NISTory of the Periodic Table

Cesium:
The frequency of microwave radiation from this atom in atomic clocks such as the NIST-F2 (2014), is used to define the second.
Image Credit: NIST

Sodium:
NIST scientists used lasers to cool a gas of these atoms to more than theoretically expected to temperatures even closer to absolute zero. (Nobel Prize 1997)
Image Credit: A. Palm-Heller/NIST

Rubidium:
These atoms were used by researchers at JILA (NIST-CU Boulder) to create the first Bose-Einstein condensate (Nobel Prize 2001).
Image Credit: NIST/JILA/CU-Boulder

Potassium and Rubidium:
JILA researchers married these elements into an ultracold gas of molecules and demonstrated striking predictions of quantum physics by hitting the atoms with "rulers of light" known as frequency combs (Nobel Prize 2005) and trapping them in webs of light known as optical lattices.
Image Credit: Steven Burrows and Nuta group/JILA

Krypton:
Wavelengths of light from this atom, measured by NIST researchers, defined the official meter until 1983.
Image Credit: Ben Tucker/Wikipedia

Deuterium:
This rare heavy isotope of hydrogen was concentrated at NIST and then identified by Columbia University's Harold Urey (Nobel Prize 1934). On the left is a deuterium lamp; the light on the right comes from the NIST SURF III Synchrotron Ultraviolet Radiation Facility.
Image Credit: Jake Ely/NIST

Beryllium and Aluminum:
Individual ions of these atoms were probed in a NIST trap to create "quantum logic" clocks that measured the second more precisely than before and tested Einstein's general theory of relativity. Such quantum manipulations were recognized in the 2012 Nobel Prize.
Image Credit: J. Amini/NIST

1931 — **1960** — **1967** — **1988** — **1995** — **2008** — **2010/2011**

Design by: Natasha Hanacek/NIST

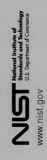

NIST National Institute of Standards and Technology
U.S. Department of Commerce
www.nist.gov

Undergraduate Instrumental Analysis

Analytical instrumentation is crucial to research in molecular biology, medicine, geology, food science, materials science, forensics, and many other fields. *Undergraduate Instrumental Analysis, 8th Edition*, provides the reader with an understanding of all major instrumental analyses, and is unique in that it starts with the fundamental principles, and then develops the level of sophistication that is needed to make each method a workable tool for the student. Each chapter includes a discussion of the fundamental principles underlying each technique, detailed descriptions of the instrumentation, and a large number of applications. Each chapter includes an updated bibliography and problems, and most chapters have suggested experiments appropriate to the technique.

This edition has been completely updated, revised, and expanded. The order of presentation has been changed from the 7th edition in that after the introduction to spectroscopy, UV-Vis is discussed. This order is more in keeping with the preference of most instructors. Naturally, once the fundamentals are introduced, instructors are free to change the order of presentation. Mathematics beyond algebra is kept to a minimum, but for the interested student, in this edition we provide an expanded discussion of measurement uncertainty that uses elementary calculus (although a formula approach can be used with no loss of context). Unique among all instrumental analysis texts we explicitly discuss safety, up front in Chapter 2. The presentation intentionally avoids a finger-wagging, thou-shalt-not approach in favor of a how-to discussion of good laboratory and industrial practice. It is focused on hazards (and remedies) that might be encountered in the use of instrumentation. Among the new topics introduced in this edition are:

- Photoacoustic spectroscopy.
- Cryogenic NMR probes and actively shielded magnets.
- The nature of mixtures (in the context of separations).
- Troubleshooting and leaks in high vacuum systems such as mass spectrometers.
- Instrumentation laboratory safety.
- Standard reference materials and standard reference data.

In addition, the authors have included many instrument manufacturer's websites, which contain extensive resources. We have also included many government websites and a discussion of resources available from National Measurement Laboratories in all industrialized countries. Students are introduced to standard methods and protocols developed by regulatory agencies and consensus standards organizations in this context as well.

Caption for cover art:

Photographs: R: Prof. April Hill of Metropolitan State University of Denver instructs a class in analytical instrumentation (courtesy of Metropolitan State University of Denver); L: Prof Lisa Ott of California State University, Chico, works with a research student in the instrumentation laboratory (courtesy of California State University, Chico).

Spectra, clockwise from top: X-ray photoelectron spectrum of lunar soil; mass spectrum of ultralow sulfur diesel fuel; infrared spectrum of rocket propellant 1; gas chromatogram of aviation kerosene; fully coupled NMR spectrum of sucrose in D2O; hypothetical TGA and DTA curves.

Undergraduate Instrumental Analysis

Eighth Edition

Thomas J. Bruno, James W. Robinson,
Eileen M. Skelly Frame, and George M. Frame II

CRC Press
Taylor & Francis Group
Boca Raton London New York

CRC Press is an imprint of the
Taylor & Francis Group, an **informa** business

Eighth edition published 2024
by CRC Press
2385 NW Executive Center Drive, Suite 320, Boca Raton FL 33431

and by CRC Press
4 Park Square, Milton Park, Abingdon, Oxon, OX14 4RN

CRC Press is an imprint of Taylor & Francis Group, LLC

© 2024 Taylor & Francis Group, LLC

First edition published by Routledge 1970
Seventh edition published by Routledge 2014

Reasonable efforts have been made to publish reliable data and information, but the author and publisher cannot assume responsibility for the validity of all materials or the consequences of their use. The authors and publishers have attempted to trace the copyright holders of all material reproduced in this publication and apologize to copyright holders if permission to publish in this form has not been obtained. If any copyright material has not been acknowledged please write and let us know so we may rectify in any future reprint.

Except as permitted under U.S. Copyright Law, no part of this book may be reprinted, reproduced, transmitted, or utilized in any form by any electronic, mechanical, or other means, now known or hereafter invented, including photocopying, microfilming, and recording, or in any information storage or retrieval system, without written permission from the publishers.

For permission to photocopy or use material electronically from this work, access www.copyright.com or contact the Copyright Clearance Center, Inc. (CCC), 222 Rosewood Drive, Danvers, MA 01923, 978-750-8400. For works that are not available on CCC please contact mpkbookspermissions@tandf.co.uk

Trademark notice: Product or corporate names may be trademarks or registered trademarks and are used only for identification and explanation without intent to infringe.

Library of Congress Cataloging-in-Publication Data
Names: Bruno, Thomas J., author. | Robinson, James W., 1923–2018, author. | Frame, George M., II, author. | Frame, Eileen M. Skelly, 1953– author.
Title: Undergraduate instrumental analysis / Thomas J. Bruno, James W. Robinson, Eileen M. Skelly Frame, and Dr. George M. Frame II.
Description: Eighth edition. | Boca Raton, FL : CRC Press, 2023. | Revised edition of: Undergraduate instrumental analysis / James W. Robinson, Eileen Skelly Frame, George M. Frame II. Seventh edition. 2014. | Includes bibliographical references and index. |
Identifiers: LCCN 2022041820 | ISBN 9781032036915 (hbk) | ISBN 9781032036953 (pbk) | ISBN 9781003188544 (ebk)
Subjects: LCSH: Instrumental analysis–Textbooks.
Classification: LCC QD79.I5 R6 2023 | DDC 543–dc23/eng20230111
LC record available at https://lccn.loc.gov/2022041820

ISBN: 9781032036915 (hbk)
ISBN: 9781032036953 (pbk)
ISBN: 9781003188544 (ebk)

DOI: 10.1201/9781003188544

Typeset in Times
by Newgen Publishing UK

Access the Support Material: www.routledge.com/9781032036953

Contents

Preface xxiii
Acknowledgments xxvii
Authors..xxix

Chapter 1
Concepts of Instrumental Analytical Chemistry 1

1.1 Introduction: What Is Analytical Chemistry?..1
    1.1.1 The Difference between Weight and Mass 2
1.2 Analytical Approach 3
    1.2.1 Defining the Problem...3
        1.2.1.1 Qualitative Analysis 4
        1.2.1.2 Quantitative Analysis 5
    1.2.2 Designing the Analytical Method 10
    1.2.3 Sampling...11
        1.2.3.1 Gas Samples..13
        1.2.3.2 Liquid Samples...14
        1.2.3.3 Solid Samples...14
    1.2.4 Storage of Samples 14
1.3 A Conceptual Discussion about Uncertainty in Measurements 15
1.4 Basic Statistics and Data Handling..17
    1.4.1 Significant Figures 17
    1.4.2 Accuracy and Precision 18
    1.4.3 Types of Errors...19
        1.4.3.1 Determinate Error 20
        1.4.3.2 Indeterminate Error 22
    1.4.4 Definitions for Statistics 23
    1.4.5 Quantifying Random Error..23
        1.4.5.1 Confidence Limits...26
        1.4.5.2 Variance...27
    1.4.6 Detection of Outliers..27
    1.4.7 Expressing Uncertainty..28
1.5 Sample Preparation 29
    1.5.1 Acid Dissolution and Digestion...29
    1.5.2 Fusions 31
    1.5.3 Dry Ashing and Combustion...31
    1.5.4 Extraction...32
        1.5.4.1 Solvent Extraction...32
        1.5.4.2 Solid-Phase Extraction..35
        1.5.4.3 QuEChERS 35
        1.5.4.4 Solid-Phase Microextraction 36
1.6 Performing the Measurement 38
    1.6.1 Signals and Noise 39
    1.6.2 Plotting Calibration Curves 41
1.7 Assessing the Data 42
    1.7.1 Limit of Detection..43
    1.7.2 Limit of Quantitation 43
1.8 Reference Materials and Reference Data 44
    1.8.1 Reference Materials...44
    1.8.2 Reference Data...44
Problems 45
Bibliography 46

Chapter 2
Safety in the Instrumentation Laboratory 49

- 2.1 Electrical Hazards and Safety 50
  - 2.1.1 Electrical Lines, Receptacles, and Cables 50
  - 2.1.2 PAT Testing Procedures in the U.K. 55
  - 2.1.3 Electrical Shocks 55
  - 2.1.4 Batteries 55
  - 2.1.5 Portable Generators 56
  - 2.1.6 A Brief Discussion about Ground 57
- 2.2 Instrument Cooling 58
- 2.3 Compressed Gas Cylinder Safety 58
  - 2.3.1 Compressed Gas Regulators and Manifolds 59
  - 2.3.2 Gas Cylinder Markings 60
- 2.4 Cryogenic Fluid Safety 61
  - 2.4.1 Cryogenic Fluids 61
  - 2.4.2 Cryogenic Liquid Containers 63
- 2.5 The Use of Fume Hoods in the Instrumentation Laboratory 64
- 2.6 Laser Safety 66
- 2.7 Compressed Air Line Safety 67
- 2.8 Ladder Safety 67
  - 2.8.1 Step Ladders 67
  - 2.8.2 Straight, Combination, and Extension Ladders 68
- 2.9 Hearing Protection 68
- 2.10 Selection of Laboratory Gloves 69
- 2.11 Selection of Respirator Cartridges and Filters 69
- 2.12 Safety with Nanomaterials 69
- 2.13 Relative Dose Ranges from Ionizing Radiation 73
- 2.14 Magnetic Field Safety 75
- 2.15 Miscellaneous Hazardous Conditions and Fire Safety 76
  - 2.15.1 Fire Extinguishers 77
  - 2.15.2 Hot Work Permits 78
- 2.16 Stop Work Orders 78

References 80

Chapter 3
Introduction to Spectroscopy 81

- 3.1 Interaction between Electromagnetic Radiation and Matter 81
  - 3.1.1 What Is Electromagnetic Radiation? 81
  - 3.1.2 How Does Electromagnetic Radiation Interact with Matter? 82
- 3.2 Atoms and Atomic Spectroscopy 86
- 3.3 Molecules and Molecular Spectroscopy 87
  - 3.3.1 Rotational Transitions in Molecules 87
  - 3.3.2 Vibrational Transitions in Molecules 88
  - 3.3.3 Electronic Transitions in Molecules 88
- 3.4 Absorption Laws 89
  - 3.4.1 Deviations from Beer's Law 91
- 3.5 Methods of Calibration 92
  - 3.5.1 Calibration with Standards 92
  - 3.5.2 Method of Standard Additions 93
  - 3.5.3 Internal Standard Calibration 96
  - 3.5.4 Errors Associated with Beer's Law Relationships 98
- 3.6 Optical Systems Used in Spectroscopy 100
  - 3.6.1 Radiation Sources 101
  - 3.6.2 Wavelength Selection Devices 102

		3.6.2.1	Filters	102
		3.6.2.2	Monochromator	102
		3.6.2.3	Resolution Required to Separate Two Lines of Different Wavelengths	104
	3.6.3	Optical Slits		108
	3.6.4	Detectors		110
	3.6.5	Single-Beam and Double-Beam Optics		110
	3.6.6	Dispersive Optical Layouts		112
	3.6.7	Fourier Transform Spectrometers		113
		3.6.7.1	Advantages of FT Systems	114
3.7	Spectroscopic Technique and Instrument Nomenclature			115

Suggested Experiments 115
Problems 116
Bibliography 118

Chapter 4
Visible and Ultraviolet Molecular Spectroscopy 119

4.1	Introduction		119
	4.1.1	Electronic Excitation in Molecules	122
	4.1.2	Absorption by Molecules	124
	4.1.3	Molar Absorptivity	125
	4.1.4	Shape of UV Absorption Curves	125
	4.1.5	Solvents for UV/VIS Spectroscopy	128
4.2	Instrumentation		129
	4.2.1	Optical System	129
	4.2.2	Radiation Sources	130
	4.2.3	Monochromators	132
	4.2.4	Detectors	132
		4.2.4.1 Barrier Layer Cell	133
		4.2.4.2 Photomultiplier Tube	134
		4.2.4.3 Semiconductor Detectors: Diodes and Diode Array Systems	135
		4.2.4.4 Diodes	136
		4.2.4.5 Diode Arrays	137
	4.2.5	Sample Holders	138
		4.2.5.1 Liquid and Gas Cells	138
		4.2.5.2 Matched Cells	139
		4.2.5.3 Flow-Through Samplers	140
		4.2.5.4 Solid Sample Holders	140
		4.2.5.5 Fiber-Optic Probes	141
	4.2.6	Microvolume, Nanovolume, and Handheld UV/VIS Spectrometers	141
		4.2.6.1 Microvolume Systems	142
		4.2.6.2 Variable Path Length Slope Spectroscopy™ System	143
		4.2.6.3 Handheld Visible Spectroscopy System	147
4.3	UV Absorption Spectra of Molecules		150
	4.3.1	Solvent Effects on UV Spectra	150
		4.3.1.1 Bathochromic or Red Shift	150
		4.3.1.2 Hypsochromic or Blue Shift	151
4.4	UV Spectra and the Structure of Organic Molecules		152
	4.4.1	Conjugated Diene Systems	152
	4.4.2	Conjugated Ketone Systems	154
	4.4.3	Substitution of Benzene Rings	156
4.5	Analytical Applications		156
	4.5.1	Qualitative Structural Analysis	156
	4.5.2	Quantitative Analysis	157
	4.5.3	Multicomponent Determinations	162
	4.5.4	Other Applications	162

		4.5.4.1	Reaction Kinetics	162
		4.5.4.2	Spectrophotometric Titrations	163
		4.5.4.3	Spectroelectrochemistry	163
		4.5.4.4	Analysis of Solids	164
	4.5.5	Measurement of Color		164
4.6	Accuracy and Precision in UV/VIS Absorption Spectrometry			166
4.7	Nephelometry and Turbidimetry			166
4.8	Molecular Emission Spectrometry			167
	4.8.1	Fluorescence and Phosphorescence		167
	4.8.2	Relationship between Fluorescence Intensity and Concentration		169
4.9	Instrumentation for Luminescence Measurements			170
	4.9.1	Wavelength Selection Devices		170
	4.9.2	Radiation Sources		171
	4.9.3	Detectors		172
	4.9.4	Sample Cells		173
	4.9.5	Spectrometer Systems		173
4.10	Analytical Applications of Luminescence			173
	4.10.1	Advantages of Fluorescence and Phosphorescence		175
	4.10.2	Disadvantages of Fluorescence and Phosphorescence		175
	Suggested Experiments			176
	Problems			177
	Bibliography			180

Chapter 5
Infrared, Near-Infrared, Raman, and Photoacoustic Spectroscopy ... 183

5.1	Absorption of IR Radiation by Molecules			184
	5.1.1	Dipole Moments in Molecules		184
	5.1.2	Types of Vibrations in Molecules		185
	5.1.3	Vibrational Motion		187
5.2	IR Instrumentation			190
	5.2.1	Radiation Sources		190
		5.2.1.1	Mid-IR Sources	191
		5.2.1.2	NIR Sources	193
		5.2.1.3	Far-IR Sources	193
		5.2.1.4	IR Laser Sources	193
	5.2.2	Monochromators and Interferometers		194
		5.2.2.1	FT Spectrometers	194
		5.2.2.2	Interferometer Components	197
	5.2.3	Detectors		198
		5.2.3.1	Bolometer	198
		5.2.3.2	Thermocouples	199
		5.2.3.3	Thermistors	199
		5.2.3.4	Golay Detector	199
		5.2.3.5	Pyroelectric Detectors	199
		5.2.3.6	Photon Detectors	200
	5.2.4	Detector Response Time		200
5.3	Sampling Techniques			201
	5.3.1	Techniques for Transmission (Absorption) Measurements		201
		5.3.1.1	Solid Samples	201
		5.3.1.2	Liquid Samples	202
		5.3.1.3	Gas Samples	204
	5.3.2	Background Correction in Transmission Measurements		205
		5.3.2.1	Solvent Absorption	205
		5.3.2.2	Air Absorption	205
	5.3.3	Techniques for Reflectance and Emission Measurements		206

		5.3.3.1	Attenuated Total Reflectance	206
		5.3.3.2	Specular Reflectance	208
		5.3.3.3	Diffuse Reflectance	209
		5.3.3.4	IR Emission	209

5.4 FTIR Microscopy 210
5.5 Nondispersive IR Systems 214
5.6 Analytical Applications of IR Spectroscopy 214
    5.6.1 Qualitative Analyses and Structural Determination by Mid-IR Absorption Spectroscopy 215
        5.6.1.1 Hydrocarbons 218
        5.6.1.2 Organic Compounds with C–O Bonds 222
        5.6.1.3 Nitrogen-Containing Organic Compounds 227
        5.6.1.4 Functional Groups Containing Heteroatoms 228
    5.6.2 Quantitative Analyses by IR Spectrometry 233
5.7 NIR Spectroscopy 236
    5.7.1 Instrumentation 236
    5.7.2 NIR Vibrational Bands, Spectral Interpretation, and Calibration 236
        5.7.2.1 NIR Calibration: Chemometrics 238
    5.7.3 Sampling Techniques for NIR Spectroscopy 238
        5.7.3.1 Liquids and Solutions 239
        5.7.3.2 Solids 240
        5.7.3.3 Gases 240
    5.7.4 Applications of NIR Spectroscopy 240
5.8 Raman Spectroscopy 242
    5.8.1 Principles of Raman Scattering 242
    5.8.2 Raman Instrumentation 244
        5.8.2.1 Light Sources 244
        5.8.2.2 Dispersive Spectrometer Systems 245
        5.8.2.3 FT-Raman Spectrometers 246
        5.8.2.4 Fiber-Optic-Based Modular and Handheld Systems 246
        5.8.2.5 Samples and Sample Holders for Raman Spectroscopy 249
    5.8.3 Applications of Raman Spectroscopy 249
    5.8.4 Resonance Raman Effect 254
    5.8.5 Surface-Enhanced Raman Spectroscopy 254
    5.8.6 Raman Microscopy 255
5.9 Chemical Imaging Using NIR, IR, and Raman Spectroscopy 256
5.10 Photoacoustic Spectroscopy 262
Inset: Alexander Graham Bell 265
Suggested Experiments 266
Problems 268
Bibliography 273
Spectral Databases 275

Chapter 6
Magnetic Resonance Spectroscopy 277

6.1 Introduction to Nuclear Magnetic Resonance Spectroscopy 277
    6.1.1 Properties of Nuclei 278
    6.1.2 Quantization of ^{1}H Nuclei in a Magnetic Field 279
        6.1.2.1 Saturation and Magnetic Field Strength 281
    6.1.3 Width of Absorption Lines 284
        6.1.3.1 Homogeneous Field 284
        6.1.3.2 Relaxation Time 284
        6.1.3.3 The Chemical Shift 285
        6.1.3.4 Magic Angle Spinning 285
        6.1.3.5 Other Sources of Line Broadening 286
6.2 FT-NMR Experiment 287

6.3	Chemical Shifts		287
6.4	Spin–Spin Coupling		291
6.5	Instrumentation		302
	6.5.1	Sample Holder	302
	6.5.2	Sample Probe	305
		6.5.2.1 Cryogenically Cooled Probes	305
	6.5.3	Magnet	305
	6.5.4	RF Generation and Detection	308
	6.5.5	Signal Integrator and Computer	309
6.6	Analytical Applications of NMR		309
	6.6.1	Samples and Sample Preparation for NMR	309
	6.6.2	Qualitative Analyses: Molecular Structure Determination	310
		6.6.2.1 Relationship between the Area of a Peak and Molecular Structure	310
		6.6.2.2 Chemical Exchange	311
		6.6.2.3 Double-Resonance Experiments	312
	6.6.3	Interpretation of Proton Spectra	313
		6.6.3.1 Aliphatic Alkanes, Alkenes, Alkynes, and Alkyl Halides	315
		6.6.3.2 Aromatic Compounds	319
		6.6.3.3 Oxygen-Containing Organic Compounds	321
		6.6.3.4 Nitrogen-Containing Organic Compounds	321
	6.6.4	^{13}C NMR	324
		6.6.4.1 Heteronuclear Decoupling	326
		6.6.4.2 Nuclear Overhauser Effect	326
		6.6.4.3 ^{13}C NMR Spectra of Solids	328
		6.6.4.4 Interpretation of ^{13}C Spectra	330
	6.6.5	2D NMR	332
	6.6.6	Qualitative Analyses: Other Applications	335
	6.6.7	Quantitative Analyses	338
6.7	Hyphenated NMR Techniques		342
6.8	NMR Imaging and MRI		343
6.9	Time Domain NMR		347
	6.9.1	Solid Fat Content Determination by TD-NMR	348
	6.9.2	Field Homogeneity and Spin Echo	349
	6.9.3	Relaxation Time Distribution	351
	6.9.4	TD-NMR Applications	352
6.10	Low-Field, Portable, and Miniature NMR Instruments		353
6.11	Limitations of NMR		353
6.12	Electron Spin Resonance Spectroscopy		353
	6.12.1	Instrumentation	356
		6.12.1.1 Samples and Sample Holders	357
	6.12.2	Miniature ESR Spectroscopy	358
	6.12.3	ESR Spectra and Hyperfine Interactions	360
	6.12.4	Applications	361
		6.12.4.1 Spin Labels and Spin Traps	363
	6.12.5	CW-ENDOR	364
Suggested Experiments			366
Problems			366
Bibliography			372
Spectral Databases			373

Chapter 7
Atomic Absorption Spectrometry 375

7.1	Absorption of Radiant Energy by Atoms		375
	7.1.1	Spectral Linewidth	377
	7.1.2	Degree of Radiant Energy Absorption	377

7.2	Instrumentation		377
	7.2.1	Radiation Sources	378
		7.2.1.1 Hollow Cathode Lamp	378
		7.2.1.2 Electrodeless Discharge Lamp	379
	7.2.2	Atomizers	380
		7.2.2.1 Flame Atomizers	380
		7.2.2.2 Electrothermal Atomizers	382
		7.2.2.3 Other Atomizers	384
	7.2.3	Spectrometer Optics	384
		7.2.3.1 Monochromator	384
		7.2.3.2 Optics and Spectrometer Configuration	384
	7.2.4	Detectors	386
	7.2.5	Modulation	386
	7.2.6	Commercial AAS Systems	386
		7.2.6.1 High-Resolution Continuum Source AAS	387
7.3	Atomization Process		387
	7.3.1	Flame Atomization	387
	7.3.2	Graphite Furnace Atomization	391
7.4	Interferences in AAS		392
	7.4.1	Nonspectral Interferences	392
		7.4.1.1 Chemical Interference	392
		7.4.1.2 Matrix Interference	393
		7.4.1.3 Ionization Interference	394
		7.4.1.4 Nonspectral Interferences in GFAAS	394
		7.4.1.5 Chemical Modification	395
	7.4.2	Spectral Interferences	397
		7.4.2.1 Atomic Spectral Interference	397
		7.4.2.2 Background Absorption and Its Correction	397
		7.4.2.3 Two-Line Background Correction	397
		7.4.2.4 Continuum Source Background Correction	397
		7.4.2.5 Zeeman Background Correction	399
		7.4.2.6 Smith–Hieftje Background Corrector	400
		7.4.2.7 Spectral Interferences in GFAAS	401
7.5	Analytical Applications of AAS		402
	7.5.1	Qualitative Analysis	402
	7.5.2	Quantitative Analysis	402
		7.5.2.1 Quantitative Analytical Range	402
		7.5.2.2 Calibration	403
	7.5.3	Analysis of Samples	404
		7.5.3.1 Liquid Samples	404
		7.5.3.2 Solid Samples	405
		7.5.3.3 Gas Samples	407
		7.5.3.4 Cold Vapor Mercury Technique	407
		7.5.3.5 Hydride Generation Technique	408
		7.5.3.6 Flow Injection Analysis	408
		7.5.3.7 Flame Microsampling	409
Suggested Experiments			409
Problems			410
7.A Appendix			411
7.B Appendix			415
Bibliography			418

Chapter 8
Atomic Emission Spectroscopy 421

8.1	Flame Atomic Emission Spectroscopy	421

		8.1.1	Instrumentation for Flame OES	422

- 8.1.1 Instrumentation for Flame OES 422
  - 8.1.1.1 Burner Assembly 422
  - 8.1.1.2 Wavelength Selection Devices 423
  - 8.1.1.3 Detectors 424
  - 8.1.1.4 Flame Excitation Source 424
- 8.1.2 Interferences 425
  - 8.1.2.1 Chemical Interference 425
  - 8.1.2.2 Excitation and Ionization Interferences 426
  - 8.1.2.3 Spectral Interferences 426
- 8.1.3 Analytical Applications of Flame OES 427
  - 8.1.3.1 Qualitative Analysis 427
  - 8.1.3.2 Quantitative Analysis 427
- 8.2 Atomic OES 429
  - 8.2.1 Instrumentation for Emission Spectroscopy 430
    - 8.2.1.1 Electrical Excitation Sources 431
    - 8.2.1.2 Sample Holders 435
    - 8.2.1.3 Spectrometers 435
    - 8.2.1.4 Detectors 437
  - 8.2.2 Interferences in Arc and Spark Emission Spectroscopy 441
    - 8.2.2.1 Matrix Effects and Sample Preparation 441
    - 8.2.2.2 Spectral Interference 442
    - 8.2.2.3 Internal Standard Calibration 442
  - 8.2.3 Applications of Arc and Spark Emission Spectroscopy 443
    - 8.2.3.1 Qualitative Analysis 443
    - 8.2.3.2 Raies Ultimes 443
    - 8.2.3.3 Quantitative Analysis 443
- 8.3 Plasma Emission Spectroscopy 446
  - 8.3.1 Instrumentation for Plasma Emission Spectrometry 446
    - 8.3.1.1 Excitation Sources 446
    - 8.3.1.2 Spectrometer Systems for ICP, DCP, and MP 451
    - 8.3.1.3 Sample Introduction Systems 453
  - 8.3.2 Interferences and Calibration in Plasma Emission Spectrometry 458
    - 8.3.2.1 Chemical and Ionization Interference 459
    - 8.3.2.2 Spectral Interference and Correction 459
  - 8.3.3 Applications of ICP, DCP, and MP Atomic Emission Spectroscopy 462
  - 8.3.4 Chemical Speciation with Hyphenated Instruments 464
- 8.4 GD Emission Spectrometry 465
  - 8.4.1 DC and RF GD Sources 465
  - 8.4.2 Applications of GD Atomic Emission Spectrometry 466
    - 8.4.2.1 Bulk Analysis 466
    - 8.4.2.2 Depth Profile Analysis 466
- 8.5 Atomic Fluorescence Spectroscopy 467
  - 8.5.1 Instrumentation for AFS 469
  - 8.5.2 Interferences in AFS 470
    - 8.5.2.1 Chemical Interference 470
    - 8.5.2.2 Spectral Interference 470
  - 8.5.3 Applications of AFS 470
    - 8.5.3.1 Graphite Furnace Laser-Excited AFS 471
    - 8.5.3.2 Mercury Determination and Speciation by AFS 471
    - 8.5.3.3 Hydride Generation and Speciation by AFS 472
- 8.6 Commercial Atomic Emission Systems 472
  - 8.6.1 Arc and Spark Systems 472
  - 8.6.2 ICP, DCP, and MP Systems 472
  - 8.6.3 GD Systems 473
  - 8.6.4 AFS Systems 473

8.7	Laser-Induced Breakdown Spectroscopy	473
	8.7.1 Principle of Operation	473
	8.7.2 Instrumentation	474
	8.7.3 Applications of LIBS	475
	8.7.3.1 Qualitative Analysis and Semiquantitative Analysis	476
	8.7.3.2 Quantitative Analysis	477
	8.7.3.3 Remote Analysis	477
	8.7.4 Commercial LIBS Systems	480
8.8	Atomic Emission Literature and Resources	480
8.9	Comparison of Atomic Spectroscopic and ICP-MS Techniques	480

Suggested Experiments ... 480
Problems ... 481
8.A Appendix ... 483
8.B Appendix: Comparison of Atomic Spectroscopic Analytical Techniques ... 484
Bibliography ... 486

Chapter 9
X-Ray Spectroscopy ... 489

Contributing authors: Alexander Seyfarth and Eileen Skelly Frame

9.1	Origin of X-Ray Spectra	489
	9.1.1 Energy Levels in Atoms	489
	9.1.2 Moseley's Law	494
	9.1.3 X-Ray Methods	494
	9.1.3.1 X-Ray Absorption Process	495
	9.1.3.2 X-Ray Fluorescence Process	496
	9.1.3.3 X-Ray Diffraction Process	496
9.2	X-Ray Fluorescence	498
	9.2.1 X-Ray Source	499
	9.2.1.1 X-Ray Tube	499
	9.2.1.2 Secondary XRF Sources	501
	9.2.1.3 Radioisotope Sources	501
	9.2.2 Instrumentation for Energy-Dispersive X-Ray Spectrometry	504
	9.2.2.1 Excitation Source	504
	9.2.2.2 Primary Beam Modifiers	505
	9.2.2.3 Sample Holders	508
	9.2.2.4 EDXRF Detectors	511
	9.2.2.5 Multichannel Pulse Height Analyzer	514
	9.2.2.6 Detector Artifact Escape Peaks and Sum Peaks	515
	9.2.3 Instrumentation for Wavelength-Dispersive X-Ray Spectrometry	516
	9.2.3.1 Collimators	516
	9.2.3.2 Analyzing Crystals	516
	9.2.3.3 Detectors	521
	9.2.3.4 Electronic Pulse Processing Units	525
	9.2.3.5 Sample Changers	527
	9.2.4 Simultaneous WDXRF Spectrometers	527
	9.2.5 Micro-XRF Instrumentation	528
	9.2.5.1 Micro-X-Ray Beam Optics	528
	9.2.5.2 Micro-XRF System Components	529
	9.2.6 Total Reflection XRF	530
	9.2.7 Comparison between EDXRF and WDXRF	530
	9.2.8 XRF Applications	531
	9.2.8.1 Analyzed Layer	531
	9.2.8.2 Sample Preparation Considerations for XRF	533
	9.2.8.3 Qualitative Analysis by XRF	534

		9.2.8.4	Quantitative Analysis by XRF	539
		9.2.8.5	Applications of Quantitative XRF	541
9.3	X-Ray Absorption			542
	9.3.1	EXAFS		546
9.4	X-Ray Diffraction			546
	9.4.1	Single-Crystal X-Ray Diffractometry		548
	9.4.2	Crystal Growing		549
	9.4.3	Crystal Structure Determination		549
	9.4.4	Powder X-Ray Diffractometry		551
	9.4.5	Hybrid XRD/XRF Systems		553
		9.4.5.1	Compact and Portable Hybrid Systems	553
	9.4.6	Applications of XRD		555
		9.4.6.1	Analytical Limitations of XRD	556
9.5	X-Ray Emission			557
	9.5.1	Electron Probe Microanalysis		557
	9.5.2	Particle-Induced X-Ray Emission		558
9.6	Commercial X-Ray Instrument Manufacturers			558
Suggested Experiments/Tutorials				559
Problems				561
9.A Appendix: Characteristic X-Ray Wavelengths (Å) and Energies (keV)				566
9.B Appendix: Absorption Edge Wavelengths and Energies				571
Bibliography				573

Chapter 10
Mass Spectrometry I: Principles and Instrumentation 575

10.1	Principles of MS			575
	10.1.1	Resolving Power and Resolution of a Mass Spectrometer		579
10.2	Instrumentation			580
	10.2.1	Sample Input Systems		580
		10.2.1.1	Gas Expansion	580
		Inset: A Brief Digression on Units of Measure – Vacuum Systems		580
		10.2.1.2	Direct Insertion and Direct Exposure Probes	581
		10.2.1.3	Chromatography and Electrophoresis Systems	581
		10.2.1.4	Chromatography and Infrared Spectrometry Systems Coupled with Mass Spectrometry	582
	10.2.2	Ionization Sources		582
		10.2.2.1	Electron Ionization	582
		10.2.2.2	Chemical Ionization	583
		10.2.2.3	Atmospheric Pressure Ionization Sources	584
		10.2.2.4	Desorption Ionization	586
		10.2.2.5	Ionization Sources for Inorganic MS	592
	10.2.3	Mass Analyzers		593
		10.2.3.1	Magnetic and Electric Sector Instruments	593
		10.2.3.2	Time-of-Flight Analyzer	596
		10.2.3.3	Quadrupole Mass Analyzer	601
		10.2.3.4	MS/MS and MS^n Instruments	603
		10.2.3.5	Quadrupole Ion Trap	605
		10.2.3.6	Fourier Transform Ion-Cyclotron Resonance	606
		10.2.3.7	Orbitrap MS	607
	10.2.4	Detectors		608
		10.2.4.1	Electron Multiplier	609
		10.2.4.2	Faraday Cup	610
		10.2.4.3	Array Detectors	611

10.3	Ion Mobility Spectrometry	611
	10.3.1 Handheld DMS JUNO® Chemical Trace Vapor Point Detector	613
	10.3.2 Excellims HPIMS-LC System	613
	10.3.3 Photonis Ion Mobility Spectrometer Engine	613
	10.3.4 SYNAPT G2-S Multistage MS System Incorporating the TriWAVE Ion Mobility Stage	613
10.4	Leaks in Mass Spectrometer Systems	615
Problems		616
Bibliography		617

Chapter 11
Mass Spectrometry II: Spectral Interpretation and Applications 619

11.1	Interpretation of Mass Spectra: Structural Determination of Simple Molecules	619
	11.1.1 Molecular Ion and Fragmentation Patterns	621
	11.1.2 Nitrogen Rule	623
	11.1.3 Molecular Formulas and Isotopic Abundances	623
	11.1.3.1 Counting Carbon Atoms	624
	11.1.3.2 Counting Carbon, Nitrogen, and Sulfur Atoms	625
	11.1.3.3 Counting Oxygen Atoms	625
	11.1.4 Compounds with Heteroatoms	626
	11.1.5 Halogen Isotopic Clusters	627
	11.1.6 Rings Plus Double Bonds	630
	11.1.7 Common Mass Losses on Fragmentation	631
11.2	Mass Spectral Interpretation: Some Examples	631
	11.2.1 Mass Spectra of Hydrocarbons	635
	11.2.2 Mass Spectra of Other Organic Compound Classes	640
	11.2.2.1 Alcohols	640
	11.2.2.2 Ethers	642
	11.2.2.3 Ketones and Aldehydes	642
	11.2.2.4 Carboxylic Acids and Esters	645
	11.2.3 Compounds Containing Heteroatoms	646
	11.2.3.1 Nitrogen-Containing Compounds	646
	11.2.3.2 Sulfur-Containing Compounds	647
11.3	Applications of Molecular MS	649
	11.3.1 High-Resolution Mass Spectrometry	649
	11.3.1.1 Achieving Higher Mass Accuracy (but Not Resolution) from Low-Resolution MS Instruments	650
	11.3.1.2 Improving the Quantitation Accuracy of Isotope Ratios from Low-Resolution MS Instrument Data Files	651
	11.3.2 Quantitative Analysis of Compounds and Mixtures	652
	11.3.3 Protein-Sequencing Analysis (Proteomics)	654
	11.3.4 Gas Analysis	655
	11.3.5 Environmental Applications	655
	11.3.6 Other Applications of Molecular MS	657
	11.3.7 Limitations of Molecular MS	657
11.4	Atomic MS	657
	11.4.1 ICP-MS	657
	11.4.2 Applications of Atomic MS	660
	11.4.2.1 Geological and Materials Characterization Applications	661
	11.4.2.2 Speciation by Coupled Chromatography-ICP-MS	663
	11.4.2.3 Applications in Food Chemistry, Environmental Chemistry, Biochemistry, Clinical Chemistry, and Medicine	664
	11.4.2.4 Coupled Elemental Analysis MS	666
	11.4.3 Interferences in Atomic MS	666
	11.4.3.1 Matrix Effects	666

		11.4.3.2 Spectral (Isobaric) Interferences	667
	11.4.4	Instrumental Approaches to Eliminating Interferences	670
		11.4.4.1 High-Resolution ICP-MS (HR-ICP-MS)	670
		11.4.4.2 Collision and Reaction Cells	670
		11.4.4.3 MS/MS Interference Removal	671
	11.4.5	Limitations of Atomic MS	672
11.5	Common Spurious Effects in Mass Spectrometry		674
Problems			674
11.A Appendix			682
Bibliography			684

Chapter 12
Principles of Chromatography ... 687

12.1	Introduction to Chromatography	687
12.2	The Nature of Mixtures	688
12.3	What Is the Chromatographic Process?	691
12.4	Chromatography in More than One Dimension	692
12.5	Visualization of the Chromatographic Process at the Molecular Level: Analogy to "People on a Moving Belt Slideway"	693
12.6	Digression on the Central Role of Silicon–Oxygen Compounds in Chromatography	696
12.7	Basic Equations Describing Chromatographic Separations	698
12.8	How Do Column Variables Affect Efficiency (Plate Height)?	701
12.9	Practical Optimization of Chromatographic Separations	702
12.10	Extra-Column Band Broadening Effects	703
12.11	Qualitative Chromatography: Analyte Identification	704
12.12	Quantitative Measurements in Chromatography	704
	12.12.1 Peak Area or Peak Height: What Is Best for Quantitation?	705
	12.12.2 Calibration with an External Standard	705
	12.12.3 Calibration with an Internal Standard	706
12.13	Examples of Chromatographic Calculations	707
Problems		708
Bibliography		709

Chapter 13
Gas Chromatography ... 711

13.1	Historical Development of GC: The First Chromatographic Instrumentation	711
13.2	Advances in GC Leading to Present-Day Instrumentation	712
13.3	GC Instrument Component Design (Injectors)	713
	13.3.1 Syringes	713
	13.3.2 Autosamplers	714
	13.3.3 SPME	714
	13.3.4 Split Injections	715
	13.3.5 Splitless Injections	715
13.4	GC Instrument Component Design (the Column)	716
	13.4.1 Column Stationary Phase	716
	13.4.2 Selecting a Stationary Phase for an Application	719
	13.4.3 Effects of Mobile Phase Choice and Flow Parameters	721
13.5	GC Instrument Operation (Column Dimensions and Elution Values)	722
13.6	GC Instrument Operation (Column Temperature and Elution Values)	724
13.7	GC Instrument Component Design (Detectors)	728
	13.7.1 Thermal Conductivity Detector	730
	13.7.2 Flame Ionization Detector	730
	13.7.3 Electron Capture Detector	732
	13.7.4 Electrolytic Conductivity Detector	734

	13.7.5 Sulfur–Phosphorus Flame Photometric Detector	734
	13.7.6 Sulfur Chemiluminescence Detector	734
	13.7.7 Nitrogen–Phosphorus Detector	735
	13.7.8 Photoionization Detector	735
	13.7.9 Helium Ionization Detector	736
	13.7.10 Atomic Emission Detector	737
13.8	Hyphenated GC Techniques (GC-MS, GC-IR, GC-GC, or 2D GC)	737
	13.8.1 GC-MS	737
	13.8.2 GC-IR Spectrometry	740
	13.8.3 Comprehensive 2D Gas Chromatography (GC × GC or GC^2)	740
13.9	Retention Indices (A Generalization of Relative R_t Information)	743
13.10	Scope of GC Analyses	744
	13.10.1 GC Behavior of Organic Compound Classes	744
	13.10.2 Derivatization of Difficult Analytes to Improve GC Elution Behavior	744
	13.10.3 Gas Analysis by GC	745
	13.10.4 Limitations of GC	746
Problems		746
13.A Appendix: GC Internet Web Resources		748
Bibliography		749

Chapter 14
Chromatography with Liquid (and Fluid) Mobile Phases 751

14.1	High-Performance Liquid Chromatography	751
	14.1.1 HPLC Column and Stationary Phases	751
	14.1.1.1 Support Particle Considerations	753
	14.1.1.2 Stationary-Phase Considerations	755
	14.1.1.3 Chiral Phases for Separation of Enantiomers	755
	14.1.2 Effects on Separation of Composition of the Mobile Phase	756
	14.1.2.1 New HPLC-Phase Combinations for Assays of Very Polar Biomolecules	757
	14.1.2.2 Ion-Exchange Chromatography	758
	14.1.2.3 Hydrophilic Interaction Liquid Chromatography	758
	14.1.2.4 Polar-Embedded Long Organic Chains on Silica	758
	14.1.2.5 Polar-Endcapped Short Hydrocarbon Chains on Silica	758
	14.1.3 Design and Operation of an HPLC Instrument	758
	14.1.4 HPLC Detector Design and Operation	761
	14.1.4.1 Refractive Index Detector	762
	14.1.4.2 Evaporative Light-Scattering Detector	763
	14.1.4.3 UV/VIS Absorption Detectors	765
	14.1.4.4 Fluorescence Detectors	767
	14.1.4.5 Electrochemical Detectors	768
	14.1.4.6 Conductometric Detectors	770
	14.1.4.7 Charge Detector (QD) [Thermo Scientific]	770
	14.1.4.8 Summary Comparison of Six Major HPLC Detectors	771
	14.1.5 Derivatization in HPLC	771
	14.1.6 Hyphenated Techniques in HPLC	773
	14.1.6.1 Interfacing HPLC to Mass Spectrometry	773
	14.1.7 Applications of HPLC	777
	14.1.7.1 Biochemical Applications in Proteomics	778
14.2	Chromatography of Ions Dissolved in Liquids	782
	14.2.1 Ion Chromatography	784
	14.2.1.1 Single-Column IC	787
	14.2.1.2 Indirect Detection in IC	787
14.3	Affinity Chromatography	788
14.4	Size-Exclusion Chromatography	789
14.5	Supercritical Fluid Chromatography	791

		14.5.1	Operating Conditions	792
		14.5.2	Effect of Pressure	792
		14.5.3	Stationary Phases	792
		14.5.4	Mobile Phases	792
		14.5.5	Detectors	792
		14.5.6	SFC versus Other Column Methods	792
		14.5.7	Applications	793
		14.5.8	Ultra Performance Convergence Chromatography: A New Synthesis	793
	14.6	Electrophoresis		794
		14.6.1	Capillary Zone Electrophoresis (CZE)	794
		14.6.2	Sample Injection in CZE	798
		14.6.3	Detection in CZE	799
		14.6.4	Modes of CE	800
			14.6.4.1 CZE	800
			14.6.4.2 Capillary Gel Electrophoresis	801
			14.6.4.3 Capillary Isoelectric Focusing	801
		14.6.5	Capillary Electrochromatography	803
			14.6.5.1 Micellar Electrokinetic Capillary Chromatography	803
	14.7	Planar Chromatography and Planar Electrophoresis		805
		14.7.1	Thin-Layer Chromatography	805
			14.7.1.1 Parallel 1D TLC Separations	805
			14.7.1.2 Detection in TLC	806
			14.7.1.3 2D Separations Using TLC (2D-TLC)	808
		14.7.2	Planar Electrophoresis on Slab Gels	808
			14.7.2.1 1D Planar Gel Electrophoresis	808
			14.7.2.2 2D Planar Gel Electrophoresis	808
Problems and Exercises				810
14.A Appendix LC/CE/TLC Internet Web Resources				811
Bibliography				812

Chapter 15
Electroanalytical Chemistry 815

15.1	Fundamentals of Electrochemistry		815
15.2	Electrochemical Cells		816
	15.2.1 Line Notation for Cells and Half-Cells		819
	15.2.2 Standard Reduction Potentials		819
		15.2.2.1 Standard Hydrogen Electrode	819
	15.2.3 Sign Conventions		821
	15.2.4 Nernst Equation		821
	15.2.5 Activity Series		822
	15.2.6 Reference Electrodes		823
		15.2.6.1 Saturated Calomel Electrode	823
		15.2.6.2 Silver/Silver Chloride Electrode	824
	15.2.7 Electrical Double Layer		824
15.3	Electroanalytical Methods		825
	15.3.1 Potentiometry		826
		15.3.1.1 Indicator Electrodes	827
		15.3.1.2 Instrumentation for Measuring Potential	833
		15.3.1.3 Analytical Applications of Potentiometry	835
	15.3.2 Coulometry		843
		15.3.2.1 Electrogravimetry	845
		15.3.2.2 Instrumentation for Electrogravimetry and Coulometry	845
		15.3.2.3 Applied Potential	845
		15.3.2.4 Analytical Determinations Using Faraday's Law	846
	15.3.3 Conductometric Analysis		850

		15.3.3.1	Instrumentation for Conductivity Measurements	852
		15.3.3.2	Analytical Applications of Conductometric Measurements	853
	15.3.4	Polarography		855
		15.3.4.1	Classical or DC Polarography	856
		15.3.4.2	Half-Wave Potential	859
		15.3.4.3	Normal Pulse Polarography	860
		15.3.4.4	Differential Pulse Polarography	861
	15.3.5	Voltammetry		863
		15.3.5.1	Instrumentation for Voltammetry	865
		15.3.5.2	Stripping Voltammetry	865
		15.3.5.3	pHit Sensor for pH by Cyclic Voltammetry on Carbon	866
		15.3.5.4	Applications of Anodic Stripping Voltammetry	867
15.4	Liquid Chromatography Detectors			868
	15.4.1	Voltammetric Detection		868
	15.4.2	Conductometric Detection		869
15.5	Spectroelectrochemistry			870
15.6	Quartz Crystal Microbalance			872
	15.6.1	Piezoelectric Effect		872
	15.6.2	Electrochemical Quartz Crystal Microbalance		876
15.7	Electrochemistry at the Nanoscale			876
Suggested Experiments				878
Problems				879
15.A Appendix: Selected Standard Reduction Potentials at 25 °C				880
Bibliography				880

Chapter 16
Surface Analysis ... 883

16.1	Introduction			883
16.2	Electron Spectroscopy Techniques			883
	16.2.1	X-Ray Photoelectron Spectroscopy		885
		16.2.1.1	Instrumentation for XPS	886
		16.2.1.2	Sample Introduction and Handling for Surface Analysis	890
		16.2.1.3	Analytical Applications of XPS	891
	16.2.2	Auger Electron Spectroscopy		898
		16.2.2.1	Instrumentation for AES	899
		16.2.2.2	Applications of AES	900
16.3	Ion Scattering Spectroscopy			902
16.4	Secondary Ion Mass Spectrometry			905
	16.4.1	Instrumentation for SIMS		906
		16.4.1.1	Primary Ion Sources	906
		16.4.1.2	Mass Spectrometers	907
	16.4.2	Analytical Applications of SIMS		907
		16.4.2.1	Quantitative Analysis	908
		16.4.2.2	Ion Microprobe Mass Spectrometry	909
16.5	Electron Microprobe (Electron Probe Microanalysis)			910
Problems				910
Bibliography				911

Chapter 17
Thermal Analysis ... 913

17.1	Thermogravimetry		914
	17.1.1	TGA Instrumentation	916
	17.1.2	Analytical Applications of Thermogravimetry	920
	17.1.3	Derivative Thermogravimetry	924
	17.1.4	Sources of Error in Thermogravimetry	926

17.2 Differential Thermal Analysis 926
    17.2.1 DTA Instrumentation 927
    17.2.2 Analytical Applications of DTA 929
17.3 Differential Scanning Calorimetry..930
    17.3.1 DSC Instrumentation 930
    17.3.2 Applications of DSC...934
        17.3.2.1 Pressure DSC 936
        17.3.2.2 Modulated DSC 937
17.4 Hyphenated Techniques 938
    17.4.1 Hyphenated Thermal Methods...938
    17.4.2 Evolved Gas Analysis 938
17.5 Thermometric Titrimetry 941
    17.5.1 Applications of Thermometric Titrimetry..944
17.6 Direct Injection Enthalpimetry 944
17.7 Microcalorimetry 945
    17.7.1 Micro-DSC Instrumentation 945
    17.7.2 Applications of Micro-DSC..946
    17.7.3 Isothermal Titration Calorimetry 946
    17.7.4 Microliter Flow Calorimetry...948
17.8 Thermomechanical Analysis and Dynamic Mechanical Analysis 949
    17.8.1 Instrumentation 952
        17.8.1.1 TMA Equipment 952
        17.8.1.2 DMA Equipment 953
    17.8.2 Applications of TMA and DMA...954
17.9 Optical Thermal Analysis 957
    17.9.1 Heating Microscope..957
    17.9.2 Optical Dilatometers...958
    17.9.3 Optical Thermal Analyzers with Differential Thermal Analysis 960
17.10 Summary..962
Suggested Experiments 963
Problems 964
Bibliography 967

Appendix: Abbreviations, Acronyms, and Symbols 969
Subject Index 977
Chemical Index 987

Preface to the Eighth Edition

Analytical chemistry today is almost entirely instrumental analytical chemistry, performed by many scientists, engineers, and technicians, many of whom are not chemists. Analytical instrumentation is crucial to research in molecular biology, medicine, geology, food science, materials science, forensics, and a myriad of other fields. While it is true that it is no longer necessary to have almost artistic skills to obtain accurate and precise analytical results using instrumentation, the instruments cannot be considered "black boxes" by those using them. "Garbage in, garbage out" holds true for analytical instrumentation as well as computers. The intent of this book is to provide users with an understanding of their instruments.

In keeping with the earlier editions of this text, the book is designed for teaching undergraduates and those with no instrumental analytical chemistry background how modern analytical instrumentation works and what the uses and limitations of analytical instrumentation are. Mathematics beyond algebra is kept to a minimum. Prior editions touted the need for no background in calculus, physics, or physical chemistry, but we recognize that some familiarity in these areas is helpful. Indeed, it is unrealistic to assume that a student taking analytical instrumentation will have no such background and doing so is a disservice to them. Here, we do not sell them short; we give them the opportunity to build upon their background.

The major fields of modern instrumentation are covered, including applications of each type of instrumental technique. Each chapter includes a discussion of the fundamental principles underlying each technique, detailed descriptions of the instrumentation, and a large number of applications. Each chapter includes an updated bibliography and practice problems, and most chapters have suggested experiments appropriate to the technique.

This edition has been completely updated, revised, and expanded. To match the preference of most instructors, the order of presentation has been changed from that of the seventh edition. Now, after the introduction to spectroscopy, UV-Vis is discussed, and naturally, instructors are free to change the order of presentation once the fundamentals have been introduced. The topics of chromatography and mass spectrometry have been greatly expanded to better reflect the predominance of chromatography and mass spectrometry instrumentation in modern laboratories. The equally important topic of NMR, expanded in the last edition to focus on FTNMR, ^{13}C, and 2D NMR spectral interpretation, now includes time domain NMR (relaxometry) and an overview of low-field, benchtop, and miniature instrumentation. The topic of electron spin resonance spectroscopy (ESR, EPR) has been added due to the recent availability of small, low-cost ESR instrumentation and its impact on materials characterization and bioanalysis.

A unique feature of this text is the combination of instrumental analysis with organic spectral interpretation (IR, NMR, and MS). The NMR, IR, and MS spectra, courtesy of Bio-Rad Laboratories, Informatics Division (IR, NMR), Aldrich Chemical Company (NMR), Agilent Technologies, Inc., and the NIST Chemistry Webbook were obtained on modern instruments to reflect what students will encounter in modern laboratories. Additional NMR spectra have been provided by Bruker Corporation and picoSpin LLC (now Thermo Fisher Scientific). The use of spreadsheets for performing calculations has been introduced with examples. Reflecting the ubiquitous nature of the Internet, we have included a large number of instrument manufacturers' websites and government websites, which contain extensive resources for interested students. This includes an expanded discussion of standard reference materials (available from National Measurement Laboratories) and standard reference data.

Sampling, sample handling, and storage and sample preparation methods are extensively covered, and modern methods such as accelerated solvent extraction, solid-phase microextraction (SPME), QuEChERS, and microwave techniques are included. Instrumentation, the analysis of liquids and solids, and applications of NMR are discussed in detail. A section on hyphenated NMR techniques is included, along with an expanded section on MRI and advanced imaging. The IR instrumentation section is focused on FTIR instrumentation, and a discussion of photoacoustic spectroscopy is now included. Absorption, emission, and reflectance spectroscopy are discussed, as is FTIR microscopy. ATR has been expanded. Near-IR instrumentation and applications are presented, and the topic of chemometrics is introduced. Coverage of Raman spectroscopy includes resonance Raman, surface-enhanced Raman, and Raman microscopy. Chemical imaging is described, including confocal Raman imaging. UV and visible spectroscopy includes innovations such as flow-through sample holders and fiber-optic probes, as well as instruments for analysis of submicroliter volumes and nondestructive analysis for nucleic acid and protein determinations. UV absorption spectral interpretation for organic molecules is covered in depth. Applications described include nucleic acid and protein measurements, spectrophotometric titrations, and new applications in forensic chemistry. Nephelometry, turbidimetry, fluorescence, and phosphorescence are described in detail, including instrumentation and applications. The measurement of color using the CIE system is described with examples.

Mass spectrometry is treated as two chapters and covers both organic and inorganic MS instrumentation and applications. We dispel the often confusing mix of units that describe vacuum systems, a topic that is applicable to other chapters as well. GC-MS and LC-MS are described along with MS^n instruments for organic and inorganic MS, including new triple quad ICP-MS instrumentation. Modern ionization methods such as electrospray and MALDI are also described. Surface scanning and ionization sources, such as desorption electrospray ionization (DESI) and laser ablation electrospray ionization (LAESI), and the DART atmospheric pressure ionization source are introduced. More advanced high-resolution systems, such as the Orbitrap, and long-path time of flight (JEOL SpiralTOF and LECO Citius-LC and Pegasus GC-HRT) are described. The use of MS peak profile correction software to improve mass and spectral accuracy from lower resolution MS instrumentation is described, and important cautions in the use of search software are discussed. Uses of HRMS receive increased attention. A new section on ion mobility spectrometry is added, with a discussion of four systems ranging from handheld vapor monitors to stages enabling characterization of different conformational isomers in top-end proteomics analyzers. Organic mass spectral interpretation is covered with many examples and new spectra. Organic and inorganic applications focus on speciation using GC-MS, LC-MS, and hyphenated ICP-MS, with an emphasis on proteomics, biomolecules, and species of environmental interest. Examples of atomic MS applications using new simultaneous and "triple quad" ICP-MS systems and glow discharge surface sampling and hydrogen in metals profiling are added. All major modern atomic absorption and emission techniques and instrumentation are covered, including new MP-AES and triple quadrupole ICP-MS instruments.

Chromatographic separations and instrumentation are treated in three chapters which begin with a fundamental description of mixtures from a molecular view. The first of these chapters covers the nature of the chromatographic process, starting with the nature of mixtures. A minimum of complex formulas is introduced; instead, extensive description and analogy are employed to give the student an intuitive grasp of the mechanisms giving rise to separation, resolution, and detection of separated components. Additional graphics have been added to clarify the concepts. In line with current practice, GC is treated as a method almost completely employing open tubular (capillary) columns. We treat SPME injection, low-bleed stationary phases, ionic liquid stationary phases, compound-selective detectors, and especially interfacing to spectrometric detectors and the use of computerized data processing. So-called hyphenated techniques such as GC-MS, GC-IR, GC-IR-MS, and comprehensive GC-GC are included, and new sections on retention indices, derivatization to improve detectability and volatility, and analysis of gases in air or water have been added. The requirements for implementing and instrumentalizing HPLC are developed from the preceding discussion of GC. The student gains an appreciation of the difficulties that caused this instrumentation to lag behind GC. Once overcome, liquid chromatographic instrumentation is seen to have wider applicability, especially in the burgeoning subdisciplines of bioanalysis, proteomics, and genomics. Coverage includes the use of superficially porous, core–shell–particle stationary phases, which enable much higher resolution and faster analysis. Such UHPLC methods now constitute more than half of the LC methods in operation. Detailed discussions of instrumental design and the operation of new detectors, such as the charge detector for ion chromatography, the corona discharge detector, pulsed amperometric detector for saccharides, and the evaporative light-scattering detector for spectrally inert analytes have been added. Major sections on the design and operation of HPLC interfaces to mass spectrometers (ESI and APCI) have been added, and examples of their use in protein or peptide sequencing and identification are included. Tables of amino acid structures and nomenclature have been included so that students can follow these descriptions. Separate sections on ion chromatography, affinity chromatography, size exclusion chromatography, and supercritical fluid extraction and chromatography (SFE/C) have been expanded. The revival of analytical SFC by the very recent redesign of instrumentation to enable ultra-performance convergence chromatography (UPC^2) is introduced, and examples of the improvements it offers are given. Planar and capillary electrophoresis are described in detail in Chapter 13, despite not being strictly defined as chromatographic methods. Examples are given of the use of capillary electrophoresis with fluorescence-derivatization detection to gene sequencing in genomics and of 2D slab-gel, isoelectric focusing/SDS-PAGE electrophoresis for protein peptide mapping in proteomics. New appendices provide links to websites containing examples of thousands of chromatographic separations and encourage the student to learn how to utilize the resources of commercial column vendors to find a solution to particular separation or measurement problems.

The discussion of XPS has been revised to reflect the current state of commercial instrumentation. Spectroelectrochemistry and nanoelectrochemistry topics have been expanded, with commercial instrumentation and examples. Probes based on quartz crystal microbalance response are covered. The radical new pH meter design using calibration-free, easily stored, extremely robust, voltammetric-based probes with surfaces of anthraquinone-bonded multiwalled carbon nanotubes is described. The important bioanalytical technique of microcalorimetry has been added to Chapter 17, Thermal Analysis, and TMA and DMA have been greatly expanded, thanks to the excellent materials and invaluable assistance from TA Instruments. Optical thermal instrumentation is now covered.

We are pleased that in this edition we have included both a subject index and a chemical index. While the chemical index is not exhaustive, it will allow the reader to rapidly find analytical metrology for many compounds and elements.

While the authors are extremely grateful to the many experts listed in the acknowledgments, who have provided graphics, technical advice, rewrites, and reviews of various sections, any errors that are present are entirely the responsibility of the authors.

Thomas J. Bruno
James W. Robinson
Eileen M. Skelly Frame
George M. Frame II

TO THE STUDENT OR READER:

This book is unique among texts on analytical instrumentation in that the discussion of each technique starts with the underlying atomic or molecular picture and then develops the level of sophistication that is needed to make each method work for you. Yes, you have heard this before, but this book actually delivers. What does this mean? Each student approaches the study of analytical instrumentation with some knowledge from prior course work, but the completeness of that preparation will depend on the interests of your instructors. If, for example, your organic instructor was not interested in ultraviolet spectroscopy, they likely spent little time on it. Here, on the other hand, you will start with the molecular infrastructure, and the development and application of the method will therefore make sense to you.

This is a book that you will want to keep in your professional toolbox rather than selling back to the bookstore when the course is done. We appreciate that your textbooks are costly, but because of our approach in this book, you will be able to go back to it in a few years, pick it up, and in short order be up to speed. If you go on to an advanced degree, this aspect will be indispensable to you. But, even if you go into medicine or law or some other profession, the case may arise in which an NMR spectrum or a mass spec will have bearing, and you will want to revisit the topic. This book will get you there quickly because we start simple and don't assume anything. The added chemical index will allow you to rapidly find analytical metrology and procedures for many compounds and elements.

Prior editions of this book touted minimization of mathematics and absence of calculus, and yes, you can use this book with only a bit of rudimentary algebra. But honestly, it is not really possible to appreciate uncertainty analysis without taking a derivative now and then, so in the greatly expanded discussion of uncertainty in the eighth edition, sensitivity factors are expressed with derivatives. Don't be intimidated: simple calculus is a lot like cooking — you simply follow a recipe.

Unique among all instrumental analysis texts we explicitly discuss safety, up front in Chapter 2. The presentation intentionally avoids a finger-wagging, thou-shalt-not approach in favor of a how-to discussion of good laboratory and industrial practice. It is focused on hazards (and remedies) that might be encountered in the use of instrumentation; we don't even mention chemical safety (which should have been emphasized for you before now). The information in Chapter 2 will not only help you complete the course safely, it will also be valuable material for you as a life skill. This chapter might not be discussed in class, but please read it anyway.

TO THE INSTRUCTOR:

We realize you are busy so we will make this short and sweet.

Past editions of this book touted the need for minimal mathematics knowledge and skills. In the greatly expanded discussion of uncertainty presented in the eighth edition, sensitivity factors are expressed as derivatives (typically from introductory calculus). Understand that it is still possible to approach error and uncertainty with a formulae approach, and sidestep expanded uncertainty, but here you can use either approach.

The chapter on safety is meant to augment your institution's safety program and standard operating procedures. If you do not discuss this chapter in lecture, we ask that you assign it as required reading. Or simply highlight the points relevant to your program. If you are a new teaching assistant doing this for the first time, you will have to know this material in your sleep; it is simply part of the job.

You will note that the order of presentation has been changed in the eighth edition such that UV-Vis and vibrational spectrometry follows introductory spectroscopy, then NMR and AAS and AES. We think this is a more natural progression, but feel free to modify the presentation to suit your students' needs.

Acknowledgments

The following people are gratefully acknowledged for their assistance in the successful completion of this edition. They provided diagrams, photographs, application notes, spectra, chromatograms, and many helpful comments and suggestions regarding the text. Thanks are due, in no specific order, to Birke Götz, **BÜCHI Labortechnik AG**; the entire **TA Instruments** team with special thanks to Roger Blaine, Fred Wiebke, Charles Potter, and Terry Allen; Gwen Boone, Doug Shrader, Ed McCurdy, Pat Grant, Laima Baltusis, Dan Steele, Jim Simon, Paul Canavan, Amy Herlihy, Eric Endicott, Valerie Lopez, William Champion, and Matt Nikow, **Agilent Technologies**; Dr. A. Horiba, Atsuro Okada, Juichiro Ukon, Mike Pohl, Philippe Hunault, Patrick Chapon, Phil Shymanski, Diane Surine, Joanne Lowy, Andrew Whiteley, Christophe Morin, and David Tuschel, **HORIBA Scientific**; Peter Brown, Dave Pfeil, and Dr. Manuel Almeida, **Teledyne Leeman Labs, Inc.**; Steve Sauerbrunn, **Mettler Toledo, Inc.**; Mark Mabry, Bob Coel, Ed Oliver, Lara Pryde, Jim Ferrara, John Flavell, Chuck Douthitt, Keith Bisogno, Mary Meegan-Litteer, Jackie Lathos-Markham, Todd Strother, Joseph Dorsheimer, Marty Palkovic, Dr. Julian Phillips, Wendy Weise, Carl Millholland, Michael Bradley, Janine O'Rourke, Eric Francis, Art Fitchett, Dr. Stephan Lowry, Timothy O. Deschaines, Fergus Keenan, Simon Nunn, Ryan Kershner, Elizabeth Guiney, Russell Diemer, Michael W. Allen, Bill Sgammato, Dr. John Wolstenholme, Dr. Joachim Hinrichs, Carolyn Carter, Kathy Callaghan, Allen Pierce, Arthur Fitchitt, Fraser McLeod, and Frank Hoefler, **Thermo Fisher Scientific**; Mary Allen, **Allen Design**; Todd Maxwell, **CETAC Technologies**; David Coler, **PANalytical, Inc.**; William Ickes, **Buchi Corporation**; Kenneth Krewson, **Supercritical Fluid Technologies, Inc.**; Jim McKinley, Dale Edcke, Bob Anderhalt, **AMETEK**; Mike Hurt and Luis Moreno, **Hitachi High Technologies America**; Michael Nelson, **Rigaku Corporation**; Robert Bartek, **Applied Rigaku Technologies, Inc.**; Dr. Tom Dulski, **Carpenter Technologies**; Dr. Alex Medek, **Pfizer Central Research and Development**; Dr. James Roberts, **Lehigh University**; Dr. Thorsten Thiel, Kodi Morton, Catherine Fisk, Andrew Hess, Armin Gross, Sarah Nelson, and Jerry Sooter, Mark Chaykovsky, **Bruker Corporation**; Pat Wilkinson and James Beier, **Bruker BioSpin**; Professor F. X. Webster, **State University of New York College of Environmental Science and Forestry**; Pat Palumbo, Bill Strzynski, Lorne M. Fell, and Veronica Jackson, **LECO Corporation**; Dr. Cedric Powell, Dr. William M. Haynes, Dr. Megan Harries, **NIST**; Jill Thomas and Michael Monko, **Supelco**; Michael Garriques and Karen Anspach, **Phenomenex, Inc.**; A. Audino and Kerry Scoggins, **SGE**; Dr. Ronald Starcher, **BURLE Electro-Optics, Inc.**; Merrill Loechner, **Milestone, Inc.**; Dr. Mike Collins, **CEM Corporation**; Professor Gary Siudzak, **Scripps Research Institute Center for Mass Spectrometry**; Dr. Robert Kobelski, **Centers for Disease Control**, Atlanta, Georgia; Professor David Hercules, **Vanderbilt University**; Dr. S.E. Stein and Anzor I. Mikaia, **NIST Mass Spec Data Center**; Volker Thomsen, **NITON Corp.**; Dr. Gilles Widawski and Fumi Akimaru, **NETZSCH Instruments North America, LLC**; Dr. Thomas Rampke and Stephan Knappe, **NETZSCH-Gerätebau GmbH**; Michael Fry, Dr. Chip Cody, Pam Mansfield, Masaaki Ubukata, and Patricia Corkum, **JEOL, Inc.**; Andy Rodman, Giulia Orsanigo, Christopher Tessier, Sarah Salbu, and Danielle Hawthorne, **PerkinElmer, Inc.**; Nancy Fernandes, **Newport Corporation**; Ralph Obenauf, **SPEX CertiPrep, Inc.**; Rolf Schlake, **Applied Separations, Inc.**; Paul Smaltz and Anne Logan, **Avantor Performance Materials, Inc.**; Donald Bouchard, **Anasazi Instruments, Inc.**; John Hammond, Rosemary Huett, and Keith Hulme, **Starna Scientific, Inc.**; John Price, Dean Antic, and Chuck Miller, **picoSpin, LLC**; Jenni L. Briggs and Z. Stanek, **PIKE Technologies**; Dawn Nguyen, **Enwave Optronics, Inc.**; Alicia Kimsey, Claire Dentinger, **Rigaku Raman Technologies, Inc.**; Steve Pullins and Eric Bergles, **BaySpec, Inc.**; Evan Friedmann, **Hellma USA, Inc.**; Eric Shih, Mark Salerno, **C Technologies**; Monique Claverie, Dr. Joseph E. Johnson, and Dr. David Wooton, **Microspectral Analysis, LLC**; Keith Long, Mark Talbott, Mark Taylor, and Kevin McLaughlin, Herb Pottoff, **Shimadzu Scientific Instruments, Inc.**; Steve Buckley and Edwin Pickins, **TSI, Inc.**; Chris Beasley, Burak Uglut, **Gamry Instruments**; Mark Handley, **International Crystal Manufacturing, Inc.**; Eoin Vincent, Samuel Machado, and John Nikitas, **Olympus NDT**; Dr. John C. Edwards, **Process NMR Associates, LLC**; Mary Jo Wojtusik and Michele Giordiano, **GE Healthcare Life Sciences**; Bryan C. Webb, **Norton Scientific Inc.**; Ryan Brennan, **Glass Expansion, Inc.**; Heather Grael, **Implen, Inc.**; Ray Wood, **AstraNet Systems, Ltd.**; Jamie Huffnagle, Len Morrisey, and Brent Cleveland, **ASTM International**; Brian J. Murphy and Dave DePasquale, **Waters Corp.**; Joseph H. Kennedy, **Prosolia**; Carol Morloff, **Excellims**; Margaret M. Cooley, **Photonis**; Stanton Loh, **Palisade**; Mike Festa, **IonSense Inc.**; Yongdong Wang and Ming Gu, **CERNO Inc.**; Jeff Okamitsu, **Chemring Inc.**

Special thanks go to the late Professor Milton Orchin, **University of Cincinnati**, for permission to use material from his and the late H.H. Jaffe's classic text, *Theory and Applications of Ultraviolet Spectroscopy*, Wiley, New York, 1962. Long-term collaboration with Professor Paris D.N. Svoronos of QCC of City University of New York is acknowledged. The cheerful cooperation, patience, and invaluable assistance of Marie Scandone, **Bio-Rad Informatics Division**, in providing the Bio-Rad NMR,

UV, and IR spectra used in Chapters 3 through 5 and of Wes Rawlins, Cindy Addenbrook, Sean Battles, Dean Llanas, Chris Wozniak, and Chris Lein, **Sigma-Aldrich**, in providing the Aldrich NMR spectra in Chapter 3 and the Supelco graphics in several chapters deserve a very, very special thank-you. Herk Alberry, Mike Farrell, and Danny Dirico of **Albany Advanced Imaging**, Albany, New York, are thanked for the MRI images and technical assistance they provided for Chapter 3. To Dr. Christian Bock, **Alfred-Wegner-Institute for Polar and Marine Research**, Bremerhaven, Germany, for the use of his MRI images in Chapter 3, Vielen Dank. Julie Powers, **Toshiba America Medical Systems**, is gratefully acknowledged for the medical MRI system photos used in Chapter 3. Special thanks go to Dr. Frank Dorman, **Restek**, for providing their EZ-Chrom Chromatography Simulation Program, used to create several of the figures in Chapter 12, and to Dr. Jack Cochran and Pam Decker, **Restek**, for invaluable assistance. Dr. Z. Harry Xie, **Bruker Optics, Inc.**, is gratefully acknowledged for providing the idea for and a large portion of Section 3.9 on TD-NMR. Amy Herlihy, **Agilent Technologies**, provided the information and example for SWIFT NMR. Todd Strother, **Thermo Fisher Scientific**, provided invaluable assistance with NIR in Chapter 4, both with graphics and an excellent summary of chemometrics. Timothy O. Deschaines, formerly with **Thermo Fisher Scientific**, is thanked for his assistance with updating the Raman section, especially SERS. Harald Fischer, **WITec GmbH**, contributed the sections and provided graphics for 3D confocal Raman imaging and true surface microscopy in Chapter 4. Thanks to Dr. Anne Lynn Gillian-Daniel, NMRFAM, **University of Wisconsin**, Madison (nmrfam.wisc.edu), for the 900 MHz NMR photos in Chapter 3. The **Bruker BioSpin** website was invaluable for its EPR (ESR) tutorial, written by Dr. Ralph Weber. Dr. Weber contributed a large portion of Section 6.12, with many helpful comments. Dr. Alexander Seyfarth, **Bruker AXS**, who significantly revised Chapter 8, is initiating interactive exercises to be available online in the near future and is now the coauthor of that chapter. Dr. James R. White, Christopher J. White, and Colin T. Elliott, **Active Spectrum, Inc.**, are thanked for their graphics and technical assistance with Micro-ESR. Dr. Andrzej Miziolek, **US Army Research Laboratory**, and Dr. Andrew Whitehouse, **Applied Photonics Ltd.**, provided technical advice, literature, application notes, and graphics for the section on LIBS. Drs. Doug Shrader and Ed McCurdy, **Agilent Technologies**, provided the graphics and examples for the description of their newly introduced ICP-MS triple quadrupole instrument. Alan Merrick, **SPECTRO Analytical Instruments**, provided the SPECTRO MS graphics. Dr. John F. Moulder, **Physical Electronics USA, Inc.**, provided many expert suggestions and revision of a large portion of Chapter 14, as well as most of its graphics. Dr. Hans Joachim Schaefer and C.-A. Schiller, **ZAHNER-Elektrik GbmH & Co.KG**, and Professor Stephen P. Best, **University of Melbourne**, Australia, provided graphics and examples for the discussion of spectroelectrochemistry. Dr. Chris Beasley, **Gamry Instruments, Inc.**, contributed, reviewed, and provided the majority of graphics in the section on the quartz crystal microbalance in Chapter 15. Dr. Christin Choma, is thanked for her invaluable help with understanding the applications of microcalorimetry to the study of proteins.

Thanks go to Irwin Schreiman, **CambridgeSoft Corporation**, who provided the authors with a complete copy of ChemDraw Ultra 12.0.

The authors wish to thank the following colleagues for their helpful comments and suggestions: Dr. O. David Sparkman, **Pacific University**; Professor Ronald Bailey, **Rensselaer Polytechnic Institute**; Professor Peter Griffiths, **University of Idaho**; Dr. Elizabeth Williams; Professor Julian Tyson, **University of Massachusetts, Amherst**. Professor Emeritus Robert Gale, **LSU**, who coauthored Chapter 15 for the fifth edition, is acknowledged for his substantive contributions to and review of that chapter.

The periodic table on the inside front cover is used with permission of Dragoset, R.A., Musgrove, A., Clark, C.W. and Martin, W.C., Periodic Table: Atomic Properties of the Elements, NIST SP 966 handout, August, 2019 (National Institute of Standards and Technology, Gaithersburg, MD, www.nist.gov).

Permission for the publication herein of Sadtler Spectra has been granted by Bio-Rad Laboratories, Informatics Division. All Sadtler Spectra are Copyright Bio-Rad Laboratories, Informatics Division, Sadtler Software and Databases 2014. All rights reserved.

J.T.Baker® and ULTREX® are trademarks of **Avantor Performance Materials, Inc**.

Last but certainly not least, Barbara Glunn Knott, senior editor—analytical, industrial, and physical chemistry; Alexis O'Brien, editor, Sherry Thomas, senior editorial assistant, Fiona Macdonald, senior publisher, Jonathan Pennell, cover design manager, and **all of CRC Press/Taylor & Francis Group**, are thanked for their incredible patience with and invaluable assistance to the authors.

Authors

Thomas J. Bruno was group leader in the Applied Chemicals and Materials Division at NIST, Boulder, Colorado, before retiring in 2019. He received his B.S. in chemistry from Polytechnic Institute of Brooklyn, and his M.S. and Ph.D. in physical chemistry from Georgetown University. He first served as a National Academy of Sciences-National Research Council postdoctoral associate at NIST. Dr. Bruno has researched fuels, explosives, forensics, and environmental pollutants. He was also involved in research on chemical separations, development of novel analytical methods, and novel detection devices for chromatography. Among his inventions are the Advanced Distillation Curve method (for fuel characterization), and PLOT-cryoadsorption (for vapor sampling). He has published over 270 research papers, eight books, and has been awarded 10 patents. He serves as associate editor of the *CRC Handbook of Chemistry and Physics*, and associate editor for *Fuel Processing Technology*. Bruno was awarded the Department of Commerce Bronze Medal in 1986 for his work "on the thermophysics of reacting fluids", and the Department of Commerce Silver Medal in 2010 for "a new method for analyzing complex fluid mixtures for introduction of new fuels into the U.S. energy infrastructure." He was Distinguished Finalist for the 2011 CO-Labs Governor's Award for High Impact Research, and received the American Chemical Society Colorado Section Research Award in 2015. He has served as a forensic consultant and/or an expert witness for the U.S. Department of Justice (notably during the federal trial of Terry Nichols for the Oklahoma City bombing), various United States Attorney's offices, and various offices of the U.S. Inspector General. He received a letter of commendation from the Department of Justice for these efforts in 2002. He is currently working to develop science and technology education programs for the American judiciary. In this capacity he serves on the boards of the National Courts and Science Institute, and the Bryson Institute for Judicial Education. In addition, he was chairman of the Board of Trustees of the Mamie Doud Eisenhower Library for nine years. In his spare time, he is an avid and accomplished woodworker (and a less-than-accomplished welder and metal worker), and has hand-built all the furniture in his house.

The late James W. Robinson earned his B.S. (Hons), Ph.D., and D.Sc. from the University of Birmingham, England. He was professor emeritus of chemistry, Louisiana State University, Baton Rouge, Louisiana. A fellow of the Royal Society of Chemistry, he is the author of 250 professional papers, book chapters, and several books including *Atomic Absorption Spectroscopy* and *Atomic Spectroscopy*, first and second editions. He was editor in chief of *Spectroscopy Letters* and the *Journal of Environmental Science and Health* (both Marcel Dekker, Inc.); executive editor of *Handbook of Spectroscopy Vol. 1* (1974), *Vol. 2* (1974), *Vol. 3* (1981); and *Practical Handbook of Spectroscopy* (1991) (all CRC Press). He served on the National University Accreditation Committee from 1970–1971. He was a visiting distinguished professor at University of Colorado in 1972 and University of Sydney, Australia in 1975. He served as the Gordon Conference Chairman in Analytical Chemistry in 1974. Professor Emeritus James W. Robinson passed away in November, 2018, at 95 years of age.

The late Eileen M. Skelly Frame was adjunct professor, Department of Chemistry and Chemical Biology, Rensselaer Polytechnic Institute (RPI), Troy, NY, and head of Full Spectrum Analytical Consultants. Dr. Skelly Frame was the first woman commissioned from the Drexel University Army ROTC program. She graduated from Drexel *summa cum laude* in chemistry. She served as Medical Service Corps officer in the U.S. Army from 1975 to 1986, rising to the rank of Captain. For the first five years of her military career, Dr. Skelly Frame was stationed at the 10th Medical Laboratory at the U.S. Army Hospital in Landstuhl Germany. Thereafter, she was selected to attend a three-year Ph.D. program in Chemistry at Louisiana State University. She received her doctorate in 1982 and became the first female chemistry professor at the U.S. Military Academy at West Point. Following her military service, she joined the General Electric Corporation (now GE Global Research) and supervised the atomic spectroscopy laboratory. In addition to her duties at RPI, she was Clinical and Adjunct Professor of Chemistry at Union College in Schenectady, NY. She was well known for her expertise in the use of instrumental analysis to characterize a wide variety of substances, from biological samples and cosmetics to high-temperature superconductors, polymers, metals, and alloys. She was an active member of the American Chemical Society for 45 years, and a member of several ASTM committees. Dr. Skelly Frame passed away in January of 2020.

George M. Frame II is a retired scientific director, Chemical Biomonitoring Section of the Wadsworth Laboratory, New York State Department of Health, Albany. He has a wide range of experience in analytical chemistry and has worked at the GE Corporate R&D Center (now GE Global Research), Pfizer Central Research, the US Coast Guard R&D Center, the Maine Medical Center, and in the US Air Force Biomedical Sciences Corps. He is a member of the American Chemical Society. Dr. Frame earned his AB in chemistry from Harvard College, Cambridge, Massachusetts, and his Ph.D. in analytical chemistry from Rutgers University, New Brunswick, New Jersey.

CHAPTER 1

Concepts of Instrumental Analytical Chemistry

1.1 INTRODUCTION: WHAT IS ANALYTICAL CHEMISTRY?

Perhaps the most functional definition of analytical chemistry is that it is "the qualitative and quantitative characterization of matter." The word "characterization" is used in a very broad sense, and this is intentional because it can encompass material properties as well as composition. It may mean the identification of the chemical compounds or elements present in a sample to answer questions such as "Is there any vitamin E in this shampoo as indicated on the label?" or "Is this white tablet an aspirin tablet?" or "Is this piece of metal iron or nickel?" This type of characterization, to tell us *what* is present, is called qualitative analysis. **Qualitative analysis** is the identification of one or more chemical species present in a material. Characterization may also mean the determination of how much of a particular compound or element is present in a sample, to answer questions such as "How much acetylsalicylic acid is in this aspirin tablet?" or "How much nickel is in this steel?" This determination of *how much* of a species is present in a sample is called quantitative analysis. **Quantitative analysis** is the determination of the exact amount of a chemical species present in a sample. The chemical species may be an element, compound, or ion. The compound may be organic or inorganic. Characterization can refer to the entire sample (*bulk analysis*), such as the elemental composition of a piece of steel, or to the surface of a sample (*surface analysis*), such as the identification of the composition and thickness of the oxide layer that forms on the surface of most metals exposed to air and water. The characterization of a material may go beyond chemical analysis to include structural determination of materials, the measurement of physical properties of a material, and the measurement of physicochemical parameters such as reaction kinetics. Indeed, to the extent that properties depend on the chemical identity, it can be argued that all properties of matter are the result of the chemical composition.

Examples of such physicochemical measurements are the degree to which a polymer is crystalline as opposed to amorphous, the temperature at which a material loses its water of hydration, how long it takes for antacid "brand A" to neutralize stomach acid, and how fast a pesticide degrades in sunlight. These diverse applications make analytical chemistry one of the broadest in scope of all scientific disciplines. Analytical chemistry is critical to our understanding of biochemistry; medicinal chemistry; geochemistry; environmental science; atmospheric chemistry; the behavior of materials such as polymers, metal alloys, and ceramics; and many other scientific disciplines.

For many years, analytical chemistry relied on chemical reactions to identify and determine the components present in a sample. These types of classical methods, often called "wet chemical methods," usually required that a part of the sample be taken and dissolved in a suitable solvent if necessary and the desired reaction carried out. The most important analytical fields based on this approach were volumetric and gravimetric analyses. Acid–base titrations, oxidation–reduction titrations, and gravimetric determinations, such as the determination of silver by precipitation as silver chloride, are all examples of wet chemical analyses. These types of analyses require a high degree of skill and attention to detail on the part of the analyst if accurate and precise results are to be obtained. They are also time consuming, and the demands of today's high-throughput pharmaceutical development labs, forensic labs, commercial environmental labs, and industrial quality control labs often do not permit the use of such time-consuming methods for routine analysis. In addition, it may be necessary to analyze samples without destroying them. Examples include evaluation of valuable artwork to determine if a painting really is by a famous "old master" or is a modern forgery, as well as in forensic analysis, where the evidence may need to be preserved. For these types of analyses, **nondestructive analysis** methods are needed, and wet chemical analysis will not do the job. Wet chemical analysis is still used in specialized areas of analysis, and is sometimes cost effective, but many of the volumetric methods have been transferred to automated instruments. Classical analysis and instrumental analysis are similar in many respects, such as in the need for proper sampling, sample preparation, assessment of accuracy and precision, and proper record keeping. Some of the topics discussed briefly in this chapter are covered at greater length in more general texts on analytical chemistry and quantitative analysis. Several of these types of texts are listed in the bibliography; the book by Bruno and Svoronos has a comprehensive treatment including flow charts.

Most analyses today are carried out with specially designed electronic instruments controlled by computers. These instruments make use of the interaction of electromagnetic radiation and matter, or of some physical property of matter, to characterize the sample being analyzed. Often these instruments have automated sample introduction, data processing, and even sample preparation. To understand how the instrumentation operates and what information it can provide requires knowledge of chemistry, physics, mathematics, and engineering. The fundamentals of common analytical instruments and how measurements are performed with these instruments are the subjects of the following chapters on specific instrumental techniques. The analytical chemist must not only know and understand analytical chemistry and instrumentation but must also be able to serve as a problem solver to colleagues in other scientific areas. This means that the analytical chemist may need to understand materials science, metallurgy, biology, pharmacology, agricultural science, food science, geology, and other fields. The field of analytical chemistry is advancing rapidly. To keep up with the advances, the analytical chemist must understand the fundamentals of common analytical techniques and their capabilities and shortcomings. The analytical chemist must understand the problem to be solved, select the appropriate technique or techniques to be used, design the analytical experiment to provide relevant data, and ensure that the data obtained are valid. Merely providing data to other scientists is not enough; the analytical chemist must be able to interpret the data and communicate the meaning of the results, together with the accuracy and precision (the reliability) of the data, to scientists who will use the data. In addition to understanding the scientific problem, the modern analytical chemist often must also consider factors such as time limitations and cost limitations in providing an analysis. Whether one is working for a government regulatory agency, a hospital, a private company, or a university, analytical data must be legally defensible. It must be of known, documented quality. Record keeping, especially computer record keeping; assessing accuracy and precision; statistical handling of data; documenting; and ensuring that the data meet the applicable technical standards are especially critical aspects of the job of modern analytical chemists.

Analytical chemistry uses many specialized terms that may be new to you. The definitions of these terms, usually shown in boldface, must be learned. The units used in this text are, for the most part, the units of the Système International d'Unités (SI system). The SI system is used around the world by scientists and engineers. The tables inside the textbook covers give the primary units of measurement in the SI system. A comprehensive list of SI units, SI-derived units and definitions, as well as non-SI units may be found at the US National Institute of Standards and Technology (NIST) website at http://physics.nist.gov.

Many analytical results are expressed as the concentration of the measured substance in a certain amount of sample. The measured substance is called the **analyte**. Commonly used concentration units include molarity (moles of substance per liter of solution), mass percent (grams of substance per gram of sample $\times$ 100%), and units for trace levels of substances. One *part per million* (ppm) by mass is one microgram of analyte in a gram of sample, that is, 1×10^{-6} g analyte/g sample. One *part per billion* (ppb) by mass is one nanogram of analyte in a gram of sample or 1×10^{-9} g analyte/g sample. For many elements, the technique known as inductively coupled plasma mass spectrometry (ICP-MS) can detect *parts per trillion* of the element, that is, picograms of element per gram of sample (1×10^{-12} g analyte/g sample). To give you a feeling for these quantities, a million seconds is ~12 days (11.57 days, to be exact). One part per million in units of seconds would be 1 s in 12 days. A part per billion in units of seconds would be 1 s in ~32 years, and one part per trillion is 1 s in 32,000 years. Today, lawmakers set environmental levels of allowed chemicals in air and water based on measurements of compounds and elements at part per trillion levels because instrumental methods can detect part per trillion levels of analytes. It is the analytical chemist who is responsible for generating the data that these lawmakers rely on. A table of commonly encountered constants, multiplication factors, and their prefixes is found inside the textbook cover (a comprehensive listing can be found in the book by Bruno and Svoronos). The student should become familiar with these prefixes, since they will be used throughout the text.

1.1.1 The Difference between Weight and Mass

The terms weight and mass are often used interchangeably, and that is usually fine on planet Earth. Mass is the amount of substance (an inertial property, the sum of all the electrons, protons and neutrons) in an object, while weight is the gravitational interaction between objects that have mass. A (stationary) electronic balance measures the gravitational force of an object (i.e., its weight) by use of a load cell, and converts it to mass mathematically (weight = mass $\times$ gravitation, $W = mg$, where g is roughly 9.81 N/kg). Use that same balance on Mars in the rover and you will not get the correct mass. Use that same balance in a moving elevator and, likewise, you will not get the correct mass. Some highly precise measurements require the measurement of g locally to obtain a correction factor, but for analytical measurements (and especially for weighing by difference) the conversion to mass used by electronic balances is adequate. Buoyancy corrections were commonly required when two pan analytical balances were common (and are still required when weighing large objects such as gas cylinders), but for electronic balances, this consideration is typically only required for a calibration. So, what does this mean for our analytical measurements? It means that our balance will measure weight, and provide us with mass. It is therefore more correct to say: mass. And while we are on the topic, remember that the instrument that you use to measure the weight and read out the mass is called a balance, not a scale. Scientists use a balance; butchers use a scale.

1.2 ANALYTICAL APPROACH

A major personal care products manufacturer receives a phone call from an outraged customer whose hair has turned green after using their "new, improved shampoo." The US Coast Guard arrives at the scene of an oil spill in a harbor and finds two ship's captains blaming each other for the spill. A plastics company that sells bottles to a water company bottling "pure, crystal-clear spring water" discovers that the 100,000 new empty bottles it is ready to ship are slightly yellow in color instead of crystal clear. A new, contagious disease breaks out, and people are dying of flu-like symptoms. What caused the problem? How can it be prevented in the future? Who is at fault? Can a vaccine or drug treatment be developed quickly? These sorts of problems and many more occur daily around the world, in industry, in medicine, and in the environment. A key figure in the solution of these types of problems is the analytical chemist. The analytical chemist is first and foremost a problem solver and, to do that, must understand the analytical approach, the fundamentals of common analytical techniques, their uses, and their limitations.

The approach used by analytical chemists to solve problems may include the following steps:

1. Defining the problem and designing the analytical method.
2. Sampling and sample storage.
3. Sample preparation.
4. Performing the measurement.
5. Assessing the data and the corresponding uncertainty.
6. Method validation.
7. Documentation.

General sample preparation will be discussed in this chapter, but instrument-specific sample preparation is included in the appropriate chapter on each technique. Method validation and documentation will not be covered as the focus of this text is on instrumentation. The text by Christian cited in the bibliography has an excellent introduction to validation and documentation for the interested student.

Although the steps in solving analytical problems usually follow the listed order, knowledge of basic statistics is useful not just for handling the data and method validation but is required for proper sampling and selection of an analytical method. The statistics and definitions needed to understand what is meant by **accuracy**, **precision**, **error**, and so on are covered in Section 1.3. Steps 1 and 2 are covered in this section, while steps 3 through 5 are discussed in the sections following Section 1.3.

1.2.1 Defining the Problem

The analytical chemist must find out what information needs to be known about the sample, material, or process being studied, how accurate and precise the analytical information must be, how much material or sample is available for study, and if the sample must be analyzed without destroying it. Is the sample organic or inorganic? Is it a pure material or a mixture? Does the customer want a bulk analysis or information about a particular fraction of the sample, such as the surface? Does the customer need to know if the sample is homogeneous or heterogeneous with respect to a given analyte? Does the customer need elemental information or information about the chemical species (ionic or molecular, particular oxidation states) present in the sample? The answers to such questions will guide the analyst in choosing the analytical method. Of course, sometimes the answers to some of the questions may be part of the problem. If the sample is an unknown material, the analyst must find out if it is organic or inorganic, pure or a mixture, as part of solving the problem. The analyte is the substance to be measured; everything else in the sample is called the **matrix**. Of course, there may be more than one analyte in a given sample. The terms *analysis* and *analyze* are applied to the sample under study, as in "this water was analyzed for nitrate ion" or "an analysis of the contaminated soil was performed." Water and soil are the samples being analyzed. The terms *determine* and *determination* are applied to the measurement of the analyte in the sample, as in "nitrate ion was determined in the water sample," "a determination of lead in blood was made because the symptoms indicated lead poisoning," or "an analysis of the soil was performed and cyanide levels were determined." Nitrate ion, lead, and cyanide are the analytes being determined. Other components in the sample matrix may interfere with the measurement of the analyte; such components are called **interferences**.

A sample may be **homogeneous**, that is, it has the same chemical composition everywhere within the sample. Plain vanilla ice cream, a milk chocolate bar, and salt water are examples of homogeneous materials. Many samples are **heterogeneous**; the composition *varies* from region to region within the sample. Vanilla ice cream with raisins in it and a chocolate bar with whole almonds in it are heterogeneous; you can see the composition difference. In most real samples, the heterogeneity may not be visible to the human eye. The variation in composition can be *random*, or it can be *segregated* into regions of distinctly different compositions.

A significant part of defining the problem is the decision between performing a qualitative analysis and a quantitative analysis. Often the problem is first tackled with a qualitative analysis, followed by a quantitative analysis for specific analytes. The analyst needs to communicate with the customer who is requesting the analysis. Two-way communication is important, to be certain that the problem to be solved is understood and to be sure that the customer understands the capabilities and limitations of the analysis.

1.2.1.1 Qualitative Analysis

Qualitative analysis is the branch of analytical chemistry that is concerned with questions such as "What makes this water smell bad?" "Is there gold in this rock sample?" "Is this sparkling stone a diamond or cubic zirconia?" "Is this plastic item made of polyvinyl chloride, polyethylene, or polycarbonate?" "What is this white powder?"

Some methods for qualitative analysis are nondestructive, that is, they provide information about what is in the sample without destroying the sample. These are often the best techniques to begin with, because the sample can be used for subsequent analyses if necessary. To identify what elements are present in a sample nondestructively, a *qualitative elemental analysis* method such as X-ray fluorescence (XRF) spectroscopy can be used. Modern XRF instruments, discussed in Chapter 9, can identify all elements from sodium to uranium, and some instruments can measure elements from beryllium to uranium. The sample is usually not harmed by XRF analysis. For example, XRF could easily distinguish a diamond from cubic zirconia. Diamond is, of course, a crystalline form of carbon; most XRF instruments would see no elemental signal from the carbon in a diamond but would see a strong signal from the element zirconium in cubic zirconia, a crystalline compound of zirconium and oxygen. *Qualitative molecular analysis* will tell us what molecules are present in a material. The nondestructive identification of molecular compounds present in a sample can often be accomplished by the use of nuclear magnetic resonance (NMR) spectroscopy, discussed in Chapter 6, or by infrared (IR) spectroscopy, discussed in Chapter 5. IR spectroscopy can provide information about organic functional groups present in samples, such as alcohols, ketones, carboxylic acids, amines, and thioethers. If the sample is a pure compound such as acetylsalicylic acid (the active ingredient in aspirin), the IR spectrum may be able to identify the compound exactly, because the IR spectrum for a compound is unique, like a fingerprint. Qualitative identification of polymers for recycling can be done using IR spectroscopy, for example. NMR gives us detailed information about the types of protons, carbon, and other atoms in organic compounds and how the atoms are connected. NMR can provide the chemical structure of a compound without destroying it.

Many methods used for qualitative analysis are destructive; either the sample is consumed during the analysis or must be chemically altered in order to be analyzed. The most sensitive and comprehensive elemental analysis methods for inorganic analysis are ICP atomic emission spectrometry (ICP-AES or ICP optical emission spectrometry [ICP-OES]), discussed in Chapter 8, and ICP-MS, discussed in Chapters 10 and 11. These techniques can identify almost all the elements in the periodic table, even when only trace amounts are present, but often require that the sample be in the form of a solution. If the sample is a rock or a piece of glass or a piece of biological tissue, the sample usually must be dissolved in some way to provide a solution for analysis. We will see how this is done later in this chapter. The analyst can determine accurately what elements are present, but information about the molecules in the sample is often lost in the sample preparation process. The advantage of ICP-OES and ICP-MS is that they are very sensitive; concentrations at or below 1 ppb of most elements can be detected using these methods.

If the sample is organic, that is, composed primarily of carbon and hydrogen, qualitative analysis can provide chemical and structural information to permit identification of the compound. The IR spectrum will provide identification of the class of compound, for example, ketone, acid, or ether. NMR spectroscopy and mass spectrometry (MS), as we shall see in the appropriate chapters, provide detailed structural information, often including the relative molecular mass (RMM, or MW) of the compound. Use of IR, NMR, and MS, combined with quantitative elemental analysis to accurately determine the percentage of carbon, hydrogen, oxygen, and other elements, is the usual process by which analytical chemists identify organic compounds. This approach is required to identify new compounds synthesized by pharmaceutical chemists, for example. In a simple example, elemental analysis of an unknown organic compound might provide an **empirical formula** of C_2H_5. An empirical formula is the simplest whole number ratio of the atoms of each element present in a molecule. For any given compound, the empirical formula may or may not coincide with the **molecular formula**. A molecular formula contains the total number of atoms of each element in a single molecule of the compound. The results from IR, NMR, and MS might lead the analytical chemist to the molecular formula C_4H_{10} and would, in addition, indicate which of the following two different structures shown our sample was:

$$CH_3-CH_2-CH_2-CH_3 \qquad H_3C-\underset{\underset{CH_3}{|}}{\overset{\overset{H}{|}}{C}}-CH_3$$

n–Butane   2-Methyl-propane (or iso-butane)

These two structures are two different compounds with the same molecular formula. They are called **isomers**. Elemental analysis cannot distinguish between these isomers, but NMR and MS usually can distinguish isomers. Another example of a more difficult qualitative analysis problem is the case of the simple sugar, erythrose. The empirical formula determined by elemental analysis is CH_2O. The molecular formula, $C_4H_8O_4$, and some of the structure can be obtained from IR, NMR, and MS, but we

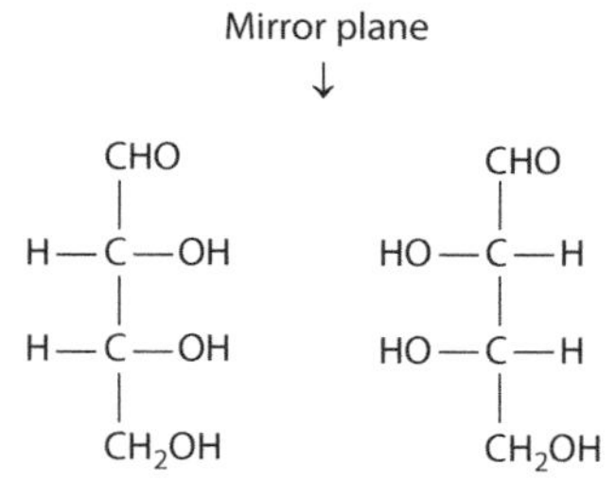

Figure 1.1 Isomers of erythrose.

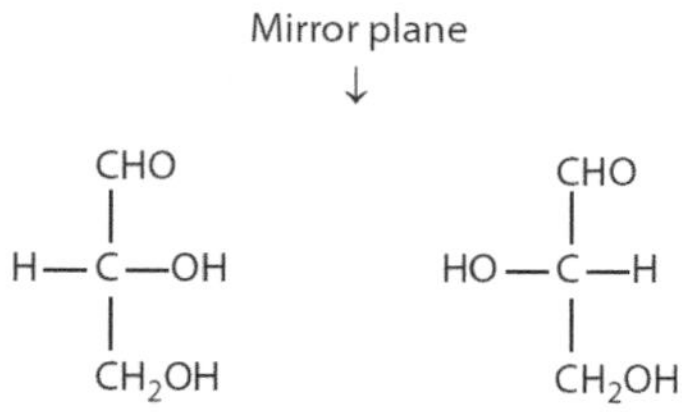

```
      Mirror plane
           ↓
   CHO              CHO
    |                |
H—C—OH         HO—C—H
    |                |
H—C—OH         HO—C—H
    |                |
  CH₂OH            CH₂OH

D-(−)-Erythrose   L-(+)-Erythrose
```

Figure 1.2 Enantiomers of glyceraldehyde.

```
      Mirror plane
           ↓
   CHO              CHO
    |                |
H—C—OH         HO—C—H
    |                |
  CH₂OH            CH₂OH

D-(+)-Glyceraldehyde   L-(−)-Glyceraldehyde
```

cannot tell from these techniques which of the two possible isomers shown in Figure 1.1 is our sample.

These two erythrose molecules are **chiral**, that is, they are nonsuperimposable mirror-image isomers, called **enantiomers**. Imagine sliding the molecule on the left *in the plane of the paper*, through the "mirror plane" indicated by the arrow, over the molecule on the right; the OH groups will not be on top of each other. Imagine turning the left molecule *in the plane of the paper* upside down and then sliding it to the right; now the OH groups are lined up, but the CHO and CH$_2$OH groups are not. That is what is meant by nonsuperimposable. You can do whatever you like to the two molecules except remove them from the plane of the paper; no matter how you move them, they will not be superimposable. They have the same molecular formula ($C_4H_8O_4$); the same IR spectrum, mass spectrum, and NMR spectrum; and many of the same physical properties such as boiling point and refractive index (RI). Such chiral compounds can be distinguished from each other by interaction with something else that possesses chirality or by interaction with plane-polarized light. Chiral compounds will interact differently with other chiral molecules, and this interaction forms the basis of chiral chromatography. Chiral chromatography (Chapter 14) can be used to separate the two erythrose compounds shown. Chiral compounds also differ in their behavior toward plane-polarized light, and the technique of polarimetry can be used to distinguish them. One of the erythrose enantiomers rotates plane-polarized light to the right (clockwise); this compound is *dextrorotatory* and is given the symbol (+) as part of its name. The other enantiomer rotates the plane of polarization to the left (counterclockwise); this compound is *levorotatory* and is given the symbol (−) in its name. Such compounds are said to be optically active. Chiral compounds are very important because biochemical reactions are selective for only one of the two structures and only one of the two enantiomers is biologically active. Biochemists, pharmaceutical chemists, and medicinal chemists are very interested in the identification, synthesis, and separation of only the biologically active compound. The letters D and L in the name of the sugar refer to the position of the alcohol group on the carbon closest to the bottom primary alcohol. There is no relationship between the D and L configuration and the direction of rotation of plane-polarized light. Figure 1.2 shows the simplest sugar, glyceraldehyde. It also has two enantiomers, one D and one L, but the D enantiomer of glyceraldehyde rotates light in the opposite direction from D-erythrose.

If organic compounds occur in mixtures, separation of the mixture often must be done before the individual components can be identified. Techniques such as gas chromatography (GC), liquid chromatography (LC), and capillary electrophoresis (CE) are often used to separate mixtures of organic compounds prior to identification of the components. These methods are discussed in Chapters 12 through 14.

Table 1.1 lists some common commercially available instrumental methods of analysis and summarizes their usefulness for qualitative elemental or molecular analysis. Table 1.2 gives a very brief summary of the use of the methods. Analyte concentrations that can be determined by common methods of instrumental analysis are presented in Table 1.3. The concentration of analyte that can be determined in real samples will depend on the sample and on the instrument, but Table 1.3 gives some indication of the sensitivity and working range of methods.

1.2.1.2 Quantitative Analysis

When qualitative analysis is completed, the next question is often "How much of each or any component is present?" or "Exactly how much gold is this rock?" or "How much of the organochlorine pesticide dieldrin is in this drinking water?" The determination of how much is quantitative analysis. Analytical chemists express how much in a variety of ways, but often in terms of *concentration*, the amount of analyte in a given amount of sample. Concentration is an expression of the quantity of analyte in a given volume or mass of sample. Common concentration units include molarity, defined as moles of analyte per liter of sample and symbolized by M or mol/L; percent by mass, defined as grams of analyte per gram of sample × 100 %, symbolized as % or % (mass/mass); and parts per million, defined as micrograms of analyte per gram of sample (ppm, μg/g). For dilute aqueous solutions, 1 mL of solution has

Table 1.1 Instrumental Methods of Analysis

Method	Qualitative		Quantitative	
	Elemental	Molecular	Elemental	Molecular
Atomic absorption spectrometry	No	No	Yes	No
Atomic emission spectrometry	Yes	No	Yes	No
Capillary electrophoresis	Yes	Yes	Yes	Yes
Electrochemistry	Yes	Yes	Yes	Yes
Gas chromatography	No	Yes	No	Yes
ICP-mass spectrometry	Yes	No	Yes	No
Infrared spectroscopy	No	Yes	No	Yes
Ion chromatography	Yes	Yes	Yes	Yes
Liquid chromatography	No	Yes	No	Yes
Mass spectrometry	Yes	Yes	Yes	Yes
Nuclear magnetic resonance	No	Yes	No	Yes
Raman spectroscopy	No	Yes	No	Yes
Thermal analysis	No	Yes	No	Yes
UV/VIS spectrophotometry	Yes	Yes	Yes	Yes
UV absorption	No	Yes	No	Yes
UV fluorescence	No	Yes	No	Yes
X-ray absorption	Yes	No	Yes	No
X-ray diffraction	No	Yes	No	Yes
X-ray fluorescence	Yes	No	Yes	No

Table 1.2 Principal Applications of Instrumental Methods of Analysis

Molecular Analysis

Nuclear magnetic resonance (NMR) spectroscopy

Qualitative analysis: NMR is one of the most powerful methods available for determining the structure of molecules. It identifies the number and type of protons and carbon atoms in organic molecules, for example, it distinguishes among aromatic, aliphatic, alcohols, and aldehydes. Most importantly, it also reveals the positions of the nuclei in the molecule relative to each other. For example, NMR will distinguish between $CH_3—CH_2—CH_2OH$ and $CH_3—CHOH—CH_3$. It does not provide the RMM of the compound. NMR is also applied to compounds containing heteroatoms such as sulfur, nitrogen, fluorine, phosphorus, and silicon.
Quantitative analysis: NMR is useful at % concentration levels, but trace levels (ppm) are becoming attainable with reasonable accuracy.

Infrared (IR) spectroscopy and Raman spectroscopy

Qualitative analysis: IR readily identifies organic functional groups present in molecules including groups containing heteroatoms—O, S, N, Si, halides. The IR spectrum is a fingerprint for a given compound, making it a very useful qualitative method. Classical mid-IR spectroscopy cannot be done on aqueous solutions. It does not give the RMM of the compound. Raman spectroscopy complements IR spectroscopy and is useful for aqueous samples.
Quantitative analysis: IR is used routinely for the quantitative analysis of organic compounds, particularly at % concentration levels. It is used for liquid, solid, and gaseous samples. The related field of Raman spectroscopy complements IR.

Ultraviolet (UV) absorption spectroscopy

Qualitative analysis: UV absorption can be used for identifying functional groups and the structures of molecules containing unsaturated bonds (π electrons), such as

$$\begin{array}{c} H \\ \diagdown \\ \end{array} C=C \begin{array}{c} H \\ | \\ \end{array} - \begin{array}{c} H \\ | \\ \end{array} C=C \begin{array}{c} H \\ \diagup \\ \end{array}$$

CONCEPTS OF INSTRUMENTAL ANALYTICAL CHEMISTRY

Table 1.2 (Continued)   Principal Applications of Instrumental Methods of Analysis

and aromatics and lone-pair electrons, such as those in pyridine:

lone pair

It does not indicate RMM or give useful information on saturated bonds (σ bonds). NMR and IR have almost entirely replaced UV absorption spectroscopy for organic compound identification.

Quantitative analysis: UV absorption is used routinely for the quantitative determination of unsaturated compounds such as those found in natural products. The method is subject to spectral overlap and therefore interference from other compounds in the sample.

UV fluorescence

Qualitative analysis: UV fluorescence is used for the determination of unsaturated compounds, particularly aromatics. It does not indicate RMM but gives some indication of the functional groups present. It is much more sensitive than UV absorption.
Quantitative analysis: UV fluorescence is a very sensitive method of analysis (10^{-8} g/g or 10 ppb), but it is subject to many kinds of interference, both from quenching effects and from spectral overlap from other compounds.

UV and visible (UV/VIS) spectrophotometry

Qualitative analysis: Organic or inorganic reagents are used for specific tests for many elements or compounds by forming a compound that absorbs at specific wavelengths. The products may or may not be colored. If the compounds are colored, analysis may be carried out visually (colorimetric analysis by eye), but the use of a spectrometer is more accurate.
Quantitative analysis: Sensitive and selective methods have been developed for most elements and many functional groups. It is used extensively in routine analysis of water, food, beverages, industrial products, etc.

X-ray diffraction (XRD)

Qualitative analysis: XRD is used for the measurement of crystal lattice dimensions and to identify the structure and composition of all types of crystalline inorganic and organic materials.
Quantitative analysis: XRD is used for the determination of percent crystallinity in polymers, the composition of mixtures, mixed crystals, soils, and natural products.

X-ray absorption spectroscopy

Qualitative analysis: X-ray absorption reveals the contours and location of high atomic mass elements in the presence of low atomic mass matrices or holes in the interior of solid samples (voids). Examples are bone locations in the human body, the contents of closed suitcases, old paintings hidden under new painting on a canvas, and voids in welded joints and opaque solid objects.

Organic mass spectrometry (MS)

Qualitative analysis: MS can be used to identify the RMM of organic and inorganic compounds, from very small molecules to large polymers and biological molecules (>100,000 Da). MS is a powerful tool in the determination of the structure of organic compounds. Fragmentation patterns can reveal the presence of substructure units within the molecule.
Quantitative analysis: MS is used extensively for the quantitative determination of the organic components of solid, liquid, and gas samples.

Thermal analysis (TA)

Qualitative analysis: TA is used to identify inorganic and some organic compounds using very small quantities of sample. It is also used to identify phase changes, chemical changes on heating, heats of fusion, melting points, boiling points, drying processes, decomposition processes, and the purity of compounds.
Quantitative analysis: TA can be used for the quantitative determination of the components of an inorganic sample, particularly at high concentration levels.

Gas chromatography (GC)

Qualitative analysis: GC can be used to separate the components of complex mixtures of gases or of volatile compounds. By comparison with known standards, it can identify components based on retention time.
Quantitative analysis: GC is an accurate method for quantitative analysis based on the area of the peak and comparison with standards. It is used extensively in organic, environmental, clinical, and industrial analysis. GC with MS detection (GC-MS) is a routine and powerful tool for quantitative analysis of organic compounds in environmental and biological samples.

(continued)

Table 1.2 (Continued) **Principal Applications of Instrumental Methods of Analysis**

Liquid chromatography (LC, high-performance LC [HPLC])

Qualitative analysis: LC is used for the identification of components of liquid mixtures, including polar compounds, ions, high-MW components, and thermally unstable compounds. Identification is based on retention time and comparison with standards.
Quantitative analysis: LC is used for the quantitative determination of components in mixtures, especially for high-MW or thermally unstable compounds. It is particularly useful for separating complicated mixtures such as natural products derived from plants or animals and biological samples such as urine and blood. Ion chromatography (IC) is used routinely in water analysis. LC with MS detection (LC-MS) is a routine and powerful tool for quantitative analysis of organic compounds in environmental and biological samples.

Capillary electrophoresis (CE)

Qualitative analysis: Used for the separation and identification of ions and neutral molecules in mixtures. Can be used for ions in aqueous solution and for organic ions.
Quantitative analysis: Quantitative determination of ions can be accomplished following separation, as in IC and LC.

Elemental analysis

Atomic emission spectrometry (AES), optical emission spectrometry (OES)

Qualitative analysis: AES is an almost comprehensive method for qualitative elemental analysis for metals, metalloids, and nonmetals with the exception of some of the permanent gases. Its sensitivity range is great, varying from ppb to percent levels. Many elements can be detected simultaneously. Spectral overlap is the major limitation.
Quantitative analysis: AES is used extensively for the quantitative determination of elements in concentrations from percent levels down to ppb. Liquids, slurries, and solids can be analyzed using the appropriate instrumentation.

Flame photometry (flame atomic emission spectrometry)

Qualitative analysis: Flame photometry is particularly useful for the determination of alkali metals and alkaline-earth metals. It provides the basis for flame tests used in qualitative analysis schemes.
Quantitative analysis: Flame photometry is used for the quantitative determination of alkali metals and alkaline-earth metals in blood, serum, and urine in clinical laboratories. It provides much simpler spectra than those found in other types of AES, but its sensitivity is much reduced.

Atomic absorption spectrometry (AAS)

Qualitative analysis: AAS is not used routinely for qualitative analysis, since with most instruments, it is only possible to test for one element at a time.
Quantitative analysis: AAS is a very accurate and sensitive method for the quantitative determination of metals and metalloids down to absolute amounts as low as picograms for some elements. It cannot be used directly for the determination of nonmetals.

X-ray fluorescence (XRF)

Qualitative analysis: XRF is useful for elements with atomic numbers greater than 4, including metals and nonmetals. For qualitative analysis, no sample preparation is required, and the method is generally nondestructive.
Quantitative analysis: XRF is used extensively for quantitative determination of elements in alloys and mineral samples, particularly of elements with high atomic masses. Sample preparation is complex for quantitative analysis.

Inorganic mass spectrometry (MS)

Qualitative analysis: Inorganic MS can identify elements, isotopes, and polyatomic ions in solutions and solid samples.
Quantitative analysis: Inorganic MS can determine elements at ppt concentrations or below. It is used for simultaneous multielement analysis for metals and nonmetals. Inorganic MS provides the isotope distribution of the elements. Special mass spectrometers are used for accurate isotope ratio measurements used in geology and geochemistry.

a mass of 1 g (because the density of water is 1 g/mL), so solution concentrations are often expressed in terms of volume. A part per million of analyte in dilute aqueous solution is equal to one microgram per milliliter of solution (µg/mL), for example.

The first quantitative analytical fields to be developed were for quantitative elemental analysis, which revealed how much of each element was present in a sample. These early techniques were not instrumental methods, for the most part, but relied on chemical reactions, physical separations, and weighing of products (gravimetry), titrations (titrimetry or volumetric analysis), or production of colored products with visual estimation of the amount of color produced (colorimetry). Using these methods, it was found, for example, that dry sodium chloride, NaCl, always contained 39.33% Na and 60.67% Cl. The atomic theory was founded on early quantitative results such as this, as were the concept of valence and the determination of atomic masses. Today, quantitative inorganic elemental analysis is performed by atomic absorption spectrometry (AAS), AES of many sorts, inorganic MS (such as ICP-MS), XRF, ion chromatography (IC), and other techniques discussed in detail in later chapters.

In a similar fashion, quantitative elemental analysis for carbon, hydrogen, nitrogen, and oxygen enabled the chemist to determine the empirical formulas of organic compounds.

CONCEPTS OF INSTRUMENTAL ANALYTICAL CHEMISTRY

Table 1.3 Analytical Concentration Ranges for Common Instrumental Methods

Technique	Destructive	Ultratrace (<1 ppm)	Trace (1 ppm–0.1%)	Minor (0.1%–10%)	Major (>10%)
X-ray diffraction	No	No	No	Yes	Yes
Nuclear magnetic resonance	No	No	Yes	Yes	Yes
X-ray fluorescence	No	No	Yes	Yes	Yes
Infrared spectroscopy	No	No	Yes	Yes	Yes
Raman spectroscopy	No	No	Yes	Yes	Yes
UV/VIS spectrometry	No	No	Yes	Yes	Yes
Colorimetry	No	Yes	Yes	Yes	No
Molecular fluorescence spectrometry	No	Yes	Yes	Yes	Yes
Atomic absorption spectrometry	Yes	Yes	Yes	Yes	No
Atomic emission spectrometry	Yes	Yes	Yes	Yes	Yes
Atomic fluorescence spectrometry	Yes	Yes	Yes	No	No
ICP-mass spectrometry	Yes	Yes	Yes	Yes	No
Organic mass spectrometry	Yes	Yes	Yes	Yes	Yes
GC-MS	Yes	Yes	Yes	Yes	Yes
LC-MS	Yes	Yes	Yes	Yes	Yes
Potentiometry	No	Yes	Yes	Yes	Yes
Voltammetry	No	Yes	Yes	Yes	Yes
Gas chromatography	May be	Yes	Yes	Yes	Yes
High-performance liquid chromatography	May be	Yes	Yes	Yes	Yes
Ion chromatography	May be	Yes	Yes	Yes	Yes
Capillary electrophoresis	No	Yes	Yes	Yes	Yes
Thermal analysis	May be	No	No	Yes	Yes

Note: The destructive nature of the instrumental method is characterized. A sample may be destroyed by a nondestructive instrumental method, depending on the sample preparation required. The chromatographic techniques may be destructive or nondestructive, depending on the type of detector employed. The nondestructive detectors generally limit sensitivity to "trace." Molecular fluorescence is not destructive if the molecule is inherently fluorescent. It may be if the molecule requires derivatization. A method with "yes" for ultratrace and "no" for major concentrations reflects the linear working range. Such methods can measure "majors" if the sample is diluted sufficiently.

An empirical formula is the simplest whole number ratio of the atoms of each element present in a molecule. For any given compound, the empirical formula may or may not coincide with the molecular formula. A molecular formula contains the total number of atoms of each element in a single molecule of the compound. For example, ethylene and cyclohexane have the same empirical formula, CH_2, but molecular formulae of C_2H_4 and C_6H_{12}, respectively. The empirical formula of many sugars is CH_2O, but the molecular formulae differ greatly. The molecular formula of glucose is $C_6H_{12}O_6$, fructose is $C_6H_{12}O_6$, erythrose is $C_4H_8O_4$, and glyceraldehyde is $C_3H_6O_3$. An example of a molecule whose empirical formula is the same as the molecular formula is tetrahydrofuran (THF), an important organic solvent. The molecular formula for THF is C_4H_8O; there is only one oxygen atom, so there can be no smaller whole number ratio of the atoms. Therefore, C_4H_8O is also the empirical formula of THF.

Empirical formulas of organic compounds were derived mainly from combustion analysis, where the organic compound is heated in oxygen to convert all of the carbon to CO_2 and all of the hydrogen to H_2O. The CO_2 and H_2O were collected and weighed, or the volume of the gas was determined by displacement of liquid in a measuring device. To distinguish between butane (C_4H_{10}), which contains 82.76% C and 17.24% H, and pentane (C_5H_{12}), which contains 83.33% C and 16.66% H, required great skill using manual combustion analysis. Today, automated analyzers based on combustion are used for quantitative elemental analysis for C, H, N, O, S, and the halogens in organic compounds. These analyzers measure the evolved species by GC, IR, or other techniques. These automated analyzers require only microgram amounts of sample and a few minutes to provide data and empirical formulas that used to take hours of skilled analytical work. Quantitative elemental analysis cannot distinguish between isomers, which are compounds with the same molecular formula but different structures. Glucose and fructose have the same molecular formula, but glucose is a sugar with an aldehyde group in its structure, while fructose is a sugar with a ketone group in its structure. They cannot be distinguished by elemental analysis but are easily distinguished by their IR and NMR spectra.

Quantitative molecular analysis has become increasingly important as the fields of environmental science, polymer chemistry, biochemistry, pharmaceutical chemistry, natural products chemistry, and medicinal chemistry have grown. Techniques such as GC, LC, or HPLC, CE, MS, fluorescence spectrometry, IR, and XRD are used to determine the amount of specific compounds, either pure or in mixtures. These techniques have become highly automated and extremely sensitive, so that only micrograms or milligrams of sample are needed in most cases. The chromatography techniques, which can separate mixtures, have been "coupled" to techniques like MS, which can identify and quantitatively measure the components in a mixture. Such techniques, like GC-MS and LC-MS, are called hyphenated techniques. Many hyphenated instruments are commercially available. These types of instruments for use in the pharmaceutical industry have been designed to process samples in very large batches in a completely automated fashion. The instruments will analyze the samples, store the data in computer files, "pattern-match" the spectra to identify the compounds, and calculate the concentrations of the compounds in the samples. Even then, more than one instrument is required to keep up with the need for characterization of potential drug candidates. As an example, one research department in a major pharmaceutical company bought its first LC-MS in 1989. By 1998, that one department had more than 40 LC-MS instruments running on a daily basis to support drug metabolism studies alone.

Instrumental methods differ in their ability to do quantitative analysis; some methods are more *sensitive* than others. That is, some methods can detect smaller amounts of a given analyte than other methods. Some methods are useful for wide ranges of analyte concentrations; other methods have very limited ranges. We will discuss the reasons for this in the chapters on the individual techniques, but Table 1.3 shows the approximate useful concentration ranges for common instrumental techniques. Table 1.3 is meant to serve as a guide; the actual sensitivity and useful concentration range (also called the working range) of a technique for a specific analysis will depend on many factors.

1.2.2 Designing the Analytical Method

Once the problem has been defined, an analytical procedure, or method, must be designed to solve the problem. The analytical chemist may have to design the method to meet certain goals, such as achieving a specified accuracy and precision, using only a limited amount of sample, or performing the analysis within a given cost limit or "turnaround time." Turnaround time is the time elapsed from receipt of a sample in the lab to delivery of the results to the person who requested the analysis. This length of time may need to be very short for clinical chemistry laboratories providing support to hospital emergency rooms, for example. A common goal for modern analytical procedures is that they are "green chemistry" processes, that is, that the solvents used are of low toxicity or biodegradable, that waste is minimized, and that chemicals used in the analysis are recycled when possible.

Designing a good analytical method requires knowing how to obtain a *representative sample* of the material to be analyzed, how to store or preserve the sample until analysis, and how to prepare the sample for analysis. The analyst must also know how to evaluate possible *interferences* and *errors* in the analysis and how to assess the accuracy and precision of the analysis. These topics will be discussed subsequently, and specific interferences for given instrumental methods are discussed in the following chapters.

There are many analytical procedures and methods that have been developed and published for a wide variety of analytes in many different matrices. These methods may be found in the chemical literature; in journals such as *Analytical Chemistry*, *The Analyst*, *Analytical and Bioanalytical Chemistry* (formerly *Fresenius' Journal of Analytical Chemistry*), and *Talanta*; and in journals that focus on specific analytical techniques, such as *Applied Spectroscopy*, *Journal of Separation Science* (formerly *Journal of High Resolution Chromatography*), *Journal of the American Society for Mass Spectrometry*, and *Thermochimica Acta*. Compilations of "standard" methods or "official" methods have been published by government agencies, such as the US Environmental Protection Agency (EPA), and private standards organizations, such as the American Association of Official Analytical Chemists (AOAC), the American Society for Testing and Materials (ASTM), and the American Public Health Association (APHA), among others. Similar organizations and official methods exist in many other countries. These standard methods are methods that have been tested by many laboratories and have been found to be reproducible, with known accuracy and precision. The bibliography lists several of these books on analytical methods. It is always a good idea to check the chemical literature first, so that you don't waste time designing a procedure that already exists.

Note: Throughout this book we mention the standards and procedures of ASTM International. Prior to 2001, the organization was known as the American Society for Testing and Materials (hence, ASTM). First formed in 1898 under the leadership of Charles Benjamin Dudley, the organization first addressed railroad component quality. Today, it is the primary international standards organization in the United States; similar bodies reside in nearly all major industrialized countries. ASTM International presides over the adoption of *consensus standards* by supporting thousands of volunteer technical committees, with international membership. Typical committee make up includes members from industry (suppliers, service providers, original equipment manufacturers, users), members from government laboratories, and members from universities. ASTM International technical committees and subcommittees maintain approximately 12,000 standard procedures and practices, published periodically in the 82 volume "Annual Book of Standards." The deliberative nature and ballot process of ASTM International is intentionally long and careful, sometimes giving rise to criticism of the organization as reactive rather than proactive.

If there are no methods available, then the analytical chemist must develop a method to perform the analysis. For very challenging problems, this may mean inventing entirely new analytical instruments or modifying existing instruments to handle the task.

The design of the method also requires the analyst to consider how the method will be shown to be accurate and precise. This requires knowledge of how we assess accuracy and precision, discussed in Section 1.3. The analyst must evaluate interferences. Interference is anything that (1) gives a response other than the analyte itself or (2) changes the response of the analyte. Interferences may be other compounds or elements present in the sample or that form on degradation of the sample. Interfering compounds or elements may respond directly in the instrumental measurement to give a false analyte signal, or they may affect the response of the analyte indirectly by enhancing or suppressing the analyte signal. Examples will be given in the chapters for each instrumental technique. The analyst must demonstrate that the method is reliable and *robust*. These will be covered in greater detail in Sections 1.5 and 1.6.

There are some fundamental features that should be part of every good analytical method. The method should require that a **blank** be prepared and analyzed. A blank is used to ascertain and correct for certain interferences in the analysis. In many cases, more than one type of blank is needed. One type of blank solution may be just the pure solvent used for the sample solutions. This will ensure that no analyte is present in the solvent and allows the analyst to set the baseline or the "zero point" in many analyses. A **reagent blank** may be needed; this blank contains all of the reagents used to prepare the sample but does not contain the sample itself. Again, this assures the analyst that none of the reagents themselves contribute analyte to the final reported value of analyte in the sample. Sometimes a **matrix blank** is needed; this is a blank that is similar in chemical composition to the sample but without the analyte. It may be necessary to use such a blank to correct for an overlapping spectral line from the matrix in AES, for example.

All instrumental analytical methods except coulometry (Chapter 15) require **calibration standards**, which have known concentrations of the analyte present in them. (In isotope dilution MS [IDMS], there must be a known amount of an isotope, usually of the analyte.) These calibration standards are used to establish the relationship between the analytical signal being measured by the instrument and the concentration of the analyte. Once this relationship is established, unknown samples can be measured and the analyte concentrations determined. Analytical methods should require some sort of **reference standard** or check standard. This is also a standard of known composition with a known concentration of the analyte. This check standard is not one of the calibration standards and should be from a different lot of material than the calibration standards. It is run as a sample to confirm that the calibration is correct and to assess the accuracy and precision of the analysis. Reference standard materials are available from government and private sources in many countries. Examples of government sources are the NIST in the United States, the National Research Council of Canada (NRCC), and the LGC (formerly Laboratory of the Government Chemist) in the United Kingdom.

1.2.3 Sampling

One of the most important single steps in an analysis is collecting the sample of the material to be analyzed. Often, this step is carried out not by the personnel performing the analyses, but rather by field technicians. Real materials are usually not homogeneous, so the sample must be chosen carefully to be **representative** of the real material. A representative sample is one that reflects the true value and distribution of the analyte in the original material. If the sample is not taken properly, no matter how excellent the analytical method or how expert the analyst, the result obtained will not provide a reliable characterization of the material. Other scientists, law enforcement officials, and medical professionals often collect samples for analysis, sometimes with no training in how to take a proper sample. The analytical chemist ideally would be part of the team that discusses collection of samples before they are taken, but in reality, samples often "show up" in the lab. It is important that the analyst talks with the sample collector before doing any analyses; if the sample has been contaminated or improperly stored, the analysis will be not only a waste of time but can also lead to erroneous conclusions. In clinical chemistry analysis, this could lead to a misdiagnosis of a disease; in forensic analysis, this could lead to a serious miscarriage of justice.

The amount of sample taken must be sufficient for all analyses to be carried out in duplicate or triplicate, if possible. Of course, if only a small quantity of sample is available, as may be the case for forensic samples from a crime scene or rocks brought back from the moon, the analyst must do the best job possible with what is provided.

A good example of the problems encountered in sampling real materials is collecting a sample of a metal or metal alloy. When a molten metal solidifies, the first portion of solid to form tends to be the most pure (remember freezing point depression from your general chemistry class?). The last portion to solidify is the most impure and is generally located in the center or *core* of the solidified metal. It is important to bear this in mind when sampling solid metals. A sample is often ground from a representative cross section of the solid, or a hole is drilled through a suitable location and the drillings mixed and used as the sample.

Samples have to be collected using some type of collection tool and put into some type of container. These tools and containers can often contaminate the sample. For example, stainless steel needles can add traces of metals to blood or serum samples. Metal spatulas, scissors, drill bits, glass pipettes, filter paper, and plastic and rubber tubing can add unwanted inorganic and organic contaminants to samples. To avoid iron, nickel, and chromium contamination from steel, some implements like tongs and tweezers can be purchased with platinum or gold tips.

The discussion of sampling, which follows, refers to the traditional process of collecting a sample at one location (often called "collection in the field") and transporting the sample to the laboratory at a different location. Today, it is often possible to analyze samples *in situ* or during the production of the material (online or process analysis) with suitable instrumental probes, completely eliminating the need for "collecting" a sample. Examples of *in situ* and online analysis and field portable instruments will be discussed in later chapters.

The process of sampling requires several steps, especially when sampling bulk materials such as coal, metal ore, soil, grain, and tank cars of oil or chemicals. First, a *gross representative sample* is gathered from the *lot*. The lot is the total amount of material available. Portions of the gross sample should be taken from various locations within the lot to ensure that the gross sample is representative. For very large lots of solid material such as coal or ore, the *long pile and alternate shovel* method can be used. The material is formed into a long rectangular pile. It is then separated into two piles by shoveling material first to one side and then to the other, creating two piles. One pile is set aside. The remaining pile may be reduced in size by repeating the process, until a sample of a size to be sent to the laboratory remains. The *cone and quarter* method is also used to collect a gross sample of solid materials. The sample is made into a circular pile and mixed well. It is then separated into quadrants. A second pile is made up of two opposite quadrants, and the remainder of the first pile discarded. This process is shown in Figure 1.3. This process can be repeated until a sample of a suitable size for analysis is obtained. This sample can still be very large. Ferroalloys, for example, are highly segregated (i.e., inhomogeneous) materials; it is not uncommon for the amount required for a representative sample of alloy in pieces about 2 in. in diameter to be 1 ton (0.9 Mg) of material from the lot of alloy.

A computer program that generates random numbers can choose the sampling locations and is very useful for environmental and agricultural sampling. If the lot is a field of corn, for example, the field can be divided into a grid, with each grid division given a number. The computer program can pick the random grid divisions to be sampled. Then a smaller, homogeneous *laboratory sample* is prepared from the gross composite sample. If the sample is segregated (i.e., highly inhomogeneous), the representative sample must be a composite sample that reflects each region and its relative amount. This is often not known, resulting in the requirement for very large samples. The smaller laboratory sample may be obtained by several methods but must be representative of the lot and large enough to provide sufficient material for all the necessary analyses. After the laboratory sample is selected, it is usually split into even smaller *test portions*. Multiple small test portions of the laboratory sample are often taken for replicate analyses and for analysis by more than one technique. The term **aliquot** is used to refer to a quantitative amount of a *dissolved* test portion; for example, a 0.100 g test portion of sodium chloride may be dissolved in water in a volumetric flask to form 100.0 mL of test solution. Three 10.0 mL aliquots may be taken with a volumetric pipette for triplicate analysis for chloride using an ion-selective electrode, for example.

As the total amount of the sample is reduced, it should be broken down to successively smaller pieces by grinding, milling, chopping, or cutting. The 1 ton sample of ferroalloy, for example, must be crushed, ground, and sieved many times. During the process, the sample size is reduced using a sample splitter called a riffle. After all this and then a final drying step, a 1 lb (454 g) sample remains. The sample must be mixed well during this entire process to ensure that it remains representative of the original. The grinding equipment used must not contaminate the sample. For example, boron carbide and tungsten carbide are often used in grinding samples because they are very hard materials, harder than most samples. However, they can contribute boron or tungsten to the ground sample, so would not be used if boron or tungsten must be measured at low concentrations. Zirconium oxide ball mills can contribute Zr and Hf to a sample. Stainless steel grinders are a source of Fe, Cr, and Ni. Some cutting devices use organic fluids as lubricants; these must be removed from the sample before analysis.

It is also possible for the grinding or milling step to cause erroneously low results for some analytes. Malleable metals like gold may adhere to the grinding or milling surface and

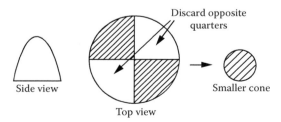

Figure 1.3 The cone and quarter method of sampling bulk materials.

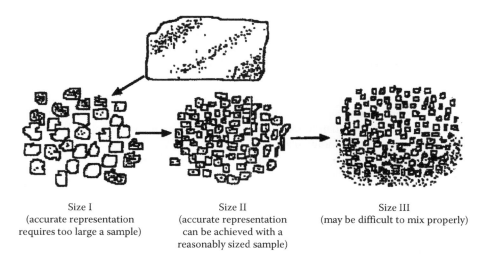

Size I
(accurate representation
requires too large a sample)

Size II
(accurate representation
can be achieved with a
reasonably sized sample)

Size III
(may be difficult to mix properly)

Figure 1.4 Sampling of a segregated material with a problematic component like gold. (Extracted from Dulski, T.R., *A Manual for the Chemical Analysis of Metals*, ASTM International, West Conshohocken, PA, 1996. With permission. Copyright 1996 ASTM International.)

be removed from the sample in the process, an undesirable effect. An example of sampling a segregated material with a problematic component like gold is illustrated in Figure 1.4. The rectangular piece at the top is a hypothetical piece of gold-bearing quartz. The gold is represented as the dark flecks. You can see that the gold appears in bands within the quartz, separated by bands of pure quartz (the white area). If the rock is crushed to Size I, the gold particles have not been liberated from the quartz; some pieces have gold flecks and many large pieces are pure quartz. At this size, it is difficult to remove a sample of the rock pieces and expect it to be representative. If the rock is crushed to a smaller size, Size II, it is evident that a representative small sample can be obtained. If the rock is crushed to Size III, the gold particles are freed from the quartz matrix. If this sample could be mixed perfectly, a smaller sample could be taken to represent the whole than the sample needed at Size II. (Why would this be desirable? The smaller the analytical sample, the less gold is used up by analysis. This is an important consideration with valuable analytes and valuable samples.) But the gold particles and the quartz particles have different densities and shapes and will be difficult to mix well. As mentioned earlier, gold is soft and malleable. If it is broken out of the quartz, it may become embedded in the grinder or smeared onto surfaces in the grinding equipment, so some gold may actually be lost from the sample particles ground to Size III. Size II will give a more representative sample than either of the other sizes.

Sampling procedures for industrial materials, environmental samples, and biological samples are often agreed upon, or *standardized*, by industry, government, and professional societies. Standard sampling procedures help to ensure that the samples analyzed are representative and are not contaminated or changed during the sampling process. Standard sampling procedures for many materials can be found in the *Annual Book of ASTM Standards*, for example. Sampling procedures for soil, water, and air are established by the US EPA in the United States and similar government organizations in other countries. Procedures for sampling of water and wastewater can be found in *Standards Methods for the Analysis of Water and Wastewater*; the AOAC publishes procedures for food products. The bibliography provides some examples of these publications. A good analytical chemist will consult the literature before sampling an unfamiliar material. Some general guidelines for sampling different classes of materials are discussed here.

1.2.3.1 *Gas Samples*

Gas samples are generally considered homogeneous, but gas mixtures may separate into layers of differing density. Samples that have been collected and allowed to settle will need to be stirred before a portion is taken for analysis. Gas samples can be taken at a single point in time (called a **grab** sample) or can be collected over a period of time or from different locations to provide an *average* or **composite** sample. Gas samples can be collected using gas-tight syringes, balloons, plastic bags, or containers made of metal or glass that can be evacuated. Sampling of toxic, flammable, or corrosive gases should be done with great care using appropriate safety equipment.

The containers used to collect the samples must not contaminate the sample with analyte. Plastic bags and balloons may leach volatile organic compounds into the gas sample, while glass may adsorb components of the sample onto the surface of the glass.

Certain components of gas samples, such as organic vapors in air, may be collected by pulling the air through activated charcoal. The organic gases are adsorbed onto the charcoal, while the majority of the air (oxygen, nitrogen, etc.) passes through. This has the advantage of *preconcentrating* the analytes of interest and reducing the physical size of the sample. Many liters of air can be pulled through an activated charcoal bed that is no bigger than a ball-point pen. It is much easier to transport the analytes trapped on the charcoal to the laboratory than to transport hundreds of liters of air. The process of trapping an analyte out of the gas phase is called "scrubbing." Scrubbing a gas sample can also be done by bubbling the gas through a liquid that will absorb the analytes of interest.

Gas samples may contain particles of solid material that need to be removed by filtration. The filter material must be chosen so that it does not adsorb analytes or add contaminants to the gas. Filters are available that will remove particles as small as 0.2 μm in diameter from a gas stream.

1.2.3.2 Liquid Samples

Liquid samples can also be collected as grab samples or as composite samples. Sampling liquids can be quite difficult; it is not always as straightforward as "pouring some" out of a bottle or dipping a bucket into a fluid. Only a few comments with respect to general sampling of liquids can be made here. It is usual to stir liquid samples adequately to obtain a representative sample; however, there may be occasions when stirring is not desired. If the analyst is only interested in identifying an oily layer floating on water, stirring the sample is not needed; the oily layer may be pulled off with a pipette or an eyedropper, for example. Samples must be collected at locations remote from sources of contamination if a representative sample is desired. For example, if a sample of "normal" river water is desired, the sample should be collected away from riverbanks, floating froth, oil, and discharges from industrial and municipal waste treatment sites. Sampling of rivers, lakes, and similar bodies of water may require samples from different depths or different distances from shore. Such samples may be analyzed individually or blended to obtain an average composition.

Liquid samples may contain particles of solid material that need to be removed by filtration or centrifugation. The filter material must be chosen so that it does not adsorb analytes or contaminate the liquid. Some samples that are mostly liquid contain suspended solid material; orange juice and liquid antacids are examples. In these types of samples, the liquid and its associated solids may need to be sampled for analysis without removing the solids. It may be difficult to obtain a representative sample from these suspensions; a standard sampling procedure is needed to ensure that results can be compared from one day to the next. Liquid samples may consist of more than one layer because they contain two or more **immiscible** liquids. Examples include samples of oil and water from an oil spill at sea, oil and vinegar salad dressing, or cream at the top of a bottle of milk. The layers may need to be *emulsified* to provide a representative sample, but it may be more useful to sample each layer separately.

Sampling of hot molten materials such as metals, alloys, and glasses is a form of liquid sampling, but one requiring very specialized equipment and techniques.

1.2.3.3 Solid Samples

Solid samples are often the most difficult to sample because they are usually less homogeneous than gases or liquids. Large amounts of solid sample cannot be conveniently "stirred up." Moreover, unlike the situation with fluids, there are no diffusion or convection currents in solids to ensure mixing. Solids must often be ground or drilled or crushed into smaller particles to homogenize the sample. There are many types of manual and automated grinders and crushers available; the choice depends on the hardness of the material to be ground. Soft materials also pose a challenge in grinding because they often just deform instead of being reduced in size. Polymer pellets may be ground in an electric coffee grinder with a small amount of liquid nitrogen added to the grinder (called cryogrinding). The liquid nitrogen freezes the polymer, making the pellets brittle and capable of being easily powdered. Other soft solids such as foods can be handled the same way. Commercial cryomills that prevent the user from coming into contact with liquid nitrogen are available. Many solid materials must be oven-dried before sampling to remove adsorbed water in order to obtain a representative sample. There are numerous published standard methods for sampling solid materials such as cement, textiles, food, soil, ceramics, and other materials. Examples of the wide variety of analytical pulverizing, grinding, and blending equipment available can be found at the following websites: SpexCertiprep (www.spexcsp.com), Retsch (www.retsch.com), and Netzsch (www.netzsch.com), among others.

1.2.4 Storage of Samples

When samples cannot be analyzed immediately, they must be stored. The composition of a sample may change during storage because of reactions with air, light, or interaction with the container material. The container used for collection and storage of the sample and the storage conditions must be chosen to minimize changes in the sample.

Plastic containers may leach organic components such as plasticizers and monomers into a sample. Plastic containers may also introduce trace metal impurities such as Cu, Mn, or Pt from the catalysts used to make the polymer or elements such as Si, Ti, Sb, Br, and P from inorganic fillers and flame retardants. Glass surfaces both adsorb and release trace levels of ionic species, which can dramatically change the trace element and trace ion concentrations in solutions. It has been observed that trace metals will "plate out" of solution along strain lines in glass. Such strain lines are not reproducible from one container to another; therefore, the loss of trace metals cannot be estimated accurately for one container by measuring the loss in a similar but different container. All containers require appropriate cleaning before use. Containers for organic samples are usually washed in solvent, while containers for samples for trace metal analysis are soaked in acid and then in deionized (DI) water.

Precautions such as freezing biological and environmental samples or displacing the air in a container by an inert gas will often extend the storage life of a sample. Samples should not be stored any longer than is absolutely necessary prior to analysis and should not be stored under conditions of high heat or high humidity. Some samples require storage in the dark to avoid photolytic (light-induced) changes in composition; water samples to be analyzed for silver are a good example. Such samples must be stored in dark plastic bottles to avoid the photolytic formation of colloidal silver, which will precipitate out of the sample. Many samples for environmental analysis require the addition of preservatives or adjustment of pH to prevent the sample from deteriorating. Water samples for trace metals determination must be acidified with high-purity nitric acid to keep the trace metals in solution, for example. Blood samples often require collection in tubes containing an anticoagulant to keep the blood sample fluid, but the anticoagulant must not interfere in the analysis. For example, a sample collected to measure a patient's sodium level cannot be collected in a tube that contains the sodium salt of ethylenediaminetetraacetic acid (EDTA) as the anticoagulant. Other biological samples may need to be collected in sterile containers.

Sample containers must be labeled accurately and in such a way that the label does not deteriorate on storage; do not use water-soluble marking pen on samples to be put in a freezer, for example. The label should clearly identify the sample and any hazards associated with the sample. Many analytical laboratories have computer-based sample tracking systems that generate adhesive bar-coded labels for samples, exactly like the bar codes used on retail items in stores. These computer-based systems are called Laboratory Information Management Systems (LIMS) and catalog and track not only the samples but also the analytical data generated on the samples.

As a student, you should get into the habit of labeling all your containers in the laboratory with your name, date, and the contents. There is nothing worse than finding four beakers of colorless liquid on the lab bench and not knowing which one is yours or what is in the other beakers! That situation would be a serious safety violation in an industrial laboratory. Academic labs in the United States are now required to follow the same safety regulations followed in industry and something as simple as beakers with "stuff" in them, but no proper labels can result in large monetary fines for your school. The cost of chemical waste disposal is very high, and it is not legal to dispose of unidentified chemicals, so unlabeled containers are a very expensive problem. The material must be analyzed to identify it just so that it can be disposed of properly.

1.3 A CONCEPTUAL DISCUSSION ABOUT UNCERTAINTY IN MEASUREMENTS

Before we get into the details about determining and expressing uncertainty in analytical measurements, it is worthwhile to appreciate that uncertainty is a concept that is inherent in the scientific enterprise as a whole. As we will encounter later in this chapter, uncertainty in analytical measurements can take on a very important if not critical character when such measurements are required in legal or regulatory venues, and especially when analytical results are presented in the courtroom.

In the context of the laboratory, uncertainty describes the confidence we have in the measurement we made. For example, we may measure the temperature of the room as 20 °C, but we know that (for reasons we will discuss in detail in this section) the uncertainty of the measurement is due to the thermometer we've used being only capable of giving the temperature to within 1 °C. This might be stated succinctly in the following manner:

$$T_{room} = 20\ °C \pm 1\ °C$$

We can read this as 20 degrees plus or minus one degree. Note that there are some instances in which the plus or minus symbol, ±, is itself problematic. We intuitively understand that uncertainty can never be a negative number, so one may sometimes find that the uncertainty is expressed without the symbol in a separate column in a table, or simply written out in a text passage. If we seek out dictionary definitions of uncertainty, a wide variety of suggestions arise:

1. Not definitely ascertainable or fixed, as in time of occurrence, number, dimensions, or quality.
2. Not confident, assured, or free from hesitancy: an uncertain smile.
3. Not clearly or precisely determined; indefinite; unknown: a manuscript of uncertain origin.
4. Vague; indistinct; not perfectly apprehended: an abstruse novel with uncertain themes.
5. Subject to change; variable; capricious; unstable: a person of uncertain opinions.
6. Ambiguous; unreliable; undependable: her loyalties are uncertain.
7. Dependent on chance or unpredictable factors; doubtful; of unforeseeable outcome or effect.
8. Unsteady or flickering, as light; of changing intensity or quality.

We can see that the term uncertainty has both general language meanings, but also more well defined scientific or technical meanings. If we look at the first definition, we see "not definitely ascertainable or fixed." What does this mean to a scientist? Not much, really, since to a scientist, NOTHING we can measure is "definitely ascertainable or fixed." Sure, we often define terms with no uncertainty. We define the molecular mass of the carbon atom as being 12 grams per mole, period. No uncertainty. This, and other quantities we call **physical constants**, might get revised from time to time, but they are fixed because *we say they are fixed*, not because we measured them so well that they are presented with metaphysical infallibility.

It is worth repeating that NOTHING we can measure will ever give us the "true" value, whether it be a temperature, pressure, or concentration of gasoline in fire debris. We may get close to the true value of a measured quantity, as science and instrumentation and resources advance, but we can never get all the way there. It's similar to approaching absolute zero in temperature; you can get close but never right there. This is important because, in science, our business consists of two elements and two elements only: measurement and uncertainty. If you present a measurement without uncertainty assessed and expressed properly, the job is only half done. A corollary to this idea is a concept that might at first glance be disconcerting. In science, our business is not seeking "truth"; we leave that to religious professionals. As Nobel laureate (Physiology or Medicine, 1996) Peter Doherty says: "If you want absolutes speak to a politician or a pope."

If we look at the sixth definition above, we find a conflict between (American) conversational English and the corresponding scientific usage. Uncertainty as applied to assessing a measurement is not at all related to ambiguity, unreliability, or undependability. A thermometer may be highly precise (and therefore correspondingly expensive) or imprecise (perhaps because it is cheaply made), but the only way it can give an ambiguous result would be if it read a random number. Put simply, scientific results that are ambiguous, unreliable, or undependable are those for which the uncertainties are poorly understood or expressed.

Indeed, this general philosophy was expressed very well by former NIST Director Dr. John W. Lyons in 1994: "It is generally agreed that the usefulness of measurement results, and thus much of the information that we provide as an institution, is to a large extent determined by the quality of the statements of uncertainty that accompany them." And, let's dispel the myth that uncertainty is bad. If you were taught that uncertainty in your lab results is naughty, embarrassing, makes you less of a person, or whatever, get over it! True, if your uncertainty, despite your best effort, is too high, your conclusions may be limited, but that's the job. Live with it; no, embrace it, or sell real estate. As is implied by Dr. Lyons, embracing the understanding of uncertainty is a virtue that will advance science and nearly any human endeavor.

The dictionary definitions mentioned above are confusing, but even in the scientific literature, we find that the language is not particularly good. In fact, it might be worse in terms of the words used to describe uncertainty, because we think we know what we are doing. If you read or edit scientific journals you will encounter such terms as:

- Accuracy.
- Absolute accuracy.
- Error.
- Absolute error.
- Uncertainty.
- Absolute uncertainty.
- Maximum uncertainty.
- Reproducibility.
- Repeatability.
- Precision.
- Known to within…
- Confidently given to…
- Measured to…
- Reliable to…
- +/−…
- Believed to be reproducible to…
- Personally guaranteed to be...
- Aleatoric.
- Don't give a damn, and never gave a damn.

Some of these might be worthy of explanation; they are taken directly from the experiences of one of the authors (TJB) who is a journal editor, and they are all real. Take the last three entries, for example. We have encountered papers in which authors insist that their measurements are accurate, and that they "personally guarantee" them. How one exercises or collects on this guarantee is a mystery. Aleatoric actually means bad luck, as in: "the uncertainty in my measurements is due to bad luck". You can't make this up! Another author, when asked to provide uncertainty in a revision of his paper, actually replied that he did not give a damn about uncertainty and never did give a damn about it! No, his paper was not accepted by the journal. Admittedly, these are extremely rare instances. We emphasize that it is never acceptable for a scientist to use such imprecision (or worse) in describing uncertainty.

Other than the last three, which are spectacular, the other terms on the list are all too common. The problem with them is that none of these terms have any quantitative meaning, and a quantitative meaning is what you want when describing the uncertainty in, say, a temperature reading from a thermometer.

The definitive guide for expressing uncertainty in any measured quantity is to be found in a consensus document called the GUM, *Guide for Expressing Uncertainty in Measurements*. It was published in 1993 (with slight revisions periodically) and resulted from an agreement among National Metrology Institutes (such as the National Institute of Standards and Technology, NIST, in the U.S.)

that are established in all industrialized countries. This document provides the scientific community with the recipe for assessing and reporting the uncertainty in all measurements, calibrations, and results.

1.4 BASIC STATISTICS AND DATA HANDLING

How many samples do we need to collect for a given analysis? How large must the sample be to ensure that it is representative? These types of questions can be answered by statistics. We also need to have a basic knowledge of statistics to understand the limitations in the other steps in method development, so we will now briefly introduce the statistical concepts and calculations used by analytical chemists.

In order to design the correct experiment to answer the analytical question being asked, statistical methods are needed to select the size of the sample required, the number of samples, and the number of measurements that must be performed to obtain the needed accuracy and precision in the results generated by the experiment. Statistical analysis of data is also used to express the uncertainty in measured values, so that the users of the data understand the limitations associated with results. Every measured result has uncertainty associated with it; it is part of the analyst's job to express that uncertainty in a way that defines the range of values that are reasonable for the measured quantity. Uncertainty, properly evaluated and reported with the measured value, indicates the level of confidence (CL) that the value actually lies within the range reported.

1.4.1 Significant Figures

The result of an analytical measurement is a number with associated units such as 50.1 % iron, 10 ppm parathion (a pesticide), or 25 mg isopropanol/L. To deal with numbers that result from measurements and calculations involving these numbers, the concept of **significant figures** must be understood. All measurements have uncertainty, so the result should be reported as a number that tells us about the uncertainty. The numbers reported from a measurement should be all of the digits known with certainty plus the first uncertain digit. These digits, with uncertainty only in the last digit of the number, are called significant figures. This gives a result that reflects the precision of the measurement. For example, the number 50.1 % means that the percentage is closer to 50.1 than to 50.2 or 50.0, but it does not mean that the percentage is exactly 50.1. In short, we are sure of the "50" part of the number, but there is some uncertainty in the last figure reported. If we were to analyze two samples containing 50.08 % and 50.12 % of a component by using an instrument accurate to 0.1 %, we would not be able to distinguish the difference in the compositions of the samples but would report them both as 50.1 %.

The number 50.1 has three significant figures (5, 0, 1). Since the measurement is no better than 0.1 %, the last digit in 50.1 is uncertain by at least ± 1. The last significant figure reported should reflect the precision of the measurement. There is no point in reporting any more figures because they would have no meaning, even though they might be obtainable mathematically. For example, scientific calculators generally display eight or more digits; that does not mean that all of the displayed digits are meaningful.

The reporting of figures implies that all the numbers are significant and only the last number is in doubt, even if that number is zero. For example, 1.21×10^6, which has three significant figures (1, 2, 1), implies that the number is closer to 1.21×10^6 than to 1.22×10^6 or 1.20×10^6. But writing 1,210,000, with seven significant figures, implies that the number is closer to 1,210,000 than to 1,210,001 or 1,209,999. Terminal zeros can be confusing because they may not be significant if they are only used to show the decimal place. Furthermore, the number 50.10 implies 10× lower uncertainty than 50.1. A zero can be a significant figure or it can be used to show the decimal place. If a zero occurs between two nonzero significant figures, it is significant, as in 12,067, which has five significant figures. In a number such as 0.024, the two initial zeros just show the decimal place. Initial zeros are never significant, and the number 0.024 has only two significant figures, the 2 and the 4. One way to confirm that the zeros are only placeholders is to write the number in scientific notation, 2.4×10^{-2}, which clearly has only two significant figures. If a zero is written after a decimal point, it is significant; for example, 24.0 would have three significant figures. It is important to write a zero at the end of a number (a terminal zero) only if it is significant. For example, in a number like 54,300 or 1,210,000, the zeros at the end of each number may or may not be significant; it is impossible to tell by looking at the number. In cases like these, scientific notation should be used to indicate exactly how many figures are significant. The number 5.4300×10^4 has five significant figures; 5.430×10^4 has four significant figures; and 5.43×10^4 has three significant figures. To be clear about the number 1,210,000, it should be written in scientific notation: 1.210000×10^6 is the unambiguous way to show that there are seven significant figures.

Numbers that represent discrete objects have no uncertainty. If five measurements were made, and we calculate the average by adding the five results and dividing by 5, the number 5 has no effect on the number of significant figures in the answer. There is no uncertainty in the number of measurements made, so the 5 can be considered to have an infinite number of significant figures.

The following are some rules that should be observed when reporting results:

1. In enumerating data, report all significant figures, such that only the last figure is uncertain. Reject all other

figures, rounding off in the process. Rounding off should be done at the end of the calculation. When rounding off, look at all the digits beyond the last digit to be retained. If the digits beyond the last digit to be retained are more than halfway to the next digit, round up.

2. If a number such as 1.37286 must be reported to four significant figures, look at the digits after the fourth place (the 2). If the following digits are greater than 5, increase the fourth figure by 1 and drop the remaining digits. If the following digits are less than five, the fourth digit remains unchanged and the remaining digits are dropped. In the special case where the fifth digit is exactly 5, round to the nearest even digit. For example, to four significant figures, 1.3245 becomes 1.324; 1.3235 becomes 1.324. This results in half of the numbers being rounded up and half being rounded down, avoiding systematic error. Table 1.4 shows some examples of rounding to four significant figures using these rules.

3. In reporting results obtained by addition and subtraction, the figures in each number are significant only as far as the first uncertain figure of any one of the numbers to be added or subtracted. The result of an addition or subtraction should have the same absolute uncertainty as the number in the calculation with the largest uncertainty. The number with the largest uncertainty is the one with the fewest significant figures. For example, the sum of the numbers in the set 21.1, 3.216, and 0.062 is reliable to the first decimal point, because 21.1 is uncertain in the tenths place. Therefore, the sum is only known to the tenths place. Round off all calculations at the end. When adding or subtracting numbers with exponents, such as numbers written in scientific notation, the numbers should be adjusted so that the exponents are all the same. For example, to add $3.25 \times 10^{-2} + 3 \times 10^{-6}$, express 3×10^{-6} as 0.0003×10^{-2}. The sum is then 3.2503×10^{-2}, which is rounded off to 3.25×10^{-2}.

4. For multiplication and division, the number of significant figures in the answer should be no greater than that of the term with the least number of significant figures. In the case of multiplication and division, the answer will have a relative uncertainty of the same order of magnitude as the number with the largest relative uncertainty. For example, $(1.236 \times 3.1 \times 0.18721 \times 2.36) = 1.692860653$, according to a scientific calculator that allows for 10 digits to be expressed. The figures being multiplied have four, two, five, and three significant figures, respectively. The term 3.1 has the least number of significant figures, two. The answer is therefore rounded off to two significant figures and would be reported as 1.7.

5. The **characteristic** of a **logarithm** indicates an order of magnitude; it has no uncertainty and does not affect the number of significant figures in a calculation. The **mantissa** should contain no more significant figures than are in the original number. For example, the number 12.7 has a logarithm of 1.1038. The mantissa is rounded off to three significant figures and log 12.7 is reported as 1.104. The pH of a solution equals the negative logarithm of the hydrogen ion concentration. Therefore, in the following calculation, pH = $-\log(3.42 \times 10^{-2})$ = 1.4659 = 1.466. The mantissa has three significant figures, as did the original number 3.42.

6. If several analyses have been obtained for a particular sample (**replicate analysis**), it should be noted at what point there is doubt in the significant numbers of the result. The final answer should be reported accordingly, and we will see how this is done in the next section. For example, given the triplicate results 11.32, 11.35, and 11.32, there is no doubt about 11.3, but there is uncertainty in the hundreds place (the fourth figure). The average should be reported as 11.33, with four significant figures (i.e., [11.32 + 11.35 + 11.32] ÷ 3). Remember that the number 3 (the divisor) indicates the exact number of measurements and has no uncertainty, so it does not affect the number of significant figures in the answer.

If a calculation involves a combination of multiplication (or division) and addition (or subtraction), the steps must be treated separately. When using a calculator or spreadsheet program, the best approach is to keep all significant figures throughout the calculation and round off at the end.

Why place all this emphasis on significant figures? When you know the limitations of a measurement in terms of its uncertainty, you can design an analytical method efficiently. If you can only read the absorbance of light in a spectrometric measurement to three significant figures (a typical value), it is a waste of time to weigh the sample to five significant figures.

1.4.2 Accuracy and Precision

It is very important to understand the definitions of accuracy and precision and to recognize the difference between precision and accuracy. ***Accuracy*** is a measure of how close a measured analytical result is to the true answer. For nearly all analytical work, the "true answer" is not known. We often work with an "accepted" value or "accepted reference value." Accuracy is evaluated by analyzing known, standard samples. NIST (formerly the National Bureau of Standards), has well-characterized **standard reference materials** of many types that can serve as the known

Table 1.4 Rounding Off to Four Significant Figures

Number	Four Significant Figures
1.37286	1.373
1.37243	1.372
1.3735	1.374
1.3725	1.372
1.37251	1.373

sample. Similar **certified reference materials** are available from government standards agencies in Canada, the United Kingdom, Europe, and other countries, as well as from a wide variety of commercial standards suppliers. We will discuss reference materials in more detail later.

Another way of assessing accuracy is to *spike* a sample with a known amount of the pure analyte. The sample is analyzed and the amount of added analyte *recovered* is reported. A spike recovery of 100% would indicate that all of the added analyte was measured by the analytical method, for example. Accuracy can be expressed by reporting the difference between the measured value and the accepted value (with the appropriate confidence level, CL, as discussed later) or by reporting the spike recovery as a percentage of added analyte.

Precision is a measure of how close replicate results on the same sample are to each other. Precision is often discussed in terms of *reproducibility* and *repeatability*, two of the less-than-precise terms we have encountered above. A common analogy used to envision the difference between accuracy and precision is to imagine a bull's-eye target used by an archer. If all the arrows hit in the bull's-eye, the archer is both accurate (has hit the center) and precise (all the arrows are close together). If the archer puts all the arrows into the target close together (a "tight shot group") but to the upper left of the bull's-eye, the archer is precise but not accurate. If the arrows hit the target in many locations—top, bottom, center, left, and right of the center—the archer is neither precise nor accurate. The difference between precision and accuracy is illustrated in Table 1.5. There are several ways to express precision mathematically; Table 1.5 uses **standard deviation** (to be defined shortly) as a measure of precision. A superficial examination of the results provided by Analyst 2 could be misleading. It is very easy to be deceived (by the closeness of the results) into believing that the results are accurate. The closeness, expressed as the standard deviation, shows that the results of Analyst 2 are *precise* and not that the analysis will result in obtaining the answer. The latter must be discovered by an independent method, such as having Analyst 2 analyze a sample of known composition. The accepted true value for the determination is 10.1 % ± 0.2 % according to the table footnote, so the determination by Analyst 2 is not accurate. Analyst 3 is both inaccurate and imprecise. It is very unlikely that an imprecise determination will be accurate. Precision is required for accuracy, but does not guarantee accuracy.

It is important for students to realize that the inability to obtain the correct answer does not necessarily mean that the analyst uses poor laboratory techniques or is a poor chemist. Many causes contribute to poor accuracy and precision, some of which we will discuss in this chapter as well as in later chapters. Careful documentation of analytical procedures, instrument operating conditions, calculations, and final results are crucial in helping the analyst recognize and eliminate errors in analysis.

The quantitative analysis of any particular sample should generate results that are precise and accurate. The results should be reproducible, reliable, and truly representative of the sample. For analytical results to be most useful, it is important to be aware of the reliability of the results. To do this, it is necessary to understand the sources of uncertainty and to be able to recognize when they can be eliminated and when they cannot.

The student will often find confusion in the distinction between uncertainty and other common terms such as accuracy, precision, and error. Error is the difference (by subtraction) between the reference result (or accepted true result) and the measured result of an analysis. This is an important concept; if the error in an analysis is large, serious consequences may result. A patient may undergo expensive and even dangerous medical treatment based on an incorrect laboratory result, or an industrial company may implement costly and incorrect modifications to a plant or process because of an analytical error. There are numerous sources of error and several types of errors, some of which are described here. The uncertainty of a measurement is a bit different (as we will illustrate later in this chapter) in that it describes an interval within which any subsequent measurement will fall. That interval is called the confidence limit (CL, mentioned earlier), and it is chosen not by whimsy but rather considerations of purpose. Will the measurement be used to bolster a theory, ensure regulatory compliance, or determine a medical treatment?

1.4.3 Types of Errors

There are two principal types of error in analysis: **determinate** or **systematic** error and **indeterminate** or **random** error.

Table 1.5 Replicate Determinations of Analyte in a Sample[a]

	% Analyte		
	Analyst 1[b]	Analyst 2[c]	Analyst 3[d]
	10.0	8.1	13.0
	10.2	8.0	10.2
	10.0	8.3	10.3
	10.2	8.2	11.1
	10.1	8.0	13.1
	10.1	8.0	9.3
Average (%)	10.1	8.1	11.2
Absolute error[e]	0.0	2.0	1.1
Standard deviation	0.089	0.13	1.57

[a] Accepted "true" answer is 10.1 % ± 0.2 % (obtained independently).
[b] Results are precise and accurate.
[c] Results are precise but inaccurate.
[d] Results are imprecise and inaccurate.
[e] Absolute error = |"true" value − measured value|.

1.4.3.1 Determinate Error

Broadly speaking, **determinate errors** are caused by faults in the analytical procedure or the instruments used in the analysis. The name determinate error implies that the cause of this type of error may be found out and then either avoided or corrected. Determinate errors are *systematic errors*; that is, they are not random. This can also be called bias. A particular determinate error may cause the analytical results produced by the method to be always too high; another may render all results too low. Sometimes the error is *constant*; all answers are too high (or too low) by the same amount. If the true results for three samples are 25, 20, and 30 mg/L of analyte, but the measured (or determined) results are 35, 30, and 40 mg/L, respectively, the analysis has a *constant error* of 10 mg/L. Since these results are all too high, the constant error is positive; a constant negative error of 10 mg/L would result in the three measured results being 15, 10, and 20 mg/L of analyte, respectively. Sometimes the determinate error is proportional to the true result, giving rise to *proportional errors*. For example, if the measured results for the same three earlier samples are 27.5, 22.0, and 33.0 mg/L analyte, respectively, the measured results are too high by 10% of the true answer. This error varies in proportion to the true value. Other determinate errors may be variable in both sign and magnitude, such as the change in the volume of a solution as the temperature changes. Although this variation can be positive or negative, it can be identified and accounted for. Determinate errors can be additive or they can be multiplicative. It depends on the error and how it enters into the calculation of the final result.

If you look again at the results in Table 1.5 for Analyst 2, the results produced by this analyst for the repetitive analysis of a single sample agree closely with each other, indicating high precision. However, the results are all too low (and therefore inaccurate), given that Table 1.5 states the true value of the sample to be 10.1 % ± 0.2 % analyte. There is a negative determinate error in the results from Analyst 2. This determinate error could be the result of an incorrectly calibrated balance. If the balance is set so that the zero point is actually 0.5 g too high, all masses determined with this balance will be 0.5 g too high. If this balance was used to weigh out the potassium chloride used to make the potassium standard solution used in the clinical laboratory, the standard concentration will be erroneously high, and all of the results obtained using this standard will be erroneously low. The error is reported as the absolute error, the absolute value of the difference between the true and measured values. However, there is not enough information provided to know if this is a constant or a proportional error. It can be seen that close agreement between results (i.e., high precision) does not rule out the presence of a determinate error.

Determinate errors arise from some faulty step in the analytical process. The faulty step is repeated every time the determination is performed. Whether a sample is analyzed 5 times or 50 times, the results may all agree with each other (good precision) but differ widely from the true answer (poor accuracy). An example is given in Table 1.6. Although the replicate results are close to each other, that tells us nothing about their accuracy. We can see from the true value given in Table 1.6 that the experimental results are too high; there is a determinate error in the procedure. An analyst or doctor examining the measured analytical results in Table 1.6 might be deceived into believing that the close agreement among the replicate measurements indicates high accuracy and that the results are close to the true potassium concentration. (Potassium in adult human serum has a normal range of 3.5–5.3 mmol potassium/L serum. Assume that the true value given is for this particular patient.)

In the example in Table 1.6, the true value was 4.0 mmol potassium/L and the average measured result was 5.2 mmol potassium/L in the patient's serum. However, the analyst and the doctor do not know the true value of an unknown serum sample. The measured result is in the normal range for adult human serum potassium concentrations, so neither the analyst nor the doctor is likely to be suspicious of the results.

If a faulty analytical procedure is used to analyze five different patients' serum samples and the results shown in Table 1.7 are obtained, it can be seen that in all cases the error is +1.2 mmol/L. This indicates a constant, positive determinate error. As you can see, this faulty procedure would result in one patient being misdiagnosed with a false high serum K level and a patient with a truly low serum K level being misdiagnosed as "normal."

Table 1.6 **Potassium Concentration in a Single Serum Sample**

	Measured Value (mmol/L)	True Value[a] (mmol/L)
	5.2	4.0
	5.1	
	5.3	
	5.1	
	5.1	
Average	5.2	4.0

[a] Normal range for potassium in serum: 3.5–5.3 mmol/L.

Table 1.7 **Potassium Concentrations in Patients' Serum**

Patient	Measured Value[a] (mmol/L)	True Value (mmol/L)
A	5.3	4.1
B	4.8	3.6
C	6.3	5.1
D	5.0	3.8
E	4.1	2.9

[a] Constant error of +1.2 mmol/L.

Table 1.8 Potassium Concentration in Serum

Patient	Measured Value (mmol/L)	True Value[a] (mmol/L)
A	5.8	4.8
B	4.3	3.6
C	7.4	6.2
D	3.5	2.9
E	6.6	5.5

[a] Results indicate a positive proportional error of 20 % of the true value.

An analyst working at a different hospital with different instrumentation obtains the results shown in Table 1.8. Examination of these analytical results shows they are all ~20 % greater than the true answer. The error is *proportional* to the true concentration of the analyte. Such information as to the nature of the error is useful in the diagnosis of the source of the determinate error.

Systematic error is under the control of the analyst. It is the analyst's responsibility to recognize and correct for these *systematic errors* that cause results to be *biased*, that is, offset in the average measured value from the true value. Indeed, in the development of standard analytical methods (such as those published by ASTM International), bias is a mandatory section that must be included in the method. How are determinate errors identified and corrected? Two methods are commonly used to identify the existence of systematic errors. One is to analyze the sample by a completely different analytical procedure that is known to involve no systematic errors. Such methods are often called "standard methods"; they have been evaluated extensively by many laboratories and shown to be accurate and precise. If the results from the two analytical methods agree, it is reasonable to assume that both analytical procedures are free of determinate errors. The second method is to run several analyses of a reference material of known, accepted concentration of analyte. The difference between the known (true) concentration and that measured by analysis should reveal the error. If the results of analysis of a known reference standard are consistently high (or consistently low), then a determinate error is present in the method. The cause of the error must be identified and either eliminated or controlled if the analytical procedure is to give accurate results. In the earlier example of potassium in serum, standard serum samples with certified concentrations of potassium are available for clinical laboratories. Many clinical, forensic, and environmental analytical laboratories participate in proficiency testing programs, where "unknown" standard samples are sent to the laboratory on a regular basis. The results of these samples are sent to the government or professional agency running the program. The unknowns are of course known to the agency that sent the test samples; the laboratory receives a report on the accuracy and precision of its performance.

Determinate errors can arise from uncalibrated balances, improperly calibrated volumetric flasks or pipettes, malfunctioning instrumentation, impure chemicals, incorrect analytical procedures or techniques, and analyst error.

Analyst error: The person performing the analysis causes these errors, which are really mistakes as distinct from the errors we have been considering. Analyst errors are sometimes called human errors, a term that is vague and should not be part of the scientist's vocabulary. They may be the result of inexperience, insufficient training, or being "in a hurry." An analyst may use the instrument incorrectly, perhaps by placing the sample in the instrument incorrectly each time or setting the instrument to the wrong conditions for analysis. Consistently misreading a meniscus in a volumetric flask as high (or low) and improper use of pipettes, such as "blowing out" the liquid from a volumetric pipette, are possible analyst errors. Some other analyst-related errors are (1) *carelessness*, which is not as common as is generally believed; (2) *transcription errors*, that is, copying the wrong information into a lab notebook or spreadsheet or onto a label; and (3) *calculation errors*. Proper training, experience, and attention to detail on the part of the analyst can correct these types of errors.

Reagents and instrumentation: Contaminated or decomposed reagents can cause determinate errors. Impurities in the reagents may interfere with the determination of the analyte, especially at the ppm level or below. Prepared reagents may also be improperly labeled. The suspect reagent may be tested for purity using a known procedure, or the analysis should be redone using a different set of reagents and the results compared.

Numerous errors involving instrumentation are possible, including incorrect instrument alignment, incorrect wavelength settings, incorrect reading of values, and incorrect settings of the readout (i.e., zero signal should read zero). Any variation in proper instrument settings can lead to errors. These problems can be eliminated by a systematic procedure to check the instrument settings and operation before use. Such procedures are called standard operating procedures (SOPs) in many labs. There should be a written SOP for each instrument and each analytical method used in the laboratory.

In instrumental analysis, electrical line voltage fluctuations are a particular problem. This is especially true for automated instruments running unattended overnight. Instruments are often calibrated during the day, when electrical power is in high demand. At night, when power demand is lower, line voltage may increase substantially, completely changing the relationship between concentration of analyte and measured signal. Regulated power supplies are highly recommended for analytical instruments. This is discussed more fully in Chapter 2. The procedure for unattended analysis should include sufficient calibration checks during the analytical run to identify such problems. Many instruments are now equipped with software that can check the measured value of a standard and automatically recalibrate the instrument if that standard falls outside specified limits.

Analytical method: The most serious errors are those in the method itself. Examples of method errors include (1) incomplete reaction for chemical methods, (2) unexpected interferences from the sample itself or reagents used, (3) having the analyte in the wrong oxidation state for the measurement, (4) loss of analyte during sample preparation by volatilization or precipitation, and (5) an error in calculation based on incorrect assumptions in the procedure (errors can arise from assignment of an incorrect formula or RMM to the sample). Most analytical chemists developing a method check all the compounds likely to be present in the sample to see if they interfere with the determination of the analyte; unlikely interferences may not have been checked. Once a valid method is developed, an SOP for the method should be written so that it is performed the same way every time it is run.

Contamination: Contamination of samples by external sources can be a serious source of error and may be extremely variable. An excellent example of how serious this can be has been documented in the analysis of samples for polychlorinated biphenyls (PCBs). PCBs are synthetic mixtures of organochlorine compounds that were first manufactured in 1929 and have become of concern as significant environmental pollutants. It has been demonstrated that samples archived since 1914, before PCBs were manufactured, picked up measurable amounts of PCBs in a few hours just sitting in a modern laboratory (Erickson). Aluminum levels in the dust in a normal laboratory are so high that dust prohibits the determination of low ppb levels of aluminum in samples. A special dust-free "clean lab" or "clean bench" with a filter to remove small dust particles may be required, similar to the clean rooms needed in the semiconductor industry, for determination of traces of aluminum, silicon, and other common elements such as iron. When **trace** (<ppm level) or **ultratrace** (<ppb level) organic and inorganic analysis is required, the laboratory environment can be a significant source of contamination.

Another major source of contamination in an analysis can be the analyst. It depends on what kinds of analytes are being measured, but when trace or ultratrace levels of elements or molecules are being determined, the analyst can be a part of the analytical problem. Many personal care items, such as hand creams, shampoos, powders, and cosmetics, contain significant amounts of chemicals that may be analytes. The problem can be severe for volatile organic compounds in aftershave, perfume, and many other scented products and for silicone polymers, used in many health and beauty products. Powdered gloves may contain a variety of trace elements and should not be used by analysts performing trace element determinations. Hair, skin, and clothing can shed cells or fibers that can contaminate a sample. Having detected the presence of a determinate error, the next step is to find its source. Practical experience of the analytical method or first-hand observation of the analyst using the procedure is invaluable.

1.4.3.2 Indeterminate Error

After all the determinate errors of an analytical procedure have been detected and eliminated, the analytical method is still subject to random or indeterminate error arising from inherent limitations in making physical measurements. Each error may be positive or negative, and the magnitude of each error will vary. Indeterminate errors are not constant or biased. They are random in nature and are the cause of slight variations in results of replicate samples made by the same analyst under the same conditions. Sources of random error include the limitations of reading balances, scales such as rulers or dials, and electrical "noise" in instruments. For example, a balance that is capable of measuring only up to 0.001 g cannot distinguish between two samples with masses of 1.0151 and 1.0149 g. In one case, the measured mass is low; in the other case, it is high. These random errors cause variation in results, some of which may be too high and some too low, as we see for Analyst 1 in Table 1.5. The average of the replicate determinations is accurate, but each individual determination may vary slightly from the true value. Indeterminate errors arise from sources that cannot be corrected, avoided, or even identified, in some cases. All analytical procedures are subject to indeterminate error. However, because indeterminate error is random, the errors will follow a random distribution. This distribution can be understood using the laws of probability and basic statistics. The extent of indeterminate error can be calculated mathematically.

Let us suppose that an analytical procedure has been developed in which there is no determinate error. If an infinite number of analyses of a single sample were carried out using this procedure, the distribution of numerical results would be shaped like a symmetrical bell (Figure 1.5). This bell-shaped curve is called the **normal** or **Gaussian distribution**. The frequency of occurrence of any given measured value when only indeterminate error occurs is represented graphically by a plot such as Figure 1.5.

If only indeterminate errors were involved, the most frequently occurring result would be the true result, that is, the result at the maximum of the curve would be the true answer. In practice, it is not possible to make an infinite number of analyses of a single sample. At best, only a few analyses can be carried out, and frequently, only one analysis of a particular sample is possible. We can, however, use our knowledge of statistics to determine how reliable these results are. The basis of statistical calculations is outlined in Section 1.4.4. Statisticians differentiate between the values obtained from a finite number of measurements, N, and the values obtained from an infinite number of measurements, so we need to define these statistical terms.

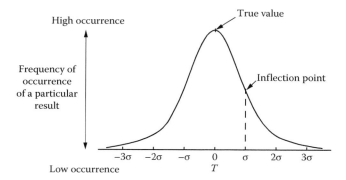

Figure 1.5 A normal or Gaussian distribution of results when only indeterminate error is present. The value that occurs with most frequency is the true value (*T*) or mean value, while the spread of the distribution is expressed in units of standard deviation from the mean, symbolized by σ. The larger the random error, the broader the distribution will be.

1.4.4 Definitions for Statistics

True value T: The true or accepted value, also symbolized by x_t

Observed value x_i: A single value measured by experiment

Sample mean $\bar{x}$: The arithmetic mean of a finite number of observations, that is,

$$\bar{x} = \frac{\sum_{i=1}^{N} x_i}{N} = \frac{(x_1 + x_2 + x_3 + \cdots + x_N)}{N} \quad (1.1)$$

where
N is the number of observations
$\sum x_i$ is the sum of all the individual values x_i

Population mean μ: The limit as N approaches infinity of the sample mean, that is,

$$\mu = \lim_{N \to \infty} \sum_{i=1}^{N} \frac{x_i}{N} \quad (1.2)$$

In the absence of systematic error, the population mean μ equals the true value T of the quantity being measured.

Error E: The difference between the true value T and either a single observed value x_i or the sample mean of the observed values, $\bar{x}$; error may be positive or negative:

$$E = x_i - T \quad \text{or} \quad \bar{x} - x_t \quad (1.3)$$

The total error is the sum of all the systematic and random errors.

Absolute error: The absolute value of E and can be defined for a single value or for the sample mean:

$$E_{abs} = |x_i - T| \quad \text{or} \quad |\bar{x} - x_t| \quad (1.4)$$

Relative error: The absolute error divided by the true value; it is often expressed as a percent by multiplying by 100:

$$E_{rel} = \frac{E_{abs}}{x_t} \quad \text{or} \quad \%E_{rel} = \frac{E_{abs}}{x_t} \times 100 \quad (1.5)$$

Absolute deviation d_i: The absolute value of the difference between the observed value x_i and the sample mean $\bar{x}$,

$$d_i = |x_i - \bar{x}| \quad (1.6)$$

Relative deviation D: The absolute deviation d_i divided by the mean $\bar{x}$,

$$D = \frac{d_i}{\bar{x}} \quad (1.7)$$

Percent relative deviation: The relative deviation multiplied by 100,

$$D(\%) = \frac{d_i \times 100\%}{\bar{x}} = D \times 100 \quad (1.8)$$

Sample standard deviation s: For a finite number of observations N, the sample standard deviation is defined as

$$s = \sqrt{\frac{\sum_{i=1}^{N} d_i^2}{N-1}} = \sqrt{\frac{\sum_{i=1}^{N} (x_i - \bar{x})^2}{N-1}} \quad (1.9)$$

Standard deviation of the mean s_m: The standard deviation associated with the mean of a data set consisting of N measurements,

$$s_m = \frac{s}{\sqrt{N}} \quad (1.10)$$

Population standard deviation σ: For an infinite number of measurements,

$$\sigma = \sqrt{\lim_{N \to \infty} \frac{\sum_{i=1}^{N} (x_i - \mu)^2}{N}} \quad (1.11)$$

Percent relative standard deviation (% RSD),

$$\%RSD = \frac{s}{\bar{x}} \times 100 \quad (1.12)$$

Variance σ^2 or s^2: The square of the population standard deviation σ or the sample standard deviation s.

1.4.5 Quantifying Random Error

If the systematic errors have been eliminated, the measured value will still be distributed about the true value owing to random error. For a given set of measurements, how close is the average value of our measurements to the true value? This is where the Gaussian distribution and statistics are used.

The Gaussian distribution curve assumes that an infinite number of measurements of x_i have been made. The maximum of the Gaussian curve occurs at $x = \mu$, the true value of the parameter we are measuring. So, for an infinite number of measurements, the population mean is the true value x_t. We assume that any measurements we make are a subset of the Gaussian distribution. As the number of measurements, N, increases, the difference between $\bar{x}$ and μ tends toward zero. For N greater than 20, the sample mean rapidly approaches the population mean. For 25 or more replicate measurements, the true value is approximated very well by the experimental mean value. Unfortunately, even 20 measurements of a real sample are not usually possible. Statistics allows us to express the random error associated with the difference between the population mean μ and the mean of a small subset of the population $\bar{x}$. The random error for the mean of a small subset is equal to $\bar{x} - \mu$.

The area under any portion of the Gaussian distribution, for example, between a value x_1 and a value x_2, corresponds to the fraction of the measurements which will yield a measured value of x between and including these two values. The spread of the Gaussian distribution, that is, the width of the bell-shaped curve, is expressed in terms of the population standard deviation σ. The standard deviation σ coincides with the point of inflection of the curve as can be seen in Figure 1.5. The curve is symmetrical, so we have two inflection points, one on each side of the maximum. The x-axis relates the area under the curve to the standard deviation. For $\pm\sigma$ on either side of the maximum, 68.3 % of the area under the curve lies in this range of x. This means that 68.3 % of all measurements of x_i will fall within the range $x = \mu \pm \sigma$. About 95.5 % of the area under the curve lies between $x = \mu \pm 2\sigma$ and 99.7 % of the area lies between $x = \mu \pm 3\sigma$.

The precision of analytical results is usually stated in terms of the standard deviation σ. As just explained, in the absence of determinate error, 68.3% of all results can be expected to fall within $\pm \sigma$ of the true value, 95.5 % of the results will fall within $\pm 2\sigma$ of the true answer, and 99.7 % of our results will fall within $\pm 3\sigma$ of the true value if we perform enough measurements (more than 20 or so replicates). It is common practice to report analytical results with the mean value and the standard deviation expressed, thereby giving an indication of the precision of the results.

Let us suppose that we know by previous extensive testing that the standard deviation σ of a given analytical procedure for the determination of Si in an aluminum alloy is 0.1 % w/w. This is the standard deviation in concentration units (0.1 g Si/100 g Al alloy), not the RSD. Assume we know that there are no determinate errors in the analysis. Also, when we analyze a particular sample using this method and perform sufficient replicate analyses, we obtain an average result of 19.6 % Si by mass. We can now report that the analytical result is 19.6 % ± 0.1 % Si with the understanding that 68.3 % of all results will fall in this range. We could also report that the result is 19.6 % ± 0.2 % (where 0.2 % = 2σ). A report stating that an analysis of a sample indicated 19.6 % Si and that 2σ for the method is 0.2 % means that we are 95.5 % certain that the true answer is 19.6 % ± 0.2 % Si. We say that the CL of the measurement is 95.5 %, with the understanding that there are no determinate errors and that we have performed a sufficient number of replicates to know both μ and σ. Notice that the more certain we are that the answer falls within the range given, the bigger the range actually is. It is important when reporting data with both mean and precision information that the analyst tells the customer what precision is reported. If Analyst A reported 19.6 % ± 0.1 % Si and Analyst B reported 19.6 % ± 0.2 % Si, without telling the customer what the 0.1 % and 0.2 % mean, the customer might think incorrectly that Analyst A is more precise than Analyst B.

However, we are usually dealing with a small, finite subset of measurements, not 20 or more; in this case, the standard deviation that should be reported is the sample standard deviation s. For a small finite data set, the sample standard deviation s differs from σ in two respects. Look at the equations given in the definitions. The equation for sample standard deviation s contains the sample mean, not the population mean, and uses $N - 1$ measurements instead of N, the total number of measurements. The term $N - 1$ is called the **degrees of freedom**.

Let us go through an example of calculating some of these statistical parameters. The equations are simple enough that the values can be calculated manually, although the calculations can be tedious for large values of N. Most scientific handheld calculators have programs that calculate mean, sample standard deviation s (sometimes marked σ_{N-1} on a calculator button), and σ (sometimes marked σ_N on a button). In addition, the calculations can be set up in a spreadsheet in programs like Microsoft Excel®. Such programs have built-in functionality that eliminates the need for coding. Two very good books on extending your applications of Excel are by the late Prof. Robert deLevie.

Suppose you have measured mercury in eight representative samples of biological tissue from herons (which eat fish that may be contaminated with mercury) using cold vapor AAS (which is discussed in Chapter 7). The values in column 2 of the following table were obtained.

Sample	Hg Content (ppb)	$(x_i - \bar{x})^2$
1	5.34	0.010
2	5.37	0.005
3	5.44	0.000
4	5.22	0.048
5	5.84	0.160
6	5.67	0.053
7	5.27	0.029
8	5.33	0.012

The mean is calculated using Equation 1.1:

$$\bar{x} = \frac{5.34 + 5.37 + 5.44 + 5.22 + 5.84 + 5.67 + 5.27 + 5.33}{8}$$

$$= \frac{43.48}{8} = 5.44 \text{ ppb Hg}$$

Since we have only eight measurements, we will calculate s, the sample standard deviation, using Equation 1.9. The standard deviation s is calculated manually in steps, with the intermediate values for $(x_i - \bar{x})^2$ shown in the preceding table in the third column. The sum of the $(x_i - \bar{x})^2$ values equals 0.317; therefore,

$$s = \sqrt{\frac{0.317}{8-1}} = 0.213$$

You could report the average concentration of Hg in the heron tissue as 5.44 ± 0.21 ppb Hg and indicate in your report that 0.21 equals 1 σ. The Hg concentration can be reported in terms of 2 σ, 3 σ, and so on, just as for the population standard deviation. Analytical results published without such precision data lose much of their meaning. They indicate only the result obtained and not the reliability of the answer.

Spreadsheets are extremely useful for performing the repetitive calculations used by analysts, displaying each step in the calculation, tabulating data, and presenting data graphically. A variety of commercial software programs are available. The example given here uses the spreadsheet program Microsoft Excel. The other commercial programs have similar capabilities but the instructions will differ for each. The following example assumes some familiarity with using Microsoft programs. If more fundamental directions are needed, the texts by Christian, Harris, Billo, or Diamond and Hanratty listed in the bibliography have excellent instructions and examples.

When an Excel spreadsheet is opened, the page consists of blank boxes, called cells, arranged in rows and columns. Each cell is identified by its column letter and row number. The individual mercury concentrations from the earlier example can be typed into separate cells, as can text such as column headings. In the sample spreadsheet page shown, text is typed into cells A1, B1, A12, and A13, while the sample numbers and data are put in as shown. The data points are in cells B3 through B10. The width of the columns can be varied to fit the contents.

	A	B	C	D
1	Sample	Hg conc. (ppb)		
2				
3	1	5.34		
4	2	5.37		
5	3	5.44		
6	4	5.22		
7	5	5.84		
8	6	5.67		
9	7	5.27		
10	8	5.33		
11				
12	Mean			
13	Std. dev.			

It is possible to write mathematical formulas and insert them into the spreadsheet to perform calculations, but Excel has many functions already built into the program. By clicking on an empty cell and then on f_x on the toolbar, these functions can be accessed. This opens the Paste Function window. Select *Statistical* in the *Function category* on the left side of the window and a list of *Function names* appears on the right side of the window. Click on the cell B12, then select AVERAGE from the Function name list. Click OK, and type in (B3:B10) in the active box. The average (mean) will appear in cell B12. Alternatively, in cell B12, you can type = AVERAGE(B3:B10) and the mean will be calculated. The standard deviation can be calculated by selecting STDEV from the Function name list and clicking on cell B13 or by typing = STDEV(B3:B10) into cell B13.

	A	B	C	D
1	Sample	Hg conc. (ppb)		
2				
3	1	5.34		
4	2	5.37		
5	3	5.44		
6	4	5.22		
7	5	5.84		
8	6	5.67		
9	7	5.27		
10	8	5.33		
11				
12	Mean	5.435		
13	Std. dev.	0.212804		
14				

The values calculated by Excel are shown in cells B12 and B13. Of course, they must be rounded off to the correct number of significant figures but are the same as the results obtained manually. Learning to use spreadsheets can save time and permit the data to be stored, processed, and presented in a variety of formats. The spreadsheets can be made part of the electronic lab notebooks that are common in industry and government laboratories as well as in many college laboratories.

The value obtained for σ is an estimate of the precision of the method. If an analyst sets up a new analytical procedure and carries out 20 determinations of a standard sample, the precision obtained is called the **short-term precision** of the method. This is the optimum value of σ because it was obtained from analyses run at the same time by the same analyst, using the same instrumentation and the same chemicals and reagents. In practice, the short-term precision data may be too optimistic. Routine analyses may be carried out for many years in a lab, such as the determination of Na and K in serum in a hospital laboratory. Different analysts, different chemicals and reagents, and even different instrumentation may be used. The analysis of a standard sample should be carried out on a regular basis (daily, weekly, etc.), and these results compiled on a regular basis. Over several months or a year, the **long-term precision** of the method can be calculated from these compiled results. This is a more realistic measure of the reliability of the analytical results obtained on a continuing basis from that laboratory.

There are a variety of terms used to discuss analytical methods that are related to the precision of the method. The **repeatability** of the method is the short-term precision of the method under the same operating conditions. **Reproducibility** refers to the ability of multiple laboratories to obtain the same results on a given sample and is determined from collaborative studies. **Ruggedness** is the degree of reproducibility of the results obtained by one laboratory under a variety of conditions (similar to the long-term precision). Repeatability, reproducibility, and ruggedness are all expressed in terms of the standard deviation (or RSD) obtained experimentally. **Robustness** is another term for the **reliability** of the method, that is, its accuracy and precision, under small changes in conditions. These changes can be in operating conditions such as laboratory room temperature, sample variables such as concentration and pH, and reagent and standard stability.

1.4.5.1 Confidence Limits

It is impossible to determine μ and σ from a limited set of measurements. We can use statistics to express the probability that the true value μ lies within a certain range of the measured average mean $\bar{x}$. That probability is called a **confidence limit** (CL, mentioned earlier) and is usually expressed as a percentage (e.g., the CL is 95 %). The term confidence limit refers to the extremes of the **confidence interval** (the range) about $\bar{x}$ within which μ is expected to fall at a given CL.

When s is a good approximation for σ, we can state a CL or confidence limit for our results based on the Gaussian distribution. The CL is a statement of how close the sample mean lies to the population mean. For a single measurement, we let $s = \sigma$. The CL is then the certainty that $\mu = x \pm z\sigma$. For example, if $z = 1$, we are 68.3 % confident that x lies within $\pm\sigma$ of the true value; if we set $z = 2$, we are 95.5% confident that x lies within $\pm 2\sigma$ of the true value. For N measurements, the CL for $\mu = \bar{x} \pm z s_m$.

Table 1.9 Student's *t* Values

Degrees of Freedom ($N - 1$)	*t* Value for CL		
	90 %	95 %	99 %
1	6.31	12.7	63.7
2	2.92	4.30	9.92
3	2.35	3.18	5.84
4	2.13	2.78	4.60
5	2.02	2.57	4.03
6	1.94	2.45	3.71
7	1.90	2.36	3.50
8	1.86	2.31	3.36
9	1.83	2.26	3.25
10	1.81	2.23	3.17
∞	1.64	1.96	2.58

In most cases, s is not a good estimate of σ because we have not made enough replicate analyses. In this case, the CL is calculated using a statistical probability parameter, *Student's t*. The parameter t is defined as $t = (x - \mu)/s$ and the CL for $\mu = \bar{x} \pm ts/\sqrt{N}$. An abbreviated set of t values is given in Table 1.9; complete tables can be found in mathematics handbooks or statistics books.

As an example of how to use Table 1.9 and CLs, assume that we have made five replicate determinations of the pesticide dichlorodiphenyltrichloroethane (DDT) in a water sample using gas chromatography. The five results are given in the spreadsheet as follows, along with the mean and standard deviation. How do we report our results so that we are 95 % confident that we have reported the true value?

	A	B	C	D
1	Replicate sample	DDT conc. (ppb)		
2				
3	1	1.3		
4	2	1.5		
5	3	1.4		
6	4	1.4		
7	5	1.3		
8				
9	Mean	1.4		
10	Std. dev.	0.08		

The number of determinations is five, so there are four degrees of freedom. The t value for the 95 % CL with $N - 1 = 4$ is 2.78, according to Table 1.9. Therefore,

$$95\ \%\ \mathrm{CL} = \bar{x} \pm \frac{ts}{\sqrt{N}} = 1.4 \pm \frac{2.78 \times 0.08}{5^{1/2}} = 1.4 \pm 0.1$$

We are 95 % confident that the true value of the DDT concentration in the water sample is 1.4 ± 0.1 ppb, assuming no determinate error is present.

The Student's t value can be used to test for systematic error (bias) by comparing means of different determinations. The CL equation is rewritten as

$$\pm t = (\bar{x} - \mu)\frac{\sqrt{N}}{s} \quad (1.13)$$

By use of a known, valid method, μ is determined for a known sample, such as a certified reference material. The values of $\bar{x}$ and s are determined for the known sample using the new method (new instrument, new analyst, etc.). A value of t is calculated for a given CL and the appropriate degrees of freedom. If the calculated t value exceeds the value of t given in Table 1.9 for that CL and degrees of freedom, then a significant difference exists between the results obtained by the two methods, indicating a systematic error. Using the DDT data earlier, assume that we know that the true value of the DDT concentration is 1.38 ppb. A new analyst runs the five determinations and obtains a mean of 1.20 ppb and a standard deviation of 0.13. If we calculate t using Equation 1.13, we get $\pm t = (1.20 - 1.38)(5)^{1/2}/0.13 = -3.09$. At 95 % CL, the absolute value of this calculated t is larger than the tabulated t for 4 degrees of freedom found in Table 1.9; that is, $3.09 > 2.78$. Therefore, a determinate error exists in the procedure as performed by the new analyst. We note, *inter alia*, that the concentration of DDT in the listing above is incomplete as ppm. In practice one must specify the basis, either mass or volume, as (mass/mass) or (vol/vol).

1.4.5.2 Variance

Variance is defined as the square of the standard deviation, σ^2. Variance is often preferred as a measure of precision because variances from m independent sources of random error are additive. The standard deviations themselves are not additive. The use of variance allows us to calculate the random error in the answer from mathematical calculations involving several numbers, each of which has its own associated random error. The total variance is the sum of the individual variances:

$$\sigma^2_{\mathrm{tot}} = \sum_{i=1}^{m} \sigma^2_i \quad (1.14)$$

For addition and subtraction, the absolute variance, σ^2, is additive. For multiplication and division, the relative variances are additive, where the relative variance is just the square of the standard deviation divided by the mean, $\sigma^2_{rel} = (\sigma/\bar{x})^2$.

The square of the standard deviation, σ^2, can also be used to determine if sets of data from two methods (analysts, instruments, etc.) are statistically significantly different from each other in terms of their precision. In this test, the variances of two sets of results are compared. The variance σ^2_2 of one set of results is calculated and compared with the variance σ^2_1 of earlier results, or the variance of a new method is compared with that of a standard method. The test is called the *F-test*. The ratio of the variances of the two sets of numbers is called the *F-function*:

$$F = \frac{\sigma^2_1}{\sigma^2_2} \quad (1.15)$$

where $\sigma^2_1 > \sigma^2_2$ (i.e., the ratio should be greater than 1).

Each variance has its associated degrees of freedom, $(N - 1)_1$ and $(N - 1)_2$. Tables of F values are found in mathematics handbooks or statistics books. An abbreviated table for the 95 % CL is given in Table 1.10. If the calculated value of F is larger than the tabulated value of F for the appropriate degrees of freedom and CL, then there is a significant difference between the two sets of data.

1.4.6 Detection of Outliers

When a set of replicate results is obtained, it may be the case that one of the results appears to be "out of line"; such a result is called a suspected **outlier**. While it is tempting to discard data that do not "look good" in order to make the analysis seem more precise, it is never a good practice unless there is justification for discarding the result. If it is known that an error was made, such as spillage of the sample, use of the wrong size pipette, incorrect dilution, or allowing the sample to boil to dryness when it should not have done so, the result should be rejected and not used in any compilation of results. In practice, if something of this sort is suspected, a good analyst will discard the sample and start over if possible.

There are a variety of statistical tests that have been used to decide if a data point should be considered an outlier, as

Table 1.10 *F* Values at 95 % CL

	$(N - 1)_1$					
$(N - 1)_2$	2	4	6	8	10	20
2	19.0	19.2	19.3	19.4	19.4	19.4
4	6.94	6.39	6.16	6.04	5.96	5.80
6	5.14	4.53	4.28	4.15	4.06	3.87
8	4.46	3.84	3.58	3.44	3.35	3.15
10	4.10	3.48	3.22	3.07	2.98	2.77
20	3.49	2.87	2.60	2.45	2.35	2.12

well as some rough guides. We emphasize that there is no test or guide that indicates that a result should be discarded, only that a result might be an outlier. The range chosen to guide the decision will limit all of these tests and guidelines. A large range will retain possibly erroneous results, while a very small range will reject valid data points. It is important to note that the outlier must be either the highest value in the set of data or the lowest value in the set. A value in the middle of a data set cannot be discarded unless the analyst knows that an error was made.

One rough guide is that if the outlier value is greater than $\pm 4\sigma$ from the mean, it should be considered an outlier. When calculating the mean and standard deviation, an outlier result should *not be* included in the calculation. After the calculation, the suspected result should be examined to see if it is more than 4σ from the mean. If it is outside this limit, this guide indicates it should be ignored; if it is within this limit, the value for σ should be recalculated with this result included in the calculation. This guide does not permit rejection of more than one result on this basis. A suspected result should not be included in calculating σ. If it is included, it will automatically fall within 4σ because such a calculation includes this number. Other reference sources recommend an even smaller range for rejection, such as a 2.5σ limit.

A statistical test called the *Q*-test can be used effectively for small data sets (see the reference by Rorabacher) to determine if a given data point should be rejected. The *Q*-test at the 90 % CL is typically used. The data are arranged in order of increasing value. The range of the data is calculated, that is, the lowest value is subtracted from the highest value. The range is $x_n - x_1$. Then the "gap" is calculated, where the gap is defined as the difference between the suspect value and the nearest value, $x_n - x_{n-1}$. The *Q* ratio is defined as Q = gap/range. Using a table of *Q* values, if *Q* observed > *Q* tabulated, the suspect value is discarded. The *Q*-test cannot be used if all but one data point are the same in a set. For example, if triplicate results are 1.5, 1.5, and 3.0, you cannot discard the 3.0 value using statistics. It ultimately falls to the analyst to make the decision about rejecting data, but it should not be done lightly. Additional discussion on outlier detection can be found in the book by Bruno and Svoronos.

1.4.7 Expressing Uncertainty

In the discussion above we have drawn the distinction between accuracy, precision, error, and uncertainty. Earlier in our discussion we realized that most often the target quantity that you are after (such as the concentration of a pollutant) cannot be measured directly, but rather it is the result of an instrument response that is itself dependent on multiple factors. The relationship is described mathematically: $y = f(x_1, x_2, \ldots x_i)$. Here, y is the target quantity, and x_i are individual measured values that determine its value.

These may include voltage, current, temperature, pressure, mass, composition, etc. Just as y contains many different contributions, similarly the uncertainty of y contains many different contributions, $u(x_i)$. Good scientific practice requires us to include in our list as many potential sources of uncertainty as possible, although we realize that there may only be a few very important ones. The list of uncertainty sources is called the *uncertainty budget*, and MUST always include contributions from any calibrations. We add the individual contributions to uncertainty in *quadrature*. This means that we square the individual contributions, multiply by a sensitivity factor, and then take the square root of the sum. If we simply add the uncertainty of the contributing factors, we will greatly overestimate the uncertainty.

Is this so bad? After all, isn't it good to be cautious, especially with our claims of uncertainty? Scientifically, doing this would be an *artificial departure from an objective assessment* of uncertainty. In applied science, it can be even more problematic; it can lead to the overdesign of bridges, for example, to the point where a bridge is too costly to build. Let's consider the example of measuring two lengths of wooden board, and then laying them end to end. We measure the length of the first board to be 500 cm with a ruler that has an uncertainty of plus or minus 3 cm. We use the same ruler to measure the second board at 300 cm. Why is not the uncertainty simply 6 cm? Well, in order for that to happen, the (random) uncertainty of the measurement of both boards would have to be at their maximum possible value AT THE SAME TIME. Sure, that might happen once in millennia, but as long as both measurements are independent (one measurement does not affect the other), this is very unlikely. We calculate uncertainty in quadrature by use of:

$$u_c^2(y) = \sum_{i=1}^{N} \left(\frac{\partial f}{\partial x_i} \right)^2 u^2(x_i) \qquad (1.16)$$

The term in the long parentheses is called the sensitivity factor, determined by calculating the partial derivative of quantity y with respect to each x_i contribution. Physically, this is the rate of change of y with respect to the contribution x_i and provides us with the relative importance of each contribution. The final term is the square of the individual uncertainty for each contribution. Taking the square root of the right-hand side of Eq. 1.16 gives us the *standard* or *propagated uncertainty*. Similar to our discussion above regarding confidence levels on precision calculations, the standard uncertainty represents the 68 % CL. If we multiply the standard uncertainty by a factor of 2, we obtain an *expanded uncertainty* at a CL of 95 %; likewise multiplying by 3 gives the 99 % CL. The multiplicative factors are called coverage factors, usually represented by the letter *h*. The use of Eq. 1.16 to determine the standard uncertainty is permissible only when the individual contributions x_i are not correlated (i.e., they do not depend on one another). If some variables are correlated, then cross-correlation terms must be added to Eq. 1.16, but this will be rare.

So, what is the difference between the uncertainty and precision metrics that we discussed earlier? A proper assessment considers the entire uncertainty budget; all factors that contribute to the measured result. This must also include calibration, whereas earlier we were dealing with the precision of individual results. In research environments, as well as critical applications such as regulatory and forensics, determination and expression of the uncertainty is required. Having described what uncertainty is, we can conversely state what uncertainty is not. Uncertainty is NOT repeatability, reproducibility, precision, deviation from a theory, an instrument maker's claim, or deviation from a literature value.

1.5 SAMPLE PREPARATION

Few samples in the real world can be analyzed without some chemical or physical preparation. The aim of all sample preparation is to provide the analyte of interest in the physical form required by the instrument, free of interfering substances, and in the concentration range required by the instrument. For many instruments, a solution of analyte in organic solvent or water is required. We have already discussed some of the sample preparation steps that may be needed. Solid samples may need to be crushed or ground, or they may need to be washed with water, acid, or solvent to remove surface contamination. Liquid samples with more than one phase may need to be extracted or separated. Filtration or centrifugation may be required.

If the physical form of the sample is different from the physical form required by the analytical instrument, more elaborate sample preparation is required. Samples may need to be dissolved to form a solution or pressed into pellets or cast into thin films or cut and polished smooth. The type of sample preparation needed depends on the nature of the sample, the analytical technique chosen, the analyte to be measured, and the problem to be solved. Most samples are not homogeneous. Many samples contain components that interfere with the determination of the analyte. A wide variety of approaches to sample preparation has been developed to deal with these problems in real samples. Only a brief overview of some of the more common sample preparation techniques is presented. More details are found in the chapters on each instrumental method.

Note: None of the sample preparation methods described here should be attempted without approval, written instructions, and close supervision by your professor or laboratory instructor. It is strongly recommended that Chapter 2 be carefully read before any laboratory work. The methods described present many potential hazards. Many methods use concentrated acids, flammable solvents, and/or high temperatures and high pressures. Reactions can generate harmful gases. The potential for "runaway reactions" and even explosions exists with the preparation of real samples. Sample preparation should be performed in a laboratory fume hood for safety. Goggles, lab coats or aprons, and gloves resistant to the chemicals in use should be worn at all times in the laboratory.

1.5.1 Acid Dissolution and Digestion

Metals, alloys, ores, geological samples, ceramics, and glass react with concentrated acids, and this approach is commonly used for dissolving such samples. Organic materials can be decomposed (digested or "wet ashed") using concentrated acids to remove the carbonaceous material and solubilize the trace elements in samples such as biological tissues, foods, and plastics. A sample is generally weighed into an open beaker, concentrated acid is added, and the beaker heated on a hot plate until the solid material dissolves. Dissolution often is much faster if the sample can be heated at pressures greater than atmospheric pressure. The boiling point of the solvent is raised at elevated pressure, allowing the sample and solvent to be heated to higher temperatures than can be attained at atmospheric pressure. This can be done in a sealed vessel, which also has the advantage of not allowing volatile elements to escape from the sample. Special stainless steel high-pressure vessels, called "bombs," are available for acid dissolution and for the combustion of organic samples under oxygen. While these vessels do speed up the dissolution, they operate at pressures of hundreds of atmospheres and can be very dangerous if not operated properly. Another sealed vessel digestion technique uses **microwave digestion**. This technique uses sealed sample vessels made of polymer, which are heated in a specially designed laboratory microwave oven. (*Never* use a kitchen-type microwave oven for sample preparations. The electronics in kitchen-type units are not protected from corrosive fumes, arcing can occur, and the microwave source, the magnetron, can easily overheat and burn out.) The sealed vessel microwave digestion approach keeps volatile elements in solution, prevents external contaminants from falling into the sample, and is much faster than digestion on a hot plate in an open beaker. Microwave energy efficiently heats solutions of polar molecules (such as water) and ions (aqueous mineral acids) and samples that contain polar molecules and/or ions. In addition, the sealed vessel results in increased pressure and increased boiling point. Commercial analytical microwave digestion systems for sealed vessel digestions are shown in Figure 1.6.

The acids commonly used to dissolve or digest samples are hydrochloric acid (HCl), nitric acid (HNO_3), and sulfuric acid (H_2SO_4). These acids may be used alone or in combination. The choice of acid or acid mix depends on the sample to be dissolved and the analytes to be measured. The purity

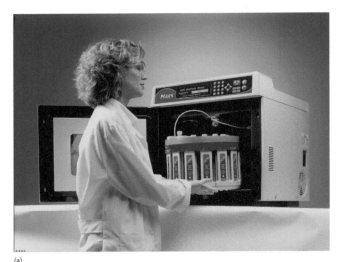

(a)

(b)

Figure 1.6 (a, b) System for sealed vessel digestions systems. (Courtesy of CEM Corporation, Matthews, NC. www.cem.com.)

of the acid must be chosen to match the level of analyte to be determined. Very-high-purity acids for work at ppb or lower levels of elements are commercially available but are much more expensive than standard reagent grade acid. For special applications, perchloric acid ($HClO_4$) or hydrofluoric acid (HF) may be required. A student should never use $HClO_4$ or HF without specific training from an experienced analytical chemist and only then under close supervision. While a mixture of HNO_3 and $HClO_4$ is extremely efficient for wet ashing organic materials, $HClO_4$ presents a very serious explosion hazard. Specially designed fume hoods are required to prevent $HClO_4$ vapors from forming explosive metal perchlorate salts in the hood ducts, and reactions of hot $HClO_4$ with organic compounds can result in violent explosive decompositions. A blast shield should be used, and the organic sample must first be heated with HNO_3 alone to destroy easily oxidized material before the $HClO_4$ is added. Concentrated HF is used for dissolving silica-based glass and many refractory metals such as tungsten, but it is extremely dangerous to work with. It causes severe and extremely painful deep tissue burns that do not hurt immediately upon exposure. However, delay in treatment for HF burns can result in serious medical problems and even death from contact with relatively small amounts of acid.

HCl is the most commonly used nonoxidizing acid for dissolving metals, alloys, and many inorganic materials. HCl dissolves many materials by forming stable chloride complexes with the dissolving cations. There are two major limitations to the universal use of HCl for dissolution. Some elements may be lost as volatile chlorides; examples of volatile chlorides include arsenic, antimony, selenium, and germanium. Some chlorides are not soluble in water; the most common insoluble chloride is silver chloride (AgCl), but mercury (I) chloride (Hg_2Cl_2), copper (1) chloride (CuCl), bismuth oxychloride, (BiOCl), and gold (III) chloride (Au_2Cl_6 dimer of $AuCl_3$) are not soluble, while lead (II) chloride ($PbCl_2$) and thallium (1) chloride (TlCl) are only partially soluble. A 3:1 mixture of HCl and HNO_3 is called *aqua regia* and has the ability to dissolve gold, platinum, and palladium. The mixture is also very useful for stainless steels and many specialty alloys.

HNO_3 is an oxidizing acid; it has the ability to convert the solutes to higher oxidation states. It can be used alone for dissolving a number of elements, including nickel, copper, silver, and zinc. The problem with the use of HNO_3 by itself is that it often forms an insoluble oxide layer on the surface of the sample that prevents continued dissolution. For this reason, it is often used in combination with HCl, H_2SO_4, or HF. A mixture of HNO_3 and H_2SO_4 or HNO_3 and $HClO_4$ can be used to destroy the organic material in an organic sample, by converting the carbon and hydrogen to CO_2 and H_2O when the sample is heated in the acid mixture. The trace metals in the sample are left in solution. This use of acids to destroy organic matter is called wet ashing or digestion, as has been noted. H_2SO_4 is a strong oxidizing acid and is very useful in the digestion of organic samples. Its main drawback is that it forms a number of insoluble or sparingly soluble sulfate salts.

HF is a nonoxidizing, complexing acid like HCl. Its most important attribute is that it dissolves silica-based substances like glass and many minerals. All or most of the silicon is volatized on heating with sufficient HF. Glass beakers and flasks cannot be used to hold or store even dilute HF. Teflon or other polymer labware and bottles are required. Commercial "heatable" Teflon beakers with graphite bottoms are available for use on hot plates. HF is used in acid mixtures to dissolve many refractory elements and minerals by forming fluoride complexes; such elements include tungsten, titanium, niobium, and tantalum. Some elements can be lost as volatile fluorides (e.g., Si, B, As, Ge, and Se). There are a number of insoluble fluoride compounds, including most of the alkaline-earth elements (Ca, Mg, Ba, and Sr) and the rare-earth elements (lanthanides). Table 1.11 gives examples of some typical acid digestions.

Table 1.11 **Common Acid Dissolutions of Metals, Alloys, and Materials for Inorganic Compositional Analysis**

Material[a]	Total Volume of Reagent (mL)	Reagent (vol:vol)
Elements		
Copper metal	20	1:1 HNO_3/H_2O
Gold metal	30	3:1 HCl/HNO_3
Iron metal	20	1:1 HCl/H_2O
Titanium metal	20	H_2SO_4, 3–5 drops HNO_3
Zinc metal	20	HCl
Zirconium metal	15	HF
Alloys		
Copper alloys	30	1:1 HNO_3/H_2O
Low alloy steels	20	3:1 HCl/HNO_3
Stainless steels	30	1:1 HNO_3/HCl
Titanium alloys	100	1:1 HCl/H_2O, 3–5 drops HNO_3
Zinc alloys	30	1:1 HCl/H_2O, dropwise HNO_3
Zirconium alloys	40	1:1 H_2SO_4/H_2O, 2 mL HF dropwise
Other materials		
Borosilicate glass	12	10 mL HF + 2 mL 1:1 H_2SO_4/H_2O
Dolomite	40	1:1 HCl/H_2O
Gypsum	50	1:1 HCl/H_2O
Portland cement	20	HCl + 3 g NH_4Cl
Silicate minerals	30	10 mL HF + 20 mL HNO_3
Titanium dioxide	15	HF
Zinc oxide	15	1:1 HCl/H_2O

Source: Extracted from Dulski, T.R., *A Manual for the Chemical Analysis of Metals*, ASTM International, West Conshohocken, PA, 1996. With permission. Copyright 1996 ASTM International.

Note: "Dropwise" means add drop by drop until dissolution is complete.

[a] 1 g test portion is used; warm to complete reaction.

Some bases, such as sodium hydroxide and tetramethylammonium hydroxide, are used for sample dissolution, as are some reagents that are not acids or bases, like hydrogen peroxide. The chemical literature contains sample dissolution procedures for virtually every type of material known and should be consulted. For elements and inorganic compounds, the *CRC Handbook of Chemistry and Physics* gives guidelines for dissolution in the tables of physical properties of inorganic compounds.

1.5.2 Fusions

Heating a finely powdered solid sample with a finely powdered salt at high temperatures until the mixture melts is called a fusion or molten salt fusion. The reacted and cooled melt is leached with water or dilute acid to dissolve the analytes for determination of elements by atomic spectroscopy or ICP-MS. Often, the molten fusion mixture is poured into a flat-bottomed mold and allowed to cool. The resulting glassy disk is used for quantitative XRF measurements. Molten salt fusions are useful for the dissolution of silica-containing minerals, glass, ceramics, ores, human bone, and many difficultly soluble materials like carbides and borides. The salts used (called "fluxes") include sodium carbonate (Na_2CO_3), borax (sodium tetraborate, $Na_2[B_4O_5(OH)_4]\cdot 8H_2O$), lithium metaborate ($LiBO_2$), and sodium peroxide ($Na_2O_2$). The fusions are carried out over a burner or in a muffle furnace in crucibles of the appropriate material. Depending on the flux used and the analytes to be measured, crucibles may be made of platinum, nickel, zirconium, porcelain, quartz, or glassy carbon. Automated "fluxers" are available that will fuse up to six samples at once and pour the melts into XRF molds or into beakers, for laboratories that perform large numbers of fusions. The drawback of fusion is that the salts used as fluxes can introduce many trace element contaminants into the sample, the crucible material itself may contaminate the sample, and the elements present in the flux itself cannot be analytes in the sample. Fusion cannot be used for boron determinations if the flux is borax or lithium metaborate, for example. Platinum crucibles cannot be used if trace levels of platinum catalyst are to be determined. Table 1.12 gives examples of typical fusions employed for materials.

1.5.3 Dry Ashing and Combustion

To analyze organic compounds or substances for the inorganic elements present, it is often necessary to remove the organic material. Wet ashing with concentrated acids has been mentioned as one way of doing this. The other approach is "dry ashing," that is, ignition of the organic material in air or oxygen. The organic components react to form gaseous carbon dioxide and water vapor, leaving the inorganic components behind as solid oxides. Ashing is often done in a crucible or evaporating dish of platinum

Table 1.12 Molten Salt Fusions of Materials

Material[a]	Dissolution Procedure
Bauxite	2 g Na_2CO_3; Pt c&l
Corundum	3 g Na_2CO_3 + 1 g H_3BO_3; Pt c&l
Iron ores	5 g Na_2O_2 + 5 g Na_2CO_3; Zr c&l
Niobium alloys	10 g $K_2S_2O_7$; fused silica c&l
Silicate minerals	10 g 1:1 Na_2CO_3:$Na_2B_4O_7$; Pt c&l
Tin ores	10 g Na_2O_2 + 5 g NaOH; Zr c&l
Titanium ores	7 g NaOH + 3 g Na_2O_2; Zr c&l
Tungsten ores	8 g 1:1 Na_2CO_3/K_2CO_3; Pt c&l

Source: Extracted from Dulski, T.R., A Manual for the Chemical Analysis of Metals, ASTM International, West Conshohocken, PA, 1996. With permission. Copyright 1996 ASTM International.

Note: c&l, crucible and lid.
[a] 1 g test portion is used.

or fused silica in a muffle furnace. Volatile elements will be lost even at relatively low temperatures; dry ashing cannot be used for the determination of mercury, arsenic, cadmium, and a number of other metals of environmental and biological interest for this reason. Oxygen bomb combustions can be performed in a high-pressure steel vessel very similar to a bomb calorimeter. One gram or less of organic material is ignited electrically in a pure oxygen atmosphere with a small amount of absorbing solution such as water or dilute acid. The organic components form carbon dioxide and water and the elements of interest dissolve in the absorbing solution. Combustion in oxygen at atmospheric pressure can be done in a glass apparatus called a *Schöniger flask*. The limitation to this technique is sample size; no more than 10 mg sample can be burned. However, the technique is used to obtain aqueous solutions of sulfur, phosphorus, and the halogens from organic compounds containing these heteroatoms. These elements can then be determined by ion-selective potentiometry, IC, or other methods.

1.5.4 Extraction

The sample preparation techniques previously discussed are used for inorganic samples or for the determination of inorganic components in organic materials by removing the organic matrix. Obviously, they cannot be used if we want to determine organic analytes. The most common approach for organic analytes is to extract the analytes out of the sample matrix using a suitable solvent. Solvents are chosen with the polarity of the analyte in mind, since "like dissolves like." That is, polar solvents dissolve polar compounds, while nonpolar solvents dissolve nonpolar compounds. Common extraction solvents include hexane, methylene chloride, methyl isobutyl ketone (MIBK), and xylene.

1.5.4.1 Solvent Extraction

Solvent extraction is based on preferential solubility of an analyte in one of two immiscible phases. There are two common situations that are encountered in analysis: extraction of an organic analyte from a solid phase, such as soil, into an organic solvent for subsequent analysis and extraction of an analyte from one liquid phase into a second immiscible liquid phase, such as extraction of PCBs from water into an organic solvent for subsequent analysis.

Liquid–liquid extraction is similar to what happens when you shake oil and vinegar together for salad dressing. If you pour the oil and vinegar into a bottle carefully, you will have two separate clear layers because the oil and vinegar are not soluble in each other (they are *immiscible*). You shake the bottle of oil and vinegar vigorously, and the "liquid" gets cloudy and you no longer see the two separate phases. On standing, the two immiscible phases separate into clear liquids again. Our two immiscible solvents will be called solvent 1 and solvent 2. The analyte, which is a solute in solvent 2, will distribute itself between the two phases on vigorous shaking. After allowing the phases to separate, the ratio of the concentration of analyte in the two phases is approximately a constant, K_D:

$$K_D = \frac{[A]_1}{[A]_2} \quad (1.17)$$

K_D is called the distribution coefficient and the concentrations of A, the analyte, are shown in solvent 1 and solvent 2. A large value of K_D means that the analyte will be more soluble in solvent 1 than in solvent 2. If the distribution coefficient is large enough, most of the analyte can be extracted quantitatively out of solvent 2 into solvent 1. The liquid containing the analyte and the extracting solvent are usually placed into a separatory funnel, shaken manually, and the desired liquid phase drawn off into a separate container. The advantages of solvent extraction are to remove the analyte from a more complex matrix, to extract the analyte into a solvent more compatible with the analytical instrument to be used, and to preconcentrate the analyte. For example, organic analytes can be extracted from water using solvents such as hexane. Typically, 1 L of water is extracted with 10–50 mL of hexane. Not only is the analyte extracted, but it is also now more concentrated in the hexane than it was in the water. The analyte percent extracted from solvent 2 into solvent 1 can be expressed as

$$\%E = \frac{[A]_1 V_1}{[A]_2 V_2 + [A]_1 V_1} \times 100 \quad (1.18)$$

where

%E is the percent of analyte extracted into solvent 1, the concentration of analyte in each solvent is expressed in molarity

V_1 and V_2 are the volumes of solvents 1 and 2, respectively.

The percent extracted is also related to K_D:

$$\%E = \frac{100 K_D}{K_D + \left(\dfrac{V_2}{V_1}\right)} \quad (1.19)$$

The percent extracted can be increased by increasing the volume of solvent 1, but it is more common to use a relatively small volume of extracting solvent and repeat the extraction more than once. The multiple volumes of solvent 1 are combined for analysis. Multiple small extractions are more efficient than one large extraction.

Liquid–liquid extraction is used extensively in environmental analysis to extract and concentrate organic compounds from aqueous samples. Examples include the extraction of pesticides, PCBs, and petroleum hydrocarbons from water samples. Extraction is also used in the determination of fat in milk. Liquid–liquid extraction can be used to separate organometallic complexes from the matrix in clinical chemistry samples such as urine. For example, heavy metals in urine can be extracted as organometallic complexes for determination of the metals by flame AAS. The chelating agent and a solvent such as methylisobutyl ketone (MIBK) are added to a pH-adjusted urine sample in a separatory flask. After shaking and being allowed to stand, the organic solvent layer now contains the heavy metals, which have been separated from the salts, proteins, and other components of the urine matrix. In addition to now having a "clean" sample, the metals have been extracted into a smaller volume of solvent, increasing the sensitivity of the analysis. An added benefit is that the use of the organic solvent itself further increases the sensitivity of flame AAS measurement (as discussed in Chapter 7).

Extraction of organic analytes such as pesticides, PCBs, and fats from solid samples such as food, soil, plants, and similar materials can be done using a Soxhlet extractor. A Soxhlet extractor consists of a round-bottom flask fitted with a glass sample/siphon chamber in the neck of the flask. On top of the sample chamber is a standard water-cooled condenser. The solid sample is placed in a cellulose or fiberglass porous holder, called a thimble; the solvent is placed in the round-bottom flask. Using a heating mantle around the flask, the solvent is vaporized, condensed, and drips or washes back down over the sample. Soluble analytes are extracted and then siphoned back into the round-bottom flask. This is a continuous extraction process as long as heat is applied. The extracted analyte concentrates in the round-bottom flask. An unfavorable aspect of extractions such as those with the Soxhlet apparatus is that large volumes of solvent are used that must eventually be disposed of properly.

As you can imagine, performing these extractions manually is time-consuming and can be hard work (try shaking a 1 L separatory funnel full of liquid for 20 min and imagine having to do this all day!). In addition, a lot of bench space is needed for racks of 1 L flasks or multiple heating mantles. There are several instrumental advances in solvent extraction that have made extraction a more efficient process. These advances generally use sealed vessels under elevated pressure to improve extraction efficiency and are classified as pressurized fluid (or pressurized solvent) extraction methods. One approach is the Accelerated Solvent Extraction (ASE®) system, from Dionex, now part of Thermo Fisher Scientific (www.thermofisher.com). This technique is used for extracting solid and semisolid samples, such as food, with liquid solvents. The technique is shown schematically in Figure 1.7. ASE uses conventional solvents and mixtures of solvents at elevated temperature and pressure to increase the efficiency of the extraction process. Increased temperature, up to 200 °C compared with the 70–80 °C normal boiling points of common solvents, accelerates the extraction rate, while elevated pressure keeps the solvents liquid at temperatures above their normal boiling points, enabling safe and rapid extractions. Extraction times for many samples can be cut from hours using a conventional approach to minutes, and the amount of solvent used is greatly reduced. Table 1.13 presents a comparison of the use of a commercial ASE system with conventional Soxhlet extraction. Dozens of application examples, ranging from fat in chocolate through environmental and industrial applications, can be found at the Thermo Fisher website. The SpeedExtractor from Büchi Corporation, introduced in 2009, also has applications available at www.mybuchi.com. The US EPA has recognized ASE and other instruments that use pressure and temperature to accelerate extraction of samples for environmental analysis by issuing US EPA Method 3545A (SW-846 series) for Pressurized Fluid Extraction of samples. The method can be found at www.usepa.gov.

A second approach also using high pressure and temperature is that of microwave-assisted extraction. The sample is heated with the extraction solvent in a sealed vessel by microwave energy, as was described for microwave digestion. The temperature can be raised to about 150 °C with the already described advantages of high temperature and high pressure. One limitation of microwave-assisted extraction is that some solvents are "transparent" to microwave radiation and do not heat; pure nonpolar solvents such as the normal alkanes are examples of such transparent solvents. Several instrument companies manufacture microwave extraction systems. Milestone Inc. (www. milestonesci.

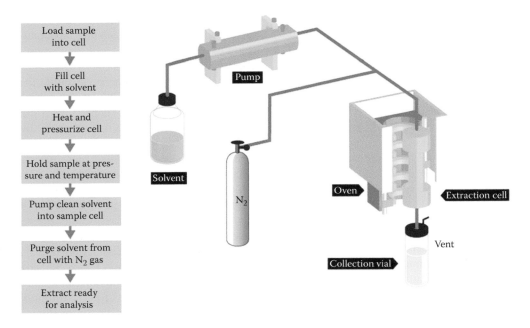

Figure 1.7 ASE technique. (© Thermo Fisher Scientific (www.thermofisher.com). Used with permission.)

Table 1.13 Comparison of Soxhlet Extraction with ASE

Extraction Method	Average Solvent Used per Sample (mL)	Average Extraction Time per Sample	Average Cost per Sample (US $)
Manual Soxhlet	200–500	4–48 h	27
Automated Soxhlet	50–100	1–4 h	16
Accelerated solvent extraction	15–40	12–18 min	14

Source: © Thermo Fisher Scientific (www.thermofisher.com). Used with permission.

Table 1.14 Comparison of Microwave-Assisted Extraction with Conventional Solvent Extraction for Herbicides in Soil Samples

Extraction Method	Time (min)	Volume of Solvent (mL)	% Recovery
Separatory funnel	15	25	42–47
Soxhlet	90	40	51–52
Microwave extraction	10 (90°C)	20	66–78

Source: Data in table courtesy of Milestone Inc., Shelton, CT, www.milestonesci.com.

com) has numerous applications on their website as well as video compact disks (CDs) of microwave extraction, ashing, and digestion available. CEM (www.cem.com) also has application notes available for microwave extraction. An example of improved performance from a microwave extraction system versus conventional extraction is shown in Table 1.14.

The third instrumental approach is the use of supercritical fluid extraction (SFE). A **supercritical fluid** is a substance at a temperature and pressure above the critical point for the substance. You may want to review phase diagrams and the critical point on the phase diagram in your general chemistry text or see the Applied Separations, Inc. website at www.appliedseparations.com. Supercritical fluids are more dense and viscous than the gas phase of the substance but not as dense and viscous as the liquid phase. The relatively high density (compared with the gas phase) of a supercritical fluid allows these fluids to dissolve nonvolatile organic molecules. Also of major importance is the diffusion of solutes in supercritical fluids, which is intermediate between gases and liquids. Carbon dioxide, CO_2, has a critical temperature of 31.3 °C and a critical pressure of 72.9 atm (see diagram); this temperature and pressure are readily attainable, making supercritical CO_2 easy to form. Supercritical fluid CO_2 dissolves many organic compounds, so it can replace a variety of common solvents; supercritical CO_2 is used widely as a solvent for extraction. The advantages of using supercritical fluid CO_2 include its low toxicity, low cost, nonflammability, and ease of disposal. Once the extraction is complete and the pressure returns to atmospheric pressure, the carbon dioxide immediately changes to a gas and escapes from the opened extraction vessel. The pure extracted analytes are left behind. Automated SFE instruments can extract multiple samples at once at temperatures up to 150 °C and pressures up to 10,000 psi (psi, pounds per square inch, is not an SI unit; 14.70 psi = 1 atm). SFE instrument descriptions and applications can be found at a number of company websites: Jasco, Inc. (www.jascoinc.com); Supercritical Fluid Technologies, Inc. (www.supercriticalfluids.com); Büchi Corporation (www.mybuchi.com); and Newport Scientific, Inc. (www.newport-scientific.com). SFE methods have been developed for extraction of analytes from environmental, agricultural, food and beverage, polymer and pharmaceutical samples, among other matrices.

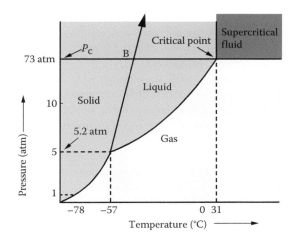

1.5.4.2 Solid-Phase Extraction

In **solid-phase extraction** (SPE), the "extractant" is not an organic liquid but a solid-phase material. Organic compounds are chemically bonded to a solid substrate such as silica beads or polymer beads. The bonded organic layer interacts with organic analytes in the sample solution and extracts them from the sample solution as it is poured through a bed or disk of the solid extractant. The excess solution is allowed to drain away, and interfering compounds are washed off the extractant bed with a solution that does not remove the target analytes. The extracted organic analytes are then **eluted** from the solid-phase extractant by passing a suitable organic solvent through the bed. The interactions that cause the analytes to be extracted are those intermolecular attractive forces you learned about in general chemistry: van der Waals attractions, dipole–dipole interactions, and electrostatic attractions.

The types of organic compounds that can be bonded to a solid substrate vary widely. They can be hydrophobic nonpolar molecules such as C_8 and C_{18} hydrocarbon chains, chains with highly polar functional groups such as cyano (—C≡N), amine (—NH_2), and hydroxyl (—OH) groups and with ionizable groups like sulfonic acid anions ($-SO_3^-$) and carboxylic acid anions ($-CO_2^-$), to extract a wide variety of analytes. The term **sorbent** is used for the solid-phase extractant. Commercial SPE cartridges have the sorbent packed into a polymer syringe body or disposable polymer pipette tip. Figure 1.8a shows commercial plastic syringe-type cartridges with a variety of sorbents, including several of those just mentioned, while Figure 1.8b shows a schematic of how the sorbent is packed and held in the cartridge. These are used only once, preventing cross-contamination of samples and allowing the cleanup of extremely small sample volumes (down to 1 μL), such as those encountered in clinical chemistry samples. Specialized sorbents have been developed for the preparation of urine, blood, and plasma samples for drugs of abuse, for example. Automated SPE systems that can process hundreds of samples simultaneously are now in use in the pharmaceutical and biotechnology industries.

SPE is used widely for the cleanup and concentration of analytes for analysis using LC, HPLC, and LC-MS, as discussed in Chapter 14. As you will see, the phases used in HPLC for the separation of compounds are in many cases identical to the bonded solid-phase extractants described here. Detailed examples and application notes are available from a number of SPE equipment suppliers: Avantor Performance Materials, Inc. (www.avantormaterials.com), Supelco (www.sigma-aldrich.com/supelco), Phenomenex (www.phenomenex.com), and Büchi Corporation (www.mybuchi.com) are a few of the companies that supply these products.

1.5.4.3 QuEChERS

A very important analytical method is the determination of pesticide residues in fruits, vegetables, and processed foods such as cereal. The method used to require complex extraction and cleanup of samples prior to HPLC and MS analysis. In 2003, US Department of Agriculture scientists (Anastassiades et al.) developed a simplified method for food samples, called QuEChERS (pronounced "ketchers" or "catchers"), which stands for quick, easy, cheap, effective, rugged, and safe. QuEChERS has now been applied to food samples, dietary supplements, and other matrices for pesticides, veterinary drugs, and other compounds and is recognized as a sample preparation method in AOAC and European standards. The QuEChERS approach uses

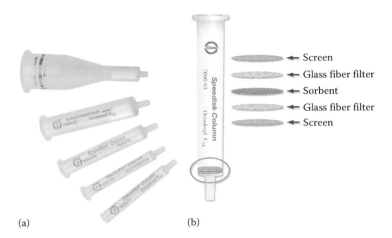

Figure 1.8 (a) Commercial plastic syringe-type cartridges and (b) schematic of sorbent packing. (Permission to reprint from Avantor Performance Materials, Inc., Center Valley, PA, formerly known as Mallinckrodt Baker, Inc., www.avantormaterials.com.)

a simple two-step approach, an extraction step followed by dispersive SPE. First, the sample is homogenized and weighed into a 50 mL centrifuge tube and appropriate organic solvent and salts are added. The salts may include magnesium sulfate and sodium acetate for drying and buffering the sample. The sample is shaken and centrifuged. The supernatant is further extracted and cleaned using dispersive SPE. In dispersive SPE, the SPE sorbent(s) is(are) in a centrifuge tube along with a small amount of $MgSO_4$. An aliquot of the supernatant from step 1 is added, the tube is shaken and centrifuged, and the sorbent removes interfering matrix materials from the sample. Sorbents include primary and secondary amine (PSA) exchange materials to remove sugars, fatty acids, organic acids, and some pigments; C18 (a common octadecyl-substituted hydrophobic LC material) is used to remove lipids and nonpolar interferences and graphitized carbon black (GBC) to remove pigments and sterols. Multiple companies now market packaged QuEChERS kits that meet regulatory standards in the United States and Europe. The websites www.quechers.com, www.restek.com, www.gerstel.com/en/applications.htm, and www.agilent.com/chem/Quechers offer a variety of tutorials, application notes, audio slideshows, webinars, and video demonstrations of the QuEChERS process.

A recent environmental application of the QuEChERS method was demonstrated by researchers from Gerstel, Inc. and the Arkansas Public Health Laboratory to deal with the huge 2010 oil well spill in the Gulf of Mexico (Whitecavage et al.). Estimates of the number of samples to be analyzed for petroleum hydrocarbons such as polyaromatic hydrocarbons (PAHs) in seafood ran as high as 10,000 samples per month. Standard regulatory assays took 1 week to analyze 14–25 samples. Using Gerstel's Twister™ stir bar sorptive extraction (SBSE), sample throughput for PAHs in seafood could be increased to 40 samples/day. The Twister™ is a glass-coated magnetic stir bar with an external layer of polydimethylsiloxane (PDMS). The stir bar is added to an aliquot of the supernatant from the QuEChERS step, and while stirring the solution, organic compounds are extracted into the PDMS phase. The Twister is removed from the sample, rinsed with DI water, dried with a lint-free cloth, and placed in a thermal desorption tube for automated direct thermal desorption into a GC-MS system. The SBSE step provides a concentration of the analytes as well as an additional cleanup of the supernatant, resulting in limits of detection 10–50 times lower than the regulatory assay while using significantly less solvent.

The SPE field is still developing, with the introduction in 2009 of an automated SPE instrument from Dionex designed to be used with large-volume samples (20 mL to 4 L) for the isolation of trace organics in water and other aqueous matrices.

1.5.4.4 Solid-Phase Microextraction

Solid-phase microextraction (SPME, pronounced "spee-mee") is a sampling technique developed first for analysis by GC; the use of SPME for GC and related applications is discussed in greater detail in Section 12.3. The solid phase in this case is a coated fiber of fused silica. The coatings used may be liquid polymers like PDMS, which is a silicone polymer. Solid sorbents or combinations of both solid and liquid polymers are also used. Figure 1.9a shows a commercial SPME unit with the coated fiber inserted into a sample vial; the coated fiber tip is shown in Figure 1.9b. No extracting solvent is used when the sample is analyzed by GC. The coated fiber is exposed to a liquid or gas sample or to the vapor over a liquid or solid sample in a sealed vial (this is called sampling the **headspace**) for a period of time. Analyte is adsorbed by the solid coating or absorbed by the liquid coating on the fiber and then thermally desorbed by insertion into the heated injection port of

 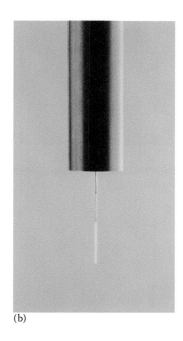

Figure 1.9 (a) A commercial SPME unit with (b) the coated fiber and inserted into a sample vial. (Reprinted with permission of Supelco, Bellefonte, PA, www.sigma-aldrich.com.)

the gas chromatograph. The process is shown schematically in Figure 1.10.

Unlike solvent extraction, the entire amount of analyte is not extracted. The amount of analyte extracted by the coated fiber is proportional to the concentration of analyte in the sample. This will be true if equilibrium between the fiber and the sample is achieved or before equilibrium is achieved if the sampling conditions are controlled carefully. SPME sampling and desorption can be used for qualitative and quantitative analyses. Quantitative analysis using external calibration, internal standard calibration, and the method of standard additions (MSA) are all possible with SPME. Calibration is discussed in Section 1.5.2 and at greater length in Chapter 3.

SPME sampling is used for a wide variety of analytes, including environmental pollutants, volatiles from botanical samples (e.g., used to identify tobacco species), explosives, and chemical agent residues. Gasoline and other accelerants in the headspace over fire debris can be sampled with SPME to determine whether arson may have caused the fire. As little as 0.01 µL of gasoline can be detected. Gas samples such as indoor air and breath have been sampled using SPME. Liquid samples analyzed by either immersion of the fiber into the sample or sampling of the headspace vapor include water, wine, fruit juice, blood, milk, coffee, urine, and saliva. Headspace samplings of the vapors from solids include cheese, plants, fruits, polymers, pharmaceuticals, and biological tissue. These examples and many other application examples are available in pdf format and on CD from Supelco at www.sigma-aldrich.com/supelco. SPME probes that are about the size of a ballpoint pen (Figure 1.11) are available for field sampling (e.g., see www.fieldforensics.com or www.sigma-aldrich.com/supelco). These can be capped and taken to an on-site mobile lab or transported back to a conventional laboratory for analysis.

While SPME started as a solvent-free extraction system for GC analysis, it can now also be used to introduce samples into an HPLC apparatus. An SPME–HPLC interface (Figure 1.12) allows the use of an SPME fiber to

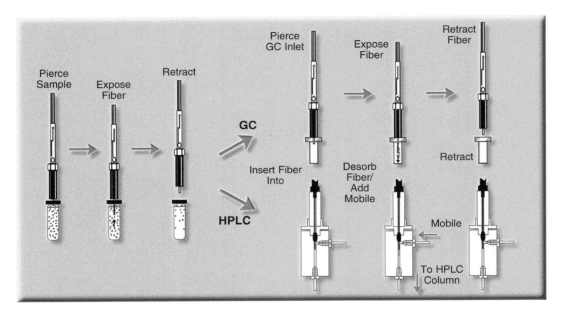

Figure 1.10 Schematic of the SPME process. (Reprinted with permission of Supelco, Bellefonte, PA, www.sigma-aldrich.com.)

(c)

Figure 1.11 An SPME probe the size of a ball-point pen. (Reprinted with permission of Supelco, Bellefonte, PA. www.sigma-aldrich.com.)

sample nonvolatile analytes such as nonionic surfactants in water and elute the analyte into the solvent mobile phase used for the HPLC analysis. The sampling process and elution are shown schematically in Figure 1.10. HPLC and its applications are covered in Chapter 14.

1.6 PERFORMING THE MEASUREMENT

To determine an analyte using an instrumental method of analysis, we must establish the relationship between the magnitude of the physical parameter being measured and the amount of analyte present in the sample undergoing the measurement. In most analyses, the measurement is made and then a calculation is performed to convert the result of the measurement into the amount of analyte present in the original sample. The calculation accounts for the amount of sample taken and dilutions required in the process of sample preparation and measurement.

In an instrumental method of analysis, a **detector** is a device that records a change in the system that is related to the magnitude of the physical parameter being measured. We say that the detector records a **signal**. If the detector is properly designed and operated, the signal from the detector can be related to the amount of analyte present in the sample through a process called **calibration**. Before we discuss calibration, we need to understand a little about what the detector is recording. A detector can measure physical, chemical, or electrical changes or signals, depending on its design. A **transducer** is a detector that converts nonelectrical signals to electrical signals (and vice versa). There are transducers used in spectroscopic instruments that convert photons of light into an electrical current, for example. Another term used in place of detector or transducer is **sensor**. The operation of specific detectors is covered in the chapters on the different instrumental methods (Chapters 3 through 17).

CONCEPTS OF INSTRUMENTAL ANALYTICAL CHEMISTRY

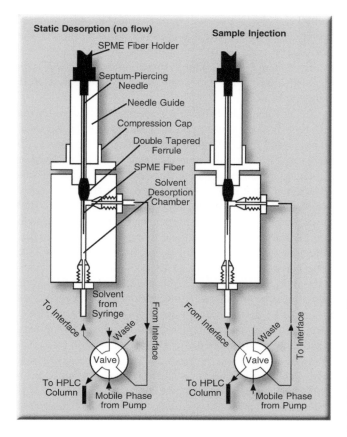

Figure 1.12 An SPME–HPLC interface. (Reprinted with permission of Supelco, Bellefonte, PA. www.sigma-aldrich.com.)

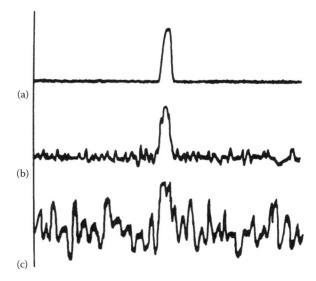

Figure 1.13 Signal versus wavelength with different noise levels: (a) no detectable noise, (b) moderate noise, and (c) high noise.

1.6.1 Signals and Noise

Instrumental analysis uses electronic equipment to provide chemical information about the sample. Older instruments used vacuum tubes and output devices like strip chart recorders, while modern instruments use semiconductor technology and computers to control the instrument, collect signals, and process and report data. Fundamentals of modern instrument electronics are covered in the text by Malmstadt et al. listed in the bibliography.

All instruments measure some chemical or physical characteristic of the sample, such as how much light is absorbed by the sample at a given wavelength, the mass-to-charge ratio of an ion produced from the sample, or the change in conductivity of a wire as the sample passes over it. A detector of some type makes the measurement and the detector response is converted to an electrical signal. The electrical signal should be related directly to the chemical or physical property being measured and that should be related to the amount of analyte present. Ideally, the signal would represent only information related to the analyte. When no analyte is present, there should be no signal. For example, in Figure 1.13, we are looking at signals from a spectrometer that is measuring the amount of light emitted by a sample at a given wavelength. The three traces in Figure 1.13 show a **peak**, which is the signal at the emission wavelength. The response on either side of the peak is called the **baseline**.

An ideal signal for intensity of light emitted by the analyte versus wavelength would be a smooth baseline when no light is emitted and a smooth peak at the emission wavelength, as shown in Figure 1.13a. In this case, when the instrument does not detect the analyte, there is no signal, represented by the flat baseline. When the instrument does detect the analyte, the signal increases. When the instrument no longer detects the analyte, the signal decreases back to the baseline. In this case, the entire signal (the peak) is attributed to the analyte. In practice, however, the recorded signal and baseline are seldom smooth but include random signals called **noise**. All measured signals contain the desired information about the analyte and undesirable information, including noise. Noise can originate from small fluctuations of various types in the instrumentation, in the power provided to the instrument, from external sources such as TV and radio stations, other instruments nearby, building vibrations, electrical motors and similar sources, and even from fundamental quantum effects that cannot be eliminated. Provided that the signal is sufficiently greater than the noise, it is not difficult to make a reliable measurement of the desired analyte signal. The **signal-to-noise (S/N) ratio** is a useful quantity for comparing analytical methods or instruments.

In Figure 1.13b, the noise level is increased over that in Figure 1.13a and is superimposed on the signal. The value of the signal is less certain as a result of the noise. The measurement is less precise, but the signal is clearly discernible. In Figure 1.13c, the noise level is as great as the signal, and it is virtually impossible to make a meaningful measurement of the latter. It is very important to be able to separate data-containing signals from noise. When the signal is very weak, as it might be for trace amounts of analyte, the noise must be reduced or the signal enhanced. Noise is random in nature; it can be either positive or negative at any given point, as can be seen in Figure 1.13b and c. Because noise is random, it can be treated statistically.

If we consider the signal S to be our measured value, the noise N is the variation in the measured value when repeat measurements of the same sample are made. That is, the noise can be defined as the standard deviation s of repeat measurements; the signal S is then the average value of the measurement, $\bar{x}$. While making the repeat measurements at the peak maximum would provide the best estimate of the S/N ratio at the exact point we want to measure, there are some difficulties associated with this approach. One is that the noise measured at the peak maximum may be hard to detect, as a small variation in a large signal is more difficult to measure than a large variation in a small signal. A second problem is that the S/N ratio will be dependent on the size of the signal if measured at the peak maximum. In practice, the noise is often measured along the baseline where the signal should be zero, not at the peak maximum. An easy way to measure the noise in Figure 1.13b is to use a ruler to measure the maximum amplitude of the noise somewhere away from the signal along the baseline. The magnitude of the noise is independent of the magnitude of the signal along the baseline. The signal magnitude is measured from the middle of the baseline to the middle of the "noisy" peak. The middle of the baseline has to be estimated by the analyst. The effect of noise on the relative error of the measurement decreases as the signal increases:

$$\frac{S}{N} = \frac{\bar{x}}{s} = \frac{Mean}{Std.\,dev.} \quad (1.20)$$

There are several types of noise encountered in instrumental measurements. The first is **white noise**, the random noise seen in Figure 1.13b and c. White noise can be due to the random motions of charge carriers such as electrons; the random motion results in voltage fluctuations. This type of white noise is called *thermal noise*. Cooling the detector and other components in an instrument can reduce thermal noise. A second type of white noise is *shot noise*, which occurs when charge carriers cross a junction in an electric circuit. **Drift** or **flicker noise** is the second major type of instrumental noise. It is inversely proportional to the frequency of the signal being measured and is most significant for low frequencies. The origin of drift or flicker noise is not well understood. The third type of instrumental noise is that due to the surroundings of the instrument, such as the line noise due to the power lines to the instrument or building vibrations. Some of this type of noise is frequency dependent and may occur at discrete frequencies.

Improvement in S/N ratio requires that the signal be different from the noise in some way. Most differences can be expressed in terms of time correlation or frequency. To increase the S/N ratio, either the noise must be reduced or the signal enhanced, or both must occur. There are a variety of hardware and software approaches to reduce noise in instruments. External sources of noise can be eliminated or reduced by proper grounding and shielding of instrument circuits and placement of instruments away from drafts, other instruments, and sources of vibration. The intrinsic noise can be reduced using a variety of electronic hardware such as lock-in amplifiers, filters, and signal modulators. Signals can be enhanced by a variety of computer software programs to perform signal averaging, Fourier transformation, filtering, and smoothing. Many of these software manipulations are applied after the data have been collected. Many of the hardware methods for reducing noise are now being replaced by computer software methods, since even simple instruments now output data in digital form. The discussion of analog-to-digital conversion and details of methods for S/N ratio enhancement are beyond the scope of this chapter; the references by Enke or Malmstadt et al. can be consulted for details.

Signal averaging is one way to improve the S/N ratio; repetitive measurements are made and averaged. In this instance, advantage is taken of the fact that noise is random but the signal is additive. If a signal, such as that shown in Figure 1.13b, is measured twice and the results added together, the signal will be twice as intense as the first measurement of the signal. If n measurements are made and added together, the signal will be n times as intense as the first signal. Because noise is random, it may be positive or negative at any point. If n measurements are added together, the noise increases only as the square root of n or $n^{1/2}$. Since S increases by a factor of n, and N increases by $n^{1/2}$, the S/N ratio increases by $n/n^{1/2} = n^{1/2}$. Averaging multiple signal measurements will improve the S/N ratio by a factor of $n^{1/2}$, as shown in Figure 1.14. Therefore, to improve the S/N ratio by 10, about 100 measurements must be made and averaged. The disadvantage to signal averaging is the time required to make many measurements. Some instrumental methods lend themselves to rapid scanning, but others such as chromatography do not.

Instruments that use Fourier transform (FT) spectroscopy, introduced in Chapter 3, collect the entire signal, an interferogram, in a second or two. Hundreds of measurements can be made, stored by a computer, and averaged by an FT instrument very quickly, greatly improving the S/N ratio using this approach. The FT approach discriminates

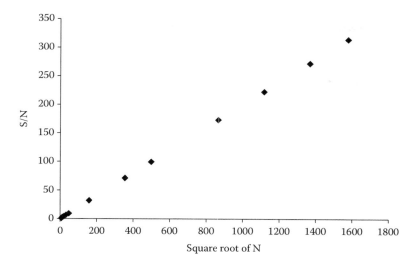

Figure 1.14 Improvement in the S/N ratio by averaging multiple signal measurements.

signal from noise based on frequency. An FT is a mathematical transformation that converts a variable measured as a function of time, $f(t)$, into a function of reciprocal time, $f(1/t)$. A function of reciprocal time is also a function of frequency. The FT permits the removal of noise that differs in frequency from the signal and also permits enhancement of frequencies associated with the signal. FT and a related algorithm, the fast FT (FFT), are now available as part of many data-handling software packages. (FFT is an algorithm for efficiently computing FT.) The use of FT is required to process data from experiments based on interferometry, such as FTIR spectroscopy (Chapter 5), from pulsed NMR experiments (FTNMR, Chapter 6), and the signal pulses from cyclotron resonance MS experiments (FTMS, Chapter 10).

The S/N ratio is a limiting factor in making analytical measurements. It is theoretically possible to measure any signal in any noise level, provided the measurements are taken over a long enough period of time or repeated enough times to accumulate the signal. In practice, there is usually a limitation on the number of measurements that can be made, owing to factors such as time, cost, amount of sample, and the inherent limitations of the instrument. If many measurements are made over too long a period of time, other sources of error may occur in the measurement and these types of errors may not be random.

1.6.2 PLOTTING CALIBRATION CURVES

Calibration is the process of establishing the relationship between the instrument signal we measure and the known concentrations of the analyte. There are many ways to establish this relationship, some very specialized. As appropriate, these will be covered in more detail in the following chapters. Samples with known concentrations or amounts of analyte used to establish this signal–concentration relationship are called *calibration standards*. For many techniques, calibration standards with a range of known concentrations are prepared and a signal is measured for each concentration. The magnitude of the signal for each standard is plotted against the concentration of the analyte in the standard, and the equation that relates the signal and the concentration is determined. Once this relationship, called a **calibration equation or calibration curve**, is established, it is possible to calculate the concentration of the analyte in an unknown sample by measuring its signal and using the equation that has been determined. The calibration curve is a plot of the sets of data points (x_i, y_i) where x is the concentration or amount of analyte in the standards and y is the signal for each standard. Many calibration curves are straight-line relationships or have a linear region, where the concentration of analyte is related to the signal according to the equation $y = mx + b$. The slope of the line is m and the intercept of the line with the y-axis is b. Determining the proper calibration curve is critical to obtaining accurate results. We have already learned that all measurements have uncertainty associated with them. The calibration curve also has uncertainty associated with it. While using graph paper and a ruler and trying to draw the best straight line through the data points manually can be used to plot calibration curves, it is much more accurate to use the statistical programs on a calculator, in a spreadsheet, or in an instrument's computer data system to plot calibration curves and determine the equation that best fits the data. The use of graphing calculators, computerized data handling, and spreadsheets also permits the fitting of nonlinear responses to higher-order equations accurately.

We will assume that the errors in the measured signal, y, are significantly greater than any errors in the known concentration, x. In this case, the best-fit straight line is the one that minimizes vertical deviations from the line. This is the line for which the sum of the squares of the deviations of the points from the line is the minimum. This fitting technique is called the **method of least squares**. For a given x_i, y_i, and the line $y = mx + b$, the vertical deviation of y_i from y equals $(y_i - y)$. Each point has a similar deviation, so the sum of the squares of the deviations is

$$D = \Sigma(y_i - y)^2 = \Sigma\left[y_i - (mx_i + b)\right]^2 \quad (1.21)$$

To obtain the slope m and the intercept b of the best-fit line requires the use of calculus, and the details will not be covered in this text. The results obtained are

$$m = \frac{n\Sigma x_i y_i - \Sigma x_i \Sigma y_i}{n\Sigma x_i^2 - (\Sigma x_i)^2} \quad (1.22)$$

$$b = \overline{y} - m\overline{x} \quad (1.23)$$

Expressions for the uncertainty in the measured value, y, in the slope, and in the intercept are similar to the expressions for the standard deviation (Equation 1.9). Two degrees of freedom are lost in the expression for the uncertainty in y, because both the slope and intercept have been defined:

$$s_y = \sqrt{\frac{\sum_{i=1}^{N} d^2}{N-2}} = \sqrt{\frac{\sum_{i=1}^{N}(y_i - mx + b)^2}{N-2}} \quad (1.24)$$

$$s_m = \sqrt{\frac{s_y^2}{\Sigma(\overline{x} - x_i)^2}} \quad (1.25)$$

$$s_b = s_y \sqrt{\frac{1}{N - (\Sigma x_i)^2 / \Sigma x_i^2}} \quad (1.26)$$

Equation 1.24 defines the uncertainty in y, s_y; Equation 1.25 gives the uncertainty in the slope, s_m; and Equation 1.26 gives the uncertainty in the intercept, s_b. The uncertainty in x_i is calculated using all the associated variances, as has been discussed in Section 1.4.4. The use of a spreadsheet program such as Microsoft Excel to plot a calibration curve eliminates the need to do manual calculations for least squares and the associated statistics. The Excel program permits the calculation of the best-fit equation using linear least squares regression (as well as many nonlinear curve-fitting routines). The program also calculates the **correlation coefficient**, r, for the line. A correlation coefficient with a value of 1 means that there is a direct relationship between x and y; the fit of the data to a straight line is perfect. A correlation coefficient of 0 means that x and y are completely independent. The range of the correlation coefficient is from 1 to -1. Most linear calibration curves should have a correlation coefficient of 0.99 or greater. Statistics programs usually calculate the square of the correlation coefficient, r^2, which is a more realistic measure of the goodness of fit of the data. It is always a good idea to plot the data graphically so that they can be looked at to ensure that the statistical calculations are not misleading about the goodness of fit.

Modern computerized analytical instruments have quantitative analysis programs that allow the analyst to specify the calibration standard concentrations, select the curve-fitting mode, and calculate the results of the samples from the calibration curve equation. Many of these programs will rerun outlier standards and samples automatically, flag suspect data, compute precision and recovery of spikes, track reference standards for quality control, and perform many other functions that used to be done manually by the analyst. An in-depth treatment of calibration and standardization can be found in the book by Bruno and Svoronos.

1.7 ASSESSING THE DATA

A good analytical method should be both precise and accurate; that is, it should be reliable or robust. A robust analytical method is one that gives precise and accurate results even if small changes are made in the method parameters. The robustness of a method is assessed by varying the parameters in the analysis, such as temperature, pH, reaction time, and so on, and observing the effect of these changes on the results obtained. The **specificity** of a method refers to the ability of the method to determine the analyte accurately in the presence of interferences. Method development should include checking the response of the method to other chemicals known to be in the sample, to possible degradation products, and to closely related compounds or elements. Ideally, the analytical method would be specific for only the analyte of interest. It is possible that analytical methods published in the literature may appear to be valid by a compensation of errors; that is, although the results appear accurate, the method may have involved errors that balanced each other out. When the method is used in another laboratory, the errors may differ and not compensate for each other. A net error in the procedure may result. This is an example of a method that is not reliable or robust. It is always prudent to run known reference samples when employing a method from the literature to evaluate its reliability.

The **sensitivity** of an analytical method can be defined as the slope of the calibration curve, that is, as the ratio of change in the instrument response with a change in the analyte concentration. Other definitions are also used. In AAS, sensitivity is defined as the concentration of analyte that produces an absorbance of 0.0044 (an absorption of 1 %), for example. When the term sensitivity is used, it should be defined.

Once the relationship between the signal and analyte concentration (i.e., the calibration curve) has been established, the linear working range of the method can be determined. The **range** is that interval between (and including) the lowest and highest analyte concentrations that have been demonstrated to be determined with the accuracy, precision, and linearity of the method. Linear working ranges vary greatly among instrumental methods and may depend on the instrument design and the detector used, among other factors. Some instruments, such as a gas chromatograph with an electron capture detector or an atomic absorption spectrometer, have short linear ranges of one to two orders of magnitude. Other instruments, like ICP-AES, may have linear ranges of five orders of magnitude, while MS may be linear over nine orders of magnitude. All results should fall within the linear range of the method. This may require dilution of samples with analyte concentrations that are higher than the highest calibration standard in order to bring the signal into the known linear response region. Extrapolating beyond the highest standard analyzed is very dangerous because many signal–concentration relationships become nonlinear at high levels of analyte. Extrapolating below the lowest standard analyzed is also very dangerous because of the risk of falling below the limit of quantitation (LOQ). If samples fall below the lowest calibration standard, they may need to be concentrated to bring them into the working range.

1.7.1 Limit of Detection

All measurements have noise and the magnitude of the noise limits the amount of analyte that can be detected and measured. As the concentration of analyte decreases, the signal decreases. At some point, the signal can no longer be distinguished from the noise, so the analyte can no longer be "detected." Because of noise, it is not possible to say that there is *no* analyte present; it is only possible to establish a **detection limit**. Detection is the ability to discern a weak signal in the presence of background noise, so reducing the noise will permit the detection of smaller concentrations of analyte. The **limit of detection** (LOD) is the lowest concentration of analyte that can be detected. Detected does not mean that this concentration can be measured quantitatively; it only specifies whether an analyte is above or below a certain value, the LOD. One common approach to establishing the LOD is to measure a suitable number of replicates (8–10 replicates is common) of an appropriate blank or low concentration standard and determine the standard deviation of the blank signal. The blank measures only the background or baseline noise. The LOD is then considered to be the concentration of analyte that gives a signal that is equal to three times the standard deviation of the blank. This is equivalent to defining the LOD as that concentration at which the S/N ratio = 3 at 99 % CL:

$$\text{LOD} = \bar{x}_{\text{blank}} \pm 3\sigma_{blank} \qquad (1.27)$$

The use of 3σ is often specified by regulatory methods such as those of the US EPA; it results in an LOD with a 99 % CL.

There are other approaches used for calculating LODs; it is important to specify exactly how an LOD has been determined. An LOD can be defined by the instrument (e.g., using pure water); this would be the instrument detection limit (IDL). One can be defined for an entire analytical method including sample preparation. This would be the method detection limit (MDL). The MDL is in general not as low as an IDL, so it is important to know how the LOD has been defined. The calculated LOD should be validated by analyzing standards whose concentrations are below, at, and above the LOD. Any "results" from samples that fall below the established detection limit for the method are reported as "not detected" or as "<LOD." They should not be reported as numerical values except as "<the numerical value of the LOD"; for example, <0.5 ppb if the LOD is 0.5 ppb for this analysis.

1.7.2 Limit of Quantitation

The precision of an analysis at or near the detection limit is usually poor compared with the precision at higher concentrations. This makes the uncertainty in the detection limit and in concentrations slightly above the detection limit also high. For this reason, many regulatory agencies define another limit, the LOQ, which is higher than the LOD and should have better precision.

The LOQ is the lowest concentration of analyte in a sample that can be determined quantitatively with a given accuracy and precision using the stated method. The LOQ is usually defined as that concentration equivalent to an S/N ratio of 10/1. The LOQ can also be determined from the standard deviation of the blank; the LOQ is 10× the standard deviation of the blank, expressed in concentration units. The LOQ is stated with the appropriate accuracy and precision and should be validated by running standards at concentrations that can confirm the ability of the method to determine analyte with the required accuracy and precision at the LOQ.

Analytical results that fall between the LOD and the LOQ should be reported as "detected but not quantifiable." These results are only estimates of the amount of analyte present, since, by definition, they cannot be determined quantitatively.

The limit of detection and the limit of quantitation are but two parameters that are called figures of merit. Many other figures of merit are commonly used, and others will be discussed throughout this book. A comprehensive listing in tabular form can be found in the book by Bruno and Svoronos.

Note: In this chapter and in the following chapters, the websites for many instrument manufacturers, government agencies, and so on are given. While there are many useful technical notes, application examples, tutorials, and some very useful photographs and videos on commercial websites, it should be understood by the student that these publications and tutorials are in most cases not peer-reviewed. The information should be treated accordingly. In addition, the student will find that there are many online forums (message boards, chat rooms) dedicated to analytical chemistry and to specific techniques in analytical chemistry. While these can be valuable, not all of the information provided on these types of sites is accurate, and students are encouraged to use the peer-reviewed literature as the first source of answers to your questions.

1.8 REFERENCE MATERIALS AND REFERENCE DATA

1.8.1 Reference Materials

In several places throughout this book, we discuss the standardization or calibration of instrumentation by use of standards available from the National Institute of Standards and Technology (NIST). Every industrialized country maintains a national measurement laboratory (NML), and in the United States, NIST satisfies this requirement by statute. Several different kinds of materials are available from NIST.

Reference Materials (RM): An RM is a material sufficiently homogeneous and stable with respect to one or more specified properties established to be fit for its intended use in a measurement process. The term RM is generic and not fixed by statute. An RM can be used for calibration, identification, procedure assessment, or quality control. An RM generally cannot be used for both validation and calibration in the same instrument or procedure.

Certified Reference Material (CRM): A CRM is a reference material characterized by a metrologically valid procedure for one or more specified properties. It is provided with a certificate that lists the value of the specified property, its associated uncertainty (which may be expressed as an expanded uncertainty or a probability statement), and a statement of traceability. The certificate lists the numerical value of the property (e.g., enthalpy, density, etc.) and the traceability.

Standard Reference Material (SRM): An SRM is a CRM that meets additional certification criteria that might differ material to material. These criteria are determined by NIST, and may include the requirement of two separate, orthogonal measurement methods for the property (and its associated uncertainty), which is listed on the accompanying certificate. An SRM is prepared and used for three main purposes: (1) to help develop accurate methods of analysis; (2) to calibrate measurement systems used to facilitate exchange of goods, institute quality control, determine performance characteristics, or measure a property at the state-of-the-art limit; and (3) to ensure the long-term adequacy and integrity of measurement quality assurance programs.

The term "Standard Reference Material" is fixed by statute and is a registered trade mark.

The uncertainty provided with a reference material can be used to set a lower limit for the uncertainty in a procedure developed with that material. The use of these materials in no way guarantees that the result obtained using the material will have the same uncertainty as that listed on the certificate. You must still develop an uncertainty budget for the method and perform an uncertainty analysis.

For more details, the student is referred to the NIST Standard Reference Materials Catalog, available as a pdf: https://nvlpubs.nist.gov/nistpubs/SpecialPublications/NIST.SP.260-176-2018.pdf (accessed October, 2020). It is of interest to note that a NIST-supplied material may very well not be the highest purity material available, it is simply a material that is provided from a consistent and large enough batch to ensure the long-term integrity of the measurement process. For example, NIST provides an SRM sample of toluene for the calibration of density measurement instruments. The toluene that is provided may have a small quantity of known and characterized impurities, however the density has been measured precisely with the best metrology available, and the uncertainty has been well characterized. That density value is listed on the certificate. Additional details on this topic can be found in the book by Bruno and Svoronos.

1.8.2 Reference Data

In many places throughout this book, we mention sources of additional information and data that will be required at decision points during the instrumental analysis process, both in your coursework and as a professional. Reference books, text books, manufactures' literature and application guides, standard reference handbooks and standard reference databases are among these media. Even legally required safety data sheets (SDS; the old terminology was material safety data sheet, MSDS) have some limited physical property information listed. As you use these various sources, you will come to realize that not all of these sources are reliable, and depending upon the application, you will have to develop the ability to discern between the good and not-so-good. Surprisingly, the physical property data provided on a chemical SDS are notoriously unreliable. We understand that often an estimate will suffice, such as an approximate boiling temperature of a fluid. Other analytical situations will require nothing but the most well determined information with well characterized uncertainty. This includes all legal venues, such as forensic and environmental analyses. In those situations, it is imperative to obtain the required data (and the corresponding uncertainty) from standard reference sources.

The data products made available by NIST in the U.S. are an important source of standard reference data. An easily accessible and reliable place to start is the NIST Chemistry

WebBook, but there are many data sources that are either free or provided at minimal cost. In recent years, standard handbooks have undertaken a reconciliation process so as to standardize data with the NIST data collections. Thus, the *CRC Handbook of Chemistry and Physics*, as well as the book by Bruno and Svoronos, provide data that are consistent with the NIST standard databases.

PROBLEMS

1.1
   a. Define determinate error and give two examples of determinate errors.
   b. In preparing a sample solution for analysis, the pipette used actually delivered 4.92 mL instead of the 5.00 mL it was supposed to deliver. Would this cause determinate or indeterminate error in the analysis of this sample?

1.2
   a. Define precision.
   b. Do determinate errors affect precision?

1.3
   a. Define accuracy.
   b. How can the accuracy of an analytical procedure be determined?

1.4
   a. Define uncertainty.
   b. How do the terms uncertainty, precision, accuracy, and error differ?

1.5
   a. What is the statistical definition of sigma (σ)?
   b. What percentage of measurements should fall within $\pm 2\sigma$ of the true value for a data set with no determinate error, assuming a Gaussian distribution of random error?

1.6 Calculate the standard deviation of the following set of measured values: 3.15, 3.21, 3.18, 3.30, 3.25, 3.13, 3.24, 3.41, 3.13, 3.42, 3.19.

1.7 The true mass of a glass bead is 0.1026 g. A student takes four measurements of the mass of the bead on an analytical balance and obtains the following results: 0.1021 g, 0.1025 g, 0.1019 g, and 0.1023 g. Calculate the mean, the average deviation, the standard deviation, the % RSD, the absolute error of the mean, and the relative error of the mean.

1.8 How many significant figures are there in each of the following numbers:
3.216, 32.1, 30, 3×10^6, 3.21×10^6, 321,000?

1.9 Round off the following additions to the proper number of significant figures:

	3.2	1.9632	1.0×10^6
	0.135	0.0013	1.321×10^6
	3.12	1.0	1.13216×10^6
	0.028	0.0234	4.32×10^6
Total	6.483	2.9879	7.77316×10^6

1.10 The following data set represents the results of replicate measurements of lead, expressed as ppm Pb. What are
   a. The arithmetic mean?
   b. The standard deviation?
   c. The 95 % CL of the data?
   2.13, 2.51, 2.15, 2.17, 2.09, 2.12, 2.17, 2.09, 2.11, 2.12
   d. Do any of the data points seem "out of line" with the rest of the data? Are there any point(s) outside the 4σ "rule of thumb" for rejecting suspect data? Should the suspect data be ignored in the calculation?

1.11
   a. Explain the importance of good sampling.
   b. Give three examples of precautions that should be taken in sample storage.

1.12
   a. Illustrate the difference between precision and accuracy.
   b. Do indeterminate errors affect precision or accuracy?

1.13 The results in Problem 1.9 were obtained for the lead content of a food sample. The recommended upper limit for lead in this food is 2.5 ppm Pb. (a) Are the results greater than 2.5 ppm Pb with 95% CL? (b) If the regulatory level is decreased to allow no more than 2.00 ppm Pb, is the Pb content of the food greater than 2.00 ppm with 95 % CL?

1.14 The determination of Cu in human serum is a useful diagnostic test for several medical conditions. One such condition is Wilson's disease, in which the serum Cu concentration is lowered from normal levels and the urine Cu concentration is elevated. The result of a single copper determination on a patient's serum was 0.58 ppm. The standard deviation σ for the method is 0.09 ppm. If the serum copper level is less than 0.70 ppm Cu, treatment should be started. Based on this one result, should the doctor begin treatment of the patient for low serum copper? Support your answer statistically. If the doctor were unsure of the significance of the analytical result, how would the doctor obtain further information?

1.15 Add the following concentrations and express the answer with the correct units and number of significant figures:
3.25×10^{-2} M, 5.01×10^{-4} M, 8×10^{-6} M

1.16 With what confidence can an analytical chemist report data using σ as the degree of uncertainty?

1.17 The mean of eight replicate blood glucose determinations is 74.4 mg glucose/100 mL blood. The sample standard deviation is 1.8 mg glucose/100 mL blood. Calculate the 95 % and 99 % CLs for the glucose concentration.

1.18 An analysis was reported as 10.0 with $\sigma = 0.1$. What is the probability of a result occurring within (a) ± 0.3 or (b) ± 0.2 of 10.0?

1.19 Name the types of noise that are frequency dependent. Which types of noise can be reduced by decreasing the temperature of the measurement?

1.20 The following measurements were obtained on a noisy instrument:
1.22, 1.43, 1.57, 1.11, 1.89, 1.02, 1.53, 1.74, 1.83, 1.62
  a. What is the S/N ratio, assuming that the noise is random?
  b. How many measurements must be averaged to increase the S/N ratio to 100?

1.21 Estimate the S/N ratio in Figure 1.13b and c. What is the improvement in S/N ratio from (c) to (b)? If the S/N ratio value in (c) is the result of one measurement, how many measurements must be made and averaged to achieve the S/N ratio value in (b)?

1.22 The tungsten content of a reference ore sample was measured both by XRF spectrometry, the standard method, and by ICP-AES. The results as mass percent tungsten are given in the following table. Are the results of the two methods significantly different at the 95 % CL? Is there any bias in the ICP method? (*Hint*: The standard method can be considered to have a mean = μ.)

Replicate Number	XRF	ICP
1	3.07	2.92
2	2.98	2.94
3	2.99	3.02
4	3.05	3.00
5	3.01	2.99
6	3.01	2.97

1.23 Assuming that the results of many analyses of the same sample present a Gaussian distribution, what part of the curve defines the standard deviation σ?

1.24 A liquid sample is stored in a clear glass bottle for the determination of trace metals at the ppm level. What factors can cause the results to be (a) too low or (b) too high?

1.25 Round off the following numbers according to the rules of significant figures: (a) $10.2 \div 3$, (b) $10.0 \div 3.967$, (c) $11.05 \div 4.1 \times 10^{-3}$, (d) $11.059 \div 4.254$, and (e) $11.00 \div 4.03$.

1.26 Nitrate ion in potable water can be determined by measuring the absorbance of the water at 220 nm in an ultraviolet–visible (UV/VIS) spectrometer. The absorbance is proportional to the concentration of nitrate ion. The method is described in Standard Methods for the Examination of Water and Wastewater listed in the bibliography. Calibration standards were prepared and their absorbances measured. The results are given in the following table. (*Note*: In the absorbance of the blank, the 0.0 mg nitrate/L "standard" was set to 0.000 in this experiment. A nonzero blank value is subtracted from all the standards before the calibration curve is plotted and the equation calculated.)

Nitrate Ion (mg/L)	Absorbance
0.00	0.000
1.00	0.042
2.00	0.080
5.00	0.198
7.00	0.281

  a. Make an x–y graph showing the experimental data with absorbance plotted on the y-axis.
  b. Determine the equation of the least squares line through the data points, with y as absorbance and x as nitrate ion concentration.
  c. Calculate the uncertainty in the slope and intercept.
  d. If you have done this using a statistics program or spreadsheet, what is the value of the correlation coefficient for the line?

1.27
  a. Name the instrumental methods that can be used for elemental qualitative analysis.
  b. Name the instrumental methods that are used for elemental quantitative analysis.

1.28
  a. Name the instrumental methods that can be used for molecular organic functional group identification.
  b. What instrumental methods provide molecular structural information, that is, indicate which functional groups are next to each other in an organic molecule?

1.29 What instrumental methods are best for quantitative analysis of (a) complex mixtures, (b) simple mixtures, or (c) pure compounds?

1.30 What instrumental methods can provide measurements of the RMM of a molecule?

1.31 What is the purpose of a blank in an analysis? What is the purpose of a reference material or reference standard in an analysis?

1.32 Equation 1.27 shows how to calculate the LOD of a method at both the 95 % and 99 % CLs. You have measured the blank for a determination of arsenic in food samples by hydride-generation atomic fluorescence spectrometry. The blank values are 0.23 ppb, 0.14 ppb, 0.16 ppb, 0.28 ppb, 0.18 ppb, 0.09 ppb, 0.10 ppb, 0.20 ppb, 0.15 ppb, and 0.21 ppb As. What is the LOD at (a) 95 % CL and (b) 99 % CL?

1.33 Based on your answer to Problem 1.32, what are the respective LOQs for As at the two CLs?

BIBLIOGRAPHY

The reader will note that occasionally multiple editions of books are cited as sources or reference works. The reason for this is that sometimes different figures are used in different editions, which may in fact be written by different authors. The sources found in the bibliography will be the most recent versions of a particular work.

The reader will note that some of the older literature speaks of statistical tests for the rejection of data. We emphasize that this terminology is not correct; there is no test that can indicate a datum should be rejected, only that a datum might be a potential outlier.

In some cases instrument manufacturers and suppliers are cited by providing a URL to the exact internet location. We have endeavored to ensure that these links are functional at the time of publication, but clearly these links change. If a link has become nonfunctional, a good approach is to use your browser's advanced search capability. For example: "company name" AND "instrument type" AND "specific descriptors" is a good place to start.

American Society for Testing and Materials. *2021 Annual Book of ASTM Standards*. ASTM International: West Conshohocken, PA, 2021.

Anastassiades, M.; Lehotay, S.J.; Stajnbaher, D.; Schenck, F.J. Fast and easy multiresidue method employing acetonitrile extraction/partitioning and "dispersive solid-phase extraction" for the determination of pesticide residues in produce *J. AOAC Int. (JAOAC), 86*, 412–413, 2003.

APHA, AWWA, WPCF. In Greenberg, A.E., Clesceri, L.S., and Eaton, A.D. (eds.), *Standard Methods for the Examination of Water and Wastewater*, 18th edn. American Public Health Association: Washington, DC, 1992.

Bruno, T.J., Svoronos, P.D.N., CRC Handbook of Basic Tables for Chemical Analysis – Data Driven Methods and Interpretation, 4th Ed., CRC Taylor and Francis, Boca Raton, FL, 2021.

Christian, G.D. *Analytical Chemistry*, 6th edn. John Wiley & Sons, Inc.: Hoboken, NJ, 2004.

deLevie, R., *Advanced Excel for Scientific Data Analysis*, 3rd ed., Atlantic Academic, 2012.

deLevie, R., *How to Use Excel in Analytical Chemistry and in General Scientific Data Analysis*, Cambridge Univ. Press (2001).

Diamond, D.; Hanratty, V. *Spreadsheet Applications in Chemistry Using Microsoft Excel*. John Wiley & Sons, Inc.: New York, 1997.

Dulski, T.R. *A Manual for the Chemical Analysis of Metals*. ASTM International: West Conshohocken, PA, 1996.

Enke, C.G. *The Art and Science of Chemical Analysis*. John Wiley & Sons, Inc.: New York, 2001.

Erickson, M.D. *Analytical Chemistry of PCBs*, 2nd edn. CRC Press: Boca Raton, FL, 1997.

Evaluation of measurement data — Guide to the expression of uncertainty in measurement (GUM)(Corrected version 2010), copyright of this document is shared jointly by the JCGM member organizations.

Harris, D.C. *Quantitative Chemical Analysis*, 5th edn. W.H. Freeman and Company: New York, 1999.

Horwitz, W. (ed.) *Official Methods of Analysis of the Association of Official Analytical Chemists*, 13th edn. Association of Official Analytical Chemists: Washington, DC, 1980.

Keith, L.H. (ed.) *Principles of Environmental Sampling*. American Chemical Society: Washington, DC, 1988.

Malmstadt, H.V.; Enke, C.G.; Crouch, S.R. *Microcomputers and Electronic Instrumentation: Making the Right Connections*. American Chemical Society: Washington, DC, 1994.

Mark, H.; Workman, J. *Statistics in Spectroscopy*. Academic Press, Inc.: San Diego, CA, 1991.

NIST Chemistry WebBook, NIST Standard Reference Database Number 69, Eds. P.J. Linstrom and W.G. Mallard, National Institute of Standards and Technology, Gaithersburg MD, 20899, https://doi.org/10.18434/T4D303, (retrieved February 17, 2022).

Rorabacher, D.B. Statistical treatment for rejection of deviant values *Anal. Chem.*, 63, 139, 1991.

Rumble, J. R., ed., CRC Handbook of Chemistry and Physics, 102nd ed. CRC Taylor and Francis, Boca Raton, FL, 2021.

Simpson, N.J.K. (ed.) *Solid-Phase Extraction: Principles, Techniques and Applications*. Marcel Dekker, Inc.: New York, 2000.

Skoog, D.A.; Holler, J.A.; Nieman, T.A. *Principles of Instrumental Analysis*, 5th edn. Saunders College Publishing; Harcourt, Brace and Company: Orlando, FL, 1998.

Tyson, J. *Analysis. What Analytical Chemists Do*. Royal Society of Chemistry: London, U.K., 1988.

US Environmental Protection Agency. Methods for the chemical analysis of water and wastes, EPA-600/4-79-020. Environmental Monitoring and Support Laboratory: Cincinnati, OH, March 1983.

US Environmental Protection Agency. Test methods for evaluating solid waste-physical/chemical methods, 3rd edn., SW-846. Office of Solid Waste and Emergency Response: Washington, DC, 1986 (most recent version available at www.epa.gov).

Wercinski, S.A.S. (ed.) *Solid Phase Microextraction: A Practical Guide*. Marcel Dekker, Inc.: New York, 1999.

Whitecavage, J.; Stuff, J.R.; Pfannkoch, E.A.; Moran, J.H. High throughput method for the determination of PAHs in seafood by QuEChERS-SBSE-GC-MS. Application Notes 2010/6a and 2010/6b. www.gerstel.com/en/applications.htm.

Zwillinger, D. (ed.) *CRC Standard Mathematical Tables and Formulae*, 32nd edn. CRC Press: Boca Raton, FL, 2011.

CHAPTER 2

Safety in the Instrumentation Laboratory

By the time the student takes a course in analytical instrumentation, they will have had chemical laboratory safety drilled into them in multiple venues. There is no doubt that, especially in your organic chemistry laboratory courses, you will have had instruction on the proper and safe use of not only reagents and solvents, but also some instruction on the safe handling of instrumentation. Why then should we have yet another discussion on safety in this book? The reason is actually quite simple: the analytical instrumentation laboratory poses some unique potential hazards that reach beyond the safe handling of reagents and solvents. To address these issues, every university, corporate entity, and government laboratory has in place a detailed, written safety plan, guidelines, and standard operation procedures (SOPs) to protect workers from hazards. In some respects, the implementation of these documents is an administrative response to potential liability; however there are many important lessons to be learned from them. We realize that it is not possible to eliminate all hazards, but it is certainly possible to build on these documents to minimize risk posed by these hazards.

We mentioned that your study of general and organic chemistry has acquainted you with chemical safety. Indeed, in this chapter we will not treat chemical hazards at all, since these are effectively treated in those other courses and by university safety programs. More extensive information on the chemical hazards that might be encountered in the analytical instrumentation laboratory can be found in the book by Bruno and Svoronos if that is needed. We will also not discuss safety glasses, food and drink in the laboratory, eyewash stations and safety showers, and the proper disposal of waste, since these topics will have already been covered in your earlier coursework and should be second nature.

The purpose of this chapter is actually two-fold. First, it is important that the student be cognizant of safety protocols and hazards in the analytical laboratory so that he/she can complete the laboratory aspects of the course in safety. Beyond this, however, is the philosophy that we hope you will carry with you through your further education, undergraduate and graduate research, career, and even in your home life. The advice in this chapter not only reflects good industrial and laboratory practice, it also reflects the accumulated practical experience of the authors, and will serve more as a how-to guide than a how-not-to guide. For example, the information provided on electrical safety will be applicable to the home or workplace; the guidelines and tips will be valuable, so please take advantage of them.

We intentionally provide this as the second chapter of the text, to follow the conceptual discussion of analytical instrumental analysis. This is because a discussion about safety is not an afterthought or a distraction. It is inherent in all the laboratory work that you will perform in the course. We urge the student to read this chapter carefully even if it is not formally discussed in lecture, and apply the guidelines each time they use an instrument. Moreover, graduate students and teaching assistants supervising laboratory classes MUST be conversant with all of the information in this chapter.

The vast majority of analytical instruments are operated by use of electric power, and that comes with power sources, power cables, and cords. These are to be treated with respect and caution, since if they are misused, they can kill you. Many instruments require the use of compressed air or compressed gases from cylinders. Have no doubt that careless handling of these items can make for a very bad outcome. Larger instruments such as NMRs often require a ladder to insert a sample. Not surprisingly, falls from ladders are among the most common causes of laboratory injuries. Some instruments such as GC detectors and some spectrometers incorporate radiation sources. These not only require caution in their use, but also adherence to federal, state, and university regulations. These are only a few examples; indeed, there are many ways to get hurt (or worse) in the analytical instrumentation laboratory, and this chapter is intended to minimize such events. In all of the following chapters, we will mention the potential hazards, referring back to this chapter, with the suggestion that the material presented in this chapter be reviewed before attempting any of the experiments.

2.1 ELECTRICAL HAZARDS AND SAFETY

Analytical instrumentation is almost universally electrically powered (some portable devices are battery powered with voltage and power requirements that vary), and for this reason we begin our treatment of safety with this topic. Introductory courses in physics typically treat the basic electrical power equations and relationships, so that discussion will not be repeated here. If the student feels the need to brush up on the relations between voltage, current, resistance, and power, please refer to your physics text or online sources.

2.1.1 Electrical Lines, Receptacles, and Cables

Analytical instruments typically utilize computer and microprocessor controls that are electrically powered through built-in power supplies, thus, the required current and voltage needed for the various instrument functions are generated internally by use of the applied power from the laboratory receptacles. In the United States, the most common receptacles in the laboratory furnish alternating current of 110 to 120 volts at 60 Hz with a capacity for 15- or 20-amp current loads. Some instruments will require a higher voltage, which in some cases will be 240 volts, but in some industrial or commercial environments, 208 volts will be provided. The latter voltage results from a service that is delivered as three-phase power instead of the more familiar two phase. In those cases, the 208 V results from tapping two of the three phases, the 208 V line being single-phase power. If a device requires a current of more than 20 amp, it will usually require one of the higher voltage lines. If an instrument requires 240 volts but only 208-volt lines are available (and vice versa), then a transformer will have to be purchased for proper operation. Yes, it is possible to jury-rig a connection between 208 and 240 V receptacles or plugs, and vice versa. The device may operate with no apparent difficulty for a time, but use of incorrect voltage will ultimately result in premature instrument failure. It is important to recognize the different receptacles and the power ratings that they represent. Note that students in the U.K. (and western Europe) will encounter circuits of 240 V AC at 50 Hz. In Figure 2.1, some of the more common configurations are shown. Students in the U.S. will note that prongs of plugs will have holes at the end of the prongs. These holes serve two purposes. First, these holes engage small detents or bumps in the outlets, to more securely hold the plug in the outlet. Second, one will occasionally see a safety lockout device in the form of a zip tie or even a small padlock in these holes, preventing the plug being inserted into an outlet. The prongs are also polarized, meaning you can only insert it into the outlet one way, with the wider prong engaging the neutral line, and the thinner one engaging the hot line. Hot and neutral lines will be discussed in more detail later in this chapter.

> Jury rig, or jerry rig, is a makeshift repair or installation typically accomplished with whatever rudimentary tools and materials may be at hand. Thus, if a sailing boat is de-masted at sea, the crew might fabricate and rig a temporary mast called a jury mast.

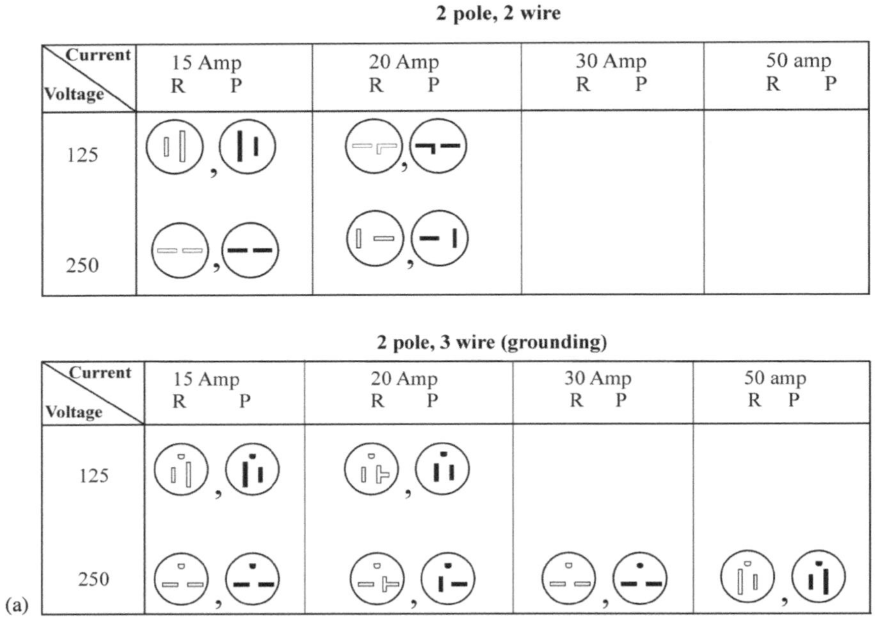

2 pole, 2 wire

Current / Voltage	15 Amp R P	20 Amp R P	30 Amp R P	50 amp R P
125	⊙,⊙	⊙,⊙		
250	⊙,⊙	⊙,⊙		

2 pole, 3 wire (grounding)

Current / Voltage	15 Amp R P	20 Amp R P	30 Amp R P	50 amp R P
125	⊙,⊙	⊙,⊙		
250	⊙,⊙	⊙,⊙	⊙,⊙	⊙,⊙

(a)

Figure 2.1a A schematic diagram of typical receptacle (R) and plug (P) configurations common in the analytical instrumentation laboratory in the United States (taken from the book by Bruno and Svoronos). These have in common one or more hot lines, a neutral line, and a ground line, which will be discussed in detail later in this section.

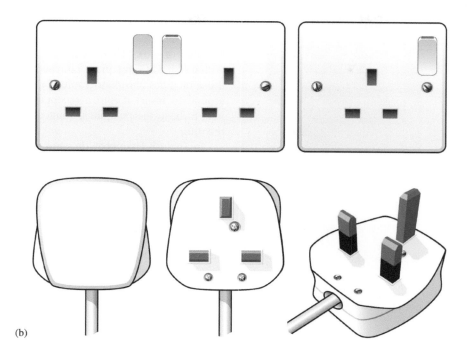

Figure 2.1b The typical outlet and plug configuration used in the U.K. Note that the two power prongs are partially insulated and the ground prong is longer than the power prongs.

Instrumentation laboratory electrical receptacles are often provided as wall mounted raceways that span the room. A photograph of such a raceway is shown in Figure 2.2. It is tempting to use these raceways as a convenient shelf for vials, pens, and other small items. This practice is usually prohibited by all safety guidelines. It is especially hazardous to have any metallic objects on the raceways; in the past one could often find spare lengths of copper tubing stored on the raceways in a laboratory. Such items can slip off the raceway and contact electrical connections, causing arcing and a possible fire. Vials containing chemicals that are placed on raceways can leak, and drip chemicals into the raceway itself. The only purpose of the raceway is to deliver electrical power. Optimally, the outlets on the raceways should be staggered among separate circuits; this means that adjoining outlets on a raceway should be on separate circuits that can repeat around the room. A simple 1-2-3-1-2-3... sequence is an example of this, with the number of circuits being determined by the total load of the room.

Students in the United Kingdom (and western Europe) will notice that the plugs are considerably larger than the ones used in the United States. First, the ground plug prong is longer than either of the power prongs, and the power prongs are partially insulated near the plug body. The longer ground plug actually serves two purposes: (1) it serves as a ground terminal (which we will discuss in detail later) and (2) it serves as a key to unlock the power in the receptacle. Power is not applied to the outlet unless the ground prong is inserted far enough to affect power release. This feature, and also the partial insulation of the power prongs, makes electrical shock from the plug less likely. Be aware that plugs in the U.K. also incorporate an internal fuse. So, in our later discussion of fuses and circuit breakers, in the U.K. you might encounter a fuse problem in the plug itself. Overall the electrical fixtures in the U.K. incorporate a greater degree of safety because of the engineering described above, but at a considerable additional cost. An unfortunate side effect of this is that shortages of outlets are often encountered.

Students may notice some unusual receptacles and fixtures installed in certain hazardous locations (examples are the unit operation laboratories in chemical engineering departments). When there is the potential of explosive vapors or dusts being present, special explosion-proof (EP) or vapor-proof fixtures must be used. While these fixtures and receptacles serve to prevent electrical sparks and arcing affecting the environment of the laboratory, the real purpose is actually quite different. These fixtures are designed such that if an explosive vapor were to be present in the laboratory and enter the fixture, an arc might cause deflagration or detonation inside the fixture. In an EP fixture, the flame front so produced would be prevented from reaching outside the fixture. This is done by the design of a suitable quench distance into the fixture. An example of an explosion-proof receptacle is shown in Figure 2.3. Explosion-proof lighting and switches, and even explosion-proof telephones are also available. Note that many older EP telephones are of the

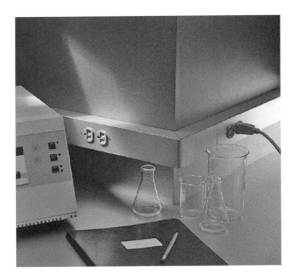

Figure 2.2 A wall-mounted electrical raceway above a laboratory bench. Reproduced with permission of Continuant Inc.

Figure 2.3 An example of an explosion-proof electrical receptacle. Reproduced with permission of Hubbell Incorporated (Delaware), Killark Division.

rotary dial type; if you are unfamiliar with this, learn how it works because you might have to use it to summon help in an emergency.

While we are not treating chemical hazards in this chapter, we will mention that refrigeration units that are used for flammable materials such as solvents will always be explosion proof in that the controls will be located outside of the chemical containment area. Occasionally, standard refrigerators are found in laboratories that store biological samples or amino acids/proteins. If you notice solvents or other volatile organic compounds stored in non-explosion-proof units, please alert your instructor.

Plugs should be fully inserted into receptacles, and they should be located so as to cause no undue stress on the plug or wire. This situation is commonly encountered with molded 20-amp plugs attached to heavy cables, such as those on a gas chromatograph. If the plug has to be twisted around to insert it into a receptacle, it would be a good idea to have the receptacle rotated to better accommodate the plug.

In Table 2.1 we provide some approximate current and power requirements of common laboratory instruments (taken from the book by Bruno and Svoronos).

The power requirements of instruments must be considered when instruments are installed, but since instruments are often moved from laboratory to laboratory, one must always match the power requirement to the power source if an instrument is relocated. It is also important to note that when a device is first energized, the current draw will typically spike to a significantly higher level than that encountered in normal operation. This should be considered if multiple instruments are connected to the same circuit. Turning on a mechanical vacuum pump, diffusion vacuum pump, and a gas chromatograph (as might be on a GC-MS instrument) should be done sequentially rather than simultaneously. Energizing all three at the same time will possibly cause the circuit breaker to trip due to an overload. When this happens, all devices plugged into the circuit protected by that circuit breaker will shut down. Note that most instruments and computers are not designed for this kind of a hard power-down. A hard power-down can result in data loss or instrument damage; at the very least, when the instrument is re-energized, it will typically present the user with an on-screen scolding while disk and buffers are checked for damage.

If the circuit breaker trips, it is important to turn off all instruments that have been affected by the shutdown, not simply the one that you were working with. While the power will be off and the devices will not be working, one must not reset the circuit breaker with all devices set in the "on" position; that will only result in another circuit breaker trip. Once all devices on the circuit are physically turned off, one may then locate the circuit breaker box (which may be in the corridor or in a separate mechanical or utility space) and find the tripped breaker. In undergraduate courses, this is best left to the instructor. Indeed, some organizations require that a tripped circuit breaker be reset only by an authorized electrician, sometimes equipped with arc-protective clothing. Before resetting the breaker, examine the breaker for evidence of electrical arcing such as soot, evidence of excessive heating, or a burning smell. If all seems well, the breaker can then be reset, but it is good practice to stand by for 10 seconds or so to make sure that the breaker does remain in the "on" position. With the breaker reset, one can then begin to energize the equipment sequentially. Note that not all instances of a circuit breaker trip are the result of an accidental overload. An internal fault such as a short circuit in one of the instruments must also

SAFETY IN THE INSTRUMENTATION LABORATORY

Table 2.1 Electrical requirements of common laboratory instruments

Instrument	Current (amperes)	Power (watts)
balance (electronic)	0.1–0.5	12–60
biological safety cabinet	15	1,800
blender	3–15	400–1,800
centrifuge	3–30	400–6,000
chromatograph	15	1,800
computer (PC)	2–4	400–600
freeze dryer	20	4,500
fume hood blower	5–15	600–1,800
furnace/oven	3–15	500–3,000
heat gun	8–16	1,000–2,000
heat mantle	0.4–5	50–600
hot plate	4–-2	450–1,400
Kjeldahl digester	15–35	1,800–4,500
refrigerator/freezer	2–10	250–1,200
stills	8–30	1,000–5,000
sterilizer	12–50	1,400–12,000
vacuum pump (mechanical)	4–20	500–2,500
vacuum pump (diffusion)	4	500

Taken from the book by Bruno and Svoronos.

be suspected. Thus, whenever a circuit breaker trips, it is important to ask oneself why this occurred and determine the cause. Students in the U.K. will encounter a different scenario, since as we discussed fuses are incorporated into device plugs. The typical circuit breaker in the U.K. is a 40 A breaker, with the protection being furnished by the plug more than the breaker.

While we are discussing the topic of circuit breakers and tripping, some additional important issues deserve mention. It is a good idea to label each electrical receptacle in the laboratory to identify the breaker. This might be something like: "power panel 2, breaker 15" printed on the receptacle cover. Proper labeling will make for easier location of the tripped breaker, and it will also help identify the equipment that is operated on a given circuit in the laboratory. Note that room lighting will typically be connected to a separate lighting panel, which will help avoid confusion. Circuit breaker panels must be unobstructed for three feet in all directions (with appropriate signage). This means no desks, cabinets, or equipment can be in the way. Since the ability of a circuit breaker to provide protection against an overload depends on its' rapid response, it is a good idea to exercise the circuit breakers yearly. This means that at least once a year, all devices on the circuit should be shut down and the circuit breaker should be clicked back and forth 2–3 times. This will dislodge oxide from the contacts in the breaker. Some organizations exercise circuit breakers through yearly building-wide shut downs, or localized shut downs that are needed for other electrical modifications or alterations.

It is best to avoid the use of electrical extension cables (extension cords) in the instrumentation laboratory, but there are circumstances in which they will be necessary, such as when power is needed for a device that is simply too far from the receptacle. The improper and unwise application of electrical extension cables is a major cause of electrical fires. A heavy reliance on extension cables in a laboratory is an indication that too few receptacles or circuits are available. The only remedy to this is the installation of additional circuits in the laboratory. Students should have no hesitation about informing instructors of such situations. Most organizations have policies and standard operating procedures that cover the use of electrical extension cables, so this section serves as a supplement to those guidelines. Most of the guidelines presented below are founded in good industrial practice and in good common sense, as well as national and local codes.

1. Use only electrical cables that are properly rated for the load. Table 2.1 provides some guidance on the approximate load of common laboratory devices, but the equipment may also be labeled with the electrical load. Always refer to this information if it is available. A discussion on cable ratings will be presented later.
2. Do not run electrical cables through walls, doorways, ceilings, or floors. If an electrical cable is obscured, it cannot be observed for problems.
3. Do not cover electrical cables with floor mats, since this can hamper heat dissipation and can obscure signs of wear.
4. If an electrical cable must be run along a floor temporarily, it must be taped securely to the floor to prevent movement and to minimize the tripping hazard. It must

be clearly visible, also to prevent a tripping hazard. Never staple or nail an electrical cable to any surface.

5. Never use an electrical cable that is damaged in any way; check it beforehand for any signs of deterioration. This includes broken or cracked insulation. Do not use electrical cables that have been "repaired" with duct tape or even electrical tape. Instead, obtain a serviceable replacement. This rule applies not only to extension cables but to the instrument cables as well. Note that the wiring near certain devices such as vacuum pumps is particularly vulnerable to deterioration and cracking because of ozone that these devices can generate.
6. Always use a grounded three-pronged extension cable, and always use a grounded outlet. Some instruments are operated from a constant voltage transformer that might have a floating ground, but in these cases the metal case of the transformer should be grounded. A more in-depth discussion of grounding will be provided later.
7. Avoid the temptation of grinding polarized plugs to make them more convenient. They are polarized to ensure that the "hot" rather than the neutral line is feeding the switch or on/off mechanism.
8. Fully insert the plug of an extension cable into an outlet. Insert and remove plugs from receptacles by holding the plug itself and not the wire. Repeatedly pulling on the wire to remove a plug will damage the insulation on the plug. Note that the plugs used in the U.K., discussed earlier, are designed with the ground wire longer than the power wires, thus the ground wire will be the last wire to detach from the plug.
9. Keep extension cables away from water. In general, if you can touch an electrical appliance or outlet with one hand, and a water source (such as a faucet) with the other, a ground fault circuit interrupter (GFCI) should be used. A GFCI should be used in all damp environments and when using a circuit outdoors.
10. Electrical extension cords must be connected such as to prevent tension at joints and terminal screws; straining a cord can cause the strands of one conductor to loosen. This was mentioned earlier in the discussion of receptacles.
11. Do not use an extension cable with such power-intensive devices as heat guns or space heaters. Indeed, such devices might be prohibited outright in your laboratory or organization.
12. Unused extension cables should be unplugged and stored indoors, preferably wrapped or tied with string or rubber band to prevent them from lying on the floor where they can be damaged.
13. Do not "daisy chain" multiple electrical extension cables.
14. Power strips should not be permanently secured to a bench or instrument. These are temporary connection devices, and any fastening implies a more permanent installation.
15. An extension cable should not be used as a disconnect means for an apparatus. To deenergize a piece of equipment, first turn off the on/off switch before unplugging the power cable or extension cable. If the power switch of a device is faulty, inform your instructor so that it can be replaced.

In Table 2.2 we provide some basic information on electrical extension cable size and nominal capacity. The AWG is the American wire gauge scale, the SWG is the British standard wire gauge. To convert to the metric gauge scale, multiply the wire diameter in mm by a factor of 10 to obtain the nominal metric gauge. Note that in AWG and SWG, the diameter increases as the gauge decreases. Additional details and guidance should be sought in your institutional safety plan or local codes. Since this information is geared for laboratory use, we present information only for 25-foot (7.6 m) lengths. As we noted earlier, some laboratory apparatus will draw large current upon start up. This must be taken into account when choosing an electrical extension cable.

Ubiquitous in most laboratories is a drawer or cabinet with unused computer cables that faculty and staff members have refused to part with. These cables typically have a standard plug for the receptacle at one end and a C13 adapter at the other. It is often tempting and expedient to cut off the C13 adapter and use the cable to hard-wire or re-wire laboratory equipment. This may be prohibited by your organization's standard operating procedures. If not prohibited, please pay close attention to the loads being served, noting that these electrical cables are almost always intended for very low loads.

We also must pay attention to the condition of the plug. Most molded plugs will function well for years and will not require maintenance, but some plugs are replaceable

Table 2.2 **Wire gauges and current ratings for electrical cables**

AWG	SWG	Diameter, in (mm)	Nominal current rating, A	Comments
12	14	0.0808 (2.052)	20	Good for most medium- and heavier-duty applications.
14	16	0.0641 (1.628)	15	Good for light- and medium-duty laboratory apparatus.
16	18	0.0508 (1.290)	10	Used in computers and monitors as well as other light-duty applications.
18	19	0.0453 (1.024)	7	Used in laptop computer power supplies and other very light-duty applications.

and are fastened with screws. It is important that the screws be tight and the insulation at the plug entry is intact. The prongs themselves will become covered with oxide and general grime over years of use. It is a good idea to clean the prongs with emery cloth and wipe them clean to bare metal.

2.1.2 PAT Testing Procedures in the U.K.

All laboratories in the United Kingdom are covered by the Portable Appliance Testing requirement, which covers all electrically operated laboratory equipment. Each year, an inspection must be performed by a designated duty staff member that includes a visual inspection of the cable and plug, and also a insulation resistance test and a ground test. The exception is computer equipment, which requires this inspection every two years. If the device is deemed suitable for use a color-coded sticker is affixed, bearing the date of the next required inspection. If the device is deemed condemned (that is, it fails either the visual or insulation or ground test), the plug is removed and the staff member responsible for the device is informed. A new instrument or device less than a year old does not require a PAT sticker. Current, reliable PAT records are required in all universities.

2.1.3 Electrical Shocks

Our discussion of electrical issues must consider the effect of current and voltage on humans. In Table 2.3 we show the effects of electrical exposure and shock. We list the current in milliamperes, which is an indication of how little current is required to do harm. The current difference between a barely noticeable shock and a lethal shock is only a factor of 100. In individuals with cardiac problems, the threshold may be lower. The voltage is an important consideration as well because of the relationship with resistance: $I = V/R$, where I is the current, V is the voltage, and R is the resistance. The presence of moisture can significantly decrease the resistance of the human skin, and thereby increase the hazard of an electrical shock.

2.1.4 Batteries

A final caution regarding electrical hazards applies to the battery-operated devices that were mentioned at the start of this section. Some instruments are meant to be taken out into the field for use. This includes some refractometers, colorimeters, sampling devices, even some spectroscopic and chromatographic devices. There are other devices that are wearable, such as monitors and cameras; these are typically used while sampling for environmental compliance and for forensic analyses. The study of battery safety is actually very large, with conferences and journals dedicated to the topic. We will present only a brief discussion here, but be assured that failure to appreciate battery hazards will impact health and safety as well as the environment.

There are several types of batteries used in portable analytical instruments. These include the lead-acid battery, variants of the alkaline battery, the nickel-cadmium battery, and the lithium-ion battery. The lead-acid battery is often used in computer uninterruptable power supplies (UPS units), but they are also used in field instrumentation. Their purpose is not really to provide power to the device (power is applied through the electrical mains in the laboratory, or a generator in the field), but rather they are a means to allow "gentle" power-down of devices in the event of a power failure; immediate loss of power to these devices is a sure way to cause instrument damage or data loss. These batteries are rechargeable, and are often maintained connected to the electrical power source during operation. When these batteries fail, they must be disposed of through a recycler since improper disposal (i.e., dumping) will cause lead contamination in the environment. The casing of these batteries, although sealed, will often show a bulge when failure has occurred or is imminent. Indeed, it might be difficult to get

Table 2.3 **Effects of electrical current on the human body**

Current (milliamperes)	Reaction
1	Perception level, a faint tingle.
5	Slight shock felt; disturbing but not painful. Average person can let go. However, vigorous involuntary reactions to shocks in this range can cause accidents.
6 to 25 (women) 9 to 30 (men)	Painful shock, muscular control is lost. Called freezing or "let-go" range.*
50 to 150	Extreme pain, respiratory arrest, severe muscular contractions, individual normally cannot let go unless knocked away by muscle action. Death is possible.
1,000 to 4,300	Ventricular fibrillation (the rhythmic pumping action of the heart ceases). Muscular contraction and nerve damage occur. Death is most likely.
10,000–	Cardiac arrest, severe burns, and probable death.

* The person may be forcibly thrown away from the contact if the extensor muscles are excited by the shock. Note that the gender difference listed in the table is far from conclusive and likely to depend on the route of entry; however, it appears that males may respond to shock with significant changes in systolic blood pressure while females may respond with significant respiration changes. This possible difference is mentioned to alert students and instructors of specific symptoms to check in the event of a shock.

Figure 2.4 The incorrect application of AAA alkaline batteries in a device designed to accept AA alkaline batteries.

the battery out of the housing if the bulge is pronounced. There will often be a choice of current capacities for these batteries, and experience shows that the use of a battery of the highest current capacity for a given application gives the best results, despite the somewhat higher cost. When replacing a lead-acid battery in a piece of equipment, it is critical that the user not touch the terminals and complete a circuit, or a potentially serious shock can result.

The alkaline battery is convenient and readily available. These batteries are not rechargeable, and must be disposed of when depleted (Figure 2.4). Most organizations require the recycling of these batteries. Under normal use, these batteries will not leak, but if left unused for a long period, or if the contact with the instrument terminals is compromised, leakage can occur. If this has happened to a device you are using, take care not to touch the leaked material; use a paper towel or wood applicator to remove the battery, and wear gloves. Before replacing with a fresh set of batteries, clean any scale from the instrument contacts. Use only batteries intended for the instrument, even if the voltage is the same. This may seem like common sense advice, but it is surprising how many home-made adapters for AA batteries have been installed in a device requiring D batteries!

The nickel-cadmium battery (NiCad) is still found in some devices, such as combustible gas detectors that are based on thermal conductivity. These batteries, based on a nickel oxide hydroxide chemistry, are rechargeable. This battery is becoming far less common since the introduction of the lithium-ion battery. Depleted nickel-cadmium batteries must be recycled and not disposed in the trash.

By far the most common type of battery that will be encountered in portable analytical instrumentation is the lithium battery. There are two types of lithium battery: the lithium-metal battery and the lithium-ion battery (Lion). The lithium-metal batteries are the small coin-type batteries found in calculators, penlights, automobile key fobs, temperature data loggers, and some defibrillators. These batteries are not rechargeable. Lithium-ion batteries, on the other hand, are usually rechargeable without removing them from the device. They are used in cellular telephones, power tools, and many portable analytical instruments.

Lithium batteries are generally safe to use if handled properly and if no defects are present from the manufacturing process. Failure of lithium batteries will present both fire/explosion hazards and chemical exposure hazards. Of course, physical impacts to these batteries (such as dropping the device to the floor) can damage the cells as well as the case in which the cells are contained. Exposure to temperatures in excess of 54 °C (130 °F) will damage any lithium battery regardless of type. While such exposure temperatures might not be common in the laboratory, it will not be uncommon in some locations if a device is left in a hot vehicle. Moreover, recharging of a lithium-ion battery at temperatures below 0 °C will cause damage. One must avoid overcharging or charging too rapidly; some devices have a recharging control (based on cycling power on and off) built in that extends the recharge time over several hours. This also extends battery life. The most spectacular failure mechanism reported for lithium batteries is fire and explosion (sometimes reported as a thermal runaway). Although uncommon, an internal short circuit, or exposure to external heat can certainly cause a thermal runaway. A thermal runaway in one cell of a battery will damage the entire battery. During failure, the casing will bulge and the release of carbon monoxide, carbon dioxide, hydrogen, and hydrocarbons is possible. In some batteries, the release of fluorine and hydrofluoric acid is possible. If a lithium-ion battery begins to smoke or make a hissing sound, it is best to assume that a fire and chemical release is possible. If it is possible to do safely, turn the instrument off. The proper response is the same as in the case of a fire; evacuate the area and summon emergency personnel.

2.1.5 Portable Generators

There are some portable or field instruments for which a battery is impractical or not of sufficient capacity for the application. While emergency vehicles and fire apparatus will usually have provision for electrical equipment operation, it is usually preferable to use a separate portable engine-generator device called a "genset." These devices are operated with a four-stroke reciprocating piston engine that can operate on gasoline or diesel fuel, and some units are dual fuel (gasoline or propane). For powering instrumentation, it is critical to only use a genset that is equipped with an inverter generator; use of a non-inverter will produce a widely varying voltage that will harm instrumentation. Moreover, the inverter type will operate more quietly and with less fuel consumption. The inverter-type gensets are more expensive initially but are well worth the investment. If an inverter genset is not available, it is important to use a device to condition the power to the extent possible. An autotransformer, commonly called a variac, can be used for this and is commonly available in laboratories. The output of the generator will be clearly marked in kW, and it is important not to exceed this capacity. It is also critical to operate the generator outdoors, since it will produce carbon monoxide during operation.

2.1.6 A Brief Discussion about Ground

We will complete our treatment of electrical hazards with a discussion about ground (sometimes called earth ground), grounding, and static charge. This discussion is important because the term is often misunderstood. We have mentioned earlier the need to only use grounded outlets and cables. The reason for this is actually two-fold. First, the earth serves as a nearly infinite sink for charge. This means that the earth can absorb a large (infinite is a bit of an overstatement) amount of current without any measurable change in its potential. The second reason is that for measurement purposes, this nearly infinite capacity for charge allows the earth to serve as a reference point of zero potential. Using a 120 V electrical (U.S.) receptacle as an example, we note that a circuit has a "hot" wire (usually with black insulation), a neutral wire (usually with white insulation), and a ground wire (usually with green insulation). The hot wire will be 120 V above ground potential, while the neutral will be (theoretically) at zero volts. In practice, however, the neutral may be a few V above zero potential. The ground wire, on the other hand, will ultimately be connected to earth, and will be at zero potential. The chassis, case, or shrouding of an instrument is connected to the ground wire. The purpose of having a ground wire is that in the event a hot wire in an instrument becomes dislodged and comes into contact with the instrument chassis or cover, the ground wire will safely direct the current to the earth and away from personnel. The term ground is often misapplied to electronic devices that do not have a separate grounding lead. In these cases, the device actually has a "common" lead that may not actually be connected to ground. Cellular phones are an example.

In the discussion above we referred to the white, black, and green insulation on the conductors. Please note that in many industrial and laboratory environments, there may be many different colors and even combinations of colors on conductor insulation. This is to facilitate the installation of multiple circuits in these locations. Indeed, the student might recall seeing wire carts with many different-colored reels of wire around areas undergoing renovation or construction.

In the first paragraph above, the possibility of the neutral wire being slightly above ground potential was mentioned. This is termed "neutral-to-ground" voltage. In many analytical instruments, the maximum allowable neutral-to-ground voltage is specified, and it is typically 5 V. Voltages above this level can cause the instrument to malfunction, which will be manifest in an error message, such as a buffering overload error. If the instrument repeatedly throws this type of otherwise unexplainable error, it will often be difficult to pinpoint the cause. Many an instrument manufacturer's service technician has been baffled as a result, replacing circuit board after circuit board with no success. The only way to troubleshoot the problem is to place a line monitor on the circuit, which will record the electrical characteristics of the circuit (including current, voltage quality, loading, harmonic content, and grounding). Neutral-to-ground departures can be intermittent and are typically the result of "dirty power," that is, interference from other devices and equipment in the facility. Note that this can include the motors that operate the building elevators and HVAC (heating, ventilation, and air conditioning) equipment. Nearly every instance of dirty power comes from within the building and not from the power delivered to the mains or service entry. The engineered solution to this problem is the installation of "clean" (that is, isolated) power circuits, while an expedient would be the installation of a constant voltage transformer (or isolation transformer) to operate the affected instrument.

In some instrumentation laboratories, you may observe a copper rail that is below the electrical utility receptacles, an example of which is shown in Figure 2.5. This is a grounding bar that is available for use directly as a safe earth ground. It can be used to minimize instances of static buildup on devices by bleeding off charge to ground; simply run a wire from the static-prone instrument to the grounding bar. We mentioned earlier that some laboratories will have explosion-proof fixtures. In such laboratories, it is common to have conductive flooring as well. The floor will be made from an epoxy-based composite into which metal fibers are embedded. This will aid in dissipating static charge that may cause a spark and initiate an explosion. In these laboratories, the floor itself will be connected to ground.

So how is grounding actually accomplished? It is actually as simple as it sounds. A metallic rod (sometimes copper, sometimes low-cost rebar) called the grounding electrode is placed into the earth beneath the building, usually during construction but often as part of a renovation. This rod may extend below the lowest level of the building by between 30 and 50 feet. All circuits and ground rails are then connected to this grounding bar; everything should be on the same ground to prevent so-called floating grounds.

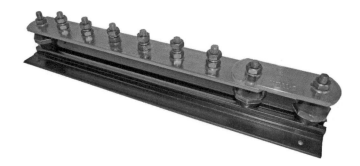

Figure 2.5 A grounding bar that is connected to an earth ground. Reproduced with permission of A.N. Wallis & Co Ltd, Nottingham, U.K..

In most cases, the cold-water piping of the building is also connected to ground. Thus, if a grounding rail is unavailable in the laboratory, one can often use a nearby cold-water pipe; be sure to determine if the institution allows this as part of the safety plan. Of course, this will not work if the cold-water line is made from non-conductive PEX (cross-linked polyethylene).

2.2 INSTRUMENT COOLING

Most of the instrumentation encountered in the laboratory generates heat; this heat can be directly associated with the function (such as with calorimeters and TGA instruments) or it may be waste heat that must be dissipated. Nearly all modern instruments incorporate some means to dissipate waste heat, usually a small fan or fans. It is important that the fans not be obstructed, and that they be free of dirt and lint buildup. It is also important that one not operate an instrument with the cover or shrouding open. The covers of instrumentation and computers are deliberately designed to direct incoming cooling air over the most heat-sensitive components such as power transistors. Instruments are actually more likely to overheat if the cover is removed. The intake louvers or screens should be clean and free from lint.

Gas chromatographs typically provide column temperature control inside of a forced air oven by use of a fan at the rear of the oven. Cool-down sequences in these instruments are typically achieved by opening a small door or doors at the rear of the instrument to dissipate hot air from the oven. The air that is blown from the back of a gas chromatograph after a high-temperature program will be very hot and should not be allowed to interfere with other instruments or flammable materials.

2.3 COMPRESSED GAS CYLINDER SAFETY

Compressed gases are used on many analytical instruments, and they involve hazards that must be minimized by good engineered controls and good laboratory practice. All laboratories that use compressed gas cylinders will have written rules and guidelines that must be followed. Much of the content of these documents deals with topics that will likely not concern the student in an instrumental analysis course, such as transporting, storing, or even connecting cylinders to instruments. This will have been done before you use an instrument. In this section, we will stress those aspects of compressed gas cylinder handling that most directly concern the student during operation of an instrument.

The compressed gas cylinder must be labeled as to the contents. There are some color conventions and, in the U.K., these have been standardized (by the British Compressed Gas Association, BCGA), while in the U.S. they can vary by manufacturer. The important point here is never to rely on the color of the cylinder for anything but decorative purposes; always check the label provided by the supplier.

The compressed gas in a cylinder contains a great deal of energy. A full cylinder can typically contain gas at a pressure of 17.3 MPa (2500 psig). Even if the gas is inert, this energy alone requires utmost respect. Add to this the fact that the gas can be toxic, flammable, or both. Therefore, first and most important, a compressed gas cylinder must be securely restrained to a laboratory bench by either a strap clamp or a chain. It is usually permissible to use a chain around two cylinders side by side, but only one can be fixed with an individual strap clamp. Chaining more than two cylinders with a single chain is prohibited by both standard operating procedures and codes. Older floor-mounted stands for gas cylinders are not used any longer. Cylinder restraint is important at all times, but it is most critical when a pressure-reducing regulator is installed on the cylinder. If a cylinder should fall over with the regulator attached, the force of the fall can break the cylinder valve and rupture the cylinder at the neck. The cylinder then becomes a missile in every sense of the word. Take a few moments to view the video from Georgetown University (https://ehs.georgetown.edu/occupational/physical-hazards/compressed-gas-cylinders/#, accessed February 2021).

SAFETY IN THE INSTRUMENTATION LABORATORY

This video mentions an incident in which a ruptured cylinder penetrated two metal floors of a building, then penetrated the roof to become airborne to 140 ft, and then crashed back down to penetrate back through the roof a second time, all in a matter of seconds. You will not have time to react in such a situation.

Gas cylinders are most commonly made from steel, and large cylinders can have an empty weight of up to 66 kg (145 lbs). Some cylinders, designed to be weighed in the laboratory, are made from aluminum, and have a somewhat lower pressure rating. Because of the weight, cylinders can be awkward and difficult to handle.

2.3.1 Compressed Gas Regulators and Manifolds

In the paragraphs above we mentioned the regulator; this device is actually called a pressure regulating valve (PRV), used to decrease the very high pressure of the cylinder to a level at which it is usable in an instrument. Most regulators encountered in an analytical instrumentation laboratory are two-stage units. This refers to the two pressure zones whose pressures are indicated on gauges typically mounted on top of the round valve housing of the regulator. The first-stage gauge (nearest the cylinder valve) indicates the pressure inside the cylinder, while the second-stage gauge (nearest the outlet) indicates the lower regulated pressure applied to the instrument. The PRV itself is typically a spring-operated diaphragm valve. Screwing the valve handle clockwise (tightening the spring) will increase the outlet pressure, and vice versa. It is important, however, not to turn the valve counter-clockwise (loosening the spring) while there is pressure on the outlet side; this can rupture the diaphragm, which is an especially dangerous situation if the gas is flammable or toxic. If it is necessary to decrease the outlet pressure by loosening the spring tension, first bleed off the outlet pressure into the instrument, hood, or vent.

Connection from the PRV to the gas cylinder is done with gas-specific fittings defined by the Compressed Gas Association in the United States. More information on these fittings can be found in the book by Bruno and Svoronos. Note that flammable gases will usually be connected with a fitting that has a left-handed (or reverse) thread; the presence of a reverse threaded fitting is indicated on the fitting nut with a series of engraved lines. An example of a two-stage pressure-regulating valve is shown in Figure 2.6.

The student might sometimes encounter a single-stage valve connected to a gas cylinder. This is a valve with a pressure gauge that indicates the pressure inside the cylinder, and no pressure reduction is provided. They are used when the full cylinder pressure is required, such as for some reactors and extractors. These types of valves are simply called manifolds. If the cylinder contains a liquid under a high-pressure gas (such as a carbon dioxide cylinder used for supercritical fluid extraction instruments), the cylinder might be equipped with a dip tube that directs liquid into the manifold, rather than gas. Cylinders such as these will usually be marked as DT, and they will not use a PRV, rather a manifold will be used.

When not in use, the cylinder valve should be closed and if the device is to be shut down for an appreciable period, the regulator should be removed and the cylinder cap should be re-installed. If cylinders must be changed out or moved, it is imperative to remove the regulator and install the cap. Then, one must use a gas cylinder cart or hand truck that allows the cylinder to be chained or restrained during the move. If the cylinder must be moved to another floor, the cylinder should be (preferably) placed in a designated freight elevator and personnel should not ride with the cylinder. The cylinder

Figure 2.6 An example of a two-stage pressure regulating valve for gas cylinders. Note the grooves on the fitting nut at the right; this indicates a flammable or toxic gas application, and is left-hand threaded.

should be met at the destination floor by a staff member or instructor. If the cylinder must travel more than one floor, it should be marked with a "Cylinder Transfer—DO NOT ENTER Elevator" placard to ensure that nobody gets into the elevator between the origin and destination floors.

It is good practice to replace a cylinder before all the gas is depleted, leaving, for example, 50 psig (note that the definitions of pressure unit conventions are provided on page 64) of gas still in the cylinder. This will prevent accidental backflow of air into the cylinder, causing contamination. In the past, it was common when installing a new cylinder of an inert gas (such as nitrogen or helium) to crack open the cylinder valve briefly, to blow out any dust in the valve body. This was done to avoid dust and dirt contamination from entering the PRV, before installing the PRV to the cylinder. If this is done, it is important to warn all in the laboratory of the loud noise that will result. Nowadays, cylinders usually come with the cylinder valve in a protective shrink wrap, eliminating the need for the blow-out. Needless to say, a blow-out is not done with toxic or flammable gas cylinders. Certainly, the presence of flammable or toxic gases in the cylinder is reason for extra caution. For example, compressed cylinders of hydrogen gas are used to operate flame ionization detectors (FID) on gas chromatographs. Although the hydrogen cylinder operating the FID will last a long time due to the relatively low flow rate needed, it will require occasional replacement.

2.3.2 Gas Cylinder Markings

Compressed gas cylinders are subject to periodic hydrostatic (or non-destructive imaging) testing for pressure integrity. Although this will not be a major concern to the student, it is nonetheless important for the student to be able to recognize test markings and certifications on gas cylinders. All those permanent stamped markings on the cylinder have a purpose, as illustrated in Figure 2.7. Note that all of these markings will pertain to the cylinder itself rather than the contents.

The owner of the cylinder is typically found on the collar at the neck of the cylinder. This is usually not the name of the university or laboratory, but rather the name of the supplier. In some organizations, cylinders may be owned in-house, but this is rare. Below the collar on the upper body of the cylinder are four fields with permanent stamped markings.

Field 1 lists the cylinder specifications. DOT stands for the United States Department of Transportation, the agency that regulates the transport and specification of gas cylinders in the United States. The next entry, for example, 3AA, is the specification for the type and material of the cylinder. The most common cylinders are 3A, 3AA, 3AX, 3AAX, 3T, and 3AL. All but the last refer to steel cylinders,

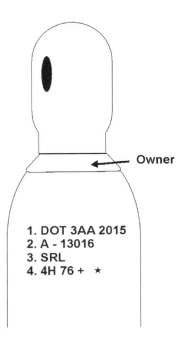

Figure 2.7 Gas cylinder stamped markings.

while 3AL refers to aluminum cylinders. The individual specifications differ mainly in the chemical composition of the steel, and the gases that are approved for containment and transport. The 3T specification only deals with large bundles of tube trailer cylinders, not cylinders that are encountered in the laboratory. Gas cylinders are also made from composite materials but these are not found in laboratories. The next (numerical) entry is the service pressure, in psig (in English units rather than SI units, as a matter of convention). Note that the service pressure is not the pressure at which the cylinder is tested; the test pressure is 5/3 times the service pressure stamped on the cylinder.

Field 2 is the serial number, a unique number assigned by the manufacturer.

Field 3 contains the manufacturer identification symbol. The manufacturer identifying symbol historically can be a series of letters or a unique graphical symbol. In recent years, the DOT has standardized this identification with the "M" number, for example, M1004. This is a number issued by DOT that identifies the cylinder manufacturer.

Field 4 shows the manufacturing information for the cylinder. The date of manufacture is provided as a month and year. With this date is the inspector's official mark, for example, H. In recent years, this letter has been replaced with an "IA" number, for example IA02, pertaining to an independent agency that is approved by DOT as an inspector. If "+" is present, the cylinder qualifies for an overfill of 10 %

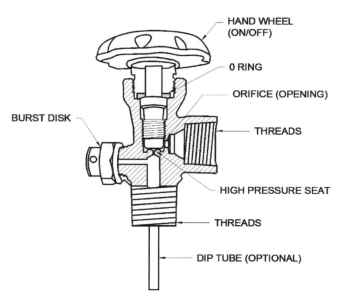

Figure 2.8 A cutaway diagram of a gas cylinder valve showing the component parts including the burst (rupture) disk nut. Reproduced with permission of Cylinder Training Services, Edmonds, WA, USA.

in service pressure. If a star is present, the cylinder qualifies for a 10-year rather than a 5-year retest interval. Also stamped on the cylinder will be the retest dates. A cylinder must have a current (that is, within 5 or 10 years) test stamp. Let your instructor know if this is not the case. A cylinder that has an expired test date is not necessarily a hazard, but it cannot be refilled until the retest is completed.

One final aspect on compressed gas cylinder safety deals with the cylinder valve. If you look closely at the valve, you will note that on the side opposite the connection for the PRV there is another nut that usually has holes in it. This is the rupture disk nut; it holds in place a thin metal sheet (called a rupture or burst disk) that is designed to burst if the cylinder pressure gets beyond a safe limit. This is usually the test pressure, which as we mentioned above, is 5/3 times the stamped service pressure. Overpressure of a cylinder can be caused by heat (such as exposure to fire) or it can be the result of accidental overfilling. When a disk ruptures, the cylinder content is released very rapidly through the vents of the rupture disk nut. For some flammable and reactive gases, the vent holes are plugged with a fusible metal that will melt on exposure to heat and cause pressure release. The gas release that occurs upon a ruptured disk is quite dramatic and can itself be a cause of damage and injury. Early in the career of one of the authors (TJB), an accidentally overfilled cylinder containing nitrogen underwent a rupture disk burst in a laboratory (luckily) during the overnight hours. Upon arrival the next day, staff found the laboratory in an astounding state of disarray, with laboratory furniture atop gas chromatographs, other equipment overturned, and papers scattered everywhere. At first, intentional mayhem was suspected and the police were summoned. After a great deal of head-scratching and investigation, the culprit was found to be an apparently overfilled nitrogen cylinder with a burst rupture disk (Figure 2.8).

So, what are the lessons from this? First, never attempt to remove the rupture disk fitting from a cylinder valve. Second, while it is probably rare that you should want to clean a cylinder valve (as mentioned earlier, gas cylinders usually are delivered with the valves shrink wrapped for cleanliness), do so with nothing but mild soap and water, and absolutely never with an ammonia-based cleaner; some of the disks are made from copper. Many rupture disk failures have occurred in hospitals because of the understandable desire to clean and disinfect (oxygen) cylinders, and the ammonium hydroxide solution used initiated corrosion of the copper disk. Third, if you hear gas escaping from the nut, have the cylinder replaced.

2.4 CRYOGENIC FLUID SAFETY

Cryogenic fluids (or cryogens, liquified gases, or cryogenic liquids) are used extensively in analytical instruments. A cryogen is typically defined as a liquefied gas with a normal boiling temperature of no higher than −90 °C (−130 °F, different sources cite differing upper limits). Dry ice (solid carbon dioxide, which sublimes at 194.65 K, −78.5 °C) is also used to achieve low temperatures in the laboratory, but is not considered to be a cryogen. The hazards associated with the use of cryogens are actually twofold. There are hazards associated with the cryogenic fluids themselves, and there are hazards associated with the containers used to store and transport the cryogenic fluids; we will treat these issues separately.

2.4.1 Cryogenic Fluids

The major hazards associated with the use of cryogens stem from their low temperatures, their high liquid-to-gas expansion ratios, toxicity, and air displacement. Some important properties germane to the handling of common cryogens are provided in Table 2.4. The gas-to-liquid expansion ratios are on the basis of 1:R, and have a variability of up to ± 5 depending on the local ambient temperature. The boiling temperatures are at atmospheric pressure (101.325 kPa).

Clearly, the main purpose of the use of a cryogen stems from the low temperatures at their boiling points. This leads to the potential of severe frostbite burn if a cryogen comes into contact with skin. While a thin layer of vapor formation will protect the skin initially, if a cryogen is allowed to pool (such as in clothing), the frostbite danger is severe. Protective clothing is essential (Figure 2.9) and should include cryogen gloves and safety glasses (a full-face shield

Table 2.4 Some important properties of cryogenic fluids

Cryogen Fluid	Expansion Ratio R (approximate)	Boiling Temperature K	Boiling Temperature °C
Argon	847	87.302	−185.84
Helium (^{4}He)	757	4.2238	−268.92
Hydrogen	850	20.271	−252.87
Nitrogen	696	77.355	−195.79
Oxygen	860	90.188	−182.96

Figure 2.9 Examples of protective clothing for the handling of cryogenic fluids showing long gauntlet gloves, apron, and face shield. Reproduced with permission of National Safety Apparel, Inc.

is recommended). In addition, canvas shoes are discouraged since pooling can occur in the case of spillage. Pooling can also occur in cuffs on pant legs, which should also be avoided. Pants should not be tucked into shoes. A lab coat or shirt cuffs should be tucked under glove gauntlets. If skin should come in contact with a cryogen, it should be rinsed with cold water; do not apply dry heat to the affected area. If clothing has frozen to an individual due to cryogen exposure, cold water should be used to free the clothing, and emergency personnel must be summoned.

Another hazard due to the low temperature results from the ability of these fluids to embrittle materials including hoses, floor mats, and other surfaces in the laboratory. Of particular concern is the embrittlement of electrical insulation, which can lead to a fire hazard.

Since the boiling temperature of liquid nitrogen is below that of liquid oxygen, it is possible for oxygen to condense on any surface or vessel cooled by liquid nitrogen. Liquid oxygen is an oxidizer that can enhance the flammability characteristics of liquids and solids that it contacts. The liquid air that is seen dripping from lines transferring liquid nitrogen can be up to 50 % oxygen. If a blue tint is observed in a vessel being used with liquid nitrogen, the presence of liquid oxygen must be assumed. Additional hazards of liquid oxygen will be discussed below.

The high liquid-to-gas expansion ratios listed in the table above show that when a cryogen is vaporized, it has the potential to displace air in a laboratory. While most cryogens are not toxic *per se*, they can act as simple asphyxiants. Laboratories that use cryogens should be adequately ventilated. In labs with large containers of cryogens (such as those found with NMR instruments), it is important to have an oxygen monitor with an audible alarm. Personnel, including rescue workers, should not enter areas where the oxygen concentration is below 19.5 % (vol/vol), unless provided with a self-contained breathing apparatus or air-line respirator. The safe range as indicated on an oxygen monitor is between 19.5 % and 23 % (vol/vol). Personnel in an area of low oxygen concentration may be unaware of the condition, thus instrumented monitoring is critical.

In Chapter 6, we discuss the particular (and spectacular) effects of an NMR magnet quench. Clearly, in the event of an accidental quench, it is imperative to evacuate the laboratory immediately and summon help. If a person seems to become dizzy or loses consciousness while working with cryogens, they should be moved to a well-ventilated area immediately. If breathing has stopped, apply CPR and summon emergency personnel.

Related to the liquid-to-gas expansion ratio is the overpressure that can result if a cryogen is allowed to warm within an enclosure. Indeed, no cryogen can remain liquid within a container; some venting must be provided. If a vent becomes disabled or is not present, warming cryogen will vaporize and produce very high pressures based on the pressure–volume–temperature (PVT) surface of the fluid.

Some cryogens pose specific hazards or handling requirements based on their chemistry:

Liquid oxygen (LOx) cannot be permitted to contact organic materials; some common cautions include solvents and vacuum pump oil. Organic materials can be readily ignited by spark or shock after exposure to LOx, including fingerprints on a surface. One of the authors (TJB) has seen demonstrations in which a thumbprint was intentionally placed on a steel beam, and a stream of LOx was directed against the print. The print was then tapped gently with a hammer, resulting in a small detonation. Clothing saturated with oxygen is readily ignitable and will vigorously burn. If LOx spills on an asphalt surface, do not walk over or roll equipment over that surface for one hour. While not having specific toxicity issues, if LOx is exposed to high energy electromagnetic radiation it can produce ozone, which will solidify at LOx temperatures. Solid ozone is unstable, toxic (upon vaporization), and can explode if disturbed.

Liquid hydrogen handling requires all of the precautions used for hydrogen gas. Liquid hydrogen should not be transferred in an atmosphere of air since it will readily condense in the liquid hydrogen, resulting in a potentially explosive mixture. Liquid hydrogen must be transferred by helium pressurization in properly designed vacuum-insulated transfer lines pre-purged with helium or gaseous hydrogen. Liquid hydrogen, like liquid helium, can solidify air, which can block vents and safety relief devices. Dewars and other containers made of glass should not be used for liquid hydrogen service. Breakage makes the possibility of explosion too hazardous to risk.

2.4.2 Cryogenic Liquid Containers

Several different types of cryogenic liquid containers are encountered in the laboratory during routine chemical analyses, and each of them has its own associated hazards and precautions. It is common parlance to refer to all of these containers as Dewars, but this is imprecise. The small portable containers used to assemble laboratory cold baths and the small transport containers (with loose-fitting lids and carry handles) are Dewars. These containers are used at ambient pressure. The larger supply containers (from which Dewars are filled) are called liquid cylinders. These containers are pressurized, with different pressure ratings available. In Figure 2.10 we show a variety of cryogenic liquid cylinders.

Liquid cylinders are large heavy containers with integrated casters (or a separate dolly) to facilitate movement. At least two of these casters should be equipped with a braking mechanism. Liquid helium cylinders, which often incorporate a liquid nitrogen jacket, are usually heavier than those used for the common cryogens. Moreover, there is a larger variety of available sizes, ranging from 50 to 500 L, whereas for other cryogens the capacity is usually between 160 and 230 L. The weight of these cylinders can make them

Figure 2.10 A variety of cryogenic liquid cylinders. Reproduced with permission of Cryofab, Inc.

challenging to handle. The personal protective equipment discussed above (for cryogens) must be used when handling liquid cylinders. Cylinders should be moved by pushing, not pulling, to reduce the potential of an upset. In locations of frequent transport of liquid cylinders, bottom door sills should be removed to eliminate the potential of bouncing or rough handling. If a cylinder must be transported by elevator, a freight elevator is preferred. Personnel should not ride in the elevator car with the cylinder. The cylinder should be transported in the elevator with no personnel, and be met at the receiving floor. A placard reading "Cryogen Transfer—DO NOT ENTER Elevator" should be posted facing the door if the elevator is to travel more than one story.

If liquid cylinders must be transported between buildings, it is critical to use ramps and to ensure that there are no large cracks in paving that must be traversed. Note also that the casters commonly found on liquid cylinders are not rated for travel along long distances of pavement.

In the event of loss of control of a liquid cylinder during transport, if the cylinder begins to fall, it is usually best to simply let it go and summon qualified help as defined in the organization standard operating procedures.

Liquid cylinders are pressurized and can contain up to 350 psi (2411 kPa), depending on the cylinder specifications. One commonly encounters pressure specifications (in English units rather than SI units) on liquid cylinders that may sometimes be confusing. The commonly encountered specifications are:

- psia (pounds-force per square inch absolute) gauge pressure plus local atmospheric pressure.
- psid (psi difference) difference between two pressures, specified on the cylinder label.
- psig pounds (force) per square inch, gauge.
- psi-vg (psi vented gauge) difference between the measuring point and the local pressure.

- psi-sg (psi sealed gauge) difference between a chamber of air sealed at atmospheric pressure and the pressure at the measuring point.

Pressure-relief devices are integral to all liquid cylinders and must remain un-obstructed with frost. If the outlet fitting on a pressure-relief valve is facing the same direction as the liquid- or gas-dispensing valve, a fitting directing vented gas away from users should be added to protect personnel. Over-pressurization of liquid cylinders is a serious hazard; cylinders can rupture if a pressure-relief valve becomes impaired or inoperative.

Dewar flasks or simply, Dewars, are small cryogen containers used at atmospheric pressure, with or without a loose-fitting cover or cap. Smaller Dewars are usually made from an evacuated silvered glass insert set into a metal jacket. Any exposed glass should be taped to prevent flying glass in the event of a catastrophic rupture. A variety of Dewar flasks are shown in Figure 2.11.

When filling a small Dewar from a liquid cylinder, it is best to pre-cool the interior of the Dewar with a small amount of cryogen first, swirling it around for a few seconds, before completing the fill. Boiling and splashing generally occur when filling a warm container, so stand clear and wear appropriate personal protective equipment as discussed above. The flask should be clean and dry before filling. Surprisingly, the temperature of the cryogen in the Dewar must be given time to equilibrate in order to achieve the uniformity required by most instruments.

A Dewar flask should not be filled to beyond 80 % of its capacity. Over filling increases the risk of splashing and spillage. A beverage thermos bottle is not a substitute for a Dewar flask in the laboratory. When carrying a small Dewar flask, make sure it is the only item you are carrying. Hold the Dewar flask as far away from the face as possible. Be aware of other personnel in the area.

Small Dewar flasks with liquid nitrogen are often used as cold traps in the instrumentation laboratory. When instrument components are placed in the filled Dewar cold bath, it is important to avoid splashing and excessive boiling by inserting components slowly. Surprisingly, it takes some time for the cryogenic liquid in a Dewar flask to stabilize in temperature. Indeed, some instruments such as isotherm measurement instruments require a wait time of 15–30 min after inserting the samples to be cooled.

If Pyrex or glass wool insulation is placed around the flask, do not let the wool dip into the cryogen or become a vapor barrier. If the liquid nitrogen acquires a blue tint, it has become contaminated with liquid oxygen, and the discussion of LOx hazards above applies.

If a Dewar flask cold trap is used in association with a vacuum pump, it is important to periodically empty the trap carefully to avoid exposure to toxic chemicals and to prevent over-pressurization should the trap run dry. Note also that venting liquid nitrogen near a vacuum pump v-belt can embrittle the belt and shorten its service life. Likewise, if the venting is near electrical cables. Embrittlement of the electrical insulation can result in a fire hazard.

2.5 THE USE OF FUME HOODS IN THE INSTRUMENTATION LABORATORY

Engineered safety equipment is preferred over the reliance on personal protective equipment, and the fume hood is one such engineered safety device that is nearly ubiquitous in laboratories. Laboratories concerned with biological specimens (microbes, spores) also commonly are equipped with biological safety cabinets (BSCs). Occasionally, an instrument may be located or installed in a fume hood for safe operation. Most of the chemical fume hoods consist of a cabinet or enclosure set at waist level (above a table or storage cabinet) that is connected to a blower located above the hood or external to the hood through a duct system. The cabinet has an open side (or sides) to allow a user to perform work within. A movable transparent sash separates the user from the work. Most chemical fume hoods have a sill that functions as an airfoil at the work surface below the sash. The walk-in hood is a chemical fume hood that is mounted directly on the laboratory floor or a slightly raised chemical resistance platform. It is used for the ventilation of larger pieces of equipment, with the advantage that these pieces of equipment can be wheeled in and out of the walk-in hood. The walk-in hood typically uses two separate sashes. Additional types of hood systems include snorkels and canopies that are portable. Each type must be understood to be operated most efficiently within specifications. Additional information on fume hood types and biological safety cabinets can be found in the book by Bruno and Svoronos. Some common configurations of fume hoods are shown in Figure 2.12.

While the operation of chemical fume hoods is straightforward, safe operating practices must be observed. The face velocities should be optimized at between 80 and 120

Figure 2.11 Examples of Dewar flasks used with cryogenic fluids. Reproduced with permission from Pope Scientific Co.

Figure 2.12 Some common configurations of laboratory fume hoods. Reproduced by permission of Labconco Inc.

fpm. Face velocities in excess of 125 fpm can cause turbulent flow that can actually cause outflow of contaminants from the hood and potentially expose the user to hazards. Indeed, instruments located in chemical fume hoods can be a cause of turbulence as well. Face velocities are checked periodically by facilities managers or safety personnel. Ideally, chemical fume hoods are located in low traffic areas of the laboratory, away from entry doors. Safe operation of chemical fume hoods requires observance of the following practices:

1. Before using the chemical fume hood, ensure that it is in working order. A simple air flow test can be done with a laboratory wipe, or one can observe the flow monitor if one is installed on the hood. These monitors will produce a loud alarm (similar to a smoke alarm) if the air flow is insufficient. Be aware of clattering sounds that might indicate a broken (or soon to be broken) drive belt, or screeching sounds that might indicate a failed or failing bearing on a motor.
2. The motor of the chemical fume hood should be running at all times, except for maintenance. If an on/off switch is located inside the laboratory, it should be equipped with a lockout device to protect maintenance staff. Unfortunately, more than a few technicians have lost fingers because somebody turned the motor on while it was being serviced.
3. Users should perform all work at least six inches inside the plane of the sash, and they should avoid placing their heads inside the chemical fume hood, beyond the plane of the sash.
4. Laboratory occupants should avoid or minimize traffic in front of the chemical fume hood to minimize turbulence.
5. Users should avoid rapid movements in front of the chemical fume hood; this includes rapidly raising and lowering the sash.
6. Keep the sash closed down as far as possible. There will be a sticker on the side of the hood indicating the maximum height at which the sash can be operated. This level is measured yearly by safety personnel by use of an anemometer. If an air flow monitor is installed on the hood, operating the hood with the sash above the indicated maximum height will typically trigger the alarm.
7. The baffles in the rear or sides of the chemical fume hood should be free of obstruction, and one must avoid allowing bits of paper such as laboratory wipes to get sucked into the baffle.
8. If a heating device such as a hot plate is being used in the hood, the interior air flow can be dramatically changed by convection. In those instances, the lower baffles of the hood should be closed or minimized and the middle baffles (if present) fully opened to accommodate the convection.
9. As mentioned earlier, equipment located inside the cabinet can potentially disrupt the air flow; such equipment should be raised above the level of the lower baffles by a small shelf or blocks. Cinder blocks are often used and provide stability because of their weight.
10. Hoses and power cords that must be run into the chemical fume hood from the outside should be run through the airfoil beneath the sash.
11. Some chemical fume hoods have a small cup sink located inside the cabinet. The main purpose of this cup sink is for the discharge of cooling water. Water should be run into this sink periodically to maintain it free of obstruction and to keep the P-trap full and thereby prevent sewer gas back up. It is also critical not to allow discharge of any chemicals into this cup sink.
12. Some chemical fume hoods have compressed air lines that allow the use of air-operated equipment such as vortex tubes (for heating and cooling). When using vortex tubes inside a fume hood, the sash should be fully closed because additional air flow inside the hood is present.
13. In the event of a power outage, lower the sash to within an inch of the fully closed position.
14. Evaporations and digestions with perchloric acid must only be done in a specifically designed and designated perchloric acid chemical fume hood. The same considerations apply for the use of radionuclides and infectious agents.

The air movement inside of a fume hood with the sash open and closed is shown schematically in Figure 2.13.

As a final note on fume hood operations and safety, we will discuss the distinction between the devices discussed above and similar safety cabinets: the laminar flow hood and the biological safety cabinet. Both of these latter devices are very different in operation and purpose to the typical laboratory fume hood discussed above, and while they find extensive use in laboratories, it is relatively rare to encounter them in analytical instrumentation laboratories. They will only be discussed for completeness. The laminar flow hood is essentially an opposite of the

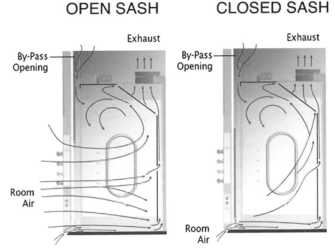

Figure 2.13 The air movement inside a fume hood with the sash open and closed. Reproduced by permission of Labconco, Inc.

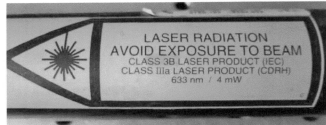

Figure 2.14 An example of an identification label on a laser used in analytical instrumentation. Reproduced with permission of Niagara College, Prof. Mark Csele.

fume hood. It directs a uniform, relatively slow stream of HEPA (high-efficiency particulate air) filtered air downwards toward the work surface and toward the user and out into the laboratory. This device does not provide protection from chemically hazardous or infectious materials such as particulate, volatile hazards and aerosols. The main use for the laminar flow hood is in the clean manipulation of crystals, electronic devices, pharmaceuticals, etc. Biological safety cabinets, as distinct from chemical fume hoods, are enclosed ventilated cabinets intended to provide both a clean and a safe working environment for aerosols and biological hazards. All exhaust air is HEPA-filtered as it exits the biosafety cabinet, removing harmful bacteria and viruses.

2.6 LASER SAFETY

Safety in the use of lasers is a very large topic that is treated in many books and online sources. Lasers in the analytical instrumentation laboratory are usually embedded inside of instruments and users are protected by shielding, barriers, and interlocks that will de-energize the laser if an instrument cover is removed. We will therefore present only a brief discussion here, relevant to devices most commonly encountered in instrumentation, since exposure to laser light is possible when changing samples or performing adjustments.

It is difficult to state *a priori* what laser hazards will be present in a given instrument; however each instrument will have a tag or label indicating the type of laser within (an example of a tag on a laser is shown in Figure 2.14). Thus, it is possible to discuss safety guidelines on the basis of general laser types, and no doubt your instructor will have specific information regarding the equipment that you will be using. The following is a summary for the laser classes following the ANSI guidelines used in the United States (note that the International Electrotechnical Commission, IEC, publishes a different classification scheme).

Class I
Class I lasers are inherently safe with no possibility of eye damage under conditions of normal use. The safety can result from a low output power (in which case eye damage is impossible even after prolonged exposure), or due to an enclosure preventing user access to the laser beam during normal operation, such as in CD players, laser printers, surveying transits, or analytical measurement instruments.

Class II
With this class of laser, the blink reflex of the human eye will prevent eye damage, unless the person deliberately stares into the beam for an extended period. Thus, a Class II laser can cause some eye damage if this is done. Output power may be up to 1 mW. This class includes only lasers that emit visible light. Some laser pointers are in this category.

Class IIIa
Lasers in this class are mostly dangerous in combination with certain optical instruments that change the beam diameter or power density. Output power does not exceed 5 mW. Beam power density may not exceed 2.5 mW/cm^2. Many laser sights for firearms and some laser pointers are in this category.

Class IIIb
Lasers in this class may cause damage if the beam enters the eye directly. This generally applies to lasers powered from 5–500 mW. Lasers in this category can cause permanent eye damage with exposures of 1/100th of a second or less depending on the strength of the laser. A diffuse reflection (on paper or from a matte surface) is generally not hazardous but a specular reflection from a highly reflective surface can be just as dangerous as direct exposures. Protective eyewear is recommended when direct beam viewing of Class IIIb lasers may occur. Lasers

at the high-power end of this class may also present a fire hazard and can lightly burn skin.

Class IV
Lasers in this class have output powers of more than 500 mW in the beam and may cause severe, permanent damage to eye or skin without being magnified by optics of eye or instrumentation. Diffuse reflections of the laser beam can be hazardous to skin or eye within the nominal hazard zone (NHZ). Many industrial, scientific, military, and medical lasers are in this category.

We mentioned laser pointers in several categories above. Laser pointers, ubiquitous at meetings, shows, and in the classroom, deserve separate consideration because of recent work on actual observed power output. The human light aversion response can generally protect against the light from these lasers; however, this response is less sensitive in the near-infrared range (700–1400 nm). Thus, to prevent retinal burns, laser pointers must not emit hazardous levels of infrared radiation, and must be Class 1 compliant in terms of accessible emission level (AEL) in that wavelength range. In a testing program undertaken at NIST, it was found that of laser pointers randomly chosen and tested, all but two pointers failed to comply by more than 15 % of the specified AEL, and 48 % emitted more than twice the specified AEL at one or more specified wavelength, indicating a much higher potential risk.

2.7 COMPRESSED AIR LINE SAFETY

Compressed air piped into analytical instrument laboratories is used routinely in the operation of many devices. One commonly sees utility pods with laboratory valves or cocks for air, hot and cold water, natural gas, lab vacuum, and occasionally steam. Compressed air may be hard-plumbed into instrumentation for valve actuation, vortex tube chilling, and as a source of air for flame ionization detectors and spinning NMR samples. There are specific hazards associated with the use of compressed air sources and lines in laboratories, and safety precautions must be observed.

All pipes, hoses, and fittings must have at least the rating of the maximum pressure of the compressor. Compressed air pipelines should be identified as such with labeling affixed to the pipe (high-pressure air or plant air are also acceptable labels). Flexible hoses used for compressed air delivery must be appropriately rated, must be kept free from grease and corrosives, and must be secured to fixtures with hose clamps. The hoses themselves must be secured to prevent hose-whip in the event of a rupture. Care must be taken to prevent kinking of flexible air hoses. The maximum working pressure should be known and not exceeded. Typically, the pressure of laboratory compressed air lines is no more than 125 psig (862 kPa), and reduced pressures are provided by diaphragm regulators (similar to the PRVs used on gas cylinders).

Compressed air must not be used to clean dirt and dust from clothing or personnel. Laboratory compressed air used for cleaning should be regulated down to 15 psi (103 kPa) unless the blowguns used are equipped with diffuser nozzles to lower exit pressure and velocity. Static electricity can be generated through the use of pneumatic devices and tools. This type of equipment must be grounded or bonded if it is used where fuel, flammable vapors, or explosive atmospheres are present. The use of the laboratory grounding bars, discussed earlier in the treatment of electrical hazards, is recommended for this purpose.

All sources of laboratory or plant compressed air will contain water vapor, even if coalescence filters and particulate filters are provided in-line. This water vapor can condense into liquid or ice, and can cause hazards if the air contacts electrical or electronic devices. If the laboratory air supply is dried with a refrigerated conditioner, the dew point is still typically only dropped to 40 °F (4 °C). In some older facilities, compressed air systems can be severely contaminated with oil (usually from the compressor), rust (from steel piping), and liquid water. In such installations, it is important to provide coalescence filters and particulate filters in-line.

2.8 LADDER SAFETY

It might seem unusual to find a section of an analytical instrumentation text to have a discussion about ladders, but it is common for students and personnel to require access to above-bench heights and to areas behind instruments. Often, a ladder is employed to gain access to such places. Indeed, one often finds ladders in place semi-permanently in some laboratories, such as in front of NMR magnets. As common as ladders are in the laboratory, one of the most common type of lab injury is the result of falling from a ladder.

2.8.1 Step Ladders

When possible, it is best to use a step ladder, which is a portable, self-supporting, A-frame ladder with two front side rails and two rear side rails (note that some older tripod-type ladders are still in use). Generally, there are steps mounted between the front side rails and bracing between the rear side rails, but sometimes there are steps on both sides. While aluminum ladders are lightweight and are easy to carry, they pose a hazard when used near electrical equipment and electronics. It is generally better to use the heavier and more costly fiberglass ladders. Different ladder types are provided in Table 2.5.

When climbing or descending from a step ladder, it is important to maintain three points of contact: two hands and

Table 2.5 Ladder types and ratings

Type	Duty Rating	Use	Load lbs (kg)
1AA	Special duty	Rugged industrial	375 (170)
1A	Extra heavy duty	Industrial	300 (136)
I	Heavy duty	Industrial	250 (113)
II	Medium duty	Commercial	225 (102)
III	Light duty	Household	200 (91)

a foot, two feet and one hand. Stand on the middle of the step, only use the ladder on a level surface, and be cautious of other personnel that may be present. The semi-permanent installations, around NMR magnets for example, should be set off with barricades. If you carried tools to the top of the ladder to work on an instrument, be sure to take them down with you before folding and removing the ladder. Never try to move the ladder while standing on it, by trying to "walk" the legs; never have more than one person on the ladder, and never stand on the top step of a ladder.

2.8.2 Straight, Combination, and Extension Ladders

While step ladders are to be preferred in the lab, sometimes a straight or combination ladder is needed. Combination ladders are a hybrid of the step ladder and straight ladder, and can fold or unfold, sometimes furnishing a small scaffold. An extension ladder is essentially two locking units of straight ladder sections. All of the guidelines for step ladders apply to straight and combination ladders, with some additional considerations. The angle of the straight ladder is of critical importance for safe operation. The proper angle for setting up a straight or combination ladder is to place its base a quarter of the working length of the ladder from the wall or other vertical surface. When accessing an elevated surface with a straight or combination ladder, the ladder must extend at least three feet above that surface. Never stand on the three top rungs of a straight, combination, or extension ladder. Finally, ladders must be free of slippery coatings such as oils and greases, and must be used on a level surface unless appropriate adjustment jacks are provided.

It should be obvious without much discussion in this book that the cover or shrouding of a laboratory instrument is not to be used as a ladder to access high places. Instruments today have cabinets that are usually plastic and are designed only to cover internal components. Although this seems obvious, one of the authors (TJB), while a graduate student, was asked by an accomplished and revered professor to assist in moving a benchtop Cary 14 UV-Vis spectrophotometer to the center of an instrumentation laboratory. The professor then climbed atop the instrument to adjust the damper on a fume hood duct (the Cary 14 was built like a battle tank; https://en.wikipedia.org/wiki/Cary_14_Spectrophotometer). Once used as a ladder, this workhorse instrument was returned to its proper location. Such an occurrence would (should) be unheard of today.

2.9 HEARING PROTECTION

There are many sources of noise in the instrumentation laboratory, and in the professional laboratory workplace, employers are required to identify employees exposed to noise in excess of 85 decibels (dB) averaged over eight working hours. Students in a laboratory course on analytical instrumentation are not likely to be exposed to excessive noise for long periods of time, but cumulative exposures will cause eventual hearing loss over time. It is therefore a good idea to begin early to protect hearing while in a noisy environment. Table 2.6 provides a range of sound levels (in decibels, dB) for common laboratory equipment.

It is best to apply engineering and administrative controls to mitigate these noise exposures. For example, if the NMR that you use is equipped with a cryogenic probe, there will be a cryocooler (located in a utility space or outside the building) with transfer lines to the instrument for liquid helium. You will note that these lines are not only vacuum insulated (for the liquid helium cryogen) but also are encased in foam insulation. This layer of foam insulation is to mitigate noise from the cryocooler, not to protect the thermal integrity. Without this insulation, the noise level would be uncomfortable if not illegal. Another example would be high-power probe-type sonicators which are almost always used inside of a sound-insulated cabinet.

If, after applying engineering and administrative controls, sound levels in the local environment are higher

Table 2.6 Noise levels from some common equipment in the laboratory

Equipment	Sound level range (dB)
Air compressor	75–95
Shaker table	75–85
Pressure-relief valves	75–120
Pressurized vortex tubes	75–85
Sonicator (probe type)	70–110
Fume hood	30–70
Vacuum pump	50–70
Room air conditioner	30–65

than regulatory or recommended levels, hearing protection devices (HPDs) should be used to reduce noise exposure. In addition to the noise generated by various devices in the laboratory, additional noise can be generated by devices not directly related to the function of the lab. Noise from radios, piped-in music, HVAC systems, and telephones is also present in many laboratories.

It is not necessarily optimal to use hearing protection that simply maximizes the sound attenuation; it remains important for students and workers to communicate with one another. Thus, devices that prevent intelligible speech can be problematic. Moreover, one must be able to recognize laboratory equipment alarms (such as low flow alarms on fume hoods, temperature alarms on ovens, and in some environments, back-up alarms on forklifts, etc.). In many cases, personnel will remove hearing protection to communicate with one another or to listen for an alarm, even though removal of an HPD in a high-noise environment can substantially reduce hearing protection. By choosing the right hearing protection, removal of the HPD by personnel can be minimized. In workplaces in the United States and other industrialized countries, sound levels must be determined so that the correct noise reduction rating may be determined, and training must be provided by industrial hygienists or other appropriate personnel.

Clearly, hearing protection can only be effective if it is actually used, so user preference (between ear plugs and earmuffs, for example) must be considered. Since the anatomy of the ear varies with each individual, the effectiveness of plugs might not be universal, especially if inserted incorrectly. Moreover, in some environments, personnel inserting and removing ear plugs through the course of a workday can introduce dirt and bacteria, and cause ear infections. When a higher noise reduction rating (NRR) is needed, sometimes using two styles of HPDs simultaneously is effective, e.g., earmuffs and disposable foam earplugs, although the NRR is not simply additive. Table 2.7 provides guidance in choosing the appropriate HPD. Note that in addition to the equipment shown below, there are combination devices available. One can obtain, for example, a passive earmuff combined with a protective hard hat and various types of face shields.

2.10 SELECTION OF LABORATORY GLOVES

Much of the discussion about laboratory gloves presented here might be a review since hand protection is usually covered in general and organic chemistry courses in the context of dealing with hazardous chemicals. The reason for presenting a discussion in this text is that in the analytical instrumentation laboratory, you will encounter a wider variety of hand protection scenarios. In the instrumentation laboratory, you will encounter hazards with both organic and inorganic reagents, aqueous solutions, heat and cold. We have already discussed the use of cryogen gloves earlier in this chapter. Please review this material before working with cryogenic liquids.

The following tables provide guidance in the selection of protective gloves for laboratory use. If protection from more than one class of hazard or chemical is required, double gloving should be considered. Table 2.8 covers general hand protection from scrapes, burns, ergonomic issues, cuts, and abrasions, while Table 2.9 specifically deals with chemical hazards. In selecting the appropriate hand protection, one should identify the hazard, and where possible use engineering controls (drip controls on bottles, elimination of sharps, the use of ergonomically designed tools, etc.).

2.11 SELECTION OF RESPIRATOR CARTRIDGES AND FILTERS

Respirators are sometimes desirable or required when performing certain tasks in the instrumental analysis laboratory. This might involve field work or the sampling of environmental samples. If your laboratory work requires the use of a respirator, you will be fitted for the respirator by your instructor. Part of the fitting process will likely be a test for fit integrity by testing with an appropriate chemical such as isoamyl acetate (which has the distinctive smell of banana, to test for vapor integrity) or saccharine (which has a sweet taste, to test for particulate protection). Facial hair will cause any respirator to leak; a full beard, for example, will not allow for the proper use of a respirator.

There is a standardized color code system used by all manufacturers for the specification and selection of the cartridges and filters that are used with respirators. Table 2.10 provides guidance in the selection of the proper cartridge using the color code.

In addition to the cartridges specified in the table, particulate filters are available that can be used alone or in combination.

Note that the respirators discussed above are distinct from the face masks that have become ubiquitous for the period following the release of the CV-19 (severe acute respiratory syndrome coronavirus 2, SARS-CoV-2) virus. Mandated protection against airborne droplet infection has included face masks, face coverings, and the N95 mask. The N95 mask, commonly (and mistakenly) called a respirator, is designed to remove 95 % of all particulates from the air stream, achieved by use of a fine mesh of nonwoven polypropylene fibers. These masks can even provide protection against nanoparticles. The N95 is not a respirator in the sense of those devices discussed above that provide protection against chemical exposure. Indeed, the N95 mask will not protect against oils or organic vapors.

2.12 SAFETY WITH NANOMATERIALS

In the past decade, research into nanomaterials has burgeoned. Nanomaterials have been applied in polymers,

Table 2.7 Selection of hearing protection devices

Type	Illustration	Notes
Passive earmuff		Easy to fit; good for intermittent noise exposure, multiple headband types are available to accommodate other protective equipment such as hard hats; may be uncomfortably hot or heavy; more expensive than earplugs.
Disposable foam earplug		Cooler than earmuffs, and often more comfortable for extended use; can provide most attenuation; multiple sizes for different ear canals; attenuation depends on proper fit; proper fitting may be difficult to achieve or learn; hygiene issues in dirty environments; can absorb perspiration; single use only.
Pre-molded, reusable earplug		Comfortable for extended use; washable and reusable; does not absorb perspiration; multiple sizes for different ear canals; attenuation depends on proper fit; slightly more expensive per unit than the disposable earplug above; must be cleaned between uses, may be loosened with talking and chewing.
Canal caps/semi-insert earplug		Convenient for intermittent use; various noise reduction ratings available; hangs around neck when not in use; lower attenuation than most earplugs; lower noise reduction ratings than muffs or earplugs; wearer's voice sounds louder (occlusion effect).
Custom earplug		Comfortable with proper insertion; active sound management; variability in attenuation; comfort and fit variable with changes in ear canal; higher initial cost than other ear plugs, quality can be variable.
Combination earmuff		Easy to fit; face shield or face screen available; might be less adjustable than individual components; may be uncomfortably hot or heavy; more expensive than earmuffs or earplugs, often more expensive than individual components.

catalysis, chemical processing, medicine, energy storage, electronics, and many other fields. Indeed, nanoscience research groups are among the most sought after in graduate-level education in the chemical allied fields. We treat nanomaterials safety in this chapter even though many of the issues are related to chemical hazards, and not specifically to hazards associated with the analytical instrumentation laboratory. The reason for this is that many of the engineered and personal protective safety devices already discussed in this chapter and elsewhere in this book are used in the handling of nanomaterials.

The classification of nanomaterials includes nano-objects and nanoparticles; nano-objects are materials with at least one dimension (length, width, height, and/or

SAFETY IN THE INSTRUMENTATION LABORATORY

Table 2.8 General hand protection glove selection criteria

Glove	Application Examples
Cotton	Weighing, glass handling (avoiding contamination with skin oils); note that these gloves can be used as a first layer when also using other gloves such as for chemical hazards.
Leather	Moderate hot or cold material handling; moving equipment.
Gel-filled (anti-vibration)	Operation of vibrating equipment.
Kevlar or fine mesh	Work with sheet metal, glass, or heavy cutting; note that these will not protect against punctures.
Chemical resistant	See Table 2.9 for specifics.
Insulated for heat	Furnace work; handling hot glass or metal.
Insulated for cold	Cryogenic work, filling Dewars, replenishing NMR magnets, etc.

Table 2.9 Hand protection for chemical hazards

Glove Material	Resistant to:
Viton	PCBs, chlorinated solvents, aromatic solvents
Viton/Butyl	Acetone, toluene, aromatics, aliphatic hydrocarbons, chlorinated solvents, ketones, amines, aldehydes
SilverShield and 4H (PE/EVAL)	Morpholine, vinyl chloride, acetone, ethyl ether, many toxic solvents and caustics
Barrier	Wide range of chlorinated solvents, aromatic acids
PVA	Ketones, aromatics, chlorinated solvents, xylene, MIBK, trichloroethylene; DO NO USE WITH WATER/AQUEOUS SOLUTIONS
Butyl	Aldehydes, ketones, esters, alcohols, most inorganic acids, caustics, dioxane
Neoprene	Oils, grease, petroleum-based solvents, detergents, acids, caustics, alcohols, solvents
PVC	Acids, caustics, solvents, solvents, grease, oil
Nitrile	Oils, fats, acids, caustics, alcohols
Latex	Body fluids, blood, acids, alcohols, alkalis
Vinyl	Body fluids, blood, acids, alcohols, alkalis
Rubber	Organic acids, some mineral acids, caustics, alcohols; not recommended for aromatic solvents, chlorinated solvents

Table 2.10 Respirator cartridge color code

Color Code	Application
Gray	Organic vapors, ammonia, methylamine, chlorine, hydrogen chloride, and sulfur dioxide or hydrogen sulfide (for escape only) or hydrogen fluoride or formaldehyde.
Black	Organic vapors, not to exceed regulatory standards.
Yellow	Organic vapors, chlorine, chlorine dioxide, hydrogen chloride, hydrogen fluoride, sulfur dioxide, or hydrogen sulfide (for escape only).
White	Chlorine, hydrogen chloride, hydrogen chloride, hydrogen fluoride, sulfur dioxide, or hydrogen sulfide (for escape only).
Green	Ammonia and methylamine.
Orange	Mercury and/or chlorine.
Purple	Solid and liquid aerosols and mists.
Purple + gray	Organic vapors, ammonia, methylamine, chlorine, hydrogen chloride, and sulfur dioxide or hydrogen sulfide (for escape only) or hydrogen fluoride or formaldehyde; solid and liquid aerosols and mists.
Purple + black	Organic vapors, and solid and liquid aerosols and mists.
Purple + yellow	Organic vapors, chlorine, chlorine dioxide, hydrogen chloride, hydrogen fluoride, sulfur dioxide, or hydrogen sulfide (for escape only); solid and liquid aerosols and mists.
Purple + white	Chlorine, hydrogen chloride, hydrogen chloride, hydrogen fluoride, sulfur dioxide, or hydrogen sulfide (for escape only); solid and liquid aerosols and mists.
Purple + green	Ammonia, methylamine, and solid and liquid aerosols and mists.

diameter) that is between 1 nm and 100 nm (1×10^{-9} m), nanoparticles are materials in which all three dimensions are on this scale. [Note that the ASTM definition allows for two dimensions between 1 nm and 100 nm.] To add perspective, a human hair is approximately 50,000 nm in diameter. Beyond scale, nanomaterials can be classified as natural, incidental, and engineered, depending on origin. Natural nanomaterials include volcanic products, viruses, sea spray, and mineral aerosols, and are ubiquitous in nature at appreciable concentrations. Incidental nanomaterials include metal vapors produced during welding, sandblasting dust and other industrial effluent, cooking smoke, and diesel engine particulates. The environmental health and safety aspects of natural nanomaterials have received some study, and among the incidental nanomaterials, welding vapors and diesel fuel particulates have received extensive study. It is important to note that the nomenclature might be imprecise at times. For example, welding vapors are often referred to as fumes; the implication of historical usage speaks to the particle sizes encountered in welding operations.

In recent years, however, there has been a great emphasis on engineered nanomaterials, and it is this class, which includes metal nanoparticles, nanorods, nanowires, nanotubes, buckeyballs, nanocapsules, and quantum dots that are the main concern here. Study of the environmental health and safety risks of engineered nanomaterials remains an active area of research that is receiving increasing attention due to the widespread use of these materials in numerous applications. While much is still unknown regarding the fate and toxicity of this class of materials, here, we provide some general guidelines for the safe handling of nanomaterials. We begin with some simplified definitions or terms used in nanotechnology, needed for understanding of these safety guidelines as well as those provided elsewhere.

Definitions

- Aerodynamic diameter: An indirect measure of particle diameter defined as the diameter of a sphere with a density of 1000 kg/m^3, having the same settling velocity of a particle of interest.
- Agglomerate: A group of particles (which may include nanoparticles) held together in a loose cluster by weak forces that may include van der Waals forces, surface tension, and electrostatic forces. Agglomerates are often re-suspendable.
- Aggregate: A heterogeneous particle held together with relatively strong forces such that the particle is not easily disassembled. Aggregates are typically not re-suspendable.
- Buckeyballs: Spherical carbon (C_{60}) fullerenes.
- Fullerenes: Molecules composed entirely of carbon, usually in the form of a hollow sphere, ellipsoid, or tubes.
- Graphene: A one atom thick sheet of carbon.
- Multi-walled carbon nanotube: Multiple sheets of sheet graphene wrapped into a tube of nanoscale dimensions.
- Nanoaerosol: A collection of nanomaterials suspended in a gas.
- Nanocolloid: A nanomaterial suspended in a gel or other semi-solid substance.
- Nanocomposite: A solid material composed of two or more nanomaterials having different physical characteristics.
- Nanohydrosol: A nanomaterial suspended in a solution.
- Nanotube: A seamless tube with a diameter on the order of nanometers.
- Nanowire: A wire of dimensions on the order of nanometers.
- Quantum dot: A nanomaterial or nanoparticle that confines the motion of conduction band electrons, valence band holes, or excitons (pairs of conduction band electrons and valence band holes) in all three spatial directions.
- Single-walled carbon nanotube: A single sheet graphene wrapped into a tube of nanoscale dimensions.
- Ultrafine particle: A (usually) airborne particle with a diameter less than 100 nm.

The unique safety issues posed by nanomaterials result from the potential of deep penetration into tissue, the potential of passing through the blood–brain barrier, and the possible ability to translocate between organs. Biological effects result from the size, shape, polarity/charge, adsorptive capacity, and surface composition, and the ability to bind biological proteins and receptors. Nanomaterials have a higher reactivity than the parent compounds, often having catalytic effects and often presenting greater flammability or explosion risks. For example, bulk elemental gold is considered inert, but gold nanoparticles below 5 nm are catalytic toward a number of oxidation reactions.

The most obvious exposure route of nanomaterials is respiratory; particles depositable in the air exchange region of the lungs are considered respirable. Ingestion can occur from unintentional hand-to-mouth transfer. Finally, nanoparticles can be absorbed through skin or cuts/abrasions to the skin. The safe handling of nanomaterials will generally follow the usual laboratory safety grid:

Elimination: a change in the experimental design to avoid the hazard.
Substitution: the use of a surrogate of lower hazard.
Engineering controls: the use of enclosures, fume hoods, etc.
Administrative controls: adherence to standard procedures and protocols.
Personal protective equipment (PPE): the last line of safety, including gloves, clothing, respirators, etc.

Clearly, elimination and substitution are most useful for nanoparticles of incidental origin. Research with engineered nanoparticles must make use of engineering controls, administrative controls, and personal protective equipment.

The use of non-regenerating general ventilation systems, such as fume hoods, and high-efficiency particulate air (HEPA) dust collection systems, are critical to safe handling. Where possible, installation of ultra-low particulate air (ULPA) filters should be used, since they are widely viewed as being more effective for engineered nanomaterials. The lab in which nanomaterials are handled should be under negative pressure relative to the surroundings (corridor, service galley, etc.). The entry on fume hoods and/or biological safety cabinets in this section provides additional information on fume hood selection and operation.

Where possible, manipulations should be conducted in solution (or in a liquid phase) to minimize the potential of aerosol formation.

Solution-phase nanomaterials should be handled wearing gloves. Gloves should be compatible with the solvent used to disperse the nanomaterials in solution. In general, nitrile gloves are recommended, and double gloving is advisable for heavy or prolonged usage. Gloves with cuffs, clothing with full sleeves to protect wrists, or a laboratory coat are recommended. Liquid- or solution-phase nanomaterial manipulations are best conducted in a fume hood or biological safety cabinet, especially when employing nanomaterials dispersed in open containers, in solvents with known health risks, or when higher risk activities such as sonication, agitation, and vortex mixing are involved. Manipulations conducted in closed containers need not be performed in a fume hood or biological safety cabinet, however the container should be opened inside of such an enclosure since aerosols could be released upon opening. Disposable bench covers should be used where spillage is possible. Any spillage should be cleaned up immediately. Contaminated gloves should be removed and replaced immediately.

Manipulation of dry nanomaterials must be performed in a fume hood or biological safety cabinet. Transport of dry nanomaterials from place to place in the lab must be done in closed containers. For manipulations of air-sensitive nanomaterials, a glove box or glove bag is required. Hand washing must be done after manipulations of nanomaterials. Work areas should be cleaned after completion of tasks. Adequate consideration should be given to tasks involved with the maintenance of equipment or instrumentation used in work on nanomaterials. Such maintenance should be done with the assumption of the presence of nanomaterials.

Though the fate and toxicity of nanomaterials remains largely unknown and is still an area of active investigation, nanomaterials and any by-products from their synthesis should be treated as potentially hazardous waste. Nanomaterials should be properly disposed of based on their nanomaterial and solvent compositions. Nanomaterials containing heavy metals should be treated accordingly and separately from other waste streams. When possible, it is advisable to collect or fully dissolve nanomaterials that are present in solution to limit the volume of waste generated. Often nanomaterials can be aggregated or precipitated from solution with an anti-solvent and filtered off to be recycled or collected as solid waste. This is called "crashing out" in laboratory vernacular. Alternatively, adsorbents such as activated charcoal can often be used to remove certain types of nanomaterials from solution upon filtration and collection of the adsorbent following exposure to the nanomaterial-containing solution. Other types of nanomaterials such as metal oxides can often be dissolved completely with strong acids.

2.13 RELATIVE DOSE RANGES FROM IONIZING RADIATION

The study of the effects of ionizing radiation on the human body is the province of the field called health physics. This nomenclature can be misleading since it covers not only aspects of physics but also chemistry and biology as well. This is a very large field, beginning at the close of World War II and the dawn of the atomic age. Nuclear power, nuclear medicine, sensors, and, of course, nuclear weaponry all make use of sources of ionizing radiation; the analytical instrumentation laboratory is also a beneficiary. In the United States, every radiation source must be licensed in the U.S. by the Nuclear Regulatory Commission (NRC), and every source must have a designated custodian or responsible person. In addition, the institution or university will have a radiation safety officer (RSO) to oversee administration, compliance with rules and regulation, and periodic testing of all sources possessed by the university. While the details of source licensing are outside the scope of this text, some discussion is in order to provide some perspective on the responsibilities of users.

The most commonly encountered source in the instrumentation laboratory is a ^{63}Ni beta radiation source used in gas chromatographic electron capture detectors (ECDs) and in ion mobility spectrometers (IMSs). Indeed, this is the only source we will be concerned with in this text (x-ray tubes are not considered sources). For both instruments, the source is *sealed* and has a radioactivity of 15 mCi and a maximum emission of 65.87 keV. We list the radioactivity in units of mCi because that is the unit that will be listed on the sources themselves. The appropriate SI unit is the becquerel (Bq). One Bq is one event of radiation emission (such as a disintegration) per second. It is related to the older unit by: 1 Ci = 3.7 × 10^{10} Bq (a complete discussion of radiation safety units is provided in the book by Bruno and Svoronos). The ^{63}Ni sources themselves are made by electroplating the radioactive material in the interior of the detection cavity. It is a solid at all temperatures likely to be encountered, with a melting temperature of 1453 °C. The half-life of this source is 101.1 years.

In the United States, one can obtain such devices under either a *general* or a *specific* license, the terms of which are different. Note that the terms sealed, general, and specific are all terms of art with legal meanings and consequences. Many universities will possess sources with both types of license. By far the most common radiation sources that will be encountered in the instrumentation laboratory will be under a general license. Under a general license, the user cannot open, modify, or solvent-clean the source (specified thermal cleaning is permitted). The user cannot transfer or move the source; this must be done by the RSO, who will also conduct periodic wipe tests of the outside of the source. This is done by wiping the exterior of the source with a fiber pad dampened with ethanol, whereupon the fiber is "counted" by use of an appropriate, calibrated counter for radiation leakage. Users with a source under a specific license have a bit more flexibility in terms of transferring and cleaning, but they also typically have more responsibility. Under either type of license, the source must have attached to it a tag of identification that must not be removed. Note that any notations or warnings on the tag will refer to exposures under normal operation; it does not cover exposures if the device is dismantled or allowed to overheat.

It is important to place in perspective the relative ionizing radiation dose acquired in common laboratory settings with that of the natural background. Dose is expressed in the SI system in units called Sieverts (Sv), which describes the effect of radiation on human tissue, rather than the physical effects of the radiation alone. It is unlikely that you will encounter radiation dose described with this unit, however. Most RSOs use the more common unit, the rem, so it is important to be conversant with this unit. It is related to the Sv by use of this relation: 1 rem = 0.01 Sv. This relationship is for x-rays, gamma rays, and beta rays (note that this conversion will not hold for neutrons or alpha particles).

Natural radiation background consists of the highly variable sum of all ubiquitous sources of ionizing radiation encountered on the planet. Background in general can be divided into the four major contributions shown in Table 2.11.

In the United States, the average natural background ionizing radiation level is 311 mrem. This is variable due primarily to differences in altitude and primordial radionuclides and their daughters. For example, the averages in the United Kingdom and Finland are 200 mrem and 700 mrem, respectively. Higher levels are found at higher altitudes and regions with a larger radon budget. Within the United States, for example, the background in Denver, Colorado, is approximately 450 mrem, while in most of Florida, it is closer to 230 mrem. The terrestrial contribution primarily arises from radionuclides of potassium, uranium, and thorium, and their daughters. The cosmic contribution arises primarily from muons, neutrons, and electrons, and varies with the terrestrial magnetic field and altitude. The internal contribution results from consuming radionuclides of potassium and carbon in food and water. By far the largest contribution is from radon and radon daughters. The radon budget results from terrestrial sources of uranium. Within the United States, the action level requiring indoor radon mitigation is reached when a measurement results in 4 pCi/L (150 Bq/m^3) or higher. This level of radioactivity results in a dose of between 300 mrem and 700 mrem/year, assuming 80 % indoor occupancy. The range cited results from different dose conversion coefficients and dosimetric models used by different agencies. Radon testing is familiar to many because a radon test is usually a required part of home inspections during real estate transactions, and companies provide testing (and mitigation if necessary).

Exposure to ionizing radiation in the laboratory results in a dose level in excess of the background levels discussed above. The following charts place in perspective the additional dose received above background for some common exposures. Since in Figure 2.15 the ranges are dwarfed by tobacco use, Figure 2.16 is presented with this contribution removed. This is significant because the tobacco dose is specific to the lungs, and not the whole body. A more relevant comparison among the items in Figure 2.15 would be obtained by multiplying the listed 8000 mrem by the tissue weighting factor for the lungs, 0.12.

The high incremental level associated with tobacco use (1 mrem/hr while smoking) results from the accumulation of ^{210}Po and ^{210}Pb (radon daughters that are alpha and gamma emitters) on tobacco leaves.

The incremental dose accrued by air travel is dependent on altitude, with higher levels associated with higher altitudes, and will range between 0.3 mrem/hr and 0.5 mrem/hr. The incremental dose due to medical imaging or radiation treatment can be misleading. For many individuals this dose can be close to zero (for example, if you do not have any x-ray imaging in a given year), but in the case of radiation treatment (such as for cancers), it can be much higher. Indeed, patients given certain radiation treatments (such as brachytherapy) become incremental sources themselves, resulting in incremental dosage above background to attending medical personnel, caregivers, family members, and the general public. The listing for clocks (in both Figures 2.15 and 2.16) is for older (even antique) clocks that have dials coated with Ra/ZnS paint to provide illumination.

Table 2.11 Average Contributions of Natural Background Radiation

Contribution	Average Dose, mrem/year, United States
Terrestrial contribution	21
Cosmic contribution	33
Airborne radon (and daughter) contribution	228
Internal consumption contribution	29
Total Natural Background	**311**

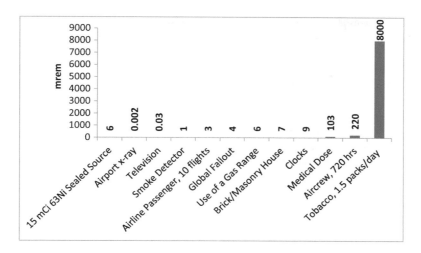

Figure 2.15 A comparison of increments to natural background levels, explicit for the 15 mCi ^{63}Ni sources used in electron capture detectors and ion mobility spectrometers. Taken from the book by Bruno and Svoronos.

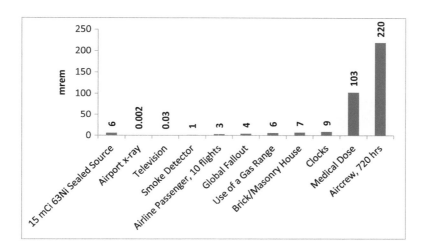

Figure 2.16 The same comparison as in Figure 2.1, but with tobacco use removed. Thus, the data shown above are all whole-body dosages instead of targeted lung-body dosages. Taken from the book by Bruno and Svoronos.

2.14 MAGNETIC FIELD SAFETY

In Chapter 6 we deal with nuclear magnetic resonance spectrometry and the high magnetic fields that are required to define the appropriate energy transitions. Most of the magnets that you will encounter in an NMR laboratory are superconducting, although some large electromagnets remain. The magnets used in NMR are very strong. It is common practice to rope off an area around the magnet where the field is about 5 gauss (0.0005T) and to stay outside of this area. It is very important to keep all metallic objects, like tools, as well as any magnetic records (including credit cards) away from the magnet. People with heart pacemakers should not go near the magnetic field. Modern actively shielded magnets are far less prone to field leakage; it is possible to place a credit card against a modern actively shielded magnet with no consequence. The use of steel tools near the magnet can be dangerous even with active shielding, however. In such magnets, there is still a powerful field below the magnet. Indeed, during probe change (which is done from the bottom of the magnet), a small wrench can be pulled from one's hand, especially if it is accidentally placed near the bottom of the magnet. Tools flying across the room is not unheard of. In Chapter 6 we discuss the sequence of events that may occur with a superconducting magnet quench. This is a spectacular event that is accompanied by a rapid release of the cryogenic liquid refrigerant in the magnet. As we have discussed above, this release is hazardous in that the vaporizing liquid helium and nitrogen will act as a simple asphyxiant to rapidly displace breathable air. An additional caution is that ferromagnetic objects such as tools can be drawn toward the core of the magnet during a quench.

2.15 MISCELLANEOUS HAZARDOUS CONDITIONS AND FIRE SAFETY

In this section we will discuss some general hazards that are not necessarily limited to the instrumentation laboratory, or any workplace. The maxim "if you see something, say something" applies here. Become familiar with the odor of fuel gas (natural gas and liquefied petroleum gas) in your area. For example, the odorant in natural gas, a mixture of t-butyl mercaptan (TBM) and other sulfur compounds, is described as having a rotten eggs odor. Well, not quite, rather it is an odor that is "gassy," characteristic of a natural gas leak. But the odorant mixture used in the northern climates is different than the odorant used in southern climates. The odorant used in Europe is typically a thiophene mixture that smells more skunky than gassy. Learn the difference if you are living in a new place; some utilities provide free *scratch and sniff* cards for familiarization. If the room that you enter smells like natural gas odorant, the prudent thing to do is to leave and inform the safety or security office. The same applies to strong burning smells and visible smoke. If any of these conditions are pervasive, widespread, or unusually strong (and always in the case of an accidental fire), then find a fire pull station and sound the alarm. Be advised, a fire pull station may require a good bit of force to activate; they are designed to avoid accidental activation (Figure 2.17).

Activation of the pull station will cause an audible and flashing light (strobe) alarm through the building by use of wall-mounted annunciators, and will also inform the safety and the security offices as well as the local fire department of the alert. Some fire annunciators have a public address loudspeaker that will allow specific information to be announced. Typical response will include police, an engine company and/or ladder company, as well as an ambulance. University police will typically meet responding fire crews and escort them to the location, but often local fire departments have firefighting plans already prepared for university buildings, especially buildings that house laboratories.

Evacuation must be done rapidly but not in panic. Many institutions will have designated assembly areas for evacuees to meet, and designated individuals to marshal evacuees. These designations differ in U.S. institutions, while in the U.K. they are typically known as fire stewards or fire wardens. It is critical that you assemble in these areas and remain there; the purpose is to facilitate a head-count. If students and staff members leave the area or wonder about, and a head-count comes up short, first responders must assume that these persons are left inside the facility. They will then enter the facility in an attempt to affect rescue, possibly endangering themselves in the process. The person who sounded the alarm will evacuate the building with everyone else, but it will be your responsibility to identify yourself to emergency personnel as the person who noticed the hazard and activated the alarm. Give them the details that you have and then let them do their jobs; entry will only be permitted when emergency personnel give the all-clear. **You will not get into trouble for summoning help when you believe a hazard exists.**

(a)

(b)

Figure 2.17 Photographs of a typical fire pull station (a) and emergency annunciator (b) equipped with a horn, strobe light, and public address loudspeaker.

*For the Instructor: Your organization standard operating procedure or safety training manual will make your responsibilities clear, but some general guidelines are important. In the event of an evacuation of the laboratory in response to an alarm or verbal alert, it will be your responsibility to ensure that all students and staff under your purview leave the laboratory. You will have to double check and triple check, by **counting heads**, that the room has been evacuated. It is also critical that private areas such as restrooms or lactation rooms be checked and evacuated. This may be the responsibility of the designated fire martial, fire warden, or safety officer. Once the evacuation is complete, you will **close the door** to the laboratory if that can be done safely. The door will be typically fire-rated, and will slow the spread of fire, smoke, or fumes. It will not provide this measure of protection if you do not use it, however. If you were unable to close the door because of a hazardous condition, or if you or another designated person were unable to check private areas, you must alert first responders of that fact.*

2.15.1 Fire Extinguishers

Near the door of the instrumentation laboratory (or in some cases in the corridor just outside the door), you will probably notice one or more fire extinguishers on hooks above the floor (at approximately hand level). In this section we will discuss the basics of fire extinguisher use, types, and guidelines to follow in the event of a laboratory fire. It is likely that the instructor will have been trained in the use of a fire extinguisher, so it is usually best that this be left to them. Unforeseen circumstances may intervene to make this your responsibility, however. We have deliberately placed this discussion to follow the earlier discussions about evacuations and alarms, because the use of a fire extinguisher should not even be considered unless the following conditions are met:

1. An alarm has been issued by use of a pull station or other means to summon first responders.
2. The laboratory room evacuation has been completed and verified by a head count.
3. The fire is small, involving a single device or location, and is not spreading.
4. The person considering use of a fire extinguisher is uninjured and is physically able to lift and aim the fire extinguisher (many are heavy and awkward).
5. The person considering the use of the fire extinguisher has a clear escape route and can avoid harmful fumes or smoke.

Noted above is that the fire extinguishers are located near the door of the laboratory; this location is intentional. It is only after the evacuation is completed, and emergency personnel notified, that a person may look back into the laboratory from the door to determine if the use of a fire extinguisher is feasible.

The types of fire extinguishers in use derive from the classification of the types of fires:

- Type A: Fire involving general combustibles such as paper, wood, cloth, etc.
- Type B: Fire involving solvents, flammable liquids such as gasoline, liquid and gaseous fuels, paints, etc.
- Type C: Fire involving active electrical transmission lines, motors, wiring.
- Type D: Fire involving burning metals such as sodium, lithium, magnesium, metallic nanomaterials.
- Type K: Fires involving animal and vegetable fats in commercial cooking facilities.

The most commonly encountered laboratory fires will be of the A, B, and C types, although with the exponential increase in research on nanoparticles, type D fires are not unknown. Following this classification of fires are the different types of common fire extinguishers, shown in Figure 2.18.

Pressurized water fire extinguishers can only be used on Type A fires involving paper, wood, or cloth. It is not likely that these fire extinguishers will be placed near laboratory doors, rather these are more often seen in office environments. The carbon dioxide fire extinguisher can be used on Types B and C fires; they should not be used on fires involving paper because the release of the pressurized carbon dioxide can blow burning materials away from the source of the fire to surrounding areas. The multipurpose dry chemical fire extinguisher, called ABC, can be used on fire types A, B, and C. The dry chemical in an ABC fire extinguisher is typically based on monoammonium phosphate (propelled with nitrogen gas), which, while not toxic, can be irritating. There is often a preference for the carbon dioxide fire extinguisher in the instrumentation laboratory because of the clean-up required (if not instrument damage) after the deployment of the ABC unit. An emergency is no time for hairsplitting; however, if an ABC is all that is available, use it. Operation of the fire extinguisher is depicted in Figure 2.19, showing the PASS method. *P*ull the pin, *a*im the hose or nozzle at the base of the fire, *s*queeze the trigger, and *s*weep side to side. Once a fire extinguisher is deployed, it must be removed from service and refilled or replaced, not returned to the wall hook.

We have not discussed the fire extinguishers that are required for Type D fires, involving burning metals. The student in analytical instrumentation is unlikely to encounter such a device, but we mention it here for completeness. Type D fires cannot be extinguished with any of the units shown in Figure 2.18; these fire extinguishers will make the situation worse. A special dry chemical based on graphite powder, sodium chloride, or copper is used to absorb heat from a metal fire. It must be understood that Type D fires are extremely dangerous and difficult to handle in the best of circumstances. Indeed, a metal fire at sea is typically handled by pushing the burning object(s) overboard with a deck dozer. The fire extinguishers for Type D fires can never be used on any other type of fire. We also mention type K fires for completeness, although this is clearly not relevant to the analytical instrumentation laboratory. Type K fires in commercial kitchens must be extinguished by wet chemical, or chemical foam fire extinguishers; as with Type D fires, use of other units will only make things worse.

We mentioned above that the fire extinguisher(s) may be located on a hook in the corridor. It is well to mention some aspects of a safe corridor in a laboratory setting. Corridors in a laboratory building must be free of obstructions such as desks, bookcases, cabinets, etc. This is critical for several reasons. First, in an emergency, they can be accidently overturned and impede evacuation, especially if the corridor is smoky due to a fire. Second, such obstructions will be problematic if fire crews must run a hose through the corridor. Hose lines can get hung up on desks and cabinets, making the response that much more difficult. The lesson here is that if you observe such obstructions in the corridor of the laboratory building,

Figure 2.18 A photograph showing a pressurized water, carbon dioxide, and ABC dry chemical fire extinguisher. Reproduced with permission of the University of South Florida.

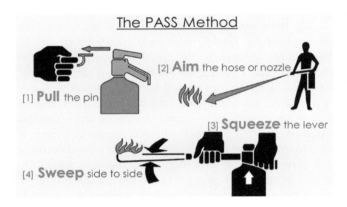

Figure 2.19 The PASS method of operating a fire extinguisher. Reproduced with permission of the National Fire Prevention Association (NFPA).

you must tell your instructor. These must be removed and relocated to a safer location.

2.15.2 Hot Work Permits

There are occasions in which laboratory work and routine repairs/maintenance to instrumentation will require soldering, brazing, or welding of parts. Ideally, these operations should be done in a designated shop area that is provided with proper engineered safety equipment, but sometimes performing such operations cannot be done remotely, but rather must be done in the laboratory. The use of these techniques will inevitably produce some smoke or fumes, and while proper ventilation should be provided for the work, it might not be possible to avoid triggering smoke alarms. In such instances, one can typically apply to the institution safety office or physical plant office for a hot work permit. Granting a hot work permit includes an intentional disabling of local smoke alarms in the area, and carries with it the requirement to post signage warning of the work in progress. It also requires the observation of a designated staff member as "fire watch." In the U.K., this person would be called the fire warden. The person on fire watch is responsible for observing the work and ensuring that no errant spark or flame contacts any flammable materials. Once the hot work is completed, the area should be ventilated and the smoke alarms activated.

2.16 STOP WORK ORDERS

We will end this chapter with a discussion that should rightly be uncomfortable, because it conveys a great deal of power and a great deal of responsibility on everybody. Throughout this chapter, we have discussed safety guidelines and provided advice on the safe conduct of work in the analytical instrumentation laboratory. Implied throughout this discussion is the responsibility of the individual to follow safety guidelines. We must now discuss our responsibility with respect to the conduct and behavior of others in the laboratory. In this context, others include classmates, instructors, staff members, teaching assistants, all the way up to Nobel laureate faculty, deans, university presidents, custodians, tradespeople, and outside contractors.

If you observe an unsafe activity or condition, it is your responsibility to speak up and prevent a potential accident. This can take the form of a gentle reminder: "don't pull on the wire, pull on the plug." It matters not at all who you must say this to. The plug-pulling example given seems almost trivial in its import, but it is nonetheless important.

If that wire becomes damaged by repeated pulling, you or your colleague might be the next person to touch it, only to get a nasty shock. Other situations might not be so seemingly trivial.

If you see another person, regardless of their position in the organization, moving an unrestrained gas cylinder with the regulator attached (or for that matter with the cap not in place), do not hesitate to act. You must clearly and firmly insist that such action cease immediately and be done properly. The same applies if you see someone standing on the top step of a ladder, or engaging in horseplay with a compressed air hose, etc.

Now comes the hard part. If Professor Knowitall tells you "don't worry, I know what I am doing," you must then call out firmly and clearly: STOP WORK. You have now issued a stop work order, which is something that anyone in an organization is legally authorized to do when confronted with a hazardous situation. A stop work order has meaning legally, although that meaning may differ by jurisdiction. [The meaning in U.S. government research facilities is nearly universal, while in other institutions and in the U.K., alternate designations are used.] The person engaged in the action must now cease; the hazard must be contained or mitigated until the event is investigated and the appropriate responsible party gives the all-clear. Failing to obey a stop work order is a violation of law in most jurisdictions. Issuance of a stop work order may require the safety officer be summoned to assess the incident. Your university or organization's safety program or standard operating procedure will make clear what procedure is to be followed. In almost every instance, the person issuing the stop work order must remain at the scene until the resolution of the incident. Have no doubt that this will be an uncomfortable situation, but you will have averted what could have become a very bad day.

We must not overlook the responsibility that comes with a stop work order; it is not something to be done without sufficient justification. You will require courage even bordering on cockiness, but if your intent is to earnestly prevent accident or injury among your fellows, you will come out OK. Be assured that incidents such as this will be very uncommon. One of the authors (TJB) was a research scientist for some 44 years (in university and government laboratories) and issued a stop work order only once; an electrical contractor was standing on the top step of a ladder in the corridor and refused requests to obtain a more suitable ladder. As another example of what not to do, earlier in this chapter we related the story of one of the authors (TJB) who assisted a professor in the (mis)use of a spectrophotometer as a ladder. Today, it would be incumbent to prevent such an action by all means possible including the issuance of a stop work order. Making this situation particularly difficult would have been that this professor was not only renowned in his field but also a favorite among all the students, and it would have indeed been uncomfortable to insist this revered person find a better way.

QUESTIONS

1. A fire pull station should be activated when:
    a. A large uncontrolled fire is observed in the laboratory
    b. A small uncontrolled fire is observed in the laboratory
    c. A strong smell of natural gas is noted
    d. All of the above
2. What are the conditions under which one may consider the use of a fire extinguisher in the laboratory?
3. Why are nanomaterials treated differently from ordinary reagents?
4. What is the action level (that is, the level at which a regulation requires corrective or remedial action) for noise exposure in the workplace? What can be done to minimize deleterious effects on hearing in a noisy environment?
5. All chemical laboratories are equipped with fire extinguishers. Which one of the four types of extinguishers most commonly used should not be found in a chemical laboratory?
    a. Water
    b. Carbon dioxide
    c. Dry chemical
    d. Type D
6. When operating a fire extinguisher, remember the mnemonic PASS. PASS represents the steps used to properly operate the extinguisher and it stands for which of the following?
    a. Pray, aim, sweep, pass
    b. Pull, aim, squeeze, sweep
    c. Pick, ask, squeeze, sweep
    d. Pull, aim, step, sweep
7. What does a bluish tint to the liquid in a Dewar flask indicate?
8. When in use, compressed gas cylinders should be
    a. Stored laying on their side
    b. Stored standing in the corner of the room
    c. Stored secured to a bench
9. If you want to decrease the pressure of gas being delivered by a compressed gas cylinder, you should
    a. Close the big valve on top
    b. Close the needle valve
    c. Turn the big knob on the regulator clockwise
    d. Turn the big knob on the regulator counterclockwise after bleeding off pressure
10. When working in a fume hood,
    a. Store your reagents near the instrument or equipment in the hood
    b. Close the sash at least as far as the safety marking on the hood
    c. Open the sash as far as is necessary for you to get your head in the hood
    d. Run electrical cords over the sash airfoil
11. List five of the most common laboratory instruments that require the most electrical power.
12. What electrical current is required to cause a painful shock?

13. What are the meanings of a "+" sign and a star stamped on a compressed gas cylinder?
14. What are the major hazards associated with the use of cryogenic fluids?
15. Laser pointers are always of low enough power to be safe:
    a. True
    b. False
16. Every laboratory radiation source is generally overseen by two people. How are these two people designated by the Nuclear Regulatory Commission in the United States?
17. What is the typical background radiation level in the United States?
18. While working in the NMR room, a loud hissing sound and a huge vapor cloud is noticed coming from the magnet. What is the appropriate action?
    a. Take out your phone and record this one in a lifetime event
    b. Normal boil-off of cryogen is occurring; go on about your work
    c. A magnet quench has occurred; leave the lab (and close the door if it is safe to do so).
    d. Put hearing protection on and go on about your work.
19. What are the three points of contact that must be maintained to minimize hazards associated with the use of ladders in the laboratory?
20. Now that I have successfully completed this list of questions, I am an expert on safety in the instrumentation laboratory:
    a. True
    b. False

DISCUSSION QUESTIONS

1. Discuss the importance of grounding in laboratory instrumentation electrical connections.
2. What is meant by turbulence and why is it a problem when using a laboratory fume hood? What can cause turbulence?
3. Discuss the relative magnitude of the various ionizing radiation hazards commonly encountered in the instrumentation laboratory, and how these hazards are related to the normal background radiation.
4. This chapter discussed hazards specific to the instrumentation laboratory. There are other hazards associated with the handling of chemicals, samples and reagents that have been covered in your general and organic chemistry classes. Now is a good time to review all of these guidelines and discuss them.
5. It is clear that safety in the laboratory is a team effort. Discuss how each of us has a role in ensuring that the work and learning is done in safety and everyone goes home in one piece.

REFERENCES

The reader will note that occasionally multiple editions of books are cited as sources or reference works. The reason for this is that sometimes different figures are used in different editions which may in fact be written by different authors. The sources found in the bibliography will be the most recent versions of a particular work.

In some cases instrument manufactures and supplies are cited by providing a URL to the exact internet location. We have endeavored to ensure that these links are functional at the time of publication, but clearly these links change. If a link has become nonfunctional, a good approach is to your browser's advanced search capability For example: "company name " AND "instrument type AND "specific descriptors" is a good place to start.

BIBLIOGRAPHY

1. Bruno, T.J., Svoronos, P.D.N., *CRC Handbook of Basic Tables for Chemical Analysis – Data Driven Methods and Interpretation*, 4th Ed., CRC Taylor and Francis, Boca Raton, FL, 2021.
2. Furr, A.K., *CRC Handbook of Laboratory Safety*, 5th Ed., CRC Taylor and Francis, Boca Raton, FL, 2000.
3. National Fire Protection Association, www.nfpa.org.
4. Nuclear Magnetic Resonance Safety Tips, www2.lbl.gov/ehs/safety/nir/assets/docs/NMRsafetytips.
5. Compressed Gas Association, www.cganet.com.
6. British compressed Gas Association, https://bcga.co.uk.

CHAPTER 3

Introduction to Spectroscopy

3.1 INTERACTION BETWEEN ELECTROMAGNETIC RADIATION AND MATTER

We know from our observation of rainbows that *visible light* (white light) is composed of a **continuum** of colors from violet to red. If a beam of white light is passed through a beaker of water, it remains white. If potassium permanganate is added to the water, the white light appears purple after it passes through the solution. The permanganate solution allows the red and blue components of white light to pass through but absorbs the other colors from the original beam of light. This is one example of the interaction of **electromagnetic radiation**, or light, with *matter*. In this case, the electromagnetic radiation is visible light and we can see the effect of absorption of some of the light with our eyes. However, interactions between electromagnetic radiation and matter take place in many ways and over a wide range of *radiant energies*. Most of these interactions are not visible to the human eye but can be measured with suitable instruments.

The interaction of electromagnetic radiation and matter is not haphazard but follows well-documented rules with respect to the wavelengths of light absorbed or emitted and the extent of absorption or emission. The subject of **spectroscopy** is the study of the interaction of electromagnetic radiation and matter.

3.1.1 What Is Electromagnetic Radiation?

The nature of electromagnetic radiation baffled scientists for many years. At times, light appears to behave like a wave; at other times, it behaves as though it were composed of small particles. While we now understand the "wave–particle duality" of all matter, including electromagnetic radiation, in terms of quantum mechanics, it is still convenient to consider electromagnetic radiation as having the properties of waves in many cases.

Light waves can be represented as oscillating perpendicular electric and magnetic fields. The fields are at right angles to each other and to the direction of propagation of the light. The oscillations are sinusoidal in shape, as shown in Figure 3.1. We can easily and accurately measure the **wavelength** λ defined as the crest-to-crest distance between two successive maxima. The standard unit of wavelength is the SI unit of length, the meter (m), but smaller units such as centimeter (cm), micrometer (μm), and nanometer (nm) are commonly used. The **amplitude** of the wave is defined as the maximum of the vector from the origin to a point displacement of the oscillation. An example of the electric field portion of a light wave propagating along only one axis is shown in Figure 3.1. Such a wave, confined to one plane, is called **plane-polarized light**. The wave shown represents only a single wavelength, λ. Light of only one wavelength is called **monochromatic light**. Light that consists of more than one wavelength is called **polychromatic** light. White light is an example of polychromatic light.

The **frequency** v of a wave is the number of crests passing a fixed point per second. One crest-to-crest oscillation of a wave is called a cycle. The common unit of frequency is the hertz (Hz) or inverse second (s^{-1}); an older term for frequency is the cycle per second (cps). One hertz equals one cycle per second.

The wavelength of light, λ, is related to its frequency, v, by the following equation:

$$c = \lambda v \tag{3.1}$$

where

c is the speed of light in a vacuum, 2.997×10^8 m/s
v is the frequency of the light in inverse seconds (Hz)
λ is the wavelength (m)

In a vacuum, the speed of light is a maximum and does not depend on the wavelength. The frequency of light is determined by the source and does not vary. When light passes through a material other than a vacuum, its speed is decreased. Because the frequency cannot change, the wavelength must decrease. If we calculate the speed of light in air, it only differs by a very small amount from the speed of light in vacuum. In general, we use 3.00×10^8 m/s (to three significant figures) for the speed of light in air or vacuum.

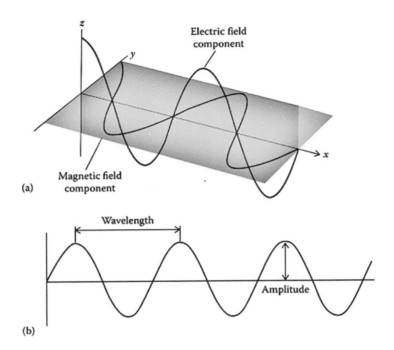

Figure 3.1 (a) A plane-polarized light wave in the x-direction, showing the mutually perpendicular electric and magnetic field components. (b) The wavelength and amplitude of a wave.

In some cases, it is more convenient to consider light as a stream of particles. Particles of light are called **photons**. Photons are characterized by their energy, E. The energy of a photon is related to the frequency of light by the following equation:

$$E = h\nu \quad (3.2)$$

where
E is the energy in joules (J)
h is Planck's constant, 6.626×10^{-34} J s
ν is the frequency in inverse seconds (Hz)
From Equations 3.1 and 3.2, we can deduce that

$$E = \frac{hc}{\lambda} \quad (3.3)$$

We can see from the relationship in Equations 3.2 and 3.3 that the energy of electromagnetic radiation is directly proportional to its frequency and inversely proportional to its wavelength. Electromagnetic radiation ranges from very low-energy (long-wavelength, low-frequency) radiation, like radiowaves and microwaves, to very high-energy (short-wavelength, high-frequency) radiation, like X-rays. The major regions of the electromagnetic spectrum of interest to us as analytical chemists are shown in Figure 3.2. It is clear from this figure that visible light, that portion of the electromagnetic spectrum to which the human eye responds, is only a very small portion of all radiant energy. Table 3.1 presents some common units and symbols for various types of electromagnetic radiation. Note that although still in common use, the Angstrom is not an accepted SI unit any longer.

3.1.2 How Does Electromagnetic Radiation Interact with Matter?

Spectroscopy is the study of the interaction of radiant energy (light) with matter. We know from quantum mechanics that energy is really just a form of matter and that all matter exhibits the properties of both waves and particles. However, matter composed of molecules, atoms, or ions, which exists as solid or liquid or gas, exhibits primarily the properties of particles. Spectroscopy studies the interaction of light with matter defined as materials composed of molecules or atoms or ions.

In a gas, atoms or molecules are widely separated from each other; in liquids and solids, the atoms or molecules are closely associated. In solids, the atoms or molecules may be arranged in a highly ordered array, called a **crystal**, as they are in many minerals, or they may be randomly arranged, or **amorphous**, as they are in many plastics. Atoms, molecules, and ions are in constant motion whatever their physical state or arrangement. For molecules, many types of motion are involved. Molecules can rotate, vibrate, and translate (move from place to place in space). Interaction with radiant energy can affect these molecular motions. Molecules that absorb IR radiation vibrate with greater amplitude; interaction with UV/VIS light can move

INTRODUCTION TO SPECTROSCOPY

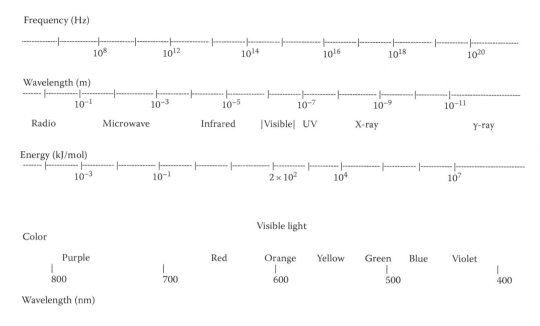

Figure 3.2 The electromagnetic spectrum. The visible light region is expanded to show the colors associated with wavelength ranges.

Table 3.1 Common wavelength symbols and units for electromagnetic radiation

Unit	Symbol	Length (m)	Type of Radiation
Angstrom	Å	10^{-10}	X-ray
Nanometer	nm	10^{-9}	Ultraviolet/visible (UV/VIS)
Micrometer	μm	10^{-6}	Infrared (IR)
Millimeter	mm	10^{-3}	IR
Centimeter	cm	10^{-2}	Microwave
Meter	m	1	Radio

Table 3.2 Some types of transitions studied by spectroscopy

Type of Transition	Spectroscopic Method	Wavelength Range
Spin of nuclei in a magnetic field	Nuclear magnetic resonance (NMR) spectroscopy	0.5–10 m
Rotation of molecules	Microwave spectroscopy	4–25 mm
Vibration of molecules	Raman and IR spectroscopy	0.8–300 μm
Bonding electron energy, valence electron energy	UV/VIS spectroscopy	180–800 nm
Core electron energy	X-ray spectroscopy	0.1–100 Å

Note: This is a very limited list of the types of transitions and spectroscopic methods in current use.

bonding electrons to higher energy levels in molecules. A change in any form of motion or electron energy level involves a change in the energy of the molecule. Such a change in energy is called a **transition**; we have the possibility of vibrational transitions, rotational transitions, and electronic transitions in molecules. We have some of the same kinds of motion in atoms and ions. Atoms can move in space, and their electrons can move between energy levels, but atoms and monoatomic ions cannot rotate or vibrate. The chemical nature of matter (its composition), its physical state, and the arrangement of the atoms or molecules in the physical state with respect to each other affect the way in which any given material interacts with electromagnetic radiation. Table 3.2 lists some of the important types of transitions studied by spectroscopy. We will cover these techniques in detail in later chapters. There are literally hundreds of types of transitions and types of spectroscopy used to investigate matter. Only the most common types of analytical spectroscopy will be covered in this text.

When light strikes a sample of matter, the light may be absorbed by the sample, transmitted through the sample, reflected off the surface of the sample, or scattered by the sample. Samples can also emit light after absorbing incident light; such a process is called **luminescence**. There are different kinds of luminescence, called **fluorescence** or **phosphorescence**, depending on the specific process that occurs; these are discussed in detail in Chapter 4. Emission of light may also be caused by processes other than absorption of light. There are spectroscopic methods based on all of these interactions. Table 3.3 summarizes the major types of interaction of light with matter and gives examples of the common spectroscopic techniques based on these interactions. For the moment, we will focus on the absorption, transmission, and emission of light by matter.

If we pass white light (i.e., visible light) through blue glass, the emerging light is blue. The glass has absorbed the

Table 3.3 Some interactions of light and matter

Interaction	Radiation Measured	Spectroscopic Method
Absorption and transmission	Incident light, I_o Transmitted light, I	Atomic absorption Molecular absorption
Absorption then emission	Emitted light, I'	Atomic fluorescence Molecular fluorescence Molecular phosphorescence
Scattering	Scattered light, I_s	Turbidimetry Nephelometry Raman
Reflection	Reflected light I_R or relative reflected I_R	Attenuated total reflection Diffuse reflection IR (the term *reflectance* is also used for these methods)
Emission	Emitted light, I_e	Atomic emission Molecular emission Chemiluminescence

other colors, such as red and yellow. We can confirm this absorption by shining red light through the blue glass. If the absorption is strong enough, all of the red light is absorbed; no light emerges from the glass and it appears black. How can this be explained?

The interaction of electromagnetic radiation and matter conforms to well-established quantum mechanical laws. Atoms, ions, and molecules exist only in certain discrete states with specific energies. The same quantum mechanical laws dictate that a change in state requires the absorption or emission of energy, ΔE, exactly equal to the difference in energy between the initial and final states. We say that the energy states are quantized. A change in state (change in energy) can be expressed as

$$\Delta E = E_{final} - E_{initial} = h\nu \quad (3.4)$$

Since we know that $c = \lambda\nu$, then

$$\Delta E = h\nu = \frac{hc}{\lambda} \quad (3.5)$$

These equations tell us that matter can absorb or emit radiation when a transition between two states occurs, but it can absorb or emit only the specific frequencies or wavelengths that correspond to the exact difference in energy between two states in which the matter can exist. Absorption of radiation increases the energy of the absorbing species ($E_{final} > E_{initial}$). Emission of radiation decreases the energy of the emitting species ($E_{final} < E_{initial}$). So the quantity ΔE can have either a positive sign or a negative sign, but when using ΔE to find the wavelength or the frequency of radiation involved in a transition, only the absolute value of ΔE is used. Wavelength, frequency, and the speed of light are always positive in sign.

A specific molecule, such as hexane, or a specific atom, such as mercury, can absorb or emit only certain frequencies of radiation. All hexane molecules will absorb and emit light with the same frequencies, but these frequencies will differ from those absorbed or emitted by a different molecule, such as benzene. All mercury atoms will absorb the same frequencies of incident light, but these will differ from the frequencies of light absorbed by atoms of lead or copper. Not only are the frequencies unique, but also the degree to which the frequency is absorbed or emitted is unique to a species. This degree of absorption or emission results in light of a given **intensity**. The uniqueness of the frequencies and amount of each frequency absorbed and emitted by a given chemical species are the basis for the use of spectroscopy for identification of chemicals. We call the set of frequencies and the associated intensities at which a species absorbs or emits its **spectrum**. In some cases, species will also emit radiation; we will encounter many examples when we discuss fluorescence spectroscopy.

The lowest energy state of a molecule or atom is called the **ground state**. All higher energy states are called **excited states**. Generally at room temperature, molecules and atoms exist in the ground state.

If we think about our example of blue glass and its ability to absorb red and yellow light, we can deduce a simple picture of the energy states in the blue glass. We will assume that the glass is in its ground state before we shine any light through it since we have performed this experiment at room temperature. We will call the ground-state energy E_1. If the glass is capable of absorbing red light, there must be an excited state such that the difference in energy between the ground state and this excited state is equivalent to the energy of a wavelength of red light. If we look at Figure 3.2, we can choose a representative wavelength in the red region of the visible spectrum, such as 653 nm. If a wavelength $\lambda = 653$ nm is absorbed by the glass, we can calculate the frequency of this light by rearranging Equation 3.1:

$$\nu = \frac{c}{\lambda} = \frac{2.997 \times 10^8 \text{ m/s}}{(653 \text{ nm}) \times (10^{-9} \text{ m/nm})}$$

$$\nu = 4.59 \times 10^{14} \text{ s}^{-1}$$

From the frequency, we are able to calculate the difference in energy between the ground state and this excited state, which we will call E_2:

$$\Delta E = E_2 - E_1 = h\nu$$
$$\Delta E = (6.626 \times 10^{-34} \text{ J S})(4.59 \times 10^{14} \text{ s}^{-1})$$
$$\Delta E = 3.05 \times 10^{-19} \text{ J}$$

INTRODUCTION TO SPECTROSCOPY

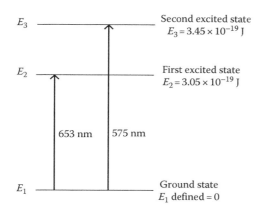

Figure 3.3 Simplified energy diagram for the absorption of visible light by blue glass. Two possible excited energy states are shown: one corresponding to the absorption of 653 nm (red) light and a higher state corresponding to the absorption of 575 nm (yellow) light. If the ground-state energy is defined as $E = 0$, the relative energies of the excited states can be determined.

So, there is one excited state with an energy that is 3.05×10^{-19} J higher than the ground state in the glass. We do not know the exact energy of the ground state itself. The glass also absorbs yellow light, so we can pick a representative wavelength of yellow light, such as 575 nm, and repeat the preceding calculation. The frequency of light corresponding to a wavelength of 575 nm is 5.21×10^{14} Hz, so there must be an excited state E_3 such that

$$\Delta E = E_3 - E_1 = h\nu$$
$$\Delta E = (6.626 \times 10^{-34} \text{ J s})(5.21 \times 10^{14} \text{ s}^{-1})$$
$$\Delta E = 3.45 \times 10^{-19} \text{ J}$$

We can now construct a simplified energy diagram for the blue glass, such as the one shown in Figure 3.3.

This diagram shows two excited states: one corresponding to the ability of the glass to absorb light with a wavelength of 653 nm and one corresponding to the ability of the glass to absorb light with a wavelength of 575 nm. Because the glass does not absorb blue light (it transmits the blue light portion of white light), there would be no energy states with a difference in energy equal to any frequency of blue light. This diagram is very oversimplified. "Red," "yellow," and "blue" light span a range of wavelengths, as can be seen from Figure 3.2. There are actually many different energy levels associated with the transitions occurring in glass. Absorption of red light occurs from 620 to 750 nm, yellow light from 550 to 595 nm, and so on. So what is the molecular reason for this "broadband" absorption observed in spectroscopic experiments with visible light? The absorption of visible light by glass is due to excitation of bonding electrons in the molecules or, in other words, due to electronic transitions. Electronic transitions require more energy than rotational or vibrational transitions. For a molecule, the relative energy of transitions is rotational < vibrational < electronic. A more realistic energy-level diagram for glass (and for molecules in general) is presented in Figure 3.4. For every electronic state E_n, there are many associated rotational and vibrational sublevels. Each sublevel has a slightly different energy, with the result that a transition from one energy level E_n to a higher energy level is not a single energy but a range of closely spaced energies, because the electron can end up in any one of the many sublevels. For this reason, absorption of red light by molecules occurs over a range of wavelengths, not at a single discrete wavelength.

Excited states are energetically unfavorable; the molecule or atom wants to return to the lowest-energy ground state by giving up energy, often by emitting light. Because they are energetically unfavorable, excited states are usually short-lived, on the order of 10^{-9} to 10^{-6} s. Emission of light therefore occurs rapidly following excitation. One notable exception is the process of phosphorescence, described in Chapter 4; for this process, the excited state lifetime can be as long as tens of seconds.

Absorption spectra are obtained when a molecule or atom absorbs radiant energy that satisfies the equation $\Delta E = h\nu = hc/\lambda$. The **absorption spectrum** for a substance shows us the energies (frequencies or wavelengths) of light absorbed as well as how much light is absorbed at each frequency or wavelength. The nature of the molecule or atom dictates the amount of light absorbed at a given energy. The complete spectrum of a substance consists of the set of energies absorbed and the corresponding intensity of light absorbed. A graph of the intensity of light amplitude change on the y-axis versus the frequency or wavelength on the x-axis is constructed. It is this graph of intensity versus energy that we call a spectrum. The IR absorption spectrum for polystyrene is shown in Figure 5.1. The UV absorption spectrum of benzene is shown in Figure 4.1.

Emission spectra are obtained when an atom or molecule in an excited state returns the ground state by emitting radiant energy. An emission spectrum can result from many different ways of forming an excited state. Atoms and molecules can be excited not only by absorption of electromagnetic radiation but also by transfer of energy due to collisions between atoms and molecules, by addition of thermal energy, and by addition of energy from electrical discharges. Different excitation methods are used in several types of emission spectroscopy and will be discussed in detail in later chapters. A special term is used for the emission of electromagnetic radiation by either atoms or molecules following excitation by absorption of electromagnetic radiation. Such emission is called luminescence. In other words, if light is used as the source of excitation energy, the emission of light is called luminescence; if other excitation sources are used, the emission of light is called simply emission.

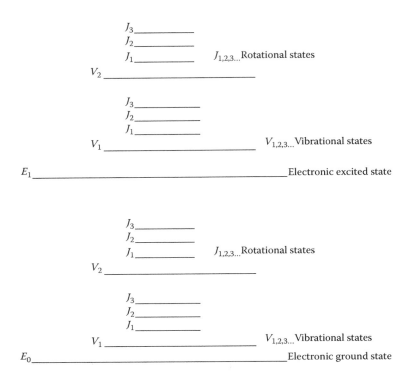

Figure 3.4 Schematic energy-level diagram for molecules. Each electronic energy level $E_{0,1,2,...}$ has associated vibrational sublevels $V_{1,2,3,...}$ and rotational sublevels $J_{1,2,3,...}$.

3.2 ATOMS AND ATOMIC SPECTROSCOPY

An atom consists of a nucleus surrounded by electrons. Every element has a unique number of electrons equal to its atomic number for a neutral atom of that element. The electrons are located in atomic orbitals of various types and energies, and the electronic energy states of atoms are quantized. The lowest-energy, most stable electron configuration of an element is its ground state. The ground state is the normal electron configuration predicted from the "rules" for filling a many-electron atom, which you learned in general chemistry. These rules are based on the location of the atom in the periodic table, the *aufbau* principle, the Pauli exclusion principle, and Hund's rule. (You may want to review your general chemistry text on the structure of the atom.) For example, the ground-state electron configuration for sodium, atomic number 11, is $1s^2 2s^2 2p^6 3s^1$ based on its position in the third row, first group of the periodic table, and the requirement to account for 11 electrons. The ground-state electronic configuration for potassium is $1s^2 2s^2 2p^6 3s^2 3p^6 4s^1$, vanadium is $1s^2 2s^2 2p^6 3s^2 3p^6 4s^2 3d^3$, and so on. If energy of the right magnitude is provided to an atom, the energy may be absorbed and an outer (valence) electron promoted from the ground-state orbital it is in to a higher-energy orbital. The atom is now in a higher-energy, less-stable, excited state. Because the excited state is less stable than the ground state, the electron will return spontaneously to the ground state. In the process, the atom will emit energy; this energy will be equivalent in magnitude to the difference in energy levels between the ground and excited states (and equivalent to the energy absorbed initially). The process is shown schematically in Figure 3.5. If the emitted energy is in the form of electromagnetic radiation, Equations 3.4 and 3.5 directly relate the wavelength of radiation absorbed or emitted to the electronic transition that has occurred:

$$\Delta E = E_{final} - E_{initial} = h\nu = \frac{hc}{\lambda} \quad (3.6)$$

Each element has a unique set of permitted electronic energy levels because of its unique electronic structure. The wavelengths of light absorbed or emitted by atoms of an element are characteristic of that element. The absorption of radiant energy by atoms forms the basis of atomic absorption spectrometry (AAS) discussed in Chapter 7. The absorption of energy and the subsequent emission of radiant energy by excited atoms form the basis of atomic emission spectroscopy and atomic fluorescence spectroscopy, discussed in Chapter 8.

In practice, the actual energy-level diagram for an atom is derived from the emission spectrum of the excited atom. Figure 3.6 shows an energy-level diagram for mercury atoms. Notice that there are no rotational or

INTRODUCTION TO SPECTROSCOPY

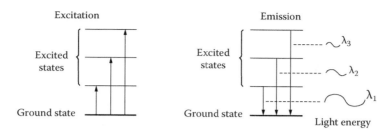

Figure 3.5 Energy transitions in atoms. Atoms may absorb energy and move a ground-state valence electron to higher-energy excited states. The excited atom may relax back to the ground state by emitting light of a wavelength equal to the difference in energy between the states. Three such emissions are shown. [© 1993–2014 PerkinElmer, Inc. All rights reserved. Printed with permission. (www.perkinelmer.com).]

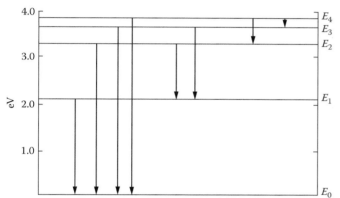

Figure 3.6 Energy levels in mercury atoms (units of energy are electron volts [eV]).

vibrational sublevels in atoms! A free gas phase atom has no rotational or vibrational energy associated with it. When an electron is promoted to a higher atomic excited state, the change in energy is very well defined, and the wavelength absorbed (or emitted on **relaxation** to the ground state) can be considered monochromatic. The wavelengths of light involved in valence electronic transitions in atoms fall in the visible and UV regions of the spectrum. This region is often called the UV/VIS region for short. The energy-level diagrams for all elements have been determined, and the tables of wavelengths absorbed and emitted by atoms are available. Appendix 6.A lists the absorption wavelengths used to measure elements by AAS, discussed in Chapter 7.

Knowing what wavelengths of light are absorbed or emitted by a sample permits qualitative identification of the elements present in the sample. Measuring the intensity of light absorbed or emitted at a given wavelength gives us information about how much of a given element is present (quantitative elemental analysis). All of the atomic spectroscopy methods—absorption, fluorescence, and emission—are extremely sensitive. As little as 10^{-12} to 10^{-15} g of an element may be detected using atomic spectroscopy.

It is possible for atoms to absorb higher-energy radiation, in the X-ray region; such absorption may result in the inner shell (core) electrons being promoted to an excited state, with the subsequent emission of X-ray radiation. This process forms the basics for qualitative and quantitative elemental analysis by X-ray fluorescence (XRF) spectroscopy, as well as other X-ray techniques, discussed in Chapter 9.

3.3 MOLECULES AND MOLECULAR SPECTROSCOPY

The energy states associated with molecules, like those of atoms, are also quantized. There are very powerful spectroscopic methods for studying transitions between permitted states in molecules using radiation from the radiowave region to the UV region. These methods provide qualitative and quantitative information about molecules, including detailed information about molecular structure.

3.3.1 Rotational Transitions in Molecules

The ability of a molecule to rotate in space has associated rotational energy. Molecules may exist in only discrete (quantized) rotational energy states. Absorption of the appropriate energy causes transitions from lower-energy rotational states to higher-energy rotational states, in which the molecule rotates faster. This process gives rise to rotational absorption spectra. The rotational energy of a molecule depends on its angular velocity, which is variable. Rotational energy also depends on the molecule's shape and mass distribution, which change as bond angles change. While a change in shape is restricted in diatomic molecules such as O_2, molecules with more than two atoms, such as hexane, C_6H_{14}, have many possible shapes and therefore many possible rotational energy levels. Furthermore, the presence of more than one natural isotope of an atom in a molecule generates new sets of rotational energy levels. Such is the case with carbon, where a small percentage of the carbon atoms in a carbon-containing molecule are ^{13}C instead of ^{12}C. Consequently, even simple molecules have complex rotational absorption spectra. The energies involved in rotational changes are very small, on the order of 10^{-24} J per molecule. The radiation absorbed is therefore in the radiofrequency (RF) and microwave regions of the spectrum. Microwave spectroscopy has been largely unexploited in analytical chemistry because of the experimental difficulties involved and the complexity of the spectra produced. The technique is limited to the gas phase and has been used by radioastronomers to detect the chemical species in interstellar clouds.

3.3.2 Vibrational Transitions in Molecules

For the purposes of basic understanding of this branch of optical spectroscopy, molecules can be visualized as a set of weights (the atoms) joined together by springs (the chemical bonds). The atoms can vibrate toward and away from each other, or they may bend at various angles to each other, as shown in Figure 3.7. Each such vibration has characteristic energy associated with it. The vibrational energy states associated with molecular vibration are quantized. Changes in the vibrational energy of a molecule are associated with absorption of radiant energy in the IR region of the spectrum. While absorption of IR radiation causes changes in the vibrations of the absorbing molecule, the increase in vibrational energy is also usually accompanied by increased molecular rotation. Remember, the rotational energy levels are sublevels of the vibrational energy levels, as we saw in Figure 3.4. So, in practice, absorption of IR radiation corresponds to a combination of changes in rotational and vibrational energies in the molecule. Because a molecule with more than two atoms has many possible vibrational states, IR absorption spectra are complex, consisting of multiple absorption bands. Absorption of IR radiation by molecules is one of the most important techniques in spectroscopy.

Through IR absorption spectroscopy, the structure of molecules can be deduced, and both qualitative identification of molecules and quantitative analysis of the molecular composition of samples can be performed. IR spectroscopy is discussed in Chapter 5.

3.3.3 Electronic Transitions in Molecules

When atoms combine to form molecules, the individual atomic orbitals combine to form a new set of molecular orbitals. Molecular orbitals with electron density in the plane of the bonded nuclei, along the axis connecting the bonded nuclei, are called sigma (σ) orbitals. Those molecular orbitals with electron density above and below the plane of the bonded nuclei are called pi (π) orbitals. Sigma and pi orbitals may be of two types: bonding orbitals or antibonding orbitals. Bonding orbitals are lower in energy than the corresponding antibonding orbitals. When assigning electrons in molecules to orbitals, the lowest-energy bonding orbitals are filled first. For a review of molecular orbital theory, see your general chemistry text.

Under normal conditions of temperature and pressure, the electrons in the molecule are in the ground-state configuration, filling the lowest-energy molecular orbitals available. Absorption of the appropriate radiant energy may cause an outer electron to be promoted to a higher-energy excited state. As was the case with atoms, the radiant energy required to cause electronic transitions in molecules lies in the visible and UV regions. And as with atoms, the excited state of a molecule is less stable than the ground state. The molecule will spontaneously revert (relax) to the ground-state emitting UV/VIS radiant energy. Unlike atoms, the energy states in molecules have rotational and vibrational sublevels, so when a molecule is excited electronically, there is often a simultaneous change in the vibrational and rotational energies. The total energy change is the sum of the electronic, rotational, and vibrational energy changes. Because molecules possess many possible rotational and vibrational states, absorption of UV/VIS radiation by a large population of molecules, each in a slightly different state of rotation and vibration, results in absorption over a wide range of wavelengths, called an **absorption band**. The UV/VIS absorption spectra of molecules usually have a few broad absorption bands and are usually very simple in comparison with IR spectra. Molecular absorption and emission spectroscopy can be used for qualitative identification of chemical species. This technique used to be one of the major methods for structural determination of organic molecules but has been replaced by the more powerful and now commonly available techniques of NMR, IR spectroscopy, and mass spectrometry (MS). UV/VIS absorption spectroscopy is most often used for quantitative analysis of the molecular composition of samples. Molecular fluorescence spectroscopy is an extremely high-sensitivity method, with the

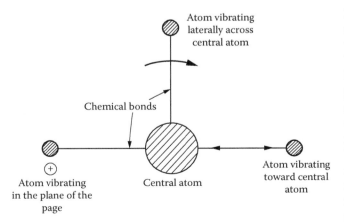

Figure 3.7 Some possible vibrations of bonded atoms in a molecule.

INTRODUCTION TO SPECTROSCOPY

ability to detect single molecules! We will learn the laws governing absorption, which permit quantitative analysis by UV/VIS spectroscopy, in this chapter. The use of UV/VIS molecular spectroscopy will be discussed at greater length in Chapter 4.

3.4 ABSORPTION LAWS

The **radiant power** P of a beam of light is defined as the energy of the beam per second per unit area. A related quantity is the **intensity** I, the power per unit solid angle. Both power and intensity are related to the square of the amplitude of the light wave, and the absorption laws can be written in terms of either power or intensity. We will use intensity I but you may see the same laws written with a P for power in other literature.

When light passes through an absorbing sample, the intensity of the light emerging from the sample is decreased. Assume the intensity of a beam of monochromatic (i.e., single wavelength) radiation is I_0. This beam is passed through a sample that can absorb radiation of this wavelength, as shown in Figure 3.8. The emerging light beam has an intensity equal to I, where $I_0 \geq I$. If no radiation is absorbed by the sample, $I = I_0$. If any amount of radiation is absorbed, $I < I_0$. The **transmittance** T is defined as the ratio of I to I_0:

$$T = \frac{I}{I_0} \qquad (3.7)$$

The transmittance is the fraction of the original light that passes through the sample. Therefore, the range of allowed values for T is from 0 to 1. The ratio I/I_0 remains relatively constant even if I_0 changes; hence, T is independent of the actual intensity I_0. To study the quantitative absorption of radiation by samples, it is useful to define another quantity, the **absorbance** A where

$$A = \log\left(\frac{I_0}{I}\right) = \log\left(\frac{1}{T}\right) = -\log T \qquad (3.8)$$

When no light is absorbed, $I = I_0$ and $A = 0$. Two related quantities are also used in spectroscopy, the **percent transmittance**, $\%T$, which equals $T \times 100$, and the **percent absorption**, $\%A$, which is equal to $100 - \%T$.

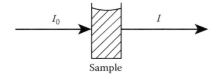

Figure 3.8 Absorption of radiation by a sample.

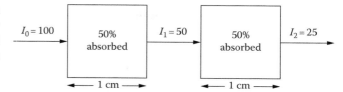

Figure 3.9 Absorption of radiation by two identical sample cells.

Suppose we have a sample of an aqueous solution of an absorbing substance in a rectangular glass sample holder with a length of 1.0 cm, as shown in Figure 3.9. Such a sample holder is called a **sample cell**, or **cuvette**, and the length of the cell is called the **path length** b. The incident light has intensity, I_0, equal to 100 intensity units. If 50 % of the light passing through the sample is absorbed, then 50 % of the light is transmitted. The emerging light beam has an intensity denoted as $I_1 = 50$ intensity units. So the $\%T = 50$, and therefore,

$$T = \frac{\%T}{100} = \frac{I_1}{I_0}$$

$$T = \frac{50}{100} = 0.50$$

From T, we calculate that absorbance equals

$$A = -\log T = -\log(0.50) = 0.30$$

If a second identical cell with the same solution is placed in the path of beam I_1, 50 % of the incident radiation, I_1, will be absorbed, and we have a new emerging beam, I_2. The intensity of I_1 is 50 intensity units, and therefore, I_2 must be 25 intensity units. So the transmittance for the second cell is

$$T = \frac{I_2}{I_1} = \frac{25}{50} = 0.50$$

The absorbance for just the second cell is $A = -\log 0.50 = 0.30$. The two cells are identical in their absorbance of light. Identical or "optically matched" cells are required for accurate quantitative analysis using spectroscopy in many cases.

Now suppose we put the two cells back to back and consider them together. For our purpose, we will assume that these will behave as if there were no glass walls between the two cells; in other words, we have one "cell" that is 2.0 cm long. The path length for this experiment is now 2.0 cm. The incident light beam has intensity I_0, with $I_0 = 100$ intensity units. We know from passing light through the two cells (as described earlier) that the emerging light beam has intensity $I = 25$ intensity units. So we see that for a path length of 2.0 cm, T now equals 25/100 or 0.25 and

$A = -\log 0.25 = 0.60$. If we put three cells in line (path length = 3.0 cm), the emerging beam has $I = 12.5$, $T = 0.125$, and $A = 0.90$. Four cells in line will give $I = 6.25$, $T = 0.063$, and $A = 1.20$. If we plot intensity I versus the number of cells (i.e., the path length in cm) as shown in Figure 3.10a, it is clear that I decreases exponentially with increasing path length. If we plot absorbance A versus path length, shown in Figure 3.10b, absorbance increases linearly with increasing path length. As analytical chemists, we find it better to work with linear equations rather than with exponential equations, because as you can see from Figure 3.10, it is easier to interpolate and read data from a linear plot. A linear plot has a constant slope that greatly simplifies calculations. That is why the absorbance is such a useful quantity—it results in a linear relationship with quantities important to analytical chemists. This proportional relationship between sample thickness (the path length) and absorbance at constant concentration was discovered by P. Bouguer in 1729 and J. Lambert in 1760.

If we perform a similar experiment keeping the path length constant by using only one cell but change the **concentration** of the absorbing species, we find the same relationship between I, A, and concentration as we found for path length. A linear relationship exists between the absorbance A and the concentration c of the absorbing species in the sample, a very important quantity! Because A is linear with respect to path length b and concentration c, we can write the following equation:

$$A = abc \qquad (3.9)$$

The term "a" is a proportionality constant called the **absorptivity**. The absorptivity is a measure of the ability of the absorbing species in the sample to absorb light at the particular wavelength used. Absorptivity is a constant for a given chemical species at a specific wavelength. If the concentration is expressed in molarity (mol/L or M), then the absorptivity is called the **molar absorptivity** and is given

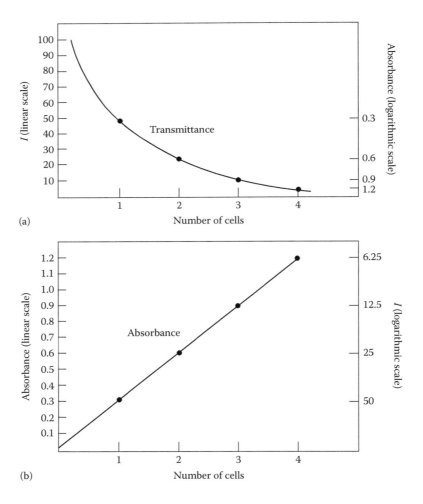

Figure 3.10 (a) Exponential relationship between intensity (or transmittance) and increasing number of cells (increasing path length). (b) L near relationship between absorbance and increasing number of cells (increasing path length). $I_0 = 100$ for both plots.

the symbol ε. The usual unit for path length is centimeters, cm, so if the concentration is in molarity, M, the unit for molar absorptivity is $M^{-1}\,cm^{-1}$. The units for absorptivity a when concentration is expressed in any units other than molarity (e.g., ppm and mg/100 mL) must be such that A, the absorbance, is dimensionless. A. Beer discovered the proportional relationship between concentration and absorbance at constant path length in 1852.

Equation 3.9, which summarizes the relationship between absorbance, concentration of the species measured, sample path length, and the absorptivity of the species, is known as the **Beer–Lambert–Bouguer law** or, more commonly, as **Beer's law**.

Since $A = -\log T$, we have the following equivalent expressions for Beer's law:

$$abc = -\log T \tag{3.10}$$

$$-\log\left(\frac{I}{I_0}\right) = abc \tag{3.11}$$

$$\log\left(\frac{I_0}{I}\right) = abc \tag{3.12}$$

$$\frac{I_0}{I} = 10^{abc} \tag{3.13}$$

$$\frac{I}{I_0} = 10^{-abc} \tag{3.14}$$

where a, the absorptivity, is equal to ε, the molar absorptivity, in all of the equations if c, the concentration of the absorbing species, is expressed as molarity.

Beer's law shows mathematically, based on observed experimental facts, that there is a linear relationship between A and the concentration of an absorbing species if the path length and the wavelength of incident radiation are kept constant. This is an extremely important relationship in analytical spectroscopy. It forms the basis for the quantitative measurement of the concentration of an analyte in samples by quantitative measurement of the amount of absorbed radiation. The *quantitative* measurement of radiation intensity is called **spectrometry**. Beer's law is used in all quantitative optical absorption spectrometry – IR absorption spectrometry, AAS, UV/VIS absorption spectrometry, and so on.

3.4.1 Deviations from Beer's Law

Beer's law is usually followed at low concentrations of analyte for homogeneous samples. Absorbance is directly proportional to concentration for most absorbing substances when the concentration is less than about 0.01 M.

Deviations from linearity are common at high concentrations of analyte. There are several possible reasons for deviation from linearity at high concentrations. At low concentrations in a solution, the analyte would be considered the **solute**. As the solute concentration increases, the analyte molecules may begin to interact with each other, through intermolecular attractive forces such as hydrogen bonding and other van der Waals forces. Such interactions may change the absorptivity of the analyte, resulting in a nonlinear response as concentration increases. At extremely high concentrations, the solute may actually become the **solvent**, changing the nature of the solution. If the analyte species is in chemical equilibrium with other species, as is the case with weak acids or weak bases in solution, changes in concentration of the analyte may shift the equilibrium (Le Chatelier's principle). This may be reflected in apparent deviations from Beer's law as the solution is diluted or concentrated.

Another source of deviation from Beer's law may occur if the sample scatters the incident radiation. Solutions must be free of floating solid particles and are often filtered before measurement. The most common reason for nonlinearity at high analyte concentrations is that too little light is available to be absorbed. At low levels of analyte, doubling the concentration doubles the amount of light absorbed, say from 25 % to 50 %. If 99 % of the light has already been absorbed, doubling the concentration still doubles the amount of remaining light absorbed, but the change is only from 99 % to 99.5 %. This results in the curve becoming flat at high absorbance.

It can be seen from Equation 3.8 that $A = \log(I_0/I)$. If $I_0 = 100$ and $A = 1.0$, then $I = 10$. Only 10 % of the initial radiation intensity is transmitted. The other 90 % of the intensity is absorbed by the sample. If $A = 2.0$, $I = 1.0$, indicating that 99 % of the incident light is absorbed by the sample. If $A = 3.0$, 99.9 % of the incident light intensity is absorbed. As we shall see, the error in the measurement of A increases as A increases (or as I decreases). In practice, Beer's law is obeyed for absorbance values less than or equal to 1.0.

We will end this section with a brief discussion of the limitations of all types of absorption spectroscopy. We have seen that absorption is not measured directly but rather as a ratio of intensity of the incident to the transmitted beams. Because of this, it is difficult if not impossible to differentiate between actual absorption and scattering (for example, from suspended particulates in the sample). Moreover, the signal is inversely proportional to the concentration of the absorbing species in the beam path. This results in high signals for low concentrations of analyte, and low signals for high concentrations of analyte. While the latter can be compensated by dilution, the former is a limitation. The demands placed on instrumentation are inherently high because of higher noise levels at higher signal levels.

3.5 METHODS OF CALIBRATION

Calibration is the process of establishing the relationship between the signal we measure (such as absorbance) and known concentrations of analyte. Once this relationship is established, it is possible to calculate the concentration of the analyte in an unknown sample by measuring its signal. The calibration methods discussed subsequently are applicable to most of the analytical instrumental methods discussed in this text, not just spectroscopic measurements. Historically, calibration data were collected manually by reading a meter or measuring a peak height with a ruler and transcribing all the numbers into a lab notebook. The relationship between signal and concentration was plotted manually on graph paper. Modern instruments have software packages that fit the data points to the best-fit line or curve statistically, display the results on the computer screen, and send them to the printer. Computerized data collection and processing greatly reduces *transcription error*, the copying of the wrong figures into a notebook. The use of linear regression and other curve-fitting approaches provides greater accuracy and precision in the calibration equation and in sample results calculated from the equation.

3.5.1 Calibration with Standards

In order to use Beer's law to determine the concentration of analyte in an unknown, it is necessary to establish the relationship between absorbance at a given wavelength and the concentration of the analyte. Solutions containing known concentrations of analyte are called **standard solutions** or, more simply, **standards**. For some types of analyses, the standards may be in the form of solids or gases. Standards must be prepared accurately from high-purity materials so that the concentration of analyte is known as accurately as possible. A series of standards covering an appropriate concentration range is prepared. The standards should include one solution with no added analyte; the concentration of analyte in this standard is zero. This solution is called the reagent blank and accounts for absorbance due to impurities in the solvent and other reagents used to prepare the samples. It also accounts for the instrumental baseline. The absorbance of the reagent blank and each standard is measured. The absorbance of the reagent blank is subtracted from the absorbances of the other standards before any calculations are performed. The absorbances from which the blank absorbance has been subtracted are called "**corrected absorbances**." A plot is made of corrected absorbance on the y-axis versus the known concentration of the standard on the x-axis. Such a plot used to be constructed manually on graph paper; now, plots are generated by computer software. A typical calibration curve of this type is shown in Figure 3.11. This calibration curve shows the relationship between the absorbance of n-hexadecane, $CH_3(CH_2)_{14}CH_3$, a hydrocarbon found in petroleum, at 3.41 µm in the IR region, and concentration of solutions of n-hexadecane in tetrachloroethylene, C_2Cl_4. This measurement of the absorbance of solutions of n-hexadecane at 3.41 µm is a method used for determining petroleum contamination in water, soil, and other environmental samples, because most hydrocarbons absorb at this wavelength. It is an official method, Method 8440, developed by the US Environmental Protection Agency (EPA) (www.epa.gov) and relies on Beer's law to permit the measurement of petroleum hydrocarbons in unknown samples. It is used to measure environmental contamination from oil spills, illegal dumping of oil, and leaking underground oil storage tanks.

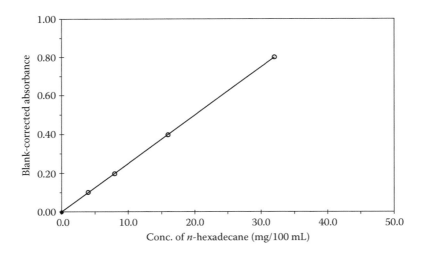

Figure 3.11 A typical external calibration curve for quantitative absorption spectrometry. The calibration standards follow Beer's law since the relationship between concentration and absorbance is linear.

Table 3.4 Calibration data for measurement of petroleum hydrocarbons by IR absorption spectrometry

Concentration of n-Hexadecane (mg/100 mL Solution)	Absorbance at 3.41 μm	Corrected Absorbance
0.0	0.002	0.000
4.0	0.103	0.101
8.0	0.199	0.197
16.0	0.400	0.398
32.0	0.804	0.802

Table 3.5 Calibration data for measurement of petroleum hydrocarbons by IR absorption spectrometry (higher concentration standards added)

Concentration of n-Hexadecane (mg/100 mL Solution)	Absorbance at 3.41 μm	Corrected Absorbance
0.0	0.002	0.000
4.0	0.103	0.101
8.0	0.199	0.197
16.0	0.400	0.398
32.0	0.804	0.802
60.0	1.302	1.300
100.0	1.802	1.800

Table 3.4 gives the values of the concentration and the measured and corrected absorbance for each standard. It is clear from Figure 3.11 that the relationship between absorbance and concentration for this measurement follows Beer's law since the plot results in a straight line ($y = mx + b$, where y is the absorbance and x is the concentration). Once the points have been plotted, the best straight line is fitted through the data points. As we learned in Chapter 1, the best-fit straight line is determined by linear regression (linear least squares) using a statistical program on your calculator or using a computer spreadsheet program such as Excel. It is, of course, possible to estimate the concentration that corresponds to a given absorbance visually from a calibration curve, but for accurate work, the equation of the best-fit straight line must be determined.

Performing a linear regression on the data in Table 3.4 provides us with the exact Beer's law relationship for this method: $A = 0.0250x - 0.001$, where x is the concentration of n-hexadecane in mg/100 mL. From the equation for the calibration curve, the concentration can be determined for any measured absorbance. For example, an unknown sample of contaminated soil is prepared according to US EPA Method 8440, and the absorbance of the sample solution is measured. The measured absorbance is 0.302, so the corrected absorbance would be $0.302 - 0.002 = 0.300$. From our calibration curve, we can see visually that this corresponds to a concentration of ~12.0 mg n-hexadecane/100 mL. The exact concentration can be calculated from the linear regression equation and is found to be 11.96 mg/100 mL or 12.0 mg/100 mL rounded to three significant figures.

Suppose that, in addition to the standards listed in Table 3.4, we also prepared n-hexadecane standards containing 60.0 mg n-hexadecane/100 mL solution and 100.0 mg n-hexadecane/100 mL solution. If we measure the absorbances for these standards, we obtain the data shown in Table 3.5. The additional points, shown as open circles, are plotted along with the original standards in Figure 3.12. Clearly, we now see a deviation from Beer's law at these high concentrations. The points no longer fit a straight line; the measured absorbances are lower than they should be if Beer's law was followed at these concentrations. This shows why it is never a good idea to *extrapolate* (extend) a calibration curve beyond the range of the measured standards. If we had a sample with an absorbance of 1.30 (just like our 60.0 mg/100 mL standard) and had used our original calibration curve extrapolated to higher absorbances as shown by the black squares in Figure 3.12, we would calculate a concentration of 51.9 mg/100 mL for the unknown, which would be erroneously low. What should we do if we have absorbances that are above the absorbance of our highest linear standard (in this case, above $A = 0.802$)? The best approach is to dilute the samples. Let us dilute our 60.0 mg/100 mL standard by a factor of 10, by taking 10.0 mL of the solution and diluting it with pure solvent to a total volume of 100 mL in a volumetric flask. If we measure the absorbance of the diluted solution, we find that the corrected absorbance = 0.153. The corresponding concentration is determined to be 6.07 mg/100 mL. Multiplied by 10 to account for the dilution, our original solution is calculated to contain 60.7 mg/100 mL, and we know that this is a reasonably accurate answer because we made up the standard to contain 60.0 mg/100 mL. So if we take our unknown sample solution with an absorbance of 1.30, dilute it by a factor of 10, and measure the absorbance of the diluted solution, we should get an accurate result for the sample as well.

Preparing a calibration curve as just described, by making a series of standards of known concentrations of analyte, is called external calibration or calibration with external standards. No calculation of uncertainty has been included in this discussion. That subject is covered in Chapter 1. We never report analytical results without also reporting the uncertainty.

3.5.2 Method of Standard Additions

An alternate method of calibration is the Method of Standard Additions (MSA) calibration. This calibration method requires that known amounts of the analyte be added directly to the sample, which contains an unknown amount of analyte. The increase in signal due to the added analyte (e.g., absorbance, emission intensity) permits us to calculate the amount of analyte in the unknown. For this method

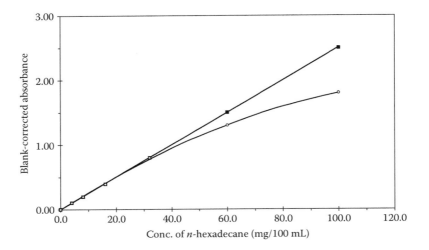

Figure 3.12 A calibration curve showing deviation from Beer's law at high concentrations (high absorbance values). The open circles are the measured absorbance values for the standards and clearly deviate from linearity above $A = 0.8$. The black squares show the expected absorbance values if all standards followed Beer's law. These points were obtained by extrapolation of the linear portion of the curve.

of calibration to work, there must be a linear relationship between the concentration of analyte and the signal.

MSA is often used if no suitable external calibration curve has been prepared. There may be no time to prepare calibration standards—for example, in an emergency situation in a hospital, it may be necessary to measure sodium rapidly in a patient's serum. It may not be possible to prepare a valid set of calibration standards because of the complexity of the sample matrix or due to lack of sufficient information about the sample. Industries often require the analysis of "mystery" samples when something goes wrong in a process. MSA calibration is very useful when certain types of interferences are present in the sample matrix. MSA permits us to obtain accurate results without removing the interferences by performing the calibration in the presence of the interferences. It is often used when only one sample must be analyzed and the preparation of external standards would be inefficient.

A typical example of the use of MSA is the determination of sodium by atomic emission spectrometry in an industrial plant stream of unknown composition. A representative sample of the plant stream is taken and split into four aliquots of 100 mL each. The first aliquot is left untreated; this is called the "no add" or "zero add" sample. To the second aliquot, 100 μg Na is added to the 100 mL sample in such a way as to not change the volume significantly. This can be done by adding a 10 μL volume of a 10,000 ppm Na solution to the sample. A 10,000 ppm Na solution contains 10,000 μg Na/mL, so a 10 μL portion contains 100 μg Na as shown:

$$\left(\frac{10{,}000 \; \mu g \; Na}{mL \; solution}\right)\left(\frac{1 \; mL}{1000 \; \mu L}\right) \times 10 \; \mu L = 100 \; \mu g \; Na$$

Table 3.6 MSA calibration

Sample Aliquot	Emission Intensity (Intensity Units)[a]	Corrected Intensity (Intensity Units)[a]	ppm Na Added
1	2.9	2.4	0.0
2	4.2	3.7	1.0
3	5.5	5.0	2.0
4	6.8	6.3	3.0
Background[b]	0.5	0.0	

[a] Intensity units = 1000 counts/s.
[b] Flame only; no sample aspirated.

The second sample aliquot now contains an additional 1.0 ppm Na, since 100 μg Na/100 mL equals 1.0 ppm Na. To the third aliquot, we add 0.020 mL of the 10,000 ppm Na solution; the third aliquot now contains an additional 2.0 ppm Na. To the fourth aliquot, an addition of 0.030 mL of the 10,000 ppm Na solution results in an additional 3.0 ppm Na in the sample aliquot. The maximum change in volume caused by the addition of Na solution is only 0.03%, an insignificant amount. It is important not to change the volume of the aliquots because a change in volume will cause a change in the concentration–signal relationship. All of the aliquots, untreated and the ones to which additions have been made, must have the same composition, or MSA calibration will not produce accurate results. The concentrations of Na added and the sample aliquot numbers are listed in Table 3.6. The emission intensity for each of the four sample aliquots is measured at the 589.0 nm sodium emission line, using a flame atomic emission spectrometer or a flame photometer. The intensities measured are also listed in Table 3.6. In addition, the emission intensity from the flame is measured with no sample present. This

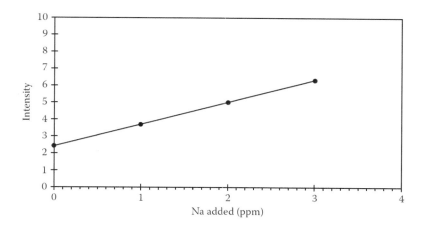

Figure 3.13 A typical MSA calibration curve. The y-axis shows corrected emission intensity.

measures "**background emission**"—a positive signal from the flame not due to the sample. The background emission signal must be subtracted from the sample intensities, just as a reagent blank is subtracted as we discussed earlier, to obtain the corrected intensities shown in Table 3.6. We will learn more about background emission from flames in Chapters 7 and 8.

The corrected emission intensity is plotted versus added Na concentration (Figure 3.13). It can be seen in Figure 3.13 that the relationship between concentration added and signal is linear over the range examined. The quantity ($\Delta I_{emission}/\Delta$ppm Na added) is the **slope** of the addition calibration line and is obtained by a linear regression calculation. In this case,

$$\frac{\Delta I_{emission}}{\Delta ppm\,Na\,added} = \frac{1.3}{1.0\,ppm\,Na}$$

The emission intensity increases by 1.3 units for every 1.0 ppm Na present. Therefore, the concentration of Na in the untreated sample is calculated from the following equation:

$$ppm\,Na_{sample} = \frac{R}{slope} = R(slope)^{-1}$$

where
R = the corrected intensity measurement for sample aliquot 1
slope = ($\Delta I_{emission}/\Delta$ppm Na added)

For our example, the Na concentration in the plant stream is

$$ppm\,Na_{Sample} = 2.4\,\text{intensity units} \times \frac{1.0\,ppm\,Na}{1.3\,\text{intensity units}}$$

$$PPm\,Na_{Sample} = 1.85\,ppm$$

Alternatively, the addition calibration curve may be extrapolated to the intercept on the negative x-axis using the linear regression equation determined. The concentration of sodium in the sample is equal to the absolute value of the negative x intercept. The extrapolation is shown in Figure 3.14 for the data in Table 3.6. The $-x$ intercept occurs at -1.85; therefore, the concentration of Na in the sample is $|-1.85\,ppm| = 1.85\,ppm\,Na$.

The MSA is a very powerful tool for obtaining accurate analytical results when it is not possible to prepare an external calibration curve. In some cases, it permits accurate results to be obtained even in the presence of interfering substances. MSA will not correct for background emission or background absorption or other **spectral interferences**. These interferences and methods for correcting for background will be discussed in the appropriate later chapters.

If the amount of sample is limited, as it may very well be for a patient's serum, a one-point standard addition technique may be used. The technique is outlined for a sodium determination in serum by flame AES. The emission intensity from the sample is measured, and then a known concentration of Na is added, again without significantly changing the volume of the sample. The emission intensity for the addition sample is measured, as is the flame background. Because the emission signal is directly proportional to the sodium concentration, we can write

$$\frac{I_{unknown}}{I_{added}} = \frac{ppm\,Na_{unknown}}{ppm\,Na_{unknown} + ppm\,Na_{added}}$$

where
$I_{unknown}$ is the corrected intensity for the original sample
I_{added} is the corrected intensity for the sample with added Na
ppm Na_{added} is the concentration of the Na added to the original sample

For example, measurement of a serum sample gives a corrected intensity of 2.4 intensity units. Sufficient sodium

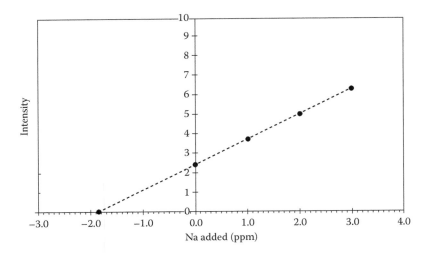

Figure 3.14 Determination of the concentration of analyte by extrapolation of the MSA curve to the negative x intercept. The y-axis shows corrected emission intensity.

is added to the serum sample to increase the concentration by 3.0 ppm, and the corrected intensity of this "addition" solution is 6.3. Setting x = ppm Na$_{unknown}$,

$$\frac{3.7}{6.3} = \frac{x \text{ ppm}}{(x+3.0) \text{ ppm}}$$

Solving for x,

$$x = 4.3 \text{ ppm Na}$$

While the use of a one-point addition may be required if sample is limited, it is generally good practice to make at least two additions whenever possible, to confirm that the additions are within the linear range of the analysis method. Again, results would need to be reported with the associated uncertainty.

3.5.3 Internal Standard Calibration

An **internal standard** is a known amount of a nonanalyte element or compound that is added to all samples, blanks, and standard solutions. Calibration with internal standardization is a technique that uses the signal from the internal standard element or compound to correct for interferences in an analysis. Calibration with internal standardization improves the accuracy and precision of an analysis by compensating for several sources of error.

For determination of analyte A, an internal standard S, which must not be present in the samples, is selected. The same concentration of S is added to all samples, standard solutions, and blanks. The signals due to both A and S are measured. The ratio of the signal due to the analyte A to the signal due to the internal standard S is calculated. The signal ratio, signal$_A$/signal$_S$, is plotted against the concentration ratio of A/S in the standards. The equation of the calibration curve, which should be linear for best results, is obtained by linear regression. The equation permits the calculation of the concentration ratio A/S in any unknown samples by measuring the signals of A and S in the sample. The relationship between concentration and signal may be expressed as follows:

$$\frac{\text{Concentration ratio}(A/S) \text{ in sample}}{\text{Concentration ratio}(A/S) \text{ in standard}} = \frac{\text{Signal ratio}(A/S) \text{ in sample}}{\text{Signal ratio}(A/S) \text{ in standard}}$$

The method of internal standardization is widely used in spectroscopy, chromatography, MS, and other instrumental methods. The use of internal standards can correct for losses of analyte during sample preparation, for mechanical or electrical "drift" in the instrument during analysis, for volume change due to evaporation and other types of interferences. The internal standard must be chosen carefully, usually so that the chemical and physical behavior of the internal standard is similar to that of the analyte. The internal standard must not interact chemically or physically with the analyte. Whatever affects the signal from the analyte should affect the signal from the internal standard in the same way. The ratio of the two signals will stay constant, even if the absolute signals change; this provides more accuracy and precision than if no internal standard is used.

The determination of lead in drinking water by inductively coupled plasma MS (ICP–MS) demonstrates how an internal standard can correct for a problem such as "instrumental drift," that is, a change in the signal over a period of time. The signal for the Pb-208 isotope (^{208}Pb) is monitored to determine lead. Two lead calibration standards are prepared, each containing 10.0 ppb Pb. One standard also contains 20.0 ppb bismuth as an internal standard. The Bi signal is measured at the Bi-209 isotope, along with the Pb-208 signal. Both standards are measured several times during the day, and the resulting signals are listed in Table 3.7. As can be seen from Table 3.7, the signals for both Pb and Bi fluctuate during the day; such fluctuations could be due to changes in electrical line voltage to the instrument, for example. However, as the last column shows, the ratio of the analyte signal to the internal standard signal stays constant throughout the day. If the ICP-MS were calibrated without using an internal standard at 8 AM, samples run at 1 PM would give erroneously high Pb concentrations, because the signal for a given amount of lead has increased by a factor of 2.5. At 3 PM, samples would be approximately 30% higher than the true value. If 20.0 ppb Bi had been added to all the standards and samples, the ratio of the Pb signal/Bi signal would have remained constant and an accurate lead concentration would be determined. For example, the ICP-MS is calibrated at 8 AM with the 10.0 ppb Pb only standard. The signal obtained for 10.0 ppb Pb is 12,050 counts. This is the calibration factor: 12,050 counts/10.0 ppb Pb. The second solution containing 10.0 ppb Pb and 20.0 ppb Bi is measured; the Pb counts are 12,050 (the same as the standard) and the counts for Bi are 60,000, as shown in Table 3.7. A drinking water sample to which 20.0 ppb Bi has been added as internal standard is measured throughout the day. The signals obtained are given in Table 3.8. At 8 AM, the Pb signal is equal to 6028 counts, so the concentration of Pb in the water may be calculated from our calibration factor for lead with no internal standard:

$$\text{ppb Pb in sample} = (6.028 \text{ counts}) \frac{10.0 \text{ ppb Pb}}{12{,}050 \text{ counts}}$$

$$\text{ppb Pb in sample} = 5.00 \text{ ppb Pb}$$

This is the true value for lead in the water sample. The rest of the concentrations using the lead signal only are calculated the same way, and the results are shown in the third column of Table 3.8. Due to the instrument "drift," the results obtained at 1 and 3 PM are clearly in error. For example, the 1 PM sample has a signal equal to 15,010 counts for Pb. The concentration is calculated using the lead calibration factor:

$$\text{ppb Pb in sample} = (15{,}010 \text{ counts}) \frac{10.0 \text{ ppb Pb}}{12{,}050 \text{ counts}}$$

$$\text{ppb Pb in sample} = 12.5 \text{ ppb Pb}$$

The percent error in this result due to instrumental drift is calculated:

$$\%\text{Error} = \frac{\text{Measured value} - \text{True value}}{\text{True value}} \times 100$$

$$\%\text{Error} = \frac{12.5 - 5.00}{5.00} \times 100$$

$$\%\text{Error} = +150$$

Table 3.7 Use of internal standard in calibration

Time of Measurement	Signal Counts[a] 10 ppb Pb, m/e = 208	Signal Counts 20 ppb Bi, m/e = 209	Ratio of Counts Pb/Bi
8 AM	12,050	60,000	0.2008
10 AM	12,100	60,080	0.2013
1 PM	30,000	149,200	0.2010
3 PM	15,750	78,400	0.2009

[a] Both standards give the same signal for Pb, so only one column is shown. The counts for bismuth are from the second standard only.

Table 3.8 Lead in water with and without internal standard

Time of Measurement	Pb Counts, m/e = 208	ppb Pb, No Internal Standard	Pb Counts, m/e = 208	Bi Counts, m/e = 209	Ratio of Counts: Pb/Bi	ppb Pb, with Internal Standard
8 AM	6,028	5.00	6,028	60,010	0.1004	5.00
10 AM	6,063	5.03	6,063	60,075	0.1009	5.02
1 PM	15,010	12.5	15,010	149,206	0.1006	5.01
3 PM	7,789	6.46	7,789	78,398	0.0994	4.95
Mean		7.25				4.99
SD		3.57				0.03
% RSD		49.2				0.60
True value		5.00				5.00
% Error		45.0				−0.2

Note: SD, standard deviation; % RSD, % relative standard deviation = (SD/mean) × 100.

If the Bi internal standard signal is used and the ratio of the Pb signal to the Bi signal is calculated, the internal standard calibration factor is obtained:

$$10.0 \text{ ppb Pb} = \frac{12{,}050 \text{ counts Pb}}{60{,}000 \text{ counts Bi}} = 0.2008$$

The sample run at 8 AM also contained Bi as an internal standard and the counts for Pb and Bi are shown in Table 3.8. The ratio of the sample Pb counts to the sample Bi counts is 6,028 counts Pb/60,010 counts Bi = 0.1004. The Pb concentration in the sample is obtained as follows:

$$\text{ppb Pb in sample} = (0.1004)\frac{10.0 \text{ ppb Pb}}{0.2008}$$

$$\text{ppb Pb in sample} = 5.00 \text{ ppb Pb}$$

This is the same result obtained without an internal standard. If the internal standard ratio is used for all of the samples run during the day, the instrument "drift" is taken into account and the correct results are obtained. For example, at 1 PM, the sample Pb counts are 15,010 and the sample Bi counts are 149,206. These are clearly much higher than the counts obtained at 8 AM, but the ratio of Pb counts to Bi counts is 15,010/149,206 = 0.1006, and the calculated Pb concentration in the 1 PM sample is

$$\text{ppb Pb in sample} = (0.1006)\frac{10.0 \text{ ppb Pb}}{0.2008}$$

$$\text{ppb Pb in sample} = 5.01 \text{ ppb Pb}$$

The percent error in this result is

$$\%\text{Error} = \frac{5.01 - 5.00}{5.00} \times 100$$

$$\%\text{Error} = 0.2$$

An error of 0.2 % is well within the expected accuracy for an instrumental method.

Compare this correct result to the 12.5 ppb Pb result at 1 PM calculated with no internal standard correction and the importance of using an internal standard when possible is clear.

3.5.4 Errors Associated with Beer's Law Relationships

All spectrometric measurements are subject to indeterminate (random) error, which will affect the accuracy and precision of the concentrations determined using spectrometric methods. A very common source of random error in spectrometric analysis is instrumental "noise." Noise can be due to instability in the light source of the instrument, instability in the detector, and variation in placement of the sample in the light path, and is often a combination of all these sources of noise and more. Because these errors are random, they cannot be eliminated. Errors in measurement of radiation intensity lead directly to errors in measurement of concentration when using calibration curves and Beer's law.

We can evaluate the impact of indeterminate error due to instrumental noise on the information obtained from transmittance measurements. The following discussion applies to UV/VIS spectrometers operated in regions where the light source intensity is low or the detector sensitivity is low and to IR spectrometers where noise in the thermal detector is significant.

From Beer's law, it can be shown that

$$\frac{\Delta c}{c} = \frac{0.434 \Delta T}{T \log T} \tag{3.15}$$

where
$\Delta c/c$ is the relative error in concentration
ΔT is the error in measurement of the transmittance

Table 3.9 Relative concentration error from 1 % spectrometric error

Transmittance (T)	Relative Error in Concentration ($\Delta c/c$) × 100 (%)
0.02	12.8
0.08	4.9
0.15	3.5
0.30	2.8
0.37	2.7
0.45	2.8
0.65	3.6
0.80	5.6
0.97	33.8

Note: $\Delta T = 0.01$; $\Delta c/c = (0.434 \Delta T)/(T \log T)$.

The value of ΔT can be estimated from a large number ($n > 20$) of replicate measurements of the same solution. If we assume that we have a constant error of 1 % in the measurement of T, or $\Delta T = 0.01$, the relative error in concentration can be calculated using Equation 3.15. Table 3.9 presents the relative error in concentration for a wide range of transmittance measurements when a constant error of 1 % T is assumed. It can be seen from Table 3.9 that the relative error in concentration is high when T is very low or very high; significant errors

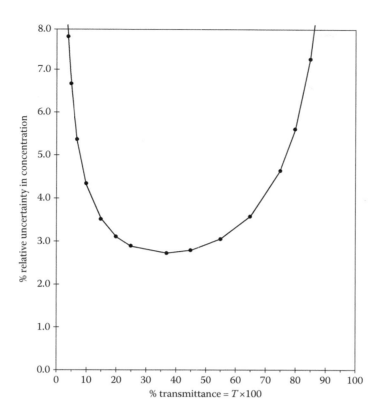

Figure 3.15 Relative uncertainty in measured concentration due to random error in spectrometric measurements due to some types of instrument noise. The data shown are for a constant 1 % error in transmittance. The curve will have the same shape for other values of error in T, but the magnitude of the uncertainty will change.

result when using Beer's law at very low concentrations of analyte (high %T) and at very high concentrations of analyte (low %T).

We can plot the relative error data in Table 3.9 as a function of transmittance. The resulting plot is shown in Figure 3.15. It can be seen from this plot that the minimum relative error occurs at $T = 0.37$ (37 % T), although satisfactory results can be obtained over the range of 15 %–65 % T. This range corresponds to an absorbance range of 0.82–0.19. For the greatest accuracy in quantitative absorption measurements, it is advisable to determine concentration from samples with absorbances between 0.82 and 0.19. Samples that are too concentrated ($A > 0.82$) should be diluted to bring their absorbance values below 0.8. Samples that are too dilute ($A < 0.19$) should be concentrated by evaporation or solvent extraction. If it is not possible to alter the sample solution, the analyst must be aware that the relative error will be large for very dilute or very concentrated samples when using an instrument with the limitations described.

Modern UV/VIS spectrometers are generally limited by "shot noise" in the photon detector as electrons cross a junction. In this case, the plot of relative uncertainly due to indeterminate instrument error looks very different from Figure 3.15. For good-quality, shot noise-limited instruments, the relative error is high for very low values of A (high %T), but absorbance values from 0.2 to above 2.0 have approximately the same low (<1 %) relative uncertainty. In other words, modern spectrometers are much more accurate and precise than older ones because of improvements in instrument components.

Another way of plotting spectrometric data is to use the Ringbom method in which the quantity $(100 - \%T)$ is plotted against the logarithm of the concentration. The resulting S-shaped curve is called a Ringbom plot. A Ringbom plot for absorption by Mn as permanganate ion is shown in Figure 3.16. The Ringbom plot shows the concentration range where the analysis error is minimal; this is the steep portion of the curve where the slope is nearly linear. The plot also permits the evaluation of the accuracy at any concentration level. From Beer's law, it can be shown that the percent relative analysis error for a 1 % transmittance error is given by

$$\frac{\% \text{ relative analysis error}}{1\% \text{ transmittance error}} = \frac{100 \Delta c / c}{1} = \frac{230}{\Delta T / \Delta \log c} \quad (3.16)$$

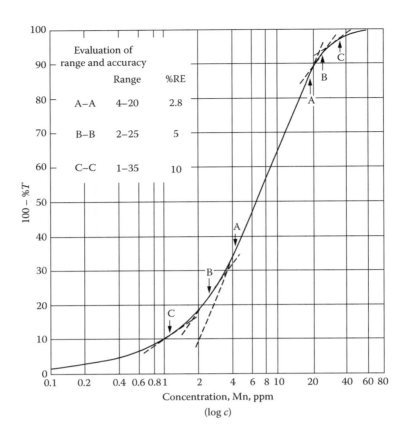

Figure 3.16 Ringbom plot of permanganate solution measured at 526 nm in a 1.00 cm cell. %RE = percent relative error in concentration for a 1 % error in transmittance. The magnitude of %RE is shown for three ranges of Mn concentration. Mn concentration is in units of ppm Mn in solution (1 ppm = 1 μg Mn/mL solution).

Because the quantity ($\Delta T/\Delta \log c$) is the slope of the curve, the relative analysis error per 1 % transmittance error at any point on the curve is equal to 230 divided by the slope at that point. The slope can be determined by constructing a tangent to the curve at the desired concentration. The difference in y for a 10-fold difference in x is calculated. This value, divided into 230, is the percent relative analysis error per 1 % transmittance error. For example, the slope between the two points labeled A in Figure 3.16 is determined by drawing a tangent that extends through the concentrations 2 and 20 ppm (a 10-fold change) as shown. The values of y from the plot are 9 % at 2 ppm and 90 % at 20 ppm. The difference in y values is $90 - 9 = 81$; therefore, $230/81 = 2.8$. The relative analysis error is 2.8 % over this range. Other ranges and their respective errors are shown in Figure 3.16 in the inset box at the top left. Of course, this calculation can be done more accurately using a computer than by manually drawing tangent lines.

A practical application of the Ringbom plot is the determination of the concentration range over which the percent relative analysis error will not exceed a specified value. This sort of limit is often set for industrial analyses, where specifications for "good" product are established for upper and lower limits of product composition based on spectrometric measurements. A product falling outside of these specifications would be rejected as being not in compliance.

Interpretation of the Ringbom plot leads to the same conclusions we deduced from Figure 3.15. The error is lowest at approximately $100 - \%T = 63$, or 37 % T. The relative analysis error per 1 % transmittance is about 2.8 %. The error is not significantly greater over the range $(100 - \%T)$ of 40 %–80 % T or between 20 % and 60 % T; this is the steep, nearly linear portion of the Ringbom plot. At very low and very high values of $100 - \%T$, the slope of the Ringbom plot approaches zero, and therefore, the % relative analysis error approaches infinity for spectrometers with the limitations noted previously.

Table 3.10 provides a summary of the nomenclature, symbols, and definitions commonly used in spectroscopy. We will use these symbols throughout many of the later chapters.

3.6 OPTICAL SYSTEMS USED IN SPECTROSCOPY

In optical analytical spectroscopy, the absorption or emission of radiation by a sample is measured.

INTRODUCTION TO SPECTROSCOPY

Table 3.10 **Nomenclature and definitions for spectroscopy**

Term	Symbol	Definition
Transmittance	T, where $T = I/I_0$	Ratio of light intensity after passing through sample, I, to light intensity before passing through sample, I_0
Absorbance	A	$-\log T = abc$
Absorptivity	a	The proportionality constant a in Beer's law, $A = abc$, where A is absorbance, c is concentration, and b is path length
Molar absorptivity	ε	The proportionality constant ε in Beer's law, $A = \varepsilon bc$, where A is absorbance, b is path length, and c is concentration of the absorbing solution in molarity (M)
Path length	b	Optical path length through the sample
Sample concentration	c	Amount of sample (usually in terms of the absorbing species) per unit volume or mass. Typical units are g/mL, mol/L, ppm,%
Absorption maximum	λ_{max}	Wavelength at which greatest absorption occurs
Wavelength	λ	Distance between consecutive wave crests
Frequency	ν	Number of oscillations of a wave per second; the number of wave crests passing a given point per second
Wavenumber	$\bar{\nu}$	$1/\lambda$ or the number of waves per centimeter

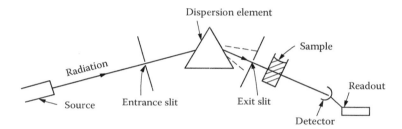

Figure 3.17 Schematic diagram of a single-beam absorption spectrometer.

The instrumentation designed to measure absorption or emission of radiation must provide information about the wavelengths that are absorbed or emitted and the intensity (I) or absorbance (A) at each wavelength. The instrumentation for spectroscopic studies from the UV through the IR regions of the spectrum is very similar in its fundamental components. For the moment, the term **spectrometer** will be used to mean an instrument used for optical spectroscopy. More specific terms for instruments will be defined after the components are discussed.

Instruments for analytical spectroscopy require a radiation source, a wavelength selection device such as a monochromator, a sample holder transparent to the radiation range being studied, a detector to measure the intensity of the radiation and convert it to a signal, and some means of displaying and processing the signal from the detector. Fourier transform (FT) spectrometers, discussed subsequently, do not require a wavelength selection device. If emitted radiation is being measured, the sample, excited by some means, is the radiation source. If absorption, fluorescence, phosphorescence, or scattering of light is measured, an external radiation source is required. The specific arrangement of these components is referred to as the **optics** or **optical configuration** or **optical layout** of the instrument. The optical layout of a simple single-beam absorption spectrometer is shown schematically in Figure 3.17. The placement of the sample holder and the wavelength selector may be inverted; in UV/VIS absorption spectrometry, the sample holder is usually placed after the wavelength selector, so that monochromatic light falls on the sample. For atomic absorption, IR, and fluorescence spectroscopy, the sample is usually placed in front of the wavelength selector.

3.6.1 Radiation Sources

An ideal radiation source for spectroscopy should have the following characteristics:

1. The source must emit radiation over the entire wavelength range to be studied.
2. The intensity of radiation over the entire wavelength range must be high enough so that extensive amplification of the signal from the detector can be avoided.
3. The intensity of the source should not vary significantly at different wavelengths.
4. The intensity of the source should not fluctuate over long time intervals.
5. The intensity of the source should not fluctuate over short time intervals. Short time fluctuation in source intensity is called "flicker."

The design of the radiation source varies with the wavelength range for which it is used (e.g., IR, UV, visible), and details of specific sources will be discussed in the appropriate instrumentation chapter. Most sources will have their intensities change exponentially with changes in voltage, so in all cases, a reliable, steady power supply to the radiation source is required. Voltage regulators (also called line conditioners) are available to compensate for variations in incoming voltage. A doublebeam optical configuration may also be used to compensate for variations in source stability, as described in Section 3.6.5.

There are two major types of radiation sources used in analytical spectroscopy: **continuum sources** and **line sources**. Continuum sources emit radiation over a wide range of wavelengths, and the intensity of emission varies slowly as a function of wavelength. Typical continuum sources include the tungsten filament lamp that produces visible radiation (white light) and high-pressure xenon arc lamps, the deuterium lamp for the UV region, high-pressure mercury or xenon arc lamps for the UV region, and heated solid ceramics or heated wires for the IR region of the spectrum. Continuum sources are used for most molecular absorption and fluorescence spectrometric instruments. Line sources, in contrast, emit only a few discrete wavelengths of light, and the intensity is a strong function of the wavelength. Typical line sources include hollow cathode lamps and electrodeless discharge lamps, used in the UV and visible regions for AAS and atomic fluorescence spectrometry, sodium or mercury vapor lamps (similar to the lamps now used in streetlamps) for lines in the UV and visible regions, and lasers. Lasers are high-intensity **coherent** line sources; lasers are available with emission lines in the UV, visible, and IR regions. They are used as sources in Raman spectroscopy and molecular and atomic fluorescence spectroscopies.

3.6.2 Wavelength Selection Devices

3.6.2.1 Filters

The simplest and most inexpensive way to select certain portions of the electromagnetic spectrum is with a filter. There are two major types, absorption filters and interference filters. Absorption filters can be as simple as a piece of colored glass. In Section 3.1, we discussed how blue glass transmits blue wavelengths of the visible spectrum but absorbs red and yellow wavelengths. This is an example of an absorption filter for isolating the blue region of the visible spectrum. Colored glass absorption filters can be purchased that isolate various ranges of visible light. These filters are stable, simple, and cheap, so they are excellent for use in portable spectrometers designed to be carried into the field. The biggest limitation is that the range of wavelengths transmitted is broad compared with prisms and gratings that are also devices used to select a narrow wavelength range from a broadband polychromatic source. The transmission range may be 50–300 nm for typical absorption filters. Absorption filters are limited to the visible region of the spectrum and the X-ray region.

The second type of filter is the interference filter, constructed of multiple layers of different materials. The filter operates on the principle of constructive interference to transmit selected wavelength ranges. The wavelengths transmitted are controlled by the thickness and refractive index of the center layer of material. Interference filters can be constructed for transmission of light in the IR, visible, and UV regions of the spectrum. The wavelength ranges transmitted are much smaller than for absorption filters, generally 1–10 nm, and the amount of light transmitted is generally higher than for absorption filters.

3.6.2.2 Monochromator

A **monochromator** consists of a dispersion element, an entrance slit and an exit slit, plus lenses, and mirrors for collimating and focusing the beam of radiation. The function of the dispersion element is to spread out in space, or *disperse*, the radiation falling on it according to wavelength. The two most common types of dispersion elements are prisms and gratings. You are probably already familiar with the ability of a glass prism to disperse white light into a rainbow of its component colors.

The entrance slit allows light from the source to fall on the dispersion element. The dispersed light falls on the exit slit of the monochromator. The function of the exit slit is to permit only a very narrow band of light to pass through to the sample and detector. One way to accomplish this is to rotate the dispersion element to allow dispersed light of different wavelengths to fall on the exit slit in sequence. For example, a white light source is dispersed into violet through red light by a prism or grating. The dispersion element is rotated slowly, allowing first violet light through the exit slit, then blue light, and so on, all the way to red light. In this way, the monochromator sorts polychromatic radiation from a source into nearly monochromatic radiation leaving the exit slit.

Prisms: Prisms are used to disperse IR, visible, and UV radiation. The most common prisms are constructed of quartz for the UV region, silicate glass for the visible and near-IR (NIR) region, and NaCl or KBr for the IR region. Prisms are shaped like bars with triangular cross sections. The 60°–60°–60° triangle, called a Cornu prism and shown in Figure 3.17, is widely used, but other geometries are available. Polychromatic light passing through the entrance slit is focused on a face of the prism such that *refraction*, or bending, of the incident light occurs. Different wavelengths of light are refracted to different degrees, and the spatial separation of wavelengths is therefore possible. The refractive index of prism materials varies with wavelength. A quartz prism has a higher index of refraction for

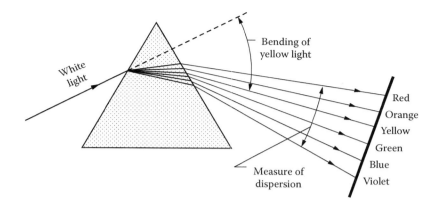

Figure 3.18 Dispersion of visible light by a prism.

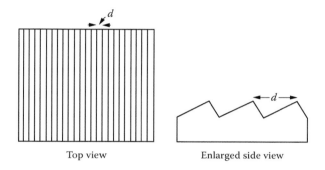

Figure 3.19 Highly magnified schematic view of a diffraction grating.

short-wavelength radiation than for long-wavelength radiation; therefore, short-wavelength radiation is bent more than long-wavelength radiation. In the visible region of the spectrum, red light would be bent less than blue light on passing through such a prism, as shown in Figure 3.18. Prisms were historically the most used dispersion devices in monochromators, but they have been replaced by diffraction gratings or by FT systems.

Diffraction gratings: UV, visible, and IR radiation can be dispersed by a diffraction grating. A diffraction grating consists of a series of closely spaced parallel grooves cut (or *ruled*) into a hard glass, metallic, or ceramic surface (Figure 3.19). The surface may be flat or concave and is usually coated on the ruled surface with a reflective coating. A grating for use in the UV and visible regions will contain between 500 and 5000 grooves/mm, while a grating for the IR region will have between 50 and 200 grooves/mm. Traditionally, the grooves in a grating were cut mechanically with a diamond-tipped tool, a time-consuming and expensive operation. Such a grating is called a *master* and is used as a mold for casting *replica* gratings out of polymer resin. The replica gratings duplicate the grooves in the master and are coated with a reflective coating, such as aluminum, for use. Most instruments use replica gratings because they are much less expensive than a master grating. Most gratings are now produced by a holographic technique. The grating is made by coating the grating substrate with an optically flat photosensitive polymer film. The film is exposed to the interference pattern from laser beams, and the interference pattern is "burned" into the film. The grooves from the interference pattern are then etched into the substrate to make the master grating, using chemical or ion etching to shape the grooves to the desired shape. The use of a laser interference pattern to form the grooves results in more perfect gratings at lower cost than mechanically ruled master gratings. These gratings, called holographic gratings, can be used in instruments directly or can serve as master gratings for the manufacture of replica holographic gratings. Holographic gratings can be made in many shapes other than the traditional plane or concave shape, and the grooves may be uniform or nonuniform in spacing, depending on the application. The size of a typical diffraction grating varies from about 25 × 25 to 110 × 110 mm.

Dispersion of light at the surface of a grating occurs by diffraction. Diffraction of light occurs because of constructive interference between reflected light waves. The path of one wave is shown in Figure 3.20. Parallel waves can be envisioned on adjacent grooves. Constructive interference or diffraction of light occurs when

$$n\lambda = d(\sin i \pm \sin \theta) \qquad (3.17)$$

where
n is the order of diffraction (must be an integer 1, 2, 3, ...)
λ is the wavelength of the radiation
d is the distance between grooves
i is the angle of incidence of the beam of light
θ is the angle of dispersion of the light of wavelength λ

The angle of incidence i and the angle of dispersion θ are both measured from the normal to the grating. For a given value of n, but different values of λ, the angle of dispersion θ is different. Separation of light occurs because light of different wavelengths is dispersed (diffracted) at different angles.

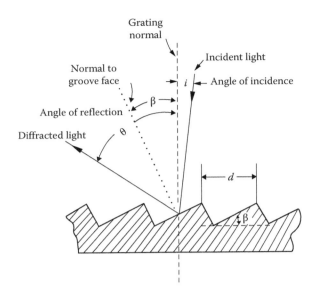

Figure 3.20 Cross-sectional diagram of a diffraction grating showing diffraction of a single beam of light. The symbols used are as follows: i = angle of incidence, θ = angle of diffraction (or reflectance), β = blaze angle of the grating, d = grating spacing. (Modified from Dean, J.A. and Rains, T.C., eds., *Flame Emission and Absorption Spectrometry*, Vol. 2, Marcel Dekker, Inc., New York, 1971. Used with permission.)

One problem with gratings is that light with several different wavelengths may leave the grating at the same angle of dispersion. For example, suppose that a beam of radiation falls on a grating at an angle i. From Equation 3.17 for a given angle of dispersion θ, the product $n\lambda$ is a constant. Any combination of n and λ that equals this constant will satisfy the equation. Assume $\lambda = 600$ nm and $n = 1$ gives an angle of dispersion = θ; then, if $\lambda = 200$ nm and $n = 3$, the angle is also θ, and so on. In practice, radiation with each of these wavelengths is dispersed at an angle θ and travels down the same light path, as illustrated in Figure 3.21. Wavelengths of light that are related in this way are said to be different *orders* of diffracted radiation. They are not separated by gratings, as is seen in Figure 3.21. The wavelengths of radiation traveling the same path after dispersion are related by the number n, which may take the value of any whole number. On high-quality spectrometers, different orders are separated by using a small prism or a filter system as an *order sorter* in conjunction with the grating (Figures 3.22 and 3.23). It is common for IR instruments to use filters as order sorters. As the grating rotates to different wavelength ranges, the filters rotate to prevent order overlap, and only one wavelength reaches the detector. Excellent tutorials on diffraction gratings and the optics of spectroscopy are available on the Internet from the instrument company Horiba Scientific (www.horiba.com) by typing "optics of spectroscopy tutorial" into the search box on the home page.

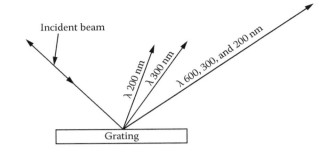

Figure 3.21 The angle of diffraction of light from a grating depends not only on the wavelength but also on the *order of diffraction, n*. Wavelengths of 200, 300, and 600 nm are diffracted at different angles in first order ($n = 1$), but 200 nm light in third order ($n = 3$) and 300 nm light in second order ($n = 2$) diffract at the same angle as 600 nm light in first order. The three wavelengths overlap.

3.6.2.3 Resolution Required to Separate Two Lines of Different Wavelengths

Monochromator wavelength accuracy: Before discussing the ability of a system to separate two wavelengths, it is important that the wavelength scale of the system be assessed. For the UV/VIS region of the spectrum, this can be done by scanning either a holmium or didymium glass or commercially available sealed quartz cuvettes filled with a stable solution of holmium perchlorate or didymium perchlorate. One source of such wavelength reference standards is Starna Cells, Inc. (www.starnacells.com). These compounds give a series of sharp peaks throughout the UV/VIS region, as seen in Figure 3.24, and should be run on a regular basis to ensure that the wavelength reading from the monochromator system is accurate. Similar types of standards are available to check wavelength accuracy in the far UV and NIR regions.

Resolution of a monochromator: The ability to disperse radiation is called *resolving power*. Alternative designations include *dispersive power* and *resolution*. For example, in order to observe an absorption band at 599.9 nm without interference from an absorption band at 600.1 nm, we must be able to resolve, or separate, the two bands.

The resolving power R of a monochromator is equal to $\lambda/\delta\lambda$, where λ is the average of the wavelengths of the two lines to be resolved and $\delta\lambda$ is the difference in wavelength between these lines. In the present example, the required resolution is

$$R = \frac{\lambda}{\delta\lambda} \qquad (3.18)$$

$$R = \frac{\text{Average of 599.9 and 600.1}}{\text{Absolute difference between 599.9 and 600.1}}$$

$$= \frac{600}{0.2} = 3000$$

INTRODUCTION TO SPECTROSCOPY

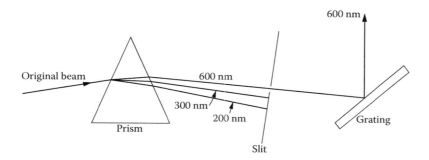

Figure 3.22 Prism used as an order sorter for a grating monochromator.

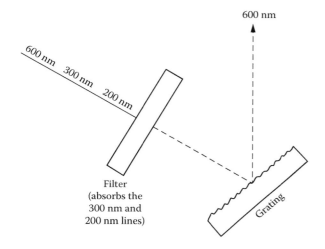

Figure 3.23 A filter is used as an order sorter to prevent higher-order wavelengths from reaching the grating.

Resolution of a prism: The resolving power R of a prism is given by

$$R = t \frac{d\eta}{d\lambda} \qquad (3.19)$$

where
t is the thickness of the base of the prism
$d\eta/d\lambda$ is the rate of change of dispersive power (or *refractive index*) η of the material of the prism with wavelength

For the resolution of two beams at two wavelengths λ_1 and λ_2, it is necessary that the refractive index of the prism be different at these wavelengths. If it is constant, no resolution occurs. The resolving power of a prism increases with the thickness of the prism. Resolution can be maximized for a given wavelength region by choosing the prism material to maximize $d\eta/d\lambda$. For example, glass prisms disperse visible light better than quartz prisms. For maximum dispersion, a prism is most effective at wavelengths close to the wavelengths at which it ceases to be transparent. At longer wavelengths, the resolving power decreases.

Resolution of a grating: The resolving power of a grating is given by

$$R = nN \qquad (3.20)$$

where
n is the order
N is the total number of grooves in the grating that are illuminated by light from the entrance slit

Therefore, longer gratings, smaller groove spacing, and the use of higher orders ($n > 1$) result in increased resolution. Suppose that we can obtain a grating with 500 grooves/cm. How long a grating would be required to separate the sodium D lines at 589.5 and 589.0 nm in first order?

We know from Equation 3.18 that the required resolution R is given by

$$R = \frac{589.25}{0.5} = 1178.5$$

The resolution of the grating must therefore be at least 1179 (to four significant figures). But R (for a grating) $= nN$; therefore, $1179 = nN$. Since we stipulated first order, $n = 1$; hence, N, the total number of grooves required, is 1179. The grating contains 500 grooves/cm. It must be (1179/500) cm long, or 2.358 cm. This assumes that the entire grating surface is illuminated during use.

In a separate example, we may ask how many grooves per centimeter must be cut on a grating 3.00 cm long to resolve the same sodium D lines, again assuming that the entire grating is illuminated. The required resolution is 1179, and for first order, $nN = N = 1179$ total grooves are required. Therefore, the number of grooves needed per cm is

$$N/\text{cm} = 1179/3.00 \text{ cm} = 393 \text{ lines/cm}$$

It is not possible to cut a fraction of a groove or to illuminate a fraction of a groove; hence, N must be a whole number in all calculations. This may require rounding off a calculated answer to the nearest whole number.

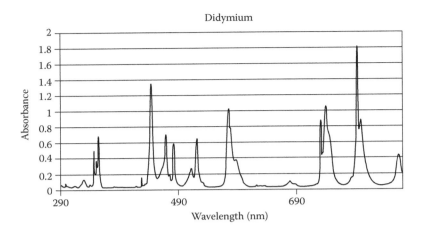

Figure 3.24 Didymium oxide UV/VIS wavelength reference spectra, from a sealed quartz cell containing didymium oxide in perchloric acid solution. (Courtesy of Starna Cells, Inc., Atascadero, CA, www.starnacells.com.)

In practice, the resolution of a grating monochromator system in the UV is usually assessed by measuring benzene vapor, which gives rise to a number of close but sharp absorption lines. Sealed quartz cells containing benzene vapor are commercially available for performing this assessment, which is normally done at several bandwidths, as shown in Figure 3.25.

Dispersion of a grating monochromator: The resolution of a monochromator measures its ability to separate adjacent wavelengths from each other. Resolution is related to a useful quantity called the reciprocal dispersion, or reciprocal linear dispersion, D^{-1}:

$$D^{-1} = \frac{d\lambda}{dy} \quad (3.21)$$

where the reciprocal linear dispersion equals the change in wavelength, $d\lambda$, divided by dy, the corresponding change in y, the distance separating the wavelengths along the dispersion axis. Units for D^{-1} are usually nm/mm. The reason D^{-1} is useful is that the **spectral bandpass** or **spectral bandwidth (SBW)** of the light exiting a monochromator is directly related to D^{-1} and the slit width of the monochromator:

$$\text{SBW} = sD^{-1} \quad (3.22)$$

where s is the slit width of the monochromator. The SBW represents the width of the wavelength range of 75% of the light exiting the monochromator. For a monochromator that uses a grating as the dispersion device, the reciprocal linear dispersion is

$$D^{-1} = \frac{d}{nF} \quad (3.23)$$

where
d is the distance between two adjacent grooves on the grating
n is the diffraction order
F is the focal length of the monochromator system

The reciprocal dispersion for a grating-based system is therefore essentially constant with respect to wavelength. SBW in the UV region can be determined by measuring a solution of 0.02 % toluene in hexane using a matched hexane reference cell. The absorbance is measured at both 268.7 and 267.0 nm. The ratio of absorbances ($A_{268.7}/A_{267.0}$) is calculated and is directly related to the SBW, as shown in Table 3.11 (Figure 3.26).

The dispersion of a prism-based monochromator is a more complex calculation and will not be covered, due to the predominance of gratings in even inexpensive modern instruments.

Echelle monochromator: From Figures 3.19 and 3.20, you can see that the cuts shown on the surface of the gratings are not symmetrical v-shapes. Each cut has a short face and a long face. This type of grating is known as a blazed grating. The conventional blazed diffraction grating uses the long face of the groove, as seen in Figure 3.20, and the angle of the cut, called the blaze angle β, is generally optimized for first-order diffraction. It is possible to rule a grating with a much higher blaze angle and to use the short side of the groove for diffraction; this type of grating is called an **echelle** grating. The angle of dispersion θ is much higher from an echelle grating than from a conventional grating. The echelle system improves dispersion by this increase in θ and by the use of higher orders (larger values of n). The result is a 10-fold improvement in resolution over a conventional grating monochromator of the same focal length. Because of the multiple high orders diffracted, it is necessary to use a second dispersing element to sort the

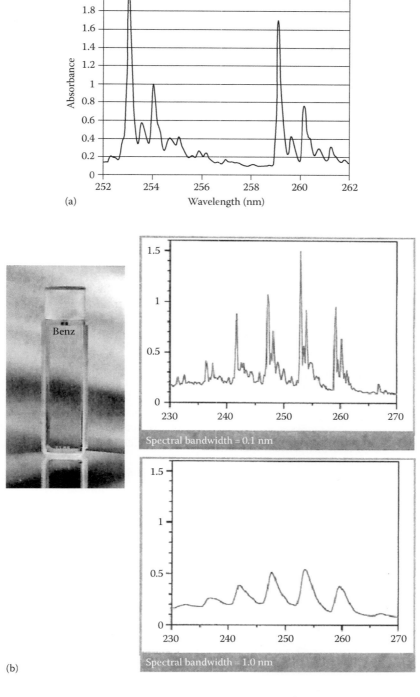

Figure 3.25 (a) Spectrum of benzene vapor. (b) Spectra of benzene vapor taken at two different bandwidths for testing resolution of a grating-based instrument. (Courtesy of Starna Cells, Inc., Atascadero, CA, www.starnacells.com.)

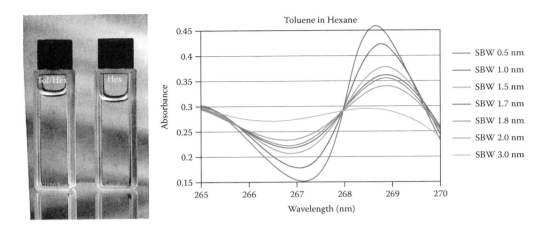

Figure 3.26 SBW measurements using a sealed cell containing 0.02 % toluene in hexane and a hexane blank cell. (Courtesy of Starna Cells, Inc., Atascadero, CA, www.starnacells.com.)

Table 3.11 Bandwidth from wavelength ratios

Ratio	2.5	2.1	1.6	1.4	1.0
SBW	0.5	1.0	1.5	2.0	3.0

Source: Courtesy of Starna Calls, Inc., Atascadero, CA, www.starnacells.com.

overlapping orders. The second dispersing element, called a cross disperser, is arranged to sort the light at right angles to the grating, so a 2D spectrum results. An echelle optical layout for ICP optical emission spectroscopy (ICP-OES) is shown in Figure 3.27. An example of the 2D output, with wavelength plotted on the *y*-axis and diffraction order on the *x*-axis, is shown in Figure 3.28. Commercial echelle spectrometers will be discussed in Chapter 8.

3.6.3 Optical Slits

A system of slits (Figure 3.17) is used to select radiation from the light beam both before and after it has been dispersed by the wavelength selector. The jaws of the slit are made of metal and are usually shaped liked two knife edges. They can be moved relative to each other to change the mechanical width of the slit as desired. For the sake of simplicity, Figure 3.17 does not show the system of lenses or mirrors used in a monochromator to focus and collimate the light as needed.

The *entrance slit* permits passage of a beam of light from the source. Radiation from the light source is focused on the entrance slit. Stray radiation is excluded. After being passed through the entrance slit, the radiation is collimated into a parallel beam of light, which falls onto and completely illuminates one side of the prism or the entire grating. The prism or grating disperses the light in different directions according to wavelength. At the setting selected for the dispersion device, one wavelength is refocused onto the exit slit. Light of other wavelengths is also focused, but not onto the exit slit. The emerging light is redirected and focused onto the detector for intensity measurement.

Lenses or front-faced mirrors are used for focusing and collimating the light. In the IR, front-faced mirrors are always more efficient than lenses and do not absorb the radiation. They are also easily scratched, since the reflecting surface is on the front and not protected by glass, as is the case with conventional mirrors. Back-faced mirrors are not used because the covering material (e.g., glass) may absorb the radiation. One type of monochromator system using mirrors for focusing, collimation, and a grating for dispersion is presented in Figure 3.29, with the entrance and exit slits shown.

The physical distance between the jaws of the slit is called the *mechanical slit width*. Instruments used to have a micrometer scale attached so that one read off the mechanical slit width directly; modern computer-controlled instruments set and read the slit width through the software that controls a stepper motor operating the slit mechanism. In UV absorption spectroscopy, mechanical slit widths are of the order of 0.3–4 μm. In IR spectroscopy, slit widths between 0.1 and 2.0 mm are common for dispersive instruments. There are no slits in FTIR spectrometers.

The wavelength range of the radiation that passes through the exit slit is called the *spectral bandpass* or *SBW* or *spectral slit width*. As noted previously, this bandpass can be measured by using the absorption spectra of toluene in hexane. It can also be measured by passing an emission line of very narrow width through the slits to the detector. By rotating the dispersion element, we can record the wavelength range over which the response occurs. After correcting for the actual width of the emission line, we can calculate the spectral bandpass. For example, to measure the spectral bandpass for a monochromator system used as an AAS, we can use a cadmium hollow cathode lamp, which produces

INTRODUCTION TO SPECTROSCOPY

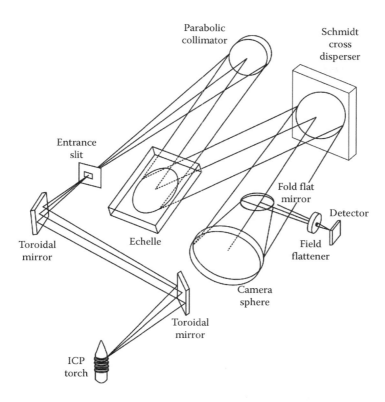

Figure 3.27 An echelle spectrometer optical layout. The echelle grating disperses the light to a second wavelength selector, called a cross disperser. The cross disperser may be a prism or a conventional grating. [© 1993–2014 PerkinElmer, Inc. All rights reserved. Printed with permission. (www.perkinelmer.com).]

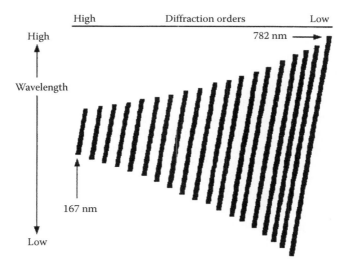

Figure 3.28 Illustration of the 2D array of dispersed light produced by the echelle spectrometer. [© 1993–2014 PerkinElmer, Inc. All rights reserved. Printed with permission. (www.perkinelmer.com).]

very narrow atomic emission lines from cadmium. One of those lines occurs at 228.8 nm. We move our dispersion device and monitor the signal at the detector. The emission line from cadmium gave a signal at all wavelengths from 228.2 to 229.4 nm. This means that the cadmium emission line reached the detector over a wavelength range that was 1.2 nm wide. Therefore, 1.2 nm is the spectral bandpass of this monochromator system. In this example, no correction was made for the actual width of the cadmium 228.8 nm line, which is negligible in this case. The signal that is measured in the aforementioned experiment has a Gaussian peak shape. The SBW in this type of measurement is usually

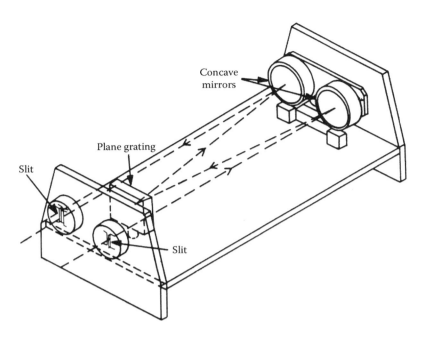

Figure 3.29 A grating monochromator showing the optical slits. The entrance slit is on the right and the exit slit on the left. (From Dean, J.A. and Rains, T.C., eds., *Flame Emission and Absorption Spectrometry*, Vol. 2, Marcel Dekker, Inc., New York, 1971. Used with permission.)

defined as the width of the signal peak at one-half of the maximum peak height, called the full width at half maximum (FWHM). Spectral bandpasses are normally on the order of 0.3–4 nm. Note that the SBW is three orders of magnitude smaller than the physical slit width, nm versus µm.

If the mechanical slit width were made wider, the spectral bandpass would simultaneously increase and vice versa. The spectral bandpass is one of the components of the spectrometer that affects resolution, as seen in Figure 3.25. For example, with the mechanical slit settings described, it would not be possible to resolve an emission line at 229.0 nm from the 228.8 nm Cd line, because both would pass through the slits. In practice, the slits are kept as narrow as possible to ensure optimum resolution; however, they must be wide enough to admit sufficient light to be measured by the detector. The final choice of slit width is determined by the analyst based on the particular sample at hand. A good rule of thumb is to keep the slits as narrow as possible without impairing the functioning of the detector or the ability to detect a specified amount of analyte.

By rotating the prism or grating (or by moving the exit slit across the light beam from the monochromator), the wavelength range passing through the exit slit can be changed. By continuously rotating the dispersion element from one extreme to another, the complete spectrum can be scanned.

3.6.4 Detectors

The detector is used to measure the intensity of the radiation that falls on it. Normally, it does this by converting the radiation energy into electrical energy. The amount of energy produced is usually low and must be amplified. The signal from the detector must be steady and representative of the intensity of radiation falling on it. If the signal is amplified too much, it becomes erratic and unsteady; it is said to be *noisy*. The degree of random variation in the signal is called the *noise level*. Amplifying the signal from the detector increases its response. In practice, the response can be increased until the noise level of the signal becomes too great; at this point, the amplification is decreased until the noise level becomes acceptable.

There are a number of different types of photon detectors, including the photomultiplier tube, the silicon photodiode, the photovoltaic cell, and a class of multichannel detectors called charge transfer devices. Charge transfer detectors include photodiode arrays, charge-coupled devices (CCDs), and charge-injection devices (CIDs). These detectors are used in the UV/VIS and IR regions for both atomic and molecular spectroscopies.

In addition to photon detectors, there are several important detectors that measure heat. These heat detectors or thermal detectors are particularly useful in the IR region, where the energy of photons is very low. The detectors will be discussed at length in the following chapters on specific techniques, for example, IR detectors in Chapter 5, photomultiplier detectors and photodiodes in Chapter 4, and CCDs and CIDs in Chapter 8.

3.6.5 Single-Beam and Double-Beam Optics

Single-beam optics, shown schematically in Figure 3.17, are used for all spectroscopic *emission* methods. In emission procedures, the sample is put where the source is located in

Figure 3.17. In spectroscopic *absorption* studies, the intensity of radiation before and after passing through the sample must be measured. When single-beam optics are used, any variation in the intensity of the source while measurements are being made may lead to analytical errors. Slow variation in the average signal with time is called *drift*, as displayed in Figure 3.30. Drift can cause a direct error in the results obtained. As shown in Figure 3.30, a signal has been set to zero at time 0 with no analyte present. As time increases toward time 1, the signal with no analyte present (called the baseline signal) increases due to drift. At time 1, a sample is measured and gives an increased signal due to analyte present (the peak shown above the baseline). The total signal, sample plus baseline, at time 1 is 5 units. The baseline continues to drift upward and, at time 2, the sample is measured again. As can be seen in the figure, the peak for sample above the baseline is the same height as the peak at time 1, but the total signal (peak plus baseline) is now 10 units. If the baseline drift were not accounted for, the analyst would conclude that the sample at time 2 has twice as much analyte as the sample at time 1 – a direct error.

There are numerous sources of drift. The *radiation source* intensity may change because of line voltage changes, the source warming up after being recently turned on, or the source deteriorating with time. The *monochromator* may shift position as a result of vibration or heating and cooling causing expansion and contraction. The line voltage to the *detector* may change, or the detector may deteriorate with time and cause a change in response. Errors caused by drift lead to an error in the measurement of the emission signal or the absorption signal compared with the standards used in calibration. The problem can be reduced by constantly checking the light intensity or by using a standard solution measured at frequent intervals during the analysis.

Single-beam optics are particularly subject to errors caused by drift. However, the problems associated with drift can be greatly decreased by using a double-beam system.

The double-beam system is used extensively for spectroscopic absorption studies. The individual components of the system have the same function as in the single-beam system, with one very important difference. The radiation from the source is *split* into two beams of approximately equal intensity using a *beam splitter*, shown in Figure 3.31. One beam is termed the *reference beam*; the second beam, which passes through the sample, is called the *sample beam*. The two beams are then recombined and pass through the monochromator and slit systems to the detector. This is illustrated schematically in Figure 3.31. In this schematic, there is a cell in the reference beam that would be identical to the cell used to hold the sample. The reference cell may be empty or it may contain the solvent used to dilute the sample, for example. This particular arrangement showing the monochromator after the sample is typical of a dispersive IR double-beam spectrophotometer. There are many commercial variations in the optical layout of double-beam systems. While many have become archaic, they still find applications.

As shown in Figure 3.32a, the beam splitter may be a simple mirror plate into which a number of holes are drilled. Light is reflected by the mirror plate and passes down the sample beam path. An equal portion of light passes through the holes in the plate and forms the reference beam. Another convenient beam splitter is a disk with opposite quadrants removed (Figure 3.32b). The disk rotates in front of the radiation beam and the mirrored surface reflects light into the sample path. The missing quadrants permit radiation to pass down the reference path. Each beam of light is intermittent and arrives at the detector in the form of an alternating signal. When no radiation is absorbed by the

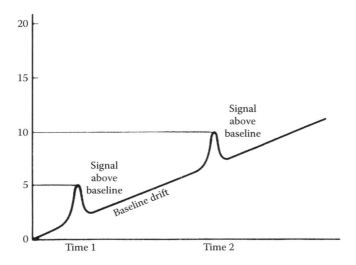

Figure 3.30 Error caused by baseline drift in a spectroscopic measurement.

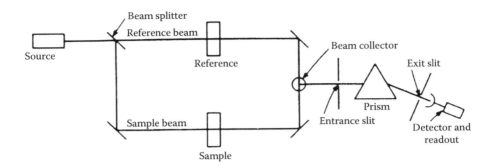

Figure 3.31 Schematic diagram of a double-beam optical system.

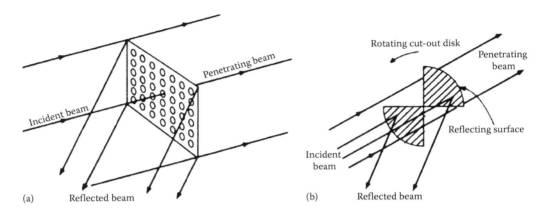

Figure 3.32 (a) Plate beam splitter. (b) Rotating disk beam splitter (or *chopper*).

sample, the two beams are equal, recombine, and form a steady beam of light. However, when radiation is absorbed by the sample, the two beams are not equal, and an alternating signal arrives at the detector. This is illustrated in Figure 3.33.

Using the double-beam system, we can measure the *ratio* of the reference beam intensity to the sample beam intensity. Because the ratio is used, any variation in the intensity of radiation from the source during measurement does not introduce analytical error. This advantage revolutionized absorption spectroscopy. If there is a drift in the signal, it affects the sample and reference beams equally. The recombined beam will continue to give accurate signal information unless the drift is very great, in which case correction is not complete. Absorption measurements made using a double-beam system are virtually independent of drift and therefore more accurate.

3.6.6 Dispersive Optical Layouts

Configurations of the common components of dispersive spectroscopy systems are shown in the following for the most used types of spectroscopy. Arrows show the light path in each layout. In layouts 1 and 3, an external source of radiation is required, but for 3, the source is generally oriented at right angles to the sample. Emission, layout 2, does not require an external radiation source; the excited sample is the source. For absorption, fluorescence, phosphorescence, and scattering, the source radiation passes through the dispersive device, which selects the wavelength, and into the sample. The selected wavelength passes through the sample and reaches the detector, where the intensity of the signal is converted to an electrical signal. In emission, the radiation that emanates from the sample passes through the dispersive device, which selects one wavelength at a time to reach the detector. The detector signal in all types of spectroscopy is processed (amplified, smoothed, derivatized, or otherwise transformed), and an output is obtained. The output may be graphical (e.g., a plot of intensity vs. wavelength or what we call a *spectrum*), tabular, or both. In some absorption spectrometers, the position of the sample and the dispersive device may be reversed. In AAS, for example, the sample is positioned between the source and the dispersive device for reasons discussed in Chapter 7:

1. Absorption spectroscopy:
   Source → Dispersive device → Sample → Detector → Data output

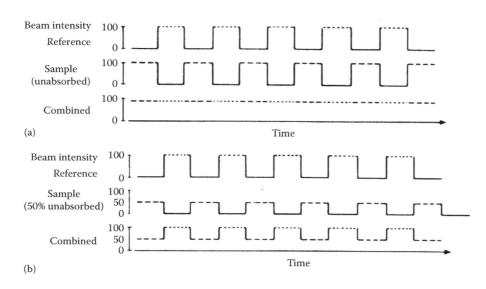

Figure 3.33 Radiation intensity reaching the detector using double-beam optics and a rotating disk beam splitter. (a) No absorption by the sample. (b) A 50% absorption by the sample.

2. Emission spectroscopy:
   Sample → Dispersive device → Detector → Data output
3. Fluorescence, phosphorescence, and scattering spectroscopies:
   Sample → Dispersive device → Detector → Data output
   ↑
   Dispersive device
   ↑
   Source

3.6.7 Fourier Transform Spectrometers

The aforementioned dispersive spectroscopy systems separate light into its component wavelengths and spread them into a spectrum. In these systems, the intensity can be measured at each point along a path where wavelength is proportional to position. The intensity over a narrow region around each point in the spectrum can be determined by slowly moving the grating so that each region of the dispersed spectrum passes by a single fixed detector. Alternatively, a system can measure all wavelength regions simultaneously with a continuous array of detectors. The latter approach acquires more information in less time. It is achieved in the UV/VIS region by employing one-dimensional photodiode arrays or 2D CCDs, similar to those found in modern digital cameras. These will be discussed in Chapters 4 through 8.

Detectors for the less-energetic IR wavelengths cannot be as easily miniaturized, so dispersive IR operates with the slow scanning approach. To obtain high wavelength resolution with scanning instruments requires restricting the wavelength region reaching the detector to a very narrow window. This in turn requires scanning the spectrum slowly to achieve a desired sensitivity.

FT spectrometers take a very different approach to acquiring a spectrum than that used in dispersive systems. It would be ideal to measure the light at all wavelengths simultaneously in a manner that will permit reconstruction of the intensity versus wavelength curve (i.e., the spectrum). If the wavelength information is encoded in a well-defined manner, such as by modulation of the light intensity using an interferometer, mathematical methods allow the information to be interpreted and presented as the same type of spectrum obtained from a dispersive instrument. An instrument that does this without a dispersive device is called a **multiplex** instrument. If all of the wavelengths of interest are collected at the same time without dispersion, the wavelengths and their corresponding intensities will overlap. The resulting overlapping information has to be sorted out in order to plot a spectrum. A common method of sorting or "deconvoluting" overlapping signals of varying frequency (or wavelength) is a mathematical procedure called Fourier analysis. The example presented here is of IR spectroscopy, its first application in instrumental analytical chemistry, but the principle is also employed with other techniques in which analytical data are displayed as a spectrum of response versus frequency, for example, NMR and ion cyclotron resonance MS.

Fourier analysis permits any continuous curve, such as a complex spectrum of intensity peaks and valleys as a function of wavelength or frequency, to be expressed as a sum of sine or cosine waves varying with time. Conversely, if the data can be *acquired* as the equivalent sum of these sine and cosine waves, they can be *Fourier transformed* into the spectrum curve. This requires data acquisition in digital form, substantial computing power, and efficient

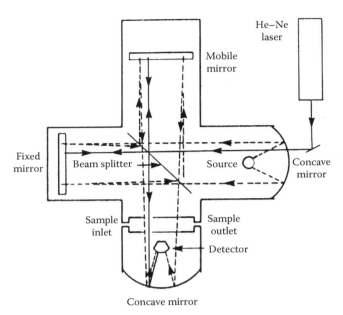

Figure 3.34 Schematic of a Michelson interferometer-based FTIR spectrometer.

software algorithms, all now readily available at the level of current-generation personal computers, laptops, and handheld devices. The computerized instruments employing this approach are called FT spectrometers—FTIR, FTNMR, and Fourier transform mass spectrometry (FTMS) instruments, for example.

FT optical spectroscopy uses an **interferometer** similar in design to that of the Michelson interferometer shown schematically in Figure 3.34. To simplify the discussion, we will initially consider a source that produces monochromatic radiation of wavelength λ. The source radiation strikes the beam splitter, which transmits half of the light to the fixed mirror and reflects the rest to a mobile mirror. The mobile mirror can be programmed to move at a precisely controlled constant velocity along the path of the beam. The beams are reflected from the mirrors back to the beam splitter. Half of each ray is directed through the sample holder to the detector. The other halves travel back in the direction of the source and do not need to be considered further. If the fixed and mobile mirrors are at exactly equal distances from the beam splitter, the two half beams will combine in phase. The combined wave will have twice the amplitude of each half, and the detector signal will be at its maximum. If the mobile mirror then moves a distance equal to $\lambda/4$, the two half beams will combine 180° (i.e., $\lambda/2$) out of phase. The beams interfere destructively and the detector registers no signal. For all other values of path difference between the mirrors, partial destructive interference occurs. As the mobile mirror moves at constant speed, the signal reaching the detector cycles regularly through this repetitive pattern of constructive and destructive interference. It maximizes when the path difference δ is an integral multiple of λ and

goes to zero when δ is a half-integral multiple of λ. In FTIR, δ is called the *retardation*. For monochromatic light, a plot of the signal (power, intensity) versus δ is called an **interferogram** and has the form of a simple pure cosine curve:

$$P(\delta) = B(\overline{u})\cos(2\pi\delta\overline{u}) \quad (3.24)$$

where

$P(\delta)$ is the amplitude of the signal reaching the detector
$B(\overline{u})$ is a frequency-dependent constant that accounts for instrumental variables such as detector response, the amount of light transmitted or reflected by the beam splitter, and the source intensity

The wavenumber $\overline{u}$ is equal to $1/\lambda$. The interferogram is the record of the interference signal reaching the detector. It is actually a "time-domain" spectrum; it records how the detector response changes with time. If the sample absorbs light at a specific frequency, the amplitude of that frequency changes. For a continuum source (a source with a continuously variable output of all wavelengths over the region of interest), the interferogram is a complex curve that can be represented as the sum of an infinite number of cosine waves and different amplitudes that reflect absorption of light by the sample. Although complex, if the interferogram is sampled at a sufficient number of points, modern computers using fast Fourier transform (FFT) can process the interferogram and identify enough of the cosine waves to permit deconvolution of the data into the IR spectrum plot of intensity versus wavelength. This will be discussed at greater length in Chapter 5 for FTIR and in Chapter 6 for FTNMR.

3.6.7.1 Advantages of FT Systems

Compared with dispersive systems, FT spectrometers produce better S/N ratios. This results from several factors. FT instruments have fewer optical elements and no slits, so the intensity of the radiation reaching the detector is much higher than in dispersive instruments. The increase in signal increases the S/N ratio. This is called the *throughput* advantage or Jacquinot advantage. All available wavelengths are measured simultaneously, so the time needed to collect all the data to form a spectrum is drastically reduced. An entire IR spectrum can be collected in less than 1 s. This makes practical the collection and signal averaging of hundreds of repetitions of the spectrum measurement. The theoretical improvement in S/N from signal averaging is proportional to the square root of the number of spectra averaged, $(n)^{1/2}$. This advantage is called the *multiplex* or Fellgett advantage.

FT spectrometers have high wavelength reproducibility. In an FTIR spectrometer, the position of the mobile mirror during its motion is continuously calibrated with extreme precision by parallel interferometry through the optical system using highly monochromatic light from a visible range laser, as seen in Figure 3.34. This accurate position measurement translates into accurate and reproducible analytical wavelength measurements after Fourier transformation of the interferogram. This accurate position measurement permits the precise addition of multiple spectra to achieve the multiplex advantage.

It should be noted that FT spectrometers are single-beam instruments. The background must be collected separately from the sample spectrum. The ratio of the two spectra results in a background-corrected spectra, similar to that obtained from a double-beam instrument. While the sample and background spectra are not collected at exactly the same time, because the spectra can be collected rapidly and processed rapidly, background spectra can be collected regularly to avoid the problems encountered with a single-beam dispersive instrument.

3.7 SPECTROSCOPIC TECHNIQUE AND INSTRUMENT NOMENCLATURE

Spectroscopy and spectroscopic instrumentation have evolved over many years. It is not surprising that the terminology used has also evolved and is in fact constantly evolving. Scientific terms are often defined by professional organizations, and sometimes these organizations do not agree on the definitions, leading to the use of the same term to mean different things. Scientists (and students) have to keep up to date on the meaning of terms in order to communicate effectively but must also know the older usage of terms in order to understand the literature. The term **spectroscopy** means the study of the interaction of electromagnetic radiation and matter. The term **spectrometry** is used for quantitative measurement of the intensity of one or more wavelengths of electromagnetic radiation. **Spectrophotometry** is a term reserved for absorption measurements, where the ratio of two intensities (sample and reference) must be measured, either simultaneously in a double-beam system or sequentially in a single-beam system. The term is gradually being replaced by spectrometry; for example, *atomic absorption spectrometry* is now more common than *atomic absorption spectrophotometry*.

The terms used for instruments generally distinguish how wavelengths are selected or the type of detector used. An optical spectrometer is an instrument that consists of a prism or grating dispersion device, slits, and a photoelectric detector to measure transmittance or absorbance. However, the term spectrometer is now also applied to IR interferometer-based FT systems that are nondispersive and have no slits. Spectrophotometer used to mean a double-beam spectrometer; however, the term is now used for both single-beam and double-beam dispersive spectrometers used for absorption measurements. A **photometer** is a spectroscopic instrument that uses a filter to select the wavelength instead of a dispersive device. A **spectrograph** is an instrument with a dispersive device that has a large aperture instead of a tiny exit slit and uses either photographic film for detection (now obsolete) or a solid-state imaging detector.

SUGGESTED EXPERIMENTS

3.1 You will need a UV/VIS spectrophotometer for this experiment and plastic or glass sample holders, either cuvettes or test tubes, depending on your instrument.
  a. Prepare suitable standard solutions of (1) 0.1 g $KMnO_4$/L of water, (2) 1.0 g $K_2Cr_2O_7$/L of water, and (3) water-soluble red ink diluted 50% with water.
  b. Measure the absorption from 700 to 350 nm and determine the wavelength of maximum absorption for each solution.
  c. Measure the transmittance I/I_0 at the wavelength of maximum absorption determined for each solution.

3.2
  a. Choose one of the standard solutions prepared in Experiment 3.1(a) and measure the transmittance at the wavelength where maximum absorption occurs. Take 50 mL of this solution (solution A) and transfer it to a 100 mL volumetric flask; make up to volume with deionized water. This is solution B. Measure and record the transmittance of solution B. Dilute solution B by 50% to obtain solution C. Measure the transmittance of solution C. Repeat this process to produce solutions D, E, and F.
  b. Prepare a graph correlating transmittance T and the concentrations of solutions A, B, C, D, E, and F. *Note*: You can use most commercial spreadsheet programs (e.g., Excel) or the software package on many spectrophotometers to do the curve fitting and graph.
  c. From the data obtained in step (a), calculate A, the absorbance of each solution. Prepare a graph correlating A, the absorbance, with the concentrations of solutions A, B, C, D, E, and F.
  d. Is one graph preferred over the other for use in obtaining information about concentrations of unknown samples? Why or why not?

3.3
  a. Add 10.0 mL of the $K_2Cr_2O_7$ solution prepared in Experiment 3.1 to 10.0 mL of the $KMnO_4$ solution prepared in Experiment 3.1. Mix well and measure the absorption spectrum.
  b. Add 10.0 mL of the $K_2Cr_2O_7$ solution prepared in Experiment 3.1 to 10.0 mL deionized water. Mix well and measure the absorption spectrum.
  c. Add 10.0 mL of the $KMnO_4$ solution prepared in Experiment 3.1 to 10.0 mL of deionized water. Mix well and measure the absorption spectrum. Using the wavelength of maximum absorption for

each compound, answer the following questions. Is there a change in the absorbance at the wavelengths of maximum absorption for the solution containing both compounds compared with the solutions containing a single compound? Is the total amount of light absorbed by the single solutions equal to the amount absorbed by the mixture? Would this change in absorbance (if any) be a source of error? Is the error positive or negative? Can you think of ways to correct for this error if you have to measure a mixture of potassium permanganate and potassium dichromate?

3.4 Measure the absorbance of a freshly prepared aqueous solution of $KMnO_4$ at its wavelength of maximum absorption. The concentration of the solution should be such that the absorbance is about 0.6–0.8. Make the solution in a volumetric flask and make your first measurement by pouring the solution into the sample holder directly from the flask. Now, pour the solution from the flask into a beaker or wide-mouth jar (you want to maximize the surface area). Leave the container open to the atmosphere for 5 min and then measure the absorbance of the solution again. Repeat measurements at 5 min intervals. (If no change is seen, cap the sample and shake it well, then uncap and allow it to sit in the air.) Plot the measured absorbance against the time exposed to the air. The change is caused by the chemical instability of the $KMnO_4$ (it reacts with the air). If it were being used as a standard solution for calibration, this change would be a source of error. Many standard solutions are subject to this sort of error to a greater or lesser extent, and precautions must be taken to prevent and avoid this source of trouble. Suggest two ways to avoid this problem with $KMnO_4$.

PROBLEMS

3.1 A molecule absorbs radiation of frequency 3.00×10^{14} Hz. What is the energy difference between the molecular energy states involved?

3.2 What frequency of radiation has a wavelength of 500.0 nm?

3.3 Describe the transition that occurs when an atom absorbs UV radiation.

3.4 Arrange the following types of radiation in order of increasing wavelength: IR, radiowaves, X-rays, UV, and visible light.

3.5 For a given transition, does the degree of absorption by a population of atoms or molecules depend on the number in the ground state or the excited state? Explain.

3.6 For a given transition, does the intensity of emission by a population of atoms or molecules depend on the number in the ground state or the excited state? Explain.

3.7 Briefly describe three types of transitions that occur in most molecules, including the type of radiation involved in the transition.

3.8 State the mathematical formulation of the Beer–Lambert–Bouguer law and explain the meaning of each symbol in the equation.

3.9 (a) Define transmittance and absorbance.
(b) What is the relationship between concentration and (1) transmittance, (2) absorbance?

3.10 Using Figure 3.16, calculate the slope of the tangent drawn through the lower point marked B by extending the line to cover a 10-fold difference in concentration. Confirm that the range shown for B–B for 1 % relative error (% RE) is correct by finding where on the upper portion of the curve you have a slope equal to the one you just calculated. Repeat the calculation for point C and confirm the C–C range.

3.11 The following data were obtained in an external standard calibration for the determination of iron by measuring the transmittance, at 510 nm and 1.00 cm optical path, of solutions of Fe^{2+} reacted with 1,10-phenanthroline to give a red-colored complex.

Fe Concentration (ppm)	%T	Fe Concentration (ppm)	%T
0.20	90.0	3.00	26.3
0.40	82.5	4.00	17.0
0.60	76.0	5.00	10.9
0.80	69.5	6.00	7.0
2.00	41.0	7.00	4.5

(a) Calculate A, the absorbance, for each solution, and plot A against concentration of iron. (You can do this using a spreadsheet program very easily.) Does the system conform to Beer's law over the entire concentration range? (b) Calculate the average molar absorptivity of iron when it is determined by this method. (c) Plot $(100 - \%T)$ against log concentration (Ringbom method). (1) What are the optimum concentration range and the maximum accuracy (% RE per 1 % transmittance error) in this range? (2) Over what concentration range will the relative analysis error per 1 % transmittance error not exceed 5 %?

3.12 The following data were obtained in a standard calibration for the determination of copper, as $Cu(NH_3)_4^{2+}$, by measuring the transmittance using a filter photometer.

Cu Concentration (ppm)	%T	Cu Concentration (ppm)	%T
0.020	96.0	0.800	27.8
0.050	90.6	1.00	23.2
0.080	84.7	1.40	17.2
0.100	81.4	2.00	12.9

Cu Concentration (ppm)	%T	Cu Concentration (ppm)	%T
0.200	66.7	3.00	9.7
0.400	47.3	4.00	8.1
0.600	35.8		

Calculate A, the absorbance, for each solution, and plot A against the concentration of copper. (You can do this using a spreadsheet program very easily.) Does the system, measured under these conditions, conform to Beer's law over the entire concentration range? Is any deviation from the law of small or of large magnitude? Suggest a plausible cause for any deviation.

3.13 An amount of 0.200 g of copper is dissolved in nitric acid. Excess ammonia is added to form $Cu(NH_3)_4^{2+}$ and the solution is made up to 1 L. The following aliquots of the solution are taken and diluted to 10.0 mL: 10.0, 8.0, 5.0, 4.0, 3.0, 2.0, and 1.0 mL. The absorbances of the diluted solution were 0.500, 0.400, 0.250, 0.200, 0.150, 0.100, and 0.050, respectively. A series of samples was analyzed for copper concentration by forming the $Cu(NH_3)_4^{2+}$ complex and measuring the absorbance. The absorbances were (a) 0.450, (b) 0.300, and (c) 0.200. What were the respective concentrations in the three copper solutions? If these three samples were obtained by weighing out separately (a) 1.000 g, (b) 2.000 g, and (c) 3.000 g of sample and dissolving and diluting to 10.0 mL, what was the original concentration of copper in each sample?

3.14 Describe the standard addition method for measuring the concentration of an unknown. What are the advantages of this method of calibration?

3.15 Describe the use of an internal standard for calibration. What characteristics must a species possess to serve as an internal standard? What are the advantages of the internal standard method?

3.16 Describe what you would do for samples whose absorbances fell above the absorbance of your highest calibration standard.

3.17 What range of % transmittance results in the smallest relative error for an instrument limited by (a) noise in the thermal detector of an IR spectrometer? (b) shot noise?

3.18 What is A if the percentage of light absorbed is (a) 90 %, (b) 99 %, (c) 99.9 %, and (d) 99.99 %?

3.19 What is the purpose of having and measuring a reagent blank?

3.20 An optical cell containing a solution was placed in a beam of light. The original intensity of the light was 100 units. After being passed through the solution, its intensity was 80 units. A second similar cell containing more of the same solution was also placed in the light beam behind the first cell. Calculate the intensity of radiation emerging from the second cell.

3.21 The transmittance of a solution of unknown concentration in a 1.00 cm cell is 0.700. The transmittance of a standard solution of the same material is also 0.700. The concentration of the standard solution is 100.0 ppm; the cell length of the standard is 4.00 cm. What is the concentration of the unknown solution?

3.22 A solution contains 1.0 mg of $KMnO_4$/L. When measured in a 1.00 cm cell at 525 nm, the transmittance was 0.300. When measured under similar conditions at 500 nm, the transmittance was 0.350. (a) Calculate the absorbance A at each wavelength. (b) Calculate the molar absorptivity at each wavelength. (c) What would T be if the cell length were in each case 2.00 cm? (d) Calculate the absorptivity if concentration is in mg/L for the solution at each wavelength.

3.23 A series of standard ammoniacal copper solutions was prepared and the transmittance measured. The following data were obtained:

Cu Concentration	Transmittance	Sample	Transmittance
0.20	0.900	1	0.840
0.40	0.825	2	0.470
0.60	0.760	3	0.710
0.80	0.695	4	0.130
1.00	0.635		
2.00	0.410		
3.00	0.263		
4.00	0.170		
5.00	0.109		
6.00	0.070		

Plot the concentration against absorbance (use your spreadsheet program). The transmittance of solutions of copper of unknown concentrations was also measured in the same way, and the sample data in this table were obtained. Calculate the concentration of each solution. What is missing from this experiment? List two things a good analytical chemist should have done to be certain that the results are accurate and precise.

3.24 List the components of a single-beam optical system for absorption spectroscopy. List the components of single-beam optical system for emission spectroscopy.

3.25 Describe the components in a grating monochromator. Briefly discuss the role of each component.

3.26 State the equation for the resolution of a grating.

3.27 (a) Define mechanical slit width.
(b) Define spectral bandpass or bandwidth.

3.28 What is the effect of mechanical slit width on resolution?

3.29 Write the expression for resolution of a grating ruled to be most efficient in second order. To resolve a given pair of wavelengths, will you need more or fewer grooves if the grating were ruled in first order?

3.30 What resolution is required to separate two lines λ_1 and λ_2?

3.31 What resolution is required to resolve the Na D lines at 589.0 and 589.5 nm in first order?

3.32 How many grooves must be illuminated on a grating to separate the Na D lines in second order?

3.33 Briefly describe how a holographic grating is produced. What is the advantage of this approach over a mechanically ruled grating?

3.34 A grating contains 1000 grooves. Will it resolve two lines of λ 500.0 and 499.8 nm in first order if all 1000 grooves are illuminated?

3.35 What are the components of a double-beam system? Describe two types of beam splitters.

3.36 How does a double-beam system correct for drift? Draw the alternating signal output from a doublebeam system for a sample that absorbs 25% of the incident light.

3.37 Give an example of an absorption filter. Over what wavelength range do absorption filters function as wavelength selectors?

3.38 What are the advantages of absorption filters as wavelength selectors compared with gratings? What are the disadvantages?

3.39 Light of 300.0 nm is diffracted from a grating in first order at an angle of incidence normal to the grating (i.e., $i = 0°$). The grating contains 1180 grooves/mm. Calculate the angle of diffraction, θ, for this wavelength.

3.40 If an emission line for magnesium appears at 285.2 nm in first order, where will it appear in second order? Where will it appear in third order? If you needed to measure a first-order iron emission line at 570 nm, will the presence of magnesium in the sample cause a problem? What can you do to solve the problem if one exists?

3.41 What are the major differences between an FT system and a dispersive system for spectroscopy?

3.42 Define the throughput advantage. How does it arise?

3.43 Define the multiplex advantage. How does it arise?

3.44 Draw two cosine waves of the same amplitude in phase. Draw the resulting wave if the two waves are combined. Draw two cosine waves 180° out of phase. Draw the resulting wave if these two waves are combined.

3.45 What is the difference between a spectrometer and a photometer? What is the difference between a spectrometer and a spectrograph?

BIBLIOGRAPHY

The reader will note that occasionally multiple editions of books are cited as sources or reference works. The reason for this is that sometimes different figures are used in different editions, which may in fact be written by different authors. The sources found in the bibliography will be the most recent versions of a particular work.

In some cases instrument manufacturers and suppliers are cited by providing a URL to the exact internet location. We have endeavored to ensure that these links are functional at the time of publication, but clearly these links change. If a link has become nonfunctional, a good approach is to use your browser's advanced search capability. For example: "company name" AND "instrument type" AND "specific descriptors" is a good place to start.

Ayres, G.H. *Quantitative Chemical Analysis*, 2nd edn. Harper & Row: New York, 1968.

Beaty, R.D.; Kerber, J.D. *Concepts, Instrumentation and Techniques in Atomic Absorption Spectrophotometry.* PerkinElmer, Inc.: Norwalk, CT, 1993.

Boss, C.B.; Fredeen, K.J. *Concepts, Instrumentation and Techniques in Inductively Coupled Plasma Optical Emission Spectrometry*, 2nd edn. PerkinElmer, Inc.: Norwalk, CT, 1997.

Dean, J.A.; Rains, T.C. (eds.) *Flame Emission and Absorption Spectrometry*, Vol. 2. Marcel Dekker, Inc.: New York, 1971.

Harris, D.C. *Quantitative Chemical Analysis*, 5th edn. W.H. Freeman and Company: New York, 1999.

Hollas, J.M. *Modern Spectroscopy*. John Wiley & Sons, Ltd.: Chichester, U.K., 1996.

Ingle, J.D.; Crouch, S.R. *Spectrochemical Analysis*. Prentice Hall, Inc.: Englewood Cliffs, NJ, 1988.

Koenig, J.L. Industrial problem solving with molecular spectroscopy *Anal. Chem.*, 66(9), 515A, 1994.

Meehan, E.J. *Optical Methods of Analysis. Treatise on Analytical Chemistry*, 2nd edn. John Wiley & Sons, Ltd.: Chichester, U.K., 1981.

Settle, F.A. (ed.) *Handbook of Instrumental Techniques for Analytical Chemistry*. Prentice Hall PTR: Upper Saddle River, NJ, 1997.

Skoog, D.A.; Holler, F.J.; Nieman, T.A. *Principles of Instrumental Analysis*, 5th edn. Harcourt, Brace and Company: Orlando, FL, 1998.

Willard, H.H.; Merrit, L.L.; Dean, J.A.; Settle, F.A. *Instrumental Methods of Analysis*, 7th edn. Van Nostrand: New York, 1988.

Zumdahl, S.S.; Zumdahl, S.A. *Chemistry*, 5th edn. Houghton Mifflin: Boston, MA, 2000.

CHAPTER 4

Visible and Ultraviolet Molecular Spectroscopy

4.1 INTRODUCTION

One of the first physical methods used in analytical chemistry was based on the quality of the color in colored solutions. The first things we observe regarding colored solutions are their *hue* or color and the color's *depth* or *intensity*. These observations led to the technique historically called *colorimetry*; the color of a solution could identify species (qualitative analysis), while the intensity of the color could identify the concentration of the species present (quantitative analysis). This technique was the first use of what we now understand to be absorption spectroscopy for chemical analysis. When white light passes through a solution and emerges as red light, we say that the solution is red. What has actually happened is that the solution has allowed the red component of white light to pass through, whereas it has absorbed the complementary colors: yellow and blue. The more concentrated the sample solution, the more yellow and blue light is absorbed and the more intensely red the solution appears to the eye. For a long time, experimental work made use of the human eye as the detector to measure the hue and intensity of colors in solutions. However, even the best analyst can have difficulty comparing the intensity of two colors with slightly different hues, and there are of course people who are color blind and cannot see certain colors. Instruments have been developed to perform these measurements more accurately and reliably than the human eye. While the human eye can only detect visible light, this chapter will focus on both the ultraviolet (UV) and the visible portions of the spectrum.

The wavelength range of UV radiation starts at the blue end of visible light (about 400 nm) and ends at approximately 200 nm for spectrometers operated in air. The radiation has sufficient energy to excite valence electrons in many atoms and molecules; consequently, UV radiation is involved with electronic excitation. Visible light, considered to be light with wavelengths from 800 to 400 nm, acts in the same way as UV light. It is also considered part of the electronic excitation region. For this reason, we find commercial spectroscopic instrumentation often operates with wavelengths between 800 and 200 nm. Spectrometers of this type are called UV/visible (UV/VIS) spectrometers. The vacuum UV region of the spectrum extends below 200 nm to the X-ray region of the spectrum at ~100 Å. It is called the vacuum UV region because oxygen, water vapor, and other molecules in air absorb UV radiation below 200 nm, so the spectrometer light path must be free of air to observe wavelengths <200 nm. The instrument must be evacuated (kept under vacuum) or purged with an appropriate non-UV-absorbing gas such as helium for this region to be used. Vacuum UV radiation is also involved in electronic excitation, but the spectrometers are specialized and not commonly found in undergraduate analytical laboratories. For our purposes, the term UV will mean radiation between 200 and 400 nm, unless stated otherwise. The major types of analytical spectroscopy operating within this wavelength range are listed in Table 4.1. This chapter will focus on molecular spectroscopy—the absorption and emission of UV and visible radiation by molecules and polyatomic species. We will also look at the use of scattering of visible light to provide information about macromolecules and particles. Atomic absorption spectroscopy is covered in Chapter 7 and atomic emission spectroscopy in Chapter 8.

The interaction of UV and visible radiation with matter can provide qualitative identification of molecules and polyatomic species, including ions and complexes. Structural information about molecules and polyatomic species, especially organic molecules, can be acquired. This qualitative information is usually obtained by observing the UV/VIS spectrum, the absorption of UV and visible radiation as a function of wavelength by molecules. Typical UV absorption spectra of some organic compounds are shown in Figures 4.1a through c. The spectrum may be plotted as wavelength versus absorbance, transmittance, or molar absorptivity, ε. The molar absorptivity is defined subsequently. In Figure 4.1, the absorption spectrum of pyridine dissolved in ethanol is plotted as log ε versus wavelength in ångstroms (Å). Figures 4.1b and c are plotted as absorbance versus wavelength in nm. The older wavelength unit seen in these spectra was µm; this unit has been replaced by the Systeme International d'Unites (SI) unit nm. Note the need for dilution in Figures 4.1b and c. In Figure 4.1b, the solution was diluted 1:10 to bring the peak at 275 nm on scale. In Figure 4.1c, the absorbance peaks at 230 and 222 nm are

Table 4.1 Spectroscopy using UV and visible light

Function	Analytical Field	Analytical Application
Atomic spectroscopy		
Absorption of UV/VIS radiation	Atomic absorption spectrometry (AAS)	Quantitative elemental analysis
Emission of UV/VIS radiation	Flame photometry, atomic emission spectroscopy (AES)	Qualitative and quantitative multielement analysis
Emission of UV/VIS radiation	Atomic fluorescence spectroscopy (AFS)	Quantitative elemental analysis of ultratrace concentrations (sub-ppb)
Molecular spectroscopy		
Absorption of UV/VIS radiation	UV/VIS molecular absorption spectroscopy, spectrophotometry	Qualitative and quantitative determinations of aromatic and unsaturated organic compounds, including natural products; direct and indirect quantitative determination of inorganic ions, organic molecules, and biochemicals
Emission of UV/VIS radiation	Molecular fluorescence, molecular phosphorescence	Detection of small quantities (<ng) of certain aromatic compounds and natural products, analysis of gels and glasses, determination of organic and inorganic species by "tagging"

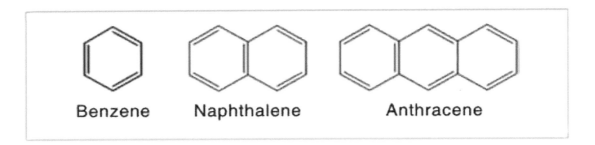

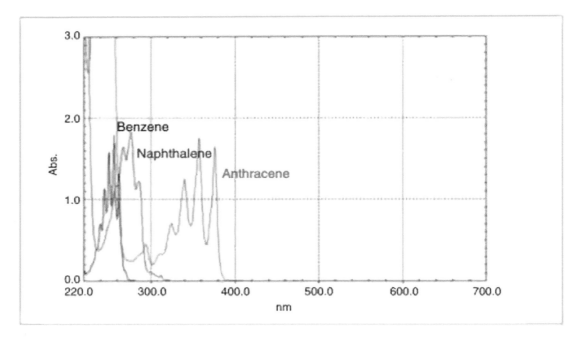

Figure 4.1 Typical UV absorption spectra for several organic molecules in solution. At the top are the structures for benzene, naphthalene, and anthracene, and below are the UV absorption spectra of those compounds dissolved in ethanol. (Reprinted with permission of Shimadzu Inc.)

VISIBLE AND ULTRAVIOLET MOLECULAR SPECTROSCOPY

602 UV

BENZONITRILE

IR 2255

Mol. Form.	C_7H_5N			
Mol. Wt.	103.12	B. P. 188-189°C		
Source	The Matheson Co., Inc., E. Rutherford, N. J.			

		A	B	C	D	E
Methanol	Conc. g/L	0.140	0.140	0.140	0.0380	0.0380
	Cell mm	5	5	5	2	2
	a_m	932	1010	804	9100	10600
	λ Max. mμ	277	269.5	263	230	222

Methanol KOH	Conc. g/L					
	Cell mm			Checked - no change		
	a_m					
	λ Max. mμ					

Methanol HCl	Conc. g/L					
	Cell mm			Checked - no change		
	a_m					
	λ Max. mμ					

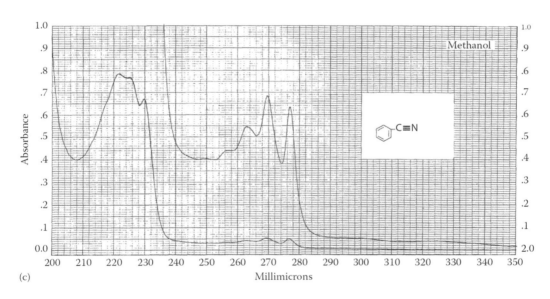

Figure 4.1 (Continued) Typical UV absorption spectra for several organic molecules in solution. (c) The UV spectrum of benzonitrile dissolved in methanol. (b and c: Copyright Bio-Rad Laboratories, Informatics Division, Sadtler Software and Databases, 2022. All rights reserved.)

off scale at a concentration of 0.140 g/L; the solution was diluted to a concentration of 0.038 g/L and run in a smaller path length cell.

Quantitative information can also be obtained by studying the absorption or emission of UV and visible radiation by molecules or polyatomic species. As a very simple example, we can look at the absorption spectrum of a red solution such as red ink in water (Figure 4.2). It can be seen that with a colorless sample of pure water, shown as the dotted line, all wavelengths of white light, including all of the red wavelengths, are transmitted through the sample. If we add one drop of red ink to water, to make a solution that appears pale red, the spectrum shows that some of the blue and some of the yellow light have been absorbed, but all of the red light has been transmitted. If we add more red ink to the water to make a dark red solution, most of the blue and yellow light has been absorbed, but again all of the red light has been transmitted. The amount of red light falling on the

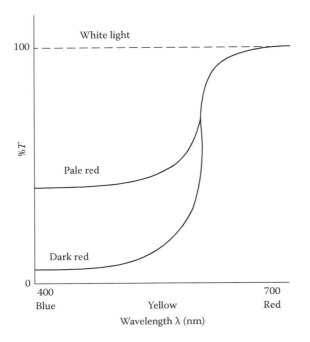

Figure 4.2 Visible absorption spectra of a colorless sample of pure water (dotted line), a pale red solution of red ink in water, and a dark red solution of red ink in water. Note that none of the samples absorbs red light.

eye or the detector is the same in each case; the amount of ink in the solution is related to the blue and yellow light absorbed, not to the color transmitted. We could construct a series of known amounts of red ink in water and quantitatively measure other ink solutions by measuring the amount of light absorbed at, for example, 450 nm. Concentrations of species in samples, especially solutions, are often measured using UV/VIS absorption spectrometry or fluorescence spectrometry. The measurement of concentrations or changes in concentrations can be used to calculate equilibrium constants, reaction kinetics, and stoichiometry for chemical systems. Quantitative measurements by UV/VIS spectrometry are important in environmental monitoring, industrial process control including food and beverage manufacturing, pharmaceutical quality control, and clinical chemistry, to name just a few areas. The emission of radiation by molecules may occur in several ways following excitation of the molecule; two processes are fluorescence and phosphorescence. These processes will be discussed in Sections 4.8 through 4.10.

4.1.1 Electronic Excitation in Molecules

Molecules are composed of atoms that are held together by sharing electrons to form chemical bonds. Electrons in molecules move in molecular orbitals at discrete energy levels as defined by quantum theory. When the energy of the electrons is at a minimum, the molecules are in the lowest energy state, or ground state. The molecules can absorb radiation and move to a higher-energy state, or *excited state*. When the molecule becomes excited, an outer shell (valence) electron moves to an orbital of higher energy. The process of moving electrons to higher-energy states is called *electronic excitation*. For radiation to cause electronic excitation, it must be in the visible or UV region of the electromagnetic spectrum.

The frequency absorbed or emitted by a molecule and the energy of radiation are related by $\Delta E = h\nu$. The actual amount of energy required depends on the difference in energy between the ground state E_0 and the excited state E_1 of the electrons. The relationship is described by

$$\Delta E = E_1 - E_0 = h\nu \tag{4.1}$$

where
E_1 is the energy of the excited state
E_0 is the energy of the ground state

You may want to review the topics of bonding, molecular orbitals, Lewis structures, and organic chemistry in your general chemistry textbook or in the texts by Chang or Zumdahl listed in the bibliography to help you understand the material discussed subsequently. The discussion will focus on organic molecules, as the bonding is relatively easy to understand. Inorganic molecules also undergo absorption and emission of UV and visible radiation, as do complexes of organic molecules with metal ions, but the bonding in inorganic molecules and complexes of the transition metals and heavier elements is complicated due to electrons in the d and f orbitals.

Three distinct types of electrons are involved in valence electron transitions in molecules. First are the electrons involved in single bonds, such as those between carbon and hydrogen in alkanes. These bonds are called *sigma* (σ) *bonds*. The amount of energy required to excite electrons in σ bonds is usually more than UV photons of wavelengths >200 nm possess. For this reason, alkanes and other saturated compounds (compounds with only single bonds) do not absorb UV radiation and are therefore frequently very useful as transparent solvents for the study of other molecules. An example of such a nonabsorbing compound is the alkane hexane, C_6H_{14}.

Next, we have the electrons involved in double and triple (unsaturated) bonds. These bonds involve a *pi* (π) *bond*. Typical examples of compounds with π bonds are alkenes, alkynes, conjugated olefins, and aromatic compounds (Figure 4.3). Electrons in π bonds are excited relatively easily; these compounds commonly absorb in the UV or visible region.

Electrons that are not involved in bonding between atoms are the third type of electrons in molecules. These

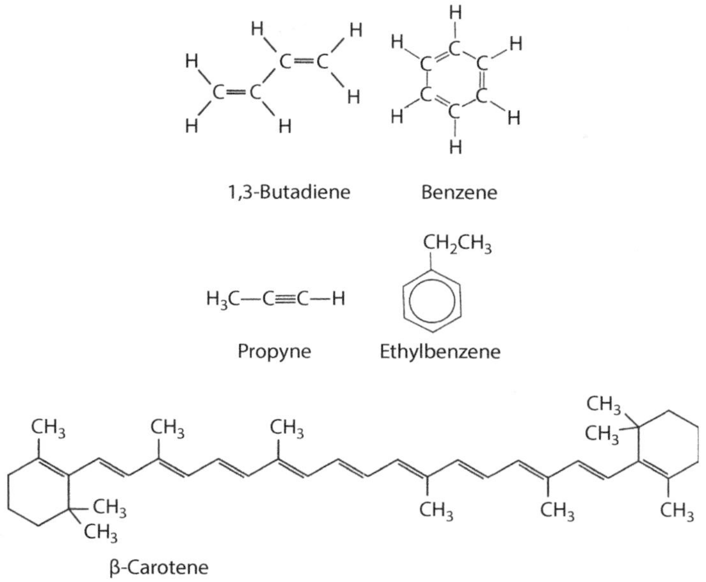

VISIBLE AND ULTRAVIOLET MOLECULAR SPECTROSCOPY

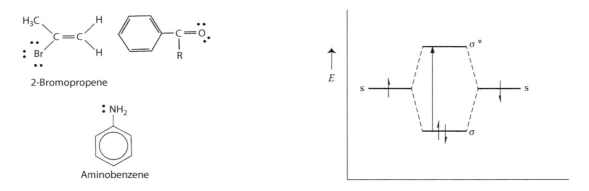

Figure 4.3 Examples of organic molecules containing π bonds. Note that benzene rings can be drawn showing three π bonds (the Kekulé structure) or with a circle inside the ring, as has been done for ethylbenzene, to more accurately depict the delocalized nature of the π electrons in aromatic compounds.

Figure 4.4 Examples of organic molecules with nonbonding electrons. The n electrons are represented as pairs of dots around the atom on which they are located. For the carbonyl compound, if R = H, the compound is an aldehyde; if R = an organic group, the compound is a ketone.

Figure 4.5 Schematic energy diagram of two s orbitals on adjacent atoms forming a σ bonding orbital and a σ* antibonding orbital.

are called n electrons, for nonbonding electrons. In saturated hydrocarbons, the outer shell electrons of carbon and hydrogen are all involved in bonding; hence, these compounds do not have any n electrons. Organic compounds containing nitrogen, oxygen, sulfur, or halogens, however, frequently contain electrons that are nonbonding (Figure 4.4). Because n electrons are usually excited by UV or visible radiation, many compounds that contain n electrons absorb UV/VIS radiation.

A schematic energy diagram of two s electrons in atomic orbitals on adjacent atoms combining to form a σ bond is shown in Figure 4.5. Orbitals are conserved; therefore, two molecular orbitals are formed: a sigma bonding orbital and a higher-energy sigma antibonding orbital. The antibonding orbital is denoted σ* (called a sigma star orbital). The energy difference between σ and σ* is equal to ΔE, shown by the large arrow. Remember that each atom has three 2p atomic orbitals. One of those p orbitals can overlap with a p orbital on an adjacent atom to form a second set of sigma orbitals. Sideways overlap of the other two p orbitals is possible, resulting in pi bonding and antibonding orbitals. The schematic energy diagram for the formation of one set of π orbitals is shown in

A relative energy diagram of σ, π, and n electrons is shown in Figure 4.8, although there are exceptions to this general order. It can be seen that the energy required to excite an electron from a σ to a σ* orbital is considerably greater than that required to excite an electron from a π to a π* orbital or an n electron to either a σ* or a π* orbital. As a consequence, the energy necessary to excite σ electrons to σ* orbitals is greater than that available in the UV region, but usually UV radiation is sufficient to excite electrons in π orbitals to π* antibonding orbitals or n electrons to π* or σ* antibonding orbitals.

4.1.2 Absorption by Molecules

Quantum mechanics provides a theoretical basis for understanding the relative energy levels of molecular orbitals and how they vary with structure. Quantum mechanics also generates a set of "selection rules" to predict what transitions occur in molecules. The transitions that occur in molecules are governed by these quantum mechanical selection rules. Some transitions are "allowed" by the selection rules, while others are "forbidden." The selection rules are beyond the scope of this text but may be found in most physical chemistry texts or in the text by Ingle and Crouch listed in the bibliography. As is often the case with rules, there are exceptions, and many forbidden transitions do occur and can be seen in UV/VIS spectra.

When molecules are electronically excited, an electron moves from the highest occupied molecular orbital to the lowest unoccupied orbital, which is usually an antibonding orbital. Electrons in π bonds are excited to antibonding π^* orbitals, and n electrons are excited to either σ^* or π^* orbitals.

Both organic and inorganic molecules may exhibit absorption and emission of UV/VIS radiation. Molecular groups that absorb visible or UV light are called *chromophores*, from the Greek word *chroma*, color. For example, for a $\pi \rightarrow \pi^*$ transition to occur, a molecule must possess a chromophore with an unsaturated bond, such as C=C, C=O, and C=N. Compounds with these types of chromophores include alkenes, amides, ketones, carboxylic acids, and oximes, among others. The other transition that commonly occurs in the UV/VIS region is the $n \rightarrow \pi^*$ transition, so organic molecules that contain atoms with nonbonded electrons should be able to absorb UV/VIS radiation. Such atoms include nitrogen, oxygen, sulfur, and the halogen atoms, especially Br and I. Table 4.2 presents some typical organic functional groups that serve as chromophores. Table 4.3 lists types of organic compounds and the wavelengths of their absorption maximum, that is, the wavelength at which the most light is absorbed. Some

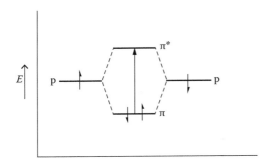

Figure 4.6 Schematic energy diagram of two p orbitals on adjacent atoms forming a π bonding orbital and a π^* antibonding orbital.

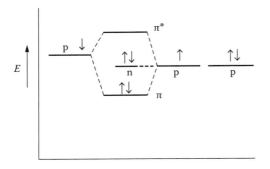

Figure 4.7 The relative energy levels of the π, π^*, and nonbonding (n) orbitals formed from p orbitals on adjacent atoms.

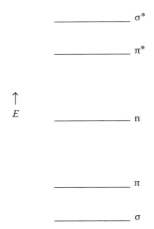

Figure 4.8 The relative energy levels of a set of σ, π, and n orbitals and the associated antibonding orbitals.

Figure 4.6, and the energy difference between the π orbital and the antibonding π^* (pi star) orbital is shown by the large arrow. If a p orbital is filled with a pair of electrons, it will have no tendency to form a bond. Figure 4.7 shows that a filled atomic p orbital (in the atom on the right) may form a nonbonding n orbital that is not shifted in energy from the atomic orbital, while the partially filled p orbitals on each atom overlap to form a pair of pi bonding and antibonding orbitals.

Table 4.2 Organic functional groups that can absorb UV/VIS radiation

Functional Group	Chemical Structure	Electronic Transitions
Acetylenic	–C≡C–	$\pi \rightarrow \pi^*$
Amide	–CONH$_2$	$\pi \rightarrow \pi^*$, $n \rightarrow \pi^*$
Carbonyl	>C=O	$\pi \rightarrow \pi^*$, $n \rightarrow \pi^*$
Carboxylic acid	–COOH	$\pi \rightarrow \pi^*$, $n \rightarrow \pi^*$
Ester	–COOR	$\pi \rightarrow \pi^*$, $n \rightarrow \pi^*$
Nitro	–NO$_2$	$\pi \rightarrow \pi^*$, $n \rightarrow \pi^*$
Alkene	>C=C<	$\pi \rightarrow \pi^*$
Organoiodide	R–I	$n \rightarrow \sigma^*$
Thiol	R–SH	$n \rightarrow \sigma^*$

Note: R = any organic group (e.g., CH$_3$, C$_2$H$_5$, C$_6$H$_5$).

Table 4.3 Absorption wavelengths of typical organic functional groups

Chromophore	System	Wavelength of Absorption Maximum, λ_{max} (nm)
Amine	$-NH_2$	195
Bromide	$-Br$	208
Iodide	$-I$	260
Thioketone	$>C=S$	460
Thiol	$-SH$	220
Ester	$-COOR$	205
Aldehyde	$-CHO$	210
Carboxylic acid	$-COOH$	200–210
Nitro	$-NO_2$	210–270
Nitrile	$-C \equiv N$	<160
Azo	$-N=N-$	285–400
Conjugated olefins	$(-HC=CH-)_2$	210–230
	$(-HC=CH-)_3$	260
	$(-HC=CH-)_5$	330
	$(-HC=CH-)_{10}$	460
Benzene	⌬	198
		255
Naphthalene	⌬⌬	210
		220
		275

compounds have more than one absorption peak, so several "maxima" are listed. Compounds such as alkanes (also called paraffins) contain only σ bonds, which do not absorb radiation in the visible or UV region.

Transition metal compounds are often colored, indicating that they absorb light in the visible portion of the spectrum. This is due to the presence of unfilled d orbitals. The exact wavelength of the absorption band maximum depends on the number of d electrons, the geometry of the compound, and the atoms coordinated to the transition metal.

4.1.3 Molar Absorptivity

Beer's law, which relates absorbance of a sample to the path length and concentration of absorbing species, was covered in Chapter 3. The proportionality constant, a, in Beer's law is the absorptivity of the absorbing species.

The *absorptivity*, a, of a molecule defines how much radiation will be absorbed by that molecule at a given concentration and at a given wavelength. If the concentration is expressed in molarity (mol/L, M), the absorptivity is defined as the *molar absorptivity*, ε. The absorptivity can be calculated directly from the measured absorbance using Beer's law:

$$A = abc = \varepsilon bc \qquad (4.2)$$

where
A is the absorbance
b is the path length
c is the concentration of the absorbing species

If b is in units of cm and c has units of molarity, then the proportionality constant is the molar absorptivity (sometimes called the extinction coefficient, or molar extinction coefficient) and is given the symbol ε, with units of $M^{-1}cm^{-1}$. Commonly $\varepsilon \approx 10^4$–10^5 $M^{-1}cm^{-1}$ for an allowed transition and is on the order of 10–100 for a forbidden transition. The magnitude of the absorptivity is an indication of the probability of the electronic transition. High values of ε give rise to strong absorption of light at the specified wavelength; low values of ε result in weak absorption of light. Both a and ε are constants for a given wavelength and are physical properties of the molecule. The molar absorptivity may be specified for any wavelength but is usually tabulated for the wavelength at which maximum absorption of light occurs for a molecule. The wavelength of maximum absorption is symbolized by λ_{max}, and the associated ε is symbolized as ε_{max}. Table 4.4 presents typical values for λ_{max} and ε_{max} for some common organic molecules. Note that some chromophores have more than one maximum, with a different molar absorptivity. We also note that while UV-VIS spectra are not as useful for organic structure determination, this table can provide valuable help as a correlation table similar to those for other spectroscopic techniques.

The absorptivity is not a direct measure of the probability that a given electronic transition will occur. This is because absorbance is measured over a wavelength range that is much smaller than the width of the absorption band. The absorptivity will differ at different wavelengths over the band profile. There are several fundamental quantities that are directly related to the transition probability; these include Einstein coefficients and the oscillator strength, f. The text by Ingle and Crouch presents the derivation of these quantities for the interested student.

4.1.4 Shape of UV Absorption Curves

Figure 4.1 shows "typical" UV absorption spectra for several organic molecules in solution. Each spectrum appears to be very simple, with a broad absorption "band" over a wide wavelength range instead of the numerous, narrower absorptions seen in infrared (IR) spectra (Chapter 5). The absorption bands are broad because each electronic energy level has multiple vibrational and rotational energy levels associated with it. Excitation from the ground electronic state can occur to more than one vibrational level and to more than one rotational level. A schematic representation of an electronic transition with vibrational and rotational sublevels is shown in Figure 4.9. In the ground state,

Table 4.4 Typical absorption maxima and molar absorptivities for common chromophores

Chromophore	Functional Group	λ_{max} nm	ε_{max}	λ_{max} nm	ε_{max}	λ_{max} nm	ε_{max}
Ether	-O-	185	1000				
Thioether	-S-	194	4600	215	1600		
Amine	-NH$_2$-	195	2800				
Amide	-CONH$_2$	<210	—				
Thiol	-SH	195	1400				
Disulfide	-S-S-	194	5500	255	400		
Bromide	-Br	208	300				
Iodide	-I	260	400				
Nitrile	-C≡N	160	—				
Acetylide (alkyne)	-C≡C-	175–180	6000				
Sulfone	-SO$_2$-	180	—				
Oxime	-NOH	190	5000				
Azido	>C=N-	190	5000				
Alkene	-C=C-	190	8000				
Ketone	>C=O	195	1000	270–285	18–30		
Thioketone	>C=S	205	strong				
Esters	-COOR	205	50				
Aldehyde	-CHO	210	strong	280–300	11–18		
Carboxyl	-COOH	200–210	50–70				
Sulfoxide	>S→O	210	1500				
Nitro	-NO$_2$	210	strong				
Nitrite	-ONO	220–230	1000–2000	300–4000	10		
Azo	-N=N-	285–400	3–25				
Nitroso	-N=O	302	100				
Nitrate	-ONO$_2$	270 (shoulder)	12				
Conjugated hydrocarbon	-(C=C)$_2$- (acyclic)	210–230	21,000				
Conjugated hydrocarbon	-(C=C)$_3$-	260	35,000				
Conjugated hydrocarbon	-(C=C)$_4$-	300	52,000				
Conjugated hydrocarbon	-(C=C)$_5$-	330	118,000				
Conjugated hydrocarbon	-(C=C)$_2$- (alicyclic)	230–260	3000–8000				
Conjugated hydrocarbon	C=C-C≡C	219	6500				
Conjugated system	C=C-C=N	220	23,000				
Conjugated system	C=C-C=O	210–250	10,000–20,000			300–350	weak
Conjugated system	C=C-NO$_2$	229	9500				
Benzene		184	46,700	202	6900	255	170
Diphenyl				246	20,000		
Naphthalene		220	112,000	275	5600	312	175
Anthracene		252	199,000	375	7900		
Pyridine		174	80,000	195	6000	251	1700

Table 4.4 (Continued) Typical absorption maxima and molar absorptivities for common chromophores

Chromophore	Functional Group	λ_{max} nm	λ_{max}	λ_{max} nm	λ_{max}	λ_{max} nm	λ_{max}
Quinoline		227	37,000	270	3600	314	2750
Isoquinoline		218	80,000	266	4000	317	3500

Source: taken from the book by Bruno and Svoronos.

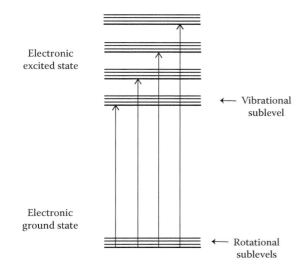

Figure 4.9 An electronic transition occurs over a band of energy due to the multiple vibrational and rotational sublevels associated with each electronic state. This schematic depicts four of the many possible transitions that occur. The length of the arrow is proportional to the energy required for the transition, so a molecular electronic transition consists of many closely spaced transitions, resulting in a band of energy absorbed rather than discrete line absorption.

only the lowest vibrational level is shown, with four rotational sublevels. At room temperature, most molecules are in the ground state in the lowest vibrational state. In the excited state, four vibrational sublevels are shown, slightly separated, with four rotational sublevels in each. Only four of the many possible transitions are shown; each arrow represents an absorption wavelength. The electronic transition consists of a large number of wavelengths that overlap to give the "continuous" absorption band observed. Even though each separate transition is quantized, the close energy spacing of the vibrational levels and the even more closely spaced rotational sublevels cause the electronic transition to appear as a broad band. This is shown schematically in Figure 4.10.

An absorption band is characterized by its shape, that is, by its width and intensity. The shape of the band is determined primarily by the vibrational energy-level spacing and the intensity of each vibrational transition. The intensity distribution is related to the probability of the transition to a given vibrational sublevel. The transition probabilities can be determined using the Franck–Condon principle. The texts by Hollas and Lambert et al. may be consulted for details. Suffice it to say that if we have a million molecules, even if they are mostly in the ground vibrational state before excitation, they may be in various vibrational states after excitation. The radiant energy required to cause electronic excitation to each vibrational energy level is slightly different and is further modified by rotational energy changes. For this reason, when UV radiation falls on the million molecules, it is absorbed at numerous wavelengths. The total range of the absorption wavelengths may stretch over 100 nm. The effect of the rotational energy of the molecule is to add even more absorption lines to the single band. The increased number of absorption lines makes the lines even closer together, but it does not appreciably increase the total range of the band. This is because the energy involved in rotation is very small compared to vibrational energy and extremely small compared to electronic excitation energy. UV radiation is therefore absorbed in *absorption bands* rather than at discrete wavelengths.

In some cases, the UV/VIS spectra will show the different energies associated with the vibrational sublevels. For example, simple molecules in the gas phase often show the vibrational levels superimposed on the electronic transitions, as seen in Figure 4.11, the gas-phase spectrum for benzene. The sharp peaks on top of the broad bands are called vibrational "fine structure." Molecules in solution (such as the spectrum shown in Figure 4.1) usually do not exhibit vibrational structure due to interactions between the solvent and the solute molecules. Compare the gas-phase spectrum of benzene (Figure 4.11) to the solution spectrum for benzene (Figure 4.12), and note the loss of much of the fine structure in solution. Fine structure due to rotational sublevels is never observed in routine UV/VIS spectra;

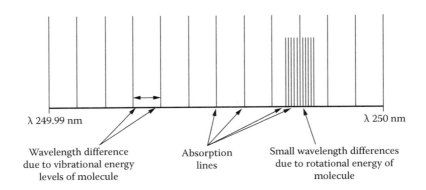

Figure 4.10 Illustration of a UV absorption band greatly expanded.

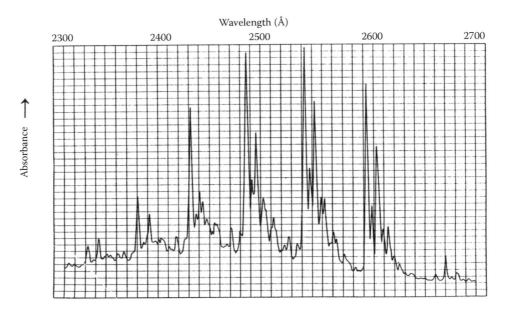

Figure 4.11 Gas-phase absorption spectrum of benzene. (Reprinted with the permission of the late Professor Milton Orchin. Jaffé, H.H. and Orchin, M., *Theory and Applications of Ultraviolet Spectroscopy*, John Wiley & Sons, New York, 1962.)

the resolution of commercial instrumentation is not high enough to separate these lines.

4.1.5 Solvents for UV/VIS Spectroscopy

Many spectra are collected with the absorbing molecule dissolved in a solvent. The solute must be soluble in the solvent and the solvent must be transparent over the wavelength range of interest. A molecule will dissolve in a solvent if the formation of a solution leads to a lower-energy system. The intermolecular attractive forces between the solute and solvent must be greater than solute–solute and solvent–solvent attractive forces. The forces involved in solution formation are dipole–dipole attraction, hydrogen bonding, and van der Waals forces. Polarity plays a major role and gives rise to the "like dissolves like" rule. Polar substances dissolve more readily in polar solvents than in nonpolar solvents. It is important for the solute to be dissolved completely; undissolved particles can scatter light from the light source. This can result in serious errors in qualitative and quantitative analyses.

The solvent may affect the appearance of the spectrum, sometimes dramatically. Polar solvents generally wipe out the vibrational fine structure in a spectrum. Solvents may also shift the position of the absorption band, as will be discussed. For the visible region of the spectrum, any colorless solvent can be used in which the sample is soluble.

The common solvents used in UV/VIS spectroscopy are listed in Table 4.5, along with their wavelength cutoff and dielectric constants. The dielectric constants are presented since this describes polarity, as discussed above. A solvent with a high dielectric constant (such as water) will be highly

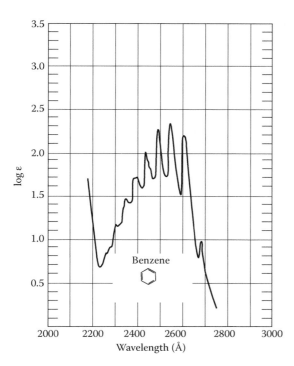

Figure 4.12 Absorption spectrum of benzene in solution. The solvent is cyclohexane. Note the loss of fine structure when compared to Figure 4.11. (Reprinted with the permission of the late Professor Milton Orchin. Jaffé, H.H. and Orchin, M., *Theory and Applications of Ultraviolet Spectroscopy*, John Wiley & Sons, New York, 1962.)

Table 4.5 Common UV solvents with their wavelength cutoffs and dielectric constants

Solvent	Wavelength Cutoff, nm	Dielectric Constant (20 °C)
Acetic acid	260	6.15
Acetone	330	20.7 (25 °C)
Acetonitrile	190	37.5
Benzene	280	2.284
sec-Butyl alcohol (2-butanol)	260	15.8 (25 °C)
n-Butyl acetate	254	
n-Butyl chloride	220	7.39 (25 °C)
Carbon disulfide	380	2.641
Carbon tetrachloride	265	2.238
Chloroform[1]	245	4.806
Cyclohexane	210	2.023
1,2-Dichloroethane	226	10.19 (25 °C)
1,2-Dimethoxyethane	240	
N,N-Dimethylacetamide	268	37.8
N,N-Dimethylformamide	270	36.7
Dimethylsulfoxide	265	4.7
1,4-Dioxane	215	2.209 (25 °C)
Diethyl ether	218	4.335
Ethanol	210	24.30 (25 °C)
2-Ethoxyethanol	210	

Table 4.5 (Continued) Common UV solvents with their wavelength cutoffs and dielectric constants

Solvent	Wavelength Cutoff, nm	Dielectric Constant (20 °C)
Ethyl acetate	225	6.02 (25 °C)
Methyl ethyl ketone	330	18.5
Glycerol	207	42.5 (25 °C)
n-Hexadecane	200	2.06 (25 °C)
n-Hexane	210	1.890
methanol	210	32.63 (25 °C)
2-Methoxyethanol	210	16.9
Methyl cyclohexane	210	2.02 (25 °C)
Methyl isobutyl ketone	335	
2-Methyl-1-propanol	230	
N-Methyl-2-pyrrolidone	285	32.0
n-Pentane	210	1.844
n-Pentyl acetate	212	
n-Propyl alcohol	210	20.1 (25 °C)
2-Propanol (sec-propyl alcohol, isopropyl alcohol)	210	18.3 (25 °C)
Pyridine	330	12.3 (25 °C)
Tetrachloroethylene[2]	290	
Tetrahydrofuran	220	7.6
Toluene	286	2.379 (25 •C)
1,1,2-Trichloro-1,2,2-trifluoroethane	231	
2,2,4-Trimethylpentane	215	1.936 (25 °C)
o-Xylene	290	2.568
m-Xylene	290	2.374
p-Xylene	290	2.270
Water		78.54 (25 °C)

Source: taken from the book by Bruno and Svoronos.

polar. At wavelengths shorter than the cutoff wavelength, the solvent absorbs too strongly to be used in a standard 1 cm sample cell. The cutoff is affected by the purity of the solvent. For spectroscopy, the solvents should be of spectral or spectrochemical grade, conforming to purity requirements set by the American Chemical Society.

4.2 INSTRUMENTATION

4.2.1 Optical System

Spectrometers are instruments that provide information about the intensity of light absorbed or transmitted as a function of wavelength. Both single-beam and double-beam optical systems (see the schematics in Chapter 3) are used in molecular absorption spectroscopy. Single-beam systems and their disadvantages were discussed in Chapter 3. Most commercial instruments for absorption spectrometry are double-beam systems, so these will be reviewed.

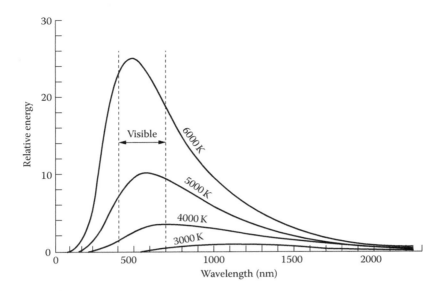

Figure 4.13 Emission intensity of blackbody radiation at various temperatures as a function of wavelength: 3000 K is equivalent to a tungsten filament lamp (an incandescent lamp); 6000 K is equivalent to a xenon arc lamp.

In the double-beam system, the source radiation is split into two beams of equal intensity. The two beams traverse two light paths identical in length; a *reference* cell is put in one path and the *sample* cell in the other. The intensities of the two beams after passing through the cells are then compared. Variation in radiation intensity due to power fluctuations, radiation lost to the optical system (e.g., cell surfaces and mirrors), radiation absorbed by the solvent, etc., should be equal for both beams, correcting for these sources of error. A dispersive spectrometer used for absorption spectroscopy that has one or more exit slits and photoelectric detectors that ratio the intensity of two light beams as a function of wavelength is called a spectrophotometer.

Commercial UV/VIS spectrometers are designed to operate with air in the light path over the range of 200–800 nm. Purging the spectrometer with dry nitrogen may permit wavelengths as low as 175 nm to be observed. For lower wavelengths, as mentioned, the spectrometer must be put under vacuum or purged with a nonabsorbing gas. Analytically, the vacuum UV region has been of minor importance for routine analysis because of the difficulties and expense inherent in instrumentation requiring a vacuum.

Simple optical systems using filters for wavelength selection and a photoelectric detector are called photometers. Photometers are used for both the visible and the UV region. For example, UV photometers were commonly used as detectors in high-performance liquid chromatography (HPLC) but have been superseded by photodiode arrays (PDAs). HPLC detectors will be discussed in greater detail in Chapter 14.

All spectrometers for absorption measurements require a light source, a wavelength selection device, a sample holder, and a detector. Complete UV/VIS and UV/VIS/near-IR (NIR) systems are available from many commercial instrument companies, including Agilent Technologies, PerkinElmer, Shimadzu Scientific Instruments, and Thermo Fisher Scientific. These large companies each offer from five to eight versions of their systems with varied capabilities. Their websites offer applications, videos on using the instruments, tutorials, and a host of other useful information.

4.2.2 Radiation Sources

Radiation sources for molecular absorption measurements must produce light over a continuum of wavelengths. Ideally, the intensity of the source would be constant over all wavelengths emitted. Traditionally, the two most common radiation sources for UV/VIS spectroscopy were the tungsten lamp and the deuterium discharge lamp. The *tungsten lamp* is similar in functioning to an ordinary incandescent light bulb. It contains a tungsten filament heated electrically to white heat and generates a continuum spectrum. It has two shortcomings: the intensity of radiation at short wavelengths (<350 nm) is low; furthermore, to maintain a constant intensity, the electric current to the lamp must be carefully controlled. However, the lamps are generally stable, robust, and easy to use. Typically, the emission intensity varies with wavelength, as shown in Figure 4.13. The shape of these curves is typical of the continuum output of a solid heated to incandescence. An incandescent solid that produces a curve of this type is called a blackbody radiator. The continuum emission is due to thermally excited transitions in the solid, in this case, the tungsten filament. The intensity versus wavelength plot for

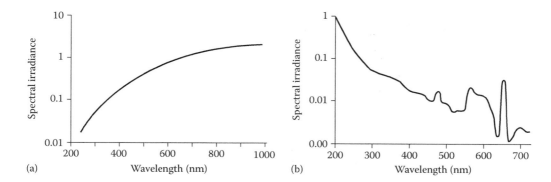

Figure 4.14 (a) Emission spectrum of a commercial tungsten-halogen lamp. (Courtesy of Agilent Technologies, Inc., Santa Clara, CA, www.agilent.com.) (b) Emission spectrum of a commercial deuterium arc lamp. (Courtesy of Agilent Technologies, Inc., Santa Clara, CA, www.agilent.com.)

a blackbody radiator is dependent on the temperature of the emitting material, not on its chemical composition. The tungsten lamp is most useful over the visible range and is therefore commonly used in spectrophotometry, discussed subsequently. Because it is used only in the visible region, the bulb (i.e., the lamp envelope) can be made of glass instead of quartz. Quartz is required for the transmission of UV light. The tungsten-halogen lamp, similar to the lamp in modern auto headlights, has replaced the older tungsten lamp in modern instruments. The tungsten-halogen lamp has a quartz bulb, primarily to withstand the high operating temperatures of the lamp. This lamp is much more efficient than a W lamp and has a significantly longer lifetime. The wavelength/intensity output of a tungsten-halogen lamp is presented in Figure 4.14.

The *deuterium arc lamp* consists of deuterium gas (D_2) in a quartz bulb through which there is an electrical discharge. The molecules are excited electrically and the excited deuterium molecule dissociates, emitting UV radiation. The dissociation of the deuterium molecule into atoms results in UV photon emission over a continuous range of energies from zero up to the energy of excitation of the molecule. This causes the lamp to emit a continuum (broadband) UV spectrum over the range of 160–400 nm rather than a narrow line atomic emission spectrum. The lamps are stable, robust, and widely used. The use of deuterium (D_2) instead of hydrogen gas results in an increase in the emission intensity by as much as a factor of three at the short-wavelength end of the UV range. Deuterium is more expensive than hydrogen but is used to achieve the high intensity required of the source. Figure 4.14b presents the emission spectrum of a deuterium arc lamp.

Xenon arc lamps operate in a manner similar to deuterium lamps. A passage of current through xenon gas produces intense radiation over the 200–1000 nm range. They provide very high radiation intensity and are widely used in modern UV/VIS instruments because they require no warm-up period. This lamp is used in fluorescence spectrometry and the lamp schematic and spectrum are shown in Section 4.9.2. Many UV/VIS spectrometers today use xenon flash lamps, which only flash when a reading is taken. They require no warm-up time and use very little electrical energy. Photodegradation is eliminated with this type of lamp, as the sample is not exposed to high amounts of radiation or heat.

Important recently introduced light sources are light-emitting diodes (LEDs). An LED consists of a chip of semiconducting material doped with impurities to create a p–n junction (see Figure 4.21). Current flows from the p side, the anode, to the n side, the cathode. The charge carriers, electrons and holes, flow into the junction from electrodes with different voltages. When an electron meets a hole, it falls into a lower energy level and emits a photon. The wavelength of light emitted (its color) depends on the bandgap energy (Figures 4.15a and 4.20) of the materials forming the p–n junction. An LED schematic is shown in Figure 4.15b.

Commercial LEDs are now available in a variety of visible colors as well as IR and UV-emitting LEDs. UV-emitting LEDs are often made of boron nitride (BN), aluminum nitride (AlN), aluminum gallium nitride (AlGaN), and aluminum gallium indium nitride (AlGaInN). To create a white light source for visible spectroscopy, two approaches are used. One is to use a combination of red, green, and blue LEDs mixed to create white light. Red LEDs, the first practical LED and the first to be commercialized, are made from gallium phosphide (GaP), gallium arsenide phosphide (GaAsP), aluminum gallium arsenide (AlGaAs), and aluminum gallium indium phosphide (AlGaInP). Green LEDs are based on indium gallium nitride (InGaN), aluminum gallium phosphide (AlGaP), and similar materials. Blue LEDs are made from indium gallium nitride (InGaN), zinc selenide (ZnSe), and other materials. The second approach, made practical after the development of high-intensity blue LEDs, is to use a blue LED and a phosphor coating based on materials like yttrium aluminum oxide (also called YAG). The phosphor produces a yellow light that appears white

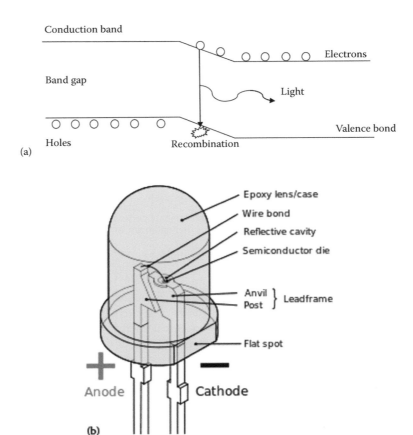

Figure 4.15 (a) Light production by recombination of an electron and hole in an LED. (b) Schematic of an LED.

when combined with blue light, very similar to the way a standard fluorescent light bulb works.

4.2.3 Monochromators

The purpose of the monochromator is to disperse the radiation according to wavelength and allow selected wavelengths to illuminate the sample. Diffraction gratings are used to disperse light in modern instruments, as discussed in Chapter 3. The monochromators in modern systems, such as the Cary spectrophotometers from Agilent Technologies, can scan at rates up to 2000–3000 nm/min, with slew rates (the time to move between wavelengths without taking measurements) as high as 16,000 nm/min to accommodate the high-throughput measurements needed in pharmaceutical and biotechnology laboratories.

New diffraction grating technology permits measurements that were impossible just a few years ago. Diffraction gratings, discussed in Chapter 3, produced by ion beam or laser beam photolithography, can produce stray light (light at unwanted wavelengths) due to the surface roughness of the surface. Today, new proprietary manufacturing methods developed for optimizing the etching process used with holographic technology result in greatly improved diffraction gratings, such as the "Lo-Ray-Ligh" gratings from Shimadzu Scientific Instruments (www.ssi.shimadzu.com). The Lo-Ray-Ligh gratings have ultralow stray light values of as little as 0.00005%, permitting absorbance measurements at up to 8.5 absorbance units with complete linearity. Figure 4.16a compares the absorbance spectra obtained on a UV-2700 with a Lo-Ray-Ligh grating with those obtained on a conventional spectrometer. Figure 4.16b demonstrates the linearity over 8.5 absorbance units. This allows the measurement of highly absorbing materials, such as sunglass polarizers and welding facemasks. Some examples will be given in Section 4.5.

4.2.4 Detectors

The earliest detector used for visible light spectroscopy was the human eye. There are still *spectroscopes* and *color comparators* designed for visual observation of color and intensity.

Most modern instruments rely on **photoelectric transducers**, detection devices that convert photons into an electrical signal. Photoelectric transducers have a surface that can absorb radiant energy. The absorbed energy either causes the emission of electrons, resulting in a photocurrent,

VISIBLE AND ULTRAVIOLET MOLECULAR SPECTROSCOPY

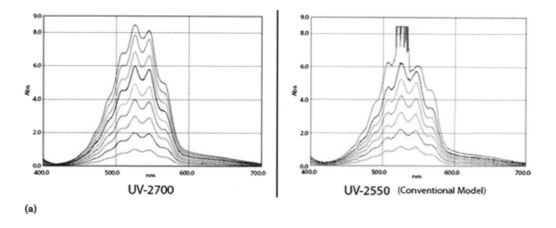

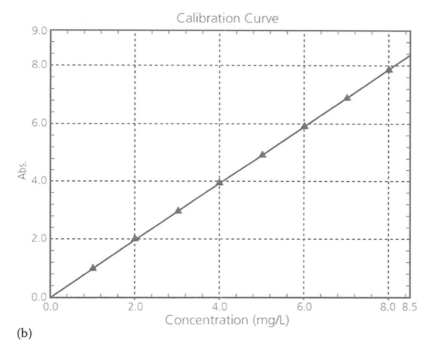

Figure 4.16 (a) Left: spectra obtained on a UV-2700 spectrometer equipped with a Lo-Ray-Ligh grating; right: the same spectra collected on a conventional spectrometer. Note that the signals on the conventional system go over the operating range. (b) Calibration curve from the UV-2700 showing linearity up to absorbance = 8.5. (© 2014 Shimadzu Scientific Instruments, www.ssi.shimadzu.com. Used with permission.)

or moves electrons into the conduction band of a solid semiconductor, resulting in an increase in conductivity. There are several common forms of these detectors including barrier layer cells, photomultiplier tubes (PMTs), and semiconductor detectors.

4.2.4.1 Barrier Layer Cell

In a barrier layer cell, also called a photovoltaic cell, a current is generated at the interface of a metal and a semiconductor when radiation is absorbed. For example, silver is coated onto a semiconductor such as selenium (see Figure 4.17) that is joined to a strong metal base, such as iron. To manufacture these cells, the selenium is placed in a container and the air pressure reduced to a vacuum. Silver is heated electrically, and its surface becomes so hot that it melts and vaporizes. The silver vapor coats the selenium surface, forming a very thin but evenly distributed layer of silver atoms. Any radiation falling on the surface generates electrons and holes at the selenium–silver interface. A barrier seems to exist between the selenium and the iron that prevents electrons from flowing into the iron; the electrons flow to the silver layer and the holes to the iron. The electrons are collected by the silver. These collected electrons migrate through an external circuit toward the holes. The photocurrent generated in this manner is proportional to the number of photons striking the cell.

Barrier layer cells are used as light meters in cameras and in low-cost, portable instruments. The response range of these cells is 350–750 nm. These detectors have two main disadvantages: they are not sensitive at low light levels and they show *fatigue*, that is, the current drops gradually under constant exposure to light. On the plus side, they require no external electrical power and they are very rugged.

4.2.4.2 Photomultiplier Tube

A very common detector is the PMT. A PMT is a sealed, evacuated transparent envelope (quartz or glass) containing a *photoemissive cathode*, an anode, and several additional electrodes called *dynodes*. The photoemissive cathode is a metal coated with an alkali metal or a mixture of elements (e.g., Na/K/Cs/Sb or Ga/As) that emits electrons when struck by photons. The PMT is a more sophisticated version of a vacuum phototube (Figure 4.18a), which contained only a photoemissive cathode and an anode; the photocurrent was limited to the electrons ejected from the cathode. In the PMT (Figure 4.18b), the additional dynodes "multiply" the available electrons. The ejected electrons are attracted to a dynode that is maintained at a positive voltage with respect to the cathode. Upon arrival at the dynode, each electron strikes the dynode's surface and causes several more electrons to be emitted from the surface. These emitted electrons are in turn attracted to a second dynode, where similar electron emission and more multiplication occur. The process is repeated several times until a shower of electrons arrives at the anode, which is the collector. The number of electrons falling on the collector is a measure of the intensity of light falling on the detector. In the process, a single photon may generate many electrons and give a high signal. The dynodes are therefore operated at an optimum voltage that gives a steady signal. A commercial PMT may have nine or more dynodes. The gain may be as high as 10^9 electrons per photon. The noise level of the detector system ultimately limits the gain. For example, increasing the voltage between dynodes increases the signal, but if the voltage is made too high, the signal from the detector becomes erratic or *noisy*. In practice, lower gains and lower noise levels may be preferable for accuracy.

PMTs are extremely sensitive to UV and visible radiation. In fact, they are so sensitive that care must be taken not to expose PMTs to bright light, to avoid damage. There are a wide variety of photoemissive surfaces available, which respond to different wavelength ranges. A plot of detector signal versus wavelength is called a *response curve*. Figure 4.19 displays some response curves for commercial PMTs. The PMT detector should be chosen so that it has maximum response to the wavelength range of interest. For example, the IP28 is not useful at 800 nm, but the R136 and Ga/As PMT detectors respond in this range. PMTs have very fast response times, but they are limited in sensitivity by their *dark current*. Dark current is a small, constant signal from the detector when no radiation is falling on it. Dark current can be minimized or eliminated by cooling the

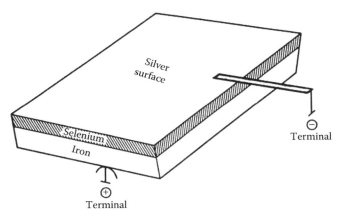

Figure 4.17 Barrier layer cell.

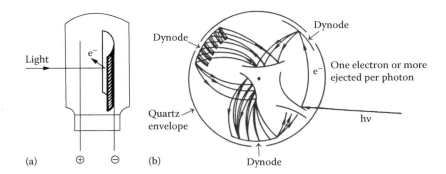

Figure 4.18 (a) A vacuum phototube. (b) Schematic of a PMT, looking down through the tube. Impinging photons pass through the quartz envelope and liberate electrons from the light-sensitive cathode. The electrons are accelerated to the first dynode, where each electron liberates several electrons on impact. The process is repeated at the other dynodes, resulting in a cascade of electrons for every photon hitting the PMT.

VISIBLE AND ULTRAVIOLET MOLECULAR SPECTROSCOPY

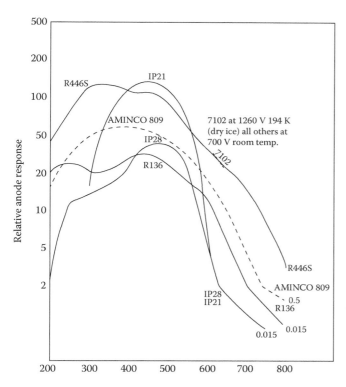

Figure 4.19 Response curves for various commercial PMTs. Note the variable response among different models and the sharp drop-off in response outside the usable range.

detector housing. Cooling devices are available commercially for this purpose.

4.2.4.3 Semiconductor Detectors: Diodes and Diode Array Systems

Solid semiconducting materials are extremely important in electronics and instrumentation, including their use as radiation detectors. To understand the behavior of a semiconductor, it is necessary to briefly describe the bonding in these materials.

When a large number of atoms bond to form a solid, such as solid silicon, the discrete energy levels that existed in the individual atoms spread into *energy bands* in the solid. The valence electrons are no longer localized in space at a given atom. The width of the energy bands increases as the interatomic spacing in the solid decreases. The highest band that is at least partially occupied by electrons is called the *valence band*; the energy band immediately above the valence band is called the *conduction band*. The valence and conduction bands are separated by a forbidden energy range (forbidden by quantum mechanics); the magnitude of this separation is called the *bandgap*, E_g. A set of energy bands and the bandgap are shown schematically in Figure 4.20. If the valence band of a solid is completely filled at a temperature of 0 K, the material is a

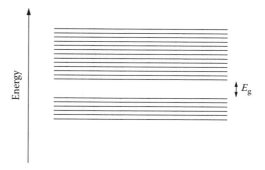

Figure 4.20 Schematic of energy bands in a solid material separated by a bandgap of energy E_g.

semiconductor or an **insulator**. The difference between a semiconductor and an insulator is defined by the size of the bandgap. If $E_g \leq 2.5$ eV, the material is a semiconductor; if $E_g > 2.5$ eV, the material is an insulator. The third type of material, a **conductor**, has a partially filled valence band at 0 K.

The two elements most used for semiconductor devices are silicon and germanium; both are covalently bonded in the solid state and both belong to group 4A of the periodic table. (This group is also called group 14 in the new International Union of Pure and Applied Chemistry [IUPAC] nomenclature and group IVA in some texts.) Other semiconductors include GaAs, CdTe, InP, and other inorganic and organic compounds. Most semiconductors are covalently bonded solids. Bandgap energies for semiconductors are tabulated in the *CRC Handbook of Chemistry and Physics*.

Silicon has the valence electronic structure $3s^2 3p^2$. The partially filled p orbitals might lead one to suppose that silicon has a partially filled valence band and would therefore be an electrical conductor. Because silicon is covalently bonded, the two 3s electrons and the two 3p electrons occupy sp^3 hybrid orbitals. This results in a solid with two electron energy bands, each with four closely spaced sublevels, one for each electron in the valence shell of Si. The four electrons occupy and fill the valence band at 0 K and are therefore nonconducting. However, at temperatures above 0 K, a few electrons can be thermally promoted from the valence band into the conduction band; there, they become conductors of electricity. When an electron leaves the valence band, it leaves behind a positive hole that is also mobile, thus producing an electron–hole pair. Both the electron and the hole are charge carriers in a semiconductor. Semiconductors such as Si and Ge are called intrinsic semiconductors; their behavior is a result of the bandgap and band structure of the pure material.

The conductivity can be increased by doping either one of these elements with a group 5A element, such as arsenic or antimony, or a group 3A element, such as indium or gallium. Doping means to add another species to the *host* material; the added species is referred to as the *dopant*. The electrons associated with the dopant atom do not have the same energy levels as the host and may lie at energies forbidden to the host. Conductivity caused by addition of a dopant is called *extrinsic conduction*. A group 5A element has an extra electron (or extra negative charge). This electron is not held as tightly as the covalently bonded electrons of the host and requires less energy to move it into the conduction band. This is an n-type semiconductor. Similarly, adding a group 3A element leads to "missing" electrons; this can be considered to be the generation of extra positive holes. These positive holes from the dopant atom can accept electrons from the valence band. The energy needed to move an electron into an acceptor hole is less than the energy needed to move an electron into the conduction band. This is a p-type semiconductor. In an n-type semiconductor, the electron is mobile, and in the p-type, the positive hole is mobile. In an intrinsic semiconductor, two charge carriers are formed for every excitation event. In extrinsic semiconductors, either n- or p-type, only one charge carrier is formed per excitation event.

Semiconductors can be used as detectors for electromagnetic radiation. A photon of light with $E > E_g$ is sufficient to create additional charge carriers in a semiconductor. Additional charge carriers increase the conductivity of the semiconductor. By measuring the conductivity, the intensity of the light can be calculated. Selection of a material with the appropriate bandgap can produce light detectors in the UV, visible, and IR regions of the spectrum.

4.2.4.4 Diodes

A diode or rectifier is an electronic device that permits current to flow in only one direction. If we put together a p-type semiconductor and an n-type semiconductor, the junction between the two types is a p–n junction, as shown in Figure 4.21. It is formed from a single piece of semiconductor by doping one side to be a p-type and the other side to be an n-type semiconductor. The junction is formed where the two types meet. Before any potential is applied to the device, holes will be the major charge carriers on the p side and electrons will be the major charge carriers on the n side. If we apply a positive potential to the p-type side and a negative potential to the n-type side, as shown in Figure 4.22, positive charges (holes) flow from the p region to the junction and negative charges flow from the

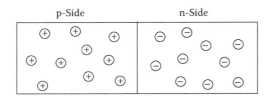

Figure 4.21 A p–n junction with no applied electrical potential.

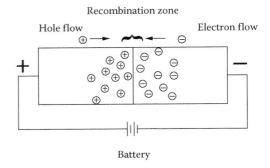

Figure 4.22 A p–n junction under forward bias. The positive battery terminal is connected to the p side.

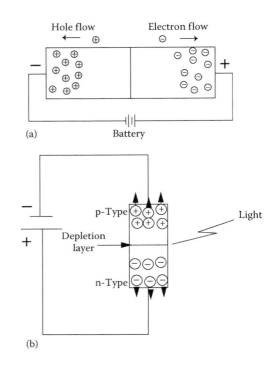

Figure 4.23 (a) A p–n junction under reverse bias. The positive battery terminal is connected to the n side. A depletion layer forms along the junction. (b) Diagram of a photodiode showing light incident upon the depletion layer. (From Brown, C., In *Analytical Instrumentation Handbook*, 4th edn., Ewing, G.W., ed., CRC Taylor and Francis, Boca Raton, FL, 2018. With permission.)

n region to the junction. At or near the junction, the holes and electrons recombine and are annihilated. This is called a *forward bias*, and under these conditions, current flows easily across the semiconductor. The annihilation of the electron–hole pair produces energy. If this energy is in the form of light, we have the LED, discussed previously.

However, if the applied voltage were in the reverse direction, the flow of carriers would be in the opposite direction, as shown in Figure 4.23a. These are the conditions of *reverse bias*. The junction region is depleted of mobile

VISIBLE AND ULTRAVIOLET MOLECULAR SPECTROSCOPY

charge carriers, recombination cannot occur, and no significant flow of current occurs. There is always a small flow of current due to the intrinsic conductivity. In short, the p–n junction acts as a rectifier and permits significant current flow only under forward bias.

If a diode is held under reverse bias and photons of energy greater than the bandgap energy fall on the diode junction as shown in Figure 4.23b, electron–hole pairs are formed in the depleted region. These carriers will move through the diode, producing a current that is proportional to the intensity of the light falling on the diode. These detectors cover spectral ranges from the UV (about 190 nm) into the NIR (about 1000 nm) but are not as sensitive as PMTs. They have limited dynamic range compared to PMTs, and when they deviate from linearity, they do so precipitously.

4.2.4.5 Diode Arrays

In UV/VIS spectroscopy, a complete absorption spectrum can be obtained by scanning through the entire wavelength range and recording the spectrum with a PMT, one wavelength at a time. This takes time using a conventional scanning monochromator system, although modern instruments are much faster than old instruments. The absorption at one wavelength is measured at a different time from that at another wavelength.

There are two conditions under which scanning optical systems do not work very well. The first is when there is a rapid chemical reaction taking place and conventional scanning is too slow to follow the reaction. The second is when the sample is available only for a limited time and complete scanning is not possible. Examples of the latter include the eluent from a liquid chromatographic separation, the flowing stream in a flow-injection system, or a process stream in a chemical or pharmaceutical production plant. In cases like this, many wavelengths need to be monitored simultaneously. Ideally, intensity over the entire spectral range of interest should be measured at the same instant. Today, a third condition often pushes the need for fast analysis—the need for high sample throughput in many industries—often 24/7 unattended operation of instruments, which improves analytical "turnaround" time and impacts costs.

The linear PDA (LPDA) is a transducer developed to enable simultaneous measurement of light intensity at many wavelengths. The diode array consists of a number of semiconductors embedded in a single crystal in a 1D linear array. A common procedure is to use a single crystal of doped silicon that is an n-type semiconductor. A small excess of a group 3A element, such as arsenic, is embedded into the surface at regular intervals. This creates local p-type semiconductors. The semiconductor device ideally has a cross section such as that shown in Figure 4.24. The surface contains a linear series or array of p–n junctions, each of which is a photodiode. The individual diodes are called *elements*, *channels*, or *pixels*.

The PDA is arranged as part of a circuit. A reverse bias is created across the p–n junction by charging a capacitor in the circuit. Radiation falling on the array creates charge carriers in both p and n regions. The electrons will then flow to the nearest p-type semiconductor, and the holes are collected in the p-type region. The current flow partially discharges the capacitor. At the end of the measurement cycle, the capacitor is recharged; the charging current results in a voltage that is directly related to the light intensity. The number of charge carriers formed in a particular element depends on the light intensity falling on the array in the immediate vicinity of that particular element. By measuring the charges on each individual element, it is possible to get an instantaneous measurement of light intensity versus wavelength of the whole spectral range but in discrete elements. This is tantamount to a digital UV absorption spectrum.

The optical layout of a commercial multichannel instrument is shown schematically in Figure 4.25. In this system, radiation from a source, which may be a deuterium lamp or other UV/VIS light source, passes through the sample to a holographic grating, where the radiation is separated by wavelengths and directed to the diode array detector. No exit slit is used. The entire spectral region is measured in much less than 1 s. In practice, the spectrum is usually acquired

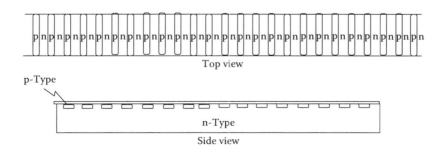

Figure 4.24 A PDA. The top view shows the face that the light would fall on. The side view shows that the p-type elements are embedded in a continuous layer of n-type semiconductor. (From Brown, C., In *Analytical Instrumentation Handbook*, 4th edn., Ewing, G.W., ed., Taylor and Francis, Boca Raton, FL, 2018. With permission.)

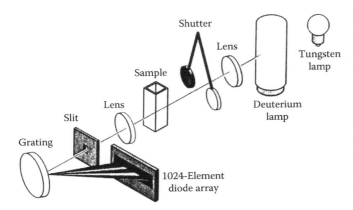

Figure 4.25 Optical diagram of a diode array spectrophotometer, showing both UV and visible light sources and a 1024-element PDA detector. (Courtesy of Agilent Technologies, Inc., Santa Clara, CA, www.agilent.com.)

for more than 1 s and stored by the computer. This practice improves the S/N ratio for the measurement. By acquiring multiple measurements, the signal can be accumulated and sensitivity considerably increased.

PDAs are available to cover the wavelength range between 190 and 1100 nm. Simultaneous use of the entire wavelength range provides the multiplex advantage and improves the resolution of the system. The resolution of the system is limited by the number of diode elements involved. Typical diode spacing is 0.025 mm. Each one can be thought of as covering a finite spectral range. Detectors have been developed with as many as 4096 elements in the array, although 1024 is probably the most common number.

The most important applications for diode array systems are in molecular spectroscopy, since in general they do not have the resolution necessary for atomic spectroscopy. In molecular spectroscopy, the most useful areas of application are for (1) scanning fast reactions to determine kinetics, (2) applications involving low light levels because spectra can be stored and added to each other, increasing the intensity, and (3) detectors for HPLC and capillary electrophoresis (CE). HPLC and CE are discussed in Chapter 14.

4.2.5 Sample Holders

Samples for UV/VIS spectroscopy can be solids, liquids, or gases. Different types of holders have been designed for these sample types. As will be discussed in Section 4.2.6, a new class of spectrometers designed for nanoliter sample volumes does not use the standard sample cells described in the following.

4.2.5.1 Liquid and Gas Cells

The cells or *cuvettes* (also spelled *cuvets*) used in UV absorption or emission spectroscopy must be transparent to UV radiation. The most common materials used are quartz and fused silica. Quartz and fused silica are also chemically inert to most solvents, which makes them sturdy and dependable in use. (*Note*: Solutions containing hydrofluoric acid or very strong bases, such as concentrated NaOH, should never be used in these cells. Such solutions will etch the cell surfaces, making them useless for quantitative work.) Quartz and fused silica cells are also transparent in the visible and into the NIR region, so these could be used for all work in the UV and visible regions. These are also the most expensive cells, so if only the visible portion of the spectrum is to be used, there are cheaper cell materials available, such as Pyrex®.

Some typical cell types are shown in Figure 4.26. Cells are available in many sizes. The standard size for spectrophotometry had long been the 1 cm path length rectangular cell, which holds about 3.5 mL of solution, shown in the upper left of Figure 4.26. There are microvolume cells (second from the left in the top row of Figure 4.26) with volumes as small as 40 μL, flow-through cells for process streams or

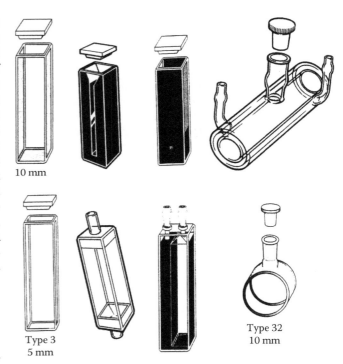

Figure 4.26 Cells for liquid samples, showing just a few of the wide variety of types and sizes available. From left to right, top row: standard 1 cm spectrophotometer cuvette with two optical faces and two frosted faces, semimicro 0.7 mL cuvette, 10 μL submicro cell, constant temperature cell with a jacket for circulating a temperature-controlling fluid. From left to right, bottom row: 5 mm fluorometer cuvette (all four faces are optically clear), in-line continuous flow cell for process monitoring (sample flow is from bottom to top), 10 mm flow cell, and cylindrical cell. (Courtesy of Starna, Inc., Atascadero, CA, www.starna.com.)

the routine analysis of large numbers of samples, microflow cells for chromatographic systems, and larger path length/volume cells for gases and highly dilute solutions. Two flow-through cells are shown in the middle of the bottom row in Figure 4.26. In general, gas cells are long path cells, such as the one shown on the upper right of Figure 4.26 and must be able to be closed once filled with a gas sample. New innovations in cell technology include the development of nanovolume (i.e., sub-µL volume) cells, such as the Starna Demountable Micro-Volume (DMV)-Bio Cell, an ultralow volume cell for use in most UV/VIS spectrometers. The DMV-Bio is available in 0.5, 0.2, and 0.125 mm path lengths, with nominal sample volumes of 2.5, 1.0, and 0.6 µL, respectively. Another nanovolume cell is the TrayCell, from Hellma Analytics (www.hellmausa.com), with path lengths of 0.2 or 1.0 mm, allowing volumes as low as 0.5 µL to be measured and recovered (see Figure 4.31).

For spectrophotometric analysis in the visible region of the spectrum, glass or disposable plastic cells may be used. These are less expensive than quartz or fused silica but cannot be used at short UV wavelengths. Polystyrene is often used for visible range cuvettes (340–800 nm), while acrylic polymer cuvettes can be used down to 285 nm. Plastic cells cannot be used with any organic solvent in which the plastic is soluble. Disposable plastic cells are not suitable for accurate quantitative work. Price differences are significant between the materials. For example, a high-quality 1 cm quartz cuvette for use in the UV range costs about $80, while a 1 cm Infrasil® Quartz cell (from Starna, Inc.) with a transmission range of 220–3800 nm costs about $130; the same size glass cuvette for use in the visible region costs about $35, and a 1 cm plastic disposable cuvette costs about 10–20 cents. Microvolume cells, flow cells, and other specialty cells are expensive, with costs of $200–$500 per cell. Some spectrometers are designed to use ordinary glass test tubes as the "cells." These test tube "cells" should not be used for accurate quantitative work. Transparencies of some typical cell materials are presented in Figure 4.27.

It is important that cells be treated correctly in order to achieve the best results and to prolong their lifetime. To that end, the analyst should (1) always choose the correct cell for the analysis; (2) keep the cell clean and check for stains, etch marks, or scratches that change the transparency of the cell; (3) hold cells on the nontransparent surfaces if provided; (4) clean cells thoroughly before use and wash out the cell with a small amount of the sample solution before filling and taking a measurement; (5) not put strongly basic solutions or hydrofluoric acid (HF) solutions into glass, quartz, or fused silica cells; (6) check for solvent compatibility with disposable plastic cells before putting them into the spectrometer; (7) for nondisposable cells, always dry carefully and return to their proper storage case; and (8) never wipe the optical surfaces with paper products, only lens-cleaning paper or cloth material recommended by the manufacturer. At all times when not in use, cells should

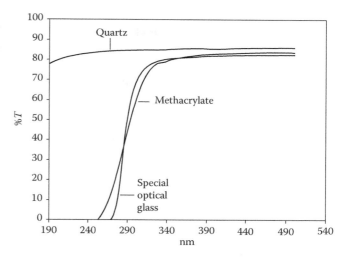

Figure 4.27 Transparencies of materials used to make sample cells for UV/VIS spectroscopy.

be kept clean and dry and stored so that the optical surfaces will not become scratched.

4.2.5.2 Matched Cells

When double-beam instrumentation is used, two cells are needed: one for the reference and one for the sample. It is normal for absorption by these cells to differ slightly. This causes a small error in the measurement of the sample absorption and can lead to analytical error. For most accurate quantitative work, *optically matched cells* are used. These are cells in which the absorption of each one is equal to or very nearly equal to the absorption of the other. Large numbers of cells are manufactured at one time and their respective absorptivities measured. Those with very similar absorptivities are designated as optically matched cells. Matched cells are usually etched near the top with an identification mark and must be kept together. It is important for the analyst to understand that even closely matched cells will show small differences in absorption due to differences in raw material characteristics. (The transmission of matched cells will also change due to normal use, so a new cell of the same "match code" will not necessarily match an older, used cell.) Less commonly used cells (other than the 1 cm type) can be supplied in matched sets of two or four cells. The proper use of matched cells is to fill both the sample and the reference cells with the solvent and run a baseline spectrum, which is stored by the instrument computer system. The sample cell is then cleaned and sample solution put into it, while the reference cell and its solvent are left in place. After measuring the sample spectrum, the baseline is subtracted from the sample spectrum by the computer. This approach will correct for small differences in the cells. It is also important that the sample cell be reinserted into the spectrometer facing in the same direction it was facing when the background was obtained. The etch mark on the top of the cell helps to facilitate this.

Modern cell manufacturing practices have improved greatly, and high-quality cell manufacturers such as Starna are now producing cells with tolerances for window flatness, parallelism of windows, polish, and path length precision better than older "matched" cells. These modern cells are in effect optically matched by the nature of the manufacturing process, and standard cells like the 1 cm cell are available in large quantities. Tolerances for modern cells can be found on the Starna website (www.starna.com). It is still necessary for the analyst to check the cells on a routine basis by measuring a dilute solution of an absorbing material in all cells. This will identify any possible problems with microscopic scratches, residual film or deposits on the windows, and so on. Matched cells are not needed for qualitative analysis, such as obtaining the spectrum of a compound to help identify its structure.

4.2.5.3 Flow-Through Samplers

For routine analysis of large numbers of samples, the filling, cleaning, and emptying of large numbers of cells is time-consuming. A flow cell and a peristaltic pump can be used to measure sample solutions directly from their original containers. Flow cells for sample volumes as small as 80 μL are available. This eliminates the need for sample handling and can minimize errors from sample handling, as well as eliminating the need to wash many cuvettes. Flow-through samplers for routine analysis are commercially available. Dedicated flow-injection systems and segmented flow systems are available for specific routine analyses, such as nitrate, sulfate, and fluoride in drinking water using methods specified by regulatory agencies around the world, such as those found in *Standard Methods for the Examination of Water and Wastewater*, listed in the bibliography. These systems are automated to take the sample, add and mix the reagents, and send the absorbing solution through a fixed wavelength spectrometer for completely unattended quantitative analysis. One such stand-alone system is the OI Analytical Flow Solution™ FS 3100 (www.oico.com).

Also available from multiple instrument companies are stopped-flow cell accessories for mixing reagents and measuring the kinetics of short-lived reactions. A stopped-flow rapid kinetics accessory from Agilent Technologies, Inc., has an empirical dead time of less than 8 ms and can monitor reaction rates up to 100 s^{-1} when used with their Cary spectrophotometers, using as little as 350 μL per reagent. Similar rapid mixing accessories are available from other instrument companies. The Rapid Mix accessory from Thermo Fisher Scientific mixes two reactants and fills cells using as little as 120 μL of reactant, with a dead time of 8 ms. The flow circuit and drive syringes can be water thermostated for temperature control. Sample surfaces are chemically inert and biocompatible. Reactant ratios are changed by changing the syringe diameters. The pneumatic drive mechanism and software triggering allows precise zero time measurement and can measure reaction rates from millisecond to minutes (see Figure 4.29a).

4.2.5.4 Solid Sample Holders

The absorption spectrum of thin films of transparent solid materials, such as polymers, can be obtained by using a film holder. The simplest type of holder can be a paper slide mount with the sample taped to the slide mount. However, producers of films, gels, and other sheet materials are often interested in the homogeneity of the film or sheet. The film holder accessory for the Cary line of spectrometers (Agilent Technologies, Inc.) allows samples up to 160 mm in length to be mounted. The absorption spectrum can be collected and then the sample moved automatically to produce a plot of absorption versus position along the sample length.

Gel electrophoresis is an important technique for separating high relative molecular mass (RMM) biological molecules, such as deoxyribonucleic acid (DNA), lipoproteins, immunoglobulins, and enzyme complexes. The classic method for visualizing the separated molecules is to stain them with colored dyes. Slices of electrophoresis gels can be mounted in a holder (Figure 4.28) and the absorption spectrum collected as a function of distance along the gel using the same device used to move film samples. The holder can be moved in increments of 0.25 mm, and gels up to 100 mm long can be analyzed in this fashion.

As described in Chapter 5 for Fourier transform IR (FTIR) and NIR, solid samples can be measured by diffuse reflectance or specular reflectance with the appropriate accessories. Solid and powder samples as small as 3 mm can be measured over the entire wavelength range of the spectrometer using a diffuse reflectance device called the Praying Mantis™ (Harrick Scientific Products, www.harricksci.com). It consists of two large hemispherical mirrors that collect the light from very small samples,

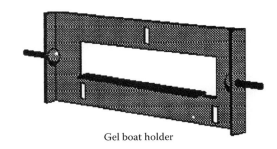

Gel boat holder

Figure 4.28 Gel holder for UV/VIS absorption spectroscopy of electrophoresis gels. (Courtesy of Agilent Technologies, Inc., Santa Clara, CA, www.agilent.com.)

VISIBLE AND ULTRAVIOLET MOLECULAR SPECTROSCOPY

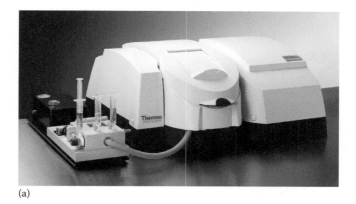

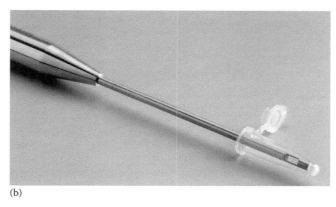

Figure 4.29 (a) The Evolution™ 300 with the Rapid Mixing accessory for fast kinetic studies in the UV/VIS region. (b) A fiber-optic microprobe for UV/VIS determinations, shown in a polymerase chain reaction (PCR) tube; 125 μL of sample can be measured. (© Thermo Fisher Scientific (www.thermofisher.com). Used with permission.)

eliminating the need for solvent dissolution. For materials science and nanomaterials, a sample of powder, 1–2 mm thick and 3 mm in diameter, is placed in the purgeable chamber, so oxygen- and water-sensitive materials can be studied. Diffuse reflectance measurements can be obtained at varying conditions: temperatures from −150 °C to 600 °C and pressures from 10^{-6} torr to 3 atm.

4.2.5.5 Fiber-Optic Probes

In all the cells described earlier, the sample had to be brought to the spectrometer and placed in the light path (or pumped into the light path). Modern fiber-optic probes enable the spectrometer to be brought to the sample. Using a fiber-optic probe such as the one shown in Figure 4.29b, the absorption spectrum can be collected from a very small sample volume in a microcentrifuge tube. These probes can cover the range from 200 to 1100 nm and operate at temperatures up to 150 °C. Fiber-optic probes can be used to collect a spectrum from inside almost any container—an open beverage can, a 55 gallon drum of material, a tanker truck, or railroad car full of liquid. Probes are made in various path lengths, just as cells are, but eliminate the need to collect a sample and put it into a cell for measurement. This is especially useful for unknown and possibly hazardous samples. Fiber-optic reflectance and fixed-angle specular reflectance probes are available for remote measurement of solid samples. They are generally made from 316 stainless steel for ease of cleaning and relative inertness.

4.2.6 Microvolume, Nanovolume, and Handheld UV/VIS Spectrometers

The intense interest in recent years in determining DNA, ribonucleic acid (RNA), proteins, and the like in limited amounts of biological samples, as well as the drive in chemical laboratories to reduce sample and reagent volumes, thereby reducing waste, and the "need for speed" in many laboratories has led to the development of a variety of microvolume UV/VIS instruments, many of which do not require a sample cell and some of which permit recovery of almost all of the sample. In addition, as in IR and Raman, handheld devices have now appeared, permitting analyses in the field or manufacturing plant. Examples of some of these instruments are presented, and while not comprehensive, attention has been paid to demonstrating the wide variety of

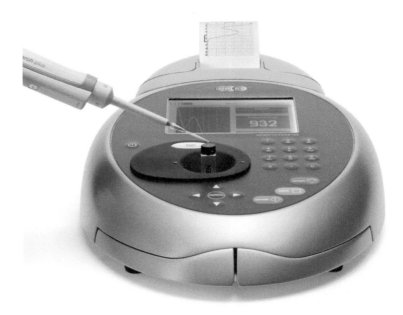

Figure 4.30 The NanoPhotometer™ Pearl. (Courtesy of Implen GmbH, Munich, Germany, www.implen.com. With permission.)

innovation in these products. The instruments themselves will be described in this section, while applications will be discussed in Section 4.5.

The advantages of directly measuring small-volume samples include no dilution errors, no contamination, no or decreased sample preparation time, and an increase in sample throughput, especially if no cuvettes are required. As the path length decreases, the measured absorbance decreases (Beer's law); decreasing the path length is therefore equivalent to dilution of the sample—a "virtual dilution." By using very short path lengths, highly absorbing samples can be measured directly without dilution. Many of these instruments permit recovery of the sample if desired.

4.2.6.1 Microvolume Systems

The NanoPhotometer™ Pearl (from Implen, GmbH; www.implen.com) is a compact dual-channel Czerny Turner grating spectrometer with a 1024 pixel charge-coupled detector (CCD) array detector and xenon flash lamp source (Figure 4.30).

Its wavelength range is 190–1100 nm with the ability to scan from 200 to 950 nm in 3.5 s. The instrument needs no warm-up time and with sealed optics and no moving parts, the need for recalibration is eliminated. Up to 81 methods can be stored in the system, and methods for nucleic acids (DNA, RNA), protein quantification, cell density, and others are predefined and come with the system. The instrument can use cuvettes but also permits cuvetteless measurement of sample volumes as small as 0.3 μL. This is accomplished by Sample Compression Technology™ (www.implen.com/nanophotometer/how-it-works.php). A sample as small as 0.3 μL is pipetted directly onto the spectrometer window, and a cap is placed over the drop. The cap squeezes the sample to an exactly defined path length that is independent of surface tension and prevents evaporation, as shown in Figure 4.31.

Five different "virtual dilution" lids are available to provide exact path lengths corresponding to dilutions of 1/5, 1/10, 1/50, 1/100, and 1/250. Using fiber optics, the light source is directed up through the sample and reflected from the mirror in the cap to the detector. After measurement, the sample can be retrieved with a micropipette or the sample is simply wiped off the mirror and window. Videos of the complete operation of the instrument can be seen at www.implen.de by clicking on applications under the instrument tab. (*Note*: The cell patent is jointly held by Implen and Hellma Analytics. Only Implen offers the 0.3 μL option and the five virtual dilution lids. Hellma Analytics, the manufacturer of these cells, offers a twofold virtual dilution version, the TrayCell, which is available directly from Hellma and also from a number of instrument companies, although often under different names.)

Thermo Fisher Scientific has a family of nanovolume spectrometers, the Thermo Scientific NanoDrop™ series (Figure 4.32).

The NanoDrop™ 2000c is a compact (14 × 20 cm) spectrometer with a range of 190–840 nm, <5 s measurement time for absorbance at a single wavelength, cuvette capability, and heating and stirring capability for kinetics and cell cultures (Figure 4.32). A 1–2 μL of sample can be measured over the full spectral range. It also

VISIBLE AND ULTRAVIOLET MOLECULAR SPECTROSCOPY

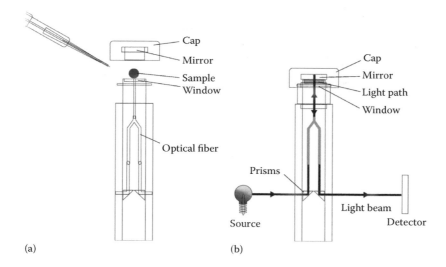

Figure 4.31 (a) The sample is pipetted onto the cell window. (b) The cap compresses the sample into a fixed path length. (Graphic courtesy of Hellma Analytics, Müllheim, Germany, www.hellmausa.com.)

permits cuvetteless measurement of as little as 0.5 μL sample volumes. Using the NanoDrop™, the sample is pipetted directly onto the measurement surface, the cover is closed, and a "column" is formed by the sample's surface tension (Figures 4.33a and b) between the upper and lower pedestals. The exact volume is not critical, only that a column forms. The sample is pipetted, the absorbance read, and the surfaces wiped clean in 15 s. A video of the operation is available at www.nanodrop.com/nd-1000-video.html and nanodrop.com/HowItWorks.aspx.

An eight-sample version is available for high-throughput laboratories (Figures 4.34a through c) using 96-well plates. Using an eight-channel pipette as shown, 96 samples can be run in about 6 min. A new palm-sized version, the NanoDrop™ Lite, was introduced in 2012 for dedicated nucleic acid applications by measuring absorbance at 260 and 280 nm. (This method will be discussed in Section 4.5.) Most of the systems have built-in methods for the life sciences, including nucleic acids, protein, and colorimetric protein assays like the Bradford and Lowry assays.

The AstraGene iCF UV/VIS spectrophotometer (AstraNet Systems, Ltd., www.astranetsystems.com; www.norsci.ca) is a CCD array spectrophotometer with a xenon light source and fiber-optic coupling of the source and detector through the sample holder, which is a micropipette tip (Figures 4.35a and b). A 2 μL sample is taken up in a UV transparent, disposable pipette tip. The polymer pipette tips transmit UV light down to 230 nm and have a fixed 1 mm path length. The absorbance measurement is made through the pipette tip (Figure 4.35c), the sample returned to the vial, and the tip disposed of. The advantages of this "through-the-tip" system include easy sample handling, no sample carryover or contamination, no cleaning of the optics, and complete recovery of the sample. The system is designed for routine analysis of nucleic acids and protein using the absorbance measured at 260 and 280 nm, but the system can be used as a scanning UV/VIS spectrophotometer or in microarray mode to measure absorbance, concentration, and ratio at two selected wavelengths. Measurement for nucleic acids or proteins takes approximately 2 s.

4.2.6.2 Variable Path Length Slope Spectroscopy™ System

Absorption spectroscopy is usually performed using a fixed path length and dilutions of standards and samples, to prepare a calibration curve and keep the absorbance readings in the linear range, as discussed in Chapter 3. Dilutions are subject to error and can be a problem when diluting from a concentrated solution of high viscosity, since displacement pipettes are usually not calibrated for high-viscosity solutions.

A new approach to UV/VIS/NIR spectroscopy is taken by C Technologies, Inc., with the SoloVPE variable path length extension, used in conjunction with an Agilent Technologies, Inc., Cary 60 spectrophotometer (Figure 4.36a). The SoloVPE leverages the path length term in Beer's law by allowing accurate measurements in the same sample by varying the path length over a range of 0.005 mm (5 μm)–15,000 mm (1.5 cm) with a path length resolution of 5 μm.

A fiber-optic cable carries light from the spectrophotometer to the fiber platform in the SoloVPE. At the end of the fiber platform is a disposable single piece of optical fiber, called a Fibrette™. The light passes through the Fibrette™ into the sample solution (Figure 4.36b). The optical fiber can be moved up and down relative to the bottom of the sample vessel, changing the available path length. An example of

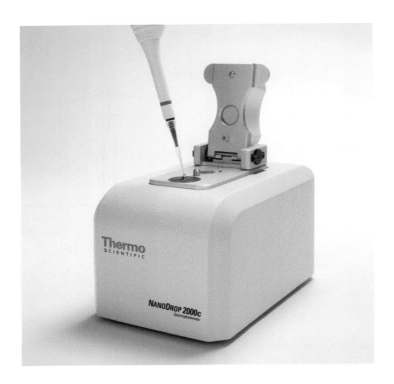
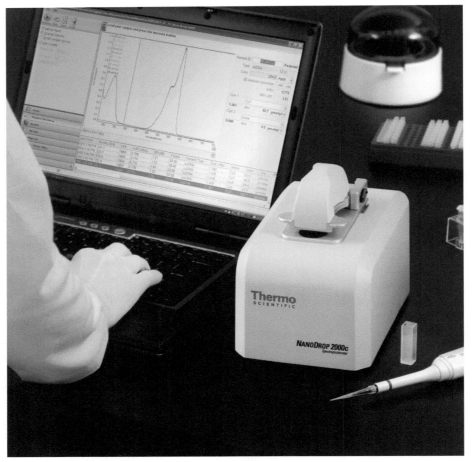

Figure 4.32 Two views of the Thermo Scientific NanoDrop™ 2000c. (© Thermo Fisher Scientific (www.thermofisher.com). Used with permission.)

Figure 4.33 (a) A 0.5 μL sample pipetted onto the measurement surface of the NanoDrop™ 2000c. (b) Cuvetteless sample column formed between the pedestals by closing the cover. (© Thermo Fisher Scientific (www.thermofisher.com). Used with permission.)

the control over the path length is shown in Figure 4.36c. A single sample of myoglobin is measured at 20 different path lengths from 0.01 to 0.2 mm. The decrease in absorbance with decreasing path length demonstrates the "virtual dilution" of the sample simply by moving the Fibrette™.

The Fibrette™ and vessels are made of quartz, giving a wavelength range of 190–1100 nm; plastic vessels are also available for either the UV or VIS–NIR range. Minimal sample volumes (1–3 μL) can be measured and the sample recovered.

The power of this technique lies in the plot that is made. At any given wavelength, a plot of absorbance versus path length can be made, called a wavelength cross-sectional plot. As is the case with a standard calibration curve, linear

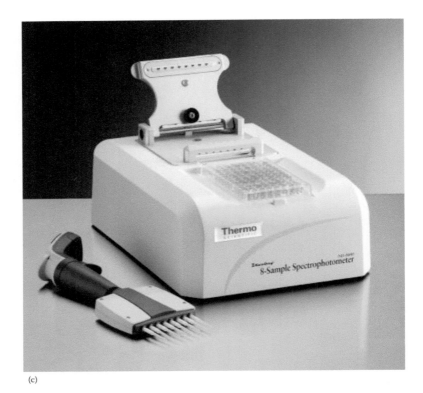

(c)

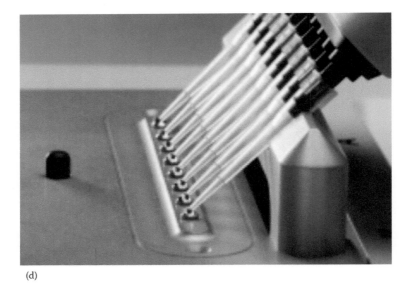

(d)

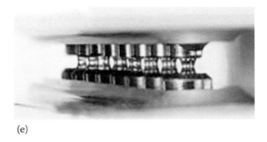

(e)

Figure 4.34 (a) NanoDrop™ 8000 with eight-channel pipette. (b) Sample pipetted onto the eight measuring surfaces. (c) Eight sample columns formed simultaneously. (© Thermo Fisher Scientific (www.thermofisher.com). Used with permission.)

VISIBLE AND ULTRAVIOLET MOLECULAR SPECTROSCOPY

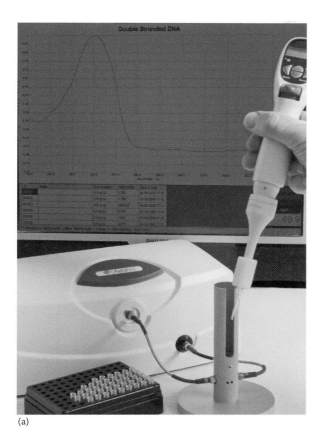

(a)

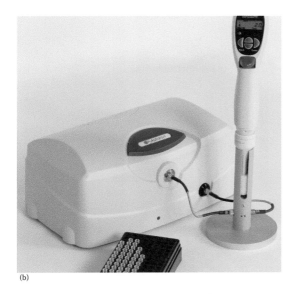

(b)

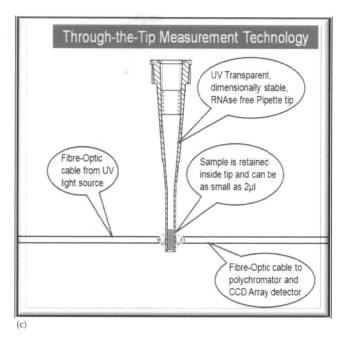

(c)

Figure 4.35 (Continued)

Figure 4.35 (a) The AstraGene iCF UV/VIS spectrophotometer with the pipette being placed in the stand. (b) The pipette containing the sample in place for measurement. (c) Schematic of the "through the tip" technology. (Courtesy of AstraNet Systems, Ltd., Cambridgeshire, U.K., www.astranetsystems.com, www.norsci.ca. With permission.)

regression is used to provide the regression equation. The slope term of the regression equation for the wavelength cross-sectional plot, m, with units of Abs/mm, is used in a rearranged version of Beer's law and is called the *Slope Spectroscopy™ Equation*: $m = A/L = \varepsilon c$, where L is the path length. Using the *Slope Spectroscopy™ equation*, if the molar absorption coefficient of a sample is known, the concentration is obtained from $c = m/\varepsilon$. Likewise, if the concentration is known, the molar absorption coefficient can be readily obtained at not just one but many wavelengths. The slopes from such plots can be used to rapidly quantify dilution ratios and to construct complete extinction spectra while using a single recoverable sample with no dilution.

4.2.6.3 Handheld Visible Spectroscopy System

The i-LAB® Visible Hand Held Analyzing Spectrometer from Microspectral Analysis, LLC (www.microspectralanalysis.com) is a portable miniaturized spectrometer powered by 3 AA batteries that covers the 400–700 nm region and weighs only 200 g (Figure 4.37). The instrument has a bandwidth of 4–7 nm, spectral resolution of 1.4 nm, and no moving parts.

The i-LAB® uses spectrally balanced LEDs to provide a uniform white light source, with a linearized PDA detector. The source light passes through a liquid sample and reflects back (Figure 4.38) or reflects directly from a solid sample (Figure 4.37) and onto a linear variable filter attached to a multiple diode array detector. In this sensor, the signals are "binned" or separated by wavelength and transformed

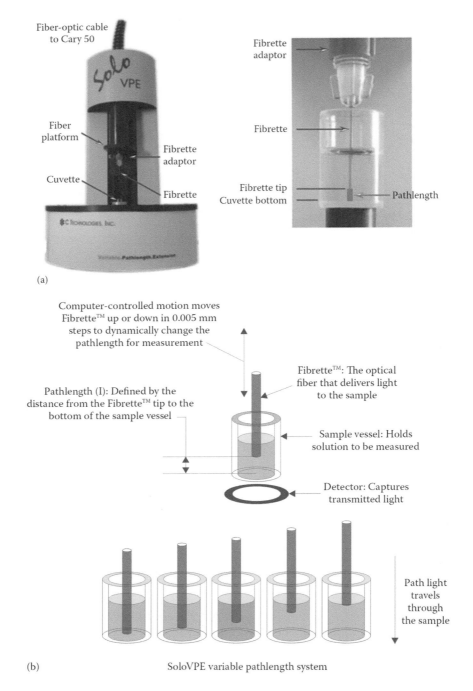

Figure 4.36 (a) Left, the SoloVPE. Right, close up of the cuvette and Fibrette™. (b) Principle of operation of the SoloVPE. (Top) Fiber optic delivers light though sample solution to a detector below the sample vessel. The path length is defined as the distance between the tip of the Fibrette™ and the bottom of the vessel. (Bottom) Accurate control of the Fibrette™ position results in multiple path lengths available.

from analog to digital signals. A spectrum is recorded and automatically saved. The data can be transformed to provide the quantitative concentration using a Beer's law calculation, absorption, transmission, area under the curve, peak maximum, peak ratios, spectral matching, and color measurement using the L*, a*, b*, or Red-Green-Blue (RGB) systems, to be described in applications. A new system, the i-LAB® LITE meter, does not use LEDs, but sunlight or artificial light as the source, and can be used in greenhouses and for the study of commercial light sources.

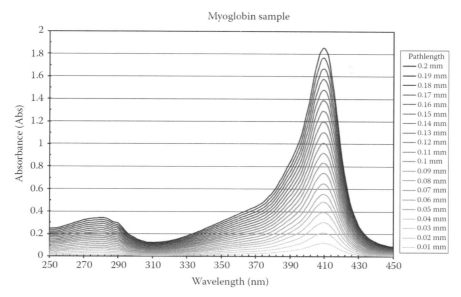

Figure 4.36 (Continued) (c) Absorbance spectrum of myoglobin collected at 20 different path lengths, with the highest absorbance at the longest path length. (Courtesy of C Technologies, Inc., Bridgewater, NJ, www.solovpe.com. With permission.)

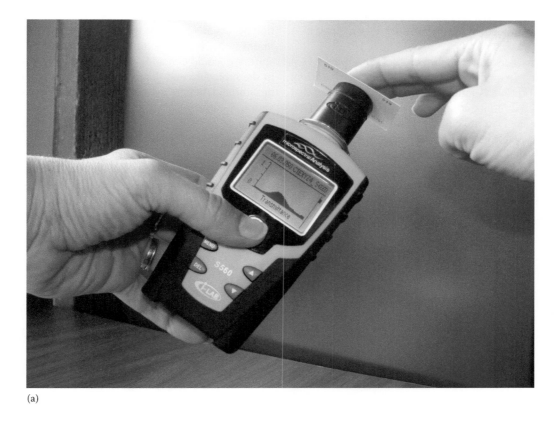

(a)

Figure 4.37 The i-LAB® Visible Hand Held Analyzing Spectrometer measuring a solid sample. (Courtesy of Microspectral Analysis, LLC, Wilton, ME, www.microspectralanalysis.com.)

The system is available with a variety of sample adaptors, including a cuvette holder, a round vial holder, and custom vials for liquids and a surface reader for solids. The software allows users to create and transfer customized measurement methods to the i-LAB. Applications will be discussed in Section 4.5.

4.3 UV ABSORPTION SPECTRA OF MOLECULES

As has been discussed, the shape and intensity of UV/VIS absorption bands are related to the electronic structure of the absorbing species. This discussion will focus on the relationship of the absorption to the structure of simple organic molecules. Tables 4.3 and 4.4 list the approximate absorption maxima of common organic chromophores, functional groups that absorb UV and/or visible light. Of course, the chromophore is not an isolated unit; it is part of a molecule. The molecule is often dissolved in a solvent to acquire the spectrum. We will look at how we can use the absorption maximum of a chromophore and a set of guidelines to predict the position of the absorption maximum in a specific molecule. We will also consider how the solvent affects the spectrum of some molecules. As a reminder, the transitions that give rise to UV/VIS absorption by organic molecules are the $n \rightarrow \sigma^*$, $\pi \rightarrow \pi^*$, and $n \rightarrow \pi^*$ transitions.

Some terms need to be defined. A **chromophore** is a group of atoms (part of a molecule) that gives rise to an electronic absorption. An **auxochrome** is a substituent that contains unshared (nonbonding) electron pairs, such as OH, NH, and halogens. An auxochrome attached to a chromophore with π electrons shifts the absorption maximum to longer wavelengths. A shift to *longer* wavelengths is called a **bathochromic shift** or **red shift**. A shift to *shorter* wavelengths is called a **hypsochromic shift** or **blue shift**. An increase in the *intensity* of an absorption band (i.e., an increase in ε_{max}) is called **hyperchromism**; a decrease in intensity is called **hypochromism**. These shifts in wavelength and intensity come about as a result of the structure of the entire molecule or as a result of interaction between the solute molecules and the solvent molecules.

4.3.1 Solvent Effects on UV Spectra

4.3.1.1 Bathochromic or Red Shift

There is a general observation that many molecules that absorb radiation due to a $\pi \rightarrow \pi^*$ transition exhibit a shift in the absorption maximum to a longer wavelength when the molecule is dissolved in a polar solvent compared to a nonpolar solvent. The shift to a longer wavelength is called a bathochromic or red shift. This does not mean that the solution turns red or that the absorption occurs in the red portion of the visible spectrum, merely that the wavelength is shifted toward the red or longer wavelength end of the spectrum. This observation can be used to confirm the presence of $\pi \rightarrow \pi^*$ transitions in a molecule. The confirmation is carried out by dissolving a sample in two different solvents. For example, one solution can be made in a nonpolar solvent such as hexane and the second in a polar solvent such as alcohol. The spectra of the two solutions are recorded. If the absorption maximum in the alcohol solution occurs at a longer wavelength than the absorption maximum in the hexane solution, the compound exhibits a red shift. We would conclude that a $\pi \rightarrow \pi^*$ transition is present in the molecule and that the molecule must therefore have unsaturated bond(s). One major exception to this observation is the $\pi \rightarrow \pi^*$ transition in dienes and other hydrocarbon polyene molecules; the spectra of these molecules are not shifted by solvent polarity.

The reason for the shift in wavelength is related to the energy level of the excited state. The excited π^* state is more affected by attractive dipole–dipole interactions and hydrogen bonding than the unexcited π state. Therefore, if a molecule is dissolved in a polar solvent, the energy level

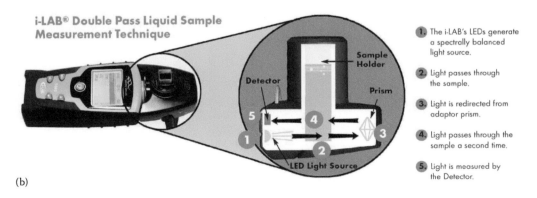

(b)

Figure 4.38 Schematic of the i-LAB® liquid sample measurement technique. (Courtesy of Microspectral Analysis, LLC, Wilton, ME, www.microspectralanalysis.com.)

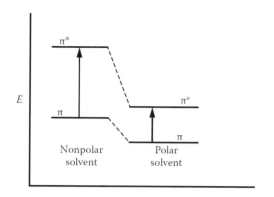

Figure 4.39 The energy difference between the π and π* levels is decreased in the polar solvent. The absorption wavelength therefore increases. This is a bathochromic or red shift.

of the π* antibonding orbital will decrease more than the energy level of the π bonding orbital. This is illustrated in Figure 4.39. The energy difference in the polar solvent is less than the energy difference when the molecule is in a nonpolar solvent. As a consequence, the absorption maximum occurs at a longer wavelength in a polar solvent.

If the π* energy level is decreased by attractive forces in polar solvents, it should be expected that the n → π* transition will also show a red shift in polar solvents. It does, but there is a much more important interaction that overcomes the red shift for the n → π* transition.

4.3.1.2 Hypsochromic or Blue Shift

The intermolecular attractive force known as hydrogen bonding can occur when a molecule contains hydrogen covalently bound to oxygen or nitrogen. The O–H or N–H bond is highly polarized; the electrons in the covalent bond are pulled toward the electronegative atom, while the hydrogen can be thought of as having a partial positive charge. This hydrogen is able to form a "hydrogen bond" with an atom containing a pair of nonbonded electrons in an adjacent molecule. The hydrogen bond is the strongest of the intermolecular attractive forces. Examples of solvent molecules capable of hydrogen bonding include water, alcohols such as ethanol and methanol, and molecules such as amines that contain an N–H bond.

The n electrons in a molecule are highly affected by hydrogen bond formation. The energy levels of n electrons decrease significantly in a solvent that has the ability to form hydrogen bonds. The result is an increase in the energy difference between the n orbital and the π* orbital. This causes a shift in the absorption maximum of an n → π* transition to shorter wavelengths, a blue shift, by as much as 25–50 nm. The decrease in the energy of the n electrons is almost equal to the energy of the hydrogen bond formed. The n → π* transition also shows a hypsochromic shift as solvent polarity increases even in non-hydrogen-bonding solvents. This is thought to be due to the increased solvation of the n electrons; the energy level of the solvated electrons is lowered. The energy involved in the transition n → π* when the solvent is nonpolar is less than the energy involved in the same transition when the molecule is in a polar or hydrogen-bonding solvent.

The hypsochromic shift effect on n electrons is much larger than the n → π* bathochromic shift due to lowering of the π* orbital described earlier. As a consequence, the absorption maximum for the n → π* transition in a molecule that contains a lone pair of electrons will move to a shorter wavelength when dissolved in ethanol versus hexane. Again, blue shift does not mean the absorption becomes blue, merely that the absorption maximum wavelength becomes shorter. This information can be used to confirm the presence of n electrons in a molecule. The sample is dissolved in a non-hydrogen-bonding solvent such as hexane and also dissolved in a hydrogen-bonding solvent such as ethanol. If the absorption spectrum of the ethanol solution exhibits a blue shift (absorption maximum at a shorter wavelength) compared to the spectrum in hexane, n electrons are present in the sample molecule. Both n → π* and n → σ* transitions are blue-shifted by hydrogen-bonding solvents. However, there are few n → σ* transitions above 200 nm to begin with; a hypsochromic shift puts these absorptions further into the vacuum UV, so they are rarely observed in routine analysis.

A hypothetical example of how we can use this solvent effect information follows. A compound with molecules that contain both π and n electrons may exhibit two absorption maxima and may show both a red shift and a blue shift with a change in solvent polarity. In general, π → π* transitions absorb approximately 10 times more strongly than n → π* transitions. A molecule that contains both π and n electrons and is dissolved in a nonpolar solvent such as hexane might have an absorption spectrum similar to the spectrum in Figure 4.40. By comparing the relative intensities of the two peaks, we would suspect that the absorption at 250 nm was due to a π → π* transition and the absorption at 350 nm was due to an n → π* transition. This is because the absorption coefficient of the π → π* transition is considerably greater than that of the n → π* transition, thus generating a higher degree of absorption. Also, in general, n → π* transitions occur at longer wavelengths than π → π* transitions because the energy difference for the n → π* transition is lower.

If we now needed to confirm these assignments, we would put the compound in a solvent such as ethanol, which is both polar and capable of hydrogen bonding. The polar nature of the solvent would induce a red shift in the π → π* transition, and hydrogen bonding would induce a blue shift in the n → π* transition. If our assignments were correct, then the absorption spectrum of the compound in ethanol might be as shown in Figure 4.41. The combined evidence

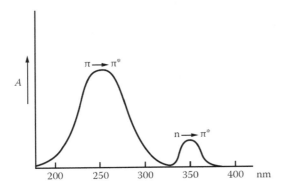

Figure 4.40 Expected absorption spectrum of a molecule in a nonpolar solvent exhibiting $\pi \rightarrow \pi^*$ and $n \rightarrow \pi^*$ transitions. Note the difference in absorptivity of the two transitions.

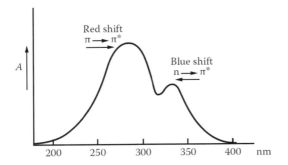

Figure 4.41 Expected absorption spectrum for the molecule in Figure 4.40 dissolved in a polar solvent such as ethanol. There is a bathochromic (red) shift in the $\pi \rightarrow \pi^*$ transition and a hypsochromic (blue) shift in the $n \rightarrow \pi^*$ transition.

of the relative intensity (degree of absorption) and the blue and red shifts occurring in ethanol strongly supports the idea that the molecule contains both π bonds and n electrons.

4.4 UV SPECTRA AND THE STRUCTURE OF ORGANIC MOLECULES

Empirical rules based on thousands of laboratory observations have been developed over the years relating the wavelengths of the UV absorption maxima to the structures of molecules. R. B. Woodward in 1941 and then L. Fieser and M. Fieser developed rules for predicting the absorption maxima of dienes, polyenes, and conjugated ketones by studying terpenes, steroids, and other natural products. The rules are known as Woodward's rules or the Woodward–Fieser rules.

There are essentially four organic molecular systems of interest. The principal parent chromophore systems are (1) conjugated dienes, (2) monosubstituted benzene rings, (3) disubstituted benzene rings, and (4) conjugated carbonyl

Table 4.6 Empirical rules for calculating the absorption

a) Rules of Diene Absorption

Base value for diene:	214 nm
Increments for (each) (in nm):	
Heteroannular diene	+ 0
Homoannular diene	+39
Extra (extension) double bond	+30
Alkyl substituent or ring residue	+ 5
Exocyclic double bond	+ 5
Diene system within a ring	+36
Polar groups:	
–OOCR	+ 0
–OR	+ 6
–S–R	+30
halogen	+ 5
–NR$_2$	+60
λ Calculated	= Total

Rules for disubstituted benzene derivatives
Parent Chromophore (benzene): 250 nm

systems. The method of calculation is to identify a parent system and assign an absorption maximum. The parent system is then modified by the presence of other systems within the molecule. From these modifications, the absorption maximum of a specific molecular structure can be calculated.

4.4.1 Conjugated Diene Systems

The parent diene of conjugated diene systems is C=C–C=C. This system in a hexane solvent absorbs at 217 nm. If the conjugated system is increased, the wavelength of the absorption maximum is increased by 30 nm for every extension. Similarly, an alkyl group attached to the conjugated system increases the absorption maximum by 5 nm. Other substitutions, such as O-alkyl, cause an increase in the absorption maximum, as does inclusion within a ring or the exocyclic character of a double bond. These shifts are listed in Table 4.6. It should be emphasized that these results are empirical and not theoretical. They result from experimental observations, and they are often referred to as Woodward's Rules, after Nobel laureate Robert B. Woodward. In this table, R refers to an alkyl group.

Substituent	o–	m–	p–
–R	+3	+3	+10
–COR	+3	+3	+10
–OH	+7	+7	+25
–OR	+7	+7	+25
–O$^-$	+11	+20	+78 (variable)
–Cl	+0	+0	+10
–Br	+2	+2	+15
–NH$_2$	+13	+13	+58

Substituent	o–	m–	p–
–NHCOCH$_3$	+20	+20	+45
–NHCH$_3$	—	—	+73
–N(CH$_3$)$_2$	+20	+20	+85

[1] Stabilized with ethanol to avoid phosgene formation.
[2] Stabilized with thymol (esopropyl meta-cresol).

Source: taken from the book by Bruno and Svoronos.

Some applications of these rules are demonstrated. The first compound is shown in Example 4.1. We have a diene (the two double bonds) in a ring with an alkyl group attached to one of the diene carbon atoms. The predicted absorption maximum is calculated to be 268 nm. How is this done? Since we have a diene, we use Table 4.6. In Example 4.1, the value assigned to the parent diene is 217 nm. This diene is within a ring (a diene within a ring is called a "homoannular" diene); therefore, we add 36 nm to the absorption maximum. It is not so clear that there are *three* alkyl substituents on the diene. One is obviously the ethyl group, C$_2$H$_5$, but the "groups" in positions A and B are also alkyl substituents. In each case, carbons at A and B are attached to the diene system and therefore contribute to the electron density and to the shift to a longer wavelength. The fact that carbons A and B are attached to each other does not change their effect on the shift of the absorption maximum. Each alkyl group adds 5 nm to the absorption maximum, so the peak should appear at 214 + 36 + 3(5) = 265 nm.

Example 4.1

Diene	214 nm
Within a ring	36
Alkyl substituent (3)	15
Predicted λ_{max}	265 nm

Example 4.2 is a little more complicated. The parent diene is at 217 nm. The diene system is within a ring, as in Example 4.1, so we add 36 nm. There is one double bond extension to the system. As you can see, the conjugated system consists of three double bonds rather than two double bonds.

Example 4.2

Diene	214 nm
Within a ring	36
Double bond extension	30
Exocyclic double bond	5
Alkyl substituents (5)	25
λ_{max}	310 nm

This adds 30 nm to the absorption maximum, from Table 4.6. There is one exocyclic double bond between carbons C and D, which adds 5 nm to the absorption maximum. This double bond is *touching* the adjacent ring and is within the conjugated system. The double bond between carbons E and F is also within a ring, but neither carbon *touches* the other ring, so this double bond is *not* exocyclic to a ring. The same thing is true for the double bond between carbons A and B; since neither carbon touches the adjacent ring, this bond is not exocyclic to a ring. There are five alkyl substitutes as follows: two on carbon A, one on carbon B, one on carbon D, and one on carbon F. The alkyl substituents on carbons D and F are designated by asterisks. The five alkyl substituents add another 25 nm to the absorption maximum. So we predict the absorption maximum to be (214 + 36 + 30 + 5 + 25) = 310 nm.

Examples 4.3 and 4.4 are molecules with very similar formulas but different structures. The double bonds are in different positions. The wavelength of the absorption maximum for Example 4.3 is 309 nm, which is 54 nm longer than in Example 4.4. The major difference between the two compounds is that the molecule in Example 4.3 has two double bonds within one ring and a conjugated system of three double bonds, that is, the double bonds are each separated by one single bond. That results in the addition of 36 nm for the diene in a ring and 30 nm for the double bond extension. In Example 4.4, there are three double bonds, but the system is not conjugated. The double bond farthest to the right (on the upper part of the ring containing the two double bonds) is separated by two single bonds from the parent diene and is not part of a conjugated system.

Example 4.3

Parent diene	217 nm
Within a ring (1)	36
Double bond extension (1)	30
Exocyclic double bond (1)	5
Alkyl substituents 3 × 5	15
–OCH$_3$	6
λ_{max}	309 nm

The only contributions to the absorption maximum in Example 4.4 are the exocyclic double bond (the diene bond on the right at the bottom of the ring containing the two double bonds) and the alkyl substituents. These two compounds can readily be distinguished by a simple UV absorption spectrum because of the large difference in λ_{max}.

Example 4.4

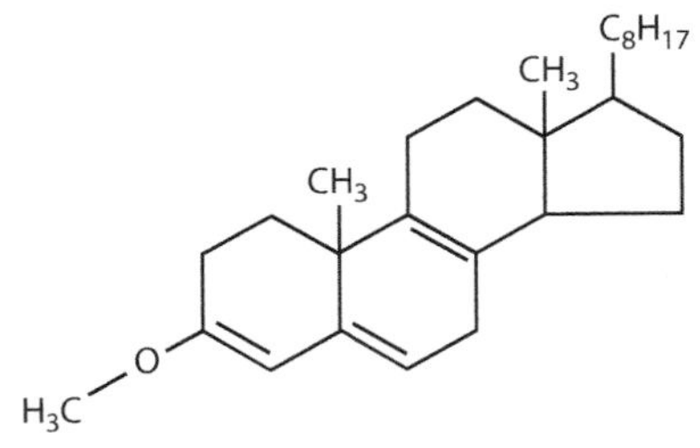

Parent diene	214 nm
Within a ring (0)	—
Double bond extension (0)	—
Exocyclic double bond (1)	5
Alkyl substituents (3)	15
–OCH$_3$	6
λ_{max}	240 nm

With the rules in Table 4.6, the absorption maximum in Example 4.5 would be calculated as shown in the following example.

Example 4.5

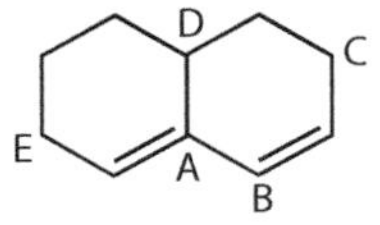

Parent diene	214 nm
One exocyclic double bond (A)	5
Three alkyl substituents (C, D, E)	15
Calculated λ_{max}	234 nm

The experimentally observed value is 234 nm. It should be noted that there is no homoannular diene system because the complete diene system is not contained in a single ring. Also, double bond B is not exocyclic to the ring, because it is not attached to a carbon that is part of another ring.

4.4.2 Conjugated Ketone Systems

The parent system is

$$C=C-C=C-C=O$$
$$\delta \quad \gamma \quad \beta \quad \alpha$$

The absorption maximum assigned to this parent system is 215 nm. In a manner similar to that for conjugated dienes, the wavelengths of the absorption maxima for conjugated ketones are modified by extension of the double bond substitution and position relative to rings and relative to the carbonyl group. The carbons are labeled α, β, γ, and δ, and substitutions in these positions change the shift of the absorption maximum. The empirical values used for calculating the absorption maxima of different compounds are shown in Table 4.7.

An example of the calculation of the wavelength of the absorption maximum using Table 4.7 is shown in Example 4.6. In this example, the calculations are similar to those used in the conjugated diene system. Note that in order to affect the absorption maximum, the substituents must be attached to the parent or conjugated system. The β-carbon has two alkyl substituents and therefore increases the absorption maximum by 36 nm. The double bond between the γ- and δ-carbons is not an exocyclic double bond: it is within two different rings but is not exocyclic to either of them. The OH group is *not* attached to the conjugated system and therefore does *not* contribute to its spectrum.

Example 4.6

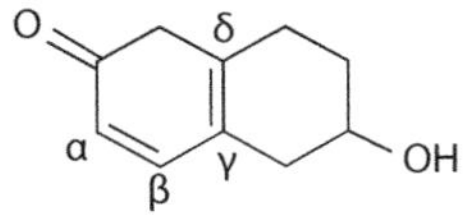

Parent ketone	215 nm
Alkyl substituent γ(1)	18
δ(2)	36
Within a ring (1)	39
Double bond extension	30
Calculated λ_{max}	338 nm

An isomer of the molecule used in Example 4.6 is shown in Example 4.7. The calculated absorption maximum is 286 nm. This spectrum is different in several ways from Example 4.6. For example, the carbon at the δ-position has only one alkyl substitution. The double bond between the γ- and δ-positions is exocyclic to a ring in this case and therefore increases the wavelength of the absorption maximum. The conjugated system is not within a ring, and therefore, contributions from ring currents are not observed. These two isomers could be distinguished from each other by their UV spectra.

Example 4.7

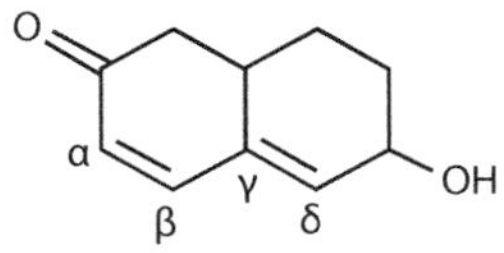

Table 4.7 **Rules for α,β-unsaturated ketone (enone) and aldehyde absorptions**

b) Rules for Enone Absorption*

$$\underset{|}{\overset{\delta}{C}}=\underset{|}{\overset{\gamma}{C}}-\underset{|}{\overset{\beta}{C}}=\underset{|}{\overset{\alpha}{C}}-\underset{\underset{O}{\|}}{C}-$$

Base value for acyclic (or six-membered) α,β-unsaturated ketone: 215 nm
Base value for five-membered α,β-unsaturated ketone: 202 nm
Base value for α,β-unsaturated aldehydes: 210 nm
Base value for α,β-unsaturated esters or carboxylic acids: 195 nm
Increments for (each) (in nm):

Heteroannular diene	+0
Homoannular diene	+39
Double bond	+30
Alkyl group:	
α–	+10
β–	+12
γ– and higher	+18
Polar groups:	
–OH	
α–	+35
β–	+30
δ–	+50
–OOCR	
α,β,γ,δ	+6
–OR	
α–	+35
β–	+30
γ–	+17
δ–	+31
–SR	
β–	+85
–Cl	
α–	+15
β–	+12
–Br	
α–	+25
β–	+30
–NR$_2$	
β–	+95
Exocyclic double bond	+5
δ Calculated =	Total

* Solvent corrections should be included. These are: water (–8), chloroform (+1), dioxane (+5), ether (+7), hexane (+11), cyclohexane (+11). No correction for methanol or ethanol.

Source: taken from the book by Bruno and Svoronos.

Parent ketone	215 nm
Alkyl substituents γ(1)	18
δ(1)	18
Within a ring	0
Double bond extension	30
Exocyclic to a ring	5
Calculated λ_{max}	286 nm

The calculation of the absorption wavelength maximum for Example 4.8 is also shown to be 286 nm. The calculation follows the steps used in previous examples. It should again be emphasized that the parts of the molecule that do not touch the conjugated or parent system do not shift the absorption maximum. This is true even for complex molecules, as some of these examples demonstrate. This makes prediction of the absorption maximum easier, but the absorption maximum gives no information as to the structure of the entire molecule.

Example 4.8

Parent ketone	215 nm
Alkyl substituents γ(1)	18
δ(1)	18
Exocyclic to a ring	5
Double bond extension	30
Calculated λ_{max}	286 nm

The idea behind being able to predict absorption maxima based on structure is the hope that the absorption spectrum will tell us about the structure of an unknown. Examples 4.7 and 4.8 demonstrate the impossibility of using the UV absorption maximum to deduce molecular structure. These are two very different compounds, easily distinguished by mass spectrometry (MS) or nuclear magnetic resonance (NMR), but each gives the same UV absorption maximum. In fact, one could draw many different organic molecules that would give the same predicted absorption maximum. The UV absorption spectrum is only one of many tools needed to elucidate the structure of an organic molecule. While the UV spectrum does not enable absolute identification of an unknown, it is often used to confirm the identity of a compound through comparison with a reference spectrum.

4.4.3 Substitution of Benzene Rings

Benzene is a strong absorber of UV radiation and particularly in the gas phase shows considerable fine structure in its spectrum (Figure 4.11). Substitution on the benzene ring causes a shift in the absorption wavelengths. It is not uncommon for at least two bands to be observed and frequently several more. The observed wavelengths of the absorption maxima of some substituted benzene rings are given in Tables 4.8 and 4.9. These are experimental data and may be insufficient to completely identify unknown compounds. If the benzene ring is disubstituted, then calculations are necessary to predict the absorption maximum, because a list containing all the possible combinations would be very long and unwieldy and would need further experimental supporting evidence. There are some rules that help us understand the disubstitution of benzene rings. These are as follows:

1. An electron-accepting group, such as NO_2, and an electron-donating group, such as OH, situated ortho or para to each other tend to cancel each other out and provide a spectrum not very different from the monosubstituted benzene ring spectrum (Table 4.8).
2. Two electron-accepting groups or two electron-donating groups para to each other produce a spectrum little different from the spectrum of the monosubstituted compound.
3. An electron-accepting group and an electron-donating group para to each other cause a shift to longer wavelengths. A para-disubstituted benzene ring illustrating this is shown in Example 4.9.

Example 4.9

Parent system λ_{max} 230 nm	
Parent system	230 nm
Para NH_2	58
Calculated λ_{max}	288 nm

To become an expert in the interpretation of UV spectra requires reading of more detailed texts, as well as practice. It can be seen that the absorption maxima can be calculated with reasonable accuracy if the structure is known, usually within 5 nm or so. A complete study of this subject is beyond the scope of this book. More detailed treatments can be found in the texts by Creswell and Runquist, Pavia, Lampman and Kriz, or Silverstein, Bassler, and Morrill (5th edition or earlier) listed in the bibliography.

4.5 ANALYTICAL APPLICATIONS

4.5.1 Qualitative Structural Analysis

As described at the beginning of this chapter, the types of compounds that absorb UV radiation are those with nonbonded electrons (n electrons) and conjugated double-bond systems (π electrons) such as aromatic compounds and conjugated olefins. Unfortunately, such compounds absorb over similar wavelength ranges, and the absorption spectra overlap considerably. As a first step in qualitative analysis, it is necessary to purify the sample to eliminate absorption bands due to impurities. Even when pure, however, the spectra are often broad and frequently without fine structure. For these reasons, UV absorption is much less useful for the qualitative identification of functional groups or particular molecules than analytical methods such as MS, IR, and NMR. UV absorption is rarely used for organic

Table 4.8 Absorption maxima of monosubstituted benzene rings Ph–R

	λ_{max} (nm) (Solvent H_2O or MeOH)	
R	Band 1	Band 2
–H	203	254
–NH_3^+	203	254
–Me	206	261
–I	207	257
–Cl	209	263
–Br	210	261
–OH	210	270
–OMe	217	269
–SO_2NH_2	217	264
–CO_2^-	224	268
–CO_2H	230	273
–NH_2	230	280
–O^-	235	287
–NHAc	238	238
–COMe		245
–$CH=CH_2$		248
–CHO		249
–Ph		251
–OPh		255
–NO_2		268
–CH′=$CHCO_2H$		273
–CH=CHPh		295

Source: Jaffé, H.H. and Orchin, M., *Theory and Applications of Ultraviolet Spectroscopy*, John Wiley & Sons, New York, 1962. Reprinted with the permission of the late Professor Milton Orchin.

Note: Me, methyl; Ph, phenyl.

structural elucidation today in modern laboratories because of the ease of use and power of NMR (Chapter 6), IR and Raman (Chapter 5), and MS (Chapters 10 and 11).

When UV spectra are used for qualitative identification of a compound, the identification is carried out by comparing the unknown compound's absorption spectrum with the spectra of known compounds. Compilations of UV absorption spectra in electronic formats can be found from commercial sources such as the Informatics Division, Bio-Rad Laboratories (www.bio-rad.com), or the American Petroleum Institute (API) indices. Computer searching and pattern matching are the ways spectra are compared and unknowns identified in modern laboratories. Some academic libraries still maintain the printed spectra collections, which must be searched manually.

One area where qualitative UV/VIS spectroscopy is still very useful is in the rapid screening of samples for UV-absorbing compounds. In a high-throughput environmental lab, UV absorption can be used qualitatively to screen for samples that may have high levels of organic compounds, to avoid contaminating a sensitive instrument or to evaluate dilution factors needed for quantitative analysis. In today's very high-throughput pharmaceutical laboratories, UV absorption spectra can provide a quick survey of synthesized compounds, to screen out those that do not have the expected absorbance and therefore probably do not have the desired structure.

Another approach that can be taken is the use of derivative spectra, that is, plotting the first, second, or even higher derivatives of the absorbance spectra. Derivative spectra can enhance the differences among spectra, resolve overlapping bands, and reduce the effects of interference from other absorbing compounds. The number of bands increases with higher orders of the derivative. The increased complexity of the derivative spectrum may aid in compound identification. For example, the absorbance spectrum of testosterone shows a single broad peak centered around 330 nm; the second-derivative spectrum has six distinct peaks (Owen).

4.5.2 Quantitative Analysis

UV and visible absorption spectrometry is a powerful tool for quantitative analysis. It is used in chemical research, biochemistry, chemical analysis, and industrial processing. Quantitative analysis is based on the relationship between the degree of absorption and the concentration of the absorbing material. Mathematically, it is described for many chemical systems by Beer's law, $A = abc$, as discussed in Chapter 3. The term applied to quantitative absorption spectrometry by measuring intensity ratios is **spectrophotometry**. The use of spectrophotometry in the visible region of the spectrum used to be referred to as *colorimetry*. (The term colorimetry appears in much of the older literature, but the term is also used for an unrelated measurement, the specification of color in samples using, for example, the Commission Internationale de l'Eclairage (CIE) L*a*b* system, described in the following.) To avoid confusion, the term spectrophotometry should be used for both UV and visible regions when quantitative determination of an analyte species is meant. While qualitative analysis is most useful for organic molecules and some transition and rare-earth compounds, quantitative UV/VIS spectrophotometry is useful for the determination of organic molecules, inorganic molecules, metal and nonmetal ions, and organometallic complexes.

UV/VIS spectrophotometry has found use everywhere in the world for research, clinical analysis, industrial analysis, environmental analysis, and many other applications. Some typical applications of UV absorption spectroscopy include determination of the concentrations of phenol, nonionic surfactants, sulfate, sulfide, phosphates, fluoride, nitrate, a variety of metal ions, and other chemicals in drinking water in environmental testing, natural products such as steroids or chlorophyll, dyestuff materials, and vitamins, proteins, DNA, and enzymes in biochemistry.

Table 4.9 Absorption maxima of disubstituted benzene derivatives

Substituent	Orientation	Shift for Each Substituent, γ, in EtOH (nm)
X = alkyl or ring residue		246
X = H		250
X = OH or O-alkyl		230
R = alkyl or ring residue	o-, m-	3
	p-	10
R = OH, O–Me, O-alkyl	o-, m-	7
	p-	25
R = O⁻	o-	11
	m-	20
	p-	78
R = Cl	o-, m-	0
	p-	10
R = Br	o-, m-	2
	p-	15
R = NH$_2$	o-, m-	13
	p-	58
R = NH–Ac	o-, m-	20
	p-	45
R = NH–Me	p-	73
R = NMe$_2$	o-, m-	20
	p-	85

Source: Jaffé, H.H. and Orchin, M., *Theory and Applications of Ultraviolet Spectroscopy*, John Wiley & Sons, New York, 1962. Reprinted with the permission of the late Professor Milton Orchin.

Note: o, ortho; m, meta; p, para; Me, methyl.

Quantitative UV/VIS spectrophotometry has been used for the determination of impurities in organic samples, such as in industrial plant streams using flow-through cells. For example, it can be used to determine traces of conjugated olefins in simple olefins or aromatic impurities in pure hexane or similar paraffins. It has also been used in the detection of possible carcinogenic materials in foods, beverages, cigarette smoke, and air. In the field of agriculture, UV/VIS spectrophotometry is used for the determination of nitrogen- and phosphorus-containing fertilizers. In the medical field, it is used for the determination of enzymes, vitamins, hormones, steroids, alkaloids, and barbiturates. These measurements are used in the diagnosis of diabetes, kidney damage, and myocardial infarction, among other ailments. In the pharmaceutical industry, it can be used to measure the purity of drugs during manufacture and the purity of the final product. For example, aspirin, ibuprofen, and caffeine, common ingredients in pain relief tablets, all absorb in the UV region and can be determined easily by spectrometry. Methods for many pharmaceuticals can be found in the US Pharmacopeia (USP) and are called USP methods.

For example, the purity of acetaminophen (paracetamol), $C_8H_9NO_2$, can be determined by measuring the absorbance of an aqueous solution of the drug at 244 nm and comparing it to a solution of acetaminophen of known purity and concentration. See the suggested experiments at the end of this chapter for details.

Foods and beverages are often analyzed by UV/VIS spectrophotometry both to ensure quality and to determine impurities. An example of such an analysis is the determination of hydroxymethylfurfural (HMF) in honey. The method is described in the reference by Keppy et al. and the reference by White. HMF is an aldehyde that is generated by the decomposition of fructose in acidic conditions; it has a chromophore that absorbs at 284 nm. While HMF occurs naturally over time in most honeys, high levels of HMF may indicate poor storage conditions, adulteration with sugar additives, or high heat. Levels in honey are regulated by the European Union (EU) and the Korean Food Code.

Olive oil is a major agricultural product for many countries. It is more valuable than other vegetable oils, especially high grades such as extra-virgin olive oil (EVOO), but in the past 20 years, numerous cases around the world have found EVOO to be adulterated with cheaper oils, such as sunflower, canola, and soybean oil (Claverie and Johnson and references therein). The standard method for evaluating olive oil is a titration of the acidity using phenolphthalein. It is a slow method and can give "passing" results to adulterated samples. A rapid, accurate test for EVOO was clearly needed. Olive oil comes in a range of green-gold colors, due to the pigments found in olives, primarily chlorophylls and carotenoids. Using a handheld i-LAB visible spectrometer, Claverie and Johnson measured 12 cooking oils using a 5 cm cuvette over the range 400–700 nm. The EVOO samples were found to have major absorption peaks at 417, 455, 478, and 667 nm (Figure 4.42), while the canola and blended oils had no such absorbance peaks. Three of the four spectra with absorbance less than 0.45 at 400 nm in Figure 4.42 are the known non-EVOO samples. However, one of the four spectra, clearly not containing the characteristic absorption peaks, was from a sample labeled as EVOO. The visible spectra, as can be seen, provide a clear and quick ID of olive oil in about 10 s. In addition, the i-LAB provided the CIE tristimulus color values, discussed in the following, at the same time as the spectra were taken.

UV spectrophotometric determination of nucleic acids and proteins is a major application, and as we have seen, systems are now able to perform this determination on sample volumes of 0.3–2 μL, often nondestructively. This is critical for DNA microarrays, PCR, and forensic DNA measurements. The nitrogenous bases, purines, and pyrimidines in nucleotides absorb at a λ_{max} of 260 nm. Using known extinction coefficients, concentrations of double-stranded DNA (dsDNA), single-stranded DNA, and RNA can be determined. Using the microvolume systems described earlier, dsDNA can be measured over a range of approximately 2–18,000 ng/μL. Proteins and peptides absorb strongly at 280 nm, due to the aromatic amino

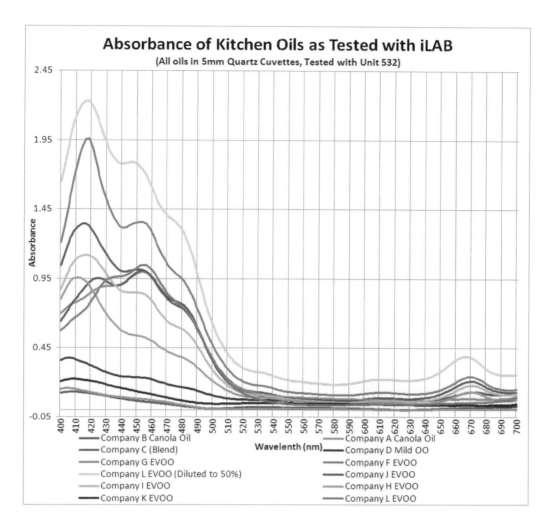

Figure 4.42 Absorbance spectra for a variety of vegetable oils using the i-LAB visible spectrometer. EVOO is clearly differentiated from other vegetable oils by the strong absorption peaks in the 400–480 nm region and the peak at 667 nm. (From Claverie, M. and Johnson, J.E. The determination of EVOO from other oils by visible spectroscopy, Courtesy of Microspectral Analysis, LLC. Wilton, ME, www.microspectralanalysis.com. With permission.)

acids tryptophan, tyrosine, phenylalanine, and, to a smaller extent, disulfide bonds. The absorbance of a DNA sample at 280 nm gives an estimate of the protein concentration of the sample, and the ratio of the absorbance at 260 nm to the absorbance at 280 nm is a measure of the purity of the DNA sample. The ratio should be between 1.65 and 1.85. The measurement of proteins and peptides at 280 nm is an important application in itself, given the increasing use of these as therapeutic agents for the treatment of diseases. Extinction coefficient assays, the determination of the extinction coefficient over the linear working range, can be used to identify proteins and peptides. There are also a number of classic spectrophotometric protein assays based on the formation of a complex between a dye and proteins in solution. The named assays are referenced in the bibliography. The Bradford assay is based on the shift of Coomassie Brilliant Blue G-250 from 465 to 595 nm. This shift is due to the stabilization of the anionic form of the dye upon binding to protein. The bicinchoninic acid (BCA) assay (Smith et al.) is based on the reaction of proteins and alkaline copper ion with BCA to form a purple complex with an absorbance maximum at 562 nm.

Many clinical chemistry assays are based on UV/VIS spectrophotometry, often by reaction of the chemical of interest, such as glucose, with an enzyme and dye to create a colored complex. A novel method for determining the age of dried bloodstains at a crime scene has been developed by researchers at the National Center for Forensic Science at the University of Central Florida using the Implen NanoPhotometer™ Pearl (Hanson and Ballantyne). The researchers discovered a previously unidentified hypsochromic shift (a blue shift to shorter wavelengths) in the Soret band of hemoglobin (λ_{max} = 412 nm) that had a high correlation to the time since deposition of dried bloodstains. The extent of the shift made it possible to distinguish among bloodstains that were deposited minutes, hours, days, and weeks prior to recovery and analysis. The test can be made on bloodstains as small as 1 µL, and the spot can be confirmed as blood by collecting the hemoglobin UV/VIS spectrum on the NanoPhotometer.

Spectrophotometry is used routinely to determine the concentrations of metal and nonmetal ions in a wide variety of samples, although many of the classic spectrophotometric measurements are being replaced by instrumental methods such as ion chromatography (IC) (Chapter 14), atomic spectroscopy (Chapters 7 and 8), and inorganic MS (Chapters 10 and 11). Spectrophotometry in the UV region of the spectrum is used for the direct measurement of many organic compounds, especially those with aromatic rings and conjugated multiple bonds. There are also colorless inorganic species that absorb in the UV region. A good example is the nitrate ion, NO_3^-. A rapid screening method for nitrate in drinking water is performed by measuring the absorbance of the water at 220 and at 275 nm. Nitrate ion absorbs at 220 but not at 275 nm; the measurement at 275 nm is to check for interfering organic compounds that may be present. Spectrophotometric analysis in the visible region can be used whenever the sample is colored. Many materials are inherently colored without chemical reaction (e.g., inorganic ions such as dichromate, permanganate, cupric ion, and ferric ion) and need no further chemical reaction to form colored compounds. Colored organic compounds, such as dyestuffs, are also naturally colored. Solutions of such materials can be analyzed directly. The majority of metal and nonmetal ions, however, are colorless. The presence of these ions in a sample solution can be determined by first reacting the ion with an organic reagent to form a strongly absorbing species. If the product of the reaction is colored, absorbance can be measured in the visible region; alternatively, the product formed may be colorless but absorb in the UV. The majority of spectrophotometric determinations result in an increase in absorbance (darker color if visible) as the concentration of the analyte increases. However, there are analyses that cause a bleaching of color (decrease in absorbance) with increasing concentration of analyte.

As was mentioned in the Introduction, the color we observe in a colored solution is the color that is transmitted by the solution. The color absorbed is the complementary color. The relationship between the color of light absorbed and the color observed is given in Table 4.10.

There are thousands of possible compounds and complexes that can be formed by reacting analyte species with organic reagents. Ideally, the reagent chosen should be selective; that is, it should react with only one ion or molecule under the conditions present. Second, the reagent should cause an abrupt color change or absorbance change when mixed with the analyte. This imparts high sensitivity to the method. Third, this intensity of color or UV absorbance should be related to the concentration of analyte

Table 4.10 Absorbed and observed colors

Color Absorbed	Color Observed
Red	Blue-green
Orange	Green-blue
Yellow	Blue
Yellow-green	Violet
Green	Violet-red
Blue-green	Red
Green-blue	Orange
Blue	Yellow
Violet	Yellow-green

in the sample. Spectrophotometric reagents have been developed for almost all metal and nonmetal ions and for many molecules or classes of molecule (i.e., for functional groups). Many of these reactions are both sensitive and selective. Several examples of these reagents and their uses are given in Table 4.11. The books by Boltz and by Sandell and Onishi listed in the bibliography are classic reference sources but may be hard to locate. The *CRC Handbook of Chemistry and Physics* contained a set of tables by Ackerman, Sommer, and Burns that in 2012 underwent a major revision. Subsequent editions of the handbook (beginning with the 93rd) have an extensive compilation that is revised each year. This compilation can also be found in the book by Bruno and Svoronos. The analytical literature contains thousands of direct and indirect methods for quantitative analysis of metals and nonmetals. A good summary of methods with literature references for most metal and nonmetal ions may be found in the handbook by Dean listed in the bibliography.

Quantitative analysis by absorption spectrophotometry requires that the samples be free from particulates, that is, free from **turbidity**. The reason for this is that particles can scatter light. If light is scattered by the sample away from the detector, it is interpreted as an absorbance. The absorbance will be erroneously high if the sample is turbid. We can make use of the scattering of light to characterize samples as discussed in Section 4.7, but particulates must be avoided for accurate absorbance measurements.

Quantitative analysis by spectrophotometry generally requires the preparation of a calibration curve, using the same conditions of pH, reagents added, and so on, for all of the standards, samples, and blanks. It is critical to have a reagent blank that contains everything that has been added to the samples (except the analyte). The absorbance is measured for all blanks, standards, and samples. The absorbance of the blank is subtracted from all other absorbances, and a calibration curve is constructed from the standards. The concentrations of analyte in the samples are determined from the calibration curve. The highest accuracy results from working in the linear region of the calibration curve. These quantitative methods can be quite

Table 4.11 Typical reagents used in spectrophotometry

Aluminon (also called ammonium aurintricarboxylate)

This compound reacts with aluminum in a slightly acid solution (pH 4–5) to form an intense red color in solution. It detects 0.04–0.4 μg/mL (ppm) of Al. Other elements, such as Be, Cr, Fe, Zr, and Ti, also react with aluminon. These elements must be removed if a sample is being analyzed for Al. The absorbance of the red solution is measured at 525 nm. The red color is the result of formation of a metal–dye complex called a "lake."

4-Aminophenazone (also called 4-aminoantipyrine)

This compound reacts with a variety of phenols to give intensely colored compounds and will detect 0.02–6.4 ppm of phenol in water. Drinking water is steam-distilled to separate the volatile phenols from interfering compounds. The distillate is treated with the reagent, and the colored complex is extracted into $CHCl_3$. The absorbance of the chloroform solution is measured at 460 nm. The reagent does not react with some para-substituted phenols, such as paracresol. This reaction is an example of the determination of organic compounds by spectrophotometric analysis following reaction with a color-producing reagent.

Thiourea

H_2NCSNH_2

Thiourea will react with osmium, a very toxic element, in sulfuric acid solution to form a colored product. The absorbance is measured at 460 nm with a detection range of 8–40 ppm Os. The only interferences are Pd and Ru. Compared with the other reagents in this table, the sensitivity of the reagent is low. Interestingly, under different analytical conditions (different acid, pH), thiourea reacts with Bi. The absorbance of this product is measured at 322 nm and detects 0.06–0.6 ppm of Bi. A number of elements such as Ag, Cu, Hg, and Pb interfere. This is an example of a reagent that works under different chemical conditions to produce a low-sensitivity determination for one element and a high-sensitivity determination for another.

Chloranilic acid

Chloranilic acid forms solutions that are intensely red. The addition of calcium to the solution precipitates the chloranilic acid, and the intensity of the red diminishes. The change (loss) in color is a measure of the quantity of calcium added. Numerous other elements interfere with the procedure. This is an example of spectrophotometric analysis by loss of color after addition of the sample.

Quinalizarin

This reagent gives intensely colored solutions in aqueous solutions. In 93% w/w H_2SO_4/H_2O, the color is red. The presence of borate causes the color to become blue. Numerous other ions, such as Mg^{2+}, Al^{3+}, and Be^{3+}, also react with quinalizarin. This is an example of a change of color of the reagent after reaction with the sample.

Curcumin

Curcumin is a sensitive reagent for boron, detecting 0.01–0.1 ppm of B by absorbance at 555 nm. Fluoride, nitrate, and nitrite interfere but can be eliminated by separating the boron from the sample by distillation of B as a methyl borate ester.

SPADNS [also known as 4,5-dihydroxy-3-(2-hydroxy-5-sulfophenylazo)-2,7-naphthalenedisulfonic acid]

SPADNS (pronounced "spa-dens") is used to determine fluoride ion in drinking water. The SPADNS dye reacts with zirconium to form a dark red Zr–dye "lake." The F^- ion reacts to dissociate the Zr–dye complex and form $\left(ZrF_6^{2-}\right)$, which is colorless. The color of the solution decreases with increasing fluoride ion concentration. The absorbance is measured at 570 nm, with a range of 0.2–1.40 ppm F^-. There are both positive and negative interferences from chlorine, chloride, phosphate, sulfate, and other species in drinking water. This is another example of a reaction where the color is "bleached" with increasing concentration of analyte.

Source: Examples were extracted from *Standard Methods for the Examination of Water and Wastewater*, and the references by Dean and Dulski.

complicated in the chemistry involved and the number of steps required (extraction, back extraction, pH adjustment, precipitation, masking, and many other types of operations may be involved in a method), and the analyst must pay attention to all the details to achieve accurate and precise results. Both science and art are involved in performing many of these analyses. Many standard or regulatory methods (e.g., from *Standard Methods for the Examination of Water and Wastewater*, American Society for Testing and Materials [ASTM], and Environmental Protection Agency [EPA]) have published precision and accuracy data in the methods. These are the precisions and accuracies that can be achieved by an experienced analyst.

4.5.3 Multicomponent Determinations

It has been seen that UV/VIS absorption peaks are generally broad, so if there are two compounds, X and Y, in solution, it is likely that they will not be completely resolved from each other. That is, both X and Y contribute to the absorbance at most wavelengths. It is possible to calculate the concentrations of X and Y from a series of measurements. Measurements must be made at a number of wavelengths equal to the number of components in the mixture. In this case, there are two components, so two wavelengths are needed. Caveats for this approach are that the two components each follow Beer's law and that the two components must not interact in solution; their absorbances must be additive.

Four calibration curves need to be prepared: X at λ_1, X at λ_2, Y at λ_1, and Y at λ_2. All calibration curves should be blank corrected to pass through the origin. The absorbance of the sample mixture is measured at λ_1 and at λ_2.

Two equations can be written:

$$A_1 = C_X S_{X1} + C_Y S_{Y1}$$
$$A_2 = C_X S_{X2} + C_Y S_{Y2} \quad (4.3)$$

where
A_1 is the absorbance of the unknown at λ_1
A_2, the absorbance of the unknown at λ_2
C_X, the concentration of X in the unknown
C_Y, the concentration of Y in the unknown
S_{X1}, the slope of the calibration curve for X at λ_1
S_{X2}, the slope of the calibration curve for X at λ_2
S_{Y1}, the slope of the calibration curve for Y at λ_1
S_{Y2}, the slope of the calibration curve for Y at λ_2

The absorbances and slopes are known; this leaves us with two equations and two unknowns: C_X and C_Y. The equations can be solved for the concentrations of X and Y in the unknown mixture. Dulski (see bibliography) gives an example of this approach with a method for the simultaneous determination of niobium and titanium by reaction with hydroquinone and measurement at 400 and 500 nm. Another example is the analysis of a common nasal spray that contains L-phenylephrine (PEH) and chlorpheniramine maleate (PAM). Each compound in aqueous solution gives a broad absorption peak in the UV. Choose two wavelengths: preferably the λ_{max} for each, if possible. The amounts of the two ingredients in the nasal spray can be found by preparing a calibration curve of each compound, obtaining the slopes of the lines at two wavelengths, for example, at 266 and 272 nm. This gives the four S terms as shown previously. Then, the sample is measured at the two wavelengths, giving the A terms. The experiment is outlined in the suggested experiments at the end of this chapter.

This same approach can be used for a mixture of three components. More complex mixtures can be unraveled through computer software that uses an iterative process at multiple wavelengths to calculate the concentrations. Mathematical approaches used include partial least squares (PLS), multiple least squares, principal component regression, and other statistical methods. Multicomponent analysis using UV absorption has been used to determine how many and what type of aromatic amino acids are present in a protein and to quantify five different hemoglobins in blood.

4.5.4 Other Applications

4.5.4.1 Reaction Kinetics

In common with other spectroscopic techniques, UV spectroscopy can be used to measure the kinetics of chemical reactions, including biochemical reactions catalyzed by enzymes. For example, suppose that two compounds, A and B, react to form a third compound C. If the third compound absorbs UV radiation, its concentration can be measured continuously. The original concentrations of A and B can be measured at the start of the experiment. By measuring the concentration of C at different time intervals, the kinetics of the reaction A + B → C can be calculated. Enzyme reactions are important biochemically and also analytically; an enzyme is very selective, even specific, for a given compound. The compound with which the enzyme reacts is called the substrate. If the enzyme assay is correctly designed, any change in absorbance of the sample will result only from reaction of the substrate with the enzyme. The rate of an enzyme reaction depends on temperature, pH, enzyme concentration and activity, and substrate concentration. If conditions are selected such that all of the substrate is converted to product in a short period of time, the amount of substrate can be calculated from the difference between the initial absorbance of the solution and the final absorbance. This approach is called an endpoint assay. Alternatively, in a rate assay, the other experimental variables are controlled so that the rate of the enzyme reaction is directly proportional to the substrate concentration.

From a practical point of view, many reactions occur so rapidly that manual mixing of reagents is not practical. Mixing dead time must be small compared to the reaction half-life. Temperature must be accurately controlled, since reaction rates are temperature dependent. For accurate determination of start times, electronic triggering is necessary. Sophisticated accessories are available with most major instruments. The example used here is from Dr. Michael Allen, Thermo Fisher Scientific, using the Rapid Mixing accessory described in Section 4.2.5.3.

The dichromate ion, $Cr_2O_7^{2-}$, contains the toxic Cr(VI) ion and is usually removed from chemical waste by oxidation to Cr(III). The oxidation of dichromate with hydroxide ion proceeds in two steps, the first step slow and the second fast:

$$Cr_2O_7^{2-} + OH^- \rightarrow HCrO_4^- + CrO_4^{2-} \quad \text{slow}$$

$$CrO_4^- + OH^- \rightarrow CrO_4^{2-} + H_2O \quad \text{fast}$$

At low hydroxide concentration, both the fast and slow steps can be measured (Figure 4.43).

4.5.4.2 Spectrophotometric Titrations

Many titration procedures in volumetric analysis use an indicator that changes color to signal the endpoint of the titration. For example, acid–base titrations are often performed with indicators such as phenolphthalein. Figure 4.44 shows the structure of phenolphthalein in an acid solution and in a basic solution. As can be seen, the loss of protons results in a change in the structure of the molecule. As we know, this should result in a change in the energy levels in the molecule. In phenolphthalein, the energy-level difference gives rise to the absorption of visible radiation when it is in an alkaline solution, but not in an acid solution. Phenolphthalein appears red in basic solution but colorless in acidic solution. Such structure changes and energy-level changes are the basis of many acid–base indicators. The use of the human eye to detect the color change at the end of a titration is subject to the problems described at the beginning of this chapter. Each analyst may "see" the endpoint slightly differently from other analysts, leading to poor precision and possible errors. The use of a spectrophotometer to detect the color change is more accurate and reproducible. The use of the spectrophotometer also permits any change in absorbance in the UV or visible region by the titrant, analyte, or product to be used to determine the endpoint of the titration, so the method is not limited to reactions that use a colored indicator.

Spectrophotometric titrations have been used for redox titrations, acid–base titrations, and complexation titrations. The spectrophotometer can be used in a light-scattering mode to measure the endpoint for a precipitation titration by turbidimetry. Spectrophotometric titrations can be easily automated.

4.5.4.3 Spectroelectrochemistry

Oxidation–reduction reactions of inorganic and organic compounds can be studied by using a combination of electrochemistry (Chapter 15) and spectroscopy. Diode array systems are usefully employed when transparent thin electrodes are used to study these reaction mechanisms. By taking the absorption spectra in rapid succession and accumulating the data, it is possible to detect and measure intermediates formed in complex reactions. This is much more reliable than using absorption at a single wavelength to measure the

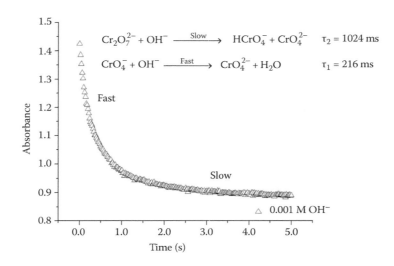

Figure 4.43 Measurement of the fast and slow steps in the oxidation of dichromate ion by spectroscopy using a Rapid Mix accessory. (© Thermo Fisher Scientific (www.thermofisher.com). Used with permission.)

Figure 4.44 The structure of phenolphthalein in acidic solution and basic solution. The change in structure results in a change in the light absorbed by the molecule.

reactions, since the choice of the single wavelength is often made with the assumption that the intermediates and end products are well known and suitable absorption wavelengths are therefore easily chosen. This is often not the case. Using the diode array system, the complete UV absorption spectra can be obtained, and much more information on the identity and concentration of species is therefore available.

4.5.4.4 Analysis of Solids

Solids can be measured in transmission or reflection (reflectance) modes. Both specular reflection and diffuse reflection are used. Diffuse reflection accessories include the Praying Mantis™ from Harrick Scientific Products, Inc., and a variety of integrating spheres available from most major instrument companies. Specular reflection is used for highly reflective materials: diffuse reflectance for powders and rough surfaced solids. Materials characterization relies heavily on techniques like these.

The Praying Mantis™ and its high-temperature reaction chamber have been used for studying temperature-induced wavelength changes in Thermal Liquid Crystal Paint, analyzing gas–solid reactions such as heterogeneous catalysis, and routine analysis of powders and solids. Application notes can be found at www.harricksci.com.

Using the Shimadzu ISR-2600 Plus integrating sphere on their UV-2600 spectrophotometer, it is possible to make transmission measurements of solids such as polycrystalline silicon. Figure 4.45a shows the transmission characteristics of the bandgap region near 1000 nm. A reflection measurement on the same system of an antireflective film (Figure 4.45b) clearly shows the suppressed reflectance in the visible region.

4.5.5 Measurement of Color

Color is a very important parameter for many manufactured products including paint, cosmetics, foods, beverages, pharmaceutical liquids and tablets, textiles, and so on. If the calcium supplements in a bottle are supposed to be pink, all the tablets in the bottle should be the same color. All of the bottles in a lot should be the same color. If they are not, consumers may think there is something wrong with the tablets. Both the generation and sensation of color are complex and depend on factors such as the spectrum of the illuminant and the surface structure of solid samples, for

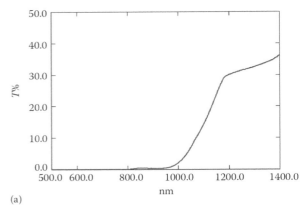

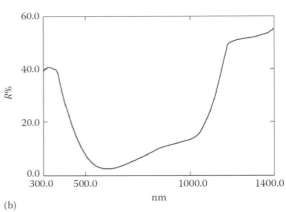

Figure 4.45 (a) Transmission measurement of polycrystalline silicon. Since the UV-2600 system is capable of measurement to 1400 nm, the transmission characteristics of the bandgap region near 1000 nm are evident. (b) Reflection measurement of an antireflective film. The suppressed reflectance in the visible region is clearly seen. (© 2014 Shimadzu Scientific Instruments, www.ssi.shimadzu.com. Used with permission.)

example. The most widely used international scale for color is the CIE L*a*b* color space introduced in 1976. CIE stands for *Commission internationale de l'eclairage*, French for International Commission on Illumination. The CIE L*a*b* color space describes all colors visible to the human eye and is device independent. Most spectrophotometers, equipped with the appropriate software, can be used to measure color.

The three coordinates, L*, a*, and b*, are organized like the x-, y-, and z-axes in a cube or sphere; they are orthogonal to each other. The L*-axis runs from top to bottom (like the z-axis) and represents the lightness of the color. L* = 0 represents pure black; L* = 100 is pure diffuse white. The a*-axis represents the color position between green and red, with negative a* values indicating greener colors and positive a* indicating redder colors. The b*-axis is the blue–yellow axis, with negative b* values on the blue end and positive b* values toward yellow. Because there are three coordinates, this is also called a tristimulus color system. Delta values associated with each coordinate indicate how much a standard and sample differ from each other and are often used for quality control or formula adjustment. An overall measure of color difference is ΔE^*. ΔE^* is equal to the square root of the sum of the squares of the ΔL^*, Δa^*, and Δb^* values. The values for the coordinates are calculated from a mathematical combination of visible absorbance spectra combined with standard functions for the observation angle and illumination source. Assuming that our spectrophotometer has the appropriate software, we'll look at some examples of how colors of products are measured.

Earlier, we discussed the use of visible absorbance spectra to compare olive oils and to detect non-olive oil vegetable oils. Using the same instrument, the i-LAB (Claverie and Johnson), the L*a*b* values were also calculated, using as the illumination source the CIE D65 function, which represents standard daylight. The observer function was the CIE 10° standard angle. For all the oils measured, the L* value correlated well with visual lightness observations. Olive oils have very definite colors ranging from light yellow or green to darker green and golden colors. Most of the oils had a* values of −2 to −14 (light green to dark green) and b* values > +7, indicating light yellow to golden yellow. All of the EVOOs had L* < 92.3, a* > (−10), and b* > 50; the canola oils had L* values > 94, meaning they were lighter than the EVOOs, with a* values in the −2 to −4 range and b* values < 10. All measurements were made in less than 10 s. The tristimulus values can easily be used to check batch-to-batch variability of vegetable oils.

The color of beer is often the first thing people notice about it (Johnson et al.). Beer color ranges from pale yellow for a wheat beer to opaque black for stout. The white paper by Johnson et al. is a highly recommended reading, for both a history of color measurement using spectrophotometry and also the color graphics. The i-LAB visible spectrometer was used to measure the color of beer according to the American Society of Brewing Chemists (ASBC) protocol but can also calculate the tristimulus values as noted earlier and the European Brewing Convention color, which uses a slightly different protocol. The ASBC color measurement defines beer color as 10 times the absorbance at 430 nm for a calculated 0.5 in. path length of decarbonated beer samples. Typical absorbance curves for four beers are shown in Figure 4.46.

Several important measures of wine quality can be evaluated by mathematical combination of absorbance values at multiple wavelengths (Bain). Wine *color inten*sity, a measure of how dark the wine is, is calculated from the

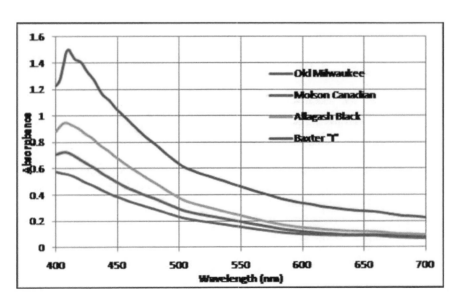

Figure 4.46 Visible absorbance spectra of selected beers from 400 to 700 nm. The lowest curve is the Old Milwaukee, with Molson Canadian, Allagash Black, and Baxter increasing in absorbance, respectively. The Allagash Black was diluted 1:10 and the Baxter 1:2 with distilled water. (From Johnson, J.E. et al., The color and turbidity of beer. White Paper, Courtesy of Microspectral Analysis, LLC, 2011. www.microspectralanalysis.com. With permission.)

Table 4.12 Color and color differences in wines

Wine	L*	a*	b*	ΔL*	Δa*	Δb*	CIE ΔE*
Standard	99.9702	−0.0047	−0.074	—	—	—	—
Red 1	56.392	36.660	27.775	−43.578	36.664	27.849	63.394
Red 2	65.478	31.896	17.716	−34.492	31.901	17.790	50.238
Red 3	70.463	27.620	12.179	−29.508	27.625	12.253	42.237
Red 4	75.233	25.744	9.695	−24.737	25.749	9.769	37.018
Red 5	74.303	23.064	15.526	−25.667	23.068	15.600	37.873
White 1	100.185	−0.144	0.819	0.215	−0.479	0.893	1.037
White 2	99.258	−0.099	1.621	−0.713	−0.095	1.695	1.842

Source: Modified from Bain, G., Wine color analysis using the Evolution™ array UV-visible spectrophotometer, Thermo Fisher Scientific Application Note 51852. Application note and data © Thermo Fisher Scientific (www.thermofisher.com). Used with permission.

sum of the absorbances at 420, 520, and 620 nm. The wine *hue* is a measure of the appearance of the wine and is calculated from the ratio of absorbance at 420 nm to absorbance at 520 nm. The Thermo Fisher Scientific software on the Evolution Array UV/VIS spectrophotometer can calculate the intensity, hue, and the CIE L*a*b* values, as well as the color difference values (the delta values) compared to a standard. Red wines exhibit an absorbance between 400 and 650 nm, centered at about 500 nm, due to the presence of anthocyanin. No such peak appears in the spectra of white wines. The CIE color measurements are carried out in transmittance mode. Table 4.12 shows results for the color and color difference measurements.

You can see from the L* values that the white wines are lighter than the reds; in fact, one of the white wines has an L* value >100. The L* = 100 value is for diffuse white; samples that give specular reflection may be higher than 100. You can also see that the a* values for the red wines are positive, as would be expected, while the white wines are negative (more greenish than red in color).

4.6 ACCURACY AND PRECISION IN UV/VIS ABSORPTION SPECTROMETRY

There are three major factors that affect the accuracy and precision of quantitative absorption measurements: the instrument, the skill of the analyst, and the method variables. Instruments vary in the quality of their optical, mechanical, and electrical systems and also in their data processing. Each instrument has fixed limitations; these must be understood by the analyst and optimized when possible. Wavelength calibration must be checked routinely using recognized wavelength standards. Holmium oxide standards are commonly used for this purpose. Stray light, transmittance, resolution, and other instrument parameters should be checked regularly. The analyst must optimize slit widths if the instrument is equipped with variable slits; too narrow a slit width may result in errors due to a low S/N ratio, while too wide a slit width causes both loss of resolution and negative deviations from Beer's law. Sample cells are very often the cause of error; they must be cleaned and handled properly for the best accuracy. Method variables include the quality of the reagents used, pH, temperature control, color stability, reaction kinetics, and stoichiometry. It may be necessary to remove interferences, to buffer the sample, to control exposure to air and light, and to perform other chemical manipulations to achieve accurate results. The analyst must be trained to operate the instrument and to perform all the chemical manipulations required. Attention to detail; accurate recordkeeping; routine use of replicates, spiked samples, or reference materials; and the preparation and measurement of appropriate blanks and standards are the analyst's responsibility. The accuracy of a method will depend on the analyst, the method specificity, the removal or control of interferences, and finally on the spectrophotometer itself.

Spectrophotometric analyses are capable of being performed with relative standard deviations (RSDs) as low as 0.5 %. Detection limits depend on the molar absorptivity of the transition being measured but are often sub-ppm for many analytes. The linear working range for spectrophotometry used to be only one to two orders of magnitude. This is a short linear range compared to fluorescence, but as has been discussed, new systems are capable of much higher linear working ranges.

4.7 NEPHELOMETRY AND TURBIDIMETRY

Much of the theory and equipment used in spectrophotometry applies with little modification to **nephelometry** and **turbidimetry**. These fields involve the **scattering** of light by nontransparent particles suspended in a liquid; examples of such particles include fine precipitates and colloidal suspensions. In *nephelometry*, we measure the amount of radiation *scattered* by the particles; in *turbidimetry*, we measure the amount of light *not scattered* by the particles.

VISIBLE AND ULTRAVIOLET MOLECULAR SPECTROSCOPY

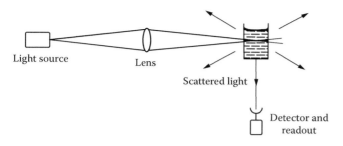

Figure 4.47 Schematic optical system for nephelometry.

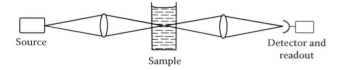

Figure 4.48 Schematic optical system for turbidimetry.

These processes are illustrated in Figures 4.47 and 4.48. The applications of nephelometry include the estimation of the clarity of drinking water, beverages, liquid pharmaceuticals, and other products where the transparency is important and in the determination of species that can be precipitated, such as calcium or barium by precipitation as the phosphate or sulfate insoluble salt. The quantity of calcium or barium present is measured by the amount of radiation scattered by the precipitated compound. From the intensity of scattered radiation, the original concentration of calcium or barium can be determined. Conversely, sulfate and phosphate can be determined by precipitation as the barium compound. Process analyzers using nephelometry or turbidimetry can be used to monitor the clarity of a plant stream or water treatment facility stream on a continuous basis.

When using nephelometry or turbidimetry for quantitative analysis, standard suspensions or standard turbid solutions are required for calibration. The precipitate or suspension standards must be prepared under rigidly controlled conditions. This is essential because the scattering of light depends on the *size*, *shape*, and *refractive index* (RI) of the *particles* involved, as well as on the concentration of particles. Some particles also absorb light, which will cause an error in the turbidity measurement. It is necessary for a given solution to produce the same number of particles of the same size and other properties listed for the degree of light scattering to be meaningful. Interferences include dirty sample cells and any colored or absorbing species that will remove light from the light path. Any absorbance of light will result in an erroneously high turbidity, just as turbidity results in an erroneously high absorbance.

The wavelength of the light scattered most efficiently depends on the physical size of the scattering particles. From this, it can be reasoned that the size of the scattering particle may be determined if the wavelength of scattered light is accurately known. This type of light scattering forms the basis for the measurement of polymer RMMs from the size of polymer molecules.

For water analysis, the formulation of turbid standards is very difficult, so most water laboratories use a synthetic polymer suspension as a standard. The formazin polymer suspension is easy to make and more stable and reproducible than adding clay or other particles to water to prepare standards. Alternatively, suspensions of polymer beads of the appropriate size can be used as scattering standards. (See *Standard Methods for the Examination of Water and Wastewater* for details.)

In the determination of a given species by a precipitation reaction, it is critical to control the experimental conditions. Two identical samples of equal concentration of analyte will scatter light equally only if they form the same number and size distribution of particles when they are precipitated. This depends on many experimental conditions, including the sample temperature, the rate at which the precipitant and the sample are mixed, the degree of agitation or stirring, and the length of time the precipitates are allowed to stand before measurement. Procedures usually call for the use of a stopwatch to make all measurements at the same point in time, such as 60 s after the reagent was added. Interferences include other particles and colored or absorbing species. Sulfate in drinking water can be determined turbidimetrically by precipitation as barium sulfate over the range of 1–40 mg/L sulfate, with precision of about 2 % RSD and accuracy, estimated by recovery of spiked samples, of about 90 %.

Other working definitions of turbidity can be used. We discussed measuring the color of beer earlier, using the ASBC protocol. Turbidity in beer is defined by them as a function of the difference in the absorbances of decarbonated beer at 700 nm and at 430 nm (Johnson et al.). If the absorbance at 700 nm for a calculated 0.5 in. path length is less than or equal to 0.039 times the absorbance for the same path length at 430 nm, the beer is said to be free of turbidity. If the value at 700 nm is $>0.039 \times \text{Abs}_{430\,nm}$, the beer is turbid and would need to be clarified by centrifugation or filtration in order to get a true color measurement.

4.8 MOLECULAR EMISSION SPECTROMETRY
4.8.1 Fluorescence and Phosphorescence

If "black light" (UV light) illuminates certain paints or certain minerals in the dark, they give off visible light. These paints and minerals are said to *fluoresce*. An energy diagram of this phenomenon is shown in Figure 4.49. For fluorescence to occur, a molecule must absorb a photon and be promoted from its ground state to an excited vibrational state in a higher electronic state. There are actually *two* possible electronic transitions. Electrons possess the

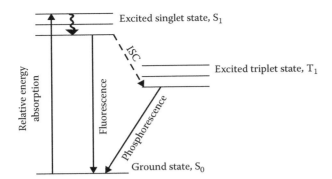

Figure 4.49 Schematic diagram of the ground state, excited singlet state, and excited triplet state, of a molecule. The wavy line denotes a radiationless transition from a higher vibrational level in the excited singlet state to the lowest vibrational level in the excited singlet state. The dotted arrow marked intersystem crossing (ISC) shows the radiationless intersystem crossing from the excited singlet state to the excited triplet state. The length of the solid arrows denotes the relative energy of the transitions: absorption > fluorescence > phosphorescence. This results in $\lambda_{phosphorescence} > \lambda_{fluorescence} > \lambda_{absorption}$.

property of spin; we can think of this simplistically as the electron rotating either clockwise or counterclockwise. For two electrons to occupy the same orbital, their spins must be opposite to each other; we say that the spins are paired. If one electron is raised to the excited level without changing its spin, the electron in the excited level is still opposite in spin to the electron left behind in the valence level. This excited state of the molecule in which electron spins are paired is called a *singlet* state. If the electron spins are parallel (both spinning in the same direction as a result of the excited electron reversing its spin), the excited state is called a *triplet* state. Each "excited state" has both a singlet and corresponding triplet state. Singlet state energy levels are higher than the corresponding triplet state energies. Singlet states are designated S_1, S_2, S_3, and so on; triplet states are designated T_1, T_2, T_3, and so on. The ground state is a singlet state, S_0. Figure 4.49 shows a ground state with the first excited singlet and triplet states. Some vibrational sublevels of the excited states are also shown.

The molecule absorbs energy and an electron is promoted to one of the higher vibrational levels in the singlet state; this is a vibrationally excited electronic state. The vibrationally excited molecule will rapidly "relax" to the lowest *vibrational* level of the electronic excited state S_1. This relaxation or loss of energy is a radiationless process, shown by the wavy arrow. Energy decreases but no light is emitted. Now the molecule can return to the ground state by emitting a photon equal to the energy difference between the two levels. This is the process of *fluorescence*: excitation by photon absorption to a vibrationally excited state, followed by a rapid transition between two levels with the same spin state (singlet to singlet, in this case) that results in the emission of a photon. The emitted photon is of lower energy (longer wavelength) than the absorbed photon. The wavelength difference is due to the radiationless loss of vibrational energy, depicted by the wavy line in Figure 4.49. This type of fluorescence, emission of a longer wavelength than was absorbed, is what is usually seen in solutions; it is called *Stokes* fluorescence. The lifetime of the excited state is very short, on the order of 1–20 ns, so fluorescence is a virtually instantaneous emission of light following excitation. However, the lifetime of the fluorescent state is long enough that time-resolved spectra can be obtained with modern instrumentation. A molecule that exhibits fluorescence is called a *fluorophore*.

The transition from the singlet ground state to a triplet state is a forbidden transition. However, an excited singlet state can undergo a radiationless transition to the triplet state by reversing the spin of the excited electron. This is an energetically favorable process since the triplet state is at a lower energy level than the singlet state. This radiationless transition, shown schematically in Figures 4.49 and 4.50, is called intersystem crossing (ISC). The molecule can relax to the ground state from the triplet state by emission of a photon. This is the process of *phosphorescence*: excitation by absorption of light to an excited singlet state and then an ISC to the triplet state, followed by emission of a photon due to a triplet–singlet transition. The photon associated with phosphorescence is even lower energy (longer wavelength) than the fluorescence photon, as seen from the relative energy levels in Figures 4.49 and 4.50. Because the triplet–singlet transition is forbidden, the lifetime of the triplet excited state is long, up to 10 s in some cases. The sample will "glow" for some time after the excitation light source is removed. "Glow in the dark" paint is an example of phosphorescent material.

Fluorescence and phosphorescence are both types of luminescence. They are specifically types of *photoluminescence*, meaning that the excitation is achieved by absorption of light. There are other types of luminescence. If the excitation of a molecule and emission of light occurs as a result of chemical energy from a chemical reaction, the luminescence is called *chemiluminescence*. The "glow sticks" popular at rock concerts and Halloween are an example of chemiluminescence. The light emitted by a firefly is an example of *bioluminescence. Electroluminescence* is induced by current. Both photoluminescence and electroluminescence are exhibited by semiconductor quantum dots and solid-state sources of single photons, made from materials such as InAs.

As shown in Figure 4.50, there are other ways for molecules to return to the ground state. Excited molecules may collide with other molecules; it is possible for energy to be transferred during this collision. The molecule returns to the ground state but does not emit radiation. This

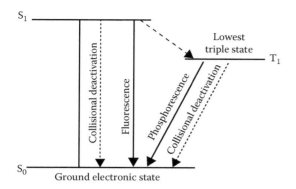

Figure 4.50 Processes by which an excited molecule can relax to the ground state. (Adapted from Guilbault, G.G., ed., *Practical Fluorescence*, 2nd edn., Marcel Dekker, Inc., New York, 1990. With permission.)

is called collisional deactivation or *quenching*. Quenching occurs in solution by collision of the excited analyte molecule with the more numerous solvent molecules. Quenching in fluorescence is often a serious problem, but with care, it can be minimized. Quenching by collision with the solvent molecules can be reduced by decreasing the temperature, thus reducing the number of collisions per unit time. The same result can be achieved by increasing the viscosity—for example, by adding glycerin. Dissolved oxygen is a strong quenching agent. It can be removed by bubbling nitrogen through the sample. Phosphorescence is very susceptible to quenching; the molecule in a triplet state has an extended lifetime in the excited state, so it is quite likely that it will collide with some other molecule and lose its energy of excitation without emitting a photon. Phosphorescence is almost never seen in solution at room temperature because of collisional deactivation. Low temperatures must be used and the analyte must be constrained from collision. This can be done for fluorescence and phosphorescence by converting the sample into a gel (highly viscous state) or glass, or by adsorption of the analyte onto a solid substrate. "Organized" solvents such as surfactant micelles have been used successfully to observe room-temperature phosphorescence and to greatly enhance fluorescence by reducing or eliminating collisional deactivation. Even with the appropriate experimental care, only a small fraction of available analyte molecules will actually fluoresce or phosphoresce, since radiationless transitions are very probable.

4.8.2 Relationship between Fluorescence Intensity and Concentration

The intensity of fluorescence F is proportional to the amount of light absorbed by the analyte molecule. We know from Beer's law that

$$\frac{I_1}{I_0} = e^{-abc} \quad (4.4)$$

so, subtracting each side of the equation from 1 gives

$$1 - \frac{I_1}{I_0} = 1 - e^{-abc} \quad (4.5)$$

We multiply each side by I_0:

$$I_0 - I_1 = I_0\left(1 - e^{-abc}\right) \quad (4.6)$$

Since, $I_0 - I_1$ = amount of light absorbed, the fluorescence intensity, F, may be defined as

$$F = \left(I_0 - I_1\right)\Phi \quad (4.7)$$

where Φ is the quantum efficiency or quantum yield. The quantum yield, Φ, is the fraction of excited molecules that relax to the ground state by fluorescence. The higher the value of Φ, the higher the fluorescence intensity observed from a molecule. A nonfluorescent molecule has $\Phi = 0$.

Therefore, fluorescence intensity is equal to

$$F = I_0\left(1 - e^{-abc}\right)\Phi \quad (4.8)$$

From Equation 4.8, it can be seen that fluorescence intensity is related to the concentration of the analyte, the quantum efficiency, the intensity of the incident (source) radiation, and the absorptivity of the analyte. Φ is a property of the molecule, as is the absorptivity, a. A table of typical values of Φ for fluorescent molecules is given in Table 4.13. The absorptivity of the compound is related to the fluorescence intensity

Table 4.13 Fluorescence quantum yields, Φ

Compound	Solvent	Φ
9-Aminoacridine	Ethanol	0.99
Anthracene	Hexane	0.33
9,10-Dichloroanthracene	Hexane	0.54
Fluorene	Ethanol	0.53
Fluorescein	0.1 N NaOH	0.92
Naphthalene	Hexane	0.10
1-Dimethylaminonaphthalene-4-sulfonate	Water	0.48
Phenol	Water	0.22
Rhodamine B	Ethanol	0.97
Sodium salicylate	Water	0.28
Sodium sulfanilate	Water	0.07
Uranyl acetate	Water	0.04

Source: Guilbault, G.G., ed., *Practical Fluorescence*, 2nd edn., Marcel Dekker, Inc., New York, 1990. With permission.

Note: Solutions are 10^{-3} M, temperatures 21°C–25°C.

(Equation 4.8). Molecules like saturated hydrocarbons that do not absorb in the UV/VIS region do not fluoresce.

The fluorescence intensity is directly proportional to the intensity of the source radiation, I_0. In theory, the fluorescence intensity will increase as the light source intensity increases, so very intense light sources such as lasers, mercury arc lamps, or xenon arc lamps are frequently used. There is a practical limit to the intensity of the source because some organic molecules are susceptible to photodecomposition.

When the term abc is <0.05, which can be achieved at low concentrations of analyte, the fluorescence intensity can be expressed as

$$F = I_0 abc\Phi \qquad (4.9)$$

That is, F, total fluorescence, $= kI_0 c$, where k is a proportionality constant. At low concentrations, a plot of F versus concentration should be linear. But only a portion of the total fluorescence is monitored or measured; therefore,

$$F' = Fk' \qquad (4.10)$$

where F' is the measured fluorescence and

$$F' = k'I_0 c \qquad (4.11)$$

where k' is another proportionality constant.

A plot of F versus c is shown in Figure 4.51. It is linear at low concentrations. The linear working range for fluorescence is about five orders of magnitude, from 10^{-9} to 10^{-4} M.

At higher concentrations, the relationship between F and c deviates from linearity. The plot of F versus c rolls over, as seen in Figure 4.52. It can be seen that at higher concentrations, the fluorescence intensity actually decreases because the molecules in the outer part of the

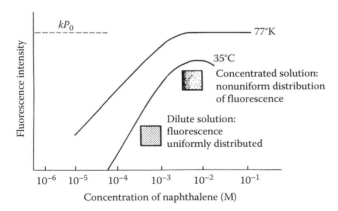

Figure 4.51 Dependence of fluorescence on the concentration of the fluorescing molecule. (From Guilbault, G.G., ed., *Practical Fluorescence*, 2nd edn., Marcel Dekker, Inc., New York, 1990. With permission.)

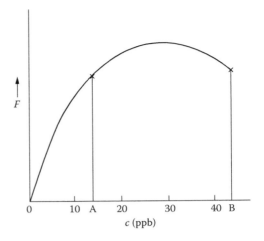

Figure 4.52 Fluorescence intensity at high concentrations of analyte. Note the reversal of fluorescence at high concentration. Concentrations A and B give the same fluorescence intensity and could not be distinguished in a single measurement.

sample absorb the fluorescence generated by those in the inner part of the sample. This is called the "inner cell" effect or self-quenching. In practice, it is necessary to recognize and correct for this effect. It is impossible to tell directly if the fluorescence measured corresponds to concentration A or concentration B, as shown in Figure 4.52. Both concentrations would give the same fluorescence intensity. Diluting the sample slightly can solve the dilemma. If the original concentration were A, then the fluorescence intensity would sharply decrease on dilution. On the other hand, if the concentration were B, then the fluorescence should increase on slight dilution of the sample.

4.9 INSTRUMENTATION FOR LUMINESCENCE MEASUREMENTS

A schematic diagram of a spectrofluorometer is shown in Figure 4.53.

4.9.1 Wavelength Selection Devices

Two monochromators are used: the primary or excitation monochromator and the secondary or fluorescence monochromator. These are generally grating monochromators, although filters can be used for specific analyses. The excitation monochromator selects the desired narrow band of wavelengths that can be absorbed by the sample. The sample emits light in all directions. The second monochromator is placed at 90° to the incident light beam. The second monochromator is set to pass the fluorescence wavelength to the detector. The 90° orientation of the second monochromator is required to avoid the detector "seeing" the intense

VISIBLE AND ULTRAVIOLET MOLECULAR SPECTROSCOPY

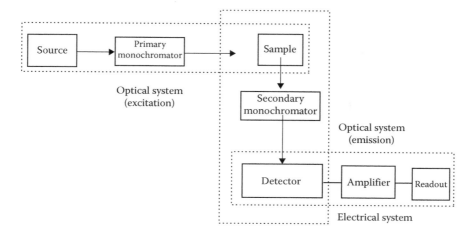

Figure 4.53 Block diagram of the optical components of a typical fluorometer. (From Guilbault, G.G., ed., *Practical Fluorescence*, 2nd edn., Marcel Dekker, Inc., New York, 1990. With permission.)

incident light, thus eliminating the background caused by the light source. Unlike absorption spectrophotometry, the measurement is not of the small difference between two signals but of a signal with essentially no background. This is one reason for the high sensitivity and high linearity of fluorescence. Most fluorescence instruments are single-beam instruments. This means that changes in the source intensity will result in changes in the fluorescence intensity. To compensate for changes in the source intensity, some instruments split off part of the source output, attenuate it, and send it to a second detector. The signals from the two detectors are used to correct for drift or fluctuations in the source.

The 90° geometry is the most common orientation for measuring fluorescence and works very well for solution samples that do not absorb strongly. Other angles are used in specific applications. For strongly absorbing solutions or for solid samples such as thin-layer chromatography plates, fluorescence is measured from the same face of the sample illuminated by the source. This is called front-surface geometry. It is shown schematically for a solid sample in Figure 4.54.

4.9.2 Radiation Sources

The fluorescence intensity is directly proportional to the intensity of the light source. Therefore, intense sources are preferred. Excitation wavelengths are in the UV and visible regions of the spectrum, so some of the same sources used in UV/VIS absorption spectrometry are used for fluorescence. The optical materials will of course be the same—quartz for the UV and glass for the visible region.

Mercury or xenon arc lamps are used. A schematic of a xenon arc lamp is given in Figure 4.55. The quartz envelope is filled with xenon gas, and an electrical discharge through the gas causes excitation and emission of light. This lamp

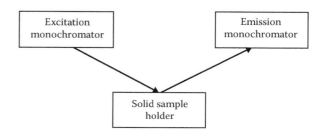

Figure 4.54 Front-surface fluorescence geometry for a solid sample. (From Froelich, P.M. and Guilbault, G.G., In *Practical Fluorescence*, 2nd edn., Guilbault, G.G., ed., Marcel Dekker, Inc., New York, 1990. With permission.)

Figure 4.55 Compact xenon arc lamp used in fluorometers. The quartz envelope is filled with xenon gas. The lamp is ignited by a 10–20 kV pulse across the electrodes. (From Froelich, P.M. and Guilbault, G.G., In *Practical Fluorescence*, 2nd edn., Guilbault, G.G., ed., Marcel Dekker, Inc., New York, 1990. With permission.)

emits a continuum from 200 nm into the IR. The emission spectrum of a xenon arc lamp is shown in Figure 4.56. Current fluorescence spectrometers use xenon flash lamps. The xenon lamps require a particular start up procedure that must be observed when first energizing the lamp. These will differ with manufacturer but must be observed to prevent lamp failure. Also, it is important to avoid sudden shut down of the instrument, such as during a power failure. An uninterruptible power supply might be a good addition to the instrument. Mercury lamps under high pressure can be used to provide a continuum, but low-pressure Hg lamps, which emit a line spectrum, are often used with filter

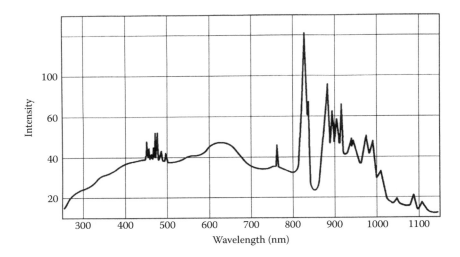

Figure 4.56 Spectral output of a compact xenon arc lamp. (From Froelich, P.M. and Guilbault, G.G., In *Practical Fluorescence*, 2nd edn., Guilbault, G.G., ed., Marcel Dekker, Inc., New York, 1990. With permission.)

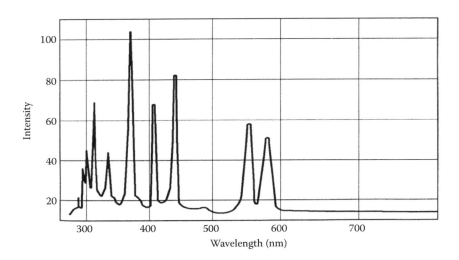

Figure 4.57 Spectral output of a mercury arc lamp, used as a source in fluorometers. (From Froelich, P.M. and Guilbault, G.G., In *Practical Fluorescence*, 2nd edn., Guilbault, G.G., ed., Marcel Dekker, Inc., New York, 1990. With permission.)

fluorometers. The spectrum of a low-pressure Hg lamp is presented in Figure 4.57.

Because of their high intensity, laser light sources are ideal sources for fluorescence. The laser must exhibit a wide range of emission wavelengths, so tunable dye lasers have been the only choice until recently. The dye lasers are generally pumped by an Nd:YAG laser. Nd stands for neodymium and YAG is yttrium aluminum garnet. These pumped dye laser systems are very expensive and complicated to operate. They have much greater intensity output than lamps and so enable lower detection limits to be achieved. Recent advances in solid-state lasers have made small, less expensive visible wavelength lasers available. Solid-state UV lasers are available but do not have the required intensity for use as a fluorescence source.

4.9.3 Detectors

The most common detector is the PMT. The operation of a PMT was described earlier in this chapter. Because the signal is small due to the low concentrations of analyte used, the PMT is often cooled to subambient temperature to reduce noise. The limitation of the PMT is that it is a single-wavelength detector. This requires that the spectrum be scanned. As we have discussed, scanning takes time and is not suitable for transient signals such as those from a chromatographic column. Diode array detectors are now available to collect the entire spectrum at once instead of scanning. The CCD, a 2D array detector, is another alternative to scanning in fluorescence spectrometry. Both PDA and CCD detectors can be used with LC or CE systems for separation and detection of fluorescent compounds or "tagged" compounds in mixtures. LC and CE are discussed in Chapter 14.

4.9.4 Sample Cells

The most common cell for solutions is a 1 cm rectangular quartz or glass cuvette with **four** optical windows. For extremely small volumes, fiber-optic probes, microvolume cells, and flow cells are available. Gas cells and special sample compartments for solid samples are commercially available. Microplate readers are available for most instruments, permitting analysis of up to 384 samples.

4.9.5 Spectrometer Systems

Spectrometer systems for luminescence measurements are available from the major instrument companies, including Agilent, Horiba, PerkinElmer, Shimadzu, and Thermo Fisher Scientific. Only a few examples will be discussed; the company websites provide information, applications, notes, and photos of their instrumentation. Luminescence spectrometers range from high-end research instruments to dedicated application instruments.

For example, the Agilent Cary Eclipse spectrophotometer is a versatile instrument that allows fluorescence, phosphorescence, chemiluminescence, bioluminescence, and time-resolved phosphorescence measurements. It uses a xenon flash lamp, red-sensitive PMT, captures a data point every 12.5 ms, and can scan at 24,000 nm/min from the UV to 900 nm. Sample volumes can range from <0.5 mL to large, odd-sized samples using fiber-optic probe technology and accommodate a 384-well microplate reader. On the other end of the spectrum, there are instruments like the Thermo Fisher Scientific Quantech® Life Science UV Filter fluorometer, featuring a phosphor-coated Hg lamp for excitation in the 270–315 nm range, a red-sensitive PMT, and suited for nucleic acid and intrinsic protein fluorescence measurements. Versions of this instrument are available with a quartz halogen lamp and extended spectral range. Also from Thermo Fisher Scientific is a fluorescence version of the NanoDrop, the NanoDrop 3300 Fluorospectrometer, which uses UV, blue, and white LED excitation sources covering a range from 400 to 750 nm, a 2048-element linear silicon CCD array detector, and uses as little as 1 µL of sample.

4.10 ANALYTICAL APPLICATIONS OF LUMINESCENCE

Fluorescence occurs in molecules that have low-energy $\pi \rightarrow \pi^*$ transitions; such molecules are primarily aromatic hydrocarbons and polycyclic aromatic compounds. Examples include those in Table 4.13, as well as compounds like indole and quinoline. Molecules with rigid structures exhibit fluorescence; the rigidity evidently decreases the probability of a radiationless deactivation. Some organic molecules increase their fluorescence intensity on complexation with a metal ion. The resulting complex structure is more rigid than the isolated organic molecule in solution. Molecules that fluoresce can be measured directly; the number of such molecules is estimated to be between 2000 and 3000 from the published literature. There are several compounds that exhibit strong fluorescence; these can be used to derivatize, complex, or "tag" nonfluorescent species, thereby extending the range of fluorescence measurements considerably. Other analytes are very efficient at quenching the fluorescence of a fluorophore; there are quantitative methods based on fluorescence quenching.

A fluorometric analysis results in the collection of two spectra: the excitation spectrum and the emission spectrum. The excitation spectrum should be the same as the absorption spectrum obtained spectrophotometrically. Differences may be seen due to instrumental factors, but these are normally small, as seen in Figure 4.58, which shows the absorption and excitation spectra for Alizarin garnet R, a fluorometric reagent for aluminum ion and fluoride ion. The longest wavelength absorption maximum in the excitation spectrum is chosen as the excitation wavelength; this is where the first monochromator is set to excite the sample. It would seem reasonable to choose the wavelength that provides the most intense fluorescence as the excitation wavelength, but often short wavelengths from the high-intensity sources used can cause a compound to decompose. The emission spectrum is collected by the second monochromator. The emission or fluorescence spectrum for Alizarin garnet R is shown in Figure 4.58. Similar excitation and emission spectra are shown in Figure 4.59 for quinine and anthracene. Note, especially for anthracene, that the fluorescence (emission) spectrum is almost a mirror image of the excitation spectrum. The shape of the emission spectrum and wavelength of the fluorescence maximum do not depend on the excitation wavelength. The same fluorescence spectrum is obtained for any wavelength the compound can absorb. However, the intensity of the fluorescence is a function of the excitation wavelength.

Fluorometry is used in the analysis of clinical samples, pharmaceuticals, natural products, and environmental samples. There are fluorescence methods for steroids, lipids, proteins, amino acids, enzymes, drugs, inorganic electrolytes, chlorophylls, natural and synthetic pigments, vitamins, and many other types of analytes. The detection limits in fluorometry are very low. Detection limits of 10^{-9} M and lower can be obtained. Single-molecule detection has been demonstrated under *extremely* well-controlled conditions. This makes fluorometry as one of the most sensitive analytical methods available. Therefore, the

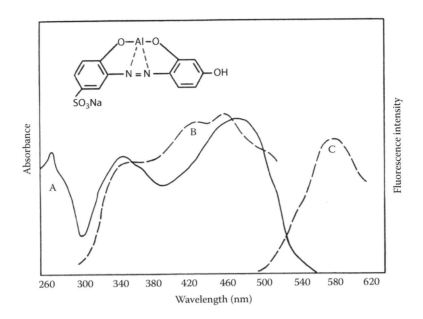

Figure 4.58 Absorption and fluorescence spectra of the aluminum complex with acid Alizarin garnet R (0.008%): Curve A, the absorption spectrum; Curve B, the fluorescence excitation spectrum; and Curve C, the fluorescence emission spectrum. (From Guilbault, G.G., ed., *Practical Fluorescence*, 2nd edn., Marcel Dekker, Inc., New York, 1990. With permission.)

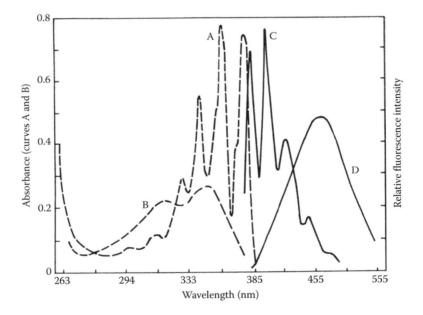

Figure 4.59 Excitation and fluorescence spectra of anthracene and quinine: Curve A, anthracene excitation; Curve B, quinine excitation; Curve C, anthracene fluorescence; and Curve D, quinine fluorescence. (From Guilbault, G.G., ed., *Practical Fluorescence*, 2nd edn., Marcel Dekker, Inc., New York, 1990. With permission.)

technique is widely used in quantitative trace analysis. For example, Table 4.10 indicates that Al^{3+} could be detected spectrophotometrically using aluminon at about 0.04 ppm in solution and fluoride could be detected at 0.2 ppm with SPADNS. Using fluorometry and Alizarin garnet R, whose structure is shown in Figure 4.60, Al^{3+} can be determined at 0.007 ppm and F^- at 0.001 ppm. The strongly fluorescent compounds like fluorescein can be detected at part per trillion levels (ng/mL in solution), so use of such a compound as a "tag" can result in a very sensitive analytical method for many analytes. The table mentioned earlier in the book by Bruno and Svoronos contains a large number

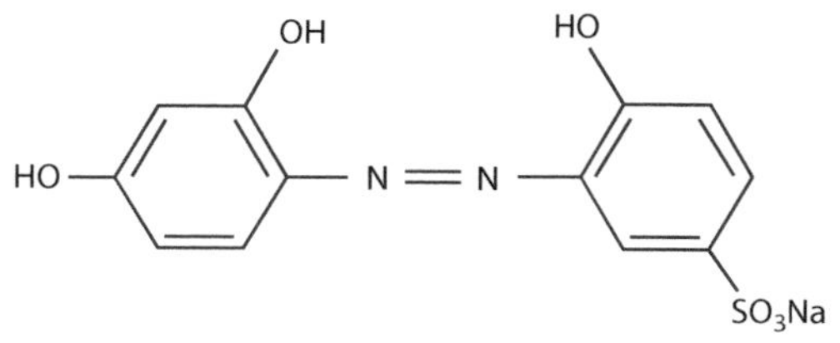

Figure 4.60 Structure of Alizarin garnet R.

of reagents and analytes that are amenable to analysis by this method.

The analysis of solid samples, especially using fiber-optic probes, has opened up new applications for fluorescence spectroscopy. Some of these include the examination of rocks and minerals to identify the presence of oil in shale, evaluation of gemstone quality and origin, and the identification of minerals at potential mining sites. Materials characterization of solar cell materials, semiconductors, organic LEDs, and polymers by fluorescence is another emerging area of study. Solid samples give much sharper fluorescence peaks than liquid samples due to the elimination of nonradiative relaxation paths. Therefore, high-resolution spectrometers are needed. Solar cell researchers need the ability to adjust the position of the sample in the excitation beam to mimic the movement of the sun, and fiber-optic probes eliminate the need to cut up large samples to fit into sample compartments. For example, stalactites often fluoresce due to the presence of humic and fulvic acids that precipitate along with the calcite. Use of a fiber-optic probe allows for nondestructive analysis of such samples in order to understand the environmental conditions under which they formed (McGarry and Baker).

Many materials benefit from analysis at cryogenic temperatures. At 77 K, enhanced fluorescence and phosphorescence are seen, and low-temperature studies are used for elucidation of the mechanisms of photochemical reactions, characterizing bandgap changes in semiconductors and other applications.

An interesting application of the use of chemiluminescence is the measurement of a ruthenium complex upon reaction with codeine (Purcell and Barnett). A 1 mM solution of tris(2,2′-bipyridyl) Ruthenium(II) in 0.05 M H_2SO_4 was oxidized to the 3+ state and then mixed with 1 mM codeine. The codeine is oxidized and the Ru reduced to the 2+ state, producing an excited-state intermediate that returns to the ground state by emission of a photon centered at about 610 nm. In this case, the excited state is produced by a chemical reaction; the emission of light is due to chemiluminescence. No light source was required for excitation. The same Ru^{2+} complex can also be excited using an external light source; the same emission spectrum results, but this would be due to phosphorescence (Girardi et al.)

4.10.1 Advantages of Fluorescence and Phosphorescence

The advantages of fluorescence and phosphorescence for analyses of molecules include extremely high sensitivity, high specificity, and a large linear dynamic range. The sensitivity is a result of the direct measurement of the fluorescence or phosphorescence signal against a zero background signal, as described. Specificity is a result of two factors: First, not all molecules fluoresce; therefore, many molecules are eliminated from consideration. Second, two wavelengths, excitation and emission, are used in fluorometry instead of one in spectrophotometry. It is not likely that two different compounds will emit at the same wavelength, even if they absorb the same wavelength and vice versa. If the fluorescing compounds have more than one excitation or fluorescent wavelength, the difference in either the emission spectrum or the excitation spectrum can be used to measure mixtures of compounds in the same solution. In Figure 4.59, for example, the excitation spectra of quinine and anthracene overlap, but they do not emit at the same wavelengths, so the two compounds could be measured in a mixture. The linear dynamic range in fluorometry is six to seven orders of magnitude compared to one to two orders of magnitude that can be achieved in spectrophotometry.

4.10.2 Disadvantages of Fluorescence and Phosphorescence

Other compounds that fluoresce may need to be removed from the system if the spectra overlap. This can be done, for example, by column chromatography. Peaks may appear in the fluorescence spectrum which are due to other emission and scattering processes; Rayleigh, Tyndall, and Raman scattering may be seen because of the high intensity of the light source used. Peaks due to fluorescent impurities may occur.

Reversal of fluorescence intensity or self-quenching at high concentrations is a problem in quantitative analysis but can be eliminated by successive dilutions. Quenching by impurities can also occur and can cause significant problems in analysis. Changes in pH can frequently change structure, as we saw with phenolphthalein in Figure 4.44, and thereby change fluorescence intensity; pH must therefore be controlled. Temperature and viscosity need to be controlled as well for reproducible results.

Photochemical decomposition or photochemical reaction may be induced by the intense light sources used. In general, the approach of using the longest excitation wavelength possible and the shortest measurement time possible will minimize this problem.

SUGGESTED EXPERIMENTS

4.1 Add a drop of toluene to a UV absorption cell, and cap or seal the cell. Record the absorption spectrum of toluene vapor over the UV range (220–280 nm) several times, varying the slit widths but keeping the scan speed constant. For example, slit widths of 0.1, 0.5, 1, and 5 nm can be used. Explain what happens to the spectral resolution as the slit width is changed.

4.2 Record the absorption spectrum of a solution of pure octane. Record the absorption spectrum of a 0.02 % v/v toluene in octane solution from 220 to 280 nm. For a double-beam spectrometer, pure octane can be put into the reference cell. Compare with Experiment 4.1. Explain your observations. Change the slit width as in Experiment 4.1, and observe what happens to the resolution.

4.3 Record the absorption spectrum between 400 and 200 nm of (a) 1-octene, (b) 1,3-butadiene, and (c) a nonconjugated diolefin (e.g., 1,4-pentadiene). What is the effect of a conjugated system on the absorption spectrum? What does the spectrum tell you about the relative energy of the molecular orbitals in each compound?

4.4 Record the absorption spectrum of a polynuclear aromatic compound such as anthracene and of a quinonoid such as benzoquinone. How does the structure of the compound affect the spectrum?

4.5 From Experiment 4.2, choose a suitable absorbance wavelength (or wavelengths) for toluene. Based on the maximum absorbance for your 0.02 % solution, prepare a series of toluene in octane solutions of higher and lower concentrations (e.g., 0.1 %, 0.05 %, and 0.01 % v/v toluene in octane might be suitable). Measure the absorbance of each solution (using octane as the reference) at your chosen wavelength.

4.6 Using the absorbance data obtained from Experiment 4.5, plot the relationship between the absorbance A and concentration c of toluene at each wavelength chosen. Indicate the useful analytical range for each wavelength.

4.7 Prepare a standard solution of quinine by dissolving a suitable quantity of quinine in water. Record the UV absorption spectrum of the solution between 500 and 200 nm. Record the absorption spectra of several commercial brands of quinine water (after allowing the bubbling to subside). Which brand contained the most quinine? (Tonic water contains quinine.)

4.8 Prepare ammonium acetate buffer by dissolving 250 g of ammonium acetate in 150 mL of deionized (DI) water and then adding 700 mL of glacial acetic acid. Prepare a 1,10-phenanthroline solution by dissolving 100 mg 1,10-phenanthroline monohydrate in 100 mL DI water to which two drops of conc. HCl have been added. One milliliter of this reagent will react with no more than 100 µg of ferrous ion, Fe^{2+}. Prepare a stock ferrous iron solution containing 1 g/L of ferrous sulfate. By taking aliquots of the stock solution and diluting, prepare four standard solutions and a blank in 100 mL volumetric flasks as follows: Pipet 0, 100, 200, 300, and 400 µg of Fe^{2+} into 100 mL flasks, then add 2 mL conc. HCl to each flask. Add 10 mL of ammonium acetate buffer solution and 4 mL of 1,10-phenanthroline solution. Dilute to the mark with DI water. Mix completely and allow to stand for 15 min for color development. Measure the absorbance at 510 nm. Correlate the absorbance with the concentration of iron in the solutions (remember to subtract the blank), and prepare a calibration curve. *Note*: All reagents used should be low in iron or trace metal grade. The use of this method for determining total iron, ferric iron, and ferrous iron in water may be found in *Standard Methods for the Examination of Water and Wastewater*. Sample preparation is required for real water samples, as the reagent only reacts with ferrous ion.

4.9 Prepare a concentrated stock solution of acetaminophen (paracetamol) by dissolving 20 mg, accurately weighed, of dried acetaminophen in 2 mL methanol. Bring this solution to a total volume of 100 mL with DI water in a volumetric flask. Dilute 3 mL of the stock standard to 100 mL with DI water in a second flask to prepare the working standard. Make a test sample of acetaminophen in exactly the same way. Measure the absorbance of the working standard and the test sample at 244 nm, using a 1 cm quartz cuvette. Measure the absorbance of the DI water as a blank, and subtract as needed.

Quantity of acetaminophen in the test sample =
Standard sample wt $(mg) \times \left(\dfrac{Abs_{224} \text{ test sample}}{Abs_{244} \text{ standard}} \right)$

$\%\text{Purity} = 100 \times \dfrac{\text{Calculated mass of test sample}}{\text{Actual mass of test sample}}$

4.10 The nasal spray Dristan® used to contain L-phenylephrine (PEH) and chlorpheniramine maleate (PAM). (Current formulation may differ; in that case, your instructor can make up a "synthetic unknown" mixture of the two.) Prepare the following solutions: (a) 1 L of 0.10 M HCl; (b) 1 L PEH stock solution containing 200 (±10) mg of PEH in 0.010 M HCl and record the actual concentration; and (c) 1 L PAM stock solution containing 80 (±5) mg PAM in 0.010 M HCl and record the concentration. Record the UV spectrum of each stock solution from 230 to 300 nm. Choose the best wavelength for PEH and the best wavelength for PAM. Using the PEH stock solution and 10 mL volumetric flasks, make up six standard solutions covering the Abs range from 0 to 1 by diluting your stock solution. Each solution should be 0.010 M in HCl. So what is your blank? Record the concentration of each standard solution. Do the same for your PAM stock solution. Record absorbance measurements on all solutions (blank and six standards for each compound) at the two wavelengths you have chosen. Use a graphing program (or graph paper) to obtain Beer's law plots for PEH and PAM at both wavelengths. Are they linear? Using a least squares program, obtain the slopes, intercepts, and correlation coefficients for the Beer's law plots. Next, prepare three solutions containing both PEH and PAM. Record the absorbance at both wavelengths. Are the absorbance values additive? If you have the nasal spray product, dilute it 1 in 200 using 0.010 M HCl by weighing 0.5 g Dristan on an analytical balance. The easiest way to do this is to squirt the spray into a tared 10 mL beaker on the balance. Pour the sample into a 100 mL volumetric, and then rinse the beaker multiple times with 0.010 M HCl, adding the washings to the volumetric and making it up to the mark. This should be done in triplicate. Record the absorbance of the three samples at the two wavelengths. Set up and solve the simultaneous equations. Calculate the average % PEH and PAM in Dristan from the diluted samples. Determine the standard deviation. Report results to the correct number of significant figures. (This experiment is courtesy of Prof. T.C. Werner, Dept. of Chemistry, Union College, Schenectady, NY.)

Concentration (ppm)	Absorbance
1.2	0.24
2.5	0.50
3.7	0.71
5.1	0.97
7.2	1.38
9.8	1.82

4.9 Several samples of monochlorobenzene were brought to the laboratory for analysis using the calibration curve in Problem 4.8. The absorbance of each sample is listed in the following table.

Sample	Absorbance
A	0.400
B	0.685
C	0.120
D	0.160
E	3.0

(a) What are the respective concentrations of monochlorobenzene in samples A through D?
(b) What is the problem with sample E? How could the analysis of sample E be obtained?

PROBLEMS

4.1 What types of molecules are excited by UV radiation? Why?

4.2 Indicate which of the following molecules absorb UV radiation, and explain why: (a) heptane, (b) benzene, (c) 1,3-butadiene, (d) water, (e) 1-heptene, (f) 1-chlorohexane, (g) ethanol, (h) ammonia, and (i) n-butylamine.

4.3 Draw a schematic diagram of a double-beam spectrophotometer. Briefly explain the function of each major component.

4.4 List the principal light sources used in UV/VIS spectrometry.

4.5 Radiation with a wavelength of 640 nm is dispersed by a simple grating monochromator at an angle of 20°. What are the other wavelengths of radiation that are dispersed at the same angle by this grating (lowest wavelength 200 nm)?

4.6 Explain the operating principle of the PMT.

4.7 What are the limitations of UV absorption spectroscopy as a tool for qualitative analysis?

4.8 (a) Plot a calibration curve for the determination of monochlorobenzene from the data listed in the following table.
(b) Three samples of monochlorobenzene were brought in for analysis. The samples transmitted (1) 90 %, (2) 85 %, and (3) 80 % of the light under the conditions of the calibration curve just prepared. What was the concentration of monochlorobenzene in each sample?

4.10 Which of the following absorb in the UV region? (a) N_2, (b) O_2, (c) O_3, (d) CO_2, (e) CH_4, (f) C_2H_4, (g) I_2, (h) Cl_2, (i) cyclohexane, and (j) C_3H_6.

4.11 Why does phenolphthalein change color when going from an acid to a basic solution?

4.12 Why do UV absorption spectra appear as broad bands?

4.13 What causes the blue shift and the red shift in spectra?

4.14 Why do D_2 lamps emit a continuum and not line spectra?

4.15 How does a p–n diode work?

4.16 Describe a diode array.

4.17 Describe the processes of UV molecular fluorescence and phosphorescence.

4.18 What is the relationship between fluorescence and excitation light intensity I_0?

4.19 Explain the reversal of fluorescence intensity with an increase in analyte concentration. How is this source of error corrected?

4.20 Draw a schematic diagram of the instrumentation used for measuring UV fluorescence intensity.

4.21 (a) What interferences are encountered in UV fluorescence?
(b) Why is phosphorescence not used as extensively as fluorescence for analytical measurements?

4.22 From the emission spectra of quinine and anthracene (Figure 4.59), pick a wavelength that will permit you to determine quinine in a mixture of quinine and anthracene. Do the same for anthracene. Can you use the excitation spectra to distinguish between the two compounds? Explain.

Calculate the wavelength of the absorption maximum of the following compounds.

4.23

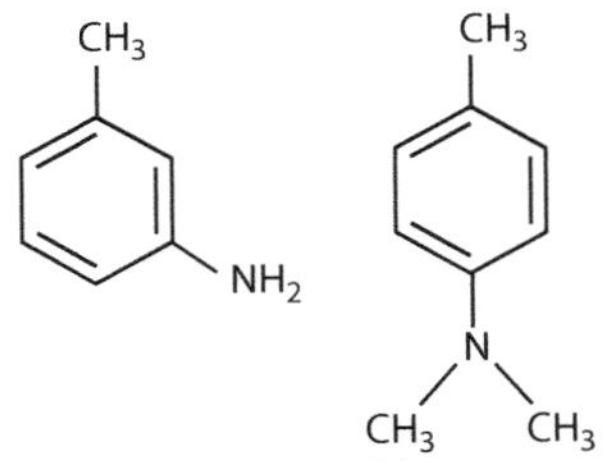

(a) 3-methylaniline (m-CH₃, NH₂ on benzene)

(b) 4-methyl-N,N-dimethylaniline (p-CH₃, N(CH₃)₂ on benzene)

4.24

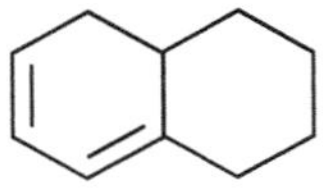

4.25

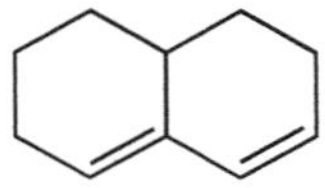

4.26

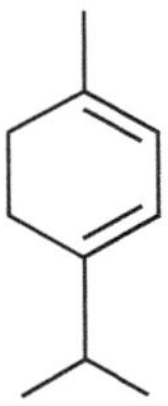

4.27

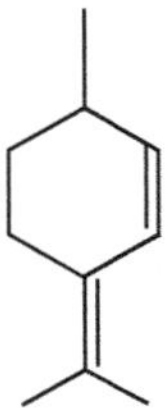

4.28

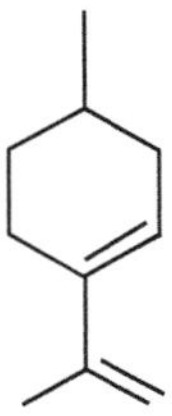

4.29

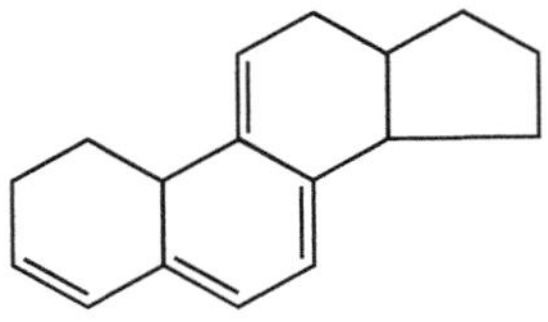

4.30

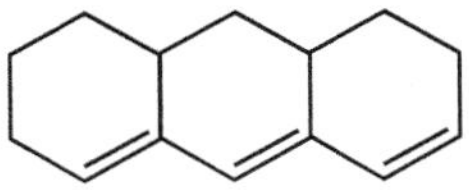

4.31

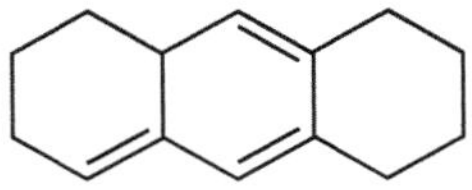

4.32

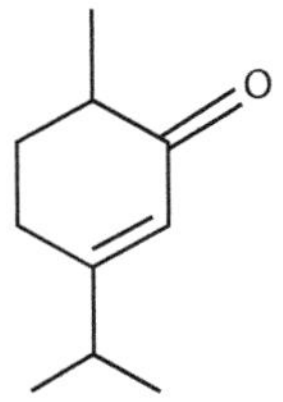

4.33

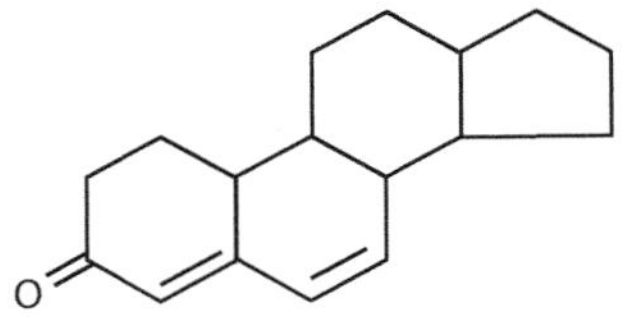

4.34

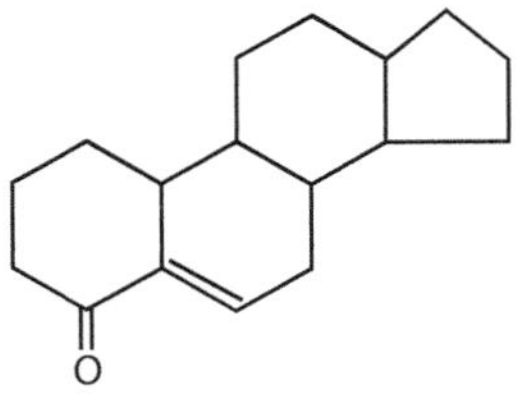

4.35

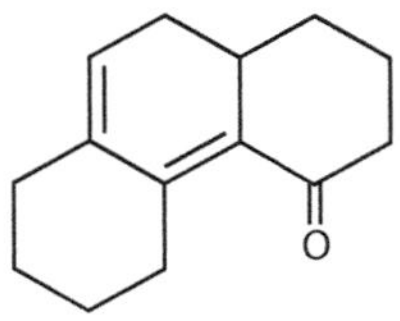

VISIBLE AND ULTRAVIOLET MOLECULAR SPECTROSCOPY

4.36

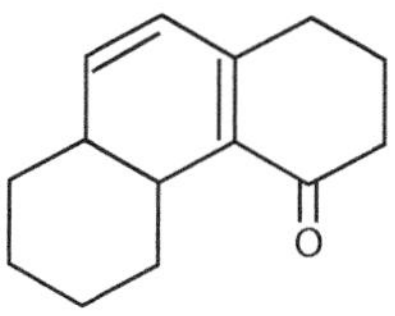

4.37

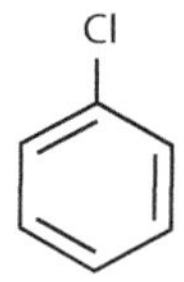

4.38

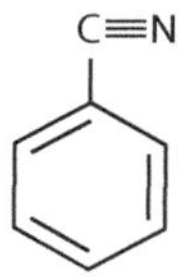

4.39

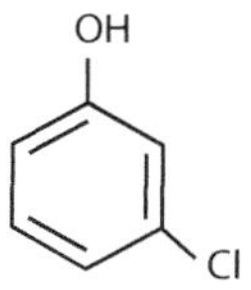

4.40

4.41

4.42 If a solution appears blue when a white light is passed through it, what color(s) has the solution absorbed?

4.43 State Beer's law. What conditions must be met for Beer's law to apply? Complete the following table by providing the values for T and A:

Solution	Absorption (%)	T	A	Concentration (ppm)
1	1			1
2	13			6
3	30			15
4	55			34
5	80			69

4.44 Assuming the data obtained in Problem 4.43 were for a calibration curve and the same cell was used for all measurements, complete the following table:

Solution	Absorption (%)	T	Concentration (ppm)
A	30		
B	3		
C	10		
D	50		
E	70		

4.45 What is the relationship between the absorption cell length b and the absorbance A? Complete the following table (the concentration c was equal in all cases):

Sample	Path Length b (cm)	A
1	0.1	0.01
2	0.5	
3	1.0	
4	2.0	
5	5.0	

4.46 Name three reagents used for quantitative UV/VIS spectrometric analysis and the elements they are used to determine.

4.47 Name two fluorometric reagents. What are the structural characteristics that make a molecule fluoresce?

4.48 Below are the absorption spectra of naphthalene and anthracene (from Jaffé and Orchin, with permission). Their structures are shown on the spectra. These molecules are polycyclic aromatic hydrocarbons, formed by fusing together benzene rings.

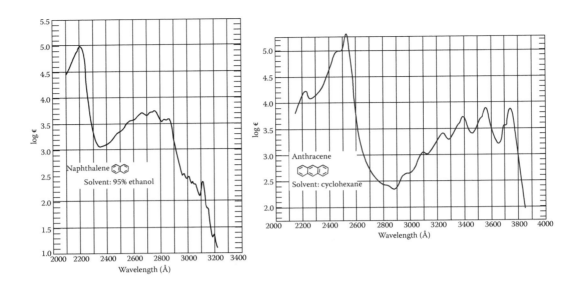

(a) Tabulate the wavelengths for the absorption maxima for these two compounds and in the spectrum of benzene (Figure 4.12 in the text). What trend do you observe?

(b) What transition is causing the peaks observed in these compounds?

(c) Explain the trend you observe in the absorption maxima.

(d) The next larger molecule in this family is naphthacene, a four-ring compound, with the structure shown here:

Predict where the absorption maxima will occur for naphthacene. Explain your prediction.

4.49 The UV/VIS absorption spectrum shown here is the spectrum of holmium oxide glass (from Starna Cells, Inc., www.starna.com, with permission). It is a rare-earth oxide glass and is available in high purity.

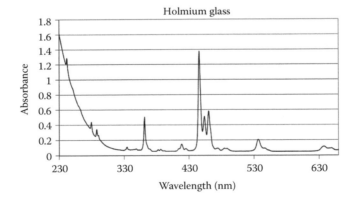

(a) Qualitatively, what differences do you see between this spectrum and the spectrum of an organic molecule such as pyridine (Figure 4.1)? Why do you think they are different in appearance?

(b) Consider the spectrum. Think of how you might use holmium oxide to check on the operation of your UV/VIS spectrometer. What could you check?

BIBLIOGRAPHY

The reader will note that occasionally multiple editions of books are cited as sources or reference works. The reason for this is that sometimes different figures are used in different editions, which may in fact be written by different authors. The sources found in the bibliography will be the most recent versions of a particular work.

In some cases instrument manufacturers and suppliers are cited by providing a URL to the exact internet location. We have endeavored to ensure that these links are functional at the time of publication, but clearly these links change. If a link has become nonfunctional, a good approach is to use your browser's advanced search capability. For example: "company name" AND "instrument type" AND "specific descriptors" is a good place to start.

Agilent Technologies, *Fundamentals of Modern UV–Visible Spectroscopy,* Publication Number 5980-1397E and the companion *Fundamentals of Modern UV–Visible Spectroscopy Workbook*, Publication 5980-1398E, Agilent Technologies, 2000 (Both publications may be accessed as pdf files at www.chem.agilent.com).

American Public Health Association. *Standard Methods for the Examination of Water and Wastewater*, 18th edn. American Public Health Association: Washington, DC, 1992.

Bain, G. Wine color analysis using the evolution array UV-visible spectrophotometer. Thermo Fisher Scientific Application Note 51852, Thermo Fisher Scientific, Inc., 2013. www.thermoscientific.com.

Boltz, D.F. (ed.) *Colorimetric Determination of Nonmetals*. Interscience Publishers: New York, 1958.

Bradford, M.M. A rapid and sensitive method for the quantitation of microgram quantities of protein utilizing the principle of protein-dye binding. *Anal. Biochem.*, 72, 248, 1976.

Brown, C. Ultraviolet, visible and near-infrared spectrophotometers. In *Analytical Instrumentation Handbook*, 4th edn. Ewing, G.W. (ed.), Taylor and Francis, Boca Raton, FL, 2018.

Bruno, T.J., Svoronos, P.D.N., *CRC Handbook of Basic Tables for Chemical Analysis – Data Driven Methods and Interpretation*, 4th Ed., CRC Taylor and Francis, Boca Raton, FL, 2021.

Burgess, C.; Knowles, A. (eds.) *Techniques in Visible and Ultraviolet Spectrometry*. Chapman & Hall: London, U.K., 1981.

Callister, W.D., Jr. *Materials Science and Engineering: An Introduction*, 5th edn. John Wiley & Sons: New York, 2000.

Chang, R. *Essential Chemistry*, 2nd edn. McGraw-Hill Companies, Inc: New York, 2000.

Claverie, M.; Johnson, J.E. The determination of extra virgin olive oil from other oils by visible spectroscopy. White Paper, Microspectral Analysis, LLC, 2011. www.microspectralanalysis.com.

Creswell, C.J.; Runquist, O. *Spectral Analysis of Organic Compounds*. Burgess: Minneapolis, MN, 1970.

Dean, J.A. *Analytical Chemistry Handbook*. McGraw-Hill, Inc.: New York, 1995.

Dulski, T.R. *A Manual for the Chemical Analysis of Metals*. American Society for Testing and Materials: West Conshohocken, PA, 1996.

Ewing, G.W. (ed.) *Analytical Instrumentation Handbook*, 2nd edn. Marcel Dekker, Inc.: New York, 1997.

Froelich, P.M.; Guilbault, G.G. In *Practical Fluorescence*, 2nd edn. Guilbault, G.G. (ed.) Marcel Dekker, Inc.: New York, 1990.

Gerardi, D.; Barnett, N.W.; Lewis, S.W. *Anal. Chim. Acta, 378*, 1, 1999.

Guilbault, G.G. (ed.) *Practical Fluorescence*, 2nd edn. Marcel Dekker, Inc.: New York, 1990.

Hanson, E.K.; Ballantyne, J. *PLoS One, 5*(9), e12830, 2010.

Hollas, J.M. *Modern Spectroscopy*, 3rd edn. John Wiley & Sons, Ltd.: England, U.K., 1996.

Huber, L.; George, S.A. (eds.) *Diode Array Detection in HPLC*. Marcel Dekker, Inc.: New York, 1993.

Ingle, J.D., Jr.; Crouch, S.R. *Spectrochemical Analysis*. Prentice Hall, Inc.: Englewood Cliffs, NJ, 1988.

Jaffé, H.H.; Orchin, M. *Theory and Applications of Ultraviolet Spectroscopy*. John Wiley & Sons: New York, 1962.

Johnson, J.E.; Claverie, M.; Wooton, D. The color and turbidity of beer. White Paper, Microspectral Analysis, LLC, 2011. www.microspectralanalysis.com.

Keppy, N.K.; Allen, M.W. The determination of HMF in honey with an Evolution™ array UV-visible spectrophotometer, Thermo Fisher Scientific Application Note 51864, Thermo Fisher Scientific, Inc., 2013. www.thermoscientific.com.

Lambert, J.B.; Shurvell, H.F.; Lightner, D.; Cooks, R.G. *Introduction to Organic Spectroscopy*. Macmillan Publishing Company: New York, 1987.

Lowry, O.H. et al. Protein measurement with the Folin phenol reagent, *J. Biol. Chem.*, 193, 265, 1951.

McGarry, S.F.; Baker, A. Integrating U-Th, 13C, 210Pb methods to produce a chronologically reliable isotope record of the Belize River Valley Maya from a low uranium stalagmite. *Quaternary Sci. Rev.*, 19, 1087, 2000.

Meehan, E.J. Optical methods of analysis. In Elving, P.J., Meehan, E., Kolthoff, I.M. (eds.), *Treatise on Analytical Chemistry*, Vol. 7. John Wiley & Sons: New York, 1981.

Owen, A. Tutorial on UV-visible spectroscopy, Agilent Technologies, 2013. www.chem.agilent.com.

Pavia, D.L.; Lampman, G.M.; Kriz, G.S. *Introduction to Spectroscopy: A Guide for Students of Organic Chemistry*, 3rd edn. Harcourt College Publishers: Fort Worth, TX, 2001.

Pisez, M.; Bartos, J. *Colorimetric and Fluorometric Analysis of Organic Compounds and Drugs*. Marcel Dekker, Inc.: New York, 1974.

Purcell, S.; Barnett, N. Using the Agilent Cary Eclipse to measure the chemiluminescence of a ruthenium complex. Application Note, www.chem.agilent.com.

Rumble, J. R., (ed.), CRC Handbook of Chemistry and Physics, 102nd ed. CRC Taylor and Francis, Boca Raton, FL, 2021.

Sandell, E.B.; Onishi, H. *Colorimetric Determination of Traces of Metals*, 4th edn. Interscience: New York, 1978.

Schaffer, J.P.; Saxena, A.; Antolovich, S.D.; Sanders, T.H., Jr.; Warner, S.B. *The Science and Design of Engineering Materials*, 2nd edn. WCB/McGraw-Hill: Boston, MA, 1999.

Scott, A.I. *Interpretation of the Ultraviolet Spectra of Natural Products*. Pergamon: Oxford, U.K., 1964.

Settle, F.A. (ed.) *Handbook of Instrumental Techniques for Analytical Chemistry*. Prentice Hall, Inc.: Upper Saddle River, NJ, 1997.

Shackelford, J.F. *Introduction to Materials Science for Engineers*, 4th edn. Prentice Hall, Inc.: Upper Saddle River, NJ, 1996.

Silverstein, R.M.; Bassler, G.C.; Morrill, T.C. *Spectrometric Identification of Organic Compounds*, 5th edn. John Wiley

& Sons: New York, 1991. (*Note*: The most recent edition of this text [Silverstein, R.M.: Webster, F.X., 6th edn., John Wiley and Sons: New York, 1998] has eliminated entirely the topic of UV spectroscopy.)

Smith, P.K. et al. *Anal. Biochem.*, *150*, 76, 1985.

United States Pharmacopeia and National Formulary, Rockville, MD, 2014.

White, J.W. Spectrophotometric method for hydroxymethylfurfural in honey. *J. Assoc. Off. Anal. Chem.*, *62*(3), 509, 1979.

Zumdahl, S.S.; Zumdahl, S.A. *Chemistry*, 5th edn. Houghton Mifflin Co.: Boston, MA, 2000.

CHAPTER 5

Infrared, Near-Infrared, Raman, and Photoacoustic Spectroscopy

Infrared (IR) radiation was first discovered in 1800 by Sir William Herschel, who used a glass prism with blackened thermometers as detectors to measure the heating effect of sunlight within and beyond the boundaries of the visible spectrum. Coblentz laid the groundwork for IR spectroscopy with a systematic study of organic and inorganic absorption spectra. Experimental difficulties were immense. Since each point in the spectrum had to be measured separately, it could take 4 hours to record the full spectrum. But from this work came the realization that each compound had its own unique IR absorption pattern and that certain functional groups absorbed at about the same wavelength even in different molecules. The IR absorption spectrum provides a "fingerprint" of a molecule with covalent bonds. This can be used to identify the molecule. Qualitative identification of organic and inorganic compounds is a primary use of IR spectroscopy. In addition, the spectrum provides a quick way to check for the presence of a given functional group such as a carbonyl group in a molecule.

IR spectroscopy and spectrometry as used by analytical and organic chemists is primarily *absorption* spectroscopy. IR absorption can also be used to provide quantitative measurements of compounds. IR spectroscopy became widely used after the development of commercial spectrometers in the 1940s. Double-beam monochromator instruments were developed, better detectors were designed, and better dispersion elements, including gratings, were incorporated. These conventional spectrometer systems have been replaced by Fourier transform IR (FTIR) instrumentation. This chapter will focus on FTIR instrumentation and applications of IR spectroscopy. In addition, the related techniques of near-infrared (NIR) spectroscopy and Raman spectroscopy will be covered, as well as the use of IR and Raman microscopy.

The wavelengths of IR radiation of interest to chemists trying to identify or study organic molecules fall between 2 and 20 μm. These wavelengths are longer than those in the red end of the visible region, which is generally considered to end at about 0.75 μm. IR radiation therefore is of lower energy than visible radiation but of higher energy than radio waves. The entire IR region can be divided into the *NIR*, from 0.75 to 2.5 μm, the *mid-IR*, from about 2.5 to 20 μm, and the *far-IR*, from 20 to 200 μm. Visible radiation (red light) marks the upper energy end or minimum wavelength end of the IR region; the maximum wavelength end is defined somewhat arbitrarily; some texts consider the far-IR to extend to 1000 μm. The IR wavelength range tells us the IR frequency range from the equation (introduced in Chapter 3)

$$v = \frac{c}{\lambda} \qquad (5.1)$$

where
v is the frequency
c is the speed of light
λ is the wavelength

We also know that $\Delta E = hv$. When the frequency is high, λ is short and the energy of the radiation is high.

It is common to use *wavenumber*, symbolized by either $\tilde{v}$ or $\bar{v}$, with units of cm^{-1}, in describing IR spectra. The unit cm^{-1} is called a *reciprocal centimeter*. The wavenumber is the reciprocal of the wavelength. The wavenumber is the number of waves of radiation per centimeter, $1/\lambda$; frequency is the number of waves per second, c/λ. Wavelength and wavenumber are related by

$$\text{Wavelength } (\mu m) \times \text{wavenumber } (cm^{-1}) = 10,000 = 1 \times 10^4 \qquad (5.2)$$

Both wavenumbers and wavelengths will be used throughout the chapter, so it is important to be able to convert between these units. The older IR literature used the term *micron* and the symbol μ for wavelength in micrometers (μm).

5.1 ABSORPTION OF IR RADIATION BY MOLECULES

Molecules with covalent bonds may absorb IR radiation. This absorption is quantized, so only certain frequencies of IR radiation are absorbed. When radiation (i.e., energy) is absorbed, the molecule moves to a higher energy state. The energy associated with IR radiation is sufficient to cause molecules to rotate (if possible) and to vibrate. If the IR wavelengths are longer than 100 µm, absorption will cause excitation to higher rotational states in the gas phase. If the wavelengths absorbed are between 1 and 100 µm, the molecule will be excited to higher vibrational states. Because the energy required to cause a change in rotational level is small compared to the energy required to cause a vibrational level change, each vibrational change has multiple rotational changes associated with it. Gas-phase IR spectra therefore consist of a series of discrete lines. Free rotation does not occur in condensed phases. Instead of a narrow line spectrum of individual vibrational absorption lines, the IR absorption spectrum for a liquid or solid is composed of broad vibrational absorption **bands**. A typical IR spectrum for a condensed phase (liquid or solid) is shown in Figure 5.1. This is the spectrum of a thin film of polystyrene; note that the absorption band at about 2950 cm^{-1} is more than 100 cm^{-1} wide at the top. The individual vibration–rotation lines can be seen in gas-phase IR spectra. These narrow lines are clearly seen in Figure 5.2, the gas-phase spectrum of hydrogen chloride, HCl.

Molecules absorb radiation when a bond in the molecule vibrates at the same frequency as the incident radiant energy. After absorbing radiation, the molecules have more energy and vibrate at increased amplitude. The frequency absorbed depends on the masses of the atoms in the bond, the geometry of the molecule, the strength of the bond, and several other factors. Not all molecules can absorb IR radiation. The molecule must have a change in **dipole moment** during vibration in order to absorb IR radiation.

5.1.1 Dipole Moments in Molecules

When two atoms with different electronegativities form a bond, the electron density in the bond is not equally distributed. For example, in the molecule hydrogen fluoride, HF, the electron density in the bond shifts away from the H atom toward the more electronegative fluorine atom. This results in a partial negative charge on F and a partial positive charge on H. The bond is said to be **polar** when such charge separation exists. The charge separation can be shown as

$$\overset{\delta^+}{H} - \overset{\delta^-}{F}$$

where δ indicates a partial charge. The dipole in the bond is indicated by a crossed arrow placed with the point on the more negative end of the bond as shown here:

The HF molecule is a linear diatomic molecule with one polar bond; therefore, the molecule is polar and has a dipole moment. The dipole moment µ is the product of the charge Q and the distance between the charges, r:

$$\mu = Q \times r \tag{5.3}$$

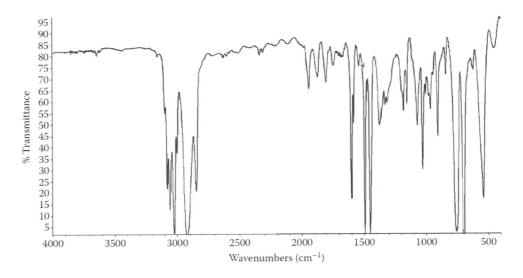

Figure 5.1 FTIR spectrum of a thin film of polystyrene. The *y*-axis unit is %*T*; the *x*-axis is in wavenumbers (cm^{-1}). Collected on a Thermo Scientific 6700 FTIR spectrometer with a deuterated triglycine sulfate (DTGS) detector. (© Thermo Fisher Scientific, www.thermofisher.com. Used with permission.)

INFRARED, NEAR-INFRARED, RAMAN, AND PHOTOACOUSTIC SPECTROSCOPY

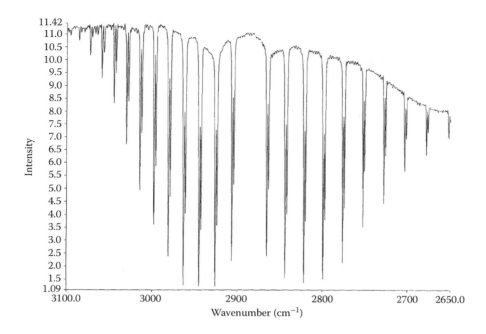

Figure 5.2 Vapor-phase FTIR spectrum of hydrogen chloride, HCl. Spectrum was collected in a 10 cm gas cell with NaCl windows on a Paragon 1000 FTIR spectrometer. [© 1993–2014 PerkinElmer, Inc. All rights reserved. Printed with permission. (www.perkinelmer.com).]

The partial positive and negative charges in HF must be equal in magnitude but opposite in sign to maintain electrical neutrality. We can imagine that the HF molecule vibrates by having the bond stretch and compress over and over, just as a spring can be stretched and compressed by first pulling and then squeezing the spring. On vibration of the HF bond, the dipole moment changes because the distance between the charges changes. Because the dipole moment changes in a repetitive manner as the molecule vibrates, it is possible for HF to absorb IR radiation. It follows that diatomic molecules containing atoms of the same element such as H_2, N_2, O_2, and Cl_2 have no charge separation because the electronegativity of each atom in the bond is the same. Therefore, they do not have dipole moments. Since a change in dipole moment with time is necessary for a molecule to absorb IR radiation, symmetrical diatomic molecules do not absorb IR radiation. Diatomic molecules made up of different atoms such as HCl and CO do have dipole moments. They would be expected to absorb IR radiation or to be "IR active."

Molecules with more than two atoms may or may not have permanent dipole moments. It depends on the geometry of the molecule. Carbon dioxide has two equal C=O bond dipoles, but because the molecule is linear, the bond dipoles cancel and the molecule has no net dipole moment:

Water, H_2O, has two equal H–O bond dipoles. Water has a bent geometry and the vector sum of the bond dipoles does not cancel. The water molecule has a permanent net dipole moment:

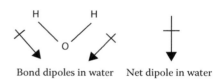

Bond dipoles in water   Net dipole in water

Predicting the IR absorption behavior of molecules with more than two atoms is not as simple as looking at diatomic molecules. It is not the net dipole moment of the molecule that is important, but any change in the dipole moment on vibration. We need to understand how molecules vibrate. This is relatively simple for diatomic and triatomic molecules. Large molecules cannot be evaluated simply because of their large number of vibrations and interactions between vibrating atoms. It can be said that most molecules do absorb IR radiation, which is the reason this technique is so useful.

5.1.2 Types of Vibrations in Molecules

The common molecular vibrations that are excited by IR radiation are **stretching** vibrations and **bending** vibrations. These are called **modes** of vibration. Stretching involves a change in bond lengths resulting in a change in interatomic

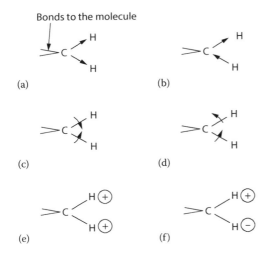

Figure 5.3 Principal modes of vibration between carbon and hydrogen in an alkane: (a) symmetrical stretching, (b) asymmetrical stretching and the bending vibrations, (c) scissoring, (d) rocking, (e) wagging, and (f) twisting.

distance. Bending involves a change in bond angle or a change in the position of a group of atoms with respect to the rest of the molecule. For a group of three or more atoms, at least two of which are the same type of atom, there are two stretching modes: symmetrical stretching and asymmetrical stretching. These two modes of stretching are shown, respectively, in Figure 5.3 for a CH_2 group. In Figure 5.3a, the two H atoms both move away from the C atom—a symmetrical stretch. In Figure 5.3b, one H atom moves away from the C atom and one moves toward the C atom—an asymmetrical stretch. Bending modes are shown in Figures 5.3c–f. Scissoring and rocking are in-plane bending modes—the H atoms remain in the same plane as the C atom (i.e., in the plane of the page). Wagging and twisting are out-of-plane (oop) bending modes—the H atoms are moving out of the plane containing the C atom. The + sign in the circle indicates movement above the plane of the page toward the reader, while the − sign in the circle indicates movement below the plane of the page away from the reader. Bends are also called deformations and the term antisymmetric is used in place of asymmetric in various texts.

For the CO_2 molecule, we can draw a symmetric stretch, an asymmetric stretch, and a bending vibration, as shown here:

$$\overset{\leftarrow\;\;\;\rightarrow}{O=C=O} \qquad \overset{\rightarrow\;\;\;\rightarrow}{\underset{\leftarrow}{O=C=O}} \qquad \overset{\uparrow\;\;\;\uparrow}{\underset{\downarrow}{O=C=O}}$$

The CO_2 molecule on the left is undergoing a symmetric stretch, the one in the middle an asymmetric stretch, and the one on the right an in-plane bend. The symmetric stretching vibration does *not* change the dipole moment of the molecule. This vibrational mode does not absorb IR radiation—it is said to be *IR inactive*. However, the other two modes of vibration do change the dipole moment—they are *IR active*. The asymmetric stretching frequency occurs at 2350 cm^{-1} and the bending vibration occurs at 666 cm^{-1}.

For diatomic and triatomic molecules, it is possible to work out the number and kind of vibrations that occur. To locate a point in 3D space requires three coordinates. To locate a molecule containing N atoms in three dimensions, $3N$ coordinates are required. The molecule is said to have $3N$ degrees of freedom. To describe the motion of such a molecule, translational, rotational, and vibrational motions must be considered. Three coordinates or degrees of freedom are required to describe translational motion and three degrees of freedom are required to describe rotational motion about the molecule's center of gravity. This leaves $3N - 6$ degrees of freedom to describe vibrational motion. There are $3N - 6$ possible normal modes of vibration in a molecule of N atoms. For example, the water molecule contains three atoms, so it has $3 \times 3 = 9$ degrees of freedom and $(3 \times 3) - 6 = 3$ normal modes of vibration. For the water molecule, these normal modes of vibration are a symmetric stretch, an asymmetric stretch, and a scissoring (bending) mode. Linear molecules cannot rotate about the bond axis. As a result, only two degrees of freedom are needed to describe rotation, so linear molecules have $3N - 5$ normal modes of vibration. If we look at the CO_2 molecule given earlier, three modes of vibration are shown, but $3N - 5 = 4$ normal modes of vibration, so one is missing. The fourth is a bending mode equivalent to that shown, but oop:

$$\overset{+\;\;\;-\;\;\;+}{O=C=O}$$

where
+ indicates movement toward the reader
− indicates movement away from the reader

The oop bending mode and the in-plane bending mode already shown both occur at 666 cm^{-1}. They are said to be *degenerate*. Both are IR active, but only one absorption band is seen since they both occur at the same frequency.

For simple molecules, this approach predicts the number of fundamental vibrations that exist. The use of the dipole moment rule indicates which vibrations are IR active, but the IR spectrum of a molecule rarely shows the number of absorption bands calculated. Fewer peaks than expected are seen due to IR-inactive vibrations, degenerate vibrations, and very weak vibrations. More often, additional peaks are seen in the spectrum due to overtones and other bands.

The excitation from the ground state V_0 to the first excited state V_1 is called the *fundamental transition*. It is the most likely transition to occur. Fundamental absorption bands are strong compared with other bands that may appear in the spectrum due to overtone, combination, and difference bands. *Overtone* bands result from excitation from the ground state to higher energy states V_2, V_3, and so on. These absorptions occur at approximately integral multiples of the frequency of the fundamental absorption. If the fundamental absorption occurs at frequency v, the overtones will appear at about 2v, 3v, and so on. Overtones are weak bands and may not be observed under real experimental conditions. Vibrating atoms may interact with each other. The interaction between vibrational modes is called *coupling*. Two vibrational frequencies may couple to produce a new frequency $v_3 = v_1 + v_2$. The band at v_3 is called a *combination* band. If two frequencies couple such that $v_3 = v_1 - v_2$, the band is called a *difference* band. Not all possible combinations and differences occur; the rules for predicting coupling are beyond the scope of this text.

The requirements for the absorption of IR radiation by molecules can be summarized as follows:

1. The natural frequency of vibration of the molecule must equal the frequency of the incident radiation.
2. The frequency of the radiation must satisfy $\Delta E = hv$, where ΔE is the energy difference between the vibrational states involved.
3. The vibration must cause a change in the dipole moment of the molecule.
4. The amount of radiation absorbed is proportional to the square of the rate of change of the dipole during the vibration.
5. The energy difference between the vibrational energy levels is modified by coupling to rotational energy levels and coupling between vibrations.

5.1.3 Vibrational Motion

A molecule is made up of two or more atoms joined by chemical bonds. Such atoms vibrate about each other. A simple model of vibration in a molecule can be made by considering the bond to be a spring with a weight on each end of the spring (Figure 5.4). The stretching of such a spring along its axis in a regular fashion results in simple harmonic motion. Hooke's law states that two masses joined by a spring will vibrate such that

$$v = \frac{1}{2\pi}\sqrt{\frac{f}{\mu}} \quad (5.4)$$

where
v is the frequency of vibration
f is the force constant of the spring (a measure of the stiffness of the spring)
μ is the reduced mass

The reduced mass is calculated from the masses of the two weights joined by the spring:

$$\mu = \frac{M_1 M_2}{M_1 + M_2} \quad (5.5)$$

where
M_1 is the mass of one vibrating body
M_2 is the mass of the other

From Equation 5.4, it can be seen that the natural frequency of vibration of the harmonic oscillator depends on the force constant of the spring and the masses attached to it but is independent of the amount of energy absorbed. Absorption of energy changes the amplitude of vibration, not the frequency. The frequency v is given in hertz (Hz) or cps. If one divides v in cps by c, the speed of light in cm/s, the result is the number of cycles per cm. This is $\bar{v}$, the wavenumber:

$$\bar{v} = \frac{v}{c} \quad (5.6)$$

Dividing both sides of Equation 5.4 by c, we get

$$\bar{v} = \frac{1}{2\pi c}\sqrt{\frac{f}{\mu}} \quad (5.7)$$

where
$\bar{v}$ is the wavenumber of the absorption maximum in cm^{-1}
c is the speed of light in cm/s
f is the force constant of the bond in dyn/cm
μ is the reduced mass in g

The term *frequency of vibration* is often used when the vibration is expressed in wavenumbers, but it must be remembered that wavenumber is directly proportional to the frequency v, not identical to it. Equation 5.7 tells us the frequency of vibration of two atoms joined by a chemical bond, where the force constant f is a measure of the

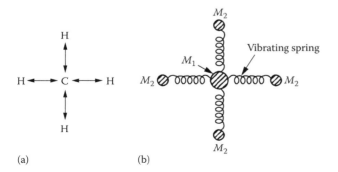

Figure 5.4 (a) The atoms and chemical bonds in methane, CH_4, presented as (b) a system of masses and springs.

strength of the chemical bond and μ is the reduced mass of the vibrating atoms. The term *f* varies with bond strength; a rough but useful approximation is that a triple bond between two atoms is three times as strong as a single bond between the same two atoms, so *f* would be three times as large for the triple bond. A double bond is roughly twice as strong as a single bond. The force constant *f* is directly proportional to the bond order and depends on the electronegativity of the vibrating atoms and the mean distance between them. These are all physical constants and properties of the molecule. Since *f* and μ are constant for any given set of atoms and chemical bonds, the frequency of vibration ν is also constant. *The radiation absorbed by the system has the same frequency and is constant for a given set of atoms and chemical bonds*, that is, for a given molecule. The absorption spectrum is therefore a physical property of the molecule. Average values of the force constant for single, double, and triple bonds are given in Table 5.1. Using these values of *f* and the masses of given atoms, it is possible to estimate the wavenumber for fundamental stretching vibrations of given bonds as discussed. Table 5.2 presents frequencies for common vibrations.

From Equation 5.7, we can deduce that a C–C bond vibrates at a lower wavenumber than a C=C bond, because the force constant for the C–C bond is smaller than that for C=C. For example, C–C vibrates at 1200 cm^{-1} and C=C vibrates at 1650 cm^{-1}. In general, force constants for bending vibrations are lower than stretching vibrations. Resonance and hybridization in molecules also affect the force constant for a given bond.

Table 5.1 Average values for bond force constant

Bond Order	Average Force Constant *f* (N/m)[a]
1 (Single bond)	500
2 (Double bond)	1000
3 (Triple bond)	1500

[a] N/m is the Systeme International d'Unites (SI) unit; dyn/cm is the cgs unit. The single-bond force constant in dyn/cm is about 5×10^5.

Table 5.2 Molecular vibrations and approximate absorption frequencies in wavenumbers (cm^{-1})

Molecular Group	Structure	Stretching		Bending	
C–H bonds		C → H		C ↓ H	
Methylene (alkane)	R–CH$_2$–R	2929, 2850		1460	780–760
Methyl (alkane)	R–CH$_2$–R	2960, 2870		1450	1375
Alkene (terminal)	R–CH=CH$_2$	3080, 2990		1410	890
Aromatic	C$_6$H$_6$	3050		1200	680–900
Aldehyde	R–CHO	2820, 2710		1390	
Alkyne (terminal)	R–C≡CH	3300		615–680	
Nitrogen bonds		C → N	N → H	N ↓ H	
Primary amine	R–NH$_2$	1065	3390, 3330	1610	

INFRARED, NEAR-INFRARED, RAMAN, AND PHOTOACOUSTIC SPECTROSCOPY

Table 5.2 (Continued) Molecular vibrations and approximate absorption frequencies in wavenumbers (cm^{-1}).

Molecular Group	Structure	Stretching			Bending	
Nitrogen bonds		$C \rightarrow N$	$N \rightarrow H$		$N \downarrow H$	
Secondary amine	$R_1\text{-NH-}R_2$	1150, 850	3330		1610	
Tertiary amine	$R_1,R_2,R_3\text{-N}$	780				
Oxygen bonds		$O \rightarrow H$	$C \Rightarrow O$	$C \rightarrow O$		$C\text{-}O \downarrow H$
Primary alcohol	$R\text{-CH}_2\text{-OH}$	3635		1050, 850		1300
Secondary alcohol	$R_2\text{CH-OH}$	3625		1100, 830		1300
Tertiary alcohol	$R_1R_2R_3\text{C-OH}$	3615		1150, 780		1300
Phenol	Ph-OH	3600		1220		1360
Carboxylic acid	$R\text{-COOH}$	2500–3000	1710			1420, 925
Oxygen bonds		$O \rightarrow H$	$C \Rightarrow O$	$C \rightarrow O$		$C\text{-}O \downarrow H$
Ketone	$R_1R_2\text{C=O}$		1715			
Ether	$R\text{-CH}_2\text{-O-CH}_2\text{-R}$			1100, 860		
Aldehyde	$R\text{-CHO}$		1730			
C–C bonds		$C \rightarrow C$	$C \Rightarrow C$	$C \Rightarrow C$		
Alkane	$R\text{-CH}_2\text{-CH}_2\text{-R}$	800–1200				
Alkene	$R\text{-CH-C=CH}_2$		1650			
Alkyne	$R\text{-C}\equiv\text{CH}$			2120		
Aromatic	Ph-CH_3		1590, 1450			
Chloroalkanes	$R\text{-CHCl-R}$	$C \rightarrow Cl$ 550–850				

Note: A single arrow ($\rightarrow$) denotes a single-bond stretching vibration, a double arrow ($\Rightarrow$) denotes a double-bond stretching vibration, and so on. A vertical arrow ($\downarrow$) denotes a bending vibration.

5.2 IR INSTRUMENTATION

Until the early 1980s, most mid-IR spectrometer systems were double-beam dispersive grating spectrometers, similar in operation to the double-beam system for ultraviolet/visible (UV/VIS) spectroscopy described in Chapter 3. These instruments have been replaced almost entirely by FTIR spectrometers because of the advantages in speed, signal-to-noise ratio, and precision in determining spectral frequency that can be obtained from a modern multiplex instrument. There are NIR instruments that are part of double-beam dispersive UV/VIS/NIR systems, but many NIR instruments are stand-alone grating instruments.

The first requirement for material used in an IR spectrometer is that the material must be transparent to IR radiation. This requirement eliminates common materials such as glass and quartz for use in mid-IR instruments because glass and quartz are not transparent to IR radiation at wavelengths longer than 3.5 μm. Second, the materials used must be strong enough to be shaped and polished for windows, sample cells, and the like. Common materials used are ionic salts, such as potassium bromide, calcium fluoride, sodium chloride (rock salt), and zinc selenide. The final choice among the compounds is determined by the wavelength range to be examined; for example, sodium chloride is transparent to radiation between 2.5 and 15 μm.

This wavelength range was therefore termed the *rock salt region* when an ionic salt prism was used as the wavelength dispersion device in early instruments. Potassium bromide or cesium bromide can be used over the range of 2.1–2.6 μm, and calcium fluoride in the range of 2.4–7.7 μm. The wavelength ranges of some materials used for IR optics and sample holders are given in Table 5.3. A table with more extensive property data (including thermal properties) can be found in the book by Bruno and Svoronos. Knowing the thermal properties of the optics is important when doing temperature studies in which expansion might affect sample volumes. It must also be noted that some of these materials such as thallium bromide are highly toxic and must be handled with utmost care.

The major problem with the use of NaCl, KBr, and similar ionic salts is that they are very soluble in water. Any moisture, even atmospheric moisture, can dissolve the surface of a polished salt crystal, causing the material to become opaque and scatter light. Optics and sample containers made of salts must be kept desiccated. This limitation is one of the reasons salt prisms are no longer used in dispersive IR spectrometers.

5.2.1 Radiation Sources

A radiation source for IR spectroscopy should fulfill the requirements of an ideal radiation source, namely, that

Table 5.3 Typical materials used in mid-IR optics

Material	Transmission Range (μm)	Solubility (g/100 g Water)	Refractive Index (RI)	Comments
Sodium chloride (NaCl)	0.25–16	36	1.49	Most widely used; reasonable range and low cost.
Potassium chloride (KCl)	0.30–20	35	1.46	Wider range than NaCl; used as a laser window.
Potassium bromide (KBr)	0.25–26	65	1.52	Extensively used; wide spectral range.
Barium fluoride (BaF$_2$)	0.2–11	0.1	1.39	Extremely brittle.
Cesium iodide (CsI)	0.3–60	160	1.74	Transmits to 60 μm.
Cesium bromide (CsBr)	0.3–45	125	1.66	
Thallium bromide/iodide eutectic (KRS-5)	0.6–40	<0.05	2.4	For internal reflection in the far IR when moisture is a problem.
Strontium fluoride (SrF$_2$)	0.13–11	1.7×10^{-3}		Resistant to thermal shock.
Silver chloride (AgCl)	0.4–25	1.5×10^{-4}	2.0	Darkens under UV light.
Silver bromide (AgBr)	0.5–35	1.2×10^{-5}	2.2	Darkens under UV light.
Germanium	2–11	Insoluble	4.00	
Fused silica	0.2–4.5	Insoluble	1.5 at 1 μm	Most useful for NIR.
Magnesium fluoride (MgF$_2$)	0.5–9	Insoluble	1.34 at 5 μm	This and the next five materials are known commercially at Irtran® 1 through 6.
Zinc sulfide (ZnS)	0.4–14.5	Insoluble	2.2	
Calcium fluoride (CaF$_2$)	0.4–11.5	Insoluble	1.3	
Zinc selenide (ZnSe)	0.5–22	Insoluble	2.4	
Magnesium oxide (MgO)	0.4–9.5	Insoluble	1.6 at 5 μm	
Cadmium telluride (CdTe)	0.9–31	Insoluble	2.7	

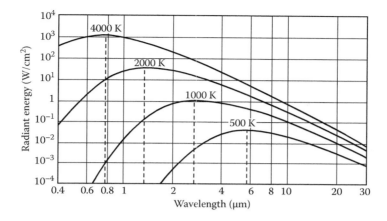

Figure 5.5 Radiant energy distribution curves for a blackbody source operated at various temperatures. (From Coates, J., Vibrational spectroscopy, in Ewing, G.W., ed., *Analytical Instrumentation Handbook*, 2nd edn., Marcel Dekker, Inc., New York, 1997. With permission.)

the intensity of radiation (1) be continuous over the wavelength range used, (2) cover a wide wavelength range, and (3) be constant over long periods of time. The most common sources of IR radiation for the mid-IR region are *Nernst glowers, Globars,* and *heated wires*. All of these heated sources emit continuous radiation, with a spectral output very similar to that of a blackbody radiation source. Spectral curves for blackbody radiators at several temperatures are shown in Figure 5.5. The normal operating temperatures for IR sources are between 1100 and 1500 K. The range of light put out by mid-IR sources extends into both the NIR and far-IR regions, but intensity is at a maximum in the mid-IR region from 4000 to 400 cm^{-1}.

5.2.1.1 Mid-IR Sources

The two main types of sources for mid-IR radiation are electrically heated rigid ceramic rods and coiled wires.

The Nernst glower is a cylindrical bar composed of zirconium oxide, cerium oxide, and thorium oxide that is heated electrically to a temperature between 1500 and 2000 K. The source is generally about 20 mm long and 2 mm in diameter. The rare-earth oxide ceramic is an electrical resistor; passing current through it causes it to heat and glow, giving off continuous IR radiation. The Nernst glower requires an external preheater because of the negative coefficient of electrical resistance; it only conducts at elevated temperature. In addition, the Nernst glower can easily overheat and burn out because its resistance decreases as the temperature increases. The circuitry must be designed to control the current accurately. A related source, the Opperman, consists of a bar of rare-earth ceramic material with a Pt or other wire running coaxially through the center of the ceramic. Electrical current through the wire heats the wire, which heats the ceramic, providing a source similar to the Nernst glower without the preheating requirement. The Globar is a bar of sintered silicon carbide, which is heated electrically to emit continuous IR radiation. The Globar is a more intense source than the Nernst glower. These rigid cylinders were designed so that their shape matched the shape of the slit on a classical dispersive spectrometer. Modern FTIRs do not have slits, so the geometry of the source can now be made more compact. Commercial ceramic IR sources are available in a variety of sizes and shapes, as seen in Figure 5.6. Typical spectral outputs from these commercial ceramic sources are compared with a blackbody radiator in Figure 5.7.

Electrically heated wire coils, similar in shape to incandescent light bulb filaments, have also been used successfully as a light source. Nichrome wire is commonly used, although other metals such as rhodium are used as well. These wires are heated electrically in air to a temperature of ~1100°C. The main problem with these wire coils is "sagging" and embrittlement due to aging, resulting in fracture of the filament, exactly the way a light bulb filament "burns out." Some coiled wire sources are wound around a ceramic rod for support; this results in a more uniform light output over time than that from an unsupported coil.

Modern sources for the mid-IR region are variants of the incandescent wire source or the Globar, but generally in a compact geometry. Commercial furnace ignitors and diesel engine heaters such as the silicon carbide tipped "glow plug" have been adapted for use as IR sources because of their robustness, low operating voltage, and low current requirements.

Sources are often surrounded by a thermally insulated enclosure to reduce noise caused by refractive index gradients between the hot air near the source and cooler air in the light path. Short-term fluctuations in spectral output are usually due to voltage fluctuations and can be overcome by use of a stabilized power supply. Long-term changes

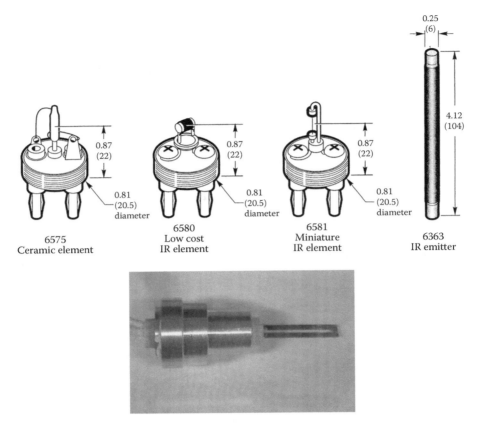

Figure 5.6 Commercial IR radiation sources. (Top) A variety of designs. Dimensions given are in inches and (mm). (Courtesy of Newport Corporation, Irvine, CA, www.newport.com.) (Bottom) FTIR source element used in the PerkinElmer Spectrum One instrument. It is made of a proprietary ceramic/metallic composite and is designed to minimize hot spots to the end of the element. Only the last 5 mm on the end lights up. [© 1993–2014 PerkinElmer, Inc. All rights reserved. Printed with permission. (www.perkinelmer.com).]

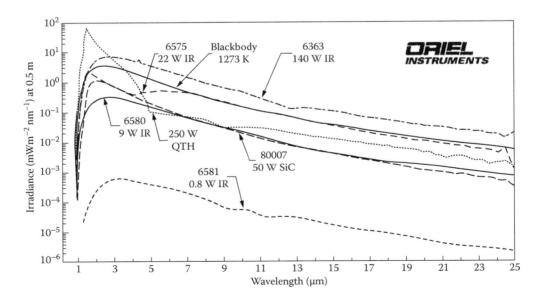

Figure 5.7 Spectral output of a variety of commercial IR radiation sources, including a silicon carbide source (dashed line marked SiC) and an NIR quartz tungsten-halogen lamp (the dotted line marked QTH). A blackbody curve at 1273 K is included for comparison. (Courtesy of Newport Corporation, Irvine, CA, www.newport.com.)

occur as a result of changes in the source material due to oxidation or other high-temperature reactions. These types of changes may be seen as hot or cold "spots" in the source and usually require replacement of the source.

5.2.1.2 NIR Sources

As can be seen in Figure 5.5, operating a mid-IR source at higher temperatures (>2000 K) increases the intensity of NIR light from the source. Operation at very high temperatures is usually not practical due to the excessive heat generated in the instrument and premature burnout of the source. For work in the NIR region, a quartz-halogen lamp is used as the source. A quartz-halogen lamp contains a tungsten wire filament and iodine vapor sealed in a quartz envelope or bulb. In a standard tungsten filament lamp, the tungsten evaporates from the filament and deposits on the lamp wall. This process reduces the light output as a result of the black deposit on the wall and the thinner filament. The halogen gas in a tungsten-halogen lamp removes the evaporated tungsten and redeposits it on the filament, increasing the light output and source stability. The intensity of this source is very high compared to a standard tungsten filament incandescent lamp. The range of light put out by this source is from 25,000 to 2,000 cm^{-1}. Figure 5.8 shows typical commercial quartz tungsten-halogen lamps and a plot of the spectral output of such a source.

5.2.1.3 Far-IR Sources

While some of the mid-IR sources emit light below 400 cm^{-1}, the intensity drops off. A more useful source for the far-IR region is the high-pressure mercury discharge lamp. This lamp is constructed of a quartz bulb containing elemental Hg, a small amount of inert gas, and two electrodes. When current passes through the lamp, mercury is vaporized, excited, and ionized, forming a plasma discharge at high pressure (>1 atm). In the UV and visible regions, this lamp emits atomic Hg emission lines that are very narrow and discrete, but it emits an intense continuum in the far-IR region.

5.2.1.4 IR Laser Sources

A laser is a light source that emits very intense monochromatic radiation. Some lasers, called tunable lasers, emit more than one wavelength of light, but each wavelength emitted is monochromatic. The combination of high intensity and narrow linewidth makes lasers ideal light sources for some applications. Two types of IR lasers are available: gas phase and solid state. The tunable carbon dioxide laser is an example of a gas-phase laser. It emits discrete lines in the 1100–900 cm^{-1} range. Some of these lines coincide with the narrow vibrational–rotational lines of gas-phase analytes. This makes the laser an excellent source for measuring gases in the atmosphere or gases in a production process. Open path environmental measurements of atmospheric hydrogen sulfide, nitrogen dioxide, chlorinated hydrocarbons, and other pollutants can be made using a carbon dioxide laser.

Tunable gas-phase lasers are expensive. Less expensive solid-state diode lasers with wavelengths in the NIR are available. Commercial instruments using multiple diode lasers are available for NIR analyses of food and fuels. Because of the narrow emission lines from a laser system, laser sources are often used in dedicated applications for specific analytes. They can be ideal for process analysis and product quality control (QC), for example, but are not as flexible in their applications as a continuous source or a tunable laser.

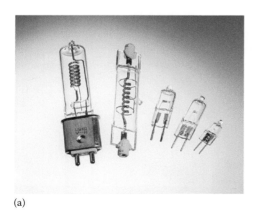
(a)

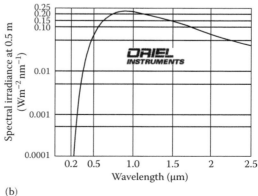
(b)

Figure 5.8 (a) Commercial quartz tungsten-halogen lamps for use in the NIR region. The lamps are constructed of a doped tungsten coiled filament inside a quartz envelope. The envelope is filled with a rare gas and a small amount of halogen. (b) The spectral output of a model 6315 1000 W quartz tungsten-halogen lamp. The location and height of the peak depend on the model of lamp and the operating conditions. (Courtesy of Newport Corporation, Irvine, CA, www.newport.com.)

5.2.2 Monochromators and Interferometers

The radiation emitted by the source covers a wide frequency range. However, the sample absorbs only at certain characteristic frequencies. In practice, it is important to know what these frequencies are. To obtain this information, we must be able to select radiation of any desired frequency from our source and eliminate that at other frequencies. This can be done by means of a monochromator, which consists of a dispersion element and a slit system, as discussed in Chapter 3. This type of system is called a *dispersive* spectrometer or spectrophotometer. Double-beam spectrophotometers were routinely used because both CO_2 and H_2O present in air absorb IR radiation. Changes in humidity and temperature would cause an apparent change in the source intensity if single-beam optics were used, resulting in error in recording the spectrum. A double-beam system automatically subtracts the absorption by species in the air and also can subtract absorption by solvent if the sample is dissolved in a solvent. Double-beam systems for the mid-IR required that the optics be transparent to IR radiation. Lenses are rarely used because of the difficulty of grinding lenses from the ionic salts that are IR transparent. Salt prisms and metal gratings are used as dispersion devices. Mirrors are generally made of metal and front surface polished. The IR spectrum is recorded by moving the prism or grating so that different frequencies of light pass through the exit slit to the detector. The spectrum is a plot of transmission intensity, usually as percent transmittance, versus frequency of light. A dispersive system is said to record a spectrum in the *frequency domain*. It was estimated (Coates, 1997) that no more than 5% of the IR spectrometers in use in 1997 were dispersive instruments and that figure has undoubtedly dropped today. Therefore, the discussion will focus on the FTIR based on a Michelson interferometer.

5.2.2.1 FT Spectrometers

If two beams of light of the same wavelength are brought together in phase, the beams reinforce each other and continue down the light path. However, if the two beams are out of phase, destructive interference takes place. This interference is at a maximum when the two beams of light are 180° out of phase (Figure 5.9). Advantage is taken of this fact in the FT instrument. The FT instrument is based on a Michelson interferometer; a schematic is shown in Figure 5.10. The system consists of four optical arms, usually at right angles to each other, with a *beam splitter* at their point of intersection. Radiation passes down the first arm and is separated by a beam splitter into two perpendicular beams of approximately equal intensity. These beams pass down into other arms of the spectrometer. At the ends of these arms, the two beams are reflected by mirrors back to the beam splitter, where they recombine and are reflected

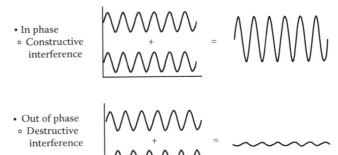

Figure 5.9 Wave interactions. (Top) Constructive interference occurs when both waves are in phase. (Bottom) Destructive interference occurs when both waves are out of phase. (© Thermo Fisher Scientific, www.thermofisher.com. Used with permission.)

together onto the detector. One of the mirrors is fixed in position; the other mirror can move toward or away from the beam splitter, changing the path length of that arm.

It is easiest to discuss what happens in the interferometer if we assume that the source is monochromatic, emitting only a single wavelength of light. If the side arm paths are equal in length, there is no difference in path length. This position is shown in Figure 5.10 as the zero path difference (ZPD) point. For ZPD, when the two beams are recombined, they will be in phase, reinforcing each other. The maximum signal will be obtained by the detector. If the moving mirror is moved from ZPD by 1/8 of a wavelength, the total path difference on recombination is $[2 \times (1/8)\lambda]$ or $(1/4)\lambda$ and partial interference will occur. If the moving mirror is moved from ZPD by 1/4 of a wavelength, then the beams will be one-half of a wavelength out of phase with each other; that is, they will destructively interfere with each other such that a minimum signal reaches the detector. Figure 5.11 shows the signal at the detector as a function of path length difference for monochromatic light. In practice, the mirror in one arm is kept stationary and that in the second arm is moved slowly. As the moving mirror moves, the net signal falling on the detector is a cosine wave with the usual maxima and minima when plotted against the travel of the mirror. The frequency of the cosine signal is equal to

$$f = \frac{2}{\lambda}(v) \qquad (5.8)$$

where
f is the frequency
v is the velocity of the moving mirror
λ is the wavelength of radiation. (Note: *v* is an italic "v," not the Greek letter "nu.")

The frequency of modulation is therefore proportional to the velocity of the mirror and inversely proportional to

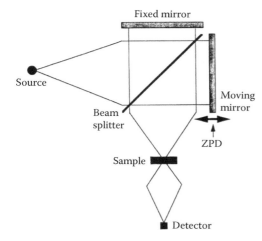

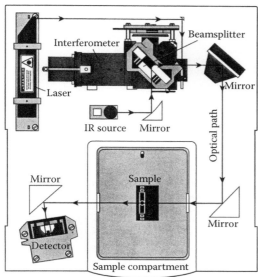

Figure 5.10 (Top) Schematic diagram of a Michelson interferometer. ZPD stands for zero path length difference (i.e., the fixed mirror and moving mirror are equidistant from the beam splitter). (From Coates, J., Vibrational spectroscopy, in Ewing, G.W., ed., *Analytical Instrumentation Handbook*, 2nd edn., Marcel Dekker, Inc., New York, 1997. With permission.) (Bottom) A simple commercial FTIR spectrometer layout showing the He–Ne laser, optics, the source, interferometer, sample, and detector. [© Thermo Fisher Scientific (www.thermofisher.com). Used with permission.]

the wavelength of the incident radiation. The frequency is therefore also proportional to the wavenumber of the incident radiation, as we know from the relationship between wavelength and wavenumber.

Real IR sources are polychromatic. Radiation of all wavelengths generated from the source travels down the arms of the interferometer. Each wavelength will generate a unique cosine wave; the signal at the detector is a result of the summation of all these cosine waves. An idealized interferogram from a polychromatic source is shown in Figure 5.12. The "centerburst" is located in the center of the interferogram because modern FTIR systems scan the moving mirror symmetrically around ZPD. The interferogram holds the spectral information from the source (or sample) in a *time domain*, a record of intensity versus time based on the speed of the moving mirror. The spectral information about the sample is obtained from the wings of the interferogram.

If the unique cosine waves can be extracted from the interferogram, the contribution from each wavelength can be obtained. These individual wavelength contributions can be reconstructed to give the spectrum in the frequency domain, that is, the usual spectrum obtained from a dispersive spectrometer. A Fourier transform is used to convert the time-domain spectrum into a frequency-domain spectrum, hence, the term FTIR spectrometer for this type of system.

It is mechanically difficult to move the reflecting mirror at a controlled, known, steady velocity, and position variations due to temperature changes, vibrations, and other environmental effects must be corrected for. The position of the moving mirror must be known accurately. The position and the velocity are controlled by using a helium–neon (He–Ne) laser beam that is directed down the light path producing an interference pattern with itself. The cosine curve of the interference pattern of the laser is used to adjust the moving mirror in real time in many spectrometers. The He–Ne laser is also used to identify the points at which the interferogram is sampled for the FT, as shown schematically in Figure 5.13.

There are a number of advantages to the use of FTIR over dispersive IR. Because the sample is exposed to all source wavelengths at once, all wavelengths are measured simultaneously in less than 1 s. This is known as the *multiplex* or *Fellgett's advantage*, and it greatly increases the sensitivity and accuracy in measuring the wavelengths of absorption maxima. This multiplex advantage permits collection of many spectra from the same sample in a very short time. Many spectra can be averaged, resulting in an improved signal-to-noise ratio. An FTIR is considerably more accurate and precise than a monochromator (*Connes' advantage*). Another advantage is that the intensity of the radiation falling on the detector is much greater because there are no slits; large beam apertures are used, resulting in higher energy throughput to the detector. This is called the *throughput* or *Jacquinot's advantage*. Resolution is dependent on the linear movement of the mirror. As the total distance traveled increases, the resolution improves. It is not difficult to obtain a resolution of 0.5 cm^{-1}. A comparison between FTIR and dispersive IR is given in Table 5.4.

The signal-to-noise improvement in FTIR comes about as a result of the multiplex and throughput advantages. These permit a rapid spectrum collection rate. A sample spectrum can be scanned and rescanned many times in a few seconds and the spectra added and averaged electronically. The IR signal (S) accumulated is additive, but the

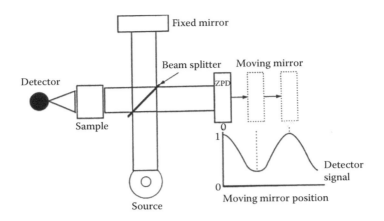

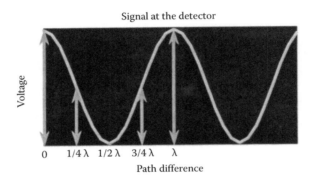

Figure 5.11 (Top) A simplified schematic showing the generation of an interferogram from monochromatic light by displacement of the moving mirror. (Modified from Coates, J., Vibrational spectroscopy, in Ewing, G.W., ed., *Analytical Instrumentation Handbook*, 2nd edn., Marcel Dekker, Inc., New York, 1997. With permission.) (Bottom) An enlarged view of the signal at the detector as a function of path difference between the moving and fixed mirrors for monochromatic light of wavelength λ. [© Thermo Fisher Scientific (www.thermofisher.com). Used with permission.]

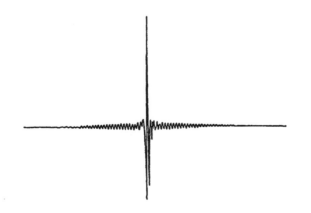

Figure 5.12 An idealized interferogram. (From Coates, J., Vibrational spectroscopy, in Ewing, G.W., ed., *Analytical Instrumentation Handbook*, 2nd edn., Marcel Dekker, Inc., New York, 1997. With permission.)

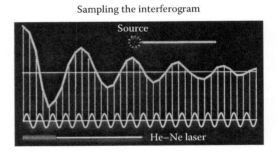

Figure 5.13 Schematic diagram showing how the interferogram is digitized by sampling it at discrete points based on the He–Ne laser signal, shown at the bottom. Each vertical line represents a sampling point. [© Thermo Fisher Scientific (www.thermofisher.com). Used with permission.]

noise level (N) in the signal is random. The S/N ratio therefore increases with the square root of the number of scans (i.e., if 64 scans are accumulated, the S/N ratio increases eight times). FTIR has the potential to be many orders of magnitude more sensitive than dispersive IR. However, in practice, the sheer number of scans necessary to continue to improve the sensitivity limits the improvement. For example, 64 scans improve sensitivity eight times. It would require 4096 scans to increase the S/N 64-fold. A practical limit of one to two orders of magnitude sensitivity increase

Table 5.4 **Comparison of dispersive IR and FTIR instruments**

Dispersive IR	FTIR
Many moving parts result in mechanical slippage. Calibration against reference spectra required to measure frequency.	Only mirror moves during an experiment. Use of laser provides high frequency precision (to 0.01 cm^{-1}) (*Connes' advantage*).
Stray light within instrument causes spurious readings.	Stray light does not affect detector, since all signals are modulated.
In order to improve resolution, only a small amount of IR beam is allowed to pass through the slits.	Much larger beam aperture used; higher energy throughput (*throughput* or *Jacquinot's advantage*).
Only narrow-frequency radiation falls on the detector at any one time.	All frequencies of radiation fall on detector simultaneously; improved S/N ratio obtained quickly (*multiplex* or *Fellgett's advantage*).
Slow scan speeds make dispersive instruments too slow for monitoring systems undergoing rapid change (e.g., gas chromatography [GC] effluents).	Rapid scan speeds permit monitoring samples undergoing rapid change.
Sample subject to thermal effects from the source due to length of scan time.	Short scan times; hence, sample is not subject to thermal effects.
Any emission of IR radiation by sample will fall on detector due to the conventional positioning of the sample before the monochromator.	Any emission of IR radiation by sample will not be detected.
Double-beam optics permit continuous real-time background subtraction.	Single-beam optics; background spectrum collected separately in time from sample spectrum. Can result in error if background spectra not collected frequently.

is therefore normal unless circumstances merit the additional time. Of course, the ability to process many spectra rapidly is a result of the improvement in computer hardware and software that has occurred over the past decade or so. Inexpensive powerful computers and commercially available user-friendly software allow this technology to be used in undergraduate laboratories as well as in industrial and academic research labs.

5.2.2.2 Interferometer Components

The schematic interferometer diagrams given do not show most of the optics, such as beam collimators and focusing mirrors. Mirrors in an FTIR are generally made of metal. The mirrors are polished on the front surface and may be gold coated to improve corrosion resistance. Commercial FTIRs use a variety of flat and curved mirrors to move light within the spectrometer, to focus the source onto the beam splitter, and to focus light from the sample onto the detector.

The beam splitter can be constructed of a material such as Si or Ge deposited in a thin coating onto an IR-transparent substrate. The germanium or silicon is coated onto the highly polished substrate by vapor deposition. A common beam splitter material for the mid-IR region is germanium and the most common substrate for this region is KBr. Both the substrate and the coating must be optically flat. KBr is an excellent substrate for the mid-IR region because of its transparency and its ability to be polished flat. Its major drawback is that it is hygroscopic; this limits the use of KBr as a substrate for field or process control instruments, where environmental conditions are not as well controlled as laboratory conditions. Germanium on KBr is also used for the long-wavelength end of the NIR region, while Si coated on quartz can be used for the short-wavelength end of the NIR region. A thin film of Mylar is used as a beam splitter for the far-IR region. Other combinations of coatings and substrates are available, including complex multilayer materials, especially for applications where moisture may limit the use of KBr.

Ideally, the beam splitter should split all wavelengths equally, with 50% of the beam being transmitted and 50% reflected. This would result in equal intensity at both the fixed and moving mirrors. Real beam splitters deviate from ideality.

As noted earlier and discussed subsequently under background correction in IR spectroscopy, CO_2 and H_2O absorb IR radiation (Figure 5.14). To reduce the spectral background from CO_2 and H_2O and increase the light intensity in the regions where these gases absorb, many spectrometers have sealed and desiccated optical systems. Only a small air path in the sample compartment remains. Alternately, some spectrometers allow the optics and the sample path to be purged with dry nitrogen or other dry gas, decreasing the H_2O and CO_2 in the light path.

IR spectrometers must be calibrated for wavelength accuracy. FTIRs are usually calibrated by the manufacturer and checked on installation. Wavelength calibration can be checked by the analyst by taking a spectrum of a thin film of polystyrene, which has well-defined absorption bands across the entire mid-IR region, as seen in Figure 5.1. Polystyrene calibration standard films are generally supplied with an IR instrument or can be purchased from any instrument manufacturer. Recalibration of the spectrometer should be left to the instrument service engineer if required.

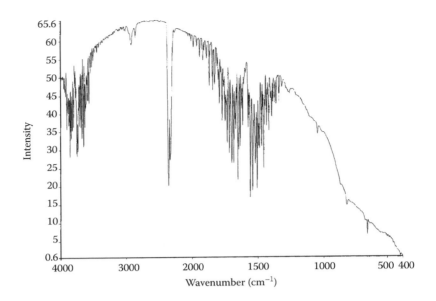

Figure 5.14 A background spectrum of air, showing the absorption bands due to water vapor and carbon dioxide. Collected on a Paragon 1000 FTIR spectrometer. [© 2012–2014 PerkinElmer, Inc. All rights reserved. Printed with permission. (www.perkinelmer.com).]

5.2.3 Detectors

Detectors for IR radiation fall into two classes: thermal detectors and photon-sensitive detectors. Thermal detectors include *thermocouples*, *bolometers*, *thermistors*, and *pyroelectric* devices. Thermal detectors tend to be slower in response than photon-sensitive semiconductors. The most common types of detectors used in dispersive IR spectroscopy were bolometers, thermocouples, and thermistors, but faster detectors are required for FTIR. FTIR relies on pyroelectric and photon-sensitive semiconducting detectors. Table 5.5 summarizes the wavenumber ranges covered by commonly used detectors.

5.2.3.1 Bolometer

A bolometer is a very sensitive electrical-resistance thermometer that is used to detect and measure weak thermal radiation. Consequently, it is especially well suited as an IR detector. The bolometer used in older instruments consisted of a thin metal conductor, such as platinum wire. Incident IR radiation heats up this conductor, which causes its electrical resistance to change. The degree of change of the conductor's resistance is a measure of the amount of radiation that has fallen on the detector. In the case of platinum, the resistance change is 0.4 %/°C. The change in temperature depends on the intensity of incident radiation and on the thermal capacity of the detector. It is important to use a small detector and to focus the radiation on it. The *rate* at which the detector heats up or cools down determines how fast the detector responds to a change in radiation intensity

Table 5.5 Detectors for IR spectroscopy

NIR (12,000–3800 cm^{-1}; 0.8 μm)	Mid-IR (4000–400 cm^{-1}; 2.5–25 μm)	Far-IR (400–20 cm^{-1}; 25–500 μm)
InGaAs (12,000–6000 cm^{-1})		
PbSe (11,000–2000 cm^{-1})		
InSb (11,500–1850 cm^{-1})		
Mercury cadmium telluride (MCT) (11,700–400 cm^{-1})	MCT (11,700–400 cm^{-1})	
DTGS/KBr (12,000–350 cm^{-1})	DTGS/KBr (12,000–350 cm^{-1})	
	Photoacoustic (10,000–400 cm^{-1})	
	DTGS/CsI (6400–200 cm^{-1})	
		DTGS/PE (700–50 cm^{-1})
		Si bolometer (600–20 cm^{-1})

Source: © Thermo Fisher Scientific (www.thermofisher.com). Used with permission.

Note: KBr, CsI, and PE (polyethylene) are the window materials for the DTGS detectors. The MCT detector can vary in its spectral range depending on the stoichiometry of the material.

as experienced when an absorption band is recorded. This constitutes the *response time* of the detector. For these older types of bolometers, the response time is long, on the order of seconds. Consequently, a complete mid-IR scan using a dispersive instrument and bolometer could take 20 min.

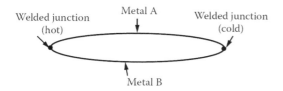

Figure 5.15 Schematic diagram of a thermocouple.

Modern bolometers are micromachined from silicon. This type of bolometer is only a few micrometers in diameter and is usually placed in one arm of a Wheatstone bridge for measurements. The modern microbolometer has a fast response time and is particularly useful for detecting far-IR radiation (600–20 cm^{-1}).

5.2.3.2 Thermocouples

A thermocouple is made by welding together at each end two wires made from different metals (Figure 5.15). If one welded joint (called the *hot junction*) becomes hotter than the other joint (the *cold junction*), a small electrical potential develops between the joints. In IR spectroscopy, the cold junction is carefully screened in a protective box and kept at a constant temperature. The hot junction is exposed to the IR radiation, which increases the temperature of the junction. The potential difference generated in the wires is a function of the temperature difference between the junctions and, therefore, of the intensity of IR radiation falling on the hot junction. The response time of the thermocouple detector is slow; thermocouples cannot be used as detectors for FTIR due to their slow response.

5.2.3.3 Thermistors

A thermistor is made of a fused mixture of metal oxides. As the temperature of the thermistor increases, its electrical resistance decreases (as opposed to the bolometer). This relationship between temperature and electrical resistance allows thermistors to be used as IR detectors in the same way as bolometers. The thermistor typically changes resistance by about 5 % per °C. Its response time is also slow.

5.2.3.4 Golay Detector

The Golay detector was a pneumatic detector, a small hollow cell filled with a nonabsorbing gas such as xenon. In the center of the cell was a blackened film. Radiation was absorbed by the blackened film, causing an increase in temperature. In turn, the film heated the enclosed gas. Thermal expansion of the gas caused the internal pressure of the cell to increase. One wall of the cell was a thin convex mirror that was part of the optical system. As the pressure inside the cell increased, the mirror bulged. A change in radiation intensity falling on the Golay detector caused a change in the readout from the detector. An important advantage of this detector was that its useful wavelength range was very wide. The response was linear over the entire range from the UV through the visible and IR into the microwave range to wavelengths as long as 7.0 mm. The Golay detector's response time was about 10^{-2} s, much faster than that of the bolometer, thermistor, or thermocouple. The detector was very fragile and subject to mechanical failure.

5.2.3.5 Pyroelectric Detectors

Pyroelectric materials change their electric polarization as a function of temperature. These materials may be insulators (dielectrics), ferroelectric materials, or semiconductors. A dielectric placed in an electrostatic field becomes polarized with the magnitude of the induced polarization depending on the dielectric constant. The induced polarization generally disappears when the field is removed. Pyroelectric materials, however, stay polarized and the polarization is temperature dependent.

A pyroelectric detector consists of a thin single crystal of pyroelectric material placed between two electrodes. It acts as a temperature-dependent capacitor. Upon exposure to IR radiation, the temperature and the polarization of the crystal change. The change in polarization is detected as a current in the circuit connecting the electrodes. The signal depends on the rate of change of polarization with temperature and the surface area of the crystal. These crystals are small; they vary in size from about 0.25 to 12.0 mm^2.

The most common pyroelectric material in use as an IR detector is DTGS. The formula for triglycine sulfate is $(NH_2CH_2COOH)_3 \cdot H_2SO_4$; replacement of hydrogen with deuterium gives DTGS. DTGS with a cesium iodide window covers the 6400–200 cm^{-1} range, which includes part of the NIR, all of the mid-IR, and some of the far-IR regions. With the use of a polyethylene window, a DTGS detector can be used as a far-IR detector (700–50 cm^{-1}). Other pyroelectric detectors for the mid-IR region include lithium tantalate, $LiTaO_3$, and strontium barium niobate.

Pyroelectric materials lose their polarization above a temperature called their Curie point. For DTGS, this temperature is low. DTGS detectors are cooled by thermoelectric cooling to between 20 °C and 30 °C to prevent loss of polarization. Lithium tantalate has a much higher Curie temperature and does not require cooling, but is less sensitive than DTGS by about an order of magnitude. Lithium tantalate has a high linear response range, unlike DTGS. DTGS does not respond linearly over the IR frequency range. Its response is inversely proportional to the modulation frequency of the source, resulting in lower sensitivity at the high-frequency end of the spectral range than at the low-frequency end.

5.2.3.6 Photon Detectors

Semiconductors are materials that are insulators when no radiation falls on them but become conductors when radiation falls on them. Exposure to radiation causes a very rapid change in their electrical resistance and therefore a very rapid response to the IR signal. The response time of a semiconductor detector is the time required to change the semiconductor from an insulator to a conductor, which is frequently as short as 1 ns. The basic concept behind this system is that absorption of an IR photon raises an electron in the detector material from the valence band of the semiconductor across a band gap into the conduction band, changing its conductivity greatly. In order to do this, the photon must have sufficient energy to displace the electron. IR photons have less energy than UV or visible photons. The semiconductors chosen as IR detectors must have band gaps of the appropriate energy. The band gap of the detector material determines the longest wavelength (lowest wavenumber) that can be detected.

Materials such as lead selenide (PbSe), indium antimonide (InSb), indium gallium arsenide (InGaAs), and mercury cadmium telluride (HgCdTe, also called MCT) are intrinsic semiconductors commonly used as detectors in the NIR and mid-IR regions. Cooling of these detectors is required for operation. MCT requires operation at 77 K and must be cooled with liquid nitrogen; other detectors such as InGaAs can operate without cooling but show improved S/N if cooled to −30 °C or so with thermoelectric cooling. For the far-IR region, extrinsic semiconductors such as Si and Ge doped with low levels of copper, mercury, or other dopants are used. The dopants provide the electrons for conductivity and control the response range of the detector. These doped germanium or silicon detectors must be cooled in liquid helium but are sensitive to radiation with wavelengths as long as 200 μm. The spectral response curves of some semiconductor detectors are shown in Figure 5.16. The MCT material used is nonstoichiometric and can be represented as $Hg_{(1-x)}Cd_xTe$. The actual spectral range of an MCT detector can be varied by varying the Hg/Cd ratio.

Semiconductor detectors are very sensitive and very fast. The fast response time has permitted rapid-scan IR to become a reality, as is needed in FT spectrometers and coupled techniques such as GC-IR that generate transient peaks. The sensitivity of these detectors has opened up the field of microsampling, IR microscopy, and online IR systems for process control.

5.2.4 Detector Response Time

The length of time that a detector takes to reach a steady signal when radiation falls on it is called its *response time*. This varies greatly with the type of detector used and has a significant influence on the design of the IR instrument.

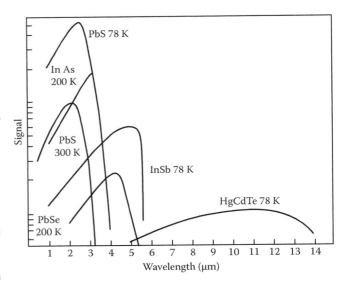

Figure 5.16 Spectral response of various semiconductor detectors. The operating temperature in kelvin is given next to the material.

Thermal detectors such as thermocouples, thermistors, and bolometers have very slow response times, on the order of seconds. Consequently, when a spectrum is being scanned, it takes several seconds for the detector to reach an equilibrium point and thus give a true reading of the radiation intensity falling on it. If the detector is not exposed to the light long enough, it will not reach equilibrium and an incorrect absorption trace will be obtained. It was normal for dispersive IR instruments with older-style thermal detectors to take on the order of 15 min to complete an IR scan. Attempts to decrease this time resulted in errors in the intensity of the absorption bands and recording of distorted shapes of the bands.

The slow response time of thermal detectors is due to the fact that the detector temperature must change in order to generate the signal to be measured. When there is a change in radiation intensity, the temperature at first changes fairly rapidly, but as the system approaches equilibrium, the change in temperature becomes slower and slower and would take an infinitely long time to reach the true equilibrium temperature. It should also be remembered that when an absorption band is reached, the intensity falling on the detector decreases and the response depends on how fast the detector cools.

Semiconductors operate on a different principle. When radiation falls on them, they change from a nonconductor to a conductor. No temperature change is involved in the process; only the change in electrical resistance is important. This takes place over an extremely short period of time. Response times of the order of nanoseconds are common. This enables instruments to be designed with very short scanning times. It is possible to complete the scan in a few seconds using such detectors. These kinds of instruments are very valuable when put onto the end of a GC and used to obtain the IR spectra of the effluents. Such scans must be made in a few seconds and be completely recorded before the next component emerges from the GC column.

Response time is not the only detector characteristic that must be considered. Linearity is very important in the mid-IR region where wide variations in light intensity occur as a result of absorption by a sample. The ability of the detector to handle high light levels without saturating is also important. The MCT detectors saturate easily and should not be used for high-intensity applications; DTGS, on the other hand, while not as sensitive as MCT, does not saturate readily. DTGS can be used for higher intensity applications than MCT.

5.3 SAMPLING TECHNIQUES

IR spectroscopy is one of the few analytical techniques that can be used for the characterization of solid, liquid, and gas samples. The choice of sampling technique depends upon the goal of the analysis, qualitative identification, or quantitative measurement of specific analytes, upon the sample size available, and upon sample composition. The water content of the sample is a major concern, since the most common IR-transparent materials are soluble in water. Samples in different phases must be treated differently. Sampling techniques are available for transmission (absorption) measurements and, since the advent of FTIR, for several types of reflectance (reflection) measurements. The common reflectance measurements are attenuated total reflectance (ATR), diffuse reflectance (DR) or DR-IR-FT spectroscopy (DRIFTS), and specular reflectance. The term reflection may be used in place of reflectance and may be more accurate; specular reflection is actually what occurs in that measurement, for example. However, the term reflectance is widely used in the literature and will be used here.

5.3.1 Techniques for Transmission (Absorption) Measurements

These are the oldest and most basic sampling techniques for IR spectroscopy and apply to both FTIR and dispersive IR systems. Transmission analysis can handle a wide range of sample types and can provide both qualitative and quantitative measurements. Transmission analysis provides maximum sensitivity and high sample throughput at relatively low cost. Substantial sample preparation may be required.

The sample and the material used to contain the sample must be transparent to IR radiation to obtain an absorption or transmission spectrum. This limits the selection of container materials to certain salts, such as NaCl or KBr, and some simple polymers. A final choice of the material used depends on the wavelength range to be examined. A list of commonly used materials is given in Table 5.3. If the sample itself is opaque to IR radiation, it may be possible to dissolve it or dilute it with an IR-transparent material to obtain a transmission spectrum. Other approaches are to obtain IR reflectance spectra or emission spectra from opaque materials.

5.3.1.1 Solid Samples

Three traditional techniques are available for preparing solid samples for collection of transmission IR spectra: mulling, pelleting, and thin-film deposition. First, the sample may be ground to a powder with particle diameters less than 2 μm. The small particle size is necessary to avoid scatter of radiation. A small amount of the powder, 2–4 mg, can be made into a thick slurry, or *mull*, by grinding it with a few drops of a greasy, viscous liquid, such as Nujol (a paraffin oil) or chlorofluorocarbon greases. The mull is then pressed between two salt plates to form a thin film. This method is good for qualitative studies, but not for quantitative analysis. To cover the complete mid-IR region, it is often necessary to use two different mulling agents, since the mulling agents have absorption bands in different regions of the spectrum. The spectrum of the mulling agents alone should be obtained for comparison with the sample spectrum.

The second technique is the KBr pellet method, which involves mixing about 1 mg of a finely ground (<2 μm diameter) solid sample with about 200 mg powdered dry potassium bromide. Vials with 200 mg pre-measured KBr are available into which one only needs to place the sample. The mixture is compressed under very high pressure (>50,000 psi) in a vacuum to form a small disk about 1 cm in diameter. An evacuable die is designed for use in a hydraulic press. A die consists of a body and two anvils that will compress the powdered mixture. The faces of the anvils are highly polished to give a pressed pellet with smooth surfaces. A schematic of an evacuable die is shown in Figure 5.17. When done correctly, the KBr pellet looks like glass. The disk is transparent to IR radiation and may be analyzed directly by placing it in a standard pellet holder. There are small, hand-operated presses available for making KBr pellets, but the quality of the pellet obtained may not be as good as that obtained with an evacuable die. The pellet often will contain more water, which absorbs in the IR region and may interfere with the sample spectrum. There are several types of handheld presses available. A common design consists of two bolts with polished ends that thread into a metal block or nut. The nut serves as the body of the die and also as the sample holder. One bolt is threaded into place. The KBr mix is added into the open hole in the nut so that the face of the inserted bolt is covered with powdered mix. The second bolt is inserted into the nut. Pressure is applied using two wrenches, one on each bolt. The bolts are then removed; the KBr pellet is left in the

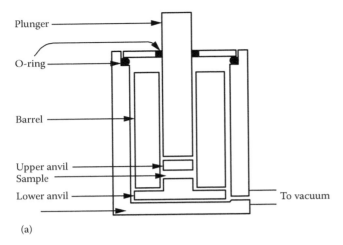

Figure 5.17 Schematic drawing of a typical IR pellet die showing the arrangement of the major components. (Reprinted from Aikens, D.A. et al., *Principles and Techniques for an Integrated Chemistry Laboratory*, Waveland Press, Inc., Prospect Heights, IL, 1978, Reissued 1984. All rights reserved. With permission.)

nut and the nut is placed in the light path of the spectrometer. The pellet should appear clear; if it is very cloudy, light scattering will result, giving a poor spectrum. The pellet is removed by washing it out of the nut with water. One disadvantage of this type of die is that the pellet usually cannot be removed from the nut intact; if pellets need to be saved for possible reanalysis, a standard die and hydraulic press should be used. Micropellet dies are available that produce KBr pellets on the order of 1 mm in diameter and permit spectra to be obtained on a few micrograms of sample. A beam condenser is used to reduce the size of the IR source beam at the sampling point.

It is critical that the KBr be dry; even then bands from water may appear in the spectrum because KBr is so hygroscopic. The KBr used should have its IR spectrum collected as a blank pellet; reagent-grade KBr sometimes contains nitrate, which has IR absorption bands. IR-grade KBr should be used when possible. The quality of the spectrum depends on having small particle size and complete mixing. A mortar and pestle can be used for mixing, but better results are obtained with a vibrating ball mill such as the Wig-L-Bug®. This is a device that has been used extensively in dentistry for the preparation of amalgam fillings and many used Wig-L-Bugs are available from the dental community. To use this device, a small vial contains the sample along with 200 mg of KBr, and a small polished steel ball bearing. The vial is vibrated for a minute; it is important not to vibrate the vial longer than is required because KBr particles that are too fine can cause scattering and distort the spectrum.

It is also very important that the polished faces of the anvils not be scratched. The anvils should never have pressure applied to them unless powdered sample is present to avoid scratching the polished faces.

In the third method, the solid sample is deposited on the surface of a KBr or NaCl plate or onto a disposable "card" by evaporating a solution of the solid to dryness or allowing a thin film of molten sample to solidify. IR radiation is then passed through the thin layer deposited. It is difficult to carry out quantitative analysis with this method, but it is useful for rapid qualitative analysis. The thin-film approach works well for polymers, for example. It is important to remove all traces of solvent before acquiring the spectrum. Disposable salt "cards" are available for acquiring the IR spectrum of a thin film of solid deposited by evaporation. These cards have an extremely thin KBr or NaCl window mounted in a cardboard holder but are manufactured so that atmospheric moisture does not pose a storage problem (Real Crystal™ IR cards, International Crystal Laboratories, Garfield, NJ). Water can even be used as the solvent for casting films of polar organic molecules on these cards. As an alternative to the cards, fine stainless-steel screens have been used to hold samples, either solid nanoparticles or from solution.

A new approach to collecting transmission spectra of solids is the use of a *diamond anvil cell*. Diamond is transparent through most of the mid-IR region, with the exception of a broad absorption around 2000 cm^{-1}. A solid sample is pressed between two small parallel diamond "anvils" or windows to create a thin film of sample. A beam condenser is required because of the small cell size. Very high pressures can be used to compress solid samples because diamonds are very hard materials. As a result, the diamond anvil cell permits transmission IR spectra to be collected of thin films of very hard materials. Hard materials cannot be compressed between salt windows because the salt crystals are brittle and crack easily.

In general, spectra from solid samples are used for qualitative identification of the sample, not for quantitative analysis. The spectrum of a solid sample is generally collected when the sample is not soluble in a suitable IR-transparent solvent. There are some problems that can occur with spectra from solid samples. Many organic solids are crystalline materials. The mull and pellet approaches result in random orientation of the finely ground crystals; deposition of thin films by evaporation may result in a specific crystal orientation with respect to the light beam. Hence, thin-film spectra may look different from the spectrum of the same material collected as a mull or a pellet. When possible, spectra of known materials obtained by the same sample preparation method should be compared when trying to identify an unknown. The use of a high-pressure hydraulic press for KBr pellets may cause crystal structure changes in some materials; again, standards and samples should have the same sample preparation method used if spectra are to be compared.

5.3.1.2 Liquid Samples

The easiest samples to handle are liquid samples. Many liquids may be analyzed "neat," that is, with no sample preparation. Neat liquids that are not volatile are analyzed by pressing a drop of the liquid between two flat salt plates to form a very thin film. The salt plates are held together by capillary action or may be clamped together. NaCl, KBr, AgCl, and similar salts are used as the plates. Volatile liquids may be analyzed neat by using a pair of salt plates with a thin spacer in a sealed cell. The path length of these cells depends on the spacer thickness. For neat liquids, very small path lengths, less than 0.05 mm, must be used to avoid complete absorption of the source beam. Sample sizes used for the collection of neat spectra are on the order of a few milligrams of material.

The use of dilute solutions of material for IR analysis is the preferred choice for several reasons. Solutions give more reproducible spectra, and dilution in an IR-transparent solvent eliminates the problem of total absorption by the strong bands in a neat sample. Solvents commonly used for IR spectroscopy include carbon tetrachloride, carbon disulfide, methylene chloride, and some alkanes such as hexane. No one solvent is transparent over the entire mid-IR region, so the analyst must choose a solvent that is transparent in the region of interest. Figure 5.18 shows the IR-transparent regions for some common solvents. An extensive table of information for solvents of importance in infrared spectrometry can be found in the book by Bruno and Svoronos. This table provides not only the spectra, but also solubility parameters and safety information.

Liquid cells for solutions are sealed cells, generally with a path length of 0.1–1 mm and two salt windows. The path length is fixed by a spacer placed between the two salt windows. Some cells come with a single fixed path length; other cells can be purchased with a variety of spacers. These cells can be disassembled and the path length changed by inserting a different spacer (Figure 5.19a). The windows and spacer are clamped into a metal frame that has two ports: one inlet and one outlet port. The cell is filled by injecting sample solution with a syringe into one port and allowing it to flow until the solution exits the other port. Solution concentrations of 1 %–10 % sample are used for most quantitative work. Solvent absorption peaks are compensated for in a double-beam dispersive IR by using matched cells. One cell is used to contain the sample solution, and the other cell to contain the solvent used to make the solution. Matched cells have the same window material, window thickness, and path length. In the single-beam FTIR, solvent absorption bands are corrected for by obtaining a blank spectrum of the solvent and subtracting the blank spectrum from the sample solution spectrum, just as the background is subtracted. In this case, the same cell is used for both the blank spectrum and the sample spectrum.

Most IR cells must be protected from water because the salt plates are water soluble and hygroscopic. Organic liquid samples should be dried over an appropriate drying agent before being poured into the cells; otherwise, the cell surfaces become opaque and uneven. Such etching of the internal window surfaces is frequently a serious problem, particularly when quantitative analyses are to be performed. Light scattering will occur; the path length within the cell becomes uneven and erroneous quantitative results may be obtained.

It will be remembered that Beer's law indicates that the absorbance = abc, where b is the path length through the sample or, in this case, the width of the empty cell. In order for quantitative data to be reliable, b must be a constant or at least measurable and correctable. A measurement of b may be performed by using a procedure based on interference fringes. An empty and dry cell is put into the light path, and the interferogram collected (or a suitable wavelength range is scanned if a dispersive instrument is used). Partial reflection of the light takes place at the inner surfaces, forming two beams. The first beam passes directly through the sample cell, and the second beam is reflected twice by the inner surfaces before proceeding to the detector. The second beam therefore travels an extra distance $2b$ compared with the first beam. If this distance is a whole number of wavelengths ($n\lambda$), then the two emerging beams remain in phase and no interference is experienced. However, if $2b = (n + 1/2)\lambda$, interference is experienced and the signal reaching the detector is diminished. The interference signal generated is a sine wave, and each wave indicates an interference fringe. The path length of the sample holder can be measured by using the following formula:

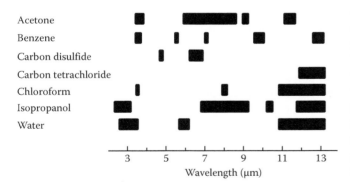

Figure 5.18 IR absorption characteristics of some common solvents. Regions of strong IR absorbance in 0.1 mm cells (except water, 0.01 mm cell) are shown as shaded areas. Longer cell paths will broaden the regions of absorption and in some cases introduce new regions where absorption is significant. (Reprinted from Aikens, D.A. et al., *Principles and Techniques for an Integrated Chemistry Laboratory*, Waveland Press, Inc., Prospect Heights, IL, 1978, Reissued 1984. All rights reserved. With permission.)

$$b(\mu m) = \frac{n}{2\eta}\left(\frac{\lambda_1 \lambda_1}{\lambda_2 - \lambda_1}\right) \quad (6.9)$$

where
n is the number of fringes
η is the RI of the sample (or air, if empty light path)

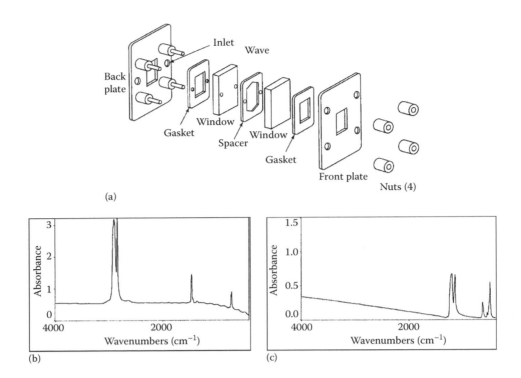

Figure 5.19 (a) Standard demountable cell for liquid samples, shown in an "exploded" view. The spacer is of Teflon or metal. The width of the spacer used determines the path length of the assembled cell. The nuts screw onto the four threaded posts to seal the assembled cell. Once the cell is assembled, it is filled via syringe. The inlet port on the back plate is equipped with a fitting for a syringe (not shown); the outlet port is the hole in the back plate opposite the inlet hole. Sample is injected until the liquid appears at the top of the outlet port. Plugs are put into both inlet and outlet ports to seal the cell. [© 2000–2014 PerkinElmer, Inc. All rights reserved. Printed with permission. (www.perkinelmer.com).] (b) and (c) show the absorbance spectra for two commercial disposable IR cards with polymer film windows. The choice of polymer depends on the region of the spectrum to be studied. Polytetrafluoroethylene (PTFE) (c) would be used if the C–H stretch region needs to be measured, while clearly polyethylene (b) is not suited for that use. [© Thermo Fisher Scientific (www.thermofisher.com). Used with permission.]

λ_1 and λ_2 are the wavelengths between which the number of fringes is measured

If λ is measured in μm, b also has units of μm. For example, if $n = 14$, $\lambda_1 = 2$ μm, and $\lambda_2 = 20$ μm, b can be calculated as

$$b = \frac{14}{2}\left(\frac{2 \times 20}{20 - 2}\right) = 15.5 \text{ μm (assuming that } \eta = 1)$$

For quantitative analysis, it is necessary to measure the path length in order to use calibration curves obtained with the same cell but at different times. If the cell becomes badly etched, the interference pattern becomes noisy and the cell windows have to be removed and repolished.

IR spectra of samples containing water can be accomplished using special cells with windows of barium fluoride, silver chloride, or KRS-5. These materials are not very water soluble (see Table 5.3). However, a more useful technique is to measure attenuated total reflection (Section 5.3.3.1).

Disposable IR cards with a thin polymer film window are available for the qualitative analysis of liquids. (These cards were originally manufactured by 3M® but are now available from International Crystal Laboratories, Garfield, NJ, and other suppliers.) Two polymer substrates are available: PTFE for the 4000–1300-wavenumber region and polyethylene for the lower-wavenumber region. The absorption spectra for these two materials are displayed in Figures 5.19b and c. A thin film can be deposited onto the polymer window by evaporation from solution or by smearing the liquid onto the polymer. A major advantage of these cards is that the polymer films do not dissolve in water; therefore, samples containing water can be analyzed. Absorption bands from the polymer substrate are subtracted from the sample spectra by running a blank card spectrum.

Microcells are available for the analysis of as little as 0.5 μL of liquid sample. These microcells also require a beam condenser as described for solid microsamples.

5.3.1.3 Gas Samples

Gas sample cells have windows of KBr and cell bodies made of glass or metal, along with two ports with valves for evacuating the cell and filling the cell from an external gas source. Gases are much more dilute than liquids or solids; a gas has many fewer molecules per unit volume than does a condensed phase. To compensate for the small concentration of sample molecules in a gas (the c term in Beer's law), the gas cells have longer path lengths (b is increased). The sample cavity of an IR spectrometer is generally about 10 cm long. There are gas cells with a single-pass 10 cm path length, but most gas cells make use of multiple reflections to increase the effective path length. Commercial gas cells with effective path lengths of 2, 10, 20, 40, and up to 120 m are available. The IR beam is reflected multiple times from internal mirrors in the cell. Such a cell is shown schematically in Figure 5.20, where the multiple reflections make the effective path length five times longer than the actual physical length of the cell. A single-pass 10 cm cell requires about 50 torr of sample pressure to obtain a good IR spectrum. However, multiple reflection cells with long path lengths permit the analysis of ppm concentrations of gases. Gas cells are also used to obtain the vapor-phase spectrum of highly volatile liquids. A drop or two of liquid is placed in the cell, the valves are closed, and the sample is allowed to come to equilibrium. The vapor-phase spectrum of HCl (Figure 5.2) was collected by placing a few drops of concentrated hydrochloric acid in a 10 cm gas cell with a glass body and KBr windows. The gas sample must not react with the cell windows or surfaces. Temperature and pressure must also be controlled if quantitative measurements are being made.

A second type of gas sampling FTIR is the open-path instrumentation used for the analysis of ambient air for environmental monitoring, chemical agent, and toxic vapor detection. These open-path systems are specifically designed for use in the field and are "hardened" against mechanical shock, vibration, temperature, and humidity changes. These systems use interferometry and either a passive mid-IR detector which measures IR signatures naturally emitted by chemicals or an active detector measuring absorption signatures using transmitted light from an active IR source. The systems can measure chemical agents and pollutants at distances of from one mile to several kilometers. They are used for smoke stack emission measurements, air pollution measurements, emissions from waste disposal sites, chemical accidents, or toxic agents deliberately released into the air. The Bruker website, www.bruker.com, has pictures and technical details of its HAWK First Responder and EM27 Open Path FTIR instruments, for example.

5.3.2 Background Correction in Transmission Measurements

The two main sources of background absorption (i.e., absorption from material other than the sample) are the solvent used for liquid solutions and the air in the optical light path. In a conventional double-beam dispersive system, comparing the sample beam to the reference beam and recording the difference spectrum in real time automatically eliminate absorption from air and solvent. If the sample is a liquid solution, a matching liquid cell with pure solvent is placed in the reference beam. The absorption from the solvent and from the air is measured simultaneously and subtracted from the sample beam signal. However, FTIR is a single-beam system and both air and solvent contribute to the signal, so corrections must be made in several steps.

5.3.2.1 Solvent Absorption

The solvent absorption spectrum is measured by putting pure solvent in the liquid sample cell and recording its spectrum. This spectrum is stored by the computer under an assigned file name (e.g., Spectrum A). The cell (or an identical cell) is then filled with the sample solution in that solvent, its spectrum taken, recorded, and stored under a file name (e.g., Spectrum B). Spectrum A is then subtracted from Spectrum B, giving the net spectrum of the sample. Of course, in this approach, any absorption by the air is also measured in both Spectrum A and Spectrum B, so the absorption by air is also corrected for.

5.3.2.2 Air Absorption

Gaseous CO_2 and H_2O vapor are both strong IR absorbers. Both occur in air in the optical light path and contribute to any IR absorption signal measured. This background signal may be corrected for in one of two ways. First, the air spectrum may be recorded by running a spectrum with no sample present. This constitutes the "blank" spectrum and is recorded and stored as a file (usually called a BLANK). Samples of any type—mulls, pellets, or neat liquids—may be run and their total spectrum (air + sample) recorded and stored. The blank (air) spectrum is then subtracted by the computer, leaving the net sample spectrum. Any suspected changes in humidity or CO_2 content can be corrected by updating the blank spectrum at regular intervals. This is an easy and rapid measurement for an FTIR, and in routine analysis, the background spectrum should be collected and the file updated on a regular basis. The second method, *purging the optical path*, is more difficult. The optical system

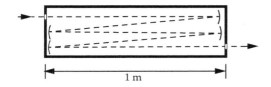

Figure 5.20 Schematic gas absorption cell. Reflection of the light beam through the cell makes the effective path length longer than the cell length.

can be purged with dry N_2 or argon, removing CO_2 and H_2O in the process. This eliminates the necessity of correcting for the blank signal derived from impurities in the air if done effectively. However, the ease of collection and subtraction of the background with modern FTIR systems and the difficulty of purging the sample compartment completely makes the first option the more common approach.

A typical background spectrum of air taken by an FTIR spectrometer is shown in Figure 5.14. The bands above 3000 cm^{-1} and between 1400 and 1700 cm^{-1} are due to water; the main CO_2 band is the band at about 2350 cm^{-1}. The FTIR is a single-beam system; this background spectrum is collected and stored for subtraction from all subsequent sample spectra. However, the absorption of the source intensity by carbon dioxide and water reduces the energy available in the regions where they absorb. To reduce the spectral background from carbon dioxide and water and increase the light intensity in these regions, many spectrometers have sealed and desiccated optical systems or a means of purging the optical path to remove the air.

5.3.3 Techniques for Reflectance and Emission Measurements

The sample techniques just described are designed for collection of transmission (absorption) spectra. This had been the most common type of IR spectroscopy, but it was limited in its applications. There are many types of samples that are not suited to the conventional sample cells and techniques just discussed. Thick, opaque solid samples, paints, coatings, fibers, polymers, aqueous solutions, samples that cannot be destroyed such as artwork or forensic evidence samples, and hot gases from smokestacks—these materials posed problems for the analytical chemist who wanted to obtain an IR absorption spectrum. The use of reflectance techniques provides a nondestructive method for obtaining IR spectral information from materials that are opaque, insoluble, or cannot be placed into conventional sample cells. In addition, IR emission from heated samples can be used to characterize certain types of samples and even measure remote sources such as smokestacks. In reflectance and emission, the FTIR spectrometer system is the same as that for transmission. For reflectance, the sampling accessories are different and in some specialized cases contain an integral detector. The heated sample itself provides the light for emission measurements; therefore, there is no need for an IR source. There may be a heated sample holder for laboratory emission measurements.

5.3.3.1 Attenuated Total Reflectance

ATR or internal reflectance uses an optical element of high RI. This optical element is called the internal reflection element (IRE) or the ATR crystal. Light traveling in a high-RI material is reflected when it encounters an interface with a material of a lower RI. The amount of light reflected depends upon the angle of incidence at the interface. When the angle of incidence is greater than the *critical angle*, which depends on the ratio of the two RIs, complete reflection of light occurs at the interface (i.e., total internal reflection). If a sample of material, such as a squishy polymer or rubber or a liquid, is placed directly against the IRE, an interface is formed. The light beam traveling in the IRE will be completely reflected at the interface if the critical angle condition is met, but the beam of light actually penetrates a very short distance (generally less than 2 µm) into the lower RI material (in this case, the sample). This penetrating beam is called an **evanescent wave**. If the sample cannot absorb any of the light, all of the intensity is reflected. If the sample can absorb part of the light, the light beam is **attenuated**, that is, reduced in intensity, at the frequencies where the sample absorbs. This is the basis of the ATR sampling technique. A schematic representation of a multiple reflection ATR crystal is shown in Figure 5.21. The interaction of the evanescent wave with the sample essentially provides an IR absorption spectrum of the sample.

The selection of the ATR crystal material for a specific sample analysis depends on the crystal RI, its spectral range, and its chemical and physical properties. The crystal should have a higher index of refraction than the sample. The majority of organic samples have RIs of approximately 1.5. ATR crystals span a range of 2.4–4.0 in RI. If the RI ratio is not appropriate, spectral features may be distorted, with diminished peak symmetries, sharp baseline/peak shoulder transitions, and derivative-like features in the spectrum.

Typical ATR crystal materials are listed in Table 5.6. Samples must be in actual intimate physical contact with the ATR crystal. The first ATR systems were designed to analyze solids that could be pressed against the surface of the crystal: polymers, films, moldable resins, textiles, canvas paintings, and the like. Little or no sample preparation is required. For example, the IR spectrum of a valuable painting could be obtained without destroying the

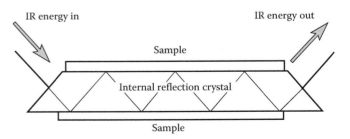

Figure 5.21 Schematic ATR sampling accessory. The internal reflection crystal permits multiple reflections. At each reflection or bounce, a small amount of IR energy penetrates the sample, and absorption occurs at the vibrational frequencies for the sample. (Courtesy of Pattacini Associates, LLC, Danbury, CT.)

Table 5.6 Common ATR IRE materials

Source: taken from the book by Bruno and Svoronos.

Material	Frequency Range (cm⁻¹)	Index of Refraction	Characteristics
Thallium iodide-thallium bromide (KRS-5)	16,000–250	2.4	Relatively soft, deforms easily; warm water, ionizable acids and bases, chlorinated solvents, and amines should not be used with this ATR element.
Zinc selenide (Irtran-4)	20,000–650	2.4	Brittle; releases H_2Se, a toxic material, if used with acids; water insoluble; electrochemical reactions with metal salts or complexes are possible.
Zinc sulfide (Cleartran)	50,000–770	2.2	Reacts with strong oxidizing agents; relatively inert with typical aqueous, normal acids and bases and organic solvents; good thermal and mechanical shock properties; low refractive index causes spectral distortions at 45 °C.
Cadmium telluride (Irtran-6)	10,000–450	2.6	Expensive; relatively inert; reacts with acids.
Silicon	9,000–1550, 400–33	3.5	Hard and brittle; useful at high temperatures to 300 °C; relatively inert.
Germanium	5000–850	4.0	Hard and brittle; temperature- opaque at 125 °C.

painting. This is essential in examining artwork and in other applications such as forensic science, archeology, and paleontology. Very hard materials such as minerals could not be pressed against traditional ATR crystals because the IRE would be damaged. Designs of ATR probes include cylindrical probes used for analysis of liquids and diamond ATR probes for hard materials. The diamond ATR probes permit analysis of hard, rigid samples and probes with inert diamond tips are available for direct insertion into process streams. The ATR crystal chosen must be chemically and physically compatible with the sample. Some crystal materials may react with certain samples, damaging the crystal and, in some cases, producing dangerous side effects. For example, ZnSe cannot be used in acidic solutions, because solutions with pH < 5 will etch the crystal and very strong acids will generate toxic hydrogen selenide. The crystal KRS-5 is not only soft but somewhat soluble in water and should be used only in the pH range of 5–8. Other crystals are susceptible to temperature and pressure changes.

ATR can be used to monitor organic reactions and processing of organic materials. For example, if an ATR probe is put into a mixture of reacting organic compounds, one particular wavelength can be monitored to indicate the disappearance of one of the reactants or the appearance of a product as the reaction proceeds. This eliminates the need to remove samples from the reaction vessel or process line in order to obtain an IR spectrum and permits continuous monitoring of the reaction without disturbing the system. ATR systems are also available with heaters to monitor processes at elevated temperatures and to study reaction kinetics and thermal degradation. Making quality chocolate is an example of a process that can be monitored by ATR at elevated temperature. ATR can be applied to the study of fossils. IR spectra can be obtained from the surface of fossilized plants or animals. The method is nondestructive, and the samples need not be removed from the fossil surface. The method is of particular interest to paleontologists and archeologists. Fossilized leaves, amber, bone, fish, trilobites, teeth, and many other sample types have been examined.

Remote mid-IR probes such as the PIKE FlexIR™ Hollow Waveguide Accessory (Pike Technologies, www.piketech.com) permit *in vivo* studies, such as investigations of chemical diffusion through the skin, residual chemicals on skin from personal care type products, analysis of large painted panels, and similar studies which could not be done in the sampling compartment of an FTIR. The hollow waveguide consists of hollow silica tubing coated on the inside with a highly reflective, smooth Ag/AgI dielectric layer and on the outside with a polymer coating for strength (Figures 5.22a and b). The spectral range spanned by hollow wave guides is from 11,000 to 400 cm⁻¹ and remote probes are available for ATR, specular reflectance, and DR measurements. The ATR crystals available are ZnSe, Ge, and diamond/ZnSe

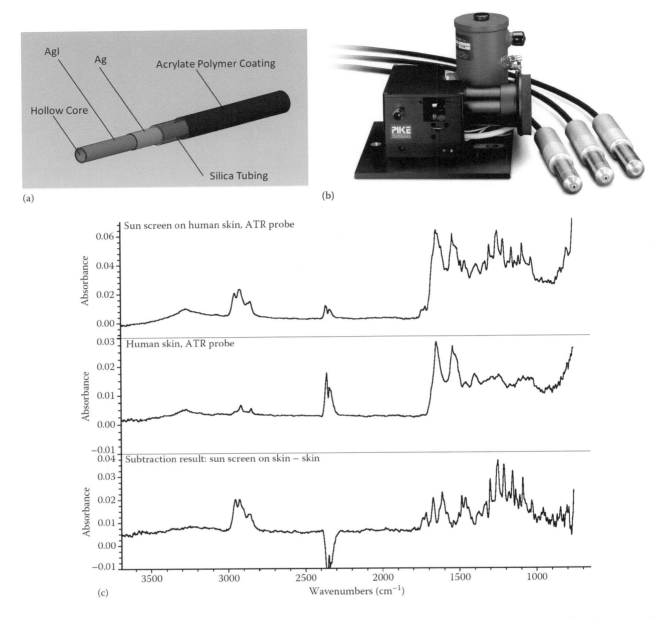

Figure 5.22 (a) Hollow wave guide schematic and (b) mid-IR FlexIR™ with remote probes. (c) ATR spectra of human skin, skin treated with sunscreen and the result of spectral subtraction of the skin spectrum (middle spectrum) from the top spectrum. (Courtesy of PIKE Technologies, www.piketech.com. Used with permission.)

composite. As an example, the FlexIR™ ZnSe ATR probe was used to study the effect of sunscreen on skin. Untreated skin measured directly with the probe showed the amide I and amide II bands at 1650 and 1550 cm^{-1}. Spectral subtraction of the untreated skin spectrum from the sunscreen treated skin permitted the residual chemicals from the sunscreen to be studied (Briggs; www.piketech.com).

For more extreme environments, such as reactors that operate at high temperature and pressure, and in supercritical fluid extraction devices, special ATR probes suitable for high pressure (up to 30 MPa, 4350 psi) and temperature (up to 200 °C) are available. These probes are typically small (thus, with only a small number of bounces) to minimize the mechanical loading caused by the pressure. The housing containing the ATR elements and sealing is typically 316 stainless-steel for corrosion resistance.

5.3.3.2 Specular Reflectance

When light bounces off a smooth polished surface, specular reflection occurs. By specular reflection, we mean that the angle of reflected light is equal to the angle of incident light just as happens with a mirror. Specular reflectance is a nondestructive way to study thin films on smooth, reflective surfaces. The measurement is a combination of absorption and reflection. The IR or NIR beam passes through the thin coating where absorption can occur. The beam is reflected from the polished surface below the coating and then passes through the coating again on its way out. Spectra can be obtained from inorganic and organic coatings from submicrometer to 100 μm in thickness. An angle of incidence of 45° from the normal is typically used for thin films. Ultrathin films, as thin as 20 Å, may be studied using a much larger angle of incidence, such as 80° from normal. This technique is called grazing angle analysis.

The thin films or coatings can be studied nondestructively, with no sample preparation other than deposition on a polished metal surface if necessary. Specular reflectance has been used to study lubricant films on computer disks, oxide layers on metal surfaces, paint curing as a function of time, and molecules adsorbed on surfaces. For example, the IR absorption spectrum of proteins adsorbed onto a polished gold surface can be studied. This spectrum from an adsorbed layer can form the basis of sensors for compounds that will bind to the proteins and change the spectrum. The use of a polarizer in conjunction with grazing angle analysis can provide information about the orientation of molecules adsorbed onto surfaces.

5.3.3.3 Diffuse Reflectance

DR, also called DRIFTS, is a technique used to obtain an IR or NIR spectrum from a rough surface. The rough surface may be a continuous solid, such as a painted surface, fabric, an insect wing or a piece of wood, or it may be a powder that has just been dumped into a sample cup, not pressed into a glassy pellet. The incident light beam interacts with the sample in several modes. Specular reflectance from the surface can occur; this is not desired and samples may need some preparation or dilution with KBr to minimize the specular component. The desired DR occurs by interaction of the incident beam with the sample. Ideally, the beam should penetrate about 100 μm into the sample and the reflected light is scattered at many angles back out of the sample. A large collecting mirror or, for NIR, an integrating sphere is used to collect the scattered radiation.

DRIFTS works very well for powdered samples. The sample powder is generally mixed with loose KBr powder at dilutions of 5%–10% and placed into an open sample cup. A commercial DR arrangement for the mid-IR region is shown in Figures 5.23a and b. Other types of probes, including fiber-optic probes, are available for DR measurements in the NIR region. A commercial IR integrating sphere for NIR DR measurements is diagrammed in Figure 5.23c.

The DR experiment requires that the incident beam penetrate into the sample, but the path length is not well defined. The path length varies inversely with the sample absorptivity. The resulting spectrum is distorted from a fixed path absorbance spectrum and is not useful for quantitative analysis. Application of the Kubelka–Munk equation is a common way of making the spectral response linear with concentration.

The Kubelka–Munk relationship is

$$f(R_\infty) = \frac{(1-R_\infty)^2}{2R_\infty} = K'C \qquad (5.10)$$

where
R_∞ is the ratio of the sample reflectance spectrum at infinite sample depth to that of a nonabsorbing matrix such as KBr
K' is a proportionality constant
C is the concentration of absorbing species

The Kubelka–Munk equation gives absorbance-like results for DR measurements, as can be seen by comparing it to Beer's law, $A = abc = Kc$ for a fixed path length. In Beer's law, K is a proportionality constant based on the absorption coefficient and the path length. K' is also a proportionality constant, but based on the ratio of absorption coefficient to scattering coefficient. The term $f(R^\infty)$ can be considered a "pseudoabsorbance."

5.3.3.4 IR Emission

Some samples are not amenable to transmission/absorption or reflectance spectroscopy. Samples can be characterized by their IR emission spectrum under certain conditions. If the sample molecules are heated, many will occupy excited vibrational states and will emit radiation upon returning to the ground state. The radiation emitted is characteristic of the vibrational levels of the molecule, that is, the IR spectrum, and can be used to identify the emitting sample. The IR emission from the sample is directed into the spectrometer instead of the usual IR light source. Very small samples can have their IR emission collected with an IR microscope, as discussed later in this chapter. Large, physically remote samples can be imaged with a telescope arrangement and the emitted light directed into the spectrometer.

IR emission can be used in the laboratory to study heated samples. Most modern research-grade instruments offer an emission sampling port as an option. One significant advantage of IR emission spectrometry is that the sample can be remote—such as the emission from a flame or smokestack. Some typical applications of IR emission include analysis of gases, remote flames and smokestacks or other hot discharges,

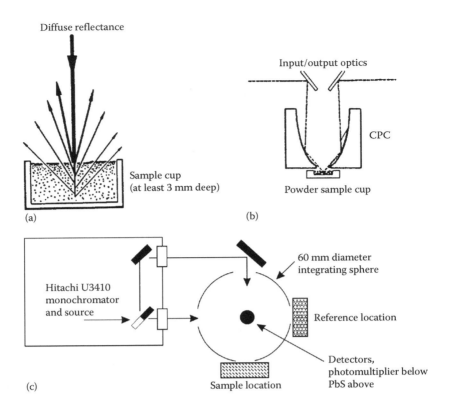

Figure 5.23 (a) Schematic diagram of DR from a powdered sample in a cup, showing the depth of penetration of the incident and reflected beams. Ideally, specular reflectance should be minimized to prevent distortion of the DR spectrum. (From Coates, J., Vibrational spectroscopy, in Ewing, G.W., ed., *Analytical Instrumentation Handbook*, 2nd edn., Marcel Dekker, Inc., New York, 1997. With permission.) (b) A DRIFTS sampling accessory with a compound parabolic concentrator (CPC) design. The CPC design minimizes specular reflection from the sample surface, reduces sample packing and height effects, and avoids damage to the optics from sample spills by placement of the sample below the optics. [© Thermo Fisher Scientific (www.thermofisher.com). Used with permission.] (c) Schematic diagram of an NIR integrating sphere for DR. The sphere is placed in the sample compartment of a dispersive UV/VIS/NIR spectrophotometer. The sphere design permits only DR to reach the detector; the specular component is reflected out through the same opening the light enters. (Courtesy of Hitachi High Technologies America, Inc., Pleasanton, CA, www.hitachi-hhta.com.)

process measurements, photochemical studies, and the analysis of thin films and coatings. IR emission measurements are nondestructive and do not suffer from atmospheric background problems since the room-temperature water and carbon dioxide in air do not emit radiation. The major limitation is that thick samples cannot be measured due to reabsorption of the emitted radiation by cool parts of the sample.

5.4 FTIR MICROSCOPY

FTIR instruments with sensitive MCT detectors have permitted the development of the IR microscope, which extends IR spectroscopy to the examination of very small samples with detection limits up to two orders of magnitude better than can be achieved with dispersive instruments. An IR microscope uses two light beams, one visible and the other IR, that travel through the microscope optics to the sample following identical paths, as shown in Figure 5.24. The sample is viewed optically and the exact region to be studied is centered and focused using the microscope controls. In some microscope designs, the visible beam is then moved out of the light path and the IR beam is moved in. Microscopes designed with dichroic optics allow both beams to reach the sample so that the analyst can view the sample while the IR spectrum is collected. It is possible to collect an IR spectrum in either transmission or reflectance mode from an area as small as 10 μm in diameter. The IR signal from the sample passes to a dedicated MCT detector designed for small samples.

To obtain a transmission spectrum, the sample must be prepared. A microtome is used for cutting a very thin slice of the sample through which radiation can penetrate. Sample thickness must be in the range of 15 μm and the sample must be flat. The quality of the spectrum depends on the sample preparation. All of the reflectance modes are available for microscopy, including ATR and grazing angle analysis. These generally require little or no sample preparation. The sensitivity obtainable is subnanogram quantities of analyte.

Modern FTIR microscopes are available with computer-controlled microscope stages and video imaging systems that

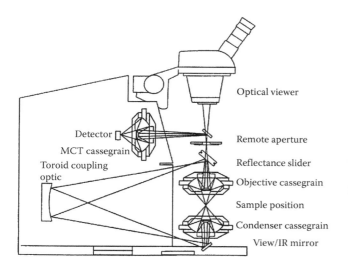

Figure 5.24 IR microscope schematic with the detector integrated into the microscope. The microscope is usually coupled to a light port on the side of the FTIR spectrometer. The FTIR spectrometer supplies a modulated, collimated beam of light to the microscope. [© 1997–2014 PerkinElmer, Inc. All rights reserved. Printed with permission. (www.perkinelmer.com).]

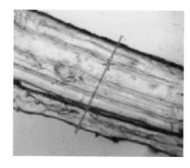

Figure 5.25 Micrograph of a polymer laminate, showing two broad layers and a narrow middle layer. The sample was mounted in a NaCl compression cell and spectra were collected automatically in 2 μm steps along the line indicated. A Centaurμs Analytical FTIR Microscope System from Thermo Fisher Scientific was used for the automatic data collection and results. [© Thermo Fisher Scientific (www.thermofisher.com). Used with permission.]

permit a 2D picture of the sample to be displayed, and areas containing a specified functional group to be highlighted using "false color" to show differences in composition with respect to position in the sample. Microscopes are available that allow the use of polarized light for imaging and that can obtain fluorescence images. These are useful to improve the contrast in samples that lack features under normal illumination.

A prime example of the use of FTIR microscopy is in the examination of polymers, a very important class of engineering materials. The physical properties of polymers are very dependent on their molecular structure. The presence of impurities, residual monomers, degree of crystallinity, size, and orientation of crystalline regions (the microstructure of a polymer) greatly affects their mechanical behavior. FTIR microscopy can identify polymers and additives and determine the presence of impurities.

Food-packaging materials may be made up of several layers of different polymers, called a laminate, to provide a single plastic sheet with the desired properties. Typical layers are between 10 and 200 μm thick. Using an automated FTIR microscope, it is possible to obtain acceptable spectra from each layer and identify the polymers involved. As an example, a cross section of a polymer laminate, compressed between NaCl plates, is shown in Figure 5.25. Three layers were seen under magnification. The sample was moved in a straight line, as shown, and IR spectra were collected every 2 μm across the sample. The spectra collected from the laminate can be displayed in a variety of formats, such as the "waterfall display" presented in Figure 5.26. This display gives the analyst a very clear picture of the differences in the three layers. If we look at the band on the left, between 3200 and 3400 cm^{-1}, we see that it is high in the layer plotted at the front of the display and as we move back (along the sample), we reach the thin middle layer. Note that the band is still there but much less intense. Then, moving back into the third layer, the band disappears. The same thing happens to the intense band at about 1700 cm^{-1}. Three distinct IR spectra were obtained, one from each layer (Figure 5.27). The front layer was identified as a polyamide polymer by matching its spectrum to a known spectrum. The back layer is identified as polyethylene—note that this spectrum does not show the bands seen in the polyamide spectrum at 3400 and 1700 cm^{-1}. The middle layer was not immediately identified. A search of a computerized IR spectral library matched the spectrum of the middle layer to a urethane alkyd, as shown in Figure 5.28. Figure 5.28 shows what a spectral search routine does—it picks a series of possible "fits" to the unknown from its database and assigns a goodness-of-fit or match number. In this case, the urethane alkyd spectrum has the highest match number, 77, of the spectra in this database.

In forensic science, FTIR microscopy has been used extensively. Rapid chemical imaging of documents and paper currency allows an analyst to distinguish between various inks and the paper itself using ATR mode, which minimizes absorption from cellulose, enabling identification of counterfeit currency. In addition, modern currency often has small security fibers in the paper. These can be visually identified and chemically imaged to confirm authenticity. FTIR microscopy is used to examine paint chips from automobile accidents. An example of a paint chip spectrum is shown in Figure 5.29. The spectral region below 1000 cm^{-1} is where pigments absorb. Hit-and-run drivers frequently leave traces of paint on cars, structures, and victims with which they collide. Identification of the paint can help to identify the car. Both transmission and ATR sampling can

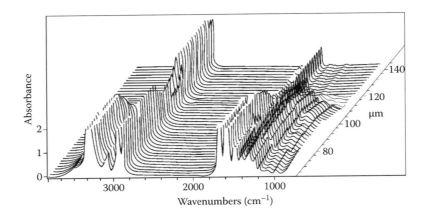

Figure 5.26 The spectra collected from the polymer laminate are displayed as a function of position along the sample, in a "waterfall display." The chemical differences in the layers are clearly seen. For example, the top layer has a large broad peak at about 3400 cm^{-1} (the peak on the far left); that peak disappears as the middle and bottom layers are scanned. [© Thermo Fisher Scientific (www.thermofisher.com). Used with permission.]

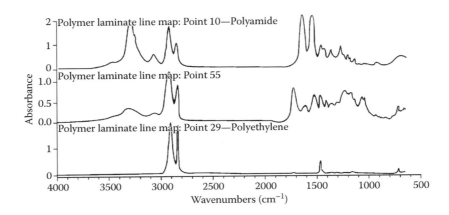

Figure 5.27 One spectrum from each layer is displayed. The spectrum from the top layer matches that of a polyamide; that on the bottom is the spectrum of polyethylene. The middle spectrum has not yet been identified. [© Thermo Fisher Scientific (www.thermofisher.com). Used with permission.]

be used in conjunction with optical microscopy and FTIR imaging. A typical paint chip will have layers of coatings, base coat, and binder, as well as possibly the substrate (plastic, fiberglass). In a real case, a jogger was intentionally struck by a car and killed. The driver of the car was convicted of murder based on the FTIR microscopy data from a tiny paint chip found on the victim's clothing. The tiny sample from the victim and a sample of paint from the car were mounted in paraffin and cross sectioned with a microtome. IR spectra were collected from five layers in each paint sample. Based on the data, the paints were shown to be identical and the hit-and-run driver found guilty.

Other uses of an IR microscope in forensic analysis include the examination of fibers, drugs, and traces of explosives. For example, oxidation of hair can occur chemically or by sunlight; oxidation of cystine to cysteic acid can be seen in hair fibers by FTIR microscopy (Robotham and Izzia). Excellent examples in full color of FTIR imaging microscopy can be found on the websites of companies like PerkinElmer and Thermo Fisher Scientific. Our limitations in use of gray scale make many of the examples unsuited for reproduction in the text. A novel IR microscope combined with atomic force microscopy, the nanoIR™ platform from Anasys Instruments (www.anasysinstruments.com), permits nanoscale IR spectroscopy, AFM topography, nanoscale thermal analysis, and mechanical testing.

IR microscopy is used in the characterization of pharmaceuticals, catalysts, minerals, gemstones, adhesives, composites, processed metal surfaces, semiconductor materials, fossils, and artwork. Biological samples such as plant leaves and stems, animal tissue, cells, and similar samples can be imaged. Frequently, such information cannot be obtained by any other means. A microscope that combines both IR and Raman measurements will be discussed in the section on Raman spectroscopy.

INFRARED, NEAR-INFRARED, RAMAN, AND PHOTOACOUSTIC SPECTROSCOPY

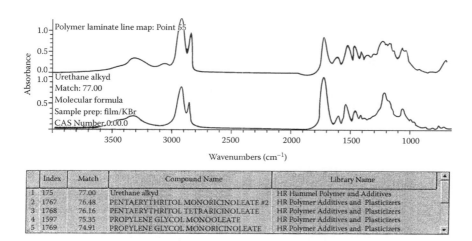

Figure 5.28 The results of a library search of the spectrum from the middle layer of the laminate. The top spectrum is that collected from the sample; the bottom spectrum, urethane alkyd, is the best match found in the search of a polymer database. Other possible compounds are suggested by the search routine and listed in the box below the spectra. Note the match number—the higher the number, the better the agreement between the sample spectrum and the library spectrum. [© Thermo Fisher Scientific (www.thermofisher.com). Used with permission.]

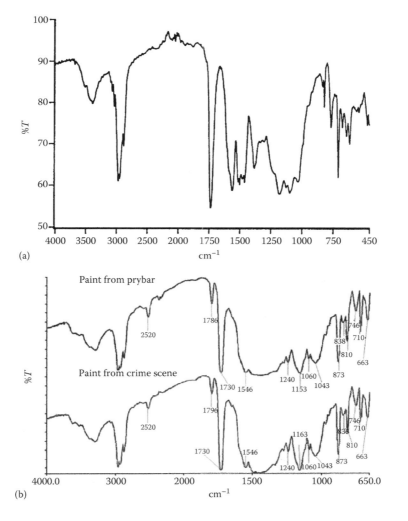

Figure 5.29 (a) Transmission spectrum of a blue paint chip from an American car measured using a miniature diamond anvil cell. (b) Comparison of microscopic paint chips taken from a crowbar and compared to paint from a window at the site of a burglary. [© 2000–2014 PerkinElmer, Inc. All rights reserved. Printed with permission. (www.perkinelmer.com).]

5.5 NONDISPERSIVE IR SYSTEMS

In industry, it is often necessary to monitor the quality of a product on a continuous basis to make certain the product meets its specifications. This online, real-time approach to analysis is called **process analysis**. IR spectroscopy is often the method of choice for process monitoring of organic chemical, polymer, and gas production. It is usually not feasible to use laboratory IR instruments under production conditions because they are too delicate, too big, and too expensive. Nondispersive systems have therefore been developed that are much sturdier and can be left running continuously. Many nondispersive systems have been designed for the NIR region. These will be discussed in Section 5.7. The mid-IR region is used mainly for monitoring gas streams.

Nondispersive IR spectrometers may use filters for analysis of gaseous substances. Each filter is designed to measure a specific compound. Figure 5.30a presents a commercial filter photometer for the mid-IR region with a filter wheel containing multiple narrow bandpass filters. The compound measured is selected by turning the wheel to put the proper filter in the light path. Other photometers have been designed for dedicated measurement of a single gaseous species. A schematic diagram of such a dedicated nondispersive IR instrument is shown in Figure 5.30b. The system consists essentially of the radiation source and two mirrors that reflect two beams of light, which pass through the sample and reference cells, respectively, to two detectors. These detectors are transducers similar in design to the Golay detector; each contains the gas phase of the compound being determined. The detector is therefore selective for the compound. For example, imagine that the two detectors are filled with gas-phase carbon tetrachloride. If there is no carbon tetrachloride vapor in the sample cell, detectors A and B will absorb the IR radiation equally and consequently their temperatures will increase. The temperature difference between the two detectors is measured and is at a minimum when there is no sample in the light path. When carbon tetrachloride vapor is introduced into the sample cell, it absorbs radiation. The light falling on detector A is therefore decreased in intensity and the temperature of the detector decreases. The temperature difference $T_2 - T_1$ between the two detectors increases. As the concentration of the sample increases, the temperature of detector A decreases and $T_2 - T_1$ increases. The relationship between $T_2 - T_1$ and sample concentration is positive in slope. The reference cell may be a sealed cell containing N_2 gas, which does not absorb in the IR, or a flow-through cell that is purged with a nonabsorbing gas. Clearly, the system can be made specific for any IR-absorbing gas by putting that vapor in the detectors. The response of the system is usually better if the light beams are chopped, and it is common to chop the light as it leaves the source so that the sample and reference beams are chopped equally. This provides an ac signal that helps correct for changes in room temperature during operation, instrument drift, and other types of noise in the system.

A problem may be encountered if an interfering material is present in the sample that has absorption bands overlapping those of the sample. This will result in a direct interference in the measurement. The problem can be overcome by placing a cell containing a pure sample of the interfering material in the sample arm. In this fashion, all the absorbable radiation at a common wavelength is absorbed, and this eliminates any variation in absorption due to the impurity in the sample because all the light at this wavelength has been removed.

Nondispersive IR systems are good for measuring concentrations of specific compounds under industrial and other similar circumstances. As can be readily understood, they are not generally used as research instruments and do not have scanning capabilities. However, they are robust and enduring and can be used for the continuous monitoring of specific compounds. Nondispersive systems are very common in NIR applications, as discussed in Section 5.7.

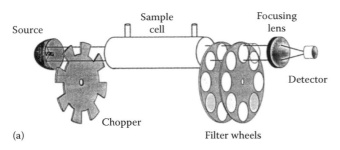

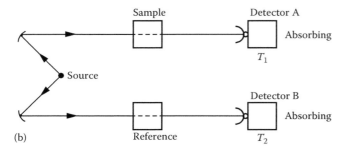

Figure 5.30 (a) Schematic of a two-filter wheel, multiwavelength filter photometer. (From Coates, J., Vibrational spectroscopy, in Ewing, G.W., ed., *Analytical Instrumentation Handbook*, 2nd edn., Marcel Dekker, Inc., New York, 1997. With permission.) (b) Schematic of a positive-filter nondispersive IR system.

5.6 ANALYTICAL APPLICATIONS OF IR SPECTROSCOPY

The two most important analytical applications of IR spectroscopy are the qualitative and quantitative analyses of organic compounds and mixtures. We pointed out at the beginning of this chapter that the frequencies of radiation absorbed by a given molecule are characteristic of the molecule. Since different molecules have different IR spectra that depend on the structure and mass of the component atoms, it is possible, by matching the absorption spectra of unknown samples with the IR spectra of known compounds, to identify the unknown molecule. Moreover, functional groups, such as $-CH_3$, $-C=O$, $-NH_2$, and $-OH$, act almost as separate groups and have characteristic absorption frequencies relatively independent of the rest of the molecule they are part of. This enables us to identify many of the functional groups that are important in organic chemistry and provides the basis for **qualitative structural identification** by IR spectroscopy. By examining the absorption spectrum of an unknown sample and comparing the bands seen with the characteristic absorption frequencies of known functional groups, it is possible to classify the sample as, say, a ketone or a carboxylic acid very quickly, even if it is not possible to identify the compound exactly. In some cases, the structure of an unknown can be deduced from its IR spectrum, but this requires much practice and is not always possible, even for an expert. Computerized libraries of spectra are now commonly used to identify compounds from their IR spectra. In addition to identifying a molecule, or its functional groups, we can acquire information about structural and geometrical isomers from IR spectroscopy. IR spectroscopy has been coupled to chromatography to identify separated compounds and to thermogravimetric analyzers to identify compounds or degradation products that volatilize as a sample is heated.

We can measure the extent of absorption at a specific frequency for an analyte of known concentration. If now we were to measure the extent of absorption at the same frequency by a sample solution of unknown concentration, the results could be compared. We could determine the sample's concentration using Beer's law. Thus, as a *quantitative tool*, IR spectroscopy enables us to measure the concentration of analytes in samples.

The introduction and widespread use of FTIR has resulted in considerable extension of the uses of IR in analytical chemistry. With regard to wavelength assignment, speed of analysis, and sensitivity, FTIR has opened new fields of endeavor. Some of these uses are described.

Typical analyses include the detection and determination of paraffins, aromatics, olefins, acetylenes, aldehydes, ketones, carboxylic acids, phenols, esters, ethers, amines, sulfur compounds, halides, and so on. From the IR spectrum, it is possible to distinguish one polymer from another or determine the composition of mixed polymers or identify the solvents in paints. Atmospheric pollutants can be identified while still in the atmosphere. Another interesting application is the examination of old paintings and artifacts. It is possible to identify the varnish used on the painting and the textile comprising the canvas, as well as the pigments in the paint. From this information, fake "masterpieces" can be detected.

Modern paints and textiles use materials that were not available when many masterpieces were painted. The presence of modern paints or modern synthetic fabrics confirms that the painting must have been done recently. In a similar manner, real antiques can be distinguished from modern imitations.

As already discussed, paints and varnishes are measured by *reflectance analysis*, a process wherein the sample is irradiated with IR light and the reflected light is introduced into an IR instrument. The paint, or other reflecting surface, absorbs radiation in the same manner as a traversed solution. This technique can be used to identify the paint on appliances or automobiles without destroying the surface. Tiny scraps of paint from automobiles involved in accidents can be examined. From the data obtained, the make and year of the car may be able to be determined and matched to that involved in a hit and run, for example, or to a car involved in a crime.

Forensic science makes use of IR spectroscopy and IR microscopy, not only for paint analysis but for analysis of controlled substances. IR can be used to detect the active compounds in hallucinogenic mushrooms, for example. IR is often used to confirm the identity of controlled substances such as cocaine. It has the advantage of being able to differentiate isomers that cannot be distinguished by mass spectrometry (MS), for example, ephedrine and pseudoephedrine.

In industry, IR spectroscopy has important uses. It is used to determine impurities in raw materials. This is necessary to ensure good products. It can be used for QC by checking the composition of the product, either in batch mode or continuously (online or process analysis). Online IR analyzers can be used to control the process in real time, a very cost-effective way of producing good products. IR spectroscopy is used in the identification of new materials made in industrial research laboratories and in the analysis of materials made or used by competitors (a process called "reverse engineering"). New handheld portable FTIR systems are available and can be used in manufacturing plants to identify incoming raw materials and finished products on site, to identify coatings (composition, thickness, homogeneity), in the field to identify minerals, to characterize materials, to evaluate surface cleanliness, and to determine cure times for polymer coatings in real time. A handheld instrument can evaluate pieces too large or too valuable to be analyzed in a laboratory setting. The handheld instrument in Figure 5.31 is a Michelson interferometer system, which weighs 7 lb; is battery powered; has both DR and single-reflection diamond ATR sampling heads and a DTGS detector; covers the 4000–650 cm^{-1} range, is water, shock, and vibration resistant; and can operate from 32 °F to 120 °F (0 °C–50 °C).

5.6.1 Qualitative Analyses and Structural Determination by Mid-IR Absorption Spectroscopy

Qualitative analysis of unknown samples is a major part of the work of an analytical chemist. Since it is better

Figure 5.31 The Agilent 4100 ExoScan FTIR, used (a) to characterize minerals in the field and (b) to evaluate rubber tire condition in real time. (© 2013 Agilent Technologies, Inc., www.agilent.com/chem. Used with permission.)

to give no answer than an incorrect answer, most analytical chemists perform qualitative analysis using an array of techniques that overlap and confirm each other, providing in the sum more information than could be obtained with the separate individual methods. For qualitative analysis of an unknown organic compound, the most commonly used spectroscopic methods are as follows: IR spectroscopy to see which functional groups are present, nuclear magnetic resonance (NMR) to indicate the relative positions of atoms in the molecule and the number of these atoms, and MS to provide the relative molecular mass (RMM) of the unknown and additional structural information. UV spectroscopy was used in the past to study unsaturated or substituted compounds; it has been almost entirely replaced for qualitative structural information by NMR and MS, which are commonly available in undergraduate chemistry labs. Each technique provides an abundance of valuable information on molecular structure, but a combination of methods is used to ensure more reliable identification. In addition to spectroscopy, real samples may be submitted to chromatography to determine if the unknown is a pure substance or a mixture, to determine the number of compounds present, and to separate and purify the compound of interest.

The value to qualitative analysis of prior knowledge about the sample cannot be overemphasized. Before trying to interpret an IR spectrum, it is important to find out as much as possible about the sample, For example, to identify the products of an organic reaction, it is very valuable to have information about the materials that were present before the reaction started, the compounds the reaction was expected to produce, the possible degradation products that may come about after the reaction, and so on. Armed with as much of this information as possible, we may be able to identify the molecules in the sample from their IR spectra.

The general technique for qualitative analysis is based on the characteristics of molecular structure and behavior mentioned at the beginning of the chapter. That is, the frequency of vibration of different parts of a molecule depends on the mass of the vibrating atoms (or groups) and the bond strength. Many groups can be treated as isolated harmonic oscillators and their vibrational frequencies calculated. More commonly, vibrational frequencies for functional groups are identified by the collection of spectra from hundreds of different compounds containing the desired functional group. These characteristic group vibrational frequencies are tabulated in correlation tables or correlation charts. Table 5.2 is a short list of functional groups and their relevant vibrational frequencies. Since the absorption frequency is the same as the vibration frequency, the presence of absorption at a given frequency is an indication that the functional group may be present. More tables are found later and very detailed tables are found in the references listed in the bibliography by Bruno and Svoronos, Silverstein and Webster, Pavia et al., Lambert et al., Colthup et al., Dean, Robinson, and in the *CRC Handbook of Chemistry and Physics*.

Qualitative analysis is carried out by matching the wavelengths of the absorption bands in the spectrum of the sample with the wavelengths of functional groups listed in a correlation table. Before a positive identification can be made, all the absorption bands typical of the functional group must be observed. More importantly, the lack of an absorption band where one should be can be used to rule out certain functional groups. For example, as we will see later, if there is no strong absorption at about 1700 cm^{-1} due to the C=O stretch in a pure unknown compound, we can state with certainty that the compound does not contain a C=O group and therefore is not a ketone, aldehyde, amide, ester, or carboxylic acid.

Because the IR spectrum of each compound is unique, matching the IR spectrum of an unknown peak for peak to a reference spectrum of a known material is a very good way to identify the unknown. This is often done with the aid of computerized spectral libraries and search routines, as we saw for the polymer laminate (Figure 5.28). A number of companies, instrument manufacturers, government agencies, and other sources publish collections of reference spectra in electronic format and in hardcopy. These spectral databases may contain spectra of more than 200,000 compounds, with subsets of the database available for specific fields of endeavor, such as environmental chemistry, pharmaceuticals, polymers, and forensic science. The unknown spectrum or some predetermined number of the strongest absorption bands from the unknown spectrum may be entered into a computerized search routine, which compares the unknown with stored spectra. It then retrieves all compounds from the database that may match the unknown spectrum, assigning a goodness of fit or probability to the suggested matches. The analyst then identifies the spectrum of the unknown based on spectral matching and chemical knowledge of the sample to rule out improbable compounds suggested by the search routine. A short list of reference spectra suppliers is located at the end of the bibliography. Most large spectral databases are expensive to buy. Many small companies or individuals can now access these by a "pay for what you use" approach. The KnowItAll™ system from the Informatics Division, Bio-Rad Laboratories (www.bio-rad.com) and the FTIR/Raman system from Thermo Fisher Scientific (www.ftirsearch.com) are two examples of this very new and cost-effective approach to spectral matching. In addition, there are some free databases that allow the user to view spectra of known compounds. These sources include FTIR and FTNMR spectra from Sigma-Aldrich (www.sigma-aldrich.com), gas-phase IR spectra from National Institute of Standards and Technology (NIST) in the United States (www.nist.gov), and a comprehensive spectral database, including IR spectra, from the Japanese National Institute of Advanced Industrial Science and Technology (https://aist.go.jp/index_en.html). If a database or spectral library is not available or the spectra do not match exactly, the analyst must try to identify the compound from its spectrum. Even when electronic databases are available, it is useful for the analyst to understand how to look at and interpret an IR spectrum. This process is described subsequently for common classes of organic compounds. Only a limited number of examples of the most common types of compounds are discussed here. For detailed basic IR spectral interpretation, the texts by Colthup et al., Pavia et al., Lambert et al., or Silverstein and Webster should be consulted. The ability to interpret an IR spectrum and deduce molecular structure requires a great deal of practice and experience as well as detailed correlation tables and knowledge of organic chemistry and molecular geometry. But even experienced analysts never try to assign every peak in an IR spectrum!

The first region to look at is the 4000–1300 cm^{-1} region, called the **functional group region**. This is the region where strong absorptions due to stretching from the hydroxyl, amine, carbonyl, and CH$_x$ groups occur. The region also has areas of weak absorptions that are nonetheless very informative. The 2000–1660 cm^{-1} region will show a set of weak overtone/combination bands if an aromatic ring is present. The intensity pattern of these weak bands can identify how the ring is substituted (i.e., *ortho*, *meta*, or *para*). Weak absorptions from triple bonds occur in this region, identifying alkynes (C≡C), cyano groups (C≡N), and diazonium salts (N≡N); also occurring in the region are absorptions by single-bonded heteroatom groups such as S–H, Si–H, and P–H.

The region 1300–910 cm^{-1} is called the **fingerprint region** because the complex absorption patterns are really what make the IR spectrum unique—a molecular "fingerprint." These absorptions are not easily interpreted because they arise from interactions between vibrations. There are some very important bands in this region, especially the C–C–O band of alcohols and the C–O–C band of esters. These should be confirmed in conjunction with the appropriate bands (OH or C=O) in the functional group region of the spectrum.

The low-frequency end of the spectrum, 910–650 cm^{-1}, is sometimes called the aromatic region. If there are no strong absorptions in this region, the structure is probably not aromatic. Strong absorptions in this region are due to oop bending of aromatic ring C–H bonds. Broad absorption bands in this region are usually due to nonaromatic amines and amides or due to carboxylic acid dimers. Below 800 cm^{-1}, absorption due to C–Cl and C–Br bonds occurs.

Before trying to interpret an IR spectrum, there are some things the analyst should note. The method of collecting the spectrum should be stated—mull, thin film, KBr pellet, solution, and the solvent—because the appearance of the spectrum may change, as has been discussed earlier. The analyst should compare reference spectra collected under the same conditions when possible. Older spectra were printed with the spectrum displayed linear in wavelength (on the *x*-axis); modern grating and FTIR spectra are generally plotted linear in wavenumber. The two plots look very different for the same spectrum; for example, bands will appear to have expanded or contracted depending on their position in the spectrum. It is important that the analyst pay attention to the scale when comparing spectra from the older literature. It should also be noted that in grating IR spectra, the scaling changes at 5 μm. The spectrum should be of a pure compound (at least 95% pure when possible), should have adequate intensity and resolution, and should have been collected on a wavelength-calibrated instrument in order for the interpretation to be useful. The *y*-axis units should also be noted. Until recently, the *y*-axis for IR absorption spectra was "transmittance" or "% transmittance," with 100%*T* at the top of the spectrum. Transmittance is the ratio of radiant power transmitted by the sample to the radiant power incident on the sample, P/P_0 or I/I_0. 100%*T* is the transmittance multiplied by 100. Transmittance ranges from 0 to 1.0; %*T* ranges from 0 to 100. The absorption peaks therefore are pointing toward the bottom of the spectrum as printed. The *y*-axis could be given in *A*, where *A* is defined as $-\log T$. If 0.00 absorbance is at the top of the *y*-axis, the spectrum is similar to the standard %*T* plot, but the contrast between strong and weak bands is not as good because *A* ranges from infinity to 0. However, it is becoming more common to see IR spectra plotted with *A* on the *y*-axis and with 0.00 *A* at the bottom of the *y*-axis, resulting in the peaks pointing up to the top of the plot. This is the inverse of the traditional %*T* spectrum. One reason for this is that Raman spectra are plotted with peak intensity increasing from the bottom to the top of the plot; having the complementary IR spectrum in the same format may make comparisons easier.

5.6.1.1 Hydrocarbons

Alkanes, alkenes, and alkynes: Figure 5.32 shows the IR spectrum of *n*-hexane, a typical straight-chain hydrocarbon with only single C–C and C–H bonds. A hydrocarbon with all single C–C bonds is called an alkane or paraffin. All of the absorption bands in the spectrum of hexane or any other alkane must be due to the stretching or bending of C–H and C–C bonds, since these are the only types of bonds in the

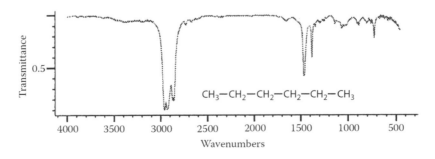

Figure 5.32 Hexane, C$_6$H$_{14}$, a normal alkane. (Copyright Bio-Rad Laboratories, Informatics Division, Sadtler Software and Databases, 2014. All rights reserved.)

INFRARED, NEAR-INFRARED, RAMAN, AND PHOTOACOUSTIC SPECTROSCOPY 219

Table 5.7 Alkanes

Functional Group	Vibrational Mode and Strength	Wavenumber (cm^{-1})
Methyl (–CH$_3$)	C–H stretching, s	2870 (sym); 2960 (asym)
Methylene (–CH$_2$–)	C–H stretching, s	2860 (sym); 2930 (asym)
Methyl (–CH$_3$)	Bending, m	1375 (sym); 1450 (asym)
Methylene (–CH$_2$–)	Bending, m (scissoring, rocking)	1465; 720 (if four or more CH$_2$ groups in an open chain); weak bands in the 1100–1350 region
gem-Dimethyl[a]	Bending, m	Doublet (two peaks); 1380–1390 and 1365–1370
Aldehyde (–CHO)[b]	Aldehydic C–H stretching, m	2690–2830 (two peaks often seen)
Aldehyde (–CHO)[b]	Aldehydic C–H bending, m	1390

[a] The term gem-dimethyl refers to two methyl groups attached to the same carbon atom, such as in an isopropyl group.
[b] An aldehyde group has a hydrogen bonded to a carbon atom that also has a double bond to the oxygen atom.

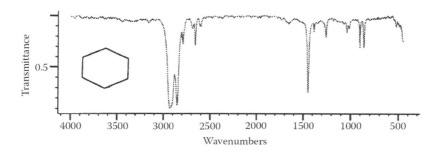

Figure 5.33 Cyclohexane, C$_6$H$_{12}$, a cyclic alkane. The structure shown on the spectrum is the typical shorthand notation—each point of the hexagon is a CH$_2$ group and the sides of the hexagon are the single covalent bonds between the six CH$_2$ groups. (Copyright Bio-Rad Laboratories, Informatics Division, Sadtler Software and Databases, 2014. All rights reserved.)

molecule. The C–C stretching vibrations are weak and are widely distributed across the fingerprint region; they are not useful for identification. The C–C bending vibrations generally occur below 500 cm^{-1}; they also are not useful for identification. Therefore, in the IR spectrum of an alkane, all of the bands observed are due to stretching and bending of C–H bonds. The approximate frequencies of some common C–H vibrational modes are given in Table 5.7. The notation sym for symmetrical and asym for asymmetrical will be used in the charts. A horizontal arrow between atoms indicates a stretch, with C → H symbolizing a C–H stretch; a vertical arrow between two atoms indicates a bending mode, with C ↓ H used for a C–H bend. (The arrows do not specify the type of bend or the symmetry of the vibrational mode.) The relative strengths of the bands are not the same for all vibrations and the intensity of the absorptions provides valuable information about structure. The abbreviations for intensity used in the tables are as follows: s, strong; m, medium; and w, weak. Peaks not marked are of variable intensity depending on the structure of the molecule.

We can assign the peaks in the n-hexane spectrum by looking at the table. The band between 2840 and 3000 cm^{-1} contains peaks from the symmetric and asymmetric stretching of the methyl C–H and methylene C–H bonds. The band at about 1460 cm^{-1} is due to the overlapped asymmetric methyl bend and the methylene bending mode called scissoring. The symmetric methyl bend is located at 1375 cm^{-1}, while the peak at 725 cm^{-1} is due to the methylene "rocking" bending mode. This band only appears if there are four or more adjacent methylene groups in an acyclic chain. These peaks are seen in the spectrum of any straight-chain alkane, regardless of the number of C–C and C–H bonds. Mineral oil, also known as Nujol®, is a paraffin oil often used for preparing mulls of solids for IR analysis. The peaks seen in the IR spectrum of mineral oil are virtually identical to those in hexane. The other regions of the IR spectrum are "clear," so that other functional groups can be observed without interference.

Figure 5.33 shows the spectrum of cyclohexane, a cyclic alkane that has no methyl groups, only methylene groups. Cyclic alkanes show similar absorption peaks to the straight-chain (acyclic) alkanes. Note the differences in this spectrum compared with that of hexane. The C–H stretching band appears narrower (no broadening due to overlap of CH$_3$ stretches with the methylene stretches), the two peaks clearly match the methylene peaks listed in the table (lower wavenumber than the corresponding methyl peaks), there is no peak at 1375 cm^{-1}, which confirms the absence of methyl groups, and there are two peaks at about 900 and 860 cm^{-1} due to ring deformation. The band at 725 cm^{-1} is also missing, because there is not an open long chain of methylene groups.

Aliphatic hydrocarbons containing one or more C=C double bonds are called alkenes or olefins. The IR spectra of alkenes contain many more peaks than those of alkanes. The major peaks of interest are due to the stretching and bending of the C–H and C=C bonds shown in Table 5.8, but the position and intensity of these vibrations is affected by the substituents on the C=C carbons. This includes the geometry of substitution, *cis* versus *trans*, for example. In addition to the vibrations themselves, many of the strong vibrations give rise to overtones, adding to the complexity of the spectrum. A strong bending mode at 900 cm^{-1} may give rise to an overtone at about 1800 cm^{-1}.

The C=C stretch in alkenes occurs in the region noted in the table. If the molecule is symmetrically substituted about the C=C bond, no change in dipole moment occurs and the band will not appear in the IR spectrum. Molecules that are close to symmetrical, such as those with symmetrically disubstituted C=C bonds, may show only a weak absorption for the C=C stretch. The position of the C=C stretch shifts if the bond is in a ring or exocyclic to a ring, with the shift depending on the size of the ring. Clearly, the spectrum of an alkene can provide many clues to the structure of the molecule. The oop bending mode for the =C–H bond is often the strongest band in the spectrum of an alkene.

The spectrum shown in Figure 5.34 is that of an olefin, 1-decene. Looking at the C–H stretch region, it should be noted that the weak but sharp absorption above 3000 cm^{-1} is due to the olefinic =C–H stretch. The olefinic C–H bends occur at 910 and 990 cm^{-1} and the C=C stretch at about 1640 cm^{-1} is typical of a monosubstituted olefin.

Figure 5.35 is the spectrum of an alkyne, 1-hexyne. Alkynes contain at least one carbon–carbon triple bond, denoted as C≡C. The carbon triple bond peak appears in the region 2100–2200 cm^{-1}, depending on its position in the molecule. Very few other functional groups absorb in this region; nitrile, C≡N, is probably the most common. Other less common groups that can absorb in the 2000–2500 cm^{-1} range are isocyanates and other

Table 5.8 Alkenes and alkynes

Functional Group	Vibrational Mode and Strength	Wavenumber (cm^{-1})
C=C → H	Stretching, w	3000–3040
C=C ↓ H	Bending, s (oop)	650–1000
C=CH$_2$	Vinyl stretching, m	2990 (sym), 3085 (asym)
C=CH$_2$	Vinyl bending, s	910, 990
C≡C → H	Stretching, s	3285–3320
C≡C ↓ H	Bending, s	610–680
C=C	C=C stretching, s[a]	1600–1680
C≡C	C≡C stretching, m-w[a]	2120 (terminal); 2200 (nonterminal)

[a] The C=C and C≡C stretching modes are weak or absent in symmetric molecules.

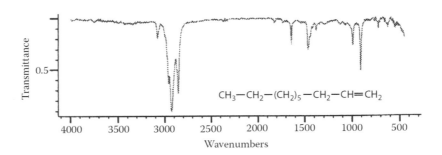

Figure 5.34 1-Decene, C$_{10}$H$_{20}$, a linear alkene. (Copyright Bio-Rad Laboratories, Informatics Division, Sadtler Software and Databases, 2014. All rights reserved.)

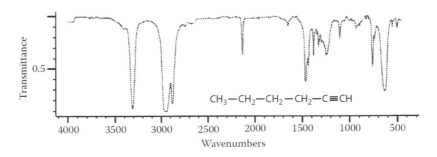

Figure 5.35 1-Hexyne, C$_6$H$_{10}$, an alkyne. (Copyright Bio-Rad Laboratories, Informatics Division, Sadtler Software and Databases, 2014. All rights reserved.)

INFRARED, NEAR-INFRARED, RAMAN, AND PHOTOACOUSTIC SPECTROSCOPY

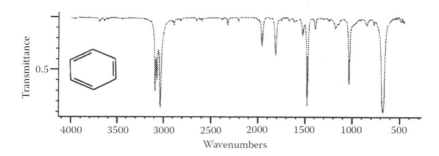

Figure 5.36 Benzene, C_6H_6, an aromatic hydrocarbon. The structure shown on the spectrum is one way of representing the delocalized electrons; they can be shown as three double bonds in one of two resonance structures. A more accurate way of representing them is to draw a circle inside the hexagon, with the circle representing the six delocalized electrons in the molecular orbital. (Copyright Bio-Rad Laboratories, Informatics Division, Sadtler Software and Databases, 2014. All rights reserved.)

nitrogen-containing conjugated double-bond systems and molecules with P–H or Si–H bonds. As is the case with the C=C stretch, symmetrical or nearly symmetrical substitution on the C≡C carbons will result in a weak or missing band. The other important peak in the spectrum of a terminal alkyne is the ≡C–H stretch, occurring near 3300 cm^{-1}.

Aromatic hydrocarbons: Aromatic compounds are those in which some valence electrons are delocalized over the molecule, as opposed to aliphatic compounds with localized electron pairs in covalent bonds. A typical example of an aromatic compound is benzene, a six-membered carbon ring. All the carbon atoms have sp^2 hybridization, giving a planar ring. The six remaining p orbitals overlap above and below the plane of the ring to give a "ring" of delocalized electrons. This can be contrasted with cyclohexane, also a six-membered carbon ring where all the carbons are sp^3 hybridized. The ring is not planar, and the electrons can be considered to be localized in pairs in the covalent bonds between the atoms. Cyclohexane is aliphatic while benzene is aromatic. Cyclohexane is often shown as a plain hexagon (Figure 5.33); aromatic rings such as benzene may be shown either with a circle in the middle of the hexagon to denote the ring of delocalized electrons or in the older "resonance" format. Benzene can be drawn, as it is in Figure 5.36, with three double bonds to account for the six delocalized electrons. The three bonds can be shifted by one position each clockwise in the ring, giving an equivalent structure. The two equivalent structures, which differ only in the location of the double bonds, are called resonance structures. The student should remember that there are no real double bonds in an aromatic ring; the drawing is just a convention.

Aromatic hydrocarbons have many IR active vibrations, resulting in complex spectra. The C → H aromatic absorption occurs above 3000 cm^{-1}, but in the same region as that for alkenes. The aromatic bands are often weak or appear as shoulders on the aliphatic bands. More useful for structural identification are the aromatic C=C ring stretching absorptions in the 1400–1600 cm^{-1} region that often appear as doublets and the aromatic C ↓ H oop bands in the 690–900 cm^{-1} region. These oop bending peaks are very strong

Table 5.9 Aromatic C–H ring deformation modes and benzene ring substitution

Substitution Positions	Wavenumber (cm^{-1})	Band Strength
Monosubstitution	730–770	s
	680–720	s
1,2 Disubstitution (*ortho*)	740–770	s
1,3 Disubstitution (*meta*)	860–890	m
	760–810	s
	680–710	s
1,4 Disubstitution (*para*)	800–855	s
1,2,3 Trisubstitution	740–790	s
	680–720	m
1,2,4 Trisubstitution	860–900	m
	790–830	s
1,3,5 Trisubstitution	820–910	s
	670–700	m

and result in weak overtone and combination bands between 1660 and 2000 cm^{-1}. The exact frequencies and intensities of these peaks are very dependent on substitution of the ring. Table 5.9 lists these ring deformation modes used to identify the positions substituted on a single benzene ring. The IR spectrum of benzene itself is shown in Figure 5.36.

Figure 5.37 depicts the overtone peaks in the 1660–2000 cm^{-1} region for substituted benzene rings. The number of peaks and peak intensities are very characteristic for the substitution on the ring and are used in conjunction with the oop bending information in a spectrum to confirm the location of substituents. Table 5.9 and Figure 5.37 may not hold for aromatic rings with very polar groups such as acids and nitro groups attached directly to the ring.

Spectra for several aromatic compounds are shown in Figures 5.38 through 5.41. Note the aromatic C–H stretch in these spectra—a sharp peak just above 3000 cm^{-1}. Just to the right, in the 2900 cm^{-1} region, are the methyl C–H stretching bands. Toluene clearly shows the pattern of four peaks in the 1650–2000 cm^{-1} region and the pair of strong

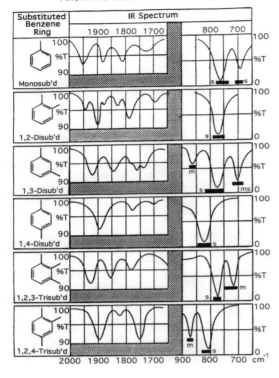

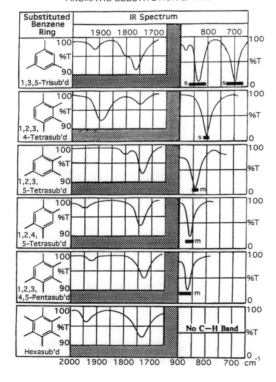

Figure 5.37 Characteristic absorption bands in the 1700–2000 cm^{-1} region for various substituted benzene rings. Source: Taken from the book by Bruno and Svoronos.

peaks for the oop C–H bend at about 690 and 740 cm^{-1} expected for a monosubstituted benzene ring. *o*-Xylene shows only a single strong oop C–H bend at about 745 cm^{-1}, as expected from the table, and the pattern in the 1650–2000 cm^{-1} region matches that in Figure 5.37 for an *ortho* substitution pattern. The student should compare the *m*-xylene and *p*-xylene spectra with Figure 5.37 and Table 5.9 as has been done for toluene and *o*-xylene. Can you distinguish all four compounds based on their IR spectra?

Figure 5.1, the first spectrum in this chapter, is that of polystyrene. Polystyrene is a polymer containing a monosubstituted benzene ring in each repeating unit. There is a pair of strong peaks between 700 and 800 cm^{-1}, characteristic of the monosubstituted oop bending pattern. The overtone pattern of four peaks in the 2000–1650 cm^{-1} region confirms that the ring is monosubstituted. The sharp peaks on the far left of the CH stretch area occur at >3000 cm^{-1}, indicative of a bond between hydrogen and the sp^2 hybridized carbon found in aromatic and alkene compounds. The peaks in the 1400–1500 cm^{-1} region are the aromatic C=C stretching bands.

5.6.1.2 Organic Compounds with C–O Bonds

There are many classes of organic compounds that contain carbon–oxygen bonds, including alcohols, carboxylic acids, ethers, peroxides, aldehydes, ketones, esters, and acid anhydrides. Not all classes of compounds are included in the following discussion, so the interested student is advised to look at the more detailed references already mentioned for additional information.

Alcohols and carboxylic acids: Alcohols are organic compounds that contain an OH group bonded to carbon. Aromatic alcohols, with an OH group substituted on a benzene ring, are called phenols. Carboxylic acids contain an OH group bonded to a carbon that also has a double bond to a second oxygen atom; the functional group is written as COOH, but it must be remembered that the OH group is attached to the carbon atom, not to the other oxygen atom:

$$-\underset{\underset{O}{\|}}{C}-O-H$$

Carboxylic acids also show a very distinctive absorption band due to the carbonyl C=O group. Typical absorption frequencies for the OH group and related C–O vibrations in alcohols and carboxylic acids are shown in Table 5.10. Carbonyl frequencies are given in Table 5.11.

The appearance and frequency of the OH stretching band is very dependent on hydrogen bonding. Hydrogen bonding is an attractive force between a hydrogen atom bonded to either oxygen or nitrogen and a nearby electronegative oxygen or nitrogen atom. Hydrogen bonding may occur between molecules (intermolecular) or within the

INFRARED, NEAR-INFRARED, RAMAN, AND PHOTOACOUSTIC SPECTROSCOPY

Table 5.10 OH and related C–O group frequencies for alcohols and carboxylic acids

Functional Group	Vibrational Mode and Strength	Wavenumber (cm⁻¹)
Aliphatic alcohols		
C–O → H	OH stretch (free), s	3600–3640
	OH stretch (hydrogen bonded), s	3200–3400
C–O ↓ H	C–O–H bend, w	1200–1400
C–C → O	C–C–O stretch (asym), s	1000–1260
Phenols		
C–O → H	OH stretch (free), s	3600
	OH stretch (hydrogen bonded), s	3100–3300
C–O ↓ H	C–O–H bend, s	1300–1400
C–C → O	C–C–O stretch (oop, asym), s	1160–1300
Carboxylic acids		
C–O → H	OH stretch (free)	Not usually observed
	OH stretch (hydrogen bonded), s	2500–3200
C–O ↓ H	C–O–H bend (in plane), m	1400–1440
	C–O–H bend (dimer, oop), m	900–950
C–C → O	C–C–O stretch (dimer), m	1200–1320

Table 5.11 Carbonyl stretching frequencies for selected compound classes

Functional Group	Vibrational Mode and Strength	Wavenumber (cm⁻¹) (±10 cm⁻¹)
Aldehydes		
Aliphatic C ⇒ O	C=O stretch, s	1730
Aromatic C ⇒ O	C=O stretch, s	1690
Amides		
C ⇒ O	C=O stretch, s	1690
Carboxylic acids		
Aliphatic C ⇒ O	C=O stretch (dimer), s	1710
Aromatic C ⇒ O	C=O stretch, s	1690
Esters		
Aliphatic C ⇒ O	C=O stretch, s	1745
Aromatic C ⇒ O	C=O stretch, s	1725
Ketones		
Aliphatic C ⇒ O	C=O stretch, s	1710
Aromatic C ⇒ O	C=O stretch, s	1690

same molecule if the geometry permits (intramolecular). The force is weaker than a covalent bond but is the strongest of the intermolecular attractive forces (van der Waals forces). A hydrogen bond is depicted as O–H···N or O–H···O, where the dotted line indicates the hydrogen bond and the solid dash a covalent bond. Alcohols and carboxylic

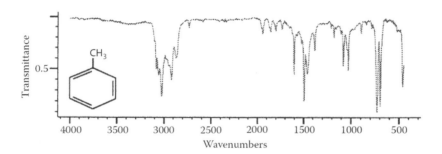

Figure 5.38 Toluene. (Copyright Bio-Rad Laboratories, Informatics Division, Sadtler Software and Databases, 2014. All rights reserved.)

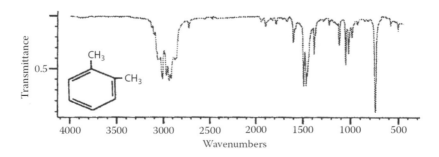

Figure 5.39 *o*-Xylene. (Copyright Bio-Rad Laboratories, Informatics Division, Software and Databases, 2014. All rights reserved.)

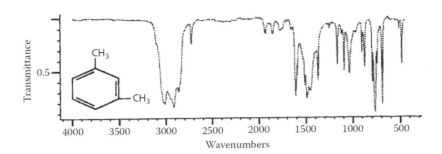

Figure 5.40 *m*-Xylene. (Copyright Bio-Rad Laboratories, Informatics Division, Sadtler Software and Databases, 2014. All rights reserved.)

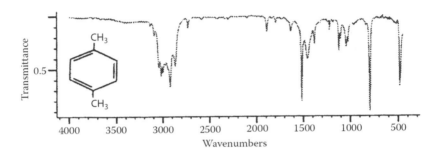

Figure 5.41 *p*-Xylene. (Copyright Bio-Rad Laboratories, Informatics Division, Sadtler Software and Databases, 2014. All rights reserved.)

acids are capable of intermolecular hydrogen bonding unless prevented by steric hindrance. The OH band in neat (pure) aliphatic alcohols appears broad and centered at about 3300 cm^{-1} due to intermolecular hydrogen bonding. In very dilute solutions of these compounds in a nonpolar solvent or in the gas phase, there is minimal hydrogen bonding; the OH band then appears "free" as a sharp peak at about 3600 cm^{-1}. The free OH peak shifts to slightly lower frequency, by about 10 cm^{-1}, in the order 1° > 2° > 3° for aliphatic alcohols. The free OH band in phenols appears slightly below that for a 3° aliphatic alcohol, again by about 10 cm^{-1}. Carboxylic acids generally exist as dimers in all but extremely dilute solutions; most acids show only the broad hydrogen-bonded band listed in Table 5.10. The H-bonded OH band in phenols is about 100 cm^{-1} lower than that for aliphatic alcohols and even lower for carboxylic acids, as seen in Table 5.10. The C–O–H bend is strong and broad in phenols, as is the C–O stretch.

The C–O stretch is also very useful for structural diagnosis of alcohols. The band is due to the C–O stretch coupled to the C–C stretch and is called an oop or asymmetric stretch in some references. The band shifts position in aliphatic alcohols depending on the substitution on the OH-bearing carbon. The band increases in frequency by about 50 cm^{-1} in the order 1° < 2° < 3°. As noted, the band is strong and broad in phenols and is shifted about 50 cm^{-1} higher than a 3° aliphatic alcohol. This C–O band is shifted to a lower frequency by about 100–200 cm^{-1} in aromatic carboxylic acids from the position given in Table 5.10.

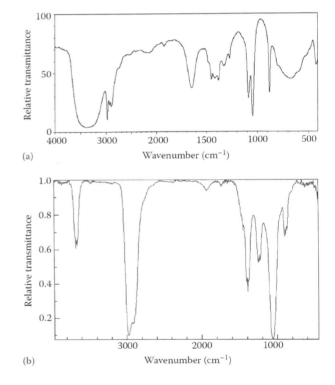

Figure 5.42 (a) IR spectrum of ethanol, collected as a liquid film between salt plates. (Courtesy of http://aist.go.jp/RIODB/SDBS.) (b) IR spectrum of ethanol vapor. (Courtesy of NIST, http://webbook.nist.gov/chemistry.) Note the difference in the position and appearance of the OH band in the hydrogen-bonded liquid versus the non-hydrogen-bonded gas phase.

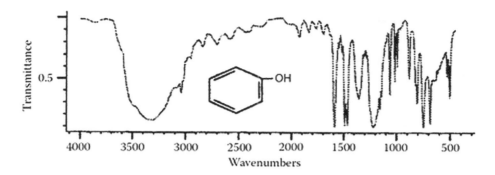

Figure 5.43 Phenol. (Copyright Bio-Rad Laboratories, Informatics Division, Sadtler Software and Databases, 2014. All rights reserved.)

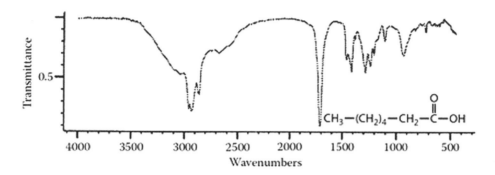

Figure 5.44 Heptanoic acid. (Copyright Bio-Rad Laboratories, Informatics Division, Sadtler Software and Databases, 2014. All rights reserved.)

Note that the characteristic bands for a carboxylic acid arise from its existence as a dimer due to strong intermolecular hydrogen bonding.

The spectrum of neat ethanol is shown in Figure 5.42a. This spectrum was collected from a thin liquid film of ethanol between salt plates. The broad hydrogen-bonded OH stretch covers the region from 3100 to 3600 cm^{-1}, centered at about 3350 cm^{-1}. The strong C–C–O stretch at 1048 cm^{-1} is characteristic of a primary alcohol. Coupling of the weak OH bend to the methylene bending mode and overlap of the methyl bending vibrations gives the broad, weak peaks between 1200 and 1500 cm^{-1}. Figure 5.42b is the spectrum of ethanol in the gas phase. Note the dramatic change in the OH stretch peak. It is shifted to 3650 cm^{-1} and is now very sharp, because there is no significant hydrogen bonding in the gas phase.

Figure 5.43 shows the absorption spectrum of phenol. Phenol has one OH group substituted on a benzene ring, so you need to consider not only the peaks in Table 5.10 but also the peaks in Table 5.9 and the earlier discussion of aromatic hydrocarbons. Note that the CO → H is very broad due to hydrogen bonding and the C → H at about 3050 cm^{-1}, on the right side of the OH band, is a short sharp peak typical of aromatics. The C–C → O band appears at about 1225 cm^{-1}, shifted up from the position in ethanol, a primary alcohol, as expected. The broad, moderately strong band at about 1350 cm^{-1} is the C–O–H bend. The sharp peaks between 1450 and 1600 cm^{-1} are from stretching of the aromatic ring carbon bonds; note that there is a doublet, as previously discussed. Characteristic bands at 745 and 895 cm^{-1} and the pattern of four peaks between 1650 and 2000 cm^{-1} indicate a monosubstituted benzene.

Figure 5.44 is the spectrum of heptanoic acid, an aliphatic carboxylic acid. As expected from the table, the broad OH band appears below that of phenol, centered at about 2900 cm^{-1}. The C–H stretching bands from the methyl and methylene groups occur in the same region; they are the peaks sticking out at the bottom of the broad OH band. The appearance of this region is very characteristic of an aliphatic carboxylic acid. The strong band at 1710 cm^{-1} is due to the C=O stretch, discussed in the next section. The peak at 1410 cm^{-1} is the in-plane C–O–H bend and that at about 930 cm^{-1} is the oop C–O–H bend. The C–C–O stretch from the dimer is at 1280 cm^{-1}.

Carboxylic acids, esters, ketones, and aldehydes: These compounds, along with acid anhydrides, acid halides, amides, and others, all possess a double bond between carbon and oxygen, C=O, called a carbonyl group. The IR spectra of these compounds are easily recognized from the very strong C ⇒ O stretching band between 1650 and

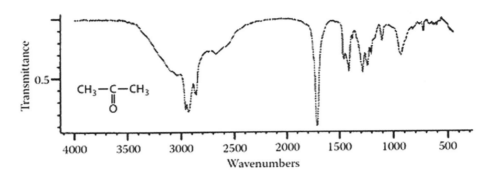

Figure 5.45 Acetone. (Copyright Bio-Rad Laboratories, Informatics Division, Sadtler Software and Databases, 2014. All rights reserved.)

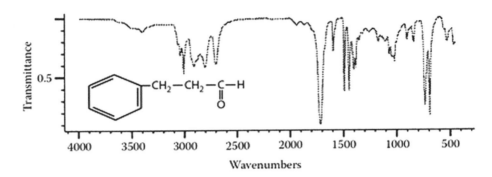

Figure 5.46 3-Phenylpropionaldehyde (or hydrocinnamaldehyde). (Copyright Bio-Rad Laboratories, Informatics Division, Sadtler Software and Databases, 2014. All rights reserved.)

1850 cm^{-1}. The position of this band is very characteristic of the class of compound.

The carbonyl stretch position depends on its environment; electronegative substituents, the position of substituents relative to the C=O bond, position in a ring and the size of the ring, resonance, conjugation, and both intermolecular and intermolecular hydrogen bonding affect the vibrational frequency. The band is usually extremely strong. The use of related bands can narrow the possible chemical class of an unknown; for example, an ester will also show a characteristic C–O–C stretch, an acid halide will show the C–X band, a carboxylic acid will show an OH band, and amides will show the NH stretch bands. Only a few examples of the huge variety of carbonyl-containing compounds will be discussed.

The spectrum for acetone is shown in Figure 5.45. Acetone is a ketone. The most intense peak in the spectrum is the C=O stretching band at 1710 cm^{-1} (Table 5.11). The peak at 1200 cm^{-1} is due to bending of the carbonyl group. The peaks between 1350 and 1450 cm^{-1} and the peaks around 2900 cm^{-1} are the C–H bend and stretch modes, respectively. The C–H peaks tell us something more. The low intensity of the CH stretch band and the lack of a "long-chain" band tell us we have no long aliphatic chains and few methyl groups in the molecule. What is the small, sharp peak at 3400 cm^{-1} due to? Remember that it is possible to see overtones (generally approximately whole number multiples of a given frequency) and combinations of strong bands in a spectrum. The peak at 3400 cm^{-1} is an overtone of the intense C=O band at 1710 cm^{-1}. The analyst needs to keep these possibilities in mind when interpreting a spectrum.

The spectrum for 3-phenylpropionaldehyde (Figure 5.46) shows a variety of features and is quite complex. This compound is an aldehyde (hence its name) and contains an aromatic ring and a short aliphatic chain. The aldehyde functional group, – CHO, has the structure $-\overset{\overset{\displaystyle}{}}{\underset{\underset{\displaystyle O}{\|}}{C}}-H$, as shown on the spectrum.

The most intense band is the C=O stretch, at 1730 cm^{-1}, indicating that the aldehyde carbonyl is part of the aliphatic portion of the molecule. Looking at the C–H stretching region, the sharp aldehydic C–H bands appear at 2725 and 2825 cm^{-1}. The doublet appears due to resonance with an overtone of the aldehydic C–H bend at 1390 cm^{-1}. This is called Fermi resonance and is always seen in aldehydes. The sharp peaks above 3000 cm^{-1} are the aromatic C–H stretching peaks and the peaks between the aromatic and aldehydic peaks are due to the aliphatic C–H stretches. The sharp peaks in the 1400–1600 cm^{-1} region are due to carbon–carbon aromatic ring stretches. The strong peak near 700 cm^{-1} is attributed to ring bending and is seen

Table 5.12 Frequencies of selected nitrogen-containing functional groups

Functional Group	Vibrational Mode and Strength	Wavenumber (cm^{-1})
Aliphatic amines		
Primary R–NH$_2$	N → H stretch (sym and asym), m	3300–3500 doublet
Secondary R$_2$NH	N → H stretch, w	3300 singlet
Tertiary R$_3$N	No NH	
Primary R–NH$_2$	N ↓ H bend, s	1550–1650; 800
Secondary R$_2$NH	N ↓ H bend, s	700
Tertiary R$_3$N	No NH	
Aromatic amines		
Primary Ar–NH$_2$	N → H stretch, m	3500
Secondary Ar$_2$NH	N → H stretch, m	3400
Tertiary Ar$_3$N	No NH	
Primary Ar–NH$_2$	N ↓ H bend, s-m	1620, 650
Nitriles, isonitriles		
C–C≡N	C≡N stretch, w-m	2220–2260
C–N≡C	N≡C stretch, w-m	2110–2180
Nitro compounds		
Aliphatic RNO$_2$	NO$_2$ stretch, asym, s	1530–1580
	NO$_2$ stretch, sym, s	1350–1390
Aromatic ArNO$_2$	NO$_2$ stretch, asym, s	1500–1550
	NO$_2$ stretch, sym, s	1285–1350
Nitrates RONO$_2$	NO$_2$ stretch, asym, s	1620–1650
	NO$_2$ stretch, sym, s	1280
Azides	N≡N stretch, m	2140

in other substituted benzenes, such as in Figure 5.51, nitrobenzene.

The spectrum for heptanoic acid (Figure 5.44) was already examined, but the student should confirm that the position of the C=O stretch is consistent with that for an aliphatic carboxylic acid.

5.6.1.3 Nitrogen-Containing Organic Compounds

There are many classes of nitrogen-containing organic compounds, including amines, nitriles, pyridines, and azides. In addition, there are a variety of classes of compounds containing both nitrogen and oxygen, such as amides, oximes, nitrates, and nitrites. Absorptions from NH, CN, NN, and NO vibrational modes result in many IR spectral bands. The peaks listed in Table 5.12 are just a small sample of some of the bands in nitrogen-containing compounds.

Primary amines have two modes of "bending" of the NH$_2$ group, a scissoring mode and the low-frequency wagging mode. Secondary amines have only one H on the nitrogen; they cannot scissor, but display the NH "wagging" band also at low frequency. Tertiary amines have no NH bands and are characterized by C–N stretching modes in the regions 1000–1200 cm^{-1} and 700–900 cm^{-1}. Primary, secondary, and tertiary amides are similar to their amine counterparts with the addition, of course, of the C=O stretching band. In primary and secondary amides, the C=O stretch is often called the Amide I band, while the adjacent NH combination band is called the Amide II band. Go back to Figure 5.22c and look for the Amide I and II bands in human skin.

Figure 5.47 shows the spectrum of a primary aliphatic amine, represented generally as RNH$_2$. This is the spectrum of propylamine, C$_3$H$_7$NH$_2$. The N–H stretch doublet characteristic of a primary amine is seen at 3370 and 3291 cm^{-1}; the primary NH bends occur at 1610 and about 800 cm^{-1}. The C–N stretching band around 1100 cm^{-1} is weak and not very useful. The rest of the bands are from C–H stretching and bending, as expected. Figure 5.48 is the spectrum of a related compound, dipropylamine, a secondary amine, (C$_3$H$_7$)$_2$NH or R$_2$NH. Note the single NH stretch characteristic of a secondary amine at 3293 cm^{-1} and the expected shift to lower frequency of the broad NH bend, from about 800 cm^{-1} in propylamine to 730 cm^{-1}. The spectrum of tripropylamine, a tertiary amine with the formula (C$_3$H$_7$)$_3$N, is presented in Figure 5.49. Note the absence of the NH stretch and bend bands since there is no N–H group in a tertiary amine, R$_3$N. The peaks observed are due to C–H stretches and bends, both methyl and methylene, and a C–N stretch at about 1100 cm^{-1}.

The spectrum in Figure 5.50 is that of butyronitrile, CH$_3$CH$_2$CH$_2$C≡N. Note the characteristic sharp C≡N stretch at 2260 cm^{-1}. Figure 5.51 is the spectrum of nitrobenzene, a monosubstituted benzene ring. The nitro group is very polar and attached directly to the ring; the normal

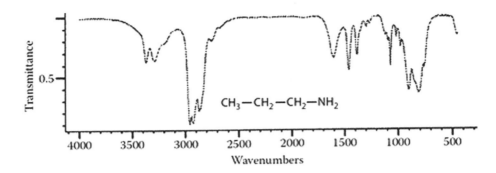

Figure 5.47 Propylamine. (Copyright Bio-Rad Laboratories, Informatics Division, Sadtler Software and Databases, 2014. All rights reserved.)

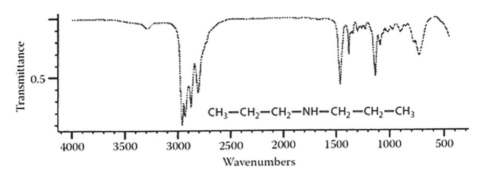

Figure 5.48 Dipropylamine. (Copyright Bio-Rad Laboratories, Informatics Division, Sadtler Software and Databases, 2014. All rights reserved.)

"rules" for substitution are not useful, as was mentioned earlier. The peaks due to the nitro group are located at 1350 and 1510 cm^{-1}, and the position of the sharp aromatic C–H stretching peaks above 3100 cm^{-1} is typical of an aromatic ring, but the overtone bands in the 1700–2000 cm^{-1} region do not show the expected monosubstitution pattern.

Amino acids: These are the building blocks of proteins and as such are very important in biochemistry and natural products chemistry. As their name implies, amino acids contain both an amine group and a carboxylic acid group and are represented as RCH(NH$_2$)COOH. Amino acids exist as *zwitterions*, that is, the H ion from the acid protonates the basic amine group, resulting in the formation within a single molecule of a positively charged ammonium cation and a negatively charged carboxylate anion. The result is that the IR absorption spectrum of an amino acid has an appearance related to salts of acids and salts of amines. Important absorption bands are given in Table 5.13.

The bands are broad and overlap with each other and with the CH bands. A broadband at 2100 cm^{-1} often appears in these spectra. The spectrum of valine, Figure 5.52, shows the broad NH stretch overlapping the CH and OH bands (2600–3200 cm^{-1}); trying to sort out the methyl and CH bands from the OH and NH bands is not practical. The carbonyl stretch is typical of the carboxylate ion, shifted to a

Table 5.13 Amino acid absorptions

Functional Group	Vibrational Mode and Strength	Wavenumber (cm^{-1})
Amine cation, RNH$_3^+$	N → H, s	2700–3100
Amine cation, RNH$_3^+$	N ↓ H, s	1600 (asym); 1500 (sym)
Carboxylate ion	CO$_2^-$ ion stretch, s	1600 (asym); 1400 (sym)

lower wavenumber than a typical carbonyl stretch, broader and overlapped by the NH bend. The symmetric NH bend and symmetric carboxylate ion stretch can be seen at 1500 and 1400 cm^{-1}, respectively.

5.6.1.4 Functional Groups Containing Heteroatoms

Halogenated compounds: The halogens F, Cl, Br, and I are usually represented in organic formulas by the letter X, as in CHX$_3$ for chloroform, CHCl$_3$, and bromoform, CHBr$_3$. The C → X stretch absorption bands occur in the fingerprint and aromatic regions and are quite strong (Table 5.14). Several halogen atoms on the same C atom cause an increase in absorption intensity, more absorption peaks, and a shift

Table 5.14 Halogen-containing functional groups

Functional Group	Vibrational Mode and Strength	Wavenumber (cm^{-1})
Aliphatic halide	C → F, s	720–1400
	C → Cl, s	530–800
	C → Br, s	510–690
	C → I, s	>600
Aromatic halide[a]	C → F, s	1100–1250
	C → Cl, s	1000–1100
	C → Br, s	1000–1090

[a] The halide atom is substituted directly on a carbon atom in an aromatic ring.

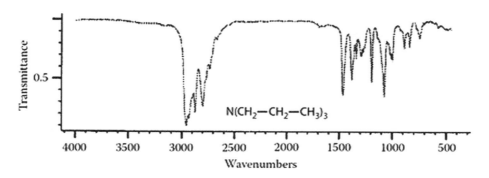

Figure 5.49 Tripropylamine. (Copyright Bio-Rad Laboratories, Informatics Division, Sadtler Software and Databases, 2014. All rights reserved.)

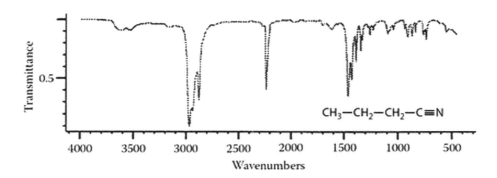

Figure 5.50 Butyronitrile. (Copyright Bio-Rad Laboratories, Informatics Division, Sadtler Software and Databases, 2014. All rights reserved.)

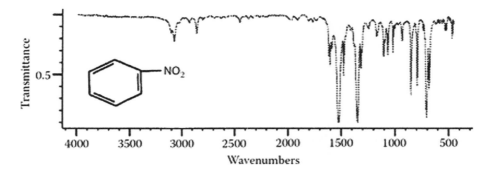

Figure 5.51 Nitrobenzene. (Copyright Bio-Rad Laboratories, Informatics Division, Sadtler Software and Databases, 2014. All rights reserved.)

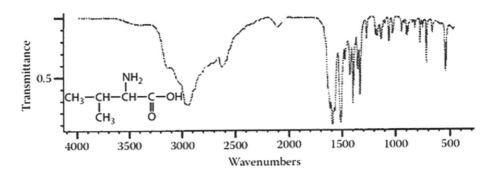

Figure 5.52 Valine. (Copyright Bio-Rad Laboratories, Informatics Division, Sadtler Software and Databases, 2014. All rights reserved.)

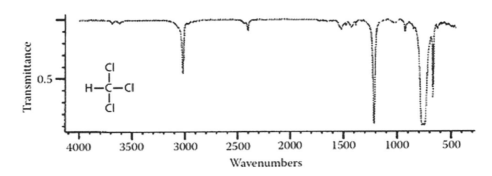

Figure 5.53 Chloroform, CHCl₃. (Copyright Bio-Rad Laboratories, Informatics Division, Sadtler Software and Databases, 2014. All rights reserved.)

to higher wavenumbers (higher frequency or shorter wavelength) for the C–X stretch.

Compounds containing the –CH$_2$X group, where X = Cl, Br, or I, have a C–H bending mode that occurs in the 1150–1280 cm^{-1} region. Absorption due to this bend can be quite strong. As is to be expected, the overtones of the very strong stretch and bend modes may often be observed, at twice the wavenumber of the fundamental band. The spectrum shown in Figure 5.53 is of chloroform, CHCl$_3$. The bands for C → H at 3000 cm^{-1}, C ↓ H (from the –CH$_2$X group) at 1200 cm^{-1}, and the very strong, broad C → Cl at 760 cm^{-1} can be seen. Overtones (weak) at about 1510 and 2400 cm^{-1} are also visible.

Sulfur, phosphorus, and silicon compounds: Some absorption bands of the heteroatoms S, P, and Si commonly seen in organic compounds are listed in Tables 5.15 through 5.17. There are many classes of compounds containing S, P, and Si atoms; detailed discussions of these compounds may be found in the texts already recommended.

Little information is obtained on sulfides or disulfides in the IR region. The compound carbon disulfide, S=C=S, is a thiocarbonyl, but atypical. It shows an intense band at about 1500 cm^{-1}, as seen in Figure 5.54, but the rest of the IR region is clear, making CS$_2$ a useful IR solvent. Figure 5.55 is the spectrum of benzylmercaptan and shows the SH stretch between 2500 and 2600 cm^{-1}, along with the expected aromatic ring peaks and the characteristic monosubstituted benzene pattern between 1600 and 2000 cm^{-1}. While the SH stretch is weak in the IR spectrum, it is strong in the Raman spectrum, indicating the complementary nature of the two techniques, as discussed later in the chapter.

Table 5.15 Frequencies in select sulfur compounds

Functional Group	Vibrational Mode and Strength	Wavenumber (cm^{-1})
Mercaptan, RSH	S → H, w	2575
Sulfide, R$_2$S	C → S, w	>700; not useful
Disulfide, RSSR	S → S	>500; not observed
Thiocarbonyl, R$_2$C=S	C ⇒ S, w	1000–1250
Sulfoxide, R$_2$S=O	S ⇒ O, s	1050
Sulfone, R$_2$SO$_2$[a]	O=S=O, s (note that there are two other single bonds, not shown, from the sulfur atom to the two R groups)	1325 (asym); 1140 (sym)

[a] Many classes of compounds contain the O=S=O group found in sulfones. These include sulfonyl chlorides, sulfonates, sulfonamides, and sulfonic acids and their salts. The exact positions of the symmetric and asymmetric stretches vary for each class.

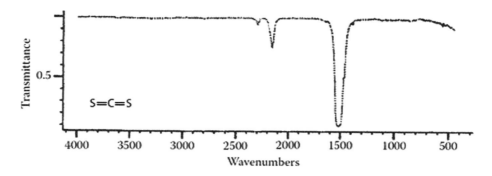

Figure 5.54 Carbon disulfide, CS_2. (Copyright Bio-Rad Laboratories, Informatics Division, Sadtler Software and Databases, 2014. All rights reserved.)

Table 5.16 Frequencies in select phosphorus compounds

Functional Group	Vibrational Mode and Strength	Wavenumber (cm⁻¹)
Phosphines, RPH_2	P → H, m	2275–2350
	P ↓ H, m	890–990
	↓ PH_2, m	1080; 900
Phosphine oxides	P ⇒ O, s	1150–1200
Phosphate esters	P ⇒ O, s	1250–1290
	P → O, m	720–850

Table 5.17 Frequencies in select silicon compounds

Functional Group	Vibrational Mode and Strength	Wavenumber (cm⁻¹)
Si–X	Si → F	800–1000
	Si → Cl	>650
Si–H	Si → H, m	2100–2200
	Si ↓ H	850–950
Si–C	Si → C, m	700–820
Si–O	Si → OR, s	1000–1100
	Si → O → Si, s	1000–1100
	Si → OH, s	820–920

One reason for concern about the IR absorption of organosilicon compounds is that silicone polymers are widely used in many laboratories and in many consumer products. Silicone lubricants, greases, caulks, plastic tubing, and o-rings may dissolve in the solvents used for IR absorption or may contaminate samples extracted in an apparatus that uses silicone polymer o-rings or lubricants. Hand cream, moisturizers, hair care products, and lubricants in protective gloves and similar products often contain silicones, so the analyst may contaminate the sample. Many common silicones are poly(dimethylsiloxanes), polymers with long Si–O–Si–O chains and two methyl groups on each Si atom. The methyl CH stretch is usually very sharp, the major peak is a strong broad band (doublet) between 1000 and 1100 cm⁻¹ due to the Si–O–Si stretch and the Si–C stretch appears at about 800 cm⁻¹.

Now that you have a basic understanding of the information that is available in an IR absorption spectrum, it is possible, with some additional information, to work out the likely structure of an unknown for small molecules (RMM < 300). For unknowns where the RMM has been determined by MS, the analyst generally takes the following approach.

Using the IR spectrum:

1. Identify the major functional groups present from the strong absorption peaks in the spectrum, for example, C=O, C–OH, NH_2, C–O, NO_2, and C≡N. Note their formula weight (FW).
2. Identify if the compound is aromatic (look for the out of plane bends in the region below 900 cm⁻¹ and the sharp but weak CH stretch above 3000 cm⁻¹). Remember that the "aromatic region" also may have bands due to the halogens Cl and Br—these are important to identify.
3. Subtract the FW of all functional groups identified from the given RMM of the compound.
4. Look for the other unique CH bands, such as aldehyde and vinyl. Look for the C≡C and C=C stretching peaks.
5. Accommodate the difference between step 3 and the RMM as aliphatic or aromatic hydrocarbon components, depending on your answers from steps 2 and 4.

For example, an unknown has been determined to have a RMM of 94. The IR spectrum shows absorptions due to OH and a monosubstituted aromatic ring pattern. The OH group has an FW of 17. Subtracting 17 from the RMM of 94 leaves 77 mass units to be accounted for. The FW of a benzene ring minus a hydrogen atom is 77, which is consistent with the IR spectrum. The unknown structure is a benzene ring with one hydrogen replaced by an OH group—therefore, it is probably phenol. (As an analyst, what other tests can you do to prove that the unknown is phenol?)

Another IR spectrum of a compound with RMM = 60 shows C ⇒ O and OH peaks consistent with a carboxylic acid group, COOH. The FW of COOH = 45. Then, RMM − FW = 60 − 45 = 15. This does not give us too many

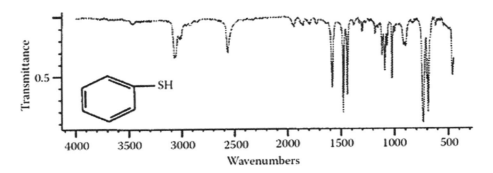

Figure 5.55 Benzylmercaptan. (Copyright Bio-Rad Laboratories, Informatics Division, Sadtler Software and Databases, 2014. All rights reserved.)

possibilities—the 15 mass units are due most likely to a CH₃ group. Hence, CH$_3$ and COOH give us acetic acid, CH$_3$COOH.

The third sample has RMM = 147. The IR spectrum shows a very strong peak due to an aromatic C–Cl stretch, and the rest of the IR spectrum is consistent with an aromatic hydrocarbon and nothing else. A benzene ring minus one hydrogen has FW = 77. One Cl atom FW = 35.5. That adds up to 112.5, leaving a difference of 34.5. This is an example that demonstrates a limitation of IR spectroscopy. You know there is Cl present from the IR spectrum; you do not know how many Cl atoms there are. What if there are two Cl atoms substituted on the benzene ring? We need to subtract another hydrogen, giving us FW = 76 for the benzene ring minus 2 hydrogen atoms. Two Cl atoms = 2 × 35.5 = 71.71 + 76 = 147, the RMM of the unknown. So the unknown is probably a dichlorobenzene. We need to look at the 1600–2000 cm^{-1} region to determine if it is *ortho*, *meta*, or *para* substituted.

These are very simple examples but provide a plan of attack for the problems at the end of the chapter. It is highly unlikely that an analyst can identify a complete unknown by its IR spectrum alone (especially without the help of a spectral library database and computerized search). For most molecules, not only the RMM but also the elemental composition (empirical formula) from combustion analysis and other classical analysis methods, the mass spectrum, proton and ^{13}C NMR spectra, possibly heteroatom NMR spectra (P, Si, and F), the UV spectrum, and other pieces of information may be required for identification. From these data and calculations such as the unsaturation index, likely possible structures can be worked out.

An application of using FTIR to determine the source oil in biodiesel is described from an application note from the PerkinElmer, Inc., website (www.perkinelmer.com). Biodiesel is a renewable fuel produced from a wide range of naturally occurring fats and oils by a transesterification reaction in which the triglycerides are broken down and fatty acid methyl esters (FAMEs) are formed. Biodiesel is allowed in fossil fuel up to 5 % in the United States and Europe without the need for labeling. Concerns include the oxidative stability of these fuels on long-term storage. The fatty acid distribution in the original oil is retained in the biodiesel, so the physical and chemical properties of the biodiesel have some dependence on the feedstock. One important property particularly influenced by the feedstock is the cloud point, the temperature at which solid crystals start to form, creating a suspension that can block fuel filters. For operation in cool climates, a low cloud point is needed. A sample of palm oil biodiesel may have a cloud point of 15 °C, while rapeseed oil may be around −10 °C. FTIR spectra collected using a PerkinElmer® Spectrum™ 100 FTIR spectrometer with a diamond ATR accessory provided a rapid and simple way to check the provenance of a biodiesel sample using the bands related to the alkene functional groups, for example, the alkene C=C–H stretch, the C=C stretch, and the CH=CH oop band at about 1400 cm^{-1}. The intensity of the bands increase in the order: palm < rapeseed < soy. Palm oil is more than 60% saturated, while the other two have high proportions of unsaturated fatty acid chains, as seen in Table 5.18. Consistent with the FTIR data, the ratio of double bonds to average chain length (a quantity approximately proportional to the volume density of double bonds) increases in the order: palm < rapeseed < soy. Acquiring the spectrum of a biodiesel sample takes only a few seconds and contains readily accessible information about the extent of unsaturation in the fatty acid chains. This information is directly related to both the source oil used and to properties such as the cloud point.

FTIR spectroscopy, as well as Raman spectroscopy, can be used to great advantage in the analysis of gemstones, to detect treatment of gemstones and to distinguish natural gemstones from synthetic stones. Emeralds are among the most valuable gemstones. They are a type of beryl, composed of aluminum and beryllium silicates with traces of chromium creating the color. There are a number of processes for creating synthetic emeralds, including the hydrothermal process and the flux process. The top spectrum in Figure 5.56 is of a natural emerald (green stone).

Table 5.18 Typical fatty acid distribution in three oils

Chain	Length	Double Bond	Palm Oil % Chains	Soy % Chains	Rapeseed Oil % Chains
Caprylate	8	0	7	0	0
Caprate	10	0	5	0	0
Laurate	12	0	48	0	0
Myristate	14	0	15	0	1
Palmitate	16	0	7	6	4
Stearate	18	0	3	3	3
Arachidate	20	0	0	3	3
Bhenate	22	0	0	0	3
Lignocerate	24	0	0	0	3
Oleate	18	1	12	35	45
Erucate	22	1	0	0	20
Linoleate	18	2	3	50	15
Linolenate	18	3	0	3	3
Average chain length			13.28	17.94	19.04
Average double bonds per chain			0.18	1.44	1.04
Double bond fraction			1.4%	8.0%	5.5%

Source: [© 1993–2014 PerkinElmer, Inc. All rights reserved. Printed with permission. (www.perkinelmer.com).]
Note: Fatty acid distribution from www.accustandard.com. The bottom three rows are weighted averages over the distributions for each oil.

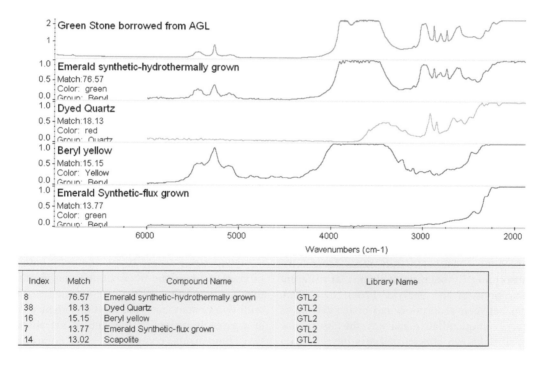

Figure 5.56 FTIR spectra of natural and synthetic gemstones. [From Lowry, S., *Using FT-IR Spectroscopy to Analyze Gemstones*. © Thermo Fisher Scientific (www.thermofisher.com). Used with permission.]

The next lower spectrum is from a hydrothermally grown synthetic emerald and the library match between the two spectra is quite good. It is obvious from the rest of the spectra that the natural emerald can easily be distinguished from a flux-grown synthetic emerald.

5.6.2 Quantitative Analyses by IR Spectrometry

The quantitative determination of various compounds by mid-IR absorption is based on the measurement of the concentration of one of the functional groups of the analyte compound. For example, if we have a mixture of hexane and hexanol, the hexanol may be determined by measuring the intensity of absorption that takes place near 3300 cm^{-1} by the OH band. The spectrum of pure hexanol would be used to determine the exact wavenumber corresponding to the absorption maximum. Alternatively, the intensity of absorption due to the C → O at about 1100 cm^{-1} could be used. From this, the concentration of alcohol can be calculated, once the intensity from a set of hexanol standards of known concentrations has been measured at the same wavenumber. Whenever possible, an absorption band unique to the sample molecule should be used for measuring purposes. This reduces the problem of overlapping bands, although there are mathematical approaches for deconvoluting overlapping peaks. The quantitative calculation is based on Beer's law

$$A = abc \tag{5.11}$$

where
A is the absorbance
a is the absorptivity of the sample
b is the internal path length of the sample cell
c is the concentration of the solution

If the same sample cells are used throughout, b is a constant. Also, the absorptivity a is a property of the molecular species being determined and can be taken as a constant. Therefore, A is proportional to c. The intensity of the peak being measured in each sample must be measured from the same baseline. Usually, a straight baseline is drawn from one side of the peak to the other or across multiple peaks or a given wavenumber range to correct for sloping background. Quantitative measurements are generally made on samples in solution, because scattered radiation results in deviations from Beer's law. KBr pellets have been used for quantitative work, but the linear working range may be small due to light scatter if the pellets are not well made. A hydraulic press is recommended for preparation of pellets for quantitative analysis.

To find the proportionality constant between A and c, calibration curves are constructed from measurements of intensity of absorption by solution standards at different known concentrations of the analyte. From these data, the relationship between absorption and concentration is obtained. To carry out quantitative analysis, the absorbance A of a sample of unknown concentration is measured. In comparison with the calibration curve, the sample composition can be determined. Quantitative analysis using mid-IR absorption is not as accurate as that using UV or visible absorption. The sample cells for mid-IR absorption have much smaller path lengths than those for UV and must be made with material transparent to IR radiation, such as NaCl or KBr. These materials are very soft and easily etched by traces of water or distorted by clamping the cell together; as a result, the sample path length may vary significantly from one sample to the next. Any change in b between samples or between samples and standards causes an error in the calculation of c, the concentration of the analyte. In addition, the nonlinear response is more of a problem in the IR when dispersive spectrometers are used than in the UV/VIS, due to stray radiation, external IR radiation, slit width (really the spectral bandpass) in regions of low source intensity, or low detector response, among other causes. Nonlinear response means that Beer's law is not obeyed, especially at high absorbance. Samples must often be diluted to keep the absorbance below 0.8 for quantitative measurements with a dispersive instrument. FTIR instruments do not suffer from stray or external radiation and they do not have slits, so it might be expected that accurate absorbance measurements can be made on solutions with absorbance values greater than 1.0. However, the detector and the apodization function limit accuracy in FTIR instruments. It is still advisable to stay below 1.0 absorbance units for FTIR quantitative work. An external calibration curve covering the range of concentrations expected in the samples should be used for the best accuracy, instead of relying on a calculation of the proportionality factor from one standard. An example of quantitative analysis by ATR is the determination of water in a polyglycol. The wavelength measured, peak intensities, and Beer's law plot are displayed in Figure 5.57.

Hydrocarbon contamination from oil, grease, and other sources in drinking water and wastewater is determined quantitatively by extraction of the hydrocarbons and measurement of the CH stretching band. The water is extracted with trichlorotrifluoroethane or other suitable solvent, the IR spectrum of the extracted solution is obtained, and the absorbance is measured at 2930 cm^{-1} after a baseline is drawn as described earlier. A series of standards of known oil diluted in the same solvent is prepared as the external calibration curve. The method can measure as little as 0.2 g oil/L of water, with a precision of about 10% relative standard deviation (RSD). Precisions of 5%–10% RSD are typical for IR measurements. The entire method can be found in *Standard Methods for the Examination of Water and Wastewater*, listed in the bibliography.

Quantitative analysis of multiple components in a mixture can be done by assuming that Beer's law is additive for a mixture at a given frequency. For a mixture with two components, the total absorbance of the mixture at a given frequency is the sum of the absorbance of the two components, compound F and compound G, at that frequency:

$$A_{\text{total}} = A_F + A_G = a_F b c_F + a_G b c_G \tag{5.12}$$

The absorptivity of F, a_F, and the absorptivity of G, a_G, are determined at two different wavenumbers, $\bar{v}_1$ and $\bar{v}_2$, from a series of mixtures with known amounts of F and G. This results in four absorptivity values, a_{F1}, a_{F2}, a_{G1}, and a_{G2}; the symbol for wavenumber is eliminated for simplicity.

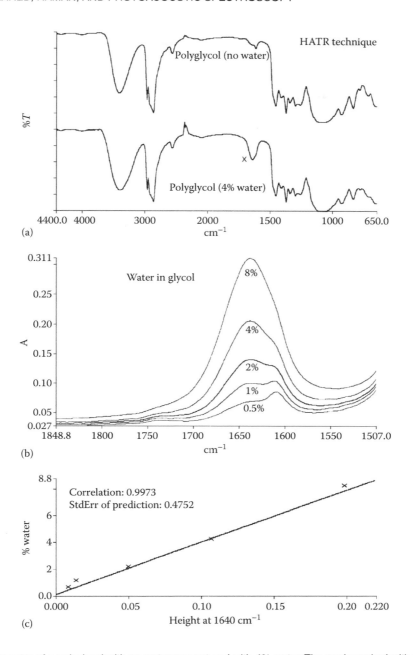

Figure 5.57 (a) The IR spectra of a polyglycol with no water present and with 4% water. The peak marked with an X is the O–H bend at 1640 cm^{-1} that will be used for quantitation. (b) Standards of the polyglycol containing 0.5%–8% water were scanned and the spectra saved. The spectra of the standard solutions were collected from a horizontal ATR (HATR) liquid sample cell using a ZnSe crystal. All spectra were collected on a Perkin Elmer Paragon 1000 FTIR spectrometer at 8 cm^{-1} resolution. (c) The data from (b) were input into the Spectrum® Beer's law quantitative analysis software. The calibration curve is shown. (Spectrum® is a registered trademark of PerkinElmer, Inc. All rights reserved. Data and spectra courtesy of Pattacini Associates, LLC, Danbury, CT.)

The concentrations of F and G in an unknown mixture can be calculated from two absorbance measurements at $\bar{v}_1$ and $\bar{v}_2$. Two simultaneous linear equations with two unknowns are constructed and solved for c_F and c_G:

$$A \text{ at } \bar{v}_1 = a_{F1}bc_F + a_{G1}bc_G$$
$$A \text{ at } \bar{v}_2 = a_{F2}bc_F + a_{G2}bc_G$$

This approach can be used for multicomponent mixtures by applying matrix algebra. This is generally done with a software program and even nonlinear calibrations can be handled with statistical regression methods.

In addition to the quantitative analysis of mixtures, measurement of the intensity of an IR band can be used to determine reaction rates of slow to moderate reactions. The reactant or product has to have a "clean" absorption band and the absorbance–concentration relationship must be determined from calibration standards. The reaction cannot be extremely fast because the band has to be scanned and even FTIR spectrometers are not instantaneous. FTIR spectrometers have permitted the determination of the kinetics of reactions much more rapid than those that could be handled by dispersive instruments.

Many applications of FTIR spectroscopy and microscopy can be found by searching the websites of major instrument manufacturers.

5.7 NIR SPECTROSCOPY

The near-infrared (NIR) region covers the range from 0.75 μm (750 nm or 13,000 cm^{-1}) to about 2.5 μm (2,500 nm or 4,000 cm^{-1}). This is the range from the long-wavelength end of visible light (red) to the short-wavelength side of the mid-IR region. The NIR region bridges the region between molecular vibrational spectroscopy and electronic spectroscopic transitions in the visible region. The bands that occur in this region are generally due to OH, NH, and CH bonds. The bands in this region are primarily overtone and combination bands; these are less intense than the fundamental bands in the mid-IR region. While weaker absorptions and limited functional group information might seem to limit the usefulness of the NIR region, there are some inherent advantages to working in the range. High-intensity sources such as tungsten-halogen lamps give strong, steady radiation over the entire range. Very sensitive detectors, such as lead-sulfide photodetectors can be used. These detectors need not be operated in liquid N_2. The third important advantage is that quartz and fused silica can be used both in optical systems and as sample containers. Fiber-optic probes and long path length cells can be used to compensate for the weaker absorptions and can be used to advantage for process analysis applications. The NIR region is used primarily for quantitative analysis of solid and liquid samples for compounds containing OH, NH, and CH bonds, such as water, pharmaceutical compounds, and proteins. This is in contrast to the mid-IR region, which is used primarily for qualitative analysis. Correlations charts for the NIR region can be found in the book by Bruno and Svoronos.

Advantages of NIR include speed, little or no sample preparation, solvents and reagents are rarely needed, high dynamic concentration range, and deep sample penetration. NIR obeys the same selection rules as mid-IR. Only molecules that exhibit a dipole moment will absorb NIR radiation.

5.7.1 Instrumentation

NIR instrumentation is very similar to UV/VIS instrumentation, discussed in Chapters 3 and 4. Quartz and fused silica optics and tungsten-halogen lamps identical to those

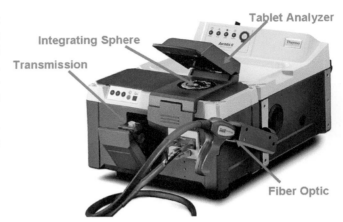

Figure 5.58 Antaris II ™ FTNIR analyzer, Thermo Scientific, equipped with a fiber-optic probe for remote measurement, transmission, and integrating sphere and a tablet analyzer for direct analysis of pharmaceuticals. [© Thermo Fisher Scientific (www.thermofisher.com). Used with permission.]

used in UV/VIS systems may be used. Commercial research-grade double-beam dispersive grating spectrometers are available that cover the optical range from the UV (180 or 200 nm) through the visible up to the NIR long wavelength limit of 2500 or 3000 nm. In addition, many customized portable filter-based NIR instruments, FTNIR instruments, and other designs for dedicated applications are commercially available. Dispersive, nonscanning spectrographs are also used in NIR spectroscopy, with a silicon photodiode array (PDA) detector. These are used for dedicated applications, such as moisture analyzers. Absorption, transmission, and reflection measurements are made in NIR spectrometry. A modern NIR analyzer with a variety of measurement modes is shown in Figure 5.58.

Quartz optical fibers are transparent to NIR radiation and are often used to interface the sample and spectrometer or spectrograph. The low-OH-content quartz fiber used for many NIR applications is single filament, between 100 and 600 μm in diameter. Fibers can be bundled together when it is necessary to collect light over a large area. Fiber-optic probes can be used for remote sampling and continuous monitoring of bulk flowing streams of commercial products as well as for simple "dip" probes. This greatly simplifies sample preparation and, in many cases, eliminates it completely. Dilute liquid solutions can be analyzed in the standard 1 cm quartz cuvettes used for UV/VIS spectrometry.

5.7.2 NIR Vibrational Bands, Spectral Interpretation, and Calibration

As seen in Figure 5.59a, an absorbance at 3,000 cm^{-1} will produce overtone bands at 6,000, 9,000, and 12,000 cm^{-1} as well as combination bands. The NIR absorptions are

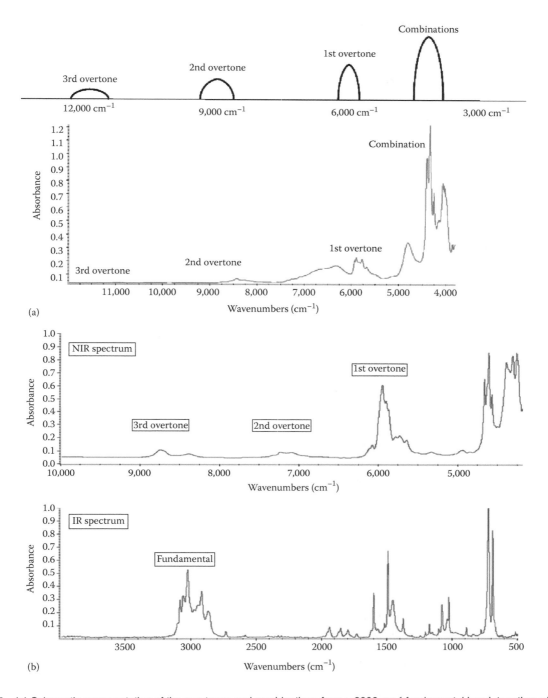

Figure 5.59 (a) Schematic representation of the overtones and combinations from a 3000 cm^{-1} fundamental band, together with the NIR spectrum. (b) NIR and IR spectra of the same compound with the bands marked. [© Thermo Fisher Scientific (www.thermofisher.com). Used with permission.]

generally not as strong as the fundamental IR band. Figure 5.59b shows the IR and NIR spectra of the same compound for comparison.

The primary absorption bands seen in the NIR region are as follows:

C–H bands between 2100 and 2450 nm and between 1600 and 1800 nm

N–H bands between 1450 and 1550 nm and between 2800 and 3000 nm

O–H bands between 1390 and 1450 nm and between 2700 and 2900 nm

Bands for S–H, P–H, C≡N, and C=O also appear in the NIR region. Water has several distinct absorption peaks at

1400, 1890, 2700, and 2750 nm. These bands enable the determination of hydrocarbons, amines, polymers, fatty acids, proteins, water, and other compounds in a wide variety of materials.

There is some correlation between molecular structure and band position for certain bands, but because these are often overtone and combination bands, their positions are not as structure dependent as the fundamental bands in the mid-IR. For example, primary amines, both aliphatic and aromatic, have two absorption bands, one at about 1500 nm and the second at about 1990 nm. Secondary amines have only one band at about 1500 nm. As expected, a tertiary amine has no NH band. Amides with an $-NH_2$ group can be distinguished from R–NH–R' amides by the number and position of the N–H bands. The reference by Goddu has a detailed table of NIR structure–wavelength correlations.

The molecular absorption coefficients (molar absorptivities) for NIR bands are up to three orders of magnitude lower than the fundamental bands in the mid-IR. This results in reduced sensitivity. Greater sample thickness can be used to compensate for this, giving more representative results with less interference from trace contaminants. Sample path lengths of 0.1 mm to 10 cm are common.

5.7.2.1 NIR Calibration: Chemometrics

NIR analysts often use a statistical methodology called **chemometrics** to "calibrate" an NIR analysis. Chemometrics is a specialized branch of mathematical analysis that uses statistical algorithms to predict the identity and concentration of materials. Chemometrics is heavily used in NIR spectral analysis to provide quantitative and qualitative information about a variety of pure substances and mixtures. NIR spectra are often the result of complex, convoluted, and even unknown interactions of the different molecules and their environment. As a result, it is difficult to pick out a spectral peak or set of peaks that behave linearly with concentration or are definitive identifiable markers of particular chemical structures. Chemometrics uses statistical algorithms to pick out complex relationships between a set of spectra and the material's composition and then uses the relationship to predict the composition of new materials. Essentially, the NIR system, computer, and associated software are "trained" to relate spectral variation to identity and then apply that training to new examples of the material.

This inherently requires a good training set of standards to ensure accurate prediction of the identity or concentrations of new examples of the material. An adequate training set that has enough examples and describes the variation expected from test samples is critical for effective chemometric modeling. The concentrations and identity of the standards must first be confirmed by some other primary means, such as chromatography, gravimetric analysis, and titration. This information is entered as part of the training set for development of the chemometric model. For example, if an NIR method is being developed to measure primary amines, it may be necessary to actually measure the standards (or samples that will then be used as standards) by a method such as the Kjeldahl titration method. The quantitative nitrogen results from the titration method are used as the concentrations for the training set. Once the standards have been analyzed and the chemometric model developed, it can be employed to automatically predict the identity and composition of new samples. When a model has been completed and validated, it may be used for long periods of time and transferred to other NIR instruments.

There are a variety of chemometric models for both qualitative and quantitative analysis. A common quantitative model is called partial least squares (PLSs) (or projection to latent structure). PLS condenses the thousands of individual data points across the whole spectrum in all of the standards to a few sources of variation that contrast broad bands of frequencies. These condensed sources of variation are called **factors** and are selected as part of the model in order of importance in accounting for the variation. The first factor of a good model will account for a large part of the variation and subsequent factors will account for the remaining variation. In practice, only a limited number of factors is required to account for most of the variation.

A practical example of the chemometrics approach is described in the reference by Heil. In order to develop a PLS calibration model for moisture, oil, protein, and linolenic acid in soybeans, a total of 207 calibration and 50 validation samples were used to build and validate the calibration model. While this may seem like a lot of work, 64 coaveraged scans were collected using a Thermo Scientific Antaris™ II FTNIR instrument in 30 s with a resolution of 8 cm^{-1}. No processing or grinding of the raw soybeans was required, and once the method is validated, no additional calibration is required. The wet chemical methods to measure these four parameters would require grinding, digestion, extraction, and multiple analytical techniques.

5.7.3 Sampling Techniques for NIR Spectroscopy

NIR spectroscopy has a real advantage over many spectroscopic techniques in that many plastic and glass materials are transparent to NIR radiation at the common thicknesses encountered for films, packaging materials, and coatings. It is practical to take an NIR spectrum of a sample without even opening the sample container in many cases; the spectrum is collected through the plastic or glass bottle or through the plastic film on food or the plastic bubble packs used for pharmaceutical tablets. In many cases, no sample preparation is required at all. When DR measurements are made, not only is no sample preparation required, but the method is also nondestructive, as Figure 5.60 shows.

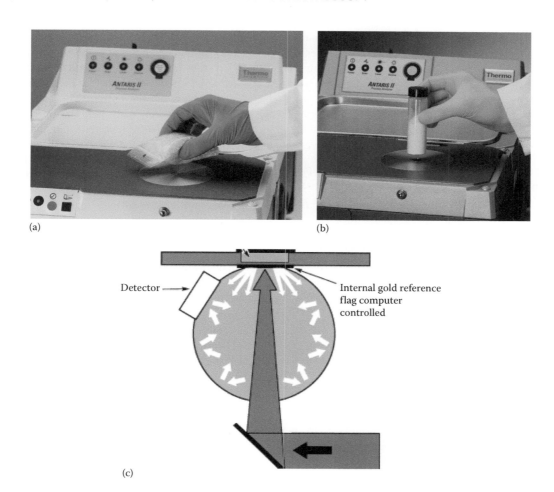

Figure 5.60 (a) NIR measurement directly through a plastic bag. (b) NIR measurement through a glass vial. Both require no sample preparation. Instrument shown is a Thermo Scientific Antaris™ II FTNIR analyzer equipped with an integrating sphere and a remote NIR probe, shown on the side of the instrument. (c) The integrating sphere (schematic), the bag, or glass vial is placed on top of the instrument (at the top of the schematic). [© Thermo Fisher Scientific (www.thermofisher.com). Used with permission.]

5.7.3.1 Liquids and Solutions

With a fiber-optic dip probe, many liquids and solutions can be analyzed with no sample preparation. One such sampling accessory for most commercial spectrometers is the FlexIR NIR fiber-optic accessory (Pike Technologies, www.piketech.com). The wand tip is inserted into the liquid sample and the spectrum collected. The FlexIR NIR probe is directly coupled to an integrated InGaAs detector to eliminate energy loss due to transfer optics and beam divergence. The use of a dip probe for transmission measurements requires that the liquid or solution be free from small particles or turbidity. Suspended particles scatter light and reduce the sensitivity of the measurement. The already low absorptivity of NIR bands makes transmission measurements of limited use for liquid samples that are not clear.

There are a number of solvents that can be used to prepare solutions for NIR measurements. Carbon tetrachloride and carbon disulfide are transparent over the entire NIR range. Many other organic solvents are transparent up to 2200 nm, with only a short region between 1700 and 1800 nm obscured by the solvent. Solvents as varied as acetonitrile, hexane, dimethylsulfoxide (DMSO), and dibutyl ether fit into this category. Methylene chloride and chloroform can be used up to 2600 nm, with short "gaps" at about 1700 and 2300 nm.

Liquids, gels, and solutions can be poured into cuvettes for transmission measurements in either a scanning NIR instrument or an FTNIR instrument. Modern FTNIR transmission instruments permit the analysis of liquid samples in the cylindrical sample vials used for liquid chromatography (LC) or high-performance LC (HPLC) (discussed in Chapter 14). For example, the 7 mm disposable glass vials used for LC can be used for transmission NIR measurements. Pharmaceutical laboratories often need to collect the NIR spectrum as well as performing HPLC or HPLC–MS on samples. The ability to fill one type of sample vial and use it for multiple measurements is very valuable. It increases sample throughput by eliminating transfer of samples to multiple sample holders and by eliminating washing of sample cells or cuvettes. It also conserves sample, decreases waste, and saves money.

5.7.3.2 Solids

For reflectance measurements, most solids require no sample preparation. Powders, tablets, textiles, solid "chunks" of material, food, and many other solids are analyzed "as is." Samples can be analyzed through plastic bags, glass vials, or in sample cups (see Figure 5.60). Examples will be discussed in the applications section. NIR reflectance spectra of solids are often plotted as the second derivative of the spectrum. This format shows small differences between samples more clearly and eliminates scatter and slope in the baseline. An example of this is seen is Figure 5.61.

Polymer films can be measured in either reflectance or transmission mode. For transmission, the film may be taped across an IR transmission card or cardboard slide mount.

Remote analysis of solids is possible through the use of fiber-optic probes such as the FlexIR NIR fiber-optic accessory described in the previous section. The probe tip is simply touched to the sample, in drums or through packaging, and the spectrum collected. This eliminates sample preparation, speeds up analyses, and minimizes exposure of the analyst to large amounts of materials.

5.7.3.3 Gases

Gas samples are handled by filling the same type of gas cell used for mid-IR gas analysis, except that the cell windows are of quartz instead of salt.

5.7.4 Applications of NIR Spectroscopy

The most important bands are overtones or combinations of the stretching modes of C–H, O–H, and N–H. These bands enable the quantitative characterization of polymers, chemicals, foods, and agricultural products for analytes such as water, fatty acids, and proteins. In many cases, the use of NIR reflectance spectroscopy has been able to replace time-consuming, classical "wet" chemical analyses, such as the Kjeldahl method for protein nitrogen and Karl Fischer titration for water content. The NIR region has been used for qualitative studies of hydrogen bonding, complexation in organometallic compounds, and solute–solvent interactions because the NIR absorptions are sensitive to intermolecular forces.

Polymer characterization is an important use of NIR spectrometry. Polymers can be made either from a single monomer, as is polyethylene, or from mixtures of monomers, as are styrene–butadiene rubber from styrene and butadiene and nylon 6-6, made from hexamethylenediamine and adipic acid. An important parameter of such copolymers is the relative amount of each present. This can be determined by NIR for polymers with the appropriate functional groups. Styrene content in a styrene–butadiene copolymer can be measured using the aromatic and aliphatic C–H bands. Nylon can be characterized by the NH band from the amine monomer and the C=O band from the carboxylic acid monomer. Nitrogen-containing polymers such as nylons, polyurethanes, and urea formaldehyde resins can be measured by using the NH bands. Block copolymers, which are typically made of a "soft block" of polyester and a "hard block" containing aromatics, as is polystyrene, have been analyzed by NIR. These analyses have utilized the C=O, aromatic, and aliphatic C=O bands. NIR is used to measured hydroxyl content in polymers with alcohol functional groups (polyols), both in final product and online for process control.

Fibers and textiles are well suited to NIR reflectance analysis. Analyses include identifying the type of fiber, the uptake of dyes, the presence of processing oil in polyester yarns, and the presence of fabric "sizing" agents.

Proteins in foods can be measured by NIR reflectance spectrometry with no sample preparation. This has replaced the standard Kjeldahl protein nitrogen determination, which required extensive sample preparation to convert protein nitrogen to ammonia, distillation of the released ammonia, and subsequent titration of the ammonia. The replacement of the Kjeldahl method for routine analysis by NIR has permitted online measurement of protein in food and beverage products. The Kjeldahl method is required for assaying the materials used to calibrate the NIR and for method validation.

The determination of ppm amounts of water in many chemicals is critical. Organic solvents used for organic synthesis may have to be very dry so that traces of water do not interfere with the desired reaction, for example. The classic method for measuring water at low levels in organic solvents and other chemicals is the Karl Fischer titration. Using the O–H bands characteristic of water, NIR has been used to measure water quantitatively in materials from organic solvents to concentrated mineral acids.

The use of NIR for process analysis and real-time analysis of complex samples is impressive. Grains such as wheat and corn can be measured with no sample preparation for protein content, water content, starch, lipids, and other components. In the reference by Heil, whole soybean samples were analyzed for water, protein, fatty acid, and oil by placing whole beans in a 12 cm sample cup spinner accessory (Thermo Scientific) combined with the integrating sphere in the FTNIR for diffuse reflection measurements. Spinning the sample allows average, representative spectra to be collected from a bulk sample. Measurement of the

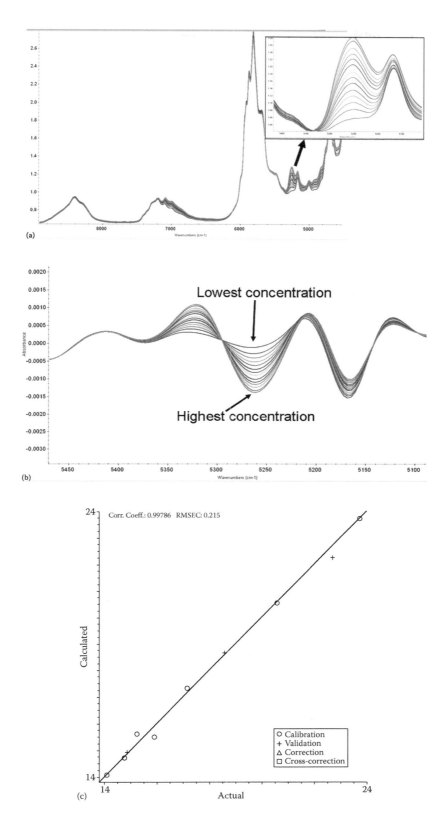

Figure 5.61 (a) NIR spectra of 14 samples of nitrocellulose solutions differing in concentration, with the region around 5200 cm^{-1} expanded in the inset (upper right) clearly showing the intensity difference due to compositional variation. (b) The derivative spectra of the 5200 cm^{-1} region, showing an improvement in peak shape (more nearly Gaussian) and elimination of the baseline shift on the low wavenumber side seen in the inset in (a). (c) Comparison of the chemometrics-based calculated results versus the actual concentrations. [© Thermo Fisher Scientific (www.thermofisher.com). Used with permission.]

four parameters permits NIR to perform multiple steps in the processing of soybeans, from grading of incoming soybeans, monitoring extraction efficiency, to ensuring the correct protein content of final product such as soy protein concentrate.

An NIR spectrometer is now commercially available on grain-harvesting machines to measure protein, moisture, and oil as the grain is being harvested (von Rosenburg et al., 2000). The analysis is performed in real time, as the farmer is driving around the field harvesting the grain. The data are coupled with a global positioning unit on the harvester, resulting in a map of the farmer's field, showing the quality of the grain harvested from different points in the field. This gives the farmer important information about where more fertilizer or water is needed. The real-time field analyzer gives excellent accuracy and precision, as shown by comparison to standard laboratory analyses.

Pharmaceutical tablets packaged in plastic "blister packs" can be analyzed nondestructively by NIR spectrometry right through the package. In the QC laboratory, this permits rapid verification of product quality without loss of product. In forensic analysis, unknown white powders in plastic bags seized as evidence in a drug raid can be identified by obtaining the NIR spectrum nondestructively right through the bag, eliminating the need to open the bag. This avoids possible contamination or spillage of the evidence.

The cost savings provided by the use of NIR instead of titrations for water and protein are enormous, not just in labor cost savings but in the cost of buying and then properly disposing of expensive reagents. NIR permits increased efficiency and increased product quality by online and at-line rapid analysis in the agricultural, pharmaceutical, polymer, specialty chemicals, and textile industries, among others.

5.8 RAMAN SPECTROSCOPY

Raman spectroscopy is a powerful, noninvasive technique for studying molecular vibrations by light scattering. Raman spectroscopy complements IR absorption spectroscopy because some vibrations, as we have seen, do not result in absorptions in the IR region. A vibration is only seen in the IR spectrum if there is a change in the dipole moment during the vibration. For a vibration to be seen in the Raman spectrum, a change in polarizability is necessary. That is, a distortion of the electron cloud around the vibrating atoms is required. Distortion becomes easier as a bond lengthens and harder as a bond shortens, so the polarizability changes as the bound atoms vibrate. As we learned earlier in the chapter, homonuclear diatomic molecules such as Cl_2 do not absorb IR radiation, because they have no dipole moment. The Cl–Cl stretching vibration is said to be *IR inactive*. Homonuclear diatomic molecules do change polarizability when the bond stretches, so the Cl–Cl stretch is seen in Raman spectroscopy. The Cl–Cl stretching vibration is said to be *Raman active*. Some molecular vibrations are active in IR and not in Raman, and *vice versa*; many modes in most molecules are active in both IR and Raman. Looking at CO_2 again, shown in the following diagrams, the mode on the left is the IR-inactive symmetric stretch, while the other two asymmetric vibrations are both IR active. The symmetric stretch is Raman active:

$$\begin{array}{ccc} \leftarrow \quad \rightarrow & \rightarrow \quad \rightarrow & \uparrow \quad \uparrow \\ O=C=O & O=C=O & O=C=O \\ & \leftarrow & \downarrow \end{array}$$

In general, symmetric vibrations give rise to intense Raman lines; nonsymmetric ones are usually weak and sometimes unobserved.

Raman spectroscopy offers some major advantages in comparison to other analytical techniques. Because it is a light-scattering technique, there are few concerns with sample thickness and little interference from ambient atmosphere. Therefore, there is no need for high-vacuum systems or instrument purge gas. Glass, water, and plastic packaging have weak Raman spectra, allowing samples to be measured directly inside a bottle or package, thereby minimizing sample contamination. Aqueous samples are readily analyzed. No two compounds give exactly the same Raman spectra and the intensity of the scattered light is proportional to the amount of material present. Raman spectroscopy is therefore a qualitative and quantitative technique.

5.8.1 Principles of Raman Scattering

When radiation from a source is passed through a sample, some of the radiation is scattered by the molecules present. For simplicity, it is best to use radiation of only one frequency and the sample should not absorb that frequency. The beam of radiation is merely dispersed in space. Three types of scattering occur. They are called *Rayleigh scattering*, *Stokes scattering*, and *anti-Stokes scattering*. Most of the scattered radiation has the same frequency as the source radiation. This is Rayleigh scattering, named after Lord Rayleigh, who spent many years studying light scattering. Rayleigh scattering occurs as a result of elastic collisions between the photons and the molecules in the sample; no energy is lost on collision. However, if the scattered radiation is studied closely, it can be observed that slight interaction of the incident beam with the molecules occurs. Some of the photons are scattered with less energy after their interaction with molecules and some photons are scattered with more energy. These spectral lines are called *Raman lines*, after Sir C.V. Raman, who first observed them in 1928. Only about 1 photon in a million will scatter with a shift in wavelength. The Raman–Stokes lines are from those photons scattered with less energy than the incident radiation; the Raman–anti-Stokes lines are from the photons

INFRARED, NEAR-INFRARED, RAMAN, AND PHOTOACOUSTIC SPECTROSCOPY

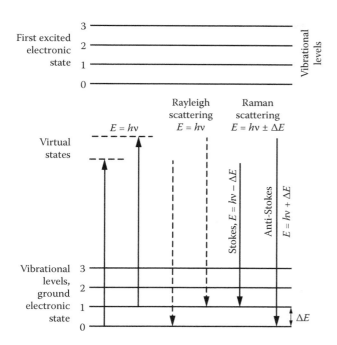

Figure 5.62 The process of Rayleigh and Raman scattering. Two virtual states are shown, one of higher energy. Rayleigh and Raman scattering are shown from each state. Normal IR absorption is shown by the small arrow on the far right marked ΔE, indicating a transition from the ground-state vibrational level to the first excited vibrational level within the ground electronic state.

scattered with more energy. The slight shifts in energy and therefore slight shifts in the frequencies of these scattered photons are caused by inelastic collisions with molecules. The differences in the energies of the scattered photons from the incident photons have been found to correspond to vibrational transitions. Therefore, the molecules can be considered to have been excited to higher vibrational states, as in IR spectroscopy, but by a very different mechanism. Figure 5.62 shows a schematic diagram of the Rayleigh and Raman scattering processes and of the IR absorption process.

The energy of the source photons is given by the familiar expression $E = h\nu$. If a photon collides with a molecule, the molecule increases in energy by the amount $h\nu$. This process is not quantized, unlike absorption of a photon. The molecule can be thought of as existing in an imaginary state, called a virtual state, with an energy between the ground state and the first excited electronic state. The energies of two of these virtual states are shown as dotted lines in Figure 5.62. The two leftmost arrows depict increases in energy through collision for a molecule in the ground state and a molecule in the first excited vibrational state, respectively. The arrows are of the same length, indicating that the interacting photons have the same energy. If the molecule releases the absorbed energy, the scattered photons have the same energy as the source photons. These are the Rayleigh

scattered photons, shown by the two middle arrows. The molecules have returned to the same states they started from, one to the ground vibrational state and the other to the first excited vibrational state. The arrows are the same length; therefore, the scattered photons are of the same energy.

If the molecule begins to vibrate with more energy after interaction with the photon, that energy must come from the photon. Therefore, the scattered photon must decrease in energy by the amount equal to the vibrational energy gained by the molecule. That process is shown by the second arrow from the right. Instead of returning to the ground vibrational state, the molecule is now in the first excited vibrational state. The energy of the scattered photon is $E - \Delta E$, where ΔE is the difference in energy between the ground and first excited vibrational states. This is Raman scattering, and the lower energy scattered photon gives rise to one of the Stokes lines. Note that ΔE is equal to the frequency of an IR vibration; if this vibration were IR active, there would be a peak in the IR spectrum at a frequency equal to ΔE. In general, the Raman–Stokes lines have energies equal to $E - \Delta E$, where ΔE represents the various possible vibrational energy changes of the molecule. This relationship can be expressed as

$$E - \Delta E = h(\nu - \nu_1) \qquad (5.13)$$

where
ν is the frequency of the incident photon
ν_1 is the *shift* in frequency due to an energy change ΔE

Several excited vibrational levels may be reached, resulting in several lines of energy $h(\nu - \nu_1)$, $h(\nu - \nu_2)$, $h(\nu - \nu_3)$, and so on. These lines are all shifted in frequency from the Rayleigh frequency. The Stokes lines, named after Sir George Gabriel Stokes, who observed a similar phenomenon in fluorescence, are shifted to lower frequencies than the Rayleigh frequency. The Raman shifts are completely independent of the wavelength of the excitation source. Sources with UV, visible, and NIR wavelengths are used, and the same Raman spectrum is normally obtained for a given molecule. There are exceptions due to instrumental variations and also if a resonance or near-resonance condition applies at certain wavelengths (Section 5.8.4).

Less commonly, the molecule *decreases* in vibrational energy after interacting with a photon. This might occur if the molecule is in an excited vibrational state to begin with and relaxes to the ground vibrational state. In this case, the molecule has given energy to the scattered photon. The photon is shifted to a frequency higher than the incident radiation. These higher-frequency lines, the Raman–anti-Stokes lines, are less important to analytical chemists than the Stokes lines because of their lower intensity. One exception to this is for samples that fluoresce strongly. Fluorescence interferes with the Stokes lines but to a much lesser extent with the anti-Stokes lines.

It is convenient to plot the Raman spectrum as intensity versus shift in wavenumbers in cm^{-1}, because these can be related directly to IR spectra. The Raman shift in cm^{-1} is identical to the IR absorption peak in cm^{-1} for a given vibration, because both processes are exciting the same vibration. That is, if a vibration is both Raman and IR active, it will be seen at the same position in wavenumbers in both spectra, although with different intensity. The Raman spectra for benzene and ethanol are shown in Figure 5.63, along with the related IR transmission spectra.

5.8.2 Raman Instrumentation

A Raman spectrometer requires a light source, a sample holder or cell, a wavelength selector (or interferometer), and a detector, along with the usual signal processing and display equipment. Since Raman spectroscopy measures scattered radiation, the light source and sample cell are usually placed at 90° to the wavelength selector, as shown schematically in Figure 5.64. The radiation being measured in Raman spectroscopy is either visible or NIR; therefore, spectrometer optics, windows, sample cells, and so on can be made of glass or quartz. It is critical in Raman spectroscopy to completely exclude fluorescent room lights from the spectrometer optics. Fluorescent lights give rise to numerous spurious signals.

5.8.2.1 Light Sources

Monochromatic light sources are required for Raman spectroscopy. The light sources used originally were simple UV light sources, such as Hg arc lamps; however, these were weak sources and only weak Raman signals were observed. The Raman signal is directly proportional to the

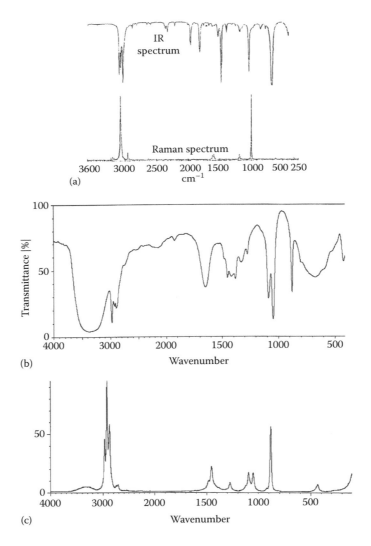

Figure 5.63 (a) The IR spectrum and Raman spectrum of benzene. (b) The IR spectrum of ethanol. (c) The Raman spectrum of ethanol. (b) and (c) are not plotted on the same scale. (The ethanol spectra are courtesy of http://aist.go.jp/RIODB/SDBS.)

power of the light source, which makes the laser, which is both monochromatic and very intense, a desirable light source. It was the development in the 1960s of lasers that made Raman spectroscopy a viable and useful analytical technique. Modern Raman instruments use a laser as the light source. The use of these intense light sources has greatly expanded the applications of Raman spectroscopy, because of the dramatically increased intensity of the signal and a simultaneous improvement of the signal-to-noise ratio. Lasers and excitation wavelengths commonly used for Raman instruments include visible wavelength helium/neon lasers and ion lasers such as the argon ion laser (488 nm) and the krypton ion laser (532 nm). The intensity of Raman scattering is proportional to the fourth power of the excitation frequency or to $1/\lambda^4$, so the shorter wavelength blue and green ion lasers have an advantage over the red helium/neon laser line at 633 nm. The disadvantage of the shorter wavelength lasers is that they can cause the sample to decompose on irradiation (photodecomposition) or fluoresce, an interference discussed subsequently. NIR lasers, such as neodymium/yttrium aluminum garnet (Nd/YAG) with an excitation line at 1064 nm, and the 785 nm NIR diode laser are used to advantage with some samples, such as biological tissue, because they do not cause fluorescence or photodecomposition. However, longer integration time or a higher-powered laser may be needed to compensate for the decrease in scattering efficiency of NIR lasers. Because of the wavelength-dependent nature of fluorescence excitation, a Raman spectrometer that integrates multiple laser sources and makes changing lasers easy should be considered if a diverse sample load is expected.

5.8.2.2 Dispersive Spectrometer Systems

Traditional Raman spectrometers used a monochromator with two or even three gratings to eliminate the intense Rayleigh scattering. The optical layout is similar to that for the UV/VIS single-grating monochromators discussed in Chapters 3 and 4. Holographic interference filters, called *super notch filters*, have been developed that dramatically reduce the amount of Rayleigh scattering reaching the detector. These filters can eliminate the need for a multiple grating instrument unless spectra must be collected within 150 cm^{-1} of the source frequency. Dispersive systems generally use a visible laser as the source. The low-end cutoff of the Raman spectrum is determined by the ability of the filters to exclude Rayleigh scattering. Since inorganic compounds have Raman bands below 100 cm^{-1}, modern instruments should provide a 50 cm^{-1} low-end cutoff.

The traditional detector for these systems was a photomultiplier tube. Multichannel instruments with PDA, charge injection device (CID), or charge-coupled device (CCD) detectors are commonly used today. All three detectors require cryogenic cooling. Room-temperature InGaAs detectors are also available. The PDA has the advantage of having the fastest response but requires more complicated optics than the other detectors. The CID has the advantage over both the PDA and CCD of not "blooming." Blooming means having charge spill over onto adjacent pixels in the array, which would be read in error as a signal at a frequency where no signal exists. CCDs are the slowest of the three array detectors because they have to be read out by transferring the stored charge row by row, but they are also the least expensive of the detector arrays. Sensitivity is improved in newer CCD designs as well. A dispersive Raman spectrometer with a CCD detector is shown schematically in Figure 5.65.

As described in Chapter 3, spectral resolution determines the amount of detail that can be seen in the spectrum. If the resolution is too low, it will be impossible to distinguish

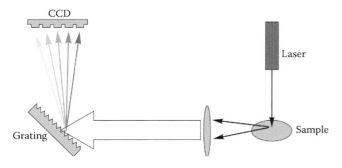

Figure 5.64 Idealized layout of a Raman spectrometer.

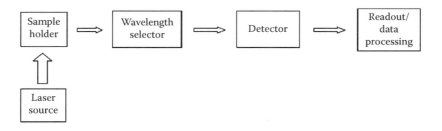

Figure 5.65 Schematic of a dispersive Raman spectrometer. [Reprinted from Weesner, F. and Longmire, M., *Spectroscopy*, 16(2), 68, 2001. With permission from Advanstar Communications, Inc.]

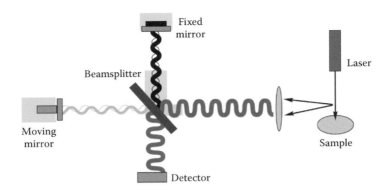

Figure 5.66 Schematic of an FT-Raman spectrometer. [Reprinted from Weesner, F. and Longmire, M., *Spectroscopy*, 16(2), 68, 2001. With permission from Advanstar Communications, Inc.]

between spectra of closely related compounds; if the resolution is too high, noise increases without any increase in useful information. Spectral resolution is determined by the diffraction grating and by the optical design of the spectrometer. With a fixed detector size, there is a resolution beyond which not all of the Raman wavelengths fall on the detector in one exposure. Ideally, gratings should be matched specifically to each laser used. A dispersive Raman echelle spectrometer from Perkin Elmer covers the spectral range 3500–230 cm^{-1} with a resolution better than 4 cm^{-1}.

CCDs used in modern instruments are generally Si-based 2D arrays of light-sensitive elements, called pixels. Each pixel, typically <30 μm, acts as an individual detector. Each dispersed wavelength is detected by a different pixel. CCD detectors commonly respond over the 400–1100 nm range, but specialized detectors can extend the range over 1100 nm and down into the UV range. The longest excitation wavelength that can be used with a silicon CCD detector is about 785 nm. Dispersive Raman, with its high sensitivity and high spatial resolution, is the best choice for small particle analysis and minor component analysis.

5.8.2.3 FT-Raman Spectrometers

FT-Raman systems generally use an NIR laser source, such as the Nd/YAG laser, and a Michelson interferometer. A schematic FT-Raman spectrometer is shown in Figure 5.66. The NIR laser source line is at 1064 nm, so the Raman–Stokes lines occur at longer wavelengths. This is beyond the detection range of the materials used in array detectors. The detector for an NIR-laser-based FT-Raman system is a liquid nitrogen-cooled photoconductive detector such as Ge or InGaAs. InGaAs detectors that do not require cooling are also available.

FT-Raman has many of the advantages of FTIR. There is high light throughput, simultaneous measurement of all wavelengths (the multiplex advantage), increased signal-to-noise ratio by signal averaging, and high precision in wavelength due to the internal interferometer calibration provided by the built-in He–Ne laser. A major advantage is in the use of the NIR laser excitation source, which dramatically reduces fluorescence in samples. Fluorescence occurs when the virtual states populated by excitation overlap excited electronic states in the molecule. Then, the molecule can undergo a radiationless transition to the lowest ground state of the excited electronic state before emitting a fluorescence photon on relaxation to the ground state. The fluorescence photon is of lower energy than the exciting radiation, and so fluorescence occurs at longer wavelengths, interfering with the Stokes scattering lines. The NIR laser is of low energy and does not populate virtual states that overlap the excited electronic states, as higher energy visible lasers can. As an example, the Raman spectrum of cocaine is shown in Figure 5.67. The spectrum in Figure 5.67a was collected with an FT-Raman spectrometer using an NIR laser, while that in Figure 5.67b was collected with a dispersive Raman system and a visible laser. Figure 5.67b shows a large fluorescence band that obscures most of the Raman spectrum below 2000 cm^{-1}. With appropriate mathematical "smoothing" algorithms and multipoint baseline correction, it is possible to extract a useable Raman spectrum from samples that exhibit strong fluorescence, as shown in Figure 5.68. One consideration in FT-Raman is that the laser line at 1064 nm is very close to a water absorption band. While this does not prevent aqueous solutions from being studied by FT-Raman, aqueous solutions cannot be studied as easily as they can with dispersive Raman. FT-Raman is the better choice for samples that fluoresce or contain impurities that fluoresce. FT-Raman is widely used in the analysis of illicit drugs and pharmaceuticals, as many of these compounds fluoresce strongly at visible wavelengths.

5.8.2.4 Fiber-Optic-Based Modular and Handheld Systems

Low-resolution Raman systems based on fiber optics have been developed to provide miniaturized, low-cost, rugged systems for lab, process, and field testing. This

INFRARED, NEAR-INFRARED, RAMAN, AND PHOTOACOUSTIC SPECTROSCOPY

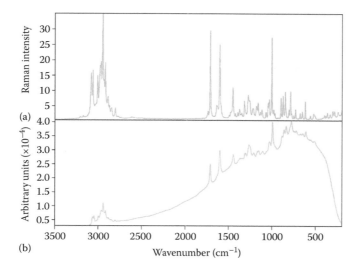

Figure 5.67 Analyses of crack cocaine using (a) FT-Raman and (b) 785 nm dispersive Raman. Note the lack of fluorescence in the FT-Raman spectrum (flat baseline vs. the big "hump" due to fluorescence in the lower spectrum) and the rich spectral information in the upper spectrum. This information is partially obscured by the fluorescence band in the dispersive spectrum. [Reprinted from Weesner, F. and Longmire, M., *Spectroscopy*, 16(2), 68, 2001. With permission from Advanstar Communications, Inc.]

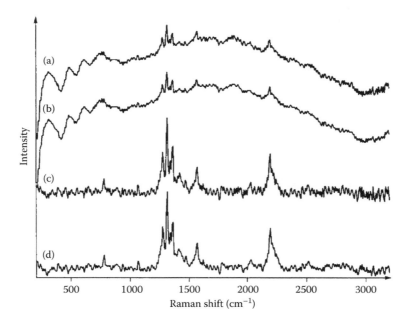

Figure 5.68 It is possible to extract data from a Raman spectrum that exhibits fluorescence interference by the application of mathematical data treatments. The data treatments were applied sequentially to give the final spectrum shown in (d). (a) is the raw spectrum of a weak scatterer with a fluorescence background, (b) is the spectrum after Savitzky–Golay smoothing, (c) after a multipoint baseline correction performed by the analyst, and (d) after a Fourier smoothing. [Reprinted from Kawai, N. and Janni, J.A., *Spectroscopy*, 15(10), 32, 2000. With permission from Advanstar Communications, Inc.]

development has occurred in the last 10 years, due to improved, low-cost semiconductor lasers, cheap and highly sensitive solid-state detectors, and cheaper, faster computer chips. Portable and handheld Raman systems are now available from several companies, including Rigaku Raman Technologies, Inc. (www.rigakuraman.com), BaySpec, Inc. (www.bayspec.com), Thermo Fisher Scientific (www.thermo.com), Enwave Optronics, Inc. (www.enwaveopt.com), B&W Tek, Inc. (www.bwtek.com), and Ocean Optics (www.oceanoptics.com) for screening of incoming raw materials, hazardous materials, forensic applications, in-line pharmaceutical processing, gem and semiprecious stone quality, authentication and anticounterfeiting, and QC (Figure 5.69). These field and portable instruments weigh

(a)

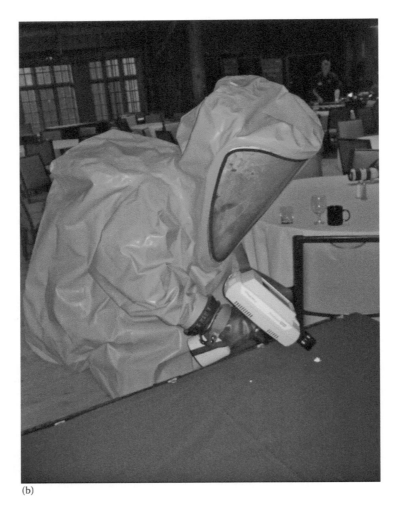

(b)

Figure 5.69 (a) Handheld TruScan® Raman instrument collecting a spectrum from a powder sample directly through the plastic bag. [© Thermo Fisher Scientific (www.thermofisher.com). Used with permission.] (b) Hazardous material identification by a first responder using the handheld FirstGuard™ 1064 nm advanced Raman system. (Used with permission of Rigaku Raman Technologies, Inc., www.rigakuraman.com.)

between 2 and 5 lb, and can be configured with one or more lasers and different detectors. The field-portable EZRaman-I series instruments from Enwave Optronics, Inc., for example, are lithium-battery-powered, compact instruments with a built-in laptop; one of three lasers (or configured with two different lasers); a high-sensitivity CCD spectrograph, with the CCD thermoelectrically cooled to −50 °C; spectral ranges from 100 to 3300 cm^{-1}; and average resolution of 6–7 cm^{-1}.

Handheld and portable systems may have less resolution and spectral range than laboratory systems but can be customized for specific applications with customized spectral libraries. Specialized applications for these instruments include measuring the "freshness" of fish by monitoring the concentration of dimethylamine, which is directly related to the age and temperature of processed frozen fish (Herrero et al., 2004). Fat concentration, composition, and saturation index allows a handheld Raman instrument to determine in less than 1 min if edible oils such as olive oil have been adulterated or if cheaper oil is being sold as more expensive olive oil.

Another critical food safety application is the determination of melamine in human and pet food. Melamine has been found in milk and pet food, deliberately added to watered-down or inferior products to boost the apparent protein concentration. The standard wet chemical method for "protein" in food is the Kjeldahl method, which actually measures nitrogen and is an indirect measure of protein. Melamine responds to the Kjeldahl method just like protein, but is not a protein and is toxic; it combines with uric acid to form kidney stones. The USDA has set a maximum allowable limit of 2.5 ppm melamine for adults. Using their portable Nunavut™ Raman System, BaySpec, Inc., scientists demonstrated the measurement of melamine at levels down to 3 ppm in the field; the analysis takes less than 10 s, is accurate, is repeatable, and is nondestructive.

5.8.2.5 Samples and Sample Holders for Raman Spectroscopy

Because the laser light source can be focused to a small spot, very small samples can be analyzed by Raman spectroscopy. Samples of a few microliters in volume or a few milligrams are sufficient in most cases. Liquid samples can be held in beakers, test tubes, glass capillary tubes, 96-well plates, or NMR tubes. Aqueous solutions can be analyzed since water is a very weak Raman scatterer. This is a significant advantage for Raman spectroscopy over IR. Other solvents that can be used for Raman studies include chloroform, carbon tetrachloride, acetonitrile, and carbon disulfide. Solid powders can be packed into glass capillary tubes, NMR tubes, plastic bags, or glass vials for analysis. The spectra are obtained through the glass. Solid samples can also be mounted at the focal point of the laser beam and their spectra obtained "as is" or pressed into pellets. Gas samples do not scatter radiation efficiently, but can be analyzed by being placed into a multipath gas cell, with reflecting mirrors at each end. The body of the gas cell must be of glass to allow collection of the scattered light at 90°.

The sample must be placed at the focal point of an intense laser beam, and some samples may be subject to thermal decomposition or photodecomposition. Accessories that spin the sample tube or cup are available, to distribute the laser beam over the sample and reduce heating of the sample. Spinning or rotating the sample minimizes thermal decomposition, but does not stop photodecomposition. Sample spinning is required for *resonance Raman spectroscopy*, discussed later.

Raman spectroscopy does not suffer interference from atmospheric water vapor or carbon dioxide, as does IR. Gases do not scatter well, so even though Raman-active bands occur for these gases, the contribution to the Raman signal from air in the optical path is insignificant. Materials in the optical path outside of the laser focus also have negligible scattering.

Sample cells have been designed for low- and high-temperature operation, well-plate accessories allow high-throughput analyses, and remote probes and user-friendly video stages are available for many systems.

5.8.3 Applications of Raman Spectroscopy

Quantitative and qualitative analyses of inorganic and organic compounds can be performed by Raman spectroscopy. Raman spectroscopy is used for bulk material characterization, online process analysis, remote sensing, microscopic analysis, and chemical imaging of inorganic, organic, and organometallic compounds, polymers, biological systems, art objects, carbon nanomaterials, and much more. Forensic science applications include identification of illicit drugs, explosives, and trace evidence like hair, fibers, and inks. Raman spectra have fewer lines and much sharper lines than the corresponding IR spectra, as seen in Figure 5.70. This makes quantitative analysis, especially of mixtures, much simpler by Raman spectroscopy than by IR spectroscopy.

Quantitative analysis had not been as common as in IR spectroscopy until recently, due to the high cost of Raman instruments. With prices for Raman systems dropping below $40,000, and even as low as $10,000, the use of Raman spectroscopy for quantitative analysis is increasing. Quantitative analysis requires measurement of the intensity of the Raman peaks and the use of a calibration curve to establish the concentration–intensity relationship. The intensity of a Raman peak is directly proportional to the concentration:

$$I = KJ\nu^4 c \qquad (5.14)$$

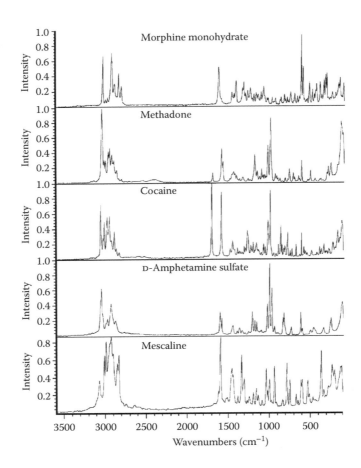

Figure 5.70 Raman spectra of common drugs of abuse, showing how sharp the Raman spectral lines are. [© Thermo Fisher Scientific (www.thermofisher.com). Used with permission.]

where

I is the intensity of Raman signal at frequency v
K is the proportionality constant including instrument parameters such as laser power
J is the scattering constant for the given Raman peak
v is the scattering frequency of the Raman peak
c is the concentration of analyte

K, J, and v are all constants for a given sample measurement. The frequency and intensity of the desired Raman lines are measured, and the intensity of an unknown compared to the calibration curve. It is common to use an internal standard for Raman analysis, because of the dependence of the signal on the laser power (in the K term). Without an internal standard, the laser power, sample alignment, and other experimental parameters must be carefully controlled. If an internal standard is used, the intensity of the internal standard peak is also measured, and the ratio of intensities plotted versus concentration. Upon division, this reduces Equation 5.14 to

$$\frac{I_{unk}}{I_{std}} = K'c \quad (5.15)$$

The use of the internal standard minimizes the effect of changes in instrumental parameters and can result in better accuracy and precision. Newer Raman systems such as the Thermo Scientific DXR series incorporate a laser power regulator that delivers reproducible laser power to the sample and compensates for laser aging and variability, minimizing the need for internal standards.

Two other major factors have impacted Raman spectroscopy for both qualitative and quantitative work. Older systems had problems with reproducibility due to wavelength dependence of the silicon CCD detector response and x-axis nonreproducibility, a result of inadequate spectrograph and laser calibration. Modern systems use automated, software-driven multipoint calibration of the laser and spectrograph and automatic intensity correction, dramatically improving reproducibility. Wavelength accuracy can be checked by using a high-purity mercury vapor or argon light source. Laser power output can be measured in a variety of ways, including the use of NIST Raman relative intensity standards, fluorescent glasses sensitive to specific laser wavelengths.

Quantitative analyses that can be done by Raman spectroscopy include organic pollutants in water, inorganic oxyanions and organometallic compounds in solution, aromatic/aliphatic hydrocarbon ratios in fuels, antifreeze concentration in fuel, and concentration of the active pharmaceutical ingredient in the presence of excipients such as microcrystalline cellulose. Other common applications include raw materials identification, forensic testing of illegal drugs, explosives, and other hazardous materials, cancer screening, and dialysis monitoring. Mixtures of compounds in pharmaceutical tablets can be determined quantitatively, without dissolving the tablets. Raman sensitivity varies greatly, depending on the sample and the equipment. In general, analyte concentrations of at least 0.1–0.5 M (or 0.1%–1%) are needed to obtain good signals.

Another use of Raman spectroscopy for quantitative analysis is the determination of percent crystallinity in polymers. Both the frequency and intensity of peaks can shift on going from the amorphous to the semicrystalline state for polymers. The percent crystallinity can be calculated with the help of chemometrics software.

Qualitative analysis by Raman spectroscopy is very complementary to IR spectroscopy and in some cases has an advantage over IR spectroscopy. The Raman spectrum is more sensitive to the organic framework or backbone of a molecule than to the functional groups, in contrast to the IR spectrum. IR correlation tables are useful for Raman spectra, because the Raman shift in wavenumbers is equal to the IR absorption in wavenumbers for the same vibration. Raman spectral libraries are available from commercial and government sources, as noted in the bibliography. These are not yet as extensive as those available for IR, but are growing rapidly. Table 5.19 gives common functional groups and vibration regions and compares the strength of the Raman and IR bands.

The same rules for the number of bands in a spectrum apply to Raman spectra as well as IR spectra: $3N - 6$ for nonlinear molecules and $3N - 5$ for linear molecules. There may be fewer bands than theoretically predicted due to degeneracy and nonactive modes. Raman spectra do not usually show overtone or combination bands; they are much weaker than in IR. A "rule of thumb" that is often true is that a band that is strong in IR is weak in Raman and *vice versa*. A molecule with a center of symmetry, such as CO_2, obeys another rule: if a band is present in the IR spectrum, it will not be present in the Raman spectrum. The reverse is also true. The detailed explanation for this is outside the scope of this text, but the rule "explains" why the symmetric stretch in carbon dioxide is seen in the Raman spectrum, but not in the IR spectrum, while the asymmetric stretch appears in the IR spectrum but not in the Raman spectrum.

Since the physical properties of the sample do not greatly impact the Raman spectrum, it is not able to differentiate between materials that differ only in physical form, for example, powder versus tablet. This does have an advantage in that a Raman spectral library can be built using only a single reference for each substance. Raman is an excellent tool for substances that differ chemically, even only a little. Raman can differentiate polymorphs, isomers, anhydrates from hydrates, monohydrates from polyhydrates, L and D forms, and other chemically similar forms. A good example is the ease of distinguishing between pseudoephedrine and ephedrine, diastereomers that differ only in the position of

Table 5.19 **Raman band frequencies**

Functional Group/Vibration	Region (cm^{-1})	Raman	Infrared
Lattice vibrations in crystals, LA modes	10–200	Strong	Strong
δ(CC) aliphatic chains	250–400	Strong	Weak
v(Se–Se)	290–330	Strong	Weak
v(S–S)	430–550	Strong	Weak
v(Si–O–Si)	450–550	Strong	Weak
v(X–O)	150–450	Strong	Med weak
v(C–I)	480–660	Strong	Strong
v(C–Br)	500–700	Strong	Strong
v(C–Cl)	550–800	Strong	Strong
v(C–S) aliphatic	630–790	Strong	Medium
v(C–S) aromatic	1080–1100	Strong	Medium
v(O–O)	845–900	Strong	Weak
v(C–O–C)	800–970	Medium	Weak
v(C–O–C) asym	1060–1150	Weak	Strong
v(CC) aliphatic chain vibrations	600–1300	Medium	Medium
v(C=S)	1000–1250	Strong	Weak
v(CC) aromatic ring chain vibrations	1580, 1600	Strong	Medium
	1450, 1500	Medium	Medium
	1000	Strong/Medium	Weak
δ(CH$_3$)	1380	Medium	Strong
δ(CH$_2$) δ(CH$_3$) asym	1400–1470	Medium	Medium
δ(CH$_2$) δ(CH$_3$) asym	1400–1470	Medium	Medium
v(C–(NO$_2$))	1340–1380	Strong	Medium
v(C–(NO$_2$)) asym	1530–1590	Medium	Strong
v(N=N) aromatic	1410–1440	Medium	—
v(N=N) aliphatic	1550–1580	Medium	—
δ(H$_2$O)	~1640	Weak broad	Strong
v(C=N)	1610–1680	Strong	Medium
v(C=C)	1500–1900	Strong	Weak
v(C=O)	1680–1820	Medium	Strong
v(C≡C)	2100–2250	Strong	Weak
v(C≡N)	2220–2255	Medium	Strong
v(–S–H)	2550–2600	Strong	Weak
v(C–H)	2800–3000	Strong	Strong
v(=(C–H))	3000–3100	Strong	Medium
v(≡(C–H))	3300	Weak	Strong
v(N–H)	3300–3500	Medium	Medium
v(O–H)	3100–3650	Weak	Strong

Source: Courtesy of HORIBA Scientific, www.horiba.com/scientific.
Notes: X, metal atom; asym, asymmetric; v, stretching, δ, bending.

a hydroxyl group, or the ease of distinguishing between methamphetamine (*N*-methyl-1-phenylpropan-2-amine, a drug widely abused) and phentermine (a diet pill, 2-methyl-1-phenylpropan-2-amine). The two compounds have the same RMM and, as we'll see in Chapter 11, are very difficult to distinguish based on their mass spectra. The Raman spectra (Figure 5.71a) are clearly different; even without any knowledge of Raman spectroscopy, one can see that these are not the same spectra. The sharpness of the Raman bands is the key to using this technique for identification.

Structural identification of an unknown is usually done with both IR and Raman spectra or by matching Raman spectra to spectral libraries of known compounds. The subtraction of known spectra from the spectrum of an unknown mixture to identify the components of the mixture works better for Raman spectra than for IR spectra, because there

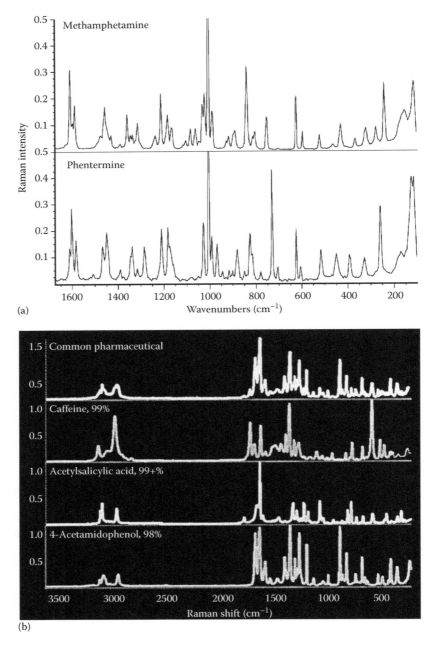

Figure 5.71 (a) The Raman spectra of methamphetamine and phentermine. (b) The analysis of a common multicomponent pharmaceutical tablet by Raman spectroscopy with spectral subtraction. The top spectrum is the entire tablet. The spectra for acetylsalicylic acid and caffeine were subtracted, resulting in the spectrum of the third component, identified as 4-acetamidophenol. [© Thermo Fisher Scientific (www.thermofisher.com). Used with permission.]

are fewer Raman peaks, the peaks are sharp, and their position and shape are not affected by hydrogen bonding. For example, it is possible to identify the components of a commercial pain relief tablet by spectral subtraction from the Raman spectrum of the intact tablet, as shown in Figure 5.71b.

The sharpness of Raman spectra also makes library searching very reliable, as seen in Figures 5.72a and b. The spectrum of an unknown white powder collected through an evidence bag, as in Figure 5.69, is compared to spectra of a variety of drugs. The match to methamphetamine is significantly better than the other possibilities (83 vs. 62–65 for the other compounds).

Raman spectroscopy is particularly useful for studying inorganic and organometallic species. Most inorganic, oxyanionic, and organometallic species have vibrations in the far-IR region of the spectrum, which is not easily accessible with commercial IR equipment. These metal–ligand and metal–oxygen bonds are Raman active and are easily studied in aqueous solutions and in solids. Raman spectroscopy is used in geology and gemology for identification and analysis and in art restoration and identification and verification of cultural objects.

As noted earlier, fused silica optical fiber is used for remote NIR measurements. The same type of fiber-optic probe can be used for Raman spectroscopy and enables remote measurement of samples and online process measurements. *In situ* reaction monitoring by Raman spectroscopy has been used to study catalytic hydrogenation, emulsion polymerization, and reaction mechanisms. It is a powerful tool for real-time reaction monitoring, allowing feedback for reaction control. Remote sensing of molecules in the atmosphere can be performed by Raman scattering measurements using pulsed lasers.

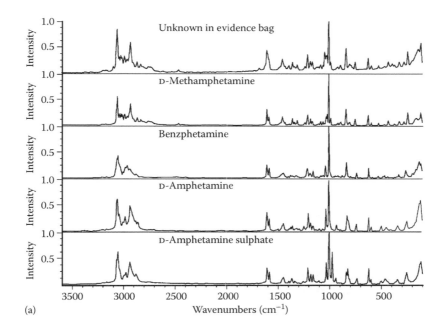

Figure 5.72 (a) Raman spectra of an unknown material (top) and four related drug compounds. (b) The library search match numbers (second column) show that the unknown is very probably methamphetamine. Note that the entire analysis was performed in 30 s. [© Thermo Fisher Scientific (www.thermofisher.com). Used with permission.]

5.8.4 Resonance Raman Effect

When monochromatic light of a frequency that cannot be absorbed by the sample is used, the resulting Raman spectrum is the **normal** Raman spectrum. Normal Raman spectroscopy, as has been noted, is an inefficient process resulting in low sensitivity and it suffers from interfering fluorescence in many samples. If a laser excitation wavelength is used that falls within an excited electronic state of the molecule, the intensity of some Raman lines increases by as much as 10^3–10^6 over the normal Raman intensity. This is known as the *resonance Raman effect*, and the technique is called resonance Raman spectroscopy. The molecule must possess a *chromophore* that can absorb visible or UV radiation (discussed in Chapter 4). The process that occurs is shown schematically in Figure 5.73. The resulting spectra are even simpler than normal Raman spectra because only totally symmetric vibrations associated with the chromophore are enhanced in intensity. This makes resonance Raman a very selective probe for specific chromophores. Resonance Raman spectroscopy has been used to study low concentrations of biologically important molecules such as hemoglobin.

Lasers that have wavelengths in the UV and visible regions of the spectrum are used for resonance Raman spectroscopy. Tunable dye lasers are often used; these lasers can be set to a selected wavelength over the UV/VIS range of 200–800 nm. This permits maximum flexibility in the choice of excitation wavelength.

5.8.5 Surface-Enhanced Raman Spectroscopy

Surface-enhanced Raman spectroscopy or surface-enhanced Raman scattering (SERS) is another technique for obtaining strong Raman signals from surfaces and interfaces, including species adsorbed onto surfaces. Fleischmann and coworkers developed SERS in 1974. The SERS technique requires adsorption of the species to be studied onto or in close proximity to a prepared "rough" metal or metal colloid surface, where the roughness is at the atomic level. The surface roughness features are much smaller than the wavelength of the laser. For example, roughness features or particle sizes in the 20–100 nm range are used for laser wavelengths in the 532–780 nm range. The Raman excitation laser produces *surface plasmons* (coherent electron oscillations) on the surface of the metal. These surface plasmons interact with the analyte to greatly enhance the Raman emission. The process is shown schematically in Figure 5.74.

Inorganic and organic species adsorbed onto such surfaces show enhancement of Raman signals by up to 10^6 over normal Raman signals. The reasons for the enhancement are not yet well understood, although to achieve the most effective enhancement, there must be resonance between the laser and the metal. Correct pairing of substrate and laser is of critical importance. Metals used as

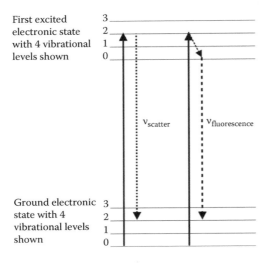

Figure 5.73 The resonance Raman process.

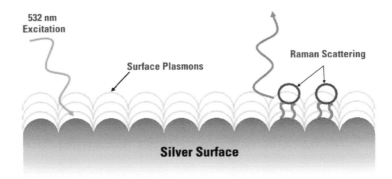

Figure 5.74 The SERS process using 532 nm excitation of a silver substrate with adsorbed species (the circles). Surface plasmons interact with the adsorbed species to enhance the Raman emission, shown on the right side of the diagram. [© Thermo Fisher Scientific (www.thermofisher.com). Used with permission.]

surfaces include gold, silver, copper, platinum, and palladium. Metal electrodes, metal films, and metal colloids have been used. The adsorbed or deposited analyte molecule must be less than 50 Å from the surface for the enhancement to be observed. Samples have been deposited electrolytically onto electrode surfaces or mixed with colloidal metal suspensions. Samples separated on thin-layer chromatography (TLC) plates have been sprayed with metal colloid solution. One simple approach involves mixing a few microliters of a colloidal substrate (metal nanoparticles of 20–100 nm dimensions suspended in solution) with an equal amount of the analyte. This is deposited onto a glass slide, is allowed to dry, and is then ready for analysis (Deschaines and Wieboldt, 2010 and references therein). It should be noted that preparing colloids is not simple; glassware must be extremely clean and ultrapure water must be used. Reaction conditions must be carefully controlled so that particles of the correct size and shape form.

Preparing metal surfaces requires expertise and can be expensive. Commercial sources for SERS substrates, colloids, and sample holders are available from companies such as Thermo Fisher Scientific, Renishaw, and Real-Time Analyzers (www.rta.biz), which focuses on high-throughput needs with SERS 96-well plates.

The enhancement leads to the ability to detect extremely small amounts of material, making SERS an effective tool for a variety of problems, including corrosion studies, detection of chemical warfare agents, bacteria on food, trace evidence in forensic science, blood glucose, research into infectious diseases, and many more. Detection limits for SERS are in the nanogram range. SERS can be used to study the way in which an analyte interacts with or binds to a surface. SERS can be used to obtain information on very dilute solutions and small amounts of material that cannot be obtained by regular Raman spectroscopy. Current research in SERS includes practical biomedical applications such as detecting disease states based on deoxyribonucleic acid (DNA) signatures.

5.8.6 Raman Microscopy

Raman spectroscopy coupled to a microscope permits the analysis of very small samples nondestructively. The use of Raman microscopy allows the characterization of specific domains or inclusions in heterogeneous samples with very high spatial resolution. With dispersive Raman microscopy, the spatial resolution is often better than 1 μm. FT-Raman microscopy is limited to spot sizes of about 2–10 μm, but with no interference from fluorescence. The use of a confocal microscope (Figure 5.75) allows only the light at the sample focus to pass into the detector; all other light is blocked. This permits nondestructive depth profiling of samples without the need for cross sectioning of the sample. For example, confocal Raman microscopy can identify the polymers in complex

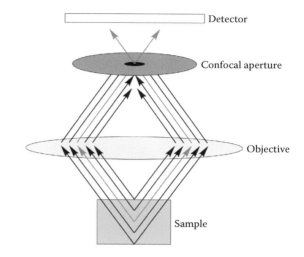

Figure 5.75 A simplified illustration of confocal microscopy. A spectrum is collected only from the area of the sample denoted by the middle arrow; only that light exits the confocal aperture. Information from all other sample areas is blocked. Raising or lowering the sample will allow the detector to collect Raman scatter from levels in the sample lower or higher than the area in the middle. [Reprinted from Weesner, F. and Longmire, M., *Spectroscopy*, 16(2), 68, 2001. With permission from Advanstar Communications, Inc.]

layered structures, such as multilayer films used for food packaging. Characterization of heterogeneous materials includes inclusions in minerals, the pigments, and other components in cosmetics and the study of pigments, resins, and dyes in art and archeological objects. Using Raman microscopy, it is possible to identify if a red pigment in a painting is expensive cinnabar (HgS), cheaper hematite (Fe_2O_3), a mixture of the two, or an organic dye. The article by Edwards provides a detailed table of Raman bands from common minerals used in art and an overview of the use of Raman microscopy for the study of art objects. Raman microscopy is particularly useful for art and archeological objects, because there is no sample preparation required and the natural water present in paintings, manuscripts, and ancient textiles does not interfere, as it does in IR spectroscopy.

Since spatial resolution is diffraction limited, short-wavelength lasers are optimal for analyzing small sample features. In order to achieve micron-level spatial resolution, the alignment of the Raman microscope is critical. The visual light path, the excitation laser beam path, and the Raman scatter beam path from the sample to the detector must all be targeted precisely on the same spot.

Raman microscopes are available from a number of instrument companies including Bruker, HORIBA Scientific, Jasco, Thermo Fisher Scientific, WITec, and Renishaw plc, among others.

FTIR microscopy was discussed earlier in the chapter, and given the complementary nature of IR and Raman, it is reasonable that laboratories performing IR microscopy might

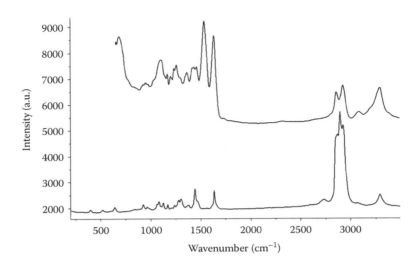

Figure 5.76 Raman and FTIR spectra of a 10 μm fiber of a nylon 6-PEG block copolymer. Spectra were collected at exactly the same spot on the fiber. (Courtesy of HORIBA Scientific, www.horiba.com.)

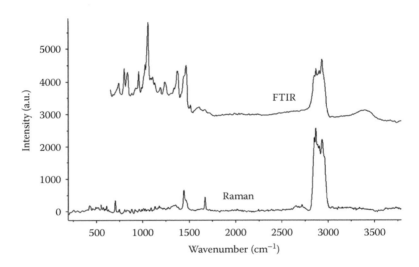

Figure 5.77 Raman and FTIR spectra of a gallstone. The spectra were recorded with the LabRAM-IR microscope using a 532 nm laser. The FTIR spectrum is more sensitive to the OH bands, while the Raman spectrum starts below 600 cm^{-1} and shows details of the cholesteric species and the C=C bands. (Courtesy of HORIBA Scientific, www.horiba.com.)

well need Raman microscopy and *vice versa*. Two microscope systems were required and the sample had to be moved from one system to the other. The difficulty of relocating the exact spot to be sampled can be imagined. A combination dispersive Raman and FTIR microscopy system, the LabRAM-IR2 (HORIBA Scientific, Edison, NJ), allows both Raman and IR spectra to be collected at exactly the same location on the sample. The resolution depends on the wavelength observed, because resolution is limited by diffraction, but is <1 μm for the Raman spectrum and 10–20 μm for the IR spectrum. Examples of the type of data that can be obtained with this combination microscopy system are shown in Figures 5.76 and 5.77.

5.9 CHEMICAL IMAGING USING NIR, IR, AND RAMAN SPECTROSCOPY

The most recent breakthrough in the use of vibrational spectroscopy for chemical analysis is in the area of *chemical imaging*. Chemical imaging is the use of 2D or 3D detectors to collect spectral data from a large number of locations within a sample and then using the variations in the spectral data to map chemical differences within the sample. The chemical differences are often displayed as a false-color image of the sample, called a chemigram. The use of chemical imaging technology in NMR has been described in Chapter 6; it is the technique more commonly called magnetic resonance imaging (MRI).

FTIR imaging has been commercially available since 1996. The usual "detector" is an MCT array detector, called a focal plane array (FPA) detector, used in conjunction with an FTIR microscope. A 64 × 64 FPA detector has 4096 detector elements and allows 4096 interferograms to be collected simultaneously. Because each pixel in the detector array generates a spectrum, there are three dimensions in the data set. These data sets are often referred to as data cubes or image cubes. The x and y coordinates of the cube are the spatial positions while the z coordinate represents the wavelength, as shown in Figure 5.78. The data can be handled in many ways, including the use of library searching, principal component analysis, and more, making this a powerful technique. Principal component analysis is a technique that sorts the image spectra into an independent set of subspectra (the principal components) from which the image spectra can be reconstructed. If there are five layers in a polymer laminate, for example, and there are 1000 image spectra, in theory, five subspectra would be sufficient to describe all 1000 image spectra. In practice, more than five subspectra are needed, due to variations in baseline, instrumental conditions, impurities, and so on. The amounts of the principal components in the original image are calculated at each pixel and the resulting scores are used to enhance the IR image contrast. See the reference by Sellors for more detail and color images that cannot be easily reproduced in gray scale, as well as the websites of the major instrument companies for full-color application notes. As a simple example, the intensity of the carbonyl-stretching band in each pixel can be "reassembled" into the visual image seen through the microscope, giving a distribution of carbonyl-containing material in the sample. Such an image is shown in Figure 5.79, collected with a Spectrum Spotlight FTIR Imaging System (www.perkinelmer.com). Samples can be measured in reflectance mode or transmission mode. Solid samples such as biological tissues may need to be sliced into thin sections for transmission analysis. FTIR imaging can be performed on a wide variety of sample matrices, including polymers, pharmaceutical tablets, fibers, and coatings.

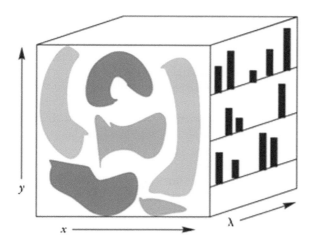

Figure 5.78 Schematic representation of a 3D spectral data cube. Sample spatial positions are represented by the x and y coordinates; the spectral wavelength is represented by the λ axis. (Reprinted from McLain, B.L. et al., *Spectroscopy*, 15, 28, 2000. With permission from Advanstar Communications, Inc.)

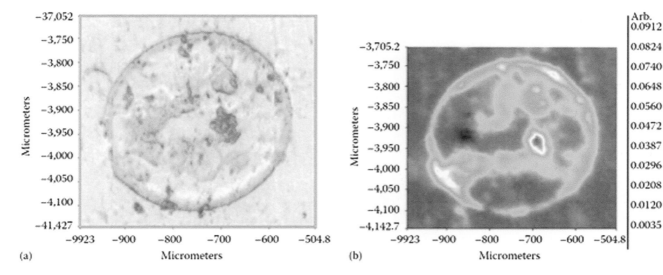

Figure 5.79 Use of FTIR imaging. (a) The visible light image of an inclusion in polypropylene film. Sample size is approximately 450 × 450 μm. (b) The false-color IR image of the same inclusion showing the distribution of a carbonyl contaminant in the inclusion. The relative intensity of the gray scale is related to concentration, which is actually displayed in full color by the instrument. The image contains more than 5000 spectra at 6.25 μm pixel size. The total data collection time was 90 s. Collected on a Spectrum™ Spotlight FTIR Imaging System. [© 2004–2014 PerkinElmer, Inc. All rights reserved. Printed with permission. (www.perkinelmer.com).]

Commercial Raman and NIR imaging microscope systems are also available. Raman imaging of polymer blend surfaces and Raman and NIR imaging of silicon integrated circuits, of whole pharmaceutical tablets, and of tooth enamel and counterfeit documents are just a few applications. The FALCON™ Raman Chemical Imaging Microscope (ChemIcon, Pittsburgh, PA, www.chemimage.com) can be equipped with fiber-optic probes for remote monitoring and for high-temperature remote monitoring, such as in a heated process stream. 3D confocal Raman imaging is a powerful technique available from a number of companies, including WITec GmbH. A confocal Raman microscope is a combination of a Raman spectrometer with a confocal microscope, so that the resolution of the optical microscope is combined with the analytical power of Raman spectroscopy. The sample is scanned point by point and line by line, and at every image pixel, a complete Raman spectrum is taken. This process is also called hyperspectral imaging. These multispectrum files are then analyzed to display the distribution of chemical sample properties. By taking a stack of images with different focal positions, the geometry of samples can be reconstructed in 3D. Figure 5.80 shows such a 3D Raman image of a fluid inclusion in garnet (60 × 60 × 30 μm) with the fluid inclusion (water) displayed in blue and the garnet in red. Typical integration times in a confocal Raman microscope are between 700 μs and 100 ms per spectrum, so that a complete Raman image of 10,000 spectra takes between a few seconds and 20 min.

The ChemIcon CONDOR™ macro chemical imaging system can do NIR absorption imaging as well as fluorescence and visible emission imaging. PerkinElmer also makes an FTNIR version of is Spectrum™ Spotlight imaging system. It can be used to show the inhomogeneous distribution of active pharmaceutical ingredients in tablets that otherwise were identical in bulk chemical composition and whose bulk FTNIR spectra were identical. Such a difference could point to a problem in the manufacturing process or to the use of a different manufacturing process by a company making counterfeit tablets.

A fast NIR Raman imaging microscope system (NIRIM) is described by McLain and coauthors that uses a fiber-optic bundle and CCD detector to collect a complete 3D Raman data cube from a sample in 1 s or less. A schematic of the NIRIM is shown in Figure 5.81. The system has been used to image inorganic inclusions in crystals, mixtures of metal oxides, and amino acid mixtures and to perform surface-enhanced Raman imaging of catalyst and nanoparticle

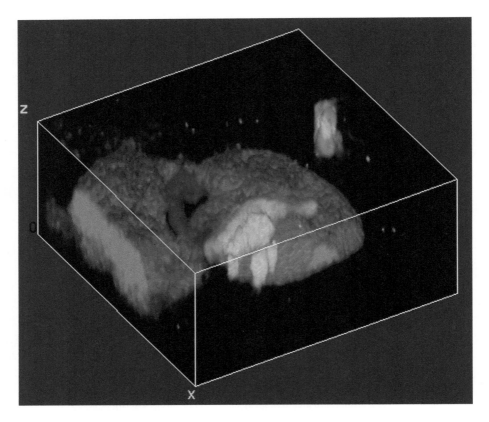

Figure 5.80 3D confocal Raman image of a fluid inclusion in garnet (60 × 60 × 30 μm; red, garnet; blue, water; green, calcite; turquoise, mica). In the gray scale used, the garnet areas are the large areas on the left and right; water is the diamond-shaped dark area in the center (follow the line from the corner x up to the center of the image); calcite is the separate bright spot on the far right. (Courtesy of WITec GmbH. Used with permission.)

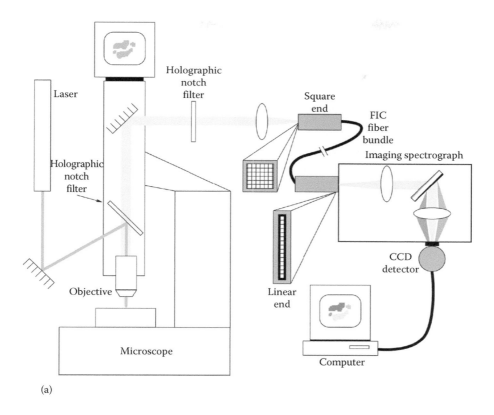

Figure 5.81 Schematic of the NIRIM. (Reprinted from McLain, B.L. et al., *Spectroscopy*, 15, 28, 2000. With permission from Advanstar Communications, Inc.)

surfaces, among other studies. Two examples are shown in Figures 5.82 and 5.83.

Silicon deposited on glass or silicon carbide is widely used in manufacturing photovoltaic cells. It is critical to monitor both the proportion and distribution of amorphous and crystalline silicon in such materials. Raman microscopy is ideal for this application as the two forms are readily distinguishable and permit simple quantification using Beer's law (Deschaines, 2009). Raman spectroscopy using a Thermo Scientific DXR Raman microscope with a 532 nm laser permitted the authors to quantify the relative amounts of amorphous and crystalline silicon in thin-layer deposits. Crystalline silicon has a sharp band at 521 cm^{-1}, while amorphous silicon has a broadband centered at 480 cm^{-1}, as seen in Figure 5.84.

A simple Beer's law calculation based on the ratio of the respective peak heights permits quantitation. In addition, a 2D map was collected, showing the distribution of the two forms of silicon. This can be used as a QC check during manufacturing. (The interested student can find the full-color image in the technical note listed in the bibliography at www.thermo.com; it does not convert to gray scale very well.) One point noted by the authors is the need for a laser power regulator (present in the DXR Raman microscope) because too high a laser power can convert amorphous Si to crystalline Si, which would result in an inaccurate analysis.

A second combination of two techniques provides a solution to the problem of obtaining the Raman spectrum from a rough surface. On rough or inclined samples, a high level of confocality can cause part of the scanned area to be out of focus. This is especially true for larger scans on the mm scale because it can become quite a challenge to achieve a flatness of only a few μm over such a large area. Decreasing the confocality and thus making the system less sensitive to variations due to a lower depth resolution is a solution if the topographic changes are small (i.e., <10 μm), but even in those cases, it is not desirable due to the loss in surface sensitivity. Especially when looking at intensity changes of peaks or when working with transparent samples, this option should be avoided. However, without such a compromise or if the topographic variation is too large, it is only possible to image a slice of the sample in one scan in conventional confocal Raman imaging systems (Figure 5.85, top). For samples with high optical density for the wavelength of choice (=low penetration depth), one could in this fashion acquire many slices and then reconstruct the surface, but this would be a very time-consuming procedure. In contrast, confocal Raman imaging in combination with True Surface Microscopy (WITec GmbH, www.witec.de) allows the user to remain in focus on the surface at every point and thus acquire the complete image in one scan (Figure 5.85, bottom).

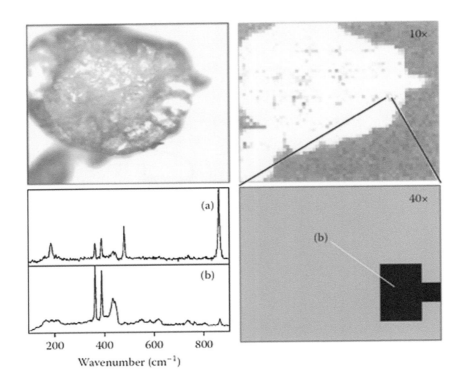

Figure 5.82 Identification of a small molybdenum sulfide inclusion (b) in a larger boric acid crystal (a) using the NIRIM system. (Reprinted from McLain, B.L. et al., *Spectroscopy*, 15, 28, 2000. With permission from Advanstar Communications, Inc.)

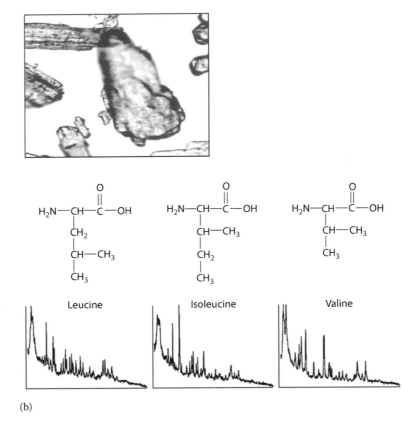

Figure 5.83 Imaging and identification of three different amino acid crystals in a mixture using the NIRIM system. (Reprinted from McLain, B.L. et al., *Spectroscopy*, 15, 28, 2000. With permission from Advanstar Communications, Inc.)

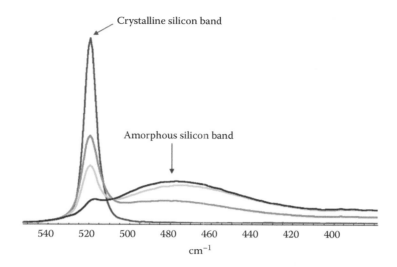

Figure 5.84   Raman spectra of silicon samples ranging from pure crystalline to one containing primarily amorphous silicon. The spectra show the sharp band at 521 cm^{-1} from crystalline silicon and the much broader band centered at approximately 480 cm^{-1} from amorphous silicon. Spectra were collected on a Thermo Scientific DXR Raman microscope using a 532 nm excitation laser. [From Deschaines, T.O. et al., Characterization of amorphous and microcrystalline silicon using Raman spectroscopy, Technical Note 51735, © Thermo Fisher Scientific, 2009 (www.thermofisher.com). Used with permission.]

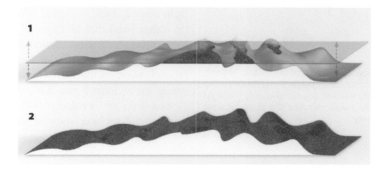

Figure 5.85   Top: (1) Confocal Raman imaging on large samples (mm size) delivers an optical cross section through the sample. If the sample is not transparent, only the intersection of focal plane and sample surface will give a Raman signal. Depending on the objective used, the focal plane might have a thickness of less than 1 μm. Bottom: (2) The True Surface Microscopy option enables the precise tracing of the surface while acquiring Raman imaging data, resulting in a true surface Raman image. (Courtesy of WITec GmbH, www.witec.de. Used with permission.)

True Surface Microscopy uses the principle of a confocal chromatic sensor as shown in Figure 5.86. A white-light point source is focused onto the sample with a hyperchromatic lens assembly, a lens system that has a good point mapping capability, but a strong linear chromatic error. Every color has therefore a different focal distance. The light reflected from the sample is collected with the same lens and focused through a pinhole onto a spectrometer. As only one specific color is in focus at the sample surface, only this light can pass through the confocal pinhole. The detected wavelength is therefore related to the surface topography. Scanning the sample in the XY plane reveals a topographic map of the sample. Such a map can be followed in a subsequent confocal Raman image scan so that the Raman excitation is always kept in focus with the sample surface (or at any chosen distance below the surface).

In order to illustrate the advantage of such a sensor, a highly inclined and very rough geological sample was investigated with True Surface Microscopy. A photograph of the sample (Josefsdal Chert 99SA07, courtesy of Frances Westall, CNRS Orleans, France) can be seen in Figure 5.87.

The scanned area was 2 × 2 mm with a maximum topographic range in the scanned area of approximately 1.3 mm. If such an area is scanned in confocal Raman imaging mode without correction of the topography, only a very limited range can be recorded, which is illustrated in Figure 5.88a. The same area was also scanned with the topographic data taken into account in order to adjust the Z-position at each image pixel such that the surface is in focus. The results (Figure 5.88b) clearly show much more detailed information of the sample surface.

The colors in Figure 5.88 match the colors of the corresponding Raman spectra shown in Figure 5.89. Quartz (blue) and kerogen (green) as well as anatase (red) could be found in this sample. In the gray scale used for Figure 5.88b, the brightest spots, especially the lower front right side are yellow; the green areas are along the bottom left and upper right sides, while the medium gray in the center area is quartz and the lighter spots in the center are anatase. Areas marked in yellow indicate high levels of fluorescence. Chemical imaging provides a nondestructive and noninvasive way to map the chemical composition of a wide variety of samples very rapidly and at the microscopic level. Applications range from real-time process monitoring and biomonitoring to materials characterization, engineering materials fabrication, and failure analysis. Chemical imaging systems are now available for about $150,000, making this technology feasible for even routine analysis.

5.10 PHOTOACOUSTIC SPECTOSCOPY

We will conclude this chapter with a discussion of photoacoustic spectroscopy (PAS), based on the photoacoustic effect, discovered by Alexander Graham Bell, who is recognized as the inventor of the telephone. A brief description of the life of this remarkable man is provided in the inset. This topic is treated in this chapter because instrumentation for PAS is typically provided as an accessory for vibrational spectrophotometers.

The photoacoustic effect is the observation that a sound wave could be induced in matter by exposure to a pulsating light source. Bell's original observation in 1880 was made on a thin solid that he exposed to sunlight that was pulsed by an opaque, rotating, slotted disk. The sound wave was the result of the rapid heating of the solid that induced a pressure (sound) wave in the surrounding air. Bell initially used the sun as the incident beam, a foot-operated device to rotate the slotted disk, and a microphone to sense the acoustic waves produced. Bell later showed the same effect with infrared

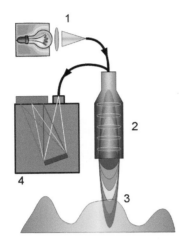

Figure 5.86 Schematic operation of a confocal chromatic sensor. White light (1) is focused through a hyperchromatic lens system (2) onto the sample (3) and only the light scattered in its focal plane reaches the spectrometer (4) with high intensity.

Figure 5.87 A 2 × 2 mm area at the indicated position was investigated on this inclined and rough rock sample while the sample was positioned under the microscope exactly as shown in the image. The height profile was measured by the integrated optical profilometer. (Courtesy of Frances Westall, CNRS Orleans, France and WITec GmbH. Used with permission.)

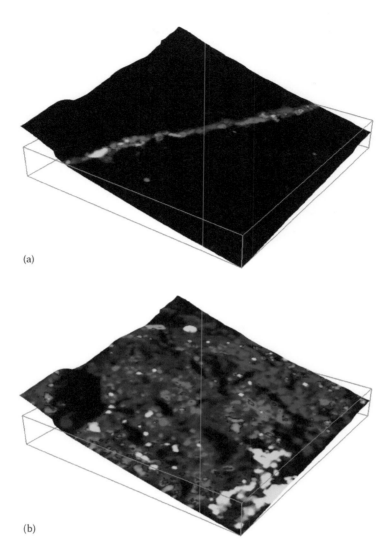

Figure 5.88 (a) Field of view of a confocal Raman imaging measurement without correcting for the topography and (b) true surface Raman image as a result of the height profile overlaid with the topography-traced confocal Raman image. (Courtesy of Frances Westall, CNRS, Orleans, France and WITec GmbH. Used with permission.)

and ultraviolet radiation. As with any spectroscopic method, one can vary the wavelength of the incident radiation and follow the acoustic signature produced by solids, liquids, and gases. The technique differs conceptually from other spectroscopic methods in that one does not study the effect of matter on the incident beam (such as through absorption), but rather one studies the effect on the sample itself, directly. The incident beam is converted into translational energy in the sample which then causes the local heating. Modulating or pulsing the incident beam or the absorption by the sample produces the pressure wave modulation. A conceptual schematic of the method is provided in Figure 5.90.

PAS differs operationally from absorption spectroscopic techniques in that typical absorption spectra are measured indirectly. One measures the ratio of incident vs. transmitted radiation in normal spectroscopic methods. Because of this, it is not possible to discriminate between absorption and scattering. Since at low concentrations of analytes, the transmitted beam is intense, and for concentrated analytes, the transmitted beam is weak, the requirements for traditional spectroscopy are more demanding because of the inherent noise. PAS is not encumbered with these difficulties, and is more akin to calorimetry.

It was the general availability of lasers that made for a revolution in the application of PAS. The acoustic signal is proportional to the incident beam power, the spectral brightness of the laser has allowed for detection of very low acoustic signals and the measurement of trace constituents. While PAS is not as prevalent in the analytical arena as the other approaches covered in this book, it is on the verge of real commercial impact for the analysis of gases, liquids, solids, and biological and living tissue. In the future, we

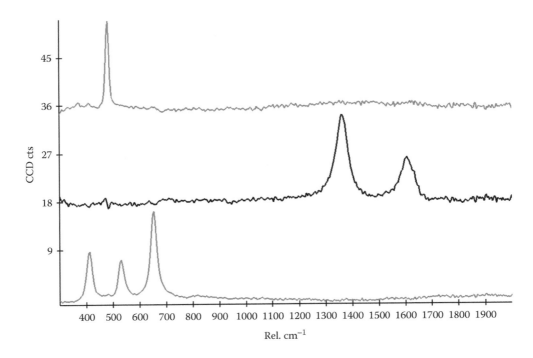

Figure 5.89 Top spectrum in blue, quartz; middle spectrum in green, kerogen; bottom spectrum in red, anatase. (Courtesy of Frances Westall, CNRS, Orleans, France and WITec GmbH. Used with permission.)

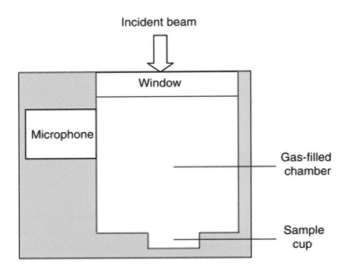

Figure 5.90 A conceptual diagram showing the principle of photoacoustic spectroscopy. Reproduced with permission of Elsevier.

anticipate that PAS will require not merely a short section in a chapter on absorption spectroscopy, but rather a fully dedicated chapter. Here we will summarize the principal features focusing only on gas analysis.

Continuous light sources can be used to study any optoacoustic effect (of which PAS is one), but the detection of modulated or pulsed effects is easier to process. Modulation is achieved by use of the chopper described above with Bell's early experiments, and used in many forms of spectroscopy. Note here that this is intensity modulation, whereas it is also possible to vary the wavelength (wavelength modulation). Focusing on gaseous samples, modulated light is passed into a chamber or cell containing the stationary or flowing gas. This light will change (excite) the populations of vibrational, rotational, and electronic states of the gas molecules, and the excited states will decay by radiative emissions, chemical reactions, and collisional processes, with radiative emission and chemical reaction being unlikely. Most of the energy released will cause enhanced collisions, and therefore heat. The resulting pressure wave will be detected as sound. To make this process more sensitive, the gas cell is actually an acoustic resonator, and the chopper matches the resonant frequency of the resonator. An example of this can be seen in Figure 5.91.

We have mentioned the two approaches of modulation and pulsing but have yet to discuss what this means. In Figure 5.91, we see that the chopper is modulated or tuned to the resonant frequency of the cell. If, instead, we use a pulsed tunable laser, tuned to the resonant frequency of the cell, then pulse mode becomes possible. This is far more sensitive since the decay process becomes essentially adiabatic. Coupling pulsed excitation with the electronics allows PAS to be used in severe environments such as refinery streams and even in test engines under full load. The noise that would plague traditional absorption spectra is avoided. It is possible to measure trace gas concentrations to the part per trillion (vol/vol) level with PAS. It is clear that in the future, PAS will play more important roles in industry and clinical settings.

Alexander Graham Bell (1847–1922) was a Scottish-American inventor and applied scientist who is best known for first patenting the telephone. His accomplishments went far beyond this, however. He did seminal work in acoustics, optics, hydrofoil design, and aeronautics. His inventiveness began in childhood, when with a friend he invented a machine with nails and wooden paddles that could remove the husk from wheat, a device that was used in a flour mill for several years. His early years were influenced by his family who were interested in elocution and speech (his mother and wife were deaf). He had a great sense of humor as well, training the family dog to growl while the young Bell manipulated the dog's lips to make it appear that the dog could talk! Bell later completed his formal education at University College of London, and obtained a teaching position at Boston University. In addition to his work on the telephone, Bell developed an early metal detector, high-speed hydrofoils, and the first acoustic spectrometer. Interestingly, he had no love for the telephone himself and considered it to be an intrusion; he did not have one in his office. His work led to the formation of the Bell Telephone Company in 1887, and he was the co-founder of the American Telephone and Telegraph Company (AT&T) in 1885. He was the second president of the National Geographic Society and one of the founders of *Science* magazine. He died in 1922 from complications of diabetes, and on the day of his burial, all telephone service in the U.S. was suspended.

Alexander Graham Bell at the opening of long-distance telephone service between New York City and Chicago in 1892. Prints and Photographs Division, Library of Congress. Reproduction Number LC-G9-Z2-28608-B.

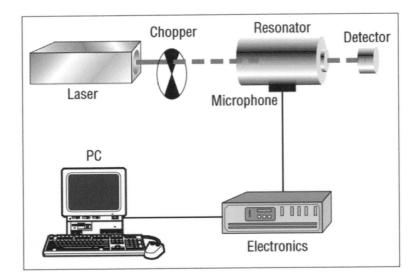

Figure 5.91 An example of a gas phase PAS instrument with a continuous wave laser modulated with a chopper. The chopper is tuned to the resonant frequency of the resonator. The associated electronics includes a lock in amplifier to reduce noise. Reproduced with permission of the American Chemical Society.

SUGGESTED EXPERIMENTS

5.1 Record the IR absorption spectrum of hexane. Identify the absorption bands caused by the C–H stretching frequency, the C–H bending frequency, and the C–C stretching frequency.

5.2 Record the IR spectrum of heptane. Note the similarity with the spectrum obtained in Experiment 5.1. Would it be possible to distinguish between these compounds based on their IR spectra?

5.3 Record the IR spectrum of *n*-butanol. Note the O → H stretching peak and the C → OH peak. Repeat with *iso*-butanol and *tert*-butanol. Note any changes in the positions of these peaks. Is there a peak that can be used to distinguish among primary, secondary, and tertiary alcohols?

5.4 Record the IR spectrum of *n*-butylamine. Note the N → H and C → N stretching peaks. Repeat with *sec*-butylamine and *tert*-butylamine. Compare the spectra to what you expect from your reading.

5.5 Place a few drops of a volatile organic liquid (toluene, xylene, carbon tetrachloride, etc.) in a 10 cm gas cell and close the valves. Allow the cell to sit for several minutes. Record the gas phase spectrum of the compound. Compare the spectrum to the liquid phase spectrum and explain your observations.

5.6 Make up several different solutions of known concentrations of ethanol in carbon tetrachloride over the concentration range of 1 %–30 %. Measure the intensity of the C–OH absorption band. Plot the relationship between absorbance and the concentration of ethanol. (Alternatively, run the quantitative "oil and grease" method from Standard Methods for the Examination of Water and Wastewater or the similar EPA method 413.2 or 418.1.)

5.7 Repeat Experiment 5.6 using acetone as the sample. Note the sharp C ⇒ O stretching band and use the absorbance of this band for quantitative studies as suggested in Experiment 5.6.

5.8 Take a sample of unsaturated cooking oil and record the IR absorption spectrum. Note the C ⇒ C stretching frequency. Repeat with several brands of cooking oil. Based on the IR absorption spectrum, which brand was the most unsaturated?

5.9 Check the calibration of your instrument by comparing the wavelengths measured against a standard spectrum, provided below:

Polystyrene Wavenumber Calibration

The following are wavenumber readings assigned to the peaks on the spectrum:

1	–	3027.1	8	–	1583.1
2	–	2924.0	9	–	1181.4
3	–	2850.7	10	–	1154.3
4	–	1944.0	11	–	1069.1
5	–	1871.0	12	–	1028.0
6	–	1801.6	13	–	906.7
7	–	1601.4	14	–	698.9

Film Thickness: 50 μm

INFRARED, NEAR-INFRARED, RAMAN, AND PHOTOACOUSTIC SPECTROSCOPY

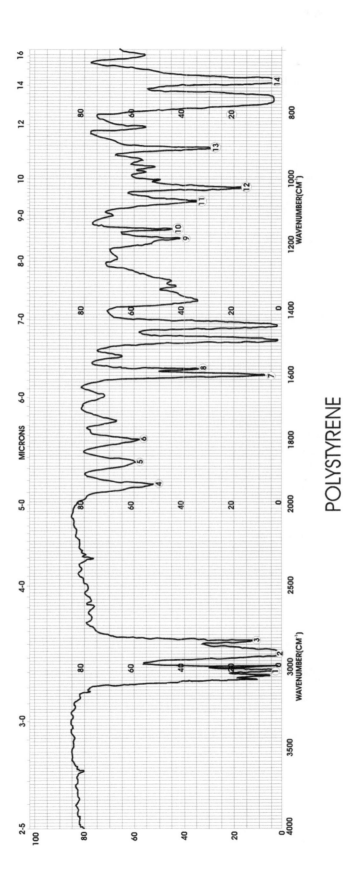

POLYSTYRENE

PROBLEMS

5.1 What types of vibrations are encountered in organic molecules?

5.2 What materials are used for making the cell windows used in IR spectroscopy? Why must special precautions be taken to keep these materials dry?

5.3 Why do organic functional groups resonate at characteristic frequencies?

5.4 How can primary, secondary, and tertiary alcohols be distinguished by their IR absorption spectra?

5.5 Indicate which C–H and C–C stretching and bending vibrations can be detected in an IR absorption trace over the range of 2.5–16 μm. What would be the frequency of each vibration?

5.6 In discussing quantitative analysis using the mid-IR region, it was suggested that either the OH stretching band or the C–O stretching band could be used to measure hexanol in a mixture of hexanol and hexane. Which band would you choose to give more accurate results? Why?

5.7 A solution is known to contain acetone and ethyl alcohol. Draw the expected IR absorption curve for each compound separately. Which absorption band could be used to identify the presence of acetone in the mixture?

5.8 What requirements must be met before a molecule will absorb IR radiation?

5.9 In preparing a calibration curve for the determination of methyl ethyl ketone (MEK), solutions with different concentrations of MEK were prepared in chloroform. The absorbance at the C=O stretching frequency was measured. The measured absorbance A for each solution is given. A blank (the pure solvent) had an absorbance = 0.00:

MEK Concentration (%)	Absorbance
2	0.18
4	0.36
6	0.52
8	0.64
10	0.74

(a) Does the relationship between A and sample concentration deviate from Beer's law? (b) Several unknown samples containing MEK were measured at the same wavelength as that used for the calibration curve. The results were as follows:

Sample	Absorbance
A	0.27
B	0.44
C	0.58
D	0.69

What were the concentrations of MEK in the solutions A–D?

5.10 Explain what is meant by the Fellgett advantage.

5.11 What are (a) fundamental and (b) overtone vibrational bands?

5.12 Is FTIR a single-beam or double-beam technique? How is background correction achieved?

5.13 How does a semiconductor detector such as an MCT detector compare to a thermocouple detector for use in IR spectroscopy?

5.14 Describe the components of an FTIR spectrometer. Which detectors are used for FTIR?

5.15 List the advantages of FTIR over dispersive IR spectroscopy.

5.16 Describe how ATR works to give an IR absorption spectrum.

5.17 Give two examples of the use of FTIR microscopy for chemical analysis.

5.18 What wavelength range is covered by NIR? What bands occur here?

5.19 What is the advantage of using NIR compared with mid-IR? What are the disadvantages?

5.20 What is meant by the term "virtual state?"

5.21 Diagram the processes that give rise to Rayleigh, Stokes, and anti-Stokes scattering.

5.22 What are the requirements for a molecule to be Raman active?

5.23 Explain why fluorescence is a problem in normal Raman spectroscopy. Give two examples of how the fluorescence interference in Raman spectroscopy can be minimized or eliminated.

5.24 Describe the resonance Raman process and its advantages.

For Problems 5.25 through 5.41, deduce a reasonable structure for a compound consistent with the IR spectrum and the other information provided. (Spectra for these problems are Copyright Bio-Rad Laboratories, Informatics Division, Sadtler Software and Databases, 2014. All rights reserved.)

5.25 The RMM of the unknown = 86.

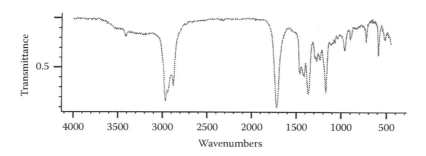

5.26 The RMM of the unknown = 128.

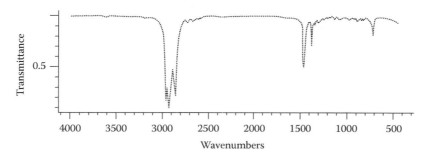

5.27 The RMM of the unknown = 46.

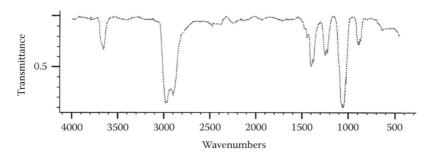

5.28 The RMM of the unknown = 154.

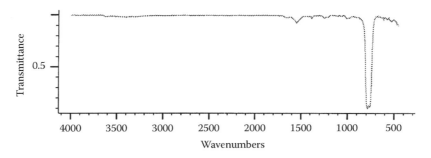

5.29 The RMM of the unknown = 106.

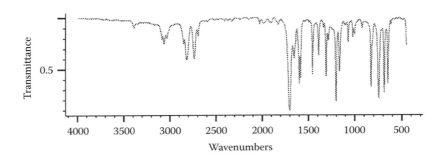

5.30 The RMM of the unknown = 93 (Hint: note that the RMM is an odd number).

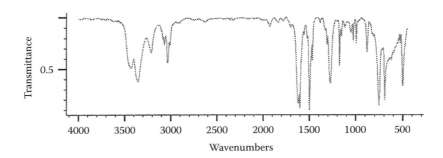

5.31 The RMM of the unknown = 284.

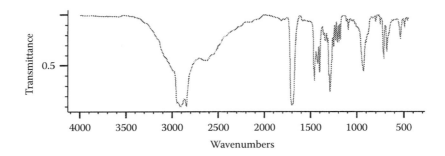

5.32 The RMM of the unknown = 107.

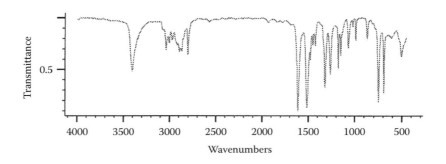

5.33 The RMM of the unknown = 104. Molecular formula from elemental analysis is $C_3H_6NO_3$.

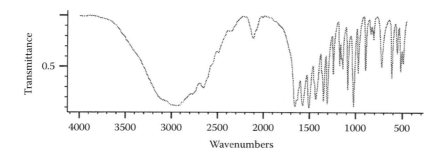

5.34 The RMM of the unknown = 153. Molecular formula from elemental analysis is $C_8H_{11}NO_2$.

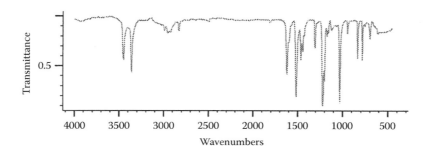

5.35 The RMM of the unknown = 60.

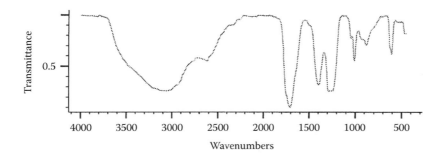

5.36 The RMM of the unknown = 124 (Hint: the molecule contains a heteroatom).

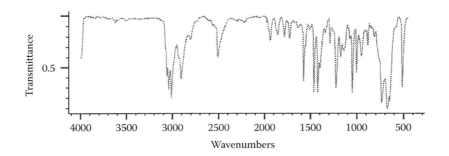

5.37 The RMM of the unknown = 87.

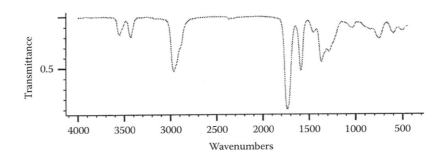

5.38 The RMM of the unknown = 126.

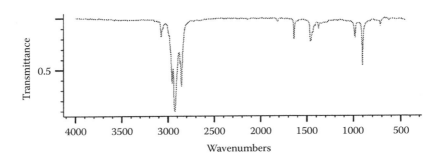

5.39 The RMM of the compound = 149.5. There are two types of heteroatoms (atoms other than C and H) in the compound.

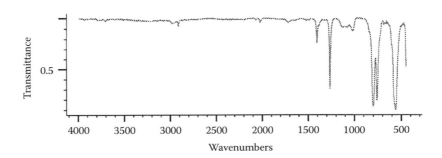

5.40 The RMM of the unknown = 160 (Hint: there are more C atoms than H atoms).

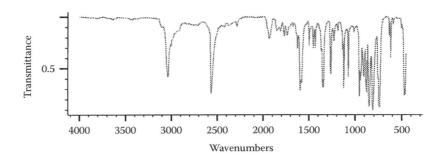

5.41 You have three compounds, A, B, and C, giving rise to three spectra, a, b, and c, respectively. Note that spectrum a and spectrum b look very similar. (But they are not identical; an easy way to see what the differences are is to make a copy of each spectrum, overlay them, and hold the pages up to a light. Align the baselines so that they match, and spectral differences will be more apparent.) The RMM of Compound A is 102. The RMM of Compound B is 114. Compound C also has RMM = 114 and has the exact same molecular formula as Compound B. However, spectrum b and spectrum c are very different. Draw three plausible structures for Compounds A, B, and C.

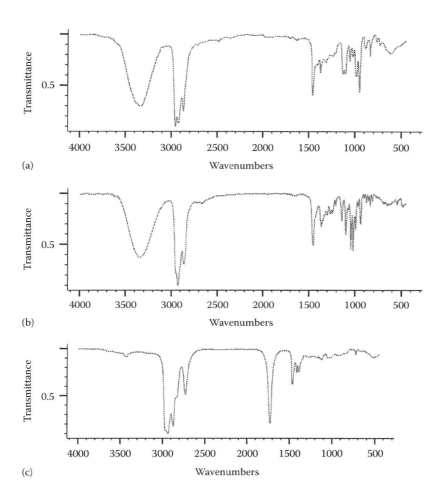

5.42 What types of acoustical resonators are commonly encountered? Hint: ask a classmate who plays a musical instrument. What are the laboratory applications of acoustic resonators? Hint: investigate the measurement of heat capacity of fluids, and the most accurate measurement of the universal gas constant, R.

BIBLIOGRAPHY

The reader will note that occasionally multiple editions of books are cited as sources or reference works. The reason for this is that sometimes different figures are used in different editions, which may in fact be written by different authors. The sources found in the bibliography will be the most recent versions of a particular work.

In some cases instrument manufacturers and suppliers are cited by providing a URL to the exact internet location. We have endeavored to ensure that these links are functional at the time of publication, but clearly these links change. If a link has become nonfunctional, a good approach is to use your browser's advanced search capability. For example: "company name" AND "instrument type" AND "specific descriptors" is a good place to start.

Aikens, D.A.; Bailey, R.A.; Moore, J.A.; Giachino, G.G.; Tomkins, R.P.T. *Principles and Techniques for an Integrated Chemistry Laboratory*. Waveland Press, Inc.: Prospect Heights, IL, 1978. (Reissued 1984).

Asher, S.A. UV resonance Raman spectroscopy for analytical, physical and biophysical chemistry, Pts.1 and 2. *Anal. Chem.*, 65(2), 59A–66A and 201A–210A, 1993.

Briggs, J.L. Hollow waveguides: The next generation of mid-IR remote sampling accessories. *FT-IR Technology for Today's Spectroscopists*, August 2010, a supplement to Spectroscopy, 2010.

Bruno, T.J., Svoronos, P.D.N., *CRC Handbook of Basic Tables for Chemical Analysis – Data Driven Methods and Interpretation*, 4th Ed., CRC Taylor and Francis, Boca Raton, FL, 2021.

Burns, D.R.; Ciurczak, E.W. (eds.) *Handbook of Near Infrared Analysis*. 4th ed., CRC Press, Boca Raton, 2021.

Carroll, J.E.; He, H.; Landa, I. *The NIR Desk Reference*. LT Industries, Inc.: Rockville, MD, 1998.

Coates, J. Vibrational spectroscopy. In Ewing, G.W. (ed.), *Analytical Instrumentation Handbook*, 4th edn. CRC Press, Boca Raton, 2018..

Colthup, N.B.; Daly, L.H.; Wiberley, S.E. *Introduction to Infrared and Raman Spectroscopy*, 3rd edn. Academic Press: New York, 1990.

Cooke, P.M. Chemical microscopy. *Anal. Chem.*, 68(12), 339R, 1996.

De Blase, F.J.; Compton, S. IR emission spectroscopy: A theoretical and experimental review. *Appl. Spectrosc.*, 45(4), 611, 1991.

Deschaines, T.O.; Hodkiewicz, J.; Henson, P. Characterization of amorphous and microcrystalline silicon using Raman spectroscopy. Technical Note 51735, Thermo Fisher Scientific, Inc.: Madison, WI, 2009.

Deschaines, T.O.; Wieboldt, D. Practical applications of surface-enhanced Raman scattering (SERS). Technical Note 51874, Thermo Fisher Scientific, Inc.: Madison, WI, 2010.

Dieing, T.; Hollricher, O.; Toporski, J., (eds.) *Confocal Raman Microscopy*, Springer Series in Optical Sciences, Vol. 158, 1st edn. Springer Verlag: Berlin/Heidelberg, Germany, 2011.

Edwards, H.G.M. Raman microscopy in art and archeology. *Spectroscopy*, 17(2), 16, 2002.

Ferraro, J.R.; Nakamoto, K. *Introductory Raman Spectroscopy*. Academic Press: New York, 1994.

Fleischmann, M.; Hendra, P.J.; McQuillan, A.J. *Chem. Phys. Lett.*, 26, 163, 1974.

Garrell, R.L. Surface-enhanced Raman spectroscopy. *Anal. Chem.*, 61(6), 401A, 1989.

Goddu, R.F. Near infrared spectrophotometry. In Reilly, C.N. (ed.), *Advances in Analytical Chemistry and Instrumentation*, Vol. 1. John Wiley & Sons, Inc.: New York, 1960.

Grasselli, J.G.; Bulkin, B.J. (eds.) *Analytical Raman Spectroscopy*. John Wiley & Sons, Inc.: New York, 1991.

Grasselli, J.G.; Snavely, M.K.; Bulkin, B.J. *Chemical Applications of Raman Spectroscopy*. John Wiley & Sons, Inc.: New York, 1981.

Greenberg, A.E.; Clesceri, L.S.; Eaton, A.D. (eds.) *Standard Methods for the Examination of Water and Wastewater*, 18th edn. American Public Health Association: Washington, DC, 1992.

Gremlich, H.-U.; Yan, B. (eds.) *Infrared and Raman Spectroscopy of Biological Materials*. Marcel Dekker, Inc.: New York, 2001.

Griffiths, P.R.; de Haseth, J.A. *Fourier Transform Infrared Spectrometry*. Chemical Analysis, Vol. 83. Wiley-Interscience: New York, 1986.

Heil, C. Rapid, multi-component analysis of soybeans by FT-NIR spectroscopy. Thermo Fisher Scientific application note, www.thermo.com, ©2010 Thermo Fisher Scientific, Madison, WI (www.thermofisher.com).

Herrero, A.M.; Carmona, P.; Careche, M. Raman spectroscopic study of structural changes in Hake (Merluccius merluccius L.) muscle proteins during frozen storage. *J. Agric. Food Chem.*, 52(8), 2147, 2004.

Hsu, C.-P.S. Infrared spectroscopy. In Settle, F.A. (ed.), *Handbook of Instrumental Techniques for Analytical Chemistry*. Prentice Hall PTR: Upper Saddle River, NJ, 1997.

Humecki, H.J. (ed.) *Practical Guide to Infrared Microspectroscopy*. Practical Spectroscopy Series, Vol. 19. Marcel Dekker, Inc.: New York, 1995.

Ingle, J.D., Jr.; Crouch, S.R. *Spectrochemical Analysis*. Prentice Hall, Inc.: Englewood Cliffs, NJ, 1988.

Kawai, N.; Janni, J.A. Chemical identification with a portable Raman analyzer and forensic spectral database. *Spectroscopy*, 15(10), 32, 2000.

Kneipp, K.; Moskvits, M.; Kneipp, H. (eds.) *Surface-Enhanced Raman Scattering: Physics and Applications*. Topics in Applied Physics, Vol. 103. Springer: New York, 2006.

Lambert, J.B.; Shurvell, H.F.; Lightner, D.; Cooks, R.G. *Introduction to Organic Spectroscopy*. Macmillan Publishing Company: New York, 1987.

Lowry, S. *Using FT-IR Spectroscopy to Analyze Gemstones*: Application Note 51123. Thermo Fisher Scientific: Madison, WI, 2006.

Marbach, R.; Kosehinsky, T.; Gries, F.A.; Heise, H.M. Noninvasive blood glucose assay by NIR diffuse reflectance spectroscopy of the human inner lip. *Appl. Spectrosc.*, 47(7), 875, 1983.

McLain, B.L.; Hedderich, H.G.; Gift, A.D.; Zhang, D.; Jallad, K.; Haber, K.S.; Ma, J.; Ben-Amotz, D. Fast chemical imaging. *Spectroscopy*, 15(9), 28, 2000.

Metzel, D.L.; LeVine, S.M. In-situ FTIR microscopy and mapping of normal brain tissue. *Spectroscopy*, 8(4), 40, 1993.

Mirabella, F.M. (ed.) *Internal Reflection Spectroscopy: Theory and Applications*. Practical Spectroscopy Series, Vol. 15. Marcel Dekker, Inc.: New York, 1992.

Morris, M.D. (ed.) *Microscopic and Spectroscopic Imaging of the Chemical State*. Marcel Dekker: New York, 1993.

Nakamoto, K. *Infrared and Raman Spectra of Inorganic and Coordination Compounds*, 5th edn. John Wiley & Sons, Inc.: New York, 1996.

Pattacini, S. *Solving Analytical Problems Using Infrared Spectroscopy Internal Reflectance Sampling Techniques*. Pattacini Associates, LLC: Danbury, CT, 1995.

Pavia, D.L.; Lampman, G.M.; Kriz, G.S. *Introduction to Spectroscopy*, 3rd edn. Harcourt College Publishers: New York, 2001.

Raman, C.V. *Indian J. Phys.*, 2, 387, 1928.

Richard, S.; Tang, P.L. Identification and evaluation of coatings using hand-held FTIR, Publication number 5990-8075EN. Agilent Technologies, Inc., Santa Clara, CA, 2011.

Robinson, J.W. (ed.) *Handbook of Spectroscopy*, Vol. II. CRC Press: Boca Raton, FL, 1974.

Robinson, J.W. *Practical Handbook of Spectroscopy*. CRC Press: Boca Raton, FL, 1991.

Robotham, C.; Izzia, F. *FT-IR Microspectroscopy in Forensic and Crime Lab Analysis*. Application Note 51517, Thermo Fisher Scientific, Inc.: Madison, WI, 2008.

Rumble, J. R., ed., CRC Handbook of Chemistry and Physics, 102nd ed. CRC Taylor and Francis, Boca Raton, FL, 2021.

Schultz, C.P. Precision infrared spectroscopic imaging. *Spectroscopy, 16*(10), 24, 2001.

Sellors, J. ATR imaging of laminates. *Am. Lab.*, *40*(8), 24, 2008.

Silverstein, R.M.; Webster, F.X. *Spectrometric Identification of Organic Compounds*, 6th edn. John Wiley & Sons, Inc.: New York, 1998.

Smith, A.L. *Infrared Spectroscopy, Treatise on Analytical Chemistry*, Part 1, Vol. 7. John Wiley & Sons, Inc.: New York, 1981.

Strommen, D.P. Raman spectroscopy. In Settle, F.A. (ed.), *Handbook of Instrumental Techniques for Analytical Chemistry*. Prentice Hall PTR: Upper Saddle River, NJ, 1997.

von Rosenberg, Jr., C.W.; Abbate, A.; Drake, J.; Mayes, D.M. A rugged near-infrared spectrometer for the realtime measurement of grains during harvest. *Spectroscopy, 15*(6), 35, 2000.

Weesner, F.; Longmire, M. Dispersive and Fourier transform Raman. *Spectroscopy, 16*(2), 68, 2001.

SPECTRAL DATABASES

This list is by no means complete. Many instrument manufacturers offer spectral databases packaged with their IR and Raman instruments, with specialized subsets of spectra for pharmaceutical, geological, forensics, law enforcement, hazardous materials, and other applications. Many government agencies, including the US NIST (www.nist.gov and webbook.nist.gov), the US EPA (www.usepa.gov), and the Georgia State Crime Lab publish general and specialized spectral databases. The NIST Chemistry WebBook has IR, Raman, and mass spectra for a wide variety of compounds accessible for free. The Canadian government publishes a forensic spectral database.

Bio-Rad Laboratories, Informatics Division, Philadelphia, PA (www.bio-rad.com), publishes the Sadtler IR and Raman spectra collections of over 240,000 spectra. They are available in electronic format and in a variety of specialized subsets, including ATR-IR, with collections of controlled and prescription drugs, nutraceuticals, forensics, polymers, and more. Sigma-Aldrich Chemical Company (www.sigma-aldrich.com) publishes over 18,000 FTIR spectra. National Institute of Advanced Industrial Science and Technology, Tsukuba, Ibaraki, Japan, publishes a free spectral database system of organic compounds. The spectra include IR, Raman, NMR, and MS for most compounds. The database may be found at www.aist.go.jp/RIODB/SDBS.

CHAPTER 6

Magnetic Resonance Spectroscopy

6.1 INTRODUCTION TO NUCLEAR MAGNETIC RESONANCE SPECTROSCOPY

Nuclear magnetic resonance (NMR) spectroscopy is one of the most powerful techniques available for studying the structure of molecules. The NMR technique has developed very rapidly since the first commercial instrument, a Varian HR-30, was installed in 1952 at the Humble Oil Company in Baytown, Texas. These early instruments with small magnets were useful for studying protons (^{1}H) in organic compounds, but only neat liquids or solutions with a high concentration of analyte. That has now changed—much more powerful magnets are available. NMR instruments and experimental methods are now available that permit the determination of the 3D structure of proteins as large as 900,000 Da. "Magic angle" NMR instruments are commercially available for studying solids such as polymers, and ^{13}C, ^{19}F, ^{31}P, ^{29}Si, and other nuclei are measured routinely. NMR imaging techniques under the name *magnetic resonance imaging* (MRI) are in widespread use in noninvasive diagnosis of cancer and other medical areas. NMR instruments coupled with liquid chromatographs and mass spectrometers for separation and characterization of unknowns are commercially available. Resolution and sensitivity have both increased; detection and identification of ppm concentrations of substances with NMR is easily achieved using modern instruments, and detection limits are approaching nanogram levels. NMR detection is being coupled with liquid chromatographic separation in high-performance liquid chromatography (HPLC)–NMR instruments for identification of components of complex mixtures in the flowing eluent from the chromatograph, and NMR is now used as a nondestructive detector combined with mass spectrometry (MS) and chromatography in HPLC–NMR-MS instruments, an extremely powerful tool for organic compound separation and identification. In short, the field has broadened greatly in scope, especially since the 1970s, and gives every indication of continuing to advance for many years to come.

The NMR phenomenon is dependent upon the magnetic properties of the nucleus. Certain nuclei possess spin angular momentum, which gives rise to different spin states in the presence of a magnetic field. Nuclei with $I = 0$, where I is the spin quantum number, including all nuclei with both an even atomic number and an even mass number, such as ^{16}O and ^{12}C, do not have a magnetic moment and therefore do not exhibit the NMR phenomenon. In theory, all other nuclei can be observed by NMR.

NMR involves the absorption of radio waves by the nuclei of some combined atoms in a molecule that is located *in a magnetic field*. NMR can be considered a type of absorption spectroscopy, not unlike ultraviolet/visible (UV/VIS) absorption spectroscopy. Radio waves are low-energy electromagnetic radiation with frequencies on the order of 10^7 Hz. The SI unit of frequency, 1 Hz, is equal to the older frequency unit 1 cycle per second (cps) and has the dimension of inverse seconds, s^{-1}. The energy of radiofrequency (RF) radiation can therefore be calculated from

$$E = h\nu$$

where Planck's constant h is 6.626×10^{-34} J s and ν (the frequency) is between 4 and 1000 MHz (1 MHz = 10^6 Hz).

The quantity of energy involved in RF radiation is very small. It is too small to vibrate, rotate, or electronically excite an atom or a molecule. It is great enough to affect the nuclear spin of atoms in a molecule. As a result, spinning nuclei of *some* atoms in a molecule *in a magnetic field* can absorb RF radiation and change the direction of the spinning axis. In principle, each chemically distinct atom in a molecule will have a different absorption frequency or **resonance** if its nucleus possesses a *magnetic moment*. The analytical field that uses absorption of RF radiation by such nuclei in magnetic fields to provide information about a sample is NMR spectroscopy.

In analytical chemistry, NMR is a technique that enables us to study the shape and structure of molecules. In particular, it reveals the different chemical environments of the NMR-active nuclei present in a molecule, from which we can ascertain the structure of the molecule. NMR provides information on the spatial orientation of atoms in a molecule. If we already know what types of compounds are present, NMR can provide a means of determining how much of each is in the mixture. It is thus a method for both qualitative and quantitative analysis, particularly of organic compounds. In addition, NMR is used to study chemical equilibria, reaction kinetics, motion of molecules, and intermolecular interactions.

Three Nobel Prizes have been awarded in the field of NMR. The first was in 1952 to the two physicists, E. Purcell and F. Bloch, who demonstrated the NMR effect in 1946. In 1991, R. Ernst and W. Anderson were awarded the Nobel Prize for developing pulsed Fourier transform (FT)-NMR and 2D NMR methods between 1960 and 1980. FT-NMR and 2D experiments form the basis of most NMR experiments run today, even in undergraduate instrumental analysis laboratories. We will use the acronym NMR to mean FT-NMR, since this is the predominant type of NMR instrument currently produced. The 2002 Nobel Prize in Chemistry was shared by three scientists for developing methods to use NMR and MS (MS is discussed in Chapters 10 and 11) in the analysis of large biologically important molecules such as proteins. K. Wüthrich, a Swiss professor of molecular biophysics, received the prize for his work in determining the 3D structure of proteins using NMR.

Since the 1970s, the technology associated with NMR has advanced dramatically. The theory, instrument design, and mathematics that make NMR so powerful are complex; a good understanding of quantum mechanics, physics, and electrical engineering is needed to understand modern NMR experiments. Fortunately, we do not need to completely understand the theory in order to make use of the technique. This chapter will address NMR in a simplified approach using a minimum of mathematics.

6.1.1 Properties of Nuclei

To understand the properties of certain nuclei in an NMR experiment, we must assume that nuclei rotate about an axis and therefore have a **nuclear spin**, represented as I, the spin quantum number. In addition, nuclei are charged. The spinning of a charged body produces a **magnetic moment** along the axis of rotation. For a nucleus to give a signal in an NMR experiment, it must have a nonzero spin quantum number and must have a magnetic dipole moment.

As a nucleus such as ^1H spins about its axis, it displays two forms of energy. Because the nucleus has a mass and because that mass is in motion (it is spinning), the nucleus has spin angular momentum and therefore mechanical energy. So, the first form of energy is *mechanical energy*. The formula for the mechanical energy of the hydrogen nucleus is

$$\text{Spin angular momentum} = \frac{h}{2\pi}\sqrt{I(I+1)} \quad (6.1)$$

where I is the spin quantum number. For example, $I = 1/2$ for the proton ^1H.

The spin quantum number I is a physical property of the nucleus, which is made up of protons and neutrons. From observations of the nuclear spins of known nuclei, some empirical rules for predicting the spin quantum numbers can be tabulated. These rules are summarized in Table 6.1. For example, ^{12}C has atomic mass 12 and atomic number 6. Hence, it has six protons (atomic number = 6) and six neutrons (atomic mass − atomic number = no. of neutrons, so 12 − 6 = 6 neutrons). Since the mass and the number of protons are both even numbers, Table 6.1 predicts that the net spin quantum number of ^{12}C is zero, denoting no spin. Therefore, the spin angular momentum (Equation 6.1) is zero and ^{12}C does not possess a magnetic moment. Nuclei with $I = 0$ do not absorb RF radiation when placed in a magnetic field and therefore do not give an NMR signal. NMR cannot measure ^{12}C, ^{16}O, or any other nucleus with both an even atomic mass and an even atomic number.

For ^{13}C, on the other hand, the atomic mass is 13 (i.e., $P + N = 13$), an odd number, and the atomic number is 6, an even number. From Table 6.1, we predict that I for ^{13}C is therefore a half integer; like ^1H, ^{13}C has $I = 1/2$. So, NMR can detect ^{13}C, and although ^{13}C represents only 1.1 % of the total C present in an organic molecule, ^{13}C NMR spectra are very valuable in elucidating the structure of organic molecules.

The physical properties predict whether the spin number is equal to zero, a half integer, or a whole integer, but the actual spin number—for example, 1/2 or 3/2 or 1 or 2—must be determined experimentally. All elements in the first six rows of the periodic table have at least one stable isotope with a nonzero spin quantum number, except Ar, Tc, Ce, Pm, Bi, and Po. It can be seen from Table 6.1 and Appendix 11.A that many of the most abundant isotopes of common elements in the periodic table cannot be measured by NMR, notably those of C, O, Si, and S, which are very important components of many organic molecules of interest in biology, the pharmaceutical industry, the polymer industry, and the chemical manufacturing industry. Some of the more important elements that can be determined by NMR and their spin quantum numbers are shown in Table 6.2. The two nuclei of most importance to organic chemists and biochemists, ^{13}C and ^1H, both have a spin quantum number = 1/2.

Table 6.1 Rules predicting spin number of nuclei

Mass ($P + N$) (Atomic Mass)	Charge (P) (Atomic Number)	Spin Quantum Number (I)
Odd	Odd or even	1/2, 3/2, 5/2,...
Even	Even	0
Even	Odd	1, 2, 3

MAGNETIC RESONANCE SPECTROSCOPY

Table 6.2 **NMR-active nuclei and their quantum numbers**

Element Isotope	I	Element Isotope	I
^{13}C	1/2	^{35}Cl	3/2
^{17}O	5/2	^{37}Cl	3/2
^{1}H	1/2	^{79}Br	3/2
^{2}H (deuterium)	1	^{81}Br	3/2
^{3}H (tritium)	1/2	^{125}I	5/2
^{19}F	1/2	^{129}I	7/2
^{31}P	1/2	^{14}N	1
^{29}Si	−1/2	^{15}N	1/2
^{33}S	3/2	^{10}B	3
^{35}S	3/2	^{11}B	3/2

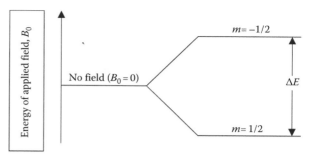

Figure 6.1 In the presence of an applied magnetic field, a nucleus with $I = 1/2$ can exist in one of two discrete energy levels. The levels are separated by ΔE. The lower energy level ($m = 1/2$) has the nuclear magnetic moment aligned with the field; in the higher energy state ($m = -1/2$), the nuclear magnetic moment is aligned against the field.

The second form of nuclear energy is *magnetic*. It is attributable to the electrical charge of the nucleus. Any electrical charge in motion sets up a magnetic field. The nuclear magnetic moment µ expresses the magnitude of the magnetic dipole. The ratio of the nuclear magnetic moment to the spin quantum number is called the **magnetogyric** (or gyromagnetic) **ratio** and is given the symbol, where $\gamma = (2\pi/h)(\mu/I)$. This ratio has a different value for each type of nucleus. The magnetic field of a nucleus that possesses a nuclear magnetic moment can and does interact with other local magnetic fields. The basis of NMR is the study of the response of such magnetically active nuclei to an external applied magnetic field.

6.1.2 Quantization of ^{1}H Nuclei in a Magnetic Field

When a nucleus is placed in a very strong, uniform external magnetic field B_0, it tends to become lined up in definite directions relative to the direction of the magnetic field. Each relative direction of alignment is associated with an energy level. Only certain well-defined energy levels are permitted; that is, the energy levels are **quantized**. Hence, the nucleus can become aligned only in well-defined directions relative to the magnetic field B_0. (*Note*: The symbol B is the SI symbol for magnetic field; many texts still use the symbols H and H_0 for magnetic field.)

The *number of orientations* or number of magnetic quantum states is a function of the physical properties of the nucleus and is numerically equal to $2I + 1$:

$$\text{Number of orientations} = 2I + 1 \quad (6.2)$$

In the case of ^{1}H, $I = 1/2$; hence, the number of orientations is $2 \times (1/2) + 1 = 2$. The permitted values for the magnetic quantum states, symbolized by the magnetic quantum number, m, are $I, I-1, I-2, \ldots, -I$. Consequently, for ^{1}H, only two energy levels are permitted, one with $m = +1/2$ and one with $m = -1/2$. The splitting of these energy levels in a magnetic field is called nuclear Zeeman splitting. (This is analogous to the electronic Zeeman effect, the splitting of electronic energy levels in a magnetic field. The electronic Zeeman effect is discussed in Section 6.12 and in Chapter 7.)

When a nucleus with $I = 1/2$, such as ^{1}H, is placed in an external magnetic field, its magnetic moment lines up in one of two directions, with the applied field or against the applied field. This results in two discrete energy levels, one of higher energy than the other, as shown in Figure 6.1. The *lower* energy is that where the magnetic moment is aligned *with* the field. The lower energy state is energetically more favored than the higher energy state, so the population of nuclei in the lower energy state will be higher than the population in the higher energy state. The difference in energy between levels is proportional to the strength of the external magnetic field. The axis of rotation also rotates in a circular manner about the external magnetic field axis, like a spinning top, as shown in Figure 6.2. This rotation is called **precession**. The direction of precession is either *with* the applied field B_0 or *against* the applied field.

So, we have nuclei, in this case, protons, with two discrete energy levels. In a large sample of nuclei, more of the protons will be in the lower energy state. The basis of the NMR experiment is to cause a transition between these two states by absorption of radiation. It can be seen from Figure 6.1 that a transition between these two energy states can be brought about by absorption of radiation with a frequency that is equal to ΔE according to the relationship $\Delta E = h\nu$.

The difference in energy between the two quantum levels of a nucleus with $I = 1/2$ depends on the applied magnetic field B_0 and the magnetic moment µ of the nucleus. The relationship between these energy levels and the frequency ν of absorbed radiation is calculated as follows. E is the expression for a given nuclear energy level in a magnetic field:

$$E = -m\left(\frac{\mu}{I}\right)B_0 = -m\left(\gamma \frac{h}{2\pi}\right)B_0 \quad (6.3)$$

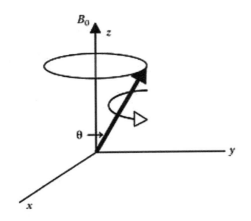

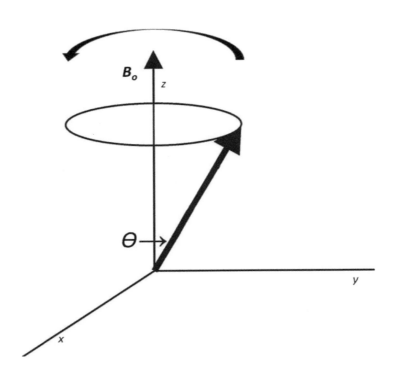

Figure 6.2 In the presence of an applied magnetic field, B_0, shown parallel to the +z-axis, a spinning nucleus precesses about the magnetic field axis in a circular manner. The nucleus is shown spinning counterclockwise (arrow with white head). The bold arrow with the black head represents the axis of rotation, which traces the circular path shown.

where
m is the magnetic quantum number
μ is the nuclear magnetic moment
B_0 is the applied magnetic field
I is the spin quantum number
γ is the magnetogyric ratio
h is Planck's constant

Equation 6.3 is the general equation for a given energy level for all nuclei that respond in NMR. However, if we confine our discussion to the ^{1}H nucleus, then $I = 1/2$. Therefore, there are only two levels.

$$\Delta E = E_{-\frac{1}{2}} - E_{+\frac{1}{2}} = h\nu \tag{6.4}$$

and

$$\Delta E = h\nu = \gamma \frac{h}{2\pi} B_0 = \frac{\mu}{I} B_0 = 2\mu B_0 \tag{6.5}$$

Therefore, the absorption frequency that can result in a transition of ΔE is

$$\nu = \gamma \frac{B_0}{2\pi} \qquad (6.6)$$

In NMR, it is common to express frequency in radians/s, ω, versus ν in Hz. Since one cycle = 2π radians, $\omega = 2\pi\nu$, therefore $2\pi\nu = \gamma B_0$. More commonly, Equation 6.6 is written as

$$\omega = \gamma B_0 \qquad (6.7)$$

where ω is the frequency in units of rad/s.

Equation 6.7 is the **Larmor equation**, which is fundamental to NMR. It indicates that for a given nucleus, there is a direct relationship between the frequency ω of RF radiation absorbed by that nucleus and the applied magnetic field B_0. This relationship is the basis of NMR.

The absorption process can be understood in terms of a classical approach to the behavior of a charged particle in a magnetic field. The spinning of the charged nucleus (Figure 6.2) produces an angular acceleration, causing the axis of rotation to move in a circular path with respect to the applied field. As already noted, this motion is called precession. The frequency of precession can be calculated from classical mechanics to be $\omega = \gamma B_0$, the Larmor frequency. Both quantum mechanics and classical mechanics predict that the frequency of radiation that can be absorbed by a spinning charged nucleus in a magnetic field is the Larmor frequency.

As shown in Figure 6.2, the axis of rotation of the precessing hydrogen nucleus is at an angle θ to the applied magnetic field. The energy of the precessing nucleus is equal to $E = \mu B_0 \cos\theta$. When energy in the form of RF radiation is absorbed by the nucleus, the angle θ must change. For a proton, absorption involves "flipping" the magnetic moment from being aligned with the field to being aligned against the applied field. When the rate of precession equals the frequency of the RF radiation applied, absorption of RF radiation takes place. The nucleus then becomes aligned *opposed* to the magnetic field and is in an excited state. To measure organic compounds containing protons by NMR, the sample is first put into a magnetic field and then irradiated with RF radiation. When the frequency of the radiation satisfies Equation 6.7, the magnetic component of the radiant energy becomes absorbed. If the magnetic field B_0 is kept constant, we may plot the absorption against the frequency ν of the RF radiation. The same experiment could be done by holding the RF constant and varying B_0. The resulting spectrum for absorption at a single frequency is shown schematically in Figure 6.3a; an actual NMR spectrum with an absorption at a single frequency is shown in Figure 6.3b. NMR spectra of the compound toluene are shown in Figure 6.4. Figure 6.4b is the 300 MHz proton spectrum; Figure 6.4c is the 60 MHz proton spectrum. Figure 6.4a is the ^{13}C NMR spectrum.

When a nucleus absorbs energy, it becomes excited and reaches an excited state. It then loses energy and returns to the unexcited state. Then it reabsorbs radiant energy and again enters an excited state. The nucleus alternately becomes excited and unexcited and is said to be in a state of *resonance*. This is where the term resonance comes from in NMR spectroscopy.

Magnetic field strengths are given in units of tesla (T) or gauss (G). The relationship between the two units is 1 T = 10^4 G. If the applied magnetic field is 14,092 G (or 1.41 T), the frequency of radiation absorbed by a proton is 60 MHz. The nomenclature "60 MHz proton NMR" indicates the RF for proton resonance and also defines the strength of the applied magnetic field if the nucleus being measured is specified. For example, the ^{13}C nucleus will also absorb 60 MHz RF radiation, but the magnetic field strength would need to be 56,000 G. Similarly, a 100 MHz proton NMR uses 100 MHz RF and a magnetic field of 14,092 × 100/60 = 23,486 G (2.35 T) for ^{1}H measurements. At that field strength, a ^{13}C nucleus would absorb at 25.1 MHz due to its very different magnetogyric ratio. If a frequency is specified for an NMR instrument without specifying the nucleus, the proton is assumed (e.g., a 500 MHz NMR would be assumed to refer to an instrument with a magnetic field strength such that a proton will absorb at 500 MHz).

6.1.2.1 Saturation and Magnetic Field Strength

The energy difference ΔE between ground-state and excited-state nuclei is very small. The number of nuclei in the ground state is the number lined up with the magnetic field B_0. The ratio of excited nuclei to unexcited nuclei is defined by the Boltzmann distribution:

$$\frac{N^*}{N_0} = e^{-\frac{\Delta E}{kT}} = e^{-\frac{\gamma \hbar B_0}{2\pi kT}} \qquad (6.8)$$

where
N^* is the number of excited nuclei
N_0 is the number of unexcited (ground-state) nuclei

For a sample at 293 K in a 4.69 T magnetic field, the ratio $N^*/N_0 = 0.99997$. There are almost as many nuclei in the excited state as in the ground state because the difference between the two energy levels is very small. Typically, for every 100,000 nuclei in the excited state, there may be 100,003 in the ground state, as in this case. This is always the case in NMR; the Boltzmann ratio is always very close to 1.00. For this reason, NMR is inherently a low-sensitivity technique.

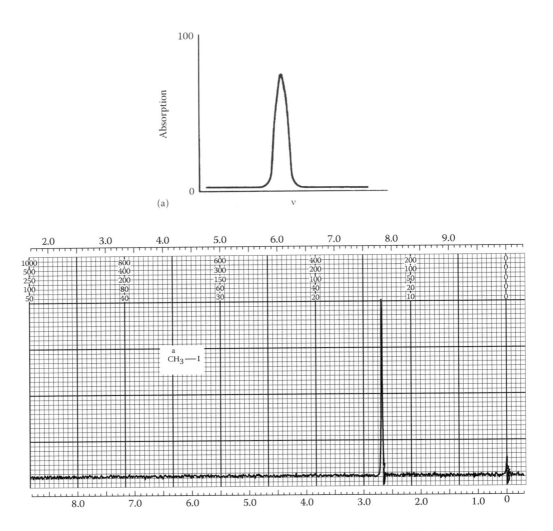

Figure 6.3 (a) Absorption versus frequency of RF radiation. (b) A 60 MHz proton NMR spectrum of methyl iodide, CH_3I. (From *The Sadtler Handbook of Proton NMR Spectra*, W.W. Simons, ed., Sadtler Research Laboratories, Inc., Philadelphia, PA, 1978. [Now Bio-Rad Laboratories, Inc., Informatics Division, Sadtler Software and Databases. Copyright Bio-Rad Laboratories, Informatics Division, Sadtler Software and Databases, 2014. All rights reserved.])

A system of molecules in the ground state may absorb energy and enter an excited state. A system of molecules in an excited state may emit energy and return to the ground state. A signal can be seen only if there is an excess of molecules in the ground state. If the number of molecules in the ground state is equal to the number in the excited state, the net signal observed is zero and no absorption is noted.

The excess of unexcited nuclei over excited nuclei is called the *Boltzmann excess*. When no radiation falls on the sample, the Boltzmann excess is maximum, N_x. However, when radiation falls on the sample, an increased number of ground-state nuclei become excited and a reduced number remain in the ground state. If the RF field is kept constant, a new equilibrium is reached and the Boltzmann excess decreases to N_s. When $N_s = N_x$, absorption is maximum. When $N_s = 0$, absorption is zero. The ratio N_s/N_x is called Z_0, the **saturation factor**.

If the applied RF field is too intense, all the excess nuclei will be excited, $N_s \rightarrow 0$, and absorption $\rightarrow 0$. The sample is said to be **saturated**. The saturation factor Z_0 is

$$Z_0 = \left(1 + \gamma^2 B_1^2 T_1 T_2\right)^{-1} \quad (6.9)$$

where
γ is the magnetogyric ratio
B_1 is the intensity of RF field
T_1 and T_2 are, respectively, the longitudinal and transverse relaxation times (discussed in Section 6.1.3.2)

As a consequence of this relationship, the RF field must not be very strong so as to avoid saturation. However, under certain experimental conditions, saturation of a particular nucleus can provide important structural information (Section 6.3).

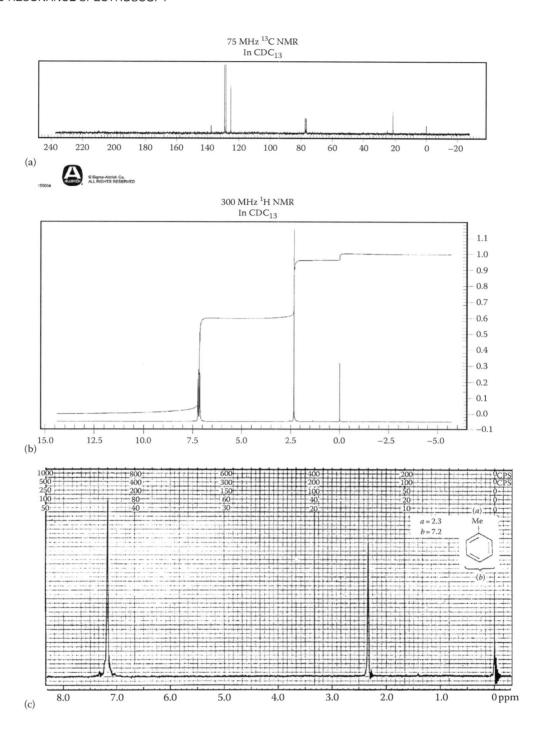

Figure 6.4 (a) The ¹³C NMR spectrum of toluene; (b) the 300 MHz proton spectrum of toluene. Both reprinted with permission of Aldrich Chemical Company, Inc. (c) The 60 MHz proton spectrum of toluene. The structure of toluene is shown, with Me indicating a methyl group, –CH₃. Two groups of absorption peaks are seen, one due to the protons of the methyl group and the other to the aromatic ring protons. The spectrum is discussed in Section 6.4.

From Equation 6.8, we can derive the expression

$$\frac{N^*}{N_0} = 1 - \frac{\gamma h B_0}{2\pi k T} = 1 - \frac{2\mu B_0}{kT} \quad (6.10)$$

which shows that the relative number of excess nuclei in the ground state is related to B_0. As the magnetic field strength increases, the NMR signal intensity increases. This is the driving force behind the development of high-field-strength magnets for NMR.

6.1.3 Width of Absorption Lines

The resolution or separation of two absorption lines depends on how close they are to each other and on the absorption linewidth. The width of the absorption line (i.e., the frequency range over which absorption takes place) is affected by a number of factors, only some of which we can control. These factors are discussed next.

6.1.3.1 Homogeneous Field

An important factor controlling the absorption linewidth is the applied magnetic field B_0. It is very important that this field be constant over all parts of the sample, which may be 1–2 in. long. If the field is not homogeneous, B_0 is different for different parts of the sample, and therefore the frequency of the absorbed radiation will vary in different parts of the sample, as described by Equation 6.7. This variation results in a wide absorption line. For molecular structure determination, wide absorption lines are very undesirable, since we may get overlap between neighboring peaks and loss of structural information. The magnetic field must be constant within a few ppb over the entire sample and must be stable over the time required to collect the data. This time period is short for routine ^{1}H and ^{13}C measurements, on the order of 5–30 min, but may be hours or days for complex analyses. Most magnets used in NMR instruments do not possess this degree of stability. Several different experimental techniques are used to compensate for field inhomogeneity, such as spinning the sample holder in the magnetic field.

Most instruments require magnetic field shimming to improve (or at least fine tune) the magnetic field homogeneity. This is done by the use of multiple shim coils that are electronically energized to produce slight magnetic fields. These coils can be controlled by the operator, manually, or by the console computer. In most cases, the analyst will allow the computer to take care of shimming automatically. Indeed, in a well shimmed magnetic field, it may not be necessary (or desirable) to spin the sample.

6.1.3.2 Relaxation Time

Another important feature that influences the absorption linewidth is the length of time that an excited nucleus stays in the excited state. The **Heisenberg uncertainty principle** tells us that

$$\Delta E \Delta t = \text{constant} \quad (6.11)$$

where
ΔE is the uncertainty in the value of E
Δt is the length of time a nucleus spends in the excited state.
Since $\Delta E \Delta t$ is a constant, when Δt is small, ΔE is large

But we know that $\Delta E = h\nu$ and that h is a constant. Therefore any variation in E will result in a variation in ν. If E is not an exact number but varies over the range $E + \Delta E$, then ν will not be exact but will vary over the corresponding range $\nu + \Delta \nu$. This can be restated as

$$E + \Delta E = h(\nu + \Delta \nu) \quad (6.12)$$

We can summarize this relationship by saying that when Δt is small, ΔE is large and therefore $\Delta \nu$ is large. If $\Delta \nu$ is large, then the frequency range over which absorption takes place is wide and a wide absorption line results.

The length of time the nucleus spends in the excited state is Δt. This lifetime is controlled by the rate at which the excited nucleus loses its energy of excitation and returns to the unexcited state. The process of losing energy is called **relaxation**, and the time spent in the excited state is the **relaxation time**. There are two principal modes of relaxation: longitudinal and transverse. **Longitudinal relaxation** is also called spin–lattice relaxation; **transverse relaxation** is called spin–spin relaxation.

Longitudinal relaxation T_1: The entire sample in an NMR experiment, both absorbing and nonabsorbing nuclei, is called the **lattice**. An excited-state nucleus (said to be in a *high-spin* state) can lose energy to the lattice. When the nucleus drops to a lower energy (*low-spin*) state, its energy is absorbed by the lattice in the form of increased vibrational and rotational motion. A very small increase in sample temperature results in spin–lattice (longitudinal) relaxation. This process is quite fast when the lattice molecules are able to move quickly. This is the state of affairs in most liquid samples. The excitation energy becomes dispersed throughout the whole system of molecules in which the analyte finds itself. No radiant energy appears; no other nuclei become excited. Instead, as numerous nuclei lose their energy in this fashion, the temperature of the sample goes up very slightly. Longitudinal relaxation has a relaxation time, T_1, which depends on the magnetogyric ratio and the lattice mobility. In solids or viscous liquids, T_1 is large because the lattice mobility is low.

Transverse relaxation T_2: An excited nucleus may transfer its energy to an unexcited nucleus nearby. In the process, a proton in the nearby unexcited molecule becomes excited and the previously excited proton becomes unexcited. There is no net change in energy of the system, but the length of time that one nucleus stays excited is shortened because of the interaction. The average excited-state lifetime decreases and line broadening results. This type of relaxation is called transverse relaxation or spin–spin relaxation, with a lifetime T_2.

It is found in practice that in liquid samples, the net relaxation time is comparatively long and narrow absorption lines are observed as in Figure 6.4. In solid samples, however, the transverse relaxation time T_2 is very short. Consequently,

ΔE and therefore Δν are large. For this reason, solid samples generally give wide absorption lines. As we will see, solid samples require a different set of experimental conditions than liquids to give useful analytical information from their NMR spectra. One approach is to make the solid behave more like a liquid. For example, solid polymer samples normally give broad NMR spectra. But if they are "solvated," narrower lines are obtained and the spectra are more easily interpreted. A sample is "solvated" by dissolving a small amount of solvent into the polymer. The polymer swells and becomes jellylike but does not lose its chemical structure. The solvating process greatly slows down transverse relaxation and the *net* relaxation time is increased. The linewidth is decreased and resolution of the spectrum for structural information is better.

6.1.3.3 The Chemical Shift

The movement of electrons in the sample, close to the nucleus being excited, will produce small magnetic fields that will affect the field "felt" by that nucleus. These small magnetic fields will counteract the applied magnetic field of the instrument. This effect is called shielding. Because of this shielding, less energy is required to excite the nucleus from one spin state to another. Since this energy is proportional to the frequency, ν, the absorption will be observed at a lower value of ν. This change is called the chemical shift, and it is usually expressed as a difference with some reference molecule. For proton NMR, the typical reference is tetramethylsilane, or TMS. Since this concept is critical to structure determination, especially in organic molecules, you will have no doubt studied it in organic chemistry, and a review of that material might be helpful. Moreover, a more in-depth discussion of chemical shift is presented later in this chapter.

It should be noted that TMS is not the only standard used in NMR against which chemical shifts are measured. In Table 6.3 we list the most common standard compounds used for various nuclei.

6.1.3.4 Magic Angle Spinning

A problem with the examination of solids is that the nuclei can be considered to be frozen in space and cannot freely line up in the magnetic field. The NMR signals generated are dependent, among other things, on the orientation of the nuclei to the magnetic field. Since the orientation of nuclei in solids is fixed, each nucleus (even chemically identical nuclei) "sees" a different applied magnetic field, resulting in broad NMR spectra. The effective magnetic field seen by a nucleus depends on the chemical environment in which the nucleus finds itself; the position at which a given nucleus resonates is its chemical shift, as discussed above. Chemical shift due to **different environments in a**

Table 6.3 Reference standards for selected nuclei in NMR spectroscopy. Taken from the book by Bruno and Svoronos

Nucleus	Name	Formula
^{1}H	tetramethylsilane [TMS]	$(CH_3)_4Si$
	3-(trimethylsilyl)-1-propanesulfonic acid, sodium salt [DSS][a]	$(CH_3)_3Si(CH_2)_3SO_3Na$
	3-(trimethylsilyl)-propanoic acid, d_4, sodium salt [TSP]	$(CH_3)_3Si(CD_2)_3CO_2Na$
^{2}H	deuterated chloroform [chloroform-d]	$CDCl_3$
^{11}B	boric acid	H_3BO_3
	boron trifluoride etherate	$(C_2H_5)_2O \cdot BF_3$
	boron trichloride	BCl_3
^{13}C	tetramethylsilane [TMS]	$(CH_3)_4Si$
^{15}N	ammonium nitrate	NH_4NO_3
	ammonia	NH_3
	nitromethane	CH_3NO_2
	nitric acid	HNO_3
	tetramethylammonium chloride	$(CH_3)_4NCl$
^{17}O	water	H_2O
^{19}F	trichlorofluoromethane [Freon 11, R-11]	CCl_3F
	hexafluorobenzene	C_6F_6
^{31}P	trimethylphosphite [methyl phosphite]	$(CH_3O)_3P$
	phosphoric acid (85%)	H_3PO_4
^{35}Cl	sodium chloride	$NaCl$
^{59}Co	cobalt (III) hexacyanide anion	$[Co(CN)_6]^{-3}$
^{119}Sn	tetramethyltin	$(CH_3)_4Sn$
^{195}Pt	platinum (IV) hexacyanide	$[Pt(CN)_6]^{-2}$
	dihydrogen platinum (IV) hexachloride	H_2PtCl_6
^{183}W	sodium tungstate (external)	Na_2WO_4

molecule is the key to structure determination by NMR. The phenomenon in solids of nuclei having different chemical shifts *as a result of orientation in space* is called **chemical shift anisotropy**.

The chemical shift due to magnetic anisotropy is directly related to the angle between the sample and the applied magnetic field. It has been shown theoretically and experimentally that by spinning the sample at an angle of 54.76° to the magnetic field, the **magic angle**, rather than the usual 90° for liquid sample analysis, the chemical shift anisotropy is eliminated and narrow line spectra are obtained. Figure 6.5 demonstrates the difference in ^{13}C NMR spectra of a solid with and without magic angle spinning (MAS). The spectrum in Figure 6.5a was taken without spinning (i.e., static) on a crystalline powder sample of l-DOPA. The spectrum has broad, unresolved lines and does not provide much useful information. The same sample is then spun at 9.6 kHz at the magic angle, 54.76°. The spectrum obtained is shown in Figure 6.5b. The linewidths are dramatically decreased and seven carbon peaks are resolved, providing significant structural information about the compound. The spinning is carried out at very high frequencies (5–15 kHz) for optimum performance. This spinning gives better resolution and improved measurement of chemical shift and spin–spin splitting. The functional groups and their positions relative to each other in the solid sample molecule can be determined, as will be discussed.

Special probes have been developed for solid-state NMR that automatically position the sample at the magic angle. Modern instruments with MAS make the analysis of solid samples by NMR a routine analytical procedure. MAS, combined with two RF pulse techniques called cross-polarization and dipolar decoupling (discussed in Section 6.6.4), permits the use of the low-abundance nuclei ^{13}C and ^{29}Si to analyze insoluble materials by NMR, including highly cross-linked polymers, glasses, ceramics, and minerals.

6.1.3.5 Other Sources of Line Broadening

Any process of *deactivating* or relaxing an excited molecule results in a decrease in the lifetime of the excited state. This in turn causes line broadening. Other causes of deactivation include the following: (1) the presence of ions (the large local charge deactivates the nucleus), (2) paramagnetic molecules such as dissolved O_2 (the magnetic moment of electrons is about 10^3 times as great as nuclear magnetic moments and this high local field causes line broadening), and (3) nuclei with quadrupole moments. Nuclei in which $I > 1/2$ have quadrupole moments, which cause electronic

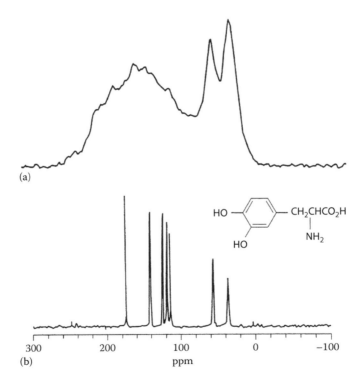

Figure 6.5 (a) The ^{13}C NMR spectrum of solid-powdered L-DOPA obtained without spinning the sample. (b) The ^{13}C spectrum of the same sample obtained with MAS at a frequency of 9.6 kHz. Note the dramatic decrease in linewidth and increase in resolution when using MAS to obtain the NMR spectrum of a solid sample. (Spectra provided courtesy of Dr. James Roberts, Department of Chemistry, Lehigh University, Bethlehem, PA.)

interactions and line broadening; one important nucleus with a quadrupole field is ^{14}N, present in many organic compounds such as amines, amides, amino acids, and proteins.

6.2 FT-NMR EXPERIMENT

The time required to record an NMR spectrum by scanning either frequency or magnetic field is Δ/R, where Δ is the spectral range scanned and R the resolution required. For ^{1}H NMR, the time required is only a few minutes because the spectral range is small, but for ^{13}C NMR, the chemical shifts are much greater; consequently, the spectral range scanned is much greater and the time needed to scan is long. For example, if the range is 5 kHz and a resolution of 1 Hz is required, the time necessary would be (5000 s)/1 or 83 min, an unacceptably long time for routine analysis and an impossible situation if rapid screening of thousands of compounds is needed, as it is in the development of pharmaceuticals.

This problem and quite a few other problems with the NMR experiment were overcome with the development of FT-NMR. The fundamentals of FT spectroscopy and the advantages gained through the use of FT spectroscopy were discussed in Chapter 3 for optical spectroscopy. In FT-NMR, the RF is applied to the sample as a short, strong "pulse" of radiation. The experiment is shown schematically in Figure 6.6. Because there are slightly more nuclei lined up with the field, the excess nuclei in the ground state can be thought of as having a single magnetic moment lined up with the external field, B_0. This net magnetization is shown in Figure 6.6d as M_z. The net magnetization behaves as a magnet. An electric current applied to a coil of wire surrounding a magnet will cause the magnet to rotate. An RF pulse through a coil of wire around the sample is used to generate a second magnetic field, B_1, at right angles to B_0; this provides the excitation step in the NMR experiment. In a modern NMR instrument, the field B_1 is applied as a **pulse** for a very short time, on the order of 10 µs, with a few seconds between pulses. The net magnetic moment of the sample nuclei is shifted out of alignment with B_0 by the pulse (Figure 6.6e). Most often, a 90° pulse is used; the net magnetization vector is shifted 90° (from the z-axis to the y-axis). Such a 90° change gives the largest signal possible. The pulse is discontinued and the excited nuclei precess around the applied magnetic field at an angle to B_0. This "rotating magnet" induces a current in the wire coil. This induced current is the NMR signal and is picked up by the coil very quickly after the B_1 pulse ends. The signal undergoes free induction decay (FID); the current decreases with time as the freely precessing nuclei relax back to the ground state, as shown in Figure 6.6f. This FID signal is a time domain signal and must be processed using the FT (or other mathematical transform) to produce the usual frequency-domain spectrum. Because all frequencies are excited simultaneously, the FID signal consists of one exponentially decaying cosine wave for each frequency component in the spectrum. This type of pattern for a ^{13}C experiment is shown in Figure 6.6g as the dense black multiple line signal at the top of the diagram, along with the Fourier transformation of the FID, resulting in the frequency-domain ^{13}C spectrum shown at the bottom of the diagram.

One advantage of the FT-NMR experiment is that the entire spectrum is taken in a single pulse. This occurs because pulsing the RF field broadens the frequency distribution of the RF source. All resonances within several kHz of the source frequency are excited simultaneously. While the intensity of the FID signal is very low, the process may be repeated many times very rapidly, for example, 8192 times in 0.8 s. The signal increases linearly, but the noise increases only as the square root of the number of readings. The net effect is a significant improvement in the S/N ratio, as seen in Figure 6.7. This directly improves the sensitivity of the method. Signals can be obtained that are orders of magnitude more sensitive than those obtained with old "continuous wave" (CW) NMR. FT-NMR is now the major type of NMR instrumentation being manufactured; it is critical to obtaining ^{13}C NMR spectra and in performing 2D NMR experiments, both discussed later in the chapter.

6.3 CHEMICAL SHIFTS

From the Larmor equation, it would seem that all protons in a sample in a given applied magnetic field would absorb at exactly the same frequency. NMR would therefore tell us only that protons were present in a sample. The same would be true for ^{13}C; NMR would tell us only that carbon was present. If NMR were suitable only for detecting and measuring the presence of hydrogen or carbon in organic compounds, it would be a technique with very limited usefulness. There are a number of other fast, inexpensive methods for detecting and measuring hydrogen and carbon in organic compounds. Fortunately, this is not the case. In the NMR experiment, protons in different chemical environments within a molecule absorb at slightly different frequencies. This variation in absorption frequency is caused by a slight difference in the electronic environment of the proton as a result of different chemical bonds and adjacent atoms. The absorption frequency for a given proton depends on the chemical structure of the molecule. This variation in absorption frequency is called the **chemical shift**. The same type of chemical shift occurs for carbon in different chemical environments within a molecule. The physical basis for the chemical shift is as follows.

Suppose that we take a molecule with several different "types" of hydrogen atoms, such as the molecule ethanol,

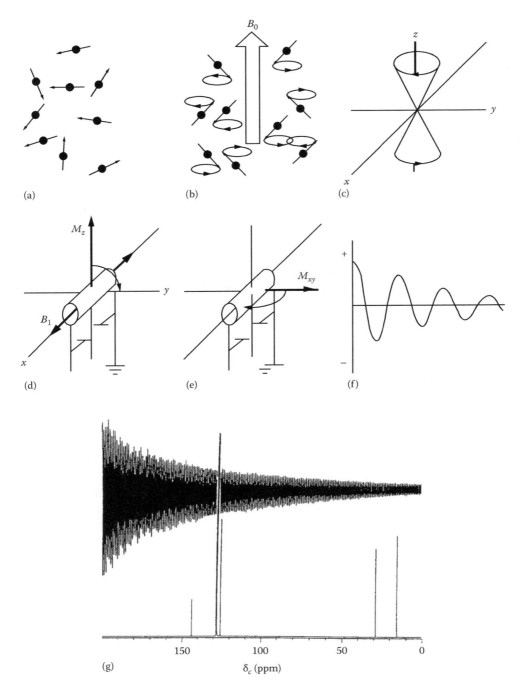

Figure 6.6 The pulsed NMR experiment. (a) NMR-active nuclei have magnetic dipole moments. (b) In the presence of an external static field, B_0, the dipoles precess about the field axis, each with an average component either parallel or antiparallel to the field. (c) In this case, there are two spin populations precessing in opposite directions represented by the two cones. At equilibrium, the top cone (the dipoles with components parallel to the field) has a slightly greater population. (d) The difference between the two populations is represented by the net magnetization, M_z. The NMR signal derives from M_z, which can be observed by rotating it into the plane of an RF coil using a resonant RF field, B_1. (e) B_1 is usually gated on just long enough to rotate M_z into the x–y plane. After B_1 is gated off, the net magnetization M_{xy} begins to rotate about the z-axis at a frequency equal to the difference in precession frequencies for the two populations. (f) This rotating M_{xy} induces an emf in the RF coil along the x-axis. The signal appears as a cosine wave decaying in time following the pulse and is referred to as free induction decay (FID). (From Petersheim, M., Nuclear magnetic resonance, in Ewing, G.A., ed., *Analytical Instrumentation Handbook*, 2nd edn., Marcel Dekker, Inc., New York, 1997. Used with permission.) The pulsed NMR experiment. (g) FID and resulting ^{13}C frequency spectrum after Fourier transformation for an organic compound containing multiple absorption frequencies due to chemically different carbon nuclei. (From Williams, E.A., Polymer molecular structure determination, in Lifshin, E., ed., *Characterization of Materials*, Part I, VCH Publishers, Inc., New York, 1992. Copyright Wiley-VCH Verlag GmbH & Co. KGaA. Used with permission.)

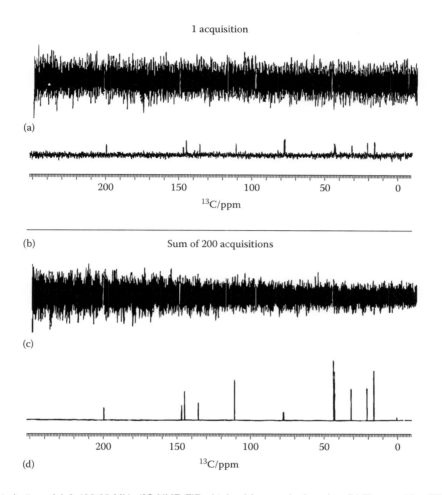

Figure 6.7 From top to bottom: (a) A 100.62 MHz ^{13}C NMR FID obtained from a single pulse. (b) The resulting FT. (c) The resulting FID after adding together 200 FIDs. (d) The resulting FT of the sum of 200 FIDs, showing the improvement in signal [compare peak heights in (b) and (d)] and the reduction in noise [compare the baselines in (b) and (d)]. (From Akitt, J.W. and Mann, B.E., *NMR and Chemistry*, 4th edn., CRC/Taylor & Francis, Boca Raton, FL, 2000. With permission.)

CH$_3$CH$_2$OH. This molecule has hydrogen atoms in three different chemical environments: the three hydrogen atoms in the terminal CH$_3$, the two hydrogen atoms in the CH$_2$ group, and the one in the OH group. Consider the nuclei of the different types of hydrogen. Each one is surrounded by orbiting electrons, but the orbitals may vary in shape and the bonds vary in electron density distribution. This changes the length of time the electrons spend near a given type of hydrogen nucleus. Let us suppose that we place this molecule in a strong magnetic field B_0. The electrons associated with the nuclei will be rotated by the applied magnetic field B_0. This rotation, or *drift*, generates a small magnetic field σB_0, which opposes the much larger applied magnetic field B_0. The nuclei are *shielded* slightly from the applied magnetic field by the orbiting electrons. The extent of the shielding depends on the movement of the electrons caused by the magnetic field (not by the simple orbiting of the electrons). If the extent of this shielding is σB_0, then the nucleus is not exposed to the full field B_0, but to an *effective* magnetic field at the nucleus, B_{eff}:

$$B_{\text{eff}} = B_0 - \sigma B_0 \tag{6.13}$$

where
B_{eff} is the effective magnetic field at the nucleus
B_0 is the applied field
σB_0 is the shielding by the drift of local electrons
σ is the *screening constant* or *diamagnetic shielding constant*

In the case of ethanol, σB_0 is different for each type of hydrogen; therefore, the effective field B_{eff} is different for each type of hydrogen. In order to get absorption at frequency ν, we must compensate for this variation by varying B_0. In other words, resonance of the hydrogen atoms in different chemical environments takes place at slightly different *applied* magnetic fields. Another way of expressing this relationship is that if the applied field is kept constant, the nuclei in different chemical environments resonate at slightly different frequencies ν. A shielded nucleus resonates or absorbs at a lower frequency than an

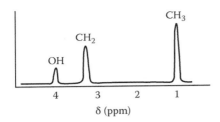

Figure 6.8 Schematic low-resolution proton NMR absorption spectrum of ethanol, CH_3CH_2OH.

unshielded nucleus. This change in frequency of absorption because of shielding is the chemical shift. Instead of one absorption signal for the protons in ethanol, we would predict three absorption signals at slightly different frequencies. Figure 6.8 is a schematic low-resolution proton NMR spectrum of ethanol; indeed, three absorption peaks are seen. Looking at the structure of ethanol, we can also predict what the ^{13}C NMR spectrum should look like. There are two chemically different C atoms in the molecule, the methyl carbon and the methylene carbon; two peaks should be seen in the ^{13}C NMR spectrum of ethanol. It is important to note that B_{eff} and the peak separations resulting from the chemical shift differences are directly proportional to the applied magnetic field strength. This is another reason for the development of high-field-strength magnets: better resolution and more structural information result.

The chemical shifts of nuclei are measured (and defined) relative to a standard nucleus. A popular standard for proton NMR is tetramethylsilane (TMS), which has the chemical formula $Si(CH_3)_4$. In this compound, all 12 hydrogen nuclei are chemically equivalent; that is, they are all exposed to the same shielding and give a single absorption peak. The chemical shift for other hydrogen nuclei is represented as follows:

$$\text{Chemical shift} = \frac{\nu_S - \nu_R}{\nu_{NMR}} \quad (6.14)$$

where
ν_S is the resonant frequency of a specific nucleus
ν_R is the resonant frequency of the reference nucleus
ν_{NMR} is the spectrometer frequency

The chemical shift is the difference in the observed shift between the sample and the reference divided by the spectrometer frequency. It is a dimensionless number. Typical values for the frequency difference between a nucleus and the reference are in the Hz or kHz range; the spectrometer frequency is in the MHz range. In order to make these numbers easier to handle, they are usually multiplied by 10^6 and then expressed in ppm. (This unit should not be confused with the concentration expression ppm used in trace analyses.) The symbol for the chemical shift is δ. It is expressed as

$$\delta = \frac{\nu_S - \nu_R}{\nu_{NMR}} \times 10^6 \text{ ppm} \quad (6.15)$$

Look at Figure 6.13b. The x-axis has units of chemical shift in ppm. The TMS peak is the single peak located at 0.0 ppm, by definition. It is a convention that NMR spectra are presented with the magnetic field increasing from left to right along the x-axis. Diamagnetic shielding therefore also increases from left to right. A nucleus that absorbs to the right-hand side of the spectrum is said to be *more shielded* than a nucleus that absorbs to the left side of the spectrum. One reason TMS is used as the reference peak is that the protons in common organic compounds are generally deshielded with respect to TMS; the TMS peak appears at the far right of the spectrum. (Look at the proton spectra in Figures 6.4b and c again—the x-axis is in ppm, although it is not marked as such in the 300 MHz spectrum, and both show the TMS peak at 0.0 ppm.) The student should be able to recognize that this means that δ decreases from left to right along the x-axis in proton NMR spectra. This is not the way one normally plots numbers; they are usually plotted increasing from left to right. An alternative scale was developed, where the chemical shift was plotted as τ, with τ defined as

$$\tau = 10.00 - \delta(\text{ppm}) \quad (6.16)$$

This tau scale is no longer used, but can be found in the literature through the 1970s, as can the now unused terms upfield and downfield to refer to resonance positions. The terminology that should be used is deshielded (higher δ) or shielded (lower δ). Shielding by the drifting electrons is modified by other nuclei in their vicinity. These in turn are affected by the chemistry, geometry, and electron density of the system. Consequently, some deshielding takes place in most molecules; the end result is that chemically identical nuclei are shifted from chemically different nuclei. Deshielded nuclei would be moved to the left on the NMR plot. As a result, we are able to distinguish different functional groups in a molecule, even ones that contain the same atoms; methyl groups are separated from methylene groups, as seen in the ethanol spectrum. Two examples of how shielding and deshielding occur based on the geometry of the molecule are shown in Figure 6.9. Figures 6.9a and b show the molecule acetylene. When the long axis of the molecule is aligned with the external magnetic field, the circulation of the π-electrons in the triple bond, shown by the partially shaded curved arrow in Figure 6.9a, induces a magnetic field along the molecular axis that opposes the

MAGNETIC RESONANCE SPECTROSCOPY

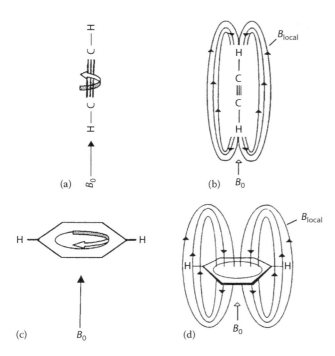

Figure 6.9 (a) and (b) The circulation of the π-electrons in the C≡C bond in acetylene is shown by the partially shaded arrow in (a). The direction of the induced magnetic field, B_{local}, at the hydrogen atoms is shown by the arrows in (b). The direction of the applied field, B_0, is shown. The two protons are shielded from the applied field by the induced local field. (c) and (d) The circulation of the π-electrons in the benzene ring, shown by the partially shaded arrow in (c), induces the magnetic field, B_{local}, shown in (d). Arrows show the induced and the applied magnetic field directions. All six protons in the plane of the benzene ring are deshielded by the induced local magnetic field. [(b) and (d) are from McClure, C.K., in Bruch, M.D., ed., *NMR Spectroscopy Techniques*, 2nd edn., Marcel Dekker, Inc., New York, 1996. Used with permission.]

applied field. The induced magnetic field lines are shown in Figure 6.9b. The arrows show the direction of the induced magnetic field lines at the positions of the hydrogen atoms. Both of the protons in acetylene are shielded from the applied magnetic field by the induced magnetic field. The field these protons experience is $B_0 - B_{local}$, which is smaller than B_0. Therefore, a higher frequency is needed to attain resonance. Figures 6.9c and d show the molecule benzene, a planar molecule with one proton on each carbon atom. The protons are located on the outside of the carbon ring in the plane of the ring. Only two of the six protons attached to the ring are shown here. The delocalized π-electrons are depicted as the circle inside the carbon hexagon. When the benzene ring is perpendicular to the applied magnetic field, the delocalized electrons circulate as shown by the partially shaded curved arrow in Figure 6.9c and generate an induced magnetic field as depicted in Figure 6.9d. In this case, the induced magnetic field shown reinforces the applied magnetic field outside the ring; all of the protons

on the benzene ring feel a stronger field than B_0. They are deshielded by the induced field caused by the *ring current*. The field these protons experience is $B_0 + B_{local}$, so a lower frequency needs to be applied for resonance to occur. The term anisotropic effect or chemical shift anisotropy means that different chemical shifts occur in different directions in a molecule.

It is the chemical shift that allows us to identify which functional groups are present in a molecule or a mixture. Chemical shift enables us to ascertain what types of hydrogen nuclei are present in a molecule by measuring the shift involved and comparing it with known compounds run under the same conditions. For example, it is easy to distinguish between aromatic and aliphatic hydrogen atoms, alkenes and alkynes, or hydrogen bonded to oxygen or nitrogen, and hydrogen on a carbon adjacent to ketones, esters, etc. Examples of typical chemical shifts for protons are given in Table 6.4. More extensive tables of chemical shifts can be found in the book by Bruno and Svoronos listed in the bibliography.

Figure 6.8 shows schematically the low-resolution spectrum of ethyl alcohol (or ethanol). This absorption spectrum, which is historically significant as one of the first NMR spectra to be recorded (Arnold et al., 1951), discloses that there are three types of hydrogen nuclei present in the ethanol molecule. Because NMR spectra provide valuable information about a molecule's structure, they are one of the most powerful tools available for characterizing unknown compounds or compounds for which we know the empirical formula but not the structure.

However, most proton NMR spectra show more peaks than one would predict based on chemical equivalency. With improved instrumentation resulting in higher resolution, the absorption spectrum of ethyl alcohol has been found to be more complex than Figure 6.8 indicates, as seen in Figures 6.19 and 6.23, discussed in detail later. When examined under *high* resolution, each peak in the spectrum can be seen to be composed of several peaks. This **fine structure** is brought about by **spin–spin splitting** or **spin–spin coupling**.

6.4 SPIN–SPIN COUPLING

As we have already seen, the hydrogen nuclei of an organic molecule are spinning, and the axis of rotation may be with or against the applied magnetic field. Since the nucleus is magnetic, it exerts a slight magnetic field, which may be either with or against the applied magnetic field.

Suppose that we have a molecule such as the aldehyde shown, 1-butanal, also called butyraldehyde:

$$CH_3-CH_2-\underset{\underset{H}{|}}{\overset{\overset{H}{|}}{C}}-\underset{}{\overset{\overset{H}{|}}{C}}=O$$

Table 6.4 Approximate chemical shifts of protons in common organic compounds

(a) Type of Hydrogen	Chemical Shift (ppm)			
	a	b	c	d
TMS $CH_3-Si(CH_3)_3$ with CH_3		0		
$CH_3CH_2CH_2CH_3$ a b b a	1.0	1.2		
$CH_3CH_2-C(CH_3)_2-CH_2CH_3$ a b b a CH_3(c)	1.0	1.2	2.0	
CH_3CH_2Br a b	1.2	3.48		
CH_3CH_2I a b	1.8	3.2		
CH_3CH_2OH a b c	1.32	3.7	0.5–4.0	
$CH_3C=O$ a | OH (b)	2.1	11.4		
$CH_3CH_2CH_2NH_2$ a b c d	0.92	1.5	2.7	1.1
$CH_3CH_2CH=CH_2$ a b c d	.9	2.0	5.8	4.9
$CH_3CH_2CH=CHCH_2CH_3$ a b c c b a	1.0	1.3	5.4	
$CH_3CH_2CH-CH_3$ a b |c d NO_2	1.0	1.9	4.2	1.5
$CH_3C-CH_2CH_3$ a || b c O	2.1	2.5	1.1	
$CH_3CH_2S-CH_3$ a b c	1.3	2.53	2.1	
$CH_3CH_2C\equiv CH$ a b c	1.0	2.2	2.4	
$CH_3CH_2CH_2C(=O)H$(d) a b c	1.0	1.7	2.4	9.7

The low-resolution NMR spectrum for 1-butanal is shown in Figure 6.10. The protons on a given carbon atom are chemically equivalent to each other, but each group of protons is different from the rest. There are four chemically different groups of protons, labeled a–d in Figure 6.10. The low-resolution spectrum shows four peaks, as expected from the structure.

We will focus on the type c and type d protons first and ignore the rest of the molecule for the moment. The type c protons are the two methylene protons adjacent to the aldehyde group; the type d proton is the single proton on the aldehyde carbon. The type d proton on the aldehyde group, represented as CHO, may be spinning either *with* or *against* the applied magnetic field. The spinning of the nucleus creates a small magnetic field, either with or against the applied magnetic field. This changes the effective field felt by the two protons of the adjacent CH_2 group. The CH_2 protons next to a CHO proton that is spinning *with* the field absorb at a slightly different frequency from that of the CH_2 protons next to a CHO proton spinning *against* the field. Statistically, a sample will contain as many protons spinning with the field as against it; so both groups will be equally represented. The single spinning proton of the CHO group splits the absorption peak (peak c in Figure 6.10) for the adjacent protons in the CH_2 group into two peaks absorbing at slightly different frequencies but of equal intensity (1:1 peak area ratio). The spin–spin splitting is shown schematically in Figure 6.11 (peak c). The two peaks are moved from the original frequency, one to a slightly higher frequency and one to a slightly lower frequency, but they are symmetrically located about the original frequency.

MAGNETIC RESONANCE SPECTROSCOPY

Table 6.4 (Continued) Approximate chemical shifts of protons in common organic compounds

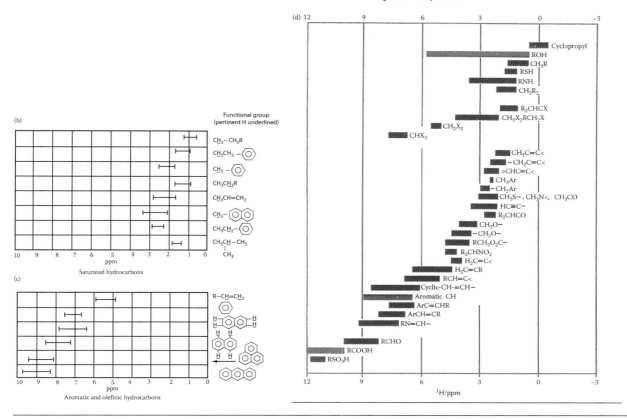

Notes: Chart of approximate ^{1}H chemical shift ranges of different types of protons in organic compounds. X = halogen; R = organic substituent; Ar = aryl substituent.

Source: (d) With kind permission from Springer Science+Business Media: *Multinuclear NMR*, 1987, Akitt, J.W., in Mason (ed.), Plenum Press, New York, 1987.

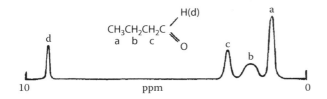

Figure 6.10 Low-resolution proton NMR spectrum of 1-butanal, showing four absorption peaks. The *x*-axis is chemical shift in ppm; the *y*-axis is absorption. The four groups of chemically nonequivalent protons are labeled a–d on the chemical structure. The peak corresponding to each type of proton is marked with the corresponding letter. Table 6.4 should be consulted to confirm the approximate chemical shifts for each type of proton.

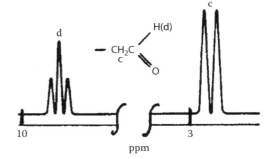

Figure 6.11 A portion of the expected 1-butanal NMR spectrum showing spin–spin coupling between the aldehyde proton and the adjacent methylene group. The *x*-axis is chemical shift in ppm; the *y*-axis is absorption. Peaks c and d are shown, with peak d split into a triplet by the adjacent –CH$_2$–group and peak c split into a doublet by the aldehyde proton. (The rest of the molecule is ignored for the moment.)

At the same time, the two protons of the CH$_2$ group are also spinning, and they affect the frequency at which the neighboring CHO proton absorbs. In this case, each of the protons of the CH$_2$ group can spin with or against the applied field. Several spin combinations are possible. These combinations may be depicted as in Figure 6.12, where the arrows indicate the directions of the magnetic fields created by the spinning nuclei. In a typical NMR sample, many billions of molecules are present, and each spin combination is represented equally by number. The three

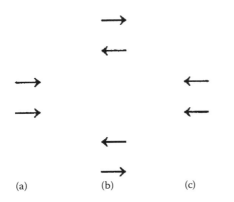

Figure 6.12 Possible alignments of the magnetic field of pairs of protons: (a) both aligned with the applied field (B_0 appears to be increased in magnitude), (b) one aligned with the field and one against the field (there are two possible arrangements with equal probability. The net effect is that B_0 is unchanged), and (c) both aligned against the field (B_0 appears to be reduced in magnitude).

combinations, (a), (b), and (c) in Figure 6.12, modify the applied field felt by the CHO to three different degrees, and absorptions occur at three frequencies for the CHO proton. It can be seen that there are two ways in which the combination of Figure 6.12 (b) can exist, but only one way for the combination of Figure 6.12 (a) or (c). There will therefore be twice as many nuclei with a magnetic field equal to the combination (b) as of either (a) or (c). As a result, the CH_2 group will cause the neighboring H of the CHO group to absorb at three slightly different frequencies with relative intensities in the ratio 1:2:1, as shown schematically in Figure 6.11 (peak d).

The relative peak intensities in a multiplet can be worked out from the possible combinations of spin states, as was done in Figure 6.12; the result is that the relative peak intensities in multiplets match the coefficients in the binomial expansion. Pascal's triangle can be used to rapidly assign peak intensities

```
                    1
                 1     1
              1     2     1
           1     3     3     1
        1     4     6     4     1
     1     5    10    10     5     1
```

and so on. From the triangle relationship, it is seen that a quartet (fourth line from the top of the triangle) will have a 1:3:3:1 intensity ratio and a sextuplet (sixth line from the top) a 1:5:10:10:5:1 ratio.

The proton NMR spectrum of 1-butanal, $CH_3CH_2CH_2CHO$, is shown in Figure 6.13. Figure 6.13a is the predicted first-order spectrum of 1-butanal, Figure 6.13b is an actual 60 MHz proton spectrum, and Figure 6.13c is an actual 90 MHz proton spectrum. Looking at Figure 6.13a, the aldehydic proton, peak d, appears as the 1:2:1 triplet just

predicted, but peak c is not the 1:1 doublet predicted from the splitting by the aldehyde proton alone. We of course ignored for the simple argument earlier that the CH_2 group adjacent to the CHO group is also adjacent to another CH_2 group on the opposite side.

Spin–spin splitting or coupling is the transmission of spin state information between nuclei through chemical bonds. Spin–spin splitting is quite strong between protons on *adjacent* carbons, but is generally negligible between protons farther removed than this. It is important to remember that spin–spin splitting by a given nucleus causes a change in the fine structure of peaks for the *adjacent* protons in the molecule. For example, the CH_2 splits the H of the adjacent CHO group, and the H of the CHO group splits the adjacent CH_2, but the CH_2 *does not* split itself because that interaction is forbidden by quantum theory.

The number of peaks in the fine structure is termed the **multiplicity** of the peak. In Figure 6.11, the multiplicity of the aldehyde proton peak is 3 and the multiplicity of the methylene peak is 2; these are also referred to as a triplet and a doublet, respectively. The multiplicity of proton peaks due to spin–spin splitting is

$$2nI + 1 \qquad (6.17)$$

where

n is the number of *equivalent hydrogen atoms* on the carbon atoms adjacent to the one whose proton peak is being examined

I is the spin quantum number

For hydrogen, $I = 1/2$ and Equation 6.17 simplifies to $n + 1$.

If two different adjacent groups cause splitting, the multiplicity is given by $(2nI + 1)(2n'I + 1)$, where n and n' are the numbers of equivalent protons on each group.

Using Equation 6.17 and the structure of butanal, we can work out the splitting patterns expected. The predicted spectrum is Figure 6.13a, which can be compared with the experimentally obtained spectra in Figures 6.13b and c. We predicted that the aldehyde proton should be split into a triplet by the adjacent CH_2 protons; the peak at 9.7 ppm is a triplet. We predicted that the aldehyde proton would split the adjacent CH_2 group into a doublet, but we did not take into account the protons on the other side of this CH_2 group. There are two methylene protons (the "b" protons) next to the "c" methylene protons. This is a case of having two different groups, b and d, adjacent to the "c" protons we are interested in. As stated earlier, the multiplicity is calculated from $(2nI + 1)(2n'I + 1)$, so we have $(2 + 1)(1 + 1) = 6$. The multiplicity of the peak for the "c" protons should be 6. We predict that peak c will consist of six peaks; in Figure 6.13b, the peak is found at 2.4 ppm and looks like a triplet with each triplet peak split into a doublet for a total of six peaks. The methyl protons (the "a" protons located at 1.0 ppm) are split into a triplet by the adjacent "b" methylene protons.

MAGNETIC RESONANCE SPECTROSCOPY

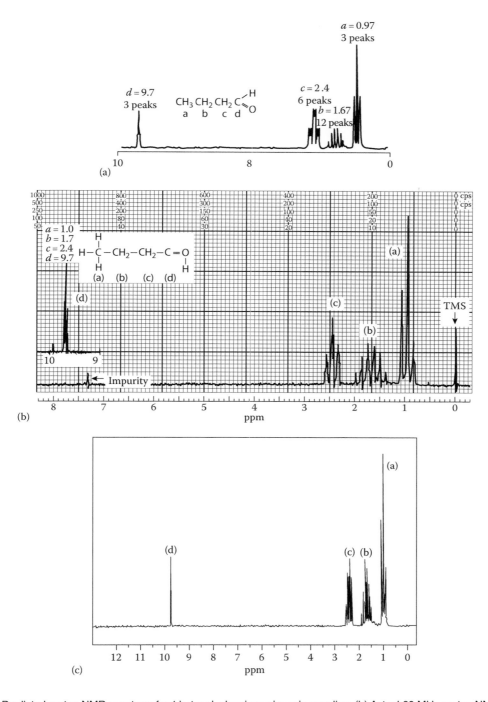

Figure 6.13 (a) Predicted proton NMR spectrum for 1-butanal, showing spin–spin coupling. (b) Actual 60 MHz proton NMR spectrum of 1-butanal. Note the peak for TMS and the presence of a small impurity peak at about 7 ppm; 60 MHz instruments required that chemical shifts greater than 8 ppm be plotted offset from the 0 to 8 ppm baseline; the aldehyde proton shift is plotted at the far left above the first baseline. (c) Actual 90 MHz proton NMR spectrum of 0.04 mL 1-butanal dissolved in 0.5 mL CDCl$_3$. SDBS No. 1925HSP-04-071. (Courtesy of National Institute of Industrial Science and Technology, Japan, SDBSWeb: http://aist.go.jp/RIOBD/SDBS. Accessed on December 13, 2002.)

The "b" methylene protons are split by both the methyl group (three equivalent protons) and the "c" methylene protons (two equivalent protons), so the multiplicity of the "b" proton peak should be $(3 + 1)(2 + 1) = 12$. While all 12 lines cannot be distinguished without expanding the scale around the peak at 1.7 ppm, it is clear, especially from the 90 MHz spectrum shown in Figure 6.13c, that there are more than six peaks present for the "b" peak.

A few more examples of spin–spin coupling should help you to predict splitting patterns for simple compounds.

The structure of 1,1-dichloroethane is

$$\mathrm{Cl-\underset{\underset{Cl}{|}}{\overset{\overset{H}{|}}{C}}-CH_3}$$

The single proton on the carbon containing the two Cl atoms will be split into (3 + 1) peaks by three equivalent protons on the adjacent methyl group. The multiplicity of this peak is 4 and it is called a quartet. Each peak in the quartet will be separated by exactly the same frequency; they will be evenly spaced. We will get back to this in a bit. The relative peak intensities will be 1:3:3:1, which can be worked out as was done in Figure 6.12. The peak due to the three methyl protons will be split into (1 + 1) peaks, a doublet with a multiplicity of 2, by the adjacent single proton. The spacing between the two peaks in the doublet will be exactly the same as the peak spacing in the quartet and the ratio of peak intensities in the doublet will be 1:1.

If we replace one of the H atoms in the CH_3 group with Cl, we have the compound 1,1,2-trichloroethane:

$$\mathrm{Cl-\underset{\underset{Cl}{|}}{\overset{\overset{H}{|}}{C}}-CH_2Cl}$$

and the single proton on the left-hand carbon will split the peak due to the two protons on the methylene carbon into a doublet, and the single proton peak will appear as a triplet, since it is now split by two equivalent protons.

If we have the compound 1,2-dichloroethane, with one Cl on each carbon,

$$\mathrm{Cl-CH_2CH_2-Cl}$$

the two methylene groups are chemically equivalent and the protons should not be split, that is, they will appear as a singlet.

We will look at the spectra for these compounds in Section 6.6.3 (see Figures 6.29b–d).

The compound benzene, C_6H_6, has six chemically equivalent protons. We would predict that a single absorption peak should be seen at a chemical shift of about 7 ppm (from Table 6.4). The 300 MHz proton spectrum of benzene (Figure 6.14) confirms a single peak located at 7.3–7.4 ppm. We cannot tell anything about the number of protons giving rise to this peak; the relative area is obviously equal to 1 when only one peak is seen. Looking back at Figure 6.4, the proton spectrum of toluene, the structure shows a methyl group substituting for one of the hydrogen

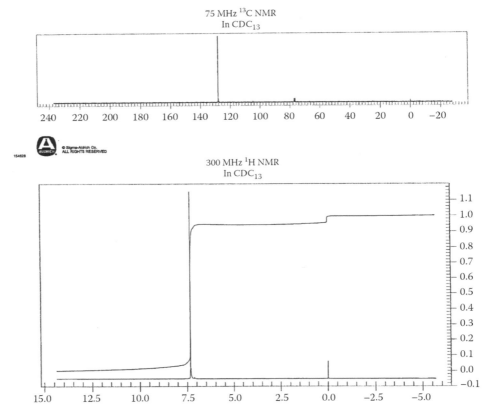

Figure 6.14 The 300 MHz proton NMR spectrum of benzene, C_6H_6, obtained in deuterated chloroform. The ^{13}C spectrum is also shown. (Reprinted with permission of Aldrich Chemical Co., Inc.)

atoms on a benzene ring. We should expect a singlet peak for the methyl group, since there are no protons adjacent to it. The chemical shift of the peak will be about 2 ppm (from Table 6.4). There is a singlet in that position in the spectrum. We also see a multiplet at about 7.3 ppm, where we would expect the aromatic ring protons to absorb. The five *ortho*, *meta*, and *para* protons on the ring are not chemically equivalent (unlike the six protons on a benzene ring), and at 300 MHz, we can tell this. In older spectra obtained at lower field strengths, such as the 60 MHz spectrum shown in Figure 6.4b, the electronic environments of the ortho, meta, and para protons are so similar as to cause the peaks to overlap, resulting in a single broadened peak. The overlap of peaks is often a problem in proton NMR. The expected area ratio of the toluene peaks should be 5/3 and can be checked by the stepped line shown on the spectrum, as will be explained in Section 6.6.2. The aromatic compound naphthalene and its proton NMR spectrum are shown in Figure 6.15a; there are two types of protons, marked A and B on the structure. Each A proton is adjacent to a single B proton and vice versa. The spectrum shows the expected two doublets in the aromatic region of the spectrum. The relative area ratio of the doublets should be 1:1, since there are equal numbers of A and B protons. Figure 6.15b is also the spectrum of naphthalene, with the relative peak areas shown by the stepped line. The two steps are equal in height, so there are equal numbers of A protons and B protons.

The spectra can be predicted for the alkanes butane and isobutane (or 2-methylpropane). The peaks should appear in the 1–1.5 ppm chemical shift region according to Table 6.4. Butane, $CH_3CH_2CH_2CH_3$, has two types of protons as noted in Figure 6.16a. Isobutane also has two types of protons, as shown in Figure 6.16b. Therefore, both spectra should have two absorption peaks. In butane, the methyl protons should be split by the adjacent methylene protons into a triplet; the methylene protons would be split by the methyl protons into a quartet. We would predict that the proton NMR spectrum of butane would look like the schematic spectrum in Figure 6.16a, with the relative peak areas shown. Isobutane would show a very different splitting pattern. There are nine chemically equivalent protons (marked "b" on the structure) on the three methyl groups; the peak for these nine protons will be split into a doublet by the single "a" type proton on the middle carbon. The peak for the single proton will be split into a $(9 + 1) = 10$ peak multiplet by the "b" type protons, with the relative peak areas shown schematically in Figure 6.16b.

If we replace one of the methyl groups in isobutane with any other substituent, the resulting compound will contain an isopropyl group, $-CH(CH_3)_2$:

$$CH_3 - \underset{\underset{H}{|}}{\overset{\overset{R}{|}}{C}} - CH_3$$

where R represents a substituent. An example is the molecule cumene, which contains a phenyl ring as R. The structure of cumene and its proton NMR spectrum are presented in Figure 6.17. The ring protons appear at about 7 ppm. The isopropyl group shows its very characteristic splitting pattern: a doublet equivalent to six protons located at about 1 ppm and a septuplet equivalent to one proton. The chemical shift of the septuplet depends on the nature of the R group, but the isopropyl splitting pattern is an easy one to spot in a spectrum.

If our substituent R is NO_2, we have 2-nitropropane (structure in Figure 6.18a). This is also a compound containing a substituted isopropyl group. A similar structure is seen for 2-chloropropane (Figure 6.18b). Figure 6.18a is a 60 MHz spectrum of 2-nitropropane. Figure 6.18b is the 300 MHz spectrum of 2-chloropropane. The single proton on the carbon atom containing the "R" group will be split into $(6 + 1)$ peaks, a septuplet, by the six equivalent methyl protons. The two methyl groups are chemically equivalent and will absorb at the same frequency, so the single peak for all six protons will be split into a doublet by the single proton on the central carbon. You can easily compare the step heights in both of these examples. We see the characteristic isopropyl splitting pattern in the cumene, 2-nitropropane, and 2-chloropropane spectra, although the chemical shift of the peaks varies with R.

The NMR spectrum of very dry ethanol is presented in Figure 6.19a. The CH_3 protons are split by the CH_2 group into $(2 + 1) = 3$ peaks. The CH_2 is split by the OH proton and by the CH_3 protons, giving $(3 + 1)(1 + 1) = 8$ peaks. Finally, the OH proton is split into three peaks by the adjacent CH_2 group. Remember that n is the number of equivalent protons taking part in spin–spin splitting. For example, on a methyl group, $-CH_3$, there are three equivalent protons; on the methylene group, $-CH_2-$, there are two equivalent protons. Also, the number of peaks due to spin–spin coupling equals $(n + 1)$ only for nuclei where $I = 1/2$, which is the case for ^1H.

The magnitude of the separation between peaks in the fine structure is a measure of the strength of the magnetic interaction between the nuclei involved in the coupling. The magnitude of the separation is given the symbol J. The value of J is the **coupling constant** or **spin–spin coupling constant** between two nuclei and is given the symbol J_{AB}, with A and B representing the coupled nuclei. The value of J is measured directly from the NMR spectrum by measuring the peak separation in the fine structure and is usually expressed in Hz or cps, as shown at the top of Figure 6.19a. Figure 6.19a is a proton spectrum collected at 300 MHz; the separation between the different peaks is improved over lower-field instruments, but in order to see the splitting patterns, the NMR spectroscopist must select and "expand" each peak. The expanded peak plots are shown either above the spectral peaks or just off to the side of the peak. In very dry ethanol, the hydroxyl proton can be considered fixed on

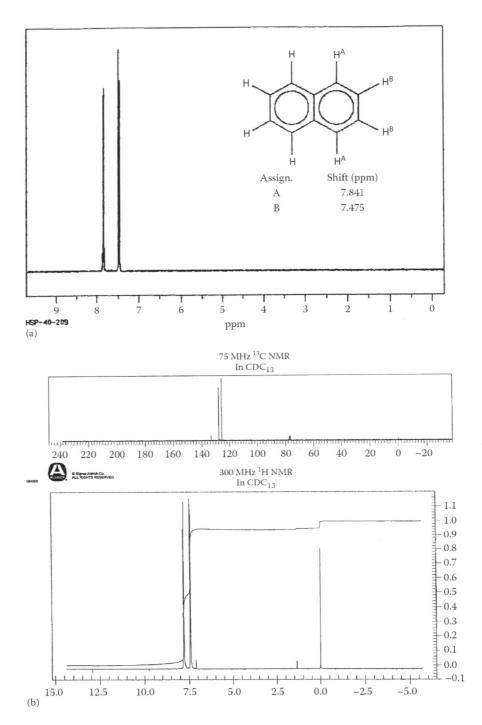

Figure 6.15 (a) The 400 MHz proton NMR spectrum of naphthalene obtained by dissolving 10.5 mg naphthalene in 0.5 mL deuterated chloroform. SDBS No. 1350HSP-40-209. (Courtesy of National Institute of Industrial Science and Technology, Japan, SDBSWeb: http://aist.go.jp/RIOBD/SDBS. Accessed on May 11, 2002.) (b) The 300 MHz proton spectrum and associated ^{13}C spectrum of naphthalene. The proton spectrum shows the peak integration. (Reprinted with permission of Aldrich Chemical Co., Inc.)

the oxygen atom. (This is not the case in "normal" ethanol, which contains water. We will learn more about this later.) Therefore, the hydroxyl proton will be split by the adjacent methylene protons into a triplet, which appears at about 4.3 ppm. Looking at the expanded peak (inset above the hydroxyl peak), it can be seen that the triplet peaks occur at 1291, 1296, and 1301 Hz; that is, they are spaced 5 Hz apart. The J coupling constant between the methylene and

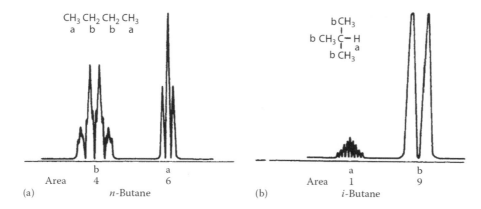

Figure 6.16 (a) The structure and predicted proton NMR spectrum for butane showing relative peak area and spin–spin coupling. The schematic spectrum is not to scale. (b) The structure and predicted proton NMR spectrum for isobutane showing relative peak area and spin–spin coupling. The schematic spectrum is not to scale. The chemical shift of the peaks can be predicted from Table 6.4.

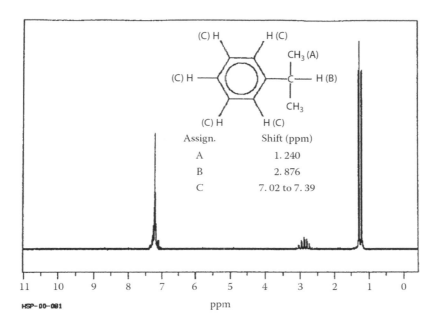

Figure 6.17 The 90 MHz proton NMR spectrum of cumene, showing the characteristic isopropyl group splitting pattern. SDBS No. 1816HSP-00.081. (Courtesy of National Institute of Industrial Science and Technology, Japan, SDBSWeb: http://aist.go.jp/RIOBD/SDBS. Accessed on December 14, 2002.)

hydroxyl protons, J_{AB}, is 5 Hz. The methyl group protons appear between 1.0 and 1.1 ppm; the peak is split into a triplet by the adjacent methylene protons. The J coupling constant between the methyl protons and the methylene protons is 7 Hz, as you can tell by measuring the distance in Hz between the peaks shown in the expanded inset just to the left of the methyl peak. Since we have used A for the hydroxyl proton, we would represent the methyl–methylene J as J_{BC}. We know that the methylene protons, at about 3.5 ppm, are split into eight peaks by the adjacent methyl and hydroxyl protons. Because the methyl–methylene constant, J_{BC}, is larger than the methylene–hydroxyl constant, J_{AB}, the methylene peak will be split first into a quartet, with each peak in the quartet separated by 7 Hz, as shown schematically in Figure 6.19b. Then each peak in the quartet will be split into a doublet, with a peak spacing of 5 Hz. The resulting pattern is shown in Figure 6.19a and in detail in Figure 6.19b.

In a compound such as chloroethane, CH_3CH_2Cl, there are only two groups of equivalent protons. The methyl protons split the methylene protons and vice versa. The coupling constant is given the symbol J_{AB}, where A represents the methyl protons and B the methylene protons. J_{AB} must equal J_{BA} and the spacing between the peaks in the quartet

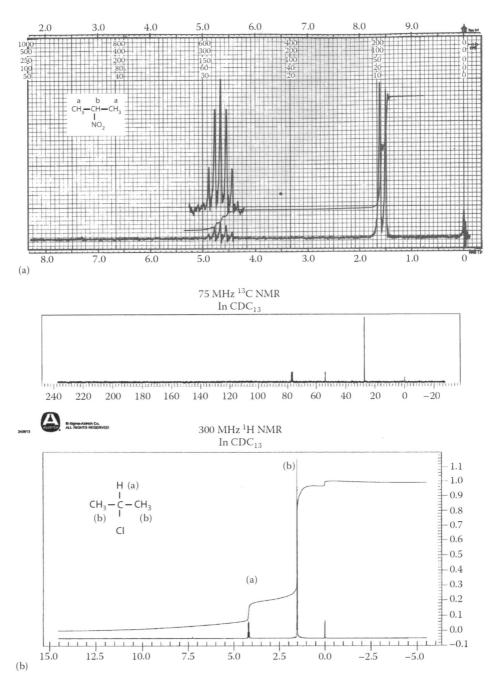

Figure 6.18 (a) The 60 MHz proton spectrum of 2-nitropropane. (From *The Sadtler Handbook of Proton NMR Spectra*, W.W. Simons, ed., Sadtler Research Laboratories, Inc., Philadelphia, PA, 1978. [Now Bio-Rad Laboratories, Inc., Informatics Division, Sadtler Software and Databases. Copyright Bio-Rad Laboratories, Informatics Division, Sadtler Software and Databases, 2014. All rights reserved.]) (b) The 300 MHz proton NMR spectrum of 2-chloropropane. (Reprinted with permission of Aldrich Chemical Co., Inc.) Both spectra show the characteristic isopropyl splitting pattern with the 1:6 peak area ratio. Compare the position of the septuplets in this figure with Figure 6.17.

and the peaks in the triplet will be identical. Measurement of the coupling constants by measuring the peak spacing tells us which protons are splitting each other; this helps in deducing the structure of an unknown from its NMR spectrum. J coupling constants provide valuable information to physical chemists and to organic chemists interested in molecular interactions. The magnitude of J does not change if the applied magnetic field changes, unlike the chemical shift.

The magnitude of the J coupling constant between protons on adjacent single-bonded carbons is between 6 and 8 Hz. The magnitude of J decreases rapidly as the protons

MAGNETIC RESONANCE SPECTROSCOPY

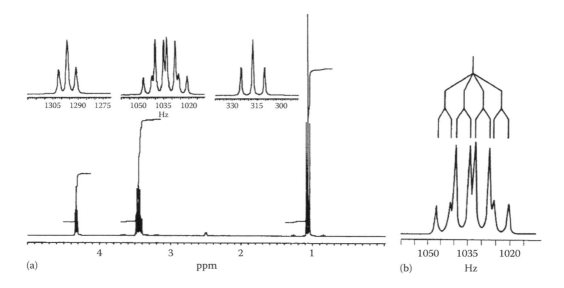

Figure 6.19 (a) The 300 MHz proton NMR spectrum of dry ethanol dissolved in deuterated dimethylsulfoxide (DMSO). (b) Detailed splitting pattern of the methylene protons. The *J* value for the methyl–methylene coupling is 7 Hz, which splits the methylene protons into a quartet. The *J* coupling constant for the hydroxyl–methylene coupling is 5 Hz, so each peak in the quartet is split into a doublet. (From Silverstein, R.M. and Webster, F.X., *Spectrometric Identification of Organic Compounds*, 6th edn., John Wiley & Sons, Inc., New York, 1998. This material is used by permission of John Wiley & Sons, Inc.)

move farther apart in a compound containing saturated C–C bonds. The introduction of multiple bonds or aromatic rings changes the value of *J* and also permits longer-range coupling. There is no coupling between protons on the same carbon. For example, in a compound such as butane,

$$\underset{A}{CH_3} - \underset{B}{CH_2} - \underset{C}{CH_2} - \underset{D}{CH_3}$$

the coupling between the protons on carbon A is zero; the coupling between the protons on carbon A and those on carbon B is about 6–8 Hz (this is written $J_{AB} = 6$–8). The coupling between the protons on carbon A and those on carbon C is never more than 1 Hz, that is, $J_{AC} \leq 1$, and J_{AD} is very small and usually negligible. If one of the C–C bonds is replaced by a C=C, the adjacent protons on the carbon atoms would have a coupling constant of 7–10 Hz if the protons are on the same side of the double bond (*cis*); the coupling constant for *trans* protons is even larger, 12–19 Hz. The magnitude of the coupling constant therefore provides structural information about the compound. Look back at Figure 6.13b, the spectrum of butanal. Peak a, the terminal methyl peak, shows the triplet expected from the adjacent methylene group, but each peak in the triplet is also split by a much smaller longer-range coupling.

The ^{13}C nucleus is also an $I = 1/2$ nucleus; it would be expected that two adjacent ^{13}C nuclei would split the peak for each carbon into a doublet. Given that ^{13}C is only 1.1 % of naturally occurring carbon, the chance of finding more than one ^{13}C in a low-molecular mass (RMM) organic molecule is very small. The chance of finding two adjacent ^{13}C nuclei is even smaller. Consequently, C–C spin–spin coupling is not usually observed in carbon NMR spectra, unless special techniques or isotopically enriched molecules are studied. Values of *J* for carbon–carbon coupling have been measured using these special approaches and range from about 20 to 200 Hz.

Spin–spin interaction is also possible between ^{13}C and protons and can be seen in both proton NMR and ^{13}C NMR spectra under the appropriate conditions. Owing to the low abundance of ^{13}C, the peaks due to ^{13}C–^1H interactions are very small. In the proton spectrum of a neat liquid, they can be observed as weak satellite peaks, one on either side of the central proton peak.

Interpretation of NMR spectra can be very difficult if we consider all the multiplicities that are possible from the interactions of all the nuclei. If we confine ourselves to spectra in which the chemical shift between interacting groups is large compared with the value of *J*, the splitting patterns and the spectra are easier to interpret. This is called utilizing *first-order spectra*. The more complicated systems resulting in second-order and higher-order spectra will not be dealt with in this text. The term second-order spectrum must not be confused with 2D NMR spectra, which will be discussed later in the chapter.

The capital letters of the alphabet are used to define spin systems that have strong or weak coupling constants (actually large or small values of $\Delta v/J$ where Δv is the difference in chemical shift between the nuclei, in Hz). For example, a system A_2B_3 indicates a system of two types of nuclei interacting strongly together of which there are two of type A and three of type B. For an AB system, the value of $\Delta v/J$ will be small; a value of 8 is an arbitrary limit. A break in the alphabetical lettering system indicates weak or no coupling, resulting in a large value for $\Delta v/J$. The system A_2X would indicate two protons of type A that weakly couple with one proton of type X. An AX spin system will give a first-order splitting pattern. As the value of $\Delta v/J$ decreases, the splitting pattern becomes more complex. The rules for interpreting first-order proton spectra can be summarized as follows:

1. A proton spin-coupled to any equivalent protons will produce $(n + 1)$ lines separated by J Hz, where J is the coupling constant. The relative intensities of the lines are given by the binomial expansion $(r + 1)^n$, where n is the number of equivalent protons. Splitting by one proton yields two lines of equal intensity; splitting by two protons yields three lines with a ratio of intensities 1:2:1. Three protons give four lines with a ratio of intensities 1:3:3:1. Four protons yield five lines with a ratio of intensities 1:4:6:4:1, and so on.
2. If a proton interacts with two different sets of equivalent protons, then the multiplicity will be the product of the two sets. For example, a proton split by both a methyl (CH_3) and a methylene (CH_2) group will be split into four lines by the methyl group. Each line will be split into three other lines by the methylene group. This will generate a total of 12 lines, some of which will probably overlap each other. The exact pattern (e.g., quartet of triplets vs. triplet of quartets) is determined by the magnitude of the coupling constants.
3. Equivalent protons do not split each other, the transition being forbidden. In practice, however, interactions do take place and can be seen in second-order spectra.

Interpretation of simple NMR spectra will be discussed in Sections 6.6.2 and 6.6.3. It is not possible in this text to give complete instructions on the interpretation of NMR spectra, but it is hoped that by looking at some simple examples, it can be understood that NMR spectra give important and detailed structural information about molecules. For example, the chemical shift indicates the functional groups that are present, such as aromatics, halides, ketones, amines, alcohols, and so on. Spin–spin splitting indicates which groups are coupled to each other and therefore close to each other in the molecular structure. The multiplicity indicates how many equivalent protons are in the adjacent functional groups. The peak area gives us relative numbers of each type of nucleus. The 3D geometry of even large, complex molecules can be determined.

6.5 INSTRUMENTATION

For complex organic structural determination, a high-resolution NMR is generally required and this type of instrument is discussed first. Low-resolution instruments are discussed in Sections 6.9 and 6.10. The original low-field (e.g., 60 MHz) NMR instruments used a fixed RF, a permanent magnet or electromagnet with a set of Helmholtz coils in the pole faces of the magnet, shown schematically in Figure 6.20b. These coils could be adjusted to vary the applied magnetic field slightly by passing a current through them, causing each chemically different nucleus to come into resonance sequentially. Such instruments were called *continuous wave* (CW) or *field sweep* instruments. The sample holder was placed in the magnetic field. Two RF coils surrounded the sample so as to be orthogonal to each other and to the applied magnetic field. One coil applied a constant RF; the second coil detected the RF emission from the excited nuclei as they relaxed. These systems were simple and rugged, but limited in resolution and capability. Modern NMRs, even 60 MHz online NMR systems, are FT-NMR systems. The most important parts of an FT-NMR instrument are the magnet, the RF generator, and the sample chamber or probe, which not only houses the sample but also the RF transmission and detection coils. In addition, the instrument requires one or more pulse generators, an RF receiver, lots of electronics, and a computer for data processing. Block diagrams of NMRs are shown in Figure 6.20.

The increased sensitivity of FT-NMR is so critical to the measurement of ^{13}C and other less-abundant nuclei as well as to increased proton sensitivity that all modern NMR spectrometers are FT instruments. The term NMR spectrometer therefore is used in the remainder of the chapter to mean a pulsed FT system.

Modern commercial NMR systems are shown in Figure 6.21 and show the range of sizes and magnets available. NMR spectrometers cost from about $20,000 for a miniature 45 MHz NMR to $200,000 for a 300 MHz instrument to $1,000,000 for a high-field wide-bore instrument for solids. MRI and NMR research imaging instruments cost over $3 million dollars in 2023.

6.5.1 Sample Holder

The sample holder in NMR is normally a cylindrical tube shape and is therefore called the sample tube. The tube must be transparent to RF radiation, durable, and chemically inert. Glass or Pyrex tubes are commonly used. These are sturdy, practical, and cheap. They are usually about 6–7 in. long and ~1/8 in. in diameter, with a plastic cap to contain the sample. This type of tube is used for obtaining spectra of bulk samples and solutions. For high-resolution NMR, it is critical that the sample tube be of uniform wall thickness to avoid field distortions. Sample tubes range in size from this "standard" size down to tubes designed to hold 40 μL of sample. Flow-through cells are used for hyphenated techniques such as HPLC–NMR and online analysis.

MAGNETIC RESONANCE SPECTROSCOPY

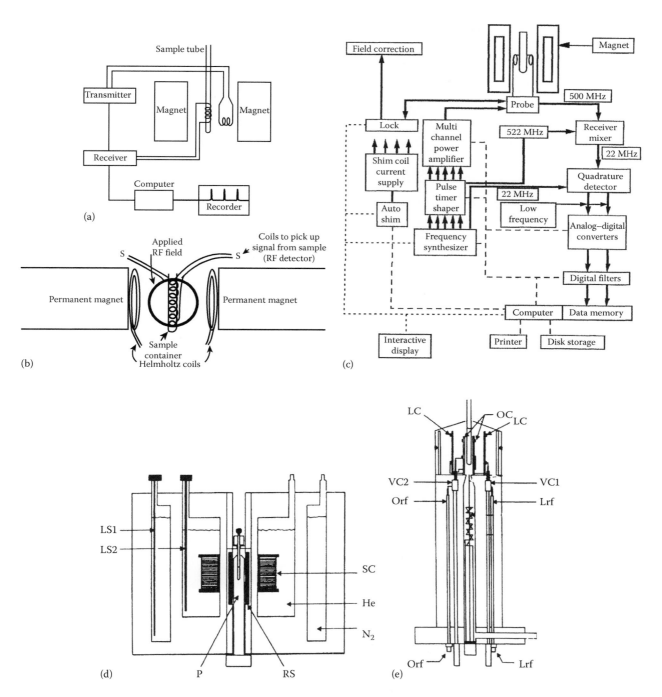

Figure 6.20 (a) Basic components of an NMR spectrometer. (From Williams, E.A.: *Characterization of Materials*, Part I, 1992. Copyright Wiley-VCH Verlag GmbH & Co. KGaA. Used with permission.) (b) Schematic diagram of a CW NMR spectrometer with a permanent magnet. (c) Schematic diagram of an FT-NMR spectrometer. (From Akitt, J.W. and Mann, B.E., *NMR and Chemistry*, 4th edn., CRC/Taylor & Francis, Boca Raton, FL, 2000. Used with permission.) (d) Magnet and assembly for a modern NMR spectrometer. The superconducting coils (the primary solenoid and the shim coils, marked SC) are submerged in a liquid helium Dewar (He), which is suspended in an evacuated chamber. A liquid nitrogen Dewar (N_2) surrounds that to reduce the loss of the more expensive He. The levels of cryogenic fluids are measured with level sensors, marked LS. The room-temperature shim coils (RS) and probe (P) are mounted in the bore of the magnet. A capped sample tube, shown inserted into the probe at the top, is introduced and removed pneumatically from the top of the bore. (e) A schematic probe assembly. A sample tube (uncapped, with a liquid sample as indicated by the meniscus) is shown inserted at the top of the probe. The RF coil for the observed nucleus (OC) is mounted on a glass insert closest to the sample volume. The RF coil for the lock signal (LC) is mounted on a larger glass insert. Variable capacitors (VC) are used to tune the appropriate circuit (lock, observe) to be in resonance with the appropriate RF. Only a small portion of the RF circuitry in the probe is shown. (From Petersheim, M., Nuclear magnetic resonance, in Ewing, G.A., ed., *Analytical Instrumentation Handbook*, 2nd edn., Marcel Dekker, Inc., New York, 1997. Used with permission.)

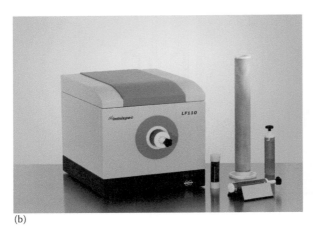
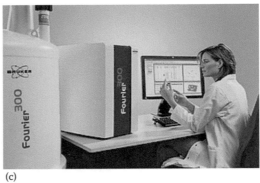
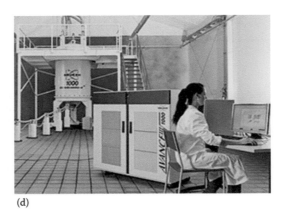

Figure 6.21 (a) Two views of a 900 MHz NMR at the National Magnetic Resonance Facility at Madison (NMRFAM), University of Wisconsin. (Photos by Robin Davies, MediaLab, University of Wisconsin-Madison, used with permission.) (b) Benchtop NMR, (c) 300 MHz FT-NMR, and (d) 1 GHz NMR (1000 MHz). [Photos (b–d) courtesy of Bruker BioSpin. www.bruker.com.]

6.5.2 Sample Probe

The sample chamber into which the sample holder is placed is called the **probe** in an NMR spectrometer. The probe holds the sample fixed in the magnetic field, contains an air turbine to spin the sample holder while the spectrum is collected, and houses the coil(s) for transmitting and detecting NMR signals. A schematic of a probe is presented in Figure 6.20.

The probe is the heart of the NMR system. The most essential component is the RF transmitting and receiving coil, which is arranged to surround the sample holder and is tuned to the precession frequency of the nucleus to be measured. Modern NMR probes use a single wire coil to both excite the sample and detect the signal. The coil transmits a strong RF pulse to the sample; the pulse is stopped and the same coil picks up the FID signal from the relaxing nuclei.

For maximum sensitivity, a fixed-frequency probe is needed. This means that a separate probe is required for each nucleus to be studied: ^{1}H, ^{13}C, ^{19}F, and so on. A high-end probe costs on the order of $140,000 in 2023. Changing probes is a time-consuming and delicate procedure. Each time a probe is changed, it must be retuned and shimmed. In most laboratories, a given probe is operated for an optimal period of time to complete as many measurements as possible. Then, if a probe change is needed, for example to accommodate another nuclei set, the probe is changed and maintained in place for the duration of any measurements that might require it. In this way, probe change operations are minimized. Some variable frequency probe designs are available but have decreased power, sensitivity, and spectral quality compared with fixed-frequency probes. Probes for double-resonance experiments require two concentric coils for the two RF sources. Triple-resonance probes with many gradient options for liquids, solids, and flow experiments are available. Probes usually have variable temperature control to run experiments at temperatures selected by the analyst. In some systems, samples can be heated up to 80 °C. New probe designs with flowthrough sample holders are commercially available for use in coupled HPLC–NMR instruments, HPLC–NMR-MS instruments, and online process control analyzers. These hyphenated instruments and process instruments are discussed under applications later in the chapter.

The probe is installed in the spectrometer magnet so that the coils are centered in the magnet. The sample tube is inserted into the top of the probe and is moved by an air column through the magnet bore and centered among the probe coils. The tube exits the spectrometer at the top of the probe, again moved by an air column from the bore.

6.5.2.1 Cryogenically Cooled Probes

During the past 15 years or so, the emergence of cryogenically cooled probes has dramatically improved the sensitivity and resolution of NMR measurements. This development was the result of nearly 20 years of engineering. Recall that the sensitivity of NMR depends on the ratio between the signal and the noise of the entire receiver system. The signal is the voltage sensed by the receiver coil, while the noise is due to thermal noise in the coil, thermal noise in the preamplifier, and noise in the sample. Thus, there are two ways to increase signal to noise. One can increase the signal by increasing the sample concentration, increasing the field strength, or otherwise changing the experiment. Thermal noise can be reduced by reducing the stochastic motion of electrons in the receiver coil; at very low temperature, common metals have a residual constant resistance that is temperature independent. Thus, cooling the receiver coil to near absolute zero results in a dramatic improvement in sensitivity. This is done by playing a very cold stream of He gas on the coil and preamplifier electronics.

The sensitivity gain provided by a cryoprobe is below the theoretical value because of the necessary thermal insulation between the sample and the receiver coil. A cold receiver coil reduces the thermal noise by a factor of about two, and the cold preamplifier adds another factor of two. Thus, the overall reduction of thermal noise in a cryogenic probe is typically about a factor of four. This also provides a reduction in data collection time of a factor of approximately 20, with the proton and ^{13}C channels available at this enhanced performance.

The cryoprobe system consists of a compressor for the cryocooler, helium circulation regulation and temperature control units, and the probe itself. A schematic diagram of a cryoprobe system is shown in Figure 6.22, and a typical installation is pictured in Figure 6.23.

Although all components are pictured in Figure 6.23 as being in the same location, it is common to locate the compressor and cryocooler remotely, in a utility space, outside the building or even on a roof above the laboratory. This is because these units are noisy; there is thick insulation covering the helium gas lines, not for thermal insulation (that is provided by a vacuum jacket of the lines) but to dampen the vibrations. The helium gas required for the cryoprobe must be of the highest purity, or contaminants will damage this very expensive device. Typically, the helium cylinder, which must run continuously while the probe is installed, needs to be changed every two or three months.

6.5.3 Magnet

The magnet in an NMR spectrometer must be strong and stable and produce a homogeneous field. Homogeneous in this context means that the field does not vary in strength or direction from point to point over the space occupied by the sample. The field must be homogeneous to a few ppb within the sample area. It is common to express the magnetic field strength in terms of the equivalent proton frequency from the Larmor equation. A field strength of 1.4 T is equivalent to a proton frequency of 60 MHz. Commercial magnets range from 60 MHz (1.4 T) to 700 MHz (16.4 T) and higher. Varian, Inc., the same company that introduced the first commercial NMR instrument in 1952, introduced a 900 MHz NMR instrument in 2001. Two views of this instrument are shown in Figure 6.21a. The magnet is so large that this instrument comes with its own staircase so that the analyst can insert and remove the sample tube from the top of the probe. This instrument costs approximately 4.5 million U.S. dollars. (Varian is now Agilent Technologies, Inc.) Other instrument companies have introduced similar high-frequency instruments, with the trend now heading to GHz instruments, as seen in Figure 6.21. Indeed, in 2020, Bruker introduced a 1.2 GHz instrument, pictured in Figure 6.24.

Modern NMR spectrometers can use electromagnets or permanent magnets (see Section 6.10), but most use superconducting solenoid magnets, as shown schematically in Figure 6.20. The magnet consists of a main field coil made of superconducting Nb/Sn or Nb/Ti wire with a dozen or more superconducting **shim coils** wound around the main coil to improve field homogeneity. The superconducting coils must be submerged in liquid helium. The magnet and liquid helium reservoir are encased in a liquid nitrogen reservoir to decrease the evaporative loss of the more expensive liquid helium. As can be seen in Figure 6.20, there is an open bore in the middle of the solenoid. The sample probe is mounted in the bore along with a set of room-temperature shim coils. These coils are adjusted with every new sample placed in the probe to compensate for sample composition, volume, and temperature. Adjusting the field homogeneity, called "shimming," used to be a time-consuming task. Now, computers with multivariate optimization procedures perform this task automatically. The magnetic field strength is held constant by a "frequency locking" circuit. The circuit is used to monitor a given nucleus, such as deuterium used in the solvent for liquid samples, and to adjust the magnetic field strength to keep this nucleus at a constant resonant frequency.

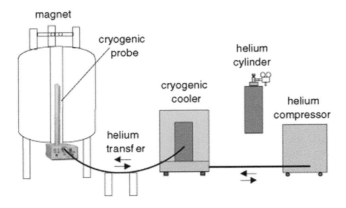

Figure 6.22 A schematic diagram of a NMR cryoprobe system. Reproduced with permission of Bruker, Inc.

Figure 6.23 A photograph of a cryoprobe installed on a 700 MHz NMR instrument. Reproduced with permission of Bruker, Inc.

MAGNETIC RESONANCE SPECTROSCOPY

Figure 6.24 A photograph of the Bruker 1.2 GHz NMR instrument, installed at the University of Florence, Italy. Reproduced with permission of Bruker, Inc.

The bore also contains air conduits for pneumatic sample changing and spinning of the sample holder in the magnetic field. The size of the bore determines how large a sample can be introduced into the magnetic field. Conventional analytical NMRs generally have bore diameters of 5–10 cm, with the larger diameters used for NMR of solids by MAS; the wider bore is needed to accommodate the instrumentation required. Field homogeneity is better in narrow-bore magnets. A very large bore size is that used in human whole-body MRI systems, where the bore is large enough to accommodate a table and the patient.

To see how an NMR magnet is constructed, the interested student should visit the JEOL website (www.jeolusa.com) and look for NMR Magnet Destruction in the links under NMR. The file also can be accessed by typing "JEOL magnet destruction" into your web browser. A JEOL scientist, Dr. Michael Frey, cut open a 270 MHz magnet layer by layer. The pictures and commentary give a good appreciation for the complexity of the magnet construction. In this particular magnet, over 12 miles of superconducting wire were used for the main coil alone.

The cryogenic fluids must be replenished on a regular basis, usually every three or four weeks for the liquid nitrogen and every four to five months for the liquid helium. The superconducting coils must be kept cold; if permitted to warm up, they stop functioning. The magnet is then said to have "quenched." The magnetic field will be maintained by the low temperature, with no external power source. Indeed, if the laboratory or site power is cut, the magnetic field will be unaffected as long as the cryogenic fluid is present. In some instances, it might be necessary to manually quench the magnet. If the instrument is decommissioned or if it must be moved, for example, it must be quenched before moving. A magnet quench is typically accompanied by a rapid release of vaporized helium and nitrogen from the top of the magnet, and must be done safely because of the asphyxiation hazard posed by the oxygen displacement. The quench process is an impressive sight, and on-line videos show what occurs during this process (see, for example: www.youtube.com/watch?v=d-G3Kg-7n_M, accessed October, 2020). Once a magnet is quenched, in order to use it again it must be reenergized (remagnetized) with a powerful, dedicated power supply, usually by the manufacturer. Accidental quenching must be avoided since this is always considered the fault of the user and is not covered by service contracts.

The magnets used in NMR are very strong. It is common practice to rope off an area around the magnet where the field is about 5 G (0.0005 T) and to stay outside of this area. It is very important to keep all metallic objects, like tools, as well as any magnetic records (including credit cards) away from the magnet. People with heart pacemakers should not go near the magnetic field. Modern actively shielded magnets are far less prone to field leakage; it is possible to place a credit card against a modern actively shielded magnet with no consequence. The use of steel tools near the magnet can be dangerous even with active shielding. In such magnets, there is still a powerful field below the magnet. Indeed, during probe change, a small wrench can be pulled from one's hand, especially if it is accidentally placed near the bottom of the magnet. Tools flying across the room is not unheard of.

6.5.4 RF Generation and Detection

The RF radiation is generated by using an RF crystal oscillator. The output of the oscillator is amplified, mixed, and filtered to produce essentially monochromatic RF radiation for an NMR instrument. The RF radiation delivered to the sample is pulsed. A simple pulse might be a rectangular pulse of 500 MHz frequency for a 10 μs duration. The process of pulsing actually widens the RF band, providing a range of frequencies able to excite all nuclei whose resonances occur within the band of frequencies. All resonances within the band are excited simultaneously. An example of this is shown in Figure 6.25. A 10 μs, 500 MHz rectangular pulse leads to a power distribution around the 500 MHz frequency as shown. A proton spectrum occurs over a chemical shift range of ~10 ppm, which corresponds to ±2.5 kHz at 500 MHz. As seen in Figure 6.25, all of the protons in the sample would see 98 %–100 % of the power of the 500 MHz radiation delivered and all would be excited simultaneously. A pulse programmer is used to control the timing and shape of the RF pulses used to excite the sample. Square wave pulses are commonly used, but multipulse experiments and 2D NMR experiments with other pulse shapes are performed. There are hundreds of pulse sequences and 2D experiments that have been developed, with curious names like attached proton test (APT), DEPT, INEPT, INADEQUATE, COSY, and many more, some of which will be discussed later in the chapter. Each pulse sequence provides specific and unique NMR responses that enable the analyst to sort out the NMR spectrum and deduce the chemical structure of a molecule.

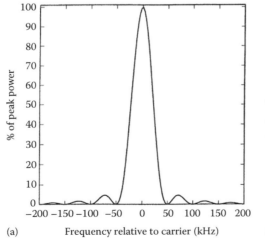 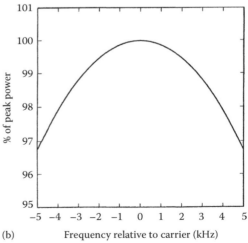

Figure 6.25 The power distribution for a 10 μs rectangular pulse of 500 MHz RF radiation. (a) Distribution over a 400 kHz range. (b) The power level drops to 97 % of the maximum at 5 kHz (10 ppm) on both sides of the center of the spectrum. This can affect the accuracy of integrals for resonances in different regions of the spectrum. (From Petersheim, M., Nuclear magnetic resonance, in Ewing, G.A., ed., *Analytical Instrumentation Handbook*, 2nd edn., Marcel Dekker, Inc., New York, 1997. Used with permission.)

All signals are collected simultaneously. The RF pulse delivered is generally on the order of watts, while the NMR signal collected is on the order of microwatts. The FID signal in the time domain must be converted to a frequency-domain spectrum by application of a Fourier transformation or other mathematical transformation. Commercial instruments generally use quadrature phase-sensitive detection to avoid spectrum artifacts resulting from the mathematical transformation. The operational details of the electronics required to provide the strong pulse and detect the very weak signals are complex and beyond the scope of this text. The interested student should consult the texts by Fukushima and Roeder and Akitt and Mann or the references cited by Petersheim for more details.

6.5.5 Signal Integrator and Computer

All NMR spectrometers are equipped with a signal integrator to measure the area of peaks. The area often appears on the NMR spectrum as a step function superimposed on the spectrum, as seen in Figure 6.19a, but is also printed out in a data report.

All NMR spectrometers require a computer to process the data and much of the instrument is under computer control. Multitasking or networked computer workstations are normally used, so that data can be processed while long experiments are running on the instrument. A typical single NMR data file can be processed in less than 1 s. Two- and three-dimensional NMR experiments generate megabytes to gigabytes of data that must be processed for each experiment. 2D NMR plot generation in the early 1980s required days of operator and computer time. The same 2D plots are now generated in minutes with today's improved software and faster computers.

Computer control of the room-temperature shim currents that control the field homogeneity, of sample spinning rates, autosamplers, pulse sequences, and many other aspects of the instrument operation is a necessity.

6.6 ANALYTICAL APPLICATIONS OF NMR

NMR spectroscopy is used for both qualitative and quantitative analyses. The applications of NMR are very diverse and only a very few examples can be given here. The student interested in NMR applications is advised to look at journals such as *Analytical Chemistry*, published by the American Chemical Society; *Applied Spectroscopy*, published by the Society for Applied Spectroscopy; and *Journal of Magnetic Resonance* (Elsevier) for an overview of the wide range of uses for NMR. Particularly useful is the annual issue of *Analytical Chemistry* dedicated to Applications Review (odd-numbered years) or Fundamental Reviews (even-numbered years). A list of specialty NMR journals can be found at www.spincore.com/nmrinfo. Many of the instrument company websites listed, such as JEOL, Bruker, Agilent Technologies, Anasazi (www.aiinmr.com), Oxford Instruments (www.oxinst.com), and Thermo Scientific (www.thermoscientific.com/picoSpin), have lots of application notes available to the public.

6.6.1 Samples and Sample Preparation for NMR

Liquid samples are the simplest samples to analyze by NMR. Neat low to moderate viscosity liquids are run "as is" by placing about 0.5 mL of the liquid in a glass NMR tube. Liquids can be mixed in a suitable solvent and run as solutions; the analyte concentration is generally about 2 %–10 %. For the examination of liquid samples, the sensitivity is sufficient to determine concentrations down to about 0.1 %. NMR has not been considered a "trace" analytical technique, but that is changing as instrumentation continues to improve. Microtubes with as little as 1 μg of sample in 100 μL of solvent are now in use. Samples in solution and neat liquids are degassed to remove oxygen and filtered to remove iron particles: both O_2 and iron are paramagnetic and cause undesired line broadening.

Soluble solid samples are dissolved in a suitable solvent, usually deuterated, for analysis. A typical sample size is 2–3 mg dissolved in 0.5 mL of solvent. Some solid polymer samples may be run under "liquid" conditions, that is, without MAS, by soaking the solid in solvent and allowing it to "swell." This gives enough fluidity to the sample that it behaves as a liquid with respect to the NMR experiment. Other solid samples must be run in an instrument equipped with MAS, as has been discussed. Sample solutions are generally filtered through alumina to remove any particles.

NMR does not have sufficient sensitivity to analyze gas-phase samples directly. Gases must usually be concentrated by being absorbed in a suitable solvent, condensed to the liquid phase, or adsorbed onto an appropriate solid phase. Recent experiments have shown that one can in fact measure the vapor liquid equilibria of a sample (that is, the coexisting vapor and liquid concentrations) in selected cases. NMR tubes capable of containing elevated pressure are available for some gas analyses.

A suitable solvent for NMR should meet the following requirements: (1) be chemically inert toward the sample and the sample holder, (2) have no NMR absorption spectrum itself or a very simple spectrum, and (3) be easily recovered, by distillation, for example, if the original sample is required for other testing. The best solvents for proton NMR contain no protons and therefore give no proton NMR signals. Carbon tetrachloride and carbon disulfide fall into this category. Replacing protons, ^1H, with deuterium, ^2H or D, will remove most of the proton signal for the solvent from the spectrum. Deuterated chloroform, $CDCl_3$; deuterated water, D_2O; and many other deuterated solvents are commercially available for use. Deuterated solvents have two drawbacks: they are expensive and they generally contain a small amount of ^1H, so some small signal from the solvent may be seen. A spectrum of the solvent (the blank spectrum) should be run regularly and whenever a new lot of solvent is used. Deuterated chloroform is the solvent used for most of the ^{13}C spectra shown in the chapter, and the small signal from the coupling of D with ^{13}C in the sample is visible in these spectra. An example will be pointed out when discussing these spectra.

6.6.2 Qualitative Analyses: Molecular Structure Determination

The primary use of NMR spectroscopy is for the determination of the molecular structure of compounds. These may be organic compounds synthesized or separated by organic chemists, pharmaceutical chemists, and polymer chemists; organic compounds isolated by biologists, biochemists, and medicinal chemists; and organometallic and inorganic compounds synthesized by chemists and materials scientists. The importance of NMR spectroscopy in deducing molecular structure cannot be overstated.

While some features of the proton NMR spectrum have been discussed earlier, we will go through some examples of how to use a proton NMR spectrum to work out the structure of some simple organic molecules. We need to understand a few more aspects of the NMR spectrum first.

6.6.2.1 Relationship between the Area of a Peak and Molecular Structure

As we learned in Section 6.4, the multiplicity of a given peak tells us the number of adjacent equivalent protons. The multiplicity tells us nothing about the number of protons that give rise to the peak itself. That information comes from the peak area. The total area of an absorption peak is directly proportional to the number of protons that resonate at the frequency. By total area, we mean the area of all the peaks in the multiplet, if the peak is a multiplet, or the total area of the peak if the peak is a singlet.

In a sample of dry ethanol, Figure 6.19a, the methyl group, CH_3, is split by the methylene group to give three peaks. The area corresponding to the methyl group is the area enclosed by all three peaks of the triplet measured from a baseline. The two protons of the methylene group are split by the methyl group and by the proton of OH into a total of eight peaks. The area contributed by the methylene protons is that enclosed by all eight peaks measured from a baseline on the spectrum. The total area of the methyl group, CH_3, will be equal to (3/2) × (the total area of the CH_2 group). The OH proton is split into a triplet by the methylene protons; the area corresponding to that proton is the entire area of its triplet measured from a baseline; the area corresponding to the hydroxyl proton should be one-third that of the methyl group.

In practice, the peak areas are measured by integrating the signal area automatically. The computer prints out the relative area. Relative areas are often shown as step function traces on the spectrum printout, as has been done for the spectrum in Figure 6.19a; each peak has a line traced over it with a baseline at the bottom and at the top. The peak areas are **relative**; it is not possible to assign a molecular formula to an unknown from the NMR spectrum alone. For the ethanol example given earlier, the relative areas of the peaks should be 3:2:1 for the peaks as shown from right to left, that is, A:B:C. For ethanol, the relative area of the peaks is also the absolute area. We know this *only* because we know the sample is ethanol and we know the molecular formula for ethanol. The areas are shown as the stepped lines drawn across the spectrum. In the rare case that the integrator is not functioning, the relative areas can be obtained the old-fashioned way, by measuring the height of each step with a ruler. Take a ruler (marked in any units as long as they are small enough, such as 1/16 in., cm, or mm divisions) and measure the height of the step for peak C, on the far left of the spectrum. It may help to extend the bottom and top plateaus using your ruler and a pencil. Measure the height of this step in mm and write down the value. Do the same for the other two steps for peaks B and A. Now, it is necessary to divide each of the measurements by the smallest measurement to get the relative peak areas. For the methyl peaks, divide the height of step A by the height of step C; the value should be equal to 3. Therefore, the area of peak A is 3 relative to peak C. As you can see by doing the other two ratios, the relative areas are 3:2:1 for peaks A, B, and C, respectively. It is important to measure and read the step heights as accurately as you can for this method to work well.

From this relative area calculation, we can state that there are twice as many protons in the group giving rise to peak B as there are in the group giving rise to peak C and that there are three times as many protons in the peak A group as in the peak C group. We *cannot* say that peak C is due to one proton and peak A due to three protons without more information. The empirical formula can be determined if we have elemental analysis information and the molecular formula can be calculated if the RMM is known from another measurement such as MS. Confirm that the isopropyl groups in Figure 6.18 fit the predicted 1:6 ratio using this method.

It is also important to keep in mind that the smallest peak may be equal to more than one proton. You can deduce this if your ratios look like 1.9:1.5:1, for example. Clearly, you cannot have a molecule with 1.5 protons on a carbon atom. Now, you need to multiply everything through by the same common factor, until you get whole number values. If you multiply 1.9, 1.5, and 1 by 2, you get 4, 3, and 2 protons, respectively, which is a reasonable set of whole numbers to work with in building a structure. A good practice example of this type is the spectrum of butylamine (see Figure 6.41).

6.6.2.2 Chemical Exchange

In a solution of methanol and water, or ethanol and water, the hydrogen of the alcohol OH group physically exchanges with hydrogen in the water. Such a proton is said to be *labile*. If this physical exchange rate is greater than the change in resonance frequency for the nuclei involved, the nearby nuclei see only the average position of the nucleus, and spin–spin splitting due to the labile proton disappears. The exchange rate is affected by temperature, increasing with increased temperature. In addition, the proton on the OH group can participate in hydrogen bonding, which is also temperature dependent, concentration dependent, and very solvent dependent. The spectra of alcohols are therefore affected by any traces of water in the sample and by temperature. Figure 6.26 is the 60 MHz spectrum of normal reagent grade ethanol, which contains water. Compare this spectrum with that of dry ethanol in Figure 6.19a. There are two dramatic differences. The hydroxyl peak at 4.40 ppm appears as a singlet, not a triplet, because the exchange rate is so fast that the hydroxyl proton is not split by the methylene protons. For the same reason, the methylene protons are not split by the hydroxyl proton, only by the methyl protons. Therefore, the methylene peak at 3.58 ppm appears as a quartet in normal ethanol. The peak at 3.32 ppm is due to an impurity and is not part of the methylene quartet. The spacing between the peaks in the triplet and the peaks in the quartet should be equal since $J_{AB} = J_{BA}$.

For another example, the spectra of CH_3OH at 20 °C and −40 °C are shown schematically in Figure 6.27a, demonstrating the effect of temperature on the spectrum. The rate of exchange is so rapid at the higher temperature that no spin–spin splitting is seen. Changes in hydrogen bonding will dramatically affect the chemical shift position of the alcohol proton. The same is true for any proton capable of hydrogen bonding, which means that any compound with an NH or OH proton has a spectrum dependent on temperature, concentration, and solvent polarity.

If the chemical exchange frequency is lower than the spectral frequency for the nuclei of interest, there will be a discrete set of resonances for each state. Processes with intermediate rates are studied by physical, organic, and inorganic chemists using NMR because the positions and shapes of the peaks can be used to estimate reaction kinetics and the lifetimes of reactants and products in a reaction.

Chemical exchange is not limited to the exchange of hydrogen-bonded protons in solution. Chemical exchange includes conformational changes as well as actual bond-breaking and bond-forming changes that result in new chemical shifts for the nuclei involved. Examples include dimer formation and tautomerism. The important point is that chemical exchange can alter the appearance of an NMR spectrum, as is clear from the alcohol example. This must be remembered in interpreting NMR spectra.

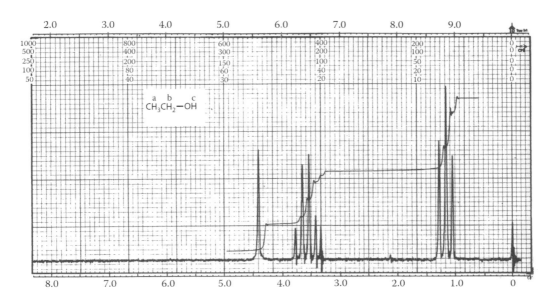

Figure 6.26 The 60 MHz proton NMR spectrum of normal reagent grade ethanol, which always contains water. Compare this spectrum with that in Figure 6.19a. Note that in the presence of water, the spin–spin coupling due to the OH proton disappears. (From *The Sadtler Handbook of Proton NMR Spectra*, W.W. Simons, ed., Sadtler Research Laboratories, Inc., Philadelphia, PA, 1978. [Now Bio-Rad Laboratories, Inc., Informatics Division, Sadtler Software and Databases. Copyright Bio-Rad Laboratories, Informatics Division, Sadtler Software and Databases, 2014. All rights reserved.])

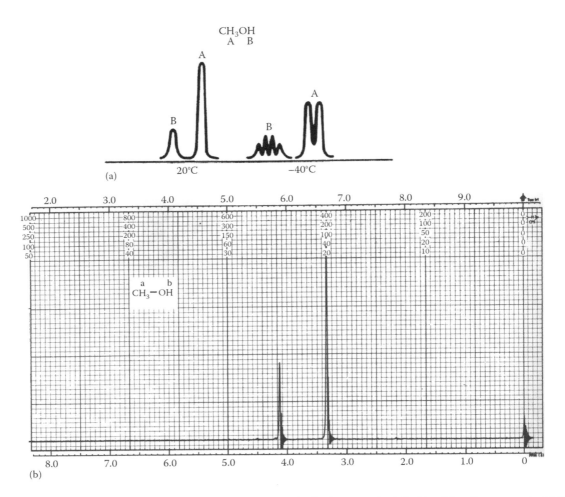

Figure 6.27 (a) Schematic proton spectra of methanol obtained at 20 °C, shown on the left side of the diagram, and −40 °C, on the right of the diagram. The fine structure (spin–spin splitting) is not observed at 20 °C due to the rapid exchange rate of the hydroxyl proton. Cooling the sample to −40 °C slows the exchange rate sufficiently that the hydroxyl proton can be considered fixed in place; spin–spin coupling is then observed. (b) The 60 MHz proton spectrum of reagent grade methanol at room temperature, with no fine structure observed. (From *The Sadtler Handbook of Proton NMR Spectra*, W.W. Simons, ed., Sadtler Research Laboratories, Inc., Philadelphia, PA, 1978. [Now Bio-Rad Laboratories, Inc., Informatics Division, Sadtler Software and Databases. Copyright Bio-Rad Laboratories, Informatics Division, Sadtler Software and Databases, 2014. All rights reserved.])

6.6.2.3 Double-Resonance Experiments

When we first examine an NMR spectrum of an unknown sample, it is often not easy to tell which nuclei are coupled, especially if the splitting patterns are complex. Double-resonance experiments employ two different RF sources and a variety of pulse sequences to sort out complex splitting patterns. It was discussed in Section 6.1.2 that a system of protons could become saturated by applying a strong RF field to the sample. When saturation occurs, the saturated protons do not give an NMR signal and they do not couple with and split the peaks for adjacent protons. The saturated proton is said to have been "decoupled." Advantage is taken of this phenomenon to simplify complicated NMR spectra. In a double-resonance **spin-decoupling** experiment for proton NMR, one RF source is scanned as usual. The second RF source selectively saturates one resonance frequency (one group or type of protons). This causes the collapse of all the splitting patterns of nuclei to which that group is coupled. Irradiating each resonance frequency in turn and observing which peaks have their fine structure disappear permits identification of all the coupled protons. For example, a peak that had been a triplet will collapse into a singlet if the methylene group that is causing the spin–spin splitting is saturated. A simple example of the use of spin decoupling by double resonance is shown in Figure 6.25. The proton spectrum of ethylbenzene is shown in Figure 6.28a. If the methyl resonance is saturated, the NMR spectrum obtained is that shown in Figure 6.28b. The signal for the methyl group at 1.2 ppm disappears, and the methylene quartet at 2.6 ppm collapses into a singlet, because the methyl protons no longer split the methylene protons. If in a second experiment the methylene resonance

MAGNETIC RESONANCE SPECTROSCOPY

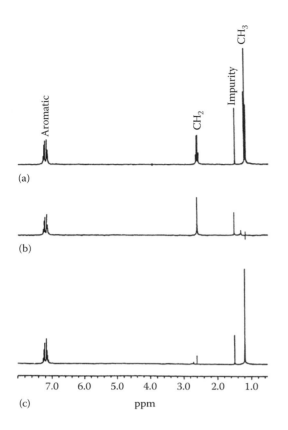

Figure 6.28 Homonuclear decoupling experiment. A 250 MHz ^{1}H NMR of ethylbenzene in deuterated chloroform obtained (a) without decoupling, (b) with irradiation of the methyl resonance, and (c) with irradiation of the methylene resonance. (From Bruch, M.D. and Dybowski, C., Spectral editing methods for structure elucidation, in Bruch, M.D., ed., *NMR Spectroscopy Techniques*, 2nd edn., Marcel Dekker, Inc., New York, 1996. Used with permission.)

is saturated, the spectrum in Figure 6.25c is obtained. The methylene signal at 2.6 ppm disappears and the methyl triplet collapses into a singlet. From these two experiments, it can be deduced that the methyl and methylene groups are coupled and therefore are adjacent to each other.

A more complex spectrum and decoupling experiments are shown for sucrose in Figure 6.26. The structure of sucrose is shown in Figure 6.29. Ignoring the hydroxyl protons, each methine proton and methylene group is chemically different, giving rise to 11 resonance peaks in the spectrum. The fully coupled spectrum for sucrose dissolved in deuterated water, D$_2$O, is shown in (c), and the exchange rate is fast enough that no coupling is seen from the hydroxyl protons. Look at the top spectrum (a) and compare it with spectrum (c). Saturation of the triplet at 4.05 ppm, which is due to the proton marked (a) in the structure, causes the triplet to disappear from spectrum (a). It also collapses the doublet at 4.22 ppm, indicating that proton (a) is coupled to the proton causing the absorbance at 4.22 ppm. If we know that the peak at 4.05 is due to proton (a), it follows that the peak at 4.22 ppm must be from the proton marked (b). [Why? Why can it not be the proton on the carbon to the right of proton (a)? Hint: think about the splitting.] Now look at spectrum (b). Saturation of the doublet at 5.41 ppm, due to proton (c), causes the doublet to disappear from spectrum (b) and also collapses the doublet of doublets at 3.55 ppm, leaving a doublet. This tells us that the resonance at 3.55 ppm is due to proton (d). [Why? Why does proton (d) appear as a doublet of doublets in the fully coupled spectrum?] The singlet peak at 4.8 ppm in each spectrum in Figure 6.26 is due to residual HDO in the D$_2$O solvent.

By using double-resonance experiments, one can greatly simplify the spectrum. Coupling between different types of nuclei is confirmed by both the disappearance of the peak for the saturated nuclei and the collapse of the fine structure of the coupled nuclei. Nuclei of the same type can be decoupled, as in the proton–proton example given earlier; this is called **homonuclear decoupling**. It is of course possible to decouple unlike nuclei, such as ^{1}H–^{13}C decoupling; this is called **heteronuclear decoupling**. Both of these are examples of spin decoupling.

In modern instruments, it is not uncommon to use double- and triple-resonance experiments to simplify the spectrum sufficiently for interpretation. One interesting result of the use of double resonance, the **nuclear Overhauser effect** (NOE), is very important in ^{13}C NMR and will be discussed in Section 6.6.4. Table 6.5 lists a few commonly used double-resonance, multipulse, and 2D NMR experiments by name. Several of these experiments will be explained in more detail subsequently.

6.6.3 Interpretation of Proton Spectra

NMR absorption spectra are characterized by the chemical shift of peaks and spin–spin splitting of peaks. Recall that the chemical shift is caused by the drifting, not orbiting or spinning, of nearby electrons under the influence of the applied magnetic field. It is therefore a constant depending on the applied field (i.e., if the field is constant, the chemical shift is constant). The chemical shift therefore identifies the functional group, such as methyl, methylene, aldehydic H, aromatics, and so on (see Table 6.3). All proton spectra shown have TMS as the reference, with the TMS absorbance set at 0.0 ppm. The student should note that all of the 300 MHz proton NMR spectra provided by Aldrich Chemical Company, Inc., also include the 75 MHz ^{13}C spectrum at the top. ^{13}C NMR spectra are discussed in Section 6.6.4.

Spin–spin splitting is caused by adjacent nuclei and is transmitted through the bonds. It is independent of the applied field. The multiplicity is therefore a function of the number of equivalent ^{1}H nuclei in the adjacent functional groups. Numerically, it is equal to $(2nI + 1)$, where n is the number of equivalent ^{1}H and I is the spin number (in this

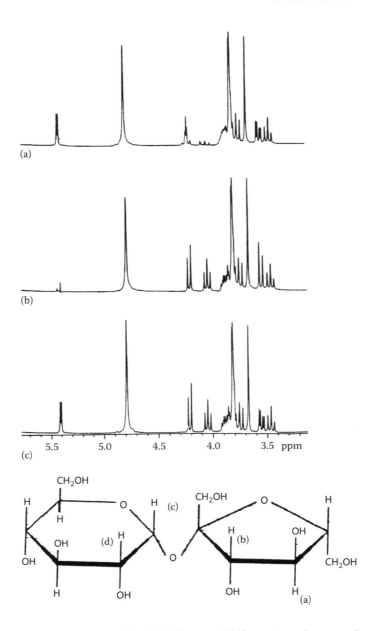

Figure 6.29 Homonuclear decoupling experiments of the 300 MHz proton NMR spectrum of sucrose dissolved in D_2O. The fully coupled spectrum is shown in (c). (a) Selective saturation of the triplet at 4.05 ppm collapses the doublet at 4.22 ppm, showing the coupling between the positions of protons a and b marked on the sucrose structure. (b) Saturation of the doublet at 5.41 ppm collapses the doublet of doublets. The experiment shows the coupling between the protons marked c and d on the structure. (The spectra are from Petersheim, M., Nuclear magnetic resonance, in Ewing, G.A., ed., *Analytical Instrumentation Handbook*, 2nd edn., Marcel Dekker Inc., New York, 1997. Used with permission. The sucrose structure is that of D-(+)-sucrose, obtained from the SDBS database, courtesy of the National Institute of Industrial Science and Technology, Japan, http://aist.go.jp/RIOBD/SDBS. Accessed on May 11, 2002.)

case, $I = 1/2$). For two adjacent groups, the number is $(2nI + 1)(2n'I' + 1)$ where n and n' are the numbers of ^1H nuclei in each separate group and I and I' each equal 1/2. It can readily be seen from the real spectra we have already looked at that the intensity ratios in multiplets are often not the symmetrical intensities predicted from Pascal's triangle. Look, for example, at the peak (a) triplet in Figures 6.13b and c. It is certainly not 1:2:1 or symmetrical. This deviation from theory actually provides us with more structural information, as we will see.

Another piece of information is obtained from the relative area of the absorption peaks in the spectrum, which tells us the relative number of protons in each group. So, a proton NMR spectrum should be examined for (1) the number of proton resonances, which tells you how many different types of protons are in the molecule; (2) the

Table 6.5 Double-resonance, multipulse, and 2D NMR experiments

Acronym	Experiment Description
APT: Attached Proton Test	Multipulse sequence used to distinguish even and odd numbers of protons coupled to ^{13}C through one bond. Even numbers of bound protons give positive peaks; odd numbers give negative peaks.
DEPT: Distortionless enhancement by polarization transfer	Multipulse; conditions are chosen so that only ^{13}C nuclei with the same number of bound protons have enhanced resonances. Four experiments must be performed, but DEPT is more definitive than APT.
INADEQUATE: Incredible natural abundance double quantum transfer	Multipulse; allows observation of natural abundance $^{13}C-^{13}C$ coupling. Only 1 carbon atom in 10^4 carbon atoms is a ^{13}C bonded to another ^{13}C, hence the name "incredible."
COSY: Correlated spectroscopy	Homonuclear 2D experiment; plot of chemical shift versus chemical shift identifies spin-coupled resonances. Many variations permit measurement of J coupling constants, long-range connectivity, suppression, and enhancement of selected resonances.
HETCOR: Heteronuclear chemical shift correlated experiment	Heteronuclear 2D experiment; usually to connect ^1H resonances with ^{13}C resonances or ^1H-X, where X is another NMR-active nucleus. Plot is ^{13}C chemical shift (or X chemical shift) versus ^1H chemical shift.
NOESY: Nuclear Overhauser effect spectroscopy	Identifies dipolar-coupled nuclei within certain distances (e.g., within 0.4 nm for first-order coupling) and identifies connectivities through cross-relaxation.

Source: Modified from Petersheim, M., Nuclear magnetic resonance, in Ewing, G.A. ed., *Analytical Instrumentation Handbook*, 2nd edn., Marcel Dekker Inc., New York, 1997. With permission.

chemical shift of the resonances, which identifies the type of proton; (3) the multiplicity of the resonances, which identifies the adjacent equivalent protons; and (4) the intensity (area) of the resonances, which tells you the relative number of each type of proton. Some examples of common classes of organic compounds are discussed. Not every type of compound is covered in this brief overview. Students needing more detailed spectral interpretation should consult the book by Bruno and Svoronos.

6.6.3.1 Aliphatic Alkanes, Alkenes, Alkynes, and Alkyl Halides

Alkane protons absorb in the region between 0.6 and 1.7 ppm. The spectrum of octane, $CH_3(CH_2)_6CH_3$, Figure 6.30, is typical of straight-chain alkanes. The methyl groups absorb at 0.88 ppm and the methylene groups at 1.26 ppm. In long-chain alkanes, the methylene protons often overlap and spin–spin splitting cannot be resolved. The relative area ratio of the methylene peak/methyl peak is expected to be 12:6 or 2:1, which can be confirmed by measuring the relative height of the steps shown on the spectrum. The peak at 0.88 ppm due to the methyl groups is split into a triplet by the adjacent methylene protons as expected.

Substitution of an electronegative halide atom, F, Cl, Br, or I, on an alkane will deshield any protons attached to the same carbon atom. The deshielding is a result of the electronegativity of the halogen atom. Fluorine shows the largest deshielding effect, iodine the smallest. Two halogen atoms on a single carbon have a greater deshielding effect than a single halogen atom. Since ^{19}F is an $I = 1/2$ nucleus, fluorine-substituted hydrocarbons will show H–F and F–F spin–spin coupling. The other halogens do not couple.

Figure 6.31a is the 60 MHz proton spectrum of bromoethane (or ethyl bromide). The structure and chemical shifts are shown. The methyl peak is split into a triplet by the methylene group and the methylene group is split into a quartet by the methyl group. The peak a/peak b area ratio is expected to be 3:2. The position of the methylene peak is shifted to a significantly higher resonance frequency by the deshielding due to Br. It occurs at about 3.4 ppm compared with the normal position of 1.2–1.4 ppm for methylene protons in an alkane such as octane (Figure 6.30). The electronegative Br also affects the resonance position of the adjacent methyl protons, moving them to a higher resonance frequency as well, 1.7 ppm versus the normal alkane position of about 0.9 ppm. Now look at the triplet; you will note that it is not symmetrical. The intensity of the peak on the left side of the triplet is higher than that of the peak on the right. Look at the quartet; you should note that the two peaks on the right of the quartet are higher than the two peaks on the left side. The triplet appears to be "leaning" toward the quartet and the quartet is "leaning" toward the triplet. This is a clue that the two peaks are coupled together. The J coupling constant should be identical for the triplet and the quartet (i.e., $J_{AB} = J_{BA}$). You should be able to measure it from this spectrum. Figure 6.31b is the 300 MHz proton spectrum of bromoethane and its ^{13}C spectrum. You can confirm the 3:2 area ratio from the height of the steps plotted on the 300 MHz spectrum.

The compound chloroethane (or ethyl chloride), CH_3CH_2Cl, would show the same splitting pattern as bromoethane; the position of the methylene peak would be slightly more deshielded (3.5 vs. 3.4 ppm) due to the greater electronegativity of Cl. In both of these compounds, the electronegative halogen has a deshielding effect on the adjacent methyl protons as well; they absorb at higher chemical

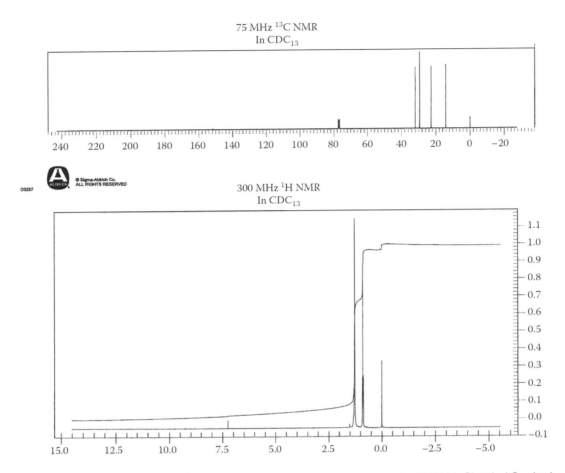

Figure 6.30 Proton and ^{13}C NMR spectra of octane, $CH_3(CH_2)_6CH_3$. (Reprinted with permission of Aldrich Chemical Co., Inc.)

shift (1.2–1.5) than they would in an alkane (0.7–1.0 ppm). Figure 6.32a shows the effect of two chlorine atoms on the same carbon; the proton on that carbon is moved to 5.9 ppm by the deshielding. Compare the position of this proton and the methyl protons to the chemical shifts given for chloroethane and shown for bromoethane (Figure 6.31). You will see that the two chlorine atoms deshield the adjacent methyl protons as well and by a larger amount than a single chlorine atom. The methyl group is split into a doublet by the adjacent proton, while the single proton (the peak at 5.9 ppm) is split into a quartet by the methyl group. In Section 6.4, we hypothesized what spin–spin splitting would be seen for ethane substituted with two and three chlorine atoms. The 60 MHz spectra are given in Figures 6.32b–d. You should confirm that the predictions in Section 6.4 were correct. As a final example of this type of compound, Figure 6.33 shows the 60 MHz spectrum of 1,3-dichloropropane. The "a" methylene protons are split into five peaks by the four equivalent "b" methylene protons. The "a" protons are also deshielded; they absorb at 2.2 ppm instead of the 1.2–1.4 ppm position in an alkane. The "b" methylene protons are split into a triplet by the two "a" protons and are more strongly deshielded since the Cl atom is on the same carbon atom as the "b" protons. The expected area ratio of peak a/peak b is 1:2. There is asymmetry in the multiplets; they are "leaning" toward one another as expected.

Alkenes have two characteristic types of protons. The vinyl protons are those attached to a double bond. Vinyl protons are deshielded by the double bond and generally appear at chemical shifts between 4.5 and 6 ppm. The spin–spin coupling of vinyl protons is complicated because they are generally not chemically equivalent due to the lack of free rotation about the double bond. Allylic protons are those located on carbon atoms adjacent to the double bond; these are slightly deshielded by the double bond and appear between 1.2 and 2.5 ppm. The presence of the double bond allows long-range coupling, so the allylic protons may couple to the protons on the far end of the double bond as well as to the adjacent protons. Figure 6.34 shows the spectrum of 1-octene, $CH_3(CH_2)_5CH=CH_2$ or C_8H_{16}. The structure of 1-octene is shown on the spectrum. Note the shift of the allylic protons (the "c" protons) to 2 ppm by deshielding from the double bond and the strong deshielding of the vinyl protons, "d, e, and f." All three vinyl protons are nonequivalent, resulting in the complex splitting pattern

MAGNETIC RESONANCE SPECTROSCOPY

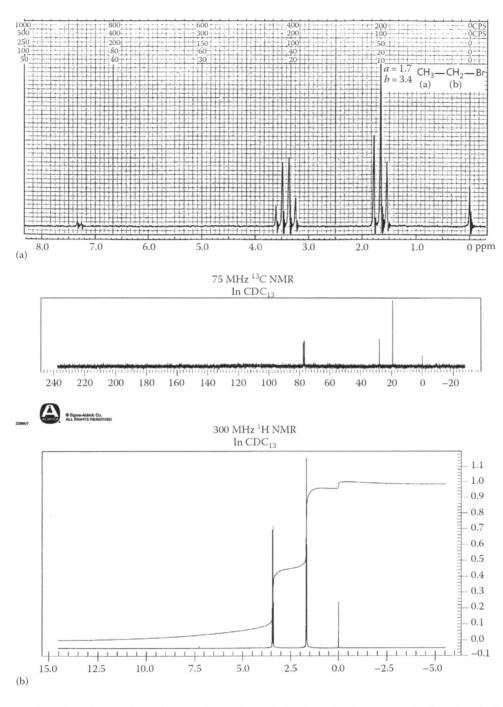

Figure 6.31 (a) The 60 MHz proton spectrum of bromoethane. The methyl protons at 1.7 ppm are split into a triplet by the two methylene protons, since $(2nI + 1) = (2)(2)(1/2) + 1 = 3$. The methylene protons are deshielded by the Br and moved to 3.4 ppm. They are split into a quartet by the adjacent methyl protons, $(2nI + 1) = (2)(3)(1/2) + 1 = 4$. (b) The 300 MHz proton spectrum and ^{13}C spectrum for bromoethane. The 3:2 proton area ratio can be measured from this spectrum. (Reprinted with permission of Aldrich Chemical Co., Inc.)

observed in the 4.9–5.8 ppm region. The terminal methyl group peak is a triplet, but the methylene groups in the middle of the chain overlap to give the broad peak from 1.0 to 1.5 ppm. The relative peak areas, from left to right, should be 1:2:2:8:3.

An alkyne with a triple bond at the end of a chain is called a terminal alkyne and the hydrogen atom at the end of the triple bond is referred to as an acetylenic hydrogen. This terminal proton is shielded by the anisotropy of the triple bond π-electrons, as shown in Figure 6.9, and so absorbs

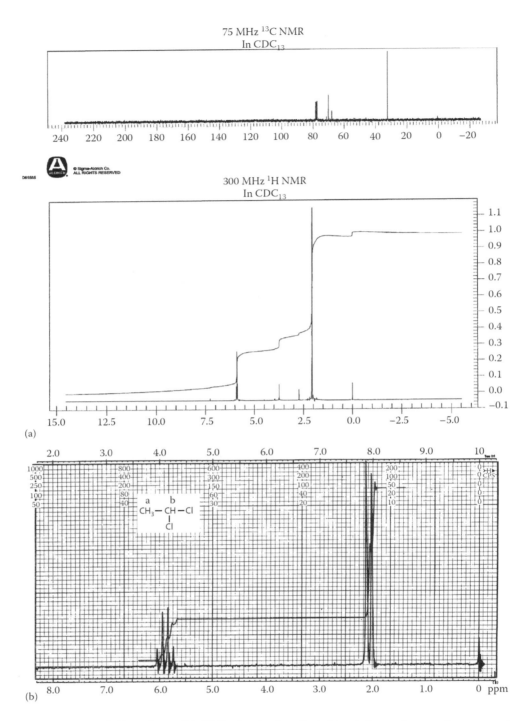

Figure 6.32 (a) Spectrum of 1,1-dichloroethane, CH_3CHCl_2, showing the effect of two halogen atoms on the same carbon on the resonance frequencies of the protons. The CH_3 group absorbs at 2.06 ppm, while the proton on the chlorine-containing carbon is highly deshielded and absorbs at 5.89 ppm. The J_{AB} coupling constant is 6.0 Hz. (Reprinted with permission Aldrich Chemical Co., Inc.) (b) The 60 MHz proton spectrum of 1,1-dichloroethane, CH_3CHCl_2. (c) The 60 MHz proton spectrum of 1,1,2-trichloroethane, $CHCl_2CH_2Cl$.

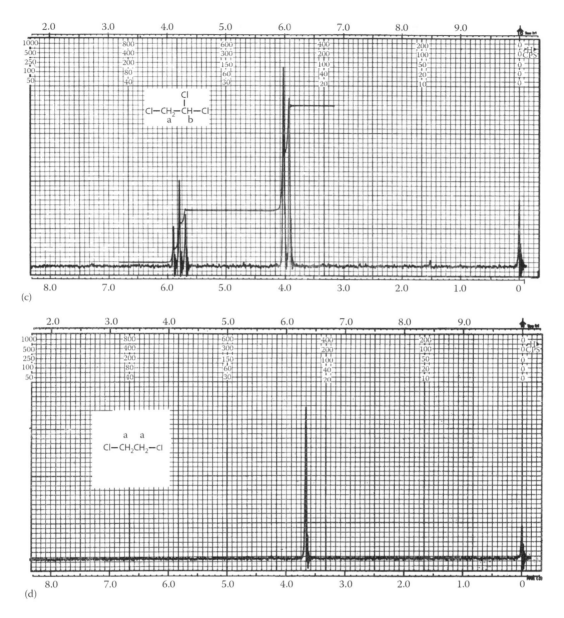

Figure 6.32 (Continued) (d) The 60 MHz proton spectrum of 1,2-dichloroethane, CH_2ClCH_2Cl. (Spectra [b–d] from *The Sadtler Handbook of Proton NMR Spectra*, W.W. Simons, ed., Sadtler Research Laboratories, Inc., Philadelphia, PA, 1978. [Now Bio-Rad Laboratories, Inc., Informatics Division, Sadtler Software and Databases. Copyright Bio-Rad Laboratories, Informatics Division, Sadtler Software and Databases, 2014. All rights reserved.])

at about 1.8 ppm. The protons on the carbon next to the triple bond are affected in the same way as allylic protons in alkenes and absorb in the same chemical shift range.

6.6.3.2 Aromatic Compounds

The proton absorbances used to identify aromatic compounds are the protons on the aromatic ring itself, the *ring protons*, and the protons on the carbon atoms adjacent to the ring. The latter are called *benzylic protons*. Protons attached directly to an aromatic ring are highly deshielded by the π-electrons, as shown in Figure 6.9. Ring protons therefore absorb between 6.5 and 8.0 ppm. There are few other protons that absorb in this region, so an absorbance in this region is characteristic of aromatic compounds. The spectrum of toluene, Figure 6.4, shows this characteristic absorbance; the ring protons give rise to the signal at 7.2 ppm. This peak shows a complex splitting pattern, because the ring protons are not equivalent; they are ortho, meta, and para to the substituent. The splitting pattern and *J* values can identify the number and position of substituents on the ring. A few examples will be shown but not discussed in detail. The patterns are second order and the details are beyond the scope of this text, but

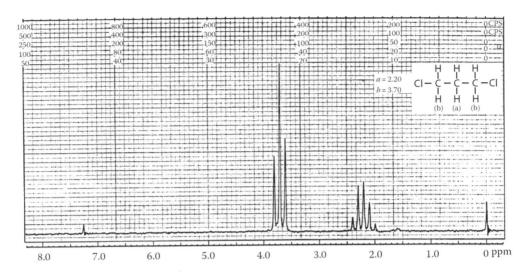

Figure 6.33 The 60 MHz proton spectrum of 1,3-dichloropropane.

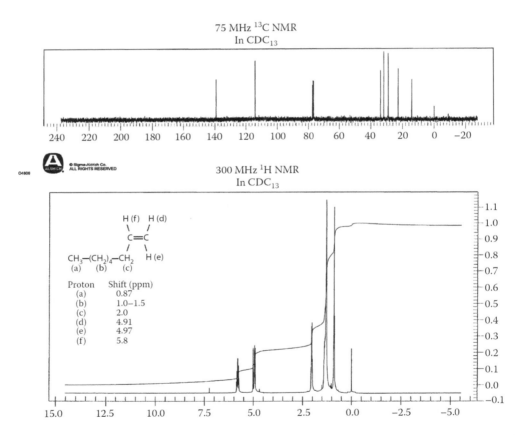

Figure 6.34 Spectra for 1-octene. The chemical shift of the terminal protons on the C=C bond is between 5 and 6 ppm. Protons (d) and (e) are not equivalent because the double bond does not permit rotation of these protons. (Reprinted with permission of Aldrich Chemical Co., Inc. with the structure and chemical shift information added by the authors.)

the interested student can consult the texts by Silverstein and Webster, Pavia et al., or Lambert et al. listed in the bibliography. One typical pattern is seen in Figure 6.35, the proton spectrum of 1-bromo-4-chlorobenzene, a para-disubstituted ring. The molecule contains a plane of symmetry, so that there are two sets of equivalent protons on the ring, as shown:

$$Cl-\underset{H_{(b)}\ H_{(a)}}{\underset{H_{(b)}\ H_{(a)}}{\bigcirc}}-Br \qquad \text{Plane of symmetry}$$

This results in a characteristic pattern of four lines in the aromatic region.

If both para substituents were the same, as in 1,4-dibromobenzene, all four protons would be equivalent and a singlet peak would be seen (Figure 6.35b). For an ortho-substituted ring, a plane of symmetry is also seen, as in o-dichlorobenzene:

$$\underset{\underset{\text{Plane of symmetry}}{H_{(b)}\ H_{(b)}}}{\overset{Cl\ \ Cl}{\underset{H_{(a)}}{\bigcirc}}H_{(a)}}$$

The spectrum shown in Figure 6.35c shows a complex second-order pattern for the protons. The same is true of meta-substituted rings (Figure 6.35d). Compounds such as o-dichlorobenzene and m-dichlorobenzene have a plane of symmetry, but as seen in Figure 6.32, the splitting patterns can be complex.

The *benzylic protons* on carbon atoms adjacent to an aromatic ring are also deshielded by the ring but to a lesser extent than the ring protons. The characteristic absorbance values for benzylic protons are in the 2.2–2.8 ppm range. An example is ethylbenzene, $C_6H_5-CH_2-CH_3$, whose spectrum is shown in Figure 6.36. The compound is a monosubstituted aromatic ring, so the ring protons at 7.2 ppm show a complex splitting pattern. The benzylic protons are the methylene protons, on the carbon attached to the ring. The quartet appears at ~2.6 ppm as expected. The methyl triplet is the peak at 1.2 ppm.

6.6.3.3 Oxygen-Containing Organic Compounds

Alcohols are organic compounds containing the hydroxyl group, –OH, attached to a carbon atom. As we have discussed and already seen in Figures 6.19, 6.26, and 6.27, the chemical shift of the hydroxyl proton depends on variables such as temperature, solvent, and concentration. The range of chemical shifts covers 0.5–5.0 ppm for aliphatic alcohols. The hydroxyl proton on an –OH group attached to an aromatic ring is deshielded by the ring, as can be seen in the spectrum of phenol, C_6H_5OH (Figure 6.37). Any other protons on the carbon to which the hydroxyl group is attached, $-CH_xOH$, are shifted due to the electronegative oxygen atom. These protons absorb between 3.1 and 3.8 ppm. In Figure 6.26, for example, we see these protons located at a chemical shift of ~3.58 ppm; in Figure 6.27b, they absorb at 3.34 ppm.

Aldehydes have a terminal $-\overset{\overset{O}{\|}}{C}-O$ group. We have discussed the spectrum of butanal, shown in Figures 6.13b and c. The aldehydic proton occurs at ~9.0–10 ppm. Protons on the carbon adjacent to the C=O are deshielded slightly and occur at about 2–2.5 ppm. Ketones contain a nonterminal —C=O group, so a ketone is represented by

$$RCH_{(x)}-\underset{R'}{\overset{\overset{O}{\|}}{C}}=O$$

The protons on the carbon adjacent to the C=O group are in the same environment as those in an aldehyde. If you replace the R′ with H, you have the formula for an aldehyde. Therefore, the protons also absorb at about 2–2.5 ppm due to slight deshielding by the C=O. The spectrum of cyclohexanone, $C_6H_{10}O$, is presented in Figure 6.38. The four equivalent protons on the two carbon atoms adjacent to the carbonyl group appear at 2.4 ppm.

Carboxylic acids contain the functional group $-\overset{\overset{O}{\|}}{C}OH$, also written –COOH. The acidic proton, the one on the COOH group, is strongly deshielded and absorbs at 10–13 ppm. A peak in this position is a good indication of a carboxylic acid. As is the case for ketones and aldehydes, the protons on the carbon atom adjacent to the –COOH group are slightly deshielded and absorb in the 2.1–2.5 pm range. A simple example is acetic acid, CH_3COOH, whose spectrum is given in Figure 6.39. The acid proton occurs at about 11.8 ppm, while the methyl protons, adjacent to the COOH group, appear at 2.1 ppm. The 90 MHz spectrum of benzoic acid (Figure 6.40) shows the structure and assignments. The complex splitting pattern for a monosubstituted benzene ring is clearly seen in this spectrum. The ortho protons are more deshielded than either the meta or para protons; this is typical when a carbonyl or like group is bonded to the ring and is seen in benzaldehyde, nitrobenzene, and similar compounds.

6.6.3.4 Nitrogen-Containing Organic Compounds

Proton NMR spectra of compounds such as amines, RNH_2, and amides, $RCONH_2$, with protons bonded to the nitrogen atom, are complicated by several factors. The protons on nitrogen in these compounds, like the hydroxyl proton in alcohols, exhibit a very variable chemical shift. These protons can hydrogen-bond, so the chemical shift depends on temperature, solvent, and concentration. The ^{14}N nucleus is a spin = 1 nucleus, so in theory, a proton on a nitrogen atom should be split into $2(1) + 1 = 3$ peaks. Similarly, a proton on a carbon atom adjacent to a nitrogen

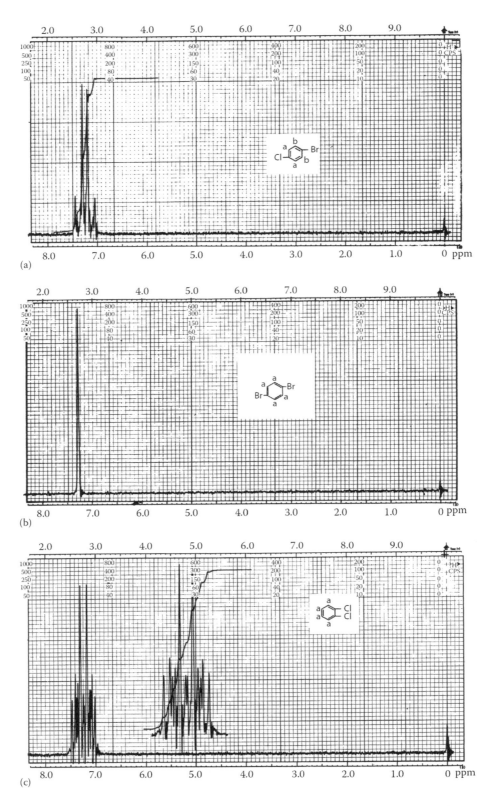

Figure 6.35 (a) The 60 MHz spectrum of 1-bromo-4-chlorobenzene (the only absorbance seen is due to the aromatic ring protons). The plane of symmetry in this molecule results in the characteristic "four-line" splitting pattern in the aromatic ring signal. (b) The 60 MHz spectrum of 1,4-dibromobenzene, showing that all protons are equivalent. (c) The 60 MHz spectrum of o-dichlorobenzene.

MAGNETIC RESONANCE SPECTROSCOPY

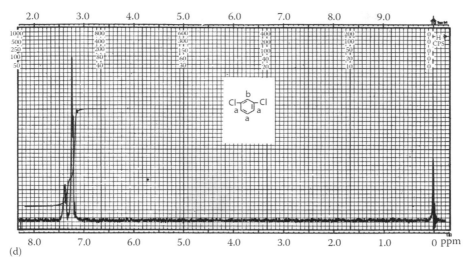

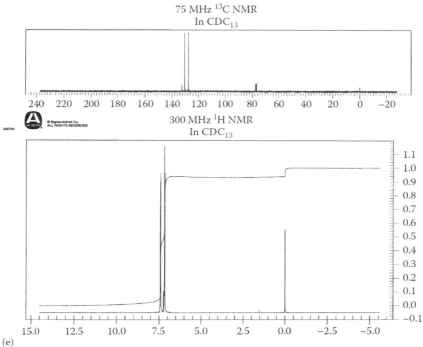

Figure 6.35 (Continued) (d) The 60 MHz spectrum of *m*-dichlorobenzene. (a–d: From *The Sadtler Handbook of Proton NMR Spectra*, W.W. Simons, ed., Sadtler Research Laboratories, Inc., Philadelphia, PA, 1978. [Now Bio-Rad Laboratories, Inc., Informatics Division, Sadtler Software and Databases. Copyright Bio-Rad Laboratories, Informatics Division, Sadtler Software and Databases, 2014. All rights reserved.]) (e) The 300 MHz proton spectrum of 1,2-dichlorobenzene (*o*-dichlorobenzene, compare with (c) earlier). (Reprinted with permission of Aldrich Chemical Co., Inc.)

atom should be split into three peaks. This is usually not seen for two reasons. Protons on electronegative nitrogen can undergo exchange (just like alcohols), and the rate of exchange will determine if the proton is coupled to the nitrogen. In aliphatic amines and amides, the exchange rate is fast enough that no splitting is seen. In addition, nitrogen has an electrical quadrupole moment. This quadrupole moment interacts in such a way as to broaden N–H peaks, resulting in no observed splitting. The N–H peak in amines can vary in chemical shift from 0.5 to 4.00 ppm, that in amides from 5 to 9 ppm. The peaks can be sharp singlets or weak broadened signals. Coupling between N–H, N–CH, and –HC–NH– is usually not seen because of proton exchange and quadrupole interaction. The protons on the carbon atom adjacent to the nitrogen atom in amines are shifted to 2.3–3 ppm; the protons on the carbon atom adjacent to the amide group are shifted into the 2–2.5 ppm region, which is the same region where protons adjacent to a C=O absorb.

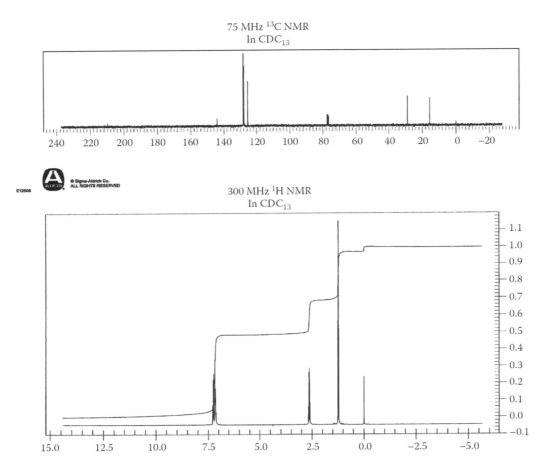

Figure 6.36 The spectra of ethylbenzene, showing the characteristic aromatic ring and benzylic proton absorbances typical of alkyl-substituted aromatic compounds. (Reprinted with permission of Aldrich Chemical Co., Inc.)

The chemical shift position of the proton on nitrogen in amines is not a good diagnostic tool because of its variability. In general, the peak due to the amine proton will be a sharp singlet in aliphatic amines because of fast proton exchange. A typical aliphatic amine spectrum is that of butylamine, $CH_3(CH_2)_2CH_2NH_2$ (Figure 6.41), with the structure and peak assignments shown. The amine proton peak is peak d, the sharp singlet at 1.1 ppm. The relative peak areas from the step heights are, from left to right, 2:4:1:3. This is a good spectrum on which to practice the manual step height measurement technique described earlier. The proton spectrum of propionamide, $CH_3CH_2CONH_2$, is shown in Figure 6.49 and should be looked at. There are two very small broad peaks located at 6.3 and 6.6 ppm. These are due to the two protons on the nitrogen atom; the peaks are broad and therefore of low intensity due to the nitrogen quadrupole interaction. The fact that there are two peaks means that these protons are not equivalent. But the formula for the molecule seems to show that the two protons on the nitrogen atom are chemically the same. We need to think back to general chemistry and Lewis structures. The two protons are not equivalent, because the nitrogen atom has an unshared pair of electrons on it. The unshared pair of electrons on nitrogen interacts with the unshared pairs of electrons on the oxygen. The interaction restricts rotation about the C–N bond, making the two protons on nitrogen nonequivalent; this results in two separate peaks. The observation of these two peaks is typical of primary amides, that is, amides with an –NH_2 group.

Compounds containing a nitro group, –NO_2, have a characteristic shift for the protons on the carbon atom adjacent to the nitro group. These protons absorb between 4 and 4.7 ppm, as seen in the spectrum of nitropropane, $CH_3CH_2CH_2NO_2$ (Figure 6.42), or 2-nitropropane (Figure 6.18a). The methylene group next to the nitro group gives the triplet peak at 4.3 ppm in nitropropane, with the splitting due to the adjacent methylene group, not to the nitro group. In 2-nitropropane, $(CH_3)_2CHNO_2$, the methine proton is a septuplet shifted to 4.65 ppm.

6.6.4 ^{13}C NMR

All nuclei with an odd mass number have a fractional (1/2, 3/2, etc.) spin number. Also, all nuclei with an even mass number and an odd atomic number have a unit (1, 2,

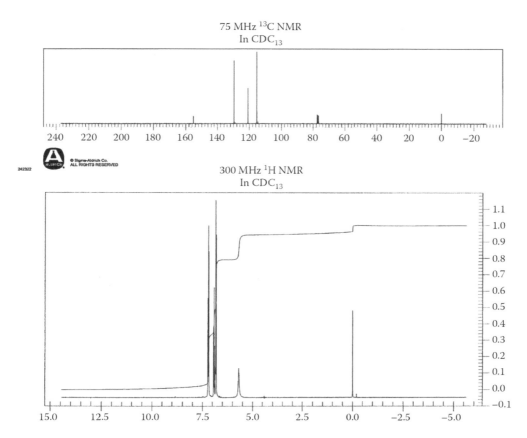

Figure 6.37 The spectra of phenol, C_6H_5–OH, an aromatic alcohol. The hydroxyl proton is deshielded by the aromatic ring and appears at a chemical shift of 5.6 ppm. (Reprinted with permission of Aldrich Chemical Co., Inc.)

3, etc.) spin number. In fact, only those nuclei with an even mass and an even atomic mass have a zero spin number and therefore give no NMR signal. Unfortunately, this includes ^{12}C and ^{16}O—two important nuclei in organic chemistry. Carbon is the underlying "backbone" of organic molecules, and knowledge of carbon atom locations in molecules is crucial to structural determination.

However, it can be seen readily that essentially all other elements have at least one isotope that can be examined by NMR. Examples were given in Table 6.1, and some additional isotopes are listed in Table 6.6, together with their sensitivity by NMR compared to the proton.

There has been great incentive to develop ^{13}C NMR because carbon is the central element in organic chemistry and biochemistry. However, useful applications were not forthcoming until the 1970s because of the difficulties in developing instrumentation. Two major problems were involved. First, the ^{13}C signal was very weak because the natural abundance of ^{13}C is low, only 1.1 % of the total carbon present in a sample. Also, the ratio γ and the sensitivity are low compared to that for ^1H. The net result was a carbon NMR signal only 0.0002 as intense as a comparative ^1H signal. Second, the chemical shift range was up to 200 ppm using TMS as a standard. This increased range precluded a simple "add on" to the ^1H NMR instruments already commercially available. However, the incentive to develop such sensitive instruments was great. The introduction of FT-NMR, which allowed the excitation of all ^{13}C nuclei simultaneously, made ^{13}C readily determinable by NMR. ^{13}C NMR is now a routine technique, providing important structural information about the carbon backbone of organic molecules.

There are several advantages to obtaining ^{13}C NMR spectra for structural identification of organic molecules. First, the wide range over which chemical shift occurs, 200 ppm for carbon compared with only 10 ppm for protons, greatly diminishes overlap between carbons in different chemical environments. The spectra are less crowded and a peak is usually seen for each unique carbon nucleus. Second, adjacent ^{12}C atoms do not induce spin–spin splitting, and the probability of two ^{13}C atoms being adjacent to each other is very low. Therefore, spin–spin coupling between ^{13}C nuclei is not seen and the spectra are much simpler than proton spectra. ^{13}C NMR spectra are included with all of the Aldrich proton NMR spectra examples we have used. Some of these carbon NMR spectra will be discussed in detail.

Coupling between ^{13}C and ^1H nuclei does occur, but there are techniques available to decouple these nuclei.

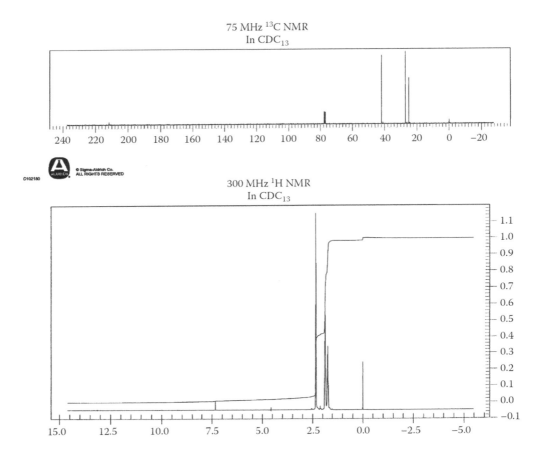

Figure 6.38 The spectra of cyclohexanone. The only characteristic absorbance for ketones in the proton NMR spectrum is that of the protons adjacent to the carbonyl group, which are shifted to 2.4 ppm by slight deshielding. (Reprinted with permission of Aldrich Chemical Co., Inc.)

Consequently, the ^{13}C NMR spectra are very simple, with a singlet seen for each chemically distinct carbon atom. This facilitates interpretation of the spectra. And, as we will see later, the comparison of ^{13}C and ^{1}H spectra leads to data that can be interpreted with a high degree of confidence, thereby elucidating the structure of even very complicated molecules. Coupling also occurs between ^{13}C and D, which can be seen when deuterated solvents are used. Look at the ^{13}C spectra in Figures 6.39 and 6.40, for example. Both spectra were acquired in CDCl$_3$. The triplet at 77 ppm is due to the natural ^{13}C in the solvent split by the D nucleus. In fact, the signal from the deuterated solvent, such as CDCl$_3$, is often used as the frequency lock for the NMR. Since D is an $I = 1$ nucleus, the signal from the single C atom is split into a *triplet* by the single deuterium nucleus, not a doublet, because $2I + 1 = [(2 \times 1) + 1] = 3$.

The chemical shift of a ^{13}C nucleus is determined by its chemical environment (electronegativity and anisotropy) as for protons, but in a more complex manner. Chemical shifts for some ^{13}C functional groups are shown in Figure 6.43. More detailed chemical shift tables can be found in the references on spectral interpretation listed in the bibliography.

6.6.4.1 Heteronuclear Decoupling

Coupling between ^{13}C and ^{1}H occurs and results in complex spectra with overlapping multiplets, but in practice, it is eliminated by broad band decoupling. The sample is irradiated with a wide RF range to decouple all the protons at once. Each chemically different carbon atom should then give a single NMR resonance peak. Figure 6.44 shows the ^{13}C NMR spectrum of sucrose before proton decoupling (top) and after proton decoupling (bottom). As can be seen in the bottom spectrum, each of the 12 carbon atoms in sucrose gives rise to a single discrete absorption peak in the spectrum. Two of the resonances are fairly close (the ones at ~26.5 ppm), but they can be distinguished. The structure of sucrose is given in Figure 6.29; you should confirm for yourself that each carbon atom is chemically nonequivalent.

6.6.4.2 Nuclear Overhauser Effect

When broad band decoupling is used simplify to the ^{13}C spectrum, it is noted that peak areas increase more than is

MAGNETIC RESONANCE SPECTROSCOPY

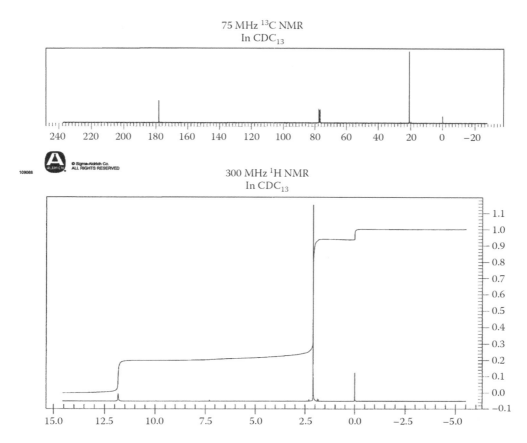

Figure 6.39 The spectra of acetic acid. The acid proton absorbance at 11.8 ppm is characteristic of carboxylic acids. (Reprinted with permission of Aldrich Chemical Co., Inc.)

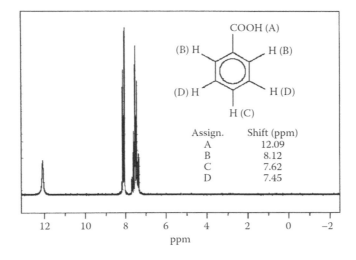

Figure 6.40 The 90 MHz proton spectrum of benzoic acid. (Courtesy of National Institute of Industrial Science and Technology, Japan, SDBSWeb: http://aist.go.jp/RIOBD/SDBS. Accessed on May 11, 2002.)

expected from elimination of the peak splitting. This is the nuclear Overhauser effect (NOE). Direct dipolar coupling between a saturated nucleus and an unsaturated nucleus results in a change in the ground-state population of the unsaturated nucleus. This change in ground-state population is a result of quantum transitions resulting in cross-relaxation between the nuclei; a transition in one nucleus induces a transition in the second nucleus. In ^{13}C NMR, the

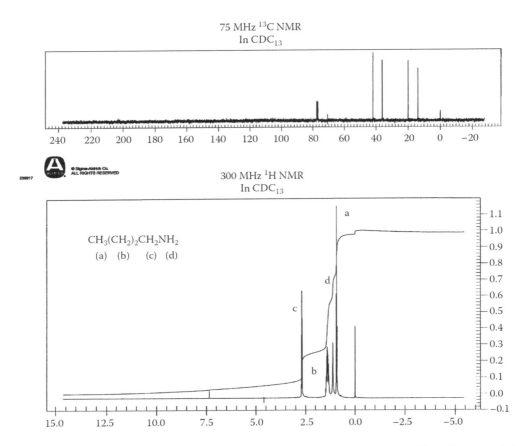

Figure 6.41 The spectra of butylamine. The amine protons give rise to the sharp singlet at 1.1 ppm. The adjacent methylene protons are shifted to about 2.7 ppm by the electronegative nitrogen, while the other alkyl methylene and methyl protons appear at their usual chemical shift positions. (Reprinted with permission of Aldrich Chemical Co., Inc., modified by addition of the structure and peak identification.)

result of the NOE is that the signal for the low-abundance ^{13}C nucleus is increased dramatically on proton decoupling. Under ideal conditions, the NOE can double the ^{13}C peak intensity; this decreases the time needed to collect a spectrum by a factor of 4. The maximum NOE between two isotopes can be expressed as

$$\text{NOE} = \frac{\gamma_{obs}}{2\gamma_{sat}} + 1 \qquad (6.18)$$

where
γ_{obs} is the magnetogyric ratio for the nucleus being measured
γ_{sat} is the magnetogyric ratio for the nucleus being saturated

The major disadvantage of the NOE is that the relationship between peak area and number of carbon atoms giving rise to the peak is lost. An example of this is seen in Figure 6.45, where the peaks for the protonated aromatic carbons (peaks 2 and 3) are more than twice the height of the peaks for the unprotonated aromatic carbons (peaks 1 and 4), even though each peak is due to a single carbon atom. The NOE can be eliminated experimentally, which must be done if quantitative analysis of the carbon spectral data is required. The NOE is also seen in homonuclear spin decoupling and can be used to determine the distances between nuclei, providing more structural information about a molecule.

6.6.4.3 ^{13}C NMR Spectra of Solids

Solid samples present a number of problems in ^{13}C NMR. Line broadening arises from chemical shift anisotropy, because of the many orientations the different carbon atoms have in a solid sample relative to the applied magnetic field. The chemical shift anisotropy can be eliminated by MAS at rotation frequencies ~5–15 kHz around an axis forming an angle of 54.76° (the magic angle) with the applied magnetic field. This technique increases the resolution observed in the spectrum for a solid by averaging the chemical shift anisotropies to their isotropic values. Figure 6.5 shows how dramatically the use of MAS reduces line broadening in a solid sample spectrum. Line broadening also occurs as a result of interaction between

MAGNETIC RESONANCE SPECTROSCOPY

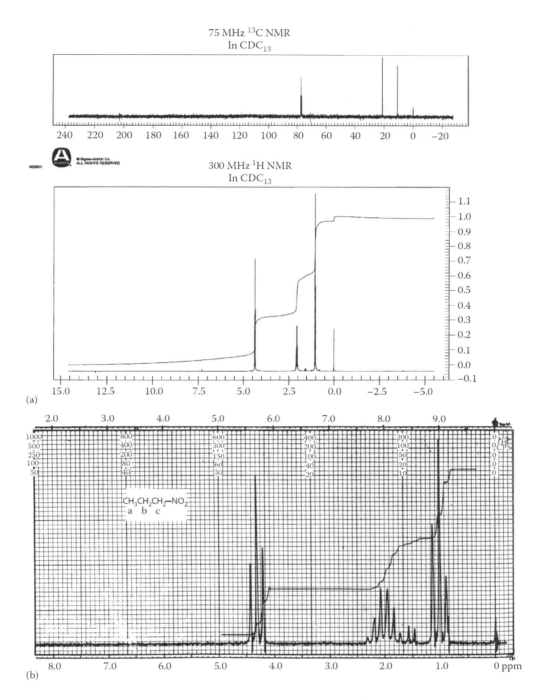

Figure 6.42 (a) The 300 MHz spectra of nitropropane. (Reprinted with permission of Aldrich Chemical Co., Inc.) (b) The 60 MHz spectrum of 1-nitropropane, clearly showing the triplet due to the methylene protons adjacent to the nitro group much more deshielded at 4.31 ppm than the second set of methylene protons at 2.0 ppm (small peaks at 1.48 and 1.59 ppm are impurities). (From *The Sadtler Handbook of Proton NMR Spectra*, W.W. Simons, ed., Sadtler Research Laboratories, Inc., Philadelphia, PA, 1978. [Now Bio-Rad Laboratories, Inc., Informatics Division, Sadtler Software and Databases. Copyright Bio-Rad Laboratories, Informatics Division, Sadtler Software and Databases, 2014. All rights reserved.])

^{13}C and ^{1}H; decoupling of the dipolar interaction in a manner similar to spin decoupling also reduces the linewidth.

The spin–lattice relaxation time for ^{13}C in solids is very long (several minutes). Since the nuclei have to relax before another excitation pulse can be sent, this requires hours of instrument time in order to collect a spectrum of reasonable intensity. A pulse technique called **cross-polarization** can be used to reduce this effect by having the protons interact with the carbon nuclei, causing them to relax more rapidly. FT-NMR systems for solid samples include the hardware

Table 6.6 Natural abundance, spin, and sensitivity of selected NMR-active nuclei

Nucleus	Natural Abundance (%)	Spin	Sensitivity Relative to ^{1}H
^{1}H	99.98	1/2	1.0
^{7}Li	92.6	3/2	0.3
^{13}C	1.1	1/2	0.0002
^{14}N	99.6	1	0.001
^{17}O	0.037	5/2	0.03
^{19}F	100	1/2	0.83
^{23}Na	15.9	3/2	0.09
^{25}Mg	10.1	5/2	0.003
^{27}Al	15.6	5/2	0.206
^{29}Si	4.7	1/2	0.008
^{31}P	24.3	1/2	0.07
^{33}S	0.8	3/2	0.002

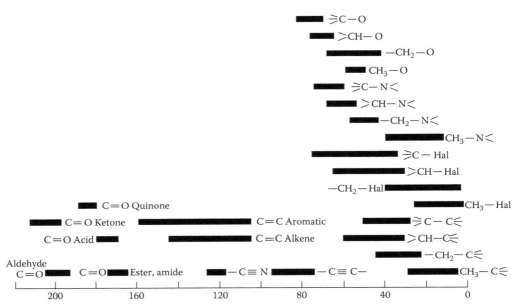

Figure 6.43 Chemical shifts for ^{13}C using TMS as the 0.0 ppm reference. The abbreviation Hal stands for halogen (i.e., Cl, Br, or I.). (From Leyden, D.E. and Cox, R.H., *Analytical Applications of NMR*, John Wiley & Sons, Inc., New York, 1977. Used with permission of John Wiley & Sons, Inc.)

and software to produce narrow line spectra from solid samples in a reasonable amount of time using high-power dipolar decoupling, MAS, and cross-polarization.

6.6.4.4 Interpretation of ^{13}C Spectra

A few examples of the interpretation of ^{13}C spectra will be discussed. Figure 6.43 provides the chemical shift information for a variety of organic compounds and should be used to follow the discussion. Remember that you need to look at the carbon atoms in the structures, not the protons. From Figure 6.43, we can see that alkane carbons are found between 0 and 75 ppm, aromatic and alkene carbons in the 100–160 ppm region, and the carbon of a carboxylic acid C=O group in a very narrow region between 170 and 180 ppm, for example.

Looking at the ^{13}C spectrum of acetic acid (at the top of Figure 6.39), there are two single peaks: one at 21 ppm and one at 178 ppm. The small triplet at 77 ppm is due to the CDCl$_3$ solvent. The structure of acetic acid is CH$_3$COOH; it has two distinct carbon atoms, one alkyl carbon and one carboxylic acid carbon. From Figure 6.43, the peak at 178 ppm is due to the acid carbon and the peak at 21 ppm is due to the CH$_3$ carbon. The peak for the alkyl carbon is much higher than the acid carbon peak. Proton decoupling has enhanced the protonated methyl carbon signal, while the

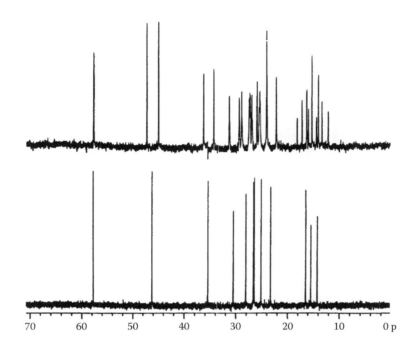

Figure 6.44 Heteronuclear decoupling of the 75 MHz ^{13}C spectrum of sucrose. The fully coupled spectrum is at the top. The bottom spectrum is the broadband ^{1}H decoupled spectrum. The structure of sucrose was given in Figure 6.29. The molecule contains 12 nonequivalent carbon atoms; the decoupled spectrum clearly shows 12 single peaks, one for each nonequivalent C atom. (From Petersheim, M., Nuclear magnetic resonance, in Ewing, G.A., ed., *Analytical Instrumentation Handbook*, 2nd edn., Marcel Dekker Inc., New York, 1997. Used with permission.)

intensity of the unprotonated acid carbon is not changed. The relative size of the peaks cannot be used to estimate the numbers of atoms because of the NOE.

The ^{13}C spectrum of benzene, C_6H_6, is given in Figure 6.46. All six carbon atoms are chemically equivalent, so the spectrum consists of a single peak at 128 ppm, in the aromatic carbon region. Figure 6.47 shows the spectrum of cyclohexanol, $C_6H_{11}OH$, a cyclic aliphatic alcohol. There are six carbon atoms in the ring, but they are not all equivalent. The structure and assignments are shown on the spectrum. The carbon to which the hydroxyl group is attached is unique and gives the deshielded peak at 70 ppm. The carbon at the opposite end of the ring from the hydroxyl is also unique; it gives the peak at 25 ppm. There are two equivalent "b" carbons and two equivalent "d" carbons due to the symmetry of the molecule. The solvent triplet at 77 ppm is seen.

Monosubstituted benzene rings have the same pattern of symmetry as does cyclohexanol; the ortho carbons are equivalent, the meta carbons are equivalent, while the substituted carbon and the para carbon are each unique. Therefore, a monosubstituted benzene ring should show four carbon peaks in the aromatic region. Benzaldehyde, C_6H_5CHO, is an example of this pattern. The ^{13}C spectrum of benzaldehyde in Figure 6.48 shows the expected four peaks in the aromatic region between 130 and 140 ppm. The smallest of the four peaks is the substituted carbon, the next largest is the para carbon, and the two tallest peaks are from the ortho and meta carbons. Although the NOE does not permit the exact area/number of nuclei calculations, the height of a peak is still a function of the number of nuclei and the number of protons on the carbon. All else being equal, a single unprotonated carbon will give a smaller peak than a peak from multiple carbon nuclei bearing multiple protons. There is an additional signal in the spectrum from the carbon in the aldehyde group. From Figure 6.43, we would expect to find the aldehydic carbon in the 190–120 ppm range; it appears in this spectrum at 192 ppm.

Figure 6.49 shows the NMR spectra for an amide, a class of compounds with $-NH_2$ substituted for the hydroxyl group of a carboxylic acid. The compound here is propionamide, $CH_3CH_2CONH_2$. There are three unique carbon atoms, the carbonyl carbon in the amide and two alkyl carbons. From Figure 6.43, the carbonyl carbon in an amide is expected to absorb in the 165–175 ppm range; the peak occurs at 177 ppm in this spectrum. The methyl carbon is located at 10 ppm, while the methylene carbon appears at 29 ppm. The proton spectrum was discussed earlier, but it is worth noting again the two small broad peaks due to the nonequivalent amide protons at 6.3 and 6.6 ppm.

The student is encouraged to look at the ^{13}C spectra presented in the earlier figures in the chapter and try to work through the number and assignment of the peaks using Figure 6.43 as a guide.

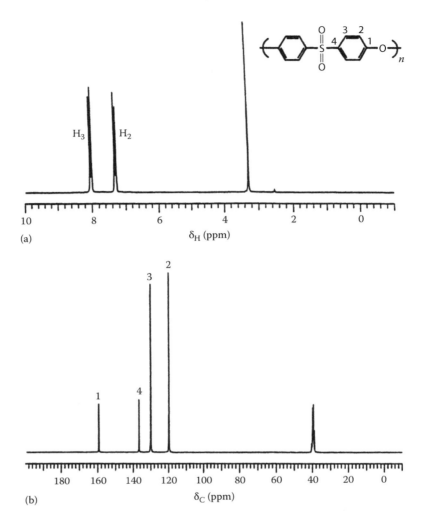

Figure 6.45 (a) ^{1}H and (b) ^{13}C NMR spectra of poly(1,4-phenylene ether sulfone) in deuterated dimethyl-sulfoxide, DMSO-d_6. Spectrum (b) shows the NOE that occurs on proton decoupling of the ^{13}C spectrum. The peaks for the protonated carbon atoms 2 and 3 are enhanced by the NOE over the signals from the unprotonated carbon atoms 1 and 4. The peak in the proton spectrum at 3.3 ppm is due to water in the solvent; the peak at 40 ppm in the ^{13}C spectrum is due to natural ^{13}C in the solvent. (From Williams, E.A.: *Characterization of Materials*, Part I., 1992. Copyright Wiley-VCH Verlag GmbH & Co. KGaA. Used with permission.)

6.6.5 2D NMR

High-resolution NMR spectra of organic compounds can be complex, with overlapping resonances and overlapping spin–spin couplings. The use of 2D NMR experiments and even 3D and 4D experiments extends the information obtained into a second (or third or fourth)-frequency dimension. The spectrum becomes easier to interpret and much more structural information is usually provided. Two- and higher-dimensional experiments rely on the selective manipulation of specific nuclear spins, followed by interaction between nuclear spins. A series of such experiments can provide the entire molecular structure including the stereochemistry of the molecule.

A 2D experiment generally consists of the following: a pulse, followed by a time interval t_1, then a second pulse, followed by a time interval t_2. The first time interval t_1 is called the evolution period; t_2 is the acquisition period. It is during the evolution period that nuclear spins interact. A nucleus detected during the acquisition period has been frequency-modulated by the nuclei it has interacted with during t_1. By varying the evolution period, t_1, in increments, and collecting the resulting FIDs, two frequencies are generated from a double Fourier transformation of the data. It is common to collect 1024 or more FIDs. Each one is Fourier-transformed to give a frequency axis v_2; a second FT is performed at right angles to the first one, resulting in a frequency axis v_1 related to t_1. One frequency axis is the nucleus detected during the acquisition period; the other axis can be the same nucleus (a homonuclear experiment such as COSY) or a different nucleus (a heteronuclear experiment such as HETCOR). The data are plotted as frequency versus frequency, usually presented as a contour plot.

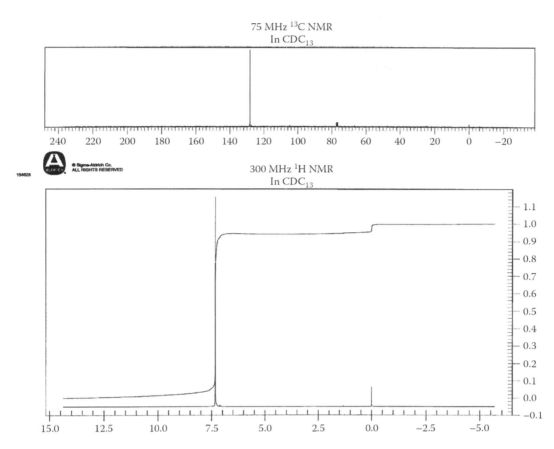

Figure 6.46 NMR spectra of benzene, C_6H_6. The six carbon atoms are equivalent, resulting in a single peak in the ^{13}C spectrum in the aromatic region at 128 ppm. The six protons are also equivalent, resulting in a single peak in the proton spectrum. (Reprinted with permission of Aldrich Chemical Co., Inc.)

For example, a homonuclear COSY experiment (see Table 6.5) is used to map the proton–proton J coupling in a molecule. Consider a simple system in which two protons are coupled to each other, CH—CH. The basic pulse sequence for a 2D COSY experiment consists of a relaxation or preparation period to establish spin equilibrium. A 90° pulse is applied and the spins precess at their characteristic frequencies during the evolution period, t_1. After the evolution period, a second 90° pulse, called the mixing pulse, is applied. This second pulse causes exchange of magnetization between J-coupled spins. The normal 1D proton spectrum is plotted on both the x-axis and the y-axis. For the simple system we are considering, the spectrum would show two doublets, one centered at a chemical shift of δ_A for proton A and one centered at a chemical shift of δ_X for proton X. The plot is shown schematically in Figure 6.50. If the magnetization undergoes identical modulation during t_1 and t_2, the resulting frequencies will be the same. A plot v_1 versus v_2, where the two frequencies are identical, results in a point along the diagonal of the x–y graph. The contour peaks (points in the schematic diagram) that appear along the diagonal are the resonances in the "normal" spectrum and provide no additional information. The four points labeled 1–4 on the diagonal are just the frequencies of the four peaks (i.e., the two doublets) in the 1D NMR spectrum. Peaks 5–8 are called autocorrelation peaks and will not be discussed. What we are interested in are those results where the magnetization exhibits one frequency during t_1 and a different frequency during t_2, that is, peaks that have been frequency-modulated by interaction. In this case, peaks appear *off* the diagonal in the x–y plot; such peaks are called cross-correlation peaks, or cross-peaks. It is the cross-peaks that provide the additional information we are looking for. In the homonuclear COSY experiment, the cross-peaks tell us which protons are coupled to each other. For this simple example, magnetization exchange results in eight peaks that appear as symmetric pairs off the diagonal. We can draw a connection between the two diagonal peaks from A and X and the symmetric pair of off-diagonal peaks, as shown in Figure 6.51, proving that protons A and X are coupled. Any pair of diagonal peaks that can be connected through symmetric pairs of off-diagonal peaks are spin-coupled; in this way, the coupling throughout a complex spectrum can be traced. As shown in the figure, the fine structure or spacing between the peaks gives us J_{AX}. A COSY spectrum usually looks more like the simulated example presented in

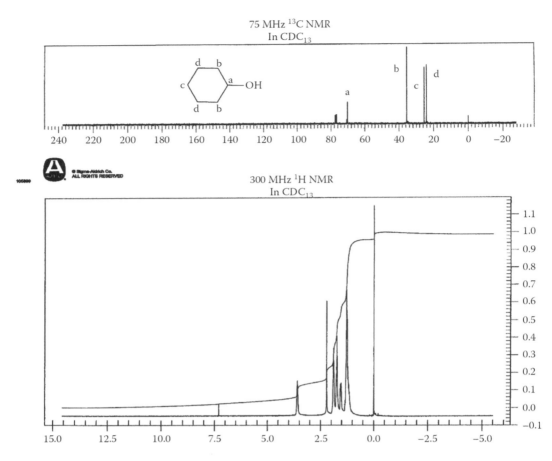

Figure 6.47 NMR spectra of cyclohexanol, $C_6H_{11}OH$. (Reprinted with permission of Aldrich Chemical Co., Inc., modified by addition of the structure and peak identification by the authors.)

Figure 6.52a, where the peaks are represented as contour plots. In this simulated example, there are six protons shown along the diagonal. The connecting lines in Figure 6.52b indicate that protons 1 and 2 are spin-coupled, protons 3 and 4 are spin-coupled, but protons 5 and 6 are not spin-coupled to any other protons. It is this type of information that helps to deduce the structure of an unknown; for this molecule, protons 5 and 6 cannot be adjacent to methyl, methylene, or methine protons, for example. The COSY spectrum of sucrose is shown in Figure 6.53. The connecting lines in Figure 6.53 confirm the couplings worked out earlier in the decoupling experiment (Figure 6.29). COSY spectra for large molecules can be complex and difficult to interpret. Figure 6.54 shows just a small portion of the proton NMR spectrum (from 3.8 to 5.4 ppm) and the related complex COSY plot of a large glucopyranoside molecule.

Another popular 2D experiment is the HETCOR experiment, which identifies which protons are directly bonded to which ^{13}C nuclei. HETCOR stands for heteronuclear chemical shift correlation. The ^{13}C spectrum is plotted on one axis and the ^1H spectrum on the other axis. The HETCOR spectrum shows spots of intensity. If a straight line drawn from a carbon signal and a straight line drawn from a proton signal intersect at a spot on the HETCOR plot, the protons are attached to that carbon. The result of a 2D HETCOR experiment for sucrose is shown in Figure 6.55. The glucose ring nuclei are marked with a G and the fructose ring nuclei with an F on both spectra; the structure of sucrose was given in Figure 6.29. Look at the peak in the carbon spectrum marked G1 at 93 ppm. By drawing a vertical line from G1 in the carbon spectrum, we reach an intensity spot in the lower left-hand portion of the HETCOR plot. A straight horizontal line from that point to the proton spectrum indicates that the proton(s) that absorb at 5.4 ppm are bonded to the G1 carbon. A vertical line from the F2 carbon does not intersect any spots; therefore, the F2 carbon has no protons bonded to it. You should be able to identify the F2 carbon in the sucrose structure in Figure 6.26 based on this knowledge.

The INADEQUATE experiment gives us the couplings between ^{13}C nuclei attached to each other and provides the carbon backbone of a molecule. As was mentioned in Table 6.5, the sensitivity of this experiment is low because of two factors: the low abundance of ^{13}C and the low probability that two ^{13}C nuclei are bonded to each other.

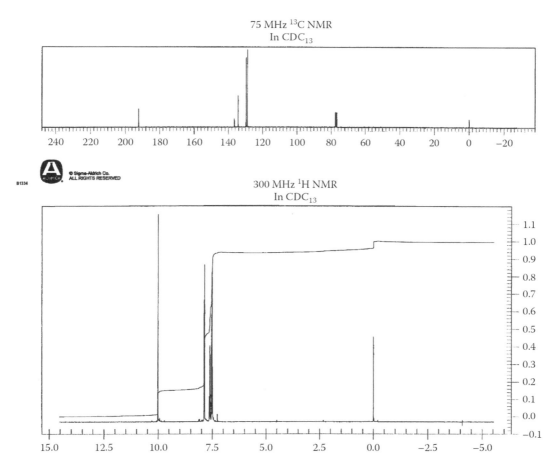

Figure 6.48 NMR spectra of benzaldehyde, showing the typical pattern of a monosubstituted benzene ring in the ^{13}C spectrum and the characteristic signal from an aldehydic carbon. (Reprinted with permission of Aldrich Chemical Co., Inc.)

6.6.6 Qualitative Analyses: Other Applications

NMR spectra can be used to identify unknown compounds through spectral pattern matching. A number of companies, instrument manufacturers, government agencies, and other sources publish collections of reference spectra in electronic format and in hard copy. These spectral databases may contain spectra of more than 200,000 compounds. The unknown spectrum or some predetermined number of the strongest absorption bands from the unknown spectrum may be entered into a computerized search routine, which compares the unknown with stored spectra. It then retrieves all compounds from the database that may match the unknown spectrum, assigning a "goodness of fit" or probability to the suggested matches. The analyst then identifies the spectrum of the unknown based on spectral matching and chemical knowledge of the sample to rule out improbable compounds suggested by the search routine. A short list of reference spectra suppliers is located at the end of the bibliography. Most large spectral databases are expensive, but the amount of work required to compile these databases is considerable. Aldrich Chemical Company (www.sigma-aldrich.com) provides the Aldrich Spectral Library, which is the source of many of the spectra used in this chapter, and Bio-Rad Laboratories (www.bio-rad.com) provides the Sadtler spectra collection. More than 20 vendors market the National Institute for Standards and Technology (NIST) spectral database. A comprehensive spectral database, including ^{13}C and ^{1}H NMR spectra, from the Japanese National Institute of Advanced Industrial Science and Technology (www.aist.go.jp/RIODB/SDBS) can be searched for no charge as of the time of this writing. ChemNMR, part of the ChemDraw/ChemBioDraw suite from CambridgeSoft® Corporation (Cambridge, MA, www.cambridgesoft.com), permits one to predict the proton and carbon NMR shifts based on a chemical structure. CambridgeSoft is now part of PerkinElmer (www.perkinelmer.com).

In addition to identification of unknowns, NMR can be used for conformational and stereochemical analyses. This includes the determination of **tacticity** in polymers, that is, whether the side chains are arranged regularly (isotactic and syndiotactic) or randomly (atactic) along the polymer backbone. Fundamental studies of bond distances from dipolar coupling and molecular

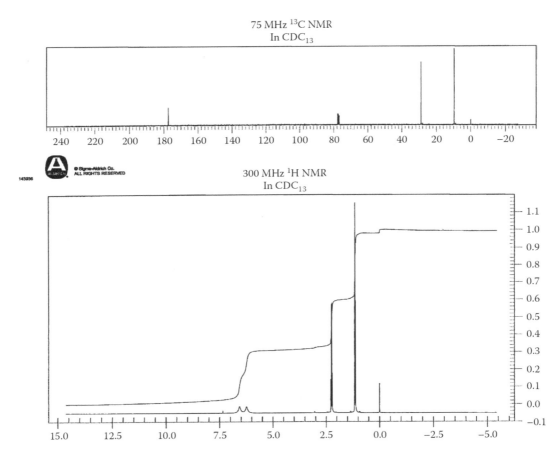

Figure 6.49 NMR spectra of propionamide, $CH_3CH_2CONH_2$. (Reprinted with permission of Aldrich Chemical Co., Inc.)

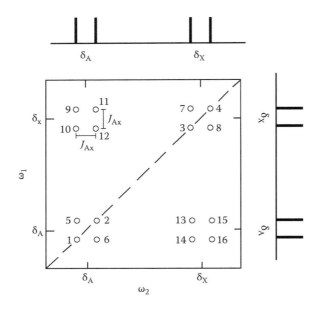

Figure 6.50 A simulated AX proton–proton COSY plot. (Modified from Bruch, used with permission.)

motion from relaxation time measurements are used by physical chemists and physical organic chemists. In biology and biochemistry, the use of isotope-labeled compounds and NMR can be used to study metabolism and understand metabolic pathways in living organisms. Chemical reaction rates can be measured and studied as a function of temperature directly in the NMR spectrometer.

Polymers of many chemical compositions are used to make materials and composites for use in appliances, electrical equipment, telecommunications equipment and fiber optics, computers, aircraft, aerospace, automobiles, food and beverage packaging, potable water delivery systems, and medical devices, to name a few applications. Adequate knowledge of the polymer structure and its relationship to the performance of these polymeric materials is crucial to their successful use. NMR is extremely useful in polymer characterization. ^{13}C and ^{1}H are the elements most commonly examined, followed by ^{29}Si, ^{19}F, ^{31}P, and ^{15}N. NMR can be used to determine the monomer sequence, branching, and end groups in polymers of many types. For example, Figure 6.56 shows the ^{29}Si NMR spectrum of a polydimethylsiloxane (PDMS) polymer with trimethylsilyl end groups. PDMSs are a major class of silicone polymers, used in many products from silicone caulk to shampoo.

The NMR spectrum enables the analyst to identify the trimethylsilyl end groups (peak A) and the repeat unit of the chain (Peak B), the dimethyl-substituted SiO unit shown in parentheses, from the chemical shifts. The degree of polymerization can be determined from the areas of the peaks; the integrations are shown as the stepped lines on the spectrum, so quantitative information about the polymer is also obtained in this case.

Polymer composites or fiber-reinforced plastics (FRPs) have a wide range of applications in the aerospace and automotive industries. In these composite systems, fibers of carbon or silica are embedded in a polymer matrix to provide desirable physical properties, such as strength, while maintaining the low mass of a polymer. For example, polymer composite turbine blades are replacing heavier metal alloy blades in jet engines. Organic coupling agents are used to treat fiber surfaces to improve bonding between the fibers and the polymer matrix. High-resolution cross-polarization MAS (CP–MAS) NMR is used to observe the structure, orientation, and interactions of coupling agents bound to surfaces.

Chemists and materials scientists are working to develop new materials with specific properties such as high strength and high modulus, resistance to temperature extremes, corrosion resistance, demanding optical or electrical properties, and the like. This requires detailed knowledge of composition of the material, orientation of molecules, crystalline state, homogeneity, and other parameters. NMR is one tool that can provide much of the structural information necessary and often in a nondestructive manner. NMR is useful for studying both amorphous and crystalline materials, unlike X-ray diffraction (XRD) (Chapter 9), which requires a crystalline sample.

The chemical shift differences between reactants and products permit NMR to be used to follow the course of a reaction and to choose the optimum reaction conditions. In Figure 6.57, the ^{29}Si NMR spectra show that NMR can follow the process of making β-SiC, a refractory ceramic, from polymethylvinylsilane and silicon metal. The NMR spectrum of the product silicon carbide (top spectrum) is clearly different from the spectrum of the starting mixture (bottom spectrum). In the bottom spectrum, the resonance at −18 ppm is due to the organosilane; the resonance at −82 ppm is the elemental silicon signal. All reactant and product signals are well separated in chemical shift, so any unreacted starting

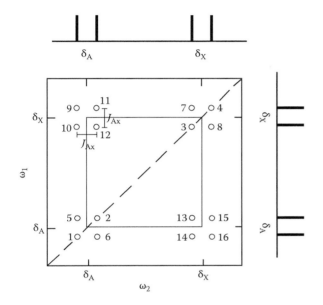

Figure 6.51 The connection between the cross-peaks and the A and X peaks on the diagonal proves that A and X are spin-coupled and also provides a measurement of J_{AX}. (Modified from Bruch, used with permission.)

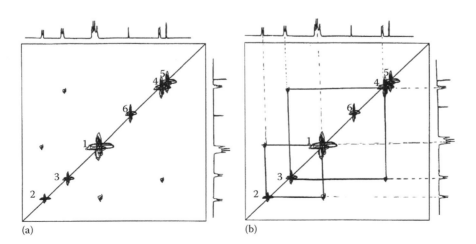

Figure 6.52 (a) Simulated ^{1}H–^{1}H COSY plot of an unknown compound. (b) Connecting lines indicate which protons are spin-coupled to each other. The COSY plot indicates that protons 5 and 6 are not coupled to any other protons. Any postulated structure for the unknown must be consistent with the coupling information from the COSY experiment.

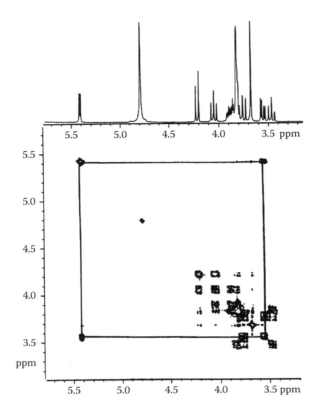

Figure 6.53 COSY profile and contour spectra of sucrose in D$_2$O. Note the strong contour connecting the diagonal peak for the proton at 5.41 ppm with the doublet of doublets at 3.35 ppm. In addition, the triplet at 4.05 ppm has strong off-diagonal contours indicating it is coupled to the doublet at 4.22 ppm. Compare with Figure 6.29. (Modified from Petersheim, M., Nuclear magnetic resonance, in Ewing, G.A., ed., *Analytical Instrumentation Handbook*, 2nd edn., Marcel Dekker Inc., New York, 1997. Used with permission.)

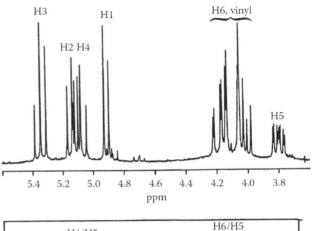

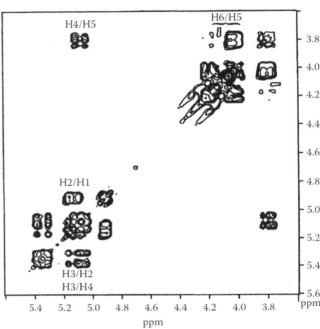

Figure 6.54 A small portion of an actual 2D-COSY spectrum of a glucopyranoside. This is an expansion of the region from 3.6 to 5.4 ppm. (Modified from Bruch, used with permission.)

material can be measured in the product and the production process can be optimized.

Solid-state NMR is proving to be a powerful technique for the study of reactions at surfaces. For example, NMR has been used in catalysis studies for determining the structure of chemisorbed molecules and for monitoring changes occurring in those structures as a function of temperature.

The need to determine the structures of large biological molecules like proteins is driving a new revolution in NMR. Extremely fast multidimensional NMR and new mathematical approaches, such as G-matrix FT-NMR, are being developed to rapidly collect and process 4D and 5D NMR experiments on biological macromolecules. Articles on these developments can be found in *Chemical and Engineering News*, December 23, 2002, p. 7 and January 27, 2003, p. 15.

6.6.7 Quantitative Analyses

Both high- and low-resolution NMR are used for quantitative analyses of mixtures, quality control (QC) of both incoming raw materials and finished products, determination of percent purity of pharmaceuticals and chemicals, quantitative determination of fat and water *in vivo* in animals, and many other applications.

A significant advantage of NMR is that data can be obtained under experimental conditions where the area under each resonance is directly proportional to the number of nuclei contributing to the signal. No response factors are necessary to obtain quantitative results. A universal reference standard can be used for the analyses of most materials because the NMR response can be made the same for all components. This is a significant advantage over other methods of analysis. In cases where proton decoupling is used, such as in the analyses of polymers and organic compounds containing ^{13}C, ^{29}Si, ^{19}F, and other spin 1/2

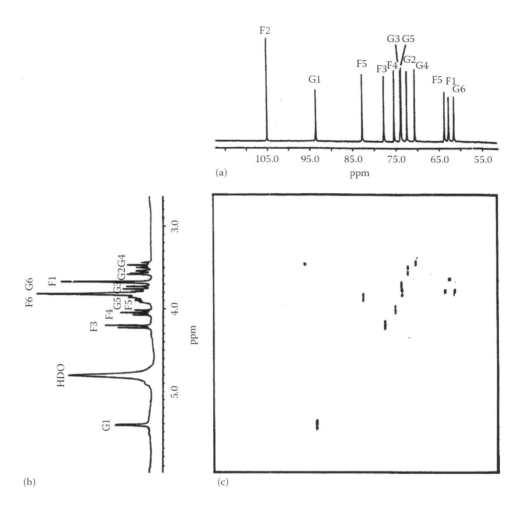

Figure 6.55 A 2D-HETCOR experiment. (a) A 75 MHz ^{13}C spectrum of 1 M sucrose in D_2O. (b) A 300 MHz ^{1}H spectrum of 1 M sucrose in D_2O. In both spectra, the labels G and F refer to the glucose ring and the fructose ring, respectively. (c) The 2D-HETCOR spectrum of sucrose. Carbon F2 has no protons directly bonded to it because there is no spot of intensity in the HETCOR plot in line with the F2 chemical shift.

nuclei, the NOE must be eliminated to reestablish the peak area to number of nuclei relationship. This is done by using gated decoupling (i.e., turning the decoupler off during the delay between RF pulses). This permits the return of normal equilibrium populations and results in the peak intensity (area) again relating to the number of nuclei giving rise to the resonance.

In polymers and ceramics, NMR can be used to determine quantitatively the amounts of amorphous and crystalline material in a sample. Molecules in amorphous regions "move" more than molecules in crystalline regions, so the relaxation times of molecules in these different environments are different. NMR can measure the difference in relaxation times and relate that to the percent crystallinity of the sample. Whether a sample is amorphous, crystalline, or a combination of both directly impacts the material's physical properties and behavior. It is an important piece of information for materials scientists and polymer chemists to know.

One type of organic compound can be determined quantitatively in the presence of a different type, such as the percentage of alcohols in alkanes, amines in alcohols, aromatics and aliphatics in petroleum, olefins in hydrocarbon mixtures, organic halides and organometallic compounds in other organic compounds, or the number of side chains in a hydrocarbon. The method is not limited in the number of components that can be identified as long as there is at least one peak in the spectrum that is unique to each component.

NMR can be used to provide determination of chemical purity and quantitative measurements of impurities in materials. The accuracy and precision of quantitative NMR measurements are comparable to other instrumental analyses methods. Major components can be accurately determined with precisions better than 1 % relative standard deviation (RSD), while impurities in materials may be quantified at 0.1 % or lower (Maniara et al.). The most common nuclei used for quantitative analyses are ^{1}H, ^{13}C, and ^{31}P. Quantitative

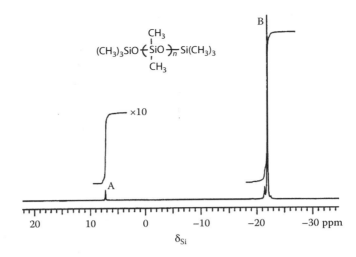

Figure 6.56 The use of ^{29}Si NMR to characterize a silicone polymer. The polymer end groups can be identified as trimethylsilyl groups and the measurement of peak areas permits the calculation of the average degree of polymerization. (Williams, E.A., Polymer molecular structure determination, in Lifshin, E. ed., *Characterization of Materials*, Part I, VCH Publishers, Inc., New York, 1992. Copyright Wiley-VCH Verlag GmbH&Co. KGaA. Used with permission.)

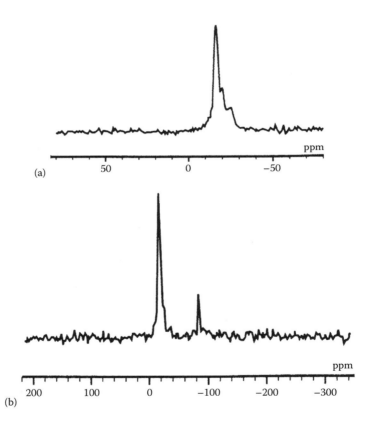

Figure 6.57 (a) The ^{29}Si NMR spectrum of the pyrolyzed product, β-SiC. (b) The ^{29}Si NMR spectrum of polymethylvinylsilane and elemental silicon, the precursors to silicon carbide, SiC. (From Apple, T.M., *Appl. Spectrosc.*, 49(6), 12A, 1995. Used with permission.)

^{31}P NMR has been used to determine phospholipids, inorganic phosphorus, and organophosphorus compounds, such as phosphorus-based insecticides (Maniara et al.).

As an example of a simple quantitative analysis, suppose that we have a mixture of *n*-octane and 1-octene. The structure of *n*-octane, C_8H_{18}, which contains 18 protons, is composed of methyl and methylene groups. All atoms are joined by single bonds:

$$CH_3-CH_2-CH_2-CH_2-CH_2-CH_2-CH_2-CH_3$$

The structure of 1-octene, C_8H_{16}, which contains one terminal C=C bond, is

$$CH_3-CH_2-CH_2-CH_2-CH_2-CH_2-CH=CH_2$$

It can be seen that in 1-octene, there are two protons on the *terminal* carbon that are olefinic in nature. These olefinic protons will absorb at about 5.0 ppm (Table 6.4). Quantitatively, they constitute 2 of 16 protons in 1-octene. If we measure the area of the peaks at 5.0 ppm (call it area *A*), then the total area in the whole spectrum due to the presence of 1-octene is equal to area $A \times 8$:

$$\text{Area due to } 1-\text{octene} = \text{Area } A \times 8$$

and

$$\text{Area due to octane} = \text{Total area} - (1-\text{octene area})$$

However, one molecule of octane contains 18 protons and one molecule of 1-octene contains 16 protons. Hence, if these compounds were present in equimolar proportions, the ratio of the relative area of the NMR absorption curves would be 18:16. A correction must be made for this difference in the final calculation. The mole ratio of octane to 1-octene in the mixture would therefore be obtained as

$$\frac{\text{Area due to octane}/18}{\text{Area due to octene}/16}$$

that is, the mole ratio of octane/octene equals

$$\frac{\left[\text{Total area} - (\text{Area } A \times 8)\right]/18}{(\text{Area } A \times 8)/16}$$

Sample calculation: In an actual experiment involving a mixture of octane and 1-octene, it was found that the total area of all peaks = 52 units. The area of peaks at 5.0 ppm (area *A*) = 2 units. From the data and the relationships previously set out, we find that the area due to 1-octene protons = 2 × 8 = 16 units. The area due to octane protons = 52 − 16 = 36 units:

$$\text{Ratio of octane to octene} = \frac{36/18}{16/16} = \frac{2}{1}$$

Therefore, the mole ratio of octane to 1-octene in this sample is 2:1. Similar calculations can be made for many other combinations of compounds.

NMR can be used to determine the composition of mixtures, polymers, and other materials as long as there is at least one peak representing each different component. For example, the degree of bromination of a brominated polymer can be determined from the ^{13}C NMR spectrum. The degree of bromination can be calculated from a direct comparison of the integrated areas labeled *g* and *h* in Figure 6.58, which represent the unbrominated and brominated aromatic rings or by using peak *d* to represent the unbrominated ring and the sum of peaks *e* and *f* to represent the brominated ring.

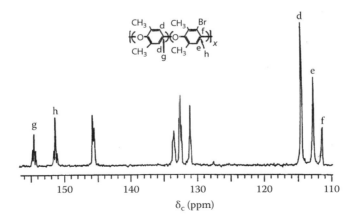

Figure 6.58 The 75 MHz ^{13}C NMR spectrum of the aromatic region of a 57 mol % brominated poly(2,6-dimethyl-1,4-phenylene oxide) polymer. Assignments are shown for peaks used in determining the degree of bromination. (From Williams, E.A. et al., *Appl. Spectrosc.*, 44, 1107, 1990. With permission.)

The end groups in a polymer are extremely important to the chemical and mechanical behavior of the polymer, but they are present at only low concentrations compared with the polymer chain repeat units. End groups may be less than 1 mol % of the total polymer. An example of end group identification in a silicone polymer was given earlier (Figure 6.57). In cases where the polymer end groups are hydrocarbons, the less sensitive ^{13}C spectrum must often be used to identify the end groups because there is less likelihood of spectral overlap. The proton spectrum, with a chemical shift range of only 10 ppm, often has the proton signals overlapping and impossible to sort out for the low concentration of end groups in a large polymer. Under ideal conditions, ^{13}C NMR may be used to determine end group concentrations as low as 0.01 mol %.

NMR can be used to determine rates of reaction and reaction kinetics. In a chemical reaction such as

$$C_8H_{18} \xrightarrow[\text{catalyst}]{\text{Pd}} C_8H_{16} + H_2$$

one type of proton (paraffinic) is consumed and another type (olefinic) is formed. Samples of a reacting mixture can be taken at frequent intervals during the reaction, and the rate of disappearance or formation of the different types of protons can be measured. The results can be used to calculate the rate of the chemical reaction and the kinetics involved. With a modern FT-NMR, many reactions can be monitored in the probe without the need for physically taking samples at intervals. Spectra are collected every few seconds, and the rate and order of the reaction are calculated after the data are processed.

The octane number of gasoline is a measure of the tendency of the gasoline to resist "knock." The octane number is determined using a standardized single-cylinder engine. The method requires the proper test engine, calibration with standard fuels, and then operation of the test engine with the sample gasoline. It has been shown that the octane number can be determined directly on the liquid gasoline by NMR (Ichikawa et al.). In this procedure, various types of protons are measured quantitatively and the volume percentages of the components calculated. The octane number of the gasoline can be calculated using an empirical formula. A similar approach using infrared (IR) spectroscopy (Chapter 5) has also been demonstrated, although none are legally accepted for determining the merchantability of the product.

6.7 HYPHENATED NMR TECHNIQUES

Many samples of interest to researchers in biochemistry, pharmaceutical chemistry, medical chemistry, forensic chemistry, and industrial chemistry are not pure compounds. It is often necessary to go through complex extraction and separation procedures to isolate the compounds of interest before they can be studied. These separation procedures can be time-consuming and labor-intensive. Costly high-purity solvents are used, which then must be disposed of (costing even more money). In the pharmaceutical industry, for example, **combinatorial chemistry** approaches are used to synthesize thousands of new compounds a day. These syntheses do not result in pure products; in fact, many products and unreacted starting materials may be present in the sample to be analyzed. Various types of instrumental **chromatography** have been developed to expedite the separation and detection of compounds in complex mixtures. These techniques are discussed in detail in Chapters 12 through 14. One of the most important types of chromatography, especially for pharmaceutical, biochemical, and clinical chemists, is HPLC, covered in Chapter 14. HPLC is used for the separation of nonvolatile molecular compounds; there are methods available to separate compounds with low MWs, high MWs, low polarity, high polarity, and everything in between. Detectors for HPLC do not provide molecular structure or RMM, except mass spectrometers (Chapter 14).

We have seen that NMR can provide detailed molecular structure information. It is possible to join together or "couple" an HPLC instrument with an NMR spectrometer. The HPLC performs the separation of a complex mixture, and the NMR spectrometer takes a spectrum of each separated component to identify its structure. We now have a "new" instrument, an HPLC–NMR instrument. We call a coupled instrument like this a **"hyphenated"** instrument. The coupling of two instruments to make a new technique with more capabilities than either instrument alone provides results in a **hyphenated technique** or hybrid technique. HPLC–NMR is made possible with a specially designed flow probe instead of the standard static probe. For example, Bruker Corporation (www.brukerbiospin.com) has a flow probe for proton and ^{13}C NMR with a cell volume of 120 µL. Complex mixtures of unknown alkaloids extracted from plants have been separated and their structures completely characterized by HPLC–NMR using a variety of 2D NMR experiments (Bringmann et al.). Only a simple aqueous extraction of the plant leaves, lyophilization, and dissolution in D_2O were required for injection into the HPLC–NMR, saving time and resources. HPLC–NMR has been used for the analyses of metabolites in body fluids; body fluid samples are very complex mixtures and are usually of limited volume (Lindon et al.).

HPLC–NMR and another more powerful instrument, HPLC–NMR-MS (the MS stands for mass spectrometry) are used in pharmaceutical research and development. These hyphenated techniques identify not only the structures of unknowns but, with the addition of MS, the RMM of unknown compounds. The HPLC–NMR-MS instrument separates the sample on the HPLC column, takes the NMR spectra as the separated components flow through the probe, and then acquires the mass spectrum of each separated component to determine the RMM and additional structural information from the mass spectral fragmentation pattern.

The MS must be placed after the NMR, since MS is a destructive technique. MS is covered in Chapters 10 and 11.

6.8 NMR IMAGING AND MRI

NMR is extensively used in imaging solid objects in a nondestructive, noninvasive manner. The most important application of this imaging is in medicine, where humans benefit tremendously from the power of NMR imaging. NMR imaging and **magnetic resonance imaging** (MRI) are the same technique. The term "nuclear" was dropped from instrumentation used for medical imaging so that patients would not mistakenly think this was a procedure that involved radioactive materials or gamma rays (X-rays). We will use the term MRI for NMR imaging applications in general.

MRI has found valuable applications in medical imaging of the human body. In this technique, a highly uniform magnetic field is used, but a linear magnetic field gradient is superimposed in three orthogonal directions using auxiliary coils. ^{1}H nuclei therefore respond at different frequencies at different physical locations, thus locating the physical position of the nuclei in three dimensions. By observing the NMR signal from numerous different angles simultaneously, the physical outline of the various body tissue components can be revealed. Abnormalities such as fractures or cancerous growths can then be located and measured in 3D at high resolution. The magnets used must have a large bore, large enough to accommodate a human patient lying on a table, or must use an open magnet design. MRI magnet designs are very different from NMR instrument magnet designs. Open magnet or short-bore designs are used to avoid inducing claustrophobia in patients undergoing MRI. Field strengths of MRI medical imaging units range from 0.8 to 3 T, much lower than NMRs for chemical analyses. The cost of medical MRI units is extremely high compared with chemical analysis instrumentation; about $1.5 million is typical for a 1.5 T state-of-the-art MRI unit in 2023. Photographs of two medical MRI systems from Toshiba Medical Systems, a high-resolution long-bore design and an open magnet (four-pole) design, are shown in Figures 6.59a and b. The Toshiba Medical Systems website at http://medical.toshiba.com has numerous MRI images taken with its systems available for viewing.

X-rays have been used extensively for noninvasive medical imaging studies in the past, but the contrast between different forms of soft tissue is low and therefore abnormalities in soft tissues are hard to visualize using X-rays. Also, X-rays are ionizing radiation and can cause tissue damage at high exposure levels. In contrast, MRI, also a noninvasive procedure, has essentially no side effects and can more readily visualize very small differences in soft tissue. MRI can monitor *in vivo* concentrations of biologically important molecules like adenosine triphosphate (ATP) in a noninvasive manner; this permits studies of the effect of drugs on metabolism, for example. MRI has proven to be a very valuable noninvasive medical tool for studying cancer, stroke, epilepsy, heart problems, arthritis, and many other conditions. Figure 6.60 illustrates the use of MRI to locate a brain tumor. The image in Figure 6.60a shows the tumor, the gray oval object on the front left side of the brain behind the eye. The image in Figure 6.60b is of the same tumor after a "contrast agent" containing gadolinium was given to the patient. The tumor absorbs more of the contrast agent because of its fast growth and high blood supply. The image shown in Figure 6.60c, that of one of the authors (TJB), is far less definitive in its etiology. It can be a (surprising and unusual) metastasis from a previous stage 1 kidney cancer, a new brain tumor, or a harmless capillary telangiectasia. It is too small for any attempt at biopsy, and if it were a capillary telangiectasia, biopsy would be dangerous. But, never being

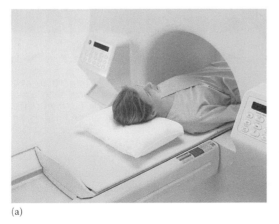
(a)

(b)

Figure 6.59 Two commercial medical MRI systems. (a) Female patient being placed in the bore of the Toshiba EXCELART™ MRI system. (b) The four-pole open magnet Toshiba OPART™ MRI system. (Photos courtesy of Toshiba America Medical Systems, http://medical.toshiba.com.)

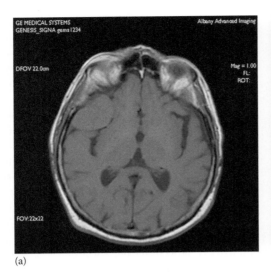
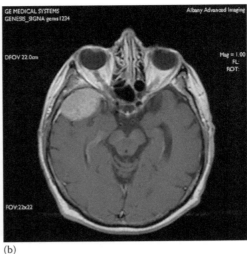

Figure 6.60 (a) An unenhanced MRI image of a human brain tumor. The tumor is the gray circular object on the left side of the brain behind the eye. (b) MRI image of the same tumor after administration of a gadolinium-containing contrast agent. Note the brightness of the tumor in this image compared to (a). (Images courtesy of H.T. Alberry, D. Derico, and M. Farrell, Albany Advanced Imaging, Albany, NY.)

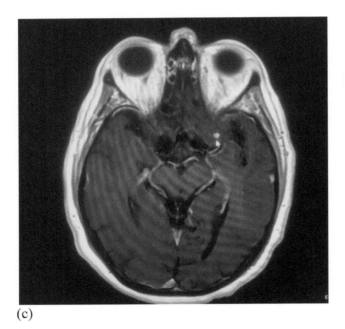
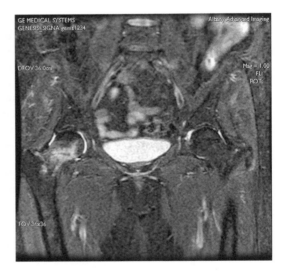

Figure 6.60 (Continued) Image (c) shows an enhancing nodule (indicated by the arrow) with an etiology that is far from certain, unlike the images in (a) and (b).

Figure 6.61 An MRI image of a hip fracture. The fracture in the hip on the left side of the image shows up as a bright area due to bleeding. The fracture was too small to be detected by conventional X-ray absorption (XRA). (Image courtesy of H.T. Alberry, D. Derico, and M. Farrell, Albany Advanced Imaging, Albany, NY.)

one to leave well enough alone, the author had it treated via a 7-minute stereotactic radio surgery. This is presented here to demonstrate that although great strides have been made in medical imaging by magnetic resonance, significant uncertainty remains and one often must make decisions based on the best information available.

The Gd ion is paramagnetic and changes the relaxation times of nuclei in its vicinity resulting in signal enhancement; the tumor now appears bright or "lit up" against the surrounding tissue. A brief discussion of gadolinium MRI contrast agents can be found in the reference by Skelly Frame and Uzgiris. However, recent advances in technology now permit MRI visualization of soft tissues and flowing blood even without the use of contrast agents. Some impressive examples can be found at http://medical.toshiba.com.

Another example of the use of MRI is seen in Figure 6.61. This is an image of a fractured hip; the

MAGNETIC RESONANCE SPECTROSCOPY

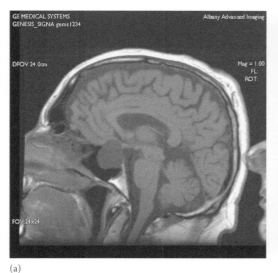
(a)

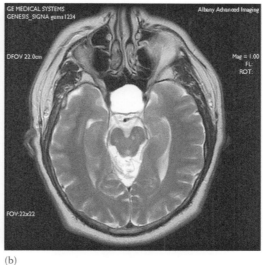
(b)

Figure 6.62 (a) Side view of a pituitary gland adenoma. (b) Top view (slice) of the adenoma. (Images courtesy of H.T. Alberry, D. Derico, and M. Farrell, Albany Advanced Imaging, Albany, NY.)

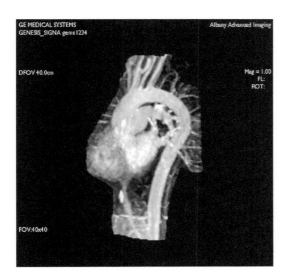

Figure 6.63 MRI image of a heart and associated blood vessels. (Image courtesy of H.T. Alberry, D. Derico, and M. Farrell, Albany Advanced Imaging, Albany, NY.)

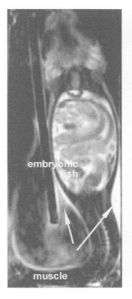 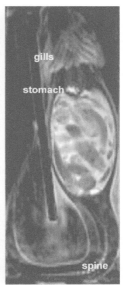

Figure 6.64 *In vivo* MR images of a North Sea fish, the eelpout. This fish is pregnant and embryonic fish are visible inside the uterus. Other tissues are also visible as marked. This fish was free-swimming in a salt water-filled flow-through chamber. (Courtesy of Dr. Christian Bock, Alfred-Wegener-Institute for Polar and Marine Research, Bremerhaven, Germany; www.awi.de.)

fracture is in the hip bone on the left side of the image and appears as the bright area in the region of the hip bone. Compare the right hip, which appears dark in this location. The brightness is due to blood in the fractured bone area; this fracture was too small to see by X-ray absorption (XRA). Two MRI views of a pituitary adenoma are shown in Figure 6.62. Figure 6.62a is a side view of the head; the adenoma is the dark gray circle in the sinus area, behind the eyes. A view from the top of the head shows the large bright adenoma centered in front of the brain and behind the eyes. A final example of human soft tissue imaging is the image of a heart and its blood vessels (Figure 6.63).

MRI instruments can be equipped to study polar ice and marine organisms in their saltwater environments, permitting the imaging of new marine species. The references by Bock and the website of the Alfred-Wegener-Institute for Polar and Marine Research offer amazing MRI pictures of ice microstructures and marine species *in vivo* (www.awi.de). A few examples follow. Figures 6.64 and 6.65 are *in vivo* MRI images and the NMR spectra of embryos

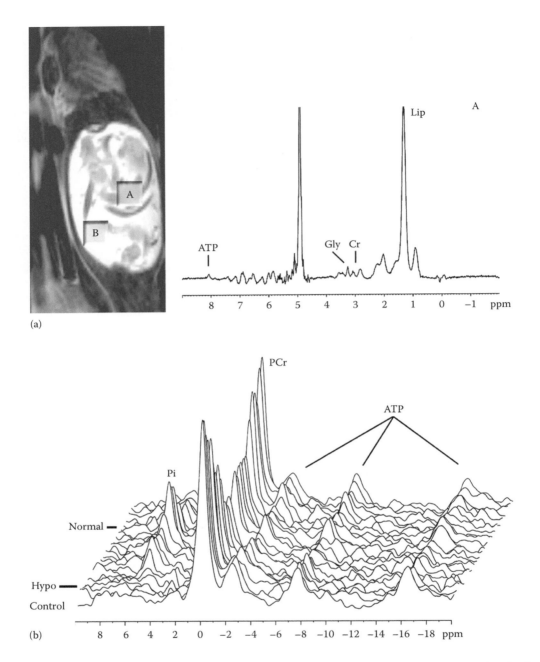

Figure 6.65 (a) *In vivo* localized proton NMR spectrum collected from the same fish embryos seen in Figure 6.64. At location A on the image, signals identified in the embryo spectrum include ATP, glycine (Gly), and creatine (Cr) as well as large signals from lipids (Lip). (Courtesy of Dr. Christian Bock, Alfred-Wegener-Institute for Polar and Marine Research, Bremerhaven, Germany; www.awi.de.) (b) Stack plot of *in vivo* ^{31}P NMR spectra from the muscle of a living codfish. Each spectrum was acquired over 5 min. When hypoxia (hypo) was induced in the fish by decreasing its oxygen supply, the inorganic phosphate levels (Pi) increased while phosphocreatine (PCr) decreased. As conditions return to normal, the Pi signal disappears, the PCr signal increases, and the fish recovered its energy. (Courtesy of Dr. Christian Bock, Alfred-Wegener-Institute for Polar and Marine Research, Bremerhaven, Germany, www.awi.de.)

in a live pregnant fish. Figure 6.65b shows a stack plot of phosphorus-containing compounds in a living codfish that undergoes hypoxia (lack of oxygen). As hypoxia is induced, looking from the front of the plot to the back, the inorganic phosphate signal on the left increases. At the same time, the PCr signal decreases. As conditions return to normal, the inorganic phosphate disappears and the PCr returns to normal (as did the fish). These are all ^{31}P NMR spectra. The Bruker BioSpin® website at www.bruker-biospin.com offers a wide variety of MRI images, including a movie of a beating rat heart with a myocardial infarction imaged *in vivo*, an MRI of a living newly discovered saltwater fish, and false color images showing temporal differences in brain activity during an epileptic seizure in a rat.

MRI permits the noninvasive imaging of the interior of solid objects. This has been successfully utilized in the study of extruded polymers and foams and the study of spatial distributions of porosity in porous materials. The structure of ice in polar ice cores has been studied as noted earlier; unlike optical imaging, MRI is nondestructive. Ice has sufficient mobile protons to be imaged with conventional MRI.

While conventional MRI has been used effectively for imaging soft tissues, it had not been used for studying hard tissues in the body due to its inability to capture the rapidly decaying signals of these tissues. In addition, conventional MRI cannot image around air or metallic objects. As anyone who has undergone an MRI can attest, the measurement process produces loud noises that cause some patients discomfort. New MRI technology from Agilent Technologies overcomes these problems and expands the applicability of MRI to materials and tissues previously invisible to MRI. The technique is called sweep imaging with Fourier transform (SWIFT). SWIFT uses broadband frequency-swept RF excitation with nearly simultaneous acquisition, allowing the imaging of objects and materials with extremely fast transverse relaxation rates. Applications include bones and teeth (Figure 6.66).

In conventional MRI, RF excitation and signal acquisition are separated by an echo time greater than 1 ms. By the time this echo delay is completed, hard materials with fast transverse relaxation times have already decayed and cannot be measured. In SWIFT, the echo time approaches zero because the acquisition begins within a few microseconds of excitation.

SWIFT can image semisolid materials such as macromolecules and quadrupolar nuclei, such as oxygen-17, sodium-23, and potassium-39. With the SWIFT technique, acoustic noise is all but eliminated. The technique tolerates static magnetic field inhomogeneity, making it possible to image near and around air (e.g., imaging of lungs) and metal, such as orthopedic hardware and iron nanoparticles. The article by Amy Herlihy provides more detail on this breakthrough imaging technology.

6.9 TIME DOMAIN NMR

There is a class of available physically small, benchtop NMR instruments, useful for dedicated quantitative analysis. These instruments are pulsed, time-domain NMRs (TD-NMRs). TD-NMR is also called relaxometry. TD-NMR does not deal with spectroscopy or magnetic resonance (MR) images. The TD-NMR instrument is usually a small benchtop or handheld, low-resolution and low-field analyzer designed to detect hydrogen or fluorine nuclei. TD-NMR analysis is quantitative and rapid (normally within seconds or minutes). It is also nondestructive and noninvasive. Thanks to these advantages and its ease of use, it is widely used for routine analysis in agriculture, food science, polymer, chemical, petroleum, and pharmaceutical and medical industries.

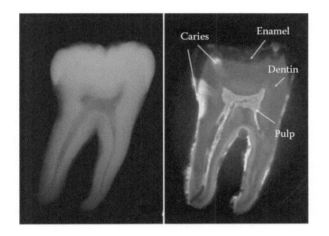

Figure 6.66 Comparison of an X-ray image (left) with a sweep imaging with Fourier transform (SWIFT) MRI image (right) shows that SWIFT expands MRI to hard substances such as teeth. (Used with permission of Agilent Technologies, www.chem.agilent.com. X-ray imaging will be discussed in Chapter 9.)

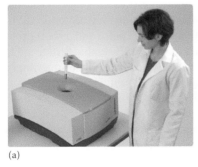
(a)

(b)

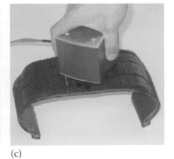
(c)

Figure 6.67 (a) A modern TD-NMR instrument, with a sample being inserted for quantitative analysis. (b) and (c) Single-sided TD-NMR, the minispec ProFiler, inspecting yogurt samples and a piece of rubber tire, respectively. (Courtesy of Bruker Optics, Inc., www.bruker.com.)

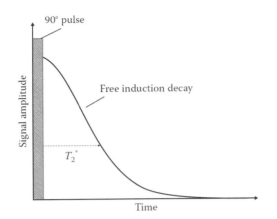

Figure 6.68 A typical TD-NMR FID from a water sample. (Courtesy of Bruker Optics, Inc., www.bruker.com.)

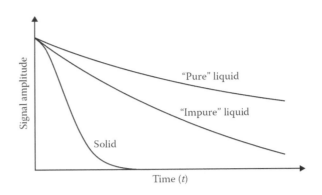

Figure 6.69 FID signals from solid and liquid samples, showing the dramatic difference in decay time for solids versus liquids and between one liquid and another. (Courtesy of Bruker Optics, Inc., www.bruker.com.)

TD-NMR instruments are similar to the high-resolution instruments described in Section 6.5 except the magnet is much smaller and made of permanent magnetic materials. Small permanent magnets have relatively low magnetic field strength and homogeneity. The magnetic field strength of commercial TD-NMR instruments is usually in the range of 1 MHz (0.023 T) to 60 MHz (1.4 T). The sample probe sizes vary from 5 mm up to 150 mm in diameter. The cost of TD-NMR instruments is between $30,000 and $150,000 (in 2023) depending on instrument configuration and applications. TD-NMR instruments are available from several suppliers, including Oxford Instruments (www.oxinst.com) and Bruker (www.bruker.com). A modern benchtop TD-NMR instrument is shown in Figure 6.67a. The handheld minispec ProFiler (Bruker Corporation) is seen in Figures 6.67b and c.

The time evolution of the transient signal produced after an RF excitation pulse is the FT of the absorption spectra. When the FT is applied to the time evolution of the transient signal, an NMR spectrum is obtained as was shown in Figure 6.6g. However, if the time evolution of the transient signal, that is, the time domain signal, is from a low-resolution NMR instrument, the frequency spectrum after the Fourier transformation would be a single broad line. This does not provide any useful spectroscopic information. Therefore, instead of Fourier transforming time domain signals, TD-NMR only deals with raw time domain signals directly.

TD-NMR instruments are designed in such a way that amplitudes of NMR signals are acquired and processed. Figure 6.68 shows a typical FID from a water sample. In general, the total signal intensity of a FID is directly proportional to the number of nuclei in the sample, which is the basis of quantitative analysis by TD-NMR. On the other hand, within the magnetic homogeneity limit, the decay rate of a FID, denoted by a decay time constant, is associated with molecular mobility. In general, the more mobile a molecule is, the longer its FID decay time will be. That means a FID from a solid sample decays much faster than a FID from a liquid sample. Figure 6.69 illustrates FIDs from different forms of samples.

6.9.1 Solid Fat Content Determination by TD-NMR

The simplest and earliest application of TD-NMR was the determination of solid fat content (SFC) in fats or oils. The traditional method for the determination of the solids content in edible oils was dilatometry, based on a change in volume as a chilled oil sample is warmed. Dilatometry is not a direct measure of SFC and is highly dependent on the analyst's skill. In addition, fat samples containing emulsifiers do not give accurate results. There was a strong impetus to develop an alternative method. As early as the 1970s, Bruker GmbH and its partners decided to explore the use of the TD-NMR signal for the determination of solid-to-liquid ratios in edible oil compositions. A **Bruker minispec** TD-NMR system was used for the following example. The minispec employs the "pulse" method to perform the NMR experiment. A short intense burst of RF energy is applied to the sample in the static magnetic field to cause excitation of the hydrogen nuclei. The time evolution of the NMR response is recorded and the resulting signal contains the information from which the ratio of protons in the solid state to protons in the liquid state may be deduced. As illustrated in Figure 6.70, after a 90° pulse and a short probe ringdown time, or dead time (usually under 8 μs), a FID signal is acquired. The FID is composed of superimposed signals from both solid and liquid components in the sample. After the pulse, the signal is detectable for milliseconds to seconds. The NMR signal decays because of relaxation. In solid matter, oscillations are heavily damped and the signal decays relatively quickly. In a liquid, the surroundings are more mobile, resulting in less damping and a slower decay of the signal. The portion of the signal marked "phase a" represents the combined signals from solid and liquid components; "phase b" represents the pure liquid signal.

MAGNETIC RESONANCE SPECTROSCOPY

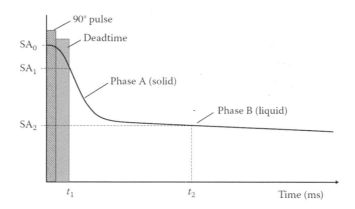

Figure 6.70 TD-NMR FID of an edible oil to determine solid/liquid fat ratio, using a Bruker minispec instrument. Phase a is the combined solid + liquid signal; phase b is due to liquid alone. Amplitudes at t_1 and t_2 can be used to calculate the ratio of solid/liquid or the amount of solid phase. (Courtesy of Bruker Optics, Inc., www.bruker.com.)

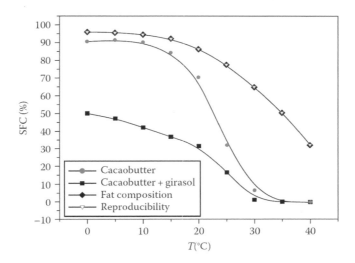

Figure 6.71 SFC melting curves for two samples of coconut oil with the minispec TD-NMR system. (Courtesy of Bruker Optics, Inc., www.bruker.com.)

Point SA_1 is proportional to the total protons in the liquid and solid phases. Point SA_2 is proportional to the total protons in the liquid phase alone. Correction of the instrumental conditions such as the dead time and field homogeneity results in a ratio of S1 to S2 that is proportional to the real solid/liquid ratio or SFC of the sample.

In the standard SFC application in the dairy industry or the edible oil industry, the SFC of a fat composite sample is measured as a function of sample temperature to obtain melting curves as shown in Figure 6.71. A typical direct solid fat measurement by TD-NMR uses enough sample to fill a 10 mm tube to a height of about 4 cm. The actual NMR measurement takes only 6 s. The accuracy of the SFC determination using TD-NMR is ca. 0.5 %.

Because it is a direct measurement with good accuracy and precision, the TD-NMR method for determination of SFC has been made a standard method by a number of international organizations including International Standards Organization (ISO), American Oil Chemists' Society (AOCS), and International Union of Pure and Applied Chemistry (IUPAC).

The same principle of SFC measurement and method can be applied to other industries or other products. For instance, the product behavior of a lipstick after application on the lips can be characterized by measuring its melting curve, as shown in Figure 6.72.

6.9.2 Field Homogeneity and Spin Echo

All NMR instruments need a uniform magnetic field, and all the nuclei in the sample should experience the same magnetic field strength so all nuclei can have one resonance frequency according to Equation 6.7. It is almost impossible to produce a perfectly homogeneous magnetic field from a magnet that has a limited size of pole faces. Most TD-NMR instruments use magnets with pole faces less than 150 mm in

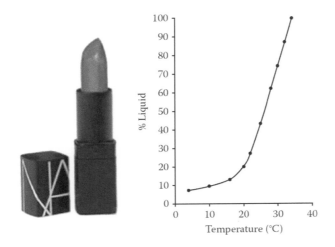

Figure 6.72 A melting curve from a commercial lipstick sample. The % liquid is plotted as a function of sample temperature. (Courtesy of Bruker Optics, Inc., www.bruker.com.)

diameter. Such an inhomogeneous magnet shortens the decay of NMR FID signals. T_2^* is used to denote the time constant of a FID, which is the observed transverse relaxation time, T_2, as described in Section 6.1.3.2 and can be expressed as

$$\frac{1}{T_2^*} = \frac{1}{T_2} + \frac{1}{T_2^M}$$

where
T_2 is the true transverse relaxation time of the sample
T_2^M is an expression of magnetic field homogeneity in terms of relaxation time

A typical permanent magnet in a 20 MHz TD-NMR instrument has a T_2^M of 1 ms in a cylindrical volume of 10 mm in diameter and 10 mm long, which corresponds to a field homogeneity of about 15 ppm. It can be seen from the earlier equation that if $T_2 \gg T_2^M$, that is, in the case of measuring liquid samples that have a T_2 of hundreds of milliseconds or longer, then T_2^* is a direct measure of the magnet homogeneity. On the other hand, if $T_2 \ll T_2^M$, that is, in the case of measuring solid samples that have a T_2 of 100 μs or shorter, then T_2^* is the true T_2 of the solid sample.

It is obvious that magnetic field inhomogeneity in TD-NMR instruments can prevent using a FID to measure the true T_2 of liquid samples. The spin echo or Hahn echo pulse sequence has solved this problem. Using the spin echo pulse sequence, a premature decrease in signal intensity of a liquid sample due to magnetic field inhomogeneity can be recovered with a second refocusing pulse (the 180° pulse) after a short time gap, as shown in Figure 6.76. Repeating the refocusing pulse produces a train of echoes (the CPMG echo train, named after the developers Carr, Purcell, Meiboom, and Gill), and the envelope of the echo peaks forms a decay whose time constant is the true T_2 of a liquid sample.

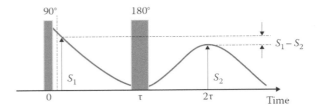

Figure 6.73 A diagram of the spin echo pulse sequence. The FID is observed after the first 90° pulse. The 180° pulse is applied at time t and the echo is then formed at $2t$. (Courtesy of Bruker Optics, Inc., www.bruker.com.)

One example of the use of the spin echo technique is to measure moisture and oil content simultaneously in many agriculture and food products such as oilseeds, nuts, chocolate, milk powder, cheese, cookies, and sausages. The total oil and moisture analyses using TD-NMR for seeds and seed residues have become international standard methods (ISO CD 10565 and ISO CD 10632). In seeds, oil and water molecules all have hydrogen and they all contribute to NMR signals. In dried seeds, the water molecules are bound, and the mobility of the water molecules is much more restricted than that of oil molecules, which remain free in the seeds. Therefore, the water in the seeds has a short relaxation time, T_2, while the oil has a long relaxation time. Using the spin echo method as shown in Figure 6.73 the FID signal amplitude S_1 is measured at 0.05 ms after a 90° pulse, and the echo amplitude S_2 is measured at $2t = 7.0$ ms, where $t = 3.5$ ms is the delay between the 90° and 180° pulses. S_1 represents the sum of the moisture and oil signals, S_2 is proportional to the oil content, and the difference, $S_1 - S_2$, is proportional to the moisture content.

Of course, this method cannot be used for samples with high moisture content because the relaxation time of free water is as long as that of the oil and the NMR echo contains signals from both oil and free water. The echo signal is no longer proportional to the oil content. Such samples (fresh olives, sausages, or meats) must be predried to remove the free water. Then the earlier method can be applied and will provide precise oil or fat content.

The use of low-resolution TD-NMR for oil, fat, and moisture is often easier and more accurate than the use of near IR (NIR) (Chapter 5) because it is less dependent on the product matrix, so it is possible to calibrate an oil determination against 100 % oil. In addition, NMR acquires the signal from the whole sample, not the surface. In some cases, if water is <10 %, oil and moisture may be able to be determined simultaneously. Other applications of TD-NMR can be found on the websites of Oxford Instruments and Bruker and include oils in snack foods, chocolate and cocoa, milk powder, and so on.

Another example of using the spin echo technique in an industrial application is the measurement of the surface coating or "spin finish" on artificial fibers (Rodgers, 1994). At the end of the manufacturing process for artificial fibers,

MAGNETIC RESONANCE SPECTROSCOPY

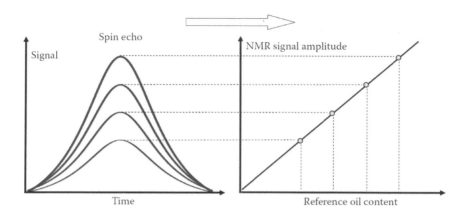

Figure 6.74 Calibration curve for oil content (right side of the figure) constructed from the spin echo signals of standards of known oil content (left side of Figure). (Courtesy of Bruker Optics, Inc., www.bruker.com.)

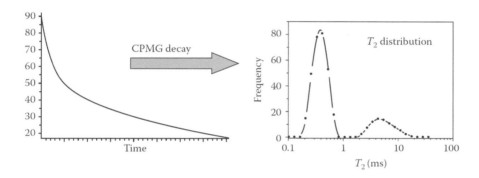

Figure 6.75 The inverse Laplace transformation is applied to a T_2 decay signal (the CPMG echo train) to obtain a continuous distribution of T_2. (Courtesy of Bruker Optics, Inc., www.bruker.com.)

the fibers are sprayed with an oil-based coating to lubricate, to control static, and to improve handling during subsequent use of the fibers in textile manufacturing. The oil coating is usually called spin finish. The amount of coating must be strictly controlled to ensure proper performance of the fibers, so samples must be regularly and frequently tested in QC laboratories for the spin finish level. The use of TD-NMR to determine spin finish levels directly on the fibers is ideal for the application, replacing the costly and time-consuming solvent extraction of the finish and subsequent determination by IR spectroscopy or gravimetry.

Similar to the determination of oil content in oilseeds, the spin echo technique is employed and the echo signal is used to measure the oil content in fiber, while the signals from the residual water and the fiber itself decay away before the refocusing pulse in the measurement. The TD-NMR analysis involves weighing a fiber sample of 2–5 g, putting it into a test tube, and placing the tube into the instrument. The measurement takes a few seconds to a few minutes with a precision of ±0.02 % in a spin finish range of 0.06 %–1.00 %.

Any analytical instrument needs to be calibrated for quantitative analysis. TD-NMR is no exception. In the calibration process, samples with known amounts of oil content are measured. The spin echo amplitudes are plotted against oil content and a calibration curve is then obtained (Figure 6.74). The echo signal amplitude of a sample is converted directly to oil content based on the calibration curve.

6.9.3 Relaxation Time Distribution

It can be seen from the discussion earlier that NMR relaxation T_2 can be treated as an indicator of molecular mobility. In some complex sample systems such as foodstuffs, the NMR relaxation mechanism is very complicated, and it is impossible to use a small number of discrete relaxation times to describe molecular mobility. A new approach has been explored to use a "continuous distribution of T_2" to study the thermodynamics of complex systems. An inverse Laplace transformation is applied to the echo train to resolve a number of different exponentials. The result is expressed in a form of T_2 distribution as illustrated in Figure 6.75.

The T_2 distribution has applications as diverse as petroleum geology and bread making. Its distribution has been applied in the petroleum industry for many years to characterize rock cores to obtain pore size distributions in

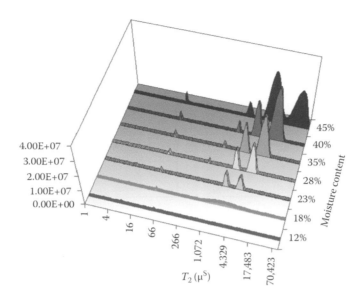

Figure 6.76 Continuous distribution of T_2 in dough samples of different moisture contents. (From Ruan, R. and Chen, L., *Water in Foods and Biological Materials—A Nuclear Magnetic Resonance Approach*. Technomic Publishing Co., Inc., Lancaster, PA, and Basel, Switzerland, 1998. Used with permission.)

well-logging operations. Rock cores from oil wells are filled with water or oil. The NMR CPMG echo train is acquired in a TD-NMR instrument and the T_2 distribution is obtained. This is essentially a mirror image of pore size distribution, as water in small pores is more restricted; it is less mobile (short T_2). Water in large pores has more freedom to move (long T_2).

The T_2 distribution method can be employed to study water behavior in complex food systems. The example in Figure 6.76 shows changes in dough samples when moisture content changes.

6.9.4 TD-NMR Applications

Typical examples of other TD-NMR analyses include extractables in polyethylene and polypropylene, the determination of tacticity in polymers, simultaneous determination of fat, moisture, and protein content in meats or meat products, and body composition determination of live animals. For example, the Bruker Optics Minispec NMR analyzer (www.minispec.com) has been optimized for the determination of whole-body fat, lean tissue, and fluid in live mice. The pulse sequence and relaxation times distinguish protons in water from protons in fat. These measurements are critical to pharmaceutical and medical studies of obesity and the development of drugs to treat obesity. The mice are restrained but not anesthetized and placed in the magnetic field. Although the magnet is low strength, the field penetrates the entire volume of the mouse. Total body fat, lean tissue, and fluid are measured in less than 2 min. The mice are not harmed during the NMR measurement, which makes this approach valuable for long-term clinical studies.

The same mice can be studied for the entire duration of the clinical trial. The NMR method is more precise than methods currently in use for these measurements.

Dedicated benchtop NMR analyzers for a variety of applications are available. These include an analyzer to determine fluoride in toothpaste quantitatively, such as the MQC from Oxford Instruments, a 23 MHz benchtop NMR, and another to determine water droplet size distribution in oil/water emulsions. Fluoride is often added to toothpaste as sodium fluoride or sodium monofluorophosphate to prevent tooth decay. The fluorine analyzer can determine fluorine at the level of a few hundred ppm. Toothpaste is squeezed into a glass sample tube and the quantitative determination of fluorine takes less than 1 min. The NMR method uses no solvents or reagents and is independent of the sample color and clarity, unlike the colorimetric methods and other instrumental methods such as ion chromatography (IC) that are used for this purpose. In the water droplet size distribution analyzer, droplets as small as 0.25 μm can be measured. The shelf life and palatability of products such as margarine, mayonnaise, salad dressings, and soft cheese depend on the size of water droplets in the water–oil emulsion. For example, products with multiple small droplets are less susceptible to bacterial growth than products with large droplet sizes. No sample preparation or dilution is required, the method is nondestructive and noninvasive, and the NMR results agree with conventional methods such as laser light scattering.

Studies of rock cores from oilfields are used in the petroleum industry to assess the hydrocarbon content of rock strata and the ease with which the oil can be recovered. The GeoSpec™ NMR Rock Core Analyser, a 2 MHz TD-NMR from Oxford Instruments, can provide information on the

Figure 6.77 The miniature Thermo Scientific™ picoSpin™ proton NMR. (© Thermo Fisher Scientific, www.thermo-fisher.com.) Used with permission.

fluids in oil- and water-saturated rock including oil viscosity and clay-bound water, porosity, fluid distribution, and permeability. Rock cores can be studied at high temperatures and pressures and the instrument can provide diffusion and imaging information.

6.10 LOW-FIELD, PORTABLE, AND MINIATURE NMR INSTRUMENTS

While it is true that the original 60 MHz CW NMRs are no longer being made, there are a number of low-field instruments with permanent magnets or electromagnets commercially available. These instruments are aimed at academic institutions where low resolution is adequate for educational purposes or where cost and space are issues. Installations where temperature fluctuations or environmental vibrations occur are more suitable for a rugged permanent magnet or electromagnet than a superconducting magnet.

FT-NMR instruments of 60 and 90 MHz are available from Anasazi Instruments (www.aiinmr.com) in configurations for proton, carbon, fluorine, phosphorus, and silicon studies, with common research capabilities such as 2D (COSY, HETCOR) and DEPT. The proton spectrum and peak assignments for ibuprofen and a good comparison of S/N, resolution, and second-order effects in the 60 MHz spectrum versus a 90 MHz spectrum for ibuprofen can be found in their application note at http://aiinmr.com/applications/60-vs-90mhz-sample-2m-ibuprofen.

A novel miniature 45 MHz proton FT-NMR is available from Thermo Scientific (www.thermo-scientific.com/picoSpin). The NMR, seen in Figure 6.77, is based on a permanent magnet, weighs 7 lbs, and needs only 20 μL of sample. The permanent magnet does not require cryogenic cooling. The very small footprint (8" × 11", less than a standard sheet of paper) and low cost allow dedicated NMR systems to be placed in manufacturing operations, QC labs, and classrooms. The instrument can be used for a variety of educational studies and process control applications, such as monitoring transesterification and condensation reactions, QC of food and beverage components, and so on. Essential oils extracted from natural products such as fruit, herbs, and spices are often used in foods, perfumes, and personal care products. NMR can be used to evaluate purity of the oils and to identify different lots of product or even different geographical sources for a given oil. The major component of both orange oil and lemon oil is (+)-limonene, but analysis of these two oils using the picoSpin-45 shows small peaks in the spectrum of the lemon oil that are due to other trace compounds extracted from the raw material (Figure 6.78).

Another example of the use of NMR for pharmaceutical QC is the screening of final products for active pharmaceutical ingredients (APIs) as well as binders, excipients, and contaminants. In softgel tablets, a major component is polyethylene glycol (PEG), seen in the NMR spectrum of an over-the-counter ibuprofen softgel product (Figure 6.79a).

6.11 LIMITATIONS OF NMR

There are two major limitations to NMR: (1) It is limited to the measurement of nuclei with magnetic moments and (2) it is often less sensitive than other spectroscopic and chromatographic methods of analysis. As we have seen, although most elements have at least one nucleus that responds in NMR, that nucleus is often of low natural abundance and may have a small magnetogyric ratio, reducing sensitivity. The proton, ^{1}H, and fluorine, ^{19}F, are the two most sensitive elements.

Elements in the ionic state do not respond in NMR, but the presence of ions in a sample contributes to unacceptable line broadening. Paramagnetic contaminants such as iron and dissolved oxygen also broaden NMR lines. Nuclei with quadrupole moments, such as ^{81}Br, broaden the NMR signal. Line broadening in general reduces the NMR signal and hence the sensitivity.

6.12 ELECTRON SPIN RESONANCE SPECTROSCOPY

Electron spin resonance (ESR) spectroscopy, also called electron paramagnetic resonance (EPR) spectroscopy and

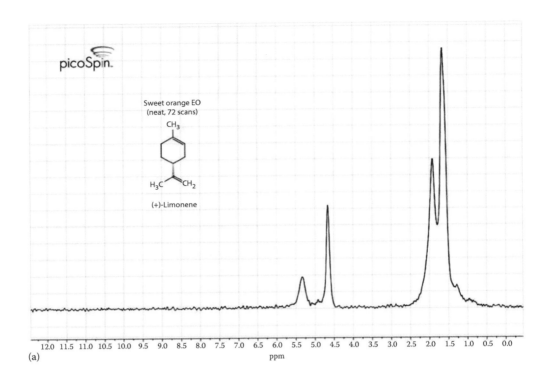

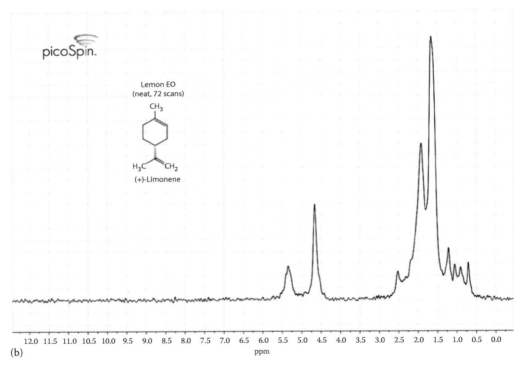

Figure 6.78 (a) Proton NMR spectrum of sweet orange essential oil. (b) Proton NMR spectrum of lemon essential oil. Collected neat using the picoSpin-45. The lemon oil shows other trace compounds not found in the orange oil. (© Thermo Fisher Scientific, www.thermofisher.com.) Used with permission.

MAGNETIC RESONANCE SPECTROSCOPY

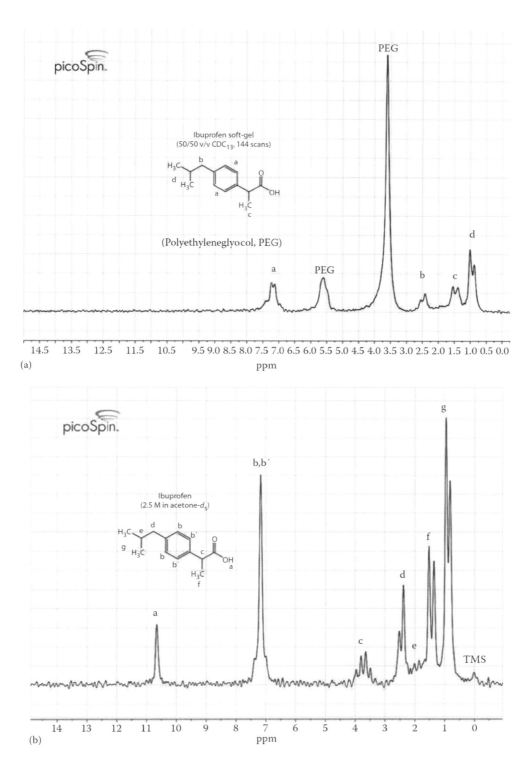

Figure 6.79 (a) Ibuprofen softgel, showing the active ingredient as well as the PEG component. Compare to (b) pure ibuprofen. (© Thermo Fisher Scientific, www.thermofisher.com.) Used with permission.

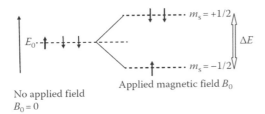

Figure 6.80 In the presence of an applied magnetic field, electrons with $S = 1/2$ can exist in one of two discrete energy levels. The levels are separated by ΔE. The lower energy level ($m_s = -1/2$) has the magnetic moment aligned with the field; in the higher energy state ($m_s = +1/2$), the magnetic moment is aligned against the field.

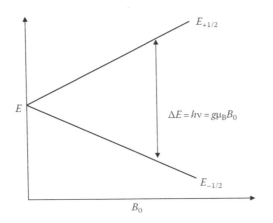

Figure 6.81 Variation of spin state energies as a function of the applied magnetic field, B_0.

electron magnetic resonance (EMR) spectroscopy, is an MR technique for studying species with one or more unpaired electrons. These species include organic and inorganic free radicals and transition metals ions. Unpaired electrons can also exist as defects in materials. Unpaired electrons possess spin, with a spin quantum number $S = 1/2$, and therefore also possess a spin magnetic moment. Just like protons in an NMR experiment, in an applied external magnetic field of B_0, electrons with magnetic moment $m_s = +1/2$ will align antiparallel to the field, while those with $m_s = -1/2$ will align parallel to the field, as shown in Figure 6.80. Just like protons in an NMR experiment, absorption (resonance) will occur under certain combinations of frequency and applied magnetic field. The important point to remember is that we are now talking about electrons, not nuclei. The energy difference or Zeeman splitting seen in ESR is the electronic Zeeman effect.

As shown in Figure 6.81, the energies of the spin states diverge linearly as the magnetic field increases. An unpaired electron can move between the two energy levels by absorbing or emitting electromagnetic radiation such that the energy equals $\Delta E = h\nu$.

The fundamental ESR equation is

$$\Delta E = h\nu = g\mu_B B_0 = g\beta B_0 \quad (6.19)$$

where

g is the g-factor, a proportionality constant approximately equal to 2 for most samples, but which varies depending on the electronic configuration of the radical or ion
μ_B is the Bohr magneton (the electron's spin magnetic moment, sometimes denoted as β)
B_0 is the magnetic field strength

This equation is analogous to the Larmor equation in NMR. Because of mass differences between nuclei and electrons, the electronic magnetic moment is about three orders of magnitude larger than the nuclear magnetic moment. This means that frequencies for ESR are generally in the 9–10 GHz range, much higher than the usual 60–300 MHz range for NMR. The generation of these high GHz frequencies, previously a major challenge for electronics, benefited greatly from the development of radar for World War II.

In principle, an absorption spectrum can be obtained by applying a constant magnetic field and scanning the frequency of the electromagnetic radiation. Alternatively, the frequency of the electromagnetic radiation can be held constant and the magnetic field scanned. Owing to limitations with microwave electronics, the frequency is normally held constant and the magnetic field varied. When the magnetic field tunes the two spin states so that their energy difference matches the frequency of the applied radiation, unpaired electrons can move between their two spin states; they are now in resonance. This value of the magnetic field is called the *field for resonance*. An absorption spectrum results as electrons in the lower state absorb energy to move to the upper state, as shown in Figure 6.82.

For a free electron, $g = g_e = 2.0023$. Using $g = 2$, a frequency of 9.75 GHz, and Equation 6.19, the predicted field for resonance $B_0 = 3480$ G $= 0.3480$ T, and a single absorption peak at this field value would be observed. It is customary for ESR spectra to be plotted as first derivative spectra, as will be seen in the following.

6.12.1 Instrumentation

ESR/EPR spectrometers use a wide variety of magnetic field strengths and microwave radiation sources.

CW ESR, pulsed ESR, and FT ESR systems are available. Table 6.7 provides a sample of frequency and field for resonance commonly used in ESR systems.

Spectrometers in the 1 GHz range are called "L-band" spectrometers; those in the 100 GHz range, "W-band" spectrometers; and so on. A schematic of a commercially available ESR/EPR spectrometer is given in Figure 6.83, along with a photo of a commercial instrument.

MAGNETIC RESONANCE SPECTROSCOPY

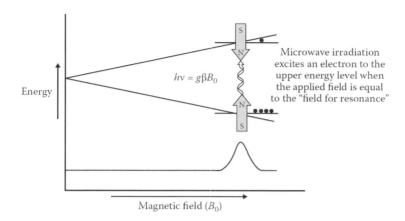

Figure 6.82 An absorption peak is seen (lower portion of Figure) as microwave irradiation excites an electron to the upper energy level when the applied field is equal to the field for resonance. (Courtesy of Bruker Corporation, www.bruker-biospin.com. This and all subsequent Bruker EPR schematics are the work of Dr. Ralph T. Weber, Bruker BioSpin, and are used with permission.)

Table 6.7 Field for resonance, B_{res}, for a $g = 2$ signal at selected frequencies

Microwave Band	Frequency (GHz)	B_{res} (G)
L	1.1	392
S	3.0	1,070
X	9.75	3,480
Q	34.0	12,000
W	94.0	34,000

Source: Courtesy of Bruker Corporation, www.bruker-biospin.com/cwtheory.html.

Classical sources for microwave generation of GHz and THz frequencies were klystrons or magnetrons. A magnetron microwave source is exactly what is present in your kitchen microwave. In the magnetron, electrons travel from a wire cathode to the anode walls of an evacuated chamber, forced into a circular path by a magnetic field. The anode wall contains hollow resonator cavities. The resonant frequency of the magnetron is determined by the resonator cavity dimensions. In your kitchen microwave, the frequency is tuned to a rotational absorption band in water, so water-containing samples absorb this frequency and heat up.

The microwave source in Figure 6.83 is located in the area marked "bridge" in the diagram. The use of phase-sensitive detection results in the absorption signal being presented as the first derivative. The absorption maximum is the point where the first derivative passes through zero.

The sample is placed in a microwave cavity (Figure 6.83). A microwave cavity is a metal box or cylinder that resonates, that is, stores, microwave energy. At the resonance frequency of the cavity, all microwaves remain inside the cavity. A consequence of resonance is that there will be a standing electromagnetic wave inside the cavity, with electric and magnetic field components exactly out of phase. Samples are placed in the cavity at the point where the electric field is at a minimum and the magnetic field is at a maximum to obtain the biggest absorption signals and highest sensitivity. The magnetic field is swept using a field controller, as seen in Figure 6.83. When the field matches (Figure 6.82), a spectrum is generated. (For the student interested in the details of instrument control and signal generation, see www.bruker-biospin.com and the EPR theory links.)

6.12.1.1 Samples and Sample Holders

For a sample to be EPR/ESR active, it must have one or more unpaired electrons. Stable free radicals, paramagnetic metal ions, and irradiated materials are some examples of such materials. The amount of sample required depends on the type of spectrometer (what band is used) and on the type of experiment—CW, pulsed, and double-resonance experiments like electron nuclear double resonance (ENDOR) (discussed in the following)—but in general, liquid and solid samples can be measured. Volumes of sample required range from 20 μL to 1 mL at concentrations of 10 nM to 20 mM for most experiments.

Sample holders are small-diameter or capillary tubes or flat cells made of clear fused quartz. High-purity quartz is required instead of glass because most glass contains traces of iron oxide, which is paramagnetic and gives rise to large background signals. Specialty cells include flowthrough cells, tissue-holding cells, and cells for electrochemical studies. Wide varieties of sample holders for EPR (and NMR) are offered by Wilmad-LabGlass and may be viewed at www.wilmad-labglass.com. Other vendors include Vitrocom and Bruker.

Organic nonpolar solvents can normally be used at all temperatures. Aqueous samples present difficulties at temperatures above 0 °C, due to the interaction of the large electric dipole of water with the standing wave in the

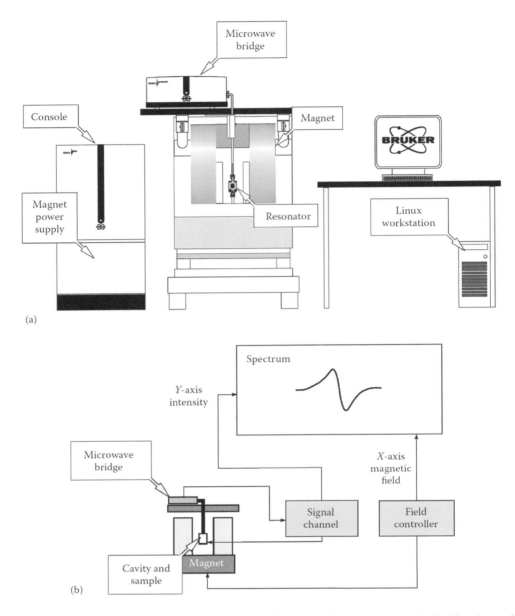

Figure 6.83 (a) and (b) A schematic ESR/EPR spectrometer. (Courtesy of Bruker Corporation, www.bruker-biospin.com.)

resonator cavity. To minimize this interaction, thin, flat cells are used for aqueous samples in X-band systems and very narrow quartz capillaries are used in Q-band systems.

Organic radicals can often be measured at room temperature. For many metals, the spin–lattice relaxation time broadens their EPR signals significantly at room temperature. In order to narrow the EPR linewidth, such samples are often run at cryogenic temperatures (liquid N_2, 77 K, or liquid He, 4 K). The mechanism for the relaxation time is interaction of the orbital angular momentum with local vibrational modes. Metal ions with little orbital angular momentum, such as Cu^{2+} and VO^{2+}, can usually be observed at room temperature.

6.12.2 Miniature ESR Spectroscopy

The standard commercial instrumentation described earlier is of large size and high cost and generally requires highly trained personnel. Recent advances in miniaturization of sensor technology and in microwave electronics have resulted in the development of smaller ESR systems. Several companies now offer benchtop systems. For example, a miniature ESR system, the Micro-ESR™ (Active Spectrum, Inc., www.activespectrum.com), brings ESR spectroscopy to the benchtop and to online process monitoring. The benchtop instrument and online sensor are

MAGNETIC RESONANCE SPECTROSCOPY

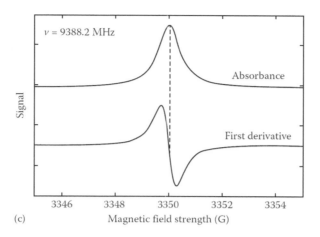

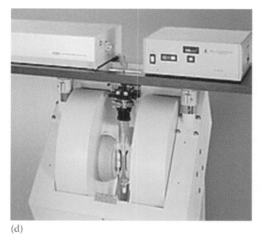

Figure 6.83 (continued) (c) Simulated absorbance spectrum and its first derivative. ESR spectra are plotted as first derivative spectra. (d) Bruker ELEXSYS-II EPR. (Courtesy of Bruker Corporation, www.bruker-biospin.com.)

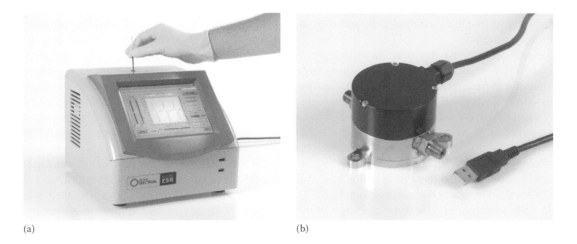

Figure 6.84 (a) Benchtop Micro-ESR™ spectrometer with integrated touch screen. The dimensions are 11.25" × 10" × 10". (b) Online Micro-ESR™ sensor with 24V DC power input, USB interface, and 2.5" diameter. (Active Spectrum, Inc., www.activespectrum.com. With permission.)

shown in Figure 6.84. Note the first derivative form of the spectrum shown on the instrument screen.

The S-band system uses a samarium cobalt permanent magnet, with a fixed field of 1180 G that is swept by a coil over a narrow range of ± 135 G. This corresponds to a frequency of 3.5 GHz. The system (Figure 6.84) includes an integrated touch screen, internal computer, thermoelectric cooler, and power supply. The instrument is available in an X-band version (9.5 GHz; 3480 ± 200 G).

Samples are introduced in capillary tubes, 2.0 mm or 5.8 mm inner-diameter tubes. Solid samples and liquid samples in any solvent may be analyzed, with only 25–50 μL sample volume required for the S-band system and 100 μL sample volume minimum for the X-band system.

6.12.3 ESR Spectra and Hyperfine Interactions

The g-factor for a free electron is $g_e = 2.0023$. The excitation frequencies in ESR spectra depend on the total magnetic moment. If we obtain a spectrum at known frequency and magnetic field, we can calculate the g-factor for our sample. The energy levels of a bound, unpaired electron differ from those of a free electron due primarily to electron orbital angular momentum. If g does not equal g_e, the electron giving rise to the signal is not a free unpaired electron. The magnitude of the change in the g-factor gives information about the atomic or molecular orbital containing the unpaired electron. Organic radicals have g values from 1.99 to 2.01, while transition metal compounds have a wider range of g values, 1.4–3.0, due to factors like spin–orbit coupling. The g-factor helps us characterize the type of sample we are measuring. It can identify a specific metal ion, its oxidation state, spin state, and coordination environment. For example, cytochrome oxidase, a metalloprotein with more than one metal center, displays an ESR spectrum with multiple g-factors that are used to identify and characterize the different centers.

ESR is a sensitive probe for the local environment of a transition metal. Orbital angular momentum can be very large for the d orbitals of transition metals. As an example, if a transition metal is surrounded by six identical ligands bound symmetrically by the d orbitals, a single transition results (i.e., a single line in the spectrum). As the symmetry is lowered either by substitution of one or more ligands or distortion of the symmetry (Jahn–Teller distortion), anisotropy appears in the ESR signal. ESR can be used to determine the oxidation state and coordination of transition metal centers in compounds.

Since the ESR spectrum results from a change in the electron's spin state (with or against the field), it might be supposed that all ESR spectra would consist of a single transition. Fortunately, as with NMR, this is not the case. Bound unpaired electrons are very sensitive to the nuclei around them, through magnetic moment interaction, shown in Figure 6.85.

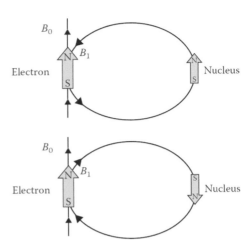

Figure 6.85 Local magnetic field at the electron, B_I, due to a nearby nucleus. (Courtesy of Bruker Corporation.)

The interaction between an unpaired electron and nearby nuclei gives rise to additional allowed energy levels; this interaction is called the hyperfine interaction or hyperfine coupling. Hyperfine coupling can tell us the identity and number of atoms in a radical or complex as well as their distances from the electron. The interactions with neighboring nuclei can be described by Equation 6.20:

$$E = g\mu_B B_0 M_s + aM_s m_I \qquad (6.20)$$

where
a is the hyperfine coupling constant
m_I is the nuclear spin quantum number

The value of a is measured as the distance between the centers of two signals (Figure 6.86a).

The magnetic moment of a nearby nucleus produces a magnetic field, B_I, at the electron, which may oppose or add to the magnetic field from the instrument magnet. When B_I adds to the magnetic field, the field for resonance is lowered by the amount B_I; when B_I opposes the magnetic field, the field for resonance is raised by an amount B_I. For a spin 1/2 nucleus, such as the proton, the single ESR signal splits into two signals, each B_I away from the original signal, as seen in Figure 6.86b.

Hyperfine interactions follow the same selection rules as NMR. For isotopes with $I = 0$ (atomic number and mass number both even), no EPR or NMR spectra are seen. For isotopes with odd atomic number and even mass number, the value for I will be an integer, such as ^{14}N with $I = 1$. For isotopes with odd mass numbers, I values will be fractional, such as ^1H, $I = 1/2$.

For one adjacent nucleus with spin = I, the number of lines seen in the ESR spectrum is $2I + 1$. (This equation should be compared with the NMR equivalent). For N

MAGNETIC RESONANCE SPECTROSCOPY

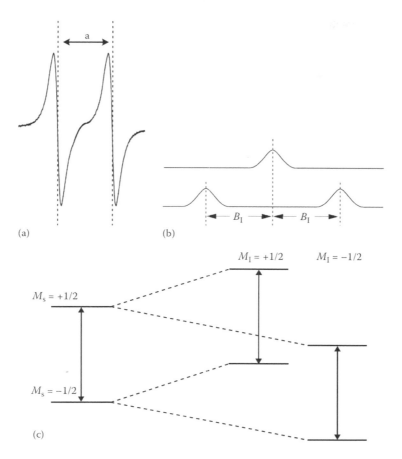

Figure 6.86 (a) Hyperfine splitting of an ESR signal by neighboring nuclei. The distance *a* is the hyperfine coupling constant. (b) Splitting of the ESR signal due to the local magnetic field of a nearby nucleus. (Courtesy of Bruker Corporation.) (c) Energy levels resulting from splitting by a single spin 1/2 nucleus. (From Ebsworth, E. et al., *Structural Methods in Inorganic Chemistry*, CRC Press, Boca Raton, FL, 1987. Used with permission.)

nuclei, the number of lines = 2NI + 1. For example, an adjacent nitrogen atom, I = 1, would give a three-line ESR spectrum, while Mn, I = 5/2, would give a six-line spectrum. Figure 6.87a shows the simulated spectra expected from N adjacent nuclei with spin = 1/2. The intensity patterns follow the binomial distribution (Pascal's triangle) as in NMR. Figure 6.87b shows similar spectral patterns for I = 1, but the relative intensities differ. For hyperfine coupling to 1N atom, for example, the pattern would be 1:1:1, 2N atoms, 1:2:3:2:1, 3N atoms, 1:3:6:7:6:3:1, and so on.

As in NMR, if there are two adjacent protons, each line would be split into two additional lines and these would overlap to give a 1:2:1 triplet pattern. Four protons would give a five-line spectrum with intensities predicted (from Pascal's triangle) to be 1:4:6:4:1.

In general, for a radical having M_1 equivalent nuclei each with a spin of I_1 and a group of M_2 nuclei, each with a spin of I_2, the number of lines expected is $N = (2M_1I_1 + 1)(2M_2I_2 + 1)$. The simulated ESR spectra of a methyl radical and a methoxymethyl radical are shown in Figure 6.88.

From the earlier equations (and having seen similar splitting patterns in NMR), it is clear that in Figure 6.88a, the three equivalent ^{1}H nuclei give 2(3)(1/2) + 1 lines in a 1:3:3:1 ratio; (b) shows a triplet of quartets due to the 1:2:1 splitting from the two methyl protons followed by each of these lines splitting into a quartet by the CH$_3$ protons. A question the student should be able to answer from having studied NMR is "why doesn't the ESR spectrum show a quartet of triplets instead of the triplet of quartets?"

If we have an electron delocalized over a six-membered ring, as in the pyrazine anion, C$_4$H$_4$N$_2^-$, the electron is coupled to two equivalent N atoms (I = 1) and then couples to four equivalent protons (I = 1/2). The spectrum should show a quintet with intensities of 1:2:3:2:1, with each of those lines split into a quintet with intensities of 1:4:6:4:1 (Figure 6.89).

6.12.4 Applications

ESR is used in chemistry, biochemistry, biology, archeology, geology, and physics to detect, identify, and study

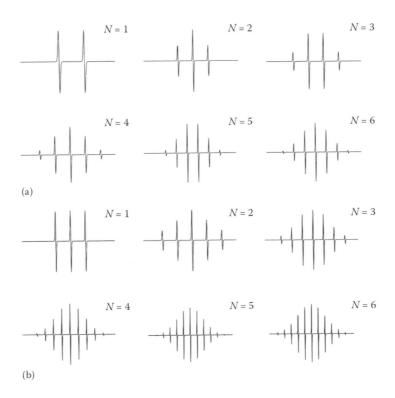

Figure 6.87 (a) Spectral patterns expected for N interacting nuclei with $I = 1/2$. (b) Spectral patterns expected for N interacting nuclei with $I = 1$.

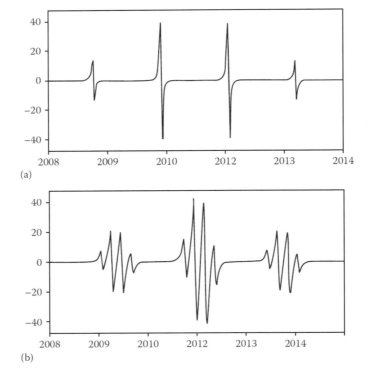

Figure 6.88 Simulated ESR spectra of (a) $H_3C\cdot$ radical and (b) $(CH_3O)CH_2\cdot$ radical. (The unpaired electron is on the carbon atom in both cases.)

Figure 6.89 ESR spectrum of the pyrazine anion.

Table 6.8 Applications of ESR

Chemistry
- Kinetics of radical reactions
- Polymerization reactions
- Spin trapping
- Organometallic compounds
- Catalysis
- Petroleum research
- Oxidation and reduction processes

Physics
- Measurement of magnetic susceptibility
- Transition metal, lanthanide, actinide ions
- Conduction electrons in semiconductors, conductors
- Defects in crystals (e.g., color centers in alkali halides)
- Optical detection of MR, excited states
- Crystal fields in single crystals

Biology and medicine
- Spin-label and spin-probe techniques
- Spin trapping
- Free radicals in living tissue and fluids
- Antioxidants, radical scavengers
- Oximetry
- Enzyme reactions
- Photosynthesis
- Structure of metalloprotein active sites
- Photochemical generation of radicals
- NO in biological systems

Source: Courtesy of Dr. Ralph T. Weber, Bruker BioSpin, Bruker Corporation.

free radicals and paramagnetic compounds and materials. Table 6.8 summarizes some of the uses for ESR.

It is the only direct method available to measure free radicals. ESR is used to study both the radicals and the reactions, chemical and physical, which create or modify free radicals. Examples of reactions that can create or change free radicals are electrochemical reactions, exposure to UV radiation, and exposure to ionizing radiation, among others. ESR samples can be liquid, such as oil, blood, or saliva, solid, or gaseous, such as measurement of soot particulates in air. ESR is both a qualitative and quantitative technique, measuring the type and concentration of free radicals in a sample. Qualitative information is obtained from the g value of a peak, while quantitative results are obtained by integrating the area under the peaks.

ESR has been used to measure engine oil oxidation. As the oil oxidizes, the free radical signal increases in intensity. Measurement of the population of stable free radicals generated by breakdown of the oil is a direct measurement of oxidation. The intensity of the organic radical signal increases dramatically when the oil approaches the end of its useful life. This application has been demonstrated using the Active Spectrum online Micro-ESR™ for real-time monitoring of engine oil. Many types of oil can be analyzed for the amount of oxidation, the amount of soot in the oil, the percentage of both free and dissolved water, and certain metal and sulfur compounds. Crude oils also can be monitored online to determine both organic radicals and metals like vanadium (as the vanadyl radical) (White et al.).

Free radicals are known to play a major role in the damage caused to biological systems by ionizing radiation, UV-induced cancer, poisoning by CCl_4 and some herbicides, and in oxygen toxicity in premature infants treated with hyperbaric oxygen (www.niehs.nih.gov). ESR is used by medical and biological scientists studying the involvement of free radicals in disease. For example, EPR can be used to study free radicals in saliva, which can cause periodontal disease (the inflammation and/or loss of connective tissue surrounding the teeth).

In biochemistry, metal ions are involved in many reactions. All metal atoms with unpaired electrons (paramagnetic ions) can theoretically be studied by ESR, but the experimental conditions vary widely. Atoms with important paramagnetic states that can be measured by ESR include Cu, Fe, Mn, Gd, and V, among others. Oxidation state and coordination can be determined in proteins, for example.

Radiation damage over time creates free radicals in materials. In the case of teeth, free radicals in tooth enamel have been measured to date teeth in archeology studies and to evaluate exposure of people to ionizing radiation after several major nuclear reactor leaks. Radiation dosimetry and analysis of irradiated foods, gamma-irradiated polymers, and other solids such as ceramics, bone, and coal are common applications of ESR.

6.12.4.1 Spin Labels and Spin Traps

A spin label is a stable paramagnetic molecule that contains an atom or group of atoms with an unpaired electron spin that can be bonded to another molecule. In this way, one can detect molecules that otherwise would not give an ESR spectrum. Most spin labels are nitroxide compounds. For example, 2,2,6,6-tetramethylpiperidine-*N*-oxyl (known as TEMPO) is a common spin label with a characteristic three-line ESR spectrum, seen in Figure 6.90.

An advantage of spin-label ESR studies is that they can be carried out at temperatures relevant to physiology; they

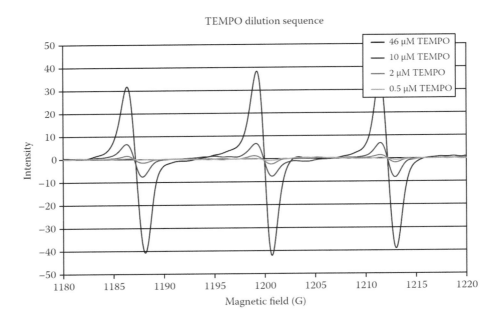

Figure 6.90 Micro-ESR sensitivity for TEMPO in deionized water solutions at a concentration between 0.5 and 46 µM. (Active Spectrum, Inc., www.activespectrum.com. Used with permission.)

are therefore sensitive to molecular motion. Spin labeling permits the study of the dynamics, folding, and structure of proteins by ESR.

Many free radicals of interest, especially in biological systems, are extremely short-lived and are of low abundance. Spin trapping is a technique for measuring short-lived free radicals by causing an addition reaction to occur that results in a longer-lived radical. The compound "trapping" the radical is called a spin trap; the addition product is called a spin adduct. The spin adduct is detected by ESR and its spectrum provides a signature for the type of radical that was trapped. Spin traps are often nitrone or nitroso compounds such as 5,5-dimethyl-pyrroline-*N*-oxide (DMPO). Thousands of published spin-trapping experiments can be found in the NIEHS database. (See Spectral Databases.) Unstable free radicals such as superoxide, hydroxyl, and xenobiotic radicals can be studied in this way.

An interesting application of spin trapping, provided courtesy of Dr. Ralph Weber, is the study of beer. Specifically, ESR is used in the study of beer "staling," that is, the development of oxidized compounds like aldehydes, ketones, and alcohols that impart a stale, undesirable flavor to the beer. Long distribution channels make fresh beer a challenge to breweries, because excess oxygen and improper storage or processing of beer promote free radical oxidation. Beer contains antioxidants intended to prevent premature staling. If the antioxidant levels get too low, oxidation of beer components increases and stale flavor results. Breweries use ESR to study the free radical oxidation process in beer using spin traps to capture unstable radicals such as peroxyl radicals in order to better optimize their antioxidant levels and formulations.

6.12.5 CW-ENDOR

NMR and ESR/EPR are the two MR techniques used to study the structure, dynamics, and electronic structure of radicals, metal ions, and defects. While each of these techniques provides crucial and exclusive data, each suffers from certain limitations. ESR/EPR is highly sensitive with detection limits approaching nanomolar concentrations, but linewidths of signals may at times obscure important spectral information. NMR has superior resolution but much less sensitivity than ESR.

The unpaired electron, which gives the ESR spectrum, is very sensitive to its local surroundings. The nuclei of the atoms in a radical or complex often have a magnetic moment, which produces a local magnetic field at the electron. The electron also produces a local field at the nucleus. The interaction between the electron and the nucleus is called the hyperfine interaction. The hyperfine splittings provide information such as the number and identity of atoms, their distances from the unpaired electron, and electron spin densities at the nuclei. The goal of the ENDOR experiment is to obtain information (the hyperfine splittings) from the individual nuclei. Figure 6.91 outlines the interaction between an electron and four equivalent nuclei.

ENDOR is a double-resonance technique that combines the high resolution and nuclear sensitivity of an NMR experiment with the sensitivity of an ESR/EPR experiment by observing NMR signals as changes in the intensity of

MAGNETIC RESONANCE SPECTROSCOPY

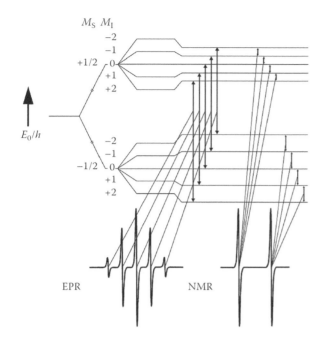

Figure 6.91 Energy level diagram for the interaction of an electron and four equivalent nuclei. (Courtesy of Bruker Corporation.)

ESR signals. The ENDOR experiment is performed by monitoring the ESR signal while sweeping an RF signal to drive the NMR transitions. An ESR spectrum is acquired in order to determine the magnetic field values at which the ESR signals occur. The magnetic field is then fixed at the top of the ESR absorption signal. The ESR signal is then saturated by high-power microwaves. The final step is to sweep the RF while monitoring the ESR signal. When the RF drives an NMR transition, the saturated ESR signal desaturates and increases in intensity, resulting in an ENDOR signal. The ENDOR experiment provides superior resolution of the hyperfine coupling constants and hyperfine splitting information compared to ESR. A more detailed explanation of the ENDOR experiment can be found at http://bruker-biospin.com/cwendor.html.

A schematic diagram of an ENDOR spectrometer is shown in Figure 6.92.

As with NMR, many pulsed ESR techniques, where the ESR signal is obtained as a function of time, are now routinely providing data (and the proliferation of acronyms such as DEER, ESEEM, and so on). The discussion of these is beyond the scope of this text.

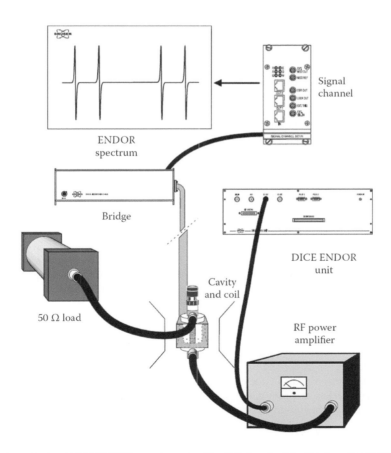

Figure 6.92 Configuration for a Bruker E 560 ENDOR spectrometer. The RF signal originates from the DICE ENDOR system. (Courtesy of Bruker Corporation.)

SUGGESTED EXPERIMENTS

6.1 (For the instructor) Demonstrate to students in the laboratory the principal components of an NMR instrument. Indicate the steps taken in tuning the instrument.

6.2 Obtain the NMR spectrum of a straight-chain alkane, such as *n*-octane. Identify the methyl and methylene peaks. Note the chemical shift and the spin–spin splitting. Measure the total area of the methyl and methylene peaks and correlate this with the number of methyl and methylene protons in the molecule.

6.3 (a) Obtain the NMR spectrum of ethyl alcohol (1) wet and (2) very dry.
(b) Identify the methyl, methylene, and alcohol protons. Note the chemical shift and spin–spin splitting.
(c) Measure *J*, the coupling constant, between the methyl and the methylene protons. Note the effect of water on the alcohol peak. Explain this phenomenon.

6.4 Integrate the peak areas obtained in Experiment 6.3. Measure the ratios of the areas and the numbers of hydrogen nuclei involved in the molecules. What is the relationship between the area and the number of hydrogen nuclei?

6.5 Obtain the NMR spectrum of a mixture of toluene and hexane. Integrate the peak areas of the different parts of the spectrum. From the ratio of the aromatic protons to the total protons, calculate the percentage of toluene in the mixture.

6.6 Obtain the NMR spectra of organic halides, olefins, aromatics, and substituted aromatics. Correlate the peaks with the different types of protons present.

6.7 Identify an unknown compound from its NMR spectrum. The difficulty of the problem should match the level of competence of the student.

6.8 Obtain the NMR spectra of (a) hexane and (b) heptane. Integrate the areas under the peaks in each spectrum. Would it be possible to distinguish between these compounds based on their NMR spectra? Compare with IR spectroscopy when Chapter 5 is covered and with MS (Chapter 11).

6.9 Record the NMR spectrum of a sample of cooking oil. Measure the ratio of hydrogen in unsaturated carbon and that in saturated carbons. Compare the degrees of unsaturation of various commercial cooking oils.

6.10 Repeat Experiment 6.9 using margarine as the sample. First, dissolve a known amount of margarine in CCl_4 and obtain the NMR spectrum. Compare the degrees of unsaturation of different brands of margarine.

6.11 Obtain the NMR spectra, proton and ^{13}C, for commonly available headache tablets and pain relievers. Most will dissolve in deuterated chloroform or acetone. Inert fillers may need to be filtered out of the solution. Read the labels—some are "pure" compounds such as aspirin, others contain more than one ingredient. Obtain the spectra of the pure components—acetylsalicylic acid, acetaminophen, and caffeine are common ingredients of these tablets. (The "pure" material can often also be purchased as a commercial tablet. Just be aware that you may see some impurity peaks if you are not using reagent grade materials.) Correlate the peaks with the structure of the compounds. In those tablets with more than one ingredient, note any spectral overlaps. Estimate the purity of different commercial brands of aspirin using your spectra.

6.12 Obtain the proton and ^{13}C spectra of geraniol. Perform the following 2D experiments: COSY, HETCOR, and INADEQUATE. Using the 2D data, work out the structure of geraniol.

6.13 Repeat Experiment 6.12 with caffeine, acetaminophen, or butyl acetate.

6.14 The reaction of 2,2-dimethoxypropane with acid to form methanol and acetone can be followed by both proton and ^{13}C NMR. Into a dry NMR tube, add 0.6 mL of 2,2-dimethoxypropane. Obtain the NMR spectrum and integration. Then add 0.1 mL of 1 M HCl and mix well for about 5 min. Obtain the NMR spectrum again. (a) Write out the reaction, using structures. Predict how many proton and/or carbon peaks you should see in the first spectrum. Predict how many peaks should be seen in the second spectrum. (b) Look up the chemical shifts of the protons (and ^{13}C, if obtained) in the tables in the chapter and predict where each proton or carbon will be observed relative to TMS. (c) Confirm the identity of the peaks in your spectra.

PROBLEMS

6.1 Define the term chemical shift. Why does it occur? How is it measured for protons? For ^{13}C?

6.2 (a) Explain why spin–spin coupling occurs.
(b) A methylene group (CH_2) is adjacent to a CH group. Into how many peaks is the CH_2 peak split by the adjacent single hydrogen?

6.3 What does a *J* coupling constant tell you? Proton A is coupled to proton B with $J_{AB} = 9$. Proton A is also coupled to proton C with $J_{AC} = 2$. Draw the predicted splitting pattern for proton A.

6.4 Draw a schematic proton NMR spectrum for each of the following compounds. Indicate chemical shift and peak multiplicity.

(a) *n*-Butane

```
 H  H  H  H
 |  |  |  |
H-C--C--C--C-H
 |  |  |  |
 H  H  H  H
```

(b) 2,2-dimethylpropane (also called neopentane)

```
      CH_3
       |
CH_3 - C - CH_3
       |
      CH_3
```

6.5 Draw a schematic proton NMR spectrum for each of the following compounds. Indicate chemical shift and peak multiplicity.

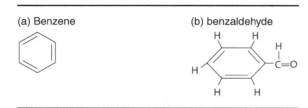

(a) Benzene   (b) benzaldehyde

6.6 The total area of the peaks in 1-butene (shown) is 80 units. What will be the area of the peaks caused by (a) the CH_2 group on carbon b and (b) the CH group on carbon c?

```
    H H H        H
    | | |       /
  H-C-C-C=C
    | |         \
    H H          H
    a b c d
```

6.7 Draw a schematic block diagram of an FT-NMR instrument.

6.8 Define magnetogyric ratio.

6.9 What is the "magic angle"? Explain why MAS is used to acquire the NMR spectrum of a solid sample.

6.10 Explain why liquid samples are spun in the magnetic field while acquiring an NMR spectrum.

6.11 Aldehydic protons occur at about 9–10 ppm, meaning that they are highly deshielded compared with other protons. Using Figure 6.9 as an example show why the aldehyde proton is deshielded.

6.12 What are the rules for determining the spin number of an element?

6.13 Which of the following have a spin number = 0? ^{12}C, ^{13}C, ^{16}O, ^{17}O, ^{1}H, ^{2}H, ^{19}F, ^{28}Si, ^{35}Cl, ^{108}Ag, ^{96}Mo, ^{66}Zn, ^{65}Zn

6.14 What is the spin number for (a) ^{1}H and (b) ^{2}H? What is the number of orientations for (a) ^{1}H and (b) ^{2}H in a magnetic field (spin number = $2I + 1$)?

6.15 What information does the COSY experiment provide? Explain how you interpret a COSY plot.

6.16 What information does the HETCOR experiment provide? Explain how you interpret a HETCOR plot.

6.17 What information does the INADEQUATE experiment provide? What is plotted on the x- and y-axes in an INADEQUATE plot? What is the major limitation of this experiment?

6.18 What are the requirements for a standard reference material, such as TMS, in NMR?

6.19 Draw the proton NMR spectrum you would expect for the compound propylamine, $CH_3CH_2CH_2NH_2$. Include the chemical shifts and spin–spin splitting. Explain your rationale.

6.20 What would be the spin–spin splitting caused by an NH group on an adjacent CH group? Why?

6.21 What would be the spin–spin splitting caused by an adjacent (a) CH_3, (b) CH_2, (c) CF_3, (d) CF_2, (e) CH, and (f) CFH on a CH group?

6.22 What are (a) transverse relaxation and (b) longitudinal relaxation?

6.23 List the causes of relaxation in NMR.

6.24 How does the relaxation time affect linewidth?

6.25 What is the NOE? What advantages and disadvantages does the NOE result in for ^{13}C NMR?

6.26 What is the chemical exchange rate at the temperature at which multiplicity is lost?

6.27 What is the effect of viscosity on T_1?

6.28 A proton appears at a chemical shift of 7.0 ppm (vs. TMS) in an NMR spectrometer operated at 60 MHz. What is the resonance frequency of the proton in Hz?

6.29 What is meant by saturation of a nucleus?

6.30 Explain why the spectrum of naphthalene (Figure 6.15) exhibits a "four-line" splitting pattern for the aromatic protons. Do the same for the proton spectrum of the polymer in Figure 6.45a.

6.31 In the quantitative analysis example of a mixture of octane and 1-octene discussed in the text, the terminal olefinic protons that absorbed at 5.0 ppm were used to quantify the 1-octene. Could you have used any other resonance? (Look at Table 6.4.) If so, show how the calculation would change from the sample calculation in the text. Is there any advantage to one calculation over the other?

6.32 What are the requirements for solvents used for NMR studies of dissolved organic compounds?

6.33 If a disubstituted benzene ring has the same two substituents, for example, dibromobenzene, the number of unique carbon atoms in the ring can be 2, 3, or 4. The ^{13}C NMR spectrum depends on where the bromine atoms are located on the ring.
(a) Draw the three possible dibromobenzene molecules. (b) Using the dichlorobenzene example in Section 6.6.3, draw a mirror plane in each of your dibromobenzene molecules. A "mirror plane" is a plane of symmetry that gives you identical halves of the molecule on each side of the plane, as was done for dichlorobenzene. Identify how many unique carbon atoms there are in each dibromobenzene molecule and predict how many peaks you will see in the ^{13}C NMR spectrum of each molecule.

Use the proton and ^{13}C spectra provided and the chemical shift tables in the chapter for Problems 6.34 through 6.48, and draw a reasonable structure for the compound, given the formula or RMM. All spectra given here are reprinted with permission of the Aldrich Chemical Co. In some cases, the 300 MHz proton spectra have had peak "expansions" added to clarify the spin–spin splitting. (These "expansions" are actually 60 MHz signals.) All proton spectra have a peak for TMS at 0.0 ppm and all ^{13}C spectra have a triplet from the $CDCl_3$ solvent at 77 ppm.

6.34 $C_{12}H_{10}$

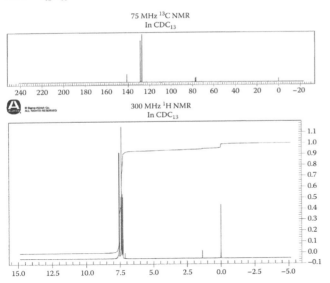

6.35 C_4H_9I

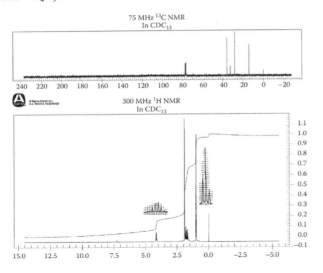

6.36 C_4H_8O

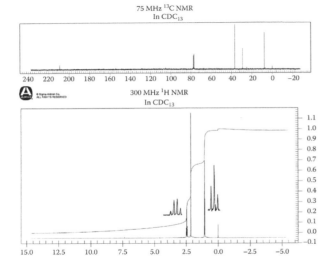

6.37 RMM = 120.97. The molecular formula contains a halogen atom.

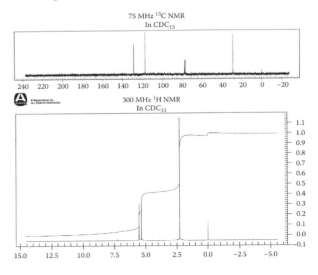

6.38 RMM = 156.23

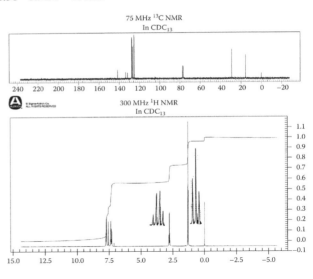

6.39 C_3H_7O

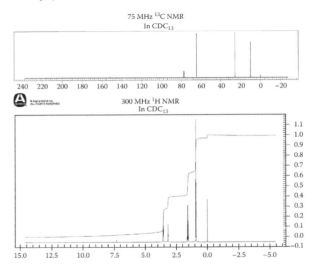

6.40 C$_3$H$_5$O$_2$Cl

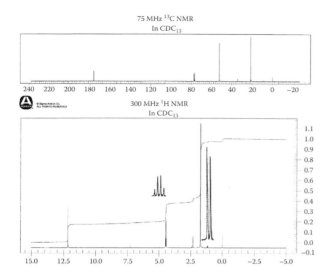

6.41 C$_3$H$_6$Cl$_2$

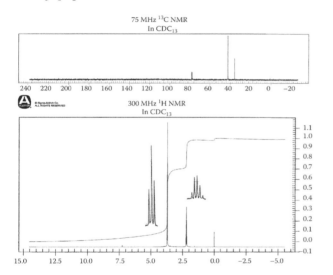

6.42 C$_3$H$_6$Cl$_2$

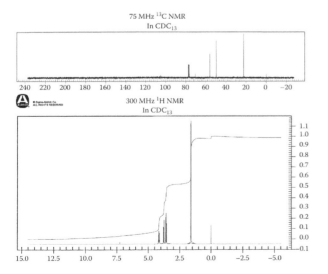

6.43 RMM = 106.17

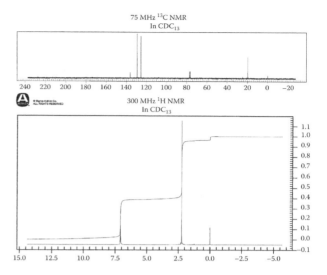

6.44 RMM = 106.17

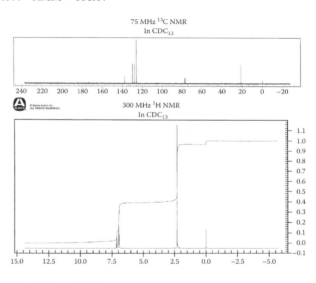

6.45 RMM = 106.17

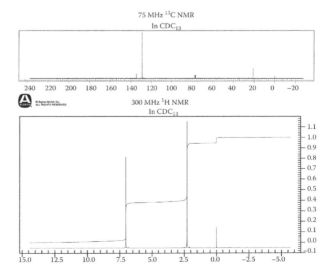

6.46 RMM = 108.14

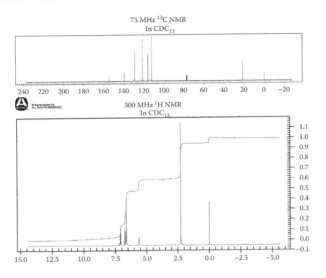

6.47 RMM = 84.16. The compound is a hydrocarbon.

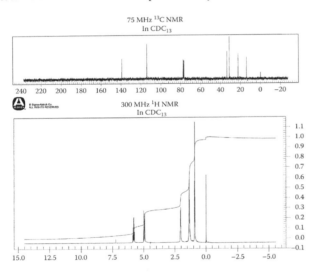

6.48 C$_4$H$_{11}$N

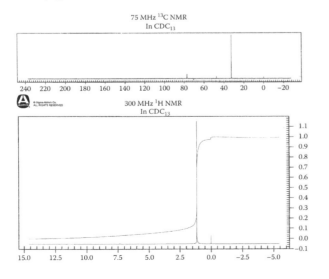

Spectra 6.49 through 6.52 are from *The Sadtler Handbook of Proton NMR Spectra*, W.W. Simons, ed., Sadtler Research Laboratories, Inc., Philadelphia, PA, 1978. [Now Bio-Rad Laboratories, Inc., Informatics Division, Sadtler Software and Databases. Copyright Bio-Rad Laboratories, Informatics Division, Sadtler Software and Databases, 2014. All rights reserved.])

6.49 RMM = 84.93. Compound is a substituted hydrocarbon.

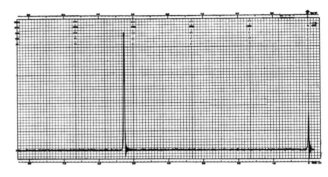

6.50 RMM = 78.11. The compound is a hydrocarbon.

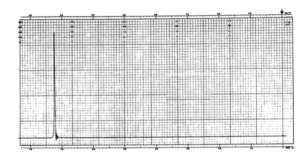

6.51 C$_7$H$_6$ClF.

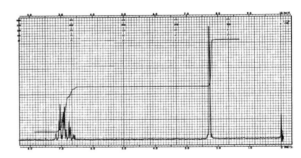

6.52 C$_3$H$_8$O.

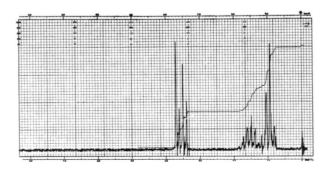

6.53 The fluorine NMR spectrum of a substituted benzene ring, RMM = 186. (© Thermo Fisher Scientific, www.thermofisher.com.) Used with permission.

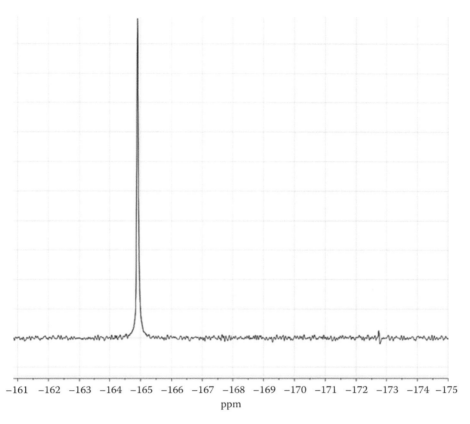

6.54 $C_{10}H_{12}O$. (© Thermo Fisher Scientific, www.thermofisher.com.) Used with permission. The compound is the essential oil of anise.

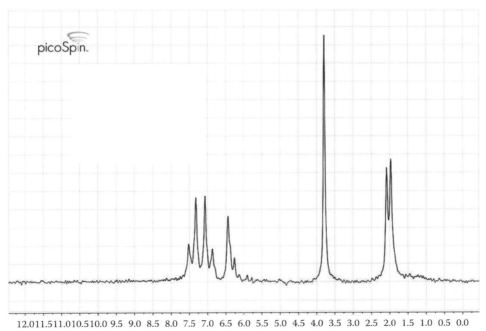

6.55 $C_8H_8O_3$. (© Thermo Fisher Scientific, www.thermofisher.com.) Used with permission. The compound is the essential oil of wintergreen.

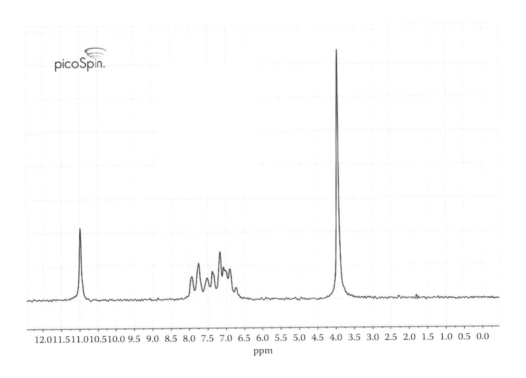

6.56 What species are able to be measured using ESR/EPR spectroscopy?

6.57 Superoxide free radicals are extremely short-lived. Explain what can be done to allow these free radicals to be measured.

6.58 How are quantitative results obtained by ESR?

6.59 What are the advantages and disadvantages of ESR versus NMR?

6.60 In the compound vanadyl acetylacetonate, $VO(acac)_2$, how many lines and of what intensity would you expect to see in the ESR spectrum. For vanadium, $I = 7/2$.

BIBLIOGRAPHY

The reader will note that occasionally multiple editions of books are cited as sources or reference works. The reason for this is that sometimes different figures are used in different editions, which may in fact be written by different authors. The sources found in the bibliography will be the most recent versions of a particular work.

In some cases instrument manufacturers and suppliers are cited by providing a URL to the exact internet location. We have endeavored to ensure that these links are functional at the time of publication, but clearly these links change. If a link has become nonfunctional, a good approach is to use your browser's advanced search capability. For example: "company name" AND "instrument type" AND "specific descriptors" is a good place to start.

Akitt, J.W. In Mason, J. (ed.), *Multinuclear NMR*. Plenum Press: New York, 1987.

Akitt, J.W.; Mann, B.E. *NMR and Chemistry*, 4th edn. CRC/Taylor & Francis: Boca Raton, FL, 2000.

Ando, I.; Yamanobe, T.; Asakura, T. *Prog. NMR Spectrosc., 22*, 349, 1990.

Apple, T.M. NMR applied to materials analysis. *Appl. Spectrosc., 49*(6), 12A, 1995.

Arnold, J.T.; Dharmetti, S.S.; Packard, M.E. *J. Chem. Phys., 19*, 507, 1951.

Atherton, N.M. *Principles of Electron Spin Resonance*, Ellis Horwood: Chichester, U.K., 1993.

Bell, A.T.; Pines, A. (eds.), *NMR Techniques in Catalysis*, Marcel Dekker, Inc.: New York, 1994.

Bock, C.; Frederic, M.; Witting, R.M.; Pörtner, H.-O. *Magn. Reson. Imaging, 19*, 1113–1124, 2001.

Bock, C.; Sartoris, F.-J; Pörtner, H.-O. In vivo MR spectroscopy and MR imaging on nonanaesthetized marine fish: Techniques and first results. *Magn. Imaging, 20*, 165–172, 2002.

Bovey, F.A. *Nuclear Magnetic Resonance*, 2nd edn. Academic Press: New York, 1988.

Bringmann, G.; Günther, C.; Schlauer, J.; Rückert, M. HPLC-NMR on-line coupling including the ROESY technique: Direct characterization of naphthylisoquinoline alkaloids in crude plant extracts. *Anal. Chem., 70*, 2805, 1998.

Bruch, M.D.; Dybowski, C. Spectral editing methods for structure elucidation. In Bruch, M.D. (ed.), *NMR Spectroscopy Techniques*, 2nd edn. Marcel Dekker, Inc.: New York, 1996.

Bruno, T.J., Svoronos, P.D.N., CRC Handbook of Basic Tables for Chemical Analysis – Data Driven Methods and Interpretation, 4th Ed., CRC Taylor and Francis, Boca Raton, FL, 2021.

Ebsworth, E.; Rankin, D.; Craddock, S. *Structural Methods in Inorganic Chemistry*. CRC Press: Boca Raton, FL, 1987.

Farrar, T.C.; Becker, E.D. *Pulsed and Fourier Transform NMR*. Academic Press: New York, 1971.

Fukushima, E.; Roeder, S.B.W. *Experimental Pulse NMR: A Nuts and Bolts Approach*. Addison-Wesley: Reading, MA, 1981.

Fyfe, C.A. *Solid State NMR for Chemists*. CFC Press: Guelph, Ontario, Canada, 1985.

Herlihy, A. SWIFT MRI: Silent MRI that images the "invisible". *Access Agilent eNewsletter*, February 2011, www.chem.agilent.com/en-US/Newsletters/Accessagilent/2011/feb/pages/swift_mri.aspx.

Ichikawa, M.; Nonaka, N.; Amano, H.; Takada, I.; Ishimori, S.; Andoh, H.; Kumamoto, K. *Appl. Spectrosc.*, *46*, 1548, 1992.

Lambert, J.B.; Shurvell, H.F.; Lightner, D.; Cooks, R.G. *Introduction to Organic Spectroscopy*. Macmillan Publishing Company: New York, 1987.

Laupretre, F. *Prog. Polym. Sci.*, *15*, 425, 1990.

Leyden, D.E.; Cox, R.H. *Analytical Applications of NMR*. John Wiley & Sons, Inc.: New York, 1977.

Lindon, J.C.; Nicholson, J.K.; Wilson, I.D. *Advances in Chromatography*, Vol. 36. Marcel Dekker, Inc.: New York, p. 315, 1996.

Maniara, G.; Rajamoothi, K.; Rajan, S.; Stockton, G.W. Method performance and validation for quantitative analysis by ^{1}H and ^{31}P NMR spectroscopy. Applications to analytical standards and agricultural chemicals. *Anal. Chem.*, *70*(23), 4921, 1998.

Mathias, L.J. *Solid State NMR of Polymers*. Plenum Press: New York, 1991.

McClure, C.K. In Bruch, M.D. (ed.), *NMR Spectroscopy Techniques*, 2nd edn. Marcel Dekker, Inc.: New York, 1996.

Pavia, D.L.; Lampman, G.M.; Kriz, G.S. *Introduction to Spectroscopy*, 3rd edn. Harcourt College Publishers: New York, 2001.

Petersheim, M. Nuclear magnetic resonance. In Ewing, G.A. (ed.), *Analytical Instrumentation Handbook*, 2nd edn. Marcel Dekker Inc.: New York, 1997.

Robinson, J.W. (ed.) *Handbook of Spectroscopy*, Vol. II. CRC Press: Boca Raton, FL, 1974.

Rodgers, J.E. Wide line nuclear magnetic resonance in measurement of finish-on-fiber of textile products. *Spectroscopy*, *9*(8), 40, 1994.

Ruan, R.; Chen, L. *Water in Foods and Biological Materials—A Nuclear Magnetic Resonance Approach*. Technomic Publishing Co., Inc.: Lancaster, PA, 1998.

Shoolery, J.J. NMR spectroscopy in the beginning. *Anal. Chem.*, *65*(17), 731A, 1993.

Silverstein, R.M.; Webster, F.X. *Spectrometric Identification of Organic Compounds*, 6th edn. John Wiley & Sons, Inc.: New York, 1998.

Simons, W.W. (ed.) *The Sadtler Handbook of Proton NMR Spectra*. Sadtler Research Laboratories, Inc.: Philadelphia, PA, 1978. (Now Bio-Rad Laboratories, Informatics Division, Sadtler Software and Databases.)

Skelly Frame, E.M.; Uzgiris, E.E. The determination of gadolinium in biological samples by ICP-AES and ICP-MS in evaluation of the action of MRI agents. *Analyst*, *123*, 675–679, 1998.

Skloss, T.W.; Kim, A.J.; Haw, J.F. High-resolution NMR process analyzer for oxygenates in gasoline. *Anal. Chem.*, *66*, 536, 1994.

Smith, M.E.; Strange, J.H. NMR techniques in materials physics: A review. *Meas. Sci. Tech.*, *7*, 449, 1996.

Wendesch, D.A.W. A practical guide to geometrical regulation. *Appl. Spectrosc. Rev.*, *28*(3), 165, 1993.

White, C.J.; Elliott, C.T.; White, J.R. Micro-ESR: Miniature electron spin resonance spectroscopy. *Am. Lab.*, *43*, 18–23, April 2011.

Williams, E.A. Polymer molecular structure determination. In Lifshin, E. (ed.) *Characterization of Materials*, Part I. VCH Publishers, Inc.: New York, 1992.

Williams, E.A.; Skelly Frame, E.M.; Donahue, P.E.; Marotta, N.A.; Kambour, R.P. Determination of bromine levels in brominated polystyrenes and poly(2,6-dimethyl-1,4-phenylene oxides). *Appl. Spectrosc.*, *44*, 1107, 1990.

Yu, T.; Guo, M. Recent developments in 13C solid state high-resolution NMR of polymers. *Prog. Polym. Sci.*, *15*, 825, 1990.

SPECTRAL DATABASES

This list is not complete. Many instrument manufacturers offer spectral databases packaged with their instruments. The publishers listed in the following offer their databases in electronic and some hardcopy formats, with CD versions and electronic versions becoming increasingly popular. The major drawback of the high-resolution databases for the beginner learning NMR spectral interpretation is the lack of peak expansion and area integrations on the spectra in many cases. The authors are deeply indebted to Aldrich Chemical Co., Agilent Technologies, Inc., Bio-Rad Informatics Division, and AIST for their permission to use their spectra in this chapter.

Aldrich Chemical Company (www.sigma-aldrich.com) publishes 12,000 high-resolution ^{13}C and proton spectra, in the volumes by Pouchert, C.J., Behnke, J. *The Aldrich Library of ^{13}C and ^{1}H FT-NMR Spectra, 300 MHz*, Aldrich Chemical Company, Milwaukee, WI, 1993. *The Aldrich/ACD Library of FTNMR Spectra, Pro Version*, is available on CD with 15,000 ^{13}C and ^{1}H 300 MHz spectra.

Bio-Rad Laboratories, Informatics Division, Philadelphia, PA (www.bio-rad.com), publishes the electronic Sadtler spectra collections of high-resolution proton and ^{13}C NMR spectra, and specialty NMR databases for metabolites, monomers, and polymers. Bio-Rad offers a powerful informatics tool called KnowItAll®, which permits spectral processing, search, analysis, and prediction, among other tools. As of 2012, the proton and carbon NMR databases contained over 560,000 spectra, and the "other elements" NMR databases such as B, P, N, and Si had over 90,000 spectra.

CambridgeSoft® publishes ChemDraw and ChemBioDraw, with the ability to predict NMR proton and carbon shifts based on structures, using their ChemNMR application (CambridgeSoft Corporation, Cambridge, MA, www.cambridgesoft.com, now part of PerkinElmer Informatics).

National Institute of Advanced Industrial Science and Technology, Tsukuba, Ibaraki, Japan, publishes a free spectral database system for organic compounds. The spectra include IR, Raman, NMR, and MS for most compounds. The database may be accessed at www.aist.go.jp/RIODB/SDBS.

NIST, the United States, publishes a comprehensive database of NMR, IR, and MS spectra. The NIST database is available for sale through 21 commercial distributors, such as Bio-Rad Laboratories.

The National Institute of Environmental Health Sciences (NIEHS) publishes a database of over 10,000 published spin-trapping ESR experiments at http://tools.niehs.nih.gov/stdb/index.cfm and an ESR software database at http://tools.niehs.nih.gov/stdb/index.cfm/spintrap/epr_home.

CHAPTER 7

Atomic Absorption Spectrometry

The basis of atomic absorption spectrometry (AAS) is the absorption of discrete wavelengths of light by ground-state, gas-phase free atoms. Free atoms in the gas phase are formed from the sample by an "atomizer" at high temperature. AAS was developed in the 1950s by Alan Walsh and rapidly became a widely used analytical tool. AAS is an **elemental analysis** technique capable of providing quantitative information on ~70 elements in almost any type of sample. As an elemental analysis technique, it has the significant advantage in many (but not all) cases of being practically independent of the chemical form of the element in the sample. A determination of cadmium in a water sample is a determination of the total cadmium concentration. It does not matter whether the cadmium exists as the chloride, sulfate, or nitrate, or even if it exists as a complex or an organometallic compound, if the proper analysis conditions are used. Concentrations as low as ppt levels of some elements in solution can be determined and AAS is used routinely to determine ppb and ppm concentrations of most metal elements. Another principal advantage is that a given element can be determined in the presence of other elements, which do not interfere by absorption of the analyte wavelength. Therefore, it is not necessary to separate the analyte from the rest of the sample (the matrix). This results in rapid analysis times and eliminates some sources of error. This is not to say that AAS measurements are completely free from interferences; both chemical and spectral interferences do occur and must be compensated for, as will be discussed. The major disadvantages of AAS are that no information is obtained on the chemical form of the analyte (no "speciation") and that often only one element can be determined at a time. The latter makes AAS of very limited use for qualitative analysis. AAS is used almost exclusively for quantitative analysis of elements, hence the use of the term "spectrometry" in the name of the technique instead of "spectroscopy."

7.1 ABSORPTION OF RADIANT ENERGY BY ATOMS

AAS is based on the absorption of radiant energy by free gas-phase atoms. In the process of absorption, an atom changes from a low energy state to a higher energy state as discussed in Chapter 3. Gas-phase atoms do not vibrate in the same sense that molecules do. Also, they have virtually no rotational energy. Hence, no vibrational or rotational energy is involved in the electronic excitation of atoms. As a result, atomic absorption spectra consist of a few very narrow absorption lines, in contrast to the wide bands of energy absorbed by molecules in solution.

Each element has a specific number of electrons "located" in an orbital structure that is unique to each element. The lowest energy electronic configuration of an atom is called the ground state. The ground state is the most stable electronic state. If energy ΔE of exactly the right magnitude is applied to a free gas-phase atom, the energy will be absorbed. An outer electron will be promoted to a higher energy, less stable excited state. The frequencies and wavelengths of radiant energy capable of being absorbed by an atom are predicted from $\Delta E = h\nu = hc/\lambda$. The energy absorbed, ΔE, is the difference between the energy of the higher energy state and the lower energy state. As shown schematically in Figure 7.1, this atom has four electronic energy levels. E_0 is the ground state, and the other levels are higher energy excited states. If the exact energies of each level are known, the three wavelengths capable of being absorbed can be calculated as follows:

$$\Delta E' = \frac{hc}{\lambda_1} E_1 - E_0$$

$$\Delta E'' = \frac{hc}{\lambda_2} E_2 - E_0$$

$$\Delta E''' = \frac{hc}{\lambda_3} E_3 - E_0$$

The calculated wavelengths λ_1, λ_2, and λ_3 all arise from transitions from the ground state to excited states. Absorption lines due to transitions from the ground state are called **resonance lines**. It is possible for an electron in an excited state to absorb radiant energy and move to an

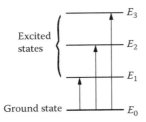

Figure 7.1 Schematic electronic energy levels in a free atom.

even higher excited state; in that case, we use the ΔE values for the appropriate energy levels involved. As we will see, in AAS, most absorptions do arise from the ground state.

Quantum theory defines the electronic orbitals in an atom and predicts the lowest energy configuration (from the order of filling the orbitals). For example, the 11 electrons in sodium have the configuration $1s^2 2s^2 2p^6 3s^1$ in the ground state. The inner shells (principal quantum number, $n = 1$ and 2) are filled and there is one electron in the $n = 3$ shell. It is this outer-shell electron that is involved in atomic absorption transitions for sodium. Ultraviolet (UV) and visible (VIS) wavelengths are the range of radiant energies absorbed in AAS. UV/VIS radiation does not have sufficient energy to excite the inner-shell electrons. Only the electrons in the outermost (valence) shell are excited. This is true of all elements: only the outermost electrons (valence electrons) are excited in AAS.

The number of energy levels in an atom can be predicted from quantum theory. The actual energy differences of these levels have been deduced from studies of atomic spectra. These levels have been graphed in *Grotrian diagrams*, which are plots for a given atom showing energy on the y-axis and the possible atomic energy levels as horizontal lines. An extensive compilation of Grotrian diagrams can be found in the book by Bashkin and Stoner. Although this book is relatively old, the compilation is valuable.

A partial Grotrian diagram for sodium is shown in Figure 7.2. The energy levels are split because the electron itself may spin one way or another, resulting in two similar energy levels and therefore two possible absorption lines rather than a single line (a singlet). For the transition from the ground state to the first excited state of sodium, the electron moves from the 3s orbital to the empty 3p orbital. The latter is split into two levels, designated $^2P_{1/2}$ and $^2P_{3/2}$, by the electron spin, so two transitions are possible. The levels differ very slightly in energy because of the interaction of the electron spin and the orbital motion of the electron. The wavelengths that are associated with these transitions are 589.5 and 589.0 nm, respectively, the well-known sodium D lines that give rise to the yellow color of sodium vapor street lights.

Under the temperatures encountered in the atomizers used in commercial AAS systems, a large majority of the atoms exist in their lowest possible energy state, the ground

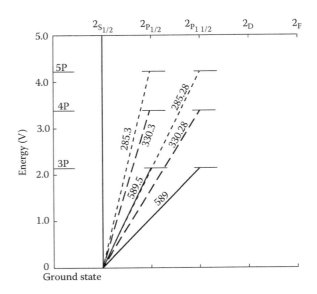

Figure 7.2 Partial Grotrian diagram for sodium.

state. Very few atoms are normally in the higher energy states. The ratio of atoms in an upper excited state to a lower energy state can be calculated from the Maxwell–Boltzmann equation (also called the Boltzmann distribution):

$$\frac{N_1}{N_0} = \frac{g_1}{g_0} e^{-\frac{\Delta E}{kT}} \qquad (7.1)$$

where
N_1 is the number of atoms in the upper state
N_0 is the number of atoms in the lower state
g_1, g_0 are the number of states having equal energy at each level 0, 1, etc. (g is called the degeneracy of the level)
ΔE is the energy difference between the upper and lower states (J)
k is the Boltzmann constant = 1.380649×10^{-23} J/K, exact
T is the absolute temperature (K)

For example, it can be calculated from the Boltzmann distribution that if zinc vapor (Zn^0 gas) with resonance absorption at 213.9 nm is heated to 3000 K, there will be only one atom in the first excited state for every 10^{10} atoms in the ground state. Zinc atoms need a considerable amount of energy to become excited. On the other hand, sodium atoms are excited more easily than the atoms of most other elements. Nevertheless, at 3000 K, only one sodium atom is excited for every 1,000 atoms in the ground state. In a normal atom population, there are very few atoms in states E_1, E_2, E_3, and higher. The total amount of radiation absorbed depends, among other things, on how many atoms are available in the lower energy state to absorb radiation and become excited. Consequently, the total amount of radiation absorbed is greatest for absorptions from the ground state. Excited-to-excited state transitions are very rare, because there are so few excited atoms; only the ground-state resonance lines are useful analytically in AAS.

This greatly restricts the number of absorption lines that can be observed and used for measurement in atomic absorption. Quite frequently only three or four useful lines are available in the UV/VIS spectral region for each element, and in some cases, fewer than that. The wavelengths of these absorption lines can be deduced from the Grotrian diagram of the element being determined but are more readily located in AAS instrument methods manuals (called "AAS cookbooks") available from the major instrument manufacturers. A list of the most intense absorption wavelengths for flame AAS (FAAS) determination of elements is given in Appendix 7.A.

AAS is useful for the analysis of approximately 70 elements, almost all of them metal or metalloid elements. Grotrian diagrams correctly predict that the energy required to reach even the first excited state of *nonmetals* is so great that they cannot be excited by normal UV radiation (>190 nm). The resonance lines of nonmetals lie in the vacuum UV region. Commercial AAS systems generally have air in the optical path, and the most common atomizer, the flame, must operate in air. Consequently, using flame atomizers, atomic absorption cannot be used for the direct determination of nonmetals. However, nonmetals have been determined by indirect methods, as will be discussed in the applications section.

7.1.1 Spectral Linewidth

According to the Bohr model of the atom, atomic absorption and emission linewidths should be infinitely narrow, because there is only one discrete value for the energy of a given transition. However, there are several factors that contribute to line broadening. The natural width of a spectral line is determined by the Heisenberg uncertainty principle and the lifetime of the excited state. Most excited states have lifetimes of 10^{-8}–10^{-10} s, so the uncertainty in the energy of the electron slightly broadens the spectral line. This is called the *natural* linewidth and is on the order of 10^{-4} Å (1.0 Å = 1.0 × 10^{-10} m).

Collisions with other atoms in the atomizer lead to *pressure (Lorentz) broadening*, on the order of 0.05 Å. *Doppler broadening*, due to random kinetic motion toward and away from the detector, results in broadening of the spectral line on the order of 0.01–0.05 Å. Doppler and collisional broadening are temperature dependent. In an atomization source with high concentrations of ions and electrons (such as in a plasma, discussed in Chapter 8), *Stark broadening* occurs as a result of atoms encountering strong local electrical fields. In the presence of a magnetic field, *Zeeman splitting* of the electronic energy levels also occurs. Localized magnetic fields within atomizers from moving ions and electrons are negligibly small and their effects are generally not seen. However, by adding an external magnetic field, we can use Zeeman splitting to assist in the correction of background absorption. The width of atomic absorption lines is on the order of 0.002 nm. These are very narrow lines, but not infinitely narrow.

7.1.2 Degree of Radiant Energy Absorption

The fraction of incident light absorbed by atoms at a particular wavelength is proportional to the number of atoms, N, that can absorb the wavelength and to a quantity called the oscillator strength f. The oscillator strength f is a dimensionless quantity whose magnitude expresses the transition probability for a specific transition. The **oscillator strength** is a constant for a particular transition; it is an indicator of the probability of absorbing the photon that will cause the transition. N is the number of ground-state atoms in the light path, since most atoms are in the ground state at normal atomizer temperatures.

As discussed in Chapter 3, the fraction of incident light absorbed by a species can be expressed as the absorbance, A. The relationship between absorbance and the amount of analyte, in this case, atoms, in the light path is given by Beer's law:

$$A = abc = (\text{constant} \times f)(b)\left(\frac{N}{\text{cm}^3}\right) \quad (7.2)$$

The proportionality constant a is called the absorptivity and includes the oscillator strength f. The term b is the length of the light path and c is the concentration of ground-state atoms in the light path (i.e., atoms/cm^3).

The amount of radiation absorbed is only slightly dependent on temperature, as shown by Equation 7.1. This is an advantage for AAS over atomic emission spectrometry and flame photometry (Chapter 8) where the signal intensity is highly temperature dependent. Although the temperature does not affect the *process* of absorption by atoms, it does affect the efficiency with which free atoms are produced from a sample and therefore indirectly affects the atomic absorption signal. This effect can sometimes be significant. Some atoms, particularly those of the alkali metals, easily ionize. Low ionization energies and high temperatures result in the formation of ions rather than atoms. Ions do not absorb at atomic absorption wavelengths. Atoms that become ionized are effectively removed from the absorbing population, resulting in a loss of signal.

7.2 INSTRUMENTATION

A schematic block diagram of the instrumentation used for AAS is shown in Figure 7.3. The components are similar to those used in other spectroscopic absorption methods as

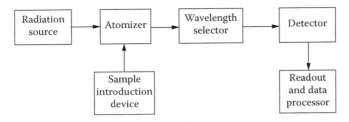

Figure 7.3 Block diagram of the components of an AAS.

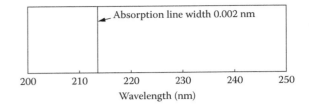

Figure 7.4 Width of an atomic absorption line (Zn 213.9 nm line), greatly exaggerated, compared with the emission bandwidth from a continuum source such as a deuterium lamp.

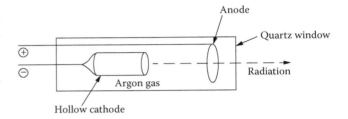

Figure 7.5 Schematic diagram of an HCL. The anode and cathode are sealed in a glass cylinder filled with argon or neon gas at low pressure. The window must be transparent to the emitted radiation.

discussed in Chapters 3, 4, and 5. Light from a suitable source is directed through the **atomizer**, which serves as the sample cell, into a wavelength selector and then to a detector. The detector measures how much light is absorbed by the sample. The sample, usually in solution form, is introduced into the atomizer by some type of introduction device. The atomizer converts the sample to gas-phase ground-state atoms that can absorb the incident radiation.

7.2.1 Radiation Sources

Two radiation sources are commonly used in commercial AAS instruments, the hollow cathode lamp (HCL) and the electrodeless discharge lamp (EDL). Both types of lamps are operated to provide as much intensity as possible while avoiding line-broadening problems caused by the collision processes described earlier.

7.2.1.1 Hollow Cathode Lamp

As we have already mentioned, atomic absorption lines are very narrow (about 0.002 nm). They are so narrow that if we were to use a continuous source of radiation, such as a hydrogen or deuterium lamp, it would be very difficult to detect any absorption of the incident radiation at all. Absorption of a narrow band from a continuum is illustrated in Figure 7.4, which shows the absorption of energy from a deuterium lamp by zinc atoms absorbing at 213.9 nm. The width of the zinc absorption line is exaggerated for illustration purposes. The wavelength scale for the deuterium lamp in Figure 7.4 is 50 nm wide and is controlled by the monochromator bandpass. If the absorption line of Zn were 0.002 nm wide, its width would be $0.002 \times 1/50 = 1/25{,}000$ of the scale shown. Such a narrow line would be detectable only under extremely high resolution (i.e., very narrow bandpass), which is not encountered in commercial AAS equipment.

With the use of slits and a good monochromator, the bandwidth falling on the detector can be reduced to about 0.2 nm (much less than the earlier 50 nm example). If the light source is continuous, the entire 0.2 nm bandwidth contributes to the signal falling on the detector. If an absorption line 0.002 nm wide were absorbed from this light source, the signal reaching the detector would be reduced by only 1 % from the original signal. Since this is about the absorption linewidth of atoms, even with complete absorption of the radiation at 213.9 nm by Zn atoms, the total signal from a continuous light source would change by only 1 %. This would result in an insensitive analytical procedure of little practical use.

What is needed for a light source in AAS is a source that produces very narrow emission lines at the exact wavelengths capable of being absorbed by analyte atoms. The problem was solved by the development of the HCL, shown schematically in Figure 7.5. The cathode is often formed by hollowing out a cylinder of pure metal or making an open cylinder from pure metal foil. The metal used for the cathode is the metal whose spectrum will be emitted by the lamp. If we want to determine Cu in our AAS experiment, the lamp cathode must be a copper cylinder; if we want to determine gold, the cathode must be a gold cylinder, and so on. The cathode and an inert anode are sealed in a glass cylinder filled with Ar or Ne at low pressure (the "filler gas"). A window of quartz or glass is sealed onto the end of the lamp; a quartz window is used if UV wavelengths must be transmitted. Most HCLs have quartz windows, because most elements have emission and absorption lines in the UV. Glass can be used for some elements, such as sodium, where all the strong absorption lines are in the visible region of the spectrum.

The HCL emits narrow, intense lines from the element that forms the cathode. Applying a high voltage across

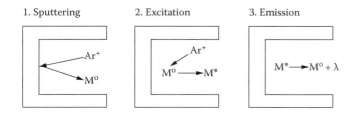

Figure 7.6 The HCL process, where Ar⁺ is a positively charged argon ion, M⁰ is a sputtered ground-state metal atom, M* is an excited-state metal atom, and λ is emitted radiation at a wavelength characteristic for the sputtered metal. [© 1993–2014 PerkinElmer, Inc. All rights reserved. Printed with permission. (www.perkinelmer.com).]

the anode and cathode creates this emission spectrum. Atoms of the filler gas become ionized at the anode and are attracted and accelerated toward the cathode. The fast-moving ions strike the surface of the cathode and physically dislodge some of the surface metal atoms (a process called "sputtering"). The displaced atoms are excited by collision with electrons and emit the characteristic atomic emission spectrum of the metal used to make the cathode. The process is shown in Figure 7.6. The emitted atomic lines are extremely narrow. Unlike continuum radiation, the narrow emission lines from the HCL can be absorbed almost completely by unexcited atoms. Using this light source, atomic absorption is easily detected and measured. Narrow line sources such as the HCL provide not only high sensitivity but also specificity. If only Cu atomic emission lines are produced by the Cu HCL, there are few species other than Cu atoms that can absorb these lines. Therefore, there is little spectral interference in AAS. The emitted spectrum consists of all the emission lines of the metal cathode, including many lines that are not resonance absorption lines, but these other lines do not interfere in the analysis. A tabulation of some relative intensities can be found in the book by Bruno and Svoronos.

Each hollow cathode emits the spectrum of metal used in the cathode. For this reason, a different HCL must be used for each different element to be determined. This is an inconvenience in practice and is the primary factor that makes AAS a technique for determining only one element at a time. The handicap is more than offset, however, by the advantage of the narrowness of the spectral lines and the specificity that results from these narrow lines.

It is possible to construct a cathode from more than one element. These are called "multielement" HCLs and can be used for the determination of all the elements in the cathode. This can be done sequentially but without having to change the lamp, which saves some time. In general, multielement cathodes do not perform as well for all of the elements in the cathode as do single HCLs for each element. The multielement cathode may have reduced intensity for one or more of the elements. Each of the elements present will emit its atomic emission spectrum, resulting in a more complex emission than from a single-element lamp. This may require that a less-sensitive absorption line be chosen to avoid a spectral interference. The obvious reason to use a multielement lamp is in the hope that more than one element can be measured simultaneously, making AAS a multielement technique. In fact, there are a few commercial simultaneous multielement AAS systems available for measuring up to eight elements or so. Most use a bank of single-element lamps all focused on the atomizer rather than multielement cathodes. The disadvantage with this approach is that only one set of conditions in the atomizer can be used, and this set of atomization conditions may not be optimum for each element.

HCLs have a limited lifetime, usually due to loss of filler gas atoms through several processes. Adsorption of filler gas atoms onto the lamp surfaces causes decreased sputtering and decreased intensity of emission; eventually, the number of filler gas atoms becomes so low that the lamp will not "light." The sputtering process causes atoms to be removed from the cathode; these metal atoms often condense elsewhere inside the lamp, trapping filler gas atoms in the process and decreasing lamp life. This is particularly a problem for HCLs of volatile metals like Cd and As. HCLs operated at currents higher than recommended will have shorter lifetimes than those operated according to the manufacturer's recommendation. Operating at higher currents results in more intensity in the lamp output but also may increase noise, which impacts both precision and limit of detection (LOD). Since we are measuring the *ratio* of light absorbed to incident light, there is little to be gained by increasing the lamp current.

Single-element HCLs cost between $400 and $700 per lamp, while multielement lamps cost between $500 and $700 each for most elements in 2023. Particularly rare elements like Rh, Ir, and Os may cost four times as much as a single element lamp; if the user has a service contract for the instrument, one new lamp per year may be included, however, replacement of a burned out lamp is not included as a routine claim on a service contract.

7.2.1.2 Electrodeless Discharge Lamp

It is difficult to make stable hollow cathodes from certain elements, particularly those that are volatile, such as arsenic, germanium, or selenium. The HCLs of these elements have short lifetimes and low intensities. An alternative light source has been developed in the EDL. A commercial EDL design is shown in Figure 7.7. A small amount of metal or a salt of the element whose spectrum is desired is sealed into a quartz bulb with a low pressure of Ar gas. The bulb is shown centered inside the coils in Figure 7.7. The coils are part of a self-contained radio-frequency (RF) generator. When power is applied to the coils, the RF field

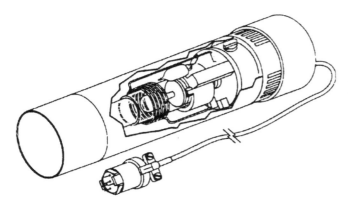

Figure 7.7 EDL. [© 1993–2014 PerkinElmer, Inc. All rights reserved. Printed with permission. (www.perkinelmer.com).]

generated will "couple" with the metal or salt in the quartz bulb. The coupled energy will vaporize and excite the metal atoms in the bulb. The characteristic emission spectrum of the metal will be produced.

EDLs are very intense, stable emission sources. They provide better detection limits (DLs) than HCLs for those elements that are intensity limited either because they are volatile or because their primary resonance lines are in the low-UV region. Some elements like As, Se, and Cd suffer from both problems. For these types of elements, the use of an EDL can result in a LOD that is two to three times lower than that obtained with an HCL. EDLs are available for many elements, including antimony, arsenic, bismuth, cadmium, germanium, lead, mercury, phosphorus, selenium, thallium, tin, and zinc. Older EDLs required a separate power supply to operate the lamp. Modern systems are self-contained. EDLs cost more than the comparable HCLs.

Most modern AAS systems have "coded" lamps that are recognized by the spectrometer software, which can then set up the analysis parameters automatically. The instrument "knows" that a copper lamp or a calcium lamp has been inserted and can apply the default analytical conditions without analyst intervention.

Having explained why narrow line sources are needed for AAS, in Section 7.2.6, a new high-resolution AAS using a continuum source will be introduced.

7.2.2 Atomizers

The atomizer is the sample cell of the AAS system. The atomizer must produce the ground-state free gas-phase atoms necessary for the AAS process to occur. The analyte atoms are generally present in the sample as salts, molecular compounds, or complexes. The atomizer must convert these species to the reduced, free gas-phase atomic state. The atomizer generally does this via thermal energy and some chemistry. The two most common atomizers are flame atomizers and electrothermal (furnace) atomizers.

7.2.2.1 Flame Atomizers

To create a flame, we need to mix an oxidant gas and a fuel gas and light the mixture. In modern commercial FAAS, two types of flames are used. The first is the air–acetylene flame, where air is the oxidant and acetylene is the fuel. The second type of flame is the nitrous oxide–acetylene flame, where nitrous oxide is the oxidant and acetylene is again the fuel. The fuel and oxidant gases are mixed in a burner system, called a premix burner. An exploded view of one type of commercial flame atomic absorption burner is shown in Figure 7.8a. In this design, the fuel gas is introduced into the mixing chamber through one inlet (not shown), while the oxidant gas is introduced through the sidearm on the *nebulizer*. The premix burner design generates laminar gas flow and this results in a very steady flame. The steady flame generates less noise due to "flame flicker"; this improves precision. Mixing the gases in the mixing chamber eliminates the safety hazard of having a combustible gas mixture piped through the laboratory. The flame burns just above the burner head, along the slot shown in the figure.

The sample is introduced into the burner in the form of a solution. The solution is aspirated into the nebulizer, which is basically a capillary tube. The nebulizer sprays the solution into the mixing chamber in the form of a fine aerosol. Three nebulizer designs are shown schematically in Figure 7.8b. The term "to nebulize" means to convert to a fine mist, like a cloud. The solution exits the capillary tube at high velocity and breaks into tiny droplets as a result of the pressure drop created. Kinetic energy transfer from the nebulizer gas overcomes the surface tension and cohesive forces holding the liquid together. The fuel and oxidant gases carry the sample aerosol to the base of the flame. In the flame, the sample aerosol is desolvated, vaporized, and atomized to form free gas-phase atoms of the analyte. The process will be discussed in detail later.

When the sample solution passes through the nebulizer, an aerosol is formed, but the droplets are of different sizes. As a droplet enters the flame, the solvent (water or organic solvent) must be vaporized, the residue must vaporize, and the sample molecules must dissociate into atoms. The larger the droplet, the more inefficient this process is. The two devices shown in Figure 7.8a are used to overcome this problem. The impact bead is made of glass, quartz, Teflon, or ceramic and is placed directly in front of the nebulizer spray inside the mixing chamber. The impact bead improves nebulization efficiency by breaking larger droplets into smaller ones through collision of the spray with the bead. The flow spoiler is a piece of polymer or other corrosion-resistant material machined into three or more vanes. The flow spoiler is placed in the mixing chamber, about midway between the end cap and the burner. It physically removes larger droplets from the aerosol through collision, while

ATOMIC ABSORPTION SPECTROMETRY

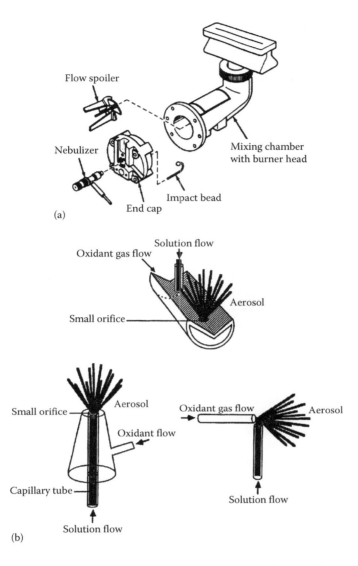

Figure 7.8 (a) Premix burner system. [© 1993–2014 PerkinElmer, Inc. All rights reserved. Printed with permission. (www.perkinelmer.com).] (b) Schematic nebulizer designs. (Top) modified Babington type; (left) concentric (the most common in FAAS); (right) cross-flow type. (From Parsons, M.L., Atomic absorption and flame emission, in Ewing, G.A., ed., *Analytical Instrumentation Handbook*, 2nd edn., Marcel Dekker, Inc., New York, 1997. Used with permission.)

smaller droplets pass through the openings between the vanes. The larger droplets drain from the mixing chamber through a drain opening (not shown). The aim of this system is to produce an aerosol with droplets about 4 μm in diameter. The two devices may be used alone or in combination. The drain opening is connected to a liquid waste container with a length of polymer tubing. It is extremely important that there be a trap between the drain opening and the waste container to prevent the free flow of flame gases out of the burner assembly. The presence of the trap helps to eliminate "flashback," discussed subsequently. The trap is often a simple water-filled loop in the drain tubing itself.

The nebulizer described is a self-aspirating, pneumatic nebulizer and is the one shown schematically on the lower left of Figure 7.8b. The nebulizer capillary is usually made of stainless steel, but other materials such as Pt, Ta, and polymers may be used for corrosive solvents when contamination from the elements in steel must be avoided. Other nebulizer designs have been developed for specific applications but are not commonly used in AAS. These other nebulizers are often used in atomic emission spectrometry and will be described in Chapter 8.

The burner head is constructed either of stainless steel or titanium. Different sizes and geometries of burner heads are used for various flames. The single-slot burner head shown in Figure 7.8a with a 10 cm long slot is used for air–acetylene flames. The 10 cm long flame is the sample path length for the AA spectrometer in this case. A smaller 5 cm long single-slot burner head with a narrower slot is used for nitrous oxide–acetylene flames. Some burner head

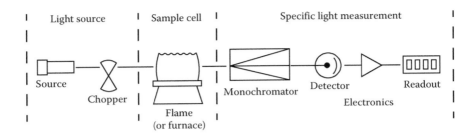

Figure 7.9 Basic AAS system. [© 1993–2014 PerkinElmer, Inc. All rights reserved. Printed with permission. (www.perkinelmer.com).]

designs use Ti and multiple fins for both flames. Usually, the slot is oriented parallel to the light beam from the radiation source, so the path length is as long as possible to achieve the highest sensitivity (remember Beer's law), as shown in the AAS layout in Figure 7.9. The modern burner head design coupled with the use of a liquid-filled trap between the mixing chamber drain and the waste container prevents the possibility of a flashback. The gas mixture is ignited above the burner head and the flame, a highly energetic chemical reaction between the fuel and oxidant, propagates rapidly. The flame is supposed to propagate up from the burner head. It will do so if the linear gas flow rate through the burner slot is higher than the **burning velocity** of the flame. Burning velocity is a characteristic of the flame type; both nitrous oxide–acetylene and air–acetylene flames have low burning velocities. If the premixed gas flow rate is less than the burning velocity, a flashback can occur when the gas mixture is ignited. Flashback is the very undesirable and extremely hazardous propagation of the flame below the burner head slot and back into the mixing chamber. A flashback results in an explosion in the mixing chamber; it can destroy the burner assembly, create flying debris, rupture the fuel and oxidant lines thereby releasing combustible gases into the laboratory, and cause injury to the analyst or other people in the laboratory. Early burner designs and high-burning-velocity flames such as those using pure oxygen as oxidant were often prone to flashback. It is imperative that AAS systems be operated according to the manufacturer's directions, that only the correct gases, properly regulated, and the correct burner head be used, and that the trap and all safety interlocks be in place and functioning.

One reason for the long path length used in FAAS (compared with the typical 1 cm path length in UV/VIS or infrared [IR] absorption spectrometry) is that the premix burner and nebulizer system used is very inefficient and wasteful of sample. Sample solution is aspirated into the nebulizer at ~5 mL/min but only a small amount (<5 %) of that solution reaches the burner. The path length has to be as long as possible to compensate for the loss of sample in the mixing chamber. FAAS is very popular because it is fast, has high element selectivity, and the instruments are easy to operate, but the inefficiency of the sample introduction

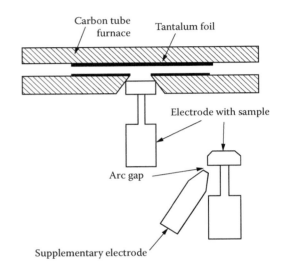

Figure 7.10 High-temperature furnace atomizer designed and used by L'vov.

system combined with noise inherent in the system restricts DLs in flames to the ppm range for most elements.

7.2.2.2 Electrothermal Atomizers

In order to measure ppb concentrations of metals, a different type of atomizer is needed. A furnace or electrothermal atomizer (ETA) was developed less than 10 years after the technique of AAS was developed. In 1961, B.V. L'vov built a heated carbon tube atomizer, illustrated in Figure 7.10. The system used a carbon tube heated by electrical resistance. The tube was lined with Ta foil and purged with argon gas. After the tube or furnace reached an elevated temperature, the sample, on a carbon electrode, was inserted at the bottom, as shown in the figure. Atomization took place and the analyte atoms absorbed the light beam passing through the carbon tube. His system was orders of magnitude more sensitive than flame atomizers but was difficult to control. Other workers in the field, particularly West, Massman, and Robinson, refined carbon rod and furnace atomizers. Most commercial instruments use some version of a carbon tube atomizer for electrothermal

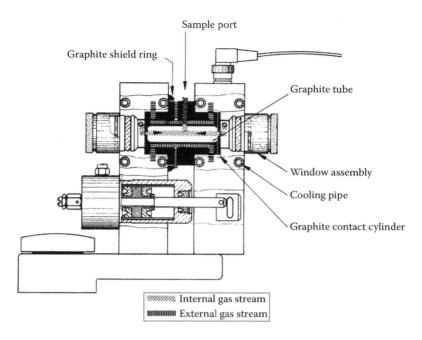

Figure 7.11 Graphite furnace atomizer. This is a longitudinally heated furnace design. The graphite tube is shown in greater detail in Figure 7.21. [© 1993–2014 PerkinElmer, Inc. All rights reserved. Printed with permission. (www.perkinelmer.com).]

atomization. The most common atomizer of this type uses a tube of graphite coated with pyrolytic graphite and heated by electrical resistance; therefore, this commercial ETA is called a **graphite furnace atomizer**. The acronym GFAAS, which stands for graphite furnace AAS, tells the reader that the atomizer used is a graphite furnace, as opposed to a flame.

A commercial graphite furnace atomizer is illustrated in Figure 7.11. This atomizer consists of a graphite tube, approximately 6 mm in diameter and 25–30 mm long, the electrical contacts required to heat the tube, a system for water-cooling the electrical contacts at each end of the tube, and inert purge gas controls to remove air from the furnace. An inert gas is used to prevent the graphite from being oxidized by air during the heating process. Quartz windows at each end of the furnace assembly permit the light from the HCL or EDL to pass through the furnace and out to the spectrometer. A small amount of sample solution, between 5 and 50 μL, is dispensed into the graphite tube through a small hole in the top of the tube. The furnace is heated in a series of programmed temperature steps to evaporate the solvent, decompose (ash, char) the sample residue, and finally to atomize the sample into the light path. Details of the graphite furnace atomization process and temperature program will be discussed later. The diagram in Figure 7.11 shows a longitudinally heated graphite furnace. The electrical contacts are at each end of the tube and they must be water-cooled. This longitudinal heating results in a temperature gradient in the heated furnace—the ends are cooler than the center. This may result in recondensation of vaporized species at the ends of the tube. This can be

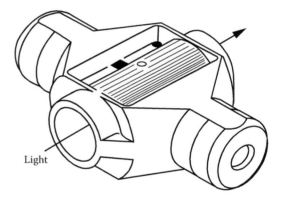

Figure 7.12 A graphite tube for a transversely heated furnace. [© 1993–2014 PerkinElmer, Inc. All rights reserved. Printed with permission. (www.perkinelmer.com).]

a problem for the next sample analysis if material from the previous sample is still in the furnace. This problem is called "carryover" or a "memory effect" and can result in poor accuracy and precision. To overcome this problem, new graphite furnaces that are heated transversely have been developed. A transverse graphite tube is shown in Figure 7.12. The electrical contacts are transverse to the light path, and the tube is heated across the circumference. This results in even heating over the length of the furnace and reduces the carryover problem significantly.

Modern graphite furnace atomizers have a separate power supply and programmer that control the electrical power, the temperature program, the gas flow, and some spectrometer functions. For example, the spectrometer can be programmed to "read," that is, collect absorbance data, only when the furnace reaches the atomization temperature. This saves data storage space and data processing time.

Researchers have developed other types of ETAs over the years, including filaments, rods, and ribbons of carbon, tantalum, tungsten, and other materials, but the only commercial ETA available is the graphite furnace atomizer.

7.2.2.3 Other Atomizers

Two additional commercially available atomizers (really analysis techniques with unique atomizers) must be discussed, because they are extensively used in environmental and clinical analysis. They are the **cold vapor AAS (CVAAS) technique** for determination of the element mercury, Hg, and the **hydride generation AAS (HGAAS) technique** for several elements that form volatile hydrides, including As, Se, and Sb. These elements are toxic; federal and state laws regulate their concentrations in drinking water, wastewater, and air, so their measurement at ppb concentrations is very important. Because the CVAAS and HGAAS "atomizers" are analysis techniques, they will be discussed under applications of AAS.

An atomizer based on glow discharge (GD) techniques is commercially available for the analysis of solid metal samples by AAS. It will be discussed in the applications sections under solid sample analysis.

7.2.3 Spectrometer Optics

7.2.3.1 Monochromator

A monochromator is required to separate the absorption line of interest from other spectral lines emitted from the HCL and from other elements in the atomizer that are also emitting their spectra. Because the radiation source produces such narrow lines, spectral interference is not common. Therefore the monochromator does not need high resolution.

A typical monochromator is shown in Figure 7.13. The most common dispersion element used in AAS is a diffraction grating. The grating disperses different wavelengths of light at different angles, as discussed in Chapter 3. The grating can be rotated to select the wavelength that will pass through the exit slit to the detector. All other wavelengths are blocked from reaching the detector. The *angle of dispersion* at the grating is a function of the density of lines ruled on the grating. The more lines/mm on the grating, the higher is the dispersion. Higher dispersion means greater separation between adjacent lines. A high-dispersion grating permits the use of wider entrance and exit slits on the monochromator to achieve the same resolution. Wider slits allow more light to pass through the system to the detector.

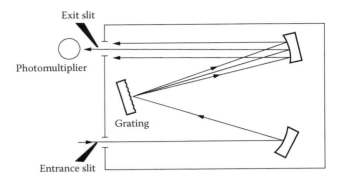

Figure 7.13 Typical grating monochromator layout. [© 1993–2014 PerkinElmer, Inc. All rights reserved. Printed with permission. (www.perkinelmer.com).]

Large high-quality gratings with high dispersion are expensive, but they offer better energy throughput than cheaper low-dispersion gratings. In addition to the number of lines ruled on the grating, the blaze angle affects the intensity of diffracted wavelengths. Gratings can be blazed for any wavelength; the farther away a given wavelength is from the wavelength for which the grating is blazed, the more light intensity will be lost in diffraction. The analytically useful wavelength range for commercial AAS is from 190 to 850 nm. If a grating is blazed for the middle of this range, there will be loss in intensity at both extremes. That is particularly bad for elements such as Se, As, P, Cd, and Zn, with resonance lines at the low-UV end, and for the alkali metals, with resonance lines at the high end of the visible region. To overcome this problem, the monochromator can be equipped with two gratings, one blazed for the UV and one for the visible. Modern spectrometers automatically control the movement and alignment of the grating or gratings, so that the correct one is used for the wavelength being measured.

Most commercial AAS systems have the monochromator, optics, and detector designed for the measurement of one wavelength at a time; they are single-element instruments. There are a few systems available that do perform multielement determinations simultaneously, using an echelle spectrometer (discussed in Chapter 3) and a bank of HCLs all focused on the atomizer. The limitation to this approach is not the sources or the spectrometer or the detector, but the atomizer. The atomizer can only be at one set of conditions, and those conditions will not necessarily be optimum for all of the elements being measured. There will be a trade-off in DLs for some of the elements.

7.2.3.2 Optics and Spectrometer Configuration

The spectrometer system for AAS can be configured as a single-beam system, as shown in Figure 7.8; as a double-beam system, shown in Figure 7.14; or as a

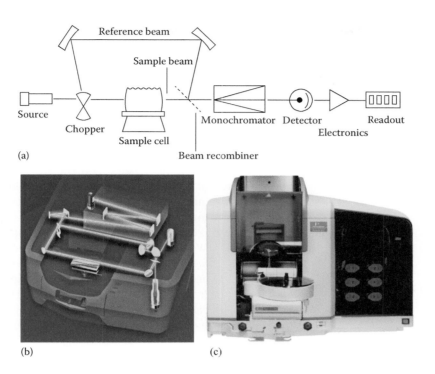

Figure 7.14 (a) Schematic double-beam AAS configuration. [© 1993–2014 PerkinElmer, Inc. All rights reserved. Printed with permission. (www.perkinelmer.com).] (b) The optical path of the Shimadzu Scientific Instruments AA-7000 superimposed over the actual instrument body. On the right, from front to back is the HCL, the beam splitter (slanted rectangle) and the deuterium lamp. In the center, from front to back, we have the burner head with the sample beam passing over it, the reference beam, and the monochromator at the center top. On the left is the mirrored chopper and in the back, the detector. (c) A photograph of a modern AA-7000 instrument. (Courtesy of Shimadzu Scientific Instruments, Inc. www.shimadzu.com.)

pseudo-double-beam system, which will not be discussed. (See the reference by Beaty and Kerber for a description of this system.) Note that in AAS, the sample cell is placed in front of the monochromator, unlike UV/VIS spectrometers for molecular absorption or spectrophotometry, where the sample is placed after the monochromator.

A single-beam system is cheaper and less expensive than a double-beam system, but cannot compensate for instrumental variations during analysis. In a double-beam system, part of the light from the radiation source is diverted around the sample cell (flame or furnace atomizer) to create a *reference beam*, as shown in Figure 7.14. The reference beam monitors the intensity of the radiation source and electronic variations (noise, drift) in the source. The signal monitored by the detector is the ratio of the sample and reference beams. This makes it possible to correct for any variations that affect both beams, such as short-term changes in lamp intensity due to voltage fluctuations in the power lines feeding into the instrument. Compensation for these variations is performed automatically in modern double-beam spectrometers. Even though part of the source radiation is directed to the reference beam, modern double-beam instruments have the same signal-to-noise ratio as single-beam instruments with the advantage of more accurate and precise absorbance measurements.

Lenses or mirrors are used to gather and focus the radiation at different parts of the optical system. This avoids losing too much signal as a result of the light beams being nonparallel and focuses the light beam along the flame so that as much light as possible passes through the sample. Any light that does not pass through the sample cannot be absorbed and results in a loss in sensitivity. Quartz lenses have been used for this purpose, but most instruments today use front-surfaced concave mirrors, which reflect and focus light from their faces and do not lose much radiation in the process. The monochromator has two slits, an entrance slit and an exit slit. The entrance slit is used to prevent stray radiation from entering the monochromator. Light passes from the entrance slit to the grating. The entrance slit should be as wide as possible to permit as much light as possible to fall on the grating but must be narrow enough to isolate the wavelength of interest. After dispersion by the grating, the radiation is directed toward the exit slit. At this point, the desired absorption line is permitted to pass, but the other lines emitted from the source and atomizer are blocked from reaching the detector by the monochromator exit slit. The system of slits and grating enables the analyst to choose the wavelength of radiation that reaches the detector. There is rapid loss in sensitivity as the mechanical slit width, and hence, the spectral bandpass (spectral slit width) is increased.

Commercial AAS instrumentation may be purchased with fixed slits or with variable slits. Fixed mechanical slit widths are available so that the resolution and sensitivity are acceptable for most analytical purposes at lower cost than instruments with variable slit widths. Variable slit widths are desirable for maximum flexibility, especially if samples are varied and complex. Instruments that have both flame and graphite furnace atomizers often have separate sets of slits of different heights for each atomizer. The furnace slits are usually shorter to avoid having emission from the small-diameter incandescent furnace reach the detector. In general, the analyst should use the widest slit widths that minimize the stray light that reaches the detector while spectrally isolating a single resonance line for the analyte from the HCL.

7.2.4 Detectors

The common detector for AAS is the photomultiplier tube (PMT). The construction and operation of a PMT has been described in Chapter 4. While PMTs are the most common detectors, solid-state single- and multichannel detectors such as photodiode arrays (PDAs) (discussed in Chapter 4) and charge-coupled devices (CCDs) (discussed in Chapters 4 and 8) are increasingly being used in AAS spectrometers. Many small systems, particularly those dedicated to one element such as a dedicated CVAAS mercury analyzer, use solid-state detectors instead of PMTs. Multielement simultaneous AAS systems also use multichannel solid-state detectors to measure more than one wavelength at a time.

7.2.5 Modulation

Many metals, when atomized in a flame or furnace, emit strongly at the same wavelength at which they absorb. The emission signal can cause a serious error when the true absorption is to be measured. This problem is illustrated in Figure 7.15. Emission by the metal in the atomizer is at precisely the same wavelength as the absorption wavelength of the metal because the same electronic transition is involved. Better resolution cannot improve the situation.

Furthermore, since we are trying to measure I_1, the interference by the emission from the flame will result in a direct error. Unless a correction is made, the signal recorded will be $(I_1 + E)$, where E is the emission intensity.

This problem is overcome most simply by the modulation of the radiation source. **Modulation** means that the source radiation is switched on and off very rapidly. This can be done by using a rotating mechanical *chopper* placed directly in front of the source. A chopper is shown in Figures 7.9 and 7.14. The mechanical chopper is a circle of metal sheet with opposite quadrants cut out. Every quarter turn of the chopper alternately blocks and passes the source radiation. Another way to modulate the source is by pulsing the power to the lamp at a given frequency. When modulated, the signal from the source is now an alternating current (AC) signal. Therefore, the signal for both I_0 and I_1 is AC but the sample emission E is not modulated. It is a direct current (DC) signal. The detector electronics can be "locked in" to the frequency of the AC signal. The modulated absorption signal will be measured and the DC emission signal will be ignored. This eliminates emission from the atomizer as a source of error. Modulation of the source is required for accurate results in AAS. All modern commercial instruments have some means of source modulation in the instrument.

7.2.6 Commercial AAS Systems

AAS is considered a mature analysis technique with decades-long production of instruments by a large number of companies around the world. The biggest change in most modern AA systems is in the size of the instruments and the capabilities of the software. Instruments today are much smaller than older designs. Some systems have automated switching between flame and furnace atomizers, eliminating what had been a labor-intensive and potentially error-prone job; some have the ability to run both flame and furnace at the same time. Many have multiple lamp turrets, with lamps already warmed up to avoid waiting when switching from one element to the next. They incorporate features such as video cameras to observe what is happening in the furnace to enable better optimization and method development.

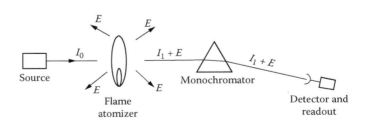

Figure 7.15 Emission by excited atoms in the atomizer can occur at the resonance wavelength, resulting in a direct error in the AAS measurement.

Intelligent or smart software is found on most modern systems that will recognize the lamp installed, set the optimum instrument parameters or user-specified method parameters, and will even validate the performance of an instrument, all without any intervention by the analyst.

Complete systems are available from a large number of instrument companies. These include but are not limited to Agilent Technologies, Aurora Biomed, GBC, PerkinElmer, Shimadzu Scientific Instruments, which makes the only instrument with a vibration sensor to shut down the flame in case of earthquakes, and Thermo Fisher Scientific. There are numerous smaller companies, like Buck Scientific and companies in China, India, and elsewhere who only distribute locally. A completely new AAS concept is now available from Analytik Jena AG, as described in the next section.

CV- and hydride-AAS systems for mercury, As, Se, and other hydride-forming elements, discussed in Sections 7.5.3.4 and 7.5.3.5, are available from the major AAS manufacturers and also from companies making dedicated analyzers, including Teledyne Leeman Labs, Teledyne Cetac Technologies, and Milestone, Inc.

7.2.6.1 High-Resolution Continuum Source AAS

AAS, as developed in the 1950s by Alan Walsh, required a narrow line source of light and modulation of that source. Recently, advances in source and detector technology, combined with a high-resolution double monochromator, have enabled the use of a continuum source for AAS (Welz and Heitmann; Welz et al.). The instruments, from Analytik Jena AG (www.analytik-jena.de), are the contrAA® high-resolution continuum source AAS (HR-CS AAS) systems. The source is a small xenon arc lamp, which provides a continuous source of radiation with high intensity, especially in the UV range. The systems use a high-resolution double monochromator consisting of a prism pre-monochromator followed by an echelle grating monochromator. The detector is a UV-sensitive CCD linear array detector. No modulation is required. Each pixel of the CCD functions as an independent detector, with all pixel information shifted simultaneously into a readout register. This configuration means that the analytical line does not need to be separated from the adjacent emission of the continuum source by an exit slit. There is no exit slit since the CCD array itself represents a large number of high-resolution detectors. Each individual detector accumulates a spectral range of about 2 pm; the entire array comprises a spectral range of about 0.4 nm.

Only 1–3 pixels are used for registering the analyte absorption, so all the remaining pixels are available for correction purposes. Both fluctuations in lamp emission and transmission of the atomizer are accounted for, making the system a simultaneous double-beam system of high precision with simultaneous background correction.

Since the intensity of the source does not affect sensitivity, but does influence the noise, the high-intensity source increases the DLs of this system by a factor of 2–5 over conventional AAS. The linear dynamic range of the system is 5–6 orders of magnitude, much broader than conventional AAS and similar to inductively coupled plasma-optical emission (ICP-OES) (discussed in Chapter 8).

The high-intensity continuum source offers a number of advantages over narrow line sources. Secondary lines (nonresonance lines) can be used in order to reduce sensitivity, thereby avoiding dilutions. All spectral lines are available, including lines for which no HCL or EDL sources are available, such as fluorine and chlorine. In addition, molecular absorptions, such as those of diatomic fragments like CS or PO, can be used to measure sulfur and phosphorus (Huang et al.)

The CCD detector permits not only simultaneous background correction but complete correction of spectral interferences such as structured background from molecular species such as OH and NO. The resolution permits collection of the spectra from these species and removal of them from the absorbance of the sample. Direct-line overlap can also be corrected, if the matrix has an additional absorption line within the registered spectral range of the detector. Direct-line overlap is rare and usually is due to line-rich matrices such as Fe, so the system can reliably correct for the direct overlap of Fe on Zn, for example.

We discussed earlier the desire to speed up AAS by having simultaneous multielement measurement and the compromises that current HCL/EDL systems make. The HR-CS AAS technique is not a simultaneous multielement system, but since it uses a single source, the "change" of elements, that is, the reading of a different line, is extremely fast, especially for FAAS. The system therefore functions as a fast sequential FAAS with the advantage that every element can be measured under automatically set optimum conditions, rather than the compromise conditions a bank of HCLs or a multielement HCL uses. The transient signals from a graphite furnace are still problematic due to the large amount of data collected with the concomitant increase in detector readout times.

7.3 ATOMIZATION PROCESS

7.3.1 Flame Atomization

Most samples we want to examine by AAS are solid or liquid materials. Examples of solid samples are soil, rock, biological tissues, food, metal alloys, ceramics, glasses, and polymers. Examples of liquid samples are water, wastewater, urine, blood, beverages, oil, petroleum products, and organic solvents. For FAAS and most GFAAS determinations, the sample must be in the form of a solution. This requires that most samples be prepared by acid

1. Nebulization	$M^+ + A^-$	(Solution)
	↓	
	$M^+ + A^-$	(Aerosol)
2. Desolvation	↓	
	M A	(Solid)
3. Liquefaction	↓	
	M A	(Liquid)
4. Vaporization	↓	
	M A	(Gas)
5. Atomization	↓	
	$M^0 + A^0$	(Gas)
6. Excitation	↓	
	M^*	(Gas)
7. Ionization	↓	
	$M^+ + e^-$	(Gas)

Figure 7.16 The processes occurring in a flame atomizer. M^+ is a metal cation; A^- is the associated anion. M^0 and A^0 are the ground-state free atoms of the respective elements. [© 1993–2014 PerkinElmer, Inc. All rights reserved. Printed with permission. (www.perkinelmer.com).]

Table 7.1 Temperatures Obtained in Various Premixed Flames (°C)

	Oxidant	
Fuel	Air	N_2O
H_2	2000–2100	
Acetylene	2100–2400	2600–3000
Propane (natural gas)	1700–1900	

digestion, fusion, ashing, or other forms of sample preparation (Chapter 1) to give us an aqueous, acidic solution or a solution in a combustible organic solvent. We will look at aqueous acidic solutions, since they are the most common form of sample for FAAS. Metals are present in aqueous acidic solutions as dissolved ions; examples include Cu^{2+}, Fe^{3+}, Na^+, and Hg^{2+}.

To measure an atomic absorption signal, the analyte must be converted from dissolved ions in aqueous solution to reduced gas-phase free atoms. The overall process is outlined in Figure 7.16. As described earlier, the sample solution, containing the analyte as dissolved ions, is aspirated through the nebulizer. The solution is converted into a fine mist or aerosol, with the analyte still dissolved as ions. When the aerosol droplets enter the flame, the solvent (water, in this case) is evaporated. We say that the sample is "desolvated." The sample is now in the form of tiny solid particles. The heat of the flame can melt (liquefy) the particles and then vaporize the particles. Finally, the heat from the flame (and the combustion chemistry in the flame) must break the bonds between the analyte metal and its anion, and produce free M^0 atoms. This entire process must occur very rapidly, before the analyte is carried out of the observation zone of the flame. After free atoms are formed, several things can happen. The free atoms can absorb the incident radiation; this is the process we want. The free atoms can be rapidly oxidized in the hostile chemical environment of the hot flame, making them unable to absorb the resonance lines from the lamp. They can be excited (thermally or by collision) or ionized, making them unable to absorb the resonance lines from the lamp. The analyst must control the flame conditions, flow rates, and chemistry to maximize production of free atoms and minimize oxide formation, ionization, and other unwanted reactions. While complete atomization is optimum and will yield the highest signal, it is much more important that the atomization process be consistent. If the atomization is not complete, any variation in the fractional atomization will result in errors in analysis.

The flame is responsible for production of free atoms. Flame temperature and the fuel/oxidant ratio are very important in the production of free atoms from compounds. Many flame fuels and oxidants have been studied over the years, and temperature ranges for some flames are presented in Table 7.1. In modern commercial instruments, only air–acetylene and nitrous oxide–acetylene flames are used, however data on additional flames can be found in the book by Bruno and Svoronos.

Lower temperature flames are subject to interferences from incomplete atomization of the analyte. The air–acetylene flame is useful for many elements, but the hotter and more chemically reducing nitrous oxide–acetylene flame is needed for refractory elements and elements that are easily oxidized, such as Al and Si. Appendix 7.A lists the usual flame chosen for each element. This appendix also gives the fuel/oxidant ratio conditions used for maximum sensitivity for each element. Flames are classified as oxidizing (excess oxidant) or reducing (excess fuel). Air–acetylene flames can be used in either an oxidizing mode or a reducing mode; nitrous oxide–acetylene flames are usually run in a reducing mode. In general, excess oxidant helps to destroy organic material in samples. However, excess oxidant can react with elements that exist as stable oxides to form oxide molecules. These oxide molecules cannot undergo atomic absorption. Elements that form stable oxides, such as aluminum, boron, molybdenum, and chromium, are therefore determined using reducing flame conditions, usually with the high-temperature nitrous oxide–acetylene flame, to prevent the formation of oxide molecules. The chemistry that occurs depends on what part of the flame is observed; the base of the flame differs from the inner core of the flame and both differ from the outer mantle of the flame.

If we measure the absorbance signal of an atom with respect to the height of the signal above the burner, we arrive at a relationship called a flame profile. The burner assembly can be moved up and down with respect to the light source, usually by turning a knob manually or through the software that controls the burner position. The beam of light from the source is fixed in position. The flame profile is determined by aspirating a solution of an element into a flame with the burner height set so that the light beam from the lamp is at the base of the flame, just above the burner head. Then the burner head is slowly lowered, so that the light beam passes through higher and higher regions of the flame. The absorbance is plotted versus the height above the burner. Flame profiles for Cr, Mg, and Ag in an oxidizing air–acetylene flame are shown in Figure 7.17. Let us look at each of these profiles. The signal for magnesium, Mg, starts off low at the base of the flame, because free atoms have not yet formed from the sample. The absorbance increases with increasing height in the flame (above the burner head), to a maximum. Moving farther up in the flame causes the absorbance to decrease. This curve is brought about by the complicated reactions that take place in a flame, including atomization, oxidation, and ionization. All of these processes compete in the flame. The maximum sensitivity for Mg will be obtained when the light beam is at the position of maximum absorbance, above the burner head, but not very high up in the flame. The decrease in absorbance as we go higher up in the flame is due to the formation of oxide molecules. Many elements have flame profiles similar to Mg, but not all, as can be seen in Figure 7.17. Chromium forms very stable oxides that are hard to atomize. As can be seen from its profile, the absorbance by Cr atoms is actually highest at the base of the flame. Cr atoms combine with oxygen in air almost as soon as they are formed, and the absorbance decreases rapidly with height above the burner head (i.e., with increasing exposure to air). Chromium is an element that would be better determined in a fuel-rich (reducing) flame to minimize the formation of chromium oxide. Silver, on the other hand, starts with low absorbance at the base of the flame just like Mg. Then, the absorbance increases with increasing height in the flame, indicating that silver atoms are stable once formed; there is no decrease in absorbance due to oxide formation or ionization.

In practice, the optimum position for taking absorbance measurements is determined in exactly this way. A solution of the element to be determined is aspirated, the absorbance is monitored continually, and the burner head is moved up and down (and forward and back) with respect to the light beam until the position of maximum absorbance is located. Older instruments have a ruler mounted along the burner compartment to indicate where the burner head is with respect to the light beam to assist in adjusting the height of the burner. Modern systems can control burner adjustment through the software. Figure 7.17b shows the 2D distribution of atoms in an air–acetylene flame for several elements. These are side views of the flames, as would be seen by the detector. The contour lines differ by 0.1 in absorbance, with the maximum absorbance at the center of the profile.

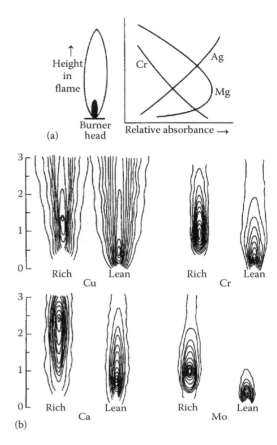

Figure 7.17 (a) Flame profiles for Cr, Mg, and Ag demonstrating differences in free atom formation as a function of height above the burner in an oxidizing air–acetylene flame. (b) 2D atom distributions in a 10 cm air–acetylene flame. Fuel-rich (reducing) and fuel-lean (oxidizing) results are shown. The scale on the left is the height above the burner head (in cm). Maximum absorbance is shown as the smallest circle in the plot with the contour lines separated by 0.1 in absorbance. For Cr, the fuel-lean plot shows maximum absorbance close to the burner, in accordance with the results shown in (a). The fuel-rich flame shows that the maximum absorbance for Cr moves up above the burner head by approximately 1 cm. Molybdenum is even more dramatically affected by the fuel/oxidant ratio. (The diagrams in (b) are adapted and reprinted with permission from Rann, C.S. and Hambly, A.N., *Anal. Chem.*, 37, 879. Copyright 1965 American Chemical Society.)

Fuel-rich (reducing) and fuel-lean (oxidizing) conditions are displayed. As we discussed, Cr shows the highest absorbance in a fuel-lean flame just above the burner head; in a fuel-rich flame, the maximum absorption position is moved up in the flame. Molybdenum shows the same response, but even more strongly. Calcium can be determined in either a fuel-rich or fuel-lean flame, but the position of maximum absorbance is very different for each flame. In the fuel-rich flame, the optimum position is more than 2 cm above the burner, while in an oxidizing flame, the position is between 0.5 and 1 cm above the burner. The need to optimize the horizontal position of the burner with respect to the light source can be seen from the width of the intervals of the atom distributions.

The flame profile results from the complex physical and chemical processes occurring in the flame. The sample, introduced into the base of the flame, goes through the process of evaporation, forming solid particles, which are then decomposed by the hot flame to liberate free atoms. The latter can then oxidize, forming metal oxides in the upper and outer parts of the flame. The formation of free atoms depends on the flame temperature and on the chemical form of the sample. Chemical species with small dissociation energies at high temperatures will dissociate to form free atoms. For a given flame temperature, free atom formation depends on the chemical species and its dissociation constant. If the metal exists in the sample in a stable chemical form (small dissociation constant), it may be difficult to decompose, and **atomization efficiency** is low. On the other hand, if the metal is in a chemical form that is easily decomposed, the number of atoms formed is high and the atomization efficiency is high. The atomization efficiency is a measure of how many free atoms are formed from all the possible species containing the atom. It can be expressed as the fraction of the total element in the gaseous state that is present as free atoms at the observation height in the atomizer or as the fraction of the total element present as both free atoms and ions. By total element, we mean all atomic, molecular, and ionic species containing the atom of interest, M (e.g., M, MX, MO, M$^+$). The atomization efficiency can vary from 0 to 1, and the value will depend on where in the atomizer the observation of the fraction M/(total M-containing species) is made. The interested student should consult the more detailed discussions in the *Handbook of Spectroscopy*, Vol. 1, edited by Robinson, or the references by Dean and Rains (Vol. 1) or Ingle and Crouch listed in the bibliography.

The loss of free atoms in the atomizer is also a function of the chemistry of the sample. If the oxide of the analyte element is readily formed, the free atoms will form oxides in the flame and the population of free atoms will simultaneously decrease. This is the case with elements such as chromium, molybdenum, tungsten, and vanadium. On the other hand, some metal atoms are stable in the flame and the free atoms exist for a prolonged period. This is particularly so with the noble metals platinum, gold, and silver. Adjusting the fuel/oxidant ratio can change the flame chemistry and atom distribution in the flame, as shown in Figure 7.17b. Atoms with small ionization energies will ionize readily at high temperatures (and even at moderate temperatures). In an air–acetylene flame, it can be shown that moderate concentrations of potassium are about 50 % ionized, for example. Ions do not absorb atomic lines.

The maximum absorbance signal depends on the number of free atoms in the light path. These free atoms are in dynamic equilibrium with species in the flame; they are produced continuously by the flame and lost continuously in the flame. The number produced depends on the original concentration of the sample and the atomization efficiency (Table 7.2). The number lost depends on the formation of oxides, ions, or other nonatomic species. The variation of atomization efficiency with the chemical form of the sample is called *chemical interference*. It is the most serious interference encountered in AAS and must always be taken into account. Interferences are discussed in detail in Section 7.4. Table 7.2 can be used as a guide for choosing

Table 7.2 Efficiencies of Atomization[a] of Metals in Flames

	Flame	
Metal	Air–C_2H_2	N_2O–C_2H_2
Ag	1.0	0.6
Al	<0.00005	0.2
Au	1.0	0.5
B	<0.0005	0.004
Ba	0.001	0.2*
Be	0.00005	0.1
Bi	0.2	0.4
Ca	0.07	0.5*
Cd	0.5	0.6
Co	0.3	0.3
Cu	1.0	0.7
Cr	0.07	0.6
Cs*	0.7	—
Fe	0.4	0.8
Ga	0.2	0.7
In	0.6	0.9
K*	0.4	0.1
Li*	0.2	0.4
Mg	0.6	1.0
Mn	0.6	0.8
Na*	1.0	1.0
Pb	0.7	0.8
Rb*	1.0	—
Si	—	0.06
Sn	0.04	0.8
Sr	0.08	0.03
Ti	—	0.2
Tl	0.5	0.56
V	0.01	0.3
Zn	0.7	0.9

Source: Modified from Robinson, J.W. (ed.), *Handbook of Spectroscopy*, Vol. 1, CRC Press, Boca Raton, FL, 1975.

[a] The efficiency of atomization in the flames has been measured as the fraction of total element in the gaseous state present as free neutral atoms or ionized atoms at the observation height in the atomizer. That is, efficiency of atomization = (neutral atoms + ions)/total element. Entries marked * obtained using ionization suppression, discussed in Section 7.4.1. The data were obtained under a variety of conditions by multiple researchers and may not be directly comparable between elements. The reference should be consulted for details.

which flame to use for AAS determination of an element. For example, Al will definitely give better sensitivity in a nitrous oxide–acetylene flame than in air–acetylene, where hardly any free atoms are formed; the same is true of Ba. But for potassium, clearly, an air–acetylene flame is a better choice in terms of sensitivity than the hotter nitrous oxide–acetylene flame. This is because K is very easily ionized at high temperatures and ions do not give atomic absorption signals. Based on ionization energy trends, you should expect that Cs and Rb would also be more sensitive in the cooler air–acetylene flame. As you can see, there are no entries in the table for Cs or Rb in a nitrous oxide–acetylene flame; they ionize to such an extent that the nitrous oxide–acetylene flame cannot be used for determining Cs or Rb by FAAS, just as Si and Ti are too refractory to be determined in an air–acetylene flame.

7.3.2 Graphite Furnace Atomization

There are many differences in the atomization process in a flame and in a graphite furnace. One very important difference to keep in mind is that in FAAS, the sample solution is aspirated into the flame continuously for as long as it takes to make the absorbance measurement. This is usually not long—about 30 s once the flame has stabilized after introducing the sample solution, but it is a continuous process. GFAAS is not a continuous process; the atomization step produces a transient signal that must be measured in less than 1 s. We will again consider an aqueous acidic solution of our sample.

A small volume of solution, between 5 and 50 μL, is injected into the graphite tube via a micropipette or an autosampler. The analyte is once again in the form of dissolved ions in solution, and the same process outlined in Figure 7.16 must occur for atomic absorption to take place. The graphite furnace tube is subjected to a multistep temperature program. The program controls the temperature ramp rate, the final temperature at each step, the length of time the final temperature is held at each step, and the nature and flow rate of the purge gas through the furnace at each step. A typical graphite furnace program consists of six steps: (1) dry, (2) pyrolyze (ash, char), (3) cool, (4) atomize, (5) clean out, and (6) cool down.

A generic temperature program for GFAAS might look like that in Figure 7.18. The process of atomization is extremely fast and must be rigidly controlled. The temperature program is therefore very carefully controlled, both with respect to the times used for each section of the heating program and the temperature range involved in each step. It is vital to avoid loss of sample during the first two steps but it is also extremely important to eliminate as much organic and other volatile matrix material as possible.

The "dry" step is used to remove the solvent. The solvent must be removed without splattering the sample by rapid

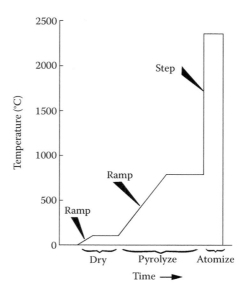

Figure 7.18 Typical temperature program for a graphite furnace atomizer. [© 1993–2014 PerkinElmer, Inc. All rights reserved. Printed with permission. (www.perkinelmer.com).]

boiling, which would result in poor precision and accuracy. A slow temperature ramp from room temperature to about 110 °C is used for aqueous solutions. The upper temperature is held for about a minute. The purge gas during this step is the normal inert gas (nitrogen or argon) at its maximum flow of about 300 mL/min to remove the solvent vapor from the furnace. The "pyrolyze" step is also called the ashing step or the "char" step. Its purpose is to remove as much of the matrix as possible without volatilizing the analyte. The "matrix" is everything except the analyte; it may consist of organic compounds, inorganic compounds, or a mixture of both. The sample is again subjected to a temperature ramp. The upper temperature is chosen to be as high as possible without losing the analyte and held for a short time. The gas flow is normally 300 mL/min and is usually the inert purge gas. With some organic sample matrices, switching to air is done in this step to help oxidize the organic materials.

A cool-down step before the atomization step is used for longitudinally heated furnaces. This has been shown to improve sensitivity and reduce peak "tailing" for some refractory elements. This improves the accuracy of the measurement for these elements. The cool-down step is not used in transversely heated furnaces.

The atomization step must produce gas-phase free analyte atoms. The temperature must be high enough to break molecular bonds. In general, the temperature is raised very rapidly for this step, as shown in Figure 7.18, and the purge gas flow is stopped to permit the atoms to remain in the light path. Stopping the purge gas flow increases the sensitivity of the analysis. The atomization occurs very rapidly and the signal generated is a transient

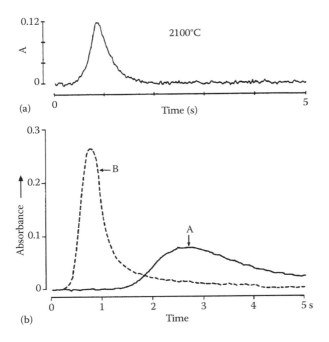

Figure 7.19 (a) An ideal Gaussian-shaped atomization signal from a graphite furnace. (b) Atomization signals for molybdenum in a graphite furnace atomizer. A is the signal resulting from an uncoated (normal) graphite tube atomizer. B is the signal from a graphite tube coated with pyrolytic graphite. The absorbance signal is transient in GFAAS. [© 1993–2014 PerkinElmer, Inc. All rights reserved. Printed with permission. (www.perkinelmer.com).]

signal; as the atoms form, the absorbance increases. As the atoms diffuse out of the furnace, the absorbance decreases, resulting in a nearly Gaussian peak-shaped signal. An idealized signal is presented in Figure 7.19a. Atomization signals for molybdenum from two different types of graphite furnace tubes are shown in Figure 7.19b. The spectrometer system is programmed to begin to acquire the absorbance data as soon as the atomization step begins. The integrated area under the absorbance peak is used for quantitative measurements.

Finally, the furnace is taken to a temperature higher than the atomization temperature to burn out as much remaining residue as possible; this is the "clean out" step. The furnace is allowed to cool before the next sample is injected. The entire program for one replicate of one sample is usually about 2 min long. Modern GFAAS systems may allow for temperature programs of up to 20 separate steps. GFAAS systems are now available with high-definition video cameras in the furnace, so that the analyst can watch the sample behavior as part of method development. The camera could show, for example, if the sample spatters during the dry step, permitting the analyst to lower the temperature to avoid loss of sample.

Early GFAAS instruments were operated by having the sample pipetted onto the bottom wall of the tube. The tube was heated in a programmed fashion as has been described, and atomization occurred "off the wall" of the tube. Poor precision and a variety of interferences were serious problems with this approach. Modern graphite furnace atomizers make use of a pyrolytic graphite platform, first introduced by L'vov, inserted into the graphite tube or fabricated as an integral part of the tube. The sample is pipetted onto the platform and is atomized from the platform, not from the wall of the tube. The reasons for use of the L'vov platform will be discussed in Section 7.4. In addition, modern GFAAS methods add one or more chemicals to the sample in the furnace. These chemicals are called "modifiers" and are used to control interferences.

Some safety considerations should be mentioned concerning all furnaces and burners. The potential for exposure to heavy metal contamination due to the vaporized compounds is real, and therefore it is best to have a snorkel in place above the burner or furnace. This snorkel should be connected to the laboratory fume hood system or connected to a dedicated exhaust blower. Details on these types of safety devices can be found in Chapter 2 of this book and more details can be found in the book by Bruno and Svoronos.

7.4 INTERFERENCES IN AAS

Interferences are physical or chemical processes that cause the signal from the analyte in the sample to be higher or lower than the signal from an equivalent standard. Interferences can therefore cause positive or negative errors in quantitative analysis. There are two major classes of interferences in AAS, spectral interferences and nonspectral interferences. Nonspectral interferences are those that affect the formation of analyte free atoms. Nonspectral interferences include chemical interference, ionization interference, and solvent effects (or matrix interference). Spectral interferences cause the amount of light absorbed to be erroneously high due to absorption by a species other than the analyte atom. While all techniques suffer from interferences to some extent, AAS is much less prone to spectral interferences and nonspectral interferences than atomic emission spectrometry and X-ray fluorescence (XRF), the other major optical atomic spectroscopic techniques.

7.4.1 Nonspectral Interferences

7.4.1.1 Chemical Interference

A serious source of interference is *chemical interference*. Chemical interference occurs when some chemical component in the sample affects the atomization efficiency of the sample compared with the standard solution. The result is either an enhancement or a depression in the analyte signal from the sample compared with that from the standard. This effect is associated most commonly with the predominant anions present in the sample. The anion affects the stability of the metal compound in which the analyte is bound, and this, in turn, affects the efficiency with which the atomizer produces metal atoms. For example, a solution of calcium chloride, when atomized in an air–acetylene flame, decomposes to calcium atoms more readily than a solution of calcium phosphate. Calcium phosphate is more thermally stable than calcium chloride. A solution of calcium chloride containing 10 ppm Ca will give a higher absorbance than a solution of calcium phosphate containing 10 ppm Ca. If phosphate ion is added to a solution of calcium chloride, the absorbance due to Ca will decrease as the concentration of phosphate increases. This is a chemical interference. It occurs in the atomization process. Chemical interference is a result of having insufficient energy in the flame or furnace to break the chemical bonds in molecules and form free atoms.

There are three ways of compensating for chemical interference. The first approach is to match the matrix of the standards and samples; that is, to have the same anion(s) present in the same concentrations in the working standards as in the samples being analyzed. This supposes that the samples have been thoroughly characterized and that their composition is known and constant. This may be the case in industrial production of a material or chemical, but often, the sample matrix is not well characterized or constant.

A second approach is to add another metal ion that forms an even more stable compound with the interfering anion than does the analyte ion. Such an ion is called a "releasing agent" because it frees the analyte from forming a compound with the anion and permits it to atomize. For example, lanthanum forms a very thermally stable phosphate, more stable than calcium phosphate. To determine Ca in solutions that contain an unknown or variable amount of phosphate, such as those from biological samples, the analyst can add a large excess of lanthanum (as the chloride or nitrate salt) to all standards and samples. The lanthanum "ties up" the phosphate by forming lanthanum phosphate. If all of the phosphate is now present as lanthanum phosphate, this eliminates the dependence of the formation of Ca atoms on the phosphate concentration. The exact concentration of phosphate does not have to be measured; it is only necessary to add enough La to completely react with the phosphate in the solution to be analyzed. Usually 500–2000 ppm La is sufficient for most types of samples. The same amount of La must be added to all the solutions, including the blank. The releasing agent should be of the highest purity possible.

The third approach is to eliminate the chemical interference by switching to a higher-temperature flame, if possible. For example, when a nitrous oxide–acetylene flame is used, there is no chemical interference on Ca from phosphate, because the flame has sufficient energy to decompose the calcium phosphate molecules. Therefore, no lanthanum addition is required.

A fourth possible approach is the use of the method of standard additions (MSA), discussed in Chapter 3. This approach can correct for some chemical interferences but not all. For example, in the graphite furnace, if the analyte is present as a more volatile compound in the sample than the added analyte compound, MSA will not work. The analyte form in the sample is lost prior to atomization as a result of volatilization, while the added analyte compound remains in the furnace until atomization; therefore, the standard additions method will not give accurate analytical results.

7.4.1.2 Matrix Interference

Other potential sources of interference are the sample **matrix** and the solvent used for making the sample solution. The sample matrix is anything in the sample other than the analyte. In some samples, the matrix is quite complex. Milk, for example, has a matrix that consists of an aqueous phase with suspended fat droplets and suspended micelles of milk protein, minerals, and other components of milk. The determination of calcium in milk presents matrix effects that are not found when determining calcium in drinking water. Sample solutions with high concentrations of salts other than the analyte may physically trap the analyte in particles that are slow to decompose, interfering in the vaporization step and causing interference. Differences in viscosity or surface tension between the standard solutions and the samples, or between different samples, will result in interference. Interference due to viscosity or surface tension occurs in the nebulization process for FAAS because different volumes of solution will be aspirated in a given period of time and nebulization efficiency will change as a result of the solvent characteristics. Metals in aqueous solutions invariably give lower absorbance readings than the same concentration of such metals in an organic solvent. This is due in part to enhanced nebulization efficiency of the organic solvent. In aqueous acidic solutions, higher acid concentrations often lead to higher solution viscosity and a decrease in absorbance due to decreased sample uptake. Matrix and solvent effects are often seen in GFAAS as well as FAAS and may be more severe in furnaces than in flames. The presence of matrix interference can be determined by comparing the slope of an external calibration curve with the slope of an MSA curve (discussed in Chapter 3). If the slopes of the two calibrations are the same (parallel to each other), there is no matrix interference; if the slopes are different (not parallel), interference exists and must be corrected for.

The solvent may interfere in the atomization process. If the solvent is an organic solvent, such as a ketone, alcohol, ether, or a hydrocarbon, the solvent not only evaporates rapidly but may also burn, thus increasing the flame temperature. The atomization process is more efficient in a hotter flame. More free atoms are produced and a higher absorbance signal is registered from solutions in organic solvents than from aqueous solutions, even though the metal concentration in the two solutions is equal.

Matching the solutions in the working standards to the sample solutions can compensate for matrix or solvent interferences. The type of solvent (water, toluene, methyl isobutyl ketone [MIBK], etc.), amount and type of acid (1 % nitric, 5 % HCl, 20 % sodium chloride, etc.), and any added reagents such as releasing agents must be the same in calibration standards and samples for accurate results.

Alternatively, the MSA may be used to compensate for matrix interferences. This calibration method uses the sample to calibrate in the presence of the interference. Properly used, MSA will correct for the solvent interference but care must be taken. The assumption made in using MSA is that whatever affects the rate of formation of free atoms in the sample will affect the rate of formation of free atoms from the added analyte spike in the same way. MSA *will not* correct for spectral interference or for ionization interference. MSA can correct for interferences that affect the slope of the curve but not for interferences that affect the intercept of the curve.

7.4.1.3 Ionization Interference

In AAS, the desired process in the atomizer should stop with the production of ground-state atoms. For some elements, the process continues as shown in Figure 7.16 to produce excited-state atoms and ions. If the flame is hot enough for significant excitation and ionization to occur, the absorbance signal is decreased because the population of ground-state atoms has decreased as a result of ionization and excitation. This is called ionization interference. Ionization interferences are commonly found for the easily ionized alkali metal and alkaline-earth elements, even in cool flames. Ionization interferences for other elements may occur in the hotter nitrous oxide–acetylene flame, but not in air–acetylene flames.

Adding an excess of a more easily ionized element to all standards and samples eliminates ionization interference. This addition creates a large number of free electrons in the flame. The free electrons are "captured" by the analyte ions, converting them back to atoms. The result is to "suppress" the ionization of the analyte. Elements often added as ionization suppressants are potassium, rubidium, and cesium. For example, in the AAS determination of sodium, it is common to add a large excess of potassium to all samples and standards. Potassium is more easily ionized than sodium. The potassium ionizes preferentially and the free electrons from the ionization of potassium suppress the ionization of sodium. The DL of the sodium determination thereby decreases. The ionization suppression agent, also called an *ionization buffer*, must be added to all samples,

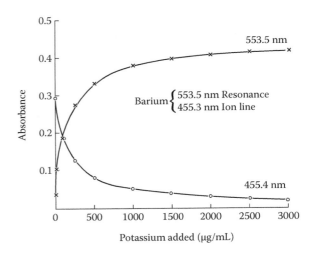

Figure 7.20 The suppression of barium ionization in a flame atomizer by addition of the more easily ionized element potassium. [© 1993–2014 PerkinElmer, Inc. All rights reserved. Printed with permission. (www.perkinelmer.com).]

standards, and blanks at the same concentration for accurate results. An example of the use of ionization suppression is shown in Figure 7.20. Absorbance at a barium resonance line (atomic absorption) and absorbance at a barium ion line (by barium ions in the flame) are plotted as a function of potassium added to the solution. As the potassium concentration increases, barium ionization is suppressed; the barium stays as barium atoms. This results in increased atomic absorption at the resonance line and a corresponding decrease in absorbance at the ion line. The trends in absorbance at the atom and ion lines very clearly show that barium ion formation is suppressed by the addition of 1,000 ppm of the more easily ionized potassium.

7.4.1.4 Nonspectral Interferences in GFAAS

Graphite furnace atomizers experience significant nonspectral interference problems, some of which are unique to the furnace. Compensation or elimination of these interferences is different than what is done in flame atomizers.

Some analytes react with the graphite surface of the atomizer at elevated temperatures to form thermally stable carbides. These carbides do not atomize, thereby decreasing the sensitivity of the analysis. Carbide formation also results in poor precision and poor accuracy, because the amount of carbide formation depends on the condition of the graphite atomizer surface. Elements that tend to form carbides include W, Ta, B, Nb, and Mo, among others. The use of a dense pyrolytic graphite coating on the tube wall is highly recommended to minimize carbide formation. Normal graphite has a very porous structure that permits solution to soak into the graphite tube, thereby increasing the graphite/solution contact area and allowing more carbide formation at high temperatures. Pyrolytic graphic forms an impervious surface that prevents the solution from entering the graphite structure, decreasing carbide formation. Comparing the atomization profiles for Mo in Figure 7.19b, it is apparent that the atomization from an uncoated tube (peak A) appears at a much later time than that from a pyrolytic graphite-coated tube. This time delay is due to the need for Mo to diffuse out of the porous uncoated graphite. The result is not only a time delay but also a very broad absorbance signal that is difficult to integrate accurately. As you can see, the peak has not returned to the baseline, even after 5 min at the atomization temperature. The signal from the pyrolytic graphite-coated tube rises rapidly and returns to baseline rapidly, evidence that no significant diffusion of Mo into the graphite has occurred. This results in a higher absorbance signal and a peak that can be integrated accurately. It also permits a much shorter atomization step and increases tube life; both contribute to higher sample throughput.

When an analyte is atomized directly from the wall of the graphite tube, chemical and matrix interferences can be severe. The analyte, either as atoms or gas-phase molecules, is released from the hot wall at the atomization temperature into a cooler inert gas atmosphere inside the tube. The atmosphere is heated by conduction and convection from the tube walls, so the temperature of the atmosphere lags behind that of the walls. The atoms or molecules enter this cooler atmosphere where a number of processes may occur. For example, the atoms may recombine with matrix components into molecules or the temperature may be low enough that vaporized analyte-containing molecules fail to atomize. In either case, nonspectral interference occurs. A solution to this problem is the use of a platform insert in the tube, shown schematically in Figure 7.21. The platform surface is pyrolytic graphite and is deliberately designed to be in poor contact with the tube. The platform is heated by radiation from the tube walls. When an analyte is placed on the platform, it does not atomize when the wall reaches the right temperature, but later, when the platform and the atmosphere are both at the same, higher temperature. The difference in temperature of the gas phase inside the atomizer for wall and platform atomization is shown in Figure 7.22. This higher temperature improves the atomization of molecules and prevents recombination of free atoms. The result is a significant reduction in interferences and a significant improvement in precision.

7.4.1.5 Chemical Modification

Chemical modification, also commonly called **matrix modification**, is the addition of one or more reagents to the graphite tube along with the sample. The use of these chemical modifiers is to control nonspectral interferences by altering the chemistry occurring inside the furnace. The reagents are chosen to enhance the volatility of the matrix or to decrease the volatility of the analyte or to modify the surface of the atomizer. The use of a large amount of chemical modifier may, for example, convert all of the analyte into a single compound with well-defined properties. The end result of the use of chemical modifiers is to improve the

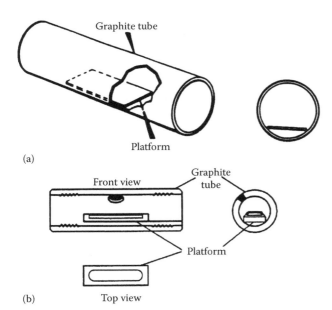

Figure 7.21 (a) Schematic of a L'vov platform inserted into a standard longitudinal graphite tube. The left diagram is a cutaway view of the platform inside the tube. The right diagram is an end-on view of the tube and platform looking along the light path. (b) Commercial platforms have a shallow depression into which the sample is pipetted through the opening in the top of the tube. The opening is shown as a dark area on the front and end-on views. [© 1993–2014 PerkinElmer, Inc. All rights reserved. Printed with permission. (www.perkinelmer.com).]

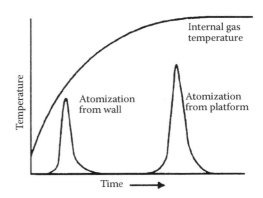

Figure 7.22 Tube wall and platform temperature profiles. [© 1993–2014 PerkinElmer, Inc. All rights reserved. Printed with permission. (www.perkinelmer.com).]

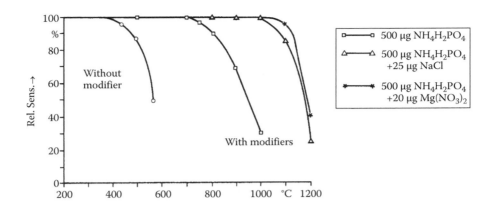

Figure 7.23 Effect of matrix modification on pyrolysis temperatures for cadmium. Without modification, Cd begins to volatilize out of the furnace at temperatures below 500 °C. By adding modifiers or combinations of modifiers, Cd is retained in the furnace at much higher temperatures. [© 1993–2014 PerkinElmer, Inc. All rights reserved. Printed with permission. (www.perkinelmer.com).]

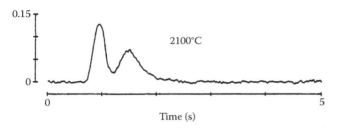

Figure 7.24 Multiple atomization signals caused by different chemical forms of analyte in the sample. [© 1993–2014 PerkinElmer, Inc. All rights reserved. Printed with permission. (www.perkinelmer.com).]

accuracy and precision of the analysis by permitting the use of the highest possible pyrolysis temperature. This permits removal of the matrix with no loss of analyte.

Cadmium is one of the heavy metal elements whose concentrations in the environment are regulated by law due to their toxicity. Drinking water, blood, urine, and environmental samples are analyzed routinely for Cd at low ppb concentrations by GFAAS. Many common cadmium compounds, such as cadmium chloride, are relatively volatile. With no modifiers, Cd starts to volatilize out of the furnace at pyrolysis temperatures as low as 400 °C, as seen in Figure 7.23. By adding either single compounds or mixtures of compounds, the pyrolysis temperature can be increased significantly, to over 1,000 °C. The addition of the modifier(s) converts cadmium in the sample to a much less volatile form. In general, this improves both the precision and accuracy of the analysis. If, for example, cadmium had been present in the sample in two different chemical forms, such as an organometallic Cd compound and an inorganic Cd compound, these two compounds might atomize at different rates and at different temperatures. The result would be multiple atomization signals, as shown in Figure 7.24. This can cause a serious error in the measurement of the signal. Multiple analyte peaks can usually be eliminated through the use of an appropriate matrix modifier that converts the analyte to one common species.

Another example of a modifier used to stabilize the analyte is the use of nickel nitrate in the determination of selenium in biological tissues. Addition of nickel nitrate as a modifier retains Se in the furnace as nickel selenide while allowing the pyrolysis and removal of the organic matrix. Without the addition of nickel nitrate, the pyrolysis temperature must be 350 °C or less to prevent loss of Se. With the use of nickel nitrate, temperatures up to 1100 °C can be used.

In some cases, the matrix itself can be made more volatile through chemical modification. For example, in the determination of traces of elements in seawater the matrix in the graphite tube after drying would be primarily NaCl. NaCl is a high-melting ionic compound and requires relatively high temperatures to volatilize it. The NaCl molecule stays intact on volatilization, increasing the background absorbance (a spectral interference). If all the NaCl could be volatilized out of the furnace, very volatile analyte elements like Hg, As, Se, Pb, and many more would have been lost in the process. By adding an excess of ammonium nitrate as a *matrix modifier* to the sample in the graphite tube, the NaCl matrix can be converted to the much more volatile compounds ammonium chloride and sodium nitrate. Most of the matrix can be removed at a temperature below 500 °C, which substantially reduces background absorption without loss of volatile analytes.

A generic "mixed modifier" of palladium and magnesium nitrate improves the GFAAS determination of many elements. The AAS cookbook of methods provided with commercial instruments should contain the recommendations for matrix modification for standard GFAAS determination of each element. Of course, it is imperative that very high-purity reagents be used for matrix modification and blank determinations that include the modifiers must be run.

7.4.2 Spectral Interferences

Spectral interferences occur when absorption of the hollow cathode resonance line occurs by species other than the element being determined. For example, the Pb 217.0 nm line may be absorbed by the components of a flame even though no lead is present in the sample or in the flame. This is a spectral interference.

7.4.2.1 Atomic Spectral Interference

The resonance absorption lines of the various elements are very narrow, on the order of 0.002 nm, and at discrete wavelengths. Direct overlap between absorption lines of different elements is rare and can usually be ignored as a source of error. Absorption by the wings of the absorption lines of interfering elements present in high concentrations has been observed, but this is also a rare occurrence. A table of reported atomic spectral overlaps can be found in the handbook by Dean cited in the bibliography. The only cures for direct atomic spectral interference are (1) to choose an alternate analytical wavelength or (2) to remove the interfering element from the sample. Extracting the interfering element away from the analyte or extracting the analyte away from the interfering element can accomplish the last option. There are many successful methods in the literature for the separation of interferences and analytes, but care must be taken not to lose analyte or contaminate the sample in the process. The extraction approach also permits the analyst to concentrate the analyte during the extraction, improving the accuracy and sensitivity of the analysis when performed correctly.

7.4.2.2 Background Absorption and Its Correction

A common occurrence that results in spectral interference is absorption of the HCL radiation by molecules or polyatomic species in the atomizer. This is called "background absorption"; it occurs because not all of the molecules from the sample matrix and the flame gases are completely atomized. This type of interference is more commonplace at short wavelengths (<250 nm) where many compounds absorb, as discussed in Chapter 4. Incomplete combustion of organic molecules in the atomizer can cause serious background interference problems. If a flame atomizer is used, incomplete combustion of the solvent may take place, particularly if the flame is too reducing (fuel rich). The extent of the interference depends on flame conditions (reducing or oxidizing), the flame temperature used, the solvent used, and the sample aspiration rate. Background interference is much more severe when graphite furnace atomizers are used because the pyrolysis step is limited to a maximum temperature that does not volatilize the analyte. Consequently, many matrix molecules are not thermally decomposed. They then volatilize into the light path as the higher atomization temperature is reached and absorb significant amounts of the source radiation.

We have seen that atoms absorb over a very narrow wavelength range and that overlapping absorption by other atoms of the resonance lines of the analyte is extremely unlikely. However, molecular absorption occurs over broad bands of wavelengths and is observed in flame and graphite furnace atomizers. The molecular absorption may come from hydroxyl radicals generated from water in the flame, incomplete combustion of organic solvent, stable molecular residues, metal oxides, and so on. Solid particles in flame atomizers may scatter light over a wide band of wavelengths; this is not absorption but results in less light reaching the detector. Scattering of light therefore causes a direct error in the absorption measurement. If these broad absorption and scattering bands overlap the atomic absorption lines, they will absorb the resonance line from the hollow cathode and cause an increase in absorbance, a spectral interference. There are several ways of measuring this background absorption and correcting for it.

7.4.2.3 Two-Line Background Correction

An early manual method of measuring background and applying a correction for it used the absorption of a nearby nonresonance line. The emission spectrum from a hollow cathode is quite rich in emission lines and contains the resonance lines of the element of interest plus many other nonresonance emission lines from the element and the filler gas. The analyte atoms do not absorb these nonresonance lines; however, the broad molecular background absorbs them. Two measurements are made. The absorbance at a nonresonance line close to the analyte resonance line but far enough away that atomic absorption does not occur is measured. Then the absorbance at the resonance line of the analyte is measured. The absorbance at the nonresonance line is due *only* to background absorption. The absorbance at the resonance line is due to *both* atomic and background absorption. The difference in absorbance between the resonance and nonresonance lines is the net atomic absorption. The two-line method of background correction was convenient when all AAS instruments were manual instruments, since no change in light source or complicated instrumentation was necessary. However, this method is not always accurate. Unless the nonresonance line is very close to the atomic absorption line, inaccurate correction of the background will result. Modern computerized instrumentation has made this technique obsolete.

7.4.2.4 Continuum Source Background Correction

Let us remind ourselves why we do not use a broadband, continuum lamp as a light source in AAS. The

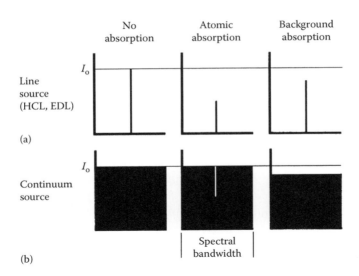

Figure 7.25 (a) Atomic and background absorption with a line source (HCL, EDL). (b) Atomic and background absorption with a continuum source (deuterium lamp, tungsten filament lamp). The width (x-axis) of each diagram is the spectral slit width. [© 1993–2014 PerkinElmer, Inc. All rights reserved. Printed with permission. (www.perkinelmer.com).]

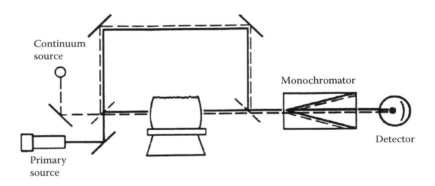

Figure 7.26 Continuum background corrector in a double-beam AAS system. [© 1993–2014 PerkinElmer, Inc. All rights reserved. Printed with permission. (www.perkinelmer.com).]

normal monochromator and slit system results in a spectral slit width about 0.2 nm wide. If we have a continuum source, the light from the source will fill the entire spectral window, as shown in the lower left-hand side drawing in Figure 7.25. When the continuum emission passes through the flame, the atoms can only absorb that portion of it exactly equal to the resonance wavelength. The atomic absorption line has a total width of about 0.002 nm. Consequently, if the atoms absorb all of the radiation over that linewidth, they will absorb only 1 % of the radiation from the continuum lamp falling on the detector, as shown in the lower center drawing of Figure 7.25. All light within the 0.20 nm bandwidth, but not within the 0.002 nm absorption line, will reach the detector and not be absorbed by the sample. In other words, 99 % or more of the emitted light reaches the detector. Consequently, the effect of atomic absorption on the continuum lamp is negligible and we can say that atoms do not measurably absorb the continuum lamp emission. It is this observation that permits the use of a continuum light source to measure and correct for broadband molecular background absorption.

In the continuum source background correction method, when an HCL source is used, the absorption measured is the total of the atomic and background absorptions. When a continuum lamp source is used, only the background absorption is measured. The continuum lamp, which emits light over a range or band of wavelengths, is placed into the spectrometer system as shown in Figure 7.26. This setup allows radiation from both the HCL and the continuum lamp to follow the same path to reach the detector. The detector observes each source alternately in time, either through the use of mirrored choppers or through pulsing of the lamp currents. When the HCL lamp is in position as the source, the emission line from the HCL is only about 0.002 nm wide; in other words, it fills only about 1 % of the spectral window as shown in the upper left-hand-side drawing in Figure 7.25. When the HCL radiation passes through the flame, both the free atoms at the resonance line and the broadband absorbing molecules will absorb the line. This results in significant attenuation of the light reaching the detector, as shown in the upper center and right-hand-side drawing in Figure 7.25. When the continuum source is in place, the continuum lamp emission fills the entire spectral window as shown in the lower left-hand-side drawing in Figure 7.25. Any absorption of the radiation from the continuum lamp observed is broadband background absorption, since it will be absorbed over the entire 0.2 nm as seen in the lower right-hand-side drawing in Figure 7.25. We showed earlier that any *atomic* absorption of the continuum lamp is negligible. The absorbance of the continuum source is therefore an accurate measure of background absorption. An advantage of this method is that the background is measured at the same nominal wavelength as the resonance line, resulting in more accurate correction.

This correction method is totally automated in modern AAS systems. The HCL or EDL and the continuum lamp are monitored alternately in time; instrument electronics compare the signals and calculate the net atomic absorption.

The lamps used as continuum sources are a deuterium lamp (D_2) for the UV region and a tungsten filament lamp for the visible region. Commercial instruments using continuum background correction normally have both of these continuum sources installed to cover the complete wavelength range; the continuum lamp used is computer-selected based on the analyte wavelength.

Continuum background correction is used for FAAS and can be used for furnace AAS. There are limitations to the use of continuum background correction. If another element (not the analyte) in the sample has an absorption line within the spectral bandpass used, especially if this element is present is large excess, it can absorb radiation and result in inaccurate background correction. The background correction is made using two different sources. The correction can be inaccurate if the sources do not have similar intensities and if they are not aligned properly to pass through exactly the same region of the atomizer. This is especially true when the background levels are high. Continuum sources can generally correct for background absorbance up to $A = 0.8$. A problem can arise when the background absorption spectrum is not uniform over the spectral bandpass; such background absorption is said to have "fine structure." Absorption with fine structure may not be corrected properly using a continuum lamp corrector; the correction may be too large or too small. These limitations are particularly problematic in GFAAS, which is prone to high background levels.

7.4.2.5 Zeeman Background Correction

Atomic absorption lines occur at discrete wavelengths because the transition that gives rise to the absorption is between two discrete energy levels. However, when a vapor-phase atom is placed in a strong magnetic field, the electronic energy levels split. This gives rise to several absorption lines for a given transition in place of the single absorption line in the absence of a magnetic field. This occurs in all atomic spectra and is called **Zeeman splitting** or the Zeeman effect. In the simplest case, the Zeeman effect splits an absorption line into two components. The first component is the π-component, at the same wavelength as before (unshifted); the second is the σ-component, which undergoes both a positive and negative shift in wavelength, resulting in two equally spaced lines on either side of the original line. This splitting pattern is presented in Figure 7.27. The splitting results in lines that are separated by approximately 0.01 nm or less depending on the field strength. The strength of the magnetic field used is between 7 and 15 kG. Background absorption and scatter are usually not affected by a magnetic field.

The π- and σ-components respond differently to polarized light. The π-component absorbs light polarized in the direction parallel to the magnetic field. The σ-components absorb only radiation polarized 90° to the applied field. The combination of splitting and polarization differences can be used to measure total absorbance (atomic

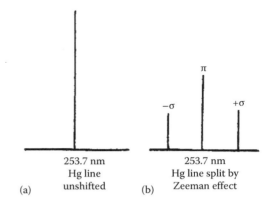

Figure 7.27 Zeeman effect causing shifts in electronic energy levels. (a) Hg resonance line in the absence of a magnetic field occurs as a single line at 253.7 nm. (b) Hg resonance line in the presence of a magnetic field. The π-component is at the original, unshifted wavelength (253.7 nm); the ±σ-components are shifted equally away from the original line to higher and lower wavelengths.

plus background) and background only, permitting the net atomic absorption to be determined.

A permanent magnet can be placed around the furnace to split the energy levels. A rotating polarizer is used in front of the HCL or EDL. During that portion of the cycle when the light is polarized parallel to the magnetic field, both atomic and background absorptions occur. No atomic absorption occurs when the light is polarized perpendicular to the field, but background absorption still occurs. The difference between the two is the net atomic absorption. Such a system is a DC Zeeman correction system.

Alternately, a fixed polarizer can be placed in front of the light source and an electromagnet can be placed around the furnace. By making absorption measurements with the magnetic field off (atomic plus background) and with the magnetic field on (background only), the net atomic absorption signal can be determined. This is a transverse AC Zeeman correction system. The AC electromagnet can also be oriented so that the field is along the light path (a longitudinal AC Zeeman system) rather than across the light path. No polarizer is required in a longitudinal AC Zeeman system.

The use of a polarizer in either DC or AC Zeeman systems cuts the light throughput significantly, adversely affecting sensitivity and precision. DC Zeeman systems require less power but have poorer linear working range and sensitivity than AC systems. AC Zeeman systems are more expensive to operate but have higher sensitivity and larger linear working ranges than DC systems.

Zeeman correction can also be achieved by having an alternating magnetic field surround the hollow cathode, causing the emission line to be split and then not split as the field is turned on and off. By tuning the amplifier to this frequency, it is possible to discriminate between the split and unsplit radiation. A major difficulty with the technique is that the magnetic field used to generate Zeeman splitting also interacts with the ions in the hollow cathode. This causes the emission from the hollow cathode to be erratic, which in turn introduces imprecision into the measurement. Most commercial instruments with Zeeman background correctors put the magnet around the atomizer. Zeeman correction systems can be used with flame atomizers and are commercially available; one disadvantage to using this approach is that the magnet blocks the analyst's view of the flame. This makes it impossible to visually check that the flame conditions are correct, since the analyst cannot see the color of the flame. The analyst cannot see if the burner slot is partially blocked or if charred material is building up on the burner when organic solvents are run. These unobserved problems might result in errors in the analysis. The significant advantage of Zeeman background correction is to compensate for the high background absorption in graphite furnace atomizers.

The advantages of Zeeman background correction are numerous. Only one light source is required. Since only one light source is used, there is no need to match intensity or align multiple sources. The physical paths of the split and unsplit light are identical, as opposed to the use of a continuum lamp, where the radiation may not pass along the identical path as the HCL radiation. The background correction is made very close to the resonance line and at the same absorption bandwidth as the atomic absorption signal. Therefore, the correction is generally very accurate, even for background with fine structure and for high background levels, up to absorbance = 2.0. The Zeeman correction system can be used for all elements at all wavelengths. One limitation to the use of Zeeman background correction is that it shortens the analytical working range. The relationship between absorbance and concentration becomes nonlinear at lower absorbances than in a non-Zeeman corrected system, and then the absorbance–concentration relationship "rolls over." That is, as concentration increases, absorbance first increases linearly, then nonlinearly until a maximum absorbance is reached. After the maximum, absorbance actually *decreases* with increasing concentration. It is possible for two completely different concentrations, one on either side of the absorbance maximum, to exhibit the same absorbance value. The rollover point must be established by carefully calibrating the instrument and unknown samples may need to be run at more than one dilution factor to ensure that an unknown is not on the "wrong side" of the absorbance maximum. If this well-understood potential problem is kept in mind, the use of Zeeman background correction with modern GFAAS instrumentation and methodology provides accurate and precise GFAAS results. The use of Zeeman correction causes significant loss in sensitivity compared to continuum background correction for some elements, notably Li, Cs, Tl, Be, Hg, B, and V. Systems with Zeeman background correction are more expensive than those with only continuum background correction.

7.4.2.6 Smith–Hieftje Background Corrector

It will be remembered that the HCL functions by the creation of excited atoms that radiate at the desired resonance wavelengths. After radiating, the atoms form a cloud of neutral atoms that, unless removed, will absorb resonance radiation from other emitting atoms.

If the HCL is run at a high current, an abundance of free atoms form. These free atoms absorb at precisely the resonance lines the hollow cathode is intended to emit; an example is shown in Figure 7.28, with the free atoms absorbing exactly at λ over an extremely narrow bandwidth (Figure 7.28b). The result is that the line emitted from the HCL is as depicted in Figure 7.28c, instead of the desired emission line depicted in Figure 7.28a. The emitted lines are broadened by the mechanisms discussed in Section 7.1.1. The phenomenon of absorption of the central portion of the emission line by free atoms in the lamp is called

ATOMIC ABSORPTION SPECTROMETRY

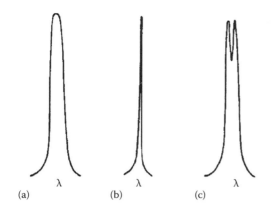

Figure 7.28 Distortion of spectral line shape in an HCL due to self-reversal. (a) Shape of the spectral line emitted by the HCL. (b) Shape of the spectral energy band absorbed by cool atoms inside the HCL. (c) Shape of the net signal emerging from an HCL, showing self-absorption of the center of the emission band.

"self-reversal." Such absorption is not easily detectable, because it is at the very center of the emitted resonance line and very difficult to resolve. It is, of course, exactly the radiation that is most easily absorbed by the atoms of the sample. In practice, if the HCL is operated at too high a current, the self-reversal decreases the sensitivity of the analysis by removing absorbable light. It also shortens the life of the lamp significantly.

The Smith–Hieftje background corrector has taken advantage of this self-reversal phenomenon by pulsing the lamp, alternating between high current and low current. At low current, a normal resonance line is emitted and the sample undergoes normal atomic absorption. When the HCL is pulsed to a high current, the center of the emission line is self-absorbed, leaving only the wings of the emitted resonance line; the emitted line would look similar to Figure 7.28c. The atoms of the sample do not significantly absorb such a line. The broad background, however, absorbs the wings of the line. Consequently, the absorption of the wings is a direct measurement of the background absorption at the wavelength of the atomic resonance absorption. The Smith–Hieftje technique can be used to automatically correct for background. In practice, the low current is run for a fairly short period of time and the absorbance measured. The high current is run for a very short, sharp burst, liberating intense emission and free atoms inside the hollow cathode and the absorbance is again measured. The difference between the measurements is the net atomic absorption. There is then a delay time to disperse the free atoms within the lamp before the cycle is started again.

The advantages of the Smith–Hieftje technique are that it can be used in single-beam optics and it is not necessary to align the beam measuring background absorption and atomic absorption since it is the same beam. In addition, the electronics are much simpler than that used in a Zeeman background correction system and no expensive magnet is needed. There are some disadvantages to the technique. It requires a special type of HCL that can operate both at low and high currents and the life of the special HCLs is significantly less than that of standard HCLs (150–500 h depending on the element vs. 5000 h for standard HCLs). The loss in sensitivity for this technique is about 50 % of that obtained using a continuum correction system, the dynamic range is shorter, and the technique cannot correct for structured background. The Smith–Hieftje background correction system is commercially available in Shimadzu Scientific AAS instruments, where it is known as rapid self-reversal background correction. All lamps for the Shimadzu instruments are self-reversing lamps that can be run at low current for D_2 background correction work. A more detailed comparison of the methods used for background correction is given in the reference by Carnrick and Slavin listed in the bibliography.

7.4.2.7 Spectral Interferences in GFAAS

Graphite furnace atomizers have their own particular spectral interference problems. Any emission from the light source or atomizer that cannot be absorbed by the analyte should be prevented from reaching the detector. Usually, the emission from flames is low. At atomization temperatures in excess of 2200 °C, the graphite furnace is a white-hot emission source. If this intense visible radiation reaches the PMT, noise increases, which decreases precision. The emission from the hot graphite is essentially blackbody radiation, extending from 350 to 800 nm and varying in intensity with wavelength. The extent of emission interference caused by this radiation falling on the detector varies with analyte wavelength. Elements with resonance lines in the 400–600 nm region like Ca and Ba can be seriously affected. The control of emission interference lies in spectrometer design, which is not controlled by the analyst and in the correct installation, alignment, and maintenance of the furnace and optical windows, which are very definitely controlled by the analyst.

Background absorption can be severe in GFAAS. Decreasing the sample volume and choosing alternate wavelengths can be used to reduce background levels in some cases. Control of the chemistry in the furnace is key to reducing background. The pyrolysis step must be designed to volatilize as much of the matrix as possible without loss of analyte. Ideally, the matrix would be much more volatile than the analyte, so that 100 % matrix removal could occur in the pyrolysis step with 100 % retention of analyte in the furnace. This situation rarely occurs. It is possible to control the volatility of the matrix and analyte to some extent through matrix modification as discussed earlier. The use of matrix modifiers seems to control background (a spectral interference) by controlling nonspectral interferences.

A potential problem with any automatic background corrector is that the analyst may be unaware of the extent to which the background is present. It should be remembered that when absorbance $A = 2$, the transmittance $T = 1/100$, and the absorbed radiation is 99 % of the available radiation. Therefore, the total signal falling on the detector is only 1 % of I_0. If quantitative analysis is carried out using this very small signal, major errors may result. A 1 % error in measuring the background may result in a 100 % error in measuring the atomic absorption by the sample. Instrument manufacturers claim quantitative atomic absorption measurements while correcting background absorption in excess of 99 % using Zeeman background correction. Analysts should be aware of the magnitude of the background signal, because a small error in measuring the background becomes a major error in the net atomic absorption measurement. Steps such as the use of L'vov platforms, matrix modification, alternate wavelengths, reduction in sample volume, and other techniques to minimize background should be taken to keep background absorbance as low as possible to obtain accurate results.

7.5 ANALYTICAL APPLICATIONS OF AAS

AAS is a mature analytical technique. There are thousands of published methods for determining practically any element in almost any type of sample. There are books and journals devoted to analytical methods by AAS and other atomic spectrometry techniques. The bibliography provides a list of some texts on AAS. Journals such as *Analytical Chemistry*, *Applied Spectroscopy*, *Journal of Analytical Atomic Spectroscopy*, *The Analyst*, *Spectroscopy Letters*, *Spectrochim. Acta Part B*, and others are sources of peer-reviewed articles, but many applications articles can be found in specialized journals on environmental chemistry, food analysis, geology, and so on. The applications discussion here is necessarily limited, but the available literature is vast.

AAS is used for the determination of all metal and metalloid elements. Nonmetals cannot be determined directly because their most sensitive resonance lines are located in the vacuum UV region of the spectrum. Neither flame nor furnace commercial atomizers can be operated in a vacuum. It is possible to determine some nonmetals indirectly by taking advantage of the insolubility of some compounds. For example, chloride ion can be precipitated as insoluble silver chloride by adding a known excess of silver ion in solution (as silver nitrate). The silver ion remaining in solution can be determined by AAS and the chloride ion concentration calculated from the change in the silver ion concentration. Similar indirect approaches for other nonmetals or even polyatomic ions like sulfate can be devised.

7.5.1 Qualitative Analysis

The radiation source used in AAS is an HCL or an EDL, and a different lamp is needed for each element to be determined (except for the new continuum source system discussed earlier). Because it is essentially a single-element technique, AAS is not well suited for qualitative analysis of unknowns. To look for more than one element requires a significant amount of sample and is a time-consuming process. For a sample of unknown composition, multielement techniques such as XRF, ICP-mass spectrometry (ICP-MS), ICP-OES, and other atomic emission techniques are much more useful and efficient.

7.5.2 Quantitative Analysis

Quantitative measurement is one of the ultimate objectives of analytical chemistry. AAS is an excellent quantitative method. It is deceptively easy to use, particularly when flame atomizers are utilized.

7.5.2.1 Quantitative Analytical Range

The relationship between absorbance and concentration of the analyte being determined in AAS follows Beer's law (discussed in Chapter 3) over some concentration range. There is an optimum linear analytical range for each element at each of its absorption lines. The *minimum* of the range is a function of the DL of the element under the operating conditions used. The ultimate limiting factor controlling the DL is the noise level of the instrument being used. The LOD in AAS is defined by the International Union of Pure and Applied Chemistry (IUPAC) and by many regulatory agencies as the concentration giving a signal equal to three times the standard deviation of a blank solution measured under the same operating conditions being used for the analysis. However, the LOD is never used as the minimum for reporting quantitative data. The limit of quantitation (LOQ) is defined as that concentration giving a signal equal to 10 times the standard deviation of a blank solution measured under the operating conditions for the analysis. This is the lowest quantitative concentration that should be reported. Concentrations between the LOD and the LOQ should be reported as "detected but not quantifiable," and concentrations below the LOD should be reported as "not detected." The *maximum* of the analytical range is determined by the sample. The linear working range for AAS is small for most systems, generally only one to two orders of magnitude at a given wavelength. The calibration curve deviates from linearity, exhibiting a flattening of the slope at high absorbance values. With a flat slope, changes in concentration of the sample produce virtually no changes in absorption. Hence, it is impossible to measure concentrations accurately at extremely high absorbance values. Many calibration curves deviate from linearity below $A = 0.8$, and Zeeman background correction shortens the analytical working range even more. With modern computerized instruments, nonlinear curve fitting can be used to some advantage to measure concentrations beyond the linear range. However, it is advisable to keep the absorbance values of samples and standards below 0.8, as was discussed in Chapter 3, unless your system can handle a higher linear working range. When a concentration range is quoted for an analytical method, it is important to know how the lower and upper limits were determined. Approximate upper limits for the linear range of elements determined by FAAS are listed in Appendix 7.A. DLs for elements by FAAS, GFAAS, CV Hg and HGAAS, and other atomic spectroscopic techniques discussed in later chapters are given in Appendix 7.B. For AAS, it had been customary to report *method sensitivity*, where sensitivity was defined as the concentration of analyte that gives an absorbance of 0.0044 (equal to 1 % absorption), so this term may be encountered in the literature. Sample sensitivities for FAAS and GFAAS are tabulated in Appendix 7.B. Both the LOD and the sensitivity are highly dependent on the sample matrix, operating conditions, the particular instrument used, and the way in which the data are processed.

Modern AAS instruments have computerized data collection and data processing systems. These computer-based systems have the ability to fit different types of calibration curves to the data using a variety of equations, including linear, quadratic, and higher-order polynomial functions. Such systems are capable of calculating concentration results from nonlinear calibration curves. The accuracy of such calculated results depends on the equations used for calibration and on the number and concentrations of the standards used to provide the calibration data. It is not advisable to use a nonlinear calibration curve without verification that the results are accurate. When in doubt, dilute the samples into the linear region, especially if using Zeeman background correction.

7.5.2.2 Calibration

Calibration of AAS methods can be performed by the use of an external calibration curve or by MSA; both calibration methods were presented in Chapter 3. Internal standards are not used in AAS because it is usually a single-element technique; we cannot measure an internal standard element at the same time we measure the analyte.

External calibration curves are prepared from solutions of known concentrations of the sample element. High-purity metals dissolved in high-purity acids are used to make the stock standard solution. For AAS, stock standard concentrations are either 1,000 or 10,000 ppm as the element. Working standards are diluted from the stock standard. For example, if we wanted to make a calibration curve for copper determination in the range of 2–20 ppm, we would make a stock standard solution of 1,000 ppm Cu by dissolving Cu metal in nitric acid. Then we would dilute this solution to make a series of working standards. The concentrations of the working standards must cover the complete range of 0–20 ppm. A calibration blank (the 0.0 ppm standard) is of course required to correct for any analyte present in the water or reagents used to prepare the standard solutions. In this example, the calibration blank would be prepared from deionized water and the same high-purity nitric acid used to make the standard solutions. Three to five working standards are typical, so we might make Cu solutions of 0, 2, 5, 10, 15, and 20 ppm. These standard solutions would then be aspirated into the flame and the absorbance of each standard would be measured. The absorbance would be plotted versus concentration. A calibration curve like that in Figure 7.29 results. It should be noted that the relationship between absorbance and concentration is linear over the range of 0–10 ppm Cu, but at higher concentrations, the relationship deviates from linearity. Provided that the slope of the curve is fairly steep, it may be possible to distinguish between concentrations outside the linear range, such as between 12 and 14 ppm. Above 16 ppm, the curve becomes so flat that accurate concentration results are difficult or impossible to obtain.

When a quantitative analysis is to be performed, the sample is atomized and the absorbance measured under exactly the same conditions as those used when the calibration standards were measured. The calibration standards are usually measured first, and the sample solutions should be measured immediately following calibration. If the sample has been prepared using reagents such as acids, it is necessary to prepare a method blank that is carried through the same sample preparation steps as the sample. The method blank is used to correct for any contamination that may have occurred during sample preparation. The absorbance of the method blank is subtracted from the sample absorbance to give the corrected or net sample absorbance. The

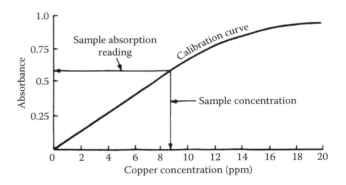

Figure 7.29 AAS calibration curve for the copper 324.8 nm line showing linear and nonlinear regions of the curve.

concentration of the unknown copper solution is determined from the calibration curve. For example, suppose that the corrected absorbance reading of the sample solution was 0.58. Using the calibration curve shown in Figure 7.29, it can be deduced that the concentration of the solution is about 8.7 ppm.

Stock standard solutions of single elements at 1,000 and 10,000 ppm can be purchased from a number of companies making these solutions. Purchasing the stock solutions saves a great deal of time and effort and is often more accurate and less expensive than making these solutions "in-house." The US National Institute for Standards and Technology (NIST) also sells standard reference material (SRM) solutions of many elements, certified for their concentration (https://nist.gov/srm). NIST SRM solutions are expensive but are often used as verification standards to confirm that the analyst has made the working standard solutions correctly as part of a method quality control process. Recall that a description of NIST standards was presented in Chapter 1 of this book; please review this material. Stock standards of most elements have a shelf life of at least 1 year; commercial solutions will have the expiration date marked on the bottle. Working standards should be prepared fresh daily or as often as needed as determined by a stability test. Low concentrations of metal ions in solution are not stable for long periods of time; they tend to adsorb onto the walls of the container. In addition, evaporation of the solvent over time will change the concentration of the solution, making it no longer a "standard."

It is most important in preparing external calibration curves that the solution matrix for samples and standards be as similar as possible. To obtain reliable quantitative data, the following should be the same for the samples and standards:

1. The same solvent (e.g., water, 5 % nitric acid, alcohol, methyl ethyl ketone [MEK]); same matrix modifier if used.
2. The same predominant anion (e.g., sulfate, chloride) at the same concentration
3. The same type of flame (air–acetylene or nitrous oxide–acetylene) or the same graphite tube/platform.
4. Stable pressure in the flame gases during the analysis.
5. Absorbance measured at the same position in the flame/furnace.
6. Background correction carried out on each sample, blank, and standard using the same correction technique.

Sample solutions should be measured immediately following calibration. If the instrument is shut down for any reason (the gas tank runs out, the power fails in a thunderstorm, the lamp burns out, the nebulizer clogs up and needs to be cleaned, the graphite tube cracks, the analyst goes to lunch, etc.), the calibration must be repeated when the instrument is turned back on to be sure that items 3–6 are exactly the same for samples and standards. For GFAAS, the platform and tube must be the same for the calibration curve and the samples. If a tube cracks during a run, a new tube and platform are inserted, conditioned per the manufacturer's directions and the calibration standards rerun before samples are analyzed. For extremely complex sample matrices, it may not be possible to make external standards with a similar matrix. In this case, the MSA should be used for quantitative analysis.

The signal in FAAS is a continuous, steady-state signal; as long as sample is aspirated into the flame, the absorbance stays at a constant value. Measurement of the absorbance in FAAS is fairly simple. In GFAAS, the signal is transient and the shape and height of the peak depend on the rate of atom formation. That rate can be affected by interferences from the sample matrix. The *integrated peak area* is independent of the rate of atom formation. For quantitative analysis by GFAAS, the concentration should be plotted against the peak area, not the peak height, for the most accurate results. Modern instruments are equipped with integrators to measure absorbance over the atomization period.

The accuracy can be excellent for both FAAS and GFAAS. The precision of FAAS is usually less than 1 % relative standard deviation (RSD). The precision of GFAAS can be in the 1–2 % RSD range but is often 5–10 % for complex matrices.

7.5.3 Analysis of Samples

7.5.3.1 Liquid Samples

Liquid solutions are the preferred form for sample introduction into flame and furnace atomizers. Frequently, liquid samples can be analyzed directly or with minimal sample preparation, such as filtration to remove solid particles. Typical samples that have been analyzed directly include urine, electroplating solutions, petroleum products, wine, fruit juice, and, of course, water and wastewater. If the samples are too concentrated, they may be diluted prior to analysis. If they are too dilute, the solvent may be evaporated or the analyte concentrated by solvent extraction or other methods.

Milk may be analyzed for trace metals by FAAS by adding trichloroacetic acid to precipitate the milk proteins. The precipitate is removed by filtration or centrifugation. The standards should also contain trichloroacetic acid to match the samples. Trace metals can be determined in seawater, urine, and other high-salt liquid matrices by complexing the metals with a chelating agent and extracting the metal complexes into organic solvent. For example, ammonium pyrrolidine dithiocarbamate (APDC) will complex Cu, Fe, Pb, and other metals. Complexation allows the metals to be extracted into MIBK. The standards are made in MIBK and aspirated directly into the flame. This permits not just extraction but preconcentration; for example, 1 L of seawater can be extracted into 50 mL of MIBK, resulting in a 20-fold concentration factor. Seawater can be analyzed for the major cations by preparing an artificial seawater solution for making calibration standards in a matrix similar to the samples. Artificial seawater contains calcium carbonate, magnesium oxide, potassium carbonate, and sodium chloride at levels that reflect the cation and anion levels found in real seawater. Oils may be diluted with a suitable organic solvent such as xylene and the soluble metals determined directly. Organometallic standards soluble in organic solvent are commercially available for preparing calibration curves for these types of analyses.

There are liquid sample matrices, such as blood, serum, very viscous oils, and the like, that require sample preparation by acid digestion or ashing prior to FAAS determination. In many cases, these matrices can be analyzed without digestion or prior ashing by GFAAS.

7.5.3.2 Solid Samples

Solid samples are usually analyzed by dissolving the sample to form a liquid solution that can be introduced into the flame or furnace. Dissolution can be accomplished by mineral acid digestion ("wet ashing"), fusion of solids with molten salts and dissolution of the fusion bead, dry ashing of organic solids with acid dissolution of the residue, combustion in oxygen bombs, and other procedures too numerous to mention. Dissolution is time consuming, even with fast microwave digestion and ashing systems, automated fusion fluxers, and other automated sample preparation devices now available. In many cases, solid samples have to be ground into smaller particles before digestion, sometimes in a multistep process requiring the use of a variety of grinding tools or mixers to ensure a homogeneous sample. Care must be taken that these tools do not contain analytes of interest, such as heavy metals. Grinding generates heat and volatile elements may be lost; cooling the grinder with liquid nitrogen or dry ice is often used to prevent loss of volatiles. Dissolution entails the possibility of introducing impurities or losing analyte, causing error. This is particularly problematic at ppb and lower levels of analyte. However, hundreds of types of solid samples are dissolved for AAS analysis on a daily basis; it is still the most common approach to AAS analysis of solids. Types of samples that have been analyzed after dissolution include metals, alloys, soils, animal tissue, plant material, foods of all types, fertilizer, ores, cements, polymers, cosmetics, pharmaceuticals, coal, ash, paint chips, and many solid industrial chemicals. Solid particulates in air, gas, and fluid streams can be collected by filtration and analyzed by digesting the filter and the collected particulates.

For example, grains, plant tissues, and many other organic materials can be digested on a hot plate in a mixture of nitric and other mineral acids to destroy the organic matrix. The cooled, clear solution is diluted to volume and analyzed against aqueous acidic calibration standards. Dry ashing of food, plant, and biological tissue is performed by placing 10–50 g of the sample in a suitable crucible and heating in a muffle furnace for several hours to burn off the organic material. The ash is dissolved in mineral acid (nitric, HCl, or acid mixtures) and diluted to volume with deionized water. An aqueous acidic calibration curve would be used. Inorganic materials, ceramics, and geological materials often require fusion in molten salts at high temperatures to convert them to soluble forms. For example, bauxite, an aluminum ore, can be fused in a Pt crucible with a mixture of sodium carbonate and sodium borate. The mixture is heated over a Bunsen burner to form a clear melt. The fusion salts convert the aluminum, silicon, and titanium compounds in bauxite to salts that will dissolve in aqueous HCl. The calibration standards must contain HCl, sodium chloride, and sodium borate at levels that match those in the diluted fused samples. It is critical that a blank be prepared in the same manner as samples are prepared, to correct for any traces of analytes that might be present in the acids, salts, and other reagents used and for contamination from "the environment." Dust in the air is often a major source of contamination of samples by "the environment," especially for analytes such as Al and Si.

The analysis of solids is time consuming and is prone to many sources of error. Analytical chemists have been trying for many years to analyze solid samples directly, without having to dissolve them. For some types of samples, this can be done by AAS. Solids can be analyzed using *a glow discharge (GD) atomizer*, by inserting small pieces or particles of sample directly into the flame or furnace or by the use of *laser ablation*.

A unique direct solid analysis approach for graphite furnace AAS, solid AA®, is the combination of the SSA 600 autosampler from Analytik Jena AG with their ContrAA systems (Figure 7.30). The SSA 600 is a fully automated solid sampler that will analyze up to 84 samples. All that is required is to place the solid samples onto the sample carriers. The autosampler weighs the sample using an integrated microbalance with a precision of 1 µg, transports it to the graphite furnace, and returns the reusable carrier back to the sample tray. Sample sizes are on the order of 100 µg–10 mg. Samples can be in the form of powders, granulates, fibers, and paste-like materials such as creams, sludge, and viscous oils.

Some of these approaches are also used in atomic emission spectrometry and in ICP-MS, and are discussed in greater detail in Chapters 8, 10, and 11 and in the references by Sneddon, Robinson, and Skelly Frame and Keliher listed in the bibliography.

One approach to the direct analysis of solids is to form a *slurry* of the sample in a suitable solvent and introduce the slurry directly into a graphite furnace atomizer. A slurry

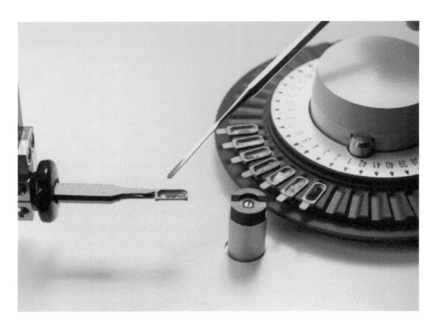

Figure 7.30 The SS 600® solids autosampler for direct solids analysis by graphite furnace AAS. (© 2013 Analytik Jena, www.analytik-jena.de. Used with permission.)

is a suspension of fine solid particles in a liquid. Slurry preparation requires that the sample either is in the form of a fine powder or can be ground to a fine powder without contamination from the grinding process. This can be done successfully for many types of samples, such as foods, grains, pharmaceuticals, and sediments. A key development in achieving reproducible results from slurry samples introduced into graphite furnace atomizers was to keep the slurry "stirred" with an ultrasonic agitator during sampling, as discussed in the reference by Miller-Ihli listed in the bibliography.

A commercial GD source, the AtomSource® (Teledyne Leeman Labs, Inc., Hudson, NH, www.teledyneleemanlabs.com), is available for the analysis of solid metals and metal alloys. A GD is more commonly used as an atomic *emission* source, but this source acts as a unique atomizer for conductive samples. The source is shown in Figure 7.31a. The atomizer cell is a vacuum chamber with the light path along the axis. The conductive sample is positioned as shown and is bombarded with six streams of ionized argon gas. This process sputters the sample, releasing free atoms into the light path, as shown in Figure 7.31b. The method does not rely on a high temperature for atomization, so refractory metals such as boron, tungsten, zirconium, niobium, and uranium can be atomized easily. In addition to ground-state atoms, the source also produces excited atoms. The emission from the excited atoms can be used to measure elements not able to be measured by AAS. Carbon in steels can be determined by emission, for example, while the other steel components are determined by absorption. The

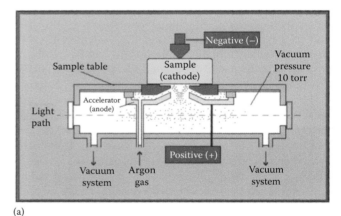

(a)

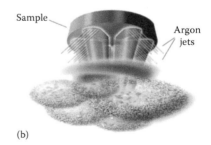

(b)

Figure 7.31 (a) The AtomSource GD atomizer for conductive solid samples. (b) Schematic of the sputtered atom cloud produced by the bombardment of the sample surface by ionized argon. (Courtesy of Teledyne Leeman Labs, Inc., Hudson, NH, www.teledyneleemanlabs.com.)

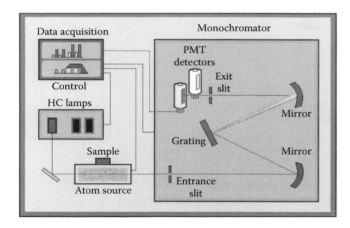

Figure 7.32 The AtomSource GD atomizer incorporated into an analyzer for solid metals and metal alloys (the pulsar system). (Courtesy of Teledyne Leeman Labs, Inc., Hudson, NH, www.teledyneleemanlabs.com.)

AtomSource atomizer is shown incorporated into the entire spectrometer system in Figure 7.32.

7.5.3.3 Gas Samples

There are some metal-containing compounds that are gaseous at room temperature and in principle the gas sample can be introduced directly into a flame atomizer. More commonly, the gas is bubbled through an appropriate absorbing solution and the solution is analyzed in the usual manner.

The introduction of a gas-phase sample into an atomizer has significant advantages over the introduction of solids or solutions. The transport efficiency may be close to 100 %, compared to the 5–15 % efficiency of a solution nebulizer. In addition, the gas-phase sample is homogeneous, unlike many solids. There are two commercial analysis systems with unique atomizers that introduce a gas-phase sample into the atomizer. They are the CV technique for mercury and the HGAAS technique. Both are used extensively in environmental and clinical chemistry laboratories.

7.5.3.4 Cold Vapor Mercury Technique

Mercury is unusual in that it exists as gas-phase free atoms at room temperature. Elemental mercury is a liquid at room temperature, but a liquid with a very high vapor pressure. So it is not necessary to apply heat from a flame or furnace to measure mercury vapor by AAS.

The CV or CVAAS technique requires the *chemical reduction* of mercury in a sample solution to elemental Hg. This is usually done with a strong reducing agent such as sodium borohydride or stannous chloride in a closed reaction system external to the AA spectrometer. The Hg atoms are then purged out of solution (sparged) by bubbling an inert gas through the solution or pumping the postreduction solution through a gas/liquid separator. The gas stream containing the free Hg atoms is passed into an unheated quartz tube cell sitting in the AAS light path. The cell is usually clamped into the position normally occupied by the burner head. Mercury atoms will absorb the HCL or EDL Hg wavelength and the measurement is made. The cell is sometimes heated to prevent water condensation in the cell, but no heat is required to atomize the Hg. The mercury atomization process is a chemical reduction reaction. CVAAS can be performed manually or can be automated using a technique called flow injection, described below. Because Hg determinations are required in many environmental samples including drinking water, instrument companies have built small AAS CV systems just for Hg measurements. Figure 7.33 shows a schematic of an automated dedicated mercury analyzer that uses a gas–liquid separator, a mercury lamp, a long optical path quartz cell, and a solid-state detector. Samples are pumped from the autosampler tray and mixed automatically with the stannous chloride reductant. Stannous chloride will reduce mercuric ion, Hg^{2+}, to Hg^0, but will not reduce organomercury compounds or mercurous ion. Samples are normally predigested with a strong oxidizing agent to ensure that all Hg in the sample is in the form of mercuric ion before adding the stannous chloride (Figure 7.33).

The sensitivity of the CVAAS technique for Hg is about 0.02 ppb, much better than that obtained from FAAS or GFAAS. The reasons are simple; the sampling is 100 % efficient and a large sample volume is used. All of the reduced Hg atoms are sent into the light path of the spectrometer, unlike the 5 % or so efficiency of a flame nebulizer system. Unlike GFAAS, the CV technique uses large sample volumes (10–100 mL of sample) compared with the microliter sample volume put into a graphite furnace. Incorporating a gold amalgamation accessory can increase the sensitivity of the CVAAS technique even further. The mercury vapor can be trapped and concentrated in the gold as an amalgam at room temperature and then released into the optical path as a concentrated "plug" of vapor by rapidly heating the gold.

For example, the PerkinElmer CVAAS systems are available as flow injection accessories for their AAS systems or dedicated flow injection mercury systems (FIMS), with or without amalgamation capability. Flow injection is described in Section 7.5.3.6. The DL for mercury varies with the system. Using a flow injection atomic spectroscopy (FIAS)-100 or 400 system with amalgamation, the DL for Hg is 0.009 ppb; without an amalgamation system, the DL is 0.2 ppb with a HCL and 0.05 ppb with an EDL. The DL using dedicated FIMS-100 or FIMS-400 systems is <0.005 ppb without an amalgamation and <0.0002 ppb with amalgamation. Dedicated Hg analyzers for the analysis of solid samples are also available, for measuring total Hg in hair, biological tissues, coal, whole blood, food, soil, and the like, using sample pyrolysis. Samples are usually

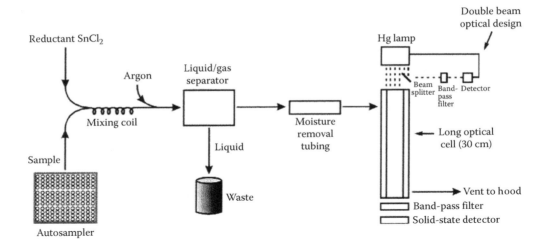

Figure 7.33 Automated CV mercury analyzer. Mercury in the sample is reduced to Hg. The Hg vapor is separated from the solution by a gas–liquid separator and carried to the optical cell. The Hg vapor absorbs the 253.7 nm emission line from the Hg lamp and the amount of atomic absorption is measured using a solid-state detector. (Courtesy of Teledyne Leeman Labs, Inc., Hudson, NH, www.teledyneleemanlabs.com.)

weighed into quartz sample boats and placed into a solid sample autosampler. The DMA-80 from Milestone, Inc. (www. milestonesci.com) has a 40-position autosampler, the ability to store specific sample methods and unattended operation. Mercury results can be obtained in about 6 min with no sample preparation. This can result in faster, cheaper analyses with less hazardous waste to dispose of.

7.5.3.5 Hydride Generation Technique

The HGAAS technique is similar in many ways to CVAAS. The atomizer is a quartz tube cell sitting in the light path of the AA spectrometer. In the HGAAS technique, the cell must be heated. Having the cell clamped above the burner head and lighting an air–acetylene flame accomplishes this. The flame surrounds the cell and heats it. Alternatively, some systems have electrically heated cells.

Sample solutions are reacted with sodium borohydride in a closed reaction system external to the AA spectrometer, as in CVAAS. Some elements will react with the sodium borohydride to form volatile hydride compounds. Arsenic, for example, forms arsine gas, $AsH_3(g)$. The volatile hydrides are sparged from solution by an inert gas and sent to the heated quartz cell. Arsine is a molecule, not an atom, so to measure atomic absorption by arsenic, the hydride species must be decomposed (by breaking bonds) and reduced to free atoms. That is done in the heated quartz cell, and the absorption signal measured as usual. The elements that form volatile hydrides include As, Bi, Ge, Pb, Sb, Se, Sn, and Te. The HGAAS technique can also be operated in manual mode or automated through the use of flow injection techniques. The sensitivity for As, Se, and other elements is excellent because of the efficient transport of the hydride to the atomizer and the larger sample volumes used. DLs for CVAAS and HGAAS are given in Appendix 7.B. Similar analyzers based on atomic fluorescence will be discussed in Chapter 8.

7.5.3.6 Flow Injection Analysis

The term **flow injection analysis** (FIA) describes an automated continuous analytical method in which an aliquot of sample is injected into a continuously flowing carrier stream. The carrier stream usually contains one or more reagents that react with the sample. A transient signal is monitored as the analyte or its reaction product flows past a detector. We have just described CVAAS, where a reduction reaction occurs and the analyte must be swept in a carrier stream into the optical path. This results in a transient signal. All of the components for an FIA are present—sample, reagent, carrier stream, and transient signal measurement. The same is true of HGAAS. The implementation of FIA for CVAAS and HGAAS permits these determinations to be carried out completely automatically, from sample uptake to printout of concentrations measured.

An FIA system consists of four basic parts: a pump or pumps for regulation of flow, an injection valve to insert sample volumes accurately and reproducibly into the carrier stream, a manifold, and a flow-through detector. A manifold is the term used for the tubing, fittings, mixing coils, and other apparatus used to carry out the desired reactions. The flow-through detector in AAS is the atomizer/detector combination in the spectrometer.

FIA techniques for AAS are described in detail in the reference by Tyson. In addition to automated CVAAS and HGAAS, FIA has been used to automate online dilution for the preparation of calibration curves, online matrix matching of solutions, online preconcentration and extraction for GFAAS, automated digestion of samples, and much more. Commercial FIA systems are available for most AAS instruments.

7.5.3.7 Flame Microsampling

Most FAAS determinations are performed using a continuous flow of solution taken up by the types of nebulizers described in Section 7.2.2.1. Uptake rates in these types of nebulizers can be as high as 5 mL/min, so a significant amount of sample solution is needed for a measurement.

The Shimadzu Scientific Instruments AA-7000 has a flame microsampling option that uses a sampling port, which allows 50–100 µL of sample solution to be injected into the flame for a measurement. This microsampling option can be used when sample is in limited supply. In addition, it minimizes salt buildup on the burner and minimizes waste. The use of such small volumes allows autocalibration using one standard and also permits "autodilution" of the sample, simply by injecting less volume into the flame. More details are available at www.shimadzu.com.

SUGGESTED EXPERIMENTS

For the following experiments, prepare at least 25–100 mL of each solution in acid-cleaned volumetric flasks. If the solutions must be stored before analysis, be certain that the solutions are acidified and that an acid blank is prepared, stored, and analyzed with the samples.

7.1 Prepare solutions of zinc chloride in deionized water containing 1 % (v/v) HNO_3. The solutions should contain known concentrations of zinc over the range of 0.5–50.0 ppm and an acid blank of 1 % (v/v) nitric acid containing no added zinc should be prepared. Aspirate each solution in turn into the burner. Measure the absorbance of the solutions at 213.9 nm, the zinc resonance line. Plot the relationship between the absorbance and the zinc concentrations. Note the deviation from Beer's law. Indicate the useful analytical range of the calibration curve. If your instrument permits it, try different curve-fitting algorithms and evaluate how they affect the working range of the analysis.

7.2 (a) Look up the wavelength, preferred flame conditions, and linear range for calcium in Appendix 7.A or your instrument "cookbook" and set up the AAS to measure Ca. Prepare 1 L of a stock standard solution of 1,000 ppm calcium using calcium phosphate. Prepare 1 L of a 10,000 ppm La standard using lanthanum chloride. Make two sets of calcium working standards. The two sets should have the same concentrations of Ca, but to one set of standards (including the blank), add enough of the La stock solution to make each of these solutions 2000 ppm in La. The La must be added before the Ca standards are made to volume. For example, to make a 10 ppm Ca standard, take 1 mL of the 1,000 ppm Ca stock and dilute it to 100 mL in a volumetric flask. To make the 10 ppm Ca standard with added La, take 1 mL of the Ca stock solution and add it to an empty 100 mL volumetric flask. Add 20 mL of the stock La solution to the same flask and then add deionized water to bring the solution to volume. Optimize the burner position for maximum absorbance. Run both sets of solutions and plot absorbance vs. concentration of Ca. Explain your observations.

(b) Using one of the Ca working standards with La and the same concentration solution without La, aspirate each solution in turn while moving the burner position. Start with the HCL beam just above the burner head (at the base of the flame) and then lower the burner in increments, making note of the absorbance. Construct a flame profile for Ca with and without La by plotting absorbance versus height above the burner. Are the two profiles different? Give a reasonable explanation for the observed flame profiles, based on your reading of the chapter.

(c) You can use the 1–10 ppm Ca working standards with La to determine the amount of Ca in commercial multivitamin tablets. Grind 1–2 coated multivitamin tablets into a fine powder using a mortar and pestle. For tablets containing about 200 mg Ca, accurately weigh out approximately 200 mg of the powder into a beaker and add 20 mL of concentrated HCl (analytical or trace metal grade). Stir well. Quantitatively transfer to a 100 mL volumetric flask that should already contain some deionized water and dilute to the mark with deionized water (Solution A). Pour this solution through coarse filter paper into a beaker. Dilute 10.0 mL of the filtrate to 100 mL with deionized water (Solution B). Test solution: Put 10.0 mL of Solution B and 20.0 mL of La solution into a 100 mL volumetric flask and dilute to the mark with deionized water. Using your instrument cookbook parameters (usually 422.7 nm, air–acetylene flame), run the calibration standards and the test solution. Using the instrument software, determine the amount of Ca in the test solution. The % calcium in the multivitamin is calculated taking into account the two dilutions:

$$\%\text{Ca in tablet} = \left[\frac{(10)(10)(X \text{ mg Ca/L})(0.1000 \text{ L})}{\text{mg of tablet powder}} \right] \times 10^2$$

For tablets containing more than 200 mg Ca per tablet, use a higher dilution factor.

Why should there be some deionized water in the volumetric flask before you transfer the HCl solution?

Is there undissolved material in the first volumetric flask? If so, the analysis would be described as determining "leachable" calcium.

How does your measured %Ca compare to the amount of Ca listed on the label of the vitamin? If your measured amount is less than expected, explain why. If your measured amount matches what is on the label, why?

7.3 Prepare a series of standard solutions containing sodium at concentrations of 1, 3, 5, 7, 9, and 10 ppm. Measure the absorbance of each solution and plot the absorbance against the sodium concentration. Take water samples from various sources, such as tap water, bottled drinking water, distilled water, river water, distilled water stored in a polyethylene bottle, and distilled water stored in a glass bottle. Determine the sodium concentration in each sample. (If you choose any carbonated bottled waters as samples, allow the sample to "go flat" or loosen the cap and shake it slightly to release the dissolved gases to avoid poor precision in your measurements. If you want to make the comparison, analyze a carbonated water sample that is not flat. The reproducibility of the absorbance measurements should be better in the "flat" sample.) All samples should be at room temperature when measured.

7.4 Prepare a solution containing 20 ppm of Pb as Pb$(NO_3)_2$. Prepare similar solutions containing 20 ppm of Pb as (a) $PbCl_2$, (b) lead oxalate, and (c) lead acetate. Measure the absorbance by the solutions of the Pb resonance line at 283.3 nm. Note the change in absorbance as the compound changes. This is chemical interference. Find a literature method for the elimination of chemical interference using excess ethylenediaminetetraacetic acid (EDTA). Following the literature method, see if the use of EDTA does eliminate the chemical interference you observed.

7.5 Prepare an aqueous 1 % HNO_3 solution containing 5 ppm of NaCl; also prepare five separate aqueous 1 % HNO_3 solutions (100 mL each) containing 500 ppm of (a) Ca, (b) Mg, (c) Fe, (d) Mn, and (e) K. Take an aliquot of each of these solutions (10–25 mL, but use the same volume for all of the solutions). Spike each aliquot with 5 ppm Na. Be sure to prepare an acid blank and use the same acid for all solution preparations. Measure the absorbance at the Na resonance line at 589.0 nm by all solutions. Compare the absorbance of the Na-spiked solutions (a) through (e) with the aqueous acidic Na solution (i.e., compare all the 5 ppm Na solutions in all matrices). Are any of the signals enhanced compared to the aqueous acidic solution? Which ones? Explain. Are any suppressed? Again, explain. For the solutions (a), (b), (c), (d), and (e) with and without Na, plot two-point standard addition calibration curves and calculate the amount of Na present in the unspiked solutions. Any sodium present was caused by an Na impurity in the original (500 ppm) solutions. Look at the slopes of all five plots. Are they the same? If not, why not?

7.6 For laboratories with a graphite furnace atomizer, the following experiments can be run. The instrument should be set up for the element according to the instrument "cookbook." (a) Dilute the lead solutions from Experiment 7.4 to give 10 ppb lead solutions (or some other concentration within the linear range of your furnace). Using the "cookbook" furnace program and background correction but *no matrix modifiers*, run triplicate injections of each solution (20 µL is a usual injection volume) and observe the peak shape, appearance time of the peak maximum, the peak height, and area. Do the peak maxima appear at the same time? Do all peaks have the same shape? If not, why not? Repeat using a lower pyrolysis temperature. If the background absorbance is printed out, compare the background absorbance values at both sets of conditions. How is the background affected by pyrolysis temperature? Does it depend on the compound? (b) Using background correction *and the matrix modifiers* and furnace program recommended by the "cookbook" for Pb, rerun the 10 ppb solutions of lead compounds. Compare the peak appearance time, shape, and so on with and without matrix modification. Comment on the approach you would use to obtain accurate results, based on your observations. (c) If your system has the ability to run atomization from the tube wall and from a L'vov platform, run the lead nitrate solution both ways (wall and platform). Perform 20 replicate injections at each condition and compare the precision of wall vs. platform measurements. Explain your results.

PROBLEMS

7.1 Why are atomic absorption lines very narrow?

7.2 Why must HCLs or EDLs be used as the radiation source for most AAS instruments? Illustrate schematically an HCL for lead (Pb). How would you make the cathode?

7.3 Why is "modulation" of the line source necessary for accurate results? How is modulation achieved?

7.4 If the radiation source is not modulated, will emission from the analyte in the atomizer result in a positive or negative error? Show your calculation to support your answer.

7.5 Why is atomic absorption not generally used for qualitative analysis?

7.6 What causes chemical interference in AAS? Give three examples.

7.7 How are solid samples analyzed by AAS?

7.8 Several standard solutions of copper were prepared. These were aspirated into a flame and the absorption measured with the following results. Prepare a calibration curve from the data.

Sample Concentration (ppm)	Absorbance
Blank (0.0)	0.002
0.5	0.045
1.0	0.090
1.5	0.135
2.0	0.180
2.5	0.225
3.0	0.270

7.9 Samples of polymer were brought to the lab for a determination of their copper content. They were ashed, the residue was dissolved in mineral acid, and diluted to a known volume. The absorbance of each solution and a digestion blank was measured. Using the calibration curve from Problem 7.8, calculate the copper concentration in each solution and the blank and fill in the following table.

Sample	Absorbance	Concentration (ppm)
Blank	0.006	
A	0.080	
B	0.105	
C	0.220	
D	0.250	

Exactly 2.00 g of each sample were digested and the final solution volume for each sample and the blank was 100.0 mL. Calculate the concentration of copper in the original polymer for each sample.

7.10 What is shown in a Grotrian diagram?

7.11 Why can nonmetal elements not be determined directly by AAS?

7.12 What is the relationship between the amount of light absorbed and the oscillator strength of the transition involved?

7.13 How can the population distribution of atoms in various energy levels be calculated?

7.14 What is the basis for concluding that at temperatures up to 3000 K the great majority of an atom population is in the ground state?

7.15 Describe the two major light sources used in AAS—the HCL and the EDL.

7.16 Why are EDLs used? List the elements that benefit from use of an EDL instead of an HCL.

7.17 What is meant by a flame profile?

7.18 Describe the atomization process that takes place in a flame.

7.19 How does the rapid formation of a stable oxide of the analyte affect its flame profile?

7.20 Define chemical interference. Give an example of how this interference is corrected.

7.21 What is the source of background absorption?

7.22 How is the background corrected?

7.23 Describe the operation of a Zeeman background corrector. Discuss its advantages and disadvantages compared with the use of a deuterium lamp for background correction.

7.24 Describe the operation of a deuterium (D_2) lamp background corrector.

7.25 Draw a schematic of a typical graphite furnace atomization tube.

7.26 Describe the L'vov platform.

7.27 What is the advantage of the L'vov platform?

7.28 What are the advantages of graphite furnace atomizers over flames?

7.29 What are the disadvantages of graphite furnace atomizers versus flames?

7.30 Define ionization interference and give an example. How is this interference corrected?

7.31 You need to determine potassium in serum samples by FAAS. What will you add to correct for ionization interference in the determination?

7.32 Distinguish between spectral and nonspectral interference in AAS.

7.33 The indirect determination of chloride ion by precipitation as silver chloride was described. What metal element could you use to determine sulfate ion in water indirectly by AAS? Describe how you would do this experiment and give an example calculation for a sulfate solution containing 200 mg SO_4^{2-}/L. (You may need to consult some external references on this question.)

7.34 Briefly explain the advantages of the newly developed HR-CS AAS compared to line source instruments.

7.A APPENDIX

Absorption Wavelengths, Preferred Flames, and Characteristic Concentrations for Elements by FAAS

Abbreviations Used

A.A.	Air–acetylene flame
Char. conc.	Characteristic concentration in ppm (mg/L) in aqueous solution equivalent to an absorbance = 0.2.
EDL	Electrodeless discharge lamp
Linear range	The upper concentration limit in ppm of the linear region of the concentration–absorbance plot
N.A.	Nitrous oxide–acetylene flame
ox.	Oxidizing flame (excess oxidant, either air or nitrous oxide). An oxidizing A.A. flame is clear and blue in color. Oxidizing conditions are rarely used for N.A. flames
red.	Reducing flame (excess fuel, acetylene.) A reducing A.A. flame is yellow in color; a reducing N.A. flame is red in color

Absorption Wavelength (nm)	Preferred Flame	Char. Conc. (ppm)	Linear Range (ppm)
Aluminum			
309.3	N.A. (red.)	1.1	100.0
396.2	N.A. (red.)	1.1	150.0
308.2	N.A. (red.)	1.5	150.0
394.4	N.A. (red.)	2.2	—
237.3	N.A. (red.)	3.3	—
236.7	N.A. (red.)	4.8	—
257.5	N.A. (red.)	6.7	—

(continued)

Absorption Wavelength (nm)	Preferred Flame	Char. Conc. (ppm)	Linear Range (ppm)	Absorption Wavelength (nm)	Preferred Flame	Char. Conc. (ppm)	Linear Range (ppm)
Antimony				216.5	A.A. (ox.)	0.17	20.0
217.6	A.A. (ox.)	0.55	30.0	222.6	A.A. (ox.)	1.1	50.0
206.8	A.A. (ox.)	0.85	50.0	249.2	A.A. (ox.)	5.8	100.0
231.2	A.A. (ox.)	1.3	50.0	224.4	A.A. (ox.)	14.0	—
Arsenic[a]				244.2	A.A. (ox.)	24.0	—
189.0	A.A. (ox.)	0.78	180.0	Dysprosium			
193.7	A.A. (ox.)	1.0	100.0	421.2	N.A. (red.)	0.70	20.0
197.2	A.A. (ox.)	2.0	250.0	404.6	N.A. (red.)	0.97	50.0
Barium				418.7	N.A. (red.)	1.0	—
553.5	N.A. (red.)	0.46	20.0	419.5	N.A. (red.)	1.3	—
350.1	N.A. (red.)	5.6	—	Erbium			
Beryllium				400.8	N.A. (red.)	0.068	40.0
234.8	N.A. (red.)	0.025	1.0	415.1	N.A. (red.)	1.2	150.0
Bismuth				389.3	N.A. (red.)	2.3	150.0
223.1	A.A. (ox.)	0.45	20.0	Europium			
306.8	A.A. (ox.)	1.3	100.0	459.4	N.A. (red.)	0.67	200.0
206.2	A.A. (ox.)	3.7	—	462.7	N.A. (red.)	0.79	300.0
227.7	A.A. (ox.)	6.1	—	466.2	N.A. (red.)	0.98	—
Boron				Gadolinium			
249.7	N.A. (red.)	13.0	400.0	407.9	N.A. (red.)	19.0	1,000
208.9	N.A. (red.)	27.0	—	368.4	N.A. (red.)	19.0	1,000
Cadmium				405.8	N.A. (red.)	22.0	—
228.8	A.A. (ox.)	0.028	2.0	Gallium			
326.1	A.A. (ox.)	11.0	—	287.4	N.A. (red.)	1.3	100.0
Calcium				294.4	N.A. (red.)	1.1	50.0
422.7	A.A. (ox.)	0.092	5.0	417.2	N.A. (red.)	1.5	100.0
239.9	A.A. (ox.)	13.0	800.0	403.3	N.A. (red.)	2.8	400.0
Cerium				Germanium			
520.0	N.A. (red.)	30	—	265.1	N.A. (red.)	2.2	200.0
Cesium				259.3	N.A. (red.)	5.2	300.0
852.1	A.A. (ox.)	0.21	15.0	271.0	N.A. (red.)	5.0	400.0
894.4	A.A. (ox.)	0.40	15.0	275.5	N.A. (red.)	6.1	—
455.5	A.A. (ox.)	25.0	—	Gold			
459.3	A.A. (ox.)	94.0	—	242.8	A.A. (ox.)	0.33	50.0
Chromium				267.6	A.A. (ox.)	0.57	60.0
357.9	A.A. (red.)	0.078	5.0	274.8	A.A. (ox.)	210.0	—
359.3	A.A. (red.)	0.10	7.0	312.8	A.A. (ox.)	210.0	—
360.5	A.A. (red.)	0.14	7.0	Hafnium			
425.4	A.A. (red.)	0.20	7.0	286.6	N.A. (red.)	11.0	500.0
Cobalt				294.1	N.A. (red.)	14.0	—
240.7	A.A. (ox.)	0.12	3.5	307.3	N.A. (red.)	16.0	—
242.5	A.A. (ox.)	0.15	2.0	Holmium			
241.2	A.A. (ox.)	0.22	3.0	410.4	N.A. (red.)	0.87	50.0
252.1	A.A. (ox.)	0.28	7.0	405.4	N.A. (red.)	1.0	50.0
352.7	A.A. (ox.)	3.2	—	416.3	N.A. (red.)	1.4	70.0
Copper				Indium			
324.8	A.A. (ox.)	0.077	5.0	304.9	A.A. (ox.)	0.76	80.0
327.4	A.A. (ox.)	0.17	5.0	325.6	A.A. (ox.)	0.80	60.0

Absorption Wavelength (nm)	Preferred Flame	Char. Conc. (ppm)	Linear Range (ppm)	Absorption Wavelength (nm)	Preferred Flame	Char. Conc. (ppm)	Linear Range (ppm)
410.5	A.A. (ox.)	2.5	200.0	Nickel			
Iridium				232.0	A.A. (ox.)	0.14	2.0
264.0	A.A. (red.)	12.0	1,000.0	231.1	A.A. (ox.)	0.20	5.0
266.5	A.A. (red.)	13.0	—	352.5	A.A. (ox.)	0.39	20.0
285.0	A.A. (red.)	15.0	—	341.5	A.A. (ox.)	0.40	—
Iron				Niobium			
248.3	A.A. (ox.)	0.11	6.0	334.4	N.A. (red.)	15.0	600.0
252.3	A.A. (ox.)	0.18	10.0	334.9	N.A. (red.)	15.0	600.0
248.8	A.A. (ox.)	0.19	10.0	408.0	N.A. (red.)	20.0	—
302.1	A.A. (ox.)	0.40	10.0	412.4	N.A. (red.)	26.0	—
296.7	A.A. (ox.)	0.81	20.0	353.5	N.A. (red.)	42.0	—
Lanthanum				374.0	N.A. (red.)	47.0	—
550.1	N.A. (red.)	48.0	1,000.0	Osmium			
418.7	N.A. (red.)	63.0	2,000.0	290.9	N.A. (red.)	1.0	10–200
495.0	N.A. (red.)	72.0	3,000.0	305.9	N.A. (red.)	1.6	—
Lead				263.7	N.A. (red.)	1.8	—
217.0	A.A. (ox.)	0.19	20.0	301.8	N.A. (red.)	3.0	—
283.3	A.A. (ox.)	0.45	20.0	Palladium			
205.3	A.A. (ox.)	5.4	—	244.8	A.A. (ox.)	0.22	10.0
202.2	A.A. (ox.)	7.1	—	247.6	A.A. (ox.)	0.25	15.0
Lithium				276.3	A.A. (ox.)	0.74	—
670.8	A.A. (ox.)	0.035	3.0	340.5	A.A. (ox.)	0.72	—
323.3	A.A. (ox.)	10.0	—	Phosphorus[a]			
610.4	A.A. (ox.)	150.0	—	178.3	N.A. (red.)	5.0	2,000.0
Lutetium				213.6	N.A. (red.)	290.0	10,000.0
336.0	N.A. (red.)	6.0	500.0	214.9	N.A. (red.)	460.0	20,000.0
331.2	N.A. (red.)	11.0	—	Platinum			
337.7	N.A. (red.)	12.0	—	265.9	A.A. (ox.)	2.2	60.0
451.9	N.A. (red.)	66.0	—	306.5	A.A. (ox.)	3.2	—
Magnesium				262.8	A.A. (ox.)	4.2	—
285.2	A.A. (ox.)	0.0078	0.50	283.0	A.A. (ox.)	5.4	—
202.6	A.A. (ox.)	0.19	10.0	293.0	A.A. (ox.)	6.1	—
Manganese				Potassium			
279.5	A.A. (ox.)	0.052	2.0	766.5	A.A. (ox.)	0.043	2.0
279.8	A.A. (ox.)	0.067	5.0	769.9	A.A. (ox.)	0.083	20.0
280.1	A.A. (ox.)	0.11	5.0	404.4	A.A. (ox.)	7.8	600.0
Mercury[a]				Praseodymium			
253.7	A.A. (ox.)	4.2	300.0	495.1	N.A. (red.)	39.0	5,000.0
Molybdenum				513.3	N.A. (red.)	61.0	—
313.3	N.A. (red.)	0.67	40.0	492.5	N.A. (red.)	79.0	—
317.0	N.A. (red.)	1.1	30.0	Rhenium			
319.4	N.A. (red.)	1.4	60.0	346.0	N.A. (red.)	14.0	500.0
390.3	N.A. (red.)	2.9	—	346.5	N.A. (red.)	24.0	500.0
Neodymium				345.2	N.A. (red.)	36.0	1,000.0
492.4	N.A. (red.)	7.3	200.0	Rhodium			
463.4	N.A. (red.)	11.0	200.0	343.5	A.A. (ox.)	0.20	15.0
471.9	N.A. (red.)	19.0	—	369.2	A.A. (ox.)	0.35	20.0

(continued)

Absorption Wavelength (nm)	Preferred Flame	Char. Conc. (ppm)	Linear Range (ppm)	Absorption Wavelength (nm)	Preferred Flame	Char. Conc. (ppm)	Linear Range (ppm)
339.7	A.A. (ox.)	0.45	15.0	Technetium			
350.3	A.A. (ox.)	0.65	20.0	261.4	A.A. (red.)	3.0	70.0
Rubidium				260.9	A.A. (red.)	12.0	1,000.0
780.0	A.A. (ox.)	0.11	5.0	429.7	A.A. (red.)	20.0	1,000.0
794.8	A.A. (ox.)	0.19	5.0	Tellurium			
420.2	A.A. (ox.)	8.7	—	214.3	A.A. (ox.)	0.43	20.0
421.6	A.A. (ox.)	19.0	—	225.9	A.A. (ox.)	4.4	—
Ruthenium				238.6	A.A. (ox.)	18.0	—
349.9	A.A. (ox.)	0.66	20.0	Terbium			
372.8	A.A. (ox.)	0.86	20.0	432.6	N.A. (red.)	5.9	200.0
379.9	A.A. (ox.)	1.6	—	431.9	N.A. (red.)	6.8	200.0
392.6	A.A. (ox.)	7.5	—	390.1	N.A. (red.)	9.5	400.0
Samarium				406.2	N.A. (red.)	11.0	200.0
429.7	N.A. (red.)	6.7	400.0	Thallium			
476.0	N.A. (red.)	12.0	—	276.8	A.A. (ox.)	0.67	40.0
511.7	N.A. (red.)	14.0	—	377.6	A.A. (ox.)	1.6	100.0
Scandium				238.0	A.A. (ox.)	3.7	200.0
391.2	N.A. (red.)	0.30	25.0	258.0	A.A. (ox.)	13.0	—
390.8	N.A. (red.)	0.40	25.0	Thorium			
402.4	N.A. (red.)	0.41	25.0	324.6	N.A. (red.)	500	—
408.2	N.A. (red.)	2.1	25.0	Thulium			
Selenium[a]				371.8	N.A. (red.)	0.45	60.0
196.0	A.A. (ox.)	0.5	50.0	410.6	N.A. (red.)	0.66	50.0
204.0	A.A. (ox.)	1.5	—	374.4	N.A. (red.)	0.74	60.0
206.3	A.A. (ox.)	6.0	—	409.4	N.A. (red.)	0.80	100.0
207.5	A.A. (ox.)	20.0	—	Tin[a]			
Silicon				224.6	N.A. (red.)	1.7	300.0
251.6	N.A. (red.)	2.1	150.0	235.5	N.A. (red.)	2.2	—
251.9	N.A. (red.)	3.0	200.0	286.3	N.A. (red.)	3.2	400.0
250.7	N.A. (red.)	5.9	—	Titanium			
252.9	N.A. (red.)	6.1	—	364.3	N.A. (red.)	1.8	100.0
252.4	N.A. (red.)	7.0	—	365.4	N.A. (red.)	1.9	100.0
Silver				320.0	N.A. (red.)	2.0	200.0
328.1	A.A. (ox.)	0.054	4.0	363.5	N.A. (red.)	2.4	200.0
338.3	A.A. (ox.)	0.11	10.0	Tungsten			
Sodium				255.1	N.A. (red.)	9.6	500.0
589.0	A.A. (ox.)	0.012	1.0	294.4	N.A. (red.)	13.0	1,500.0
330.2	A.A. (ox.)	1.7	—	268.1	N.A. (red.)	12.0	1,000.0
Strontium				272.4	N.A. (red.)	13.0	500.0
460.7	N.A. (red.)	0.11	5.0	Uranium			
407.8	N.A. (red.)	2.0	20.0	351.5	N.A. (red.)	110.0	5,000.0
Sulfur				358.5	N.A. (red.)	47.0	—
180.7	A.A. (ox.)	9.0	—	356.7	N.A. (red.)	76.0	—
Tantalum				Vanadium			
271.5	N.A. (red.)	11.0	1,200.0	318.4	N.A. (red.)	1.9	200.0
260.9	N.A. (red.)	23.0	—	306.6	N.A. (red.)	4.6	200.0
277.6	N.A. (red.)	24.0	—	306.0	N.A. (red.)	4.7	400.0

ATOMIC ABSORPTION SPECTROMETRY

Absorption Wavelength (nm)	Preferred Flame	Char. Conc. (ppm)	Linear Range (ppm)
305.6	N.A. (red.)	6.2	500.0
320.2	N.A. (red.)	13.0	—
390.2	N.A. (red.)	13.0	—
Ytterbium			
398.8	N.A. (red.)	0.12	15.0
346.4	N.A. (red.)	0.45	15.0
246.4	N.A. (red.)	1.0	—
267.2	N.A. (red.)	5.0	—
Yttrium			
410.2	N.A. (red.)	1.6	50.0
407.7	N.A. (red.)	1.9	50.0
412.8	N.A. (red.)	1.9	50.0
362.1	N.A. (red.)	2.9	100.0
Zinc			
213.9	A.A. (ox.)	0.018	1.0
307.6	A.A. (ox.)	79.0	—
Zirconium			
360.1	N.A. (red.)	7.0	600.0
303.0	N.A. (red.)	11.0	600.0
301.2	N.A. (red.)	11.0	600.0
298.5	N.A. (red.)	13.0	—
362.4	N.A. (red.)	17.0	—

Source: The data are from Analytical Methods for Atomic Absorption Spectrometry, PerkinElmer, Inc., Shelton, CT, 1994. [© 1993–2014 PerkinElmer, Inc. All rights reserved. Printed with permission. (www.perkinelmer.com).]

Note: At least the three most intense lines (where available) are listed for each element. Some elements (e.g., Hg, S, P) have their most intense lines in the vacuum UV region (<190 nm); these lines are not accessible on most commercial AAS systems. Data not from the aforementioned reference are courtesy of the late Dr. Fred Brech.

[a] The Char. conc. obtained by using EDL.

7.B APPENDIX

Table B.1 Representative Atomic Spectroscopy DLs (µg/L)

Element	Flame AA	Hg/Hydride	Graphite Furnace	ICP Emission	ICP-MS
Ag	1.5		0.005	0.6	0.002
Al	45		0.1	1	0.005[a]
As	150	0.03	0.05	1	0.0006[b]
Au	9		0.15	1	0.0009
B	1,000		20	1	0.003[c]
Ba	15		0.35	0.03	0.00002[d]
Be	1.5		0.008	0.09	0.003
Bi	30	0.03	0.05	1	0.0006
Br					0.2
C					0.8[e]
Ca	1.5		0.01	0.05	0.0002[d]
Cd	0.8		0.002	0.1	0.00009[d]
Ce				1.5	0.0002
Cl					12
Co	9		0.15	0.2	0.0009
Cr	3		0.004	0.2	0.0002[d]
Cs	15				0.0003
Cu	1.5		0.014	0.4	0.0002[c]
Dy	50			0.5	0.0001[f]
Er	60			0.5	0.0001
Eu	30			0.2	0.00009
F					372
Fe	5		0.06	0.1	0.0003[d]
Ga	75			1.5	0.0002
Gd	1,800			0.9	0.0008[g]
Ge	300			1	0.001[h]

(continued)

Table B.1 (Continued) Representative Atomic Spectroscopy DLs (μg/L)

Element	Flame AA	Hg/Hydride	Graphite Furnace	ICP Emission	ICP-MS
Hf	300			0.5	0.0008
Hg	300	0.009	0.6	1	0.016[j]
Ho	60			0.4	0.0006
I					0.002
In	30			1	0.0007
Ir	900		3.0	1	0.001
K	3		0.005	1	0.0002[d]
La	3,000			0.4	0.0009
Li	0.8		0.06	0.3	0.001[c]
Lu	1,000			0.1	0.00005
Mg	0.15		0.004	0.04	0.0003[c]
Mn	1.5		0.005	0.1	0.00007[d]
Mo	45		0.03	0.5	0.001
Na	0.3		0.005	0.5	0.0003[c]
Nb	1,500			1	0.0006
Nd	1,500			2	0.0004
Ni	6		0.07	0.5	0.0004[c]
Os	(120)			6	
P	75,000		130	4	0.1[a]
Pb	15		0.05	1	0.00004[d]
Pd	30		0.09	2	0.0005
Pr	7,500			2	0.00009
Pt	60		2.0	1	0.002
Rb	3		0.03	5	0.0004
Re	750			0.5	0.0003
Rh	6			5	0.0002
Ru	100		1.0	1	0.0002
S				10	28[j]
Sb	45	0.15	0.05	2	0.0009
Sc	30			0.1	0.004
Se	100	0.03	0.05	2	0.0007[b]
Si	90		1.0	10	0.03[a]
Sm	3,000			2	0.0002
Sn	150		0.1	2	0.0005[a]
Sr	3		0.025	0.05	0.00002[d]
Ta	1,500			1	0.0005
Tb	900			2	0.00004
Te	30	0.03	0.1	2	0.0008[k]
Th				2	0.0004
Ti	75		0.35	0.4	0.003[l]
Tl	15		0.1	2	0.0002
Tm	15			0.6	0.00006
U	15,000			10	0.0001
V	60		0.1	0.5	0.0005
W	1,500			1	0.005
Y	75			0.2	0.0002
Yb	8			0.1	0.0002[m]
Zn	1.5		0.02	0.2	0.0003[d]
Zr	450			0.5	0.0003

Table B.1 (Continued) Representative Atomic Spectroscopy DLs (μg/L)

Source: [© 1993–2014 PerkinElmer, Inc. All rights reserved. Printed with permission. (www.perkinelmer.com).]

Notes:
All DLs are given in micrograms per liter and were determined using elemental standards in dilute aqueous solution. All DLs are based on a 98 % confidence level (3 standard deviations or 3 SD). All atomic absorption DLs were determined using instrumental parameters optimized for the individual element, including the use of System 2 EDLs where available. All Optima ICP-OES DLs were obtained under simultaneous multielement conditions with the axial view of a dual-view plasma using a cyclonic spray chamber and a concentric nebulizer. CV mercury DLs were determined with a FIAS-100 or FIAS-400 flow injection system with amalgamation accessory. Hydride DLs shown were determined using an MHS-15 Hg/hydride system. GFAA DLs were determined on a PerkinElmer AA using 50 μL sample volumes, an integrated platform, and full stabilized temperature platform furnace (STPF) conditions. Unless otherwise noted, ICP-MS DLs were determined using a PerkinElmer ICP-MS equipped with Ryton™ spray chamber, Type II cross-flow nebulizer, and nickel cones. All DLs were determined using 3s integration time and a minimum of eight measurements.

Letters following ICP-MS DLs refer to the use of specialized conditions or a different model instrument: [a]Run on a PerkinElmer ICP-MS in standard mode using Pt cones and quartz sample introduction system. [b]Run on a PerkinElmer ICP-MS in dynamic reaction cell (DRC) mode using Pt cones and a quartz sample introduction system. [c]Run on a PerkinElmer ICP-MS in standard mode in a Class-100 clean room using Pt cones and quartz sample introduction. [d]Run on a PerkinElmer ICP-MS in DRC mode in a Class-100 clean room using Pt cones and quartz sample introduction. [e]Using C-13. [f]Using Dy-163. [g]Using Gd-157. [h]Using Ge-74. [i]Using Hg-202. [j]Using S-34. [k]Using Te-125. [l]Using Ti-49. [m]Using Yb-173.

Osmium FAAS data are provided by the authors, not by PerkinElmer.

Table B.2 Sensitivity (1 % Absorption) in FAAS

Element	λ (nm)	Sensitivity (ppm)	Element	λ (nm)	Sensitivity (ppm)
Al	309.2	1.0	Nd	463.4	10.0
Sb	217.6	0.1	Ni	232.0	0.1
As	193.7	1.0	Nb	334.9	20.0
Ba	553.5	0.2	Pd	247.6	0.5
Be	234.9	0.1	Pt	265.9	1.0
Bi	223.1	0.1	K	766.5	0.1
B	249.7	30.0	Pr	495.1	10.0
Ca	442.7	0.05	Re	346.0	15.0
Cs	852.1	0.1	Rh	343.5	0.1
Cr	357.9	0.1	Rb	780.0	0.1
Co	240.7	0.1	Ru	349.9	0.8
Cu	324.7	0.1	Sm	429.7	10.0
Dy	421.2	1.0	Sc	391.2	1.0
Er	400.8	1.0	Se	196.1	1.0
Eu	459.4	2.0	Si	251.6	0.8
Gd	368.4	20.0	Ag	328.1	2.5
Ga	287.4	1.0	Na	589.0	2.5
Ge	265.2	2.0	Sr	460.7	0.1
Au	242.8	1.0	Ta	471.4	10.0
Hf	307.2	10.0	Te	214.3	0.5
Ho	410.4	2.0	Th	377.6	0.4
In	304.0	0.1	Sn	235.4	0.5
Fe	248.3	0.1	Ti	364.3	1.0
La	392.8	75.0	W	400.9	1.0
Pb	217.0	0.05	U	351.5	100.0
Li	670.7	0.03	V	318.4	1.0
Mg	285.2	0.001	Y	398.8	2.0
Mn	279.5	0.05	Zn	218.9	0.01
Hg	253.7	1.0	Zr	360.1	50.0
Mo	313.3	0.1			

Note: Data include air–acetylene and nitrous oxide–acetylene flames as appropriate for the element (see Appendix 7.A for standard flame).

Table B.3 Sensitivity (1 % Absorption) in the Graphite Tube Furnace (HGA-70)

Element	Absolute Sensitivity (g × 10^{-12})	20 µL Solution (ng/µL)
Al	150	0.007
As	160	0.008
Be	3.4	0.0002
Bi	280	0.014
Ca	3.1	0.05
Cd	0.8	0.00004
Co	120	0.006
Cr	18	0.01
Cs	71	0.004
Cu	45	0.02
Ga	1,200	0.06
Hg	15,000	1.5
Mn	7	0.01
Ni	330	0.10
Pb	23	0.001
Pd	250	0.013
Pt	740	0.02
Rb	41	0.002
Sb	510	0.15
Si	24	0.10
Sn	5,500	0.2
Sr	31	0.0015
Ti	280	0.5
Tl	90	0.1
V	320	0.2
Zn	2.1	0.0001

Source: [© 1993–2014 PerkinElmer, Inc. All rights reserved. Printed with permission. (www.perkinelmer.com).]

BIBLIOGRAPHY

The reader will note that occasionally multiple editions of books are cited as sources or reference works. The reason for this is that sometimes different figures are used in different editions, which may in fact be written by different authors. The sources found in the bibliography will be the most recent versions of a particular work.

In some cases instrument manufacturers and suppliers are cited by providing a URL to the exact internet location. We have endeavored to ensure that these links are functional at the time of publication, but clearly these links change. If a link has become nonfunctional, a good approach is to use your browser's advanced search capability. For example: "company name" AND "instrument type" AND "specific descriptors" is a good place to start.

Analytical Methods for Atomic Absorption Spectrometry. PerkinElmer, Inc.: Shelton, CT, 1994.

Bashkin, S., Stoner, J.A., *Atomic Energy Levels and Grotrian Diagrams*, North Holland/American Elsevier, New York, 1975.

Beaty, R.D.; Kerber, J.D. *Concepts, Instrumentation and Techniques in Atomic Absorption Spectrophotometry.* PerkinElmer, Inc.: Shelton, CT, 1993.

Bruno, T.J., Svoronos, P.D.N., *CRC Handbook of Basic Tables for Chemical Analysis – Data Driven Methods and Interpretation*, 4th Ed., CRC Taylor and Francis, Boca Raton, FL, 2021.

Burguera, J.L. (ed.) *Flow Injection Atomic Spectroscopy.* Marcel Dekker, Inc.: New York, 1989.

Carnrick, G.R.; Slavin, W. The determination of trace elements in natural waters using the stabilized temperature platform furnace. *Appl. Spectrosc.*, 37(1), 1, 1983.

Dean, J.A. *Analytical Chemistry Handbook.* McGraw-Hill, Inc.: New York, 1995.

Dean, J.A.; Rains, T.C. (eds.) *Flame Emission and Atomic Absorption Spectrometry*, Vol. 1. Marcel Dekker, Inc.: New York, 1969 (1971; Vol. 2; 1975; Vol. 3).

Huang, M.D.; Becker-Ross, H.; Florek, S.; Heitmann, U., Okruss, M., Determination of phosphorus by molecular absorption of phosphorus monoxide using a high-resolution continuum source absorption spectrometer and an air–acetylene flame. *J. Anal. Atom. Spec.*, 21, 338, 2006.

Ingle, J.D. Jr.; Crouch, S.R. *Spectrochemical Analysis.* Prentice Hall: Englewood Cliffs, NJ, 1988.

Jackson, K.W. (ed.) *Electrothermal Atomization for Analytical Atomic Spectrometry.* John Wiley & Sons, Inc.: New York, 1999.

Kirkbright, G.F.; Sargent, M. *Atomic Absorption and Fluorescence Spectroscopy.* Academic Press: New York, 1974.

Lajunen, L.H.J. *Spectrochemical Analysis by Atomic Absorption and Emission.* Royal Society of Chemistry: Cambridge, U.K., 1992.

L'vov, B.V. *Atomic Absorption Spectrochemical Analysis.* Elsevier: New York, 1970.

Maugh, T.H., A new way to correct background in AA. *Science*, 220, 183, 1983.

Miller-Ihli, N. J. Sample preparation for multi-element AAS. *Anal. Atomic Spectrosc.*, 3, 73, 1988.

Parsons, M.L. Atomic absorption and flame emission. In Ewing, G.A. (ed.), *Analytical Instrumentation Handbook*, 2nd edn. Marcel Dekker, Inc.: New York, 1997.

Rann, C.S.; Hambly, A.N. *Anal. Chem.*, 37, 879, 1965.

Robinson, J.W. (ed.) *Handbook of Spectroscopy*, Vol. 1. CRC Press: Boca Raton, FL, 1975.

Robinson, J.W. (ed.) *Atomic Spectroscopy.* Marcel Dekker, Inc.: New York, 1990.

Schlemmer, G.; Radziuk, B. *A Laboratory Guide to Graphite Furnace Atomic Absorption Spectroscopy.* Birkhauser Verlag: Basel, Switzerland, 1999.

Skelly Frame, E.M.; Keliher, P.N. Atomic spectrometry. In Cahn, R.W., Haasen, P., and Kramer, E.J. (eds.), *Materials Science and Technology*, Vol. 2A. VCH Publishers, Inc.: New York, 1992.

Slavin, W. *Graphite Furnace Source Book*. PerkinElmer, Inc.: Shelton, CT, 1984.

Smith, S.B. Jr.; Hieftje, G.M. A new background correction method for atomic absorption spectroscopy. *Appl. Spectrosc.*, *37*, 419, 1983.

Sneddon, J. (ed.) *Sample Introduction in Atomic Spectroscopy*. Elsevier: Amsterdam, the Netherlands, 1990.

Sneddon, J. (ed.) *Advances in Atomic Spectroscopy*, Vol. 1. JAI Press: Greenwich, CT, 1992 (1994; Vol. II).

Tyson, J.F. Flow injection techniques for atomic spectrometry. In Sneddon, J. (ed.), *Advances in Atomic Spectroscopy*, Vol. I. JAI Press: Greenwich, CT, 1992.

Walsh, A. *Spectrochim. Acta,* *7*, 108, 1955.

Welz, B.; Becker-Ross, H.; Florek, S.; Heitmann, U. *High Resolution Continuum Source AAS*. Wiley-VCH: Weinheim, Germany, 2005.

Welz, B.; Heitmann, U. 50 Years after Alan Walsh-AAS redefined. Technical Note, Analytik Jena, Jena, Germany, 2005 (www.analytik-jena.de).

Welz, B.; Sperling, M. *Atomic Absorption Spectrometry*, 3rd edn. John Wiley & Sons: New York, 2002.

Welz, B., Vale, M.R., in Grinberg, N., Rodriguez, S., eds., Ewings Analytical Instrumentation Handbook, 4th Ed., CRC Press Taylor and Francis, Boca Raton, FL, 2018.

# CHAPTER 8

Atomic Emission Spectroscopy

Atomic emission spectroscopy is the study of the radiation emitted by excited atoms and monatomic ions. Excited atoms and ions relax to the ground state, as discussed in Chapter 3 and shown schematically in Figure 8.1. Relaxation often results in the emission of light, producing line spectra in the visible (VIS) and ultraviolet (UV) regions of the spectrum, including the vacuum UV region. These emitted atom and ion lines can be used for the qualitative identification of elements present in a sample and for the quantitative analysis of such elements at concentrations ranging from low ppb to percent. Atomic emission spectroscopy has relied in the past on flames and electrical discharges as excitation sources, but these sources have been overtaken by plasma sources, such as the inductively coupled plasma (ICP) source and the microwave plasma (MP) source. Atomic emission spectroscopy is a multielement technique with the ability to determine metals, metalloids, and some nonmetal elements simultaneously. The major difference between the various types of atomic emission spectroscopy techniques lies in the source of excitation and the amount of energy imparted to the atoms or ions (i.e., the excitation efficiency of the source). For many years, the abbreviation AES was used for atomic emission spectroscopy; however, the same abbreviation is used for Auger electron spectrometry, discussed in Chapter 16. To minimize confusion, the term *optical emission spectroscopy* (OES) is now usually used for atomic emission spectroscopy.

Flame sources impart relatively low quantities of energy to the atoms produced in the flame. Electrons are promoted to only a few low-energy excited states in flames; this results in simple emission spectra. In practice, flame emission spectroscopy is most useful for the easily excited alkali metals and alkaline earth metals. Electrical discharges such as arcs and sparks provide significantly more energy than flames. Most elements can be atomized and ionized in these discharges; the higher energy input causes electrons to populate many higher energy levels. This results in spectra with many emission lines. Plasma sources, such as the ICP, MP, and the direct current plasma (DCP), are high-energy sources that permit the excitation of most elements, both metals and nonmetals, and result in very line-rich spectra. This is one of the major advantages of emission spectroscopy; many elements, as atoms and ions, emit multiple wavelengths simultaneously. The analyst has a choice of several wavelengths for each element and the ability to measure multiple elements concurrently. The disadvantage is that as the number of emission wavelengths increases, so does the *spectral interference* from overlapping lines. This mandates the use of high-resolution spectrometers, which are more expensive than the spectrometers needed for flame emission systems or for atomic absorption (Chapter 7). Atomic emission spectroscopy provides elemental analysis and can measure most metals, metalloids, and nonmetals in a wide variety of liquid, solid, and gaseous samples.

The related subject of atomic fluorescence spectroscopy (AFS), the emission of photons by excited gas-phase atoms following excitation by absorption of photons, is also covered in this chapter.

Some safety considerations should be mentioned concerning all flames and burners. The potential for exposure to heavy metal contamination due to the vaporized compounds is real, and therefore it is best to have a snorkel in place above the burner or furnace. This snorkel should be connected to the laboratory fume hood system or connected to a dedicated exhaust blower. Details on these types of safety devices can be found in Chapter 2 of this book and more details can be found in the book by Bruno and Svoronos.

8.1 FLAME ATOMIC EMISSION SPECTROSCOPY

Flame atomic emission spectroscopy, also called flame photometry, is based on the measurement of the emission spectrum produced when a solution containing metals or some nonmetals such as halides, sulfur, or phosphorus is introduced into a flame. In early experiments, the detector used was the analyst's eye. Those elements that emitted visible light could be identified qualitatively, and these "flame tests" were used to confirm the presence of certain elements in the sample, particularly alkali metals and alkaline-earth metals. A list of visible colors emitted by elements in a flame is given in Table 8.1.

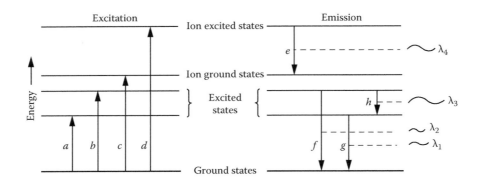

Figure 8.1 Schematic diagram of the excitation and emission process. The energies *a* and *b* represent atomic excitation, *c* represents ionization, and *d* represents ionization and excitation. Four possible emission energies and their respective wavelengths are shown; *e* is ionic emission and *f*, *g*, and *h* are atomic emission. The emission wavelength and energy are related by $\Delta E = hc/\lambda$. [© 1993–2014 PerkinElmer, Inc. All rights reserved. Printed with permission. (www.perkinelmer.com).]

Table 8.1 Flame Colors from Atomic Emission

Flame Color	Flame Color through Blue Glass[a]	Element
Carmine red	Purple	Lithium
Dull red	Olive green	Calcium
Crimson	Purple	Strontium
Golden yellow	(Absorbed)	Sodium
Greenish yellow	Bluish green	Barium, molybdenum
Green		Copper, P as phosphate, B as borate
Blue		Copper
Violet	Violet red	Potassium

[a] The blue glass is used to absorb the intense yellow emission from sodium, which is a common constituent of most samples.

The human eye is a useful detector for qualitative analysis but not for quantitative analysis. Replacing the human eye with a spectrometer and photon detector such as a photomultiplier tube (PMT) or charge-coupled device (CCD) permits more accurate identification of the elements present because the exact wavelengths emitted by the sample can be determined. In addition, the use of a photon detector permits quantitative analysis of the sample. The wavelength of the radiation indicates what element is present, and the radiation intensity indicates how much of the element is present. Flame AES is particularly useful for the determination of the elements in the first two groups of the periodic table, including sodium, potassium, lithium, calcium, magnesium, strontium, and barium. The determination of these elements is often called for in medicine, agriculture, and animal science. Remember that the term *spectrometry* is used for quantitative analysis by the measurement of radiation intensity.

8.1.1 Instrumentation for Flame OES

Flame OES can be performed using most modern atomic absorption spectrometers (discussed in Chapter 7). No external lamp is needed since the flame serves as both the atomization source and the excitation source. A schematic diagram of a flame emission spectrometer based on a single-beam atomic absorption spectrometer is shown in Figure 8.2. For measurement of the alkali metals in clinical samples such as serum or urine, only a low-resolution filter photometer is needed because of the simplicity of the spectra. The filter photometer is discussed in Section 8.1.1.2. Both instruments require a burner assembly, a flame, a wavelength selection device, and a detector.

8.1.1.1 Burner Assembly

The central component of a flame emission spectrometer is the burner assembly. The assembly has a device to nebulize the sample and then introduce the sample aerosol into the flame. In the flame, free atoms are formed and then excited, which causes them to emit radiant energy. For analytical purposes, it is essential that the emission intensity be steady over reasonable periods of time (1–2 min). The Lundegardh or premix burner is the most commonly used and is depicted in Figure 7.8a.

In the premix burner, the sample, in solution form, is first aspirated into a nebulizer where it forms an aerosol or spray. An impact bead or flow spoiler is used to break the droplets from the nebulizer into even smaller droplets. Larger droplets coalesce on the sides of the spray chamber and drain away. Smaller droplets and vapor are swept into the base of the flame in the form of a cloud. An important feature of this burner is that only a small portion (about 5 %) of the aspirated sample reaches the flame. The droplets that reach the flame are, however, very small and easily decomposed. This results

ATOMIC EMISSION SPECTROSCOPY

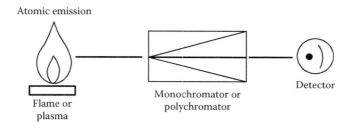

Figure 8.2 Schematic atomic emission spectrometer. [© 1993–2014 PerkinElmer, Inc. All rights reserved. Printed with permission. (www.perkinelmer.com).]

Table 8.2 Process Generating Atomic Emission in Flames

	Liquid sample enters nebulizer	→	Small droplets of liquid enter flame	→	Evaporation of droplets
Oxidation of atoms	← Formation of excited atoms and emission of radiation from atoms as relaxation occurs	←	Decomposition of particles to free atoms	←	Formation of fine solid particles

in an efficient atomization of the sample in the flame. The high atomization efficiency leads to increased emission intensity and increased analytical sensitivity compared with other burner designs. The process that occurs in the burner assembly and flame is outlined in Table 8.2. This process is identical to the atomization process for atomic absorption spectrometry (AAS), but now, we want the atoms to progress beyond ground-state free atoms to the excited state.

The common nebulizers used in flame emission spectroscopy are the pneumatic nebulizer and the cross-flow nebulizer, just as in AAS.

8.1.1.2 Wavelength Selection Devices

Two wavelength selectors are used in flame OES, monochromators, and filters.

Monochromators. These consist of slits and dispersion elements and are described in Chapters 3 and 7. The common dispersion element in modern flame atomic absorption and emission spectrometers is a diffraction grating.

Filters. The alkali metals in a low-temperature flame emit only a few lines and therefore have a simple emission spectrum. In this case, wider wavelength ranges may be allowed to fall on the detector without causing errors, so an optical filter may replace the more expensive diffraction grating. Filters are built with materials that are transparent over a narrow spectral range. The transparent spectral range is designed to be the one in which emission from a given element occurs. When a filter is placed between the flame and the detector, radiation of the desired wavelength from the sample is allowed to reach the detector and be measured. Other radiation is absorbed by the filter and is not measured. Therefore, a separate filter is required for each element to be measured. For example, a filter transparent to radiation with a wavelength of 766 nm is used to determine potassium, while a filter transparent to radiation of 589 nm permits the determination of sodium. The potassium filter absorbs sodium emission and the sodium filter absorbs potassium emission, so the two elements can be measured in the same sample without interference from the other element. Instruments that use filters as wavelength selectors are very convenient for simple repetitive analysis but are limited with regard to the number of elements for which they can be used unless a large number of filters are employed. Most flame photometers are single-channel instruments; one element at a time is determined, then the filter is changed and a new element can be determined. Multichannel instruments have been designed where the emission from the flame falls on two or more filters. Each filter transmits the radiation for which it has been designed, and the transmitted radiation falls on a PMT behind the filter. The intensity of the radiation is measured for calibration standards, blanks, and samples. A calibration curve of concentration versus emission intensity is made for the standards and the concentration of the sample is determined by comparison to the calibration curve. This multichannel approach permits the use of an internal standard (IS) to improve precision. For example, a multichannel instrument with filters for lithium and sodium may be used for the determination of lithium in serum. Lithium compounds are used in drugs to treat depression. Patients taking such a drug need to be monitored to be certain that the prescribed dose is effective, so the Li concentration in serum samples from patients is measured using flame photometry. Alternatively, the Li channel could be used as an IS channel; lithium is added to all samples and standards to be analyzed for Na. The lithium IS corrects for instrument drift, changes in sample uptake due to viscosity differences, and other interferences, as described in Chapter 3. Filter photometers designed for use in hospital or veterinary clinical laboratories often have autosamplers and autodilutors attached, permitting the unattended analysis of many samples per hour.

8.1.1.3 Detectors

The detectors in common use for these systems are the PMT or solid-state detectors such as CCDs and charge injection devices (CIDs). PMTs and photodiode arrays (PDAs) are discussed in Chapter 4. More detailed discussion of solid-state detectors is covered in Section 8.2.

8.1.1.4 Flame Excitation Source

A flame is the result of the exothermic chemical reaction between two gases, one of which serves as the fuel and the other as the oxidant. The reaction is an oxidation–reduction reaction, with the oxidant oxidizing the fuel. In the process, the reaction generates a great deal of heat. Common oxidants for modern AAS/OES systems are air and nitrous oxide. The only fuel used in modern AAS/OES systems is acetylene, although commercial filter photometer systems can use propane, natural gas, or butane as fuel. When a liquid sample is introduced into a flame, a complex process to produce excited-state atoms occurs.

Spectral emission lines are generated by the excited atoms formed during the process of combustion in a flame. Emission lines are characterized by wavelength and intensity. The wavelengths emitted depend on the atoms present. Each element has a different set of quantized energy levels (such as those shown schematically in Figure 8.1) and will emit different, characteristic wavelengths of light. The intensity of the emission depends on several factors including (1) the concentrations of the elements in the sample and (2) the rate at which excited atoms are formed in the flame. The latter depends on (3) the rate at which the sample is introduced into the flame, (4) the composition of the flame, and (5) the temperature of the flame. The intensity–concentration relationship is the basis for quantitative analysis by flame OES.

Flame temperature is probably the most important single variable in flame photometry. The type of fuel and oxidant used controls the temperature. In general, an increase in flame temperature causes an increase in emission intensity for most elements. This does not happen with elements that ionize easily, such as sodium, potassium, and lithium. If these elements are heated at too high a temperature, they become ionized; the outer electrons move to higher and higher energy states until they leave the atom completely, forming an ion. If the atoms ionize, the valence electrons are lost and therefore cannot return to the ground state and emit atomic radiation in the process. A loss of atomic emission intensity results. These elements must be determined in low-temperature flames to minimize ionization.

The ratio of the number of atoms in an upper excited state to the number of atoms in a lower energy state can be calculated from the Maxwell–Boltzmann equation (also called the Boltzmann distribution):

$$\frac{N_1}{N_0} = \frac{g_1}{g_0} e^{-\frac{\Delta E}{kT}} \quad (8.1)$$

where
N_1 is the number of atoms in the upper state
N_0 is the number of atoms in the lower state
g_1 and g_0 are the number of states having equal energy at each level 0, 1, etc. (g is called the *degeneracy* of the level)
ΔE is the energy difference between the upper and lower states (J)
k is the Boltzmann constant = 1.380649×10^{-23} J/K, exact
T is the absolute temperature (K)

The Boltzmann distribution assumes the system is in thermal equilibrium. The emission intensity is related to the number of atoms in the higher excited state, N_1, since we are looking at emission as the atom relaxes from a higher state to a lower state.

Using the Boltzmann equation, we can calculate the ratio of the number of excited-state atoms at two different temperatures. For potassium atoms, the major atomic emission line occurs at 766.5 nm. The energy of this transition in joules is

$$\begin{aligned}\Delta E &= \frac{hc}{\lambda} \\ &= (6.626 \times 10^{-34} \text{ J s})(2.998 \times 10^8 \text{ m/s}) / \\ &\quad (766.5 \text{ mm})(1 \times 10^{-9} \text{ m/nm}) \\ &= 2.59 \times 10^{-19} \text{ J}\end{aligned}$$

The temperature in a typical air–acetylene flame is about 2200 °C or 2473 K. The ratio of excited-state potassium atoms at 2498 versus 2473 K is calculated by dividing the Boltzmann equation at 2498 K by that at 2473 K. N_0 and the degeneracy terms cancel and we are left with

$$\begin{aligned}\frac{N_{2498}}{N_{2473}} &= \frac{\exp[-2.59 \times 10^{-19}(\text{J})/1.38 \times 10^{-23}(\text{J/K})(2498 \text{ K})]}{\exp[-2.59 \times 10^{-19}(\text{J})/1.38 \times 10^{-23}(\text{J/K})(2473 \text{ K})]} \\ &= \frac{5.458 \times 10^{-4}}{5.059 \times 10^{-4}} = 1.08\end{aligned}$$

This tells us that a 25 K increase in flame temperature results in an 8 % increase in the excited-state population of potassium atoms that give rise to this emission line. The intensity of the emission line is directly proportional to the excited-state population, even for systems not in thermal equilibrium. The relationship between emission intensity, S, and excited-state population can be expressed as

$$S = kN \qquad (7.2)$$

where

S is the intensity

k is a proportionality constant that includes a number of factors such as the transition probability and energy of the emitted photon

N is the excited-state atom population

S is related directly to the number of atoms in the excited state. As this number increases, the intensity of radiation increases. As the absolute temperature increases, the number of atoms in the excited state increases; therefore, the emission intensity increases. Even a small change in temperature results in a significant change in excited-state population, as just shown. OES is very sensitive to changes in temperature. The temperature in the atomizer must be carefully controlled for quantitative measurement of emission intensity. This is not the case in AAS, where transitions from the ground state are measured. The ground-state population is relatively unaffected by small changes in temperature. Table 8.3 lists representative maximum temperatures for some common flames.

As the energy required to cause excitation increases, it is more difficult to excite the atoms, and the number of atoms in the excited state decreases. As a consequence of the relationship $\Delta E = h\nu = hc/\lambda$, a decrease in the wavelength of the emission line indicates that more energy is required to excite the atom. The process becomes more difficult, fewer atoms are excited, and the intensity of radiation decreases. Consequently, elements with emission lines in the short-wavelength part of the spectrum give weak emission signals in low-temperature flames. For these elements, the high-temperature nitrous oxide–acetylene flame is favored, or the high-energy electrical or plasma excitation sources discussed later in the chapter should be used.

Another factor that influences emission intensity is the ratio of fuel to oxidant in the flame. The highest flame temperature is obtained when a stoichiometric mixture of the two is used. In a stoichiometric flame, the number of moles of fuel and oxidant present react completely and there is no excess of either after combustion. Any excess of oxidant or fuel results in a decrease in the temperature of the flame. However, some atoms are unstable in certain kinds of flames. Aluminum atoms oxidize very quickly to aluminum oxide in a stoichiometric flame or a flame with an excess of oxidant. Aluminum oxide emits molecular radiation that is not at the same wavelength as the line emission associated with aluminum atoms. This results in a direct loss of atomic emission intensity. To prevent the formation of aluminum oxide, the flame is usually run in a "reducing state," that is, with an excess of fuel. The excess fuel "mops up" free oxidant and minimizes oxidation of the aluminum in the flame. Some elements emit more strongly in oxidizing flames. Consequently, an oxidizing flame is recommended for these elements. In addition, the excess oxygen helps decompose other materials present in the sample and reduces the molecular background. Manufacturers of flame emission instruments provide a list of recommended flame compositions for elements measured by OES. Table 8.A.1 in Appendix 8.A provides flame OES detection limits (DLs) for many elements and identifies the flame used.

8.1.2 Interferences

The radiation intensity measured may not represent the concentration of analyte in the sample accurately because of the presence of interferences. Interferences fall into two categories: spectral and nonspectral. Three principal sources of interference are encountered in flame OES; they are the same interferences that occur in flame AAS (FAAS).

8.1.2.1 Chemical Interference

If the analyte element is present in the sample with anions with which it combines strongly, it will not decompose to free atoms easily. If the anions present in solution combine only weakly with the analyte element, decomposition and formation of free atoms will be easier. For example, if calcium is to be determined, a given concentration of calcium in the presence of phosphate ion will give a lower signal than the same amount of calcium in a chloride solution. It is easier to decompose calcium chloride into free calcium atoms than it is to decompose calcium phosphate in the same flame. This effect is called *chemical interference*. Chemical interference is a nonspectral interference. It can be reduced or eliminated in a number of ways. The analyte ion may be extracted away from the sample matrix by using a chelating agent or the interfering anion may be removed by ion exchange (discussed in Chapter 14). The standards can be made from the same salt (phosphate, chloride, etc.) present in the sample, compensating for the effect of the interference. A *releasing agent* may be added to the sample

Table 8.3 Maximum Flame Temperatures (°C)

Fuel	Oxidant		
	Air	O_2	N_2O
H_2	2100	2900	
Acetylene	2200	3100	3200
Propane	1900	2800	

solution. Releasing agents reduce or minimize chemical interferences, often by forming a more stable salt with the anion than is formed by the analyte metal. For example, adding a large excess of lanthanum releases calcium from the effect of phosphate ion by preferentially forming lanthanum phosphate.

8.1.2.2 Excitation and Ionization Interferences

Excitation and ionization interferences are nonspectral interferences. When a sample is aspirated into a flame, the elements in the sample may form neutral atoms, excited atoms, and ions. These species exist in a state of dynamic equilibrium that gives rise to a steady emission signal. If the samples contain different amounts of elements, the position of equilibrium may be shifted for each sample. This may affect the intensity of atomic emission. For example, if sodium is being determined in a sample that contains a large amount of potassium, the potassium atoms may collide with unexcited sodium atoms in the flame, transferring energy in the collision and exciting the sodium atoms in the process. This results in an increased number of excited sodium atoms and increased atomic emission signal compared to a solution of sodium atoms at the same concentration with no potassium atoms in the solution. This is *excitation interference* and is generally restricted to the alkali metals. It can be overcome by matrix matching the samples and standards; the standards must be made up so that the standard solutions contain concentrations of the nonanalyte elements similar to those present in the sample. Matrix matching is often done in routine industrial analysis where the sample composition is known. It cannot be done for samples of unknown or highly variable composition. In cases where matrix matching is not practical, the methods of standard additions (MSA) can be used. An example of the MSA is given in Section 8.1.3.

A related problem is *ionization interference*. If the analyte atoms are ionized in the flame, they cannot emit atomic emission wavelengths, and a reduction in atomic emission intensity will occur. For example, if potassium is ionized in the flame, it cannot emit at its atomic emission line at 766.5 nm and the sensitivity of the analysis will decrease. If a large amount of a more easily ionized element (EIE), such as cesium, is added to the solution, the cesium will ionize preferentially and suppress the ionization of potassium. The potassium ions will capture the electrons released by the cesium, reverting to neutral potassium atoms. The intensity of emission at 766.5 nm will increase for a given amount of potassium in the presence of an excess of cesium. The added cesium is called an *ionization suppressant*. Ionization interference is a problem with the EIEs of groups 1 and 2. The use of ionization suppressants is recommended for the best sensitivity and accuracy when determining these elements. Of course, as ionization increases, *ion emission* line intensity increases. It may be possible to use an ionic emission line instead of an atomic emission line for measurements.

8.1.2.3 Spectral Interferences

There are two types of spectral interferences in flame OES: (1) background radiation and (2) overlapping line emission from different elements. Excited molecules and radicals in the flame emit over broad regions of the spectrum; it is this broad *band emission* that we call background radiation. The molecules and radicals may be combustion products from the flame gases, oxides, and hydroxides formed from elements in the sample and similar species. For example, species such as BaOH, CaOH, SrOH, MnOH, CaO, CN, and the rare earth oxides all emit band spectra when introduced into a flame. These band spectra can in fact be used to identify these species in a gas-phase form of molecular UV/VIS emission spectroscopy but interfere when atomic emission is to be measured. An example of background emission from an air–acetylene flame is illustrated in Figure 8.3. Atomic emission lines of magnesium, nickel, sodium, and potassium are superimposed on the broad molecular background. In order to accurately determine the intensity emitted by the atom of interest, the intensity due to the background emission must be measured and subtracted from the total intensity. A blank solution (no analyte present) is aspirated and the intensity at the analyte wavelength is measured. This intensity is due to the background emission, since no analyte is present. Then the sample and standard solutions are aspirated and their intensity measured. The sample or standard intensity is due to the sum of the background emission plus the analyte emission. Subtraction of the blank intensity from the sample or standard intensity results in the intensity due to analyte. A sample calculation is given in Section 8.1.3.

All samples and standards must be corrected for background emission. A second method for measuring background intensity is to measure the background intensity of the actual sample at a wavelength very close to the analyte emission line, either slightly higher in wavelength or slightly lower in wavelength than the emission line. This intensity is then subtracted from the intensity of the sample measured at the analyte emission wavelength. This second approach can be less accurate than the first method if the background spectrum is not "flat" in the region of interest. For example, look at Figure 8.3 in the region of the Mo emission line. The background slopes sharply in the region of the Mo line; it is much higher on the high-wavelength side than on the low-wavelength side. The use of an "off-line" background correction for Mo could result in a high degree of error because of the rapid change in background intensity with wavelength in this region. On the other hand, the background region around the potassium emission

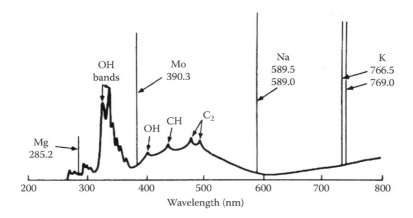

Figure 8.3 Emission spectrum of a flame. The broad background emission bands from polyatomic species such as OH and CH are seen. Superimposed on the broad background emission are the very narrow atomic emission lines from the elements Mg, Mo, Na, and K.

line is relatively flat; the method of "off-line" background correction would result in accurate background correction for potassium.

The other type of spectral interference is the emission by another element of the same wavelength as the analyte element. This is a direct source of error. The analyst may choose another analyte wavelength, extract the interfering element, or apply a correction factor if the concentration of the interfering element is known. This is not much of a problem in flame OES, due to the low temperature of flames and the relatively few excited states that are populated, but it is a major source of interference in high-temperature excitation sources such as electric discharges and plasmas. By "same wavelength" we mean that the instrument we are using cannot resolve the interfering line from the analyte line, even if the actual wavelengths are slightly different. The spectral bandpass is such that both wavelengths pass through the exit slit of the system to the detector. For this reason, high-resolution spectrometers are needed for nonflame OES.

8.1.3 Analytical Applications of Flame OES

8.1.3.1 Qualitative Analysis

Unlike AAS, atomic emission spectroscopy is an excellent qualitative method for determining multiple elements in samples. Flames are not the optimum emission source for most elements because of the limited temperatures available but they can be used. The presence of elements in a sample is determined qualitatively by observing emission at the wavelength characteristic of the element. Table 8.A.1 in Appendix 8.A lists common characteristic emission lines for flame OES, and a more comprehensive list can be found in the *Handbook of Spectroscopy* (Robinson) cited in the bibliography. Emission at 766.5 nm indicates that potassium is present in a sample, while the bright yellow emission of sodium at 598 nm indicates its presence. While some elements can be detected visually as we noted in Table 8.1, it is much more accurate to use a spectrometer to identify the wavelengths emitted. The use of a flame source is not as reliable as the use of higher energy excitation sources where multiple wavelengths can be observed for each element, thereby providing more proof of the presence of a given element. For this reason, plasma sources have become the preferred atomic emission source for most elements. Flame OES is a fast, simple method for the qualitative identification of the group 1 and 2 elements and can be used for any element that emits radiation in a flame, provided care is taken to discriminate the emission lines from any spectral interference.

8.1.3.2 Quantitative Analysis

Flame OES can be used to determine the concentrations of elements in samples. The sample usually must be in solution form. Generally, one element is determined at a time if using an AAS system in emission mode. Multichannel instruments are available for the simultaneous determination of two or more elements. DLs can be very low as seen in Appendix 8.A, Table 8.A.1. DLs for the alkali metals are in the ppt concentration range when ionization suppression is used. One ppt in an aqueous solution is 1 pg of analyte per mL of solution or 1×10^{-12} g/mL. Most elements have DLs in the high ppb to low ppm range.

An important factor in quantitative analysis by flame OES is the solvent used for the samples and standards. When water is the solvent, the process of atomization is endothermic and relatively slow. If the solvent is organic, the reactions in the flame are often exothermic and atomization is rapid. Other things being equal, more free atoms are liberated and emission intensity increased in organic solvent than in aqueous solution. For accuracy, it is therefore critical that the samples and standards be prepared in the same solvent.

For a quantitative measurement, the sample is aspirated into the flame and the intensity of radiation is measured at the emission wavelength of the analyte element to be determined. The concentration of the element in the sample is calculated from comparison with an external calibration curve or by the standard addition method. Both of these methods have been discussed in Chapter 3. A typical external calibration curve for lithium is shown in Figure 8.4. The calibration curve was constructed by aspirating Li standard solutions of 5 and 10 ppm Li and the blank solution (0.0 ppm Li). Each intensity was measured and a plot of intensity versus concentration made. Modern systems use a statistical curve-fitting program to construct the calibration curve and to calculate the concentrations of unknowns. Note that an emission signal was detected in the standard that contained no lithium (the zero concentration standard or calibration blank). This is called the *background signal* or *blank emission signal*. The source of the background signal was discussed in Section 8.1.2.2. In this case, the background emission was not subtracted from each standard but was plotted as the 0.0 ppm Li standard. To use this curve, an unknown sample is aspirated, its emission intensity is measured and the concentration read (or calculated by the computer) from the calibration curve. The assumption in this case is that the samples and standards all have the same background emission as the blank. If the blank is subtracted from all the standards and the calibration curve then plotted, the line will go through the origin. To use that type of calibration curve, the blank signal must be subtracted from the unknown sample signals. In one way or another, the background emission signal must be corrected for in the final calculation of analyte concentration in the sample. The relationship between emission intensity and lithium concentration is linear in this example. The calibration curve follows Beer's law at low concentrations of analyte. Deviations from linearity at high concentrations are commonly observed in flame emission calibration curves.

Linear regression equation: $y = mx + b = 200x + 450$.

As an example, assume that we have the calibration curve in Figure 8.4. The data points for this curve and the equation of the calibration curve determined by linear regression are given in Table 8.4.

If we aspirate an unknown sample solution and measure an intensity of 1626 counts, we can use the equation to calculate the Li concentration in the unknown. Inserting 1626 as y in the equation and solving for x, the concentration x is calculated to be 5.88 ppm Li in the unknown solution.

If we subtracted the blank first, the calculation would be as shown in Table 8.5. Now in order to use the blank-corrected calibration curve, we have to subtract the background intensity from our sample intensity. The corrected intensity is shown in Table 8.5. Inserting $y = 1176$ into the equation shown and solving for x gives us 5.88 ppm Li. As expected, the results are the same since the background was accounted for in both approaches (and assumed to be the same in all samples and standards).

Suppose we need to determine Li in human serum to monitor a lithium-based antidepressant medication. We would not expect the background from a serum sample to be the same as the background from an aqueous standard

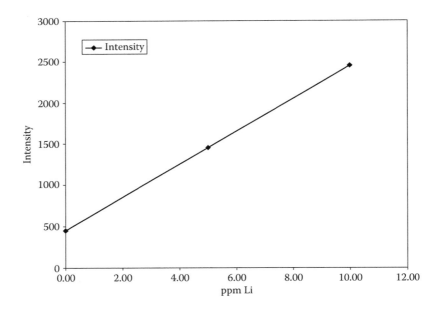

Figure 8.4 Calibration curve for lithium by flame emission using external calibration standards. Note the nonzero intercept for the blank due to background emission from the flame.

Table 8.4 Calibration Curve for Li by Flame OES

Li Concentration (ppm)	Intensity
0.00	450
5.00	1450
10.00	2450

Table 8.5 Blank-Corrected Calibration Curve for Li by Flame OES

Li Concentration (ppm)	Intensity	Blank-Corrected Intensity
0.00	450	0
5.00	1450	1000
10.00	2450	2000
Unknown	1626	1176

Linear regression equation: $y = mx + b = 200x + 0 = 200x$.

Table 8.6 Data for Li in Serum by MSA

Standard Addition, ppm Li Added	Intensity	Blank-Corrected Intensity
0	1432	1247
5.0	2117	1932
10.0	2802	2617
Blank	185	0

Linear regression equation: $y = mx + b = 137x + 1247$.
Correlation coefficient = 1.00.

solution. We might make our calibration standards in lithium-free serum, but if we only have one sample to run, that is not efficient in terms of cost or time. This is a good example of when to use the MSA. In MSA, the calibration curve is made up in the sample itself.

The sample is split into three aliquots of equal volume. Two aliquots have known amounts of Li added, while the third is just the sample aliquot itself or the "zero addition" sample. As we discussed in Chapter 3, the Li is added in such a way that the total volume of the three aliquots does not change significantly. The intensity of emission for all three solutions is measured, as presented in Table 8.6 and the intensity plotted against added Li, as shown in Figure 8.5. The concentration of Li in the original sample can be calculated using the calibration curve and the background-corrected sample intensity. The equation for the line obtained is included in Table 8.6. The sample concentration is the absolute value of y at $x = 0.0$ (no added Li). The calculated value for x at $y = 0.0$ is -9.1, so the sample Li concentration is 9.1 ppm Li.

The sample concentration can also be read from the intercept of the line with the negative x-axis from a graphical plot of the calibration curve. In Figure 8.5, the sample can be seen to contain 9.1 ppm Li, because the intercept occurs at -9.1 ppm. It is critical to measure the background emission from the flame and to subtract the background emission signal from all the aliquot signals before plotting the standard addition curve. If this is not done, a positive error in the reported concentration will result.

In order to obtain a steady emission signal of constant intensity, each step in the process shown in Table 8.2 must be controlled. Modern equipment is able to do this with little attention from the analyst. The skillful analyst, however, is always aware of the chemical and physical processes involved and is thus able to recognize and correct any problems that may arise. Variations in emission intensity can be caused by several factors, including (1) blockage in the nebulizer or burner, preventing the flow of sample; (2) viscosity differences in samples or between samples and standards; (3) change of solvent in the sample; (4) change in the fuel or oxidant flow rate to the burner; and (5) change of the position of the burner in the instrument, causing the radiation to be displaced from the light path. The operating conditions must be optimized for the analysis, including the selection of the flame to use, the wavelength to monitor, the fuel/oxidant ratio, the sample uptake rate, the slit widths, and the position of the burner in the light path.

The major use of flame emission spectroscopy is in the determination of Na, K, Li, Ca, Mg, and Ba in biological tissues and fluids by medical laboratories. The same elements are often measured in plant tissue, animal feed and forage, and food. While flame OES is a simple, rapid method for the determination of the group 1 and 2 elements, it has been replaced for the determination of most other elements in liquid samples by ICP-OES. A typical commercial flame photometer for clinical determination of Na, K, Li, and other elements is capable of DLs as shown in Table 8.7. Buck Scientific, Inc. (www.bucksci.com), makes a specialized flame photometer, their Model 300, for the measurement of ions in power plant feedwater, with DLs of 5–15 ppt for Na, K, and Li, with a linear working range from 0.1 to 20 ppb. The high sensitivity comes from an optimized low-temperature hydrogen flame. For quantitative analysis, the precision, as expressed by the percent relative standard deviation (RSD), is about 1 %. There are many flame emission applications published in the literature before plasma emission systems became dominant in the early 1980s.

8.2 ATOMIC OES

Radiation from excited atoms and ions is emitted by a sample when it is introduced into an electrical discharge, a glow discharge (GD), or a plasma. Because these excitation sources are of higher energy than a flame source, all metallic and metalloid elements can be detected in low concentration,

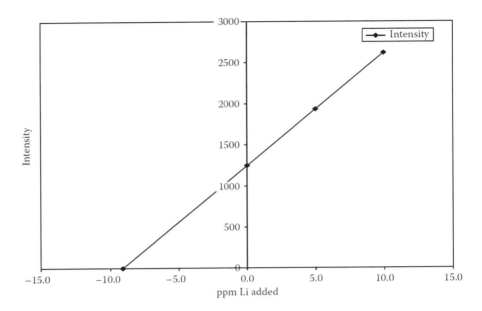

Figure 8.5 Determination of lithium in a sample by the MSA. The background emission from a blank reading was subtracted from all sample aliquots before plotting the data.

Table 8.7 DLs for Flame Photometry in Commercial Instruments

Element	LOD (ppm) Model PFP-7	Range[a] (mmol/L) or LOD[b] (ppm) Model PFP-7C	LOD (ppt) Model 300
Na	≤0.2	120–190[a]	8
K	≤0.2	0–10.0[a]	15
Li	≤0.25	≤0.25[b]	5
Ca	≤15	≤15[b]	
Ba	≤30	≤30[b]	

Source: Data courtesy of Buck Scientific, Inc., East Norwalk, CT, www.bucksci.com.

Note: LOD, limit of detection; ppt, parts per trillion; or ng/L. The Model PFP-7C is designed specifically for clinical chemistry and displays the Na and K concentrations in mmol/L. The Model 300 is designed for high-sensitivity analysis of power plant feedwater.

[a] Indicates range.
[b] Indicates LOD.

including refractory elements such as boron, tungsten, tantalum, and niobium. Nonmetals can be measured, including C, N, H, Cl, Br, and I. Liquids and solids can be analyzed quite easily using a plasma source; electrical and GD sources are used for the analysis of solids only. Pure gas-phase elements, such as hydrogen, helium, neon, and mercury vapor, can be identified by the emission spectra generated by an electrical discharge through the gas confined in a sealed quartz tube. This same process generates the light in "neon" signs and mercury vapor street lights.

Because the temperatures of electrical discharges and plasmas are much higher than temperatures that can be achieved in flames, the emission spectra from these excitation sources are very complex. Two types of line spectra are generated from these systems. First are the *atomic emission spectra* from neutral atoms. These lines are designated in tables of emission lines by a Roman numeral I following the wavelength. Under spark and plasma conditions, ions are often formed. A second electron in the ion may then become excited and enter one of the higher energy states. From these states, the ion relaxes and emits a photon in the process. The energy levels of the ions are not the same as the energy levels of the atoms. They therefore emit different emission lines, called *ion lines*. Lines from singly charged ions are designated by a Roman numeral II and lines from doubly charged ions by a Roman numeral III following the wavelength. Several reference books that list tables of emission lines are listed in the bibliography.

Ion lines are less sensitive than atomic lines but are not subject to reversal due to self-absorption, discussed in Section 8.2.3.2. They are used for quantitative analyses when a sufficient concentration of the analyte is available. Ion lines are seldom used for qualitative analysis because of their lack of sensitivity.

8.2.1 Instrumentation for Emission Spectroscopy

In simple terms, an emission spectrometer with an electrical discharge source (Figure 8.6) functions as follows. An electrical source produces an electrical discharge in a spatial "gap" between two electrodes, the *sample electrode* and the *counterelectrode*. A solid machined piece of the metal to be analyzed serves as the sample electrode; the counterelectrode is an "inert" electrode of tungsten or

ATOMIC EMISSION SPECTROSCOPY

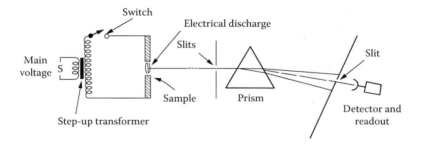

Figure 8.6 Schematic diagram of an emission spectrometer with an electrical discharge excitation source. The prism is meant only to illustrate a dispersive device; a diffraction grating is used in all modern spectrometers with a single dispersive device. Echelle spectrometers use two dispersive devices, either a prism and a grating or two gratings.

graphite. Material from the sample electrode is introduced into the discharge, where it becomes vaporized and excited. The excited atoms emit radiation, which is detected and measured by the detector readout system. The wavelengths of the emitted lines identify the elements present and the intensity of the emission at each wavelength can be used to determine the amount of each element present. The individual components of the instrumentation are described in the following sections.

8.2.1.1 Electrical Excitation Sources

Historically, several types of electrical excitation sources have been used since the early part of the 1900s, among them are the direct current (DC) arc, the alternating current (AC) arc, and the AC spark. Commercial instruments using electrical excitation sources became available about 1940 with PMT detectors; prior to this, emission instruments used a photographic plate or film as the detector. Modern instruments are either DC arc emission instruments or high-voltage spark emission instruments capable of operating under "spark-like" or "arc-like" conditions as necessary. A significant advantage of these electrical excitation sources is that conductive solids can be analyzed directly, avoiding the need for tedious sample dissolution and the degradation of DLs by having to dissolve and therefore dilute the sample. Nonconductive solid powders can often be analyzed by mixing them with conductive graphite, which does dilute the sample to some extent, but not as much as dissolution.

DC arc. In this source, a DC voltage is applied across a pair of graphite or metal electrodes separated by a small gap. The typical voltage for a DC source is 200 V with currents of 3–20 A. The arc must be initiated by the formation of ions, usually by a low-current spark. Ions are formed in the gap, electrical conduction occurs, and the motion of electrons and ions formed by thermal ionization sustains the arc. A *plasma* forms into which the sample is vaporized by electrical and thermal energies. The term *burning* is used to describe the process, although no

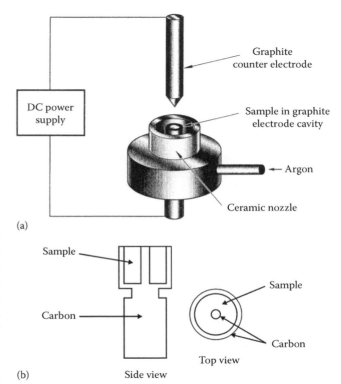

Figure 8.7 (a) Diagram of a direct current arc source. The solid sample is packed into the cupped end of the lower graphite electrode. The graphite counterelectrode is also shown. (From Hareland, W., Atomic emission spectroscopy, in Ewing, G.W. ed., *Analytical Instrumentation Handbook*, 2nd edn., Marcel Dekker, Inc., New York, 1997. With permission.) (b) A schematic of the lower electrode used to hold the sample.

combustion is involved. The sample, in the form of solid powder, chips, or filings, is generally packed into the cupped end of a graphite electrode (Figures 8.7a and b). This electrode would be at the bottom of the gap, as seen in Figure 8.6. The top electrode or *counterelectrode* is a solid graphite electrode. A photograph of the electrode assembly in a commercial DC arc emission spectrometer is shown in Figure 8.8a. If the sample is a solid piece of conductive

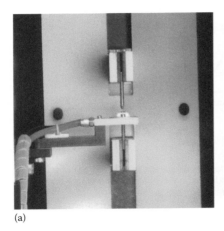

Figure 8.8 (a and b) Photographs of the electrode assembly in commercial DC arc emission spectrometers. (c) Photograph of the plasma formed by the DC arc in operation. The discharge is established in the gap between the electrodes.

metal, it may be machined into the proper shape and serve as the bottom electrode. In DC arc emission, the entire sample packed into the electrode cup is usually vaporized or "burned" over a period of several minutes. The arc is a *continuous* discharge, and the emission intensity is collected and integrated over the burn period. An operating DC arc is shown in Figure 8.8b.

The temperature of the arc depends upon the composition of the plasma and varies with the nature of the sample. If the sample is made of material with low ionization energy, the temperature of the plasma will be low; if the ionization energy of the material is high, the temperature will be high. In addition, the temperature is not uniform in either the axial or radial directions. This results in matrix effects and self-absorption. Arc temperatures are on the order of 4500 K with a range of 3000–8000 K. Emission spectra from arc sources contain primarily atomic lines with few ion lines. The DC arc can excite more than 70 elements.

A significant problem with DC arc sources is the poor stability of the emission signals. Generally, the plasma will link to a point or projection on the surface of the solid sample and will continuously erode the sample, vaporizing it in the process. A pit is formed in the sample by the erosion, and finally, the pit becomes so large that the discharge can no longer be maintained at that point. The discharge then "wanders" to another nearby point on the sample surface. In practice, the wandering of the discharge is quite fast and causes the signals to fluctuate. Numerous such discharge points are created at the same time. Local hot spots are formed, which cause the sample in the immediate vicinity to evaporate rapidly and become excited easily; however, these hot spots are not continuous and soon die away, and the local emission intensity is reduced. As a consequence of this behavior, the total emission intensity is somewhat erratic and the signals produced lack precision. On the other hand, the signals are very intense, and the analytical sensitivity is better than when other types of electrical discharges are used.

Another problem with the DC arc is that volatile elements will selectively vaporize and enter the plasma before the less volatile elements in a sample. The electrode temperature increases slowly from the initiation of the arc. If the sample is the anode, which heats more rapidly and to a higher temperature than the cathode, volatile elements will rapidly enter the arc. For example, tungsten powder for light bulb wire contains added potassium and silicon. In a DC arc analysis, the low-melting potassium would volatize first, followed later (possibly several minutes later) by the high-melting silicon and tungsten. This can cause serious problems with quantitative analysis if standards are used that do not match the volatilization behavior of the sample matrix. The problem may be overcome by mixing the samples and standards with an excess of a "spectrochemical buffer." This is a salt or mixture of salts such as NaF and Na_2CO_3, often mixed with graphite. The "buffer" increases conductivity and provides a uniform matrix that keeps the arc temperature constant. Mixing with pure graphite is also used to give a common matrix. The purity of the graphite or other buffer materials used is very important, given the ppb DLs that can be reached.

There is a positive aspect to the selective volatility of low-melting elements, in that spectral interferences are likely to be less at the beginning of a burn (low temperature) than at high temperatures when line-rich elements such as Fe start to vaporize. In modern DC arc sources, this problem of selective volatility is addressed by having the polarity of the electrodes reversed for the first minute or so of the burn. With the sample as the cathode, the temperature is lower and the rate of heating lower. This slows the rate of evolution of volatile species and improves sensitivity and precision for the volatile elements. Sensitivity for the non-volatile elements is somewhat reduced when the sample is the cathode. A common sequence for analysis would consist of a burn of 60 s with the sample as cathode (reverse polarity) at 8–10 A, a 60 s burn of normal polarity (sample as anode) at 8–10 A, and then 60 s at 15 A to volatilize the very refractory elements.

The DC arc is used primarily for semiquantitative and qualitative analysis of solids, including powdered solids, chips, and filings. Most elements can be determined. The spectrometer operates from about 190–800 nm, which covers the UV and visible range in which most atomic emission occurs. The precision attainable in a modern DC arc spectrometer is 5–10 %, with DLs in the 10–100 ppb range in solid samples. DLs for trace elements in high-purity copper by DC arc emission using a CID detector are given in Table 8.8. These DLs are comparable to the instrumental DLs of graphite furnace AAS (GFAAS) and ICP-OES, both of which generally require dissolution of the solid and dilution to form a solution. The effective DLs in the solid sample therefore would be poorer by GFAAS or ICP-OES than the DC arc DLs. The limitation to DC arc is the requirement that the solid sample be electrically conductive. The linear working range of a DC arc source is about three orders of magnitude, limited by the relatively low energy of the source.

The uranium oxide data in Table 8.8 were collected on a Teledyne Leeman Labs Prodigy DC Arc. The uranium oxide was mixed with a small amount of gallium oxide/cobalt oxide buffer and 125 mg of the buffered sample were analyzed. The preburn time was 4 s, and the elements were determined in a 4–45 s integration time. The details are available in an application note available from Teledyne Leeman Labs, along with other examples of the use of DC arc emission for materials analysis.

Spark source. A schematic diagram of a high-energy spark source is shown in Figure 8.9. When the switch is closed, the capacitor is charged. When the switch is opened, the capacitor discharges through the inductor and the resistor and produces an electrical discharge in the spark gap. The voltage in the circuit shown in a modern instrument is between 400 and 1000 V. This voltage is not high enough to cross the 3–4 mm gap between the sample and the counterelectrode. A separate ignition circuit, similar to those used in automobiles, is used to create a very high voltage (10 kV). This high voltage initiates electrical excitation across the gap. Once the electrical path is initiated, the operating voltage sustains the discharge. The spark source is an interrupted discharge, unlike the continuous DC arc discharge.

In a modern spark source, the discharge is only in one direction, from the counterelectrode to the sample. This is done by including a diode across the capacitor so that current flows in only one direction. The switch is opened and closed electronically many times per second. This is the source frequency and is equal to the number of discharges per second. Spark source frequencies are generally in the 200–600 Hz range. A modern spark source can be used in either spark-like or arc-like mode. Spark-like refers to a high-current discharge that comes in short bursts. Arc-like refers to a lower current, almost continuous discharge. A diagram of current in the gap versus time for the two types of discharges shows the difference between the two modes (Figure 8.10). The spark frequency is hundreds of sparks per second. The sparks strike the sample surface randomly, resulting in an averaging of the composition and improved precision. The spark source is flushed with argon gas to eliminate oxidation of samples and to permit measurement of wavelengths below 200 nm. The flow of argon through the spark stand is as high as 250 L/h when intensity measurements are taken.

The temperature in a spark-like discharge is much higher than the temperature in a DC arc discharge. Temperatures are not stable in either time or space in the spark, but may be as high as 40,000 K. This permits excitation of most elements, including oxygen, nitrogen, the halogens, and the

Table 8.8 DLs for Elements in High-Purity Copper and High-Purity Uranium Oxide Using DC Arc Emission Spectrometry

Element	High-Purity Copper DC Arc with PMT Detection (ppm)	High-Purity Copper DC Arc with CID Detection (ppm)	High-Purity Uranium Oxide Current-Controlled DC Arc
Ag	0.2	0.15	0.12
Al			0.028
As	1.0	0.20	
B			0.083
Be			0.015
Bi	0.1	0.012	0.50
Ca			4.7
Cd			0.11
Cr			4.8
Cu			0.32
Fe	1.0	0.15	19.5
Mn	0.1	0.06	1.2
Ni	0.1	0.02	2.7
Pb	0.1	0.008	0.39
Sb	1.0	0.25	
Sn	0.5	0.02	0.13
Te	1.0	0.15	
Zn	1.0	0.02	0.33

Copper data: The DC arc/CID data were collected using a commercial system no longer in production. © Thermo Fisher Scientific (www.thermofisher.com). Used with permission.

Uranium oxide data: collected on a Prodigy DC Arc spectrometer. DLs based on analysis of a certified uranium oxide standard and are therefore higher than those obtainable using a high-purity uranium oxide blank. Courtesy of Teledyne Leeman Labs, Inc., Hudson, NH, www.teledyneleemanlabs.com. Used with permission.

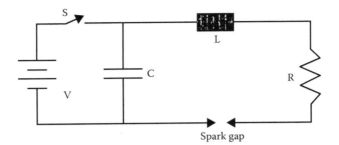

Figure 8.9 Schematic spark excitation source. S is a switch, C is a capacitor, L is an inductor, R is a resistor, and V is a voltage source. (From Thomsen, V.B.E., *Modern Spectrochemical Analysis of Metals: An Introduction for Users of Arc/Spark Instrumentation*, ASM International, Materials Park, OH, 1996. With permission from ASM International®.)

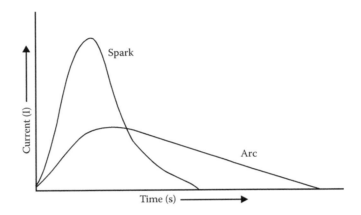

Figure 8.10 Current–time relationship for a spark discharge and an arc discharge. (From Thomsen, V.B.E., *Modern Spectrochemical Analysis of Metals: An Introduction for Users of Arc/Spark Instrumentation*, ASM International, Materials Park, OH, 1996. With permission from ASM International®.)

group 18 gases (Ne, Ar, etc.). The higher temperature also results in many ion emission lines being produced. Spark spectra contain ion lines for singly ionized elements and for doubly ionized elements. Spark spectra therefore are more complex than DC arc spectra. The spark-like discharge is more reproducible than the DC arc; this makes the spark a better choice for quantitative analysis. However, signal intensity is lost with this technique, with a consequent loss of analytical sensitivity. Spark-like conditions are used for the best precision; arc-like conditions for the best sensitivity.

While the primary function of the electrical excitation source is to generate a discharge that will vaporize and excite sample atoms, the source serves a second function. It provides a high-energy prespark (HEPS), a high-current spark discharge that homogenizes the sample surface by essentially melting it. The emission during the HEPS period (formerly called a preburn) is not integrated. After the HEPS, the sample is more homogeneous and the integration of the signals from the sparking period results in higher precision. A typical spark cycle for a metal sample would include a 10–20 s HEPS period, a 5 s measurement under spark-like conditions for major and minor constituents, and then a 5 s measurement under arc-like conditions for trace elements. Typical excitation source parameters are given in Table 8.9.

The SPECTROLAB spark emission system from SPECTRO Analytical Instruments GmbH, using microintegrators to process PMT photon current, enables µs readings

Table 8.9 Typical Spark Excitation Source Parameters

Parameter	HEPS	Spark	Arc
R (Ω)	1	1	15
L (µhenry)	30	130	30
C (µfarad)	12	2	12
V (V)	400	400	400
f (Hz)	300	300	300
I (A)	130	60	20
t (µs)	100	90	600

Source: From Thomsen, V.B.E., *Modern Spectrochemical Analysis of Metals: An Introduction for Users of Arc/Spark Instrumentation*, ASM International, Materials Park, OH, 1996. Used by permission of ASM International®.

to be obtained. A single spark can be divided into 100 or more steps, allowing for optimization of dynamic range and S/N. Inclusions, segregations, and correlations between elements can be detected and evaluated. For example, a correlation in time between Ti and N indicates the presence of TiN in the sample; a similar correlation between Mn and S would indicate MnS.

8.2.1.2 Sample Holders

The function of the sample holder is to introduce the sample into the electrical discharge. There are two types of sample holders, those for solid samples and those for liquid samples. More than 50 types and sizes of electrodes for holding samples are commercially available.

Solid sample holders. Solid samples, especially nonmetallic materials, may be analyzed as powders. High-purity carbon electrodes are available with a small well drilled in one end to hold powdered samples, such as the electrode shown in Figure 8.7b. Powdered samples may be loaded into the carbon electrode and packed or tamped down. In some cases, the sample powder is mixed with a large amount of a high-purity matrix powder, such as alumina or silica mixed with graphite. This ensures a constant matrix and therefore a constant plasma temperature, which improves the precision of the analysis. Another approach is to mix the sample powder with a large excess of a conducting powder, such as high-purity graphite or copper, and then to press a pellet of the mixture to serve as the electrode. The conducting powder also serves as a constant matrix, improving the precision of the measurements.

Metallic or electrically conductive samples, such as sheets, wires, or rods of alloys or pure metals, may be used directly as one or both of the electrodes. Some preparation may be required, such as milling the metal into the desired shape or melting the metal and casting it into a mold. The end of the electrode is either polished flat or tapered at one end. The formed electrode must be cleaned to remove any surface contamination before analysis. The most common approach used in modern spark instruments is to have a polished flat metal sample serve as the cathode and a tungsten electrode as the anode, shown schematically in Figure 8.11.

Liquid sample holders. Oil samples may be analyzed directly by spark emission spectrometry using a rotating disk electrode. The rotating disk electrode or rotrode (Figure 8.12) is a graphite disk that rotates through the liquid sample, which wets the surface of the disk. The rotating wet surface carries the sample into the discharge at a steady rate. Calibration data must be obtained under identical conditions and an identical number of revolutions. This approach had been used routinely for the determination of metals in lubricating and fuel oils. The analysis of oil for metal content indicates wear of the engine and other lubricated parts and is an important quality control measure in commercial and military aircraft and vehicle maintenance. The rotrode was used with spark sources, not with DC arc systems. Many modern oil analysis laboratories have switched to DCP or ICP instruments for these determinations, because of the improved precision and lower DLs that can be obtained.

8.2.1.3 Spectrometers

Simultaneous multichannel spectrometers are required for most arc and spark emission work because the instability of the source or the interrupted nature of the discharge requires that emission signals be integrated for 10 s or longer. A scanning spectrometer is therefore not practical. There are three types of spectrometers in use, spectrographs, polychromators, and CID-based or CCD-based systems,

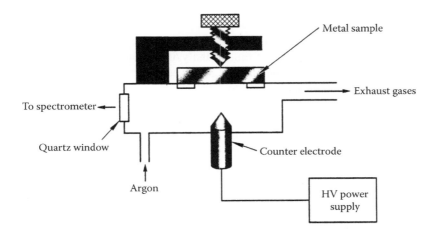

Figure 8.11 A schematic spark source, showing a flat metal sample as the cathode and a tungsten counterelectrode as anode. (From Hareland, W., Atomic emission spectroscopy, in Ewing, G.W. ed., *Analytical Instrumentation Handbook*, 2nd edn., Marcel Dekker, Inc., New York, 1997. With permission.)

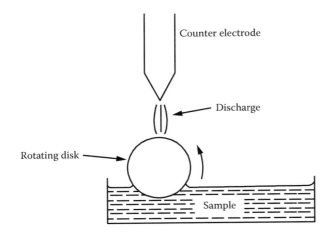

Figure 8.12 A schematic rotating disk electrode (or rotrode) used for the introduction of liquid samples into an electrical discharge. The system is used primarily for the determination of metals in oils.

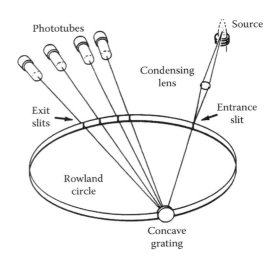

Figure 8.13 A Rowland circle polychromator. This configuration is called a Paschen–Runge mount. The source shown is an ICP, but any of the electrical excitation sources can be substituted. Multiple photomultiplier detectors are placed at the appropriate positions on the circumference of the circle to measure dispersed wavelengths. [© 1993–2014 PerkinElmer, Inc. All rights reserved. Printed with permission. (www.perkinelmer.com).]

also called electronic spectrographs. The basic components of spectrometers were discussed in Chapter 3 and more detail on atomic spectrometers was covered in Chapter 7. A significant difference between systems for AAS and AES is in the spectral band width (SBW) or spectral bandpass used. The bandpass in AAS is generally in the range of 0.2–0.5 nm, while for atomic emission, it is usually less than 0.1 nm. The reason for this difference is left as a problem for the student at the end of the chapter.

Spectrographs. Spectrographs are spectrometers that use photographic film or photographic plates to detect and record emitted radiation. Spectrographs were introduced in the 1930s and served as the fundamental elemental analysis instrument in industries for many years, especially in the production of steel, other metals, and alloys. Some of these instruments were still in use as recently as 2000. No new spectrographs of this type have been built for years and spare parts are becoming very hard to find. The photographic plates are expensive because they are now a specialty product and must be ordered a year or more in advance. Because of this and because of the hours of time and the skill required to record spectra, develop the plates, and analyze the film images, most industries have turned to CID, CCD, and PMT polychromator systems.

Polychromators. Polychromators are multichannel spectrometers with PMTs as detectors. Instruments with up to 108 PMTs are commercially available. These instruments generally use a concave diffraction grating as the dispersion device, as shown schematically in Figure 8.13. A concave grating focuses light of different wavelengths, $\lambda_1, \lambda_2, \lambda_3$, and so on, at different points on the circumference of a circle.

For a given radius of curvature of the grating, the entrance slit, the surface of the grating, and the focal points of the dispersed wavelengths lie on a circle called the *Rowland circle*. The *diameter* of the Rowland circle is equal to the radius of curvature of the surface of the concave grating. Photomultiplier detectors are put behind fixed slits at the focal points of up to 64 different wavelengths, as illustrated in Figure 8.13, which shows only four slits and detectors. The limitation on the number of channels (wavelengths monitored) is set by the physical size of the PMTs in a conventional Rowland circle system. The instrument buyer chooses the wavelengths, so each instrument is custom built. The advantage of this system is that routine multielement analysis for many elements can be performed rapidly. Once chosen, the wavelengths, and therefore the elements detected, are fixed, so the instrument lacks flexibility if nonroutine samples need to be analyzed. The precision of these instruments with a spark source is excellent, often better than 1 % RSD, and quantitative analysis can be performed rapidly. If a DC arc source is used, rapid semiquantitative analysis with ppb DLs can be performed.

In order to measure 70 or 80 elements simultaneously, it is necessary to cover the entire range of wavelengths from the vacuum UV through the visible region (120–800 nm). Using a single Rowland circle, the number of PMTs is limited by the size of the detectors. The spark spectrometer shown schematically in Figure 8.14 uses fiber optics to carry the light from the excited sample to up to four separate polychromators, each optimized for a specific wavelength region. This greatly increases the number of wavelengths that can be measured; this system has a CCD detector in one spectral module and up to 108 separate PMTs can be arranged in the second spectral module. The UV

ATOMIC EMISSION SPECTROSCOPY

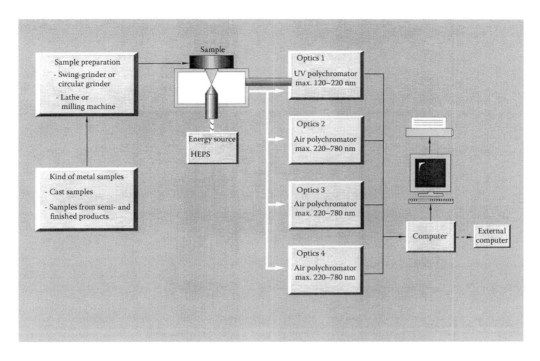

Figure 8.14 Schematic diagram of the SPECTROLAB spark emission spectrometer system. (Courtesy of SPECTRO Analytical Instruments GmbH, a division of AMETEK, Inc., Berwyn, PA, www.spectro.com, www.ametek.com.)

polychromator has a sealed optical system filled with argon, permitting wavelengths as short as 120 nm to be measured without the need for vacuum pumping. A steel sample can be completely analyzed quantitatively, including determination of C and N in the steel, in less than 30 s using this system. A sealed purged optical system, continuous inert gas purging, or a vacuum spectrometer is required to measure wavelengths in the 120–190 nm range.

Echelle monochromators. In the past 10 years, echelle monochromators have become increasingly common in multielement emission systems. The echelle design uses two dispersion devices in tandem, a prism and an echelle grating, or two gratings. The design of the echelle grating is discussed in Chapter 3. The two dispersion devices are placed so that the prism disperses the light in one plane. This light falls upon the grating, which disperses in a plane at right angles to the plane of dispersion of the prism. This is illustrated in Figure 8.15a. The final dispersion takes place over a 2D plane rather than along the single line of the Rowland circle optics. The radiation intensity can be measured using either PMT detectors or array detectors. As shown in Figure 8.15b, wavelengths can be isolated by the use of an aperture grid, which is a 2D array of multiple individual exit slits. Numerous photomultipliers can be located behind the grid. More commonly, the entire 2D "echellogram" is imaged using a 2D solid-state detector, such as the CID or CCD discussed later. Figure 8.16 shows a commercial echelle spectrometer with a CID detector. A schematic representation of the resulting echellogram,

wavelength versus diffraction order, is shown in Figure 8.17. With new array detectors, up to 5000 different wavelengths can be monitored simultaneously, many more than could possibly be detected using PMTs. An additional advantage of an echelle system with a solid-state detector is the ability to correct for background emission in real time (i.e., at exactly the same time as the analyte intensity is measured). These echelle systems are electronic versions of the old photographic imaging systems, and while they are usually called spectrometers, the term electronic spectrograph can be applied as well.

8.2.1.4 Detectors

PMTs and linear PDA detectors are discussed in detail in Chapter 4. This section will cover the 2D array detectors used in arc/spark and plasma emission spectrometers. In order to take advantage of the 2D dispersion of wavelengths from an echelle spectrometer, a 2D detector is required. The detector should consist of multiple individual detectors positioned (arrayed) so that different wavelengths fall on each individual detector. Such an array detector is called a multichannel detector, whether the detector units are arranged linearly or in two dimensions. The charge-injection device (CID) and charge-coupled device (CCD) are both silicon-based sensors that respond to light. They belong to a broader class of devices called charge transfer devices (CTDs). The two detectors discussed can be considered to

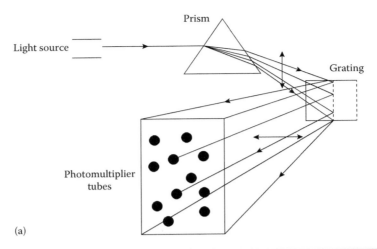

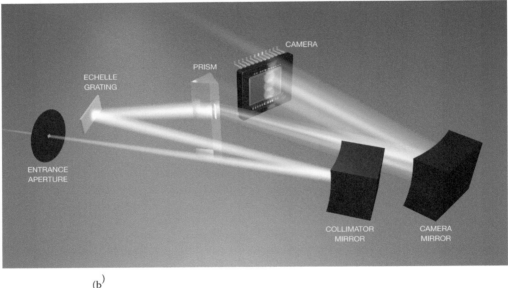

Figure 8.15 (a) A schematic diagram of an echelle spectrometer. Two dispersive devices are used. The prism disperses light in the vertical plane; the grating disperses the vertically dispersed light in the horizontal plane. The physical arrangement of the grating and prism can be reversed as shown in (b) or two gratings may be used. (b) Schematic of a commercial echelle spectrometer system, the Profile ICP spectrometer, which uses an aperture grid to isolate the 2D dispersed wavelengths. This system uses PMT detection (not shown). (Diagram (b) courtesy of Teledyne Leeman Labs, Inc., Hudson, NH, www.teledyneleemanlabs.com.)

be 2D arrays of silicon photodiode detectors. CTDs store the charge created when electromagnetic radiation strikes them and permit the integrated charge to be read. The amount of charge measured is proportional to the intensity of the light striking the detector units.

An insulating layer of SiO_2 is grown onto a crystalline silicon substrate, and thin conducting electrodes are placed on top of the silica layer to create metal–oxide–semiconductor (MOS) capacitors. The crystalline silicon has Si–Si bonds arranged in a 3D lattice. A bond can be broken by a photon of sufficient energy, releasing an electron into the silicon lattice and creating a "hole" simultaneously. This is exactly analogous to the formation of an electron–hole pair in a photodiode, described in Chapter 4. In the presence of an applied electric field across the device, the electrons will move in one direction and the holes will move in the opposite direction, leaving an area depleted of positive charge (a depletion region or potential well). The electrons accumulate and are stored in the depletion region. The more photons absorbed by the silicon, the more electrons are accumulated in the potential well. Wells can hold as many as 10^6 charges before "leaking" or "blooming" into an adjacent storage unit. The CTD storage units, or pixels, are arranged in a 2D array on a silicon wafer. The number of pixels varies from 512×512 to 4096×4096. The difference between the CID and the CCD is the manner in which the charge is obtained, accessed, and stored. Both are multichannel detectors, permitting multielement analysis using several hundred wavelengths. A significant advantage to these multichannel detectors is that the background is measured simultaneously with the analyte emission, providing more accurate background correction and faster analysis times.

CID. Each pixel in a CID can be accessed randomly to determine the amount of charge accumulated during a given period of time. This time period is called the integration time. Positive charges from the photon production of an electron–hole pair are stored in a potential well beneath a negatively biased capacitor. Reading the amount of charge does not change the amount of charge; the CID readout is nondestructive. The charge can be accessed even while the pixel is exposed to light, that is, during the integration time. This helps eliminate blooming, the spilling over of charge from one pixel to another. The random access and nondestructive readout are advantages for the CID. Another advantage is that in its operation, the CID provides complete wavelength coverage, just like a photographic plate. It is possible to measure more than 250,000 lines simultaneously. Background correction and spectral interference correction are performed by processing the data after they are collected. The CID has a high noise level and must be cooled to liquid nitrogen temperature for operation. CIDs are available to cover the UV and visible regions of the spectrum (190–800 nm). The dynamic range for the CID used in the spectrometer shown in Figure 8.16 is eight orders of magnitude, significantly more than the one to two orders of magnitude offered by photographic film. A picture of this CID and a real echellogram image collected by the CID are shown in Figure 8.18. A megapixel large-format programmable array detector (L-PAD), a CID design, used by Teledyne Leeman Labs in their Prodigy High-Dispersion ICP Spectrometer, has an area of 28 mm^2, the largest active area solid-state detector currently in use. The L-PAD provides continuous wavelength coverage from 165 to 1100 nm with a linear working range of over six orders of magnitude. For a given analytical wavelength, the Prodigy uses a 3 × 15 pixel subarray centered on the wavelength of interest. The analytical peak and background correction points are defined in each subarray with pixel position and width values. For example, to measure lead in paint samples using the 220.353 nm Pb line, a typical subarray setup might use the first five pixels to measure the background on the left of the emission line. Pixels 7–9 are used for integrating the analytical peak. If a background measurement is needed on the right side of the analytical line, pixels 10–15 could be used. Details of the measurement can be found in Teledyne Leeman Labs application note 1056 on the analysis of lead-based paint (www.teledyneleemanlabs.com).

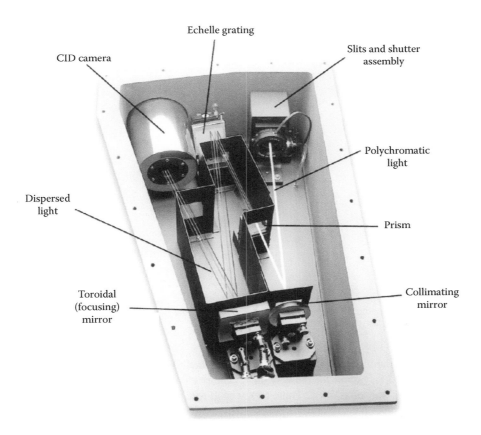

Figure 8.16 Photograph of the interior of a commercial echelle system with a CID detector. (© Thermo Fisher Scientific (www.thermofisher.com). Used with permission.)

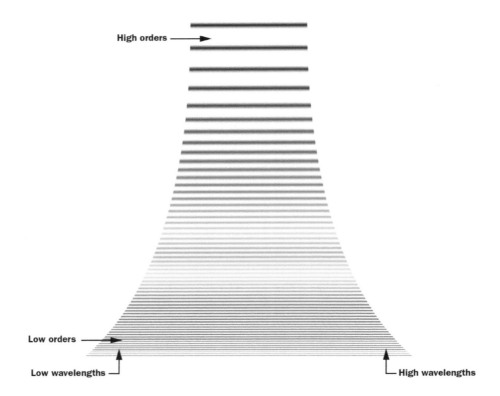

Figure 8.17 Exit plane representing the 2D array of wavelengths produced by the echelle configuration. [© 1993–2014 PerkinElmer, Inc. All rights reserved. Printed with permission. (www.perkinelmer.com).]

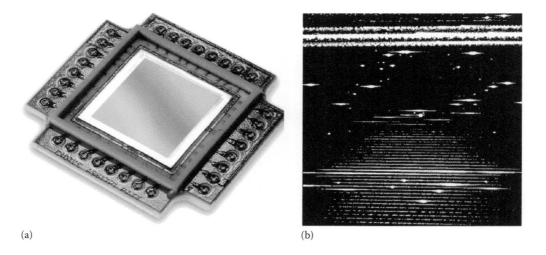

Figure 8.18 (a) Photograph of a CID. This solid-state array detector captures an electronic image of the complete emission spectrum. (b) The echellogram of a sample of zinc oxide collected by a CID detector from a DC arc source. (© Thermo Fisher Scientific (www.thermofisher.com). Used with permission.)

CCD. The CCD operates in a manner identical to a linear PDA detector and similar to the CID, except for the way the individual detectors are accessed and controlled. The CCD depicted in Figure 8.19 shows the depletion region (potential well) and stored negative charge from photoelectrons generated when light hits a pixel. In a CCD, the accumulated charge must be read sequentially. In the process of reading it, the charge is destroyed. The charge is transferred from one pixel to the next in a row in a "bucket brigade" manner. The readout "buckets" are at the edge of the CCD chip. As each row is read, the charge from the next row is shifted to the corresponding pixels in the empty row, a process called parallel charge transfer. The readout electronics know the location in the original array of the pixel being dumped into the end bucket and match the measured charge with that position. This allows the CCD to provide intensity–wavelength

ATOMIC EMISSION SPECTROSCOPY

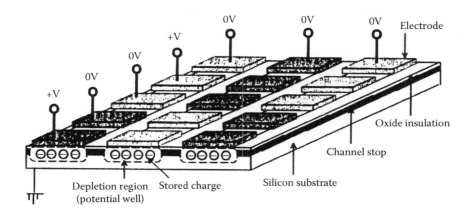

Figure 8.19 Schematic of a CCD. In this CCD, three electrodes define a pixel. (Courtesy of HORIBA, Inc., HORIBA Scientific, Kyoto, Japan, www.horiba.com/scientific.)

correlation even though the charge has been moved. The readout is very fast and has low noise associated with it. CCDs can be cooled by liquid nitrogen but also by more convenient thermoelectric cooling devices and still have good S/N ratios. They are more susceptible to blooming than CIDs at high levels of illumination; this limits their dynamic range but makes them good detectors for weak sources.

While PMT detectors are more sensitive and operate with lower S/N than CID or CCD detectors, they cannot match the multichannel capacity of the CTD array detectors. The references by Harnly and Fields, Sweedler et al., Epperson et al., and Pilon et al. provide more detail on CID and CCD detectors.

8.2.2 Interferences in Arc and Spark Emission Spectroscopy

In most modern laboratories, arc and spark emission is limited to the analysis of solid samples, because liquids are much more easily analyzed by ICP, MP, or DCP emission spectroscopy.

Arc and spark emission spectroscopy is widely used for the qualitative, semiquantitative, and quantitative determination of elements in geological samples, metals, alloys, ceramics, glasses, and other solid samples. Quantitative analysis of more than 70 elements at concentration levels as low as 10–100 ppb can be achieved with no sample dissolution. The determination of critical nonmetal elements in metal alloys, such as oxygen, hydrogen, nitrogen, and carbon, can be performed simultaneously with the metal elements in the alloys.

8.2.2.1 Matrix Effects and Sample Preparation

The matrix, or main substance, of the sample greatly affects the emission intensity of the trace elements present. For example, the emission intensity from 5 ppm of iron in a sample of pure aluminum may be very different from the intensity of the same 5 ppm of iron in a sample of aluminum–copper alloy. This is because the matrix (i.e., the aluminum or the mix of aluminum and copper) has different properties such as thermal conductivity, crystal structure, and boiling point (or heat of vaporization). The matrix affects the temperature of the plasma and the heat buildup in the electrode, which affect the rate of vaporization of the iron into the electrical discharge. The rate of vaporization into the discharge affects the number of iron atoms excited per unit time and therefore the emission intensity from the iron. If the iron volatilizes from the sample matrix faster than it does from the matrix of the iron standards used for calibration, a positive error will result. If the iron volatilizes more slowly from the sample than from the iron standards, a negative error will result. There are other causes for this matrix or *interelement effect*, including the physical state of the analyte element. Some elements exist in solid solution in metals; others occur as precipitated elements or compounds in the sample. Some alloys consist of two or more solid phases, and the proportion of these phases will change with how a metal alloy has been treated or worked. The physical state of the analyte in the sample will often affect the emission intensity. The text by Slickers discusses this in detail, especially for metals. Interelement effects result in changes in the slope of the calibration curve. This differs from the effect of spectral interference from an overlapping line. In the case of line overlap, the entire calibration curve is shifted parallel to a calibration curve with no spectral interference; no change in slope occurs.

To minimize this problem, it has been common practice to convert all samples to a common matrix, especially for arc emission analysis. This may be done by converting the sample to an oxide powder through either wet ashing with oxidizing acid or dry ashing. The sample oxide powder is then mixed with a large excess of a common inert matrix, such as graphite powder. Organic compounds, such as animal tissue, body fluids, plant materials, plastics, paper, and textiles, are analyzed in this manner. Metal powders are often oxidized before analysis. Mixing with a matrix powder allows the sample to enter the electrical discharge at a steady rate. It avoids sudden "flares" of emission from the sample, which give rise to erratic results. The common matrix permits the accurate and precise determination of the elements present.

Bulk solid samples must be machined or polished into the correct shape, and this must be done without contaminating the surface from the grinding material or altering the surface composition by polishing out a soft phase, for example. The surface must be cleaned of any lubricant or oil used in the machining process before analysis.

Advances in understanding of the spark and arc sources and improvement in electronics have led to redesign of the sources and spectrometers, so that in modern instruments, structure-induced interelement effects can be minimized by proper choice of excitation and measurement conditions. This has reduced the need for a common matrix and has permitted, in some cases, the use of one "global" calibration curve for multiple materials. However, if an interelement effect does exist, correction factors can be calculated to compensate for the effect.

8.2.2.2 Spectral Interference

As discussed in the section on flame emission spectrometry, there are two types of spectral interference: background interference and line overlap. Background in arc/spark emission can be due to thermal radiation (light emitted by a heated object), emission by molecular or polyatomic species such as CN from atmospheric nitrogen, or the edges of adjacent emission lines reaching the detector. Background from the edges of adjacent emission lines will vary depending on the elements present. In general, the background for atomic lines from an arc source is lower than the background for the same lines in a spark source. Background must be measured as close to the analyte emission line as possible and the analyte intensity corrected for accurate results.

Spectral line overlap is a serious problem with high-temperature excitation sources. Thousands of wavelengths are emitted from metal alloys such as steels and from heavy elements such as molybdenum and tungsten. Overlapping lines are unavoidable. The resolution of the spectrometer needs to be as high as possible to separate analytical lines of interest. Even with high resolution, spectral overlap will occur. The analyst needs to use published tables of emission lines, such as the Massachusetts Institute of Technology (MIT) wavelength tables (see the bibliography), to choose an analytical line that is not directly overlapped by another element in the sample. For example, Mn has an intense emission line at 293.306 nm; the MIT tables indicate that uranium emits at exactly the same wavelength, so clearly, this line cannot be used to measure Mn in uranium or U in manganese. There are five emission lines within ±0.008 nm of the Mn 293.306 nm line. None of the lines is from Cu or Al, so this Mn line could be used to measure Mn in copper alloys or Al alloys, but not in U, Th, W, or Ir.

If there is no suitable analytical line free of spectral overlap, corrections can be made for the interfering element, but only if the overlapping element's intensity is less than 10 % of the analyte intensity. Less than 1 % is even better, if possible. Corrections cannot be made if the overlapping line is one of the major elements in the sample (e.g., overlap from an iron line on Cr in a steel sample cannot be corrected mathematically).

8.2.2.3 Internal Standard Calibration

Apart from matrix effects, which must be controlled to obtain accurate analyses, such factors as physical packing and the size of particles cause a variation in the quantity of sample vaporized into the discharge from one sample to the next. Instrument drift, slight changes in the source energy, and other system problems can cause variations in intensity from measurement to measurement. Because arc/spark emission spectroscopy is a multielement technique, these problems may be compensated for by using *IS calibration*. In this calibration method, described in Chapter 3, the ratio of analyte intensity at its chosen wavelength to the intensity of an IS element at its emission wavelength is measured. The intensity ratio is plotted versus the concentration ratio of analyte to IS. The use of the ratio should compensate for some variations if the IS element and its wavelength are chosen correctly. If something causes the analyte intensity to increase, the IS intensity should also increase and the ratio will remain constant.

For bulk solid samples such as metals and alloys, the easiest way to do this is to choose an emission line of the major matrix element as the IS line. For example, an iron emission line would be used in steels, a Cu line in copper alloys, and so on. If the sample is a powder, either the major element can be used or a separate element added to all standards and samples.

The following guidelines are used when choosing an element and wavelength to serve as an IS:

1. The element used as an IS should be similar in properties such as enthalpy of vaporization and ionization energy to the element to be determined.
2. The wavelength of the emission line from the IS should be *homologous* with the wavelength of the analyte. Homologous means that the lines should behave similarly with respect to excitation. Atomic IS lines should be used for atomic analyte lines and ion IS lines for ion analyte lines. (The Roman numerals after the wavelengths chosen should be the same. Examples of homologous line pairs for elements in steel are shown in Table 8.10.)
3. The concentration of the IS added should be similar to that of the analyte measured.
4. For spectrometers with more than one polychromator, the IS line and analytical line must be measured on the same polychromator.

The guidelines should be adhered to as far as possible to obtain the best accuracy. It is often not possible to meet all of the guidelines with one IS element. For example, if the

Table 8.10 Some Homologous Line Pairs in Steel

Analyte Line (nm)	Fe II 273.0 nm Internal Std. Line	Fe I 281.3 nm Internal Std. Line
Mn II 293.3	×	
Cr II 298.9	×	
Ni II 376.9	×	
Mo II 281.6	×	
Si I 251.6		×
Ni I 352.4		×

Source: Adapted from Thomsen, V.B.E., *Modern Spectrochemical Analysis of Metals: An Introduction for Users of Arc/Spark Instrumentation*, ASM International, Materials Park, OH, 1996. With permission of ASM International®.

Note: The Roman numerals denote the type of line (atomic, ionic), *not* the valence state.

analyst is required to measure 23 trace elements in a given sample, including transition metals; REEs; nonmetals such as B, S, and P; and refractory metals such as Nb, W, and Zr, it is not likely that one element chosen as an IS will meet all of the guidelines. More than one IS element may be added or more than one matrix wavelength chosen, but the possibility of spectral interference must be kept in mind. The more elements present, the more emission lines there will be and the greater the possibility of spectral overlap.

8.2.3 Applications of Arc and Spark Emission Spectroscopy

8.2.3.1 Qualitative Analysis

Qualitative analysis is performed by recording the emission spectrum of the sample. The sample elements are then identified by comparing their emission spectra with previously recorded spectra from known elements. Generally, for trace amounts, the *raies ultimes*, or *RU lines*, must be present and identified in the emission spectra as discussed in the following text. For higher concentrations, the RU lines are present together with many other emission lines. The emission spectra encountered in arc and spark spectroscopy are rich and intense for most elements in the periodic table, in contrast to those observed in flame photometry. As a result, these techniques are widely used in elemental qualitative analysis of solid samples. Most elements can be detected at low concentrations (ppm or ppb) with a high degree of confidence. In practice, emission spectra are now "matched" to identify the elements present using a computer database of emission lines contained in the instrument software.

8.2.3.2 Raies Ultimes

When an element is excited in an electrical discharge, it emits a complex spectrum that consists of many lines, some strong, some weak. When the concentration of the element is decreased, the weaker lines disappear. Upon continuing dilution of the element, more and more lines disappear until only a few are visible. The final lines that remain are called the RU lines, although this term is not used in many modern publications. The RU lines are the most intense emission lines and invariably involve an electronic transition from an excited state to the ground state. These lines can be detected even when small concentrations of the element are present. Conversely, if the element is present in a sample, whatever other spectral lines are emitted from the sample, the RU lines should always be detectable. Some elements and their RU lines are listed in Table 8.11. Generally, for confirmation of the presence of a trace amount of a given element, two or three RU should be detected. Confirmation based on a single RU line is unreliable because of the possibility of spectral interference from other elements.

Because the plasma is not uniform in temperature throughout the discharge, there will be unexcited atoms of the analyte element in the cooler regions of discharge. These free atoms absorb radiation at the same wavelength as the emitted RU lines from other excited atoms of the same element since they involve the same transition. The greater the concentration of a given element, the greater the number of free atoms in the discharge. As a result, such emission lines are reabsorbed by neighboring unexcited atoms in a process called self-absorption or self-reversal. (This is the same process of self-reversal that occurs in hollow cathode lamps [HCLs] and serves as the basis for the Smith–Hieftje background correction technique discussed in Chapter 7.) The result is seen as curvature in the emission calibration curve for the element at high concentrations. If the concentration is high enough, the self-absorption is so effective that the net emission intensity actually decreases as shown in Figure 8.20 for the beryllium 313.11 nm line, an RU line. A plot of emission intensity against concentration will go through a maximum and then decrease. The intensity therefore "reverses" at higher concentrations. Such lines are said to be *reversible lines*. They cannot be used for quantitative analysis at higher concentrations, but they are useful for measuring very small concentrations, where reversal is not a problem. In fact, these lines are usually the most sensitive; the Be 313 nm line can be used for determining ppb levels of Be, a highly toxic element, in biological samples. Ion lines do not suffer from self-reversal, because the ion energy states are not the same as the atomic energy states.

8.2.3.3 Quantitative Analysis

Quantitative analysis is carried out by measuring the intensity of one emission line of the spectrum of each element to be determined. The choice of the line to be measured depends on the expected concentration of the element to be determined and the sample matrix. For

Table 8.11 RU Lines of Some Common Elements

Element	Wavelength of RU Lines (nm)	Element	Wavelength of RU Lines (nm)	Element	Wavelength of RU Lines (nm)
Ag	328.07	Co	344.36	Mo	379.63
	520.91		240.73		386.41
Al	396.15	Cr	399.86	Na	589.59
	394.40		425.43		568.8
As	193.76	Cs	852.11	Ni	341.48
	189.04		894.35		352.45
	286.04	Cu	324.75	P	253.57
Au	242.80		327.40		255.33
	267.60	Fe	385.99	Pb	405.78
B	249.77		371.99		368.35
	249.68	Ge	265.12	Pt	265.95
Ba	265.05		303.90		292.98
	553.55	Hg	185.0	Se	196.03
Be	234.86		253.65		206.28
	313.11		435.84		203.99
Bi	306.77	K	766.49	Si	288.16
	289.80		769.90		251.61
C	193.09	La	624.99	Sn	286.33
	247.86		579.13		284.00
Ca	422.67	Li	670.78	Sr	460.73
	445.48		610.36		483.21
Cd	228.80	Mg	285.21	Ti	498.18
	326.11		383.83		399.86
Ce	569.92	Mn	403.08	V	318.40
	404.0		403.31		437.92

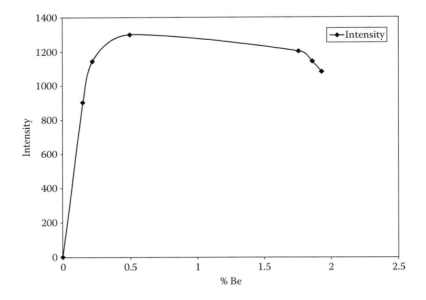

Figure 8.20 A plot of emission intensity versus % Be at the 313.11 nm Be line. This is a reversible line for Be. Note that the emission intensity reaches a maximum and then decreases as the Be concentration increases. Such lines cannot be used for high concentrations of analyte but are excellent for low concentrations. (Modified from Thomsen, V.B.E., *Modern Spectrochemical Analysis of Metals: An Introduction for Users of Arc/Spark Instrumentation*, ASM International, Materials Park, OH, 1996. With permission from ASM International®.)

extremely low concentrations, the RU lines must be used. These are the most intense lines and must be used when high sensitivity is required. These lines are reversible, and their intensity does not increase linearly with concentration. For high concentrations, a nonreversible line must be used in order to stay within the linear calibration range.

When a sample is introduced into an electrical discharge, there is a lag in time before the emission signal becomes steady. In the first few seconds of discharge, the intensity of emission is erratic and difficult to control. In order to obtain reproducible quantitative results, it is advisable to ignore these erratic signals. This is done by not recording the emission until the signal is judged stable. By plotting emission intensity from a standard against time, this period can be measured. For example, for tin, this period may be 15 s. When a sample is analyzed for tin, the first 15 s of the discharge are not recorded. This period is called the *preburn time*. The term used for modern spark instruments is HEPS and is usually set for 10–20 s. In modern DC arc systems, the preburn is performed under conditions of reverse polarity for about 60 s. A preburn time is always used if accurate results are required. In modern spectrometers, the preburn time is set in the software and performed automatically.

The relative accuracy and precision obtained by arc and spark emission spectroscopy is commonly about 5 %, but may be as poor as 20–30 %. Arc emission is much more prone to matrix effects than spark emission due to the lower temperature of the discharge. Both arc and spark excitation may require matrix matching of sample and standards for accurate analyses and usually require the use of an IS.

In practice, the sample is excited in the electrical discharge. Intensities of emission lines are measured using the detector(s) in the spectrometer, and concentrations of elements are determined by comparison to the intensities obtained from standard calibration curves. External calibration (discussed in Chapter 3) can be used for many materials, with the IS being the major matrix element, as has been described. Certified metal alloy standards in flat polished forms and bulk solid pieces are available from government organizations such as the US National Institute for Standards and Technology (NIST) and from private standards companies. It is possible to purchase sets of solid steel standards, brass standards, nickel alloy standards, and so on for use as calibration standards and quality control check samples. High-purity metal powders are available from commercial suppliers as well. The accuracy of the analysis will be only as good as the accuracy of the standards used for calibration of the instrument. Using commercially available automatic readout systems, the simultaneous determination of the elements present in a sample can be carried out in a minute or less.

Emission spectroscopy with arc and spark excitation has been used since the 1930s for many industrial analyses. In metallurgy, for example, the presence in iron and steel of the elements nickel, chromium, silicon, manganese, molybdenum, copper, aluminum, arsenic, tin, cobalt, vanadium, lead, titanium, phosphorus, and bismuth have been determined on a routine basis. Modern instruments can also measure oxygen, nitrogen, and carbon in metals, which used to require separate measurements with dedicated high-temperature analyzers. Frequently, the analysis can be carried out with very little sample preparation, apart from physically shaping the sample to fit in the discharge unit. Alloys of aluminum, magnesium, zinc, and copper (including brass and bronze), lead and tin alloys (including solders), titanium alloys, high-purity metals such as gold, platinum, palladium, and many others are routinely analyzed during production by arc and spark spectrometry. Alloy identification for metal scrap sorting and recycling has become so important that portable, handheld spark emission spectrometers have been developed. These units are battery operated and contain a built-in computer database of alloy compositions, so that a metal recycling facility can immediately separate aluminum alloys from nickel alloys, for example, even when the metal pieces are out in a field or as they are unloaded from a truck.

Production and quality control of uranium fuel rods used in nuclear power plants are monitored by DC arc emission spectroscopy. Trace elements in high-purity metal powders are measured for quality control purposes. Tungsten powder can be analyzed for trace elements by arc/spark emission spectroscopy without the need to dissolve the tungsten; this eliminates the use of expensive and hazardous hydrofluoric acid (HF).

In the oil industry, lubricating oils can be analyzed for iron, nickel, chromium, manganese, iron, silicon, copper, aluminum, and other elements. These metals get into the lubricant when bearings and pistons wear. Such analyses give the engine designer valuable information on the parts of the system that are likely to fail or need protection from wear and corrosion. The information is used in aircraft and railway engine maintenance to warn of bearing wear in the engine. The bearings may be replaced based on lube oil analysis rather than taking the engine apart for physical inspection, which is a long and expensive procedure. Oil feedstocks to petroleum refinery catalytic cracking units are analyzed for iron, nickel, and vanadium, which can poison the catalyst and interfere in the production of the branched hydrocarbons desired in fuel. Many of the oil analysis applications are now performed by ICP or DCP emission spectrometers, since these systems handle liquids more easily.

Arc and spark excitation sources are still widely used in the analysis of solid materials by emission spectroscopy, especially if the solids are difficult to dissolve or rapid analysis for quality control and production is required. They are still the mainstay of analysis in foundries, where samples are easily cast into electrodes and the range of compositions analyzed is well known.

8.3 PLASMA EMISSION SPECTROSCOPY

A *plasma* is a form of matter that contains a significant percentage (>1 %) of electrons and ions in addition to neutral species and radicals. Plasmas are electrically conductive and are affected by a magnetic field. The plasmas used in emission spectroscopy are highly energetic, ionized inert gases. The most common plasma in commercial use is the argon ICP. Other commercial plasma sources are the DCP, also usually supported in argon, the recently introduced nitrogen MP, and the helium microwave-induced plasma (MIP). The temperature of a plasma excitation source is very high, from 6500 to 10,000 K, so almost all elements are atomized or ionized and excited to multiple levels. The resulting emission spectra are very line rich, which necessitates the use of high-resolution spectrometers to avoid spectral overlap.

8.3.1 Instrumentation for Plasma Emission Spectrometry

8.3.1.1 Excitation Sources

Radio-frequency (RF) ICP source. In ICP-OES, the sample is usually introduced to the instrument as a stream of liquid. The sample solution is nebulized and the aerosol transported to the plasma. In the plasma, the sample undergoes the same process outlined in Table 8.2. The argon plasma serves to atomize, ionize, and excite the elements in the sample. The emitted radiation is sorted by wavelength in a spectrometer and the intensity is measured at each wavelength. A schematic of a typical ICP-OES system is presented in Figure 8.21.

In Figure 8.21, the argon plasma is the flame-like object at the top of the ICP torch, above the coils (the dark lines) from the RF generator. Figure 8.22 depicts a cross section of a typical ICP torch. This torch contains three concentric tubes for argon flow and sample aerosol introduction. The two outer tubes are normally made of quartz. The inner tube, called the injector tube, may be made of quartz, alumina, sapphire, or other ceramic. Surrounding the torch is a water-cooled copper load coil or induction coil, which acts as an antenna to transmit power to the plasma gas from the RF generator. The power required to generate and sustain an argon plasma ranges from 700 to 1500 W. RF generators on some commercial instruments operate at 27.12 MHz; many modern instruments operate at 40.68 MHz, which results in better coupling efficiency and lower background emission intensity. The plasma initiation sequence is depicted in Figure 8.23. When RF power is applied to the load coil, an AC oscillates within the coil at the frequency of the generator. The oscillating electric field induces an oscillating magnetic field around the coil. Argon is swirled through the torch and slightly ionized using a Tesla coil. The few ions and electrons formed are immediately affected by the magnetic field. Their translational motion changes rapidly from one direction to the other, oscillating at the same frequency as the RF generator. The rapid movement in alternating directions is *induced* energy; adding energy in this manner is called *inductive coupling*. The high-energy electrons and ions

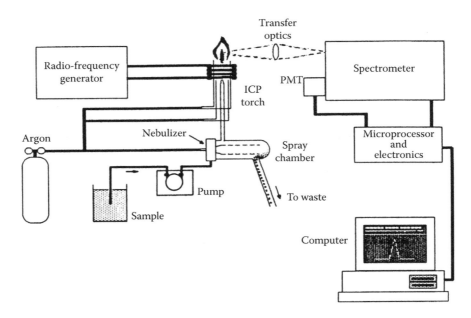

Figure 8.21 Major components and layout of a typical ICP-OES instrument. [© 1993–2014 PerkinElmer, Inc. All rights reserved. Printed with permission. (www.perkinelmer.com).]

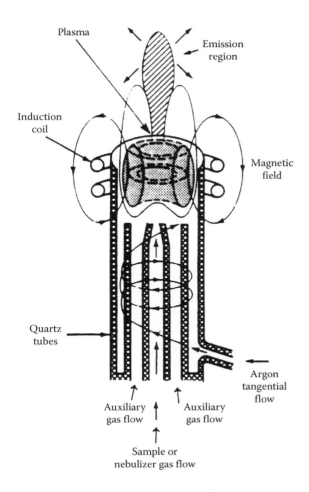

Figure 8.22 Cross section of an ICP torch. The emission is viewed in the region above the coils when the torch is in a radial position.

collide with other argon atoms and cause more ionization. This continues in a rapid chain reaction to convert the argon gas into a plasma of argon ions, free electrons, and neutral argon atoms.

The ICP torch is designed with narrow spacing between the two outermost tubes, so that the gas emerges at a high velocity. The outer tube is designed so that the argon flow in this tube, called the plasma flow, follows a tangential path as shown in Figure 8.22. This flow keeps the quartz tube walls cool and centers the plasma. A typical flow rate for the plasma flow is 7–15 L argon/min. The argon flow in the middle channel is called the auxiliary flow and can be 0–3 L argon/min. The auxiliary gas flow serves several purposes, including that of reducing carbon deposits at the injector tip when organic solvents are being analyzed. The gas flow that carries the sample aerosol into the plasma goes through the center or injector tube. It is called the nebulizer flow or sample flow and is typically about 1 L/min. The tangential or radial flow spins the argon to create a toroidal or doughnut-shaped region at the base of the plasma through which the sample aerosol passes. The temperatures for various regions of the plasma are shown in Figure 8.24a. Immediately above the load coil, the background emission is extremely high. The background signal drops with distance from the load coil, and emission is usually measured slightly above the load coil, where the optimum signal-to-background ratio is achieved. This area is called the "normal analytical zone," shown in Figure 8.24b.

The advantage of the argon ICP as an excitation source lies in its high temperature and its stability. The gas temperature in the center of the plasma is about 6800 K, which permits the efficient atomization, ionization, and excitation of most elements in a wide range of samples. In addition, the high temperature reduces or eliminates many of the chemical interferences found in lower-temperature electrical sources and flames, making the ICP relatively free from matrix effects. Another important advantage of the ICP is that the sample aerosol is introduced through the center of the plasma and is exposed to the high temperature of the plasma for several milliseconds, longer than in other excitation sources; this contributes to the elimination of matrix effects. Elements are atomized and excited simultaneously. The stability of the ICP discharge is much better than arc or spark discharges, and precision of less than 1 % RSD is easily achieved. The dynamic range of an ICP source is approximately four to six orders of magnitude. It is often possible to measure major, minor, and trace elements in a single solution with an ICP source. Many ICP systems permit the injection of air or oxygen into the plasma when running organics, to prevent carbon buildup and reduce background.

DCP source. The design of the DCP source is shown in Figure 8.25. Two jets of argon issue from graphite anodes. These join and form an electrical bridge with the cathode, which is made of tungsten. In operation, the three electrodes are brought into contact, voltage is applied, and the electrodes are then separated. Ionization of the argon occurs and a plasma is generated, forming a steady discharge shaped like an inverted letter Y.

As shown in Figure 8.25, the region that is monitored for analytical measurements is at the junction of the anode argon streams. This small region gives good emission intensity from analyte atoms and has a background lower than the regions in the immediate vicinity. The sample is injected into this region with a stream of argon from a separate injection system. In the excitation region, the effective electronic temperature is about 5000 K. This results in spectra that are simpler than ICP emission spectra, with fewer ions lines. Sensitivities are within an order of magnitude of those reported for ICP sources for most elements and for the alkali metals and alkaline-earth elements are as good as ICP. The power supply is less expensive and the quantity of argon required for DCP operation is much less than what is necessary to operate an ICP. A DCP can be operated for several days on a single cylinder of argon gas; an ICP would use the cylinder up in less than 8 h. Because of the way

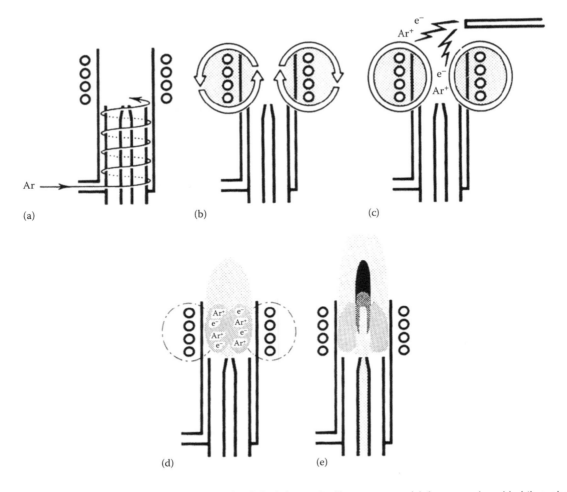

Figure 8.23 Cross section of an ICP torch and the load coil depicting an ignition sequence. (a) Argon gas is swirled through the torch. (b) RF power is applied to the load coil. (c) A spark produces some free electrons in the argon. (d) The free electrons are accelerated by the RF field causing more ionization and formation of a plasma. (e) The nebulizer flow carries sample aerosol into the plasma. [© 1993–2014 PerkinElmer, Inc. All rights reserved. Printed with permission. (www.perkinelmer.com).]

the sample is introduced into the plasma, the DCP can analyze solutions with high dissolved solids content better than some ICP sample introduction systems. There are several disadvantages to the DCP source. The graphite electrodes must be replaced frequently and the wearing away of the electrodes during analysis contributes to long-term drift in the signal. The dynamic range of a DCP is about three orders of magnitude, less than that of an ICP. The residence time of the sample in the plasma is short because the plasma is small compared to an ICP source, making it difficult to atomize and excite highly refractory elements.

MP source. In 2011, Agilent Technologies, Inc., introduced a unique excitation source for AES. The 4100 MP-AES uses a magnetically excited nitrogen MP (Figure 8.26a). The instrument uses an industrial magnetron, similar to that used in kitchen microwave ovens, and its magnetic field, rather than an electric field, to couple the microwave energy into the plasma. The magnetron is air-cooled and operates at 2450 MHz. It creates a concentrated axial magnetic field around a conventional torch and produces a toroidal plasma. The plasma is ignited using a brief flow of Ar and then automatically switches to nitrogen, either from a cylinder, Dewar (liquid N_2), or produced from air by a nitrogen generator. An external gas control module allows for injection of air into the plasma when analyzing organics to prevent carbon buildup and reduce background.

MIP source. An atmospheric pressure helium MIP is generated using a 2.45 GHz microwave generator and an electromagnetic cavity resonator, called a Beenakker cavity. The helium gas is passed through a discharge tube placed in the cavity, as seen in Figure 8.26b. The plasma is initiated by a spark from a Tesla coil. The electrons produced by the spark oscillate in the microwave field and ionize the helium gas by collision, producing a plasma. The microwave energy is coupled to the gas stream in the discharge tube by the external cavity. The plasma is centered in the discharge tube and is represented in Figure 8.26b by the shaded spheroid shape.

ATOMIC EMISSION SPECTROSCOPY

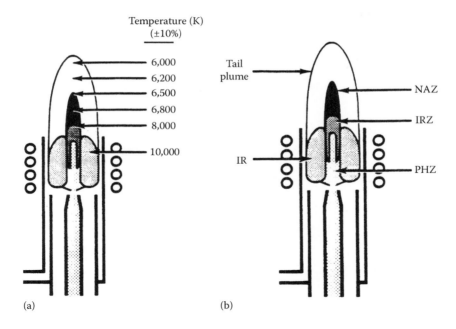

Figure 8.24 (a) Temperature regions in a typical argon ICP discharge. (b) Zones of the ICP discharge. IR, induction region; PHZ, preheating zone; IRZ, initial radiation zone; NAZ, normal analytical zone. [© 1993–2014 PerkinElmer, Inc. All rights reserved. Printed with permission. (www.perkinelmer.com).]

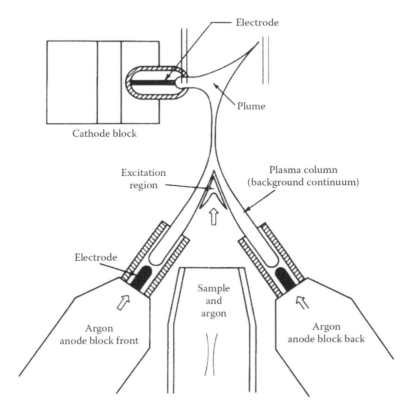

Figure 8.25 The inverted Y configuration of a DCP jet. (© Thermo Fisher Scientific (www.thermofisher.com). Used with permission.)

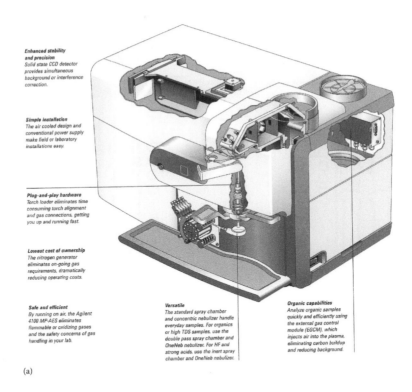

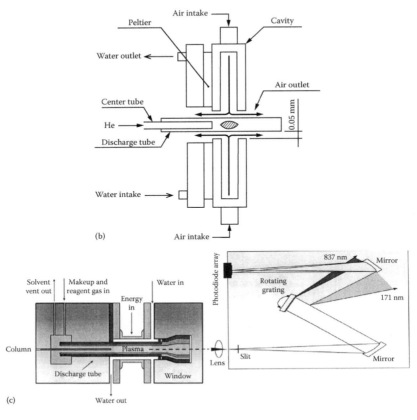

Figure 8.26 (a) The 4100 MP-AES nitrogen-based atomic emission spectrometer. (Courtesy of Agilent Technologies, Inc., Santa Clara, CA, www.agilent.com.) An updated version is the 4120 MP-AES with a more advanced software. (b) A helium MIP. (Courtesy of Yokogawa Electric Company, Musashino, Japan.) (c) Atomic emission detector for GC using a helium MIP as the excitation source. It uses a scanning diffraction grating spectrometer and PDA detector to measure atomic emission from samples excited in the MIP. The wavelength range covered is 171–837 nm. (Courtesy of Joint Analytical Systems GmbH, Moers, Germany, www.jas.de.)

The MIP operates at lower power than the ICP and at microwave frequencies instead of the RFs used for ICP. Because of the low power, an MIP cannot desolvate and atomize liquid samples. Therefore, MIPs have been limited to the analysis of gaseous samples or very fine (1–20 μm diameter) particles. Helium is the usual plasma gas for an MIP source. Electronic excitation temperatures in a helium MIP are on the order of 4000 K, permitting the excitation of the halogens, C, N, H, O, and other elements that cannot be excited in a flame atomizer. The lower temperature results in less spectral interference from direct-line overlap than in ICP or high-energy sources, but also causes more chemical interference.

A helium MIP has been used as an element-specific detector for gas chromatography (GC). This detector is shown in Figure 8.26c and the use of it for GC will be covered in greater detail in Chapter 13. The effluent from the GC column consists of carrier gas and separated gas-phase chemical compounds. The separated compounds flow through the plasma contained in the discharge tube shown. A compound in the plasma is decomposed, atomized, and excited and emits the wavelengths characteristic of the elements present. The light from the plasma is sent to a grating monochromator with a PDA detector, as shown. Another unique commercial instrument that uses a helium MIP as the excitation source is a particle analysis system (Yokogawa Electric Company, Japan) designed to both count and identify the chemical composition of particles. Particles that have been collected on a filter are "vacuumed" into the He MIP source, where they are atomized and excited and emit the characteristic radiation from the elements present. Graphite furnaces (GFs), hydride generation (HG) instruments, and other devices have been used to generate gas-phase samples or desolvated particles for introduction into MIPs.

8.3.1.2 Spectrometer Systems for ICP, DCP, and MP

Spectrometer systems for ICP and DCP include all of the dispersive devices and designs discussed in Chapter 3 and earlier in this chapter. Scanning monochromators of all optical designs are used in *sequential* spectrometer systems. In a sequential instrument, one wavelength at a time is measured, and the grating must be moved if more than one element is to be determined. Scanning rates can be very rapid in commercial instruments, but sequential instruments are slower than *simultaneous* spectrometers. Simultaneous systems contain either a polychromator or an echelle spectrometer and measure multiple wavelengths at the same time, as has been described for arc/spark instruments. *Combination instruments* are available, with both polychromators and a monochromator, to take advantage of the speed of the simultaneous system for routine work but to have the flexibility of the sequential system if a nonroutine wavelength needs to be accessed. An example of the optical layout of a combination instrument is shown in Figure 8.27.

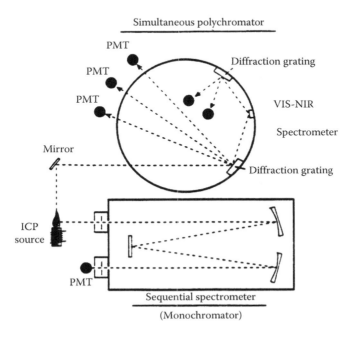

Figure 8.27 A combination sequential–simultaneous ICP emission spectrometer. Such a combination permits rapid multielement analysis using the polychromator and preselected wavelengths. The monochromator adds the flexibility to monitor additional elements or alternate wavelengths in case of spectral interferences. (Courtesy of HORIBA Scientific, Inc., HORIBA Scientific, Kyoto, Japan, www.horiba.com/scientific.)

The detectors used in ICP and DCP systems include PMTs, CCDs, and CIDs. One variation of the CCD used for ICP is the segmented array CCD or segmented charge detector (SCD). The SCD has individual small subarrays positioned on a silicon substrate in a pattern that conforms to the echellogram pattern (Figure 8.28). More than 200 subarrays are used; they cover approximately 236 of the most important wavelengths for the 70 elements routinely measured by ICP emission spectrometry. The SCD differs from a standard CCD in that the individual subarrays can be rapidly interrogated in random order (much like a CID). The SCD detector responds from 160 to 782 nm.

The DCP source is oriented in only one configuration within the spectrometer. The ICP source had traditionally been viewed from the side of the plasma, as shown in Figure 8.21. This configuration is called "radial viewing." However, looking along the plasma axis instead of across the diameter provides a longer optical path length and that results in lower DLs. Unfortunately, the longer optical path includes more temperature zones, so more chemical interference can result. Radial viewing was chosen to avoid the problems of increased spectral interference, matrix interferences, and self-absorption in the cooler tip of the plasma that would result from axial viewing. Axial viewing of the plasma does, however, provide improved DLs for many elements. Axial or end-on viewing of the plasma was made practical by use of a flow of argon, called the shear gas, to cut off the cool tip of the plasma, as shown in Figure 8.29. This reduces self-absorption and matrix or chemical interferences. There are commercial ICP systems

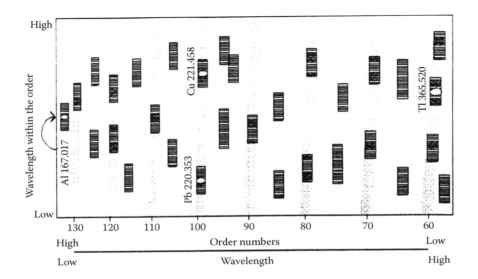

Figure 8.28 SCD detector. [© 1993–2014 PerkinElmer, Inc. All rights reserved. Printed with permission. (www.perkinelmer.com).]

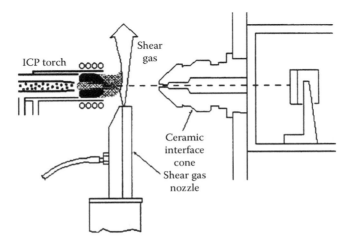

Figure 8.29 Axial (end-on) ICP torch position showing how the shear gas cuts off the cool plasma tail plume. Light emitted from the plasma passes through the ceramic interface into the spectrometer on the right side of the diagram (not shown). [© 1993–2014 PerkinElmer, Inc. All rights reserved. Printed with permission. (www.perkinelmer.com).]

ATOMIC EMISSION SPECTROSCOPY

available with only radial viewing, with only axial viewing, with simultaneous radial and axial viewing (dual view), and with the ability to flip the torch into either radial or axial position. A comparison of axial and radial DLs for some elements on a specific instrument, as well as a comparison of various nebulizers is presented in Table 8.12.

In the 4100 MP-AES instrument (Agilent Technologies), the detector is a UV-sensitive solid-state CCD detector, cooled to 0 °C using a thermoelectric Peltier device and provides simultaneous background and interference correction. The spectrometer is a fast-scanning, high-resolution sequential Czerny–Turner design that can be purged with air or nitrogen. The wavelength range is 178–780 nm. The plasma is vertically oriented but viewed end-on (axially) by using a 25 L/min flow of compressed air to protect the preoptics.

8.3.1.3 Sample Introduction Systems

Liquids are the most common form of sample to be analyzed by plasma emission. These are usually introduced with a nebulizer and spray chamber combination. An aerosol is formed and introduced into the plasma by the nebulizer gas stream through the injector tube. Nebulizers and spray chambers come in a variety of designs to handle aqueous solutions, high-salt (high total dissolved solids) solutions, HF-containing solutions, and organic solvents.

There are sample introduction systems that can handle slurries of particles suspended in liquids. Powders can be injected directly into the plasma for analysis. Lasers, sparks, and GFs (exactly the same as AAS GFs) are used to generate gaseous samples from solids for introduction into the

Table 8.12 DL Comparison of Radial and Axial ICP with Pneumatic and Ultrasonic Nebulization

Element	Wavelength (nm)	Detection Limits (µg/L)			
		Radial[a]		Axial[b]	
		Pneumatic Nebulizer	Ultrasonic Nebulizer	Pneumatic Nebulizer	Ultrasonic Nebulizer
Ag	328.068	2	0.3	1	0.03
Al	396.153	9	0.5	2	0.06
As	193.69	18	2		
As	188.979			3	0.7
Ba	233.527			0.5	0.01
Be	313.107	0.3	0.02	0.1	0.009
Bi	233.061			2	0.2
Ca	317.933			2	0.03
Cd	214.43	2	0.2		
Cd	228.802			0.1	0.02
Co	228.616	5	0.3	0.2	0.02
Cr	267.716	4	0.2	0.2	0.01
Cu	324.754	2	0.3	0.6	0.02
Fe	238.204			0.1	0.02
Mg	285.213			0.5	0.06
Mn	257.610			0.1	0.03
Mo	202.031	8	0.3	0.6	0.3
Ni	231.604	10	0.4	0.4	0.06
Pb	220.353	27	1	2	0.2
Sb	206.836	12	3	2	0.3
Se	196.026	20	1.3	3	0.5
Sn	189.927			1	0.4
Ti	334.940			0.2	0.006
Tl	190.801	22	5	2	0.5
V	292.40	2	0.2		
V	290.880			2	0.02
Zn	213.857	2	0.2	0.2	0.03

Source: Data courtesy of Teledyne CETAC Technologies, Omaha, NE, www.cetac.com.

Note: The ultrasonic nebulizer (USN) used in both cases was the CETAC U-5000AT+.

[a] Radial DLs based on 3σ, 10s integration time.
[b] Axial DLs based on 3σ, 20s integration time.

plasma. HG for As and Se and cold-vapor (CV) Hg introduction are used for ICP as for AAS; these two techniques were discussed in Chapter 7.

Liquid sample introduction. Solutions are introduced into the plasma through a nebulizer. The nebulizer is usually connected to a spray chamber. While some nebulizers discussed later can draw in the solution by self-aspiration, like the nebulizers used in AAS, it is common to pump the sample solution at a fixed rate into all nebulizers for ICP, DCP, and MP measurements. A pump, nebulizer, and spray chamber arrangement are shown schematically in Figure 8.21. The use of a pump eliminates sample uptake changes due to viscosity differences and permits rapid washing out of the nebulizer and spray chamber using a fixed fast flow rate in between samples. The type of pump used in most ICP systems is a peristaltic pump. This pump uses a series of rollers that push the sample solution through tubing. Only the tubing comes into contact with the sample. Peristaltic pump tubing must be compatible with the sample being pumped and most instrument manufacturers provide different types of tubing to handle highly acidic solutions, various organic solvents, highly basic solutions, and so on. The analyst should check the chemical compatibility of a given polymer tubing with the samples to be run through it and should also check the tubing for leachable elements by running blanks before analyzing samples. Some polymers have traces of metal catalysts that might cause erroneous results to be obtained if the catalyst metal is one of the analytes. Silicone rubber tubing should not be used if the determination of silicon in samples is required.

Nebulizers. The nebulizer converts a liquid into an aerosol of small droplets that can be transported to the plasma. Two approaches for aerosol formation are used commercially for ICPs, pneumatic nebulization and ultrasonic nebulization. Pneumatic nebulizers use high-speed gas flows to create the aerosol. There are three commonly used pneumatic nebulizers: the concentric nebulizer, the cross-flow nebulizer, and the Babington nebulizer. Pneumatic nebulizers produce a range of aerosol droplet sizes. Large droplets cannot be efficiently desolvated, so the aerosol is passed through a *spray chamber* to remove large droplets. As is the case with AAS, less than 5 % of the original sample liquid actually reaches the plasma.

The *concentric nebulizer* works by having the liquid solution introduced through a capillary tube into a region of low pressure created by a concentric gas flow past the tip of the capillary at high speed. Concentric means that the flow of liquid and flow of gas are parallel to each other. The liquid breaks up into a fine mist as it exits the capillary tip. The typical nebulizer gas flow rate is about 1 L/min. The concentric nebulizers have very small orifices, which result in highly efficient aerosol formation; this results in excellent sensitivity and stability. The most common glass concentric nebulizer is the Meinhard® nebulizer, shown in Figure 8.30a. The glass Meinhard design is easily clogged by solutions containing as little as 0.1 % dissolved solids; the solids dry on the capillary tip and block it. The glass

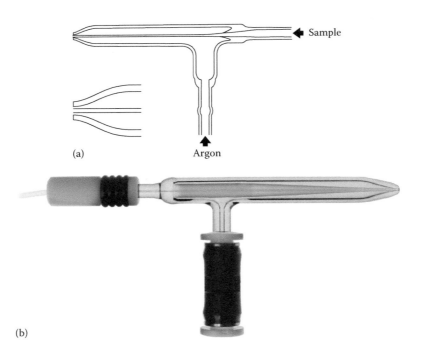

Figure 8.30 (a) Meinhard glass concentric nebulizer, showing a close-up of the tip. Courtesy of Meinhard®, Golden, CO. (www.meinhard.com). (b) Sea Spray nebulizer. Courtesy of Glass Expansion, Pty, Ltd. Melbourne, Victoria, Australia, www.geicp.com.

nebulizer cannot be used with HF-containing solutions, since HF will dissolve the glass. High-solids concentric nebulizers are available that can run solutions with up to 20 % dissolved solids without clogging. An example is the SeaSpray from Glass Expansion Pty, Ltd. (www.geicp.com), seen in Figure 8.30b, which uses an extremely smooth self-washing tip and a uniform VitriCone capillary, with a zero dead volume connector. Nebulizers made of polymer are also available for HF solutions. The sample uptake rate of a standard concentric nebulizer is about 1–3 mL solution/min.

The *cross-flow nebulizer* uses a high-speed gas flow perpendicular to the sample flow capillary tip, as shown in Figure 8.31a, to form the aerosol. Droplets formed by cross-flow nebulizers are larger than those from concentric nebulizers, so the cross-flow nebulizer is not as efficient. However, the sample capillary is a larger diameter and this minimizes clogging of the nebulizer by salt deposits. The cross-flow nebulizer has an inert polymer body and sapphire and ruby tips on the capillaries; it is more rugged and chemically resistant than a glass nebulizer.

The *Babington nebulizer* was originally developed to aspirate fuel oil. A variation of the Babington nebulizer, called a *V-groove nebulizer*, is commonly used for ICP sample introduction. A V-groove nebulizer is shown in Figure 8.31b. The liquid sample flows down a smooth-surfaced groove in which there is a small hole. The high flow of gas from the hole shears the film of liquid into small drops. This nebulizer is the least prone to clogging and is used for viscous solutions and high-salt solutions and can be made of polymer for use with HF solutions.

Microconcentric nebulizers have been developed that put solution directly in the ICP torch sample capillary, thereby eliminating the need for a spray chamber. These nebulizers, also called direct insertion nebulizers (DINs), can be used for very small (microliter) sample volumes and for direct coupling of liquid chromatographs to the ICP as an element-specific detector.

USNs operate by having the liquid sample pumped onto an oscillating quartz plate. The frequency of oscillation of the quartz plate for a commercial USN is 1.4 MHz, for example. The rapid oscillations of the quartz plate break the liquid film into an aerosol. This aerosol is of very fine, uniform droplets, unlike the aerosol from a pneumatic nebulizer. The fine aerosol contains a significant amount of solvent that must be removed before the aerosol reaches the plasma. The aerosol flows through a desolvation unit consisting of a heated section to vaporize the solvent and

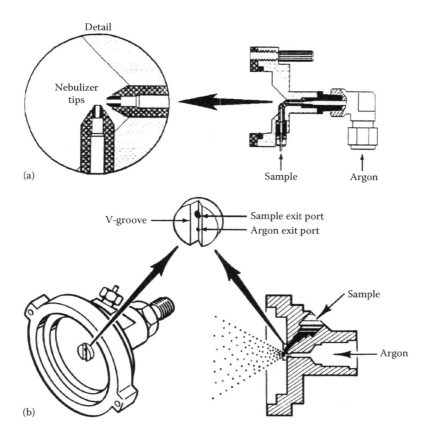

Figure 8.31 (a) A cross-flow nebulizer, with a close-up of the nebulizer tips. (b) A V-groove nebulizer. [© 1993–2014 PerkinElmer, Inc. All rights reserved. Printed with permission. (www.perkinelmer.com).]

then a chiller to condense the solvent so that a dry aerosol reaches the plasma. The efficiency of a USN is 5–10 times greater than that of a pneumatic nebulizer. This results in improved sensitivities and better DLs because more sample reaches the plasma. USNs have good stability but are prone to carryover or memory effects from one sample to the next. Carryover is caused by one sample not being washed out of the nebulizer completely before the next sample is introduced. It is often related to the ease of adsorption of certain elements on nebulizer surfaces and is a problem if the concentrations of analyte in samples vary from trace to major. USNs cannot handle high total dissolved solids solutions and are not HF resistant. They are very good at removing solvent from the sample before it reaches the plasma. Organic solvents can present problems in analysis by ICP because they tend to extinguish the plasma. Also, molecular species and radicals emit broadband spectra, which give rise to high background radiation. The USN removes the solvent by cryogenic trapping in the chiller. A commercial USN is shown schematically in Figure 8.32. Comparative DLs for some elements determined using this nebulizer are presented in Table 8.12.

Spray chambers. Pneumatic nebulizers generally require a spray chamber placed in between the nebulizer and the plasma. The spray chamber is designed to remove large droplets from the aerosol, because large droplets are not efficiently desolvated, atomized, and excited. In addition, the spray chamber smoothes out pulsations in the aerosol flow that arise from the pumping action of the peristaltic pump. Droplets with diameters of 10 μm or less generally pass through the spray chamber to the plasma. Only about 5 % of the sample solution makes it into the plasma; the rest is drained to waste. The spray chamber shown in Figure 8.21 is a Scott double-pass spray chamber. New more efficient cyclonic spray chambers are now available. The spray chamber can be made of glass or inert polymer depending on the application. Water-cooled spray chambers may be used for volatile organic solvents, to decrease the amount of solvent vapor entering the plasma. Figure 8.33a shows the individual parts needed for aqueous liquid sample introduction. A three-part demountable torch of quartz is shown, with a Tracey cyclonic spray chamber and a glass concentric nebulizer. Figure 8.33b shows the components assembled for use. The fittings are for argon gas connection to provide the auxiliary and plasma gas flows. The demountable quartz torch parts are held in a polymer torch body. Other torch designs are constructed in a single piece but still require a nebulizer and spray chamber.

Solid sampling systems. The direct analysis of solid materials is the goal of every analyst who must deal with solid samples. Dissolution of solid samples is time consuming and error prone. There have been devices designed by researchers to place solid samples directly into the argon plasma of an ICP for analysis but these have not been commercialized with the exception of the slurry nebulizer. A slurry is a suspension of fine particles in liquid. Samples that are powders or can be finely ground into powder can be suspended in water or an organic solvent and aspirated directly into the plasma. Slurry nebulizers are available in both concentric and V-groove types. Two of these available from Glass Expansion Pty. Ltd. (www.geicp.com) are the Slurry® concentric nebulizer for analysis of samples with particles up to 150 μm in diameter and the VeeSpray® nebulizer, a V-groove type in either quartz or HF-resistant ceramic that can handle particulates up to 300 μm in diameter. Slurry introduction is used for analysis of suspensions of cement powder and milk powder, for example.

The other commercial solid sampling systems all vaporize a portion of the solid sample first; the vapor or fine particulate is transported to the plasma where it is atomized and excited. The three techniques commercially available for solid sampling are electrothermal vaporization (ETV), spark ablation, and laser ablation. The advantage of separating the sampling step (using the ETV, spark, or laser) from the atomization and excitation step (in the ICP) is that each process can be optimized and controlled separately. The disadvantage of direct solid sampling is calibration of the system to achieve quantitative results. It is often necessary to purchase a series of well-characterized solids similar to the sample matrix as standards, stainless steel alloys for stainless steel samples, nickel alloys for nickel alloy samples, certified glass standards for glass samples, and so on. There are limits as to the kinds of certified solid reference materials available and they are generally very costly.

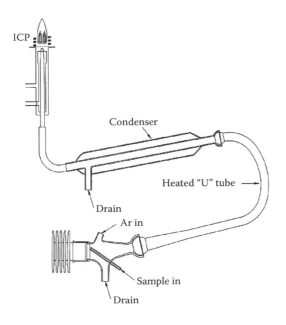

Figure 8.32 Schematic of a commercial USN, the CETAC U-5000AT+ and desolvation system. (Courtesy of Teledyne CETAC Technologies, Omaha, NE, www.cetac.com.)

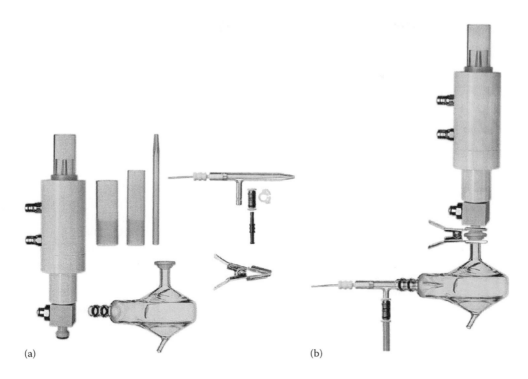

Figure 8.33 (a) The components of a liquid sample introduction system for ICP: glass concentric nebulizer, demountable three-piece quartz torch, torch body, and glass cyclonic spray chamber. (b) The components assembled for use. This torch design is typical of that used in HORIBA Scientific ICP instruments. (Courtesy of Glass Expansion Pty, Ltd., Melbourne, Victoria, Australia, www.geicp.com.)

ETV. ETV techniques adapted from GF AAS have been successfully applied as a means of introducing the vapor from a solid sample into plasma. The technique is particularly useful for organic matrices such as polymers and biological tissues since it eliminates the need to digest the sample. The electrothermal vaporizer follows the same type of temperature program used in AAS, drying, ashing, and "atomizing" the sample in sequence, although in this case, vaporization is really all that needs to be accomplished since the ICP also serves as an atomizer. The vapor formed is swept into the ICP by a flow of gas, where it is atomized and excited. The method can be used with low-volume liquid samples and solid samples. About 10–15 μL of liquid or a few milligrams of solid sample or slurry can be analyzed in this fashion. DLs for ETV-ICP-OES are given in Table 8.A.2, Appendix 8.A. The same table compares GFAAS DLs and the use of ETV with ICP-mass spectrometry (MS). The ICP-MS instrumentation will be discussed in Chapters 10 and 11.

Spark ablation. Spark ablation solid sampling uses the same type of spark source already described. The function of the spark in this case is to vaporize the solid sample; the ICP plasma can atomize any nonatomic vapor reaching it. Spark ablation is limited to the analysis of solids that conduct electricity. It is very useful for metals and alloys because it eliminates time-consuming sample dissolution and costly high-purity acids.

Laser ablation. Another technique for direct sampling of solids is laser ablation. A high-energy laser beam is directed at the solid sample. The photon energy is converted to thermal energy on interaction with the sample. This thermal energy vaporizes surface material and the vaporized material is swept into the plasma. Lasers can vaporize most types of solid surfaces, not just conductive materials, so laser ablation can be used to sample metals, glass, ceramics, polymers, minerals, and other materials. The sample does not have to be of any particular shape or size, as long as it fits into the sample chamber and a flat polished surface is not required. Laser beams can be focused to small spot sizes, so microscopic inclusions in a surface can be analyzed. The laser beam can be moved across the surface (a process called "rastering") for bulk analysis of a material. The laser beam can be focused on one position on the sample and the emission signals measured as material is vaporized from deeper and deeper below the surface, creating a *depth profile* of the sample composition. A depth profile is useful for the analysis of coatings on surfaces and multilayered composite materials, for example. Gas-filled excimer lasers such as XeCl and ArF, which emit in the UV region, can be used for laser ablation, but excimer lasers are costly and complex. Recent advances in solid-state laser design have resulted in a commercial Nd:yttrium aluminum garnet (YAG) solid-state laser. This solid-state laser can be operated at its primary wavelength, 1064 nm, which is in the infrared (IR) region, or can be frequency doubled or quadrupled to make lasers that operate at 532 or 266 nm. The UV solid-state laser shows efficient coupling of energy with most types of materials, high sensitivity, good precision, and controlled ablation characteristics. With ICP-OES, the laser ablation technique is used primarily for bulk analysis of solid samples, although depth profiling of materials is possible. The analysis of inclusions, multilayered materials, and microscopic phases in solids is often done by laser ablation and the more sensitive technique of ICP-MS (Chapters 10 and 11). A schematic of a commercial UV Nd:YAG laser system for laser ablation is shown in Figure 8.34.

8.3.2 Interferences and Calibration in Plasma Emission Spectrometry

Calibration is achieved by the same techniques used for AAS: external calibration with standard solutions or MSA. Plasma emission does not suffer from significant chemical interference because of the high temperatures achieved in the plasma, but it is good practice to matrix match the standard solutions and calibration blank to the samples to be analyzed. Matrix matching can correct for differences in viscosity and suppression or enhancement of analyte signal due to interelement effects. Matrix matching enables accurate background corrections to be made. The same solvent should be used in standards and samples, the same concentration of acid, and when possible, the same major chemical species in solution. For example, in the analysis of tungsten metal for trace elements, if the sample solutions contain 100 mg of tungsten, the standards and calibration blank should contain 100 mg of high-purity tungsten. Standards for the analysis of steels should contain the appropriate amount of high-purity iron, and so on.

Most samples for analysis by plasma emission spectrometry are dissolved in acids, because metal elements are more soluble in acids. The acidic nature of the solution prevents elements from adsorbing onto the surface of the glassware. Some bases, such as tetramethylammonium hydroxide (TMAH), are used to prepare solutions, but care needs to be taken to avoid the formation of insoluble metal hydroxides. Organic solvents may be used to dissolve organometallic compounds directly or to extract chelated metals from aqueous solution into the solvent as part of separation and preconcentration steps in sample preparation.

Calibration standards are made by dissolving very high-purity elements or compounds accurately in aqueous acid solution. In general, a concentrated stock standard is prepared and then serially diluted to prepare the working calibration standards. Many vendors now supply 1000 or 10,000 ppm single-element standards in aqueous acidic solution, with certified concentration values. One new product recently introduced is a series of stable 1 ppm stock solutions from SPEX CertiPrep (www.spexcertiprep.com). As instrument DLs continue to improve, less dilution is needed from a 1 ppm stock than from a 1000 ppm stock to make working standards. It is easier, more accurate, and more efficient to purchase these stock solutions from a high-quality supplier than to make them from scratch. Because emission is a multielement technique, the analyst combines single-element standards to make the necessary working standings that contain all the analytes of interest. If the same

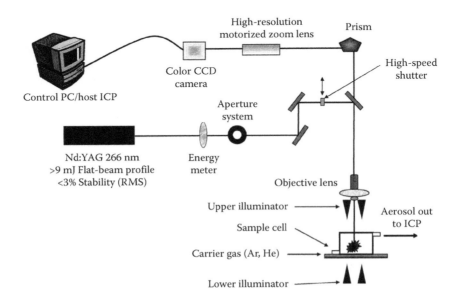

Figure 8.34 A commercial laser ablation system for solid samples, the Teledyne CETAC Technologies LSX-500. (Courtesy of Teledyne CETAC Technologies, Omaha, NE, www.cetac.com.)

set of multielement standards is used routinely, these can be ordered from the standards vendors. Multielement solutions for environmental analyses are stocked by vendors, for example. Custom mixtures can be ordered for routine analyses in industry.

For most analyses, standard solutions are aqueous solutions containing several percent of a mineral acid such as HCl or nitric acid. When mixing different elements in acids, it is important to remember basic chemistry and solubility rules for inorganic compounds. The elements must be compatible with each other and soluble in the acid used so that no precipitation reactions occur. Such reactions would change the solution concentration of the elements involved in the reaction and make the standard useless. Combinations to be avoided are silver and HCl, barium and sulfuric acid, and similar combinations that result in insoluble precipitates. The use of certain reagents, such as HF or strong bases, can etch glass and contaminate solutions prepared in glass with Si, B, and Na. HF is required to keep certain elements such as tungsten and silicon in solution, so the analyst needs to be aware that purchased stock solutions of these types of elements contain small amounts of HF that can etch glassware and nebulizers.

Because emission is inherently a multielement technique, accuracy and precision can be improved through the use of an IS. The IS is an element that is not present in the samples, that is soluble in the sample and standard solutions, and that does not emit at a wavelength that interferes with the analyte emission lines. Ideally, the IS element should behave similarly to the analyte in the plasma; matching the ionization energies of the analyte and IS is a good approach. The use of IS ion lines to match analyte ion lines and IS atom lines to match analyte atom lines is also a good practice. In a multielement analysis, it may be necessary to use more than one IS element to achieve this "ideal" behavior. The IS element is added in the same concentration to all blanks, standards, and samples. The emission intensities of the analyte elements and the IS element are measured. The calibration curve is plotted as described in Section 8.5.3. The use of the ratio of intensities corrects for instrument drift, sample uptake rate changes, and other instrumental sources of error. The use of an IS does not correct for background interferences or spectral interferences.

8.3.2.1 Chemical and Ionization Interference

Chemical interferences are rare in plasma emission spectroscopy because of the efficiency of atomization in the high-temperature plasma. For example, in FAAS, there is a severe suppression of the calcium signal in samples containing high amounts of Al, as seen in Figure 8.35. The same figure shows the lack of chemical interference in the ICP emission signal for Ca in solutions containing Al up

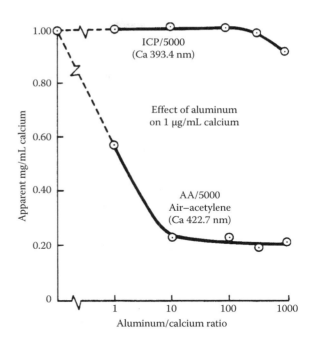

Figure 8.35 A comparison of the chemical interference effect of Al on the determination of Ca by FAAS and ICP-OES. The FAAS signal for Ca is severely depressed by small increases in the Al/Ca ratio, while the emission signal from the high-temperature plasma is not significantly affected. [© 1993–2014 PerkinElmer, Inc. All rights reserved. Printed with permission. (www.perkinelmer.com).]

to very high Al/Ca ratios. In the few cases where chemical interference is found to occur, increasing the RF power to the plasma and optimizing the Ar flow rates usually eliminates the problem.

The alkali metals Li, Na, and K are easily ionized in flames and plasmas. Concentrations of an EIE greater than 1000 ppm in solution can result in suppression or enhancement of the signals from other analytes. The EIE effect is wavelength dependent; ion lines are affected differently than atom lines. The EIE effect can be minimized by optimizing the RF power and plasma conditions, by matrix matching, by choosing a different wavelength that is not subject to the EIE effect, or by application of mathematical correction factors.

8.3.2.2 Spectral Interference and Correction

Spectral interference is much more common in plasmas than in flames due to the great efficiency of excitation in plasmas. Elements such as Fe, Mn, Ta, Mo, W, and U emit thousands of lines in a plasma source. Ideally, the analyte wavelength chosen should have no interference from other emission lines, but this is often not possible.

The most common type of spectral interference is a shift in the background emission intensity between

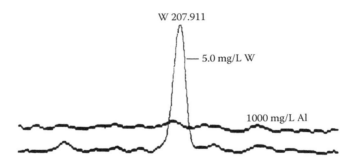

Figure 8.36 A simple background shift caused by 1000 ppm Al at the tungsten 207.911 nm line. A 1000 ppm Al solution would give an erroneous positive reading for tungsten due to the increased background from the aluminum solution. (The baseline of the tungsten solution on either side of the W emission line can be assumed to be zero intensity.) [© 1993–2014 PerkinElmer, Inc. All rights reserved. Printed with permission. (www.perkinelmer.com).]

samples and standards. Figure 8.36 shows a background shift that is a constant broad emission, from a solution containing only 1000 ppm Al, measured at the tungsten 207.911 nm emission line. The tungsten emission spectrum shown is from an aqueous acidic solution containing no Al. The signal intensity increases from bottom to top in the figure. The spectral window being examined is 0.25 nm wide, centered on the tungsten emission line. This type of background emission difference can result in two positive errors. If we were trying to measure W in an Al alloy and used aqueous W standards (no Al), the continuum emission from the Al solution would result in a positive error in a determination of tungsten, if not corrected. We would be measuring both the background and the tungsten signal and assuming it was all due to tungsten. Another error could occur if we were determining tungsten in mineral samples and did not know that some samples had a high Al content. Even if these Al-containing samples had no tungsten in them, we would erroneously report that tungsten was present, because we were measuring the background intensity inadvertently. Because the background is shifted uniformly in this example, the background signal can be corrected for by picking a wavelength near but not on the tungsten emission line, as shown in Figure 8.37. The background correction point could be on either side of the tungsten peak in this case. The intensity at this correction point is measured and subtracted from the total intensity measured at the tungsten peak. Computer software is used to set the background point and automatically make the correction.

A more complicated background is seen in Figure 8.38. The Cd 214.438 nm emission line is overlapped by a nearby broad Al emission line, so the background intensity is not the same on both sides of the Cd peak. If Cd were to be measured in an aluminum alloy, this asymmetrical background would result in a positive error. When the background emission intensity is asymmetrical, two background correction points are needed, one on either side of the emission peak, as shown in Figure 8.39. The computer software uses the two chosen points to draw a new baseline for the peak and automatically corrects for the background emission.

Figure 8.40 is an example of a very complex background pattern at the Au 267.595 nm emission line caused by closely spaced emission lines from a line-rich matrix, tungsten. The two-point approach will not work here, because of the severe overlap by the intense tungsten line on the left side of the Au line. Modern instrument software is available with mathematical programs that will model or "fit" the background and then deconvolute complex backgrounds. If such spectral deconvolution software is not available, the first choice of the analyst should be to find an alternate gold emission line with less interference. If there are no lines without significant interference, it may be necessary to separate the analyte from the matrix using classical "wet" chemistry, chelation and extraction techniques, ion exchange methods, or electrochemical methods. The chemical literature should be consulted for advice, since there are thousands of published papers on analysis by ICP-OES.

Figure 8.41 shows a case of direct spectral overlap between two emission lines, one from Pt and one from Cr. A high-resolution spectrometer will limit the number of direct spectral overlaps encountered, but not eliminate them entirely, as in this case. If the analyst has a choice of wavelengths, the first thing to do is to look for alternate emission lines for Pt and Cr. If no other analyte wavelength is available, the technique of IEC may be used. The emission intensity of the interfering element at another wavelength is measured, and a calculated correction factor is applied to the intensity at the wavelength that has the interference. The correction factor has to be determined through a series of prior experimental measurements. The use of the technique requires one interference-free emission line for each interfering element in the sample. This used to be quite complicated and time consuming to set up, requiring the preparation of known solutions, obtaining the wavelength

ATOMIC EMISSION SPECTROSCOPY

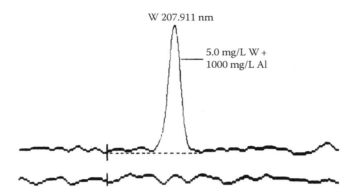

Figure 8.37 An example of single-point background correction for the simple background shift caused by Al at the tungsten emission line. A point on the left side of the tungsten emission is chosen and the intensity at this point subtracted from the intensity at the tungsten peak. This will have the effect of subtracting the increased background due to the Al matrix. The new baseline, shown as the dotted line, will return to the same baseline intensity (shown as the lower line) given by the tungsten solution in Figure 8.36. Because the baseline is symmetric with respect to the emission line, a background point on the right side of the tungsten peak could have been chosen instead; the result would be the same. [© 1993–2014 PerkinElmer, Inc. All rights reserved. Printed with permission. (www.perkinelmer.com).]

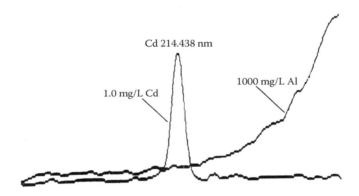

Figure 8.38 An aluminum background spectrum representing a sloping background shift at the Cd 241.438 nm line. This asymmetric background cannot be corrected using the one-point correction shown in Figure 8.37, because the background intensity is different on each side of the emission line. [© 1993–2014 PerkinElmer, Inc. All rights reserved. Printed with permission. (www.perkinelmer.com).]

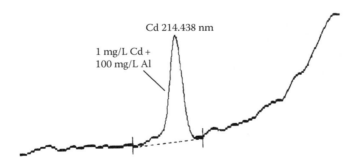

Figure 8.39 The sloping background shown in Figure 8.38 requires the use of a two-point background correction, with one correction point on each side of the Cd emission line. The peak is corrected using a straight line fit between the background correction points as shown by the dotted line (the new baseline). [© 1993–2014 PerkinElmer, Inc. All rights reserved. Printed with permission. (www.perkinelmer.com).]

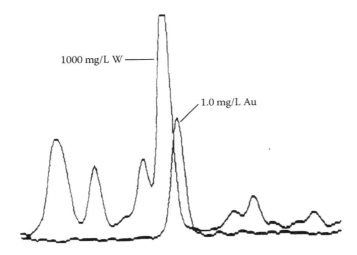

Figure 8.40 The complex, line-rich emission spectrum of tungsten severely overlaps the Au emission line at 267.595 nm. The overlap is very severe on the left of the Au peak, so a two-point approach will not work for accurate determination of trace amounts of Au in tungsten. Finding a gold emission line with less interference or extracting the gold from the tungsten matrix is a possible solution to this problem. [© 1993–2014 PerkinElmer, Inc. All rights reserved. Printed with permission. (www.perkinelmer.com).]

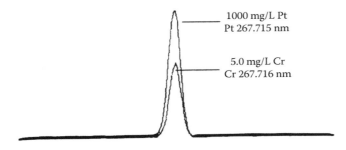

Figure 8.41 Direct spectral overlap of Pt and Cr emission lines. No background correction technique can solve this problem. A line with no interference must be found, an interelement correction (IEC) factor must be applied, or the elements must be separated chemically. [© 1993–2014 PerkinElmer, Inc. All rights reserved. Printed with permission. (www.perkinelmer.com).]

8.3.3 Applications of ICP, DCP, and MP Atomic Emission Spectroscopy

The applications of plasma emission spectrometry are very broad. The technique is used for clinical chemistry; biochemistry; environmental chemistry; geology; specialty and bulk chemical production; materials characterization of metal alloys, glasses, ceramics, polymers, and composite materials; atmospheric science; forensic science; conservation and restoration of artworks by museums; agricultural science; food and nutrition science; industrial hygiene; and many other areas. The versatility of plasma emission spectrometry comes from its ability to determine a large number of elements rapidly in a wide variety of sample matrices and over a wide linear dynamic range.

The strength of plasma emission spectrometry is rapid, multielement qualitative and quantitative analyses. ICP emission spectrometry has a linear dynamic range of about six to ten orders of magnitude, meaning that elements from trace level to major constituents can usually be analyzed in one run. This eliminates the need for dilution of samples in many cases. By contrast, the linear range of AAS is usually one to two orders of magnitude. Precision of ICP measurements is usually better than 2 % RSD and, with a minor amount of optimization, can often be better than 1 % RSD. Aqueous solutions, organic solvent solutions, and high-salt solutions can be analyzed with virtually the same precision and accuracy. In addition to the peer-reviewed literature, thousands of application notes and methods can be found on the websites of the major instrument manufacturers.

Plasma emission is used in agricultural science to study elements in soils, plant tissue, animal tissue, fertilizers, and

profile at each wavelength at different concentrations, and finally either finding an interference-free line or calculating a correction factor, but it is useful for the repetitive analysis of the same sample type when an interference-free analyte line cannot be found. Most modern instruments have powerful software programs that assist the analyst in this type of correction and may take different mathematical approaches. As an example, the HORIBA Scientific Ultima 2 high-resolution sequential ICP uses a software called Collection of Line Intensity Profiles (CLIP), which calculates the profile of each line based on the instrument configuration: focal length, slit combination, diffraction grating, and order used. Only a few minutes are needed to select lines for every element.

Alternatively, the analyte or the interfering element may be extracted from the sample, thereby eliminating the interference.

feed. Samples are usually acid digested in a heating block or by using a microwave digestion system or ashed in a muffle furnace and the ash dissolved in acid. The samples may be extracted instead of being totally dissolved; extraction results in the measurement of "available" elements. Such extracted available elements are usually considered to be biologically available to plants and animals, while digestion or ashing gives total element content. Applications include the determination of metals in beer, wine, infant formula, and fruit juice. It is possible to identify the country of origin of some food products based on their trace elemental "fingerprint."

Plasma emission is used to study elements in environmental and geological samples. The analysis of soil, water, and air for industrial pollutants is a common application. Contaminated soil and water can be analyzed as well as soil and water that have been treated to remove heavy metals, for example. This will verify if the treatment worked and provides the data engineers need to optimize and improve their removal processes. The soil and water content of naturally occurring potentially toxic metals such as As and Se can be studied as well as the uptake of As and Se by plants and animals grazing on the plants or living in the water. Particulates in air can be trapped on filters and the entire filter digested in acid or ashed prior to dissolution of the residue. This approach is used to study coal fly ash and Pb in house dust from lead-based paint. Geologists use ICP to measure major, minor, and trace components of minerals and rocks, to identify sources of metal ores, to study marine geochemistry, and to identify the origins of rocks. A picture of a mineral specimen, garnet, that has been sampled by laser ablation is shown in Figure 8.42. Laser ablation permits the spatial distribution of elements in minerals to be studied. The surface of the garnet was studied by rastering the laser across the face of the sample. Two lines of 50 µm diameter laser spots can be seen.

Biological and clinical chemistry applications of plasma emission spectrometry include determinations of those metals required for proper functioning of living systems, such as Fe, Cu, K, Na, P, S, and Se, in urine, blood, serum, bone, muscle, and brain tissue. Aluminum exposure was suspected of playing a role in Alzheimer's disease and Al concentrations in blood and tissue can be determined by emission spectrometry. No link between exposure to aluminum and Alzheimer's disease has been found. In addition to essential elements, plasma emission spectrometry can monitor metal-based therapeutic drugs in patients. Gadolinium compounds are used in patients undergoing some forms of magnetic resonance imaging (MRI) (discussed in Chapter 6) to enhance the contrast between normal tissue and tumors. The optimal dose of the Gd "contrast agent" can be determined from studies where the tissues are analyzed for Gd by ICP-OES (Skelly Frame and Uzgiris). ICP-OES and ICP-MS are going to play a major role in the analysis of pharmaceutical products with the implementation of stringent new limits on the permitted daily exposure to approximately 16 elements (generally 5–2500 mg/day, which can translate into low to sub-ppm levels in individual drugs). An example

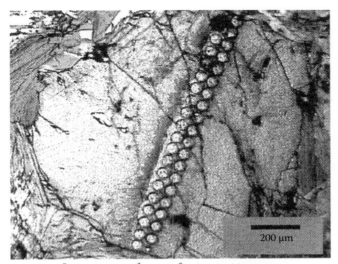

Screen capture of garnet after two spot traverses.
50 µm spots, 4 Hz, 200 shots. Crossed polarized light.

Figure 8.42 A garnet sample analyzed by laser ablation to determine the spatial distribution of elements in the mineral. Two traverses of the laser beam can be seen. Each ablated spot is 50 µm in diameter. (Courtesy of Teledyne CETAC Technologies, Omaha, NE, www.cetac.com.)

Table 8.13 **Mean Percentages of Elemental Oxides in Various Types of Glass**

Glass Usage	Al$_2$O$_3$	CaO	Fe$_2$O$_3$	MgO	NaO
Structural window	0.158	8.50	0.123	3.65	13.53
Auto windshield	0.150	8.10	0.555	3.93	12.90
Headlamp	1.370	0.017	0.060	0.017	5.40
Container	1.416	8.266	0.117	0.283	11.73

Courtesy of Tom Catalano, NYPD Police Laboratory, Jamaica, NY; Reprinted with permission from DeForest, P.R., Petraco, N., and Kobilinsky, L., Chemistry and the challenge of crime, in Gerber, S.M. (ed.), *Chemistry and Crime*, American Chemical Society, Washington, DC. Copyright 1983 American Chemical Society.

of the use of ICP-OES for this type of application is the reference by Cassap.

Plasma emission spectroscopy is used in forensic science. By measuring the elemental composition of glass fragments, it is possible to distinguish among automobile windshield glass, headlamp glass, and glass used for containers and bottles, as seen in Table 8.13.

Another forensic application is in the analysis of gunshot residue. Gunshot residue is the mixture of organic and inorganic compounds originating from the discharge of a gun and deposited on the person or clothing of an individual in close proximity to the weapon when it is discharged. The target elements looked at are barium and antimony found together above a certain baseline as well as copper and lead. Plasma emission spectroscopy, GFAAS, and methods including X-ray (Chapter 9) and neutron activation analysis have been used to study such residues. The goal is to place the discharged weapon into the hands of the shooter. This goal has not yet been attained. Other examples may be found in the text by James and Nordby.

Plasma emission spectrometry is widely used to study metals, alloys, ceramics, glass, polymers, and engineered composite materials. The major difficulty with metals and alloys is the possibility of spectral interference from the rich emission spectra of the major metal elements, but high-resolution instruments permit the routine determination of trace elements in steels, nickel alloys, copper alloys, tungsten, and other refractory metals. Most metals and alloys can be dissolved in one or more acids for analysis. Elements like tungsten and silicon require HF for dissolution; it is therefore critical to use a polymeric sample introduction system for analysis, not a quartz nebulizer and spray chamber. Glass and ceramic samples are often fused in molten salt for dissolution; the resulting bead of molten salt is dissolved in acid and water for analysis. This fusion is done at high temperatures over a burner in a suitable crucible, often a Pt or Pt/Au crucible. There are commercial automatic fusion fluxers that will fuse up to six samples at once.

Polymers, oils, and other organic materials can often be analyzed by dissolution in an organic solvent. Lubricating oil and petroleum feedstock analysis was discussed in Section 8.2.3.3. Most oils and fuels are now analyzed by ICP, DCP, or MP instead of arc or spark emission instruments, due to the better precision that can be obtained. Lead in gasoline is determined by dilution of the gasoline in organic solvent and analysis of the solvent solution. Metals in cooking oils can be determined by dilution with organic solvent and analysis of the organic solution. Organophosphates and organochlorine compounds from environmental and agricultural samples can be determined after extraction into organic solvents. The silicon in silicone polymers can be measured by dissolution of the silicone in organic solvent and analysis of the solvent solution. Such a method can be used to measure the amounts of silicone used on the two components of self-stick labels; both the adhesive on the label and the release coating on the paper backing are organosilicon polymers. In general, the ICP is very tolerant of organic solvents. Toluene, xylene, hexane, tetrahydrofuran (THF), acetonitrile, methylene chloride, and many other solvents work perfectly well in this type of analysis. In some cases, it may be necessary to increase the RF power to the plasma or to add a small amount of oxygen to the plasma to help with combustion of the organic matrix.

ICP-OES, along with ICP-MS and X-ray fluorescence (XRF), is used for the analysis of the materials in electronic equipment. The EU has established directives on the disposal of waste electrical and electronic equipment (WEEE) and the restriction of the use of hazardous substances (RoHS) in electronic equipment sold in, into, and out of the EU. The maximum allowable quantities in electrical equipment of the following hazardous substances are 0.1 % by mass for Pb, hexavalent chromium (CrVI), mercury, and polybrominated biphenyl and polybrominated diphenyl ethers (PBDEs) and 0.01 % by mass for Cd. Pb, Hg, Cd, and total Cr can be measured by ICP-OES, while the determination of hexavalent chromium requires a separation step in order to determine the oxidation state. This can be done using a hyphenated instrument, described in the following. Total bromine can also be measured by ICP-OES, but the determination of the PBDEs is generally done by GC-MS, described in Chapter 13. The WEEE/RoHS requirements have led many instrument manufacturers to have an installed method template for such analyses in their software.

8.3.4 Chemical Speciation with Hyphenated Instruments

Plasma emission spectroscopy is an elemental analysis technique; that is, it provides no information on the chemical form or oxidation state of the element being determined. The identification of the chemical state of an element in a sample is called *speciation*. For example, in environmental samples, mercury may exist in a variety of species: mercuric ion, mercurous ion, methylmercury compounds, and the extremely toxic compound dimethylmercury. Determination of mercury by ICP-OES results in total mercury concentration; chemical speciation would tell us how much mercury is present in each of the different forms. Arsenic is another element of environmental and health interest because of its toxicity. Arsenic, like mercury, exists in multiple organoarsenic compounds and multiple oxidation states as inorganic arsenic ions. Why is speciation so important? The most common arsenic compound found in shellfish is not metabolized by humans; therefore, it is not toxic. Estimating toxicity from a determination of total arsenic concentration would vastly overestimate the danger of eating shellfish, for example. In order to perform chemical speciation, we need to separate the different species in the sample and measure each of them. Chromatography is a separation method; it permits different chemical compounds in a sample to be separated from one another by the processes and instruments described in Chapters 12 through 14. The use of chromatography to separate chemical species and then introduction of the separated species into an ICP sequentially can identify different chemical forms of the same element. An instrument combination formed by the connection of two separate types of instrumentation is called a hyphenated instrument. A very common type of hyphenated instrument is one that couples a separation technique such as chromatography with a spectroscopic technique such as nuclear magnetic resonance (NMR), Fourier transform IR (FTIR), AES, or MS. GC, high-performance liquid chromatography (HPLC), ion chromatography (IC), and molecular exclusion chromatography instruments have all been interfaced to plasma emission spectrometers to identify chemical species in a sample.

For example, an HPLC coupled to a sequential ICP set to measure a silicon emission line has been used to separate and detect the small organosilicon molecules that result from the hydrolysis of high relative molecular mass (RMM+) silicone polymers (Dorn and Skelly Frame). The chromatograph separates the compounds in solution by the selective interaction of the compounds with the chromatographic *column stationary phase* and the *liquid mobile phase* that carries the sample through the column. The mobile phase with the now separated compounds flows from the column into the ICP. The ICP serves as the detector for this system; it detects and measures any silicon-containing compound that enters the plasma. This permits the study of silicone polymer degradation in a variety of environments, including degradation inside the human body. Leaking silicone polymer implants in humans have been suspected of causing serious health problems. The coupling of HPLC and ICP is one of the very few ways to separate and measure trace quantities of organosilicon compounds. This is only one of many applications of coupled chromatography-AES. There is a website dedicated to speciation methods at www.speciation.com.

The use of a helium MIP as a detector for compounds separated by GC has been mentioned and will be discussed in Chapter 13, but has been used for the determination of thiophene in benzene at ppb levels; F, C, N, and S in gasoline; organolead and organomanganese compounds in gasoline; and elements in sediment and coal. The website at www.jas.de provides a number of application notes.

8.4 GD EMISSION SPECTROMETRY

The glow discharge (GD) is a reduced-pressure gas discharge generated between two electrodes in a tube filled with an inert gas such as argon. The sputtered atom cloud in a GD source consists of excited atoms, neutral atoms, and ions. The emission spectrum can be used for emission spectrometry in the technique of GD-OES, but the GD source can also be used for AAS, AFS, and MS. The source can be used with any of the types of spectrometers discussed for plasma emission: sequential monochromator, Rowland circle polychromator, echelle spectrometer, or combination sequential–simultaneous designs. The detectors used are the same as described for plasma emission spectrometry: PMTs, CCDs, or CIDs.

8.4.1 DC and RF GD Sources

A schematic DC GD source is shown in Figure 8.43. The argon gas is present at a pressure of a few torr. The DC GD source is operated with a negative DC potential of -800 to -1200 V applied between the electrodes. The sample is in electrical contact with and serves as the cathode, as seen in Figure 8.43. The applied potential causes spontaneous ionization of the Ar gas to Ar^+ ions. The Ar^+ ions are accelerated to the cathode and a discharge is produced by argon ions colliding with argon atoms. The resulting plasma is called a GD. The electric field accelerates some Ar^+ ions to the sample surface where impact causes neutral sample atoms to be *sputtered* or uniformly removed from the sample surface. Therefore, the GD source is both the sample introduction system and the atomizer. Dissociation, atomization, and excitation of the sputtered cathode material in the GD occur by collision. The emission spectrum of the cathode, that is, of the sample, is produced. The sample must be machined to have a flat surface and must be electrically conductive. The Grimm GD source (also called a GD lamp) was designed so that a flat conductive sample served as the cathode. The cathode could be easily changed for analysis of different samples. The Grimm design results in lower self-absorption and low material redeposition on the sample surface.

A DC GD source can only be used to sputter conductive samples. Some applications to nonconductive materials

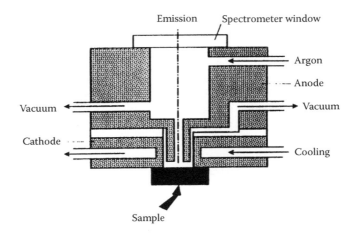

Figure 8.43 A DC GD source. A flat conducting sample serves as part of the cathode. (Courtesy of HORIBA Scientific, Inc., HORIBA Scientific, Kyoto, Japan, www.horiba.com/scientific.)

were made by mixing the nonconductive sample powder with graphite or pure copper powder and pressing flat pellets. In the 1990s, a new GD source based on RF excitation was developed (Marcus et al.). This RF GD source permits the sputtering and excitation of nonconductive materials, such as glass and ceramics, as well as conductive samples. This source still has low Ar pressure and two dissymmetric electrodes, a planar analytical sample as the cathode, and a tubular copper anode facing the sample. The sample is subjected to an RF of 13.56 MHz. If it is an insulator, the sample acts as a capacitor with respect to the alternating RF voltage. This results in an alternating charge on the sample surface through each cycle. When the sample surface is negatively charged (a very rapid process), the positive Ar^+ ions are accelerated to the negative surface and the sample material is sputtered into the GD. The charge alternates very rapidly, so the sputtering process is very efficient. Excitation and emission occur as described for the DC source.

8.4.2 Applications of GD Atomic Emission Spectrometry

8.4.2.1 Bulk Analysis

The first application of GD-OES using a DC source was for direct multielement analysis of solid metals and alloys, much like a spark source. The bulk composition of the sample was determined. The DC GD source for analysis of conductive solids has several advantages over spark emission spectrometry. The spectra produced by the GD exhibit lower background and narrower emission lines than spark sources due to less Doppler broadening. Emission lines that appear in the spectrum are almost exclusively atom lines. Linear calibration curves are achieved over a wide concentration range, which reduces the number of standards required for calibration. The GD source exhibits much lower levels of interelement and matrix interferences than arc or spark sources, since the sputtering step and the excitation step are separate events. This allows the use of one set of calibration standards for different families of materials, because there is less need to matrix match for accurate results. DLs are on the order of 0.1–10 ppm for most elements, with a precision of 0.5–2 % RSD.

Applications now include the analysis of nonconductive materials such as polymers, ceramics, and glasses using the RF Marcus-type source. The advantages are similar to those just discussed, with a major improvement in DLs and a decrease in sources of error. With a DC GD source, a nonconductive material had to be diluted with a conductive powder; this decreased the amount of analyte that could be detected in the sample. Use of an excess of conductive powder and the process of blending and pressing always introduced the possibility of contamination of the sample. This source of error has been eliminated by direct analysis of nonconductive materials.

8.4.2.2 Depth Profile Analysis

Because the GD sputtering process removes uniform layers of materials from the sample surface, it can be used to examine coatings and multilayered materials. The analysis of composition of a sample as a function of depth from the surface is called *depth profiling*. Depth profiling is a major application for GD instruments. Uniform removal of the sample can be seen in Figure 8.44, which shows a GD sputtered spot in a steel sample. Layers that are only nanometers thick can be analyzed quantitatively. Applications of depth profiling are many. Paint layers on automobile bodies, the thickness of the Zn layer on galvanized steel, carbon

ATOMIC EMISSION SPECTROSCOPY

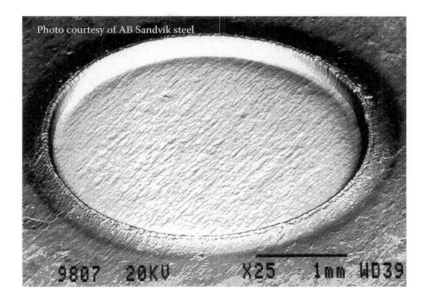

Figure 8.44 A GD "sputter spot" in a steel sample showing the uniform removal of the surface. It is this uniform sputtering process that permits accurate depth profiling of layered materials by GD-OES. (Courtesy of Sandvik Materials Technology, Sandvikem, Sweden, www.smt.sandvik.com.)

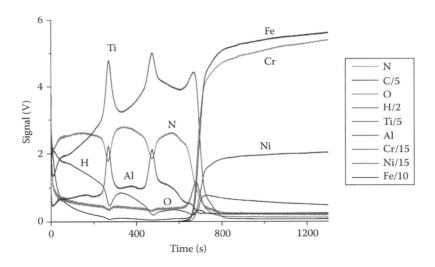

Figure 8.45 A GD depth profile of a multilayered coating on stainless steel. The *y*-axis is concentration in atomic%; the *x*-axis is time, equivalent to depth in micrometers from the surface. (Courtesy of HORIBA Scientific, Inc., HORIBA Scientific, Kyoto, Japan, www.horiba.com/scientific.)

and nitrogen concentrations as a function of depth in nitrocarburized steel, the structure of multilayer semiconductor materials, nonconductive coatings of nitrides or carbides on alloys, thermal barrier coatings on alloys, and many other complex samples can be easily characterized by depth profiling with GD-OES. Several examples are shown in Figures 8.45 through 8.47. Many other examples and application notes can be found on the HORIBA Scientific website at www.horiba.com//scientific and the LECO website at www.leco.com.

8.5 ATOMIC FLUORESCENCE SPECTROSCOPY

The process of atomic fluorescence involves the emission of a photon from a gas-phase atom that has been excited by the absorption of a photon, as opposed to excitation by thermal or electrical means. Figure 8.48 shows a few of the various types of fluorescence transitions that can occur. The solid arrows are photon absorptions or emissions; the dotted arrows are transitions that do not involve a photon. The lowest energy level represents the ground state. Two *resonance fluorescence*

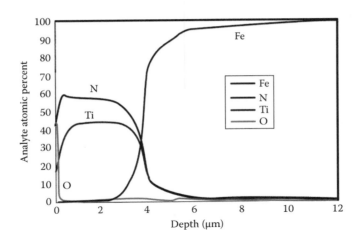

Figure 8.46 A GD depth profile of a titanium nitride coating on steel. TiN is a hard, brittle material often used to modify the surface of steel. The quantitative depth profile software on this system can verify the stoichiometry and thickness of the coating layer. (Courtesy of LECO Corporation, St. Joseph, MI, www.leco.com.)

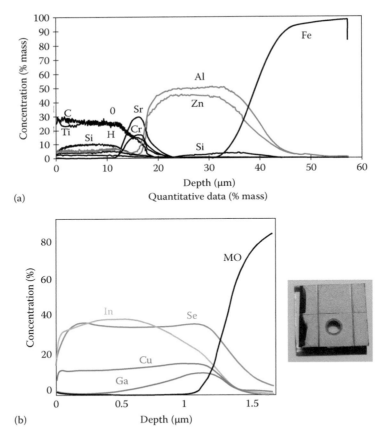

Figure 8.47 (a) A quantitative GD depth profile of a painted, galvanized steel sheet. This profile shows an initial layer containing Ti, Si, C, H, and O. The paint is white, so titania and silica are probable ingredients and latex paints would certainly contain hydrocarbons. An intermediate layer containing Sr and Cr is seen, then the third layer (with the Zn peak) shows the galvanization and the increasing Fe signal shows where the steel base begins. (b) A GD depth profile of a photovoltaic cell, seen on the right, composed of the following layered materials: CuInGaSe coating/MO/glass. MO stands for metal oxide. (Courtesy of HORIBA Scientific, Inc., HORIBA Scientific, Kyoto, Japan, www.horiba.com/scientific.)

ATOMIC EMISSION SPECTROSCOPY

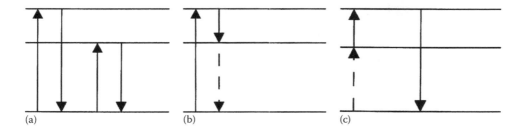

Figure 8.48 Example of atomic fluorescence transitions: (a) resonance fluorescence, (b) Stokes direct-line fluorescence, and (c) anti-Stokes direct-line fluorescence.

transitions are shown in Figure 8.48a, where the initial and final energy levels involved in the absorption and emission steps are the same. The absorption and emission wavelengths are the same in resonance fluorescence. Figure 8.48b depicts *Stokes direct-line fluorescence*, while Figure 8.48c shows an *anti-Stokes direct-line fluorescence* transition. There are other transitions that can occur, including stepwise transitions and multiphoton processes.

We know from the Boltzmann distribution that most atoms are in the ground state even at the temperatures found in GFs. Therefore, most analytical AFS uses ground-state resonance fluorescence transitions because these have the greatest transition probability and thus result in the highest fluorescent yield. Even so, the fluorescent yield is only a fraction of the total excitation source power, because not all the excitation source power is absorbed, not all absorbed photons result in fluorescence, and only a small amount of the total fluorescence is measured. A major disadvantage of the use of resonance fluorescence lines is that scattered light from the source has exactly the same wavelength as the fluorescence emission and is a direct interference.

The calculation of fluorescence yields for AFS are similar to those for molecular fluorescence (Chapter 4). Ingle and Crouch present an extensive discussion of the theory of atomic fluorescence. Given the limited commercial applications of AFS, the theory will not be covered here. It is sufficient to understand that for a resonance transition and low analyte concentration, the fluorescence signal is proportional to the analyte concentration and to the intensity of the source. This assumption is valid for sources that do not alter the population of the analyte states. Intense laser sources can deplete the population of lower energy states, including the state from which excitation occurs. This condition is called *saturation* and is discussed under applications of AFS in Section 8.5.3.

8.5.1 Instrumentation for AFS

A block diagram for an AFS spectrometer is shown in Figure 8.49. The fluorescence signal is generally measured

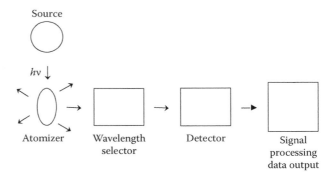

Figure 8.49 Schematic AFS spectrometer system showing the typical 90° orientation between the source and the wavelength selector and detector to minimize scattered radiation.

at an angle of 90° with respect to the excitation source to minimize scattered radiation from the source entering the wavelength selector, as was done in the measurement of molecular fluorescence (Chapter 4). While the fluorescence radiation is emitted in all directions from the atomizer, only a small fraction of it is collected and sent to the detector. This combined with the low fluorescent yield results in a very small signal in most cases.

The atomizer used can be a flame, a plasma, a GF, or a quartz cell atomizer used for HG and CVAAS systems. The wavelength selector can be any of the devices discussed in earlier chapters, from grating monochromators to nondispersive filter systems. The detector commonly used is the PMT because of its high sensitivity. The fluorescence signal is directly proportional to the intensity of the excitation source at low concentration of analyte, so intense stable light sources are needed for AFS. The most common sources are line sources and include electrodeless discharge lamps (EDLs) and HCLs (described in Chapter 7); the HCLs are generally modified to be high-intensity lamps. Laser excitation is ideal for AFS since lasers provide extremely intense monochromatic radiation and permit the use of less-sensitive nonresonance fluorescence transitions. The primary disadvantage to lasers is the cost and complexity of operation of some types of lasers. In all cases, the source is pulsed or mechanically chopped to eliminate the interference from emission by the analyte caused by thermal or other means of excitation. Resonance emission occurs at the same wavelength as resonance fluorescence; however, the fluorescence signal will be modulated at the source frequency, while the emission signal (from thermal or other excitation) will be a steady signal. A lock-in amplifier set to the source modulation frequency will measure only the fluorescence signal.

Commercial instruments are available for the determination of mercury by CV AFS and for the determination of As, Se, and other hydride-forming elements by HG-AFS. The chemistry of these methods is the same as for their AAS counterparts (discussed in Chapter 7). The instruments use either a cold quartz cell for Hg or a heated quartz cell for As, Se, Sb, Bi, Te, Sn, and other hydride-forming elements. PMT detectors with high sensitivity in the UV region are used, since the most sensitive fluorescence lines for Hg and the hydride-forming elements are in the 180–260 nm region. These commercial instruments are equipped with autosamplers and automated sample/reagent mixing systems. Flow injection systems exactly as described for AAS in Chapter 7 are used for some systems. The Hg systems are often equipped with an amalgamation trap to preconcentrate the mercury vapor on gold. The concentrated Hg vapor is released by heating the gold amalgamation trap. A major research laboratory tool (not commercially available) is the use of a GF atomizer with a laser excitation source; this technique is called GF laser-excited AFS (GF-LEAFS).

8.5.2 Interferences in AFS

AFS would be expected to suffer from chemical interferences and spectral interferences, as do the other atomic emission techniques we have discussed. The focus will be on the commercial instruments available and on the GF-laser system.

8.5.2.1 Chemical Interference

Chemical interference in the CV Hg AFS method is not a significant problem. The chemistry of the reaction is well understood and the interferences are both few and rare. Large amounts of gold and several other metals interfere in the release of Hg atomic vapor, but the interferences are not encountered commonly in the biological and environmental samples routinely analyzed for mercury. Chemical interference in HG is also well understood. The elements must be converted to the appropriate oxidation state, for example. For routine analysis of biological and environmental samples using standard analytical protocols, AFS determination of As and Se does not suffer from significant chemical interference. GF-LEAFS using modern furnace techniques has not been reported to suffer from serious chemical interference.

8.5.2.2 Spectral Interference

Spectral interferences can be significant in AFS and some of them are unique to AFS. The total fluorescence signal at the detector can include light scattered from the source, fluorescence from nonanalyte atoms and molecules, background emission, analyte emission, and analyte fluorescence. To measure only analyte fluorescence, the other spectral interferences must be eliminated or corrected for. As noted, if the source is modulated, analyte emission is not measured. Fluorescence from nonanalyte atoms and molecules or background fluorescence is rare. The multistep nature of the fluorescence process makes it unlikely that nonanalyte species will absorb and fluoresce at the same wavelengths as the analyte. Any background fluorescence that occurs can be compensated for by having a suitable blank and subtracting the blank signal from the measured total signal. Background emission and scattered light are the major sources of spectral interference in AFS. Scattered light and background emission can both be compensated for by subtraction of a suitable blank. Scattered light is more significant in resonance fluorescence, where the excitation and fluorescence wavelengths are the same, than in nonresonance fluorescence. Wavelength selection can distinguish between the source scattering and the analyte signal in nonresonance fluorescence because the fluorescence wavelength is not the same as the excitation wavelength. The advantage to the use of a laser as the excitation source is that less-sensitive nonresonance transitions can be used and the scattered light eliminated as a source of interference by wavelength selection.

Spectral interferences are minimal for CV mercury AFS and for HG-AFS, due to the chemical separation step that is used prior to excitation. Background signals in GF-LEAFS are smaller than those from GFAAS, but Zeeman background correction, described in Chapter 7, has been used for GF-LEAFS to provide accurate background correction.

8.5.3 Applications of AFS

The AFS phenomenon has been studied extensively since the late 1950s. An overview of the research into AFS can be found in the book by Ingle and Crouch, the chapter by Michel, and in the handbook by Settle (see the bibliography). AFS is primarily used for quantitative analysis. The accuracy and precision of AFS depend on the atomizer used but are generally comparable to CV AAS for mercury or HG for As and Se, on the order of 1–2 % RSD. The use of a GF as the atomizer results in precisions typical of GFAAS, on the order of 5 % RSD for solutions to 10–30 % RSD for

ATOMIC EMISSION SPECTROSCOPY

solid samples. Linear ranges vary from three to eight orders of magnitude.

8.5.3.1 Graphite Furnace Laser-Excited AFS

The use of a GF atomizer and a laser excitation source combine to give the technique called GF-LEAFS. While this is a research technique limited at present to a few laboratories, the technique has dramatic advantages over other atomic spectroscopic methods. Small solid samples, such as polymer pieces, can be analyzed directly, as described by Lonardo et al., with DLs that are two to four orders of magnitude better than GFAAS and one to two orders of magnitude better than ICP-MS. There are definite advantages to direct solid sampling, such as the elimination of costly high-purity acids for dissolution, reduction in contamination of the sample, minimization of loss of volatile elements during preparation, and faster sample analysis. Direct solid sampling has been used to determine a number of elements, including Pb, Tl, and Te in alloys, while many other elements have been determined on dissolved samples. References to these applications, from the research group of Professor Robert Michel, can be found at http://chemweb.chem.uconn.edu.

The use of an intense laser source can change the population of states in the analyte. In particular, the lower state, from which excitation occurs, can be depleted and an excited state populated using high-intensity sources such as pulsed dye lasers. For a simple two-level system (ground state and first excited state), complete saturation is defined as having one half of the atoms in the excited state. As we know from the Boltzmann distribution, this is not the usual distribution of atoms. If we graph the logarithm of the relative fluorescence signal versus the logarithm of the source intensity, we obtain the *saturation curve*, shown schematically in Figure 8.50. In the region of low source intensity, which would occur with low laser power or with conventional EDL or HCL sources, the signal is linearly related to the source intensity. This region demonstrates the need for high-intensity conventional sources. The higher the intensity, the higher is the signal in the linear region of the curve for conventional sources. With very-high-intensity sources, the signal becomes independent of the source intensity because the population of ground-state and excited-state atoms has been altered. The intense laser source has altered the population of atoms and caused the rates of absorption and fluorescence to become equal. This condition is called saturation or optical saturation. Saturated fluorescence gives several advantages. The fluorescence signal is the maximum obtainable signal, so the DLs are the lowest obtainable. The signal is independent of fluctuations in the source intensity. Saturation eliminates quenching of the fluorescence signal by collisional deactivation and improves linearity by decreasing self-absorption of the fluorescence signal. In practice, sufficient laser power is used to just saturate a given transition. This provides the advantages just listed while minimizing scattered radiation.

8.5.3.2 Mercury Determination and Speciation by AFS

Commercial CV AFS systems for mercury are available because of the dramatic improvement in DLs for Hg using AFS. DLs for Hg on the Teledyne Leeman Labs Hydra II$_{AF}$ Gold instrument using amalgamation to preconcentrate the mercury are reported to be less than 0.05 ppt with a dynamic range of 0.1 ppt to 250 ppb. That is at least 10–100 times more sensitive than CVAAS. This system is completely

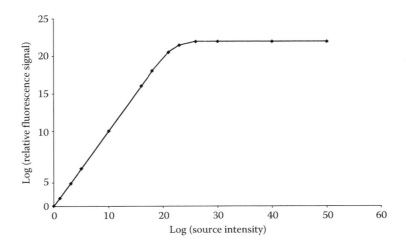

Figure 8.50 Schematic saturation curve for AFS. At low intensity, the signal is linear with source intensity. At high source intensity, the signal is independent of the source intensity. This is the saturation region.

automated, performs an analysis every 3 min with amalgamation, and has an autosampler that can hold 180 samples. These instruments are used to determine low levels of Hg in biological fluids and tissues, environmental samples, air, and gases. The commercial instruments can be coupled to an LC or GC for speciation of mercury compounds in samples, as can be seen at www.psanalytical.com. The P S Analytical GC-based Hg speciation system consists of a GC coupled to an AFS with an integrated pyrolysis control unit. The detector is a PMT with a low-pressure Hg lamp as the excitation source. Application notes and system diagrams are available from Teledyne Leeman Labs (www.teledyneleemanlabs.com) and P S Analytical Ltd., United Kingdom (www.psanalytical.com).

8.5.3.3 Hydride Generation and Speciation by AFS

The hydride-forming elements, including As, Se, Sb, Bi, Te, Ge, Sn, and others, can be determined routinely using commercial HG-AFS instruments. (Note that upper case HG means hydride generation, not mercury, Hg.) A number of these elements, like Hg, are important analytes because of their toxicity. DLs for most of these elements are in the low ppt (ng/L) range, with linear ranges up to 10 ppm. The Lumina 3300 AFS from Aurora Biomed, Inc. (www.aurorabiomed.com), states DLs for Hg and Cd of 1 ppt, while most other hydride-forming elements have DLs of about 10 ppt, for example, with <1 % RSD. The DLs are, in general, 5–10 times better than the equivalent HG AAS DLs. Speciation for these elements can be accomplished by coupling an LC or GC with the AFS system. Mercury, arsenic, and selenium are found in both inorganic and organic forms, with the toxicity of the forms varying widely. For example, hydrogen selenide is extremely toxic, while selenomethionine is an essential nutrient. Selenium in fact has the smallest range between essentiality and toxicity of any essential element. Inorganic arsenic species are toxic, while compounds such as arsenobetaine and arsenocholine, found in shellfish, are nontoxic. Using an LC attached to a P S Analytical Millennium Excalibur, their HG-AFS system, P S Analytical has demonstrated the separation and detection of five selenium species, five arsenic species, and the two inorganic oxidations states of Sb (not simultaneously). Details and chromatograms can be found at www.psanalytical.com.

8.6 COMMERCIAL ATOMIC EMISSION SYSTEMS

Providing lists of commercial manufacturers and types of instruments is a dangerous business, as instrument company takeovers, sell-offs, and the usual yearly introduction of "new" instruments, sometimes with only minor changes but with new names, occur on a frequent basis. The instrument company websites should be consulted for current details. There are also many companies who manufacturer instruments with limited geographic distribution who are not included in the discussion. This list is not inclusive or comprehensive, but is meant to give the student a feel for the types of instruments commercially available at the present time.

8.6.1 Arc and Spark Systems

Commercial DC arc emission spectrometers and commercial spark spectrometers are available as stand-alone laboratory instruments, as sample stands for an existing spectrometer, and as small, mobile field units. This list is not comprehensive, but examples include the Prodigy DC Arc system from Teledyne Leeman Labs (www.teledyneleemanlabs.com); a combined arc/spark probe, the Q4 Mobile, from Bruker AXS (www.bruker.com); and polychromator spark systems with PMT and/or CCD detectors from Bruker AXS, PANalytical (www.panalytical.com), SPECTRO Analytical Instruments (www.spectro.com), Thermo Fisher Scientific (www.thermo.com), and HORIBA Scientific (www.horiba.com/scientific). Many instrument companies make portable spark instruments, some small enough to be handheld, for use in sorting metal alloys at scrap yards, confirming the identity of alloys received at loading docks, and for at-line quality control in production. Examples include the Q2 Ion and the Q4 Mobile, from Bruker AXS. Instrument manufacturers also offer an interchangeable spark ablation stand for analysis of solid samples using an ICP spectrometer. An example is the Thermo Fisher Scientific SSEA.

8.6.2 ICP, DCP, and MP Systems

Sequential and simultaneous ICP systems of many types are available from numerous instrument manufacturers. Most are available in radial or axial view or dual view (radial and axial). Most spectrometers can be purged to reach 160 nm and some can reach 120 nm. Examples of some of the companies who make ICPs and DCPs are given. HORIBA Scientific makes a line of PMT-based high-resolution sequential systems, simultaneous systems, and combination sequential–simultaneous systems, with 120–800 nm spectral range and ten orders of magnitude linear working range. PerkinElmer, Inc., makes a scanning ICP with CCD detection and a line of simultaneous CCD-based echelle systems, with newer instruments using 50 % less Ar than older instruments. Teledyne Leeman Labs makes an echelle ICP spectrometer with an aperture grid available as a sequential system with rapid-scanning PMT detection, a simultaneous system with a PMT array, and a combination system, with a spectral range of 170–780 nm and an option for high-throughput analysis of two samples per minute. Agilent Technologies makes a sequential ICP with PMT detection and a simultaneous system with CCD detection. SPECTRO makes several instruments including simultaneous systems covering 175–770 nm and a simultaneous CCD Paschen–Runge axial or radial system covering 130–770 nm. Thermo Fisher Scientific offers ICP echelle spectrometers with CID detection and a sequential diffraction grating instrument with PMT detection.

Spectrometer systems are now available with nitrogen-purged sealed optics for routine analysis from 120 to 800 nm. ICP systems that have this capability include instruments from HORIBA Scientific and SPECTRO Analytical Instruments. There is one commercial DCP available from Thermo Fisher Scientific, an echelle spectrometer with CID detection. The ability to use lines in the "vacuum UV" permits determination of the halogens and other elements whose most intense lines occur at <160 nm.

Most systems, especially for use in industry, include an autosampler that allows unattended, overnight operation of the instrument. Computer software in commercial instruments controls the autosampler and the instrument, collects the data, performs the calculations, prints out the results, and shuts down the instrument when the analysis is completed. Most instruments have computer-controlled safety interlocks that shut down the plasma or the instrument and autosampler in the event a problem is detected.

The only MP-based atomic emission spectrometer as of this writing, leaving aside the specialty GC MIP detector, is the Agilent Technologies 4100 MP-AES (www.agilent.com). This instrument is meant primarily to compete with FAAS, providing a multielement technique without the need for multiple light sources or flammable gases and with DLs in the ppt–ppb range for most elements.

8.6.3 GD Systems

Commercial GD spectrometers with RF sources are available from HORIBA Scientific (www.horiba.com/scientific), LECO (www.leco.com), and Spectruma Analytik (www.spectruma.de); DC sources are available from LECO and from SPECTRO Analytical Instruments (www.spectro.com), among others.

8.6.4 AFS Systems

Teledyne Leeman Labs (www.teledyneleemanlabs.com) makes Hg AFS systems; Aurora Biomed, Inc. (www.aurorabiomed.com) makes an AFS system for Hg and all hydride-forming elements; and P S Analytical Ltd. (www.psanalytical.com) makes commercial AFS systems for Hg and HG, as well as systems for speciation of Hg, As, Se, and related elements using coupled chromatography-AFS.

8.7 LASER-INDUCED BREAKDOWN SPECTROSCOPY

Laser-induced breakdown spectroscopy (LIBS) is a relatively new atomic emission spectroscopy technique that uses a pulsed laser as the excitation source. LIBS is also referred to as laser spark spectroscopy (LASS) and laser-induced plasma spectroscopy, with the unfortunate acronym of LIPS. The technique was developed in the early 1960s, after the invention of the laser, but the high cost and large size of lasers and spectrometers made this a specialized research tool until the 1990s. The early development of LIBS is covered in the reference by Myers et al. Recent advances in miniature fiber-optic spectrometers and small high-peak-power lasers have led to the development of compact, portable analytical systems for laboratory and field use. There is, as of August 2012, a LIBS system in operation on Mars, on board the Mars rover named Curiosity.

8.7.1 Principle of Operation

The output of a pulsed laser, such as a compact Q-switched Nd:YAG laser, is focused onto the surface of the material to be analyzed, using simple focusing lenses. For the duration of the laser pulse, typically 10 ns, the power at the surface of the material can exceed 1 GW/cm^2. This creates an induced dielectric breakdown or plasma spark. At these high power densities, a very small amount of material, less than a microgram, is ejected from the surface by the process of laser ablation. A short-lived but highly luminous plasma with instantaneous temperatures reaching 10,000 °C is formed at the surface of the material. In LIBS, the laser itself creates the plasma. Within the hot plasma, the ejected material becomes atomized, ionized, and excited. At the end of the laser pulse, the plasma quickly cools by expansion at supersonic speeds. The excited atoms and ions emit their characteristic emission lines as they return to lower energy levels. Using collecting lenses and fiber-optic connections to an optical spectrometer, the elemental composition of the material can be determined.

LIBS can be used to analyze any material—solids, liquids, gases, slurries, and the like—for all elements. Many elements can be detected at <1–100 ppm levels, with a smaller group of elements having DLs in the 100–500 ppm range. The Applied Photonics Ltd. website at www.appliedphotonics.co.uk has a periodic table color-coded to show LIBS DLs. Due to the extremely small amount of sample ablated, on the order of nanograms to micrograms of material per laser shot, the method is considered nondestructive. No sample preparation is usually required. Because LIBS is an optical technique, remote analysis is possible, as will be seen. Heating of the sample is negligible as the average power incident on the sample is generally less than 1 W.

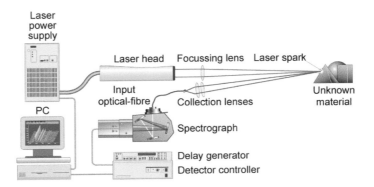

Figure 8.51 LIBS schematic showing the major components. (Image reproduced courtesy of Applied Photonics Ltd., Yorkshire, U.K., www.appliedphotonics.co.uk.)

8.7.2 Instrumentation

A schematic of a modern LIBS system is shown in Figure 8.51. The components required are a high-peak-power Q-switched laser, optics to focus the laser onto the sample surface, laser pulse emission and detection timing circuitry, optics to gather the emitted light and transfer it to the spectrometer, and computer software for spectral analysis.

Different lasers have been used for LIBS, but the most common is a Q-switched Nd:YAG laser operated at its fundamental wavelength of 1064 nm, at the frequency doubled wavelength at 532 nm, or higher harmonics at 355 and 266 nm. The requirements for the laser are that it has a short pulse duration or pulse width (1–20 ns) and a minimum of 10 mJ per pulse. Laser pulses in LIBS are typically 10–100 mJ. Neodymium-doped YAG is the material used in this optically pumped solid-state laser. A Q-switch is an optical device that controls the laser's ability to oscillate, in this case by waiting for a maximum population inversion in the Nd ions. By keeping the laser Q-factor low, the laser is able to store higher levels of energy. The stored energy is then released in the form of extremely short pulses of high power. (It is important to note that these high-power lasers constitute a safety hazard to people and a potential fire risk and should only be operated by trained, experienced analysts with the appropriate safety equipment.)

Time-gated detectors that allow the optical emission from the laser plasma to be recorded at some time delay after the laser pulse are required to accurately capture the emission spectra. For the first few microseconds after the "ignition" of the laser spark, the plasma emits a strong white light continuum (also called bremsstrahlung), which decays as the plasma cools; the characteristic atomic and ionic emission lines only appear as the plasma cools. A detector delay on the order of several microseconds after the laser pulse is used to eliminate interference from the continuum radiation. The principle is demonstrated in Figure 8.52.

Spectrometer designs are similar to those already discussed, including broadband diode array, echelle, Czerny–Turner, and Paschen–Runge spectrometers, with miniature echelle and CCD array systems most suited to portable LIBS systems. The LIBSCAN 25+ (Applied Photonics Ltd.) uses up to six compact CCD array spectrometers covering the 185–900 nm range. Figure 8.53 shows the field-portable LIBSCAN 25+.

The TSI ChemReveal Desktop LIBS comes in a variety of configurations, including four Czerny–Turner spectrographs with CCD detectors, and has a wavelength range of 190–950 nm. PMT detectors can be used, especially with multiple PMTs, with the drawback being that the wavelengths that can be monitored are fixed and limited to the number of detectors.

The spectra are processed by computer software that now often includes advanced chemometric methods (see the discussion of chemometrics in Section 5.7.2.1). Methods such as partial least squares (PLS) and principal component analysis are used by instrument software to construct calibration curves and provide quantitative analysis. LIBS software on many commercial systems supports library matching, sample classification, qualitative, semiquantitative, and quantitative analysis including preloaded libraries and calibrations, and customer-built libraries and calibrations.

A low-pressure Ar purge of the sample surface is used to prevent particulates from the laser ablation of the sample from dirtying the system optics, to generate a stronger emission signal and to displace air from the LIBS spark to enable the determination of N and O from samples.

Instruments can be run as open-beam systems by focusing the laser onto a sample that is not in an enclosure, or as benchtop enclosed systems, and as standoff systems where the laser propagates through the atmosphere over significant distances and the emission is collected by a telescope.

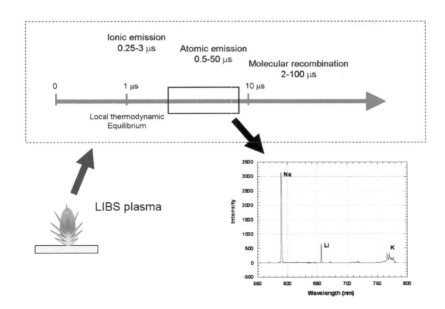

Figure 8.52 Schematic time sequence for atomic emission after initiation of the LIBS plasma. (Courtesy of Photon Machines, Inc., Bozeman, MT, www.photon-machines.com.)

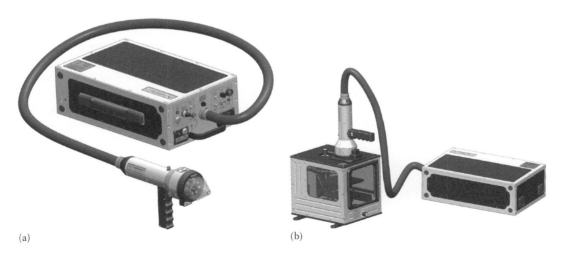

Figure 8.53 (a) LIBSCAN 25+ showing the handheld probe that contains the compact flashlamp-pumped laser, the laser beam optics, the UV/VIS and VIS-near-IR (NIR) plasma light collection optics, and an optional miniature color video camera in the transparent head of the probe. The optical spectrometers, laser power supply, Li-ion battery, and other electrical components are located in the instrument "box." The flexible umbilical contains the optical fibers used to transmit the plasma light to the spectrometers and electrical cabling for the head. (b) The LIBSCAN 25+ with an optional sample chamber, allowing operation of the system as a Class 1 laser system. (Images reproduced courtesy of Applied Photonics Ltd., Yorkshire, U.K., www.appliedphotonics.co.uk.)

8.7.3 Applications of LIBS

There are significant advantages to LIBS when compared with other atomic emission techniques discussed earlier and XRF techniques, discussed in Chapter 9. LIBS is capable of detecting all elements, including low atomic number elements like H, Be, B, C, N, and O, at DLs in the ppm range. Some of these elements cannot be detected by most atomic emission instrumentation or by XRF, especially at such low levels. Little or no sample preparation is required. If the sample surface is coated with an oxide layer, paint, oil, or other contaminants, the laser may be used to rapidly clean the surface to expose the underlying material. Coatings several hundred microns thick can be rapidly removed in this way. Since the laser can remove surface coatings, this allows LIBS to be used for depth profiling of coated and multilayer materials, as long as the sample can be ablated at the laser power available although not with the resolution of a GD system. Bulk simultaneous determination of all elements can be done in approximately 20 s. LIBS can rapidly provide an elemental fingerprint of a sample.

LIBS is used in industry, biomedical applications, analysis of materials in hazardous environments, such as inside nuclear reactors, analysis of materials underwater, and for the remote detection of hazardous materials, including explosives. Samples may be solids, liquids, gases, slurries, sludges, and other mixtures. Applications for LIBS (courtesy of Applied Photonics Ltd.) include the following:

- Remote, noninvasive characterization and identification of materials.
- Remote detection and elemental analysis of hazardous materials, including radioactive, high-temperature, and chemically toxic substances.
- *In situ* detection of radioactive contamination.
- *In situ* compositional analysis of steel components in difficult-to-access environments, such as nuclear reactor pressure vessels.
- Rapid identification of metals, alloys, and plastics during scrap recycling. LIBS can be used to identify pieces moving at high speed on a conveyor belt.
- Positive metal identification of critical components during manufacturing and assembly.
- Online compositional analysis of liquid metals, alloys, and glasses for process control.
- *In situ* identification of materials submerged in water.
- Depth profiling and compositional analysis of surface coatings (e.g., galvanized steel).
- Online monitoring of particles in air (stack emission monitoring).
- Compositional analysis of objects with complex shapes.

Other applications of LIBS are in archeology; geology; gemology; art restoration and conservation; military and homeland security analyses of explosives, landmines, chemical, or biological warfare agents; bioaerosols and biohazards; forensics and biomedical studies of bones, hair, teeth, cancer tissue, bacteria, and other biological materials; and real-time environmental monitoring of water, air, exhaust gases, and industrial waste.

8.7.3.1 Qualitative Analysis and Semiquantitative Analysis

Qualitative analysis is performed by identifying the specific atomic emission lines present in a LIBS spectrum. Commercial instrument manufacturers now include atomic spectral libraries and automatic line identification software.

Since all elements can be detected by LIBS, qualitative elemental "fingerprints"—lines and intensities—can be used to match samples. This can be used for quality control, for example. Biological materials will give a characteristic fingerprint of multiple lines and intensities from oxygen, nitrogen, hydrogen, carbon, and perhaps heteroatoms like S and P. An interesting example is the use of LIBS spectra to fingerprint bacteria and easily distinguish between similar bacteria, for example, distinguishing between anthrax (*Bacillus anthracis*) and the very common, related, but not deadly, *Bacillus thuringiensis*, *Bacillus cereus*, and others. Based on demonstrated LIBS capabilities, it may soon be possible to perform *in vivo* analysis of tissue and diagnosis of pathogens in real time. The article by Rehse et al. presents an excellent overview of biomedical applications of LIBS.

Gemstones are very valuable materials, especially rare colored gemstones of certain types, such as deep blue diamonds like the Hope Diamond. Artificial stones and treated, dyed, and coated gemstones are readily available, often being sold as "real" gemstones. In the evaluation of gemstones, using a UV 355 nm frequency-tripled Nd:YAG laser produces a smaller focal point on the gemstone and is only absorbed at the surface, causing an inconsequentially small area of laser "damage" while still producing useful spectra. Using this approach with a TSI LIBS system, applications scientists were easily able to identify beryllium-treated sapphires and cobalt coatings of topaz, treatments used to change the color of the stones.

Diamonds, gold and gold-bearing ores, and minerals that are major sources of tin, niobium, tantalum, and tungsten are mined in various locations around the world. Profits from some of these "conflict minerals" may be used to finance wars. LIBS is being used to study the mineral "fingerprints" with the goal of identifying the geological source of these materials (Arnaud, 2012; Hark et al., 2012; Harmon et al., 2011). In addition, researchers have used LIBS to trace the origins of Neolithic artifacts made of obsidian, a volcanic glass (Remus et al.).

The determination of composition without having to use calibration standards is a goal of every spectroscopic technique. There are a number of spectral modeling approaches available for various techniques that often permit semiquantitative analysis. (How "semiquantitative" the results are is often a matter of how the models have been developed—"semiquant" can mean just about anything from qualitative to quantitative. We'll see in Chapter 9 that a powerful standard-less calibration approach is available for XRF.) For LIBS, a software is available that uses a database of atomic emission lines to create a theoretical spectrum using defined plasma parameters (Yaroshchyk et al.), permitting composition to be determined in some matrices without calibration standards with results that are in "good agreement," according to the authors, with certified values.

8.7.3.2 Quantitative Analysis

Quantitative analysis usually requires the use of standards and/or certified reference materials (CRMs), the selection of an appropriate elemental optical emission line, and, in most cases, the selection of a normalization line, used as an IS. Calibration curves are then constructed using the normalized peak area versus concentration, as previously described for calibration using an IS. When there is a dominant matrix component for which the concentration will remain approximately constant across the calibration set, it is best to use an emission line from that matrix element for normalization. This approach helps minimize effects due to changes in plasma conditions caused by shot-to-shot fluctuations in laser intensity. Alternatively, chemometric correlation analysis of the entire observed spectrum with the concentration of the analyte can be used to construct calibration curves automatically. In general, RSDs of 5–10 % are readily achievable. To improve quantitation, sample preparation methods such as pressing pellets may improve results for soils and sediments and fusion with salts to convert the sample into a glass bead can eliminate matrix effects. Fusion was discussed in Chapter 1 and is used extensively in XRF analysis (Chapter 9).

As discussed earlier, the RoHS directives of the EU limit the amount of certain elements, including Pb, in electronic devices and equipment. Scientists at Applied Spectra, Inc. (www.AppliedSpectra.com), used their RT-100 series LIBS system to demonstrate the quantitative determination of Pb in tin-plated semiconductor lead-frame packages without the need for sample preparation. The RT-100 used a 50 mJ laser at 1064 nm, with a laser spot size of 120 μm to determine lead over the range 20–1000 ppm Pb using the Pb 405.7 nm emission line. Quantification was done by calibration with a set of four standards obtained from NIST and MBH Analytical Ltd. The LIBS analysis showed excellent agreement with analyses by ICP-OES and laser ablation-ICP-MS, performed by third-party laboratories. The advantages to the LIBS method are fast measurement time, the ability to measure the Pb content at various stages in manufacturing, and the elimination of toxic waste and expensive, high-purity acids, making the method cheaper and safer than methods like AAS and ICP-OES.

For the quantitative determination of boron in borosilicate glass and the boron-containing ores colemanite and ulexite, which are used as feedstocks in the glass industry, samples were ground to <60 μm, pressed into pellets using a hydraulic press, and analyzed using the TSI, Inc., LIBS system. Using the oxygen emission line at 777 nm for normalization, B was measured at the 206.723 nm emission line. The analysis took 20 s. The DL for boron was estimated to be 0.001 % and quantitative results from 5 % to 50 % calculated as B_2O_3 were obtained with RSDs of 1–2%.

The lanthanide elements, $Z = 58–71$, are widely used in industrial materials. Fluorescent lamp phosphors, nuclear reactor control rods, coloring agents in glass and ceramics, and as constituents in laser materials and solid-state devices are some examples of materials that use lanthanide elements. The current global supply of lanthanides is controlled by China; as a result, many countries are evaluating the reopening of mines where low levels of lanthanide-containing ores may be found. Acid digestion or fusion for determination by ICP-OES or ICP-MS is tedious and often requires the addition of ISs and the use of expensive high-purity acids and fluxes. XRF, which can analyze solid samples directly, suffers from overlap of the lanthanide fluorescence lines with those of the transition elements. LIBS can readily measure the lanthanide elements in ores at levels at or below 10 ppm and the ability to do this in the field allows geologists to quickly determine if the ore is worth mining.

8.7.3.3 Remote Analysis

Because LIBS is an optical technique, only optical access to the material is required to carry out an analysis. Remote measurements using an instrument equipped with a telescope for a direct line of sight method allows analysis of objects up to distances of about 10 m, while fiber-optic probes permit analyses at up to 100 m. This is extremely useful for analysis of hazardous or high-temperature materials and for deployment in hostile environments. An exciting use of LIBS in the harsh environment of the planet Mars is currently underway. The Mars rover named Curiosity is equipped with a LIBS system called the Chemistry and Camera (ChemCam) instrument, capable of determining the elemental composition of rocks and soil up to 23 ft (7 m) away and through the thin dust layer that covers the Martian surface. This system uses an IR laser, a telescopic lens, and a trio of spectrometers with over 6000 UV, visible, and IR channels covering a range of 240–850 nm. This LIBS system was a collaboration between the French space agency Centre National d'Études Spatiales (CNES) and the US Los Alamos National Laboratory (LANL). The first test of the ChemCam system occurred on August 19, 2012, when the laser was fired at a rock about 10 ft away. Articles and pictures (Figure 8.54) can be found on the websites of the National Aeronautics and Space Administration (NASA), LANL, and CNES, and magazines such as *Sky and Telescope* or *Astronomy Magazine*. The ChemCam-equipped rover is still active as of July 10, 2023, having been functional for 3985 days: 10 years, 328 days.

LIBS has an advantage on Mars, in that the atmospheric pressure is only about 1 % of Earth's atmospheric pressure at sea level, allowing the plasma to expand and become brighter (Figure 8.55).

Figure 8.54 (a) Artist's conception of the Mars rover Curiosity, showing the raised mast head. (b) Close-up of the actual mast head with the ChemCam. (Image credits: Courtesy of NASA/JPL-Caltech/LANL, Pasadena, CA.)

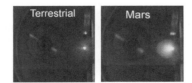

Figure 8.55 LIBS plasmas on Earth and Mars. Studies at LANL show LIBS plasmas at typical atmospheric pressures on Earth (a) and Mars (b). At the lower atmospheric pressure on Mars, the plasma is larger and brighter. (Image credit: Courtesy of LANL, Pasadena, CA.)

The first LIBS spectrum from the ChemCam was collected from a rock near the rover. The rock (Figure 8.56a), called "Coronation," is about 3" (7.6 cm) across and located about nine feet (2.7 m) from ChemCam on the mast. The spectrum is shown in Figure 8.56b. The plot is a composite of spectra taken over 30 laser shots at a single 0.016" (0.4 mm) diameter spot on the rock. Major elements are identified and the upper left inset shows an expansion of the 398–404 nm region, identifying minor elements Ti and Mn. An inset on the right shows the hydrogen and carbon

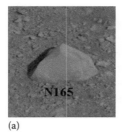

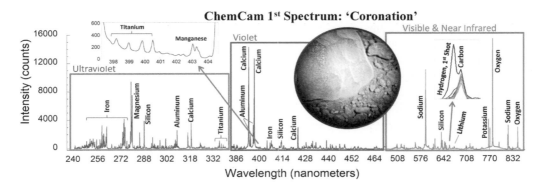

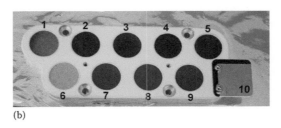

Figure 8.56 (a) "Coronation," the first rock studied on Mars by LIBS. (b) The first ChemCam spectrum of the rock, showing major and minor elements present. (c) The calibration target for ChemCam. (Image credits: courtesy of NASA/JPL-Caltech/LANL/CNES/IRAP, Pasadena, CA.)

peaks. Carbon is present in carbon dioxide in the Martian atmosphere. The hydrogen peak was present only on the first laser shot, indicating that the element was only on the very surface of the rock. The heights of the peaks do not directly indicate relative abundances of the elements, since some emission lines are more easily excited than others. A preliminary analysis of the spectrum indicates that it is consistent with basalt, a type of volcanic rock known to be abundant on Mars from previous explorations.

The ChemCam is calibrated using the target shown in Figure 8.56c, consisting of nine materials representing those expected on Mars. Circles 1–4 are glass samples representing igneous rocks, #5 is graphite, and circles 6–9 are ceramic samples representing sedimentary rocks. The square plate of Ti, #10, is used for laser diagnostics and wavelength calibration of the ChemCam.

Remote (standoff) LIBS systems have been built by Applied Photonics Ltd. with a range capability of >100 m for defense departments. They also built a transportable standoff LIBS with a 20 m range to characterize radioactive materials in a hot cell at the Sellafield, UK, high-level nuclear waste vitrification plant by directing the laser beam through the lead-glass window of the cell. LIBS is an excellent tool for remote and *in situ* detection of uranium oxide fuel located in hard-to-reach places, underwater or covered in sludge, for example, at a damaged nuclear reactor site such as the Fukushima Daiichi plant in Japan, because it is essentially immune to the effects of ionizing radiation. Conventional gamma spectroscopy would be swamped by the high gamma field from fission products such as Cs-137. LIBS, therefore, can be used to measure quantities of low specific activity radionuclides such as U-235, U-238, Pu-239, Pu-241, and Tc-99 in high gamma field environments.

8.7.4 Commercial LIBS Systems

Companies that make commercial LIBS systems or custom-made devices include Applied Photonics Ltd. (www.appliedphotonics.co.uk); Applied Spectra, Inc. (www.appliedspectra.com); Bertin Technologies (www.bertin.fr); Energy Research Co. (www.er-co.com); IVEA (www.ivea-solution.com/libs/); LTB Lasertechnik Berlin (www.ltb-berlin.de); Marwan (www.marwan-technology.com); Ocean Optics, Inc. (www.oceanoptics.com); TSI, Inc. (www.tsi.com), Photon Machines, Inc. (www.photon-machines.com); and StellarNet (www.stellarnet-inc.com). Applications notes, videos, and technical information on LIBS are available on most of these companies' websites.

8.8 ATOMIC EMISSION LITERATURE AND RESOURCES

There are many journals that publish articles on atomic emission spectroscopy. *Analytical Chemistry*, *Applied Spectroscopy*, *Spectrochimica Acta Part B*, and *The Analyst* publish articles on atomic emission spectroscopy as well as other analytical methods. The *Journal of Analytical Atomic Spectrometry* is a more focused journal, as the name implies. Applications articles that use atomic emission spectroscopy for analysis of specific materials may be found in journals related to the field of application, such as geology, agriculture, food science, pharmaceutical science, polymer science, and the like.

Methods and applications can be found online from government, academic, and instrument manufacturer sites. The US Environmental Protection Agency (EPA) methods that use AES for analysis of environmental samples can be found at www.usepa.gov, for example. Most of the instrument companies whose websites are given in the chapter have applications notes and methods on their websites. Many also have tutorials and videos on the various techniques.

Tables of atomic spectral lines may be found in print in the books and handbooks by Boumans, Harrison, Robinson, and Zaidel listed in the bibliography. Atomic spectral wavelengths can be found at the US NIST website, www.nist.gov. Most commercial instrument software contains a library of emission lines.

8.9 COMPARISON OF ATOMIC SPECTROSCOPIC AND ICP-MS TECHNIQUES

The atomic spectroscopic techniques discussed in Chapters 7 and 8 and ICP-MS discussed in Chapters 10 and 11 each have advantages and disadvantages in the determination of elements in real samples. A summary of the capabilities of each technique and a comparison of the techniques is given in Appendix 8.B. Most major instrument companies have "selection guides" on their websites that compare their atomic spectroscopy instruments by DL, analytical working range, sample throughput, interferences, cost, and so on.

SUGGESTED EXPERIMENTS

Flame Emission

8.1 Warm up the flame photometer or flame atomic absorption spectrometer in emission mode, following the manufacturer's directions. Using an air–acetylene burner, set the flow rates of air–acetylene to (a) oxidizing flame, (b) stoichiometric flame, and (c) reducing flame, following the manufacturer's directions. Note the change in flame color, shape, and size. With each flame, measure the emission at 589.0 and 589.5 nm. This is from sodium contamination in the flame gases and from dust in the atmosphere. Aspirate into the flame (a) deionized (DI) water from a plastic container, (b) DI water from a soda-lime glass container (your typical "glass" jar or bottle), and (c) DI water from a borosilicate glass container (e.g., Pyrex® or Kimax®). (Allow water to sit in each container overnight.) Compare with freshly drawn DI water. Compare the relative sodium contamination of the water by the containers (emission intensity at 589.0 nm).

8.2 Prepare aqueous solutions containing amounts of potassium varying from 1.0 to 100.0 ppm of K^+. Aspirate each sample into the flame. Measure the intensity of emission at 766.0 and 404.4 nm. Plot the relationship between the emission intensity of each line and the concentration of the solution aspirated into the burner.

8.3 Repeat Experiment 8.2, but add 500 ppm of Cs to each solution. Note the change in the emission intensity. Explain.

8.4 Prepare solutions containing 20 ppm of Mg, Ca, Zn, Na, Ni, and Cu. Measure the emission intensity of Mg at 285.2 nm, Ca at 422.7 nm, Zn at 213.9 nm, Na at 589.0 nm, Ni at 341.4 nm, and Cu at 324.7 nm. Note the wide divergence in emission intensity, even though the metal concentration was constant. Compare the sensitivity of your results with that of the results of Experiments 8.1 and 8.2. (This experiment can be done with an ICP-OES spectrometer as well as a flame emission spectrometer.) If both a flame emission system and an ICP system are available, do Experiments 8.1 through 8.4 on both instruments and compare the results. Explain your observations in terms of the excitation source temperature, type of nebulizer, and any other instrument variables that apply.

8.5 Take a sample containing an unknown quantity (about 3 ppm) of sodium. Split it into four 10 mL aliquots. To the first aliquot, add 10 mL of standard sodium solution containing 2 ppm Na. To the other aliquots, add solutions containing 4 and 6 ppm of Na (10 mL of each, one per aliquot). To the remaining aliquot, add 10 mL of DI water so that all solutions now have a volume of 20 mL. Measure the intensity of Na emission at 589.0 and 589.5 nm. Using the MSA, calculate the Na concentration in the original sample.

8.6 Determine the sodium content of the drinking water in your chemistry department or city.

Arc/Spark or ICP Emission

For those of you who still own working emission spectrographs with photographic plate detectors, a series of experiments is suggested. Some of these experiments can be modified to suit other emission sources for solid samples—GD, spark, DC arc with modern PMT, or CCD/CID detectors. Experiments 8.11 through 8.14 can be made suitable for ICP-OES by dissolution of the solid samples in acid. For example, the point of Experiment 8.10 for an emission spectrograph is that the matrix can be a problem. Does that problem exist for ICP-OES if the samples are all in solution? Matrix effects (Experiment 8.10) can be studied in ICP-OES by running one of the copper salts in 5 %, 10 %, and 20 % nitric acid, then 5 % nitric acid plus varying amounts of NaCl (e.g., 5–20 % NaCl) or by running Experiment 8.12 using external calibration standards in acid alone and then using the MSA for brass, bronze, and so on to see if there is a difference in slope that would indicate a matrix effect. Note that if you want to dissolve SiO_2 (Experiment 8.10), you need to use HF and have an HF-resistant sample introduction system in your ICP. HF is EXTREMELY toxic, causes severe burns that can result in death if not treated immediately, and must only be used with the permission and under the supervision of your instructor. Alternatively, go through a molten salt fusion or substitute another material.

8.7 Take several 1 g portions of dry graphite powder to be used as the matrix. Add powdered anhydrous copper sulfate to make a powder containing 0.01 % Cu. Load this into the graphite electrode of an emission spectrograph and fire under spark conditions for 3 min. Record the spectrum on the photographic plate.

8.8 Repeat Experiment 8.7, but record the spectrum after exposures of 1, 3, 5, 7, 10, 15, and 20 min. Measure the intensity of the emission lines at 327.4 and 324.7 nm. Plot the intensity of each line and a background intensity at a suitable position against the exposure time. Note that after a certain period of time, the intensity of the line does not increase, but the background does.

8.9 Using graphite powder and anhydrous copper sulfate powder, prepare powders that contain 0.01 %, 0.05 %, 0.1 %, 0.2 %, 0.5 %, 0.7 %, and 1.0 % Cu. Insert each powder separately into the emission spectrograph and expose for 3 min. Measure the intensity of the lines at 324.7 and 327.4 nm. Plot the relationship between intensity and the concentration of the copper. Note that the intensity of the lines goes through a maximum. These are reversible lines.

8.10 Repeat Experiment 8.9, but use (a) MgO, (b) SiO_2, and (c) powdered Al metal as the matrix powder. Note that the changes in the matrix compound cause a variation in the emission from the copper, even though other conditions are constant. This is the matrix effect.

8.11 Repeat Experiment 8.9, but use (a) $CuCl_2$, (b) $Cu(NO_3)_2$, (c) Cu acetate, (d) CuO, and (e) CuS as the source of copper. Note the variation in the intensity of the copper emission when different salts are used, even though other conditions are constant. What causes this effect?

8.12 Determine the percentage of copper in (a) brass, (b) bronze, (c) a quarter (or other suitable coin), and (d) a nickel (or other suitable coin). Record the emission spectra of these samples.

8.13 Record the emission spectra of (a) Zn, (b) Sn, (c) Ni, (d) Fe, (e) Co, and (f) Cd.

8.14 Based on the emission spectra obtained in Experiments 8.12 and 8.13, what elements are present in (a) brass, (b) bronze, (c) a US quarter coin, and (d) a US nickel coin (or other suitable coins)?

ICP Emission

8.15 Determine the elements present in powdered infant formula by weighing a known amount of formula, dissolving the formula in DI water, and aspirating it into an ICP. Aqueous multielement standards can be used for calibration. Compare your results to the label. Compare your results from this dissolution procedure to a sample of formula you prepared by wet ashing or dry ashing.

8.16 Determine the elements present in a multivitamin tablet prepared by digestion in concentrated nitric acid and dilution to a suitable volume in DI water. You may need to filter out a white insoluble powder (or allow it to settle) after digestion. Pharmaceutical tablets often contain silica or titania, which will not dissolve in nitric acid. Aqueous multielement standards in the same concentration of nitric acid should be used for calibration.

PROBLEMS

8.1 What is the difference between a monochromator and a polychromator?

8.2 When a sample is introduced into a flame, what processes occur that lead to the emission of radiant energy?

8.3 Does a spectral interference in the standards for atomic emission cause a positive or negative intercept in the calibration curve? Briefly discuss the possibilities.

8.4 List three elements that are commonly analyzed by flame photometry. What is the major use of flame photometry?

8.5 In preparing a calibration curve for the determination of potassium by AES, the following data were obtained:

Concentration of K (ppm)	Emission Intensity
0.1	10
0.1	410
0.2	800
0.4	1620
0.6	2450
0.8	3180
1.0	3600
1.2	4850

(a) Plot the results using a spreadsheet. Determine the equation of the calibration curve using linear regression. (b) Which of the results appear(s) to be in error? (c) What should the emission intensity of these point(s) have been assuming the relationship between concentration and intensity is linear? (d) What are you going to do with the data point(s) that appear to be in error? Explain your reasoning (you may want to reread Chapter 1).

8.6 The emission intensities from five samples were, respectively, 1020, 2230, 2990, 4019, and 3605 units. The blank prepared with the samples has an intensity of 398. Using the final calibration curve prepared in Problem 8.5 after you have decided what to do with any suspect data, calculate the potassium concentration of the five samples.

8.7 The following results were obtained using the standard addition method:

	Emission Intensity from Calcium
Sample alone	5.1
Sample + 0.1 ppm Ca	6.2
Sample + 0.2 ppm Ca	7.3
Sample + 0.5 ppm Ca	10.6
Flame alone	0.7

What was the concentration of calcium in the original sample?

8.8 Can a given nonanalyte element present in a sample cause both spectral and nonspectral interferences in AES? Explain, with diagrams as necessary.

8.9 In preparing a calibration curve for the determination of lithium, the following data were obtained:

Standard Concentration (ppm Li)	Radiation Intensity
0.1	360
0.2	611
0.4	1113
0.6	1602
0.8	2091
1.0	2570

Plot the curve. What was the radiation intensity of the flame with no lithium present (i.e., what is your y intercept)? The radiation intensity from a sample containing lithium was measured three times; the intensity readings were 1847, 1839, and 1851. What was the lithium concentration in the sample? Express the result including the uncertainty.

8.10 Spectral emission lines in flames are the same width as atomic absorption lines in flames, on the order of 0.01 nm. Why is the spectral bandpass in an atomic emission spectrometer much smaller than that in an AAS?

8.11 What causes background emission in AES? Consider all of the excitation sources discussed.

8.12 What is meant by chemical interference? Compare flame photometry, arc/spark, ICP, and GD sources in the amount of chemical interference exhibited by each technique.

8.13 How does emission spectrometry differ from flame photometry? How does this difference affect the number of elements that are detectable by each method? Explain.

8.14 Describe and illustrate a graphite electrode used for analysis of powdered samples by DC arc emission spectrometry.

8.15 What effect does the matrix have on the intensity of the emission signal in arc/spark emission, ICP emission, and GD emission?

8.16 Define the preburn time. Why is it used? Would you expect that (a) gallium and (b) tungsten could be satisfactorily determined in steel using the same preburn time? Explain your answer.

8.17 Sketch the components of a Rowland circle polychromator. What is the advantage of this design over a scanning monochromator?

8.18 Describe the components and operation of an echelle monochromator. What are its advantages over a Rowland circle design?

8.19 What is the advantage to using a CCD or CID detector for an echelle spectrometer instead of a PMT or multiple PMTs?

8.20 What is meant by matrix effect in AES? Give at least two examples.

8.21 How is an ICP formed? What gas is used to form the plasma? What other kinds of plasmas are there?

8.22 Draw a cross-flow nebulizer and explain its operation. What are its advantages and disadvantages when compared with a USN?

8.23 Describe how laser ablation is used for sample introduction. What kinds of samples would you run by laser ablation? Explain your answer.

8.24 What is the dynamic analytical range of a typical ICP? How does this compare to AAS? Is there any advantage to using ICP based on its dynamic range?

8.25 What are the advantages of using a GD source over a spark source for analysis of solids?

8.26 Explain what is meant by a depth profile and how it can be used in analysis of materials.

8.27 Compare atomic absorption and atomic emission for qualitative analysis. What are the advantages and disadvantages of each approach?

8.28 Compare atomic absorption and atomic emission for quantitative analysis. What are the advantages and disadvantages of each approach?

8.29 Draw the processes that occur for (a) resonance fluorescence, (b) Stokes direct-line fluorescence, and (c) anti-Stokes direct-line fluorescence.

8.30 What are the advantages of using resonance fluorescence? What is the major disadvantage to its use?

8.31 Why are high-intensity line sources preferred for AFS over less intense or broadband sources?

8.32 What are the advantages of using a laser as the source for AFS? What are the disadvantages?

8.33 Briefly describe the spectral interferences in AFS and how they are eliminated or accounted for.

8.34 What is meant by saturated fluorescence? Why is saturated fluorescence used for analytical AFS?

8.35 Compare atomic fluorescence with atomic absorption and atomic emission. What are the advantages and disadvantages of each technique?

8.36 Describe how LIBS works. What are the advantages of LIBS versus other atomic emission techniques?

8.A APPENDIX

Table 8.A.1 Flame Atomic Emission DLs

Element	Wavelength (nm)	Type of Flame	Flame Conditions	DLs (ppm)
Aluminum	396.2	NA	R	0.005
Antimony	259.8	NA	R	1.0
Arsenic	235.0	NA	Ox	2.2
Barium	553.6	NA	R	0.001
Beryllium	234.9	NA	R	
Bismuth	223.1	NA	R	6.4
Boron	249.7	NA	R	7
Cadmium	326.1	NA	R	0.5
Calcium	422.7	NA	R	0.0001
Cerium	569.9	NA	R	30
Cesium	852.1	NA	Ox	0.5
Chromium	425.4	NA	R	0.005
Cobalt	345.4	NA	R	0.5
Copper	327.4	NA	R	0.01
Dysprosium	404.6	NA	R	0.07
Erbium	400.8	NA	R	0.04
Europium	459.4	NA	R	0.0006
Gadolinium	440.2	NA	R	1.0
Gallium	403.3	NA	R	0.01
Germanium	265.1	NA	R	0.5
Gold	267.6	NA	R	0.5
Hafnium	368.2	NA	R	
Holmium	405.4	NA	R	0.02
Indium	451.1	NA	R	0.002
Iridium	380.0	NA	R	30
Iron	371.9	NA	R	0.05
Lanthanum	579.1	NA	R	0.1
Lead	405.8	NA	R	0.2
Lithium	670.8	AA	Ox	5×10^{-6a}
Lutetium	451.9	NA	R	1.0
Magnesium	285.2	NA	R	0.005
Manganese	403.1	NA	R	0.005
Mercury	253.7	NA	R	2.5
Molybdenum	390.3	NA	R	0.2
Neodymium	492.4	NA	R	0.2
Nickel	341.5	NA	R	0.1
Niobium	405.9	NA	R	1.0
Osmium	442.0	NA	R	
Palladium	363.5	NA	R	0.1
Phosphorus	253.0	NA	R	1.0
Platinum	265.9	NA	R	2.0
Potassium	766.5	AA	Ox	15×10^{-6a}
Praseodymium	495.1	NA	R	1.0
Rhenium	346.0	NA	R	0.3
Rhodium	369.2	NA	R	0.1
Rubidium	780.0	AA	Ox	0.001
Ruthenium	372.8	NA	R	0.02
Samarium	476.0	NA	R	
Scandium	402.0	NA	R	0.012
Silicon	251.6	NA	R	4.0
Silver	328.1	NA	R	0.02
Sodium	589.0	AA	Ox	8×10^{-6a}
Strontium	460.7	NA	R	0.001
Tantalum	481.3	NA	R	5
Tellurium	238.3	NA	R	2.0
Terbium	431.9	NA	R	0.4
Thallium	535.0	NA	R	0.6
Thorium	570.7	NA	R	0.02
Thulium	371.8	NA	R	0.02
Tin	284.0	NA	R	0.5
Titanium	399.9	NA	R	0.2
Tungsten	400.9	NA	R	0.5
Uranium	591.5	NA	R	
Vanadium	437.9	NA	R	0.01
Ytterbium	398.8	NA	R	0.002

(continued)

Table 8.A.1 (Continued) Flame Atomic Emission DLs

Element	Wavelength (nm)	Type of Flame	Flame Conditions	DLs (ppm)
Yttrium	362.1	NA	R	0.3
Zinc	213.9	NA	R	77
Zirconium	360.1	NA	R	3

Note: AA, air–acetylene flame; NA, nitrous oxide–acetylene flame; R, reducing flame conditions; Ox, oxidizing flame conditions. Wavelengths and flame conditions from Analytical Methods for Atomic Absorption Spectrometry, PerkinElmer Inc., Shelton, CT, 1994 [© 1993–2014 PerkinElmer, Inc. All rights reserved. Printed with permission. (www.perkinelmer.com)]; and Robinson, used with permission.

[a] Data from Model 300 flame photometer, courtesy of Buck Scientific Company, Norwalk, CT.

Table 8.A.2 Approximate DLs for Electrothermal Atomizer (ETA) Techniques

	ETA-ICP-MS		GFAAS		ETA-ICP-OES	
	ng/mL		ng/mL		ng/mL	
Elements	(2 µL)	pg	(10 µL)	pg	(10 µL)	pg
^{107}Ag	0.08	0.16	0.01	0.1	0.1	1
^{27}Al	0.03	0.05	0.2	2	1.5	15
^{75}As	0.05	0.1	1	10	20	200
^{44}Ca	0.7	1.4	0.5	5	0.002	0.02
^{114}Cd	0.15	0.3	0.01	0.1	0.2	2
^{52}Cr	0.1	0.2	1	10	0.3	3
^{39}K	1.5	3	0.1	1	110	1100
^{23}Na	0.2	0.4	1	10	80	800
^{58}Ni	0.47	0.93	1	10	0.9	9
^{208}Pb	0.1	0.3	0.2	2	2	20
^{78}Se	5.7	11.4	1	10	6	60
^{28}Si	2.7	5.4	5	50	10	100
^{64}Zn	0.2	0.4	0.05	0.5	0.05	0.5

Note: This table presents a comparison of GF AAS with the use of a GF (ETA) as the sample vaporization step in conjunction with ICP-MS and ICP-OES. The solution DL, the sample volume, and the absolute DL in picograms are given for each technique. The isotope measured is specified for the ICP-MS technique; the isotope number has no meaning for the AAS or ICP-OES results.

8.B APPENDIX: COMPARISON OF ATOMIC SPECTROSCOPIC ANALYTICAL TECHNIQUES

Atomic spectroscopy is the most widely used technique for determining elements in almost every conceivable matrix. The techniques commercially available and in wide use are FAAS, GFAAS, and ICP-OES. Commercial AFS is limited to the determination of Hg, As, Se, and a few other elements using the CV and hydride chemistry discussed in Chapter 7. Elemental MS using an argon ICP as the ionization source (ICP-MS) is now also widely used for ultratrace determination of elements and is discussed in Chapters 10 and 11. Although ICP-MS complements atomic spectroscopy, it is not a spectroscopic method; mass/charge ratio is determined, not radiant energy.

Metal and metal alloy producers and materials research laboratories routinely use spark and arc emission spectrometry, and GD spectrometry has been gaining in popularity because it can analyze nonconducting solids directly. It is difficult to make direct sensitivity and accuracy comparisons between AAS, ICP, ICP-MS, and the arc/spark/GD methods, since the latter techniques have sensitivities based on solid samples (limits of detection [LODs] in µg/g) and the former are solution techniques (LODs in µg/mL). While the solution DLs of GF or ICP may appear to be "better" than arc/spark or GD, the actual dilution factor for a real solid sample needs to be considered to compare the techniques for analysis of solids.

Atomic absorption, atomic fluorescence, and AES have the same goal—the qualitative or quantitative determination of elements, particularly metals and semimetals. These methods have a number of things in common and a number of differences. The choice of method for a particular analysis depends on the type of sample to be analyzed and the elements to be determined. While some generalizations can be made, it is important for the student to understand that the generalized statements below about sensitivity and other parameters may not hold true for all elements in all matrices and may not hold true for all instrument manufacturers of a given type of instrument. DLs based on the S/N ratio for AAS, ICP, and ICP-MS instruments, primarily from one manufacturer, are listed in Appendix 8.A. It can be seen that ICP-MS holds a distinct advantage over the other techniques as far as sensitivity is concerned for most elements in solution. LODs for other atomic spectroscopic techniques are given in Table 8.12, and Tables 8.A.1 and 8.A.2. The literature should be searched for specific applications when needed. The precision of these techniques is similar when operating in the linear region and above the limit of quantitation; that is, above 10× the 3σ DL. The relative precision is 1–2 % for FAAS, 1–2 % for ICP-AES, 1–10 % for furnace AAS, and 1–2 % for ICP-MS. It is possible to achieve precisions less than 1 % for most of these techniques with simple matrices and proper optimization of the operating parameters.

Table 8.B.1 lists several of the parameters that are important in the selection of the instrumental method to be used for an analysis.

8.B.1 Flame Atomic Absorption

FAAS is an excellent choice for a small laboratory. The instrumentation is relatively inexpensive and easy to operate. FAAS is a mature technique with many published textbooks, applications, and comprehensive guides for analysis and

ATOMIC EMISSION SPECTROSCOPY

Table 8.B.1 Analytical Parameters

Analytical accuracy
Analytical precision
Analytical concentration range
Amount of sample available
Total analytical time per sample
Total analytical cost per sample
Bulk analysis or region specific analysis
Elemental information or speciation
Instrument cost (initial and operating costs)
Degree of automation
Skills required to develop methods and interpret data
Analytical interferences
Difficulties with contamination

method development (the "cookbooks" from instrument manufacturers). Spectral interferences are few and well documented. "Flame" AAS instrumentation is also a platform for CV Hg and HG determination of As, Se, and Tl by AAS, so the instrumentation is a staple of environmental laboratories. On the negative side, the refractory elements like boron, tungsten, zirconium, and tantalum are difficult to determine because the flame is not hot enough to atomize these analytes. The low temperature of the flame results in chemical interference in many samples. Elimination of chemical interference may require the addition of reagents or the use of the slower MSA calibration or even extraction of the analyte from the interfering substances. If more than two or three analytes must be determined in each sample, even automated AAS is a much slower process than ICP-OES. Several milliliters of solution are required per analyte, so multiple analytes per sample may need 25 mL or more of solution. The operating cost of a FAAS is low; the gases used are relatively inexpensive. The cost of HCLs and EDLs may be significant, depending on what elements are required, but most lamps last for years if operated correctly.

8.B.2 GFAAS

The analytical advantages of furnace AAS are that its DLs are about two orders of magnitude better than either FAAS or ICP-OES and that very small sample volumes can be analyzed. The entire analytical process is slow but is completely automated in modern instruments. Environmental labs rely on automated unattended GFAAS analysis of the priority pollutant metals in drinking water and wastewater. Like FAAS, it usually determines a single element at a time. Furnace AAS can suffer from severe interferences, including high background signals, possible loss of volatile analytes, and formation of refractory carbides by analytes such as B, W, Ta, and Nb. The use of matrix modifiers, L'vov platforms, and transverse heating designs for the atomizer has alleviated some of these problems. Slurries can be analyzed directly and automatically by GFAAS, eliminating the sample preparation time for some classes of materials. A GFAAS instrument costs more than an FAAS instrument; a modern instrument with both flame and GF atomizers costs about the same as an echelle ICP-OES instrument, approximately $60,000–70,000. Operating costs for furnace AAS are significant due to the high cost of the graphite parts, especially the tubes and platforms, which must be replaced frequently.

8.B.3 ICP-OES

The emission signal can be recorded for all elements of interest either simultaneously or in rapid sequence, thus making ICP-OES much faster than AAS techniques if many elements must be determined in each sample and requires less sample volume in the process. The linear working range for ICP-OES is much greater than that of AAS, so trace to major components of a sample can often be determined at the same time. However, the measurement of analyte emission at high temperature requires a monochromator of better resolution than required for AAS; thus, ICP-OES instrumentation is more expensive than FAAS. The high-temperature plasma is very efficient at atomizing and exciting all of the metal and semimetal elements, as well as some nonmetal elements. The high temperature of the plasma virtually eliminates chemical interferences of the type found in AAS. The sensitivity and S/N ratio are better than those for FAAS. Because of the high-temperature source, ICP-OES may suffer from spectral interference due to the large number of emission lines associated with each element. This is particularly troublesome when materials to be analyzed have very complex spectra, like iron, nickel, uranium, tungsten, molybdenum, and the rare earths. With a fixed polychromator system, it may be necessary to measure and apply correction factors if an interference-free line cannot be found for an analyte. In most sequential systems and in the simultaneous echelle systems with solid-state detectors, it is usually possible to find a suitable analyte line; if not, the interference must be determined and corrected for. Operating costs are high; the instruments require a large amount of argon and may require a high-purity purge gas; the torches and quartz nebulizers are expensive and fragile, and pump tubing must be replaced at least daily.

8.B.4 ICP-MS

ICP-MS is discussed in Chapters 10 and 11, but it is a powerful trace elemental analysis method and many analytical texts cover it with atomic spectroscopy. Its advantages and disadvantages are compared here for completeness.

For most elements, the ICP ionizes a significant amount of the elemental vapor. This serves as the ionization source for the MS. Coupled together, ICP-MS measures the analyte concentration by the mass spectrum, that is, by the ratio of mass/charge for separated ions rather than by the emission of light. The instrumentation is more expensive than AAS and slightly more expensive than high-end ICP-OES instrumentation. ICP-OES instruments generally cost $65,000–130,000; ICP-MS instruments cost $160,000–210,000 as of this writing. Much of the cost is due to the requirement for a high-vacuum system. ICP-MS possesses the fast, multielement capabilities of ICP-OES and better DLs than the GF. The linear working range is very broad, and DLs are in the ppt range for most elements. The extreme sensitivity of ICP-MS means that "environmental" contamination of samples is a serious problem; sample preparation areas and the instrument itself may need to be housed in a "clean lab" with filtered air. The ability of the mass spectrometer to measure the elements' individual isotopes with great accuracy adds some very special applications for ICP-MS, such as calibration by isotope dilution and the use of stable isotopes as tracers instead of more hazardous radioactive isotopes. Specific applications are discussed in Chapter 11. ICP-MS is not suitable for the analysis of solutions high in dissolved solids, unlike ICP-OES. Sensitivities for some elements are not as good as those by ICP-OES; S, Si, C, and Ca are examples. Calcium suffers from a direct isobaric interference from Ar at its most abundant isotope. Isobaric polyatomic Ar-containing interferences can limit LODs for As, Se, and some other elements, depending on the design of the ICP-MS. Operating costs are similar to those for ICP-OES; additional very costly expense items are the skimmer and sampler cones, especially if Pt cones are required. The installation and upkeep costs of a metal-free clean lab are very high.

8.B.5 LIBS

A relative newcomer to the atomic emission instruments commercially available, LIBS has some significant advantages over the other emission techniques and over ICP-MS and XRF. It is virtually nondestructive and often can be used with no sample preparation. It has the ability to detect all of the elements in virtually any sample type. Versions are available that are field-portable, and versions are available for remote/standoff analysis. Instrumentation is relatively inexpensive. The major disadvantage is the lack of LIBS spectral libraries for easy fingerprint matching, but that is being remedied as the instrumentation becomes more common.

BIBLIOGRAPHY

The reader will note that occasionally multiple editions of books are cited as sources or reference works. The reason for this is that sometimes different figures are used in different editions, which may in fact be written by different authors. The sources found in the bibliography will be the most recent versions of a particular work.

In some cases instrument manufacturers and suppliers are cited by providing a URL to the exact internet location. We have endeavored to ensure that these links are functional at the time of publication, but clearly these links change. If a link has become nonfunctional, a good approach is to use your browser's advanced search capability. For example: "company name" AND "instrument type" AND "specific descriptors" is a good place to start.

American Society for Testing and Materials. *Annual Book of ASTM Standards*, Vols. 3.05 and 3.06. ASTM: West Conshohocken, PA, 2021.

Arnaud, C.H. Fingerprinting conflict minerals. *C&E News*, April 30, 2012, www.cen-online.org.

Beaty, R.D.; Kerber, J.D. *Concepts, Instrumentation and Techniques in Atomic Absorption Spectrophotometry*. PerkinElmer Instruments: Shelton, CT, 1993.

Berneron, R. A study of passive films formed on pure ferritic steels. *Corros. Sci.*, 20, 899–907, 1980; Pergamon Press Ltd.: Great Britain.

Boss, C.B.; Fredeen, K.J. *Concepts, Instrumentation and Techniques in Inductively Coupled Plasma Optical Emission Spectrometry*. PerkinElmer Instruments: Shelton, CT, 1999.

Boumans, P.W.J.M. *Line Coincidence Tables for Inductively Coupled Plasma Emission Spectrometry*, Vols. I and II. Pergamon Press: New York, 1984.

Boumans, P.W.J.M. *Inductively Coupled Plasma Emission Spectroscopy*, Parts 1 and 2. Wiley: New York, 1987.

Broekaert, J.A.C. Glow discharge atomic spectroscopy. *Appl. Spectrosc.*, 49(7), 12A, 1995.

Bruno, T.J., Svoronos, P.D.N., *CRC Handbook of Basic Tables for Chemical Analysis – Data Driven Methods and Interpretation*, 4th Ed., CRC Taylor and Francis, Boca Raton, FL, 2021.

Busch, K.W.; Busch, M.A. *Multielement Detection Systems for Spectrochemical Analysis*. Wiley: New York, 1990.

Cassap, M. Analysis of elemental impurities in liquid pharmaceutical products using the Thermo Scientific iCAP 6000 Series ICPOES, Application note 43086, Thermo Fisher Scientific, Waltham, MA, www.thermo.com.

Chandler, C. *Atomic Spectra*, 2nd edn. D. Van Nostrand Co. Inc.: Princeton, NJ, 1964.

Cremers, D.A.; Radziemski, L.J. *Handbook of Laser-Induced Breakdown Spectroscopy: Fundamentals and Applications*. John Wiley & Sons Ltd.: Chichester, U.K., 2006.

Dean, J.A. *Flame Photometry*. McGraw-Hill: New York, 1960.

Dean, J.A.; Rains, T.E. *Flame Emission and Atomic Absorption Spectroscopy*, Vols. 1–3. Marcel Dekker, Inc.: New York, 1965–1971.

DeForest, P.R.; Petraco, N.; Kobilinsky, L. Chemistry and the challenge of crime. In Gerber, S.M. (ed.), *Chemistry and Crime*. American Chemical Society: Washington, DC, 1983.

Dorn, S.B.; Skelly Frame, E.M. Development of a high-performance liquid chromatographic-inductively coupled plasma method for speciation and quantification of silicones: From silanols to polysiloxanes. *Analyst*, 119, 1687–1694, 1994.

Dulski, T.R. *A Manual for the Chemical Analysis of Metals*. ASTM: West Conshohocken, PA, 1996.

Epperson, P.M.; Sweedler, J.V.; Billhorn, R.B.; Sims, G.R.; Denton, M.B. A linear charge coupled device for spectroscopy. *Anal. Chem.*, 60, 327A, 1988.

Hahn, D.W.; Omenetto, N. Laser-induced breakdown spectroscopy (LIBS), part II: review of instrumental and methodological approaches to material analysis and applications to different fields. *Appl. Spectrosc.*, *64*, 355A, 2010.

Hahn, D.W.; Omenetto, N. Laser-Induced Breakdown Spectroscopy (LIBS), Part II: Review of Instrumental and Methodological Approaches to Material Analysis and Applications to Different Fields. *Appl. Spectrosc.*, *66*, 347A, 2012.

Hareland, W. Atomic emission spectroscopy. In Ewing, G.W. (ed.), *Analytical Instrumentation Handbook*, 2nd edn. Marcel Dekker, Inc.: New York, 1997.

Hark, R.R.; Remus, J.J.; East, L.J.; Harmon, R.S.; Wise, M.A.; Tansi, B.M.; Shughrue, K.M.; Dunsin, K.D.; Liu, C. Geographical analysis of "conflict minerals" utilizing laser-induced breakdown spectroscopy. *Spectrochim. Acta B*, *74*, 131–136, 2012.

Harmon, R.S.; Shughrue, K.M.; Remus, J.J.; Wise, M.A.; East, L.J.; Hark, R.R. Laser-Induced Breakdown Spectroscopy: LIBS instrumentation is sufficiently compact, simple and robust to facilitate the recent development of commercial handheld LIBS analyzers. *Anal. Bioanal. Chem.*, *400*, 3377, 2011.

Harnly, J.M.; Fields, R.E. Solid-state array detectors for analytical spectrometry. *Appl. Spectrosc.*, *51*(9), 334A, 1997.

Harrison, G.R. *MIT Wavelength Tables*. The MIT Press: Cambridge, MA, 1969.

Ingle, J.D., Jr.; Crouch, S.R. *Spectrochemical Analysis*. Prentice Hall: Englewood Cliffs, NJ, 1988.

James, S.H.; Nordby, J.J. *Forensic Science, An Introduction to Scientific and Investigative Techniques*, 2nd edn. CRC Press: Boca Raton, FL, 2005.

Kirkbright, G.F.; Sargent, M. *Atomic Absorption and Fluorescence Spectroscopy*. Academic Press: London, U.K., 1974.

Lajunen, L.H.J. *Spectrochemical Analysis by Atomic Absorption and Emission*. The Royal Society of Chemistry: Cambridge, U.K., 1992.

Lieberman, M.A.; Lichtenberg, A.J. *Principles of Plasma Discharges and Materials Processing*. Wiley: New York, 1994.

Lonardo, R.F.; Yuzefovsky, A.; Yang, K.X.; Michel, R.G. Electrothermal atomizer laser-excited atomic fluorescence spectrometry for the determination of phosphorus in polymers by direct solid analysis and dissolution. *JAAS*, *4*, 279, 1996.

Marcus, R.K. (ed.) *Glow Discharge Spectroscopies*. Plenum Press: New York, 1993.

Michel, R.G., Skelly-Frame, E., Barren, J. Atomic fluorescence spectrometry. In Riordan, J.F.; Vallee, B.L. (eds.), *Metallobiochemistry*, Part A, Vol. 158. Academic Press: San Diego, CA, 1988.

Miziolek, A.W. Progress in fieldable laser induced breakdown spectroscopy (LIBS), *SPIE-DSS*, Baltimore, MD, April 2012.

Miziolek, A.W.; Palleschi, V.; Schechter, I. (eds.) *Laser Induced Breakdown Spectroscopy*. Cambridge University Press: U.K., 2006.

Montaser, A.; Golightly, D.W. (eds.) *Inductively Coupled Plasmas in Analytical Atomic Spectrometry*, 2nd edn. VCH: New York, 1992.

Myers, M.J.; Myers, J.D.; Myers, A.G. Laser induced breakdown spectroscopy. In Lackner, M. (ed.), *Lasers in Chemistry*. Wiley-VCH: Weinheim, Germany, 2008.

Parsons, M.L.; Foster, A. *An Atlas of Spectral Interferences in ICP Spectroscopy*. Plenum Press: New York, 1980.

Payling, R.; Nelis, T.; Browner, R.; Chalmers, J. *Glow Discharge Optical Emission Spectroscopy: A Practical Guide*. Royal Society of Chemistry: Cambridge, U.K., 2003.

Pilon, M.J.; Denton, M.B.; Schleicher, R.G.; Moran, P.M.; Smith, S.B. Evaluation of a new array detector atomic emission spectrometer for inductively coupled plasma atomic emission spectrometry. *Appl. Spectrosc.*, *44*, 1613, 1990.

Rehse, S.J.; Salimnia, H.; Miziolek, A.W. Laser-induced breakdown spectroscopy (LIBS): An overview of recent progress and future potential for biomedical applications. *J. Med. Engr. Technol.*, *36*, 77–89, 2012 (http://informahealthcare.com/toc/jmt/current).

Remus, J.J.; Harmon, R.S.; Hark, R.R.; Potter, I.K.; Bristol, S.K.; Baron, D.; Haverstock, G.; East, L.J. Advanced signal processing analysis of laser-induced breakdown spectroscopy data for the discrimination of obsidian sources. *Appl. Optics*, *51*, B65, 2012.

Robinson, J.W. (ed.) *Handbook of Spectroscopy*, Vol. 1. CRC Press: Boca Raton, FL, 1974.

Robinson, J.W. *Atomic Spectroscopy*. Marcel Dekker: New York, 1990.

Sacks, R.D. Emission spectroscopy. In Elving, P.J.; Meehan, E.; Kolthoff, I. (eds.), *Treatise on Analytical Chemistry*, Vol. 7. Wiley: New York, 1981.

Settle, F. (ed.) *Handbook of Instrumental Techniques for Analytical Chemistry*. Prentice Hall PTR: Upper Saddle River, NJ, 1997.

Skelly Frame, E.M.; Suzuki, T.; Takamatsu, Y. Particle characterization by helium microwave induced plasma spectrometry. *Spectroscopy*, *11*, 1, 1996.

Skelly Frame, E.M.; Uzgiris, E.E. The determination of gadolinium in biological samples by ICP-AES and ICP-MS in evaluation of the action of MRI agents. *Analyst*, *123*, 675–679, 1998.

Slickers, K.A. *Automatic Emission Spectroscopy*, 2nd edn. Bruehlische University Press: Giessen, Germany, 1993.

Sneddon, J. (ed.) *Sample Introduction in Atomic Spectroscopy*. Elsevier: Amsterdam, the Netherlands, 1990.

Sneddon, J. (ed.) *Advances in Atomic Spectroscopy*, Vols. I and II. JAI Press: Greenwich, CT, 1992, 1994.

Sweedler, J.V.; Jalkian, R.D.; Denton, M.B. Crossed interferometric dispersive spectroscopy. *Appl. Spectrosc.*, *43*, 953, 1989.

Syty, A. Flame photometry. In Elving, P.J.; Meehan, E.; Kolthoff, I. (eds.) *Treatise on Analytical Chemistry*, Vol. 7. Wiley: New York, 1981.

Thomsen, V.B.E. *Modern Spectrochemical Analysis of Metals: An Introduction for Users of Arc/Spark Instrumentation*. ASM International: Materials Park, OH, 1996.

Winchester, M.R.; Lazik, C.; Marcus, R.K. Comparison between direct current and radio frequency gas jet boosted glow discharge atomic emission spectrometry for the analysis of steel. *Spectrochim. Acta*, *46B*, 483, 1991.

Winchester, M.R.; Payling, R. Radio frequency glow discharge spectrometry: A critical review. *Spectrochim. Acta, 59B*, 607, 2004.

Yaroshchyk, P.; Body, D.; Morrison, R.J.S.; Chadwick, B.L. Quantitative measurements of loss on ignition in iron ore using laser induced breakdown spectroscopy and partial least squares regression analysis. *Spectrochim. Acta, 61B*, 200, 2006.

Zaidel, A.N.; Prokofev, V.K.; Raiskii, S.M.; Slavnyi, V.A.; Shreider, E.Y. *Table of Spectral Lines*, 3rd edn. IFI/Plenum Press: New York, 1970.

CHAPTER 9

X-Ray Spectroscopy[1]

Contributing authors: Alexander Seyfarth[2] and Eileen Skelly Frame

9.1 ORIGIN OF X-RAY SPECTRA

X-rays were discovered in 1895 by Wilhelm Conrad Röntgen who received the first Nobel Prize in Physics, awarded in 1901, for his discovery. X-ray absorption, emission, and fluorescence spectra are used in the qualitative and quantitative determination of elements in solid and liquid samples. XRA is used in the nondestructive evaluation of flaws in objects, including voids or internal cracks in metals, cavities in teeth, and broken bones in humans, a technique called radiography or X-ray fluoroscopy. This same technique is used to perform security screening of baggage at airports. A computerized version of radiography, computed tomography (CT) scanning or computed axial tomography (CAT), provides a powerful, high-resolution medical diagnostic tool by giving a 3D cross-sectional image of body tissues. Diffraction of X-rays by crystalline materials, a technique called X-ray crystallography, provides crystal structure identification, orientation of atomic planes in materials, and other physical information about samples. X-ray astronomy uses cosmic X-rays to study the universe, and X-ray spectrometers have been sent to the moon and Mars to study the surface rocks *in situ*. This chapter will focus primarily on X-ray fluorescence (XRF) spectrometry and X-ray diffractometry (XRD), the techniques of most use to analytical chemists.

X-rays consist of electromagnetic radiation with a wavelength range from 0.005 to 10 nm (0.05–100 Å). X-rays have shorter wavelengths and higher energy than ultraviolet (UV) radiation. X-rays are generated in several ways, by deceleration of electrons in matter or by electronic transitions of inner core electrons.

9.1.1 Energy Levels in Atoms

An atom is composed of a nucleus and electrons. The electrons are arranged in shells around the nucleus with the valence electrons in the outer shell. The different shells correspond to the different principal quantum numbers of the possible quantum states. The principal quantum number, n, can have integral values beginning with 1. The shells are named starting with the shell closest to the nucleus, which is called the K shell. The K shell is the lowest in energy and corresponds to the quantum level with $n = 1$. The shells moving out from the nucleus are named the L shell, M shell, and so on alphabetically. The letters used for the two lowest shells are historical; K is from the German word *kurz*, meaning short, and L is from the German word *lang*, meaning long. An atom is shown schematically in Figure 9.1a, with Φ_K, Φ_L, and Φ_M representing the energy of the K, L, and M shells, respectively. A partial list of elements and their electron configurations is given in Table 9.1. For example, a sodium atom contains filled K and L shells and one electron in the M shell.

When an X-ray photon or a fast-moving electron collides with an atom, its energy may be absorbed by the atom. If the X-ray photon or electron has sufficient energy, it knocks an electron out of one of the atom's inner shells (e.g., the K shell), and the atom becomes ionized as shown in Figure 9.1b. An electron from a higher-energy shell (e.g., the L shell) then falls into the position vacated by the dislodged inner electron, and an X-ray photon is emitted as the electron drops from one energy level to the other (Figure 9.1c). The wavelength (energy) of this emitted X-ray photon is characteristic of the element being bombarded.

A fourth process can also occur, as shown in Figure 9.1d. Instead of emitting an X-ray photon, the energy released knocks an electron out of the M shell. This electron is called an Auger electron. This Auger process is the basis for a sensitive surface analysis technique. Auger electron spectroscopy (AES) and the related method of X-ray photoelectron spectroscopy, based on the measurement of the emitted electron shown in Figure 9.1b, are discussed in Chapter 16.

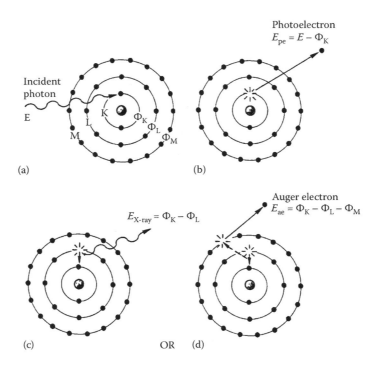

Figure 9.1 (a, b) A schematic atom showing the steps leading to the emission of an X-ray photon (c) or an Auger electron (d). (From Jenkins, R. et al., *Quantitative X-Ray Spectrometry*, Marcel Dekker, Inc., New York, 1981. With permission.)

If we plot the energy levels of the K, L, and M shells for a given element, we get a diagram similar to Figure 9.2.

Note that the K shell has only one energy level, while the higher shells have sublevels within each shell. If an electron is dislodged from the K shell, an electron from an L or an M shell may replace it. The resulting ion emits radiation with energy E equal to the energy difference between the electronic energy levels, such as

$$E_{X-ray} = \Phi_L - \Phi_K \quad (9.1)$$

where Φ_L is the energy of the electron in a specific electronic state within the L shell that "drops" to the K shell. Similar equations may be written for other transitions, such as between an M shell sublevel and an L shell sublevel, using the appropriate energy of the electron in the M shell sublevel that drops into the L shell and so on. As we know from Chapter 3,

$$E = h\nu = \frac{hc}{\lambda}$$

This relationship relates the energy of the emission to the wavelength and can be used to convert wavelength and energy. Most X-ray systems express wavelength in angstroms and energy in keV. To convert between these units, Equation 9.2 gives

$$\text{Energy (keV)} = \frac{12.4}{\lambda}(\text{Å}) \quad (9.2)$$

Therefore, for the X-ray photon released when an L electron in a specific sublevel drops down to fill a vacancy in the K shell,

$$h\nu = \frac{hc}{\lambda} = \Phi_L - \Phi_K$$

Hence the frequency of the emitted X-ray is

$$\nu = \frac{\Phi_L - \Phi_K}{h} \quad (9.3)$$

The frequency or wavelength for transitions between other sublevels and shells is calculated in the same manner. Transitions are not possible between all available energy levels.

As in all forms of spectroscopy, transitions are governed by quantum mechanical *selection rules*. Some transitions are allowed by these rules while others are forbidden. For a brief discussion of the selection rules, the interested student should consult the texts by Jenkins or Bertin listed in the bibliography.

X-ray emission lines from electron transitions **terminating** in the K shell are called K lines, lines from transitions

Table 9.1 Electron Configurations of Various Elements

Element	Z	K	L		M			N	
		1s	2s	2p	3s	3p	3d	4s	4p
H	1	1							
He	2	2							
Li	3	2	1						
Be	4	2	2						
B	5	2	2	1					
C	6	2	2	2					
N	7	2	2	3					
O	8	2	2	4					
F	9	2	2	5					
Ne	10	2	2	6					
Na	11	Neon core (10)			1				
Mg	12				2				
Al	13				2	1			
Si	14				2	2			
P	15				2	3			
S	16				2	4			
Cl	17				2	5			
Ar	18				2	6			
K	19	Argon core (18)						1	
Ca	20							2	
Sc	21						1	2	
Ti	22						2	2	
V	23						3	2	
Cr	24						5	1	
Mn	25						5	2	
Fe	26						6	2	
Co	27						7	2	
Ni	28						8	2	
Cu	29						10	1	
Zn	30	Cu+ core (28)						2	
Ga	31							2	1
Ge	32							2	2
As	33							2	3
Se	34							2	4
Br	35							2	5
Kr	36							2	6

terminating in the L shell are called L lines, and so on. There are three L levels differing by a small amount of energy and five M levels. These sublevels are different quantum states, as shown in Figure 9.2; the quantum numbers and states will not be discussed here in detail. An electron that drops from an L shell sublevel to the K shell emits a photon with the energy difference between these quantum states. This transition results in a K_α line. There are two possible K_α lines for atoms with atomic number >9: $K_{\alpha 1}$ and $K_{\alpha 2}$, which originate in different sublevels of the L shell. The K_α lines are often not resolved, and only one peak is seen. These lines are illustrated in Figure 9.2. The use of a Greek letter and numerical subscript to identify an X-ray emission line is called the *Siegbahn* notation. For the purposes of this text, the notation is just a "name" for the peak. An electron that drops from an M shell sublevel to the K shell generates a K_β X-ray. There is more than one K_β line, but the energy differences are so small between $K_{\beta 1}$ and $K_{\beta 3}$ that only a single K_β line is seen unless a high-resolution spectrometer is used. If an electron is ejected from an L shell, an electron from an M shell may fall into its place and emit an X-ray of characteristic wavelength with energy equivalent to the difference between the L and M shell sublevels. These are designated as L lines. A number of L lines are possible, as indicated by Figure 9.2. Table 9.2 indicates the actual transition that gives rise to selected X-ray emission lines. Electrons originating in an N or O shell and falling into the L shell also generate L lines. The energy levels of the K, L, M, and higher shells are characteristic of the element being examined, and the sharp emission lines resulting from electronic transitions are called *characteristic lines* or characteristic radiation. A schematic X-ray emission spectrum obtained under certain conditions by bombarding a solid metal, such as rhodium or lead or silver, with high-energy electrons is shown in Figure 9.3. The characteristic lines are shown as sharp peaks on a broad continuous background. The characteristic K X-ray emission lines from some elements are given in Table 9.3. A more comprehensive table of K and L lines for the elements is found in Appendix 9.A and in handbooks such as the *CRC Handbook of Chemistry and Physics* and the *CRC Handbook of Spectroscopy*, Vol. 1. Not all X-ray lines have a Siegbahn designation, so the International Union of Pure and Applied Chemistry (IUPAC) established a new identification system for X-ray lines in 1991. Appendix 9.A contains a list of the Siegbahn notation for lines and the IUPAC designation for these lines. For all new publication purposes, the IUPAC designation should be used.

The wavelengths and energies of the characteristic lines depend only on the element, because the inner electrons do not take part in bonding. Therefore, the lines are independent of oxidation state and bonding and physical state, making the use of the characteristic lines an *elemental* analysis technique. No molecular information is obtained from these lines.

The broad continuous "background" emission of X-radiation seen in Figure 9.3 is due to a second process that occurs when high-energy electrons strike a solid. The continuous radiation results from the collision of electrons with the atoms of the solid. At each collision, the electron loses some energy and decelerates, with the production of an X-ray photon. The energy of the photon is equal to the kinetic energy difference of the electron as a result of the collision. Each electron generally undergoes a series of collisions with each collision resulting in a photon of slightly different energy. The result of these many collisions is emission of a continuum of X-rays over a wide wavelength range. This continuous radiation is called *Bremsstrahlung*, *white radiation*, or *braking radiation*.

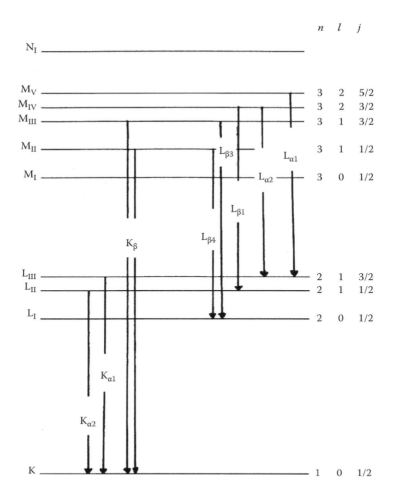

Figure 9.2 Atomic energy levels and symbols used for some common X-ray transitions. (Modified from Parsons, M.L., X-ray methods, in Ewing, G.W. (ed.), *Analytical Instrumentation Handbook*, 2nd edn., Marcel Dekker, Inc., New York, 1997. Used with permission.)

Table 9.2 Electron Transitions for Selected X-Ray Emission Lines

Siegbahn Line Designation	Electron Transition	Siegbahn Line Designation	Electron Transition
$K_{\alpha 1}$	$L_{III} \to K$	$L_{\beta 1}$	$M_{IV} \to L_{II}$
$K_{\alpha 2}$	$L_{II} \to K$	$L_{\beta 3}$	$M_{III} \to L_I$
$K_{\beta 1}$	$M_{III} \to K$	$L_{\beta 4}$	$M_{II} \to L_I$
$K_{\beta 3}$	$M_{II} \to K$	L_{η}	$M_I \to L_{II}$
$K_{\beta 5}$	$M_{IV,V} \to K$	$L_{\gamma 2,3}$	$N_{II,III} \to L_I$
$K_{\beta 2}$	$N_{II,III} \to K$	$L_{\beta 6}$	$N_I \to L_{III}$
$K_{\beta 4}$	$N_{IV,V} \to K$	$M_{\xi 1}$	$N_{III} \to M_V$
$L_{\alpha 1}$	$M_V \to L_{III}$	$M_{\xi 2}$	$N_{II} \to M_{IV}$
$L_{\alpha 2}$	$M_{IV} \to L_{III}$		

Note: Not all lines are seen for all elements, and many of the lines are not resolved with standard X-ray spectrometers. Many $M \to M$, $N \to M$, $O \to L$, and $O \to M$ transitions have no Siegbahn notation associated with them.

When all the energy of the impinging electrons is turned into X-rays, as would occur if the electrons transferred all their energy in one collision, the wavelength of the emitted photons is the shortest attainable. This is termed the minimum λ or λ_{min}. The radiation with the highest energy (and therefore the shortest wavelength) is deduced as follows. When all the energy of the electrons is converted to radiant energy, then the energy of the electrons equals the energy of the radiation. The energy of the radiation is given by $E = h\nu$, whereas the energy of the electrons is given by $E = eV$. When they are equal, $h\nu = eV$, where e is the charge of the electron; V, the applied voltage; and ν, the frequency of the radiation. But

$$\nu = \frac{c}{\lambda}$$

where
c is the speed of light
λ is the wavelength of radiation

X-RAY SPECTROSCOPY

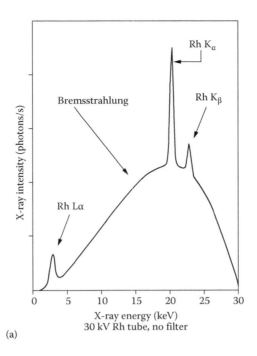

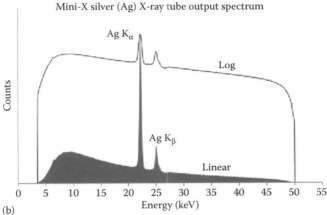

Figure 9.3 X-ray emission spectrum obtained by bombarding (a) rhodium metal (Rh) and (b) silver metal (Ag) with electrons. Both a broad continuum (Bremsstrahlung) and sharp characteristic emission lines from the metal are seen. (a: © Thermo Fisher Scientific. www.thermofisher.com. Used with permission; b: Amptek Mini-X silver (Ag) XRF tube output spectrum; Courtesy of Amptek, Inc., Bedford, MA, www.amptek.com.)

Therefore,

$$h\nu = \frac{hc}{\lambda} = eV \quad (9.4)$$

Rearranging, we get

$$\lambda = \frac{hc}{eV} \quad (9.5)$$

When all the energy of the electron is converted to X-radiation, the wavelength of the radiation is a minimum, and we achieve minimum λ conditions:

$$\lambda_{min} = \frac{hc}{eV} \quad (9.6)$$

Inserting the values for h, c, and e, which are constants, we have the **Duane–Hunt law**,

$$\lambda_{min} = \frac{(6.626 \times 10^{-34} \text{ J s})(3.00 \times 10^8 \frac{\text{m}}{\text{s}})(10^{10} \frac{\text{Å}}{\text{m}})}{(1.60 \times 10^{-19} \text{ C}) \times V}$$

$$= \frac{12{,}400}{V}$$

$$(9.7)$$

Table 9.3 Wavelengths of Absorption Edges and Characteristic Emission Lines of Various Elements

Element	K Absorption Edge (Å)	Emission (Å)	
		$K_{\beta 1,3}$[a]	$K_{\alpha 1,3}$[b]
Mg	9.512	9.559	9.890
Ti	2.497	2.514	2.748, 2.752
Cr	2.070	2.085	2.290, 2.294
Mn	1.896	1.910	2.102, 2.106
Ni	1.488	1.500	1.658, 1.662
Ag	0.4858	0.4971, 0.4977	0.5594, 0.5638
Pt	0.1582	0.1637, 0.1645	0.1855, 0.1904
Hg	0.1492	0.1545, 0.1553	0.1751, 0.1799

[a] When more than one number is listed, $K_{\beta 1}$ is listed first.
[b] When more than one number is listed, $K_{\alpha 1}$ is listed first.

where
h is Planck's constant
c is the speed of light
e is the charge of an electron
V is the applied voltage (in volts)
λ_{min} is the shortest wavelength of X-rays radiated (in angstroms)

The continuum radiation spectrum from a solid metal therefore has a well-defined short-wavelength limit. This limit is a function of the accelerating voltage, but not of the solid metal. The same λ_{min} would be obtained by bombardment of lead or tungsten or rhodium at the same accelerating voltage.

An X-ray emission spectrum is similar for all elements, in that K_α, K_β, and L_β lines may be seen, if the element possesses enough electrons to populate the appropriate levels. However, the actual wavelengths of these lines vary from one element to another, depending on the *atomic number* of the particular element. A mathematical relationship was discovered between the wavelengths of the K series and the atomic number of the element, and similar relationships were found for the L lines, and others.

9.1.2 Moseley's Law

Henry Moseley, a young graduate student working at Cambridge, UK, in 1913, discovered the relationship between wavelength for characteristic X-ray lines and atomic number. After recording the X-ray spectra from numerous elements in the periodic table, he deduced the mathematical relationship between the atomic number of the element and the wavelength of the K_α line. A similar relationship was found between the atomic number and the K_β line, the L_α line, and so on. The relationships were formulated in **Moseley's law**, which states that

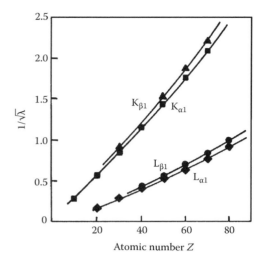

Figure 9.4 Partial Moseley's law plots for selected K and L lines, showing the relationship between the X-ray emission wavelength and atomic number of the element. Using this relationship, it was possible to predict undiscovered elements and to correctly assign atomic numbers to known elements. (From Helsen, L.A. and Kuczumow, A., in Van Griekin, R.E.; Markowicz, A.A. (eds.), *Handbook of X-Ray Spectrometry*, 2nd edn. Marcel Dekker, Inc.: New York, 2002. Used with permission.)

$$\nu = \frac{c}{\lambda} = a(Z-\sigma)^2 \quad (9.8)$$

where
c is the speed of light
λ is the wavelength of the X-ray
a is the constant for a particular series of lines (e.g., K_α or L_α lines)
Z is the atomic number of the element
σ is the screening constant that accounts for the repulsion of other electrons in the atom

A partial Moseley's law plot for the K_α, K_β, L_α, and L_β emission lines is shown in Figure 9.4. Shortly after this monumental discovery, Moseley was killed in action in World War I. The impact of Moseley's law on chemistry was substantial, in that it provided a method of unequivocally assigning an atomic number to newly discovered elements, of which there were several at that time. In addition, it clarified disputes concerning the positions of all known elements in the periodic table, some of which were still in doubt in the early part of the twentieth century.

9.1.3 X-Ray Methods

There are several distinct fields of X-ray analysis used in analytical chemistry and materials characterization, namely, XRA, XRD, XRF, and X-ray emission. The basic principles of each are described in the following. X-ray emission is generally used for microanalysis, with either an electron microprobe (Chapter 16) or a scanning electron microscope (SEM).

9.1.3.1 X-Ray Absorption Process

The absorption spectrum obtained when a beam of X-rays is passed through a thin sample of a pure metal is depicted in Figure 9.5.

As is the case with other forms of radiation, some of the intensity of the incident beam may be absorbed by the sample while the remainder is transmitted. We can write a Beer's law expression for the absorption of X-rays by a thin sample:

$$I(\lambda) = I_0(\lambda) e^{-(\mu_m)\rho x} \quad (9.9)$$

where
$I(\lambda)$ is the transmitted intensity at wavelength λ
$I_0(\lambda)$ is the incident intensity at the same wavelength
μ_m is the **mass absorption coefficient** (cm²/g)
ρ is the density of the sample (g/cm³)
x is the sample thickness (cm)

The mass absorption coefficient is a constant for a given element at a given wavelength and is independent of both the chemical and physical state of the element. An updated compilation of mass absorption coefficients can be found online at the NIST website: www.nist.gov/pml/data/xraycoef/index.cfm (J.H. Hubbell and S.M. Seltzer) or in the text by Bertin or handbooks listed in the bibliography or references in the literature.

Of course, most samples do not consist of a single pure element. The total mass absorption coefficient for a sample can be calculated by adding the product of the individual mass absorption coefficients for each element times the mass fraction of the element present in the sample. That is, for a metal alloy like steel,

$$\mu_{total} = w_{Fe}\mu_{Fe} + w_{Cr}\mu_{Cr} + w_{Ni}\mu_{Ni} + \cdots \quad (9.10)$$

where
w_{Fe} is the mass fraction of iron
μ_{Fe} is the mass absorption coefficient for pure iron
w_{Cr} is the mass fraction of chromium, and so on for all the elements in the alloy

For accurate quantitative work, the *mass attenuation coefficient* is used in place of the mass absorption coefficient. The mass attenuation coefficient takes into account both absorption and scattering of X-rays by the sample.

The amount of light absorbed increases as the wavelength increases (and the energy decreases). This is reasonable since longer wavelengths have less energy, and a less energetic photon has less "penetrating power" and is more likely to be absorbed. Only a few absorption peaks are seen in an X-ray absorption spectrum, but there is a very distinct feature in these spectra. An abrupt change in absorptivity (or the mass absorption coefficient) occurs at the wavelength of the X-ray necessary to eject an electron

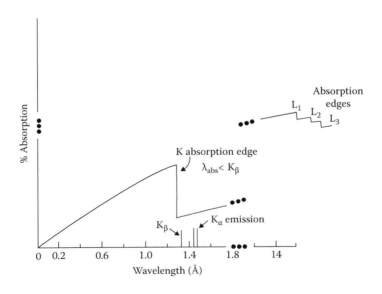

Figure 9.5 The X-ray absorption spectrum of a pure metal. Longer wavelengths are more readily absorbed than shorter wavelengths. The absorption spectrum is characterized by absorption edges, which are abrupt increases in absorption at energies sufficient to eject an electron from one of the atomic shells. The K absorption edge occurs at an energy sufficient to eject an electron from the K shell of the given metal.

from an atom. These abrupt changes in X-ray absorptivity are termed **absorption edges**. Looking at Figure 9.5, it can be seen that radiation with a wavelength of 1.8 Å has a certain percent absorption value. As the wavelength of the X-ray *decreases*, its energy increases, its penetrating power increases, and the percent absorption decreases. This can be seen by the downward slope of the absorption trace, moving to the left along the *x*-axis from a wavelength of 1.8 Å. As the wavelength decreases further, the X-ray eventually has sufficient energy to displace electrons from the K shell. This results in an abrupt *increase* in absorption. This is manifested by the **K absorption edge**. After the absorption edge, the penetrating power continues to increase as the wavelength decreases further until finally the degree of absorption is extremely small at very small wavelengths. At wavelengths less than 0.2 Å, penetrating power is extremely high, and we are approaching the properties of interstellar radiation such as cosmic rays, which have extremely high penetrating power. Wavelengths shorter than the K absorption edge have sufficient energy to eject K electrons; the bombarded sample will exhibit both continuum radiation and the characteristic K lines for the sample. This process is called XRF and will be discussed in detail. Wavelengths just slightly longer than the K absorption edge do not have enough energy to displace K electrons. The absorption spectrum is unique for each element; portions of the absorption spectrum showing the position of the K absorption edge for several pure elements are shown in Figure 9.6.

Another way of looking at X-ray absorption is to plot the mass absorption coefficient as a function of wavelength or energy. For a thin sample of pure metal and a constant incident intensity, Equation 8.9 indicates that if the percent absorption changes as a function of wavelength, it must be that μ_m changes. A plot of μ_m versus X-ray energy for the element lead is shown in Figure 9.7. The K, L, and M absorption edges are seen.

The wavelengths of the absorption edges and of the corresponding emission lines do not quite coincide, as seen in Figure 9.6b. This is because the energy required to dislodge an electron from an atom (the absorption edge energy) is not quite the same as the energy released when the dislodged electron is replaced by an electron from an outer shell (emitted X-ray energy). The amount of energy required to displace the electron must dislodge it from its orbital and remove it completely from the atom. This is more than the energy released by an electron in an atom falling from one energy level to another. A few absorption edge values are given in Table 9.3. Figure 9.6b shows that the energy of the K absorption edge is greater than the energy of the K emission lines. As opposed to emission spectra, only one K absorption edge is seen per element, since there is only one energy level in the K shell. Three absorption edges of different energies are observed for the L levels, five for the M levels, and so on; these can be seen in Figure 9.7.

A comprehensive table of absorption edge wavelengths is located in Appendix 9.B.

9.1.3.2 X-Ray Fluorescence Process

X-rays can be emitted from a sample by bombarding it with electrons, alpha particles, or with other X-rays. When electrons or alpha particles are used as the excitation source, the process is called X-ray emission or particle-induced X-ray emission (PIXE). This is the basis of X-ray microanalysis using an electron microprobe (Chapter 16) or an SEM. An alpha-particle X-ray spectrometer (APXS) is currently on the Mars Curiosity rover collecting data on Martian rock composition.

When the excitation source is a beam of X-rays, that is, photons, the process of X-ray emission is called fluorescence. This is analogous to molecular fluorescence discussed in Chapter 5 and atomic fluorescence discussed in Chapter 8, because the wavelength of excitation is shorter than the emitted wavelengths. The beam of exciting X-rays is called the *primary* beam; the X-rays emitted from the sample are called *secondary* X-rays. The use of an X-ray source to produce secondary X-rays from a sample is the basis of XRF spectroscopy. The primary X-ray beam must have a λ_{min} that is shorter than the absorption edge of the element to be excited.

9.1.3.3 X-Ray Diffraction Process

Crystals consist of atoms, ions, or molecules arranged in a regular, repeating 3D pattern called a crystal lattice. This knowledge came from the pioneering work of German physicist Max von Laue and the British physicists W.H. Bragg and W.L. Bragg. Max von Laue demonstrated in 1912 that a crystal would diffract X-rays, just as a ruled grating will diffract light of a wavelength close to the distance between the ruled lines on the grating. The fact that diffraction occurs indicates that the atoms are arranged in an ordered pattern, with the spacing between the planes of atoms on the order of short-wavelength electromagnetic radiation in the X-ray region. The diffraction pattern could be used to measure the atomic spacing in crystals, allowing the determination of the exact arrangement in the crystal, the *crystal structure*. The Braggs used von Laue's discovery to determine the arrangement of atoms in several crystals and to develop a simple 2D model to explain XRD.

If the spacing between the planes of atoms is about the same as the wavelength of the radiation, an impinging beam of X-rays is reflected at each layer in the crystal, as shown in Figure 9.8. Assume that the X-rays falling on the crystal are parallel waves that strike the crystal surface at angle θ. That is, the waves O and O' are in phase with each other and reinforce each other. In order for the waves to emerge as a reflected beam after scattering from atoms at points B and

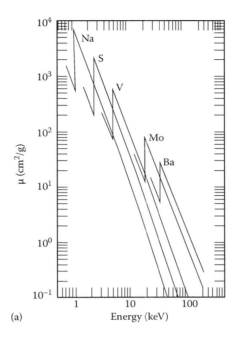

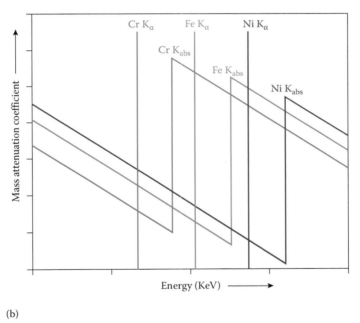

Figure 9.6 (a) Energies of the K absorption edges for several pure elements. (Courtesy of ORTEC [Ametek]. www.ortec-online.com; From Jenkins, R. et al., *Quantitative X-Ray Spectrometry*, Marcel Dekker, Inc., New York, 1981. Used with permission.) (b) Energies of the K absorption edge plotted schematically for the main elements in steel. (Courtesy of the University of Western Ontario XRF Short Course, London, Ontario, Canada.)

D, they must still be in phase with each other, that is, waves Y and X must still be parallel and coherent. If the waves are completely out of phase, their amplitudes cancel each other; they are said to interfere destructively and no beam emerges. In order to get reinforcement, it is necessary that the two waves stay in phase with each other after diffraction at the crystal planes.

It can be seen in Figure 9.8 that the lower wave travels an extra distance AB + BC compared with the upper wave. If AB + BC is a whole number of wavelengths, the emerging beams Y and X will remain in phase and reinforcement will take place. From this deduction, we can calculate the relationship between the wavelengths of X-radiation, the distance d between the lattice planes, and the angle at which

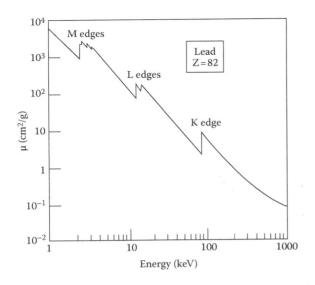

Figure 9.7 The mass absorption coefficient for Pb as a function of energy. The K, L, and M absorption edges are seen. (Courtesy of ORTEC [Ametek]. www.ortec-online.com; From Jenkins, R. et al., *Quantitative X-Ray Spectrometry*, Marcel Dekker, Inc., New York, 1981. Used with permission.)

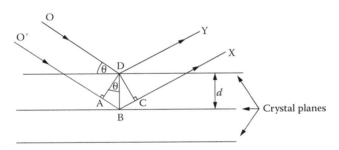

Figure 9.8 Diffraction of X-rays by crystal planes.

a diffracted beam can emerge. We employ the following derivation.

X-ray waves O and O′ are parallel. The extra distance traveled by wave O′ in traveling through the crystal is AB + BC. For diffraction to occur, it is necessary that this distance be a whole number of wavelengths, n; that is,

$$\text{distance } AB + BC = n\lambda \quad (9.11)$$

but

$$AB + BC = 2AB \quad (9.12)$$

and

$$AB = DB \sin \theta \quad (9.13)$$

where θ is the angle of incidence of the X-ray beam with the crystal; therefore

$$AB = d \sin \theta \quad (9.14)$$

where d is the distance between the crystal planes, called the interplanar distance. (Note from Figure 9.8 that $d = $ DB.) Therefore

$$AB + BC = n\lambda = 2AB = 2d \sin \theta$$

or

$$n\lambda = 2d \sin \theta \quad (9.15)$$

The equation $n\lambda = 2d \sin\theta$ is known as the **Bragg equation**. The important result of this equation is that at any particular angle of incidence θ, only X-rays of a particular wavelength fulfill the requirement of staying in phase and being reinforced and are therefore diffracted by the crystal. Diffraction of X-rays by crystals forms the basis of XRD for crystal structure determination and is also the reason XRF spectrometry is possible, as will be seen.

9.2 X-RAY FLUORESCENCE

Instrumentation for X-ray spectrometry requires a source, a wavelength (or energy) selector, a detector, and beam conditioners. Component parts of the instrument are similar for XRF, XRD, and the other fields, but the optical system varies for each one. One major point to note is that some systems have the source located above the sample (the sample is faceup to the X-ray beam); other systems have the source located below the sample, with the sample facedown to the beam. There are advantages and disadvantages to both designs, as will be discussed later in this chapter.

Generally, we can distinguish between two main techniques in XRF, energy-dispersive XRF (EDXRF) and wavelength-dispersive XRF (WDXRF), as can be seen in Figure 9.9. EDXRF is based on a detector, which can discriminate the various photon energies and count them individually, whereas WDXRF is based on a set of analyzer crystals, which only diffract one energy/(wavelength) into a detector that counts all the photons. The subsequent instrumentation section will illustrate the components and setups used in today's commercially available instrumentation. EDXRF is sometimes called EDX or energy-dispersive spectroscopy (EDS). It must be stressed that all X-ray systems use ionizing radiation, which poses significant health hazards. X-ray systems of all types are regulated, and commercial systems have the proper shielding and safety interlocks to comply with regulatory requirements and prevent accidental exposure to radiation.

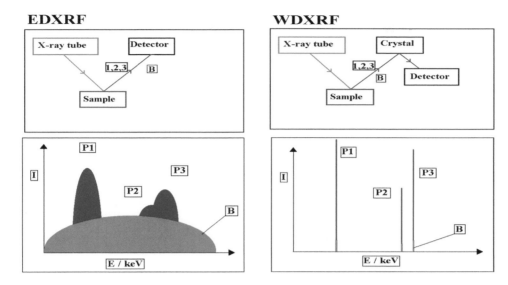

Figure 9.9 Schematic comparison of EDXRF and WDXRF. The upper squares show the differences in instrument configuration, while the lower squares show the resulting spectra. B indicates the background radiation, while 1, 2, and 3 (P1, P2, P3) are photons of different energies.

9.2.1 X-Ray Source

Three common methods of generating X-rays for analytical use in the laboratory or the field are:

1. Use of a beam of high-energy electrons to produce a broadband *continuum* of X-radiation resulting from their deceleration upon impact with a metal target as well as element-specific X-ray *line* radiation by ejecting inner-core electrons from the target metal atoms; this is the basis of the X-ray tube, the most common source used in XRD and XRF.
2. Use of an X-ray beam of sufficient energy (the primary beam) to eject inner-core electrons from a sample to produce a secondary X-ray beam (XRF).
3. Use of a radioactive isotope, which emits very-high-energy X-rays (also called gamma radiation) in its decay process.

A fourth method of producing X-rays employs a massive, high-energy particle accelerator called a synchrotron. These are available at only a few locations around the world, such as the Brookhaven National Laboratory or the Stanford Accelerator Center in the United States, and are shared facilities servicing a large number of clients. X-rays may be generated when alpha particles or other heavy particles bombard a sample; this technique is called PIXE and requires a suitable accelerator facility. The use of an electron beam to generate X-rays from a microscopic sample as well as an image of the sample surface is the basis of X-ray microanalysis using an electron microprobe or scanning electron microscopy.

These different X-ray sources may produce either broadband (continuum) emission or narrow line emission, or both simultaneously, depending on how the source is operated. Figure 9.3 displays the simultaneous generation of both a broad continuum of X-ray energies with element-specific lines superposed upon it, obtained by bombarding rhodium metal with electrons in an **X-ray tube**.

9.2.1.1 X-Ray Tube

A schematic X-ray tube is depicted in Figure 9.10. The X-ray tube consists of an evacuated glass or ceramic envelope containing a wire filament cathode and a pure metal anode. A thin beryllium window sealed in the envelope allows X-rays to exit the tube. The envelope is encased in lead shielding and a heavy steel jacket with an opening over the window, to protect the tube. When a cathode (a negatively charged electrode) in the form of a metal wire is electrically heated by the passage of current, it gives off electrons, a process called thermionic emission. If a positively charged metallic electrode (called an anode) is placed near the cathode *in a vacuum*, the negatively charged electrons will be accelerated toward the anode. Upon striking it, the electrons give up their energy at the metallic surface of the anode. If the electrons have been accelerated to a high enough velocity by a sufficiently high voltage between the cathode and anode, energy is released as radiation of short wavelength (0.1–100 Å), called X-radiation or X-rays. X-ray tubes are generally operated at voltage differentials of 2–100 kV between the wire filament cathode and the anode.

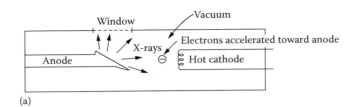

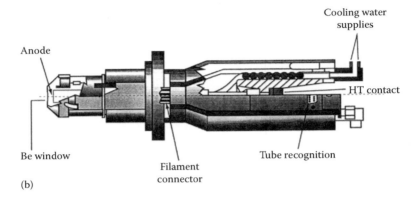

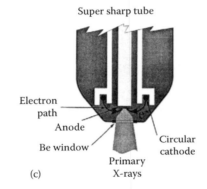

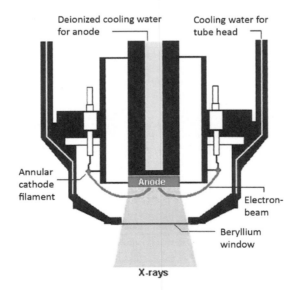

Figure 9.10 (a) Schematic diagram of a side-window X-ray tube. (b) Schematic of the 4.0 kW ceramic end-window X-ray tube, called the Super Sharp Tube™. (c) Close-up of the window end of the Super Sharp Tube, showing the circular cathode design of this tube. (b and c: Courtesy of PANalytical, Inc., The Netherlands, www.panalytical.com.). (d) Schematic side view of an end-window X-ray tube showing the Be window and cooling water. (© 2013 Bruker, Inc., www.bruker.com. Used with permission.)

The cathode is normally a tungsten-based wire filament. The anode is called the target. The X-ray tube is named for the anode; a copper X-ray tube has a copper anode, a rhodium tube has a rhodium anode, a tungsten tube has a tungsten anode; all have tungsten-based wire cathodes. The target is usually not a solid metal but a copper slug coated with a layer of the desired target material. Numerous metals have been used as target materials, but common target elements are titanium, copper, chromium, molybdenum, rhodium, gold, silver, palladium, and tungsten. The target material determines the characteristic emission lines and affects the intensity of the continuum. The voltage between the anode and cathode determines how much energy the electrons in the beam acquire, and this in turn determines the overall intensity of the wide range of X-ray intensities in the continuum distribution and the maximum X-ray energy (shortest wavelength or λ_{min}, as shown by the Duane–Hunt law, Equation 9.7). Figure 9.11 shows how both the intensity of the continuum radiation and the short-wavelength/high-energy cutoff vary as the applied voltage to this silver (Ag) X-ray tube varies. This illustration also shows that the Ag L line emission varies as a function of the excitation.

In choosing the element to be used for the target, it should be remembered that it is necessary for the energy of the X-rays emitted by the source to be greater than that required to excite the element being irradiated in an XRF analysis. As a simple rule of thumb, the target element of the source should have a greater atomic number than the elements being examined in the sample. This ensures that the energy of radiation is more than sufficient to cause the sample element to fluoresce. This is not a requirement in X-ray absorption or XRD, where excitation of the analyte atoms is not necessary. In XRD, the target is chosen to avoid the excitation of characteristic X-ray (XRF) emissions from the sample. For example, copper (Cu) target-based XRD units can excite the iron (Fe) commonly found in geological samples, and the fluorescence from the sample will increase the background of the measurement. Using a cobalt (Co) target will avoid this.

In many high-power (>200 W) X-ray tube designs, the anode, or target, gets very hot in use because it is exposed to a constant stream of high-energy electrons, with most of the energy being dissipated as heat on collision. This problem is overcome by water-cooling the anode. Modern low-power and compact X-ray tubes have been designed to operate at lower voltages and do not require any liquid-based cooling of the anode.

The exit window of the X-ray tube is usually made of beryllium, which is essentially transparent to X-rays. The Be window is thin, generally 0.3–0.5 mm thick, and is very fragile. The window may be on the side of the tube, as shown in Figure 9.10a, or in the end of the tube (Figures 9.10b through d). Side-window tubes are common, but end-window tubes permit the use of a thinner beryllium window. This makes end-window tubes good for low-energy X-ray excitation by improving the low-energy output of the tube. To minimize the tube size, a transmission target can be used. The target is a layer of the target element on the beryllium window. For example, a silver transmission target can be constructed from a 0.75 μm thick layer of Ag on a 127 μm thick Be window. Figure 9.12 shows both the schematic of a side and transmission target rhodium X-ray tube.

X-ray tubes must provide adequate intensity over a relatively wide spectral range for XRF in order to excite a reasonable number of elements. In some applications, monochromatic or nearly monochromatic X-rays are desired; that is accomplished by using filters or a monochromator as described later or by using a secondary fluorescent source, described subsequently. The tube output must be stable over short time periods for the highest precision and over long time periods for accuracy without frequent recalibration. The X-ray emission lines from the anode element must not interfere with the sample spectrum. Tube lines can be scattered into the detector and be mistaken for an element present in the sample.

9.2.1.2 Secondary XRF Sources

If it is necessary to prevent the continuum emission from an X-ray tube from scattering on a sample, a standard tube can be used to excite another pure metal target. The resulting XRF from the secondary target is used as the source of X-ray excitation for the sample. Such an example is shown in Figure 9.13. A standard tungsten X-ray tube is used to produce the emission spectrum on the left, with the tungsten characteristic lines superimposed on the continuum radiation. The radiation from this tube is used to strike a secondary pure copper target. The resulting emission from the copper is the copper XRF spectrum on the right.

This source emits very little or no continuum radiation but does emit quite strongly at the copper K and L lines. Of course, the metal used in the target of the first source must have a higher atomic number than copper to generate fluorescence. The Cu lines then can be used as an excitation source, although the intensity of the secondary source is much less than that of a Cu X-ray tube. However, when monochromatic or nearly monochromatic radiation is required, the loss of intensity is more than offset by the low background from the secondary source.

9.2.1.3 Radioisotope Sources

X-radiation is a product of radioactive decay of certain isotopes. The term gamma ray is often used for an X-ray resulting from such a decay process. Alpha and beta decay and electron capture (EC) processes can result in the release of gamma rays. Table 9.4 lists some common radioisotopes used as XRF sources.

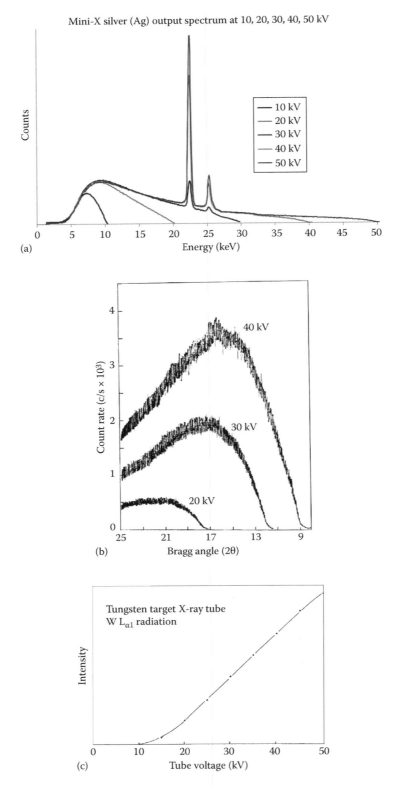

Figure 9.11 (a) Amptek Mini-X silver output spectrum at various voltages. Ten kV starts on the left side of the diagram; the curves increase in voltage moving from left to right. Below 30 kV, only continuum radiation is seen. The high-energy cutoff increases as the voltage increases, and the Ag L line emission intensity increases with voltage. (Courtesy of Amptek, Inc., Bedford, MA, www.amptek.com.) (b) The intensity of the continuum radiation from an X-ray tube and the short-wavelength cutoff vary as the applied voltage varies. This plot is of a tungsten X-ray tube. (c) The characteristic radiation from an X-ray tube also varies as the applied voltage varies. Below 20 kV, the intensity of the tungsten $L_{\alpha 1}$ line is very low. The intensity of this characteristic line increases as the applied voltage increases.

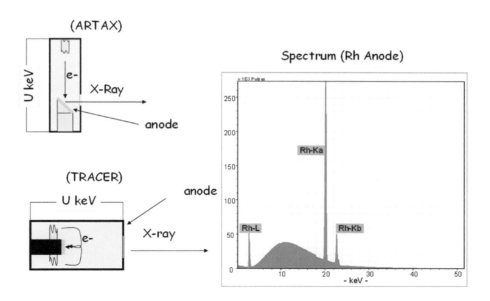

Figure 9.12 Schematic diagrams of a Rh target side-window tube (ARTAX system) and a modern handheld transmission target design (TRACER system). (Courtesy of Roald Tagle, BRUKER NANO GmbH, Berlin, Germany. © 2013 Bruker, Inc. Used with permission.)

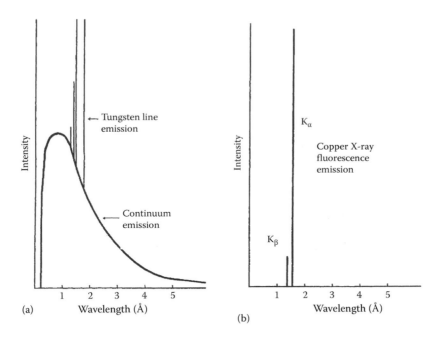

Figure 9.13 (a) The primary X-ray emission from a tungsten target. (b) The secondary emission from a copper target. Note the removal of the continuum radiation and the nearly monochromatic output from the secondary target. (From Parsons, M.L., X-ray methods, in Ewing, G.W. [ed.], *Analytical Instrumentation Handbook*, 2nd edn., Marcel Dekker, Inc., New York, 1997. Used with permission.)

The advantages of radioisotope sources are that they are small, rugged, and portable and do not require a power supply. They are ideal for obtaining XRF spectra from large samples or field measurements while not requiring any power, enabling small compact units with long operation times. Examples include the analysis of bulky objects that cannot have pieces cut from them, such as aircraft engines, ship hulls, and art objects. The disadvantage is that the intensity of these sources is weak compared with that of an X-ray tube, the source cannot be optimized by changing

Table 9.4 Characteristics of Radioisotope Sources for XRF Spectrometry

Isotope	Primary Decay Mode	Half-Life (Years)	Useful Photon Energies Emitted	% Theoretical Yield (Photons per 100 Decay Transformations)	Typical Activity (mCi)
^{55}Fe	EC[b]	2.7	5.9, 6.4 keV Mn K X-rays	28.5	5–100
^{109}Cd	EC	1.3	22.2–25.5 keV Ag K X-rays	102	0.5–100
			88.2 keV γ-ray	4	
^{241}Am	Alpha	458	14–21 keV Np L X-rays	37	1–50
			59.6 keV γ-ray	36	
^{57}Co	EC	0.74	6.4, 7.1 keV Fe K X-rays	51	1
			14.4 keV γ-ray	8.2	
			122 keV γ-ray	88.9	
			136 keV γ-ray	8.8	
^{3}H[a]	Beta	12.3	Bremsstrahlung source, endpoint at 18.6 keV		3000–5000
^{147}Pm	Beta	2.6	Bremsstrahlung source, endpoint at 225 keV		500

Source: Table from Jenkins, R. et al., *Quantitative X-Ray Spectrometry*, Marcel Dekker, Inc., New York, 1981. Used with permission.
[a] Radioactive tritium gas is adsorbed on nonradioactive metal foil, such as titanium foil.
[b] Electron capture.

voltage as can be done with an X-ray tube, and the source intensity decreases with time depending on the half-life of the source. In addition, the source cannot be turned off. This requires care on the part of the analyst to avoid exposure to the ever-present ionizing radiation. Isotope sources are more regulated than tube-based sources, and although isotope sources are convenient for portable XRF systems, tube-based portable and handheld devices are more common. Isotope source uses are currently limited to K line excitation approaches as well as for economical dedicated analyzers, such as for Pb in paint.

9.2.2 Instrumentation for Energy-Dispersive X-Ray Spectrometry

In EDXRF spectrometry, there is no physical separation of the fluorescence radiation from the sample. There is no dispersing device prior to the detector. All the photons of all energies arrive at the detector simultaneously. Any EDXRF is designed around the following components—a source of primary excitation, a sample holder, and a detector, seen in Figure 9.14. In this figure the sample compartment shown has a six sample automatic sampler. Alternatively it is possible to insert a single sample (of the same size as the six shown), or one large sample.

All EDXRF systems are able to modify the primary signal to control the excitation of the sample. This is achieved by different means, which result in three groups of systems: direct excitation, secondary or 3D excitation, and total reflectance XRF (TXRF).

In direct excitation systems, the source is directed toward the sample, thus exciting the sample directly with the radiation from the target or isotope source (Figure 9.15). To modify the excitation or attenuate undesired parts of the excitation spectra, primary beam filters are used. To reduce the spot size and enable "micro" spot analysis, pinhole masks are in use as well. For micro spots below 100 μm in certain units, capillary optics are used. Typical direct excitation units are all handheld units, isotope-based units, the majority of low-power and medium-power benchtop EDXRF systems, as well as the micro-XRF and dedicated plating thickness XRF units.

In secondary excitation units, the source is used to illuminate a selectable secondary target, which is made of specific material to either scatter the beam or to act as a new "source." The geometric arrangement is such that the signal from the secondary target illuminates the sample and also ensures that the X-ray beam is polarized (Figure 9.16).

TXRF systems use a secondary target and the principle of Bragg diffraction to create a more monochromatic beam. This beam is directed at very low incident angle to the sample to achieve "total reflectance." The signal from the sample, which has to be prepared as very thin film or micropowder, will exhibit a very low background and thus an excellent signal-to-noise ratio. These systems are quite compact, as seen in Figure 9.17b.

9.2.2.1 Excitation Source

Most tubes in EDXRF systems are end-window or transmission target designs. For secondary target and mobile XRF units, side-window tubes are used. Tube voltage and excitation is limited to usually 50 or 60 kV for handheld, portable, and benchtop devices. High voltages

X-RAY SPECTROSCOPY

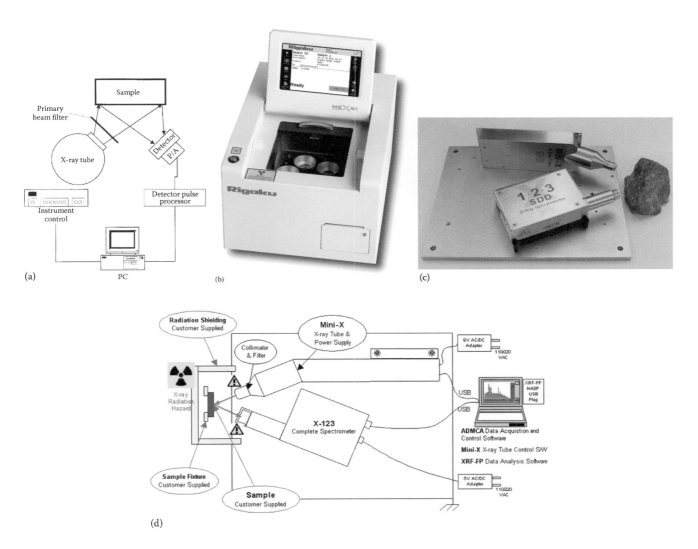

Figure 9.14 (a) A schematic EDXRF system with an X-ray tube source. There is no dispersion device between the sample and the detector. Photons of all energies are collected simultaneously. (From Ellis, A.T., in Van Griekin, R.E.; Markowicz, A.A. [eds.], *Handbook of X-Ray Spectrometry*, 2nd edn., Marcel Dekker, Inc.: New York, 2002. Used with permission.) (b) A commercial benchtop EDXRF spectrometer, the Rigaku NEX QC, setup to measure sulfur in crude oil with a six position automatic sampler. (Courtesy of Applied Rigaku Technologies, Inc., Austin, TX, www.rigakuedxrf.com.) (c) Component-based compact XRF kit, showing relative positions of the X-ray tube (top), the sample (right), and the X-ray spectrometer (bottom). (d) The compact component kit as part of a complete XRF system. (c and d: Courtesy of AMPTEK, Inc. www.amptek.com.)

(≥ 100 kV) require more shielding and therefore require large floor units.

9.2.2.2 Primary Beam Modifiers

Primary beam filters, beam masks, and other devices are used to modify the excitation from the source. One of the problems with using an X-ray tube is that both continuum and characteristic line radiation are generated at certain operating voltages, as seen in Figure 9.3. For many analytical uses, only one type of radiation is desired. Filters of various materials can be used to absorb unwanted radiation but permit radiation of the desired wavelength to pass by placing the filter between the X-ray source and the sample.

Primary beam filters are used to modify the excitation spectra by making use of characteristic, selective absorption. The nature of the material and thickness of the material are the parameters used to tune the primary excitation. Filters are customized based upon the target material and application. Manual filters are inserted by the user into the beam path, whereas automatic filters are generally arranged in a wheel-like fashion and controlled by the instrument software (see Figure 9.15).

A simple example of how a filter is used is shown in Figure 9.18. The solid line spectrum is the output of a Rh tube operated at 20 kV with no filter between the tube and

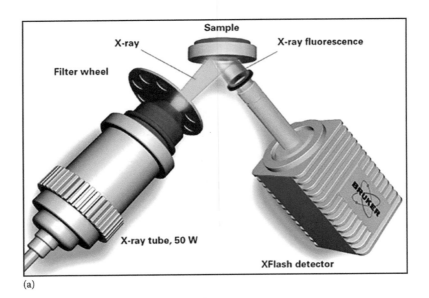

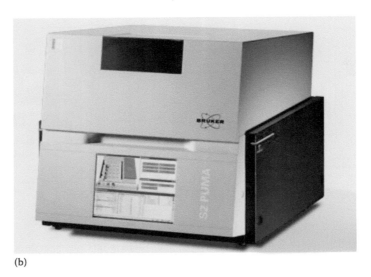

Figure 9.15 (a) Schematic 3D view of a direct excitation EDXRF system. (© 2013 Bruker, Inc. www.bruker.com. Used with permission.) (b) A photograph of an instrument capable of measurement from carbon to americium, used for quality control. (Bruker, Inc. www.bruker.com. Used with permission.)

the detector. The Rh L_α line at 2.69 keV is seen, along with a broad continuum of X-rays from 4 to 19 keV. If the Rh L_α line gets scattered into the detector, as it can from a crystalline sample, it can be mistaken for an element in the sample or may overlap another line, causing spectral interference. Placing a cellulose filter over the tube window causes the low-energy Rh characteristic line to be absorbed; only the continuum radiation reaches the detector, as shown by the dotted line spectrum.

Alternatively, when monochromatic radiation is desired, a filter is chosen with its absorption edge between the K_α and the K_β emission lines of the target element or between the continuum and the characteristic emission lines of the target. Using an Al filter in front of a silver X-ray tube removes most of the continuum radiation (Bremsstrahlung) between 5 and 15 keV; the light reaching the sample is essentially the characteristic line or lines of the target (Figure 9.19a). As seen in Figure 9.19b, using the correct filter greatly improves the signal-to-background ratio of the sample spectrum, which will reduce uncertainty in the measurement of these lines and improve quantitative analysis.

Filters are commonly thin metal foils, usually pure elements, but some alloys such as brass and materials like cellulose are used. Varying the foil thickness of a filter is used to

X-RAY SPECTROSCOPY

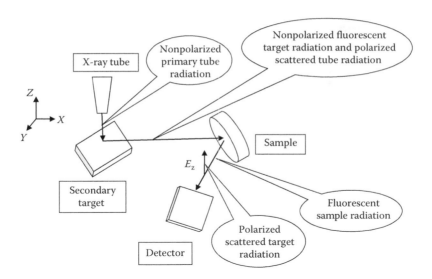

Figure 9.16 Schematic drawing of a 3D beam of a secondary target EDXRF system. (Modified from a diagram provided by Bruno Vrebos, 209, PANalytical Inc., The Netherlands. www.panalytical.com.)

(a)

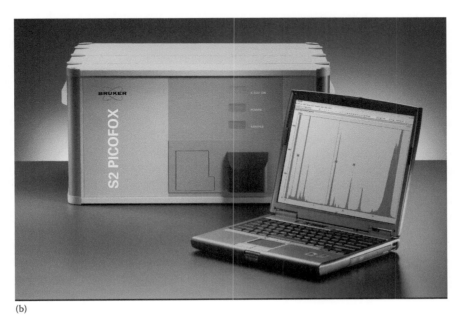

(b)

Figure 9.17 (a) Schematic drawing of a TXRF system. From left to right is the X-ray source, the synthetic multilayer monochromator (Section 9.2.3.2), the thin sample on a support, the silicon drift detector (SDD) above the sample, and finally a beam stop. (b) A commercial TXRF spectrometer, the Bruker S2 PICOFOX with its laptop computer. The displayed spectrum shows the high S/N ratio. (© 2013 Bruker, Inc. www.bruker.com. Used with permission.)

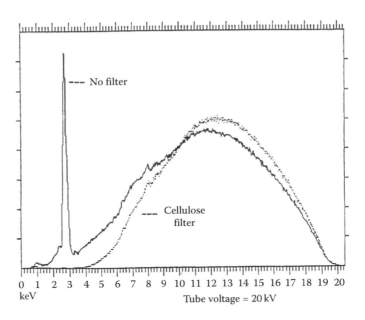

Figure 9.18 The use of a filter to remove unwanted radiation from entering the spectrometer is demonstrated. A cellulose filter placed between a Rh X-ray tube and the sample removes the Rh L_α line at 2.69 keV and allows only the continuum radiation to excite the sample. (© Thermo Fisher Scientific. www.thermofisher.com. Used with permission.)

optimize peak-to-background ratios. Commonly used filters for various targets are listed in Tables 9.5 and 9.6.

Applications of primary beam filters include:

- Attenuation or suppression of the characteristic tube lines (e.g., Rh K_α) to enable analysis for the target element or neighboring elements (Cd, Ag).
- Attenuation of elemental ranges (e.g., removal of all energies below 10 kV).
- Optimization of signal-to-noise (signal-to-background) levels.
- Use as "secondary" targets to emit the target radiation in addition to the nonabsorbed source radiation.

In addition, the primary beam filter units can be equipped with a beam-reducing device referred to as a mask or pinhole collimator. The device is based on the complete absorption of the primary beam except what passes through the opening. The smaller the reduced beam diameter, the more signal is being eliminated, thus reducing the signal from the sample. Units with small spot capability therefore are usually higher power (e.g., 50 W) to compensate for the decreased signal.

Beam restrictors and concentrators using capillary optics concentrate the beam on a small spot without the loss of signal from masks. Commercial units can achieve spot sizes of 25 μm with suitable intensity. Masks and beam restrictors are used to observe microscopic regions within a sample and, using raster techniques, can provide elemental image maps of samples that are heterogeneous.

9.2.2.3 Sample Holders

XRF is used for the analysis of solid and liquid samples, and similar sample holders and autosamplers are used for both EDXRF and WDXRF. Sample preparation and other considerations will be discussed in the applications section. For quantitative analysis, the surface of the sample must be as flat as possible, as also will be discussed in the applications section. There are two classes of sample holders: cassettes for bulk solid samples and cells for loose powders, small drillings, and liquids. A typical cassette for a flat bulk solid such as a polished metal disk, a pressed powder disk, and a glass or polymer flat is shown in Figure 9.20a. The cassette is a metal cylinder, with a screw top and a circular opening or aperture, where the sample will be exposed to the X-ray beam. The maximum size for a bulk sample is shown. The sample is placed in the cassette. For a system where the sample is analyzed facedown, the cassette is placed with the opening down and the bulk sample sits in the holder held in position by gravity, as shown in Figure 9.20b. If the system requires the sample faceup, the body of the cassette must be filled with an inert support (often a block of wood) to press the sample surface against the opening. These cassettes are available with a variety of apertures, usually from 8 to 38 mm in diameter, to accommodate samples of different diameters. Other types of solid samples, such as coatings on a solid substrate, can be placed directly in this type of cassette.

The analysis of liquids, loose powders, or small pieces requires a different holder. The cells for these types of

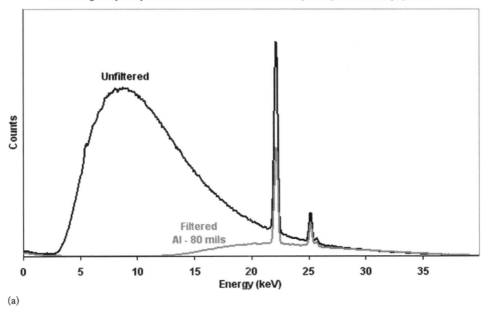

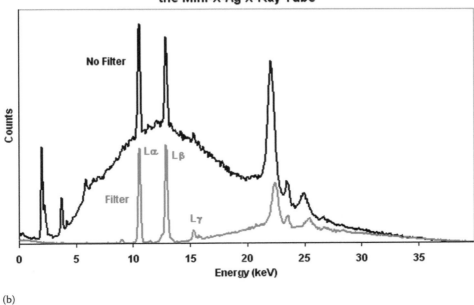

Figure 9.19 (a) Output spectra of the mini-X silver X-ray tube at 40 kV unfiltered and filtered with 2 mm Al. The filter removes the majority of the continuum radiation. (b) A Pb sample analyzed with and without a filter on the Ag X-ray tube. With no filter, the Pb characteristic X-rays are superimposed on a large background of scattered X-rays. With the Al filter, the signal-to-background ratio for Pb is greatly improved. Note that the Pb L_γ is much more visible in the filtered spectrum. (Courtesy of Amptek, Inc. www.amptek.com.)

Table 9.5 Filters for Commonly Available X-Ray Tubes

Target	Target K_α (Å)	Target K_β (Å)	Filter Element	Filter K Edge (Å)	Foil Thickness (µm)	% K_β Absorbed
Cr	2.289	2.085	V	2.269	15.3	99.0
Fe	1.936	1.757	Mn	1.896	12.1	98.7
Co	1.789	1.621	Fe	1.743	14.7	98.9
Ni	1.658	1.500	Co	1.608	14.3	98.4
Cu	1.541	1.393	Ni	1.488	15.8	97.9
Mo	0.709	0.632	Zr	0.689	63.0	96.3
Ag	0.559	0.497	Pd	0.509	41.3	94.6
W	0.209	0.185				

Source: From Parsons, M.L., X-ray methods, in Ewing, G.W. (ed.), *Analytical Instrumentation Handbook*, 2nd edn., Marcel Dekker, Inc., New York, 1997. Used with permission.

Note: No suitable filter is available for tungsten.

Table 9.6 Typical Primary Beam Filters and Range of Use in EDXRF Systems

Filter	Thickness (µm)	X-Ray Tube	Range (kV)	Elements Comments
None		4–50	Na–Ca	Optimum for light elements, 4–8 kV excitation[a]
Cellulose		5–10	Si–Ti	Suppresses tube L lines (see Figure 9.18)[a]
Al, thin	25–75	8–12	S–V	Removes tube L lines[a]
Al, thick	75–200	10–20	Ca–Cu	Used for transition elements
Anode element, thin	25–75	25–40	Ca–Mo	Good for trace analysis[b]
Anode element, thick	100–150	40–50	Cu–Mo	Trace analysis with heavy element L lines[b]
Cu	200–500	50	>Fe	Suppresses tube K lines

Source: Modified from Ellis, A.T., in Van Griekin, R.E.; Markowicz, A.A. (eds.), *Handbook of X-Ray Spectrometry*, 2nd edn., Marcel Dekker, Inc.: New York, 2002. With permission.

[a] He purge or vacuum path needed to avoid attenuation of low-energy lines.
[b] Serves as a secondary fluorescence source, also called a monochromatizing filter; preferentially transmits tube K lines.

samples are multipart plastic holders, shown in Figure 9.21a, and require squares or circles of thin polymer film to hold the sample in the cell. The body of the cell is a cylinder open on both ends. One end of the cylinder is covered with the plastic film (or even clear plastic adhesive tape), and the film or tape is clamped into place by a plastic ring. The cell is placed with the film down, and the sample of liquid, powder, or filings is added. The film surface should be completely covered, as uniformly as possible. A plastic disk that just fits into the cell is inserted and pressed against the sample to obtain as flat a surface as possible, and a top cap is screwed or pressed on. For liquid samples, a vented top is used to avoid pressure buildup from heating of the sample by the X-ray beam. This assembled cell may be used "as is" or may be inserted into a standard cassette, as shown in Figure 9.21b in a facedown configuration. As you can imagine, if the thin polymer film breaks, samples of loose powder, chips, or liquid will spill into the interior of the spectrometer, contaminating the system and possibly breaking the Be window of the X-ray tube, if the tube is below the sample. It is for this reason that liquid sample cells are vented and a vacuum is not used. This is the main disadvantage of the facedown configuration; for anything other than bulk samples, there is a risk of contaminating the instrument if the film covering the sample ruptures. Figure 9.21b shows that in the facedown configuration, a liquid naturally assumes a flat surface. Imagine what the liquid sample would look like faceup. An air bubble will form at the film surface if a sealed cell is used and not filled completely. A bubble may form at the surface by heating of the sample in the X-ray beam if the cell is filled completely. If this occurs, the intensity of XRF from the sample will drop dramatically, and the possibility of film rupture as the pressure in the cell builds increases dramatically. So, if liquid samples must be analyzed, the facedown configuration gives better quantitative results, even at the risk of contaminating the spectrometer.

The sample cassette is moved into position, either manually or with an automatic sample changer. In position, the sample is spun, generally at about 30 rpm, to homogenize the surface presented to the X-ray beam.

Polymer films used to cover the cell opening must be low in trace element impurities, strong enough to hold the sample without breaking, thermally stable, and chemically

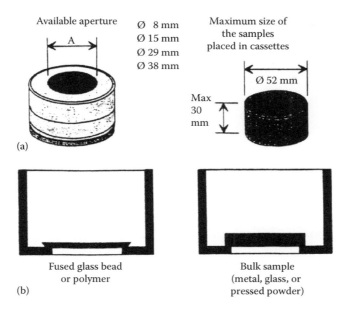

Figure 9.20 (a) Sample cassette for bulk solid samples. (b) Position of a bulk solid in the facedown configuration used in many spectrometers. (© Thermo Fisher Scientific. www.thermofisher.com. Used with permission.)

inert. They certainly must not be soluble in any liquid samples to be analyzed. Films of polyester (Mylar®), polyimide (Kapton®), polycarbonate, polypropylene, and fluoropolymer (Teflon®) are commonly used, with film thickness ranging between about 3–8 μm. Films of different composition and thickness transmit X-rays to varying degrees (Figure 9.22), and the film chosen must transmit the wavelengths for the elements to be measured in the sample.

Manual XRF units are designed for the operator to place the sample into the measurement position, and this is done one sample at a time. Manual systems also include handheld XRF (HH XRF) analyzers, where the sample is pressed against the unit's "nose," with no sample holder (or preparation). The sample position and beam path from source to sample and detector accommodate different atmospheres, including vacuum, helium, nitrogen, and air.

Portable and handheld units can accommodate either vacuum or helium purges in their instrument nose. Sample changers or autosamplers are used to enable units to operate unattended, and the number of samples in the autosampler varies with instrument make and model.

Random access sample changers are more complex than standard autosamplers and allow the operator to add and remove samples while the unit is analyzing another sample in the measurement position. Batch sample changers are usually based on a circular tray or a conveyer belt with multiple sample positions. Samples are rotated or moved over the measurement position. This design does not allow a sample change while the measurement sequence is started since it is connected to an interlock and/or purged with the user-selected atmosphere. This design though is technically less complex and can be used for small units.

9.2.2.4 EDXRF Detectors

The detector is the most crucial component of the EDXRF unit since it detects and sorts the incoming photons originating from the sample. The detector type and associated electronics determine the performance with respect to count rate, resolution, and detection efficiency.

The count rate is the total of *all* photons detected and counted by the detector over the energy range being detected.

Two types of detectors are used in commercially available units: proportional detectors and semiconductor detectors such as silicon PIN, Si(Li), Ge(Li), and silicon drift detectors. The detectors used in EDXRF have very high intrinsic energy resolution. In these systems, the detector resolves the spectrum. The signal pulses are collected, integrated, and displayed by a multichannel analyzer (MCA).

Semiconductor detectors. When an X-ray falls on a solid-state semiconductor, it generates an electron (e^-) and a hole (e^+). Based on this phenomenon, semiconductor detectors have been developed and are now of prime importance in EDXRF and scanning electron microscopy. The total ionization caused by an X-ray photon striking the detector is proportional to the energy of the incident photon. A formerly common semiconductor detector for laboratory EDXRF systems was the *lithium-drifted silicon diode*, represented as Si(Li) and called a "silly" detector. A schematic diagram of a silicon lithium-drifted detector is shown in Figure 9.23.

A cylindrical piece of pure, single-crystal silicon is used. The size of this piece is 4–19 mm in diameter and 3–5 mm thick. The density of free electrons in the silicon is

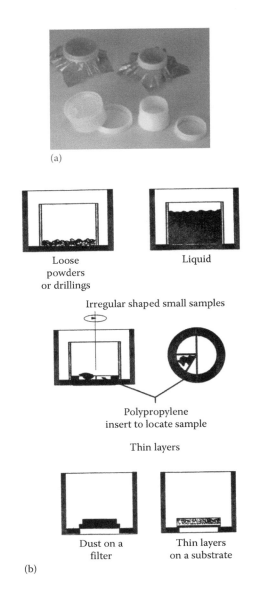

Figure 9.21 Cells for liquid samples, loose powders, and chips. (a) Two types of disposable polyethylene sample cups for liquids and loose samples, consisting of a cup and snap ring to hold the polymer film cover. The cells with polymer film in place are shown at the top of the photograph. The disassembled cup and ring pieces are shown at the bottom of the photograph. (Courtesy of SPEX Certiprep, Inc. Metuchen, NJ. www.spexcsp.com.) (b) Liquid and other loose samples in cells such as those shown in the photo, and then inserted into a sample cassette of the type shown in (a) for a facedown configuration spectrometer. As shown, dust sampled on impact filters or thin-layer samples may be inserted directly into the sample cassette. (© Thermo Fisher Scientific, www.thermofisher.com. Used with permission.)

very low, constituting a p-type semiconductor. If the density of free electrons is high in a semiconductor, then we have an n-type semiconductor. Semiconductor diode detectors always operate with a combination of these two types.

The diode is made by plating lithium onto one end of the silicon. The lithium is drifted into, that is diffused into, the silicon crystal by an applied voltage. The high concentration of Li at the one end creates an n-type region. In the diffusion process, all electron acceptors are neutralized in the bulk of the crystal, which becomes highly nonconducting. This is the *intrinsic* material. The lithium drifting is stopped before reaching the other end of the silicon crystal, leaving a region of pure Si (p-type), as shown in Figure 9.23. Submicron gold layers are applied at each end as electrical contacts. The detector is reverse-biased, removing any free charge carriers from the intrinsic region. Under this condition, no current should flow since there are no charge carriers in the intrinsic region. However, the bandgap between the valence band and the conduction band is small, only 1.1 eV for Si(Li). At room temperature, thermally generated charge carriers cross this barrier easily and become conductive even with no X-ray photons striking the detector. This causes a high noise level. To decrease this noise and increase the sensitivity of the detector, the temperature of the system must be decreased significantly. This is accomplished by cooling the detector to 77 K with liquid nitrogen, which must be replenished regularly. Peltier effect-based electronic cooling is used with temperatures of as low as −90 °C on benchtop EDXRF units. A Peltier cooling device, also known as a thermoelectric cooling device, uses the Peltier effect to create a heat flux between the junction of two different materials. A Peltier cooler is a solid-state heat pump, which transfers heat from one side of the device to the other, depending on the direction of the electric current. The solid-state nature of the cooler means no moving parts, compact size, and no maintenance. These devices are commonly used to cool electronic components.

In exactly the same fashion, germanium, also in group IV of the periodic table, can be used instead of silicon, making a Ge(Li)-drifted detector. (You might guess this is called a "jelly" detector.) The Ge(Li) detector also requires liquid nitrogen or electronic cooling, since its bandgap is only 0.66 eV.

An X-ray photon striking the detector produces multiple electron–hole pairs in the intrinsic region (Figure 9.23b). The number of electron–hole pairs produced is proportional to the photon energy. The energy required to make an electron–hole pair is 3.86 eV in Si(Li), so the number of electron–hole pairs formed is approximately

$$n = \frac{E}{\varepsilon} = \frac{E}{3.86 \text{ eV}} \quad (9.16)$$

where
n is the number of electron–hole pairs
E is the energy of the incident X-ray photon (eV)
ε is the energy to form an electron–hole pair (eV)

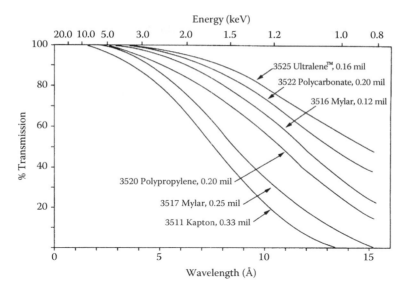

Figure 9.22 The X-ray transmission characteristics of some thin polymer films used as sample holder covers for liquid, loose powder, and similar samples. (Courtesy of SPEX Certiprep, Inc., Metuchen, NJ. www.spexcsp.com.)

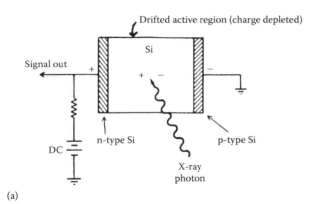

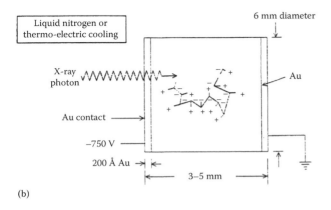

Figure 9.23 The Si(Li) semiconductor detector. (a) Schematic shows the n-type Si region on one end of the Si crystal, a central charge-depleted intrinsic region and p-type Si on the other end. (b) The actual detector has 200 Å layers of gold as electric contacts on each end of the crystal. An X-ray photon striking the intrinsic region generates electron–hole pairs within the diode. (b: © Thermo Fisher Scientific. www.thermofisher.com. Used with permission.)

For a similar Ge lithium-drifted detector, the energy required for ionization is 2.96 eV. This is much less than the energy required for ionization in a proportional counter or a NaI(Tl) scintillation detector.

Under the influence of an applied voltage, the electrons move toward the positive end and the holes toward the negative end of the detector. The total charge collected at the positive contact is

$$Q = nq_e \quad (9.17)$$

where
Q is the total charge in coulombs (C)
n is the number of electron–hole pairs = E/ε
q_e is the charge on one electron = 1.69×10^{-19} C/electron

The collection of charge results in a voltage pulse. Since the total charge is proportional to the energy of the incident photon, the amplitude of the voltage pulse produced is also directly proportional to the energy of the incident photon. The voltage pulses are amplified and "shaped" electronically and sent to a *multichannel pulse height analyzer* (Section 9.2.2.5) to be sorted by pulse height and counted.

Silicon PIN diode detectors. These consist essentially of a couple of mm thick silicon junction type p-i-n diode with a bias of 1000 V across it. The heavily doped central part forms the nonconducting i-layer (intrinsic layer), where the doping compensates the residual acceptors that would otherwise make the layer p-type. When an X-ray photon passes through, it causes a swarm of electron–hole pairs to form, and this causes a voltage pulse. To obtain sufficiently low conductivity, the detector must be maintained at low temperature with a Peltier device. Continuous improvement of the pulse processing results in resolution as low as 125 eV. Count rate and resolution are strongly correlated; the resolution is best at a low count rate. The yield of this detector for higher energies is good although less than for the Si(Li) detector (Figure 9.24).

Silicon drift detector. Commercially available silicon drift detectors (SDDs) are based on the drift chamber principle. The detector crystal is moderately cooled by vibration-free thermoelectric coolers. A monolithically integrated on-chip field effect transistor (FET) acts as a signal amplifier and controls energy resolution. The sideward depletion of the active detector volume in connection with the integrated drift structure provides an extremely small detector capacitance that enables the use of fast signal-processing techniques, thus enabling high count rate processing. The processing electronics enable high count rate and the lowest possible resolution (Figure 9.25).

The unique property of this type of detector is the extremely small value of the anode capacitance, allowing the FET to be either integrated on the chip or connected to it by a short metal strip. Using very elaborate and proprietary processing technology in the fabrication reduces the leakage current level to such low levels that the detector can be operated with moderate cooling. This cooling can be readily achieved by Peltier cooling and is in the vicinity of −15 to −20 °C.

Figure 9.24 Schematic of a silicon PIN detector. (Courtesy of Amptek, Inc. www.amptek.com.)

9.2.2.5 Multichannel Pulse Height Analyzer

A multichannel pulse height analyzer, also called a multichannel analyzer (MCA) or digital pulse processor, collects, integrates, and displays the signal pulses from the detector. The operation of an MCA can be modeled in a simple fashion. Assume we have a handful of coins of different denominations and a coin-sorting device to put each coin into a separate stack (Figure 9.26a). The stacks will have different heights, depending on the number of coins of each type (Figure 9.26b, second plot from the bottom). We can plot "counts" or number of coins versus the height of the stack. Indeed, old pulse height analyzers actually had a readout that mimicked coin counting, or otherwise a series of spinning "electric eyes" that showed the progressive incrementation. A few of these old-style units are still in use but most have been replaced by digital processors, although the name is retained. An MCA does the same thing with photons of different energies. Assume that we have a pulse height analyzer of a given total voltage range with the ability to change the voltage in small increments. As an example, the total voltage range is 10 V, and the interval of change is 0.1 V. X-rays of short wavelengths (high energies) must be separated from X-rays of long wavelengths (low energies). That is what a pulse height analyzer does; it rejects energy signals that are higher or lower than a selected energy window. If the analyzer window can be changed in small energy increments, only photons with that energy will pass through. Those photons are counted and stored in that energy window location in the analyzer memory. Each energy window location is called a *channel*. Then the energy window (voltage) is changed by 0.1 V, and only photons corresponding to the new energy window will pass through and be counted and stored in a second channel. Sweeping the voltage range in steps of 0.1 V permits us to distinguish between X-rays of various energies. If the X-ray photons are counted by energy, we can obtain I, the X-ray intensity at given energy. This permits us to plot I versus wavelength (energy), which gives us an energy spectrum

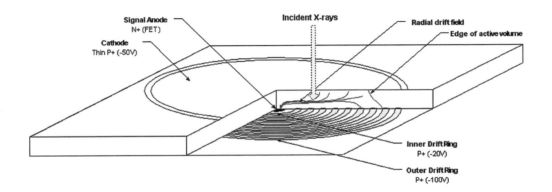

Figure 9.25 Schematic of an SDD. (Courtesy of Amptek, Inc. www.amptek.com.)

X-RAY SPECTROSCOPY

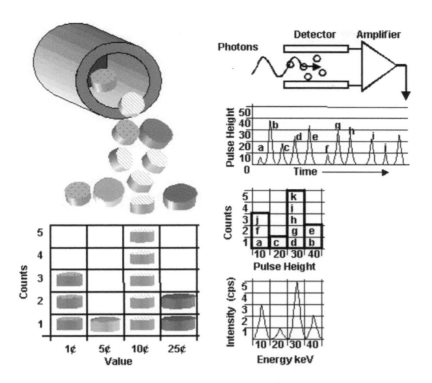

Figure 9.26 Schematic of the pulse processing of an MCA. (a) The "coins" (photons) are separated by denomination and binned. (b) Second from the bottom: The number of coins gives a "stack height" (counts). The typical EDXRF output is the lower-right plot of counts per second versus energy.

of the XRF from the sample. An EDXRF spectrum is in the form of a histogram, usually plotted as "counts" on the y-axis, where counts means the number of photons counted in a given channel, versus energy on the x-axis. In practice, an EDXRF is equipped with a pulse height analyzer with many channels and complicated signal processing circuitry. A typical multichannel pulse height analyzer may have 2048 channels, each corresponding to a different energy interval.

Resolution in a semiconductor detector EDXRF system is a function of both the detector characteristics and the electronic pulse processing. The energy resolution of semiconductor detectors is much better than either proportional counters or scintillation counters (SCs). Their excellent resolution is what makes it possible to eliminate the physical dispersion of the X-ray beam; without the energy resolution of semiconductor detectors, EDXRF would not be possible. Resolution is generally defined as the smallest energy difference observable between peaks. In EDXRF, the energy resolution is defined as the FWHM of the Mn K emission peak. Resolution is dependent on the energy of the detected photon and due to the pulse processing also depends on the total number of photons counted (total input count rate). The area of the detector as well as the electronic parameters of the pulse processor affect resolution as well and need to be included in a complete system comparison.

The most common detectors in benchtop EDXRF units are SDDs of <10 mm² area with a resolution of <150 eV for the Mn K line at 100,000 counts/s.

In handheld or portable instrumentation, Si-PIN detectors of <10 mm² area are common, with a resolution of <170 eV at 40,000 cps. For the detection of light elements (Mg to S), SDDs are more suitable and achieve a resolution of <190 eV at 90,000 cps with detector areas of ≥25 mm².

9.2.2.6 Detector Artifact Escape Peaks and Sum Peaks

Spectrum artifacts may appear in the energy-dispersive (ED) spectrum. These are peaks that are not from elements in the sample, but are caused by interaction between the sample and the detector material. For example, when measuring pure iron or steel, some of the Fe photon energy is transferred to the Si detector atoms; the amount of energy absorbed by a Si atom has *escaped* from the Fe photon. This type of peak, which may appear in the spectrum, is called an escape peak. The Si escape peak, from the Si K_α line, results in an artifact peak 1.74 keV lower than the parent peak when any silicon-based detector is used. Similar escape peaks at different energies appear for Ge if a Ge detector is used. Table 9.A.2 in the appendix gives the keV values for the K and L lines of all the elements and can be used to calculate where an escape peak might appear in the spectrum.

Sum peaks in the EDXRF spectrum occur when two high-intensity photons arrive so close in time that the signal-processing electronics cannot separate them. A single peak is registered at an energy that is the sum of the two peaks. Sum peaks can be avoided by reducing the current and thus the signal on the detector. Major elements in the sample, such as Fe in steel, are generally the source of the sum peaks. Most EDXRF systems come with software that automatically can correct for escape peaks and some also for sum peaks. Figure 9.27 shows some examples of these artifact peaks in EDXRF spectra. Table 9.A.2 in the appendix can help with identification of these artifacts. In Figure 9.27a, the instrument software "marks" a peak and identifies it as the element francium. In reality, the peak at 2.77 keV is the Ti K_α escape peak. Note the position of the marker line—it is not at the center of the emission peak, and francium is not a very likely element in most samples. Also check Table 9.A.2. The peak energy does not match Fr. This should alert the analyst to an artifact in the spectrum.

For the major elements in a sample at high count rates, it is possible that two photons are detected at the same time. These two photons are "summed" in the detector output. Figure 9.27b shows the Fe emission spectrum on the left with three observed sum peaks, calculated as follows: 6.40 keV + 6.40 keV = 12.8 keV; 6.40 keV + 7.05 keV = 13.4 keV; and 7.05 keV + 7.05 keV = 14.1 keV.

Figure 9.27c is the spectrum of pure iron, measured with a Si(Li) detector, which shows both escape peaks and sum peaks.

9.2.3 Instrumentation for Wavelength-Dispersive X-Ray Spectrometry

There are three types of WDXRF instruments: **sequential spectrometers**, which use a **goniometer** and sequentially measure the elements by scanning the wavelength; **simultaneous spectrometers**, which use multiple channels, with each channel having its own crystal/detector combination optimized for a specific element or background measurement; and **hybrid spectrometers**, which combine sequential goniometers or scanners with fixed channels as well as XRD channels and goniometers. Hybrid instruments will be discussed in Section 9.4 with XRD instruments.

A WDXRF spectrometer can be generally divided into four major components based on their functionality: the generator, spectrometer, electronic pulse processing unit, and sample changer. The **generator** supplies the current and high voltage for the X-ray tube to produce the tube radiation (primary X-rays). The generators available today can provide a maximum output of around 4 kW with a maximum voltage of up to 70 kV and up to 170 mA current.

The **spectrometer** contains the X-ray tube, primary beam filters, collimators, analyzing crystals, and detectors. Parts of the spectrometer and the sample measurement position are generally under vacuum, but other atmospheres (purge gases) can be used based on the desired application. X-ray tubes and primary beam filters have been discussed earlier. Most commercially available WDXRF units today use end-window X-ray tubes, which are water-cooled. The window thickness is around 75 μm or less (see Figure 9.10).

The primary beam filter wheel is equipped with a selection of absorbing foils, commonly Al and Cu foils of various thicknesses. It is located between the tube and the sample, filtering out undesirable or interfering components of the tube radiation to increase the peak-to-background ratio.

9.2.3.1 Collimators

The X-rays emitted by the anode in an X-ray tube are radially directed. As a result, they form a hemisphere with the target at the center. In WDXRF spectroscopy or XRD structural determination, the spectrometer's analyzing crystal or the crystalline substance undergoing structure determination requires a nearly parallel beam of radiation to function properly. A narrow, nearly parallel beam of X-rays can be made by using two sets of closely packed metal plates or blades separated by small gaps. This arrangement absorbs all the radiation except the narrow beam that passes between the gaps (Figure 9.28).

Decreasing the distance between the plates or increasing the total length of the gaps decreases the divergence of the beam of X-rays (i.e., it collimates or renders them parallel). The use of a collimator increases the wavelength resolution of a spectrometer's analyzing crystal, cuts down on stray X-ray emission, and reduces background (Figure 9.29).

Commercial instruments use multiple tube or multiple slit collimator arrangements, often both before the analyzing crystal (the primary collimator) and before the detector (the secondary collimator). The collimator positions in a sequential WDXRF spectrometer are shown schematically in Figure 9.30. In many wavelength-dispersive (WD) instruments, two detectors are used in tandem, and a third auxiliary collimator may be required. Such an arrangement is shown in Figure 9.31.

Collimators are not needed for curved crystal spectrometers where slits or pinholes are used instead, nor are they needed for ED spectrometers. **Collimator masks** are situated between the sample and collimator and serve the purpose of cutting out the radiation coming from the edge of the sample cup aperture (Figure 9.32).

9.2.3.2 Analyzing Crystals

The analyzing crystals are the "heart" of a WDXRF spectrometer. As we have discussed, a crystal is made up

X-RAY SPECTROSCOPY

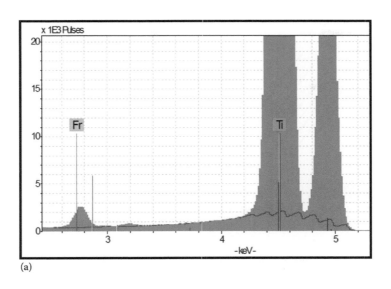

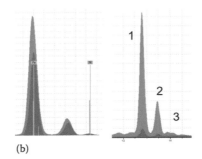

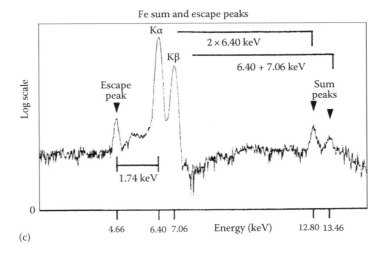

Figure 9.27 (a) The peak at 2.77 keV with the Fr emission line marker shown is in reality the Ti K_α escape peak: 4.51 keV (Ti K_α) − 1.74 (Si K_α) = 2.77 keV. (b) Fe emission shown on the left (K_α at 6.40 keV); sum peaks shown on the right as described earlier. (a and b: Data courtesy of Bruker Elemental © 2013 Bruker, Inc. www.bruker.com. Used with permission.) (c) Iron spectrum showing both escape and sum peaks. The escape peak is lower in energy than the Fe K_α peak by an amount exactly equal to the energy of the Si K_α line. Sum peaks also appear in EDXRF spectra when two intense photons arrive at the detector simultaneously. A sum peak from two K_α photons is shown along with a sum peak from one K_α and one K_β photon. (c: © Thermo Fisher Scientific. www.thermofisher.com. Used with permission.)

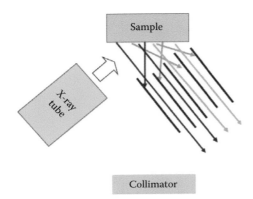

Figure 9.28 The function of a collimator is shown. All radiation is absorbed except for the photons passing through the gaps, forming a narrow, nearly parallel beam of X-rays. (© 2013 Bruker, Inc. www.bruker.com. Used with permission.)

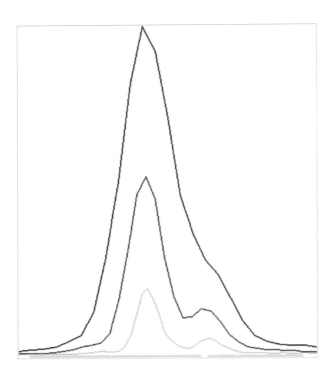

Figure 9.29 Peak resolution and intensity as a function of finer and finer divergence (blade spacing). Finer divergence increases resolution but decreases intensity. (© 2013 Bruker, Inc. www.bruker.com. Used with permission.)

of layers of ions, atoms, or molecules arranged in a well-ordered system or lattice. If the spacing between the layers of atoms is about the same as the wavelength of the radiation, an impinging beam of X-rays is reflected at each layer in the crystal (Figure 9.8). Bragg's law (Equation 9.15) indicates that at any particular angle of incidence θ, only X-rays of a particular wavelength fulfill the requirement of staying in phase and being reinforced, and are therefore diffracted by the crystal. If an X-ray beam consisting of a range of wavelengths falls on the crystal, the diffracted beams of different wavelengths emerge at different angles. The incident beam is thus split up by the crystal into its component X-ray wavelengths, just as a prism or grating splits up white light into a spectrum of its component colors. The analyzing crystal is an X-ray monochromator, with the small lattice distances found in natural or synthetic crystals acting exactly like the ruled diffraction gratings used in the UV/visible (VIS) region. Because X-ray wavelengths are so small, it is not possible to mechanically rule a grating for use in the X-ray region. The crystal separates X-rays of different wavelengths by diffracting them at different angles. Bragg's law fixes the spectral range of a given crystal. Since the maximum value of sinθ is 1.00, the upper spectral limit $\lambda_{max} = 2d$. The diffraction efficiency and the resolution depend on the purity and perfection of the crystal lattice. The crystal should be as perfect as possible, so that the d-spacing for a given plane will be constant in all parts of the crystal. The principle is illustrated in Figure 9.33a. Figure 9.33a shows schematically that two detectors placed at the proper locations could detect two diffracted wavelengths simultaneously. Alternatively, the detector or analyzing crystal could move, allowing each wavelength to be detected sequentially. Both types of spectrometers are commercially available. Some crystals in common use for dispersion of X-rays in commercial XRF spectrometers are listed in Table 9.7. Note that the Miller indices listed in the table are plane identifiers and will be discussed in more detail later in the chapter. Most crystals are natural inorganic or organic compounds. A serious limitation in XRF was the lack of natural crystals with d-spacings large enough to diffract the low-energy X-rays from low atomic number elements. That limitation has been overcome by the synthesis of multilayer "pseudocrystals." The desired lattice plane d-spacing is produced from alternating layers of materials with high and low optical densities, such as Si and W or Ni and BN, deposited on a silicon or quartz flat. Figure 9.33b shows how such a synthetic multilayer crystal functions. The PX-3 multilayer is made from B_4C alternating with Mo, for example. These engineered multilayers are stable and commercially available and have revolutionized the determination of light elements, permitting elements as light as Be to be detected.

When applying Bragg's law, the d-spacing of a specific analyzing crystal limits its detectable element range. The shorter the d-spacing of a crystal, the better the separation of two adjacent or overlapping peaks. For example, looking at vanadium (V) and chromium (Cr), the V $K_{\alpha 1}$ and Cr $K_{\alpha 1}$ lines are farther apart when measured with LiF-220 ($2d = 0.2848$ nm) than when measured with LiF-200 crystal ($2d = 0.4028$).

As is clear from Table 9.7, different crystals are needed to measure different elements. Commercial sequential XRF systems have a computer-controlled multiple-crystal holder

X-RAY SPECTROSCOPY

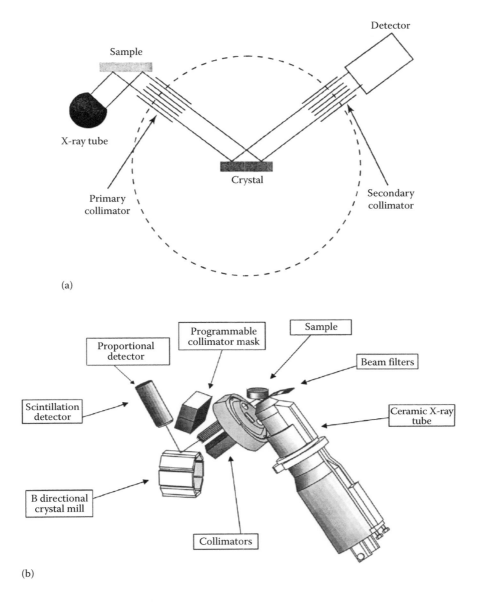

Figure 9.30 (a) Schematic of the optical path in a WD sequential spectrometer, showing the positions of the collimators. (b) Schematic layout of a commercial sequential WDXRF system. (Courtesy of PANalytical, Inc., The Netherlands. www.panalytical.com.)

inside the spectrometer, with positions for as many as 8–10 crystals in some instruments, as seen in Figure 9.32.

The analyzing crystal is mounted on a turntable that can be rotated through θ degrees (see the arrow marked θ on the lower left side of Figure 9.31). The detector(s) are connected to the crystal turntable so that when the analyzing crystal rotates by θ degrees, the detector rotates through 2θ degrees, as shown by the marked arrow. Therefore, the detector is always in the correct position (at the Bragg angle) to detect the dispersed and diffracted fluorescence. This crystal positioning system is called a **goniometer**. Figure 9.34 shows the turntable and the concentric circles made by the crystal and the detector. In most systems, the maximum diffraction angle attainable is 75° θ (or 150° 2θ).

In some systems, the rotation of the crystal and the detector is mechanically coupled with gears. Other systems have no mechanical coupling but use computer-controlled stepper motors for the crystal and the detector. The newest systems use optical position control by optical sensors or optical encoding devices. Optical position control permits very high angular precision and accuracy and very fast scanning speeds.

The analyzing crystals are interchanged by rotation of the crystal holder, but the same goniometer is used to select the diffraction angle, meaning that only one wavelength can be measured at a time with a sequential system.

The analyzing crystal(s) shown schematically in Figures 9.30 through 9.32 have a flat surface. Flat crystals are used

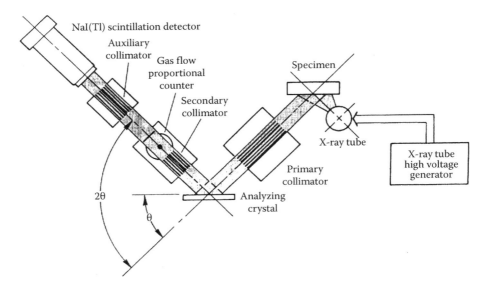

Figure 9.31 A sequential spectrometer with two tandem detectors, showing the placement of the collimators in the optical path. (From Jenkins, R. et al., *Quantitative X-Ray Spectrometry*, Marcel Dekker, Inc., New York, 1981. Used with permission.)

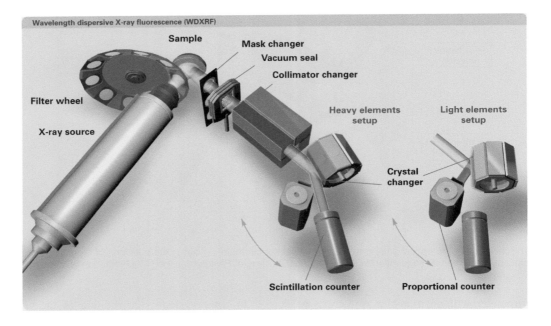

Figure 9.32 Schematic commercial WDXRF system showing the position of a collimator mask changer, as well as the layout of X-ray tube, filter wheel, sample, collimator, analyzing crystal changer, and detector. (© 2013 Bruker, Inc. www.bruker.com. Used with permission.)

in scanning (sequential) spectrometers. Curved crystals, both natural and synthetic multilayers, are used in simultaneous spectrometers, electron microprobes, and for synchrotron X-ray spectrometry. The advantage to a curved crystal is that the X-rays are focused and the collimators replaced by slits, resulting in much higher intensities at the detector than with flat crystal geometry. This makes curved crystals excellent for analysis of very small samples. The use of a curved crystal and slits in a simultaneous spectrometer is illustrated schematically in Figure 9.35. The curved crystal spectrometer geometry should remind you of the Rowland circle geometry for optical emission spectrometers discussed in Chapter 8. Curved crystals are used in the "fixed" channel design of most simultaneous (multichannel) spectrometers.

X-RAY SPECTROSCOPY

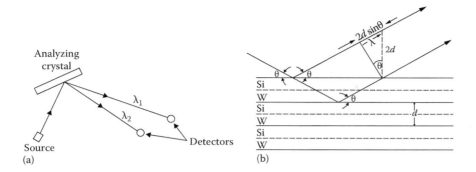

Figure 9.33 (a) The analyzing crystal as a monochromator. (b) A synthetic crystal made from multiple alternating layers of Si and W, showing the *d*-spacing and the ability to diffract X-rays. (b: © 2013 Bruker, Inc., www.bruker.com. Used with permission.)

Table 9.7 Analyzing Crystals Used in Modern X-Ray Spectrometers

Crystal	Name and Orientation (Miller Indices)	2d-Spacing (nm)	Element Range
LiF-420	Lithium fluoride (420)	0.1891	$\geq NiK_{\alpha 1}$
LiF-220	Lithium fluoride (220)	0.2848	$\geq VK_{\alpha 1}$
LiF-200	Lithium fluoride (200)	0.4028	$\geq KK_{\alpha 1}$
Ge	Germanium (111)	0.6530	P, S, Cl
InSb	Indium antimonide (111)	0.7481	Si
PET[a]	Pentaerythritol (002)	0.8740	Al–Ti, Kr–Xe, Hf–Bi
ADP	Ammonium dihydrogen phosphate	1.064	Mg
TlAP[b]	Thallium acid phthalate (100) (Thallium hydrogen phthalate)	2.576	Fe–Na
XS55, OV055, AX06, PX-1, PX-3, and others	Multilayered synthetic crystal[c]	5.0–19	Light elements from Be to N, N–Al, Ca–Br depending on the *d*-spacing

[a] The most heat-sensitive crystal and the only completely organic material used as an analyzing crystal.
[b] Toxic.
[c] The designations for these multilayered synthetic crystals are commercial trade names from different instrument manufacturers.

9.2.3.3 Detectors

X-ray detectors transform photon energy into electrical pulses. The pulses (and therefore, the photons) are counted over a period of time. The *count rate*, usually expressed as counts per second, is a measure of the intensity of the X-ray beam. Operating the detector as a photon counter is particularly useful with low-intensity sources, as is often the case with X-radiation.

There are three major classes of X-ray detectors in commercial use: gas-filled detectors, scintillation detectors, and semiconductor detectors. Semiconductor detectors are used in EDXRF and handheld systems and were discussed earlier with EDXRF instrumentation. Both WDXRF and EDXRF detection make use of a signal processor called a pulse height analyzer or selector in conjunction with the detector.

WDXRF systems commonly use one or more of the following detectors: gas flow proportional counter (flow counter [FC]), sealed gas-proportional counter, and scintillation counters (SC).

Gas-filled detectors. Suppose we take a metal cylinder, fit it with X-ray transparent windows, place in its center a positively charged wire, fill it with inert filler gas, such as helium, argon, or xenon, and seal it. If an X-ray photon enters the cylinder, it will collide with and ionize a molecule of the filler gas by ejecting an *outer shell electron*, creating a *primary ion pair*. With helium as a filler gas, the ion pair would be He$^+$ and a photoelectron e$^-$. A sealed gas-filled detector of this type is illustrated in Figure 9.36. The interaction

$$h\nu + He \rightarrow He^+ + e^- + h\nu$$

takes place inside the tube. The electron is attracted to the center wire by the applied potential on the wire. The positive charge causes the wire to act as the anode, while the positive ion, He$^+$ in this case, migrates to the metal body (the cathode). The ejected photoelectron has a very high kinetic energy. It loses energy by colliding with and ionizing many additional gas molecules as it moves to the center wire. A plot of the number of electrons reaching the wire versus the applied potential is given in Figure 9.37.

With no voltage applied, the electron and the positive ion (He$^+$) recombine and no current flows. As the voltage is slowly increased, an increasing number of electrons

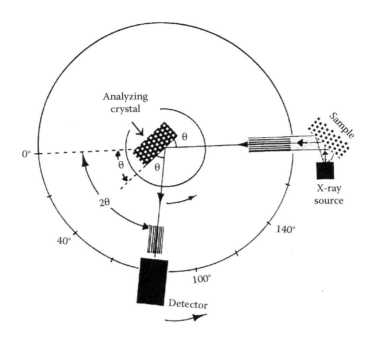

Figure 9.34 Goniometer layout for a sequential XRF spectrometer.

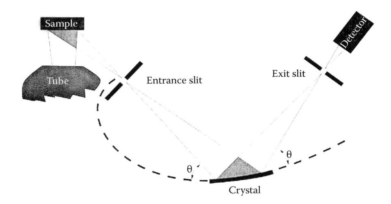

Figure 9.35 Schematic of the optical path in a curved crystal spectrometer. (Courtesy of PANalytical, Inc., The Netherlands. www.panalytical.com.)

reach the anode, but not all of them; recombination still occurs. This is the sloping region marked A in Figure 9.37. At the plateau marked B in Figure 9.37, all the electrons released by a single photon reach the anode and the current is independent of small changes in the voltage. A detector operating under these voltage conditions is known as an *ionization counter*. Ionization counters are not used in X-ray spectrometers because of their lack of sensitivity.

As the voltage increases further, the electrons moving toward the center wire are increasingly accelerated. More and more electrons reach the detector as a result of an avalanche of secondary ion pairs being formed, and the signal is greatly amplified. In the region marked C in Figure 9.37, the current pulse is proportional to the energy of the incoming X-ray photon. This is the basis of a *proportional counter*. In X-ray spectrometry, gas-filled detectors are used exclusively in this range, that is, as proportional counters. The amplification factor is a complex function that depends on the ionization potential of the filler gas, the anode potential, the mean free path of the photoelectrons, and other factors. It is critical that the applied potential, filler gas pressure, and other factors be kept constant to produce accurate pulse amplitude measurements. There are two main types of proportional counter: flow proportional counters and sealed proportional counters.

As shown in Figure 9.37, if the voltage is further increased, electrons formed in primary and secondary ion pairs are accelerated sufficiently to cause the formation

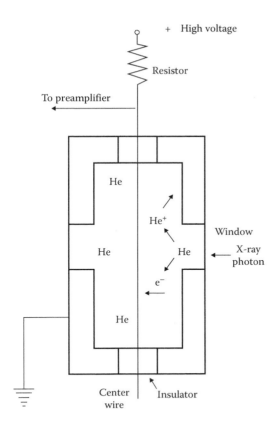

Figure 9.36 Schematic diagram of a gas-filled X-ray detector tube. He filler gas is ionized by X-ray photons to produce He⁺ ions and electrons, e⁻. The electrons move to the positively charged center wire and are detected. (Modified from Parsons, M.L., X-ray methods, in Ewing, G.W. [ed.], *Analytical Instrumentation Handbook*, 2nd edn., Marcel Dekker, Inc., New York, 1997. Used with permission.)

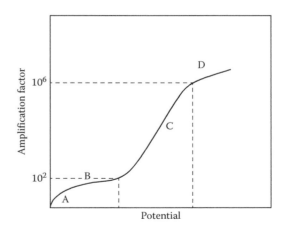

Figure 9.37 Gas-filled detector response versus potential. A detector operating at the plateau marked B is an ionization counter. A proportional counter operates in the sloping region marked C where the response is proportional to the energy of the incoming photon. The plateau marked D represents the response of a Geiger counter. (Modified from Helsen, L.A. and Kuczumow, A., in Van Griekin, R.E.; Markowicz, A.A. [eds.], *Handbook of X-Ray Spectrometry*, 2nd edn., Marcel Dekker, Inc., New York, 2002. Used with permission.)

of more ion pairs. This results in huge amplification in electrons reaching the center wire from each X-ray photon falling on the detector. The signal becomes independent of the energy of the photons and results in another plateau, marked D. This is called the Geiger–Müller plateau; a detector operated in this potential range is the basis of the *Geiger counter* or *Geiger–Müller* tube. It should be noted that a Geiger counter gives the highest signal for an X-ray beam without regard to the photon energy. However, it suffers from a long *dead time*. The dead time is the amount of time the detector does not respond to incoming X-rays. It occurs because the positive ions move more slowly than the electrons in the ionized gas, creating a *space charge*; this stops the flow of electrons until the positive ions have migrated to the tube walls. The dead time in a Geiger counter is on the order of 100 μs, about 1000 times longer than the dead time in a proportional counter. Due to the long dead time compared with other detectors, Geiger counters are not used for quantitative X-ray spectrometry. They are, however, very important portable detectors for indicating the presence or absence of X-rays. Portable radiation detectors equipped with Geiger counters are used to monitor the operation of equipment that creates or uses ionizing radiation to check for leaks in the shielding.

As was the case with EDXRF, escape peaks that are detector artifacts occur in WDXRF gas-filled detectors. Ionization of the filler gas by an X-ray photon usually results in the ejection of an outer shell electron. However, it is possible for ionization to occur by ejection of an inner shell electron. When this happens, the incoming X-ray photon is absorbed, and the filler gas emits its characteristic K or L lines. This will result in peaks appearing at an energy E' equal to

$$E' = E(incoming\ X-ray) - E(filler\ gas\ characteristic\ X-ray) \quad (9.18)$$

As an example, if the detector filler gas is Ar, the Ar K line has energy of about 3 keV (or a wavelength of 3.87 Å). If an incoming X-ray has a wavelength shorter than 3.87 Å, it can eject an argon K electron. Assuming that the incoming X-ray is the Fe K_α line at 6.4 keV, a peak will appear at (6.4 − 3) keV or about 3.4 keV. This peak at 3.4 keV is called an **escape peak** and can be called either the Fe K_α escape peak or the argon escape peak. An escape peak appears at a constant distance from the parent fluorescence X-ray (in this case, Fe K_α) on the low-energy side. Escape peaks can often be very intense and can be useful in identifying elements.

FC. This detector consists of a metallic cylinder with a thin (counting) wire mounted in the middle (Figure 9.38).

The cylinder is filled with continuous flow of P-10 gas (10 % methane [CH_4] and 90 % Ar). A high positive voltage (+1400 to +1900 V) is applied to the thin wire. The cylinder is sealed with a 0.3–1 μm very thin foil window permeable to X-ray photons with low energies and is thus suitable for measuring light elements, with $Z < 27$. The thin windows allow the filler gas to leak out; therefore, a supply of filler gas is constantly provided to the detector through the inlet, as shown in Figure 9.38a. The pressure, flow, and temperature of the gas must be precisely controlled for an accurate detector response.

X-ray photons interact with the inert gas in the detector by ionizing it. This ionization is based on the ejection of an outer shell electron and creates a primary ion pair. When argon is used, we will observe both an Ar^+ ion and an electron (e^-). The electron is attracted to the wire, which is positively charged, and the Ar^+ will be attracted and migrate to the metallic body. Each primary ion pair produces 10–10,000 electrons due to an "avalanche" of secondary pairs produced as the initial electrons are accelerated toward the wire (Figure 9.39). The principle of operation is that the number of electron–ion pairs created is proportional to the energy of X-rays that entered the detector. The pulse voltage as recorded by the counting electronics is proportional to the energy of the "counted" photon.

Sealed gas-proportional counter. This detector is a sealed or closed system with a fixed volume of filler gas (Figure 9.38b). The filler gas used in a sealed proportional counter may be Ne, Kr, or Xe. The detector is sealed with thicker windows than those used in the FC and therefore does not leak. Window materials include polymers, mica, aluminum, and beryllium. The window thickness generally prevents the sealed proportional counter from being used for the measurement of light elements from Be to Na. It is used for analyzing elements from Al to Ti.

Multiple proportional counters are used in simultaneous X-ray spectrometers, described later, while one proportional counter is often used in tandem with an SC in a sequential system. It is for this reason that the detector has two windows as shown in Figure 9.38. X-ray photons pass

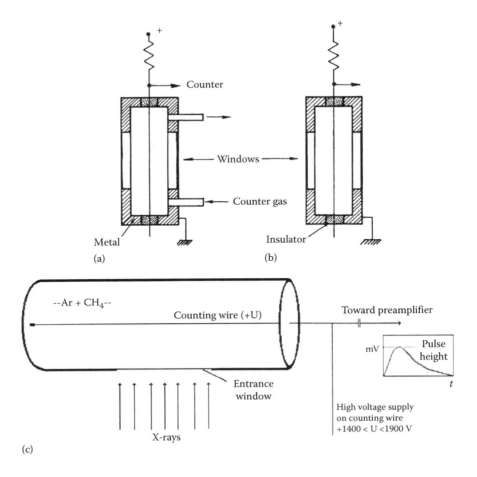

Figure 9.38 Schematics of (a) a flow proportional counter and (b) a sealed proportional counter. (From Helsen, L.A. and Kuczumow, A., in Van Griekin, R.E.; Markowicz, A.A. [eds.], *Handbook of X-Ray Spectrometry*, 2nd edn., Marcel Dekker, Inc., New York, 2002. Used with permission.) (c) Schematic view of a flow proportional counter.

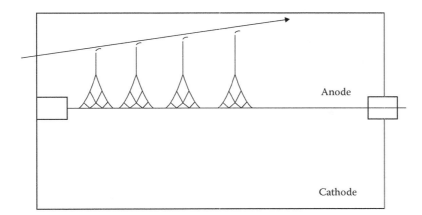

Figure 9.39 Four primary ions pairs produce four "avalanches," all of which contribute to a single pulse. (Copyright Oak Ridge Associated Universities, www.orau.org.)

through the proportional counter to the SC located behind it, as illustrated in Figure 9.31, and signals are obtained from both detectors. It should be noted that this tandem arrangement does not permit independent optimization of both detectors. There are sequential spectrometer systems available with independent proportional and scintillation detectors.

SC. Photomultiplier detectors, discussed in Chapter 4, are very sensitive to visible and UV light, but not to X-rays, to which they are transparent. In a *scintillation detector*, the X-radiation falls on a compound that absorbs X-rays and emits visible light as a result. This phenomenon is called *scintillation*. A PMT can detect the visible light scintillations. The scintillating compound or phosphor can be an inorganic crystal, an organic crystal, or an organic compound dissolved in solvent.

The most commonly used commercial scintillation detector has a thallium-doped sodium iodide crystal, NaI(Tl), as the scintillating material. A single crystal of NaI containing a small amount of homogeneously distributed Tl in the crystal lattice is coupled to a PMT, shown in Figure 9.40.

When an X-ray photon enters the crystal, it causes the interaction

$$I^- \rightarrow I^0 + e^-$$

and the ejection of photoelectrons, as in the gas-filled detector. The ejected photoelectrons cause excited electronic states to form in the crystal by promotion of valence band electrons. When these excited electrons drop back to the ground state, flashes of visible light (scintillations) are emitted. The excited state lies about 3 eV above the ground state, so the emitted light has a wavelength of 410 nm. The intensity of the emitted light pulse from the crystal is proportional to the number of electrons excited by the X-ray photon. The number of electrons excited is proportional to the energy of the X-ray photon; therefore, the scintillation intensity is proportional to the energy of the X-ray.

The scintillations (visible light photons) from the crystal fall on the cathode of the PMT, which is made of a photoemissive material such as indium antimonide. Photoemissive materials release electrons when struck by photons. Electrons ejected from the cathode are accelerated to the first dynode, generating a larger number of electrons. The electron multiplication process occurs at each successive dynode, resulting in approximately 10^6 electrons reaching the anode for every electron that strikes the cathode. The amplitude of the current pulse from the photomultiplier is proportional to the energy of the X-ray photon causing the ionization in the crystal.

To summarize, the scintillation detector works by (1) formation of a photoelectron in the NaI(Tl) crystal after an X-ray photon hits the crystal, (2) emission of visible light photons from an excited state in the crystal, (3) production of photoelectrons from the cathode in the photomultiplier, and (4) electron multiplication.

The NaI(Tl) scintillation detector is most useful for short-wavelength X-rays, <2 Å (Z > 27), so it complements the proportional counter. It also has the potential for escape peaks caused by the iodine K line (about 30 keV or 0.374 Å). Incoming X-rays with wavelengths less than 0.374 Å will result in escape peaks about 30 keV lower in energy than the true energy. The major disadvantage of the NaI(Tl) detector is that its resolution is much worse than that of the proportional counter. This is due to the wider pulse height distribution that occurs in the output pulse because of the multiple steps involved in the operation of this detector.

9.2.3.4 Electronic Pulse Processing Units

In the detector, a photon generates a number of ion pairs, that is, a current pulse with a certain magnitude or pulse height. The pulse height (voltage) in an XRF detector

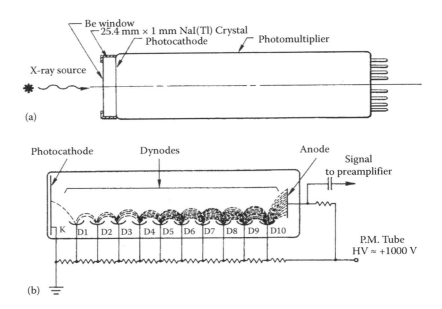

Figure 9.40 Schematic NaI(Tl) scintillation detector. (a) The assembled detector. (b) Schematic representation of the photomultiplier and its circuitry. (Courtesy of ORTEC (Ametek). www.ortec-online.com; From Jenkins, R. et al., *Quantitative X-Ray Spectrometry*, Marcel Dekker, Inc., New York, 1981. Used with permission.)

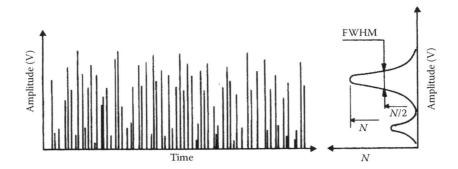

Figure 9.41 Amplitude or pulse height and time record of signals from the detector is on the left. Transformation of the data into a pulse height distribution is shown on the right. The FWHM measurement is shown for the higher peak. (From Helsen, L.A. and Kuczumow, A., in Van Griekin, R.E.; Markowicz, A.A. [eds.], *Handbook of X-Ray Spectrometry*, 2nd edn., Marcel Dekker, Inc., New York, 2002. Used with permission.)

depends on the energy of the photon. Unfortunately, the height of the current pulse that results is not exactly the same for photons of the same energy. Formation of ion pairs and secondary ion pairs is a statistically random process, so a Gaussian distribution of pulse heights centered on the most probable value results.

A series of pulses and their heights is shown in Figure 9.41. On the left side, this figure shows a series of current pulses from photons of two different energies counted over a period of time. If the pulses are plotted by height (amplitude), the result is a Gaussian **pulse height distribution**, shown on the right side of the figure. Two Gaussian pulse height distributions are seen since we had two photons of different energies reaching the detector. The width of the distribution is measured at half of the maximum height; this is called the FWHM. The FWHM is a measure of the energy resolution of a detector. Energy resolution is best in semiconductor detectors and worst in scintillation detectors, with gas-filled proportional counters in the middle.

In a WD instrument, the analyzing crystal separates the wavelengths falling on it, but as Bragg's law tells us, it is possible for higher-order ($n > 1$) lines of other elements to reach the detector. A higher-order line from a different element will have a very different energy and will result in a second pulse height distribution centered at a different energy reaching the detector. This would result in an error, since the signal would be misinterpreted as coming from just one element. This problem is eliminated by the use of a **pulse height discriminator**. The pulse height discriminator sets a lower and an upper pulse height threshold. Only the pulse heights that lie within these limits are counted. It thus reduces the background noise from the electronics and eliminates the interference of higher-order reflections. **A sine amplifier** ensures that a discriminator window, once set for a crystal, will be applicable for all detectable energies.

Pulse processing also incorporates **dead time correction**. Dead time results from the inability of the detector electronics to process the pulses fast enough to match the volume of input signals. Therefore, the greater the incident intensity, the greater the losses would be during measurement. Dead time is typically 300–400 ns for modern spectrometers.

9.2.3.5 Sample Changers

Commercial instruments are equipped with automatic sample changers. These are of various types and sizes, permitting either controlled loading of samples with sequential access or X–Y random access.

9.2.4 Simultaneous WDXRF Spectrometers

A simultaneous WDXRF system, also called a multichannel system, uses multiple channels, optimized for a specific element or background measurement. Each individual channel consists of a crystal, detector, and electronics module, and is dedicated to a specific wavelength. Instruments with as many as 38 fixed crystal/detector channels or as few as two are available. These systems are designed for specific applications, such as the analysis of steel in a production facility where hundreds of samples must be analyzed very quickly (e.g., in less than 60 s) for the same suite of elements every day.

It is difficult to arrange a large number of channels in close proximity to the sample. Therefore, curved crystal monochromators with slit optics are used (Figure 9.42). Two commonly used curvatures are those that follow a logarithmic spiral, shown in Figure 9.42, and the Johansson curvature, which results in a Rowland circle monochromator (Figure 9.43) with the curve of the crystal matching the curve of the Rowland circle. The curvature of the crystal is selected so that the X-ray entrance slit is focused by the crystal onto the exit slit. This allows for higher intensities in a space-saving geometric arrangement.

As seen in Figure 9.42, the X-ray beam is applied from above the sample in these systems. The detector is located behind the exit slit. Scintillation or gas-proportional counters are used depending on the wavelength. All the monochromators are located in a large vacuum chamber. All channels are measured simultaneously, so the generator setting must provide the best compromise for all of the elements measured. The measurement time depends on the statistical accuracy requirements of the element with the lowest intensity; the time is typically 20–60 s.

When measuring trace and major elements simultaneously, the generator is normally set so that the trace elements can be measured with the highest possible intensity. Absorbers or attenuators are then needed for the major element channels to reduce their intensities so that they are in the operational range of the detectors.

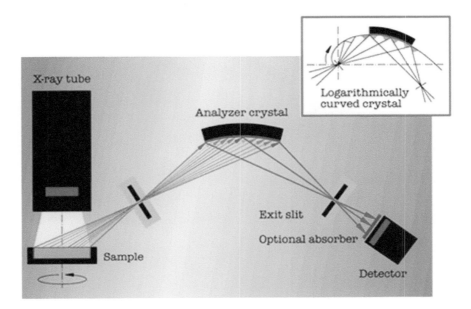

Figure 9.42 Monochromator schematic with a logarithmic spiral curved crystal. (© 2013 Bruker, Inc. www.bruker.com. Used with permission.)

While the fixed channels are used exclusively for quantitative analysis, a scanner or movable channel may be installed in the vacuum chamber to provide qualitative analysis ability and some flexibility in what is otherwise a system with a fixed suite of elements.

A single crystal, usually LiF200 or pentaerythritol (PET), is used, and the scanner's 2θ angular range is limited to 30°–120°. Therefore, some elements must be measured in second order. The scanner works on the principle of the Rowland circle, in which the crystal and the detector move in such a way that the entrance slit, exit slit, and crystal lie on a fixed radius circle that changes in position, as shown in Figure 9.43. A typical commercial instrument is shown in Figure 9.44.

9.2.5 Micro-XRF Instrumentation

Classically, XRF has been considered a "bulk" analysis technique because standard EDXRF and WDXRF systems have analysis spot sizes with a diameter in the mm–cm range, depending on the system. This requires a relatively large volume of sample, with inhomogeneous materials requiring a great deal of sample preparation, discussed in Section 9.2.7. Developments in X-ray optics now permit the analysis of discrete microscopic particles, and the creation of elemental maps of a sample with high spatial resolution. The systems are variously called micro-XRF spectrometers, μ-XRF systems, X-ray analytical microscopes (not to be confused with electron microscopes equipped with EDS detectors discussed in Section 9.5.1), and X-ray microanalyzers.

9.2.5.1 Micro-X-Ray Beam Optics

In order to decrease the analysis spot size, it is necessary to generate an intense, narrow X-ray beam of the necessary diameter. One way to do this is to pass the X-ray beam through an aperture (in effect, a small collimator) with diameters of a few mm to a few hundreds of micrometers. In 1984, Russian scientist Muradin Kumakhov discovered that X-rays were efficiently reflected from glass surfaces when the angle of incidence is low. This total external reflection from a smooth solid surface could be used to focus, collimate, and guide X-ray beams and has led to a new generation of micro-X-ray optics. The production of carefully shaped glass capillary optics allows X-ray beam diameters of 10 μm to 1 mm. A monocapillary optic is shown in Figure 9.45a. The solid angle from the X-ray source to the capillary is relatively large, which allows efficient coupling. Very-high-intensity X-rays can be channeled into the capillary, and high-intensity microbeams as small as 10 μm in diameter are formed by total external reflection.

Capillary optics can also be bundled together to form polycapillary optics, also called Kumakhov lenses. These lenses can be bundles of several thousand capillaries or a monolithic polycapillary structure. Beams produced using polycapillary optics (Figure 9.45b) are highly intense and strongly focused. In comparison to a pinhole collimator, the

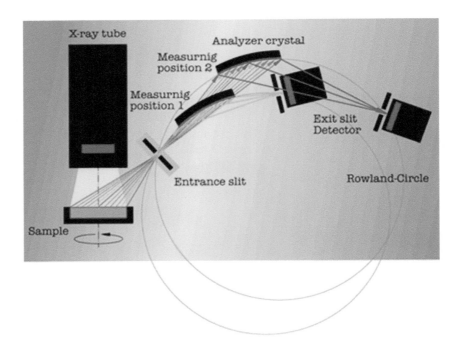

Figure 9.43 Rowland circle type monochromator schematic with a Johansson curved crystal used as a scanning channel. (© 2013 Bruker, Inc. www.bruker.com. Used with permission.)

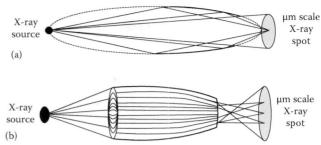

Figure 9.45 (a) Schematic of a monocapillary X-ray optic for producing microbeams of X-rays. (b) Schematic of polycapillary X-ray optics for producing microbeams of X-rays. (© 2013 HORIBA, Ltd. www.horiba.com. All rights reserved.)

Table 9.8 Evolution of Kumakhov Capillary X-Ray Optics

Lens Parameter	1985	2003
Channel size	1 mm	<1 μm
Lens length	1000 mm	10 mm
No. of channels	Max. 1000	>1,000,000
Lens diameter	100 mm	1 mm
Focal distance	Min. 50 mm	≤1 mm
Spot size	500 μm	3 μm

Source: © 2013 Unisantis Europe GmbH. All rights reserved.

9.2.5.2 Micro-XRF System Components

Commercial systems consist of an optical microscope for observation of the sample area and an ED-based XRF system with a low-power, microfocused X-ray source using Kumakhov optics, a filter wheel, and a detector (Si[Li], Si-PIN, SDD). The X-ray source is generally air-cooled, and some systems allow a choice of beam diameters. Systems are available that enable He purge gas for light elements, atmospheric pressure analysis, and low- or high-vacuum analysis.

Systems generally have large sample chambers allowing for analysis of complete circuit boards, for example, Figure 9.46, and open beam systems are available to bring the measurement head anywhere, permitting investigation of very large samples as well as uneven or structured surfaces. The chamber contains an automated motor-driven sample stage that allows the sample to be moved in the x-, y-, and z-directions for single-point, raster, or mapping analyses. A laser beam is used for reproducible sample positioning.

Many systems have proprietary mapping software and a charge-coupled device (CCD) camera for visual observation of the area being measured. Applications of micro-XRF spectroscopy will be discussed in Section 9.2.8.

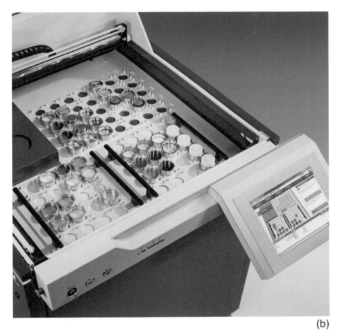

Figure 9.44 (a) The S8 TIGER, a floor-mounted commercial WDXRF system. (b) The sample compartment of the S8 TIGER, showing the multiposition autosampler with typical samples such as polished metal flats on the left and pressed pellets on the right. (© 2013 Bruker, Inc. www.bruker.com. Used with permission.)

fluorescence intensity of a polycapillary lens is increased by a factor of more than 1000. Focused beam diameters of 40–50 μm are possible. Table 9.8 shows how rapidly these X-ray optics have evolved.

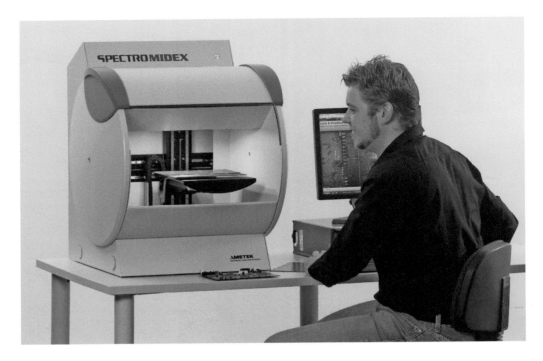

Figure 9.46 A micro-XRF system with a large sample chamber, the SPECTRO MIDEX. This is used for analysis of large objects, such as the complete circuit board shown to the left of the analyst. (© AMETEK, Inc. www.ametek.com. Used with permission.)

9.2.6 Total Reflection XRF

Total reflection (or reflectance) XRF (TXRF) is a technique that looks at thin film samples or thin-layer samples with the ability to detect elements from Na to U. A schematic TXRF spectrometer and a commercial system can be seen in Figures 9.17a and b. The Bruker S2 PICOFOX consists of an air-cooled Mo anode metal–ceramic X-ray tube, a multilayer monochromator, and an XFlash® SDD, and can be equipped with a 25 position autosampler.

The beam incident on the sample is at a very shallow angle, close to 0°, while the fluorescent radiation is detected at 90° from the sample disk, as seen in Figure 9.17a. Because the samples are very thin, matrix effects are negligible. Elemental composition can be quantified using an internal standard (IS).

Samples for TXRF must be prepared on a reflective medium. Polished quartz, glass, or polyacrylic glass disks are used. The samples must be in the form of a thin layer, thin film, or microparticulates. Sample sizes are on the order of microliters for liquids and milligrams for solids. Liquid samples are pipetted directly onto the disk and then dried using heat or vacuum. Solid samples can be analyzed directly if thin enough or prepared by suspension. Solid materials are ground to a fine particle size, weighed, and transferred to a test tube. The sample powder is suspended in a detergent solution, an IS is added, and the suspension is homogenized. An aliquot is pipetted onto the disk, dried under heat or vacuum, and is ready for analysis. TXRF is suitable for the analysis of clinical chemistry samples for trace elements, biological tissues, fluids, plant material, and the like. Other applications can be found at www.bruker-axs.com.

9.2.7 Comparison between EDXRF and WDXRF

The major difference between EDXRF and WDXRF is in the achievable energy (spectral) resolution, shown in Figure 9.47. WDXRF systems can routinely provide working resolutions between 5 and 20 eV, while EDXRF systems typically provide resolutions of 150–300 eV, depending on the type of detector used.

The higher resolution of WDXRF provides several advantages: reduced spectral overlaps, allowing complex sample matrices to be more accurately characterized, and reduced background, which improves detection limits (DLs). However, the additional optical components of a WDXRF instrument, including the collimators and analyzing crystals, reduce the efficiency of the instrument. This can be compensated for by the use of high-power X-ray sources, but the increased complexity of the instrument and the requirements for water-cooling, counting gases, and the like increase the cost of the instrument. WDXRF systems are larger (often floor mounted such as the system in Figure 9.44) and more expensive than EDXRF systems.

Another significant difference is in spectral acquisition. With an EDXRF system, the entire spectrum is acquired

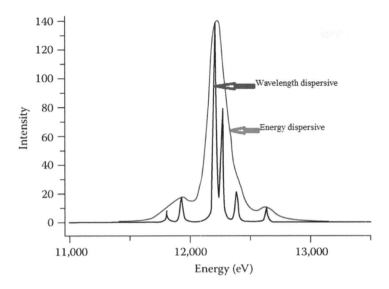

Figure 9.47 Comparison of the resolution of EDXRF (outer spectrum) and WDXRF (inner spectrum) systems. WDXRF provides higher resolution than EDXRF. (© 2013 HORIBA, Ltd. www.horiba.com. All rights reserved.)

simultaneously within a few seconds. In a WDXRF system, the spectrum is acquired point by point, which requires more time. The alternative, a simultaneous WD system such as has been described, is very expensive.

9.2.8 XRF Applications

When a sample is placed in a beam of X-rays, atoms in the sample are excited by the X-rays and emit X-rays of *characteristic* wavelengths (energies) from the elements in the sample. This process is called *XRF*. Since the wavelength (energy) of the fluorescence is characteristic of the element being excited, measurement of this *wavelength or energy* enables us to *identify* the fluorescing element. Tables of X-ray lines are given in the appendices of this chapter. The *intensity* of the fluorescence depends on *how much* of that element is in the sample. For most laboratory XRF equipment, the energy of the emitted X-ray is independent of the chemical state of the element; therefore, XRF is generally considered to be an elemental analysis method. Solid and liquid samples can be analyzed directly.

Modern WDXRF instruments permit the determination of all elements from fluorine ($Z = 9$) to uranium ($Z = 92$). Some WDXRF systems allow measurement of elements from Be to U. Benchtop EDXRF instruments can determine elements from sodium ($Z = 11$) to uranium ($Z = 92$); with a special SDD detector, it is possible to measure from fluorine to uranium. Handheld EDXRF units with SDD detectors can measure from magnesium ($Z = 12$) to uranium, and Si-PIN units can do the same with vacuum or He purge.

Elements with atomic numbers between Mg ($Z = 12$) and U ($Z = 92$) can be analyzed in air. Elements with atomic numbers 3 (beryllium) through 11 (sodium) fluoresce at long wavelengths (low energy), and air absorbs the fluorescence. Analysis of this group must be carried out in a vacuum or in a helium atmosphere.

WDXRF units usually operate primarily under vacuum and use a helium atmosphere for loose powder samples and liquid samples. Benchtop EDXRF and micro-XRF systems can operate under vacuum, helium, or air, based on the configuration. Handheld or portable units generally operate in air at atmospheric pressure.

The DLs are in the ppm range for most elements with a wide linear working range. XRF thus permits multielement analysis of alloys and other materials for major, minor, and trace elements. Sensitivity is poorest for the low Z elements and best for the high Z elements. XRF capabilities are dependent on the technique (ED or WD used) as well as the power and other design variables of the actual instrument employed.

9.2.8.1 Analyzed Layer

Both solid and liquid samples can be analyzed by XRF as described earlier in the chapter. With the exception of micro-XRF, XRF is considered to be a "bulk" analysis technique. This means that the analysis represents the elemental composition of the entire sample, assuming the sample is homogeneous. The term "bulk" analysis is used to distinguish such techniques from "surface analysis" techniques (Chapter 16), which look at only a very thin layer at the sample surface. But there are conditions that must be considered in XRF in order to obtain accurate results. The limiting factor for direct XRF analysis or the analysis of prepared samples is that the signal of the characteristic

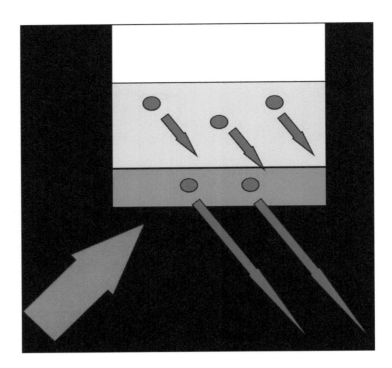

Figure 9.48 Illustration of the analyzed layer from the sample: The incoming primary radiation excites the lower and middle parts of the sample as shown but does not excite the upper part since the sample is infinitely thick for the primary radiation. The excited characteristic radiation from the element in the middle part is reabsorbed within the sample. Only the signal from the layer near the surface is able to be detected by the detector.

radiation from the sample originates from different layers within the sample.

The excited radiation inside the sample will need to travel out of the sample before it can be detected by the detector. Assuming a constant density for the sample, lower-energy characteristic radiation will only exit from relatively close to the surface, whereas higher-energy radiation will be detectable from increasingly deeper layers of the sample. Figure 9.48 illustrates the concept.

XRF is a *surface-sensitive* technique because the depth of penetration and the thickness of the analyzed layer depend on the exciting radiation and the sample composition (atomic number).

The analyzed layer is a function of the characteristic energy of the emission line being measured as well as the density and composition of the sample. The effect is based on absorption of the excited radiation. The effective analyzed layer is also dependent on the geometry of the instrument. The effective analyzed layer is usually defined as the layer from which either 50 % or 90 % of the signal originates.

To better understand the implications for accurate XRF determinations of composition, Table 9.9 shows selected characteristic K_α emission lines and shows the analyzed layer depth (90 % signal) for various matrices, which increase in density from the left to the right. Note that the table uses a typical "shortcut" for line notation: KA1

Table 9.9 X-Ray Emission Lines and Depth of the Analyzed Layer in Various Matrices

Line	Energy	Graphite	Glass	Iron	Lead
Cd K_α 1	23.17 keV	14.46 cm	8.20 mm	0.70 mm	77.30 μm
Mo K_α 1	17.48	6.06	3.60	0.31	36.70
Cu K_α 1	8.05	5.51 mm	0.38	36.40 μm	20.00
Ni K_α 1	7.48	4.39	0.31	29.80	16.60
Fe K_α 1	6.40	2.72	0.20		11.10
Cr K_α 1	5.41	1.62	0.12	104.00	7.23
S K_α 1	2.31	116.0 μm	14.80 μm	10.10	4.83
Mg K_α 1	1.25	20.00	7.08	1.92	1.13
F K_α 1	0.68	3.70	1.71	0.36	0.26
N K_α 1	0.39	0.83	1.11	0.08	0.07
C K_α 1	0.28		0.42	0.03	0.03
B K_α 1	0.18	4.19	0.13	0.01	0.01

instead of $K_{\alpha 1}$ and the energies of the lines are expressed in keV. Some examples of the use of this table are presented.

Considering the S K_a emission line, in a graphite matrix, the emission is detectable from a depth of 116 μm in the sample, but that depth is reduced to 14.8 μm in a silicate glass matrix and reduced even more in the dense matrices of iron and lead. This means that if the sulfur in the glass sample lies much deeper than 14.8 μm from the surface of the sample, it will not be detected because the radiation cannot escape

Table 9.10 Analyzed Layer Thickness of NIST 88B Dolomite Sample for Various Elements*

Compound	Line	Concentration (%, mass/mass)	Energy (keV)	Layer Thickness (μm)
Fe_2O_3	Fe K_α 1	0.722	6.40	174
MnO	Mn K_α 1	0.016	5.89	139
TiO_2	Ti K_α 1	0.016	4.51	66
CaO	Ca K_α 1	30.12	3.69	104
K_2O	K K_α 1	0.103	3.31	77
SO_3	S K_α 1	0.000	2.31	27
P_2O_5	P K_α 1	0.004	2.01	19
SiO_2	Si K_α 1	1.130	1.74	13
Al_2O_3	Al K_α 1	0.277	1.49	8
MgO	Mg K_α 1	21.03	1.25	7
Na_2O	Na K_α 1	0.029	1.04	4
CO_2		46.37		

* Note that the concentration data are for an older lot of the NIST SRM 88b, and this material is currently no longer available.

from the matrix. Conversely, this means that to measure sulfur in glass accurately, the top 14.8 μm of the sample must be homogeneous and representative of the sulfur concentration throughout the glass; in graphite, the top 116 μm of the sample must be homogeneous. The XRF analyst needs to ensure that the analyzed layer is *representative* of the entire sample. This might require intensive sample preparation to ensure that in a powdered sample, the "grain" or particle size is well below the size of the analyzed layer. Table 9.10 shows this issue for a limestone sample with no binder.

The preparation of the sample needs to be designed to ensure that the analyzed layer is representative of the sample. This has implications mainly for solid samples such as minerals where the sample needs to be pulverized to be made homogeneous with the resulting grain size less than the analyzed layer of the lightest element to be determined. As can be seen in Table 9.10, the magnesium emission originates from a depth of only 7 μm. To grind a sample to particle diameters substantially below 7 μm while keeping the sample homogeneous is extremely difficult.

A coal sample (which we can approximate by using the graphite data in Table 9.9) would need to be about 14.5 cm thick to measure 90 % of the Cd K_α signal. Since most benchtop or laboratory units can only load samples with a maximum thickness of 5 cm, one has to ensure that all samples, standards, and reference samples are prepared with identical thickness.

9.2.8.2 Sample Preparation Considerations for XRF

While general sample preparation approaches were covered in Chapter 1, as seen from the examples earlier, XRF sample preparation requires some thought as well as an understanding of the chemical and physical properties of the sample material. Samples must be representative of the material to be analyzed. They must fit the sample holder being used and completely cover the opening in the sample holder or be larger than the measurement spot size being used.

Very flat surfaces are required for quantitative analysis, as discussed subsequently. Liquid samples flow naturally into flat surfaces, but cannot be run under vacuum. Liquid samples are usually run in a polymer cup with a sleeve or ring to hold a polymer film in place over the liquid. The film needs to be mechanically stable, chemically inert, and as thin as possible to allow transmission of radiation, especially for the low Z elements. Elements with atomic numbers less than Na cannot be detected through polymer films. As discussed in Section 9.2.2.3, liquid samples and loose powders are best run in a facedown configuration for greatest accuracy, that is, the source should be below the sample. The best solvents for samples that must be dissolved for analysis are H_2O, HNO_3, hydrocarbons, and oxygenated carbon compounds, because these compounds contain only low atomic number elements. Solvents such as HCl, H_2SO_4, CS_2, and CCl_4 are undesirable because they contain elements with higher atomic numbers; they may reabsorb the fluorescence from lower Z elements and will also give characteristic lines for Cl or S. This will preclude identification of these elements in the sample. Organic solvents must not dissolve or react with the film used to cover the sample.

Solid samples may be prepared using several techniques. Solid samples that can be cut and polished to give a flat surface can be analyzed after polishing. Care must be taken not to contaminate the sample with the cutting tool or polishing compound. For example, cutting a flat piece of polymer with a steel razor blade can result in iron being detected in the polymer sample. Other solids should be ground to a powder, preferably using a ball mill or similar device to pulverize the sample. Again, the grinding tools must not contaminate the sample, so boron carbide is often used to contain and grind samples. The sample powder may be pressed "as is" or mixed with lithium borate salt, borax, wax, or other suitable "binder" and formed into a briquette or disk. This procedure provides a sample that can be easily handled and has the advantage that the borate salts and hydrocarbon binders provide a standard matrix for the sample. Furthermore, the matrix is composed of low atomic mass elements, which interact only slightly with the X-ray beam. In addition, the disks or pellets are pressed to a uniform thickness, as is required for accurate analysis. Figure 9.49 presents a summary of the various sample preparation techniques used for XRF analysis and shows some types of sample holders used.

Minerals in particular present challenges to grinding and achieving a uniform grain size due to the complex composition of mineral samples, a problem known as the mineralogical effect.

Figure 9.49 Sample preparation techniques and sample holders for XRF analysis.

It may not be possible to grind a sample to achieve ideal grain size. The next best approach is to ensure that the grinding process results in a repeatable grain size in the samples to be analyzed. The sample grain size must be similar to the grain size in standards and reference materials used in the analysis. This may require different grinding procedures for different materials. Integrating over a larger sample by spinning the sample over the beam or making multiple measurements enables the analyst to average the analyzed layer and obtain more repeatable data. These approaches will create methods that work but may limit the accuracy of the measurement and applicability of the method.

The ultimate preparation is to remove the "grains" from the sample to enable accurate and efficient analysis of multiple types of materials. The best approach is to mix the sample and salts such as lithium metaborate or lithium tetraborate and heat the mixture in a Pt crucible over high heat. The process is called fusion and is shown in the photographs in Figures 9.50a through c. The heat melts the sample and salt, and the molten mixture is poured into a Pt/Au mold with a flat bottom. When the melt is allowed to cool, a glassy flat "fused bead" is formed, suitable for quantitative XRF analysis with the advantage that the matrix and sample thickness are now the same for all samples. Fusion is usually required for analysis of geological samples to eliminate mineralogical effects, for example. Calibration standards are prepared by fusing known amounts of the analyte with the borate salt and casting standard beads. Automated fusion devices, called fluxers, are available, as are a wide variety of devices for grinding and powdering samples. A handbook of XRF sample preparation methods and equipment, with pictures of the various fluxers, crucibles, molds, and grinders, is available from Spex Certiprep at www.spexcsp.com. Videos and links to online sample preparation resources from Claisse® (www.claisse.com) are available on their website. A short video of automatic sample fusion using the Claisse M4 fluxer can be seen at: www.youtube.com/watch?v=pSMAbFMZ1Ko. The use of six position commercial fluxers allows the preparation of 20–30 fusion beads/h, faster and more accurately than the manual method shown in Figure 9.50.

Since the physics of X-ray emission is governed by the sample and not by the instrumentation, special care needs to be applied to the use of handheld and portable XRF in the direct analysis of field samples that have not been "prepared." Especially for the low Z elements, one needs to understand that these only emit from very close to the sample surface. The calibration of the analyzer must be tuned to the preparation or presentation of the sample. If not done correctly (or not done at all), the values obtained from the analyzer will be incorrect.

9.2.8.3 Qualitative Analysis by XRF

Each element fluoresces at its own characteristic wavelengths (energies); the fluorescing element can be identified from a table of wavelengths (energies) such as those in Appendix 9.A. An example of qualitative analysis is shown in Figure 9.51. The EDXRF spectrum is that of a zinc-coated iron washer. The spectra are much simpler than those from atomic emission spectrometry. Fe gives hundreds of strong emission lines in an inductively coupled plasma (ICP) or direct current (DC) arc emission experiment; here, only two lines are seen for Fe in this XRF spectrum. The coated washer also shows Zn, which is fully resolved.

Figure 9.50 Preparing a fusion bead for analysis. (a) The sample and a salt such as lithium tetraborate are heated until liquid in a Pt crucible over a burner. The crucible tongs are tipped with Pt to avoid contamination of the crucible. (b) The molten mixture is poured into a mold, in this case, a ring placed on a smooth marble slab. (c) Cooled beads removed from the molds. The flat bottom surface, shown turned up, is the side used for analysis. (From Jenkins, R. et al., *Quantitative X-Ray Spectrometry*, Marcel Dekker, Inc., New York, 1981. Used with permission.)

Even with the relatively simple spectra, spectral interference does occur in XRF. The major sources of spectral interference are scattered radiation from the tube, higher-order lines diffracted in a WD system, and L lines of higher atomic number elements overlapping K lines of lower atomic number elements. Two examples are given in Figure 9.52. The peaks in each spectrum are labeled, but you should use the tables in Appendix 9.A to confirm that they have been labeled correctly and to give yourself some practice in figuring out what elements are present from the energy positions of the peaks. The top spectrum shows the overlap of the spectral lines from pure Fe and pure Mn; the spectral interference occurs between the Fe K_α and Mn K_β peaks. The peak maxima are not at exactly the same energy, so in a mixture or alloy of the two elements, the Fe K_α line can be identified, and if the Mn content of a mixture or alloy were high, the Mn K_β peak might appear as a shoulder on the Fe K_α peak. The overlap is not critical in a two-component system; as you can see, the unobstructed Mn K_α, the strong Fe K_α, and the unobstructed Fe K_β peaks can easily identify the presence of both elements.

The ratio between the K_α and K_β line of an emission is fixed and tabulated for every element. A K_β emission is *always* accompanied by a K_α emission. Similarly, an L_β emission is always accompanied by an L_α emission. Keep these facts in mind when examining an XRF spectrum.

The bottom spectrum in Figure 9.52 is that of a more complex alloy containing Cr, Mn, Fe, Co, Ni, and W. Qualitatively, it is easy to see that Cr, Fe, W, and Co are present. Cr and W have unobstructed lines that tell us they are in the alloy. Fe is clearly present, because if there is any Mn, it must be in low concentration compared to the Cr and Fe. The Mn K_α peak overlaps the Cr K_β peak; we know there is Cr present from its K_α peak, so some (at least) of the intensity of the small peak at 5.8 keV is due to Cr. If Mn is present but in low concentration, then most of the intensity at about 6.4 keV is from Fe, since the Mn K_β peak is much smaller than its K_α peak, as seen in the top spectrum. Looking at the peak to the right of the Fe K_α peak, we see a much bigger peak than would be expected from Fe K_β alone. (Compare the Fe peak heights in the top spectrum.) Therefore, most of the peak at about 6.9 keV must be from Co K_α. That indicates that Co is present in the alloy. From the Mn and Fe peak ratios in the top spectrum, it is reasonable to assume that the Co K_β peak is less intense than the Co K_α peak. But the peak at approximately 7.5 keV, about where we would expect the Co K_β peak, is actually more intense than the Co K_α peak (and the "Co K_α" peak intensity also includes some interfering Fe K_β intensity). Therefore, the high intensity of the 7.5 keV peak tells us we have another element, Ni, in the alloy. The one element that is not clearly present is Mn; the intensity ratio for the two Cr lines needs to be ascertained. If the "Cr K_β" peak is too high relative to the K_α peak, then Mn is probably in the alloy as well. So even though only two elements, Cr and W, have lines with no spectral interference, knowledge of the

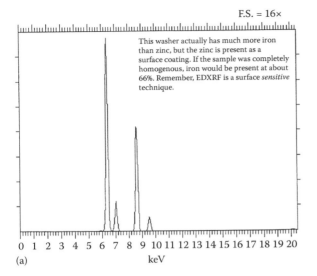

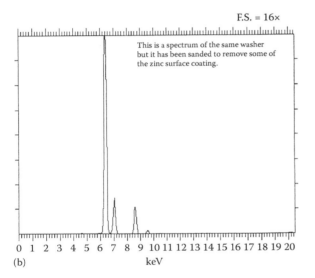

Figure 9.51 Qualitative EDXRF spectrum of a zinc-coated iron washer. (a) The coated washer. The peaks present are, from left to right, Fe K_α, Fe K_β, Zn K_α, and Zn K_β. (Confirm these peaks by looking in Table 9.A.2.) (b) The same washer, but after sanding to remove some of the surface. Note the decrease in the intensity of the Zn peaks and the increase in the Fe peaks. These spectra were collected for 50 s with an Rh tube at 30 kV and 0.01 mA, under vacuum, using a filter to remove the X-ray tube lines. (© Thermo Fisher Scientific. www.thermofisher.com. Used with permission.)

relative intensities of lines allows us to state with certainty that Fe, Co, and Ni are present, and that Mn may be present in low concentration. We will come back to this spectrum in the discussion of quantitative analysis by XRF. Modern instrument software generally displays the position and relative heights of the lines of each element, superimposed on the collected spectrum.

The spectrum shown in Figure 9.53 is a more complex spectrum of a duplex steel sample, containing Cr, Mn, Fe, Cu, Ni, and Mo. Note that the instrument software automatically shows element line positions and identity. The emission line from the Rh tube can be seen in the spectrum. There is also an emission line from argon. (Look at the figure caption. Can you suggest where the argon comes from?)

To analyze a spectrum to determine what elements are present, identify the strongest (most intense) line first and then continue with the next most intense lines. Look for the tube emission line and any lines that may come from air. Lines that do not match an element line may be scatter lines from the tube, an escape peak, or a sum peak.

The XRF method is nondestructive, an important feature when the sample is available in limited amounts or when it is valuable or even irreplaceable, as in the case with works of art, antiques, rocks from the moon, or forensic samples. The nondestructive nature of XRF coupled with the fact that sample preparation may not be required means that direct multielement analysis can be performed *in situ*. Portable HH XRF analyzers are used in the field by geologists to study soil and minerals, to sort metal scrap, glass, and plastics at recycling centers, for quality control in steel plants, and to check incoming lots of material before unloading a ship or a truck (Figure 9.54).

Another large handheld application segment is in positive material identification (PMI) where, similar to scrap sorting, the sample is identified as a specific alloy or material. PMI ensures that, for example, the repair of a crucial pipe in a hazardous material-containing process is only performed using the correct and certified material. After a catastrophic accident in a Texas refinery, PMI inspection of components carrying hazardous products was made mandatory. This mandate is easily accomplished on site using HH XRF instrumentation.

Antiques, jewelry, gems, and art objects can be characterized. Original art and copies of masterpieces can be distinguished from each other based on elemental composition and layering of pigments. Natural gems can be distinguished from synthetic gems in many cases, because the laboratory synthesis often adds specific elements not usually found in the natural gems or may even produce gems with fewer trace elements than real gems. Plated jewelry can be distinguished from solid gold or solid silver jewelry. Museums rely heavily on this method for examining works of art. The sample is usually unaffected physically or chemically by the analytical process. Usually low-power EDXRF or micro-XRF units are used for this application, and some units have dedicated software for particular applications. It is possible for some sensitive materials like paper or plastics to turn yellow or brown upon prolonged exposure to X-rays, so low power and short analytical times are used for these materials.

Forensic samples are often very small and inhomogeneous. As discussed later, it is difficult to perform quantitative

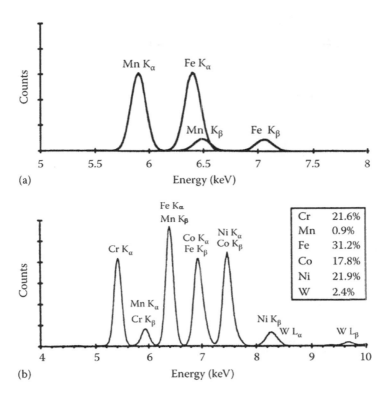

Figure 9.52 Spectral overlap in EDXRF spectra. (a) A simple overlap demonstrated by overlaying the spectra of pure Fe and pure Mn. There is overlap of the Fe K_α line with the Mn K_β line. (b) The spectrum of a multielement alloy showing multiple spectral interferences due to line overlaps. Despite the overlaps, knowledge of the relative line intensities permits qualitative identification of most of the elements present.

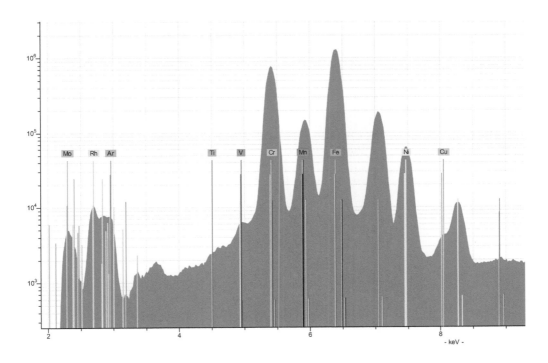

Figure 9.53 XRF spectrum of a duplex steel sample. Handheld XRF (HH XRF) measurement under air, 15 s, using a Rh tube at 15 kV. (Courtesy of Bruker Elemental, © 2013 Bruker, Inc. www.bruker.com. Used with permission.)

(a)

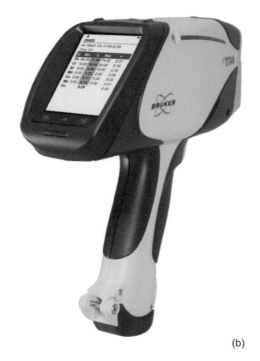

(b)

Figure 9.54 (a) Scrap metal sorting using the TITAN, an HH XRF analyzer, to identify elemental composition. (b) A photograph showing the readout of the S1 TITAN Handheld XRF analyzer. (Courtesy of Bruker Elemental, ©2013 Bruker, Inc. www.bruker.com. Used with permission.)

analysis on such samples. However, it is possible to use qualitative XRF to good advantage. Using the spectral "fingerprint" of a sample does not require exact concentrations to be determined. By proving that a spectrum of soil found in a shoe matches the spectrum of soil from the location of a crime scene, it is possible to place the object at that scene.

Examples of the use of nondestructive spatial investigation of element distribution of artworks, archeological items, pharmaceutical products, and other samples using micro-XRF are available on several instrument companies' websites, including Bruker, Horiba Scientific, SPECTRO, and Unisantis. A few examples will be discussed.

An altarpiece from 1424 AD, the Göttingen Barfüßer altarpiece, was examined using a Bruker ARTAX® µ-XRF spectrometer (Bruker Nano GmbH, www.bruker.com). Thin layers of gold and silver approximately 1 µm thick were examined by single-point measurements with a W target tube and a 0.65 mm collimator. Layers of pure silver, historical gold alloys like "green gold" containing 30 % Cu, and layers of Au and Ag hammered together ("Zwischgold") were identified. Studies of two colored copperplate engraving copies of a famous work by Albrecht Dürer using the ARTAX® µ-XRF system showed that the pigments used in one copy were consistent with it being painted in the sixteenth century, while the second copy contained pigments that were not in use until the 1800s (Gross; Hahn et al.). With the same system equipped with a Mo source and polycapillary lens, 2D mapping of a 2 × 2.5 mm piece of fabric with gunshot residue permitted identification of particles ≥10 µm in diameter. The particle distribution pattern across the fabric allowed the determination of the angle of incidence of the gun shot.

A key area that benefits from micro-XRF is the pharmaceutical industry where tablet imaging can determine active pharmaceutical ingredient (API) distribution, mixing uniformity, contamination, and interbatch concentration fluctuations. The ability to pinpoint single microscopic particles and identify them is very important in ocular and intravenous (IV) formulations, which must meet stringent particulate concentration limits. An example of the use of micro-XRF in mapping a pharmaceutical tablet is shown in Figure 9.55.

Other pharmaceutical applications for micro-XRF can be found at www.horiba.com/scientific/products/X-ray-fluorescence-analysis/applications/pharmaceutics/.

The Unisantis XMF-104 X-ray microanalyzer (Unisantis S.A., www.unisantis.com) was used by researchers at the Institute for Roentgen Optics, Moscow, Russian Federation, to examine nondestructively the composition of ancient coins from the fourth century BC through the second century AD. The fourth century BC coins were found to be an alloy of 82 % Ag/18 % Cu, but areas of pure Ag showed the inhomogeneity of the alloy. A drachma coin depicting Alexander was composed of 99 % Ag/1 % Cu. The

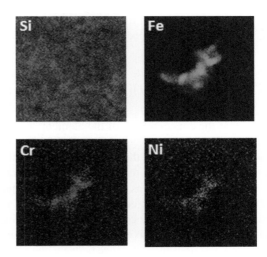

Figure 9.55 Mapping across a pharmaceutical tablet allows a buried steel particle to be visualized and identified. Contaminants such as this can be introduced from the manufacturing process and are typically caused by wear debris from mechanical mixing/compressing equipment. (© 2013 HORIBA Scientific. www.horiba.com/scientific. All rights reserved.)

XMF-104 system had a 50 W Mo tube, a two-stage Peltier-cooled compact Si-PIN detector and polycapillary focused X-ray beam with a 50–250 µm focal spot. Spectra, images of the coins, and details are available at www.unisantis.com, application note 605.

9.2.8.4 Quantitative Analysis by XRF

Quantitative analysis can be carried out by measuring the intensity of fluorescence at the wavelength characteristic of the element being determined. The method has wide application to most of the elements in the periodic table, both metals and nonmetals and many types of sample matrices. It is comparable in precision and accuracy to most atomic spectroscopic instrumental techniques. The sensitivity limits are on the order of 1–10 ppm, although sub-ppm DLs can be obtained for the heavier Z elements under appropriate conditions. Accuracy depends on a thorough understanding of the sample heterogeneity, the sample matrix, and the absorption and fluorescence processes.

X-rays only penetrate a certain distance into a sample, and the fluorescing X-rays can only escape from a relatively shallow depth (otherwise, they would be reabsorbed by the sample) as we have seen. While XRF is considered a "bulk" analysis technique, the X-rays measured are usually from no deeper than 1000 Å from the surface. True surface analysis techniques such as Auger spectroscopy only measure a few angstroms into the sample, so in that respect, XRF does measure the bulk sample, but only if the surface is homogeneous and representative of the entire sample composition.

The relationship between the measured X-ray intensity for a given peak and the concentration of the element can be written

$$C = K(I)(M)(S) \qquad (9.19)$$

where C is the concentration of the element. The factor K is a function of the spectrometer and the operating conditions. I, the intensity of the measured signal, is the net intensity after subtraction of the background. M represents interelement effects such as absorption and enhancement effects. Interelement effects must be corrected for or minimized to obtain accurate concentration values. S is the sample preparation term (also called the sample homogeneity term) and describes the difference between the reference sample with which K was obtained and the sample being measured. The classical approach to quantitative analysis is to use an external calibration curve. A set of standards at various concentrations is prepared in a manner similar to the sample preparation, intensity is carefully measured for all standards and samples using the same instrumental conditions, and a plot of C versus I is made. M and S are considered to be constant (a good assumption only if the standards reflect the matrix and particle size distribution of the samples). Precision depends only on careful measurement of intensity; accuracy depends on the elimination or evaluation of M, the interelement effects, and the elimination of S by using the same preparation of standards and unknowns.

Peak and background intensities are measured by counting photons at the appropriate wavelengths/energies. The net intensity is obtained via background correction. The process of background correction differs between EDXRF and WDXRF. In WDXRF, a fixed position close to the emission line to be measured needs to be identified. This position needs to be free of other emission lines, which could occur in the sample. It is usual to perform a scan where the signal as a function of diffraction angle 2θ is obtained. From this scan (spectrum), a background is mathematically obtained by fitting a numerical function (e.g., a polynomial) using a user-defined minimization based on least squares. Once the background is determined, the remainder of the spectrum is the net intensity from the sample. In EDXRF, the area of the peak is used for quantification; in WDXRF, the peak height is used. In ED systems, background subtraction is usually performed. The background is very matrix dependent in XRF, so background correction is more difficult than in other types of spectroscopy.

Calibration methods include the use of external standards and may include the use of ISs to improve precision. The IS may be an added element or elements not present in the sample, but this is a very time-consuming endeavor, both in making the standards and samples and in measuring all the lines. Frequently, the scattered radiation from the X-ray tube is used as an IS; this is a reasonable approach because the scattered radiation (background) is very dependent on the sample matrix. The ratio of analyte line to scattered radiation results in better precision than the measurement of the analyte line alone. XRF calibration used to require a set of accurate standards, as well as certified reference materials (CRMs) for verification of the accuracy and precision of the analysis. Depending on the range of materials that require analysis, laboratories can spend a great deal of money on the highly pure elements, compounds, and reference materials required for calibration and sample preparation.

A variety of mathematical approaches to the correction of absorption-enhancement effects and calibration is now in use, due to the availability of inexpensive, powerful computers and powerful software programs. One approach is the **fundamental parameters (FP) method**.

Using a set of pure element samples or well-certified multielement reference materials, the M factor for each element (Equation 9.19) can be calculated using the fundamental parameters describing the primary spectral distribution (tube spectrum), effective mass absorption of the primary and secondary radiation, and the fluorescent yield for each element. The M factor represents the effect the "other" elements have on the analyte element. The parameters used assume that the sample has a planar surface and is homogeneous and infinitely thick (no transmission of X-rays through the sample). This approach enables wide-ranging "universal" calibrations to be established for metals and alloys, such as a FeCrNiCo base curve for WDXRF or dedicated calibrations for cement, glass, and so on.

These universal calibrations or standardless calibrations can now be found on any commercial XRF. They are still based on a calibration of a larger set of samples, but this is usually performed at the instrument manufacturer's factory and the calibration transferred to the unit. From the user's point of view, quantitative analysis is standardless since no calibration standards or standard reference materials (SRMs) are required. These calibrations allow for good accuracy if the samples are homogeneous, flat and infinitely thick, or described in terms of thickness and mass to allow for corrections. Universal calibrations will fail when the sample is inhomogeneous. Results become more "semiquantitative" as samples become smaller and less homogeneous. Forensic samples and failure analysis samples may be small, nonplanar, and not particularly homogeneous. For forensic analysis, one does not compare concentrations but the ratio of the net signal, or one tries to match the spectral "fingerprint."

Most commercial XRF systems have a fundamental parameters software program as part of their data processing package. The advantage to the FP approach is that many industrial materials do not have readily available matrix-matched calibration standards commercially available. Preparation of good, stable calibration standards takes a lot of time and money even when it is possible to make such standards. Even when well-defined standards are available commercially, as they are for steels, brasses, bronzes, nickel-based superalloys, and many other types of alloys, as well as some glasses and ceramics, the standards are expensive and must be handled carefully to avoid scratching or contaminating their surfaces. Sources for prepared XRF standards include government standards agencies such as NIST in the United States, as well as commercial firms. Although this standardless software approach has improved dramatically in accuracy over the years, the analyst should, when possible, verify the accuracy of any mathematical approach with known materials that reflect the samples being analyzed. More information on these approaches can be found in several of the introductory XRF texts (e.g., those by Bertin, Jenkins et al., and Herglotz and Birks listed in the bibliography).

Modern WDXRF systems can measure the entire periodic table in less than 20 min and calculate the composition of a complete unknown using an FP program. Using a "fast scan" approach, a semiquantitative analysis can be obtained in less than 2 min of measurement time. Handheld EDXRF systems with SDD detector technology require less than 5 s to identify a standard alloy and about 20–30 s to identify an aluminum or magnesium alloy.

When quantitative analysis is needed on complex samples with peak overlaps, a method of extracting the individual intensities of each element is needed. These mathematical corrections are now performed in instrument data processing software and may include overlap corrections, interelement corrections, matrix corrections, corrections for non-Gaussian peak shapes, and different statistical fitting routines. A combination of these corrections was used to determine the concentration of the elements in the alloy shown in Figure 9.53. The HH measurement of a 2205 type duplex steel sample took less than 15 s.

9.2.8.5 Applications of Quantitative XRF

Quantitative XRF is used in virtually every industry for almost any type of liquid or solid sample. XRF is used daily to analyze minerals, metals, paper, textiles, ceramics, cement, polymers, wood, environmental samples, food, forensic samples, cosmetics and personal care products, and more. Only a few examples will be given here.

The petroleum industry uses XRF to measure sulfur in fuels, the elemental composition of catalysts used to "crack" petroleum hydrocarbons, and lubricating oil additives. The sulfur in oil or fuel application is driven by the US federal government clean air regulations. Diesel fuel, for example, is required to contain less than 15 ppm S. Refineries, pipeline operators, and commercial and government labs need to be able to certify sulfur content to ensure fuel classification and compliance. There are procedures published by the American Society for Testing and Materials (ASTM) to ensure minimum performance for accuracy and precision: ASTM D2622 for WDXRF, ASTM D4294 for direct excitation EDXRF, and D7220 for polarized (3D) EDXRF. The performance and DLs for the methods are different, but all enable the user to be able to comply with the regulatory requirements. Refineries often choose WDXRF due to the higher accuracy and precision achieved in about 3 min of measurement when compared to EDXRF.

Petroleum hydrocarbon cracking catalysts usually contain more than 12 elements, including transition metals, rare earth elements (REEs), and alkali and alkaline earth elements at concentrations varying from 0.005 to 35 mass %. The composition of the catalyst can be determined accurately after fusion into borate beads; the total analysis takes up to 1 h. The procedure and approach are described in ASTM D7085. Lubrication oil blenders use EDXRF to check their blends and ensure that control samples from "lube" shops are what the products claim to be (ASTM D6481 outlines the procedure).

Used engine oil used to be monitored routinely for metal content, since the metal content indicated how parts of the engine were wearing away in use. The wear metals analysis indicated what parts needed to be replaced during maintenance. Analysis also enables extension of the oil life by verifying that there is still enough P, Ca, or Zn in the oil to function correctly. This analysis could be performed by XRF or by atomic emission spectrometry (e.g., ICP or DC plasma [DCP] emission). However, filters in modern engines are now so efficient at trapping wear metal particles that the analysis of oil is no longer a good indicator of engine wear. Micro-XRF is now used to measure the metals on the filters themselves to predict engine failure and part replacement by quantifying and analyzing the individual particles from the filter.

In metallurgy, alloy composition can be rapidly determined and unknown samples identified rapidly. XRF has an advantage over wet chemistry in that all the components can be measured due to the wide dynamic range of XRF. For example, in the analysis of nickel alloys, a wet chemical approach would measure all the other elements and calculate the Ni content as the balance. With XRF, the major element, nickel, as well as the minor and trace components, can be measured accurately. For high-grade steel and alloys with multiple major components, WDXRF achieves better accuracy and repeatability than optical emission spectroscopy (OES).

A limitation for XRF in a metallurgical application is the measurement of carbon in steel. The problem with carbon in this application is due to the small analyzed layer and the inhomogeneous distribution of carbon in steel. To ID the correct steel grade for low-carbon steels, OES or combustion analysis is required.

The areas of restricted/prohibited elements and consumer product safety have major applications where XRF can be used successfully. Inspection of consumer goods for regulated elements such as Br (from flame retardants), Hg, Pb, Cr, and Cd as part of government regulations such as restriction of the use of hazardous substances (RoHS), waste electrical and electronic equipment (WEEE), and the Consumer Protection and Safety Improvement Act 2008 (CPSIA) use XRF (especially portable, handheld, and stationary microspot XRF) as screening tools. This is achieved by factory calibrations, which adjust and switch automatically for common matrices: high-density (HD) and low-density (LD) polymers and polyvinyl chloride (PVC) and metals. These matrices are the ones found in toys and jewelry. Controlled elements can be measured rapidly using such "dedicated" systems. Pb concentrations as low as 10 ppm can be determined in soil samples with a portable HH XRF system. XRF is used to measure the amount of phosphorus-based flame retardant in textiles and the amount of antimony-based flame retardant in plastics. For textiles, a piece of fabric is cut into a square piece and stretched across a standard sample cup and held in place by the sample container ring. This results in a flat specimen for analysis. Phosphorus levels in the range of 0.3–3 % P can be measured in a sample in as little as 30 s. (Details of the method are available from SPECTRO Analytical Instruments [www.spectro-ai.com].)

The ceramics industry routinely measures 6–12 elements quantitatively in both pressed pellet and fusion bead form for quality control using either WD or ED XRF. The elements (reported as oxides) vary in concentration from 0.01 to 70 mass %. Using a modern WDXRF spectrometer, aluminum (reported as Al_2O_3) can be determined at the 1 mass % level in a magnesium oxide-based ceramic with a standard deviation of 0.006 and a DL of about 8 ppm.

The cement industry requires the close monitoring of at least 10 major elements as oxides (Na_2O, MgO, Al_2O_3, SiO_2, SO_3, P_2O_5, K_2O, CaO, MnO_2, and Fe_2O_3) and determination of adverse trace elements (Zn, Cr, Cl). Based on the requirements of time to result, either WDXRF (simultaneous or sequential) or benchtop EDXRF can be used.

Most mining companies use XRF for the analysis of their process streams and monitor the "separation" of metals from the ore. XRF is faster than any wet or atomic spectroscopy technique since it can measure the sample as a solid. Trace elements in soil and sediment can be measured to collect geological, agricultural, and environmental data both in the lab and in the field, using portable or benchtop EDXRF analyzers.

Online XRF analyzers are available for monitoring process streams for metals in plating baths, metals in hydrocarbons, silicon in adhesive coatings on paper, acid leaching solutions, effluent discharge monitoring, and similar applications. The thickness of coatings can be measured on a wide variety of materials, such as paint on steel for the automotive industry and silicone release coating on paper (the shiny nonstick backing paper you peel adhesive labels from is coated with a silicone polymer called a "release coating"). Solar panels are routinely checked with XRF either online or at line.

Due to the extremely high current price for gold, many jewelers, pawnshops, and the like are buying scrap gold and old jewelry. Handheld or portable systems with dedicated software to determine % Au and carat weight, silver and other precious metals are now available for accurate evaluation of the content of scrap jewelry, such as the Thermo Fisher Scientific Niton DXL precious metal analyzer. This is a countertop unit that delivers fast XRF analysis for all precious metals and has an integrated program called AuDIT™, which automatically identifies gold-plated items, based on the thickness of the Au layer. Units with dedicated software for determining plating thickness of other metal coatings on substrates are also common.

XRF can be superior to atomic spectroscopy when calibration and consumable costs are compared, assuming the DLs are comparable. XRF units are calibrated usually either at the factory or on site when the source is changed; calibration is adjusted or checked with a set of solid check samples. Consumables for XRF are limited to those for sample preparation (fusion fluxes, disposable sample holders, polymer film) and helium, when liquid samples are measured. A significant advantage for XRF over atomic spectroscopy is the ability to measure major, minor, and trace elements simultaneously and directly in the solid or liquid; for most atomic spectroscopy techniques, major and minor components require serial dilutions and multiple analyses. Most solid samples for atomic spectroscopy must be dissolved for analysis. AAS, atomic emission systems, and ICP-mass spectrometry (MS) systems need to be calibrated daily before use and when sample introduction systems are changed, and when parts are cleaned or replaced. They all require large amounts of gases, calibration standards for each element, large volumes of high-purity acids, disposable autosampler cups or tubes, graphite furnace (GF) components, torches, nebulizers, disposable pump tubing, and so on.

9.3 X-RAY ABSORPTION

If the wavelength of an X-ray beam is short enough (high energy), it will excite an atom that is in its path. In other words, the atom absorbs X-rays that have enough energy to cause it to become excited. As a rule of thumb, the X-rays emitted from a particular element will be absorbed by elements with a lower atomic number. The ability of each element to absorb increases with atomic number.

Beer's law states that

$$\log\left(\frac{P_0}{P_x}\right) = \mu_x t \qquad (9.20)$$

where
μ_x is the linear absorption coefficient
t is the path length through the absorbing material
P_0 is the X-ray power before entering sample
P_x is the X-ray power leaving sample

However,

$$\left(\frac{\mu_x}{\rho}\right) = \mu_m \qquad (9.21)$$

where
μ_m is the mass absorption coefficient
ρ is the density

But

$$\mu_m = \left(\frac{CN_0 Z^4 \lambda^3}{A}\right) \qquad (9.22)$$

where
C is a constant
N_0 is Avogadro's number
Z is the atomic number
A is the atomic mass
λ is the wavelength of the radiation

It can be seen that at a given wavelength, μ_m is proportional to Z^4 divided by the atomic mass. This relationship is shown in Figure 9.56. The setup for X-ray absorption is slightly different than that for XRF, as seen in Figure 9.57. The sample is placed directly in line with the X-ray tube, in a configuration very similar to UV/VIS absorption spectrometry. The parameter measured is the decrease in intensity of the incident beam after passing through the sample. The same detectors described for XRF may be used for X-ray absorption

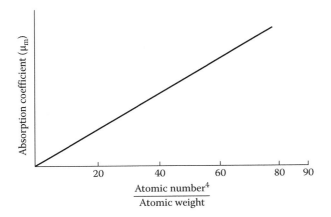

Figure 9.56 Relationship between atomic number and X-ray absorption coefficient.

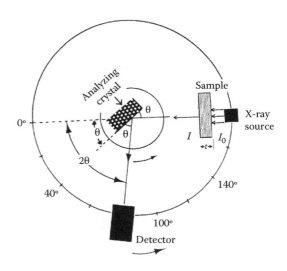

Figure 9.57 X-ray absorption. The sample is placed between the X-ray tube and the detector. The intensity of the source is I_0. The intensity of light reaching the detector after passing through a sample of thickness t is I. I will be less than I_0 if absorption occurs.

spectrometry. Older systems and some current medical systems use photographic film for detection (radiography).

The most familiar example (and the oldest use) of X-ray absorption does not provide chemical information, but rather physical information. That is the use of X-ray absorption in medical radiography, but it is based on the relationship between absorption coefficient and atomic number. For example, the human arm consists of flesh, blood, and bone. The flesh or muscle is made up primarily of carbon, nitrogen, oxygen, and hydrogen. These are all low atomic number elements, and their absorptive power is very low. Similarly, blood, which is primarily water, consists of hydrogen and oxygen, plus small quantities of sodium chloride and trace materials. Again, the absorptive power of blood is quite low. In contrast, bone contains large quantities of calcium and phosphorus, primarily as calcium phosphate. The atomic numbers of these elements are considerably higher than those in tissue and blood, and so the absorptive power is considerably higher. When an X-ray picture is taken of a hand (Figure 9.58), the X-radiation penetrates the muscle tissue and blood quite readily, but is absorbed significantly by the bone. A photograph of this absorption indicates the location of the bones in the hand. The procedure is routinely used in medicine to detect broken bones, fractures, and arthritic changes in joints.

Another application of X-ray absorption in medicine is to define the shapes of arteries and capillaries. Normally, the blood absorbs only poorly; however, it is possible to inject a solution of strongly absorbing cesium iodide into the veins. The material is then swept along with the blood and follows the contours of the arteries. An X-ray video is recorded as the highly absorbing cesium iodide flows through the arteries, showing the contours of the arteries. This can be used to identify breaks in the veins or arteries

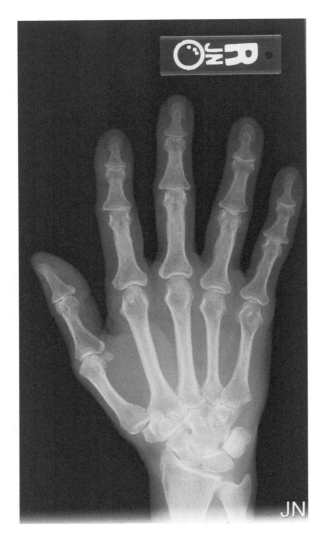

Figure 9.58 X-ray photograph of a human hand.

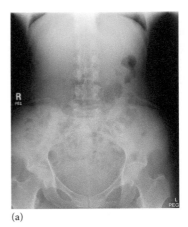

(a)

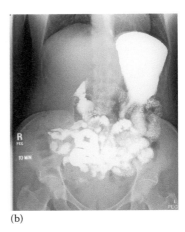

(b)

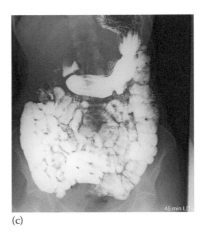

(c)

Figure 9.59 X-ray photographs of a human colon before and after ingestion of a barium-containing liquid. (a) Before ingestion of the barium solution, the colon cannot be seen at all. (b) Ten min after ingestion, the highly absorbing barium has entered the colon. (c) After 45 min, the entire colon can be clearly seen.

that could cause internal bleeding. Such internal bleeding can be the cause of a stroke. The technique may also be used to indicate a buildup of coating (plaque) on the inside of the veins. This is particularly dangerous in the heart, where deposits of cholesterol restrict the flow of blood through the heart. If this is left unchecked, a heart attack will result. X-ray absorption can be used to diagnose this problem and to locate exactly the position of deposits. Surgery is made much easier by this technique. In a similar fashion, a barium-containing liquid ingested by a patient permits the detailed observation of the colon since barium, a high Z element, absorbs very strongly (Figures 9.59a through c).

In the field of metallurgy, applications of X-ray absorption include the detection of voids or the segregation of impurities, such as oxides, in welds and other joints. Figure 9.60 shows an idealized X-ray absorption photograph of a mechanical weld that contains voids or internal holes. Such holes indicate that the weld is mechanically weak and might break in use. If the weld is weak, it must be strengthened to form a sufficiently strong joint. This type of nondestructive testing is used to check the manufacturing

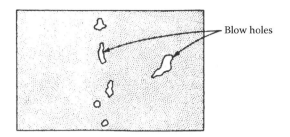

Figure 9.60 Schematic X-ray absorption photograph of a mechanical weld.

Table 9.11 Molar Absorptivities of Elements in Petroleum

Element	Molar Absorptivity (cm²/g) at 21 keV
H	0.37
C	0.41
O	0.79
S	5.82

Source: Courtesy of Applied Rigaku Technologies, Inc. www.rigakuedxrf.com.

quality of ships, aircraft, bridges, and buildings. It is also used to check these structures during routine maintenance. X-ray absorption is routinely used for measuring the thickness of thin metal films.

The technique of X-ray absorption can also be used to determine the levels of liquids in enclosed vessels or pipes without opening or breaking them. The same process can be used to detect metal supports or metal fillings inside constructed objects as diverse as buildings and small works of art. A major advantage is that X-ray techniques are usually nondestructive. Sometimes artists paint over old paintings, using the canvas for their own work and covering unrecognized masterpieces in the process. Using X-ray absorption, it is possible to reveal the covered painting without removing the top painting. When used to examine a metal horse sold for several million dollars as an ancient Greek art piece, X-ray absorption showed that the horse contained internal metal supports and was therefore a fake. This was done without destroying the art piece, in case it had been authentic.

For quantitative elemental analysis, X-ray absorption is not particularly useful except in the case of a single heavy element in a light matrix. As we saw in Equation 9.10, the mass absorption coefficient needed for Beer's law calculation (Equation 9.20) must be calculated from the mass fractions of elements present in the sample. The mass fractions are usually unknown. Quantitative analysis by X-ray absorption is usually only used for the determination of a high atomic number element in a matrix of lower atomic number elements. Examples include the determination of lead or sulfur in hydrocarbon fuels and the determination of Pt catalyst in polymers, where the difference in mass absorption coefficients between analyte and matrix is large. One approach to quantitative analysis using X-ray absorption is based on the measurement of the intensities of two or more monochromatic X-rays passed through the sample. This is called X-ray preferential absorption analysis or dual-energy transmission analysis. The analysis depends on the selective absorption of the transmitted X-rays by the analyte compared with absorption by the rest of the sample (the matrix). The sensitivity of the analysis also depends on the difference in mass absorption coefficients of the analyte and sample matrix for the transmitted X-rays; a big difference results in a more sensitive analysis. The analyte concentration calculation in any absorption method requires that the thickness and density of the sample be known and requires a homogeneous matrix for accurate quantitative results.

An example of the use of X-ray absorption (transmission) is the determination of sulfur in heavy hydrocarbon process streams. Accurate determination of sulfur is required for assaying and blending of crude oil, marine oil, and bunker fuel. The term used by the petroleum industry is X-ray transmission (XRT) or absorption (XRA) gauging. XRT/XRA gauging involves measuring the attenuation of a monochromatic X-ray beam at 21 keV. This energy is chosen because at 21 keV, the molar absorptivity for S is 14 times larger than the molar absorptivity of the hydrocarbon matrix and seven times larger than oxygen (Table 9.11).

As seen in Table 9.11, the absorptivities of C and H are almost identical. The method is therefore insensitive to changes in the C:H ratio and is sensitive only to the sulfur content. In practice, a process stream passes through a flow cell where sulfur in the hydrocarbon matrix absorbs X-rays transmitted between the source and detector (Figure 9.61a). The recorded X-ray intensity is inversely proportional to the sulfur concentration; that is, X-ray transmission decreases as sulfur concentration increases. The commercial process analyzer is shown in Figure 9.61b. The process analyzer can withstand pressures up to 1480 psig and temperatures to 200 °C.

Transmission of X-rays through the flow cell is given by the following equation:

$$T = \frac{I}{I_0} = \exp - dt \left[\mu_m (1 - C_s) + \mu_s C_s \right] \quad (9.23)$$

where

I is the measured X-ray intensity (after the flow cell) (photons/s)

I_0 is the initial X-ray intensity (before the flow cell) (photons/s)

d is the density of the hydrocarbon stream

t is the thickness of the flow cell (cm)

μ_m is the molar absorptivity of the hydrocarbon matrix at 21 keV (cm²/g)

μ_s is the molar absorptivity of sulfur at 21 keV (cm²/g)

C_s is the mass fraction of sulfur (% mass/mass)

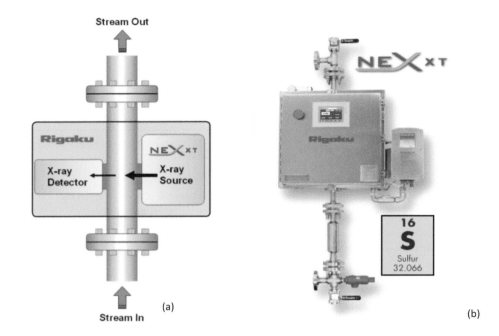

Figure 9.61 (a) Schematic of the X-ray transmission NEX XT system showing the flow cell, source, and detector. (b) The Rigaku NEX XT on-line sulfur analyzer. (Courtesy of Applied Rigaku Technologies, Inc., TX. www.rigakuedxrf.com.)

9.3.1 EXAFS

A recent development in the use of X-ray absorption is a technique called EXAFS, extended X-ray absorption fine structure spectroscopy. A sample is placed in a beam of X-rays, and the incident and transmitted intensities are measured as the energy of the X-ray beam is varied. A plot of the absorption versus energy gives us the position (energy) and exact shape of the absorption edge for the element being measured. The exact energy of the absorption edge and its shape, the "fine structure," does change slightly depending on the oxidation state of the element and the number and type of nearest neighbor atoms. This change in position of the absorption edge is analogous to the chemical shift seen in nuclear magnetic resonance (NMR). EXAFS does provide oxidation state information and molecular structure information, unlike normal X-ray absorption or XRF, which are strictly elemental analysis techniques. Details of the spectral interpretation of EXAFS are beyond the scope of this book. The interested student can consult the text by Teo and Joy listed in the bibliography.

9.4 X-RAY DIFFRACTION

XRD is a technique that is useful for the analysis of solid crystalline or semicrystalline materials. Most organic and inorganic compounds, minerals, metals, and alloys, and many types of polymers form crystals and can be analyzed by XRD. XRD can provide the exact crystal structure of a pure single-crystal material. In addition, XRD can provide the qualitative

Figure 9.62 An electron micrograph of the (110) plane in crystalline silicon.

and quantitative identification of the molecules present in pure crystalline powders or mixtures of crystalline powders.

The ions or molecules that make up a crystal are arranged in well-defined positions, called a crystal lattice. Figure 9.62 is an electron micrograph of the (110) plane of crystalline silicon. Three coordinates, called Miller indices, identify the plane in space; the Miller indices for this plane are 1, 1, and 0. The light spots are individual Si atoms. As can be seen, they are arranged in a very regular pattern in the 2D plane. The dark area is the empty space or interstitial space between the atoms in the lattice. A crystal is a 3D well-ordered array of atoms. An illustration of a typical crystal structure, greatly magnified, is shown in Figure 9.63a. As we examine the structure of the crystal, we see that the ions or atoms or molecules form planes in three dimensions. You can imagine stacking identical planes of Si atoms on top of each other to create a 3D crystal, for example.

X-RAY SPECTROSCOPY

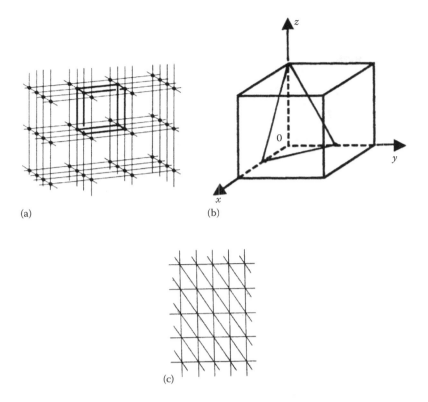

Figure 9.63 (a) A portion of a 3D crystal lattice. The unit cell, or basic repeating unit, of the lattice is shown in heavy outline. The black dots represent the atoms or ions or molecules that make up the crystal. (b) A cubic unit cell, with the corners of the cell located at 1 unit from the origin (O). The triangular plane drawn within the unit cell intersects the x-axis at 1/2, the y-axis at 1/2, and the z-axis at 1. This plane has Miller indices of (221). (c) A family of planes shown in a 2D lattice.

The **unit cell**, shown in heavy outline in the lattice, can be moved in three dimensions to recreate the entire crystal lattice. The unit cell is the smallest volume that can be used to describe the entire lattice. A Cartesian coordinate system is used to locate points, directions, and planes in a crystal lattice. A unit cell has its origin at the intersection of the three axes and is designated by its edge lengths in the x-, y-, and z-directions and by three angles. An atom (molecule or ion) in the crystal lattice is a point, identified by its x, y, and z coordinates. A plane is identified by its Miller indices, the reciprocals of the intersection points of the plane with the x, y, and z axes. A triangular plane is shown within the unit cell (Figure 9.63b). The plane intersects the x-axis at 1/2, the y-axis at 1/2, and the z-axis at 1; it has intercepts of 1/2, 1/2, and 1. The reciprocals are 2, 2, and 1, so the Miller indices for this plane are (221). A plane that is parallel to a given axis has an intercept of infinity; the reciprocal of infinity is 0. A crystal lattice will have many parallel planes, each uniformly spaced from each other. Such groups of planes are called families of planes and will have related Miller indices [e.g., the (110), (220), (330), and (440) planes are a family of planes]. These planes in a given family are all parallel, as shown in Figure 9.63c, just at different distances from the origin specified for the coordinate system. The (110) plane is the farthest from the origin, and the (440) plane is the closest to the origin of the set of planes (110), (220), (330), and (440). Miller indices for commonly used analyzing crystals were given in Table 9.7.

If a monochromatic X-ray beam falls on such a crystal, each atomic plane reflects the beam. Each separate reflected beam interacts with other reflected beams. If the beams are not in phase, they destroy each other and no beam emerges. Other beams reinforce each other and emerge from the crystal. The net result is a **diffraction pattern** of reinforced beams from many planes. It is the atomic planes that are important in XRD. It is of course possible to draw an infinite number of planes in three dimensions, but only those planes with electron density on them reflect X-rays.

In Figure 9.64, radiation from the source falls on the crystal, some on the top atomic plane and some on the second plane. Since the two beams are part of the same original beam, they are in phase on reaching the crystal. However, when they leave the crystal, the part leaving the second plane has traveled an extra distance ABC. If ABC is a whole number of wavelengths, the two beams leaving the crystal will be in phase and the light is coherent. If ABC is not a whole number of wavelengths, the two beams come together out of phase, and by destructive interference, the light is destroyed.

As we derived earlier, $n\lambda = 2d \sin\theta$. This is the Bragg equation, which states that coherence occurs when $n\lambda = 2d \sin\theta$. It can be used to measure d, the distance between

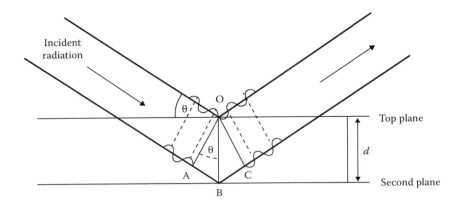

Figure 9.64 Reinforcement of light diffracted from two crystal planes.

planes of electron density in crystals, and is the basis of **X-ray crystallography**, the determination of the crystal structure of solid crystalline materials. Liquids, gases, and solids such as glasses and amorphous polymers have no well-ordered structure; therefore, they do not exhibit diffraction of X-rays.

For any given crystal, d is constant for a given family of planes; hence, for any given angle θ and a given family of planes, $n\lambda$ is constant. Therefore, if n varies, there must be a corresponding change in λ to satisfy the Bragg equation. For a given diffraction angle, a number of diffracted lines are possible from a given family of planes; n is known as the **order** of diffraction. As an example, if $2d \sin\theta$ equals 0.60, each of the conditions of Table 9.12 will satisfy the Bragg equation. Radiation of wavelength 0.60, 0.30, 0.20, or 0.15 Å will diffract at the same angle θ in first, second, third, or fourth order, respectively, as seen in Figure 9.65. This is called order overlap and can create difficulty in interpretation of crystal diffraction data.

It should be noted that radiation of 0.30 Å would also be diffracted at a different angle in first order from the same family of planes (same d value), as shown in Figure 9.65. Wavelengths corresponding to low orders such as first and second order give observable diffraction lines. Consequently, a single plane will generate several diffraction lines for each wavelength. Each of the planes in the three dimensions of the crystal will give diffraction lines. The sum total of these diffraction lines generates a diffraction pattern. From the diffraction pattern, it is possible to deduce the different distances between the planes as well as the angles between these planes in each of the three dimensions. Based on the diffraction pattern, the physical dimensions and arrangement of the atomic planes in the crystal can be identified.

9.4.1 Single-Crystal X-Ray Diffractometry

The schematic layout of a single-crystal diffractometer is given in Figure 9.66. This system uses an X-ray tube,

Table 9.12 Order of Diffraction for Values of n, λ

n	λ (Å)	nλ	Order
1	0.60	0.60	First
2	0.30	0.60	Second
3	0.20	0.60	Third
4	0.15	0.60	Fourth

a sample specimen, and a detector that rotates in an arc described by a Rowland circle. Note that the single-crystal sample takes the place of the analyzing crystal in a WDXRF analyzer. The goniometer mounting for a single-crystal diffractometer is very complex, because the crystal must be moved in three dimensions to collect data from many planes. A commercial benchtop single-crystal XRD instrument is shown in Figure 9.67.

For any given experiment, λ is the known wavelength of the monochromatic X-ray beam; θ is controlled and varied by the goniometer. From this information, d can be calculated. By rotating the goniometer and examining various sides of the crystal, hundreds (or thousands) of diffracted X-rays are collected. These data are processed to identify the positions of the planes and atoms in the crystal in three dimensions. Modern single-crystal diffractometers use computers to control the goniometer and to process the data. The diffraction data are usually converted to a 2D electron density map by Fourier transformation. The electron density map shows the location of atoms. A 2D electron density map is produced for each angle. The computer program uses the 2D maps plus the rotation angle data to generate the 3D coordinates for atoms (molecules, ions) in the crystal. The mathematical treatment of the experimental data to produce a crystal structure from an unknown single-crystal diffraction pattern is complicated and beyond the scope of this text, but an example of the results will be shown in the following.

The diffraction pattern of a single crystal of an inorganic salt is shown in Figure 9.68; an actual diffraction pattern

X-RAY SPECTROSCOPY

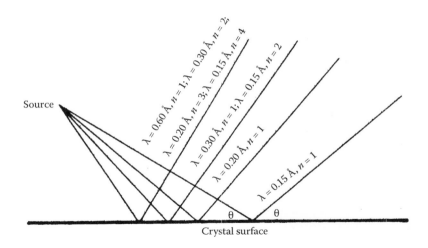

Figure 9.65 Diffraction of radiation of different wavelengths. Overlap can occur when different orders are diffracted at the same angle.

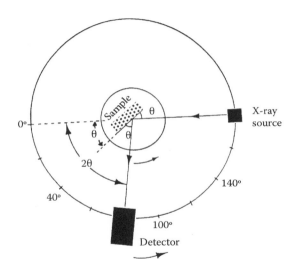

Figure 9.66 Schematic layout of a single-crystal XRD.

is shown in Figure 9.69. A given molecule always gives the same diffraction pattern, and from this pattern, we can determine the spacing between planes and the arrangement of planes in the crystal. Also, qualitative identification can be obtained by matching this pattern to previously identified patterns. Modern instruments are equipped with 2D imaging detectors such as CCDs.

9.4.2 Crystal Growing

In order to determine the structure of a single crystal, such a crystal must be grown from the material to be studied. The crystal quality determines the quality of diffraction results obtained. One limitation of XRD is that data are obtained on only one single crystal of the bulk material, unlike other techniques where multiple replicates are usually analyzed. The growth of single crystals of materials often is not easy. Simple inorganic salts and small organic molecules can be crystallized as single crystals by very slow evaporation of a supersaturated solution of the salt or compound. Once one tiny single crystal forms, it will grow in preference to the formation of more small crystals. Sublimation under vacuum can be used. Proteins and other biomolecules are more difficult to grow as single crystals because they are complex. One method that often works is to suspend a drop of protein solution over a reservoir of buffer solution. Water diffusion from the drop often results in single-crystal growth. Different techniques are required to form metal crystals. The interested student can find many references and resources on the Internet, by searching the term "X-ray crystallography." The single crystal then has to be mounted; various mount types are available depending on the instrument and goniometer being used.

9.4.3 Crystal Structure Determination

From the diffraction pattern, the crystal structure can be determined mathematically and the compound identified. While the discussion of the details is beyond the scope of this text, small, benchtop automated single-crystal XRD systems have come into use. These instruments possess a variety of powerful data-processing programs and libraries that permit the determination of high-resolution crystal structures of small molecules rapidly and automatically. Systems include the Rigaku XtaLAB mini™ and the Bruker X2S. In the past, crystal structure determination for one complex crystal could form the basis of a doctoral dissertation.

The XtaLAB mini™ software shows the approach used by these automated systems. Figure 9.70 lists the steps involved in obtaining a structure from the diffraction

Figure 9.67 (a) A benchtop single-crystal XRD system, the Rigaku XtaLAB mini™. (b) Close-up of the goniometer, X-ray tube, and detector in the XtaLAB mini. The single-crystal sample is mounted at the tip of what looks like a sewing needle in the center of the photo. (Used by permission of Rigaku Corporation, www.rigaku.com.)

pattern. Once the single crystal of the analyte has been mounted, a "project" is opened and the data collected. Then, the data are evaluated and the structure solved using one or more of the stored mathematical approaches. Some of the various methods and programs can be seen on the computer screen. A model is created and refined, and a report with the crystal structure, lattice parameters, and molecular identification is produced. Figure 9.71 shows the refinement results for a particular crystal obtained using the program SHELX. The results after five cycles of refinement (which

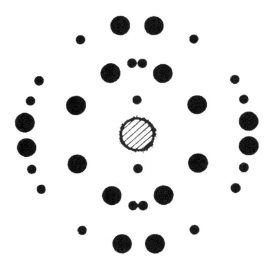

Figure 9.68 Diffraction pattern in two dimensions of a single crystal of an inorganic salt.

Figure 9.69 Actual single-crystal diffraction pattern of a small molecule collected with a Rigaku XtaLAB mini™. (Used by permission of Rigaku Corporation. www.rigaku.com.)

requires over 2000 observations and 231 variables) give a "goodness of fit" of 1.066, an indication that the result is highly probable. Perfect fit would be 1.000. The number of cycles, observations, and variables makes it clear that without modern computer processing, determination of a structure from an XRD pattern would take days or weeks of work, as it used to.

Measurement times for crystals of small molecules using such a system vary and depend on the complexity of the molecule. Potassium tetrachloroplatinate II, K_2PtCl_4, required less than 2 h of measurement time using the XtaLAB mini™, while a structure like raffinose, $C_{18}H_{32}O_{16} \cdot 5H_2O$, required 5 h and 30 min. The overall results from an analysis of the diffraction pattern would include the chemical structure, the name of the compound, the formula mass, the space group to which the crystal belongs, and the crystal lattice constants.

9.4.4 Powder X-Ray Diffractometry

Powdered crystalline samples can also be studied by XRD. The sample is loaded onto the specimen holder, which is placed in the X-ray beam in a setup similar to that used for single-crystal XRD. The sample must be powdered by hand or by mechanical grinding and is pressed into a sample holder to form a flat surface or packed into a thin glass or polymer capillary tube. After mounting, the specimen is rotated relative to the X-ray source at a rate of (degrees θ)/min. Diffracted radiation comes from the sample according to the Bragg equation. The detector is simultaneously rotated at (degrees 2θ)/min, using traditional θ–2θ geometry. Alternatively, instruments such as the Thermo Fisher Scientific ARL X'TRA powder diffraction system are available using a θ–θ geometry (Bragg–Bretano geometry).

A powdered crystalline material contains many thousands of tiny crystals. These crystals are oriented in all possible directions relative to the beam of X-rays. Hence, instead of the sample generating only single diffraction spots, it generates cones of diffracted X-rays, with the point of all the cones at the sample (Figure 6.72). Each family of planes will have a different circular diameter, so the result is a series of concentric cones radiating from the sample. Imagine that inside the Rowland circle opposite the sample, we have a strip of X-ray film as shown in Figure 6.72 on the right. The circular cones of X-rays will hit the film, resulting in a series of curved lines on the film (Figure 9.72 left). A typical diffraction pattern from a powdered sample collected on film is shown in Figure 9.73. These are called Laue photographs, after Max von Laue, the German scientist who developed the technique. Film has been replaced by automated scanning with a standard X-ray detector as discussed for XRF or by the use of imaging detectors such as CCDs or image plates to give a 2D image. A cylindrical image plate detector used by Rigaku Corporation has an active area of 454 mm × 256 mm, a pixel size of 100 μm × 100 μm, extremely rapid readout, and a sensitivity of 1 X-ray photon per pixel for Cu $K_α$ radiation. One X-ray photon per pixel is a quantum efficiency of 100%. The major advantage of these imaging detectors is that the images can be stored and manipulated electronically without the need for a photographic film-developing lab. The Thermo Fisher Scientific ARL X'TRA uses a Peltier-cooled Si(Li) solid-state detector, which can be cooled to about −100 °C, resulting in extremely low internal noise. This solid-state detector allows the user to electronically select photons based on their energy signature, which eliminates the need for beta filters or diffracted beam monochromators to remove Bremsstrahlung and sample fluorescence.

In order to obtain accurate XRD spectra, the sample must be ground finely and pressed, so that there is sufficient

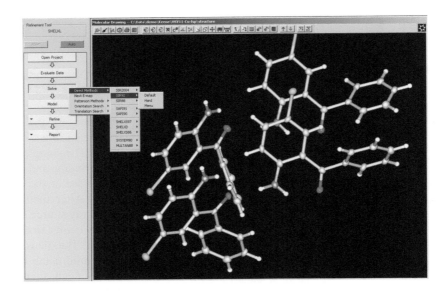

Figure 9.70 Computer screen of the XtaLAB mini™ showing the automated program steps for solving a crystal structure, including a list of some of the mathematical methods available and the molecular structure resulting from the data processing. (Used by permission of Rigaku Corporation. www.rigaku.com.)

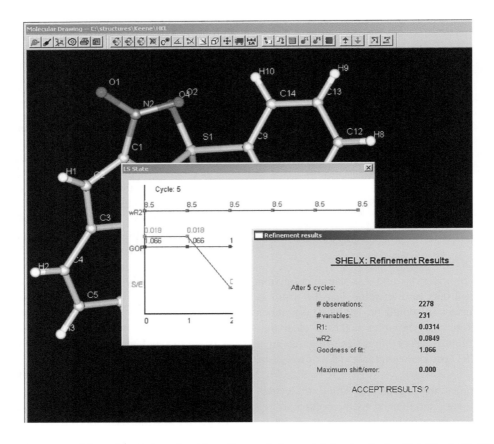

Figure 9.71 Computer screen of the automated results of the refinement of a crystal structure by the XtaLAB mini™. (Used by permission of Rigaku Corporation. www.rigaku.com.)

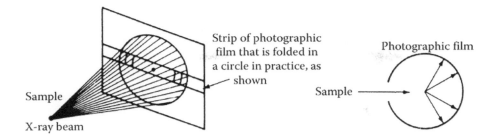

Figure 9.72 Schematic of diffraction from a powdered crystalline sample. The powdered sample generates the concentric cones of diffracted X-rays because of the random orientation of crystallites in the sample. The X-ray tube exciting the sample is not shown in this diagram. The cones of diffracted light intersect X-ray film curved to fit the diameter of the Rowland circle. The result is a series of curved lines on the X-ray film.

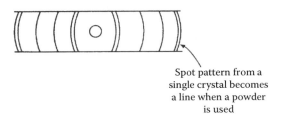

Figure 9.73 A typical diffraction photograph, called a Laue photograph, from a powdered crystalline sample.

random orientation of the crystals in the sample. Nonrandom orientation (preferred orientation) will result in a distorted spectral pattern, like a broken line or series of spots instead of a complete curved line. Some materials exhibit preferred orientation, either naturally or by design, and can be identified through these distortions. Powder diffractometers come in a variety of sizes, including small benchtop units and field-portable units (Section 9.4.5.1).

9.4.5 Hybrid XRD/XRF Systems

Hybrid systems are designed to combine the speed and flexibility of XRD and XRF systems in one spectrometer. Such systems permit more complete characterization of a given crystalline sample. These systems are varied: some are XRF systems with a few powder XRD-based channels to identify compounds, such as a system designed with a CaO channel for the cement industry. The combination can go as far as including a complete powder diffractometer and XRF spectrometer for flexible compound identification and quantification. These types of systems are generally used in industries, including metal and alloy production, cement production, mining, and refractory materials production, for both R&D and process control.

Thermo Fisher Scientific's ARL 9900 series hybrid systems offer a variety of configurations of the XRF and powder XRD components for R&D, fast process analysis, qualitative and quantitative elemental analysis, and phase analysis. The system can be configured in many ways: with up to 32 monochromators for fast elemental analysis, up to three goniometers for analysis of specific elements (quantitative and standardless analysis), scanning the XRF spectrum (qualitative and semiquantitative analysis), a compact XRD system for process control, or a full powder XRD system. With the XRF unit, up to 83 elements (B to U) can be determined from ppm levels to 100 %. The full XRD goniometer, called the NeXRD, provides qualitative and quantitative phase analysis. Full-pattern quantitative phase analysis results can be obtained in 5 min using automatic interpretation of the XRD pattern. The 9900 series has options for a variety of automated sample changers (12 or 98 position) and can be integrated with the ARL robotic sample preparation systems, allowing unattended continuous operation of the instrument. This system is a floor-mounted laboratory unit. The advantages of a hybrid unit for the laboratory are that it provides a single user interface for both XRF and XRD techniques, minimizes occupied floor space, merges elemental and phase analysis into a single report, and permits rapid, precise analysis of solid samples. One limitation of such systems is that they can only handle solid samples. Pictures and details are available at www.thermo.com./xray.

9.4.5.1 Compact and Portable Hybrid Systems

A revolution in powder XRD/XRF systems has resulted from the development of portable, compact, rugged systems designed originally for rock and mineral analysis in the field. The Terra Mobile XRD/XRF system (from Innov-X and Olympus [www.olympus-ims.com; www.inxitu.com]) is a high-performance combination 2D powder diffraction/XRF system that is battery powered, completely self-contained, very rugged, and field portable (Figures 9.74a and b).

The Terra provides full phase ID of major, minor, and trace components in rocks and minerals and an XRF system capable of scanning from Ca to U. It has a sample handling system that can sieve powders to <150 μm and requires only 20 mg of sample. To obtain accurate powder XRD spectra,

(a)

(b)

Figure 9.74 (a, b) Two images of geologists using the Terra Mobile XRD/XRF system in real mining applications in field conditions. (Courtesy of Olympus NDT. www.olympus-ims.com; www.inxitu.com.)

it is generally necessary to finely grind a sample and press a pellet, to ensure sufficient random orientation of the crystals in the sample. To avoid the problems of pressing a pellet, which is difficult to do in the field or on Mars, the Terra has a patented integrated sample vibration chamber. By vibrating the 15–20 mg sample, all orientations of the crystal structure are presented to the instrument optics. This virtually eliminates the problem of preferred orientation induced by poor sample preparation. The system provides images of the XRD pattern and permits identification of preferred orientation and particle influences on the spectrum. The XRF uses energy discrimination to eliminate fluorescence, scatter, and other background and contains an XRF library for pattern confirmation. Peak identification in XRD and XRF is as accurate as that obtained with laboratory-based systems. The Terra system is now onboard the Mars Science Laboratory rover. Full-color images of the Curiosity rover and Terra XRD/XRF data from Mars are available at www.nasa.gov/mission_pages/msl.

Based on the Terra design is the BTX Benchtop XRD/XRF system (www.olympus-ims.com; www.inxitu.com). This system is a one-button operation compact XRD/XRF unit suitable for educational use as well as dedicated quality control powdered sample applications. It is easy to use, fast, small, and compact and requires minimal sample preparation and provides the high-quality results seen in many large laboratory units. A sample spectrum from the BTX system is shown in Figure 9.75.

9.4.6 Applications of XRD

The analytical applications of XRD are numerous. The method is nondestructive and gives detailed information on the structure of crystalline samples. Comparing powder diffraction patterns from crystals of unknown composition with patterns from crystals of known compounds permits the identification of unknown crystalline compounds. The number of peaks or lines, intensities of peaks or lines, and the angular positions of peaks or lines (in terms of 2θ) are used to identify the material (Figure 9.75).

Diffraction patterns are unique for each compound and serve as a fingerprint for a crystalline material. For example, as shown in Figures 9.76a and b, pure crystals of compound A and pure crystals of compound B give different diffraction patterns. A *mixture* containing both A and B will show diffraction peaks from both pure compounds (Figure 9.76c). If we had a mixture of 15 % KCl and 85 % NaCl, the diffraction pattern would show strong NaCl peaks with a weak pattern of KCl intermixed. A mixture containing 15 % NaCl and 85 % KCl would show the diffraction pattern of KCl with a weak pattern of NaCl. Such a mixture is a *multiphase material*, and interpretation of multiphase patterns is more difficult than for single-phase (pure) materials. If, on the other hand, the crystal were a *mixed crystal* of sodium potassium chloride, in which the sodium and potassium ions are in the *same crystal lattice*, there would be changes in the crystal's lattice size from that of pure NaCl or pure KCl. However, the mixed crystal is a single-phase material, resulting in a unique diffraction pattern, as shown in Figure 9.76d. X-ray powder diffraction therefore can be used to distinguish between a *mixture of crystals*, which would show both diffraction patterns, and a *mixed crystal*, which would give a separate unique diffraction pattern. The exact crystallographic lattice constants can be measured using XRD. Powder diffraction pattern matching to identify unknowns is now done with a computer, software, and a powder diffraction pattern spectral library or database

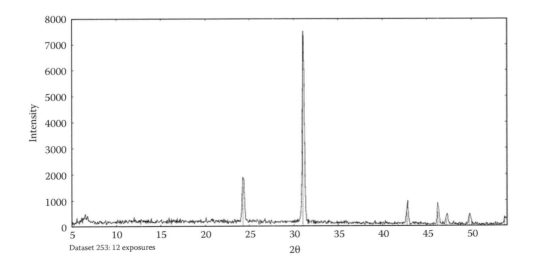

Figure 9.75 Powder XRD spectrum of pure quartz, taken with the BTX Benchtop system using a Co target and 12 exposures (100 s). (Courtesy of Olympus NDT. www.olympus-ims.com; www.inxitu.com.)

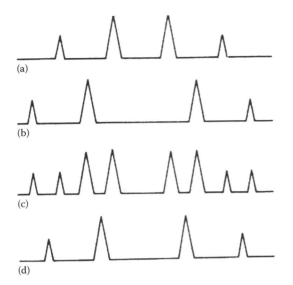

Figure 9.76 Schematic X-ray powder diffraction patterns for simple inorganic salts. (a) Pure salt A. (b) Pure salt B. (c) A physical mixture of salt A and salt B. Note that the peaks for both pure salts can be seen in the mixture; every peak matches a peak in either (a) or (b). (d) A diffraction pattern for a mixed crystal containing the same elements present in both A and B, but chemically combined in the same crystal lattice. The diffraction pattern for the mixed crystal (a unique structure) is unique; it does not match either pure A or pure B.

that can be searched by computer. An example is shown in Figure 9.77. The International Centre for Diffraction Data (ICDD), located in Newtown Square, PA, United States, maintains a database of more than 50,000 single-phase powder XRD patterns (www.icdd.com).

In polymer characterization, it is possible to determine the degree of crystallinity of semicrystalline polymers. The noncrystalline (amorphous) portion simply scatters the X-ray beam to give a continuous background, whereas the crystalline portion gives diffraction lines. A typical schematic diffraction spectrum of a semicrystalline polymer is shown in Figure 9.78. The ratio of the area of diffraction peaks to scattered radiation is proportional to the ratio of crystalline to noncrystalline material in the polymer. The ultimate quantitative analysis must be confirmed using standard polymers with known percent crystallinity and basing the calculation on the known ratio of crystalline diffraction to amorphous scattering.

Using an XRD pattern and looking at the intensity of peaks, it is possible to determine if the crystals of a polymer or a metal are oriented in any particular direction. "Preferred orientation" can occur after the material has been rolled out into a sheet, for example. This is sometimes a very undesirable property, since the material may be very weak in one direction and strong in another, with the result that it tears easily in one direction. Sometimes, however, this is a desirable property, as, for example, in packaging material that we may wish to tear easily in one direction to open the package.

Powder XRD is used for phase analysis and compound identification in a variety of industries, especially mineral, mining, and metal production, and for materials such as rocks, minerals, oxide materials, and products. Typical examples include the levels of Fe phases such as FeO, Fe_2O_3, Fe_3O_4, and FeC; determination of free lime in clinker and slags in the cement industry; phases related to the electrolysis of Al; and CaO, $CaCO_3$, and $Ca(OH)_2$ content.

XRD at different temperatures can be used to study phase transitions between different crystallographic forms of a material (e.g., tetragonal vs. monoclinic forms of yttria-stabilized zirconia). This approach can be used to measure thermal expansion coefficients and to study crystalline-to-amorphous transitions in materials.

A property of metals that can be determined by XRD is the state of anneal. Well-annealed metals are in an ordered crystal form and give sharp diffraction lines. If the metal is subjected to drilling, hammering, or bending, it becomes "worked" or "fatigued"; the crystal structure and the diffraction pattern change. Working a metal initially increases its strength, but continued deformation (fatigue) weakens the metal and can result in the metal breaking. (Try bending a paper clip slowly—you should note that it becomes harder to bend after one or two workings, but then if you continue to bend it at the same point, it will eventually become fatigued and break.) Disorder in metals and alloys can also result from rapid cooling from the molten state. Disorder results in lower density and a more brittle material. Electrical conductivity of metals is also affected by the order in the metal crystal. XRD can be used to distinguish between ordered and disordered materials and provide valuable information on the suitability of a material for a given use.

A very important use of XRD is in the determination of the structure of single crystals, that is, identifying the exact position in 3D space of every atom (molecules, ions) in the crystal. Single-crystal XRD was a major tool in elucidating the structure of ribonucleic acid (RNA) and deoxyribonucleic acid (DNA), insulin, vitamins, and proteins. Single-crystal diffractometry is used for structural determination of biomolecules, natural products, pharmaceuticals, inorganic coordination complexes, and organometallic compounds.

9.4.6.1 Analytical Limitations of XRD

Amorphous materials cannot be identified by XRD, so many polymeric and glassy materials cannot be studied. XRD is not a trace analysis technique. A component should be present at 3–5 % by mass at a minimum in order for diffraction peaks to be detected. The unique pattern from a mixed crystal, where one atom has replaced another in the lattice, was discussed earlier. A mixed crystal is analogous to a solid solution of a contaminant in a pure material. The contaminated

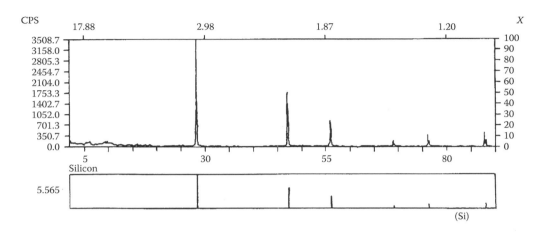

Figure 9.77 The X-ray powder diffraction pattern of an unknown material is shown in the upper spectrum. A search of a computerized database identified the unknown as silicon, based on the match to the stored spectrum for silicon (lower spectrum).

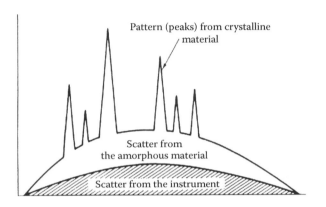

Figure 9.78 Schematic diffraction pattern from a semicrystalline polymer, showing how both crystalline and amorphous phases may be detected. The amorphous portion results in broad scattering, while the crystalline portion shows a typical diffraction peak pattern. A totally amorphous polymer would show no diffraction peaks. (There are no 100 % crystalline polymers.) The units on the x-axis are degrees θ.

material will appear to be a single phase, since it is a solid solution, but the lattice of the pure material will be expanded or contracted as the result of contaminant atoms of the "wrong" size in the lattice. There will be a unique diffraction pattern, but such a pattern may be hard to match. Mixtures may be difficult to identify because of overlapping peaks.

9.5 X-RAY EMISSION

In XRF, characteristic X-rays from an X-ray source are used to excite characteristic X-rays from a sample. X-ray emission differs in that particles such as electrons and alpha particles are the sources used to generate characteristic X-rays from a sample. Several types of X-ray emission systems will be described.

9.5.1 Electron Probe Microanalysis

A beam of electrons striking a target results in the emission of characteristic X-rays. This is the basis of the X-ray tube, as was discussed earlier in this chapter. A beam of electrons striking a sample will also generate characteristic X-rays from the sample. The use of a small diameter electron beam, on the order of 0.1–1.0 μm, to excite a sample is the basis of **electron probe microanalysis**. An electron probe microanalyzer is an X-ray emission spectrometer. The small-diameter electron beam excites an area of the surface of the sample that is about 1 μm in diameter. Elemental composition and variation of composition on a microscopic scale can be obtained.

Two different instruments are available for microanalysis. The electron microprobe analyzer (EMA) uses high electron beam currents to provide elemental analysis of samples, with moderate spatial resolution and low magnification of the sample. The intensity of emitted X-rays from the sample is high. The scanning electron microscope (SEM) is designed to provide high-resolution, high-magnification images of a sample. The SEM operates at low electron beam currents; the characteristic X-rays are of lower intensity than in the EMA but still provide elemental composition. In addition to X-ray spectra, other information may be obtained as a result of interactions of the electron beam with the sample. Some samples will exhibit UV or visible fluorescence (called *cathodoluminescence*); secondary electrons may be ejected, such as Auger electrons, and backscattered electrons are used to provide information about the sample topography and composition.

Both systems operate on similar principles. A tungsten filament emits electrons, which are focused by an electron optical system. The electron beam can scan the sample surface and can provide composition at a point, along a line or over a rectangular area, by *rastering* (moving) the beam across the surface in a series of parallel lines. The sample is mounted on a stage that can be accurately moved in the *x*- and *y*-direction and in the *z*-direction, normal to the plane of the sample. The system has an optical microscope, to permit alignment of the sample and selection of the area or feature of the sample to be analyzed.

The X-ray analysis system for the EMA is a WD spectrometer with gas-proportional counter detectors. In the SEM, an ED X-ray spectrometer with a Si(Li) detector is used. The entire electron and X-ray optical systems are operated under a vacuum of about 10^{-5} torr. Modern systems are completely automated with computer control of the instrument parameters, specimen stage movement, data collection, and data processing.

Samples must be solid and may be in almost any form. Thin films, bulk solids, particles, powders, machined pieces, and small objects (including biological specimens) can be analyzed. All elements from beryllium ($Z = 4$) to uranium can be determined at concentrations of about 100 ppm or greater. For qualitative analysis, the surface finish of the sample is not important. For quantitative analysis, the surface of the specimen must be flat. A common method for achieving a flat surface for an SEM sample is to embed the sample in epoxy and then carefully polish the hardened epoxy to expose a flat surface of the sample. Calibration standards should have flat surfaces as well, and the composition of the standards should be similar to that of the samples. Alternatively, an FP approach using pure element standards can be used. For nonconductive samples, a thin coating of osmium is deposited on the surface of the sample.

Applications of electron probe microanalysis are numerous in analytical chemistry, metallurgy, geology, medicine, biology, and materials science. In materials science, not only can the composition of the material be determined, but the location of elements in the solid structure can be visualized. Segregation of a particular element in grain boundaries between crystals can be seen, for example. Such segregation may have a dramatic effect on the properties of a material. Pure tungsten wire is too brittle to make a useful light bulb. The wire fractures easily as the light bulb is turned on (wire heats up) and off (wire cools down) because the crystals separate along the grain boundaries. By adding certain elements such as potassium to tungsten, an impure (doped) tungsten wire that is much more ductile results; this ductile wire can be heated and cooled many times without fracturing. Analysis of the doped tungsten using SEM shows that the added elements are located in the boundaries between the crystals. The elements in the grain boundaries cause the tungsten crystals to slide along the grain boundaries, permitting the tungsten wire to expand and contract on heating without fracturing. Thin films and multilayer materials can be analyzed and film thickness measured. Small particles filtered from air and water can be analyzed, and both particle count and particle size determined. The number, size, and type of particles in air are measures of air quality; this is very important in the manufacturing of semiconductor devices.

The distribution of elements in biological samples, mineral samples, soils, and other heterogeneous materials can be determined. Many other applications can be found in the literature. This is one of the few techniques that can provide spatial variation of composition at the micrometer scale. Other related techniques are discussed in Chapter 16.

9.5.2 Particle-Induced X-Ray Emission

The use of a radioactive isotope as a source for XRF was discussed, and a list of common radioactive sources was given in Table 9.4. The use of one that generates "particles"—electrons or alpha particles—gives rise to the term PIXE. The 2012 NASA Martian Space Laboratory rover, Curiosity, has on board a PIXE instrument called an APXS. The APXS uses a Curium-244 source to excite characteristic X-rays from Martian rocks. The spectrometer includes a calibration target with multiple elements. The target is a well-characterized rock slab brought from Earth to enable accurate determination of the composition of Martian rocks. The calibration target spectrum is shown in Figure 9.79.

Figure 9.80 shows a Martian rock known as "Jake Matijevic" that has been studied using the APXS and the ChemCam laser-induced breakdown spectroscopy (LIBS) instrument (described in Chapter 8). The two black circles on the rock face show the areas studied by the APXS. The black dots are the laser pits from the LIBS instrument. The results of the APXS analysis are shown in Figure 9.81.

The spectra from the rock and target were collected for 1 h. Results showed that, compared to previous Martian rocks, the Jake rock is low in Fe and Mg, but high in Na, Al, Si, and K, which are often found in feldspar minerals. The results point to an igneous or volcanic origin for the rock.

9.6 COMMERCIAL X-RAY INSTRUMENT MANUFACTURERS

Commercial XRF systems include WDXRF and EDXRF laboratory-based instruments from floor models to benchtop systems, handheld X-ray analyzers based on EDXRF, field-portable XRF systems, and micro-XRF systems. Commercial XRD systems include single-crystal diffractometers, powder XRD systems, and field-portable XRD systems. Hybrid XRD/XRF systems are available in laboratory and field-portable versions. Companies include but are not limited to Bruker, Horiba, PANalytical, Rigaku MSC and Applied Rigaku Technologies, Olympus (formerly Innov-X), SPECTRO, Thermo Fisher Scientific, EDAX, and Unisantis GmbH. The websites for most of these companies contain a wealth of information, including application notes, tutorials, videos, and on-demand webinars.

X-RAY SPECTROSCOPY   559

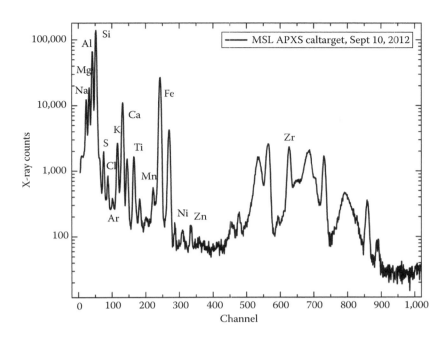

Figure 9.79 Spectrum of the calibration target from the APXS on the Martian rover "Curiosity." (Image credit: NASA/JPL-Caltech/LANL.)

Figure 9.80 The "Jake Matijevic" rock on Mars, showing the circles where the APXS instrument obtained data, as well as the laser pits (black dots) from the LIBS instrument. (Image credit: NASA/JPL-Caltech/LANL. Full-color images are available at www.nasa.gov/mission_pages/msl.)

SUGGESTED EXPERIMENTS/TUTORIALS

Safety note: The student is reminded that X-radiation poses significant health hazards. Safety is a very important issue with X-ray methods. X-radiation is ionizing radiation, a significant health hazard because it can damage or kill living cells and produce genetic mutations. Some instrumentation uses radioactive isotopes as sources; these sources are always producing ionizing radiation and cannot be "turned off." Modern commercial instrumentation is equipped with

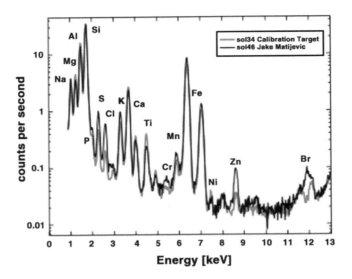

Figure 9.81 Spectrum from "Jake" and the calibration target acquired by the APXS system. (Image credit: NASA/JPL-Caltech/LANL.)

the proper shielding and safety interlocks that will shut off the ionizing radiation automatically or protect the analyst from exposure when operated properly. Operators of X-ray instrumentation usually must undergo specialized safety training and wear radiation dosimeters to ensure that they are not being exposed to damaging X-rays. Only commercial equipment equipped with safety interlocks should be used and only after appropriate training and under the guidance of your instructor or radiation safety official at your facility. Under no circumstances should safety interlocks be tampered with. Experiments should never be performed with unconfined or unshielded X-ray sources. (An example of an unshielded X-ray source is the setup shown in Figure 9.14c.)

X-Ray Absorption

9.1 Take several metal disks made of different pure elements, each about 1 in.² in area and 1/8 in. in thickness. It is important that the thickness of each disk be constant. The disks should be made from (a) Mg, (b) Al, (c) Fe, (d) Ni, (e) Cu, (f) Sn, and (g) Pb. Expose them to an X-ray beam simultaneously on a single sheet of X-ray photographic film. Develop this film. Note that the absorption of X-rays by each disk is proportional to the atomic mass of the particular element.

9.2 Take a piece of glass known to contain an entrapped bubble of air. Place it in front of X-ray film and expose it to X-rays. Note that the bubble can be detected on the film as a dark patch. Similar holes in metal castings (or other metal objects) can be detected by XRA, even though the holes are not visible to the naked eye.

X-Ray Fluorescence

9.3 Using the metal disks used in Experiment 9.1, record the fluorescence spectra of the different metals. Identify the K and L lines for each element. Plot the relationship between the wavelengths of these lines and the atomic numbers of the metals using a spreadsheet program such as Excel. Explain your results.

9.4 Record the fluorescence spectrum of metal disks of unknown composition. By measuring the wavelength of the K_α line, identify the major components. Use disks of brass, bronze, stainless steel, and aluminum and coins of various denominations, for example.

9.5 Using silver powder and copper powder, prepare mixtures with various ratios of these metals. Record the XRF spectrum of each mixture. Relate the silver/copper ratio to the ratio of the Ag K_α and Cu K_α lines. Measure the Ag/Cu ratio of various US silver coins minted (a) before 1964 and (b) after 1964.

9.6 Record the fluorescence spectrum of sulfur. Locate the S K_α line. Record the fluorescence spectra of various grades of motor oil. Compare the sulfur contents of the oils by the intensities of the S K_α lines. The oil may be held in a polymer "zipper lock" bag during analysis if no liquid sample cups are available. Run a blank spectrum of the polymer bag as well to be certain it contains no detectable sulfur.

9.7 Analyze small chips of various rocks, or samples of sand or soil from different geographic locations. Identify the elements present in each sample. Can you distinguish between sands or soils from different locations by their XRF spectra? Look for any trace or minor elements that may differentiate the samples. (Remember that peak intensities will vary due to surface roughness, so particle size differences between samples may account for peak height differences in the spectra.)

9.8 Take the spectrum of various pieces of jewelry, supported on cellophane tape or a suitable heavy polymer film so that they do not fall into the spectrometer. Identify the elements present. Is your gold ring solid gold or gold-plated?

X-Ray Diffraction

9.9 Use a tungsten or copper X-ray source and a powder diffractometer. Load crushed NaCl powder into the sample holder. Measure the diffraction pattern of the NaCl. Repeat with powdered KCl. Note the difference in spectra. Make mixtures by varying the amounts of NaCl and KCl. Can you distinguish the mixtures and the relative amounts of each component from the powder patterns?

9.10 For the student interested in crystal growth, potassium aluminum sulfate, also called alum, is a salt available in most pharmacies. It will form single crystals relatively easily from a supersaturated solution. Alum crystals are octahedral in shape and colorless. The related compound, chromium aluminum sulfate dodecahydrate, called chrome alum, forms dark-purple octahedral crystals. Mixing various proportions of supersaturated alum and chrome alum solutions allows the growth of mixed crystals that resemble amethysts in color. (The more alum in the mix, the lighter the purple color of the crystal.) Take the powder patterns of crushed pure alum and crushed pure chrome alum, and a mixture of the two compounds. After you have grown some mixed crystals, crush some and take their powder pattern. Is it different from the two pure materials and from the mixture? Explain your observations.

PROBLEMS

9.1 Draw a schematic diagram of an X-ray tube, label the major parts, and describe its operation.

9.2 What is the origin of the K_α X-ray lines? The K_β lines? The L lines?

9.3 State Moseley's law. What was the importance of this law to chemistry in general?

9.4 What are the three major analytical fields of X-ray spectroscopy? State three analytical applications of each field.

9.5 The intensity of XRF is weak compared to some other spectroscopic techniques. What instrumental changes are made to maximize the fluorescence signal?

9.6 What elements can be detected by XRF? What elements cannot be detected by XRF in air? Explain.

9.7 Can X-rays from a tungsten target be used to excite copper atoms? Can X-rays from a copper target be used to excite tungsten atoms? Explain.

9.8 Look at Figures 9.3a and 9.18. The spectrum in Figure 9.3 was obtained at a tube voltage of 30 kV; that in Figure 9.18 with *no* filter was obtained at 20 kV. Explain the difference in the spectra.

9.9 What is the relationship between the wavelengths of the absorption edges and the related emission lines? Explain.

9.10 Why can XRF not distinguish between Cu^{2+} and Cu^{1+} in a sample?

9.11 Explain the difference between the two spectra in Figure 9.18. Why is a filter useful?

9.12 Derive the Bragg equation using reflection from two parallel planes as the model.

9.13 (a) Plot the signal-to-voltage relationship in an ionization counter and proportional counter.
(b) Which is the most sensitive? Why?

9.14 Diagram a scintillation detector and describe its principle of operation.

9.15 Describe how a Si(Li)-drifted detector operates.

9.16 (a) What is meant by the mass absorption coefficient?
(b) What is the relationship between the mass absorption coefficient, atomic number, and atomic mass at the same wavelength?

9.17 Describe the equipment used to collect XRD patterns from a single crystal.

9.18 Why are X-ray powder diffraction patterns useful in analytical chemistry? What advantage does a powdered sample have over a single crystal in terms of identifying an unknown sample?

9.19 Describe the basic components of a goniometer.

9.20 Look up the Ge line energies in Appendix 9.A and predict where you would find a Ge escape peak in an EDXRF spectrum of pure Cu using a Ge(Li) detector. Where would the escape peak occur if a Si(Li) detector were used for the same sample?

9.21 Predict what sum peaks you might see as artifacts in the EDXRF spectrum of pure Cu.

9.22 Looking at the cell in Figure 9.63b with the origin in the back lower left corner (marked O), explain why the (440) plane is closer to the origin than the (110) plane. Remember that the Miller indices are the reciprocals of the intercepts. Draw the planes. (You may find drawing a 2D picture is easier than a 3D one.)

9.23 What axis is the (110) plane in Si parallel to? Draw the (110) plane in a cubic unit cell such as the one shown in Figure 9.63b.

To interpret Problems 9.24 through 9.31, students should use the following table of tube voltage/filter combinations optimized for element ranges as well as Table 9.A.2. These spectra were collected on a Thermo Scientific EDXRF using different voltages, filters, and atmospheres. The conditions are given in each spectrum. Check the voltage and filter combinations used for the elements that can be detected using that combination.

Filter	Tube Voltage (kV)	Elements Optimized	Other Elements Detected
None	8–15	Na–S (K lines)	K–Fe (K lines)
		Zn–Mo (L lines)	Tc–Ce (L lines)
Cellulose	8–15	Cl–Sc (K lines)	Al–Zn (K lines)
		Tc–Cs (L lines)	Zr–Gd (L lines)
Al, 0.127 mm	10–20	Ti–Mn (K lines)	Cl–Br (K lines)
		Ba–Sm (L lines)	Ag–Hf (L lines)
Thin (0.05 mm Rh)	20–30	Fe–Ge (K lines)	K–Mo (K lines)
		Eu–Au (L lines)	Ba–U (L lines)
Five (0.127 mm Rh)	30–45	As–Mo (K lines)	Ti–Mo (K lines)
		Hg–U (L lines)	
Six (0.63 mm Cu)	45–50	Tc–La (K lines)	Zn–La (L lines)

9.24 Identify the binary inorganic salt that gives the following spectrum.

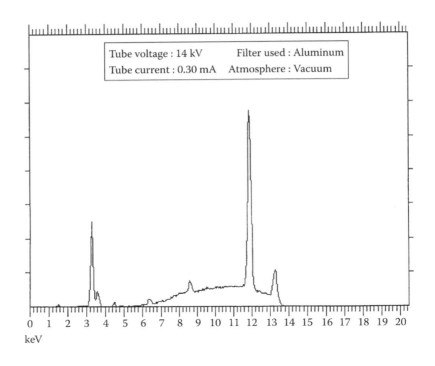

X-RAY SPECTROSCOPY

9.25 Identify the elements present in this spectrum.

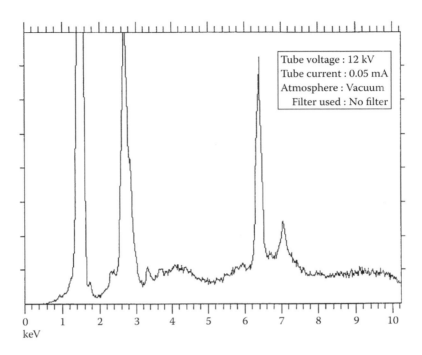

9.26 Identify the elements present in this spectrum.

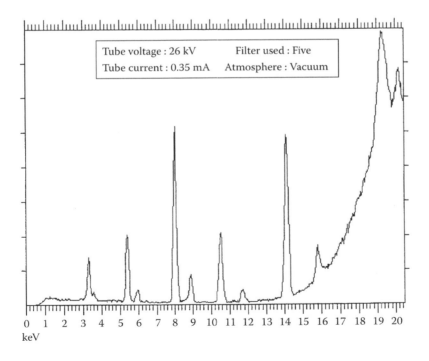

9.27 Identify the elements present and label the lines.

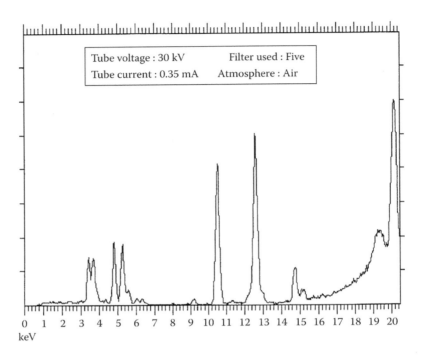

9.28 Identify the elements present—there are at least seven. (Hint: Many are high Z elements.)

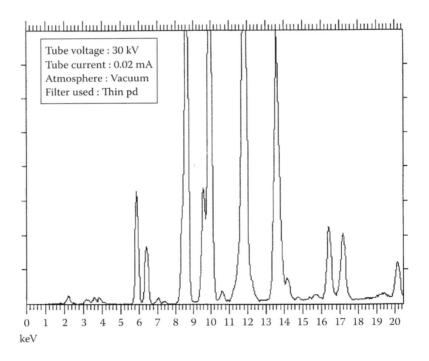

9.29 More high Z elements to identify. Label the lines.

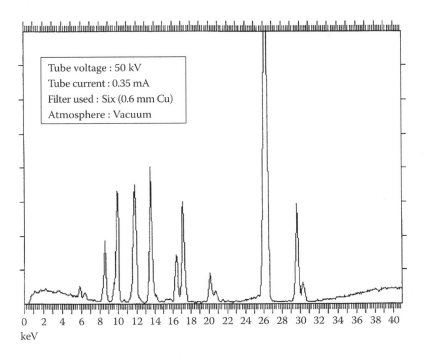

9.30 This is the spectrum of the X-ray tube. The line spectrum was obtained with NO filter in the X-ray tube path. The dotted spectrum was obtained using a 0.63 mm Cu filter. (1) Identify the X-ray tube anode. (2) What do we call the radiation emitted by the filtered X-ray tube? (3) What elements will be excited using the filtered tube radiation? (4) Why do you want the tube voltage so high? (50 kV is high voltage.) (5) Why not use the unfiltered X-ray tube spectrum to excite your sample? Can you list two elements that could not be determined using the unfiltered X-ray tube, other than the anode element itself?

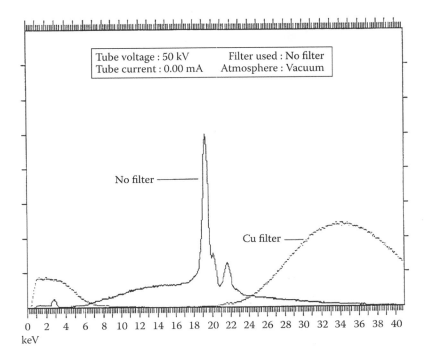

9.31 A sample with many elements. Identify as many as you can. This sample is loaded with stuff! GOOD LUCK!!!

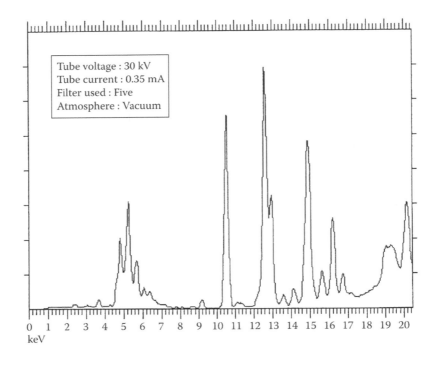

9.A APPENDIX: CHARACTERISTIC X-RAY WAVELENGTHS (Å) AND ENERGIES (KEV)*

Table 9.A.1 K Series Lines (Å)

Line		$\alpha_{1,2}$	α_1	α_2	β_1		β_3
Approximate Intensity		150	100	50	15		
Element	**Z**						
Li	3	230					
B	4	113					
B	5	67					
C	6	44					
N	7	31.603					
O	8	23.707					
F	9	18.307					
Ne	10	14.615			14.460		
Na	11	11.909			11.574		11.726
Mg	12	9.889			9.559		9.667
Al	13	8.339	8.338	8.341	7.960		8.059
Si	14	7.126	7.125	7.127		6.778	
P	15	6.155	6.154	6.157	5.804		
S	16	5.373	5.372	5.375	5.032		
Cl	17	4.729	4.728	4.731	4.403		
Ar	18	4.192	4.191	4.194	3.886		
K	19	3.744	3.742	3.745	3.454		
Ca	20	3.360	3.359	3.362	3.089		
Sc	21	3.032	3.031	3.034	2.780		

Table 9.A.1 (Continued) K Series Lines (Å)

Line		$\alpha_{1,2}$	α_1	α_2	β_1	β_3
Approximate Intensity		150	100	50	15	
Element	Z					
Ti	22	2.750	2.749	2.753	2.514	
V	23	2.505	2.503	2.507	2.285	
Cr	24	2.291	2.290	2.294	2.085	
Mn	25	2.103	2.102	2.105	1.910	
Fe	26	1.937	1.936	1.940	1.757	
Co	27	1.791	1.789	1.793	1.621	
Ni	28	1.659	1.658	1.661	1.500	
Cu	29	1.542	1.540	1.544	1.392	1.393
Zn	30	1.437	1.435	1.439	1.296	
Ga	31	1.341	1.340	1.344	1.207	1.208
Ge	32	1.256	1.255	1.258	1.129	1.129
As	33	1.177	1.175	1.179	1.057	1.058
Se	34	1.106	1.105	1.109	0.992	0.993
Br	35	1.041	1.040	1.044	0.933	0.933
Kr	36	0.981	0.980	0.984	0.879	0.879
Rb	37	0.927	0.926	0.930	0.829	0.830
Sr	38	0.877	0.875	0.880	0.783	0.784
Y	39	0.831	0.829	0.833	0.740	0.741
Zr	40	0.788	0.786	0.791	0.701	0.702
Nb	41	0.748	0.747	0.751	0.665	0.666
Mo	42	0.710	0.709	0.713	0.632	0.633
Tc	43	0.676	0.675	0.679	0.601	
Ru	44	0.644	0.643	0.647	0.572	0.573
Rh	45	0.614	0.613	0.617	0.546	0.546
Pd	46	0.587	0.585	0.590	0.521	0.521
Ag	47	0.561	0.559	0.564	0.497	0.498
Cs	48	0.536	0.535	0.539	0.475	0.476
In	49	0.514	0.512	0.517	0.455	0.455
Sn	50	0.492	0.491	0.495	0.435	0.436
Sb	51	0.472	0.470	0.475	0.417	0.418
Te	52	0.453	0.451	0.456	0.400	0.401
I	53	0.435	0.433	0.438	0.384	0.385
Xe	54	0.418	0.416	0.421	0.369	0.360
Cs	55	0.402	0.401	0.405	0.355	0.355
Ba	56	0.387	0.385	0.390	0.341	0.342
La	57	0.373	0.371	0.376	0.328	0.329
Ce	58	0.359	0.357	0.362	0.316	0.317
Pr	59	0.346	0.344	0.349	0.305	0.305
Nd	60	0.334	0.332	0.337	0.294	0.294
Pm	61	0.322	0.321	0.325	0.283	
Sm	62	0.311	0.309	0.314	0.274	0.274
Eu	63	0.301	0.299	0.304	0.264	0.265
Gd	64	0.291	0.289	0.294	0.255	0.256
Tb	65	0.281	0.279	0.284	0.246	0.246
Dy	66	0.272	0.270	0.275	0.237	0.238
Ho	67	0.263	0.261	0.266	0.230	0.231
Er	68	0.255	0.253	0.258	0.222	0.223
Tm	69	0.246	0.244	0.250	0.215	0.216
Yb	70	0.238	0.236	0.241	0.208	0.208

(continued)

Table 9.A.1 (Continued) K Series Lines (Å)

Line		$\alpha_{1,2}$	α_1	α_2	β_1	β_3
Approximate Intensity		150	100	50	15	
Element	Z					
Lu	71	0.231	0.229	0.234	0.202	0.203
Hf	72	0.224	0.222	0.227	0.195	0.196
Ta	73	0.217	0.215	0.220	0.190	0.191
W	74	0.211	0.209	0.213	0.184	0.185
Re	75	0.204	0.202	0.207	0.179	0.179
Os	76	0.198	0.196	0.201	0.173	0.174
Ir	77	0.193	0.191	0.196	0.168	0.169
Pt	78	0.187	0.158	0.190	0.163	0.164
Au	79	0.182	0.180	0.185	0.159	0.160
Hg	80	0.177	0.175	0.180	0.154	0.155
Tl	81	0.172	0.170	0.175	0.150	0.151
Pb	82	0.167	0.165	0.170	0.146	0.147
Bi	83	0.162	0.161	0.165	0.142	0.143
Po	84	0.185	0.156	0.161	0.138	
At	85		0.152	0.157	0.134	0.135
Rn	86		0.148	0.153	0.131	0.132
Fr	87		0.144	0.149	0.127	0.128
Ra	88		0.144	0.149	0.127	0.128
Ac	89		0.140	0.145	0.124	0.125
Th	90	0.135	0.133	0.138	0.117	0.118
Pa	91		0.131	0.136	0.115	0.116
U	92	0.128	0.126	0.131	0.111	0.112

* Appendices 9.A and 9.B are modified from Markowicz, used with permission. The data in these tables were obtained from the following sources: Clark (1963), Burr (1974), and Jenkins et al. (1991).

Table 9.A.2 Energies of Principal K and L X-Ray Emission Lines (keV)

Z	Element	$K_{\beta 2}$	$K_{\beta 1}$	$K_{\alpha 1}$	$K_{\alpha 2}$	$L_{\gamma 1}$	$L_{\beta 2}$	$L_{\beta 1}$	$L_{\alpha 1}$	$L_{\alpha 2}$
3	Li				0.052					
4	Be				0.110					
5	B				0.185					
6	C				0.282					
7	N				0.392					
8	C				0.523					
9	F				0.677					
10	Ne				0.851					
11	Na			1.067	1.041					
12	Mg			1.297	1.254					
13	Al			1.553	1.487	1.486				
14	Si			1.832	1.40	1.739				
15	P			2.136	2.015	2.014				
16	S			2.464	2.309	2.306				
17	Cl			2.815	2.622	2.621				
18	Ar			3.192	2.957	2.955				
19	K			3.589	3.313	3.310				
20	Ca			4.012	3.691	3.688			0.344	0.341
21	Sc			4.460	4.090	4.085			0.399	0.395
22	Ti			4.931	4.510	4.504			0.458	0.492

Table 9.A.2 (Continued) Energies of Principal K and L X-Ray Emission Lines (keV)

Z	Element	$K_{\beta2}$	$K_{\beta1}$	$K_{\alpha1}$	$K_{\alpha2}$	$L_{\gamma1}$	$L_{\beta2}$	$L_{\beta1}$	$L_{\alpha1}$	$L_{\alpha2}$
23	V		5.427	4.952	4.944			0.519	0.510	
24	Cr		5.946	5.414	5.405			0.581	0.571	
25	Mn		6.490	5.898	5.887			0.647	0.636	
26	Fe		7.057	6.403	6.390			0.717	0.704	
27	Co		7.647	6.930	6.915			0.790	0.775	
28	Ni	8.328	8.264	7.477	7.460			0.866	0.849	
29	Cu	8.976	8.904	8.047	8.027			0.943	0.928	
30	Zn	9.657	9.571	8.638	8.615			1.032	1.009	
31	Ga	10.365	10.263	9.251	9.234			1.122	1.096	
32	Ge	11.100	10.981	9.885	9.854			1.216	1.166	
33	As	11.863	11.725	10.543	10.507			1.517	1.282	
34	Se	12.651	12.495	11.221	11.181			1.419	1.379	
35	Br	13.465	13.290	11.923	11.877			1.526	1.480	
36	Kr	14.313	14.112	12.648	12.597			1.638	1.587	
37	Rb	15.184	14.960	13.394	13.335			1.752	1.694	1.692
38	Sr	16.083	15.834	14.164	14.097			1.872	1.806	1.805
39	Y	17.011	16.736	14.957	14.882			1.996	1.922	1.920
40	Zr	17.969	17.666	15.774	15.690	2.302	2.219	2.124	2.042	2.040
41	Nb	18.951	18.261	16.614	16.520	2.462	2.367	2.257	2.166	2.163
42	Mo	19.964	19.607	17.478	17.373	2.623	2.518	2.395	2.293	2.290
43	Tc	21.012	20.585	18.410	18.328	2.792	2.674	2.538	2.424	2.420
44	Ru	22.072	21.655	19.278	19.149	2.964	2.836	2.683	2.558	2.554
45	Rh	23.169	22.721	20.214	20.072	3.144	3.001	2.834	2.696	2.692
46	Pd	24.297	23.816	21.175	21.018	3.328	3.172	2.990	2.838	2.833
47	Ag	25.454	24.942	22.162	21.988	3.519	3.348	3.151	2.994	2.978
48	Cd	26.641	26.093	23.172	22.982	3.716	3.528	3.316	3.133	3.127
49	In	27.859	27.274	24.207	24.000	3.920	3.713	3.487	3.287	3.279
50	Sn	29.106	28.483	25.270	25.042	4.131	3.904	3.662	3.444	3.436
51	Sb	30.387	29.723	26.357	26.109	4.347	4.100	3.543	3.605	3.595
52	Te	31.698	30.993	27.471	27.200	4.570	4.301	4.029	3.769	3.758
53	I	33.016	32.292	28.610	28.315	4.800	4.507	4.220	3.937	3.926
54	Xe	34.446	33.644	29.802	29.485	5.036	4.720	4.422	4.111	4.098
55	Cs	35.819	34.984	30.970	30.623	5.280	4.936	4.620	4.286	4.272
56	Ba	37.255	35.376	32.191	31.815	5.531	5.156	4.828	4.467	4.451
57	La	38.728	37.799	33.440	33.033	5.789	5.384	5.043	4.651	4.635
58	Ce	40.231	39.255	34.717	34.276	6.052	5.613	5.262	4.840	4.823
59	Pr	41.772	40.746	36.023	35.548	6.322	5.850	5.489	5.034	5.014
60	Nd	43.298	42.269	37.359	36.845	6.602	6.090	5.722	5.230	5.208
61	Pm	44.955	43.945	38.649	38.160	6.891	6.336	5.956	5.431	5.408
62	Sm	46.553	45.400	40.124	39.523	7.180	6.587	6.206	5.636	5.609
63	Eu	48.241	47.027	41.529	40.877	7.478	6.842	6.456	5.846	5.816
64	Gd	49.961	48.718	42.983	42.280	7.788	7.102	6.714	6.039	6.027
65	Tb	51.737	50.391	44.470	43.737	8.104	7.368	6.979	6.275	6.241
66	Dy	53.491	52.178	45.985	45.193	8.418	7.638	7.249	6.495	6.457
67	Ho	55.292	53.934	47.528	46.686	8.748	7.912	7.528	6.720	6.680
68	Er	57.088	55.690	49.099	48.205	9.089	8.188	7.810	6.948	6.904
69	Tm	58.969	57.576	50.730	49.762	9.424	8.472	8.103	7.181	7.135
70	Yb	60.959	59.352	52.360	51.326	9.779	8.758	8.401	7.414	7.367
71	Lu	62.946	61.282	54.063	52.959	10.142	9.048	8.709	7.654	7.604
72	Hf	64.936	63.209	55.757	54.579	10.514	9.346	9.021	7.898	7.843
73	Ta	66.999	65.210	57.524	56.270	10.892	9.649	9.341	8.145	8.087
74	W	69.090	67.233	59.310	57.973	11.283	9.959	9.670	8.396	8.333

(continued)

Table 9.A.2 (Continued) Energies of Principal K and L X-Ray Emission Lines (keV)

Z	Element	$K_{\beta 2}$	$K_{\beta 1}$	$K_{\alpha 1}$	$K_{\alpha 2}$	$L_{\gamma 1}$	$L_{\beta 2}$	$L_{\beta 1}$	$L_{\alpha 1}$	$L_{\alpha 2}$
75	Re	71.220	69.298	61.131	59.707	11.684	10.273	10.008	8.651	8.584
76	Os	73.393	71.404	62.991	61.477	12.094	10.596	10.354	8.910	8.840
77	Ir	75.605	73.549	64.886	63.278	12.509	10.918	10.706	9.173	9.098
78	Pt	77.866	75.736	66.820	65.111	12.939	11.249	11.069	9.441	9.360
79	Au	80.165	77.968	68.794	66.980	13.379	11.582	11.439	9.711	9.625
80	Hg	82.526	80.258	70.821	68.894	13.828	11.923	11.823	9.987	9.896
81	Tl	84.904	82.558	72.860	70.320	14.288	12.268	12.210	10.266	10.170
82	Pb	87.343	94.922	74.957	72.794	14.762	12.620	12.611	10.549	10.448
83	Bi	89.833	87.335	77.097	74.805	15.244	12.977	13.021	10.836	10.729
84	Po	92.386	89.809	79.296	76.868	15.740	13.338	13.441	11.128	11.014
85	At	94.976	92.319	81.525	78.956	16.248	13.705	13.873	11.424	11.304
86	Rn	97.616	94.877	83.800	81.080	16.768	14.077	14.316	11.724	11.597
87	Fr	100.305	97.483	86.119	83.243	17.301	14.459	14.770	12.029	11.894
88	Ra	103.048	100.136	88.485	85.446	17.845	14.839	15.233	12.338	12.194
89	Ac	105.838	102.846	90.894	87.681	18.405	15.227	15.712	12.650	12.499
90	Th	108.671	105.592	93.334	89.942	18.977	15.620	16.200	12.966	12.808
91	Pa	111.575	108.408	95.851	92.271	19.559	16.022	16.700	13.291	13.120
92	U	114.549	111.289	98.428	94.648	20.163	16.425	17.218	13.613	12.438
93	Np	117.533	114.181	101.005	97.023	20.774	16.837	17.740	13.945	12.758
94	Pu	120.592	117.146	103.653	99.457	21.401	17.254	18.278	14.279	14.082
95	Am	123.706	120.163	106.351	101.932	22.042	17.667	18.829	14.618	14.411
96	Cm	126.875	123.235	109.098	104.448	22.699	18.106	19.393	14.961	14.743
97	Bk	130.101	126.362	111.896	107.023	23.370	18.540	19.971	15.309	15.079
98	Cf	133.383	129.544	114.745	109.603	24.056	18.980	20.562	15.661	15.420
99	Es	136.724	132.781	118.646	112.244	24.758	19.426	21.166	16.018	15.764
100	Fm	140.122	136.075	120.598	114.926	27.475	19.879	21.785	16.379	16.113

Note: The conversion equation between energy in keV and wavelength in Å is $E \text{ (keV)} = 12.4/\lambda \text{ (Å)}$.

Table 9.A.3 Correspondence between Old Siegbahn and New IUPAC Notation X-Ray Diagram Lines

Siegbahn	IUPAC	Siegbahn	IUPAC	Siegbahn	IUPAC
$K_{\alpha 1}$	$K-L_3$	$L_{\alpha 1}$	L_3-M_5	$L_{\gamma 1}$	L_2-N_4
$K_{\alpha 2}$	$K-L_2$	$L_{\alpha 2}$	L_3-M_4	$L_{\gamma 2}$	L_1-N_2
$K_{\beta 1}$	$K-M_3$	$L_{\beta 1}$	L_2-M_4	$L_{\gamma 3}$	L_1-N_3
$K_{\beta_2^I}$	$K-N_3$	$L_{\beta 2}$	L_3-N_5	$L_{\gamma 4}$	L_1-O_4
$K_{\beta_2^{II}}$	$K-N_2$	$L_{\beta 3}$	L_1-M_3	$L_{\gamma 4}'$	L_1-O_2
$K_{\beta 3}$	$K-M_2$	$L_{\beta 4}$	L_1-M_2	$L_{\gamma 5}$	L_2-N_1
$K_{\beta_4^I}$	$K-N_5$	$L_{\beta 5}$	$L_3-O_{4,5}$	$L_{\gamma 6}$	L_2-O_4
$K_{\beta_4^{II}}$	$K-N_4$	$L_{\beta 6}$	L_3-N_1	$L_{\gamma 8}$	L_2-O_1
$K_{\beta 4x}$	$K-N_4$	$L_{\beta 7}$	L_3-O_1	$L_{\gamma 8}'$	$L_2-N_{6(7)}$
$K_{\beta_5^I}$	$K-M_5$	$K_{\beta 7}$	$L_3-N_{6,7}$	L_η	L_2-M_1
$K_{\beta_5^{II}}$	$K-M_4$	$L_{\beta 9}$	L_1-M_5	L_l	L_3-M_1
		$L_{\beta 10}$	L_1-M_4	L_s	L_3-M_3
		$L_{\beta 15}$	L_3-N_4	L_t	L_3-M_2
		$L_{\beta 17}$	L_2-M_3	L_u	$L_3-N_{6,7}$
				L_v	$L_2-N_{6(7)}$

X-RAY SPECTROSCOPY

Table 9.A.3 (Continued)   Correspondence between Old Siegbahn and New IUPAC Notation X-Ray Diagram Lines

Siegbahn	IUPAC	Siegbahn	IUPAC	Siegbahn	IUPAC
				$M_{\alpha 1}$	M_5–N_7
				$M_{\alpha 2}$	M_5–N_6
				M_β	M_4–N_6
				M_γ	M_3–N_5
				M_ζ	$M_{4,5}$–$N_{2,3}$

9.B APPENDIX: ABSORPTION EDGE WAVELENGTHS AND ENERGIES

Table 9.B.1   Critical Absorption Wavelengths and Critical Absorption Energies

Atomic Number	Element	K Edge Å	K Edge keV	L_1 Edge Å	L_1 Edge keV	L_2 Edge Å	L_2 Edge keV	L_3 Edge Å	L_3 Edge keV	M_4 Edge Å	M_4 Edge keV	M_5 Edge Å	M_5 Edge keV
1	H	918	0.014										
2	He	504	0.025										
3	Li	226.953	0.055										
4	Be	106.9	0.116										
5	B	64.6	0.192										
6	C	43.767	0.283										
7	N	31.052	0.399										
8	O	23.367	0.531										
9	F	18.05	0.687										
10	Ne	14.19	0.874	258	0.048	564	0.022	564	0.022				
11	Na	11.48	1.08	225	0.055	365	0.034	365	0.034				
12	Mg	9.512	1.303	197	0.063	248	0.050	253	0.049				
13	Al	7.951	1.559	143	0.087	170	0.073	172	0.072				
14	Si	6.745	1.837	105	0.118	125	0.099	127	0.098				
15	P	5.787	2.142	81.0	0.153	96.1	0.129	96.9	0.128				
16	S	5.018	2.470	64.2	0.193	75.6	0.164	79.1	0.163				
17	Cl	4.397	2.819	52.1	0.238	61.1	0.203	61.4	0.202				
18	Ar	3.871	3.202	43.2	0.287	50.2	0.247	50.6	0.245				
19	K	3.437	3.606	36.4	0.341	41.8	0.297	42.2	0.294				
20	Ca	3.070	4.037	30.7	0.399	35.2	0.352	35.5	0.349				
21	Sc	2.757	4.495	26.8	0.462	30.2	0.411	30.8	0.402				
22	Ti	2.497	4.963	23.4	0.530	27.0	0.460	27.3	0.454				
23	V	2.269	5.462	20.5	0.604	23.9	0.519	24.2	0.512				
24	Cr	2.070	5.987	18.3	0.679	21.3	0.583	21.6	0.574				
25	Mn	1.896	6.535	16.3	0.762	19.1	0.650	19.4	0.639				
26	Fe	1.743	7.109	14.6	0.849	17.2	0.721	17.5	0.708				
27	Co	1.608	7.707	13.3	0.929	15.6	0.794	15.9	0.779				
28	Ni	1.488	8.329	12.22	1.015	14.2	0.871	14.5	0.853				
29	Cu	1.380	8.978	11.27	1.100	13.0	0.953	13.3	0.933				
30	Zn	1.283	9.657	10.33	1.200	11.87	1.045	12.13	1.022				
31	Ga	1.196	10.365	9.54	1.30	10.93	1.134	11.10	1.117				
32	Ge	1.117	11.100	8.73	1.42	9.94	1.248	10.19	1.217				
33	As	1.045	11.860	8.107	1.529	9.124	1.358	9.39	1.32				
34	Se	0.980	12.649	7.506	1.651	8.416	1.473	8.67	1.43				
35	Br	0.920	13.471	6.97	1.78	7.80	1.59	8.00	1.55				
36	Kr	0.866	14.319	6.46	1.92	7.21	1.72	7.43	1.67				

(continued)

Table 9.B.1 (Continued) Critical Absorption Wavelengths and Critical Absorption Energies

Atomic Number	Element	K Edge Å	K Edge keV	L₁ Edge Å	L₁ Edge keV	L₂ Edge Å	L₂ Edge keV	L₃ Edge Å	L₃ Edge keV	M₄ Edge Å	M₄ Edge keV	M₅ Edge Å	M₅ Edge keV
37	Rb	0.816	15.197	5.998	2.066	6.643	1.865	6.89	1.80				
38	Sr	0.770	16.101	5.583	2.220	6.172	2.008	6.387	1.940				
39	Y	0.728	17.032	5.232	2.369	5.755	2.153	5.962	2.079				
40	Zr	0.689	17.993	4.867	2.546	5.378	2.304	5.583	2.220				
41	Nb	0.653	18.981	4.581	2.705	5.026	2.467	5.223	2.373				
42	Mo	0.620	19.996	4.298	2.883	4.718	2.627	4.913	2.523				
43	Tc	0.589	21.045	4.060	3.054	4.436	2.795	4.632	2.677				
44	Ru	0.561	22.112	3.83	3.24	4.180	2.965	4.369	2.837				
45	Rh	0.534	23.217	3.626	3.418	3.942	3.144	4.130	3.001				
46	Pd	0.509	24.341	3.428	3.616	3.724	3.328	3.908	3.171				
47	Ag	0.486	25.509	3.254	3.809	3.514	3.527	3.698	3.351				
48	Cd	0.464	26.704	3.085	4.018	3.326	3.726	3.504	3.537				
49	In	0.444	27.920	2.926	4.236	3.147	3.938	3.324	3.728				
50	Sn	0.425	29.182	2.777	4.463	2.982	4.156	3.156	3.927				
51	Sb	0.407	30.477	2.639	4.695	2.830	4.380	3.000	4.131				
52	Te	0.390	31.800	2.511	4.937	2.687	4.611	2.855	4.340				
53	I	0.374	33.155	2.389	5.188	2.553	4.855	2.719	4.557				
54	Xe	0.359	34.570	2.274	5.451	2.429	5.102	2.592	4.780				
55	Cs	0.345	35.949	2.167	5.719	2.314	5.356	2.474	5.010				
56	Ba	0.331	37.399	2.068	5.994	2.204	5.622	2.363	5.245	15.56	0.7967	15.89	0.7801
57	La	0.318	38.920	1.973	6.282	2.103	5.893	2.258	5.488				
58	Ce	0.307	40.438	1.889	6.559	2.011	6.163	2.164	5.727				
59	Pr	0.295	41.986	1.811	6.844	1.924	6.441	2.077	5.967	13.122	0.9448	13.394	0.9257
60	Nd	0.285	43.559	1.735	7.142	1.843	6.725	1.995	6.213	12.459	0.9951	23.737	0.9734
61	Pm	0.274	45.207	1.665	7.448	1.767	7.018	1.918	6.466				
62	Sm	0.265	46.833	1.599	7.752	1.703	7.279	1.845	6.719	11.288	1.0983	11.552	1.0732
63	Eu	0.256	48.501	1.536	8.066	1.626	7.621	1.775	6.981	10.711	1.1575	11.013	1.1258
64	Gd	0.247	50.215	1.477	8.391	1.561	7.938	1.710	7.250				
65	Tb	0.238	51.984	1.421	8.722	1.501	8.256	1.649	7.517				
66	Dy	0.231	53.773	1.365	9.081	1.438	8.619	1.579	7.848				
67	Ho	0.223	55.599	1.317	9.408	1.390	8.918	1.535	8.072				
68	Er	0.216	57.465	1.268	9.773	1.338	9.260	1.482	8.361	8.601	1.4415	8.847	1.4013
69	Tm	0.209	59.319	1.222	10.141	1.288	9.626	1.433	8.650			8.487	1.4609
70	Yb	0.202	61.282	1.182	10.487	1.243	9.972	1.386	8.941				
71	Lu	0.196	63.281	1.140	10.870	1.199	10.341	1.341	8.239				
72	Hf	0.190	65.292	1.100	11.271	1.155	10.732	1.297	9.554				
73	Ta	0.184	67.379	1.061	11.681	1.114	11.128	1.255	9.874	6.87	1.804	7.11	1.743
74	W	0.178	69.479	1.025	12.097	1.075	11.533	1.216	10.196	6.59	1.880	6.83	1.814
75	Re	0.173	71.590	0.990	12.524	1.037	11.953	1.177	10.529	6.33	1.958	6.560	1.890
76	Os	0.168	73.856	0.956	12.968	1.001	12.380	1.140	10.867	6.073	2.042	6.30	1.967
77	Ir	0.163	76.096	0.923	13.427	0.967	12.817	1.106	11.209	5.83	2.126	6.05	2.048
78	Pt	0.158	78.352	0.893	13.875	0.934	13.266	1.072	11.556	5.59	2.217	5.81	2.133
79	Au	0.153	80.768	0.863	14.354	0.903	13.731	1.040	11.917	5.374	2.307	5.584	2.220
80	Hg	0.149	83.046	0.835	14.837	0.872	14.210	1.008	12.3	5.157	2.404	5.36	2.313
81	Tl	0.415	85.646	0.808	15.338	0.843	14.695	0.979	12.655	4.952	2.504	5.153	2.406
82	Pb	0.141	88.037	0.782	15.858	0.815	15.205	0.950	13.041	4.757	2.606	4.955	2.502
83	Bi	0.137	90.420	0.757	16.376	0.789	15.713	0.923	13.422	4.572	2.711	4.764	2.603
84	Po	0.133	93.112	0.732	16.935	0.763	16.244	0.897	13.817				
85	At	0.130	95.740	0.709	17.490	0.739	16.784	0.872	14.215				
86	Rn	0.126	98.418	0.687	18.058	0.715	17.337	0.848	14.618				
87	Fr	0.123	101.147	0.665	18.638	0.693	17.904	0.825	15.028				

Table 9.B.1 (Continued) Critical Absorption Wavelengths and Critical Absorption Energies

Atomic Number	Element	K Edge Å	K Edge keV	L₁ Edge Å	L₁ Edge keV	L₂ Edge Å	L₂ Edge keV	L₃ Edge Å	L₃ Edge keV	M₄ Edge Å	M₄ Edge keV	M₅ Edge Å	M₅ Edge keV
88	Ra			0.645	19.229	0.671	18.478	0.803	15.439				
89	Ac	0.116	106.759	0.625	19.842	0.650	19.078	0.782	15.865				
90	Th	0.113	109.741	0.606	20.458	0.630	19.677	0.761	16.293	3.557	3.485	3.729	3.325
91	Pa	0.110	112.581	0.588	21.102	0.611	20.311	0.741	16.731	3.436	3.608	3.618	3.436
92	U	0.108	115.610	0.56	21.764	0.592	20.938	0.722	17.160	3.333	3.720	3.497	3.545
93	Np	0.105	118.619	0.553	22.417	0.574	21.596	0.704	17.614				
94	Pu	0.102	121.720	0.537	23.097	0.557	22.262	0.686	18.066				
95	Am	0.099	124.816	0.521	23.793	0.540	22.944	0.669	18.525				
96	Cm	0.097	128.088	0.506	24.503	0.525	23.640	0.653	18.990				
97	Bk	0.094	131.357	0.491	25.230	0.509	24.352	0.637	19.461				
98	Cf	0.092	134.683	0.477	25.971	0.494	25.080	0.622	19.938				
99	Es	0.090	138.067	0.464	26.729	0.480	25.824	0.607	20.422				
100	Fm	0.088	141.510	0.451	27.503	0.466	26.584	0.593	20.912				

NOTES

1. Dedicated to the memory of Dr. Ron Jenkins and Dr. Eugene Bertin, outstanding X-ray spectroscopists and excellent teachers.
2. Dr. Seyfarth is Sr. Product Manager, HHXRF, Bruker Elemental, Kennewick, WA.

BIBLIOGRAPHY

The reader will note that occasionally multiple editions of books are cited as sources or reference works. The reason for this is that sometimes different figures are used in different editions, which may in fact be written by different authors. The sources found in the bibliography will be the most recent versions of a particular work.

In some cases instrument manufacturers and suppliers are cited by providing a URL to the exact internet location. We have endeavored to ensure that these links are functional at the time of publication, but clearly these links change. If a link has become nonfunctional, a good approach is to use your browser's advanced search capability. For example: "company name" AND "instrument type" AND "specific descriptors" is a good place to start.

ASTM E1621-13, Standard Guide for Elemental Analysis by Wavelength Dispersive X-Ray Fluorescence Spectrometry American Society for Testing and Materials. *Annual Book of ASTM Standards*. ASTM: West Conshohocken, PA, 2021.

Bertin, E.P. *Principles and Practice of X-Ray Spectrometric Analysis*, 2nd edn. Plenum Press: New York, 1975.

Bertin, E.P. *Introduction to X-Ray Spectrometric Analysis*. Plenum Press: New York, 1978.

Burr, A. In Robinson, J.W. (ed.), *Handbook of Spectroscopy*, Vol. 1. CRC Press: Boca Raton, FL, 1994.

Clark, G.L. (ed.) *Encyclopedia of X-Rays and Gamma Rays*. Reinhold Publishing: New York, 1963.

Criss, J.W.; Birks, L.S. *Anal. Chem.*, 40, 1080–1086, 1968.

Ellis, A.T. In Van Griekin, R.E.; Markowicz, A.A. (eds.), *Handbook of X-Ray Spectrometry*, 2nd edn. Marcel Dekker, Inc.: New York, 2002.

Formica, J. X-ray diffraction. In Settle, F. (ed.), *Handbook of Instrumental Techniques for Analytical Chemistry*. Prentice-Hall, Inc.: Upper Saddle River, NJ, 1997.

Gross, A. The characterization of historic pigments by μXRF spectrometry. Bruker Nano Lab Report XRF 422, www.bruker.com. 2010.

Hahn, O.; Oltrogge, D.; Bevers, H. *Archaeometry*, 2003.

Havrilla, G.J. X-ray fluorescence spectrometry. In Settle, F. (ed.), *Handbook of Instrumental Techniques for Analytical Chemistry*. Prentice-Hall, Inc.: Upper Saddle River, NJ, 1997.

Helson, L.A.; Kuczumov, A. In Van Griekin, R.E.; Markowicz, A.A. (eds.), *Handbook of X-Ray Spectrometry*, 2nd edn. Marcel Dekker, Inc.: New York, 2002.

Herglotz, H.K.; Birks, L.S. *X-Ray Spectrometry*. Marcel Dekker, Inc.: New York, 1978.

Jenkins, R. X-ray fluorescence. *Anal. Chem.*, 36, 1009A, 1984.

Jenkins, R. *X-Ray Fluorescence Spectrometry*, 2nd edn. John Wiley & Sons, Inc.: New York, 1999.

Jenkins, R.; Gould, R.W.; Gedcke, D. *Quantitative X-Ray Spectrometry*. Marcel Dekker, Inc.: New York, 1981.

Jenkins, R.; Manne, R.; Robin, J.; Senemaud, C. Nomenclature, symbols, units and their usage in spectrochemical analysis - VIII. Nomenclature system for X-ray spectroscopy (Recommendations 1991). *Pure Appl. Chem.*, 63(5), 785, 1991.

Jenkins, R.; Snyder, R.L. *Introduction to X-Ray Powder Diffractometry*. John Wiley & Sons, Inc.: New York, 1996.

Lubhofsky, H.A.; Schweikert, E.A.; Myers, E.A. *Treatise on Analytical Chemistry*, 2nd edn. Wiley Interscience: New York, 1986.

Markowicz, A.A. In Van Griekin, R.E., Markowicz, A.A. (eds.), *Handbook of X-Ray Spectrometry*, 2nd edn. Marcel Dekker, Inc.: New York, 2002.

Parsons, M.L. X-ray methods.

Robinson, J.W. *Handbook of Spectroscopy*, Vol. 1. CRC Press: Boca Raton, FL, 1974.

Robinson, J.W. *Practical Handbook of Spectroscopy*. CRC Press: Boca Raton, FL, 1991.

Stout, G.H.; Jensen, L.H. *X-Ray Structure Determination*. Macmillan Publishing Co., Inc.: New York, 1968.

Teo, B.K.; Joy, D.C. (eds.) *EXAFS Spectroscopy: Techniques and Applications*. Plenum Press: New York, 1981.

Van Griekin, R.E.; Markowicz, A.A. (eds.) *Handbook of X-Ray Spectrometry*, 2nd edn. Marcel Dekker, Inc.: New York, 2002.

Variankaval, N,. in Grinberg, N., Rodriguez, S., eds., Ewing's Analytical Instrumentation Handbook, 4th Ed., CRC Press Taylor and Francis, Boca Raton, FL, 2018.

CHAPTER 10

Mass Spectrometry I
Principles and Instrumentation

Mass spectrometry (MS) is a technique for creating gas-phase ions (or radical ions) from the molecules or atoms in a sample, separating the ions according to their mass-to-charge ratio, *m/z*, and measuring the abundance of the ions formed. MS is currently one of the most rapidly advancing fields of instrumental analysis. It has developed from an inorganic method used to prove that most elements exist as isotopes of differing masses to one of the cornerstone techniques used to elucidate the structure of biomolecules, especially high relative molecular mass (RMM) proteins. MS provides the analyst with information as diverse as the structure of complex organic and biomolecules to the quantitative determination of ppb concentrations of elements and molecules in samples.

MS is an analytical technique that provides qualitative and quantitative information, including the mass of molecules and atoms in samples as well as the molecular structure of organic and inorganic compounds. MS can be used as a qualitative analytical tool to identify and characterize different materials of interest to the chemist or biochemist. MS is used routinely for the quantitative analysis of mixtures of gases, liquids, and solids. MS is used in many fields in addition to identifying and quantifying organic and biological molecules. These include atomic physics, physical chemistry, and geology.

This chapter will focus on MS principles and instrumentation for both organic and inorganic analysis. Chapter 11 will cover applications of MS for organic and inorganic analyses as well as interpretation of simple mass spectra for structural identification of organic compounds.

10.1 PRINCIPLES OF MS

The mass spectrometer is an instrument that separates gas-phase ionized atoms, molecules, and fragments of molecules by the difference in their mass-to-charge ratios. The mass-to-charge ratio is symbolized by *m/z*, where the mass *m* is expressed in **unified atomic mass units** and *z* is the number of charges on the ion. The statement found in many texts that MS separates ions based on *mass* is only true when the charge on the ion is a single + or −; in this case, *m/z* is numerically equal to the mass of the ion. While some MS methods do generate mostly +1 charged ions, many new techniques generate ions with multiple charges.

The mass of an ion is given in **unified atomic mass units, u**. One unified atomic mass unit is equal to 1/12 of the mass of the most abundant, stable, naturally occurring isotope of carbon, ^{12}C. The mass of ^{12}C is defined as exactly 12 u. The abbreviation **amu**, for atomic mass unit, is now considered obsolete but may still be encountered in the literature. A synonym for the unified atomic mass unit is the dalton (Da); 1 u = 1 Da. In the Systeme International d'Unites (SI) unit of mass, 1 u = 1.665402 × 10^{-27} kg. Table 10.1 presents the exact masses for some common isotopes encountered in organic compounds.

The term *z* symbolizes the *number of charges* on the ion; this number may be positive or negative, such as +1, −1, +2, and +10. The number of charges is not the same as the total charge of the ion in coulombs (C). The total charge $q = ze$, where *e* is the magnitude of the charge on the electron, 1.6 × 10^{-19} C.

For simplicity, the following discussion will focus on what happens to an organic molecule in one type of MS experiment. One method of forming ions from sample molecules or atoms is to bombard the sample with electrons. This method is called **electron ionization** (EI, note that one will often encounter this process referred to as electron impact as well as electron ionization):

$$M + e^- \rightarrow M^{+\bullet} + 2e^- \quad (10.1)$$

where
M is the analyte molecule
e$^-$ is the electron
M$^{+\bullet}$ is the ionized analyte molecule; this species is called the **molecular ion**

Table 10.1 Comparison of Atomic Masses and Measured Accurate Isotope Masses for Some Common Elements in Organic Compounds*

Element	Atomic Mass	Isotope	Mass[a]	% Abundance
Hydrogen[b]	1.00794(7)	^{1}H	1.007825	99.99
		^{2}H (or D)	2.01410	0.01
Carbon[b]	12.0107(8)	^{12}C	12.000000 (defined)	98.91
		^{13}C	13.00336	1.1
Nitrogen[b]	14.0067(2)	^{14}N	14.0031	99.6
		^{15}N	15.0001	0.4
Oxygen[b]	15.9994(3)	^{16}O	15.9949	99.76
		^{17}O	16.9991	0.04
		^{18}O	17.9992	0.20
Fluorine	18.99840	F	18.99840	100
Phosphorus	30.97376	P	30.97376	100
Sulfur[b]	32.065(5)	^{32}S	31.9721	95.02
		^{33}S	32.9715	0.76
		^{34}S	33.9679	4.22
Chlorine[b]	35.453(2)	^{35}Cl	34.9689	75.77
		^{37}Cl	36.9659	24.23
Bromine	79.904(1)	^{79}Br	78.9183	50.5
		^{81}Br	80.9163	49.5

[a] Many of the isotope masses have been determined by MS accurately to seven or more decimal places. Numbers in parentheses in the atomic masses represent the first uncertain figure.
[b] The atomic masses for these elements and a few others are now given as ranges due to the variability of values in natural terrestrial materials. (Atomic masses are based on the 2009 values in *Pure Appl. Chem.*, 2011, 83, 359–396. With permission.) Newer tables now provide a range of isotope abundances of some elements due to variations in Earthly sources. Other celestial bodies (e.g., Moon, Mars, asteroidal meteorites) display even more different isotope variations, which enables MS measurements to identify the particular source of some meteorites.
* Additional nuclei can be found in the book by Bruno and Svoronos.

In many cases, only ions with a single positive charge are formed, that is, the number of charges on the ion, z, equals + 1. The mass spectrometer separates ions based on the mass-to-charge ratio, m/z; for ions with a single positive charge, m/z equals the mass of the ion in unified atomic mass units. For a molecular ion, m/z is related to the relative molecular mass of the compound. The symbol M$^{+\bullet}$ indicates that a molecular ion from an organic compound is a radical cation formed by the loss of one electron. In most cases, molecular ions have sufficient energy as a result of the ionization process to undergo *fragmentation* to form other ions of lower m/z. All these ions are separated in the mass spectrometer according to their m/z values, and the abundance of each is measured.

A **magnetic sector** mass spectrometer is shown schematically in Figure 10.1. This instrument is a *single-focus* mass spectrometer. The gas-phase molecules from the sample inlet are ionized by a beam of high-energy (i.e., high-velocity) electrons passing closely among them. The rapidly changing electric field produced by the passage of an energetic electron can both eject electrons from the atom or molecule (ionization) and transfer sufficient energy to the molecule to cause its bonds to rupture (fragmentation). Only a very small percentage (0.1 %–0.001 %) of the molecules is ionized. A permanent magnet is often used to produce a magnetic field, which causes the beam electrons to follow a spiral trajectory of longer path length as they transit the ion source, thus increasing the ionization efficiency and improving sensitivity. However, too many ions in the ion source, especially when present from a high excess of unseparated matrix components, will add to and distort the instrument's operating fields from this **space charge**. The ions are then accelerated in an electric field at a voltage V. If each ion has a single positive charge, the magnitude of the charge is equal to +1. The energy of each ion is equal to the charge z times the accelerating voltage, zV. The energy acquired by an ion on acceleration is kinetic energy. It is important to note that the kinetic energy of an ion accelerated through a voltage V depends only on the charge of the ion and the voltage, not on the mass of the ion. The translational component of the kinetic energy is equal to $1/2mv^2$. The kinetic energy of all singly charged ions is the same for a given accelerating voltage; therefore, those ions with small masses must travel at higher velocity than ions with larger masses. That is, for ions with a single positive charge,

$$\frac{1}{2}mv^2 = zV \tag{10.2}$$

hence,

$$v = \left(\frac{2zV}{m}\right)^{\frac{1}{2}} \tag{10.3}$$

where
m is the mass of the ion
v is the velocity of the ion
z is the charge of the ion
V is the accelerating voltage

As m varies, the velocity v changes such that $1/2mv^2$ remains a constant. This relation can be expressed for two different ions as follows:

$$zV = \frac{1}{2}m_1v_1^2 \text{ (ion 1)}$$
$$zV = \frac{1}{2}m_2v_2^2 \text{ (ion 2)} \tag{10.4}$$
$$\frac{1}{2}m_1v_1^2 = \frac{1}{2}m_2v_2^2$$

where
m_1 is the mass of ion 1
v_1 is the velocity of ion 1
m_2 is the mass of ion 2
v_2 is the velocity of ion 2

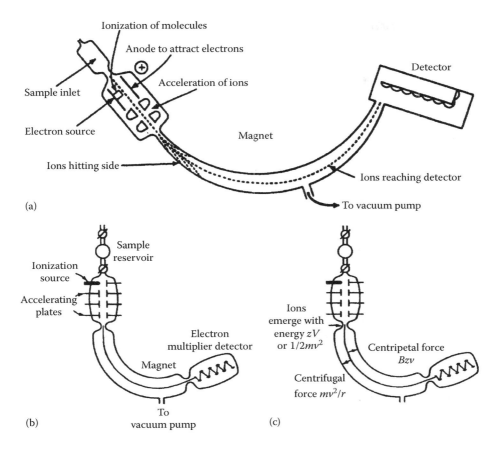

Figure 10.1 (a) Schematic magnetic sector mass spectrometer. (b) Schematic magnetic sector mass spectrometer showing a gas sample reservoir with fill and inlet valves and an electron multiplier (EM) detector. (c) Forces and energies associated with ions traveling through a magnetic sector mass spectrometer. For an ion to reach the detector, Bzv must equal mv^2/r.

The velocity of an ion depends on its mass; the velocity is inversely proportional to the square root of the mass of the ion.

After an applied voltage has accelerated the charged ions, they enter a curved section of a magnet of homogeneous magnetic field B and a fixed radius of curvature. This magnetic field acts on the ions, making them move in a circle. The attractive force of the magnet equals Bzv. The centrifugal force on the ion is equal to mv^2/r, where r is the radius of the circular path traveled by the ion. If the ion is to follow a path with the radius of curvature of the magnet, these two forces must be equal:

$$\frac{mv^2}{r} = Bzv \qquad (10.5)$$

or

$$\frac{1}{r} = \frac{Bzv}{mv^2}$$

and

$$r = \frac{mv}{zB} \qquad (10.6)$$

Substituting for v (Equation 10.3), we get

$$r = \frac{m}{zB}\left(\frac{2zV}{m}\right)^{1/2}$$

Squaring both sides, we have

$$r^2 = \frac{m^2}{z^2 B^2}\frac{2zV}{m} = \frac{m2V}{zB^2} \qquad (10.7)$$

which can be rearranged to give

$$\frac{m}{z} = \frac{B^2 r^2}{2V} \qquad (10.8)$$

That is, the radius of the circular path of an ion depends on the accelerating voltage V, the magnetic field B, and the ratio m/z. When V and B are kept constant, the radius of the circular path depends on the m/z value of the ionized molecule. Ions of different m/z travel in circles with different radii; this is the basis of the separation by m/z. Ions with different paths are shown as the dotted lines in Figure 10.1a; only one particular m/z has the right r to pass through the mass spectrometer under a given V and B. The others, as shown, follow paths that cause them to hit the sides of the instrument and be lost. By varying V or B, we can select which m/z will pass through the system. Voltage scanning is cheaper and was used in early instruments with permanent magnets. Scanning the magnetic field strength is more sensitive and is easily done with modern electromagnets. (In reality, a curved magnet with a path of significant width could accommodate a range of different m/z values, and it is the width and position of the exit slit that determine precisely which ions finally get through to the detector in a real scanning magnetic sector mass spectrometer.)

In all modern magnetic sector mass spectrometers, the applied magnetic field B of the electromagnet is varied, while the accelerating voltage V is held constant. The radius of curvature of the magnetic sector of a given instrument is a constant, so only ions with a trajectory of radius r pass through. For a particular magnetic field strength, then, only ions with an m/z value that satisfies Equation 10.8 will exit the chamber. Consequently, for different values of B, ions with different mass-to-charge ratios will pass through the instrument to the detector, and by varying B, we can scan the range of mass-to-charge ratios of the sample ions and measure their abundance with a detector that either counts and sums individual ions or measures a current produced by their impact on the detector. A plot of abundance versus m/z is called a **mass spectrum**. The mass spectral data can be shown as a histogram plot or as a table. A digitized mass spectrum of benzene, C_6H_6, is shown in Figure 10.2a; the plot is shown as a histogram. The most abundant peak in the spectrum, called the **base peak**, is scaled to 100, so the y-axis represents the **relative abundance** of the ions of each m/z value. The nominal mass of the benzene molecular ion is equal to the sum of the nominal masses of the C and H isotopes, so $6(12) + 6(1) = 78$; the most abundant peak in this benzene mass spectrum is the molecular ion peak at m/z 78. A number of fragment ions at lower values of m/z are also seen in this spectrum. Table 10.2 presents the same mass spectral data for benzene as shown in the plot. Figure 10.2b is a mass spectrum of cocaine, which exhibits many fragment ions. This example shows that the molecular ion is not always the most abundant ion. The base peak in the cocaine mass spectrum is the fragment ion at m/z 82. The m/z values and the fragmentation pattern enable the analyst to determine the RMM and structure of organic compounds by MS. Detailed interpretation of simple organic mass spectra will be covered in Chapter 11.

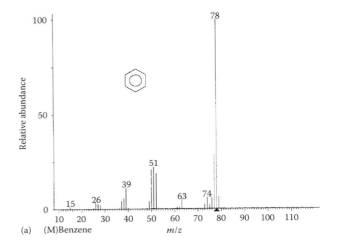

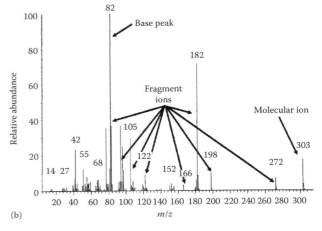

Figure 10.2 (a) A mass spectrum of benzene, C_6H_6 (RMM = 78), plotted as relative abundance versus m/z, with the most abundant peak set to a relative abundance of 100. In this spectrum, the most abundant peak (or base peak) is the molecular ion peak at m/z 78, marked by the dark triangle on the x-axis. Some fragment ion m/z values are also marked on the spectrum. (b) A mass spectrum of cocaine. In this spectrum, the molecular ion peak is not the base peak; the fragment ion at m/z 82 is the most abundant ion. (The cocaine spectrum is courtesy of Dr. Robert Kobelski, CDC, Atlanta, GA.)

Table 10.2 Mass Spectral Data for Benzene

m/z	Relative Abundance	m/z	Relative Abundance
37	4	53	0.80
37.5	1.2	63	2.9
38	5.4	64	0.17
38.5	0.35	73	1.5
39	13	74	4.3
39.5	0.19	75	1.7
40	0.37	76	6.0
48	0.29	77	14
49	2.7	78	100
50	16	79	6.4
51	18	80	0.18
52	19		

There is a difference between the mass of an atom (or ion) and what chemists are used to thinking of as the atomic mass or RMM of a species. Atomic mass is the average mass of all the isotopes of an element and RMM is calculated from these average atomic masses. This is the value we would need to use if we wished to weigh out an exact number of moles of a substance whose component atoms have a distribution of different isotopes, which is usually the case. Each stable isotope (nuclide) of an element has an exact mass (more properly called a measured accurate mass), which has been determined accurately by MS. MS separates compounds containing ^{12}C atoms from compounds with ^{13}C atoms precisely because of the difference in mass between the isotopes. Table 10.1 gives a few examples of the atomic masses of some of the elements found in organic compounds as well as the measured accurate mass and abundance of the isotopes of the element. In a mass spectrum, a given ion is monoisotopic, not a weighted average. For example, the RMM of acetone, C_3H_6O, is calculated from the atomic masses of the elements to be $(3 \times 12.011) + (6 \times 1.00797) + (1 \times 15.9994) = 58.0795$ g/mol. The molecular ion of acetone as measured by MS has a mass that consists of only contributions from the one most abundant isotope of each element, that is, ^1H, ^{12}C, and ^{16}O; its mass can be calculated from the formula $^{12}C_3^1H_6^{16}O$ to be $(3 \times 12.000) + (6 \times 1.00783) + (1 \times 15.9949) = 58.0419$ u or 58.0419 g/mol if we had a mole of the monoisotopic compound.

The term m/z is the correct term to use for the mass-to-charge ratio of an ion. Older literature used the term m/e; however, the symbol e is used for the charge on the electron in coulombs and is *not* what goes into the divisor when the mass is given in unified atomic mass units, u. Two terms used in the older literature that are no longer acceptable for use in MS are "parent ion" for molecular ion and "daughter ion" for fragment ions. Ions do not have gender. The terms molecular ion and fragment ion should be used; "precursor ion" and "product ion" are used for tandem MS/MS experiments described later in this chapter.

10.1.1 Resolving Power and Resolution of a Mass Spectrometer

The **resolving power** of a mass spectrometer is defined as its ability to separate ions of two different m/z values. Numerically, the resolution is equal to the mass of one singly charged ion, M, divided by the difference in mass between M and the next m/z value that can be separated. For example, to separate two singly charged ions of 999 and 1001 Da requires a resolving power of $999/(1001 - 999) = 500$. That is,

$$\text{Resolving power} = \frac{M}{\Delta M} \qquad (10.9)$$

In practice, it is found that if we wish to distinguish between ions of 600 and 599 Da, the resolving power required is 600, or 1 Da in 600 Da. As a rule of thumb, if we wish to distinguish between ions differing by 1 Da in the 600 Da mass range, we need a resolving power of 600. If we need to distinguish between ions differing by 1 Da in the 1200 Da range, we need a resolving power of at least 1200. This is not a very high resolution. Some isotopes of different elements or molecular fragment ions composed of different combinations of atomic isotopes adding up to close to the same mass may produce ions differing by *much less* than 1 Da. To distinguish between these types of species, high-resolution MS, with resolution in the range of 20,000–100,000 or higher, is required.

Both the magnetic sector and other designs of MS instruments control ion movement and trajectories by controlling the voltage fields from charge (electrons) on conductive metal components, often described as *lenses*, which focus and concentrate ions, or *extractors*, which move them between stages, or multipole rods with applied radio-frequency (RF) alternating voltages and polarities, which induce controlled resonant spiral orbits to select ions of a given m/z, etc. All these components must be interactively set to obtain the optimum MS ion peak shapes, ideally a Gaussian profile with the top of the symmetrical peak at a location on the mass axis corresponding to the actual m/z value. Early magnetic sector or quadrupole MS instruments used a great array of knobs and an oscilloscope readout to manually "tune" the instrument to this optimum (analogous to the much simpler process of tuning a radio to a particular station). As these components age or become contaminated, the voltages on the surfaces will shift, the MS peak shapes will deteriorate, and the instrument will require retuning. Depending on usage conditions, this might be required at the beginning of each day's runs. Most modern computerized instruments employ "autotune" algorithms, whereby the computer observes selected peaks from an autotune compound vapor (often perfluorotributylamine [PFTBA]) continuously infused into the ion source and iteratively adjusts all these interlocking voltages, checking selected MS peak shapes and positions, until an optimum set of tuning conditions is achieved. When optimal tuning is required at specific RMM ranges, one can manually tune on other peaks of the PFTBA mass spectrum. A complete listing of the mass spectrum of PFTBA can be found in the book by Bruno and Svoronos. In addition, other compounds used for tuning are provided in this source.

The resolving power is determined by actual measurement of the mass spectral peaks obtained. The method for calculating ΔM must also be specified. Two methods are commonly used to indicate the separation between peaks, and these are shown in Figure 10.3. One definition is that the overlap between the peaks is 10 % or less for two peaks of equal height; that is, the height of the overlap should not

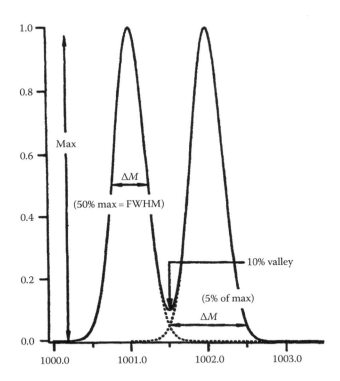

Figure 10.3 Illustration of the peaks used to calculate the resolving power of a mass spectrometer showing the location of the FWHM, the 10 % valley and the 5 % valley. (From Sparkman, O.D, *Mass Spectrometry Desk Reference*, Global View Publishing, Pittsburgh, PA, 2000. Used with permission.)

be more than 10 % of the peak height. A second method is to use the full width at half maximum (FWHM) as a measure of ΔM. The FWHM method results in a resolving power twice that of the 10 % overlap method, so it is important to state how the calculation was performed.

Resolving power for commercial mass spectrometers depends on the instrument design and can range from 500 to more than 1×10^6. In general, the higher the resolving power, the greater the complexity and cost of the MS instrument.

Resolution is the value of ΔM at a given M and is often expressed in ppm. For the spectrometer with a resolving power of 600 earlier described, the resolution would be 1 part in 600 parts. To distinguish between $^{14}N_2^+$ with an exact mass of 28.0061 Da and $^{12}C^{16}O^+$ with an exact mass of 27.9949 Da, we would need a resolution of (28.0061 − 27.9949) = 0.0112 Da in 28 Da. To convert this resolution to ppm, divide 0.0112 by 28 and multiply by 1×10^6; a resolution of 400 ppm is required. The resolving power needed would be 27.9949/0.00112 = 2500. **Mass accuracy** describes how well the top of the MS peak can be defined and kept stable and its position identified on the mass axis. Mass axis calibration is usually established by a separate portion of the tuning algorithm, which identifies the centroids of several mass peaks of precisely known mass in the spectrum of a suitable calibration compound and adjusts tuning parameters to place these at the appropriate positions on the mass axis. Like resolution, it can be expressed as a percentage or ppm value of the calibration mass values.

10.2 INSTRUMENTATION

All mass spectrometers require a sample input system, an ionization source, a mass analyzer, and a detector. All of the components with the exception of some sample input systems or ion source volumes are under vacuum (10^{-6}–10^{-8} torr for that portion where ions are separated by mass, i.e., the **analyzer**, or 10^{-4}–10^{-5} torr in some **ion sources**, where the ions are initially formed), so vacuum pumps of various types are required. Other ion sources, such as the direct analysis in real time (DART) (discussed in Section 10.2.2.3), electrospray ionization (ESI) (Sections 10.2.2.3 and 14.1.6.1), or chemical ionization (CI) (Section 10.2.2.2), operate at atmospheric pressure and use extraction lenses set to a polarity opposite that of the ions to draw them into subsequent stages of the MS instrument. Modern mass spectrometers have all of the components under computer control, with a computer-based data acquisition and processing system. A block diagram of a typical mass spectrometer is shown in Figure 10.4.

10.2.1 Sample Input Systems

10.2.1.1 Gas Expansion

This method of sample introduction is useful for gases and for liquids with sufficiently high vapor pressures. The gas or vapor is allowed to expand into an evacuated, heated vessel. The sample is then "leaked" into the ionization source through pin holes in a gold foil seal. This is sometimes termed a **molecular leak inlet**. Vacuum pumps control the system so that the pressure in the ionization source is at the required 10^{-6}–10^{-8} torr.

A Brief Digression on Units of Measure—Vacuum Systems

In this book, we express vacuum in units of torr (named after the Italian physicist Evangelista Torricelli). This unit is not part of the SI system, nevertheless it is still widely used in discussion of mass spectrometry and in the presentation of measurements. Understanding this unit is critical to the interpretation of the literature. One torr is approximately one millimeter of mercury in a mercury manometer operated at 0 °C. It is converted to the SI unit, the Pascal, using the factors below:

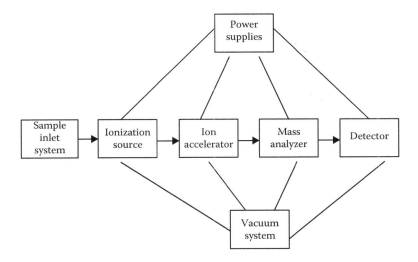

Figure 10.4 Block diagram of a mass spectrometer.

torr	Pa	mbar	micron	psi	mmHg (0 °C)
1	133.322	1.33322	1000	0.0193368	0.99999984

The unit micron is commonly used in the vacuum industry; it is derived from the unit torr where one milliTorr is equal to one micron. A complete list of conversion factors can be found in the book by Bruno and Svoronos.

10.2.1.2 Direct Insertion and Direct Exposure Probes

A **direct insertion probe** is used for introduction of liquids with high boiling points and solids with sufficiently high vapor pressure. The sample is put into a glass capillary that fits into the tip of the probe shown in Figure 10.5. The probe is inserted into the ionization source of the mass spectrometer and is heated electrically, vaporizing sample into the electron beam where ionization occurs. Direct insertion probes can be retrofitted onto existing mass spectrometers that are made for other sample introduction methods. A problem with this type of sample introduction is that the mass spectrometer can be contaminated because of the relatively large sample volume introduced in a small quartz crucible that is heated directly.

A **direct exposure probe** usually has a rounded glass tip. The sample is dissolved in solvent, a drop of the solution is placed on the end of the probe, and the solvent is allowed to evaporate. A thin film of sample is left on the glass tip. The tip is inserted into the ion source and heated in the same manner as the direct insertion probe. Much less sample is introduced into the ion source and the spectrometer is less likely to be contaminated as a result.

10.2.1.3 Chromatography and Electrophoresis Systems

The appropriate chromatographic instrument can separate mixtures of gases and liquids, and the separated components are then introduced sequentially into a mass spectrometer for detection. Mass spectrometrists consider the chromatograph to be a "sample inlet" for their spectrometers, while chromatographers consider the mass spectrometer to be "a detector" for their chromatographs. The truth is that these hyphenated (or coupled) chromatography-MS systems are extremely powerful analytical techniques, much more powerful than either instrument alone. The major problem to be overcome in coupling these two types of instruments is that the chromatography systems run at atmospheric pressure with large amounts of "carrier gas" or solvent. The mass spectrometer operates at very low pressures as noted earlier. The carrier gas or solvent must be removed without losing the analyte before the analyte can be introduced into the evacuated ionization source or analyzer regions. Alternatively, one may use an ionization source that can ionize the target analytes under the conditions of higher pressure produced by the accompanying carrier fluid and then selectively extract the ions into a lower-pressure region while diverting and discarding the vast majority of that vaporized fluid. Interfaces for these systems have been developed and are described in the chapters on the individual chromatographic techniques.

Gas chromatography (GC) coupled with MS is known as GC-MS. It is a well-established method for separating gases or volatile compounds; the mass spectrum of each component in a mixture can be obtained and the components measured quantitatively. The interfacing, operation, and applications of GC-MS are discussed in Chapters 12 and 13.

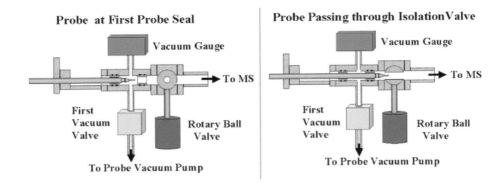

Figure 10.5 Schematic diagram of a direct insertion probe for solids and high boiling liquids. The sample is placed into the cavity (or in a quartz insert called a crucible) in the tip of the probe. When not in use, the cutoff valve is closed, Used with permission of Adaptas Scientific Instrument Services.)

Several types of liquid chromatography (LC) and one nonchromatographic separation system for liquids have been interfaced with MS. High-performance LC (HPLC) is widely used to separate nonvolatile organic compounds of all polarities and RMMs. Coupled to a mass spectrometer, the technique is called LC-MS. Supercritical fluid chromatography (SFC) and the nonchromatographic separation technique of capillary electrophoresis (CE) are also used with mass spectrometric detection. The interfacing, ionization sources, operation, and applications of these hyphenated methods are covered in Chapters 12 and 13.

10.2.1.4 Chromatography and Infrared Spectrometry Systems Coupled with Mass Spectrometry

A multiple hyphenated (or tandem) technique that is worth mentioning is the combination of in-line FT-IR spectrophotometry before a typical quadrupolar mass spectrometer. The desire to couple IR spectrophotometry ahead of mass spectrometry arose from the limited capability of MS to distinguish between structural isomers and provide direct functional group information. This was usually implemented as a high-resolution gas chromatograph in which the column effluent is first passed into a "light pipe," which is a quartz capillary with IR optical materials (typically salt plates) at either end. Two approaches were used to achieve this: a post column split between the two detectors or a true piggyback in which the MS follows the IR measurement. If a split is used, typically 90–99 % of the flow is directed into the much less sensitive IR section, with the rest being sent to the more sensitive MS analyzer. The piggyback has been the more common approach; it allows one to obtain infrared spectra of the separated compounds before they are sent into the (usually) EI source of the mass spectrometer. The insertion of the column flow into the light pipe results in the loss of some chromatographic resolution, but with careful control of the chromatographic conditions, one can usually gain additional information about the sample that the mass spectrum cannot alone provide. The application of GC-ITR-MS had been most prominent in forensic analysis of drug samples, especially heroin tars. It has also been used in fuels analysis, especially coal-derived liquid fuels with large numbers of naphthalenic compounds. GC-IR-MS systems are costly and delicate, with a high maintenance overhead.

10.2.2 Ionization Sources

10.2.2.1 Electron Ionization

The EI source is a commonly used source for organic MS. Electrons are emitted from a heated tungsten or other metal filament. The electrons are accelerated by a potential of 50–100 V toward the anode (Figure 10.6). A standard value of 70 V is often used to produce comparable fragmentation patterns. Modern EI sources have a small permanent magnet that produces a field aligned with the electron beam. This causes the electrons to follow a spiral path, increasing the number of target atoms or molecules encountered. Still, only 1 in 10^5 or 10^6 of them are ionized. As shown in this figure, the paths of the electrons and sample molecules meet at right angles. Ionization of the sample molecules and fragmentation into smaller ions occur as a result of interaction with the high-energy electrons. Ions are accelerated out of the center of the source into the mass analyzer by an accelerating voltage of about 10^4 V.

The EI source forms both positive and negative ions, so it can be used as a source for negative ion MS. Negative ions form from molecules containing acid groups or electronegative atoms. The high energy imparted to the ions by the EI source causes significant fragmentation of organic molecules. This type of high-energy ionization source is referred to as a *hard* ionization source. The fragmentation of the molecular ion into smaller ions is very useful in

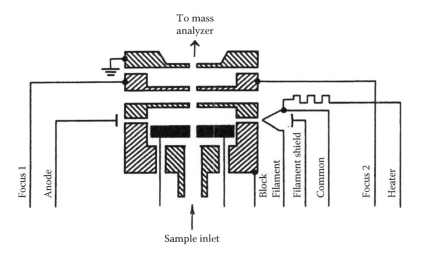

Figure 10.6 Cross section of an EI source. The filament and anode define the electron beam. The ions are formed in the space above the two repellers (the solid color blocks). A positive charge on the repellers together with a negative potential on the focus electrodes cause positive ions to be accelerated upward in the diagram, into the mass analyzer. (Modified from Ewing, G.W., Mass spectrometry, in Ewing, G.W. [ed.], *Analytical Instrumentation Handbook*, 2nd edn., Marcel Dekker, Inc., New York, 1997. Used with permission.)

deducing the structure of a molecule. However, EI fragmentation can be so significant for some types of molecules that either no molecular ions remain or they are so few that they cannot be reliably distinguished from the ions from background contamination. This means that the molecular mass of the compound cannot be determined from the spectrum. This is a critical loss of information, as deduction of the structure of an unknown compound is greatly facilitated by knowing its molecular mass, and this provides the starting point for assigning the mass losses to fragment ions formed.

Collisions between ions and molecules in the source can result in the formation of ions with higher m/z values than the molecular ion. A common ion–molecule reaction is that between a proton, H^+, and an analyte molecule, M, to give a protonated molecule, MH^+ or $(M + H)^+$. Such a species has a +1 charge and a mass that is 1 u greater than that of the molecule and is called a proton **adduct**, often represented as (M + 1). In LC-MS experiments, sodium salts from the LC buffer solution often produce a mass (M + 23) Na-adduct ion. One reason for keeping the sample pressure low in the EI ionization source is to prevent reactions between ions and molecules that would complicate interpretation of the mass spectrum.

The EI source used to be called the electron impact source and the term EI meant electron impact ionization. The use of the term *impact* is now considered archaic. Ionization and fragmentation are more often induced by the close passage of the energetic electron and the consequent large fluctuation in the electric field as opposed to a physical direct "impact," and an energetic electron may well perform this function multiple times on different molecules. As the student will note throughout this chapter, MS terminology has changed in recent years as a result of agreements by professional scientific organizations to standardize definitions and usage of terms to avoid confusion. Not all organizations have agreed to the same terms and definitions. The recommendations from Sparkman (see the bibliography) have been followed, but even current literature will be found that uses "archaic" terminology. The old or alternate terms will be provided when necessary.

10.2.2.2 Chemical Ionization

A CI source is considered a *soft* ionization source; it results in less fragmentation of analyte molecules and a simpler mass spectrum than that resulting from EI. Most importantly, the molecular ion is much more abundant using CI, allowing the determination of the RMM. Since proton adduct ions $(M + 1)^+$ form more easily in CI-MS, one needs to exercise care in assigning the molecular ion mass (M^+). If the CI process is "soft" enough, the spectrum may consist almost entirely of only the molecular ion or an adduct ion. Such a lack of fragmentation provides less structural information than an EI spectrum. Concentration of the charge on mostly this one ion in CI-MS improves the sensitivity of detection when the method of selected ion monitoring (SIM) is employed for quantitative measurements in GC-MS (Section 13.8.1). If a fragmented EI spectrum is missing a molecular ion, then combining data from a CI spectrum containing a strong molecular ion will greatly assist interpretation of an unknown compound's spectra. The two modes complement one another for identification and quantitation of unknowns.

In CI, a large excess of **reagent gas** such as methane, isobutane, or ammonia is introduced into the ionization region. The pressure in the ion source is typically several orders of magnitude higher than in EI ion sources. A CI source design will be more enclosed with smaller orifices for the source vacuum pump to remove the reagent gas, allowing the higher source pressure to be maintained. The mixture of reagent gas and sample is subjected to electron bombardment. The reagent gas is generally present at a level 1000–10,000× higher than the sample; therefore, the reagent gas is most likely to be ionized by interaction with the electrons. Ionization of the sample molecules occurs indirectly by collision with ionized reagent gas molecules and proton or hydride transfer. A series of reactions occurs. Methane, for example, forms CH_4^+ and CH_3^+ on interaction with the electron beam. These ions then react with additional methane molecules to form ions as shown:

$$CH_4^+ + CH_4 \rightarrow CH_5^+ + CH_3^{\cdot}$$

$$CH_3^+ + CH_4 \rightarrow C_2H_5^+ + H_2$$

Collisions between the ionic species CH_5^+ or $C_2H_5^+$ and a sample molecule M cause ionization of the sample molecule by proton transfer from the ionized reagent gas species to form MH⁺ or by hydride (H⁻) transfer from the sample molecule to form (M − H)⁺, also written as (M − 1)⁺:

$$M + CH_5^+ \rightarrow MH^+ + CH_4$$

$$M + C_2H_5^+ \rightarrow MH^+ + C_2H_4$$

$$M + C_2H_5^+ \rightarrow (M-H)^+ + C_2H_6$$

$$M + CH_5^+ \rightarrow (M-H)^+ + CH_4 + H_2$$

Hydride transfer from M occurs mainly when the analyte molecule is a saturated hydrocarbon. In addition, the ionized reagent gas can react with M to form, for example, an (M + C_2H_5)⁺ ion with m/z = (M + 29). The presence of such an *adduct* ion of mass 29 Da above a candidate molecular ion in a methane CI mass spectrum is a good confirmation of the identity of the molecular ion. A comprehensive listing of reagent gases, for positive as well as negative ions, along with their common mass spectral peaks, is provided in the book by Bruno and Svoronos. This compilation also lists the proton affinity of the reagent gases (note that the older literature will use units of kcal/mol instead of the SI preference of kJ/mol), as well as the proton affinity of common simple molecules.

Many commercial sources are designed to switch from EI to CI rapidly to take advantage of the complementary information obtained from each technique. The main advantage of CI is that fragmentation of the sample molecule is greatly reduced and significant peaks at m/z = (M + 1) or (M − 1) are seen, permitting the identification of the RMM of the analyte.

It is possible to introduce a sample directly into the CI source on a tungsten or rhenium wire. A drop of sample in solution is applied to the wire, the solvent is allowed to evaporate, and the sample inserted into the CI source. The sample molecules are desorbed by passing a current through the wire, causing it to heat. The analyte molecules then ionize by interaction with the reagent gas ions as has been described. This technique is called **desorption CI** and is used for nonvolatile compounds.

10.2.2.3 Atmospheric Pressure Ionization Sources

There are two major types of ionization sources that operate at atmospheric pressure, ESI, and atmospheric pressure CI (APCI). A modified version of the ESI source is the ion spray source. These sources are described in detail in Section 14.1.6.1, because they are used to interface LC with MS for the separation and mass spectrometric analysis of mixtures of nonvolatile high RMM compounds, especially in the fields of pharmaceutical chemistry, biochemistry, and clinical biomonitoring. ESI will be described briefly so that its use may be demonstrated, but more detail will be found in Chapter 14.

When a strong electric field is applied to a liquid passing through a metal capillary, the liquid becomes dispersed into a fine spray of positively or negatively charged droplets, an electrospray. The electric field is created by applying a potential difference of 3–8 kV between the metal capillary and a counter electrode. The highly charged droplets shrink as the solvent evaporates until the droplets undergo a series of "explosions" due to increasing coulombic repulsion of the electrons as their droplet surface density increases. Each "explosion" forms smaller and smaller droplets. When the droplets become small enough, the analyte ions desorb from the droplets and enter the mass analyzer. A schematic *ESI source* is shown in Figure 10.7. The ESI source is at atmospheric pressure. The droplets and finally the analyte ions pass through a series of orifices and skimmers. These serve to divert and exclude unevaporated droplets and excess vaporized solvent from the higher vacuum regions where analyte ions are accelerated and analyzed by m/z. A flow of gas such as nitrogen or argon serves to desolvate the droplets and to break up ion clusters. The skimmers act as velocity filters. Heavier ions have less velocity from random thermal motions transverse to the direction of voltage acceleration through the orifices than lighter ions and continue in a straight path to the mass analyzer while the lighter ions (and solvent vapors and gases) are pumped away, permitting the

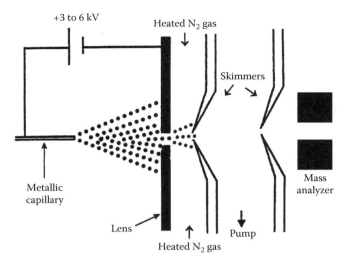

Figure 10.7 Schematic diagram of an ESI source. This source is shown with a lens system and skimmers to focus the ions and heated nitrogen gas to desolvate the ions.

pressure to be reduced without affecting the ion input to the mass analyzer. Liquid flow through the metal capillary is in the range of 1–10 μL/min for the standard ESI design. For the increasingly important HPLC-MS instrumentation used in analysis of biomolecules, orthogonal spray ESI interfaces operate at 1 mL/min, and by addition of jets of heated gas (e.g., N_2) to increase droplet evaporation rates, they can handle flow of up to 4–8 mL/min from monolithic HPLC columns.

The advantage of ESI lies in the fact that large molecules, especially biomolecules like proteins, end up as a series of multiply charged ions, M^{n+} or $(M + nH)^{n+}$ with little or no fragmentation. For example, a given analyte M might form ions of M^{9+}, M^{10+}, M^{11+}, and so on. If the mass of the analyte M is 14,300 Da, then peaks would appear in the mass spectrum of this analyte at m/z values of (14,300/9) = 1588.9, (14,300/10) = 1430.0, and (14,300/11) = 1300.0. These ions are at much lower m/z values than would be the case if we had a singly charged M^+ ion at m/z 14,300. One advantage to having multiply charged ions with low m/z values is that less-expensive mass analyzers with limited mass range can be used to separate them. Another is that high m/z ions such as high RMM biomolecules with only a single charge leave a CI source with low velocities; these low velocities result in poor resolution due to dispersion and other processes in the mass spectrometer. Ions with low m/z values due to high charge are easily resolved.

Examples of mass spectra of biological molecules obtained with ESI are shown in Figures 10.8 and 10.9. In reality, the analyst does not know the numerical charge on the peaks in the mass spectrum, but the successive peaks often vary by 1 charge unit. Computer-based algorithms have been developed for deconvoluting the sequence of m/z values of the multiply charged ions into the equivalent mass of single charged ions; such a deconvolution has been done in Figure 10.8. This permits identification of the RMM of the analyte. Applications of LC-ESI-MS are described in Section 14.1.6.1. Dr. John Fenn, one of the inventors of ESI, received the Nobel Prize in Chemistry in 2002.

10.2.2.3.1 Direct Analysis in Real Time (the DART Source)

This source was originally developed by Japan Electron Optics Laboratory (JEOL)—a major MS vendor—and is now owned by Ion Sense. This ion production process reacts electronically excited atoms or vibrationally excited molecules to form energetic metastable species M^*. When M^* collides with a sample, energy is transferred to the analyte molecule A, causing it to lose an electron and become a radical cation $A^{+•}$. This process is called Penning ionization. In Figure 10.10 for the DART source, we see a gas stream of N_2 or Ne, which carries the ions thus formed to the inlet cone of an ambient pressure MS source. When He is used as a carrier, a different ionization process occurs. The He first collides at atmospheric pressure with H_2O to form $H_2O^{+•}$, which undergoes several more transformations to become a protonated water cluster. This cluster eventually transfers the proton to the analyte molecule to produce MH^+. The excitation in either case is due to an applied potential of +1 to +5 kV, which generates the ionized gas or metastable species. A potential of 100 V on an electrode lens removes other charged particles from the gas stream, which permits only excited species to continue on. The excited species can react with solid, liquid, or gas samples to desorb and ionize the analytes. After entry into the ambient pressure sampling cone of the MS inlet, ions in the gas stream are pulled out at an angle by a charged lens into the MS analyzer stage inlet, while the obliquely directed gas stream containing neutral contaminants is directed to a trapping region and pumped away. This helps greatly to keep the ion optics free from the major contaminants of the outside world. DART spectra are simple, mainly protonated MH^+ in positive ion mode or $[M - H]^-$, and even sometimes just M^+ for some polynuclear aromatic hydrocarbons (PAHs). One never observes multiply charged or alkali metal cation adducts, although ammonia and chloride may form $[M + NH_4]^+$ or $[M + Cl]^-$ adducts. The DART source's strength is in the simplicity of its operation: one simply presents the sample to the source entrance as one might bring something to one's nose to sniff, and M^+, MH^+, or $[M - H]^-$ ions are picked up for everything present above the detection limits. The use of a high-resolution MS with a library of likely vapors to be found for the environment being tested can provide almost instant identification, hence the designation DART. Major hits not in the library will become "elephants in the room" begging for further consideration. The source can "sniff" a variety of

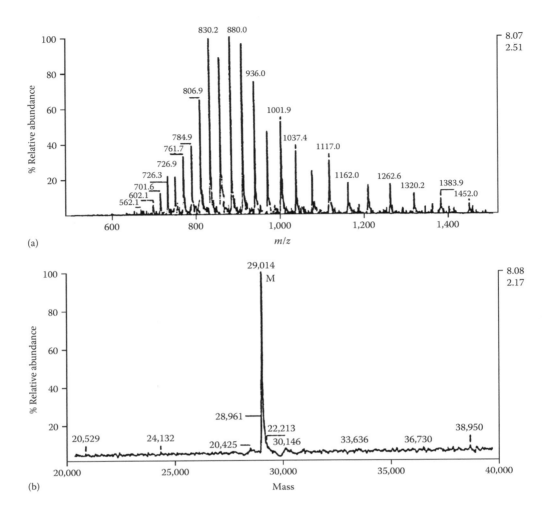

Figure 10.8 ESI mass spectrum (a) and deconvoluted mass spectrum (b) of carbonic anhydrase. (© Thermo Fisher Scientific. www.thermofisher.com. Used with permission.)

simple sample collection media, for example, wipe papers or cloths, thin-layer chromatography (TLC) plate spots, and even "stains"—just present them to the input and stand back (figuratively). A tale told by the DART developers was that an employee walked into the lab after passing a construction site where blasting was underway and on a whim waved his necktie in front of the sampling orifice, and the MS immediately reported a nitrate-explosive MS signature.

10.2.2.4 Desorption Ionization

Large molecules, such as proteins and polymers, do not have the thermal stability to vaporize without decomposing. Desorption ionization (DI) sources permit the direct ionization of solids, facilitating the analysis of large molecules. There are several types of desorption sources in which solid samples are adsorbed or placed on a support and then ionized by bombardment with ions or photons. Desorption CI, one form of DI, has already been described. The important technique of **secondary ion MS (SIMS)** is used for surface analysis as well as characterization of large molecules; SIMS is covered in detail in Section 15.4. Several other important desorption sources are described subsequently.

10.2.2.4.1 Laser Desorption and Matrix-Assisted Laser Desorption Ionization

The use of a pulsed laser focused on a solid sample surface is an efficient method of ablating material from the surface and ionizing the material simultaneously. A variety of lasers have been used; examples include infrared (IR) lasers such as the CO_2 laser ($\lambda = 10.6$ μm) and ultraviolet (UV) lasers such as Nd:YAG ($\lambda = 266, 355$ nm). YAG stands for yttrium aluminum garnet. Selective ionization is possible by choosing the appropriate laser wavelength. The laser can be focused to a small spot, from submicron to several microns in diameter. This permits the investigation of inclusions and multiple phases in solids as well as

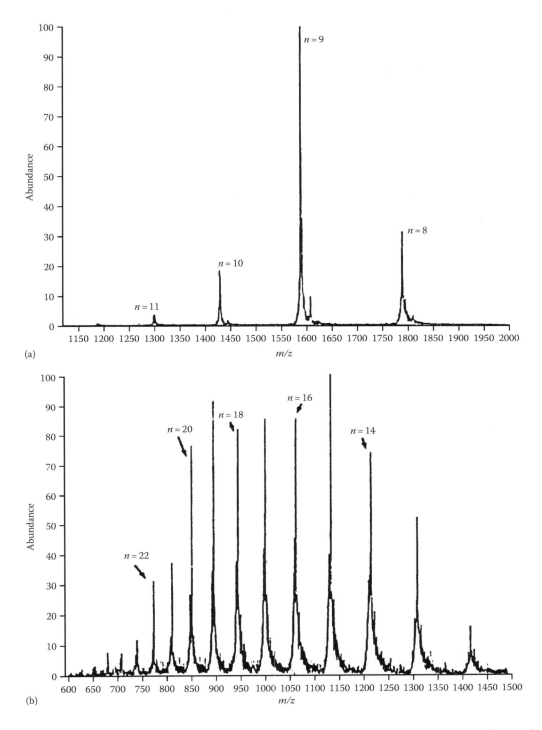

Figure 10.9 ESI mass spectra of two proteins: (a) hen egg white lysozyme and (b) equine myoglobin. n indicates the number of charges on the ion, M^{n+}.

bulk analysis. The laser pulses generate transient signals, so a simultaneous detection system or a time-of-flight (TOF) mass analyzer or a Fourier transform (FT) mass spectrometer is required. The laser provides large amounts of energy to the sample. This energy must be quickly dispersed within the molecule without fragmenting the molecule. Until the development of matrix-assisted laser desorption, the use of a laser resulted in fragmentation of biological molecules with molecular masses above about 1000 Da.

By mixing large analyte molecules with a "matrix" of small organic molecules, a laser can be used to desorb and ionize analyte molecules with RMMs well over 100,000

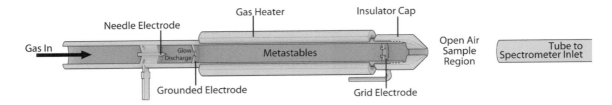

Figure 10.10 Diagram of the DART source. (Used with permission from Ion Sense. www.ionsense.com.)

Table 10.3 **MALDI Experimental Conditions**

Matrix	Wavelength		
Nicotinic acid	266 nm	2.94 μm	10.6 μm
2,5-Dihydroxybenzoic acid	266 nm	337 nm	355 nm
	2.8 μm	10.6 μm	
Succinic acid	2.94 μm	10.6 μm	
Glycerol (liquid)	2.79 μm	2.94 μm	10.6 μm
Urea (solid)	2.79 μm	2.94 μm	10.6 μm

Da with little fragmentation. The function of the matrix is to disperse the large amounts of energy absorbed from the laser, thereby minimizing fragmentation of the analyte molecule. This technique of matrix-assisted laser DI has revolutionized the mass spectrometric study of polymers and large biological molecules such as peptides, proteins, and oligosaccharides. The actual process by which ions are formed using the MALDI approach is still not completely understood.

Typical matrices and optimum laser wavelengths are shown in Table 10.3. A matrix is chosen that absorbs the laser radiation but at a wavelength at which the analyte absorbs moderately or not at all. This diminishes the likelihood of fragmenting the analyte molecule.

A typical sample is prepared for MALDI by mixing 1–2 μL of sample solution with 5–10 μL of matrix solution. A drop (<2 μL) of the mixture is placed on the MALDI probe and allowed to dry at room temperature. The solvent evaporates and the now-crystallized solid is placed into the mass spectrometer. The analyte molecules are completely separated from each other by the matrix, as shown in Figure 10.11.

The matrix material must absorb strongly at the wavelength of the laser used for irradiation. In addition, the matrix must be stable in a vacuum and must not react chemically. Matrix compounds used for MALDI include 2,5-dihydroxybenzoic acid, 3-hydroxypicolinic acid, and 5-chlorosalicylic acid for the UV region of the spectrum and carboxylic acids, alcohols, and urea for the IR region of the spectrum. Intense pulses of laser radiation are aimed at the solid on the probe. The laser radiation is absorbed by the matrix molecules and causes rapid heating of the matrix. The heating causes desorption of entire analyte molecules along with the matrix molecules. Desolvation and ionization of the analyte occur; several processes have been suggested for the ionization, such as ion–molecule reactions, but the MALDI ionization process is not completely understood. A useful, if simplistic, analogy is to think of the matrix as a mattress and the analyte molecules as dinner plates sitting on the mattress. The laser pulses are like an energetic person jumping up and down on the mattress. Eventually, the oscillations of the mattress will cause the dinner plates to bounce up into the air without breaking. The plates (i.e., molecules) are then whisked into the mass analyzer intact.

MALDI acts as a soft ionization source and generally produces singly charged molecular ions from even very large polymers and biomolecules, although a few multiple-charge ions and some fragment ions and cluster ions may occur (Figure 10.12).

10.2.2.4.2 Fast Atom Bombardment

Fast atom bombardment (FAB) uses a beam of fast-moving neutral inert gas atoms to ionize large molecules. In this technique, the sample is dissolved in an inert, nonvolatile solvent such as glycerol and spread in a thin layer on a metal probe. The probe is inserted into the mass spectrometer through a vacuum interlock. The beam of fast atoms is directed at the probe surface (the target). Argon is commonly used as the bombarding atom, although xenon is more effective (and more expensive).

Argon ions are produced in a heated filament ion source or gun, just as in SIMS, a surface analysis technique discussed in Chapter 16. The ions are accelerated through a cloud of argon atoms under an electrostatic field of 3–8 kV toward the target. The fast-moving ions exchange their charge with slow-moving argon atoms but lose no kinetic energy in the process. This results in a beam of fast-moving argon atoms and slow-moving argon ions. The latter are repelled and excluded from the system using a negatively charged deflection plate. The fast-moving atoms now strike the target, liberating molecular ions of the sample from the solvent matrix. The process may again be visualized using the analogy of dinner plates atop a mattress. Instead of responding to repetitive laser pulses, the matrix (mattress) absorbs, moderates, and transfers impact energy from the heavy fast atoms to the analyte molecules (dishes). If the analytes have surfactant character, they will preferentially

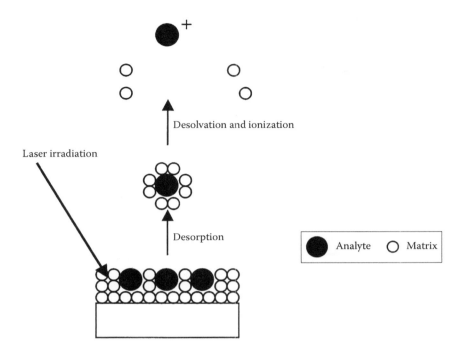

Figure 10.11 The MALDI process. Isolated analyte molecules are desorbed from a bed of matrix molecules by laser irradiation of the matrix. Subsequent desolvation and ionization of the analyte molecules occur by processes that are not completely understood.

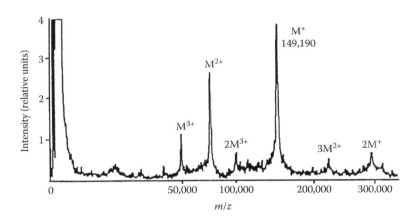

Figure 10.12 The MALDI mass spectrum of a mouse monoclonal antibody. The matrix used was nicotinic acid; the laser radiation used was 266 nm. (Reprinted from *Trends Anal. Chem.*, 9, Karas, M. and Bahr, U., 323, Copyright 1990, with permission from Elsevier.)

concentrate at the liquid matrix surface, in a location optimal for being lofted into the vapor state. Positively charged $(M + H)^+$ or negatively charged $(M - H)^-$ ions may be produced, so positive ion or negative ion mass spectra may be collected. The process is shown schematically in Figure 10.13.

There are several advantages to the FAB technique. The instrumentation is simple and the sensitivity is high. Analytes such as surfactants have been measured quantitatively at concentrations as low as 0.1 ppb. It is difficult to get very large molecules into the gas phase because of their low volatility, and it is difficult to ionize large molecules and retain the molecular ion in many ionization sources. The FAB process works at room temperature; volatilization is not required, so large molecules and thermally unstable molecules can be studied. The duration of the signal from the sample is continuous and very stable over a long period.

Sample fragmentation is greatly reduced in the FAB process, resulting in a large molecular ion, even with somewhat unstable molecules. This provides information on the RMM of the molecule, which is particularly important in biological samples such as proteins. Spectra from molecules with RMMs greater than 10,000 have been obtained. Although a strong molecular ion is obtained with

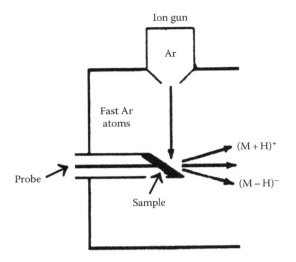

Figure 10.13 Schematic FAB ionization source. The sample, dissolved in solvent, is spread in a thin film on the end of a metal probe and bombarded by fast-moving argon atoms. Both positive and negative ions are produced.

FAB, fragmentation patterns are also obtained, providing structural information on biologically important molecules such as proteins.

An important advantage of FAB is that only those analyte molecules sputtered from the glycerol are lost; the remainder can be recovered for other analyses. Consequently, samples as small as 1 μg can be placed on the probe, and after the mass spectrum is obtained, a significant amount of the sample can be recovered.

A modification of the FAB technique is **continuous flow FAB (CFFAB)**. In this approach, the sample in solution is introduced into the mass spectrometer through a fused silica capillary. The tip of the capillary is the target. The solution is bombarded by fast atoms produced as described earlier. Solvent is flowing continuously, and the liquid sample is introduced by continuous flow injection (Figure 10.14). The mass spectrum produced has the same characteristics as that from conventional FAB, but with low background. Typically, the solvent used is 95 % water and 5 % glycerol. The ability to inject aqueous samples is an enormous advantage in biological and environmental studies. Very frequently, these types of samples are aqueous in nature, such as blood, urine, and other body fluids; water; and wastewater.

The sensitivity at the lower RMM range (1500 Da) is increased by two orders of magnitude over conventional FAB. Further, the background is reduced because of the reduced amount of glycerol present. In addition, when the solvent alone is injected, a background signal can be recorded. This can be subtracted from the signal due to sample plus solvent, and the net signal of the sample is obtained. This is especially valuable for trace analysis; amounts as low as 10^{-12} g have been detected using CFFAB.

The CFFAB system can be incorporated into LC-MS systems. The mobile phase is the solvent used. The effluent from the LC is transported directly into the mass spectrometer and the MS obtained by CFFAB. This provides a mass spectrum of each separated peak in mixtures. (See Chapter 14 for a detailed discussion of LC-MS.)

In summary, FAB and CFFAB have greatly increased the potential of MS by increasing the RMM range of molecules whose molecular ion can be determined. The system can be directly attached to LC, permitting identification of the components of a solution. Also, trace analysis is possible. The method can be applied to the important research areas of the health sciences, biology, and environmental science, as well as to chemistry.

10.2.2.4.3 Desorption Electrospray Ionization Source

While ESI, APCI, FAB, etc., operate from flowing chromatographic effluents or dipped point sources, other techniques introduce ions from 2D planar surfaces such as tissue thin sections or TLC plates (cf. Section 14.7.1). Desorption ESI (DESI) combines ESI and DI. Under ambient atmospheric conditions, an electrically charged cloud of droplets is directed at a small spot on a sample surface only a few mm distant. This charged mist is pulled to the surface by a voltage of opposite polarity applied to it. The droplet charge is transferred to analytes on the surface, and the resultant ions travel into an atmospheric pressure interface similar to that in APCI or ESI. For high RMM targets such as large protein molecules, multiple charges can be transferred with the subsequent behavior of the droplets essentially the same as the series of coulomb explosions and desolvation steps seen in ESI. By contrast, for low RMM analytes, the ionization occurs by charge transfer of either an electron or a proton (thus yielding MH$^+$ ions, but not adducts such as M + Na$^+$), similar to the APCI or DART ionization mechanisms. The geometry of the DESI source is illustrated in Figure 10.15. The key features are the two angles α and β, d (*distance tip to surface*), and the unmarked distance (*surface to center of MS inlet*). Optimal conditions for high RMM analytes are high α = 70°–90° and short d = 1–3 mm. For low RMM analytes, conditions are reversed: α = 35°–50° and d = 7–10 mm. This is consistent with and predicted for the two different ionization mechanisms described earlier. The spray emitter and the sample surface holder are each attached to a 3D stage that moves with three orthogonal degrees of freedom, which allows variation of all four of those geometric parameters. One can take full-scan mass spectra from a raster of points covering the whole 2D surface to discover and identify all unknown molecules within the MS capability or set the MS to map the distribution of a specific analyte or analytes

MASS SPECTROMETRY I: PRINCIPLES AND INSTRUMENTATION

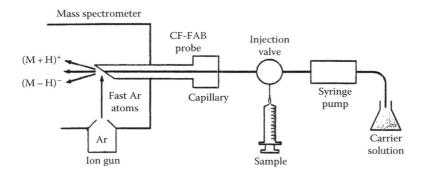

Figure 10.14 Schematic of CFFAB MS operated in the flow-injection mode.

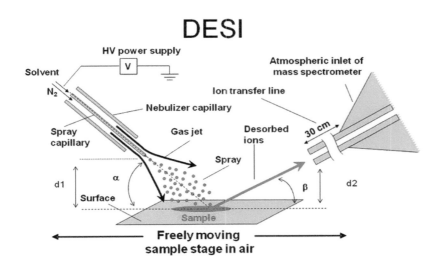

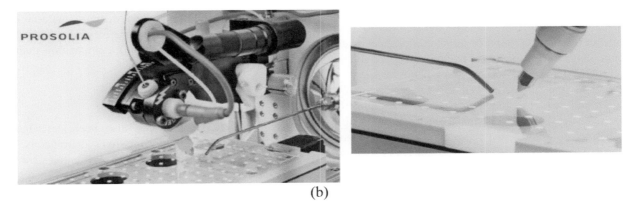

Figure 10.15 (a) Prosolia DESI Omni Spray ion source geometry and operation diagram. The family of different (Manual, 1D, and 2D) ion sources can be used on MS equipment from Agilent, Thermo, Waters, ABSciex, Bruker, and LECO. (b) A photograph of this device, with a close-up view of the source. (Used with permission from Prosolia, Inc. www.prosolia.com.)

and process the data to be displayed as an image of their location(s). One of the authors (GMF) worked in the drug metabolism and distribution department of Pfizer Central Research in the 1980s, where the only practical way to achieve this was to synthesize radiolabeled analogs of a drug in question and feed them to test animals, which were then killed. Tissue slices were obtained from quick frozen (in liquid N_2) target organs and stored for weeks in a freezer, pressed against X-ray film, which was finally developed to produce similar images. These showed only the tissue location of the drug and/or perhaps some unknown metabolites retaining the radiolabel. This was achieved only with much time and at great expense. With automated 2D DESI-MS, the imaging portion of such a study can now be completed in an hour or so, using unlabeled drug, and with all the drug-related materials in the image yielding positive MS identification.

10.2.2.4.4 Laser Ablation Electrospray Ionization

This method of ionization of samples from a surface and introduction for MS profiling is similar in some ways to DESI. An electrospray plume containing charged droplets is created over the sample surface, and a small spot on this *water-containing or water-wetted* surface is hit with a short (5 ns) intense pulse from a mid-IR laser tuned to the 2.94 µm absorption line of water. Four steps then ensue:

1. A small hemispherical plume rises over the spot without ionization.
2. The plume collapses, ejecting a sample jet from the surface, still with little ionization.
3. The ESI spray plume hits the jet and transfers charge to molecules within it.
4. The ionized molecules (some from ESI droplet shrinkage) are swept into the MS for analysis.

Steps 1 and 2 are the laser ablation process and steps 3 and 4 are the ESI-MS analytical finish.

The sample surface stage can be positioned and rastered in the *x*–*y* plane as in the DESI process to enable rapid production of an image or to sample arrays of 96 or 384 microplate wells containing microliters of aqueous solutions. Each analysis from a laser shot is completed in 2 s, so throughput is high, and use of internal standards and calibration curves permits absolute quantitation of targeted molecules over a dynamic range of four orders of magnitude. Laser ablation ESI (LAESI) has been used to image plants, tissues, cell pellets, even single cells, as well as historical documents and untreated urine and blood spots. Analytes range from small molecules, saccharides, lipids, and metabolites up to large biomolecules such as peptides and proteins of mass >100,000 Da. For single-cell experiments and analyses, etched optical fibers are used to shrink the laser spot sizes to less than 50 µm: a 4× improvement on the imaging resolution. LAESI is an immensely powerful technique.

10.2.2.5 Ionization Sources for Inorganic MS

Mass spectrometry as applied to inorganic compounds is very similar in most respects as for organic compounds, however there is little need to study masses above 300 Da. The following ionization sources are used mainly in inorganic (atomic) MS, where the elemental composition of the sample is desired. The glow discharge (GD) and spark sources are used for solid samples, while the inductively coupled plasma (ICP) is used for solutions. All three sources are also used as atomic emission spectroscopy sources; they are described in more detail with diagrams in Chapter 8.

10.2.2.5.1 GD Sources and Spark Sources

The GD source and spark source are both used for sputtering and ionizing species from the surface of solid samples and have been discussed in Chapter 8 for use as atomic emission sources. As MS ionization sources, they are used primarily for atomic MS to determine the elements present in metals and other solid samples. The direct current (DC) GD source has a cathode and anode in 0.1–10 torr of argon gas. The sample serves as the cathode. When a potential of several hundred volts is imposed across the electrodes, the argon gas ionizes forming a plasma. Electrons and positive argon ions are accelerated toward the oppositely charged electrodes. The argon ions impact the cathode surface, sputtering off atoms of the cathode material. The sample atoms are then ionized in the plasma by electrons or by collision with excited argon (Penning ionization). The sample ions are extracted from the plasma into the mass analyzer by a negatively charged electrode with a small aperture. The DC GD source is used for the analysis of conductive samples including metals, alloys, and semiconductors. The sample must be conductive to serve as the cathode. RF GD sources have been developed that enable the sputtering of electrically nonconductive samples such as ceramics and other insulators. Spark sources also can be used for sputtering of solids, but the GD source produces a more stable ion beam with better signal-to-noise ratio. The GD source sputters more material from a sample and gives more representative and more quantitative results of the elemental bulk composition than the spark source.

10.2.2.5.2 ICP Source

The argon ICP source has also been described in Chapter 8 for use with atomic emission spectrometers. This source is used primarily for liquids. Solid samples are generally dissolved in acid and diluted for analysis; organic solvents can also be used. The source produces ions from the elements introduced into the plasma as well as radiation; these ions can be extracted into a mass analyzer. The ICP torch is usually mounted horizontally with the tip of the plasma at the entrance to the mass analyzer as shown in Figure 10.16. Most of the plasma gas is deflected by a metal cone with a small orifice in its center, called the sampling cone. The gas that enters through the orifice expands into an evacuated region. The central portion passes through another metal cone, the skimmer cone, into the evacuated mass analyzer. Singly charged positive ions are formed

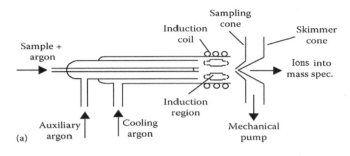

Figure 10.16 (a) Argon ICP torch used as an ionization source for ICPMS. (From Ewing, G.W., Mass spectrometry, in Ewing, G.W. [ed.], *Analytical Instrumentation Handbook*, 2nd edn., Marcel Dekker, Inc., New York, 1997. Used with permission.) (b) Photo of the plasma in a SPECTRO MS system (described in Section 10.2.3). (Courtesy of SPECTRO Analytical Instruments, Inc., AMETEK® Materials Analysis Division. www.spectro.com, www.ametek.com. Used with permission.)

from most elements, metallic and nonmetallic. The ICP has a high ionization efficiency, which approaches 100 % for most of the elements in the periodic table. The mass spectra are very simple and elements are easily identified from their m/z values and their isotope ratios observed. Background ions from the solvent and from the argon gas used to form the plasma are usually observed. Such ions include Ar^+, ArH^+, ArO^+, and polyatomic ions from water and the mineral acids used to dissolve most samples.

10.2.3 Mass Analyzers

The mass analyzer is at the core of the mass spectrometer. Its function is to differentiate among ions according to their mass-to-charge ratio. There are a variety of mass analyzer designs. *Magnetic sector* mass analyzers employing narrow metal slits to isolate individual m/z ions and *quadrupole* mass analyzers are scanning instruments; only ions of a given mass-to-charge ratio pass through the analyzer at a given time. The m/z range is scanned over time. Other mass analyzers allow simultaneous transmission of all ions; these include *TOF*, *ion trap*, and *ion cyclotron resonance (ICR)* mass analyzers as well as *dispersive magnetic* mass analyzers. Tandem mass spectrometers are instruments with several mass analyzers in sequence; these allow the selection of one ion in the first analyzer (the precursor ion), collisional fragmentation or decomposition of that ion or its reaction with a vapor reagent in a second element using four, six, or eight rods to focus and confine the precursor ions in the second stage (often called a "collision cell") and mass determination of these products in a third quadrupole, TOF, or magnetic sector analyzer stage.

10.2.3.1 Magnetic and Electric Sector Instruments

The principle of operation of a simple **single-focusing magnetic sector** mass analyzer was described briefly in Section 10.1. An ion moving through a magnetic field B will follow a circular path with radius r (Equation 10.6). Changing B as a function of time allows ions of different m/z values to pass through the fixed radius flight tube sequentially. This scanning magnetic sector sorts ions according to their masses, assuming that all ions have a +1 charge and the same kinetic energy. A schematic of a 90° sector instrument is shown in Figure 10.17. A variety of other magnetic mass spectrometers are shown in Figure 10.18; some of these will be discussed later. The sector can have any apex angle, but 60° and 90° are common. It can be demonstrated that a divergent beam of ions of a given m/z will be brought to a focus by passing through a sector-shaped magnetic field, as shown by the three ion paths in Figure 10.17.

A **dispersive magnetic sector** mass analyzer does not use a flight tube with a fixed radius. Since all ions with the same kinetic energy but different values of m/z will follow paths with different radii, advantage can be taken of this. The ions will emerge from the magnetic field at different positions and can be detected with a position-sensitive detector such as an array detector. Examples of dispersive magnetic sector systems are shown in Figures 10.18c and d.

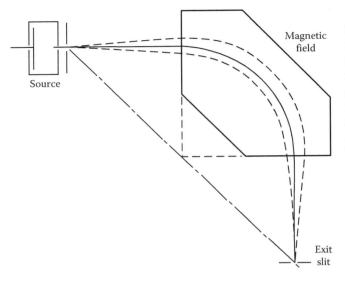

Figure 10.17 A 90° magnetic sector mass spectrometer. (From Ewing, G.W., Mass spectrometry, in Ewing, G.W. [ed.], *Analytical Instrumentation Handbook*, 2nd edn., Marcel Dekker, Inc., New York, 1997. Used with permission.)

A single-focusing instrument such as the system shown in Figure 10.18b has the disadvantage that ions emerging from the ion source do not all have exactly the same velocity. This is due to several factors. The ions are formed from molecules that have a Boltzmann distribution of energies to begin with. The ion source has small variations in its electric field gradient, causing ions formed in different regions of the source to experience different acceleration. Also, when fragmentation occurs, kinetic energy is released. This results in a distribution of velocities and adversely affects the resolution of the instrument by broadening the signal at the detector.

However, ions in a radial electrostatic field also follow a circular trajectory. The electrostatic field is an *electric sector* and separates ions by kinetic energy, not by mass (Figure 10.19). The ion beam from the source can be made much more homogeneous with respect to velocities of the ions if the beam is passed through an electric sector before being sent to the mass analyzer. The electric sector acts as an energy filter; only ions with a very narrow kinetic energy distribution will pass through.

Most magnetic sector instruments today combine both an electric sector and a magnetic sector. Such instruments

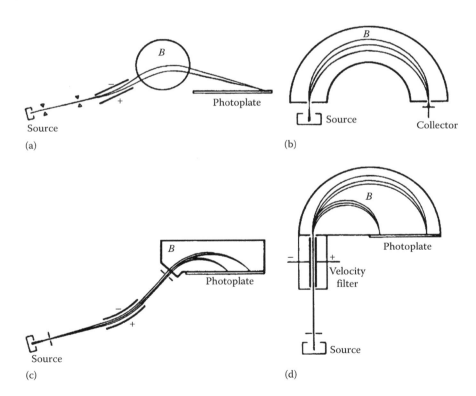

Figure 10.18 Early mass spectrometer designs. (a) Aston, 1919; (b) Dempster, 1918; (c) Mattauch–Herzog, 1935; (d) Bainbridge, 1933. In each case, *B* signifies the magnetic field. Spectrometers (c) and (d) are dispersive mass spectrometers; (c) the Mattauch–Herzog design is also a double-sector instrument, using an electric sector before the magnetic field. Modern versions of these spectrometers are still in use, with the photoplate replaced by one of the modern detectors discussed subsequently.

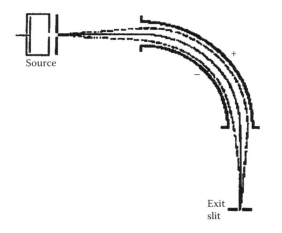

Figure 10.19 A cylindrical electrostatic-sector energy filter. (From Ewing, G.W., Mass spectrometry, in Ewing, G.W. [ed.], *Analytical Instrumentation Handbook*, 2nd edn., Marcel Dekker, Inc., New York, 1997. Used with permission.)

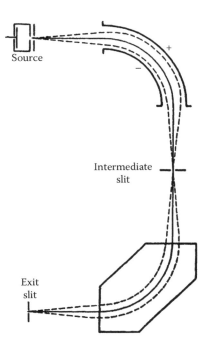

Figure 10.20 A Nier–Johnson double-focusing mass spectrometer. (From Ewing, G.W., Mass spectrometry, in Ewing, G.W. [ed.], *Analytical Instrumentation Handbook*, 2nd edn., Marcel Dekker, Inc., New York, 1997. Used with permission.)

are called double-focusing mass spectrometers. One common commercial double-focusing design is the Nier–Johnson design (Figure 10.20), introduced in 1953; this design is also used with the ions traveling in the opposite direction in "reverse" Nier–Johnson geometry. A second common design using two sectors is the Mattauch–Herzog dispersive design, shown in Figure 10.23.

Mass ranges for magnetic sector instruments are in the m/z 1–1400 range for single-focusing instruments and m/z 5000–10,000 for double-focusing instruments. Very high mass resolution, up to 100,000, is possible using double-focusing instruments.

Double-focusing sector field instruments of the Nier–Johnson and Mattauch–Herzog designs are used in ICP-MS instruments (ICP-sector field MS or ICP-SFMS, also referred to as high-resolution ICP-MS or HR-ICP-MS) as well as in organic MS instruments. In atomic MS, the ICP-SFMS instruments offer high sensitivity, high resolution, very low noise, and very low detection limits—on the order of $<1 \times 10^{-15}$ g/L (i.e., less than 1 fg/mL or less than one part per quadrillion [ppq]) for most elements. The Thermo Fisher Scientific Element XR, introduced in 2005 (Figure 10.21), uses a reverse Nier–Johnson geometry with the magnetic sector first for directional focusing and then an electric sector for energy focusing of the ion beam. The instrument is a single-ion collector and is available as an ICP-MS or GDMS. The instrument has three mass resolution settings: low, $m/\Delta m \approx 300$; medium, $m/\Delta m \approx 3,000$; and high, $m/\Delta m \approx 10,000$. Medium resolution and high resolution can separate isobaric interferences of analyte ions from polyatomic ions. Examples include $^{80}Se^+$ from $^{40}Ar_2^+$ and $^{56}Fe^+$ from $^{40}Ar^{16}O^+$. Interferences and examples are discussed in detail in Chapter 11. The Element is equipped with two detectors, a secondary EM and a Faraday detector for a linear dynamic range of 12 orders of magnitude (Figure 10.22). This permits detection of sub-part-per-quadrillion to % concentrations. The detectors are described in Section 10.2.4.

In 2011, SPECTRO Instruments GmbH (www.spectro.com, www.ametek.com) introduced the first simultaneous ICP-MS, the SPECTRO MS, based on Mattauch–Herzog geometry and using a novel direct charge detector (Section 10.2.4) placed in the focal plane of the magnet occupying the location of the photoplate shown in Figure 10.18c. The instrument consists of a novel ion optic (Figure 10.23a) and a double-focusing Mattauch–Herzog spectrometer (Figure 10.23b) with an entrance slit (which defines resolution), a 31.5° electrostatic analyzer (ESA), a drift length with energy slit, and a 90° magnetic field sector with the detector attached at the focal plane. The assembled instrument is shown in Figure 10.24.

The novel ion optic consists of a curved ESA (not to be confused with the ESA in the mass spectrometer), an Einzel lens and a quadrupole doublet. The ESA separates ions of a defined kinetic energy, which have been extracted by a prefilter, from neutrals, particles, and photons. The ions are focused by the Einzel lens into a quadrupole doublet, which changes the shape of the ion beam from round to rectangular to match the entrance slit of the mass spectrometer.

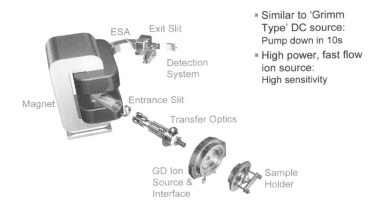

Figure 10.21 The reverse Nier–Johnson geometry of the Thermo Scientific Element with a GD source. (© Thermo Fisher Scientific. www.thermofisher.com. Used with permission.)

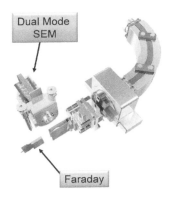

Figure 10.22 The dual detectors on the Thermo Scientific Element MS. (© Thermo Fisher Scientific. www.thermofisher.com. Used with permission.)

The ESA sector reduces the ion beam energy band width to achieve high-resolution *m/z* separation in the magnetic field. The Mattauch–Herzog geometry of the mass spectrometer focuses all the ions, separated by *m/z*, onto the focal plane at the exit of the magnet using a fixed setting for the ESA and a permanent magnet (see Section 10.2.4).

This enables the use of a flat-surfaced array detector. No scanning is required, resulting in a fully simultaneous determination of all the elements (5–240 amu) and fully utilizing the continuous ion beam produced by the ICP. Because the entire spectrum is recorded at once, it is possible to go back and retrieve concentration data for any element in any sample at any time.

The major disadvantage of double-focusing MS systems for elemental analysis is their high cost compared to quadrupole-based systems.

10.2.3.2 Time-of-Flight Analyzer

A TOF analyzer does not use an external force to separate ions of different *m/z* values. Instead, pulses of ions are accelerated into an evacuated field-free region called a drift tube. If all ions have the same kinetic energy, then the velocity of an ion depends on its mass-to-charge ratio, or on its mass, if all ions have the same charge. Lighter ions will travel faster along the drift tube than heavier ions and are detected first. The process is shown schematically in Figure 10.25.

A schematic TOF mass spectrometer is shown in Figure 10.26. The drift tube in a TOF system is approximately 1–2 m in length. Pulses of ions are produced from the sample using pulses of electrons, secondary ions, or laser pulses (e.g., MALDI). Ion pulses are produced with frequencies of 10–50 kHz. The ions are accelerated into the drift tube by a pulsed electric field, called the ion extraction field, because it extracts (or draws out) ions into the field-free region. Accelerating voltages up to 30 kV and extraction pulse frequencies of 5–20 kHz are used.

Ions are separated in the drift tube according to their velocities. The velocity of an ion, v, can be expressed as

$$v = \sqrt{\frac{2zV}{m}} \quad (10.10)$$

where V is the accelerating voltage. If L is the length of the field-free drift tube and t is the time from acceleration to detection of the ion (i.e., the flight time of the ion in the tube):

$$v = \frac{L}{t} \quad (10.11)$$

and the equation that describes ion separation is

$$\frac{m}{z} = \frac{2Vt^2}{L^2} \quad (10.12)$$

MASS SPECTROMETRY I: PRINCIPLES AND INSTRUMENTATION

The flight time, t, of an ion is

$$t = L\sqrt{\frac{m}{2zV}} \qquad (10.13)$$

Equation 10.13 can be used to calculate the difference in flight time between ions of two different masses. Actual time separations of adjacent masses can be as short as a few nanoseconds, with typical flight times in microseconds.

TOF instruments were first developed in the 1950s but fell out of use because of the inherent low resolution of the straight drift tube design (as in Figure 10.26). The drift tube length and flight time are fixed, so resolution depends on the accelerating pulse. Ion pulses must be kept short to avoid overlap of one pulse with the next, which would cause mass overlap and decrease resolution. Interest in TOF instruments resurfaced in the 1990s with the introduction of MALDI and rapid data acquisition methods. The simultaneous transmission of all ions and the rapid flight time mean that the detector can capture the entire mass spectral range almost instantaneously.

The resolution of a TOF analyzer can be enhanced by the use of an ion mirror, called a **reflectron**. The reflectron is used to reverse the direction in which the ions are traveling and to energy-focus the ions to improve resolution. The reflectron's electrostatic field allows faster ions to penetrate more deeply than slower ions of the same m/z value. The faster ions follow a longer path before they are turned around, so that ions with the same m/z value but differing velocities end up traveling exactly the same distance and arrive at the detector together. The use of a curved field reflectron permits the focusing of ions over a broad mass range to collect an entire mass spectrum from a single laser shot. In a reflectron TOF, the ion source and the detector are at the same end of the spectrometer; the reflectron is at the opposite end from the ion source. The ions traverse the drift tube twice, moving from the ion source to the reflectron and then back to the detector. A schematic of a commercial reflectron TOF mass analyzer is shown in Figure 10.27.

The mass range of commercial TOF instruments is up to 10,000 Da. Resolution depends on the type of TOF and ranges from 1000 for instruments designed as dedicated detectors for GC-TOF-MS to 20,000 for reflectron instruments. One limitation to the use of a conventional reflectron instrument is a loss in sensitivity; about 10 % of the ions are lost with a conventional wire grid reflectron.

10.2.3.2.1 JEOL Spiral TOF

The ion flight path length of a TOF instrument can be extended to increase resolution by bending the flight path into a series of figure-8's offset or staggered along an

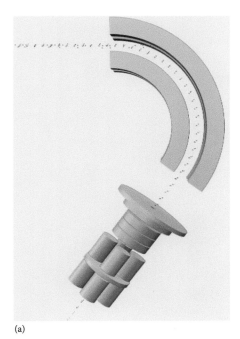

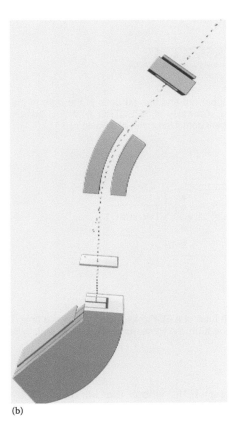

Figure 10.23 (a) The ion optic in the SPECTRO MS. Ions of a defined kinetic energy are introduced into the curved ESA (top). An Einzel lens focuses the ions (middle) into a quadrupole doublet (bottom). (b) The Mattauch–Herzog geometry of the SPECTRO MS. (Courtesy of SPECTRO Analytical Instruments, Inc., AMETEK® Materials Analysis Division. www.spectro.com, www.ametek.com. Used with permission.)

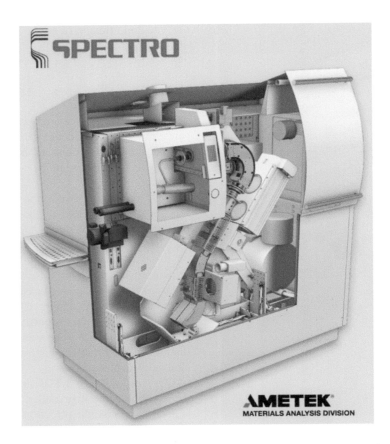

Figure 10.24 Cutaway view of the SPECTRO MS, showing the ICP torch box upper left and the ion optic to the right of the torch box, leading to the mass spectrometer on the left under the torch box. (Courtesy of SPECTRO Analytical Instruments, Inc., AMETEK® Materials Analysis Division. www.spectro.com, www.ametek.com. Used with permission.)

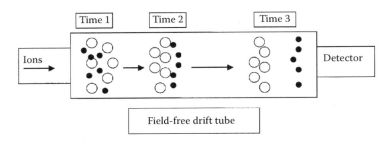

Figure 10.25 A pulse of ions of two different m/z values enters the field-free drift tube of a TOF mass spectrometer at time 1. The large white circles have m/z > the small dark circles. As they travel down the tube, the lighter ions move faster and, by time 3, have been separated from the heavier ions.

axis perpendicular the plane of the curved paths. Curved electrodes bend the paths of ion packets as illustrated in Figure 10.28 for multiple figure-8 stages. They are also focused back after every figure-8 stage before transfer to the next ones, until a total path length of 17 m is traversed without the packets diverging at the detection plane, thus achieving high mass accuracy (<1 ppm) and resolution (>60,000 FWHM) and high ion transmission. Several configurations are available: (1) MALDI-TOFMS; (2) a linear TOF, where the ion packets bypass the spiral stages and proceed directly to a detector (for high RMM analytes and those undergoing post source decay); and (3) TOF/TOF, where a highly resolved and isolated ion from the ion packets from the 17 m spiral array separation is fragmented in a collision cell and its fragment ions are passed on to a single reflectron second TOF stage for further MS characterization. All this is compacted into a space not much larger than a

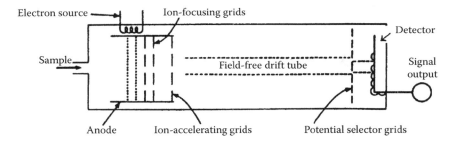

Figure 10.26 Schematic TOF mass spectrometer.

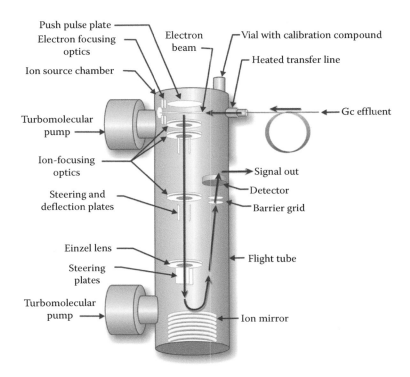

Figure 10.27 A commercial reflectron TOF mass analyzer, the Pegasus III from LECO. The analyzer is shown with sample introduction from a GC. (Diagram courtesy of LECO Corporation. www.leco.com.)

student desk. In the MALDI-TOF/TOF configuration, the long spiral path enables monoisotopic selection of precursor ions. The selected precursor ions undergo high energy fragmentation in the collision cell, and the second TOFMS incorporates a re-acceleration mechanism and an offset parabolic reflection, which permits observation of product ions from very low m/z up to that of the precursor ion. The high resolution of the system simplifies interpretation of the product ion spectra, as both they and the selected precursor ion are monoisotopic. The curved electric sectors in the spiral TOF first stage remove ions arising from post source decay, which are common in MALDI, yielding simpler product ion spectra characteristic only of the high-energy collision-induced fragmentation.

10.2.3.2.2 LECO Pegasus GC-HRT® and Citius LC-HRT™

These two new designs of high-resolution TOFMS instruments employ multiple (20–40) gridless back-and-forth reflections of the ion packets to produce flight paths of 20–40 m with minimal reflection losses, achieving resolutions of 50,000–100,000, respectively. These exceed even the 17 m in the Spiral TOF. Figure 10.29a depicts the multiple reflection stages and Figure 10.29b shows three selectable path length modes. The ultrahigh-resolution mode can cover the m/z range of 300–1200, while the other two cover 50–2500. The instruments are therefore suited to high-resolution MS characterization of "small molecules," and the scan rate of up to 200 full spectra/s makes them

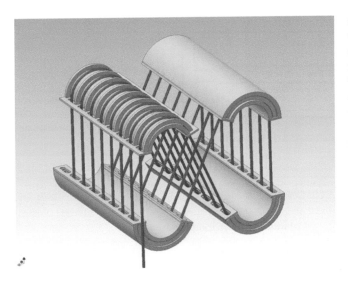

Figure 10.28 Figure-8 ion trajectories in the JEOL Spiral TOF. (Used with permission from JEOL Inc. www.jeol.com.)

a good match for the narrow peaks from high-speed ultrahigh-pressure LC (UHPLC) separations or fast, short-capillary GC separations. Figure 10.30 shows how accurate mass measurements from a Pegasus GC-high-resolution TOF (HRT) mass spectrum with sub-ppm mass accuracy enabled assignment of the correct formula of $C_{24}H_{23}NO$ to the measured m/z of 341.17749, where that empirical formula was the only one fitting the constraints of the high-resolution mass fit table. This formula in fact corresponds to that of a cannabinomimetic compound being used in the preparation of synthetic cannabis. The JEOL Spiral TOF or TOF/TOF systems with MALDI sources actually complement the performance of the LECO LC-HRT by working with higher RMM range biomolecules introduced from a MALDI source, whereas the latter is more suitable for metabolomic studies of smaller RMM metabolites and can employ DESI source introduction (cf. end of Section 10.2.2.4).

The rapid collection of the entire mass spectrum made possible by the TOF makes it ideal for interfacing with a

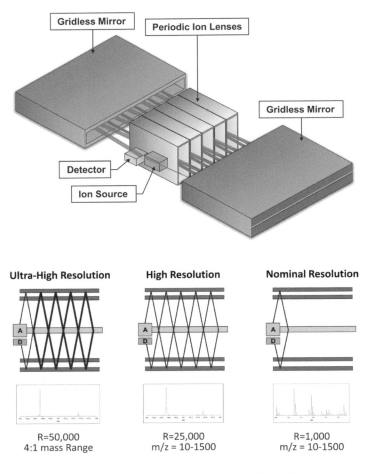

Figure 10.29 LECO Multireflectron TOF-MS: (a) Diagram of mass analyzer assembly. (b) Selection of three different TOF path lengths to give three increasingly higher resolutions. (Materials used with permission from LECO Corporation. www.leco.com.)

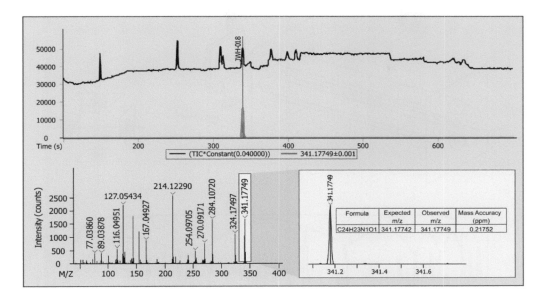

Figure 10.30 HRT mass spectrum of cannabinomimetic unknown obtained with LECO Pegasus GC-HRT. (Material used with permission from LECO Corporation. www.leco.com.)

chromatograph. It is especially useful when combined with fast GC, which requires the rapid collection of hundreds of mass spectra. For example, the LECO Pegasus 4D GC-TOFMS collects the entire mass range from 1 to 1000 Da in 170 μs and collects up to 500 mass spectra. (A detailed description of these instruments can be found at www.leco.com.)

TOF-based ICP-MS instruments for elemental analysis and speciation using coupled chromatographs are commercially available. GBC Scientific Equipment Pty Ltd. makes an orthogonal acceleration ICP-TOF-MS, the OptiMass series. Because of the speed of data collection, this instrument is well suited for interfacing with chromatographic instruments, especially HPLC or for use with other transient signal sample introduction, such as solid-phase microextraction (SPME) with thermal desorption.

Note that TOF MS systems can be part of multistage MS/MS operation configurations (Section 10.2.3.4) as are "triple quads" (QQQ or QqQ). There are "Q-TOFs" and "TOF/TOFs." An example of the latter is implemented by JEOL, where their high-resolution Spiral TOF (Section 10.2.3.2) is coupled to a single reflectron TOF to analyze complex biological mixtures from MALDI sampling.

10.2.3.3 Quadrupole Mass Analyzer

The quadrupole mass analyzer does not use a magnetic field to separate ions. The quadrupole separates ions in an electric field (the quadrupole field) that is varied with time. This field is created using an oscillating RF voltage and a constant DC voltage applied to a set of four precisely machined parallel metal rods (Figure 10.31). This results in an alternating current (AC) potential superimposed on the DC potential. The ion beam is directed axially between the four rods.

The opposite pairs of rods A and B, and C and D, are each connected to the opposite ends of a DC source, such that when C and D are positive, A and B are negative. The pairs of electrodes are then connected to an oscillating RF electrical source. They are connected in such a way that the potentials of the pairs are continuously 180° out of phase with each other. The magnitude of the oscillating voltage is greater than that of the DC source, resulting in a rapidly oscillating field. The RF voltage can be up to 1200 V, while the DC voltage is up to 200 V. The rods would ideally be hyperbolic instead of circular in cross section, with their hyperboloid axes pointed toward the center of the rod array, to provide a more uniform field. Under these conditions, the potential at any point between the four poles is a function of the DC voltage and the amplitude and frequency of the RF voltage. The shape of the rods varies with different manufacturers; cheaper circular cylindrical rods are often used instead of hyperbolic rods. Agilent Technologies manufactures the only hyperbolic quadrupole used in ICP-MS.

An ion introduced into the space between the rods is subjected to a complicated lateral motion due to the DC and RF fields. Assume that the x-direction is the line through the midpoint of the cross sections of rods A and B; the y-direction is the line through the midpoint of the cross sections of rods C and D, as shown in Figure 10.31b. The forward motion of the ion in the z-direction (along the axis between the rods) is not affected by the field. The following equations describe the lateral motion of the ion:

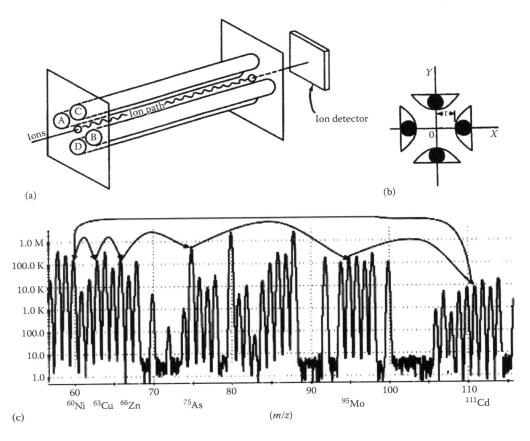

Figure 10.31 (a) Transmission quadrupole mass spectrometer. Rods A and B are tied together electrically, as are Rods C and D. The two pairs of rods, AB and CD, are connected both to a source of direct potential and a variable RF excitation such that the RF voltages are 180° out of phase. (b) The geometry of the rods. (c) Sequential detection of elements by peak hopping.

$$\frac{d^2 x}{dt^2} + \frac{2}{r^2 \left(\frac{m}{z}\right)} \left(V_{DC} + V_{RF} \cos 2\pi ft\right) y = 0 \quad (10.14)$$

$$\frac{d^2 y}{dt^2} + \frac{2}{r^2 \left(\frac{m}{z}\right)} \left(V_{DC} + V_{RF} \cos 2\pi ft\right) x = 0 \quad (10.15)$$

where
V_{DC} is the voltage of the DC signal
V_{RF} is the amplitude of the voltage of the RF field
f is the frequency of oscillation of the RF field (rad/s)
r is half the distance between the inner edges of opposing poles such as A and B as shown in Figure 10.31b
t is the time

The motion is complex because the velocity in the x-direction is a function of the position along y and *vice versa*. In order for an ion to pass through the space between the four rods, every time a positive ion is attracted to a negatively charged rod, the AC electric field must be present to push it away; otherwise, it will collide with the rod and be lost. The coordination between the oscillating (AC) field and the time of the ion's arrival at a rod surface over the fixed distance between the rods is critical to an ion's movement through the quadrupole. As a result of being alternately attracted and repelled by the rods, the ions follow an oscillating or "corkscrew" path through the quadrupole to the detector. For a given amplitude of a fixed ratio of DC to RF at a fixed frequency, only ions of a given m/z value will pass through the quadrupole. If the mass-to-charge ratio of the ion and the frequency of oscillation fit Equations 10.14 and 10.15, the ion will oscillate toward the detector and eventually reach it. If the m/z value and the frequency do not meet the conditions required by Equations 10.14 and 10.15, these ions will oscillate with an increasingly wide path until they collide with the rods or are pulled out by the vacuum system. In any case, the ions will not progress to the detector. Only a single m/z value can pass through the quadrupole at a given set of conditions. In this respect, the quadrupole acts like a filter and is often called a mass filter. One m/z value is filtered from the ion beam and passed to the detector; the elements are measured sequentially by "peak hopping" as shown in Figure 10.31c.

The separation of ions of different m/z can be achieved by several methods. The frequency of oscillation of the RF field can be held constant while varying the potentials of the DC and RF fields in such a manner that their ratio is kept constant. It can be shown mathematically that the best resolution is obtained when the ratio V_{DC}/V_{RF} is equal to 0.168. If the ratio is greater than this number, a stable path cannot be achieved for any mass number; if the ratio is lower than this number, resolution is progressively lost.

The resolution of the system is dependent on the number of oscillations an ion undergoes in the drift chamber. Increasing the rod lengths, therefore, increases resolution and extends the use of the system to higher RMM compounds. Increasing the frequency of the RF field can bring about this same improvement. The rod diameter is also important. If the diameter is increased, the sensitivity is greatly increased, but then the mass range of the system is decreased. The manufacturer must come to a compromise with these factors when designing an instrument for analytical use. The resolution achievable with the quadrupole mass spectrometer is approximately 1000; the m/z range for a quadrupole mass analyzer is 1–1000 Da. As with other mass spectrometers, the sample must be available in the gas phase and must be ionized.

Quadrupole mass analyzers are found in most commercial ICP-MS instruments, in most GC-MS instruments (Chapter 13), and in many LC-MS instruments (Chapter 14). Quadrupoles or higher number multipole rod arrays are also used in both organic and inorganic MS/MS systems as mass analyzers or collision cells, respectively. This use will be described in Section 10.2.3.4.

Although the quadrupole mass analyzer does not have the range or resolution of magnetic sector instruments, it is very fast. It can provide a complete mass spectrum in less than 100 ms. This property and its wide angle for acceptance of ions make it suitable for coupling to transient signal sources such as those from chromatography or laser ablation. In addition, the quadrupole mass analyzer is inexpensive, compact, and rugged. Most ICP-MS, GC-MS, and LC-MS instruments with quadrupoles are small enough to fit on a benchtop. Figure 10.32 shows the cutaway view of the smallest benchtop ICP-MS. Quadrupoles are the most common mass analyzer in commercial use. The term **transmission quadrupole** mass spectrometer is sometimes used for this mass analyzer to avoid confusion with the **quadrupole ion trap (QIT)** mass spectrometer discussed in Section 10.2.3.5.

10.2.3.4 MS/MS and MS^n Instruments

Many analytical questions require the mass spectrometrist to obtain more information about the structure of fragment ions or about ion–molecule reactions than can be obtained from the initial ionization of an analyte. In such cases, the technique of MS/MS, also called tandem

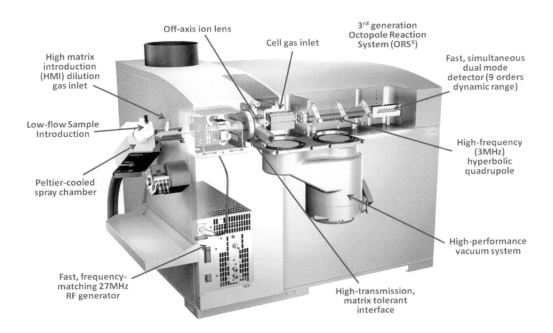

Figure 10.32 Cutaway view of the single quadrupole 7700-series benchtop ICP-MS from Agilent Technologies. The hyperbolic quadrupole is shown between the reaction cell and the detector. The sample introduction system and ICP torch are on the left of the diagram. (© 2013 Agilent Technologies. www.agilent.com/chem. All rights reserved. Used with permission.)

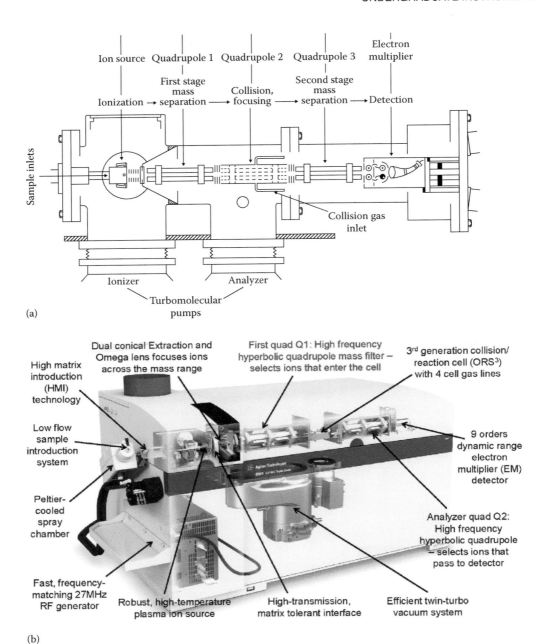

Figure 10.33 (a) A commercial quadrupole tandem MS/MS instrument for organic MS. (© Thermo Fisher Scientific. www.thermofisher.com. Used with permission.) (b) The 8800 ICP-QQQ with two hyperbolic quadrupoles separated by an octupole cell. (© 2013 Agilent Technologies. www.agilent.com/chem. All rights reserved. Used with permission.)

MS may be useful. MS/MS is a mass spectral technique that uses two (or more) stages of mass analysis combined with a process that causes a change in mass of the ion of interest, such as dissociation into lighter fragment ions by collision with an inert gas or conversion into a heavier ion by reaction with a neutral molecule. An ICP-MS/MS is now available (Figure 10.33b) and operates in a similar fashion, described in detail in Chapter 11.

The stages of mass analysis may be performed by two physically separate mass analyzers, such as two quadrupoles coupled in series on either side of a multipole collision cell; this type of arrangement for MS/MS is called "tandem in space." Figure 10.33a shows a quadrupole MS/MS instrument with three quadrupoles for "tandem in space" analysis. We reiterate again that the second "quad" may actually have more than four rods. It serves to isolate and focus individual selected ions from the first quad, allows them to collide with a collision gas, and moves the secondary fragment ions to the third quad analyzer stage. Such an arrangement is often referred to as a "triple quad" despite the different design and function of the second stage. Alternatively, ion traps, discussed in Sections 10.2.3.5 and 10.2.3.6, may be used to perform MS/MS experiments within the same mass analyzer; this type of MS/MS experiment is called "tandem in time."

Using Figure 10.33a, we will look at a simple MS/MS experiment. For example, an analyte may be ionized as usual by the ion source. One ion of a particular *m/z* value is of interest. This ion is called the **precursor** ion. The precursor ion is selected by the first quadrupole, which is operating as a mass analyzer, permitting only the precursor ion (plus any direct on-mass interferences) to pass into the second stage. The precursor ion enters the second stage. This second stage is the reaction region and acts as a collision cell and ion lens, not as a mass analyzer. An inert gas is usually added in this region to cause collision-induced fragmentation of the precursor ion into lighter product ions, or a reactive reagent gas may be introduced to form heavier product ions through ion–molecule reactions. The second stage also serves to confine and transfer the product ions; that is,

$$M^+_{precursor} \rightarrow M^+_{product}$$

where the precursor *ion* and product *ions* have different *m/z* values. The product ions then undergo mass analysis as usual in the third quadrupole. This type of design, where the first and third quadrupoles are used for mass analysis and the center quadrupole is used for collision and focusing, is often abbreviated as a QqQ design, to indicate that there are only two stages of mass analysis symbolized by the capital Q. The schematic of the ICP-MS instrument in Figure 10.33b shows the same two quadrupole analyzers, separated in this case by an octupole collision/reaction cell; it is still called a triple quadrupole instrument (by the manufacturer) and is of the QqQ type.

If we had an instrument with three mass analyzers, the fragmentation process could be repeated before final analysis. A precursor ion is selected and fragmented and a given product ion is selected and fragmented again before mass analysis of its product ions; that is,

$$M^+_{precursor} \rightarrow M^+_{product\ 1} \rightarrow M^+_{product\ 2}$$

where all three ions have different *m/z* values. This is an example of MS/MS/MS or MS^3; the number of steps can be increased to give an MS^n experiment. It is not practical to build "tandem in space" instruments with large numbers of mass analyzers; three or four is the upper limit. Commercial MS/MS instruments are limited to two mass analysis stages. Ion trap instruments are used for higher-order experiments.

In general, $n = 7$ or 8 is a practical upper limit in ion trap instruments.

Tandem mass spectrometers have been built with three quadrupoles as shown in Figure 10.33a, with hexapole or octupole reaction/collision cells (Figure 10.33b) and with other combinations of sector and TOF mass analyzers. Electric and magnetic sector analyzers have been combined with quadrupoles and with TOF analyzers. Quads are often used as the first stage of mixed-mode tandem systems, where the second stage is a high-resolution system (e.g., a TOF or Orbitrap; see Section 10.2.3.7). The second-stage resolution and performance could be degraded by the fields of extraneous ions, especially from high levels of nonanalyte contamination (the space charge interference). The analyte M$^+$ or a characteristic fragment ion can be isolated from most of these and then transferred to the next stage (e.g., Q-TOF and Q-Trap).

10.2.3.5 Quadrupole Ion Trap

An **ion trap** is a device where gaseous ions can be formed and/or stored for periods of time, confined by electric and/or magnetic fields. There are three commercial types of ion traps in use in MS, the QIT; the Orbitrap, which uses a spiral trajectory that oscillates along a 1D linear axis; and the ICR trap.

The QIT mass spectrometer is also called a **Paul ion trap, a 3D ion trap** or, more commonly, just an ion trap. This analyzer uses a quadrupole field to separate ions, so "quadrupole" is used in the name to distinguish this system from the Orbitrap and ICR traps discussed in the next section. The QIT is shown schematically in Figure 10.34.

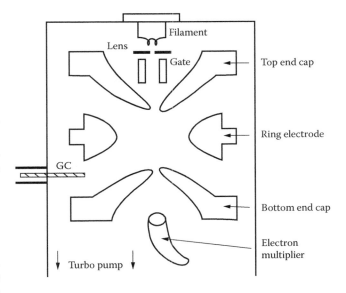

Figure 10.34 Cross section of a QIT mass spectrometer. This schematic shows a gas-phase sample introduced from a GC and ionized inside the trap by electrons from the filament. (From Niessen, W.M.A., J., *Liquid Chromatography–Mass Spectrometry*, 3rd ed, CRC Taylor and Francis, Boca Raton, FL, 2006. Used with permission.)

A ring-shaped electrode and two end cap electrodes, one above and one below the ring-shaped electrode, are used to form a 3D field. A fixed-frequency RF voltage is applied to the ring electrode while the end caps are either grounded or under RF or DC voltages. Ions are stored in the trap by causing them to move in stable trajectories between the electrodes under the application of the field. This is done by varying the potentials, so that just before an ion collides with an electrode, the potential changes sign and repels the ion. Ions with a very broad range of m/z values can be stored simultaneously in the ion trap.

Ionization of the sample can take place outside of the ion storage area of the ion trap; such external ionization is required for LC-MS using an ion trap and may be used for GC-MS. Alternatively, ionization can take place inside the ion storage area; this internal ionization approach can be used for GC-MS. Inert gas may be introduced into the trap after initial ionization for MS/MS experiments using collision-induced dissociation.

Ions are extracted from the trap by changing the amplitude of the ring electrode RF. As the amplitude increases, the trajectory of ions of increasing m/z becomes unstable. These ions move toward the end caps, one of which has openings leading to the detector. Ions of a given m/z value pass through the end cap sequentially and are detected.

The use of various RF and DC waveforms on the end caps allows the ion trap to selectively store precursor ions for MS/MS experiments or to selectively store analyte ions while eliminating ions from the matrix. This can result in improved detection limits in analysis. The ion trap has limitations. Because the stored ions can interact with each other (a space–charge effect), thereby upsetting stability of trajectories, the concentration of ions that can be stored is low. This results in a low dynamic range for ion trap mass spectrometers. Trace level signals from a target analyte ion at one mass can be destabilized by the presence of great excesses of contaminant ions, even if these are of sufficiently different mass to be well resolved from the ion of interest. Ion trap MS instruments are less forgiving of "dirty samples" than are quadrupoles, which "throw away" such unwanted ions as they are measuring the target ion. The stored ion interaction also limits the accuracy of the mass-to-charge ratio measurement. Resolution of commercial QIT mass spectrometers is on the order of 0.1–1, with an m/z range of 10–1000.

10.2.3.6 Fourier Transform Ion-Cyclotron Resonance

The ICR instrument, also called a **Penning ion trap**, uses a magnetic field to trap and store ions. As shown in Figure 10.35, six conducting plates arranged as a cube serve as the ion trap. The cubic cell is about 100 mm on a side, is under very high vacuum (<10^{-8} torr), and is located inside a strong magnetic field produced by a superconducting magnet. Sample is introduced into the cell and ionized by an external ion source such as an electron beam passing through the trap. Ions in the presence of a magnetic field move in circular orbits perpendicular to the applied field, at a frequency called the cyclotron frequency:

$$\omega_c = \frac{z}{m}(eB) = \frac{v}{r} \quad (10.16)$$

where
ω_c is the frequency of rotation of ions (rad/s)
e is the charge on electron (Coulombs)
B is the magnetic field (Tesla)
z is the charge on the ion
m is the mass of the ion
v is the velocity of the ion
r is the radius of orbit

The frequency of motion of an ion depends on the inverse of its m/z in a fixed magnetic field. Mass analysis is performed by applying an RF pulse of a few milliseconds duration to the transmitter plates. The RF pulse provides energy to the ions, causing them to move in larger circular orbits at the same frequency. For a given m/z value, a pulse at a frequency of ω_c causes all ions of that m/z value to absorb energy and increase their orbit of rotation. When the RF pulse is off, the motion of the ions is detected by current induction in the receiver plates. As a group of positive ions approaches the receiver plate, its charge attracts electrons to the inside surface of the plate. As the group recedes, the electrons are released. This induced current, called an "image current," is a sinusoidal signal with frequency ω_c. The larger the orbit, the larger is the induced current. The frequency provides the m/z information about the ion and the current amplitude depends on the number of ions of that m/z value, providing information about the concentration of ions.

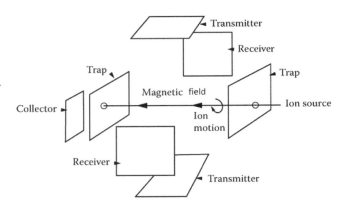

Figure 10.35 "Exploded" view of an ICR ion trap. The ICR has been the primary mass analyzer used in FTMS, both alone and in newer "hybrid" FTMS instruments.

It would be possible to scan the RF and measure the magnitude of the image current at each m/z value to obtain the mass spectral information but the process would be very slow. Instead, an RF pulse is used that contains a range of frequencies. The range of frequencies is chosen to excite the desired m/z range. When the pulse is off, all of the excited ions induce image currents in the receiver plates as they rotate. The output current, which contains all of the frequency and magnitude information from all of the ions present, can be converted mathematically to a mass spectrum by the application of the Fourier transformation. The use of an ICR ion trap and Fourier transformation is called FTICR-MS or just FTMS. As of early 2003, this was the only type of FTMS instrument commercially available.

There are several advantages to the ICR. One is that the ion detection is nondestructive. Therefore, signals can be accumulated by averaging many cycles, resulting in greatly improved S/N and signal-to-background as well as very low detection limits. Detection of attomoles of analyte is possible. Frequency can be measured very accurately, so the mass accuracy and resolution of these FTMS systems can be very high, on the order of 1 ppb for a mass of 100 Da. In order to acquire sufficient information to achieve such high resolutions by the FT process, the data must be acquired over a longer period. In order that collisions with residual gas atoms in the ICR trap do not remove the ions during this period, it must be operated at very high vacuum (e.g., $<10^{-8}$ torr), if such high resolution is to be attained. The ICR can also be used for MS/MS and MS^n experiments, by storing precursor ions and fragmenting them in the trap using a collision gas, lasers, or ion beams. An advantage of the FTMS system is that it is nondestructive, so ions at all stages of an MS^n experiment can be measured. A QIT instrument expels ions to be analyzed, so only ions in the final step can be measured.

The major disadvantages of the ICR are a limited dynamic range due to the same space charge effect described for the QIT, a more complex design, and high instrument cost, resulting from the need for extremely high vacuum operation and high field superconducting magnets.

Despite the high cost of the FTICR instrument, new "hybrid" FTMS instruments costing significantly more than 1 million US dollars were introduced commercially in 2003 because of their ability to determine the structure of proteins. Protein structure determination is critical to fundamental biology, genomics, proteomics, and the understanding of drug–biomolecule interactions for development of pharmaceuticals. "Hybrid" FTMS instruments combining either an ion trap or quadrupole(s) on the front end with the FTICR on the back end exhibit both high sensitivity and high resolution. The diversion of nontarget ions in the earlier stages greatly reduces their space charge effects in the ICR trap, enabling it to operate at its maximum capability. Ion trap and FTICR instruments are not available for ICP-MS due to the large Ar^+ intensity (from the plasma) relative to the analyte ions. Only a limited number of ions can be stored in the ion trap, resulting in poor sensitivity for the analytes compared to other ICP mass spectrometer designs.

10.2.3.7 Orbitrap MS

The final and most recent MS design combines features of the Paul Trap and the FTICR trap described earlier, with vastly greater resolution and mass range than the former, and much better data acquisition rate and resistance to space–charge disruption than the latter. The operation is of greater complexity than can be addressed in full here, but some general points can be described. Figure 10.36 displays the heart of the instrument, a cross

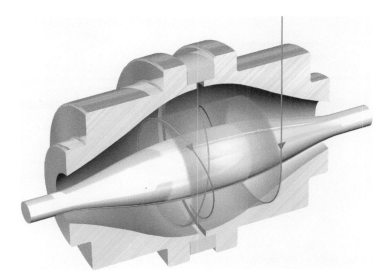

Figure 10.36 The Orbitrap mass analyzer. (© Thermo Fisher Scientific. www.thermofisher.com. Used with permission.)

section of the Orbitrap itself, with part of the trajectory of a single ion represented. Ions resonantly orbit around the central spindle, in circles perpendicular to its axis. The outer casing converges slightly toward the inner spindle, creating a converging field along the long axis, which causes the circling ions to oscillate back and forth from one end to the other, with differing frequencies related to their *m/z* values. The same principle of image charge detection described earlier in the FTICR section is employed. Here, this is done at the end of the spindle and reflects the frequencies of the longitudinal back-and-forth motion of the circling ions, while in the ICR instruments, the varying image current is induced by the transverse passage of the circling, cyclotron-resonant ions. In both spectrometers, the very complex signal of summed ion frequencies is deconvoluted by fast Fourier transformation (FFT) software to give the amplitudes and frequencies of all the ions, which translate to intensity and *m/z* values, respectively. The use of FFT data acquisition rather than scanning conveys the same Fellgett's advantage as was discussed in the sections on FTIR and FT-nuclear magnetic resonance (NMR). In the Orbitrap geometry, the frequencies of the ions' axial oscillations are independent of their energy and spatial spread, leading to very high resolutions (up to 240,000 in the latest models). There is a large space charge capacity (i.e., less interference) due both to the independence of the trapping potential on *m/z* and the larger trapping volume compared to FTICR and Paul traps. Mass accuracy is good (2–5 ppm), and *m/z* range is at least 6000, with a dynamic range greater 1000. The complete high-resolution mass spectra can be acquired in less than 1 s. These specifications are from the initial introduction model in 2005 and have been improved with subsequent models. The Orbitrap analyzer stage illustrated in Figure 10.36 is incorporated with more MS source units: lenses and extra MS stages (quads, high-energy collisional dissociation [HCD], or electron transfer dissociation [ETD]) to produce higher-quality MS/MS spectra, and detectors. A particular element in the Orbitrap systems is the C-Trap, an element that transforms incoming ion beams into very short pulses more suitable for Orbitrap mass analyzer operation. All this can be engineered for very fast data acquisition in high-end systems, which allows characterization of narrow peaks from fast UHPLC separations. A typical such train is illustrated in Figure 10.37. The combination of all these improvements at a better price has severely diminished the market for FTICR systems, except when the highest possible resolution is needed and samples are clean and data can be sampled for long periods. An extensive online resource is the Thermo site at www.planetorbitrap.com. This is billed as "All Things Orbitrap," and the links under "technology" provide more detailed descriptions of the design and operation of the Orbitrap than there is space for in this text. If you register on the site, you can get access to a vast array of applications notes and product literature for the Orbitrap MS family and even view discussion boards.

10.2.4 Detectors

Most mass spectrometers measure one *m/z* value at a time. A single-channel ion detector is used for these instruments, either an **electron multiplier (EM)** or a **Faraday cup**. TOF, ion trap, and FTICR mass spectrometers have the ability to extract ions of many *m/z* values simultaneously, so simultaneous detection of these ions is desirable. One approach

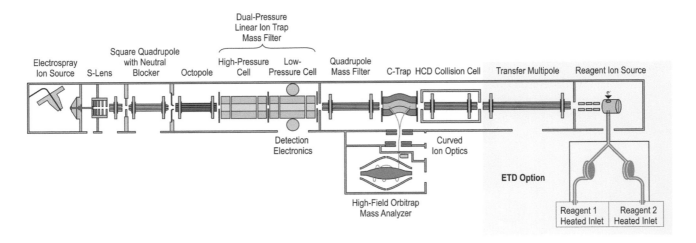

Figure 10.37 Diagram of a complete Hybrid Orbitrap-based MS system. (© Thermo Fisher Scientific. www.thermofisher.com. Used with permission.)

to multiple ion detection has been to use multiple detectors. Multiple detectors are also used for high-resolution magnetic sector MS instruments designed for very precise isotope ratio determination and for quantitative analysis using isotope dilution. Instruments with multiple detectors are called "multicollectors." New detector developments in array detectors now permit simultaneous *m/z* measurement over a wide mass range, such as the SPECTRO MS instrument.

10.2.4.1 Electron Multiplier

The most common detector used for ions in mass spectrometers is the EM. The EM is very similar in concept to the photomultiplier tube for optical detection. It is very sensitive and has a fast response. The EM is based on the dynode, which has a surface that emits electrons when struck by fast-moving electrons, positive ions, negative ions, or neutrals. A **discrete-dynode EM** uses a series of 12–24 dynodes, each biased more positively than the preceding dynode. A collision releases several electrons from the dynode surface. These electrons are then accelerated to a second such surface, which, in turn, generates several electrons for each electron that bombards it. This process is continued until a cascade of electrons (an amplified current) arrives at the collector. The process is shown schematically in Figure 10.38. Typically, one ion can produce 10^5 electrons or more; this ratio of electrons measured per ion is referred to as the gain. The gain of the detector can be adjusted, with operating gains of 10^4–10^8 used, depending on the application. An animation of the operation of an EM can be viewed at www.sge.com/products/electron_multipliers. Figure 10.38b shows a commercial discrete-dynode EM. A **continuous dynode EM**, also called a channel EM (CEM), uses a continuous glass tube, either lead-doped or coated on the inside with a conductive surface of high electrical resistance, such as those shown in Figure 10.39. A potential difference is applied across the tube ends so that the potential varies in a linear manner along the tube. Each incident ion releases electrons that are accelerated and strike the tube again, resulting in the same cascade effect seen in the discrete-dynode EM. The curved or coiled form is designed to reduce electrical noise by preventing positive ions from returning upstream.

A disadvantage to dynode-based detectors is that the number of secondary electrons released depends on the type of incident primary particle, its angle, and energy. The dependence of the number of secondary electrons emitted on incident energy is shown for electron impact in Figure 10.38c; the same plot for ion impact would be similar. Therefore, they can exhibit *mass discrimination* due to differences in ion velocity. Heavy ions from quadrupole mass analyzers and from QIT mass analyzers impact

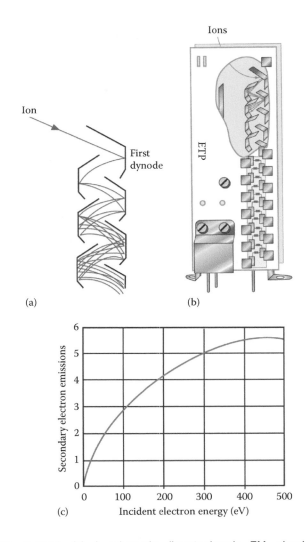

Figure 10.38 (a) A schematic discrete-dynode EM, showing the electron gain at each successive dynode after impact of an ion on the first dynode surface. The electron cascading process results in gains of up to 10^8 being achieved with approximately 21 dynodes. (b) An ETP EM schematic showing the position of the dynodes in the detector. (c) Dependence of the number of secondary electrons emitted on impact energy. (Images courtesy of SGE, Inc., Austin, TX, and ETP Electron Multipliers Pty Ltd, a division of SGE, Sydney, Australia. www.etspci.com and www.sge.com.)

the dynode surface at lower velocities than light ions. EM detectors for these instruments must be designed to overcome the difference in velocities, often by accelerating the ions prior to them striking the first electron-emitting dynode. An excellent source of information on how discrete dynode EMs work is the SGE website at www.sge.com, which describes their ETP EMs. Similarly, the Photonis website at www.photonis.com provides technical information on their Channeltron® continuous-dynode EM.

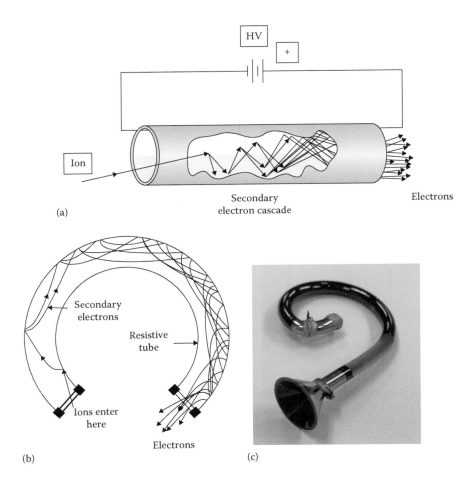

Figure 10.39 (a) A schematic CEM, consisting of a glass or interior-coated ceramic tube that emits secondary electrons upon ion impact. (b) A schematic curved CEM. The curved shape minimizes ion feedback noise. (c) Photo of the Channeltron® EM, showing the curved glass tube without the associated electronics. (Courtesy of Photonis USA, Fiskdale, MA, www.photonisusa.com.)

10.2.4.2 Faraday Cup

The least-expensive ion detector is the Faraday cup, a metal or carbon cup that serves to capture ions and store the charge. The resulting current of a few microamperes is measured and amplified. The cup shape decreases the loss of electrons from the metal due to ion impact. The Faraday cup is an absolute detector and can be used to calibrate other detectors. The current is directly proportional to the number of ions and to the number of charges per ion collected by the detector. Unlike dynode-based detectors, the Faraday cup does not exhibit mass discrimination. The detector does have a long response time, which limits its utility. The Faraday cup detector is used for making very accurate measurements in isotoperatio MS, where the ion currents do not change rapidly. The Faraday cup detector has no gain associated with it, unlike dynode-based detectors. This limits the sensitivity of the measurement.

This low sensitivity can be an advantage when combined with a high-sensitivity discrete dynode secondary EM (SEM) detector. The SEM detector can cover a linear dynamic range of nine orders of magnitude, but at a concentration range equivalent to ppq to ppm. If concentrations of elements in the sample are high, the detector will saturate. If one wants to analyze matrix elements (usually at % levels) as well as trace and ultratrace levels, as is often needed in geochemical analysis and with laser ablation analysis, the sample usually must be diluted and/or run multiple times under several sets of conditions. By adding a Faraday detector to an ICP-MS system, levels of all elements, % to ultratrace, can be run at one time (still with sequential detection of the elements). Such a high-resolution system with a dual-mode discrete dynode-type detector and a single Faraday detector is marketed by Thermo Fisher Scientific as the ELEMENT XR. The system automatically switches between detection modes with <1 ms delay (Figures 10.21 and 10.22).

High-precision isotope ratio mass spectrometers are designed with combinations of multiple Faraday cup detectors and multiple miniature EMs (used as ion counters) for simultaneous isotope measurement. These instruments are called multiple ion collector mass spectrometers (e.g., multiple collector [MC]-ICP-MS). For example, the TRITON and NEPTUNE MC-ICP-MS systems from Thermo Scientific can be configured with up to nine Faraday cups and eight ion counters to detect 17 ion beams simultaneously. Details of these instruments can be found at www.thermo.com. Other high-precision double-focusing sector field MC-ICP-MS systems are available from Nu Instruments and GV Instruments, both located in the UK. The use of multiple ion collector instruments improves precision by two to three orders of magnitude over a single collector magnetic sector instrument, and this high precision is needed for accurate isotope ratio measurements.

10.2.4.3 Array Detectors

The microchannel plate is a spatially resolved array detector formed of $10^5–10^7$ continuous-dynode EMs,

(a)

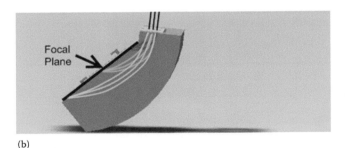

(b)

Figure 10.40 (a) The 4800-channel array detector in the SPECTRO MS. (b) The detector mounted into the focal plane of the magnet. (Courtesy of SPECTRO Analytical Instruments, Inc., AMETEK® Materials Analysis Division, www.spectro.com, www.ametek.com. Used with permission.)

each only 10–100 μm in diameter. This detector is used in focal plane mass spectrometers as a replacement for photograph plate detectors and is used in some TOFMS instruments.

A new 12 cm array detector with 4800 separate channels (detectors) is employed in the SPECTRO MS simultaneous double-focusing sector field ICP-MS (Figure 10.40). Each channel consists of two separate detectors with different signal amplification electronics (high and low gain). A nondestructive read-out algorithm adjusts the integration time as needed in real time, permitting low- and high-intensity signals to be measured simultaneously.

Each detector in high and low gain can cover 4.5 orders of magnitude in signal range, with an overlap of about 1 order of magnitude, resulting in about 8 orders of dynamic range. The nondestructive read-out algorithm also works on transient signals, such as those from HPLC or laser ablation. As seen schematically for two different *m/z* ions in Figure 10.40b, each *m/z* ratio is focused and detected on a different part of the detector. This permits fully simultaneous detection from Li to U with no scanning.

The *focal plane camera* (FPC), still in initial development, consists of an array of 31 Faraday cups, each 145 μm wide. Up to 15 *m/z* values can be measured simultaneously. This detector shows improved precision compared with single-channel detectors and has the ability to measure fast transient signals such as those from laser ablation. The detector design is described in the references by Barnes et al. and Knight et al. cited in the bibliography.

10.3 ION MOBILITY SPECTROMETRY

A final variation of MS is a method that separates ions (primarily unfragmented molecular ions) by a combination of their *masses*, *charges*, *sizes*, and *shapes*, using their **mobility** in a gas atmosphere in a **drift tube** under the influence of an applied electric field. Unlike classical MS, where the separation by *m/z* only occurs under high vacuum conditions such that the residual gas molecules have no significant effect on the ion trajectories, **ion mobility spectroscopy (IMS)** separates ions under low but significant (several mm Hg to atmospheric pressure) gas pressures. The resistance of this gas to the ion's motion varies with the previous parameters, so the ions arrive at a detector plate at times determined by their mobility in the gas, rather than their initial velocity out of an ion source as in TOF-MS, which also often uses "drift tubes" of varying geometries. The ion's drift velocity v_d is proportional to the applied field E by the mobility factor K, that is, $v_d = KE$. K is often reported as a reduced mobility (its value at standard temperature and pressure [STP], where $T = 273°K$ and $P = 1013$ mbar). The ions are sampled into the drift tube in narrow pulses with minimal velocity distributions, and they migrate against the gas, arriving at the detector plate at different times. The physics is vaguely analogous to the electric-field-generated migration currents of ions in liquid solutions seen in polarography (cf. Section 16.3.4), and the ion mobility differences that determine the dissolved ion flow rates and directions in capillary zone electrophoresis (CZE, cf. Section 14.6.1). The sample molecules are usually ionized by corona discharge, atmospheric pressure photoionization (APPI), ESI, or by radioactive sources such as ^{63}Ni or ^{241}Am, as are used in the GC ECD detector (cf. Section 13.7.3). A drift tube's resolving power R can be calculated as

$$R = \frac{t}{\Delta t} = \left(\frac{LEQ}{16kT \ln 2}\right)^{\frac{1}{2}} \qquad (10.17)$$

where
L is the drift tube length
Q is the ion charge
E is the electric field strength
k is the Boltzmann constant
T is the drift gas temperature (K)

The simplest form of IMS is TOFIMS, where the speeds of ion travel through the drift tube are registered as the time of arrival at the detector from the fastest to the slowest. In simple systems this is just a Faraday plate to collect the ions, but in more advanced systems it may be a mass spectrometer where both size and mass information may be acquired simultaneously. Operation at ambient pressure is best for portable handheld or field-deployable monitors, of which more than 50,000 have been used by the US Army alone. Ambient pressure is also used in IMS detectors for GC, LC, and SFC (see Chapters 13 and 14). Reduced pressure (several torr) IMS is more complex to operate, but enables ion focusing for more sensitivity and flexibility, and easier interfacing to MS systems. **Differential mobility spectrometers (DMS)** vary the field strength E at high values for different lengths of time. As ion mobility K depends on E, the result is that only ions of a narrow mobility range survive passage through the drift tube, and the DMS can be operated as a scannable ion filter, with improved sensitivity and selectivity.

IMS units can vary from handheld devices for ambient detection of explosives and volatile toxic agents on the battlefield or around pipelines or manufacturing and refining plants, to stages in high-end MS systems which can add shape or size discrimination to the mass information in molecular structure determination or isolation. This helps to resolve some isomer characterization problems, and even to let MS distinguish among different tertiary structure folded forms of proteins of the same mass and amino acid sequence. Such identifications may be important in unraveling the causes of diseases such as Alzheimer's, which is suspected to arise from protein misfolding. Four examples of IMS instruments for these different applications are discussed here.

Figure 10.41 JUNO® handheld DMS vapor detector. (Used with permission from Chemring Sensors and Electronic Systems, www.chemringds.com.)

10.3.1 Handheld DMS JUNO® Chemical Trace Vapor Point Detector

Figure 10.41 displays a DMS vapor detector, battery powered, weighing only 2 lb. Use of DMS mode improves sensitivity and allows it to be tuned to detect specific compounds, improving selectivity and allowing it to detect trace levels with fewer false alarms. It can monitor most chemical warfare agents, particularly the five major nerve agents, and toxic industrial vapors such as HCN, Cl_2, H_2S, HNO_3, and HF, while it is not affected by common interferents including paint, Windex®, engine exhaust, or N,N-diethyl-meta-toluamide (DEET) (insect spray). The particular DMS mode used in this instrument employs two high-field, high-frequency asymmetric potential waveforms superposed on a low potential DC field. Ions that have similar mobilities under the low-field conditions generally used in IMS instruments are distinguished by the nonlinear dependence of mobility under high-field conditions. DMS also enables simultaneous detection of both positive and negative ions.

10.3.2 Excellims HPIMS-LC System

High-performance ion mobility spectrometry (HPIMS™) can be combined with and is "orthogonal" (i.e., measures different aspects of a molecule) to MS (distinguishing by m/z ratio) and chromatography (separation based on polarity differences). A compact IMS unit is marketed by Excellims for interfacing ("hyphenation") with other manufacturers' separation instrumentation (HPLC or MS). A diagram of the unit, about the size of a small paperback novel, together with an illustration of the action on three different compounds in the ion desolvation and drift regions, is shown in Figure 10.42. The desolvation step is required as this is a LC-ESI-interfaced system. The unit can be "bolted on" to its parent instrument, either before (MS) or after (HPLC) that instrument's operation.

10.3.3 Photonis Ion Mobility Spectrometer Engine

This foot-long unit is diagrammed in Figure 10.43 and is sold for incorporation with other manufacturers' MS components. It can operate with DESI, DART, electrospray, corona, or radiation ion sources. It features a 30 mm diameter × 250 mm drift tube, a Bradbury–Nielsen (B-N) design ion gate that has been photoetched, and is therefore much simpler and easier to maintain than the traditional wire grids. It requires line power, has a heater capable of 125 °C, operates with drift voltages up to 15 kV, and achieves high IMS resolving powers of 64–150 depending on the ions in question. Key components are constructed from novel resistive glass to replace conventional lens assemblies, providing one-piece construction that produces minimal ion divergence and a uniform counterflow of drift gas without a need for additional confinement. This is an example of a product that is not sold as a complete instrument but rather for incorporation with other components and software; hence, its description as an "engine." Here, Photonis is operating as an original equipment manufacturer (OEM), but this engine will likely be marketed as part of a more complex system under other names.

10.3.4 SYNAPT G2-S Multistage MS System Incorporating the TriWAVE Ion Mobility Stage

Our final high-end application of IMS is as a stage in a very powerful (and expensive) system for characterizing large biomolecules from Waters Corporation. In Figure 10.44, the ion mobility (IM) section is labeled by the trade name TriWAVE® and consists of three tandem multiring electrode assemblies. The first one traps ions exiting a quadrupole analyzer stage and

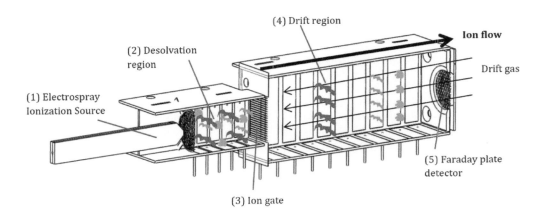

Figure 10.42 Excellims HPIMS device operation diagram. (Used with permission from Excellims Corp. www.excellims.com.)

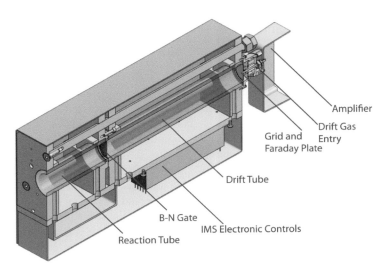

Figure 10.43 Diagram of Photonis ion mobility spectrometer engine. (Used with permission from Photonis USA, Fiskdale, MA, www.photonis.com.)

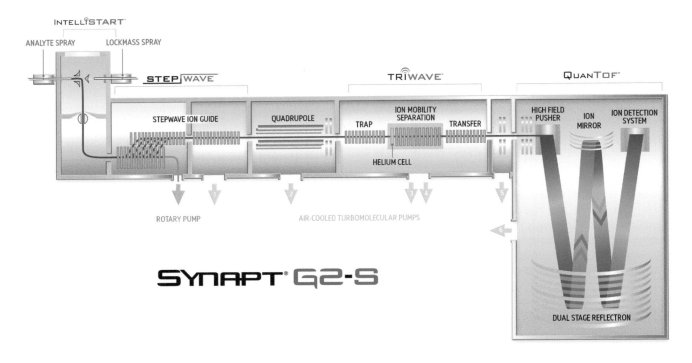

Figure 10.44 Waters SYNAPT G2-S System with TriWAVE® IMS Stage. (Used with permission from Waters Corp. www.waters.com.)

feeds them slowly to the second; the second is the IMS itself, which separates individual ions isolated by the preceding Q stage by their mobility, adding shape and size discrimination to the analysis; and the third then transfers the sequence of IMS-selected ions to a HRT MS (QuanTOF) for accurate mass determination. Inspection of the diagram will reveal other components whose function we have discussed elsewhere in this chapter. If this were coupled to an LC, one might abbreviate the whole train as [HPLC-ESI-Q-IMS-TOF MS] and at that point one really might need to consult the Acronym Index at the end of this book! As is the case with military units, these mix-and-match MS systems are more often discussed by their acronyms than by their full names. Another excellent vendor online resource is "The Mass Spectrometry Primer" from the supplier of the SYNAPT system, Waters Corporation, accessible at www.waters.com//primers. This 80-page, extensively illustrated booklet contains brief descriptions of all significant MS instrument designs and, in particular, describes in greater detail the operation of the IMS TriWave® stage in the SYNAPT system illustrated in Figure 10.44. In general, the websites of major MS manufacturers such as Agilent, Bruker Daltonics, JEOL, LECO, PerkinElmer, SPECTRO, Thermo Fisher Scientific, and Waters can be accessed through Internet search engines like Google. They will provide elaborate menu trees of instruments, operation tips, troubleshooting tips, and a host of applications and sample spectra. Smaller companies mentioned in the credit lines of figures in this chapter also provide more limited and specialized information.

10.4 LEAKS IN MASS SPECTROMETER SYSTEMS

Earlier in this chapter we discussed the use of molecular leaks as a sample introduction system. There, a leak was used to our advantage. The fact is, however, that mass spectrometers that operate under high vacuum are often plagued with leaks, that is, unwanted entry of air or other contaminants from the laboratory. A leak would be manifest with an inability to tune the instrument, high background (often described as grassy ion syndrome, since the display looks like grass), or spurious peaks that are present with all samples. The first step to solving a suspected leak problem is to determine if there really is a leak. Instrument malfunctions can occur that may at first seem like a vacuum system leak. The grassy ion display can be a malfunction in the mother board, and a poor vacuum measurement by an ion gauge can simply result from an ion gauge that has reached the end of its useful life. Moreover, what seems to manifest as a leak can simply be months (or years) of accumulated crud in the analyzer or source (crud is a technical term to describe all manner of detritus). In this case, there is no alternative but to open the instrument up for an overhaul and cleaning. The last section of the next chapter has some additional information about contamination and spurious signals.

After a column change or cleaning, the instrument will need to be pumped back down to vacuum. During this process, you will note peaks for nitrogen to oxygen (at 28 and 32) showing a ratio of approximately 4:1. If these peaks diminish during pump down, but poor behavior persists, then there is probably not an external leak but rather a dirty instrument. If, however, you have checked and have

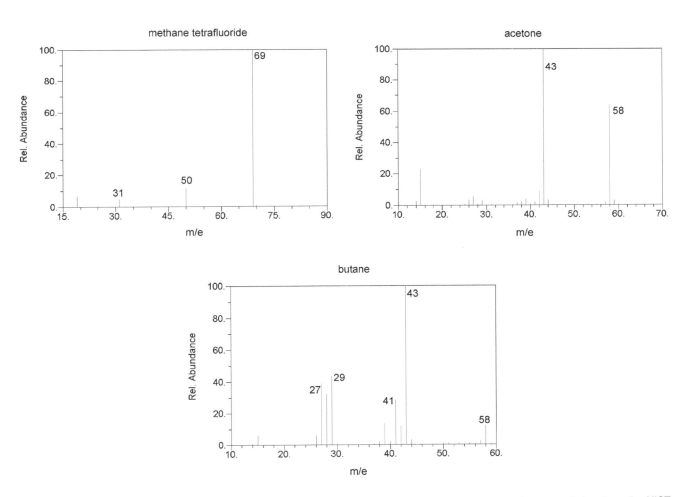

Figure 10.45 The mass spectra methane tetrafluoride, 1,1,1,2-tetrafluoroethane (R-134a), n-butane and acetone (taken from the NIST Chemistry WebBook).

confidence in the ion gauge and the instrument is clean, you must suspect a leak. Some laboratories possess a helium leak detector, which is actually a separate mass spectrometer that is tuned to just detect He. Many older units are still in use; you will recognize them because they look like a washing machine, and are equally as heavy and cumbersome. Modern He leak detectors are portable and benchtop, and are easy to use. The He leak detector will need to be somehow connected to the vacuum system of the mass spectrometer via a port or fitting, usually provided as an auxiliary fitting. Once connected, the operator plays a gentle flow of He (from a compressed gas cylinder; see Chapter 2 for safety guidelines) on each fitting, flange, or joint. There will be an audible alarm and a visual display on the He leak detector indicative of a leak, allowing necessary repairs to be made.

If it is safe to operate the instrument (that is, if the leak is not so severe that operation will not burn out the source), one can employ other fluids (gases and liquids) and use the instrument as its own leak detector. In these cases, the fluids are gently applied to the various suspected components. One must be patient and wait for a response, and not jump from place to place. A stream of the gas sulfur hexafluoride (SF_6) will show a peak at a mass of 127 in EI mode, for example. Not all labs will have SF_6 on hand but there are other fluids that are useful. Moreover, SF_6 is a large molecule and will move much more slowly through the system than He. In Figure 10.45 we present the mass spectra of methane tetrafluoride, 1,1,1,2-tetrafluoroethane (R-134a), n-butane, and acetone. Methane tetrafluoride, 1,1,1,2-tetrafluoroethane, and n-butane are handled as gases, while acetone is handled as a liquid. Typically, n-butane is dispensed from a disposable lighter, and acetone is dispensed from a dropper. Care must be taken when using acetone or a butane lighter for leak checking because of the flammability of these fluids. One then looks for the more abundant peaks in these mass spectra.

An important take away from the foregoing discussion is that if a leak is suspected, approach the problem systematically and do not begin dismantling the instrument without good cause; this will likely only make things worse. And, clearly, it is far better to avoid leaks in mass spectral instruments than to have to locate and fix them. It is good practice to replace fitting flanges and gaskets on the recommended schedule rather than waiting for a problem. While replacing a gasket or O-ring, it is essential to keep everything clean; a hair on an O-ring will make for a leak path. Never try to reuse a Vespel® or graphite ferrule or seal; get a new one.

PROBLEMS

10.1 What is meant by "molecular ion"? What is the importance of identifying it to a chemist?

10.2 How are resolving power and resolution defined for a mass spectrometer?

10.3 What is the difference between hard ionization and soft ionization?

10.4 Describe how an EI source forms ions from analyte molecules. Is this a hard or soft ionization source? What are the advantages and disadvantages of this source?

10.5 Describe how a CI source forms ions from analyte molecules. Is this a hard or soft ionization source? What are the advantages and disadvantages of this source?

10.6 Draw a block diagram of a typical mass spectrometer.

10.7 Consider an ion with $z = 1$. Calculate the kinetic energy added to the ion by acceleration through a potential of 1000 V in an EI source. Does the kinetic energy acquired depend upon the mass of the ion? Explain.

10.8 In an EI source, does the mass of the ion affect its velocity? Explain.

10.9 Describe two methods for ionizing liquid samples for mass spectral analysis.

10.10 Diagram a transmission quadrupole mass analyzer. What are the advantages and disadvantages of this mass analyzer compared with a double-focus mass spectrometer?

10.11 What resolution is necessary to differentiate between
(a) An ion with $m/z = 84.0073$ and one with $m/z = 84.0085$?
(b) $ArCl^+$ and As^+?
(c) $C_4H_6O_2^+$ ($m/z = 86.0367$) and $C_5H_{10}O^+$ ($m/z = 86.0731$)?

10.12 How would you expect the EI and CI mass spectra of stearyl amine (RMM = 269) to differ? If the CI mass spectrum is collected using methane as a reagent gas, would you expect to see any peaks in the spectrum at $m/z > 269$? Explain.

10.13 Diagram the process that occurs in MALDI. What are the advantages of this technique?

10.14 Diagram a simple TOF mass spectrometer and explain how mass separation is achieved in a TOF mass analyzer. What is the advantage to a reflectron TOF over a TOF with a straight drift tube?

10.15 Explain the steps in an MS/MS experiment. Why are MS^n experiments valuable?

10.16 Diagram an ESI source and describe its uses.

10.17 Diagram a discrete-dynode EM and describe its operation.

10.18 Diagram a CEM and describe its operation.

10.19 What is meant by "gain" in an EM?

10.20 Compare the advantages and disadvantages of the Faraday cup detector and EMs.

10.21 What is meant by FTMS? What is the most common type of FTMS instrument?

10.22 Compare the principles of operation, advantages, and disadvantages of the QIT and the ICR. How are they used in MS^n experiments?

10.23 Describe or illustrate the ion trajectories in an Orbitrap analyzer. Why do you think it is less susceptible to space–charge resolution degradation than other trap designs? Compare the acquisition of image currents in the Orbitrap and an FTICR instrument. Why does the latter's longer full-scan acquisition period for maximum resolution require higher vacuum operation? Which is more suited for interfacing with UHPLC sample introduction? Why?

10.24 What fundamental properties of ions are the bases of separations in IMS versus all other forms of "classical" MS. Which is better for small metabolite molecule characterization and which for misfolded protein discrimination?

10.25 Why is a low-mass IMS spectrometer easier to engineer for a cheap, simple, handheld scanner for detecting target toxic vapors than a quadrupole, trap, TOF, or permanent magnetic sector MS?

10.26 If the source, analyzer, and detector stages of the latter MS designs in Problem 10.25 could be ultraminiaturized using silicon chip manufacture techniques (a project well under way as of this writing), how might handheld, field-portable versions have performance superior to the IMS designs. [Speculate!]

10.27 The long-standing problem of how to achieve atmospheric pressure vapor introduction or even more demanding LC effluent interfacing was eventually solved by the APCI, ESI, DART, etc., source designs. These employ sprays, skimmer cones, and "differential pumping" to separate the analytes from the matrix and present it to the higher vacuum environment required for the analyzer stage(s). Outline how these features function. Why is this less necessary for IMS spectrometers?

BIBLIOGRAPHY

The reader will note that occasionally multiple editions of books are cited as sources or reference works. The reason for this is that sometimes different figures are used in different editions, which may in fact be written by different authors. The sources found in the bibliography will be the most recent versions of a particular work.

In some cases instrument manufacturers and suppliers are cited by providing a URL to the exact internet location. We have endeavored to ensure that these links are functional at the time of publication, but clearly these links change. If a link has become nonfunctional, a good approach is to use your browser's advanced search capability. For example: "company name" AND "instrument type" AND "specific descriptors" is a good place to start.

Alexander, M.L.; Hemberger, P.H.; Cesper, M.E.; Nogar, N.S. Laser desorption in a quadrupole ion trap: Mixture analysis using positive and negative ions. *Anal. Chem.*, 65, 1600, 1993.

Barker, J. *Mass Spectrometry: Analytical Chemistry by Open Learning*, 2nd edn. Wiley: Chichester, U.K., 1999.

Barnes, J.H., IV; Hieftje, G.M.; Denton, M.B.; Sperline, R.; Koppenaal, D.W.; Baringa, C. A mass spectrometry detector array that provides truly simultaneous detection. *Am. Lab.*, 35, 15, October 2003.

Becker, J.S. *Inorganic Mass Spectrometry: Principles and Applications*. John Wiley & Sons, Ltd.: Chichester, U.K., 2007.

Bruno, T.J., Svoronos, P.D.N., *CRC Handbook of Basic Tables for Chemical Analysis – Data Driven Methods and Interpretation*, 4th Ed., CRC Taylor and Francis, Boca Raton, FL, 2021.

Buchanan, M.V.; Hettich, R.L. Fourier transform mass spectrometry of high-mass biomolecules. *Anal. Chem.*, 65(5), 245, 1993.

Caprioli, R.M. Continuous flow fast atom bombardment mass spectrometry. *Anal. Chem.*, 62(8), 177A, 1990.

Cody, R.N.; Tamura, L.; Musselman, B.D. Electrospray ionization/magnetic sector MS calibration and accuracy. *Anal. Chem.*, 64, 1561, 1992.

Cole, R.B. *Electrospray and MALDI Mass Spectrometry*. Wiley: New York, 2010.

Dass, C. *Fundamentals of Contemporary Mass Spectrometry*. Wiley: New York, 2007.

De Hoffman, E. *Mass Spectrometry—Principles and Applications*, 3rd edn. Wiley: New York, 2007.

Duncan, W.P. *Res. Dev.*, 57, April 1991.

Eiceman, G.A. *Ion Mobility Spectrometry*, 3rd edn. Taylor & Francis: Boca Raton, FL, 2013.

Fenn, J.B.; Mann, M.; Meng, C.K.; Wong, S.E.; Whitehouse, C. Electrospray ionization for mass spectrometry of large biomolecules. *Science, 64*, 246, 1989.

Goldner, H.J. Electrospray excites scientists with its amazing potential. *R&D Mag.*, 10, 43, 1993.

Gross, J.H. *Mass Spectrometry—A Textbook*. Springer VCH: Berlin, Germany, 2011.

Karas, F.H.; Beavis, R.C.; Chait, B.T. Matrix-assisted laser desorption ionization mass spectrometry of biopolymers. *Anal. Chem.*, 63(24), 1193A, 1991.

Karas M, Bahr U, Ingendoh A, Nordhoff E, Stahl B, Strupat K, Hillenkamp F. Principles and applications of matrix-assisted UV-laser desorption/ionization mass spectrometry. *Analytica Chimica Acta*. 241: 175–185.

Knight, A.K.; Sperline, R.P.; Hieftje, G.M. et al. The development of a micro-Faraday array for ion detection. *Int. J. Mass Spectrom.*, 215, 131, 2002.

Lambert, J.B.; Shurvell, H.F.; Lightner, D.; Cooks, R.G. *Introduction to Organic Spectroscopy*. MacMillan Publishing Company: New York, 1987.

Mahoney, J.; Perel, J.; Taylor, S. Primary ion source for FABMS. *Am. Lab.*, 92, March 1984.

March, R.E. *Quadrupole Ion Trap Mass Spectrometry*, 2nd edn. Wiley: New York, 2005.

McLafferty, F.W.; Tureček, F. *Interpretation of Mass Spectra*, 4th edn. University Science: Mill Valley, CA, 1993.

Montaser, A. (ed.) *Inductively Coupled Plasma Mass Spectrometry*. VCH: Berlin, Germany, 1998.

Niessen, W.M.A., J., *Liquid Chromatography–Mass Spectrometry*, 3rd ed, CRC Taylor and Francis, Boca Raton, FL, 2006.

Silverstein, R.M.; Webster, F.X. *Spectrometric Identification of Organic Compounds*, 6th edn. Wiley: New York, 1981.

Smith, R.D.; Wahl, J.H.; Goodlett, D.R.; Hofstadler, S.A. Capillary Electrophoresis-Mass Spectrometry. *Anal. Chem.*, 65(13), 574A, 1993.

Smith, R.M.; Busch, K.L. *Understanding Mass Spectra—A Basic Approach*. Wiley: New York, 1999.

Sparkman, O.D. *Mass Spectrometry Desk Reference*. Global View Publishing: Pittsburgh, PA, 2000.

Su, Y., Yu, L-R., Conrads, T.P., Veenstra, T.D., Grinberg, N., Rodriguez, S., eds., Ewing's Analytical Instrumentation Handbook, 4th Ed., CRC Press Taylor and Francis, Boca Raton, FL, 2018.

Svartsburg, A.A. *Differential Ion Mobility Spectrometry*. Taylor & Francis: Boca Raton, FL, 2008.

Thomas, R. *Practical Guide to ICP-MS*, 2nd edn. Taylor & Francis: Boca Raton, FL, 2011.

Voreos, L. Electrospray mass spectrometry. *Anal. Chem.*, 66(8), 481A, 1993.

Wilkins, C.L. *Ion Mobility Spectrometry—Mass Spectrometry*. Taylor & Francis: Boca Raton, FL, 2010.

CHAPTER 11

Mass Spectrometry II
Spectral Interpretation and Applications

J.J. Thompson in 1913 first used mass spectrometry (MS) to demonstrate that neon gas consisted of a mixture of nonradioactive isotopes, ^{20}Ne and ^{22}Ne. The atomic mass of neon listed in a modern periodic table is 20.18. Thomson obtained two peaks in the mass spectrum of neon, at masses of 20 and 22 with a roughly 10:1 intensity ratio, but no peak at mass 20.18. This work was revolutionary because it demonstrated that elements existed as isotopes with different atomic masses and simultaneously explained why the apparent atomic mass of an element based on chemical reactions was not a whole number. Neon in fact has three natural isotopes, but ^{21}Ne is present in much smaller amounts than the other two isotopes. In 1923, Francis W. Aston used a higher-resolution instrument he designed to determine the atomic masses of the elements and the isotope ratios of each particular element. This was extremely useful to inorganic chemists and helped solve many of the problems concerning the position of elements in the periodic table at that time. During World War II, Nier at the University of Minnesota developed the high-resolution double-focusing instrument that permitted the analysis and separation of ^{235}U from ^{238}U, aiding in the development of the atomic bomb.

In the 1940s, the first commercial mass spectrometers were developed for petroleum analysis. Subsequent instrument developments, many of them only in the past decade, have led to the widespread use of MS in many branches of science. It has been estimated (Busch) that a billion mass spectra are recorded daily.

MS is a powerful analytical tool with vast applications in organic, inorganic, environmental, polymer, and physical chemistry, physics, geology, climatology, paleontology, archeology, materials science, biology, and medicine. Advances in MS instrumentation have made possible major advances in our understanding of the human genome, protein structure, and drug metabolism. For example, intact viruses of millions of daltons have been analyzed by MS using electrospray ionization (ESI) with retention of viral activity and structure (Fuerstenau et al.). Commercial hyphenated gas chromatography (GC)-MS^n and liquid chromatography (LC)-MS^n systems permit rapid, sensitive biomonitoring of humans for exposure to chemicals, including chemicals used by terrorists. It is impossible to cover all applications of MS in one chapter, but examples of important uses of both molecular and atomic MS will be presented. In addition, this chapter introduces the interpretation of simple mass spectra for the identification of molecules.

11.1 INTERPRETATION OF MASS SPECTRA: STRUCTURAL DETERMINATION OF SIMPLE MOLECULES

The major reasons for learning to interpret mass spectra of molecules are so that the structure of an unknown compound can be deduced and an unknown molecule can be completely and unambiguously identified. For even fairly small organic molecules, a thorough knowledge of structural organic chemistry and lots of practice are required to do this. Fortunately, simple mass spectra often can be interpreted using basic arithmetic and some topics you learned in general chemistry: chemical bonding, valence, isotopes, atomic masses, and relative molecular mass (RMM) calculations. For completely unknown molecules, more than one mass spectral technique may need to be combined with other analytical techniques to positively identify a molecule. As will be seen, there are some types of molecules that cannot be identified using MS alone.

A mass spectrum is a plot or table of the mass-to-charge ratio, m/z, of detected ions versus their relative abundance (relative concentration). A typical mass spectral plot for a small organic molecule, benzene, is presented in Figure 11.1. The m/z values are plotted on the x-axis; relative abundance is plotted on the y-axis. The *most abundant peak* in the spectrum is called the **base peak**. The base peak is assigned an abundance of 100 % and the other peak heights are plotted as percentages of that base peak. A tabular form of benzene mass spectral data is given in Table 11.1. The tabular data have the advantage that very low abundance

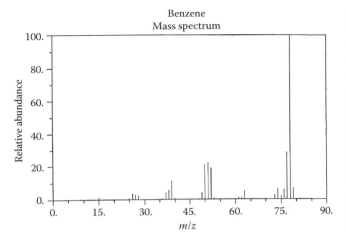

Figure 11.1 A mass spectrum of benzene, C_6H_6. (From the NIST Mass Spectrometry Data Center accessed 2022 via http://webbook.nist.gov. U.S. Secretary of Commerce for the United States of America. All rights reserved. Used with permission.)

Table 11.1 Mass Spectral Data for Benzene

m/z	Relative Abundance	m/z	Relative Abundance
37	4.0	53	0.80
37.5	1.2	63	2.9
38	5.4	64	0.17
38.5	0.35	73	1.5
39	13	74	4.3
39.5	0.19	75	1.7
40	0.37	76	6.0
48	0.29	77	14
49	2.7	78	100
50	16	79	6.6
51	18	80	0.18
52	19		

ions can be listed, such as the ions at $m/z = 64$ and 80, which are too small to be seen on the normalized plot.

The mass spectra we will study have been obtained by electron ionization (EI) since other techniques enhance the molecular ion but greatly reduce fragmentation, making structural analysis difficult. Consequently, the following discussion is concerned mostly with EI spectra of pure compounds. The ions that appear in these spectra are all positively charged ions. These are also low-resolution mass spectra; the m/z values are measured only to the units place. (We will explain the half-integral values in Table 11.1 later in this chapter.) Exact RMMs, which are not integer values, are obtained by high-resolution MS, also discussed later in this chapter.

There are two ways to interpret such spectra. The first is to compare the spectrum you have with those in a searchable computerized mass spectral database. The second is to evaluate the spectrum using the interpretation procedure described

subsequently. In either case, once an unknown compound has been "identified" from its mass spectrum, the pure compound should be obtained and analyzed under the same conditions as the sample for confirmation. Over 10 million chemical compounds have been identified. No mass spectral database contains spectra for every possible compound, although mass spectral databases of over 719,000 compounds are available, such as the Wiley Registry of Mass Spectral Data, 10th edition (www.wiley.com/go/databases), ~$8000 on DVD-ROM. The mass spectral database from the US National Institute for Standards and Technology (NIST) (www.nist.gov/srd/nista.cfm) contains EI spectra for 306,869 compounds in the 2020 release (up to date as of May, 2023) and may be purchased from a number of licensed vendors (this database is not distributed directly). This database is unique in that it also contains 112,253 compounds with both mass spectral data but also chromatographic retention index data to make identification more reliable. Many mass spectra from NIST are also available online in the NIST Chemistry WebBook (http://webbook.nist.gov). Commercial vendors and publishers offer specialized mass spectral libraries of compounds, such as environmental compounds, pharmaceuticals, natural products, and oil industry compounds. As discussed in the following, many compounds of the same empirical formula may exist as multiple structural isomers, with substituents such as halogen or alkyl groups attached at different locations. For example, there are 46 possible different pentachlorobiphenyls. Such isomers may have essentially indistinguishable EI mass spectra. As we will see in Chapter 13, these may be separable by GC coupled to an MS instrument and the isomers distinguished by their GC retention index (RI) (cf. Section 13.9) values. Many MS spectral databases also provide some GC-RI values for this purpose.

In practice, the analyte spectrum is entered into the computer, which compares it to the spectra in the stored database using a search algorithm. There are a number of algorithms currently available, including probability-based matching, designed by Professor F. W. McLafferty and coworkers at Cornell University; the INCOS dot-product algorithm; and the NIST library search algorithm. These algorithms use pattern matching, peak intensities, weighting factors, and other information from the spectrum to compare the candidate spectrum to spectra in the library database. The search will result in a list of possible candidate compounds with a probability attached to the "match." This match quality is based on intensity and on the ion mass, with greater weight being given to ions with higher mass. The analyst should visually compare the candidate spectra to that of the analyte. Using knowledge, judgment, and experience, the analyst then chooses which of the candidate compounds matches the unknown compound. This spectral matching method in theory requires little training on the part of the analyst to identify the compound but requires pure compounds, a good mass spectrum of the sample, and a comprehensive mass spectral database. Users must be cautioned not simply to accept the highest "quality" match; it may in fact be incorrect. The database is no substitute for critical thinking. Use all information available, including chromatographic retention times and indices, when possible. When in doubt, the measurement of a pure component mass spectrum is an important technique. The inclusion of chromatographic retention indices in the NIST database is especially valuable when dealing with ambiguous situations.

The technique of identifying a pure compound by comparing its spectrum with known spectra works well if we already know something about the compound (e.g., odor, color, melting point, functional groups present from an infrared [IR] spectrum, and elements present from combustion analysis) and if the compound's mass spectrum is in our database. This enables us to make an informed deduction as to the identity of the compound by direct comparison with the library spectrum. Unfortunately, this is not always the case; more frequently, the nature of the sample is not known. Furthermore, the sample may not be pure, and therefore, direct comparison of spectra will not constitute a valid confirmation of the compound's identity. It is possible that spectra in a database are not identified correctly. It is not prudent to rely completely on a library match, especially for a complete unknown, since there are many cases of multiple compounds with very similar mass spectra; for example, polychlorinated biphenyls (PCBs), dioxins, and polychlorinated diphenyl ethers (PCDEs) with the same number of Cl substituents each having many structural isomers, and these often have nearly indistinguishable mass spectra. Indeed, it can be difficult to distinguish two such simple species as carbon dioxide and propane based on the mass spectra. Refer also to Figure 11.30 at the end of Section 11.2, which displays an unusual example of two radically different molecular formulas and structures yielding essentially identical mass spectra. In these circumstances, it is necessary to take the spectrum and deduce the information from it. Therefore, an analyst should understand how to identify the compound from the information in the mass spectrum; this ability will also enable the analyst to evaluate the validity of the spectral database search results. The procedure for interpreting a mass spectrum consists of the following steps:

1. Identify the molecular ion if present.
2. Apply the "nitrogen rule."
3. Evaluate for "A + 2" elements.
4. Calculate "A + 1" and "A" elements.
5. Look for reasonable loss peaks from the molecular ion.
6. Look for characteristic low-mass fragments.
7. Postulate a possible formula.
8. Calculate "rings + double bonds."
9. Postulate a reasonable structure.

We will define the terms in quotation marks as we go through the process. Interpretation of mass spectra is much like solving a jigsaw puzzle: all or most of the pieces are there, but putting them together involves a lot of educated guesswork followed by confirmation. Already mentioned above is the use of retention data as an aid. In addition, one should not neglect simple thermophysical properties. In the analysis of fuels, for example, one might encounter hundreds of hydrocarbon species that are isomeric, and a chromatographic column used for GC-MS will control in large measure how these are delivered to the mass spectrometer. Boiling temperature progressions, and interactions with polar or mildly polar stationary phases all provide guidance.

11.1.1 Molecular Ion and Fragmentation Patterns

When a molecule is ionized by EI, a **molecular ion** forms by loss of an electron from the molecule. This radical cation, symbolized as $M^{+\bullet}$, has the same mass as the neutral molecule, because the loss in mass of one electron is too small to measure. The molecular ion generally has absorbed excess energy in the ionization process. The excess energy causes the molecular ion to break apart. The fragments formed may be ions, neutral molecules, radicals, and the like. Fragments may undergo more fragmentation into even smaller pieces. The ions that appear in the mass spectrum are called fragment ions. If the ionization conditions are kept constant, a given molecule will always produce the same fragments in the same proportions. The mass and abundance of these fragments are called the molecule's **fragmentation pattern**. Each molecule has its own characteristic fragmentation pattern under particular ionization conditions.

Using low-resolution MS, the molecular ion, if present, will provide the RMM of the compound to unit mass accuracy, a very valuable piece of information. High-resolution spectra with high mass accuracy can reduce the number of possible elemental formulas for the molecular ion. The identification of the molecular ion is important because from the RMM, we can immediately conclude a great deal about the compound. For example, for an organic compound with an RMM = 54, we would be able to conclude that the compound can only have a limited number of empirical formulas, such as C_2NO, $C_2H_2N_2$, C_3H_2O, C_3H_4N, or C_4H_6. Some of these possibilities can be eliminated by inspection, because it is not possible to have a neutral molecule with certain empirical formulas. For example, C_2NO cannot exist as a neutral molecule. Thus, if we know the RMM, we have considerable information about the possible empirical formula of the molecule. MS is one of the most rapid and reliable methods of determining the RMM of compounds. Unfortunately, in EI spectra, the molecular ion is often of low abundance and is not always observed, or it cannot be distinguished from ions originating from low-level, unrelated background contaminants. In short, the m/z value of the molecular ion indicates the RMM of the molecule; the fragmentation pattern is used to deduce the structure of the molecule. The molecular ion is not

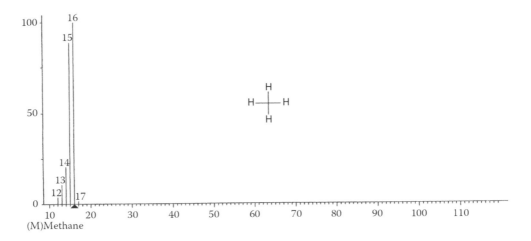

Figure 11.2 A mass spectrum of methane, CH_4.

necessarily the base peak in the spectrum; particularly in EI spectra, the base peak is usually not the molecular ion.

In this chapter, we shall emphasize compounds with molecular ions that can be identified or deduced with reasonable certainty. If the molecular ion is present, it must have the highest m/z in the spectrum, excluding the effects of isotopes. Examples are shown for methane (Figure 11.2, Table 11.2), methanol (Figure 11.3, Table 11.3), and benzene (Figure 11.4, Table 11.1). In each case, the molecular ion was very abundant and not difficult to identify. This is not always the case, as will be seen in later examples. The student should note that in most of the mass spectra used as examples, the molecular ion m/z value is marked by a black triangle on the x-axis. The x-axis is in units of m/z, while the y-axis is relative abundance, even though these units are not marked on the spectra. The most intense peak is set to 100 %, and the rest of the peaks are normalized to that peak. The structure of the compound is also shown on the spectrum, using a shorthand method that does not show the hydrogen atoms or the carbon atoms. The methanol spectrum (Figure 11.3) demonstrates clearly that the molecular ion is not the base peak in the spectrum; the fragment ion at $m/z = 31$ is the most abundant ion.

When a molecule is ionized by electron impact, it undergoes the reaction

$$M + e^- \rightarrow M^{+\cdot} + e^- + e^-$$

Thus, a molecular ion is always a radical cation, usually with a single positive charge. It is evident that if an organic molecule loses an electron, it must be left with an unpaired electron (i.e., it is a radical ion), as shown by the dot representing the unpaired electron. This radical ion with an unpaired electron is termed an *odd electron* (OE) *ion*. The molecular ion is always an OE ion. For example, in the reaction $C_2H_6 + e^- \rightarrow C_2H_6^{+\cdot} + e^- + e^-$, the $C_2H_6^{+\cdot}$ is the molecular ion.

Table 11.2 **Mass Spectral Data for Methane**

m/z	Relative Abundance
1	3.1
2	0.17
12	1.0
13	3.9
14	9.2
15	85
16	100
17	1.11

However, the molecule may fragment in such a way as to leave a pair of electrons behind on one ion. This is an *even electron* (EE) *ion*. Such an ion arises by *fragmentation* and *cannot* be the molecular ion. Hence, the molecular ion is *never* an EE ion. As an example, CH_3OH may ionize and fragment as follows:

$$CH_3OH + e^- \rightarrow CH_3OH^{+\cdot} (m/z = 32) + e^- + e^-$$

$$CH_3OH^{+\cdot} \rightarrow CH_2OH^+ (m/z = 31) + H^\cdot$$

The CH_2OH^+ ion is an EE ion and cannot be the molecular ion.

While a detailed discussion of fragmentation patterns and how to identify a molecular ion peak is beyond the scope of this chapter, there are certain classes of organic compounds that tend to give stable molecular ions and classes that do not. Aromatic compounds and conjugated hydrocarbons give more intense molecular ion peaks than alkanes; aliphatic alcohols, nitrates, and highly branched compounds tend not to give peaks for the molecular ion. Texts listed in the bibliography such as those by Silverstein and Webster, McLafferty and Tureček, Pavia et al., or Lambert et al. should be consulted for more details.

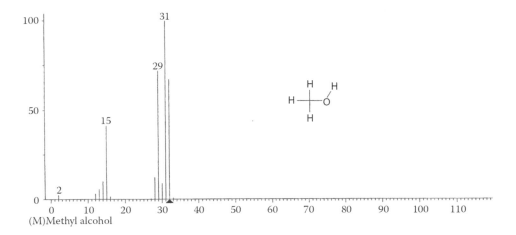

Figure 11.3 A mass spectrum of methanol, CH₃OH.

Table 11.3 Mass Spectral Data for Methanol

m/z	Relative Abundance
12	0.3
13	0.7
14	2.4
15	13.0
16	0.2
17	1.0
28	6.3
29	64.0
30	3.8
31	100
32	66
33	1.0
34	0.14

11.1.2 Nitrogen Rule

The "nitrogen rule" can be used to help decide if a peak is the molecular ion peak. Because of the atomic masses and valences of elements commonly present in organic molecules, it transpires that the *m/z* value of the molecular ion *is always an even number if the molecular ion contains either no nitrogen atoms or an even number of nitrogen atoms*. If the molecular ion contains an odd number of nitrogen atoms, the *m/z* value (and the RMM) must be an odd number. This is a very important rule to remember, and it is valid for organic compounds containing C, H, N, O, S, the halogens, P, Si, and many other elements. If the highest *m/z* value in a mass spectrum is an odd number, the ion must contain an odd number of nitrogen atoms if it is the molecular ion. To use this rule, RMMs are calculated using the most abundant isotope and unit atomic masses. This calculation is discussed in the next section. Some examples of the rule follow:

Compound	Number of Nitrogen Atoms	RMM	
CH₄	0	16	Even
HCN	1	27	Odd
H₂NNH₂	2	32	Even
C₂H₅OH	0	46	Even
C₆H₆	0	78	Even
C₂H₅NH₂	1	45	Odd
C₆H₅NH₂	1	93	Odd
C₆H₄(NH₂)₂	2	108	Even
C₉H₇N	1	129	Odd

11.1.3 Molecular Formulas and Isotopic Abundances

If you look at the mass spectrum for methane (Figure 11.2) and the corresponding Table 11.2, you note from the marked peak on the spectrum that the molecular ion is at *m/z* = 16. The RMM to the units place for methane is, of course, equal to the atomic mass of carbon plus four times the atomic mass of hydrogen, or (1)(12) + (4)(1) = 16. The table shows the presence of an additional small peak at *m/z* = 17. The peak at 17 exists because the element carbon has more than one isotope. The two stable isotopes of carbon are ^{12}C and ^{13}C. ^{12}C constitutes 98.90 % of all naturally occurring carbon atoms; ^{13}C is 1.10 % of all naturally occurring carbon. (A third carbon isotope, ^{14}C, is radioactive and does not need to be accounted for in organic structure determination but is very important for other applications, such as radiocarbon dating.) Therefore, in a compound containing one carbon atom, 98.90 % of the molecules will have a ^{12}C atom, while 1.10 % will have a ^{13}C atom.

In every organic compound (natural or synthesized from natural sources), the ^{13}C will result in a peak that is one

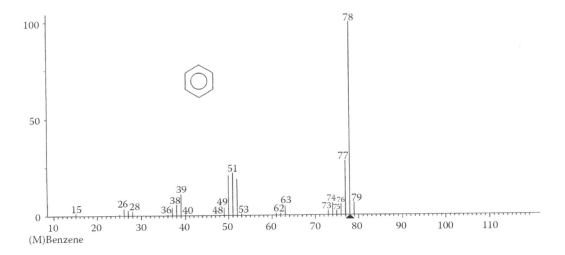

Figure 11.4 Another mass spectrum of benzene, C_6H_6. Compare with Figure 11.1.

mass number greater than the mass of the molecular ion. This ^{13}C-containing peak is generally designated as (M + 1). For example, with methane, CH_4, the RMM is 16, assuming that carbon has an atomic mass of 12 (i.e., $^{12}C^1H_4$ = 12 + 4 = 16). Let the abundance of this peak be 100. But the abundance of ^{13}C is 1.1 % of ^{12}C; therefore, $^{13}C^1H_4$ = 13 + 4 = 17, and its abundance is 1.1 % of $^{12}C^1H_4$, as shown in the following:

Peak Designation	Isotopic Formula	RMM	Rel. Abundance
M	$^{12}C^1H_4$	16	100
M + 1	$^{13}C^1H_4$	17	1.1

Similarly, for ethane, C_2H_6, which has two carbon atoms, we can designate the MS pattern as follows:

Peak Designation	Isotopic Formula	RMM	Rel. Abundance
M	$^{12}CH_3\ ^{12}CH_3$	30	100
M + 1	$^{12}CH_3\ ^{13}CH_3$	31	1.1
M + 1	$^{13}CH_3\ ^{12}CH_3$	31	1.1
Total M + 1		31	2.2
M + 2	$^{13}CH_3\ ^{13}CH_3$	32	0.012

With C_2H_6, we have $^{12}C_A\ ^{12}C_B\ ^1H_6$ for an RMM of 30 and a corresponding m/z = 30 for the molecular ion. But we also will have ions of $^{13}C_A\ ^{12}C_B\ ^1H_6$ (m/z = 31) and $^{12}C_A\ ^{13}C_B\ ^1H_6$ (m/z = 31). Both combinations are equally possible. If the abundance at m/z = 30 is 100, then $^{13}C\ ^{12}C\ ^1H_6$ is 1.1 % of that and $^{12}C\ ^{13}C\ ^1H_6$ is also 1.1 % of that, so the total abundance at m/z = 31 is 2.2 % of m/z = 30. The small peak at two mass units above the molecular ion peak occurs when both carbon atoms are ^{13}C atoms.

If our molecule contains only carbon and hydrogen, since there is 1.1 % relative natural abundance of ^{13}C compared to 100 % ^{12}C, and our mass spectrum shows a 1.1 % abundance of (M + 1) to M, only one carbon atom can be present in the molecule. That is, if we are dealing with a hydrocarbon and only one carbon atom is present, then the ratio (M + 1)/M = 1.1 %. If two carbons are present in the molecule, as there are in ethane, the probability of ^{13}C being present is twice as great and (M + 1)/M = 2.2 %.

11.1.3.1 Counting Carbon Atoms

There is a definite relationship between the peaks at (M + 1) and M that is directly related to the (maximum) number of carbons present in hydrocarbon molecules:

$$\frac{M+1}{M} = 1.1\ \% \times \text{Number of C atoms in the molecule} \quad (11.1)$$

This relationship gives the maximum number of carbon atoms but not necessarily the actual number; nonetheless it is very valuable in characterizing an unknown compound. The calculation just presented is for hydrocarbons (compounds containing only carbon and hydrogen) and ignores the contribution from the ^2H isotope of hydrogen, called deuterium, due to its low relative abundance of 0.016 %. In rigorous calculations, the contribution from deuterium would be taken into account.

For benzene, C_6H_6 (Table 11.1), the molecular ion m/z = 78, and the (M + 1) m/z = 79. The relative abundance of the (M + 1) peak is 6.6 %; dividing 6.6 % by 1.1 % confirms that there are six carbons in the molecule.

Using a similar process for C_3H_8, RMM = 44, the abundance of an (M + 1) peak at m/z = 45 will be predicted to be

Table 11.4 Relative Natural Isotope Abundances of Common Elements in Organic Compounds

Isotope	Mass	Relative Abundance	Mass	Relative Abundance	Mass	Relative Abundance
H	1	100	2	0.016		
C	12	100	13	1.11		
N	14	100	15	0.37		
O	16	100	17	0.04	18	0.20
F	19	100				
Si	28	100	29	5.07	30	3.36
P	31	100				
S	32	100	33	0.78	34	4.39
Cl	35	100			37	32.7
Br	79	100			81	97.5
I	127	100				

3×1.1 % of the abundance at $m/z = 44$. The same calculation can be made for any hydrocarbon.

A list of common isotopes found in organic compounds and their relative abundance in nature is given in Table 11.4. A complete table of the relative abundance of natural isotopes for all elements is located in Appendix 11.A. As can be seen in Table 11.4, the natural abundance of deuterium is only 0.016 % of the ^{1}H abundance, so it can usually be ignored in calculations.

However, there are a number of common isotopes that we cannot ignore for many organic molecules, such as those of oxygen, nitrogen, sulfur, and the halogens. This gives rise to the question of what atomic masses to use when calculating RMMs for MS so that we have the correct mass for the molecular ion.

There are three expressions for mass that are important to know. The **integer mass** of the most abundant naturally occurring stable isotope is the **nominal mass** of the element. The **exact mass** of an isotope is determined by high-resolution MS; the exact mass of ^{35}Cl is 34.9689 Da and that for ^{37}Cl is 36.9659 Da. The **atomic mass** is the **average mass** of the element. For example, if we have a compound containing chlorine, most periodic tables list the atomic mass of chlorine as 35.45 or 35.5. The RMM of CH_2Cl_2 is therefore equal to $12 + 2(1) + 2(35.5) = 85$. But a mass spectrum of dichloromethane does not have an intense peak at $m/z = 85$. The "atomic mass" of chlorine in the periodic table is a weighted average of the exact masses of ^{35}Cl and ^{37}Cl, and ^{35}Cl is about three times as abundant as ^{37}Cl. If chlorine were analyzed by MS, there would be a peak at $m/z = 35$ and a peak about one-third as intense at $m/z = 37$, but no peak at $m/z = 35.5$. For MS, RMMs are calculated using the most abundant isotope of the element with its mass rounded to the nearest integer value; that is, the nominal mass is used. For most organic compounds, the most abundant isotope is the lowest mass isotope. Therefore, the RMM of dichloromethane would be calculated as $12 + 2(1) + 2(35) = 84$. There is a significant peak at $m/z = 84$ in the mass spectrum of dichloromethane. At what higher values of m/z would you expect to see peaks in this spectrum based on the isotopes present?

11.1.3.2 Counting Carbon, Nitrogen, and Sulfur Atoms

From Table 11.4, it is clear that not only carbon but also other elements contribute to the (M + 1) peak intensity, notably nitrogen and sulfur when they are present. Silicon obviously will impact the (M + 1) peak in organosilicon compounds such as silicone polymers. A more general equation for the ratio of the (M + 1) to M peak is

$$\frac{M+1}{M} = 1.1(\#\,C\,atoms) + 0.016(\#\,H\,atoms) + 0.37(\#\,N\,atoms) + 0.78(\#\,S\,atoms) + \cdots$$

(11.2)

where #C atoms is the (maximum) number of carbon atoms in the molecule and so on. The general expression for compounds containing only C, H, N, and O would use only the terms for carbon and nitrogen. This equation would also apply to CHNO compounds containing fluorine, phosphorus, and/or iodine, since these have only one isotope. It can be seen that the equation is of most use in confirming molecular formulas, since we need to know what elements are present to use the equation effectively. We may also have previous information about what elements are present from combustion analysis; that type of analysis would indicate oxygen, nitrogen, or sulfur in the molecule, for example.

11.1.3.3 Counting Oxygen Atoms

Similarly, the (M + 2) peak contains intensity contributions from oxygen, Cl, Br, and multiple heavy isotopes of

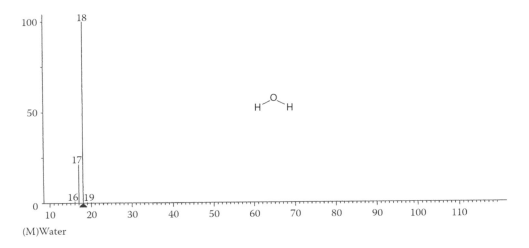

Figure 11.5 A mass spectrum of water, H_2O.

carbon, hydrogen, and so on. A general formula for compounds that contain only C, H, N, O, F, P, and I can be used to ascertain the number of oxygen atoms, in theory. Examination of Table 11.4 shows that oxygen has two important isotopes, ^{16}O and ^{18}O, separated by 2 Da, with a relative abundance $^{18}O/^{16}O$ of 0.2 %. The number of oxygen atoms in a molecular ion can be calculated from

$$\frac{M+2}{M} = 0.20(\#\text{O atoms}) + \frac{(1.1(\#\text{C atoms}))^2}{200} \quad (11.3)$$

This equation again ignores the contribution from hydrogen. Compounds containing the halogens Cl and Br will be discussed later in this chapter. Iodine does not affect the equation, since iodine is monoisotopic.

For example, the mass spectrum of methanol, CH_3OH, is shown in Figure 11.3, and the mass spectral data for methanol are shown in Table 11.3. Figure 11.3 notes with the triangle that the molecular ion is at $m/z = 32$. (This can be deduced with practice; aliphatic alcohols often fragment by loss of a proton or water. Is there an M − 18 peak in the methanol spectrum corresponding to loss of water?) The ratio of (M + 1)/M, the $m/z = 33$ peak to that at $m/z = 32$, is 1.0/66 or 1.5 %, indicating not more than one carbon atom. Also, the (M + 2)/M ratio, m/z 34/32, is 0.21 %; inserting one carbon atom and 0.21 for the ratio into Equation 11.3 shows us that only one oxygen atom is present in the molecular ion.

There are some other interesting points in the methanol spectrum. For instance, the most abundant peak (the base peak) is at $m/z = 31$, indicating that methanol very easily loses one hydrogen atom and forms CH_3O. Another abundant peak is at $m/z = 29$, indicating that methanol loses three hydrogen atoms to form COH. The next most abundant peak is at 15, which is equivalent to a CH_3 ion, indicating breaking of the OH bond to the methyl group. The fragment that would correspond to the hydroxyl ion would be at $m/z = 17$; it is not abundant, indicating that if OH ions are formed, they

Table 11.5 Mass Spectral Data for H_2O

m/z	Relative Abundance
1	<0.1
16	1.0
17	21
18	100
19	0.08
20	0.22

do not reach the detector. Since we are detecting positively charged ions in EI MS, is this surprising?

The mass spectrum of water is shown in Figure 11.5 and Table 11.5. To calculate the number of carbon and oxygen atoms in a molecule, we assume that only the elements C, H, N, O, F, P, and I are present and that there are no interferences from other ions. The molecular ion $^1H_2^{16}O$ is at $m/z = 18$, but there is also a peak listed in the table at $m/z = 20$, the (M + 2) peak caused by $^1H_2\ ^{18}O$. Its abundance is approximately 0.2 % of the abundance of 18 ($^1H_2\ ^{16}O$), confirming that only one oxygen atom is present in the molecular ion. The (M + 1) peak is only 0.08 % of the M peak, which does not allow for a carbon atom in the molecular ion.

While Equation 11.3 can be used, in theory, to calculate the number of oxygen atoms in a compound, in reality, the ^{18}O isotope is of low abundance, and oxygen-containing compounds often fragment so that the molecular ion is of low intensity or not detected. In practice, the observation of isotope information for oxygen is often difficult.

11.1.4 Compounds with Heteroatoms

The elements shown in Table 11.4 can be classified into three categories, following the recommended nomenclature of McLafferty. "A" elements are those that are monoisotopic,

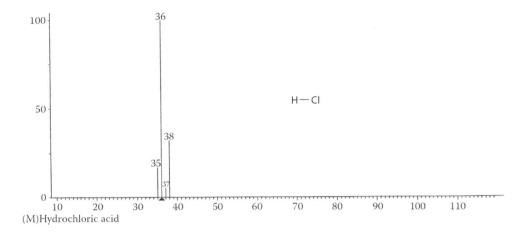

Figure 11.6 A mass spectrum of hydrogen chloride, HCl, showing the characteristic 3:1 ratio of the M/(M + 2) peaks for a compound containing a single Cl atom.

such as F, P, and I. Hydrogen is also classified as an A element, because the natural abundance of deuterium is so low. "A + 1" elements are those with two isotopes, one heavier than the most abundant isotope by 1 Da. Carbon and nitrogen are A + 1 elements. "A + 2" elements have an isotope that is 2 Da heavier than the most abundant isotope. Cl, Br, and O are the three most common of these; S and Si are also A + 2 elements that must be considered if their presence is detected by elemental combustion analysis or other analytical methods.

The A + 2 elements are the easiest to recognize, so they should be looked for first when presented with an unknown mass spectrum. Oxygen has already been discussed, so look at the isotope ratios of Cl and Br from Table 11.4. Cl has two isotopes, ^{35}Cl and ^{37}Cl, in a ratio of 100:33. A compound that contains one Cl atom would have an (M + 2) peak approximately one-third as intense as the M peak. Br has two isotopes, ^{79}Br and ^{81}Br, in a ratio of ~1:1. A compound with one Br atom would have approximately equal M and (M + 2) peaks. These are very distinctive patterns.

Figure 11.6 and Table 11.6 show two strong peaks at $m/z = 38$ (M + 2) and 36 (M) in an abundance ratio of ~1/3. (We know that the peak at $m/z = 36$ is the M peak because it is marked by the triangle.) This is typical of the pattern seen when one Cl atom is present. If a chlorine atom is present, then $m/z = 36$ arises from ^{35}Cl plus one ^1H atom (only ^1H could increase the mass from 35 to 36). Then $m/z = 38$ would be due to $^{37}Cl^1H$. This is the mass spectrum of hydrochloric acid, HCl. The peaks at $m/z = 35$ and 37 are of course due to the ionized chorine isotopes that have lost the hydrogen through fragmentation. The ratio of the m/z 37/35 peaks is also 1/3, supporting the identification of a single Cl atom.

For organochlorine compounds with one Cl atom, such as CH_3Cl, we will expect to see a peak at (M + 2) because there will be ^{37}Cl present. The ratio (M + 2)/M will be the

Table 11.6 Mass Spectral Data for HCl

m/z	Relative Abundance
35	12
36	100
37	4.1
38	33
39	0.01

same as the natural ratio of ^{37}Cl to ^{35}Cl, that is, 32.7 %, resulting in the same one-third pattern seen in HCl.

Looking at Figure 11.7 and Table 11.7, the almost equal intensity of the two peaks at $m/z = 94$ and 96 is characteristic of the presence of one bromine atom. The presence of one Br atom is confirmed by the peaks at $m/z = 79$ and 81 in approximately equal abundance. These two peaks can reasonably be attributed to ^{79}Br and ^{81}Br ions. If the RMM is 94 as indicated by the identification of $m/z = 94$ as the molecular ion (note the triangle), a mass of ^{79}Br subtracted from 94 shows that atoms equivalent to a group of mass 15 must be present. This is most likely a methyl group, CH_3-, confirmed by an abundant peak at $m/z = 15$. Based on the m/z 16/15 ratio, which is 0.62/59 = 1.05 %, the ion of $m/z = 15$ contains 1 C and is probably CH_3. Consecutive peaks at masses of 14, 13, and 12 confirm three hydrogen atoms, ending in a C ion with no H attached. Hence, the molecule is most likely composed of a methyl group and one Br atom: Br + CH_3 or CH_3Br. This is bromomethane (also called methyl bromide).

11.1.5 Halogen Isotopic Clusters

If more than one chlorine atom or bromine atom is present in a molecule, distinctive "isotope cluster patterns" are seen in the mass spectrum. Such compounds are commonly

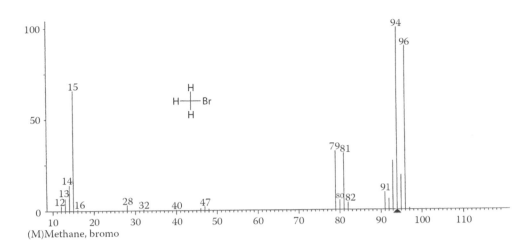

Figure 11.7 A mass spectrum of bromomethane (or methyl bromide), CH_3Br, showing the characteristic 1:1 ratio of the M/(M + 2) peaks for a compound containing a single Br atom.

Table 11.7 Mass Spectral Data for CH_3Br

m/z	Relative Abundance	m/z	Relative Abundance
12	1.2	48	0.95
13	1.4	79	10
14	3.8	81	10
15	59	91	4.2
16	0.62	92	2.4
39.5	0.19	93	6.8
40.5	0.20	94	100
46	1.3	95	0.9
46.5	0.30	96	96
47	2.3	97	1.1
47.5	0.28		

encountered in environmental sample analyses. Figure 11.8 gives a graphical representation of the isotope peak intensity patterns for Cl and Br. The numerical values for the peak ratios are given in Table 11.8. The patterns arise as follows.

If we have two Cl atoms present in an ion, the distribution of masses in each ion may be $^{35}Cl^{35}Cl$, $^{37}Cl^{35}Cl$, $^{35}Cl^{37}Cl$, or $^{37}Cl^{37}Cl$. The $^{35}Cl:^{37}Cl$ abundance is 100:33, so the probability of these isotope distributions occurring is approximately 100:66:11.

Consequently, any molecule RCl_2 containing two Cl atoms will exhibit masses of R + 70, R + 72, and R + 74 with a relative abundance due to Cl isotopes of 100:66:11, due to the presence of $R^{35}Cl^{35}Cl$, $R^{37}Cl^{35}Cl$, $R^{35}Cl^{37}Cl$, and $R^{37}Cl^{37}Cl$. This pattern in the mass spectrum is a Cl_2 *isotope cluster*. The same pattern will be seen not only for the molecular ion but also for any fragment ion that contains both Cl atoms. Similarly, if three chlorine atoms are present (RCl_3), a cluster will occur with masses R + 105, R + 107, R + 109, and R + 111 with approximate abundances of 100:98:32:3.

Similar sets of clusters can be determined for bromine compounds and for compounds containing both Br and Cl. Bromine has two isotopes, ^{79}Br and ^{81}Br, and they are approximately equal in abundance. A compound RBr containing one Br atom will have masses of R + 79 and R + 81, and they will be about equal in abundance. If two Br atoms are present, as in RBr_2, there will be masses of R + 158, R + 160, and R + 162 from $R^{79}Br^{79}Br$, $R^{79}Br^{81}Br$, $R^{81}Br^{79}Br$, and $R^{81}Br^{81}Br$, and their relative abundances will be 51:100:49, based purely on probability.

The mass spectrum in Figure 11.9 shows a molecular ion at $m/z = 130$, with a pattern of peaks at m/z values of 130, 132, 134, and 136. Looking at the patterns shown in Figure 11.8, this group of four peaks most closely resembles the pattern from three Cl atoms. Using a ruler to measure the peak heights and dividing all four peak heights by the largest (the peak at 130) give a 100:95:31:3 ratio, very close to the values predicted in Table 11.8. There is another cluster of peaks at $m/z = 95$, 97, and 99. Using Figure 11.8, it is apparent that this cluster is due to a fragment with only two Cl atoms, and indeed, the difference between 130 and 95 is equal to the mass of one ^{35}Cl atom. Similarly, the pair of peaks at m/z 60 and 62 is due to a fragment containing only one Cl atom, as shown by the 3:1 peak height ratio. Three ^{35}Cl atoms have a mass of 105, leaving a mass of 130 − 105 = 25 to be accounted for in the molecule. The RMM is an even number, so there are no nitrogen atoms in the molecule (two nitrogen atoms would have a mass of 28, more than is possible based on the molecular ion mass). The peak ratio m/z 131/130 is about 2.3 %, indicating two carbon atoms at most. Two carbon atoms (total mass of 24) would leave room for only one H atom. Putting together two C atoms, one H atom, and three Cl atoms gives a molecular formula of C_2HCl_3. A reasonable structure is $HClC=CCl_2$; the structure is shown in Figure 11.9 (the hydrogen atom is not shown but would be bonded to

MASS SPECTROMETRY II: SPECTRAL INTERPRETATION AND APPLICATIONS

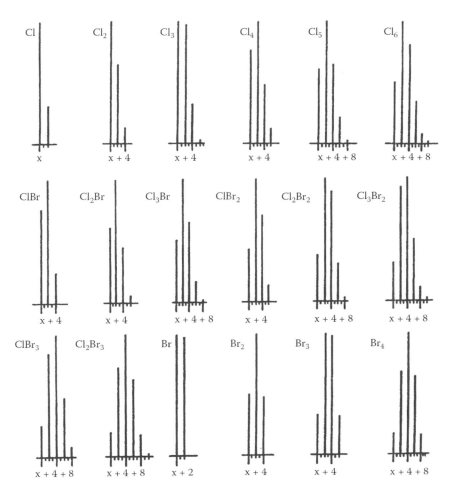

Figure 11.8 Graphical representation of relative isotope peak intensities for any given ion containing the indicated number of chlorine and/or bromine atoms. The numeric values are given in Table 11.8. (From the NIST Mass Spectrometry Data Center accessed via http://webbook.nist.gov. © 2011 U.S. Secretary of Commerce for the United States of America. All rights reserved. Used with permission.)

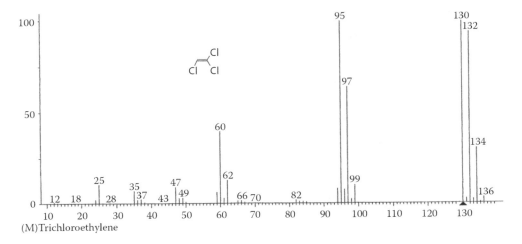

Figure 11.9 A mass spectrum of trichloroethylene, C_2HCl_3, showing the typical isotope cluster pattern for a compound containing three Cl atoms.

Table 11.8 Chlorine and Bromine Isotope Abundance Ratios

Atoms of Cl and Br	X	X + 2	X + 4	X + 6	X + 8	X + 10
Cl	100	32.5				
Cl$_2$	100	65.0	10.6			
Cl$_3$	100	97.5	31.7	3.4		
Cl$_4$	76.9	100	48.7	0.5	0.9	
Cl$_5$	61.5	100	65.0	21.1	3.4	0.2
Cl$_6$	51.2	100	81.2	35.2	8.5	1.1
ClBr	76.6	100	24.4			
Cl$_2$Br	61.4	100	45.6	6.6		
Cl$_3$Br	51.2	100	65.0	17.6	1.7	
ClBr$_2$	43.8	100	69.9	13.7		
Cl$_2$Br$_2$	38.3	100	89.7	31.9	3.9	
Cl$_3$Br$_2$	31.3	92.0	100	49.9	11.6	1.0
ClBr$_3$	26.1	85.1	100	48.9	8.0	
Cl$_2$Br$_3$	20.4	73.3	100	63.8	18.7	2.0
Br	100	98				
Br$_2$	51.0	100	49.0			
Br$_3$	34.0	100	98.0	32.0		
Br$_4$	17.4	68.0	100	65.3	16.0	

Source: From Sparkman, O.D., *Mass Spectrometry Desk Reference*, Global View Publishing, Pittsburgh, PA, 2000. Used with permission.

the carbon atom on the left). This compound is trichloroethylene, a commercial solvent.

A more comprehensive listing of ratios is presented in the book by Bruno and Svoronos, but it must be noted that in their compilation the base peak (which may or may not be the molecular ion) is always assigned a value of 100 % and the ratios are calculated on that basis. In Table 11.8 the base peak intensity varies.

Note that experimental uncertainty will change these ideal projections; many users will have an instrument capable of only unit mass resolution. In such cases, one must expect these patterns to be approximate. For example, if a molecule has no Cl and two Br, the P/P+2/P+4 ratio will be roughly 100/200/100 (P is assumed to be 100, and is not listed in the table). To confirm the presence of two bromine atoms, look for a peak at P-79, attributed to the loss of one bromine atom. Losing one bromine would lead to a "doublet" (roughly 1:1 ratio), which is indicative of a fragment containing only one bromine atom. Furthermore, at P-158 one would see that the 1:1 pattern has disappeared. These two observations will confirm the presence of two bromine atoms in the original P+ base peak ion.

If a molecule has two Cl and no Br present, the P/P+2/P+4 ratio will be roughly 100/67/11 in a similar fashion. To confirm consider P-35 area of the spectrum (loss of one chlorine). Losing one chlorine would lead to a doublet (roughly 3:1 ratio). In general, no lost fragments of 35 or 79 are indicative of any other atoms or fragments besides Cl and Br. Occasionally a halogen is lost as the hydrogen halide which will lead to the loss of 80 (HBr) or 36 (HCl), respectively, but the doublet pattern P/P+2 will be the same.

The spectrum becomes much more complicated if more than two bromine, or two chlorine atoms, or a combination of both of them is involved, as one can observe in the table. The rule of thumb however is that one ought to always look at the (P-79) or (P-35) to alert the presence of a bromine or chlorine, respectively. We also note that the statistical likelihood of encountering molecules with high multiples of Cl and Br will be low outside of environmental analyses.

11.1.6 Rings Plus Double Bonds

The formula for *n*-hexane is C_6H_{14}. The formula for cyclohexane is C_6H_{12}, the same as the formula for hexene. The presence of a ring or double bond results in a change in the ratio of carbon to hydrogen. This change is a general property, and based on it, the following formula is derived.

The number of rings + double bonds in a molecule of formula $C_xH_yN_zO_m$ is

$$x - \frac{1}{2}y + \frac{1}{2}z + 1 \quad (11.4)$$

For example, for *n*-hexane, $x = 6$, $y = 14$, so the number of rings + double bonds is $6 - (1/2 \times 14) + 0 + 1 = 0$. For cyclohexane, $x = 6$, $y = 12$; therefore, rings + double bonds = $6 - (1/2 \times 12) + 0 + 1 = 1$. For benzene, C_6H_6, rings

+ double bonds = 6 − (1/2 × 6) + 0 + 1 = 4, that is, three double bonds plus one ring. Acetylene, C_2H_2, contains a triple bond between the two carbon atoms. Equation 11.4 applied to acetylene yields 2 − (1/2 × 2) + 0 + 1 = 2. A triple bond is equivalent to two double bonds. The equation does not tell us whether we are dealing with a ring or a double bond or two double bonds versus a triple bond, or, as is the case with benzene, a delocalized molecular orbital that can be represented for bonding purposes as three double bonds. We need to use the equation in conjunction with the molecular formula (and any other information we have from IR, nuclear magnetic resonance [NMR], etc.) to postulate a reasonable structure.

When atoms other than CHNO are present, the heteroatoms are matched according to their lowest absolute valence. Si takes four bonds, like C; P is equivalent to N, while the halides, with a lowest absolute valence of 1, act like H. Hence, the general equation is

$$\text{Rings and double bonds} = x - \frac{1}{2}y + \frac{1}{2}z + 1 \quad (11.5)$$

where
x = C, Si
y = H, F, Cl, Br, I
z = N, P

11.1.7 Common Mass Losses on Fragmentation

When a molecule ionizes, it follows the reaction M + e⁻ → M⁺· + e⁻ + e⁻, as described earlier. When an OE molecular ion fragments, it can do so in a variety of ways, and resulting OE fragment ions can also fragment in similar ways. Simple cleavage results in the expulsion of a neutral fragment that is an OE radical (OE·) and formation of an EE ion (EE⁺) as shown in the following:

$$M^{+\cdot} \rightarrow EE^+ + OE^\cdot$$

Multicentered fragmentation results in the expulsion of an EE neutral species and formation of an OE ion (OE⁺·):

$$M^{+\cdot} \rightarrow EE + OE^{+\cdot}$$

In the first case, we would detect the EE⁺ ion and, in the second case, the OE⁺· ion. The neutral species lost on fragmentation can be identified by the difference between the mass of M⁺· and the fragment ion. Common losses from the molecular ion are given in Table 11.9. Tables 11.10 and 11.11 list more common neutral fragments expelled by simple cleavage and by multicentered fragmentations.

Thus far we have considered fragmentation from specific moieties, but it is useful to recast this on the basis of chemical families as well. In Table 11.12 we provide some fragmentation guidelines on the basis of major chemical families. The discussion of fragmentation patterns based on chemical families will be expounded later in the discussion of mass spectral interpretation.

Since many of the samples that are presented for mass spectral analysis are dissolved in a liquid solvent, it is also critical that the student be able to distinguish between sample peaks and solvent peaks. Even if the sample has been recovered from a solvent by recrystallization or vaporization, trace residual solvent may be present. In Table 11.13 we present the most common solvents used in mass spectrometry with their expected fragmentation patterns.

11.2 MASS SPECTRAL INTERPRETATION: SOME EXAMPLES

The general appearance of the mass spectrum depends on the type of compound analyzed. Seeing the patterns that distinguish, say, a normal alkane from an aromatic hydrocarbon requires practice and lots of it. The following

Table 11.9 Common Neutral Losses from the Molecular Ion

M − 1	Loss of hydrogen radical	M − ·H
M − 15	Loss of methyl radical	M − ·CH$_3$
M − 29	Loss of ethyl radical	M − ·CH$_2$CH$_3$
M − 31	Loss of methoxy radical	M − ·OCH$_3$
M − 43	Loss of propyl radical	M − ·CH$_2$CH$_2$CH$_3$
M − 45	Loss of ethoxy radical	M − ·OCH$_2$CH$_3$
M − 57	Loss of butyl radical	M − ·CH$_2$CH$_2$CH$_2$CH$_3$
M − 2	Loss of hydrogen	M − H$_2$
M − 18	Loss of water	M − H$_2$O
M − 28	Loss of CO or ethylene	M − CO; M − C$_2$H$_4$
M − 32	Loss of methanol	M − CH$_3$OH
M − 44	Loss of CO$_2$	M − CO$_2$
M − 60	Loss of acetic acid	M − CH$_3$CO$_2$H

Table 11.10 Some Neutral Fragments Expelled by Simple Cleavage

Mass	Fragment (OE·)	Mass	Fragment (OE·)
1	·H	46	·NO$_2$
15	·CH$_3$	47	·CH$_2$SH
16	·NH$_2$	57	·C$_4$H$_9$, ·C$_3$H$_5$O
17	·OH, ·NH$_3$	58	·C$_3$H$_8$N
19	·F	61	·C$_2$H$_5$S
26	·CN	69	·CF$_3$
29	·C$_2$H$_5$, ·CHO	71	·C$_4$H$_7$O
30	·CH$_3$NH, ·CH$_2$NH$_2$	79	·Br, ·C$_6$H$_7$
31	·CH$_3$O, ·CH$_2$OH	81	·C$_6$H$_9$, ·C$_5$H$_5$O
35	·Cl	91	·C$_7$H$_7$
45	·C$_2$H$_5$O, ·COOH	127	·I

Table 11.11 Some Neutral Fragments Expelled by Multicentered Fragmentations

Mass	Fragment (EE⁺)	Mass	Fragment (EE⁺)
2	H_2	45	C_2H_7N
17	NH_3	46	C_2H_6O, $H_2O + C_2H_4$
18	H_2O	48	CH_4S
20	HF	54	C_4H_6
27	HCN	56	C_4H_8, C_3H_4O
28	CO, C_2H_4	58	C_3H_6O
30	CH_2O	59	C_3H_9N
31	CH_5N	60	C_3H_8O, $C_2H_4O_2$
32	CH_4O	62	C_2H_6S
34	H_2S	74	$C_3H_6O_2$
36	HCl	76	C_6H_4
42	C_3H_6, C_2H_2O	78	C_6H_6
44	CO_2	80	HBr

examples include simple gas molecules and simple compounds representative of a variety of organic chemicals.

Looking at the spectrum in Figure 11.10 (data in Table 11.14), assume that $m/z = 44$ is the molecular ion, since it has the highest m/z (excepting isotope peaks) in the spectrum. Since the ratio of m/z 45/44 = 1.2 %, it appears that not more than one carbon atom is present. M − 12 = 32 Da is to be accounted for. The peak ratio m/z 46/44 = 0.42 %, which is consistent with two oxygen atoms, for an additional 2(16) = 32 Da. 12 + 32 equals the assumed RMM; therefore, a reasonable suggestion is that the compound is CO_2. Looking at the rest of the spectrum, the peak at $m/z = 28$ could be due to the formation of a CO ion by loss of one oxygen atom from the molecular ion, while the peak at $m/z = 16$ can be attributed to an ionized oxygen atom. Using the rings + double bonds equation, the compound is predicted to have two double bonds, as it does.

Table 11.12 Fragmentation Patterns for Major Organic Compound Families (taken from the book by Bruno and Svoronos)

Family	Molecular Ion Peak	Common Fragments; Characteristic Peaks
Acetals		Cleavage of all C–O, C–H and C–C bonds around the original aldehydic carbon
Alcohols	Weak for 1° and 2°; not detectable for 3°; strong for benzyl alcohols	Loss of 18 (H_2O—usually by cyclic mechanism); loss of H_2O and olefin simultaneously with four (or more) carbon-chain alcohols; prominent peak at $m/z = 31(CH_2\ddot{O}H)^+$ for 1° alcohols; prominent peak at $m/z = (RCH\ddot{O}H)^+$ for 2° and $m/z = (R_2C\ddot{O}H)^+$ for 3° alcohols
Aldehydes	Low intensity	Loss of aldehydic hydrogen (strong M − 1 peak, especially with aromatic aldehydes); strong peak at $m/z = 29(HC\equiv O^+)$; loss of chain attached to alpha carbon (beta cleavage); McLafferty rearrangement via beta cleavage if gamma hydrogen is present
Alkanes a) Chain b) Branched c) Alicyclic	 Low intensity Rather intense 	Loss of 14 mass units (CH_2) Cleavage at the point of branch; low-intensity ions from random rearrangements Loss of 28 mass units ($CH_2=CH_2$) and side chains
Alkenes (olefins)	Rather high intensity (loss of π-electron) especially in case of cyclic olefins	Loss of units of general formula C_nH_{2n-1}; formation of fragments of the composition C_nH_{2n} (via McLafferty rearrangement); retro Diels-Alder fragmentation
Alkyl halides a) Fluorides b) Chlorides c) Bromides d) Iodides	Abundance of molecular ion F<Cl<Br<I; intensity decreases with increase in size and branching Very low intensity Low intensity; characteristic isotope cluster Low intensity; characteristic isotope cluster Higher than other halides	Loss of fragments equal to the mass of the halogen until all halogens are cleaved off Loss of 20 (HF); loss of 26 (C_2H_2) in case of fluorobenzenes Loss of 35 (Cl) or 36 (HCl); loss of chain attached to the gamma carbon to the carbon carrying the Cl Loss of 79 (Br); loss of chain attached to the gamma carbon to the carbon carrying the Br Loss of 127 (I)
Alkynes	Rather high intensity (loss of π-electron)	Fragmentation similar to that of alkenes
Amides	Rather high intensity	Strong peak at $m/z = 44$ indicative of a 1° amide (O=C=NH⁺$_2$); base peak at $m/z = 59$ ($CH_2=C(OH)N^+H_2$); possibility of McLafferty rearrangement; loss of 42 (C_2H_2O) for amides of the form $RNHCOCH_3$ when R is aromatic ring
Amines	Hardly detectable in case of acyclic aliphatic amines; high intensity for aromatic and cyclic amines	Beta cleavage yielding >C=N⁺<; base peak for all 1° amines at $m/z = 30$ ($CH_2=N^+H_2$); moderate M − 1 peak for aromatic amines; loss of 27 (HCN) in aromatic amines; fragmentation at alpha carbons in cyclic amines
Aromatic hydrocarbons (arenes)	Rather intense	Loss of side chain; formation of RCH=CHR' (via McLafferty rearrangement); cleavage at the bonds beta to the aromatic ring; peaks at $m/z = 77$ (benzene ring; especially mono-substituted), 91 (tropyllium); the ring position of alkyl substitution has very little effect on the spectrum

Table 11.12 (Continued) Fragmentation Patterns for Major Organic Compound Families (taken from the book by Bruno and Svoronos)

Family	Molecular Ion Peak	Common Fragments; Characteristic Peaks
Carboxylic acids	Weak for straight-chain monocarboxylic acids; large if aromatic acids	Base peak at $m/z = 60$ ($CH_2=C(OH)_2$) if *-hydrogen is present; peak at $m/z = 45$ (COOH); loss of 17 (–OH) in case of aromatic acids or short-chain acids
Disulfides	Rather low intensity	Loss of olefins (m/z equal to $R-S-S-H^+$); strong peak at $m/z = 66$ ($HSSH^+$)
Phenols	Highly intense peak (base peak* generally)	Loss of 28 (C=O) and 29 (CHO); strong peak at $m/z = 65$ ($C_5H_5^+$)
Sulfides (thioethers)	Rather low intensity peak but higher than that of corresponding ether	Similar to those of ethers (–O– substituted by –S–); aromatic sulfides show strong peaks at $m/z = 109$ ($C_6H_5S^+$); 65 ($C_5H_5^+$); 91 (tropyllium ion)
Sulfonamides	Rather intense	Loss of $m/z = 64$ ($SONH_2$) and $m/z = 27$ (HCN) in case of benzenesulfonamide
Esters	Rather weak intensity	Base peak at m/z equal to the mass of $R-C\equiv O^+$; peaks at m/z equal to the mass of $^+O\equiv C-OR'$, the mass of OR' and R'; McLafferty rearrangement possible in case of: a) presence of a beta hydrogen in R' (peak at m/z equal to the mass of $R-C(^+OH)OH$, and b) presence of a gamma hydrogen in R (peak at m/z equal to the mass of $(CH_2=C(^+OH)OR)$; loss of 42 ($CH_2=C=O$) in case of benzyl esters; loss of ROH via the ortho effect in case of o-substituted benzoates
Ketones	Rather high-intensity peak	Loss of R–groups attached to the >C=O (alpha cleavage); peak at $m/z = 43$ for all methyl ketones (CH_3CO^+); McLafferty rearrangement via beta cleavage if gamma hydrogen is present; loss of $m/z = 28$ (C=O) for cyclic ketones after initial alpha cleavage and McLafferty rearrangement
Mercaptan (thiols)	Rather low intensity but higher than that of corresponding alcohol	Similar to those of alcohols (–OH substituted by –SH); loss of $m/z = 45$ (CHS) and $m/z = 44$ (CS) for aromatic thiols
Nitriles	Unlikely to be detected except in case of acetonitrile (CH_3CN) and propionitrile (C_2H_5CN)	M + 1 ion may appear (especially at higher pressures); M − 1 peak is weak but detectable ($R-CH=C=N^+$); base peak at $m/z = 41$ ($CH_2=C=N^+H$); McLafferty rearrangement possible; loss of HCN is case of cyanobenzenes
Nitrites	Absent (or very weak at best)	Base peak at $m/z = 30$ (NO^+); large peak at $m/z = 60$ ($CH_2=O^+NO$) in all unbranched nitrites at the alpha carbon; absence of $m/z = 46$ permits differentiation from nitrocompounds
Nitrocompounds	Seldom observed	Loss of 30 (NO); subsequent loss of CO (in case of aromatic nitro- compounds); loss of NO_2 from molecular ion peak
Sulfones	High intensity	Similar to sulfoxides; loss of mass equal to RSO_2; aromatic heterocycles show peaks at M-32 (sulfur), M-48 (SO), M-64 (SO_2)
Sulfoxides	High intensity	Loss of 17 (OH); loss of alkene (m/z equal to $RSOH^+$); peak at $m/z = 63$ ($CH_2=SOH)^+$; aromatic sulfoxides show peak at $m/z = 125$ ($^+S-CH=CHCH=CHC=O$), 97($C_5H_5S^+$), 93(C_6H_5OH); aromatic heterocycles show peaks at M-16 (oxygen), M-29(COH); M-48(SO)

* The base peak is the most intense peak in the mass spectrum, and is often the molecular ion peak, M^+.

Table 11.13 Important Peaks in the Mass Spectra of Solvents (taken from the book by Bruno and Svoronos)

Solvents	Formula	M+	Important Peaks (m/z)
Water	H_2O	18 (100 %)	17
Methanol	CH_3OH	32	31 (100 %), 29, 15
Acetonitrile	CH_3CN	41 (100%)	40, 39, 38, 28, 15
Ethanol	CH_3CH_2OH	46	45, 31 (100 %), 27, 15
Dimethylether	CH_3OCH_3	46 (100%)	45, 29, 15
Acetone	CH_3COCH_3	58	43 (100 %), 42, 39, 27, 15
Acetic acid	CH_3CO_2H	60	45, 43, 18, 15
Ethylene glycol	$HOCH_2CH_2OH$	62	43, 33, 31 (100 %), 29, 18, 15

(continued)

Table 11.13 (Continued) Important Peaks in the Mass Spectra of Solvents (taken from the book by Bruno and Svoronos)

Solvents	Formula	M+	Important Peaks (m/z)
Furan	C_4H_4O	68 (100 %)	42, 39, 38, 31, 29, 18
Tetrahydrofuran	C_4H_8O	72	71, 43, 42 (100%), 41, 40, 39, 27, 18, 15
n-Pentane	C_5H_{12}	72	57, 43 (100%), 42, 41, 39, 29, 28, 27, 15
Dimethylformamide (DMF)	$HCON(CH_3)_2$	73 (100 %)	58, 44, 42, 30, 29, 28, 18, 15
Diethylether	$(C_2H_5)_2O$	74	59, 45, 41, 31 (100 %), 29, 27, 15
Methylacetate	$CH_3CO_2CH_3$	74	59, 43 (100 %), 42, 32, 29, 28, 15
Carbon disulfide	CS_2	76 (100 %)	64, 44, 32
Benzene	C_6H_6	78 (100 %)	77, 52, 51, 50, 39, 28
Pyridine	C_5H_5N	79 (100 %)	80, 78, 53, 52, 51, 50, 39, 26
Dichloromethane	CH_2Cl_2	84	86, 51, 49 (100 %), 48, 47, 35, 28
Cyclohexane	C_6H_{12}	84	69, 56, 55, 43, 42, 41, 39, 27
n-Hexane	C_6H_{14}	86	85, 71, 69, 57 (100 %), 43, 42, 41, 39, 29, 28, 27
p-Dioxane	$C_4H_8O_2$	88 (100 %)	87, 58, 57, 45, 43, 31, 30, 29, 28
Tetramethylsilane (TMS)	$(CH_3)_4Si$	88	74, 73, 55, 45, 43, 29
1,2-Dimethoxy ethane	$(CH_3OCH_2)_2$	90	60, 58, 45 (100 %), 31, 29
Toluene	$C_6H_5CH_3$	92	91 (100 %), 65, 51, 39, 28
Chloroform	$CHCl_3$	118	120, 83, 81 (100 %), 47, 35, 28
Chlorodorm-d_1	$CDCl_3$	119	121, 84, 82 (100 %), 48, 47, 35, 28
Carbon tetrachloride	CCl_4	152 (not seen)	121, 119, 117 (100 %), 84, 82, 58.5, 47, 35, 28
Tetrachloroethene	$CCl_2=CCl_2$	164 (not seen)	168, 166 (100 %), 165, 164, 131, 129, 128, 94, 82, 69, 59, 47, 31, 24

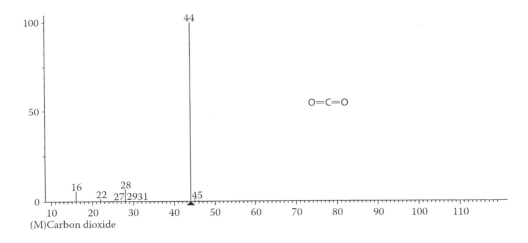

Figure 11.10 A mass spectrum of carbon dioxide, CO_2.

Table 11.14 Mass Spectral Data for CO_2

m/z	Relative Abundance
12	8.7
16	9.6
22	1.9
28	9.8
29	0.13
30	0.02
44	100
45	1.2
46	.42

Now consider the mass spectrum in Figure 11.11 and compare it to the carbon dioxide spectrum. The spectrum in Figure 11.11 is one that every mass spectrometrist should know. It is the spectrum of air, a mixture of components. (Why should a mass spectrometrist recognize this spectrum?) The peak at $m/z = 44$ is from carbon dioxide, while the other peaks are due to water (18), nitrogen (28), and oxygen (32). The peak at m/z 28 in the CO_2 spectrum is due to a CO ion, while in the air spectrum, it is due to nitrogen plus any carbon monoxide that is in the sample. At this resolution, it is not possible to distinguish between N_2 and CO. It is important to remember that not all mass spectra are of a single pure component. In the real world, mixtures and impure samples are much more commonly encountered.

The actual proportions of the major gases in air are N_2 = 78 %, O_2 = 21 %, argon (mass 40) = 1%, and CO_2 = ~0.039 %. Water varies widely with relative humidity. Careful measurement of the relative bar heights in Figure 11.11 does not quite reflect these values. Recall that this is not a single molecule but a mixture. The ionization efficiency (i.e., number of ions formed under operating conditions) will vary with the substance (e.g., the noble gas argon is probably not greatly ionized under these conditions, so no mass 40 peak is to be seen). The degree of fragmentation and division of charge among fragments will also vary among polyatomic gases. Precise quantitation will therefore vary with instrument settings and stability. Hence, the very precise measurements of increasing global atmospheric CO_2 concentration due to fossil fuel burning inputs that underlie the anthropogenic global greenhouse warming controversies are monitored by IR spectrometry rather than MS.

Consider the mass spectrum in Figure 11.12 (data in Table 11.15) and assume the molecular ion is the peak at m/z = 27.

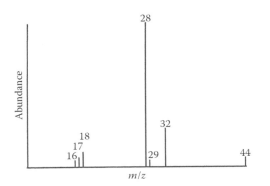

Figure 11.11   A mass spectrum of air, a mixture of gases. (Courtesy of Dr. Robert Kobelski, CDC, Atlanta, GA.)

If the RMM is 27, there must be an odd number of nitrogen atoms present in the molecule. There cannot be three, because the mass is too low. Therefore, there must be one nitrogen atom in the molecule. The ratio of the (M + 1)/M peaks at 28/27 = 1.5, indicating one carbon atom. The mass accounted for is 1 C (12) + 1 N (14) = 26 Da. Since the RMM is 27 Da, the only possibility is the presence of one hydrogen atom, giving a molecular formula of HCN, hydrogen cyanide. The mass and the (M + 2) peak are too small to allow for any oxygen atoms. The rings + double bond calculation results in 1 − 1/2 + 1/2 + 1 = 2. Hydrogen cyanide has a triple bond between the C and N, as shown in Figure 11.12. A triple bond is equivalent to two double bonds, as we learned.

11.2.1 Mass Spectra of Hydrocarbons

The general appearance of the mass spectrum depends on the type of compound analyzed. We will look first at the normal alkanes (saturated straight-chain hydrocarbons of the formula C_nH_{2n+2}, also called paraffins). Figure 11.13 shows the spectra of n-C_{16} and n-C_{30} alkanes. It can be seen that these spectra are very similar and would be hard to distinguish even though one molecule is twice as large as the other. Such problems are common in the analysis of kerosene-based liquid fuels such as diesel fuel and aviation turbine fuels. Small n-alkanes generally show an intense molecular ion peak. The molecular ion may be very small for large n-alkanes using EI. (To obtain the molecular ion, chemical ionization could be used.)

It can also be noted that with each compound, there are groups of fragments differing by 14 Da from each other. This is caused by the loss of successive numbers of CH_2 (methylene) groups as the molecules fragment by breaking the carbon–carbon bonds. For the homologous series of alkanes, peaks at m/z = 15, 29, 43, 57, etc., will be noted starting at 15 (CH_3). For other homologous series such

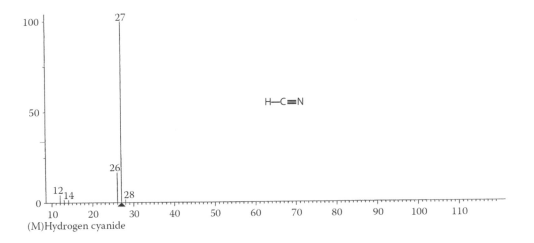

Figure 11.12   A mass spectrum of hydrogen cyanide, HCN.

as straight-chain alcohols or amines, the same general appearance is observed in the spectrum, but the *m/z* values will be displaced by the difference in mass of the functional group. Table 11.16 shows the terminal group and the first several *m/z* peaks of such series.

Branched alkanes fragment more readily than straight-chain alkanes, so branched alkanes are less likely to show a molecular ion peak than *n*-alkanes. Cycloalkanes show a strong molecular ion peak and also show the characteristic peaks separated by 14 Da. Figure 11.14 is the mass spectrum of hexane, C_6H_{14}, RMM = 86, and Figure 11.15 is the mass spectrum of cyclohexane, C_6H_{12}, RMM = 84. Cyclohexane has a stronger molecular ion peak than hexane, although both are clearly present. Cycloalkanes preferentially fragment by loss of ethene, C_2H_4. This preferred fragmentation gives rise to the base peak for cyclohexane at *m/z* = 56. Look at the mass spectrum of 2,2-dimethylpropane, $CH_3C(CH_3)_2CH_3$ or $(CH_3)_4C$, a branched alkane as shown by the structure in Figure 11.16. The molecular ion at *m/z* = 72 is too weak to be observed; the strongest peak in the spectrum is due to the loss of a methyl group.

Alkenes and alkynes show strong molecular ion peaks because the double and triple bonds are able to absorb energy. Alkenes with more than four carbon atoms often show a strong peak at *m/z* = 41 due to formation of an allyl ion. Alkynes show strong (M − 1) peaks. The mass spectra of alkene isomers are very similar, especially the mass spectra of *cis*- and *trans*-isomers, so it is not usually possible to locate the position of the double bonds in these compounds through MS. Figures 11.17 and 11.18 show the mass spectra of 1-hexene and acetylene, respectively.

Table 11.15 Mass Spectral Data for HCN

m/z	Relative Abundance
12	4.2
18	1.7
13.5	0.88
14	1.6
15	0.12
26	17
27	100
28	1.6
29	0.06

Table 11.16 Homologous Series Separated by CH_2

Functional Group	End Group	m/z Series
Amine	$-CH_2-NH_2$	30, 44, 58, …
Alcohol	$-CH_2-OH$	31, 45, 59, …
Aldehyde	$-\overset{\overset{O}{\parallel}}{C}-H$	29, 43, 57, 71, …
Ketone	$-\overset{\overset{O}{\parallel}}{C}-CH_3$	43, 57, 71, …

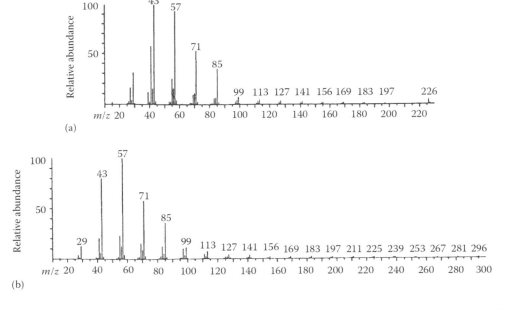

Figure 11.13 (a) Mass spectrum of *n*-hexadecane, $C_{16}H_{34}$, and (b) *n*-tridecane, $C_{30}H_{62}$. The RMM of *n*-hexadecane is 226 Da; the molecular ion is visible in spectrum (a) but small. The RMM of *n*-tridecane is 422 Da; the spectrum does not extend far enough to see the molecular ion, but it would be very small as well. Both *n*-alkanes show a similar pattern of peaks separated by 14 Da due to the loss of CH_2 groups, and both show very similar patterns at the low *m/z* end of the spectrum.

MASS SPECTROMETRY II: SPECTRAL INTERPRETATION AND APPLICATIONS

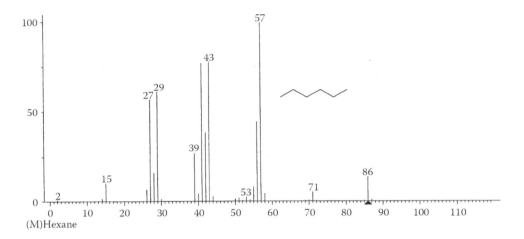

Figure 11.14 A mass spectrum of a normal alkane, hexane, C_6H_{14}.

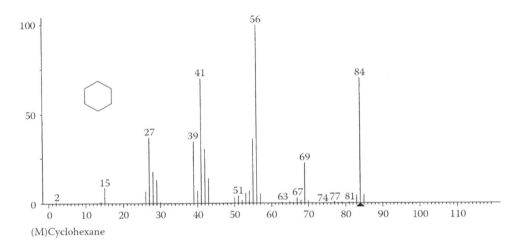

Figure 11.15 A mass spectrum of a cyclic alkane, cyclohexane, C_6H_{12}.

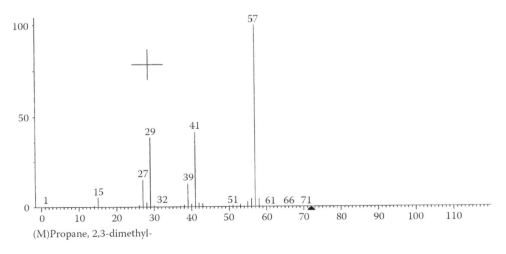

Figure 11.16 A mass spectrum of a branched alkane, 2,2-dimethylpropane, $(CH_3)_4C$. The structure is shown schematically in the figure. The hydrogen atoms are not shown.

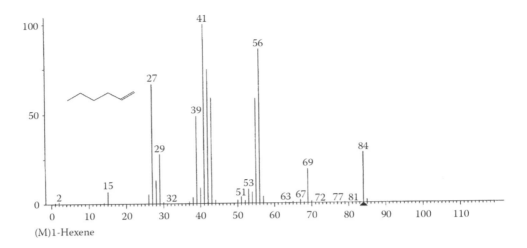

Figure 11.17 A mass spectrum of an aliphatic alkene, 1-hexene, C_6H_{12}.

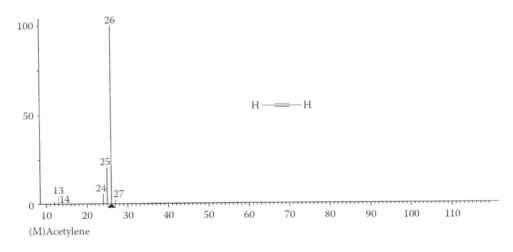

Figure 11.18 A mass spectrum of an alkyne, acetylene, C_2H_2.

Table 11.17 Mass Spectral Data for Acetylene

m/z	Relative Abundance
12	0.91
13	3.6
14	0.10
24	6.1
25	23
26	100
27	2.2
28	0.02

The mass spectral data for acetylene are given in Table 11.17. The molecular formula can be deduced as follows. If the molecular ion is at $m/z = 26$, then the ratio of (M + 1) to M is 2.2 %, indicating two carbon atoms. Their total mass is 24 Da. The missing mass of 2 Da can only be two hydrogen atoms. Hence, the formula is C_2H_2. Note that the (M + 2) peak is too small to indicate the presence of oxygen, and the even-numbered m/z value for the molecular ion indicates zero or two nitrogen atoms. Two nitrogen atoms have a mass of 28 Da, so there cannot be two nitrogen atoms in this molecule if the RMM is 26 Da. The rings + double bond calculation results in a value of two, which can only be a triple bond between the two carbon atoms. Also, the strong (M − 1) peak at $m/z = 25$, due to the loss of one hydrogen, is typical of alkynes.

Aromatic compounds present quite different spectra from alkanes, as can be seen in Figures 11.4 (benzene) and 11.19 (naphthalene). Aromatic molecules are very stable and do not fragment easily. The molecular ion peak for aromatic compounds is very intense, as seen in Figures 11.4 and 11.19.

Using naphthalene (Figure 11.19 and Table 11.16) as an example, we see an intense molecular ion peak at $m/z = 128$

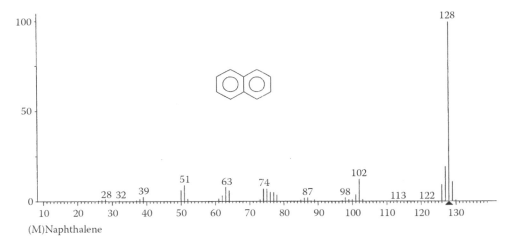

Figure 11.19 A mass spectrum of an aromatic hydrocarbon, naphthalene, $C_{10}H_8$.

and relatively little fragmentation. This should suggest an aromatic compound just from the appearance of the spectrum. The ratio (M + 1)/M is 11/100, indicating 10 carbon atoms, although a formula with nine carbon atoms cannot be completely ruled out based on isotopic abundances alone. If we assume 10 carbon atoms, their mass is 10 × 12 = 120 Da. The molecule, therefore, has 128 − 120 = 8 Da to be accounted for; eight hydrogen atoms are the most reasonable suggestion. The formula would be $C_{10}H_8$ and is probably naphthalene, whose structure is shown in the figure. Can you work out the rings + double bond calculation and explain it?

It will also be noted that in Table 11.1 for benzene and in Table 11.18 for naphthalene, peaks at noninteger m/z values are observed. For naphthalene, there are peaks at m/z values of 64.5, 63.5, and 62.5. These are not fragments with fractional masses. Rather, they are doubly charged ions. It will be remembered (Chapter 10) that

$$\frac{m}{z} = \frac{B^2 r^2}{2V} \quad (11.6)$$

and it is normally assumed for EI spectra that $z = +1$. But if the ion is doubly charged to M^{++}, it appears to have half the mass to satisfy Equation 11.6. Therefore, these ions really have masses of 64.5 × 2 = 129 Da, 63.5 × 2 = 127 Da, and 62.5 × 2 = 125 Da. Large M^{++} peaks are also typical of aromatic compounds. The apparent mass at $m/z = 42.67$ is due to a triply charged ion (i.e., M^{+++}), and its true mass is 42.67 × 3 = 128 Da. This should drive home the message that the x-axis in a mass spectrum is not mass but m/z. In order to confirm that this is the mass spectrum of naphthalene, a known sample should be run.

There is another phenomenon that gives rise to peaks with noninteger masses. Ions with lower internal energy than those that produce the "normal" product ions can

Table 11.18 Mass Spectral Data for Naphthalene

m/z	Relative Abundance	m/z	Relative Abundance
38	1.8	64.5	1.1
39	3.9	65	0.24
40	0.19	66	0.02
41	0.01	74	4.7
42	0.06	75	4.9
42.67	0.10	76	3.9
43	0.16	77	4.1
43.5	0.02	78	2.7
49.5	0.24	79	0.19
50	6.4	87	1.4
50.5	0.09	88	0.22
51	12	89	0.73
51.5	0.37	90	0.06
52	1.6	101	2.7
53	0.27	102	7.1
54	0.04	103	0.64
55	0.14	104	0.15
55.5	0.03	110	0.06
56	0.12	111	0.09
56.5	0.06	112	0.02
61	1.4	113	0.18
61.5	0.08	125	0.85
62	2.7	126	6.1
62.5	0.20	127	9.8
63	7.4	128	100
63.5	0.95	129	11
64	10	130	0.52

fragment after they leave the ion source but before entering the mass analyzer. These **metastable ions** have the velocity of the precursor ion, m_1, but the mass of the product ion, m_2. In a magnetic sector instrument, a **metastable peak** will appear at m^*, where

$$m^* = \frac{(m_2)^2}{m_1} \qquad (11.7)$$

The value of m^* is generally not an integer. The spectral interpretation texts cited in the bibliography should be consulted for more details.

The lack of significant fragmentation and the intense molecular ion illustrate how stable aromatic molecules are. As a general rule, increased unsaturation increases stability. It is harder to break double bonds than single bonds and harder to break apart the very stable, delocalized molecular orbitals in aromatic compounds than to break bonds in aliphatic compounds. Also, rings are more stable than straight chains.

Aromatic rings with aliphatic side chains will show characteristics of both types of compounds. In particular, benzene rings with an alkyl group as a substituent, such as a methyl group, often show an intense peak at $m/z = 91$, due to the formation of a seven-membered carbon ring, $C_7H_7^+$. This ion is called the **tropylium ion** and results from a rearrangement of the benzyl cation, $C_6H_6CH_2^+$. This and many other well-known rearrangement reactions are detailed in the specialized texts listed in the bibliography on interpretation of mass spectra.

11.2.2 Mass Spectra of Other Organic Compound Classes

Functional groups are of two main types: those that have a single bond to a heteroatom and those that have a multiple bond to a heteroatom. Alcohols, halides, ethers, and amines are examples of the first type, while aldehydes, ketones, and carboxylic acids are examples of the second type.

11.2.2.1 Alcohols

The mass spectra of three aliphatic alcohols, butanol, 2-butanol, and 2-methyl-2-propanol, are given in Figures 11.20 through 11.22, respectively. These three alcohols are isomers; they have the same chemical formula, $C_4H_{10}O$, also written as C_4H_9OH to emphasize the alcohol OH functional group, but different structures, as shown in the following.

$CH_3CH_2CH_2CH_2OH$ $CH_3CH_2-\underset{\underset{CH_3}{|}}{CH}-OH$ $CH_3-\underset{\underset{CH_3}{|}}{\overset{\overset{CH_3}{|}}{C}}-OH$

Butanol   2-Butanol   2-Methyl-2-propanol
(*n*-Butanol)   (*sec*-Butanol)   (*tert*-Butanol)

The names in parentheses are old names for these alcohols. The *n* stands for normal (meaning a primary alcohol; the OH group is attached to a carbon atom attached to only one other carbon atom), *sec* for secondary (note that the OH group is attached to a carbon atom attached to two other carbon atoms), and *tert* for tertiary (the OH group is on a carbon atom attached to three carbon atoms). The terms primary, secondary, and tertiary to describe the position of the functional group are still used. Table 11.19 presents relative abundances for these compounds (again remembering that the spectra and the tabular data are from different sources and may not be exactly comparable).

In the butanol spectrum, the most abundant fragment is at $m/z = 31$, corresponding to CH_2OH^+, with the structure $H_2C=OH^+$. The fragment lost was $74 - 31 = 43$, probably as a propyl radical, $C_3H_7^{\cdot}$. This indicates that fragmentation between the first and second carbon atoms is favored as shown in the following:

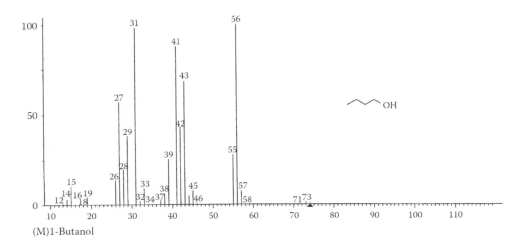

Figure 11.20 A mass spectrum of a normal (straight-chain) alcohol, butanol, $C_4H_{10}O$.

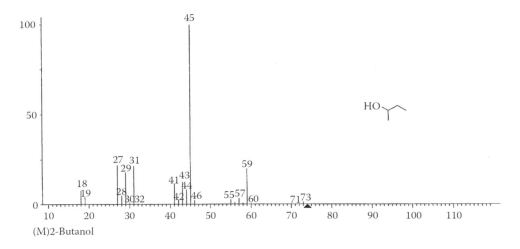

Figure 11.21 A mass spectrum of a secondary alcohol, 2-butanol, $C_4H_{10}O$.

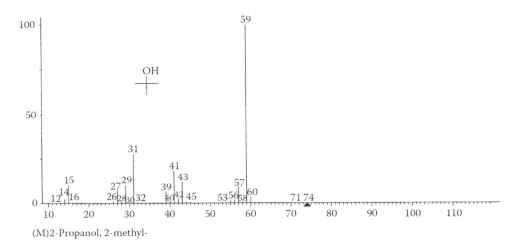

Figure 11.22 A mass spectrum of a tertiary alcohol, 2-methyl-2-propanol, $C_4H_{10}O$.

$$CH_3CH_2CH_2-CH_2OH \rightarrow CH_3CH_2CH_2{}^{\cdot} + H_2C=OH^+$$

The reasons for this favored fragmentation, a process called **alpha cleavage**, meaning cleavage at the bond adjacent to the carbon to which the functional group is attached, are discussed in detail in the text by McLafferty and Tureček. The fragmentation can be thought of as loss of an alkyl group, with loss of the largest alkyl group most likely. In butanol, the propyl group is the largest possible alkyl fragment. Alpha cleavage occurs for many functional groups in addition to alcohols, as shown schematically in the following:

Bond cleavage here
↓
$(R-CH_2-OR')^{+\bullet} \rightarrow R^{\bullet} + (CH_2=OR')^+$
↑
α carbon

As shown, cleavage occurs at the first C–C bond counting from the functional group.

However, the spectrum of butanol also shows an almost equally abundant peak at $m/z = 56$, due to loss of water $(74 - 18 = 56)$ from the molecular ion. Notice how much more likely the loss of water is than the loss of just a neutral OH fragment, which gives rise to the smaller peak at $m/z = 57$.

The butanol spectrum shows the primary decomposition paths for aliphatic alcohols: alpha cleavage and loss of water. Alcohols containing more than four carbons often lose both water and ethylene simultaneously (total loss of 18 + 28 Da). Detailed interpretation of the peaks requires extensive knowledge of the reaction mechanisms, which requires a good understanding of physical organic chemistry. For example, the peak in the butanol spectrum at $m/z = 43$ can be due to the formation of a C_3H_7 ion by loss

Table 11.19 Mass Spectral Distribution of the Fragments of Isomers of Butanol[a] Using EI at 70 eV

m/z	Butanol	2-Butanol	2-Methyl-2-propanol
15	8.4	6.80	13.3
27	50.9	15.9	9.9
28	16.2	3.0	1.7
29	29.9	13.9	12.7
31	*100	20.31	35.5
33	8.5	0	0
39	15.6	3.4	7.7
41	61.6	10.1	20.8
42	32.4	1.7	3.3
43	61.4	9.8	14.5
45	6.6	*100	0.6
55	12.3	2.0	1.6
56	99.9	1.0	1.5
57	6.7	2.7	9.0
59	0.3	17.7	*100
60	0	0.7	3.2
74	0.8	0.3	0

[a] RMM for all isomers = 74 Da. Asterisks in table body denote the most abundant fragment (the base peak). The abundance is normalized to 100 for the base peak in each pattern, but the actual abundances of these peaks probably differ from each other.

of the CH_2OH fragment as a neutral species. The observation of a peak at $m/z = 18$ (water) should be interpreted with caution, because water is a common contaminant in samples or may have come from air getting into the instrument.

With 2-butanol, the most abundant peak is at $m/z = 45$, indicating loss of a neutral fragment, $74 - 45 = 29$, probably C_2H_5. Favored fragmentation is again by alpha cleavage, with the largest alkyl group (ethyl) being lost and formation of a $CH_3CH=OH^+$ ion:

$$(CH_3CH_2-\overset{CH_3}{\underset{\uparrow}{CH}}-OH)^{+\bullet} \rightarrow CH_3CH_2^{\bullet} + CH_3CH=OH^+$$
Bond cleavage here

The loss of CH_3 from the secondary alcohol is a common fragmentation, since the peak at $m/z = 59$ is more abundant than with butanol. (Is this also alpha cleavage?) With 2-methyl-2-propanol, the most abundant peak is 59, showing a loss of $74 - 59 = 15$ Da, a methyl group. Looking at the structure in the following, it is easy to understand, since any one of the CH_3 groups may break off. All three methyl groups are equivalent, and each CH_3 group is attached to the alpha carbon, so this is again an example of alpha cleavage. Tertiary alcohols also tend to lose OH rather than water, accounting for the M − 17 peak at $m/z = 57$ in the spectrum:

$$(CH_3-\overset{CH_3}{\underset{CH_3}{C}}-OH)^{+\bullet} \rightarrow CH_3^{\bullet} + CH_3\overset{CH_3}{C}=OH^+$$

Aromatic alcohols such as phenols usually show strong molecular ion peaks and the characteristic loss of CO and CHO to give (M − 28) and (M − 29) peaks. A substituted phenol mass spectrum is given in Figure 11.23. Two features of this spectrum should stand out. There is a strong molecular ion peak (the base peak) and not much fragmentation, suggesting an aromatic compound and the almost equal heights of the peaks at $m/z = 172$ and 174 tell us there is a bromine atom present in the compound. The ratio of the (M + 1) to M peaks is about 7 %, allowing for six carbon atoms. If we assume that we have a substituted benzene ring with one Br atom as a substituent, we have $172 - 79 = 93$ Da to account for. A disubstituted benzene ring would have six carbon atoms and four H atoms, for a mass of 76 Da. The difference between 93 and 76 is 17 Da, which could be an OH group. This would suggest one of the bromophenol compounds. The rings + double bond calculation gives an answer of four, consistent with a disubstituted benzene ring containing a hydroxyl group. (Remember that Br is counted along with H for this calculation and of course the H atom in the OH group is also counted.) The spectrum should be matched to those for the possible bromophenol compounds; it is in fact 4-bromophenol.

11.2.2.2 Ethers

Ethers are compounds of the formula R–O–R'. Like alcohols, they undergo alpha cleavage as well as cleavage of the C–O bond to form a hydrocarbon ion and an oxygen-containing radical. The molecular ion is usually weak but present.

11.2.2.3 Ketones and Aldehydes

Ketones and aromatic aldehydes show strong molecular ion peaks, while aliphatic aldehydes give a weak but measurable molecular ion peak. Both ketones and aldehydes fragment by alpha cleavage. Aldehydes also fragment by beta cleavage, at the bond shown as follows:

Bond cleavage here
$$\underset{\uparrow\ \beta\ carbon}{(R-CH_2-CH_2-OR')^{+\bullet}} \rightarrow R^+ + CH_2=CH-OR'^{\bullet}$$

A very important fragmentation with rearrangement occurs with aldehydes, ketones, and other classes of compounds. This is the **McLafferty rearrangement**. The ion must be able to form a six-membered cyclic transition state. A transfer of hydrogen to the oxygen atom occurs and an alkene molecule splits off. The McLafferty rearrangement is shown for a ketone:

MASS SPECTROMETRY II: SPECTRAL INTERPRETATION AND APPLICATIONS

The rearrangement product detected is the ionic product (an enol) shown; a neutral alkene is also produced. At least one of the alkyl chains on the ketone and the alkyl chain on the aldehyde must have three or more carbon atoms for the McLafferty rearrangement to occur. The mass spectrum of 3-heptanone is shown in Figure 11.24. The peak at $m/z = 72$ is due to the McLafferty rearrangement; propene is the neutral alkene that splits off. The peak at $m/z = 57$ is due to alpha cleavage on the left side of the carbonyl group as the structure is shown in Figure 11.24 to form $(C_4H_9)^+$ or $(C_3H_5O)^+$; cleaving the bond gives two possible ions with the same m/z value. Alpha cleavage can also occur on the right side, forming $(CH_3CH_2)^+$ with $m/z = 29$ or $(C_5H_9O)^+$ with $m/z = 85$. Both ions are seen in the spectrum.

Let us look at the spectrum of an aromatic ketone in Figure 11.25 (data in Table 11.20). The general appearance of the spectrum indicates an aromatic compound because there are few peaks indicating little fragmentation. The molecular ion is at $m/z = 182$ and is strong, as expected for an aromatic ketone. The value of $(M + 1)/M$ is 12.8, indicating 12 or 13 carbon atoms. The $(M + 2)/M$ value is 1.18. Using Equation 11.3 and an assumed value of 12 C atoms based on the $(M + 1)$ peak, there is no more than one oxygen atom present. The same conclusion is drawn if 13 C atoms

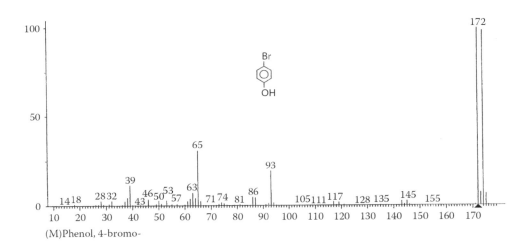

Figure 11.23 A mass spectrum of a halogen-substituted aromatic alcohol, 4-bromophenol, C_6H_5OBr.

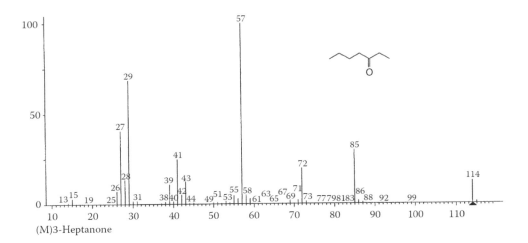

Figure 11.24 A mass spectrum of an aliphatic ketone, 3-heptanone, $C_7H_{14}O$.

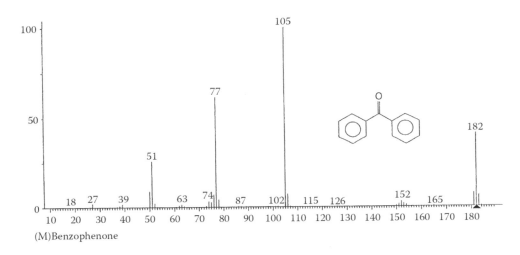

Figure 11.25 A mass spectrum of an aromatic ketone, benzophenone, $C_{13}H_{10}O$.

Table 11.20 Mass Spectral Data for Benzophenone

m/z	Relative Abundance	m/z	Relative Abundance	m/z	Relative Abundance	m/z	Relative Abundance
27	1.3	74	2.0	105	100	153	1.8
28	1.0	75	1.7	106	7.8	154	1.4
38	0.37	75.5	0.21	107	0.51	155	0.16
39	1.1	76	4.3	119	0.02	164	0.06
50	6.2	76.5	0.33	126	0.63	165	0.44
51	19	77	62	127	0.40	166	0.06
52	1.4	78	4.2	128	0.13	181	8.2
53	0.29	79	0.16	139	0.17	182	60
63	1.3	91	0.08	151	1.1	183	8.5
64	0.61	104	0.48	152	3.4	184	0.71
65	0.09						

are used in Equation 11.3. (Put in the numbers and demonstrate to yourself that this is the case.) This highlights the fact that the (M + 1) and (M + 2) peaks do not always give us a definite answer for the numbers of atoms.

The most abundant peak is $m/z = 105$, showing the loss of a neutral fragment of $182 - 105 = 77$ Da, probably a phenyl group, C_6H_5, based on the value of the $m/z = 78$ peak. The $m/z = 78$ peak can be thought of as the (M + 1) peak for the fragment at $m/z = 77$. The idea of the (M + 1) and (M + 2) peak ratios can be used to count carbon atoms and oxygen atoms in fragments, but with caution. These peaks may arise from some other fragmentation pathway, in which case, the answer will be wrong and misleading. The fragment at $m/z = 105$ probably contains seven carbon atoms, based on the value of the $m/z = 106$ peak. Using this information, calculating the ratio of peaks 107/105 and Equation 11.3 leads to the conclusion that the 105 fragment contains seven C atoms and one oxygen atom.

The peak at $m/z = 77$ is very strong. It is *not* a good assumption in general that lower m/z peaks result from sequential fragmentation of higher m/z fragments. However, in this case, it is notable that the difference between 105 and 77 Da is 28, the mass of a C=O group. CO is a commonly lost neutral species. The compound is aromatic, we were told it was a ketone, and the compound has at least one phenyl ring. What aromatic ketone can we make from 13 carbon atoms and an oxygen atom? We could postulate two phenyl rings and a carbonyl group. One possibility that would account for the spectrum is benzophenone, $C_{13}H_{10}O$. Alpha cleavage of one phenyl group would leave an ion consisting of the other phenyl ring substituted with CO, with mass = 105 Da. Loss of CO from this ion would result in a second phenyl ring fragment. To help confirm our hypothesis, we would either run a spectral library search to match the spectrum or run a sample of benzophenone on our mass spectrometer and match the spectrum or run the sample on a GC (cf. Chapter 13) and match retention times. The more the types of confirmation, the better the identification.

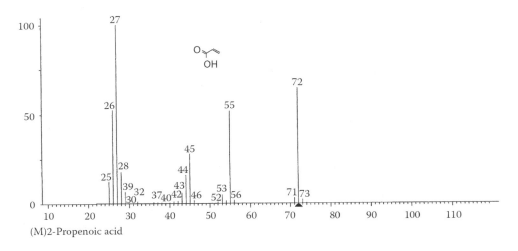

Figure 11.26 A mass spectrum of a carboxylic acid, 2-propenoic acid, $C_3H_4O_2$.

11.2.2.4 Carboxylic Acids and Esters

Carboxylic acids are compounds with the general formula

$$\underset{RC}{\overset{O}{\|}}-OH$$

while esters are compounds with the acidic hydrogen of the acid replaced by another alkyl or aromatic group to give the formula

$$\underset{RC}{\overset{O}{\|}}-OR'$$

Aliphatic carboxylic acids and small aliphatic esters (those with four to five carbon atoms) have a weak but measurable molecular ion in their mass spectra. Larger esters may show no molecular ion peak. Aromatic carboxylic acids give strong molecular ions.

The acids typically lose OH and COOH through alpha cleavage, giving fragment ions of M − 17 and M − 45. The COOH$^+$ ion at m/z = 45 also forms and is a very characteristic peak in the spectra of these acids. The mass spectrum of 2-propenoic acid, $H_2C=CH–COOH$, shown in Figure 11.26, displays the characteristic peaks of a carboxylic acid. The molecular ion occurs at m/z = 72 and is actually quite strong. The loss of the OH group from the molecular ion results in the (M − 17) ion at m/z = 55, while the loss of the COOH group from the molecular ion produces the (M − 45) ion at m/z = 27. The COOH$^+$ ion arising from loss of the rest of the molecule as a radical gives the peak at m/z = 45. A table of mass spectral data (again from a different source than the spectrum shown in Figure 11.26) for 2-propenoic acid is presented in Table 11.21. From the (M + 1) peak, three carbon atoms are likely; from the (M + 2) peak and Equation 11.3, we would deduce the presence of two oxygen atoms. Three carbon atoms plus two oxygen atoms have a mass of 68 Da, leaving 4 Da to be accounted for as hydrogen atoms. Therefore, the formula is most likely $C_3H_4O_2$. Rings + double bonds: $3 − 1/2(4) + 1 = 2$. One of the double bonds is the C=O bond in the acid; the other must be a C=C bond.

Esters undergo alpha cleavage to form an RCO$^+$ ion. Those esters with four or more carbon atoms (butyrates and higher) undergo the McLafferty rearrangement. For methyl esters, where the R' is CH_3, the ion formed as a result of the McLafferty rearrangement is

$$(CH_3O-\underset{|}{\overset{OH}{C}}=CH_2)^+$$

which has an m/z value of 74. This is a characteristic peak in the spectra of methyl esters.

Table 11.21 Mass Spectral Data for 2-Propenoic Acid

m/z	Relative Abundance	m/z	Relative Abundance
25	5.2	45	32
26	38	46	2.5
26.5	0.15	47	0.14
27	74	52	1.4
27.5	0.26	53	6.0
28	12	54	2.3
29	4.3	55	74
30	0.13	56	2.6
31	0.48	57	0.19
41	1.2	71	4.3
42	1.3	72	100
43	5.8	73	3.5
44	14	74	0.48

11.2.3 Compounds Containing Heteroatoms

Compounds containing the halogen atoms Cl and Br have been discussed in Sections 11.1.4 and 11.1.5. The two other important halogen atoms, fluorine and iodine, are monoisotopic and have no isotope cluster patterns. It is important to remember that they do need to be counted in the rings + double bonds equation. In general, iodine compounds tend to fragment by loss of the halogen, giving a weak or no molecular ion. Chlorine compounds tend to lose HCl. Fluorine compounds undergo some unique reactions with rearrangements that are beyond the scope of this text, although we will look at a fluorinated inorganic compound in the next section. One very characteristic indication of the presence of fluorine, mass = 19 Da, in a molecule is the loss of fluorine, resulting in an (M − 19) peak.

While nitrogen, sulfur, silicon, and phosphorus are important heteroatoms, we will look briefly only at nitrogen and sulfur compounds. In general, when the heteroatom is incorporated into an aromatic ring, strong molecular ions are seen. Table 11.4 should be reviewed to remind you of the isotope ratios for these heteroatoms to help in interpreting spectra.

11.2.3.1 Nitrogen-Containing Compounds

There are a number of classes of organic compounds that contain nitrogen, including amines, amides, nitro compounds, and nitriles. All of these compounds follow the *nitrogen rule*, which tells us that a compound with an odd number of nitrogen atoms has an RMM that is an odd number. This rule is very useful in determining if a compound contains nitrogen atoms. Unfortunately, many nitrogen-containing compounds give no detectable molecular ion. Aliphatic amines, nitriles, and nitro compounds rarely give detectable molecular ions. Amides, cyclic aliphatic amines, aromatic amines, nitriles, and nitro groups do give measurable molecular ions.

The primary fragmentation reaction for aliphatic amines is alpha cleavage. A primary amine, RCH_2NH_2, gives the $CH_2NH_2^+$ ion, $m/z = 30$, as a result of alpha cleavage, and this serves as a diagnostic peak. The spectrum (Figure 11.27) of a secondary amine, 2-butanamine (RMM = 73), shows the peaks expected from alpha cleavage on both sides of the amine group, an (M − 15) peak due to loss of the methyl radical at $m/z = 58$ and the intense peak at $m/z = 44$ due to cleavage and loss of the ethyl radical (M − 29). Remember that the longest alkyl group is preferentially lost as a neutral species, which explains why the peak at $m/z = 44$ is larger than at $m/z = 58$. However, the molecular ion peak ($m/z = 73$) is very small for this compound, as is typical of aliphatic amines. Its odd number value is consistent with the nitrogen rule and gives a great deal of information for so small a peak.

Amides have fragmentation patterns similar to their corresponding carboxylic acids. Nitro compounds often have the ions NO^+ ($m/z = 30$) and NO_2^+ ($m/z = 46$) in their spectra. Aromatic nitro compounds have characteristic peaks at M − 30 and M − 46, due to loss of the radicals NO• and NO_2•. Heteroaromatic nitrogen compounds like pyrrole often fragment to lose a neutral HCN molecule; this loss of HCN is also seen with aromatic amines such as aniline.

Inorganic molecules can also be determined by MS. The compound nitrogen fluoride, NF_3, exhibits characteristics of both a nitrogen-containing compound and a fluorine-containing compound in its mass spectrum (Figure 11.28). The molecular ion occurs at $m/z = 71$, an odd number,

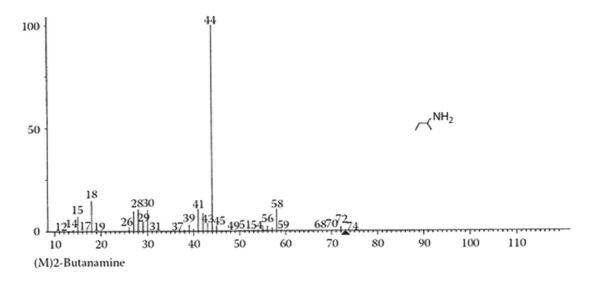

Figure 11.27 A mass spectrum of a secondary aliphatic amine, 2-butanamine, $C_4H_9NH_2$.

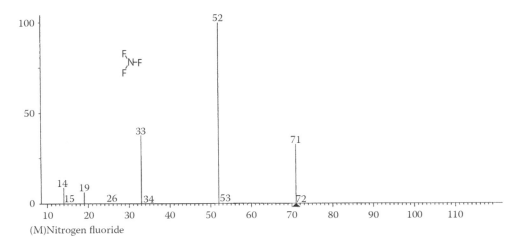

Figure 11.28 A mass spectrum of the inorganic compound, nitrogen fluoride, NF_3.

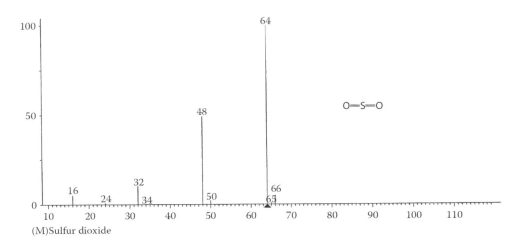

Figure 11.29 A mass spectrum of the inorganic compound, sulfur dioxide, SO_2.

indicating that the molecule contains an odd number of nitrogen atoms. The base peak shows a loss of 19 Da from the M+ peak; this is characteristic of the loss of fluorine. The peak at $m/z = 19$ confirms the presence of fluorine. The peak at $m/z = 33$ results from a loss of two fluorine atoms from the molecular ion or another fluorine atom from the $m/z = 52$ fragment. Two fluorine atoms and one nitrogen atom account for a mass of 52 Da. The difference between 71 and 52 is 19, leading to the conclusion that there are three fluorine atoms and one nitrogen atom in the compound. The formula is NF_3.

11.2.3.2 Sulfur-Containing Compounds

Thiols, also called aliphatic mercaptans, with the formula RSH, are the sulfur analogs of alcohols. Thiols generally show stronger molecular ion peaks than the equivalent alcohols. Looking back at Table 11.4, sulfur has a significant ^{34}S isotope. This gives rise to an enhanced (M + 2) peak in the mass spectrum of sulfur-containing compounds.

Primary thiols lose H_2S just as alcohols lose H_2O on fragmentation. This results in a characteristic peak at M − 34. The fragmentation patterns for thiols are similar to those for alcohols, with alpha cleavage and loss of the larger fragment being common. Heteroaromatic compounds such as thiophene, with the sulfur atom in the ring, give strong molecular ion peaks. Other sulfur-containing organic compounds show similar fragmentation patterns to their oxygen-containing counterparts, but often with stronger molecular ion peaks.

Inorganic sulfur compounds can be determined by MS, as illustrated by the spectrum of SO_2, sulfur dioxide. The spectrum is shown in Figure 11.29. The molecular ion is shown at $m/z = 64$ and is the base peak in the spectrum. Using a ruler, the (M + 1) and (M + 2) peaks can be measured, but not very accurately. Doing so gives an (M + 1)/M value that is about 0.6, much less than 1.1, and an (M + 2)/M value of 4.8. Consulting Table 11.4, there cannot be any carbon present due to the small (M + 1) peak, but both the (M + 1) and (M + 2) peaks are consistent with the presence of one sulfur atom. The (M + 2) peak is actually larger than the 4.4 due to a sulfur atom, so oxygen may also be present. The peak in the spectrum at $m/z = 48$ is an (M − 16) peak, due to the loss of an oxygen atom. The peak at $m/z = 32$ is sulfur and is also 16 Da lower than the $m/z = 48$ peak, indicating loss of another oxygen atom from that fragment. Oxygen is confirmed by the peak at $m/z = 16$. Two oxygen atoms and one sulfur atom account for the RMM of 64 needed. The formula is therefore SO_2.

We have only touched on a few examples of select organic compound classes and a few inorganic compounds. The student is advised to consult the texts on mass spectral interpretation cited in the bibliography for more details. Interpreting mass spectra requires a great deal of practice coupled with knowledge of the major fragmentation reactions and rearrangement reactions, which we cannot cover in detail in this text. It is rare that a practicing mass spectrometrist actually tries to assign an unknown structure by evaluating every peak in the mass spectrum and the isotope ratios, as we have outlined here. Matching to a spectral library plays a predominant role in modern MS with confirmation performed by analysis of a known sample. However, with much practice, knowledge of the molecular ion m/z value, fragmentation patterns, isotope ratios, and the like, it may be possible to come to well-founded conclusions about the empirical formula and the RMM of an unknown compound.

MS alone often cannot achieve confirmation of the identity of an unknown, even when a pure sample of the candidate compound is analyzed by MS. An extreme example of two totally unrelated compound structures that produce very similar EI-MS fragmentation patterns is displayed in Figure 11.30. This pair was discovered by searching an MS database of hundreds of thousands of spectra to find the two closest matches. A library search yielding the MS of either of these would not allow one to identify which one gave the spectrum. Other types of spectral or chromatographic (cf. Chapters 13 and 14) information or common sense evaluation of the likelihood of finding one or the other in the particular sample must come into play. Let this be a cautionary example against too much reliance on library searching for MS identifications.

Corroborative evidence is often necessary from IR, which identifies the presence of many functional groups, and NMR, which confirms functional groups and, by spin–spin splitting patterns, the placement of these groups. Elemental analysis to determine the C, H, N, O, and heteroatom content is usually performed on pure compounds to assist in the assignment of an empirical formula. Optical activity measurements may be needed for chiral compounds. When used in conjunction with other analytical methods, such as elemental analysis, IR, and NMR, MS makes it possible to identify unknown compounds. Combined with a separation method like chromatography, as in GC-MS or LC-MS, even impure samples and mixtures can be analyzed and components identified. GC-MS and LC-MS are described in Chapters 13 and 14, respectively.

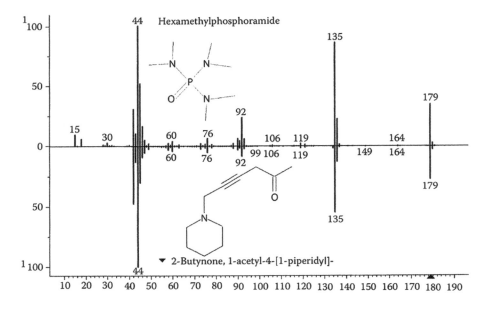

Figure 11.30 Matching MS of two radically different compound structures: one aligned and displayed inverted below the other. (From the NIST Mass Spectrometry Data Center accessed via http://webbook.nist.gov. © 2011 U.S. Secretary of Commerce for the United States of America. All rights reserved. Used with permission.)

11.3 APPLICATIONS OF MOLECULAR MS

The applications for molecular MS have grown enormously in recent years. MS (molecular and atomic) is used by clinical chemists, pharmaceutical chemists, synthetic organic chemists, petroleum chemists, geologists, climatologists, molecular biologists, marine biologists, environmental scientists, forensic chemists, food scientists, nuclear scientists, materials scientists, inorganic chemists, organometallic chemists, analytical chemists, and scientists in many other fields. MS is not just used in laboratories, but online MS analyzers and field-portable MS systems are available commercially for process control and on-site detection of volatile pollutants, explosives, and gases. Applications are as diverse as determination of RMM, molecular structure, reaction kinetics, dating of minerals, fossils, and artifacts, fundamental studies of ion–molecule reactions, and quantitative analysis of elements and compounds at sub-ppb concentrations. Inorganic and organic solids, liquids, and gases can be analyzed. Subcellular structures such as whole ribosomes can be studied. Some application examples will be discussed but the field is vast and cannot be adequately covered in this text. In addition to the scientific literature, many applications of MS can be found on the websites of the instrument manufacturers. Some of the websites are given subsequently.

11.3.1 High-Resolution Mass Spectrometry

Double-focusing magnetic sector mass spectrometers; Fourier transform (FT) mass spectrometers; long path, folded, multireflection or multiturn spiral time of flight (TOF); and Orbitrap instruments (described in Chapter 10) are capable of mass measurements with high *resolution*, which enables separate measurements of mass and signal intensity of ions of closely similar mass and high *mass accuracy*, which enables determination of the exact masses of the resolved ions. The most important use of high-resolution MS is the direct determination of molecular formulas by exact mass measurements.

The atomic masses of the individual isotopes (nuclides) of the elements are nominally whole numbers, H = 1 and 2, C = 12 and 13, N = 14 and 15, O = 16, 17, and 18, and so on. If these masses are measured with sufficient accuracy, we find that actually they are only close to being whole numbers. This is due to the "mass defect," discovered by the early MS pioneer Aston in the 1920s. The mass defect is characteristic of a given isotope. If we use ^{12}C as the standard, the atomic masses of some common isotopes are as shown in Table 11.22. As a consequence, two molecular formulas may have the same nominal unit RMM but actually differ slightly but significantly. For example, $C_{20}H_{26}N_2O$ and $C_{16}H_{28}N_3O_3$ have the same unit RMM, 310 Da, but if we take into account the exact atomic masses of the nuclei involved, the RMMs are actually 310.204513 and 310.213067, respectively. Numerous other formulas with this nominal mass can be written; they also have slightly different actual RMMs. With sufficient resolution, it is possible to distinguish among almost all reasonable formulas. More importantly, knowing the exact RMM allows us to identify the molecular formula of the sample. Not only can the formula of the molecular ion be determined but also the formulas of fragments of the molecule. This provides valuable information on large fragments of very large molecules and on fragments when no molecular ion is observed. Much of the pioneer work in this field was carried out by F. W. McLafferty at Cornell University.

Resolution was defined in Chapter 10, and a graphical example of the effect of resolution on the measured mass is shown in Figure 11.31. The compound $C_{101}H_{145}N_{34}O_{44}$ has a nominal mass of 2537 Da, calculated from the unit masses of the most abundant stable isotopes, ^{12}C, ^{1}H, ^{14}N, and ^{16}O. A spectrum with resolution of 200 (low resolution) would give the single broad peak shown. The mass measured from this peak at the peak maximum would be 2539.5 Da (or 2540 Da for unit resolution). The measured average mass is slightly higher than the nominal mass because the large number of atoms in the molecule makes it probable that there will be higher isotopes present, most likely ^{13}C. So the measured average mass at low resolution could be misleading in trying to assign the elemental composition. With a resolution of 2500, the monoisotopic peaks that contribute to the average mass are separated, allowing much more exact mass measurement as shown. If the resolution was infinite, each monoisotopic peak would be a line, as shown.

Loss of resolution (peak broadening) is largely caused by the narrow beam of ions spreading out while traveling through the sector. Similar masses may spread out enough to actually overlap each other and therefore be unresolved. Double-focusing instruments refocus the ion beam to prevent overlap of peaks. Exact mass measurements are made by comparing the position of the unknown peak with the position of an internal standard of

Table 11.22 Exact Mass of Common Isotopes

Element	Atomic Mass	Isotope (Nuclide)	Exact Mass
Hydrogen	1.00794	^{1}H	1.007825
		^{2}H (or D)	2.014102
Carbon	12.01115	^{12}C	12.000000
		^{13}C	13.003355
Nitrogen	14.0067	^{14}N	14.003074
		^{15}N	15.0001
Oxygen	15.9994	^{16}O	15.994915
		^{17}O	16.9991
		^{18}O	17.999159
Fluorine	18.998405	^{19}F	18.998405

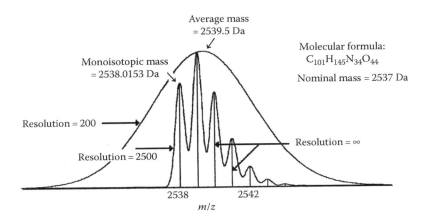

Figure 11.31 The difference between low-resolution and high-resolution MS. The nominal mass of the molecule $C_{101}H_{145}N_{34}O_{44}$ = 2537 Da. At the low resolution of 200, a single peak is measured with an average mass at the peak maximum = 2539.5 Da. A high-resolution instrument (resolution = 2500) separates the monoisotopic peaks and permits measurement of exact masses. (Courtesy of Professor Gary Siuzdak, Scripps Research Institute Center for Mass Spectrometry, https://masspec.scripps.edu/. The website has a number of tutorials on MS.)

known composition and known exact mass. The ions of the known internal standard are used to calibrate the mass scale. The magnetic sector instruments often use selected ions from these internal standards as a **lock mass** against which to reference values on the mass axis and compensate for instrumental drift in parameters over time. High-resolution MS can be used to identify the formulas of fragments of large molecules such as proteins. It can also be used to assign empirical formulas to newly synthesized organic compounds or complex molecules isolated from natural products.

11.3.1.1 Achieving Higher Mass Accuracy (but Not Resolution) from Low-Resolution MS Instruments

Review the discussion of resolving power and resolution in MS instruments in Section 10.1.1. Recall that it was noted there that the ideal Gaussian profile of the *m/z* of an ion was distorted by the necessary compromises made in tuning, where the voltages on a large number of metal elements that shape and guide the ion trajectories through the instrument must be interactively adjusted to an optimum line shape and position of the peak on the mass axis. Distortions, misalignments, and accumulation of contaminants on these elements degrade the tuning and accuracy of these values and require frequent retuning. Even then, distortions will likely remain. In the early days of scanning MS, the continuous ion signal response as a function of scanning voltage along the mass axis produced more data points than could be accommodated by the speed and memory capacities of the computerized data systems. Therefore, the signal was processed in real time and converted to individual mass values with unitary or better accuracy, and the MS line signal counts for each "unit mass" were stored with its centroid position. This gives rise to the "line or bar diagrams" so familiar to users of MS data systems and from which computerized library search routines for EI fragmentation patterns operate. These are what you have been looking at in Figures 11.1 through 11.29. The MS sees a series of distorted and misaligned Gaussian peaks and converts them to a bar graph, and you see the output of these figures: good enough for unit mass work, but "exact mass" will enable more interpretation.

Software is now available that can deconvolute the more data-intense continuous output signal from a "unit mass" or even a higher-resolution MS instrument. One such program is the CERNO MassWorks™ program from CERNO Bioscience, www.cernobioscience.com. Using stored data on the distortions in line shapes obtained by measuring the response to calibration compounds yielding ions of precisely known or calculated mass, the software can restore the exact position and theoretical shape of the observed lines of the monoisotopic ion (of lowest isotopic mass) and with the centroid of each peak measurable to a much higher accuracy (e.g., 1 ppm) than the nominal accuracy of the instrument. *Note well*: This improves measurement of the monoisotopic ion's mass **only** *if its peak is from a single ion well resolved by the instrument from any other*. The process improves single-ion *mass accuracy* but does *not* improve *resolution* between ions of closely similar mass if both are present. The software does not convert a low-resolution MS into one that can do *everything* achievable by a high-resolution MS. The application of this software improves the mass accuracy obtainable from lower-resolution instruments for the lowest mass ion of any isotope cluster, which can reduce the number of possible empirical formulas from dozens or hundreds (the possibilities increase rapidly as *m/z* increases) to perhaps only one or two. But you must ensure that you are working with the isotope of the line from the MS of a molecular ion or fragment from only one compound. In practice, this will be of more use if a high-resolution *chromatographic* separation providing complete separation of compounds is interfaced to the MS ion source. Still, the possession of data that can give the probable empirical formula of an ion, especially the molecular ion if that is identifiable, is of great value. The few remaining candidate formulas may often be excluded on grounds of their improbability in the sample or from other information in hand, such as the spectral accuracy measurements discussed in the following.

11.3.1.2 Improving the Quantitation Accuracy of Isotope Ratios from Low-Resolution MS Instrument Data Files

If we can measure the intensity of all the higher isotope masses (i.e., A + 1, A + 2, where A is the mass of the lowest mass monoisotopic ion) in an isotope cluster, we can determine which combination of different isotopes of different atoms best fits the candidate empirical formulas. To do this, we must attain high *spectral accuracy*. Another feature of the CERNO program that corrects the mass line shapes to fix the exact centroid position on the mass axis is improvement of spectral accuracy of these higher isotope cluster mass peaks, by applying the instrument calibration algorithm to restore the peaks to their ideal Gaussian shape, magnitude, and position. This enables much more precise and exclusive assignment of the best elemental composition match to the candidate mass and formula. Thus, mass accuracy and spectral accuracy from low-resolution MS instruments can be improved *in silico* (i.e., by computer processing) to values similar to those attained by high-resolution instruments, at greatly reduced cost, *if* one can be assured that the cluster MS peaks are only from one pure ion. Hence, a preliminary MS stage or chromatographic isolation of the compound in question is desirable.

An illustration of the operation of the CERNO MassWorks™ program is displayed in Figure 11.32, which shows both a heavily distorted raw MS signal of an ion isotope cluster and the restored ideal trace after the processing with a correction algorithm. As an example of the power of these algorithms, a compound of unit mass $M^+ = 400$ whose formula is $C_{25}H_{23}N_2OS$ could be measured in a low-resolution MS with a peak shape with full width at half maximum (FWHM) of 0.600 Da. There are 4110 possible formulas within a mass tolerance window of 20 mDa (51 ppm). When the spectral accuracy match for each of these, which was calculated from the data on all the higher masses

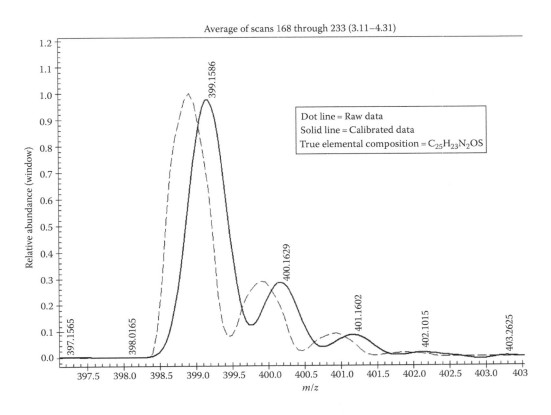

Figure 11.32 Application of CERNO MassWorks™ and CLIPS programs to MS isotope cluster raw data. (Used with permission from CERNO Bioscience, www.cernobioscience.com.)

in the isotope cluster, is added to the search parameters, the aforementioned formula is the second closest hit among the total of 4110 formulas having a spectral error of 1 %. Even applying the industry standard mass matching tolerance of 5 ppm (2 mDa), there remain 411 possible candidates. This not only demonstrates the feasibility for unknown formula determination on a unit mass resolution quadrupole MS but demonstrates that spectral accuracy is more important than mass accuracy for differentiation among large numbers of formula candidates, regardless of the MS resolving power available. These programs can be usefully applied even to data obtained on high-resolution instruments.

11.3.2 Quantitative Analysis of Compounds and Mixtures

Quantitative analysis of compounds can be performed by measuring the relative intensities of a spectral peak or peaks unique to the compound and comparing the intensity of the sample to a series of known standards of the compound. The use of **isotope dilution** for quantitative measurements in MS is common. Isotope dilution is a special case of calibration using an internal standard; the internal standard calibration approach was covered in Chapter 3. For molecular analysis, the internal standard is the analyte molecule, but with some of its atoms replaced by heavier isotopes. Such a compound has been **isotopically labeled**. Labeling with deuterium (^{2}H) and ^{13}C is common, but isotopes of heteroatoms can also be used. For example, orotic acid, $C_5H_4N_2O_4$, is a compound with two nitrogen atoms in a six-membered ring; it is a metabolite present in only trace amounts in human urine, but an increase in orotic acid in urine can signify diseases that disrupt the urea cycle (McCann et al. 1995). If both ^{14}N atoms in natural orotic acid are replaced with ^{15}N atoms, signified by calling it 1,3-[^{15}N$_2$] orotic acid, the isotopically labeled orotic acid has an RMM that is 2 Da higher than natural orotic acid but behaves in all respects like natural orotic acid during analysis. An isotopically labeled compound added in a known amount to known amounts of the natural compound is the perfect calibration standard in this respect. A quantitative mass spectral determination of orotic acid in urine by GC–MS as the trimethylsilyl derivative (Section 11.3.7 and Chapter 13) can be performed by measuring ions with $m/z = 357$ or 254 for natural orotic acid and $m/z = 359$ or 256 for the labeled acid. The ratio of $m/z = 357/359$ or 254/256 is plotted versus the concentration of orotic acid in a set of calibration standards.

Quantitative analysis can also be performed by external calibration or by the use of an internal standard that is not a labeled analyte molecule but a compound that is not present in the sample.

In some cases, components in a mixture can be determined quantitatively without prior separation if the mass spectrum of each component is sufficiently different from the others. Suppose that a sample is known to contain only the butanol isomers listed in Table 11.19. It can be seen from Table 11.19 that the peak at $m/z = 33$ is derived from butanol, but not from the other two isomers. A measurement of the $m/z = 33$ peak intensity compared to butanol standards of known concentration would therefore provide a basis for measuring the butanol content of the mixture. Also, we can see that the abundances of the peaks at $m/z = 45$, 56, and 59 vary greatly among the isomers. Three simultaneous equations with three unknowns can be obtained by measuring the actual abundances of these three peaks in the sample and applying the ratio of the abundances from pure compounds. The three unknown values are the percentages of butanol, 2-butanol, and 2-methyl-2-propanol in the mixture. The three equations can be solved and the composition of the sample determined. Computer programs can be written to process the data from multicomponent systems, make all necessary corrections, and calculate the results.

Complex mixtures of molecules such as biological fluids, natural products, foods, and beverages will result in mass spectra that are quite complicated, even if soft ionization is used to minimize fragmentation. Mixtures are more commonly analyzed by using GC-MS^n or LC-MS^n to separate the components of the mixture and to obtain mass spectral information on the separated components. This allows the analyst to obtain the pure MS spectra of each chromatographic peak by GC-MS or LC-MS. Chromatographic retention time matches will add to the confidence of the peak assignments. Standards are run, often using isotope dilution and internal standard calibration, and the peak intensities or intensity ratios of appropriately selected peaks are used to make a calibration curve from which unknown concentrations in samples can be determined.

GC-MS and LC-MS instrumentation and applications are described in more detail in Chapters 13 and 14, but a few examples will be given here. A Japan Electron Optics Laboratory (JEOL) GCMate II™ high-resolution GC-MS with a magnetic sector mass analyzer was used to analyze Scotch whiskey and tequila samples. Over 50 compounds were identified in the whiskey using exact mass measurements (see Section 11.3.1) from EI high-resolution spectra. Figure 11.33a shows the gas chromatogram for a sample of whiskey; the peaks are the separated compounds, identified by their retention time. Figure 11.33b presents the high-resolution exact mass measurements and the elemental compositions of the identified compounds. Such an analysis would permit distillers, food scientists, and regulators to compare brands, compare different batches for quality, and monitor contaminants. One of the compounds identified in the whiskey was a plasticizer that may have leached from packaging material. The high-resolution GC-MS system was also able to separate and measure trace amounts of PCBs in crude oil, quite a complicated mixture. (These and

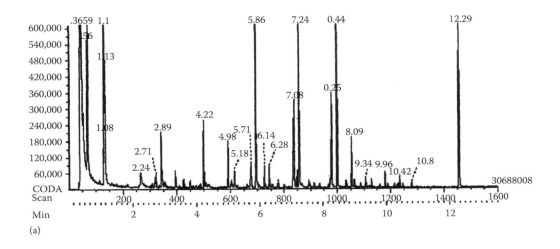

Compound	Meas. m/z	Composition	Ion	Calc. m/z	Error
ethyl acetate	88.05077	$C_4H_8O_2$	$M^{+\cdot}$	88.05243	-1.7
1-propanol, 2-methyl	74.07131	$C_4H_{10}O$	$M^{+\cdot}$	74.07317	-1.9
diethoxyethane	103.0752	$C_5H_{11}O_2$	$[M-CH_3]^+$	103.0759	-0.7
furfural	96.02313	$C_5H_4O_2$	$M^{+\cdot}$	96.02113	2
furfural	95.01513	$C_5H_3O_2$	$[M-H]^+$	95.01331	1.8
1-butanol, 3-methyl acetate	87.04336	$C_4H_7O_2$	Fragment	70.07825	-0.2
1-butanol, 3-methyl acetate	70.07805	C_5H_{10}	Fragment	87.04461	-1.2
hexanoic acid, ethyl ester	88.05026	$C_4H_8O_2$	Fragment	88.05243	-2.2
hexanoic acid, ethyl ester	99.07999	$C_6H_{11}O$	Fragment	99.081	-1
hexanoic acid, ethyl ester	101.0588	$C_5H_9O_2$	Fragment	101.0603	-1.5
hexanoic acid, ethyl ester	115.0717	$C_6H_{11}O_2$	Fragment	115.0759	-4.2
hexanoic acid, ethyl ester	117.0923	$C_6H_{13}O_2$	Fragment	117.0916	0.7
hexanoic acid, ethyl ester	144.1134	$C_8H_{16}O_2$	$M^{+\cdot}$	114.115	-1.6
phenylethyl alcohol	122.0713	$C_8H_{10}O$	$M^{+\cdot}$	122.0732	-1.9
phenylethyl alcohol	92.06038	C_7H_8	Fragment	92.0626	-1
phenylethyl alcohol	91.05377	C_7H_7	Fragment	91.05478	-1
decanoic acid	172.1481	$C_{10}H_{20}O_2$	$M^{+\cdot}$	172.1463	1.8
decanoic acid, ethyl ester	200.1776	$C_{12}H_{24}O_2$	$M^{+\cdot}$	200.1776	0
dodecanoic acid	200.1764	$C_{12}H_{24}O_2$	$M^{+\cdot}$	200.1776	-1.2
dodecanoic acid, ethyl ester	228.2078	$C_{14}H_{28}O_2$	$M^{+\cdot}$	228.2089	-1.1

(b)

Figure 11.33 (a) Gas chromatogram of compounds in a sample of whiskey analyzed on a GCMate II™ high-resolution magnetic sector GC–MS instrument. (b) The exact masses, elemental compositions, and identification of the compounds in a whiskey sample analyzed by high-resolution GC-MS. (From JEOL Applications Note MS-1126200-A. Both graphics courtesy of JEOL USA, Inc., Peabody, MA, www.jeol.com.)

the following JEOL applications, along with many others, can be found on the JEOL website at www.jeol.com.)

While we have focused on positive ion MS, there are classes of compounds that give much better mass spectral detection limits as negative ions. The use of a highly selective ionization source, a tunable-energy electron monochromator (the TEEM™ from JEOL), allows negative ions to be formed directly from analyte molecules. The electron beam energy can be tuned in the range of 0–25 eV; this permits selective ionization of the analyte, not matrix molecules. The combination of the tunable-energy EI source with the JEOL GC-MS system into an instrument called the TEEMmate™ has permitted the determination of less than 10 ppb of explosives such as trinitrotoluene (TNT), nitroglycerin, pentaerythritol tetranitrate, and RDX in complex matrices like soils by negative ion MS. An example is shown in Figure 11.34. The instrument has also been applied in the detection of chemical warfare agents, bacterial spores, and environmental contaminants. Another mode of negative ion MS operation is negative ion chemical

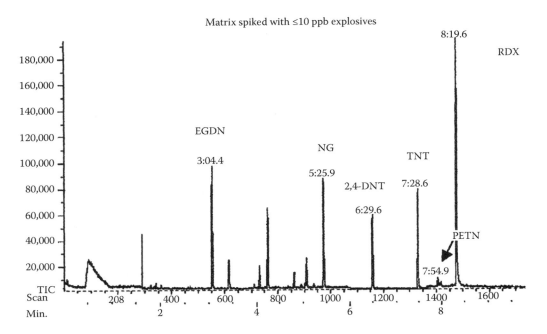

Figure 11.34 Determination of less than 10 ppb of a mixture of explosives spiked into a "dirty" matrix by negative ion MS using a JEOL TEEMmate™ with a highly selective tunable energy EI source. (Courtesy of JEOL USA, Inc., Peabody, MA, www.jeol.com.)

ionization (NICI) where the target molecules possess strong electronegativity (e.g., heavily halogen substituted) and the "chemical" is actually an electron captured from a donor molecule or low-energy electron beam, sometimes referred to as an electron capture (EC) process.

11.3.3 Protein-Sequencing Analysis (Proteomics)

Proteins make up some of the most important components of the human body and indeed all animal life. They are major components in all living cells and are therefore of great importance in all chemical studies of life sciences. It is estimated that in human blood plasma alone, there are about 100,000 proteins ranging in concentration from millimolar levels for proteins such as albumin to femtomolar levels for proteins such as tumor necrosis factor.

Proteins are made up of amino acids, which, as the name implies, include both an amino group and a carboxylic acid. There are 20 essential amino acids and the many hundreds of thousands of proteins are made up of these building blocks. What distinguishes one protein from another is the sequence in which these amino acids are arranged in the molecule. Sequencing has become a cornerstone of the research into proteomics, the study of protein structure and function. There are several approaches that can be taken to identify the sequence of amino acids in a protein by MS. The reference by de Hoffmann and Stroobant contains a large section on analysis of biomolecules by MS. 2D gel electrophoresis separation followed by matrix-assisted laser desorption ionization (MALDI)-TOF MS and digestion of the protein into peptide fragments followed by high-performance LC-MS-MS are two ways in which protein sequences and structure can be obtained. A detailed description of the HPLC-MS^n approach to protein sequencing is given in Chapter 14 (Section 14.1.7). The analysis generally requires digestion of the protein with trypsin to form short peptide fragments. Such a tryptic digest of a pure 14 kDa protein was analyzed on a TOF-based LC-MS system, the JEOL AccuTOF™ using nano-ESI. As expected for ESI, multiply charged peptides were seen in the mass spectrum. A database search of a protein library (ProFound, available on the Internet at http://prowl.rockefeller.edu/cgi-bin/ProFound) resulted in 10 possible protein matches from the peptide distributions, one of which had a 100 % probability.

MS is used to identify proteins, protein complexes with deoxyribonucleic acid (DNA), and even intact ribosomes (a multiple protein–ribonucleic acid [RNA] structure that produces proteins), but the analysis had been slow. Due to the large number of possible proteins and the wide concentration range found in samples, MS methods at the time of the sixth edition (2004) could identify only about 1000 proteins in up to ten samples a month. A recent article by Richards updates the progress in proteome mapping by MS. This article conveys the breakneck pace of progress in the technology. It is noted that "Thermo Scientific's newest Orbitrap, the Q-Exactive, adds a quadrupole mass filter which was not available on the LTQ Orbitrap in 2008. The quadrupole helps rapidly separate peptide ions of different masses for fragmentation, allowing them to be broken apart almost as quickly as they are selected by the filter, which increases speed." In 2008, using the LTQ Orbitrap, it took 3 months to generate the yeast proteome map, which included almost 4000 proteins. The Q-Exactive's advances enable the same procedure to be done in about 4 h with a relative ease of use that "puts complex proteomics experiments within the reach of MS neophytes" (see Orbitrap discussion in Section 10.2.3.7). AB SCIEX is implementing a new approach to protein mapping called "SWATH," the simultaneous acquisition of multiple "swaths" of data on all proteins in a sample (a so-called "shotgun" procedure). This is implemented on their TripleTOF 5600 MS. Richards noted that only a few years ago, a shotgun approach aimed at mapping a proteome could take a month, because it required prefractionating the proteins and running each fraction as many as eight times to avoid missing peptides. SWATH requires no prefractionation before samples are applied to LC columns, and about 2000 yeast proteins have been mapped in a 2 h period. At a conference in March of 2013, Dr. Leroy Hood, the director of the US portion of the worldwide collaboration that first sequenced the human genome at the end of the twentieth century at a cost of $3,000,000,000 and years of effort by hundreds of workers, predicted that within a decade, such powerful new MS technologies would enable individuals to have their full genome, proteome, and metabolomes characterized, completely transforming the practice of medicine. Premeds using this text in undergraduate studies should be on notice to keep an eye on the pace of progress in this area!

11.3.4 Gas Analysis

The direct measurement of gases in medicine, industry, materials, and the environment is an important use of MS.

One direct application of MS in medicine is in blood gas analysis. The speed of the analysis allows surgeons to monitor the blood of patients during operations. The concentration in the blood of compounds such as CO_2, CO, O_2, and N_2 and anesthetics such as N_2O can be controlled in this way.

Mass spectrometric analysis of gases has many industrial applications. In petroleum chemistry applications, MS is used to identify chemicals in hydrocarbon fractions, to detect and identify impurities such as ethane in polymer feedstocks such as ethylene, and to determine CO, CO_2, H_2, and N_2 in feed streams used for making alcohols. MS is also used for analysis of jet and automobile exhausts. Online quadrupole MS gas analyzers, also called residual gas analyzers (RGAs), are used to monitor noble gases, solvent vapors, corrosive acid vapors, inorganic gases, and other species in semiconductor and optoelectronics manufacturing processes. The composition of feed, exhaust and stack gases, evolved gas, and headspace gas in many industries can be determined using these types of mass spectrometers. Systems such as the MKS Instruments, Inc. (Andover, MA), Mini-Lab™ online analyzers can track gas concentrations over a wide dynamic range (ppb to percent levels) in seconds. These systems are used to monitor trace contamination in bulk gases, to study catalysis, fuel cells, and semiconductor wafer processing. For example, in the deposition of epitaxial silicon layers on a silicon wafer, a Mini-Lab™ quadrupole mass analyzer was used to measure oxygen and water vapor, which can interfere with deposition on the wafer surface. The unit was also used to study the deposition of Si. The deposition of Si occurs by hydrogen reduction of $SiHCl_3$, a gas, to form solid Si. The $SiHCl_3$, gaseous intermediates such as $SiCl_2$, and gaseous products such as HCl can all be measured using the mass spectrometer. (This and other applications as well as a useful tutorial with pictures of the quadrupole and other parts of the mass spectrometer can be found on the MKS Instruments, Inc., website at www.mksinst.com). Even smaller quadrupole mass analyzer systems are available as field-portable gas analyzers; quadrupoles in these field-portable systems are as small as 0.5 inches in length.

11.3.5 Environmental Applications

Molecular MS, especially GC-MS, is widely used in the identification and quantitative measurement of compounds that are of concern in the environment. PCBs, dioxins, dibenzofurans, and pesticides can be measured in the tissues of shellfish, fish, birds, and other animals, as well as in water, soil, sediments, and materials using MS, generally coupled to GC.

Brominated flame retardants are under increasing scrutiny for their environmental impact. Common brominated flame retardants are the polybrominated diphenyl ethers (PBDEs), of which there are 209 individual compounds (congeners) just as there are for PCBs. High-resolution GC-MS is the method of choice for determining PBDEs in the environment. PCBs or PBDE congeners with the same number of halogen substituents (from one to ten halogens are possible) are isomers with identical RMMs and very similar mass spectra. The retention time information from the GC portion of the GC-MS will be needed to distinguish among such isomers. The isomers cannot be differentiated on the basis of the mass spectra alone.

GC-MS is the basis for most US Environmental Protection Agency (EPA) official analytical methods for organic pollutants in water and wastewater. Detection limits in the ppb to ppt ranges are common. Accuracy of ±20 % relative standard deviation (RSD) is usual.

For example, EPA Methods 8260 and 8270 use GC-MS to determine the pollutant 1,4-dioxane in water. Using

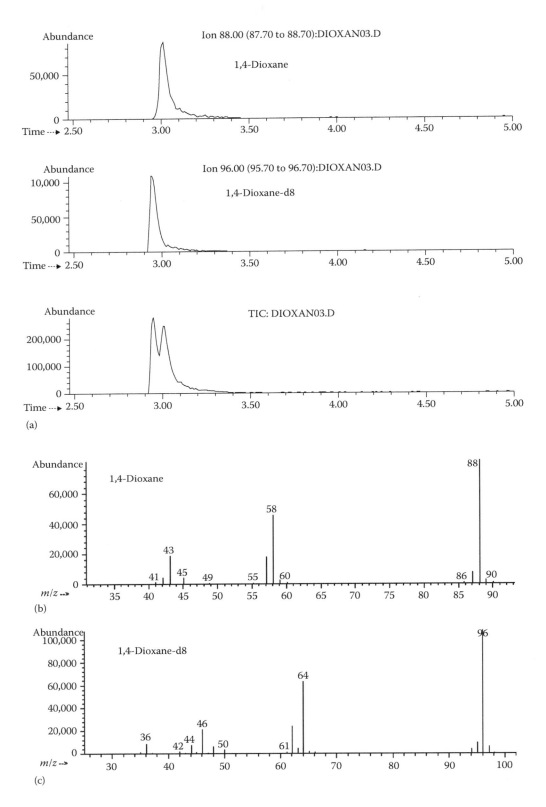

Figure 11.35 (a) Separation of labeled and unlabeled dioxane by GC; the labeled dioxane elutes first in this case. (b) Mass spectrum of unlabeled 1,4-dioxane ($m/z = 88$). (c) Mass spectrum of deuterium-labeled 1,4-dioxane-d_8 ($m/z = 96$). The ions of the known internal standard are used to calibrate the mass scale.

deuterium-labeled 1,4-dioxane-d_8 and isotope dilution GC-MS, it can be demonstrated that the percent recovery of ppb-level spikes improves from 60 % to >90 % when corrected using the isotope dilution technique. The two compounds differ slightly in retention time and can be separated on the GC column and measured quantitatively. Figure 11.35 shows the chromatogram and the mass spectra of the labeled and unlabeled 1,4-dioxane.

11.3.6 Other Applications of Molecular MS

Biomolecules of all types can be analyzed by MS. The RMM and sequence of bases in nucleotides can be determined by MALDI-TOF-MS or by MS-MS. This allows the comparison of normal and mutated genes, for example. Analysis of nucleotides is important in medicine, agriculture, environmental science, and molecular biology. Oligosaccharides, long chains of sugar units, are extremely complex in structure, more so than proteins or nucleotides. These compounds can be linear or branched, and the sugar units have isomers and specific configurations at the bonds between units. All of this structural information can be obtained from mass spectral analysis of the oligosaccharides. FT ion cyclotron resonance (ICR) and MS^n instruments are valuable in determining the structure of these complex molecules. Lipids such as fatty acids and steroids can be determined by MS. Drugs, toxins, and other compounds often undergo metabolic changes in the human body; MS is used to measure metabolites of compounds as well as the compounds themselves. As discussed in Chapters 13 and 14, the whole families of "-omics" studies have depended on highly automated, large-scale use of hyphenated chromatographic MS systems to profile the hundreds or thousands of DNA nucleotide sequences, peptide sequences in proteins, lipids, low-RMM metabolites, etc., that characterize the molecular biology and disease states of life forms ranging from bacteria to human beings.

In organic and inorganic chemistry, MS can be used to identify reaction products and byproducts. Impurities at concentrations as low as parts per quadrillion can be detected; MS is widely used for this purpose. In inorganic chemistry, with special inlet techniques, the elemental composition of materials as diverse as crystals and semiconductors can be determined. Reaction kinetics and ion–molecule reactions can be studied using MS.

Polymers are routinely characterized by MS, with MALDI now being the most common form of sampling and ionization of polymeric materials, using MALDI-TOF-MS or MALDI-FTMS. Synthetic polymers actually have a distribution of chain lengths and so have a distribution of RMMs. Polymer RMMs are reported as the average RMM (the center of the RMM distribution). The average mass of polymers can be determined more accurately by MS than by the more commonly employed method of gel permeation chromatography (Section 14.4).

11.3.7 Limitations of Molecular MS

The major limitation to the use of MS is that compounds must be volatile or must be able to be put into the gas phase without decomposing. A variety of ionization sources have been developed to handle materials that are volatile, semivolatile, nonvolatile, or thermally labile, as described in Chapters 10 and 14. For compounds that are not volatile, chemical derivatization to a more volatile form can be performed to make them suitable for MS or GC-MS, as described in Chapter 13. Carboxylic acids can be converted to the corresponding volatile methyl esters, for example; trimethylsilane is another common derivatizing agent used to make volatile ether derivatives.

Certain isomers cannot be distinguished by MS alone. Some isomers, such as the PCBs and PBDE isomers discussed in Section 11.3.5, may be able to be separated by chromatography prior to determination by MS.

11.4 ATOMIC MS

The original use of MS was for the detection and determination of elements. The elements have different masses, so MS provided a method of determining atomic masses. The various elements as well as their isotopes could be separated from each other with this technique. This made it possible to obtain the isotope distribution of pure elements (Appendix 11.A). The application of MS to the determination of atomic masses and isotope distribution was crucial in the development of atomic chemistry and physics. While there are several types of ionization sources for atomic MS, such as the glow discharge (GD) and spark source (described in Chapters 8 and 10) used for atomic mass spectrometric analysis of solids, it is the development of the inductively coupled plasma (ICP) and the quadrupole mass analyzer that has caused a recent enormous increase in the use of atomic MS as an analytical tool.

11.4.1 ICP-MS

ICP-MS permits determination of most of the elements in the periodic table at very low concentration levels. More than 60 different elements from lithium to uranium can be determined in a few seconds using an ICP-MS with a quadrupole mass analyzer. This technique has great advantages over other techniques for elemental analysis in that ICP-MS can be used to determine most elements at high sensitivity and over a wide range of concentrations. In addition, isotope ratios can be obtained, providing geochemical and

Table 11.23 **Complementary Aspects of MS and ICP-OES: Why ICP-MS Evolved**

ICP-OES	MS
Efficient but mild ionization source (produces mainly singly charged ions)	Ion source required
Sample introduction for solutions of inorganic salts is rapid and convenient	Sample introduction can be difficult for inorganic samples (generally not volatile). Thermal ionization, spark source, GD, and lasers restricted to solid samples and are time consuming
Sample introduction is at atmospheric pressure. Efficiency is poor (5 % of sample into plasma; rest to waste using conventional nebulizer)	Requires reduced-pressure sample introduction
Matrix or solvent interelement effects are observed, but relatively large amounts of dissolved solids can be tolerated	Limited to small quantities of sample and low amounts (<1 %) of dissolved solids
Complicated spectra with frequent spectral overlaps	Relatively simple spectra
Detectability is limited by relatively high-background continuum over much of the useful wavelength range	Very low background level throughout a large section of the mass range
Moderate sensitivity (ppm to ppb range)	Excellent sensitivity (ppb to ppt range)
Isotope ratios cannot be determined	Isotope ratio determinations are possible

geochronological information, for example, that cannot be obtained by other elemental analysis techniques.

There are several properties of the mass spectrometer as an analyzer and the argon ICP as an ionization source that make ICP-MS an attractive combination. The ICP has a high ionization efficiency, which approaches 100 % for most of the elements in the periodic table, and it produces mainly singly charged positive ions. The mass spectra therefore are very simple, and elements are easily identified from the masses and the isotope ratios easily measured. ICP-MS in this respect overcomes the major problem of ICP-optical emission spectroscopy (OES) (i.e., the line-rich complicated optical emission spectrum with high and variable background). In addition, the mass spectrometer is very sensitive, with detection limits up to three orders of magnitude better than ICP-OES (see Appendix 9.A), and is linear over a wide dynamic range of up to five orders of magnitude. Table 11.23 presents some of the aspects of ICP-OES and MS that complement each other. The ability to rapidly analyze solutions of dissolved inorganic materials (metals, rocks, bones, ceramics, ash, and the like), combined with the simple mass spectra and low background and excellent sensitivity, led to the rapid spread in the use of ICP-MS in many fields.

The radio-frequency (RF)-ICP source was described in Section 8.3.1, and the quadrupole mass analyzer was described in Chapter 10. The interfacing between the ICP source and the quadrupole mass analyzer will be looked at with a little more detail to explain how ICP-MS provides the information it does.

Simplistically, the instrument is an ICP interfaced with a quadrupole MS as shown in Figure 11.36a. This ICP source is horizontal, with the argon plasma concentric to the mass spectrometer inlet. Figure 11.36b shows the layout of a new ICP-MS, the PerkinElmer NexION 300, with the plasma vertical (right side of diagram) to the mass spectrometer. This instrument has a quadrupole ion deflector (QID) above the torch. The QID turns positively charged ions from the plasma 90° to the left, into the mass analyzer, while nonionized material flows straight up and out of the system. While the instrument has a QID and a quadrupole reaction cell, it has only one analyzing quadrupole, not a QqQ system as described earlier. ICPs operate at atmospheric pressure and at a temperature of about 10,000 K. On the other hand, the mass spectrometer requires a high vacuum (10^{-4}–10^{-6} torr) and operates at room temperature. Interfacing of the two systems is therefore the critical problem to be overcome. Most ICP-MS systems have an interface similar to the one shown in Figure 11.37, but the NexION 300 shown in Figure 11.36b has three cones. In Figure 11.37, the argon ICP plasma is on the right side of the diagram. Ions from the plasma enter into the mass spectrometer through a two-stage interface. The plasma is centered on the *sampler cone*, and ions and plasma gas pass through the orifice in the cone into a vacuum-pumped region. Most of the argon gas is pumped away in this region. The remaining ions pass through the *skimmer cone* into the mass spectrometer. The skimmer cone is a sharper-angled cone with an orifice of about 0.9 mm in diameter. The design of this cone restricts the flow of ions into the mass spectrometer to the central part of the flow initially coming from the plasma. The region behind the skimmer cone is evacuated to a pressure of about 10^{-4} torr by a turbomolecular pump. This region can be isolated from the higher pressure of the interface region by a gate valve. This permits the sampler and skimmer cones to be removed for cleaning without breaking the vacuum in the mass spectrometer. Both sampler and skimmer cones are made of either nickel or platinum and are water cooled by contact with water flowing within the interface chamber. The hyper skimmer cone in the PerkinElmer instrument is designed to focus the ions even more tightly, with the aim of increasing stability and eliminating drift.

A series of ion-focusing elements (ion lenses) similar to those developed for double-focusing mass spectrometers have been utilized to introduce the ions into the quadrupole. Interference effects due to the presence of large numbers of photons reaching the detector are eliminated by photon stops in older designs or off-axis ion lenses in newer instruments.

MASS SPECTROMETRY II: SPECTRAL INTERPRETATION AND APPLICATIONS

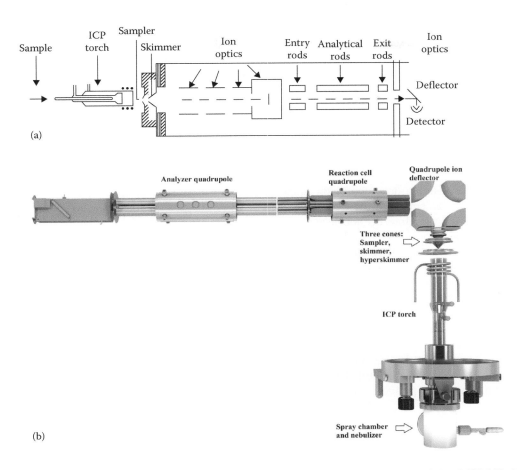

Figure 11.36 (a) Schematic diagram of an ICP-MS system with a quadrupole mass analyzer. (b) NexION 300 ICP-MS showing its three quadrupoles, three cones, and the ICP torch and sample introduction system. [© 1993–2014 PerkinElmer, Inc. All rights reserved. Printed with permission. (www.perkinelmer.com).]

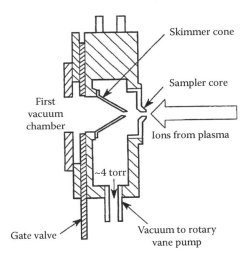

Figure 11.37 Close-up of an ICP-MS interface, showing the sampler and skimmer cones. [© 1993–2014 PerkinElmer, Inc. All rights reserved. Printed with permission. (www.perkinelmer.com).]

Background signals have been largely eliminated in modern ICP-MS instruments.

Other mass analyzers are used with the ICP ionization source, including high-resolution magnetic sectors and TOFs. ICP-TOF mass analyzer systems as well as other designs have the advantage of simultaneous measurement of ions. Simultaneous measurement is critical for accurate determination of isotope abundances as well as for accurate quantitative work using isotope dilution. The article by Blades contains an excellent review of atomic MS instruments and an extensive bibliography. However, the instrumentation is constantly improving. In 2003, Varian, Inc., introduced a new quadrupole-based ICP-MS design with an ion mirror that reflects the analyte ions through 90°, while the neutrals and photons pass through. This results in exceptionally high S/N ratio and extremely low detection limits compared with older quadrupole instrument designs. The instrument uses an all-digital extended range detector that gives nine orders of linear dynamic range, meaning that concentrations from ppt to hundreds of ppm can be measured reliably. (Varian was acquired by Agilent Technologies. The Varian ICP-MS line is now owned by Analytik Jena.) In 2012, Agilent Technologies introduced the first triple quadrupole ICP-MS, the 8800 ICP-QQQ. This instrument has tandem quadrupoles separated by an octopole reaction system, making it a QqQ design as discussed earlier. The advantages to this design will be described later.

11.4.2 Applications of Atomic MS

The high-temperature plasma decomposes the sample into its elements. A high percentage of these elements are ionized in the plasma and therefore do not need to pass through an additional ionization source. ICP-MS is particularly useful for rapid multielement determination of metals and nonmetals at concentrations of ppq, ppt, and ppb.

Only unit mass (low) resolution is required to discriminate between different elements, because isotopes of different elements differ by 1 unit mass as can be seen in Appendix 11.A. There are only a few isotopic overlaps between elements, so one can usually find an isotope to measure for any given element. In fact, there is only one element that cannot be definitely identified by ICP-MS, the element indium. One of the problems at the end of this chapter asks you to explain why. The abundance of each isotope is a quantitative measure of that element's concentration in the original sample. The isotope patterns for the elements are shown in Figure 11.38.

The applications discussed are from many forms of atomic MS, including ICP-MS, GDMS, and coupled chromatography-ICP-MS. The websites of the major instrument manufacturers (Agilent, PerkinElmer, SPECTRO, and Thermo Fisher Scientific, to name a few, for ICP-MS) are an excellent source for applications.

Some of the many uses for ICP-MS include analysis of environmental samples for ppb levels of trace metals and nonmetals; the analysis of body fluids for elemental toxins such as lead and arsenic; determination of trace elements in geological samples, metals, and alloys; determination of isotope ratios; analysis of ceramics and semiconductors; analysis of pharmaceutical and cosmetic samples; determination of platinum group catalysts in polymers; forensic analysis; elemental analysis in the petroleum and chemical industries; and metal determinations in clinical chemistry and food chemistry.

The most common samples analyzed by ICP-MS are aqueous solutions. The sample is dissolved in acid, digested or fused in molten salt (all described in Chapter 1), and then diluted to volume with water. All acids, bases, reagents,

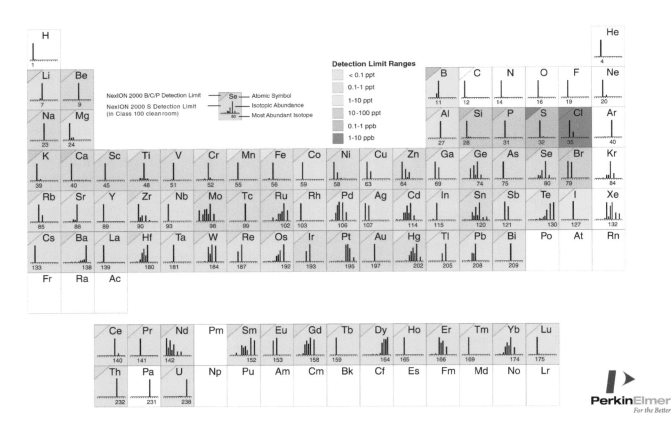

Figure 11.38 Element isotope patterns. [© 1993–2014 PerkinElmer, Inc. All rights reserved. Printed with permission. (www.perkinelmer.com).]

Table 11.24 Approximate Detection Limits for ETA-ICP-MS

Elements	ETA–ICP-MS		GFAAS	
	ng/mL (2 µL)	pg	ng/mL (10 µL)	pg
^{107}Ag	0.08	0.16	0.01	0.1
^{27}Al	0.03	0.05	0.2	2
^{75}As	0.05	0.1	1	10
^{44}Ca	0.7	1.4	0.5	5
^{114}Cd	0.15	0.3	0.01	0.1
^{52}Cr	0.1	0.2	1	10
^{39}K	1.5	3	0.1	1
^{23}Na	0.2	0.4	1	10
^{58}Ni	0.47	0.93	1	10
^{208}Pb	0.1	0.3	0.2	2
^{78}Se	5.7	11.4	1	10
^{28}Si	2.7	5.4	5	50
^{64}Zn	0.2	0.4	0.05	0.5

Note: This table presents a comparison of GFAAS with the use of a graphite furnace (ETA) as the sample vaporization step in conjunction with ICP-MS. The solution detection limit, the sample volume (µL), and the absolute detection limit in picograms are given for each technique. The isotope measured is specified for the ICP-MS technique; the isotope number has no meaning for AAS.

and water must be of extremely high purity, given the sensitivity of the ICP-MS technique. "Ultratrace metal" grade acids, solvents, and deionized water systems are all commercially available. The aqueous solution is introduced into the plasma using a peristaltic pump, nebulizer, and spray chamber system identical to those used for ICP-OES (Chapter 8).

Solid samples can be analyzed by laser ablation ICP-MS or by coupling an electrothermal vaporization (ETV) or electrothermal atomizer (ETA) (identical to a graphite furnace used in atomic absorption spectrometry [AAS] [Chapter 7]) to the ICP-MS. Both sample introduction systems are described in Chapter 8 and operate the same way they do for ICP-OES. The direct analysis of solid samples is certainly desirable as it eliminates sample preparation and the errors that go along with sample preparation. Table 11.24 compares the detection limits of ETV-ICP-MS with those of graphite furnace AAS (GFAAS). Solids can also be analyzed by GDMS and spark source MS. Laser ablation ICP-MS is used to measure elements in fluid inclusions in rocks, microscopic features in heterogeneous materials, and individual crystals in samples such as granite. It can be used for analysis of artworks and jewelry, since the small laser spot size (<10 µm) available with modern lasers results in minimal damage to the object.

Quantitative analysis by ICP-MS is usually done with external calibration standards and the addition of internal standards to all standards and samples. When a large number of elements across the periodic table are to be determined, it is usual to add Li, Y, In, Tb, and Bi and measure the ions ^{6}Li, ^{89}Y, ^{115}In, ^{159}Tb, and ^{209}Bi as internal standards (unless you need to measure one of these elements as an analyte). Not all of the internal standard elements are used to quantitate every analyte. The internal standard that is most closely matched in first ionization potential to the analyte is generally used, since this will compensate for ionization interferences in matrices containing easily ionized elements such as Na. Results obtained using this approach are generally very accurate and precise. Table 11.25 presents typical spike recovery and precision data for ICP-MS determination of 25 elements in a certified reference material (CRM) "Trace Elements In Drinking Water" from High-Purity Standards, Charleston, SC.

11.4.2.1 Geological and Materials Characterization Applications

Atomic MS has the ability to measure the isotope ratios of the elements. It is of interest to note that this was the original application of MS both by Aston and by Thompson. Both quantitative measurement of trace elements in minerals and accurate isotope ratio determinations are extremely important in geology, climatology, and earth science, among other fields. Most elements have fixed ratios because their isotopes are stable, but some elements, such as uranium, carbon, lead, and strontium, have isotopes that vary in their abundance because of natural radioactive decay processes. The isotope ratios of elements such as these can be used for estimating the age of a sample or for identifying the source of the element. The application of atomic MS to geological and fossil samples for accurate isotope ratios permits the dating of samples (geochronology), as well as applications in paleothermometry, marine science, and climatology.

The incorporation rates of the elements uranium, magnesium, and strontium into corals appear to be a function of water temperature. Accurate measurements of the elements and their ratios in fossil corals can provide information to climatologists about historical sea temperatures (paleothermometry). An ICP-MS method for determining these ratios accurately using internal standard isotope dilution is described by Le Cornec and Correge. As these authors note, the precision of the ICP-MS instrument must be very high, because the rate of change of the Sr/Ca ratio is only 0.9 % per °C and yearly variations in the sea surface temperature are only 5 °C–6 °C.

A long-standing method for thermometry has been to measure the stable oxygen isotope ratios in carbonate rocks using MS. Uranium/lead isotope ratios, determined by MS, are used to estimate the age of the Earth.

Accurate isotope ratios often require high-precision mass spectrometers. Instruments capable of such high precision are either ICP-TOF-MS or high-resolution magnetic sector ICP-MS instruments of either the single detector or

Table 11.25 Precision and Recovery Data for "Trace Metals in Drinking Water" CRM by ICP-MS

Analyte	Isotope	Average Measured Conc (µg/L)	Std. Dev.	Rel.% Dif.	High Purity Certified Value	Recovery of Certified Value (%)	Spike Level (µg/L)	Average Spike Recovery (%)	Std. Dev. of Spike Rec.	Rel.% Dif.
Be	9	17.17	0.13	0.75	20	85.8	50	92.1	1.0	1.1
Na	23	5,711.45	113.71	1.99	6,000	95.2	—	—	—	—
Mg	24	8,544.22	11.45	0.13	9,000	94.9	—	—	—	—
Al	27	113.60	0.37	0.33	115	98.8	50	106.7	3.8	3.5
K	39	2,421.85	22.01	0.91	2,500	96.9	—	—	—	—
Ca	44	33,871.41	611.84	1.81	35,000	96.8	—	—	—	—
V	51	28.92	0.08	0.28	30	96.4	50	101.0	1.3	1.2
Cr	52	18.99	0.04	0.22	20	94.9	50	103.2	5.6	5.4
Mn	55	33.58	0.73	2.18	35	95.9	50	99.0	2.6	2.6
Co	59	23.71	0.81	3.41	25	94.8	50	100.9	2.2	2.2
Ni	60	60.46	0.26	0.43	60	100.8	50	102.6	3.9	3.8
Cu	63	19.32	0.51	2.64	20	96.6	50	101.1	4.6	4.5
Zn	66	68.98	1.29	1.87	70	98.5	50	101.2	3.2	3.2
As	75	78.67	0.54	0.68	80	98.3	50	108.4	0.0	0.0
Se	82	9.21	0.15	1.61	10	92.1	50	91.7	0.6	0.7
Mo	98	94.93	3.23	3.40	100	94.9	50	105.4	1.1	1.0
Ag	107	2.34	0.04	1.78	2.5	93.6	50	99.4	1.0	1.0
Cd	111	11.35	0.01	0.12	12	94.6	50	96.3	1.1	1.2
Sb	121	9.55	0.03	0.31	10	95.5	50	99.8	1.0	1.0
Ba	135	48.49	0.73	1.51	50	97.0	50	117.5	0.3	0.2
Hg	202	0.11	0.03	27.27	Not available	—	1.5	100.7	0.04	0.04
Tl	205	10.21	0.14	1.34	10	102.1	50	106.3	2.1	2.0
Pb	208	34.85	0.73	2.10	35	99.6	50	109.7	4.8	4.4
Th	232	0.01	0.01	79.75	Not available	—	50	107.6	0.8	0.8
U	238	10.16	0.20	1.97	10	101.6	50	106.7	2.7	2.5

Source: Wolf, R. et al., *EPA Method 2008 for the Analysis of Drinking Waters, Application Note ENVA-300*; [© 1995–2014 PerkinElmer, Inc. All rights reserved. Printed with permission. (www.perkinelmer.com).]

Note: The CRM is from High-Purity Standards, Charleston, SC.

multicollector type. An excellent fundamental discussion of high-precision isotope ratio measurements using the LECO Renaissance™ TOF-based instrument is the cited technical brief by Allen and Georgitis. Despite the high cost of the magnetic sector instruments (approximately two to six times the cost of a quadrupole ICP-MS), it is estimated that in 2003, there were more than 350 magnetic sector single-detector ICP-MS instruments and more than 100 multicollector magnetic sector ICP-MS units in operation around the world, testifying to the need for high-precision, high mass resolution measurements. The high mass resolution permits analysis of complex minerals and other matrices. The high sensitivity of these instruments makes them especially suitable for laser ablation analysis of geological specimens. (The instrument numbers and cost estimates are courtesy of Chuck Douthitt of Thermo Fisher Scientific, a manufacturer of magnetic sector-based high-resolution instruments.)

Trace levels of all of the rare earth elements (REEs) can be determined in rock samples and their distribution provides critical information to geologists. Some of the REE concentrations in basalt samples are in the sub-ppm range but can be measured accurately in basalt samples that have been fused with lithium metaborate and diluted 5000-fold. The high dilution is necessary to keep the total dissolved solids <2000 mg/L, as the ICP-MS instrument does not tolerate high salt solutions well. The application is described by Ridd. The use of ICP-MS for the REEs gives spectra that are simpler than those obtained by ICP-OES and all of the REEs can be measured, which is not the case with techniques such as neutron activation analysis.

An interesting overlap of geology and synthetic materials characterization is the result of improvements in recent years in the ability to synthesize gem-quality diamonds in large quantities and in rare and therefore valuable colors. In addition, it is now possible to chemically diffuse trace amounts of metals like Ti and Be into low-quality sapphires and rubies, dramatically improving their color and value. Rubies and sapphires are both Al_2O_3 minerals, and as few as five Be atoms in the Al_2O_3 crystal lattice can result in dramatic color change in the gem. In most cases, these synthetic diamonds and chemically treated gemstones cannot be detected by traditional gemological testing. Laser ablation ICP-MS is now being used to quantify Be and other trace elements in gems to distinguish natural gemstones from treated or synthetic ones. (Figure 8.42 in Chapter 8 shows laser ablation spots in a garnet crystal, for example.) The cited article by Yarnell and the website of the Gemological Institute of America (www.gia.edu) provide details of this application and other spectroscopic techniques used to examine gems.

GDMS is used for the characterization of solid materials, especially coated and layered materials. The GD ablates the sample from the surface inward, providing the ability to depth-profile materials. Bulk analysis of solid metals with detection limits in the low ppb range is possible. Coatings and layered materials of a few nanometers thick can be characterized by GDMS. Examples are given by Broekaert. The major disadvantage to GDMS is its relatively poor precision and the severe matrix effects encountered in analysis. Matrix effects are described in Section 11.4.3.1. GDMS measurements generally have 20 %–30 % RSD compared with <5 % RSD for solutions analyzed by ICP-MS. The poor precision and severe matrix effects make GDMS suitable only for qualitative and semiquantitative work unless matrix-matched standards are available. In the experience of one of the textbook authors, GDMS determination of an easily ionized trace element in a heavy metal matrix gave results that were 100× higher by GDMS than results obtained by ICP-OES and AAS after dissolution of the sample. The GDMS results were in error because of a matrix effect. Because the analyte element was in the zero oxidation state in the metal sample but was present as an oxide in the glass used as the GDMS "standard," the analyte was ionized from the metal sample very differently than it was ionized from the glass used as the calibration standard. This is an example of really poor "matrix matching" of sample to calibration standard.

11.4.2.2 Speciation by Coupled Chromatography-ICP-MS

The coupling of chromatography (GC, LC, and ion chromatography [IC]) or capillary electrophoresis (CE) to ICP-MS allows the separation of complex mixtures and speciation by compound or oxidation state of the elements present. GC is described in Chapter 13; LC including HPLC and IC and the nonchromatographic separation method of CE are described in Chapter 14. All of these separation instruments have been interfaced to ICP-MS systems. Coupling a separation technique to a quadrupole ICP-MS may require the use of additional software that can handle transient signals; such software is commercially available, as are complete hyphenated speciation ICP-MS systems.

Chong and Houk used an argon ICP mass spectrometer as a GC detector for chemical compounds containing nitrogen, oxygen, phosphorus, sulfur, carbon, chlorine, bromine, boron, and iodine. GC-ICP-MS applications include the measurement of siloxanes, brominated flame retardants, and sulfur in fuels. Interfacing a GC to an ICP-MS is a relatively simple matter of connecting the GC outlet to the base of the ICP torch with a heated transfer line. In general, detection limits for GC-ICP-MS are 10–100 times better than GC with an atomic emission detector (AED) for those elements that can be measured. Gases such as arsine, germane, phosphine, carbonyl sulfide, hydrogen sulfide, and silane can be determined at ppb or lower levels in bulk process gases. This is becoming an important area of analysis as demands for high-purity gases in the microelectronics and polymer industries now specify impurity levels for metal hydrides and similar compounds be 10 ppb or lower (Geiger and Raynor).

Commercial coupling of IC and MS was introduced by Dionex and Thermo Fisher Scientific, who now owns Dionex. An application of ion chromatographic separation of ionic species coupled to ICP-MS is the determination of halogen oxyanions such as IO_4^-, IO_3^-, BrO_3^-, and ClO_3^-. IC generally uses a conductivity detector (described in Chapter 14), which is not a specific detector. Identification of ionic species must be made based on retention time and comparison to known standards. There is always some risk that two ionic species will have the same retention time, which could lead to misidentification of the species. Coupling the IC to an ICP-MS allows the anions to be separated, their retention times measured, and the specific halogens confirmed by ICP-MS. The mass spectrometer monitors the ^{127}I isotope, confirming that the ions giving rise to these peaks in the ion chromatogram are iodine-containing ions. The ions can be measured quantitatively using either the IC conductivity detector or the much more sensitive detector on the mass spectrometer, depending on the concentrations. For perchlorate ion, an environmental concern in drinking water at ppb concentrations, two peaks, at 99 and 101 amu, can be used to confirm perchlorate due to the relative isotopic abundances of ^{35}Cl and ^{37}Cl. The use of MS as the detector permits quantitative determination of perchlorate in drinking water at ppt (ng/L) concentrations. Many other IC-MS application examples can be found at the Thermo Fisher Scientific website (www.thermo.com).

Organic solvents and organic petroleum fractions often need to be analyzed by ICP-MS for trace element content. LC is the most common separation technique coupled to ICP-MS, and commercial systems for conventional HPLC as well as capillary and nano-LC are available from several companies. HPLC separations are often carried out in organic solvents ranging from nonpolar solvents like xylene to polar alcohol–water mixtures. Organic solvents can be analyzed by ICP-MS with the same limitations that organic solvents present in ICP-OES. The RF power, the temperature of the ICP spray chamber, the nebulizer gas flow rate, and sample uptake rate

Table 11.26 Potential Interferents in a Carbon-Rich Plasma with Oxygen Addition

Potential Interfering Species	m/z	Affected Element
$^{12}C_2$	24	Mg
$^{12}C^{13}C$	25	Mg
$^{40}Ar^{16}O$	56	Fe
$^{12}C^{16}O$	28	Si
$^{12}C^{16}O_2$	44	Ca
$^{40}Ar^{12}C$	52	Cr
$^{40}Ar^{12}C^{16}O$	68	Zn

Source: Modified from Elliott, S., *Analysis of Organic Solvents, Including Naphtha by ICP-MS*, Varian Applications Note ICP-MS-19, October 1998. All species would be in the form of ions as measured, used with permission. (Varian is now Agilent Technologies.)

may need to be adjusted depending on the volatility of the solvent to avoid extinguishing the plasma or causing carbon deposition on the sampler cone. Running organic solvents can cause specific polyatomic interferences, discussed in Section 11.4.3. Examples of polyatomic interferences are given in Tables 11.26, 11.28, and 11.29.

Speciation of compounds of environmental concern is another important application of coupled GC-ICP-MS and LC-ICP-MS. Arsenic compounds in shellfish are one example of the importance of being able to speciate arsenic-containing compounds, that is, to determine the exact chemical forms of arsenic present. Most of the arsenic in shellfish (80 %–99 %) is in the form of organoarsenic compounds, including arsenobetaine and monomethylated and dimethylated arsenic acids. Arsenobetaine is not toxic because humans do not metabolize it, while the other organoarsenic compounds are much less toxic than inorganic arsenic compounds. To assess the risk of eating shellfish based on a determination of the total amount of arsenic present would greatly overestimate the danger of eating shellfish. Inorganic arsenic is present as well, in two oxidation states, As(III) and As(V). As(III) is more toxic than As(V). The ability to separate all of these arsenic species and detect them at extremely low levels by LC-ICP-MS permits a much better assessment of As exposure and hazards from consumption of water and food than a determination of total As by elemental techniques such as ICP-OES, AAS, atomic fluorescence spectroscopy (AFS), or ICP-MS. The cited articles by Milstein et al. describe arsenic speciation using IC-ICP-MS. A website dedicated to speciation analysis can be accessed at www.speciation.net.

11.4.2.3 Applications in Food Chemistry, Environmental Chemistry, Biochemistry, Clinical Chemistry, and Medicine

Foods and beverages of many types have been analyzed by ICP-MS. Solid or semisolid samples are generally digested with mineral acid, as are some beverages. Peanut butter, commercial breakfast cereal, dried milk, fish and shellfish, wine, beer, and the like have been analyzed for trace elements such as Cu, Fe, Se, and Zn for nutritional purposes as well as for toxic metals like As and Pb. Al has been determined in many foods because dietary Al was being studied for a possible link to Alzheimer's disease; no such "link" has been demonstrated.

Whole blood and serum have been analyzed by ICP-MS for Al, Cu, Se, Zn, As, Cd, Mn, and Pb, among other elements. Blood lead (Pb) measurements are done to determine exposure to lead-based paint in infants and small children. Pb, like many of the heavy metals, is a central nervous system toxin, and ingestion of lead-based paint chips or exposure to lead-containing dust has been implicated in learning disabilities in small children. ICP-MS is sensitive enough to permit multielement quantitative analysis of blood using sample volumes of 100 μL or less. In addition, the isotopic ratio pattern of the four Pb isotopes can be used like a fingerprint to match the source of lead (home, environment, and industrial exposure) to the lead in a patient's blood (Allen and Georgitis).

The unenhanced magnetic resonance imaging (MRI) (described in Chapter 6) of tumors cannot always reliably distinguish tumors from normal tissue. However, gadolinium compounds targeted at tumors can enhance the MRI signal from tumors, as seen in Figure 11.39. Gadolinium affects the T1 relaxation time in MRI, making the tissue that has absorbed the Gd compound "stand out" from adjacent tissue. The subject in Figure 11.39 is a rat with a mammary tumor. The tumor is at the top of both images just to the left of center. Before injection with the Gd-containing contrast agent, the tumor cannot be distinguished from the surrounding tissue very well. Following injection with the contrast agent, the tumor is clearly enhanced (seen as a much brighter area) because of preferential uptake of the contrast agent by the tumor. The other bright area in the lower right of the postinjection image is the rat's liver/kidneys. The use of ICP-MS to accurately determine the amount of Gd in rat tissues, blood, plasma, and serum is discussed in the article by Skelly Frame and Uzgiris. Accurate concentrations were needed so that the dose of contrast agent could be optimized for use in humans to obtain the best image with the lowest possible dose. Blood, serum, and plasma samples of 30–100 μL and tissue samples of 0.1–1.0 g wet mass were microwave digested in high-purity nitric acid and analyzed after dilution. The Gd isotopes 156 and 158 were measured; the detection limit was ~0.005 ppb Gd.

Environmental applications of ICP-MS are numerous and include analysis of water, wastewater, soil, sediment, air particulates, and so on. A typical environmental analysis is to determine the leachable metals from soil or sediment; the solid is not dissolved but leached or extracted to determine labile elements. These labile or leachable elements are the ones that might be mobilized from a landfill into a drinking water supply, for example. Table 11.27 gives an

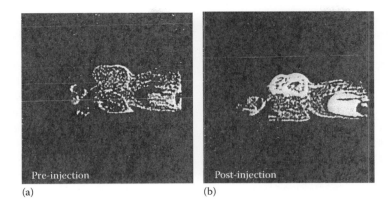

Figure 11.39 MRI of a tumor in a rat before injection with a Gd-containing contrast agent (a) and after injection (b). The tumor is at the top of each image slightly to the left of center. The contrast agent is preferentially taken up by the tumor and appears enhanced (brighter) compared to the surrounding muscle tissue. (Images courtesy of Dr. E. E. Uzgiris.)

Table 11.27 ICP-MS Results from Leaching of NIST SRM 2711 Moderately Contaminated Montana Soil

Analyte	Mass	Measured Conc (mg/kg)	Std Dev	RSD	RPD	NIST Leach Value (mg/kg)	Range Low	Range High	Spike Amount (ppb)	Spike Recovery (%)
Be	9	1.1	0.03	2.54	3.59				100.0	104
Al	27	20,066.5	621.82	3.10	4.38	18,000.0	12,000.0	23,000.0	100.0	—
V	51	48.2	1.63	3.39	4.79	42.0	34.0	50.0	100.0	98
Cr	52	23.7	0.23	0.95	1.35	20.0	15.0	25.0	100.0	97
Mn	55	493.0	14.82	3.01	4.25	490.0	400.0	620.0	100.0	110
Co	59	8.1	0.12	1.44	2.03	8.2	7.0	12.0	100.0	98
Ni	60	17.1	0.01	0.08	0.11	16.0	14.0	20.0	100.0	96
Cu	63	104.1	3.62	3.48	4.92	100.0	91.0	110.0	100.0	99
Zn	66	315.8	8.79	2.78	3.94	310.0	290.0	340.0	100.0	111
As	75	94.0	1.93	2.06	2.91	90.0	88.0	110.0	100.0	103
Se	82	2.2	0.11	4.92	6.96	NR			100.0	109
Mo	98	1.2	0.02	1.48	2.09	<2			100.0	105
Ag	107	4.3	0.03	0.67	0.94	4.0	2.5	5.5	100.0	102
Cd	111	40.0	0.91	2.27	3.20	40.0	32.0	46.0	100.0	102
Sb	123	3.9	0.14	3.71	5.25	<10			100.0	98
Ba	135	192.2	6.64	3.45	4.88	200.0	170.0	260.0	100.0	98
Tl	205	1.8	0.06	3.58	5.07				100.0	105
Pb	208	1,087.3	35.25	3.24	4.58	1,100.0	930.0	1,500.0	100.0	132
Na	23	320.3	5.82	1.82	2.57	260.0	200.0	290.0		—
Mg	24	7,726.8	201.46	2.61	3.69	8,100.0	7,200.0	8,900.0		—
K	39	5,064.4	128.65	2.54	3.59	3,800.0	2,600.0	5,300.0		—
Ca	44	20,742.0	166.68	0.80	1.14	21,000.0	20,000.0	25,000.0		—
Fe	54	21,662.2	526.25	2.43	3.44	22,000.0	17,000.0	26,000.0		—

Source: Wolf, R. et al., *RCRA SW-846 Method 6020 for the ICP-MS Analysis of Soils and Sediments, Application Note ENVA-301*; [© 1996–2014 PerkinElmer, Inc. All rights reserved. Printed with permission. (www.perkinelmer.com).]

example of determining leachable metals from an NIST standard reference material (SRM) soil sample by ICP-MS.

As is the case with isotopic labeling of molecules, enriched levels of stable isotopes of elements can be used as tracers. Isotopes of elements can be used as nutritional supplements for plants or animals to trace absorption, assimilation, and metabolism of elements (Allen and Georgitis). Processes such as biomethylation of elements like mercury and arsenic in the environment can be studied using isotopically enriched elements. In some cases, methylated metals are more toxic than the inorganic species and generally accumulate up the food chain.

Molecular MS is a widely used technique in the pharmaceutical industry for the identification and analysis of organic compounds. ICP-MS has recently become a very important tool for screening and quantitation of elements in pharmaceuticals due to recent (implemented on January 1, 2018, USP Chapters 232 and 233) changes in the United States Pharmacopeia (USP) and new proposed guidelines from the US Food and Drug Administration. The USP is the official standard-setting agency for pharmaceutical products manufactured in or sold in the United States. Similar organizations exist around the world (e.g., the European Medicines Agency, EMEA), and some have already implemented limits that require the sensitivity of ICP-MS as well as the speciation ability of coupled chromatography-ICP-MS in some cases. Inorganic impurities in pharmaceuticals can arise from impurities in raw materials or reagents, contaminants introduced during manufacturing, residues from catalysts, naturally occurring elements from plants or minerals, and leachable elements from packaging materials, among other sources. Elements, chemical form, and limits vary depending on the nature of the pharmaceutical (drug vs. dietary supplement) and the method of dosing—oral, inhalation, and parenteral (by intravenous drip or injection). Limits are set by permissible daily exposure (PDE) based on a 50-kg person, by concentration, and, for parenteral administration, by the large-volume parenteral (LVP) component limit, defined as the absolute level that may occur in any component of the solution.

For example, in dietary supplements, the USP individual component limits in μg/g are inorganic As, 1.5; Cd, 0.5; Pb, 1.0; Hg (total), 1.5; and methylmercury (as Hg), 0.2. Speciation is clearly required for As and Hg. In pharmaceuticals, the USP regulates inorganic As, Cd, Pb, inorganic Hg, Cr, Cu, Mn, Mo, Ni, Pd, Pt, V, Os, Rh, Ru, and Ir at levels varying from 5 to 2500 μg/day PDE. The platinum group metals are usually residues from catalysts, but there are several platinum cancer drugs in use. The EMEA regulates an almost identical list of elements. ICP-MS is almost the only technique available for rapidly determining all of these elements at the required sensitivity. More details are available in the white paper '"Simultaneous ICP-MS in the Pharmaceutical Industry" available from SPECTRO Analytical Instruments GmbH at www.spectro.com.

11.4.2.4 Coupled Elemental Analysis MS

While not strictly speaking "atomic" MS, mass analyzers have now been coupled with traditional elemental analyzers for trace levels of C, H, N, S, and O in metals and alloys. These elemental analyzers are also confusingly called "gas analyzers" because they are determining C, H, N, O, and S as gases released from metal samples by heat. These traditional elemental analyzers are of two basic types: oxygen-assisted combustion for carbon and sulfur and inert gas fusion for hydrogen, oxygen, and nitrogen determinations. They utilize high-temperature furnaces and either TC or IR detectors. These instruments, made by a number of instrument companies, have been used for years to analyze metals and alloys for materials development, process control, and quality assurance.

Hydrogen has a particularly deleterious effect on the physical properties of steel. Because hydrogen atoms are so small, they can easily diffuse into the metal lattice during fabrication and processing of steel. Hydrogen atoms can accumulate in the voids of the lattice and form pockets of molecular hydrogen with high internal pressure. This process is called hydrogen embrittlement. When the steel is stressed, the metal can crack at these hydrogen pockets (Oxley et al.). Bruker Corporation has introduced a series of extremely low detection limit hydrogen analyzers for diffusible hydrogen and for diffusible and total hydrogen, as well as oxygen and nitrogen by coupling their traditional elemental analyzers with a compact quadrupole mass spectrometer, the ESD 100, from InProcess Instruments (www.in-process.com). Figure 11.40 shows the Bruker G8 Galileo O/N/H analyzer with an ESD 100 mass analyzer (lower right) and an external IR furnace for diffusible hydrogen measurements (upper right). The G8 Galileo utilizes the inert gas fusion principle. The steel sample is fused in a graphite crucible at >2500 °C. Total hydrogen is released from the sample and measured with the ESD 100. The ESD has a channeltron detector optimized for the low m/z range. Using a TC detector, a steel sample containing 0.43 ppm hydrogen is at the detection limit of the TC detector. However, the mass spectrometer can detect not only 0.43 ppm hydrogen (Figure 11.41) but a steel sample with 0.043 ppm (430 ppb) hydrogen (Figure 11.42).

11.4.3 Interferences in Atomic MS

Mass spectral interferences are observed in atomic MS. These interferences are of two types: direct m/z overlap between two ions and matrix interferences that cause enhancement or suppression of analyte signals.

11.4.3.1 Matrix Effects

The introduction of the sample into the plasma suffers from the same problems in ICP-MS as in ICP emission spectrometry. These include the complicated process of nebulization and atomization. These processes occur before introduction of the sample into the mass spectrometer system.

The interface between the two systems includes the 1 torr region where deposition of products can occur. Severe suppression of signal has been observed when high concentrations of dissolved solids are present. This

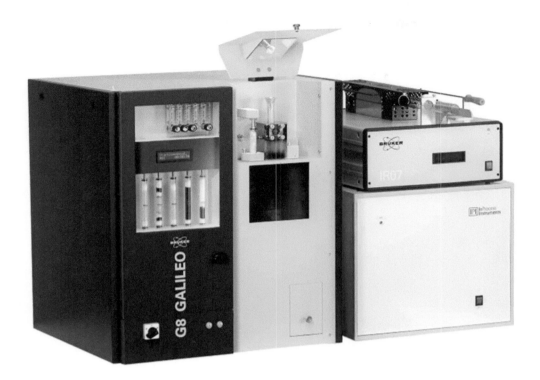

Figure 11.40 The Bruker G8 Galileo O/N/H analyzer, equipped with a quadrupole mass spectrometer detector (lower right) and an external IR furnace (upper right) for diffusible hydrogen. (© 2013 Bruker Corporation, www.bruker-axs.com. Used with permission.)

may be caused by suppression of ionization by other more easily ionized elements present in the sample. The exact cause of this interference is not clear, but the fact remains that interference does take place. The problem can be overcome to some extent by limiting the concentration in the samples to less than 0.2 % total solids. This can be a serious limitation, particularly when body fluids or fused minerals are being examined. Newer instruments such as the Agilent 8800 system employ a variety of usually proprietary sample introduction devices to enable higher matrix concentrations to be tolerated. Agilent's high matrix introduction (HMI) technology claims that sample matrices of 2 % total dissolved solids can be measured routinely with a significant reduction in matrix suppression effects.

The solvent and other elements present in the sample cause matrix effects. These affect atomization efficiency, ionization efficiency, and therefore the strength of the MS signal. This directly impacts quantitative results. Signals may be suppressed or enhanced by matrix effects. Aqueous solutions act very differently from organic solvents, which in turn act differently from each other. The problem can be overcome for the most part by matrix matching (i.e., the standards used for calibration are matched for acid concentration or solvent, major elements, viscosity, etc., to the matrix of the samples being analyzed). This is similar to AAS and atomic emission spectroscopy (AES) where the same requirement in matching solvent and predominant matrix components is required for accurate quantitative analysis. The use of internal standards will also compensate for some matrix effects and will improve the accuracy and precision of ICP-MS measurements.

11.4.3.2 Spectral (Isobaric) Interferences

In ICP-MS, interferences are seen from argon ions and argon-containing polyatomic species, since these are present in high abundance in the argon plasma. Argon is composed of three isotopes, but ^{40}Ar is 99.600 % of the total, so ^{40}Ar$^+$ is the ion of most concern. ^{40}Ar$^+$ and ^{40}Ca$^+$ have the same nominal mass, and we can say that they overlap. A single quadrupole MS cannot distinguish between the two. Such mass overlaps are called isobaric interferences; they are spectral interferences. Unfortunately for us, ^{40}Ca$^+$ is the most abundant Ca isotope, constituting 96.941 % of Ca (see Appendix 11.A). To determine Ca using a standard quadrupole ICP-MS instrument, another less abundant Ca isotope must be chosen for measurement. Polyatomic ions such as ArH$^+$, ArN$^+$, ArCl$^+$, ArOH$^+$, and Ar$_2^+$ are commonly formed in the plasma and interfere with a number of possible analytes, as shown in Tables 11.26, 11.28, and 11.29. These polyatomic spectral interferences are very problematic when the

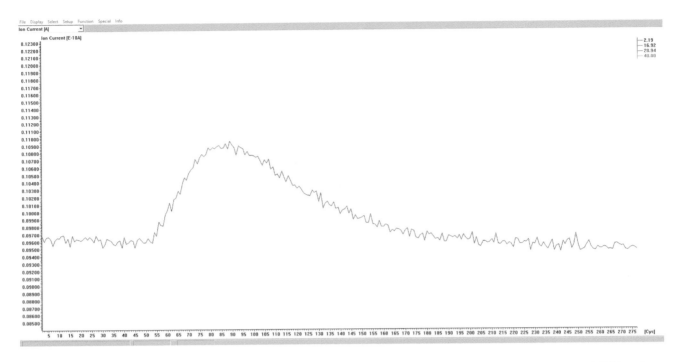

Figure 11.41 The mass spectrum of a steel sample containing 0.23 ppm hydrogen, determined with the G8 Galileo O/N/H analyzer, equipped with an ESD 100 mass spectrometer detector. (© 2023 Bruker Corporation, www.bruker-axs.com. Used with permission.)

element affected is monoisotopic, as is the case with ^{75}As and ^{55}Mn, or when the isotope affected comprises more than 90 % abundance of the element, as happens with ^{56}Fe, ^{39}K, ^{40}Ca, and ^{51}V.

Polyatomic ionic species also form from air, solvents, and matrix components, such as oxides of metals present in large amounts in the sample as well as from the acids and other reagents used to digest samples. For example, titanium has five isotopes, ^{46}Ti, ^{47}Ti, ^{48}Ti, ^{49}Ti, and ^{50}Ti, which can form Ti^{16}O oxide ions that interfere with ^{62}Ni, ^{63}Cu, ^{64}Zn, ^{65}Cu, and ^{66}Zn. Examples in addition to the argon ions mentioned earlier include CO_2^+, CO^+, SH^+, SO^{2+}, NOH^+, ClO^+, and ArS^+, which are derived from the plasma gas and from reactions of Ar with water, carbon, and other elements present in the solvent or the sample. Based on studies of these interfering ions, nitric acid is considered to be the most attractive acid to be used in preparing sample solutions. Unfortunately, not all elements are soluble in nitric acid alone. No ions occur, however, with a mass greater than 82 from any of the common acids. Therefore, the higher part of the mass range is unaffected by these interfering ions.

The analysis of organic solvents presents some unique polyatomic interferences. Tables 11.26, 11.28, and 11.29 list potential polyatomic interfering species and the affected element. Some analysts add oxygen to the plasma when running organic solvents to minimize carbon (soot) formation on the cones. The additional oxygen can not only create the polyatomic species listed in Table 11.26 but can also react with some elements to form refractory oxides, as happens in aqueous solution. Examples include ^{46}Ti^{16}O, which interferes with ^{63}Cu, and REE oxides, such as ^{143}Nd^{16}O, which interferes with ^{159}Tb.

One approach to minimize the formation of polyatomic species is to adjust plasma and operating conditions. A "cool plasma" or "cold plasma" can be formed by decreasing the power to the plasma. Using cool plasma conditions eliminates the formation of or reduces many of the Ar polyatomic ions, permitting accurate ppt-level determination of Na, K, Ca, and Fe. The drawbacks to changing the operating conditions to a cool plasma are that detection limits for some elements may be poorer than in a normal "hot plasma," matrix effects may increase, and longer analysis times may be needed. It may be necessary to run samples under two sets of operating conditions to optimize detection limits for all elements.

Appendix 11.A indicates that there are also direct spectral interferences of one element on another. ^{58}Ni overlaps with ^{58}Fe; even though ^{58}Fe is the least abundant Fe isotope at 0.28 %, trying to measure trace ^{58}Ni (the most abundant Ni isotope at 68 %) in an iron sample would be difficult. It would be better to measure ^{60}Ni, which is not interfered with by Fe. Other overlaps can

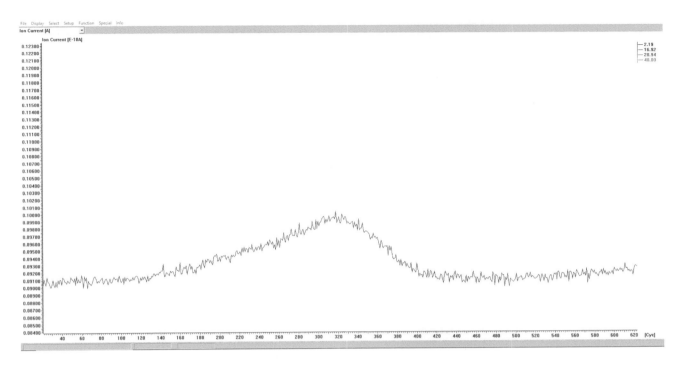

Figure 11.42 The mass spectrum of a steel sample containing 0.0074 ppm (7.4 ppb) hydrogen, determined with the G8 Galileo O/N/H analyzer, equipped with an ESD 100 mass spectrometer detector. (© 2023 Bruker Corporation (www.bru ker- axs.com). Used with permission.)

Table 11.28 Common Polyatomic Interferences in ICP-MS

Polyatomic Ion Interference	m/z	Affected Element
$^{14}N_2$	28	^{28}Si
$^{12}C^{16}O$	28	^{28}Si
$^{14}N^{16}O^{1}H$	31	^{31}P
$^{16}O^{16}O$	32	^{32}S
$^{38}Ar^{1}H$	39	^{39}K
^{40}Ar	40	^{40}Ca
$^{35}Cl^{16}O$	51	^{51}V
$^{36}Ar^{16}O$	52	^{52}Cr
$^{40}Ar^{12}C$	52	^{52}Cr
$^{40}Ar^{14}NH$	55	^{55}Mn
$^{38}Ar^{16}O^{1}H$	55	^{55}Mn
$^{40}Ar^{16}O$	56	^{56}Fe
$^{23}Na^{40}Ar$	63	^{63}Cu
$^{40}Ar^{35}Cl$	75	^{75}As
$^{40}Ar^{40}Ar$	80	^{80}Se

Table 11.29 Examples of Spectral Interferences Addressed Using a Reaction Cell

Isotope	Abundance (%)	Interfering Species	Typical Reaction Gas
$^{39}K^+$	93.3	$^{38}Ar^{1}H^+$	NH_3
$^{40}Ca^+$	96.9	$^{40}Ar^+$	NH_3
$^{51}V^+$	99.8	$^{35}Cl_{16}O^+$	NH_3
$^{52}Cr^+$	83.8	$^{36}Ar^{16}O^+$	NH_3
		$^{40}Ar^{12}C^+$	
		$^{35}Cl^{16}O^{1}H^+$	
$^{55}Mn^+$	100	$^{38}Ar^{16}O^{1}H^+$	NH_3
$^{56}Fe^+$	91.7	$^{40}Ar^{16}O+$	NH_3
$^{63}Cu^+$	69.2	$^{23}Na_{40}Ar$	NH_3
$^{75}As^+$	100	$^{40}Ar^{35}Cl^+$	H_2/Ar mix
$^{80}Se^+$	49.6	$^{40}Ar^{40}Ar^+$	CH_4

be seen in this table; Zr and Mo have three isotopes that directly interfere with each other, Lu, Hf, and Yb all have a 176 isotope, while Ta, W, and Hf all have a 180 isotope. Although most ions formed in the plasma have a single positive charge, some doubly charged ions form with the same apparent mass as a singly charged ion of another element. Remember that we are still measuring m/z in atomic MS, not mass. When the doubly charged ion is the most abundant isotope and is present in large quantities in the sample, this can cause significant problems. For example, $^{88}Sr^{2+}$ interferes with $^{44}Ca^+$, and $^{86}Sr^{2+}$ interferes with $^{43}Ca^+$; $^{112}Cd^{2+}$ interferes with $^{56}Fe^+$, and $^{138}Ba^{2+}$ interferes with $^{69}Ga^+$. An analyst needs to consider the relative abundances and nature of the sample matrix and acids or other reagents used when assessing potential isobaric interferences in ICP-MS.

Many of these interferences are well-documented, and in a well-controlled system, many can be corrected for mathematically. Most instrument companies have interference-correction software as part of their data handling software. A problem with this approach has been that sometimes the necessary data were not collected during analysis, requiring reanalysis if possible. Multiple scans or longer analysis times may have been needed. One instrument that has an advantage with respect to software corrections is the simultaneous ICP-MS, from SPECTRO, described in Chapter 10. Because the SPECTRO MS collects the complete mass spectrum simultaneously without scanning, all data are available. Having been collected at the same time, measurement-based corrections can be made more reliably.

11.4.4 Instrumental Approaches to Eliminating Interferences

There are now five primary instrumental approaches to eliminating or reducing interferences in atomic MS: (1) use of high mass resolution ICP-MS (HR-ICP-MS); (2) use of a collision cell to break apart polyatomic interferences; (3) use of gas-phase chemical reactions in a "reaction cell" to eliminate polyatomic interferences; (4) changing the instrument operating conditions to form a cool or cold plasma as discussed earlier; and (5) use of tandem quadrupoles in MS-MS mode to exclude all nontarget analyte masses.

11.4.4.1 High-Resolution ICP-MS (HR-ICP-MS)

As is the case with high-resolution molecular mass spectrometers, these magnetic sector or TOF-based instruments permit accurate mass determination (resolution <0.1 amu). Quadrupole systems are limited to unit mass resolution (1 amu). Therefore, the use of HR-ICP-MS eliminates many of the spectral interferences seen with quadrupoles because the "overlaps" do not occur; the resolution is sufficient to separate the exact masses of the polyatomic species and analyte elements. HR-ICP-MS is the only way to measure accurately low levels of elements that suffer from significant interferences, including S, P, Si, F, and Cl, using MS. HR-ICP-MS allows interferences to be characterized using exact mass measurements. Using HR-ICP-MS, it is possible to separate ^{31}P, ^{30}SiH, $^{12}C_{19}F$, $^{14}N_{17}O$, and $^{13}C_{18}O$. All these species have a nominal mass of 31 Da. The high S/N ratio of these instruments results in extremely low detection limits (parts per quintillion limits are claimed by one manufacturer). This allows direct analysis of low-concentration samples without preconcentration, allows dilution of samples, and allows the use of low flow rates, minimizing matrix effects.

11.4.4.2 Collision and Reaction Cells

A collision cell is often a small hexapole, octopole, or quadrupole inserted between the ion optics and the mass analyzer of the instrument. As the ions, including polyatomic ions, pass through the cell, they are bombarded with gas molecules, usually helium or hydrogen. While both analyte ions and polyatomic ions lose the same amount of energy every time they collide with an He atom, polyatomics are bigger and therefore collide more often and lose more energy as they transit the cell than do analyte ions. At the exit of the collision cell, the ion energies no longer overlap. A bias voltage "step" rejects the low-energy polyatomic ions. This method of removing polyatomic interferences is called **kinetic energy discrimination** (KED). A simple hexapole collision cell can reduce the interference on $^{56}Fe^+$ by the polyatomic ArO^+ ion by 10^2. Collision cells are commonly used with single quadrupole ICP-MS systems to remove polyatomic interferences in high matrix samples. Advantages include no reactive signal loss for analytes and no formation of new interferences in the cell, which can occur with reactive gases. However, He gas collision mode cannot effectively remove intense interferences on analytes such as P, S, and Se or enable ppt-level analysis of high-purity materials. He collision mode cannot remove direct isobaric interferences such as ^{40}Ar on ^{40}Ca and is not effective against doubly charged interferences such as $^{150}Sm^{2+}$ and $^{150}Nd^{2+}$ on $^{75}As^+$. For ultratrace analysis, a reactive gas is used in "reaction mode" to eliminate interferences.

One example of a collision/reaction cell is the ORS^3 octopole reaction system from Agilent Technologies. The ORS^3 can operate in collision mode with He and KED removal of interferences as seen in the following (graphic Copyright Agilent Technologies, Inc., used with permission). It can also operate in reaction mode, using specific gas-phase chemical reactions to greatly reduce polyatomic interferences. The ORS^3 is an octopole placed inside an enclosed reaction chamber. The cell is positioned between the ion optics and the mass analyzer in a single quad MS (see Figure 10.33). The gas inlet allows a small volume of reactive gas into the cell. Some of the gases used in reaction mode are ammonia, methane, oxygen, and hydrogen. The gas is selected based upon its ability to undergo a gas-phase chemical reaction with the interfering species. Examples are given in Table 11.29. The gas-phase ion–molecule reaction converts the interfering ions into noninterfering products. These reaction products are ejected from the analyte ion path either by the reaction cell or by the analyzer quadrupole. Using this approach, the ArO^+ interference on $^{56}Fe^+$ is reduced by 10^6.

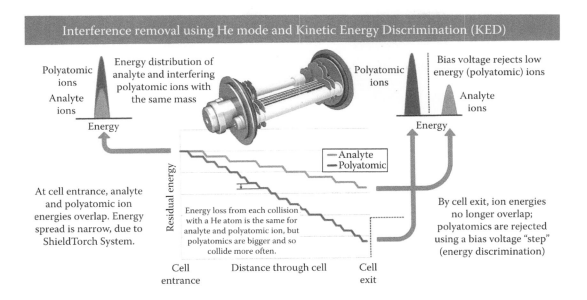

For ultratrace analysis, reaction mode may provide better interference removal than collision mode. There are some disadvantages to this approach. Each reaction gas will only remove interferences that react with that cell gas, so some interferences may remain. The analyst must know what interferences to remove in order to choose the reaction gas. This is not always possible with unknown samples. Reaction gases can react with the sample matrix and analytes to create new interferences. These cell-formed interferences are often unpredictable and may lead to erratic results. Reaction gases may react with analytes and reduce their sensitivity; severe loss of sensitivity for Cu and Ni has been reported when using H_2 and ammonia.

PerkinElmer uses a patented scanning quadrupole-based collision/reaction cell, called a *universal cell*, which can operate with collision/KED or reactive gas. The scanning quadrupole in the cell removes side reaction product and new interferences, so that only the analyte passes to the analyzing quadrupole.

11.4.4.3 MS/MS Interference Removal

Quadrupole-based reaction cells used in single quadrupole ICP-MS instruments can reject some nontarget ions by operating the ion guide as a low-mass cutoff or band-pass filter. However, these devices cannot remove all overlapping ions, so the reaction mode can be erratic except in the case of simple, well-characterized matrices. In 2012, Agilent Technologies introduced the first tandem MS instrument for ICP-MS, the 8800 ICP-QQQ. QQQ stands for triple quadrupole or "triple quad," although there are only two quadrupole mass analyzers in any "triple quad" system. Triple quad commercial instruments have been used for molecular MS (see Chapter 10) since the early 1970s. In the 8800 ICP-QQQ, a single quadrupole mass analyzer (Q1) is placed in front of the Agilent ORS³ octopole collision/reaction cell. A second quadrupole mass analyzer (Q2) follows the collision/reaction cell. This configuration is called MS/MS or tandem MS. The instrument layout was shown in Chapter 10; Figure 11.43 shows the hardware without the instrument box.

The MS/MS mode removes interferences in a number of ways, depending on the configuration of the quadrupoles. The two main modes are MS/MS on-mass mode and MS/MS mass-shift mode. (Remember, we are using mass to mean *m/z*.)

In on-mass mode, both Q1 and Q2 are set to the same target mass. In this case, Q1 controls which ions enter the collision/reaction cell. All masses except the analyte mass and any direct mass interferences are rejected by Q1; only the analyte mass and any direct on-mass interferences pass into the collision/reaction cell. The ORS³ separates the analyte from interferences by neutralizing the interference or shifting it to a new mass. Q2 then rejects any cell-formed interferences except for the target analyte ion. The on-mass mode removes the variability seen in reaction cell systems caused by different samples (ions) changing the reaction processes and product ions. The on-mass mode is used when the analyte is *not* reactive, and the interferences *are* reactive with a particular cell gas.

An example of this mode of interference removal is the use of O_2 cell gas to remove the $^{186}W^{16}O^+$ overlap on $^{202}Hg^+$. Note that both ions have the same *m/z*, 202 amu. The mercurous ion does not react with oxygen, but WO^+ does, as follows:

$$WO^+ + O_2 \rightarrow WO_2^+ \text{ and } WO_3^+$$

(b)

Figure 11.43 The Agilent Technologies 8800 ICP-QQQ. From the left, sample introduction, the ICP torch box (note the cooling coils for the torch), Q1, the ORS³ octopole collision/reaction cell, Q2, and finally, the electron multiplier detector. (© 2014 Agilent Technologies, www.agilent.com/chem. All rights reserved. Used with permission.)

Q1 and Q2 are both set to 202 amu. If Q1 is set to 202 amu, both the analyte Hg⁺ and the interference WO⁺ will pass into the ORS³ cell. WO⁺ reacts with O_2 in the cell and is converted to WO_2^+ and WO_3^+. These product ions are no longer at mass 202 and are rejected by Q2. Only the analyte ion passes through Q2 and is detected free from overlap interference. The resulting mass spectra are shown in Figure 11.44.

Another example of the use of this mode is the removal of S-based interferences on ⁵¹V. In high-purity materials, the interferences are generally well known and consistent. In the analysis of high-purity sulfuric acid for trace elements, ⁵¹V⁺ is overlapped by ³³S¹⁸O⁺ and ³⁴S¹⁶OH⁺ ions from the matrix. The sulfur-based interferences react rapidly with ammonia gas in the cell; vanadium reacts very slowly with ammonia, so the use of on-mass mode with the quadrupoles set at 51 amu removes the SO/SOH interferences by formation of S ion–ammonia cluster ions. No other ammonia cluster ions can form in the cell because only the ions at 51 amu were allowed to enter the cell. Only ⁵¹V⁺ exits the cell to Q2 and the detector. Detection limits for this type of analysis are generally one to two orders of magnitude better than those achieved with a standard He collision cell.

The second mode of interference removal is mass-shift mode, where Q1 and Q2 are set to different masses. This mode is used when the analyte *is* reactive with a particular cell gas and the interferences are *not* reactive.

In mass-shift mode, Q1 is set to a precursor ion mass, usually the analyte mass. Only the analyte mass and any on-mass interferences pass through to the ORS³. In the ORS³, the reactive analyte is separated from the nonreactive interferences by shifting the analyte to a new product ion mass.

Q2 is set to the mass of a target reaction product ion containing the analyte. Q2 rejects all cell-formed product ions except the target analyte product ion.

An example of using mass-shift mode is the conversion of ³²S⁺ to ³²S¹⁶O⁺ (at 48 amu) by reaction with oxygen gas. In a single quadruple system, ³²S⁺ is overlapped by O_2^+, while ⁴⁸SO⁺ is overlapped by ⁴⁸Ca⁺, ⁴⁸Ti⁺, and ³⁶Ar¹²C⁺. By setting Q1 to 32 amu and Q2 to 48 amu (Q1 + 16 amu in oxygen mode), only the ³²S⁺ ion enters the reaction cell and only ⁴⁸SO⁺ enters Q2 and is detected. The Ca, Ti, and ArC ions are rejected by Q1 and do not interfere. Since all of the S isotopes react with oxygen, it is possible using this MS/MS mode and oxygen cell gas to perform S isotope ratio analysis by measuring ³²S¹⁶O, ³³S¹⁶O, and ³⁴S¹⁶O ions. A similar O atom addition reaction to Ti⁺ enables all of the Ti isotopes to be measured as the respective TiO ions with no interference from on-mass ions Ni, Cu, and Zn.

11.4.5 Limitations of Atomic MS

The major limitations to atomic MS at this point are the inefficient introduction system to the ICP (a common problem with the ICP), matrix effects, and isobaric interference. In addition, quadrupoles are limited in sensitivity for trace amounts of light elements such as S, Cl, P, and Si due to significant interferences from C-, N-, and O-based ions. ICP-MS cannot measure H, N, O, C, or Ar because it is open to the atmosphere and the plasma is Ar-based. Fluorine cannot be measured because of its ionization potential.

The flow rate, gas mixture, and solvent used affect the formation of polyatomic species and therefore the degree of interference from polyatomic species including refractory oxides. For reproducible results, the operating conditions must be rigorously controlled. Plasma conditions can be chosen that minimize the formation of interferences but may cause loss of sensitivity.

The isobaric interference problem can be assessed and perhaps avoided by measuring a different isotope

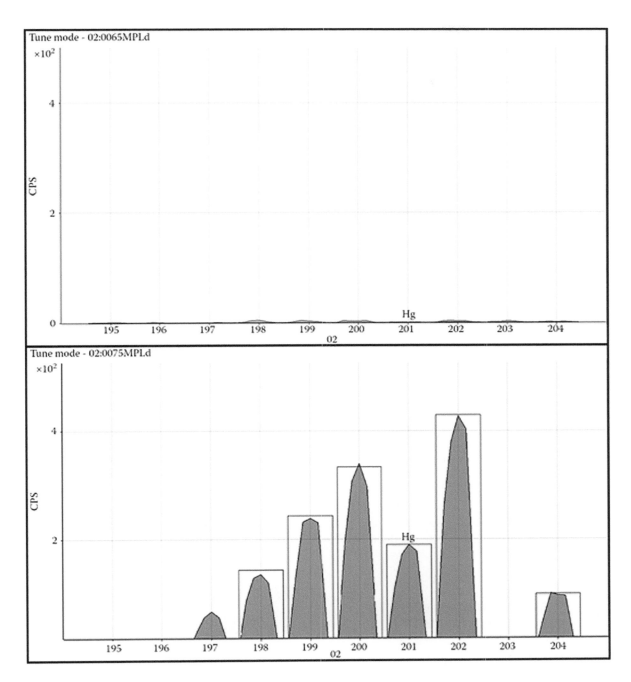

Figure 11.44 MS/MS on-mass mode. (Top) Spectrum of 5 ppm W. Peaks are from WO⁺. (Bottom) Spectrum of 2 ppb Hg on the same scale. Note that there is an order of magnitude difference in the W and Hg concentrations and that the interference from W is hardly discernible. (© 2014 Agilent Technologies, www.agilent.com/cs/library/applications/application-hg-cosmetics-8900-icp-ms-5994-0461en-us-agilent.pdf. All rights reserved. Used with permission.)

of the analyte. That may require measuring a less abundant isotope, reducing the sensitivity of the analysis. The use of high-resolution MS can eliminate the mass overlap problem, but the cost of such instruments is very high compared with quadrupole instruments. The use of collision cells and gas-phase reaction cells in a quadrupole instrument can be used to eliminate or reduce polyatomic interferences. The new simultaneous ICP-MS from SPECTRO offers more reliable software interference corrections from simultaneous acquisition of the complete

Table 11.30 Common Spurious Effects in Mass Spectrometry (taken from the book by Bruno and Svoronos)

Ions observed, m/z	Possible Compound	Possible Source
13, 14, 15, 16	Methane*	Chlorine reagent gas
18	Water*	Residual impurity; outgassing of ferrules; septa and seals
14, 28	Nitrogen*	Residual impurity, outgassing of ferrules; septa and seals; leaking seal
16, 32	Oxygen*	Residual impurity; outgassing of ferrules; septa and seals; leaking seal
44	Carbon dioxide*	Residual impurity; outgassing of ferrules; septa and seals; leaking seal; note it may be mistaken for propane in a sample
31, 51, 69, 100, 119, 131, 169, 181, 214, 219, 264, 376, 414, 426, 464, 502, 576, 614	Perfluorotributyl amine (PFTBA), and related ions	This is a common tuning compound; may indicate a leaking valve
31	Methanol	Solvent; can be used as a leak detector
41, 43, 55, 57, 69, 71, 85, 99	Hydrocarbons	Mechanical pump oil; fingerprints
43, 58	Acetone	Solvent; can be used as a leak detector
78	Benzene	Solvent; can be used as a leak detector
91, 92	Toluene	Solvent; can be used as a leak detector
105, 106	Xylenes	Solvent; can be used as a leak detector
151, 153	Trichloroethane	Solvent; can be used as a leak detector
69	Mechanical pump fluid, PFTBA	Back diffusion of mechanical pump fluid; possible leaking valve of tuning compound vial
73, 147, 207, 221, 281, 295, 355, 429	Dimethylpolysiloxane	Bleed from a column or septum, often during high-temperature program methods in GC-MS
77, 94, 115, 141, 168, 170, 262, 354, 446	Diffusion pump fluid	Back diffusion from diffusion pump, if present
149	Phthalates	Plasticizer in vacuum seals, gloves
X − 14 peaks	Hydrocarbons	Loss of a methylene group indicates a hydrocarbon sample

* It is often possible to operate the analyzer to ignore these common background impurities. They will be present to contribute to poor vacuum if these impurities result from a significant leak.

mass spectrum, while the Agilent triple quad is taking the MS/MS instrumental approach to removing interferences completely.

11.5 COMMON SPURIOUS EFFECTS IN MASS SPECTROMETRY

We will end this chapter with a discussion that can affect all of the techniques and interpretations that have been described. It is almost inevitable that artifacts will crop up in mass spectra that must be understood. Some of these are called background effects, since they are caused by the operation of a vacuum system in an air atmosphere. In Chapter 13 we discuss the interface of gas chromatography with mass spectrometry, a powerful combination but one that can introduce deceptive effects that must be understood. Direct insertion probes will have their own issues. In Table 11.30 we present the most common spurious effects that can adversely affect the reliability of interpretation and instrument operation.

PROBLEMS

11.1 Why can the element indium not be identified definitively by atomic MS?

11.2 Identify the compound and the isotopic composition of each fragment given the following mass spectral data:

m/z	1	16	17	18	19	20
Relative abundance	<0.1	1.0	21	100	0.15	0.22

11.3 Identify the compounds and the isotopic composition of each fragment given the following mass spectral data:

(a)

m/z	12	16	28	29	30	44	45	46
Relative abundance	8.7	9.6	9.8	0.13	0.02	100	1.2	0.42

(b)

m/z	35	36	37	38	39
Relative abundance	12	100	4.1	33	0.01

11.4

m/z	28	29	30	31	32	33	34
Relative abundance	6.3	64	3.8	100	66	0.98	0.14

Identify the following compounds from their mass spectra and the data provided:

11.5

m/z	14	15	16	19	31	32	33	34	35
Relative abundance	17	100	1.0	2.0	10	9.3	89	95	1.1

11.6

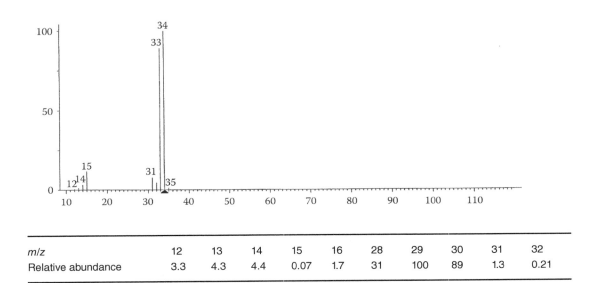

m/z	12	13	14	15	16	28	29	30	31	32
Relative abundance	3.3	4.3	4.4	0.07	1.7	31	100	89	1.3	0.21

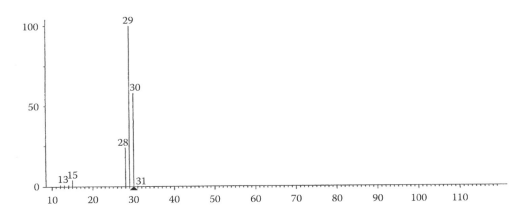

11.7

m/z	16	24	25	32	34	48	50	64	65	66	67
Relative abundance	5.2	0.8	0.04	11	0.42	49	2.3	100	0.88	49	0.04

11.8

m/z	14	19	33	52	53	71	72
Relative abundance	5.2	8.4	42	100	0.39	30	0.11

11.9

m/z	12	13	14	15	24	25	26	27	30	31	35	36	37	38	47	47.5	48	48.5
Relative abundance	2.7	3.0	0.63	0.05	4.0	15	34	1.2	0.06	0.32	7.0	1.9	2.3	0.68	6.5	0.22	5.9	0.19
m/z	49	50	51	59	60	61	62	63	64	74	95	96	97	98	99	100	101	
Relative abundance	4.2	1.5	0.31	2.6	24	100	9.9	32	0.67	0.02	3.0	67	3.3	43	1.2	7.0	0.40	

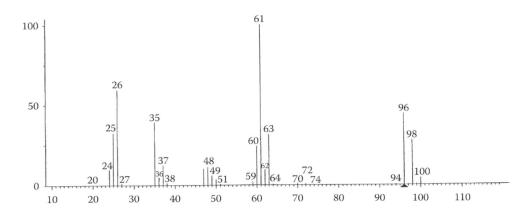

11.10

m/z	15	16	26	27	28	29	38	39	40	41	42	43
Relative abundance	6.3	0.20	1.4	15	2.4	38	1.4	13	1.4	41	2.3	1.6
m/z	51	52	53	54	55	56	57	58	59	71	72	
Relative abundance	1.0	0.27	1.2	0.20	2.8	4.3	100	4.4	0.08	0.04	0.01	

11.11

m/z	15	16	26	27	28	29	30	31	39	40	41
Relative abundance	2.8	0.03	1.4	10	5.2	6.0	0.33	4.5	5.8	1.0	7.2
m/z	42	43	44	45	46	47	57	58	59	60	61
Relative abundance	4.2	19	3.9	100	2.5	0.19	0.41	0.20	4.3	0.51	0.03

11.12

m/z	12	14	19	24	26	27	31	32
Relative abundance	13	2.1	2.0	2.7	11	0.11	22	0.28
m/z	38	50	69	70	76	77	95	96
Relative abundance	6.2	25	100	1.08	46	1.0	2.4	0.06

In the following spectra, remember that the molecular ion peak is indicated by a dark triangle. A *white triangle* indicates where the molecular ion should appear in those spectra that do not show a peak for the molecular ion.

11.13

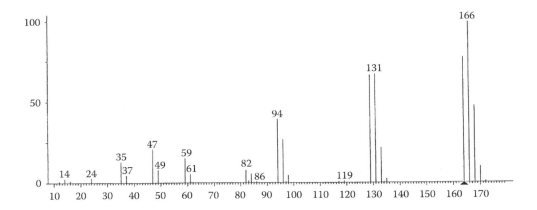

11.14

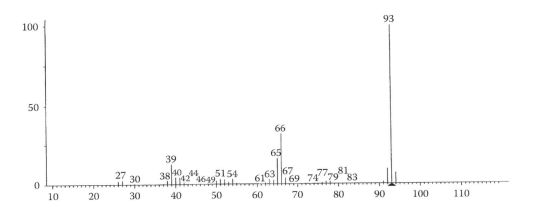

11.15

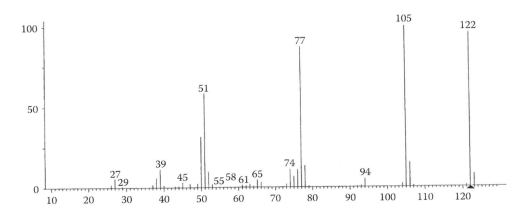

11.16 Hint: This compound contains one halogen atom.

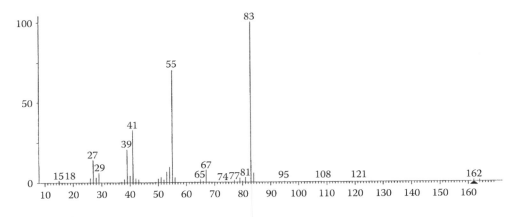

11.17 Hint: This compound contains three halogen atoms.

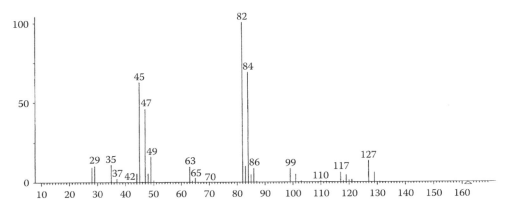

11.18

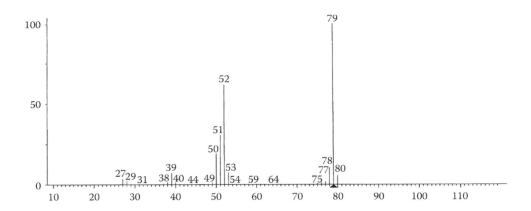

11.19 Hint: This compound is an ether.

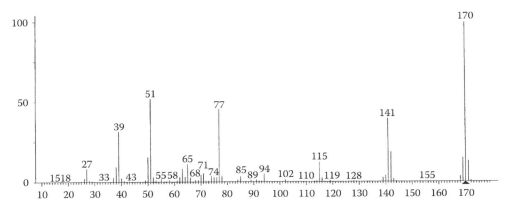

11.20 Hint: This compound is a carboxylic acid.

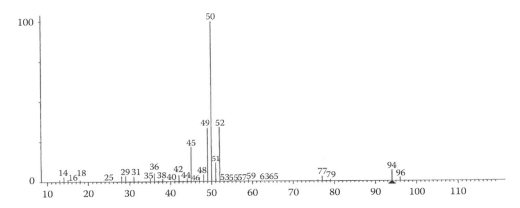

These are a little more challenging. Draw the structure and name the compound.

11.21 The empirical formula is $C_4H_{11}N$.

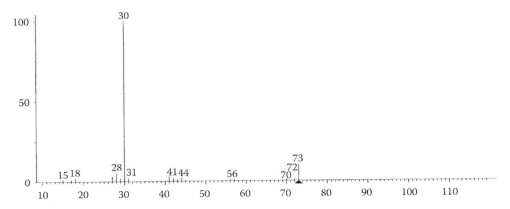

11.22 The empirical formula is $C_7H_6N_2O_4$.

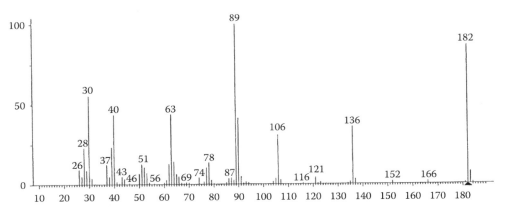

11.23 The empirical formula is $C_5H_{10}O$.

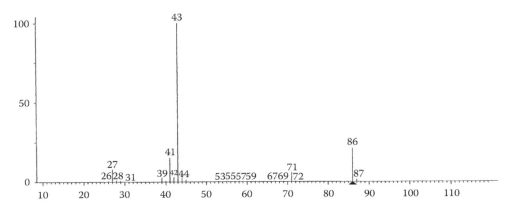

11.24 Hint: This compound has six "rings + double bonds."

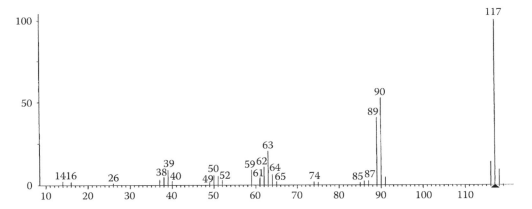

11.25

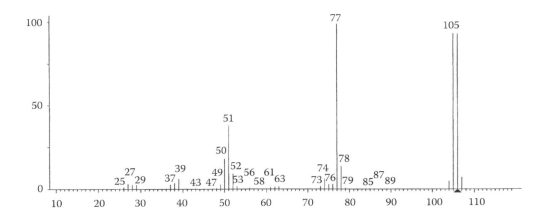

11.26 Hint: This compound is a ketone and contains three halogen atoms.

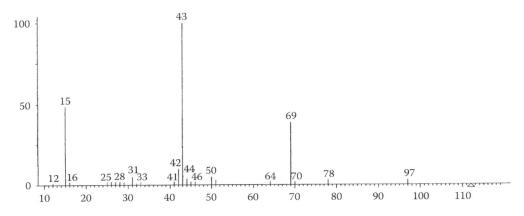

11.27 What are the advantages of high-resolution MS for molecular MS? What are the disadvantages?

11.28 What are the advantages and disadvantages of high-resolution atomic MS?

11.29 Why is it important to understand the difference between atomic mass and exact mass in interpreting mass spectra?

11.30 Explain how quantitative analysis is performed using isotope dilution. What is the advantage of an isotope-labeled compound as a calibration standard?

11.31 Diagram an ICP-MS interface. What is the purpose of the sampler and skimmer cones?

11.32 Ni and Pt are the two common materials for making ICP-MS cones. What are the advantages and disadvantages of the two materials for this application?

11.33 Which is easier to interface to a mass spectrometer, a GC or an LC? Why?

11.34 What mass analyzers are used for ICP-MS instruments?

11.35 What advantages are there to a TOF-based ICP-MS? (You may need to review the TOF mass analyzer, Chapter 10.)

11.36 Explain the two major types of interference in atomic MS.

11.37 Describe two instrumental approaches for eliminating or reducing polyatomic interferences in ICP-MS.

11.38 Explain the difference between MS/MS on-mass and mass-shift modes. When is on-mass mode used? Give an example of how this removes interferences.

11.39 Explain how mass-shift mode removes interferences and give an example.

11.40 Using the exact mass values for the monoisotopic (lowest mass ions) of each of the elements in Table 10.1, prepare a spreadsheet program using Excel or other spreadsheet program to construct an "**exact mass calculator**" for the lowest mass ion in a molecular ion cluster for any molecule of a given empirical formula containing only these atoms. Input in one line of cells the number of each kind of atom and set up the sheet to produce the exact mass in an answer cell. Use this calculator to compute the exact mass of the monoisotopic $M^+ = 2538$ for $C_{101}H_{145}N_{34}O_{44}$ illustrated in Figure 11.31. Print and submit the sheet with the cells displaying the formulas used. How close does your value come to that on the figure? Given the number and identity of the atoms in this molecule, what class of compound is it likely to be? (Hint: an important biological structure.)

11.41 Looking at the structures in Figure 11.30 of the two very different compounds with essentially the same EI mass spectrum, draw the structure of each with a ⌐ crossing the bond that breaks to form each of the major fragment ions and put their mass on each side of the break symbol.

11.42 If one of the candidate hits in Figure 11.30 had one sulfur atom, how could you identify it? Would this be easier or harder if the odd atom was Cl or Br? Suppose these hypothetical heteroatoms were at different positions in the molecules (structural isomers). Would the MS spectra always still be hard to distinguish? What information could you use to locate the position of the heteroatom on the backbone?

11.43 Using your exact mass calculator, what is the difference in exact masses of the M^+ monoisotopic ions of each candidate in Figure 11.30? What MS resolution would you require to distinguish between these?

11.44 Given the compound formula in Figure 11.32, try to devise a spreadsheet employing both the number of each of the different atoms in the formula **and** their percent abundances. Design this to calculate the exact proportions of each ion in the molecular ion cluster (i.e., M^+, $M^+ + 1$, $M^+ + 2$, $M^+ + 3$) that gives the theoretical spectrally accurate relative proportions for each ion in the cluster. The CERNO algorithm corrects and matches the acquired mass spectrum to this, in order to find the closest matches at high resolution (along with the separately calculated value of high mass accuracy for the monoisotopic, lowest mass ion in the cluster).

11.A APPENDIX

Table 11.A.1 **Relative Abundance of the Natural Isotopes**

Isotope		%		%		%	Isotope		%		%		%
1	H	99.985					51			V	99.750		
2	H	0.05					52					Cr	83.789
3			He	0.000137			53					Cr	9.501
4			He	99.999863			54	Fe	5.8			Cr	2.365
5							55			Mn	100		
6					Li	7.5	56	Fe	91.72				
7					Li	92.5	57	Fe	2.2				
8							58	Fe	0.28			Ni	68.077
9	Be	100					59			Co	100		
10			B	19.9			60					Ni	26.223
11			B	80.1			61					Ni	1.140
12					C	98.90	62					Ni	3.634
13					C	1.10	63	Cu	69.17				
14	N	99.643					64			Zn	48.6	Ni	0.926
15	N	0.366					65	Cu	30.83				
16			O	99.762			66			Zn	27.9		
17			O	0.038			67			Zn	4.1		
18			O	0.200			68			Zn	18.8		
19					F	100	69					Ga	60.108
20	Ne	90.48					70	Ge	21.33	Zn	0.6		
21	Ne	0.27					71					Ga	39.892
22	Ne	9.25					72	Ge	27.66				
23			Na	100			73	Ge	7.73				
24					Mg	78.99	74	Ge	35.94	Se	0.89		
25					Mg	10.00	75					As	100
26					Mg	11.01	76	Ge	7.44	Se	9.36		
27	Al	100					77			Se	7.63		
28			Si	92.23			78	Kr	0.35	Se	23.78		
29			Si	4.67			79					Br	50.69
30			Si	3.10			80	Kr	2.25	Se	49.61		

Table 11.A.1 (Continued) Relative Abundance of the Natural Isotopes

Isotope		%		%		%	Isotope		%		%		%
31					P	100	81					Br	49.31
32	S	95.02					82	Kr	11.6	Se	8.73		
33	S	0.75					83	Kr	11.5				
34	S	4.21					84	Kr	57.0	Sr	0.56		
35			Cl	75.77			85					Rb	72.165
36	S	0.02			Ar	0.337	86	Kr	17.3	Sr	9.86		
37			Cl	24.23			87			Sr	7.00	Rb	27.835
38					Ar	0.063	88			Sr	82.58		
39	K	93.2581					89					Y	100
40	K	0.0117	Ca	96.941	Ar	99.600	90	Zr	51.45				
41	K	6.7302					91	Zr	11.22				
42			Ca	0.647			92	Zr	17.15	Mo	14.84		
43			Ca	0.135			93					Nb	100
44			Ca	2.086			94	Zr	17.38	Mo	9.25		
45					Sc	100	95			Mo	15.92		
46	Ti	8.0	Ca	0.004			96	Zr	2.80	Mo	16.68	Ru	5.52
47	Ti	7.3					97			Mo	9.55		
48	Ti	73.8	Ca	0.187			98			Mo	24.13	Ru	1.88
49	Ti	5.5					99					Ru	12.7
50	Ti	5.4	V	0.250	Cr	4.345	100			Mo	9.63	Ru	12.6
101					Ru	17.0	151					Eu	47.8
102	Pd	1.02			Ru	31.6	152	Gd	0.20	Sm	26.7		
103			Rh	100			153					Eu	52.2
104	Pd	11.14			Ru	18.7	154	Gd	2.18	Sm	22.7		
105	Pd	22.33					155	Gd	14.80				
106	Pd	27.33	Cd	1.25			156	Gd	20.47	Dy	0.06		
107					Ag	51.839	157	Gd	15.65				
108	Pd	26.46	Cd	0.89			158	Gd	24.84	Dy	0.10		
109					Ag	48.161	159					Tb	100
110	Pd	11.72	Cd	12.49			160	Gd	21.86	Dy	2.34		
111			Cd	12.80			161			Dy	18.9		
112	Sn	0.97	Cd	24.13			162	Er	0.14	Dy	25.5		
113			Cd	12.22	In	4.3	163			Dy	24.9		
114	Sn	0.65	Cd	28.73			164	Er	1.61	Dy	28.2		
115	Sn	0.34			In	95.7	165					Ho	100
116	Sn	14.53	Cd	7.49			166	Er	33.6				
117	Sn	7.68					167	Er	22.95				
118	Sn	24.23					168	Er	26.8	Yb	0.13		
119	Sn	8.59					169					Tm	100
120	Sn	32.59	Te	0.096			170	Er	14.9	Yb	3.05		
121					Sb	57.36	171			Yb	14.3		
122	Sn	4.63	Te	2.603			172			Yb	21.9		
123			Te	0.908	Sb	42.64	173			Yb	16.12		
124	Sn	5.79	Te	4.816	Xe	0.10	174			Yb	31.8	Hf	0.162
125			Te	7.139			175	Lu	97.41				
126			Te	18.95	Xe	0.09	176	Lu	2.59	Yb	12.7	Hf	5.206
127	I	100					177					Hf	18.606
128			Te	31.69	Xe	1.91	178					Hf	27.297
129					Xe	26.4	179					Hf	13.629
130	Ba	0.106	Te	33.80	Xe	4.1	180	Ta	0.012	W	0.13	Hf	35.100
131					Xe	21.2	181	Ta	99.988				
132	Ba	0.101			Xe	26.9	182			W	26.3		

(continued)

Table 11.A.1 (Continued) Relative Abundance of the Natural Isotopes

Isotope		%		%		%	Isotope		%		%		%
133			Cs	100			183			W	14.3		
134	Ba	2.417			Xe	10.4	184	Os	0.02	W	30.67		
135	Ba	6.592					185					Re	37.40
136	Ba	7.854	Ce	0.19	Xe	8.9	186	Os	1.58	W	28.6		
137	Ba	11.23					187	Os	1.6			Re	62.60
138	Ba	71.70	Ce	0.25	La	0.0902	188	Os	13.3				
139					La	99.9098	189	Os	16.1				
140			Ce	88.48			190	Os	26.4			Pt	0.01
141					Pr	100	191			Ir	37.3		
142	Nd	27.13	Ce	11.08			192	Os	41.0			Pt	0.79
143	Nd	12.18					193			Ir	62.7		
144	Nd	23.80	Sm	3.1			194					Pt	32.9
145	Nd	8.30					195					Pt	33.8
146	Nd	17.19					196	Hg	0.15			Pt	25.3
147			Sm	15.0			197			Au	100		
148	Nd	5.76	Sm	11.3			198	Hg	9.97			Pt	7.2
149			Sm	13.8			199	Hg	16.87				
150	Nd	5.64	Sm	7.4			200	Hg	23.10				
201	Hg	13.18					220						
202	Hg	29.86					221						
203					Tl	29.524	222						
204	Hg	6.87	Pb	1.4			223						
205					Tl	70.476	224						
206			Pb	24.1			225						
207			Pb	22.1			226						
208			Pb	52.4			227						
209	Bi	100					228						
210							229						
211							230						
212							231	Pa	100				
213							232	Th	100				
214							233						
215							234	U	0.0055				
216							235	U	0.7200				
217							236						
218							237						
219							238	U	99.2745				

Source: Isotopic composition of the elements 1989. *Pure Appl. Chem.*, 63(7), 991–1002, © 1991 IUPAC.

BIBLIOGRAPHY

The reader will note that occasionally multiple editions of books are cited as sources or reference works. The reason for this is that sometimes different figures are used in different editions, which may in fact be written by different authors. The sources found in the bibliography will be the most recent versions of a particular work.

In some cases instrument manufacturers and suppliers are cited by providing a URL to the exact internet location. We have endeavored to ensure that these links are functional at the time of publication, but clearly these links change. If a link has become nonfunctional, a good approach is to use your browser's advanced search capability. For example: "company name" AND "instrument type" AND "specific descriptors" is a good place to start.

Allen, L.A.; Georgitis, S.J. High precision isotope ratio measurements by the LECO Renaissance™ TOFICP-MS, LECO Technical Brief, available at www.leco.com.

Application reviews. *Anal. Chem.* American Chemical Society, published biennially in odd years.

Barker, J. *Mass Spectrometry: Analytical Chemistry by Open Learning*, 2nd edn. Wiley: Chichester, U.K., 1999.

Blades, M.W. Atomic mass spectrometry. *Appl. Spectrosc.*, 48(11), 12A, 1994. (Reprinted in Holcombe, J.A.; Hieftje, G.M.; Majidi, V. (eds.) *Focus on Analytical Spectrometry.*

Society for Applied Spectroscopy: Frederick, MD, 1998. www.s-a-s.org.)

Borman, S.; Russell, H.; Siuzdak, G. A mass spec timeline. *Today's Chemist at Work.* September 2003, p. 47. www.tcawonline.org.

Broekaert, J.A.C. Glow discharge atomic spectroscopy. *Appl. Spectrosc., 49*(7), 12A, 1995. (Reprinted in Holcombe, J.A.; Hieftje, G.M.; Majidi, V. (eds.) *Focus on Analytical Spectrometry.* Society for Applied Spectroscopy: Frederick, MD, 1998. www.s-a-s.org.)

Bruno, T.J., Svoronos, P.D.N., *CRC Handbook of Basic Tables for Chemical Analysis – Data Driven Methods and Interpretation*, 4th Ed., CRC Taylor and Francis, Boca Raton, FL, 2021.

Buchanan, M.V.; Hettich, R.L. Fourier transform mass spectrometry of high-mass biomolecules. *Anal. Chem., 65*(5), 245, 1993.

Busch, K.L. Units in mass spectrometry. *Curr. Trends Mass Spectrometr. (Spectrosc. Mag.), 18*(5S), S32, May 2003. www.spectroscopyonline.com.

Chong, N.S.; Houk, R.S. Inductively Coupled Plasma-Mass Spectrometry For Elemental Analysis and Isotope Ratio Determinations in Individual Organic Compounds Separated by Gas Chromatography. *Appl. Spectrosc., 41*, 66, 1987.

de Hoffmann, E.; Stroobant, V. *Mass Spectrometry: Principles and Applications*, 2nd edn. John Wiley & Sons Ltd.: Chichester, U.K., 1999.

Fenn, J.B.; Mann, M.; Meng, C.K.; Wong, S.E.; Whitehouse, C. *Science, 64*, 246, 1989.

Fuerstenau, S.D. et al. *Angew. Chem. Intl. Ed., 40*, 541, 2001.

Geiger, W.M.; Raynor, M.W. ICP-MS: A universally sensitive GC detection method for specialty and electronic gas analysis. *The Application Notebook (Spectroscopy Magazine Supplement)*, February 2009. www.spectroscopyonline.com.

Holmes, J.L.; Aubry, C.; Meyer, P.M. *Assigning Structures to Ions in Mass Spectrometry.* Taylor & Francis: Boca Raton, FL, 2006.

Jarvis, K.E.; Gray, A.L.; Houk, R.S. *Handbook of Inductively Coupled Plasma Mass Spectrometry.* Blackie: London, U.K., 1992.

Karas, F.H.; Beavis, R.C; Chait, B.T. Matrix-assisted laser desorption ionization mass spectrometry of biopolymers. *Anal. Chem., 63*(24), 1193A, 1991.

Kraj, A.; Silberring, J. (eds.) *Introduction to Proteomics.* Wiley: New York, 2008.

Lambert, J.B.; Shurvell, H.F.; Lightner, D.; Cooks, R.G. *Introduction to Organic Spectroscopy.* MacMillan Publishing Company: New York, 1987.

Le Cornec, F.; Correge, T. *A New Internal Standard Isotope Dilution Method for the Determination of U/Ca and Sr/Ca Ratios in Fossil Corals by ICP-MS.* Varian ICP-MS Applications Note ICP-MS-18, September 1998. (Varian is now Agilent Technologies and the Varian Application Notes are no longer available).

Lovric, J. *Introducing Proteomics, from Concepts to Sample Separation, Mass Spectrometry and Data Analysis.* Wiley-Blackwell: Oxford, U.K., 2011.

McCann, M.T.; Thompson, M.M.; Gueron, I.C.; Tuchman, M. Quantification of orotic acid in dried filter-paper urine samples by stable isotope dilution. *Clin. Chem., 41/5*, 739–743, 1995.

McLafferty, F.W.; Tureček, F. *Interpretation of Mass Spectra*, 4th edn. University Science: Mill Valley, CA, 1993.

Milstein, L.S.; Essader, A.; Fernando, R.; Akinbo, O. Development and Application of a Robust Speciation Method for Determination of Six Arsenic Compounds Present in Human Urine. *Environ. Int., 28*(4), 277–283, 2002.

Milstein, L.S.; Essader, A.; Fernando, R.; Levine, K.; Akinbo, O. Development and Application of a Robust Speciation Method for Determination of Six Arsenic Compounds Present in Human Urine *Environ. Health Perspect., 111*(3), 293–296, 2003.

Montaser, A. (ed.) *Inductively Coupled Plasma Mass Spectrometry.* VCH: Berlin, Germany, 1998.

Niessen, W.M.A.; van der Greef, J. *Liquid Chromatography–Mass Spectrometry.* Marcel Dekker, Inc.: New York, 1992.

NIST Chemistry WebBook, NIST Standard Reference Database Number 69, Eds. P.J. Linstrom and W.G. Mallard, National Institute of Standards and Technology, Gaithersburg MD, 20899, https://doi.org/10.18434/T4D303, (retrieved February 17, 2022).

NIST Standard Reference Database 1A, NIST/EPA/NIH Mass Spectral Library with Search Program, Data Version: NIST v20, Software Version: 2.4, 2020 release.

Oxley, E.S.; Stremming, H.; Paplewski, P. Ultra-low hydrogen detection in steel. *Bruker FIRST Newsletter*, Bruker Corporation, May 2012. www.bruker-axs.com.

Pavia, D.L.; Lampman, G.M.; Kriz, G.S. *Introduction to Spectroscopy: A Guide for Students of Organic Chemistry*, 3rd edn. Harcourt College Publishers: Fort Worth, TX, 2001.

Richards, S. Proteome portraits. *The Scientist*, August 2012.

Ridd, M. *Determination of Trace Levels of Rare Earth Elements in Basalt by ICP-MS.* Varian ICP-MS Applications Note ICP-MS-2, April 1994. (Varian is now Agilent Technologies and the Varian Application Notes are no longer available).

Sevastyanov, V. *Isotope Ratio Mass Spectrometry.* Taylor & Francis: Boca Raton, FL, 2013.

Silverstein, R.M.; Webster, F.X. *Spectrometric Identification of Organic Compounds*, 6th edn. Wiley: New York, 1981.

Skelly Frame, E.M.; Uzgiris, E.E. The determination of gadolinium in biological samples by ICP-AES and ICP-MS in evaluation of the action of MRI agents. *Analyst, 123*, 675–679, 1998.

Smith, R.M.; Busch, K.L. *Understanding Mass Spectra—A Basic Approach.* Wiley: New York, 1999.

Sparkman, O.D. *Mass Spectrometry Desk Reference.* Global View Publishing: Pittsburgh, PA, 2000.

Su, Y., Yu, L-R., Conrads, T.P., Veenstra, T.D., in Grinberg, N., Rodriguez, S., eds., Ewing's Analytical Instrumentation Handbook, 4th Ed., CRC Press Taylor and Francis, Boca Raton, FL, 2018.

VanHaecke, F.; Degryse, P. (eds.) *Isotopic Analysis: Fundamentals and Applications Using ICP-MS.* Wiley-VCH Verlag & Co. KGaA: Weinheim, Germany, 2012.

Voreos, L. Electrospray mass spectrometry. *Anal. Chem., 66*(8), 481A, 1994.

Watson, J.T.; Sparkman, O.D. *Introduction to Mass Spectrometry*, 4th edn. John Wiley & Sons Ltd.: Chichester, U.K., 2007.

Wolf, R.; Denoyer, E.; Grosser, Z. EPA Method 2008 for the analysis of drinking waters, Application Note ENVA-300. PerkinElmer Life and Analytical Sciences: Shelton, CT, July 1995.

Wolf, R.; Denoyer, E.; Sodowski, C.; Grosser, Z. RCRA SW-846 Method 6020 for the ICP-MS analysis of soils and sediments, Application Note ENVA-301. PerkinElmer Life and Analytical Sciences: Shelton, CT, January 1996.

Yarnell, A. The many facets of man-made diamonds. *Chem. Eng. News, 82*(5), 26, 2004.

CHAPTER 12

Principles of Chromatography

12.1 INTRODUCTION TO CHROMATOGRAPHY

Many of the techniques of spectroscopy we have discussed in the preceding chapters are selective for certain atoms or functional groups and structural elements of molecules. Often, this selectivity is insufficient to distinguish compounds of closely related structure from each other. It may be insufficient for measurement of low levels of a compound in a mixture of others that produce an interfering spectral signal. Conversely, when we wish to measure the presence and amounts of a large number of different compounds in a mixture, even the availability of unique spectral signatures for each analyte may require laborious repeated measurements with different spectral techniques to characterize the mixture fully. Living systems have evolved large protein molecules (e.g., receptors, enzymes, immune system antibodies) with unique 3D structures that can bind strongly only to very specific organic compounds. Analytical procedures such as immunoassay, enzyme-mediated assays, and competitive binding assays employ the extreme selectivity of such protein macromolecules to measure particular biomolecules in very complex mixtures. Microarrays of thousands of these, each with unique selectivity, can rapidly screen complex mixtures for a list of expected components. However, precise quantitation of each detected component is difficult to achieve this way.

Analysis of complex mixtures often requires separation and isolation of components or classes of components. Examples include extraction, precipitation, and distillation. These procedures partition components between two phases based on differences in the components' physical properties. In liquid–liquid extraction, components are distributed between two immiscible liquids based on their similarity in polarity to the two liquids (i.e., "like dissolves like"). In precipitation, the separation between solid and liquid phases depends on relative solubility in the liquid phase. In distillation, the partition between the mixture liquid phase and its vapor (prior to recondensation of the separated vapor) is primarily governed by the relative vapor pressures of the components at different temperatures (i.e., differences in boiling points). When the relevant physical properties of the two components are very similar, their distribution between the phases at equilibrium will result in slight enrichment of each in one of the phases, rather than complete separation. To attain nearly complete separation, the partition process must be repeated multiple times and the partially separated fractions recombined and repartitioned multiple times in a carefully organized fashion. This is achieved in the laborious *batch processes* of *countercurrent liquid–liquid extraction*, *fractional crystallization*, and *fractional distillation*. The latter appears to operate continuously, as the vapors from a single equilibration chamber are drawn off and recondensed, but the equilibration in each of the chambers or "plates" of a fractional distillation tower represents a discrete equilibration at a characteristic temperature.

A procedure called **chromatography** automatically and simply applies the principles of these "fractional" separation procedures. Chromatography can separate very complex mixtures composed of many very similar components. The various types of chromatographic instrumentation that have been developed and made commercially available over the past half century now comprise a majority of the analytical instruments for measuring a wide variety of analytes in mixtures. They are indispensable for detecting and quantitating trace contaminants in complex matrices. A single chromatographic analysis can isolate, identify, and quantitate dozens or even hundreds of components of mixtures. Without their ability to efficiently characterize complex mixtures of organic biochemicals at trace levels from microscale samples, research in molecular biology as we know it today would be impossible. We will be discussing another powerful separation technology, **electrophoresis**, together with chromatography, because much of the way electrophoresis operates, and the appearance of its separations, is analogous to chromatography, even though the underlying separation principle is different. Chromatographic-type separation is at the heart of earlier generations of the instrumentation used to separate fragmented deoxyribonucleic acid (DNA) molecules and to sequence the order of the "bases" that form the genomes of all living creatures on Earth. The heroic sequencing of around three billion DNA bases in the human genome, completed in the year 2001, could never have been envisioned, let alone completed, without chromatographic-type instrumentation. This forms the basis of the new field of *genomics*. Even more extensive analytical separation, identification, and quantitation problems arise in the next new field of molecular biological research, coming in the so-called postgenomic era; namely, *proteomics*. This requires separation, identification, and quantitation of the thousands of proteins that may be produced (expressed) by the genome of each type of living cell. The contents and proportions of this mix will depend on what the cell is doing and whether it is in a healthy or diseased state. The amounts of analytes may be at the nanomolar scale or less, and unlike DNA analysis, there is no handy polymerase chain reaction (PCR) procedure to amplify the amounts of analytes. If genomics provides a complete parts list for an organism, then proteomics will be necessary to understand the assembly and operating manual. Billions of dollars of chromatography-based instrumentation, and hundreds of thousands of jobs using it, will be required for research in these fields. Environmental monitoring and regulation is already a multimillion dollar market that uses chromatography's ability to isolate, identify, and measure trace pollutants in complex environmental mixtures.

At this point, it bears mentioning that many of the chromatographic or electrophoretic separations that lay at the heart of earlier generations of the methods of delineating the large numbers of proteins, or DNA bases or other huge biological compound classes, are now superseded by faster, cheaper devices of even greater capacity. Many of these employ huge arrays of specific binding sites created on microchips that can then be scanned rapidly and automatically by a variety of detection methodologies such as laser-induced fluorescence. These support research into classes of biocompounds gathered under the rubric of "-omics," such as genomics, proteomics, metabolomics, lipidomics, and a host of other-omics. Each instrument is highly specialized for the application, with a set of reactants and purpose-built hardware, and requires powerful computer support to keep up with the torrent of data it produces. For example, metabolomics studies the complete array of relatively small polar metabolite molecules in blood circulation and urinary excretion and employs ultrafast liquid chromatography (LC) columns interfaced to very fast scanning mass spectrometry (MS) or MS/MS instrumentation monitored by fast computer matching to libraries of known metabolites (and sometimes even identifying heretofore unknown metabolites). Although this is an exception to the idea that chromatography has been superseded, the instrumentation is so specialized that it is usable only for this particular application.

12.2 THE NATURE OF MIXTURES

Most of the materials that are useful to us are mixtures, not individual pure compounds. These mixtures may be very simple, such as one solute (or sample) dissolved in a solvent of our own choosing, or quite complex, such as a potentially hazardous waste sample over which we have no control or prior knowledge. Mixture complexity increases as (a) the number of components increases and (b) the diversity in the physical properties of the components increases. The first kind of complexity is called "speciation," which describes the multiplicity of different species (compounds, ions, etc.) that might be present. This applies not only to chemicals but to anything. A bag of 100 green marbles is simple to describe and understand: it has one species: green marbles. But if you have a bag of 100 marbles, each a different color, it is much harder to describe: there are 100 different "species" that are present. This is illustrated in Figure 12.1.

Mixture speciation is a critically important subfield of analytical chemistry, especially in the environmental area. For example, it is important to understand in what form a contaminant is found in a sample. One oxidation state of a cation can be much more damaging to organisms than

Figure 12.1 Mixture speciation complexity illustrated by two collections of marbles, one all green and one of all different colors. (Adapted from Vapor Characterization: A NIST Short Course, resource book, The Behavior of Mixtures by M.E. Harries.)

Table 12.1 A Comparison of Intermolecular Interaction and Bonding Energies (Taken from the book by Bruno)

Bond/Interaction	Energy kJ/mol
Simple fluids (dispersion forces)	
$CH_4 \cdots CH_4$	0.5
$C_6H_6 \cdots C_6H_6$	1.6
Associating fluids (hydrogen bonding)	
C—H$\cdots$O	11
C—H$\cdots$N	13-21
Chemical bonds	
C—C	347
O—H	465

another, and speciation is done to sort out the forms that may be observed. This is a major focus of the atomic spectroscopies we have discussed in earlier chapters.

The second kind of complexity results from differences in the physicochemical properties of the component molecules, since these properties describe how the individual molecules interact with each other.

The nature of mixtures and how we must deal with them in a chemical analysis is ultimately determined by interactions and forces which arise between the individual molecules comprising the mixture. To affect a separation of components of a mixture, we will need to understand and take advantage of intermolecular forces and the physical properties which result from these forces. A comparison of some important intermolecular forces and their energies with representative chemical bond energies is presented in Table 12.1. The energy value next to each type of force or bond is representative of the strength of that bond—it is how much energy would be required to displace the indicated species to infinite separation.

The following discussion about intermolecular forces is likely to be a review, since you will have encountered this in both general and organic chemistry classes. The first interaction we will discuss is the dispersion forces (studied by Fritz London in 1930). These are the weakest intermolecular forces; they contribute to both attraction and repulsion among essentially nonpolar molecules. They are represented in Table 12.1 as the interactions occurring between simple fluids. These dispersion forces will be present among all molecules, regardless of the atoms they contain. Importantly, they are the *only* forces which occur between nonpolar simple molecules such as the alkanes (like methane, CH_4), the noble gases (like helium, He), or even more complex but nonpolar molecules (such as carbon tetrachloride, CCl_4). At any given instant, the electrons of such a molecule are in some configuration which will result in an instantaneous dipole moment. A dipole moment occurs in a molecule whenever the electrons and nuclei of the atoms are arranged so that, at any given instant, part of the molecule has a net negative charge and part has a net positive charge. This instantaneous dipole moment can then electrostatically induce a dipole moment in another molecule, resulting in a weak, short range net attraction. This interaction is therefore alternatively called the dipole-induced dipole interaction. The attraction can be additive over several molecules, resulting in, for example, short-lived nearest neighbor ordering. As the molecules approach one another more closely, the weak attractive force gives way to a net repulsive force. The susceptibility of a molecule to induced dipole forces depends on the polarizability of the molecule. A molecule has a high polarizability when it contains groups with large, bulky electron clouds that can be easily deformed. These dispersion forces will always be present among molecules, regardless of the constituent atoms. They will be the only forces which will occur between nonpolar simple molecules such as the alkanes or the noble gases. Molecules with more actively interacting atoms (such as oxygen, nitrogen, and in some cases the strongly electronegative halogens, for example) will display other interactions in addition to dispersion.

A compound such as formamide, CH_3NO, which has a large permanent dipole moment, will show very strong intermolecular interactions. The forces occurring between these kinds of molecules are called Keesom forces. These attractive interactions will promote self-association, causing the formation of clusters of molecules. Because of this effect, these dipole–dipole forces are considered to be orientation forces, having a significant influence on long-range order (that is, the scale is large as compared to molecular size) and structure in a fluid. Such highly polar molecules will be more strongly attracted to one another than nonpolar molecules of comparable molecular mass. Because of the strong interactions, mixtures of polar molecules can be especially difficult and challenging to separate.

If we were to change one of the chlorine atoms of CCl_4 to a hydrogen, the formula and structure of chloroform, $CHCl_3$, results (Figure 12.2), where δ^+ and δ^- represent fractional charges on the halogen atoms and on the hydrogen.

We now notice a significant change in behavior due to the dipolar interaction known as the **hydrogen bond**. The hydrogen atom, being small and not shielded by a large electron cloud can interact with nearby electron-rich groups or molecules. Thus, the hydrogen atom of chloroform will be attracted to the oxygen linkage of an ether or the carbonyl of a ketone, since they are relatively electron-rich. The hydrogen bonding potential of a particular

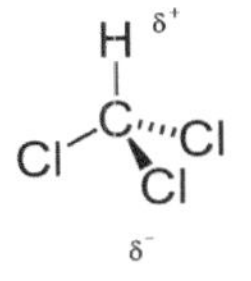

Figure 12.2 The structure of chloroform showing the net positive and negative charges distributed about the carbon atom.

hydrogen is controlled by the chemical neighborhood of the molecule. The electronegative environment set up by the three chlorine atoms on the carbon tends to unshield the hydrogen and make it "feel" more positively charged to its neighbors. The hydrogen bond between chloroform and a ketone would still be relatively weak, however. A very strong hydrogen bond system exists in water, where a relatively stable, extended structure of four to five water molecules can exist as a cluster at ambient temperature. No such stable long-range structure exists in nonpolar fluids such as CCl_4, in which dispersion forces are the predominant interaction. The hydrogen bond is relatively strong as intermolecular forces go, having energies of 4 to 11 kJ/mol. These interactions are highly directional in space, and may even be considered as being weak bonds. Hydrogen bonded systems will usually present problems in the design of separations, especially those involving chromatographic techniques. Peak asymmetry (which we will discuss in detail) and the resulting quantitation problems will usually have to be addressed in analyses of systems such as these.

The stronger intermolecular interactions described above can cause the formation of intermolecular complexes. A complex is an observable, distinct pairing between molecules which is higher in interaction energy and of a longer lifetime than that which would be expected from random molecular collisions. The complex is itself a discrete chemical entity, having its own unique properties. For example, the well-known complex formed between benzene and iodine has a unique ultraviolet spectrum, having a band not present in the electronic spectra of either pure benzene or pure iodine. This complex is called a charge-transfer complex, since the movement of an electron between the benzene π-orbital system and the iodine molecule is demonstrated. The hydrogen bonded clusters of water molecules may also be considered a kind of complex. Actually, complex formation can occur in any system containing molecules which are net negative charge donors in combination with negative charge acceptors. These complexes may be weak and short lived, as in the case of chloroform with an ether, or strong enough to actually be isolated chemically. Complexation will always have a significant effect on the properties of a mixture and our ability to separate that mixture using analytical techniques.

The forces which arise between ions in aqueous solution are both very strong and long range. They will be an important factor in many separations using liquid chromatography, especially when ion exchange columns and ion-pairing agents are used. Ions in solution will interact strongly with dipolar molecules, causing ion–dipole interactions.

The physical arrangement of atoms in a molecule (molecular geometry) can have a profound effect on the properties of a mixture, again because of intermolecular interactions. We shall consider two examples of how differences in the molecular structure or geometry of compounds can have a striking effect in one case and a very subtle effect in another. Ethanol and dimethyl ether are isomers that have the chemical formula C_2H_6O, both with a relative molecular mass of approximately 46 g/mol. Table 12.2a gives some of their relevant physical properties.

Table 12.2a Comparative Properties of Dimethyl Ether and Ethanol

Property	Dimethyl ether	ethanol
State (at ambient)	Gas	Liquid
Relative molecular mass, g/mol	46	46
Normal boiling temperature, °C	−24.9	78.4
Melting temperature, °C	−138.0	−114.2

Table 12.2b Comparative Properties of R123 and R123a (Taken from the book by Bruno)

Property	R-123, CF_3CHCl_2	R123a, $CClF_2CHClF$
Relative molecular mass, g/mol	153	153
Normal boiling temperature, °C	27.9	28
Melting temperature, °C	−107	−78

Clearly, these two chemicals have very different properties despite having the same formula. The reason for the stark differences is intermolecular interactions. The predominant interaction between molecules of dimethyl ether is the London dispersion (or the van der Waals interaction). While these interactions are attractive, the overall effect is very weak. Ethanol, on the other hand, due to the change in molecular geometry, has the specific interaction of hydrogen bonding that we discussed earlier. As a result, it actually exists in a long-range, self-associated network. The effect of the hydrogen bond is so dramatic that it causes ethanol to exist in the liquid state (at ambient temperature and pressure) rather than in the gaseous state, as is the case for dimethyl ether. We would not expect the problem of separating a mixture of ethanol and dimethyl ether to be particularly challenging on the basis of their physical properties.

Let us now turn our attention to two other isomers, each having the formula $C_2F_3Cl_2H$ and a relative molecular mass of approximately 152 g/mol. These compounds were both representatives of the intermediate alternative refrigerant fluids developed after the ban on Freons. Table 12.2b lists some relevant properties of these materials.

The slight difference in geometrical structure between these two materials has a very subtle effect on their physical properties at ambient temperature and pressure. The major intermolecular forces to be considered for both are the dispersion forces. As you may imagine, separation of these isomers for analysis does indeed cause some difficulties, especially since there is usually cross contamination of one in the other.

Table 12.3 Data from a Hypothetical Chromatogram

Peak	t_r (min)	k	w_b (s)	$w(1/2)$ (s)	Resolution
1	3.70	1.31	5.13	3.02	
2	4.00	1.50	5.54	3.26	3.37
3	4.10	1.56	5.68	3.34	1.07
4	4.30	1.69	5.96	3.51	2.06
5	4.35	1.72	6.03	3.55	0.50
6 IS	8.00	4.00	11.09	6.52	5.77
Conc. (ppm)	10	25	50	100	200
Peak area counts for five-level standard curve					
1	64,871	138,589	300,767	654,611	1,132,301
2	53,422	150,838	345,670	647,345	1,319,829
3	65,668	151,856	235,309	530,813	1,094,460
4	54,705	116,870	253,633	552,025	954,854
5	49,904	140,906	322,911	604,723	1,232,931
6 IS[a]	287,296	262,670	270,879	300,977	279,087
Peak area ratios (standard/peak 6 IS)					
1	0.23	0.53	1.11	2.17	4.06
2	0.19	0.57	1.28	2.15	4.73
3	0.23	0.58	0.87	1.76	3.92
4	0.19	0.44	0.94	1.83	3.42
5	0.17	0.54	1.19	2.01	4.42

Note: Capillary GC 30 m × 0.25 mm i.d. column; 20 psi helium mobile phase; 31.3 cm/s; t_m = 1.6 min; temperature = 200 °C; N = 30,000.

[a] The concentration of IS was 50 ppm.

Our discussion of molecular geometry and isomerism would be incomplete without a few words about stereoisomerism and chirality, although much of this will be a review from your organic chemistry courses. Here, we focus on the aspects that affect separations, an especially important topic since many of the more vexing and rigorous analytical separations performed in recent years have concerned stereoisomers. This is especially important in the study of biomolecules for the pharmaceutical industry. Stereoisomers are isomeric species which differ only in the way the individual atoms are arranged in space. This can be contrasted with the R-123 and R123a isomers described above, in which there were actually differences in the number of halogen species on each carbon. Stereoisomers can be subdivided into two groups, called enantiomers and diastereomers. Isomers which are not superimposable upon their mirror image are enantiomers.

Molecules which cannot be superimposed on their mirror images possess a stereogenic center. An older terminology is chirality (chiral, from the Greek work meaning handedness), while those that can are called "achiral." Carbon atoms which contain four different substituents are always chiral, and the carbon in question is called the chiral center. If we can isolate one enantiomer from the other, an interesting optical phenomenon may be observed. The individual enantiomers can rotate the plane of polarized light to the right or left, depending on their structure; one of the isomers will rotate the plane to the right (the dextrorotary, or "+" isomer), and the other to the left (the levorotary, "–" isomer) by the same rotation. While the physics of this interesting interaction need not concern us in the design of effective analytical separations, we must have an appreciation of the consequences of the structural features of these molecules. An equimolar enantiomeric mixture (50 percent of each isomer) will not change the plane of polarized light, since the tendency to rotate to the right and left is exactly balanced. This mixture is called a racemic mixture. If there is a greater amount of one over the other, optical activity may be observed in the mixture. The direction of polarized light rotation in such a mixture will depend on the predominating enantiomer. The physical properties of the individual enantiomers which compose the mixture will be identical. This includes density, boiling point, vapor pressure, refractive index, etc. The only difference, other than the possibility of optical activity, will be in the relative reactivity and interactions toward enantiomeric reagents. This is fortunate, since it will allow us to carry out separations of chiral materials using the chromatographic methods discussed in this book.

Diastereomers are a class of stereoisomers which are not mirror image isomers, as opposed to being simply not superimposable. These isomers are much simpler to deal with from a separation point of view. The physical properties of diastereomers are always different, in many cases strikingly so. The chemical properties of diastereomers, while similar, will also be different. It will be much easier to successfully apply the methods of chromatography to mixtures of diastereomers than to mixtures of enantiomers.

12.3 WHAT IS THE CHROMATOGRAPHIC PROCESS?

The Russian botanist Mikhail Tswett invented the technique and coined the name chromatography. Early in the twentieth century, he placed extracts containing a mixture of plant pigments on the top of glass columns filled with particles of calcium carbonate. Continued washing (**elution**) of the mixtures through the columns with additional solvent (called the **eluent**) resulted in the separation of the pigments, which appeared as colored bands adsorbed on the solid particles, moving along the column at different rates, thus appearing at different distances down the column. He named this technique "chromatography," from the Greek for "color writing." By collecting separated components as they exited the bottom of the column, one could isolate pure material. If a detector in the exiting fluid stream (called the **effluent**) can respond to some property of a separated component other than color, then neither a transparent column nor colored analytes are necessary to separate and measure materials by "chromatography."

When there is a great difference in the retention of different components on the material filling the column, a short column can be used to separate and isolate a rapidly moving component from highly retained material. This is the basis of a useful sample preparation technique called **solid-phase extraction (SPE)**.

Real instrumental chromatography employs highly engineered materials for the **stationary phase** through which the **mobile phase** fluid carrying the mixture of analytes passes, with continuous partitioning of the analytes between the two phases. The column needs to be sufficiently long and the eluent flow sufficiently slow for the process to approach equilibrium conditions. Then partitioning will be repeated a large number of times. Even very slight differences among mixture components in the ratio of the amounts that would exist in each phase at equilibrium will result in their separation into bands along the column. These bands can be separately detected or collected as they elute off the column.

The critical defining aspects of a chromatographic process are as follows:

1. Immiscible stationary and mobile phases.
2. An arrangement whereby a mixture is deposited at one end of the stationary phase.
3. Flow of the mobile phase toward the other end of the stationary phase.
4. Different rates or ratios of partitioning for each component of the mixture and many cycles of this process during elution.
5. A means of either visualizing **bands** of separated components on or adjacent to the stationary phase or detecting the eluting bands as **peaks** in the mobile phase effluent. Here, we emphasize that peaks are traces of electronic detector responses to the eluting analyte bands exiting the separation column. Often, chromatographers use the word peak to refer to the band itself. This more common usage is acceptable, and the distinction between bands and peaks is useful mainly to clarify how the former gives rise to the latter. Mostly, we observe peaks in chromatography. (Note that in some applications, the bands are also referred to as **zones**.)

The first major implementation of instrumental chromatography was based on partition between a liquid, nonvolatile, stationary phase supported on inert solid particles packed in metal columns and an inert gaseous mobile phase flowing through the column from a pressurized tank. The various partition ratios between the phases, which affected separation, were primarily related to the components' different volatilities (i.e., boiling points). The initial application of this procedure, **gas chromatography (GC)**, to petroleum hydrocarbon samples led to facile separation of many more components than could be achieved by the fractional distillation process used in petroleum refining. A schematic GC instrument is shown in Figure 12.3. The inert mobile phase (carrier gas) is in the pressurized tank on the left; the packed column is shown curled inside an oven.

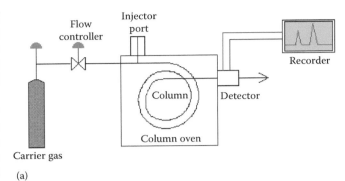

Figure 12.3 A schematic GC instrument, showing carrier gas, column, oven, injection port, and detector. Details of the instrumentation will be discussed in Chapter 13.

Sample is injected through the injection port and is carried through the column to a detector.

Instead of just petroleum "fractions," individual specific hydrocarbons could be resolved and isolated, if the GC column was long enough. The analogy to the well-studied theory of multiple plate fractional distillation columns used in refining processes led to the initial plate theory of the chromatographic process. This assumed that the separation of the bands as they migrated down the column proceeded as a series of sequential, discrete equilibrium partitionings, instead of the actual situation, which is a continuous, not-quite-in-equilibrium process. Nevertheless, this **plate theory** proved fairly accurate for most cases. It contained useful terms (e.g., N = *the number of plates* in a column as a descriptor of its separation efficiency and the **height equivalent to a theoretical plate [HETP]**). These remain central to the simplified theoretical characterization of the chromatographic process, despite the fact that continuous, nonequilibrium kinetic (or rate) theory is more accurate, and there exist no actual plates in a chromatographic column. GC was not suitable for separation and measurement of more polar and less volatile compounds or less thermally stable large biomolecules. Optimized chromatographic separations of these types of molecules at ambient temperatures were achieved later using more polar, water-miscible liquid mobile phases. The instruments and stationary phases that enabled this improvement on the traditional vertical **gravity column LC** apparatus form the basis of the even more widely used **high-performance liquid chromatography (HPLC)**.

12.4 CHROMATOGRAPHY IN MORE THAN ONE DIMENSION

In GC, the gaseous mobile phase must be confined in a column, so that a pressure gradient can cause it to flow past the stationary phase and eventually elute the separated bands out the effluent end of the column. This is inherently a 1D separation, along the column, from one end to the other. This dimensionality applies even should the column be coiled to fit in a GC oven rather than vertically straight, like Tswett's gravity flow liquid mobile phase column. However, unlike gases, liquids as mobile phases do not always require confinement to move in a desired direction or retain their volume. If they are in contact with porous beds of small particles or fiber mats, surface forces (capillary attraction) can often induce them to flow. Thus, it is possible to carry out the LC process on a stationary phase arrayed as a thin surface layer, usually a planar 2D surface. Examples include the matted cellulose fibers of a sheet of paper or a thin layer of silica gel or alumina particles on a planar support (e.g., a pane of glass).

An example of paper chromatography, beloved of grade school science fairs for its simplicity, is to place a drop of colored ink in the center of a sheet of paper. After drying, a series of drops of organic solvent are slowly added to the spot. Capillary attraction causes a radial eluent flow away from the spot in all directions. The different colored dyes in the ink adhere more or less strongly to the multiple hydroxyl groups on the cellulose molecule chains through polar or hydrogen-bonding interactions. Therefore, the capillary eluent flow carries some dyes further than others, resulting in a pattern of colored circles of different radii. This recreates Tswett's original 1D column chromatography experiment on a 2D surface using his original definition. Being on a piece of paper, it seems even more like "color writing."

The radial separation in the earlier example is not the most efficient way to perform surface chromatographic separations. A square planar **thin-layer chromatography (TLC)** plate (not to be confused with the "theoretical plates" discussed previously!) may have a line of spots containing sample mixtures and reference standard materials deposited just above one edge. Submerge that edge in solvent to a level just below the line of spots, and capillary attraction will produce a unidirectional, ascending, eluent flow that will move and separate each spot's components vertically up the plate in separate **lanes**. If they are not already colored, they may be sprayed with a reagent that will react to "develop" a color, just as a photographic plate may be developed chemically to reveal a latent image. Alternatively, the separated component spots may be visualized, or even quantified, by observing their fluorescence emission induced by exposing them to ultraviolet (UV) light or by their quenching of the natural fluorescence of a silica gel TLC layer. Note that an advantage to this form of surface chromatography is that multiple 1D (i.e., in one direction) chromatographic separations of samples or standard mixtures are carried out simultaneously in parallel.

If a mixture has enough components, and some are sufficiently similar in structure and physical properties, it is likely that chromatography in one dimension, either on a column or a surface, will not separate all of them from each other.

Many column bands or surface spots will contain multiple unseparated components. These are said to **coelute** in the chromatographic system. If we repeat the previous TLC plate separation, but this time place just one spot containing a mixture to be separated near one corner of the plate, we will again achieve a line of eluted spots in a lane along a vertical edge. Some of these spots may contain still unseparated, coeluting components. Now let us rotate the plate 90° and place this line of separated spots back in the tank just above *a new eluting solvent* with polarity or pH or some property very different from the first solvent. This will elute each of the partially separated spots from the first chromatographic separation vertically up the plate perpendicular (or orthogonal) to the initial separation dimension. If the nature of the solvent used for the second, orthogonal elution produces different stationary/mobile phase partition ratios for components that coeluted in spots from the first separation, they will now separate in the dimension perpendicular to that of the first chromatographic separation. This is an example of *true 2D chromatography*. It is most easily achieved in the planar surface mode described here. A 2D *column* chromatography instrument, necessary for 2D GC separations, requires more complex hardware. The effluent from one column may be sampled one time, or in continuous segments, by a device that collects, reconcentrates, and redirects it to a second column. The second column must have separation selectivity different from the first one. If the coeluting components in just a *single* peak isolated from the first column are separated on the second one, the procedure is **heart-cutting 2D chromatography**. If segments of the effluent from the first column are *continuously* and very rapidly separated by the second column at a pace matching that at which they are sampled, the resulting detector signal may be transformed by computer to yield **comprehensive 2D chromatograms**, displayed as a planar array of spots like those observed directly on the TLC plate. These examples of multidimensional chromatography should not be confused with the use of a spectroscopic detector (e.g., a mass spectrometer at the end of a GC column or an ultraviolet/visible (UV/VIS) diode array spectrometer monitoring the effluent of a liquid chromatograph). Such **hyphenated systems**, so called from their abbreviations such as GC-MS or LC-diode array, are said to add *a second spectral dimension* to the initial chromatographic separation. If components coeluting in a chromatographic peak have sufficiently different spectral characteristics, they may be measured separately in the peak despite their coelution.

12.5 VISUALIZATION OF THE CHROMATOGRAPHIC PROCESS AT THE MOLECULAR LEVEL: ANALOGY TO "PEOPLE ON A MOVING BELT SLIDEWAY"

Molecules undergoing chromatographic separation may be likened to people standing on a moving slideway, such as those that transport people between terminals in an airport. Imagine a very broad moving belt, capable of accepting a large number of passengers across its width, with no side rails to prevent them from stepping on and off along its length. If several ranks of people step on together at one end and rigidly hold their positions, the group will proceed to the end in the formation in which it boarded. This is like a crystalline particle suspended in a mobile phase flowing down a column. The time (t_m) taken to transit the belt or column is the product of the speed of the belt and the length of the belt. If the people are free to mill about while on the belt, like molecules dissolved in a flowing mobile phase, they will tend to spread out along the length of the belt as they transit. The narrow, compact band of people/molecules that was put on at the head of the belt/column becomes increasingly wider, and the degree of spread is proportional to the time elapsed before the exit is reached. In the case of a narrow band of molecules, this spreading occurs by a process of **random diffusion**, resulting in a **Gaussian distribution** about the mean, producing a peak with the **bell-shaped curve** familiar from statistical description. The center, and maximum, concentration of people/molecules still travels at the velocity of the belt/mobile phase. Now imagine that during their progress down the slideway/column, the people/molecules can *step on and off/partition* to the *nonmoving floor/stationary phase*. This happens continuously, rapidly, and repeatedly during the trip. During their visits to the floor/stationary phase, they are delayed from moving toward the exit, so they come off the slideway/column later than if they had spent all their time on it. We can characterize the extent of the effect of the interaction between the people/molecules and the floor/stationary phase by the **adjusted retention time**: $(t'_r) = t_r - t_m$, where t_m is the time it takes people/molecules *not* getting on and off/partitioning to go from one end to the other, and t_r is the time required when these processes take place.

Imagine we have two different categories of people (e.g., people in purple shirts and people in gold shirts) in our initial compact group at the start, and the purple shirts are slightly more inclined than gold shirts to linger on the floor when they step off the moving slideway. On average, the gold shirts will reach the end of the system sooner, and their t_r and t'_r times will be shorter. If the tendency of the purple shirts to linger is sufficiently greater than that of the gold shirts, the two groups will separately exit the slideway (Figure 12.4a). However, do not forget that both groups will mill about and spread out, mainly while on the slideway. If the distance that they spread out is close to the average difference by which they separate while moving down the system as a whole, they will still partially overlap at the end (Figure 12.4b).

Several factors affect how much separation can be achieved:

1. How rapidly does the *milling about/molecular diffusion* spread the bands along the length of the system compared

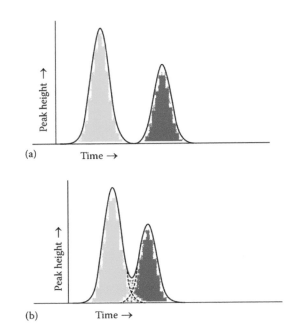

Figure 12.4 (a) Two different groups separately exiting the slideway. The first peak will be all gold shirts and the second, all purple shirts. Note the Gaussian peak shapes. (b) Two different groups exiting the slideway not completely separated. There will be all gold shirts followed by a mix of purple and gold shirts followed by all purple shirts.

with how rapidly the bands separate on it? Suppressing this diffusion improves separation (i.e., the ability of the system to *resolve* a pair of components).

2. How much time do these two processes have to compete with each other (i.e., how much time does a closely separating pair of bands spend on the system)? This depends directly on its length and inversely on the speed of the slideway/mobile phase.
3. How different is the relative attraction of the slideway to the floor, and therefore the relative times spent in each, between the purple shirts and the gold shirts? The greater the difference, the more selective and complete the separation.
4. How far along the slideway does the crowd to be separated extend when it gets on at the start? Even without any further band spreading during movement to the exit, the differential separation process must cause the bands to separate by at least this much before they exit the system to obtain complete separation.

Note that while in the slideway/mobile phase, all people/molecules move at its rate and therefore spend the same time in it. What varies is the amount of time different members of the mixture stay off it in the nonmoving environment. In the chromatography side of this analogy, the variations in the chemical interactions of the mixture components with differing stationary phases vary these times and achieve different separations. Different gas mobile phases in GC have much the same affinity for given compounds. The major variable affecting distribution between the phases is temperature, which controls vapor pressure. This largely controls the proportions in the vapor mobile phase and the "liquid" stationary phase. When the mobile phase is a liquid, a variety of chemical interactions equivalent to or even greater than those of the stationary phases can be chosen to optimize separations.

Let us stretch our moving belt slideway analogy to chromatography a bit further. Assume a very wide belt, capable of holding many people across its width (after all, molecules are very small compared to dimensions in a chromatography column). Another process that could affect band broadening is that the people randomly milling about will sometimes drift toward or away from the edge of the belt where they could step on and off. Sufficiently slow, pressure-driven fluid flow across small dimensions is **laminar** (in smooth straight parallel lines, not swirling with turbulence). In unobstructed narrow columns, its rate is variable across the width of the tube. The flow is stationary at the tube wall and increases to a maximum at the center. This is actually the design for multibelt moving slideways envisioned by science fiction writers for very high traffic loads in some distant future. One could step on and off slow-moving belts at the outer edge and step with small increases in velocity to successively faster belts by moving toward the center. In our analogy, people randomly milling about not only along the direction of travel but across it would undergo additional spreading out from traveling at a range of different forward speeds. If we use a column packed with small particles to support the stationary phase, we are less likely to achieve laminar flow, and additional band broadening factors come into play. Imagine the people on our moving slideway confronting thick obstructing columns hanging down from the ceiling. Some will dodge one way and some another, and the different paths taken during these dodges will cause some people to travel further and to be held up longer than others. In dodging, they may bump one another more than in random milling about and form swirls and eddies of humanity. This spreads the bands even more and works against maintaining the separation of populations achieved by different rates of stepping on and off the belt to the floor.

In our chromatographic analogy, separation of the people by categories is affected only by the relative rates of stepping on and off the belt, working in opposition to processes that spread out or broaden the initial band. On actual slideways, many people often walk purposefully in one direction or another (usually forward). If different categories of people walked forward or backward at specific rates, while some simply stood still, the system could achieve separation without the need for the people to step on and off to the side at different rates. If molecules in an open tubular column with flowing fluid could be induced to behave this way, we could achieve column separations without the actual chromatographic process. That process by definition requires continuous partition of analytes between a mobile and a stationary phase. If some of our analytes are positively or negatively charged species together with uncharged species in solution, the charged species can in fact be separated from the uncharged species and from each other in just this way.

In a procedure called **capillary zone electrophoresis (CZE)**, the *anions* will appear to "walk backwards" at different rates, the *cations* to "walk forward," and uncharged molecules, or those with a balance of positive and negative charged sites (**zwitterions**), will flow along with the solvent for the ride. Many of the same processes of band broadening that apply in chromatography occur here, although not all. For example, the mobile fluid flow is not pressure driven, and its speed is constant across the width of the column, removing one of chromatography's sources of band broadening. Although its mechanism is distinctly different from that of chromatography, electrophoresis as a separation method shares many of its features. Therefore, it will be discussed in Chapter 14, along with other forms of chromatography that employ a liquid mobile phase.

In a perfectly operating chromatography system or its people on a moving belt analogy, the moving people/molecules form gradually spreading, symmetrical bands advancing through the system (Figure 12.5a). One way this can fail is for there to be insufficient room on the floor to accommodate all the people ready to step off or in the stationary phase for molecules to partition into. When this happens, a portion of them will be forced to remain on the moving portion of the system, and they will advance further

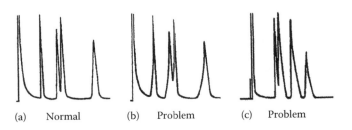

(a) Normal (b) Problem (c) Problem

Figure 12.5 (a) Normal Gaussian peaks shapes, (b) "leading" peak shapes, and (c) "tailing" peak shapes.

until they reach less crowded areas where they can resume normal partitioning. This will distort or skew the shape of the bands, with more people/molecules moving in advance of the highest concentration in the band than behind it. When the band elutes off the column, the resulting peak is said to display a **leading** shape (Figure 12.5b). The greater the **overload** of the stationary phase, the larger the percentage of people/molecules that cannot be accommodated and the more pronounced is this leading asymmetry.

Imagine that on the floor by the moving belt, there are limited numbers of comfortable chairs. Many people may step back on to the belt without approaching one, but those who get close enough to sit down remain for a much longer time than the rest. They will lag well behind people who do not sit down. Depending on the ratio of people to available chairs, the band advancing along the system will lose its symmetry in the other direction; that is, a portion of its members will lag behind. It will be said to display a **tailing** shape (Figure 12.5c). In chromatography, the analogy to our chairs is a limited number of highly retentive sites on the solid that supports the liquid stationary phase where normal equilibrium partitioning of molecules takes place. Both leading due to overloading and tailing due to "active sites" cause additional band broadening and degrade resolution of closely eluting peaks.

In some chromatography, such as Tswett's color bands on the transparent column or the spots on a TLC plate, we measure separated components directly on the separating system. More generally, in column chromatography, a detector that responds to separated components in the mobile phase effluent as they elute off the column gives a signal that is plotted as a **peak**. In our moving belt analogy, we must imagine the **detector cell** as being a continuation of the belt without any more floors for people to step off to. This is different from the end of actual moving walkways, where people step off the end of the belt and walk away. In the case of chromatography of molecules, there are two types of detector cells:

1. The **mass detector**, actually an "amount detector," gives a signal proportional to the total amount of molecules exiting the column per unit time. In our analogy, we might imagine a trap door that opens continuously at regular intervals in the detector cell and drops people in a bin to be counted. If the belt moves more rapidly, more people will be collected and the peak will be higher and narrower. Peak area is independent of mobile phase velocity. If the belt stops, people will not continue to be counted. The signal falls to zero.
2. The **concentration detector** measures the total number of people in a given volume (say that of the detector cell) at a given time. If the belt moves faster, the peak will be narrower, but its height will remain the same. Peak area depends on mobile phase velocity. If the belt stops, the people are still in the measurement volume, and the signal stays constant.

Note: The observed *peak* width depends not only on its band width but also on the rate at which the *band* corresponding to it moves through the chromatographic system (i.e., the mobile phase/stationary phase combination). This is proportional to the rate at which it comes off the column. More retained bands of people/molecules exit the system over a longer time interval, giving rise to peaks broader in time. *This is the major cause of peak broadening*, and it is separate from the process of *band broadening* due to mobile phase diffusion, nonequilibrium partitioning, multipath travel, turbulence, and so on. These latter processes affect whether or how well two closely eluting bands can be separated as resolved peaks as they elute from the system into the detector cell. Contrary to common misconception, they are not the primary reason that later eluting peaks appear broader. This confusion arises from a tendency to speak of peaks and bands interchangeably and to forget that a band has a dimension along the column of *length*, while a peak has a corresponding dimension in the detector signal of *time*.

12.6 DIGRESSION ON THE CENTRAL ROLE OF SILICON–OXYGEN COMPOUNDS IN CHROMATOGRAPHY

Elemental silicon is central to the vast industry of solid-state electronics. Appropriately doped with other elements, it forms a variety of semiconductors that constitute most transistors and integrated circuits. Other elements and compounds such as germanium or gallium arsenide have also found a niche as semiconductors in electronics, but silicon occupies the prime position. How fortunate that it is the second most abundant element in the Earth's crust. Its compounds with the most abundant crustal element, oxygen, are equally central in many different aspects of chromatography. Silica, silica gel, "glass," quartz, fused silica, and silicones all have a remarkable variety of key roles to play in chromatography. Let us familiarize ourselves with some of their relevant properties.

Crystalline silicon dioxide, SiO_2, *quartz*, is a transparent, very high melting, 3D network of silicon–oxygen tetrahedra. When melted and cooled, it forms a noncrystalline, high-melting glass known as *fused silica*. This no longer has the regular crystalline lattice of quartz. Being a glass, it can easily be drawn while molten into long fibers. If other elements, such as calcium, sodium, or boron are mixed in, the lower-melting glasses can be more easily shaped, blown, molded, and so on. Their transparency, moldability, and ability to be ground and polished lead to many applications in optics, which are well represented in optical spectroscopic instrumentation. Their resistance to heat and most chemical reagents makes them the primary material for the vessels in which chemists perform and observe reactions. In chromatography, we shall see that the lower-melting glasses with their guest atoms in the silica structure are generally less desirable than pure silica itself.

Silica can also be precipitated from aqueous solutions of silicic acid, H_4SiO_4. Quartz itself forms in nature from very slow hydrothermal crystallization from hot solutions under great pressure deep underground. A more porous open structure called *silica gel* can be precipitated quickly in the laboratory. Particles of this can be coated as the stationary phase on a glass substrate to prepare one of the most useful types of TLC plates. A vast variety of very pure, monolithic, or porous silica particles, as carefully controlled monodisperse spheres (all about the same dimension, 3–50 μm in diameter), can be synthesized for use as supports for the liquid-like stationary phase of packed columns in GC or LC. At the surface of silica, those oxygen atoms that are attached to silicon by only one bond generally have an H atom as the other bond. The resulting –Si–OH on the surface of silica interacts with water both by polar and hydrogen-bonding interactions to bind it tightly. Thus, unless vigorously heated and kept in a very dry environment, silica has a bound surface layer of adsorbed water. This may serve as the very polar stationary phase in a chromatographic system. Many early simple gravity flow liquid–solid column chromatography systems employed silica or similar particles such as *alumina* (Al_2O_3). Nonpolar compounds could be separated on such a polar stationary phase by eluting with and partitioning between a less polar, water-immiscible mobile phase. This was called **normal-phase** chromatography.

The great majority of molecules we want to separate by LC are polar organic biomolecules, which are more suitably eluted in an aqueous mobile phase against a less-polar stationary phase than "bare" silica or alumina. We can still use the array of specially tailored silica particles. We simply replace adsorbed water by chemically attaching a less-polar organic substituent, such as a long hydrocarbon chain, to the dangling –OH groups on the silica surface. Using this method, we have replaced very polar water adsorbed by hydrogen bonding to the silica surface silanol groups with nonpolar hydrocarbon. Since such columns reverse the relative polarity of the stationary and mobile phases from what used to be considered "normal" for column chromatography, this is called **reversed-phase** LC. It becomes increasingly difficult to force more of these large chains to attach to every last –OH group on the silica surface, so some are left uncovered. Analytes penetrating the less-polar stationary phase layer during the chromatographic partitioning process may burrow down to bind strongly to these "active" sites. They are the "chairs" of our slideway analogy. They can be deactivated and blocked by reacting with a small molecule reagent that attaches as a methyl group, a process called "**end capping**." Glass or silica columns, or particular supports with which a stationary phase is coated but not chemically bonded, are similarly pretreated to deactivate all such sites that could cause tailing peaks. As the pH of an aqueous mobile phase in LC increases (becomes more basic), the higher mobile phase [OH^-] concentration combines with the H on the surface of –Si–OH and at high enough pH, slowly dissolves and destroys the silica surface and its attached stationary phase. To operate at high pH, a more resistant substrate, such as *zirconia* (ZrO_2), or the use of polymer resins, is required.

Chromatographers learned that coating a stationary phase inside a long narrow tube of *capillary dimension* (i.e., internal diameter about 0.1–1.0 mm) yielded exceptionally efficient separations, especially for GC. Metal capillaries were replaced by glass, which was more inert. Glass still had some activity due to its dissolved metal ions and was brittle and broke easily when being installed into instruments. The billion-dollar fiber-optic communication industry learned to draw fibers from rods of extremely pure silica, which absorbed light much less than glass and was much more flexible when incorporated in fiber-optic cable bundles. If an open tube of such pure fused silica were drawn by similar machinery, an inert, flexible capillary perfectly suited for supporting a chromatographic stationary phase was formed (Figure 12.6). A deliberate scratch across the capillary allows it to be broken cleanly by hand to a desired length. Protection against inadvertent scratches that might induce unwanted breaks is provided by fusing a high-temperature-resistant **polyimide coating** to the outside of the capillary. As a result, instead of appearing clear, fused silica capillary tubing for chromatography has a brown to golden hue.

The separation technique of CZE has no stationary phase deposited or bonded to the interior of the capillary. As discussed in Section 12.4, it is not a chromatographic process, but it yields a separation into bands or peaks in a fashion that looks much like chromatography. The –Si–OH on the surface actually begins to lose some H^+ at pH above 2, so the silica surface becomes more and more negatively charged with –Si–O^- units as the pH rises. As we will see, in capillary electrophoresis (CE), a high electrical voltage between the ends of the capillary causes charged analytes to move in different directions at different speeds through the aqueous solvent. Because of the charge on the silica surface, the solvent as a whole moves from one end of the capillary to the other, carrying the separating charged analytes past a detector and out of the column, even those that move in a direction opposite to the solvent flow!

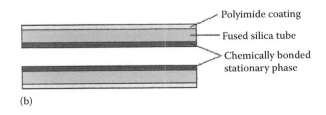

(b)

Figure 12.6 Schematic of a fused silica open tubular column for chromatography.

When charges flow through a CE column, great frictional heat is released. This could result in excessive band broadening and loss of resolution. Fused silica has much higher heat conductivity than glass, and the small diameters and thin walls of the capillaries enable this heat to be efficiently dissipated to a surrounding bath. Conversely, when used for GC, a fused silica chromatographic column rapidly equilibrates with temperature changes produced by the column oven.

Finally, chains of alternating silicon and oxygen form the backbone of a class of polymers called *silicones*, whose temperature stability and variable polarity make them the workhorse of GC stationary phases. Their general structure is

$$R_3-\underset{\underset{R_2}{|}}{\overset{\overset{R_1}{|}}{Si}}-O-\underset{\underset{R_2}{|}}{\overset{\overset{R_1}{|}}{Si}}-O-\underset{\underset{R_2}{|}}{\overset{\overset{R_1}{|}}{Si}}-O-\underset{\underset{R_2}{|}}{\overset{\overset{R_1}{|}}{Si}}-O-\underset{\underset{R_2}{|}}{\overset{\overset{R_1}{|}}{Si}}-R_3$$

where R_1, R_2, and R_3 are organic groups.

If R_1 and R_2 are both methyl groups, this is **polydimethylsiloxane (PDMS)**, the most widely used silicone polymer. If the chains are much longer than the 5-unit one illustrated here, the material is very viscous and nonvolatile. It can be coated on silica particles or on the wall of a fused silica capillary. It will dissolve small molecules and partition them to liquid or gaseous mobile phases. Changing the R-groups will affect the polarity of the polymer and thus the relative separability of pairs of molecules dissolving in it. The silicon–oxygen backbone is more resistant to higher temperatures than many other polymers used as stationary phases. At high enough temperatures, the ends of the chains can cyclize with themselves to successively shorten the chain by losing four- or five-membered rings of $-Si(R_x)_2-O$. Substituting a bulky aromatic hydrocarbon group for occasional O atoms in the chain can interrupt this decomposition and make a phase with a higher temperature limit. Reactive functional groups on some of the R-groups can cross-link the polymer chains and bond them to $-Si-OH$ on the silica support to further increase phase stability. The remaining $-Si-OH$ on the silica surface can then be reacted with reagents like trimethylchlorosilane to deactivate them to $-Si-O-Si-(CH_3)_3$.

The majority of HPLC columns are packed with very highly engineered, spherical, **monodisperse** (very narrow diameter range) silica support particles, 1.7–10 μm in diameter, with an interior network of much smaller-scale passages or **pores** permeating them (cf. Figure 14.2). These greatly increase the surface area for bonding the adsorption coatings, thereby increasing analyte capacity. The depth that analytes must travel in and out of these pores adversely affects the equilibration of these materials with the flowing liquid mobile phase and degrades column resolution. A further refinement of particle structure puts a porous layer (~20 %–40 % of its radius) on the surface above an impermeable solid core. Such **core–shell** particles have much better resolution than fully porous particles of the same diameter, while retaining adequate sample capacity. They can be operated at lower LC pump pressures than columns packed with the smaller particles while approaching the latter's performance. Columns using particles at the lowest end of the earlier diameter range or with core–shell structure give rise to such improved performance that they define *ultra*high-performance LC (UHPLC), which has already taken over more than 50% of the LC market.

12.7 BASIC EQUATIONS DESCRIBING CHROMATOGRAPHIC SEPARATIONS

We will state or derive a bare minimum of the equations most useful for understanding and describing how the most easily measured or controlled variables affect separations in chromatography. Many of the factors in these equations can be derived or calculated from more fundamental parameters, such as diffusion coefficients of analytes in the two chromatographic phases, column dimensions, or variables defined in the statistical theory of random variation. Such details are covered in more advanced texts.

The chromatographic process for a component X is governed by its **equilibrium partition ratio**, K_x (also called the *distribution ratio*). This is the ratio of its concentrations at distribution equilibrium between the two immiscible phases [(A) in Equation 12.1]. In chromatography, these immiscible phases are the stationary phase (s) and the mobile phase (m). In the chromatographic separation process, we are not interested so much in knowing the ratio of concentrations in each phase as in the ratio of the amounts. These are related to volumes of each phase at each point along the separation path [(B) and (C) in Equation 12.1]. We define the amount ratios and volume ratios in expression (C) by the new terms in (D), namely, k' and β. The ratio of the amounts distributed at equilibrium, k', is called the **capacity factor** or **retention factor**. It tells us the relative capacity of each phase in the particular chromatographic system, under the particular operating conditions (e.g., temperature in GC), for each particular compound. Instead of measuring these directly in the phases, we most often infer them from measurements of retention times, thus the use of "retention factor." The β factor is the (volume) phase ratio. Note that the terms for (s) and (m) are in opposite order in these two ratios:

$$\overset{(A)}{K_x = \frac{[X]_s}{[X]_m}} = \overset{(B)}{\frac{\text{Amt}\frac{X_s}{V_s}}{\text{Amt}\frac{X_m}{V_m}}} = \overset{(C)}{\frac{\text{Amt }X_s}{\text{Amt }X_m} \times \frac{V_s}{V_m}} = \overset{(D)}{k' \times \beta}$$

(12.1)

Manufacturers of open tubular GC columns often specify **column radius** r_c and stationary phase **film thickness** (d_f), which is usually much smaller. Simple geometry enables us to calculate the phase ratio:

$$\beta = \frac{r_c}{2d_f} \text{ (the limit when } d_f \text{ is much smaller than } r_c\text{)} \quad (12.2)$$

For packed GC columns, one might have weighed the stationary phase applied to the particulate support, calculated its volume knowing its density, and measured the gas flow rate through the column and combined it with the time for an unretained component to transit the column to get the mobile phase volume. The gas volume flow rate can be measured with a soap bubble flow meter. This is a simple handheld glass cylinder with volume calibrations along its length, similar to a burette. Instead of a stopcock at the bottom, it has a medicine dropper-like squeeze bulb, filled with soap solution that can create a flat "bubble" above the bulb and above a side port that is attached to the column gas effluent flow. The bubble rises through the tube and the time it takes to transit two volume markings is obtained with a stopwatch, and the volume flow rate is calculated. Many modern GC instruments now feature electromechanical flow controllers by which the gas flow can be set, measured and displayed, and held to a constant optimum value as the oven temperature scans through a wide range. The manual bubble meter used on simple instructional GCs is much cheaper to operate. In Section 12.7 and Figures 12.9 and 13.9, one can see the justification for controlling the (linear) mobile phase flow rate to achieve optimal column efficiency. In the case of LC columns where a molecular-scale layer of stationary phase is bonded to an unknown surface area of support, stationary phase volume is very difficult to estimate.

Figure 12.7 illustrates a very simple chromatogram with a single retained component peak. The y-axis represents detector signal response, while the x-axis has either mobile phase volume eluted or time, increasing from left to right. In chromatography, the more accessible measurement is the retention time of a component, t_r, which is measured from its introduction at one end, until elution of the peak maximum at the center of the eluting band. In the figure, point O is the start. An **unretained component** (i.e., one that does not partition to the stationary phase and so travels through the system at the speed of the mobile phase) gives a small peak at point D at time (t_m). This could be air, methane, or dichloromethane in GC depending on the detector and conditions, while in LC, it might be a large ionized dye molecule that does not interact with the stationary phase. The retention time of the component peak is t_r, at point B. Tangents at maximum slope to the peak intersect the **baseline** at points A and C. Since the peak merges imperceptibly at each end with the baseline, time segment AC can be used to define the **peak width at base** (w_b). If AB = BC, the peak has perfect Gaussian shape. If AB > BC, it displays some degree of **leading** or **fronting**, often associated with stationary phase overload. If AB < BC, it displays a **tailing** character, associated with some adsorption on active sites (cf. Figure 12.5).

How much a component is retained by the stationary phase is better described by $(t_r - t_m) = t'_r$, which is defined as the "**adjusted retention time**" (i.e., we count time from the introduction on the column, and then adjust it by subtracting the time it would take to flow through the column if no stationary phase interaction occurred). The capacity or retention factor described in Equation 12.1 can be calculated from these time measurements using Equation 12.3:

$$k' = \frac{(t'_r)}{(t_m)} = \frac{(t_r - t_m)}{t_m} \quad (12.3)$$

If a component is unretained, $k' = 0$. If $k' < 2$, all peaks are packed together too closely at the beginning of the chromatogram. If k' is between 2 and about 10, good separations are observed. If $k' > 10$, the peaks are separated, but they take too long to elute and have become wider and lower, and if close to baseline noise in magnitude, they may become difficult to distinguish from it. These wide peaks with high k' have spent the same amount of time in the mobile phase, where most band spreading takes place, as

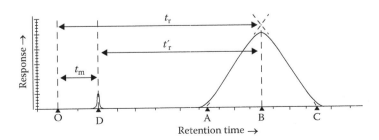

Figure 12.7 Chromatogram with a single eluted peak; peak position and shape terms.

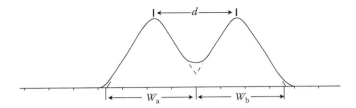

Figure 12.8 Closely resolved peaks of equal size; parameters defining resolution (R_S).

have the narrower ones eluting earlier. Their greater width is due to slower elution of their band off the column to form the peak.

The amount by which we can separate two components as peaks in time or bands in distance on a chromatographic system is determined by the selectivity of the stationary phase. It is defined for specific conditions, namely, temperature in GC or mobile phase composition in LC. Selectivity, α, is related to the ratio of partition constants of two components, A and B, or their more easily measured adjusted retention times, as illustrated in Figure 12.8 and calculated from Equation 12.4:

$$\alpha = \frac{K_B}{K_A} = \frac{(V'_R)_B}{(V'_R)_A} = \frac{k'_B}{k'_A} = \frac{(t'_r)_B}{(t'_r)_A} \quad (12.4)$$

When the selectivity is close to 1, the peaks elute closely together, as illustrated in Figure 12.8. They may even overlap one another; in which case we say that they are incompletely resolved. This concept of **peak resolution**, R_S, is put on a quantitative basis by ratioing twice the **separation d** between two closely eluting peaks (determined by the column's selectivity), to the sum of their two widths at the base, w_b (Figure 12.8):

$$R_S = \frac{2d}{(w_b)_A + (w_b)_B} \text{ or } \frac{d}{w_b} \text{ for adjacent peaks of equal area and width} \quad (12.5)$$

Resolution is a measure of the efficiency of the column. Another parameter that defines the efficiency is the **number of theoretical plates**, N. Review the discussion of plates in chromatography at the end of Section 12.2. The determination of N from measurements of retention time and peak width in time units will not be derived, but will simply be given here in Equation 12.6. An advantage of N as a measure of efficiency is that it may be calculated from measurements on a single peak. Unlike R_S, it does not require a pair of peaks and is independent of their relative selectivity α. It may be calculated using either the peak width at base, w_b, or peak width at half the peak height, w_h. The latter is sometimes better, as w_b can often be subtly broadened by overloading (leading) or the effects of adsorption (tailing):

$$N = 16\left(\frac{t_r}{w_b}\right)^2 = 5.54\left(\frac{t_r}{w_h}\right)^2 \quad (12.6)$$

The plates may be imagined as distances along the column where a complete equilibration of the sample between the two phases takes place. The greater the number of plates for a given column length, the shorter the **HETP**, H, and the more *efficiently* the column operates. This distance along the column is described as a height, since early chromatographic columns (and fractional distillation towers) were oriented vertically:

$$H = \frac{L}{N} \quad (12.7)$$

where L is the length of the column.

Since d in Equation 12.5 equals $(t_r)_B - (t_r)_A$, we can solve for w_b in Equation 12.6, insert it into the second expression of Equation 12.5, and relate resolution R_S to plate number N as follows:

$$R_S = \frac{(t_r)_B - (t_r)_A}{w_b} = \frac{(t_r)_B - (t_r)_A}{t_r} \times \frac{\sqrt{N}}{4} \quad (12.8)$$

Substitution of appropriate versions of Equations 12.3 and 12.4 into Equation 12.8 and rearrangement results in Equation 12.9:

$$R_s = \frac{\sqrt{N}}{4} \times \left(\frac{\alpha - 1}{\alpha}\right) \times \left(\frac{k'_B}{1 + k'_B}\right) \quad (12.9)$$

Equation 12.9 can be approximated by Equation 12.10:

$$R_s = \frac{\sqrt{N}}{4}(\alpha - 1) \times \left(\frac{k'}{1 + k'}\right) \quad (12.10)$$

where k' = the average of k'_A and k'_B.

PRINCIPLES OF CHROMATOGRAPHY

Equations 12.9 and 12.10 enable us to estimate the degree of resolution of two closely eluting peaks if we have information on the *column's* **selectivity** (α, Equation 12.3) and **efficiency** (N, Equation 12.6) and the *peaks'* **average capacity factor** (k', Equations 12.1 and 12.3) under the conditions of the separation.

12.8 HOW DO COLUMN VARIABLES AFFECT EFFICIENCY (PLATE HEIGHT)?

At the microscopic, molecular level, very complex theoretical equations are required to describe the chromatographic process. These include expressions for laminar or turbulent fluid flow; random walk, diffusional broadening of analyte bands in both the mobile and stationary phases; and the kinetics of near-equilibrium mass transfer between the phases. Such discussions are beyond the scope of this text.

Fortunately, Dutch scientists in the 1950s related the performance of columns in terms of H, the HETP, to a single variable, the **linear mobile phase velocity**, u. This could be calculated from column dimensions and volume flow rates or more simply measured directly using the retention time of an unretained analyte, t_m, and the measured column length, L. Three constants, the *ABC*s of chromatographic column efficiency, combine in the **Van Deemter equation** (Equation 12.11), to describe how H varies with u for a particular geometry and construction. Three terms sum together, one independent of u, one inversely proportional to it, and one directly proportional to it:

$$H = A + \frac{B}{u} + Cu \qquad (12.11)$$

Two generalized plots of the Van Deemter equation are illustrated in Figure 12.9. Each of the terms is plotted individually, and their sum is shown as the dashed line with a distorted U-shape. The important feature of the plots is that each chromatographic system will have a *minimum* value for H as a function of u. This value defines the flow conditions that will produce the highest possible resolution. When there is high selectivity of a column for all pairs of analytes, a desire to decrease analysis time may encourage operation at flows above the Van Deemter minimum. The coefficients of the three terms are governed by the following processes:

1. *The multipath term* (A): This term applies to columns packed with support particles. It becomes zero for open tubular columns when the mobile phase velocity is slow enough for the flow to be laminar (i.e., without turbulent eddies). In a packed column, the paths of individual analyte molecules will differ as they take different routes through the spaces between the particles. Thus, they will travel varying distances before they exit the column, and the difference between these distances contributes to band broadening. The relative magnitude of the multipath term depends on the particle and column dimensions. If Figure 12.9 depicts packed columns, A would be a constant value for all values of u and would appear as a horizontal line raising the curve by a constant amount. The multipath process is illustrated in Figure 12.10.

2. *The longitudinal diffusion term* (B): This term accounts for the spreading of molecules in both directions from the band center along the length of the column as a result of random walk diffusion. This occurs primarily in the mobile phase in GC, but significantly in both phases in LC (as the analytes dissolved in the stationary phase behave much as if they were in a liquid). The faster

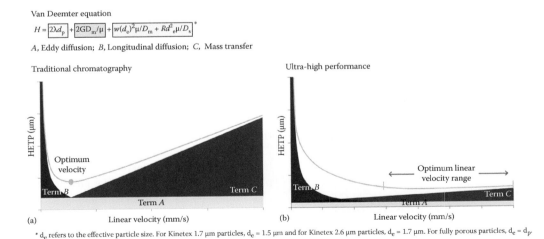

Figure 12.9 Van Deemter plots for packed particle columns as sum of three terms: A, B, and C—(a) is for HPLC with standard size packing and (b) is for UHPLC with small core–shell particles. Note the depression of both the A contribution and flatter right branch for (b). (Used with permission from Phenomenex Inc., Torrance, CA, USA, www.phenomenex.com.)

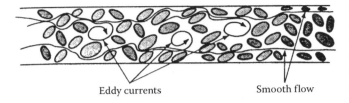

Figure 12.10 Mechanism of origin of multipath term A of Van Deemter equation.

the linear mobile phase velocity u, the less time the analytes have to spread in the mobile phase. Thus, band spreading and H are inversely proportional to u. The B term is much more significant in GC than in LC, since diffusion rates in gases are much higher than in liquids.

3. *The mass transfer term (C)*: Diffusion in both the mobile and stationary phases also occurs perpendicular to the direction of flow. If the distance molecules traveled into and back out of each phase were infinitely small, and the mobile phase flow were infinitely slow, there would be time at every point along the column for the molecules to achieve perfect equilibration between the phases. However, the faster the mobile phase moves, the more some analytes move ahead in it due to insufficient time for diffusion to allow them to come to equilibrium and conversely the further some analytes are left behind in the stationary phase as they penetrate more deeply into it. Thus, this mechanism for band spreading is directly proportional to u. Additionally, under laminar flow conditions in small channels, the flow rate will actually vary from zero at the boundary between the two phases to a maximum at the channel center. Molecules diffusing further away will be carried ahead even faster. In Figure 12.9, we can see how B and C work against each other as u varies to produce a minimum H value. The left-hand plot in Figure 12.9 is representative of packed column GC and to a lesser extent of fully porous particle packed LC. The optimal flow value is defined by the position of this pronounced minimum. The right-hand plot shows the superior performance obtained in UHPLC operation with small, superficially porous (core–shell) particles. Note that the A term contributes less and that the curve minimum is almost flat and remains much lower as flow rates increase above that at the minimum. This is due to the faster equilibration between the narrower porous shell and the mobile phase liquid, and it allows use of faster flows to attain shorter analysis times without sacrificing significant resolution or the improved sensitivity of sharper peaks.

12.9 PRACTICAL OPTIMIZATION OF CHROMATOGRAPHIC SEPARATIONS

As analysts, how do we employ the principles summarized in the equations listed in Sections 12.7 and 12.8? We want to arrive at the best trade-off between optimum resolution and separation of mixture components and the time to complete the separation. Even if time were of no concern, recall from the discussion following Equation 12.3 that components with very high retention times and thus very large k' values elute off the column slowly. This spreads out the peak in the detector, making the signal lower and more difficult to measure above the baseline, thus decreasing sensitivity of detection. On the other hand, if we operate with conditions that make k' and retention times too small, many of the earliest eluting peaks will be crowded together, and we will lose resolution. Note what happens to the value of R_S in Equation 12.10 as k' becomes much less than 2 in the last factor. The first step is to select a stationary phase (and in LC, a mobile phase solvent mixture) that optimizes selectivity for the analytes of interest. The principles governing this choice will be treated in Chapters 13 and 14 for the various types of chromatography. Catalogs and application manuals (now usually online) from suppliers of chromatography columns publish extensive descriptions of optimized separations and the most suitable stationary phases for all classes of analytes. Analysts consult these catalogs or the companies' online websites or their application help phone lines for suggested chromatographic separations of commonly encountered analytes or ones structurally similar to the ones they wish to determine. (Lists of Internet chromatography resources are found in Appendices 13.A and 14.A.)

From Equation 12.7, we see that greater efficiency of the column is indicated by lower values of H. The number of plates is directly proportional to the length L but, from Equation 12.8, the resolution is proportional only to the square roots of either N or L. The retention time increases directly with L. Thus, if all other conditions are unchanged, doubling the column length doubles the analysis time but only increases the resolution by the square root of two (i.e., 1.4 times). We could change the temperature in GC or the polarity of the mobile phase in LC to keep the analytes longer in the mobile phase and thus get them off the column faster, but as k' gets near 2 or below, the last factor in Equation 12.10 rapidly becomes much smaller than its maximum value of 1, and resolution decreases. With many analytes spread out in retention time, bringing later eluting pairs into the region of k' near 2 will force even earlier eluting pairs into the region of severely degraded resolution. Speeding up the mobile phase flow rate causes us to climb up the right-hand side of the Van Deemter curve in Figure 12.9. This results in decreased resolution by increasing the plate height, which thereby decreases the number of theoretical plates. The ability to increase the mobile phase flow rate is also limited by the instrument's ability to supply a column head pressure necessary to increase the flow rate. The more efficient columns with narrower channels for passage of the mobile phase require more pressure per unit length to maintain a given linear flow rate, and the required pressure also increases linearly with increasing column length. For these reasons, going to a longer column should be a last resort to improve resolution. A set of mobile and stationary phases with improved selectivity should be sought first, followed by optimization of k' and u (i.e., operating close to the minimum of the system's Van Deemter curve).

Increasing column temperature often improves efficiency in LC systems, because the resulting decrease of the C term is greater than the increase of the B term (i.e., improvement in mass transfer outweighs the effect of diffusive band broadening in liquids). The decrease in liquid viscosity with increasing temperature means less pressure is required to achieve optimum values of u if longer columns are used. To properly employ a liquid mobile phase at elevated temperature, one must be careful to assure that it is preheated to the desired operating temperature before entering the column. If not, a temperature and flow rate gradient will be set up across the width of the column that will act to broaden bands and reverse the desired resolution improvement. Contrary to our intuitive understanding that increasing temperature lowers the viscosity of liquids, the effect is the opposite with gases. Additionally, temperature is the primary variable that affects retention of analytes in GC, so it cannot be independently changed to affect resolution only through the terms in the Van Deemter equation that affect plate height and efficiency. In GC, increasing the temperature increases the mobile phase viscosity and thus decreases flow rate. Even though analytes will come off more quickly because k' becomes much smaller, u will have decreased. If we were originally operating with optimum efficiency at the Van Deemter minimum, the smaller u will shift operation to the steeply rising left-hand branch, resulting in a significant loss of resolution. In GC, we often employ "temperature programming" (see Section 13.6), raising the column temperature at a constant rate during the analysis to more rapidly elute mixtures of analytes with a wide range of boiling points. In these circumstances, it is advisable to start with a u value on the less steeply rising right-hand side of the Van Deemter curve, so that raising the temperature during the run causes u to shift back toward, but not much past, the position of the curve's minimum.

12.10 EXTRA-COLUMN BAND BROADENING EFFECTS

In Sections 12.7 and 12.8, we have seen how interacting parameters of column efficiency and linear flow velocity (the Van Deemter equation), selectivity, and capacity factor (related to the distribution of analytes between phases under operating conditions) affect our ability to resolve mixture components. These descriptions apply to columns generating ideal symmetrical Gaussian peaks. We know that processes of overloading or adsorption on active sites can lead to peak fronting or tailing, which degrade this ideal resolution. We have not treated these processes occurring in the column quantitatively. There are other sources of band or peak broadening that occur "outside" the column that we need to take care to avoid. Working backward from the chromatographic instrument's detector to its injector, we may lose resolution in the following ways:

1. *Mixing in too large a detector volume*: If the volume of the detector is much larger than the volume of mobile phase carrying a small fraction of the eluting band, the resulting peak will begin to broaden as the differing concentrations eluting from the band mix together in it. In an extreme case, two bands just resolved on a column could completely mix in a large-volume detector cell and all separation could be lost. A classic example of this is using a capillary GC column with very high resolution and low flow rate with a large volume detector designed to handle the much higher flows and broader peaks associated with packed column GC. This mixing can be suppressed by supplying a flow of **makeup gas** to the detector to sweep the column effluent through it rapidly before it has time to mix significantly. The resulting dilution of the effluent may lower the sensitivity of detection relative to that which could be obtained with a detector specifically designed to handle low capillary flows.

2. *Open space within a packed column*: While this is not strictly "extra-column," the effect is very similar to what happens when bands enter a large detector and lose resolution by mixing. This will happen when voids or openings in the bed of particles in a packed column are present. The problem occurs most frequently in HPLC separations, where the packing particles are of micron dimension and the bed is subject to large fluctuations in pressure during pump start-up and shutdown. If the bed is not tightly packed between supporting frits, **voids** can develop at either end or elsewhere. **Channels** may form along the walls of the column if the packing bed is not of uniform density across the entire diameter. Dissolved gases (air) may form bubbles as pressure drops along the column. Recall what happens when a pressurized carbonated drink container is opened. Even a few small **bubbles** can disrupt the packed bed. Initially, these could impede mobile phase flow, but if the gas in the bubbles redissolves in the liquid mobile phase, a void that acts as a remixing chamber within the column may remain.

3. *Injector spreading and tailing*: Mixtures deposited on the head of a column prior to elution and separation exist as a band of finite thickness. If this band is broadened beyond its ideal possible minimum when it is put on the column, maximum resolution performance from the column will not be attainable. In GC, the sample is vaporized in a hot injection port and then swept onto the column by the mobile phase flow. The analyte mixture is often dissolved in a much lower boiling solvent. If the vapor volume of solvent and analytes resulting from flash vaporization in the hot injector exceeds the capacity of the injector port, some sample may be blown back into unheated portions of the gas lines where it will continue to seep slowly back into the system. If bands of some components deposited at the head of the column move significantly down the column while more sample

is still entering the column from the injector, they will start with a larger than ideal width. Many chromatographic procedures load the sample from the injector under conditions where k' is very high. Little movement of the bands takes place until all the sample is on the head of the column, and an ideal minimum initial bandwidth is achieved. Then the temperature is raised in GC, or the mobile phase polarity in LC is changed to more closely match that of the analytes, and movement and separation of these initially narrow bands begins.

12.11 QUALITATIVE CHROMATOGRAPHY: ANALYTE IDENTIFICATION

Each compound has its own characteristic **retention time**, t_r, in a chromatographic system. This is useful in identifying compounds, but in general, it is not as specific as the compound's mass, infrared (IR), UV/VIS absorption, or nuclear magnetic resonance (NMR) spectra. The t_r depends on the compound, the stationary and mobile phase compositions, the mobile phase flow rate, and column dimensions. It is generally not practical to predict t_r for an analyte precisely from all these parameters nor to maintain reference data that will enable an analyst to identify a compound from its t_r the way we do from interpretation or comparison of spectra. Further, within the range of $k' = 2$–10, where resolution is optimal and t_r is not unreasonably long, there is space for only several hundred resolved peaks at best. There are thousands of compounds that could elute in this region, so coelutions are highly possible, and a t_r is not a unique identifier of only one compound.

We can, however, use t_r to indicate the absence of a detectable level of a compound if there is no peak at the retention time we have determined for the compound on our system by separately running a standard. How closely must the t_r of an unknown and a standard peak match for us to say that they could be the same compound? A slight difference in t_r between the two runs might be due to a shift in flow, temperature, or mobile phase composition between the runs. Spiking the standard into the unknown solution and chromatographing them together at the same time and observing a single undistorted peak eliminate this uncertainty. Another way to "factor out" instrumental variations from run to run is to coinject a **retention time internal standard (IS)** with both unknowns and standards. This compound should have a t_r different from any analytes to be measured or contaminants that might elute closely enough to be mistaken for it. **Relative retention times (RRTs)** can be calculated by ratioing the t_r of the standards to that of this IS. These RRTs are more reproducible and may even be compared between columns with the same stationary and mobile phases and operating conditions but having slightly different dimensions. Use of RRT values avoids the need to confirm identity by coelution with spiked standards when there are many peaks to be identified in a chromatogram.

Additional certainty in using t_r values to assign peak identity is obtained when retention times match on two columns with significantly different stationary phases, or a single planar stationary phase is eluted by orthogonal, sequential application of two mobile phases. This is the principle and application of 2D chromatography discussed earlier in Section 12.3. Many analyses of complex environmental mixtures specify splitting an injection to two different capillary GC columns operating in parallel to separate detectors. An analyte is considered to be measurable only if it elutes with the appropriate RRT on both columns and the measured quantities of it on each column differ by less than a specified amount. The latter condition ensures that there is no significant coelution with yet another compound on one of the columns. The second column in such a pair is called the *confirmation column*.

Even more certainty of peak identification is obtained by using detectors that are **selective** for certain classes of compounds (e.g., only those with particular heteroatoms, such as halogens, N, P, and S). These will be discussed in greater detail in Chapters 13 and 14. Even more selectivity can be obtained by the so-called hyphenated detection techniques. Here, the effluent is delivered to a spectrometer (e.g., mass spectrometer, IR spectrometer, UV/VIS spectrometer, or NMR), and full spectra or characteristic regions from them are sampled continuously and stored in computer data files as the peaks elute from the column. This provides the best of both worlds (the instrumental analytical worlds of chromatographic separation and of unique spectral fingerprints of pure separated compounds). If their spectra are sufficiently different, it is often possible to identify and quantitate compounds, which the chromatograph has failed to separate and which coelute in a single peak. Recall that there is only limited "retention space" in any chromatographic system, and there are thousands of compounds that could possibly elute and coelute there. Conversely, in MS, for example, there are isomers, such as *ortho-*, *meta-*, and *para-*xylene (dimethyl benzenes), which fragment to produce identical mass spectra, but which can be distinguished by their different t_r values in a chromatographic system. Such instrumentation is complex and expensive, and it demands computerized data handling, but it dominates instrumental analytical chemistry in the fields of biomedical and environmental monitoring and research. The techniques for interfacing chromatographs to spectrometers and acquiring and processing the resulting flood of data are discussed at the ends of Chapters 13 and 14.

12.12 QUANTITATIVE MEASUREMENTS IN CHROMATOGRAPHY

Quantitation may be done crudely on the spots separated by planar chromatographic techniques such as TLC or slab gel electrophoresis (see Chapter 14). One might compare the optical density, the fluorescence, or the degree of stationary phase fluorescence suppression by the unknown spot to a series of standards of known concentration. In contrast, the electrical signals from the variety of detectors used in various column chromatography instruments can be precisely, reproducibly, and linearly related to the amount of analyte passing through the detector cell. If all parameters of injection, separation, and detection are carefully controlled from run to run and especially if appropriate quantitative ISs are incorporated in the procedure, accuracy and precision better than ±1% may be attained.

12.12.1 Peak Area or Peak Height: What Is Best for Quantitation?

In chromatography, we measure each of the signals from a series of peaks. Unlike spectrometric measurements where a single stable signal level is proportional to the amount of analyte, the peak signal rises from the baseline (background) level to a maximum and then returns to baseline. The amount of analyte is proportional to the area of the peak. The instrument or the operator must decide when a peak starts and ends, how to estimate the position of the baseline beneath the peak signal, and then to integrate the signal level above the baseline between the start and end points.

In earlier instrumentation where the signal was displayed as a pen trace from a chart recorder, the baseline was drawn manually. The area of a symmetrical Gaussian peak could be approximated by drawing tangents to each side of the peak and calculating the area of the resulting triangle formed by these tangents and the baseline. This is illustrated in Figure 12.7. A more accurate but laborious method involved actually cutting out the peak areas from the chart paper and comparing their masses obtained on an analytical balance. Most chromatographic instrumentation now employs computerized data systems. The analog signal level is continuously sampled and digitized, and a summation of these digital "counts" during the peak's elution, after subtraction of levels calculated from a hypothetical baseline between the peak start and end points, produces a peak area value. The operator of the data system must select a "trigger level" of baseline rise that signals the start of a peak. Another one after the descent from the peak maximum level determines either that the signal has returned to a level value (baseline) or is ascending again with the elution of a following peak. These tell the system when to start and stop peak integration. At analyte levels close to the instrumental limit of detection, baseline noise (i.e., random small variations in the signal) will increasingly corrupt the proper functioning of these triggers. If they are set high enough not to be incorrectly tripped by random noise, the baseline will be drawn too high, and a significant portion of the peak will not be integrated. If they are set too low, they may be tripped by noise, and integration could be either started or cut off too early. These problems of setting parameters for computerized peak integration are often the primary source of chromatographic standard curve nonlinearity near the lower limit of quantitation.

Peak heights are easier to measure either manually or in a computerized method. The signal level at the top of the peak has the greatest S/N ratio, and if closely eluting peaks are incompletely resolved, the tops are the easiest points to locate. The problems associated with proper estimation of the baseline above which the height is to be measured are similar to the case for peak areas, but heights only have to be evaluated at the single point on the baseline beneath the peak maximum. For well-separated peaks of ideal Gaussian form, the peak height is directly proportional to the peak area. However, as the peak shape becomes more and more distorted due to leading or tailing processes, the relation no longer holds. This introduces nonlinearity into standard curves of peak height versus concentration, which would not occur if properly measured peak areas were used instead of heights. Therefore, peak area measurements are preferred whenever possible, despite their greater complexity. The only common exception may be when measuring a small peak eluting on the front or back of a larger one. In such cases, a data system may not be capable of being programmed to accurately start and stop integration and draw an appropriate baseline for the small peak. Printing the peak pairs and manually drawing and measuring baselines and peak heights of the unknown and appropriate standards may enable a quantitation calculation.

Review the description of the two different types of detector (mass detector or concentration detector) at the end of Section 12.4. Note well that the area that will be calculated for a peak measured by a mass detector is independent of mobile phase velocity, but when a concentration detector is used, it is not. When calibrating against standards using a concentration detector, it is important that the analyte peaks in both the unknown and standard runs elute at the same time and under the same mobile phase flow conditions. Conversely, if quantitation is by peak height comparison using a mass detector, the same requirements apply.

12.12.2 Calibration with an External Standard

The same principles that are outlined for calibration in Chapter 3 apply to chromatography. A series of solutions containing analytes at levels covering the concentration range to be measured are injected, peak areas are measured, and **standard curves** of peak areas versus concentrations are produced. In contrast to spectrometric measurements, a well-designed chromatographic separation will move mixture components away from the analyte's retention time, so a blank background subtraction need not be applied to peak area values. It is vital, however, to analyze a blank sample using the same injection solvent, and if possible, one processed using the same sample preparation procedure from a similar matrix not containing the analyte. This should confirm the absence of interfering peaks eluting at the same times as target analytes. If these are found, the chromatographic conditions should be adjusted to separate them from the analytes. Note that the **method of standard additions (MSA)** could only be applied if there were sufficient experience with the samples to warrant the assumption that no interferences eluting at the positions of the analytes would be encountered. Such an assumption will always have some level of uncertainty, so this method of calibration is seldom applied in quantitative chromatography. Since chromatography separates the analyte from most matrix components, its detectors do not often suffer from the matrix suppression effects that this calibration method corrects for in spectroscopy.

The standard curve should be linear over the calibration range. Samples with levels above this range should be diluted to fall within it and then be reanalyzed. Advanced data-handling systems can fit peak heights or areas to nonlinear (e.g., quadratic, logarithmic) standard curves, but best practice is to find the linear response region and to operate within it. The precision and accuracy of an externally calibrated assay are limited by the reproducibility of sample preparation recoveries, sample volume injections, and the stability of instrument detector response over the course of calibration and analysis runs. In chromatography, we may be comparing the detector response from unknowns against those from an external standard curve run hours before.

To quantitate unknown peaks against a standard curve, one plots the response (e.g., area counts from a digital integrator) on the y-axis (ordinate) against the concentration on the x-axis (abscissa). The absolute value of the y-axis intercept of a linear, least squares fit of the data to the points should be a small fraction of the value at the lowest calibrated point on the "curve" (i.e., the response extrapolates to a value close to zero at zero concentration). If so, or if we select a least squares fitting program that forces the line through zero, then we can calculate the concentration C of an unknown from its peak area A and the slope S of the calibration curve using Equation 12.12:

$$C = \frac{A}{S} \quad (12.12)$$

This is only valid for areas and concentrations within the limits of the standard curve calibration levels. While in principle one may calibrate against a standard curve that has a substantial y-axis intercept, this will yield valid results only if the background signal level causing the large intercept is the same in both the standards and unknown sample matrices. This is a dangerous and often unverifiable assumption to make, and the use of such calibration curves should be avoided.

12.12.3 Calibration with an Internal Standard

Some of the factors mentioned earlier, which introduce error when calibrating with external standards only, can be compensated for by introducing a constant amount of an IS in both the unknown and the standard calibration samples. The IS should be a compound with a chemical nature similar to that of the analytes, so that it will pass through the sample extraction and preparation procedure similarly. In general, it should elute in the chromatographic system close to the other peaks in the system, but separated from all of them, so it can be identified and measured accurately. One calculates the **ratio** of the peaks' responses to their concentrations. Then the **ratio** for each analyte peak (A) is divided by the ratio of the IS peak, to give the **relative response factor (RRF)**, for the analyte to the IS, as summarized in Equation 12.13:

$$\text{Ratio } A = \frac{\text{Area } A}{\text{Conc. } A}; \text{ Ratio IS} = \frac{\text{Area IS}}{\text{Conc. IS}}$$
$$\text{RRF} = \frac{\text{Ratio } A}{\text{Ratio IS}} = \frac{\text{Area } A \times \text{Conc. IS}}{\text{Area IS} \times \text{Conc. } A} \quad (12.13)$$

For a given constant concentration of IS added, if we plot the ratio of the analyte peak area to the IS peak area against the analyte concentration A, we should get a standard curve whose slope is the RRF. In a fashion analogous to Equation 12.12 for external standard calibration, we can determine the ratio U of analyte of unknown concentration to IS peak area and calculate the unknown concentration C using Equation 12.14:

$$C = \frac{\text{Ratio } U}{\text{RRF}} \text{ or } C = \frac{\text{Area } \dfrac{U}{\text{IS}}}{\text{RRF}} \quad (12.14)$$

The criteria for acceptable linearity of least squares fit and zero intercept when plotting ratios of analyte to IS areas versus concentration are similar to the case for external standard calibrations described earlier. More than one IS can be used, both for calculating RRTs to compensate for retention time variations as well as the RRFs for improving quantitation. The variations that a quantitation IS can compensate for depend upon the point at which it is introduced in the analysis. If it is put into the final extract prior to injection on the chromatograph, it can correct for concentration variations due to evaporative volume changes, variations in injection volume, and variations in detector response. This is called an **injection IS**. If the IS is put into the initial sample, and into calibration standards prepared in an equivalent matrix, it can additionally correct for variations in recovery during the sample preparation process. This is called a **method IS**. Combined use of separate compounds for each purpose can aid in determining the causes of peak area variability.

12.13 EXAMPLES OF CHROMATOGRAPHIC CALCULATIONS

To illustrate the operation of some of the equations and calibration principles earlier outlined, consider the data from the following hypothetical chromatogram summarized in Table 12.3 (on page 691). This is for a capillary GC column, 30 m long, with a diameter of 0.25 mm, operated at a constant temperature of 200 °C under a helium pressure of 20 psi above atmospheric pressure, resulting in an average linear flow rate of 31.3 cm/s. We measured the time for an unretained component to pass through the column as 1.6 min. We determined the typical number of theoretical plates on the column to be 30,000. (How would you easily measure and/or calculate these last two values?) If the column is 30 m, or 30,000 mm, long, it is trivial to see that the HETP is 1 mm. This is a very efficient column!

There are five analytes, giving peaks 1 through 5, plus an IS peak 6. Observe peaks 2 and 3, eluting at 4.00 and 4.10 min, respectively. Their widths at the base are somewhat more than 5.5 s, and their peaks are separated only by 0.1 min (= 6 s), so they probably overlap a bit near the bottom. This is reflected by the calculated resolution of 1.07. A rule of thumb is that we need a resolution of 1.5 or more to achieve complete baseline resolution. We note that k' is about 1.53 for this pair; a little less than the minimum preferable value of 2 or more. We can improve resolution by lowering the temperature from 200 °C to a value that will increase k' and the resolution. From the last factor in Equation 12.9, we can calculate that increasing k' to 3.06, corresponding to t_r of about 6 min, will increase the resolution from 1.07 to 1.34. Perhaps this will be sufficient. We could double the number of plates by going to a 60 m column. This improves resolution by the square root of 2, namely, 1.4. That improvement, 1.4 × 1.07, gives us a resolution of 1.5. But this is at the cost of purchasing and installing a 60 m column at nearly twice the price, requiring 40 psi pressure for the same optimum flow rate (perhaps out of our instrument's range), and IS peak 6 will now take 16 min to come off before we can run the next sample or standard. It is much easier to adjust the temperature (by lowering it) to get better resolution of the close pair of analytes 2 and 3. Speaking of close pairs, the resolution of 0.5 between peaks 4 and 5 may well cause them to overlap so much that they cannot be separately quantitated. Based on what we could do by lowering temperature or lengthening the column to increase the resolution of 1.07 for peaks 2 and 3, these remedies are unlikely to be practical for resolving 4 and 5. If we need to measure them, we should find a column with a different stationary phase that has a selectivity, α, for this pair, which is not so close to 1. Of course, we must hope that resolution for the other analyte pairs does not decrease too much with the new column. Such are the trade-offs in finding the best column and conditions for a mixture separation. One can easily imagine that the probability of problems and conflicts increases with the number of components we would like to measure.

The second segment of Table 12.3 (on page 691) shows peak area counts for each of the five analyte peaks over a standard range of five concentration levels from 10 to 200 ppm, and the areas of the 50 ppm IS peak 6 in each of those five standards. The ratios of each peak area to the IS area are calculated and displayed in the third segment. We can plot each line of data for each peak against the standard concentration and obtain the slopes and intercepts of the linear least squares fits. If you do this, you will find that

Table 12.4 Use of Retention and Quantitation IS to Correctly Identify Peaks and Correct for Sample Loss or Concentration Changes

Peak	RT (min)	RRT	Area Counts	Slope Ext. Std.	Slope Int. Std.	Ratio (U)
2	4.00	0.500		6,635	0.0235	
A	4.01	0.489	175,000			0.831
3	4.10	0.513		5,435	0.0193	
B	4.12	0.502	194,500			0.924
C	4.24	0.517	254,000			1.207
6 IS	8.00		280,182			
D IS	8.20		210,500			

	Ext. Std. Quant. (ppm)		Int. Std. Quant. (ppm)		
	Wrong ID	Right ID Uncorrected	Wrong ID	Right ID Corrected	Identification
A	26.4	Contam.	35.4	Contam.	Unknown
B	35.8	29.3	47.9	**39.3**	Cpd. 2
C	Contam.	46.7	Contam.	**62.5**	Cpd. 3

there is some small variation of the data points about the line, but the intercept is close to zero and much smaller than the response from the lowest level standard. We can therefore use the slopes, and area or area ratios for unknowns, to calculate concentrations using either Equation 12.12 or 12.14 for external or IS method calculations, respectively.

In the top segment of Table 12.4, we display data for peaks 2, 3, and IS 6 from the standard curve runs and for three unknown peaks (A, B, C) in a separate chromatographic run eluting with similar retention time (RT), plus D, which is the IS 6 peak in that run. In both runs, we use IS 6 as a retention IS to calculate RRT values displayed in the third column of the table. If we only compared RTs (i.e., t_r), we might say A is 2 and B is 3. We note, however, that IS D elutes 0.2 min later than it did in the standard runs. Something has caused a shift in retention between the runs, perhaps the flow has slowed a little, or the GC oven temperature dropped a bit. If we compare RRTs, we see that B and C are better matches to 2 and 3, respectively.

Notice that not only has the retention time of the D IS shifted, but its peak area is significantly lower than the average value from the five standard runs displayed directly above it. If D were employed as a method IS, we may have a recovery difference between the unknown sample and the standards, or perhaps the unknown solution was diluted more during the final step before injection, or perhaps a smaller volume was inadvertently injected, or possibly some combination of all these or other factors. The method IS will quantitatively correct for any or all of these, even if we do not know which may have occurred. The bottom section of Table 12.4 shows the different results that will be obtained with or without correction for either or both the retention and quantitation shifts. The bold values displayed at the lower right are obtained using the IS to calculate RRTs for peak assignment and RRFs for quantitation. Note how they differ from the uncorrected and incorrect values calculated without full benefit of IS retention and quantitation calibration.

PROBLEMS

12.1 In GC, what is the effect on the retention time of a peak (increase, decrease, remain the same, become zero) of
  (a) Raising the column temperature (if head pressure is kept constant)?
  (b) Lengthening the column?
  (c) Increasing the gas flow rate?
  (d) Increasing the volume of stationary phase in the column?
  (e) Increasing the column head pressure?
  (f) Overloading the column with sample?
  (g) Having some stationary phase slowly volatilize from column?
  (h) Forming a more volatile derivative of the analyte compound?

12.2 In LC, what is the effect on the retention time of a peak (increase, decrease, remain the same, become zero) of
  (a) Raising the column temperature (if head pressure is kept constant)?
  (b) Decreasing the particle size in packed column HPLC (at constant pressure)?
  (c) Having a bubble form in the column?
  (d) Increasing the column head pressure?
  (e) Using an open tubular column of the same length with a diameter 5× the spacing between particles in a packed column at the same head pressure?
  (f) Increasing the polarity of the mobile phase in "reversed-phase" HPLC?
  (g) Using a less-polar stationary phase in "normal-phase" HPLC?
  (h) Using a more efficient column with twice as many plates with the same length, mobile, and stationary phase compositions, operated at the same flow rate?

12.3 Can one perform planar GC? 2D GC? How?

12.4 What controls the mobile phase flow rate in ascending TLC or paper chromatography? Do you think it is constant during the separation?

12.5 Distinguish among the terms: elution, eluent, effluent.

12.6 What form of chromatography is SPE?

12.7 Acronym quiz: Identify—HETP, HPLC, TLC, CZE, GC-MS, IS, RRT, RRF.

12.8 Identify and define in words the following formula terms, and give their defining equations from Chapter 12: α, H, k', N, A, B, C, L, t_m.

12.9 For a pair of peaks, what is the effect on resolution (R_s) of increasing one of each of the terms in Problem 12.8?

12.10 What do the Van Deemter equations look like for packed column chromatography, capillary chromatography, and CE (hint: in the last of these, there is no stationary phase)?

12.11 Draw a Van Deemter curve for each of the three techniques in Problem 12.10—the answer should be "schematic," not necessarily quantitative. Explain the differences in appearance.

12.12 (a) Calculate the phase ratio for a 30 m by 0.2 mm diameter capillary column coated with a 0.5 μm film of stationary phase:
  (b) If a similar column uses a 2.0 μm film of the same stationary phase, how will k' change?
  (c) Will t_m differ between columns a and b if operated the same way?
  (d) If t_m for the column in a is 1.5 min and a peak has a t_r on it of 3.5 min, what will be the t_r of that peak on the column in b?

12.13 Under what conditions are peak heights proportional to area? When are they not?

12.14 Calculate how many theoretical plates a column must have to achieve a resolution of 1.5 (baseline) between two peaks eluting at 8 and 8.2 min.

12.15 An HPLC column received from a manufacturer was specified to have 4000 plates measured under standard conditions with a test compound. When the customer repeated the test, she found only 1500 plates, despite a symmetrical Gaussian peak shape. What might account for this?

12.16 In an experiment, an analyst injected a pure compound into a gas chromatograph in volumes increasing from 0.2 to 5.0 µL in 0.2 µL increments. At first, the amount of "tailing" observed in each successive peak decreased, but then started increasing at volumes above 3.0 µL. Account for what happened (*hint*: two different processes dominate at each end of the volume range).

Questions Based on Example in Section 12.13 and Table 12.4

12.17 Using a calculator or personal computer program, use the appropriate data from the second and third segments of Table 12.4 to calculate the slopes for all five peaks used for both the external standard and IS calibration calculation. Check your values against those reported for peaks 2 and 3 in Table 12.4.

12.18 How might you satisfy yourself that peak C and not peak B is actually component 3? Suggest two different ways. Which is easier to implement?

12.19 Describe how using separate injection and method ISs could allow you to separate analyte recovery variations from injection volume variations.

12.20 Suppose the five standard curve mixtures in this example and then two unknowns are injected at 9 min intervals. All six peaks in the five standard levels appear normal, and peaks A, B, and C appear in the first unknown as described in Table 12.4. In the second unknown, the IS is again at 8.2 min, and peaks are seen at the same times as A, B, and C in the previous chromatogram, but there is now a symmetrical Gaussian peak at 4.55 min. with a w_b of 19 s:
   (a) Should we identify and quantitate this as component 5?
   (b) What does the width of this peak tell us?
   (c) Where did it come from?
   (d) How can we confirm our hypothesis as to its origin?
   (e) How should we modify our operating procedure to deal with peaks like this?

BIBLIOGRAPHY

The reader will note that occasionally multiple editions of books are cited as sources or reference works. The reason for this is that sometimes different figures are used in different editions, which may in fact be written by different authors. The sources found in the bibliography will be the most recent versions of a particular work.

In some cases instrument manufacturers and suppliers are cited by providing a URL to the exact internet location. We have endeavored to ensure that these links are functional at the time of publication, but clearly these links change. If a link has become nonfunctional, a good approach is to use your browser's advanced search capability. For example: "company name" AND "instrument type" AND "specific descriptors" is a good place to start.

Ayres, G.H. *Quantitative Chemical Analysis*, 2nd edn. Harper & Row: New York, 1968.

Berthod, A. *Chiral Recognition in Separation Methods*. Springer VCH: Berlin, Germany, 2010.

Bouchonnet, S. *Introduction to GC-MS Coupling*. Taylor & Francis: Boca Raton, FL, 2013.

Bruno, T.J., *Chromatographic and Electrophoretic Methods*, Prentice Hall, Englewood Cliffs, NJ, 1991.

Bruno, T.J., Svoronos, P.D.N., *CRC Handbook of Basic Tables for Chemical Analysis – Data Driven Methods and Interpretation*, 4th Ed., CRC Taylor and Francis, Boca Raton, FL, 2021.

Giddings, J.C. *Unified Separation Science*. Wiley: New York, 1991.

Harris, D.C. *Quantitative Chemical Analysis*, 5th edn. W.H. Freeman: New York, 1999.

Hyötyläinen, T.; Wiedmer, S.; Smith, R.M. *Chromatographic Methods in Metabolomics*. RSC Pub.: Cambridge, U.K., 2013.

Jonsson, J.A. (ed.) *Chromatographic Theory and Basic Principles*. Marcel Dekker, Inc.: New York, 1987.

Miller, J.M. *Chromatography: Concepts and Contrasts*. Wiley: New York, 1987.

Scott, R.P.W. *Techniques and Practice of Chromatography*. Marcel Dekker, Inc.: New York, 1995.

Shalliker, R.A. *Hyphenated and Alternative Methods of Detection in Chromatography*. Taylor & Francis: Boca Raton, FL, 2011.

Skoog, D.A.; Holler, F.J.; Nieman, T.A. *Principles of Instrumental Analysis*, 5th edn. Harcourt, Brace & Company: Orlando, FL, 1998.

Willard, H.H.; Merrit, L.L.; Dean, J.A.; Settle, F.A. *Instrumental Methods of Analysis*, 7th edn. Van Nostrand: New York, 1988.

Wixon, R.L.; Gehrke, C.W. *Chromatography: A Science of Discovery*. Wiley: New York, 2010.

CHAPTER 13

Gas Chromatography

13.1 HISTORICAL DEVELOPMENT OF GC: THE FIRST CHROMATOGRAPHIC INSTRUMENTATION

In **gas chromatography (GC)** (originally called **vapor phase chromatography**), the *mobile phase* is a gas, usually nitrogen, helium, or hydrogen, introduced to the column through a pressure regulator from a cylinder of compressed gas. These mobile phase gases do not have significant solvation interactions with analyte molecules. Initially, the *stationary phases* were high boiling liquids dispersed on a particulate solid support, which was then "packed" into a column of metal tubing 1/16 to 1/4 in. in diameter, generally made of copper or stainless steel, hence the term **packed column GC** or **gas liquid chromatography (GLC)**. A typical coating liquid was squalene, with a relative molecular mass of 422.8. Liquids still do find application in specialized circumstances and are certainly encountered in the literature. A comprehensive listing of liquid stationary phases, including specialized liquid crystalline liquids, can be found in the book by Bruno and Svoronos.

The coated particles were secured by plugs of glass (or Pyrex) wool or porous metal frits at each end. The column (often wound in a coil to fit) was maintained in a thermostatted oven (or even a water bath) set to a controlled constant temperature. Higher relative molecular mass analytes, with higher boiling points, would require correspondingly higher temperatures to partition sufficiently into the gaseous mobile phase to be eluted in a reasonable time. True liquid stationary phases would eventually begin to vaporize slowly at such elevated temperatures and be carried off the column as well, a phenomenon referred to as column **bleeding**. It is interesting to note that the interaction of a solute with this liquid is a true solvation process, characterized by the same enthalpy of solution that can be measured by calorimetry. In time, such liquid stationary phases were replaced with very high molecular mass polymers, especially long-chain, substituted silicones (cf. end of Section 12.6), which would behave like liquids to the analytes but which were stable at higher temperatures. At high enough temperatures, these would eventually bleed as well, but instead of volatilization of intact molecules of the stationary phase, this was now due to stepwise loss of small fragments from the ends of the polymer chains. Mixtures of volatile liquids (e.g., the components of a petroleum oil fraction, mixed products of an organic synthesis procedure, lipid mixtures) would be sampled with a **syringe** and injected on the GC column for separation and detection of separated components. The mobile phase gas, called the **carrier gas**, would be introduced to the head of the column into an unpacked open space above the restraining frit or plug. This empty volume lay beneath a plastic or rubber **septum**, which sealed the top of the column and through which the syringe needle could be inserted. The liquid mixture was injected, vaporized at the elevated oven temperature, and swept along by the carrier gas to initially pile up as a narrow band on the beginning of the packing. Its components subsequently progressed more slowly, at different rates along the packed portion of the column, and the chromatographic separation ensued. If the carrier gas were helium or hydrogen, these very low molecular mass gases had much higher thermal conductivity than almost any analyte they might carry off the column. A simple detector at the effluent end of the column produced a signal proportional to the amount of analyte in each eluting peak. This resulted from the increase in electrical resistance of a wire placed in the effluent stream and heated by passage of an electrical current. This was initially thought to be a consequence of its increase in temperature due to the decrease in thermal conductivity and thus cooling power of the carrier gas stream, which was caused by the presence of the heavier analyte vapors. So, GC had a nearly universal detector that readily produced an electrical signal that could be plotted on a chart recorder to produce a **chromatogram**. A very simplified block diagram of a generic GC instrument was depicted in Figure 12.1.

The general concept of GC was put forward in a Nobel Prize-winning paper by Martin and Synge in 1941 and implemented by James and Martin in 1952 under the name "vapor phase chromatography." The reasons for GC being the initial realization of modern instrumental chromatographic techniques lie in the relative ease of achieving several of the key components:

1. Only low pressures (5–250 psi) of gas were needed to achieve flow rates suitable for optimal separations. Such pressures could be contained by simple metal or glass tubing and standard compression (nut and ferrule) fittings. Compressed gas cylinders using standard two-stage regulators easily provide gases at these pressures.
2. Because diffusion is so much more rapid and viscosity so much lower in gases than in liquids, mass transfer equilibrium between stationary and mobile phases in GC was easily achieved while using large-dimension packing particles, in the range of 100–1000 μm (i.e., 1 mm). It was not necessary to use complex pumps to achieve suitable pressures, as was required for the later realization of instrumental liquid chromatography (LC), which is performed in packed beds of much smaller particles (1–10 μm).
3. One parameter for controlling partition between the phases, namely, temperature, was easily varied and maintained by readily available thermostatted oven technology.
4. A simple, universal detector, the **katharometer**, or **thermal conductivity** detector (TCD), applicable to chromatography with analytes eluting in gas phase, was available to provide an electrical signal that could be displayed as a chromatogram.

13.2 ADVANCES IN GC LEADING TO PRESENT-DAY INSTRUMENTATION

If the dimensions of stationary phase coating thickness and diffusion distance to the film from the gas phase have been optimized, for further improvement of GC resolution it becomes necessary to increase the length of the column. This is seen from the simple relation of Equation 12.7, which indicates that for an optimal minimized value of H (the height equivalent to a theoretical plate [HETP]), the number of plates, N, is proportional to the length of the column. From Equations 12.8 and 12.9, we note that the resolution is proportional to the square root of N. For columns packed with particles of optimal size, and operated at the optimal linear flow rate at the minimum of the Van Deemter curve (Figure 12.7), the typical maximum pressure of ~100 psi achievable from a regulated gas cylinder requires that most packed columns be less than 4–7 m long. More typically, they are only 1–2 m in length. These considerations limit the resolution achievable in packed column GC.

In 1958, Marcel Golay demonstrated that coating the stationary phase on the inner walls of narrow-bore tubing of capillary dimension (generally in the range of 0.1–0.5 mm diameter) could allow columns up to 50 or even 100 m to be operated within these pressure ranges. When the flow rate was optimized to a minimum of the open tubular column relation, similar in form to the Van Deemter equation (Equation 12.10), the H values were similar to or even better than those obtainable from the best packed columns, so the much longer open tubular or "capillary" columns could have much larger values of N and consequently much greater resolution. Since they have no packing, the H values for capillary columns can be smaller because there is no contribution from a multipath A term as in Equation 12.10.

A number of additional advances in GC instrumentation were necessary before the advantages of capillary columns over packed columns could be widely employed:

1. The actual amount of stationary phase per unit column length on a capillary column is much less than that supported on the particles of the typical packed column. Therefore, the capacity for analytes of the capillary is much less than that of the packed column. When the capacity is exceeded, the analytes will spread out over the front of the column after injection, and the improved resolution of the capillary will be lost. This results in the phenomenon of **leading** or **fronting** peak shape described in Section 12.4. In capillary GC, one must decrease the concentrations of the injected analytes, often by dilution with solvent to levels that are at concentrations of only the part per thousand or ppm level. For this to be practical, one must have detectors orders of magnitude more sensitive than the TCD described in Section 13.1.
2. Inventors developed a variety of "ionization detectors" based on measurement of the tiny currents of ions or electrons induced or suppressed when analyte peaks eluted. One, the flame ionization detector (FID), was almost universal, giving sensitive and generally similar responses for almost all carbon-containing compounds capable of eluting through a GC system. Others were even more sensitive and displayed selectivities for certain classes of organic compounds, such as halogen or other electronegative substituent-containing compounds (electron capture detector [ECD]), nitrogen- or phosphorus-containing compounds (nitrogen–phosphorous detector [NPD]), or sulfur- and phosphorus-containing compounds (flame photometric detector [FPD]). These and several others will be discussed at length later on. The key point is that the sensitivity of these detectors was sufficient to measure the low levels of analytes that were within the column capacity of capillary columns.
3. The typical GC syringe injection volume of 0.5–2 μL of liquid solvent typically expands to 0.5–1.5 mL vapor volume, depending on injector temperature. This is a volume increase on the order of 1000×. A capillary GC column is operated at a much lower flow rate than a packed column (typically on the order of 1 mL/min instead of 20–40 mL/min). The handling of the volume of vaporized solvent and analyte(s) in capillary GC required a more complex control of gas flows in and out of the injector volume. At the typical capillary column flow rates, it may require a minute or more for all this vapor volume to be transferred to the column. In order

that the analytes (if not the solvent) should not spread out after entering the column, the column oven must be maintained at a temperature much lower than that used to vaporize the sample in the injector space. This requires that a separate "injector port" volume be independently heated to a different and higher temperature than the initial oven temperature.

If the analyte concentrations are so high that they will overload the column, an automatic diversion valve may be used to "split" the sample, allowing only a small percentage (1 %–5 %) of the injected vapor volume to enter it. On the other hand, the analytes might be present at very low concentration, and we would wish to transfer almost all of them to the column while setting the initial column oven temperature high enough to allow the more volatile solvent pass rapidly through. The vapor cloud of solvent exiting the injector port does not do so as a discrete plug. The last portions will mix and be diluted with incoming carrier gas in the injector volume and bleed slowly onto the column. In order that such a persistent "solvent tail" not overwhelm the detector and interfere with the much smaller analyte peaks, we may wish to program the **split valve** to divert the last residue of solvent vapor from continuing to enter the column after we have transferred most of the analytes to the head of the column. This procedure is designated **splitless injection.**

4. Long capillary columns must be coiled to fit in a GC thermostatic oven. This can be done easily with metal capillary tubing. Unfortunately, metal at the elevated temperatures used in GC often interacts unfavorably with many analytes. Such interactions, which could lead to complete analyte destruction, or tailing, as described in Section 12.4, will more severely impact the smaller sample levels required in capillary columns. Also, all the stationary phase is supported on the metal column instead of mostly on more inert packing particles, as is the case of GC on packed metal columns. For sensitive analytes, both capillary and packed column GC used glass columns, whose active sites leading to tailing could be deactivated by masking dangling surface hydroxyl groups by treatment with reagents such as trimethylchlorosilane (see the discussion of silica chemistry in Section 12.5). The glass capillary columns were difficult to coil, to coat uniformly with stationary phase, and to fit to the injector and detector ends of the GC instrument without breakage. Unless the greater resolution of capillaries was absolutely necessary, more robust packed glass columns were preferred.

The development of the flexible fused silica capillary GC column described in Section 12.6 resulted in the domination of the GC market by open tubular columns and instruments tailored to their use. Packed columns are relegated to a few special applications that include gas analyses. Fused silica capillaries were easy to install. They were more inert (i.e., contained fewer active sites causing tailing) than glass or metal capillaries or the particle supports used in packed column GC. Even when the resolution of packed columns was sufficient for simple separations, a capillary coated with the same stationary phase could achieve equivalent separation in a shorter run time. Packed columns, especially those with diameters larger than 2 mm, will possess much higher analyte capacity than capillaries. They are still preferred for preparative separations, where the separated peaks are captured for later processing or use. The TCD is still the most universal, responding to everything but hydrogen or helium when these are used as the carrier gas. Compounds such as carbon monoxide, oxygen, nitrogen, and small molecules not containing carbon require its use, and its low sensitivity often dictates the use of high concentrations of analytes requiring the peak capacity of packed columns. Some regulatory standard assay procedures developed in the early days of GC instrumentation still specify the use of packed columns, although most of these have been updated to capillary GC equivalents. Therefore, we shall discuss GC and GC instrumentation primarily from the perspective of open tubular column GC.

13.3 GC INSTRUMENT COMPONENT DESIGN (INJECTORS)

13.3.1 Syringes

The most common way of introducing a sample into a GC instrument is with a syringe. The typical hypodermic syringe used to administer vaccines or intravenous medications has a volume on the order of 1 mL, a plastic body and plunger, and a sharply pointed, detachable needle with an internal bore of 0.2–0.5 mm, and is often designed for one-time use. In contrast, the usual GC syringe has a narrower-bore permanently attached needle, long enough (2–3 in.) for the tip to penetrate to the center of an injector port, and a volume on the order of 5–25 µL and is composed of a glass body with a very tightly fitting metal plunger. The usual volume of liquid sample or solution injected is 0.5–2.0 µL when open tubular GC columns are used. Larger volumes might expand to a vapor volume (>1–2 mL) that is larger than the capacity of a standard heated injector port and "blow back" into the gas supply system and condense there. Such vapor volumes would require an excessive time to transfer to GC columns at typically 1–2 mL/min capillary flow rates. The GC syringe needle must be very narrow, both to minimize the volume of sample remaining in the syringe when the plunger is depressed and to pass smoothly through the deformable silicone **septum disc** that seals the entrance to the injector port to maintain the column **head pressure**. Too wide a needle diameter could cut out a plug of the septum, a process called *coring* (as in coring an apple). This results in gas leaks preventing the GC from maintaining the desired head pressure, and the **septum core** is likely to be projected into the hot injection port, where it will decompose to liberate *septum bleed* into the column or trap sensitive analytes. Syringes designed for very-small-volume injection (<1 μL) may contain the sample within the bore of the needle itself, expelling it with an internal wire plunger. Such designs place a limit on how narrow the needle may be.

Gas samples may be injected with larger, specialized *gas-tight* syringes (1 mL or larger volume), as they do not expand as much as liquids do upon vaporization. Very large gas volumes (>30 mL) may require slow injection into the larger flow rates usable with packed columns. For regular repetitive sampling of vapor streams, a rotary multiport *gas sampling valve* like that illustrated for high-performance LC (HPLC) injectors (Figure 14.6) can be set to cycle at intervals to accept a noncondensable gas sample into a metal sample loop of known volume and then insert this into the carrier gas stream for a GC separation. Sampling and injection techniques to measure trace concentrations of gases in air or water will be discussed in Section 13.10.3.

13.3.2 Autosamplers

Modern GC instruments can sequentially analyze up to 100 or more liquid samples without manual intervention. Racks or carousels of vials containing liquid samples or standard calibration solution are capped with thin septa, like miniature versions of the GC injector port septum. The vial holder moves to present each vial sequentially to a GC syringe mounted in a robotic assembly. The syringe penetrates the sample vial septum, and the plunger is pumped several times to pull up a set volume of sample and expel air bubbles. Then the syringe is withdrawn, moved over the injector port septum, and inserted smoothly and rapidly, and the plunger is depressed to complete the injection. Upon injection, a signal is sent to the GC electronics to initiate any temperature control programs needed in the various GC ovens and to the detector and recording electronics to start recording the chromatogram signal. Many refinements can be programmed into these operations. For example, a plug of air may be drawn into the syringe before the sample volume is pulled up, and after sampling, the plunger is further withdrawn in air again to pull all of the sample out of the needle bore. The sample is thus *sandwiched* between two plugs of air. When the needle is initially inserted into the hot injector port, the solvent is not present in it to start vaporizing prematurely. When the plunger is depressed, the rear air plug forces all the sample from the needle, and again, there is no boil-off of the sample from within the hot metal needle. The long thin narrow-bore needle required for these microliter injections is the weak point of the "syringe-through-septum" technique. Bad operator technique or autosampler malfunction can easily result in a severely bent ("zeed") needle. Once it is bent, the one-piece GC syringe/needle assembly is beyond repair. Typically, if a needle is misaligned or is faulty, the failure will result quickly; the needle will zee after the first or second injection. Experience has shown that a well-aligned syringe can serve for up to 1000 injections in an autosampler before metal-to-glass wear inside the barrel causes severe friction. This will result either in a malfunction of the syringe or the autosampler throwing an error.

13.3.3 SPME

A recently developed procedure called solid-phase microextraction (SPME) combines the principles of the minicolumn sample isolation and concentration technique of solid-phase extraction (SPE) and the technique of automated GC sample injection via a syringe assembly. A small amount of a solid phase into which analytes can partition from the sample is coated on one end of a fine fused silica fiber. These solid-phase materials can often be the same as GC stationary phases. The other end of the fiber is fastened to a GC syringe plunger, such that the coated portion of the fiber can be extended beyond the needle tip or be withdrawn entirely within the needle. A sample, generally liquid, is contained in a septum-capped vial. The syringe needle pierces the sample vial septum, and the fiber is extended for several minutes—either into the sample liquid or the headspace gas above its surface. Analytes partition and concentrate into the fiber coating. The volume of this solid-phase coating is very small, hence the "micro" in SPME. After equilibration, the fiber is retracted; the syringe needle is pulled from the vial, and it is then inserted through the GC septum into the heated injector port, where the fiber is immediately reextended. Analytes are thermally desorbed and are swept by the carrier gas flow to the top of the much cooler capillary GC column, where they collect. The column temperature is raised to an appropriate value and separation of analytes proceeds in the normal fashion.

The SPME injection procedure is illustrated in Figure 13.1 and Figures 1.9 through 1.12. If the sample matrix is not so dirty that it coats the fiber with high boiling material that is not thermally desorbed, the fiber may be used for many repeated sampling and injection cycles. If the analytes are quite volatile, it may be possible to sample them in the confined headspace volume above a "dirty" matrix, thereby avoiding the contamination of the fiber by high boiling contaminants. This **headspace analysis** is one of the most useful modes of SPME injection. The fiber concentrates only a small portion of the analytes, but all of the "extracted" material is injected into the GC by the thermal desorption process. Contrast this with classical liquid–liquid extraction or SPE procedures, where the extract is into a large volume (e.g., 1–10 mL), which must then be concentrated (to perhaps 25–100 μL), of which perhaps only 1–2 μL may be injected. A good application

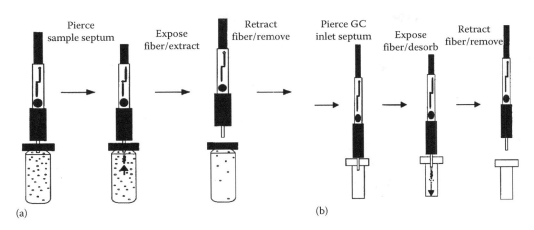

Figure 13.1 Diagrams of SPME (a) extraction sampling and (b) GC desorption injection. (Reprinted with permission of Supelco, Inc., Bellefonte, PA, www.sigmaaldrich.com.)

of headspace SPME analysis would be for volatile organic vapors (alcohols, solvents, hydrogen cyanide, etc.) from a messy whole-blood sample. The amount of analyte absorbed is based on the equilibrium concentration that accumulates in the fiber, and it is far from the total amount of analyte present in the sample. This equilibrium concentration can vary substantially with equilibration time, temperature, and other conditions. Therefore, to obtain quantitative data when employing SPME, it is necessary to calibrate against an internal standard (IS) introduced into both unknowns and calibration solutions. The IS must behave in a fashion similar to the analytes, both in its equilibration behavior and in its response to the GC detector employed. Even when using an IS, good quantitative values are difficult to obtain, unless the IS is an isotopically labeled version of the analyte, and it is used with mass spectrometric detection. The fiber coating in SPME is often the same as a GC stationary phase. Analytes are moved off it into the GC column by varying the same parameter, temperature, which controls GC elution. In the capillary GC column, the coating is on the inside of a long cylindrical tube. On the SPME fiber, it is on the outside of a short cylindrical rod, of slightly smaller diameter. For this reason, SPME sampling and injection is sometimes characterized as "inside-out GC." It is not really chromatography, because there is no continuous partitioning to the mobile phase, but otherwise, many of the principles and materials are similar.

It is often necessary to enhance the vapor-phase concentration of an analyte by **salting out**. This procedure is simply the addition of a soluble ionic compound to an aqueous matrix for which organic component headspace analysis is desired. Vapor-phase concentration enhancements ranging between a factor of 2 and 8 are possible with carefully selected salts. The book by Bruno and Svoronos has a comprehensive listing of SPME phases and salting out reagents.

13.3.4 Split Injections

The sample capacity of capillary columns is limited. GC analysis of neat mixtures or highly concentrated solutions may require that only a portion of the sample that is vaporized in the injection port enters the column. Modern capillary GC instruments often feature a **split–splitless injector**. This has a time-programmable on–off valve controlling flow to a split vent gas line from the injector port. When this valve is open, only a fraction of the carrier gas flow (and of the vaporized injected sample) passing through the injector port enters the column. The rest flows out the split vent. The relative proportions passing to the column and the vent may be adjusted by a variable needle valve (the back-pressure regulator) in the split vent line. The ratio split vent flow/column flow can typically be varied from 5 to 100 and is called the **split ratio**. When operating the GC column isothermally (at a constant temperature chosen so analytes move down the column at a reasonable rate), the sample must enter the column over a short time, if the initial chromatographic bandwidth is to be minimized. At capillary column flow rates of ~1 mL/min, the ~1 mL vapor volume from a 1 μL injection would take about a minute to load onto the column, and the sample would spread out during loading. With the much larger flow through the port with the split vent open and set to a high split ratio, a small fraction of the vapor cloud enters the column over a much shorter period. A diagram of the flows in the **split mode** of a split–splitless injector is illustrated in Figure 13.2a.

13.3.5 Splitless Injections

When sample concentrations are much lower (ppm to ppb or lower), one will wish for all the analytes to be transferred to the capillary column, and overloading will be less of a concern. As illustrated in Figure 13.2b, when the valve

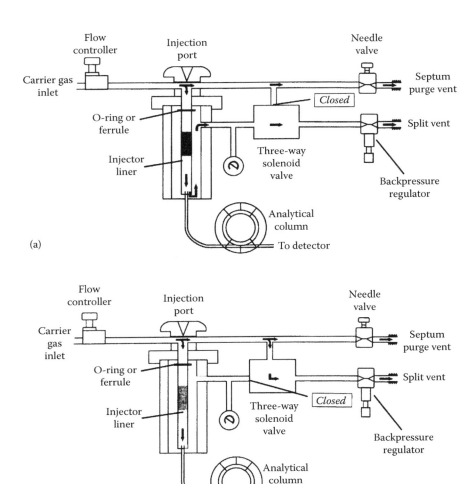

Figure 13.2 Diagram of GC flows for split and splitless injections: (a) split mode and (b) splitless mode. (Adapted with permission of Restek Inc., Bellefonte, PA, www.restek.com.)

to the split vent is closed, all of the vaporized sample must flow to the column. When this **splitless mode** of injection is used, the valve must be closed while most of the sample and its solvent vapor have time to enter the column. As mentioned in the split mode discussion, this time may be on the order of 1–2 min. In order that the analytes do not spread out on the front of the column, the initial column oven temperature must be well below that at which they begin to move down the column. The lower boiling solvent they were dissolved in will pass rapidly ahead of them. In many cases this solvent front will accumulate in the first few cm of the column, forming a quasi-stationary phase that will actually concentrate the solute species. In the splitless mode of operation, the column temperature will then be raised to temperatures (constant or steadily increasing) at which the solvent and analytes begin to move down the column and separate; the solvent moves faster than the analytes. Traces of the much higher concentration of the solvent remaining in the injection port might continue to enter the column and obscure the signals from the lower levels of analytes. Therefore, at the end of the splitless injection period, the valve is returned to the split mode position to rapidly divert the remaining solvent. A high split ratio setting enhances this.

13.4 GC INSTRUMENT COMPONENT DESIGN (THE COLUMN)

13.4.1 Column Stationary Phase

As described in Section 12.5, silicone polymers form the most common and useful class of GC stationary phases. Figure 13.3a displays a two-monomer segment of the polydimethylsiloxane (PDMS) chain. This is the fundamental silicone polymer. Various polymers used for GC phases differ in their polarity. The degree to which the polarity of the analyte is similar to that of the phase

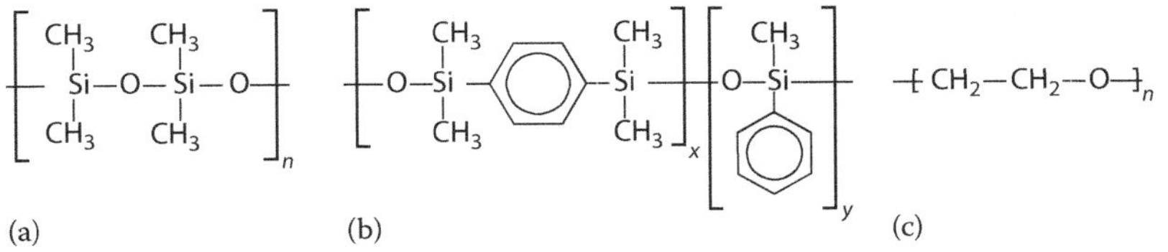

Figure 13.3 Representative monomer units for several common types of GC stationary phase polymers: (a) PDMS, (b) backbone phenylene-stabilized and phenyl-substituted PDMS (more polar π-electron character), and (c) carbowax (polyethyl ether phase).

is an indication of the strength with which it will dissolve into and be retained by the phase. This in turn depends on the analyte's polarity. The general rules of "like dissolves like" and its converse apply here. At one end of the polarity scale are pure long-chain hydrocarbons, whose intermolecular interactions are governed almost entirely by *van der Waals forces* (alternatively referred to as *London's dispersion forces*). These are the *least polar* compounds and phases. Unless pure hydrocarbons have a relative molecular mass high enough not to volatilize or decompose under the full range of GC temperatures, they make poor stationary phases. PDMS itself is of intermediate polarity, with the oxygen atoms in its backbone contributing some **dipole** and **hydrogen-bonding** character to its polarity. Substitution of the methyl groups in PDMS with longer hydrocarbon substituents (e.g., $-C_8H_{17}$ or $-C_{18}H_{35}$) yields a silicone polymer with a lower polarity more similar to hydrocarbons. The oxygen in the backbone is less accessible, and there is a higher proportion of the hydrocarbon-type van der Waals dispersive interactive surface presented to the analyte.

Figure 13.3b illustrates two other types of modification to the PDMS structure. In the y monomer, a more polar (by virtue of **π-electron bonding interaction**) phenyl group has been added. This increases the polarity of the phase and thereby the retention of analytes that have a similar functionality. In the x unit, occasionally substituting a phenyl ring for an O atom in the polymer backbone again enhances this type of polarity, but it removes the hydrogen-bonding and dipole polarity of the replaced oxygen atom. The net overall effect on the polarity interaction is difficult to predict. It will depend on the structure of the analyte. Such an inserted **spacer** is in fact not present for polarity adjustment but rather to stabilize the polymer chain against thermal decomposition at high temperatures. This suppresses the phenomenon of **column bleeding**, which results in an increased detector signal and a rising baseline signal as temperature is increased. If one wishes an even more polar column, a substituent such as $-CH_2CH_2CH_2CN$ with a very strong cyano group dipole can be substituted for a PDMS methyl. The polarity changes introduced by substitutions such as these phenyl or cyanopropyl groups can be varied by adjusting their percentage on the chain.

Occasionally, other substituents with reactive end functional groups can be introduced. After the phase is coated on the interior of the column, these groups can be reacted to bond with one another (**cross-linking**) or with free silanol groups (Section 12.5) on a capillary column's interior silica surface (to form a **bonded phase column**).

These processes further improve the thermal stability of the stationary phase and suppress degradation and bleeding. A bleeding signal may also result from decomposition of very nonvolatile material deposited on the column from prior injections of dirty samples. Cross-linked phases bonded to the walls of capillary columns can withstand rinsing with solvents to remove this kind of contamination. Recall that bare or nonbonded silanol groups on the silica wall can act as active sites to introduce tailing in eluting peaks. After coating and bonding a phase to the column wall, such remaining **active sites** may be **deactivated** by treatment with reagents that bond methyl or trimethylsilyl groups to them. With long use, these active sites may reform and can sometimes be reblocked by injection of the derivatizing agent through the column at elevated temperature. The commercial designations of capillary columns often reflect many of these design elements. For example, a widely used J&W (now Agilent) column is designated DB5-MS; the 5 indicates 5 % phenyl-substituted PDMS, the MS indicates the presence of phenyl spacers in the chain to suppress bleed that interferes particularly with mass spectrometry (MS) detection, and DB is "Durabond"; this particular manufacturer's trade designation indicates that the phase is chemically bonded to the capillary column. Other manufacturers employ a variety of very similar designations for their equivalent phase materials. Figure 13.3c displays a common nonsilicone polymer, whose monomer is characterized as a "carbowax" column. The oxygen in the chain is much less shielded than the one in silicones, so wax phases are much more polar, but they lack the high thermal stability of the silicones.

A recent development in stationary phases employs *ionic liquids*. These two properties would have seemed very inappropriate for a stationary phase, as charged ionic groups, should they be charge balanced by an inorganic counter ion, would be unstable. True liquid coatings hark back to the earliest days of GC and would be expected to vaporize and be restricted to low temperature ranges due to column bleed. The new category of **ionic liquid phases** is represented by the structure in Figure 13.4. In all cases, two cationic groups are separated by a nonpolar spacer chain

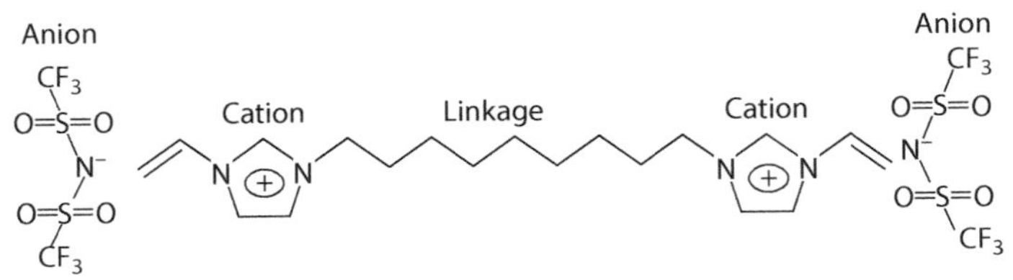

Figure 13.4 Typical ionic liquid GC stationary phase structure: Note the long nonpolar hydrocarbon "spacer" with charge-balanced organic cation–anion pairs at either end. (Reprinted with permission of Supelco, Inc., Bellefonte, PA,. www.sigmaaldrich.com.)

of variable length. Unlike the case of *zwitterions*, where there is a cationic and anionic structure at either end of the spacer, two identical charge balancing anionic groups are closely paired with the two identical cationic groups on the central chain. The charge compensation takes place over a very short range, but the fact that there is so much charge in the small molecule structure causes the boiling point of the liquid to be very high. Unlike the much higher molecular mass silicone polymers, there is no "back-biting" reaction at high temperatures to clip off the ends of the polymer chains and give rise to small volatile bleed molecules. These phases are especially suitable for separations of highly polar analytes that interact more strongly with the ionic groups, while the nonpolar spacers confer some more or less reverse-phase retention. Thus, the retention mechanism is "mixed mode." These ionic liquids are so different from the common polymer phases that they often offer a good opportunity to radically shift the selectivity term α in Equation 12.9 or 12.10, when variation of other terms will not achieve sufficient resolution of some analytes. Figure 13.5 displays the difference in elution of 22 fatty acid methyl esters (FAMEs) on a traditional silicone-based GC stationary phase and an ionic liquid. Note the achievement of separation of several geometric isomers about double bonds that coelute with other FAMEs on most traditional GC phases. The initial commercialization of these phases is from Supelco, and more details can be obtained on their website listed in Appendix 13.A.

The descriptions just given highlight a few examples from the major categories of polymer stationary phase coatings. In the modern practice of GC, these are almost always employed in open tubular or capillary columns. Any assay that might employ them in packed column mode can be implemented more efficiently or faster or both, using open tubular columns, and a vast selection of these are now readily available from commercial vendors. Because many very simple instructional GCs only operate with packed columns, the undergraduate student may still encounter them in the instructional laboratory. They may lack separately heated injector or detector regions and be capable only of isothermal separations. The student should be aware that in industrial and research practice, the analysis time and resolution advantages of open tubular columns make them the overwhelmingly dominant choice for most applications. Packed column GC is relegated to large-scale **preparative GC**, where the greater sample capacity of wide-bore, packed columns is required, or to a few "niche applications."

One example of the latter is the separation of **fixed (or permanent) gases** (e.g., H_2, He, argon, O_2, N_2, CH_4, CO, CO_2). One scarcely needs a column oven for these—a column refrigerator would seem more in order. Unfortunately, when most polymer phases are frozen solid, gas molecules do not partition efficiently into them. Packed particles of a **molecular sieve** material can separate these small ultra-volatile molecules. These materials, sometimes produced from the catalyst minerals called zeolites, have pores of the dimension of the fixed gas molecules. Larger molecules cannot enter, and they will remain with the mobile phase gas stream. The small molecules vary in the extent to which they can diffuse in and out of the pores and through the sieve and are thereby separated. Figure 13.6 displays such a separation. The principle of excluding the larger molecules is similar to what happens in "size exclusion chromatography" described in Chapter 14. For those molecules that can enter the openings of the sieve particles, the factors governing relative retention are more complex. It is not easy to rationalize the elution order seen in Figure 13.6.

Another type of packed column particle useful for relatively small molecule separations is a **porous polymer** (trade name: **Porapack®** or **HaySep®**). These are like the molecular sieve but made of organic polymers or resins with larger-sized pores. Highly volatile molecules larger than the fixed gases, but of a size containing only a few atoms, are retained by these materials at temperatures high enough to be controlled by standard GC column ovens. These porous polymers have now been engineered into coatings for open tubular GC, with its inherent advantages, and are sold as **porous layer open tubular (PLOT)** columns. A representative separation of small molecules on a PLOT column is displayed in Figure 13.7. An advantage of these porous polymer phases is that they will even tolerate and separate water, albeit with poor tailing peak shape. This ultrapolar and strong hydrogen-bonding liquid is usually totally incompatible with and destructive of other polymer stationary phase coatings. PLOT columns are susceptible to releasing the small particles coated onto the walls of capillary tubing,

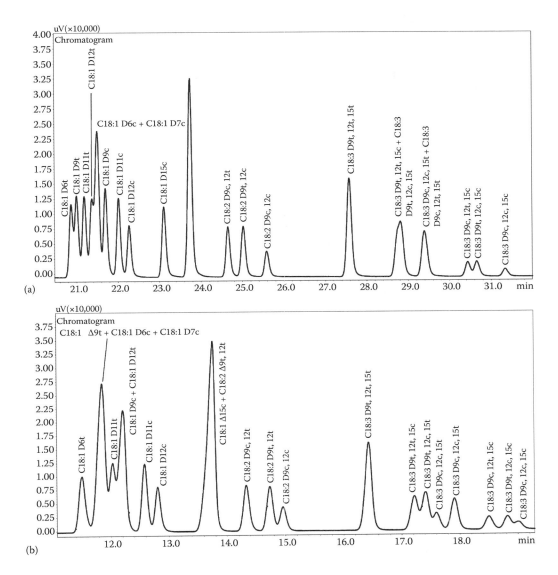

Figure 13.5 Chromatograms of GC separations of a mix of 22 18-carbon FAMEs on (a) SP-2560 and (b) ionic liquid SLB-IL 100. Note particularly the retention shifts of C18:1 D9c + C18:1 D12t, and many others. C18:X represents X number of double bonds in the 18-carbon chain, and the other D characterization represents their position and geometric orientation along the FAME chain. The ionic liquid responds to these differences more radically than traditional silicone-based polymers. (Reprinted with permission of Supelco, Inc., Bellefonte, PA, www.sigmaaldrich.com.)

and these can clog the column flow through the column or of narrow-bore components downstream of it.

Note that porous polymers are not the only PLOT columns available. Various sizes of both silica and alumna PLOT columns are also used extensively, as well as familiar molecular sieve phases. Less common PLOT columns include monolithic carbon and cyclodextrin phases. The specific separations that can be affected by use of PLOT columns are listed in the book by Bruno and Svoronos. Interestingly, short lengths of alumina PLOT columns can be used as headspace sampling traps at room temperature and especially at reduced temperatures, the latter referred to as PLOT-cryoadsorption (Bruno, 2009).

13.4.2 Selecting a Stationary Phase for an Application

With thousands of possible analytes, even more different combinations of these analytes, and upward of 100 stationary phase materials generally available, how should one go about selecting the right column for the job? In earlier times, analysts compiled tables of so-called McReynolds or Rohrschneider constants of characteristic compound types for different phase materials. If the data for the compounds of interest were available from these tables, one could employ these values with calculation algorithms to estimate the relative elution orders and resolution of the components

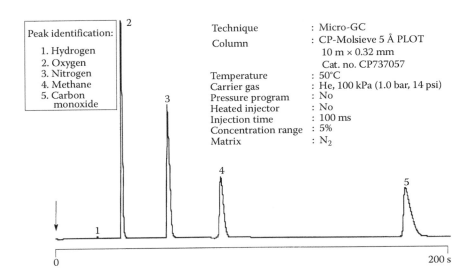

Figure 13.6 Molecular sieve PLOT column separation of fixed gases with TCD detection.

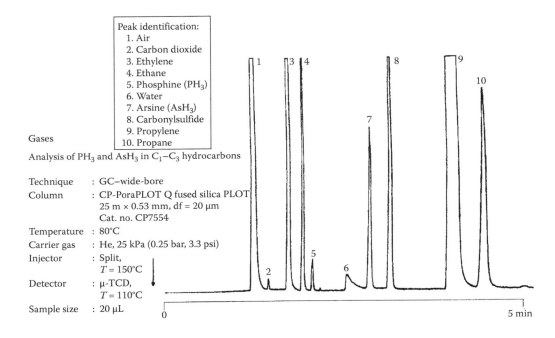

Figure 13.7 Porous polymer PLOT column separation of low-molecular-mass gases with TCD detection: analysis of PH_3 and AsH_3 in C_1–C_3 hydrocarbons.

on different columns. This approach has fallen into disuse. A modern commercial capillary column with all the features described in Section 13.4.1 is a very highly engineered product, not easily created or duplicated in the analyst's lab. The column manufacturers have amassed extensive databases and application notes for optimum separations using the most suitable columns from their stables. The best place to start when looking for a column and phase to achieve separations for a group of analytes is in the book by Bruno and Svoronos or manufacturers' **applications literature**. Their catalogs, in hardcopy or on searchable CD-ROMs, and their websites, with search programs for the analyst to enter the compound type or application, illustrate thousands of optimized separations. It is beyond the scope of this chapter to attempt to summarize this information, however, in general, the best stationary phase to start with in approaching a new or unknown analysis is the 5 % phenyl substituted polydimethylsiloxane phase, unless prior knowledge of the sample indicates otherwise. Appendix 13.A lists some selected web resources. The student is encouraged to access them, or catalog data if available, to gain a sense of the scope and organization of this information. This is where one should now start when confronted with a GC separation problem.

Once a reasonably suitable column phase has been identified, the dimensions of the column employing it should be selected, and then the separation is optimized by varying the flow and temperature parameters. The column dimension decisions will depend on the difficulty of achieving the necessary peak resolutions, which is likely to increase with the number of analytes to be measured. The column's analyte capacity depends on these dimensions as well. The column dimensions, carrier gas flow, and column oven temperature parameters thus chosen will determine the time required to perform each GC run.

It may be useful to attach a length of uncoated capillary tubing to the front of the coated column. This *retention gap* serves two purposes. It allows space for the injected sample and solvent vapors to collect on the column without being significantly retained until they encounter the coated portion. Dirty, poorly volatile contaminant residues can accumulate in this region, and front portions of the gap section can be cut away to remove these without affecting the separating and retaining functions of the column. When the gap region becomes too short, it can be replaced by another. Figure 13.8 illustrates the processes that occur following sample and solvent injection on a capillary column using a similar procedure, **direct on-column injection**. Here, an extremely fine syringe needle actually enters the capillary column, bypassing the injector port (whose heater should be turned off), and deposits the liquid sample solution directly in the gap region of the column, usually at a low starting temperature near or below the solvent's boiling point. This is a delicate process, required for thermally unstable analytes which might be destroyed by vaporizing in a hot injector port. After vaporization, the retention gap functions as illustrated, whether the injection is on-column or through a splitless injector.

13.4.3 Effects of Mobile Phase Choice and Flow Parameters

A column with a particular stationary phase and dimensions has these as fixed parameters. Changing them means obtaining or purchasing another column and installing it. New columns require a significant **conditioning** or "breaking in" period after installation. This conditioning process should be done with the column connected to the heated injection port but not to the detector port. This will prevent residual monomer and other impurities from contaminating the detector.

Another fixed parameter is the choice of mobile phase (i.e., the carrier gas to be used). Three are commonly employed: helium (He), hydrogen (H_2), or nitrogen (N_2). As will be discussed in Section 13.7, proper operation of the GC detector being used will sometimes dictate the use of a

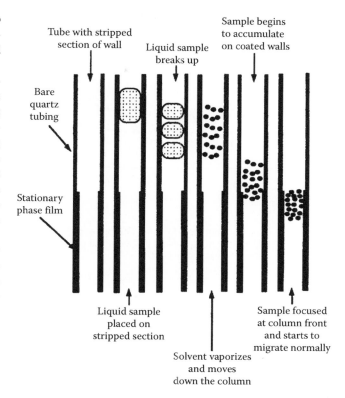

Figure 13.8 Illustration of on-column injection with a retention gap. (From Cazes, J., *Encyclopedia of Chromatography*, Marcel Dekker Inc., New York, 2001. Used with permission.)

particular carrier gas. If the detector tolerates any of these gases, then their flow and diffusion characteristics govern how well each attains optimum chromatographic resolution. This is best illustrated by their respective **Van Deemter curves** (cf. Sections 12.8 and 12.9, Equation 12.10, Figure 12.7). These are displayed in Figure 13.9. N_2 has the deepest **minimum H** (i.e., HETP) value (i.e., greatest efficiency) but only over a limited range of linear flow rates, u. Lighter gases He and H_2 do not quite attain such low H values, but they stay lower over a much wider range of u than does N_2. One would like to operate the column close to the flow rate of the Van Deemter minimum for the gas. All modern chromatographs are equipped with automatic flow controllers that can maintain the flow as other parameters change. In the absence of such capability, if one is raising the column temperature over the course of an analysis, it is better to use the light gases and to start at a flow rate above the Van Deemter minimum. As column temperature goes up, gas viscosity increases, and the flow will decrease toward the minimum position of H. Low values of H are more difficult to maintain over the steep-sided valley of the N_2 curve. In Section 13.7, we will see that such rising **temperature programs** are often required to optimize separations of mixtures of large numbers of compounds.

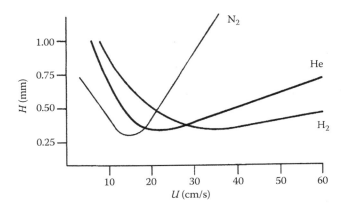

Figure 13.9 Van Deemter plots for three GC carrier gases: N_2, He, H_2. H (height equivalent to theoretical plate) versus u (linear flow velocity, cm/s).

Helium is more expensive to purchase than N_2 or H_2 (which can be generated on-site from water electrolysis). Shortfalls in the availability of He are becoming more common, even in the United States. Some large facilities that also have NMR instrumentation have helium recapture systems that can be used with chromatographs. Hydrogen forms explosive mixtures with air, and a catastrophic column rupture or leak might allow enough into the oven to contact hot heating coils and trigger an explosion. The United States held a near monopoly on He (the huge helium depot near Amarillo, TX, stored helium separated from natural gas), so GC users there preferred it to hydrogen on grounds of safety. In Europe and elsewhere, chromatographers learned to safely use H_2, since He was not available (recall that the Hindenburg dirigible was forced to employ hydrogen, to its ultimate demise). The capillary GC resolution performance of H_2 exceeds even that of He. As flow controllers become more available on GC instruments, and lower flow capillary systems become the norm, one could program them to recognize a sudden catastrophic H_2 leak and command the shutdown of carrier gas flow before the H_2 concentration reaches dangerous values. This facilitates the use of superior H_2 as a carrier gas. While it is simple to use with FIDs, which add in H_2 for the flame, or TCDs, which require either He or H_2 as mobile phase gas, care must be taken when interfacing a column end directly into the vacuum of an MS system. If an insufficiently long or narrow column is so interfaced, retention times and orders may shift significantly from values observed in non-vacuum interfaced systems, and the increasingly common GC electronic flow controllers may be forced unsuccessfully to try to set the exit pressure to negative values. Too high an H_2 pressure in the MS ion source may result in its reaction with analytes to produce a nonstandard electron ionization (EI) mass spectrum, with molecular ions adding an H atom. These problems may be circumvented by installing a flow restrictor at the column end.

When it is necessary to analyze for the gases that are commonly used as carrier gases, it is possible to use a special mixed carrier gas (8.5 % hydrogen in helium, mol/mol) that has a minimum in the thermal conductivity vs. temperature curve. It is also possible to detect and measure hydrogen in a helium carrier stream if the polarity of the TCD is momentarily reversed during the passage of the hydrogen peak. Note that this is not the most sensitive method to determine hydrogen, however. Additional information on carrier gases for GC in general and for TCD use in particular can be found in the book by Bruno and Svoronos.

13.5 GC INSTRUMENT OPERATION (COLUMN DIMENSIONS AND ELUTION VALUES)

The column temperature is the major instrumentally controllable variable that can be used to affect the speed and resolution of the GC separation. Much precision engineering goes into designing ovens that can precisely control and reproduce the temperature profile that the column encounters during each analysis.

Before we move to a description of how temperature affects peak elution on GC columns, let us review how some of the column dimension and stationary phase composition factors interact. In Figure 13.10a, we see the effect on retention time and resolution of increasing the stationary phase **film thickness** d_f and thereby lowering the **phase ratio** (cf. Section 12.7, Equation 12.2). Thicker films are especially effective at retarding faster-eluting components to elute at suitable k' values (cf. Equation 12.3) while providing modest improvements in the resolution of close pairs eluting in the middle of the run. The primary value of the thicker film is to keep very volatile components from eluting too fast before they have time to separate well. It also confers greater column capacity against overloading, which would result in leading peaks and degrade resolution. Another way to increase resolution is to use a longer column, as illustrated in Figure 13.10b. Note that for the illustrated pair separation, the longest column takes a little longer and does not achieve quite as good resolution of this pair as did the thicker film column; diffusion broadening of peaks is more of a problem as the column length is increased. The longer column would probably result in a much higher price and require higher pressures to achieve a flow to operate it at the Van Deemter optimum. Thus, longer columns are not always the best way to achieve better separation. Going to a longer column should be the last resort. Recall from Equation 12.8 that resolution increases only as the square root of the number of plates, N, or the column length, L. Note that it is not necessarily the thicker film that provides the better resolution, it is lowering of the phase ratio. This could also be achieved with a thinner film and a correspondingly narrower-bore column to achieve the same low phase

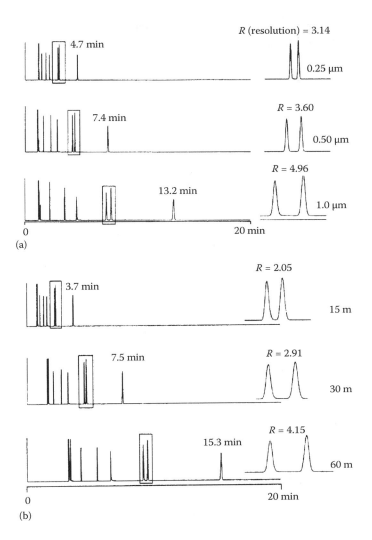

Figure 13.10 Effects of stationary phase film thickness or GC column length on retention time (R_t) and resolution (R). (a) Film thicknesses (d_f) of 0.25, 0.50, and 1.0 μm (30 m × 0.25 mm PDMS column) and (b) column lengths (L) of 15, 30, and 60 m (0.25 mm × 0.5 μm PDMS column). (Used with permission from SGE Analytical Science, Pty Ltd, Melbourne, Australia, www.sge.com.)

ratio. With a thinner film, equilibration in the stationary phase would be faster as well, leading to even lower values of H (plates/m) and a more efficient column. However, the narrow-bore thin-film column has less capacity and is more easily overloaded, and it requires higher head pressures to be operated at or above the Van Deemter optimum. The pressure limits of the GC may limit the usable length of such columns.

In Figure 13.11a, we see how a shorter column, with film thickness and diameter reduced to keep the phase ratio constant, can achieve nearly the same separation performance in half the time as a column of more normal dimensions. If overloading can be avoided, this is the best approach to achieving **fast GC**. For some detectors, the lower carrier gas flow is a benefit; for others designed for higher flow levels, special adjustments (e.g., the use of **makeup gas**; cf. Section 13.7, item 4) may be required. If none of these resolution-enhancing tricks and no adjustments of the column temperature will achieve resolution of critical analyte pairs, it is time to fall back and consider finding a column with a different stationary phase to do the job. A classic example is the analysis of BTEX (benzene, toluene, ethylbenzene, and three isomeric xylenes) single-ring aromatic hydrocarbons. Many silicone-based phases fail to separate the *meta*-(1,3-dimethyl) and *para*-(1,4-dimethyl) benzenes (xylenes). This is illustrated in Figure 13.11b, where the SolGel-1MS (a PDMS-type phase) fails, but the BTEX compounds are all separated on a SolGel-WAX (carbowax-type phase). Note that the other xylene isomer, the *ortho*-(1,2-dimethyl), is more strongly retained on either phase. In general, one cannot predict relative retention for similar structures precisely enough to guess which peak is which or whether a given phase will separate them or not. This is where recourse to the manufacturer's databases

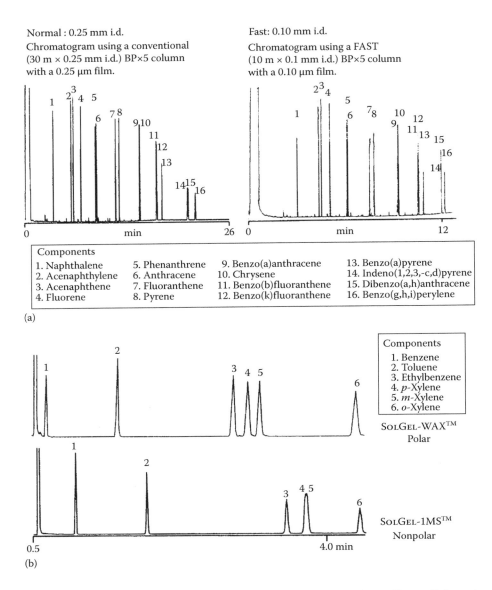

Figure 13.11 (a) Comparison of polycyclic aromatic hydrocarbon (PAH) separations on a column of "normal" dimensions and on a column designed for "fast" GC. (b) Comparison of BTEX compound separations on carbowax and PDMS GC phases. (Used with permission from SGE Analytical Science, Pty Ltd, Melbourne, Australia, www.sge.com.)

on separations and phases (and the NIST Chemistry Web Book) is invaluable. The take-away here is that database lookups trump theory for getting elution orders. The caution regarding sources of information that was provided in Chapter 1 applies here; please go back and review this.

13.6 GC INSTRUMENT OPERATION (COLUMN TEMPERATURE AND ELUTION VALUES)

Finally, we reach the discussion of the most easily adjusted variable controlling GC peak elution. Recall from Chapter 12 that for a given stationary phase, the temperature of the column is the primary determinant of the equilibrium partition ratio between the stationary and mobile phases. The larger the percentage of time the analyte spends in the gas flow of the mobile phase, the more quickly it elutes from the column. A modern GC instrument is designed to precisely and reproducibly control the temperature of the oven compartment in which the coiled-up column resides. This will promote reproducible retention times to enable peak identification. Note that temperature variations outside of the coil can be pronounced; heated injector and detector fittings inside the oven can cause departures from the indicated oven temperature of a few degrees.

In a very simple GC analysis of a group of similar analytes that would elute close together, the analyst chooses an appropriate stationary phase, optimizes its film thickness,

GAS CHROMATOGRAPHY

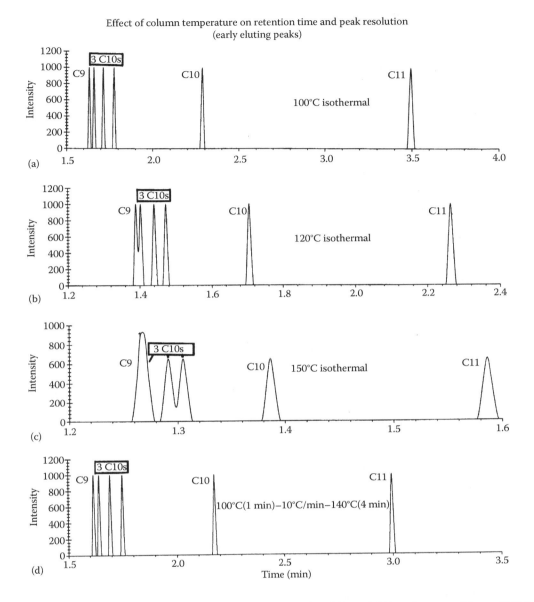

Figure 13.12 Separations of three branched-chain C10 hydrocarbons and C9, C10, and C11 *n*-alkanes on 30 m × 0.25 mm × 0.25 µm PDMS column at (a) 100 °C, (b) 120 °C, (c) 150 °C, and (d) programmed 100 °C/1 min–10 °C/min–140 °C/4 min early-eluting peaks. (Separations simulated using EZ-Chrom GC®, Restek Corp. Inc., Bellefonte, PA, www.restek.com.)

and selects the necessary column diameter and length, all as outlined in Section 13.5. Then the oven is set to a column temperature where k' for these analytes is between 2 and 10 (cf. Equation 12.3). The highest value of temperature T at which adequate resolution of the closest eluting pair is observed will yield the fastest analysis. Such operation is called **isothermal GC**. It is the simplest for the temperature controllers to maintain, and as soon as the last component has eluted, the instrument is ready to accept injection of the next sample.

Most often, more complex mixtures of components varying widely in polarity and volatility must be analyzed. In Figures 13.12 and 13.13, we examine the problems these sorts of mixtures present for the development of a simple isothermal GC analysis, and we describe how to separate them all in a single GC run. The two figures illustrate the **early-eluting** and **later-eluting** peaks of a single analysis. The column parameters have been selected: a 30 m long, 0.25 mm i.d., 0.25 µm film PDMS column; the archetypical standard capillary GC column. It uses H_2 carrier flow controlled at 46.3 cm/s, the minimum of the H_2 Van Deemter curve. The column "dead time," t_m, was determined to be 1.08 min. The analytes consist of four groups:

1. Normal straight-chain hydrocarbons, labeled C9 through C15, for the number of carbons in the chain.

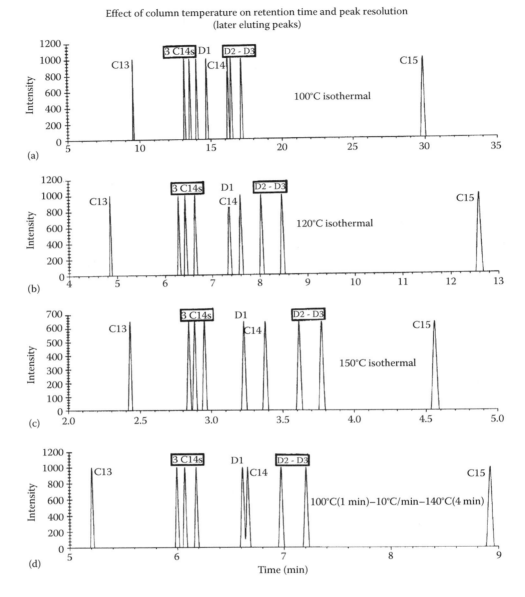

Figure 13.13 Separations of three branched-chain C14 hydrocarbons, three dimethylnaphthalene isomers, and C13, C14, and C15 *n*-alkanes on 30 m × 0.25 mm × 0.25 μm PDMS column at (a) 100 °C, (b) 120 °C, (c) 150 °C, and (d) programmed 100 °C/1 min–10 °C/min–140 °C/4 min, late-eluting peaks. (Separations simulated using EZ-Chrom GC®, Restek Corp. Inc., Bellefonte, PA, www.restek.com.)

2. A group of three branched-chain C10 hydrocarbon isomers, labeled 3 C10s.
3. A group of three branched-chain C14 hydrocarbon isomers, labeled 3 C14s.
4. Three dimethylnaphthalene isomers, labeled D1, D2, and D3.

There are several points of caution to note when interpreting these figures:

1. These "chromatograms" were not obtained on an actual GC instrument. They are accurate simulations of retention data from a computer program using data on the column and the analytes.
2. The retention time axes *differ* for each run, A through D (i.e., the peaks do not line up with one another on a single elution time axis).
3. Only retention time and peak width data are simulated to illustrate analysis time and resolution. The peaks are always displayed at the same height (except when they merge), while in a real chromatogram, as peaks having the same amount of analyte with equal detector response were eluted more slowly, their height would decrease as their width increased, thus keeping their area constant.

Sections A, B, and C of each figure illustrate three isothermal GC analyses, at 100 °C, 120 °C, and 150 °C, respectively. At 100 °C isothermal, the three early-eluting branched-chain C10s elute sufficiently far from t_m to be well resolved from each other and from C9. As the temperature is increased, they begin to elute too close to t_m (i.e., k' becomes too much less than 2) and resolution is lost. On the other hand, if we look at the later-eluting sections in Figure 13.13, we see that at 100 °C isothermal, C15 requires more than 30 min to elute. If we could run the separation at 150 °C isothermal, the last peak C15 would be out in only 4.5 min and we would still be able to resolve all the earlier-eluting peaks. By contrast, the poor separation of C9 and three C10s in Figure 13.12 (Section C) reveals that this temperature would not do for the early eluters' separation. The solution to the problem is to change the temperature as the run proceeds. Start at 100 °C, and when the initial group of peaks has been resolved and eluted, increase the temperature linearly to around 150 °C to speed up the elution of the less volatile components. This is what was done in the chromatograms depicted in D. The procedure is called programmed temperature GC. The "temperature program" illustrated is abbreviated as "100 °C (1 min)–10 °C/min–140 °C (4 min)." Translated, this states that the oven was held at 100 °C isothermal for 1 min and then programmed to increase steadily at a rate of 10 °C/min to 140 °C, where it is held for 4 min. Now all peaks are resolved and the entire analysis takes only 9 min (plus enough time for the oven to be rapidly cooled back to the starting temperature to repeat the program for the next sample). The cool-down time is short (a minute or two) unless the start temperature is close to room temperature, where the cooling rate becomes very slow. If such is the case, liquid CO_2 or even liquid N_2 injection to the column oven may be employed to speed up the last stage of cooling.

An alternative to the injection of these fluids into the oven is the application of the vortex tube to speed up cool down. A vortex tube is a device that separates a stream of compressed laboratory air into a hot flow and a cold flow, with the cold flow reaching as low as –40 °C. Vortex tube cooling in many aspects of chromatography was introduced by Bruno, and includes not only the column but syringes, extractors/concentrators, and sample conditioners (Bruno, 1992).

The simplest and most common temperature program is the single linear ramp illustrated earlier. More complex programs, with multiple ramps interrupted by isothermal intervals may be designed to optimize critical separations and minimize overall analysis time. Environmental and biological samples often contain very slowly eluting heavier background contaminants of no interest to the analysis, which must be cleared off the column before the next injection, lest they elute with and interfere with analyte peaks from subsequent injections. Such **late eluters** can be recognized because their peaks appear much broader than those of the faster-moving analyte peaks they coelute with. The cure for these is to rapidly program the column oven to a temperature close to the column stability limit after the last analyte of interest elutes, and to hold it there until all late eluters are cleared. Close comparison of isothermal runs A, B, and C with programmed temperature run D will reveal that in the former, peaks become broader with increasing retention time, while in the latter, the peaks have very similar widths. This can be understood because in the programmed temperature runs, each peak (more precisely each band) exits the column at nearly the same rate and has been in the mobile phase for the same length of time. This may seem unobvious, but it can be rationalized if one has a clear understanding of the chromatographic process (e.g., the discussion in Section 12.5), and the explanation is left as an exercise for the student (or the instructor). Figure 13.14 illustrates another aspect of this point in a different manner. Here, the same temperature program is applied to the separation of many FAMEs on equivalent columns differing in length by a factor of two. Although the small scale of the reproduction prevents it from being seen, the resolution of closely eluting peaks on the longer column is improved by a factor of about 1.4, just the amount predicted by the square root of N relation in Equation 12.8. Note, however, that the analysis time has *not* been extended by a factor of two, but only from 22 to a little more than 24 min! Under temperature programming conditions, the interval during which the temperature passes through the range where each component moves at a significant rate is only several minutes. This is the order of magnitude of the extra time it takes to complete the analysis on the longer, more highly resolving column.

When analyzing a series of related samples, such as a group of aviation kerosenes, it is often best not to vary the temperature programs between analyses. This is especially important if peaks are to be identified by a mass spectrometer (which we will discuss later). The use of the same temperature program for each sample allows the use of "reconciliation" between samples. For example, a common constituent of aviation kerosene is n-hexadecane, and this will be found at the same retention time if the same conditions are used for all analyses.

In Sections 13.5 and 13.6, we have seen that elution and resolution behavior of analytes depend in a complex interactive way upon a number of different factors:

1. The *composition* of the stationary phase (*strongly*: effects on both absolute and relative retention) and the mobile phase gas (*less strongly*: effects on resolution via differences in the parameters that control the Van Deemter constants, but essentially no effect on retention).
2. Column *dimensions* (length L) and phase ratio (column i.d. and film thickness d_f).
3. Column *gas flow* (depends on gas, pressure drop across column, and temperature).
4. Column *temperature* (the major controllable variable, can be very complex if programmed).

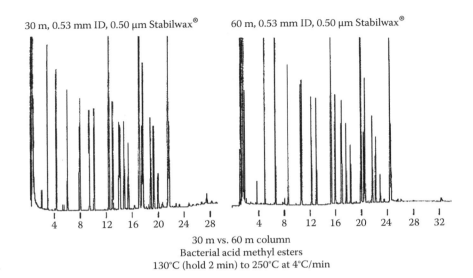

Figure 13.14 Effect of column length on elution time in programmed temperature GC of bacterial long-chain acid methyl esters. (Used with permission from Restek Corp., Bellefonte, PA, www.restek.com.)

Most of these factors interact with one another. One could specify all the equations for these interactions and require a student to master calculations using them to predict the results on a GC separation, varying these parameters. An even more challenging task would be to use them all to optimize a given separation. This is beyond the scope of this introductory chapter, but the need for it remains. The solution is to avail oneself of interactive computer programs that perform the linked calculations almost instantaneously. One such was used to create Figures 13.12 and 13.13. Several are available to download from the World Wide Web at no charge. The **Agilent pressure-flow calculator** allows one to specify all the parameters 2 to 4 and instantly see the effect of varying any one at a time. The **Agilent method translator** allows retention information from a particular method (with a column described by factor 1) to be used to quickly predict how a separation will change with variation in any of the other three categories of factors. Appendix 13.A provides instructions to access these programs. The student is encouraged to download them and experiment with the results of varying parameters. The use of free interactive GC separation optimization programs beats working through a tangle of equations whose variables interlock with one another.

13.7 GC INSTRUMENT COMPONENT DESIGN (DETECTORS)

The third and final major component analytes encounter after the injector and the column is a detector. This produces an electrical signal (usually analog, but often converted to digital) that is proportional to either the concentration or the mass-flow rate of the analyte molecules in the effluent stream. The signal is displayed as a chromatogram on a chart recorder, or more often these days, on the screen of a desktop computer data system. Retention times are automatically calculated, heights of peaks are measured, or they are automatically integrated to obtain their areas, and peaks can be identified by their elution within a retention time window, and quantitated by comparison to the areas or heights of a quantitative standard. We will not discuss the details of the operation of this signal processing equipment, but will describe only the operation and characteristics of the most useful GC detectors.

There are around a dozen GC detectors in common use. Spectroscopic instruments can be interfaced to the effluent of a GC and act as a form of detector that has the compound identification power of a spectroscopic measurement. This mating of a separation instrument to a spectroscopic instrument is called a **hyphenated technique**. The acronyms for the two classes of instruments are separated by a hyphen (or sometimes a slash [/]), as in GC-MS. These will be discussed later in Section 13.8. Some of the general characteristics of GC detectors that need to be considered are the following:

1. *Universality versus selectivity*: If a detector responds with similar sensitivity to a very wide variety of analytes in the effluent, it is said to be universal (or at least almost so—no GC detector is absolutely universal). Such detectors are valuable when one needs to be sure that no components in the separated sample are overlooked. At the other extreme, a selective detector may give a significant response to only a limited class of compounds: those containing only certain atoms (e.g., atoms other than the ubiquitous C, H, and O atoms of the majority of organic compounds) or possessing certain types of functional

groups or substituents that possess certain affinities or reactivities. Selective detectors can be valuable if they respond to the compounds of interest while not being subject to interference by much larger amounts of coeluting compounds for which the detector is insensitive.

2. *Destructive versus nondestructive*: Some detectors destroy the analyte as part of the process of their operation (e.g., by burning it in a flame, fragmenting it in the vacuum of a mass spectrometer, or by reacting it with a reagent). Others leave it intact and in a state in which it may be passed on to another type of detector for additional characterization.

3. *Mass flow versus concentration response*: These two modes of detector response are described in Section 12.4. In general, destructive detectors are mass-flow detectors. If the flow of analyte in effluent gas stops, the detector quickly destroys whatever is in its cell, and the signal drops to zero. A nondestructive detector does not affect the analyte, and the concentration measurement can continue for as long as the analyte continues to reside in the detector cell, without decline in the signal. Some types of nondestructive detectors (e.g., the ECD) measure the capture of an added substance (e.g., electrons). The "saturation" of this process results in a loss of signal to analyte proportionality, and they are mass-flow detectors.

4. *Requirement for auxiliary gases*: Some detectors do not function well with the carrier gas composition or flow rates from a capillary column effluent. **Makeup gas**, sometimes the same as the carrier gas, may be required to increase flow rates through the detector to levels at which it responds better and/or to suppress **detector dead volume** degradation of resolution achieved on the column. Some detectors require a gas composition different from that used for the GC separation. Some detectors require both air and hydrogen supplied at different flow rates than the carrier to support an optimized flame for their operation. Makeup flow dilutes the effluent but does not change the detection mechanism from concentration to mass-flow detection.

5. *Sensitivity and linear dynamic range*: Detectors (both universal and of course selective ones) vary in their sensitivity to analytes. Sensitivity refers to the lowest concentration of a particular analyte that can be measured with a specified **signal-to-noise ratio**. The more sensitive the detector, the lower this concentration. The range over which the detector's signal response is **linearly proportional** to the analyte's concentration is called the *linear dynamic range*. Some exquisitely sensitive detectors have limited linear dynamic ranges, so higher concentrations of analytes must be diluted to fall within this range. Another less than satisfactory solution to a limited linear dynamic range is to calibrate against a multilevel nonlinear standard curve. This requires injections of more standards and is more prone to introduce quantitative error. Dilution will not work satisfactorily if there is a wide range of concentrations in the sample. A very insensitive detector will perforce have a more limited dynamic range, and multilevel standard curves or dilution will be of no avail with it.

Figure 13.15 compares the sensitivities and dynamic ranges of several of the most common types of GC detectors.

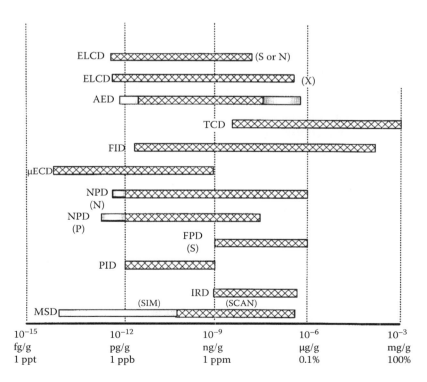

Figure 13.15 Approximate limits of detection (left end of bar) and dynamic ranges for 12 GC detectors from 1 μL sample injected.

The further to the left the range bar extends, the more sensitive the detector. The longer the bar, the greater the dynamic range. Each vertical dotted line denotes a span of three orders of magnitude (1000×), so the overall ranges covered are very large. The values along the x-axis assume an injected volume of 1 μL (1/1000 of an mL or cm^3) of solution. The ranges are approximate; their exact endpoints will depend on the design and model of the detector and optimization of its operating conditions. They provide a good general comparison. The infrared detector (IRD) and mass selective detector (MSD) at the bottom are hyphenated method spectroscopic detectors, operable in several modes, and will be discussed in Section 13.8. Note that the "original" GC detector, the TCD, is the least sensitive but also the only one suitable for handling neat (i.e., 100 % pure single component) samples. The FID has the greatest single-mode dynamic range, while the "micro-ECD" is more sensitive, but with a more limited dynamic range. Let us proceed to describe the operation, characteristics, and applications for each of these nonhyphenated method detectors.

13.7.1 Thermal Conductivity Detector

Universal; nondestructive; concentration detector; no auxiliary (aux.) gas; insensitive; limited dynamic range.

The TCD was the first widely commercially available GC detector, in the era when all the columns were packed, and samples were neat (i.e., not diluted solutions) mixtures to be separated. It measured differences in the thermal conductivity and/or specific heat of highly thermally conductive (either H_2 or He) carrier gas when diluted by small concentrations of much less conductive analyte vapor (anything else). A current through a thin resistive wire heated the wire in the detector flow cell. The thermal conductivity of the flowing carrier gas cooled the wire. When analytes were in the stream, their lower thermal conductivity produced less cooling, which caused the wire's temperature to rise and its resistance to increase. This wire resistor was in a **Wheatstone bridge** circuit (an arrangement of four resistors on the sides of a square). One of the other resistors was in a matching TCD cell connected to a matching column and flow with no analyte passing through. In isothermal GC, carrier flows and temperatures would remain constant, but with temperature programming of the column, the temperature would increase and the flow would decrease, independently affecting the conductivity. The matching *reference cell* would compensate for these effects on the resistance changes. A fixed and a variable resistance constituted the other two legs of this bridge. A voltage sensor was connected across opposite points on the diagonal of the square array. The variable resistance was used to null (or "zero" the signal from) the voltage sensor. Once the bridge circuit was thus balanced, passage of analyte changed the resistance of the wire in one leg of the bridge, throwing the bridge out of balance and producing a voltage signal proportional to the resistance change. The designs of two TCD cells are illustrated in Figures 13.16a and b. This design of the TCD cell was also referred to as a **katharometer** cell, a name for a gas thermal conductivity measurement device.

The high thermal conductivity body shown in Figure 13.16a is typically a stainless steel, for corrosion resistance toward the analytes. Stainless steels are not particularly high in thermal conductivity among metals but compared to the gases that pass through the cell, they provide sufficient thermal mass. Note that different sensing wires are available for specific analyte streams, with gold-sheathed tungsten being the most common. More information on this can be found in the book by Bruno and Svoronos. In some applications, thermistors are substituted for the sensing wires. Thermistors constructed using micro-electromechanical systems (MEMS) technology are useful for lower temperature separations and for corrosion resistance. These older style TCDs find application in custom or DIY gas chromatographic systems.

Modern gas chromatography instruments are usually equipped with a thermal conductivity detector in which a column and reference flow are directed over a single sensing wire by a pneumatically operated switching valve set for a frequency of 5 Hz. The body of the detector is a ceramic module that can be easily replaced if needed.

Note the application of the TCD for the detection of **fixed gases** in the chromatogram of Figure 13.6. These are generally not seen by other detectors (except hyphenated GC-MS). Note the low sensitivity signal for H_2, whose thermal conductivity is the only one to closely match that of the He carrier gas used. If only H_2 were being measured, it would be better to use N_2 as the carrier. This would yield a peak signal in the negative direction (due to a change in signal polarity). Alternatively, a mixture of hydrogen in helium (8.5 % hydrogen, mol/mol) can be used to avoid this polarity change. In Figure 13.7, note that the detector is described as a μ-TCD. This must employ miniaturized cells to be compatible with the low carrier flows of the open tubular PLOT column and the need for small detector cell volumes to avoid extra-column band broadening.

Related to the thermal conductivity detector is the gas density balance, a device in which the analyte never comes in contact with the sensing elements. Instead of responding to a difference in thermal conductivity, this device essentially responds to a difference in relative molecular mass. It has been used to analyze streams in which hydrogen is present, especially in petroleum refining. This detector is optimized by the use of sulfur hexachloride as a carrier, not a very common gas used in chromatography.

13.7.2 Flame Ionization Detector

Nearly universal (all carbon compounds except CO, CO_2, and HCN, but not many inorganic gases); destructive; mass-flow detector; needs H_2 and air or O_2 aux. gas; sensitive with wide dynamic range.

GAS CHROMATOGRAPHY

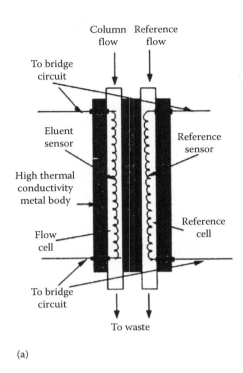

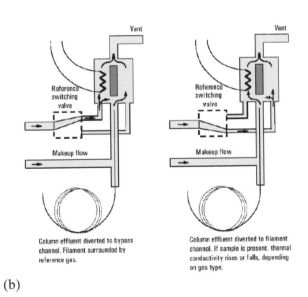

Figure 13.16 Diagrams of two types of TCDs. (a) An older style detector with two separate column flows, one for the analyte and one as a reference. (From Cazes, J., *Encyclopedia of Chromatography*, Marcel Dekker Inc., New York, 2001. Used with permission.). (b) A modern frequency-modulated μ-TCD (Agilent Technologies, used with permission).

The FID is the most commonly employed detector, as it gives a response to almost all organic compounds (any compound with C–H bonds will have a response). On a molecular basis, the signal is roughly proportional to the number of carbon atoms in the molecule, a fact that can make calibration straightforward. Hydrogen and oxygen (or more commonly air) must be separately provided to fuel a flame in the detector cell. If air is used, it must be

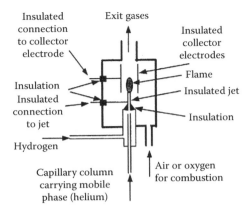

Figure 13.17 Diagram of the FID. (From Cazes, J., *Encyclopedia of Chromatography*, Marcel Dekker Inc., New York, 2001. Used with permission.)

"zero-air", meaning that it must be free from any hydrocarbon contamination. Zero air can be supplied in cylinders from a supplier or produced in the lab with a zero-air generator, a device with a heated catalytic bed that removes any hydrocarbons. As illustrated in the diagram of Figure 13.17, the H_2 is introduced and mixed with the carrier effluent from the GC column. Even if H_2 is used as capillary carrier gas, an additional separately controlled H_2 supply is necessary to adjust the appropriate **fuel** supply for the flame. The mixed gas enters the cell through a jet, where air or O_2 flows past to serve as the flame **oxidizer** supply. An electrical **glow plug** in the cell (not illustrated) can be pulsed to ignite the flame. The fuel and oxidizer flows are adjusted with needle valves to achieve a stable flame with optimal FID response, often by bleeding a volatile unretained hydrocarbon into the carrier stream to provide a reference signal. The **jet tip** is charged by several hundred volts positive relative to several **collector electrodes** or a **collector ring** surrounding the flame. In the absence of eluting organic analytes, no current flows in the jet–collector circuit. When a carbon-containing analyte elutes into the flame, the molecule breaks up into smaller fragments during the cascade of oxidation reactions. Some of these are positively charged ions, and they can carry current across the flame in the circuit. Although the ionization efficiency of the FID is low, its base current is also very low. Hence, its signal-to-noise ratio is very high. Against such a low background, even very small ionization currents can be accurately measured using modern electronics that draw very low currents (high input impedance, voltage measurement circuits), hence the good sensitivity and extraordinary dynamic range of this detector, often exceeding six orders of magnitude.

One can detect essentially any organic compound that can survive passage through the GC to the detector. Some organic compounds can be problematic to analyze, such as chlorinated compounds. Such compounds cause soot formation in the flame and the soot particles cause electronic spikes in the chromatograms.

A very wide range of analyte concentrations can be measured in a single run. The FID is the forerunner of a series of more selective and sometimes even more sensitive ionization detectors, some of which will be discussed subsequently. It led the way to GC as it is currently practiced, with analytes being present in dilute concentrations in a preparation or extraction solvent solution instead of being mixtures of the neat compounds. Precise quantitation requires calibration against reference standards of the analytes being measured. The responses on a compound mass basis are sufficiently similar so that for **approximate quantitation**, calibration for many components can be made against a single reference compound peak area. The wide dynamic range minimizes the number of calibration levels required for an adequate standard curve.

13.7.3 Electron Capture Detector

The ECD is very selective for organic compounds with halogen substituents, nitro and some other oxygen-containing functional groups; is nondestructive; is a concentration detector; needs to use N_2 or argon/CH_4 as carrier or as makeup gas if used with H_2 or He capillary carrier flow; and has extreme but highly variable sensitivity but with limited dynamic range.

If the FID brought GC into the realm of characterizing dilute solutions of organics, it was the ECD that allowed it to spark a revolution in the understanding of the threat of bioaccumulation in tissue and bioconcentration up the food chains of lipophilic, persistent organic pollutants (POPs) in the environment. The classic example of this was the discovery of the threat posed by the organochlorine pesticide dichlorodiphenyltrichloroethane (DDT) and its subsequent banning. The device was also largely responsible for the detection of atmospheric chlorofluorocarbons (CFCs) that were used as refrigerants and propellants. J.E. Lovelock (subsequently famous as the founder of the Gaia theory of the whole earth as an organism) invented this deceptively simple but exquisitely sensitive and selective detector. It is not too much of a reach to claim that observations that were only made possible by the use of this GC detector set off the revolution in environmental consciousness in the 1960s. A story is told that while applying the instrument to CFCs, the research group discovered that the use of CFC-propelled underarm deodorant was causing an unacceptable noisy ECD response. The remainder of the work had to be done by the team sans the use of deodorant!

How then does this little marvel function? Two versions are illustrated in Figure 13.18. The right-hand diagram displays features of the initial design (operation in direct current [DC] mode). The column effluent enters from the right. A nickel foil doped with radioactive ^{63}Ni (a β particle [energetic electron] emitter) constantly

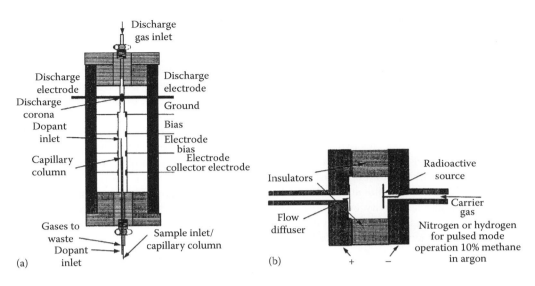

Figure 13.18 Diagrams of two types of ECDs: (a) ECD for use with pulsed electrode potential and (b) ECD for use with constant electrode potential. (From Cazes, J., *Encyclopedia of Chromatography*, Marcel Dekker Inc., New York, 2001. Used with permission.)

bombards the carrier gas, ionizing some of its molecules, creating an atmosphere of positive ions and the negative electrons that have been knocked loose. Older versions utilized a tritium source; these are no longer available. This radioactive source is chosen for its ability to withstand the high operating temperatures that GC detectors may require to prevent condensation of high boiling analytes. A *low voltage* (several volts, instead of the hundreds of volts across the FID) between the negative inlet side of the detector and the positive outlet side is sufficient to set up a current between these two electrodes. Under this low voltage, it is the more mobile, lighter free electrons that carry the so-called standing current to the positive electrode.

If an analyte with highly electronegative, highly polarizable substituents enters the cell, the electrons of the standing current can be captured by these molecules. The mobility of the captured negative charge is drastically reduced, and some may be more easily neutralized by collision with the positive ions that are generated. It is the depression of this standing negatively charged electron current that is displayed reversed and appears as a peak signal. The electron capture sensitivity varies dramatically (one to two orders of magnitude) with the nature and number of the substituents: I > Br > Cl ≫ F among the halogens and 5–8 halogen atoms > 4 > 3 ≫ 2 ≫ 1. Note that most electronegative F is surpassed by more polarizable Cl, Br, and I, increasing in that order, so it is the *polarizability* that dominates, although both are necessary for electron-capturing effectiveness. Some other groups such as $-NO_2$, $-NO$, and so on fall somewhere on the low side of these responsiveness series. Hydrocarbons are essentially unresponsive, although in great excess they can depress the sensitivity of the ECD response of ECD active compounds if they coelute with them. So, with the ECD, calibration standards for every analyte are necessary to perform quantitative work.

Note that ECDs often require special carrier or makeup gases. Most early ECD cells were designed for high packed column flow rates and will need makeup gas at higher flow rates and of different composition when He or H_2 is employed as capillary carrier gas. The use of a radioactive ionization source in the detector requires a license from the Nuclear Regulatory Commission (NRC) in the United States and a program of regular wipe tests to detect radioactivity leaks in and around the detector. Another problem is that the depression of the standing current soon "saturates." As the level of electron-capturing analytes rises, the tiny current of electrons becomes more and more depleted. An additional increase in the analyte level is not reflected by a proportional decrease in the standing current—there just are not enough electrons left to go around (the circuit). This results in severe nonlinearity of the detector response, curvature of the standard curve (it eventually levels out as concentrations continue to increase), and a severely limited linear dynamic range. Quantitative calibration becomes a tricky problem with ECDs.

The left-hand ECD diagram shows a more modern design that attempts to alleviate some of these difficulties. The ionization is achieved by the "discharge electrode" using a special discharge gas flow, thereby replacing the radioactive foil. A special dopant gas is introduced to enhance this. Instead of maintaining a constant potential across the cell, it may be intermittently pulsed at a higher voltage. Typically, a pulse at ~30 V of around 1 μs repeated at 1 ms intervals (a duty cycle of 0.1 %) collects all free electrons in the ECD cell, while during the "off period," the electrons reestablish equilibrium with the gas. The very fast collection time is enabled by admixture of 5 %–10 %

methane in the argon ECD gas, which serves to enhance the voltage-induced migration rate of the free electrons. The current during the pulsed collection is integrated to produce the standing current level. The benefit of this method of controlling the standing current is to extend the linear dynamic range, and most modern ECDs are designed to operate as pulsed-mode ECDs. Such an extended linear dynamic range is illustrated for the "μ-ECD" in Figure 13.15. The micro (μ)-designation indicates that this model is equipped with a much smaller cell volume adapted for the effluent flow rates of capillary columns. Dispensing with the need for dilution of the effluent stream with makeup gas also increases the sensitivity even more. All other things remaining equal, pushing the sensitivity even lower extends the dynamic range since it is mainly limited by the effects of saturation of the standing current at the upper end of the range. The earlier model, large volume, DC-mode ECDs often had a linear dynamic range of only 100 or less. This made measuring multiple analytes (especially if they covered a much larger range of concentration) within a valid calibration range very complicated. Many dilutions and re-assays were often required to do good quantitative work—a great contrast to operating within the huge dynamic range of the FID. On the other hand, the special selectivity and extraordinary sensitivity of the ECD for certain classes of compounds, such as multiply chlorinated DDT, were indispensable for the discovery of the bioconcentration pathways of that compound and others such as polychlorinated biphenyls (PCBs), dioxins, and polybrominated diphenyl ethers (PBDEs) (polybrominated diphenyl ether flame retardants). It is an important general principle that the detection limits for given analytes in chromatographic methods applied to real samples from environmental or biological systems are more often determined by the selectivity of the detector against background coeluting interferences than by its absolute instrumental signal-to-noise sensitivity limit.

A discussion of the ECD must include some safety precautions. The ^{63}Ni source typically employed is a sealed source, meaning that the radioactive element is not accessible to the user under normal circumstances. In most places, the acquisition of an instrument with an ECD is controlled by regulation; in the U.S., this is done under what is called a general license, which gives the owner the ability to operate the device but not to service it. It is a good idea to review the discussion on radioactive sources that was presented in Chapter 2. The main safety precaution to observe in general use, other than the obvious "don't take it apart warning," is not to allow the device to overheat.

13.7.4 Electrolytic Conductivity Detector

Heteroatom selective (for organic compounds with halogen [X] substituents or N or S atoms); destructive; mass-flow detector; requires liquid reagents; good sensitivity and linearity; response is directly proportional to the flow rate of the X, N, or S atoms.

The ELCD operates by consuming the analyte in a hydrogen–air flame and dissolving the resulting gases in an aqueous solution whose electrolytic conductivity is proportional to the amounts of any halogen (X), S, or N atoms that were present and oxidized to the corresponding acids. It represents the first example of a detector specifically and proportionately responsive to specific atoms in organic compounds. It is complex to operate and now not commonly used. Its unique value is its proportional response only to the number of heteroatoms, not to their position in the molecule. If a mixture of compounds with only one of these elements present is analyzed and additional information is available about the number of its atoms in each compound in each peak (obtainable from a GC-MS analysis), then a single standard peak response can serve to calibrate many other peaks, for which no standards exist. An example is the calibration of PCB congener peaks in commercial mixtures against the response of a chlorine compound standard. There are 209 of these PCB isomers, and their ECD responses vary widely. The TCD, FID, or GC-MS responses vary less, but still by too much, and individual standards were initially available for only a small subset. The exact atom proportionate response and good dynamic range of the ELCD permit quantitation in the absence of complete standard sets for these homologous compounds.

13.7.5 Sulfur–Phosphorus Flame Photometric Detector

Heteroatom selective (for organic compounds with S or P atoms; separately); destructive; mass-flow detector; less sensitive and shorter dynamic range than FID.

When organic compounds containing S or P atoms are burned in an FID, the flame conditions can be adjusted to produce a lower temperature that enhances the emission from S_2 fragments at 394 nm or HPO fragments at around 515 nm. The FID electrodes are omitted, and a temperature-resistant window or fiber-optic light guide in the side of the detector cell passes the emitted light to filters designed to isolate these wavelengths and pass the characteristic emission to a sensitive photomultiplier tube. Operated in S-selective mode, it is useful to characterize the organosulfur compounds in complex petroleum mixtures, as the much higher levels of coeluting hydrocarbon peaks give minimal response. Operated in P-selective mode, it is a sensitive detector for organophosphorus pesticide trace residues in complex environmental mixtures. Only the sulfur mode response range is illustrated in Figure 13.15.

13.7.6 Sulfur Chemiluminescence Detector

The sulfur chemiluminescence detector (SCD) is heteroatom selective (for organic compounds with S atoms only); destructive; mass-flow detector; more sensitive and greater dynamic range than sulfur–phosphorous flame photometric detector (SP-FPD).

This detector takes the gases from a flame produced in the same manner as in an FID (again minus the electrodes) and reacts it with ozone to induce chemiluminescence from sulfur products produced from eluting organosulfur compounds burned in the flame. Alternatively, instead of a flame, the column exit can be routed into a high-temperature ceramic tube. The intensity of this luminescence is detectable at even lower levels than the S_2 emissions produced in the SP-FPD, making it useful for monitoring trace environmental pollutant sulfur organics. This detector has an equimolar response to sulfur in sulfur compounds, thus making calibration more straightforward. Calibration can be done with any sulfur compound. In most SCD devices, the products are routed into a catalytic chamber to ensure that all residual ozone is destroyed.

13.7.7 Nitrogen–Phosphorus Detector

The NPD is heteroatom selective (for organic compounds with N or P atoms; separately); more sensitive for N and P compounds than FID, as well as selective for them; destructive; mass-flow detector; more sensitive than FID for N, more sensitive than P-FPD for P, but with less dynamic range than the FID.

The NPD is yet another variation on the workhorse of GC detectors, the FID. Comparison of Figure 13.19 for an NPD to Figure 13.17 for the FID highlights the similarities. Both are defined as **thermionic detectors**, that is, the high temperature of a flame breaks the eluting analytes into fragments, some of them positive ions, which release electrons to carry a current under the influence of the voltage between two electrodes. The NPD is operated under **fuel (i.e., H_2)-rich** conditions. Under these conditions, the normal carbon compound FID thermionic response is suppressed by orders of magnitude. The "new element" in the detector is **rubidium** in the form of a glass or ceramic **bead** doped with a rubidium salt, which is heated by immersion in the flame but is also additionally and variably heated by passing current through thin wires that support the bead in the flame. What exactly happens near the surface of the bead while organonitrogen or organophosphorus compounds are decomposing in this not-so-hot flame is complex and poorly understood. In some designs, the flame would not be self-supporting were it not for the independently heated bead embedded in it. Somehow C–N- or C–P-containing fragments interact with easily ionizable rubidium atoms to produce ions and electrons that then produce a signal like that in the FID, but one that is exquisitely sensitive and selective to either N or P atoms (depending on how the detector parameters are set). This is illustrated in Figure 13.15. The extreme sensitivity of this detector to some organophosphorus compounds explains

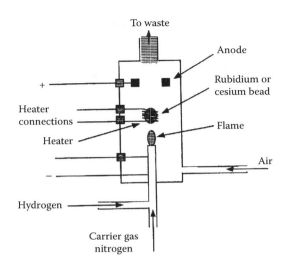

Figure 13.19 Diagram of the NPD. (From Cazes, J., *Encyclopedia of Chromatography*, Marcel Dekker Inc., New York, 2001. Used with permission.)

why it has largely supplanted the P-FPD for this application, and no range is depicted for that mode of the FPD operation in the comparison figure. The NPD produces no response to N atoms not bound to carbon in organic molecules. This is a fortunate circumstance that renders it immune from interference from ubiquitous atmospheric N_2.

13.7.8 Photoionization Detector

Compound class selective (for organic compounds with more easily ionizable π-electrons, especially aromatic compounds); nondestructive; mass-flow detector; slightly more sensitive than FID for many compounds it detects, but with less dynamic range than the FID.

Ionization detectors like the FID and the NPD have great sensitivity. Is there any other way to selectively ionize some classes of compounds to achieve a sensitive, selective ionization detector like the NPD? Let us take the "thermo" out of thermionic by dispensing with the flame and supply ionization energy instead by passing **high-energy ultraviolet (UV) radiation** into the cell. This is the reverse of measuring flame emission coming out as in the SP-FPD. The UV radiation provides enough energy to knock electrons out of some aromatic compounds and some others that have electrons in higher-energy-level π-bond orbitals. Precise selection of the UV wavelength can allow some selectivity within these compound classes. Organic compounds like alkyl hydrocarbons containing only electrons in more tightly held single bonds will be unaffected. The resulting photoionization detector (PID) is selective for aromatic hydrocarbons and other organic compounds with unsaturated bonds. The high-energy UV lamp and its power supply are much more compact and robust than the FID and its associated gas supplies.

Combined with its exceptional sensitivity to a reasonably broad-based selection of organic compounds, this operational simplicity makes the PID especially suitable as a detector for field-portable GC instruments. An additional factor supporting such applications is that it is not affected by oxygen in air and can even be used with air as a carrier gas, if the column stationary phase can withstand this at the temperature conditions employed.

13.7.9 Helium Ionization Detector

This is a universal detector (everything except neon); nondestructive; mass-flow detector; a little less sensitive than FID for FID-active compounds, and with less dynamic range than the FID.

The **helium ionization detector** (HID) is like the FID, NPD, or PID in that it is a sensitive detector that measures a current of ions produced from the analyte molecules. An analogous **argon ionization detector** employing the same principle was another early GC detector. They are sometimes also referred to as **discharge ionization detectors (DIDs)**. In the HID, a flow of He gas atoms (as He carrier or He makeup gas) is activated to an excited metastable state by **discharge electrodes**. These have sufficient energy to ionize any analyte molecules (except Ne atoms) they come in contact with, which then produce an ionization current proportional to the sample amount between two other electrodes in the detector. As with the ECD, pulsing the discharge electrode to maintain a constant current yields a design with superior dynamic range. It can be operated in series with the other universal detector, the TCD, since both are nondestructive, and their two overlapping dynamic ranges can cover the range up to 100 % of the neat compound. It is often employed in simple portable GC instruments designed to measure both fixed gases and very volatile low-molecular-mass hydrocarbons (cf. Figure 13.20).

A more modern version of the HID is the **dielectric barrier discharge ionization detector (BID)** from Shimadzu. It contains a two-stage cell (Figure 13.21). In the top stage, all sides of a He plasma discharge cell are protected by a dielectric coating, which enables a lower discharge current, preventing overheating and electrode damage. Helium flows down through this upper compartment and undergoes a series of reactions:

$$He \rightarrow He(A^1\Sigma_u^+) \rightarrow He(1S^1) ----h\nu ----\rightarrow$$
$$[M ---\rightarrow M^+ + e^-]$$

↑ Plasma   emitted He Photon at
                  17.7 eV   M = Analyte

The He plasma contains He atoms excited to high energy states as represented by the spectral term symbols in parentheses. The excited He atoms emit a very energetic

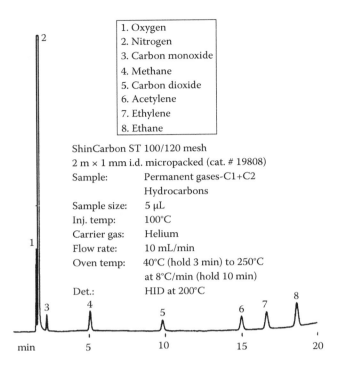

Figure 13.20 Separation and detection of fixed gases and low-molecular-mass hydrocarbons on high surface area molecular sieve using an HID. (Adapted with permission of Restek Inc., Bellefonte, PA, www.restek.com.)

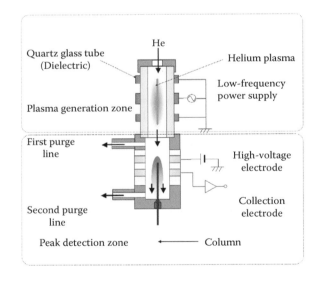

Figure 13.21 Design of the dielectric barrier discharge ionization detector (BID). (Used with permission from Shimadzu Corporation, Kyoto, Japan, www.shimadzu.com.)

photon (much higher energy than the UV photons of the PID) that can then ionize all other atoms or molecules except Ne, whose ionization potential is 21.56 eV (electron volts). The ionized target materials are collected in the lower cell, which is purged top and bottom to clear out uncollected ions and prevent them from entering the plasma generation cell. In general, this destructive detector is even more universal and orders of magnitude more sensitive than the TCD and equal or slightly more sensitive than an FID, but the dynamic range does not reach quite as high analyte percentages as these other two detectors. This is easily dealt with by diluting or reducing sample size or splitting the effluent and reinjecting or operating the BID in tandem behind a nondestructive TCD. The GC requires only a He gas supply, which has no flammability hazard. The detector can operate up to 350 °C and can in principle detect gases (H_2, H_2O, CO, CO_2, argon) up to high boiling liquids (with nearly uniform sensitivity over the range of C_8 to >C_{44} organics) in a single run, were a suitable column and elution conditions capable of handling such a wide range of molecular masses (2 to ~618) to be found! The mechanism of operation is like a combination of the PID and the FID, with more universality and no need for a UV lamp or an H_2/O_2 flame.

13.7.10 Atomic Emission Detector

Tunable, separately or simultaneously selective for compounds containing many specific atoms (e.g., C, O, N, S, P, Si, Sn, halogens, other metals); destructive; mass-flow detector; sensitivity and dynamic range vary with element measured (cf. Figure 13.15).

The AED employs a **microwave-induced He plasma** to dissociate eluted analyte molecules to their component atoms and excite them to emit at characteristic wavelengths. This is very similar to the mechanism in the argon plasma inductively coupled plasma source (cf. Section 8.3.1). A spectrometer with a diode array detector (Figures 8.26b and c) isolates and measures the intensity of sensitive emission lines unique to each element. Depending on the relative sensitivity and proportion of atoms in the molecules, separate element response channels may display peaks in several element-selective chromatograms. These data may be combined with retention information to additionally characterize the peaks, to separately quantitate coeluting peaks, and to suggest identification of unknown peaks based on their elemental composition. Like the ELCD, its response is atom selective and atom proportional, but each different element may be simultaneously detected, identified, and quantified. Agilent marketed such a system for a number of years, and after a hiatus, it was picked up by a German instrument company, joint analytical systems (lower case *sic*.; www.jas.de). It is particularly useful for detection and quantitation of the total N, F, S, and P contents of gasoline samples from a single GC injection or wear metal contents of used lubricating oils to identify failing aircraft or motor vehicle engines. Since it interfaces a chromatographic separation instrument with a multiwavelength spectrometric detector, it verges on being classified as a 2D hyphenated instrument. Such fully 2D instruments are the subject of the next section.

13.8 HYPHENATED GC TECHNIQUES (GC-MS, GC-IR, GC-GC, OR 2D GC)

Two-dimensional analytical techniques can combine two similar (e.g., 2D thin-layer chromatography [TLC] [cf. Section 12.3] or 2D gel electrophoresis [cf. Section 12.3]) separation procedures by sequentially varying the mobile phase compositions and developing with the second phase in a direction perpendicular to the first.

If we can (1) find a way to join two gas chromatography systems with differing stationary phases, (2) separately and repeatedly trap everything eluting from the first phase (column), (3) then repeatedly separate each trapped fraction on a second column, and (4) store and manipulate the data signals from a detector at the end of the second column, we can produce a comprehensive 2D chromatography system (e.g., 2D GC; cf. Section 12.3).

A final way to add a second dimension to chromatographic detection is to employ as a detector a spectroscopic instrument capable of sequentially and rapidly acquiring full spectra and storing them as a series of computer data files. In the case of GC, the two major hyphenated techniques of this type are GC-MS and GC-infrared (IR). There are four critical requirements for achieving this:

1. A spectrometric detector capable of acquiring full spectra quickly enough to sample the eluting peak at least 5–10 times over its width. This is a challenge in capillary GC, where peak widths may be as small as several seconds.
2. A method of **interfacing** the chromatographic effluent stream to the spectrometer's detection cell that presents the sample under the conditions required for proper detector functioning.
3. A computer data processing system fast enough to deal with the high information rate of such a detector, with sufficient data storage capacity for the many spectra to be acquired.
4. Efficient and intuitive software for viewing and manipulating these data files, extracting subsets of the information, and using it for qualitative identification and quantitative measurement.

13.8.1 GC-MS

If we consider their wide availability and capability, GC-MS instruments could have been said to provide the largest total amount of analytical power available to the instrumental analysis world. With the more recent spread of LC-MS instrumentation (Chapter 14) to serve the biochemical research market, they now must share this status. If analytes are volatile and thermally stable, capillary GC-MS can identify and quantitate hundreds (even thousands in GC-GC time-of-flight [TOF] MS runs) of them from a single injected mixture. The various types of mass spectrometers have been described in Chapter 10. It is the **quadrupoles** and **ion traps** that best meet the earlier-mentioned scan speed criterion. Magnetic sector instruments, with their higher mass resolution but slower scan speeds, will require that the GC peaks be broadened and slowed down. This works against the goals of faster analyses and better chromatographic resolution. On the other hand, the so-called fast GC, employing short, narrow-bore thin-film capillary columns, will require MS-scan acquisition speeds of 50–500 s^{-1}, which are attainable only by **TOFMS instruments**.

The major problem for GC-MS is interfacing. Mass spectrometers form and move their ions at highly reduced pressures, and the mass analyzer sections need to be operated at even lower pressures (10^{-5} or 10^{-6} torr). The effluent of a GC consists primarily of carrier gas around atmospheric pressure (760 torr). The MS vacuum pumps must remove this fast enough to maintain the necessary low pressure. With packed column flows of 10–40 mL/min, this was impractical. With capillary GC flows around 1 mL/min, modern diffusion pumps, or even better, powerful turbomolecular pumps, could achieve this—another reason to prefer capillary GC to packed column GC separations! Many modern GC-MS instruments simply introduce the end of the capillary column into the ion source (*direct coupling*). A separate pump evacuates the small, largely confined volume of the source, while the ions are extracted through a small orifice into the mass analyzer region, which is kept at a lower pressure by another pump. Such an MS design is called a *differentially pumped* system. To interface packed columns, a device called a **jet separator** (Figure 13.22) preferentially removed light He or H_2 carrier gas atoms or molecules from the effluent, allowing a smaller but enriched concentration of analyte molecules to enter the ion source at a lower total gas flow rate. The much higher speed of the lighter carrier gas atoms or molecules causes most of them to spread more widely in the vacuum of the separator chamber, and not to pass on through the conical orifice leading to the ion source. Jet separators are not much used or needed for GC-MS now, but we will encounter a similar design feature when we come to discuss the even more challenging problem of dealing with the much denser and larger mobile phase mass in LC-MS interfacing (Section 14.1.5).

In principle, one could monitor a GC effluent with a nondestructive detector like a TCD, pass it on through a suitable interface, and acquire and print out the mass spectra of peaks as they are eluting. With fast digital data converters and the speed and power of modern desktop PCs, it is better to simply acquire mass spectral data as a continuous sequence of spectra or selected mass fragment signals as the GC run proceeds. As described in Sections 11.3.1.1 and 11.3.1.2, improvements in computer data acquisition speed and memory storage capacity now make it possible to record and store complete MS line profiles (i.e., for all masses) instead of converting these on the fly to numbers of detector counts and calculation of the centroid of the *instrumentally distorted* peak top on the mass axis. This allows the now complete MS data acquisition files to be processed by computer programs that have stored data on MS instrument tuning distortions to correct each line to an ideal Gaussian shape whose MS peak area and position on the mass axis can be determined much more precisely. This enables "quasi" high-resolution MS determinations from "unit mass" instruments and better use of MS isotope peak areas to prune the number of possible elemental formulas.

A data file for a GC-MS analysis in computer memory consists of a sequence of MS scans ordered by retention time. There are three main modes of acquiring and employing such data files:

1. *Full scan*: If the file consists of a sequence of full-scan spectra, the total counts from ions of all masses in each scan may be summed, and these totals plotted against successive scan numbers (or equivalently, GC retention time). This display is called a **total ion chromatogram (TIC)**. Here, the mass spectrometer acts like an ionization detector, using whatever MS ion-source ionization mode was selected (EI, chemical ionization [CI], etc.). The TIC will appear similar to that obtained by an FID. One difference is that the FID output digitized by a chromatographic data system will likely define each peak at many more points across its width than the number of scans that the MS could make (unless it is the very fast scanning TOFMS; Section 10.2.3). For this reason, integration of FID peaks yields more accurate quantitation than TIC peaks. The advantage of the TIC data file is that by selecting any scan comprising a TIC peak, one can display the mass spectrum of the compound in the peak. If the peaks are sharp and narrow, the level of the analyte may vary significantly over the time of the scan, thus distorting the proportions of mass fragments in the spectrum. It would be best therefore to select a scan at the top of the peak. Even better would be to sum scans across the whole peak to get an average spectrum. Even better than that would be to select and sum an equal

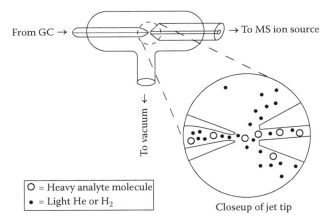

Figure 13.22 Diagram of GC-MS jet separator interface operation.

number of scans of background mass spectra near the peak where nothing else is obviously eluting. Then subtract that sum from the peak average sum, and thereby obtain a cleaner mass spectrum, with spurious fragments from column bleed eliminated. This illustrates the power and flexibility of manipulating a data file of continuous, contiguous MS scans. We have combined the great separation power of capillary GC to the great characterization and identification power of MS. This is particularly feasible because in GC the samples are in the vapor state, which is mandatory for the MS ionization and fragment identification process to proceed. The vast libraries of EI-MS spectra are based on this mode of operation. In many cases, it is possible to automate the process of peak detection, spectral selection, library spectra searching, and separated compound identification by matching to a library spectrum. A naïve person might be forgiven for wondering whether GC-MS eliminates the need for an analyst. Not quite, since the limitations are significant and were discussed earlier in the chapter on mass spectrometry. But the interfacing of vapor-separation GC with vapor-requiring MS, when combined with the fluency of PC processing of digital data files, is an excellent combination.

2. *Mass chromatograms (XIC)*: One may seek to locate and measure only certain categories of analytes in the chromatogram of a much larger and more complex mixture. These might have very characteristic mass fragments. An example would be mono-, di-, and trimethylnaphthalene isomers in a complex mixture of petroleum hydrocarbons such as a fuel oil. Their mass spectra consist mostly of a single intense molecular ion (M^+) at masses 142, 156, or 170, respectively. We can program a GC-MS data system to extract from the full-scan GC-MS files and plot only the ions of these masses in three separate chromatograms. Such plots are called **extracted ion chromatograms (XICs)**, from their mode of production, or **mass chromatograms**, since they display peaks, whose spectra contain the selected ion mass(es). If in the mass chromatogram of a petroleum sample at $M^+ = 142$, we observed a pair of large peaks close together, we might well suspect them of being from the two possible mono-methylnaphthalene isomers and could confirm our suspicion by inspecting the complete spectrum of the scans at the center of each peak. Thus, we may be able to determine the retention times of various compounds without injecting a standard. Unfortunately, mass spectra of isomers of such compounds are indistinguishable, so we will need some sort of retention information based on standards run on the particular GC stationary phase to say which isomer is which. A method of accessing such information without actually running the standards is described in Section 13.9. Such a process could be repeated at many other characteristic masses for the different classes of hydrocarbons. The data for all masses in the range scanned, at every point in the TIC chromatogram, are present in the full-scan GC-MS data file.

3. *Selected ion monitoring (SIM)*: In full-scan mode, the mass spectrometer is acquiring counts at any particular ion mass for only a brief portion of the scan. For example, if scanning from mass 50 to 550 each second, with unit mass resolution, the detector spends less than 2 ms at each mass during one scan. If we have only a few classes of analytes we wish to measure, and we know their characteristic major mass fragments, we can program the MS to acquire counts only at the selected masses, thereby increasing the dwell time at each mass, increasing the signal-to-noise ratio, and improving the sensitivity. If we wanted the three successive methylnaphthalene isomer distributions, we could monitor at only $M^+ =$ 142, 156, and 170. In 1 s, each of these three ions would be monitored for a little less than 330 ms instead of 2 ms, greatly improving the sensitivity over the 50–550 full-scan acquisition. In fact, we could cut the cycle time from 1 to 0.2 s, still acquire for 66 ms per cycle, but now be sampling and defining the GC peak shape five times per second instead of once per second. If we know that each class of isomers elutes over a unique range of retention times, we can set MS acquisition to monitor just their most characteristic and abundant ions during this period and achieve even greater sensitivity. This mode of acquisition is called SIM. SIM improves sensitivity by collecting more counts at the masses of interest, and it improves quantitative precision by enabling the GC-MS peak to be defined and integrated using more points. The extension of the linear range of measurement to lower values with SIM versus full scan is represented by the two sections of the GC-MS sensitivity range in Figure 13.15. Ultimate SIM sensitivity is a complex function of MS ionization efficiency for the particular analyte compound, number of different masses monitored, dwell time, and cycle time. Figure 13.15 indicates that in the most favorable instances, GC-MS-SIM can reach detection limits below that of the FID and in the range of the ECD. It is far more selective than any of the general GC detectors. This sort of analysis is characterized as **target compound analysis** since the system is tuned or programmed to select specific characteristic ions from specific target compounds expected to elute in specific retention time ranges.

To reiterate, the difference between XIC and SIM chromatograms is that in the former case, the desired ion masses are extracted from a full-scan data file, while in the latter case, the MS is directed to acquire the data only at those selected masses. The ability to improve sensitivity and quantitative precision by using SIM applies mainly to magnetic sector (including the use of high-resolution MS) and quadrupole MS instruments. The mode of operation of **ion-trap MS** and **TOFMS** instruments yields near-optimum sensitivity in full-scan mode. Thus, there is generally no provision for SIM acquisition on these instruments, and quantitation is done on XIC files. The MS in GC-MS is a destructive, mass-flow detector. Only with mass spectrometric detection can the analyst use the "perfect" IS, namely, the identical chemical species, labeled with stable isotopes of atoms with a higher mass (e.g., ^2H, ^{13}C, ^{15}N, ^{18}O). This procedure, **isotope dilution MS (IDMS)**, can correct for differences in sample preparation recovery, derivatization efficiency, MS ionization efficiency, and so on, for which the use of a different chemical species as an IS may not fully compensate. Review Section 12.12 for a discussion of chromatographic IS. However, IDMS IS materials are difficult to make and expensive to purchase, and commercial products are limited to only several thousand especially important target analytes.

We have really only touched the surface of the many and varied combinations of GC with mass spectral instruments. It is now possible to couple tandem mass spectrometry with GC, using combinations of quadrupolar analyzers and TOF analyzers in tandem. This discussion is beyond the scope of this book, but we provide Table 13.1 as a guide to what is possible.

The chromatographic system can be a source of spurious MS signals. These are more fully covered in Chapter 10.

13.8.2 GC-IR Spectrometry

At the concentrations eluting from GC columns, with analytes extensively diluted as vapors in the carrier gas, IR is not nearly as sensitive a detector as MS. Unfortunately, since IR spectrometry is traditionally carried out on analytes in a liquid or solid matrix, the libraries of vapor-phase IR spectra are much less extensive than those in the other phases, and the spectra differ significantly. Actually, the vapor-phase IR spectra are sharper, with greater interpretive structure than in the other phases, but comparable upper-range concentrations are unattainable. The spectra must be acquired during a rather brief passage of the peak through a detector cell. This makes acquisition by the rapid-sampling Fourier transform IR (FTIR) process (Section 5.2.2) mandatory. The beam from the FTIR passes from one end to the other of a long narrow cylindrical **light pipe** flow cell. The GC effluent enters and exits the sides of each end of this tube, while the ends are capped by flat windows of IR-transmitting material. The cell must be heated to prevent condensation of analyte vapors. The interior side walls are coated with an IR-reflective gold film, and the beam remains confined within the pipe by grazing-incidence total internal reflection.

The cell must have a volume small enough not to introduce extra-column GC band broadening. This conflicts with the need to make it as long as possible to increase the IR absorption from the dilute analyte vapor in the peak. Remember that an IRD is a nondestructive, concentration detector, and as such its absorbance signal is proportional to both concentration and path length. Something has to give, and it is sensitivity. The FTIR acquisition process could improve sensitivity if more time could be taken to acquire more scans. One way to achieve this is to stop carrier flow (best achieved by closing shut-off valves upstream and downstream of the detector) while the peak is in the cell.

For reasonable times, diffusion will not spread the band too much outside the cell. This **stopped flow analysis** is an awkward process. Chromatography of other peaks still on the column may slowly degrade during the pauses in flow, one has to know in advance exactly when to stop the flow, and the analysis time increases dramatically, especially if one wishes to do this for many peaks. The lower sensitivity range of GC-IR is reflected in Figure 13.15 (the IRD). The instrumentation is as expensive or even more so than much GC-MS.

Why then does anyone even want to do GC-IR? We have seen that MS information cannot distinguish between many isomers of the same relative molecular mass, such as the methylnaphthalenes mentioned earlier. Perhaps standards and retention data for the GC phase can help us distinguish them. If, however, several coelute, GC-MS cannot tell how much of each is in the peak or which one is present or absent. An example of this is seen in an important GC assay for "BTEX" aromatic hydrocarbons (Section 13.5, Figure 13.9b). On the PDMS column phase, *para-* and *meta-*xylenes coelute. All three xylenes have indistinguishable mass spectra. The carbowax-type phase will separate them (Figure 13.9a), but it is not as stable as PDMS if we must also elute other analytes later at much higher temperatures. If we inspect the IR spectra for xylenes in Figures 4.39 through 4.41, we can see significant differences in the positions of the large absorptions in the 1000–500 cm^{-1} region, in particular between coeluting *p-* and *m-*xylene. In the gas-phase spectra, these differences would be even sharper and more distinctive. We could monitor absorbance at these distinctive wavenumbers and separately quantify each xylene, even in the presence of a coeluting one. As with GC-MS, examination of the full spectrum of resolved GC peaks could enable identification even without retention data from standards (i.e., using spectral library matching). IR spectra will distinguish among structural isomers whose MS spectra will appear identical. Since the GC-IRD is nondestructive, after IR spectra are acquired, the peaks may be passed on to a destructive detector such as an FID or even an MS. A GC-FTIR-MS instrument is very complex and expensive but correspondingly powerful in its ability to characterize separated analytes!

13.8.3 Comprehensive 2D Gas Chromatography (GC × GC or GC^2)

At the beginning of Section 13.8, we described how an additional different GC column in series could add a second dimension to GC analyses. Now the second dimension is not a spectrum from a detector, but a very fast GC column with separation capability different from the first column. The speed of the second short fast column (cf. the discussion in Section 13.5 on "fast GC") enables it to quickly separate and elute effluent collected and briefly

GAS CHROMATOGRAPHY

Table 13.1 Combinations of Gas Chromatography with Mass Spectrometry

Method (with accepted acronyms and abbreviations)	Modes	Advantages	Limitations
Gas Chromatography Mass Spectrometry GC-MS (single quadrupole, SQ)	Scan, selected ion monitoring (SIM) and Scan/SIM	Relatively simple and relatively inexpensive; compound identification by library search; dynamic range = 10^5, but typically limited to >10^4; mass/charge range = 10^3–10^4, resolution (at m/z = 1000) 10^3–10^4 for most ions; well-developed hardware and software; used in standard protocols	Co-elution of compounds compromise library identifications; often user must interpret fragmentation patterns; SIM mode provides higher sensitivity but for target ions only, sensitivity decreases with increasing number of SIM ions; scan/SIM methods must balance scan and SIM sensitivity requirements
Gas Chromatography-Mass Spectrometry GC-MS (ion trap, IT)	Scan or selected ion monitoring (SIM); tandem MS-MS; gas or liquid chemical ionization	High sensitivity, compact design; tandem mass spectrometry is possible; dynamic range = 10^4, but typically limited to < 10^4; mass/charge range = 10^4–10^5, resolution (at m/z = 1,000) 10^4; well-developed hardware and software; used in standard protocols	Space charge effects can lead to relatively poor dynamic range, however for "clean" samples the dynamic range can be as high as single quadrupole units; specific libraries are limited; SIM is for target ions only with a sensitivity lower than that obtainable by a SIM analysis via single quadrupole, above; MS-MS analyses are limited to approximately 120–150 compounds per analysis; MS-MS analysis is typically slower than that done with a tandem MS-MS (process is done in time rather than space)
Gas Chromatography (time of flight) Mass Spectrometry, low-resolution GC-TOF	Scan or selected ion monitoring	Identification possible with good chromatographic separations, dynamic range = 10^4, mass/charge range = 10^5, resolution (at m/z = 1000) 10^3–10^4	Unit mass resolution limits identification capability; libraries are limited; if chromatographic separation is poor, comprehensive (GCXGC) might be needed; cannot distinguish neutral losses
Comprehensive Gas Chromatography (time of flight) Mass Spectrometry, low resolution GCXGC-TOF	Scan or selected ion monitoring; normal column configuration (nonpolar–polar) or reversed column configuration (polar–nonpolar)	Identification possible with good chromatographic separations in two dimensions on a nonpolar (long) and a polar (short) column; identification of families with help of principal component analysis tools; dynamic range = 10^4, mass/charge range = 10^5, resolution (at m/z = 1000) 10^3–10^4, well-developed hardware and software; used in standard protocols	Unit mass resolution limits identification capability; libraries are limited; GCXGC may not remedy all aspects of component coelution
Gas Chromatography Mass Spectrometry (triple quadrupole) GC-MS(QQQ, or QqQ)	Multiple and selected reaction monitoring	Provides product and precursor ion scans; very sensitive for target compounds or functional groups; developed to provide enhanced daughter ion resolution; relatively simple construction with straightforward scanning procedures; no high voltage arcing; dynamic range = 10^5, mass/charge range = 10^3–10^4, resolution (at m/z = 1000) 10^3–10^4	Unit mass, making identification ambiguous due to multiple structures as source of breakdown mass; empirical formula determination can be difficult, spectra and fragmentations must often be interpreted manually; full scan data can be acquired (similar to the single quad procedure) by turning off the collision cell
Gas Chromatography (time of flight) Mass Spectrometry, high-resolution GC-TOF	Scan or selected ion monitoring, tandem MS-MS	Might obviate the need for GCXGC separations with accurate mass determinations; deconvolution software can aid in identification of multiple components under peaks; sensitivity intermediate between multiple reaction monitoring and product ion scan of a QQQ; dynamic range = 10^4, mass/charge range = 10^5, resolution (at m/z = 1000) 10^3–10^4	Very large data files (currently approaching 2 Gb), software is currently developmental
Gas Chromatography Tandem Mass Spectrometry GC-QTOF	Scan or selected ion monitoring, tandem MS-MS	Might obviate the need for GCXGC separations with accurate mass determinations; isolation of parent ions and subsequent fragmentation provides identification; will detect any daughter ion passed into the TOF; sensitivity intermediate between multiple reaction monitoring and product ion scan of a QQQ dynamic range = 10^4, mass/charge range = 10^5, resolution (at m/z = 1000) 10^3–10^4	Large data files are produced; Sophisticated software is needed for processing and deconvolution; requires accurate mass and high resolution

(taken from the book by Bruno and Svoronos).

stopped from the much longer and slower separation on the first column. Because the second column uses a phase with different polarity characteristics than the first one, it may separate components coeluting in the peak trapped from the first column. What is the trick for repeatedly trapping and then transferring sequential segments of the chromatogram in the first column for separation on the second one? A length of uncoated fused silica capillary tubing connects the two. It is alternately flash cooled to **trap** and then flash heated to **desorb** segments into the second column. In some versions, the flash cooling is done by jets of liquid N_2, and the flash heating by electrically conductive paint on the outside of the bare capillary. This is the same material used for electrically heated automobile rear window defrosters. The very high heat conductivity of the thin-walled fused silica capillary facilitates these processes. This interfacing device is known as the **thermal modulator**. Data from the detector at the end of this pair of serially coupled columns are acquired and stored as a sequence of fast separations (second-column GC "scans") at equal retention time intervals from the first column. This is analogous to full-scan GC-MS data files. The computer system displays the data set as a planar 2D array of spots, looking like a developed 2D TLC plate or 2D gel electrophoresis plate. Each "spot" is a peak, and quantitation is by volume integrated over peak area, instead of area integrated over peak baseline as in 1D GC. Many coelutions are thus resolved. The technique is called **comprehensive** because all components injected are measured, and most are resolved from one another. In this sense, it differs from **heart-cutting 2D GC**, where only a selected peak is diverted by a valve to a second slow GC column for additional separation. Using comprehensive GC × GC-FID, a light fuel oil like one whose 1D separation is illustrated in Figure 13.23, has been resolved into 3000–4000 quantified peaks! The size and resolution of the textbook page is hopelessly inadequate to display this.

An additional valuable feature of the thermal modulator is that it acts in a fashion similar to the retention gap (cf. Figure 13.8). A slow-moving, late-eluting peak will be gradually accumulated off the first column during the cold trapping phase. Then it is quickly flashed back to vapor, and much more rapidly transferred to the head of the second column, where it is "refocused" to a very narrow band like that of an initial retention gap injection. The fast separation on the second column retains this very narrow peak width, so at the detector the peak height is correspondingly increased from what it had been exiting the first column, yielding a great improvement in signal to noise at the peak top. In some circumstances, the S/N improvement can be an order of magnitude or more, making it worthwhile to use a thermal modulator

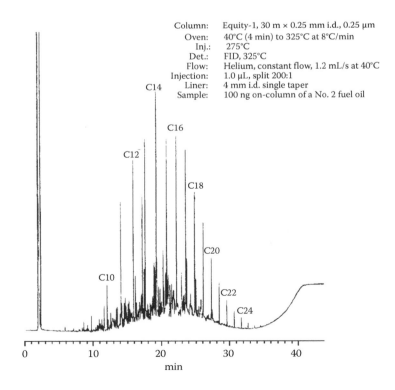

Figure 13.23 Linear programmed temperature GC-FID analysis of hydrocarbon fuel oil. (Reprinted with permission of Supelco, Inc., Bellefonte, PA,. www.sigmaaldrich.com.)

13.9 RETENTION INDICES (A GENERALIZATION OF RELATIVE R_T INFORMATION)

In the last paragraph of Section 13.8.2, we saw in the case of xylene isomers that even the great compound fingerprinting capability of full-scan MS detection could not distinguish among them. The differences in elution times on a GC column could resolve this problem, if we had run authentic standards of each isomer on the system to establish where each one elutes. What if no standards are available in the lab? We know that absolute values in minutes of R_t will depend on many factors: the column phase and exact dimensions, temperature program, and flow rate. Such numbers from someone else's lab will be useless. How about relative retention times (RRTs) (Section 12.11)? This might get us closer, but the values will still vary significantly with the temperature program used. Elution orders and RRT among compounds will differ with and be specific to the stationary phase in use. The best system devised to codify data on these parameters is a library of compound **retention index (RI)** values, specific for each stationary phase. The most common such system was devised by Kovats.

Compounds of similar structure, differing by only the number of –CH_2– units in linear hydrocarbon chain substituents, or in the number of the same substituents on aromatic rings (like the methylnaphthalenes), are called homologs. The simplest example is the sequence of **n-alkanes** (methane, C1; ethane, C2; propane, C3;… octane, C8;… hexadecane, C16; etc.). The n-alkane elution sequence on most stationary phases displays a simple regularity. Under isothermal GC conditions, their adjusted retention times t'_r (cf. Section 12.7) follow a **logarithmic** progression. In the **Kovats' RI system**, each is assigned an RI value of 100× its carbon number; thus, octane has an RI of 800 and nonane an RI of 900. By definition, these numbers are the same on any GC phase. The RI of any other compound (U) is calculated by substituting the adjusted R_t values of both the compound and the alkane pair between which it elutes into the following equation:

$$\text{RI} = 100\left[n + (N-n)\frac{\log t'_r(U) - \log t'_r(n)}{\log t'_r(N) - \log t'_r(n)}\right] \quad (13.1)$$

where n and N are the number of carbon atoms in the smaller and larger alkanes, respectively. This formulation gives the official, most accurate, thermodynamic evaluation of a compound's RI. The measurement and calculation are fairly complex. One must measure in the appropriate isothermal temperature range, use the dead time t_m to calculate t'_r values from measured t_r, and use moderately complex Equation 13.1. A shortcut in the use of Kovats' retention indices is to simply measure a series of n-alkanes on a given column (and temperature), and plot the retention time (or the adjusted retention time, subtracting for the void volume of the column) against the Kovats' retention index. For the n-alkanes, this will result in a near perfect straight line (with uncertainty only in the retention time axis since the index axis is fixed by definition). Thereafter, the Kovats' retention indices can be read off the graph by noting the corresponding point on the retention time axis. This approximation is most useful when identifying many related compounds on a chromatogram, such as in the analysis of liquid fuels. Note that methods for determining the void volume or more correctly, the gas holdup volume, are covered extensively in the book by Bruno and Svoronos.

GC runs covering a wide range of analyte boiling points will generally be done by programming temperature using a linear ramp covering the desired range of elution temperatures. Figure 13.23 illustrates a GC-FID chromatogram of a hydrocarbon fuel oil. The n-alkanes from C10 to C24 are seen under these conditions to elute at almost regularly spaced intervals. Correspondingly, a close estimation of RI can be made much more simply:

$$\text{RI} = 100\left[n + (N-n)\frac{t_r(U) - t_r(n)}{t_r(N) - t_r(n)}\right] \quad (13.2)$$

Note the use of actual rather than adjusted retention times. Although not as accurate or conforming to the operational definition of RI as defined in Equation 13.1, use of Equation 13.2 with retention times acquired over the linear ramp portion provides a useful approximate value. Essentially, the RI values of unknowns are calculated by linear interpolation between the values of the flanking alkanes. If one had the PDMS phase RI values for the three dimethylnaphthalenes illustrated in Figure 13.13, one could probably assign identity to these compounds, whose mass spectra in GC-MS would be indistinguishable. To be certain, however, one would need to have RI values for all such isomers (there are many more than these three). If there were some others whose RI values were very close to those observed in that chromatogram, it would be dangerous to rely on RI to clinch the identification. At best it would tell which were the most likely candidates, and then the confirmation standards could be selected. If RIs are within less than one unit of one another, coelution is a possibility, and resort to a different column that will resolve the candidates (as in the xylene case of Figures 13.11a and b) is indicated. Tables of RIs for the compounds on different phases will aid in selecting such an appropriate column. In general, RI values are useful for aiding in sorting out small sets of isomer elutions. There are far too many possible compounds relative to the number of reliably distinguishable RI values to make these values useful for identifying general unknowns. Kovats indices are an adjunct to identification with hyphenated GC methods rather than a substitute for them.

13.10 SCOPE OF GC ANALYSES

Appropriately configured GC instrumentation can separate and identify molecules ranging from the very smallest (H_2) to those with masses on the order of 1000 Da. This upper mass range limit applies only for compounds with great thermal stability at high temperatures and very low polarity leading to minimal boiling points for a given mass. The prototype of such compounds are the *n*-alkanes, the reference series for the Kovats RI system, which present the characteristic "picket fence" profile of a homologous series when they elute under linear temperature programs. This is illustrated in the FID chromatogram of a light fuel oil in Figure 13.23. Heavier petroleum crude oils might extend this pattern out to *n*-C60 or beyond. The limiting factors are as follows:

1. The ability to efficiently volatilize the compound.
2. A stationary phase that will not decompose at a temperature sufficient to partition the analyte into the mobile phase for a large enough percentage of the time. This will enable it to pass through the column in a reasonable time and come off the column quickly enough to produce a peak well above the noise background. Recall that very slowly eluting compounds with long retention times elute so gradually as they exit the column into the detector that their peaks spread out and eventually become not much higher than and indistinguishable from background noise and baseline drift.
3. The column tubing itself must be stable at the necessary temperature. The polyimide scratch-protective outer coating of fused silica capillaries is itself stable only to about 380 °C. Some exceptional stationary phases can exceed this limit and may require the use of metal capillaries, which for some analytes at those temperatures must be deactivated—perhaps with an inner coating of silica, which will not yield a flexible column coil.

The strength of GC relative to other chromatographic techniques, which combine separation, identification, and quantitation, is the number of compounds in a mixture that it can characterize in a single run. The illustration of capillary GC-FID analysis of the fuel oil in Figure 13.23 only suggests what has been resolved. Magnification of the smaller peak regions between successive *n*-alkane peaks would show perhaps 15–20 resolved hydrocarbons. The use of GC-MS SIM or GC-MS XIC chromatograms could tease out many times more classes of hydrocarbon peaks that are too small and overlap too much to be distinguished in the FID chromatogram. Among these would be a majority of the methylnaphthalene isomers we discussed earlier, as well as many other such extended "families" of hydrocarbons. In the comprehensive 2D GC-FID run of a similar fuel oil described in Section 13.8.3, 3000–4000 peaks were resolved and quantitated (approximately—certainly standards for many of these were not available!). This spectacular performance is a tribute both to the extreme resolution of multiple capillary GC separations and the sensitive, wide-dynamic-range FID.

13.10.1 GC Behavior of Organic Compound Classes

Hydrocarbons, aliphatic or aromatic, have low polarity and high stability, and perform well at molecular masses limited only by the three factors listed earlier. Organic compounds with only ketone groups (R–(C=O)–R′) or ester groups (R–(C=O)–O–R′) or ether groups (R–O–R′) or halide substituents (–F, –Cl, –Br, –I) are somewhat more polar and will have higher boiling points for their mass but are still relatively unreactive to components of the GC.

Compounds with strong hydrogen bonding groups or ionizable H^+, like alcohols, carboxylic acids, and phenols as well as compounds with basic functionality, including amines, amides, and aromatic ring nitrogens, may react with column active sites and produce tailing peaks. They are more prone to decomposition at higher temperatures. It requires specially engineered phases and carefully deactivated columns to handle even the lighter molecules of this sort. The heavier they become and the more such features they possess, the less likely they are to make it through the GC while exhibiting good chromatographic behavior. Really active compounds like strong acids or bases or water can destroy a column after one or two injections at elevated temperature. Ionizable compounds that form salts will also fail to elute. One must use highly specialized columns, such as the one illustrated in the chromatogram of Figure 13.7 to handle such compounds by GC.

13.10.2 Derivatization of Difficult Analytes to Improve GC Elution Behavior

One way to deal with the recalcitrant compounds described in the preceding paragraph is to modify the troublemaking functional group(s) to a more tractable form. Their reactivity and polarity are the cause of their bad behavior. We can use this reactivity to add a substituent that converts them to a less-active, less-polar group. A few examples are provided below, however a comprehensive discussion and list of derivatization reagents can be found in the book by Bruno and Svoronos.

1. We can **methylate** alcohols (–OH) to methyl ethers or **esterify** organic acids (–COOH) to esters. An excellent

reagent for this is **diazomethane**, CH_2N_2. This is a toxic gas that must be generated onsite, dissolved in ethyl ether, and handled with great care in a fume hood. Upon addition of its ether solution to a solution (not in alcohol or water) containing analytes, their alcohol or acid functionalities are instantly converted (methylated), the by-product is inert nitrogen gas, and excess reagent is quenched by adding water.

2. We can deactivate both acids and alcohols by replacing their end hydrogen with a **trimethylsilyl** ($-Si(CH_3)_3$) group and also tame the basic amines by bonding this group to their N atom. The silyl esters, ethers, or amides thus formed present a relatively nonpolar trimethyl face to the stationary phase, where earlier an active, polar irritant was present. Despite the larger mass added by this group, such derivatives are actually more volatile and will elute faster at lower temperatures. Derivatizing agents such as trimethylchlorosilane or hexamethyldisilazane or a host of others are used to accomplish this.

   Such agents function less instantaneously than diazomethane. They react faster at quite high temperatures. One way to achieve those has been to mix them with the injection solution. Upon flash vaporization in the hot injection port or the start of the column, the derivatization reaction is nearly instantaneous and complete. This procedure is known as **on-column derivatization**. Such a mode of introduction of this reagent also achieves the purpose of reacting with and deactivating the free –Si–OH, active site groups in the column, as described at the very end of Section 12.5. This treatment is called **silanization.**

3. We can **acylate** alcohols to esters by reacting them with active organic acids or anhydrides. For example, heptafluorobutyric anhydride derivatizes R–OH to R–O–(C=O)–C_4F_7. After quenching excess anhydride, the liberated heptafluorobutyric acid must be left behind in an aqueous phase by extracting the derivative into an organic layer under basic buffer conditions. The acid is extraordinarily strong and just a trace in a single injection could utterly destroy even a robust PDMS stationary phase.

   The heavy –C_4F_7 group nevertheless is so much less polar than the –OH it replaced (think "Teflon") that the derivative will elute more easily than the parent compound. An additional advantage is that all those electronegative F atoms yield a derivative with ECD sensitivity, making our analyte responsive to a more sensitive and selective detector. Derivatizing to add groups with heavier halides like Cl or Br would confer even greater ECD sensitivity, albeit with perhaps less volatility. Incorporating Cl or Br atoms into the analyte derivative would produce MS fragments with the characteristic isotope cluster patterns of these halides (cf. Chapter 11), which would make them stand out when using GC-MS detection.

To summarize, derivatization for GC can confer greater thermal stability to an analyte. It can convert the analyte to a form that is more volatile, less polar, and with less tendency to produce tailing peaks by interaction with column active sites. It can introduce elements into it that make it detectable by more sensitive or selective detectors (remember, in real-world analyses, selectivity often confers more effective sensitivity to an assay). The scope of derivatization for GC is far greater than the several examples listed here. See the bibliography (Grob, Jennings, Scott) or product literature from specialty derivatization agent vendors (Appendix 13.A) for more examples.

13.10.3 Gas Analysis by GC

Not surprisingly, GC is the method of choice for the analysis of mixtures of gases (i.e., compounds that are vapors at room temperature). The procedures for sampling and introduction to the GC instrument sometimes differ from the standard syringe injection or SPME techniques (Section 13.3) employed for neat liquids or dilute solution samples. Gases at trace levels in air or dissolved in water may be present at levels too low to measure with the volumes one might sample and inject with a syringe, even one of several milliliters volume. Certainly, one would not wish to introduce several milliliters of water into a GC column!

Trace levels of gases in air can be sampled by drawing large volumes of air (up to cubic meter volumes) with a pump through a **trap** (a tube or plate filled with a special sorbent packing) for periods up to an hour or more. One especially useful sorbent is a packed bed of **Tenax**® particles. These are beads of polyphenyl ether polymer [(...–Φ–O–Φ–O–...), where Φ = 6-carbon aromatic *para*-phenylene ring]. Some varieties have additional phenyl rings substituted on the phenylene units in the polymer backbone. At room temperature, this will absorb small organic vapor molecules while letting fixed gases and, most importantly, water vapor pass through. After these are concentrated from a large volume of air slowly sampled over time, a trap oven heats it rapidly to desorb the mixture almost instantaneously to a GC configured for low-molecular-mass gas analysis. An alternative form of trap employs **activated charcoal**. Thermal desorption is not as efficient with this, so the trap packing is emptied after sampling, and the analytes are desorbed by carbon disulfide solvent, which is then filtered, concentrated, and injected into a standard low-temperature GC system. This method works best for a range of higher-molecular-mass vapor molecules than the range covered by thermal desorption from Tenax. Figure 13.20 displays elution of both fixed gases and low-molecular-mass organics from a new GC column that can handle both while operated near room temperature. Trace level sampling for this type of analysis would probably require liquid N_2 **cryotrapping** for the fixed gases after Tenax trapping for the less volatile small organics. It is remarkable that a single GC column phase can handle separation of both classes simultaneously.

Trace levels of gases or volatiles in water can be sampled by **purging** a water sample with a cloud of gas bubbles (N_2 or He) introduced through a **frit** at the bottom of the sample vessel. The purged vapors and large amounts of entrained water vapor pass through a trap packed with Tenax or porous polymer beads that does not retain the large excess of water. After the analyte vapors are completely purged from the sample and captured on the trap, they are thermally desorbed to a GC column separation and detection system in a fashion similar to the procedure for the trace gases in air. Dedicated instruments connected to a GC are designed to sequentially perform such a procedure on multiple samples of water and are called **purge and trap samplers**.

These traps for volatiles at room temperature behave like GC columns operated at too low a temperature to have significant elution in a reasonable time. Since the sample is being loaded constantly over a long sampling time, eventually the trap may become saturated with analyte, and/or slow "elution" through the trap tube will eventually cause analytes to start coming out the downstream end of the trap. At this point, the amount on the trap reaches an equilibrium value and is no longer proportional to the time-averaged concentration in the volume sampled. **Breakthrough** is said to have occurred. Often a smaller trap will be added in series behind the main trap. If it is first separately desorbed and analyzed, and no analytes are detected, and it is then removed or left at the desorption temperature, one may be confident that breakthrough has not occurred on the primary trap and proceed to desorb and analyze its contents.

13.10.4 Limitations of GC

GC serves well for analyzing many mixtures of compounds significant for environmental pollution monitoring, checking processes and contaminants in synthetic chemical manufacture, characterizing complex petroleum hydrocarbon mixtures, and of course it shines at the low end of the compound volatility range. The huge area of application of chromatographic instrumentation in the twenty-first century is in biochemical, medical, and pharmaceutical research. The chemistry of life sciences is mostly chemistry in aqueous media, and many of the compounds are correspondingly polar to function in this very polar medium. The complexity of life processes demands large, complex molecules: enzymes, proteins, receptors, hormones, the double helix of deoxyribonucleic acid (DNA), ribonucleic acid (RNA), nucleotides, and so on. These things are generally polar, full of acidic and basic functionalities, and too large to be volatilized into a GC without suffering catastrophic thermal decomposition, even if derivatization is attempted. Chromatography in the liquid state, at temperatures closer to physiological values, will be required to separate, identify, and quantitate these complex mixtures. GC led the way in instrumentation for chromatography, but the torch is being passed to those methods that are carried out in liquids. One does not see many GC instruments in genomics or proteomics laboratories. We next focus our attention in Chapter 14 on the instruments and methods for chromatography in the liquid state.

PROBLEMS

13.1 Indicate whether packed GC columns or open tubular GC provides the best performance in each of the following categories:
1. Analyte capacity.
2. Peak resolution.
3. Number of peaks resolvable in a given time.
4. Operation of detectors without need for makeup gas flow.
5. Ease of interfacing to mass spectrometer.
6. Ease of interfacing to FTIR spectrometer.
7. Operability without need for separately heated injection port.
8. Capable of using the longest columns.
9. Capable of producing the most inert "deactivated" column.

13.2 Indicate which injection mode (split or splitless) best fits or meets the criteria in the following:
1. More sensitive detection.
2. Less solvent peak tailing.
3. Most appropriate for injection of "neat" samples.
4. Not appropriate for packed column application.
5. More likely to need initial temperature hold before temperature program begins.
6. Preferred for operation with gas sampling valve injection.
7. Works best when doing SPME desorption.

13.3 What is the effect on retention time (longer, shorter, unchanged) of the following?
1. Septum coring.
2. Loss of stationary phase by column bleeding.
3. Operating column at higher temperature.
4. Operating column at higher head pressure.
5. Clipping off front 5 % of column to remove nonvolatile contaminants.
6. The air peak's elution on a PDMS column when temperature is raised.
7. Changing carrier gas from He to N_2, if flow, temperature, and column are the same.
8. A small leak where the column is attached to the detector.
9. A small leak where the column is attached to the injector port.

13.4 Describe the sequence of events that would lead to septum bleed giving rise to spurious GC peaks.

13.5 Why does column bleed not give rise to such spurious peaks? What is its GC signature? Are you more likely to observe this signature in isothermal or programmed temperature GC? Why?

13.6 Answer the questions for each of the following analyte mixtures:
1. HCN volatilized from whole blood.
2. Pure gasoline.
3. Organophosphorus pesticides at 1 ppm each in hexane.
4. Trace volatile chlorocarbons in swimming pool water.
5. Trace inert gases in air.
6. Rapidly repeated sampling of major hydrocarbons in natural gas pipeline.
7. Dimethylbenzothiophene ($C_{12}H_8S$) isomers in crude oil.
8. <50 ppm polychlorinated biphenyl mixtures in transformer fluid.
9. Two completely unknown peaks at 0.1 % concentration each in toluene
10. BTEX vapors in refinery air.
   A. Which of the following sampling and injection methods would be best for each sample?
      a. Split injection.
      b. Splitless injection.
      c. SPME.
      d. Headspace SPME.
      e. Gas valve injection.
      f. Tenax trapping with thermal desorption.
      g. Purge and trap sampling.
   B. Select one or more detectors from Section 13.7 (and only one example from hyphenated methods, Section 13.8) that would be most suitable to analyze these samples.
   C. Which of the following columns would be best for each sample?
      a. 10 m Porapak PLOT Q.
      b. 20 m × 0.25 mm × 0.25 μm PDMS.
      c. 5 Å molecular sieve.
      d. 40 m × 0.18 mm × 0.1 μm PDMS.
      e. 10 m SolGel-WAX column.

13.7
A. Draw a representative structural formula of a portion of a PDMS polymer phase deposited on fused silica.
B. Draw the structure for such a cross-linked PDMS phase.
C. Draw the structure for such a bonded PDMS phase.
D. Draw the structure for such a bonded and cross-linked PDMS phase.

13.8 Using the chromatographic system pictured in Figures 13.10 and 13.11 to analyze two trichloronaphthalene isomers
1. Which selective detector from Section 13.7 is most suitable?
2. With N_2 carrier gas, what flow rate (cm/s) should be used?
3. With H_2 carrier gas, what flow rate (cm/s) should be used?
4. How would you measure and set these flow rates?
5. Which compound would be most suitable to use to measure the linear flow rate (Cl_2, CH_3Cl, or chlorobenzene)?
6. What auxiliary gases (if any) will be needed to operate with H_2 carrier gas?
7. If the Kovats RI of two of the analytes are known to be 915 and 920, what temperature would be suitable for an isothermal analysis to separate this pair?
8. If chloronaphthalenes with more chlorines are present as well, what temperature program should be used?
9. If the first pair cannot be separated, describe three changes in the column that might be tried to achieve this.

13.9 Which detector (from Section 13.7) is most suitable for each of the following analytes:
1. Neon.
2. Krypton and xenon in air.
3. Nitrogen in air.
4. CO_2.
5. Trace carbon tetrachloride in hexane.
6. Trace "eau de skunk," *n*-butyl mercaptan, C_4H_9SH.
7. Trace PCB mixture in solvent.
8. Trace trinitrotoluene (TNT) vapor in air.
9. Mixture of FAMEs.
10. Heavy crude oil sample.
11. Trace BTEX mixtures on handheld monitoring GC instrument.

13.10 Which hyphenated GC system detector and mode of detection (listed in Section 13.8) would be most suitable for analysis of the following samples:
1. Identification by library searching of 100 possible different pesticides in vegetation extracts.
2. Finding traces of mono-, di-, and trimethyldibenzothiophene isomers in a fuel oil sample.
3. Identifying the compounds found in (2) against standards of the authentic isomers.
4. Identifying the compounds in (2) against library spectra.
5. Identifying the maximum possible number of components in a fuel oil sample over the widest concentration range.
6. Detection of small amounts of an unanticipated class of isomers in a hydrocarbon mixture, followed by identification of the compound class, followed by quantitation of relative amounts of the peaks for this subset of compounds assuming they all had similar relative response factors (RRFs).

13.11 Retention data for three dimethylnaphthalene isomers (D1–D3) and the *n*-alkanes between which they elute in Figure 13.11 are listed in the following table:

Temperature	Compound	$R_t(n)$	$R_t(N)$	$R_t(U)$
100 °C	D1	9.55	15.20	14.80
	D2	15.20	29.80	16.40
	D3	15.20	29.80	17.20
120 °C	D1	4.84	7.60	7.37
	D2	7.60	12.60	8.03
	D3	7.60	12.60	8.47
150 °C	D1	2.43	3.38	3.23
	D2	3.38	4.57	3.61
	D3	3.38	4.57	3.78
Program	D1	5.21	6.67	6.61
	D2	6.67	8.92	6.97
	D3	6.67	8.92	7.21

The retention times are in minutes for preceding (*n*) and following (*N*) *n*-alkane peaks and for each of the three D1–D3 unknowns (*U*). As mentioned in the text, the column dead time was 1.08 min.

Using the formulas of Equations 13.1 and 13.2, set up and use a spreadsheet to calculate the RI values for each of the three compounds at each of the four temperature conditions. If you cannot program this, calculate the values with a hand calculator, which will take lots more time. How well do the RI values compare? Some of the variation is due to inaccuracies in the simulation program used to generate the data, but some reflects how these values depend on the conditions under which they are acquired.

How well do the values approximated from the programmed temperature formula compare with those of the theoretically more accurate isothermal runs? Which isothermal run do you think should provide the best RI values?

Perhaps the Ris vary more with temperature because the analytes are aromatic and the reference compounds are aliphatic. If you have the spreadsheet set up, measure the retention times in the figures and put in the appropriate values for the triplets of branched hydrocarbons C10s or C14s and see whether their Ris are more consistent with variation of temperature.

There are actually 10 possible dimethylnaphthalene isomers. See if you can draw them all. Naphthalene consists of two hexagonal rings of six carbons sharing an edge (so there are ten carbons in all). The eight carbons to which substituents can be attached are numbered clockwise from 1 to 8 around the ring starting at the top right side. Suppose we obtained a GS-MS-SIM chromatogram of a petroleum oil at mass 156, which shows a cluster of six peaks, three of which match the Ris of D1–D3, and two of these are "fatter" than the other four.

1. Why might we not be seeing all 10 isomers (more than one reason)?
2. How might we attempt to find the others?
3. How would running authentic standards of all 10 isomers help resolve the uncertainties?
4. How would having RI data on all 10 help? As much as the standards? Why or why not?
5. You find another mass 156 peak in the SIM chromatogram with an RI of 1100. Is that one of the missing isomers? What might it be? If you had acquired data with an ion-trap MS, how might you check this out?

13.12 Suppose all the compounds eluting in the chromatogram of Figures 13.10 and 13.11 had the same detector RRF and were all present at the same concentration. How would the peak heights differ from the "phony" artificial ones in the simulated chromatogram? Sketch the appearance of the actual chromatograms for the 100 °C isothermal run and the programmed temperature run using the same retention time scale. What causes the difference in peak heights, areas, and widths?

13.A APPENDIX: GC INTERNET WEB RESOURCES

The major vendors of GC instrumentation and consumables maintain elaborate websites. One can access these by putting the name of the company into a search engine (e.g., Google) and finding the main home site, then working one's way through the menu tree to find different types of instrumentation, separation media and small accessories (2D plates, separation columns, fittings, injector hardware, detectors, calibration standards, etc.), suggestions for products addressing common needs, and a host of sample chromatograms.

Major GC Instrumentation Companies: **Agilent, Bruker, PerkinElmer, SGE, Shimadzu, ThermoScientific**

Major Separation Media and Small Parts Suppliers: **Agilent, PerkinElmer, Phenomenex, Inc., Restek, SGE, Sigma-Aldrich/Supelco**

There are many smaller players in this market, some with URLs listed in figure captions, but the incomplete listing will give the student a start. There are literally thousands of separation media and application notes and sample chromatograms available online. If one has a specific problem or application, a better alternative may be to find the "Contact Us" e-mail address or phone number for their applications lab staff and pose a specific query. Free trade journals such as American Laboratory and LC/GC magazine have annual comprehensive vendor resource issues, indexed by methodology and application. Their content is easily available from their websites.

SPECIAL GC CALCULATION PROGRAMS

www.restek.com/chromatogram/ezgc. GC chromatogram modeler program (paid subscription service), free demo of operation and product description. Select your analytes; it suggests the best column and conditions, then calculates and presents a theoretical chromatogram.

www.sigma-aldrich.com/analyticalchromatography/gas-chromatography.html. GC learning center, GC applications, column selection guide, literature.

www.sigma-aldrich.com/analytical-chromatography/sample-preparation/spme/literature.html. Library/Applications/SPME Instruction and video of process.

www.chem.agilent.com/en-US/promotions/Pages/Gcapp.aspx. GC pressure-flow calculator—plug in column parameters, applied pressure, carrier gas, etc., and it will use the equations of pressure and flow in GC columns to show how variation of parameters affects these values.

www.chem.agilent.com/en-us/Technical-Support/Instrumentssystems/GasChromatography/literature/Pages/gcmethodtranslation.aspx. Calculates how elution of a set of compounds on one system changes when the parameters in the pressure-flow calculation above change.

TWO LARGE PUBLISHER CHROMATOGRAPHY WEBSITES

www.separationsnow.com. Wiley newsletter: ads, webinars, podcasts (cf. author ESF @ Pittsburgh conferences interviewing editors' award recipients). A mother lode of chromatography information.

www.chromatographyonline.com. Website of LC/GC magazine, covers all types of chromatography, search by keyword, issues/articles archive 1983–present, author–subject index.

BIBLIOGRAPHY

The reader will note that occasionally multiple editions of books are cited as sources or reference works. The reason for this is that sometimes different figures are used in different editions, which may in fact be written by different authors. The sources found in the bibliography will be the most recent versions of a particular work.

In some cases instrument manufacturers and suppliers are cited by providing a URL to the exact internet location. We have endeavored to ensure that these links are functional at the time of publication, but clearly these links change. If a link has become nonfunctional, a good approach is to use your browser's advanced search capability. For example: "company name" AND "instrument type" AND "specific descriptors" is a good place to start.

Barry, E.F.; Grob, R.L. *Columns for Gas Chromatography: Performance and Selection*. Wiley: New York, 2007.

Blumberg, L.M. *Temperature-Programmed Gas Chromatography*. Wiley: New York, 2010.

Bruno, T.J., Applications of the vortex tube in chemical analysis, *Proc. Contr. Qual.*, 3, 195–207, 1992.

Bruno, T.J., J. Simple, quantitative, headspace analysis by cryoadsorption on short alumina PLOT columns. *Chromatographic Sci.*, 47, 569, 2009.

Bruno, T.J., Svoronos, P.D.N., *CRC Handbook of Basic Tables for Chemical Analysis – Data Driven Methods and Interpretation*, 4th Ed., CRC Taylor and Francis, Boca Raton, FL, 2021.

Cazes, J. *Encyclopedia of Chromatography*. Marcel Dekker Inc.: New York, 2001.

Grob, R.L. *Modern Practice of Gas Chromatography*, 4th ed. Wiley: New York, 2004.

Hubschmann, H.-J. *Handbook of GC-MS: Fundamentals and Applications*. Wiley: New York, 2008.

Jennings, W.; Mittlefehlt, E.; Stremple, P. *Analytical Gas Chromatography*, 2nd ed. Harcourt Brace: Orlando, FL, 1997.

Koel, M. *Ionic Liquids in Chemical Analysis*. Taylor & Francis: Boca Raton, FL, 2008.

McMaster, M. *GC/MS: A Practical Users Guide*, 2nd ed. Wiley: New York, 2008.

McNair, H.; Miller, J.M. *Basic Gas Chromatography*, 2nd ed. Wiley: New York, 2009.

Niessen, W.M.A. (ed.) *Current Practice of Gas Chromatography—Mass Spectrometry*. Marcel Dekker, Inc.: New York, 2003.

NIST Chemistry WebBook, NIST Standard Reference Database Number 69, Eds. P.J. Linstrom and W.G. Mallard, National Institute of Standards and Technology, Gaithersburg MD, 20899, https://doi.org/10.18434/T4D303, (retrieved February 17, 2022).

Rood, D. *A Practical Guide to the Care, Maintenance, and Troubleshooting of Capillary Gas Chromatographic Systems*, 3rd revised ed. Wiley: New York, 1999.

Scott, R.P.W. *Chromatographic Detectors*. Marcel Dekker, Inc.: New York, 1996.

Scott, R.P.W. *Analytical Gas Chromatography*, 2nd ed. Marcel Dekker, Inc.: New York, 1998.

Sparkman, O.D.; Pentan, Z.; Kitson, F.G. *Gas Chromatography and Mass Spectrometry: A Practical Guide*, 2nd ed. Elsevier Academic Press: Oxford, U.K., 2011.

CHAPTER 14

Chromatography with Liquid (and Fluid) Mobile Phases

14.1 HIGH-PERFORMANCE LIQUID CHROMATOGRAPHY

The first chromatography was performed using liquid mobile phases to separate colored plant pigments, which caused its inventor, Tswett, to give it the name of "color writing" (Section 12.1). Crude liquid chromatography (LC) separations under gravity flow were commonly used to characterize synthetic mixtures from the organic research lab, but it was the chromatographic version of the venerable distillation process that was first instrumentalized as gas chromatography (GC). The reasons for this precedence were discussed in Section 13.1. Diffusion and equilibration in the column is much faster with gases. To attain equivalent diffusion equilibration using a liquid mobile phase requires that the dimensions for diffusion be much smaller. These dimensions will be those of interstitial space in a packed column or the diameter of an open-tubular column. Liquids are much more viscous than gases. To force liquid to flow through such narrow pathways requires that orders of magnitude greater pressures be imposed across a column capable of withstanding these pressures. Suitably small support particles need to be coated with an extremely thin layer of stationary phase, then packed tightly, and not develop resolution-destroying voids as high-pressure pumps cycle over the immense range necessary to move the liquid through the bed of microscopic particles. Detectors for analytes in the small eluted peak volumes of the liquid mobile phases needed to be designed. The prior example of GC instrumentation was required to inspire analytical chemists to invent this whole suite of improvements, resulting in an instrumental version of LC that would match or exceed the performance of GC. The instrumental version of LC became known as high-performance liquid chromatography (HPLC) and ultra-high performance liquid chromatography (UHPLC). In the early days of the method, this was often understood to stand for high-pressure LC, for it was instrumentation that could control these necessary pressures that made HPLC possible.

HPLC is now a larger field and a more important market than GC. Why has this happened? The limitations on analytes for GC outlined in Section 13.10.4 are for compounds with relatively low relative molecular mass (RMM) and low polarity, which are electrically neutral (i.e., uncharged) and thermally stable. Vast numbers of biochemically important molecules fall outside these categories. There are more parameters available to vary in liquid mobile-phase chromatography than in GC, so it is divided into a number of techniques. Some are named for the different mechanisms of separation employed (e.g., partition, adsorption, size exclusion, ion exchange, immunoaffinity). Others are described by the design of the separating technique (e.g., thin layer, high-pressure column, and electrophoresis). Electrophoresis is not a chromatographic method but has similar features such as peaks, elution times, resolution, capillary columns or 2D plates, and chromatographic-type detectors.

The major method is HPLC, operating as **partition chromatography**. This is what is commonly implied when the term HPLC is used. It is the mode we will discuss initially as we cover the design and function of the pumps, columns, stationary and mobile phases, injectors, and detectors. Recall the discussion in Section 12.5, where initially, the partitioning of analytes between nonpolar liquid mobile phases and silica particles with adsorbed water became known as "**normal-phase (NP)**" chromatography. For the larger population of more polar analytes encountered in biological systems, it became more advantageous to reverse this. A more polar, more aqueous mobile phase could carry such analytes past particles coated not with adsorbed water molecules but with covalently attached nonpolar organic groups. Thus, this received the name "**reversed-phase**" HPLC (RP-HPLC), and it became the workhorse of the field. Let us then start our discussion with RP-HPLC and contrast the design features of the instrumentation with those of GC, which were covered in Chapter 13.

14.1.1 HPLC Column and Stationary Phases

Compare GC with HPLC. The typical capillary GC column has an internal diameter around 0.25 mm (or 250 μm). A typical polymer stationary-phase coating on the inner wall of this capillary is 0.25 µm (or 250 nm). The high diffusion rate of analytes in the gas phase permits rapid near equilibration over such distances. Diffusion in a liquid is much slower. If we compensate by reducing the dimensions of a capillary LC system sufficiently to attain the desired near equilibration condition, they will be very small indeed, with a bore measured in a few microns. To retain a phase ratio similar to GC, the stationary coating must shrink to a thickness of several nanometers (nm). This is in the size range of single molecules. An open-tubular LC column with such dimensions would have a very minimal sample capacity before it overloaded, and it would be very difficult to build a detector that would work on such a small scale. Larger volume detectors would introduce severe extra-column band broadening or require extensive makeup flow that would further dilute the already miniscule concentrations of analyte eluting from such a column. Capillary HPLC is possible when samples are limited to very small amounts, but it is nowhere near as common as capillary GC.

The way to achieve HPLC is to return to packed columns, but to implement them with support particle and stationary-phase coating dimensions much smaller than those of packed column GC. Typically, one employs spherical silica particles of uniform diameters (**monodisperse**) in the range of 1.7–10 µm. The interstitial spaces between the tightly packed spheres will be of similar dimension. Unlike the much larger support particles in packed column GC, these LC particles often have a smaller-scale **secondary internal porosity** of channels permeating the particle into which analytes can diffuse. The stationary-phase coating in GC is like a true quasi-liquid polymer many molecular sizes thick—a true phase with a boundary across which molecules diffuse, dissolve, and partition from the mobile gas phase. Indeed, in GC the stationary phase behaves thermodynamically like a liquid, with the same enthalpy of solution, activity coefficients, etc., as can be measured in a bulk liquid. In contrast, the stationary phase on these porous micro-HPLC particles is an organic monomolecular layer chemically bonded to the free silanol groups on the silica surface. The most common such group is a long-chain n-C18 hydrocarbon (**octadecylsilyl group, ODS**). Check the discussion in Sections 12.4 and 12.5 and see Figure 14.1. If stretched out, such a chain is on the order of several nanometers long, which is just the thickness proposed in the previous paragraph. The structure of such coated microporous particles is diagrammed in Figure 14.2. Here are some points of contrast between HPLC and packed or capillary GC stationary phases:

1. Each LC-phase molecule is *bonded* to its solid support surface but is less often *cross-linked*.
2. The HPLC phase is so thin that it is not a fully randomized fluid, like a GC polymer coating.
3. Usually, it is impossible, due to steric hindrance or increasing crowding, to attach a molecule [like the $-O-Si-(CH_3)_2C_{18}H_{37}$ group] to every free silanol bonding site on the silica surface.
4. These unreacted silanol sites lie closer to the mobile phase than in GC and can easily act as active sites degrading the chromatography as described in Section 12.4.
5. The unreacted active sites on the support surface must be masked by reacting with a very short group [such as $-Si-(CH_3)_3$], which can penetrate to react and bond with most of the remaining silanol groups, a process called **endcapping**.
6. As the composition of the liquid mobile phase is varied, it can interact to cause alterations in the alignment of the stationary-phase molecules and thereby affect their

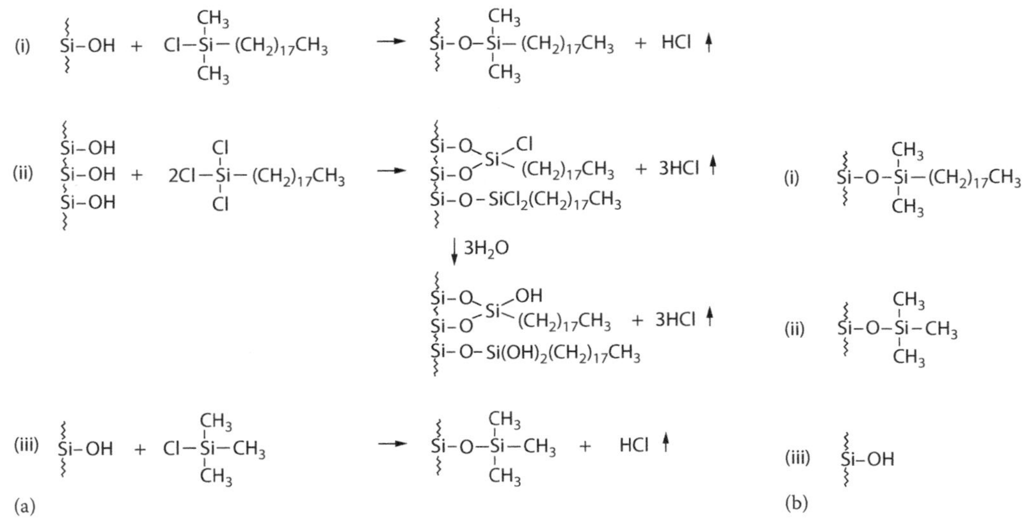

Figure 14.1 Types of ODS silica HPLC stationary phases: (a) i, synthesis of monomeric C18; ii, synthesis of polymeric C18; iii, endcapping process; and (b) i, monomeric C18 ligand; ii, endcapped silanol; and iii, residual silanol. (From Cazes, J., *Encyclopedia of Chromatography*, Marcel Dekker, Inc., New York, 2001. With permission.)

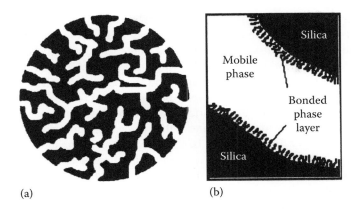

Figure 14.2 Microporous structure of bonded-phase HPLC particle. (a) A particle of column packing is honeycombed by pores. Most of the surface area is inside these pores (this drawing is "not" to scale; typical pore diameters are on the order of 1%–5% of the particle diameter). (b) Magnified cross-section of a pore, showing a layer of alkyl chains bonded to the silica surface. (Reprinted with permission of Supelco, Inc., Bellefonte, PA. www.sigmaaldrich.com.)

partition behavior with some analytes. This dependence of stationary-phase behavior upon mobile-phase composition is not found in GC, and it adds more complexity to the prediction of HPLC separation behavior.

7. The small interparticle spacing in HPLC columns requires stable, uniform, spherical particles to avoid their packing together in a fashion which will block uniform flow through these tiny interstitial passages. Mobile-phase solvents must be **filtered** to remove micron-sized particles that could plug up these interstitial volumes and block or channel mobile-phase flow.

8. Dissolved gases in the mobile phase may come out of solution and form bubbles as the pressure changes from the entrance to the exit of the column. Those will block and channel flow, degrading resolution or even blocking the column completely. Even if the bubbles redissolve, the resultant large cavity in the packing may introduce extra-column band broadening. Mobile-phase liquids must be **degassed** prior to use, by filtering under vacuum or by **sparging** with a flow of fine bubbles of poorly soluble helium gas.

9. As the instrument cycles the column through very large pressure changes as the pumps turn flow on and off, the packed bed may move or settle, yet again introducing the dead **void** and its extra-column band broadening.

10. Outside certain bounds of pH and/or temperature, the liquid mobile phase may attack and gradually dissolve the silica support particles and/or release the bonded stationary-phase molecules.

11. Because of the smaller particle size and internal porosity of LC packings, the area covered by the molecular monolayer stationary "phase" is vastly greater than that of the thick polymer film stationary phase in GC columns of similar volume. This restores the analyte capacity of these columns to a level compatible with typical chromatographic detector designs.

12. The HPLC columns must operate at pressures 500–6000 psi versus 10–50 psi—orders of magnitude higher than GC columns. The mobile phase must be pumped instead of coming from a pressurized cylinder through a reducing valve. The columns are commonly fabricated from thick-walled 304 or 316 stainless steel, with stainless steel mesh packing restraining frits, and use stainless steel or high-impact plastic nut and ferrule fittings.

The effects on analyte resolution of particle size and column length in HPLC are analogous to those of capillary column diameter and length in GC. However, one does not vary the "film thickness" of the bonded-phase molecules in HPLC. The percentage of available silanol sites to which stationary-phase groups have been bonded can vary. A larger percentage does not directly translate into a larger effective volume of stationary phase. It is the surface density of the groups that varies and, with it, the molecular interaction strength, not just the volume available for analyte partitioning. The percentage and completeness of endcapped silanol groups is an extra variable that also impacts the nature of the analyte interaction with the phase. The lesson here is that different preparations of an RP packing category, such as the workhorse silica ODS, can vary widely in their suitability for different analytes, depending on analyte size, polarity, acidity or basicity, and the characteristics of the interactions with the mobile phase. All ODS columns are not created equal! Even more so than with GC columns, comparison of manufacturer's specifications and application notes from catalogs or websites is the efficient first step in column selection.

14.1.1.1 Support Particle Considerations

Silica can be prepared as micron-sized spheres. The internal porosity can be varied. The spheres can be generated to have a narrow range of diameters, so they will pack together uniformly. The surface silanol groups make ideal attachment points for the bonded-phase molecules. At acidic values, the organosilyl bonds may hydrolyze and the phase can degrade. More densely covered and appropriately endcapped silica surfaces may tolerate lower pHs. At pH above 8, the silica itself may begin to dissolve at the surface. Recall that glass can be etched by strongly alkaline solutions. As will be discussed subsequently under mobile phases, there could be advantages to operating the column at elevated temperatures. Unfortunately, silica also becomes more soluble under these conditions. Recall that quartz crystals grow from hydrothermal solutions under pressure in the ground. HPLC column specifications need to be checked for their pH and temperature stability.

Zirconia (ZrO_2) is a new substitute for silica as an HPLC particle support, which has both greater pH range stability (1–14) and temperature stability (to 100 °C or even greater—recall that we are operating at many atmospheres pressure in HPLC and this can elevate boiling points). For some analyses, the greater cost of zirconia columns may be offset by the improvements in analysis speed and performance they can bring to HPLC.

Another new development circumvents production of the individual particles, and it casts the column as a **monolithic silica cylinder**, which possesses both the macro- and microscale porosities and analogs to the interstitial volumes of a column "packed" with micron-scale spherical silica particles. When techniques for achieving this were invented, two major new advantages ensued. First, such monolithic HPLC columns are mechanically far more stable than a packed bed of particles. There is far less possibility for the development of voids and channels. Second, the size of the macroscale void volumes is not constrained to the interstitial dimensions of a packed bed of spherical particles. Specifically, they can be larger, but the network of fused-together microporous support "particles" can still remain connected to one another. This additional degree of freedom in the column dimensions results in columns that can be operated at higher mobile-phase flow rates, eluting analytes more quickly with lower column pressures while still retaining the resolution performance of comparable packed columns.

Since the last edition of this text, a more sophisticated synthesis of the spherical silica support particles has enabled the production of columns with much higher resolutions and shorter analysis times. The narrower and taller peaks from these columns produce better S/N sensitivity improvements. These so-called **core–shell** column packings (sometimes called **superficially porous**) limit the small-scale intraparticle porosity to a shell-like layer (typically one-half to one-fifth the spherical particle radius) outside a solid impermeable core. This reduces the distance that analyte molecules can penetrate and allows them to equilibrate much faster with the flowing mobile phase (cf. the discussion in Section 12.7 and Figure 12.7). In an independent development, instrument manufacturers have redesigned the pumps, transfer lines, and connecters of the new model HPLC instruments to accommodate higher-pressure operation with *smaller, fully porous* packing particles, which also leads to the same degree of improvement, albeit with more expense and concern about column clogging and stationary-phase breakdown. These new technologies have led to the so-called ultra-high-performance liquid chromatography (UHPLC) or UPLC systems, where U stands for ultra. So many older HPLC assays have been replaced by or converted to these ultra-systems that they now comprise more than 50 % of the instrumental LC installed base, with this percentage continuing to rise. Performances similar to the ultrahigh pressure, small-particle size instrumentation can often be achieved by converting assays on older, lower-pressure instruments to those using larger core–shell phases that do not require the higher pressures needed for use with the smaller particles. This development is a reprise of the replacement of the large-particle, vertical columns, operated with gravity-driven flow, with instrumentally pumped, higher-pressure, smaller-particle HPLC. Should equivalent advances eventuate, will we someday have to speak of hyper-HPLC (HHPLC)? Almost every major HPLC column supplier offers a line of core–shell particle columns: for example, Kinetex (Phenomenex), HALO (MacMod), Accucore (Thermo Scientific), Ascentis Express (Supelco), and Poroshell (Agilent). Figure 14.3

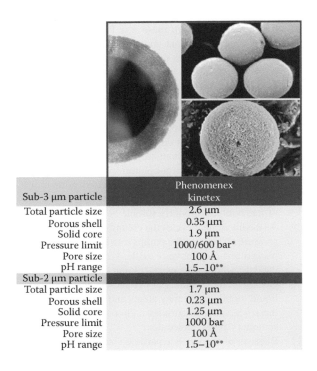

	Phenomenex kinetex
Sub-3 μm particle	
Total particle size	2.6 μm
Porous shell	0.35 μm
Solid core	1.9 μm
Pressure limit	1000/600 bar*
Pore size	100 Å
pH range	1.5–10**
Sub-2 μm particle	
Total particle size	1.7 μm
Porous shell	0.23 μm
Solid core	1.25 μm
Pressure limit	1000 bar
Pore size	100 Å
pH range	1.5–10**

Figure 14.3 Electron micrographs of core–shell UHPLC particles: on the left, a cross section showing the core–shell structure and, on the right, particle surfaces (note relative smooth, spherical surface of monodisperse particles). The superficially porous structure contributes to the flatter right-hand portion of the Van Deemter plot for UHPLC, and the particles' uniformity allows them to pack efficiently and reduce the A term contribution. (With permission from Phenomenex, Inc., Torrance, CA, www.phenomenex.com.)

displays electron micrographs of typical core–shell UHPLC particles, both surface and cross section.

For completeness, we should note that spherical, porous beads of organic polymer resin, particularly cross-linked styrene–divinylbenzene copolymer, can be used as HPLC phases. Unlike silica ODS, the whole particle is the stationary phase. These are often designated as XAD resins. Unlike hard silica particles coated with a monomolecular layer, these resin beads will take up mobile-phase solvent and swell in size. Resin bead sizes cannot be as small as the spherical silica particles, lest this swelling fill and block the interstitial macropores. The higher pressures cannot be used, and the high-resolution efficiencies of true HPLC are rarely attained. These resins are very important as the substrates for attaching and supporting ion-exchange groups to form the stationary phases for the technique of ion-exchange chromatography. These will be discussed in Section 14.2. Pure polystyrene beads can serve as supports for covalent attachment of organic groups to make phases that are stable over the pH range of 8–12, but they also suffer from the structural shortcomings of the resin beads.

14.1.1.2 Stationary-Phase Considerations

Three of the most common nonpolar organic substituents employed in stationary phases for RP-HPLC with aqueous/polar mobile phases are as follows:

$$\text{Octadecyl (C18, ODS)}: -(CH_2)_{17}CH_3$$

$$\text{Octyl (C8)}: -(CH_2)_7 CH_3$$

$$\text{Phenyl}: -(CH_2)_3 C_6H_5$$

The larger the substituent group, the more strongly and longer will it retain the nonpolar analyte, so retention will vary inversely to the order listed here.

Three examples of common polar organic substituents (incorporating functional groups to confer the desired polarity) for NP-HPLC with less-polar mobile phases are the following:

$$\text{Amino}: -(CH_2)_3 \mathbf{NH_2}$$

$$\text{Cyano}: -(CH_2)_3 - \mathbf{C \equiv N}$$

$$\text{Diol}: -(CH_2)_2 OCH_2 CH(\mathbf{OH})CH_2\mathbf{OH}$$

Here, it is the functional groups that dominate retention. The amino and diol phases will retain compounds with pronounced hydrogen-bonding potential more strongly, while the cyano phase interacts most strongly with analytes whose polar interactions result from strong dipole moments. Complex structures may interact in a combination of these mechanisms, so relative retention is less easy to predict than with the RP columns. The book by Bruno and Svoronos provides comprehensive coverage of all common and even some specialized stationary phases for HPLC.

14.1.1.3 Chiral Phases for Separation of Enantiomers

Even more specialized groups can be attached to supports to achieve particular separations. Complex cyclic carbohydrates called **cyclodextrins** exist in a single **enantiomeric form**. These are illustrated in Figure 14.4. They form a truncated conical pocket, into which different optical isomers of a compound (enantiomers) will bind with different strengths. This enables HPLC separation of these isomers, which have identical chemical structures and polarity and differ only in their stereochemical orientation of substitution at asymmetric carbons. These are described as **chiral** compounds, and use of such stationary phases enables "**chiral chromatography**." The cyclodextrin "pockets" come in a variety of sizes (α-, β-, γ-, etc.) suitable for separating enantiomer pairs of differing size. This capability is especially important for determining the relative amounts of each enantiomer that may be present. Solutions of individual enantiomers rotate the plane of oscillation of plane polarized light. At high concentrations, this may be measured by optical polarimetry instruments (not covered in this text). Polarimetric detectors also exist for HPLC. They are relatively insensitive (limit of detection [LOD] ~5 ppm in eluent solution). For trace levels, the use of even more sensitive chromatographic detectors makes resolution of enantiomers by chiral chromatography the method of choice. Often it is only one enantiomeric form of a drug compound that is active in the body. If the other form is toxic, the drug may have to be formulated to contain only the active, safe form. It then becomes critical to be able to determine whether, when, and to what extent this form may **racemize** (revert to an equilibrium mixture of both enantiomers) in the body. Chiral chromatography may be the only way to address this. A classic example is the drug thalidomide, originally marketed as a racemic mixture of *N*-phthalylglutamic acid imide. One enantiomer had the desired pharmacological activity, but it was not realized until too late that the other was a strong teratogen, capable of causing severe fetal malformations. Cyclodextrin substituent stationary phases can be used to create **chiral GC** columns as well. HPLC is better suited than GC for many chiral separations, as the elevated temperatures used in GC may induce racemization in the column. There are other chiral substituents that can be incorporated into HPLC columns to enable them to separate enantiomer pairs, some being very complex and expensive for specialized applications. The book by Bruno and Svoronos has a comprehensive listing that includes the chemical structures and separation

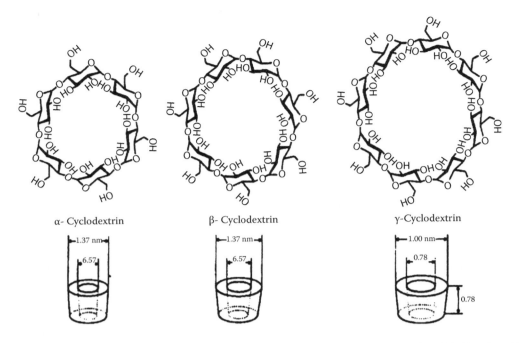

Figure 14.4 Structures of α-, β-, and γ-cyclodextrins. (From Katz, E. et al., eds., *Handbook of HPLC*, Marcel Dekker, Inc., New York, 1998. With permission.)

characteristics. Indeed, some chiral stationary phases are not commercialized but must be prepared in the lab for use in separations.

14.1.2 Effects on Separation of Composition of the Mobile Phase

In GC, elution temperature is the most easily controlled variable that affects elution and k'. In LC, the corresponding variable is mobile-phase composition. The factor that is being varied is polarity of the solvent mixture. This is more complex and less easy to define rigorously than is temperature. It is a relative value applying to the molecules to be separated and to both the mobile-phase and stationary-phase materials. Polarity interactions can be of several types (e.g., dipole–dipole interactions, π-bonding interactions, and acid–base or hydrogen-bonding interactions [proton donors or acceptors]). Both analytes and the chromatographic phases can display polarities, which are combinations of these mechanisms. In GC, there were only two polarities involved: those of the analyte and those of the stationary phase. In LC, we add the third—the mobile phase—and it is this one that we can vary most readily. A detailed discussion of how one would evaluate the polarity and interactive components of the analyte molecule, select the most appropriate stationary phase, and then adjust the solvent mixture in the mobile phase to optimize a separation is beyond the scope of this text. What follows are a few oversimplified, general "guidelines":

1. The overall polarity contributions from particular functional groups increase in the following order: hydrocarbons < ethers < esters < ketones < aldehydes < amides < amines < alcohols.
2. For partition chromatography, it is best to use a stationary phase of polarity opposite to that of the analytes and to elute with a mobile phase of different polarity.
3. In the extreme of NP-HPLC, the stationary phase is highly polar, adsorbed water on silica or alumina supports, so the eluents should be nonpolar, often hydrocarbon solvents. The mechanism of NP-HPLC is often classified as adsorption rather than partition, as the analytes are essentially binding to a highly polar surface.
4. RP partition HPLC uses less-polar bonded organic groups as the stationary phase, so the mobile phase is more polar. This is preferred for more polar biomolecules, since they have been selected by evolution to function in polar aqueous environments.
5. The most polar components of an RP-HPLC mobile phase will often be water (HOH), ACN ($CH_3-C\equiv N$), or methanol (CH_3OH). The polarity of the mobile phase can be adjusted to values between these by mixing them in different proportions. Even less-polar compounds, such as ethanol, tetrahydrofuran (THF), or diethyl ether, may be added to bring the overall eluent polarity down even more or to introduce different interaction mechanisms of polarity that may increase the separation factor α between close pairs of analytes. Being at the extreme end of the polarity scale, water is often considered the fundamental component of the mobile-phase mixture,

and the other less-polar, organic solvents that may be added are termed "**organic modifiers**."

There are several methods of quantifying the relative solvent strengths or polarities of liquids used as mobile phases in HPLC. Table 14.1 displays values from two such systems for selected solvents. The first was developed to describe the relative ability of eluents in NP adsorption LC to solvate analytes and move compounds through the column more quickly. This "**eluent strength**," ε^0, is ordered in the so-called **eluotropic series**. The values are obtained using alumina (Al_2O_3) particle columns. Its zero point is defined by n-pentane, and it goes to very high values for highly polar water. The corresponding values on silica are about 80 % of the alumina values. Clearly, water as a mobile phase will very rapidly elute anything adsorbed to the adsorbed water of these stationary phases. The second column of Table 14.1 lists values for a "**polarity index**," P', developed by L.R. Snyder. These indices are derived from solubility of the solvent in three test solvents chosen to have high or low dipole proton acceptor polarity or high dipole proton donor polarity. The polarity index of mixtures of solvents can be calculated by summing the products of their respective polarity indices and volume fractions in the mixture ($P_{AB...} = \phi_A P'_A + \phi_B P'_B + \cdots$, where ϕ is the volume fraction).

Other solvent properties are important to consider in choosing a solvent for HPLC. Viscosity is an important factor simply because of the need to pump the solvent through the bed of stationary phase; solvents of higher viscosity will pose difficulty. In addition to the Snyder polarity listed above, important factors include the solubility parameter and the dielectric constant. The solubility parameters, δ, defined fundamentally as the cohesive energy per unit volume, are calculated from vapor pressure data or estimated from group contribution methods. The dielectric constant is also a measure of polarity. As we will discuss below, buffers are often an important part of the solvent system for an HPLC separation. Preparation of buffers is made easier by use of stock solutions and making up the final volume to achieve the desired pH. A comprehensive listing of solvents for HPLC (and buffers), along with their properties, can be found in the book by Bruno and Svoronos.

Another less than obvious consideration in the selection of a solvent for HPLC is the stability of the solvent. Ethers and chlorinated alkanes usually incorporate stabilizers in the as-received solvents; these may have to be removed before use and then one must be careful of peroxide formation and phosgene formation, respectively. The peroxide concentration that is typically considered a hazard is 250 ppm (mass/mass). Ethanol and methanol often contain trace water contamination, and these solvents can become contaminated with water by exposure to the atmosphere. More information on solvent instability can be found in the book by Bruno and Svoronos.

It is possible to optimize an RP-HPLC separation of several compounds by systematically varying the proportions of three appropriate solvents, usually water and two organic modifiers. By observing the relative shifts of R_t as each change is made, and using as few as seven runs with well-chosen variations, one can create a "solvent polarity triangle" for the separations. Data from this will enable calculation of an optimum mobile-phase composition for the separation. For HPLC instruments equipped with separate mobile-phase solvent reservoirs and programmable precise mixing valves, this process can be automated, and the mobile-phase optimization can even be done without operator intervention.

The major contrast with GC is that the mobile-phase eluent strength (affecting k') and the relative selectivity between analytes (α) can both be readily varied. In GC, the carrier gas composition has practically no effect on selectivity, and varying temperature affects only k'. The large number and range of parameters to be evaluated to optimize the eluent composition makes HPLC method development more complex. The Agilent GC method translator was referenced in Appendix 13.A. Thermo Scientific offers online an LC method translator program, listed in Appendix 14.A. Experimenting with these shows the effects of parameter changes more rapidly than solving the relevant equations manually. Even more powerful is the DryLab® 4 UHPLC computer-based modeling software from the Molnar Institute for Applied Chromatography (www.molnar-institute.com, info@molnar-institute.com), which employs the principles of quality by design (QbD) to select and model the most important factors affecting the quality and robustness in the development of proposed separations. Free examples are available on their websites for informational purposes.

14.1.2.1 New HPLC-Phase Combinations for Assays of Very Polar Biomolecules

For the highly polar biomolecules, RP phases have several shortcomings, namely, low retention, often a

Table 14.1 Eluent Strengths, ε^0, and Polarity Indices, P', of Selected Mobile-Phase Solvents

Solvent	ε^0	P'
Cyclohexane	−0.2	0.04
n-Hexane	0.01	0.1
Carbon tetrachloride	0.18	1.6
Toluene	0.29	2.4
THF	0.57	4.0
Ethanol	0.88	4.3
Methanol	0.95	5.1
ACN	0.65	5.8
Ethylene glycol	1.11	6.9
Water	~∞	10.2

need for derivatization, requirement for the use of ion-pairing reagents, analysis time too short, lack of robustness, and poor quantitation. More recently, systems have been developed to separate and quantify this class of compounds.

14.1.2.2 Ion-Exchange Chromatography

If the analytes are charged cations or anions (strong or weak acids or bases), then IE may perform better. This methodology is discussed at great length in Section 14.2.

14.1.2.3 Hydrophilic Interaction Liquid Chromatography

This new combination of a column with more NP character using mobile phases with high organic content was initially used to separate peptides on strong cation-exchange (CX) columns and later extended to other polar analytes using phases like bare silica or highly polar cyano, diol, or amino coatings. Unlike classic NP-HPLC, which might use a pure nonpolar eluent like hexane, the hydrophilic interaction liquid chromatography (HILIC) system mobile phase might use a high percentage of organic solvent (e.g., >70 % ACN) but with a critical addition of miscible 3 % water. The polar stationary-phase particles will bond a small amount of water from the mostly organic stationary phase, and the separation mechanism will proceed by hydrophilic partitioning, H-bonding, and electrostatic interaction from the mobile phase to the adsorbed water layer on the stationary phase. Hydrophilic analytes will have sufficient retention, and the mode of separation will be orthogonal to RP-HPLC. Additional strong advantages accrue if the separation is hyphenated to mass spectrometry (MS) detection, where the sensitivity by HILIC separation is better than by RP separation in 89 % of cases: for example, 42 % 1–5×, 9 % 5–10×, 18 % 10–30×, and 14 % > 30× better! The highly organic mobile phase produces much less column back pressure in attaining the optimal flow rate. It therefore does not profitably use monolithic columns, and available <3 μm core–shell columns lack good column chemistry, so the best results are presently obtained with 2 μm fully porous particles that operate at pressures <400 bar under HILIC conditions. Combining HILIC and RP-HPLC orthogonally in a 2D-LC separation can be performed by offline coupling, serial coupling, or comprehensive 2D RPLC-HILIC. The high ACN content mobile phase is especially favorable when HILIC is used to characterize large mixtures of small-molecule, highly polar metabolites in a metabolomic profiling study using a UHPLC-electrospray ionization (ESI) MS/MS system.

14.1.2.4 Polar-Embedded Long Organic Chains on Silica

Instead of the classic RP C18 ODS chains on the silica, one can attach a similar chain that has a polar moiety embedded in it to produce a mixed mode of retention. Examples are the following:

Silica: $CH_2CH_2CH_2-O-C=O-CHCH_2\ldots CH_3$

Or silica: $CH_2CH_2CH_2CH_2-(3CH, 2N^+, Br^-)-CH_2CH_2CH_2\ldots CH_3$

In the second structure mentioned earlier, the moiety in parentheses is a five-membered, aromatic, positively charged heterocyclic ring attached in the chain through the 2 N atoms separated by both 1 and 2 Cs. This is the same cationic structure seen at either end of the GC ionic liquid phase depicted in Figure 13.4. This charged unit confers electrostatic and π-bonding retention, while the long hydrocarbon tail confers hydrophobic nonpolar RP retention.

14.1.2.5 Polar-Endcapped Short Hydrocarbon Chains on Silica

A long-hydrocarbon-chain RP coating such as C18 attached to the silica will require endcapping of up to 50 % of unreacted silanol sites on the silica (cf. Section 14.1.1, point 5). If instead of the short $-Si-(CH_3)_3$ cap, a longer one with a hydrocarbon spacer holding a polar end group such as $-Si-CH_2CH_2CH_2NH_2$ is used; then another mixed polarity mode stationary phase is produced, with properties and usefulness similar to the polar-embedded phase just described.

14.1.3 Design and Operation of an HPLC Instrument

Figure 14.5 is a schematic diagram of a typical HPLC instrument. We will use this figure to discuss the features of the major components and describe the construction of some of them. We will use this discussion to illustrate some of the major modes of operation of HPLC instruments. We proceed in order of the numbered items of the figure:

1. *Mobile-phase degassing*: The figure shows a setup designed to sparge four separate reservoirs with He gas. It is distributed by the fixed splitter valve to fritted glass or metal filters that disperse the gas into a fine cloud of bubbles, driving out (sparging) other dissolved gases. Removal of dissolved gases could alternatively be done by manually filtering over vacuum prior to placing in the reservoirs. In any case, solvents must be passed through a filtration membrane to remove particulates.

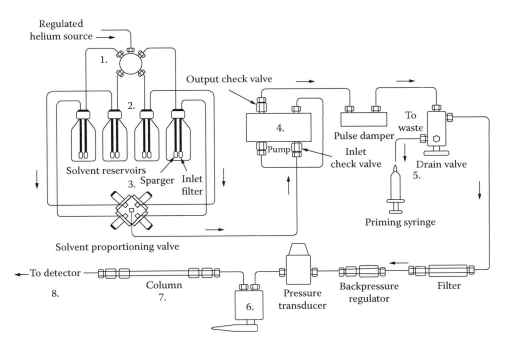

Figure 14.5 Diagram of conventional HPLC instrument. [© 1993–2014 PerkinElmer, Inc. All rights reserved. Printed with permission. (www.perkinelmer.com).]

2. *Mobile-phase storage*: The figure shows four reservoirs, one for each of up to four pure solvents (e.g., water [perhaps with pH controlling buffer], ACN, methanol, THF). Note the use of inlet filters. Alternatively, one could prepare the mobile-phase mixture to the desired composition manually and store it in a single reservoir. Operation at a single, constant mobile-phase composition is called **isocratic** HPLC elution. It is critical if a buffer is used as a solvent that it is not allowed to remain stagnant in the HPLC system; it must be flushed out after use. Most instruments have a piston seal wash procedure that can help with this. Phosphate buffers are particularly problematic. These buffers may be immiscible with or be precipitated by the presence of acetonitrile. It is good practice to use as low a concentration as possible of a phosphate buffer (or a lower concentration of acetonitrile).

3. *Mobile-phase mixing*: The solvent proportioning valve can be programmed to mix solvents from the reservoirs to produce the desired mixture composition. This could be a constant proportion mixture for isocratic operation, or one could vary the solvent strength of the mobile phase with time by changing the mixture proportions. As a simple example, the mobile-phase composition could be programmed to vary from 75 % water:25 % ACN at time zero to 25 % water:75 % ACN at the end of the run. Such a variation of mobile-phase solvent strength or polarity during a run is termed **gradient elution**. There is a "gradient" or variation in mobile-phase composition and elution strength with time. A mixture of components of widely varying polarity can thus be separated and eluted more rapidly. In an RP separation, less-polar analytes will elute later during this gradient, but the run will be completed more rapidly than would be possible with isocratic operation. Gradient elution is the HPLC analog of programmed temperature GC, and it confers all the same advantages. It is especially valuable if there are "slow eluters," which must be removed before the next sample injection. If these are very slow eluters, which need not be measured, it may be better to reverse the flow through the column and *backflush* these off the front of the column and out a drain port located between the column and the injector, using backflush valves on each end of the column (not illustrated here). It takes considerable time to re-equilibrate the stationary phase with the starting mobile-phase composition after a gradient run. Unless there is a wide polarity range of components to be eluted, isocratic elution is to be preferred. It is simpler to implement, and if R_t values of separated components do not become too long, it will permit shorter analysis cycle times. Mobile-phase mixing can be done with either high- or low-pressure approaches. Low-pressure mixing is done by use of a proportioning valve and mixing chamber before the pump. High-pressure mixing is done with a quaternary pump in which two solvents are pumped into a mixing chamber at high pressure. Interestingly, in the early days of HPLC before gradient elution was a built-in feature of the instrument, gradients were jury rigged by placing a separatory funnel containing the second solvent above the instrument solvent reservoir (which was stirred magnetically as well as He sparged). The cock of the

separatory funnel was regulated to produce a more or less constant flow into the instrument solvent reservoir, thereby producing a solvent gradient.

4. *HPLC pump*: A common type of HPLC pump is the reciprocating piston type illustrated in Figure 14.6. During *fill stroke* (Figure 14.6a), it is pulling mobile-phase liquid from the solvent side; then on *exhaust stroke* (Figure 14.6b), it is pushed through the injector and to the head of the column. This is where the high pressure that permits the high performance in HPLC is generated. The figure illustrates a *"single-piston"* design. The flow from the pump will rise during the exhaust stroke and subside during the fill stroke. The *"pulse dampener"* shown downstream attempts to smooth the flow surges to approximate a steady constant flow. The stationary-phase particle bed would become unsettled by repeated pressure surges. When using mass-flow rate detectors, variation in flow would be reflected in variation of their response. Both are undesirable consequences of pulsed flow. An improved pump design effects more uniform pressure and flow using *dual pistons* in an arrangement whereby one is in the exhaust stroke while the other is in the fill stroke. Pumps for conventional HPLC are typically constructed from inert materials (304, 316 and Nitronic stainless steels) and are capable of producing pressures up to 41 MPa (6,000 psi) and pumps for UHPLC produce 103 MPa (15,000 psi). There are many advanced pump designs that go beyond the simple piston unit shown in the figure. The most sophisticated pumping systems utilize pumps mounted in series but driven by two independent motors. The first piston in the fluid circuit has either the same or a somewhat larger volume than the second. Elaborate piston control algorithms and sensors compensate for compressibility and thermal expansion as fluid passes from the first to the second piston. This type of system provides the most precise compositional control and little pulsation. Despite the sophistication, these pumping systems require some mundane maintenance tasks. The filter frits must be changed on a regular schedule (or whenever unexpectedly high pressures or low flow rates are encountered). Occasionally the piston seals will have to be changed. The motor drive compartment will have a special lubricant that can leak if there is significant seal wear; often this will be smelled before it is seen, since it has a skunky odor. It is critical to use only the recommended replacement lubricant.

5. *Fill/drain valve*: There are many liquid transfer lines and components that the mobile phase passes through before it gets to the head of the column. The analyst needs to fill all this with each new mobile phase, and after use, one stores the system and column in a simple solvent or water rather than a complex mobile phase containing buffers, dissolved salts, and so on. A **priming syringe** pulls mobile phase through the lines of an empty system to remove air and "prime the pump." It can also be used to push mobile phase or storage solvent further downstream to clear out those lines. The **drain valve**

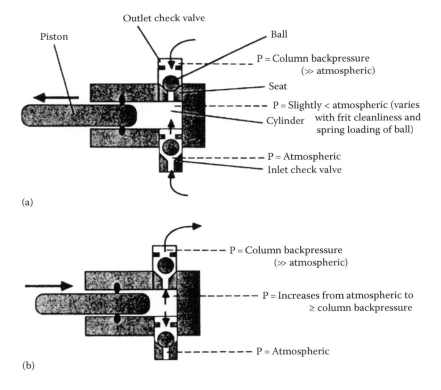

Figure 14.6 Operation of piston and check valve reciprocating pump: (a) suction stroke and (b) exhaust stroke. (From Katz, E. et al., eds., *Handbook of HPLC*, Marcel Dekker, Inc., New York, 1998. With permission.)

can be opened so the pumps can quickly rinse storage solvent through the front half of the lines without having to work against the resistance of the column. Note that the solid black lines connecting the components of the system on the high-pressure side of the pump are narrow-bore (e.g., 0.5 mm i.d.) stainless steel tubing. In more modern automated systems, the cleanout cycle can be programmed to inject the storage solvent from an autosampler vial after completion of all analyte and calibration standard injections.

6. *Rotary sample loop injector*: Figure 14.7 illustrates the most common design of HPLC injector. To introduce the sample as in GC with a syringe through a septum into the highly pressurized mobile-phase liquid flow is impractical. The top portion of the figure is a schematic end-on view of the valve, which rotates on an axis coming out of the page. There are three inlet and three outlet fittings. The valve is shown in the "*load*" position. Sample from a syringe is pushed into a **sample loop** of tubing of specified volume (e.g., 50 or 100 µL), with any excess flowing to waste. Mobile phase is flowing through one solid-line arc of the valve as shown. To *inject*, the valve is rotated 60° clockwise. Now the solid-line flow paths within the valve diagram are at the location of the dotted lines. The pump is now pushing the mobile phase and sample liquid in the loop in a direction *opposite* to that in which it was loaded, and it exits the valve to the column. All this occurs without leakage or introduction of bubbles at the high pressure maintained by the pump.

7. *The column*: HPLC columns were described in detail in Section 14.1.1. When a fused silica capillary GC column accumulates unmoving contaminants at its front end, it is a simple matter to cut off some of it and reinstall it. This is not possible with HPLC columns. Even with the most careful filtration, the fine pores may become clogged with traces of particulate matter over time. The solution is to attach in front of the "**analytical column**" a very short, inexpensive, "**guard column**," often packed with "**pellicular**" particles (i.e., without complete internal porosity; thus, the bonded stationary-phase layer is of much reduced area). When it becomes contaminated, it is detached, discarded, and replaced with a new guard column. If the stationary-phase support will tolerate it, **operation of the column at elevated temperature** can confer advantages of both speed and improved resolution. Contrary to gases, the viscosity and resistance to flow of liquids decrease with increasing temperature. Diffusion and equilibration rates in the column are improved with increasing temperature as in GC. Unlike capillary GC, overall column temperature equilibration with an oven is not as fast in HPLC columns. To avoid radial temperature gradients across the diameter of the column that will degrade resolution, it is critical to ensure that the mobile phase is preheated to the desired elution temperature before it enters a thermostated HPLC column. Even without operation at elevated temperature, thermostating an HPLC column helps make elution times more reproducible.

8. *HPLC detectors*: Like GC, HPLC has a wide variety of detectors, universal or specific, destructive or nondestructive, mass flow or concentration responsive, and with even more challenging requirements for interfacing to spectrometers in hyphenated techniques. Tubing of even smaller bore than described earlier in item 5 is necessary to connect the effluent end of the column to the detector to avoid extra-column band-broadening effects. Pressures are lower, and high-strength and -density PEEK plastic may be used in place of stainless steel.

14.1.4 HPLC Detector Design and Operation

Sections 14.1.1 and 14.1.2 described the stationary-phase column and the mobile-phase liquid, respectively. Section 14.1.3 described how these and other HPLC instrument components are connected together and operated. The last item therein, number 8, detectors, covers such a broad and important variety of devices that another subsection

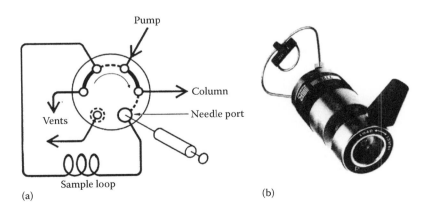

Figure 14.7 (a) Diagram of operation and (b) photo of two-position rotary injection valve. (From Katz, E. et al., eds., *Handbook of HPLC*, Marcel Dekker, Inc., New York, 1998. With permission.)

will be required to describe them. Many of the principal features of HPLC detector operation are those described for GC detectors in the introduction in Section 13.7. The student is encouraged to reread that section. Only item number 4 in Section 13.7, is peculiar to GC. An analogous feature in HPLC would be *postcolumn derivatization*, which will be treated in Section 14.1.5. As in the GC chapter, hyphenated HPLC techniques are discussed separately in Section 14.1.6.

A common feature of many HPLC detectors is a small-volume "**flow cell**" through which the column effluent passes and in which the detector measurement is made. When acquired from within a flow cell, even detector response data files that contain full spectral information are sometimes not classified as hyphenated HPLC techniques. Those usually require the presence of a special interface device. Typical HPLC flow rates are on the order of 1 mL/min (1000 μL/min). The analyte concentration in the effluent might be sampled on the order of once every second or two. It is necessary that the volume of effluent pass through the flow cell on this time scale. Thus, flow cell measurement volumes must be on the order of 5–50 μL. Many HPLC detectors rely on emission or absorption of light, so their flow cells must have transparent entrance and exit windows. The requirement for small volumes limits the length of light paths. This is one of the factors that affects the magnitude of optical spectroscopic signals that may be obtained from a given analyte concentration (e.g., the "*b*" factor in the *abc* of Beer's law [Section 2.4]).

14.1.4.1 Refractive Index Detector

14.1.4.1.1 Refractive Index (RI) Characteristics

RI is universal, nondestructive, a concentration detector, relatively insensitive, requires thermostating, and is useable only with isocratic HPLC separations.

Every transparent substance will slow the speed of light passing through it, by an amount roughly proportional to its optical density (not mass density). The difference between the optical density and the mass density (what we commonly simply call density, mass per unit volume) may seem subtle but it is really not. The term optical density describes the latency or holdup of the absorbed energy of an electromagnetic wave by matter before reemitting it as a new electromagnetic wave. The more optically dense that a material is, the slower that a wave will move through the material. Because light is a wave phenomenon, this results in a bending of the light path as it passes at an angle to the interface from a less optically dense material to a more optically dense material, or *vice versa*. The quantity that defines this effective density is called the refractive index, **RI**. The presence of analyte molecules in the mobile-phase liquid will generally change its RI by an amount almost linearly proportional to its concentration. At low concentrations, the change is small and could easily be masked by density and RI changes arising from temperature differences or slight changes in mobile-phase solvent proportions, hence the need for careful thermostating of an RI detector cell and the restriction to isocratic operation. Thermostating is typically done by use of a thermal block, making them relatively heavy instruments. Figure 14.8 diagrams one common **differential** type of RI detector. Mobile-phase mixture and analyte-containing eluent each pass through one of two transparent compartments separated by a tilted transparent plate. When the RI of the fluids in each compartment differs, the optical arrangement results in shifting of the source radiation beam's focus *location* on a photocell sensor. The photocell's response surface is arranged so that this shift in *position*, rather than a change in beam *intensity* (as in absorption spectrometry), results in a signal linearly proportional to the RI change produced by the analyte

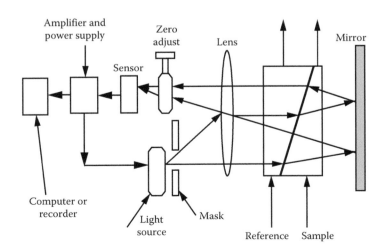

Figure 14.8 Schematic diagram of differential RI detector. (From Katz, E. et al., eds., *Handbook of HPLC*, Marcel Dekker, Inc., New York, 1998. With permission.)

concentration in the cell. The linearity is typically four orders of magnitude.

The RI detector for HPLC is somewhat analogous to the thermal conductivity detector (TCD) in GC. It is relatively insensitive (with a maximum sensitivity of approximately 10^{-7} g) but almost universal in its response, failing only in the unusual cases where the RI values of analyte and mobile phase are identical. The lack of sensitivity results from the relatively small difference in RI of many compounds with respect to the solvent stream. Both the TCD and the RI detector are more readily and stably operated without "programmed" variation of mobile-phase concentrations or GC oven temperatures. Response factors for most organic molecules vary over less than an order of magnitude. It is resorted to for analytes that give no response with other more sensitive or selective detectors. An example of analytes requiring HPLC with RI detection is mixtures of complex carbohydrates like sugars, which have no chromophores, fluorescence, electrochemical activity, and so on, and which would be difficult, even after derivatization, to get through a GC column without decomposition. One difference between RI and TCD is that RI is not very sensitive to variation in the mobile-phase flow rate, while the TCD is.

14.1.4.2 Evaporative Light-Scattering Detector

14.1.4.2.1 ELSD Characteristics

An evaporative light-scattering detector (ELSD) is a universal, destructive, mass-flow detector, and sensitive, requires auxiliary gas, and is useable with isocratic or gradient mobile phases but with *no buffer salts*, insensitive to flow variations, more expensive than RI detector, and *not* usable with mobile-phase buffers.

The ELSD is the other universal HPLC detector. Instead of absorption, stimulated emission, or refraction, it uses yet another interaction of analytes with light, namely, "**scattering**." Scattering occurs when light interacts with small particles of dimension comparable to its wavelength and is scattered, without a change in wavelength, in a direction to the side of its original path. The particles are of a phase different from the one in which they and the light are embedded (e.g., liquid droplets or solid particles in a vapor or solid particles in a liquid). How could we employ this principle to measure concentrations of molecules, which are *much* smaller than visible light wavelengths, when they are dissolved in liquid mobile phases? These do not scatter light. The trick is to *nebulize* the mobile-phase effluent into droplets and instantly *evaporate* each of these, leaving behind a residue as a tiny solid particle of the nonvolatile analyte. The mechanism of particle formation from the droplets and the relation of scattering response to particle size is complex, but empirical calibration curves can be obtained. The response is roughly proportional to the mass of analyte eluting. As long as no buffer salts are present, so all the mobile phase can be volatilized, the ELSD is insensitive to composition variations during gradient elution. It can be orders of magnitude more sensitive than RI detection. Modern instruments illuminate the particle cloud exiting a heated droplet evaporation tube with an intense monochromatic beam from a solid-state laser. Scattered light passes through a filter tuned to pass the same wavelength to a sensitive photocell that provides the detector signal.

Figure 14.9 illustrates the three steps in the operation of an ELSD: nebulization, evaporation, and light scattering. As we have seen for inductively coupled plasma (ICP)-optical emission spectroscopy (OES) (Section 8.1.3) and will see later for LC-MS (Section 14.1.6.1), careful design of nebulizers or spray chambers and thermally assisted droplet evaporation is key to many instrumental interfaces designed for introducing analytes as liquid solutions. Figure 14.10 illustrates some advantages of the ELSD as a universal HPLC detector by comparing its response to the same effluent measured with other detectors.

The ELSD has a more uniform response to analytes than an ultraviolet (UV) absorption detector, whose response

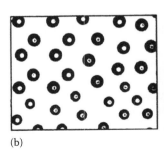
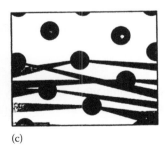

(a) (b) (c)

Figure 14.9 The ELSD detects any compound less volatile than the mobile phase to low nanogram levels using a simple three-step process: (a) Nebulization—Column effluent passes through a needle and mixes with nitrogen gas to form a dispersion of droplets, (b) Evaporation—Droplets pass through a heated "drift tube" where the mobile phase evaporates, leaving a fine mist of dried sample particles in solvent vapor, and (c) Detection—The sample particles pass through a cell and scatter light from a laser beam. The scattered light is detected, generating a signal. (Alltech 3300 ELSD, adapted with permission from Buchi Labortechnik AG, www.buchi.com.)

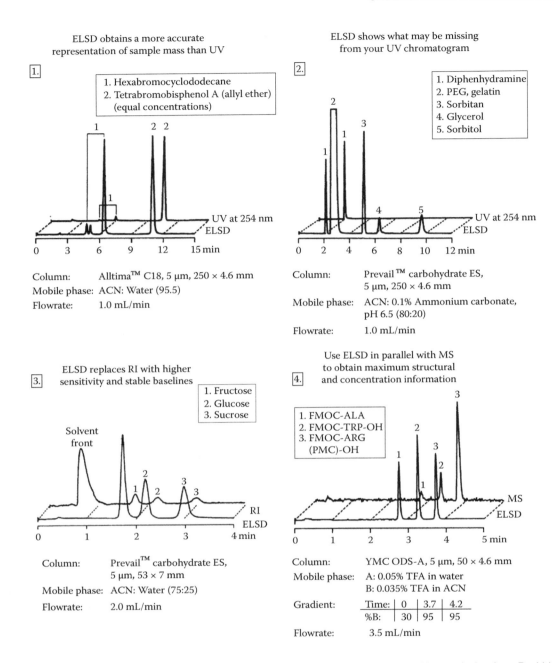

Figure 14.10 Illustrations of four advantages of ELSD versus other HPLC detectors. (Adapted with permission from Buchi Labortechnik AG, www.buchi.com.)

factors for different compounds vary widely and are wavelength dependent and which gives no response at all to many compounds, as shown in Figure 14.10(2).

Only one of five analytes eluting here (Figure 14.10[2]) is seen by the UV detector, and it misses the huge peak for PEG and gelatin, whereas all five are seen by the ELSD. The similarity of the mass proportionality responses enables one to immediately identify the major component.

ELSD has a much better S/N response to three sugars and better baseline stability than the RI detector. Note that the ELSD shows no solvent front peak, since the solvent droplets evaporate completely leaving no particle behind to scatter light (Figure 14.10[3]).

Here (Figure 14.10[4]), three derivatized amino acids or amino acid metabolites are analyzed by both LC-MS and LC-ELSD. The mass spectra in the LC-MS contain much structural information, but MS ionization efficiencies and fragmentation yields vary widely, so TICs convey the relative concentrations less accurately than do the ELSD peaks.

14.1.4.2.2 Corona Charged Aerosol Detector

This newer universal aerosol detector operates similarly to the ELSD, but instead of measuring light scattered by particles from a laser beam, it adds charge to N_2-dried vapor-borne particles formed by aerosol evaporation. As HPLC peak concentration increases, so do the sizes of the analyte particles. A second N_2 stream is positively charged by passing a high-voltage corona discharge wire, and the charge is diffusively transferred to the analyte stream particles coming from the opposite direction. The amount acquired is directly proportional to the particle size (and the analyte concentration). The charge is transferred to a collector where it is measured by an extremely sensitive electrometer (a voltmeter that operates while drawing almost no current [moving charge]). This signal is proportional to the analyte amount, and its sensitivity (mid-picogram) and wide dynamic range of 4 orders of magnitude are claimed to surpass the performance of ELSDs.

14.1.4.3 UV/VIS Absorption Detectors

14.1.4.3.1 UV/VIS Detector Characteristics

A UV/visible (VIS) detector is compound specific, nondestructive, and a concentration detector; compound sensitivities differ over a wide range; it is useable with isocratic or gradient mobile phases including buffer salts; and pre- or post-column derivatization can be used to increase the number of measurable compounds.

The principles and operation of UV/VIS spectroscopic detectors are discussed at length in Section 4.2.4. The "detector" in HPLC refers to the whole spectroscopic assembly, whose design is optimized to make rapid, continuous measurements of solutions passing through a flow cell of volume much smaller than the typical sample cell or cuvette of a conventional spectrometer. The light source, optical train, dispersion device, and light detectors are similar in design to those discussed in Chapter 4, but they are often built on a much smaller compressed scale, to match the size of the other HPLC components. The major differences from the large spectrometers are the much smaller measurement volume of the flow cell, the necessity to route continuous flow through it, and the need to focus light on smaller areas on the flow cell windows and extract it efficiently to send it on to the light "detector."

Figure 143.11 displays a simple UV/VIS flow cell. These are typically glass with UV transparent quartz windows bonded at each end enclosing volumes of 1–10 μL. Note that in this design, the incoming flow is directed against one window, to induce turbulent mixing that breaks up the parabolic laminar flow velocity profile across the diameter of small-bore tubes. The flow exits in a similar fashion. This "Z-design" flow cell geometry reduces laminar flow

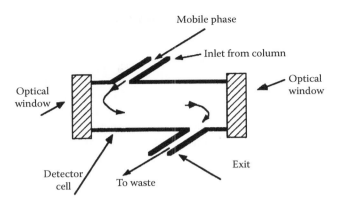

Figure 14.11 Diagram of HPLC UV–VIS detector flow cell. (From Scott, R.P.W., *Chromatographic Detectors*, Marcel Dekker, Inc., New York, 1996. With permission.)

velocity variation within the cell that could contribute to band spreading.

Compounds with aromatic rings, C=C double bonds, or with functional groups containing double bonds to some heteroatoms, particularly when these structures are multiply conjugated, may display widely varying degrees of absorption in the UV or visible regions (cf. Chapter 4). These features occur in many polar organic compounds that are separated and quantitated by RP-HPLC. In one survey, well over 70 % of compounds assayed by HPLC using nonhyphenated techniques employed UV/VIS detection and another 15 % were measured with the related UV fluorescence detector. Despite not being as universal as the previously described detectors, UV/VIS and fluorescence are the workhorse detectors of HPLC. Their position is like that of the flame ionization detector (FID) in GC. They possess the best combination of applicability to analytes of interest, compatibility with the widest range of mobile phases, and high sensitivity. The sensitivity is of course dependent on the molar absorptivity of the chromophore, but for ε between 10,000 and 20,000, the sensitivity will be approximately 10^{-9} g. Unlike the FID in GC, or the RI and ELSD detectors earlier described, these light absorption and emission detectors have compound response factors that vary widely, so calibration with standards of each analyte is mandatory. The optimum wavelengths for measurement will also vary with analyte structure.

There are three main classes of UV/VIS absorbance detectors, namely, **single-wavelength filter photometers, dispersive monochromator detectors** (DMDs), and **diode array detectors** (DADs). The first of these often uses a source lamp that emits light at wavelengths specific to the excited, vaporized element employed in its design. These lines are isolated by wavelength-selective filters. The latter two use source lamps that emit over a wide wavelength range: tungsten filament for the visible and near UV or deuterium and xenon lamps that cover a broad emission range into the nonvacuum UV (down to around 190 nm). Many designs employ a dual-beam geometry, with the reference beam and its photocell monitoring a flow cell containing only mobile phase. Sometimes the reference beam simply passes through a filter. In the first case, ratioing of the two signals can compensate for absorbance variations in the mobile phase. In both designs, variations in source intensity are compensated for. If absorbance data from a **single-beam** design instrument are stored in a computer during a run, subtraction of mobile-phase absorbance may be achieved post run, but stable source intensities must be maintained as source variations cannot be compensated.

14.1.4.3.2 Single-Wavelength Filter Photometers

This simplest of detector designs often employs a mercury vapor lamp source, with filters between the flow cell and the photocell isolating a single-line emission wavelength. The most useful such line is the intense one at 254 nm, a wavelength at which many UV-active compounds absorb strongly. Less generally useful, but sometimes more selective lines at 250, 313, 334, or 365 nm may be isolated and monitored by use of the appropriate filters. Interference filters may be used with deuterium sources to isolate specific wavelengths and measure absorbance at them.

14.1.4.3.3 Dispersive Monochromator Detector

Instead of a selected wavelength filter, a grating monochromator (Section 3.6.2.3) may be used to select a narrow wavelength band for absorption measurement. The wavelength is generally isolated *before* the beam passes through the detector flow cell. Such detectors are particularly useful when the analytes to be measured are known in advance (**target compound analysis**). The monochromator may be set to monitor the signal at the compound's wavelength of maximum absorption, and if operating under computer control, this wavelength can be varied to the optimal value for each chromatographic peak as it elutes. Conversely, if trying to use UV/VIS spectra to help identify unknown compounds, the eluent flow can be stopped while the top of the LC peak is in the flow cell, and the monochromator can be scanned across all accessible wavelengths to produce a spectrum.

14.1.4.3.4 Diode Array Detector

This detector substitutes a linear array of light-sensitive solid-state diodes for the photocell. See the extensive discussion of diode array detector (DAD) design and function in Section 4.2.4. Light from the source passes through the flow cell and is then dispersed into its component wavelengths, which are directed to the different diodes of the array. The array itself is illustrated in Figure 4.24. Each diode responds to light from a narrow wavelength range of the dispersed light, so all wavelengths are monitored simultaneously. A computer is required to read out all these parallel signals and store them in files arrayed as a function of elution time. A diagram of a diode array UV/VIS spectrometer is presented in Figure 4.25. Figure 14.12 diagrams the components of a typical DAD designed for use in HPLC. Note that the wavelength dispersion takes place *after* the beam from the source passes through the flow cell. By contrast, the dispersive monochromator detector (DMD) isolates a single wavelength *before* passage through the flow cell. In either design, there is a possibility that in some cases, the UV passing through the flow cell might excite fluorescent emission at longer wavelengths. In the DAD, this would distort the observed absorption spectrum at those longer wavelengths,

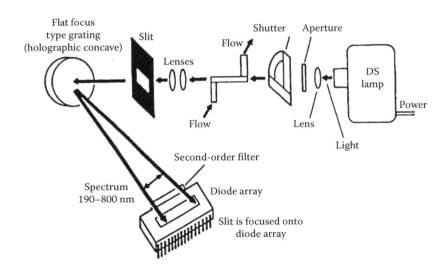

Figure 14.12 Schematic of DAD for HPLC. (From Cazes, J., *Encyclopedia of Chromatography*, Marcel Dekker, Inc., New York, 2001. With permission.)

and in either the DAD or DMD, it might add back to the absorption signal being recorded at the detector and cause the absorption value to appear less. As the DAD illuminates the cell at all wavelengths, the probability of it falling victim to distortions arising by this mechanism is greater. The DAD data set can easily provide UV/VIS spectra of everything eluting during the chromatogram. If HPLC peaks are composed of a single UV-active component, the ratios of DAD responses at several different wavelengths will remain constant across the peak. If there is an impurity with a somewhat different UV spectrum within the HPLC peak, and it elutes with its peak at a slightly different retention time from that of the major component, then these wavelength response ratios will vary across the peak. This is a powerful method for detecting the presence of small impurities "buried" under the major component HPLC peak.

14.1.4.3.5 Infrared Absorbance Detectors

For completeness, we should note that IR detectors for HPLC do exist. As is the case for GC-IR (Section 13.8.2), a continuous rapid-response HPLC-IR detector benefits from Fourier transform IR (FTIR) spectral acquisition and computerized data file storage. Such spectra as may be obtained have spectral peak widths more characteristic of a liquid matrix, instead of the more information-rich, sharp gas-phase GC-IR peaks. The major drawback to IR detection in HPLC is that most of the commonly useful HPLC mobile phases absorb strongly in many areas of the IR spectral region. Thus, HPLC-IR can be used for only a very limited set of analytes.

14.1.4.4 Fluorescence Detectors

14.1.4.4.1 Fluorescence Detector Characteristics

It is compound specific, nondestructive, and a concentration detector; compound sensitivities differ over a wide range; detection at right angles to and at wavelength different from the excitation beam results in low background noise and thus higher S/N sensitivity than UV/VIS; it is useable with isocratic or gradient mobile phases including buffer salts; and pre- or postcolumn derivatization can be used to increase the number of measurable compounds.

The general principles of fluorescence spectrometry and the design of fluorescence spectrometers are discussed in Chapter 4. The simplest detectors for HPLC excite molecules with intense UV emission lines from a mercury lamp or wavelengths isolated by filters from a xenon lamp. The fluorescent emission at longer wavelength is sampled at right angles to the excitation beam, and the desired wavelength is isolated by use of a suitable filter. A more versatile design employs one tunable monochromator to select the optimum excitation wavelength from a broadband source such as a deuterium or xenon lamp and another to select the fluorescent emission wavelength passed to the photocell or diode detector. This permits operation at optimal sensitivity for a wider range of analytes. More recent designs employ an intense monochromatic laser beam to increase the energy used for excitation. While this provides for more intense excitation, limitations of available laser wavelengths sometimes preclude excitation of analytes at their absorption maxima unless expensive tunable lasers are used. A disadvantage of this detector is that only a limited subset of analytes display substantial fluorescence emission. Its greatest advantage by comparison with the UV/VIS absorption detector is its several orders of magnitude greater sensitivity for those compounds which do display such emission. This results from the very low noise background in the measured signal. This is a consequence of measuring at right angles to the incident beam against a dark background at a wavelength longer than and isolated from that of light which might be scattered from the excitation beam. Improved sensitivity results from an improved S/N ratio.

The extreme sensitivity of fluorescence detection (approaching 10^{-11} g) encourages the use of derivatization reagents to react with a nonfluorescent analyte's functional group(s) to introduce or form a fluorescent structure in the derivative. This must be done prior to injection of the sample and HPLC separation, if the derivatization reaction is slow. More often, one attempts to quickly form derivatives of separated peaks as they elute from the column prior to passing into the detector. This procedure is called **postcolumn derivatization**, and it will be discussed in greater detail in Section 14.1.5. One common fluorescence derivatizing reagent is dansyl chloride ("dansyl" = 5-dimethylaminonaphthalene-l-sulfonyl-). Dansyl derivatives of primary and secondary amines, amino acids, or phenolic compounds form readily and are highly fluorescent

Highly sensitive detection of chromatographically (or electrophoretically) separated mixtures of fluorescent derivatives of amino acids or peptides is central to many of the instruments which support the new discipline of **proteomics**. The much greater sample processing speed and power of LC-MS or LC-MS/MS proteomic systems now has them rapidly overtaking the fluorescence-based detection systems.

A fluorescent derivatizing agent can derivatize the terminal nucleotide of a mixture of all fragment lengths produced by sequential removal of one nucleotide at a time from one end of a deoxyribonucleic acid (DNA) polymer. Chromatography or electrophoresis can separate and elute these in inverse order of their length. The smallest will elute first. There are four different nucleotides that could be at the end of each fragment, and each such derivative has a different fluorescence emission wavelength. The use of a fluorescence detector capable of discrimination among each of these emissions allows one to identify the end nucleotide in each successively longer fragment. This then gives one the **sequence** of nucleotides (i.e., the order of the "letters" of the genetic code in the particular segment of the DNA chain being analyzed). By repeating this enough times, with hundreds of automated machines operating around the clock for years, the entire human genetic code of around 3 billion nucleotides has been sequenced! An electropherogram employing this four fluorescent wavelength detection and sequencing of a chain of close to 1000 DNA "bases" (nucleotides) is displayed in Figure 14.40. The subdiscipline of **genomics** relied greatly upon the great sensitivity of fluorescent detection applied to small-volume capillary separations. As of this edition, such genomic assays are being rapidly supplanted by higher-capacity, microchip-based, purpose-built, specific binding instrumentation.

14.1.4.5 Electrochemical Detectors

14.1.4.5.1 ECD Characteristics

It is compound-specific, destructive, and a concentration or mass-flow detector depending on operation; compound sensitivities differ over a wide range; it is useable primarily with isocratic RP-HPLC including some buffer salts; and pre- or postcolumn derivatization can be used to increase the number of measurable compounds.

Electrochemical detectors (ECDs) for HPLC are summarized in the electrochemistry chapter in Section 16.4. They are divided into two classes: **voltammetric** (subdivided into amperometric, polarographic, and coulometric) and **conductometric**. The voltammetric detectors will be described in this section under the classification of ECDs (be careful not to confuse the acronym with the GC ECD, where EC denotes electron capture, not electrochemical). The conductometric detectors operate on a principle different from the voltammetric ones, so they will be treated separately in the next section. One must employ RP-HPLC aqueous or polar mobile phases capable of carrying dissolved electrolytes (i.e., the analytes measured by conductometric detectors or the supporting electrolytes to suppress migration current [cf. Section 16.4]). Thus, ECDs are not suitable for NP-HPLC separations. As distinguished from conductometric detectors, ECDs respond to substances that are oxidizable or reducible at a solid electrode surface. Their signal is the current that flows through the electrode circuit when these processes occur. If flow of analyte through the detector cell and the potentials applied to the electrodes are appropriately controlled, then the rate at which the analyte is provided to the **working electrode** surface to be consumed there to liberate electrons to the electrical circuit is proportional to its amount in the cell. The usual design of ECDs employs a three-electrode configuration consisting of:

1. **The working electrode**, where electrons are transferred in a redox reaction.
2. **An auxiliary electrode**, to which the current flows from the working electrode.
3. **A reference electrode**, which serves as a reference against which to control and maintain the potential of the working electrode relative to the auxiliary electrode.

Many different arrangements of these three electrodes are employed in the design of ECD cells. Five are depicted schematically in Figure 14.13.

If all the analyte is consumed, integration of the current determines the total charge transferred and such a voltammetric detector is classified as **coulometric** (from the coulomb unit of measurement of charge). Such complete electrolysis of the analyte in the short period of peak passage through the detector cell is difficult to achieve on a standard working electrode surface. A **porous carbon electrode** can be constructed as a frit in the detector cell through which all the effluent must pass. The pressure drop through such an electrode is small, and its huge internal surface area enables complete consumption of all analytes reacting at its potential, thus enabling coulometric rather than amperometric detection. A sequence of these electrodes (each with its accompanying auxiliary and reference electrodes) can be arrayed in series along the flow path to operate at intervals of increasing potential. Such a **multielectrode array detector** will sequentially and separately produce a signal only from analytes that react in the narrow potential interval around which each electrode is set. Analytes that would react at lower potentials have already been completely consumed by prior electrodes, and those requiring higher potentials will be measured by subsequent ones. The data acquisition and multianalyte capabilities of such an array are analogous to the DAD for UV/VIS spectral responses. An advantage of coulometric detection is that if the reaction is known, that is, the number of electrons consumed or released per mole of analyte is known, the integrated signal quantifies the analyte without need of a calibration standard.

The most common ECD for HPLC is the **amperometric detector**. Here, the analyte is completely depleted at the working electrode surface, but not in the bulk solution passing through the flow cell. An equilibrium diffusion current that is proportional to both the hydrodynamic flow parameters past the working electrode and the concentration of the oxidizable or reducible analyte is established. If steady-state effluent flow and diffusion conditions are maintained, then the amperometric current is proportional just to the analyte concentration in the cell. These conditions are more difficult to maintain during a gradient elution, so ECD detection works best with isocratic separations. There is a wide variety of modes of operation of amperometric ECDs. The working electrode may be an anode (where the analyte is oxidized). A useful material for oxidative analyses is vitreous, conductive, so-called "glassy" carbon. Reductions can also be carried out on this electrode, and various metals such as Pt, Au, and Hg are used as cathodes for reductions. These sometimes are vulnerable to buildup

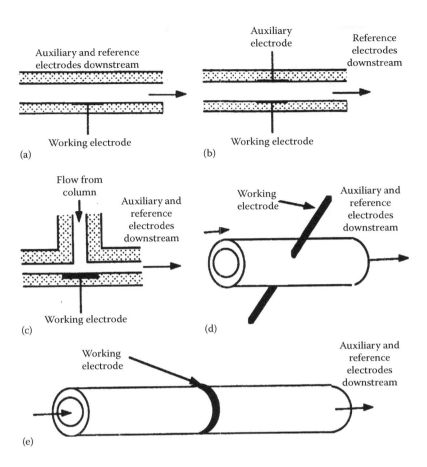

Figure 14.13 (a–e) Five HPLC ECD three-electrode arrangements. (From Scott, R.P.W., *Chromatographic Detectors*, Marcel Dekker, Inc., New York, 1996. With permission.)

of contamination by adsorbing material. A mercury electrode can be extruded from a capillary as a hanging drop, which can be easily replaced with a fresh, clean surface. As described in Chapter 15, the high overvoltage for H$^+$ reduction on Hg permits a considerable range of electrode potentials over which reductions may be induced without solvent reduction. This is the principle of polarographic analysis, and a detector with such a voltammetric working electrode is classified as a **polarographic detector**. When using such detectors, extensive degassing of the mobile phase becomes critical for the additional purpose of removing potentially interfering reduction of dissolved oxygen. A detector configured as a pair of amperometric electrode assemblies can be operated in series, with the first set to oxidize an analyte to a form suitable for subsequent reduction by the next detector in line, or *vice versa*. Optimization of the potentials for operation of each working electrode in such a chain can confer additional target analyte selectivity, which may also result in improved sensitivity by lowering background noise.

A recent refinement of such detectors is a combination of **high-performance anion-exchange (HPAE)** separation in a capillary IE (see Section 14.2) column using a capillary ECD cell employing a chemically inert gold working electrode versus a Pd/H$_2$ reference electrode operated for **pulsed amperometric detection (PAD)**. In PAD, the electrical potential is cycled up to three times a second to clean and restore the detection electrode surface after the measurement and before the next detection by oxidative desorption for amino acids. Mono- and disaccharides and higher sugars undergo adsorption followed by oxidative desorption. A prime application is the determination of mono- and disaccharides (organic carbohydrate sugars) at very high pHs at which they acid-dissociate (11–13) by using very high stationary and mobile-phase pHs. One key to this is the use of Carbopak™ columns capable of operating at extreme alkaline pH. Most other detectors give no response to this class of compounds. **HPAE-PAD** has been used to analyze hydrolysates of the polysaccharide heparin, to determine its glucosamine content and the level of impurity chondroitin sulfate, which releases galactosamine instead. The exceptional sensitivity of the PAD enables measurement of the latter sugar from very small samples, as less than 1 % of the total amount of analyte passing through the detection cell is

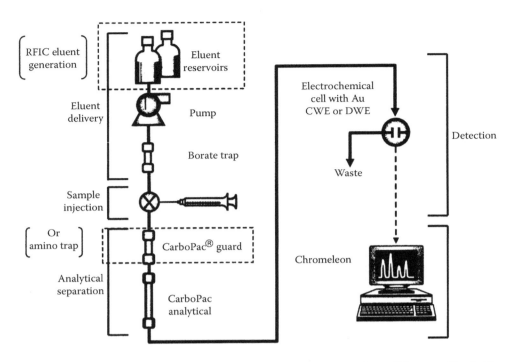

Figure 14.14 Schematic diagram of HPAE-PAD system. (© Thermo Fisher Scientific, www.thermofisher.com. Used with permission.)

consumed by the electrode reaction. (A schematic diagram of this system is displayed as Figure 14.14.)

14.1.4.6 Conductometric Detectors

14.1.4.6.1 Conductometric Detector Characteristics

It is universal for all anions or cations in solution (i.e., a "bulk property of mobile-phase" detector), nondestructive, and a concentration detector; compound sensitivities differ over an order of magnitude range, useable primarily with isocratic RP-HPLC without buffer salts (unless subsequent suppression column is used); and it is a premier detector for ion chromatography (IC).

The conductometric detector is small, simple, and easy to construct. It consists of two electrically isolated electrodes immersed in or surrounding the mobile-phase effluent flow, as illustrated in Figure 14.15. All ions in the mobile phase will participate in carrying current across the cell when an *alternating potential* (AC signal) of frequency 1000–5000 Hz is imposed between the electrodes. If a constant DC potential were imposed, the ions would migrate to the electrodes opposite to their charge, concentration polarization would occur, and current through the cell would decrease to zero. With an imposed AC voltage, this polarization is suppressed, and AC current passes through the cell. The conductance is proportional to the concentration of ions, and small changes in its inverse quantity (resistance, or with AC current, more properly impedance) may be measured directly or against a reference cell in a Wheatstone bridge arrangement (cf. discussion of GC-TCD in Section 13.7.1 and Figure 13.15).

The primary usefulness of the conductometric detector is for small anions or cations such as inorganic ions (e.g., $Cl^-, ClO_4^-, NO_3^-, PO_4^{-3}$, Na^+, NH_4^+) or small organic anions (e.g., acetate, oxalate, citrate) or cations (e.g., di-, tri-, and tetra-alkyl ammonium ions). These often possess no useful chromophore or suitable redox activity at ECDs. The difficulty with these conductometric detectors is that they are bulk property detectors responding to all ions present in the mobile phase. They cannot be made compound selective. Background ions from mobile-phase buffers or the ionic eluents used in ion-exchange chromatography (Section 14.2.1) would swamp the conductance from low levels of ionic analytes. The conductometric detector first became generally useful when the ion suppressor column was developed to remove this background, giving rise to the method termed IC (discussed in Sections 14.2.2 and 16.4).

14.1.4.7 Charge Detector (QD) [Thermo Scientific]

14.1.4.7.1 Charge Detector Characteristics

It is universal for anions and cations in solution; destructive; a concentration-sensitive detector; and higher, more linear response for weakly dissociated species; its response is proportional to both concentration and charge of analytes.

The charge detector mode of operation is illustrated in Figure 14.16. Analyte ions, cations (A^+) or anions (Y^-), in the suppressed column effluent flow through a

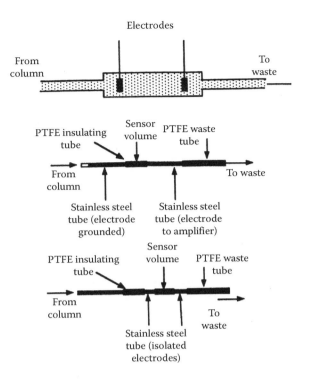

Figure 14.15 Three conductometric detector designs. (From Scott, R.P.W., *Chromatographic Detectors*, Marcel Dekker, Inc., New York, 1996. With permission.)

central channel defined by cation-exchange (CX) and anion-exchange (AX) membranes and are drawn by the applied voltage toward the electrodes of the opposite polarity. A current derived at the electrodes by the transport of the ions to the electrodes is the detector signal. The transported ions combine with electrolysis-generated ions at the electrodes (hydroxide [OH^-] or hydronium [H_3O^+] ions) to form a base (AOH) or acid (HY), respectively. The countercurrent flow of an aqueous stream in the electrode channels provides the water required for the electrolysis reaction that occurs at >1.5 V and aids in removing the base or acid products. The analyte ions are converted to these products and removed to waste. Thus, the QD is a destructive detector and must be placed at the end of a chain of multiple detectors. Advantages include stronger relative response for very weakly dissociated species such as boric acid, greater linearity than suppressed conductivity detection alone for such compounds, and the ability to combine its superior performance relative to the latter nondestructive detector by operating it in tandem behind the conductivity detector, which possesses greater sensitivity for strongly dissociated ions. Thermo Scientific now offers this small, low-footprint QC for capillary IC systems. For some applications, the tandem detector combo may provide additional information such as peak confirmation and predict coeluting peaks and trigger the need to use MS detection to identify the coeluting components in a sample.

14.1.4.8 Summary Comparison of Six Major HPLC Detectors

A comparison of the six major HPLC detectors is shown in Table 14.2.

14.1.5 Derivatization in HPLC

In Section 13.10, we saw how in GC analytes were reacted with derivatizing agents primarily to produce a compound with improved thermal stability and greater volatility by masking more polar functional groups with a less-polar substituent. Sometimes this also was done to introduce atoms that rendered the derivative detectable by more sensitive and selective GC detectors. With HPLC, analyte thermal stability is not so critical, and the workhorse RP systems perform well with very polar analytes. The primary purpose for derivatization in HPLC becomes that of introducing substituents to increase the sensitivity and selectivity of detection. One case where improvement of HPLC is a major goal is for analytes that exist as charged ions in solution (anions, cations, or zwitterions [containing both positively and negatively charged atoms]). The very important group of amino acids falls into this category. Originally, complete separation of complex mixtures of these required very lengthy runs on ion-exchange resin columns. Reaction of physiological amino acids with orthophthaldehyde produces isoindole derivatives whose polarity is compatible with much more efficient RP-C18 columns, leading to much more rapid separations. Additionally, these derivatives display intense fluorescence at 425 nm, which greatly improves the detection limits when monitoring with a fluorescence detector.

The second benefit in the earlier example illustrates the primary application of derivatization in HPLC. Inspection of Table 14.2 suggests that if we want to derivatize for improved selectivity and sensitivity, introduction of fluorescence should be our first choice. Table 14.3 lists a number of fluorescence derivatization agents and their target functionalities. A comprehensive list of derivatizing reagents can be found in the book by Bruno and Svoronos. Generally, we wish to derivatize only one group at a time.

The derivatization can be carried out in solution as a last step in sample preparation. This allows for whatever temperature and reaction time may be necessary to form the derivative. This is the standard practice for forming GC derivatives and is termed **precolumn derivatization** (or labeling). In GC, we can sometimes avail ourselves of the very high temperatures in a GC injection port during flash vaporization to perform "**on-column derivatization**" (cf. Section 13.10). While this is not possible in HPLC, we have an even better option with this latter form of chromatography; namely, **postcolumn derivatization**. Here, the labeling agent is added to the column effluent by constant rate infusion from a separate pump, passes through

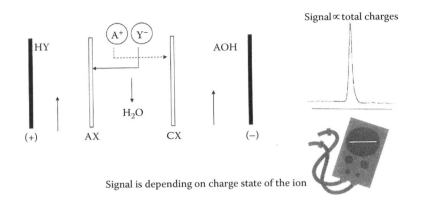

Figure 14.16 Charge detector (QD) operation diagram. (© Thermo Fisher Scientific, www.thermofisher.com. Used with permission.)

Table 14.2 Comparison of Major HPLC Detectors

Detector Type	RI	ELSD	UV/VIS	Fluorescent	ECD	Conductivity
Typical LOD (ng/injection)	100	1	1	0.01–0.1	0.01	1
Selectivity	No	No	Moderate	Very high	High	Low
Robustness	High	High	Excellent	High	Poor	High
Gradient elution	No	Yes	Yes	Yes	No	Yes[a]
Microsystem usable	No	No	Limited	High	High	High

[a] When used with suitable effluent suppression system.

Table 14.3 Reagents for Precolumn Fluorescence Labeling

Reagent	Analytes	λ_{ex} (nm)	λ_{em} (nm)
o-Phthalic anhydride (OPA)	Primary amines, thiols	340	455
Fluorescamine	Primary amines	390	475
Dansyl chloride (Dns-Cl)	Primary and secondary amines	350	530
9-Fluorenylmethyl chloroformate (FMOC)	Primary and secondary amines	260	305
Naphthyl isocyanate	Hydroxy compounds	310	350
Dansylhydrazine (Dns-H)	Aldehydes and ketones	365	505
Anthryldiazomethane (ADAM)	Carboxylic acids	365	412
N-Acridinyl maleimide	Thiols	355	465

Source: Katz, E. et al., eds., *Handbook of HPLC*, Marcel Dekker, Inc., New York, 1998.

a mixing chamber or coil, possibly at elevated temperature, for a precisely regulated and reproducible interval, and then passes into the detector cell. A block diagram of such a system is displayed in Figure 14.17. When using this mode, the purpose is clearly only to convert the analyte to a better-detected form. The postcolumn reaction needs to be fast, as too long a mixing chamber could lead to extra-column band broadening. However, the reproducibility of the mixing and its interval yields satisfactory quantitation even if the reaction does not proceed to completion. Additionally, should the derivative be unstable, it is being quickly measured at a highly reproducible interval after formation, which would not be the case when using precolumn derivatization. If the reagent is itself fluorescent in the range being monitored, the presence of the unreacted excess will interfere with the analyte measurement. If such a reagent is used in the precolumn mode and is separable from the derivatized analyte on the HPLC column, a fluorescent reagent may be usable. When properly implemented, postcolumn derivatization is actually more reproducible in its yield and quantitation than is the precolumn mode. Additionally, if more than one fluorescent product may be formed in the reaction, they will all pass through the detector cell together, while the products of a precolumn procedure may elute separately in variable proportions from the HPLC column.

Derivatization can be employed to advantage with ECD detection as well. Both precolumn or postcolumn reactions may be employed to produce electroactive compounds analogous to the formation of fluorescent derivatives. The UV-induced reaction chamber for ECDs described in Section 16.4 and the dual electrode detector oxidation–reduction series described at the end of Section 14.1.4.5 may be regarded as non-reagent instrumental forms of the principle of postcolumn derivatization.

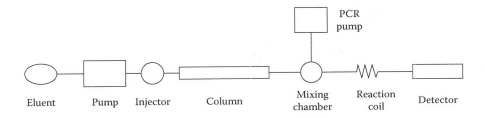

Figure 14.17 Block diagram of HPLC postcolumn derivatization system. (From Katz, E. et al., (eds.) *Handbook of HPLC*, Marcel Dekker, Inc., New York, 1998. With permission.)

14.1.6 Hyphenated Techniques in HPLC

Adding a second dimension to an HPLC separation lagged behind the application of this idea to GC. True HPLC columns and instrumentation were just becoming available in the early to mid-1970s when the first commercial computerized GC-MS systems came on the market. At the time of this writing, while comprehensive 2D-GC systems and their associated data-handling software have become widely commercially available, 2D-LC or LC-GC systems are mostly confined to development in individual academic and research laboratories. Agilent has recently introduced a 2D-LC system that combines components from its LC product line, with a novel 10-port valve having all equivalent liquid transport lengths and reverse flow for backwashing steps. When operated under computer control with Agilent software, it can provide automated heart cutting or comprehensive 2D-LC and displays of separations similar to those in comprehensive 2D-GC or even shift between these modes during the course of a run. It is easier to interface the less-dense vapor mobile phase of GC to the vacuum conditions required for MS operation. There is no equivalent in LC of GC-GC's thermal modulator (cf. Section 13.8.3) to refocus and transfer effluent peaks from one column to another continuously with few or no moving parts. The use of a DAD detector (Section 14.1.4.3) with HPLC does enable continuous acquisition of full UV/VIS spectra of peaks eluting from an HPLC column. This can be valuable in some applications; however, its utility does not begin to match that of GS-MS or GC-IR. The range of organic compounds that present useful UV/VIS spectra is much narrower than those that yield mass spectra or IR spectra (essentially all organic compounds). The UV/VIS spectra contain much less structure than mass or IR spectra and are much less easy to interpret for identifying unknowns or to match unambiguously with library searching programs. LC full-scan-DAD instruments are therefore often not considered true hyphenated techniques. The IR absorption of most HPLC mobile-phase solvents precludes use of IR detection as a full-scan second-dimension detector.

LC-nuclear magnetic resonance (NMR): One information-rich spectral technique that is more suited to the liquid mobile phase of HPLC than to the vapor phase of GC is NMR. LC-NMR has been implemented, but it has significant limitations. To obtain interpretable spectra of unknowns, concentrations in the measurement cell must be higher than with other detectors. The cell must be smaller than the usual NMR tube, so for any but the very highest concentrations of analytes, FT-NMR acquisition is preferred, with each eluted peak being retained in the measurement cell by a stopped-flow procedure similar to that employed to increase sensitivity in GC-IR (Section 13.8.2). Expensive deuterated mobile-phase solvents are required for proton NMR, which mandates the use of low mobile-phase volume flow columns: narrow-bore or even capillary HPLC. LC-NMR is expensive to implement and only just becoming available from commercial vendors at this time.

14.1.6.1 Interfacing HPLC to Mass Spectrometry

Given the clear superiority of HPLC over GC for separations of large polar biomolecules, analysts yearned to interface its effluent to MS instrumentation. A huge market in the pharmaceutical and biomedical research field and in clinical biomonitoring beckoned. The difficulty was the approximately three orders of magnitude greater mass of less-volatile mobile phase that must be prevented from entering the MS vacuum system (while still letting the analytes in), since no imaginable MS pumping system could remove it fast enough. Consider that a vapor is ~1000 times less dense than its parent liquid and that typical mobile-phase flow rates in capillary GC and packed column HPLC are both on the order of 1 mL/min. The jet separator design for GC-MS illustrated in Figure 13.20 would only work with mobile phases of very light, fast-moving atoms or molecules like He or H_2. Chromatographers attempted to deposit HPLC effluent onto a moving conveyer belt that passed through vacuum interlocks into a pumped-down chamber where the mobile phase would vaporize, leaving the analytes behind on the belt. The belt then passed through further interlocks to an even higher-vacuum ion source chamber, where it was heated in hopes of volatilizing the analytes so they could be ionized and fragmented under electron ionization (EI)-MS or chemical ionization (CI)-MS conditions as in classic GC-MS analyses. These contraptions were difficult to operate in a reproducible manner, and the prospect for volatilizing thermally sensitive large biomolecules was not much better than trying to coax them to survive passage through the high-temperature gauntlet of a GC-MS system.

14.1.6.1.1 Thermospray HPLC-MS Interface

The first really successful interface did in fact look something like the GC-MS jet separator, although its mechanism was different. The HPLC effluent from a microbore column or a fraction split from a regular diameter column was coupled to a capillary tube 10–20 cm long, nestled inside a tight-fitting cylindrical electrically heated tube. The exit end of the capillary pointed at a skimmer cone orifice leading to the first stage of a differentially pumped MS ion source. The arrangement was similar to the sample solution nebulizers feeding an ICP torch as described in Chapter 10 for ICP-MS instruments. If the temperature and heating rate from the heater were optimized to the mobile-phase flow, it would begin to boil within the capillary close to the exit and emerge from its tip as a fine spray of droplets that would evaporate, leaving a portion of the analyte molecules to enter the ion source orifice, while the rest of the solvent vapor was pumped out of the spray chamber. Any nonvolatile salts or other material would deposit around the MS orifice entrance, requiring its frequent cleaning. Mobile phases with ions in solution often gave rise to strong analyte molecular ions, while weakly ionized solutions required an electron beam to produce EI-MS spectra. Sensitivity was not as high as could be due to the need to use low-capacity HPLC columns or to split effluent flow and because of inefficient transfer from the spray through the narrow MS entrance orifice. Nonetheless, it was an effective and useful general LC-MS interface.

14.1.6.1.2 Electrospray Interface (ESI) for HPLC-MS

The thermospray interface retained some disadvantages. Heating the transfer capillary to the boiling temperature of the mobile phase was not as great a shock to thermally labile compounds as attempted volatilization in GC or from the moving belt interface. Still, even lower temperatures would be preferable. Larger molecules might clump together and "precipitate out" as the spray droplets evaporated. Only a small proportion might travel on the right trajectory to enter the orifice, and only a portion of these might enter as separate molecules instead of the precipitated solid particulates. The key to an improved spray interface was in the observation that analytes already existing as ions in the mobile phase or capable of having charge transferred to them from other ions gave exceptionally strong MS signals.

Instead of heating the inlet capillary entering the spray chamber, in the ESI, its tip is electrically charged so that as droplets emerge from it, they pick up excess positive or negative charge, which is transferred to them. A dry heated stream of highly purified nitrogen gas is directed at the emerging spray of charged droplets to accelerate evaporation. Two things happen as the charged droplets rapidly shrink by evaporation. Analyte and/or solvent molecules bearing the charge with the polarity in excess migrate to the surface under mutual repulsion. As they are forced closer together, they induce a "**coulomb explosion**" whereby the droplets fragment into multiple smaller droplets with greater total surface area to support the excess charge. This process repeats in a cascade, almost instantly producing a spray of extremely small charged droplets. As the droplets' surface electric field continues to build, analyte ions begin to be individually expelled from the microdroplets as **solvated clusters** of one analyte ion embedded in a cluster of solvent molecules. This process is illustrated in Figure 14.18a, and it is far more efficient in converting analyte molecules to ions than is the thermospray interface. The spray can be directed at right angles (**orthogonal spray**) to the ion entrance orifice and skimmer assemblies instead of directly at them. By appropriately charging the entrance orifice, the analyte ions can be electrically attracted to it and drawn in, while excess solvent droplets and particulate material pass by to waste. An example of such an ESI is illustrated in Figure 14.19. The region between the entrance orifice and capillary, and the illustrated "**skimmer cone**" seen in cross section, contains high-purity nitrogen at low pressure. Collisions with this gas result in **collision-induced dissociation (CID)**, which produces fragment ions from the original analyte molecular ions, in a process equivalent to the CID that takes place in the second stage of a tandem three-stage MS-MS instrument (cf. Section 10.2.3.4). These solvent molecules and any others pulled in through the capillary are swept away by the CID gas flow, while the analyte molecular and fragment ions are drawn on and focused through more electric field entrance lenses into the higher vacuum region of the quadrupole mass analyzer.

Originally, the electrospray interface, like the thermospray interface, was limited to use with very low flow rates of mobile phase from capillary or microbore HPLC columns or capillary electrophoretic separations. The acceleration of droplet evaporation by addition of heated gas flow enabled its use with higher flow rate separations. Some modern designs can even tolerate the 3–6 mL/min flow associated with high-speed separations on monolithic HPLC columns with large macropores. Electrospray can generate multiple charged ions from compounds that have multiple charge sites, such as peptides or oligonucleotides. This enables one to observe molecular ions or high-mass fragments of high-RMM species whose masses are greater than the nominal mass limit of the MS instrument. The multiple charges produce an ion whose mass-to-charge (m/z) ratio (cf. Section 10.1) lies within the mass range of the instrument.

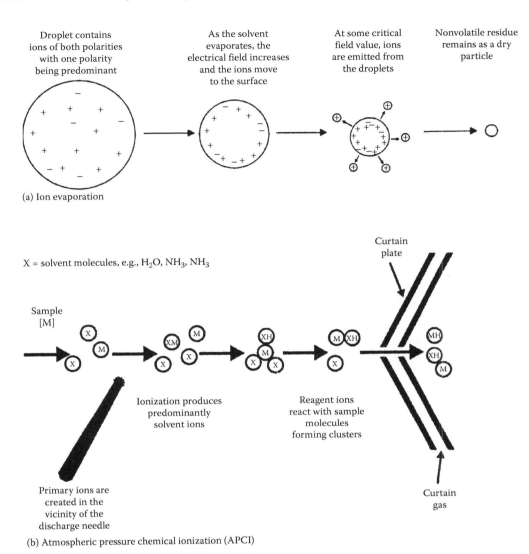

Figure 14.18 LC-ESI-MS interface ionization mechanism diagrams. (a) Coulomb explosion in ESI droplets and (b) reactions leading to molecular ion in APCI. (Adapted with permission from Applied Biosystems, San Francisco, CA; MDS Sciex, Vaughan, Ontario, Canada.)

Note that all ionization is initiated at the entrance to the spray chamber, at the tip of the sprayer. There is no electron impact beam used to ionize and fragment analytes. Large molecules are treated gently through the whole process, and if they have natural charge centers at various pHs (as is especially the case with peptide chains comprising linked amino acids), they are ionized with great ease. This method of LC-MS interfacing, along with the one to be described next, was the long-sought solution for a practical, general, and easy-to-use interface. LC-MS instruments based on these two interfaces are used by the thousands in biomedical research laboratories, and they now outsell GC-MS instrumentation. It is difficult to predict for many analytes whether negative or positive ion LC-MS operation will produce the most sensitive signal, but it is easy to shift between modes, even within a single LC-MS run, and to find out and then to program an LC-ESI-MS instrument to shift to the optimal polarity for each analyte as it elutes.

We learned earlier in Section 14.1.4.4 that fluorescence detection of end nucleotides of capillary electrophoretic separations of DNA segments is key to the instruments that perform the DNA sequencing that supports the field of **genomics**. The LC-ESI-MS instruments, with their superb ability to characterize analogous large peptide (polymer chains of amino acids) segments of protein macromolecules, likewise are key to rapid characterization of the complex protein mixtures that reflect the various patterns of gene "expression" in biological systems. This is the foundation of the discipline of **proteomics**, whose full flowering will likely entail even more massive analytical determinations

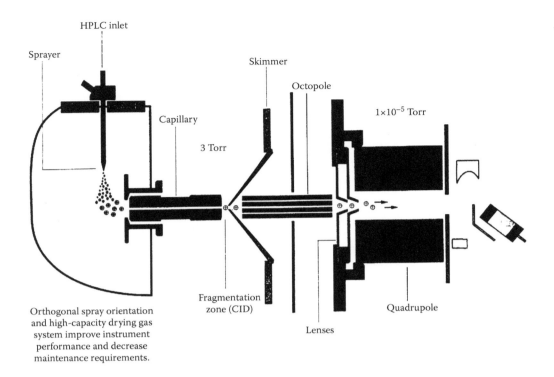

Figure 14.19 Diagram of orthogonal electrospray LC-MS interface. (Courtesy of Agilent Technologies, Santa Clara, CA, www.agilent.com/chem. With permission.)

than the huge project of sequencing selected biota genomes (e.g., the 3 billion base sequence of the human genome). Achieving a full understanding of proteomics will likely occupy much of the twenty-first century. One of the inventors of the ESI-MS interface, Dr. John Fenn, shared the 2002 Nobel Prize in Chemistry for this contribution. He entitled his lectures on the subject: "Electrospray wings for molecular elephants." For massive "-omics" studies such as proteomics, genomics, metabolomics, lipidomics, and a host of others, chromatography-ESI-MS or MS-MS systems have replaced many earlier ones using derivatization and fluorescence detection. Now, in their turn, they are beginning to be superseded by even faster, more powerful, and cheaper, purpose-built systems designed around arrays of sensors on microchips. The day of the $100 complete human genome is in sight, perhaps only two decades after the $3,000,000,000 first human genome sequence publication at the turn of the past century!

14.1.6.1.3 APCI Interface

The other highly successful HPLC-MS interface design is illustrated in Figure 14.20. This is the APCI interface. Like the ESI interface, the sample is nebulized into a spray of fine droplets, aided by and entrained in a flow of nitrogen "**sheath gas**," but without being charged by an applied potential at the spray tip. As in electrospray, heat conducted through the sheath gas is supplied from a heated ceramic tube as the droplets are carried downstream. This gently accelerates droplet evaporation. The process occurs at atmospheric pressure. At the exit of the tube, ions are produced by the **corona discharge** of electrons emitted from the high-field gradient of a sharp metal needle charged to several thousand volts. The initial ions produced by the discharge are N_2^+, O_2^+, H_2O^+, and NO^+ from air, and through a complex cascade of charge transfer, they create **reagent ions** ($[X + H]^+$ in positive mode; $[X - H]^-$ in negative mode, where X is a mobile-phase solvent molecule). The reagent ions then produce stable sample ions upon collision with sample molecules. The final step in this mechanism is the same as that occurring in classical CI-MS (cf. Section 10.2.2.2) and thus gives the name for APCI. In contrast, GC-CI-MS introduces a reagent gas maintained in a confined ion source volume at pressures also near atmospheric, but its composition is different (usually methane, isobutane, ammonia, etc.) from the normal atmospheric components that serve to form the initial primary ions in LC-APCI-MS (i.e., nitrogen, oxygen, water vapor). Contrary to experience in GC-MS, LC-APCI-MS ion sources can be opened to the atmosphere for cleaning without venting and shutting down the rest of the MS instrument. After all, they operate under normal atmospheric conditions!

The reactions leading to the production of sample molecular ions are illustrated in Figure 14.18b. Note the

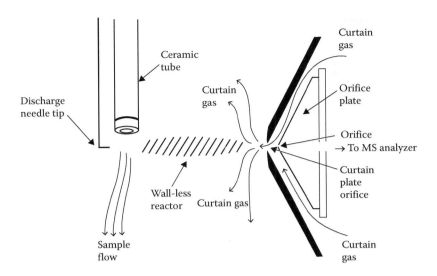

Figure 14.20 Diagram of APCI LC-MS interface. (Adapted with permission from Applied Biosystems, San Francisco, CA; MDS Sciex, Vaughan, Ontario, Canada.)

barrier to entry of solvent vapor into the vacuum pumped portions of the MS, which is produced by outflowing N_2 **curtain gas** flowing between the curtain plate and orifice plates illustrated in Figures 14.18b and 14.20. The ions are drawn through this flow by the charges applied to these plates, while uncharged molecules are swept back and away. The various charge transfer reactions forming the sample ions occur as the ions are being drawn to the entrance orifices, and this region between the corona discharge tip and the curtain plate orifice acts as an ideal "wall-less reactor." The remaining ions that do not make their way through the orifices eventually are neutralized by contact with a "wall" and play no further part in the process. While the ESI interface is especially suitable for large biomolecules with multiple charge sites, the APCI interface is more useful for smaller neutral molecules. Many analytes may be analyzed successfully using either interface. The one that produces the best sensitivity must be found by experiment. There are several modes (e.g., positive ion, negative ion, different mobile-phase salts or components that affect ionization efficiency or CI ionization) that may be evaluated. Numerous ionization voltage, orifice voltage, or evaporation temperature and flow parameters may be adjusted to maximize sensitivity. Fortunately, in LC-MS, one may easily continuously infuse into the MS interface a solution of a constant low-level concentration of analyte in a proposed LC mobile phase. The interface parameters may be quickly and automatically varied in sequence to find their optimal settings. Such optimization is not so quickly attained with a sequence of injections and elutions through a GC column in GC-MS. Many modern general-purpose HPLC-MS instruments have interchangeable ESI or APCI spray and ionization assemblies that enable rapid conversion from one mode of operation to the other without requiring venting of the MS vacuum system.

14.1.7 Applications of HPLC

In Section 13.10, we described the scope of GC. It was easy to find large important classes of compounds for which it was unsuitable, either due to lack of volatility associated with size, thermal instability, polarity, or acid–base functionality or due to existence as an ionized form. When considering HPLC, the question is reversed to define the much smaller range of analytes for which none of its manifold variations is suitable. These would be substances that cannot be put into liquid solution or be easily retained therein. Highly volatile compounds or fixed gases are one category. GC handles these well. Materials that exist only in the solid state (e.g., metals, refractory and insoluble inorganics, cross-linked polymers of indefinitely high RMM) are better handled by direct spectroscopic techniques such as X-ray fluorescence (XRF) or X-ray diffraction (XRD) or by techniques that employ selective dissolution prior to spectroscopic analysis, thermal analysis, and so on. We shall not attempt a comprehensive description of the range of compounds usefully characterized by the wide range of RP or NP-HPLC and their modifications to study small polar analytes (HILIC, polar-endcapped or polar-embedded phases [cf. Section 14.1.1.1]), adsorption chromatography, IC (Section 14.2), affinity chromatography (Section 14.3), SEC (Section 14.4), and the similar but nonchromatographic separations using electrophoresis (Section 14.6). The student is referred to the references in the bibliography and to the websites listed in Appendix 14.A for discussions of thousands of applications.

Some of the variety of categories covered include the following:

Additives	Carbohydrates	Pesticides
Aflatoxins	Condensed aromatics	Plant pigments (cf. Tswett)
Amino acids	Drugs	Phenols
Analgesics	Drug metabolites	Poisons
Antibiotics	Dyes	Propellants
Antioxidants	Estrogens	Sedatives
Artificial sweeteners	Herbicides	Steroids
Bile acids	Lipids	Surfactants
Blood alcohols	Narcotics	Urine extracts
Blood proteins		

14.1.7.1 Biochemical Applications in Proteomics

As an example of the utility of LC techniques in support of biochemical research, we shall focus on several applications to the analysis of amino acids and peptides. **Peptides** are short polymer chains consisting of **amino acid** units. Figure 14.21 illustrates schematically a peptide structure consisting of four amino acid units. The "*backbone*" of the peptide polymer chain consists of "*peptide bonds*" between the amino end ($-NH_2$) and the carboxylic acid end ($-COOH$) of each of these amino acids. Even longer amino acid polymers form **proteins**, which are crucial for a vast variety of biological functions. Proteins form enzymes, which catalyze the chemical reactions in living systems. They form structural features, such as muscles, tendons, and hair. They form compound-selective receptors on cell surfaces that receive and "interpret" messenger compounds such as hormones that regulate biological functions. Some of these messengers are themselves amino acids or peptides. Clearly, we need to characterize and measure an immense variety of these complex mixtures if we are to understand life at the molecular level. Of the thousands of possible amino acids, there are only 20 (called **"essential" amino acids**) that are incorporated in these biopolymers, each distinguished by a different "R-group" substituent structure. These are tabulated in Figure 14.22. Their R-group substituents are seen to have a wide variety of polarities; some have acidic or basic functionalities and the ability to carry both positive and negative charge. Large polymers comprising such subunits will not be separable using GC instruments. HPLC in its various forms, perhaps using derivatization, and employing a variety of detectors to enhance measurement sensitivity and selectivity, is called for. Each essential amino acid in Figure 14.22 is named, and three-letter and one-letter abbreviations for each are also listed. If amino acids 1 through 4 in the tetrapeptide in Figure 14.21 were alanine, glycine, phenylalanine, and arginine, in that order, the peptide would be designated "Ala-Gly-Phe-Arg" or more compactly "AGFR." The latter nomenclature is often encountered in the output of computerized, automated instrumentation for sequencing peptide chains.

Figure 14.23 displays chromatograms from HPLC analysis of amino acids derivatized with phenyl isothiocyanate to yield derivatives with strong absorption at the common Hg-lamp line of 254 nm. The single-letter component abbreviations are those of Figure 14.22, with some additional peaks such as ammonia (NH_3), phosphoserine (PH-S), phosphothreonine (PH-T), hydroxyproline (OH-P), galactosamine (Gal), and norleucine (NLE) internal standard (IS) and excess derivatizing reagent (Re). The fact that Re appears as a distinct peak indicates that derivatization took place in precolumn mode. The presence of altered or metabolized derivatives of the 20 essential amino acids is characteristic of actual biological samples. The top trace is of a standard mixture of 200 pM concentration amino acids, while the lower one is from fingerprint oils extracted by water and ethanol from a glass surface. This illustrates impressive sensitivity for HPLC with UV/VIS detection. A 20-fold greater sensitivity is possible when using fluorescent detectors with OPA (cf. Table 14.2) derivatives of amino acids.

A major application of HPLC interfaced with MS is in protein or peptide sequencing. This can be accomplished using a wide variety of chromatographic separations followed by different kinds of mass spectrometric characterizations.

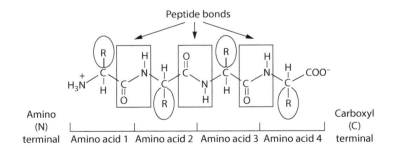

Figure 14.21 Generic structure of a four-amino-acid peptide chain. (With permission from Phenomenex, Inc., Torrance, CA, www.phenomenex.com.)

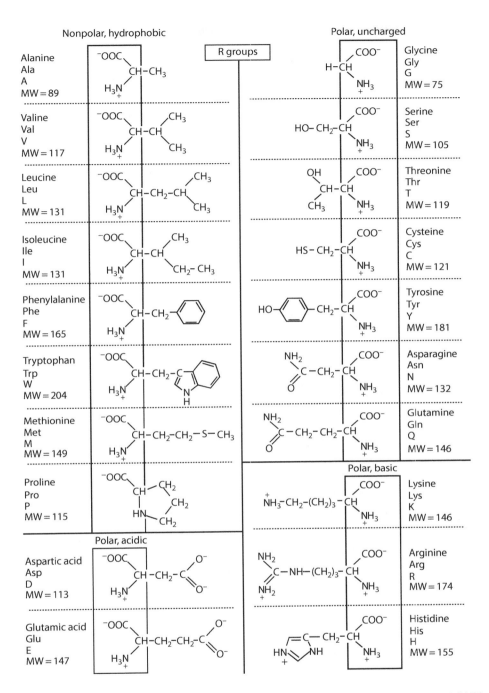

Figure 14.22 Structures of the 20 essential amino acids, names, three-letter and one-letter abbreviations, and RMMs. (With permission from Phenomenex, Inc., Torrance, CA, www.phenomenex.com.) Note that this table uses the old terminology of MW instead of the accepted RMM.

We will illustrate one example of such techniques to give an impression of the power and complexity of such analyses. A very large protein might consist of a sequence of several hundred amino acids. Such a structure would be much too difficult to unambiguously characterize with a single measurement. Even if we were to know the identity and how many of each of the component amino acids were in the protein chain, there could be an astronomical number of different ways they might be arranged. All these different possible proteins would have the same RMM. Knowing only that will not distinguish among them. Some procedures (Edman degradation) can repeatedly clip off the terminal amino acid from one end of the chain, leading to a mixture of peptide chains successively shorter by one amino acid. If we identify each of the amino acids as we repeat the procedure, we could reconstruct the order of the chain. Alternatively, we might

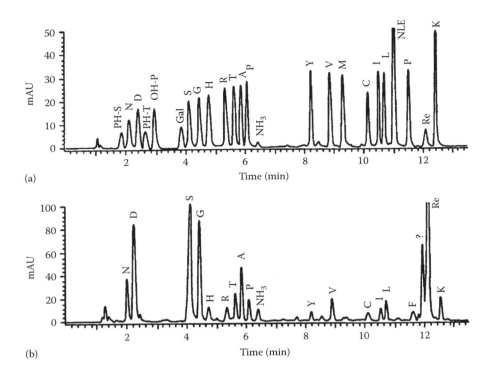

Figure 14.23 HPLC of amino acid derivatives detected by 254 nm UV absorption: (a) 200 pmol of PTC-amino acid standard, including PH-S PH-T, hydroxyproline (OH-P), galactosamine (Gal), norleucine (NLE, 1 nmol IS), excess reagent (Re), and other amino acids designated by one-letter codes listed in Figure 14.22 and (b) analysis of a human fingerprint, taken up from a watch glass using a mixture of water and ethanol. (Courtesy of National Gallery of Art and the Andrew W. Mellon Foundation; From Cazes, J., *Encyclopedia of Chromatography*, Marcel Dekker, Inc., New York, 2001. With permission.)

be able to use chromatography to separate the successively smaller peptide fragments, and if we could be confident of separating them all and measuring their masses, we could arrange them in order of descending mass and determine the identity of each successive amino acid by mass difference. One problem with this is that we cannot distinguish between pairs of amino acids I and L or Q and K, in Figure 14.22, since they have identical masses. The reliability of these sequential "clipping" approaches diminishes as the length of the peptides or protein increases.

Figure 14.24 illustrates the initial steps for evaluating the sequence of amino acids in an unknown protein chain by using LC-MS data. These data might be obtained by using only the first quad of a triple quad MS instrument (cf. Section 10.2.3.4) to measure the masses of peptide molecular ions. The first step, illustrated in Figure 14.24a, is to break the protein into smaller peptide chains by treating it with an enzyme (which is itself a catalytic protein), which cleaves the chain only between specific pairs of amino acids occurring in a particular order from the amino to the acid end of the chain. A most commonly used enzyme is **trypsin**, which cleaves bonds on the carboxyl side of K or R (lysine or arginine) to some but not all other amino acids in the protein chain. The resulting peptide fragments are only a few dozen or less amino acids in length and are easily separated by capillary LC. Figure 14.24b contrasts total ion current capillary LC-electrospray MS chromatograms of such a trypsin digest, which employ either ammonium acetate or trifluoroacetate buffers in the LC mobile phase. Note the use of buffers containing volatile salts, which are necessary to avoid buildup of salt deposits in the MS ion source. The vertical ion count scales in each chromatogram are the same, so clearly the ammonium acetate buffer gives stronger signals than trifluoroacetic acid. The separations achieved with each buffer differ. Figure 14.24c shows the single-stage ESI-MS mass spectrum of a 12-amino-acid peptide fragment from one of the chromatographic peaks in Figure 14.24b. This spectrum consists of ions with one through five positive charges on the peptide chain RAKMDRHVGKEK, whose RMM is 1453.8. Note that there are R and K amino acids within this peptide chain that were not the sites for tryptic cleavage. The number of types of peptide bonds suitable for tryptic cleavage tends to result in mixtures of peptides of the most desirable length range, neither too large nor too small to be both well separated by LC and reliably identified by their molecular masses. The fact that the four lower *m/z* values in this mass spectrum are precisely 2×, 3×, 4×, and 5× less than that of the RMM assures us that the spectrum arises only from one compound of this RMM. If the LC peak contained several unresolved

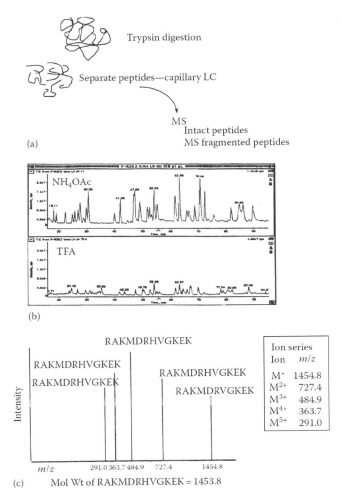

Figure 14.24 LC-MS analysis of protein peptide fragments. (a) Trypsin digestion of protein to yield peptide fragments, (b) LC-MS-total ion chromatogram (TIC) chromatograms of pepsin digest of a protein molecule, and (c) ESI-MS spectrum of the molecular ion of separated peptide fragment. (Adapted with permission from Applied Biosystems, San Francisco, CA; MDS Sciex, Vaughan, Ontario, Canada.)

to obtain data on the actual *sequence* of amino acids of an unknown peptide fragment. Figure 14.25 displays the collisionally induced fragmentation pattern of a molecular ion of the 14-amino-acid peptide EGSFFGEENPNVAR. It has been produced by isolating the peptide by LC, isolating its molecular ion using the first quad of a tandem three-stage MS, collisionally fragmenting this ion in the second collision cell multirod MS stage, and using the final quad to scan a complete mass spectrum of all the collision fragment ions. Note that all 13 fragments produced by sequentially removing amino acids one at a time from the "E" end of the chain are present and identified. Their relative proportions in the collisional fragmentation spectrum vary greatly. There are also many other fragments resulting from breaking these bonds in different orders, so it would be difficult but not impossible to obtain a definite peptide sequence from this data set. However, the MS-MS pattern is fairly unique for each such peptide and again acts as a fingerprint. Once the peptide sequence has been unraveled, its fingerprint can be filed in a database, and it can then be rapidly identified if encountered again, perhaps in a different context, by computer matching the pattern to its database entry. Thousands of such identified peptide fingerprints are submitted to international libraries. Darwinian evolution often selects and conserves certain peptide sequences, so one will often find a match for a peptide fragment MS-MS spectrum in the libraries, thus accelerating the process of sequencing the whole protein chain.

DNA nucleotide sequences for particular genes are often identical or very similar among species. Libraries of identified gene sequences are the foundation of genomics. Likewise, the corresponding libraries of peptides resulting from particular enzymatic digests and their sequences are the building blocks of analytical proteomics. Achieving a library match of some portions of these MS-MS spectra to those of the corresponding "fingerprints" of a known protein will usually be sufficient to identify it as such without the need to completely sequence it. The particular instrumental examples described here are not necessarily the only or even the primary methods used for proteomics. They illustrate the requirements for rapid and definite identification and measurement of proteins, as shown in the following:

1. A method for cutting them specifically and reproducibly into fragments (peptides) that are small enough to be separated easily by chromatographic techniques (e.g., LC, capillary electrophoresis (CE), 2D-thin-layer chromatography [TLC], electrophoresis).

2. A method for interfacing the peptides from the chromatographic separation to some form of spectroscopic detection (primarily MS, due to its superior sensitivity to underivatized peptide chains or amino acids and the high information content of mass spectra, especially when employed as multistage MS^n).

peptides, we would see several such sequences—one for each peptide, and if the coeluting mixture were sufficiently simple, we could unravel the mixture mass spectrum to identify the RMMs of each component. Note that the RMM alone does not uniquely identify the sequence of amino acids in the peptides. However, the tryptic digest of any particular protein will produce a well-defined series of LC-MS peptide peaks. The combination of the LC retention times or retention order of the different peptides, together with their molecular masses determined by ESI-MS, will produce a "fingerprint" that can be matched to libraries of protein trypsin digests and will serve to identify the protein by library matching.

Adding a second-stage mass spectral fragmentation by employing MS-MS (cf. Section 10.2.3.4) enables one

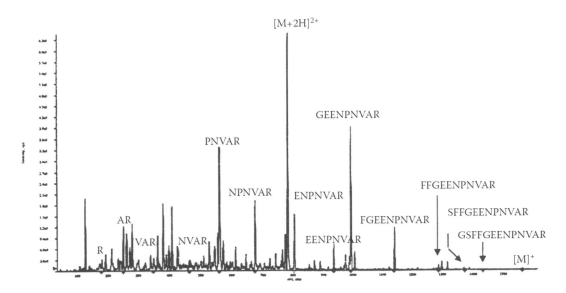

Figure 14.25 LC-MS-MS fragmentation to sequence a 14-amino acid peptide. (Adapted with permission from Applied Biosystems, San Francisco, CA; MDS Sciex, Vaughan, Ontario, Canada.)

14.2 CHROMATOGRAPHY OF IONS DISSOLVED IN LIQUIDS

Many compounds dissolve in polar and aqueous solutions to form charged species (ions). Inorganic salts dissociate (e.g., NaCl → Na$^+$ and Cl$^-$). Depending on solution pH, organic acids may dissociate (e.g., RCOOH → RCOO$^-$ and H$^+$) or bases may protonate (e.g., R$_3$N + HOH → R$_3$NH$^+$ and OH$^-$) as pH varies. Peptide chains containing acidic amino acid residues such as aspartic or glutamic acids or basic residues such as lysine, arginine, and histidine (cf. Figure 14.22) may contain both positive- and/or negative-charged side groups, with one or the other in excess. The net charge can be balanced at a pH corresponding to the **isoelectric point**, where the molecule has equal amounts of positive and negative charges, but these continue to interact with the solvent by charge–dipole interactions. Small molecules with single positive and negative charges at either end can exist as **zwitterions**. Ionic species generally do not partition smoothly between the classical mobile and stationary phases of RP or NP-HPLC. Special modifications, either to the mobile phase to suppress the ionic nature of the analytes or to the stationary phase to incorporate ions of the opposite charge to attract and retain analytes, are used to facilitate the chromatography of ionic species.

Ion-pairing chromatography (IPC): One way to render ions more suitable for separation by RP-HPLC is to incorporate a **counterion** of opposite charge into the mobile phase. If the counterion is fairly large, with relatively nonpolar organic substituents attached, then in the polar, aqueous mobile-phase characteristic of RP-HPLC, it will associate with analyte ions to form a tight "**ion pair**" with no net charge. This entity can now equilibrate and partition between the mobile and stationary phases in a fashion not possible for the parent analyte ion. The associated ion pair will often retain or even improve the spectroscopic or other detection characteristics of the original parent ion, allowing it to be measured by the standard HPLC detectors. Positively charged organic amines can be ion paired with alkyl sulfonic acids (e.g., $C_{12}H_{25}SO_3^-$), while negatively charged, dissociated carboxylic or sulfonic acids and various dye molecules are often paired with long-chain quaternary amines (e.g., tetrabutylammonium ion $(C_4H_9)_4$ N$^+$). Both are **strong** acidic or basic ions, respectively (i.e., they remain dissociated in an ionic form over a wide pH range). A comprehensive listing of ion pairing agents can be found in the book by Bruno and Svoronos. Neither of these counterions interferes with UV detection of the analyte at wavelengths above 210 nm. **IPC** thus extends the scope of RP-HPLC. It is less practical for use with NP-HPLC, as the less-polar NP mobile phases are not as capable of maintaining the counterions in solution. The ion-exchange columns to be described in the next section have difficulties separating large ionic species, as these cannot easily penetrate the cross-linked resin networks that comprise these phases. This hinders attainment of mass-transfer equilibrium and results in loss of chromatographic efficiency. Ion pairing RP-HPLC works better for such ions. Ionic surfactant molecules (e.g., phospholipids, phosphate detergents) have such high affinity for the ion-exchange resins that they are difficult to elute from them, so ion-pairing RP-HPLC is also preferred for this class of analyte.

Ion-exchange chromatography: For many years, it has been known that clays include metal ions in their mineral structure and that these can exchange with other metal ions present in solutions in contact with the clay. **Zeolites** (sodium aluminum silicates) possess similar ion-exchange properties. Natural and artificial zeolites are employed in water softeners to remove the "hardness ions," Ca^{2+} and Mg^{2+}, by binding them while releasing a charge equivalent to the number of more soluble Na^+ ions. For application of ion exchange to chromatographic separations, resin columns (cf. end of Section 14.1.1.1) can be modified by introducing charged functional groups into the cross-linked 3D polymeric structure. These groups will have either positive or negative charges opposite to those of the ions to be exchanged.

A particularly useful resin is that formed by the copolymerization of styrene (vinylbenzene) and divinylbenzene, and is illustrated in Figure 14.26. By varying the proportion of divinylbenzene incorporated in the polymerization, one can vary the extent of cross-linking of the polystyrene chains, which confers greater stability to the polymer structure and affects the porosity and tendency of the polymer to swell as it takes up mobile-phase liquid. The benzene rings in the polymer structure are ideal substrates for synthetic introduction of various charged functional groups, using standard aromatic substitution reactions:

1. *CX resins*. The functional group in CX resins is usually an acid. Sulfonation reactions can add the sulfonic acid group, $-SO_3H$, to essentially all the benzene rings of the resin. The weaker carboxylic acid group, $-COOH$, may be added instead. The former produces a "**strong acid**" CX resin of the formula Res $-(SO_3H)_n$, where Res represents the resin polymer matrix and $-(SO_3H)_n$ the numerous attached sulfonic acid groups. Introduction of the $-COOH$ group yields a "**weak acid**" exchange resin whose pK_a is more similar to that of benzoic acid. In each case, an acidic hydrogen is attached to a functional group chemically bound to the resin. The number of such exchange sites, n, in a given mass of the ion-exchange resin is a measure of its **exchange capacity**, usually expressed as milliequivalents/g (or meq/g.) To maintain charge balance in the resin, the exchange and binding of multiply charged ions require interaction with multiple singly charged sites. Thus, the equivalent value for such ions is calculated as m/z, where m is the number of moles and z is the ionic charge. If the affinity of an ion such as Na^+ in solution is greater than that of H^+ for the functional group and/or its solution concentration is sufficiently greater than the "concentration" (not to be confused with exchange capacity) of exchangeable H^+ on the resin functional groups, then exchange reactions of the following type establish an equilibrium:

$$\text{Res} - COOH + Na^+ \rightarrow \text{Res} - COONa + H^+ \quad (14.1)$$

$$\text{Res} - (SO_3H) + Na^+ \rightarrow \text{Res} - (SO_3Na) + H^+ \quad (14.2)$$

Almost any metal ion will displace hydrogen ion from the resin under these circumstances. This provides a method for removing metal ions from aqueous solutions. The solution becomes acidic only if the metal is exchanging with a hydrogen ion. A CX resin with H^+ on its functional groups is said to be in the *acid form*, and when Na^+ replaces the H^+, it is said to be in the *sodium form*. Depending on the relative affinities of the metal ions for the resin functional group, further exchanges are possible.

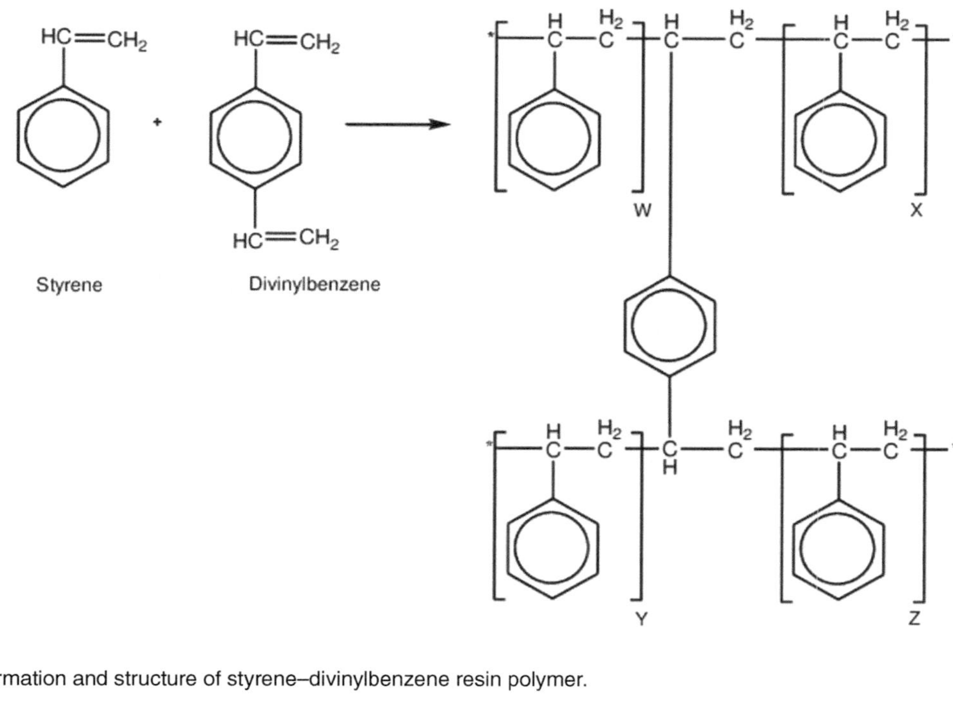

Figure 14.26 Formation and structure of styrene–divinylbenzene resin polymer.

For example, Cu^{2+} will replace the sodium in a sodium form resin according to the process

$$Res-(SO_3)_2 2Na + Cu^{2+} \rightarrow Res-(SO_3)_2 Cu + 2Na^+ \quad (14.3)$$

Note that the cupric ion is shown to be associated with two resin functional groups, releasing two sodium ions. Half the number of moles of doubly charged copper is equivalent on the basis of charge to sodium or hydrogen. If the copper were present as singly charged cuprous ions, the equivalents would contain equal numbers of moles.

2. *AX resins*. Anions can be removed from solution using the same principles that are applied to exchange cations. The functional groups of opposite positive charge are most commonly quaternary ammonium or polyalkyl amine groups, forming either **strong** or **weak base** exchangers according to the processes illustrated in Equations 14.4 and 14.5, respectively:

$$Res-CH_2N(CH_3)_3^+ OH^- + Cl^- \rightarrow Res-CH_2N(CH_3)_3^+ Cl^- + OH^- \quad (14.4)$$

$$Res-NH(-R)_2^+ OH^- + Cl^- \rightarrow Res-NH(-R)_2^+ Cl^- + OH^- \quad (14.5)$$

The rate of ion exchange is controlled by the law of mass action. At *equal* concentrations, the greater affinity of Na^+ replaces H^+ from the CX resins in the acid form. However, if a much higher concentration of strong (i.e., completely dissociated) acid is passed through the sodium form resin, it will reverse the equilibrium, displace the Na^+ ions, and convert the resin back to the acid form. In general, it is usually possible to return an ion-exchange resin column to a desired starting form by passing a large excess of the desired ion at very high concentration through the resin. Thus, one can use an ion-exchange column to retain the majority of ions of higher affinity as a less concentrated solution of them is passed through it (until most of the binding sites or functional groups have been exchanged and the exchange capacity is exceeded—a situation analogous to "overloading" in partition chromatography). Subsequent passage of a smaller volume of a very concentrated solution of the initial form, lower affinity ions restores it to that form, **regenerating** the column, and effects a *concentration* of the originally more retained ions in the regenerating wash solution effluent.

The relative affinities (i.e., selectivities) of ions for the functional groups on the resin are governed by two rules:

1. Ions with higher charge have higher affinity (e.g., $Na^+ < Ca^{2+} < Al^{3+}$ and $Cl^- < SO_4^{2-}$).
2. The ion with the greatest size (radius) and the greatest charge has the highest affinity (e.g., $Li^+ < Na^+ < K^+ < Cs^+ < Be^{2+} < Mg^{2+} < Cu^{2+}$ and $F^- < Cl^- < Br^- < I^-$).

By a careful choice of the ionic composition of the eluent, and the gradual adjustment of its strength (i.e., ion concentration and/or pH, which can control concentration of ions resulting from acid–base equilibrium) during elution using a controlled gradient, the components of a mixture of ions can be induced to separate just as do the components of a mixture separated by partition chromatography. The chromatographic principles discussed for the latter technique also apply. For ion-exchange chromatographic separations, the concentrations of the analyte ions being separated by a resin column need to be much less than those that would be present at values close to its exchange capacity.

The fundamental parameters controlling relative residence times of analyte or other eluent ions in the resin stationary phase or the ionic solution mobile phase are both the relative selectivity of the resin for the ions and their relative concentrations in each phase. In contrast, in RP or NP-HPLC, selectivity is controlled by the relative polarity interactions of different analytes with both phases. Until an overload occurs, this is independent of analyte concentration.

Another difference is that in ion exchange, the selectivity resides in relative ion-pairing interaction strengths only in the stationary phase. The effects of mobile-phase composition changes are manifested by the relative concentrations of other nonanalyte ions in the mobile phase competing for the fixed number of exchange sites on the resin.

Ion-exchange chromatography was initially developed during World War II as part of the Manhattan Project to preparatively separate the chemically very similar triply charged lanthanide rare earth and actinide series cations on the basis of slight differences in their ionic radii. A generally useful instrumental version suitable for quantitative analysis of small organic and inorganic ions awaited the development of a universal, sensitive detector. The conductometric detector (Section 14.1.4.6) would seem to fit the requirements, but the high background level of eluent ions present in traditional ion-exchange chromatography rendered it relatively useless.

14.2.1 Ion Chromatography

As described earlier, when attempting to use a conductivity detector for ion-exchange chromatography, the high electrolyte concentration required to elute analyte ions in a reasonable time tends to overwhelm the much lower conductivity contributed by low levels of analyte ions. This problem was eventually overcome by introducing an **eluent suppressor column** to convert eluent ions to nonconductive, unionized molecular forms while leaving the low concentrations of analyte ions to be measured by a conductometric detector. This mode of analytical ion-exchange chromatography goes by the name of IC. IC is defined as the analysis of ionic analytes by separation on ion-exchange stationary phases with **eluent suppression** of excess eluent ions. Detection in most IC systems is by the universal (for ionic analytes) conductivity detector (Section 14.1.4.6), and can be in the ppb (mass/mass) range or sub-ppb (mass/mass) range if large-volume injections or concentrator columns are used.

When *cations* are being exchanged on the analytical column to effect a separation, variable concentrations of hydrochloric acid are often employed as the eluent. In this case, the suppressor column is an AX column in the hydroxide form. The hydroxide on the suppressor column reacts with the hydrogen ions in the eluent exiting the analytical column to form nonconductive water, which contributes no signal to the conductivity detector. The reaction on the suppressor column is

$$H^+(aq) + Cl^-(aq) + Res^+(OH^-) \rightarrow Res^+(Cl^-) + HOH(water) \quad (14.6)$$

The cations being separated on the analytical column pass through the AX suppressor column without being retained. Conversely, for anion separations, the suppressor column is in the acid form of a CX column. The eluent contains high concentrations of a displacing anion (e.g., OH^- or HCO_3^-) that can react with the exchangeable H^+ on the suppressor column to form a largely undissociated species (e.g., H_2O or carbonic acid, H_2CO_3, respectively). An analysis of a chloride anion, Cl^-, by simple suppressor column IC is illustrated in Figure 14.27a. In this case, Na^+ from NaOH eluent and NaCl analyte displaces H^+ from the head of the suppressor column, and the liberated H^+ reacts with the OH^- to form water, leaving only the chloride ion and a balancing amount of hydrogen ion to produce a conductivity signal in the detector proportional to the small amount of eluted chloride. As more NaOH in the eluent displaces H^+ in the suppressor column, the portions that are "exhausted" increase and the "suppressing" portions decrease. Eventually, it becomes completely exhausted and then must be **regenerated** by backflowing a high concentration of strong acid solution through it, to convert it back to the acid form, and then be rinsed with water to remove residual ions. The same process with a strong base is used to regenerate AX suppressor columns employed in cation IC systems.

The need to periodically regenerate suppressor columns is an inconvenience, especially if one wishes to operate an ion chromatograph in an unattended, automated fashion over long periods. A **self-regenerating suppressor** continually regenerates the element in an ion chromatograph that performs the function of the suppressor column. It recycles the post-detector effluent of the ion chromatograph past an electrode that continuously electrolyzes water in the effluent to produce the desired H^+ or OH^- regenerant ions. The suppressor ion-exchange column is replaced by an array of microvolume ion-exchange membranes. Effluent from the analytical column passes over one side of these membranes, while the regeneration ions produced by the electrode flow on the other. The excess eluent ions pass through the membrane and are carried to waste in the eluent recycle stream from which the electrode-generated ions were produced. These latter ions pass through the membrane to maintain the charge balance, and they react to suppress the charge of the excess eluent ions, just as did the corresponding ions bound on a regenerated suppressor column. Figure 14.27b illustrates this process for the same anion IC analysis of chloride, which was illustrated in Figure 14.27a.

The employment of an electrode system to generate H^+ or OH^- on demand in the amounts required to operate a self-regenerating membrane suppressor system led to the idea of using a similar generator to also produce the desired concentrations of these ions in the eluent streams with which analytes are eluted from analytical ion-exchange columns used for IC. For this to work effectively, it is desirable to use high-efficiency, low-capacity, ion-exchange columns for the separation. These are better suited for the low levels of analytes to be separated and permit the use of lower eluent ion concentrations, which are easier to generate by these methods. Instead of the large porous resin beads, where diffusion in and out of the beads degrades chromatographic plate count and resolution, **pellicular packings** are used. An example of such an ion-exchange column "bead" is illustrated in Figure 14.28. The spherical bead consists of a rigid silica sphere of 5 μm radius, coated on the surface with a 0.1 μm layer of latex to which sulfonate groups $\left(-SO_3^-\right)$ are bonded. Large molecules containing multiple positively charged alkyl ammonium ion substituents bond electrostatically to these $-SO_3^-$ groups. The outer layer of positively charged ions serves as the site for exchange of anions with a mobile phase. Figure 14.27c displays a block diagram of an IC system that employs an eluent-generating system, EG40® (Thermo Fisher Scientific, Inc.), based on electrolysis of water. Such a system does away with the need for preparing and consuming eluent and regenerating solutions. The slogan used for such systems by their originator is "just add water."

Figure 14.29 displays an IC separation of 34 small anions. Note that both inorganic halides and oxoanions such as sulfate and bromate are measured, as well as small organic anions like formate, malonate, and phthalate. Note that pairs of ions representing different oxidation states of inorganic atoms such as nitrite/nitrate and chlorite/chlorate are readily separable. IC is particularly appropriate for speciation of inorganic elements and is a particularly useful

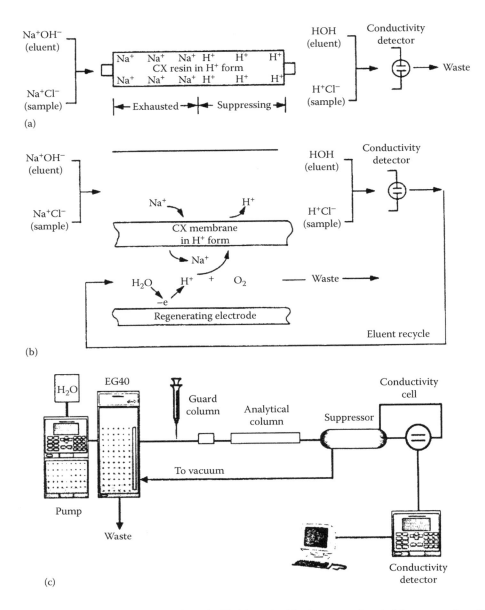

Figure 14.27 Methods of suppressing mobile-phase ions in IC: (a) packed bed suppression, batch suppression; (b) self-regenerating suppression, continuous regeneration; and (c) eluent generation from water. (© Thermo Fisher Scientific. www.thermofisher.com. Used with permission.)

mode for introduction to highly specific and sensitive detectors such as ICP-OES and ICP-MS when element-specific speciation is called for.

A final new version of ion-exchange chromatography employs arguably the most sophisticated column packing technology yet, which combines RP-HPLC and ion-exchange separation mechanisms on the same particle. This employs a novel nanopolymer–silica hybrid particle structure. Check the structure of porous silica particles illustrated in Figure 14.2. The interior pore surfaces are modified with a covalently bonded organic layer that provides both RP and AX retention properties. The outer surface is electrostatically coated with nanopolymer particles with a CX functionality. This overall structure ensures spatial separation of the AX and CX regions and allows both retention mechanisms to function simultaneously, and the retention properties of each to be varied independently. With enough design effort and experimentation with both mobile-phase and stationary-phase compositions suitable for a given separation problem, ideal selectivity may be attained for simultaneous separation of base, neutral, and acidic analytes and retention of ionic and ionizable analytes without the use of ion-pairing reagents. These "triple threat," one-pass columns are marketed by Thermo Scientific under the name

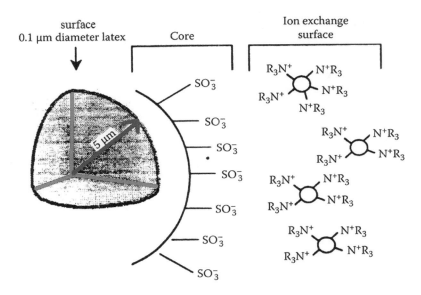

Figure 14.28 Pellicular AX bead for IC. (© Thermo Fisher Scientific. www.thermofisher.com. Used with permission.)

"**Acclaim Trinity P1 Columns**." Although sophisticated in phase design, Acclaim Trinity P1 is specially designed for simultaneous determination of pharmaceutical compounds and their counterions. Literature on the use of this column can be found at www.thermoscientific.com. To make the most of the full potential of this phase, users need to have an in-depth knowledge of the phase chemistry, master the general application development strategy (adjusting selectivity by mobile-phase ionic strength, pH, and solvent content), and most importantly, familiarize themselves with the column by collecting enough hands-on experience. Alas, there is no royal road to separation optimization with these ever more complex and powerful phase designs. Mixed-mode columns including the Acclaim Trinity provide chromatographers with unique separation tools to complement general-purpose RP columns (e.g., C18).

14.2.1.1 Single-Column IC

The development of ion-exchange columns with low exchange capacity has led to a mode of IC that dispenses with the suppressor column. When using such columns, a dilute eluent solution can be sufficient to displace ions from the column. If such an eluent can be found that has a substantially lower conductivity than the analyte ions, then the conductivity of the effluent will change (increase) as the more conductive analyte ions elute through the detector. Charge balance ensures that the number of charges (or ions of the same charge) will be constant as the exchange takes place. **Single-column anion chromatography** typically employs benzoate, p-hydroxybenzoate, or phthalate anions as eluent anions. These are either benzene mono- or dicarboxylate anions. Adjustment of eluent pH can control average eluent anionic charge and thus the eluent strength.

Single-column cation chromatography is performed with nitric acid as eluent for singly charged cations or doubly charged ethylenediammonium salts for doubly charged cations. While the instrumentation is much simpler, the sensitivity and linear dynamic range of single-column IC tends to be less than that of IC employing suppressor columns.

14.2.1.2 Indirect Detection in IC

The use of anions such as phthalate in single-column IC enables another form of detection. Many small anions will have no UV absorption in the range where the aromatic phthalate anion absorbs. If the latter is employed as an eluent, when it displaces non-UV-absorbing anions, the concentration of absorbing species passing a UV detector will decrease to the extent that the non-absorbing analyte ions replace the eluent ions in the effluent. The effect is a drop in the previously constant level of absorbance from the eluent anions, which manifests itself as a negative peak. The peak area is proportional to the concentration of the analyte anion. A suitable UV-absorbing eluent for indirect detection in cation chromatography is cupric ion, as copper sulfate. This is called **indirect detection**, since we are not measuring the analyte directly, but instead a drop in the signal produced by absorbing eluent ions that have been replaced on an exactly equivalent basis by the "invisible" analyte ions with which they have exchanged on the column. It is the equivalence of charge exchange and balance in ion-exchange chromatography that enables this method to work quantitatively. We will see later that indirect detection can also be employed in the technique of CE, where again the analytes are ions and can substitute for absorbing ions forming a background supporting electrolyte, which in this case is not an eluent.

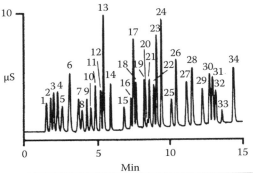

Column:	IonPac AS11			
Detection:	Suppressed conductivity, Autosuppression™ Mode			
Peaks:				
1. Isopropylethyl- phosphonate	5 ppm 5	18. Chlorate	5	
2. Quinate	1	19. Selenite	—	
3. Fluoride	5	20. Carbonate	5	
4. Acetate	5	21. Malonate	5	
5. Propionate	5	22. Maleate	5	
6. Formate	5	23. Sulfate	5	
7. Methanesulfonate	5	24. Oxalate	10	
8. Pyruvate	5	25. Ketomalonate	10	
9. Chlorite	5	26. Tungstate	10	
10. Valerate	5	27. Phthalate	10	
11. Monochloroacetate	5	28. Phosphate	10	
12. Bromate	2	29. Chromate	10	
13. Chloride	5	30. Citrate	10	
14. Nitrite	5	31. Tricarballylate	10	
15. Trifluoroacetate	3	32. Isocitrate	10	
16. Bromide	3	33. cis-Aconitate	} 10	
17. Nitrate	3	34. trans-Aconitate		

Figure 14.29 IC separation of 34 anions. (© Thermo Fisher Scientific. www.thermofisher.com. Used with permission.)

By far the most important applications of ion chromatography are for the analysis of drinking water. This includes not only naturally occurring contaminants in water but also residual disinfectant and disinfectant byproducts. Although this technique has evolved considerably since its introduction in 1975, current developments include the application of two-dimensional IC. Current thinking concerns large-diameter first columns (at 4 mm internal diameter) and a smaller (0.4 mm internal diameter) secondary column. These new techniques will make possible the determination of trace ions in more concentrated solutions of ions. An informative and very readable review of IC is provided in the chapter by Pohl in the ASM series listed in the bibliography

14.3 AFFINITY CHROMATOGRAPHY

In Section 14.1.7, we saw how chromatography interfaced with MS could separate and characterize (even sequence) peptide chains. With sufficient information of this sort, one might even reassemble the data on fragments of a tryptic digest of a large protein molecule and chart the sequence of amino acids in the whole molecule, thus completely characterizing and identifying this macromolecule. Complex biological systems, even at the level of particular types of cells, may contain hundreds, even thousands of proteins. How might these be separated and isolated from one another, so that this sort of analysis may proceed on a single protein? Indeed, we may expect that there are chromatographic procedures or electrophoresis procedures that can separate mixtures of proteins. These may not be satisfactory for isolating one out of the hundreds that might be present, especially if they are minor components of the mixture, and we do not know exactly which peak or spot is the particular molecule of interest. Many of these biological macromolecules (not all only proteins) have been tailored by natural selection to bind very selectively to other specific biomolecules to perform their function in the organism. For example, enzymes bond selectively to their substrates, antibodies to their antigens, and receptors to their messenger hormone molecules.

If we wish to isolate a single biomolecule of this sort from a complex soup of other biomolecules, the technique of **affinity chromatography** is especially powerful. The appropriate biomolecule to which the target binds (e.g., an enzyme substrate, antigen, hormone) is isolated and attached by a short molecular chain called a "**spacer arm**" to a chromatographic gel support material, which is permeable to solutions of the biological macromolecules. The "spacer arm" will have reactive functionalities (often carbamide groups) on either end, which will attach to reactive groups on both the gel and the binding biomolecule. Dangling this target away from the gel on the spacer arm reduces the steric hindrance that the large target molecule might suffer were the binding molecule attached directly to the gel.

The steps for isolating a specific biomolecule are outlined in Figure 14.30. The stationary phase might be a permeable cross-linked agarose polymer or perhaps an agarose–acrylamide copolymer gel, here represented by the coiled, looping lines in the column. The binding biomolecules (termed **ligands**) attached by the spacer arm are represented by small open circles linked to the gel polymer by a short line. A mixture containing three large target biomolecules is shown being applied to the column. Only one of these has a cavity of just the right shape to bind strongly to the ligand (**adsorption** step). This is done using a solvent designated as an **application buffer**, which is close to the pH, ionic strength, and polarity which the target and ligand experience in their natural environment. The other two nonbinding biomolecules are rinsed out of the column in the **wash** step using the application buffer. The next step is to remove the bound target compound in an **elution** step, using an **elution buffer**. If the target biomolecule to be eluted is very sensitive to changes in solvent condition (e.g., a protein structure

14.4 SIZE-EXCLUSION CHROMATOGRAPHY

SEC or gel permeation (or filtration) chromatography (gel permeation chromatography [GPC]/gel filtration chromatography [GFC]) uses a porous material as the stationary phase and a liquid as a mobile phase. The diameters of the pores of the porous material can range from 5 to 100,000 nm, which are useful for the size range of molecules having RMMs of ~500 to 1 million. The latter penetrate the pores according to their size. Small molecules penetrate more deeply into the pores than large molecules, which frequently are excluded from fully penetrating the smaller cavities of the pores. This results in a difference in the rates at which the molecules pass down the column, the larger molecules traveling faster than the smaller molecules. Ideally, there are no binding interactions between the analytes and the porous column particles, so the retention of smaller molecules is due solely to the relative ease and depth to which they can penetrate the pores by diffusion, before diffusing back again to reenter the flow of the mobile phase. The principles of this separation are illustrated in Figure 14.31. They are the same as those that are used for the GC separation of fixed gases using Zeolite molecular sieves or Porapak® materials described in Section 13.4.1. In the examples of Figures 13.6 and 13.7, it is not so easy to rationalize relative retention strictly on the basis of analyte size or mass. There are some binding interactions of the analytes with the pore materials. SEC separations have less binding interactions but are not completely immune from them. Thus, retention is more easily correlated with molecular size or mass.

There are generally two types of column packings: porous glasses or silicas and porous cross-linked organic gels such as dextrans, methacrylate-based gels, polyvinyl alcohol-based gels, and hydroxyethyl cellulose gels. GFC uses aqueous solvents and hydrophilic packings, while GPC employs nonpolar organic solvents and hydrophobic packings. The detectors used are based on UV fluorescence, UV absorption, or changes in RI. Generally, a given packing material is available in a range of pore sizes. Molecules above a given RMM are too large to enter the pores and therefore elute in the void volume. Any molecule above this **exclusion limit** will elute in the minimum possible time, and molecules with RMMs above this limit cannot be separated. Molecules small enough to freely diffuse into all portions of the pores will be the most strongly retained, spending the same portion of time trapped in the pores, so they will elute at the same maximum time, called the **permeation limit**, and will likewise fail to be separated. Thus, in contrast to partition chromatography, SEC or GPC has not only a lower limit for peak retention times but also an upper limit. This is actually an advantage, since chromatographic runs will not drag on too long with late eluters. In the RMM range between these limits (on the order of one to two orders of magnitude), the retention time of a molecule is inversely proportional to the logarithm of the RMM. A packing with

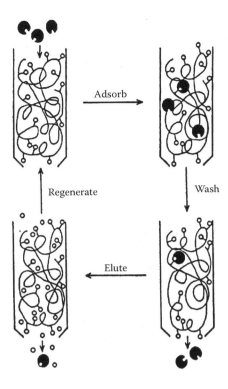

Figure 14.30 Four steps in performing affinity chromatography. (From Katz, E. et al., eds., *Handbook of HPLC*, Marcel Dekker, Inc., New York, 1998. With permission.)

that is easily denatured by changes in pH, solvent polarity), then one will not wish to break the binding by changing these buffer parameters too greatly from those of the application buffer. A more gentle elution method is **biospecific elution**, in which the elution buffer contains a similar, competing biomolecule, which *displaces* the target molecule. This can be done by adding either an agent that competes with the column ligand for the target (**normal-role elution**) or one that competes with the target for ligand-binding sites (**reversed-role elution**). While biospecific elution is very gentle, it is generally slow, and the eluted target peaks are very broad. The competing molecule may have to be removed from the eluted solution, and if analytical affinity chromatography is being performed, it must not interfere by producing a large background signal under the detection conditions being used. For analytical rather than the preparative applications for which the gentle mechanism of biospecific elution may be required, **nonspecific elution** may be employed. This employs an elution buffer of substantially different pH, ionic strength, or polarity to lower the association constant of binding to the ligand. Under such conditions, the compound of interest is eluted much more rapidly, leading to higher, narrower peaks, and consequently lower limits of detection. After the target compound is detected or isolated, the column is **regenerated** by passing the application buffer through it before injecting the next sample.

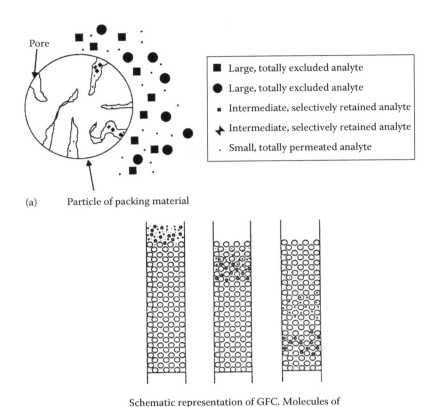

Figure 14.31 Mechanism of SEC; (a) size selectivity as a function of pore and analyte sizes and (b) separation process on a gel-filtration column. (Adapted with permission from Phenomenex, Inc., Torrance, CA, www.phenomenex.com.)

the appropriate pore size to cover the anticipated range of masses must be selected. Particles with different pore sizes can be mixed to accomplish separation of mixtures of a wider range of RMMs.

If the sample is a single polymer, the chromatogram represents the RMM distribution. This method is very valuable for determining the RMM distributions of polymers up to very high RMMs. The actual calculation is usually based on a comparison with a standard polymer material of known RMM distribution. The procedure is also used to assess the RMMs of biological compounds, such as proteins. GPC is also capable of separating different polymers from each other and, under the correct conditions, mixtures of polymers can be characterized with respect to percentage and mass range of each polymer present. GPC has been used for such important analyses as the determination of somewhat toxic aldehydes and ketones in auto exhaust. It has also been used to separate C_{60} and C_{70} fullerenes (buckyballs). Figure 14.32 illustrates a GFC separation of several large protein molecules. An additional useful application of GFC is **desalting**, wherein salts of low RMM are separated from solutions of large macromolecules in order to facilitate changing the buffer composition of their solution to one suitable for a method such as electrospray LC-MS or CE.

Care must be taken in interpreting retention data to estimate RMMs since the separation is based on the size of the molecule rather than on the actual RMM. This means that the shape of molecule has a significant effect on the results. For example, two molecules of the same RMM, one straight chained and the other highly branched, will be retained somewhat differently because each has a different penetrating power into the column particle pores. This must be taken into account when selecting standards for calibrating the RMMs.

An advantage of SEC is that it can be carried out at room temperature and the samples are not decomposed because of exposure to high temperature. This is especially important when dealing with some very high-RMM compounds that are easily broken into two or three fragments, resulting in great changes in apparent RMM. It gives a limited range of retention times that facilitates rapid sample throughput, and the peaks are narrow and therefore give good sensitivity. Since analyte interaction with the particles is minimal, sample loss and deactivation of the column are also minimized. A disadvantage of the short

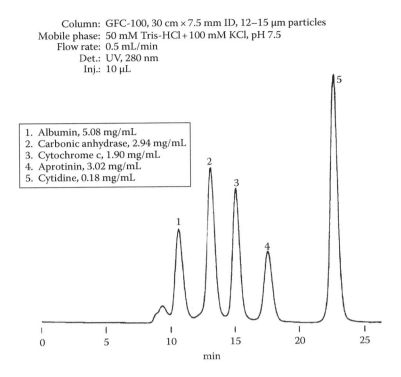

Figure 14.32 Gel filtration separation of five protein molecules. (Reprinted with permission of Supelco, Inc., Bellefonte, PA. www.sigmaaldrich.com/supelco.)

time range for peak elution between the exclusion and permeation limits is the limitation this imposes on the number of peaks that can be accommodated. The method generally requires differences on the order of ± 10 % in RMM for sample peaks to be resolved. Consequently, similar-sized variants of large biomolecules (e.g., isomers) are better analyzed by affinity chromatography, if the appropriate ligand can be obtained and bound in an affinity chromatography column.

14.5 SUPERCRITICAL FLUID CHROMATOGRAPHY

A compound such as CO_2 is a gas at normal temperature and pressure (NTP) and, like all fluids (gases and liquids) below their critical temperature, T_c, further increasing the pressure results in the formation of a liquid. The critical temperature is a fundamental thermodynamic property of any fluid; it is the temperature at a phase equilibrium end point (review simple phase diagrams in your general or physical chemistry text for more discussion on this topic). Above the critical temperature, increasing the pressure increases the density of the fluid but a distinct transition to and boundary with a liquid phase never forms. At the critical temperature and pressure, the density of the gas phase and the liquid phase are the same. This state is neither a true liquid nor a true gas but is a supercritical fluid (SCF). The use of a compound in this fluid state as a chromatographic mobile phase provides different properties than when it is used as either a gas or a liquid in GC or HPLC, as shown in Table 14.4. To date, most attention has been focused on CO_2, C_2H_6, C_3H_8, and N_2O, with critical temperatures of 31 °C, 32 °C, 96.8 °C, and 37 °C, respectively. See the phase diagram for carbon dioxide in Section 1.5.4.1. The temperatures and the necessary pressures can be accommodated in conventional GC instruments if the effluent end of the column is maintained at a sufficiently high pressure above ambient. This can be achieved by the addition of suitable flow restrictors at the column end. An important property of SCFs is their ability to dissolve poorly volatile molecules. Certain important industrial processes are based upon the high solubility of organic species in supercritical CO_2. For example, CO_2 has been employed for extracting caffeine from coffee beans to produce decaffeinated coffee and for extracting nicotine from cigarette tobacco. Supercritical CO_2 readily dissolved n-alkanes containing up to 22 carbon atoms, di-n-alkylphthalates with the alkyl groups containing up to 16 carbon atoms, and various polycyclic aromatic hydrocarbons. Discussions on the instrumentation and

Table 14.4 Typical values of the thermophysical properties of fluid phases

	Gas	Liquid	Supercritical fluid
Density, ρ, g/mL	10^{-3}	1	0.7
Diffusion Coefficient, D, cm^2/s	10^{-3}	5×10^{-6}	10^{-3}
Dynamic Viscosity, η, g/(cm·s)	10^{-4}	10^{-2}	10^{-4}

Source: Bruno, T.J., A supercritical fluid chromatograph for physicochemical studies, J. Res. Natl. Inst. Stand. Technol, 94(2), 105-112, 1989.

theory of supercritical fluid chromatography and extraction can be found in the book by Bruno and Ely.

14.5.1 Operating Conditions

To adapt equipment to supercritical applications, it is only necessary to provide an independent means for controlling the internal pressure of the system. Such systems are available as adaptations of commercial GC or HPLC instrument designs.

14.5.2 Effect of Pressure

Pressure changes in supercritical chromatography affect k'. For example, increasing the average CO_2 pressure across a packed column from about 70 to 90 bars (1 bar = 0.987 atm) decreases the elution time for hexadecane from about 25 to 5 min. This effect is general and has led to the type of gradient elution in which the column pressure is increased linearly as the elution proceeds. The results are analogous to those obtained with programmed temperature in GC and solvent-gradient elution in LC.

14.5.3 Stationary Phases

So far, most supercritical fluid chromatography has utilized column packings commonly used in LC. Capillary SFC has been performed with stationary phases of organic films bonded to capillary tubing. Because of the low viscosity of SCFs, long columns (50 m or more) can be used. This results in very high resolutions in a reasonable elapsed time. This is the key advantage of SFC: it combines the low resistance to flow of GC, enabling longer columns with more theoretical plates, while the mobile-phase density approaches that of a liquid and thus introduces a liquid-like solvating power that is totally lacking in GC gases like H_2, He, or N_2.

14.5.4 Mobile Phases

The most commonly used mobile phase for SFC is CO_2. It is an excellent solvent for many organic molecules, and it is transparent in the UV range. It is odorless, nontoxic, readily available, and inexpensive when compared with other chromatographic solvents. Carbon dioxide's critical temperature of 31 °C and its pressure of 72.9 bar at the critical point permits a wide selection of temperatures and pressures without exceeding the operating limits of modern chromatographic equipment. The effective design of separations with supercritical fluid CO_2 is facilitated by an understanding of the phase behavior of this fluid. This begins with exploitation of the P-V-T surface, which has been provided in tabular form in a number of sources (the first three editions of the book by Bruno and Svoronos, for example), but in recent years can be calculated by one of several very accurate equations of state. The most accurate available currently is implemented in the NIST Refprop program, but others are available as well.

Other substances that have served as mobile phases for SFC include ethane, pentane, dichlorodifluoromethane, diethyl ether, and THF, however specialized fluids have been used as well. The development of alternative refrigerants in the 1990s led to the large-scale synthesis of non-chlorinated heat transfer and working fluids. Many of these fluids can be used as supercritical fluid solvents. A comprehensive listing can be found in the book by Bruno, which gives the critical properties and design parameters for the solvent applications.

The most common fluid used for SFC, CO_2 (as well as C_2H_6 and C_3H_8), is nonpolar, but many analytes of interest require additional polarity for separation. While that was the reason for the less common fluids such as the alternative refrigerants, another approach is more popular. One can add a polar organic modifier to the nonpolar fluid to take advantage of the enhanced intermolecular interaction. The most common modifier is methanol, but other modifiers are used as well. The comments made earlier regarding mobile-phase instability apply to modifiers used in SFC. Sometimes the modifier is added at such a high concentration that it becomes dominant; here, the nonpolar fluid serves as a viscosity modifier to enhancement. It is worthwhile to understand that the mobile phase in SFC need not be above the critical point to provide benefits in separations. Many applications have been developed with near-critical fluids.

14.5.5 Detectors

The most widely used detectors are those found in both GC or LC (i.e., UV absorption and fluorescence, RI, flame ionization, and MS).

14.5.6 SFC versus Other Column Methods

As shown in Table 14.4, several physical properties of SCFs are intermediate between those of gases and liquids. Hence, SFC combines some of the characteristics of both GC and LC. For example, like GC, SFC is inherently faster than LC because of the lower viscosity and higher diffusion rates in SCFs. High diffusivity, however, leads to band

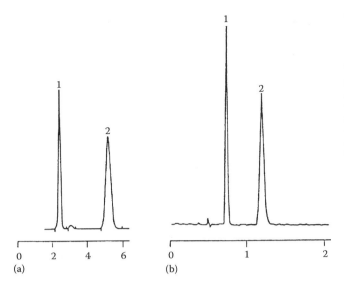

Figure 14.33 Comparison of SFC (b) and HPLC (a) separations (R_t in min).

14.5.7 Applications

SFC can handle larger molecules than GC, with higher efficiencies than LC, and its largely NP packings provide separations orthogonal to RPLC. It was also considerably easier to interface with mass spectrometers than was its liquid counterpart prior to the advent of ESI and APCI-MS interfaces. The high chromatographic resolution achievable with coated capillary columns was particularly significant. Effluent collection was especially easy when using CO_2, as the mobile phase could be readily and gently removed from collected fractions by simply opening them to atmospheric pressure, bleeding the gas off slowly to avoid the condensation and dry ice crystal formation that occur using CO_2 fire extinguishers. This feature had made **preparative SFC** and supercritical fluid extraction one of the major present applications of the technique, while **analytical SFC** was often practiced. The solubility parameter approach that was discussed earlier is particularly apt in matching fluids to analytes in such applications.

14.5.8 Ultra Performance Convergence Chromatography: A New Synthesis

Analytical SFC did not live up to its promise due to difficulties in handling the largely supercritical CO_2-based mobile phase and blending in small amounts of organic modifiers, while avoiding difficulties of using such a carrier in the injectors, columns, valves, backpressure regulators, and detectors designed for UHPLC analyses. Waters Corp. redesigned all these elements from their Acquity UPLC™ hardware to a new instrument called the Acquity UPC²™ system. This is capable of better controlling the parameters of SFC mode operation to enable greater peak shape resolution and reproducibility of retention times under gradient conditions, as well as remaining capable of standard UHPLC analyses. This new combination is named Ultra Performance *Convergence* Chromatography (UPC²), as it enables standard RP-UHPLC runs as well as greatly superior NP-SFC runs, thus achieving a convergence of these two very different LC separation methodologies. The former are developed using typically a single suitable column and varying the polar mobile-phase mixture composition to achieve optimal separation. The latter uses a standard fixed set of mobile-phase conditions (e.g., 150 bar pressure, 40 °C, and a generic gradient from 5 % to 40 % methanol in CO_2) and tests the analyte mixture against a set of columns optimized for separation of basic compounds (which are the most problematic for SFC/UPC² [e.g., a set of ethylpyridine columns]). Neutral and acidic analytes are much less problematic for NP-SFC assay development. The Acquity UPC²™ system is unique in that the dual-instrument/column/mobile-phase systems enable operation using either of the *orthogonal* RP-HPLC or NP-SFC separation techniques, with improvements in the

spreading, a significant factor with GC but not with LC. The intermediate diffusivities and viscosities of SCFs result in faster separations than are achieved with LC together with lower band spreading than encountered in GC.

Figure 14.33 compares the performance characteristics of a packed column when elution is performed with supercritical CO_2 versus a conventional HPLC mobile phase. The rate of elution is approximately four times faster using SFC. The roles of the mobile phase in GC, LC, and SFC are somewhat different. In GC, the mobile phase serves but one function–band movement. In LC, the mobile phase provides not only transport of solute molecules but also influences selectivity factors. When a molecule dissolves in a supercritical medium, the process resembles volatilization but at a much lower temperature than under normal circumstances. Thus, at a given temperature, the vapor pressure for a large molecule in an SCF may be orders of magnitude greater than in the absence of that fluid. As a result, important compounds such as high-RMM compounds, thermally unstable species, polymers, and biological molecules can be brought into a much more fluid state than that of a normal liquid solution of these same molecules. Interactions between solute molecules and the molecules of an SCF must occur to account for their solubility in these media. The solvent power is thus a function of the chemical composition of the fluid. Therefore, in contrast to GC, there is the possibility of varying retention substantially by changing the mobile phase (i.e., by adding modifiers to the SFC mobile phase, as is done in HPLC). The solvent power of an SFC is also related indirectly to the gas density and thus the gas pressure. One manifestation of this relationship is the general existence of a threshold density below which no solution of solute occurs. Above this critical value, the solubility increases rapidly and then levels off.

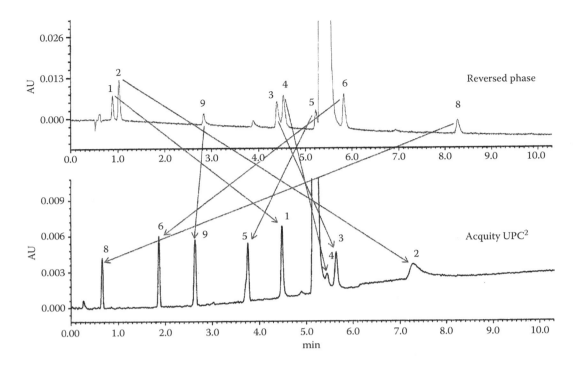

Figure 14.34 Orthogonal RP-UPLC and Acquity UPC² separations of metoclopramide and 8 contaminants. Waters BEH UPLC C18 RP-column and Acquity UPC² BEH column with UV detection. (Used with permission from Waters Corp., Milford, MA, www.waters.com/upc2.)

NP branch of this combination, which brings its performance back to parity with the dominant RP methodologies. Figure 14.34 illustrates this for a preparation of the antiemetic metoclopramide (large unlabeled peak 7) and eight other contaminants. Note the radically different elution orders and positions of these contaminants relative to the metoclopramide peak! This is a dramatic different example of the principle and use of orthogonal separations that we have already seen in 2D-GC, 2D-HPLC, and 2D-TLC. Here, the analyst must reconfigure the UPC² instrument and analyze the sample with two different separation methodologies, but it will not be long before someone links two of these instrument modes together and automates a UHPLC–UPC² system or even two different UPC² systems.

14.6 ELECTROPHORESIS

In Section 12.4, we described the mechanism of the chromatographic separation process using an extended analogy to people stepping on and off a moving slideway. We mentioned that if different classes of people stayed always on the slideway but walked forward or backward at different constant rates, they would separate themselves along the slideway and exit it at different times. If molecules could be induced to behave in a column in a similar fashion, the process would not be chromatographic, as they would not be transferring back and forth between a stationary and mobile phase. The separation process would, however, appear very much like chromatography, with a mixture being introduced at one end of a column with flowing fluid, and the components exiting the other end as separated, Gaussian-shaped peaks, to be detected and quantitated by the same type of detectors employed in HPLC.

14.6.1 Capillary Zone Electrophoresis (CZE)

If molecules in solution carry a charge (i.e., if they exist as ions in solution), application of a sufficiently high voltage between two electrodes immersed in the solution will cause positively charged cations to migrate (i.e., move through the solvent in response to the electromotive force or voltage) toward the cathode. Likewise, anions will migrate toward the anode. Such a **migration** of ions in solution under the influence of an electric field is termed **electrophoresis**. In polarography (Chapter 15), such a migration current for those analytes that exist as ions is undesirable, as we wish to measure only the diffusion current, so the migration current is suppressed by swamping it with a large excess of supporting electrolyte that does not undergo a Faradaic process (oxidation or reduction) at the electrode. In the ion source of a mass spectrometer, operating under near-vacuum conditions, electric fields of several volts are sufficient to pull ions through the entrance lenses, and then somewhat higher voltages accelerate them into the analyzer section. At a constant accelerating voltage in the vacuum, the velocity of the accelerated ions is precisely inversely proportional to their masses and directly proportional to their charge, there being no other factor or force involved than the inertia of the ions.

The situation for ions immersed in solution and migrating in an electrophoresis experiment is radically different. There is intimate interaction of the ions with the surrounding solvent as they move through it, resulting in viscous frictional drag forces that act almost instantaneously to limit the motion of the ion to a steady speed through the liquid. This speed is that at which the retarding frictional force just balances the applied electromotive force (voltage). To attain useful electrophoretic migration speeds, electrical fields of thousands of volts per meter must be applied. It is the *effective size* of the ion, *not its mass*, that governs the frictional force resisting the migration of the ion. The balance of the frictional force and the accelerating force at a given migration velocity is given by the following equation:

$$qE = fu_{ep} \qquad (14.7)$$

where
q is the charge on the ion (in C)
E is the applied field (in V/m)
f is the frictional coefficient
u_{ep} is the ion's velocity

Rearranging Equation 14.7 to 14.8, we obtain

$$u_{ep} = \left(\frac{q}{f}\right)E = \mu_{eq}E \qquad (14.8)$$

where μ_{eq}, **the electrophoretic mobility**, is defined as the constant of proportionality between the speed of the ion and the electric field. This value is proportional to the charge on the ion and inversely proportional to the frictional coefficient. That coefficient "f" is proportional to the **effective hydrodynamic radius** so the electrophoretic mobility is seen from Equation 14.8 to be inversely proportional to this "size" parameter of the ion. Note that this effective radius is not calculated from measurements of the molecular size, but is experimentally determined by measurements of μ_{eq}. A diprotic acid could bear either one or two negative charges depending on solution pH. Although the charge is exactly doubled, the effective size of the ion with its tightly bound, associated water molecules will vary with the degree of acid dissociation, so the mobility difference will not necessarily be exactly a factor of 2. Mobilities must be measured by experiment, as they are not subject to precise calculation. They will vary as the viscosity of the buffer solution changes, so they are specific for each buffer composition.

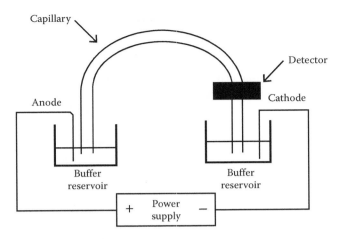

Figure 14.35 Schematic of CZE apparatus. (From Cazes, J., *Encyclopedia of Chromatography*, Marcel Dekker, Inc., New York, 2001. With permission.)

We can separate ions of different molecules by electrophoresis if they have different mobilities. Figure 14.35 is a schematic diagram of a capillary zone electrophoresis (CZE) instrument. Two reservoirs connected by a length of capillary tubing are filled with a buffer solution whose pH and electrolyte strength are adjusted to optimize the electrophoretic separation. A high voltage (typically 30–50 kV) is applied across the capillary between an anode and cathode in each of the reservoirs. Imagine that a narrow band of buffer solution containing a mixture of ions to be separated were somehow placed at the center of the capillary. Charge balance would require equal numbers of anions and cations. When the voltage was applied, anions would migrate to the anode, and cations to the cathode. Ions of greater μ_{eq} would move faster, and if the mobility differences among the components were great enough, they could be separated. Migration velocities are fairly slow, typically in the cm/min range, and if a long capillary were used, the experiment might take a long time. To measure both anions and cations, one would need a detector at each end, and ions of each charge would have only half the capillary length to achieve separation. Note that the figure displays only one detector. If we introduced the sample mixture band at the left end of the capillary and arranged for buffer to flow through it from left to right at a speed exceeding that of the greatest migration velocity of anions, then the cations would pass the detector first in order of their mobilities (fastest first). Following the cations, the anions migrating in a direction opposite to the faster buffer flow would pass the detector in inverse order of their mobilities. One detector gets them all, the sample is easily introduced at other end of the capillary, and we can choose the buffer flow to speed up or slow down the period of "elution" depending on how long we need the ions to remain migrating in the capillary to achieve a separation. One might be able to cause the buffer to flow toward the detector simply by raising the anode reservoir. We will see that long (50–100 cm) narrow-bore (50–100 μm) fused silica capillaries are generally used for CZE. To obtain a desired buffer flow rate might require pressurizing the system. Adjustment and precise control of both voltage and pressure drop across the capillary would make for a more complex instrumental design, sacrificing the elegant simplicity of the system diagrammed in Figure 14.31. It turns out that there is a different and better way to induce buffer flow through the capillary.

Toward the end of Section 12.5, we saw that one of the many useful features of fused silica capillary tubing was the ability of its surface silanol groups to lose H^+ ions and form a chemically bound layer of negatively charged groups in aqueous solutions at pH above 2. When this happens, these excess bound negative charges are partially neutralized by tightly binding positive ions from the buffer electrolyte used in the CZE analysis. The rest of the excess negative surface charge is neutralized by the presence of an excess of cations in the **diffuse double layer** very close to the silica surface. Polarizing (adding charge to) an electrode in electrochemical techniques results in the establishment of both compact and diffuse electrical double layers at the electrode surface. The diffuse double-layer cations are not tightly bound and can move in response to the applied electrophoretic voltage. Because they are in excess in this layer, more buffer cations are flowing toward the cathode than buffer anions toward the anode. This excess cationic flow in this layer drags the buffer solution as a whole toward the cathode. This motion is called the **electroosmotic flow** (EOF). Like the phenomenon of "osmotic pressure" established across semipermeable membranes, it arises from an imbalance of particles across a boundary, but the flow ends up being parallel to the boundary instead of across it, since the motion of interest is that of an excess of ions responding to a voltage gradient imposed along instead of across the surface. The thickness of the diffuse double layer that mediates this flow is inversely proportional to the concentration (more precisely the ionic strength) of the buffer. Thus, changing the electrolyte composition of the buffer will affect and allow control of the velocity of the EOF. Likewise, lowering the pH of the buffer will decrease the dissociated silanol surface charge density and decrease the velocity of the EOF. The origin of the EOF is illustrated in Figure 14.36. Note that one siloxyl group lacks a tightly bound counter cation. It is the excess of mobile countering cations in the layer just below the "plane of shear" on the diagram that provides the "handle" for the applied voltage to move the buffer through the column. This mechanism is sometimes referred to as **electrokinetic pumping**, in contrast to the pressure-driven, **hydrodynamic flow** produced by pumps in HPLC systems. For the EOF, we can write an equation similar to Equation 14.8 relating the velocity of the EOF (μ_{eo}) to the electric field E:

$$u_{eo} = \mu_{eo} E \qquad (14.9)$$

where μ_{eo} is defined as the **electroosmotic mobility**. We can then define and measure an **apparent mobility** of an ion, which is the sum of μ_{ep} and μ_{eo}:

$$\mu_{app} = \mu_{ep} + \mu_{eo} \qquad (14.10)$$

μ_{ep} is positive for cations and negative for anions, so μ_{app} will be greater or less than μ_{eo}. The apparent mobility of an ion and the electrophoretic mobility can be calculated according to

$$\mu = \frac{\frac{L_d}{t_x}}{\frac{V}{L_t}} \qquad (14.11)$$

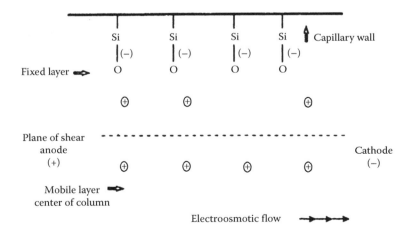

Figure 14.36 Diagram showing origin of EOF. (From Cazes, J., *Encyclopedia of Chromatography*, Marcel Dekker, Inc., New York, 2001. With permission.)

where L_d and L_t are the length of the column from injector to detector and the total length of the column, respectively. Note in CZE that the detector usually is before the end of the capillary, as illustrated in Figure 14.35. V is the total voltage across the capillary. To measure μ_{app}, the time for the ion to migrate from the injector end of the column to the detector is substituted for t_x. To measure μ_{eo}, one substitutes in the equation the time for a neutral species capable of giving a detector response (e.g., a UV-absorbing compound) to make the same journey. From these two measurements, the electrophoretic mobility μ_{ep} can be calculated by taking their difference.

There is an additional important distinction between these two modes of inducing liquid flow within a narrow capillary. Figure 14.37a illustrates the profile of flow velocity across a capillary driven by EOF. Except at points very close to the inner wall, it is uniform. In contrast, pressure-driven hydrodynamic flow is affected by shear forces varying with distance from the center to the wall, resulting in a parabolic, laminar flow velocity profile (Figure 14.37b). Band spreading resulting from analyte diffusion across this hydrodynamic flow profile will be absent in electrokinetically pumped systems. The friction with solvent molecules of ions being pulled through them by the applied voltage, which creates both the EOF and the electrophoretic separation, can raise the temperature of the buffer by a process called **Joule heating**, which is directly proportional to applied voltage, current, and time. It also depends on capillary diameter and buffer concentration. Joule heating is closely analogous to the heating produced by a current of electrons passing through a resistor in an electrical circuit. In CZE, this heat is dissipated by flowing the short distance to the wall of the capillary. If a significant temperature gradient builds up across the radius of the capillary, the higher central temperature will cause the buffer viscosity to become lower at the center than at the capillary wall, across which the heat must be dissipated. This would act to disrupt the uniform EOF flow profile, making it resemble the pressure-driven laminar flow profile, and introduce band broadening. The high thermal conductivity of fused silica helps alleviate this, and placing the capillary in a cooling bath may retard the onset of the problem, but this phenomenon places a limit on the inner diameter of capillaries that can be used to full efficiency. Typical ranges are 25–100 μm, while tubing a millimeter or more in diameter could not escape the bad effects of Joule heating.

Although we have emphasized that the fundamental principle of electrophoretic separation is different from that of chromatography, there are many similarities in the appearance of the separation. There is sample injection, flow of a "carrier fluid" along a column, similar detectors employed near the end of the column, and separated compounds being displayed as peaks emerging at characteristic times. Similar formulas can be used to describe the resolution of separations and to characterize the separation efficiency of the system in terms of theoretical plates. It is instructive to revisit the Van Deemter equation (Equation 12.10) for evaluating the impact of various processes on the height equivalent to a theoretical plate (HETP) (H) described in Section 12.8. Recall that minimizing this maximizes efficiency (i.e., permits more plates for a given column length):

$$H = A + \frac{B}{u} + Cu$$

Since CZE is an open-tubular technique, there is no multipath A term. Since CZE is actually not a chromatographic technique, there being no stationary phase, there is no mass-transfer C term. Recall that this is actually a term that combines several processes, namely, transfer in the mobile phase across the laminar flow velocity gradient, which we have shown is suppressed in a properly operated CZE system, and transfer to and from the stationary phase, of which there is none in CZE. The only remaining source of band broadening is the longitudinal diffusion term B. This is proportional to the diffusion coefficient of the analyte in the carrier fluid and varies inversely with the size or mass of the analyte molecule. Consequently, very low values of H and correspondingly high plate counts are obtained for large molecule separations. Plate counts up to 500,000 are routinely achieved for large molecule separations by

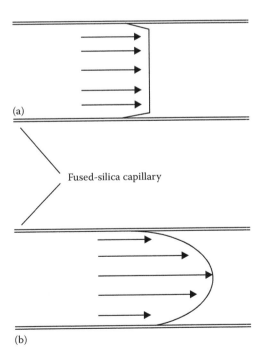

Figure 14.37 Comparison of electroosmotic and hydrodynamic flow profiles: (a) uniform EOF (electrically driven) and (b) laminar, parabolic, hydrodynamic flow (pressure driven). (From Cazes, J., *Encyclopedia of Chromatography*, Marcel Dekker, Inc., New York, 2001. With permission.)

CZE, which is an order of magnitude better than those of which HPLC is capable. Biological macromolecules such as large peptides or proteins, which have multiple sites for bearing charge as a result of adding or abstracting H+ by pH variation, may often be separated with great efficiency by CZE, provided a pH can be found at which one charge predominates and which is compatible with good electrophoretic operation. It can be demonstrated that the number of plates in a CZE separation of a given compound is related only to the voltage V across the capillary, the electrophoretic mobility, μ_{ep}, and diffusion coefficient, D, of the ion, according to the following equation:

$$N = \frac{\mu_{ep} V}{2D} \quad (14.12)$$

Contrary to our experience with chromatographic separations, N is independent of column length. Clearly, we would like to operate at higher voltages, but we will be limited by the onset of Joule heating, which is proportional to the field E. For a given applied voltage V, E is proportional to V/L. Thus, we must increase column length and analysis time to keep E below the value at which Joule heating degrades resolution. This is the reason for using longer columns. One can see that larger ions, which have lower values of D, will achieve greater plate counts. This is why CZE is often superior to HPLC for separations of charged biological macromolecules.

If the absolute value of μ_{ep} is less than μ_{eo} for all anions and cations in a CZE separation, then they may all be observed whether the applied voltage directs EOF either to the cathode or anode. If not, then those migrating in a direction opposite to the EOF at greater than its velocity will never reach the detector. It may then be necessary to do a second electrophoretic separation with the electrode polarities reversed to measure these. An alternative is to reverse the direction of the EOF while maintaining the same electrode polarity. To do this, one must replace the siloxyl negative charges fixed to the column wall with bound positive charges. This may be accomplished by adding a cationic **surfactant** such as cetyltrimethyl ammonium ion $\left[n - C_{16}H_{33}N(CH_3)_3^+ \right]$. The long C_{16} hydrocarbon tails attract one another and form a bilayer sandwich structure with the hydrophilic quaternary ammonium cation portions on the outer sides and the nonpolar organic chains grouping together on the inside. The forces driving this are twofold: cation charge repulsion and van der Waals attractions of the adjacent nonpolar hydrocarbon chains. One layer of positive charges is fixed by electrostatic attraction to the negative siloxyl groups on the silica wall, and the buffer is presented with the other layer of positive charges. Now, excess ions in the diffuse double layer will be anions instead of cations, and the EOF will move in the opposite direction. This is illustrated in Figure 14.38.

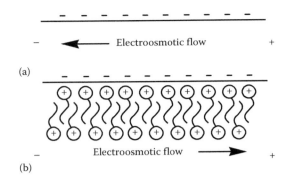

Figure 14.38 Use of cationic surfactant to reverse EOF: (a) normal EOF toward cathode (no surfactant) and (b) reversed EOF toward anode (cationic surfactant bilayer). (From Katz, E. et al., eds., *Handbook of HPLC*, Marcel Dekker, Inc., New York, 1998. With permission.)

The ability of some organic molecules to selectively bind to the silica surface in this fashion can become a problem if they are in fact the analytes one is trying to separate and measure by CZE. Such unintended "wall effects" superpose an adsorption chromatographic mechanism on the electrophoretic separation. If the binding is very tight, one may initially fail to see some components migrate to the detector, only to have them begin to appear as peaks with degraded resolution and shifting migration times, as successive injections result in saturation of such "binding sites." For some applications, it is necessary to precoat the capillary with a covalently bonded hydrophilic polymer to suppress this interference. Such complexities make developing CZE separations less simple than one might naively predict.

14.6.2 Sample Injection in CZE

In capillary GC, samples on the order of several microliters are propelled from a syringe through a septum into a complexly designed injector port where they vaporize, expanding in volume nearly 1000-fold, and then must either be split or reconcentrated in the head of the capillary. In capillary HPLC, samples are injected with a syringe into a sample loop injector that must be designed to function at high pressures. In CZE, things can be much simpler. The injection volumes are much smaller, on the order of nanoliters. One wishes to introduce a plug or zone of analytes dissolved in aqueous buffer into the injection end of the capillary. This can be done in two ways: hydrodynamic injection or electrokinetic injection. For **hydrodynamic injection**, one often simply removes the injection end of the capillary from its buffer reservoir and places it briefly in the sample buffer, which is elevated sufficiently above the detector end reservoir, and hydrostatic pressure causes sample solution to flow into the capillary. The volume of the sample flowing in will be given by

$$\text{Volume} = \frac{\Delta P \pi d^4 t}{128 \eta \, L_t} \quad (14.13)$$

where

ΔP is the pressure difference between the ends of the capillary
d is its inner diameter
L_t is its total length
t is the time it spends in the sample solution
η is the viscosity of the sample solution

If the capillary is either too narrow and/or too long, it may be necessary to apply additional pressure at the injection end or suction at the detector end in order to introduce a sufficient volume in a reasonable time. For **electrokinetic injection**, one places the injector end of the column into the sample solution and applies a voltage across the column (for this, we will need to introduce an electrode into the sample solution). The sample ions will enter the column both by migration (at rates which will vary with their electrophoretic mobilities) and by entrainment in the EOF (at a constant rate for all ions). This difference in sample amount will cause problems in quantitative CZE work, so hydrodynamic injection is preferred in that case. In the capillary mode called capillary gel electrophoresis (CGE) (discussed in Section 14.6.4), the gel in the capillary is much too viscous to employ hydrodynamic injection, so electrokinetic injection must be used.

To introduce a sufficient sample volume into a narrow CZE capillary, one might produce a band of sample much longer than the optimum bandwidth that would be implied by the number of plates calculated from Equation 14.12 operating on an ideally narrow injected band. This is an example of an "extra-column band-broadening effect." It would be desirable to have a mechanism for "focusing" the injected sample mixture into a very narrow band, as happens in chromatographic injections when the analytes riding the mobile phase into the column are arrested at its head by the stationary phase, when k' is sufficiently high. Just such a focusing effect can be achieved in CZE if one employs a buffer electrolyte concentration in the sample that is much *lower* than that in the separation buffer. The conductivity of this more dilute sample plug is much lower (thus, its resistance is higher), and since the ion current through the system must be constant, by Ohm's law ($E = IR$), the field across the sample plug must become proportionally larger. The increased field causes ions in the sample plug to migrate more rapidly to the end nearest to the electrode of opposite charge. When they reach this interface, they slow down and continue to move with the velocities characteristic of the separation buffer. This process is called **sample stacking**, and it is illustrated in Figure 14.39 for anions in a sample plug. Cations would "stack" in a similar fashion at the opposite end of the plug. The result is to quickly produce two narrow bands of either cations or anions, separated by the width of the sample injection plug, which can then experience the full separation capability of the CZE process.

14.6.3 Detection in CZE

In contrast to detectors used in GC and HPLC, those in CZE usually are placed at some point before the end dipping into the capillary reservoir, as indicated in Figure 14.35. Optical detectors are similar in design to HPLC

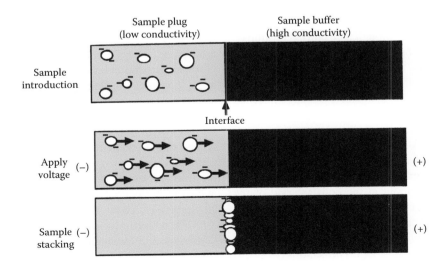

Figure 14.39 Sample stacking of anions in sample plug toward anode. (From Katz, E. et al., (eds.) *Handbook of HPLC*, Marcel Dekker, Inc., New York, 1998. With permission.)

detectors, but with modifications of the detector cell portion to fit the much smaller peak volumes encountered in CZE. For UV absorption or fluorescence detectors, the detection cell is often a portion of the capillary column itself. The polyimide protective coating of a short section of the fused silica column is removed, and the beams of the spectrometric detectors are focused with condensing optics on this small volume. This has the advantage of using a detector cell volume that is similar to the very small band volume of highly resolved CZE peaks. The very short path length for the beam across the diameter of the CZE capillary decreases sensitivity. This may be improved by bending the capillary at right angles or fitting a capillary length between two 90° fittings for a short distance approximating the bandwidth of a narrow peak and introducing the illuminating beam lengthwise down the axis of this segment through optically flat windows. Another approach to increase the light path length is to coat the outside of a short length of the capillary with reflective silver beneath the protective polyimide coating. Light entrance windows are cleared at either end on opposite sides of this segment. Light enters one window at a slight angle to perpendicular to the capillary such that it is reflected multiple times back and forth across the capillary before exiting the other window of the detector.

Miniature amperometric or conductometric detectors can be introduced into the capillary, by plating the appropriate electrodes in the inside of the capillary walls. It is often necessary to employ a porous glass or graphite joint between the capillary containing the detector electrodes and the end of the capillary in the detector end electrode reservoir, in order to isolate the detector from the high voltage used to power the electrophoretic separation.

The extremely low flow rates of buffer in a CZE capillary separation make it especially easy to interface to a mass spectrometer by using an ESI interface (cf. Section 14.1.6.1) for detection of large multiply charged biological macromolecules. Instead of a CZE detector end reservoir with its electrode, the voltage at that end is applied through a metallized coating on the inside and outside of the capillary end. This voltage also functions to charge the ions and droplets that exit the tip of the capillary end and to form the "electrospray" with its process of coulomb explosion and charge transfer, yielding molecular ions suitable for MS analysis.

Because ions in CZE migrate past spectrometric or ECDs at different rates according to their electrophoretic mobilities, they pass through the detector zone at different rates, unlike chromatographic analytes, which enter the detector cell from the column at the constant rate of the mobile-phase velocity. For concentration detectors (cf. Section 12.4), the peak areas are therefore independent of retention or elution time in chromatographic systems. However, this is **not** true in CZE. The same concentration of material giving the same response will pass more slowly through the detection region of the capillary if it has a lower migration velocity and therefore will integrate over time to a larger peak area value. If one is quantitating against an IS of different mobility, and separation conditions and "apparent migration times" shift between the time a calibration run is made and an unknown sample quantitation run is made, calibration based on peak area ratios will become invalid.

Because CZE is a method for separating and measuring ions, like IC, considerations of charge balance make it particularly suitable for use of the method of indirect detection, as described at the end of Section 14.2.1.

14.6.4 Modes of CE

There are several different modes of conducting electrophoresis in capillary columns. We have just discussed at length the basic one, CZE, which is conceptually the simplest. It is conducted with a liquid buffer of uniform composition of electrolyte concentration and pH level. There are more complex versions, in which the buffer liquid is enmeshed in a porous gel of hydrophilic polymer or the concentration or the pH of the separation buffer is not constant along the length of the column, leading to variations in the migration velocities of individual ions as the separation proceeds. Several of these will be discussed subsequently.

14.6.4.1 CZE

CZE has already been discussed in Sections 14.6.1 through 14.6.3. To summarize for purposes of comparison, it is performed using a separation buffer of constant composition. The only exception is the sample plug, which is usually introduced in a buffer of lower concentration to enable faster migration velocities within it to achieve the desirable effect of sample stacking at either end of the plug. Once migration of the sample mixture ions begins in the uniform separation buffer, the different components separate into "zones," similar conceptually to the *bands* in a chromatographic column separation. These zones migrate at an apparent constant rate, which is the sum of their electrophoretic mobility and any electroosmotic mobility, which may be induced by the particular buffer pH and its electrolyte concentration. The zones are separated by spaces of buffer. With separation buffer flow toward the cathode, cation zones will lead a wide zone (of the width of the injected sample plug) of uncharged molecules moving at the EOF velocity, followed by anion zones (if $\mu_{ep} < \mu_{eo}$ for all anions), hence the name CZE.

CZE is useful for a wide range of ion sizes and is particularly useful for small volumes of small organic and inorganic ions. Its inherent smaller sample size requirements (nL instead of μL) make it particularly useful for sampling small volumes, even ones as small as the fluids inside a single biological cell. In its simplest form, the equipment is much less expensive than comparable IC or HPLC instruments. The interaction of so many variables such as pH, electrolyte concentration, and adsorption effects, on both the electrophoretic mobility and the electroosmotic mobility makes the development and optimization of a CZE method for a particular separation less intuitive than for IC or HPLC. Once the CZE procedure is refined, it may be more cost effective to run than the corresponding chromatographic technique. As we shall see, it is even possible to separate and quantitate electrically neutral species using a CZE apparatus by combining it with a chromatographic partitioning process.

14.6.4.2 Capillary Gel Electrophoresis

As will be discussed in Section 14.7, the initial application of electrophoresis was to large biological molecule separations on a planar, unconfined surface. The separation buffer was applied to a solid planar substrate (a glass plate) mixed into a porous hydrophilic gel matrix. This mixture retards dispersion of the analyte by diffusion and convection, and such a cast **slab gel** is easier to manipulate in the instrument and to place its opposite edges into the electrode reservoirs. The gels are commonly produced by polymerizing acrylamide ($CH_2=CH–CO–NH_2$) in the presence of a cross-linking agent or using other hydrophilic polymers such as agarose or PEG. Increasing the degree of cross-linking reduces the size of the pores in the gel through which the buffer solution diffuses and the charged macromolecules and electrolyte migrate. It was found that the separation mechanism for macromolecules in such gels is actually a combination of differing migration velocities due to the charge and hydrodynamic radius of the ions and a size discrimination effect due to the **sieving action** of the pores on the ions. This sieving action is more effective than the differences in electrophoretic mobility in achieving separations of macromolecules such as oligonucleotides, DNA fragments, and proteins having similar total charges. Such gels can be formed within capillaries to confer the same advantages for certain CZE separations. This modification of CZE is called **CGE**. If the gel is cross-linked in the fashion employed for 2D slab gels, the capillary gel can be used only until clogging or other problems render it unusable, as it is nearly impossible to remove such a so-called **chemical gel** from the capillary by flushing. If the capillary is filled with a **physical gel**, formed by filling it with long linear polymer chains that produce pores by entangling randomly, this may be easily flushed and replaced by fresh physical gel material as needed.

The use of CGE separations with multiple reagent, precolumn derivatization of end nucleotides on DNA segments, and laser-induced fluorescence measurement of the four possible end nucleotides was described in Section 14.1.4.4. Figure 14.40 displays a CGE electropherogram of such a separation of a mixture of DNA fragments ranging from 3 to over 1000 nucleotides in length. These come off the column in order of their length, so the listing of each sequential end nucleotide (represented by A, G, C, T above each peak) directly yields the sequence of the nucleotides (also designated "bases") in the fragment. Automated instrumentation performs such measurements on many capillaries operating in parallel, with fluorescence excitation and emission being rapidly repeatedly scanned across the detection regions of the bank of capillaries by a single fluorescence detector spectrometer. DADs can be used to acquire multiple channels of spectral data simultaneously. The use of automatically refilled physical gels between runs enables the system to cycle repeatedly through electrokinetic injections of banks of parallel sample vials or samples in small-volume well plates, without operator intervention. The ability of large numbers of such highly automated instruments to process millions of very small-volume samples of short-length DNA segments is what made projects such as the complete assembly of the 3 billion base long human genome sequence possible. Similar massively parallel automated instrumentation for sequencing protein chains forms an important resource supporting the discipline of proteomics. As mentioned at the end of Section 14.1.6.1 in the discussion of LC-ESI-MS interfacing, these earlier-generation HPLC/CE–UV fluorescence or MS or MS/MS sequencing technologies are being supplanted by even more powerful and faster, massively parallel, chip-based instrumentation. These are so specialized that a detailed description of their operation is beyond the scope of this text, and they are likely to be supplanted by something even better in a few years. As Moore's law had the number of circuit elements on electronic microchips doubling every 18 months, so application of chip technologies to develop massively parallel analytical instruments for characterizing the vast arrays of molecules comprising the various "-omics" disciplines will engender a vicious cycle of technological obsolescence.

14.6.4.3 Capillary Isoelectric Focusing

Proteins, peptides, and their component amino acids are categorized as **amphoteric** compounds, namely, they contain both acidic (–COOH) and basic (–NH_2) functionalities in the same molecule. As the pH of their surrounding solution varies from higher and lower values, they can lose protons from the carboxylic acid groups and gain protons on the amino groups. Their net charge will be negative (excess –COO^-) on the high pH (basic) side, and positive (excess –NH_3^+) on the low pH (acidic) side, of a compound-specific **isoelectric point** (pH value), at which the compound will exist as a zwitterion with equal amounts of positive and negative charges and hence no overall net charges. Taking the simplest amino acid, glycine, as an example, we have the equilibria shown in the following:

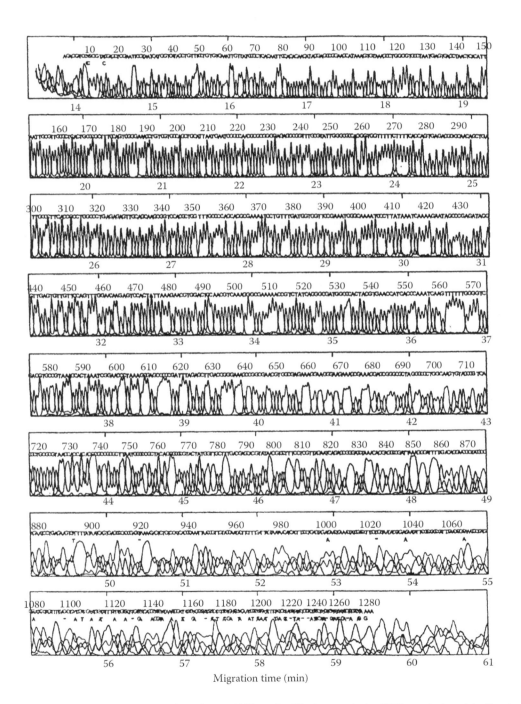

Figure 14.40 Separation and DNA base sequencing of over 1000 nucleotide fragments by CGE employing detection by fluorescence derivatization. (From Cazes, J., *Encyclopedia of Chromatography*, Marcel Dekker, Inc., New York, 2001. With permission.)

If we perform electrophoresis on such a compound, at pH below the isoelectric point, it will migrate toward the cathode, and at pH above that point, it will migrate toward the anode, and if it is in a buffer whose pH matches its isoelectric point value, it will be net neutral and undergo no migration at all. If we were to place a band or zone containing a mixture of such compounds in a capillary containing a separation buffer whose pH varied with a constant linear gradient along the length of the capillary, each amphoteric compound would migrate in one direction or the other, depending on whether the band started at a point whose pH was above or below the compound's isoelectric point. As it moved up or down the gradient, its excess charge would progressively become neutralized and its migration rate would slow until it would stop completely when it reached the point in the pH gradient that matched its isoelectric point. The mixture is separated into very sharply resolved motionless bands whose relative positions correspond to their isoelectric point values, which are characteristic for each compound, and may help serve to identify it. The process should be operated without the disruptive presence of EOF, so the capillaries are generally coated with an adsorbed organic compound that masks the surface siloxyl groups. This separation is called **capillary IEF (CIEF)**.

Two big questions remain: How do we establish and maintain a pH gradient across the column while the amphoteric analytes are migrating to their focused halt points, and how then do we move the separated and focused bands past a detector while maintaining the separation that was achieved in the first step?

To form the gradient, we employ a separating buffer that contains several other amphoteric compounds (called ampholytes). These **ampholyte** mixtures are composed of compounds whose isoelectric points cover the range of the target analytes. The cathodic end of the capillary is dipped in a reservoir solution of strong base such as NaOH, and the anodic end into a reservoir containing a strong acid. Under applied electrophoretic potential, H^+ from the anode migrates toward the cathode and OH^- from the cathode toward the anode. The buffer ampholytes will migrate toward the electrode of charge opposite their excess charge, which will be progressively neutralized by the strong acid or base ions migrating from the reservoirs in the opposite direction. The net effect of all these countering migrations and neutralizations is the establishment of an equilibrium state in which there is a uniform pH gradient along the column between the pH values of the highest and lowest isoelectric points of the separation buffer ampholytes. Note that when the migrating ampholyte of highest or lowest isoelectric point reaches that pH value, it will stop migrating, thus setting the bounds of the gradient.

Once all the amphoteric analyte bands have completed migration to and have been focused at their isoelectric points along the gradient, all net motion stops. The pattern of focused bands can then be **mobilized** along the column so that they pass the detector to be identified by their isoelectric point values and quantitated. This can be done in a straightforward manner by applying a pressure difference across the column, using the same mechanism that brings a sample plug into the column by hydrodynamic injection. Pressure-driven laminar flow profiles may cause some loss of the high resolution achieved by the isoelectric focusing step. Alternatively, either the H^+ or OH^- solutions in one of the reservoirs may be diluted with equivalent concentrations of another ion of the same charge. This will change the pH gradient, as there is a lesser concentration of one strong acid or base to neutralize the other. The gradient moves toward the electrode reservoir in which the acid–base dilution (or ion replacement) took place, and the focused bands move with it as they follow the shift in position of their isoelectric point along the length of the column as the gradient shifts.

14.6.5 Capillary Electrochromatography

One shortcoming of the very "simple" technique of CZE is that it acts only to separate ions. While it separates ions from any neutrals left behind in the sample plug, these neutrals cannot be separated from one another. A crude way of achieving this would be to introduce a chromatographic separation mechanism by packing the capillary with microparticles of RP-HPLC packing while retaining the liquid mobile-phase flow produced by a separation buffer mobilized by EOF. This becomes simply HPLC with mobile-phase flow induced by EOF instead of high-pressure pumps. The multipath A term returns to the Van Deemter equation, as well as the mass-transfer C terms. Many of the resolution enhancements seen for CZE of ions, especially for biological macromolecules, are lost. The opportunity exists to achieve both neutral separations based on chromatographic partitioning to a stationary phase, as well as ion separations based on differences in electrophoretic mobility, in the same sample mixture. The number of variables to be managed becomes even more daunting than in either HPLC or CZE alone.

14.6.5.1 Micellar Electrokinetic Capillary Chromatography

Instead of a packing composed of an organic coating on a spherical solid support acting as the stationary phase of the chromatographic component of a capillary electrochromatography (CEC) system, we may avail ourselves of something that performs the same function without blocking buffer/mobile-phase flow as dramatically as a stationary HPLC particle packing. Recall that we employed a cationic surfactant bilayer to reverse EOF, as illustrated in Figure 13.48. If we prepare a buffer solution of an anionic surfactant such as **sodium dodecylsulfate (SDS)**

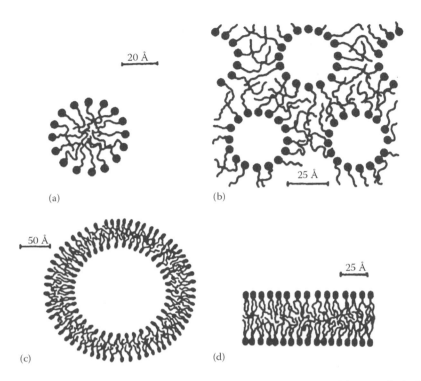

Figure 14.41 Formation of micelles, vesicles, and bilayers by surfactant molecules: (a) micelle, (b) inverted micelle, (c) vesicle, and (d) flat bilayer. (From Cazes, J., *Encyclopedia of Chromatography*, Marcel Dekker, Inc., New York, 2001. With permission.)

$(n-C_{12}H_{25}OSO_3^-Na^+)$, we can form a variety of structures, which are illustrated in Figure 14.41. In Figure 14.41a, the hydrocarbon tails coalesce together to form a small spherical **micelle**, with a charged polar exterior and an hydrophobic interior. This structure is small enough to remain in colloidal suspension in the buffer solution, and it acts in an electrophoretic system like a very large, multiply charged macromolecule. As the concentration of the surfactant in the buffer increases, it reaches a **critical micelle concentration (CMC)**, at which an equilibrium forms between a *constant* surfactant concentration and a rapidly increasing micelle concentration. If the total amount of surfactant becomes too high, an **inverted micelle** may form, with the aqueous buffer solution trapped in a multitude of spherical inverted micelles (Figure 14.41b). This would be unsuitable for electrophoresis. A spherical bilayer may form (Figure 14.41c) comprising a surface whose inside is nonpolar organic and surrounds a pocket of the aqueous buffer. This structure is called a **bilayer vesicle**. Finally, an extended **planar bilayer** (Figure 14.41d) may form, especially if there is a suitable surface to which it may adhere, as was the case for cationic surfactants used to reverse EOF.

Figure 14.42 illustrates the behavior of the anionic surfactant micelles formed in a standard CE experiment. The surfactant being anionic, it does not form a planar bilayer on the capillary wall as would a cationic surfactant. The micelles have a net negative charge and migrate slowly to

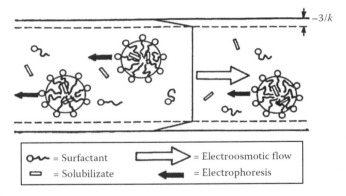

Figure 14.42 Mechanism of micellar electrokinetic chromatography. (From Cazes, J., *Encyclopedia of Chromatography*, Marcel Dekker, Inc., New York, 2001. With permission.)

the left (anode). This migration speed is slower than the induced EOF toward the cathode, so the micelles will be carried toward the detector, but at a net speed less than that of the sample plug where neutrals would ordinarily remain unseparated. The neutrals (represented as "**solubilizate**" in the figure) can partition between the polar separation buffer and the less-polar interior of the micelles, as in RP-HPLC. They will move away from the sample plug in a direction opposite the flow by that percentage of the distance to the retarded back-migrating micelles, which is simply the percentage of the time they spend in the micelle. This variable retention in elution time relative to the speed at which the buffer is moved through the capillary by the EOF results from the same mechanism that causes variable peak retention in chromatography. The micelles are acting as a "pseudostationary phase" for separating the neutrals based on differences in their partition coefficient between the moving buffer and the differently moving micelle interiors. A chromatographic process has been overlaid on the electrophoretic separation of any analyte ions present. Thus, the name of the procedure emphasizes that it is **micellar electrokinetic chromatography** (**MEKC**). The focus of the technique is on the introduction of a chromatographic separation of neutrals in a CE system. The ions themselves may interact electrostatically to some extent with the charged micelle surface, so their migration times are altered as well. The imposition of a true chromatographic process means that the mass-transfer term C in the Van Deemter equation reappears, but mass transfer in and out of micelles is so efficient that this adds little to band broadening.

Surfactants other than SDS, which have different CMCs and form different-size micelles with different interior polarity, can be used. If a cationic surfactant such as cetyltrimethylammonium bromide is used, the direction of the EOF will be reversed by the formation of a capillary surface bilayer, as illustrated in Figure 14.38. The charge on the micelle surface is also opposite that of an anionic surfactant micelle, so again the neutrals will lag the solvent plug as they are separated chromatographically by partitioning with the pseudostationary phase. The term **solubilize** for the separated analytes comes from the original use of surfactant detergents to bring nonpolar greasy material on surfaces into polar aqueous solutions in washing processes. Introduction of cyclodextrins (cf. Figure 14.3) to an MEKC separation can alter the times that enantiomers of optically active compounds spend in each phase and effect chiral separations.

14.7 PLANAR CHROMATOGRAPHY AND PLANAR ELECTROPHORESIS

If the stationary phase for a chromatographic separation, or a gel for an electrophoretic separation, is bound to a surface, it is possible to spot a sample on to one corner of such a flat plate or slab and perform a separation directed along one edge, spreading out the bands separated by chromatographic partitioning or electrophoretic migration along that edge (without eluting them off the surface or through a detector). The linear array of separated bands can then be subject to another separation process using different chromatographic mobile phase or electrophoretic buffer parameters directed at right angles to the first separation. Bands that remained unresolved in the first step may be further separated in this new direction, and the final product is a surface covered with many separated spots, which may be scanned by some detector or reacted to visualize the separated components. These basic principles were outlined in Section 12.3.

14.7.1 Thin-Layer Chromatography

14.7.1.1 Parallel 1D TLC Separations

In TLC, particles of stationary phase (e.g., silica or alumina for a NP separation or large diameter particle [10–100 µm] RP-HPLC type packing) are coated and bound as a thin layer to a square planar supporting **plate**, usually of polymer (older plates were made from glass). Alumina stationary phases can be modified by the careful addition of water on a mass percent basis to alter the separation characteristics. Modern TLC is now called high-performance TLC, HP-TLC, similar to the nomenclature of liquid chromatography. HP-TLC plates are typically smaller than the older TLC plates, and the bound particle sizes are smaller (5–6 µm) and the coatings are thinner (0.1–0.2 mm). Chiral plates are available that are composed of an RP layer coated with copper acetate and a chiral selector molecule. Such plates can be used to resolve enantiomer bases on ligand exchange mechanisms.

Sample application to the plate is a critical stem in achieving satisfactory results. Sample can be applied as a spot, or a rectangular band. Application must be in a precise initial zone, and must be done without damage (or contact) to the plate surface. Application of round spots (2–4 mm in diameter) facilitates multiple sample and standard solution applications on a given plate. Application of samples as bands facilitates high resolution and enhanced detection. Narrow repeatable bands are best applied with a device such as the Linomat 5, shown in Figure 14.43. This device sprays the sample into a precise starting band.

A liquid mobile phase is placed in the bottom of a TLC **tank** or **chamber**, a rectangular glass tank with a tightly sealing glass plate top and with dimensions that will contain the TLC plate when it is propped up in the tank at an angle inclined a little from the vertical. The tank contains an internal lip above the solvent reservoir, so that initially, the plate can be set on it out of contact with the liquid to equilibrate with the solvent vapor. The mixtures for separation are **spotted** on the plate along an origin line parallel to one edge, at a distance sufficient for the spots to lie above the level of the mobile-phase solvent when the edge with the sample spots is lowered into the reservoir. This can be done crudely by using a glass capillary to pick up the sample and transfer it to the plate. For quantitative work, a measured volume is slowly put on the plate using a microliter syringe, allowing time for solvent to evaporate during the application in order to minimize spot spreading in the thin layer. Figure 14.43 displays a side view of a TLC plate in a sealed tank. Solvent vapors from the liquid in the bottom

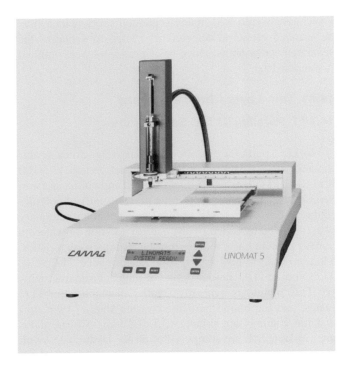

Figure 14.43 Linomat 5 instrument for the application of repeatable bands of sample on UHTLC plates. Courtesy of Camag

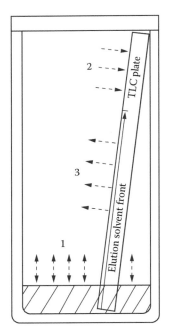

Figure 14.44 TLC plate in enclosed developing tank.

equilibrate with the headspace above it (Figure 14.43[1]). Capillary attraction to the closely spaced stationary-phase particles comprising the thin layer causes the mobile-phase solvent to move up the plate. The leading edge of this flow is called the solvent front. Above the solvent front, the thin layer remains conditioned by solvent vapor diffusing into it (Figure 14.44[2]). There will be no net evaporation from the solvent flowing up the plate (Figure 14.44[3]), because the closed system establishes an equilibrium vapor pressure. Control of these variables ensures reproducible mobile-phase flow in the system.

Multiple spots (e.g., standard mixtures, a blank, and several unknown samples) can be chromatographed in parallel. This facilitates direct comparisons of retention relative to the advance of the solvent front and comparison of separated component spot densities for estimating relative quantities. The same separation principles apply as in HPLC column chromatography. Analytes must not have significant volatility, as there is no column to contain them. Figure 14.45 displays three stages in the process of obtaining a TLC chromatogram. The left-hand-side view shows the initial, somewhat ragged line of spotted samples. If a strong elution solvent is used, initial ascent of its solvent front through the sample spots moves them to a line precisely at the solvent front and compacts the spots into narrow elongated bands. This is called **beginning development** (center panel). As **development** continues, with either the beginning solvent or a weaker one that can discriminate better among the analytes, the bands separate until the solvent front has run up most of the plate (right-hand-side panel). This is the TLC chromatogram. The various distances of the separated bands between the start point origin and the solvent front (which the analyst can see and mark when removing the plate before drying it to remove solvent) and the bandwidths are measured. The R_f values (corresponding to relative retention time [RRT] in column chromatography) and the theoretical plate number, N, are calculated as outlined in Figure 14.46. The book by Bruno and Svoronos provides a comprehensive guide to the selection of stationary phases and solvent systems for many chemical families.

14.7.1.2 Detection in TLC

The separated spots or bands may be detected in a variety of ways. In contrast to column chromatography, they remain on the stationary phase and are not eluted off it into a separate detection region. Detection is by scanning or imaging the "developed" plate. It is critical to understand that in the TLC context, development refers to the chromatographic separation process and not the subsequent process of making the separated spots visible, as is done when one "develops" a latent image on a planar photographic film. If the spots are colored or opaque, imaging may be as simple as looking at them by eye. If the analytes are colorless, but are fluorescent under UV illumination, the dried, developed plate may be placed in a darkened box, illuminated with UV, and the fluorescent spots imaged by eye. This can be done more precisely by using a **plate position scanning** (as opposed to wavelength scanning)

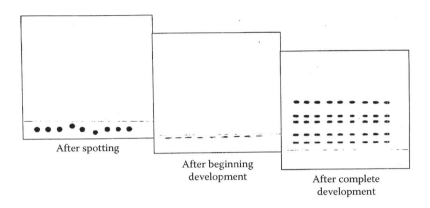

Figure 14.45 Steps in development of a TLC chromatogram.

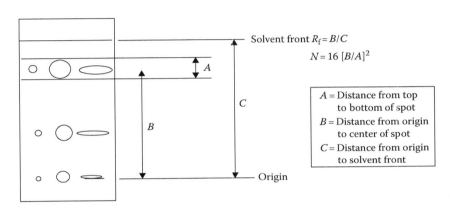

Figure 14.46 Definition and calculation of retention factor (R_f) and plate number (N) from a developed TLC chromatogram.

fluorescence spectrometer. Fluorescence can be used as an **indirect detection method** if the TLC plate contains an "**F layer**" incorporating a bound phosphor or fluorescent indicator such as zinc sulfide. Analyte spots containing compounds (such as most organic aromatic compounds and those with conjugated double bonds), which are colorless in the visible range but which absorb Hg-lamp UV at around 254 nm, will absorb the UV and **quench** the fluorescence of the plate at the position of their developed spots.

Some reagents or reactions will visualize almost all separated analytes. Examples of such **universal detection reagents** are as follows: (1) iodine vapor, which will reversibly stain almost all analyte spots except some saturated alkanes, forming a dark brown color, and (2) irreversible charring with heating in the presence of sulfuric acid or a mixture of 8 % phosphoric acid and 3 % copper sulfate. At the other end of the reagent spectrum are **selective detection reagents**, such as (1) ninhydrin for α-amino acids; (2) 2,4-dinitrophenyl-hydrazone HCl to produce orange zones by reaction with aldehydes; and (3) enzyme reactions with a substrate to form a colored product to generate uniform color over the plate except on bands containing biologically active compounds that act to block the enzyme function (e.g., cholinesterase inhibitors such as organophosphorus pesticides). These reagents are generally applied to the plates in one of two fashions: (1) **dipping** them for a set period of time in a solution of the reagent (special dipping chambers can automatically perform a reproducibly timed immersion and gentle agitation of the plate; dipping yields the best reactions for *quantitative* purposes) and (2) **spraying** solutions of the reagents as an aerosol mist from a handheld atomizer. This must often be done in a contained volume TLC spray chamber vented to a fume hood, in order to avoid exposure to hazardous solvent vapors. A comprehensive listing of spray reagents can be found in the book by Bruno and Svoronos. Intermittent spraying with intervals for drying will minimize band spreading due to mobilization and expansion of the spots from accumulated spray solvent. TLC plates and slab gels can now be sampled and interfaced to MS by desorption electrospray ionization (DESI) or laser ablation electrospray ionization (LAESI) (see end of Section 10.2.2.4), and other devices are marketed to dissolve analytes from locations on these 2D media and elute them to an MS ion source using a variety of LC-MS interfaces. All these move the plates in an X–Y plane to

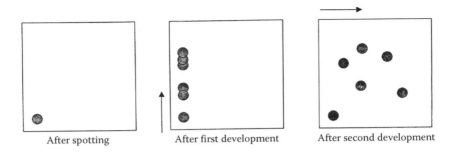

Figure 14.47 Diagram of development of a 2D TLC chromatogram.

locate selected spots or even to scan the whole plate in a raster pattern.

TLC separations rarely approach the resolution or quantitation accuracy of chromatographic instruments employing the sequence of injection, separation on a confined capillary column, and separate detection in a specially designed detector cell. For applications that do not require such precision, TLC's simplicity and the low cost of materials and equipment recommend it, especially for initial screening applications.

14.7.1.3 2D Separations Using TLC (2D-TLC)

Additional resolution of components that are incompletely separated by 1D development may be obtained by rotating the plate 90° and redeveloping the single line of partially separated components of a single mixture sample originally spotted in one corner. For the second development, a different mobile-phase solution having different selectivities for the partially separated compounds is used. This is illustrated in Figure 14.47. Only one sample can be separated on each plate. This 2D-TLC analysis is an example of a 2D separation employing a 2D stationary (i.e., planar) phase. This relatively simple trick can be contrasted with the extremely complex 2D column chromatographic separations employing modulator devices interfacing columns of two different selectivities and analysis speeds (e.g., Section 13.8.3).

14.7.2 Planar Electrophoresis on Slab Gels

14.7.2.1 1D Planar Gel Electrophoresis

Prior to the development of CZE, the primary application of electrophoresis was for large biological macromolecule separations on planar gels. The gels were supported on glass or polycarbonate plates not unlike the particles in the thin layer of a TLC plate. A schematic diagram of such an apparatus is shown in Figure 14.48. These so-called slab gels were needed to confine the separation buffer. It was easier to prepare cross-linked chemical gels on the open surface than in narrow capillaries and to remove the gel from the support plate to prepare a fresh slab gel for a new analysis. Electrophoretic separations of multiple samples in 1D could be performed and easily visualized and compared. This was the standard tool for biochemists to compare distributions of peptides from enzymatic protein digests or DNA or ribonucleic acid (RNA) fragments. Comparison of multiple lanes of 1D separations enabled easy illustration and identification of abnormal distributions or the detection of the presence of unusual fragments, which might, for example, be indicators of a mutant form or a disease state. Such patterns looked similar to the diagram of a developed TLC plate shown in the right-hand panel of Figure 14.44. Thousands of photos of such multilane slab gel electropherograms grace the pages of the molecular biology literature.

14.7.2.2 2D Planar Gel Electrophoresis

Many DNA, RNA, or protein digests can contain hundreds or even over a thousand fragments, and extracts of particular tissues or cell types may contain similar numbers of proteins. 1D planar gel electrophoresis could not possibly resolve all of these. The same principle used with 2D-TLC was applied. There are many possible combinations of separation conditions that might be applied to sequential developments of a sample on a planar gel, to produce an array of hundreds of resolved spots. One particular sequence is a standard for **protein peptide mapping**. This procedure is so well standardized, and the reagents and instrumentation so widely commercialized that such "maps" of the distribution of peptides as a particular array of separated spots on a 2D-gel are archived in international computer-searchable databases, enabling researchers to submit new patterns that they have characterized or to determine if a new pattern has already been described by others. This is analogous to the huge databases of full-scan mass spectra that help to identify unknown spectra by library searching. Similar libraries of the distributions of cell proteins are compiled by very similar procedures.

The most common standard protein or peptide mapping experiment employs first a slab gel electrophoretic

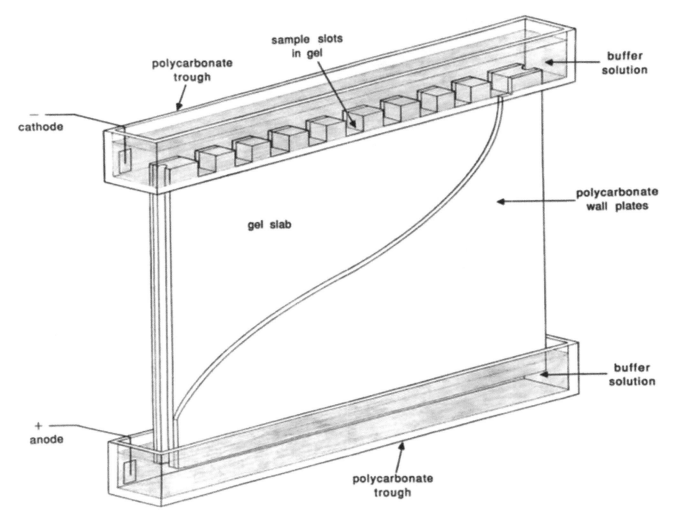

Figure 14.48 A schematic diagram of a slab-type gel apparatus showing the buffer troughs, the polyacrylamide slab gel, and the combs for the application of sample. Taken from the book by Bruno.

separation employing **isoelectric focusing** (IEF) to spread the amphoteric analytes across one axis of the plate in order of their **isoelectric point pH** (pI) values (cf. Section 14.6.4.3). These partially separated components are then subjected to **sodium-dodecylsulfate–polyacrylamide gel electrophoresis** (SDS–PAGE) separation in a direction at right angles to the initial IEF separation. The proteins are **denatured** (i.e., their specific tertiary folded structure is disrupted by elevated temperature or high ionic strength solution), and the micelle-forming SDS (at a concentration well below the CMC) will bind an amount of anionic SDS proportional to the size of a denatured protein. The resulting protein–SDS complexes have approximately constant charge-to-mass ratios and therefore identical electrophoretic mobilities. Therefore, in the sieving gel medium, the complexes migrate proportionally to their effective molecular radii and likewise to the protein RMM. Thus, the 2D planar electrophoresis separation initially sorts the proteins (or peptides) by a charge-related parameter of their isoelectric points and secondly by the size-related parameter of the SDS–PAGE sieving process.

The 2D-SDS–PAGE separation described results in a planar array of separated compounds. If these are unknown proteins (perhaps just one or a few that interrupt a standard pattern and may signal a mutant form or an abnormal biological state), the spots of interest may be cut out or dissolved out of the gel. The isolated spot can be subjected to tryptic digest, the peptide digest mixture separated by capillary HPLC, and the separated peaks characterized and, if necessary, sequenced by ESI-MS and ESI-MS-MS, as was illustrated in Figures 14.21 and 14.22. If the 2D separation is of peptides from a single protein in a peptide mapping experiment, the developed slab gel may be put on an automatically positioned stage that will place each spot at the laser target point of a matrix-assisted laser-desorption ionization (MALDI)-time of flight

(TOF)-MS (cf. Section 11.3.3), which may obtain enough information on each peptide to identify it based on archival library lookup and matching. The plate positioning engine and the MALDI-TOF operate so quickly to acquire data that for gel plates containing very large numbers of closely resolved components, it may be more efficient to sample the entire plate surface at points in a closely spaced grid, store all the data in a computer, and have it automatically processed. Pause for a moment to contemplate and marvel at the number of powerful separation and spectral characterization techniques that are being interfaced, automated, and having their output processed by complex computer algorithms in such an instrument. This melding of many instrumental techniques results in systems being successfully marketed today to support the enormous demands for analytical information in the disciplines of genomics, proteomics, and a burgeoning list of other "-omics."

PROBLEMS AND EXERCISES

14.1 Compare packed column LC, capillary column LC, and CZE (as best, intermediate, poorest, or inapplicable for the following categories):
  (a) Analyte capacity.
  (b) The highest number of theoretical plates for large molecule separations.
  (c) Suitability for separating charge neutral compounds.
  (d) Ability to separate anions and cations in a single run.
  (e) Suitability for analyzing 50 μL sample of mix of low RMM alcohols.
  (f) Suitability for analyzing 10 μL sample of mix of low RMM alcohols.
  (g) Suitability for analyzing 10 μL sample of peptides.

14.2 (a) List advantages and disadvantages of the presence of free silanol groups on silica surfaces for applications in liquid-phase separations.
  (b) How may the disadvantages be suppressed?
  (c) How may the advantages be enhanced?

14.3 Describe for the following HPLC procedures. (1) What is meant by these? (2) What is the purpose of these? (3) How is it done?:
  (a) **Degassing** the mobile phase.
  (b) **Filtering** the mobile phase.
  (c) **Modifying** the mobile phase.
  (d) **Endcapping** the stationary phase.
  (e) **Pulse-dampening** the mobile phase.
  (f) Introducing a **guard column.**
  (g) Introducing a **suppressor column.**

14.4 Describe three ways in which an HPLC support particle differs from one used in packed column GC.

14.5 Describe the primary purpose and advantages of using the following columns in HPLC:
  (a) Silica spheres with bonded organic substituents.
  (b) Monolithic columns.
  (c) Resin polymer spheres.
  (d) Polyacrylamide gels of varying pore size.
  (e) Zirconia spheres with bonded organic substituents.
  (f) Bare silica spheres.
  (g) Resin spheres with attached –COOH groups.
  (h) Silica spheres with attached latex coating with bonded –SO_3H groups.
  (i) Resin spheres with bonded –$N(CH_3)_3$ + groups.
  (j) Completely endcapped –C18 bonded silica.

14.6 Answer questions (14.1) and (14.2) for the following analyte mixtures:
  (a) A mixture of styrene–vinylbenzene polymer chain lengths.
  (b) Organophosphorus pesticides in urine.
  (c) PAHs in hexane.
  (d) REE ions in acidic aqueous solution.
  (e) Peptides from a tryptic digest.
  (f) A partially racemized, optically active, drug molecule containing phenolic groups.
  (g) Isolation of a single antibody protein from a preparation of disrupted cells.
  (h) A mixture of hydrocarbon ethers.
  (i) A mixture of polysaccharide polymers (e.g., starches or sugars).
  (j) A mixture of inorganic halides and oxyhalides.
  (k) A preparative extract of a mixture of spice essences from leaves.
    (i) Select one or more suitable liquid phase chromatographic or electrophoretic separation procedures from Sections 14.1.1, 14.2, 14.2.1, 14.3, 14.4, 14.1.5, or 14.6.
    (ii) Select one or more suitable detectors from Sections 14.1.4 and 14.1.6 and explain why.

14.7 Draw diagrams of the structures at the atomic level of the following silica surfaces:
  (a) Silica in aqueous solution at pH 1.
  (b) Silica in aqueous solution at pH 9.
  (c) An ODS silica, bonded HPLC phase (indicate no. of C atoms n, by Cn).
  (d) An endcapped, octyl-bonded silica HPLC phase.
  (e) A silica surface with an additive that reverses EOF.
  (f) A silica surface treated to suppress EOF.

14.8 Using the style of the diagram of the six-port rotary HPLC injector valve (Figure 14.6), draw two diagrams showing the flow through the injection loop, the valve, and the column during the load and injection positions.

14.9 Using the information in Sections 14.1.4 and 14.1.6, construct a table contrasting the properties of various LC and CE detectors, indicating relative sensitivity, compound selectivity, gradient compatibility, buffer salt and pH compatibility, ability to extract analyte structural information, and mass or concentration mode of detection.

14.10 Contrast guard and suppressor columns with respect to their purpose, their position relative to the separation column, and the reason that pellicular packings are often desirable for use in each.

14.11 For an NP or reversed-phase (RP) separation, predict the elution order of:
  (a) *n*-Octane, *n*-octanol, naphthalene.
  (b) Ethyl acetate, diethyl ether, nitrobutane.

14.12 In HPLC
  (a) Why is high pressure required?
  (b) Why are stationary phases bonded?
  (c) Why are supports endcapped?
  (d) Why are the buffers used usually at pH between 2 and 8?

14.13 In GC, a plot of the Van Deemter equation is shaped like a severely distorted U. In HPLC using 3 μm particles, its shape is more like the letter "L." Explain.

14.14 Describe three different modes of use for SDS to change and improve separations in chromatography and electrophoresis.

14.15 In CE, what is the effect on the EOF of:
  (a) Increasing buffer pH.
  (b) Increasing buffer concentration.
  (c) Increasing the applied voltage.
  (d) Increasing the capillary length for a constant applied voltage.
  (e) Endcapping silanol groups on the silica capillary surface.

14.16 In IC:
  (a) How does one increase the exchange capacity of the stationary phase?
  (b) Is the exchange group charge on the suppressor column the same or opposite that on the separating column? Explain why.
  (c) Why and how does one regenerate the separator column?
  (d) Why and how does one regenerate the suppressor column?
  (e) Describe how to regenerate the suppressor continuously.

14.17 Describe the instrumentation and separation media you would choose for:
  (a) Separation of amino acids by:
    (i) Ion-exchange chromatography.
    (ii) RP-HPLC of derivatives.
    (iii) CIEF.
  (b) Separation of proteins by:
    (i) Size exclusion chromatography (SEC).
    (ii) CGE.
    (iii) 2D-IEF-SDS–PAGE.

14.18 In Figure 14.21c, the RMM of the 11 amino acid peptide is given as 1453.8, but the M^+ ion is listed as 1454.8. Explain why (hint: this is an MS, not an LC question).

14.19 What is the function of:
  (a) Cross-linked polyacrylamide in SEC?
  (b) Cross-linked polyacrylamide in SDS–PAGE planar electrophoresis?
  (c) Uncross-linked polyacrylamide in rapid DNA sequencing by CGE?
  (d) Styrene–divinylbenzene copolymer resin in ion-exchange chromatography?

14.20 In what ways is SFC like LC? Like GC? What are several advantages of the use of CO_2 as eluent in SFC?

14.21 (a) How does the LC-APCI-MS interface differ from GC-CI-MS?
  (b) In an orthogonal ESI-MS interface, where does most of the LC mobile phase go?
  (c) Which LC-MS interface of (a) and (b) above works best for:
    (i) Small molecules.
    (ii) Large molecules.
    (iii) Charged molecules (i.e., ions in solution).
    (iv) Neutral (uncharged) molecules.

14.22 Sample injection in CZE is by either hydrodynamic or electrokinetic injection:
  (a) Which is better for CGE with physical gels?
  (b) Which is better for CGE with chemical gels?
  (c) How should an injection buffer differ from a separation buffer in order to achieve sample stacking? What is this, how does it work, and what are its benefits? Explain.

14.23 Explain how micellar electrokinetic chromatography combines electrophoretic and chromatographic separation principles. Why is the MEKC micelle called a "pseudostationary phase"?

14.24 If cetyltrimethylammonium ion formed micelles, and one employed it for MEKC, in which direction (toward the cathode or anode?) would:
  (a) The EOF move?
  (b) The cations move?
  (c) The anions move?
  (d) Neutrals move away from their position in the initial injection plug?

14.25 Consider the planar 2D separation illustrated in the last panel of Figure 14.42:
  (a) If this were on a TLC plate, how do you think one might identify the separated spots by fast atom bombardment (FAB)-MS? By capillary LC-ESI-MS? Describe a proposed procedure.
  (b) If it were an SDS–PAGE separation on a slab gel, how do you think you might use MALDI-TOF-MS to identify the spots? Describe a proposed procedure.

14.26 The Internet World Wide Web "Literature Research" project. Choose one of the HPLC compound class applications in Section 14.1.7 and use the links in Appendix 14.A to download manufacturer's application notes, sample chromatograms, and suggested products for their analysis. Write a detailed experimental procedure to separate your chosen compounds.

14.A APPENDIX LC/CE/TLC INTERNET WEB RESOURCES

The major vendors of LC, CE, and TLC instrumentation and consumables maintain elaborate websites. One can access these by putting the name of the company into a search engine (e.g., Google) and finding the main home site, then work one's way through the menu tree to find different types of instrumentation, separation media and small accessories (2D plates, separation columns, fittings, injector hardware, detectors, calibration standards, etc.), suggestions for products addressing common needs, and a host of sample chromatograms.

Major instrumentation companies: **LC/CE: Agilent, Shimadzu; LC: PerkinElmer, SGE, Thermo Fisher Scientific; TLC: CAMAG.**

Major separation media and small parts suppliers: **Agilent; CAMAG; PerkinElmer; Phenomenex, Inc.; Restek; SGE; Sigma-Aldrich/Supelco.**

There are many smaller players in this market; some with URLs listed in the figure captions, but the incomplete listing given earlier will give the student a start. There are literally thousands of separation media and application notes and sample chromatograms available online. If one has a specific problem or application, a better alternative may be to find the "Contact Us" e-mail address or phone number for their applications lab staff and pose a specific query. Free trade journals such as American Laboratory (www.americanlaboratory.com) or LC/GC magazine (www.chromatographyonline.com) have annual comprehensive vendor resource issues, indexed by methodology and application. Their content is accessible from their websites.

SPECIAL LC CALCULATION PROGRAMS

www.hplctransfer.com/ An online calculation tool from Thermo Fisher Scientific to enter conditions for an existing isocratic HPLC method and predict elution on a different system (with same mobile- and stationary-phase compositions, but new flows, particle sizes, column dimensions, and pressures, and injection volumes).

www.sigma-aldrich.com/Supelco—CE learning center—CE applications—Column Selection Guide—Literature.

www.molnar-institute.com—**DryLab 4 UHPLC Modeling Software**: An elaborate commercial program, 30 years in development, to help users convert an old HPLC separation to a more efficient UHPLC method. It employs QbD principles to determine the most important factors influencing the quality of a separation and then incorporate maximized robustness in the method to increase reliability and avoid need to revalidate the method. A free demo is available on the site.

TWO LARGE PUBLISHER CHROMATOGRAPHY WEBSITES

www.separationsnow.com—Wiley newsletter: Ads, Webinars, Podcasts (cf. author ESF @ Pittsburgh Conferences interviewing Editors' Award recipients. A mother lode of chromatography information.

www.chromatographyonline.com—Website of LC/GC magazine, covers all types of chromatography, search by keyword, issues/articles archive 1983–present, author–subject index.

BIBLIOGRAPHY

The reader will note that occasionally multiple editions of books are cited as sources or reference works. The reason for this is that sometimes different figures are used in different editions, which may in fact be written by different authors. The sources found in the bibliography will be the most recent versions of a particular work.

In some cases instrument manufacturers and suppliers are cited by providing a URL to the exact internet location. We have endeavored to ensure that these links are functional at the time of publication, but clearly these links change. If a link has become nonfunctional, a good approach is to use your browser's advanced search capability. For example: "company name" AND "instrument type" AND "specific descriptors" is a good place to start.

Anton, K.; Berger, C. (eds.) *Supercritical Fluid Chromatography*. Marcel Dekker, Inc.: New York, 1998.

Ardrey, R.E. *Liquid Chromatography–Mass Spectrometry, An Introduction*. Wiley: New York, 2003.

Bayne, S.; Carlin, M. *Forensic Applications of HPLC*. Taylor & Francis: Boca Raton, FL, 2010.

Berthold, A. et al. *Micellar Liquid Chromatography*. Marcel Dekker, Inc.: New York, 2000.

Bruno, T.J., A supercritical fluid chromatograph for physicochemical studies, *J. Res. Natl. Inst. Stand. Technol*, 94(2), 105–112, 1989.

Bruno, T.J., *Chromatographic and Electrophoretic Methods*, Prentice Hall, Englewood Cliffs, NJ, 1991.

Bruno, T.J., *Handbook for the Analysis and Identification of Alternative Refrigerants*, CRC Press, Boca Raton, FL, 1995.

Bruno, T.J., Ely, J.F., *Supercritical Fluid Technology*, CRC Press, Boca Raton, FL, 1991.

Bruno, T.J., Svoronos, P.D.N., *CRC Handbook of Basic Tables for Chemical Analysis – Data Driven Methods and Interpretation*, 4th Ed., CRC Taylor and Francis, Boca Raton, FL, 2021.

Byrdwell, W.; Holcapek, M.E. *Extreme Chromatography*. AOCS Press, Taylor & Francis, CRC Press: Urbana, IL, 2011.

Camilleri, P. *Capillary Electrophoresis: Theory and Practice*. CRC Press: Boca Raton, FL, 1998.

Cazes, J. *Encyclopedia of Chromatography*. Marcel Dekker Inc.: New York, 2001.

Cohen, S.A.; Shure, M.R. *Multidimensional Liquid Chromatography: Theory and Applications in Industrial Chemistry and the Life Sciences*. Wiley: New York, 2008.

Corradini, D. *Handbook of HPLC*, 2nd edn. Taylor & Francis: Boca Raton, FL, 2010.

Cunico, R.L. et al. *Basic HPLC and CE of Biomolecules*. Bay Analytical Laboratories: Richmond, CA, 1998.

Dong, M.W. *Modern HPLC for Practicing Scientists*. Wiley: New York, 2006.

Fritz, J.S.; Gjerde, D.T. *Ion Chromatography*. Wiley: New York, 2009.

Guillaume, D.; Venthey, J.L.; Smith, R.M.; Cahooter, D. *UHPLC in Life Sciences*. RSC Pub.: Cambridge, U.K., 2012.

Guzman, N.A. (ed.) *Capillary Electrophoresis Technology*. Marcel Dekker, Inc.: New York, 1993.

Hage, D.S.; Cazes, J. *Handbook of Affinity Chromatography*, 2nd edn. Taylor & Francis: Boca Raton, FL, 2005.

Hahn-Deinstrop, E. *Applied Thin-Layer Chromatography—Best Practice and Avoidance of Mistakes*, 2nd edn. Wiley: New York, 2006.

Katz, E. et al. (eds.) *Handbook of HPLC*. Marcel Dekker, Inc.: New York, 1998.

Kromidas, S. *More Practical Problem Solving in HPLC*. Wiley: New York, 2005.

Landers, J.P. *Handbook of Capillary Electrophoresis*. Marcel Dekker, Inc.: New York, 1997.

McMaster, M. *LC/MS: A Practical User's Guide*. Wiley: New York, 2005.

McMaster, M. *HPLC: A Practical User's Guide*, 2nd edn. Wiley: New York, 2006.

Meyer, V.R. *Practical High Performance Liquid Chromatography*, 5th edn. Wiley: New York, 2010.

Niessen, W.M.A. *Liquid Chromatography—Mass Spectrometry*, 3rd edn. Taylor & Francis: Boca Raton, FL, 2006.

NIST Standard Reference Database 23, Version 10, Reference Fluid Thermodynamic and Transport Properties Database, 2018.

Olsen, B.; Pack, B.W. *Hydrophilic Interaction Chromatography: A Guide for Practitioners*. Wiley: New York, 2013.

Pohl, C.A., Ion Chromatography, in ASM Handbook Volume 10, Materials Characterization, ASM International, Materials Park, OH, 2019

Pommerening, K. *Affinity Chromatography; Practical and Theoretical Aspects*. Marcel Dekker, Inc.: New York, 1985.

Scott, R.P.W. *Chromatographic Detectors*. Marcel Dekker, Inc.: New York, 1996.

Sherma, J.; Fried, B. (eds.) *Handbook of Thin Layer Chromatography*. Marcel Dekker, Inc.: New York, 1996.

Snyder, L.R.; Kirkland, J.J.; Dolan, J. *Introduction to Modern Liquid Chromatography*, 3rd edn. Wiley: New York, 2009.

Snyder, R.L. et al. *Practical HPLC Method Development*. Wiley: New York, 1997.

Spangenberg, B.; Poole, C.F.; Wiens, C. *Quantitative Thin-Layer Chromatography*. Springer VCH: Berlin, Germany, 2011.

Srinatava, M. *High Performance Thin-Layer Chromatography*. Springer VCH: Berlin, Germany, 2011.

Strengel, A. et al. *Modern Size-Exclusion Liquid Chromatography: Practice of Gel Permeation and Gel Filtration Chromatography*, 2nd edn. Wiley: New York, 2009.

Unger, K.K.; Tanaka, N.; Machtejevas, E. *Monolithic Silicas in Separation Science Concepts*. Wiley: New York, 2011.

Van Eeckhaut, A.; Michotte, Y. *Chiral Separations in Capillary Electrophoresis*. Taylor & Francis: Boca Raton, FL, 2009.

Wang, P.G.; He, W. *Hydrophilic Interaction Liquid Chromatography (HILIC) and Advanced Applications*. Taylor & Francis: Boca Raton, FL, 2011.

Wu, C.S. (ed.) *Handbook of Size Exclusion Chromatography*. Marcel Dekker, Inc.: New York, 1995.

CHAPTER 15

Electroanalytical Chemistry

Electrochemistry is the area of chemistry that studies the interconversion of chemical energy and electrical energy. Electroanalytical chemistry is the use of electrochemical techniques to characterize a sample. The original analytical applications of electrochemistry, electrogravimetry and polarography, were for the quantitative determination of trace metals in aqueous solutions. The latter method was reliable and sensitive enough to detect concentrations as low as 1 ppm of many metals. Since that time, many different types of electrochemical techniques have evolved, each useful for particular applications in organic, inorganic, and biochemical analyses.

A species that undergoes reduction or oxidation upon application of a voltage or current is known as an **electroactive** species. Electroactive species in general may be solvated or complexed, ions or molecules, in aqueous or nonaqueous solvents and even in films. Electrochemical methods are now used not only for trace metal-ion analyses but also for the analysis of organic compounds, for continuous process analysis, and for studying the chemical reactions within a single living cell. Applications have been developed that are suited for quality control of product streams in industry, *in vivo* monitoring, materials characterization, and pharmaceutical and biochemical studies, to mention a few of the myriad applications. Under normal conditions, concentrations as low as 1 ppm can be determined without much difficulty. By using electrodeposition and then reversing the current or potential, it is possible to extend the sensitivity limits for many electroactive species by three or four orders of magnitude, thus providing a means of analysis at the ppb level.

In practice, electrochemistry not only provides elemental and molecular analysis but also can be used to acquire information about equilibria, kinetics, and reaction mechanisms from research using polarography, amperometry, conductometric analysis, and potentiometry. The analytical calculation is usually based on the determination of current or voltage or on the resistance developed in a cell under conditions such that these are dependent on the concentration of the species under study. Electrochemical measurements are easy to automate because they are electrical signals. The equipment is often far less expensive than spectroscopy instrumentation. Electrochemical techniques are also commonly used as detectors for LC, as discussed in Chapter 14.

15.1 FUNDAMENTALS OF ELECTROCHEMISTRY

Electrochemistry is the study of **reduction–oxidation** reactions (called **redox** reactions) in which electrons are transferred from one reactant to another. A chemical species that loses electrons in a redox reaction is **oxidized**. A species that gains electrons is **reduced**. A species that oxidizes is also called a **reducing agent** because it causes the other species to be reduced; likewise, an **oxidizing agent** is a species that is itself reduced in a reaction. An oxidation–reduction reaction requires that one reactant gain electrons (be reduced) from the reactant that is oxidized. We can write the reduction and the oxidation reactions separately, as half reactions; the sum of the half reactions equals the net oxidation–reduction reaction. Examples of oxidation half reactions include

$$Fe^{2+} \rightarrow Fe^{3+} + e^-$$

$$Cu(s) \rightarrow Cu^{2+} + 2e^-$$

$$AsH_3(g) \rightarrow As(s) + 3H^+ + 3e^-$$

$$H_2C_2O_4 \rightarrow 2CO_2(g) + 2H^+ + 2e^-$$

Examples of reduction half reactions include

$$Co^{3+} + e^- \rightarrow Co^{2+}$$

$$(IO_3)^- + 6H^+ + 5e^- \rightarrow \frac{1}{2}I_2(s) + 3H_2O$$

$$Cl_2(g) + 2e^- \rightarrow 2Cl^-$$

$$Ag^+ + e^- \rightarrow Ag(s)$$

If the direction of an oxidation reaction is reversed, it becomes a reduction reaction; that is, if Al^{3+} accepts three electrons, it is reduced to Al(s). All of the reduction reactions are oxidation reactions if they are written in the opposite direction. Many of these reactions are reversible in practice, as we shall see.

A net oxidation–reduction reaction is the sum of the appropriate reduction and oxidation half reactions. If necessary, the half reactions must be multiplied by a factor so that no electrons appear in the net reaction. For example, the reaction between Cu(s), Cu^{2+}, Ag(s), and Ag^+ is

$$Cu(s) + 2Ag^+ \rightarrow Cu^{2+} + 2Ag(s)$$

We shall see why the reaction proceeds in this direction shortly. The net reaction is obtained from the half reactions as follows:

Oxidation reaction: $Cu(s) \rightarrow Cu^{2+} + 2e^-$

Reduction reaction: $Ag^{2+} + e^- \rightarrow Ag(s)$

Each mole of copper gives up 2 mol of electrons, while each mole of silver ion accepts only 1 mol of electrons. Therefore, the entire reduction reaction must be multiplied by 2, so that there are no electrons in the net reaction after summing the half reactions:

Oxidation reaction: $Cu(s) \rightarrow Cu^{2+} + 2e^-$
Reduction reaction: $2(Ag^{2+} + e^- \rightarrow Ag(s))$
Net reaction: $Cu(s) + 2Ag^+ \rightarrow Cu^{2+} + 2Ag(s)$

The equal numbers of electrons on both sides of the arrow cancel out.

Electrochemical redox reactions can be carried out in an **electrochemical cell** as part of an electrical circuit so that we can measure the electrons transferred, the current, and the voltage. Each of these parameters provides us with information about the redox reaction, so it is important to understand the relationship between charge, voltage, and current. The absolute value of the charge of one electron is 1.602×10^{-19} coulombs (C); this is the fundamental unit of electric charge. Since 1.602×10^{-19} C is the charge of one electron, the charge of 1 mol of electrons is

$$\left(1.602 \times 10^{-19} \frac{C}{e^-}\right)\left(6.022 \times 10^{23} \frac{e^-}{mol}\right)$$
$$= 96{,}485 \frac{C}{mol} \quad (15.1)$$

This value, 96,485 C/mol, is called the Faraday constant (F) and provides the relationship between the total charge, q, transferred in a redox reaction and the number of moles, n, involved in the reaction:

$$q = n \times F \quad (15.2)$$

In an electric circuit, the quantity of charge flowing per second is called the **current**, i. The unit of current is the ampere, A; 1 A equals 1 C/s. The potential difference, E, between two points in the cell is the amount of energy required to move the charged electrons between the two points. If the electrons are attracted from the first point to the second point, the electrons can do work. If the second point repels the electrons, work must be done to force them to move. Work is expressed in joules, J, and the potential difference, E, is measured in volts. The relationship between work and potential difference is

$$w(in\,Joules) = E(in\,volts) \times q(in\,coulombs) \quad (15.3)$$

Since the unit of charge is the coulomb, 1 V equals 1 J/C.

The relationship between current and potential difference in a circuit is expressed by Ohm's law:

$$i = \frac{E}{R} \quad (15.4)$$

where
i is the current
E is the potential difference
R is the resistance in the circuit

The units of resistance are V/A or ohms, Ω.

15.2 ELECTROCHEMICAL CELLS

At the heart of electrochemistry is the electrochemical cell. We will consider the creation of an electrochemical cell from the joining of two half-cells. When an electrical conductor such as a metal strip is immersed in a suitable ionic solution, such as a solution of its own ions, a potential difference (voltage) is created between the conductor and the solution. This system constitutes a half-cell or electrode (Figure 15.1). The metal strip in the solution is called an **electrode** and the ionic solution is called an **electrolyte**. We use the term electrode to mean both the solid electrical conductor in a half-cell (e.g., the metal strip) and the complete half-cell in many cases, for example, the standard hydrogen electrode (SHE) and the calomel electrode. Each half-cell has its own characteristic potential difference or *electrode potential*. The electrode potential measures the ability of the half-cell to do work or the driving force for the half-cell reaction. The reaction between the metal strip and the ionic solution can be represented as

$$M^0 \rightarrow M^{n+} + ne^- \quad (15.5)$$

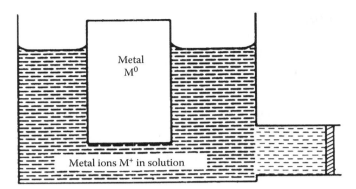

Figure 15.1 A half-cell composed of a metal electrode M^0 in contact with its ions, M^+, in solution. The salt bridge or porous membrane is shown on the lower right side.

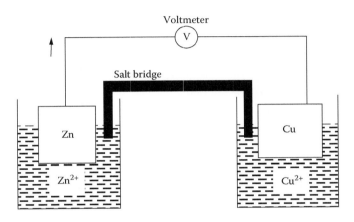

Figure 15.2 A complete Zn/Cu galvanic cell with a salt bridge separating the half-cells.

where
M^0 is an uncharged metal atom
M^{n+} is a positive ion
e^- is an electron

The number of electrons lost by each metal atom is equal to n, where n is a whole number. This is an oxidation reaction, because the metal has lost electrons. It has been oxidized from an uncharged atom to a positively charged ion. In the reaction, the metal ions enter the solution (dissolve). By definition, the electrode at which oxidation occurs is called the **anode**. We say that at the anode, oxidation of the metal occurs according to the reaction shown in Equation 15.5.

Some examples of this type of half-cell are

$$Cd(s) \rightarrow Cd^{2+} + 2e^-$$
$$Ag(s) \rightarrow Ag^+ + e^-$$
$$Cr(s) \rightarrow Cr^{3+} + 3e^-$$

Note that in normal usage, the zero oxidation state of the solid metal is understood, not shown with a zero superscript.

It has been found that with some metals the spontaneous reaction is in the opposite direction and the metal ions tend to become metal atoms, taking up electrons in the process. This reaction can be represented as

$$M^{n+} + ne^- \rightarrow M^0 \quad (15.6)$$

This is a reduction reaction because the positively charged metal ions have gained electrons, lost their charge, and become neutral atoms. The neutral atoms deposit on the electrode, a process called *electrodeposition*. This electrode is termed a **cathode**. At the cathode, reduction of an **electroactive species** takes place. An electroactive species is one that is oxidized or reduced during reaction. Electrochemical cells also contain nonelectroactive (or inert) species such as counterions to balance the charge or electrically conductive electrodes that do not take part in the reaction. Often, these inert electrodes are made of Pt or graphite and serve only to conduct electrons into or out of the half-cell.

It is not possible to measure directly the potential difference of a single half-cell. However, we can join two half-cells to form a complete cell as shown in Figure 15.2. In this example, one half-cell consists of a solid copper electrode immersed in an aqueous solution of $CuSO_4$; the other has a solid zinc electrode immersed in an aqueous solution of $ZnSO_4$. The two half-cell reactions and the net **spontaneous** reaction are shown here:

Anode (oxidation) reaction: $Zn(s) \rightarrow Zn^{2+} + 2e^-$

Cathode (reduction) reaction: $Cu^{2+} + 2e^- \rightarrow Cu(s)$

Net reaction: $Zn(s) + Cu^{2+} \rightarrow Zn^{2+} + Cu(s)$

No reaction will take place, and no current will flow, unless the electrical circuit is complete. As shown in Figure 15.2, a conductive wire connects the electrodes externally through a voltmeter (potentiometer). A salt bridge, a glass tube filled with saturated KCl in agar gel, physically separates the two electrolyte solutions. The salt bridge permits ionic motion to complete the circuit while not permitting the electrolytes to mix. The reason we need to prevent the mixing of the electrolytes is that we want to obtain information about the electrochemical system by measuring the current flow through the external wire. If we had both electrodes and both ionic solutions in the same beaker, the copper ions would react directly at the Zn electrode, giving the same net reaction but no current flow in the external circuit. In the electrolyte solution and the salt bridge, the current flow is ionic (ion motion), and in the external circuit, the current flow is electronic (electron motion). This cell and the one in Figure 15.3 show the components needed for an electrochemical cell: two electrical conductors (electrodes), suitable electrolyte solutions, and a means of allowing the movement of ions between the solutions (salt bridge in Figure 15.2, a semipermeable glass frit or membrane in Figure 15.3), external connection of the electrodes by a conductive wire and the ability for an oxidation reaction to occur at one electrode, and a reduction reaction at the other. Of course, there are counterions present in each solution (e.g., sulfate ions) to balance the charge; these ions are not electroactive and do not take part in the redox reaction. They do flow (ionic motion) to keep charge balanced in the cell. A cell that uses a spontaneous redox reaction to generate electricity is called a **galvanic cell**. Batteries are examples of galvanic cells. A cell set up to cause a nonspontaneous reaction to occur by putting electricity into the cell is called an **electrolytic cell**.

The complete cell has a potential difference, a cell potential, which can be measured by the voltmeter. The potential difference for the cell, E_{cell}, can be considered to be equal to the difference between the two electrode potentials, when both half-cell reactions are written as reductions. That is,

$$E_{cell} = E_{cathode} - E_{anode} \qquad (15.7)$$

when the potentials are **reduction potentials**. This convention is necessary to calculate the sign of E_{cell} correctly, even though the anode reaction is an oxidation. The cell potential is also called the **electromotive force** or **emf**.

While we cannot measure a given single electrode potential directly, we can easily measure the cell potential for two half-cells joined as described. So now, we have a means of measuring the **relative electrode potential** for any half-cell by joining it to a designated **reference electrode**. In real cells, there is another potential difference that contributes to E_{cell}, called a junction potential. If there is a difference in the concentration or types of ions of the two half-cells, a small potential is created at the junction of the membrane or salt bridge and the solution. Junction potentials can be sources of error. When a KCl salt bridge is used, the junction potential is very small because the rates of diffusion of K^+ and Cl^- ions are similar, so the error in measuring a given electrode potential is small.

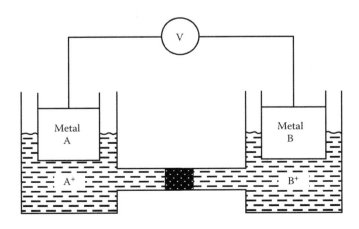

Figure 15.3 A schematic complete cell with a semipermeable frit separating the half-cells.

15.2.1 Line Notation for Cells and Half-Cells

Writing all the equations and arrows for half-cells and cells is time consuming and takes up space, so a shorthand or line notation is often used. For example, the half-cell composed of a silver electrode and aqueous 0.0001 M Ag$^+$ ion (from dissolution of silver nitrate in water) is written in line notation as

$$Ag(s) | Ag^+ (0.0001\ M)$$

The *vertical stroke or line* between Ag and Ag$^+$ indicates a *phase boundary*, that is, a difference in phases (e.g., solid | liquid, solid 1 | solid 2, or liquid | gas) in the constituents of the half-cells that are in contact with each other. The complete cell in Figure 15.2 can be represented as

$$Zn(s) | Zn^{2+} (0.01\ M) || Cu^{2+} (0.01\ M) | Cu(s)$$

The *double vertical stroke* after Zn^{2+} indicates a membrane junction or salt bridge. The double stroke shows the termination of one half-cell and the beginning of the second. This cell could also be written to show the salts used, as shown:

$$Zn(s) | ZnSO_4(aq) || CuSO_4(aq) | Cu(s)$$

It is conventional to write the electrode that serves as the anode on the left in an electrochemical cell. The other components in the cell are listed as they would be encountered moving from the anode to the cathode.

15.2.2 Standard Reduction Potentials

15.2.2.1 Standard Hydrogen Electrode

In order to compile a table of relative electrode potentials, chemists must agree upon the half-cell that will serve as the reference electrode. The composition and construction of the half-cell must be carefully defined. The value of the electrode potential for this reference half-cell could be set equal to any value, but zero is a convenient reference point. In practice, it has been arbitrarily decided and agreed upon that the **SHE** has an assigned electrode potential of exactly zero volts at all temperatures. The SHE, shown in Figure 15.4, consists of a platinum electrode with a surface coating of finely divided platinum (called a *platinized* Pt electrode) immersed in a solution of 1 M hydrochloric acid, which dissociates to give H$^+$. Hydrogen gas, H$_2$, is bubbled into the acid solution over the Pt electrode. The finely divided platinum on the electrode surface provides a large surface area for the reaction

$$2H^+ + 2e^- \rightarrow H_2(g) \quad E^0 = 0.000\ V \quad (15.8)$$

In addition, the Pt serves as the electrical conductor to the external circuit. Under standard state conditions, that is, when the H$_2$ pressure equals 1 atm and the *ideal* concentration of the HCl is 1 M, and the system is at 25 °C,

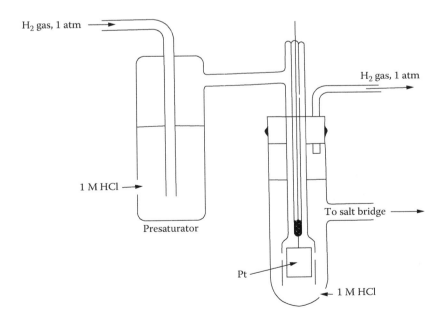

Figure 15.4 The standard hydrogen electrode (SHE). This design is shown with a presaturator containing the same 1 M HCl solution as in the electrode to prevent concentration changes by evaporation. (From Aikens et al., *Principles and Techniques for an Integrated Chemistry Laboratory*, Waveland Press Inc., Long Grove, IL, 1985. With permission.)

the reduction potential for the reaction given in Equation 15.8 is exactly 0 V. (The potential actually depends on the chemical *activity* of the HCl, not on its concentration. The relationship between activity and concentration is discussed subsequently. For an ideal solution, concentration and activity are equal.) The potential is symbolized by E^0, where the superscript zero means standard state conditions. The term **standard reduction potential** means that the ideal concentrations of all solutes are 1 M and all gases are at 1 atm; other solids or liquids present are pure (e.g., pure Pt solid). By connecting the SHE half-cell with any other standard half-cell and measuring the voltage difference developed, we can determine the standard reduction potential developed by the second half-cell.

Consider, for example, a cell at 25 °C made up of the two half-cells:

$$Zn(s) | Zn^{2+}(aq, 1M) \| H^+(aq, 1M) | H_2(g, 1\text{ atm}) | Pt(s)$$

This cell has a Zn half-cell as the anode and the SHE as the cathode. All solutes are present at *ideal* 1 M concentrations, gases at 1 atm, and the other species are pure solids, and so both half-cells are at standard conditions. The measured cell emf is +0.76 V and this is the standard cell potential, E^0, because both half-cells are in their standard states. From Equation 15.7, we can write

$$E^0_{cell} = E^0_{cathode} - E^0_{anode} \quad (15.9)$$

The total voltage developed under standard conditions is +0.76 V. But the voltage of the SHE is zero volts by definition; therefore, the standard reduction potential of the Zn half-cell is

$$+076\text{ V} = 0.000 - E^0_{Zn}$$

$$E^0_{Zn} = -0.76\text{ V}$$

Therefore, we can write

$$Zn^{2+} + 2e^- \rightarrow Zn(s) \quad E^0_{Zn} = -0.76\text{ V}$$

We have determined the Zn standard reduction potential even though the galvanic cell we set up has Zn being oxidized. By substituting other half-cells, we can determine their electrode potentials (actually, their relative potentials) and build a table of standard reduction potentials. If we set up a galvanic cell with the SHE and Cu, we have to make the SHE the anode in order for a spontaneous reaction to occur. This cell

$$Pt(s) | H_2(g, 1\text{ atm}) | H^+(aq, 1M) \| Cu^{2+}(aq, 1M) | Cu(s)$$

has a measured cell emf = +0.34 V. Therefore, the standard reduction potential for Cu is

$$+0.34\text{ V} = E^0_{Cu} - 0.000\text{ V}$$

$$E^0_{Cu} = +0.34\text{ V}$$

The quantity E^0 is the emf of a half-cell under standard conditions. A half-cell is said to be under standard conditions when the following conditions exist at a temperature of 25 °C:

1. All solids and liquids are pure (e.g., a metal electrode in the standard state).
2. All gases are at a pressure of 1 atm (760 mmHg).
3. All solutes are at 1 M concentration (more accurately, at **unit activity**).

The true electrode potential is related to the activity of the species in solution, not the concentration. For a pure substance, its mole fraction and its activity = 1. For a pure substance that is not present in the system, its mole fraction and its activity = 0. From general chemistry, you should remember that Raoult's law predicts that for an ideal solution, the mole fraction of the solute and its activity are equal. However, most solutions deviate from ideality because of interactions between the solute and solvent molecules. These deviations can be positive or negative, because the species may attract or repel each other. The amount of attraction or repulsion affects the activity of the solute. For dilute solutions, the activity is proportional to concentration (Henry's law). Activity is equal to the concentration times the activity coefficient for the species in solution. That is,

$$a_{ion} = [M_{ion}]\gamma_{ion} \quad (15.10)$$

where
a_{ion} is the activity of given ion in solution
$[M_{ion}]$ is the molar concentration of the ion
γ_{ion} is the activity coefficient of the ion

Activity depends on the ionic strength of the solution. If we compare a 1 M solution and a 0.01 M solution, the more concentrated solution may act as though it is less than 100 times more concentrated than the dilute solution. It is then said that the activity of the 1 M solution is less than unity or the activity coefficient is less than 1. For solutions with positive deviations from Henry's law, the activity coefficient will be greater than 1. For very dilute solutions (low ionic strength), the activity coefficient γ approaches 1, so concentration is approximately equal to activity for very dilute solutions. We will use concentrations in the calculations in this text instead of activities, but the approximation is only accurate for dilute solutions (<0.005 M) and for ions with single charges. Details on activity corrections can be found in most analytical chemistry texts, such as the ones by Harris or Enke listed in the bibliography.

Looking back at our cell in Figure 15.2, with the Zn half-cell as the anode and the Cu half-cell as the cathode, we can calculate the standard cell potential for this galvanic cell. In the spontaneous reaction, Zn is oxidized and Cu^{2+} is reduced; therefore, Zn is the anode and Cu is the cathode:

$$Zn^{2+} + 2e^- \rightarrow Zn^0 \quad E^0 = -0.76 \text{ V}$$

$$Cu^{2+} + 2e^- \rightarrow Cu^0 \quad E^+ = +0.34 \text{ V}$$

The standard cell potential developed is calculated from Equation 15.9:

$$+0.34 - (-0.76 \text{ V}) = +1.10 \text{ V}$$

In tables of standard potentials, all of the half-cell reactions are expressed as reductions. The sign is reversed if the reaction is reversed to become an oxidation. In a spontaneous reaction, when both half-cells are written as reductions, the half-cell with the more negative potential will be the one that oxidizes. The negative sign in Equation 15.9 reverses one of the reduction processes to an oxidation. Some standard reduction potentials for common half-cells are given in Appendix 15.A. These can be used to calculate E^0 for other electrochemical cell combinations as we have done for the Zn/Cu cell. More complete lists of half-cell potentials can be found in references such as Bard et al. listed in the bibliography.

In the reaction of our example, zinc metal dissolves, forming zinc ions and liberating electrons. Meanwhile, an equal number of electrons are consumed by copper ions, which plate out as copper metal. The net reaction is summarized as

$$Zn^0 + Cu^{2+} \rightarrow Cu^0 + Zn^{2+} \quad E^0 = +1.10 \text{ V}$$

Zinc metal is oxidized to zinc ions, and copper ions are reduced to copper metal. The copper cathode becomes depleted of electrons because these are taken up by the copper ions in solution. At the same time, the zinc anode has an excess of electrons because the neutral zinc atoms are becoming ionic and liberating electrons in the process. The excess electrons from the anode flow to the cathode. The flow of electrons is the source of external current; the buildup of electrons at the anode and the depletion at the cathode constitute a potential difference that persists until the reaction ceases. The reaction comes to an end when either all the copper ions are exhausted from the system or all the zinc metal is dissolved or an equilibrium situation is reached when both half-cell potentials are equal. If the process is used as a battery, such as a flashlight battery, the battery becomes "dead" when the reaction ceases.

When a cell spontaneously generates a voltage, the electrode that is negatively charged is the anode and the positively charged electrode is the cathode.

15.2.3 Sign Conventions

The sign of half-cell potentials has been defined in a number of different ways, and this variety of definitions has led to considerable confusion. The convention used here is in accord with the recommendations of the International Union of Pure and Applied Chemistry (IUPAC) meeting in Stockholm in 1953. In this convention, the standard half-cell reactions are written as reductions. Elements that are more powerful reducing agents than hydrogen show negative potentials, and elements that are less powerful reducing agents show positive potentials. For example, the Zn | Zn^{2+} half-cell is negative and the Cu | Cu^{2+} half-cell is positive.

15.2.4 Nernst Equation

We have seen how to use standard reduction potentials to calculate E^0 for cells. Real cells are usually not constructed at standard state conditions. In fact, it is almost impossible to make measurements at standard conditions because it is not reasonable to adjust concentrations and ionic strengths to give unit activity for solutes. We need to relate standard potentials to those measured for real cells. It has been found experimentally that certain variables affect the measured cell potential. These variables include the temperature, concentrations of the species in solution, and the number of electrons transferred. The relationship between these variables and the measured cell emf can be derived from simple thermodynamics (see any introductory general chemistry text). The relationship between the potential of an electrochemical cell and the concentration of reactants and products in a general redox reaction

$$a\text{A} + b\text{B} \rightarrow c\text{C} + d\text{D}$$

is given by the Nernst equation:

$$E = E^0 - \frac{RT}{nF} \ln \frac{[\text{C}]^c [\text{D}]^d}{[\text{A}]^a [\text{B}]^b} \quad (15.11)$$

where
E is the measured potential (emf) of the cell
E^0 is the emf of the cell under standard conditions
R is the gas constant (8.314 J/K mol = 8.314 V C/K mol)
T is the temperature (K)
n is the number of moles of electrons transferred during reaction (from the balanced half reactions)
F is the Faraday constant (96,485 C/mol e^-)
ln is the natural logarithm to base e

The logarithmic term has the same form as the equilibrium constant for the reaction. The term is called Q, the reaction quotient, when the concentrations (rigorously, the activities) are not the equilibrium values for the reaction. As in any equilibrium constant expression, pure liquids and pure solids have activities equal to 1, so they are omitted from the expression. If the values of R, T (25 °C = 298 K), and F are inserted into the equation and the natural logarithm is converted to log to the base 10, the Nernst equation reduces to

$$E = E^0 - \frac{0.05916}{n} \log \frac{[C]^c [D]^d}{[A]^a [B]^b} \qquad (15.12)$$

It should be noted that the square brackets literally mean "the molar concentration of." For example, $[Fe^{2+}]$ means "the molar concentration of ferrous ion" or moles of ferrous ion per liter of solution.

Concentrations in molarity should really be the activities of the species, but we often do not know the activities. For that reason, we define the **formal potential**, $E^{0\prime}$, as the measured potential of the cell when the species being oxidized and reduced are present at concentrations such that the ratio of the concentrations of oxidized to reduced species is unity and other components of the cell are present at designated concentrations. The use of the formal potential allows us to avoid activity coefficients, which are often unknown. This gives us

$$E = E^{0\prime} - \frac{0.05916}{n} \log \frac{[C]^c [D]^d}{[A]^a [B]^b} \qquad (15.13)$$

The Nernst equation is also used to calculate the electrode potential for a given half-cell at nonstandard conditions. For example, for the half-cell $Fe^{3+} + e^- \rightarrow Fe^{2+}$ that has an $E^0 = 0.77$ V and $n = 1$, the Nernst equation would be

$$E = 0.77 - \frac{0.05916}{1} \log \frac{[Fe^{2+}]}{[Fe^{3+}]}$$

To calculate the electrode potential, the molar concentrations of ferrous and ferric ions used to construct the half-cell would be inserted. The general form of the Nernst equation for an electrode written as a reduction is

$$E = E^0 - \frac{0.0591}{n} \log \left(\frac{[\text{red}]}{[\text{ox}]} \right) \qquad (15.14)$$

where
[red] means the molarity of the reduced form of the electroactive species
[ox] means the molarity of the oxidized form of the electroactive species

In cells where the reduced form is the metal, for example, in the Cu/Cu^{2+} half-cell, [red] = 1 because pure metals such as Cu have unit activity. There is the equivalent definition of formal potential for each half-cell.

The Nernst equation gives us the very important relationship between E, the emf of the half-cell, and the concentration of the oxidized and reduced forms of the components of the solution. Measurements using pH electrodes and ISEs are based on this relationship. If the two electrode reactions are written as reductions, the potentials can be calculated for the cathode and anode (or the electrode inserted into the positive terminal of the potentiometer, E_+ and the electrode inserted into the negative terminal of the potentiometer, E_-). The cell voltage is the difference

$$E_{\text{cell}} = E_+ - E_- \qquad (15.15)$$

15.2.5 Activity Series

The tendency for a species to become oxidized or reduced determines the sign and potential of the half-cell. The tendency is strongly related to the chemical reactivity of the species concerned in aqueous systems. Based on the potential developed in a half-cell under controlled conditions, the elements may be arranged in an order known as the activity series or electromotive series (Table 15.1). In general, the metals at the top of the activity series are most chemically reactive and tend to give up electrons easily, following the reaction $M^0 \rightarrow M^{n+} + ne^-$. The metals at the bottom of the series are more "noble" and therefore less active. They do not give up electrons easily; in fact, their cations will accept electrons from metals above them in the activity series. In the process, the cations become neutral metal atoms and plate out of solution, while the more active metals become ionic and dissolve. This is illustrated as follows:

$$\text{Metal A (active)} \rightarrow \text{metal ion } A^+ + e^-$$

$$\text{Metal ion } B^+ \text{ (noble)} + e^- \rightarrow \text{metal B}$$

$$\text{Net result: } A + B^+ \rightarrow A^+ + B$$

In short, the more active metals displace the less active metals from solution. As an example, if an iron strip is immersed in a solution of copper sulfate, some of the iron dissolves, forming iron ions, while the copper ions become metallic and copper metal plates out on the remaining iron strip. The activity series can be used to predict displacement reactions between atoms and ions in compounds of the type $A + BC \rightarrow AC + B$, where A and B are atoms. Using the activity series, any atom A will displace from a compound any element B listed below it, but will not displace any element listed above it.

This reaction is particularly important when B^+ in BC refers to a hydrogen ion solution (acid). Based on this

ELECTROANALYTICAL CHEMISTRY

Table 15.1 Activity Series of Metals

Metal	E^0 (V)	Chemical Reactivity
Li	−3.05	These metals displace hydrogen from acids and dissolve in all acids, including water.
K	−2.93	
Ba	−2.91	
Sr	−2.89	
Ca	−2.87	
Na	−2.71	
Mg	−2.37	These metals react with acids or steam.
Al	−1.66	
Mn	−1.18	
Zn	−0.76	
Cr	−0.74	
Fe	−0.44	
Cd	−0.40	
Co	−0.28	These metals react slowly with all acids.
Ni	−0.26	
Sn	−0.14	
Pb	−0.13	
H_2	0.00	
Bi	+0.32	These metals react with oxidizing acids (e.g., HNO_3).
Cu	+0.34	
Ag	+0.80	
Hg	+0.85	
Pd	+0.92	
Pt	+1.12	These metals react with aqua regia (3:1 v/v HCl/HNO_3).
Au	+1.50	

principle, all elements above hydrogen in the activity series are capable of displacing hydrogen from solution; that is, it is possible for them to be dissolved in an acid solution by reducing the H^+ ion. The metal ionizes (oxidizes) and enters the solution while the H^+ (acid) reduces to H_2 and usually bubbles off. We would predict that Al and Zn will dissolve in HCl but that Cu will not. In contrast, a simple mineral acid such as HCl cannot dissolve noble metals such as platinum and gold, because the metals will not displace hydrogen ions. These metals require an oxidizing acid such as nitric acid plus a complexing ion such as the chloride ion from HCl, as is found in aqua regia, a mixture of HCl and HNO_3, to force them into solution.

15.2.6 Reference Electrodes

In order to measure the emf of a given half-cell, it is necessary to connect it with a second half-cell and measure the voltage produced by the complete cell. In general, the second half-cell serves as a reference cell and should be one with a known, stable electrode potential. Although the SHE serves to define the standard reduction potential, in practice, it is not always convenient to use an SHE as a reference electrode. It is difficult to set up and control. Other, more convenient reference electrodes have been developed. In principle, any metal-ion half-cell could be used under controlled conditions as a reference electrode, but in practice, many metals are unsatisfactory materials. Active metals, such as sodium and potassium, are subject to chemical attack by the electrolyte. Other metals, such as iron, are difficult to obtain in the pure form. With some metals, the ionic forms are unstable to heat or to exposure to the air. Also, it is frequently difficult to control the concentration of the electrolytes accurately. As a result, only a few systems provide satisfactory stable potentials.

The characteristics of an ideal reference electrode are that it should have a fixed potential over time and temperature, long-term stability, and the ability to return to the initial potential after exposure to small currents (i.e., it should be reversible), and it should follow the Nernst equation. Two common reference electrodes that come close to behaving ideally are the saturated calomel electrode (SCE) and the silver/silver chloride electrode.

15.2.6.1 Saturated Calomel Electrode

The **SCE** is composed of metallic mercury in contact with a saturated solution of mercurous chloride, or *calomel* (Hg_2Cl_2). A Pt wire in contact with the metallic Hg conducts electrons to the external circuit. The mercurous ion concentration of the solution is controlled through the solubility product by placing the calomel in contact with a saturated potassium chloride solution. It is the saturated KCl solution that gives this electrode the "saturated" name; there are other calomel reference electrodes used that differ in the concentration of KCl solution, but all contain saturated mercurous chloride solution. A typical calomel electrode is shown in Figure 15.5. The half-cell reaction is

$$Hg_2Cl_2(s) + 2e^- \leftrightarrow 2Hg(l) + 2Cl^- \quad E^0 = +0.268 \text{ V}$$

Applying Equation 15.12, we obtain

$$E = E^0 - \frac{0.05916}{n} \log \frac{[C]^c [D]^d}{[A]^a [B]^b}$$

$$E = E^0 - \frac{0.0591}{2} \log \left(\frac{[Hg^0]^2 [Cl^-]^2}{[Hg_2Cl_2]} \right)$$

But $Hg_2Cl_2(s)$ and $Hg(l)$ are in standard states at unit activity; therefore,

$$E = E^0 - \frac{0.0591}{2} \log \left(\frac{1 \times [Cl^-]^2}{1} \right)$$

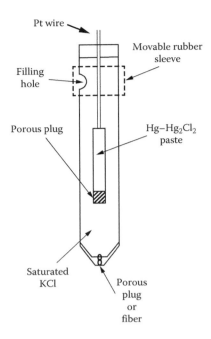

Figure 15.5 The saturated calomel electrode (SCE). (From Aikens et al., *Principles and Techniques for an Integrated Chemistry Laboratory*, Waveland Press Inc., Long Grove, IL, 1985. With permission.)

$$= E^0 - \frac{0.0591}{2}\log\left(\left[Cl^-\right]^2\right)$$

When the KCl solution is saturated and its temperature is 25 °C, the concentration of chloride ion, [Cl$^-$], is known and $E = +0.244$ V for the SCE.

For a 1 N solution of KCl, the electrode is called a normal calomel electrode (NCE) and E(NCE) = +0.280 V. The advantage of the SCE is that the potential does not change if some of the liquid evaporates from the electrode, because the Cl$^-$ concentration does not change. Remember that "saturated" for the SCE refers to the KCl concentration, not to the mercurous chloride (calomel).

15.2.6.2 Silver/Silver Chloride Electrode

Another common reference cell found in the chemical-processing industry and useful in organic electrochemistry is the silver/silver chloride half-cell. The cell consists of silver metal coated with a paste of silver chloride immersed in an aqueous solution saturated with KCl and AgCl. The half-cell reaction is

$$AgCl(s) + e^- \leftrightarrow Ag(s) + Cl^- \quad E^0 = +0.222 \text{ V}$$

The same principle applies as in the SCE, but in this case, the silver ion activity depends on the solubility product of AgCl, which is in contact with a solution of known chloride ion activity. The measured E for an Ag/AgCl electrode saturated in KCl is +0.199 V at 25 °C. A saturated KCl solution is about 4.5 M KCl. This reference electrode can also be filled with 3.5 M KCl (saturated with AgCl); the potential in this case is +0.205 V at 25 °C. The silver/silver chloride electrode provides a reproducible and reliable reference electrode free from mercury salts.

Both reference electrode designs can permit KCl to leak through the porous plug junction and contaminate the analyte solution. In addition, analytes that react with Ag or Hg ions, such as sulfides, can precipitate insoluble salts that clog the junction and impair the performance of the electrode. Electrode manufacturers have designed double-junction electrodes with an outer chamber that can be filled with an electrolyte that does not interfere with a given measurement and electrodes with large, flushable sleeve junctions that can be cleaned easily and are not prone to clogging.

15.2.7 Electrical Double Layer

While it is easy to say that we can measure the potential difference in an electrochemical cell, the reality of what we are measuring is not obvious. The potential difference between a single electrode and the solution cannot be measured. A second electrode immersed in the solution is required. The second electrode has its own potential difference between itself and the solution; this difference cannot be measured either. All we can measure are relative potential differences. The potential difference between an electrode and the solution actually exists only in a very small region immediately adjacent to the electrode surface. This region of potential difference is often called the "**electrical double layer**" and is only nanometers wide. A very simplistic view of this interface is presented. A negatively charged electrode surface will have a "layer" of excess positive ions in solution adjacent to (or adsorbed to) the surface of the electrode, hence the term "double layer." Immediately adjacent to the positive ion layer there is a complex region of ions in solution whose distribution is determined by several factors. Outside of that, and only nanometers away from the electrode surface, there is the bulk solution. A positively charged electrode surface will have a similar interfacial region with the charges reversed. The exact nature of the charged interface and the ion distribution is very complex and is not yet completely understood. Outside of this very narrow region of high potential gradient next to each electrode, there is no potential gradient in the bulk of the solution. As a consequence of the placement of electrodes in solution and the creation of charged interfaces between the electrode surface and the solution, electrochemical measurements are made in a very nonhomogeneous environment.

Any reactions that occur must occur at the electrode–solution interface and the reacting species must be brought to the electrode surface by diffusion or mass transport through stirring of the solution (convection). The ions in the bulk of the solution are not attracted to the electrodes by potential difference; there is no potential gradient in the bulk of the solution. Even when we are interested in the bulk properties of the solution, we are only analyzing an extremely small amount of the solution that is no longer homogeneous because of the formation of the complex layers of charged species at the electrode interface.

15.3 ELECTROANALYTICAL METHODS

Diverse analytical techniques, some highly sensitive, have been developed based on measurements of current, voltage, charge, and resistance in electrochemical systems. One variable is measured, while the others are controlled. Electroanalytical methods can be classified according to the variable being measured. Table 15.2 provides a summary of the more important methods. The methods are briefly defined in the following and then discussed at length in subsequent sections. A very readable discussion of all electroanalytical methods can be found in the ASM Handbook.

The Nernst equation indicates the relationship between the activity of species in solution and the emf produced by a cell involving those species. The electrochemical technique called **potentiometry** measures the potential developed by a cell consisting of an **indicator electrode** and a reference electrode. In theory, the indicator electrode response is a measure of the activity of a single component of the solution, the analyte. In practice, the indicator electrode is calibrated to respond to concentration rather than activity and is usually not specific for a single analyte. The potential, E_{cell}, is measured under negligible current flow to avoid significant changes in the concentration of the component being measured. Modern potentiometers permit measurement of potential with currents <1 pA, so potentiometry can be considered to be a nondestructive technique. Potentiometry is the basis for measurement of pH, ion-selective electrode (ISE) measurements, and potentiometric titration.

Coulometry is the term used for a group of methods based on electrolytic oxidation or reduction of an analyte. The electrolysis is performed by controlling the potential and is carried out to quantitatively convert the analyte to a new oxidation state. One form of coulometry is **electrogravimetry**, in which metallic elements are reduced, plated out onto an electrode, and weighed. The mass of the metal deposited is a measure of the concentration of the metal originally in solution. Coulometry is based on Faraday's law, which states that the extent of reaction at an electrode is proportional to the current. It is known that 1 F (96,485 C) of electricity is required to reduce (or oxidize) 1 gram-equivalent mass of an electroactive analyte. By measuring the quantity of electricity required to reduce (or oxidize) a given sample exhaustively, the quantity of analyte can be determined, provided the reaction is 100% efficient (or of known efficiency). Mass, or charge $q = i$ (A) $\times t$ (s), can be used as a measure of the extent of the electrochemical reaction.

In **voltammetry**, a controlled potential is applied to one electrode and the current flowing through the cell is monitored over time. A powerful family of techniques is available using voltammetry. **Polarography** is a technique that requires three electrodes. Polarography uses as the working electrode (WE) a dropping mercury electrode (DME) or a static (hanging) mercury drop. The auxiliary electrode (or counter electrode [CE]) is normally a Pt wire or foil. A third (reference) electrode is used as a basis for control of the potential at the WE. The current of analytical interest flows between the working and auxiliary electrodes, and the reference acts only as a high-impedance probe. As the voltage is progressively increased or decreased with time (sweep voltammetry), resultant changes in the anodic or cathodic currents occur whenever an electroactive species is oxidized or reduced, respectively. Polarography is a special case of *linear sweep voltammetry* because the

Table 15.2 Electroanalytical Techniques

Technique	Controlled Parameter	Measured Parameter
Conductometry	Voltage, V (AC)	Conductance, $G = 1/R$
Conductometric titration	Voltage, V (AC)	Titrant volume vs. conductance
Potentiometry	Current, $i = 0$	Potential, E
Potentiometric titration	Current, $i = 0$	Titrant volume vs. E
Cyclic voltammetry	Potential, E	i vs. E
Polarography	Potential, E	i vs. E
Anodic stripping voltammetry	Potential, E	i vs. E
Amperometry	Potential, E	i
Coulometry	E	Charge, q (integrated current)
Electrogravimetry	E or i	Mass deposited
Coulometric titration	i	Time, t

electrode area increases with time as each Hg drop grows and falls every 4 s or so; that is, the voltage is changing as the electrode area increases. Polarography is especially useful for analyzing and studying metal-ion and metal complex reductions and solution equilibria. Another technique for studying electrochemical reaction rates and mechanisms involves reversing the potential sweep direction to reveal the electroactive products formed in the forward sweep. In this way, it is possible to see if the products undergo reaction with other species present or with the solvent. This technique is called *cyclic voltammetry* and is usually carried out at a solid electrode. In a special case of voltammetry called *anodic stripping voltammetry*, the electrolyzed product is preconcentrated at an electrode by deposition for a reasonable time period at a fixed potential. The product is then stripped off with a rapid reverse potential sweep in the positive direction. Peak currents on the reverse sweep are used to determine the analyte concentration from a standard additions calibration. In principle, this method is applicable to both anodic and cathodic stripping.

Amperometry is the measurement of current at a fixed potential. An analyte undergoes oxidation or reduction at an electrode with a known, applied potential. The amount of analyte is calculated from Faraday's law. Amperometry is used to detect titration endpoints, as a detector for liquid chromatography (LC), and forms the basis of many new sensors for biomonitoring and environmental monitoring applications.

In **conductometry**, an alternating (AC) voltage is applied across two electrodes immersed in the same solution. The applied voltage causes a current to flow. The magnitude of the current depends on the electrolytic conductivity of the solution. This method makes it possible to detect changes of composition in a sample during chemical reactions (e.g., during a titration) although the measurement itself cannot identify the species carrying the current. Conductivity or conductance measurements are used routinely to monitor water quality and process streams. Conductivity detectors are used for measuring ion concentrations in commercial ion chromatography (IC) instruments.

A variety of electroanalytical methods are used as detectors for LC. Detectors based on conductometry, amperometry, coulometry, and polarography are commercially available.

15.3.1 Potentiometry

Potentiometry involves measurement of the potential, or voltage, of an electrochemical cell. Accurate determination of the potential developed by a cell requires a negligible current flow during measurement. A flow of current would mean that a faradaic reaction is taking place, which would change the potential from that existing when no current is flowing.

Measurement of the potential of a cell can be useful in itself, but it is particularly valuable if it can be used to measure

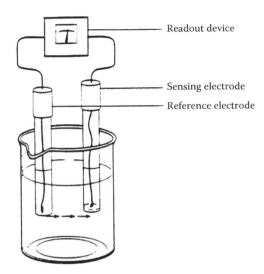

Figure 15.6 An electrochemical measuring system for potentiometry. The indicator (sensing, working) electrode responds to the activity of the analyte of interest. The potential difference developed between the reference electrode and the indicator electrode is "read out" on the potentiometer (voltmeter). (© Thermo Fisher Scientific [www.thermofisher.com]. Used with permission.)

the potential of a half-cell or **indicator electrode** (also called a sensing electrode), which responds to the concentration of the species to be determined. This can be accomplished by connecting the indicator electrode to a reference half-cell to form a complete cell as shown in Figure 15.6. When the total voltage of the cell is measured, the difference between this value and the voltage of the reference electrode is the voltage of the indicator electrode. This can be expressed as

$$E(\text{total}) = E(\text{indicator}) - E(\text{reference})$$

Compare this equation with Equations 15.7 and 15.15. By convention, the reference electrode is connected to the negative terminal of the potentiometer (the readout device). The common reference electrodes used in potentiometry are the SCE and the silver/silver chloride electrode, which have been described. Their potentials are fixed and known over a wide temperature range. Some values for these electrode potentials are given in Table 15.3. The total cell potential is measured experimentally, the reference potential is known, and therefore, the variable indicator electrode potential can be calculated and related to the concentration of the analyte through the Nernst equation. In practice, the concentration of the unknown analyte is determined after calibration of the potentiometer with suitable standard solutions. The choice of reference electrode depends on the application. For example, the Ag/AgCl electrode cannot be used in solutions containing species such as halides or sulfides that will precipitate or otherwise react with silver.

Table 15.3 Potentials of Reference Electrodes at Various Temperatures

Temperature (°C)	Potential (V) vs. SHE	
	Saturated Calomel (SCE)	Saturated Ag/AgCl
15	0.251	0.209
20	0.248	0.204
25	0.244	0.199
30	0.241	0.194
35	0.238	0.189

15.3.1.1 Indicator Electrodes

The indicator electrode is the electrode that responds to the change in analyte activity. An ideal indicator electrode should be specific for the analyte of interest, respond rapidly to changes in activity, and follow the Nernst equation. There are no specific indicator electrodes, but there are some that show a high degree of selectivity for certain analytes. Indicator electrodes fall into two classes, metallic electrodes and membrane electrodes.

15.3.1.1.1 Metallic Electrodes

A metal electrode of the **first kind** is just a metal wire, mesh, or solid strip that responds to its own cation in solution. Cu/Cu^{2+}, Ag/Ag^+, Hg/Hg^+, and Pb/Pb^{2+} are examples of this type of electrode. There are significant problems encountered with these electrodes. They have poor selectivity, responding not only to their own cation but also to any other more easily reduced cation. Some metal surfaces are easily oxidized, giving an erratic or inaccurate response unless the solution has been purged of air. Some metals can only be used in limited pH ranges because they will dissolve in acids or bases. Silver and mercury are the most commonly used electrodes of the first kind.

A metal electrode of the **second kind** consists of a metal coated with one of its sparingly soluble salts (or immersed in a saturated solution of its sparingly soluble salt). This electrode responds to the anion of the salt. For example, a silver wire coated with AgCl will respond to changes in chloride activity because the chloride ion activity is controlled by the solubility of AgCl. The electrode reaction is $AgCl(s) + e^- \leftrightarrow Ag(s) + Cl^-$, with a potential $E^0 = 0.222$ V. The Nernst equation expression for the electrode potential at 25°C is $E = 0.222 - 0.05916 \log[Cl^-]$.

A metal electrode of the **third kind** uses two equilibrium reactions to respond to a cation other than that of the metal electrode. Ethylenediaminetetraacetic acid (EDTA) complexes many metal cations, with different stabilities for the complexes but a common anion (the EDTA anion) involved in the equilibria. A mercury electrode in a solution containing EDTA and Ca will respond to the Ca ion activity, for example. The complexity of the equilibria makes this type of electrode unsuitable for complex sample matrices.

The last type of metallic electrode is the **redox indicator electrode**. This electrode is made of Pt, Pd, Au, or other inert metals and serves to measure redox reactions for species in solution (e.g., Fe^{2+}/Fe^{3+}, Ce^{3+}/Ce^{4+}). These electrodes are often used to detect the endpoint in potentiometric titrations. Electron transfer at inert electrodes is often not reversible, leading to nonreproducible potentials. Although not a metal electrode, it should be remembered that carbon electrodes are also used as redox indicator electrodes, because carbon is also not electroactive at low applied potentials.

15.3.1.1.2 Membrane Electrodes

Membrane electrodes are a class of electrodes that respond *selectively* to ions by the development of a potential difference (a type of junction potential) across a membrane that separates the analyte solution from a reference solution. The potential difference is related to the concentration difference in the specific ion measured on either side of the membrane. These electrodes do not involve a redox reaction at the surface of the electrode as do metallic electrodes. Because these electrodes respond to ions, they are often referred to as **ion selective electrodes (ISEs)**. The ideal membrane allows the transport of only one kind of ion across it; that is, it would be specific for the measurement of one ionic species only. As of this writing, there are no specific ISEs, but there are some highly selective ones. Each electrode is more or less selective for one ion; therefore, a separate electrode is needed for each species to be measured. In recent years, many different types of membrane electrodes have been developed for a wide variety of measurements.

ISEs are relatively sensitive. They are capable of detecting concentrations as low as 10^{-12} M for some electrodes. To avoid writing small concentrations in exponential form, the term **pIon** has been defined, where pIon equals the negative logarithm (base 10) of the molar concentration of the ion. That is,

$$\text{pIon} = -\log[\text{Ion}] \quad (15.16)$$

For example, pH is the term used for the negative logarithm of the hydrogen ion concentration, where H^+ is expressed as (moles of H^+)/L of solution. Concentrations of other ions can be expressed similarly: pCa for calcium ion concentration, pF for fluoride ion concentration, pOH for hydroxide ion concentration, and pCl for chloride ion concentration. If the concentration of H^+ ion in solution equals 3.00×10^{-5} M, the pH = $-\log(3.00 \times 10^{-5})$ = 4.522, for example. If the concentration is 1.00×10^{-7} M, the pH = $-\log(1.0 \times 10^{-7})$ = 7.00.

15.3.1.1.3 Glass Membrane Electrodes

The first membrane electrode to be developed was the glass electrode for measurement of pH, the concentration of hydrogen ion, H^+, in solution. The pH electrode consists of a thin hydrogen ion-sensitive glass membrane, often shaped like a bulb, sealed onto a glass or polymer tube. The solution inside the electrode contains a known concentration of H^+ ion, either as dilute HCl or a buffer solution. The solution is saturated in AgCl. The activity of H^+ inside the electrode is constant and keeps the internal potential fixed. An internal reference electrode is sealed inside the tube and is attached to one terminal of the potentiometer. The glass pH electrode is shown schematically in Figure 15.7a. The glass pH electrode is used in combination with a reference electrode, either a separate Ag/AgCl electrode or an SCE, as shown in Figure 15.7b. Both electrodes are immersed in a solution of unknown pH and the cell potential developed is a measure of the hydrogen ion concentration in the solution on the outside of the glass membrane of the pH electrode, since all other potentials are fixed.

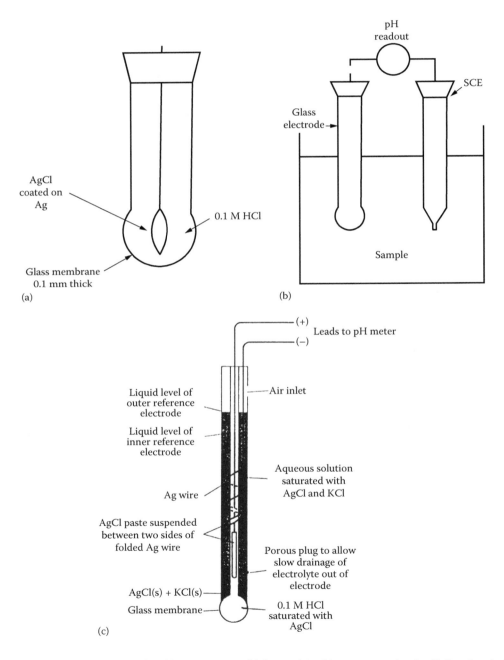

Figure 15.7 (a) A schematic glass electrode for pH measurement. (b) A complete pH measurement cell, with the glass indicator electrode and an external saturated calomel REF. (c) A commercial combination pH electrode, with built-in internal Ag/AgCl REF.

In a standard pH electrode, the glass membrane is composed of SiO$_2$, Na$_2$O, and CaO. The response of this glass to changes in hydrogen ion activity outside the membrane is complex. The glass electrode surfaces must be hydrated in order for the glass to function as a pH sensor. It is thought that a hydrated gel layer containing adsorbed H$^+$ ions exists on the inner and outer glass surfaces. It is necessary for charge to move across the membrane in order for a potential difference to be measured, but studies have proved that H$^+$ ion does not move through the glass membrane. The singly charged Na$^+$ cation is mobile within the 3D silicate lattice. It is the movement of sodium ions in the lattice that is responsible for electrical conduction within the membrane. The inner and outer glass membrane surfaces contain negatively charged oxygen atoms. Hydrogen ions from the solution inside the electrode equilibrate with the inner glass surface, neutralizing some of the negative charge. Hydrogen ions from the solution outside the electrode (our sample) equilibrate with the oxygen anions on the outer glass surface and neutralize some of the charge.

The side of the glass exposed to the higher H$^+$ concentration has the more positive charge. Na$^+$ ions in the glass then migrate across the membrane from the positive side to the negative side, which results in a change in the potential difference across the membrane. The only variable is the hydrogen ion concentration in the analyte solution. Because the electrode behavior obeys the Nernst equation, the potential changes by 0.05916 V for every 10-fold change in [H$^+$] or for every unit change in pH. The glass pH electrode is highly selective for hydrogen ions. The major interfering ion is Na$^+$. When Na$^+$ concentration is high and H$^+$ is very low, as occurs in very basic solutions of sodium hydroxide, the electrode responds to a sodium ion as if it were a hydrogen ion. This results in a negative error; the measured pH is lower than the true pH. This is called the *alkaline error*. The use of pH electrodes to measure pH is discussed under applications of potentiometry.

Commercial glass pH electrodes come in all sorts of shapes and sizes to fit into every imaginable container, including into nuclear magnetic resonance (NMR) tubes and to measure pH in volumes of solution as small as a few microliters. Some electrodes come as complete cells, with a second reference electrode built into the body of the pH electrode. These are called **combination electrodes** and eliminate the need for a second external reference electrode. Combination electrodes are required if you want to measure small volumes in small containers and have no room for two separate electrodes. A combination electrode is shown schematically in Figure 15.7c. Many glass pH electrodes have polymer bodies and polymer shields around the fragile glass bulb to help prevent breakage. A microelectrode for measuring microliter volumes in well plates is shown in Figure 15.8. Other glass compositions incorporating aluminum and boron oxides have been used in membrane electrodes to make the membrane selective for other ions instead of hydrogen ions. For example, a glass whose composition is 11% Na$_2$O, 18% Al$_2$O$_3$, and 71% SiO$_2$ is highly selective toward sodium, even in the presence of other alkali metals. For example, at pH 11, this electrode is approximately 3000× more sensitive to Na$^+$ than to K$^+$. The ratio of the response of the electrode to a solution of potassium to its response to a solution of sodium, the analyte, for solutions of equal concentration, is called the *selectivity coefficient*. The selectivity coefficient should be a very small number ($\ll 1$) for high selectivity for the analyte. An ISE has different selectivity coefficients for each ion that responds. Commercial glass ISEs are available for all of the alkali metal ions (Li, Na, K, Rb, and Cs) and for ammonium ion (NH_4^+), Ag$^+$, Fe^{3+}, Pb^{2+}, and Cu^{2+}.

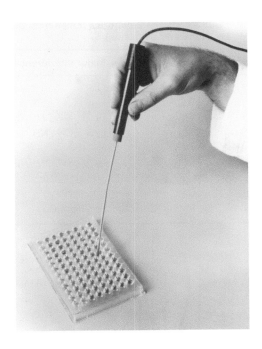

Figure 15.8 A commercial microelectrode for pH measurements in volumes as small as 10 µL. The electrode has a solid-state pH sensor and a flexible polypropylene stem, making it extremely rugged and suitable for use in the field. This electrode can be used to determine pH in single droplets of water on leaves of plants or blades of grass to study acid rain deposition. (Courtesy of Lazar Research Laboratories, Inc., Los Angeles, CA, www.lazarlab.com.)

15.3.1.1.4 Crystalline Solid-State Electrodes

Crystalline solid-state electrodes have membranes that are single-crystal ionic solids or pellets pressed from ionic salts under high pressure. The ionic solid must contain the target analyte ion and must not be soluble in the solution to be measured (usually aqueous solution). These membranes are generally about 1–2 mm thick and 10 mm in diameter. Sealing the solid into the end of a polymer tube

Table 15.4 Solid Membrane Electrodes (taken from the book by Bruno and Svoronos)

Membrane	Ion Measured	Major Interferences
LaF_3	F^-	OH^-
Ag_2S	S^{-2}, Ag^+	Hg^{+2}
$AgCl\text{-}Ag_2S$	Cl^-	Br^-, I^-, S^{-2}, CN^-, NH_3
$AgBr\text{-}Ag_2S$	Br^-	I^-, S^{-2}, CN^-, NH_3
$AgI\text{-}Ag_2S$	I^-	S^{-2}, CN^-
$AgSCN\text{-}Ag_2S$	SCN^-	Br^-, I^-, S^{-2}, CN^-, NH_3
$CdS\text{-}Ag_2S$	Cd^{+2}	Ag^+, Hg^{+2}, Cu^{+2}
$CuS\text{-}Ag_2S$	Cu^{+2}	Ag^+, Hg^{+2}
$PbS\text{-}Ag_2S$	Pb^{+2}	Ag^+, Hg^{+2}, Cu^{+2}

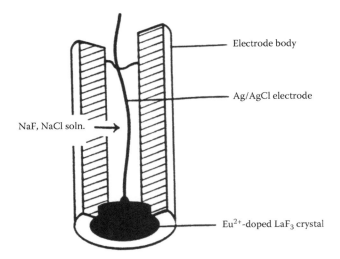

Figure 15.9 A commercial solid-state fluoride ISE. The ion-sensitive area is a solid crystal of LaF_3 doped with EuF_2. The filling solution contains NaF and NaCl. The electrical connection is made by an Ag/AgCl electrode. (© Thermo Fisher Scientific [www.thermofisher.com]. Used with permission.)

forms the electrode. Like the pH electrode, the interior of the polymer tube contains an internal electrode to permit connection to the potentiometer and a filling solution containing a fixed concentration of the analyte ion. A concentration difference in the analyte ion on the outside of the crystalline membrane causes the migration of charged species across the membrane. These electrodes generally respond to concentrations as low as 10^{-6} M of the analyte ion. In Table 15.4 we present the most common electrodes along with the ions that can be measured. Note that often the membrane is composed of a salt (listed first in the table) and a matrix (listed second).

The fluoride ISE is the most commonly used single-crystal ISE. It is shown schematically in Figure 15.9. The membrane is a single crystal of LaF_3 doped with EuF_2. The term "doped" means that a small amount of another substance (in this case, EuF_2) has been added intentionally into the LaF_3 crystal. (If the addition were not intentional, we would call the europium an impurity or contaminant!) Note that the two salts do not have the same stoichiometry. Addition of the europium fluoride creates fluoride ion vacancies in the lanthanum fluoride lattice. When exposed to a variable concentration of F^- ions outside the membrane, the fluoride ions in the crystal can migrate. Unlike the pH electrode, it is the F^- ions that actually move across the membrane and result in the electrode response. The F^- ISE is extremely selective for fluoride ion. The only ion that interferes is OH^-, but the response of the electrode to fluoride ions is more than 100 times greater than the response of the electrode to hydroxide. The hydroxide interference is only significant when the OH^- concentration is 0.1 M or higher. The electrode only responds to fluoride ions, so the pH of the solution must be kept high enough so that HF does not form.

Most of these electrode membranes are made from the corresponding silver salt mixed with silver sulfide, due to the low solubility of most silver salts in water. In addition, mixtures of silver sulfide and the sulfides of copper, lead, and cadmium make solid-state electrodes for Cu^{2+}, Pb^{2+}, and Cd^{2+} available. An advantage of the silver sulfide-based electrodes is that a direct connection can be made to the membrane by a silver wire, eliminating the need for electrolyte filling solutions. These electrodes will also respond to silver ions and sulfide ions as well as the intended analyte.

15.3.1.1.5 Liquid Membrane ISEs

Liquid membrane electrodes are based on the principle of ion exchange. Older electrodes, such as the one diagrammed in Figure 15.10a, consisted of a hydrophilic, porous membrane fixed at the base of two concentric tubes. The inner tube contains an aqueous solution of the analyte ion and the internal reference electrode. The outer tube contains an ion exchanger in organic solvent. The ion exchanger may be a cation exchanger, an anion exchanger, or a neutral complexing agent. Some of the organic phase is absorbed into the hydrophilic membrane, forming an organic ion-exchange phase separating the aqueous sample solution and the aqueous internal solution with a known concentration of analyte ion. An ion-exchange equilibrium is established at both surfaces of the membrane and the concentration difference results in a potential difference. Modern liquid membrane electrodes have the liquid ion exchanger immobilized in or covalently bound to a polymer film, shown schematically in Figure 15.10b. The selectivity of these electrodes is determined by the degree of selectivity of the ion exchanger for the analyte ion. Selective lipophilic complexing agents have been developed and this area is still an important area of research. Table 15.5 provides some common currently used liquid membrane electrodes along with the ions that can be determined and their selectivities.

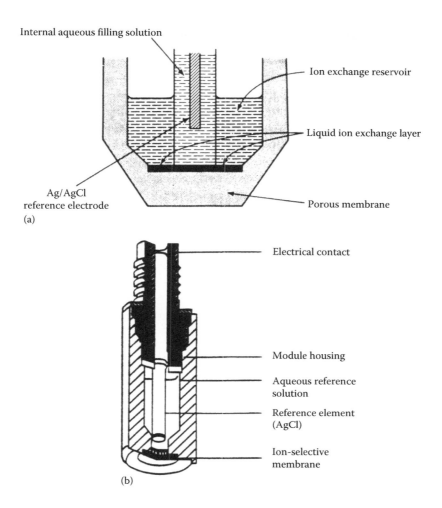

Figure 15.10 (a) A liquid membrane ISE. (b) An ISE with ion-exchange material in a solid polymer membrane. The sensed ion is exchanged across the membrane, creating the potential. This type of ISE is available for Ca^{2+}, K^+, nitrate ion, and nitrite ion determinations. (b: © Thermo Fisher Scientific [www.thermofisher.com]. Used with permission.)

As the preference of the ion exchanger increases, the selectivity increases. Note that the selectivity is also affected by the organic solvent in which the liquid ion exchanger is dissolved.

Commercial liquid membrane electrodes are available for calcium, calcium plus magnesium to measure water hardness, potassium, the divalent cations of Zn, Cu, Fe, Ni, Ba, and Sr, and anions BF_4^- and nitrate, NO_3^-, among others. In general, these electrodes can be used only in aqueous solution, to avoid attack on the membrane. Lifetimes are limited by leaching of the ion exchanger from the membrane, although newer technology (such as covalent attachment of the ion exchanger) is improving this.

Polymer film membranes are being used to coat metal wire electrodes to make miniature ISEs for *in vivo* analysis. These coated wire ISEs require no internal reference solution and have been made with electrode tip diameters of about 0.1 μm. Electrodes have been made small enough to measure ions inside a single cell, as seen in Figure 15.11.

15.3.1.1.6 Gas-Sensing Electrodes

Gas-sensing electrodes are really entire electrochemical cells that respond to dissolved gas analytes or to analytes whose conjugate acid or base is a gaseous species. Because they are complete cells, the term "probe" is often used instead of electrode. A typical gas-sensing probe consists of an ISE and a reference electrode sealed behind a gas-permeable membrane, through which the analyte gas diffuses. The ISE used can be selective for the species; for example, an Ag_2S electrode to measure hydrogen sulfide that diffuses through the gas-permeable membrane. Several gas-sensing probes use a glass pH electrode as the indicator electrode for gas species that change the pH of the internal solution. Examples include the determination of CO_2, which dissolves in the aqueous internal solution to form H^+ and HCO_3^-, and the determination of ammonia, which dissolves to form NH_4^+ and OH^-. The potential that develops in all of these cells depends on the concentration of the analyte gas in the external sample solution. Gas-sensing

Table 15.5 Liquid Membrane Electrodes. R refers to any organic radical or group (taken from the book by Bruno and Svoronos).

Ion Measured	Exchange Site	Selectivity Coefficients
K^+	Valinomycin	Na^+, 0.0001
Ca^{+2}	$(RO)_2POO^-$ (dialkylphosphate anion)	Na^+, 0.0016 Mg^{+2}, Ba^{+2}, 0.01 Sr^{+2}, 0.02 Zn^{+2}, 3.2 H^+, 10^7
Ca^{+2} and Mg^{+2}	$(RO)_2POO^-$ (dialkylphosphate anion)	Na^+, 0.01 Sr^{+2}, 0.54 Ba^{+2}, 0.94
Cu^{+2}	$RSCH_2COO^-$ (S-alkyl α-thioacetate anion)	Na^+, K^+, 0.0005 Mg^{+2}, 0.001 Ca^{+2}, 0.002 Ni^{+2}, 0.01 Zn^{+2}, 0.03
NO_3^-	[Ni(phenanthridine)$_3$]$^{+2}$ (1,10-phenanthridinium dication)	F^-, 0.0009 SO_4^{-2}, 0.0006 PO_4^{-3}, 0.0003 Cl^-, CH_3COO^-, 0.006 HCO_3^-, CN^-, 0.02 NO_2^-, 0.06 Br^-, 0.9
ClO_4^-	[Fe(phenanthridine)$_3$]$^{+2}$ (1,10-phenanthridinium dication)	Cl^-, SO_4^{-2}, 0.0002 Br^-, 0.0006 NO_3^-, 0.0015 I^-, 0.012 OH^-, 1.0

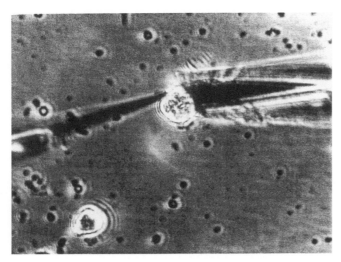

Figure 15.11 Micros copic carbon-fiber electrodes for measurements in a single cell. (From Michael, A. and Wightman, R.M., In *Laboratory Techniques in Electroanalytical Chemistry*, 2nd edn., Kissinger, P. and Heineman, W., eds., Marcel Dekker, Inc., New York, 1996. Used with permission.)

electrodes typically have long response times, on the order of a few minutes, because the gas must cross the membrane. These electrodes are highly selective and free from interference from nonvolatile species; only gases can cross the gas-permeable membrane. If a pH electrode is used as the ISE, obviously any gas that dissolves and changes the internal solution pH will be detected; therefore, some gases may interfere with other gas analytes.

15.3.1.1.7 Immobilized Enzyme Membrane Electrodes

Enzymes are biocatalysts that are highly selective for complex organic molecules of biochemical interest, such as glucose. Many enzymes catalyze reactions that produce ammonia, carbon dioxide, and other simple species for which ISEs are available as detectors. Coating an ISE with a thin layer of an enzyme holds the promise of making highly selective biosensors for complex molecules with the advantages of potentiometry—speed, low cost, and simplicity. The enzyme is generally immobilized on the electrode by incorporation into a gel, by covalent bonding to a polymer support or by direct adsorption onto the electrode

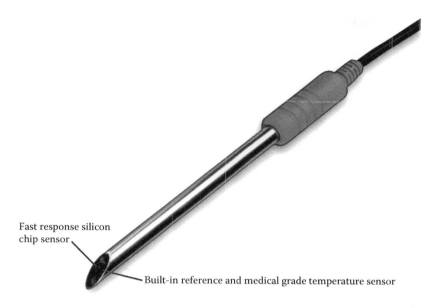

Figure 15.12 A commercial ISFET pH probe, with a silicon chip pH sensor, a built-in REF, and a built-in temperature sensor. This pH probe is housed in a stainless steel body with a slanted tip for easy insertion into meat or other soft solid samples. (Courtesy of IQ Scientific Instruments, Inc., Carlsbad, CA, www.phmeters.com.)

surface, so that a small amount of enzyme may be used for many measurements. Very few species interfere with enzyme-catalyzed reactions, although the enzyme can be inhibited. No commercial potentiometric enzyme-based electrodes are available; however, considerable research in this area is in progress.

15.3.1.1.8 Ion-Selective Field Effect Transistors

Ion-selective field effect transistors (ISFETs) are semiconductor devices related to the solid-state detectors used in spectroscopy. In this case, the surface of the transistor is covered with silicon nitride, which adsorbs H$^+$ ions from the sample solution. The degree of adsorption is a function of the pH of the sample solution and the adsorption of H$^+$ ions results in a change in the conductivity of the ion-selective field-effect transistor (ISFET) channel. The cell requires an external reference electrode. ISFET pH sensors can be made extremely small (about 2 mm^2) and are extremely rugged, unlike the fragile glass bulb pH electrode. They have rapid response times and can operate in corrosive samples, slurries, and even wet solids such as food products. The sensor can be scrubbed clean with a toothbrush, stored in a dry condition, and does not require hydrating before use. The price of an ISFET pH electrode is about half that of a standard glass electrode. A schematic of an ISFET pH electrode is shown in Figure 15.12, with the ISFET sensor embedded in a polymer body electrode.

A new (as of 2013, with recent software up-dates) solid-state pH electrode and meter, which employs voltammetric rather than potentiometric responses to measure pH, is described later in Section 15.3.5.3. This may revolutionize both lab-based and field-based pH monitoring.

15.3.1.2 Instrumentation for Measuring Potential

Understanding electrochemical instrumentation requires a basic knowledge of electricity and basic electronics. Coverage of these fundamentals is impossible in a text of this size. The student is advised to review the concepts of electricity learned in general physics and to understand the definitions of current, voltage, resistance, and similar basic terms. The texts by Kissinger and Heineman, Malmstadt et al., or Diefenderfer and Holton, listed in the bibliography, are excellent sources of information on electronics used in instrumentation. The electrochemical cell is one circuit element with specific electrical properties in the complete instrumentation circuit.

A **potentiometric cell** is a galvanic cell connected to an external potential source that is exactly equal but opposite to that of the galvanic cell. Negligible current flows in this type of cell. The potential of a complete cell can be measured with a solid-state circuit that requires negligible input current but can output the voltage to a meter, digital readout, recorder, or computer. The circuit should have high-input impedance, high-gain, low-output impedance, and ideally, no output signal when there is no input signal. **Operational amplifiers** (op-amp) have these properties. Figure 15.13 illustrates an op-amp set up as a voltage follower. The triangle represents the amplifier with two inputs on the left and an output on the right. It may contain a large number of discrete components, for example, transistors. Usually, such devices require ±15 V power lines, which are omitted in circuit diagrams. The voltage input marked with the + sign is called a *noninverting input*, and the one marked with the − sign is the *inverting input*. If A is the gain (typically 10^5), the output voltage E_{out} is

$$E_{out} = -A(E^- - E^+) \qquad (15.17)$$

where E^- and E^+ are the voltage values at the inverting and noninverting inputs, respectively. The minus sign means that the polarity of the voltage difference is reversed. Rearranging, we have

$$E^- = E^+ + \frac{E_{out}}{A}$$

and because A is very large,

$$E^- \approx E^+$$

Notice that E_{out} is fed back into the noninverting input (E^-). This is termed voltage feedback.

The function of this type of circuit is to offer a high-input impedance that can be typically 10^5–10^{13} ohms (Ω). This means that the + and − input require negligible current and can be used to monitor the voltage of a highly resistive source such as a glass electrode. However, the output current and voltage capabilities are considerable, depending on the device properties and the power supply limitations. It seems odd, at first, that the circuit just reproduces the voltage of the cell but does so in such a way as to make available much larger currents than are available from the electrochemical cell. The electrical resistance of the glass electrode for pH measurement is enormous (50–10,000 MΩ). An ordinary potentiometer or voltmeter would indicate practically no observable voltage, which is why a high-impedance solid-state measuring circuit is required. Modern instruments drain no more than 10^{-10}–10^{-12} A from the cell being measured.

Commercial solid-state potential measuring devices based on the type of op-amp described are often called **pH or pIon meters** and are designed to work with glass pH electrodes, ISEs, and other indicator electrodes described earlier. Research quality pIon meters have built-in temperature measurement and compensation, autocalibration routines for a three-point (or more) calibration curve, recognition of electrodes (so you do not try measuring fluoride ion with your pH electrode!), and the ability to download data to computer data collection programs. The relative accuracy of pH measurements with such a meter is about ±0.005 pH units. Meters are available as handheld portable devices for field or plant use; these meters have less accuracy, on the order of ±0.02 pH units, but are much less expensive. The field-portable meters are battery powered and waterproof. Commercial meters will read out in mV, pH, concentration of ion for an ISE, and mS for conductivity measurements, and many will also read out the temperature.

A second important use of op-amps and feedback circuits is to control the potential at a WE. The circuit that performs this task is called a **potentiostat**. Potentiostats and modern polarography equipment are called **three-electrode devices** because they attach to a WE, a reference electrode, and an auxiliary electrode, also called a CE (Figure 15.14). The principle of operation is that the potential between the WE and the reference electrode is maintained by a feedback circuit. The applied voltage E_{appl} may be a constant voltage

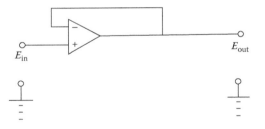

Figure 15.13 Schematic of an op-amp with voltage feedback (the voltage follower).

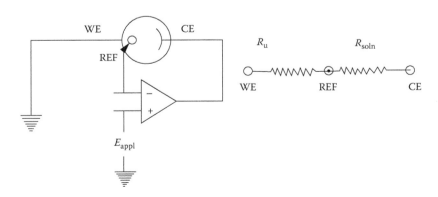

Figure 15.14 Schematic of a simple potentiostat and representation of the cell as an electrical (equivalent) circuit. WE, working electrode; REF, reference electrode; CE, counter electrode.

or some signal generator voltage, such as a linear ramp. To understand how this circuit works, consider starting with no potential at the amplifier output. The output voltage E_{out} will be given, as before, by

$$E_{out} = -A(A^- - E^+)$$

Because at time $t = 0$, $E^- = E_{appl}$, the voltage E_{out} will rise, with the same sign, until $E_{out} = E_{appl}$ at the point REF. Assuming that the uncompensated resistance R_u between WE and REF can be essentially ignored, this ensures that WE is held at E_{appl} and the potential of the CE can "float" to maintain zero differential input at the − and + inputs. Note that negligible current flows to the inverting (−) input through the REF. The measured cell current flows between the WE and the CE. In practice, the uncompensated resistance R_u is kept small by placing the REF close to WE, by using an excess of supporting electrolyte, and by keeping the net cell current small by using microelectrodes, such as a DME. Then the voltage drop in the electrolyte between WE and REF, iR_u, need not be compensated for.

On modern potentiostats, the applied voltage is given as the *actual* sign of the voltage on the WE, and there is no need to assume a potential reversal. In other words, if you set the potentiostat at −0.500 V, WE is at −0.500 V versus whatever reference electrode you have in the actual cell. If, for example, this electrode was an SCE, then on the SHE scale at 25 °C,

$$E_{WE} = -0.500 + 0.241$$
$$= -0.259 \text{ V vs. SHE}$$

If the cell had a silver/silver chloride reference (3.5 M) and the equipment were set at −0.500 V versus reference electrode, then

$$E_{WE} = -0.500 + 0.205$$
$$= -0.295 \text{ V vs. SHE}$$

In each of these cases, when we state our values on the hydrogen scale, the potential at the WE is numerically less. On the other hand, for an applied potential of +0.500 V versus SCE on the potentiostat,

$$E_{WE} = +0.500 + 0.241$$
$$= 0.742 \text{ V vs. SHE}$$

This raises an important point. It does not matter which reference electrode we use; it is best to use the one most suited for the chemical system being studied. We should always, however, quote our potentials versus the particular reference and be sure to label the axes of current–voltage curves with the appropriate reference electrode. Electrode potentials are relative values.

15.3.1.3 *Analytical Applications of Potentiometry*

Potentiometry is widely used in industrial, environmental, agricultural, clinical, and pharmaceutical laboratories to measure ions, acids, bases, and gases. Measurements may be made in a standard laboratory setting, but potentiometry is ideal for online process monitoring, *in vivo* monitoring, and field measurements using flow-through or portable instruments. Quantitative measurement of pH is extremely important and will be discussed in detail in the succeeding text. Quantitative measurement of inorganic and organic ions, acidic and basic ions, and gases and determination of ions in specific oxidation states are commonly performed by potentiometry. Potentiometry is used in research to determine stability constants of complexes, solubility product constants, to determine reaction rates and elucidate reaction mechanisms, and to study enzyme and other biochemical reactions.

Samples must be in the form of liquids or gases for analysis. Sample preparation may include buffering the sample to an appropriate pH for some ISE and gas measurements or adding ionic strength "buffer" to make all samples and standards of equivalent ionic strength. The detection limit for most common electrodes is about 10^{-6} M, while gas probes have detection limits in the low ppm range.

An understanding of stoichiometry, acid–base theory, and simple equilibrium calculations is required for the following quantitative applications. Review your introductory chemistry text or a basic quantitative analysis text such as those by Harris or Enke listed in the bibliography.

15.3.1.3.1 *Direct Measurement of an Ion Concentration*

The concentration of an ion in solution may be measured directly via the potential developed by the half-cell involved and by applying the Nernst equation. For example, a silver/silver ion half-cell plus an SCE gives the following relationship:

$$E(\text{cell}) = E(\text{Ag}) - E(\text{SCE})$$

where $E(\text{Ag})$ is the emf of the silver half-cell. But

$$E(\text{Ag}) = E^0(\text{Ag}) - \frac{RT}{nF} \ln \frac{1}{[\text{Ag}^+]}$$

Therefore,

$$E(\text{cell}) = E^0(\text{Ag}) - (0.05916) \log \frac{1}{[\text{Ag}^+]} - E(\text{SCE})$$

and at 25 °C, $E(\text{SCE}) = 0.244$ V. From Appendix 15.A, $E^0(\text{Ag}) = 0.799$ V. Substituting, we get

$$E(\text{cell}) = 0.799 - (0.05916) \log \frac{1}{[\text{Ag}^+]} - 0.244$$

This gives us $E = 0.555 + 0.05916 \log[Ag^+]$. In an experiment conducted at 25 °C, the cell potential for such a cell was measured and found to be = 0.400 V. Therefore,

$$\log[Ag^+] = \frac{0.400 - 0.555}{0.05916}$$
$$= -\frac{0.155}{0.05916}$$
$$\log[Ag^+] = -2.62$$

The concentration of Ag^+ = anti log(-26.2)
$= 2.4 \times 10^{-3}$ M

15.3.1.3.2 Potentiometric Titrations

Potentiometry is a useful way to determine the endpoint in many titrations. For example, the concentration of Ag^+ ions in solution can be used to determine the **equivalence point** in the titration of Ag^+ with Cl^-. In this titration, the following reaction takes place:

$$Ag^+ + Cl^- \rightarrow AgCl(s) \downarrow (\text{precipitation})$$

The concentration of the Ag^+ in solution steadily decreases as Cl^- is added. At the equivalence point, $[Ag^+] = [Cl^-]$. But, for the sparingly soluble salt silver chloride, $[Ag^+][Cl^-]$ is a constant called the *solubility product*, K_{sp}. In the case of the AgCl precipitation reaction, $K_{sp} = 1 \times 10^{-10}$ at 25 °C. Therefore,

$$[Ag^+][Cl^-] = 1 \times 10^{-10}$$

But, as we have seen, $[Ag^+] = [Cl^-]$ at the equivalence point; therefore,

$$[Ag^+]^2 = 1 \times 10^{-10}$$
$$[Ag^+] = \sqrt{1 \times 10^{-10}} = 1 \times 10^{-5} \text{ M}$$

From this calculation, at the equivalence point $[Ag]^+ = 1 \times 10^{-5}$ M. By applying the Nernst equation to this solution, we get

$$E(Ag) = E^0 + \frac{0.05916}{1} \log[Ag^+]$$

and

$$E^0(Ag) = +0.799$$

Therefore,

$$E(Ag) = +0.799 + \frac{0.05916}{1} \times (-5)$$
$$E(Ag) = +0.799 - 0.295$$
$$E(Ag) = +0.504$$

At the equivalence point, the emf of the half-cell is calculated to be 0.504 V versus SHE. If we make a complete cell by using the silver half-cell and an SCE half-cell (+0.244 V), the observed voltage at the equivalence point will be given by the relationship

$$E(\text{observed}) = E(Ag) - E(\text{reference})$$
$$= +0.504 - 0.244$$
$$= +0.260 \text{ V vs. SCE}$$

At the equivalence point, the measured cell emf is +0.260 V versus SCE. In the titration of silver ions, the chloride solution should be added in small portions and the emf of the cell measured after each addition. By plotting the relationship between the volume of Cl^- solution added and the voltage of the cell, we can determine the volume of Cl^- solution necessary to reach the equivalence point. A typical curve relating the volume of chloride to the voltage for titration of silver ions with 0.100 N Cl^- solution is shown in Figure 15.15. In a real titration, the endpoint is determined from the inflection point in the titration curve, not from the calculated voltage.

In the following example, the equivalence point is reached after adding 2.00 mL of 0.100 Cl^- solution to 5.00 mL of silver solution. The normality of the silver can be determined from the following equation:

$$(\text{Volume of } Cl^-) \times (\text{Normality of } Cl^-) = (\text{Volume of } Ag^+) \times (\text{Normality of } Ag^+)$$

We have (2.00 mL) (0.100 N) = (5.00 mL) (N Ag^+) or

$$\text{Normality of silver solution} = \frac{2.00 \text{ mL} \times 0.100 \text{ N}}{5.00 \text{ mL}}$$
$$= 0.0400 \text{ N}$$

From this, the mass of Ag^+ present in the solution can be calculated as follows:

1 L of 1 N Ag^+ contains 108 g Ag^+; thus,

$$\text{g Ag} = (108 \text{ g Ag/eq}) \ (0.0400 \text{ eq/L})$$
$$(1 \text{ L}/1000 \text{ mL}) \ (5.00 \text{ mL})$$

$$\text{g Ag} = 0.0216 \text{ g}$$

In this example, we used the known K_{sp} to calculate the potential at the equivalence point. Clearly, it is possible to work backward and to use the measured potential at the equivalence point to determine the K_{sp} for a sparingly soluble salt.

In Figure 15.15, the titration curve was simple and the equivalence point easily detected; however, in dilute solutions, nonaqueous solvents and titrations involving

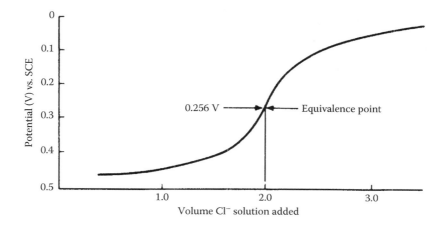

Figure 15.15 Relationship between Ag$^+$ concentration, the cell potential, and the volume of Cl$^-$ solution added.

slower reactions, the titration curve may be flatter (the inflection point hard to find) and difficult to interpret. The problem can be simplified by using equipment that records the **first and/or second derivative** of the titration curve. The curves obtained for a relationship such as that in Figure 15.15 are shown in Figure 15.16. When the first derivative of the curve is taken, the slope is measured and plotted against the volume added. As the potential changes, the slope changes. At the endpoint, the slope of the curve ceases to increase and begins to decrease, that is, the slope goes through a maximum. This can be seen in the curve of Figure 15.16b. The second derivative measures the slope of the first derivative curve and plots this against the volume added. At the endpoint, the curve in Figure 15.16b goes through a maximum. At the maximum, the slope of the curve is horizontal; that is, it has a zero slope. The second differential curve is shown in Figure 15.16c. As can be seen, there is a rapid change in the second derivative curve, whose values go from positive to negative, and at the equivalence point, the value of the second derivative equals zero. There are commercial automatic potentiometric titrators available that accurately and rapidly deliver titrant, perform the potentiometric measurements, calculate the derivatives, stop the titrant after the equivalence point is reached, and accurately calculate the concentration of the analyte. Many of these come with autosamplers and computer data-handling systems so that multiple titrations can be performed unattended.

15.3.1.3.3 pH Measurements

One of the most important applications of potentiometry is the determination of [H$^+$] or the pH of a solution, where pH is defined as the negative logarithm to the base 10 of the hydrogen ion activity (and is approximately equal to the negative logarithm to the base 10 of the hydrogen ion concentration under certain conditions discussed subsequently):

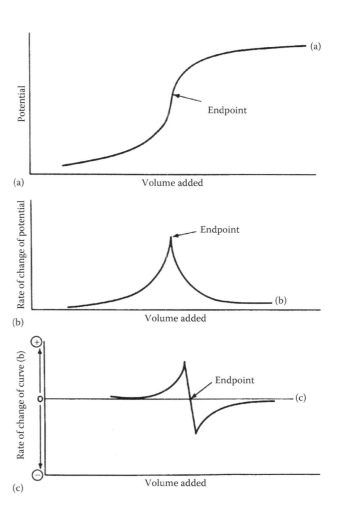

Figure 15.16 Relationship between (a) a simple potentiometric titration curve and its (b) first and (c) second derivatives.

$$\text{pH} = -\log(\text{H}^+) \approx -\log[\text{H}^+] \quad (15.18)$$

where
(H^+) stands for the activity of the hydrogen ion
$[\text{H}^+]$ is the molar concentration of the hydrogen ion

The procedure can be used for the determination of the pH of most solutions, which is important if the solution is drinking water, the water in a swimming pool, your favorite soda, or a chemical in an industrial process. Frequently, organic and inorganic synthetic reactions, biochemical reactions, and other chemical reactions carried out by research or industrial chemists are very sensitive to pH, which therefore must be measured and controlled.

The determination of $[\text{H}^+]$ in a solution is a special case of the determination of ion concentration (activity). The importance of this determination has merited special equipment designed for specific applications—field-portable, online analysis, and a wide variety of benchtop units. Hydrogen ion concentration can be determined directly by employing the Nernst equation, an SHE, and a reference electrode such as the SCE.

The potential developed by the cell is given by (if we ignore the liquid junction potential)

$$E(cell) = E(\text{H}^+) - E(\text{reference}) \quad (15.19)$$

Also,

$$E(\text{H}^+) = E^0(\text{H}) + \frac{RT}{nF}\ln[\text{H}^+]$$

$$E(\text{cell}) = -E(\text{reference}) + E^0(\text{H}) + \frac{RT}{nF}\ln[\text{H}^+]$$

$$= -E(\text{reference}) + E^0(\text{H}) + \frac{0.05916}{1}\log[\text{H}^+] \text{ at } 25°C$$

But $E(\text{H})$, the emf of a standard hydrogen cell, is zero by definition. Hence,

$$E(\text{cell}) = -E(\text{reference}) + \frac{0.0591}{1}\log[\text{H}^+]$$

The emf of an SCE = +0.241 V versus SHE; therefore,

$$E(\text{cell}) = -0.241 + 0.0591\log[\text{H}^+]$$

By rearranging, we get

$$-\log[\text{H}^+] = \frac{[E(\text{cell}) + E(\text{reference})]}{0.0591}$$
$$= \frac{[E(\text{cell}) + 0.241]}{0.0591}$$

If in a particular experiment the observed voltage of the cell is −0.6549 V versus SCE, then

$$-\log[\text{H}^+] = -\frac{(-0.6549 + 0.2412)}{0.05916} = 6.99$$

therefore,

$$[\text{H}^+] = 1.0 + 10^{-7} \text{ M}$$

SHEs are not used for routine pH determinations due to the difficulty of construction and operation of the SHE. The most common hydrogen half-cell in practical use is the glass electrode. An SCE is commonly used as the reference electrode. The cell may be formulated as

$$\text{SCE} \| [\text{H}_3\text{O}^+]_{(\text{unknown})} | \text{glass membrane} | [\text{H}_3\text{O}^+]_{(\text{internal reference})}$$

$$\| 0.1 \text{ M Cl}^-, \text{AgCl(sat.)} | \text{AgCl(s)}, \text{Ag(s)}$$

The SCE is the external reference electrode and the Ag/AgCl electrode is the internal reference electrode. Theoretically, the only unknown is the activity (concentration) of the H^+ in the sample or unknown solution. The Nernst equation tells us that the voltage of the pH electrode should change by 59.16 mV for every unit change in pH at 25 °C. However, we cannot in practice ignore the junction potential. The real equation for a cell is given by

$$E_{\text{cell}} = E_{\text{ind}} - E_{\text{ref}} + E_j \quad (15.20)$$

In addition to the junction potential E_j, membrane electrodes have asymmetry potentials of unknown magnitude. These potentials cannot be computed theoretically or easily measured individually, so direct pH measurements require that the pH electrode system be calibrated. In fact, the definition of pH used by the National Institute for Standards and Technology (NIST) and IUPAC is an *operational definition* based on calibration with a series of standard buffer solutions.

Buffer solutions are those in which the pH remains relatively constant (1) over wide ranges of concentration and (2) after the addition of small quantities of acid or base. Standard buffer solutions to calibrate pH electrode systems are commercially available from most major science supply companies, from electrode manufacturers, or may be formulated from mixtures described in chemical handbooks such as the one by Dean listed in the bibliography. Certified buffers can be purchased from NIST, but these are only needed for extremely accurate work.

The pH meter is calibrated (to compensate for asymmetry and other potentials and variations in the reference electrode) by immersing the glass electrode and the reference electrode in a buffer solution of known pH, which we will call pH_{std}. The potential of the cell generated by immersing the electrodes in the buffer is measured and is designated E_{std}. The electrodes are then placed in the sample, pH_{unk}, and the potential E_{unk} is measured. The pH of the sample is calculated from

$$pH_{unk} = pH_{std} - \frac{(E_{unk} - E_{std})}{0.05916} \quad (15.21)$$

Equation 15.21 is the operational definition of pH at 25 °C. This is our best experimental estimation of the activity of the hydrogen ion, (H$^+$), from Equation 15.18.

It is best to calibrate the instrument at a pH near to that of the sample or at two pH values that bracket the sample pH. If necessary, calibration can be performed at multiple pH levels, such as pH 4.00, 7.00, and 10.00 to check the linearity of the system. The use of a calibration curve gives the best empirical relationship between voltage and concentration, but the accuracy depends on matching the ionic strength of the sample to that of the standards to avoid differences in activity coefficients. This can be difficult to do in high-ionic-strength samples and complex matrices. In these cases, the method of standard additions (MSA), discussed subsequently, may be used to advantage.

15.3.1.3.4 Errors in pH Measurement with Glass Electrodes

There are several sources of error in the routine measurement of pH. One source of error that may occur with any pH probe, not just glass electrodes, is in the preparation of the calibration buffer or buffers. Any error in making the buffer or any change in composition on storage of the buffer will result in error in the pH measured. Common problems with buffers are bacterial growth or mold growth in organic buffers and absorption of CO_2 from air by very basic buffers (thereby making them less basic). Guidelines on storage of buffers, preservatives that can be added, and other practical advice can be found in the reference by Dean. The accuracy of the measured pH can be no more accurate than the pH of the calibration buffer.

Glass electrodes become sensitive to alkali metal ions in basic solution (pH > 11) and respond to H$^+$ *and* Na$^+$, K$^+$, and so on. This results in the measured pH being *lower* than the true pH. The magnitude of this **alkaline error** depends on the composition of the glass membrane and the cation interfering. It results from ion-exchange equilibria at the glass membrane surface between the alkali metal ions in the glass and the alkali metal ions in solutions. Special glass compositions are made for electrodes that are used in highly alkaline solutions to minimize the response to non-H$^+$ ions.

Glass electrodes also show an error in extremely acidic solutions (pH < 0.5). The **acid error** is in the opposite direction to the alkaline error; the measured pH values are too high. The cause of the acid error is not understood.

Variations in junction potential between calibration standards and samples lead to errors in pH measurement. Absolute accuracy of 0.01 pH units is difficult to obtain, but relative differences in pH as small as 0.001 pH units can be measured. The operational definition of pH is only valid in dilute solutions, with ionic strengths <0.1 and in the pH range 2–12. Both very-high-ionic-strength solutions and very-low-ionic-strength solutions (e.g., unpolluted lake water) can have serious errors in pH measurement caused by nonreproducible junction potentials.

15.3.1.3.5 pH Titrations

Direct measurement of pH can be used for quantitative acid–base titrations. For example, if NaOH is added slowly to HCl, the following reaction takes place:

Original solution: $HCl \rightarrow H^+ + Cl^-$

Addition of NaOH: $H^+ + Cl^- + Na^+ + OH^- \rightarrow H_2O + Na^+ + Cl^-$

In the process, [H$^+$] decreases and the potential of the pH electrode changes 59.16 mV for every unit change in pH, as we have seen. Near the neutralization point, the pH changes very rapidly with the addition of small amounts of NaOH. A plot of cell potential versus the volume (in mL) of NaOH added gives the relationship in Figure 15.17. The equivalence point of the titration is the inflection point of the curve.

The use of potentiometry for pH titration allows analyses to be carried out in colored or turbid solutions. Also, it solves the problem of selecting the correct indicator for a particular acid–base titration. The endpoint can be determined more accurately by using a first or second derivative curve as described earlier. It also permits pH titrations in nonaqueous solvents for the determination of organic acids and bases as described subsequently. In addition, it can be readily automated for unattended operation.

15.3.1.3.6 Titrations of Weak Acids and Bases

Most organic acids and bases and some inorganic weak acids and bases are too weak to be titrated in aqueous solution. This is a result of the **leveling effect**; the strongest acid that can exist in water is H$^+$ and the strongest base that can exist in water is OH$^-$. All acids that dissociate to give a proton are leveled to the same acid strength in water. For very weak acids and bases, the equilibrium constant for the titration reaction in water may not be large enough to give a distinct endpoint. It is common practice, therefore,

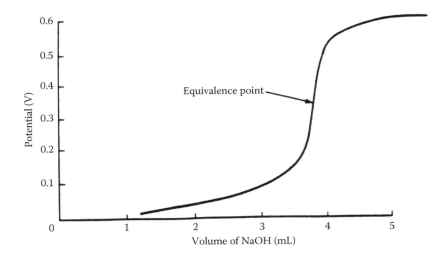

Figure 15.17 Schematic plot of the potentiometric titration of a strong acid with NaOH. A pH electrode can be used as the indicator electrode.

to choose a nonaqueous solution for these determinations. In a nonaqueous solvent, different acids are not leveled to the same extent. The nonaqueous solvents are chosen based on their acidity, dielectric constant, and the solubility of the sample in the solvent. It is important that the acidity of the solvent not be too great; otherwise, the titrant is used up in titrating the solvent rather than the sample. The same principles apply to the titration of bases, in that the solvent must not be too basic; otherwise, the reaction between the sample base and the titrant is not observed. The dielectric constant affects the relative strength of dissolved acids. For example, negatively charged acids become relatively stronger compared to uncharged acids (such as formic acid) as the dielectric constant of the solvent is decreased. It is also important that the sample be soluble in the solvent; otherwise, the reactions will be incomplete and unobservable. An advantage of using an organic solvent is that substances that are not soluble in water may be determined.

The types of cells used vary with the solvent employed. Glass and SCE electrodes may be used if the solvent is an alcohol, ketone, or acetonitrile but special modification of the salt bridge in the SCE is necessary. Platinum electrodes may be used for titrations with tetrabutylammonium hydroxide (TBAH), $(C_4H_9)_4NOH$, dissolved in a benzene–methanol mixture, or in methyl isobutyl ketone (MIBK). When TBAH is added to a sample that contains more than one acid functionality, it first reacts with the strongest acid in solution until reaction is virtually complete. It then commences reaction with the second strongest acid until the second reaction is complete, and so on. In a solution containing two different acid groups, if the K_a values differ enough, two equivalence points can be observed in a plot of emf (in mV) versus mL of titrant. These two acid groups may be on the same compound, that is, a diprotic acid, or they could be two different monoprotic acids (a mixture).

A schematic illustration of a titration of a mixture of three weak organic acids with TBAH is shown in Figure 15.18. **Nonaqueous titrations** have been used to determine amino acids, amines, alkaloids, carboxylic acids, many drugs such as barbiturates and antihistamines, and alcohols, aldehydes, and ketones.

The experimental setup for titrations in nonaqueous solvents differs from those in aqueous solution, and specialized texts (e.g., Huber) should be consulted.

15.3.1.3.7 Quantitative Determination of Ions other than Hydrogen

ISEs can be used for the analysis of aqueous and nonaqueous solutions alike and have therefore found increasing applications in organic chemistry, biochemistry, and medicine. Furthermore, in many circumstances, no sample preparation is necessary. As a result, the use of ISEs is particularly valuable for obtaining rapid results with no loss of volatile analytes. ISE potentiometry can be considered nondestructive, since the amount of contamination of the sample by leaking electrolyte is usually negligible. These electrodes can even be used for continuous analysis of waters and industrial plant streams simply by inserting the electrodes into the sample (such as the river or plant stream) and measuring the voltages generated.

It must be remembered that the cell potential is proportional to the ln(activity) of the ion rather than its concentration. The activity is a measure of the extent of thermodynamic nonideality in the solution. The activity coefficient is usually less than unity, so the activity of a solution is generally lower than the total concentration, but the values of activity and concentration approach each other with increasing dilution. If a compound is not completely ionized, the activity is further decreased. Decreased

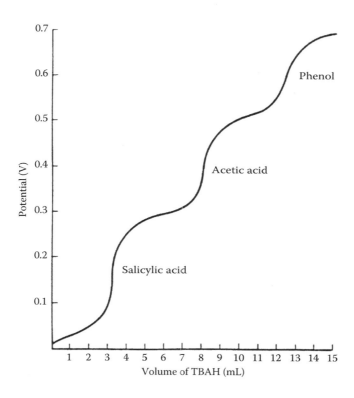

Figure 15.18 Schematic plot of the potentiometric titration of three weak acids with TBAH. Note the equivalence points after the addition of 3.5, 8.5, and 13.0 mL of titrant.

ionization can be brought about because of a weak ionization constant, chemical complexation, or a high salt concentration in the solution. Any of these factors will cause a change in the potential of the cell, even if the ion concentration is constant. In practice, it is better to determine the relationship between the cell potential and ion *concentrations* experimentally.

ISEs are calibrated in a manner similar to pH electrodes. Standard solutions of known concentrations of the ion to be measured are used and a plot of the cell emf versus concentration is made. It is important to keep the ionic strength and other matrix components of the samples and standards the same. This is often done by adding the same amount of a high-ionic-strength buffer to all samples and calibration standards. In some cases, the buffer may also adjust the pH if the form of the ion to be measured is pH dependent. The determination of fluoride ion is an example of this approach. All samples and standards have the same amount of a commercial buffer called total ionic strength adjusting buffer (TISAB) that also controls pH to insure that fluoride is present as F^-, not HF, because the fluoride electrode does not respond to HF.

Electrodes that behave in a Nernstian manner will have slopes equal to 59.16 mV per pIon unit for monovalent cations, (59.16/2) mV per pIon unit for divalent cations, −59.16 mV per pIon unit for monovalent anions, and so on, at 25 °C. pIon is defined exactly as is pH, as the negative logarithm to the base 10 of the ion concentration (activity) in M, giving pF, pCa, pAg, and so on. Many plots of concentration versus voltage for ISEs deviate from linearity due to activity coefficient effects, so an alternate approach is to use the method of standard addition (MSA) for calibration, especially for complex matrices.

15.3.1.3.8 Calibration by Method of Standard Additions

In the construction of standard addition curves for calibration, we assume that the potential is proportional to the log of the concentration. Let the original ion concentration be C_0 and the solution volume be V_0 (in L). The initial potential is given by

$$E_1 = \text{const.} + 2.303 \frac{RT}{nF} \log C_0 \quad (15.22)$$

It is usual to add a small volume V_s of a standard solution of concentration C_s from a microliter pipette such that the concentration becomes $(C_0V_0 + C_sV_s)/(V_0 + V_s)$ and the new potential E_2 is given by

$$E_2 = \text{const.} + 2.303 \frac{RT}{nF} \log\left(\frac{C_0V_0 + C_sV_s}{V_0 + V_s}\right) \quad (15.23)$$

Subtracting Equation 15.22 from Equation 15.23, we have

$$\Delta E = 2.303 \frac{RT}{nF} \log\left(\frac{C_0 V_0 + C_s V_s}{C_0(V_0 + V_s)}\right)$$

If we note that $V_0 \gg V_s$,

$$\Delta E \approx 2.303 \frac{RT}{nF} \log\left(\frac{C_0 V_0 + C_s V_s}{V_0 + V_s}\right)$$

$$\approx 2.303 \frac{RT}{nF} \log\left(1 + \frac{C_s V_s}{C_0 V_0}\right)$$

This expression, in turn, can be rearranged as

$$1 + \frac{C_s V_s}{C_0 V_0} = 10^{\frac{nF \Delta E}{2.303 RT}} \tag{15.24}$$

from which we can obtain the concentration of the unknown by plotting ($10^{nF\Delta E/2.303RT} - 1$) versus V_s for several additions of the standard. Then the slope m will have the value $C_s/C_0 V_0$ and the unknown concentration is calculated from

$$C_0 = \frac{C_s}{m V_0} \tag{15.25}$$

The method is usually very sensitive but depends upon the particular ISE. Also, special buffers may be required to avoid problems with interfering ions. In general, two additions should be made where possible, with the concentration doubled from the original concentration. It is assumed that the additions do not change the ionic strength or the junction potentials significantly.

ISEs have been used for the determination of sodium and potassium in bile, nerve and muscle tissue, kidneys, blood plasma, urine, and other body fluids. ISEs are used for the analysis of ions in sea water, river water, and industrial water and wastewater, as well as in a wide variety of commercial products, such as personal care and cosmetic products. The advantages of ISEs are that they are fast, with response times <1 min for most ISEs; they are a nondestructive technique, have a linear range of about six orders of magnitude in concentration, usually over the 10^{-6} to 1 M range, and they can be used in turbid or highly colored solutions. The disadvantages are that a different electrode is needed for each ionic species, the electrodes are selective but not specific, so interferences can occur, and the electrodes can become plugged or contaminated by components of the sample. The ionic species must be in solution and in the proper oxidation state to be detected by the electrode.

15.3.1.3.9 Oxidation–Reduction Titrations

Potentiometry can be used to follow reduction–oxidation (redox) titrations. For example, the oxidation of stannous ions by ceric ions follows the chemical reaction

$$Sn^{2+} + 2Ce^{4+} \rightarrow Sn^{4+} + 2Ce^{3+}$$

It is usually best to first identify the two half-cell reactions and to find the total number of electrons n in the reaction (n must be the same for each half-cell reaction, so it may be necessary to balance the half reactions) and to look up the standard reduction potentials in Appendix 15.A:

$$Sn^{2+} \rightarrow Sn^{4+} + 2e^- \quad E^0 = +0.13 \text{ V}$$

$$2Ce^{4+} + 2e^- \rightarrow 2Ce^{3+} \quad E^0 = +1.74 \text{ V}$$

The net $E^0_{cell} = 1.74 - 0.13 \text{ V} = +1.61 \text{ V}$. The emf $E(A)$ of the half-cell is given by the Nernst equation for a stannous/stannic half-cell, with the stannous ion oxidizing to stannic ion:

$$E(A) = E^0\left(\frac{Sn^{2+}}{Sn^{4+}}\right) + \frac{0.05916}{2} \log\left(\frac{[Sn^{4+}]}{[Sn^{2+}]}\right) \tag{15.26}$$

For the ceric/cerous half-cell, with ceric ion reduced,

$$E(B) = E^0\left(\frac{Ce^{3+}}{Ce^{4+}}\right) - \frac{0.0591}{2} \log\left(\frac{[Ce^{3+}]^2}{[Ce^{4+}]^2}\right) \tag{15.27}$$

At all times, the emf produced by each half-cell must be equal to that of the other because the mixture cannot exist at two potentials. Therefore, from the Nernst equation (Equation 15.12), we have

$$E_{cell} = E^0_{cell} - \frac{0.05916}{2} \log \frac{[Sn^{4+}][Ce^{3+}]^2}{[Sn^{2+}][Ce^{4+}]^2}$$

Since the equilibrium constant for this reaction is

$$K_{eq} = \frac{[Sn^{4+}][Ce^{3+}]^2}{[Sn^{2+}][Ce^{4+}]^2}$$

at equilibrium,

$$E^0_{cell} = \frac{0.0591}{2} \log K_{eq} \tag{15.28}$$

A similar relationship is true for all equilibrium redox reactions. With knowledge of $E^0(Sn^{2+}/Sn^{4+})$ and $E^0(Ce^{3+}/Ce^{4+})$, it is possible to calculate the equilibrium constant for the reaction.

It is possible to calculate the *equivalence point potential* by noting that we cannot ignore the small concentrations of Sn^{2+} and Ce^{4+} remaining, even though the overall reaction may be close to completion. The stoichiometry at the equivalence point demands that

$$2[Sn^{4+}] = [Ce^{3+}]$$

and

$$2[Sn^{2+}] = [Ce^{4+}]$$

If we let the potential at equivalence be E_{eq},

$$E_{eq} = E^0\left(\frac{Ce^{3+}}{Ce^{4+}}\right) + \frac{0.0591}{2}\log\left(\frac{[Ce^{4+}]^2}{[Ce^{3+}]^2}\right)$$

$$2E_{eq} = 2E^0\left(\frac{Sn^{2+}}{Sn^{4+}}\right) + \frac{0.0591}{2}\log\left(\frac{[Sn^{4+}]^2}{[Sn^{2+}]^2}\right)$$

Adding these two expressions and substituting the values for the Ce ion species in terms of Sn, the log term becomes equal to zero:

$$\log\left(\frac{2[Sn^{2+}]^2[Sn^{4+}]^2}{2[Sn^{4+}]^2[Sn^{2+}]^2}\right) = \log 1 = 0$$

Thus, $3E_{eq} = E^0(Ce^{3+}/Ce^{4+}) + 2E^0(Sn^{2+}/Sn^{4+})$. In practice, it is often easier to observe the equivalence point than to calculate it! If you must calculate the shape of a redox titration curve, the use of a spreadsheet program such as Excel is invaluable because of the multiple equilibrium equations that must be solved. (See the text by Harris for excellent examples of spreadsheet calculations for redox titrations.)

The relationship between the emf of the cell and the number of milliliters of ceric salt solution added to the stannous solution is of the same form as Figure 15.15. The equivalence point is denoted by a rapid change in potential as the ceric salt is added. If necessary, the first and second derivatives of the curve may be taken to detect the endpoint as shown in Figure 15.16. From the data obtained, calculations of solution concentrations and other variables are done in the same manner as in conventional volumetric analysis.

The same principle can be used for many redox reactions. The potential of the cell depends on the concentration of the oxidized and reduced forms of ions present. During a titration, these concentrations vary as the chemical reaction proceeds:

$$\text{red}(A) + \text{ox}(B) \rightarrow \text{ox}(A) + \text{red}(B)$$

where
red(A) is the reduced form of ion A
ox(B) is the oxidized form of ion B, and so on

At the equivalence point, there is a rapid change in potential with the addition of the titrating solution. This change makes it possible to detect the equivalence point of the reaction. Typical inorganic oxidation–reduction reactions include

$$2Fe^{2+} + Sn^{4+} \rightarrow 2Fe^{3+} + Sn^{2+}$$

$$5V^{2+} + 3MnO_4^- + 4H^+ \rightarrow 5VO_2^+ + 3Mn^{2+} + 2H_2O$$

$$IO_3^- + 8I^- + 6H^+ \rightarrow 3H_2O + 3I_3^-$$

$$I_3^- + 2S_2O_3^{2-} \rightarrow 3I^- + S_4O_6^{2-}$$

Organic compounds may be determined by redox titrimetry, including glucose and other reducing sugars, vitamin C, thiols such as cysteine, and many other organic chemicals of biological importance. There are thousands of potentiometric redox titration methods published in the chemical literature.

It may be necessary to correct for interferences to a method depending on your sample. For example, the presence of traces of other oxidizing or reducing compounds may displace the equilibrium and give an incorrect voltage for the equivalence points. Many equilibria are pH sensitive, and this too may generate inaccurate results. Care must be taken to correct for these interference effects.

15.3.2 Coulometry

It was shown earlier that when a standard cell is made from a Zn/Zn^{2+} half-cell and a Cu/Cu^{2+} half-cell, Zn dissolves, Cu^{2+} is deposited as Cu metal, and a potential of 1.100 V is developed from the spontaneous cell thermodynamics. A cell operating spontaneously is called a **galvanic cell**. It is possible to cause a cell to react in the nonspontaneous direction by application of a sufficient potential. A cell operated in this manner is called an **electrolytic cell**. The process of causing a thermodynamically nonspontaneous oxidation or reduction reaction to occur by application of potential or current is called **electrolysis**. Electrolysis is carried out for a sufficient length of time to convert a species quantitatively to a new oxidation state. There are three primary electroanalytical methods based on electrolysis; they are constant-current coulometry (coulometric titrimetry), constant-potential coulometry, and **electrogravimetry**. In electrogravimetry, the product of the electrolysis (usually a metal) is plated out on one of the electrodes and the amount is determined by weighing the electrode. The charge is measured in the **coulometric methods**. In contrast to potentiometry, where we did not want to change the concentration of the species in solution, electrolytic methods are designed to completely consume the species being measured by converting it quantitatively to a new species.

If we consider the Zn/Cu galvanic cell, we might consider how to reverse the reaction. If a sufficient voltage is applied in the opposite direction, from a power source or battery, there will be a tendency for these reactions to reverse, provided that no new reactions occur. In this case, if we try to electrodeposit Zn from aqueous solution by forcing the reaction $Zn^{2+} + 2e^- \rightarrow Zn(s)$, the evolution of hydrogen from the water will take precedence by the reaction $2H^+ + 2e^- \rightarrow H_2(g)$.

There is much more water in the cell than Zn^{2+}, so the applied voltage goes to form H_2 and we get no deposition of Zn(s). We say that the Zn/Zn^{2+} (aq) half-cell is *chemically irreversible*. On the other hand, the copper will dissolve if made an anode in an acidic solution, so the reaction $Cu(s) \rightarrow Cu^{2+} + 2e^-$ is said to be *chemically reversible*.

The differences and similarities between an electrochemical cell and an electrolysis cell are summarized in Figure 15.19. Oxidation always occurs at the anode and reduction always occurs at the cathode; notice, however, that in a spontaneous cell, the positive electrode is the one at which reduction takes place, whereas in an electrolysis cell, the negative electrode is the one at which reduction takes place. Negatively charged anions, such as Cl^-, NO_3^-, and SO_4^{2-}, are attracted to a positive electrode. Positively charged cations, such as Na^+, Ca^{2+}, and Mg^{2+}, are attracted to a negative electrode. Under the influence of the applied voltage in an electrolysis cell, the H^+ and Zn^{2+} ions are attracted to the negative electrode, the cathode; the SO_4^{2-} and OH^- ions are attracted to the positive electrode, the anode. Electrons are consumed at the cathode by H^+ ions to evolve hydrogen gas in the electrolysis cell but are consumed by Cu^{2+} ions in the spontaneous cell to deposit Cu atoms. For continuous operation of either type of cell, it is necessary for both the anode and cathode reactions to proceed with equal numbers of electrons.

Frequently, the reaction that proceeds at either the anode or cathode is the decomposition of the solvent; if water is

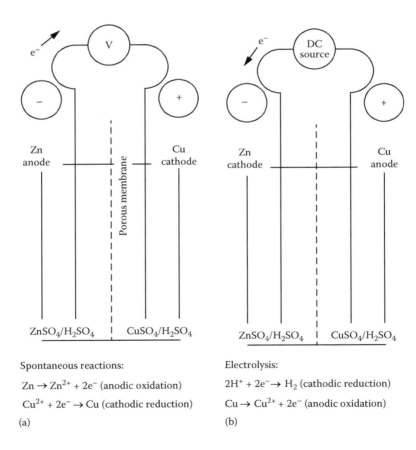

Figure 15.19 Comparison of (a) a galvanic cell and (b) an electrolysis cell.

the solvent, it may electrolyze. Note that water electrolysis depends on the pH:

For reductions: $2H^+ + 2e^- \rightarrow H_2$

$2H_2O + 2e^- \rightarrow H_2 + 2OH^-$

For oxidations: $2H_2O - 4e^- \rightarrow O_2 + 4H^+$

$4OH^- - 4e^- \rightarrow O_2 + 2H_2O$

15.3.2.1 Electrogravimetry

If a solution of a metallic ion, such as copper, is electrolyzed between electrodes of the same metal (i.e., copper), the following reaction takes place:

$Cu^0 \rightarrow Cu^{2+} + 2e^-$ (anode)

$Cu^{2+} + 2e^- \rightarrow Cu^0$ (cathode)

The net result is that metal dissolves from the anode and deposits on the cathode. The phenomenon is the basis of electroplating (e.g., chromium plating of steel). Also, it is the analytical basis of an electrodeposition method known as *electrogravimetry*. This involves the separation and weighing of selected components of a sample. Most metal elements can be determined in this manner, usually deposited as the M^0 species, although some metal elements can be deposited as oxides. The halides can be determined by deposition as the silver halide. Metals commonly determined include Ag, Bi, Cd, Co, Cu, In, Ni, Sb, Sn, and Zn.

15.3.2.2 Instrumentation for Electrogravimetry and Coulometry

The basic apparatus required is a power supply (potentiostat with a DC output voltage), an inert cathode and anode (usually platinum foil, gauze, or mesh), and arrangements for stirring. Sometimes a heater is used to facilitate the processes.

Electrogravimetry is usually carried out under conditions of controlled potential, so an auxiliary reference electrode is used together with the WE. (The WE can be either the anode or the cathode.) The auxiliary electrode–WE pair is connected to a potentiostat that fixes the potential of the WE by automatically adjusting the applied emf throughout the course of the electrolysis.

The WE should have a large surface area. Platinum is used because it is inert; Pt gauze or mesh electrodes are used to provide a large surface area. The electrode "mesh" is like wire window-screening material, with the wire diameter about 0.2 mm. The mesh is welded into an open cylinder to make the electrode. A standard size is a cylinder about 5 mm high and 5 mm in diameter, but Pt mesh electrodes of many sizes are available.

Because the deposition needs to be quantitative, the solution must be stirred during electrolysis. A magnetic stirrer is commonly used, often as a combination stirrer–hot plate. For electrogravimetry, an analytical balance of the appropriate capacity and sensitivity is needed. The electrode on which deposition will occur is weighed before the experiment is started. The electrode with the deposit is removed from solution upon completion of the electrolysis, washed, dried, and then reweighed. For coulometric methods, the charge is measured. The charge is obtained by integration of the current using an electronic integrator.

It is critical for these techniques that the current efficiency be as close to 100 % as possible. The percent current efficiency tells us how much of the applied current results in the reaction of interest. The percent current efficiency is defined as being equal to $100 \times (i_{applied} - i_{residual})/i_{applied}$, where $i_{applied}$ is the current applied to the cell and $i_{residual}$ is the "background" current. All real cells have some residual current. To achieve a 99.9 % current efficiency, the applied current must be approximately 1000 times the residual current. This is readily achieved for currents in the µA to 100 mA range used in constant-current coulometry.

15.3.2.3 Applied Potential

Electrogravimetry is generally performed under a constant, controlled potential but can also be performed under conditions of controlled current. For example, a metal alloy may contain nickel and copper. The alloy can be dissolved and the solution electrolyzed, with the result that the copper is selectively and exhaustively deposited on the cathode. The electrodeposited copper should form an adherent coating so that it can be washed, dried, and weighed. (The formation of appropriate forms of deposit has been studied thoroughly and is critical to obtaining accurate results. See the handbook by Dean for detailed information and references.) Instead of dissolution of the platinum metal taking place at the platinum anode, oxygen gas is liberated from the aqueous solution. As a consequence, platinum is a useful electrode material for redox purposes, since it is not consumed in the process. The initial reactions are as follows:

Cathode:	$Cu^{2+} + 2e^- \rightarrow Cu$	$E^0 = +0.337$ V
Anode:	$2H_2O - 4e^- \rightarrow O_2 + 4H^+$	$E^0 = -1.229$ V
Net reaction:	$2H_2O + 2Cu^{2+} \rightarrow 2Cu + O_2 + 4H^+$	$E_{net} = -0.892$ V

We may now consider on a more quantitative basis what influence the applied potential has. If the cell could behave as a thermodynamic cell and the reactants were under standard conditions, we might expect that the voltage needed would be that given by the net reaction, $E_{net} = -0.892$ V. But this would be for an infinitesimal, reversible change, and similar to any other chemical reaction, an electrochemical reaction has an *activation energy* that must be overcome for the reaction to occur. In fact, an additional voltage termed the **overpotential**, η, is required to drive the electrode reactions. About 2 V has to be applied to reach the current onset of the cell described, and this threshold is sometimes called the *decomposition voltage*. Thus, the total potential that we must apply across the cell is

$$E_{decomp} = E_{net} + IR + \eta \qquad (15.29)$$

An overpotential contribution is required for both the anodic and the cathodic reactions and, in this case, it is mainly to oxidize water at the Pt anode. As copper is deposited and the concentration of Cu^{2+} ions falls, both the ohmic potential drop, IR, across the electrolyte and the overpotential decrease. If we assume that the solution resistance R remains fairly constant, the ohmic potential drop is directly proportional to the net cell current. The overpotential increases exponentially with the rate of the electrode reaction. Once the Cu^{2+} ions cannot reach the electrode fast enough, we say that *concentration polarization* has set in. An ideally polarized electrode is one at which no faradaic reactions ensue; that is, there is no flow of electrons in either direction across the electrode–solution interface. When the potential of the cathode falls sufficiently to reduce the next available species in the solution (H^+ ions, or nitrate depolarizer), the copper deposition reaction is no longer 100% efficient.

Vigorous stirring, therefore, is necessary for several purposes. It helps to dislodge the gas bubbles that form at the electrodes. More importantly, it provides convective transport of the Cu^{2+} ions to the cathode as the solution becomes increasingly dilute in Cu^{2+} ion. At this point, if the applied voltage is raised to increase the current density, the potential could become sufficiently negative to cause the reduction of the next easiest reduced species present, in this case, H^+ ions:

$$2H^+ + 2e^- \rightarrow H_2 \quad E^0 = 0.00 \text{ V}$$

Such evolution of H_2 gas can produce unsatisfactory copper deposits, which may flake off and lead to a poor analytical determination. To prevent this, it is usual to add a cathodic depolarizer of NO_3^- ions to the solution. A *depolarizer* is a species more readily reducible than H^+ ions but one that does not complicate the cathodic deposit by occlusion or gas entrapment. In this instance, the product is ammonium ions:

$$NO_3^- + 10H^+ + 8e^- \rightarrow NH_4^+ + 3H_2O$$

If the potential becomes too negative at the cathode, the Ni^{2+} ions from our alloy sample might commence to deposit and form an alloy deposit. Note that the order of selectivity is not always that predicted by the standard potentials in the activity series. Control of the pH and the chelates or complexing agents in a solution may alter the reduction–oxidation potentials from those of an uncomplexed species.

Perhaps the most common example of overvoltage encountered in electrochemistry is that needed to reduce H^+ ions at a mercury electrode. On a catalytic Pt surface (platinized Pt, which is a large-surface-area, black Pt deposit on Pt metal), the H_2/H^+ ion couple is said to behave reversibly. This means that one can oxidize hydrogen gas, or reduce H^+ ions, at the standard reduction potential, 0.00 V, under standard conditions. At a mercury cathode, however, it takes about -1 V versus SHE to reduce H^+ ions, because hydrogen evolution is kinetically slow. This phenomenon has great importance in polarography, because it enables the analysis of many of the metals whose standard potentials are negative versus the SHE.

Controlled potential electrolysis (potentiostatic control) requires a three-electrode cell, so as not to polarize the reference electrode. Controlled potential methods enable one to be very selective in depositing one metal from a mixture of metals. If two components have electrochemical potentials that differ by no more than several hundred millivolts, it may still be possible to shift these potentials by complexing one of the species. One disadvantage of exhaustive electrolysis is the time required for analysis, and faster methods of electrochemical analysis are described.

In summary, important practical considerations in precise electrodeposition are (1) rapid stirring, (2) optimum temperature, (3) correct controlled potential or correct current density (usually expressed in A/cm^2), and (4) control of pH. The proper conditions have all been worked out for hundreds of systems; when these conditions prevail, deposits are bright and adherent and no unusual precautions are necessary in handling (washing, drying, weighing, etc.) the electrode. In commercial electroplating, additives (glue, gelatin, thiourea, etc.) are used as brighteners and for good adherence. They are not used in analytical deposition, because they would cause the mass of deposit to be erroneously high and would create an interference in the method. Electrodeposits improve the appearance, wear, and corrosion resistance of many fabricated metal parts. Electrolysis is also used to recharge batteries, study electrode reactions, extract pure metals from solutions, create shapes that cannot be machined (electroforming), and eliminate metallic impurities from solutions.

15.3.2.4 Analytical Determinations Using Faraday's Law

Electrogravimetry depends on weighing the WE before and after plating out the element under test. Therefore, it is limited to the determination of electroactive species where the product of electrolysis is a solid that forms a suitable deposit. As an alternative to weighing the deposit, the quantity of electricity used to deposit the metal can be measured. From this measurement, the quantity of metal deposited or the amount of ions reduced or oxidized can be calculated. This is an advantage if the product of electrolysis is a gas or another soluble ionic species; electrogravimetry will not work for these analytes. The calculation is based on Faraday's law, which states that equal quantities of electricity cause chemical changes of equivalent amounts of the various substances electrolyzed. The advantage of measuring the charge is that coulometry can measure electroactive species that do not form solid deposits, so it has much wider application. The accuracy is as good as gravimetric methods and coulometry is much faster than gravimetry and classical volumetric methods of analysis.

Charge is measured in either coulombs or faradays. One coulomb (C) is the amount of charge transported in 1s by a current of 1 A. One faraday (F) is the charge in coulombs of 1 mol of electrons, so 1 F = 96,485 C/mol e^-.

The relationship between charge and amount of analyte for a constant current, i, can be stated as

$$q = nFVM = \frac{nFw}{M_w} = i \times t \quad (15.30)$$

where
q is the charge (in coulombs)
n is the number of equivalents per mole of analyte
F is the Faraday constant
V is the volume (L)
M is the molar concentration of analyte (mol/L)
w is the mass of analyte (g)
M_w is the relative molecular mass of the analyte (g/mol)
i is the current (A)
t is the time (s)

This relationship involves only fundamental quantities; there are no empirically determined "calibration factors," so standardization and calibration are not required. This makes coulometry an **absolute method**; it does not require calibration with external standards. The current efficiency must be 100%. For a variable current, the charge is given by

$$q = \int_0^t i \, dt$$

Example A

A sample of stannic chloride was reduced completely to stannous chloride according to the reaction $Sn^{4+} + 2e^- \rightarrow Sn^{2}$.

The applied current was 9.65 A and the time taken for reduction was 16.0 min 40 s. What was the initial mass of stannic ion present? First convert the time to seconds. 16.0 min × 60 s/min = 960 s. Adding the additional 40s gives a total time of 1.00×10^3 s:

$$\text{Number of coulombs} = i(A) \times t(s)$$
$$= 9.65 \, A \times 1.00 \times 10^3 \, s$$
$$= 9650 \, C$$

$$\text{Number of faradays} = \frac{9,650}{96,500} = 0.100 \, F$$

But 1 F will reduce 1 g-eq mass of stannic ion:

$$1 \, g-eq \, wt. = \frac{\text{atomic wt. of tin}}{\text{valence change}} = \frac{118.69 \, g/mol}{2 \, eq/mol}$$

that is, 1 F will reduce 59.35 g of stannic ion, or 0.100 F will reduce 5.935 g of stannic ion.

Example B

If the original volume of the solution was 250.0 mL, what was the molarity of the solution?

Molarity means moles of solute contained in exactly 1 L of solution. The atomic mass of tin, the solute, is 118.69 g/mol. It was shown in Example A that 5.935 g of tin was reduced and Example A stated that the reduction was complete. The molarity of the solution is calculated from

$$\text{mol sn} = 5.935 \, g \, Sn \times \frac{1 \, mol \, Sn}{118.69 \, g}$$

$$\text{mol Sn} = 0.05000 \, mol$$

and

$$L_{\text{solution}} = 250.0 \, mL \times \frac{1.000 \, L}{1000 \, mL} = 0.2500 \, L$$

Therefore, the molarity of the solution was 0.05000 mol Sn/0.2500 L = 0.200 M stannic chloride to three significant figures.

Equation 15.30 allows us to calculate the amount of an electroactive species in either molarity or mass of material electrolyzed.

15.3.2.4.1 Controlled Potential Coulometry

Coulometry is frequently carried out under conditions of controlled potential. This is achieved by using a third electrode in the system, as described earlier. The three-electrode system maintains the potential at the WE at a constant value or permits an applied voltage pulse or ramp to be added to the WE. The reason for using controlled potential can be seen when we examine the Nernst equation. As a metal is oxidized or reduced under experimental conditions, the concentration of the remaining metal ions in the original oxidation state in solution steadily decreases. Therefore, in order to continue the deposition, the potential applied to the system must be steadily increased. When the potential is increased, different elements may begin to react or deposit and interfere with the results. Their deposition results in an increased mass of metal deposited (in electrogravimetry) and an increased number of coulombs passing through the cell in all coulometric methods. Controlled potential is therefore used to eliminate interferences from other reactions that take place at different potentials.

In theory, the equivalence point is never reached, because the current decays exponentially. Since a small amount of material always remains in solution, a correction must be made for it by measuring the current flowing at the end of the analysis. This quantity should be subtracted from the integrated signal in order to give an accurate measure of the material that has deposited or reacted during the experiment.

Controlled potential coulometry, also called bulk electrolysis, is usually used to determine the number of electrons involved in a reaction when studies are being carried out on new inorganic or organic compounds. Using a coulometer, we measure q. If we know M_w, from mass spectrometric measurements, for example, the number of moles of electrons can be calculated using Equation 15.30. This indicates how the species oxidizes or reduces; it gives us information about the reaction chemistry of new compounds. Controlled potential coulometry is also valuable in generating unstable or highly reactive substances *in situ* with good quantitative control.

15.3.2.4.2 Coulometric Titrations

An aqueous iodide sample may be titrated with mercurous ion by anodizing a mercury pool electrode. When metallic mercury is oxidized to mercurous ion by a current passing through the system, the mercurous ion reacts directly with the iodide ion to precipitate yellow Hg_2I_2:

$$2Hg - 2e^- \rightarrow Hg_2^{2+}$$

$$Hg_2^{2+} + 2I^- \rightarrow Hg_2I_2$$

The reaction continues and current passes until all the iodide is used up. At this point, some means of endpoint detection is needed. Two methods are commonly adopted. The first uses an amperometric circuit with a small imposed voltage that is insufficient to electrolyze any of the solutes. When the mercury ion concentration suddenly increases, the current will rise because of the increase in the concentration of a conducting species. The second method involves using a suitable indicator electrode. An indicator electrode may be a metal electrode in contact with its own ions or an inert electrode in contact with a redox couple in solution. The signal recorded is potentiometric (a cell voltage vs. a stable reference electrode). For mercury or silver, we may use the elemental electrodes, because they are at positive standard reduction potentials to the hydrogen/hydrogen ion couple.

Figure 15.20 illustrates an apparatus suitable for the coulometric titration (also known as constant

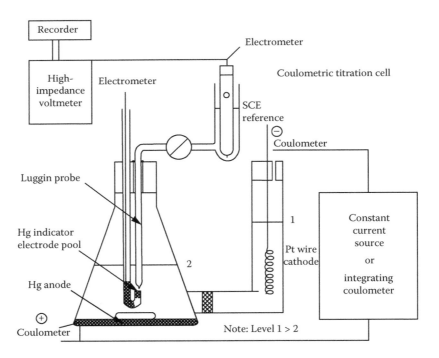

Figure 15.20 Apparatus for coulometric titration with potentiometric endpoint detection.

current coulometry) described. The anode and cathode compartments are separated with a fine porous glass membrane to prevent the anode products from reacting at the cathode, and vice versa. The porosity should be such as to allow minimal loss of titrant in the course of the experiment. The electrolysis circuit is distinct from the endpoint detector circuit. A constant-current source may be used, but it is not mandatory, provided that the current is integrated over the time of the reaction, that is,

$$q = it \,(\text{constant current})$$

or

$$q = \int_0^t i(t)\,dt \,(\text{integrator})$$

The potentiometric cell can be written as

$$\text{SCE} \| 0.1\text{ M HClO}_4 | \text{Hg}_2^{2+},\ 0.2\text{ M NaNO}_3 | \text{Hg}$$

The endpoint is sensed by recording the voltage across the potentiometric cell.

The **Luggin probe** used to monitor the potential of the Hg indicator electrode is spaced close (~1 mm) to the mercury droplet to reduce ohmic resistance. The top barrel is cleaned, wetted with concentrated $NaNO_3$ solution, and closed. Sufficient conductivity exists in this solution layer to permit the small currents necessary for potentiometric determination. When an excess of mercurous ion is generated, the potential of the Hg/Hg_2^{2+} couple varies in accordance with the Nernst equation, and a voltage follower may be used to output the voltage curve to a computer data system. For precise work, dissolved oxygen should be removed from the supporting electrolyte and efficient stirring is needed.

Coulometric titrimetry has been used for the determination of mercaptan, halide, and phosphorus compounds by using a silver electrode. The principle has also been used for many other types of titrations. Another example is the determination of ferrous ion, Fe^{2+}, in the presence of ferric ion, Fe^{3+}. This reaction can be controlled directly by coulometric analysis, but cannot be carried out to completion. As the concentration of Fe^{2+} decreases, the Nernst equation indicates that the potential necessary to continue oxidation will steadily increase until water is oxidized and oxygen is evolved. This will take place before completion of the oxidation of ferrous to ferric ions. The problem can be overcome by adding an excess of cerous ions, Ce^{3+}, to the solution. The Ce^{3+} ion acts as a *mediator* and itself undergoes no net reaction. With coulometry, the cerous ions can be oxidized to ceric ions, which are then immediately reduced by the ferrous ions present back to cerous ions, generating ferric ions. The reaction continues until all the ferrous ions have been oxidized. At this point, any new ceric ions that are formed are not reduced but remain stable in solution,

and there is a change in the current flow that signals the endpoint. The endpoint can also be detected using an indicator sensitive to ceric ions. Some additional coulometric titrations are presented in Table 15.6. Since the titrant is generated electrolytically and reacted immediately, the method gets widespread applications. The generating electrolytic concentrations need to be only approximate, while unstable titrants are consumed as soon as they are formed. The technique is more accurate than methods where visual end points are required, such as in the case of indicators. The unstable titrants (the methyl viologen radical cations) in the table below are marked with an asterisk.

$$CH_3-\overset{\oplus}{N}\!\!\!\diagdown\!\!\!\diagup\!\!\!-\!\!\!\diagdown\!\!\!\diagup\!\!\!-\overset{\oplus}{N}-CH_3 \quad CH_3-\overset{\oplus}{N}\!\!\!\diagdown\!\!\!\diagup\!\!\!-\!\!\!\diagdown\!\!\!\diagup\!\!\!-\overset{\oplus}{N}-CH_3 + e^-$$

15.3.2.4.3 Analytical Applications of Coulometry

The major advantage of coulometry is its high accuracy, because the "reagent" is electrical current, which can be well controlled and accurately measured. Coulometry is used for analysis, for generation of both unstable and stable titrants "on demand," and for studies of redox reactions and evaluation of fundamental constants. With careful experimental technique, it is possible to evaluate the Faraday constant to seven significant figures, for example.

Some typical important industrial applications of coulometry include the continuous monitoring of mercaptan concentration in the materials used in rubber manufacture. The sample continuously reacts with bromine, which is reduced to bromide. A third electrode measures the potential of Br_2 versus Br^- and, based on the measurement, automatically regulates the coulometric generation of the bromine. Coulometry is used in commercial instruments for the continuous analysis and process control of the production of chlorinated hydrocarbons. The chlorinated hydrocarbons are passed through a hot furnace, which converts the organic chloride to HCl. The latter is dissolved in water and the Cl^- titrated with Ag^+. The Ag^+ is generated by coulometry from a silver electrode, Ag^0. It is necessary for the sample flow rate to be constant at all times. Integration of the coulometric current needed to oxidize the silver to silver ion results in a measurement of the Cl^- concentration.

Another interesting application of coulometry is in the generation of a chemical reagent in solution. Acids, bases, reducing agents, and oxidizing agents can all be generated in solution and allowed to react in exactly the same manner as reagents in volumetric analysis. In effect, volumetric titrations can be carried out by generating the titrant electrochemically; the amount of titrant used is then measured directly by the amount of electricity used to generate it. This method is very exact, frequently more so than weighing the reagent. Reagents generated electrochemically are extremely pure. With coulometry, we can obtain results that are more accurate than those possible with the purest commercially available reagents.

Table 15.6 Coulometric Titrations (taken from the book by Bruno and Svoronos)

Reagent	Generator Electrode Reaction	Typical Generating Electrolyte	Substances Determined
Ag^+	$Ag \rightarrow Ag^+ + e^-$	Ag anode in HNO_3	Br^-, Cl^-, thiols
Ag^{+2}	$Ag^+ \rightarrow Ag^{+2} + e^-$		Ce^{+3}, V^{+4}, $H_2C_2O_4$, As^{+3}
*biphenyl radical anion	$(C_6H_5)_2 + e^- \rightarrow (C_6H_5)_2^{-A}$	Biphenyl/ $(CH_3)_4NBr$ in DMF	anthracene
*Br_2	$2 Br^- \rightarrow Br_2 + 2e^-$	0.2M NaBr in 0.1M H_2SO_4	As^{+3}, Sb^{+3}, U^{+4}, Tl^+, I^-, SCN^-, NH_2OH, N_2H_4, phenols, aromatic amines, mustard gas, olefins, 8-hydroxy-quinoline
*BrO^-	$Br^- + 2 OH^- \rightarrow BrO^- + H_2O + 2e^-$	1M NaBr in borate buffer, pH= 8.6	NH_3
Ce^{+4}	$Ce^{+2} \rightarrow Ce^{+4} + 2e^-$	0.1M $CeSO_4$ in 3M H_2SO_4	Fe^{+2}, Ti^{+3}, U^{+4}, As^{+3}, I^-, $Fe(SCN)_6^{-4}$
*Cl_2	$2 Cl^- \rightarrow Cl_2 + 2e^-$		As^{+3}, I^-
*Cr^{+2}	$Cr^{+3} + e^- \rightarrow Cr^{+2}$	$Cr_2(SO_4)_3$ in H_2SO_4	O_2
*$CuCl_3^{-2}$	$Cu^{+2} + 3Cl^- + e^- \rightarrow CuCl_3^-$	0.1M $CuSO_4$ in 1M HCl	V^{+5}, Cr^{+6}, IO_3^-
EDTA	$HgNH_3(EDTA)^{-2} + NH^{4+} + 2e^- \rightarrow Hg + 2NH_3 + (HEDTA)^{-3}$	0.02M Hg^{+2}/ EDTA in ammoniacal buffer, pH= 8.5, Hg cathode	Ca^{+2}, Cu^{+3}, Zn^{+2}, Pb^{+2}
EGTA	$HgNH_3(EGTA)^{-2} + NH^{4+} + 2e^- \rightarrow Hg + 2NH_3 + (HEDTA)^{-3}$	0.1M Hg^{+2}/ EGTA in triethanolamine, pH=8.6, Hg cathode	Ca^{+2} (in the presence of Mg^{+2})
Fe^{+2}	$Fe^{+3} + e^- \rightarrow Fe^{+2}$	Acid solution of $FeNH_4(SO_4)_2$	Cr^{+6}, Mn^{+7}, V^{+5}, Ce^{+4}
I_2	$2I^- \rightarrow I_2 + 2e^-$	0.2M KI in pH=8 buffer, pyridine, SO_2, CH_3OH, KI (Karl Fisher titration)	As^{+3}, Sb^{+3}, $S_2O_3^{-2}$, H_2S, H_2O
H^+	$2H_2O \rightarrow 4H^+ + O_2 + 4e^-$	0.1M Na_2SO_4 (water electrolysis)	pyridine
*Mn^{+3}	$Mn^{+2} \rightarrow Mn^{+3} + e^-$	$MnSO_4$ in 2M H_2SO_4	$H_2C_2O_4$, Fe^{+2}, As^{+3}, H_2O_2
Mo^{+5}	$Mo^{+6} + e^- \rightarrow Mo^{+5}$	0.7M Mo^{+6} in 4M H_2SO_4	$Cr_2O_7^{-2}$
*MV^{+a}	$MV^{+2} + e^- \rightarrow MV^+$		Mn^{+3} (in enzymes)
OH^-	$2H_2O + 2e^- \rightarrow H_2 + 2OH^-$	0.1M Na_2SO_4 (water electrolysis)	HCl
Ti^{+3}	$Ti^{+4} + e^- \rightarrow Ti^{+3}$ or $TiO^{+2} + 2H^+ + e^- \rightarrow Ti^{+3} + H_2O$	3.6M $TiCl_4$ in 7M HCl	V^{+5}, Fe^{+3}, Ce^{+4}, U^{+6}
U^{+4}	$UO_2^{+2} + 4H^+ + 2e^- \rightarrow U^{+4} + 2H_2O$	Acid solution of UO_2^{+2}	Cr^{+6}, Ce^{+4}

a MV^+ = methyl viologen radical cation; MV^{+2} = methyl viologen radical cation.

15.3.3 Conductometric Analysis

The ability of a solution to conduct electricity can provide analytical information about the solution. The property measured is electrical conductivity between two electrodes by ions in solution. All ions in solution contribute to the electrical conductivity, so this is not a specific method of analysis. Electrical conductivity is used to provide qualitative analysis, such as the purity of an organic solvent and relative quantitative analysis for quality control of materials, or comparison of drinking water quality in terms of total ionic contaminants.

Ohm's law states that the resistance of metal wire is given by the equation

$$R = \frac{E}{i} \tag{15.31}$$

where
E is the voltage applied to the wire (V)
i is the current of electrons flowing through the wire (A)
R is the resistance of the wire (Ω)

The resistance R depends on the dimensions of the conductor:

$$R = \frac{\rho L}{A} \tag{15.32}$$

where
ρ is the resistivity
L is the length
A is the cross-sectional area

Another valuable parameter, especially when we consider the mechanisms of current flow in solutions, is electrolytic conductivity, κ, where

$$\kappa = \frac{1}{\rho} = \frac{\left(\frac{L}{A}\right)}{R} \quad (15.33)$$

The units of electrolytic conductivity are Ω^{-1} m^{-1} (reciprocal ohm m^{-1} also called mho/m, mho m^{-1}, or S m^{-1}, where S is the siemen). The Systeme International d'Unites (SI) unit is the S/m, but practical measurement units are usually in µS/cm. **Electrolytic conductivity** is also called *specific conductance*, not to be confused with conductance. The electrolytic conductivity of a solution is a measure of how well it carries a current, in this instance by ionic carriers rather than electron transfer, and it is an intrinsic property of the solution. A related property, the conductance, G, is also used and defined as $G = 1/R$. The **conductance** is a property of the solution *in a specific cell*, at a specific temperature and concentration. The conductance depends on the cell in which the solution is measured; the units of G are siemens (S).

The charge carriers are ions in electrolyte solutions, fused salts, and colloid systems. The positive ions M$^+$ migrate through the solution toward the cathode, where they may or may not react Faradaically to pick up electrons. Anions, symbolized as A$^-$, migrate toward the anode, where they may or may not deliver electrons. The net result is a flow of electrons across the solution, but the electron flow itself stops at each electrode. Faradaic reaction of the easiest reduced and oxidized species present may occur, and hence compositional changes (reduction and oxidation) may accompany ionic conductance.

Electrolyte conductivity depends on three factors: the ion charges, mobilities, and concentrations of ionic species present. First, the number of electrons each ion carries is important, because A^{2-}, for example, carries twice as much charge as A$^-$. Second, the speed with which each ion can travel is termed its *mobility*. The mobility of an ion is the limiting velocity of the ion in an electric field of unit strength. Factors that affect the mobility of the ion include (1) the solvent (e.g., water or organic), (2) the applied voltage, (3) the size of the ion (the larger it is, the less mobile it will be), and (4) the nature of the ion (if it becomes hydrated, its effective size is increased). The mobility is also affected by the viscosity and temperature of the solvent. Under standard conditions, the mobility is a reproducible physical property of the ion. Because in electrolytes the ion concentration is an important variable, it is usual to relate the electrolytic conductivity to **equivalent conductivity**. This is defined by

$$\Lambda = \frac{\kappa}{C_{eq}} \quad (15.34)$$

where
Λ is the equivalent conductivity (Ω^{-1} cm^2/equivalent or S cm^2/equivalent)

κ is the electrolytic conductivity (mho/m)
C_{eq} is the equivalent concentration (i.e., normality of solution, where $N = M \times$ charge on ion)

Electrolyte solutions only behave ideally as infinite dilution is approached. This is because of the electrostatic interactions between ions, which increase with increasing concentration. As infinite dilution is approached, the equivalent conductance of the electrolyte Λ approaches Λ_0, where

$$\Lambda^0 = F\left(U_+^0 + U_-^0\right) = \lambda_+^0 + \lambda_-^0 \quad (15.35)$$

where
U_+^0 and U_-^0 are the cation and anion mobilities, respectively
λ_+^0 and λ_-^0 are the cation and anion equivalent limiting ion conductivities, respectively, at infinite dilution

The λ^0 values are not accessible to direct measurement, but they may be calculated from transport numbers. Kohlrausch's law of independent ionic conductivities states that at low electrolyte concentrations the conductivity is directly proportional to the sum of the n individual ion contributions, that is,

$$\Lambda^0 = \sum_{i=1}^{n} \lambda_i^0 \quad (15.36)$$

Table 15.7 shows some typical limiting ionic equivalent conductivities. From this, we can deduce that the limiting equivalent conductivity of potassium nitrate is (K$^+$) + (NO$_3^-$) = 74 + 71 = 145 and that of nitric acid is (H$^+$) + (NO$_3^-$) = 350 + 71 = 421, assuming 100% dissociation into ions. The change in conductivity of a solution upon dilution or replacement of one ion by another by titration is the basis of conductometric analysis.

The electrolytes KNO$_3$ and HNO$_3$ are strong and dissociate completely in water. Weak electrolytes do not dissociate completely. For weak electrolytes, the effects of interionic forces are less important because

Table 15.7 Equivalent Conductance of Various Ions at Infinite Dilution at 25 °C

Anions	(Ω^{-1} cm^2/eq)	Cations	(Ω^{-1} cm^2/eq)
OH$^-$	198	H$^+$	350
Cl$^-$	76	Na$^+$	50
Br$^-$	78	K$^+$	74
NO$_3^-$	71	NH$_4^+$	74
ClO$_4^-$	67	Ag$^+$	62
HCOO$^-$	55	Cu^{2+}	55
CH$_3$COO$^-$	41	Zn^{2+}	53
SO$_4^{2-}$	80	Fe^{3+}	68

the ion concentrations are lower and, in fact, the degree of ionization (α) is readily obtainable from conductance measurements:

$$\alpha \approx \frac{\Lambda}{\Lambda^0} \qquad (15.37)$$

For example, the weak acid HA partially dissociates into H^+ and A^-:

$$HA \leftrightarrow H^+ + A^- \text{ with } K' = \frac{[H^+][A^-]}{HA}$$

If the initial total molar concentration of HA = c, at equilibrium, the molar concentration of HA = $(1-\alpha)c$, and the molar concentration of $H^+ = A^- = \alpha c$.

We can express an apparent dissociation constant in terms of conductivities by substitution:

$$K' = \frac{\alpha^2 c^2}{(1-\alpha)_c} \approx \frac{\Lambda^2 c}{\Lambda^0 (\Lambda^0 - \Lambda)} \qquad (15.38)$$

This is an expression of Ostwald's dilution law, and the equation can be used to determine K', α, or Λ^0. For example, from Table 15.7, we can calculate Λ^0 for formic acid, HCOOH, to be equal to $350 + 55 = 405$ S cm^2/eq. If we measure the equivalent conductance of a 0.020 N solution of formic acid, we find it equal to 36.6 S cm^2/eq at 25°C. The degree of dissociation, α, is calculated to be

$$\alpha = \frac{\Lambda}{\Lambda^0} = \frac{36.6}{405} = 9.04 \times 10^{-2}$$

and from this, we can calculate the dissociation constant, K':

$$K' = \frac{(9.04 \times 10^{-2})^2 (0.020)}{1 - 9.04 \times 10^{-2}} = 1.8 \times 10^{-4}$$

This agrees well with the literature value for the K_a for HCOOH. Conductance measurements may also be used to find the solubility of sparingly soluble salts and complexation equilibrium constants.

15.3.3.1 Instrumentation for Conductivity Measurements

The equipment is basically a Wheatstone bridge and conductivity cell, as illustrated in Figure 15.21.

Resistance A is made up of the cell containing the sample; B is a variable resistance; resistances D and E are fixed. Resistor B and variable capacitor C may be adjusted so that the balance point can be reached. At this point,

$$\frac{R_A}{R_B} = \frac{R_D}{R_E}$$

The resistances B, D, and E can be measured, and from these measurements, the resistance and hence the conductance of the cell can be calculated. A *small* superimposed AC voltage (~20 mV peak to peak) at 1000 Hz is best as a signal, because then faradaic polarization at the electrodes is minimized. The null detector may be a sensitive oscilloscope or a tuned amplifier and meter. Stirring is often used to minimize polarization.

A wide variety of cell geometries and sizes are available for conductivity measurements, designed with two, three, or four electrodes, depending on the use. A typical dip cell (so called because it is dipped into a beaker containing the sample) usually is constructed with two parallel, platinized Pt foil electrodes, each about 1 cm^2 in area. Because the dimensions of a constructed cell are not exact, one must calibrate a particular cell. The recommended method of

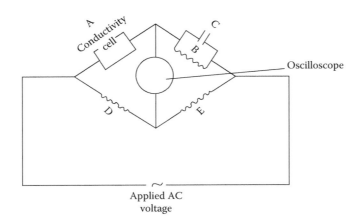

Figure 15.21 Wheatstone bridge arrangement for conductometric analysis.

calibration is to use primary standard KCl solutions of known strength. For example, at 25°C, 7.419 g of KCl in 1000 g of solution (or 0.1 molal) has a specific conductivity of 0.01286 Ω^{-1}/cm. Primary standard solutions are not always practical to make. One has to include the conductivity of the water used to prepare the standard, for example. NIST has a set of standard reference materials (SRMs) covering the range of 5–100,000 µS/cm that are satisfactory for most calibration needs. Platinizing the Pt electrodes is achieved by cleaning the Pt in hot concentrated HNO_3 and electrodepositing a thin film of Pt black from a 2% solution of platinic chloride in 2 N HCl. The Pt black is a porous Pt film, which increases the surface area of the electrodes and further reduces faradaic effects.

Cells are available for sample volumes as small as 2 mL, while standard sample size is 25–50 mL. Special cells for highly accurate conductivity measurements are available for research purposes. Flow-through cells are available for online monitoring of process streams. The reference by Berezanski depicts several cell types.

Very accurate measurements of conductivity require the use of a thermostated bath with temperature controlled to within 0.005 °C.

In addition to the laboratory meter setup described, there are handheld devices available for making conductivity measurements in the field. These are generally battery powered and are less accurate than laboratory meters.

15.3.3.2 Analytical Applications of Conductometric Measurements

Conductivity measurements can be performed on many types of solutions with no sample preparation required. The method is used to monitor solutions for their ionic content. Examples include drinking water, natural water, high-purity (DI) water, high-purity solvents, and potable beverages. Conductivity is common as a detector in IC, high-performance liquid chromatography (HPLC), and other chromatographic techniques where charged species are produced (Chapter 14). Conductivity is a powerful tool for endpoint detection in titrimetry in aqueous and nonaqueous solvents.

15.3.3.2.1 Conductimetric Titrations in Aqueous Solution

When one ion is replaced in solution by a different ion with a significantly different equivalent conductivity, a change in total conductivity occurs. As seen in Table 15.7, hydrogen ion and hydroxide ion have the highest equivalent conductivities; replacing them with less conductive ions can form the basis of conductimetric titrations for acids and bases. For example, when NaOH is added to HCl, the following reaction occurs:

Chemical reaction: $HCl + NaOH \rightarrow NaCl + H_2O$

Ionic reaction: $H^+ + Cl^- + Na^+ + OH^- \rightarrow Na^+ + Cl^- + H_2O$

The original solution contains $H^+ + Cl^-$ in water; the final solution contains $Na^+ + Cl^-$ in water. It can be seen that Na^+ replaces H^+. If we plot the conductivity measured while NaOH is being added, we observe the relationship depicted in Figure 15.22. In part A of the curve, H^+ ions are being removed by the OH^- of the NaOH added to the solution. The conductivity slowly decreases until the neutralization point C is reached. In part B of the curve, the H^+ ions

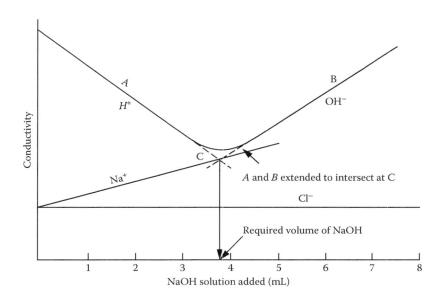

Figure 15.22 Solution conductivity during an HCl–NaOH titration. Point C is the neutralization point.

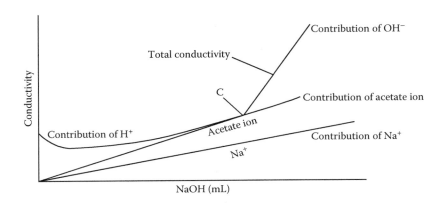

Figure 15.23 Solution conductivity during an CH₃COOH–NaOH titration. Point C is the neutralization point.

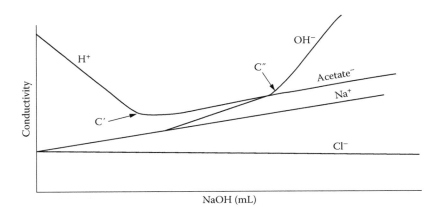

Figure 15.24 Solution conductivity during titration of a mixture of HCl and CH₃COOH with NaOH. Here, C' and C" designate the respective endpoints. HCl is titrated first and then the weaker acetic acid is neutralized.

have been effectively removed from solution. Further addition of NaOH merely adds Na⁺ and OH⁻ to the solution. An increase in conductivity results. The contribution of the Cl⁻, Na⁺, H⁺, and OH⁻ ions to the total conductivity of the solution can be seen from Figure 15.22. The volume of NaOH required to titrate the HCl can be measured by extrapolating lines A and B to the intersection C. Point C indicates the required volume of NaOH to neutralize the acid present.

Weak acids can also be titrated with NaOH and the endpoint detected by conductivity. A typical curve is shown in Figure 15.23. As with all weak acids, the H⁺ concentration is low and in equilibrium with the acetate ion. At the equivalence point, however, all the H⁺ has been neutralized. Any further addition of NaOH has the effect of adding Na⁺ and OH⁻ to the solution. A sharp increase in conductivity occurs, as shown by point C in Figure 15.23. Mixtures of weak and strong acids can be titrated with NaOH and the endpoints detected by conductivity changes. In such mixtures, the strong acid is neutralized before the weak acid. Titrations involving mixtures of acids such as these are difficult to perform using indicators because of the problem involved in detecting both endpoints separately. With conductimetric titration, the problem is simplified, as shown in Figure 15.24. It can be seen that two abrupt changes in conductivity take place (C' and C"); these correspond to the titration endpoints for the strong and weak acid, respectively. For known acids, we can determine the concentrations present in the sample quantitatively. The contributions from the Na⁺, Cl⁻, acetate⁻, OH⁻, and H⁺ to the total conductivity are noted in the figure. But remember that the conductivity measurement itself cannot distinguish individual ion contributions in a mixture. The method can be used to show how many acids are in a mixture and to indicate whether they are weak or strong acids, but not to identify them if the sample is an unknown. Conductimetric titrations can also be used to find the endpoints of reactions where precipitation takes place, as in the titration of silver solution with chloride solution. Again, the endpoint is manifested by a sharp change in conductivity.

15.3.3.2.2 Conductimetric Titrations in Nonaqueous Solvents

In nonaqueous solvents, such as alcohol, toluene, or pyridine, it is possible to titrate Lewis acids or bases that cannot be titrated in aqueous solutions. For example, it is possible to titrate phenols dissolved in an organic solvent. Phenols act as Lewis acids, releasing hydrogen ion, which can be titrated with a suitable basic material. Since the titration is carried out in a nonaqueous medium, it is necessary to use a base that is soluble in the solvent. A base commonly used is tetramethylammonium hydroxide (TMAH). This material is basic but soluble in organic solvents and will neutralize Lewis acids. Nonaqueous solvents can be used to advantage to investigate molecular species that are not soluble in water. However, interpretation of the titration curve may not be straightforward. Solvent effects, viscosity, temperature, and intermolecular attractions between solvent and solute (e.g., ion-pairing, complexation) are some of the factors that must be considered. Methods and more detail may be found in the references by Huber and Kucharsky listed in the bibliography.

15.3.3.2.3 Other Applications of Conductivity Measurements

Conductivity is used to determine the purity of drinking water and other natural waters and wastewaters. Dedicated instruments for this purpose, including handheld, portable meters, are often calibrated to read in "total dissolved solids" or TDS or in "salinity." These tell the environmental scientist, water treatment plant operator, or geochemist about the amount of ionized species (e.g., sodium chloride is assumed to be the species for salinity of seawater) in solution. It is important to remember that the conductivity measurement cannot tell if the conductivity really is due to sodium chloride. It measures any ion that contributes to conductivity. The chemist or engineer must know something about the sample or make good scientific judgments based on other information before drawing conclusions from a salinity or TDS reading on an unknown sample. Methods for conductivity measurements in potable water and wastewater may be found in *Standard Methods for the Examination of Water and Wastewater* and for all types of water, deionized to brackish, in the American Society for Testing and Materials (ASTM) standards Volumes 11.01 and 11.02. The references are listed in the bibliography.

In addition to drinking water and environmental applications, water purity is critical to many industries. Conductivity detectors are used in semiconductor and chip fabrication plants, to monitor cleanliness of pipelines in the food and beverage industry, to monitor incoming water for boilers to prevent scale buildup and corrosion. Any process stream with ions in it can be analyzed by conductometry. Conductivity detectors are part of commercial laboratory DI water systems, to indicate the purity of the water produced and to alert the chemist when the ion-exchange cartridges are exhausted. The detector usually reads out in resistivity; theoretically, completely pure water has a resistivity of 18 MΩ cm.

Conductivity detectors are widely used in IC instruments using eluent suppression detection. The detectors are inexpensive, simple, rugged, and easy to miniaturize. The same type of detectors can be used for any chromatographic process to detect charged species in a nonionic eluent.

15.3.4 Polarography

When a potential difference is applied across two electrodes immersed in a solution, even in the absence of an electroactive species of interest, a small current arises due to background reactions and dissolved impurities. If the solution contains various metal ions, these do not electrolyze until the applied negative potential exceeds the reduction potential of the metal ion, that is, it becomes more negative than the reduction potential. The difference in the current flowing through a solution under two conditions—(1) with the potential less negative than the metal-ion reduction potential and (2) with the potential more negative than the metal-ion reduction potential—is the basis of polarography. Polarography is not restricted to reductions (negative potential sweep), although these are more common. It is also possible to sweep to positive potentials and obtain oxidation curves. Metal cations, anions, complexes, and organic compounds all can be analyzed using polarography. The plot of current against applied potential for a sample solution is called a polarogram.

Polarography is the study of the relationship between the current flowing through a conducting solution and the voltage applied to a **dropping mercury electrode** (DME). It was discovered by Jaroslav Heyrovsky more than 60 years ago and has since resulted in tens of thousands of research studies. Heyrovsky's pioneering work in the field earned him the Nobel Prize in 1959. In the past two decades, many variations of Heyrovsky's classical polarographic method have evolved, principally to improve the sensitivity and resolution of this analytical method. Two important variations are described in the following, namely, normal pulse polarography and differential pulse polarography.

Classical DC polarography uses a linear potential ramp (i.e., a linearly increasing voltage). It is, in fact, one subdivision of a broader class of electrochemical methods called **voltammetry**. Voltammetric methods measure current as a function of applied potential where the WE is *polarized*. This polarization is usually accomplished by using microelectrodes as WEs; electrode surface areas are only in the μm^2 to mm^2 range. The term *polarography* is usually restricted to electrochemical analyses at the DME. Figure 15.25 illustrates the potential excitation used in DC polarography and the net current waveform relationship, usually shown as an i–E rather than an i–t curve. The net S-shaped, or sigmoidal, curve obtained when a substance is reduced (or oxidized) is called a **polarographic wave** and is the basis for both qualitative and quantitative analyses. The *half-wave potential* $E_{1/2}$ is the potential value at a current one-half of the limiting **diffusion current** i_L of the species being reduced (or oxidized). The magnitude of the limiting diffusion current i_L, obtained once the half-wave potential is passed, is a measure of the species concentration. It is standard to extrapolate the background residual current to correct for impurities that are reduced (or oxidized) before the $E_{1/2}$ potential of the species of interest is reached. Although DC polarography is no longer commonly used for analytical purposes, it provides the basis of the newer, more powerful variants of classical polarography.

15.3.4.1 Classical or DC Polarography

A modern apparatus for DC polarography is shown in Figure 15.26. The three-electrode system for aqueous work consists of the DME as the WE, a Pt wire or foil auxiliary electrode, and an SCE for reference. The potential of the WE (DME) is changed by imposition of a slow voltage ramp versus a stable reference (Figure 15.25). At the heart of polarography is the DME. A very narrow capillary is connected to a mercury column, which has a pressure head that can be raised and lowered. A drop of mercury forms at the tip of the capillary, grows, and finally falls off when it becomes too large. Typically, the level of the mercury column above the tip of the capillary is about 60 cm and the natural interval between drops is 2–6 s. The choice of mercury for the electrode is important for several reasons:

1. Each fresh drop exposes a new Hg surface to the solution. The resulting behavior is more reproducible than that with a solid surface, because the liquid drop surface does not become contaminated in the way solid electrodes can be contaminated. Organic contaminants or adsorbents must undergo re-equilibration with each new drop and are less likely to interfere.
2. As mentioned earlier, there is a high overpotential for H$^+$ ion reduction at mercury. This means that it is possible to analyze many of the metal ions whose standard reduction potentials are more negative than that of the H$_2$/H$^+$ ion couple. It is easier, too, to reduce most metals to their mercury amalgam than to a solid deposit. Conversely, however, mercury is easily oxidizable, which severely restricts the use of the DME for the study of oxidation processes.
3. Solid electrodes have surface irregularities because of their crystalline nature. Liquid mercury provides a smooth, reproducible surface that does not depend on any pretreatment (polishing or etching) or on substrate inhomogeneity (epitaxy, grain boundaries, imperfections, etc.).

In electroanalysis, diffusion currents are quite small (<100 µA), which means that the aqueous solution IR drop between the reference electrode and the DME can be neglected in all but the most accurate work. Electrolytes prepared with organic solvents, however, may have fairly large resistances, and in some instances, IR corrections must be made.

As each new drop commences to grow and expand in radius, the resulting current is influenced by two important factors. The first is the depletion by electrolysis of the electroactive substance at the mercury drop surface. This gives rise to a diffusion layer in which the concentration of the reactant at the surface is reduced. As one travels radially outward from the drop surface, the concentration increases and reaches that of the bulk homogeneous concentration. The second factor is the outward growth of the drop itself, which tends to counteract the formation of a diffusion layer. The net current waveform for a single drop is illustrated in Figure 15.27.

A mathematical description of the diffusion current, in which the current i_L is measured at the top of each oscillation just before the drop dislodges, is given by the following equation:

$$i_L = 708nCD^{\frac{1}{2}}m^{\frac{2}{3}}t^{\frac{1}{6}} \qquad (15.39)$$

where

i_L is the maximal current of the Hg drop (µA)
n is the number of electrons per electroactive species

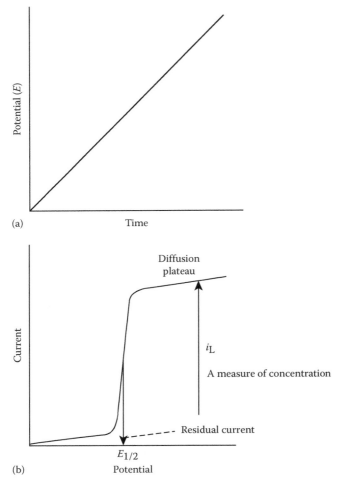

Figure 15.25 (a) Linear potential excitation and (b) the resulting polarographic wave.

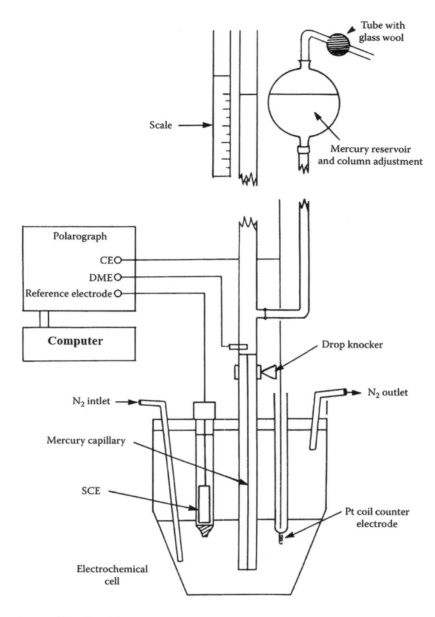

Figure 15.26 Modern polarographic cell and three-electrode circuit.

C is the concentration of electroactive species (mM)
D is the diffusion coefficient of electroactive species
m is the mercury flow rate (mg/s)
t is the drop time (s)

This is called the **Ilkovic equation**. For a particular capillary and pressure head of mercury, $m^{2/3}t^{1/6}$ is a constant. Also, the value of n and that of the diffusion coefficient for a particular species and solvent conditions are constants. Thus, i_L is proportional to the concentration C of the electroactive species, and this is the basis for quantitative analysis. The Ilkovic equation is accurate in practice to within several percent, and routinely ±1% precision is possible. It is commonplace to use standard additions to obtain a calibration curve or an internal standard. Internal standards are useful when chemical sampling and preparation procedures involve the possibility of losses. The principle is that the ratio of the diffusion currents due to the sample and the added standard should be a constant for a particular electrolyte.

Before a polarogram can be obtained from a sample, two important steps are necessary: (1) the sample must be dissolved in a suitable **supporting electrolyte** and (2) the solution must be adequately **degassed**. To understand the purpose of the supporting electrolyte, we should first review the three principal forms of transport that occur in ionic solutions: (1) diffusion, (2) convection, and (3) migration.

Diffusion is the means by which an electroactive species reaches the electrode when a concentration gradient is

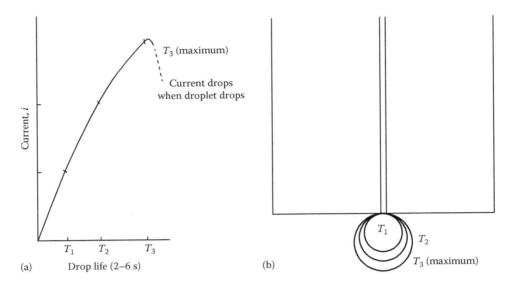

Figure 15.27 (a) Current waveform and (b) time frame of an expanding drop.

created by the electrode reaction. The electron transfer process can decrease the concentration of an electroactive species or produce a new species (not originally present in the bulk solution) that diffuses away from the electrode surface. *Convection*, that is, forced motion of the electrolyte, can arise from natural thermal currents always present within solutions, by density gradients within the solution or be produced deliberately by stirring the electrolyte or rotating the electrode. In general, natural convection must be minimized by electrode design and by making short time scale measurements (on the order of a few seconds). The third form of transport is migration of the charged species. *Migration* refers to the motion of ions in an electric field and must be suppressed if the species is to obey diffusion theory. This is done by adding an excess of inert *supporting electrolyte* to the solvent. The role of the supporting electrolyte is twofold: it ensures that the electroactive species reaches the electrode by diffusion and it lowers the resistance of the electrolyte.

Typically, supporting electrolyte concentrations are 0.1–1 M. An example is KCl solution. The K^+ ions in this solution are not easily reduced and have the added advantage that the K^+ and Cl^- ions migrate at about equal velocities in solution. In a solution of $ZnCl_2$ (0.0001 M) and KCl (0.1 M), the migration current of the zinc is reduced to a negligible amount, and we are therefore able to measure the current response due to zinc ion reduction under diffusion conditions.

Actually, in polarography, even though measurements are made in a quiescent solution (no stirring by either gas bubbling or magnetic bar is permissible), the transient currents arising with each drop depend on both diffusion and the convective motion due to the expanding mercury drop. These effects combine to produce a current that is proportional to $t^{1/6}$.

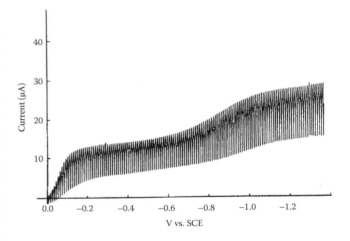

Figure 15.28 Polarogram of dissolved oxygen in an electrolyte, 1 M KNO_3 (scan rate 5 mV/s, 2 s drop time).

A further important step is the adequate degassing of the electrolyte by bubbling purified nitrogen or argon. This is necessary because dissolved oxygen from the air is present in the electrolyte and, unless removed, would complicate interpretation of the polarogram. Figure 15.28 shows an actual polarogram of a supporting electrolyte that is air saturated and without added electroactive sample. The abrupt wave at −0.1 V versus SCE is due to the reduction of molecular oxygen, and the drawn-out wave at −0.8 V versus SCE is assigned to reduction of a product from the oxygen reduction, namely, hydrogen peroxide. Once the potential exceeds the threshold for peroxide reduction, both of these reactions can occur at each new drop as it grows, and the limiting current is approximately twice as large. The first reaction is sharp and can be used for the determination of dissolved oxygen in solutions. The addition of a few drops of a dilute Triton X-100 solution is necessary to prevent the formation of maxima in the current at the diffusion plateau. Such maxima distort the waveforms and complicate the measurement of the diffusion currents. Maximum suppressants, such as gelatin and Triton X-100, are capillary-active substances, which presumably damp the streaming currents around the drop that cause the current distortions.

The relationship between current and voltage for a well-degassed solution of $Pb(NO_3)_2$ is shown in Figure 15.29. (This polarogram is actually a normal pulse polarogram, not a DC polarogram. The difference is discussed in the succeeding text.) Within the voltage range of 0 to −0.3 V versus SCE only a small *residual current* flows. It is conventional in polarography to represent cathodic (reduction) currents as positive and anodic (oxidation) currents as negative. The residual current actually comprises two components, a faradaic contribution and a charging or double-layer current contribution:

$$i_{res} = i_F + i_{ch}$$

The faradaic current i_F arises from any residual electroactive species that can be electrolyzed in this potential range, such as traces of metals or organic contaminants. Such impurities may be introduced by the supporting electrolyte, which will have to be purified for some trace analyses. The charging current i_{ch} is nonfaradaic; in other words, no electron flow occurs across the metal–solution interface, and neither redox nor permanent chemical changes result from its presence. The mercury–solution interface acts, to a first approximation, as a small capacitor and charge flows to the interface to create an electrical double layer. At negative potentials, this can be thought of as a surplus of electrons at the surface of the metal and a surplus of cations at the electrode surface. In reality, the electrical double-layer capacity varies somewhat with the potential; so the fixed capacitor analogy is not strictly accurate. An important consequence of charging currents is that they limit the analytical sensitivity of polarography. This is why pulse polarographic methods, which discriminate against charging currents, can be used to determine much lower concentrations of analyte.

Referring again to Figure 15.29, we can see that as the potential becomes increasingly negative, it reaches a point where it is sufficient to cause the electroreduction

$$Pb^{2+} + 2e^- \rightarrow Pb(Hg)$$

The Pb^{2+} concentration in the immediate vicinity of the electrode decreases as these ions are reduced to lead amalgam. With a further increase in negative potential, the Pb^{2+} concentration at the surface of the electrode becomes zero, even though the lead ion concentration in the bulk solution remains unchanged. Under these conditions, the electrode is said to be polarized. The use of an electrode of small area, such as the DME, means that there is no appreciable change of Pb^{2+} ions in the bulk of the solution from a polarographic analysis.

When the electrode has become polarized, a fresh supply of M^+ ions to its surface is controlled by the diffusion of such ions from the bulk of the solution, through the zone depleted of M^+. In other words, the current flowing is dependent on the diffusion of ions from the bulk liquid. This current is called the diffusion current and the plateau region of the curve can be used to measure the limiting diffusion current i_L. It is usual to extrapolate the residual current background and to construct a parallel line through the diffusion current plateau to correct for the residual current contribution to i_L.

15.3.4.2 Half-Wave Potential

An important point on the curve is that at which the diffusion current is equal to one-half of the total diffusion current; the voltage at which this current is reached is $E_{1/2}$, the *half-wave potential*. The half-wave potential is used to characterize the current waveforms of particular reactants. Whether a process is termed reversible or not depends on whether equilibrium is reached at the surface of the electrode in the time frame of the measurements. In other words, a process is reversible when the electron transfer reactions are sufficiently fast so that the equilibrium

$$ox + ne^- \leftrightarrow red$$

is established and the Nernst equation describes the ratio [ox]/[red] as a function of potential, that is,

$$E = E^0 + 2.303 \frac{RT}{nF} \log\left(\frac{[ox]}{[red]}\right)$$

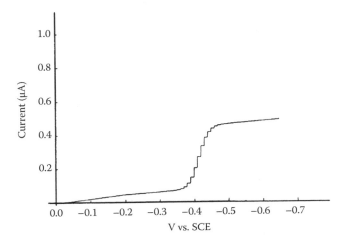

Figure 15.29 Normal pulse polarogram of Pb^{2+} ions (5 ppm) in 1 M KNO_3 (scan rate 5 mV/s, 2 s drop time).

When this is the case, it can be shown that there is a relation between the potential, the current, and the diffusion current i_L, which holds true throughout the polarographic wave:

$$E = E_{\frac{1}{2}} + \frac{0.0591}{n} \log\left(\frac{i_L - i}{i}\right) \text{ at } 25°C \quad (15.40)$$

This equation, in fact, may be practically used to test the reversibility (or Nernstian behavior) of an electroactive species. The graph of E versus $\log[(i_L - i)/i]$ will be linear with a slope of $0.0591/n$ and intercept $E_{1/2}$. Determination of the slope enables us to determine n, the number of electrons involved in the process. Substitution of $i_L/2$ in Equation 15.40 reveals that $E = E_{1/2}$ when the surface concentrations [ox] and [red] are equal.

Note, however, that the half-wave potential $E_{1/2}$ is usually similar but not exactly equivalent to the thermodynamic standard potential E^0. First, the product of reduction may be stabilized by amalgam formation in metal-ion reductions; second, there will always be a small liquid junction potential in electrochemical cells of this type that should be corrected for; and finally, it can be shown that the potential $E_{1/2}$ is the sum of two terms:

$$E_{1/2} = E_0 + \frac{0.0591}{n} \log\left(\frac{\gamma_{ox}}{\gamma_{red}}\right)\left(\frac{D_{red}}{D_{ox}}\right)^{1/2}$$

where
the γ terms are the activity coefficients
the D terms are the diffusion coefficients of the oxidized and reduced species

In most analytical work, the activity corrections are ignored and $D_{ox} \approx D_{red}$. This is because the size of the electroactive product and reactant is not greatly affected by the gain (or loss) of an electron. The value of the half-wave potential is that it can be used to characterize a particular electroactive species qualitatively. It is not affected by the analyte concentration or by the capillary constant. It can, however, be severely affected by changes in the supporting electrolyte medium. Table 15.8 lists some half-wave potentials of diverse species that may be analyzed by polarography.

15.3.4.3 Normal Pulse Polarography

Unlike classical DC polarography, normal pulse polarography does not use a linear voltage ramp; instead, it synchronizes the application of a square-wave voltage pulse of progressively increasing amplitude with the last 60 ms of the life of each drop. This is the most stable part of the drop, when its area is virtually constant. This is shown in Figure 15.30. It is necessary to use an electronically controlled solenoid to

Table 15.8 Half-Wave Potentials of Common Metal Ions

		$E_{1/2}$ vs. SCE
Ag(I)	1 M NH$_3$/0.1 M NH$_4$Cl	−0.24
Cd(II)	1 M HCl	−0.64
	0.1 M CH$_3$COONa/0.1 M CH$_3$COOH	−0.65
Cu(II)	1 M HCl	−0.22
	0.1 M CH$_3$COONa/0.1 M CH$_3$COOH	−0.07
O$_2$	0.1 M KNO$_3$	−0.05
Pb(II)	1 M HCl	−0.44
	0.1 M CH$_3$COONa/0.1 M CH$_3$COOH	−0.50
T1(I)	1 M HCl	−0.48
Zn(II)	1 M NH$_3$/0.1 M NH$_4$Cl	−1.35
	0.1 M CH$_3$COONa/0.1 M CH$_3$COOH	−1.1

Note: More complete data are available from textbooks (e.g., Bard 1980) and from polarographic equipment suppliers (e.g., EG &G Princeton Applied Research Application Briefs and Application Notes. www.princetonappliedresearch.com.)

knock the drop from the capillary at a preset time, for example, every 2 s. Precise synchronization of voltage pulses, drop time and current detection is crucial. Each drop has the same lifetime, which is shorter than its natural (undisturbed) span. The initial voltage E_{init} is chosen such that no faradaic reactions occur during most of the growth of the drop. Then, when the rate of change of the drop surface area is less than during the drop's early formation stages, the pulse is applied. As a first approximation, the drop surface area can be considered to be constant during the pulse application. If the potential is such that faradaic reaction can take place at the pulse potential, the resultant current decay transient has both a faradaic and a charging current component, as shown in Figure 15.30. The measurement of current may be made during the last 17 ms of the pulse by a sample-and-hold circuit. This outputs the average current and holds this value until the next current is ready for output. Thus, the output current is a sequence of steady-state values lasting the drop life, for example, 2 s. The i–E curve does not have the large oscillations reminiscent of classical polarography (compare Figures 15.28 and 15.29), but may provide a Nernstian waveform directly.

It is readily seen from Figure 15.30 why the current is sampled in this manner. A faradaic current i_F at an electrode of constant area will decay as a function of $t^{-1/2}$ if the process is controlled by diffusion; on the other hand, the charging current transient i_{ch} will be exponential, with a far more rapid decay. This permits better discrimination of the faradaic current and charging current contributions.

A further important advantage of this technique is that during the first 1.94s of the life of each drop, no current flows (remember, the pulse is applied in the last 0.060s of a 2s drop life). To understand how this is advantageous, we must reconsider what happens in classical DC polarography. At the start of a drop life, the potential is at a given value and increases slowly as the drop grows. If faradaic reaction can occur, this means that the electroactive species

ELECTROANALYTICAL CHEMISTRY

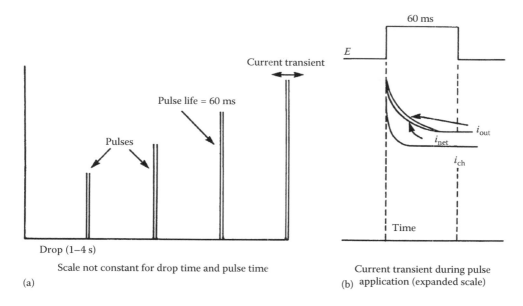

Figure 15.30 (a) Waveform for normal pulse polarography and (b) current transient response (not to scale).

is being depleted during the whole life of the drop, such that when it has grown for about 1.94s there is a large depletion region extending from the surface. In normal pulse polarography, no reaction occurs until the drop area is quite large, and this enhances the current response considerably. The analytical implication is that it is possible to detect concentration levels of about 1×10^{-6} M with the normal pulse technique, which is much better than the useful range of about 1×10^{-4}–1×10^{-2} M in classical DC polarography.

15.3.4.4 Differential Pulse Polarography

Differential pulse polarography has the most complex waveforms of the polarographic methods discussed, but it is the easiest to interpret for analytical purposes. This method provides an increase in detection limits for trace analysis to 10^{-7} to 10^{-8} mol/L. The applied voltage is a linear ramp with superimposed pulses added during the last 60 ms of the life of each drop (Figure 15.31). In this technique, however, the pulse height is maintained constant above the ramp and is termed the *modulation amplitude*. The modulation amplitude may be varied between 10 and 100 mV. As in normal pulse polarography, the current is not measured continuously, but in this technique, it is sampled twice during the mercury drop lifetime: once just prior to the imposition of a pulse and once just before each drop is mechanically dislodged. As in classical DC polarography, the presence of a potential at the start of the life of each drop reduces the analyte concentration by creating a depleted diffusion layer. However, the sampled current output is a differential measurement:

$$\Delta i_{out} = i_{after\ pulse} - i_{before\ pulse}$$

and as in normal pulse polarography, the time delay permits the charging current to decay to a very small value. As the measured signal is the difference in current, a peak-shaped i–E curve is obtained, with the peak maximum close to $E_{1/2}$ if the modulation amplitude is sufficiently small. A differential pulse waveform is shown in Figure 15.32.

To understand why the waveform is a peak, it is necessary to consider what is being measured. The output is a current difference arising from the same potential difference, that is, $\Delta i/\Delta E$. If ΔE becomes very small ($\Delta E \to dE$), we should obtain the first derivative of the DC polarogram, and the peak and $E_{1/2}$ would coincide. However, because the modulation amplitude is finite and, in fact, has to be reasonably large in order to produce an adequate current signal, the E_{peak} value of the differential pulse polarogram is shifted positive at the half-wave potential for reduction processes:

$$E_{peak} = E_{1/2} - \frac{\Delta E}{2}$$

where ΔE is the modulation amplitude. Increasing ΔE increases the current response but reduces the resolution of the waveform, and experimentally, it is best to vary the instrument parameters until a suitable combination is found for the ramp scan rate and the modulation amplitude.

The simplicity of the differential pulse polarographic method lies in the fact that the peak height is proportional to the analyte concentration (Figure 15.33). If adequate resolution exists, it is possible to analyze several ionic species simultaneously with polarography. Differential pulse polarography is especially useful for trace analyses when working close to the electrolyte background reduction or

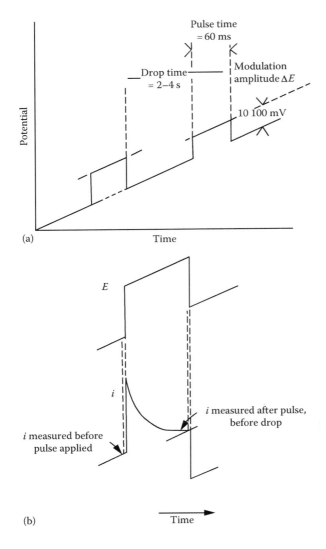

Figure 15.31 Differential pulse polarography (not to scale). (a) Voltage vs. time sequence in DPP and (b) current response vs. time during DPP pulse.

oxidation wave. Its limit of detection is typically 1×10^{-7} M or better, depending on the sample and conditions.

The addition of surfactants is not always recommended in normal and differential pulse polarography, because their presence may reduce the sensitivity of these methods in certain circumstances. Many other forms of polarography have been suggested and tested. Prominent among these are AC polarographic methods, which use sinusoidal and other periodic waveforms. Most modern instruments offer a range of techniques to the electroanalytical chemist.

The solvent used plays a very important role in polarography. It must be able to dissolve a supporting electrolyte and, if necessary, be buffered. In many cases, the negative potential limit involves the reduction of a hydrogen ion or a hydrogen atom on a molecule. It is therefore vital to control the pH of the solution with a buffer. The buffer normally serves as the supporting electrolyte. Furthermore, it must conduct electricity. These requirements eliminate the use of many organic liquids such as benzene. A polar solvent is necessary, the most popular being water. To dissolve organic compounds in water, a second solvent, such as ethanol, acetone, or dioxane, may first be added to the water. The mixture of solvents dissolves many organic compounds and can be conditioned for polarography. As an alternative to the aqueous/nonaqueous systems, a pure polar solvent may be used, especially if water is to be avoided in the electrolysis. Commonly used solvents are acetonitrile, dimethylformamide, dimethylsulfoxide (DMSO), and propylene carbonate. Tetralkylammonium perchlorates and tetrafluoroborates are useful supporting electrolytes because their cations are not readily reduced. A list of functional groups that can be determined by polarography is shown in Table 15.9.

In general, simple saturated hydrocarbons, alcohols, and amines are not readily analyzed at the DME. However, aldehydes and quinones are reducible, as well as ketones.

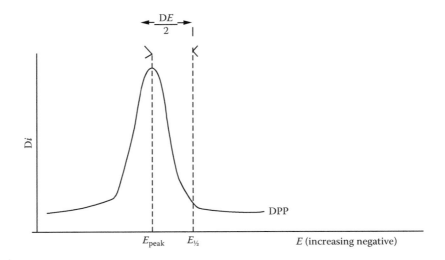

Figure 15.32 Waveform obtained in differential pulse polarography.

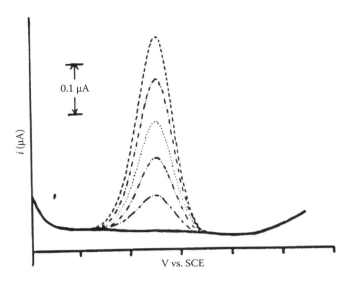

Figure 15.33 Effect of standard additions on a differential pulse polarogram.

Table 15.9 Typical Functional Groups That Can Be Determined by Polarography

Functional Group	Name	$E_{1/2}$ (V)
RCHO	Aldehyde	−1.6
RCOOH	Carboxylic acid	−1.8
RR′C=O	Ketone	−2.5
R–O–N=O	Nitrite	−0.9
R–N=O	Nitroso	−0.2
R–NH$_2$	Amine	−0.5
R–SH	Mercaptan	−0.5

Note: Electrochemical data on a large number of organic compounds are compiled in the CRC Handbook Series in *Organic Electrochemistry*.

Similarly, olefins (in aqueous solution) may be reduced according to the equilibrium

$$R-C=C-R' + 2e^- + 2H_2O \leftrightarrow R-CH_2CH_2-R' + 2OH'$$

In short, polarography can be used for the analysis of C–N, C–O, N–O, O–O, S–S, and C–S groups and for the analysis of heterocyclic compounds. Also, many important biochemical species are electroactive, such as vitamin C (ascorbic acid), fumaric acid, vitamin B factors (riboflavin, thiamine, niacin), antioxidants such as tocopherols (vitamin E), *N*-nitrosamines, ketose sugars (fructose and sorbose), and the steroid aldosterone.

15.3.5 Voltammetry

We mentioned that polarography is a form of voltammetry in which the electrode area does not remain constant during electrolysis. It is possible, however, to use electrode materials other than mercury for electroanalysis, provided that the potential "window" available is suitable for the analyte in question. Table 15.10 summarizes the accessible potential ranges for liquid mercury and for solid platinum electrodes. The precious metals and various forms of carbon are the most common electrodes in use, although a great many materials, both metallic and semiconducting, find use as analytical electrode substrates. Voltammetry is conducted using a microelectrode as the WE under conditions where polarization at the WE is enhanced. This is in sharp contrast to both potentiometry and coulometry where polarization is absent or minimized by experimental conditions. In voltammetry, very little analyte is used up in the measurement process, unlike coulometry where complete consumption of the analyte is desired.

In polarography, the DME is renewed regularly during the voltage sweep, with the consequence that the bulk concentration is restored at the electrode surface at the start of each new drop at some slightly higher (or lower) potential. At a solid electrode, the electrolysis process initiated by a voltage sweep proceeds to deplete the bulk concentration of analyte at the surface without interruption. This constitutes a major difference between classical polarography and sweep voltammetry at a solid electrode. Normal pulse polarography is a useful technique at solid electrodes, because the bulk concentration is restored by convection at the surface during the off-pulse, provided that this is at least 10 times longer than the electrolysis period. It is possible to use a mercury electrode in a static or **hanging** mode. Such a stationary Hg droplet may be suspended from a micrometer syringe capillary. Alternatively, commercially available electrodes have been developed that inject a Hg drop of varying size to the

Table 15.10 Potential Windows for Commonly Used Electrodes and Solvents

		Range (V vs. SCE[a])
Mercury		
Aqueous	1 M HClO$_4$	+0.05 to −1.0
	1 M NaOH	−0.02 to −2.5
Nonaqueous	0.1 M TEAP/CH$_3$CN[b]	+0.6 to −2.8
Platinum		
Aqueous	1 M H$_2$SO$_4$	+1.2 to −0.2
	1 M NaOH	+0.6 to −0.8
Nonaqueous	0.1 M TBABF$_4$/CH$_3$CN[c]	+2.5 to −2.5

[a] The SCE may only be used in nonaqueous systems if traces of water are acceptable.
[b] TEAP, tetraethylammonium perchlorate.
[c] TBABF$_4$, tetrabutylammonium tetrafluoroborate.

opening of a capillary. These electrodes find application for stripping voltammetry (see subsequently).

A powerful group of electrochemical methods uses *reversal* techniques. Foremost among these is **cyclic voltammetry**, which is invaluable for diagnosing reaction mechanisms and for studying reactive species in unusual oxidation states. In cyclic voltammetry, a microelectrode is used as the WE. The potential is increased linearly and the current is measured. The current increases as the potential of the electroactive material is reached. The area of the WE and the rate at which the analyte can diffuse to the electrode surface limits the current. A single voltage ramp is reversed at some time after the electroactive species reacts and the reverse sweep is able to detect any electroactive products generated by the forward sweep. This is cyclic voltammetry. A cyclic voltammogram of a typical reversible oxidation–reduction reaction is shown diagrammatically in Figure 15.34. Usually, a computer data system is used to track the voltage on the time axis so that the reverse current appears below the peak obtained in the forward sweep, but with opposite polarity. The shapes of the waves and their responses at different scan rates are used for diagnostic purposes. Substances are generally examined at concentrations around the millimolar level, and the electrode potentials at which the species undergo reduction and oxidation may be rapidly determined.

The height of the current peak of the first voltage sweep can be calculated from the Randles–Sevcik equation:

$$i_p = 2.69 \times 10^5 n^{\frac{2}{3}} A D^{\frac{1}{2}} C v^{\frac{1}{2}} \quad (15.41)$$

where
n is the number of electrons transferred
A is the electrode area (cm²)
D is the diffusion coefficient of the electroactive species (cm²/s)
C is the concentration of the electroactive species (mol/cm³)
v is the potential scan rate (V/s)

The standard potential E^0 is related to the anodic and cathodic peak potentials, E_{pa} and E_{pc}:

$$E^0 = \frac{1}{2}\left(E_{pc} + E_{pa}\right) \quad (15.42)$$

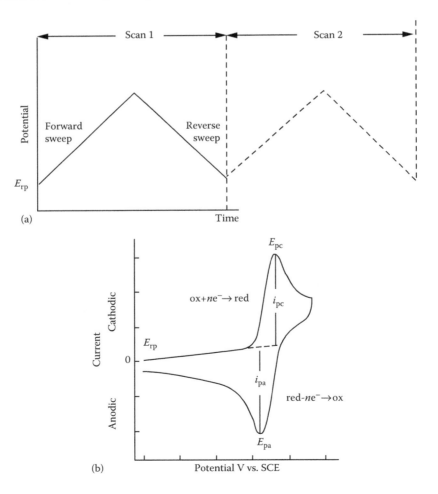

Figure 15.34 (a) Excitation waveform and (b) current response for a reversible couple obtained in cyclic voltammetry.

The peak separation is related to the number of electrons involved in the reaction:

$$E_{pc} - E_{pa} = \frac{0.059}{n} \quad (15.43)$$

assuming that the *IR* drop (resistance) in the cell is not too large, which will increase the peak separation.

Cyclic voltammetry is not primarily a quantitative analytical technique. The references at the end of this chapter provide additional guidance to its applications and interpretation. Its real value lies in the ability to establish the nature of the electron transfer reactions—for example, fast and reversible at one extreme, slow and irreversible at the other—and to explore the subsequent reactivity of unstable products formed by the forward sweep. Suffice it to say that such studies are valuable for learning the fate and degradation of such compounds as drugs, insecticides, herbicides, foodstuff contaminants or additives, and pollutants.

15.3.5.1 Instrumentation for Voltammetry

A cyclic voltammetry mode of operation is featured on many modern polarographs, or a suitable voltage ramp generator may be used in combination with a potentiostat. The three-electrode configuration is required. Pretreatment of the WE is necessary for reproducibility (unless a hanging mercury drop electrode [MDE] is used), and normally, this entails polishing the electrode mechanically with successively finer grades of abrasives.

Some confusion results in choosing an initial voltage at which to start the voltammogram. It is best to measure the open-circuit voltage between the working and reference electrodes with a high-impedance digital voltmeter. This emf is known as the **rest potential** E_{rp}. Scans may then be made in the negative and/or positive directions starting at the rest potential set on the potentiostat. By this means the potentials for reductions and oxidations, respectively, and the electrolyte limits (the "windows") are established. In other words, it is generally necessary to commence at a potential at which no faradaic reaction is possible, or else voltammograms will be distorted and irreproducible.

Scan rates typically vary from 10 mV/s to 1 V/s. The lower limit is due to thermal convection, which is always present in an electrolyte. To record faster responses, an oscilloscope or fast transient recorder used to be required; all modern potentiostats operate at fast scan voltages. The limit is determined by the sampling rate and the rise time of the op-amps. The currents in successive scans differ from those recorded in the first scan. For quantitation purposes, the first scan should always be used.

15.3.5.2 Stripping Voltammetry

The technique of *stripping voltammetry* may be used in many areas requiring trace analyses to the ppb level. It is especially useful for determining heavy metal contaminants in natural water samples or biochemical studies. Either anodic or cathodic stripping is possible in principle, but analyses by anodic stripping voltammetry are more often used. Stripping analysis is a two-step technique involving (1) the preconcentration of one or several analytes by reduction (in anodic stripping) or oxidation (in cathodic stripping) followed by (2) a rapid oxidation or reduction, respectively, to strip the products back into the electrolyte. Analysis time is on the order of a few minutes. The overall determination involves three phases:

1. Preconcentration.
2. Quiescent (or rest) period.
3. Stripping process, for example, by sweep voltammetry or differential pulse.

It is the preconcentration period that enhances the sensitivity of this technique. In the preconcentration phase, precise potential control permits the selection of species whose decomposition potentials are exceeded. The products should form an insoluble solid deposit or an alloy with the substrate. At Hg electrodes, the electroreduced metal ions form an amalgam. Usually, the potential is set 100–200 mV in excess of the decomposition potential of the analyte of interest. Moreover, electrolysis may be carried out at a sufficiently negative potential to reduce all of the metal ions possible below hydrogen ion reduction at Hg, for example. Concurrent H^+ ion reduction is not a problem, because the objective is to separate the reactants from the bulk electrolyte. In fact, methods have been devised to determine the group I metals and $\left(NH_4^+\right)$ ions at Hg in neutral or alkaline solutions of the tetraalkylammonium salts. Exhaustive electrolysis is not mandatory and 2 %–3 % removal suffices. Additionally, the processes of interest need not be 100 % faradaically efficient, provided that the preconcentration stage is reproducible for calibration purposes, which is usually ensured by standard addition.

Typical solid substrate electrodes are wax-impregnated graphite, glassy (vitreous) carbon, platinum, and gold; mercury electrodes are more prevalent in the form of either a **hanging mercury drop electrode (HMDE)** or thin-film mercury electrode (TFME). TFMEs may be electrodeposited on glassy carbon electrodes from freshly prepared $Hg(NO_3)_2$ dissolved in an acetate buffer (pH 7). There is an art to obtaining good thin films, and usually, some practice is necessary to get uniform, reproducible coverage on the carbon substrates. It is recommended that literature procedures be adhered to carefully. Some procedures recommend simultaneously pre-electrolyzing the analyte and a dilute mercury ion solution ($\sim 10^{-5}$ M) so that the amalgam is formed in a single step.

It is important to stir the solution or rotate the electrode during the preconcentration stage. The purpose of this is to increase the analyte mass transport to the electrode by convective means, thereby enhancing preconcentration.

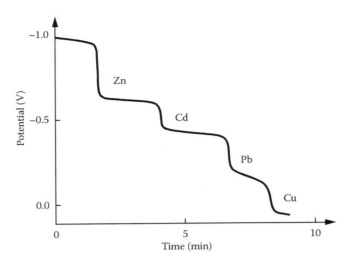

Figure 15.35 Stripping voltammogram of metal-ion species at a TFME.

In general, in electroanalysis, one seeks to obtain proper conditions for diffusion alone to permit mathematical expression of the process rate (the current). In this and controlled flow or rotation cases, it is advantageous to purposely increase the quantity of material reaching the electrode surface. Preconcentration times are typically 3 min or longer.

In the rest period or quiescent stage, the stirrer is switched off for perhaps 30 s but the electrolysis potential is held. This permits the concentration gradient of material within the Hg to become more uniform. The rest period is not obligatory for films produced on a solid substrate.

A variety of techniques have been proposed for the stripping stage. Two important methods are discussed here. In the first, in anodic stripping, the potential is scanned at a constant rate to more positive values. With this single-sweep voltammetry, the resolution of a TFME is better than that of an HMDE, because stripping of the former leads to a more complete depletion of the thin film. As illustrated in Figure 15.35, it is possible to analyze for many metal species simultaneously. The height of the stripping peak is taken to be directly proportional to the concentration. Linearity should be established in the working range with a calibration curve. The second method of stripping involves the application of a differential pulse scan to the electrode. As in polarographic methods, an increase in sensitivity is obtained when the differential pulse waveform is used.

Extremely high sensitivities in trace analyses require good analytical practice, especially in the preparation and choice of reagents, solvents, and labware. Glass cells and volumetric glassware must be soaked for 24 h in trace metal purity 6 M HNO_3. Plastic electrochemical cells are recommended if loss of sample by adsorption to the vessel walls is a likely problem. The inert gas (N_2 or Ar) used to remove dissolved oxygen should be purified so as not to introduce additional contaminants. Solid catalysts and drying agents are recommended for oxygen and water removal. A presaturator is recommended to reduce electrolyte losses by volatilization, especially when low vapor pressure organic solvents are used. A major source of contamination is the supporting electrode itself. This may be purified in part by recrystallization, but the use of sustained pre-electrolysis at a mercury pool electrode may ultimately be required. Dilute standards and samples ought to be prepared daily because of the risk of chemical (e.g., hydrolysis) or physical (e.g., adsorption) losses.

15.3.5.3 pHit Sensor for pH by Cyclic Voltammetry on Carbon

Arguably, the first commercial analytical chemical "instrument" was the glass electrode potentiometric pH meter, invented and marketed by Arnold Beckman in the 1930s. In 2004, a group at Oxford University published a means of measuring pH with special modified carbon nanotube electrodes using cyclic voltammetry with a superposed AC signal. Nearly 9 years later, after extensive research and engineering, this is being brought to market as a revolutionary handheld pH meter. The carbon electrode at its heart is composed of a multiwalled carbon nanotube material with anthraquinone molecules bonded to its surface. A narrow AC voltage signal is superposed on a DC voltage, which is swept over a broad range in which this modified surface does not reduce or oxidize water. The resulting AC current waveform is read out and appears as a peak similar to that illustrated in Figure 15.36. The key to creating a useful electrode was to find a substrate that does not react with water over a wide range of applied voltages and a molecule to attach to it that reversibly responds to H^+ concentration. The reaction controlling this current generation is the same as illustrated for the quinone couple at the end of Section 15.3.4, except

ELECTROANALYTICAL CHEMISTRY

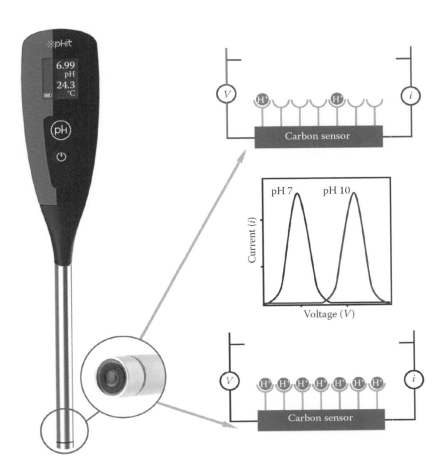

Figure 15.36 The pHit™ voltammetric pH meter. (Courtesy of Senova Systems, Sunnyvale, CA, www.senovasystems.com. Used with permission.)

that the molecule being used has fused six-member aromatic rings on either side of the quinone structure (i.e., it is anthraquinone). The reversible anthraquinone reaction is analogous and involves two H^+ ions, and the voltage value at which the AC peak top is located is precisely linearly proportional to the logarithm of the $[H^+]$, namely, the pH. This electrode is called an anthraquinone-multiwalled carbon nanotube (AQ-MWCNT). This is the first nonpotentiometric design for instrumental pH measurement in nearly 70 years.

The entire pH meter with its electrode is illustrated in Figure 15.36. It consists of two parts: a foot-long probe with the AQ-MWCNT electrode at the end and a bulb-shaped handle that contains all the electronics and digital pH readout, as well as a temperature sensor readout, for automatic temperature correction. Some of the advantages relative to glass electrodes, membrane electrodes, or ISFETs and other potentiometric systems are as follows:

1. The lack of need for a reference electrode and some type of easily fouled liquid junction. Therefore, there is no signal drift. It is calibration free. No calibration buffers required. pH response stored permanently.
2. Nonglass. Nonfragile. Robust solid-state sensor and stainless steel body.
3. Dry storage. Only needs to be rinsed or wiped clean after use.
4. Can measure inserted into meat, mud, fruit, or almost any aqueous moist solid.
5. pH range of 2–12; resolution 0.01 pH units; temperature range of 5 °C–50 °C.
6. Not responsive to high levels of background salt cations.

15.3.5.4 Applications of Anodic Stripping Voltammetry

Anodic stripping voltammetry is readily applicable for those metals that form an amalgam with mercury, for example, Ag, As, Au, Bi, Cd, Cu, Ga, In, Mn, Pb, Sb, Sn, Tl, and Zn. One important cause of interferences is intermetallic compound formation of insoluble alloys between the metals within the amalgam. For example, In/Au and Cu/Ni can form in the Hg drop and then not respond to the stripping stage, so their measured concentrations will be too low. It is imperative to carefully select the electrolyte such that possibly interfering compounds are complexed and electroinactive. This is also a means of improving resolution when there are two overlapping peaks. Some elements can be analyzed in aqueous electrolytes only with difficulty (groups I and II), but, fortunately, their determinations by atomic spectroscopy methods are sensitive and easy to carry out. Anions of carbon compounds can also be stripped by either (1) anodic preconcentration as sparingly soluble Hg salts, (2) adsorption and decomposition, or (3) indirect methods, such as displacement of a metal complex. Examples of complexes used include thiourea, succinate, and dithizone. Anions may be determined as mercurous or silver salts if these are sparingly soluble. Based on solubility determinations, it is possible to estimate some theoretical values for the minimum determinable molar concentrations of anions at mercury: Cl^-, 5×10^{-6}; Br^-, 1×10^{-6}; I^-, 5×10^{-8}; S^{2-}, 5×10^{-8}; CrO_4^{2-}, 3×10^{-9}; WO_4^{2-}, 4×10^{-7}; MoO_4^{2-}, 1×10^{-6}; and $C_2O_4^{2-}$, 1×10^{-6}. With the exception of Cl^-, the mercurous salts are less soluble than the silver salts.

In conclusion, stripping voltammetry is an inexpensive, highly sensitive analytical tool applicable to multicomponent systems; in fact, it is not recommended for metal-ion samples whose concentrations are greater than 1 ppm. Careful selection of operating conditions and especially the electrolyte buffer is necessary. The sensitivity of stripping voltammetry is less for nonmetallic and anionic species than for metals. More recently, flow-through systems have been devised for continuous monitoring purposes. Stripping voltammetry has been applied to numerous trace metal analyses and environmental studies, for example, to determine impurities in oceans, rivers, lakes, and effluents; to analyze body fluids, foodstuffs, and soil samples; and to characterize airborne particulates and industrial chemicals.

15.4 LIQUID CHROMATOGRAPHY DETECTORS

Electrochemical detectors for the determination of trace amounts of ionic and molecular components in liquid chromatographic effluents are sensitive, selective, and inexpensive. Figure 15.37 shows a schematic of an LC system with electrochemical detection. For many separations, they provide the best means of detection, outperforming, for example, ultraviolet (UV)/visible (VIS), fluorescence spectroscopy, and refractive index (RI) methods. Two major categories of electrochemical detectors will be described subsequently: voltammetric and conductometric. An efficient redox center in the ion or molecule to be detected is necessary for voltammetric methods, but this is not mandatory in the species actually undergoing separation, because many organic compounds can be derivatized with an electroactive constituent before or after column separation. Alternatively, certain classes of organic compounds can be photolytically decomposed by a post column online UV lamp into electroactive products, which can be detected. For example, organic nitro compounds can be photolyzed to nitrite and nitrate anions, which undergo electrolysis at suitable WE potentials. In the case of conductivity detectors, the method does not require species that are redox active; rather, changes in the resistivity of the electrolyte effluent are monitored. Examples of species that may be detected are simple ions such as halides, SO_4^{2-}, PO_4^{3-}, nitrate, metal cations, NH_4^+, as well as organic ionic moieties.

15.4.1 Voltammetric Detection

Various waveforms and signal presentation techniques are available, such as steady-state voltammetry with amperometric or coulometric (integrated current) outputs and sweep procedures such as differential pulse. In the simplest mode, a sufficient steady-state potential (constant E) is applied to a WE to reduce or oxidize the component(s) of interest. The output signal is a transient current peak collected and processed by a computer data system. To obtain reasonable chromatographic resolution, the electrochemical cell, positioned at the outlet of the column, must

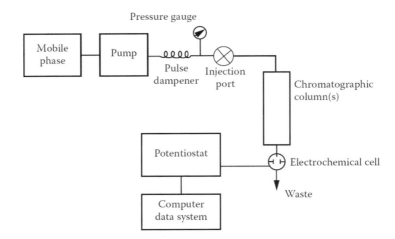

Figure 15.37 Schematic of an LC system with electrochemical detection.

be carefully designed. The volume of the detector cell must be small to permit maximum band resolution. As the reduction or oxidation is performed under hydrodynamic flow conditions, the kinetic response of the electrode/redox species should be as rapid as possible. Glassy (vitreous) carbon is a popular electrode material, and it is possible to chemically alter the electrode surface to improve the response performance. However, other substrates are available, such as Pt, Hg, and Au; these prove useful in cases where they are not passivated by the adsorption of organic reaction products. Platinum and gold electrodes are particularly prone to blockage from the oxidation products of amino acids, carbohydrates, and polyalcohols, for example. Typically, a three-electrode thin-layer cell is used, with the reference and auxiliary electrodes downstream from the WE. Dissolved oxygen may interfere in some assays and may therefore have to be removed by prior degassing.

15.4.2 Conductometric Detection

Many important inorganic and organic ions at the trace level are not easily detected by reduction–oxidation or spectroscopic methods. This has led to the development of sophisticated separation and conductometric detectors capable of ultrasensitive analyses. It is difficult to measure small changes in conductivity due to a trace analyte ion when the background level is orders of magnitude larger; consequently, procedures have been developed that are capable of suppressing high-ionic-strength backgrounds, allowing the net signal to be measured more easily. Combined with miniature conductance detectors, these procedures have led to a technique known as *IC* with eluent suppression (see Figure 14.27). Figure 15.38 shows the setup for a typical application: the analysis of the constituents of acid rain samples.

In the first ion-exchange column, whose resin exists in its carbonate form $\left(R^{2+}CO_3^{2-}\right)$, the sample species, such as Cl^- and SO_4^{2-}, are selectively separated and eluted as sodium salts in a background of sodium carbonate. Detection of the chloride and sulfate by conductivity at this stage would not be possible because of the large excess of conducting ions Na^+ and CO_3^{2-}. In the second column, the cations undergo ion exchange to produce highly conducting acids (such as the strong acids HCl and H_2SO_4, which are fully ionized) with a background of slightly ionized carbonic acid:

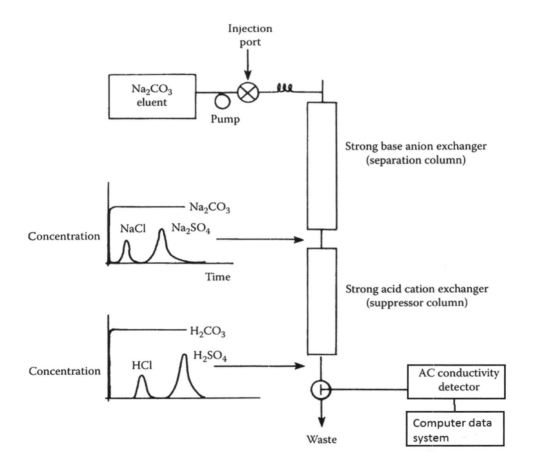

Figure 15.38 Schematic of the eluent suppression approach for IC that permits conductometric detection of analytes. (©Thermo Fisher Scientific [www.thermofisher.com]. Used with permission.)

$$H_2CO_3 \leftrightarrow 2H^+ + CO_3^{2-}$$

Now the highly ionized chloride and sulfate can be measured because the carbonic acid has a low conductance.

A major problem is that the capacity of the second column needs frequent regeneration if large amounts of Na_2CO_3 eluent are used. Various approaches have been tested to improve the practical operation of this form of separation. The first objective is to shorten the time needed to resolve the dilute sample constituents. This is done with a low-capacity separator, which decreases the resolution time and thereby the exchange consumption of the suppressor column. The next objective is to provide continuous regeneration of the suppressor column by automation or by a novel ion-exchange membrane methodology that continuously replenishes the column. For cation analyses, lightly sulfonated beads of cross-linked polystyrene have been employed; for anion analysis, composite beads are formed by electrostatically coating these beads with anion-exchanging spheres.

Liquid chromatographic electrochemical detection has been widely used for metabolite studies in complex matrices and has general applicability in many fields, for example, the pharmaceutical industry, forensic science, medicine, the explosives industry, and agriculture.

15.5 SPECTROELECTROCHEMISTRY

Spectroelectrochemistry (SEC), also called photoelectrochemistry, combines the two techniques of spectroscopy and electrochemistry, as mentioned in Chapter 4. The whole range of the electromagnetic spectrum can be employed from NMR to X-ray absorption. SEC can be used to study materials, surfaces, and inorganic, organic, and biological substances. SEC techniques allow *in situ* spectroscopic measurements of electrogenerated species, permitting the study of the chemical reversibility of reactions, the study of short-lived species, and the identification of unknown intermediate species and products in redox reactions. Because of the wide range of spectroscopy techniques that can be used, most instrumentation is custom designed by the researcher, using some commercial components such as potentiostats. We will focus on electrochemistry coupled to UV/VIS/near-infrared (NIR) and infrared (IR) systems, since these can be set up relatively inexpensively or are available commercially.

Figure 15.39 shows a schematic SEC instrument. For UV/VIS/NIR work, the light source could be a white light light-emitting diode (LED) or a tungsten light source and the spectrometer a UV/VIS/NIR diode array spectrometer.

The electrochemical cell would need quartz windows. A typical three-electrode system, with WE, CE, and reference electrode, is used. For IR work, the source and spectrometer would have to be suitable and the windows of the cell would have to be IR transparent, for example, CaF_2. Suitable materials have been discussed for both techniques in Chapters 5 and 6. Commercial systems are available from BioLogic Science Instruments, Claix, France (www.bio-logic.info), and from ZAHNER-Elektrik GmbH & Co., KG, Kronach, Germany (www.zahner.de). Their websites contain pictures and detailed instrument descriptions as well as a number of application notes and technical notes.

Optically transparent electrodes (OTEs) can be used to construct optically transparent cells for use in a conventional UV/VIS or IR spectrometer (Plieth et al. 1998). OTEs are of various types, depending on the application, and can include the following:

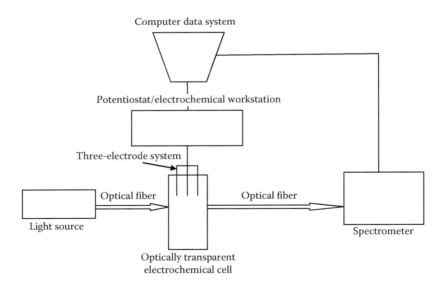

Figure 15.39 Schematic basic spectroelectrochemistry (SEC) instrument.

- A thin metal film on a transparent substrate. The thickness should not exceed 100 nm for the film to remain optically transparent. This can result in high electrical resistance.
- A glass plate with a thin film of an optically transparent, conducting material, for example, indium tin oxide (ITO).
- A gold minigrid between transparent substrates.
- A thicker freestanding metal mesh.

The optically transparent electrode (OTE) is used as a single WE or a stack of WEs and is combined with a reference electrode and a CE (also called an auxiliary electrode) in a spectrochemical cell. Variations of the OTE include thin-layer cells and long optical path length thin-layer cells. Perforated and reflective electrodes are also used as WEs.

Plots derived from SEC experiments can include, among others, absorbance as a function of potential at a constant wavelength, absorbance as a function of the wavelength at a constant potential, or absorbance as a function of time after a potential step or scan, which gives a derivative of the signal with respect to time in the shape of a cyclic voltammogram. In general, the approach taken is to set a potential, allow the system under study to equilibrate, collect a spectrum, and then repeat at different potentials.

A commercial system set up for transmission or absorption measurements in the UV/VIS/NIR range (from 245 to 1550 nm) is shown in Figure 15.40.

In the ZAHNER controlled intensity modulated photocurrent spectroscopy (CIMPS) system, the light source and the cells are aligned on an optical track. Switching between the reference and the measurement cell is automatic. A close-up of the measurement cell is shown in Figure 15.41, set up for a study of a multilayer organic solar cell model system, shown schematically on the left side. In the PECC-2 cell shown, the reference electrode is Ag/AgCl and the CE is a Pt coil. The cell construction from PTFE or similar polymers allows the use of aggressive and nonaqueous electrolytes.

By choosing "plug and play" components for the CIMPS system, more than 20 separate photoelectrochemical and spectroelectrochemical measurements can be made on processes and materials used in organic LEDs, electronic displays, and various types of solar cells. Examples of the types of data that can be obtained are shown in Figures 15.42 through 15.44.

Researchers at the University of Melbourne, Australia, have been performing IR-SEC studies of transition metal complexes, such as nickel carbonyl clusters, containing IR-active ligands like NO, CO, and CN^-. (See Best et al. and the website listed in the bibliography for photos of the system and sample spectra.) The researchers use a three-electrode system designed into a central multielectrode and an external reflectance IR-SEC cell. A thin film (10–20 μm) of sample solution is trapped between the central electrode and an IR-transparent CaF_2 window and IR spectra recorded as the sample is oxidized or reduced. Figure 15.45 shows an early IR-SEC cell design by Best's group, while Figure 15.46 shows an external reflectance cell design and how such a cell fits into an Fourier transform IR (FTIR) spectrometer. Transmission cells for IR-SEC can also be constructed, using the standard salt plate IR transmission cells described in Chapter 5.

IR-SEC is used to study electrochemically reversible reactions of a wide variety of molecules and biomolecules, including hemoglobin, myoglobin, and hydrogenase enzymes with CO or CN ligands at their Fe–Fe- or

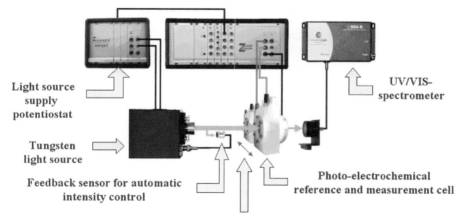

Figure 15.40 The ZAHNER CIMPS photoelectrochemical system, in one of its multiple commercially available configurations. Basics of the system include the ZENNIUM® electrochemical workstation, an XPot external potentiostat, a light source, photochemical cells (one reference cell and one measurement cell), a UV/VIS spectrometer, software, and an optical bench (not shown). (Courtesy of C.-A. Schiller, ZAHNER-Elektrik GmbH & Co. KG, ®ZAHNER-Elektrik GmbH & Co. KG, Kronach, Germany, www.zahner.de.)

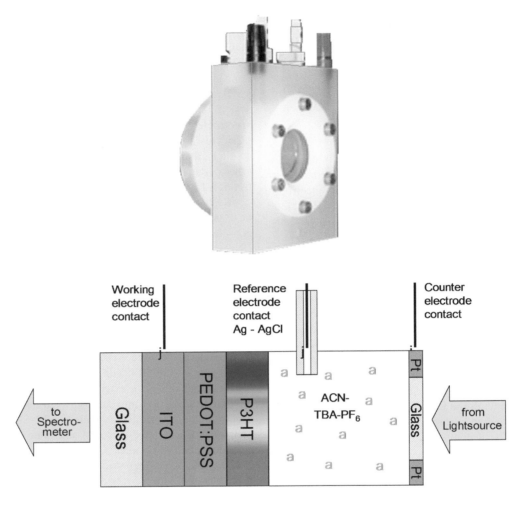

Figure 15.41 (a) A photo of the ZAHNER PECC-2 cell and (b) a schematic diagram of a model system for an organic solar cell film sample with electrode positions. The schematic is not to scale. (Courtesy of C.-A. Schiller, Zahner-Elektrik GmbH & Co. KG, ®Zahner-Elektrik GmbH & Co. KG, Kronach, Germany, www.zahner.de.)

Ni–Fe-active sites. Depending on the cell design, sample volumes of 1 µL can be studied.

"Quantum dots" are semiconductor nanocrystals with specific optical and electronic properties. Their use has grown in recent years in diodes, lasers, photovoltaic cells, and for labeling and tracking cells and other bioimaging applications. Researchers at Los Alamos National Laboratory have studied the photoluminescence behavior of single quantum dots, CdSe–CdS core–shell nanocrystals, using laser-based fluorescence microspectroscopy while manipulating the charge state of the particles in a working electrochemical cell. (See www.cen.acs.org/articles/89/2011/12/Quantum-Dots-Blink.html.) A good introduction to luminescence SEC is the article by Kirchhoff listed in the bibliography. Other luminescence phenomena, such as chemiluminescence, can be electrochemically induced and this technique is used in biochemical and clinical research, as described in the book by Bard (2004).

15.6 QUARTZ CRYSTAL MICROBALANCE

Electrochemistry, in general, is a technique that lends itself to be combined with other techniques. One such technique is gravimetry. Electrogravimetry, discussed earlier in the chapter, is a technique that relies upon the application of a current or potential to deposit an electroactive species onto an electrode. After the material is deposited, its mass is determined analytically. A quartz crystal microbalance (QCM) is an instrument that monitors mass changes in real time.

15.6.1 Piezoelectric Effect

There is a class of materials that will deform upon the application of a voltage. Conversely, the physical deformation (mechanical stress) of these materials results in a voltage. These materials exhibit what is known as the **piezoelectric effect**, that is, mechanical strain resulting in

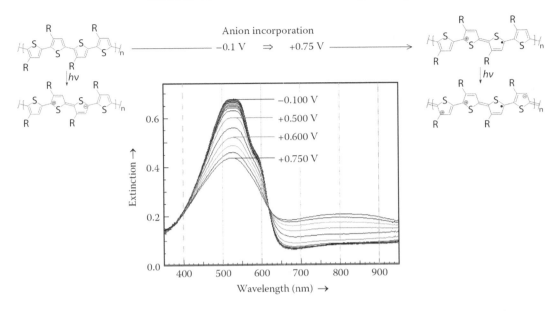

Figure 15.42 Spectra series for the model solar cell layer system from the previous figure. The plots are of extinction versus wavelength at a series of cell voltages. (Courtesy of C.-A. Schiller, Zahner-Elektrik GmbH & Co. KG, ®Zahner-Elektrik GmbH & Co. KG, Kronach, Germany, www.zahner.de.)

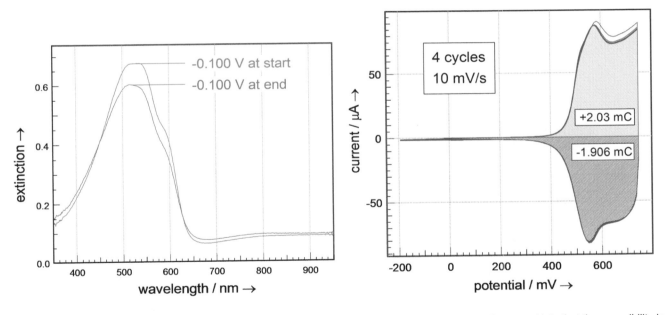

Figure 15.43 (a) Plots of stability and (b) electrochemical reversibility of the model organic solar cell system. Note that the reversibility is shown as a plot of current versus potential and has the same form as a cyclic voltammogram. (Courtesy of C.-A. Schiller, Zahner-Elektrik GmbH & Co. KG, ®Zahner-Elektrik GmbH & Co. KG, Kronach, Germany, www.zahner.de.)

the generation of an electrical potential across the material or deformation of the material as a result of an applied potential (see Curie and Curie). One common example of the piezoelectric effect is the starter in a gas grill. Pressing the starter causes a spring-loaded hammer to strike a piezoelectric material. Deformation of the material results in a voltage spark (a potential that discharges across a gap to a wire) that is used to ignite the propane. Though there are a number of materials that exhibit the piezoelectric effect, this discussion will focus only on quartz crystals.

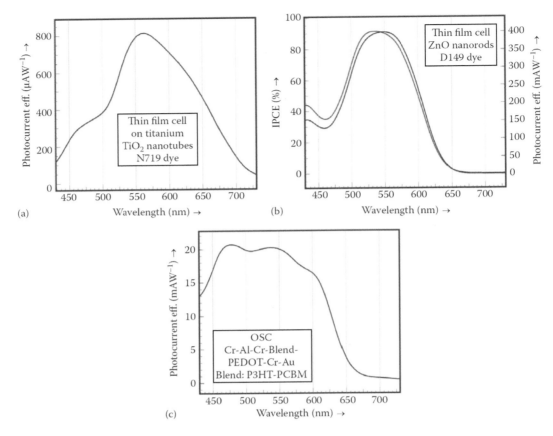

Figure 15.44 (a–c) Typical incident photon conversion efficiency (IPCE) plots for a variety of solar cells. (Courtesy of C.-A. Schiller, Zahner-Elektrik GmbH & Co. KG, ®Zahner-Elektrik GmbH & Co. KG, Kronach, Germany, www.zahner.de.)

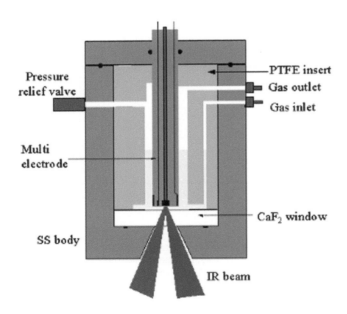

Figure 15.45 A schematic cell for IR-SEC studies. Note the IR-transparent CaF$_2$ window. (Courtesy of Professor Stephen P. Best, University of Melbourne, Parkville, Victoria, Australia.)

The application of an AC voltage to electrodes attached to a piezoelectric material causes it to oscillate. In a QCM, the piezoelectric effect is used to determine small changes in mass as a result of a change in oscillation frequency of the crystal.

Quartz crystals are cut from a block of quartz at a variety of angles in order to give the crystal certain properties. The dominant quartz crystal cut used in a QCM is the AT-cut crystal (Figure 15.47; the designation AT refers to the angle and direction of the cut—it is not an abbreviation). During oscillation in an AT-cut crystal, the top half and bottom half of the crystal move in opposite directions.

The reason for the prevalence of AT-cut crystals is simply frequency stability around room temperature. Much like a swinging pendulum stores, releases, and even loses energy during a cycle, a quartz crystal stores and loses energy as it oscillates. There is a particular frequency on the spectrum where each crystal will store a maximum amount of energy while losing a minimal amount of energy. This is called the resonant frequency and is determined by

$$f_0 = \frac{\sqrt{\dfrac{\mu}{\rho}}}{2t} \qquad (15.44)$$

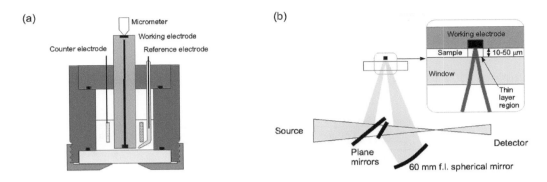

Figure 15.46 (a) Schematic diagrams of an IR-SEC cell and (b) how the cell fits into a commercial FTIR spectrometer. (From Best, S.P., Borg, S.J., and Vincent, K.A., Infrared Spectroelectrochemistry, in *Spectroelectrochemistry*, Kaim, W. and Klein, A., eds., RSC Publishing, Cambridge, U.K., 2008. Reproduced by permission of The Royal Society of Chemistry.)

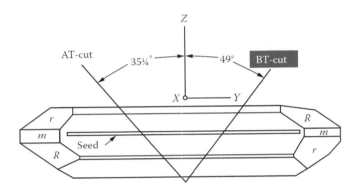

Figure 15.47 Quartz crystal showing the angle of the AT cut. (Courtesy of International Crystal Manufacturing, Inc., Oklahoma City, OK, www.icmfg.com.)

where
μ is the shear modulus
ρ is the density
t is the thickness

As can be seen from the equation, a thicker crystal results in a lower resonant frequency. Likewise, when a rigid material is added to the surface of the crystal, effectively increasing its thickness, the frequency will decrease according to the Sauerbrey equation:

$$\Delta f = \frac{2 f_0^2 m n}{\sqrt{\mu \rho}} \quad (15.45)$$

where
f_0, μ, and ρ are as defined earlier
m is the mass
n is the harmonic number

The implicit assumption in the Sauerbrey equation is that you have a thin, rigid, uniform film. Since the crystal's intrinsic properties are known, a calibration factor, C_f, can be calculated for any crystal frequency and the aforementioned equation can be reduced to

$$\Delta f = -C_f \Delta m \quad (15.46)$$

This equation shows that there is a direct, inverse correlation between a change in frequency and mass.

There are several different methods for constructing a QCM. However, common to all of these methods are the quartz crystal, oscillator, and frequency counter, as shown in Figure 15.48.

The most common frequencies for QCM crystals are 5.000 and 10.000 MHz, although crystals from 1 to >30 MHz can be produced. Electrodes for QCM crystals are usually made of gold, but a variety of metal and nonmetal electrodes are available. The International Crystal Manufacturing, Inc., website (www.icmfg.com) has detailed information on QCM crystals and their manufacture. The Gamry Instruments website (www.gamry.com) has an excellent introduction to the QCM with multiple references.

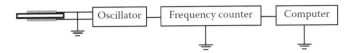

Figure 15.48 A schematic QCM. (Courtesy of Gamry Instruments, Inc., Warminster, PA, www.gamry.com.)

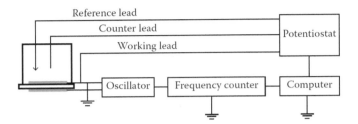

Figure 15.49 QCM interfaced to a potentiostat to create an EQCM. (Courtesy of Gamry Instruments, Inc., Warminster, PA, www.gamry.com.)

15.6.2 Electrochemical Quartz Crystal Microbalance

There are a number of applications for the QCM alone but the technique's real power lies in its ability to combine with electrochemical applications. The electrochemical QCM (EQCM) is a QCM that has been interfaced to a potentiostat such that the electrode exposed to solution doubles as the WE. In experiments such as these, the ability to use electrochemical data (from the potentiostat) combined with gravimetric data (from the QCM) gives insight into interfacial processes, chemical reactions, corrosion, and other fields of study. Interfacing a potentiostat to a QCM is shown schematically in Figure 15.49.

An example of the use of an EQCM is in the deposition of Cu^{2+} from a solution of $CuSO_4$ in 1 M H_2SO_4. Figure 15.50 is a cyclic voltammogram showing both current and change in mass versus E.

The potential of the WE was swept negatively from initial potential in order to reduce Cu^{2+} onto an Au-coated quartz crystal. A film of insoluble Cu^0 was deposited. The potential scan was then reversed, with oxidation of Cu^0 to Cu^{2+}. The soluble Cu^{2+} was removed from the electrode surface resulting in the electrode mass returning to its initial state. Plotting the change in mass versus the charge, obtained from integrating the current, results in a curve with a slope that can be converted to a molar mass by simply multiplying by the Faraday constant and the number of electrons in the reaction.

An EQCM can be used to characterize electroactive polymer films, as demonstrated in a Gamry Instruments application note. Electropolymerization is a convenient way to control film growth and the EQCM provides a way to study the redox behavior of such polymer films. Polybithiophene films were assembled by cycling an Au-coated 10 MHz quartz crystal between two voltages in 1 mM bithiophene solution. Bithiophene electropolymerizes via a two-electron oxidation at potentials greater than 1.25 V versus an Ag/Ag^+ pseudo-reference electrode. Figure 15.51 shows two cycles of film growth. In Cycle 1 (the loop formed by the middle two lines), only nonfaradaic current is seen until the potential is >1.25 V. In Cycle 2 (the loop formed by the top and bottom lines), additional faradaic current is seen beginning at about 0.75 V due to the oxidation of the polymer film.

Once the film was deposited, the cell was washed and refilled with monomer-free electrolyte. Cycling of the film (Figure 15.52) shows a broadly shaped couple at approximately 1.0 V. The mass increases during oxidation and decreases during reduction but does not return to the initial film mass.

Mass change plotted against charge for a cycle in monomer-free electrolyte provides information about mobile species during oxidation and reduction. Information about the rate of solvent ingress/egress in the film can be obtained. Details on this and other applications of the EQCM can be found at www.gamry.com.

15.7 ELECTROCHEMISTRY AT THE NANOSCALE

Since the beginnings of electrochemistry, with the development of the voltaic pile by Volta in 1799, electrochemistry (and the associated analytical methods) has been a nanoscale science. The study and exploitation of surface and interfacial properties and processes within electrical double layers has always been the study of thicknesses of nanometers and subnanometers. Technologically, scaling metallic and semiconductor electrodes to the order of 1 to 100 nm has revolutionized sensor technology and has made possible measurement of extremely low concentrations of selected molecular species. Examples have included gold nanoparticles and nanopores, and amperometric probes

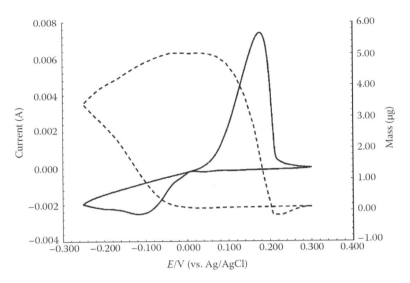

Figure 15.50 Cyclic voltammogram from an EQCM. Both the change in mass (dotted line) and change in current (solid line) versus E can be determined. (Courtesy of Gamry Instruments, Inc., Warminster, PA, www.gamry.com.)

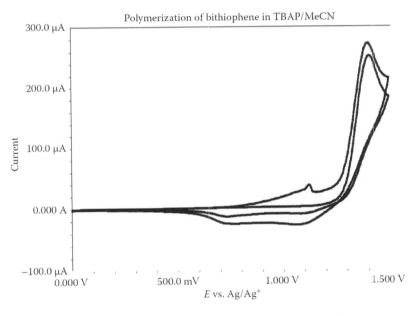

Figure 15.51 Electropolymerization of 1 mM bithiophene. Details may be found at www.gamry.com. (Courtesy of Gamry Instruments, Inc., Warminster, PA, www.gamry.com.)

that can probe single cells. This has not been a one-sided relationship; the development of nanotechnology has in large measure been enabled by synergistic application of electrodeposition and etching. This has formed the basis of major fabrication methods for nanodevices. Voltametric measurements, along with microscopy (such as scanning and transmission electron microscopy and atomic force microscopy) at the nanoscale have furnished insight in areas as diverse as quantum dots, semiconducting nanofilms, and sensors. The following summary of applications has been taken from the article by Svoronos in the ASM Handbook:

1. Determination of oxidation states and electrocatalytic "hot spots" and their impacts on analyses and catalysis.
2. Measurements of single molecular conductance in a wide variety of environments.
3. Applications of devices such as batteries, supercapacitors, fuel cells, sensors, electrochemical energy conversion and electrochromic devices.
4. Enabling surface nanostructure construction and fabrication.
5. Probing brain chemistry by the monitoring of molecular release from single cells, with spatial resolution.

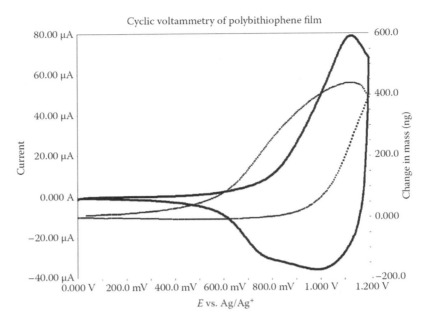

Figure 15.52 Mass (dotted line) and current (solid line) versus V for one cycle of polybithiophene. (Courtesy of Gamry Instruments, Inc., Warminster, PA, www.gamry.com.)

6. Development of bioanalytical catalysts and electrocatalysts in carbon nanotubes.
7. Measurement of transmembrane charge support and membrane potential to probe redox properties at the subcellular level.
8. Development of a large-scale homogeneous nanocup electrode array.

At this writing, the full implications of nano electrochemistry remain in development, but the promises are substantial.

SUGGESTED EXPERIMENTS

15.1 Run a DC polarogram with a known volume of 1 M KNO_3 electrolyte that has not been deaerated. Observe and identify the two reduction waves of dissolved oxygen.

15.2 Purge the electrolyte in Experiment 15.1 with purified nitrogen gas for 20 min or so. Add sufficient lead nitrate solution to make the electrolyte 1.0×10^{-4} M in Pb^{2+} ion. Record the DC polarogram and measure the limiting current and the half-wave potential ($E_{1/2}$) for the lead ion reduction wave.

15.3 Measure the mercury flow rate with the column height set as for Experiment 15.2. Collect about 20 droplets under the electrolyte (why?) and determine the drop lifetime with a stopwatch. Use the Ilkovic equation to calculate the diffusion equation for Pb^{2+} ion and compare your derived value with a literature value.

15.4 Record a normal pulse polarogram of (a) 1.00×10^{-5} M Zn^{2+}, (b) Cd ion, and (c) Cu ion in a degassed 1 M KNO_3 electrolyte. Construct a graph of potential (E) versus $\log[(i_L - i)/i]$. Determine the value of n from the slope and comment on the reversibility of this reaction. How does the electrolyte temperature affect the slope of this curve?

15.5 Record a differential pulse polarogram of a background electrolyte without and with added tap water. Which metal species are present? Use a standard additions calibration to quantify one of the metal impurities. Compare E_{peak} with $E_{1/2i}$.

15.6 Use a F^- ISE per the manufacturer's instructions to find the concentration of F^- ion in your drinking water and in your brand of (1) toothpaste and (2) mouthwash. Use a standard additions calibration procedure. Some sample preparation may be needed for the toothpaste. Several extraction procedures are available in the literature.

15.7 Collect some samples for pH analyses, for example, rainfall, snow, lake water, pool water, soft drinks, fruit juice. Standardize the pH meter with a standard buffer of pH close to the particular sample of interest. Design some experiments to change the temperature of these samples. What is the influence of temperature on pH measurements?

15.8 (a) Use a glass pH electrode combination and titrate unknown strength weak acid(s) with 0.2000 N NaOH. Stir the electrolyte continuously.
(b) Calculate the strength(s) of the acid(s) (i.e., the pK_a) and explain the shape of the curve.

15.9 Copper sulfate may be analyzed by gravimetry by exhaustive electrolysis at a weighed platinum electrode. Add 2 mL H_2SO_4, 1 mL HNO_3, and 1 g urea to 25 mL of the copper solution.

15.10 Tap waters, as well as a variety of biological fluids and natural waters, can be analyzed for heavy metals by anodic stripping voltammetry at an HMDE. Clean glassware and sample bottles by leaching for about 5 h in 6 N HNO_3. First, analyze for Cd^{2+} and Pb^{2+} ions at acidic pH (2–4). Deposition is made with gentle stirring for 2 min exactly. After a 15 s quiescent period, scan from −0.7 V SCE in the positive direction. Zn^{2+} and Cu^{2+} ions can be analyzed by raising the pH to 8–9 with 1 M NH_4Cl/1 M NH_4OH buffer. Deposit at −1.2 V SCE and strip as before. It will be necessary to adjust the deposition time and/or current output sensitivity to achieve the best results.

15.11 Cyclic voltammetry can be used to study the reversible reduction–oxidation couple of $[Fe(CN)_6]^{3-} + e^- \leftrightarrow [Fe(CN)_6]^{4-}$. You will need solutions of 1 M KNO_3, 5×10^{-3} M $K_3[Fe(CN)_6]$ in 1 M KNO_3, and 5×10^{-3} M $VOSO_4$ in 1 M KNO_3. A CV equipped with a three-electrode cell (Pt microelectrode as the WE, Pt foil electrode as the auxiliary electrode, and a calomel reference electrode) and an X–Y recorder (or computer data system) is required. The potassium nitrate solution serves as the baseline. All solutions must be purged with nitrogen to eliminate oxygen before scanning the potential. A sweep from 0.80 to −0.12 V and back versus SCE is suitable for the Fe system. The range may need to be changed for the vanadium system. Can you evaluate n for each system? Is the value what you expect for each reaction? Evaluate E^0 for each system. Do these values agree with those in the literature? (Experiment courtesy of Professor R.A. Bailey, Department of Chemistry, Rensselaer Polytechnic Institute, Troy, NY.)

PROBLEMS

15.1 Give an example of a half-cell. The absolute potential of a half-cell cannot be measured directly. How can the potential be measured?

15.2 Using the Nernst equation, complete the following table:

pH of Solution	Concentration of H+	E of Hydrogen Half-Cell
		0
1.2		
3		
	10^{-5}	
	10^{-10}	
7		
12		
14		

15.3 Describe and illustrate an SCE. Write the half-cell reaction.

15.4 How is pH measured with a glass electrode? Why does a glass electrode give pH readings lower than the actual pH in strongly basic solutions? What other errors can occur in pH measurement with a glass electrode?

15.5 Calculate the theoretical potential of the following cell. Is the cell as written galvanic or electrolytic?

$$Pt \left| Cr^{3+} \left(3.00 \times 10^{-2} \text{ M}\right), Cr^{2+} \left(4.00 \times 10^{-5} \text{ M}\right) \right| \left| Sn^{2+} \left(2.00 \times 10^{-2} \text{ M}\right), Sn^{4+} \left(2.00 \times 10^{-4} \text{ M}\right) \right| Pt$$

15.6 A salt of two monovalent ions M and A is sparingly soluble. The E^0 for the metal is +0.621 V versus SHE. The observed emf of a saturated solution of the salt is 0.149 V versus SHE. What is the solubility product K_{sp} of the salt? (Neglect the junction potential.)

15.7 Describe the process of electrodeposition and how it is used for electrogravimetric analysis.

15.8 Can two metals such as iron and nickel be separated completely by electrogravimetry? Explain your answer and state any assumptions you make.

15.9 State Faraday's law. A solution of Fe^{3+} has a volume of 1.00 L, and 48,246 C are required to reduce the Fe^{3+} to Fe^{2+}. What was the original molar concentration of iron? What is the molar concentration of Fe^{2+} after the passage of 12,061 C?

15.10 What are the three major forms of polarography? State the reasons why pulse polarographic methods are more sensitive than classical DC polarography.

15.11 What are the advantages of mercury electrodes for electrochemical measurements? What are the advantages of the DME versus a Pt microelectrode for polarography? What are the disadvantages of the DME?

15.12 Describe the method of anodic stripping voltammetry. What analytes can it be used to determine? Why is stripping voltammetry more sensitive than other voltammetric methods?

15.13 Describe the principle of an ISE. Why is the term ion-specific electrode not used?

15.14 How can conductometric measurements be used in analytical chemistry? Give two examples.

15.15 Briefly outline two types of electrochemical detectors used for chromatography.

15.16 The fluoride ISE is used routinely for measuring fluoridated water and fluoride ion in dental products such as mouthwash. A 50 mL aliquot of water containing sodium fluoride is analyzed using a fluoride ion electrode and the MSAs. The pH and ionic strength are adjusted so that all fluoride ions are present as free F^- ions. The potential of the ISE/reference electrode combination in a 50 mL aliquot of the water was −0.1805 V. Addition of 0.5 mL of a 100 mg/L F^- ion standard solution to the beaker changed the potential to −0.3490 V. Calculate the concentration of (1) fluoride ion and (2) sodium fluoride in the water sample.

15.17 Copper is deposited as the element on a weighed Pt cathode from a solution of copper sulfate in an electrolytic cell. If a constant current of 0.600 A is used, how much Cu can be deposited in 10.0 min?

(Assume no other reductions occur and that the reaction at the anode is the electrolysis of water to produce oxygen.)

15.18 Why does coulometry not require external calibration standard solutions?

15.19 Explain why a silver electrode can be an indicator electrode for chloride ions.

15.20 What are the three processes by which an analyte in solution is transported to an electrode surface? What single transport process is desired in polarography? Explain how the other transport processes are minimized in polarography.

15.21 Would you expect the half-wave potential for the reduction of copper ion to copper metal to be different at a Hg electrode from that at a platinum electrode? Explain your answer.

15.22 Sketch a schematic cyclic voltammogram for a nonreversible reduction reaction. (See Figure 15.34 for a CV of a reversible reaction.)

15.23 What is meant by SEC? What types of electromagnetic radiation can be used in this technique?

15.24 Describe two types of OTEs for use in SEC.

15.25 Briefly describe the principle of operation of a QCM.

15.26 Explain one use of the EQCM in electrochemistry.

15.A APPENDIX: SELECTED STANDARD REDUCTION POTENTIALS AT 25 °C

Half Reaction	E^0 (V)
$F_2(g) + 2e^- \rightarrow 2F^-$	+2.87
$Co^{3+} + e^- \rightarrow Co^{2+}$	+1.82
$H_2O_2 + 2H^+ + 2e^- \rightarrow 2H_2O$	+1.77
$Ce^{4+} + e^- \rightarrow Ce^{3+}$	+1.74
$PbO_2(s) + 4H^+ + SO_4^{2-} + 2e^- \rightarrow PbSO_4(s) + 2H_2O$	+1.70
$MnO_4^- + 8H^+ + 5e^- \rightarrow Mn^{2+} + 4H_2O$	+1.51
$Au^{3+} + 3e^- \rightarrow Au(s)$	+1.50
$Cl_2(g) + 2e^- \rightarrow 2Cl^-$	+1.36
$Cr_2O_7^{2-} + 14H^+ + 6e^- \rightarrow 2Cr^{3+} + 7H_2O$	+1.33
$MnO_2(s) + 4H^+ + 2e^- \rightarrow Mn^{2+} + 2H_2O$	+1.23
$O_2(g) + 4H^+ + 4e^- \rightarrow 2H_2O$	+1.23
$Br_2(l) + 2e^- \rightarrow 2Br^-$	+1.07
$NO_3^- + 4H^+ + 3e^- \rightarrow NO(g) + 2H_2O$	+0.96
$2Hg^{2+} + 2e^- \rightarrow Hg_2^{2+}$	+0.92
$Hg_2^{2+} + 2e^- \rightarrow 2Hg(l)$	+0.85
$Ag^+ + e^- \rightarrow Ag(s)$	+0.80
$Fe^{3+} + e^- \rightarrow Fe^{2+}$	+0.77
$O_2(g) + 2H^+ + 2e^- \rightarrow H_2O_2$	+0.68
$MnO_4^- + 2H_2O + 3e^- \rightarrow MnO_2(s) + 4OH^-$	+0.59
$I_2(s) + 2e^- \rightarrow 2I^-$	+0.53
$O_2(g) + 2H_2O + 4e^- \rightarrow 4OH^-$	+0.40
$Cu^{2+} + 2e^- \rightarrow Cu(s)$	+0.34
$AgCl(s) + e^- \rightarrow Ag(s) + Cl^-$	+0.22
$Cu^{2+} + 2e^- \rightarrow Cu^+$	+0.15
$Sn^{4+} + 2e^- \rightarrow Sn^{2+}$	+0.13
$2H^+ + 2e^- \rightarrow H_2(g)$	0.00
$Pb^{2+} + 2e^- \rightarrow Pb(s)$	−0.13
$Sn^{2+} + 2e^- \rightarrow Sn(s)$	−0.14
$Ni^{2+} + 2e^- \rightarrow Ni(s)$	−0.25
$Co^{2+} + 2e^- \rightarrow Co(s)$	−0.28
$Cd^{2+} + 2e^- \rightarrow Cd(s)$	−0.40
$Fe^{2+} + 2e^- \rightarrow Fe(s)$	−0.44
$Cr^{3+} + 3e^- \rightarrow Cr(s)$	−0.74
$Zn^{2+} + 2e^- \rightarrow Zn(s)$	−0.76
$2H_2O + 2e^- \rightarrow H_2(g) + 2OH^-$	−0.83
$Mn^{2+} + 2e^- \rightarrow Mn(s)$	−1.18
$Al^{3+} + 3e^- \rightarrow Al(s)$	−1.66
$Na^+ + e^- \rightarrow Na(s)$	−2.71
$Ca^{2+} + 2e^- \rightarrow Ca(s)$	−2.87
$Ba^{2+} + 2e^- \rightarrow Ba(s)$	−2.90
$K^+ + e^- \rightarrow K(s)$	−2.93
$Li^+ + e^- \rightarrow Li(s)$	−3.05

Note: All species in aqueous solution (aq) unless otherwise indicated. All dissolved species = 1 M; all gases = 1 atm.

BIBLIOGRAPHY

The reader will note that occasionally multiple editions of books are cited as sources or reference works. The reason for this is that sometimes different figures are used in different editions, which may in fact be written by different authors. The sources found in the bibliography will be the most recent versions of a particular work.

In some cases instrument manufacturers and suppliers are cited by providing a URL to the exact internet location. We have endeavored to ensure that these links are functional at the time of publication, but clearly these links change. If a link has become nonfunctional, a good approach is to use your browser's advanced search capability. For example: "company name" AND "instrument type" AND "specific descriptors" is a good place to start.

Adams, R. *Electrochemistry at Solid Electrodes*. Marcel Dekker, Inc.: New York, 1969.

Aikens, D.; Bailey, R.; Moore, J.; Giachino, G.; Torukins, R. *Principles and Techniques for an Integrated Chemistry Laboratory*. Waveland Press, Inc.: Prospect Heights, IL, 1985.

ASTM. *Annual Book of ASTM Standards, Water and Environmental Technology*, Vols. 11.01 and 11.02. ASTM: West Conshohocken, PA, 2009.

Baizer, M.M.; Lund, H. *Organic Electrochemistry*. Marcel Dekker, Inc.: New York, 1983.

Bard, A.J., ed. *Electroanalytical Chemistry*. Marcel Dekker, Inc.: New York, 1993.

Bard, A.J., ed. *Electrogenerated Chemiluminescence*. CRC Press: Boca Raton, FL, 2004.

Bard, A.J.; Faulkner, L.R. *Electrochemical Methods—Fundamentals and Applications*. John Wiley & Sons, Inc.: New York, 1980.

Bard, A.J.; Parsons, R.; Jordan, J., eds. *Standard Potentials in Aqueous Solution*. Marcel Dekker, Inc.: New York, 1985.

Bates, R.G. *Determination of pH*. John Wiley & Sons, Inc.: New York, 1973.

Berezanski, P. Electrolytic conductivity. In Settle, F.A. (ed.) *Handbook of Instrumental Techniques for Analytical Chemistry*. Prentice Hall PTR: Upper Saddle River, NJ, 1997.

Best, S.P.; Borg, S.J.; Vincent, K.A. Infrared spectroelectrochemistry. In Kaim, W.; Klein, A. (eds.) RSC Publishing: Cambridge, U.K., 2008. (See also www.chemistry.unimelb.edu.au/staff/spbest/techniques/SEC.htm.)

Bockris, J.O. *Modern Electrochemistry*, Vols. 1 and 2. Plenum Press: New York, 1970.

Bond, A.M. *Modern Polarographic Methods in Analytical Chemistry*. Marcel Dekker, Inc.: New York, 1980.

Bruno, T.J., Svoronos, P.D.N., *CRC Handbook of Basic Tables for Chemical Analysis – Data Driven Methods and Interpretation*, 4th Ed., CRC Taylor and Francis, Boca Raton, FL, 2021.

Chen, S., Liu, Y., Electrochemistry at nanosized electrodes, Phys. Chem. Chem. Phys., 16, 635, 2014. Curie, P.; Curie, J.C. *R. Acad. Sci.*, 91, 294, 1880.

Dean, J.A. *Analytical Chemistry Handbook*. McGraw-Hill, Inc.: New York, 1995.

Diefenderfer, A.J.; Holton, B.E. *Principles of Electronic Instrumentation*, 3rd edn. Saunders College Publishing: Philadelphia, PA, 1994.

Dryhurst, G. *Electrochemistry of Biological Molecules*. Academic Press: New York, 1977.

Enke, C.G. *The Art and Science of Chemical Analysis*. John Wiley & Sons, Inc.: New York, 2000.

Evans, A. *Potentiometry and Ion Selective Electrodes*. John Wiley & Sons, Inc.: New York, 1987.

Grinberg, N., Rodriguez, S., eds., Ewing's Analytical Instrumentation Handbook, 4th Ed., Chapters 16, 17, 18 and 19, CRC Press Taylor and Francis, Boca Raton, FL, 2018.

Fritz, J.S. *Acid–Base Titrations in Non-Aqueous Solvents*. G. Frederick Smith Chemical Company: Columbus, OH, 1952. (This publication had been available from GFS Chemicals upon request.)

Fry, A.J. *Synthetic Organic Electrochemistry*. Harper & Row: New York, 1972.

Gale, R.J. (ed.) *Spectroelectrochemistry*. Plenum Press: New York, 1988.

Greenberg, A.; Clesceri, L.; Eaton, A. (eds.) *Standard Methods for the Examination of Water and Wastewater*, 18th edn. American Public Health Association: Washington, DC, 1992.

Harris, D.C. *Quantitative Chemical Analysis*, 5th edn. W.H. Freeman and Co.: New York, 1999.

Hart, J.P. *Electroanalysis of Biologically Important Compounds*. Ellis Horwood: New York, 1990.

Huber, W. *Titrations in Nonaqueous Solvents*. Academic Press: New York, 1967.

Ives, D.J.G.; Janz, G.J. (eds.) *Reference Electrodes—Theory and Practice*. Academic Press: New York, 1961.

Kaim, W.; Klein, A. (eds.) *Spectroelectrochemistry*. RSC Publishing: London, U.K., 2008.

Kalvoda, R. *Operational Amplifiers in Chemical Instrumentation*. Ellis Horwood: Chichester, U.K., 1975.

Kirchhoff, J.R. *Current Separations*, 16(1), 11, 1997.

Kissinger, P.T.; Heineman, W.R. *Laboratory Techniques in Electroanalytical Chemistry*, 2nd edn. Marcel Dekker, Inc.: New York, 1996.

Koryta, J.; Stulik, K. *Ion-Selective Electrodes*, 2nd edn. Cambridge University Press: Cambridge, U.K., 1983.

Kucharsky, J.; Safarik, L. *Titrations in Nonaqueous Solvents*. Elsevier: Amsterdam, the Netherlands, 1965.

Malmstadt, H.V.; Enke, C.G.; Crouch, S.R. *Microcomputers and Electronic Instrumentation: Making the Right Connections*. American Chemical Society: Washington, DC, 1994.

Meites, L.; Zuman, P.; Rupp, E.B. (eds.) *CRC Handbook Series in Organic Electrochemistry*, Vols. 1–5. CRC Press, Inc.: Boca Raton, FL, 1982.

Michael, A.; Wightman, R.M. Microelectrodes. In Kissinger, P.; Heineman, W. (eds.) *Laboratory Techniques in Electroanalytical Chemistry*, 2nd edn. Marcel Dekker, Inc.: New York, 1996.

Plieth, W.; Wilson, G.S.; Gutierrez de la Fe, C. *Pure Appl. Chem.*, 70(7), 1395, 1998.

Sauerbrey, G. *Z. Phys.*, 155, 206, 1959.

Settle, F.A. (ed.) *Handbook of Instrumental Techniques for Analytical Chemistry*. Prentice Hall PTR: Upper Saddle River, NJ, 1997.

Streuli, C.A. Titrimetry: Acid–base titrations in non-aqueous solvents. In Kolthoff, I.M.; Elving, P.J. (eds.) *Treatise on Analytical Chemistry, Part I*, Vol. II. Wiley-Interscience: New York, 1975.

Svoronos, P.D.N, ASM Handbook, Volume 10 Materials Characterization, ASM International, Materials Park, OH, 2019.

Vydra, F.; Stulik, K.; Julakova, E. *Electrochemical Stripping Analysis*. Ellis Horwood, Chichester, U.K., 1976.

Wang, J. *Stripping Analysis*. VCH Publishers, Inc.: Deerfield Beach, FL, 1985.

Warner, M. Electrochemical detectors for liquid chromatography. *Anal. Chem.*, 66, 601A, 1994.

CHAPTER 16

Surface Analysis

16.1 INTRODUCTION

The term **surface analysis** is used to mean the characterization of the chemical and physical properties of the **surface layer** of **solid** materials. The surface layer of a solid often differs in chemical composition and in physical properties from the bulk solid material. A common example is the thin layer of oxide that forms on the surface of many metals such as aluminum upon contact of the surface with oxygen in air. The thickness of the surface layer that can be studied depends on the instrumental method. This layer may vary from one atom deep, an atomic **monolayer**, to 100–1000 nm deep, depending on the technique used. Surface analysis has become increasingly important because our understanding of the behavior of materials has grown. The nature of the surface layer often controls important material behavior, such as resistance to corrosion. The various surface analysis methods reveal the elements present, the distribution of the elements, and sometimes the chemical forms of the elements in a surface layer. Chemical speciation is possible when multiple surface techniques are used to study a sample.

Surface analysis has found great use in understanding important fields such as materials characterization of polymers, metals, ceramics, and composites; corrosion; catalysis; failure analysis; and the functioning and failure of microelectronics and magnetic storage media. In the pharmaceutical industry, it can be used to investigate the multilayered materials and coatings used in packaging and time-release products. In medicine, it has been used to study bone structure and the surface of teeth, indicating why SnF_2 fights tooth decay, for example, and to study the biocompatibility of metallic and polymeric implantable devices.

Spectroscopic surface analysis techniques are based on bombarding the surface of a sample with a beam of X-rays, particles, electrons, or other species. The bombardment of the surface by this **primary beam** results in the emission or ejection of X-rays, electrons, particles, and the like from the sample surface. This emitted beam is the **secondary beam**. The nature of the secondary beam is what provides us with information about the surface. A number of techniques have been developed for surface analysis but only the most common will be discussed in this chapter.

The names of these spectroscopic techniques and the primary and secondary beams used for each technique are listed in Table 16.1. These techniques are frequently quite different in physical approach, but all provide information about solid surfaces. Applications of these surface analysis techniques are presented in Tables 16.2 and 16.3. Very readable discussions of surface analysis methods are to be found in the ASM handbook.

The student should be aware that there is another class of surface analysis instruments based on analytical microscopy, including scanning electron microscopy, transmission electron microscopy, atomic force microscopy, and scanning tunneling microscopy. A discussion of these microscopy techniques is beyond the scope of this chapter, and while related to analytical instrumental analysis, it is more the province of material sciences. Most industrial materials characterization laboratories will have some combination of electron spectroscopy, X-ray analysis, surface mass spectrometry, and analytical microscopy instrumentation available, depending on the needs of the industry.

16.2 ELECTRON SPECTROSCOPY TECHNIQUES

In the discussion on X-rays (Chapter 9), it was mentioned that the source of an X-ray photon is an atom that is bombarded by high-energy electrons or photons. This can displace an inner shell electron, which is ejected from the atom, leaving an ion with a vacancy in an inner shell (Figure 16.1). An electron from an outer shell then drops into the inner shell and an X-ray photon is emitted simultaneously. The energy of the photon is equal to the difference between the energy of the orbital the electron was in originally and that of the one to which it descends. The energy levels of these two inner orbitals are almost independent of the chemical form of the atom, combined or otherwise. However, we know from Chapter 4 that the energy of a valence electron varies with the chemical form and chemical environment of the combined atom and provides the basis for ultraviolet (UV) absorption analysis. This variation is reflected as well in the energies of the inner shell electrons, but the changes in energy involved are very small compared with the energy of the emitted X-rays themselves. The slight differences in

Table 16.1 Selected Spectroscopic Techniques for Surface Analysis

Abbreviated Name	Full Name	Primary Beam	Secondary Beam
XPS (ESCA)	X-ray photoelectron spectroscopy (electron spectroscopy for chemical analysis)	X-rays	Electrons
AES	Auger electron spectroscopy	Electrons	Electrons
ISS	Ion scattering spectroscopy	Ions	Ions
SIMS	Secondary ion mass spectrometry	Ions	Ions
EM, EPMA	Electron microprobe, electron probe microanalysis	Electrons	X-rays

Table 16.2 Surface Analysis Applications by Industry

Industry	Application
Microelectronics	Composition of deposited layers Thickness of deposited layers Defect characterization Particle identification Process residue identification
Magnetic data storage media	Lubricant type and thickness Carbon overcoat composition Magnetic layer composition Magnetic layer thickness Defect/contamination identification Failure analysis
Optical coatings	Coating composition and thickness Adhesion layers Composition of surface defects
Automobile industry	Paint adhesion Paint weathering Catalyst poisoning Lubricant chemistry
Pharmaceutical industry	Contamination identification Distribution of active ingredients in tablets Drug distribution in time-release coatings Patent infringement monitoring
Polymer industry	Surface coatings Surface chemistry Multilayer composition and thickness
Biotechnology	Contamination identification Correlation of surface chemistry and biocompatibility Biosensor development Promotion of cell growth on surfaces

Source: Information in table provided by Physical Electronics, Inc., Eden Prairie, MN, www.phi.com.

Table 16.3 Specific Surface Analysis Applications

Type of Analysis	Applications
Failure analysis	Surface contamination Particle identification Fracture due to grain boundary impurities
Adhesion failure	Adhesive failure Cohesive failure Silicone surface contamination
Corrosion	Thickness and composition of surface oxides Identification of corrosive elements (e.g., Cl) Passivation layer composition and thickness
Surface cleanliness	Detergent residue identification Solvent residue identification Type and amount of surface impurities
Reverse engineering	Composition and identification of competitors' materials
Semiconductor wafers	Analysis of complete surface of wafer (up to 300 mm) Defect/particle identification Thin-film composition
Hot sample stage analysis (up to 700 °C)	Volatility of surface components Temperature dependence of surface chemistry Migration of bulk components to surface
Cold sample stage analysis (to −100 °C)	Surface species that are volatile at RT in ultrahigh vacuum (UHV) Hydrated surfaces Samples that outgas at RT in UHV

Source: Information in table provided by Physical Electronics, Inc., Eden Prairie, MN, www.phi.com.

Note: RT, room temperature; UHV, ultrahigh vacuum.

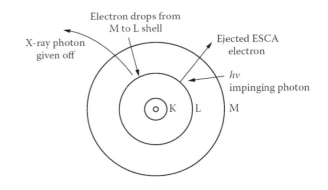

Figure 16.1 An impinging photon ejecting an inner shell XPS (or ESCA) electron. The kinetic energy of the ejected XPS (ESCA) electron, E_k, is related to the binding energy of the electron in the atom, E_b. The binding energy is characteristic of the element. An X-ray photon of energy $E_M - E_L$ is emitted in the process. An alternative process to the emission of the X-ray photon is emission of an Auger electron, which is shown in the next figure.

X-ray wavelength are extremely difficult to measure, since they are such a small fraction of the nominal wavelength of the X-rays generated. It is therefore normally accepted that the energies of the emitted X-rays are independent of the chemical form of the generating atoms and differences in energy cannot be observed except under very high resolution. X-ray fluorescence (XRF), as was discussed in Chapter 9, is therefore an elemental analysis technique. It tells us what elements are present, but does not give information on the oxidation state of the element or the chemical species present.

However, the energy E of the original electron ejected from the atom is the difference between the energy E_1 of the impinging electron (or photon) and the energy E_b required to remove the electron from the atom, that is, $E = E_1 - E_b$. The energy E_b will be slightly different depending on the chemical environment of the atom. We can determine this small difference by making the energy E_1 of the impinging electron just slightly greater than the energy E_b required to eject the electron. The residual energy E of the emitted electron will then be small, but any variation in E_b will produce a larger relative variation in the energy E of the emitted electron. In this way, small differences in E_b can be measured. For example, if the energy of the K_α line for Al is 1487 eV, the effect of the chemical environment may change this by 2 eV, resulting in Al K_α at 1485 eV. The relative shift (2/1487) is slight, and the 1485 eV X-ray line would be difficult to distinguish from the original 1487 eV line in X-ray emission and XRF. If the energy of the *impinging electrons* generating X-rays is 1497 eV, then the energy of the ejected electron is $1497 - 1487 = 10$ eV. But if the chemical environment changes the energy needed to eject the electron to 1485 eV, then the energy of the emitted electron is $1497 - 1485 = 12$ eV. It is easier to distinguish electrons with 10 eV energy from electrons with 12 eV energy than to distinguish between photons with energies of 1487 and 1485 eV. These slight changes in the energy of the ejected electron can provide information about the chemical species present and the oxidation state of the atoms present. Based on this phenomenon, the field of ESCA was developed in the 1960s by Swedish physicist Kai Siegbahn and his coworkers. More recently, the term **X-ray photoelectron spectroscopy (XPS)** has become the accepted name for the field and will be used preferentially.

A companion field, **Auger electron spectroscopy** (AES), was developed simultaneously. AES does not provide chemical species information, only elemental analysis, as we will see. Since the electrons ejected in these two techniques are of low energy and the probability of electron interaction with matter is very high, the electrons cannot escape from any significant depth in the sample. Typical escape depths for XPS and AES electrons range from 0.5 to 5 nm for materials. The phenomenon is therefore confined to a few atomic layers, combined or otherwise, which are at the surface of the sample, and provides a method of surface analysis.

16.2.1 X-Ray Photoelectron Spectroscopy

When an X-ray beam of precisely known energy impinges on a sample surface held under an ultrahigh vacuum (UHV), inner shell electrons are ejected and the energy of the ejected photoelectrons is measured. This is the phenomenon on which XPS is based. Figure 16.2 shows

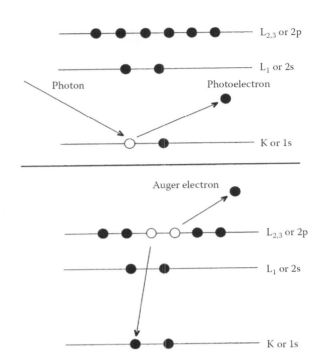

Figure 16.2 The XPS process for a model atom. The top diagram shows an incoming photon causing the ejection of the XPS photoelectron. The bottom diagram shows one possible relaxation process that follows the ejection of the photoelectron, resulting in the emission of an Auger electron. (From Moulder, J.F. et al., In *Handbook of X-Ray Photoelectron Spectroscopy*. Chastian, J. and King, R.C., eds., Physical Electronics, Inc., Eden Prairie, MN, 1995; Courtesy of Physical Electronics, Inc., Eden Prairie, MN, www.phi.com.)

the emission of both a photoelectron and an Auger electron for a model atom.

The kinetic energy of the escaping electron is designated as E_k. The binding energy of this electron is given by the equation

$$E_k = h\nu - E_b - \phi \qquad (16.1)$$

where
E_b is the binding energy of the electron
$h\nu$ is the energy of the photon (either X-ray or vacuum UV)
E_k is the kinetic energy of the escaping electron
ϕ is an instrumental constant called **the work function of the spectrometer**

The XPS spectrum is a plot of the number of emitted electrons per energy interval vs. their kinetic energy. The work function of the spectrometer can be measured and is constant for a given instrument, allowing the binding energies of the electrons to be determined.

Since E_b, the original binding energy of the emitted electron, depends on the energy of the electronic orbit and the element from which the electron is emitted, it can be used to identify the element present. In addition, the chemical form or environment of the atom affects the energy E_b to a much smaller but measurable extent. These minor variations give rise to the **chemical shift** and can be used to identify the valence of the atom and sometimes its exact chemical form. Databases of binding energies for elements and compounds are available, such as the National Institute for Standards and Technology (NIST) electronic XPS database that can be found at www.nist.gov/srd/surface or tables in reference books such as that by Moulder et al. (1995) listed in the bibliography. Quantitative measurements can be made by determining the intensity of the XPS lines of each element.

16.2.1.1 Instrumentation for XPS

A commercial XPS instrument consists of four major components housed in a UHV system with magnetic shielding: (1) the radiation source, consisting of an X-ray source and a means of providing highly monochromatic X-rays; (2) the sample holder; (3) the energy analyzer, which resolves the electrons generated from the sample by energy; and (4) an electron detector. Modern instruments have computerized data recording and processing systems. The pressure required for XPS must be very low, often less than 10^{-7} Pa, in order to prevent adsorbed residual gas from interfering with the surface analysis. This requires a UHV system. As will be seen, many of the instrument components are also used for Auger spectroscopy instruments. The sample holder will be discussed in Section 16.2.1.2.

16.2.1.1.1 Radiation Source

The radiation source used in XPS is a standard X-ray anode tube, as described in Chapter 9. Soft X-rays are used, with Al and Mg being the most common anodes. Many commercial systems offer a dual-anode X-ray tube so that the analyst can switch between excitation wavelengths. It is very important that the X-ray source be monochromatic, with a linewidth extending over as narrow an energy range as possible. Al and Mg have narrow K emission lines. The Mg K_α line has an energy of 1253.6 eV and a linewidth of 0.7 eV, while the Al K_α line has an energy of 1486.6 eV and a linewidth of 0.85 eV. Linewidths for other elements are generally about 1–4 eV. It can be seen from Equation 16.1 that any variation in the energy of the impinging X-rays will produce a similar variation in the energy of the ejected XPS electron. Most modern XPS instruments are equipped with an X-ray monochromator to narrow the bandwidth of the X-ray line. This is accomplished with the use of a quartz crystal monochromator that is constructed using the Rowland circle system shown in Figure 16.3. (If necessary, the student should review diffraction of X-rays by crystals in Chapter 9.) X-ray monochromators

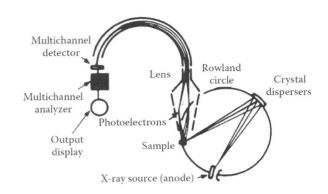

Figure 16.3 Schematic diagram of an XPS instrument using an X-ray monochromator and a multichannel detector.

can narrow the bandwidth of the Al K_α line from 0.85 eV to approximately 0.3 eV, dramatically improving the ability to detect small binding energy shifts. It should be noted that Mg cannot be used as the source with an X-ray monochromator; there are no crystals available with the proper lattice spacing.

In addition to narrowing the bandwidth of the primary X-ray beam, the monochromator eliminates X-ray satellites, radiant heat from the X-ray source, high-energy electrons, and background radiation (**Bremsstrahlung**) from the X-ray anode. This simplifies photoelectron spectra and minimizes the potential for chemical damage to the sample from the X-ray source. For nonmonochromatic X-ray sources, an aluminum window is typically placed at the end of the X-ray tube to reduce unwanted radiation from the X-ray source. The Al window must be very thin (no more than a few microns) to provide good X-ray transmission. This restriction limits the effectiveness of the Al window in reducing unwanted radiation.

Historically, XPS instruments have relied upon the energy analyzer to define the analysis area because technology was not available to finely focus the X-ray beam, limiting the spatial resolution of XPS. A commonly used approach to improve the spatial resolution is shown in Figure 16.4, where an aperture is placed in the analyzer's input lens to restrict the analysis area. Only the photoelectrons from a given small area of the sample pass through the aperture and into the analyzer. The practical lower limit for small-area XPS analysis with this approach is 25 μm.

Modern XPS instruments are typically equipped with a specially shaped crystal monochromator that, in addition to decreasing the bandwidth, also focuses the X-ray beam. These crystal monochromators are carefully thermostated so as to minimize any changes in lattice spacing. Currently, it is possible to produce X-ray beams that are less than 10 μm in diameter with this approach. XPS instruments of this type provide much higher sensitivity for small-area XPS analysis than those using an aperture to select the analysis area.

SURFACE ANALYSIS

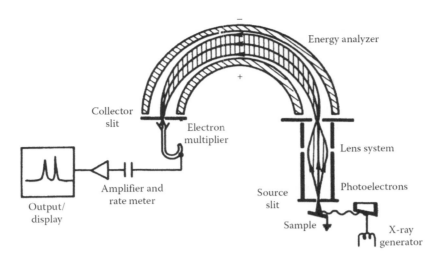

Figure 16.4 Schematic diagram of an XPS instrument with a single-channel electron multiplier detector, and an aperture for selecting the analysis area.

16.2.1.1.2 Electron Energy Analyzers

Electron energies can be filtered with energy discriminators as shown in Figure 16.5. Figure 16.5a shows two parallel charge plates with two openings. The electrons with the required energy enter and leave the holes as in a slit system. Electrons with other energies do not exit the second hole. Figure 16.5b shows a second type of energy discriminator, which is a simple grid discriminator. Electrons with insufficient energy are repelled by the second grid and do not penetrate. This system only discriminates against electrons with low energies. Any electrons that have sufficient energy penetrate the second grid. A third system uses cylindrical plates (Figure 16.5c). If the angle between the planes of the entry and exit slits is 127.17° [i.e., $\pi/(2)^{1/2}$ rad], a double-focusing effect is obtained and the intensity of the electron beam is maintained as high as possible. Electrons with the incorrect energy are lost either to the sides of the cylindrical plates or on the sides of the exit slit. Similar modifications have led to the development of a 180° cylindrical system rather than parallel plates (Figure 16.5d). This system does not provide such fine-tuning of the energy bandwidth, but it increases the intensity of the electron beam. This is another example of gaining in beam intensity but losing in energy discrimination or gaining in power and losing in resolution.

Electron energy analyzers are equivalent to the monochromators used in spectroscopy. Their function is to disperse the emitted photoelectrons based on their energies. The most commonly used electron energy analyzers incorporate an electrostatic field that is either symmetrical or hemispherical. These systems are in essence an extension of the electron energy filters shown in Figure 16.5. All electron energy analyzers require shielding from stray magnetic fields as described subsequently. One

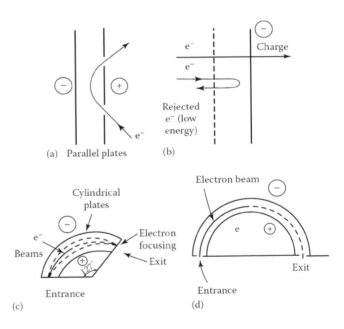

Figure 16.5 Schematic diagrams of several electron energy discriminators.

system is shown schematically in Figure 16.6; it is based on the system of Figure 16.5a and is called a cylindrical mirror analyzer (CMA). The "plates" are now two coaxial cylinders, thus providing an efficient electron-trapping system while maintaining resolution. The electron source labeled on the figure is of course the sample surface, emitting photoelectrons. A voltage (negative potential) is applied to the outer cylinder; the inner cylinder is grounded. Only photoelectrons with the appropriate energy pass through the apertures and are focused onto the detector.

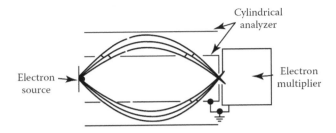

Figure 16.6 Cross section of an electrostatic cylindrical electron energy analyzer. The electron source is the excited sample surface that is emitting photoelectrons.

Photoelectrons of energy E will be focused when

$$E = \frac{KeV}{\ln\left(\dfrac{r_o}{r_i}\right)} \quad (16.2)$$

where
K is a constant
e is the charge on the electron
V is the applied voltage
r_i is the radius of the inner cylinder
r_o is the radius of the outer cylinder

The CMA as shown is used for AES. For XPS, two CMAs in series are used to obtain the required energy resolution. This design is called a double-pass CMA. The transmission of electrons through a double-pass CMA is good, but the resolution is poorer than that obtained using the concentric hemispherical analyzer (CHA) described subsequently.

The CHA, shown in Figure 16.4, is widely used in both XPS and Auger instruments. The CHA consists of an input lens assembly, two concentric hemispherical shells of differing radii, and an electron detector. The input lens focuses the photoelectrons from the sample and lowers (or retards) their energy. This permits better energy resolution. The electrons then pass through a slit into the hemispherical path. The two shells are maintained at a potential difference, ΔV, with the outer sphere negatively charged and the inner one positively charged, as shown in Figure 16.4. Only electrons with a very small energy range will pass through the hemispheres along a path of radius r_o and through the exit slit to the detector. The potential difference is varied to scan the electron energies, and a spectrum of the energies of the ejected photoelectrons is recorded. Resolution of this analyzer, $\Delta E/E$, depends on the entrance angle of the electrons into the analyzer, the slit width, and r_o, as follows:

$$\frac{\Delta E}{E} = \frac{w}{r_o} + \alpha^2 \quad (16.3)$$

where
$\Delta E/E$ is the resolution
w is the slit width
r_o is the radius of the equipotential surface within the analyzer
α is the angle at which the electrons enter the analyzer

Two approaches to retardation (deceleration) of the electrons are used. One mode of operation is called **fixed-analyzer transmission (FAT)** or **constant-analyzer transmission (CAT)**. In this mode, the electrons are retarded to a constant energy. This results in the absolute resolution of the system being independent of the kinetic energy of the incoming electrons. Energy analyzers operating in the FAT mode provide high sensitivity when the pass energy is high and high energy resolution when the pass energy is low. The other operating mode is called **fixed retardation ratio (FRR)** or **constant relative ratio (CRR)**. In this mode, the incoming electrons are decelerated by a constant ratio from their initial kinetic energies. Energy analyzers operating in the FRR mode provide relatively uniform sensitivity across the energy spectrum. However, absolute energy resolution degrades as the measured kinetic energy increases.

Other analyzers have been designed based on the magnetic deflection of electrons, but in general, these have not been very successful because of the difficulty of maintaining a uniform magnetic field. Electrostatic systems are used in all commercial instrumentation.

The instrument is calibrated regularly with known conductive standards such as gold or copper to establish the linearity of the energy scale and its position.

16.2.1.1.3 Detectors

Both single-channel and multichannel detectors are used in XPS and Auger spectroscopy. The most common single-channel detector is the channel electron multiplier. The channel electron multiplier functions much like the photomultiplier tube (PMT) used in optical spectroscopy (discussed in Chapter 4). The major difference is that electrons constitute the signal that is being amplified, not photons. The channel electron multiplier consists of a continuous dynode surface inside a tube as depicted in Figure 16.7a. Channels may be straight tubes as shown or curved. The surface is a thin-film conductor with a high resistance. When a voltage is supplied across the input and output, a potential gradient exists along the channel direction. The dynode surface and potential gradient permit electron amplification throughout the channel. An incoming electron strikes the inner wall and secondary electrons are emitted. These are accelerated by the potential gradient and travel a parabolic path until they strike the opposite wall, releasing more electrons. An ejected photoelectron

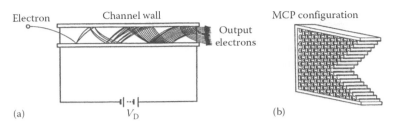

Figure 16.7 (a) A schematic channel electron multiplier. A thin-film conductive layer inside the tube serves as a continuous dynode surface. (b) A schematic microchannel plate (MCP) configuration. (Courtesy of Hamamatsu Corporation, Bridgewater, NJ, www.hamamatsu.com.)

from the sample enters at one end and an amplified pulse of electrons exits at the other end. Gains of up to 10^5 are possible with a 1 kV supply voltage. This detector is very efficient at counting electrons, even those with very low energy. Like the PMT, the detector can be saturated at high electron intensity.

Multichannel electron detectors of various types can be placed to cover the exit plane of the analyzer. Charge-coupled devices (CCDs) (discussed in Chapter 8), phosphor-coated screens, and position-sensitive detectors are some of the multichannel devices in use. One type of position-sensitive detector consists of a **microchannel plate** (MCP) electron multiplier. The MCP (Figure 16.7b) consists of a large number of very thin conductive glass capillaries, each 6–25 μm in diameter. The capillaries are fused together and sliced into a thin plate. Each capillary or channel works as an independent electron multiplier, exactly as the single-channel electron multiplier described earlier, thereby forming a 2D electron multiplier. Used in conjunction with a phosphor screen, 2D imaging of surfaces is possible.

Electrons are detected as discrete events, and the number of electrons for a given time and energy is stored and then displayed as a spectrum.

16.2.1.1.4 Magnetic Shielding

The ejected electrons have low energy. They are affected significantly by local magnetic fields, including the Earth's and those of any stray electrical impulses as generated by wiring to lights, equipment, elevators, and so on. These stray fields must be neutralized in the critical parts of the instrument in order to obtain useful data. One method is to enclose the critical regions with high-permeability magnetic alloy, which shields the sample from stray magnetic fields. Another method is to use Helmholtz coils, which produce within themselves a homogeneous field. This field may be made exactly equal and opposite to the Earth's magnetic field. A feedback system to the coils can also be used; this senses variations in local magnetic fields and varies the current in the Helmholtz coils, neutralizing transient magnetic fields as they arise. This system not only neutralizes the Earth's magnetic field but also local fields as they are generated.

16.2.1.1.5 Electron Flood Gun

In practice, when the sample is irradiated, electrons are ejected. An electron takes with it a negative charge, leaving the sample positively charged. If the sample is conductive and is connected to ground (see Chapter 2 for a discussion about earth ground), the electron "holes" are rapidly neutralized. If the conductivity of the sample is low (such as for an insulating or semiconducting material), this positive charge builds up at a steady but unpredictable rate on the surface of the sample, changing the work function of the sample itself and therefore the net energy of the ejected electrons. In practice, this would lead to a variation in the observed energy of the ejected electrons and erroneous results would be derived. It is therefore highly desirable to eliminate this variable charge.

This problem can be overcome by flooding the surface of the sample with low-energy electrons, which neutralize the positive charge built up on the surface. This is done by using an electron flood gun. Of course, these electrons in turn affect the work function of the surface to some extent, but the effect is constant and reproducible data are obtainable over an extended period of time.

16.2.1.1.6 Ion Gun

An ion gun is used in XPS and Auger instruments for two purposes: (1) to clean the sample surface of any external contamination layer and (2) to sputter atoms from the surface in order to obtain a depth profile analysis, discussed under applications. One type of ion gun uses a heated filament to ionize inert gas atoms. The ions are accelerated by a potential placed on the ionization chamber and are focused to strike the sample surface. The ions remove atoms from the sample surface by collision. The rate of removal of surface atoms is controlled by the kinetic energy of the ions and by the nature of the surface atoms; sputtering rates may be as high as 10 nm/min.

16.2.1.1.7 UHV System

Since the electrons involved in XPS are of low energy, they must not collide with other atoms or molecules in the system or they will not be detected. This is best handled by using a very high vacuum system. Pressures of 10^{-8}–10^{-9} Pa are typical in surface analysis systems. (1 pascal, Pa, is equal to 7.50×10^{-3} torr.) In order to achieve this UHV, surface analysis instruments are fabricated of stainless steel. Stainless steel structures can be evacuated to the required low pressure with the appropriate combination of vacuum pumps. The stainless steel can also be heated ("baked out") to remove adsorbed gases. Other required components such as windows and gaskets must be made of materials that do not outgas and can withstand required elevated temperatures. The vacuum pumps used in surface analysis instruments to achieve the UHV pressures required include turbomolecular pumps, ion pumps, cryopumps, and oil diffusion pumps. Oil diffusion pumps must include a liquid nitrogen trap to prevent oil vapor from entering the instrument. Details of materials used in surface analysis systems and the vacuum pumps used can be found in the handbook by Ewing listed in the bibliography and references therein.

As we will see, surface analysis equipment is very similar for several techniques, so most surface analysis instruments are configured for multiple surface analysis techniques. A schematic diagram of a commercial instrument showing the placement of the ion gun, energy analyzer, source, and so on is shown in Figure 16.8.

16.2.1.2 Sample Introduction and Handling for Surface Analysis

The sample must be kept under high vacuum during analysis. In older instruments and some modern lower-cost instruments, the main analysis chamber must be brought to atmospheric pressure in order to insert or change the sample. The system is vented with nitrogen, the sample inserted, and the entire system pumped down to UHV pressures again before analysis can begin. This presents a problem when numerous samples are to be analyzed, since the pump down can take several hours. Many instruments utilize sample prechambers, small-volume compartments that will accommodate a number of samples and can be evacuated rapidly. The samples are then transferred into the main chamber without causing an excessive rise in pressure.

Sample holders range from those that accommodate samples of 1–2 cm diameter and about 1 cm in thickness to custom-designed sample chambers that can hold much bigger samples and those that are of irregular size and

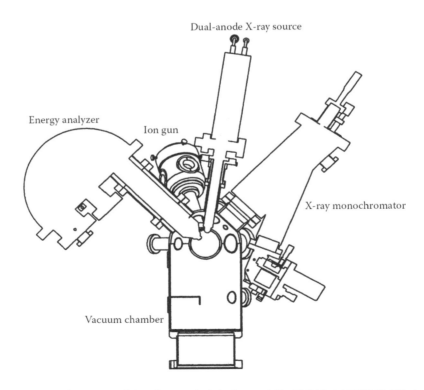

Figure 16.8 A schematic diagram of a commercial surface analysis instrument, the PHI Model 5600 MultiTechnique system. The monochromatic X-ray source is located perpendicular to the sample surface (not shown); the standard X-ray source is at 57.4° relative to the analyzer axis. (From Moulder, J.F. et al., In *Handbook of X-Ray Photoelectron Spectroscopy*. Chastian, J. and King, R.C., eds., Physical Electronics, Inc., Eden Prairie, MN, 1995; Courtesy of Physical Electronics, Inc., Eden Prairie, MN, www.phi.com.)

SURFACE ANALYSIS

shape. Chambers designed for 95 mm diameter computer hard disks are common, for example. The sample in its holder is placed on a stage that can be moved in all directions and allows the sample to be rotated and tilted for optimal positioning. The stage in a modern surface analysis instrument is under computer control. Some sample stages allow the sample to be heated or cooled as necessary. Cooling, for example, is required to analyze materials with a high vapor pressure, when volatile surface components are to be studied and for samples such as plastics that outgas at room temperature under UHV conditions.

XPS performed with commercial instruments as described is used for the analysis of the surfaces of solid samples. The solid must be in electrical contact with the spectrometer, so conducting solids are easily analyzed. Nonconductive samples can be analyzed by using a low-current electron flood gun or a combination of electron flood gun and ion gun to neutralize the surface charge of an insulator. Solid samples can be analyzed "as is," but they often must be cut to fit the sample holder. Due to the surface sensitivity of XPS and AES, great care must be taken when handing samples to prevent unwanted contamination of the sample surface. The analyst should avoid touching the sample, using solvents that may leave a residue, or solvents that remove material of interest from the surface. Dust and particulates from sample cutting or cleaving are typically removed using an inert gas such as nitrogen. Sample-handling tools should be cleaned using high-purity solvents. Analysts should wear gloves that are powder free and silicone free. Because of the surface sensitivity of these techniques, it is common to detect atmospheric or environmental contaminants on the sample surface. Common contaminants include hydrocarbons and thin oxide layers on the surface of metal samples. Sometimes, these surface layers may be of analytical interest, as they may be related to the processing used to create the sample. If they are not of interest, organic contaminants are removed with high-purity solvents, and oxide layers are removed by sputtering the sample in the instrument. An exposed chemically active surface must be kept under UHV to prevent rapid contamination of the surface.

Even under vacuum, gas molecules will reabsorb onto a clean sample surface. Hercules and Hercules describe that at pressures between 10^{-8} and 10^{-10} torr, a surface can be covered by a monolayer of adsorbed gas molecules in from 1 to 10 h, while at 10^{-6} torr, the same process occurs in a few seconds. This rapid recontamination of a cleaned surface makes clear the need for UHV conditions for accurate surface analysis.

Some surface analysis instruments are equipped with apparatus to fracture or cleave a material within the UHV chamber to expose a fresh surface for analysis. Metal samples are routinely fractured in UHV to expose grain boundaries. This permits XPS or AES detection of bulk impurities that have segregated to grain boundaries, for example. Grain boundary chemistry is of great interest to metallurgists for many reasons. As noted, even under UHV conditions, limited time is available for contamination-free analysis of freshly exposed surfaces.

16.2.1.3 Analytical Applications of XPS

Fundamental research in physical chemistry using XPS has included the study of gases and liquids to explore the energies of valence electrons, radicals, excited states, vibrational states, and more, but a discussion of this is beyond the scope of this text. Only the analysis of solid surfaces will be considered. All elements above the atomic number of Li can be studied. For qualitative elemental analysis and especially for a "first pass" at an unknown sample, a wide energy range "survey" spectrum should be collected. In order to obtain chemical state information, and for quantitative analysis of minor elements, a higher-resolution detail scan is taken of the energy regions of interest. Detail scans must be collected with a sufficient signal-to-noise ratio so that accurate peak positions and peak areas can be determined. The sampling depth for most materials is a few atomic layers, and the method is preferable when the destructive nature of ion beam or electron beam techniques presents a problem for the desired result.

Tables 16.2 and 16.3 provide examples of applications of surface analysis techniques. XPS can be used for many of these applications.

16.2.1.3.1 Qualitative Analysis

XPS is readily applied to the qualitative identification of elements present in a sample surface, to depths of 25 Å for metals and up to 100 Å for polymers, and these analyses can be done in 5 to 11 min after pump down. All elements except hydrogen and helium can be determined with the typical X-ray source. The XPS spectra for sulfur obtained with monochromated Al K_α radiation and Mg K_α radiation are presented in Figures 16.9 and 16.10, respectively. Each spectrum consists of a limited number of photoelectron peaks in the range of 0–1400 eV in binding energy for Al (0–1200 eV for Mg). XPS photoelectron peaks from the sulfur 2s (K shell) and 2p (L shell) orbitals are seen, as well as an Auger peak (marked LMM). Some small contaminant peaks from Ar and oxygen can be seen in the spectra. Auger peaks can be distinguished from photoelectron (XPS) peaks. Because Auger electron kinetic energies are independent of the ionizing radiation energy, on a plot of binding energies, the Auger lines appear to be at different positions when different X-ray sources are used. The Auger peak is shown as an inset in each of the two spectra; as you can see, the Auger peak appears at a higher binding energy when Al X-radiation is used than when Mg X-radiation is used. The XPS peaks do not change position. Figures 16.9

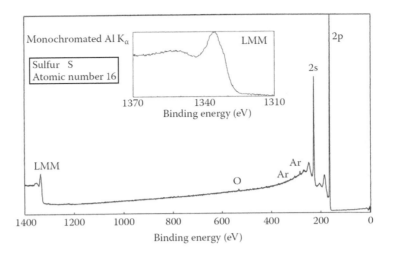

Figure 16.9 XPS spectrum of sulfur collected with monochromated Al K$_\alpha$ radiation. (From Moulder, J.F. et al., In *Handbook of X-ray Photoelectron Spectroscopy*, Chastian, J. and King, R.C., eds., Physical Electronics, Inc., Eden Prairie, MN, 1995; Courtesy of Physical Electronics, Eden Prairie, MN, Inc., www.phi.com.)

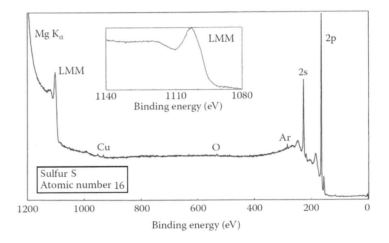

Figure 16.10 XPS spectrum of sulfur collected with Mg K$_\alpha$ radiation. (From Moulder, J.F. et al., In *Handbook of X-ray Photoelectron Spectroscopy*, Chastian, J. and King, R.C., eds., Physical Electronics, Inc., Eden Prairie, MN, 1995; Courtesy of Physical Electronics, Inc., Eden Prairie, MN, www.phi.com.)

and 16.10 are survey spectra, the type of scan that would be used for qualitative elemental analysis of an unknown.

For low atomic mass elements ($Z < 30$), energy peaks are observable corresponding to the K and L shells (the s and p electrons). For elements with atomic numbers between 35 and 70, the d electrons result in peaks in the XPS spectrum. For elements with atomic numbers greater than 70, the pattern includes f electrons. It can be seen that qualitative identification of the surface elements present is possible based on identification of the binding energies. Binding energies for all elements are tabulated and may be found in the NIST XPS database at www.nist.gov/srd/surface and in tables in references such as Moulder et al. (1995) listed in the bibliography. Typical data for S and Cu are shown in Figure 16.11. In addition, the spin doublet patterns and ratios for the lines from the p, d, and f electrons are characteristic and also tabulated. The copper XPS spectrum is shown in Figure 16.12. The spin doublet for the 2p electrons is seen ($2p_{1/2}$ and $2p_{3/2}$), and the ratio of the lines is about 1:2, which is the expected intensity ratio for these lines. XPS lines from the 2s, 3s, 3p, and 3d orbitals are also seen, along with the Auger LMM transition.

XPS spectra also exhibit some minor peaks due to a number of other processes. Small peaks called X-ray satellite peaks can appear at lower binding energies due to the nonmonochromatic nature of the X-ray source. X-ray "ghost peaks" occur due to other elements present in the

SURFACE ANALYSIS

Line positions (eV)					
Photoelectron lines					
2s	$2p_{1/2}$	$2p_{3/2}$	3s		
228	165	164	18		
Auger lines					
$L_{23}M_{23}M_{23}$					
1336	(Al)				
1103	(Mg)				
Line positions (eV)					
Photoelectron lines					
2s	$2p_{1/2}$	$2p_{3/2}$	3s	$3p_{1/2}$	$3p_{3/2}$
1097	953	933	123	77	75
Auger lines					
$L_3M_{23}M_{23}$	$L_2M_{23}M_{23}$	$L_3M_{23}M_{45}$ (^1P)			
719	712	648	(Al)		
486	479	415	(Mg)		
$L_3M_{23}M_{45}$ (^3P)	$L_2M_{23}M_{45}$ (^1P)	$L_3M_{45}M_{45}$	$L_2M_{45}M_{45}$		
640	628	568	548	(Al)	
407	395	335	315	(Mg)	

Figure 16.11 Typical XPS data tables from literature references. (Top) Data for sulfur; (bottom) data for copper. (From Moulder, J.F. et al., In *Handbook of X-Ray Photoelectron Spectroscopy*, Chastian, J. and King, R.C., eds., Physical Electronics, Inc., Eden Prairie, MN, 1995; Courtesy of Physical Electronics, Inc., Eden Prairie, MN, www.phi.com.)

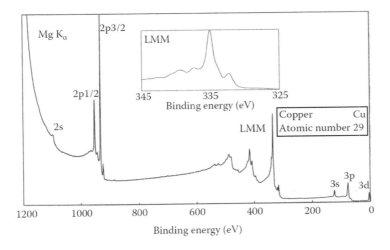

Figure 16.12 The XPS spectrum of copper collected with Mg K_α radiation. (From Moulder, J.F. et al., In *Handbook of X-Ray Photoelectron Spectroscopy*, Chastian, J. and King, R.C., eds., Physical Electronics, Inc., Eden Prairie, MN, 1995; Courtesy of Physical Electronics, Inc., Eden Prairie, MN, www.phi.com.)

X-ray source; Cu lines in spectra are an example of these ghost peaks. Copper is the base material used in the X-ray anode, so Cu lines may appear in the spectra of samples that contain no Cu. A complex photoelectric process results in "shake-up" peaks, which are peaks that occur slightly higher in binding energy than the main XPS photoelectron peak. Figure 16.13 shows the shake-up peaks in a copper oxide sample; they are the peaks (a singlet and an irregular doublet) at higher binding energy than the marked photoelectron peaks.

16.2.1.3.2 XPS Chemical Shift

There is a shift in the binding energy of the core levels of atoms brought about by the chemical environment of the atom. The shift is called a chemical shift and is on the order of 0.1–10 eV. The shift occurs as a result of changes in the electrostatic screening felt by the core electrons as a result of valence electron position. As an example, electronegative elements such as fluorine attached to the atom of interest (e.g., sulfur in SF_6) will pull valence electrons away

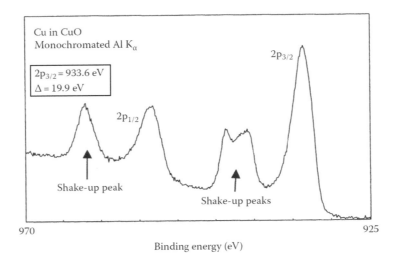

Figure 16.13 "Shake-up" peaks in the copper XPS spectrum of a copper oxide sample. These peaks are one example of several processes that result in lines other than XPS and Auger lines in an XPS spectrum. (Modified from Moulder, J.F. et al., In *Handbook of X-Ray Photoelectron Spectroscopy*, Chastian, J. and King, R.C., eds., Physical Electronics, Inc., Eden Prairie, MN, 1995; Courtesy of Physical Electronics, Inc., Eden Prairie, MN, www.phi.com.)

from the sulfur atom, resulting in a higher binding energy for the core sulfur electrons as a result of the decreased negative charge on the sulfur atom. That is, the sulfur core electrons in SF_6 are held more tightly by the positively charged sulfur nucleus due to less screening by the valence electrons than are the core sulfur electrons in elemental sulfur. Figure 16.14a shows how changes in the chemical environment cause the binding energy of the sulfur $2p_{3/2}$ electrons to shift. Elemental sulfur has a binding energy of 164.0 eV for the $2p_{3/2}$ line. Sulfur in compounds varies from this value by a few eV; some compounds are shifted to lower binding energies and some to higher binding energies. The attachment of highly electronegative fluorine atoms to the sulfur atom results in a shift in the binding energy to a higher value as just described. Figure 16.14b presents similar data for copper in a variety of compounds. The three chemically different nitrogen atoms in a cobalt complex can be identified by the nitrogen 1s XPS spectrum (Figure 16.15). In a similar fashion, the chemical environments of carbon and oxygen in a compound influence the carbon and oxygen XPS spectra as shown in Figure 16.16. This example is from Siegbahn, who with his coworkers, developed the ESCA (XPS) technique and first reported the chemical shift in copper 1s binding energies in 1957. Siegbahn was awarded the Nobel Prize in physics in 1981 for his pioneering work. Tables of chemical shifts for the oxidation states of elements are also tabulated in the literature and at the NIST XPS database site.

An interesting example of the use of XPS is the nondestructive identification of elements on the lunar surface. The Surveyor was an unmanned probe that landed on the surface of the moon. The lunar soil was exposed to magnesium K_α radiation and the XPS spectrum collected. The elemental components of the soil can be clearly identified (Figure 16.17). The advantage of a surface analysis technique for this application is obvious. It is easier to examine the surface with a remote probe (X-rays) than to devise a way of having an unmanned spacecraft take a sample for an analytical method that requires elaborate sample preparation.

A high-resolution (detail) scan of the Pb 4f electron region (Figure 16.18) for lead bromide and lead chloride using magnesium K_α radiation was obtained. In this case, 0.07 eV was the difference between the maxima for these two compounds. This resolution approaches the practical limitations of the technique.

A sample of pure lead was evaporated onto a surface, which was then exposed to air. The XPS spectra were obtained at various time intervals as indicated in Figure 16.19. The spectra show the formation of lead oxide on the surface, going through the intermediate of PbO and finally, after a number of hours, to PbO_2. This demonstrates the capability of XPS to study surface chemistry.

16.2.1.3.3 Quantitative Analysis

Quantitative analysis by XPS can be used to determine the relative concentrations of components of a sample. Either peak area or peak height sensitivity factors can be used, but the peak area method is preferred.

For a homogeneous sample (one that is homogeneous in the analysis volume), the following equation can be written for the intensity of a given spectral peak:

$$I = nf\sigma\theta\lambda yAT \qquad (16.4)$$

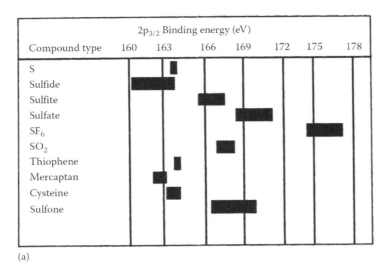

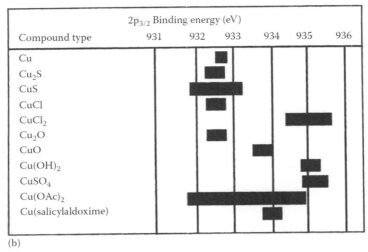

Figure 16.14 (a) Chemical shift in the sulfur $2p_{3/2}$ binding energy in various sulfur compounds. (b) Chemical shift in the copper $2p_{3/2}$ binding energy in various copper compounds. (From Moulder, J.F. et al., In *Handbook of X-Ray Photoelectron Spectroscopy*, Chastian, J. and King, R.C., eds., Physical Electronics, Inc., Eden Prairie, MN, 1995; Courtesy of Physical Electronics, Inc., Eden Prairie, MN, www.phi.com.)

where
I is the intensity (photoelectrons/s)
n is the number of atoms of the element/cm³ sample
f is the X-ray flux (photons/s cm²)
σ is the photoelectric cross section for the transition (cm²/atom)
θ is the angular efficiency factor based on the angle between the photon path and detected electron
y is the efficiency of production of primary photoelectrons
λ is the mean free path of the photoelectrons in the sample (cm)
A is the area of the sample from which photoelectrons are detected (cm²)
T is the detection efficiency for emitted photoelectrons

An atomic sensitivity factor, S, can be defined as

$$S = f\sigma\theta y AT \quad (16.5)$$

To determine two elements in a sample, an intense line from each element is measured. The relationship between the amount of each element, the intensity of the line, and the factor S for each element is

$$\frac{n_1}{n_2} = \frac{\frac{I_1}{S_1}}{\frac{I_2}{S_2}} \quad (16.6)$$

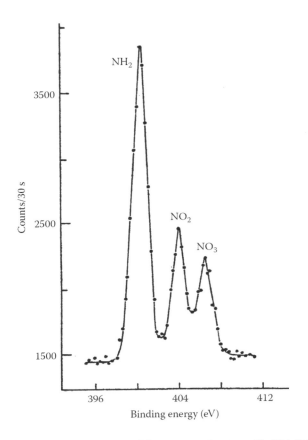

Figure 16.15 Nitrogen 1s XPS spectrum for *trans*-[Co(NH$_2$CH$_2$CH$_2$NH$_2$)$_2$(NO$_2$)$_2$]NO$_3$. The three chemically different nitrogen atoms exhibit slightly different chemical shifts. The peak intensities are an indication of the number of each type of nitrogen atom. (Reprinted with permission from Hendrickson, D.N., Hollander, J.M., and Jolly, W.L., *Inorg. Chem.*, 8, 2642. Copyright 1969 American Chemical Society.)

This equation can be used for all homogeneous samples because the ratio S_1/S_2 is for all purposes matrix independent. For any spectrometer, it is possible to develop a set of relative values for S for every element. (The instrument manufacturer usually provides such values.) To determine the atomic concentration of elements on the surface, the net peak intensity of the most sensitive line for each element detected must be measured. The fractional atom concentration, C_A, of element A is given by

$$C_A = \frac{n_A}{\Sigma n_i} = \frac{\dfrac{I_A}{S_A}}{\dfrac{\Sigma I_i}{S_i}} \qquad (16.7)$$

This use of atomic sensitivity factors will give semiquantitative results, within 10 %–20 % of the value, for most homogeneous materials. While the absolute sensitivity of XPS is high, the actual sample volume analyzed is small, so the amount of an element that can be detected is in the % range with sensitivity of about 0.5 atom %.

16.2.1.3.4 Depth Profiling and Element Location

By using an ion source to sputter or remove the outermost surface atoms, layers below the surface can be studied. Examining sample composition as a function of depth below the surface is called **depth profiling**. The sputtering of layers must be done in a controlled manner so that the depth of the layer being examined is known. This approach is very useful in determining the composition of thin multilayered

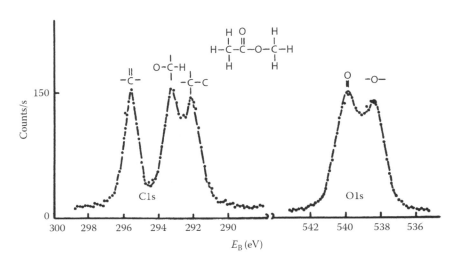

Figure 16.16 ESCA spectrum showing the chemical shifts of carbon and oxygen atoms in methyl acetate. (Reprinted from *Endeavor*, 32, Siegbahn, K., 51, Copyright 1973. With permission from Elsevier.)

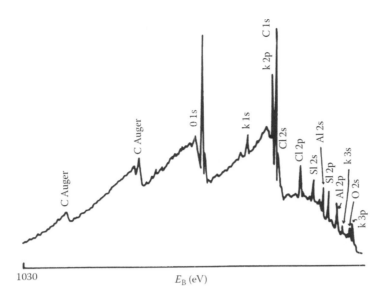

Figure 16.17 The XPS spectrum of lunar soil, taken by the Surveyor. (From Redina, J.F., *Am. Lab.*, 4(2), 1972. With permission.)

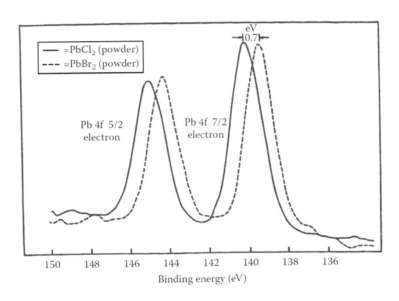

Figure 16.18 High-resolution XPS spectrum of pure PbBr$_2$ overlaid on a similarly acquired spectrum of pure PbCl$_2$. A chemical shift of 0.07 eV between the two Pb lines would enable these two compounds to be distinguished by XPS. Mg K$_\alpha$ radiation was the excitation source.

materials. One limitation of ion sputtering for depth profiling is that changes in chemical state may occur as a result of the ion bombardment. The rate of removal of atoms is material dependent; gold is removed from a surface much faster than an inorganic oxide like SiO$_2$, for example. This can pose a problem for inhomogeneous materials.

One simple way of distinguishing an element on the surface from an element in the bulk of the material using XPS is to change the angle at which the sample is mounted. A shallow angle between the plane of the sample surface and the analyzer axis results in a shallow electron takeoff angle; this will accentuate signals from the surface of the material. This shallow angle is also called a grazing angle. Samples mounted at 90° (normal) to the analyzer axis with respect to the surface plane will have the surface signal minimized with respect to the bulk signal. An example of the effect of changing the sample angle is presented in Figure 16.20.

Modern XPS systems have the ability to analyze areas with a diameter of <10 μm. This enables a surface to be mapped at high resolution and a plot of element distribution and chemical state information to be made.

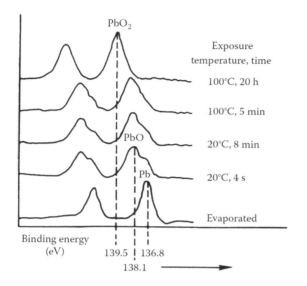

Figure 16.19 The use of ESCA (XPS) to study changes in the surface composition of Pb on exposure to air and heat. (Reprinted from *Endeavor*, 32, Siegbahn, K., 51, Copyright 1973. With permission from Elsevier.)

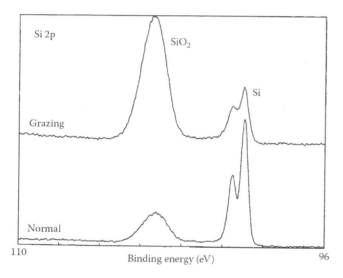

Figure 16.20 An example of the enhanced surface sensitivity achieved by varying the electron take-off angle. A thin SiO_2 oxide layer on silicon (Si) is enhanced at the shallow grazing angle. (From Moulder, J.F. et al., In *Handbook of X-Ray Photoelectron Spectroscopy*, Chastian, J. and King, R.C., eds., Physical Electronics, Inc., Eden Prairie, MN, 1995; Courtesy of Physical Electronics, Inc., Eden Prairie, MN, www.phi.com.)

16.2.2 Auger Electron Spectroscopy

The process on which AES is based is similar to that of XPS, but it is a two-step process instead of one step. As with XPS, the sample is irradiated with either accelerated electrons or X-ray photons. An inner shell electron is ejected, leaving a vacancy in the inner shell. An electron from an outer shell falls into the inner shell as in the XPS process. The ion then either emits an X-ray photon or an Auger electron. In X-ray emission, the energy balance is maintained by the emission of an X-ray photon with energy equal to the difference between the two energy levels involved (Figure 16.1). This is XRF, discussed in Chapter 9. In the Auger process, the released energy is transferred to a second electron in an outer shell, which is then emitted. This is the Auger electron. The actual process by which the energy is transferred is not clearly understood but can be represented schematically as in Figures 16.2 and 16.21.

In the examples in both figures, an electron from the K shell was ejected by bombardment. An electron from the L shell descended to the K shell, simultaneously transferring energy to a second electron in the L shell. This second electron was ejected as an Auger electron. This Auger electron is termed a KLL Auger electron. This nomenclature arises from the fact that an electron was ejected from a K shell followed by a transition from an L to a K shell, simultaneously liberating an L Auger electron. It can be deduced that other Auger transitions, such as LMM and MNN, are also possible.

The Auger electron is ejected by the energy released by relaxation, that is, by the neighboring electron that drops into the inner orbital. Note that at least three electrons are needed for the Auger process to occur, thus hydrogen and helium cannot be detected by this technique. The energy released on relaxation is a function of the particular atom and not of the energy of the source used to eject the initial electron from the inner orbital. Therefore, the kinetic energy of the Auger electron is independent of the energy range of the radiation source falling on the sample. This means that the source does not have to be monoenergetic (or monochromatic). Because the Auger peaks are independent of the source energy, unlike the XPS peaks, they can be distinguished readily from the XPS peaks. When spectra are collected with two different source energies, as shown in Figures 16.9 and 16.10, the binding energy of the Auger electron apparently changes with source energy, while the binding energies of the XPS photoelectrons do not.

The Auger process and XRF are competitive processes for the liberation of energy from bombarded atoms. In practice, it is found that the Auger process is more likely to occur with low atomic number elements; this probability decreases with increasing atomic number. In contrast, XRF is unlikely with elements of low atomic number but increases in probability with increasing atomic number. This is illustrated in Figure 16.22. The energies involved in Auger spectroscopy are similar in all respects to those of XPS, since the same atomic shells are involved. A graphical plot of Auger electron energies is shown in Figure 16.23, and tabulated values of the prominent lines are available in the literature.

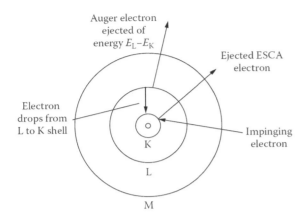

Figure 16.21 A simplified view of the process of Auger electron emission. An impinging electron ejects an inner shell electron, leaving an incomplete K shell. An electron from the L shell drops into the K shell, and simultaneously, an Auger electron is emitted from the L shell with energy $E_L - E_K$. This process competes with the process that produces X-ray photons (XRF) shown in Figure 16.1.

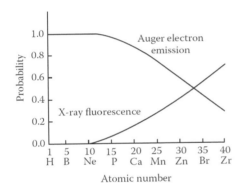

Figure 16.22 Relationship between atomic number and the probabilities of XRF and Auger electron emission. (Reprinted with permission from Hercules, D.M., *Anal. Chem.*, 42(1), 20A, 1970. Copyright 1970 American Chemical Society.)

AES is an elemental surface analysis technique that can detect elements from lithium to uranium with a sensitivity of about 0.5 atom %, in the region of 0.5 to 5 nm from the surface. Auger spectra consist of a few peaks for each element in the same energy region, 0–1000 eV, as XPS peaks. Auger peaks are less intense than XPS peaks, as seen in Figures 16.9 and 16.12. Therefore, Auger spectra are generally plotted as the first derivative of the signal, $dN(E)/dE$, vs. the Auger electron energy (Figure 16.24). The use of the first derivative of the signal is a common practice to enhance small signals and to minimize the high and sloping background from scattered electrons.

Auger chemical shift: The chemical environment of the atom also shifts the Auger lines in the same way the XPS lines are shifted. The chemical shift magnitude may be greater for the Auger lines than for the XPS chemical shift. The Auger chemical shift is very useful for the identification of chemical states. A plot of Auger electron kinetic energy vs. the XPS photoelectron binding energy is a useful tool for identifying the chemical states of an element. We can define a quantity called the Auger parameter as the difference between the kinetic energies of the Auger electron and the photoelectron:

$$AP = KE_{(A)} - KE_{(P)} = BE_{(P)} - BE_{(A)} \qquad (16.8)$$

where
AP is the Auger parameter
$KE_{(A)}$ and $KE_{(P)}$ are the kinetic energy of the Auger electron (A) and photoelectron (P), respectively
BE is the binding energy of the respective electrons

It can be shown that

$$KE_{(P)} = h\nu - BE_{(A)}$$

that leads to

$$KE_{(A)} + BE_{(P)} = AP + h\nu \qquad (16.9)$$

Plots of $KE_{(A)}$ vs. $BE_{(P)}$, with $AP + h\nu$ shown on the other ordinate (called Wagner plots) are found in the literature and in the NIST XPS database. An example for tin is presented in Figure 16.25. The Sn $3d_{5/2}$ photoelectron and the Sn MNN Auger electron binding energies are tabulated and plotted for the element Sn and a variety of tin compounds, showing how the chemical environment of the Sn affects the electron energies.

16.2.2.1 Instrumentation for AES

The instrumentation used in AES is very similar to that used in XPS including the need for ultra-high vacuum systems. The major difference is that the source used is a focused beam of electrons from an electron gun or a field emission source, not X-ray photons. A schematic diagram of an AES instrument is shown in Figure 16.26. Many instrument manufacturers provide instruments that permit both XPS and Auger spectra to be collected on one instrument.

The advantage of an electron beam is that it can be focused and deflected. The electron beam can be focused, depending on the source, from a spot size of 10 nm to a spot size of several hundred micrometers. The focused beam can be scanned over the surface. This permits compositional mapping of a surface with very high resolution and the

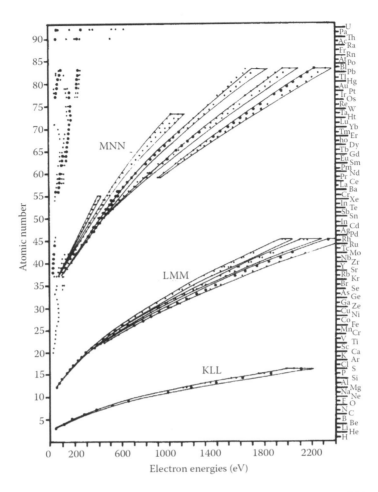

Figure 16.23 Principal Auger electron energies. (Courtesy of Physical Electronics, Inc., Eden Prairie, MN, www.phi.com.)

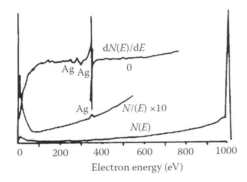

Figure 16.24 The direct Auger spectrum of silver (bottom), the direct spectrum with a 10-fold enhancement of the signal (middle), and the derivative or differential spectrum (top). The derivative spectrum significantly improves the ability to measure a small signal against a high and sloping background signal. (From Weber, R.E., *Res. Dev.*, 10, 22, 1972.)

ability to study very small features. Electron guns may use a heated tungsten filament or lanthanum hexaboride rods as the cathode. Figure 16.27 shows a schematic cross section of an electron gun with a tungsten filament.

A field emission source uses a needle-like tungsten or carbon tip as the cathode, shown in Figure 16.28. The tip is only nanometers wide, resulting in a very high electric field at the tip. Electrons can tunnel out of the tip with no input of thermal energy, resulting in an extremely narrow beam of electrons. Electron beams from heated filaments have a focal (crossover) diameter of about 50 μm, while a field emission source has a crossover diameter of only about 10 nm. Field emission sources can serve as probes of surfaces at the nanometer scale (an Auger nanoprobe).

16.2.2.2 Applications of AES

AES is primarily a surface elemental analysis technique. It is used to identify the elemental composition of solid surfaces and can be used to quantify surface components, although quantitative analysis is not straightforward. Quantitative accuracy is limited to approximately 30 % unless standards that closely resemble the sample are available. Quantitative detection sensitivity, however, can be 0.01 to 1.0 % in some cases. A complete survey spectrum can be completed in between 5 and 30 min, with longer times required when using smaller beam sizes and lower

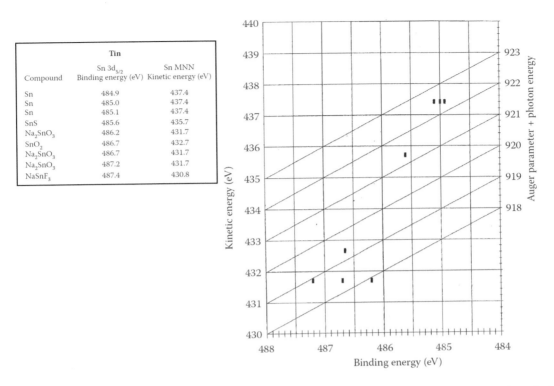

Figure 16.25 Table of values and plot of the Auger MNN electron kinetic energy for tin and Auger parameter vs. the photoelectron binding energy for the tin $3d_{5/2}$ electron in the element and a variety of compounds. The metal and different chemical states of tin in compounds can be readily distinguished. (From Moulder, J.F. et al., In *Handbook of X-Ray Photoelectron Spectroscopy*, Chastian, J. and King, R.C., eds., Physical Electronics, Inc., Eden Prairie, MN, 1995; Courtesy of Physical Electronics, Inc., Eden Prairie, MN, www.phi.com.)

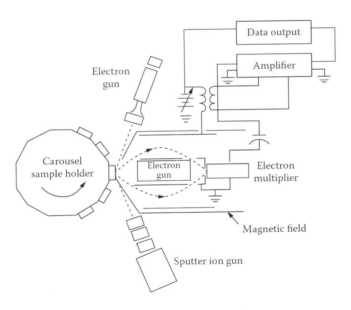

Figure 16.26 Schematic diagram of an Auger spectrometer, showing the electron gun source, an electron flood gun, and an ion gun, along with a carousel for multiple samples. The vacuum system is not shown. (Courtesy of Physical Electronics, Inc., Eden Prairie, MN, www.phi.com.)

voltages. Note that pump down time before the analysis may be on the order of an hour. The requirement of high vacuum mentioned above makes this method unsuitable for materials with significant vapor pressure or materials that may outgas. AES is a true surface analysis technique, because the low-energy Auger electrons can only escape from the first few (three to five) atomic layers or from depths of 0.5 to 5.0 nm.

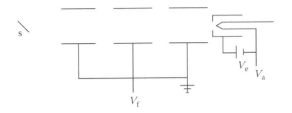

Figure 16.27 A schematic cross section of an electron gun with a tungsten filament and an Einzel lens focusing arrangement. S is the sample, V_f is the focus voltage, V_e is the emission voltage, and V_a is the acceleration voltage. (From Turner, N.H., Electron spectroscopy, in Ewing, G.W., ed., *Analytical Instrumentation Handbook*, 2nd edn., Marcel Dekker, Inc., New York, 1997. With permission.)

The electron beam can be deflected in a line across the surface of a sample to analyze multiple points or it can be rastered (moved in two dimensions) to produce a compositional map of the surface. Very small spot size and large magnifications (up to 20,000×) enable AES to study extremely small particles and intricate structures such as those used in microelectronics. Depth profiling with ion beam sputtering of the surface permits the analysis of layers and the identification of components in the grain boundaries of crystalline materials. The use of AES to study segregation of impurities in the grain boundaries of metals and alloys has played a pivotal role in understanding the embrittlement of steels, corrosion and stress corrosion cracking of alloys, and the behavior of ductile tungsten used in light bulb filaments. Very often, sulfur, phosphorus, or carbon impurities in the grain boundaries of metals and alloys are the sites at which corrosion, cracking, and failure of the alloy are initiated. Alternatively, impurities in the grain boundaries may have a beneficial effect. Potassium in the grain boundaries of ductile tungsten acts as a lubricant, allowing the tungsten wire filament to expand and contract when a light bulb is switched on and off without breaking along the grain boundaries.

Some application examples are presented. Figure 16.29 shows the Auger spectrum of stainless steel before and after heating at 750 °C for 10 min. It can be seen that the major components, iron, nickel, and chromium, are unchanged by the heating. However, the carbon that was on the surface of the stainless steel before heating disappeared after heating. This surface carbon could be from traces of lubricating oil or cutting fluid, for example. Sulfur is present on the surface after heating and the phosphorus signal has increased. This may be due to diffusion from the bulk stainless steel to the surface during heating.

The electron beam can scan a surface area systematically in what is called a raster scan. By monitoring the intensity of the Auger spectrum of a particular element during scanning, it is possible to map its distribution on the surface examined. An example of this technique is shown in Figure 16.30, which reveals the concentrations and location of sulfur and nickel on a spent catalyst surface. In another example, Figure 16.31 shows the distribution of molybdenum, gold, and oxygen on an integrated circuit. Using this technique, it is possible to map the distributions of these three elements and their positions relative to the circuit. The use of a scanning mode and small spot size is called scanning Auger microscopy (or scanning Auger microprobe), known by the initials SAM.

16.2.2.2.1 Depth Profiling

As can be done with XPS, another valuable application of AES has been developed by using an ion beam to progressively strip off the surface of a sample under controlled conditions from a sample. Spectra can be collected and the distribution of elements recorded as surface material is removed. The results show changes in distribution of different elements with depth, called a depth profile.

A schematic diagram of a sample undergoing depth profiling is shown in Figure 16.32. The ion beam sputters the surface and etches a crater into the material, while the electron beam is used as the source for AES. Figure 16.33 shows the results of such a depth profile investigation of Nichrome film on a silicon substrate. It can be seen from the plot that the Nichrome film and its outer oxide layer are about 150–175 Å thick. That can be deduced from the appearance of the Si substrate after about 175 Å of material has been sputtered off. It is also evident that the chromium at the outer surface has formed an oxide layer about 75 Å thick. Chromium readily forms a protective oxide layer when exposed to air. The oxygen signal decreases after the oxide layer has been sputtered off, revealing the Ni and Cr film itself. Information about surface oxide layers, layer thickness, and composition is critical to understanding corrosion chemistry, surface reactions, material behavior, and device fabrication.

Another example of depth profiling is shown in Figure 16.34; AES clearly shows the very localized region containing Cl at a depth of about 1100 Å from the surface of the material. The Cl-containing region is only about 100 Å wide. This extremely fine lateral resolution is one of the strengths of AES over other analytical techniques.

16.3 ION SCATTERING SPECTROSCOPY

In **ion scattering spectroscopy** (ISS), a beam of ions is directed at the sample. On collision with the surface, the ions are scattered by the sample atoms at the surface. Some of the bombarding ions are scattered after a single binary elastic collision with the surface atoms. For such

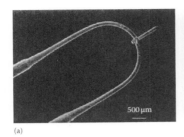

Figure 16.28 Field emitter of single-crystal tungsten is shown (a) on its supporting hairpin and (b) in a close-up view of the tip. (From Stinger, K., *Res. Dev.*, 28(9), 40, 1977.)

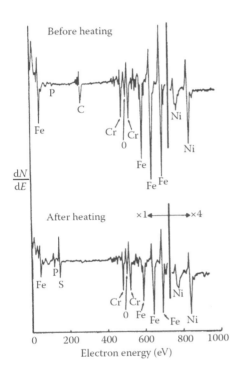

Figure 16.29 Auger spectra of stainless steel before and after heating in vacuum at 750 °C for 10 min. As a result of heating, carbon has been lost from the surface while sulfur and phosphorus have diffused from the bulk steel to the surface.

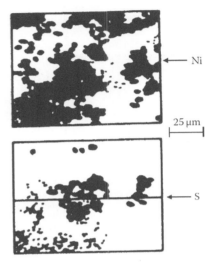

Figure 16.30 Auger element distribution maps of Ni and S on the surface of a spent catalyst, obtained by rastering the electron beam across the surface of the catalyst.

ions scattered at a given angle, conservation of momentum results in the energy of these ions being dependent only on the mass of the surface atom and the energy of the bombarding (primary) ion. By measuring the energy of the *scattered ions*, one can determine the mass of the *scattering atoms* on the surface.

For a scattering angle of 90°, the relationship between the energy of the scattered ions and the masses of the bombarding ion and scattering atom are given by the equation

$$E_S = E_0 \frac{M_S - M_{ion}}{M_S + M_{ion}} \qquad (16.10)$$

or

$$\frac{E_S}{E_0} = \frac{M_S - M_{ion}}{M_S + M_{ion}}$$

where
E_S is the energy of the scattered ion (after collision)
E_0 is the energy of the bombarding ion (before collision)
M_S is the mass of the scattering (surface) atom
M_{ion} is the mass of the bombarding ion

The process is illustrated in Figure 16.35. In this system, the bombarding ion is $^3He^+$; the scattering atoms were ^{16}O and ^{18}O. Based on Equation 16.10, the energy of the scattered helium ions can be calculated if the primary beam energy is known. The energy of the scattered helium ion will differ depending on whether it was scattered by an atom of ^{16}O or ^{18}O. By measuring the relative abundance (signal intensity) of the energies of scattered helium ions,

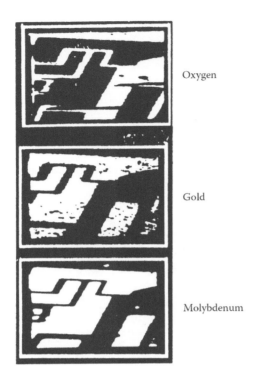

Figure 16.31 Auger element maps of an integrated circuit, showing the distribution of oxygen, gold, and molybdenum on the surface. (Courtesy of Physical Electronics, Inc., Eden Prairie, MN, www.phi.com.)

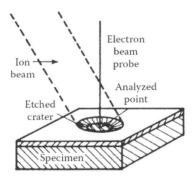

Figure 16.32 Use of ion sputtering for depth profile analysis. An ion gun is used to remove surface layers during collection of Auger spectra. (Courtesy of Physical Electronics, Inc., Eden Prairie, MN, www.phi.com.)

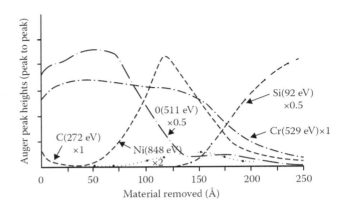

Figure 16.33 Depth profile analysis of a Nichrome film on a silicon substrate. The outermost layer shows both Cr and oxygen, probably as a stable chromium oxide layer. The Ni and Cr (the Nichrome film) signals drop off at about 175 Å while the Si substrate signal starts to appear, indicating the approximate film thickness. (From Weber, R.E., *Res. Dev.*, 10, 22, 1972.)

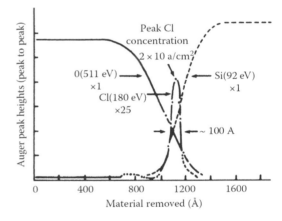

Figure 16.34 Depth profile of material with a highly localized narrow region containing Cl. The high lateral resolution of AES allows the characterization of thin layers and small features, such as this 100 Å thick Cl-containing region. (From Weber, R.E., *Res. Dev.*, 10, 22, 1972.)

the relative amounts of ^{16}O and ^{18}O on the surface of the sample can be determined.

Only exposed atoms on the surface, that is, the top monolayer of atoms, contribute to the signal in ISS. It is therefore the most surface sensitive of the surface analysis techniques. ISS is one technique that provides isotopic information on surface atoms. ISS can be used to determine all elements with an atomic number greater than that of the bombarding ion. The elemental and isotopic compositions of the surface can be determined both qualitatively and quantitatively. Sensitivity is about 1% of a monolayer for most elements.

A schematic diagram of instrumentation used for ISS is shown in Figure 16.36. This instrument is based on a CMA, discussed in Section 16.2.1.1. The instrument shown consists of an ion source, a vacuum system (not shown), an energy analyzer, and a detector. The major instrument components including the sample are under vacuum.

The most common ion guns use inert (noble) gases, particularly helium, argon, and neon, to produce monoenergetic ions. Ion guns are capable of focusing the ion beam from 100 μm to 1 mm diameter spot size and are able to control the current density of the ion beam. Ions are produced in the ion gun by electron bombardment of the gas atoms.

SURFACE ANALYSIS

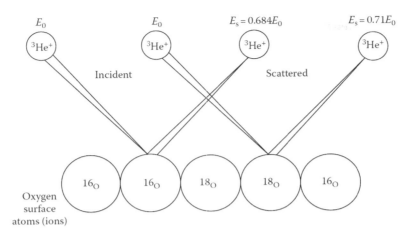

Figure 16.35 The process of ISS. The incoming helium ions are scattered with different energies from ^{16}O and ^{18}O. (From Czarderna, A.W. et al., *Ind. Res.*, January 1978.)

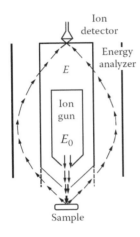

Figure 16.36 Ion scattering spectrometer based on a CMA.

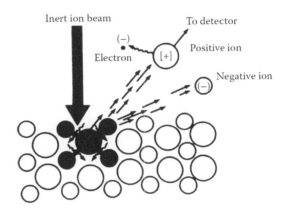

Figure 16.37 The SIMS process. Positive and negative ions are ejected from the surface by bombardment with inert ions (He$^+$).

A stream of the inert gas passes through an electron grid, which ionizes the gas by electron impact. These ions can then be accelerated to a known energy using an accelerating electrode. The accelerated ion beam is then focused onto the surface of the sample. Here, the ions are scattered, and those scattered at some predetermined angle enter an energy-analyzing system, where they are separated before reaching an ion detector.

The scattering efficiencies of different rare gas ions are different for different surfaces. Heavy surface atoms respond better to bombardment with a heavy ion such as argon rather than He. The ability of ISS to resolve atoms close in mass decreases as the difference between the bombarding ion and the target atom increases. The spatial resolution of ISS is limited to about 100 μm, so it is not as good as AES in this respect.

Applications of ISS are similar to the applications of static SIMS, described subsequently. ISS is used to study surface reaction mechanisms, catalyst behavior, and adsorption–desorption processes at surfaces. Because of its ability to discriminate among isotopes of an element, ISS can be used to study diffusion or any other reactions involving the replacement of one isotope for another.

16.4 SECONDARY ION MASS SPECTROMETRY

A more common technique than ISS is SIMS. This process is slightly different from ISS, as shown in Figure 16.37. In this instance, the surface is again bombarded with a beam of an ionized inert gas (the **primary ion beam**). Most SIMS applications to date have made use of Cs$^+$, O$_2^+$, O$^-$, Ga$^+$, and Ar$^+$ ion beams. Energies range from 1 to 20 keV. In addition to scattering, the primary ions displace (sputter) ions from the sample surface by setting up a "collision cascade." In this collision cascade, chemical bonds are ruptured in the solid by direct impact or by indirect energy transfer. Ions as well as neutral molecules are sputtered from the surface as a result of this process. The sputtered ions, called *secondary* ions, may be a complete molecular ion or an ionized fragment of a molecule. In addition, the sputtered ions may be positively or negatively charged; in practice, both types of ions are utilized in the analytical process. The mass-to-charge ratios of the sputtered ions are measured, as in other forms of MS (Chapters 10 and 11), and a spectrum of the intensity of the secondary ions vs. mass-to-charge ratio is generated. The detected secondary ions are sputtered from the top two or three monolayers of the surface, making SIMS a very surface-sensitive technique.

SIMS can be performed in two modes: **dynamic SIMS**, which uses a high-intensity primary ion beam, and **static SIMS**, which uses a very low primary ion beam intensity. Dynamic SIMS results in high sputter rates, high secondary ion production, and therefore low detection limits. The sensitivity for surface analysis by dynamic SIMS is about 1 ppm. Dynamic SIMS is a destructive technique because of the high rate of removal of surface layers. The sputtering rate depends on the ion beam intensity, the energy of the ions, and the angle of incidence, as well as the mass of the bombarding ion and the nature of the sample itself. The sample mass, crystal orientation, specific chemical composition, and other sample properties also affect the rate at which surface atoms or molecules are removed. Static SIMS can be used to study samples that are extremely thin (no deeper than 5 nm) or need to be analyzed nondestructively. The low primary beam intensity removes only a few monolayers of the surface. The sensitivity is not as good as that of dynamic SIMS because the secondary ion production is low. Detection limits are generally in the ppm–0.1 % range for static SIMS. All elements, including hydrogen, can be determined, and molecular species can also be measured. Although SIMS is considered a destructive technique, static SIMS results in very low surface damage and is often considered nondestructive.

16.4.1 Instrumentation for SIMS

In its most elementary form, a SIMS system consists of a source of primary ions, a sample holder, secondary ion extraction optics, a mass spectrometer, and an ion detector, all housed in a UHV compartment. Systems are also equipped with data processing and output systems. A schematic diagram of part of a SIMS system with a quadrupole mass analyzer is shown in Figure 16.38. The design and operation of mass spectrometers is covered in Chapter 10 and will be only briefly reviewed here.

16.4.1.1 Primary Ion Sources

SIMS instruments frequently are equipped with two ion sources: one source designed to generate positive secondary ions and the other to generate negative secondary ions. The use of an electronegative primary ion such as O_2^+ results in enhanced generation of positive secondary ions, while the use of an electropositive primary ion such as Cs^+ gives enhanced generation of negative secondary ions. The oxygen-based ion beam is generated by passing the gas through a low-voltage, low-pressure hot cathode arc, which produces a plasma of negatively and positively charged gas ions. The desired ions are extracted from the plasma, accelerated, and focused by electrostatic and magnetic fields onto the sample surface. Such ion beams can be focused to spot sizes from about 50 μm to as little as 0.5 μm. The second source is normally a cesium ion gun for

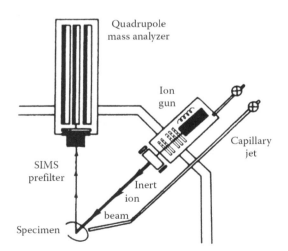

Figure 16.38 Partial schematic diagram of an SIMS instrument equipped with an ion gun source and a quadrupole mass analyzer.

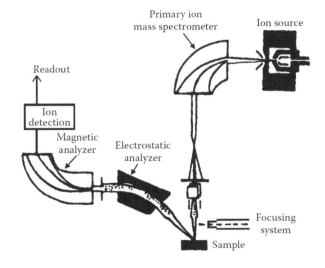

Figure 16.39 A double-focusing ion microprobe mass spectrometer with a magnetic sector to filter the primary ion beam.

the generation of negative secondary ions. A primary ion mass filter can be used to remove doubly charged ions and neutrals (Figure 16.39).

Time of flight (TOF)-SIMS instruments, described subsequently, may use two ion guns, a nanosecond pulsed Ga^+ ion gun to acquire the spectrum, and a second Cs^+ ion gun for sputtering the surface and depth profiling. The gallium and cesium ion guns have a reservoir of the metal, which is heated to form gas-phase atoms. The gas-phase atoms are ionized at a charged tungsten grid, extracted, and focused to form the ion beam.

16.4.1.2 Mass Spectrometers

The three most common types of mass analyzers in SIMS systems are (1) double-focusing magnetic sector instruments, (2) TOF mass spectrometers, and (3) quadrupole mass spectrometers. The choice of mass analyzer depends on whether dynamic or static SIMS is needed, on the requirements of mass range and resolution, and on transmission efficiency, among other factors. The mass analyzers have been discussed in Chapter 10 in detail and this chapter should be reviewed as necessary.

Double-focusing magnetic sector instruments use a magnetic field to separate the ions by mass-to-charge ratio. The mass range covered by this type of analyzer is up to 500 Da with a mass resolution, $M/\delta M$, of 10,000 and high transmission efficiency at low resolution. The advantages of this type of mass analyzer are high mass resolution, direct ion imaging capability with an imaging detector in place, and high sensitivity over the mass range. Sensitivity is lower when high mass resolution is required. When a magnetic sector instrument is used, only a single mass is measured at a time. Secondary ions are generated constantly by the continuous primary ion beam bombardment, but most of these ions are not detected. Figure 16.39 shows a schematic double-focusing ion microprobe mass spectrometer with a magnetic sector used as a primary ion mass filter.

Quadrupole mass analyzers operate by varying the applied radio-frequency (RF) and direct current (DC) voltages to the four rods that make up the quadrupole. Only ions of a given m/z ratio can pass through the rods at a given set of conditions; all other ions collide with the rods and are not detected. Direct ion imaging cannot be done with a quadrupole, although surface mapping is possible. The mass range that can be covered is up to 1000 Da, but the mass resolution is only about 200 with a transmission efficiency of about 10 %, or 30 % that of a magnetic sector instrument. The advantages of a quadrupole mass analyzer are its low cost, simple operation, its ability to switch rapidly from positive to negative ion mode, and its flexibility in handling nonconducting samples. Like the magnetic sector analyzers, only one mass is measured at a time with a quadrupole.

The TOF mass analyzer has an almost unlimited mass range, with a mass resolution of 10,000 and a transmission efficiency midway between a quadrupole and a magnetic sector. The primary ion beam is pulsed, and the pulse of secondary ions generated is mass separated in the drift tube of the TOF analyzer. Ions of light m/z reach the detector before heavy m/z ions. Unlike the other two mass analyzers, the TOF analyzer collects and analyzes all of the secondary ions. This makes the TOF system very useful for static SIMS. Advantages of the TOF analyzer are the ability to determine very large molecules such as polymers, high mass resolution with no loss of sensitivity (unlike magnetic sector analyzers), as well as high and constant sensitivity for all masses. Imaging of the surface is done by rastering the primary ion beam across the surface and collecting a mass spectrum at each pixel in a 256 × 256 array, for example.

The detectors used for SIMS are electron multipliers (discussed in Section 16.2.1.1 and in Chapter 10); array detectors are used for imaging.

16.4.2 Analytical Applications of SIMS

SIMS is used for the identification of the surface composition of solid samples by the mass spectra obtained from the surface species. Compositional information comes from the top two to three monolayers of the surface. Elements and molecular species can be determined. Surface distribution, isotope ratios, and depth profiles can be measured. SIMS, like ISS, is used to determine residues on surfaces from cleaning processes. These residues may be organic or inorganic and can be potentially damaging by causing adhesion failure of coatings, contamination of high-purity materials for semiconductor manufacture, or undesired chemical reactions. Adsorbed gases on surfaces can be studied. Element distribution in materials can provide information about homogeneity, corrosion, failure mechanisms, and the strength of materials. The distribution of deliberately added trace elements (dopants, implanted ions) in materials can be evaluated.

Both positive and negative secondary ions can be generated and analyzed. In practice, positively charged ions are most valuable for examining the elements on the left side of the periodic table, whereas the negative ions are most useful for elements on the right side of the periodic table (nonmetals).

In both SIMS and ISS, insulating samples can be analyzed if a flood gun or combination of flood gun and ion gun is used to neutralize the surface charge. This can be seen in the SIMS spectrum of a glass surface (Figure 16.40) collected without a flood gun (top spectrum) and with a flood gun (bottom spectrum). It can be seen that the spectrum has a flatter baseline and improved resolution when the flood gun is used.

The principles of ISS and SIMS are similar, in the method of bombarding the sample, the utilization of high

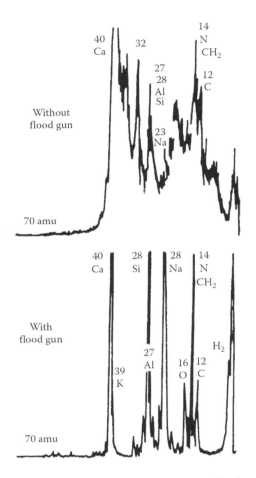

Figure 16.40 Effect of charge neutralization on the SIMS spectrum of soda lime glass. The bottom spectrum, taken with the use of a flood gun, shows improved resolution and a flat baseline compared with the top spectrum.

vacuum, and the measurement of ions. In ISS, the scattered ion beam is examined and the energies of the ions measured; in SIMS, the ions ejected from the surface are examined and their mass-to-charge ratio measured. Both techniques are very surface sensitive, with ISS obtaining information from only the top monolayer of a surface and SIMS obtaining information from two to three monolayers of the surface. Generally, the scattered and secondary ions vary considerably in mass and do not interfere with each other. The two systems can therefore be run simultaneously (ISS-SIMS), and this provides a wealth of information on surface elemental analysis, isotope ratios, and surface molecular analysis, depending on the fragmentation and ionization of the molecules involved.

SIMS can be used to obtain elemental information (so it complements other atomic mass spectrometry techniques). Examples are shown in Figure 16.41. SIMS can also be used to obtain molecular information. As an example of the type of molecular information available from a TOF-SIMS instrument, the positive ion mass spectrum of a pharmaceutical compound is shown in Figure 16.42. The M+1 ion, that is, molecular ion + H, and a variety of fragment ions are seen, permitting the relative molecular mass and structure of the compound to be worked out using the principles presented earlier in the text for interpreting mass spectra (Chapter 11). Using the TOF-SIMS, chemical images can be generated by collecting a mass spectrum at every pixel of a 256 × 256 array as the primary ion beam is rastered across the sample surface. Both molecular and elemental image maps can be produced. For example, a time-release drug tablet or coated tablet can be imaged for both elements present and the active drug compound, to identify coating thickness and evaluate homogeneity.

16.4.2.1 Quantitative Analysis

The analytical signal obtained in SIMS depends on a number of factors, namely, the abundance of the isotope examined on the surface, the properties of the surface, and the bombarding ion. The relationship is complex; calibration curves may or may not be continuous. Certified calibration standards that are matrix matched and cover the analytical range are required. For bulk analysis using dynamic SIMS, certified standards can be obtained from sources such as NIST or commercial standards firms for many metals and alloys and for some ceramics and glasses, but such standards are rarely available for surfaces, thin layers, and multilayered materials. The excellent sensitivity of SIMS allows it to distinguish between ppb and ppm concentrations of analyte in solids. SIMS is not used for quantitative analysis at the % level; XPS or methods such as XRF are better suited to this purpose. Quantitative and qualitative analysis is complicated by the sensitivity variation element to element, and the performance is often highly instrument dependent.

SIMS is used for quantitative depth profile determinations of trace elements in solids. These traces can be impurities or deliberately added elements, such as dopants in semiconductors. Accurate depth profiles require uniform bombardment of the analyzed area and the sputter rate in the material must be determined. The sputter rate is usually determined by physical measurement of the crater depth; for multilayered materials, each layer may have a unique sputter rate that must be determined. Depth profile standards are required; the use of SIMS for a true unknown is not yet possible. Government standards agencies like NIST have such standard reference materials available for a limited number of applications. For example, standard reference material (SRM) depth profile standards of phosphorus in silicon, boron in silicon, and arsenic in silicon are available from NIST for calibration of SIMS instruments. P, As, and B are common dopants in the semiconductor industry, and their accurate determination is critical to semiconductor manufacture and quality control.

SURFACE ANALYSIS

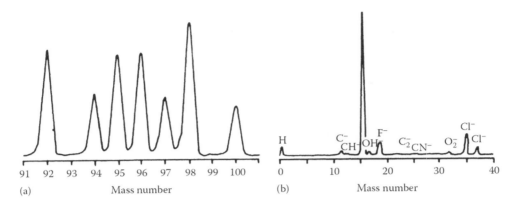

Figure 16.41 (a) Positive SIMS spectrum of molybdenum. All peaks are isotopes of Mo. (b) Negative SIMS spectrum showing both elements and polyatomic fragments of molecules. (Courtesy of Physical Electronics, Inc., Eden Prairie, MN, www.phi.com.)

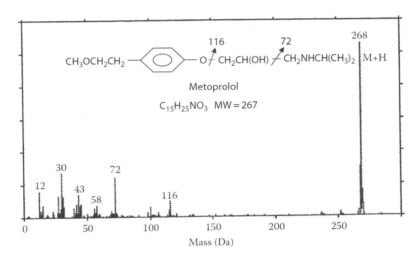

Figure 16.42 Positive SIMS spectrum of metoprolol, showing molecular and fragment peaks. (Courtesy of Physical Electronics, Inc., Eden Prairie, MN, www.phi.com.)

16.4.2.2 Ion Microprobe Mass Spectrometry

Ion microprobe mass spectrometry (IMMS) is an extension of SIMS. In this system, an ion beam is focused to a very small spot size onto the sample so that micron-sized points on the surface can be examined. The ion beam can be constricted by electrostatic lenses to a diameter of 3–10 μm, although spot sizes as small as 0.2–0.5 μm are attainable. A schematic diagram of such a system is shown in Figure 16.39. The secondary ions produced from the sample surface pass through an electrostatic field, which brings most of the ions to the same velocity. The ions then pass through a magnetic sector, which resolves them by mass-to-charge ratio. This is the same type of double-focusing mass spectrometer described in Chapter 10.

This ion beam may be used in any one of three modes: (1) it may be held stationary on the sample and the analysis at that particular spot obtained; (2) it may be rastered across the sample surface to obtain a distribution map of the elements on that surface; or (3) it may be focused on a single spot for an extended period of time, producing a crater that can be used to obtain a depth profile of the elements of interest. With this system, multielement analysis of the surface is possible. Mapping of the surface by element is also possible. In addition, the focusing potential of IMMS permits analysis of very small points on the sample surface. For example, IMMS can be used in the determination of elements on the surface of collected airborne particles and dust. Solid airborne particulates are collected by pulling a known volume of air through a small-diameter filter. The particles, often 0.5 μm or so in diameter, are trapped on the filter for analysis. Elements adsorbed on the surface of the particulates can be examined qualitatively and quantitatively. Lead, thallium, manganese, and chromium on the surface of airborne particles, all metals of environmental and toxicological interest, have been studied using this approach.

16.5 ELECTRON MICROPROBE (ELECTRON PROBE MICROANALYSIS)

The electron probe microanalyzer (EM or EPMA) uses a beam of high-energy electrons to bombard the surface of a solid sample. This results, as we have already seen, in the removal of an inner shell electron. As discussed in Section 16.2, this can result in the ejection of a photoelectron (the basis of XPS) and the emission of an X-ray photon. The X-ray photons emitted have wavelengths characteristic of the elements present. The EPMA uses either a wavelength dispersive (WD) or energy dispersive (ED) X-ray spectrometer to detect and identify the emitted X-rays. This is very much analogous to XRF spectrometry (Chapter 9), where the primary beam was of X-rays, not electrons. This instrument was discussed at the end of Chapter 9, which should be reviewed if necessary.

The instrumentation for EM uses the same type of X-ray spectrometers discussed in detail in Chapter 9, with an electron beam as the source and a UHV system that includes the sample compartment. An ED X-ray spectrometer allows the simultaneous collection and display of the X-ray spectrum of all elements from boron to uranium. The ED spectrometer is used for rapid qualitative survey scans of sample surfaces. The WD spectrometer has much better resolution and is used for quantitative analysis of elements. The WD spectrometer is usually equipped with several diffracting crystals to optimize resolution and to cover the entire spectral range. The electron beam, sample stage, spectrometer, data collection, and processing are all under computer control.

Applications of EPMA include elemental analysis of surfaces and of micron-sized features at a surface. The sensitivity of the method is about 0.2 atom %. It provides a rapid, accurate method for compositional analysis of microscopic features. Elemental mapping of the elements present at the surface can also be done, and the composition correlated to topographical maps obtained from an analytical microscopy method, allowing correlation of topographical features of a surface with elemental composition. Like XRF, EPMA is strictly an elemental analysis technique; no information on chemical speciation or oxidation state is obtained. All elements from boron to uranium can be determined. Given that X-rays can escape from depths of 1000 Å or so, EPMA has the "deepest" definition of "surface" of the techniques discussed in this chapter. In fact, like XRF, it can be considered to be a bulk analysis technique assuming the sample is homogeneous. Many surface scientists do not consider EPMA to be a surface technique at all—it depends on one's definition of surface.

The relative depth of penetration for the surface analysis techniques discussed is shown in Figure 16.43 for comparison. ISS is the most surface sensitive, while EPMA is the least surface sensitive of the techniques that have been covered.

PROBLEMS

16.1 The wavelength of XRF radiation of an element is virtually independent of the chemical form of the element, but in ESCA, the photoelectron energy is not independent of the chemical environment of the element. Explain.

16.2 Why is an electron flood gun used in ESCA, Auger, and other surface analysis instruments?

16.3 ESCA and Auger instruments are useful for the analysis of elements present in concentrations greater than 1 % although the absolute sensitivity of the method is about 10^{-10} g. Explain.

16.4 Draw a schematic diagram of an ESCA instrument.

16.5 Show the relationship between the processes that give rise to XRF, ESCA, and AES using a schematic diagram of an atom.

16.6 Why is the ESCA spectrum plotted directly as signal intensity vs. energy, but an Auger spectrum presented as the first derivative of the signal?

16.7 XRF is most sensitive for elements with a high atomic number. What is the relationship between Auger sensitivity and atomic number?

16.8 The concentration of nickel in a sample varies with depth from the surface. How may this variation be measured?

16.9 Why is it necessary to analyze surface samples under vacuum? How may several samples be examined consecutively?

16.10 Describe ISS. What depth of surface can be measured by this method? How does that compare with ESCA?

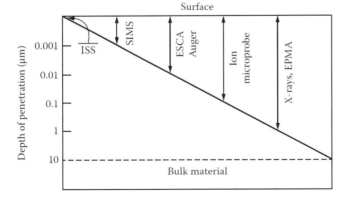

Figure 16.43 Approximate depth of sample surface probed by different surface analysis techniques.

16.11 How can the concentration and distribution of a selected metal on a plane surface be measured and displayed?

16.12 What sources are commonly used for generating ESCA and Auger spectra?

16.13 How can the effects of rapid variation in local magnetic fields on ESCA signals be eliminated? Why is it necessary to do that?

16.14 You want to identify the location, distribution, concentration, and chemical form (molecular, elemental, oxidation state) of ppm levels of potassium in tungsten wire. What techniques will you use? Explain your reasoning. Can surface analysis techniques provide all of the information needed? If not, what else will you use?

16.15 What is SIMS? Describe its analytical uses.

16.16 What are the advantages of (a) ISS and (b) SIMS over XRF?

16.17 Describe the major components of an IMMS.

16.18 Describe the detectors used in ESCA.

16.19 Describe two types of energy discriminators. What are their functions?

16.20 Describe two types of energy analyzers. What is their function?

16.21 What is meant by the ESCA chemical shift? What is meant by the Auger chemical shift?

BIBLIOGRAPHY

The reader will note that occasionally multiple editions of books are cited as sources or reference works. The reason for this is that sometimes different figures are used in different editions, which may in fact be written by different authors. The sources found in the bibliography will be the most recent versions of a particular work.

In some cases instrument manufacturers and suppliers are cited by providing a URL to the exact internet location. We have endeavored to ensure that these links are functional at the time of publication, but clearly these links change. If a link has become nonfunctional, a good approach is to use your browser's advanced search capability. For example: "company name" AND "instrument type" AND "specific descriptors" is a good place to start.

ASM Handbook, Volume 10 Materials Characterization, ASM International, Materials Park, OH, 2019.

Benninghoven, A.; Rudenauer, F.G.; Werner, H.W. *Secondary Ion Mass Spectrometry: Basic Concepts, Instrumental Aspects, Applications and Trends*. John Wiley & Sons: New York, 1987.

Briant, C.L.; Messmer, R.P. (eds.) Auger electron spectroscopy. In *Treatise on Materials Science and Technology*, Vol. 30. Academic Press, Inc.: San Diego, CA, 1988.

Brundle, C.R., Watts, J.F., Wolstenholme, J., in Grinberg, N., Rodriguez, S., eds., *Ewing's Analytical Instrumentation Handbook*, 4th Ed., CRC Press Taylor and Francis, Boca Raton, FL, 2018.

Carlson, T.A. *Photoelectric Auger Spectroscopy*. Plenum Press: New York, 1975.

Czarderna, A.W.; Metler, A.C.; Jellinek, H.H.G.; Kochi, H. *Ind. Res.* January 1978.

Davis, L.E.; MacDonald, N.C.; Palmberg, P.W.; Riach, G.E.; Weber, R.E. *Handbook of Auger Electron Spectroscopy*, 2nd edn. Physical Electronics Inc.: Eden Prairie, MN, 1976. (Now Physical Electronics, www.phi.com.)

Hendrickson, D.N.; Hollander, J.M.; Jolly, W.L. Nitrogen 1s electron binding energies. Correlations with molecular orbital calculated nitrogen charges. *Inorg. Chem.*, 8, 2642, 1969.

Hercules, D.M. Electron spectroscopy. *Anal. Chem.*, 42(1), 20A, 1970.

Hercules, D.M. Analytical Chemistry of Surfaces. *Anal. Chem.*, 58, 1177A, 1986.

Hercules, D.M.; Hercules, S.H. Analytical chemistry of surfaces. Part III. Ion spectroscopy. *J. Chem. Ed.*, 61, 403, 1984.

Lee, L.H. (ed.) *Characterization of Metal and Polymer Surfaces*, Vol. 1. Academic Press Inc.: San Diego, CA, 1974.

Lifshin, E. (ed.) Characterization of materials, parts I and II. In Cahn, R.W.; Haasen, P.; Kramer, E.J. (eds.), *Materials Science and Technology: A Comprehensive Treatment*, Vols. 2A and 2B. VCH Publishers, Inc.: New York, 1992.

Moulder, J.F.; Stickle, W.F.; Sobol, P.E.; Bomben, K.D. In *Handbook of X-Ray Photoelectron Spectroscopy*. Chastian, J.; King, R.C. (eds.), Physical Electronics, Inc.: Eden Prairie, MN, 1995 (Now Physical Electronics, www.phi.com.)

NIST X-ray Photoelectron Spectroscopy Database, NIST Standard Reference Database 20, Version 4.1, copyright 2012 by the US Secretary of Commerce on behalf of the United States of America. All rights reserved.

Perry, S.S.; Somorjai, G.A. Characterization of organic surfaces. *Anal. Chem.*, 66(7), 403A, 1994.

Rendina, J.F. Electron spectroscopy for chemical analysis. *Am. Lab.*, 4(2), 17, 1972.

Settle, F. (ed.) *Handbook of Instrumental Techniques for Analytical Chemistry*. Prentice Hall PTR: Upper Saddle River, NJ, 1997.

Siegbahn, K. Electron spectroscopy—a new way of looking into matter. *Endeavor*, 32, 51, 1973.

Stinger, K. *Res. Dev.*, 28(9), 40, 1977.

Thompson, M.; Baker, M.D.; Christie, A.; Tyson, J.F. *Auger Electron Spectroscopy*. John Wiley & Sons, Inc.: New York, 1985.

Turner, N.H. Electron spectroscopy. In Ewing, G.W. (ed.), *Analytical Instrumentation Handbook*, 2nd edn. Marcel Dekker, Inc.: New York, 1997.

Weber, R.E. Auger electron spectroscopy for thin film analysis. *Res. Dev.*, 10, 22, 1972.

Wilson, R.G.; Stevie, F.A.; Magee, C.W. *Secondary Ion Mass Spectrometry*. John Wiley & Sons Inc., New York, 1989.

CHAPTER **17**

Thermal Analysis

When matter is heated, it undergoes certain physical and chemical changes. Physical changes include phase changes such as melting, vaporization, crystallization, transitions between crystal structures, changes in microstructure in metal alloys and polymers, changes in protein structure, volume changes (expansion and contraction), adsorption, absorption, and desorption of gases, and changes in mechanical behavior. Chemical changes include reactions to form new products, oxidation, corrosion, decomposition, dehydration, chemisorption, and binding between compounds (e.g., proteins and drugs). These physical and chemical changes take place over a wide temperature range. The rates of chemical reactions vary with temperature, and the properties of some materials, such as semicrystalline polymers and metal alloys, depend on the rate at which they are cooled. Materials are used over a wide range of temperatures, from Arctic cold to desert heat, in corrosive environments, variable humidity, and under load (stress). It is necessary to characterize materials and their behavior over a range of temperatures to determine what materials are suitable for specific uses and to determine what temperature range materials or chemicals can withstand without changing. This sort of information is used in many ways: to predict safe operating conditions for products, such as which type of tire material is best for vehicles in extremely cold or extremely hot climates, the average expected lifetime of materials such as paints and polymers exposed to temperature changes, processing conditions for materials, the curing times and temperatures for dental filling material, and optimum storage conditions for food, among other uses.

The physical changes and chemical reactions a sample undergoes when heated are characteristic of the material being examined and the atmosphere in which the heating occurs. By measuring the temperature at which such reactions occur and the heat involved in the reaction, we can determine a great deal about the material. The composition of pure compounds and mixtures, including polymer blends, can be determined. The purity of pharmaceutical compounds can be determined. Rate of reaction, rate of crystallization, glass transition temperatures, decomposition temperatures, and catalysis can be studied. The effectiveness of additives and stabilizers in materials can be evaluated. Enthalpy changes (ΔH) of reactions can be measured. Percent crystallinity of polymers, which greatly affects the mechanical properties of polymers, can be determined. By heating materials under load (stress), mechanical properties such as modulus, ductility, yield point (the point at which nonpermanent elastic deformation changes to permanent plastic deformation), and volume change as a function of the load can be measured. These parameters are very important in engineering design to ensure safe and functional products.

Heat is involved in most real-life processes. This permits heat—into or out of a system—to serve as a universal detector. In many cases, the heat into or out of a system can be measured nondestructively. Heat transfer occurs in three ways: conduction, convection, and radiation. Conduction occurs between solid materials when placed in contact with each other. Convection occurs when a hot material and a cold material are separated by a fluid (gas or liquid). Radiative heat transfer involves the emission and consequent absorption of electromagnetic radiation between a hot and cold material.

The analytical techniques used to study changes in physical properties with temperature are called **thermal analysis** techniques. A very readable description of all thermal analysis methods can be found in the ASM Handbook. They include thermogravimetric analysis (TGA), differential thermal analysis (DTA), differential scanning calorimetry (DSC), thermometric titration (TT), isothermal titration calorimetry (ITC), microcalorimetry, direct injection enthalpimetry (DIE), dynamic mechanical analysis (DMA), and thermomechanical analysis (TMA). Thermal analysis techniques are used in the characterization of inorganic and organic compounds, polymers, cosmetics, pharmaceuticals, metals, alloys, geological samples, ceramics, glasses, and many manufactured products, as well as biological systems. Table 17.1 summarizes some common thermal analysis techniques and the information they provide. Table 17.2 presents a brief overview of some applications for various thermal analysis methods. Table 17.3 lists some typical sample types and some of the properties that have been measured by thermal analysis methods.

Table 17.1 Thermal Analysis Methods

Thermal Analysis Technique	Principle	Information Obtained
TGA	Measures change in mass with respect to temperature and time in an inert or reactive atmosphere. Samples from 10 to 20 mg for most applications, 50 to 100 mg for volatile samples	Thermal stability, oxidative stability, kinetics, degradation, and shelf life. EGA when coupled to gas chromatography (GC), infrared (IR), and/or mass spectrometry (MS) for chemical composition and identification. Measurement times are between 1 and 2 h
DSC	Monitors the difference in temperature between a sample and a reference material as a function of time and temperature in a specified atmosphere. Quantitatively measures heat absorbed or released by a material undergoing a physical or chemical change. Samples from 0.5 to 10 mg. Instruments have temperature range between −150 to 750 °C	Glass transition, melt and phase change temperatures, heats of reaction, heat capacity, crystallinity, aging, degradation, and thermal history. Measurements can be done quickly, between a few seconds to 30 min. DSC is usually destructive, and works best if sample is in full contact with bottom of pan
DTA	Monitors the difference in temperature between a sample and a reference material as a function of time and temperature in a specified atmosphere. Qualitative or semiquantitative measurement of heat absorbed or released by a material undergoing a physical or chemical change	Same as DSC but heat capacity and ΔH results are qualitative or semiquantitative
TMA, dilatometry	Measurement of dimensional change with temperature in a specified atmosphere; samples are typically bars, films, fibers, coatings, pellets, and cylinders	Coefficient of thermal expansion (CTE), transition temperatures, Young's modulus, and stress relaxation; polymer morphology; flow and relaxation behavior. Relatively long measurement times and careful sample preparation
DMA	Measurement of viscoelastic properties of solid materials as a function of temperature and time; samples are typically cylindrical	Glass transition, brittleness, anisotropy, crystallinity, residual stress, residual cure, and aging. Typically requires long measurement times

Table 17.2 Some Applications of Thermal Analysis Methods

Applications	TGA	DSC, DTA	TMA	DMA
Compositional analysis	×	×		
Curing studies		×		×
Glass transition		×	×	×
Heat of reaction		×		
Oxidative stability	×	×		
Corrosion	×	×		
Creep			×	×
Stress relaxation			×	×
Thermal stability	×	×		
Viscoelastic properties			×	×
Protein denaturation		×		
Shrinkage			×	×

Note: ×, sample type that has been characterized for this property.

17.1 THERMOGRAVIMETRY

Thermogravimetry or TGA measures the mass of a sample in a specified atmosphere as the temperature of the sample is programmed. The most common temperature program is a linear increase in temperature with time, although isothermal programs, stepped temperature programs, and so on can be used. In the most common TGA experiment, the sample temperature is increased linearly over a time period and the mass of the sample is constantly recorded. The output from a TGA experiment is a plot of mass (or mass %) versus temperature. The TGA plot is called a **thermal curve**. An example of a TGA thermal curve for the decomposition of calcium carbonate is shown in Figure 17.1. Mass or mass% is plotted along the y-axis and temperature (or time for a linear temperature ramp) along the x-axis. The change in mass of a sample as the temperature changes tells us several things. First, this determines the temperature at which the material loses (or gains) mass. Loss of mass indicates decomposition or evaporation of the sample. A gain in mass can indicate adsorption by the sample of a component in the atmosphere or a chemical reaction with the atmosphere, such as oxidation. Second, the temperatures at which no mass change takes place are determined, which indicate the temperature stability of the material. These mass changes at certain temperatures are physical properties of chemical compounds under the conditions of the experiment (atmosphere, heating rate). This information can be used to determine if a sample is the same as a "standard" or "good" material in a production process, for example. Knowledge of the temperatures at which a sample is unstable and subject to decomposition or chemical change is important to the engineer because

THERMAL ANALYSIS

Table 17.3 **Partial List of Sample Types and Properties Examined by Thermal Analysis**

Properties	Samples						
	Chemicals	Elastomers	Explosives	Soils	Plastics	Textiles	Metals
Identification	×	×	×	×	×	×	×
Quantitative composition	×	×	×	×	×	×	×
Phase diagram	×	×	×		×	×	
Thermal stability	×	×	×		×	×	
Polymerization	×	×			×	×	
Catalytic activity	×						×
Reactivity	×	×			×	×	×
Thermochemical constants	×	×	×	×	×		×
Reaction kinetics	×	×	×		×	×	×

Note: ×, sample type that has been characterized for this property.

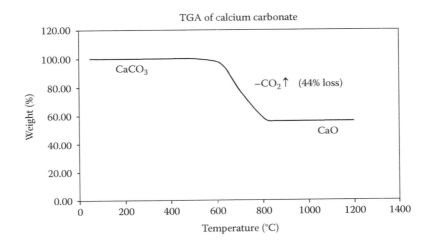

Figure 17.1 TGA thermal curve for pure anhydrous calcium carbonate, $CaCO_3$. The loss in mass is due to the loss of CO_2 (g) and the compound remaining is CaO.

it reveals the temperature above which such materials as polymers, alloys, and building materials may *not* be used, as well as the temperatures at which they may be used safely.

The *mass lost* by a sample heated to a given temperature helps the analytical chemist to determine the composition of a compound and follow the reactions involved in its decomposition. It also enables the analytical chemist to identify crystals of unknown composition or determine the percentage of a given compound in a mixture of compounds. For example, if pure calcium carbonate, $CaCO_3$, is heated to 850 °C, it loses 44 % of its mass (Figure 17.1). Also, the gas evolved can be collected and identified as CO_2. This observation virtually confirms that the reaction

$$CaCO_3(s) \rightarrow CaO(s) + CO_2(g)$$

takes place at this temperature. How do we know this? If we know our sample was pure $CaCO_3$ and CO_2 is identified as a product of the reaction, the need for mass balance tells us that CaO is a reasonable "other" product. If we had 50.0 mg of $CaCO_3$ to start with, the mass loss due to CO_2 can be calculated:

$$\text{mg } CO_2 = (50.0 \text{ mg } CaCO_3)$$
$$\times \left[\frac{1 \text{ mmol } CaCO_3}{100.00 \text{ mg } CaCO_3} \right] \times \left[\frac{1 \text{ mmol } CO_2}{1 \text{ mmol } CaCO_3} \right]$$
$$\times \left[\frac{44.00 \text{ mg } CO_2}{\text{mmol } CO_2} \right] = 22.0 \text{ mg}$$

The loss of CO_2 equals a loss in mass of (22.0 mg/50.0 mg) × 100% = 44%. From the stoichiometry, it is expected that 1 mol of CO_2 is lost for every mole of $CaCO_3$ present, which corresponds to a loss of 44% of the mass of 1 mol of

$CaCO_3$. The experimental mass loss supports the theoretical loss if the decomposition of $CaCO_3$ proceeds according to the reaction proposed. The formation of CO_2 can be verified by having the evolved gas analyzed by MS or by online IR spectroscopy. The CaO can be confirmed by analysis of the residue by X-ray diffraction (XRD) or other techniques.

The TGA technique was developed to solve problems encountered in gravimetric analysis. For example, a common precipitating and weighing form for the determination of calcium was calcium oxalate. In practice, the calcium oxalate was precipitated and filtered and the filtrate dried and weighed. The drying step was difficult to reproduce. A particular procedure might recommend drying at 110 °C for 20 min and cooling prior to weighing. The analyst would then obtain very reproducible results derived from his or her own work. A second procedure might recommend drying at 125 °C for 10 min, cooling, and weighing. As before, the analyst using this method would obtain very reproducible results, but these results might not agree with those of the first research worker. The thermal curve of hydrated calcium oxalate is shown in Figure 17.2. When we examine the thermal curve, we find that calcium oxalate loses mass over a range of temperatures from 100 °C up to 225 °C. The most logical explanation for this mass loss is that water is evaporating from the sample. The water that is driven off includes not only absorbed or adsorbed water from the precipitating solution but also water of crystallization (also called water of hydration) that is bound to the calcium oxalate as $CaC_2O_4 \cdot nH_2O$. It is only when both types of water are driven off that reproducible results are achievable. If the sample were heated to 110 °C, the absorbed water would be driven off, but a small amount of bound water (i.e., water of crystallization) would also be lost. Similarly, if the drying temperature were 125 °C, the absorbed water would be driven off and a different small amount of bound water would be lost. As long as a drying temperature was rigidly adhered to, the results would be reproducible, but results using different drying temperatures would show differences because the amount of bound water lost would vary. The TGA experiment shows that drying to a temperature above 225 °C but below 400 °C will result in a stable form suitable for gravimetric analysis.

17.1.1 TGA Instrumentation

Modern TGA equipment has a sensitive balance, usually a microbalance, for continuously measuring sample mass, a furnace surrounding a sample holder, and a purge gas system for providing inert or reactive atmospheres. A computer generally controls the furnace and the data (mass vs. sample temperature) are collected and processed by computer. Intelligent autosamplers are available for most instruments that permit the unattended analysis of samples.

Modern analytical microbalances of several different designs are commercially available—torsion balances, spring balances, and electrobalances have been used in TGA instruments. In general, the balance is designed so that a change in sample mass generates an electrical signal proportional to the mass change. The electrical signal is transformed into mass or mass loss by the data processing system and plotted on the y-axis of the thermal curve. TGA balances are available for sample masses from 1 to 1000 mg, with the usual sample weighing between 5 and 20 mg. There are specialized high-capacity TGA systems available that can accommodate samples of up to 100 g and systems that can handle microgram quantities of sample. Figures 17.3 and 17.4 show the schematic of a TGA electrobalance and a cutaway of a commercial high-capacity TGA based on the electrobalance. The usual sample size for TGA is very small, so care must be taken to obtain a homogeneous or representative sample. The balance itself must be thermally isolated from the furnace,

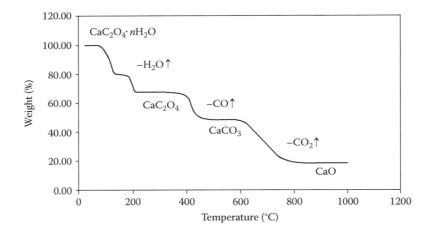

Figure 17.2 TGA thermal curve of hydrated calcium oxalate, $CaC_2O_4 \cdot nH_2O$, with both adsorbed water from the precipitation process and water of crystallization. There is a loss of adsorbed water starting at about 90 °C and loss of bound water at about 150 °C. The stable compound above 225 °C is anhydrous calcium oxalate, CaC_2O_4. This loses CO at about 450 °C to form $CaCO_3$. The calcium carbonate is stable until approximately 600 °C, when it loses CO_2 to form CaO. (Compare this step with Figure 17.1.)

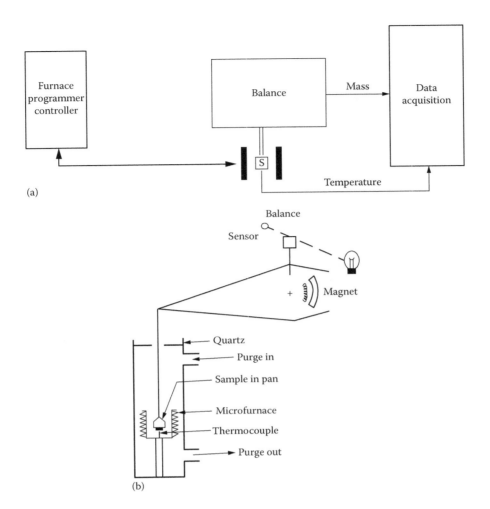

Figure 17.3 (a) Block diagram of a TGA system. S represents the sample pan hanging from the balance arm in position in the furnace (represented by the solid bars on each side). (b) Schematic of a commercial TGA, showing the purge gas inlet and outlet and the thermocouple position beneath the sample pan. (b: [© 1993–2014 PerkinElmer, Inc. All rights reserved. Printed with permission. (www.perkinelmer.com).].)

although the sample holder and sample must be in the furnace. There are two possible configurations of the balance and furnace, a horizontal furnace or a vertical furnace. Both types of configuration suffer from drift as the temperature increases. Vertical configurations suffer from buoyancy effects due to the change in gas density with temperature. The horizontal configuration was designed to minimize buoyancy effects, but horizontal configurations experience changes in the length of the quartz rod connecting the sample to the balance as the temperature changes. Buoyancy effects and changes in the quartz rod result in error in determining the mass of the sample.

The furnace surrounds the sample and sample holder. It must be capable of being programmed for a linear heating rate. Modern instruments can be heated and cooled rapidly, which increases sample throughput. Instruments that heat at rates of up to 1000 °C/min are available. One commercial instrument can heat at rates up to 200 °C/min from room temperature to about 1200 °C; cooling by forced air can be done at ~50 °C/min. There are furnaces available with upper temperatures of 1500 °C, 1700 °C, or 2400 °C; these higher-temperature instruments are useful for studying refractory materials and engineering materials. The furnace must be able to be purged with a desired gas, to provide the correct atmosphere for the experiment, and to remove gaseous products from the sample compartment. Argon or nitrogen is used when an inert atmosphere is desired. Reactive gas atmospheres can be used for certain studies. Air is often used for oxidation and combustion studies. Hydrogen gas may be used to provide a reducing atmosphere, with the appropriate precautions to prevent explosions. Modern instruments permit the purge gas to be switched automatically, so that the sample can start heating in an inert atmosphere and be switched to air or other reactive gas at high temperatures, for example.

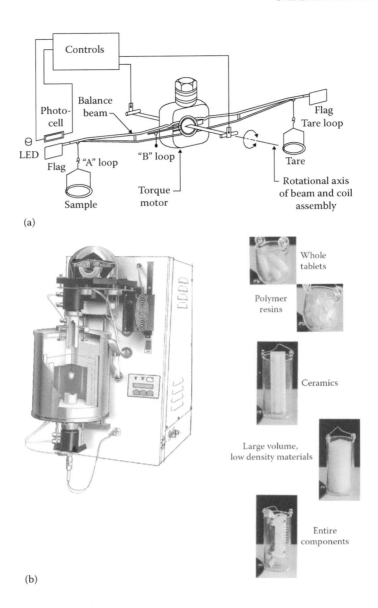

Figure 17.4 (a) Electrobalance for a TGA. As the sample mass changes, the balance arm tips, resulting in a change in the amount of light reaching the photocell. This generates a current to restore the arm position; the current is proportional to the change in mass of the sample. (b) Cutaway view of a high-capacity TGA capable of holding 100 g samples. Typical types of large samples in quartz sample holders are shown in the photographs on the right. The TGA shown is no longer produced. (© Thermo Fisher Scientific (www.thermofisher.com). Used with permission.)

The sample holder and any instrument parts inside the furnace, such as the thermocouple for measuring the temperature, must be able to withstand high temperature and be inert at these high temperatures. Quartz, platinum, and various ceramics are used for the sample holder and other parts. The sample is placed in a small pan or crucible made of Pt, aluminum, quartz, or ceramic. Pan volumes of 20, 50, 100, and 250 µL are common. Sample preparation in the pan is important to obtaining good results by TGA. It is important to have a large surface area exposed to the sample purge, and in this respect small particles are better than large chunks.

Ideally, the temperature recorded is the exact temperature of the sample. This entails measuring the temperature of the sample while the analysis is carried out. It is particularly important to measure the temperature of the sample rather than that of the furnace. This is difficult because the temperature is measured with a thermocouple that is near but not in the sample. The temperature of the sample inside the furnace is measured with a thermocouple, either a chromel/alumel thermocouple or one made of Pt alloy. The thermocouple is never inserted directly into the sample because of possible sample contamination (or

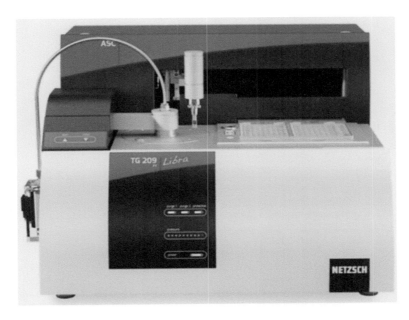

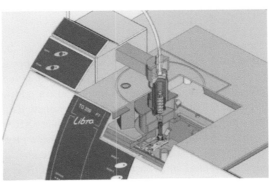

(c)

Figure 17.4 (Continued) (c) A modern TGA instrument with a cut away inset showing the sample compartment (courtesy of Netzsch Instruments).

contamination of the thermocouple resulting in errors in temperature), inadvertent initiation of a catalytic reaction, particle size effects, sample packing effects, and possible weighing errors. The thermocouple is made as small as possible and placed close to the sample holder, sometimes in contact with the bottom of the sample pan. The temperature actually recorded may be slightly different from the sample temperature; the sample temperature generally is lower than the temperature recorded by the thermocouple. This is due to factors such as the rate of heating, gas flow, thermal conductivity (TC) of the sample, and the sample holder. Modern instruments have a temperature–voltage control program in the computer software that permits reproducible heating of the furnace. With precise, reproducible heating, the thermocouple can be calibrated to provide accurate furnace temperatures, but the relationship between actual sample temperature and recorded temperature is complex. The problem is compounded by the fact that at temperatures below 500 °C, most of the heat transferred from the furnace to the sample takes place by convection and conduction, but at temperatures above 500 °C, where the furnace is red-hot, most of the energy is transferred by radiation. The switch from conduction–convection to radiative energy transfer makes choosing the position of the thermocouple to obtain accurate temperature measurements of the sample quite a complicated problem.

Temperature calibration of TGA instruments with samples of pure materials with well-characterized mass losses can be done, but often is not satisfactory. For example, one problem with this approach is that black samples such as coal behave differently from white samples such as calcium phosphate under the influence of radiant energy and the sample temperatures will

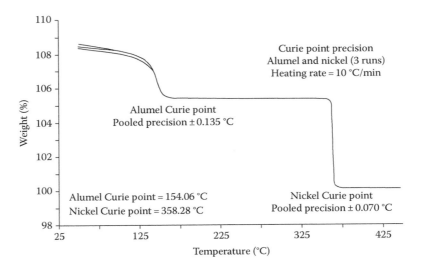

Figure 17.5 The TGA Curie point method records for each standard an apparent sharp mass change at a well-defined temperature, which corresponds to a known transformation in the standard's ferromagnetic properties at that temperature. The figure shows the relative temperature precision from three replicate calibration runs using alumel and nickel Curie point standards. (Courtesy of TA Instruments, New Castle, DE, www.tainstruments.com.)

therefore be different under the same furnace conditions. A more accurate calibration method uses the Curie temperature of various ferromagnetic standard materials. The materials undergo specific and reversible changes in magnetic behavior at their Curie temperature. Standards are available covering the temperature range of 242 °C–771 °C. A **ferromagnetic material** is magnetic under normal conditions, but at a characteristic temperature (**the Curie temperature**), its atoms become disoriented and paramagnetic and the material loses its magnetism. To take advantage of this phenomenon, we weigh the ferromagnetic material continuously, with a small magnet placed above the balance pan. The standard's apparent mass is then its gravitational weight minus the magnetic force it experiences due to the magnet immediately above. At the Curie temperature, the standard loses its magnetism and the effect of the magnet is lost. There is a change in the apparent weight of the standard, and this can be recorded. The temperatures of the Curie transitions of ferromagnetic materials are well known and therefore can be used for calibration purposes. An example of calibration using the Curie temperatures of alumel and nickel is shown in Figure 17.5.

17.1.2 Analytical Applications of Thermogravimetry

One of the first important applications of thermogravimetry was the determination of correct drying temperatures for precipitates used in gravimetric analysis. This knowledge was of vital importance if accurate and reproducible results were to be obtained from gravimetric analysis. A second important application was the identification of the gases given off while a sample's temperature is increased. In addition, the composition of the residue can be determined using techniques such as XRD and X-ray fluorescence (XRF). This information reveals the chemical decomposition process occurring when materials are heated and permits identification of the formulas of the starting materials. TGA is very important in determining the upper use temperatures of materials such as polymers by identifying the temperature at which oxidative degradation occurs on heating in air.

From Figure 17.6, it can be determined that when pure calcium oxalate monohydrate, $CaC_2O_4 \cdot H_2O(s)$, is heated, it first loses water of crystallization and forms $CaC_2O_4(s)$. The fact that the compound contains only 1 mol of water of hydration can be determined from the mass loss. The reaction is:

$$CaC_2O_4 \cdot H_2O(s) \rightarrow CaC_2O_4(s) + H_2O(g)$$
$$\text{Relative Molecular Masses (g/mol):}$$
$$146 \qquad\qquad 128 \qquad 18$$

Therefore, the %W lost due to 1 mol of water of crystallization = (18/146) × 100 = 12.3%. If our initial sample mass was 20.00 mg, the mass loss at the first step would be 2.46 mg if there is 1 mol of water of crystallization. If the mass loss actually measured was 4.92 mg, it would mean that there were 2 mol of water of crystallization and the formula would be $CaC_2O_4 \cdot 2H_2O(s)$, a dihydrate.

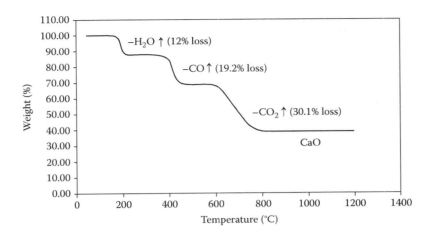

Figure 17.6 TGA thermal curve of calcium oxalate monohydrate, $CaC_2O_4 \cdot H_2O$.

Upon further heating up to about 400 °C, CO(g) is given off and the reaction that occurs is:

$$CaC_2O_4(s) \rightarrow CaCO_3(s)\ CO(g) \uparrow$$
Relative Molecular Masses (g/mol):
28

The % W lost from the initial compound = (28/146) × 100 = 19.2 %. The total % W lost at this point is the sum of the two steps: 12.3 % + 19.2 % = 31.5 %.

Finally, at even higher temperatures (about 800 °C), the $CaCO_3$ formed at 400 °C decomposes:

$$CaCO_3(s) \rightarrow CaO(s) + CO_2(g) \uparrow$$
Relative Molecular Masses (g/mol):
44

The % W lost from the original compound = (44/146) × 100 = 30.1 %. The total mass loss is the sum of all three steps: 12.3 % + 19.2 % + 30.1 % = 61.6 %. The losses correspond to what is seen in the decomposition of calcium oxalate monohydrate in Figure 17.6. The ↑ symbol indicates gas evolved from the sample and swept out of the TGA system. The gases may be identified if the TGA analyzer is connected to an IR spectrometer or a mass spectrometer, as described in Section 17.4.

TGA can be used for the identification of compounds present in mixtures of materials. When such mixtures are heated using a thermogravimetric analyzer, the thermal curve produced consists of all possible mass losses from all components superimposed on each other. Interpretation of the complete thermal curve requires that the individual thermal events be separated and identified. In many cases, the components of the mixture can be identified and a quantitative determination of each is possible from the thermal curve. An example of how this can be done is shown in Figure 17.7. The uppermost curve is the mass loss curve for pure compound A. The next lower curve is the mass loss curve for pure compound B. The bottom curve is the mass loss curve for a mixture of A and B. The amount of A present in the mixture can be determined from the mass loss between points δ and ε, while the amount of B is determined from the mass loss between points β and γ.

An illustration of the application of TGA to quantitative analysis of mixtures is the determination of the magnesium oxalate content of a mixture of magnesium oxalate and magnesium oxide, MgO. Magnesium oxalate is less stable than calcium carbonate; it decomposes to magnesium oxide, MgO, at ~500 °C. Pure MgO is stable at room temperature and to well above 500 °C; MgO does not lose any mass. The TGA curve for this mixture shows two mass losses, one at about 200 °C. The second mass loss occurs in the 397 °C–478 °C range. Thinking about what happened to calcium oxalate, it is reasonable to suppose that the first mass loss is due to loss of water, both adsorbed and water of crystallization. From the formula for $MgC_2O_4 \cdot 2H_2O$, we see that there are 2 mol of water of crystallization for every mole of magnesium oxalate. Only one other mass loss is seen, and we know MgO is stable at these temperatures. We can assume that when $MgC_2O_4 \cdot 2H_2O$ is heated to a temperature above 500 °C, it forms MgO according to the following reaction:

$$MgC_2O_4 \cdot 2H_2O(s) \rightarrow$$
$$MgO(s) + CO(g)\uparrow + CO_2(g)\uparrow + H_2O(g)\uparrow$$
Relative Molecular Masses (g/mol):
148   40   28   44   18

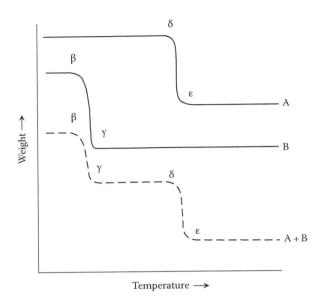

Figure 17.7 TGA thermal curves for (top) pure A, (middle) pure B, and (bottom dotted line) a mixture of A and B. Because A and B have unique temperatures at which mass is lost, the composition of the mixture may be determined.

The final mass loss is therefore %W = ([28 + 44 + 2(18)]/148) × 100 = 73 %.

We would lose 73 mg for every 100 mg of pure magnesium oxalate dihydrate we had in the mixture on heating the mixture to 500 °C. Note that the decomposition of magnesium oxalate does not follow the same process as the decomposition of calcium oxalate. Calcium oxalate first forms calcium carbonate, which only loses carbon dioxide to form CaO at temperatures above 600 °C. Magnesium oxalate appears to decompose directly to MgO in one step at about 500 °C. The mass remaining from the magnesium oxalate is equal to

$$\frac{\text{Mol wt MgO}}{\text{Mol wt MgC}_2\text{O}_2 \cdot 2\text{H}_2\text{O}} \text{ per gram of}$$
$$\text{MgC}_2\text{O}_4 \cdot 2\text{H}_2\text{O} = \left(\frac{40}{148}\right)$$
$$= 0.27 \text{ g MgO/g MgC}_2\text{O}_4 \cdot 2\text{H}_2\text{O}$$

The following data were obtained from our TGA curve:

Original mass of sample	25.00 mg
Mass of sample after heating to 500 °C	10.40 mg
Loss in mass	14.60 mg

But we have already calculated that $MgC_2O_4 \cdot 2H_2O$ loses 73 % of its mass upon heating to 500 °C; therefore, the mass of $MgC_2O_4 \cdot 2H_2O$ in the original sample was

$$\text{mg MgC}_2\text{O}_4 \cdot 2\text{H}_2\text{O} = (14.60 \text{ mg lost})\left[\frac{100}{73}\right] = 20 \text{ mg}$$

Then the concentration of $MgC_2O_4 \cdot 2H_2O$ in the original sample was

% Mg oxalate dihydrate
$$= \left(\frac{\text{mg Mg oxalate dihydrate}}{\text{Total sample mass in mg}}\right) \times 100$$

% Mg oxalate dihydrate $= \left(\dfrac{20 \text{ mg}}{25.00 \text{ mg}}\right) \times 100 = 80\%$

Thus, the other 20 % of the starting mixture is MgO.

Similarly, we can determine the magnesium oxalate content and the calcium oxalate content in a mixture of the two compounds. The TGA curve of the mixture is shown in Figure 17.8. We have already seen in Figure 17.6 that calcium carbonate decomposes to calcium oxide above 600 °C. We can deduce from the thermal curve for magnesium oxalate that magnesium carbonate is not stable because the oxalate decomposes directly to MgO. So at temperatures above 500 °C, but below 600 °C, we should have a mixture of MgO and $CaCO_3$. Above 850 °C, the $CaCO_3$ will have decomposed to CaO, so the residue is a mixture of CaO and MgO. By examining Figure 17.6 and the information we have previously on the magnesium salt, it is evident that the mass loss in Figure 17.8 above 600 °C is due to the evolution of CO_2 from the $CaCO_3$ that came from the $CaC_2O_4 \cdot H_2O$. We know one molecule of $CaC_2O_4 \cdot H_2O$ generates one molecule of CO_2. If the mass loss above 600 °C (due to CO_2 loss) was 18 %, then the mass % of $CaC_2O_4 \cdot H_2O$ in the mixture is

$$\left(\frac{\text{Mol wt CaC}_2\text{O}_4 \cdot \text{H}_2\text{O}}{\text{Mol wt CO}_2}\right) \times 10\% = \left(\frac{146}{44}\right) \times 18\% = 60\%$$

Assuming that the masses of calcium oxalate and magnesium oxalate total 100%, then it follows that the percentage of $Mg(COO)_2 \cdot 2H_2O = 100 \% - 60 \% = 40 \%$. As you can see from the examples given, TGA can be used for quantitative analysis, but not without some knowledge of the sample. If there were other components in our mixture of oxalates that we did not know about, our assumption that the two oxalates composed 100 % of the sample would be wrong. If there is another component that loses mass above 600 °C, we have another error. Without some knowledge of the sample, our calculated value for one (or both) of the components could be wrong.

TGA also provides quantitative information on organic compound decompositions and is particularly useful for studying polymers. An example is the use of TGA to determine the amount of vinyl acetate in copolymers of vinyl acetate and polyethylene. On heating, each mole of vinyl acetate present loses 1 mole of acetic acid. A TGA study of several vinyl acetate–polyethylene copolymers is presented in Figure 17.9.

TGA is very useful for providing qualitative information about samples of many types. TGA can provide qualitative information on the stability of polymers when they

THERMAL ANALYSIS

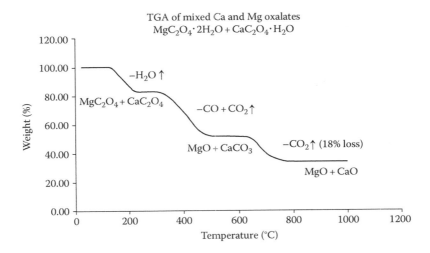

Figure 17.8 TGA thermal curve of a mixture of calcium oxalate monohydrate and magnesium oxalate dihydrate. The last mass loss is due to the loss of CO_2 only from calcium oxalate monohydrate. Therefore, the composition of the mixture can be determined, even though the other steps have combined mass losses from both compounds.

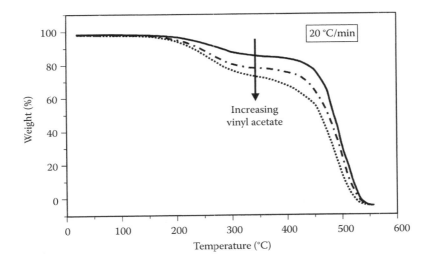

Figure 17.9 TGA thermal curves of vinyl acetate copolymers. The mass loss at 340 °C is due to loss of acetic acid and gives a quantitative measure of the amount of vinyl acetate in the polymer. (Courtesy of TA Instruments, New Castle, DE, www.tainstruments.com.)

are heated in air or under inert atmospheres. From the decomposition temperatures of various polymers heated in a TGA in an air atmosphere, the upper use temperatures of polymer materials can be determined. Figure 17.10 presents the decomposition temperatures for a variety of common polymeric materials; such a TGA comparison can be used to choose a polymer that will be stable below a certain temperature for a given application.

Another example of the application of thermogravimetry is in the characterization of coal. A TGA thermal curve for a coal sample heated in nitrogen or some other inert atmosphere indicates, in a single analysis, the percentage of volatiles present and the "fixed carbon"; if the atmosphere in the TGA is then automatically switched to air, the fixed carbon will burn and the amount of ash in the coal is determined as the residue left. This is very valuable information in the characterization of the quality of the coal and for its handling and subsequent use. Coals vary in their volatiles content and in their carbon content. Coals often contain a considerable amount of volatile material. The composition of this volatile material is variable and may contain valuable chemicals that can be used in the pharmaceutical industry, the dyestuff industry, and the chemical industry. Of course, it can be used as a fuel, and various liquefaction methods can produce both gaseous and liquid fuels; indeed, for many years South Africa had relied on coal-derived jet fuel for all aircraft departing Johannesburg. The newest form of carbon to be discovered, C_{60}, called buckminsterfullerene,

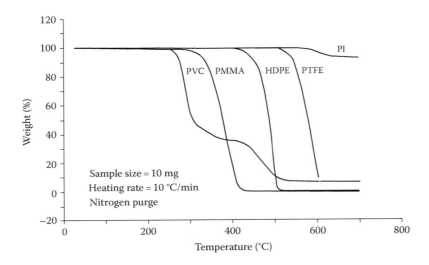

Figure 17.10 TGA thermal curves showing the decomposition temperatures of some common polymers: PVC, polyvinylchloride; PMMA, polymethylmethacrylate; HDPE, high-density polyethylene; PTFE, polytetrafluoroethylene; PI, polyimide. (Courtesy of TA Instruments, New Castle, DE, www.tain-struments.com.)

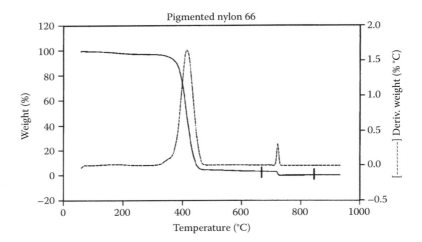

Figure 17.11 TGA determination of carbon black pigment in nylon. The sample was heated under nitrogen up to 650 °C, and then the purge gas was automatically switched to air. The polymer decomposes between 350 °C and 450 °C, leaving the carbon black pigment, which will not decompose in nitrogen, as the 3.05 % by mass residue. On switching to air, the entire residue decomposes (final mass of 0.00 %). This shows that the entire sample including the pigment was organic in nature. The solid line is the thermal curve, while the dotted line is the first-derivative thermogravimetry [DTG] plot. (Courtesy of TA Instruments, New Castle, DE, from *Thermal Analysis Applications Brief TA-122*.)

and related fullerene compounds are found in soot. TGA can be used to characterize soot for its fullerene content, as the fullerenes are more volatile than graphite. Similarly, nonvolatile additives in polymers can be measured quantitatively by TGA. The pigment additive "carbon black," which is not volatile, is measured in a sample of nylon polymer, shown in Figure 17.11. The polymer is burned off and the nonvolatile pigment determined by the mass remaining. Other polymer additives such as silica, titanium dioxide, and inorganic pigments such as cadmium red and nonvolatile flame retardants can be measured in a similar manner.

A high-precision TGA may be used under isothermal or nonisothermal conditions as a mass detector to study kinetics. Such an approach has been used to study oxidation of metals, corrosion, rates of reaction involving mass changes, and phase changes involving small changes in oxygen content for materials like high-temperature superconductors. Many applications of TGA can be found in the scientific literature and on the websites of thermal analysis instrument manufacturers. See Section 17.8.

TGA instruments can be interfaced to mass spectrometers or IR spectrometers to provide chemical information and identification of the evolved gases from samples.

17.1.3 Derivative Thermogravimetry

Examination of a TGA curve will show that a sample's mass loss associated with a particular decomposition occurs over a considerable temperature range, not at a single temperature. When TGA is used to identify an unknown compound, this wide range is a handicap because the uncertainty of identification is increased. This problem can be partially overcome by derivative thermogravimetry (DTG). In DTG, the first derivative of the TGA curve is plotted with respect to temperature. The plot that results has the change in mass with time, dw/dt, which is the rate of mass change, on the y-axis. Figure 17.11 shows the TGA and DTG curves for pigmented nylon; note how easy it is to read the temperature at the peak of the DTG curve versus trying to pick the point where the slope changes on the TGA curve. Figure 17.12a shows the TGA curve and its DTG curve for a hydrated sodium silicate, general formula $Na_xSi_yO_z \cdot nH_2O$. From the TGA curve, the temperature range over which the loss of water occurs is broad and does not have a smooth slope. The DTG curve shows that three separate steps occur in the range of 50 °C – 200 °C, all probably due to loss of water bound in different ways to the sodium silicate. Another example of the power of the derivative plot is shown in Figure 17.12b. This is a TGA and its DTG curve for a mixture of hydrated salts of barium, strontium, and calcium. From the TGA, there appears to be a single mass loss occurring between 130 °C and 210 °C. This might be misinterpreted to be a single pure compound if all we had was the TGA thermal curve. But from the very sensitive DTG, it is clear that there are three different losses of water, occurring at 140 °C, 180 °C, and 205 °C, respectively. These peaks are in fact loss of water first from the barium salt, then from the strontium salt, and finally from the calcium salt. The DTG gave us a clue that more than one event was taking place, a clue that the TGA did not provide.

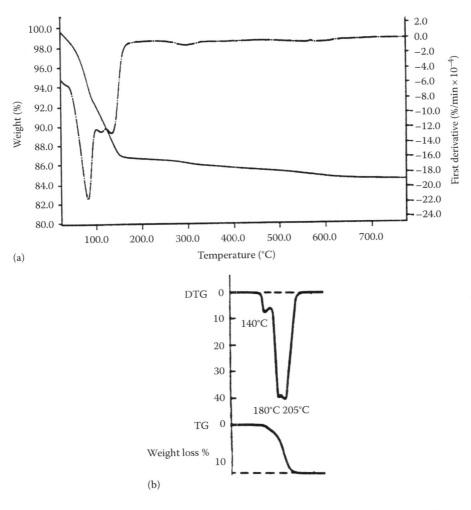

Figure 17.12 (a) TGA (solid line) and DTG (dotted line) thermal curves of a pure hydrated sodium silicate. The TGA mass loss from 50 °C to 150 °C suggests loss of water in more than one form because of the change in slope seen during the step. The first-derivative DTG plot clearly shows three separate mass losses due to water bound in different forms. [© 1993–2014 PerkinElmer, Inc. All rights reserved. Printed with permission. (www.perkinelmer.com).] (b) Partial TGA (bottom curve) and DTG (top curve) thermal curves for a mixture of hydrated barium, strontium, and calcium oxalates. Erdey et al. (1962) showed that the three hydrated oxalates lost water at three different temperatures. The TGA seems to show only one step, but the DTG clearly shows the three separate losses, one for each salt.

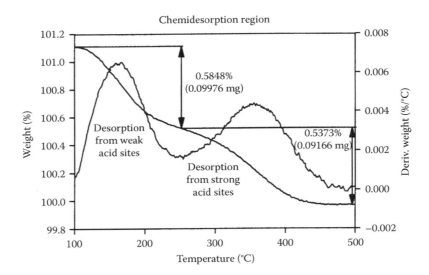

Figure 17.13 TGA and DTG thermal curves showing desorption of basic compounds from the acidic sites of a zeolite catalyst. Losses from weakly acidic sites can be distinguished from strongly acidic sites using the DTG curve, despite the fact that the overall mass loss is only about 1 %. The mass scale is on the left y-axis. (Courtesy of TA Instruments, New Castle, DE, from *Thermal Analysis Applications Brief TA-231*.)

Consequently, DTG is a valuable method of data presentation for thermal analysis. A similar example of the power of the DTG plot is shown in Figure 17.13, which shows desorption of chemisorbed basic compounds from the acidic sites of a zeolite catalyst. **Zeolites**, an important class of catalysts, are porous crystalline aluminosilicates with acidic sites. The number and relative strength of the acidic sites can be estimated by chemisorption of a basic compound, such as ammonia, and then studying its desorption by TGA/DTG. As seen in Figure 17.13, the mass change is small, about 1 % total spread over two broad steps, but the two steps are clearly evident from the DTG plot.

The decomposition of polyvinyl chloride (PVC) also demonstrates the power of the DTG plot. The TGA curve (solid line in Figure 17.14) seems to show two mass loss events, but the DTG plot (dotted line in Figure 17.14) clearly shows three steps. The loss at 280 °C is due to loss of HCl, while the mass losses at 320 °C and 460 °C are due to loss of hydrocarbons. We will learn how we know this later in the chapter.

17.1.4 Sources of Error in Thermogravimetry

Errors in thermogravimetry can lead to inaccuracy in temperature and mass data. Proper placement of the TGA instrument in the laboratory, away from sources of vibration and heat, is essential to minimize fluctuations in the balance mechanism. Older instruments suffered from an apparent gain in mass of a sample container when heated, known as the buoyancy effect. This effect, due to the decreased buoyancy of the atmosphere, changes in convection on heating, and other complex factors, has been, to a large extent, eliminated in modern TGA instruments. The buoyancy effect can be evaluated and compensated for by running a blank (empty sample container) under the same conditions of heating and gas flow used for the samples. Errors can arise due to turbulence caused by the gas flow and due to convection on heating. Gas flow rates and heating rates should be kept as low as possible to minimize these effects.

Placement of the thermocouple is critical to accurate temperature measurement. Ideally, having the thermocouple in the sample itself would give the most accurate reading of the sample temperature. However, there are problems associated with putting the thermocouple into the sample. These include reaction with the sample, reproducible sample packing, sample mass, and TC, among others. Modern instruments generally have the thermocouple in contact with the sample pan or close to the sample pan. The sample temperature is generally lower than the recorded temperature due to several factors including the finite heating rate, TC of both the sample itself and the sample container, and the gas flow rate. There is also the heat of reaction to take into account. An endothermic reaction will cause self-cooling of a sample and therefore an even greater lag in the sample temperature than would otherwise occur, while an exothermic reaction will decrease the lag in sample temperature.

17.2 DIFFERENTIAL THERMAL ANALYSIS

DTA is a technique in which the difference in temperature, ΔT, between the sample and an inert reference material is measured as a function of temperature. Both sample and reference material must be heated under carefully controlled conditions. If the sample undergoes a physical change or a chemical reaction, its temperature will change while the temperature of the reference material remains the same. That is because physical changes in a material such as phase

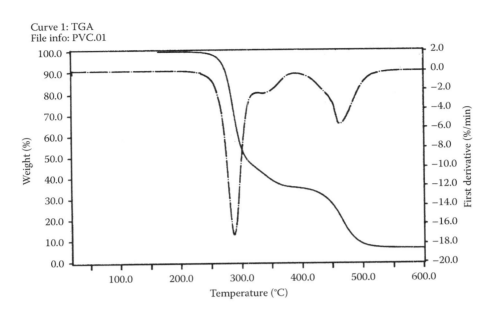

Figure 17.14 TGA (solid line) and DTG (dotted line) thermal curves of PVC polymer heated under nitrogen. The first mass loss at 280 °C is due to loss of HCl. [© 1993–2014 PerkinElmer, Inc. All rights reserved. Printed with permission. (www.perkinelmer.com).]

changes and chemical reactions usually involve changes in **enthalpy**, the heat content of the material. Some changes result in heat being absorbed by the sample. These types of changes are called **endothermic**. Examples of endothermic changes include phase changes such as melting (fusion), vaporization, sublimation, and some transitions between two different crystal structures for a material. Chemical reactions can be endothermic, including dehydration, decomposition, oxidation–reduction, and solid-state reactions. Other changes result in heat being given off by the sample. Such changes are termed **exothermic**. Exothermic changes include phase changes such as freezing (crystallization), some transitions between different crystal structures, and chemical reactions; decomposition, oxidation–reduction, and chemisorption can be exothermic. There are also physical changes that are not simple phase changes that still cause the sample temperature to change. Examples of such physical changes include adsorption and desorption of gases from surfaces and **glass transitions** in amorphous glasses and some polymers. The glass transition is a change in an amorphous material from a brittle, vitreous state to a plastic state. Glass transitions are second-order phase transitions.

DTA and the related technique of DSC to be discussed later in the chapter are capable of measuring many types of physical and chemical changes that result in enthalpy changes. It is not necessary that the sample's mass changes in order to produce a DTA response. However, if a mass change does take place, as occurs on loss of water, the enthalpy of the sample invariably changes, and a DTA response will be observed. So DTA is capable of measuring the same changes measured by TGA, plus many additional changes that TGA cannot measure because no mass change occurs. A DTA plot or thermal curve has ΔT on the y-axis and T (or time, t) on the x-axis, as shown schematically in Figure 17.15. The x-axis temperature can be the temperature of the heating block, the temperature of the sample, or the temperature of the reference, or it can be time. By convention, exothermic changes are plotted as positive, and the peaks point up, while endothermic changes are plotted as negative, and the peaks point down. (The same convention may be used for DSC; however, some instrument software uses the opposite convention. We will see examples of both conventions.) Some changes, such as the glass transition shown in Figure 17.15, do not result in a peak, but only a step change in the baseline. The reason will be discussed later in the chapter.

17.2.1 DTA Instrumentation

The equipment used in DTA studies is shown schematically in Figure 17.16. The sample is loaded into a crucible, which is then inserted into the sample well (marked S). A reference sample is made by placing a similar quantity of inert material (such as Al_2O_3) in a second crucible. This crucible is inserted in the reference well, marked R. The dimensions of the two crucibles and of the cell wells are as nearly identical as possible; furthermore, the masses of the sample and the reference should be virtually equal. The sample and reference should be matched thermally and arranged symmetrically within the furnace so that they are both heated or cooled in an identical manner. The metal block surrounding the wells acts as a heat sink. The temperature of the heat sink is slowly increased using an internal heater. The sink in

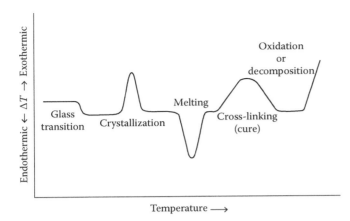

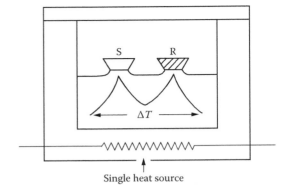

Figure 17.15 Hypothetical DTA thermal curve for a semicrystalline polymer with the ability to cross-link. The plot shows the baseline shift that occurs at the glass transition temperature, T_g, exothermic peaks for crystallization and cross-linking (or curing), an exothermic peak (off scale) for oxidative decomposition, and an endothermic peak for melting of the polymer. A similar thermal plot would be obtained by DSC analysis. (Courtesy of TA Instruments, New Castle, DE, www.tainstruments.com.)

Figure 17.16 Schematic DTA instrument. [© 1993–2014 PerkinElmer, Inc. All rights reserved. Printed with permission. (www.perkinelmer.com).]

turn simultaneously heats the sample and reference material. A pair of matched thermocouples is used. One pair is in contact with the sample or the sample container (as shown); the other pair is in contact with the reference. The output of the differential thermocouple, $T_s - T_r$ or ΔT, is amplified and sent to the data acquisition system. This allows the difference in temperature between the sample and the reference to be recorded as a function of the sample temperature, the reference temperature, or time. If there is no difference in temperature, no signal is generated, even though the actual temperatures of the sample and reference are both increasing. Operating temperatures for DTA instruments are generally room temperature to about 1600 °C, although one manufacturer makes a DTA capable of operating from −150 °C to 2400 °C. To reach the very low subambient temperatures, a liquid nitrogen cooling accessory is needed. Some low temperatures (but not −150 °C) may be reached with electrical cooling devices or with forced air cooling.

When a physical change takes place in the sample, heat is absorbed or generated. For example, when a metal carbonate decomposes, CO_2 is evolved. This is an endothermic reaction; heat is absorbed and the sample temperature decreases. The sample is now at a lower temperature than the reference. The temperature difference between the sample and reference generates a net signal, which is recorded. A typical example of a DTA thermal curve is shown in Figure 17.15. If, in the course of heating, the sample undergoes a phase transition or a reaction that results in the generation of heat, that is, an exothermic reaction, the sample becomes hotter than the reference material. In this case, the sample heats up to a temperature higher than that of the reference material until the reaction is completed. The sample then cools down or the temperature of the reference cell catches up until its temperature and that of the heat cell once again become equal. Such an effect is shown on the thermal curve as a peak that moves in a positive (upward) direction rather than in the negative direction, which allows us to distinguish between exothermic and endothermic reactions. The DTA experiment is performed under conditions of constant pressure (usually atmospheric pressure). Under constant pressure, the change in heat content of a sample (the change in *enthalpy*) is equal to the heat of reaction, ΔH. Any chemical or physical change that results in a change in ΔH gives a peak in the DTA thermal curve. There are some types of changes that do not result in a peak in the thermal curve but only a change in the baseline, as shown schematically in Figure 17.15. These types of changes do not undergo a change in ΔH, but a change in their *heat capacity*, C_p. Heat capacity is sometimes referred to as specific heat and is the amount of heat required to raise the temperature of a given amount of material by 1 K. If the amount of material is specified to be 1 mole, the heat capacity is therefore the molar heat capacity, with units of J/mol K. The most common process that gives rise to a change in baseline but not a peak in the DTA is a "glass transition" in amorphous materials such as polymers or glasses. The glass transition is discussed briefly under applications of DTA.

Modern DTA instruments have the ability to change atmospheres from inert to reactive gases, as is done in TGA. As is the case with TGA, the appearance of the DTA thermal curve depends on the particle size of the sample, sample packing, the heating rate, flow characteristics inside the furnace, and other factors. Thermal matching between the sample and the reference is often improved by diluting the sample with the inert reference, keeping the total masses in each crucible as close to each other as possible.

The peak area in a DTA thermal curve is related to the enthalpy change for the process generating the peak, so

DTA instruments must be calibrated for both temperature and for peak area. The National Institute for Standards and Technology (NIST), a US government agency, has certified high-purity metals like indium, tin, and lead with melting points known to six significant figures and enthalpies known to three or four significant figures and a series of high-purity salts with solid-state transition enthalpies that are accurately known. These materials can be used to calibrate both temperature and peak area.

Sample crucibles are generally metallic (Al, Pt) or ceramic (alumina, silica, zirconia) and may or may not have a lid. Many metal pans with lids have the lid crimped on using a special tool. Best results are obtained when the area of contact between the sample and the pan or crucible is maximized. Samples are generally in the 1–10 mg range for analytical applications. For applications where corrosive gases may evolve, boron nitride crucibles may be used. At very high temperatures, graphite and tungsten crucibles may be used but only under inert atmospheres. Tungsten may be needed in hydrogen atmospheres if the temperature exceeds 1000 °C, since Pt will react with hydrogen.

The peak area in DTA is related to the enthalpy change, ΔH; to the mass of sample used, m; and to a large number of factors like sample geometry and TC. These other factors result in the area, A, being related to the mass and ΔH by an empirically determined calibration constant, K:

$$A = K(m)(\Delta H) \quad (17.1)$$

Unfortunately, K is highly temperature dependent in the DTA experiment, so it is necessary to calibrate the peak area in the same temperature region as the peak of interest. This may require multiple calibration standards and can be time consuming. As we shall see, the calibration constant K for DSC is not temperature dependent; therefore, DTA is usually used for qualitative analysis, while DSC is used for quantitative measurements of ΔH and heat capacity.

17.2.2 Analytical Applications of DTA

DTA is based on changes of heat flow into the sample. Using DTA, we can detect the decomposition or volatilization of the sample, just as we can with TGA. In addition, however, physical changes that do not involve mass changes can be detected by DTA. Such changes include crystallization, melting, changes in solid crystal phases, and homogeneous reactions in the solid state. In each of these changes, there is a flow of heat between the sample and its surroundings caused by endothermic or exothermic transitions or by changes in the heat capacity. The main use of DTA is to detect thermal processes and characterize them as exothermic or endothermic, reversible or irreversible, but only qualitatively. DTA thermal curves can be used to determine the order of a reaction (kinetics) and can provide the information required to construct phase diagrams for materials. DTA can be used for the characterization of engineering materials; for the determination of the structural and chemical changes occurring during sintering, fusing, and heat treatments of alloys to change microstructure; for the identification of different types of synthetic rubbers; and for the determination of structural changes in polymers.

An instance of the use of transitions where no change in mass occurs is the DTA characterization of polymers. The physical properties of a polymer, such as strength, flexibility, and solubility, depend (among other things) on its degree of crystallinity. Crystalline materials are those materials that exhibit a high degree of long-range and short-range order in the arrangement of their molecules or atoms. No polymers are 100 % crystalline, but some polymers can partially crystallize; these are called **semicrystalline** polymers. A polymer is a gigantic organic molecule, or **macromolecule**, with a high relative molecular mass (RMM), typically 5,000–40,000 g/mol. Polymer molecules generally exist as long and flexible chains. The chains are capable of bending and twisting, as shown in Figure 17.17a. A bulk polymer consists of large numbers of these chains intertwined like a bowl of cooked spaghetti noodles or rumpled pieces of string. A polymer in this state, with no short-range or long-range order, is said to be **amorphous**. With some types of polymers and proper treatment, such as slow cooling from the molten state, some of these long chains can form crystal-like zones that are regularly oriented (Figure 17.17b). The DTA curves of the two samples in Figure 17.17 would appear as in Figure 17.18 for samples that had been rapidly cooled (quenched) from the liquid state. At the glass transition temperature, T_g, an amorphous material goes from a glassy, rigid state to a rubbery, flexible state as the temperature now permits large-scale molecular motion. Only amorphous materials or the amorphous regions of a semicrystalline material can exhibit a glass transition. On the molecular level, it is the temperature at which segments of 35 – 50 "backbone" atoms in the material can move in a concerted motion. At the glass transition, there is a change in the heat capacity, C_p, of the polymer, which is seen as a step change in the baseline of the thermal curve. There is no change in the enthalpy at T_g, so no peak occurs in the thermal curve; the glass transition is *not* melting. There is also a change in the rate of volume expansion of the polymer at T_g, which can be measured by TMA. On further heating, crystallization of the semicrystalline polymer occurs as the viscosity drops, molecular mobility increases, and the chains align themselves into ordered regions, resulting in an exothermic peak. The amorphous polymer shows no such exothermic peak. Additional heating will generally result in the melting of semicrystalline polymer, seen as an endothermic peak in the thermal curve. None of these transitions involves a change in mass of the sample, so none of these changes would be seen in a TGA thermal curve. Amorphous polymers do not melt but may show an endothermic decomposition peak on continued

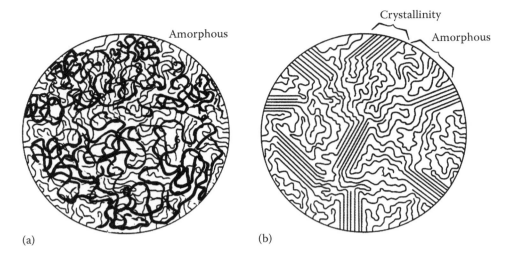

Figure 17.17 (a) Amorphous polymer structure. (b) Semicrystalline polymer structure with aligned chains in the crystalline regions, and random structure in the amorphous region.

heating. This decomposition would be seen by both DTA and TGA, since decomposition does involve mass loss. The degree of crystallinity in the semicrystalline polymer can be calculated from the heat of fusion (the area under the crystallization exotherm), but quantitative measurements of this sort are usually performed by DSC because DSC is more accurate than DTA for quantitative analysis.

Qualitative identification of materials is done by comparing the DTA of the sample to DTA thermal curves of known materials. DTA thermal curves serve as fingerprints for materials.

Theoretically, the area under each DTA peak should be proportional to the enthalpy change for the process that gave rise to the peak. Unfortunately, in a traditional simple DTA, there are many factors that are not compensated for. For example, DTA response generally decreases as the temperature increases. This makes the area under the peaks unreliable for enthalpy measurements unless the DTA has been calibrated in the temperature range of interest. Semiquantitative results for enthalpies can be obtained, but for quantitative enthalpy measurements, we turn to the DSC.

17.3 DIFFERENTIAL SCANNING CALORIMETRY

Calorimetry is the measurement of the heat of a process. The first calorimeter was developed by Lavoisier and Laplace in 1782.

In DSC, differences in heat flow into a reference and sample are measured versus the temperature of the sample. The difference in heat flow is a difference in energy; DSC is a calorimetric technique and results in more accurate measurement of changes in enthalpy and heat capacity than those obtained by DTA.

The heat flow into or out of the sample, in the absence of any transition, is

$$\frac{dQ}{dt} = \frac{dQ}{dT} \times \frac{dT}{dt} \qquad (17.2)$$

where
dQ/dt is the heat flow
dQ/dT is the heat capacity
dT/dt is the heating rate

When a transition occurs, either the heat capacity changes or some kinetic event changes the heat flow:

$$\frac{dQ}{dt} = \frac{dQ}{dT} \times \frac{dT}{dt} + f(T,t) \qquad (17.3)$$

where the first three terms are as described previously and $f(T, t)$ is the kinetic events term.

17.3.1 DSC Instrumentation

The DSC measurement requires a sample and a reference, as does DTA. Modern DSC sample and reference pans are small and usually made of aluminum, although Alodine-coated aluminum, gold, copper, platinum, graphite, and stainless steel pans are available. They may or may not have lids, and the lids may be crimped on or hermetically sealed. Crimping or hermetic sealing of pans generally requires a press for reproducible sample preparation. Sample size is generally 1–10 mg. Often the reference pan is left empty, but an inert reference material such as is used in DTA may be used. Commercial DSC equipment can operate at temperatures from −180 °C to 725 °C, with specialized instruments capable

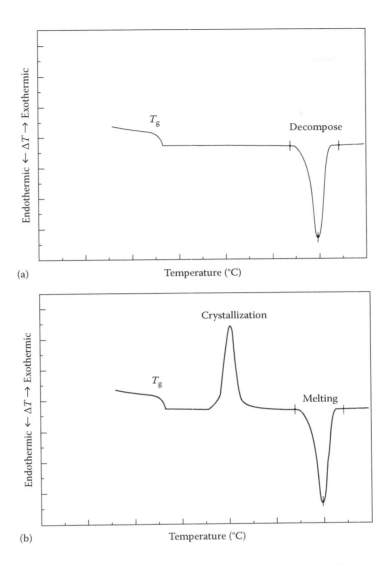

Figure 17.18 Schematic DTA thermal curves for (a) the totally amorphous polymer structure and (b) the semicrystalline polymer structure shown in Figure 17.17. Both (a) and (b) show T_g; only the semicrystalline polymer has a crystallization exotherm.

of maximum temperatures of 2000 °C. The exact temperature range depends on the model of the instrument, and most commercial thermal instrument companies offer a variety of operating ranges. The DSC must be able to be heated and cooled in a controlled manner. To achieve the very low end of the temperature range, a special liquid nitrogen cooling accessory is needed; for other cooling applications, electrical cooling or forced air cooling is used. Modern DSC instruments are available with automatic intelligent sample changers that permit the unattended analysis of as many as 50 samples or more in any order specified by the analyst.

There are two main types of DSC instrumentation, heat-flux DSC and power-compensated DSC. A schematic of a commercial **heat-flux DSC** is presented in Figure 17.19. In a heat-flux instrument, the same furnace heats both the sample and the reference. In heat-flux DSC, the temperature is changed in a linear manner, while the differential heat flow into the sample and reference is measured. The sample and reference pans sit on the heated thermoelectric disk, made of a Cu/Ni alloy (constantan). The differential heat flow to the sample and reference is monitored by area thermocouples attached to the bottom of the sample and reference positions on the thermoelectric disk. The differential heat flow into the pans is directly proportional to the difference in the thermocouple signals. The sample temperature is measured by the alumel/chromel thermocouple under the sample position. This temperature is an estimated sample temperature because the thermocouple is not inserted into the sample itself. The accuracy of this temperature will depend on the TC of the sample and its container, the heating rate, and other factors. As shown in Figure 17.19, the sample and reference pans both have lids and the reference pan is an empty pan. A schematic of a **power-compensated DSC** is presented in Figure 17.20. The major difference in power-compensated

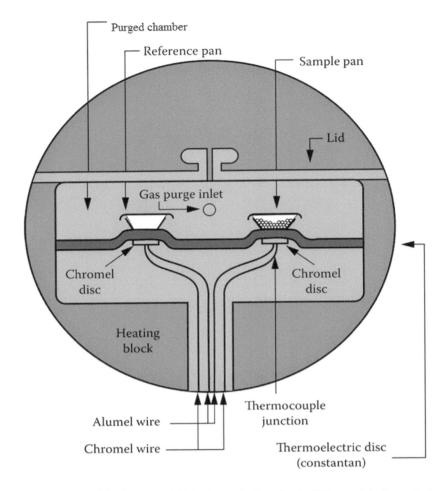

Figure 17.19 Schematic of a heat-flux DSC. (Courtesy of TA Instruments, New Castle, DE, www.tainstruments.com.)

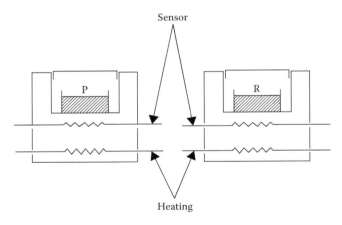

Figure 17.20 Schematic of a power-compensated DSC. [© 1993–2014 PerkinElmer, Inc. All rights reserved. Printed with permission. (www.perkinelmer.com).]

DSC instruments is that two separate heating elements are used for the sample (marked P in Figure 17.20) and the reference. A change in temperature between the sample and the reference serves as the signal to "turn on" one of the heaters so that the sample and the reference stay at the same temperature. When a phase change, reaction, glass transition, or similar event occurs in the sample, the sample and reference temperatures become different. This causes extra power to be directed to the cell at the lower temperature in order to heat it. In this manner, the temperatures of the reference and sample cells are kept virtually equal ($\Delta T = 0$) throughout the experiment. The power and temperature are measured accurately and continuously. The temperatures of the sample and reference are measured using Pt resistance sensors, shown in Figure 17.20. The difference in power input is plotted versus the average temperature of the sample and reference. Power compensation provides high calorimetric accuracy, high precision, and high sensitivity. This permits analysis of very small samples as demonstrated in Figure 17.21, showing the determination of the heat of fusion of a 6 μg sample of indium metal. Note that this figure is plotted opposite to the convention normally used for endothermic peaks. In theory, since electrical power is the quantity being measured, no calibration of the output should be necessary. In practice, these instruments are calibrated using standards to obtain the most accurate results.

THERMAL ANALYSIS

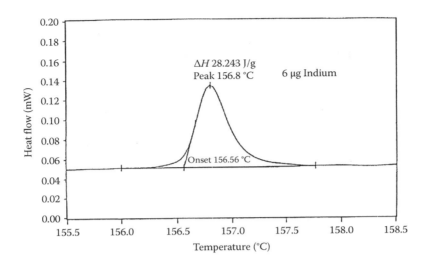

Figure 17.21 DSC thermal curve showing the determination of the enthalpy of fusion of 6 μg of indium metal. [© 1993–2014 PerkinElmer, Inc. All rights reserved. Printed with permission. (www.perkinelmer.com).]

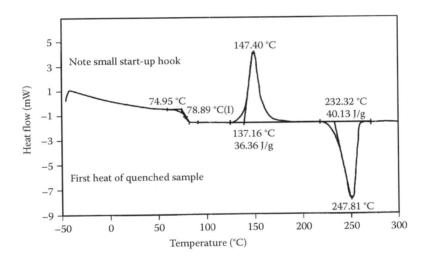

Figure 17.22 DSC thermal curve for a sample of PET. T_g is observed at 78.9 °C. Crystallization begins at 137 °C and the area under the exothermic peak is equivalent to 36.36 J/g PET. Melting begins at about 323 °C; the area under the endothermic peak is equivalent to 40.13 J/g PET. (Courtesy of TA Instruments, New Castle, DE, www.tainstruments.com.)

The DSC peak area must be calibrated for enthalpy measurements. The same types of high-purity metals and salts from NIST discussed for the calibration of DTA equipment are also used to calibrate DSC instruments. As an example, NIST standard reference material (SRM) 2232 is a 1 g piece of high-purity indium metal for the calibration of DSC and DTA equipment. The indium SRM is certified to have a temperature of fusion equal to 156.5985 °C ± 0.00034 °C and a certified enthalpy of fusion equal to 28.51 ± 0.19 J/g. NIST offers a range of similar standards. These materials and their certified values can be found on the NIST website at www.nist.gov. Government standards organizations in other countries offer similar reference materials.

The resultant thermal curve is similar in appearance to a DTA thermal curve, but the peak areas are accurate measures of the enthalpy changes. Differences in heat capacity can also be accurately measured and are observed as shifts in the baseline before and after an endothermic or exothermic event or as isolated baseline shifts due to a glass transition. Because DSC provides accurate quantitative analytical results, it is now the most used of the thermal analysis techniques. A typical DSC thermal curve for polyethylene terephthalate (PET), the polymer used in many soft drink bottles, is shown in Figure 17.22.

A third type of differential calorimeter design is the Tian–Calvet system first introduced commercially in the 1960s by Setaram Instrumentation (www.setaram.com).

The Calvet-type DSC detector is a 3D detector in a heat-flux system. Instead of two thermocouples, the sample and reference chamber are completely surrounded by an array of thermocouples that captures and measures almost all heat flow from the sample (94 % ± 1 %). A variety of Calvet detectors are available, with from 10 to 120 thermocouples surrounding the sample and reference. A 3D thermopile detector using Peltier elements is available for micro-DSC (μDSC). The advantages of the Calvet-type detector over 2D plate-type DSC detectors (Figure 17.23) are significant. The Calvet design allows the use of more sample mass with corresponding increased sensitivity.

Heat capacity, C_p, is measured using the following equation:

$$A_{\mu V} = S_{\frac{\mu V}{mV}} \times m_{kg} \times C_p \times V_{°C/min} \quad (17.4)$$

where
A is the measured signal
S is the sensitivity
m is the sample mass
V is the heating rate

By increasing both sample mass and sensitivity, the signal is increased over traditional DSC sensors. Accuracy and precision of the C_p determination are 1 % – 3 % versus 7 % – 15 % for traditional DSC.

The plate-type sensor measures heat only through the bottom of the sample, with efficiencies of 20 % – 50 %, depending on the sensor thickness and the conductivities of the crucible, sample, and purge gas. The Calvet detector is 93 % – 95 % efficient and is independent of the variables just listed, resulting in more accurate quantitative measurement. Calibration of the sensor and measurements are independent of sample shape or type, contact between sample and crucible, pressure, or flow rate. Larger sample size allows more accurate measurement on heterogeneous samples.

17.3.2 Applications of DSC

DSC is used to study all of the types of reactions and transitions that can be studied using DTA, with the added advantage of accurate quantitative measurements of ΔH and ΔC_p. Polymer chemists use DSC extensively to study percent crystallinity, crystallization rate, polymerization reaction kinetics, polymer degradation, the effect of composition on the glass transition temperature, heat capacity determinations, and characterization of polymer blends. Materials scientists, physical chemists, and analytical chemists use DSC to study corrosion, oxidation, reduction, phase changes, catalysts, surface reactions, chemical adsorption and desorption (chemisorption), physical adsorption and desorption (physisorption), and fundamental physical properties such as enthalpy, boiling point, and equilibrium vapor pressure. DSC instruments permit the purge gas to be changed automatically, so sample interactions with reactive gas atmospheres can be studied. Food chemistry, pharmaceuticals, and biological materials can be studied and some examples will be discussed. Samples can be solids, liquids, creams, gels, emulsions, and so on. Homogeneity of the sample is critical for accurate results.

For example, from Figure 17.22, the heat of crystallization is measured to be 36.36 J/g and the heat of melting (or fusion) is measured to be 40.13 J/g by integration of the respective peaks by the instrument data analysis software. For this polymer sample, the measured heat of crystallization is slightly lower than the measured heat of melting, indicating that the polymer was partly crystalline at the start of the experiment. The T_g and the specific heat can also be accurately measured from this thermal curve.

The **heat of fusion** is a useful measure of the **percent crystallinity** of polymers. Table 17.4 presents heat of fusion data and the calculated percent crystallinity for three different samples of polyethylene. The percent crystallinity is calculated from calibration with a sample of known crystallinity. The first two samples in the table are very similar in percent crystallinity and in their melting behavior; the third sample, however, has a lower percent crystallinity and a sharper melting profile (compare the melt onset and melt peak temperature columns). This tells the polymer chemist that sample 3 has been processed differently and will have different physical and mechanical properties than samples 1 and 2. The sharp melting of sample 3 is shown in Figure 17.24.

Food chemistry can be studied by DSC. Carrageenans are extracted from algae and are used in food processing for their gelation properties, forming rigid, high-strength gels used in dessert gels, pet foods, and other foods. DSC is used to study the order–disorder transitions on heating, the formation and melting behavior of these gels. The melting and solidification of cheese is studied by DSC to determine optimal thermal processing conditions and optimum storage temperatures. Dry milk powder and whey powder contain amorphous lactose. Storage of these products under high relative humidity conditions results in an unusable dense powder due to the formation of α-lactose monohydrate. The measurement of α-lactose in dry milk and whey powder is detected by DSC by measuring the dehydration of α-lactose monohydrate. Similarly, the polymorphic forms of cocoa butter, one of the main ingredients in chocolate, have been studied by DSC to determine what occurs during chocolate processing. Only one polymorph results in chocolate that melts near body temperature. The others melt too easily or at too high a temperature (see Olivon et al. 1998).

Biological and pharmaceutical applications for DSC include the study of protein denaturation, enzymatic

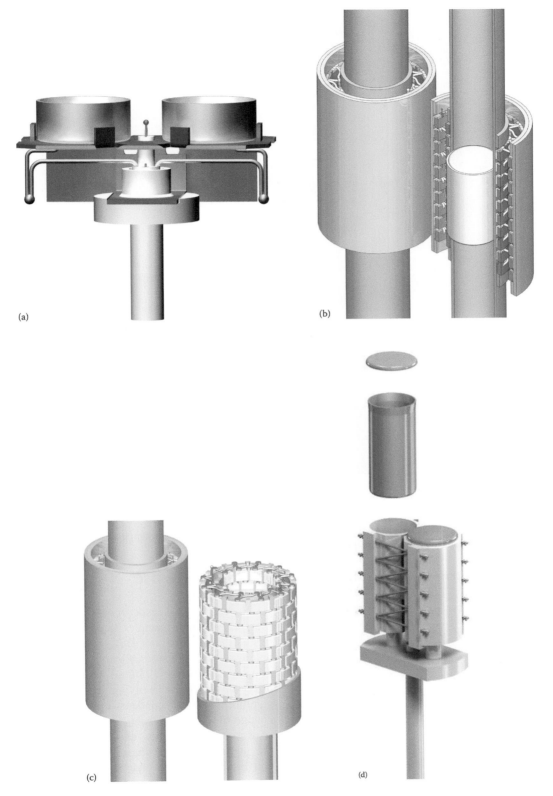

Figure 17.23 A 2D plate detector versus 3D Calvet detector designs. (a) A 2D plate detector with thermocouples under the two pans. Only the heat flow from the bottom of the pans is measured. (b) and (c) 3D Calvet calorimetric sensors with arrays of thermocouples totally surrounding both the sample and reference chamber. Approximately 94 % of the heat flow is captured and measured using these sensors. (d) A Calvet detector with 10 thermocouples for the sample and 10 for the reference. (Courtesy of Setaram Instrumentation, SA, Caluire, France, www.setaram.com. Used with permission.)

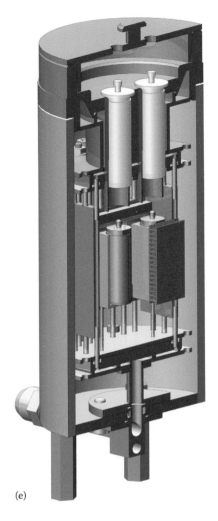
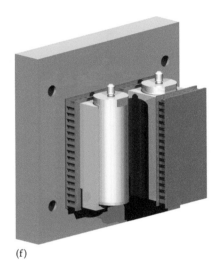

(e) (f)

Figure 17.23 (Continued) A 2D plate detector versus 3D Calvet detector designs. (e) A 3D heat flow detector using Peltier elements for the Setaram μDSC7. (f) Detail of the Peltier thermopile. (Courtesy of Setaram Instrumentation, SA, Caluire, France, www.setaram.com. Used with permission.)

Table 17.4 DSC Results for Three Samples of Polyethylene

Sample	Melt Onset (°C)	Melt Peak (°C)	Enthalpy (J/g)	% Crystallinity
1	121.9	132.9	195.9	67.6
2	121.3	132.6	194.5	67.1
3	122.3	131.6	180.1	62.1

Source: TA Instruments Applications Brief TA-123, courtesy of TA Instruments, Inc., New Castle, DE.

reactions, phase changes in lipid bilayers, and nucleic acid studies. DSC is used in the pharmaceutical and personal care industries to study phase changes in creams, gels, and emulsions. Excipient compatibility is easily studied for drug formulation. For example, aspirin mixed with magnesium stearate exhibits a large endothermic reaction, while aspirin mixed with lactose shows no reaction; magnesium stearate would not make a good excipient for an aspirin formulation.

17.3.2.1 Pressure DSC

Normally, DSC experiments are run at atmospheric pressure. Running samples under different pressures can extend the usefulness of DSC. DSC instruments are available with high-pressure cells capable of operating at up to 500 bar (50 MPa or 493 atm). Changing the pressure will affect any reactions involving gases, while not having any significant effect on condensed phases. Boiling is the phase change of a material from liquid to gas. The boiling point of a material increases as the pressure increases, while the melting point (a phase change involving only solid and liquid) does not change significantly with pressure. Therefore, changing the pressure facilitates interpretation of peaks in a DSC thermal curve. If the peak shifts in temperature as a function of pressure, a gas is involved in the reaction that gave rise to the peak. Because the pressure affects the boiling point of a liquid, a series of experiments in a pressure DSC cell at known pressures yields boiling

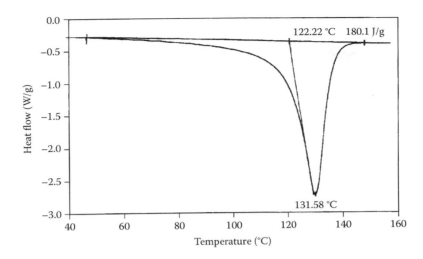

Figure 17.24 Melting point of a polyethylene sample by DSC. (Courtesy of TA Instruments, New Castle, DE, from *Thermal Analysis Applications Brief TA-123*.)

point shifts. These shifts can be used to obtain quantitative vapor pressure information on liquids.

Increasing the pressure will often increase the rate of a reaction involving gases. Adsorption of hydrogen is often used as a means of characterizing precious metal catalysts, such as the platinum catalysts used in automobile catalytic converters and the Pt and Pd catalysts used in large-scale organic chemical synthesis. A typical adsorption study may require large samples and take 6 h or more. The same type of study can be run in a pressure DSC cell in ~15 min. The high pressure of hydrogen accelerates the reaction. The adsorption and desorption of hydrogen on palladium as studied by DSC is shown in Figure 17.25. Many sample–atmosphere reactions such as oxidation of oils and greases can be accelerated by increasing the pressure of air or oxygen and can also be studied effectively by pressure DSC. Other pressure-sensitive reactions, such as volatilization, are slowed down by increased pressure. Suppression of these reactions can be a valuable tool when pressure DSC is used to study a complex system.

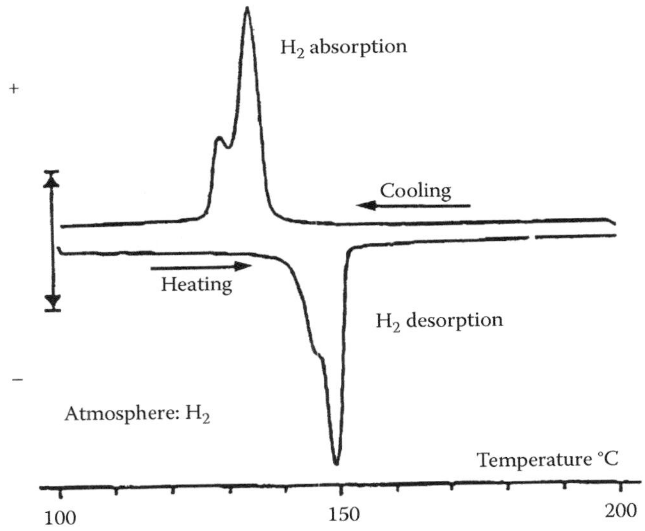

Figure 17.25 DSC thermal curve of adsorption and desorption of hydrogen from a precious metal catalyst under constant pressure.

17.3.2.2 Modulated DSC

Modulated DSC® (MDSC®) is a patented technique from TA Instruments, New Castle, DE. In MDSC, a controlled, single-frequency sinusoidal temperature oscillation is overlaid on the linear temperature ramp. This produces a corresponding oscillatory heat flow (i.e., rate of heat transfer) proportional to physical properties of the sample. Deconvolution of the oscillatory temperature and heat flow lead to the separation of the overall heat flow into heat capacity and kinetic components, called reversing and nonreversing heat flow. The overall heat flow (dQ/dt) may be described as the sum of a heat capacity term, $C_p\beta$, which is heating rate dependent, with β representing the heating rate and a kinetic term, $f(T, t)$, which is time and temperature dependent (Mohomed and Bohnsack 2013; Kraftmakher 2004).

MDSC enables the separation of the total measured heat flow into two constituents, providing increased understanding of simultaneously occurring phenomena in a sample. For example, in a polymer-like semicrystalline PET, MDSC can separate the glass transition and melting from the kinetic processes of enthalpic recovery (a measure of physical aging in polymers), cold crystallization, and crystal perfection (the process of the least perfect crystals melting below the thermodynamic melting point and then crystallizing and remelting one or more times as the temperature increases). These pieces of information cannot all be obtained by conventional DSC since several of these processes overlap. All of the processes are critical to understanding failure in polymeric materials. In pharmaceutical studies, recrystallization of a drug may be seen in the nonreversing signal only, while T_g and melting are seen in both signals.

A disadvantage to the method is that calibration is required for temperature, enthalpy, and the heat capacity, as opposed to temperature and enthalpy only for standard DSC.

17.4 HYPHENATED TECHNIQUES

17.4.1 Hyphenated Thermal Methods

While TGA provides useful data when a mass change is involved in a reaction, we have seen that many reactions do not have a change in mass associated with them. The use of both TGA and DTA or TGA and DSC provides much more information about a sample than either technique alone provides. There are several commercial thermal analysis instrument manufacturers who offer simultaneous combination systems, often called simultaneous thermal analysis (STA) systems. Simultaneous TGA-DTA and simultaneous TGA-DSC instruments are available. Instrument combinations cover a wide temperature range and come in both "analytical sample" size (1–20 mg) and high-capacity sample size. A schematic of a simultaneous TGA-DTA instrument is presented in Figure 17.26, showing the dual sample and reference pans. The instrument monitors both the mass change and the temperature difference between the sample and reference and plots both the TGA and DTA thermal curves simultaneously. A hypothetical sample might have plots like those in Figure 17.27, where two of the DTA peaks are clearly associated with the mass losses in the TGA and there are two other "events" with enthalpy changes but no mass changes.

17.4.2 Evolved Gas Analysis

The evolution of gas from a thermal analyzer such as a TGA, DTA, or DSC may be determined using *evolved gas detection* (EGD) or, if qualitative or quantitative analysis of the gas is required, EGA. These techniques are essentially a combination of thermal analysis and MS, tandem MS (MS^n), GC-MS, or other species-selective detectors, such as Fourier transform IR (FTIR). The evolved gases from the furnace are carried through a heated transfer line to the mass spectrometer or FTIR or other gas analyzer. This permits real-time identification of the gases given off by the sample during the thermal program. Figure 17.28a presents a schematic interface for capturing the evolved gases from

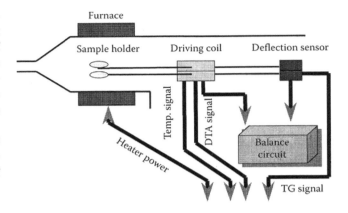

Figure 17.26 A schematic simultaneous TGA-DTA instrument. A mass change on the sample side displaces the beam and a drive coil current returns the beam to zero. The current is proportional to the mass change and serves as the TGA signal. The temperature change between the sample and reference pans is measured by thermocouples attached to the pans. The temperature differential is the DTA signal.

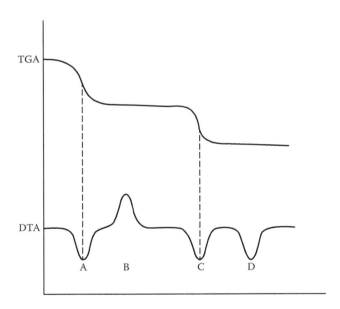

Figure 17.27 The TGA and DTA curves of a hypothetical sample. The two steps in the TGA curve result from mass losses. They correspond to the endothermic peaks A and C in the DTA plot. The DTA thermal curve also shows an exothermic event, B, and an endothermic event, D, that do not involve a change in mass.

a TGA and sending evolved gas to an FTIR or a GC-MS. This particular interface is used with a high-capacity TGA (up to 100 g sample size), so whole electronic components and fabricated parts can be analyzed. This can make failure analysis and production control very simple.

The alternative to real-time detection is to have a system for trapping the evolved gases and storing the trapped gases for analysis at a later time. Sorbents for collecting the

THERMAL ANALYSIS

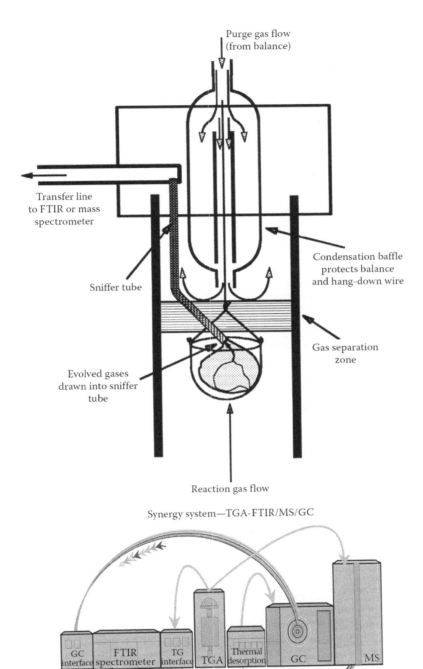

Figure 17.28 (a) Interface for EGA. The sniffer tube sits directly above the sample pan and draws evolved gases to an FTIR or a GC-MS with minimal dilution. The interface is shown in a high-capacity TGA. This system is no longer commercially available.

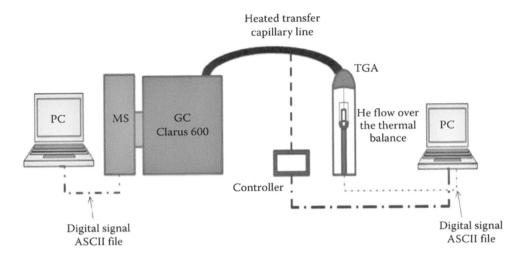

Figure 17.28 (Continued) (b) Schematic of a TGA coupled to a GC-MS through a heated transfer capillary line. (b: [© 1993–2014 PerkinElmer, Inc. All rights reserved. Printed with permission. (www.perkinelmer.com).].)

evolved gases can be liquid or solid. The sorbents can be selective for a specific gas or a class of gaseous compounds. If no adsorbent material is desired, cryogenic trapping may be used.

The most common instruments interfaced to thermal analyzers are FTIR instruments, discussed in Chapter 5, and GC-MS instruments (Figure 17.28b), discussed in Chapters 10, 11, and 13. FTIR is less sensitive and less versatile than MS but is simpler and cheaper. For hyphenated thermal analysis-FTIR, a heated transfer line from the thermal analyzer is normally connected to a heated FTIR gas cell, as shown in Figure 17.29. The interface is relatively simple, because FTIR normally operates at ambient pressure as does the thermal analyzer. IR spectra are simple to interpret, and reference libraries of gas-phase IR spectra are available for the common gases and volatile organic compounds. This makes positive identification of the evolved gas straightforward if the gas is one of the gases that often accompany the decomposition of a material, such as water vapor, CO, or HCl, for example. The decomposition of PVC polymer can be studied by TGA-FTIR. Looking back at Figure 17.14, we have the TGA thermal curve for PVC heated in nitrogen atmosphere. If the TGA is connected to an FTIR, the spectrum corresponding to the gas evolved at the first mass loss is the gas-phase spectrum of HCl (Figure 17.30), indicating that the first step in the decomposition of PVC in an inert atmosphere is loss of HCl from the polymer.

Disadvantages to the use of a separate transfer line and separate FTIR include the need for a separate heater for the transfer line, hot or cold spots in an improperly heated line, leading to the possibility of contamination due to buildup of condensable gases in the transfer line. In addition, more linear bench space is required for side-by-side instruments. New instruments are now available with no separate transfer line between the thermal analysis system and the FTIR spectrometer. In the Netzsch *Perseus* TGA-DSC-FTIR system, a Bruker Optics ALPHA FTIR spectrometer is connected directly to the thermal analyzer in a vertical arrangement (Figure 17.29b). The built-in heated gas cell of the spectrometer is directly connected to the gas outlet of the furnace. No separate heating controller is needed. The low volume of the short gas path results in rapid response and is especially useful where condensable evolved gases are present.

MS has more analytical flexibility than FTIR, but interfacing a thermal analyzer is more difficult because of the low operating pressure required for MS. MS instruments typically operate at approximately 10^{-5} torr, while thermal analyzers are usually at atmospheric pressure. One approach is to evacuate the thermal analyzer, but the common method used is a differential pumping system such as that used for GC-MS. This reduces the pressure from the thermal analyzer in several stages prior to allowing the gas flow into the mass analyzer. A commercial interface for a thermal analyzer-MS system, shown in Figure 17.31, uses a supersonic jet to skim analyte molecules into the MS, in a manner similar to the jet separators used in GC-MS. Jet separator operation is discussed in Chapter 13 under hyphenated techniques.

EGA is used for materials characterization, polymer analysis, characterization of oil shale, oxidation and reduction studies, evaluation of catalysts, and many other applications.

In the environment, persistent organic pollutants such as polyaromatic hydrocarbons have been shown by TGA-GC-MS to exhibit a six- to tenfold increased partitioning to the water phase in an octanol–water solution in the presence of TiO_2 nanoparticles, suggesting that the increasing use of engineered nanoparticles may influence the transport of pollutants and other chemicals in the environment (Sahle-Demessie et al.).

A slightly different approach to hyphenation of instruments is simultaneous analysis, where two or more

THERMAL ANALYSIS

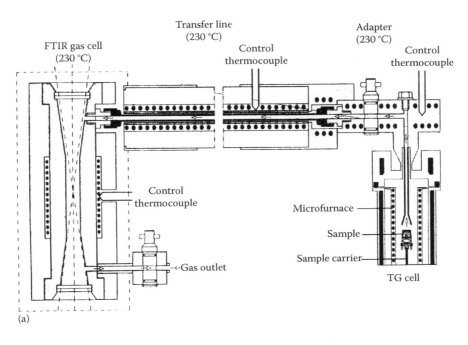

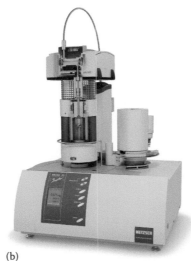

Figure 17.29 (a) Hyphenated TGA-FTIR, showing the heated transfer line and heated IR gas cell. (Courtesy of Netzsch Instruments, Inc., Burlington, MA, www.netzsch-thermal-analysis.com.) (b) The *Perseus* TGA-DSC-FTIR system with a vertical arrangement of the FTIR and thermal analysis system, with no external transfer line. (Courtesy of Netzsch Instruments, Inc., Burlington, MA, www.netzsch-thermal-analysis.com.)

methods are applied to the same sample. An example of this approach is the commercially available DSC–Raman system from PerkinElmer, Inc. (Menard et al. 2010). In this system, a three-axis positioning adapter with optics allows accurate placement of the Raman probe over the sample in the DSC. Raman spectra can be collected as the DSC experiment progresses. The authors give an example of the ability to positively identify three solid polymorphs of acetaminophen. The combination of Raman and DSC gives the precise temperatures at which the solid-state conversions from one polymorph to the next occur. The DSC provides the temperature data and the information about exothermic and endothermic changes in the material, while the polymorphs are identified from their Raman spectra. Raman spectroscopy is discussed in Chapter 5.

17.5 THERMOMETRIC TITRIMETRY

TT depends on measuring the heat generated during a chemical reaction; therefore, it is another **calorimetric**

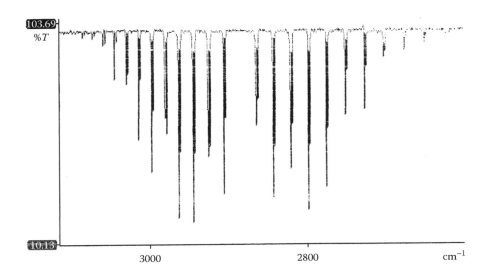

Figure 17.30 Gas-phase FTIR spectrum of HCl.

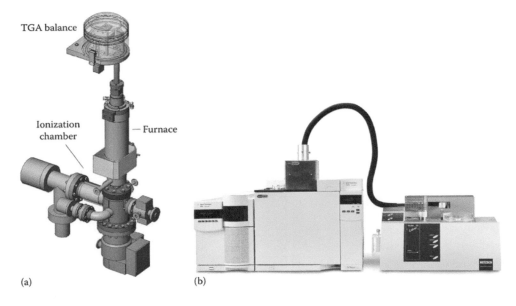

Figure 17.31 (a) Commercial "supersonic" interface between a thermal analyzer and a mass spectrometer. The mass spectrum of evolved gases (up to RMMs of 1024 Da) can be obtained to provide identification of the structure with no condensation during transfer between the thermal analyzer and the mass spectrometer. (Courtesy of Setaram Instrumentation, SA, Caluire, France, www.setaram.com.) (b) A different commercial thermogravimetry (TG)-GC-MS system, showing the Netzsch TG 209 on the right, connected to a GC-MS via an external transfer line. A photograph of the TG 209 was provided in Figure 17.4c. (Courtesy of Netzsch Instruments, Inc., Burlington, MA, www.netzsch-thermal-analysis.com.)

technique. Usually the reactions take place at room temperature under conditions such that no heat enters or leaves the titration cell except that brought in by introduction of the titrant. In this procedure, a titrant of known concentration is added to a known volume of sample. The titration reaction follows the generalized chemical equation

$$A\ (\text{sample}) + B\ (\text{titrant}) \rightarrow C\ (\text{products}) + \text{heat}$$

For any particular reaction, a mole of A titrated with a mole of B will generate a fixed quantity of heat, which is the heat of reaction. If there is half a mole of A present, half as much heat is generated, even if there is an excess of B. Heat is generated only while A and B react with each other; an excess of either one does not cause generation of heat. The reaction is usually set up in an insulated beaker, a Dewar flask, or a Christiansen adiabatic cell. This ensures a minimum of heat loss from the system during titration.

THERMAL ANALYSIS

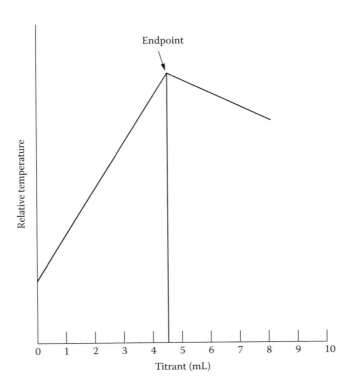

Figure 17.32 Representative TT plot. The endpoint is denoted by an abrupt change in the slope of the curve. The TT plot for HCl (a strong acid) with NaOH would look virtually identical to the TT plot for boric acid (a very weak acid) with NaOH because the ΔH of reaction is almost the same for both acids.

In practice, the temperature of the system is measured as the titrant is added to the sample. A TT plot generally has ΔT on the y-axis and volume of titrant on the x-axis. A typical plot for an exothermic reaction is shown in Figure 17.32. We will assume that the sample solution and the titrant are initially at room temperature. As titrant is added and the exothermic reaction occurs, the temperature of the solution increases. It can be seen that the temperature of the mixture increases until approximately 4.5 mL of titrant has been added to the sample. This is the equivalence point of the titration. Further addition of titrant causes no further reaction, because the entire sample A has been consumed. The temperature of the mixture steadily decreases as the cooler titrant is added to the warm mixture. The endpoint can therefore be determined as the abrupt change in slope that occurs at the equivalence point. In these circumstances, it is not necessary to add exactly the stoichiometric amount of titrant. Any excess added does not result in an increase in temperature and is therefore not a source of error as it is in conventional volumetric analysis.

For simple titrations, the equipment necessary is not particularly sophisticated; a thermometer, a burette, and an insulated beaker (or even two Styrofoam cups, one sitting in the other) will do. However, if one wants to measure specific heats or the heat of reaction, better control is required. Modern automated thermometric titrators consist of a constant delivery pump for the titrant, a temperature control system for the titrant, an insulated cell, calibration circuitry, electronic temperature sensing, and a data processing system. Most modern instruments are totally computerized, so different methods can be programmed and run unattended.

The titrant is delivered with a motorized syringe pump. This permits the volume of titrant to be calculated from the rate of delivery. These pumps are able to deliver rates of flow down to microliters per minute with high precision. The temperature control system, usually a thermostated bath for the titrant and a heater for the sample cell, is used to bring the titrant and the sample to exactly the same temperature. This is required for high-precision measurements of heat capacity and enthalpy. Modern thermostats can maintain the temperature to within 0.001 °C.

For precise measurements of reaction parameters such as rate constants, equilibrium constants, and enthalpies, a well-designed insulated titration cell is required. The cell should have minimum heat loss to the surroundings. It should have as small a mass as possible to minimize the contribution of the cell to the total heat capacity and to maximize the response of the system to temperature changes. The calibration circuitry is used to determine the heat capacity of the system, which must be known if accurate heats of reaction are to be measured. Thermistors, which are temperature-sensitive semiconductors, are used as the temperature sensors in these systems due to their small size, fast response, and sensitivity.

17.5.1 Applications of Thermometric Titrimetry

A major use of thermometric titrimetry is for the titration of very weak acids or bases. The pH for the titration of a weak acid with a strong base (or vice versa) is easily calculated and such calculations can be found in standard analytical chemistry texts. Strong acids such as HCl, when titrated with strong base, give a large change in pH as the equivalence point is reached; therefore, the quantitative determination of a strong acid using a potentiometric titration with a glass pH electrode is very simple. As the acid becomes weaker, the change in pH near the equivalence point decreases until the equivalence point becomes too shallow to detect by potentiometric titration. For example, the weak acid boric acid, with $pK_1 = 9.24$ and $pK_2 = 12.7$, does not give a good endpoint in a potentiometric titration with NaOH because the inflection is too small. However, in a TT with NaOH, boric acid gives a sharply defined endpoint and is easily determined, because the ΔH for the boric acid neutralization is large. It is very similar in magnitude to that for neutralization of HCl by NaOH. TTs can be used for quantitative analysis using neutralization reactions, oxidation–reduction reactions, and complexation reactions, as well as to determine heats of mixing and to determine equilibrium constants.

In TT, it is assumed that the rate of reaction is relatively fast and that the endpoint will occur as soon as a small excess of titrant is added. This assumption is not valid if the reaction is slow. The position of the endpoint will be distorted and the results will be inaccurate.

17.6 DIRECT INJECTION ENTHALPIMETRY

When a chemical reaction occurs at constant pressure, heat is liberated or absorbed. The heat flow into or out of a system at constant pressure is quantified using a quantity called enthalpy, H. We usually measure the change in enthalpy, ΔH, called the enthalpy of reaction. It is a reproducible physical property for a given reaction $A + B \rightarrow C + \Delta H$. Therefore, the magnitude of ΔH depends on the quantity of reactants involved in the reactions (e.g., multiplying all of the reaction coefficients by 2 means, we have to multiply ΔH by 2). Excess amounts of any of the reactants do not take part in the generation or absorption of heat.

DIE is similar in many respects to thermometric titrimetry. One essential difference is that an excess of titrant is added very rapidly to the sample and the reactants mixed vigorously. The temperature is then measured against time following the injection of the titrant, as shown

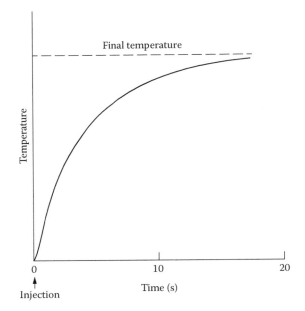

Figure 17.33 DIE titration plot. An excess of reactant B is added rapidly. The amount of analyte A is calculated from the final temperature.

in Figure 17.33. We may suppose that the following exothermic reaction takes place:

$$A(\text{analyte}) + B(\text{titrant}) \rightarrow C(\text{products}) + \text{heat}$$

The quantity of heat generated is a function of the number of moles that take part in the reaction. This, in turn, is controlled by the amounts of A and B. Any excess of either A or B does not react. In this instance, we have added an excess of B; therefore, the amount of heat generated must be a function of the amount of A present in the mixture. If the amount of A present were halved, then the amount of heat generated would be halved. Since there is an excess of B present in both cases, only the amount of A affects the amount of heat generated.

The method is quite useful, particularly if the rate of chemical reaction between A and B is slow. The final quantity of heat evolved is not a function of time, but a function of the concentration of sample. This is a distinct advantage over conventional volumetric analysis and, in some instances, TTs. As stated earlier, slow reactions give rise to errors in the endpoint determination in TTs.

The equipment used for DIE is identical to that for thermometric titrimetry. The titrant must be at the same temperature as the sample at the start of the experiment, and the syringe is emptied rapidly into the cell to deliver the titrant "instantaneously."

DIE may be used for the same applications as discussed for TTs, for example, for the volumetric analysis of materials, such as boric acid, which are virtually impossible to titrate using endpoint indicators or pH indicators. DIE can also be used in biological studies where the reaction rates may be slow. For example, proteins have been titrated with acid or base, antibodies have been titrated with antigen, and enzyme–coenzyme systems have been studied. DIE is used to determine kinetic parameters for slow reactions. The use of a large excess of one reactant (the titrant) favors the forward reaction (according to Le Chatelier's principle) even if the equilibrium constant is small, so equilibria may be studied using DIE that cannot be studied using other titrimetric methods.

A word of caution should be mentioned about possible interferences, particularly if two simultaneous reactions take place. Since both reactions may generate heat, a direct error will be involved with injection enthalpimetry methods. However, with TTs, frequently consecutive reactions take place and can be determined consecutively without prior separation. This has been demonstrated in the titration of calcium and magnesium with oxalate.

17.7 MICROCALORIMETRY

Calorimetry is the science of measuring heat changes from chemical reactions and physical events. DSC is described in Section 17.3 and has been a staple method of analysis for materials scientists. Classical DSC instruments and classical calorimetric titrimetry instruments (Sections 17.5 and 17.6) often lack the sensitivity required for the study of biological samples, where processes like the folding or unfolding of a protein may exchange only microjoules of heat. A new class of ultrasensitive microcalorimetry instrumentation has been developed primarily for studies in the life sciences, where sample amounts may be extremely limited.

Modern calorimetry has transformed and advanced to such a degree that today its acute sensitivity and flexibility have made it an indispensable tool throughout the life sciences. Ultrasensitive calorimetry now plays an integral role in biotechnology, providing researchers with essential information on the thermodynamics, structure, stability, and functionality of proteins, nucleic acids, lipids, and other biomolecules. Ultrasensitive calorimetry is a vital tool for R&D in pharmaceuticals, genetics, energy, and materials—in almost any area where the measurement and controlled manipulation of substances and interactions are required at the molecular level.

Ultrasensitive calorimeters enable researchers and product developers to gain thermodynamic information that was previously difficult or even impossible to gain with earlier models or by other techniques. The capabilities of advanced, ultrasensitive calorimeters are enormously helpful because of the amount and accuracy of the data they extract, as well as their abilities to obtain data using very small samples.

Microcalorimetry is used to study reactions involving biomolecules, interactions between molecules like ligand binding, and physical changes such as protein unfolding. The advantage of using microcalorimetry for these types of studies is that assay development is relatively simple compared to other bioanalytical approaches. No radiolabeling or derivatization of compounds is required, and since it is not an optical technique, samples without chromophores, turbid solutions, and suspensions are easily studied. Microcalorimetry instrumentation is available for micro- or nanoscale DSC, isothermal titration calorimetry (ITC), and isothermal calorimetry.

DSC measures heat as a function of changing temperature. It is typically used to discern a wide range of thermal transitions in biological systems and the thermodynamic parameters associated with these changes. ITC is typically used for monitoring a chemical reaction initiated by the addition of a binding component and has become the method of choice for characterizing biomolecular interactions.

Instruments are available from several manufacturers, including TA Instruments and Cytiva. Some companies use "micro" in their instrument names, some use "nano," and some use the Greek letter µ in their names. For the sake of simplicity, the term "micro" will be used unless a specific instrument is discussed by name.

17.7.1 Micro-DSC Instrumentation

DSCs designed for the measurement of extremely small amounts of samples must be sensitive and designed to accurately and precisely control the temperature of the sample. Sensitivity is a byproduct of sample size, S/N, and baseline stability. Today's calorimeters incorporate new control capabilities that produce baseline stability that previously has not been available. This control also makes the instrument less susceptible to external environmental influences, and with the pressure control capabilities built into the newer models, the instruments are now appropriate for the study of volumetric properties (pressure perturbation) of biopolymers.

The Nano DSC (TA Instruments) is a power-compensated DSC with a dual capillary cell design. The capillary cells (sample and reference) confer some advantage in studying proteins (discussed in Section 17.7.2), with internal surfaces that can be completely and rapidly flushed with cleaning solutions. It uses the same solid-state thermoelectric elements for heating and cooling the sample; this results in equal sensitivity for upward and downward temperature scans. Pressure control is maintained by a built-in high-pressure piston driven by a computer-controlled linear actuator. Constant pressure is applied during the DSC experiment to obtain the constant pressure heat capacity of the sample and to prevent boiling or bubble formation. The pressure can be varied up to 6 atm during "pressure perturbation" experiments to obtain compressibility and thermal expansion data. The operating temperature range is from $-10\ °C$ to $130\ °C$ or $160\ °C$. The DSC scan rate is from

0.0001 °C to 2 °C/min with a response time of 5 s. The sample cell volume is 300 µL with cells made of Pt or Au.

The MicroCal VP-Capillary DSC system from Cytiva uses a nonreactive material called Tantalum 61 for its matched capillary cells, which have an "active" cell volume of 130 µL. It has a fully automated autosampler that permits unattended analysis of 50 samples/day.

17.7.2 Applications of Micro-DSC

Minute alterations in the structure, stability, and affinity of proteins and other biological macromolecules can dramatically alter the function of the protein or biopolymer. Biopolymers can be free in solution (proteins and nucleic acids), associated with another molecule (as in a deoxyribonucleic acid [DNA]/drug complex) or part of an assembly of molecules, like lipids in a membrane. Micro-DSC is the most direct and sensitive approach for characterizing the thermodynamic parameters controlling noncovalent bond formation (and therefore stability) in proteins, other macromolecules, and even whole cells. Proteins and other biomolecules in solution are in equilibrium between the normal "folded" state and the denatured "unfolded" state, with the folded structure being thermodynamically more stable by only about 20–30 kcal/mol for a medium-sized protein or the equivalent of 5–8 hydrogen bonds. Many factors are responsible for the stability of proteins, including hydrogen bonds to the solvent and intramolecular interactions. Some proteins denature on heating, others on cooling. Correlation of thermodynamic properties with stability is needed for, among other uses, the rational design of protein-based therapeutics and small-molecule protein-specific drugs and for assessing the long-term stability of biotherapeutic formulations. DSC studies to determine the intrinsic stability of proteins in dilute solution require dilute buffer solutions with concentrations of 0.2–2 mg protein/mL needed. In a single thermal unfolding experiment, DSC can measure and permit calculation of all the thermodynamic parameters characterizing a biological molecule: the enthalpy due to thermal denaturation, ΔH; T_m, the transition midpoint, where half of the macromolecules are folded and half are unfolded; and the change in heat capacity, ΔC_p. The type of information available from a DSC experiment is shown in Figure 17.34.

Figures 17.35a and b show stability assessments of proteins. The stability is analyzed in dilute buffer solution by determining the changes in the partial molar heat capacity at constant pressure, ΔC_p. The contribution of the protein to the calorimetrically measured heat capacity is determined by subtracting a scan of a buffer blank from the sample data. As can be seen in Figure 17.35a, heating the protein solution initially produces a slightly increasing baseline, but as heating progresses, heat is absorbed by the protein and causes it to unfold over a characteristic temperature range for that protein. This gives rise to an endothermic peak. The midpoint T of the peak is denoted as the thermal transition midpoint (T_m); at this point, half the macromolecules are folded and half are unfolded. Once unfolding is completed, heat absorption decreases and a new baseline is established. Figure 17.35b shows two biomolecules in solution. The higher the thermal transition midpoint, the more stable the molecule. The solid line shows the raw data, while the dotted line is the result of statistical processing of the data as noted in the figure. The experiment provides the ΔH for each transition as well as the partial molar heat capacity of the molecule at constant pressure, ΔC_p, as well as T_m, as shown in the figure inset.

In a DSC experiment lasting about 1 h, using as little as 2 µg of material, a protein can be thermally unfolded, allowing the enthalpy and entropy of the denaturation process to be measured, as seen in Figure 17.36. The thermodynamic data collected are presented in Table 17.5. As can be seen in this table, the thermodynamic data obtained were accurate even for the 2 µg sample.

Proteins are large flexible molecules capable of recognizing and binding specific molecules such as other proteins and drugs. A ligand will bind to a protein only if the resulting complex is more stable than the original protein. Ligands can bind to native folded proteins, stabilizing the folded state, or they can bind to the denatured unfolded protein, destabilizing the folded state. Binding triggers conformational transitions and these can be measured by DSC. If a ligand binds preferentially to the native state of the protein, the temperature at which the ligand–protein complex denatures will be higher than the temperature at which the free protein unfolds. If the ligand (L) binds tightly to the native conformation of a protein (P), the resulting complex PL will have a higher thermal stability (higher T_m) than either of the free components (Jelesarov and Bosshard 1999). Figure 17.37 shows an idealized example of this behavior.

17.7.3 Isothermal Titration Calorimetry

ITC is a small-scale, small-sample-size, high-sensitivity version of the thermal titrimetry already discussed. The instrumentation consists of a highly sensitive calorimeter, capable of detecting heat effects as small as 100 nJ on 1 nmol or less of a biopolymer. Temperature stability is critical; the TA Instruments Nano ITC provides temperature stability of 0.0002 °C at 25 °C, for example. Sample cells are made of Hastelloy® or gold with a sample volume of 1–2 mL required, depending on the manufacturer. The titrant is delivered with a precision 300 µL volume syringe with a mechanical stirrer at the end. Various mixing speeds can be selected depending on the physical nature of the sample. The response time is 12–20 s, again depending on the manufacturer. Sample throughput is estimated by GE Healthcare to be 4–8 samples/8 h day for its MicroCal VP-ITC system.

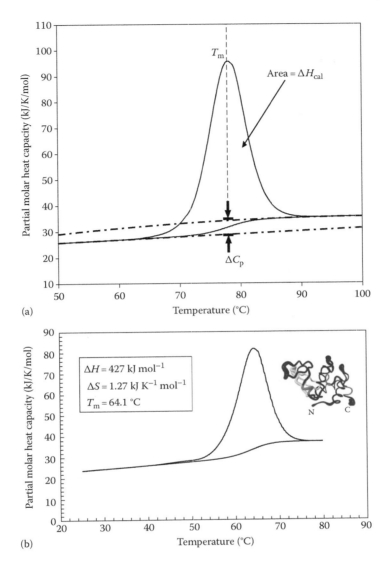

Figure 17.34 (a) A schematic DSC scan, showing how T_m, ΔH, and ΔC_p are obtained. (b) A temperature scan, converted to partial molar heat capacity for 1 mg/mL hen egg white lysozyme (shown in the upper right corner) in 0.2 M glycine buffer, pH 2.7. Data obtained at a scan rate of 1 °C/min in a Nano DSC. (Courtesy of TA Instruments, New Castle, DE, www.tainstruments.com.)

ITC is a universally applicable technique for determining the thermal effects arising from molecular interactions. It is a powerful approach for quantifying the molecular interactions between two or more proteins and macromolecules or between macromolecules such as protein and nucleic acids with small-molecule ligands such as drugs and enzyme inhibitors. Life science applications of ITC can be broadly characterized as either binding interactions or enzyme–substrate interactions. The most frequent use of ITC is in the study of binding interactions. In a single titration experiment, through the heat released or absorbed, ITC can measure binding affinity, stoichiometry, enthalpy, and entropy for nanomolar amounts of sample. Millimolar to picomolar binding constants can be measured. Common binding studies include protein–ligand binding, protein–protein interactions, biopolymer–drug binding, and lipid vesicle and liposome interactions.

To determine the affinity with which a protein binds a small ligand, such as a drug candidate molecule, a solution of the protein (1 mL of an approximately 10 µM solution) is loaded into the calorimeter sample cell and the ligand solution (approximately 100 µL) is loaded into the injector syringe connected to the sample cell. The ligand is titrated incrementally into the protein solution (Figure 17.38a). Heat is generated and registered as a deflection peak by the instrument. The peaks are integrated over the time of the experiment and plotted as area/µJ versus mole ratio of titrant to titrate

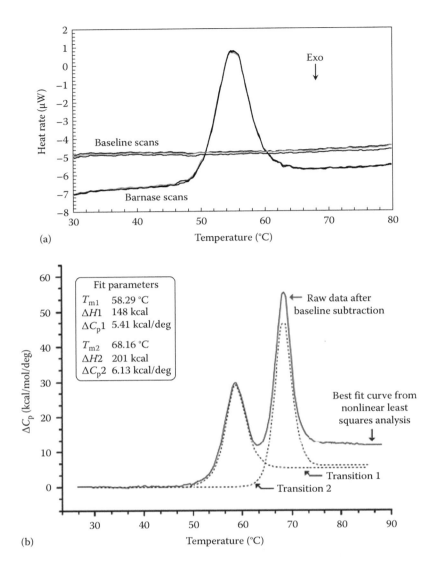

Figure 17.35 (a) Three overlaid baseline temperature scans and three overlaid scans of 60 mg barnase in 0.3 mL 20 mM phosphate buffer, pH 5.5. (Courtesy of TA Instruments, New Castle, DE, www.tainstruments.com.) (b) Stability assessment of two biomolecules using the MicroCal VP-DSC. (Courtesy of Cyvita, Inc., www.cytivalifesciences.com.)

(Figure 17.38b). ITC can characterize the kinetics of essentially any enzymatic reaction or any catalytic reaction. Both kinetic and thermodynamic data are generated. A typical fully automated ITC experiment takes about 1 h and allows determination of the association constant, binding enthalpy, stoichiometry, and reaction rate.

ITC is the most popular type of ultrasensitive calorimetry today. In the pharmaceuticals area, this technique is used in the testing of drugs that target a very specific target in the human body, possibly a particular gene or receptor. Various biomolecular interactions can be studied using ITC, such as protein–ligand, protein–protein, protein–DNA, and antibody–antigen, among others.

While ITC is particularly suitable to follow the energetics of an association reaction between biomolecules, the combination of ITC and DSC provides a more comprehensive description of the thermodynamics of an associating system. Instrumentation diagrams, videos, and applications notes for ITC and microcalorimetry may be found on the TA instruments and Cyvita websites (www.tainstruments.com; www.cytivalifesciences.com).

17.7.4 Microliter Flow Calorimetry

Microcalorimeters, as discussed earlier, are well-established instruments for protein chemistry. They typically require several hundred microliters of sample and up to 1 h for a measurement. Today's biotech industries use enzymes in biofuels, textiles, detergent, and food manufacturing. The design and selection of cost-effective,

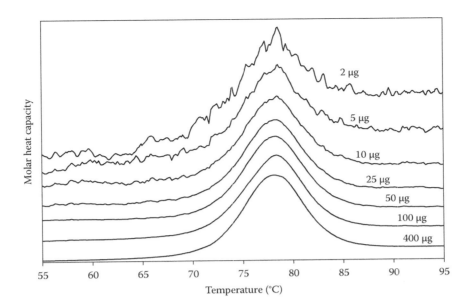

Figure 17.36 A temperature scan, converted to molar heat capacity, for hen egg white lysozyme solution in 0.2 M glycine buffer, pH 4.0. Lysozyme concentrations ranged from 400 to 2 μg in the 300 μL sample cell of the Nano DSC. Curves are offset to see them clearly. (Courtesy of TA Instruments, New Castle, DE, www.tainstruments.com.)

Table 17.5 Thermodynamic Data Collected from the Scans in Figure 17.36

Lysozyme in Cell (μg)	Calorimetric			van't Hoff	
	T_m (°C)	ΔH (kJ/mol)	ΔS (kJ/K/mol)	T_m (°C)	ΔH (kJ/mol)
400	78.0	512	1.46	78.0	515
100	78.1	512	1.46	78.0	509
50	78.1	517	1.47	77.9	513
25	78.1	513	1.46	77.8	513
10	78.1	515	1.47	78.0	515
5	78.1	490	1.40	78.1	510
2	78.1	503	1.43	77.8	499

high-specificity enzymes require high-throughput, low-volume screening methods. SPT Labtech Ltd. (Melbourn, United Kingdom, www.sptlabtech.com) has developed a flow microliter calorimeter (FMC), called the chipCAL, for enzyme studies that requires only 15 μL of sample and 8 min of analysis time. Using two syringe pumps, 15 μL each of enzyme and substrate is passed through a flow cell simultaneously. The heat evolved or absorbed is detected by a thermopile chip. The system is suitable for real-time monitoring of enzyme activity, substrate profiling to determine specificity, monitoring the metabolic activity of cell cultures, and similar processes in the biopharmaceutical, food, and fermentation industries. Enzyme kinetic parameters can be determined from a single run. The flow microcalorimeter offers a universal, label-free approach to enzyme assays with minimal sample volumes. Schematics of the system and application notes can be found on the website at www.ttplabtech.com.

17.8 THERMOMECHANICAL ANALYSIS AND DYNAMIC MECHANICAL ANALYSIS

Many materials are used under conditions that subject them to forces or loads. Examples include aluminum alloys used in airplanes, glass in windows and doors, polymers used in molded automobile bumpers, and polymer fibers in "stretch" clothing. The *mechanical* properties of these materials are very important; mechanical properties include ductility, stiffness, hardness, and strength. (For an overview of mechanical properties of materials, see an introductory materials science text, such as the one by Callister listed in the bibliography.) Many materials, such as polymers, behave both like an elastic solid and a viscous fluid; such materials are termed **viscoelastic**. See Table 17.6. The viscous or inelastic component of a material determines such properties as impact resistance. The mechanical properties of a material are a function of the applied load, the duration of the load, and environmental conditions, including temperature. The thermal properties of materials are important to understand for the proper selection of materials that will function at elevated or subambient temperatures or for materials exposed to temperature gradients or thermal cycling. The response of mechanical behavior as a function of temperature is studied using **thermomechanical methods of analysis**.

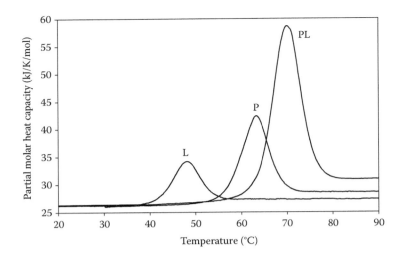

Figure 17.37 Simulated DSC data showing the higher-temperature stability of a tightly associated protein/ligand complex (PL) compared with the free protein (P) and ligand (L). (Courtesy of TA Instruments, New Castle, DE, www.tainstruments.com.)

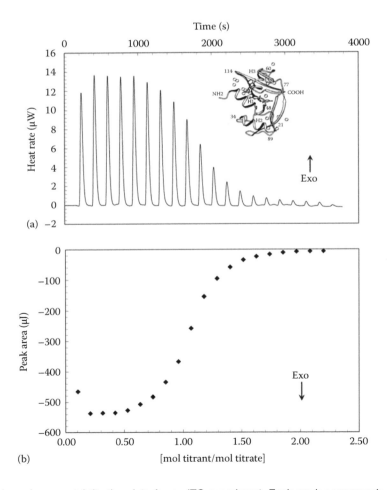

Figure 17.38 (a) Simulated raw incremental titration data for an ITC experiment. Each peak corresponds to the heat produced upon injection of a fixed amount of titrant at specific time intervals. (b) Simulated binding analysis. Area under each peak in (a) is integrated and fit to a binding model to permit calculation of affinity, enthalpy, and stoichiometry for the interaction. (Courtesy of TA Instruments, New Castle, DE. www.tainstruments.com.)

Table 17.6 Range of Material Behavior

Solid-like ↔ Liquid-like
Ideal solid ↔ Most materials ↔ Ideal fluid
Purely elastic ↔ Viscoelastic ↔ Purely viscous
Viscoelasticity: having both viscous and elastic properties

Source: Courtesy of TA Instruments, New Castle, DE, www.tainstruments.com.

Two important classes of thermomechanical methods of analysis are **TMA** and **DMA**. TMA measures changes in the physical dimensions of a sample as a function of temperature under an applied external load or stress. The dimensional change is generally measured in one direction (the length of the sample) with respect to the applied load; a 1D change is easier to measure than volume changes. The applied load is a static load in TMA. If the applied load is zero (i.e., no external load is applied), the TMA technique is called *thermodilatometry*. In the absence of an external load, changes in dimension as a function of temperature still can be measured. These changes are expressed as the thermal expansion coefficients of a material. TMA is used to measure expansion, compression, softening point, bending properties, and extension (elastic and plastic deformation) for materials, especially polymers, composites, ceramics, and glasses. DMA differs from other mechanical testing devices in two ways. Typical tensile test devices focus only on the elastic component of the material and usually work outside of the linear viscoelastic range (LVR) of the material. DMA works primarily in the viscoelastic range and measures the viscoelastic properties by applying either transient or dynamic oscillating loads to the sample. Deformation of the sample under changes in temperature and an alternating (dynamic) stress is measured as a function of time, temperature, load, and frequency. A great deal of data about the elastic and plastic behaviors of materials are obtained from DMA. Table 17.7 presents a brief overview of some parameters measured by TMA and DMA and the information obtained from these measurements.

DMA measures the viscoelastic properties of a sample using either transient or dynamic oscillatory tests. Transient tests include creep and stress relaxation. In creep, a stress is applied to the sample and held constant, while deformation is measured versus time. After some time, the stress is removed and the recovery measured. In stress relaxation, a deformation is applied to the sample and held constant; the degradation of the stress required to maintain the deformation is measured versus time. The most common test is the dynamic oscillatory test, where a sinusoidal stress (or strain) is applied to the material and the resultant sinusoidal strain (or stress) is measured. Also measured is the phase difference, δ, between the two sine waves. The phase lag will be 0° for a purely elastic material and 90° for a purely viscous material. Viscoelastic materials such as polymers will exhibit an intermediate phase difference. Since modulus equals stress divided by strain, the complex modulus, E^*, can be calculated. From E^* and δ, the storage modulus, E', the loss modulus, E'', and tan δ can be calculated:

Table 17.7 Information Obtained from TMA and DMA Experiments

Type of Measurement	Information	Results
TMA	Change in length	Expansion coefficient
Static load	Expansion	Glass transition temperature
	Compression	Shrinkage
Thermodilatometry	Penetration	Softening
No external load	Extension	Penetration with given force, T
	Bending	Strain
		Static elastic modulus (Young's modulus, i.e., the slope of the stress–strain curve in the elastic region)
		Crystalline morphological transitions
DMA	Deformation (cyclical change in length)	Complex dynamic modulus
Dynamic (alternating) load	Penetration	Storage and loss modulus
	Bending	Mechanical loss ("tan delta")
	Extension	Glass transition temperature
		Polymerization, hardening, curing
		Postpolymerization cross-linking
		Decomposition

Source: [© 1993–2014 PerkinElmer, Inc. All rights reserved. Printed with permission. (www.perkinelmer.com).]

$$E^* = \frac{\text{Stress}}{\text{Strain}}$$

$$E' = E^* \cos\delta$$

$$E'' = E^* \sin\delta$$

$$\tan\delta = \frac{E'}{E''}$$

The storage modulus is the elastic component and is related to a sample's stiffness. The loss modulus is the viscous component and is related to the material's ability to dissipate mechanical energy through molecular motion. tan δ provides information on the relationship between the elastic and inelastic components.

17.8.1 Instrumentation

17.8.1.1 TMA Equipment

A schematic of a commercial TMA instrument is shown in Figure 17.39. The instrument consists of a dimensionally stable (with temperature) sample holder and measuring probe, a programmable furnace, a **linear variable displacement transducer (LVDT)** to measure the change in length, a means of applying force (load) to the sample via the probe (core rod, push rod), and a temperature sensor (usually a thermocouple).

The sample holder is generally made of quartz or steel. A sample is placed on the holder and the selected probe is placed in contact with the sample surface. A measured load is applied and the temperature program started. As the sample expands, softens, or contracts, the position of the probe changes. The position change is measured by the LVDT, which provides a voltage proportional to the deflection of the probe.

The probe is usually also made of quartz or steel. Probes come in a variety of shapes, with the shape determining the type of dimensional change measured. Schematic diagrams of common TMA and DMA probes are presented in Figure 17.40. A flat end on the probe is used for expansion measurements and the determination of linear expansion coefficients. Compression and penetration probes have a variety of rounded or pointed ends; the hemispherical probe is used to determine the modulus of soft, rubbery materials, for example. The three-point bend apparatus is a standard mechanical test for materials. A picture of a steel three-point bending sample holder is shown in Figure 17.41. The sample holder has two knife-edge supports on which the sample is placed and a knife-edge probe through which the load is applied to the upper side of the sample. Tension or extension probes are used for measuring extension or shrinkage in thin films and fibers. The tension holder consists of a pair of grips, clamps, or hooks, one fixed and one moveable, to pull the sample in tension under load. Figure 17.39 shows in the upper left corner the placement of a sample in expansion, tension, and penetration modes.

Loads of 0.1–200 g or forces of 0.1–100 N may be selected and furnace temperatures may be programmed from −150 °C to 1000 °C, depending on the particular instrument. Samples of various sizes and shapes may be measured; lengths of 1 μm–50 mm are common, with diameters in the 10 mm range, depending on the particular probe and instrument design.

TMA instruments must be calibrated for both temperature and dimensional motion. The temperature is calibrated using the same type of melting point standards used for DSC. High-purity In, Sn, Zn, and Al are often used. The LVDT is calibrated by using a standard with a well-known linear CTE, such as Al. The initial length of the standard is determined at room temperature and then the change in

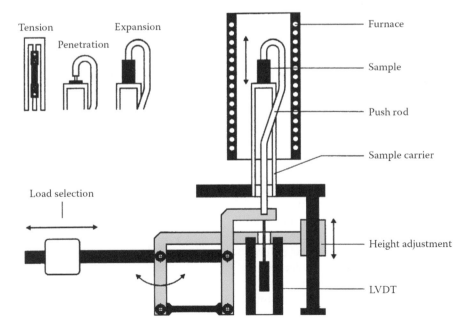

Figure 17.39 Schematic of a TMA. Three types of sample probes are shown with samples in place (dark-colored areas are the samples). The expansion probe and sample are shown as they would be located in the furnace. (Courtesy of Netzsch Instruments, Inc., Burlington, MA, www.netzsch-thermal-analysis.com.)

length of the standard with temperature is measured. The linear CTE, α, is defined by

$$\frac{L_f - L_0}{L_0} = \alpha(t_f - t_0) \qquad (17.5)$$

where
L_f is the final sample length
L_0 is the initial sample length
T_f and T_0 are the final and initial temperatures (in °C), respectively

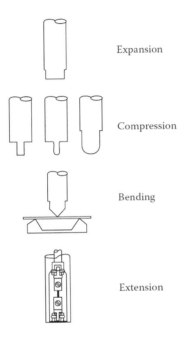

Figure 17.40 Representative sample probe shapes for TMA and DMA measurements. [© 1993–2014 PerkinElmer, Inc. All rights reserved. Printed with permission. (www.perkinelmer.com).]

The calculated value of α is compared to the literature value for the material and a correction factor is applied to the data if needed.

17.8.1.2 DMA Equipment

Commercial DMA instruments vary in their design. One commercial instrument is shown in Figure 17.42, set up for a three-point bend test under dynamic load. A different commercial instrument schematic, Figure 17.43, shows a sample clamped between two arms that are free to move about the pivot points (Figure 17.43a); the electromagnetic drive and arm/sample assembly are shown in Figure 17.43b. The electromagnetic motor oscillates the arm/sample system and drives the arm/sample system to a preselected amplitude (strain). The sample undergoes a flexural deformation as seen in Figure 17.43a. An LVDT on the driver arm measures the sample's response to the applied stress and calculates the modulus (stiffness) and the damping properties (energy dissipation) of the material. The DMA 1 from Mettler Toledo (www.mt.com) is an LVDT-based instrument that can measure changes in length over the range ±1 mm with a mean resolution of 2 nm. The instrument has a temperature range of −190 °C (liquid N_2 cooling) to 600 °C and the ability to run samples in liquids or at specific relative humidity levels. Alternatively, instead of an LVDT, some instruments are equipped with a high-resolution optical encoder to measure displacement. Based on diffraction patterns of light through gratings, one movable and one stationary, optical encoders provide exceptional resolution compared to LVDT technology. With a resolution of 1 nm, an optical encoder can measure small amplitudes very precisely.

DMA instruments have furnaces programmable from −170 °C to 600 °C. Frequency ranges can be varied from 0.001 to 1000 Hz, depending on the manufacturer. Probe

 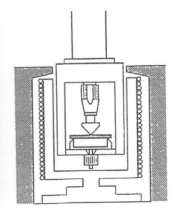

Figure 17.41 On the right is a schematic of a three-point bend setup for TMA or DMA. On the far left is a photograph of a three-point bend probe made of quartz. The center photograph is of a similar probe made of stainless steel. [© 1993–2014 PerkinElmer, Inc. All rights reserved. Printed with permission. (www.perkinelmer.com).]

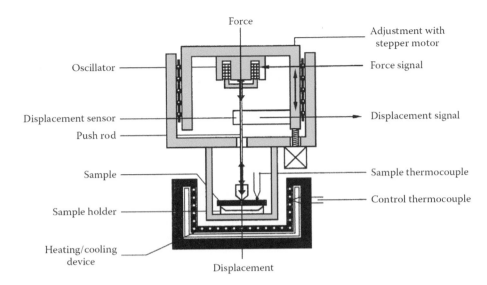

Figure 17.42 Schematic of a DMA analyzer. (Courtesy of Netzsch Instruments, Inc., Burlington, MA, www.netzsch-thermal-analysis.com.)

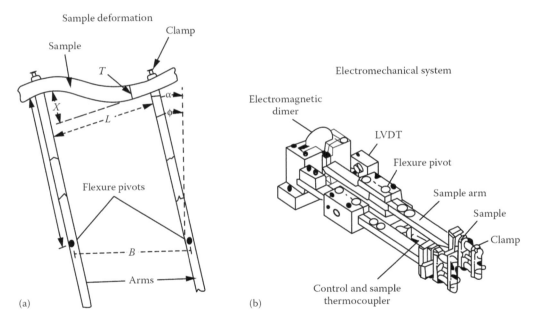

Figure 17.43 Schematic of (a) a sample holder for DMA and (b) the DMA electromechanical system. (Courtesy of TA Instruments, New Castle, DE, www.tainstruments.com.)

types and geometries are more varied than those for TMA, including probes for single/dual cantilever bending, compression, and sandwich shearing. The range of force applied is from 0.0001 to 18 N, with a force resolution of 0.00001 N. The dynamic sample deformation range for one commercial instrument from TA Instruments is ±0.5–10,000 μm. Output values for DMA experiments are given in Table 17.8.

17.8.2 Applications of TMA and DMA

TMA is used to measure expansion, compression, elastic relaxation, penetration, softening point, glass transitions, shrinkage, swelling, and elastic modulus (the stiffness or resistance of a material to elastic deformation—the higher the modulus, the stiffer the material). TMA is widely used to characterize polymers, metals, coatings, ceramics, glasses, and composite materials. TMA has been used to measure the expansion behavior of printed circuit boards. As seen in Figure 17.44, three different circuit board samples were measured using an expansion probe. Glass transition temperatures for all three can be determined from the change in slope, marked on the curves. One sample shows the melting of the tin solder used on this board, at about 180 °C. At temperatures at or above 300 °C, each board shows an abrupt expansion. This is the point at which the boards delaminate, caused by the explosive release of internal stress in the composite materials.

TMA is used to characterize polymer fibers. The characterization of polymer fibers is important in the textile industry. Processing (or texturing) the fibers by heating, spinning, twisting, and so on results in different types of stresses in the fibers. These can affect textile properties such as wrinkle resistance and stiffness, so textile chemists and polymer chemists need to evaluate the effect of processing on spun and processed fibers. Figure 17.45 compares the fiber tension in an untextured (solid line) and textured (dotted line) PET fiber. These samples were 4–5 mm long and were clamped in a fiber extension probe for testing.

Fiber tension is clearly lost during the particular texturing process for this sample. Figure 17.46 shows the behavior of three different polymer fiber samples. Two of the samples (#2 and #3) show severe shrinkage prior to melting, due to relaxation of stresses induced during spinning. Fiber sample #1 is free of induced stress (no change in length until the melting point is reached).

Penetration mode can be used to measure coating thickness, such as the polymer coating inside beverage cans and the silicone release coating on the paper backing of "peel-off" labels. Penetration can be used to measure softening points of waxes and fats used in food and cosmetics. TMA in expansion mode is used to measure shrinkage of solid-state materials and is useful in optimizing sintering conditions for materials made from compacted powders, such as ceramics and UO_2 for nuclear fuel rods. TMA can study the expansion and contraction of solid-state materials such as high-temperature superconductors to help understand the behavior of these materials. Figure 17.47 shows the difference in expansion behavior for a glass-fiber-reinforced composite. The expansion is very different (much larger) in the direction perpendicular to fiber alignment than in the direction along the fibers, an important behavior for the end user of this composite to know.

DMA provides materials scientists, polymer chemists, and design engineers with detailed information on the elastic and inelastic (plastic) deformation of materials, modulus, and damping (energy dissipation) properties of materials. The damping behavior of synthetic and natural rubbers is important in vibration and acoustic applications, for example.

The mechanical modulus can be measured under different types of load shear, flexure, and so on. The measured modulus can be expressed as the storage modulus (related to the stiffness of the material) and the loss modulus

Table 17.8 **Output Values from a DMA**

Storage modulus	Complex/dynamic viscosity	Time
Loss modulus	Creep compliance	Stress/strain
Storage/loss compliance	Relaxation modulus	Frequency
tan delta (δ)	Static/dynamic force	Sample stiffness
Complex modulus	Temperature	Displacement

Source: Courtesy of TA Instruments, New Castle, DE, www.tainstruments.com.

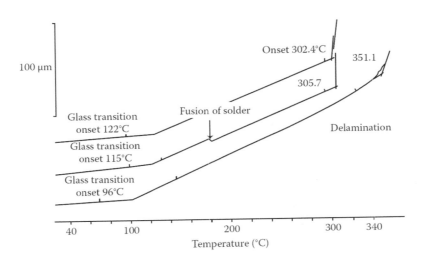

Figure 17.44 TMA of three printed circuit board samples showing the glass transitions, the melting of solder on the middle sample, and the large change in dimension (expansion) due to delamination of the composite boards at high temperature. (Courtesy of Mettler-Toledo GmbH, Analytical, Schwerzenbach, Switzerland, www.mt.com.)

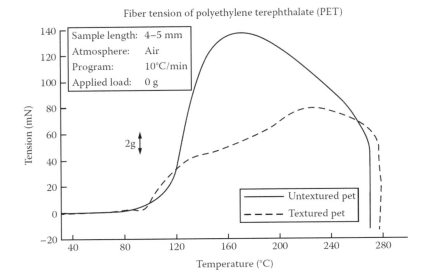

Figure 17.45 TMA curves of textured and untextured fibers measured in tension, showing the dramatic difference in behavior due to processing of the fiber. (Courtesy of TA Instruments, New Castle, DE, www.tainstruments.com.)

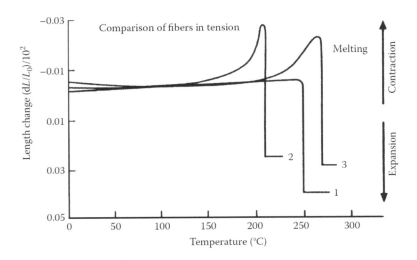

Figure 17.46 TMA curves of three types of polymer fiber, showing shrinkage due to stress relaxation before melting in polymers #2 and #3. (Courtesy of Netzsch Instruments, Inc., Burlington, MA, www.netzsch-thermal-analysis.com.)

(related to the viscous behavior or softness of the material) under the various types of loads. DMA is used to follow the curing of polymers and cross-linking of polymers and to measure glass transition temperatures.

The modes of operation of a DMA are varied. Using a multifrequency mode, the viscoelastic properties of the sample are studied as a function of frequency, with the oscillation amplitude held constant. These tests can be run at single or multiple frequencies, in time sweep, temperature ramp, or temperature step/hold experiments. In multistress/strain mode, frequency and temperature are held constant and the viscoelastic properties are studied as the stress or strain is varied. This mode is used to identify the LVR of the material. With creep relaxation, the stress is held constant and deformation is monitored as a function of time. In stress relaxation, the strain is held constant and the stress is monitored versus time. In the controlled force/strain rate mode, the temperature is held constant, while stress or strain is ramped at a constant rate. This mode is used to generate stress/strain plots to obtain Young's modulus. In isostrain mode, strain is held constant during a temperature ramp to assess shrinkage force in films and fibers.

A recent application of DMA is in the characterization of food. DMA is sensitive and can handle larger samples than DSC, so its use for inhomogeneous materials like

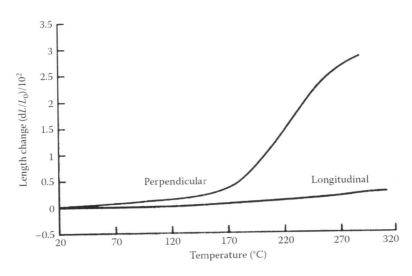

Figure 17.47 TMA curve of a fiber composite material. Thermally induced expansion differs depending on the direction of measurement because the strength of the material is related to the orientation of the fiber layers. (Courtesy of Netzsch Instruments, Inc., Burlington, MA, www.netzsch-thermal-analysis.com.)

food should have advantages. Samples of commercial white bread were studied using DMA. A fresh sample was compared with a sample exposed to the air overnight. (You know what happens to bread left "out" in air—it gets dry, stiff, and brittle.) The DMA plots of storage modulus, loss modulus, and tan δ (tan delta) for both samples are shown in Figures 17.48a and b. For the fresh sample, the changes in moduli and tan δ are postulated to be associated with changes in the physical state of water in the bread. For example, the tan δ "doublet" at about 0 °C in the fresh sample could be melting of two different forms of bound water or it could be a surface/interior water phenomenon. In any case, the "stale" bread is clearly lacking many of the features seen in the fresh sample, as we might expect if the stale sample has lost the water responsible for the features. The room temperature modulus for the fresh sample was measured to be 2 MPa, while the stale bread was 16 MPa; the increase in modulus correlates with the increased stiffness of stale bread. In a related experiment, the moduli of cooked pasta were calculated from a series of DMA experiments. The plot of the storage modulus (Figure 17.49) shows a sharp drop at about 8 min, indicating that the pasta has become "limp." The DMA data could be used by pasta manufacturers to suggest optimum cooking times to their customers. (Details of these food experiments are from *TA Instruments Thermal Analysis Applications Brief Number TA-119*, available for downloading from the website at www.tainstruments.com.)

DMA is the most sensitive of the thermal techniques for measuring T_g. The glass transition temperature can be measured from the E' onset, from the E'' peak, or from the tan δ peak. Because T_g has a kinetic component, it is affected by the frequency (rate) of deformation of a material. As the frequency of the test increases, molecular relaxation can only occur at higher temperatures; therefore, the T_g increases with increasing frequency. For a sample of PET, T_g at 0.1 Hz is 103.00 °C, and at 10 Hz, it is 113.25 °C. The suitability of a material for specific uses depends on understanding the temperature and frequency dependence of transitions.

17.9 OPTICAL THERMAL ANALYSIS

Traditional contact-based systems for TMA are limited in that they are able to follow the behavior of a material during a heating cycle only as long as the material is rigid. If the material softens or melts, the pressure imposed by the contact changes the behavior of the material. For this reason, a class of optical noncontact instruments for TMA has been designed. These optical instruments, also called thermo-optical analyzers, are used in the ceramics, enamel, glass, and steel industries, the energy sector, academia, and R&D. Especially for ceramics and similar materials, these optical instruments permit the determination of thermal expansion, sintering kinetics, melting behavior, pyroplasticity, and bending. These optical noncontact instruments make a wide variety of measurements.

17.9.1 Heating Microscope

Heating microscopes permit the direct observation of material behavior during a heating cycle. Heating microscopy techniques can be applied to direct characterization of glazes, glasses, powders for continuous casting, coal, slag, and similar materials. The heating microscope consists of three principal components mounted on an optical

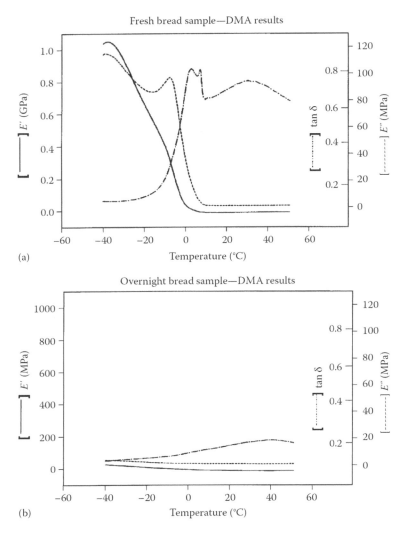

Figure 17.48 DMA plots of (a) a sample of fresh bread and (b) a sample of "stale" bread. The changes in the modulus and tan δ in the fresh bread sample are associated with water in the sample, while the overnight exposure to air of the stale sample has caused it to dry out. The room temperature modulus for the stale sample is greater than that for the fresh sample, indicating that the stale bread is stiffer than the fresh bread. (Courtesy of TA Instruments, New Castle, DE, from *Thermal Analysis Applications Brief TA-119*, www.tainstruments.com.)

bench: a halogen lamp light source, an electric furnace 100 mm long and 2 mm wide with a sample carriage, and an observation unit with a microscope and recording facility (Figure 17.50a). A sample is placed on a small alumina plate at the end of a thermocouple. Figure 17.50b shows a sample inside the furnace during a test. The microscope transfers the image of the sample in the furnace at about 5× magnification through a quartz window to the recording camera.

This heating microscope is completely automatic and can reach temperatures of 1650 °C with heating rates as high as 80 °C/min or "flash" heating, thus replicating industrial heat treatments. Images of the specimen undergoing heat treatment are stored at predetermined time or temperature intervals with the dimensional parameters measured automatically during the test. Sample sizes are small (e.g., a ceramic or glass cylinder 3 mm high and 2 mm diameter), which permit high heating rates and ensure homogeneous heating of the sample. As a ceramic or glass cylinder is heated, it begins to sinter (increase in density without changing shape), soften, and melt, changing shape through a progression such as that shown in Figure 17.51a. Figure 17.51b shows a comparison of two ceramic frits, one of which displays the "glassy" sintering, softening, and melting behavior and one which crystallizes and melts with the typical sharp melting point of a crystalline material.

17.9.2 Optical Dilatometers

Two optical noncontact dilatometers are available, a vertical dilatometer and a horizontal dilatometer. The

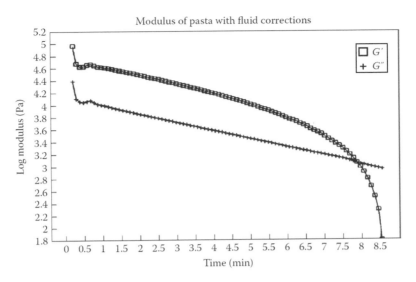

Figure 17.49 The storage modulus, G' (otherwise identified as E'), of cooked pasta measured by DMA. The modulus drops sharply at about 8 min, indicating that the pasta has lost stiffness (or become limp). (Courtesy of TA Instruments, New Castle, DE, from *Thermal Analysis Applications Brief TA-119*, www.tainstruments.com.)

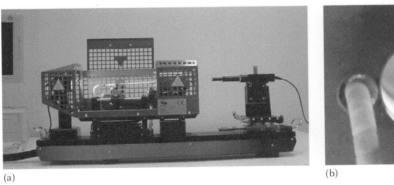
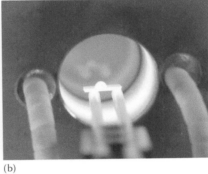

Figure 17.50 (a) A heating microscope. (b) Sample inside the furnace. (Courtesy of TA Instruments, New Castle, DE, www.tainstruments.com.)

vertical optical dilatometer (Figure 17.52a) has two high-magnification cameras, a maximum temperature of 1600 °C, and a maximum heating rate of 80 °C/min. The sample (15 mm × 5 mm × 5 mm) is placed vertically in the furnace. The first optical path is used to measure the top of the specimen and is kept aligned with the top of the sample by a stepper motor. The second optical path is focused on the sample holder and acts as a reference beam to correct for mechanical drift in the sample holder (Figure 17.52). Results are highly precise, with a resolution of <1 μm. No calibration is required. Samples can be run in air or in inert gas. The main application of this technology is in the study of ceramic sintering and sintering kinetics.

The horizontal dilatometer uses the same double optics principle, but the sample (50 mm length) is placed horizontally in the furnace and illuminated by two beams of blue light (478 nm). Two digital cameras capture the images of the last few hundred microns of each end of the sample (Figure 17.53). Measuring both ends of the sample with two beams of light simultaneously, without making contact, results in precise measurement of the CTE. The resolution for a 50 mm sample is 0.5 μm.

The study of glass and ceramic materials is an ideal application for this type of instrument, since these materials have a wide temperature range over which behavior changes from elastic through viscoelastic to viscous. The behavior can be followed through the whole thermal cycle with no contact with the specimen. Samples as thin as a few tens of microns can be studied using a special sample holder. As noted earlier, traditional contact dilatometers exert a slight pressure that modifies the behavior of the sample when it begins to soften. In addition, these traditional measuring systems also expand with temperature, requiring calibration with a known expansion sample and application of a

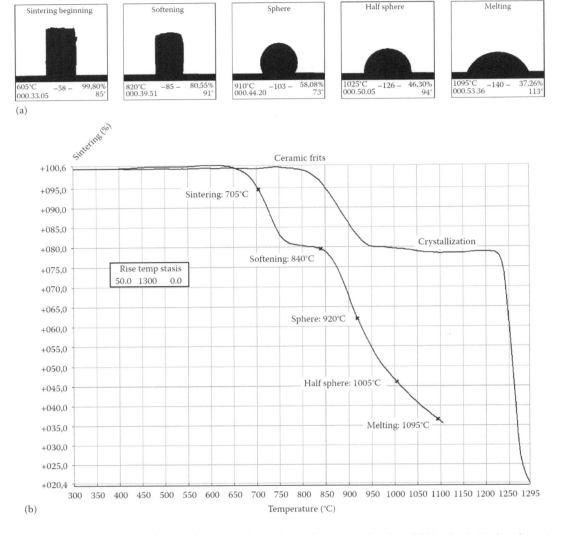

Figure 17.51 (a) Progression of shape changes for a ceramic or glass cylinder upon heating. (b) Heating behavior of two ceramic frits, one which displays the shape changes expected of glassy behavior and one which displays crystalline melting behavior. (Courtesy of TA Instruments, New Castle, DE, www.tainstruments.com.)

correction. No such corrections or calibrations are needed with noncontact optical systems.

The heating microscope and either of the dilatometers can be combined into one instrument, providing multiple pieces of information. In addition, an optical fleximeter is available. The optical fleximeter allows bending analysis of materials during a heating cycle without application of loads. A small sample bar 85 mm long is suspended between two alumina holding rods spaced 70 mm apart. A camera is centered on the sample, which moves downward or upward during the heat treatment. In the ceramics industry, the optical fleximeter is used to study pyroplasticity; that is, the tendency of a material to deform under its own mass due to a low viscosity during firing. In addition, the bending or flexion caused by differences in thermal expansion or in sintering behavior of multilayer materials can be studied. For example, solid oxide fuel cells (SOFCs) are multilayer systems, consisting of anode, cathode, solid electrolyte, and other components. SOFCs operate at temperatures up to 1000 °C. Materials used for SOFCs include layers of $LaMnO_3$, yttria-stabilized zirconia (YSZ), Ni-YSZ composite, Cu-cerium oxide composite, and Ca-doped $LaCrO_3$, for example. The thermal expansion coefficients of the components must be as close to one another as possible to minimize thermal stresses which could lead to cracking and thermal failure. Optical fleximetry is a valuable tool for studying these systems.

17.9.3 Optical Thermal Analyzers with Differential Thermal Analysis

All of the systems described are now available with an incorporated differential thermal analyzer. DTA

THERMAL ANALYSIS

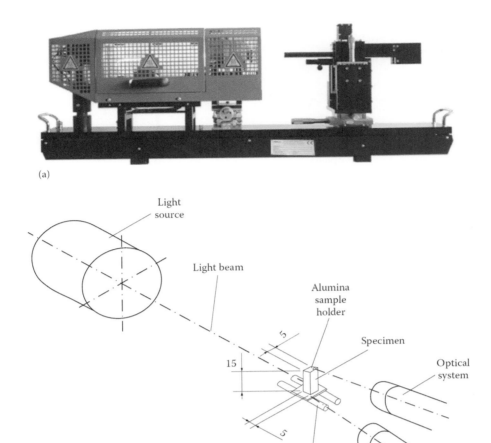

Figure 17.52 (a) Vertical optical dilatometer. (b) Schematic of the vertical optical dilatometer. (Courtesy of TA Instruments, New Castle, DE, www.tainstruments.com.)

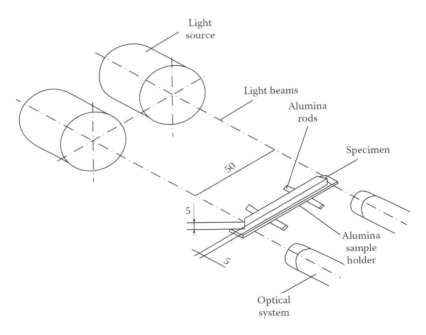

Figure 17.53 Schematic of the horizontal optical dilatometer. (Courtesy of TA Instruments, New Castle, DE, www.tainstruments.com.)

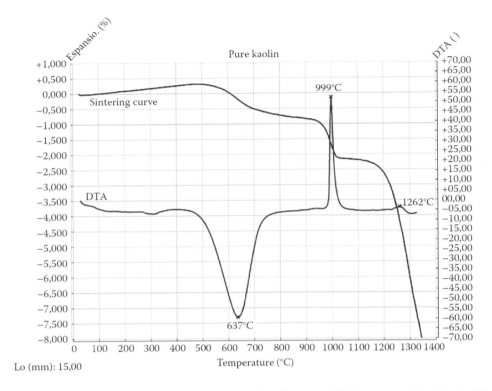

Figure 17.54   Dilatometric curve and DTA curve for a sample of pure kaolin. (Courtesy of TA Instruments, New Castle, DE, www.tainstruments.com.)

has been described in Section 17.2, so a few examples of the advantages of combining these systems will be presented. The incorporation of a DTA with one of the optical analyzers described requires the use of two thermocouples to measure the temperature difference between the sample and a reference standard (calcined kaolin or alumina). As noted, the optical noncontact instruments described are particularly useful in the glass and ceramics industries, so examples of evaluating raw materials from these industries will be described (from Paganelli and Venturelli).

Kaolin is soft white clay used to make porcelain and advanced refractory ceramics and extensively used in paper, filled rubber, paint, ink, cosmetics, and many other products. It consists primarily of the mineral kaolinite, an aluminosilicate, and varying amounts of other minerals. Figure 17.54 shows the simultaneous dilatometric curve and DTA curve for a pure kaolin from room temperature to 1400 °C. In the lower DTA curve, three peaks are observed: an endothermic peak at 638 °C due to the dehydroxylation of kaolinite, an exothermic peak at 997 °C due to crystallization of a phase of kaolinite, and a weak exothermic peak at 1263 °C due to crystallization of mullite. Mullite is also an aluminosilicate mineral with a different structure than kaolinite and is formed by high-temperature firing of kaolin. By comparing the DTA curve to the sintering (dilatometric) curve at the top of the figure, it is possible to follow the shrinkage of the material as it loses hydroxyls at 638 °C and as it crystallizes at 997 °C. However, DTA gives no information about the sintering process above 1000 °C, while the dilatometer curve shows very clearly the high degree of shrinkage that occurs above 1000 °C.

Conversely, when studying glass or glass–ceramic systems, the dilatometer provides no information on the glass transition temperature. In Figure 17.55, a sample of a calcium–magnesium–aluminosilicate clearly shows a T_g at 721 °C in the DTA curve, while the sintering curve is completely flat in this region because the sample does not change dimension.

17.10 SUMMARY

Thermometric methods of analysis have wide analytical applications due to the universal nature of heat as a detector. They can be used not only for the identification of chemical compounds but also for the quantitative analysis of mixtures of compounds and for characterizing many types of engineering materials such as polymers, glasses, ceramics, alloys, and composites. Physical and chemical properties of samples can be measured, including fundamental properties like equilibrium constants, rate constants, phase diagrams, thermal expansion coefficients, enthalpies, and specific heats. The fields of TTs and DIE have distinct

THERMAL ANALYSIS

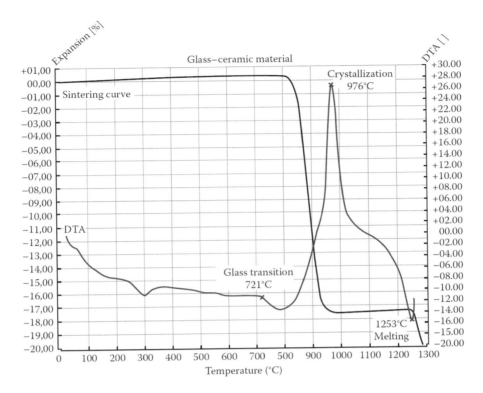

Figure 17.55 Dilatometric curve and DTA curve for a glass–ceramic material. (Courtesy of TA Instruments, New Castle, DE, www.tainstruments.com.)

advantages over conventional volumetric analysis. Micro-DSC, ITC, and microcalorimetry are used extensively in biological testing and analysis of proteins, nucleic acids, and even living organisms like bacteria. The development of new technology like the FMC is permitting analysis of less than 20 µL sample volumes with high throughput. Thermomechanical testing complements room temperature tensile testing for materials and is extremely important in the evaluation of the physical behavior of a wide variety of engineering materials.

Websites with thermal analysis instruments and applications include TA Instruments, www.tainstruments.com; Mettler Toledo, www.mt.com; PerkinElmer, www.perkinelmer.com; Netzsch Instruments, www.netzsch-thermal-analysis.com; Setaram Instrumentation, www.setaram.com; Shimadzu Scientific Instruments, www.ssi.shimadzu.com; Hitachi High Tech Science Corporation, www.hitachi-hitec-science.com; and Cytvita, www.cytivalifesciences.com.

SUGGESTED EXPERIMENTS

17.1 Using a TGA instrument, determine the TGA thermal curve for copper sulfate pentahydrate. A sample size of approximately 10 mg should be heated in nitrogen from room temperature until at least 300 °C. Explain the thermal plot in terms of loss of the water of hydration.

17.2 Repeat experiment 17.1 using KCl. Note the initial loss of free (adsorbed) water and then the loss of bound interstitial water up to temperatures above 200 °C. Could temperatures below 200 °C be used to dry KCl crystals successfully?

17.3 Using a solution of $AlCl_3$ (in acid), neutralize and precipitate the Al as $Al(OH)_3$ by the addition of NH_4OH. Filter the precipitate. Obtain the TGA thermal curve of the precipitate. Heat the precipitate until no change in mass occurs (to at least 1000 °C). Could this precipitate be dried to a constant at any temperature below 1000 °C?

17.4 Take a fresh piece of ham (about 10 mg for an analytical TGA). Obtain the TGA thermal plot in air up to 150 °C. Note the loss in mass due to water evaporation. Do any other losses occur? What is being lost, do you think? Run six additional samples of the same ham. Evaluate the accuracy and precision of the water loss measurement and explain your observations. Many commercial hams (and other meats) contain added water or broth. State regulations may specify how "water-added" products are labeled. Would there be an advantage to a high-capacity TGA if you had to determine if the amount of water in hams met the labeling regulations on a routine basis?

17.5 Obtain the TGA curve of butanol. Heat slowly. Note the slow evaporation of the butanol throughout and then the rapid loss at the boiling point.

17.6 Using about 10 mg samples of PVC powder, obtain the TGA thermal curves (to at least 700 °C) in air and in nitrogen. Compare the curves. The first loss in mass was described in the text. What is it due to? Explain any other mass losses and note any differences in the two curves.

17.7 Using DTA or DSC equipment, obtain the thermal curve of benzoic acid. In the first instance, heat the acid rapidly; in the second, heat it slowly. Note the difference in the apparent melting and boiling ranges of the benzoic acid when the different heating rates are used. Which rate gives the more accurate data?

17.8 Obtain the DTA (or DSC) and TGA curves of benzoic acid. Note that the TGA curve does not indicate the melting point but does indicate the boiling point.

17.9 Obtain the DSC curves for samples of commercial butter, margarines, lard, and solid butter substitutes. The temperature range for the DSC should be from −50 °C to 150 °C. Compare the melting behavior and explain your observations. You can also run this experiment with various types of solid chocolate.

17.10 Obtain DSC thermal curves of several semicrystalline polymers such as polymethylmethacrylate (PMMA), polystyrene, polycarbonate, high-density (HD) polyethylene, and low-density (LD) polyethylene, and look for the glass transition in these polymers. The DSC run may need to be repeated twice with rapid cooling between runs. Many "as received" polymers will show a small peak on top of the glass transition on the first run due to "relaxation effects" in the polymer. The second run should not show this "peak," but only a step change in the baseline. Compare your values of T_g to literature values. Deviations may indicate the presence of plasticizers or other additives in the polymer.

PROBLEMS

17.1 Describe the major components of a TGA instrument. Draw a schematic for a simple TGA instrument. What is measured by the TGA experiment?

17.2 A sample of mixed calcium oxalate monohydrate and anhydrous silica weighed 7.020 mg. After heating to 600 °C, the mass of the mixture was 6.560 mg. What was the mass of calcium oxalate in the original sample?

17.3 What information can be obtained by DTA that cannot be obtained by TGA?

17.4 Describe the components of a DTA instrument and discuss the differences between a DTA and a TGA instrument.

17.5 What are the advantages of using a combination of DTA and TGA to characterize a sample?

17.6 Describe the components of a DSC instrument. Draw a schematic of a heat-flux DSC. What is measured in DSC?

17.7 Make a list of the types of changes that can give exothermic peaks in DSC and DTA. Make a list of the types of changes that can give endothermic peaks in DSC and DTA.

17.8 What applications can you find for DSC that cannot be performed by TGA?

17.9 Why are Pt and quartz commonly used for sample holders in thermal analysis?

17.10 The following figure shows the TGA thermal curves of a glass-fiber-filled polymer (solid line) and the nonfilled polymer (dotted line). (a) If the original sample mass for the glass-fiber-filled polymer was 23.6 mg, what is the mass % glass fiber in the sample? (b) What would you recommend as the upper temperature limit for the use of this material, assuming you want to build in a safety factor of 20 % (i.e., you want to be no higher than 80 % of the decomposition temperature)?

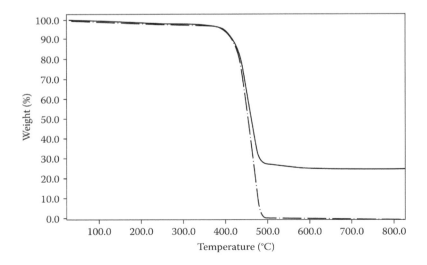

17.11 The following figure shows the TGA thermal curves for copper sulfate pentahydrate, $CuSO_4 \cdot 5H_2O$. The solid line is the TGA of the crystalline salt. (a) Assuming that all of the losses are due to water, how many different types of water are present? (b) How many moles of water are lost in each step? (c) Does the total mass loss confirm that this is $CuSO_4 \cdot 5H_2O$? The dotted line is the TGA of the crystalline salt that has been stored in a desiccator with magnesium perchlorate as the desiccant. (d) Comment on the differences observed between the two thermal curves. (e) How much water has the desiccant removed from the compound?

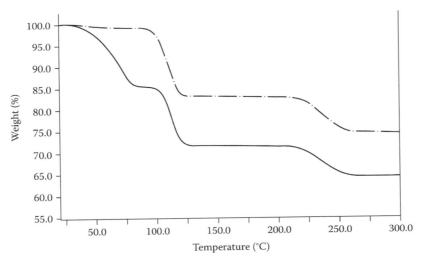

17.12 Shown in the following figure are the TGA (solid line) and the DTG (dotted line) thermal curves for a sample of hydrated citric acid, $C_6H_8O_7 \cdot nH_2O$. The formula for anhydrous citric acid is $C_6H_8O_7$. The mass of the sample used is 20.09 mg. (a) Assuming the first mass loss is the loss of water, how many moles of water of hydration are present? (What is the value of n in the formula $C_6H_8O_7 \cdot nH_2O$?) (b) What decomposition temperature would you report for anhydrous citric acid? (c) What drying temperature would you recommend for producing anhydrous citric acid from the hydrated form? Which thermal curve is more useful for assigning the temperatures in (b) and (c)?

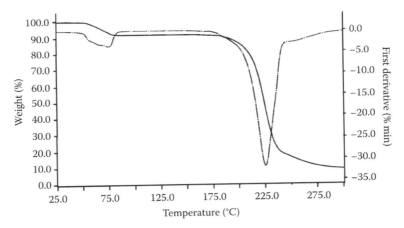

17.13 The DTA thermal curves for two polyolefins are shown in the following figure. Curve 1 is a sample of LD polyethylene. Curve 2 is a sample of stabilized polypropylene. The melting points, T_m, are shown. (a) What are the melting points for each polymer? (b) Which polymer would be better to use for hot water pipes in houses? Why? (c) What process is occurring at the high-temperature end of each thermal curve?

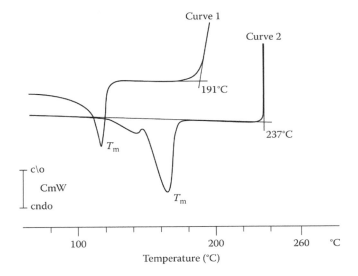

17.14 The DSC thermal curve of a sample of "light" margarine is shown. (*Note*: This plot has the endothermic peaks pointing up!) (a) What process and what substance do you think give rise to the endothermic peak at 4.60 °C? (Think about what might be added to margarine to reduce the calories.) (b) Is the endothermic peak at −24.6 °C due to the melting of the oil used to make the margarine? Explain your answer.

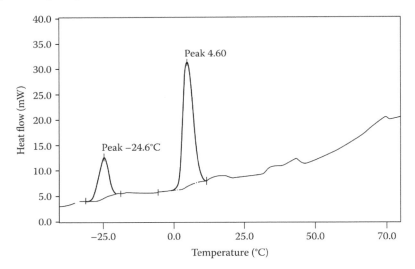

(*Note*: The figures for Problems 17.10 through 17.12 and 17.14: [© 1993–2014 PerkinElmer, Inc. All rights reserved. Printed with permission. (www.perkinelmer.com).]. The figure for Problem 17.13: Courtesy of Mettler GmbH, Switzerland.)

17.15 Describe the advantages of microcalorimetry for biological and pharmaceutical applications.
17.16 What analyses can be performed by ITC?
17.17 What are the advantages of optical noncontact dilatometers compared with traditional dilatometers?

BIBLIOGRAPHY

The reader will note that occasionally multiple editions of books are cited as sources or reference works. The reason for this is that sometimes different figures are used in different editions, which may in fact be written by different authors. The sources found in the bibliography will be the most recent versions of a particular work.

In some cases instrument manufacturers and suppliers are cited by providing a URL to the exact internet location. We have endeavored to ensure that these links are functional at the time of publication, but clearly these links change. If a link has become nonfunctional, a good approach is to use your browser's advanced search capability. For example: "company name" AND "instrument type" AND "specific descriptors" is a good place to start.

ASM Handbook, Volume 10 Materials Characterization, ASM International, Materials Park, OH, 2019.

Brown, M.E. *Introduction to Thermal Analysis: Techniques and Applications.* Chapman & Hall: New York, 1988.

Callister, W.D. *Materials Science and Engineering: An Introduction*, 5th edn. John Wiley & Sons: New York, 2000.

Connolly, M.; Tobias, B. Analysis of thermoset curve by dynamic mechanical, dielectric, and calorimetric methods. *Am. Lab., 1,* 38, 1992.

Craig, D.Q.M.; Reading, M. (eds.) *Thermal Analysis of Pharmaceuticals.* CRC Press, Taylor & Francis Group: Boca Raton, FL, 2007.

Duval, C. *Inorganic Thermogravimetric Analysis*, 2nd edn. American Elsevier: New York, 1963.

Earnest, C. Modern thermogravimetry. *Anal. Chem., November,* 1471A, 1984.

Gallagher, P.K. Thermoanalytical methods. In Cahn, R.W.; Haasen, P.; Kramer, E.J. (eds.) *Materials Science and Technology, A Comprehensive Treatment, Part 1,* Vol. 2A. VCH: Weinheim, Germany, 1992.

Groves, I.; Lever, T.; Hawkins, N. *TA Instruments Applications Brief TA-123.* TA Instruments, Ltd: U.K.

Jelesarov, I.; Bosshard, H.R. ITC and DSC as complementary tools to investigate the energetics of biomolecular recognition, *J. Mol. Recognit., 12,* 3, 1999.

Kraftmakher, Y. *Modulation Calorimetry-Theory and Applications.* Springer: New York, 2004.

Menard, K.P.; Spragg, R.; Johnson, G.; Sellman, C. Hyphenation: The next step in thermal analysis. *Am. Lab., 1,* 42, 2010.

Mohomed, K.; Bohnsack, D.A. Differential scanning calorimetry as an analytical tool in plastics failure analysis. *Am. Lab., 3,* 45, 2013.

Olivon, M.R. et al. Phase transitions and polymorphism of cocoa butter. *JAOCS, 75,* 4, 1998.

Paganelli, M.; Venturelli, C. Thermal analysis for raw materials controlling. Expert System Solutions Application Note, www.expertsystemsolutions.com.

Pazstor, A.J. Thermal analysis techniques. In Settle, F.A. (ed.) *Handbook of Instrumental Techniques for Analytical Chemistry.* Prentice Hall, Inc.: Upper Saddle River, NJ, 1997.

Sahle-Demessie, E.; Zhao, A.; Salamon, A.W. *PerkinElmer Application Note,* www.perkinelmer.com/nano. PerkinElmer, Inc., Waltham, MA.

Speyer, R.F. *Thermal Analysis of Materials.* Marcel Dekker, Inc.: New York, 1994.

Turi, E.A. (ed.) *Thermal Characterization of Polymeric Materials*, 2nd edn. Academic Press: New York, 1996.

Tyrell, H.V.; Beezer, A.E. *Thermometric Titrimetry.* Chapman & Hall: London, U.K., 1968.

Utschick, H. *Methods of Thermal Analysis: Applications from Inorganic Chemistry, Organic Chemistry, Polymer Chemistry and Technology.* Ecomed Verlagsgesellschaft AG and Co KG: Landsberg, Germany, 1999.

Venturelli, C. Heating microscopy and its applications. *Microsc. Today, 19,* 20, 2011.

Washall, J.W.; Wampler, T.P. Direct-pyrolysis Fourier transform-infrared spectroscopy for polymer analysis. *Spectroscopy,* 6(4), 38, 1992.

Wendlandt, W.W. *Thermal Methods of Analysis*, 3rd edn. John Wiley & Sons, Inc.: New York, 1986.

Wunderlich, B. *Thermal Analysis.* Academic Press: Boston, MA, 1990.

Appendix: Abbreviations, Acronyms, and Symbols

Instrumentation for analytical chemistry gives rise to many abbreviations, some forming "acronyms." These are often more encountered than the terms they abbreviate, and they appear extensively in the text. A reader new to the field may become lost or disoriented in this thicket of initials. Beginning with the 93rd edition of the *CRC Handbook of Chemistry and Physics* in 2012, the late editor William (Mickey) Haynes, together with one of the authors (TJB) developed an abbreviation/acronym list with definitions that has been updated every year. A number of recognized experts in analytical chemistry contributed to this effort, including the then NIST director Dr. Willie E. May. Beginning with this 8th edition, this more comprehensive listing is provided alphabetically, and the reader is then urged to consult the index to find the passages in which the techniques are discussed. Note that these acronyms are frequently compounded, as in UV/VIS (ultraviolet/visible) or LC-CI-TOFMS (interfaced liquid chromatograph to time-of-flight mass spectrometer operating in chemical ionization mode). The components of such compounded abbreviations are listed individually in the index, but not all the possible combinations. All acronyms are abbreviations, but the reverse is not true. For example, CLIPS, DART, DRIFTS, and COSY are acronyms because they form words or are pronounced as words; GFAAS, APCI, and FTIR are abbreviations.

Abbreviations and Symbols used in Analytical Chemistry

ABBREVIATIONS USED IN ANALYTICAL CHEMISTRY

A	peak area; surface area of solid granular adsorbent; eddy diffusion term in the Van Deemter equation
AAA	absolute activation analysis
AAD	atomic absorption detector
AAS	atomic absorption spectroscopy
AC	alternating current, affinity chromatography
ACP	alternating current plasma
ADXPS	angular dependent x-ray photoelectron spectroscopy
AED	atomic emission detector
AEM	analytical electron microscope (microscopy)
AES	Auger electron spectroscopy, atomic emission spectroscopy
AFID	alkali flame ionization detector
AFM	atomic force microscopy
AFS	atomic force spectroscopy
AIS	average of individual samples
AL	action level
AM	amplitude modulation
AMS	accelerator mass spectrometry
AN	area normalization
ANRF	area normalization with response factors
AOTF	acousto-optical tunable filter
AP	analytical pyrolysis
APCI	atmospheric pressure chemical ionization
APD	azimuthal photoelectron diffraction
API	atmospheric pressure ionization
APSTM	analytical photon scanning tunneling microscope
APT	attached proton test
ARAES	angle resolved Auger electron spectroscopy
ARF	absolute response factor
ARM	atomic resolution microscopy
ARPES	angle resolved photoelectron spectroscopy
ARUPS	angle resolved ultraviolet photoelectron spectroscopy
ASE	accelerated solvent extraction
AsFIFFF (AF4)	asymmetrical flow field flow fractionation
ATD	above-threshold dissociation
ATI	above-threshold ionization
ATR	attenuated total reflection
BB	band broadening
BE	magnetic sector—electric sector tandem mass spectrometer (note: also called a MIKE spectrometer)
BEE	magnetic sector—electric sector—electric sector mass spectrometer
BET	Brunauer-Emmett-Teller (adsorption isotherm)
BIFL	burst integrated fluorescence lifetime
BIS	Bremsstrahlung isochromat spectroscopy
BJH	Barrett-Joyner-Halenda (method)
BL	bioluminescence
BLRF	bispectral luminescence radiance factor
BQQ	magnetic sector-double quadrupole mass spectrometer BTOF
BTOF	magnetic sector—time-of-flight tandem mass spectrometer
CAD	collision-activated dissociation
CAR	continuous addition of reagent
CARS	coherent anti-Stokes Raman spectroscopy
CCC	counter-current chromatography
CCD	charge-coupled device
CCT	constant current topography
CD	circular dichroism
CE	capillary electrophoresis, counter electrode
CEC	capillary electrokinetic chromatography, capillary electrochromatography
CED	cohesive energy density
CFA	continuous How analysis
CF-FAB	continuous low-fast atom bombardment
CFM	chemical force microscopy
CGE	capillary gel electrophoresis
CHEMFET	chemical-sensing field effect transistor
CI	chemical ionization
CID	collision-induced dissociation
CIEF	capillary isoelectric focusing
CITP	capillary isotachophoresis
CL	chemiluminescence
CLLE	continuous liquid–liquid extraction
CMA	cylindrical mirror analyzer
COSY	correlation spectroscopy
CPAA	charged particle activation analysis
CP/MAS	cross polarization/magic angle spinning
CRDS	cavity ring-down spectroscopy
CRF	chromatographic response function
CRM	certified reference material
CS	carbon strip (adsorbent)
CT	cryogenic trapping
CTD	charge transfer device
CV	cyclic voltammetry
CV-ASS	cold vapor atomic absorption spectrometry
CVD	chemical vapor deposition

CW	continuous wave	EPMA	electron-probe microanalysis
CZE	capillary zone electrophoresis	EPR	electron paramagnetic resonance
DA	diode array	EPXMA	electron-probe x-ray microanalysis
DAD	diode array detector (UV-Vis)	EQL	estimated quantitation limit
DADI	direct analysis of daughter ions	ERO	elastic recoil detection
dB	de Boar t-plot	ESA	electrostatic analyzer
DBE	double bond equivalent	ESCA	electron spectroscopy for chemical analysis
DC	direct current		
DCI	desorption chemical ionization	ESEM	environmental scanning electron microscope
DCP	direct-current plasma		
DEP	differential electrolytic potentiometry	ESI	electrospray ionization
DEPT	distortionless enhancement by polarization transfer	ESP	electrospray
		ESR	electron spin resonance
		ET	electrometric titration
DETA	dielectric thermal analysis	ETA	electrothermal analyzer, emanation thermal analysis
DIN	direct injection nebulizer		
DLI	direct liquid introduction	EXAFS	extended x-ray absorption fine structure
DLS	dynamic light scattering		
DMA	dynamic mechanical analysis	FAA	flame atomic absorption
DME	dropping mercury electrode	FAAS	flame atomic absorption spectroscopy
DNMR	dynamic nuclear magnetic resonance	FABMS	fast-atom bombardment mass spectrometry
DPP	differential pulse polarography		
DRIFT	diffuse-reflectance infrared Fourier transform	FAES	flame atomic emission spectroscopy
		FAFS	flame atomic fluorescence spectroscopy
DSC	differential scanning calorimetry	FAM	field analytical method
DTA	differential thermal analysis	FAS	flame absorption spectroscopy
DTC	differential thermal calorimetry	FD	field desorption
EAES	electron-excited Auger electron spectroscopy	FD/FI	field desorption/field ionization
		FES	flame emission spectroscopy
EB	electric sector—magnetic sector tandem mass spectrometer	FFEM	freeze-fracture electron microscopy
		FFF	field-flow fractionation
EBE	electric sector—magnetic sector-electric sector tandem mass spectrometer	FFFF	flow field-flow fractionation
		FFM	friction force microscopy
		FFS	flame fluorescence spectroscopy
EC	electrochemical	FFT	fast Fourier transform
ECO	electron capture detector	FGC	fast gas chromatography
ECMS	electron capture mass spectrometry	FI	flow injection, field ionization
ECNIMS	electron capture negative ionization mass spectrometry	FIA	flow injection analysis
		FIB	focused ion beam
EDL	electrodeless discharge lamp	FID	flame ionization detector, free-induction decay
EDS	energy-dispersive spectrometer		
EDXRF	energy-dispersive x-ray fluorescence	FIM	field ion microscopy
EELS	electron energy-loss spectroscopy	FNAA	fast neutron activation analysis
EFFF	electric field flow fractionation	FOCS	fiber optic chemical sensor
EG	electrogravimetry	FPO	flame photometric detector
EGA	evolved gas analysis	FSOT	fused silica open tubular (column)
EIA	enzyme-linked immunoassay	FT	Fourier transform
EI(I)	electron impact (ionization)	FT-ICR	Fourier transform ion cyclotron resonance
EIMS	electron impact mass spectrometry		
ELCO	electrolytic conductivity detector	FT-IR	Fourier transform infrared (often "FT/IR," "FTIR," "FT IR")
ELISA	enzyme-linked immunosorbent assay		
ELSD	evaporative light scattering detector	FT-IRRAS	FT-IR reflection-absorption spectroscopy
EM	electron microscopy	FT-MS	Fourier transform mass spectrometry
EMIRS	electrochemically modulated IR spectroscopy	FWHM	full-width half-maximum
		GC	gas chromatography
EOF	electro-osmotic flow		
EPL	enhanced photoactivated luminescence	GC-IR	gas chromatography-infrared spectrometry

GCMS	gas chromatography mass spectrometry	INADEQUATE	incredible natural abundance double-quantum transfer experiment
GDL	glow discharge lamp		
GDMS	glow discharge mass spectrometry	INEPT	insensitive nuclei enhancement by polarization transfer
GE	gel electrophoresis, gradient elution		
GEMBE	gradient elution moving boundary electrophoresis	INAA	instrumental neutron activation analysis
		IP	ion pairing
GFAAS	graphite furnace atomic absorption spectroscopy	IPC	ion-pair chromatography
		IPG	immobilized pH gradient
GLC	gas–liquid chromatography	IPMA	ion probe microanalysis
GPC	gel permeation chromatography	IR	infrared (spectrophotometry)
GS	Gram-Schmidt (algorithm)	IRN	indicator radionuclide(s)
GSC	gas-solid chromatography	IRS	internal reflection spectroscopy
GSED	gaseous secondary electron detector	ISCA	ionization spectroscopy for chemical analysis
HCL	hollow cathode lamp		
HCOT	helically coiled open tubular (column)	ISE	ion selective electrode
HOC	hydrodynamic chromatography	ISP	ion spray
HETCOR	heteronuclear correlation	ISS	ion scattering spectrometry
HETP	height equivalent of (a) theoretical plate(s)	LAMMS	laser micro mass spectrometry
		LARIMS	laser atomization resonance ionization mass spectrometry
HG	hydride generation		
HIC	hydrophobic interaction chromatography	LARIS	laser atomization resonance ionization spectroscopy
HMBC	heteronuclear multiple-bond correlation		
HPAC	high-performance affinity chromatography	LASER	light amplification by stimulated emission of radiation
HPIAC	high-performance immunoaffinity chromatography		
		LBB	Lambert-Beer-Bouguer law
HPLC	high-performance liquid chromatography, high-pressure liquid chromatography	LC	liquid chromatography
		LC-LS	multidimensional liquid chromatography
HPTLC	high-performance thin-layer chromatography		
		LDMS	laser desorption mass spectrometry
HRCGC	high-resolution capillary gas chromatography	LDR	linear dynamic range
		LEAFS	laser-excited atomic fluorescence spectrometry
HRGC	high-resolution gas(-liquid) chromatography		
		LED	light-emitting diode
HS	headspace	LEED	low-energy electron diffraction
HSA	hemispherical analyzer	LEEM	low-energy electron microscopy
HSC	heteronuclear shift correlation	LEI	laser-enhanced ionization
HSQC	heteronuclear single quantum coherence	LEISS	low-energy ion scattering spectrometry
HTC	high-temperature combustion	LESS	laser-excited Shpol'skii spectroscopy
IA	isocratic analysis	LFM	lateral force microscopy
IAC	immunoaffinity chromatography	LIDAR	light detection and ranging
IAES	ion-excited Auger electron spectroscopy	LIFO	laser-induced fluorescence detection
IC	ion chromatography	LIMS	laboratory information management system
ICMS	ion chromatography mass spectrometry		
ICP	inductively coupled plasma	LLC	liquid–liquid chromatography
ICP-OES	ICP optical emission spectrometry	LLD	lower-limit detection
ICR	ion cyclotron resonance	LLE	liquid-liquid extraction
IDMS	isotope dilution mass spectrometry	LNRI	laser non-resonant ionization
IEC	ion-exchange chromatography	LO	local oscillator
IEF	isoelectric focusing	LOC	lab on a chip
IF	intermediate frequency	LOD	limit of detection
IGC	inverse gas chromatography	LPDA	linear photodiode array
IGF	inert gas fusion	LPSIRS	linear potential-sweep IR reflectance spectroscopy
ILDA	intensified linear diode array		
IMAC	immobilized metal-ion affinity chromatography	LRI	laser resonance ionization
		LRMA	laser Raman microanalysis

LSC	liquid–solid chromatography	ODS	octadecylsilane
LSE	liquid–solid extraction	OES	optical emission spectrometry, optical emission spectroscopy
LTP	low-temperature phosphorescence		
MAE	microwave-assisted extraction	OID	optoelectronic imaging device
MALDI	matrix-assisted laser desorption/ionization	OMA	optical multichannel analyzer
MAS	magic angle spinning	OPO	optical parametric oscillator
MCD	magnetic circular dichroism	OPTLC	over-pressured thin-layer chromatography
MCP	microchannel plate	ORD	optical rotary dispersion
MDGC	multidimensional gas chromatography	OTE	optically transparent electrodes
MDL	method detection limit	PA	proton affinity
MDM	minimum detectable mass	PAA	photon activation analysis
MDQ	minimum detectable quantity	PAGE	polyacrylamide gel electrophoresis
MEIS	medium-energy ion scattering	PAH	polycyclic aromatic hydrocarbon
MEKC	micellar electrokinetic chromatography	PAS	photoacoustic spectroscopy
MFM	magnetic force microscopy	PB	particle beam
MID	multiple ion detection	PC	paper chromatography
MIKE	mass analyzed ion kinetic energy mass spectrometry	PCA	principal component analysis
		PCR	polymerase chain reaction
MIP	microwave-induced plasma, mercury intrusion porosimetry	PCS	photon correlation spectroscopy
		PCSE	partially coherent solvent evaporation
MIRS	multiple internal reflection spectroscopy	PD	plasma desorption
MLC	micellar liquid chromatography	PDA	photodiode array
MLLSQ	multiple linear least squares	PDHID	pulsed discharge helium ionization detector
MMF	minimum mass fraction		
MMLLE	microporous membrane liquid–liquid extraction	PDMS	plasma desorption mass spectrometry, polydimethyl siloxane
MPD	microwave plasma detector	PED	pulsed electrochemical detection, plasma emission detector, photoelectron diffraction
MPI	multiphoton ionization		
MRDL	maximum residual disinfectant level (in water analysis)		
		PES	photoelectron spectroscopy
MRI	magnetic resonance imaging	PET	positron emission tomography
MS	mass spectrometry	PFIA	process flow injection analysis
MS-MS	tandem mass spectrometry	PGC	packed-column gas chromatography
MSPD	matric solid-phase dispersion	pH	negative logarithm of hydrogen ion concentration
MSRTP	micelle-stabilized room-temperature phosphorescence		
		PICI	positive ion chemical ionization
MWD	microwave (assisted) digestion	PID	photoionization detector
NAA	neutron activation analysis	PIXE	particle-induced X-ray emission
NCIMS	negative chemical ionization mass spectrometry	pK	negative logarithm of an equilibrium constant
NDP	neutron depth profiling	PLE	pressurized liquid extraction
NEXAFS	near edge X-ray absorption fine structure	PLOT	porous-layer open tubular
		PLOT-cryo	porous-layer open tubular (column) cryo-adsorption
NHE	normal-hydrogen electrode		
NICI	negative ion chemical ionization	PMT	photomultiplier tube
NIR	near-infrared, near-IR	ppb	parts per billion
NIRA	near-infrared reflectance analysis	ppm	parts per million
nm	nanometer	ppt	parts per thousand, parts per trillion
NMR	nuclear magnetic resonance	PSD	position sensitive detector
NOE (nOe)	nuclear Overhauser effect	PTFE	polytetrafluoroethylene
NOESY	nuclear Overhauser effect spectroscopy	PTR	proton transfer reaction (in mass spectrometry)
NPD	nitrogen-phosphorus detector, normal photoelectron diffraction		
		PTV	programmable temperature vaporizer
NPLC	normal-phase liquid chromatography	PVD	pulsed voltammetric detection, physical vapor deposition
ODMR	optically detected magnetic resonance		

QCL	quantum cascade laser	SFE	supercritical-fluid extraction
QCM	quartz-crystal microbalance	SFFF	sedimentation field flow fractionation
QFAA	quartz furnace atomic adsorption	SF-MS	sector field mass spectrometry
QIT	quadrupole ion trap	SFS	synchronous fluorescence spectroscopy
qNMR	quantitative nuclear magnetic resonance	SHE	standard hydrogen electrode
		SIA	sequential injection analysis
QQQ, QqQ, Qiq	triple quadrupole mass spectrometer	SIDA	stable isotope dilution assay
QTH	quartz tungsten halogen	SIMS	secondary ion mass spectrometry
QTOF	tandem quadrupole time-of-flight mass spectrometer	SIRIS	sputter-initiated resonance ionization spectroscopy
RAA	running annual average	SMCL	secondary maximum contaminant level (in water analysis)
RBS	Rutherford backscattering spectrometry		
REELS	reflection electron energy loss spectrometry	SMDE	static mercury drop electrode
		SMSS	spark source mass spectrometry
RES	reflection electron spectrometry	SNIFTIRS	subtractively normalized interfacial FT-IR spectroscopy
RF	radio frequency		
RHEED	reflection high-energy electron diffraction	SNMS	sputtered neutral mass spectrometry
		SPAES	spin-polarized Auger electron spectroscopy
RI	refractive index, retention index		
RIC	reconstructed ion chromatogram	SPE	solid-phase extraction
RID	refractive-index detector	SPME	solid-phase microextraction
RIMS	resonance ionization mass spectrometry	SPR	surface plasmon resonance
		SRE	stray radiant energy
RIS	resonance ionization spectroscopy, range of individual samples	SRM	standard reference material
		SSMS	spark source mass spectrometry
RM	reference material	SSRTF	solid-surface room-temperature fluorescence
RNAA	radiochemical neutron activation analysis		
		SSRTP	solid-surface room-temperature phosphorescence
ROA	Raman optical activity		
ROESY	rotating frame nuclear Overhauser effect spectroscopy	STEM	scanning transmission electron microscope
RPLC	reversed-phase liquid chromatography	STM	scanning tunneling microscope
RRDE	rotating ring-disk electrode	SVE	solvent vapor exit
RS	Raman spectroscopy	SWE	supercritical-water extraction
RSF	relative sensitivity factor	TCA	thermochemical analysis
RTIL	room-temperature ionic liquid	TCD	thermal-conductivity detector
RTP	room-temperature phosphorescence	TCT-GC-MS	thermal cold trap gas chromatography mass spectrometry thermodilatometry
SIN	signal-to-noise ratio		
SAE	sonication-assisted extraction	TD	thermodilatometry
SAM	scanning Auger microscopy, self-assembly monolayers	TDL	tunable diode laser
		TEA	thermal energy analyzer
SANS	small-angle neutron scattering	TED	thermionic emission detector
SAW	surface acoustic wave	TEELS	transmission electron energy loss spectrometry
SAXS	small-angle x-ray scattering		
SBSE	stir bar sorptive extraction	TEM	transmission electron microscope
SCE	standard calomel electrode, saturated calomel electrode	TET	thermometric enthalpimetric titration
		TFFF	thermal field flow fractionation
SCF	supercritical fluid	TGA	thermogravimetric analysis
SCOT	support-coated open tubular	TGA-IR	thermogravimetric analysis—infrared
SDD	silicon drift detector	THEED	transmission high-energy electron diffraction
SdFFF	sedimentation field flow fractionation		
SEC	size-exclusion chromatography	ThFFF	thermal field flow fractionation
SEM	scanning electron microscope	TIC	total ion current chromatogram, tentatively identified compound
SERS	surface-enhanced Raman spectroscopy		
SFC	supercritical-fluid chromatography	TIMS	thermal ionization mass spectrometry

TLC	thin-layer chromatography	A	peak asymmetry factor
TLE	thin-layer electrode	B	magnetic field strength
TLM	thermal lens microscopy	$[c]$	concentration of component c
TLV	threshold limit value	d_p	particle diameter (HPLC stationary phase)
TMA	thermomechanical analysis	D_{ab}	diffusion coefficient
TMS	tetramethylsilane	e	electron elementary charge
TOCSY	total correlation spectroscopy	ε	extinction coefficient
TOF	time-of-flight	E	energy
TOF-MS	time-of-flight mass spectrometry	E	electrode potential
TSP	thermospray	E_b	binding energy
UHV	ultrahigh vacuum	E_{ea}	electron affinity
USE	ultrasonic extraction	E_{ai}	ionization energy
UV	ultraviolet	v	frequency
UVPES, UPS	ultraviolet photoelectron spectroscopy	γ	gyromagnetic ratio
UV-VIS, UV-Vis	ultraviolet-visible	n	refractive index
VAR	variable angle reflectance	h	Planck constant
Vis	visible (radiation)	H	enthalpy, plate height
VOC	volatile organic compound(s)	I_0	incident intensity
VOX	volatile organic halogens	J	coupling constant
VUV	vacuum ultraviolet	k	coverage factor
W	Wein filter (used in mass spectrometry)	k'	capacity factor
WCOT	wall-coated open tubular	λ	wavelength, thermal conductivity
WDS	wavelength dispersive spectrometer	m/z	mass-to-elementary-charge ratio (mass spectrometry)
WWCOT	whisker-wall coated open tubular (column)	Q_{crit}	Q value (outlier test)
WWPLOT	whisker-wall coated porous layer open tubular (column)	q	quadrupole parameter (mass spectrometry)
		r	correlation coefficient
WWSCOT	whisker-wall support-coated open tubular (column)	R	resolution
		R_{00}	Rydberg constant
XAES	x-ray excited Auger electron spectroscopy, X-ray adsorption edge spectrometry	ρ	density
		s	standard deviation
		s^2	variance
XANES	X-ray absorption near-edge spectroscopy	δ	chemical shift
		δ^0	solubility parameter
XPS	X-ray photoelectron spectroscopy	r	true value of a measured quantity
XRD	X-ray diffraction	τ_{crit}	Chauvanet's criterion (outlier test)
XRF	X-ray fluorescence	$t_{1/2}$	half-life
XRFS	X-ray fluorescence spectroscopy	t_M	mobile-phase hold up
XRS	X-ray spectroscopy	t_R	retention time
ZAF	Z (element number) absorption fluorescence	t_R^0	specific retention time
		T	transmittance
		T1	spin-lattice relaxation time
		T2	spin–spin relaxation time

SYMBOLS USED IN ANALYTICAL CHEMISTRY

$\bar{u}$	carrier phase velocity		
a	Auger yield, fine structure constant	V_M	carrier hold-up volume
a_0	Bohr radius	V_R	retention volume
A	absorbance	V_R^0	specific retention volume

Subject Index

A

AAS, see atomic absorption spectrometry
Ablation 457–458
Absorbance detectors, infrared (IR) spectroscopy 767
Absorbances, corrected 92
Absorption
  by atoms 375–377
  by molecules 124–125, 184–189
  edges 571–573
  of infra-red radiation 184–189
  laws 89–91
  of radiant energy 375–377
Absorption bands 88, 184
Absorption edges 496
Absorption spectra 85
Absorptivity 90
Accelerated solvent extraction (ASE) 33
Acclaim Trinity P1 Columns 787
Accuracy
  and precision 18–19
  in UV/vis absorption spectrometry 166
Acid digestion, and dissolution 29–31
Activity series 822–823
Adduct 583
Adjusted retention time 694, 699
AES, see Auger electron spectroscopy
affinity chromatography 788–789
AFS, see atomic force spectroscopy
Agilent method translator 728
Agilent pressure-flow calculator 728
Air absorption 205
Air lines, safety 67–68
Aliquots 12
Alpha cleavage 641
Amorphous materials 82
Amperometric detectors 768
Amperometry 826
Ampholyte mixture 803
Amphoteric compounds 801
Amplitude 81
Analytical approaches 3–15
Analytical chemistry 1–2
Analytical methods, design of 10–11
Analyzers 580; see also specific types
Anode 817
Anodic stripping voltammetry 867–868
Anti-Stokes direct-line fluorescence 469
Anti-Stokes scattering 242
Antibonding orbitals 88, 123–124, 151
APCI, see atmospheric pressure chemical ionization
Application buffer 789
Applied potential 845–846
Arc and spark emission spectroscopy
  applications 443–445
  commercial systems 472
  interferences 441–443
  in matrix interference 441–442
  qualitative analysis 443
  quantitative analysis 443–445
  spectral interferences 442

Array detectors 611
ASE, see Accelerated solvent extraction
Atmospheric pressure, ionization sources 584–586
Atmospheric pressure chemical ionization (APCI) 775
  interface 776–777
Atomic absorption spectrometry (AAS)
  analytical applications 402–409
  atomizers 384
  commercial systems 386–387
  detectors 386
  instrumentation 377–387
  interferences 392–402
  liquid samples 404–405
  matrix interference 393–394
  nonspectral interferences 392–396
  qualitative analysis 402
  quantitative analysis 402–404
  radiation sources 378–384
  solid samples 405–407
  spectral interferences 397–402
Atomic emission detectors, gas chromatography 737
Atomic emission spectroscopy (AES) 480
  qualitative analysis 427
  quantitative analysis 427–429
  spectral interferences 426–427
Atomic fluorescence spectroscopy (AFS) 467–472
  applications 470–472
  commercial systems 472–473
  instrumentation 469–470
  mercury 471–472
  spectral interferences 470
  systems 473
Atomic mass 625
Atomic mass spectrometry (MS) 657–674
  applications 660–666
  instrumentation 670–672
  interferences 666–672
  limitations 672–674
  matrix effects 666–667
Atomic optical emission spectrometry (OES) 429–443
  detectors 437–441
  sample holders 435
  spectrometers 435–437
Atomic spectral interference 397
Atomic spectroscopy
  analytical techniques 484–486
  and atoms 86–87
  and ICP-MS 480
Atomization efficiency 390
Atomization process 387–392
Atomizers 380
Atoms, and atomic spectroscopy 86–87
Attenuated total reflectance 201, 206–208
Auger electron spectroscopy (AES) 885, 898–902
  applications 900–902
  instrumentation 899–900

Autosamplers 714
Auxochrome 150

B

Babington nebulizer 381, 454–455
Background absorption 397
Background correction
  spectral interferences 397
  in transmission measurements 205–206
Background emission 95
Barrier layer cell 133–134
Base peak 578, 619
Basic equations, for chromatography 698–701
Basic statistics, and data handling 17–29
Bathochromic shift 150–151
Batteries 55–56
Beam splitters 111–114, 194–195, 197, 385
Beer–Lambert–Bouguer law 91
Beer's law
  derivations 91
  errors 98–100
Bell, Alexander Graham 265
Bell-shaped curves 694
Bending vibrations 185
Bilayer vesicle 804
Biochemistry
  applications 692
  proteomics 778–782
Biospecific elution 789
Blanks 11
Bleeding 711, 717
Blue shift 151–152
Bolometer 198–199
Boltzmann constant 376, 424
Boltzmann distribution 281, 376, 424
Boltzmann ratio 281
Bonded phase column 717
Bonding orbital 88, 123, 151
Bragg equation 498
Bremsstrahlung 886
Bruker minispec 348
Bulk analysis 466
Burner assembly 422–423
Burning velocity 382

C

Calculations, significant figures 17–18
Calibration 38, 403–404
  curve plotting 41–42
  methods 92–100
Calibration standards 11, 92–93
  additions method 841–842
  external 705–706
  internal 706–707
Calorimetry
  isothermal titration 946–948
  microliter flow 948–949
Capacity factor 698
Capillary chromatography 803–805
Capillary electrochromatography 803–805

SUBJECT INDEX

Capillary gel electrophoresis 801
Capillary isoelectric focusing 801–803
Capillary zone electrophoresis (CZE) 695, 794–801
^{13}Carbon nuclear magnetic resonance (NMR) 324–331
   spectra interpretation 330–331
   spectra of solids 328–330
Carrier gas 711
Cathode 817
Cathode lamp, hollow 378–379
CE, see conducting electrophoresis
Certified reference materials 19
Changers, for samples 527
Charge detectors 770–771
Chemical exchange 311–312
Chemical imaging
   infrared (IR) spectroscopy 256–262
   spectroscopy 256–262
Chemical interference 392–393, 425–426, 470
Chemical ionization 583–584
   interference 459
Chemical modification 395–396
Chemical shift 285, 287–291
Chemical shift anisotropy 286
Chemical speciation 464–465
Chemiluminescence detectors 734–735
Chemistry, applications 664–666
Chemometrics 238
Chiral chromatography 755
Chiral phase separation, enantiomers 755–756
Chromatograms 693, 711
Chromatographic calculations 707–708
Chromatographic process 691–692
Chromatographic separations
   basic equations 698–701
   optimization 702–703
Chromatography 342, 687–688
   analyte identification 704
   dissolved ions 782–788
   and electrophoresis systems 581–582
   and (IR) infrared spectrometry systems 582
   with liquid mobile phases 751–810
   and mass spectrometry (MS) 582
   mixtures 688–691
   molecular level 693–696
   multidimensional 692–693
   principles of 687–710
   quantitative analysis 704–707
   role of silicon oxygen compounds 696–698
   silicon oxygen compounds 696–698
   size-exclusion 789–791; see also specific types
Chromatography–ICP-MS speciation 663–664
Chromophore 150
Citius LC-HRT 599–601
Classical polarography 855–859
Clinical chemistry applications 664
Cold vapor atom absorption spectrometry (CVAAS) 24, 383–384
Cold vapor mercury technique 407–408
Collector electrodes 732
Collector ring 732
Collimator masks 516
Collimators 516
Collision cells 670–671
Collision-induced dissociation (CID) 774

Column stationary phase 716–719
Columns
   band broadening effects 703–704
   variables 701–702
Combination electrodes 829
Combinatorial chemistry 342
Compounds
   with heteroatoms 626–627, 646–648
   and mixtures 652–654
Comprehensive 2D chromatograms 693
Compressed air lines, safety 67–68
Compressed gas cylinders, safety 58–61
Compressed gas regulators, and manifolds 59–60
Computed axial tomography 489
Computed tomography 489
Concentration, and fluorescence intensity 169–170
Conductance 850
Conductimetric titrations
   in aqueous solution 853–854
   in nonaqueous solvents 854
Conductimetry 826
Conducting electrophoresis (CE) 800–803
   conductivity measurements 850–855
Conductometric detection 869–870
Conductometric detectors 770
Conductometric measurements 853–855
Confidence limits 26–27
Conjugated diene systems 152–154
Conjugated ketone systems 154–156
Connes' advantage 195, 197
Constant analyzer transmission (CAT) 888
Constant relative ratio (CRR) 888
Continuous dynode electron multiplier 609
Continuous flow fast atom bombardment (CFFAB) 589–590
Continuous wave electron nuclear double resonance (CW-ENDOR) 364–365
Continuum source 102
   background correction 397–399
Controlled potential coulometry 847–848
Convergence chromatography 793–794
Corona charged aerosol detector 765
Corona discharge 776
Correlation coefficient 42
Coulomb explosion 774
Coulometric titrations 848–849
Coulometry 825, 843–860
Counterions 782
Coupled elemental analysis mass spectrometry 666
Coupling constant 297
Critical micelle concentration (CMC) 904
Cross-polarization 329
Cryogenic fluids 61–65
crystalline solid-state electrodes 829–830
Crystals 82
   analysis of 516–521
   growing 549
   structure determination 549–551
Curie temperature 199, 920
Cuvettes 89, 138–140, 142–145, 147, 173
   quartz 104, 139, 236; see also sample cells
CW-ENDOR, see continuous wave electron nuclear double resonance
Cyclic voltammetry, pHit sensor 866–867

Cylindrical mirror analyzers (CMA) 887–888, 904–905
CZE, see capillary zone electrophoresis

D

DART, see direct analysis in real time
Data
   assessment 42–44
   basic statistics 17–29
   handling 17–29
DC, see direct current
DCP, see direct-current plasma
dead time correction 527
Dead volume 729
Decoupling 312–313
Degrees of freedom 24
Depth profiling 902
   analysis 466–467
   and element location 896–898
Derivative thermogravimetry 924–926
Derivatization
   difficult analytes 744–745
   GC elution 744–745
   high-performance liquid chromatography (HPLC) 771–772
Desalting 790
Desorption electrospray ionization 586–592
   source 590–592
Detector cells 696
Detectors 38
   artifacts 515–516
   atomic absorption spectrometry (AAS) 386
   atomic OES 437–441
   dead volume 729
   diode arrays 766–767
   dispersive monochromator 766
   electrochemical 768–770
   electrolytic conductivity 734
   electron capture (ECD) 732–734, 768–770
   flame atomic emission spectroscopy 424
   infrared (IR) spectroscopy 198–200
   liquid chromatography (LC) 868–870
   luminescence measurements 172–173
   mass spectrometry (MS) 608–611
   response time 200–201
   spectroscopy 110
   supercritical fluid chromatography 792
   thermionic 735
   UV/vis molecular spectroscopy 132–138, 765–767
   wavelength-dispersive X-ray spectrometry 521–525
   X-ray photoelectron spectroscopy 888–889
Determinate error 20–22
Dielectric barrier discharge ionization detector (BID) 736
Difference band 187
Differential mobility spectrometers (DMS) 612
   JUNO detector 613
Differential pulse polarography (DPP) 861–863
Differential scanning calorimetry (DSC) 930–938
   applications 934–938
   instrumentation 930–934
   modulated 937–938

SUBJECT INDEX

Differential thermal analysis (DTA) 926–930
   analytical applications 929–930
   instrumentation 927–929
Diffraction gratings 103–104, 106, 441, 450–451, 461, 473, 518
   dispersive spectrometer systems 245–246
   electrical excitation sources 431
   monochromators 132, 384
   polychromators 436
   UV/vis 518
   wavelength selection 423
Diffraction patterns 547–549
Diffuse reflectance (DRIFTS) 201, 209, 238
Diffusion current 855
Digestion, and dissolution 29–31
Dilatometers, optical 958–960
Diode array detectors (DAD) 765–767
Diode arrays 137–138
Diodes 136–137
Dipole moments 184–185
Direct analysis in real time (DART) 580, 590, 613
   source 585–586, 588
Direct exposure probe 581
Direct injection enthalpimetry (die) 944–945
Direct insertion probe 581
Direct on-column injection 721
Direct-current (DC)
   GD sources 465–466
   polarography 855–859
Direct-current plasma (DCP)
   and microwave plasma (MP) systems 472–473
   spectrometers 451–453
Discharge electrodes 736
Discharge lamp 379–380
Discrete-dynode electron multiplier 609
Dispersive magnetic sector 593
Dispersive monochromator detectors (DMD) 765–766
Dispersive optical layouts 112–113
Dispersive spectrometer systems 245–246
Dissolution, and digestion 29–31
DMA, *see* dynamic mechanical analysis
DMS, *see* differential mobility spectrometry
Dose ranges, ionizing radiation 74–75
Double-beam optics 110–112
Double-resonance experiments 312–313
DPP, *see* differential pulse polarography
Dropping mercury electrodes 825, 855
Dry ashing 31–32
DSC, *see* differential scanning calorimetry
DTA, *see* differential thermal analysis
Duane–Hunt law 493, 501
Dynamic mechanical analysis (DMA)
   applications 954–957
   equipment 953–954
   instrumentation 953–954
   and thermomechanical analysis (TMA) 949–967

E

ECD, *see* electron capture detector
Echelle monochromators 437
EDXRF, *see* energy-dispersive X-ray fluorescence
Electrical cables 50–55
Electrical double layer 824–825
Electrical excitation sources 431–435
Electrical hazards 50–58
Electrical lines 50–55
Electrical receptacles 50–55
Electrical shocks 55
Electrochemical cells 816–825
Electrochemical detectors 768–770
Electrochemical quartz crystal microbalance 876
Electrochemistry 815–816
   nanoscale 876–878
Electrodeless discharge lamp 379–380
Electrodes
   collector 732
   combination 829
   discharge 736
   indicator 825, 827–833
   silver/silver chloride 824
   working 768
Electrogravimetry 825, 843, 845–846
Electrokinetic injection 799
Electrokinetic pumping 796
Electrolysis 843
Electrolytes 816
Electrolytic cells 843
Electrolytic conductivity 850
   detectors 734
Electromagnetic radiation 81–82
   and matter 81–86
Electromotive force (EMF) 818
Electron capture detectors (ECD) 732–734, 768–770
Electron energy analyzers 887–888
Electron flood gun 889
Electron ionization 575, 582–583
Electron microprobe 910
Electron multiplier 609–610
Electron probe microanalysis 557–558, 910
Electron spectroscopy
   qualitative analysis 891–893
   techniques 883–902
Electron spin resonance (ESR) spectroscopy 353
   applications 361–364
   instrumentation 356–358
   interactions 360–361
   sample holders 357–358
   spectra 360–361
electronic excitation 122–124
Electronic pulse processing units 525–527
Electronic transitions 88–89
Electroosmotic flow (EOF) 796
Electrophoresis 687, 794–805
Electrophoresis systems, and chromatography 581–582
Electrospray interface (ESI), for HPLC-MS 774–776
Electrothermal atomizers 382–384
Electrothermal vaporization (ETV) 457
Elemental analysis mass spectrometry 666
ELSD, *see* evaporative light scattering detector
Eluent suppression 784–785
Eluotropic series 757
Elution 725
Emission, infrared (IR) spectroscopy 209–210
Emission spectra 85
Emission spectroscopy, instrumentation 430–441
Enantiomers, chiral phase separation 755–756
End capping 697, 752
Endothermic reaction 927
Energy, characteristics 566–571
Energy levels in atoms 489–494
Energy-dispersive X-ray fluorescence (EDXRF)
   detectors 511–514
   and WDXRF spectroscopy 530–531
Energy-dispersive X-ray spectrometry
   instrumentation 504–516
   primary beam modifiers 505–508
   sample holders 508–511
Enthalpimetry, direct injection 944–945
Enthalpy 927
Environmental applications 655–657, 692
Equilibrium partition ratio 698
Equivalence point 836
Errors 19
   in thermogravimetry 926
Escape peaks 515–516, 523
ESR, *see* electron spin resonance
Evaporative light-scattering detector (ELSD) 763–765
Evolved gas analysis 938–941
Exact mass 625
EXAFS, *see* extended X-ray absorption fine structure
Excellim system 613
Excitation, and ionization interferences 426
Excitation sources 446–451, 504–505
Excited state 84
Exothermic reaction 927
Extended X-ray absorption fine structure (EXAFS) 546
Extraction 32–38

F

Far-infrared (far-IR) sources 193
Faraday cup 610–611
Faraday's law, analytical determinations 846–850
Fast atom bombardment 588–590
Fellgett's advantage 195, 197, 608
Fermi resonance 226
Fiber optics
   modular systems 246–249
   probes 141
Field homogeneity, and spin echo 349–351
Film thickness 722
Filters 102
Fine structure 291
Fingerprint region 218
Fire extinguishers 77–79
Fire safety 76–79
Fixed gases 730
Fixed retardation ratio (FRR) 888
Fixed-analyzer transmission (FAT) 889
Flame atomic absorption 484–485
Flame atomic emission spectroscopy 421–429
flame atomization 387–391
Flame atomizers 380–382
Flame excitation sources 424–425
Flame ionization detector 730–732

Flame microsampling 409
Flame optical emission spectrometry (OES) 422–429
Flicker 15, 40, 101, 380
Flow cells 762
Flow injection analysis 408–409
Flow parameters 721–722
Flow-through samplers 140
Fluorescence 83
 and phosphorescence 167–169, 175
Fluorescence detectors 767–768
Fluorescence intensity, and concentration 169–170
Food chemistry applications 692
Forensic science 215, 249
Formal potential 822
Fourier transform (FT)
 infrared (FTIR) microscopy 210–213
 ion-cyclotron resonance 606–607
 nuclear magnetic resonance (FT-NMR) 287
 spectrometers 113–115, 194–197
 systems 114–115
Fourier transform (FT)-Raman spectrometers 246
Fragmentation 621–623, 631
Frequency 81
FT, *see* Fourier transform
Fuel supply 732
Fume hoods 65
Functional group region 218
Fundamental parameters (FP) method 540
Fusions 31

G

Galvanic cells 818, 843
Gas analysis 655
 evolved 938–941
 gas chromatography (GC) 745–746
Gas cells 138–139
Gas chromatography (GC) 692
 gas analysis 745–746
 hyphenated techniques 737–743
 limitations 746
 organic compounds 744
 photoionization detector 735–736
 resources 748
 scope of analyses 744–746
 two-dimensional 740–743
Gas chromatography (GC) instrumentation 711–712
 advances 712–713
 columns 716–728
 detectors 728–737
 elution values 722–728
 injectors 713–716
Gas chromatography-infrared (GC-IR) spectrometry 740
Gas chromatography-mass spectrometry (GC-MS) 737–740
Gas cylinders 58–61
Gas expansion 580–581
Gas liquid chromatography (GLC) 711
Gas regulators, and manifolds 59–60
Gas samples 13–14, 204–205, 407
Gas-filled detectors 521–523

Gas-sensing electrodes 831–832
Gases 240
Gaussian distribution 22, 24, 694
GC, *see* gas chromatography
GC-IR spectrometry, *see* gas chromatography-infrared spectrometry
GC-MS, *see* gas chromatography mass spectrometry
GD, *see* glow discharge
Geiger–Müller tube 523
Gel boat holders 140
Generators 516
 portable 56–57
Geological materials, characterization 661–663
GFAAS, *see* graphite furnace atomic absorption spectroscopy
Glass electrodes, pH measurement errors 839
glass membrane electrodes 828–829
Glass transitions 927
Gloves 69–70
Glow discharge (GD)
 atomic emission spectrometry applications 466–467
 emission spectrometry 465–467
 source and spark source 592
 systems 473
Glow plugs 732
Golay detector 199
Goniometers 516, 519
Graphite furnace atomic absorption spectroscopy (GFAAS) 485
 atomization 391–392
 nonspectral interferences 394–395
 spectral interferences 401–402
Graphite furnace atomizer 383
Graphite furnace laser-excited atomic force spectroscopy (AFS) 471
Gravity column liquid chromatography (LC) 692
Ground 57–58
Ground state 84

H

Half-wave potential 859–860
Halogen isotopic clusters 627–630
Handheld differential mobility spectrometry (DMS), JUNO detector 613
Handheld UV/vis spectrometers 141–150
Hanging mercury drop electrode (HMDE) 865
Head pressure 713
Headspace 36, 714
Hearing protection 69
Heart-cutting 2D chromatography 693, 742
Heat-flux DSC 931
Heating microscope 957–958
Height equivalent to a theoretical plate (HETP) 692, 712, 722, 797
Heisenberg uncertainty principle 284, 317
Helium ionization detector 736–737
Heteronuclear decoupling 326
High-energy prespark (HEPS) 434
High-performance anion-exchange (HPAE) 769
High-performance liquid chromatography (HPLC) 692
 applications 777–781

 column 751–756
 detector design, and operation 761–771
 detectors 761–771
 electrospray interface (ESI) 774–776
 hyphenated 773–777
 instrumentation 758–761
 mass spectrometry (HPLC-MS) 773–777
 phase combinations 757–758
 stationary phases 751–756
High-resolution continuum source, atomic absorption spectrometry (AAS) 387
High-resolution inductively coupled plasma mass spectrometry (HR-ICP-MS) 670
High-resolution mass spectrometry 649–652
 instrumentation 650
Hollow cathode lamps 378–379
Homogeneous field 284
Hooke's law 187
Hot work permits 79
HPLC, *see* high-performance liquid chromatography
Hybrid spectrometers 516
Hybrid X-ray diffraction (XRD)/X-ray fluorescence (XRF) systems 553–555
Hydride generation atom absorption spectrometry (HGAAS) technique 384
Hydrodynamic flow 796
Hydrodynamic injection 798
Hydrophilic interactions 758
Hyperchromism 150
Hyphenated chromatography–ion mobility spectrometer-liquid chromatography (HPIMS-LC) 613
Hyphenated gas chromatography (GC) 737–743
Hyphenated high-performance liquid chromatography (HPLC) 773–777
Hyphenated instruments 464–465
Hyphenated nuclear magnetic resonance (NMR) 342–343
Hyphenated systems 693
Hyphenated thermal methods 938–941
Hypochromism 150
Hypsochromic shift 151–152

I

IC, *see* ion chromatography
ICP, *see* inductively coupled plasma
ICP-OES, *see* inductively coupled plasma optical emission spectrometry
Ilkovic equation 856
Immiscible liquids 14
Immobilized enzyme membrane electrodes 832–833
Indeterminate errors 22–23
Indicator electrodes 825, 827–833
Indirect detection, ion chromatography (IC) 787–788
Inductively coupled plasma (ICP)
 direct-current plasma (DCP) applications 462–464
 emission 481
 microwave plasma (MP) systems 472–473
 source 592–593
 spectrometers 451–453
Inductively coupled plasma (ICP) mass

SUBJECT INDEX

spectrometry (ICP-MS) 480, 485–486, 657–660
Inductively coupled plasma (ICP) optical emission spectrometry (ICP-OES) 485
Infrared (IR) spectrometry
　with mass spectrometry (MS) 582
　quantitative analysis 233–236
Infrared (IR) spectroscopy
　absorbance detectors 767
　analytical applications 214–236
　C–O bonds 222–227
　chemical imaging 256–262
　detectors 198–200
　emission 209–210
　heteroatoms 228–233
　instrumentation 190–201
　laser sources 193
　liquid samples 202–204
　nitrogen-containing compounds 227–228
　nondispersive systems 214
　pyroelectric detectors 199
　radiation sources 190–193
　sampling techniques 201–210
　solid samples 201–202
Injection, of samples 798–799
Injection internal standard (IS) 706
Inorganic mass spectrometry (MS), ionization sources 592–593
Input systems, for samples 580–582
Instrument cooling 58
Integer mass 625
Interferences 3
　arc and spark emission spectroscopy 441–443
　atomic absorption spectrometry (AAS) 392–402
　atomic force spectroscopy (AFS) 470
　atomic mass spectrometry (MS) 666–672
　flame atomic emission spectroscopy 425–427
　plasma emission spectrometry 458–462
　tandem mass spectrometry (MS/MS) 671–672
Interferometers 114
　components 197
　and monochromators 194–197
Internal standard calibration 96–98, 442–443
Ion chromatography (IC) 784–788
　indirect detection 787–788
　single column 787
Ion exchange chromatography 758
Ion gun 889
Ion microprobe mass spectrometry (MS) 909–910
Ion mobility spectrometry 611–615
Ion scattering spectroscopy 902–905
Ion selective electrode (ISE) 12, 825
　liquid membrane 830–831
Ion selective field effect transistors 833
Ion sources 580
Ionization counters 522
Ionization interference 394
　and excitation 426
Ionization sources
　atmospheric pressure 584–586
　inorganic mass spectrometry (MS) 592–593
　mass spectrometry (MS) 582–593

Ionizing radiation dose 74–75
Ions
　chromatography 782–788
　direct measurement 835–836
　quantitative analysis 840–841
IR, *see* infrared
ISE, *see* ion selective electrode
Isobaric spectral interferences 667–670
Isoelectric point 782, 801, 809
Isothermal gas chromatography (GC) 725
Isothermal titration calorimetry 946–948
Isotope dilution 652
Isotopic abundances 623–626

J

Jacquinot's advantage 195, 197
JEOL SpiralTOF mass spectrometer 597–599
Jet separators 438, 739, 940
Jet tip 732
Joule heating 797–798
JUNO detector 613

K

Katharometers 712, 730
Kinetic energy discrimination (KED) 670
Kohlrausch's law 851
Kovats retention indices 743
Kubelka–Munk equation 209

L

Ladder safety 68–69
Larmor equation 281, 287, 306, 356
Larmor frequency 281
Laser ablation 457–458
　electrospray ionization 592
Laser desorption, and MALDI 586–588
Laser safety 66–67
Laser sources, infrared (IR) spectroscopy 193
Laser-induced breakdown spectroscopy (LIBS) 473–480, 486
Lattices 294
Laue photographs 551, 553
Layers, analysis of 531–533
LC, *see* liquid chromatography
Leaks, mass spectrometers 615–616
LECO Pegasus GC-HRT 599–601
LIBS, *see* laser-induced breakdown spectroscopy
Light sources, Raman spectroscopy 244–245
Limit of detection (LOD) 43
Limit of quantitation (LOQ) 43–44
Line notation, electrochemical cells 819
Line source 102
Linear mobile phase velocity 701
Liquid cells, UV/vis molecular spectroscopy 138–139
Liquid chromatography (LC)
　detectors 868–870
　resources 811
Liquid membranes, ion selective electrode (ISE) 830–831
Liquids 14

atomic absorption spectrometry (AAS) 404–405
　infrared spectroscopy 202–204
　introduction 454
　NIR spectroscopy 239–240
Lock mass 650
LOD, *see* limit of detection
LOQ, *see* limit of quantitation
Low-resolution mass spectrometry (MS) 651–652
Luggin probe 849
Luminescence 83, 170–175
L'vov platform 392, 395, 402, 485

M

McLafferty rearrangement 632–633
Magic angle 285–286
magnetic and electric sector instrumentation 593–596
Magnetic field 1h nuclei quantitization 279–283
Magnetic fields
　safety 75–76
　strength 281–283
Magnetic resonance imaging (MRI) 343–347
Magnetic resonance spectroscopy
　instrumentation 302–309
　line broadening 286–287
　magnets 305–308
　nuclei 278–279
　probes 305
　relaxation time 284–285
　sample holders 302–304
Magnetic sector 593
Magnetic sector mass spectrometer 576
Magnetic shielding, X-ray photoelectron spectroscopy 889
Magnetogyric ratio 279
Makeup gas 723, 729
MALDI, *see* matrix-assisted laser desorption ionization
Manifolds, and compressed gas regulators 59–60
Mantissa 18
Mass 625
　and weight 2–3
Mass absorption coefficient 495
Mass accuracy, low-resolution mass spectrometry 650–651, 678
Mass analyzers 593–608
Mass losses, on fragmentation 631
Mass spectra 578
　alcohols 640–642
　aldehydes 642–644
　diene systems 152–154
　esters 645
　ethers 642
　hydrocarbons 635–640
　interpretation 619–648
　ketones 642–644
　organic compounds 640–645
Mass spectrometers
　resolution 579–580
　secondary ion mass spectrometry 907
Mass spectrometry (MS)
　carbon atom counting 624–625
　detectors 608–611

double bonds 630–631
and high-performance liquid chromatography (HPLC) 773–777
hydrocarbons 218–222
and infrared (IR) spectrometry 582
instrument data files 651
instrumentation 580–611
ionization sources 582–593
mass accuracy 678
nitrogen atom counting 625
nitrogen-containing compounds 646–647
Orbitrap 607–608
oxygen atom counting 625–626
principles 575–580
probes 581
quadrupole mass analyzer 601–603
rings 630–631
spurious effects 674
sulfur atom counting 625
sulfur-containing compounds 647–648
Matched cells 139–140
Matrix 3
Matrix interference
in arc and spark emission spectroscopy 441–442
in atomic absorption spectrometry 393–394
in atomic mass spectrometry 666–667
Matrix modification 395
Matrix-assisted laser desorption ionization (MALDI), and laser desorption 586–588
Matter, and electromagnetic radiation 81–86
Maxwell–Boltzmann equation 376, 424
Measurements
color 164–166
luminescence 170–173
performing 38–42
potential 833–835
uncertainty 15–17
vacuum systems 580
Medical applications 664–666
Membrane electrodes 827
Metallic electrodes 827
Method of standard additions (MSA) 706
Micellar electrokinetic capillary chromatography 803–805
Micro-differential scanning calorimetry (micro-DSC)
applications 946
instrumentation 945–946
Micro-X-ray beam optics 528–529
Micro-X-ray fluorescence (micro-XRF)
instrumentation 528–530
system components 529
Microcalorimetry 945–949
Microchannel plate (MCP) 889
Microliter flow calorimetry 948–949
Microvolume systems 142–143
Microvolume UV/vis spectrometers 141–150
Microwave digestion 29
Microwave plasma (MP) atomic emission spectroscopy
applications 462–464
and direct-current plasma (DCP) 472–473
Microwave plasma (MP) spectrometers 451–453
Mid-infrared (mid-IR) absorption spectroscopy 215–233

Mid-infrared (mid-IR) spectroscopy sources 191–193
Miniature electron spin resonance (ESR) spectroscopy 358–360
Mobile phases 692, 792
choice of 721–722
composition of 756–758
separation 758
Modular systems, fiber-optic-based 246–249
Modulated differential scanning calorimetry (DSC) 937–938
Modulation 386
Molar absorptivity 90, 125
Molecular emission spectrometry 167–170
Molecular formulas, and isotopic abundances 623–626
Molecular ions, and fragmentation patterns 621–623
Molecular leak inlet 580
Molecular mass spectrometry (MS)
applications 649–657
limitations 657
quantitative analysis 652–654
Molecular spectroscopy
instrumentation 129–150
and molecules 87–89
Molecular structures
and peak area 310–311
qualitative analysis 310–313
Molecules
dipole moments 184–185
electronic excitation 122–124
electronic transitions 88–89
and molecular spectroscopy 87–89
Monochromatic light 81
Monochromators 102–104, 132, 384
and interferometers 194–197
monodisperse particles 698
monolithic silica cylinder 754
Moseley's law 494
MP, see microwave plasma
MRI, see magnetic resonance imaging
MS, see mass spectrometry
Multichannel pulse height analyzer 514–515
Multicomponent determinations 162
Multiplex instruments 113

N

Near-infrared (NIR) spectroscopy
applications 240–242
calibration 236–238
chemical imaging 256–262
instrumentation 236
liquids 239–240
sampling techniques 238
solid samples 240
solutions 239–240
sources 193
spectra interpretation 236–238
vibrational bands 236–238
Nebulizers 454–456
Nephelometry 166–167
Nernst equation 821–822
NIR, see near-infrared
Nitrogen rule 623

Nitrogen–phosphorus detector 735
NMR, see nuclear magnetic resonance
NOE, see nuclear Overhauser effect
Noise
and signals 39–41
white 40
Nominal mass 625
Nondispersive infrared (IR) systems 214
Nonspectral interferences
in atomic absorption spectrometry (AAS) 392–396
in graphite furnace atomic absorption spectroscopy (GFAAS) 394–395
Normal phase chromatography 697, 751
Normal pulse polarography 860–861
Nuclear magnetic resonance (NMR) spectroscopy 277–287
aliphatic alkanes 315–319
alkenes 315–319
alkyl halides 315–319
alkynes 315–319
analytical applications 309–342
aromatic compounds 319–321
hyphenated techniques 342–343
limitations 353–365
magnetic resonance imaging (MRI) 343–347
nitrogen-containing compounds 321–324
nuclei 278–279
oxygen-containing organic compounds 321
portable instrumentation 353
qualitative analysis 335–338
quantitative analysis 338–342
samples 309–310
two-dimensional 332–335
Nuclear Overhauser effect (NOE) 326–328

O

OES, see optical emission spectrometry
One-dimensional planar gel electrophoresis 808
Operational amplifiers 833
Optical dilatometers 958–960
Optical emission spectrometry (OES) 4, 9, 520
Optical slits 108–110
Optical systems 100–115
resolution 104–108
UV/vis molecular spectroscopy 129–130
Optical thermal analyzers 957–962
differential thermal analysis 960–962
Optics 101
double-beam 110–112
single-beam 110–112
spectrometers 384–386
Orbitrap mass spectrometry (MS) 607–608
Oscillator strength 377
Ostwald's dilution law 851
Outliers, detection of 27–28
Oxidation-reduction titrations 842–843
Oxidizing agent 815

P

Packed column gas chromatography (GC) 711
Parallel 1D thin layer chromatography (TLC) 805–806

SUBJECT INDEX

Particle-induced X-ray emission 558
Partition chromatography 751
Pascal's triangle 294, 314, 361
PAT, *see* portable appliance testing
Path length 89
Paul ion trap 605
Peak 39, 696
Peak area 310–311, 705
Peak height 705
Peak resolution 700
Pellicular packings 785
pH measurements 837–839
  glass electrode errors 839
pH titrations 839
Phase ratio 722
pHit sensor 866–867
Phosphorescence 83
  disadvantages 175
Photoacoustic spectroscopy 262–264
Photoelectric transducers 132
Photoionization detectors 735–736
Photometers 115
  single-wavelength filter 766
Photomultiplier tubes (PMT) 134–135
Photon detectors 200
Photonis ion mobility spectrometer engine 613
Physical constants 16
Piezoelectric effect 872–875
Planar bilayer 804
Planar gel electrophoresis
  and planar chromatography 805–810
  on slab gels 808–810
  two-dimensional 808–810
Plane-polarized light 81
Plasma emission spectrometry
  calibration 458–462
  instrumentation 446–458
  interferences 458–462
Plasma emission spectroscopy, samples 453–458
Plate theory 692
PLOT, *see* porous open layer tubular
PLOT-cryoadsorption 719
Polarity index 757
Polarographic detector 769
Polarographic wave 855
Polarography 825, 855–863
Porous carbon electrodes 768
Porous open layer tubular (PLOT) columns 718
Porous polymers 718
Portable appliance testing (PAT) 55
Portable generators 56–57
Postcolumn derivatization 767
Potential, half-wave 859–860
Potentiometric cell 833
Potentiometry 825–843
  analytical applications 835–843
  quantitative analysis 840–841
  titrations 836–837
Potentiostat 834
Powder X-ray diffractometry 551–553
Power-compensated DSC 931
Precession 279
Precision 26
  and accuracy 18–19
  UV/vis absorption spectrometry 166

Pressure differential scanning calorimetry (DSC) 936–937
Pressure effects, supercritical fluid chromatography 792
Primary beam modifiers, energy-dispersive X-ray spectrometry 505–508
Primary ion sources, secondary ion mass spectrometry 906–907
Probes
  cryogenically cooled 305
  direct exposure 581
  direct insertion 581
  magnetic resonance spectroscopy 305
  mass spectrometry 581
  samples 305
Process analysis 214
Protein peptide mapping 808
Protein-sequencing analysis 654–655
Proteomics, biochemical applications 778–782
Proton nuclear magnetic resonance (NMR) spectra 313–324
Pulse 297
Pulse height analyzer, X-ray fluorescence 514–515
Pulse height discriminator 527
Pulse height distribution 526
Pulsed amperometric detection (PAD) 769
Purge and trap samplers 746
Pyroelectric detectors 199

Q

Quadrupole ion trap 603, 605–606, 739
Quadrupole mass analyzer 601–603
Quadrupole mass spectrometer 603
Qualitative analysis 4–5
  arc and spark emission spectroscopy 443
  atomic absorption spectrometry (AAS) 402
  atomic emission spectroscopy (AES) 427
  electron spectroscopy 891–893
  laser-induced breakdown spectroscopy 476
  mid-infrared (mid-IR) absorption spectroscopy 215–233
  molecular structures 310–313
  nuclear magnetic resonance (NMR) spectroscopy 335–338
  UV/vis molecular spectroscopy 156–157
  X-ray fluorescence 534–539
Qualitative structural identification 215
Quantifying random error 23–27
Quantitative analysis 5–10
  arc and spark emission spectroscopy 443–445
  atomic absorption spectrometry (AAS) 402–404
  atomic emission spectroscopy (AES) 427–429
  chromatography 704–707
  compounds and mixtures 652–654
  infrared (IR) spectrometry 233–236
  ions 840–841
  laser-induced breakdown spectroscopy 477
  low-resolution mass spectrometry (MS) 651–652
  molecular mass spectrometry 652–654
  nuclear magnetic resonance spectroscopy 338–342
  potentiometry 840–841

secondary ion mass spectrometry 908
UV/vis molecular spectroscopy 157–162
X-ray fluorescence (XRF) 539–542
Quartz crystal microbalance 872–876
QuEChERS 35–36
Quenching 807

R

Racemization 755
Radiant energy absorption, degree of 377
Radiant power 89
Radiation sources
  atomic absorption spectrometry (AAS) 378–384
  infrared (IR) spectroscopy 190–193
  luminescence measurements 171–172
  spectroscopy 101–102
  UV/vis molecular spectroscopy 130–132
  X-ray photoelectron spectroscopy 886
Radio frequency (RF)
  GD sources 465–466
  generation and detection 308–309
Radioisotope sources 501–504
Raies ultimes 443
Raman scattering 242–244
Raman spectroscopy
  applications 249–253
  chemical imaging 256–262
  instrumentation 244–249
  light sources 244–245
  microscopes 255–256
  sample holders 249
  samples 249
  surface enhanced 254–255
Random diffusion 694
Random error quantification 23–27
Range 43
Reaction cells 670–671
Reaction kinetics 162–163
Reagent gas 584
Red shift 150–151
Reducing agent 815
Reduction potentials 818, 880
Reduction–oxidation (redox) reaction 815
Reference data 44–45
Reference electrodes 818, 823–824
Reflectance
  and emission 206–210
  specular 208–209
Reflectron 597
Refractive index (RI) 762–763
Relative abundance 578
Relative response factor (RRF) 706
Relative retention time 704
Relaxation 294
Relaxation time
  magnetic resonance spectroscopy 284–285
  time domain nuclear magnetic resonance (NMR) 351–352
Resolution 580
  mass spectrometers 579–580
  optical systems 104–108
Resolving power 579–580
Resonance Raman effect 254
Respirators 70–72

Retention factor 698
Retention indices 743–744
Retention time 704
    internal standard 704
Reversed phase liquid chromatography 697, 751
RF, see radio frequency
Rotational transitions 87–88

S

Safety
    compressed gases 58
    cryogenic fluids 61
    electrical hazards 50–58
    fire hazards 76–79
    fume hood 65
    instrumentation 49–80
    ladders 68
    lasers 66–67
    magnetic fields 75–76
    nanomaterials 71–74
    personal protective equipment 69–71
    PPE, see personal protective equipment
    Radiation hazards 74–76
Salting out 715
Sample cells 89, 173; see also cuvettes
Sample holders
    atomic OES 435
    electron spin resonance spectroscopy 357–358
    energy-dispersive X-ray spectrometry 508–511
    magnetic resonance spectroscopy 302–304
    Raman spectroscopy 249
    UV/vis molecular spectroscopy 138–141
Sample stacking 799
Samples
    atomic absorption spectrometry (AAS) 404–409
    capillary zone electrophoresis (CZE) 798–799
    changers 527
    injection 798–799
    input systems 580–582
    nuclear magnetic resonance (NMR) 309–310
    plasma emission spectrometry 453–458
    preparation 29–38
    probes 305
    Raman spectroscopy 249
    storage 14–15
    surface analysis handling 890–891
    X-ray fluorescence (XRF) 533–534
Sampling 11–14
Sampling techniques
    infrared (IR) spectroscopy 201–210
    near-infrared (NIR) spectroscopy 238
saturated calomel electrode 823–824
Saturation, and magnetic field strength 281–283
Saturation factor 282
Scattering 763
Secondary internal porosity 752
Secondary ion mass spectrometry (SIMS) 905–910
Secondary XRF sources 501
Semiconductor detectors 135–136, 511–513

Sensitivity 42
Sensors 38
Septum core 713–714
Septum disc 713
Sequential spectrometers 516
Sign conventions 821
Signal integrators 309
Signal-to-noise ratio 39
Signals 38
    and noise 39–41
Significant figures 17
Silicon drift detectors 514
Silicon PIN diode detectors 513–514
Silver/silver chloride electrodes 824
SIMS, see secondary ion mass spectrometry
Simultaneous spectrometers 516
Simultaneous WDXRF spectrometers 527–528
Sine amplifier 527
Single-beam optics 110–112
Single-column ion chromatography (IC) 787
Single-crystal X-ray diffractometry 548–549
Single-focusing magnetic sector 593
Single-wavelength filter photometers 765–766
Size-exclusion chromatography 789–791
Skimmer cone 774
Slab gel 801
Slit widths 108
Slits, optical 108–110
Smith–Hieftje background corrector 400–401
Sodium-dodecylsulfate–polyacrylamide gel electrophoresis (SDS–PAGE) 809
Solid fat 348–349
Solid samples 14
    atomic absorption spectrometry (AAS) 405–407
    holders 140–141
    infrared spectroscopy 201–202
    near infrared (NIR) spectroscopy 240
    UV/vis molecular spectroscopy 164
Solid sampling systems 456
Solid-phase extraction (SPE) 35, 692
Solid-phase microextraction (SPME) 36–38, 714–715
Solutes 91
Solutions, NIR spectroscopy 239–240
Solvated clusters 774
Solvent absorption 205
Solvent effects, ultraviolet (UV) spectra 150–152
Solvent extraction 32–35
Solvents 91
    UV/vis spectroscopy 128–129
Sorbents 35
Space charge 576
Spacers 717
Spark ablation 457
Speciation 663–664
Specificity 42
Spectra 84–85; see also specific types
Spectral bandpass 108
Spectral interferences 95, 470
    arc and spark emission spectroscopy 442
    atomic absorption spectrometry (AAS) 397–402
    atomic emission spectroscopy (AES) 426–427

    atomic fluorescence spectroscopy 470
    background correction 397
    correction 459–462
    GFAAS 401–402
    isobaric 667–670
    plasma emission spectrometry 459–462
Spectral linewidth 377
Spectral slit width 108
Spectroelectrochemistry 163–164, 870–872
Spectrograph 115
Spectrometers 101, 516
    atomic OES 435–437
    direct current plasma (DCP) 451–453
    inductively coupled plasma (ICP) 451–453
    luminescence measurement 173
    microwave plasma (MP) 451–453
    optics 384–386
Spectrometry 115; see also specific types
Spectrophotometry
    titrations 163; see also specific types
Spectroscopy 115
    detectors 110
    radiation sources 101–102
    techniques 115
    wavelength selection devices 102–108; see also specific types
Specular reflectance 208–209
Spin echo, and field homogeneity 349–351
Spin labels 363–364
Spin traps 363–364
Spin-decoupling 312
Spin–spin coupling 291–302
Spin–spin splitting 291
Split injections 715
Split mode 715
Split ratio 715
Splitless injections 715–716
Splitless mode 716
Split–splitless injectors 715
SPME, see solid-phase microextraction
Spontaneous reactions 818
Spray chambers 456
Spurging 753
Standard additions method 93–96
    calibration 841–842
Standard curves 705
Standard hydrogen electrode 819–821
Standard reduction potentials 819–821
Standard reference materials 11, 18, 44
    calibration 92–93, 705–707
Stationary phases 692, 792
    considerations 755
    gas chromatography 719–721
Statistics, definitions 23
Stepladders 68
Stop work orders 79–80
Stretching vibrations 185
Stripping voltammetry 865–866
Sulfur–phosphorus flame photometric detector 734
Sum peaks, X-ray fluorescence 515–516
Supercritical fluid 34
Supercritical fluid chromatography (SFC) 791–794
Surface analysis 883–910
    sample handling 890–891

SUBJECT INDEX

Surface-enhanced Raman spectroscopy 254–255
Surfactants 798
SYNAPT G2-S system 613–615
Syringes 713–714

T

Tacticity 335
Tandem mass spectrometry (MS/MS)
  interferences 671–672
  and MSn instruments 603–605
TD-NMR, *see* time domain nuclear magnetic resonance
TGA, *see* thermogravimetric analysis
Thermal analysis 913–960
Thermal conductivity detectors 730
Thermal methods, hyphenated 938
Thermal modulator 742
Thermionic detectors 735
Thermistors 199
Thermocouples 199
Thermogravimetric analysis (TGA), instrumentation 916–920
Thermogravimetry 914–926
Thermomechanical analysis (TMA)
  applications 954–957
  dynamic mechanical analysis 949–967
  equipment 952–953
  instrumentation 952–953
Thermometric titrimetry 941–944
Thermospray HPLC-MS interfaces 774
Thin-layer chromatography (TLC) 693
  2D separations 808
  detection 806–808
  parallel 1D separations 805–806
Time domain nuclear magnetic resonance (TD-NMR)
  applications 352–353
  relaxation time 351–352
  solid fat 348–349
Time-of-flight (TOF) analyzer 596–601
TLC, *see* thin layer chromatography
TMA, *see* thermomechanical analysis
Total reflection X-ray fluorescence analysis (XRF) 530
Transducers 38
Transition 83
Transmission (absorption), measurement techniques 201–206
Transmission quadrupole mass spectrometer 603
Transmittance 89
Turbidimetry 166–167
Turbidity 160
Two-line background correction 397

U

Ultrahigh vacuum (UHV) system 890
Ultraperformance convergence chromatography 793–794
Uncertainty 28–29
  in measurements 15–17
Unified atomic mass units 575
Unit cell 547
UV spectra
  absorption 125–128
  ketones 154–156
  molecule structure 152–156
  solvent effects 150–152
UV/vis molecular spectroscopy
  applications 156–166
  detectors 132–138, 765–767
  gas cells 138–139
  liquid cells 138–139
  matched cells 139–140
  molar absorptivity 125
  multicomponent determinations 162
  optical systems 129–130
  qualitative analysis 156–157
  quantitative analysis 157–162
  radiation sources 130–132
  sample holders 138–141
  solid samples 164
  solvents 128–129

V

Vacuum systems, measurement in 580
Van Deemter curves 721
Van Deemter equation 701
Variable path length spectroscopy 143–147
Variance 27
Vibrations
  in molecules 88, 185–187
  motion 187–189
  transitions 88
Visible spectroscopy system, handheld 147–150
Voltammetry 825, 855, 863–868
  detection 868–869
  instrumentation 865
Vortex tube 727

W

Wavelength 81
Wavelength dispersive X-ray fluorescence spectroscopy (WDXRF)
  detectors 521–525
  and energy-dispersive XRF 530–531
  instrumentation 516–527
Wavelength selection devices
  flame OES 423
luminescence measurements 170–171
spectroscopy 102–108
Weight, and mass 2–3
Wheatstone bridge circuit 730
Width, of absorption lines 284–287
Working electrodes 768

X

X-ray absorption 542–548
  process 495–496
X-ray crystallography 548
X-ray diffraction (XRD) 546–557
  analytical limitations 556–557
  applications 555–557
  process 496–498
X-ray diffraction (XRD)/X-ray fluorescence (XRF) systems 553–555
X-ray diffractometry, single crystal 548–549
X-ray emission 557–558
  particle induced 558
X-ray fluorescence (XRF)
  applications 531–542
  methods 494–498
  process 496
  pulse height analyzer 514–515
  qualitative analysis 534–539
  quantitative analysis 539–542
  samples 533–534
  sources 499–504
  sum peaks 515–516
  tubes 499–501
X-ray instrumentation, commercial 558–559
X-ray photoelectron spectroscopy (XPS) 885–898
  analytical applications 891–899
  chemical shift 893–894
  detectors 888–889
  instrumentation 886–890
  magnetic shielding 889
  radiation sources 886
X-rays
  spectra 489–498
  wavelengths 566–571
XPS, *see* X-ray photoelectron spectroscopy
XRD, *see* X-ray diffraction
XRF, *see* X-ray fluorescence
XRS, *see* X-ray spectroscopy

Z

Zeeman background correction 399–400
Zeeman effect 279
zeolites 926
Zero air 732
Zwitterions 695, 782

Chemical Index

A

Acetals 632
4-Acetamidophenol 252
Acetate 851
Acetic acid
    conductivity 854
    fragmentation 632
    mass spectra 633
    proton NMR spectra 327
    as UV solvent 129
Acetone 129, 226, 632–633
Acetonitrile 129, 633, 757
Acetylene
    flame temperature 388, 425
    mass spectra 638
    proton NMR spectra 291
Acetylenic groups 124
Acetylide 126
Acid anhydrides 222, 225
Acid dissolution 29–31
N-Acridinyl maleimide 772
Actinium 566–570, 573
Activated charcoal 14, 745
Acylsalicylic acid 252
Adenosine triphosphate (ATP) 343, 346
Air, absorption spectra 198
Air–acetylene flames
    AAS 393–394, 404, 408
    AES 424, 426, 459, 484
    flame atomizers 380–382, 388–391
Alanine 779
Albumin 791
Alcohols 636
    fragmentation 632
    IR spectra 222–225
    mass spectra 640–642
Aldehydes
    absorption 125, 155, 188–189
    carbonyl stretching 223
    as chromophore 126
    fragmentation 632
    homologous series 636
    IR spectra 225–227
    mass spectra 642–644
    polarography 863
    vibrational mode 219
    wavenumber 219
Aliphatic alcohols 223
Aliphatic amines 227
Aliphatic halides 229
Aliphatic hydrocarbons 71, 220, 251
Alizarin garnet R 173–175
Alkanes 188–189
    fragmentation 632
    IR spectra 218–221
    NMR spectroscopy 315–319
    vibration 186, 219
    wavenumber 219
Alkenes 632
    absorption frequencies 188–189
    as chromophore 126
    IR spectra 218–221
    NMR spectroscopy 315–319
    UV/VIS radiation absorption 124
    vibrational mode 220
    wavenumber 220
Alkyl halides 632
Alkynes
    absorption frequencies 188–189
    fragmentation 632
    IR spectra 218–221
    NMR spectroscopy 315–319
    vibrational mode 220
    wavenumber 220
Alumina
    AES 435
    chromatography 693, 697, 719, 756–757, 805
    NMR spectroscopy 309
    PES 446
    thermal analysis 929, 958, 960, 962
^{27}Aluminium 330
Aluminium 823
    absorption edges 571
    absorption spectra 174
    activity 823
    AES 461
    arc emission spectrometry DLs 433
    atomic spectroscopy DLs 415
    in drinking water 662
    EDXRF system filter 510
    electron configuration 491
    electrothermal atomizer (ETA) DLs 484
    ETA-ICP-MS DLs 661
    in FAAS 417
    flame atomic emission 411, 483
    flame atomization 390
    graphite tube furnace 418
    ICP DLs 453
    ICP-MS 662, 665
    isotope abundance 682
    in Montana soil 665
    X-ray emission lines 566–570
Aluminium oxide 464, 533; see also Alumina
Aluminon 161, 174
Americium 504, 506, 566–570, 573
Amides 124, 126, 223
Amines
    absorption wavelength 125
    as chromophore 126
    fragmentation 632
    homologous series 636
    polarography 863
Amino acids
    absorptions 228
    HPLC 780
    IR spectra 228
    NIRIM system 260
    structure 778–779
9-Aminoacridine 169
4-Aminoantipyrine, see 4-Aminophenazone
Aminobenzene 123
4-Aminophenazone 161
Ammonia
    fragmentation 632
    NMR spectroscopy 285
Ammonium aurintricarboxylate, see Aluminon
Ammonium nitrate, NMR spectroscopy 285
Amorphous polymers 548, 557, 929–931
D-Amphetamine sulfate 250
Anthracene
    absorption spectra 174
    as chromophore 126
    fluorescence quantum yields 169
    UV absorption spectra 120
Anthryldiazomethane (ADAM) 772
Antimony 412
    absorption edges 572
    atomic spectroscopy DLs 416
    dissolution 30
    in drinking water 662
    electrodeless discharge lamps 380
    in FAAS 417
    flame atomic emission 412, 483
    graphite tube furnace 418
    gunshot residue 464
    ICP DLs 453
    ICP-MS 662, 665
    in Montana soil 665
    quantitative XRF 542
    in semiconductors 135
    X-ray emission lines 566–570
Aprotinin 791
Arginine 779
Argon
    absorption edges 571
    cryogenic fluid 62
    electron configuration 491
    isotope abundance 683
    X-ray emission lines 566–570
Argon ionization detectors 736
Aromatic amines 227
Aromatic compounds 319–321
Aromatic groups 188–189
Aromatic halides 229
Aromatic hydrocarbons 221–222, 632
Arsenic
    absorption edges 571
    arc emission spectrometry DLs 433
    atomic spectroscopy DLs 415
    in drinking water 662
    electron configuration 491
    electrothermal atomizer (ETA) DLs 484
    ETA-ICP-MS DLs 661
    in FAAS 417
    flame atomic emission 412, 414, 483
    graphite tube furnace 418
    ICP DLs 453
    ICP-MS 662, 665
    isotope abundance 682
    in Montana soil 665
    X-ray emission lines 566–570
Asparagine 779
Aspartic acid 779
Astatine 566–570, 572
Azo groups 125–126

B

Barium
 absorption edges 572
 activity 823
 atomic spectroscopy DLs 415
 in drinking water 662
 in FAAS 417
 flame atomic emission 412, 483
 flame atomization 390
 flame photometry DLs 430
 gunshot residue 464
 ICP DLs 453
 ICP-MS 662, 665
 isotope abundance 625, 684
 in Montana soil 665
 X-ray emission lines 566–570
Barium fluoride 190
Bauxite 32
Benzaldehyde, 13C NMR spectra 335
Benzene
 13C NMR spectra 333
 absorption spectra 129, 180
 as chromophore 126
 IR spectra 244
 mass spectra 107, 578, 620, 624, 634
 proton NMR spectra 296
 Raman spectra 244
 UV absorption spectra 120
 as UV solvent 129
Benzene groups 125
Benzene rings 157–158, 221
Benzoic acid 327
Benzonitrile 121
Benzophenone 644
Benzphetamine 253
Benzylic protons 319, 321, 324
Benzylmercaptan 232
Benzyne 632
Berkelium 566–570, 573
Beryllium
 absorption edges 571
 AES 444
 arc emission spectrometry DLs 433
 atomic spectroscopy DLs 415
 in drinking water 662
 electron configuration 491
 flame atomic emission 412, 483
 flame atomization 390
 graphite tube furnace 418
 ICP DLs 453
 ICP-MS 662, 665
 isotope abundance 682
 in Montana soil 665
 X-ray emission lines 566–570
Bismuth 412
 absorption edges 572
 activity 823
 arc emission spectrometry DLs 433
 atomic spectroscopy DLs 415
 electrodeless discharge lamps 380
 emission spectroscopy 445
 in FAAS 417
 flame atomic emission 412, 483
 flame atomization 390
 graphite tube furnace 418
 ICP DLs 453
 internal standard calibration 97
 isotope abundance 684
 X-ray emission lines 566–570
Bismuth oxychloride 30
Boric acid 260, 285
Boron
 absorption edges 571
 arc emission spectrometry DLs 433
 atomic spectroscopy DLs 415
 electron configuration 491
 in FAAS 417
 flame atomic emission 412, 483
 flame atomization 390
 isotope abundance 682
 X-ray emission lines 532, 566–570
Boron carbide 12, 533
Boron trichloride 285
Boron trifluoride etherate 285
Borosilicate glass 31
Bromides 125–126, 851
Bromine
 absorption edges 571
 atomic mass 576
 atomic spectroscopy DLs 415
 electron configuration 491
 isotopes 576, 625, 629, 682–683
 X-ray emission lines 566–570
1-Bromo-4-chlorobenzene 321–322
Bromoethane 315–317
Bromomethane 628
4-Bromophenol 643
2-Bromopropene 123
Butadiene 632
1-Butanal 299
2-Butanamine 646
Butane 299
Butanol 293, 295, 639–641
Butene 632
n-Butyl acetate 129
sec-Butyl alcohol (2-butanol) 129
n-Butyl chloride 129
Butylamine 328
Butyronitrile 227, 229

C

Cadmium
 absorption edges 572
 activity 823
 AES 461
 arc emission spectrometry DLs 433
 atomic spectroscopy DLs 415
 in drinking water 662
 electrothermal atomizer (ETA) DLs 484
 ETA-ICP-MS DLs 661
 flame atomic emission 412, 483
 flame atomization 390
 graphite tube furnace 418
 half-wave potential 860
 ICP DLs 453
 ICP-MS 662, 665
 isotope abundance 683
 in Montana soil 665
 X-ray emission lines 532, 566–570
 in XRF spectrometry 504
Cadmium sulfide 830
Cadmium telluride 190, 198, 200, 207
Caesium
 absorption edges 572
 in FAAS 417
 flame atomization 390
 isotope abundance 684
 X-ray emission lines 566–570
Caffeine 252
Calcium
 absorption edges 571
 activity 823
 arc emission spectrometry DLs 433
 atomic spectroscopy DLs 415
 in drinking water 662
 electrodes 830
 electron configuration 491
 electrothermal atomizer DLs 484
 ETA-ICP-MS DLs 661
 in FAAS 417
 flame atomic emission 412, 483
 flame atomization 389–390
 flame photometry DLs 430
 graphite tube furnace 418
 ICP DLs 453
 ICP-MS 662, 665
 isotope abundance 683
 in Montana soil 665
 X-ray emission lines 566–570
Calcium carbonate 915
Calcium oxalate 916, 921, 923
Calcium oxide
 Dolomite sample 533
 in glass 464
Calcium telluride 190
Californium 566–570, 573
Calomel electrodes 816, 823–824
Carbide formation 394–395
13Carbon 279, 330
Carbon
 absorption edges 571
 atom counting 624–625
 atomic mass 576, 649
 atomic spectroscopy DLs 415
 electron configuration 491
 gas chromatograph 719
 isotopes 576, 625, 649, 682
 petroleum molar absorptivity 545
 X-ray emission lines 532, 566–570
Carbon dioxide
 Dolomite sample 533
 extinguisher 78
 fragmentation 632
 mass spectra 634
Carbon disulfide 129, 231
Carbon monoxide 632
Carbon tetrachloride
 eluent strength 757
 mass spectra 634
 polarity index 757
 as UV solvent 129
Carbon-carbon bonds, absorption frequencies 189
Carbonic acid, polarography 863
Carbonic anhydrase
 gel filtration 791
 mass spectra 586

CHEMICAL INDEX

Carbonyl groups 124, 126
Carboxylic acid groups 124–125, 189
Carboxylic acids
    fragmentation 633
    IR spectra 222–227
    mass spectra 645
    stretching frequencies 223
Cellulose 510
Cerium
    absorption edges 572
    atomic spectroscopy DLs 415
    flame atomic emission 412, 483
    isotope abundance 684
    X-ray emission lines 566–570
Cesium
    absorption edges 572
    atomic spectroscopy DLs 415
    flame atomic emission DLs 483
    graphite tube furnace 418
    X-ray emission lines 566–570
Cesium bromide 190
Cesium iodide 190
Charcoal, activated 14, 745
Chloranilic acid 161
Chloride, conductance 851
Chlorine
    absorption edges 571
    atomic mass 576
    atomic spectroscopy DLs 415
    depth profile 904
    electron configuration 491
    isotopes 576, 625, 629, 683
    X-ray emission lines 566–570
Chloroalkanes 189
Chloroethane 299, 315–316
Chloroform 203, 228, 239, 249
    IR spectra 230
    mass spectra 634
    as UV solvent 129
2-Chloropropane 297, 300
Chromium
    absorption edge 494
    absorption edges 571
    activity 823
    AES 462
    arc emission spectrometry DLs 433
    atomic spectroscopy DLs 415
    in drinking water 662
    electron configuration 491
    electrothermal atomizer DLs 484
    emission lines 494
    ETA-ICP-MS DLs 661
    in FAAS 417
    flame atomic emission 412, 483
    flame atomization 389–390
    graphite tube furnace 418
    ICP DLs 453
    ICP-MS 662, 665
    isotope abundance 682–683
    in Montana soil 665
    steel analysis 443
    X-ray emission lines 566–570
    X-ray tube filter 510
Chromophores 126–127
Coals 12, 533, 967
    atomic absorption spectrometry 405, 407
    atomic emission spectroscopy 463, 465
    electron spin resonance 363
    fuel 582
    thermogravimetric analysis 919, 923
    X-ray fluorescence 533
Cobalt
    absorption edges 571
    activity 823
    atomic spectroscopy DLs 415
    in drinking water 662
    electron configuration 491
    in FAAS 417
    flame atomic emission 412, 483
    flame atomization 390
    graphite tube furnace 418
    ICP-MS 662, 665
    isotope abundance 682
    in Montana soil 665
    X-ray emission lines 566–570
    X-ray tube filter 510
    in XRF spectrometry 504
Cobalt (III) hexacyanide 285
Cocaine
    FT-Raman spectra 246–247, 250
    IR spectroscopy 215
    mass spectra 578
    Raman spectra 247, 250
Conjugated hydrocarbons 126
Conjugated olefins 125
Copper
    AAS 403
    absorption edges 571
    acid dissolution 31
    activity 823
    arc emission spectrometry DLs 433
    atomic spectroscopy DLs 415
    in drinking water 662
    EDXRF system filter 510
    electrodes 830
    electron configuration 491
    in FAAS 417
    flame atomic emission 412, 483
    flame atomization 389–390
    graphite tube furnace 418
    half-wave potential 860
    ICP DLs 453
    ICP-MS 662, 665
    isotope abundance 682
    in Montana soil 665
    X-ray emission lines 532, 566–570
    X-ray tube filter 510
Copper sulfide, electrodes 830
Corundum 32
Crack cocaine 247
Cumene 299
Curium 558, 566–570, 573
Cycloalkanes 636
Cyclodextrins 756
Cyclohexane
    absorption spectra 219
    eluent strength 757
    mass spectra 634, 637
    polarity index 757
    as UV solvent 129
Cyclohexanol 334
Cyclohexanone 326
Cyclopropane 632
Cyclopropanone 632
Cysteine 779
Cytidine 791
Cytochrome c 791

D

D-Amphetamine 253
Dansylchloride (Dns-Cl) 772
Dansylhydrazine (Dns-H) 772
1-Decene 220
Deuterated chloroform 285
Deuterium 131, 279
Diamond 4, 202, 307, 476, 662–663
Diazomethane 745
1,4-Dibromobenzene 322
1,2-Dichloro-1,1,2-trifluoroethane 690
9,10-Dichloroanthracene 169
Dichlorobenzene 323
Dichloroethane 129, 296, 318–319
Dichloromethane 634
1,3-Dichloropropane 320
Dichromate ion 163
Dienes 153–154
Diethyl ether 129, 634
Dihydrogen platinum (IV) hexachloride 285
4,5-Dihydroxy-3-(2-hydroxy-5-sulfophenylazo)-2,7-naphthalenedisulfonic acid, see SPADNS
2,5-Dihydroxybenzoic acid, MALDI conditions 588
1,2-Dimethoxyethane 129, 634
Dimethyl ether 633, 690
Dimethyl groups 219
Dimethyl sulfide 632
N,N-Dimethylacetamide 129
Dimethylamine 632
1-Dimethylaminonaphthalene-4-sulfonate 169
Dimethylformamide 129, 634
Dimethylnaphthalene 726
2,2-Dimethylpropane 636–637
Dimethylsulfoxide (DMSO) 129, 301
Dioxane 129, 634, 656
Diphenyl 126
Dipropylamine 228
Disulfides 126, 230, 633
DMSO, see Dimethyl sulfoxide
Dolomite 31
L-DOPA, 13C NMR spectra 286
Dysprosium
    absorption edges 572
    atomic spectroscopy DLs 415
    in FAAS 417
    flame atomic emission 412, 483
    isotope abundance 683
    X-ray emission lines 566–570

E

Einsteinium 566–570, 573
Enone 155
Erbium
    absorption edges 572
    atomic spectroscopy DLs 415

in FAAS 417
  flame atomic emission 412, 483
  isotope abundance 683
  X-ray emission lines 566–570
Erythrose 5
Ester groups 124–125
Esters
  carbonyl stretching frequencies 223
  as chromophore 126
  fragmentation 633
  IR spectra 225–227
  mass spectra 645
Ethane
  gas chromatography 720, 736, 743
  mass spectrometry 624, 655
  MR spectroscopy 316
  SFC 792
Ethanol
  eluent strength 757
  IR spectra 224, 244
  mass spectra 633
  polarity index 757
  properties 690
  proton NMR spectra 290, 301, 311
  Raman spectra 244
  as UV solvent 129
Ethene 636
Ethenone 632
Ether 126, 189, 642
2-Ethoxyethanol 129
Ethyl acetate 129
Ethyl alcohol 632
Ethylbenzene 313, 324
Ethylene 632
Ethylene glycol 633, 757
Europium 483, 830
  absorption edges 572
  atomic spectroscopy DLs 415
  in FAAS 417
  flame atomic emission 412, 483
  isotope abundance 682–683
  X-ray emission lines 566–570

F

Fatty acids, oil distribution 233
Fermium 566–570, 573
Fluorene 169
9-Fluorenylmethyl chloroformate 772
Fluorescamine 772
Fluorescein 169
Fluorides 31, 362
^{19}Fluorine 279, 330
Fluorine
  absorption edges 571
  atomic mass 576, 649
  atomic spectroscopy DLs 415
  electron configuration 491
  isotopes 576, 625, 649, 682
  X-ray emission lines 566–570
Fluorophores 168, 173
Formaldehyde 632
Formate 851
Francium 516–517, 566–570, 572
Furan 634
Fused silica 190

G

Gadolinium
  absorption edges 572
  atomic spectroscopy DLs 415
  contrast agent in MRI 665
  electron configuration 491
  in FAAS 417
  flame atomic emission 412, 483
  graphite tube furnace 418
  isotope abundance 683
  X-ray emission lines 566–570
Gallium
  absorption edges 571
  atomic spectroscopy DLs 415
  in FAAS 417
  flame atomic emission 412, 483
  flame atomization 390
  X-ray emission lines 566–570
Gallium arsenide 131, 200, 696
Gallstones 256
Garnet 258
Germanium
  absorption edges 571
  atomic spectroscopy DLs 415
  electron configuration 491
  in FAAS 417
  flame atomic emission 412, 483
  frequency range 207
  index of refraction 207
  isotope abundance 682
  mid-IR optics 190
  X-ray emission lines 566–570
Glass
  diatometric curve 963
  DTA curve 963
  SIMS spectrum 908
  thermal analysis 960
  X-ray emission lines 532
Glucopyranoside 334, 338
Glutamic acid 779
Glutamine 779
Glyceraldehyde 5
Glycerol 129, 588
Glycine 779
Gold
  absorption edges 572
  acid dissolution 31
  activity 823
  AES spectra 462
  Auger element map 904
  in FAAS 417
  flame atomic emission 412, 483
  flame atomization 390
  isotope abundance 683–684
  X-ray emission lines 566–570
Graphite
  furnace atomizer 391–392, 406, 470–471
  X-ray emission lines 532
Gypsum 31

H

Hafnium 412, 483
  absorption edges 572
  atomic spectroscopy DLs 416
  in FAAS 417
  flame atomic emission 412, 483
  isotope abundance 683
  X-ray emission lines 566–570
Halogen isotopic clusters 627–630
Halogenated compounds 228–230
Helium
  absorption edges 571
  cryogenic fluid 62
  electron configuration 491
  ion scattering spectrometry (ISS) 905
  isotope abundance 682
Helium–neon lasers 195
Heptanoic acid 225, 227
3-Heptanone 643
n-Hexadecane 129, 636
Hexafluorobenzene 285
Hexamethylphosphoramide 648
Hexane
  absorption spectra 218
  eluent strength 757
  formula 630–631
  IR spectra 218
  mass spectra 634, 637
  polarity index 757
  as UV solvent 129
Hexene 638
Hexyne 220
Histidine 779
Holmium 104, 166, 180, 412, 483
  absorption edges 572
  atomic spectroscopy DLs 416
  flame atomic emission 412, 483
  isotope abundance 683
  X-ray emission lines 566–570
Hydrides 384, 408, 663
Hydrocarbon fuel oil 742
Hydrocarbons 635–640, 725–726
Hydrofluoric acid, see Hydrogen fluoride
^{1}Hydrogen 279, 330
^{2}Hydrogen 279
^{3}Hydrogen 279
Hydrogen
  absorption edges 571
  activity 823
  atomic mass 576, 649
  cryogenic fluid 62
  DSC thermal curve 937
  electrodes 816, 819
  electron configuration 491
  flame temperature 388, 425
  fragmentation 632
  isotopes 576, 625, 649, 682
  petroleum molar absorptivity 545
  in XRF spectrometry 504
Hydrogen bromide 632
Hydrogen chloride
  AAS 401
  fragmentation 632
  FTIR spectra 185, 942
  mass spectra 627
  thermometric titrimetry (TT) plot 943
Hydrogen cyanide 632, 635–636, 715
Hydrogen fluoride 632

CHEMICAL INDEX

Hydrogen sulfide 632
Hydroxides 851
Hydroxyacetone 632

I

Ibuprofen 355
Indium
   absorption edges 572
   atomic spectroscopy DLs 416
   in FAAS 417
   flame atomic emission 412–413, 483
   flame atomization 390
   isotope abundance 683
   X-ray emission lines 566–570
Iodides 125–126
Iodine
   absorption edges 572
   atomic spectroscopy DLs 416
   isotope abundance 625, 683
   X-ray emission lines 566–570
Iridium 413, 483
   absorption edges 572
   atomic spectroscopy DLs 416
   isotope abundance 684
   X-ray emission lines 566–570
Iron
   absorption edges 571
   activity 823
   arc emission spectrometry DLs 433
   atomic spectroscopy DLs 415
   EDXRF spectra 537
   electron configuration 491
   in FAAS 417
   flame atomic emission 413, 483
   flame atomization 390
   ICP DLs 453
   ICP-MS 665
   isotope abundance 682
   in Montana soil 665
   X-ray emission lines 532, 566–570
   X-ray spectra 517
   X-ray tube filter 510
   in XRF spectrometry 504
Iron metal 31
Iron ores 32
Iron oxide
   Dolomite sample 533
   in glass 464
Isobutane 299
Isoleucine 779
Isonitriles 227
Isopropyl alcohol 632
Isoquinoline 126

K

Kaolin 962
Ketones
   absorption frequency 189
   absorption rules 155
   carbonyl stretching frequencies 223
   as chromophore 126
   fragmentation 633
   homologous series 636
   IR spectra 225–227

   mass spectra 642–644
   polarography 863
   structure 154–156
Krypton 245
   absorption edges 571
   electron configuration 491
   isotope abundance 682–683
   X-ray emission lines 566–570

L

L-Dopa 286
Lanthanum
   absorption edges 572
   atomic spectroscopy DLs 416
   in FAAS 417
   flame atomic emission 413, 483
   isotope abundance 684
   X-ray emission lines 566–570
Lanthanum trifluoride 830
Latex 71
Lead
   absorption edges 572
   activity 823
   atomic spectroscopy DLs 416
   in drinking water 662
   electrothermal atomizer (ETA) DLs 484
   ESCA spectra 898
   ETA-ICP-MS DLs 661
   flame atomic emission 413, 483
   flame atomization 390
   graphite tube furnace 418
   half-wave potential 860
   ICP DLs 453
   ICP-MS 665
   isotope abundance 683–684
   in Montana soil 665
   polarogram 859
   X-ray emission lines 532, 566–570
Lead bromide 897
Lead sulfide 830
Leucine 779
^{7}Lithium 330
Lithium
   absorption edges 571
   activity 823
   atomic spectroscopy DLs 416
   electron configuration 491
   in FAAS 417
   flame atomic emission 413, 429–430, 483
   flame atomization 390
   flame OES 429
   flame photometry DLs 430
   isotope abundance 682
   in MSA 429
   in serum 429
   X-ray emission lines 566–570
Low alloy steels 31
Lutetium 483
   absorption edges 572
   atomic spectroscopy DLs 416
   flame atomic emission 413, 483
   isotope abundance 683
   X-ray emission lines 566–570
Lysine 779
Lysozyme

   DSC scan 947, 949
   mass spectra 587

M

Magnesium
   absorption edge 494
   absorption edges 571
   activity 823
   atomic spectroscopy DLs 416
   in drinking water 662
   electrodes 830
   electron configuration 491
   emission lines 494
   in FAAS 417
   flame atomic emission 413, 483
   flame atomization 389–390
   ICP DLs 453
   ICP-MS 662, 665
   isotope abundance 682
   in Montana soil 665
   NMR active nuclei 330
   X-ray emission lines 566–570
Magnesium fluoride 190
Magnesium oxalate 923
Magnesium oxide
   Dolomite sample 533
   in glass 464
   mid-IR optics 190
Manganese
   absorption edge 494
   absorption edges 571
   activity 823
   arc emission spectrometry DLs 433
   atomic spectroscopy DLs 416
   in drinking water 662
   EDXRF spectra 537
   electron configuration 491
   emission lines 494
   in FAAS 417
   flame atomic emission 413, 483
   flame atomization 390
   graphite tube furnace 418
   ICP-MS 662, 665
   isotope abundance 682
   in Montana soil 665
   steel analysis 443
   X-ray emission lines 566–570
Manganese oxide 533
Mercaptan 633
   polarography 863
   stretching frequencies 230
Mercury
   AAS 399
   absorption edges 494, 572
   activity 823
   arc lamps 170, 172
   atomic spectroscopy DLs 416
   CV analyzer 408
   in drinking water 662
   emission lines 494
   energy levels 87
   in FAAS 417
   flame atomic emission 413, 483
   graphite tube furnace 418
   ICP-MS 662

isotope abundance 684
MS/MS 673
potential window 863
X-ray emission lines 566–570
Mescaline 250
Methadone 250
Methamphetamine 252–253
Methanamine 632
Methane 187, 622
Methanethiol 632
Methanol
  eluent strength 757
  mass spectra 623, 633
  polarity index 757
  proton NMR spectra 312
  as UV solvent 129
Methionine 779
2-Methoxyethanol 129
Methyl acetate 634, 896
Methyl alcohol 632
Methyl cyclohexane 129
Methyl ethyl ketone 129
Methyl groups 188, 219
Methyl isobutyl ketone 129
2-Methyl-1-propanol 129
2-Methyl-2-propanol 641
N-Methyl-2-pyrrolidone 129
Methylene 188, 219
2-Methylpropane 297
Metoclopramide 794
Metoprolol 909
Molybdenum
  absorption edges 571
  atomic spectroscopy DLs 416
  Auger element map 904
  in drinking water 662
  in FAAS 417
  flame atomic emission 413, 483
  flame atomization 389
  ICP DLs 453
  ICP-MS 662, 665
  isotope abundance 683
  in Montana soil 665
  SIMS spectrum 909
  steel analysis 443
  X-ray emission lines 532, 566–570
  X-ray tube filter 510
Molybdenum sulfide 260
Morphine monohydrate 250
Mylar, see polyethylene terephthalate
Myoglobin 149, 587

N

Naphthalene
  13C NMR spectra 298
  absorption wavelength 125
  as chromophore 126
  fluorescence quantum yields 169
  mass spectra 639
  proton NMR spectra 298
  UV absorption spectra 120
Naphthyl isocyanate, fluorescence 772
Neodymium
  atomic spectroscopy DLs 416
  characteristics 413

flame atomic emission 413, 483
isotope abundance 684
laser induced breakdown spectroscopy 474
radiation sources 172
Raman instrumentation 273
X-ray emission lines 566–570
Neon 491, 566–571
Neoprene 71
Neptunium 566–570, 573
Nickel
  absorption edge 494
  absorption edges 571
  activity 823
  atomic spectroscopy DLs 416
  Auger element map 903
  in drinking water 662
  electron configuration 491
  electrothermal atomizer (ETA) DLs 484
  emission lines 494
  ETA-ICP-MS DLs 661
  flame atomic emission 413, 483
  graphite tube furnace 418
  ICP DLs 453
  ICP-MS 662, 665
  isotope abundance 682
  in Montana soil 665
  steel analysis 443
  X-ray emission lines 532, 566–570
  X-ray tube filter 510
Nicotinic acid 588
Niobium
  absorption edges 571
  alloys 32
  atomic spectroscopy DLs 416
  flame atomic emission 413
  isotope abundance 683
  X-ray emission lines 566–570
Nitrates
  as chromophore 126
  conductance 851
  electrodes 830
Nitric acid 285
Nitriles 125–126
  fragmentation 633
  gloves 71
  stretching frequencies 227
Nitrites
  as chromophore 126
  fragmentation 633
  polarography 863
Nitro compounds
  as chromophore 126
  stretching frequencies 227
Nitro groups 124–125
Nitrobenzene 229
Nitrocellulose 241
Nitro compounds 633
Nitrogen
  absorption edges 571
  atom counting 625
  atomic mass 576, 649
  cryogenic fluid 62
  electron configuration 491
  isotopes 576, 625, 649, 682
  rule 623

X-ray emission lines 566–570
XPS spectra 896
Nitrogen bonds 188–189
Nitrogen fluoride 647
Nitrogen-containing compounds 321–324, 646–647
Nitrogen-containing functional groups 227
Nitromethane 285
Nitropropane 300, 329
Nitroso groups
  as chromophore 126
  polarography 863
Nitrous oxide 380
Nitrous oxide–acetylene 380–382, 388, 391, 393–394
Nobium 483
Nylon 256, 924

O

Octane 316, 342
1-Octene, proton NMR spectra 320
Organoiodide groups 124
Osmium
  absorption edges 572
  atomic spectroscopy DLs 416
  flame atomic emission 413, 483
  isotope abundance 684
  X-ray emission lines 566–570
Oxides 860
Oxidooxycarboniodidate 851
Oxime 126
Oxygen
  absorption edges 571
  atom counting 625–626
  atomic mass 576, 649
  Auger element map 904
  bonds 189
  cryogenic fluid 62
  electron configuration 491
  isotopes 576, 625, 649, 682
  NMR active nuclei 279, 330
  petroleum molar absorptivity 545
  polarogram 858
  X-ray emission lines 566–570
Oxygen-containing organic compounds 321

P

Palladium
  absorption edges 572
  activity 823
  atomic spectroscopy DLs 416
  flame atomic emission 413, 483
  graphite tube furnace 418
  isotope abundance 684
  X-ray emission lines 566–570
Palm oil 233
PDMS, see Polydimethylsiloxanes
Pentane 9, 792
n-Pentane
  chromatography 757
  mass spectra 634
  as UV solvent 129
n-Pentyl acetate 129
Peptides 781–782

CHEMICAL INDEX

Perchlorate 830
PET, see Polyethylene terephthalate
Phenol
   absorption frequency 189
   fluorescence quantum yields 169
   IR spectra 225
   proton NMR spectra 325
Phenolphthalein 164
Phenols
   fragmentation 633
   stretching frequencies 223
Phentermine 252
Phenylalanine 779
3-Phenylpropionaldehyde 226
Phosphate esters 231
Phosphine oxides 231
Phosphines 231
Phosphoric acid 285
^{31}Phosphorus 279, 330
Phosphorus
   absorption edges 571
   atomic mass 576
   atomic spectroscopy DLs 416
   electron configuration 491
   flame atomic emission 413, 483
   isotopes 576, 625, 683
   X-ray emission lines 566–570
Phosphorus compounds 230–233
Phosphorus pentoxide 533
o-Phthalic anhydride (OPA) 772
Platinum
   absorption edge 494
   absorption edges 572
   activity 823
   AES 462
   atomic spectroscopy DLs 416
   emission lines 494
   flame atomic emission 413, 483
   graphite tube furnace 418
   isotope abundance 684
   potential window 863
   X-ray emission lines 566–570
Platinum (IV) hexacyanide 285
Plutonium 566–570, 573
Polonium 532, 566–570, 572
Poly(1,4-phenylene ether sulfone) 332
Poly(2,6-dimethyl-1,4-phenylene oxide) 341
Polycarbonates 513
Polydimethylsiloxanes (PDMS) 36, 336, 698
   gas chromatography 716–717, 723–726, 740, 743, 745
Polyethylene 924, 937
Polyethylene terephthalate (PET) 513
Polyglycols 235
Polyimide 924
Polymer film 513
Polymers, see individual polymers
Polymethylmethacrylate 924
Polymethylvinylsilane 340
Polypropylene 513
Polypropylene film 257
Polystyrene 184, 222, 267
Polytetrafluoroethylene 924
Polyvinyl chloride (PVC) 71, 924, 927
Portland cement 31
Potassium
   absorption edges 571
   activity 823
   atomic spectroscopy DLs 416
   in drinking water 662
   electrodes 830
   electron configuration 491
   electrothermal atomizer DLs 484
   ETA-ICP-MS DLs 661
   flame atomic emission 413, 483
   flame atomization 390
   flame photometry DLs 430
   ICP-MS 662, 665
   isotope abundance 683
   in Montana soil 665
   in serum 20–21
   X-ray emission lines 566–570
Potassium bromide 190, 201
Potassium chloride 190
Potassium oxide 533
Praseodymium
   atomic spectroscopy DLs 416
   flame atomic emission 413, 483
   isotope abundance 684
   X-ray emission lines 566–570
Primary alcohols 189
Primary amines 188
Proline 779
Promethium 504, 566–570, 572
Propane 388, 425
2-Propanol 129
Propene 643
2-Propenoic acid 644–645
Propionamide 336
n-Propyl alcohol 129
Propylamines 228, 632
Propyne 123
Protactinium 566–570, 573
PVC, see Polyvinyl chloride
Pyrazine anion, ESR spectra 363
Pyridines 126, 129, 634

Q

Quartz 264
Quinalizarin 161
Quinate 798
Quinine 174
Quinoline 126

R

Radium 566–570, 573
Radon 74–75, 566–570, 572
Rapeseed oil 233
Rhenium 483, 584
   absorption edges 572
   atomic spectroscopy DLs 416
   flame atomic emission 413, 483
   isotope abundance 684
   X-ray emission lines 566–570
Rhodamine B 169
Rhodium
   absorption edges 571
   flame atomic emission 413–414, 483
   isotope abundance 683
   X-ray emission lines 566–570
Rock salt 190
Rubber 71
Rubidium 414
   atomic spectroscopy DLs 416
   flame atomic emission 415, 483
   flame atomization 390
   graphite tube furnace 418
   ionization interference 394
   isotope abundance 683
   nitrogen–phosphorus detectors 735
   X-ray emission lines 566–570
Ruthenium 175, 483
   absorption edges 571
   flame atomic emission 414, 483
   isotope abundance 682–683
   X-ray emission lines 566–570

S

Salt bridge 818–819, 840
Salts 556
Samarium 360, 414, 483
   absorption edges 572
   atomic spectroscopy DLs 416
   flame atomic emission 414, 483
   isotope abundance 683–684
   X-ray emission lines 566–570
Scandium 414, 483
   absorption edges 571
   atomic spectroscopy DLs 416
   electron configuration 491
   flame atomic emission 414, 483
   isotope abundance 683
   X-ray emission lines 566–570
SCD, see Sulfur chemiluminescence detectors
Secondary alcohols 189
Secondary amines 189
Selenium
   absorption edges 571
   atomic spectroscopy DLs 416
   in drinking water 662
   electron configuration 491
   electrothermal atomizer (ETA) DLs 484
   ETA-ICP-MS DLs 661
   flame atomic emission 414
   ICP DLs 453
   ICP-MS 662, 665
   isotope abundance 682–683
   in Montana soil 665
   X-ray emission lines 566–570
Serine 779
Silica 190, 752
Silicate minerals 31–32
^{29}Silicon 279, 330, 340
^{33}Silicon 279
^{35}Silicon 279
Silicon
   absorption edges 571
   atomic spectroscopy DLs 416
   depth profile 904
   electron configuration 491
   electrothermal atomizer (ETA) DLs 484
   ETA-ICP-MS DLs 661
   flame atomic emission 414, 483
   flame atomization 390
   frequency range 207

index of refraction 207
isotope abundance 625, 682
polycrystalline transmission 164
Raman spectra 261
steel analysis 443
X-ray emission lines 566–570
Silicon carbide 191–192, 259, 337, 340
Silicon compounds 230–233
Silicon dioxide 533
Silicone polymer 340
Silver
  absorption edge 494
  absorption edges 572
  activity 823
  arc emission spectrometry DLs 433
  atomic spectroscopy DLs 415
  in drinking water 662
  electrothermal atomizer (ETA) DLs 484
  emission lines 494
  ETA-ICP-MS DLs 661
  flame atomic emission 414, 483
  flame atomization 389–390
  half-wave potential 860
  ICP DLs 453
  ICP-MS 662, 665
  isotype abundance 683
  in Montana soil 665
  SERS process 254
  X-ray emission 493, 566–570
  X-ray tube filter 510
Silver bromide 190, 830
Silver chloride 190, 830
Silver iodide 830
Silver sulfide 830
Silver thiocyanate 830
^{23}Sodium 279, 330
Sodium
  absorption edges 571
  activity 823
  atomic spectroscopy DLs 416
  in drinking water 662
  electron configuration 491
  electrothermal atomizer (ETA) DLs 484
  ETA-ICP-MS DLs 661
  flame atomic emission 414, 483
  flame atomization 390
  flame photometry DLs 430
  ICP-MS 662, 665
  isotope abundance 682–683
  in Montana soil 665
  X-ray emission lines 566–570
Sodium chloride
  mid-IR optics 190
  NMR spectroscopy 285
Sodium oxide
  Dolomite sample 533
  in glass 464
Sodium salicylate 169
Sodium silicate 925
Sodium sulfanilate 169
Soy oil 233
SP-FPD, see sulfur-phosphorus flame photometric detector
SPADNS 161
Stainless steel
  acid dissolution 31

  Auger spectra 903
  GD depth profile 467–468
Steel
  mass spectra 668–669
  XRF spectra 537
Strontium
  absorption edges 571
  activity 823
  atomic spectroscopy DLs 416
  characteristics 414
  derivative thermogravimetry 925
  flame atomic emission 414, 422, 483
  flame atomization 390
  graphite tube furnace 418
  isotopes 661, 683
  X-ray emission lines 566–570
Strontium barium niobate 199
Strontium fluoride 190
Styrene–divinylbenzene resin 783
Succinic acid 588
Sucrose
  2D-HETCOR spectra 339
  13C NMR spectra 331
  contour spectra 338
  COSY spectra 338
Sulfates 851
Sulfides 230, 633
Sulfones 126, 230, 633
Sulfoxides
  as chromophore 126
  fragmentation 633
  stretching frequencies 230
^{33}Sulfur 330
Sulfur
  absorption edges 571
  atom counting 625
  atomic mass 576
  atomic spectroscopy DLs 416
  Auger element distribution 903
  electron configuration 491
  isotopes 576, 625, 683
  petroleum molar absorptivity 545
  X-ray emission lines 566–570
Sulfur dioxide 647
Sulfur trioxide 533
Sulfur-containing compounds 230–233, 647–648

T

Tantalum
  absorption edges 572
  atomic OES 430
  atomic spectroscopy DLs 416
  characteristics 414
  dissolution 31
  electrothermal atomizers 384
  flame atomic emission 414, 483, 485
  high-temperature furnace atomizers 382
  isotope abundance 683
  LIBS applications 475
  micro-DSC instrumentation 946
  X-ray emission lines 566–570
Technetium 414
  absorption edges 571
  flame atomic emission 414
  X-ray emission lines 566–570

Tellurium 483
  absorption edges 572
  atomic spectroscopy DLs 416
  flame atomic emission 414, 483
  graphite tube furnace 418
  isotope abundance 683
  X-ray emission lines 532, 566–570
Terbium 483
  absorption edges 572
  atomic spectroscopy DLs 416
  flame atomic emission 414, 483
  isotope abundance 683
  X-ray emission lines 566–570
Tertiary alcohols 189
Tertiary amines 189
Tetrachloroethene 634
Tetrachloroethylene 129
Tetrahydrofuran 129, 634, 757
Tetramethylammonium chloride 285
Tetramethylsilane (TMS) 285, 634
Tetramethyltin 285
Thallium
  absorption edges 572
  atomic spectroscopy DLs 416
  in drinking water 662
  flame atomic emission 414, 483
  flame atomization 390
  graphite tube furnace 418
  ICP-MS 662, 665
  in Montana soil 665
  X-ray emission lines 566–570
Thallium bromide/iodide eutectic (KRS-5) 190, 207
Thiocarbonyls 230
Thioether 126
Thioketone 125–126
Thiols 124–126
Thiourea 161, 846, 868
Thorium
  absorption edges 573
  atomic spectroscopy DLs 416
  characteristics 414
  in drinking water 662
  flame atomic emission 414, 483
  isotope abundance 684
  mid-IR sources 191
  radiation dose 75
  X-ray emission lines 566–570
Threonine 779
Thulium 483
  absorption edges 572
  atomic spectroscopy DLs 416
  flame atomic emission 414
  isotope abundance 683
  X-ray emission lines 566–570
Tin
  absorption edges 572
  activity 823
  atomic spectroscopy DLs 416
  flame atomic emission 414, 483
  flame atomization 390
  graphite tube furnace 418
  ICP DLs 453
  isotope abundance 683
  X-ray emission lines 566–570
Tin ores 32

CHEMICAL INDEX

Titanium
  absorption edge 494
  absorption edges 571
  atomic spectroscopy DLs 416
  electron configuration 491
  emission lines 494
  flame atomic emission 414, 483
  flame atomization 390
  graphite tube furnace 418
  isotope abundance 683
  X-ray emission lines 566–570
Titanium alloys 31
Titanium dioxide 31
Titanium metal 31
Titanium nitride 468
Titanium ores 32
Titanium oxide 533
TMS, see Tetramethylsilane
Toluene 129
  ^{13}C NMR spectrum 283
  eluent strength 757
  IR spectra 223
  mass spectra 634
  polarity index 757
1,1,2-Trichloro-1,2,2-trifluoroethane 129
Trichloroethylene 629
Trichlorofluoromethane 285
n-Tridecane 636
1,1,1-Trifluoro-2,2-dichloroethane 690
2,2,4-Trimethylpentane 129
Trimethylphosphite 285
3-(Trimethylsilyl)-1-propanesulfonic acid (DSS) 285
3-(Trimethylsilyl)-propanoic acid (TSP) 285
Tripropylamine 229
Tritium 279
Tropylium ions 632–633, 640
Trypsin 654, 780–781
Tryptophan 779
Tungsten
  absorption edges 572
  AES 460–462
  AES spectra 462
  atomic spectroscopy DLs 416
  crystal field ember 903
  filaments 902
  flame atomic emission 414, 483
  isotope abundance 683–684
  X-ray emission lines 566–570
  X-ray tube filter 510
Tungsten carbide 12
Tungsten ores 32
Tungsten oxide 673
Tungsten–halogen 131, 192–193, 236
Tyrosine 779

U

Uranium
  absorption edges 573
  atomic spectroscopy DLs 416
  in drinking water 662
  flame atomic emission 414, 483
  ICP-MS 662
  isotope abundance 684
  X-ray emission lines 566–570
Uranyl acetate 169
Urea 588

V

Valine 230, 779
Vanadium 414
  absorption edges 571
  atomic spectroscopy DLs 416
  in crystals 518
  in drinking water 662
  electron configuration 491
  ESR spectra 363
  flame atomic emission 86, 414–415, 445, 483
  flame atomization 390
  graphite tube furnace 418
  ICP DLs 453
  ICP-MS 662, 665
  isotope abundance 682–683
  in Montana soil 665
  MS/MS interference 672
  X-ray emission lines 566–570
Vegetable oil 159
Vinyl 71
Vinyl acetate copolymers 923

W

Water
  ESR spectra 364
  extinguisher 78
  fragmentation 632
  mass spectra 626, 628, 633
  NMR spectroscopy 285
  TD-NMR FID 348
  as UV solvent 129
Whiskey 653

X

Xenon
  absorption edges 572
  arc lamp 131–132, 172
  isotope abundance 683–684
  X-ray emission lines 566–570
Xylene 129, 223–224

Y

YAG, see Yttrium aluminium garnet
YSZ, see Yttria-stabilized zirconia
Ytterbium 483
  absorption edges 572
  atomic spectroscopy DLs 416
  flame atomic emission 415, 483
  isotope abundance 683
  X-ray emission lines 566–570
Yttria-stabilized zirconia (YSZ) 556, 960
Yttrium
  absorption edges 571
  atomic spectroscopy DLs 416
  characteristics 414
  flame atomic emission 415, 484
  isotope abundance 683
  radiation source 131, 172
  Raman instrumentation 245
  X-ray emission lines 566–570
Yttrium aluminium garnet (YAG) 457, 586

Z

Zeolite catalyst 926
Zeolites 718, 782–783, 789, 926
Zinc
  absorption edges 571
  activity 823
  atomic spectroscopy DLs 416
  in drinking water 662
  EDXRF spectra 536
  electron configuration 491
  electrothermal atomizer (ETA) DLs 484
  ETA-ICP-MS DLs 661
  flame atomic emission 415, 484
  flame atomization 390
  graphite tube furnace 418
  half-wave potential 860
  ICP DLs 453
  ICP-MS 662, 665
  isotope abundance 682
  in Montana soil 665
  X-ray emission lines 566–570
Zinc alloys 31
Zinc metal 31
Zinc oxide 31
Zinc selenide 207
Zinc sulfide 190, 207
Zirconia
  chromatography 697, 754
  cubic 4
  DTA 929
  X-ray diffraction 556
  yttria-stabilized (YSZ) 556, 960
Zirconium 4
  AAS 406
  absorption edges 571
  atomic spectroscopy DLs 416
  dissolution 31
  flame atomic emission 415, 484–485
  fusions 31
  isotope abundance 683
  mid-IR sources 191
  sampling 12
  in spectrometry 161
  X-ray emission lines 566–570
Zirconium alloys 31
Zwitterions 228

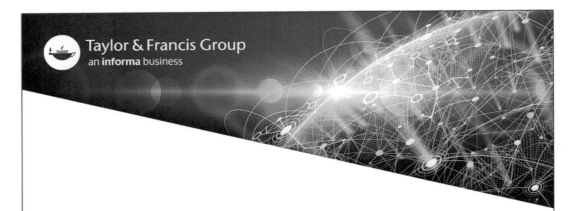

Taylor & Francis eBooks

www.taylorfrancis.com

A single destination for eBooks from Taylor & Francis with increased functionality and an improved user experience to meet the needs of our customers.

90,000+ eBooks of award-winning academic content in Humanities, Social Science, Science, Technology, Engineering, and Medical written by a global network of editors and authors.

TAYLOR & FRANCIS EBOOKS OFFERS:

- A streamlined experience for our library customers
- A single point of discovery for all of our eBook content
- Improved search and discovery of content at both book and chapter level

REQUEST A FREE TRIAL
support@taylorfrancis.com